CONTENTS

BUILDING OPERATION AND MAINTENANCE

GENERAL APPLICATIONS

INDEX

ADDITIONS AND CORRECTIONS

1995 ASHRAE HANDBOOK

Heating, Ventilating, and Air-Conditioning APPLICATIONS

SI Edition

American Society of Heating, Refrigerating and Air-Conditioning Engineers, Inc.

1791 Tullie Circle, N.E., Atlanta, GA 30329

(404) 636-8400

DEDICATED

TO THE ADVANCEMENT OF

THE PROFESSION

AND ITS ALLIED INDUSTRIES

ISBN 1-883413-23-0

CONTRIBUTORS

In addition to the Technical Committees, the following individuals contributed significantly to this volume. The appropriate chapter numbers follow each contributor's name.

Van D. Baxter (1)
Oak Ridge National Laboratory

Nick Koskolos (1)
NORDYNE, A Nortek Company

Nancy J. Banks (2, 5)
Munters DryCool

Frank Mills (2, 4)
Frank Mills Associates

John E. Wolfert (2, 5)
Melvin Simon & Associates

John C. Mentzer (3)
SSOE, Inc.

Norman A. Buckley (4, 49)
Buckley Associates

Edward D. Fitts (4, 5, 6, 7, 50)
Avanti Technologies, Inc.

Reinhold Kittler (4)
Dectron, Inc.

Charles C. Smith (4)
Colorado State University

Lynn F. Werman (4)
Henningson, Durham & Richardson, Inc.

W. Ted Ritter, Jr. (6)
Engineering Economics, Inc.

Ramnath Viswanath (6)
Etisalat

Richard D. Hermans (7)
St. Paul Public Schools

Chris P. Rousseau (7, 28)
Newcomb & Boyd

Dawood Amir Ali (8)
Eaton Corporation

David C. Allen (8)
D.C. Allen, Inc.

James Steven Brown (8)
Ford Motor Company

John Burgers (8)
Long Manufacturing Ltd.

Larry Donald Cummings (8)
Harrison Division (GMC)

Eugene A. Dianetti (8)
Parker Hannifin Corporation

William J. Hannett (8)
UTC/Carrier-Transicold

Sung Lim Kwon (8)
Thermo King Corporation

Peter Grimm Malone (8)
Eaton Corporation

John F. O'Brien (8)

Ernest W. Schumacher (8)
Fujikoki America, Inc.

Ramesh K. Shah (8)
Harrison Division (GMC)

James J. Bushnell (9)
AlliedSignal Inc.

Kim E. Linnett (9)
AlliedSignal Inc.

Thomas Nagle (9)
Douglas Aircraft Company

Neal A. Nelson (9)
Boeing Commercial Airplane Group

David R. Space (9)
Boeing Commercial Airplane Group

A. Bruce Badger (10)
Industrial Process Equipment

Arthur Bendelius (12)
Parsons Brinckerhoff

Alfred Brociner (12)
Port Authority of New York and New Jersey

Kelly A. Giblin (12)
N.J. Transit

Nils R. Grimm (12)
Sverdrup Corporation

Peter B. Gardner (13)
Schering Corporation

Louis Hartman (13)
Harley Ellington Design

Leonard H. Schwartz (13)

Alfred W. Woody (14, 24)
Giffels Associates, Inc.

Mel Crichton (15)
Eli Lilly & Company

Larry Hughes (15)
Merck & Company

Norm Maxwell (15)
Gayle King Carr & Lynch, Inc.

Phil Naughton (15)
Motorola, Inc.

Henry J. Vance (18)
Vance Professional Services

Robert F. Button (19)
Eastman Kodak Company

James W. Carty (19)
Eastman Kodak Company

Harold E. Gray (20)

Jay D. Harmon (20)
Iowa State University

Kevin A. Janni (20)
University of Minnesota

John Ogilvie (20)
University of Guelph

Gene C. Shove (21)

James A. Carlson (23)
Lawrence Livermore National Laboratory

Richard A. Evans (23)
Meier Associates

Dominic H. Flens (23)
Sargent & Lundy

Wayne A. Lawton (23, 26)
EG & G Rocky Flats

Kevin R. Scaggs (23)
Westinghouse Savannah River Company

Howard D. Goodfellow (24)
Goodfellow Consultants, Inc.

Eugene O. Shilkrot (24)
TsNIIpromzdanii

Andrey S. Strongin (24)
TsNIIpromzdanii

Alexander M. Zhivov (24, 26)
University of Illinois

Vladimir N. Posokhin (26)

Michael C. Connor (27)
McKenney's Mechanical Contractors and Engineers

Charles N. Claar (28)
International Facilities Management Association

Gary M. Elovitz (28)
Energy Economics, Inc.

Donald R. Fisher (28)
Fisher Consultants

Eliott B. Gordon (28)
American Gas Association Labs

Ronald R. Huffman (28)
LDI Manufacturing Company, Inc.

Richard M. Kelso (28)
University of Tennessee

Joseph N. Knapp (28)
McDonald's Corporation

Phil Morton (28)
Gaylord Industries

David W. Wolbrink (28)
Broan Manufacturing Company, Inc.

Steven P. Kavanaugh (29)
University of Alabama

William E. Murphy (29)
University of Kentucky

Kevin D. Rafferty (29)
Oregon Institute of Technology

Timothy J. Merrigan (30)
Florida Solar Energy Center

Gene M. Meyer (30)
Kansas State University

David L. Grumman (31)
Grumman/Butkus Associates

Charles H. Culp III (32, 37)
Emerson Electric, Inc.

Robert Fuller (32)
R H Fuller & Associates

Jeff S. Haberl (32, 33, 36, 37)
Texas A&M University

J. Michael MacDonald (32, 37)
Oak Ridge National Laboratory

Richard Mazzucchi (32)

Dick Pearson (32)
Pearson Engineering

Lawrence G. Spielvogel (32)

Cedric Trueman (32)

David G. Guckelberger (33)
The Trane Company

Carl C. Hiller (33)
EPRI

Ronald N. Jensen (33)
Jensen Engineering

CONTRIBUTORS (*Concluded*)

Dennis A. Smith (33)
Atlanta Gas Light Company

Joseph J. Watson (33)

Mark Hegberg (34)
Commonwealth Edison—Chicago

Rodney H. Lewis (34)
Rodney H. Lewis Associates, Inc.

Gaylon Richardson (34)
Engineered Air Balance, Inc.

Dennis H. Tuttle (34)
Wessels Company

John D. Roach (35)
Corporetum Management Company

T. David Underwood (35)
Isotherm Engineering, Ltd.

Frantisek Vaculik (35)
Government Services Canada

Kevin P. Cooney (36)
R.C.G./Hagler Bailly

Brian Kammers (36)
Magnetek, Inc.

Ron M. Nelson (36)
Iowa State University

Michael C.A. Schwedler (36)
The Trane Company

D. Wayne Webster (36)
Canada Mortgage & Housing

Hashem Akbari (37)
Lawrence Berkeley Laboratory

Kristin E. Heinemeier (37)
Lawrence Berkeley Laboratory

Mark P. Modera (37)
Lawrence Berkeley Laboratory

Peter R. Armstrong (38)
Battelle Pacific NW Laboratories

James E. Braun (38)
Purdue University

Zulfikar Cumali (38)
Consultants Computation Bureau

John W. Mitchell (38)
University of Wisconsin-Madison

David M. Schwenk (38)
U.S. Army CERL

Walter P. Bishop (39)
Walter P. Bishop Consulting Engineers

Peter J. Hoey (39)
Emtek, Inc.

John S. Andrepont (40)
Chicago Bridge and Iron Company

David Arnold (40)
Troup, Bywaters & Anders

William P. Bahnfleth (40)
Pennsylvania State University

Debra L. Catanese (40)
Jersey Central Power & Light Company

James L. Denkmann (40)
Denkmann Thermal Storage, Ltd.

Chad B. Dorgan (40)
Dorgan Associates, Inc.

James S. Elleson (40)
Elleson Engineering

Mark W. Fly (40)
Governair Corporation

C. Wayne Frazell (40)
TU Electric Technical Services

Kenneth L. Gillespie (40)
Pacific Gas & Electric Company

Jack Harmon (40)
H.C. Yu and Associates

David E. Knebel (40)
Mammoth, Inc.

Richard J. Kooy (40)
Chicago Bridge and Iron Company

Thomas R. Kroeschell (40)
Thomas Kroeschell, P.E.

Harold G. Lorsch (40)
Drexel University

Mark M. MacCracken (40)
Calmac Manufacturing Corporation

Edward L. Morofsky (40)
Public Works Canada

David C.J. Peters (40)
Southland Industries

George Reeves (40)
George Reeves Associates

John O. Richardson (40)
Tennessee Valley Authority

Peter Simmonds (40)
Flack & Kurtz

Martin L. Timm (40)
Henry Vogt Machine Company

Maurice W. Wildin (40)
University of New Mexico

Gemma Kerr (41)

Richard D. Rivers (41)
Environmental Quality Sciences, Inc.

Sam Silberstein (41)
NIST

Kent W. Peterson (42)
PSI Engineers, Inc.

John R. Sosoka (42)
PSI Engineers, Inc.

Ted N. Carnes (43)
Air System Components

Douglas D. Reynolds (43)
DDR, Inc.

Dirk N. Granberg (45)
RECO Industries

Edwin A. Nordstrom (45)
Viessmann Manufacturing Company

Sherwood G. Talbert (45)
Battelle Columbus Labs

Lawrence H. Chenault (46)
Hume Snow Melting Systems, Inc.

Birol I. Kilkis (46, 49)
Heatway Radiant Floor Heating

Avnit Singh (47)
Pneumafil Corporation

Patricia Thomas (47)
Munters Corporation

Thomas C. Campbell (48)
Campbell Code Consulting

John A. Clark (48)
BKBM Engineers

John H. Klote (48)
NIST

C.R. MacCluer (49)
Michigan State University

J. Marx Ayres (50)
Ayres & Ezer Associates, Inc.

Patrick J. Lama (50)
Mason Industries

Robert Simmonds (50)
Amber Booth Company

William Staehlin (50)
State of California Office of Statewide Planning and Development

ASHRAE TECHNICAL COMMITTEES AND TASK GROUPS

SECTION 1.0—FUNDAMENTALS AND GENERAL
1.1 Thermodynamics and Psychrometrics
1.2 Instruments and Measurements
1.3 Heat Transfer and Fluid Flow
1.4 Control Theory and Application
1.5 Computer Applications
1.6 Terminology
1.7 Operation and Maintenance Management
1.8 Owning and Operating Costs
1.9 Electrical Systems
1.10 Energy Resources

SECTION 2.0—ENVIRONMENTAL QUALITY
2.1 Physiology and Human Environment
2.2 Plant and Animal Environment
2.3 Gaseous Air Contaminants and Gas Contaminant Removal Equipment
2.4 Particulate Air Contaminants and Particulate Contaminant Removal Equipment
2.5 Air Flow Around Buildings
2.6 Sound and Vibration Control
TG Global Climate Change
TG Halocarbon Emissions
TG Seismic Restraint Design

SECTION 3.0—MATERIALS AND PROCESSES
3.1 Refrigerants and Brines
3.2 Refrigerant System Chemistry
3.3 Contaminant Control in Refrigerating Systems
3.4 Lubrication
3.5 Desiccant and Sorption Technology
3.6 Corrosion and Water Treatment
3.7 Fuels and Combustion

SECTION 4.0—LOAD CALCULATIONS AND ENERGY REQUIREMENTS
4.1 Load Calculation Data and Procedures
4.2 Weather Information
4.3 Ventilation Requirements and Infiltration
4.4 Thermal Insulation and Moisture Retarders
4.5 Fenestration
4.6 Building Operation Dynamics
4.7 Energy Calculations
4.9 Building Envelope Systems
4.10 Indoor Environmental Modeling
TG Cold Climate Design

SECTION 5.0—VENTILATION AND AIR DISTRIBUTION
5.1 Fans
5.2 Duct Design
5.3 Room Air Distribution
5.4 Industrial Process Air Cleaning (Air Pollution Control)
5.5 Air-to-Air Energy Recovery
5.6 Control of Fire and Smoke
5.7 Evaporative Cooling
5.8 Industrial Ventilation
5.9 Enclosed Vehicular Facilities
TG Kitchen Ventilation
TG Smoke Management Components

SECTION 6.0—HEATING EQUIPMENT, HEATING AND COOLING SYSTEMS AND APPLICATIONS
6.1 Hydronic and Steam Equipment and Systems
6.2 District Heating and Cooling
6.3 Central Forced Air Heating and Cooling Systems
6.4 In Space Convection Heating
6.5 Radiant Space Heating and Cooling
6.6 Service Water Heating
6.7 Solar Energy Utilization
6.8 Geothermal Energy Utilization
6.9 Thermal Storage

SECTION 7.0—PACKAGED AIR-CONDITIONING AND REFRIGERATION EQUIPMENT
7.1 Residential Refrigerators and Food Freezers
7.4 Unitary Combustion-Engine-Driven Heat Pumps
7.5 Room Air Conditioners and Dehumidifiers
7.6 Unitary Air Conditioners and Heat Pumps

SECTION 8.0—AIR-CONDITIONING AND REFRIGERATION SYSTEM COMPONENTS
8.1 Positive Displacement Compressors
8.2 Centrifugal Machines
8.3 Absorption and Heat Operated Machines
8.4 Air-to-Refrigerant Heat Transfer Equipment
8.5 Liquid-to-Refrigerant Heat Exchangers
8.6 Cooling Towers and Evaporative Condensers
8.7 Humidifying Equipment
8.8 Refrigerant System Controls and Accessories
8.10 Pumps and Hydronic Piping
8.11 Electric Motors—Open and Hermetic

SECTION 9.0—AIR-CONDITIONING SYSTEMS AND APPLICATIONS
9.1 Large Building Air-Conditioning Systems
9.2 Industrial Air Conditioning
9.3 Transportation Air Conditioning
9.4 Applied Heat Pump/Heat Recovery Systems
9.5 Cogeneration Systems
9.6 Systems Energy Utilization
9.7 Testing and Balancing
9.8 Large Building Air-Conditioning Applications
9.9 Building Commissioning
9.10 Laboratory Systems
TG Clean Spaces
TG Combustion Gas Turbine Inlet Air Cooling Systems
TG Tall Buildings

SECTION 10.0—REFRIGERATION SYSTEMS
10.1 Custom Engineered Refrigeration Systems
10.2 Automatic Icemaking Plants and Skating Rinks
10.3 Refrigerant Piping, Controls, and Accessories
10.4 Ultra-Low Temperature Systems and Cryogenics
10.5 Refrigerated Distribution and Storage Facilities
10.6 Transport Refrigeration
10.7 Commercial Food and Beverage Cooling, Display and Storage
10.8 Refrigeration Load Calculations
10.9 Refrigeration Application for Foods and Beverages

PREFACE

This Handbook describes heating, ventilating, and air-conditioning practices for a broad range of applications. Many chapters have been revised to reflect continued evolutionary changes and improvements in technology. These revisions reflect current requirements and design approaches.

In addition, the technical committees responsible for preparing the chapters have added information or made notable revisions to the following chapters:

- Chapter 1, Residences, includes a new section on manufactured homes.
- Chapter 4, Places of Assembly, includes a new section on atriums, information on low-intensity radiant heating, and additional information on the evaporation rate from natatoriums.
- Chapter 5, Domiciliary Facilities, includes additional information on dehumidification.
- Chapter 6, Educational Facilities, has expanded general design information for various types of school buildings ranging from preschools through colleges.
- Chapter 7, Health Care Facilities, has a revised classification system of facilities for design purposes. Also, ventilation recommendations have been expanded to include additional administrative spaces and treatment areas.
- Chapter 8, Surface Transportation, discusses the phaseout of Refrigerant 12 and its impact on vehicle air conditioning equipment.
- Chapter 9, Aircraft, includes information on air-cycle equipment from the 1988 *ASHRAE Handbook*.
- Chapter 12, has added information on bus garages and on selecting equipment for rapid transit and road tunnels.
- Chapter 13, Laboratory Systems, has been reorganized for clarity and has more information on fume hoods and exhaust systems.
- Chapter 15, Clean Spaces, now contains information on pharmaceutical clean rooms.
- Chapter 20, Photographic Materials, shows the new generation of package processing equipment and has additional information on HVAC and exhaust ventilation for photolabs.
- Chapter 24, Ventilation of the Industrial Environment, is significantly revised. The chapter now includes more specific information on thermal comfort for industrial facilities; a significant update on spot cooling; state-of-the-art ventilation design principles and the techniques for applying these principles; new sections on general comfort and dilution ventilation, natural ventilation, and air curtains; and an extensive bibliography.
- Chapter 26, Industrial Exhaust Hoods, describes new classifications for industrial hoods, and it includes additional information on jet-assisted hoods and hood design.
- Chapter 29, Geothermal Energy, has a new section on ground-source heat pumps, and the section on direct use of geothermal energy has been updated and condensed.
- Chapter 30, Solar Energy Utilization, incorporates information moved from the *Fundamentals Handbook* (Chapter 28, Energy Estimating Methods) on analyzing the performance of solar equipment.
- Chapter 31, Energy Resources, includes the most current data on energy resources and consumption, including world energy resource data. Past and projected U.S. energy consumption information has been expanded and updated.

- Chapter 33, Owning and Operating Costs, has more information in the section on owning costs. The chapter also covers innovative financing alternatives, discusses the phaseout of refrigerants, and has a revised, more useful economic analysis section.
- Chapter 34, Testing, Adjusting, and Balancing, has a revised section on temperature control verification.
- Chapter 36, Computer Applications, has been reorganized and includes new information on computer systems architecture. The section on knowledge-based systems has been updated and retitled Artificial Intelligence; and the bibliography has been substantially updated.
- Chapter 38, Building Operating Dynamics and Strategies, has been retitled to reflect its new emphasis on optimal control strategies.
- Chapter 40, Thermal Storage, has been rewritten to better inform user's unfamiliar with the subject. Also, obsolete technologies have been dropped and new technologies such as precooling inlet air for combustion turbines and international applications such as building mass thermal storage have been added. Perhaps of most importance, a section on Installation, Operation, and Maintenance has been added.
- Chapter 43, Sound and Vibration Control, has been reorganized to provide more information on acoustic design guidelines and system design requirements. Most equations have been replaced by related tables and simpler design procedures.
- Chapter 44, Corrosion Control and Water Treatment, has information on Legionnaire's disease in the section on water treatment.
- Chapter 45, Service Water Heating, includes new hot water use data and new information on refrigeration heat reclaim and heat pump water heaters.

In addition, two new chapters have been developed. They are:

- Chapter 28, Kitchen Ventilation, describes the hoods and exhaust and makeup air systems required for safe ventilation of both commercial and residential kitchens.
- Chapter 39, Building Commissioning, introduces the important topic of commissioning new HVAC systems.

Just as new topics are added, some information is deleted because technology changes and the priorities of the Society change. For these reasons, Chapter 11, Environmental Control for Survival, which is in the 1991 *Handbook*, was dropped from the Handbook series.

All Handbooks are published in two editions. One edition contains inch-pound (I-P) units of measurement, and the other contains the International System of Units (SI).

A section before the index lists additions and corrections to the 1992, 1993, and 1994 volumes of the Handbook series. Any changes to this volume will be reported in the 1996 *ASHRAE Handbook* and on ASHRAE's electronic bulletin board.

If you have suggestions on improving a chapter or would like more information on how you can help revise a chapter, write to: Handbook Editor, ASHRAE, 1791 Tullie Circle, Atlanta, GA 30329.

Robert A. Parsons
Handbook Editor

RESIDENCES

SYSTEMS

SPACE-CONDITIONING systems for residential use vary with both local and application factors. Local factors include energy source availability (both present and projected) and price, climate, socioeconomic circumstances, and the availability of installation and maintenance skills. Application factors include housing type, construction characteristics, and building codes. As a result, many different systems are selected to provide combinations of heating, cooling, humidification, dehumidification, and air filtering. This chapter emphasizes the more common systems for space conditioning of both single- (traditional site-built and modular or manufactured homes) and multifamily residences. Low-rise multifamily buildings generally follow single-family practice because constraints favor compact designs. Retrofit and remodeling construction also adopt the same systems as those for new construction, but site-specific circumstances may call for unique designs.

The common residential heating systems are listed in Table 1. Three generally recognized groups are central forced air, central hydronic, and zonal. System selection and design involve such key decisions as (1) source(s) of energy, (2) means of distribution and delivery, and (3) terminal device(s).

Climate determines the services needed. Heating and cooling are generally required. Air cleaning (by filtration or electrostatic devices) can be added to most systems. Humidification, which can also be added to most systems, is generally provided in heating systems only when psychrometric conditions make it necessary for comfort. Cooling systems generally dehumidify as well. Typical residential installations are shown in Figures 1 and 2.

Figure 1 shows a gas furnace, a split-system air conditioner, a humidifier, and an air filter. The system functions as follows: Air returns to the equipment through a return air duct (1). It passes initially through the air filter (2). The circulating blower (3) is an integral part of the furnace (4), which supplies heat during winter. An optional humidifier (10) adds moisture to the heated air, which is distributed throughout the home from the supply duct (9). When cooling is required, the circulating air passes across the evaporator coil (5), which removes heat and moisture from the air. Refrigerant lines (6) connect the evaporator coil to a remote condensing unit (7) located outdoors. Condensate from the evaporator drains away through a pipe (8).

Figure 2 shows a split-system heat pump, supplemental electric resistance heaters, a humidifier, and an air filter. The system functions as follows: Air returns to the equipment through the return air duct (1) and passes through the air filter (2). The circulating blower (3) is an integral part of the indoor unit of the heat pump (4), which supplies heat via the indoor coil (6) during the heating season. Optional electric heaters (5) supplement heat from the heat pump during periods of low ambient temperature and counteract airstream cooling during the defrost cycle. An optional humidifier (10) adds moisture to the heated air, which is distributed throughout the home from the supply

Table 1 Residential Heating and Cooling Systems

	Forced Air	Hydronic	Zonal
Most Common Energy Sources	Gas Oil Electricity Resistance Heat pump	Gas Oil Electricity Resistance Heat pump	Gas Electricity Resistance Heat pump
Heat Distribution Medium	Air	Water Steam	Air Water Refrigerant
Heat Distribution System	Ducting	Piping	Ducting Piping or None
Terminal Devices	Diffusers Registers Grilles	Radiators Radiant panels Fan-coil units	Included with product

The preparation of this chapter is assigned to TC 7.6, Unitary Air Conditioners and Heat Pumps.

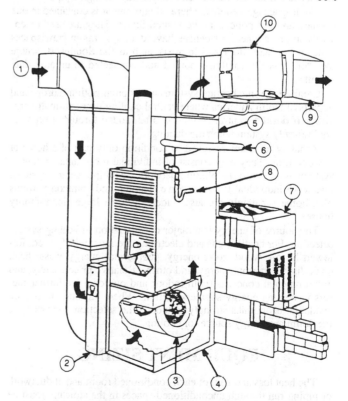

Fig. 1 Typical Residential Installation of Heating, Cooling, Humidifying, and Air Filtering System

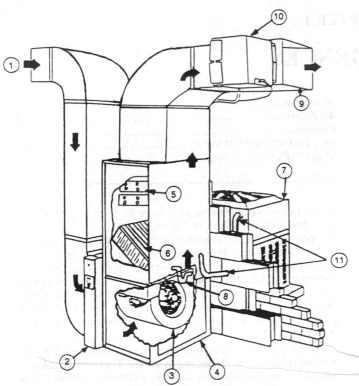

Fig. 2   Typical Residential Installation of Heat Pump Systems

duct (9). When cooling is required, the circulating air passes across the indoor coil (6), which removes heat and moisture from the air. Refrigerant lines (11) connect the indoor coil to the outdoor unit (7). Condensate from the indoor coil drains away through a pipe (8).

Single-package systems, where all equipment is contained in one cabinet, are also popular in the United States. They are used extensively in areas where residences have duct systems in crawlspaces beneath the main floor and in areas such as the Southwest, where they are typically rooftop-mounted and connected to an attic duct system.

Central hydronic heating systems are popular both in Europe and in parts of North America where central cooling is not normally provided. If desired, central cooling is often added through a separate cooling-only system with attic ducting.

Zonal systems are designed to condition only part of a home at any one time. They may consist of individual room units or central systems with zoned distribution networks. Multiple central systems that serve individual floors or serve sleeping and common portions of a home separately are also widely used in large single-family houses.

The source of energy is a major consideration in heating system selection. For heating, gas and electricity are most widely used, followed by oil, wood, solar energy, geothermal energy, waste heat, coal, district thermal energy, and others. Relative prices, safety, and environmental concerns (both indoor and outdoor) are further factors in heating energy source selection. Where various sources are available, economics strongly influence the selection. Electricity is the dominant energy source for cooling.

EQUIPMENT SIZING

The heat loss and gain of each conditioned room and of ductwork or piping run through unconditioned spaces in the structure must be accurately calculated in order to select equipment with the proper output and design. To determine heat loss and gain accurately, the floor plan and construction details must be known. The plan should include information on wall, ceiling, and floor construction and type and thickness of insulation. Window design and exterior door details are also needed. With this information, heat loss and gain can be calculated using the basic data in Chapters 22 through 28 of the 1993 *ASHRAE Handbook—Fundamentals* and the Air-Conditioning Contractors of America (ACCA) *Manual* J or similar calculation procedures. To conserve energy, many jurisdictions require that the building be designed to meet or exceed the requirements of ASHRAE *Standard* 90.2, Energy-Efficient Design of New Low-Rise Residential Buildings, or similar requirements.

Proper matching of equipment capacity to the design heat loss and gain is essential. The heating capacity of air-source heat pumps is usually supplemented by auxiliary heaters, most often of the electric resistance type; in some cases, however, fossil fuel furnaces or solar systems are used.

The use of undersized equipment results in an inability to maintain indoor design temperatures at outdoor design conditions and slow recovery from setback or set-up conditions. Grossly oversized equipment can cause discomfort due to short on-times, wide indoor temperature swings, and inadequate dehumidification when cooling. Gross oversizing may also contribute to higher energy use due to an increase in cyclic thermal losses and off-cycle losses. Variable capacity equipment (heat pumps, air conditioners, and furnaces) can more closely match building loads over specific ambient temperature ranges, usually reducing these losses and improving comfort levels; in the case of heat pumps, supplemental heat needs may also be reduced.

Recent trends are toward heavily insulated, tightly constructed buildings with improved vapor retarders and low infiltration, which may cause high indoor humidity conditions and necessitate dehumidification during the winter months. The amount of indoor air contaminants may also increase. Air-to-air heat-recovery equipment may be used to provide tempered ventilation air to tightly constructed houses. Outdoor air intakes connected to the return duct of central systems may also be used when lower installed costs are the most important factor. Minimum ventilation rates, as outlined in ASHRAE *Standard* 62, Ventilation for Acceptable Indoor Air Quality, should be maintained.

SINGLE-FAMILY RESIDENCES

HEATING EQUIPMENT

Heat Pumps

Heat pumps may be classified by thermal source and distribution medium in the heating mode. The most commonly used classes of heat pump equipment are air-to-air and water-to-air. Air-to-water and water-to-water types are also used.

Heat pump systems, as contrasted to the actual heat pump equipment, are generally described as air-source or ground-source. The thermal sink for cooling is generally assumed to be the same as the thermal source for heating. Air-source systems using ambient air as the heat source/sink are generally the least costly to install and thus the most commonly used. Ground-source systems usually employ water-to-air heat pumps to extract heat from the ground via groundwater or a buried heat exchanger.

As a heat source/sink, groundwater (from individual wells or supplied as a utility from community wells) offers the following advantages over ambient air: (1) heat pump capacity is independent of ambient air temperature, reducing supplementary heating requirements; (2) no defrost cycle is required; (3) for equal equipment rating point efficiency, the seasonal efficiency is usually higher for heating and for cooling; and (4) peak heating energy consumption is usually lower. Ground-coupled or surface-water-coupled systems offer the same advantages. However, they circulate brine or water to transfer heat from the ground via a buried or submerged heat exchanger. Direct expansion ground-source systems, using evaporators buried in the ground, are occasionally used. The total number of ground-source systems is growing rapidly, particularly of the ground-coupled type. Water-source systems that extract heat from surface water (e.g., lakes or rivers) or city (tap) water are also used where local conditions permit.

Water supply, quality, and disposal must be considered for ground-water systems. The ASHRAE publication *Design/Data Manual for Closed-Loop Ground-Coupled Heat Pump Systems* (Bose et al. 1985) provides detailed information on these subjects. Secondary coolants for ground-coupled systems are discussed in this manual and in Chapter 18 of the 1993 *ASHRAE Handbook—Fundamentals*. Buried heat exchanger configurations may be horizontal or vertical, with the vertical including both multiple-shallow- and single-deep-well configurations. Ground-coupled systems avoid water quality, quantity, and disposal concerns but are sometimes more expensive and may be less efficient than groundwater systems, if pumping power for the groundwater system is not excessive.

Heat pumps for single-family houses are normally unitary systems; i.e., they consist of single-package units or two or more factory-built modules as illustrated in Figure 2. These differ from applied or built-up heat pumps, which require field engineering to select compatible components for complete systems.

Most commercially available heat pumps (particularly in North America) are electrically powered. Supplemental heat is generally required at low outdoor temperatures or during defrost. In most cases, supplemental or backup heat is provided by electric resistance heaters, but add-on and unitary bivalent systems, which combine heat pumps with fuel-fired devices, are also used.

In add-on systems, a heat pump is added—often as a retrofit—to an existing furnace or boiler system. The heat pump and combustion device are operated in one of two ways: (1) alternately, depending on which is most cost-effective, or (2) in parallel. In unitary bivalent heat pumps, the heat pump and combustion device are grouped in a common chassis and cabinets to provide similar benefits at lower installation costs.

Extensive research and development is being conducted to develop fuel-fired heat pumps. They are just beginning to be marketed in North America.

Heat pumps may be equipped with desuperheaters (either integral or field-added) to reclaim heat for domestic water heating. Integrated space-conditioning and water-heating heat pumps with an additional full-size condenser for water heating are also available.

Furnaces

Furnaces are fueled by gas (natural or propane), electricity, oil, wood, or other combustibles. Gas, oil, and wood furnaces may draw combustion air from the indoor space or from the outdoors. If the furnace space is located such that combustion air is drawn from the outdoors, the arrangement is called an isolated combustion system (ICS). Furnaces are generally rated on an ICS basis. When outdoor air is ducted to the combustion chamber, the arrangement is called a direct vent system. This latter method is used for manufactured home applications and some mid- and high-efficiency equipment designs. Using outside air for combustion eliminates both the infiltration losses associated with the use of indoor air for combustion and the stack losses associated with atmospherically induced draft hood-equipped furnaces.

Two available types of high-efficiency gas furnaces are noncondensing and condensing. Both increase efficiency by adding or improving heat exchanger surface area and reducing heat loss during furnace off-times. The higher efficiency condensing type also recovers more energy by condensing water vapor from the combustion products. The condensate is developed in a high-grade stainless steel heat exchanger and is disposed of through a drain line. Condensing furnaces generally use PVC for vent pipes and condensate drains.

Wood-fueled furnaces are used in some areas. A recent advance in wood furnaces is the addition of catalytic converters to enhance the combustion process, increasing furnace efficiency and producing cleaner exhaust.

Hydronic Heating Systems—Boilers

With the growth of demand for central cooling systems, hydronic systems have declined in popularity in new construction, but still account for a significant portion of existing systems in the northern climates. The fluid is heated in a central boiler and distributed by piping to terminal units (fan coils, radiators, radiant panels, or base-board convectors) in each room. Most recently installed residential systems use a forced circulation, multiple zone hot water system with a series-loop piping arrangement. The equipment is described in Chapters 28 and 33 and hot water and steam heating systems in Chapters 10, 12, and 14 of the 1992 *ASHRAE Handbook—Systems and Equipment*.

Design water temperature is based on economic and comfort considerations. Generally, higher temperatures result in lower first costs because smaller terminal units are needed. However, losses tend to be greater, resulting in higher operating costs and reduced comfort due to the concentrated heat source. Typical design temperatures range from 80 to 95°C. For radiant panel systems, design temperatures range from 45 to 75°C. The preferred control system allows the water temperature to decrease as outdoor temperatures rise. Provisions for the expansion and contraction of the piping and heat distributing units and for the elimination of air from the system are essential for quiet, leaktight operation.

Fossil fuel systems that condense water vapor from the flue gases must be designed for return water temperatures in the range of 50 to 55°C for most of the heating season. Noncondensing systems must maintain high enough water temperatures in the boiler to prevent this condensation. If rapid heating is required, both terminal unit and boiler size must be increased, although gross oversizing should be avoided.

Zonal Heating Systems

Zonal systems offer the potential for lower operating costs, because unoccupied areas can be kept at lower temperatures in the winter and at higher temperatures in the summer. Common areas can be maintained at lower temperatures at night and sleeping areas at lower temperatures during the day.

One form of this system consists of individual heaters located in each room. These heaters are usually electric or gas-fired. Electric heaters are available in the following types: baseboard free-convection, wall insert (free-convection or forced-fan), radiant panels for walls and ceilings, and radiant cables for walls, ceilings, and floors. Matching equipment capacity to heating requirements is critical for individual room systems. Heating delivery cannot be adjusted by adjusting air or water flow, so greater precision in room-by-room sizing is needed.

Individual heat pumps for each room or group of rooms (zone) are another form of zonal electric heating. For example, two or more small unitary heat pumps can be installed in two-story or large one-story homes.

The multisplit heat pump consists of a central compressor and an outdoor heat exchanger to service up to eight indoor zones. Each zone uses one or more fan coils, with separate thermostatic control for each zone. Such systems are used in both new and retrofit construction.

A method for zonal heating in central ducted systems is the zone-damper system. This consists of individual zone dampers and thermostats and a zone control system. Both variable-air-volume (damper position proportional to zone demand) and on-off (damper fully open or fully closed in response to thermostat) types are available. Such systems sometimes include a provision to modulate to lower capacities when only a few zones require heating.

Solar Heating

Both active and passive solar energy systems are used to heat residences. In typical active systems, flat plate collectors heat air or water. Air systems distribute heated air either to the living space for immediate use or to a thermal storage medium (i.e., a rock pile). Water systems pass heated water through a secondary heat exchanger and store extra heat in a water tank. Due to low delivered water temperatures, radiant floor panels requiring moderate temperatures are generally used.

Trombe walls and sunspaces are two common passive systems. Glazing facing south (with overhanging eaves to reduce solar gains in the summer) and movable insulating panels reduce heating requirements.

Backup heating ability is generally needed with solar energy systems. Chapter 30 has information on sizing solar heating equipment.

AIR CONDITIONERS

Unitary Air Conditioners

In forced-air systems, the same air distribution duct system can be used for both heating and cooling. Split-system central cooling, as illustrated in Figure 1, is the most widely used forced-air system. Upflow, downflow, and horizontal airflow units are available. Condensing units are installed on a noncombustible pad outside and contain an electric motor-driven compressor, condenser, condenser fan and fan motor, and electrical controls. The condensing unit and evaporator coil are connected by refrigerant tubing that is normally field-supplied. However, precharged, factory-supplied tubing with quick-connect couplings is also common where the distance between components is not excessive.

A distinct advantage of split-system central cooling is that it can readily be added to existing forced-air heating systems. Airflow rates may need to exceed heating requirements to achieve good performance, but most existing heating duct systems are adaptable to cooling. Airflow rates of 45 to 60 L/s per kilowatt of refrigeration are normally recommended for good cooling performance. As with heat pumps, these systems may be fitted with desuperheaters for domestic water heating.

Some cooling equipment includes forced-air heating as an integral part of the product. Year-round heating and cooling packages with a gas, oil, or electric furnace for heating and an electrically driven vapor-compression system for cooling are available. Air-to-air and water-source heat pumps provide cooling and heating by reversing the flow of refrigerant.

Distribution Systems. Duct systems for cooling (and heating) should be designed and installed in accordance with accepted practice. Useful information is found in ACCA *Manuals* D and G and in Chapters 9 and 16 of the 1992 *ASHRAE Handbook—Systems and Equipment.*

Because weather is the primary influence on the load, the cooling load in each room changes from hour to hour. Therefore, the owner or occupant should be able to make seasonal or more frequent adjustments to the air distribution system to obtain improved comfort. Such adjustments may involve opening additional outlets in second-floor rooms during the summer and throttling or closing heating outlets in some rooms during the winter. Manually adjustable balancing dampers may be provided to facilitate these adjustments. Other possible refinements are the installation of a heating and cooling system sized to meet heating requirements, with additional self-contained cooling units serving rooms with high summer loads, or of separate central systems for the upper and lower floors of a house. On deluxe applications, zone-damper systems can be used.

Operating characteristics of both heating and cooling equipment must be considered when zoning is used. For example, a reduction in

the air quantity to one or more rooms may reduce the airflow across the evaporator to such a degree that frost forms on the fins. Reduced airflow on heat pumps during the heating season can cause overloading if airflow across the indoor coil is not maintained at above 45 L/s per kilowatt. Reduced air volume to a given room would reduce the air velocity from the supply outlet and could cause unsatisfactory air distribution in the room. Manufacturers of zonal systems normally provide guidelines for avoiding such situations.

Special Considerations. In split-level houses, cooling and heating are complicated by air circulation between various levels. In many such houses, the upper level tends to overheat in winter and undercool in summer. Multiple outlets, some near the floor and others near the ceiling, have been used with some success on all levels. To control airflow, the homeowner opens some outlets and closes others from season to season. Free circulation between floors can be reduced by locating returns high in each room and keeping doors closed.

In existing homes, the cooling that can be added is limited by the air-handling capacity of the existing duct system. While the existing duct system is usually satisfactory for normal occupancy, it may be inadequate during large gatherings. In all cases where new cooling (or heating) equipment is installed in existing homes, supply-air ducts and outlets must be checked for acceptable air-handling capacity and air distribution. Maintaining upward airflow at an effective velocity is important when converting existing heating systems with floor or baseboard outlets to both heat and cool. It is not necessary to change the deflection from summer to winter for registers located at the perimeter of a residence. Registers located near the floor on the inside walls of rooms may operate unsatisfactorily if the deflection is not changed from summer to winter.

Occupants of air-conditioned spaces usually prefer minimum perceptible air motion. Perimeter baseboard outlets with multiple slots or orifices directing air upwards effectively meet this requirement. Ceiling outlets with multidirectional vanes are also satisfactory.

A residence without a forced-air heating system may be cooled by one or more central systems with separate duct systems, by individual room air conditioners (window-mounted or through-the-wall), or by mini-split-room air conditioners.

Cooling equipment must be located carefully. Because cooling systems require higher indoor airflow rates than most heating systems, the sound levels generated indoors are usually higher. Thus, indoor air-handling units located near sleeping areas may require sound attenuation. Outdoor noise levels should also be considered when locating the equipment. Many communities have ordinances regulating the sound level of mechanical devices, including cooling equipment. Manufacturers of unitary air conditioners often certify the sound level of their products in an ARI program (ARI *Standard* 270). ARI *Standard* 275 gives information on how to predict the dBA sound level when the ARI sound rating number, the equipment location relative to reflective surfaces, and the distance to the property line are known.

An effective and inexpensive way to reduce noise is to put distance and natural barriers between sound source and listener. However, airflow to and from air-cooled condensing units must not be obstructed. Most manufacturers provide recommendations regarding acceptable distances between condensing units and natural barriers. Outdoor units should be placed as far as is practical from porches and patios, which may be used while the house is being cooled. Locations near bedroom windows and neighboring homes should also be avoided.

Evaporative Coolers

In dry climates, evaporative coolers can be used to cool residences. Further details on evaporative coolers can be found in Chapter 19 of the 1992 *ASHRAE Handbook—Systems and Equipment* and in Chapter 47 of this volume.

HUMIDIFIERS

For improved winter comfort, equipment that increases indoor relative humidity levels may be needed. In a ducted heating system, a central humidifier can be attached to or installed within a supply plenum or main supply duct, or installed between the supply and return duct systems. When applying supply-to-return duct humidifiers on heat pump systems, care should be taken to maintain proper airflow across the indoor coil. Self-contained humidifiers can be used in any residence. Even though this type of humidifier introduces all the moisture to one area of the home, moisture will migrate and raise humidity levels in other rooms. Overhumidification, which can cause condensate to form on the coldest surfaces in the living space (usually the windows), should be avoided.

Central humidifiers may be rated in accordance with ARI *Standard* 610. This rating is expressed in the number of liters per day evaporated by 60°C entering air. Some manufacturers certify the performance of their product to the ARI standard. Selecting the proper size humidifier is important and is outlined in ARI *Guideline F, Selection, Installation and Servicing of Residential Humidifiers.*

Since moisture migrates through all structural materials, vapor retarders should be installed near the inside surface of insulated walls, ceilings, and floors. Improper attention to this construction detail allows moisture to migrate from inside to outside, causing damp insulation, possible structural damage, and exterior paint blistering. Humidifier cleaning and maintenance schedules should be followed to maintain efficient operation and prevent bacteria buildup. Chapter 20 of the 1992 *ASHRAE Handbook—Systems and Equipment* contains more information on residential humidifiers.

AIR FILTERS

Most comfort conditioning systems that circulate air incorporate some form of air filter. Usually they are disposable or cleanable filters having relatively low air-cleaning efficiency. Higher efficiency alternatives include pleated media filters and electronic air filters. These high-efficiency filters may have high static pressure drops. The air distribution system should be carefully evaluated before the installation of such filters.

Air filters are mounted in the return air duct or plenum and operate whenever air circulates through the duct system. Air filters are rated in accordance with ARI *Standard* 680, Residential Air Filter Equipment, based on ASHRAE *Standard* 52, Method of Testing Air-Cleaning Devices Used in General Ventilation for Removing Particulate Matter. Atmospheric dust spot efficiency levels are generally less than 20% for disposable filters and vary from 60 to 90% for electronic air filters.

To maintain optimum performance, the collector cells of electronic air filters must be cleaned periodically. Automatic indicators are often used to signal the need for cleaning. Electronic air filters have higher initial costs than disposable or pleated filters, but generally last the life of the air-conditioning system. Chapter 25 of the 1992 *ASHRAE Handbook—Systems and Equipment* covers the design of residential air filters in more detail.

CONTROLS

Historically, residential heating and cooling equipment has been controlled by a wall thermostat. Today, simple wall thermostats with bimetallic strips are being replaced by microelectronic models that can set heating and cooling equipment at different temperature levels, depending on the time of day. This has led to night setback control to reduce energy demand and operation costs. For heat pump equipment, electronic thermostats can incorporate night setback with an appropriate scheme to limit use of resistance heat during recovery. Chapter 42 contains more details about automatic control systems.

MULTIFAMILY RESIDENCES

Attached homes and low-rise multifamily apartments generally use heating and cooling equipment comparable to that used in single-family dwellings. Separate systems for each unit allow individual control to suit the occupant and facilitate individual metering of energy use.

Central Forced-Air Systems

High-rise multifamily structures may also use unitary heating and cooling equipment comparable to that used in single-family dwellings. Equipment may be installed in a separate mechanical equipment room in the apartment, or it may be placed in a soffit or above a drop ceiling over a hallway or closet.

Small residential warm-air furnaces may also be used, but a means of providing combustion air and venting combustion products from gas- or oil-fired furnaces is required. It may be necessary to use a multiple-vent chimney or a manifold-type vent system. Local codes should be consulted. Direct vent furnaces that are placed near or on an outside wall are also available for apartments.

Another concept for multifamily residences (also applicable to single-family dwellings) is a combined water heating/space heating system that uses water from the domestic hot water storage tank to provide space heating. Water circulates from the storage tank to a hydronic coil in the system air handler. Space heating is provided by circulating indoor air across the coil. A split-system central air conditioner with the evaporator located in the system air handler can be included to provide space cooling.

Hydronic Central Systems

Individual heating and cooling units are not always possible or practical in high-rise structures. In this case, applied central systems are used. Hydronic central systems of the two- or four-pipe type are widely used in high-rise apartments. Each dwelling unit has either individual room units located at the perimeter or interior, or ducted fan-coil units.

The most flexible hydronic system with the lowest operating costs is the four-pipe type, which provides heating or cooling for each apartment dweller. The two-pipe system is less flexible in that it cannot provide heating and cooling simultaneously. This limitation causes problems during the spring and fall when some apartments in a complex require heating while others require cooling due to solar or internal loads. This spring/fall problem may be overcome by operating the two-pipe system in a cooling mode and providing the relatively low amount of heating that may be required by means of individual electric resistance heaters.

Through-the-Wall Units

Through-the-wall room air conditioners, packaged terminal air conditioners (PTACs), and packaged terminal heat pumps (PTHPs) give the highest flexibility for conditioning single rooms. Each room with an outside wall may have such a unit. These units are used extensively in the renovation of old buildings because they are self-contained and do not require complex piping or ductwork renovation.

Room air conditioners have integral controls and may include resistance or heat pump heating. PTACs and PTHPs have special indoor and outdoor appearance treatments, making them adaptable to a wider range of architectural needs. PTACs can include gas, electric resistance, hot water, or steam heat. Integral or remote wall-mounted controls are used for both PTACs and PTHPs. Further information may be found in Chapter 47 of the 1992 *ASHRAE Handbook—Systems and Equipment* and in ARI *Standards* 310 and 380.

Water-Loop Heat Pump Systems

Any mid- or high-rise structure having interior zones with high internal heat gains that necessitate year-round cooling can efficiently use a water-loop heat pump system. Such systems have the flexibility and control of a four-pipe system while using only two pipes. Water-source heat pumps allow for the individual metering of each apartment. The building owner pays only the utility cost (which can be prorated) for the circulating pump, cooling tower, and supplemental boiler heat. Economics permitting, solar or ground heat energy can provide the supplementary heat in lieu of a boiler. The ground can also provide a heat sink, which in some cases can eliminate the cooling tower.

Special Concerns for Apartment Buildings

Many ventilation systems are used in apartment buildings. Local building codes generally govern air quantities. ASHRAE *Standard* 62 requires minimum outdoor air values of 25 L/s intermittent or 10 L/s continuous or operable windows for baths and toilets, and 50 L/s intermittent or 12 L/s continuous or operable windows for kitchens.

In some buildings with centrally controlled exhaust and supply systems, the systems are operated on time clocks for certain periods of the day. In other cases, the outside air is reduced or shut off during extremely cold periods. If known, these factors should be considered when estimating heating load.

Buildings using exhaust and supply air systems 24 h a day may benefit from air-to-air heat recovery devices (see Chapter 44 of the 1992 *ASHRAE Handbook—Systems and Equipment*). Such recovery devices can reduce energy consumption by transferring 40 to 80% of the sensible and latent heat between the exhaust air and supply air streams.

Infiltration loads in high-rise buildings without ventilation openings for perimeter units are not controllable on a year-round basis by general building pressurization. When outer walls are pierced to supply outdoor air to unitary or fan-coil equipment, combined wind and thermal stack effects create other infiltration problems.

Interior public corridors in apartment buildings need positive ventilation with at least two air exchanges per hour. Conditioned supply air is preferable. Some designs transfer air into the apartments through acoustically lined louvers to provide kitchen and toilet makeup air, if necessary. Supplying air to, instead of exhausting air from, corridors minimizes odor migration from apartments into corridors.

Air-conditioning equipment must be isolated to reduce noise generation or transmission. The design and location of cooling towers must be chosen to avoid disturbing occupants within the building and neighbors in adjacent buildings. An important load, frequently overlooked, is heat gain from piping for hot water services.

In large apartment houses, a central panel may allow individual apartment air-conditioning systems or units to be monitored for maintenance and operating purposes.

MANUFACTURED HOMES

Manufactured homes are constructed at a factory and constitute over 7% of all housing units and about 25% of all new single-family homes sold each year. In the United States, heating and cooling systems in manufactured homes, as well as other facets of construction such as insulation levels, are regulated by HUD Manufactured Home Construction and Safety Standards. Each complete home or home section is assembled on a transportation frame—a chassis with wheels and axles—for transport. Manufactured homes vary in size from small, single-floor section units starting at 37 m^2 to large, multiple sections, which when joined together provide over 230 m^2 and have the same appearance as site-constructed homes.

Heating systems are factory-installed and are primarily forced air downflow units feeding main supply ducts built into the subfloor, with floor registers located throughout the home. A small percentage of homes sited in the far South and in the Southwest use upflow units feeding overhead ducts in the attic space. Typically there is no return duct system. Air returns to the air handler from each room

through hallways. The complete heating system is a reduced clearance type (mm) with the air-handling unit installed in a small closet or alcove usually located in a hallway. Sound control measures may be required if large forced-air systems are installed close to sleeping areas. Gas, oil and electric furnaces or heat pumps may be installed by the manufacturer of the homes to satisfy market requirements.

Gas and oil furnaces are compact direct vent types that have been approved for installation in a manufactured home. The special venting arrangement used is a vertical through-the-roof concentric pipe-in-pipe system that draws all air for combustion directly from the outdoors and discharges the combustion products through a windproof vent terminal. Gas furnaces must be easily convertible from liquefied petroleum to natural gas and back as required at the final site.

Manufactured homes may be cooled with add-on split or single-package air-conditioning systems when the supply ducts are adequately sized and rated for that purpose according to HUD requirements. The split-system evaporator coil may be installed in the integral coil cavity provided with the furnace. A high static pressure blower is used to overcome resistance through the furnace, the evaporator coil and the compact air duct distribution system. Supply air from a single-package air conditioner is connected with flexible air ducts to feed existing factory in-floor or overhead ducts. Dampers or other means are required to prevent the cooled, conditioned air from backflowing through a furnace cabinet.

A typical installation of a downflow gas or oil furnace with a split-system air conditioner is illustrated in Figure 3. Air returns to the furnace directly through the hallway (1), passing through a louvered door (2) on the front of the furnace. The air then passes through air filters (3) and is drawn into the top-mounted blower (4), which during the winter months forces air down over the heat

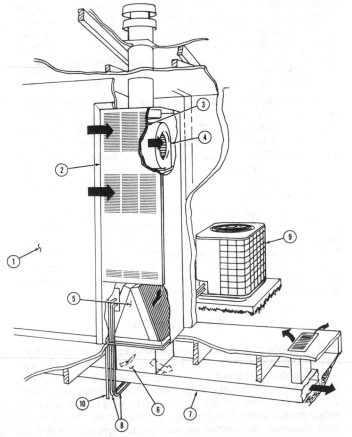

Fig. 3 Typical Installation of a Heating and Cooling System for Manufactured Homes

exchanger, where it picks up heat. For summer cooling, the blower forces air through the split-system evaporator coil (5), which absorbs heat and moisture from the passing air. During heating and cooling, the conditioned air then passes through a combustible floor base and duct connector (6), before flowing into the floor air distribution duct (7). The evaporator coil is connected via quick-connect refrigerant lines (8) to a remote air-cooled condensing unit (9). The condensate collected at the evaporator is drained by a flexible hose (10), which is routed to the outdoors through the floor construction and connected to a suitable drain.

REFERENCES

ACCA. 1970. Selection of distribution systems. *Manual* G. Air-Conditioning Contractors of America, Washington, DC.

ACCA. 1995. Duct design for residential winter and summer air conditioning and equipment selection. *Manual* D, 3rd ed. Air-Conditioning Contractors of America, Washington, DC.

ACCA. 1986. Load calculation for residential winter and summer air conditioning. *Manual* J, 7th ed. Air-Conditioning Contractors of America, Washington, DC.

ARI. 1984. Application of sound rated outdoor unitary equipment. *Standard* 275-84. Air Conditioning and Refrigeration Institute, Arlington, VA.

ARI. 1984. Sound rating of outdoor unitary equipment. *Standard* 270-84. Air Conditioning and Refrigeration Institute, Arlington, VA.

ARI. 1988. Selection, installation and servicing of residential humidifiers. *Guideline* F-1988. Air Conditioning and Refrigeration Institute, Arlington, VA.

ARI. 1989. Central system humidifiers for residential applications. *Standard* 610-89. Air Conditioning and Refrigeration Institute, Arlington, VA.

ARI. 1990. Packaged terminal air-conditioners. *Standard* 310-90. Air Conditioning and Refrigeration Institute, Arlington, VA.

ARI. 1990. Packaged terminal heat pumps. *Standard* 380-90. Air Conditioning and Refrigeration Institute, Arlington, VA.

ARI. 1993. Residential air filter equipment. *Standard* 680-93. Air Conditioning and Refrigeration Institute, Arlington, VA.

ASHRAE. 1976. Method of testing air-cleaning devices used in general ventilation for removing particulate matter. *Standard* 52-1976.

ASHRAE. 1989. Ventilation for acceptable indoor air quality. *Standard* 62-1989.

ASHRAE. 1993. Energy-efficient design of new low-rise residential buildings. *Standard* 90.2-1993.

Bose, J.E., J.D. Parker, and F.C. McQuiston. 1985. Design/Data manual for closed-loop ground-coupled heat pump systems. ASHRAE.

CHAPTER 2

RETAIL FACILITIES

THIS chapter covers the design and application of air-conditioning and heating systems for various retail merchandising facilities. Load calculations, systems, and equipment are covered elsewhere in the Handbook series.

GENERAL CRITERIA

To apply equipment properly, it is necessary to know the construction of the space to be conditioned, its use and occupancy, the time of day in which greatest occupancy occurs, the physical building characteristics, and the lighting layout.

The following must also be considered:

- Electric power—size of service
- Heating—availability of steam, hot water, gas, oil, or electricity
- Cooling—availability of chilled water, well water, city water, and water conservation equipment
- Internal heat gains
- Rigging and delivery of equipment
- Structural considerations
- Obstructions
- Ventilation—opening through roof or wall for outdoor air duct, number of doors to sales area, and exposures
- Orientation of store
- Code requirements
- Utility rates and regulations
- Building standards

Specific design requirements, such as the increase in outdoor air required for exhaust systems where lunch counters exist, must be considered. The requirements of ASHRAE ventilation standards must be followed. Heavy smoking and objectionable odors may necessitate special filtering in conjunction with outdoor air intake and exhaust. Load calculations should be made using the procedure outlined in Chapter 26 of the 1993 *ASHRAE Handbook—Fundamentals*.

In almost all localities there is some form of energy code in effect that establishes strict requirements for insulation, equipment efficiencies, system designs, and so forth, and places strict limits on fenestration and lighting. The requirements of ASHRAE *Standard* 90 should be met as a minimum guideline for retail facilities.

The selection and design of HVAC systems for retail facilities are normally determined by economics. First cost is usually the determining factor for small stores; for large retail facilities, operating and maintenance costs are also considered. Generally, decisions

about mechanical systems for retail facilities are based on a cash flow analysis rather than on a full life-cycle analysis.

SMALL STORES

The large glass areas found at the front of many small stores may cause high peak solar heat gain unless they have northern exposures. High heat loss may be experienced on cold, cloudy days. The HVAC system for this portion of the small store should be designed to offset the greater cooling and heating requirements. Entrance heaters may be needed in cold climates.

Many new small stores are part of a shopping center. While exterior loads will differ between stores, the internal loads will be similar; the need for proper design is important.

DESIGN CONSIDERATIONS

System Design

Single-zone unitary rooftop equipment is common in store air conditioning. The use of multiple units to condition the store involves less ductwork and can maintain comfort in the event of partial equipment failure. Prefabricated and matching curbs simplify installation and ensure compatibility with roof materials.

The heat pump, offered as packaged equipment, readily adapts to small-store applications and has a low first cost. Winter design conditions, utility rates, and operating costs should be compared to those for conventional heating systems before this type of equipment is chosen.

Water-cooled unitary equipment is available for small-store air conditioning, but many communities in the United States have restrictions on the use of city water for condensing purposes and require the installation of a cooling tower system. Water-cooled equipment generally operates efficiently and economically.

Retail facilities often have a high sensible heat gain relative to the total heat gain. Unitary HVAC equipment should be designed and selected to provide the necessary sensible heat removal.

Air Distribution

The external static pressures available in small-store air-conditioning units are limited, and duct systems should be designed to keep duct resistances low. Duct velocities should not exceed 6 m/s and pressure drops should not exceed 0.8 Pa/m. Average air quantities range from 47 to 60 L/s per kilowatt of cooling in accordance with the calculated internal sensible heat load.

The preparation of this chapter is assigned to TC 9.8, Large Building Air-Conditioning Applications.

Attention should be paid to suspended obstacles, such as lights and displays, that interfere with proper air distribution.

The duct system should contain enough dampers for air balancing. Dampers should be installed in the return duct and in the outdoor air duct for proper outdoor air/return air balance. Volume dampers should be installed in takeoffs from the main supply duct to balance air to the branch ducts.

Control System

Controls for small-store systems should be kept as simple as possible while still able to perform the required functions. Unitary equipment is typically available with manufacturer-supplied controls for ease of installation and operation.

Automatic dampers should be placed in the outdoor air intake to prevent outdoor air from entering when the fan is turned off.

Heating controls vary with the nature of the heating medium. Duct heaters are generally furnished with manufacturer-installed safety controls. Steam or hot water heating coils require a motorized valve for heating control.

Time clock control can limit unnecessary HVAC system operation. Unoccupied reset controls should be provided in conjunction with timed control.

Maintenance

To protect the initial investment and ensure maximum efficiency, the maintenance of air-conditioning units in small stores should be contracted out to a reliable service company on a yearly basis. The contract should clearly specify responsibility for filter replacements, lubrication, belts, coil cleaning, adjustment of controls, compressor maintenance, replacement of refrigerant, pump repairs, electrical maintenance, winterizing, system startup, and extra labor required for repairs.

Improving Operating Costs

Outdoor air economizers can reduce the operating cost of cooling systems in most climates. They are generally available as factory options or accessories with roof-mounted units.

Increased exterior insulation will generally reduce operating energy requirements and may in some cases allow a reduction in the size of the equipment to be installed. Various types of insulation are available for roofs, ceilings, masonry walls, frame walls, slabs, and foundation walls. Many codes now include minimum requirements for insulation and fenestration materials.

DISCOUNT AND OUTLET STORES

Discount and outlet stores feature a wide range of merchandise and often include a large lunch counter, an auto service area, and a garden shop. Some stores sell pets, including fish and birds. This variety of merchandise must be considered in designing an air-conditioning system.

In addition to the sales area, such areas as stockrooms, rest rooms, offices, and special storage rooms for perishable merchandise may require air conditioning or refrigeration.

The design and application suggestions for small stores also apply to discount and outlet stores. The following information should be considered in addition.

LOAD DETERMINATION

Operating economics and the spaces served often dictate the indoor design conditions for discount and outlet store air conditioning. Some stores may base summer load calculations on higher inside temperatures (e.g., 27°C db) but then set the thermostats to control at 22 to 24°C db. This reduces the installed equipment size while providing the desired inside temperature most of the time.

Special rooms for storage of perishable goods are usually designed with separate unitary air conditioners.

The heat gain from lighting will not be uniform throughout the entire area. Some areas such as jewelry and other specialty displays have lighting heat gains as high as 65 to 85 W per square metre of floor area. For the entire sales area, an average value of 20 to 40 W/m^2 is typical. For stockrooms and receiving, marking, toilet, and rest room areas, a value of 20 W/m^2 may be used. When available, actual lighting layouts rather than average values should be used for load computation.

The store owner usually determines the population density for a store based on its location, size, and past experience.

Food preparation and service areas in discount and outlet stores range from small lunch counters having heat-producing equipment (ranges, griddles, ovens, coffee urns, toasters) within the conditioned space to large deluxe installations with kitchens separate from the conditioned space. See Chapter 28, Kitchen Ventilation, for more specific information on HVAC systems for kitchen and eating spaces.

Data on the heat released by special merchandising equipment, such as amusement rides for children or equipment used for preparing speciality food items (e.g., popcorn, pizza, frankfurters, hamburgers, doughnuts, roasted chickens, cooked nuts, etc.), should be obtained from the equipment manufacturers.

Ventilation and outdoor air must be provided as required in ASHRAE standards and local codes.

DESIGN CONSIDERATIONS

Discount and outlet stores are generally constructed with a large open sales area and a partial glass storefront. The heat released by the installed lighting is usually sufficient to offset the design roof heat loss. Therefore, the interior areas of these stores need cooling during business hours throughout the year. The perimeter areas, especially the storefront and entrance areas, may have highly variable heating and cooling requirements. Proper zone control and HVAC design are essential. The location of checkout lanes in this area makes proper environmental control even more important.

System Design

The important factors in selecting discount and outlet store air-conditioning systems are (1) installation costs, (2) floor space required for equipment, (3) maintenance requirements and equipment reliability, and (4) simplicity of control. Roof-mounted units are the most commonly used.

Air Distribution

The air supply for large sales areas should generally be designed to satisfy the primary cooling requirement. In designing air distribution for the perimeter areas, the variable heating and cooling requirements must be considered.

Control System

The control system should be simple, dependable, and fully automatic because it is usually operated by personnel who have little knowledge of air-conditioning systems. The system should be as simple to operate as a residential system. Modern unitary equipment has automatic electronic controls for ease of operation.

Maintenance

Most stores do not employ trained maintenance personnel; they rely instead on service contracts with either the installer or a local service company. For suggestions on lowering operating costs, see the section on Small Stores.

SUPERMARKETS

LOAD DETERMINATION

Heating and cooling loads should be calculated using the methods outlined in the 1993 *ASHRAE Handbook—Fundamentals*. Data for calculating the loads due to people, lights, motors, and heat-producing equipment should be obtained from the store owner or manager or from the equipment manufacturer. In supermarkets, space conditioning is required both for human comfort and for proper operation of refrigerated display cases. The air-conditioning unit should introduce a minimum quantity of outdoor air. This quantity is either the volume required for ventilation or the volume required to maintain slightly positive pressure in the space, whichever is larger.

Many supermarkets are units of a large chain owned or operated by a single company. The standardized construction, layout, and equipment used in designing many similar stores simplify load calculations.

It is important that the final air-conditioning load be correctly determined. Refer to manufacturers' data for information on the total heat extraction, sensible heat, latent heat, and percentage of latent to total load for display cases. Engineers report considerable fixture heat removal (case load) variation as the relative humidity and temperature vary in comparatively small increments. Relative humidity above 55% substantially increases the load, while reduced relative humidity substantially decreases the load, as shown in Figure 1. Trends in store design, which include more food refrigeration and more efficient lighting, reduce the sensible component of the load even further.

To calculate the total load and percentage of latent and sensible heat that the air conditioning must handle, the refrigerating effect imposed by the display fixtures must be subtracted from the building's gross air-conditioning requirements.

Modern supermarket designs have a high percentage of closed refrigerated display fixtures. These vertical cases have large glass display doors and greatly reduce the problem of latent and sensible heat removal from the occupied space. The doors do, however, require heaters to prevent condensation and fogging. These heaters should be cycled based on some kind of automatic control.

DESIGN CONSIDERATIONS

Store owners and operators frequently complain about cold aisles, heating systems that operate even when the outdoor tempera-

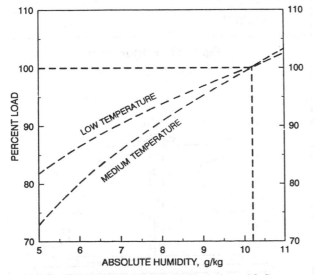

**Fig. 1   Refrigerated Case Load Variation with Store
Air Humidity**

ture is above 21°C, and air-conditioning systems that operate infrequently. These problems are usually attributed to spillover of cold air from open refrigerated display equipment.

Although refrigerated display equipment may be the cause of cold stores, the problem is not due to excessive spillover or improperly operating equipment. Heating and air-conditioning systems must compensate for the effects of open refrigerated display equipment. Design considerations include the following:

1. Increased heating requirements due to removal of large quantities of heat, even in summer.
2. Net air-conditioning load after deducting the latent and sensible refrigeration effect (Item 1). The load reduction and change in sensible-latent load ratio have a major effect on equipment selection.
3. Need for special air circulation and distribution to offset the heat removed by open refrigerating equipment.
4. Need for independent temperature and humidity control.

Each of these problems is present to some degree in every supermarket, although situations vary with climate and store layout. Methods of overcoming these problems are discussed in the following sections. Extremely high energy costs may ensue if the year-round air-conditioning system has not been designed to compensate for the effects of refrigerated display equipment.

Heat Removed by Refrigerated Display Equipment

The display refrigerator not only cools a displayed product but envelops it in a blanket of cold air that absorbs heat from the room air in contact with it. Approximately 80 to 90% of the heat removed from the room by vertical refrigerators is absorbed through the display opening. Thus, the open refrigerator acts as a large air cooler, absorbing heat from the room and rejecting it via the condensers outside the building. Occasionally, this conditioning effect can be greater than the design air-conditioning capacity of the store. The heat removed by the refrigeration equipment *must* be considered in the design of the air-conditioning and heating systems because this heat is being removed constantly, day and night, summer and winter, regardless of the store temperature.

The display cases increase the heating requirement of the building such that heat is often required at unexpected times. The following example is an indication of the extent of this cooling effect. The desired store temperature is 24°C. Store heat loss or gain is assumed to be 8 kW per K of temperature difference between outdoor and store temperature. (This value varies with store size, location, and exposure.) The heat removed by refrigeration equipment is 56 kW. (This value varies with the number of refrigerators.) The latent heat removed is assumed to be 19% of the total, leaving 81% or 45.4 kW sensible heat removed, which will cool the store 45.4/8 = 5.7 K. By constantly removing sensible heat from its environment, the refrigeration equipment in this store will cool the store 5.7 K below outdoor temperature in winter and in summer. Thus, in mild climates, heat must be added to the store to maintain comfort conditions.

The designer has the choice of discarding or reclaiming the heat removed by refrigeration. If economics and store heat data indicate that the heat should be discarded, heat extraction from the space must be included in the heating load calculations. If this internal heat loss is not included, the heating system may not have sufficient capacity to maintain the design temperature under peak conditions.

The additional sensible heat removed by the cases may change the air-conditioning latent load ratio from 32% to as much as 50% of the net heat load. Removal of a 50% latent load by means of refrigeration alone is very difficult. Normally, it requires specially designed equipment with reheat or chemical adsorption.

Multishelf refrigerated display equipment requires 55% rh or less. In the dry-bulb temperature ranges of average stores, humidity in excess of 55% can cause heavy coil frosting, product zone frosting in low-temperature cases, fixture sweating, and substantially increased refrigeration power consumption.

A humidistat can be used during summer cooling to control humidity by transferring heat from the condenser to a heating coil in the airstream. The store thermostat maintains proper summer temperature conditions. Override controls prevent conflict between the humidistat and the thermostat.

The equivalent result can be accomplished with a conventional air-conditioning system by using three- or four-way valves and reheat condensers in the ducts. This system borrows heat from the standard condenser and is controlled by a humidistat. For higher energy efficiency, specially designed equipment should be considered. Desiccant dehumidifiers have also been used.

Humidity

Cooling from the refrigeration equipment does not preclude the need for air conditioning. On the contrary, it increases the need for humidity control.

With increases in store humidity, heavier loads are imposed on the refrigeration equipment, operating costs rise, more defrost periods are required, and the display life of products is shortened. The dew point rises with the relative humidity, and sweating can become profuse to the extent that even nonrefrigerated items such as shelving superstructures, canned products, mirrors, and walls may sweat.

There are two commonly used devices for achieving humidity control. One is an electric vapor compression air conditioner, which overcools the air to condense the moisture and then reheats it to a comfortable temperature. Condenser waste heat is often used. Vapor compression holds humidity levels at 50 to 55% rh.

The second dehumidification device is a desiccant dehumidifier. A desiccant absorbs or adsorbs moisture directly from the air to its surface. The desiccant material is reactivated by passing hot air at 80 to 110°C through the desiccant base. Condenser waste heat can provide as much as 40% of the heat required. Desiccant systems in supermarkets can maintain humidity in the lower ranges of the comfort index. Lower humidity results in lower operating costs for the refrigerated cases.

System Design

The same air-handling equipment and distribution system are generally used for both cooling and heating. The entrance area is the most difficult section to heat. Many supermarkets in the northern United States are built with vestibules provided with separate heating equipment to temper the cold air entering from the outdoors. Auxiliary heat may also be provided at the checkout area, which is usually close to the front entrance. Methods of heating entrance areas include the use of (1) air curtains, (2) gas-fired or electric infrared radiant heaters, and (3) waste heat from the refrigeration condensers.

Air-cooled condensing units are the most commonly used in supermarkets. Typically, a central air handler conditions the entire sales area. Specialty areas like bakeries, computer rooms, or warehouses are better served with a separate air handler because the loads in these areas vary and require different control than the sales area.

Most installations are made on the roof of the supermarket. If air-cooled condensers are located on the ground outside the store, they must be protected against vandalism as well as truck and customer traffic. If water-cooled condensers are used on the air-conditioning equipment and a cooling tower is required, provisions should be made to prevent freezing during winter operation.

Air Distribution

Designers overcome the concentrated load at the front of a supermarket by discharging a large portion of the total air supply into the front third of the sales area.

The air supply to the space with the vapor compression system has typically been 5 L/s per square metre of sales area. This value

should be calculated based on the sensible and latent internal loads. The desiccant system typically requires due to its high moisture removal rate. In most cases, only about 40% of the circulation rate passes through the dehumidifier.

Being more dense, the air cooled by the refrigerators settles to the floor and becomes increasingly colder, especially in the first 900 mm above the floor. If this cold air remains still, it causes discomfort and serves no purpose, even when other areas of the store need more cooling. Cold floors or areas in the store cannot be eliminated by the simple addition of heat. Any reduction of air-conditioning capacity without circulation of the localized cold air would be analogous to installing an air conditioner without a fan. To take advantage of the cooling effect of the refrigerators and provide an even temperature in the store, the cold air must be mixed with the general store air.

To accomplish the necessary mixing, air returns should be located at floor level; they should also be strategically placed to remove the cold air near concentrations of refrigerated fixtures. The returns should be designed and located to avoid the creation of drafts. There are two general solutions to this problem:

1. **Return Ducts in Floor.** This is the preferred method and can be accomplished in two ways. The floor area in front of the refrigerated display cases is the coolest area. Refrigerant lines are run to all of these cases, usually in tubes or trenches. If the trenches or tubes are enlarged and made to open under the cases for air return, air can be drawn in from the cold area (see Figure 2). The air is returned to the air-handling unit through a tee connection to the trench before it enters the back room area. The opening through which the refrigerant lines enter the back room should be sealed.

If refrigerant line conduits are not used, the air can be returned through inexpensive underfloor ducts. If refrigerators have insufficient undercase air passage, the manufacturer should be consulted. Often they can be raised off the floor approximately 40 mm. Floor trenches can also be used as ducts for tubing, electrical supply, and so forth.

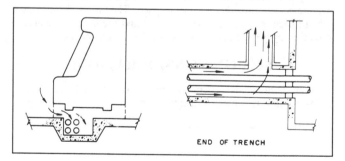

Fig. 2   Floor Return Ducts

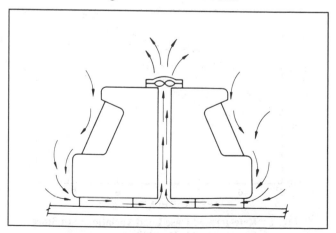

Fig. 3   Air Mixing Using Fans Behind Cases

Floor-level return relieves the problem of localized cold areas and cold aisles and uses the cooling effect for store cooling, or increases the heating efficiency by distributing the air to the areas that need it most.

2. **Fans Behind Cases.** If ducts cannot be placed in the floor, circulating fans can draw air from the floor and discharge it above the cases (see Figure 3). While this approach prevents objectionable cold aisles in front of the refrigerated display cases, it does not prevent the area with a concentration of refrigerated fixtures from remaining colder than the rest of the store.

Control Systems

The control system should be as simple as practicable so that store personnel are required only to change the position of a selector switch in order to start or stop the system or change it from heating to cooling or from cooling to heating. Control systems for heat recovery applications are more complex and should be coordinated with the equipment manufacturer.

Maintenance and Heat Reclamation

Most supermarkets, except large chains, do not employ trained maintenance personnel, but rather rely on service contracts with either the installer or a local service company. This relieves the store management of the responsibility of keeping the air-conditioning system in proper operating condition.

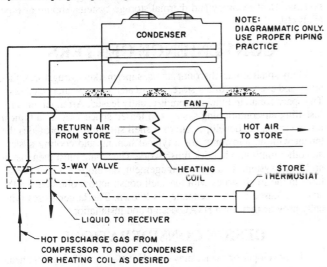

Fig. 4   Heat Reclaiming Systems

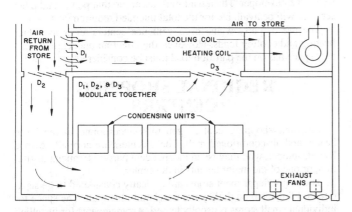

Fig. 5   Machine Room with Automatic Temperature Control Interlocked with Store Temperature Control

The heat extracted from the store and the heat of compression may be reclaimed for heating cost savings. One method of reclaiming the rejected heat is to use a separate condenser coil located in the air conditioner's air-handling system, either alternately or in conjunction with the main refrigeration condensers, to provide heat as required (see Figure 4). Another system uses water-cooled condensers and delivers its rejected heat to a water coil in the air handler.

The heat rejected by conventional machines using air-cooled condensers may be reclaimed by proper duct and damper design (see Figure 5). Automatic controls can either reject this heat to the outdoors or recirculate it through the store.

DEPARTMENT STORES

Department stores vary in size, type, and location, so air-conditioning system design should be particular to each store. An ample minimum quantity of outdoor air reduces or eliminates odor problems. Essential features of a quality system include (1) an automatic control system properly designed to compensate for load fluctuations, (2) zoned air distribution to maintain uniform conditions under shifting loads, and (3) use of outdoor air for cooling during intermediate seasons and peak sales periods. It is also desirable to adjust indoor temperatures for variations in outdoor temperatures. While the close control of humidity is not necessary, a properly designed system should operate to maintain relative humidity at 50% or below with a corresponding dry-bulb temperature of 26°C. This humidity limit eliminates musty odors and retards perspiration, particularly in fitting rooms.

LOAD DETERMINATION

Because the occupancy (except store personnel) is transient, indoor conditions are commonly set not to exceed 26°C db and 50% rh at outdoor summer design conditions, and 21°C db at outdoor winter design conditions. Winter humidification is seldom used in store air conditioning.

The number of customers and store personnel normally found on each conditioned floor must be ascertained, particularly in specialty departments or other areas having a greater-than-average concentration of occupants. Lights should be checked for wattage and type. Table 1 gives approximate values for lighting in various areas; Table 2 gives approximate occupancies.

Other loads, such as those from motors, beauty parlor and restaurant equipment, and any special display or merchandising equipment, should be determined.

Table 1   Approximate Lighting Loads for Department Stores

Area	W/m^2
Basement	30 to 50
First floor	40 to 70
Upper floors, women's wear	30 to 50
Upper floors, house furnishings	20 to 30

Table 2   Approximate Occupancy for Department Stores

Area	m^2 per person
Basement, metropolitan area	2 to 9
Basement, other with occasional peak	2 to 9
First floor, metropolitan area	2 to 7
First floor, suburban	2 to 7
Upper floors, women's wear	4.6 to 9
Upper floors, house furnishings	9 or more

The minimum outdoor air requirement should be as defined in the ASHRAE Ventilation Standards, which are generally acceptable and adequate for removing odors and keeping the building atmosphere fresh. However, local ventilation ordinances may require greater quantities of outdoor air.

Paint shops, alteration rooms, rest rooms, eating places, and locker rooms should be provided with positive exhaust ventilation, and their requirements must be checked against local codes.

DESIGN CONSIDERATIONS

Before performing load calculations, the designer should examine the store arrangement to determine what will affect the load and the system design. For existing buildings, a survey can be made of actual construction, floor arrangement, and load sources. For new buildings, an examination of the drawings and a discussion with the architect or owner will be required.

Larger stores may contain beauty parlors, restaurants, lunch counters, or auditoriums. These special areas may operate during all store hours. If one of these areas has a load that is small in proportion to the total load, the load may be met by the portion of the air-conditioning system serving that floor. If present or future operation could for any reason be compromised by such a strategy, this space should be served by a separate air-conditioning system. Because of the concentrated load in the beauty parlor, a separate air distribution system should be provided for this area.

The restaurant, because of its required service facilities, is generally centrally located. It is often used only during the noon hours. For control of odors, a separate air-handling system should be considered. Future plans for the store must be ascertained because they can have a great effect on the type of air-conditioning and refrigeration systems to be used.

System Design

Air-conditioning systems for department stores may be of the unitary or central-station type. The selection of the system should be based on owning and operating costs as well as any other special considerations for the particular store, such as store hours, load variations, and size of load.

Large department stores often use central-station systems consisting of air-handling units having chilled water cooling coils, hot water heating coils, fans, and filters. Air systems must have adequate zoning for varying loads, occupancy, and usage. Wide variations in people loads may justify consideration of variable volume air distribution systems. The water chilling and heating plants distribute water to the various air-handling systems and zones and may take advantage of some load diversity throughout the building.

Air-conditioning equipment should not be placed in the sales area; instead, it should be located in ceiling, roof, and mechanical equipment room areas whenever practicable. Maintenance and operation of the system after installation are important considerations in the location of equipment.

Air Distribution

All buildings must be studied for orientation, wind exposure, construction, and floor arrangement. These factors affect not only load calculations, but also zone arrangements and duct locations. In addition to entrances, wall areas with significant glass, roof areas, and population densities, the expected locations of various departments (e.g., the lamp department) should be considered. Flexibility must be left in the duct design to allow for future movement of departments. The preliminary duct layout should also be checked in regard to winter heating to determine any special situations. It is usually necessary to design separate air systems for entrances, particularly in northern areas. This is also true for storage areas where cooling is not contemplated.

Air curtains may be installed at entrance doorways to limit or prevent infiltration of unconditioned air, at the same time providing greater ease of entry.

Control Systems

The necessary extent of automatic control depends on the type of installation and the extent to which it is zoned. The system must be controlled so that correctly conditioned air is delivered to each zone. Outdoor air intake should be automatically controlled to operate at minimum cost.

Partial or full automatic control should be provided for the cooling system to compensate for load fluctuations. Completely automatic refrigeration plants are now practical and should be carefully considered.

Maintenance

Most department stores employ personnel for routine operations and maintenance, but rely on service and preventive maintenance contracts for refrigeration cycles, chemical treatment, central plant systems, and major repairs.

Improving Operating Costs

Outdoor air economizer systems can reduce the operating cost of the cooling system in most climates. These are generally available as factory options or accessories with the air-handling units or control systems. Heat recovery and thermal storage systems should also be analyzed.

CONVENIENCE CENTERS

Many small stores, discount stores, supermarkets, drugstores, theaters, and even department stores are located in convenience centers. The space for an individual store is usually leased. Arrangements for installing air-conditioning systems in leased space vary. In a typical arrangement, the developer builds a shell structure and provides the tenant with an allowance for a typical heating and cooling system and other minimum interior finish work. The tenant must then install an HVAC system. In another arrangement, developers install HVAC units in the small stores with the shell construction, often before the space is leased or the occupancy is known. The larger stores typically provide their own HVAC design and installation.

DESIGN CONSIDERATIONS

The developer or owner may establish standards for typical heating and cooling systems that may or may not be sufficient for the tenant's specific requirements. The tenant may therefore have to install systems of different sizes and types than originally allowed for by the developer. The tenant must ascertain that power and other services will be available for the total intended requirements.

The use of party walls in convenience centers tends to reduce heating and cooling loads. However, the effect an unoccupied adjacent space has on the partition load must be considered.

REGIONAL SHOPPING CENTERS

Regional shopping centers generally incorporate an enclosed heated and air-conditioned mall. These centers are normally owned by a developer, who may be an independent party, a financial institution, or one of the major tenants in the center.

The major department stores are typically considered to be separate buildings, although they are attached to the mall. The space for individual small stores is usually leased. Arrangements for installing air-conditioning systems in the individually leased spaces vary, but are similar to those for small stores in convenience centers.

Table 3 Typical Installed Capacity and Energy Usage in Enclosed Mall Centers[a]
Based on 1979 Data—Midwestern United States

Type of Space	Installed Cooling W/m²	Annual Consumption, MJ/m²			
		Lighting[b]	Cooling[c]	Heating[d]	Miscellaneous
Candy store	141 to 246	926 to 1147	302 to 911	372 to 221	93 to 2716
Clothing store	118 to 144	546 to 965	310 to 372	345 to 198	47 to 260
Fast food	152 to 246	647 to 1256	380 to 911	372 to 147	1476 to 2713
Game room	107 to 140	264 to 484	279 to 283	535 to 182	8 to 535
Gen. merchandise	104 to 138	512 to 903	244 to 376	244 to 217	54 to 360
Gen. service	124 to 159	608 to 674	302 to 357	465 to 229	295 to 318
Gift store	115 to 161	496 to 887	244 to 388	275 to 190	16 to 85
Grocery	219 to 272	422 to 748	205 to 345	233 to 271	721 to 814
Jewelry	169 to 209	1442 to 1740	411 to 477	217 to 140	295 to 360
Mall	95 to 151	209 to 465	322 to 531	446 to 217	8 to 58
Restaurant	126 to 167	194 to 853	302 to 477	736 to 647	667 to 814
Shoe store	116 to 160	791 to 1244	287 to 442	279 to 178	39 to 89
Center average	95 to 151	698 to 1085	310 to 581	233 to 116	39 to 116

[a]Operation of center assumed to be 12 h/day and 6.5 days/week.
[b]Lighting includes miscellaneous and receptacle loads.
[c]Cooling includes blower motor and is for unitary-type system.
[d]Heating includes blower and is for electric resistance heating.

Table 3 presents typical data that can be used as check figures and field estimates. However, this table should not be used for final determination of load, since the values are only averages.

DESIGN CONSIDERATIONS

The owner provides the air-conditioning system for the enclosed mall. The mall system may use a central plant or unitary equipment. The owner generally requires that the individual tenant stores connect to a central plant system and includes charges for heating and cooling in the rent. Where unitary systems are used, the owner generally requires that the individual tenant install a unitary system of similar design.

The owner may establish standards for typical heating and cooling systems that may or may not be sufficient for the tenant's specific requirements. Therefore, the tenant may have to install systems of different sizes than originally allowed for by the developer.

Leasing arrangements may include provisions that have a detrimental effect on conservation (such as allowing excessive lighting and outdoor air or deleting requirements for economizer systems). The designer of HVAC systems for tenants in a shopping center must be fully aware of the lease requirements and work closely with leasing agents to guide these systems toward better energy efficiency.

Many regional shopping centers now contain specialty food court areas that require special considerations for odor control, outdoor air requirements, kitchen exhaust, heat removal, and refrigeration equipment.

System Design

Regional shopping centers vary widely in physical arrangement and architectural design. Single-level and smaller centers usually use unitary systems for mall and tenant air conditioning; multilevel and larger centers usually use central plant systems. The owner sets the design of the mall system and generally requires that similar systems be installed for tenant stores.

A typical central system may distribute chilled air to the individual tenant stores and to the mall air-conditioning system and employ variable volume control and electric heating at the local use point. Some systems distribute both hot and chilled water. All-air systems have also been used that distribute chilled or heated air to the individual tenant stores and to the mall air-conditioning system and employ variable volume control at the local use point. The central

systems provide improved efficiency and better overall economics of operation. They also provide the basic components required for smoke control systems.

Air Distribution

Air distribution for individual stores should be designed for a particular space occupancy. Some tenant stores maintain a negative pressure relative to the mall for odor control.

The air distribution system should maintain a slight positive pressure relative to atmospheric pressure and a neutral pressure relative to most of the individual tenant stores. Exterior entrances should have vestibules with independent heating systems.

Smoke management systems are required by many building codes. Air distribution systems should be designed to easily accommodate smoke control requirements.

Maintenance

Methods for ensuring the operation and maintenance of HVAC systems in regional shopping centers are similar to those used in department stores. Individual tenant stores may have to provide their own maintenance.

Improving Operating Costs

Methods for lowering operating costs in shopping centers are similar to those used in department stores. Some shopping centers have successfully employed cooling tower heat exchanger economizers.

Central plant systems for regional shopping centers typically have much lower operating costs than unitary systems. However, the initial cost of the central plant system is typically higher.

MULTIPLE-USE COMPLEXES

Multiple-use complexes are being developed in most metropolitan areas. These complexes generally combine retail facilities with other facilities such as offices, hotels, residences, or other commercial space into a single site. This consolidation of facilities into a single site or structure provides benefits such as improved land use; structural savings; more efficient parking; utility savings; and opportunities for more efficient electrical, fire protection, and mechanical systems.

LOAD DETERMINATION

The various occupancies may have peak HVAC demands that occur at different times of the day or even of the year. Therefore, the HVAC loads of the various occupancies should be determined independently. Where a combined central plant is under consideration, a block load should also be determined.

DESIGN CONSIDERATIONS

Retail facilities are generally located on the lower levels of multiple-use complexes, and other commercial facilities are on upper levels. Generally, the perimeter loads of the retail portion differ from those of the other commercial spaces. Greater lighting and population densities also make the HVAC demands for the retail space different from those for the other commercial space.

The differences in HVAC characteristics for the various occupancies within a multiple-use complex indicate that separate air handling and distribution should be used for the separate spaces. However, combining the heating and cooling requirements of the various facilities into a central plant can achieve substantial savings. A combined central heating and cooling plant for a multiple-use complex also provides good opportunities for heat recovery, thermal storage, and other similar functions that may not be economical in a single-use facility.

Many multiple-use complexes have atriums. The stack effect created by atriums requires special design considerations for tenants and space on the main floor. Areas near entrances require special considerations to prevent drafts and accommodate extra heating requirements.

System Design

Individual air-handling and distribution systems should be designed for the various occupancies. The central heating and cooling plant may be sized for the block load requirements, which may be less than the sum of each occupancy's demand.

Control Systems

Multiple-use complexes typically require a centralized control system. This may be dictated by requirements for fire and smoke control, security, remote monitoring, billing for central facilities use, maintenance control, building operations control, and energy management.

CHAPTER 3

COMMERCIAL AND PUBLIC BUILDINGS

THIS chapter is organized into eight sections. The information in the first section, General Criteria, applies to all buildings and includes information on load characteristics, design concepts, and design criteria. The section on Design Criteria covers factors that may apply to all building types. The factors include comfort level; costs; local conditions and requirements; and fire, smoke, and odor control.

The remaining sections present information applicable to specific buildings: dining and entertainment centers, office buildings, libraries and museums, bowling centers, communication centers, transportation centers, and warehouses.

GENERAL CRITERIA

In theory, if properly applied, every system can be successful in any building. However, in practice, such factors as initial and operating costs, space allocation, architectural design, location, and the engineer's evaluation and experience limit the proper choices for a given building type.

Heating and air-conditioning systems that are simple in design and of the proper size for a given building generally have fairly low maintenance and operating costs. For optimum results, as much inherent thermal control as is economically possible should be built into the basic structure. Such control might include materials with high thermal properties, insulation, and multiple or special glazing and shading devices.

The relationship between the shape, orientation, and air-conditioning capacity of a building should also be considered. Since the exterior load may vary from 30 to 60% of the total air-conditioning load when the fenestration area ranges from 25 to 75% of the exterior envelope surface area, it may be desirable to minimize the perimeter area. For example, a rectangular building with a four-to-one aspect ratio requires substantially more refrigeration than a square building with the same floor area.

Proper design also considers controlling noise and minimizing pollution of the atmosphere and water into which the system will discharge. The quality of the indoor air is also a major factor to consider in design.

Retrofitting of existing buildings is also an important part of the construction industry because of increased costs of construction and the necessity of reducing energy consumption. Table 1 lists factors to consider before selecting a system for any building. The system is often chosen by the owner and may not be based on an engineering study. To a great degree, system selection is made based on the engineer's ability to relate those factors involving higher first cost or lower life-cycle cost and benefits that have no calculable monetary value.

Some buildings are constructed with only heating and ventilating systems. For these buildings, greater design emphasis should be placed on natural or forced ventilation systems to minimize occupant discomfort during hot weather. If such buildings are to provide

for future cooling, humidification, or both, the design principles are the same as those for a fully air-conditioned building.

Load Characteristics

It is important to understand the load characteristics for the building to assure that systems can respond adequately at part load as well as at full load. Systems must be capable of responding to large fluctuations in occupancy or to different occupancy schedules in different parts of the building. Areas in the building such as data or communications centers may have large 24-h equipment loads.

Any building analyzed for heat recovery or total energy systems requires sufficient load profile and load duration information on all forms of building input to (1) properly evaluate the instantaneous effect of one on the other when no energy storage is contemplated and (2) evaluate short-term effects (up to 48 h) when energy storage is used.

Load profile curves consist of appropriate energy loads plotted against the time of day. Load duration curves indicate the accumulated number of hours at each load condition, from the highest to the lowest load for a day, a month, or a year. The area under load profile and load duration curves for corresponding periods is equivalent to the load multiplied by the time. These calculations must consider the type of air and water distribution systems in the building.

Load profiles for two or more energy forms during the same operating period may be compared to determine load-matching characteristics under diverse operating conditions. For example, when thermal energy is recovered from a diesel-electric generator at a rate equal to or less than the thermal energy demand, the energy can be used instantaneously, avoiding waste. But it may be worthwhile to store thermal energy when it is generated at a greater rate than demanded. A load profile study helps determine the economics of thermal storage.

Similarly, with internal source heat recovery systems, load matching must be integrated over the operating season with the aid of load duration curves for overall feasibility studies. These curves are useful in energy consumption analysis calculations as a basis for hourly input values in computer programs (see Chapter 28 of the 1993 *ASHRAE Handbook—Fundamentals*).

Aside from environmental considerations, the economic feasibility of district heating and cooling systems is influenced by load density and diversity factors for branch feeds to buildings along distribution mains. For example, the load density or energy per unit length of distribution main can be small enough in a complex of low-rise, lightly loaded buildings located at a considerable distance from one another, to make a central heating, cooling, or heating and cooling plant uneconomical.

Concentrations of internal loads peculiar to each application are covered later in this chapter and in Chapters 25 through 27 of the 1993 *ASHRAE Handbook—Fundamentals*.

Design Concepts

If a structure is characterized by several exposures and multi-purpose use, especially with wide load swings and noncoincident

---

The preparation of this chapter is assigned to TC 9.8, Large Building Air-Conditioning Applications.

Table 1 General Design Criteria[a, b]

General Category	Specific Category	Inside Design Conditions		Air Movement	Circulation, Air Changes per Hour
		Winter	Summer		
Dining and Entertainment Centers	Cafeterias and Luncheonettes	21 to 23°C 20 to 30% rh	26°C[e] 50% rh	0.25 m/s at 1.8 m above floor	12 to 15
	Restaurants	21 to 23°C 20 to 30% rh	23 to 26°C 55 to 60% rh	0.13 to 0.15 m/s	8 to 12
	Bars	21 to 23°C 20 to 30% rh	23 to 26°C 50 to 60% rh	0.15 m/s at 1.8 m above floor	15 to 20
	Nightclubs	21 to 23°C 20 to 30% rh	23 to 26°C 50 to 60% rh	below 0.13 m/s at 1.5 mabove floor	20 to 30
	Kitchens	21 to 23°C	29 to 31°C	0.15 to 0.25 m/s	12 to 15[h]
Office Buildings		21 to 23°C 20 to 30% rh	23 to 26°C 50 to 60% rh	0.13 to 0.23 m/s 4 to 10 L/s·m²	4 to 10
Libraries and Museums	Average	20 to 22°C 40 to 55% rh		below 0.13 m/s	8 to 12
	Archival	See Special Considerations in the section on Libraries and Museums		below 0.13 m/s	8 to 12
Bowling Centers		21 to 23°C 20 to 30% rh	24 to 26°C 50 to 55% rh	0.25 m/s at 1.8 m above floor	10 to 15
Communication Centers	Telephone Terminal Rooms	22 to 26°C 40 to 50% rh	22 to 26°C 40 to 50% rh	0.13 to 0.15 m/s	8 to 20
	Teletype Centers	21 to 23°C 40 to 50% rh	23 to 26°C 45 to 55% rh	0.13 to 0.15 m/s	8 to 20
	Radio and Television Studios	23 to 26°C 30 to 40% rh	23 to 26°C 40 to 55% rh	below 0.13 m/s at 3.7 mabove floor	15 to 40
Transportation Centers	Airport Terminals	21 to 23°C 20 to 30% rh	23 to 26°C 50 to 60% rh	0.13 to 0.15 m/s at 1.8 m above floor	8 to 12
	Ship Docks	21 to 23°C 20 to 30% rh	23 to 26°C 50 to 60% rh	0.13 to 0.15 m/s at 1.8 m above floor	8 to 12
	Bus Terminals	21 to 23°C 20 to 30% rh	23 to 26°C 50 to 60% rh	0.13 to 0.15 m/s at 1.8 m above floor	8 to 12
	Garages[i]	4 to 13°C	26 to 38°C	0.15 to 0.38 m/s	4 to 6 Refer to NFPA
Warehouses		Inside design temperatures for warehouses often depend on the materials stored.			1 to 4

Table 1 General Design Criteria[a, b] (Concluded)

Noise[c]	Filtering Efficiencies[d]	Load Profile	General
NC 40 to 50[f]	35% or better	Peak at 1 to 2 p.m.	Prevent draft discomfort for patrons waiting in serving lines
NC 35 to 40	35% or better	Peak at 1 to 2 p.m.	
NC 35 to 50	Use charcoal for odor control with manual purge control for 100% outside air to exhaust ±35% prefilters	Peak at 5 to 7 p.m.	
NC 35 to 45[g]	Use charcoal for odor control with manual purge control for 100% outside air to exhaust ±35% prefilters	Peak after 8 p.m., off from 2 a.m. to 4 p.m.	Provide good air movement but prevent cold draft discomfort for dancing patrons
NC 40 to 50	10 to 15% or better		Negative air pressure required for odor control[j]
NC 30 to 45	35 to 60% or better	Peak at 4 p.m.	
NC 35 to 40	35 to 60% or better	Peak at 3 p.m.	
NC 35	35% prefilters plus charcoal filters 85 to 95% final[k]	Peak at 3 p.m.	
NC 40 to 50	10 to 15%	Peak at 6 to 8 p.m.	
to NC 60	85% or better	Varies with location and use	Constant temperature and humidity required
NC 40 to 50	85%	Varies with location and use	
NC 15 to 25	35% or better	Varies widely due to changes in lighting and people	
NC 35 to 50	35% or better and charcoal filters	Peak at 10 a.m. to 9 p.m.	Positive air pressure required in terminal
NC 35 to 50	10 to 15%	Peak at 10 a.m. to 5 p.m.	Positive air pressure required in waiting area
NC 35 to 50	35% with exfiltration	Peak at 10 a.m. to 5 p.m.	Positive air pressure required in terminal
NC 35 to 50	10 to 15%	Peak at 10 a.m. to 5 p.m.	Negative air pressure required to remove fumes
up to 75	10 to 35%	Peak at 10 a.m. to 3 p.m.	

Notes to Table 1, General Design Criteria

[a]This table shows design criteria differences between various commercial and public buildings. It should not be used as the sole source for design criteria. Each type of data contained here can be determined from the *ASHRAE Handbooks* and *Standards*.

[b]Governing codes should be consulted to determine minimum allowable requirements. Outdoor air requirements may be reduced if high-efficiency adsorption equipment or other odor- or gas-removal equipment is used, but should never be below 2.5 L/s per person (see Chapter 34). Also, see Chapter 12 of the 1993 *ASHRAE Handbook—Fundamentals* and ASHRAE *Standard* 62.

[c]Refer to Chapter 43.

[d]Average Atmospheric Dust Spot Efficiency (see ASHRAE *Standard* 52 for method of testing).

[e]Food in these areas is often eaten more quickly than in a restaurant, so that the turnover of diners is much faster. Because diners seldom remain for long periods, they do not require the degree of comfort necessary in restaurants. Thus, it may be possible to lower design criteria standards and still provide reasonably comfortable conditions. Although space conditions of 26.7°C and 50% rh may be satisfactory for patrons when it is 35°C and 50% rh outside, indoor conditions of 25.6°C and 40% rh are better.

[f]Cafeterias and luncheonettes usually have some or all of the food preparation equipment and trays in the same room with the diners. These eating establishments are generally noisier than restaurants, so that noise transmission from the air-conditioning equipment is not as critical.

[g]In some nightclubs, the noise from the air-conditioning system must be kept low so that all patrons can hear the entertainment.

[h]Usually determined by kitchen hood requirements.

[i]Peak kitchen heat load does not generally occur at peak dining load, although in luncheonettes and some cafeterias where cooking is done in the dining areas, peaks may be simultaneous.

[j]See Chapter 28, Kitchen Ventilation.

[k]Methods for removal of chemical pollutants must also be considered.

[l]Also includes service stations.

energy use in certain areas, multiunit or unitary systems may be considered for such areas, but not necessarily for the entire building. The benefits of transferring heat absorbed by cooling from one area to other areas, processes, or services that require heat may enhance the selection of such systems. Systems such as incremental closed loop heat pumps may be cost-effective.

When the cost of energy is included in the rent (rent-included) with no means for permanent or check-metering, tenants tend to consume excess energy. This energy waste raises operating costs for the owner, decreases profitability, and has a detrimental effect on the environment. While design features can minimize excess energy penalties, they seldom eliminate waste. For example, the U.S. Department of Housing and Urban Development nationwide field records for total-electric housing show that rent-included dwellings use approximately 20% more energy than those directly metered by a public utility company.

Diversity factor benefits for central heating and cooling in rent-included buildings may result in lower building demand and connected loads. However, energy waste may easily result in load factors and annual energy consumption exceeding that of buildings where the individual has a direct economic incentive to reduce energy consumption.

Design Criteria

In many applications, design criteria are fairly evident, but in all cases, the engineer should understand the owner's and user's intent, because any single factor may influence system selection. The engineer's personal experience and judgment in the projection of future needs may be a better criterion for system design than any other single factor.

Comfort Level. Comfort, as measured by temperature, humidity, air motion, air quality, noise, and vibration, is not identical for all buildings, occupant activities, or use of space. The control of static electricity may be a consideration in humidity control.

For spaces with a high population density, or with a sensible heat factor less than 0.75, a lower dry-bulb temperature reduces the generation of latent heat. Reduced latent heat may further reduce the need for reheat and save energy. Therefore, an optimum temperature should be determined.

Costs. Owning and operating costs can affect system selection and seriously conflict with other criteria. Therefore, the engineer must help the owner resolve these conflicts by considering factors such as the cost and availability of different fuels, the ease of equipment access, and maintenance requirements.

Local Conditions. Local, state, and national codes and regulations and environmental concerns must be considered in the design. Chapters 23 and 24 of the 1993 *ASHRAE Handbook—Fundamentals* give information on calculating the effects of weather in specific areas.

Automatic Temperature Control. Proper automatic temperature control maintains occupant comfort during varying internal and external loads. Improper temperature control may mean a loss of customers in restaurants and other public buildings. An energy management control system can be combined with a building automation system to allow the owner to manage energy, lighting, security, fire protection, and other similar systems from one central control system. Chapters 32 and 42 include more details.

Fire, Smoke, and Odor Control. Fire and smoke can easily spread through elevator shafts, stairwells, ducts, and other means. Although an air-conditioning system can spread fire and smoke by (1) fan operation, (2) penetrations required in walls or floors, and (3) the stack effect without fan circulation, a properly designed and installed system can be a positive means of fire and smoke control.

Knowledge of effective techniques for positive control after a fire starts is limited but increasing (see Chapter 48). Effective attention to fire and smoke control also helps prevent odor migration into unventilated areas (see Chapter 41). The ventilation system design should include careful consideration of applicable NFPA pamphlets, especially pamphlets 90A and 96.

DINING AND ENTERTAINMENT CENTERS

Load Characteristics

Air conditioning of restaurants, cafeterias, bars, and nightclubs presents common load problems encountered in comfort conditioning, with additional factors pertinent to dining and entertainment applications. Such factors include

- Extremely variable loads with high peaks, in many cases, occurring twice daily
- High sensible and latent heat gains because of gas, steam, and electric appliances, people, and food
- Localized high sensible and latent heat gains in dancing areas
- Unbalanced conditions in restaurant areas adjacent to kitchens which, although not part of the conditioned space, still require special attention
- Heavy infiltration of outdoor air through doors during rush hours
- Smoking versus nonsmoking areas

Internal heat and moisture loads come from occupants, motors, lights, appliances, and infiltration. Separate calculations should be made for patrons and employees. The sensible and latent heat load must be proportioned in accordance with the design temperature selected for both sitting and working people, because the latent to sensible heat ratio for each category decreases as the room temperature decreases.

Hoods required to remove heat from appliances may also substantially reduce the space latent loads.

Infiltration is a considerable factor in many restaurant applications because of short occupancy and frequent door use. It is increased by the need for large quantities of makeup air, which should be provided by mechanical means to replace air exhausted through hoods and for smoke removal. Wherever possible, vestibules or revolving doors should be installed to reduce such infiltration.

Design Concepts

The following factors influence system design and equipment selection:

- High concentration of food, body, and tobacco-smoke odors require adequate ventilation with proper exhaust facilities.
- Step control of refrigeration plants gives satisfactory and economical operation under reduced loads.
- Air exhausted at the ceiling removes smoke and odor.
- Building design and space limitations often favor one equipment type over another. For example, in a restaurant having a vestibule with available space above it, air conditioning with condensers and evaporators remotely located above the vestibule may be satisfactory. Such an arrangement saves valuable space, even though self-contained units located within the conditioned space may be somewhat lower in initial cost and maintenance cost. In general, small cafeterias, bars, and the like, with loads up to 35 kW, can be most economically conditioned with packaged units; larger and more elaborate establishments require central plants.
- If not required for kitchen exhaust, air required by an air-cooled or evaporative condenser may be drawn from the conditioned space. This eliminates the need for operating additional exhaust fans and also improves overall plant efficiency because the temperature of air entering the condenser is lower. Proper water treatment is necessary if an evaporative condenser is used. It may also be necessary to bypass air when air is not required.
- Smaller restaurants with isolated plants usually use direct-expansion systems.
- Mechanical humidification typically is not provided due to high internal latent loads.
- Some air-to-air heat recovery equipment can reduce the energy required for heating and cooling ventilation air. Chapter 44 of the 1992 ASHRAE Handbook—Systems and Equipment includes details. Careful consideration must be given to the potential for grease to condense on heat recovery surfaces.

Since eating and entertainment centers generally have low sensible heat factors and require high ventilation rates, fan-coil and induction systems are usually not applicable. All-air systems are more suitable. Space must be established for ducts, except for small systems with no ductwork. Large establishments are often served by central chilled water systems.

In cafeterias and luncheonettes, the air distribution system must keep food odors at the serving counters away from areas where patrons are eating. This usually means heavy exhaust air requirements at the serving counters, with air supplied into and induced from eating areas. Exhaust air must also remove the heat from hot trays, coffee urns, and ovens to minimize patron and employee discomfort and reduce air-conditioning loads. These factors often create greater air-conditioning loads for cafeterias and luncheonettes than for restaurants.

Odor Removal. Air transferred from dining areas into the kitchen keeps odors and heat out of dining areas and cools the kitchen. Outdoor air intake and kitchen exhaust louvers should be located so that exhaust air is neither drawn back into the system nor causes discomfort to passersby.

Where odors may possibly be drawn back into dining areas, activated charcoal filters, air washers, or ozonators may be used to remove odors. Kitchen, locker room, toilet, or other odoriferous air should not be recirculated unless air purifiers are used.

Kitchen Air Conditioning. If planned in the initial design phases, kitchens can often be air conditioned effectively, without excessive cost. It is not necessary to meet the same design criteria as for dining areas, but kitchen temperatures can be reduced significantly. The relatively large number of people and food loads in dining and kitchen areas produce a high latent load. Additional cooling required to eliminate excess moisture increases refrigeration plant, cooling coils, and air-handling equipment size.

Self-contained units, advantageously located, with air distribution designed so as not to produce drafts off hoods and other equipment, can be used to spot cool intermittently. The costs are not excessive, and kitchen personnel efficiency can be improved greatly.

Even in climates with high wet-bulb temperatures, evaporative cooling may be a good compromise between the expense of air conditioning and the lack of comfort in ventilated kitchens.

Special Considerations

In establishing design conditions, the duration of individual patron occupancy should be considered, because patrons entering from outdoors are more comfortable in a room with a high temperature than those who remain long enough to become acclimated. Nightclubs and deluxe restaurants are usually operated at a lower effective temperature than cafeterias and luncheonettes.

Often, the ideal design condition must be rejected for an acceptable condition because of equipment cost or performance limitations. Restaurants are frequently affected in this way, because ratios of latent to sensible heat may result in uneconomical equipment selection, unless a combination of lower design dry-bulb temperature and higher relative humidity (which gives an equal effective temperature) is selected.

In severe climates, entrances and exits in any dining establishment should be completely shielded from diners to prevent drafts. Vestibules provide a measure of protection. However, both vestibule doors are often open simultaneously. Revolving doors or local means for heating or cooling infiltration air may be provided to offset drafts.

Employee comfort is difficult to maintain at a uniform level because of (1) temperature differences between the kitchen and dining room and (2) the motion of employees, while patrons are seated. Since customer satisfaction is essential to a dining establishment's success, patron comfort is the primary consideration. However, maintenance of satisfactory temperature and atmospheric conditions for customers also helps alleviate employee discomfort.

One problem in dining establishments is the use of partitions to separate areas into modular units. Partitions create such varied load conditions that individual modular unit control is generally necessary.

Baseboard radiation or convectors, if required, should be located so as not to overheat patrons. This is difficult to achieve in some layouts because of movable chairs and tables. For these reasons, it is desirable to enclose all dining room and bar heating elements in insulated cabinets with top outlet grilles and baseboard inlets. With heating elements located under windows, this practice has the additional advantage of directing the heat stream to combat window downdraft and air infiltration. Separate smoking and nonsmoking areas should be considered. The smoking area should be exhausted or served by separate air-handling equipment. Air diffusion device selection and placement should minimize smoke migration toward nonsmoking areas.

Restaurants. In restaurants, people are seated and served at tables, while food is generally prepared in remote areas. This type of dining is usually enjoyed in a leisurely and quiet manner, so the ambient atmosphere should be such that the air conditioning is not noticed. Where specialized or open cooking is to be a feature, provisions should be made for the control and handling of cooking odors and smoke.

Bars. Bars are often part of a restaurant or nightclub. If they are establishments on their own, they often serve food as well as drinks, and they should be classified as restaurants, with food preparation in remote areas. Alcoholic beverages produce pungent vapors, which must be drawn off. In addition, smoking at bars is generally considerably heavier than in restaurants. Therefore, outdoor air requirements are relatively high by comparison.

Nightclubs. Nightclubs may include a restaurant, bar, stage, and dancing area. The bar should be treated as a separately zoned area, with its own supply and exhaust system. People in the restaurant area who dine and dance may require twice the air changes and cooling required by patrons who dine and then watch a show. The length of stay in nightclubs generally exceeds that encountered in most eating places. In addition, eating in nightclubs is usually secondary to drinking and smoking. Patron density usually exceeds that of conventional eating establishments.

Kitchens. The kitchen has the greatest concentration of noise, heat load, smoke, and odors; ventilation is the chief means of removing them and preventing these objectionable elements from entering dining areas. To assure odor control, kitchen air pressure should be kept negative relative to other areas. Maintenance of reasonably comfortable working conditions is important. For more information, see Chapter 28, Kitchen Ventilation.

OFFICE BUILDINGS

Load Characteristics

Office buildings usually include both peripheral and interior zone spaces. The peripheral zone may be considered as extending from 3 to 3.6 m inward from the outer wall toward the interior of the building and frequently has a large window area. These zones may be extensively subdivided. Peripheral zones have variable loads because of changing sun position and weather. These zone areas typically require heating in winter. During intermediate seasons, one side of the building may require cooling, while another side requires heating. However, the interior zone spaces usually require a fairly uniform cooling rate throughout the year because their thermal loads are derived almost entirely from lights, office equipment, and people. Interior space conditioning is often by systems that have variable air volume control for low or no-load conditions.

Most office buildings are occupied from approximately 8:00 a.m. to 6:00 p.m.; many are occupied by some personnel from as early as 5:30 a.m. to as late as 7:00 p.m. Some tenants' operations may require night work schedules, usually not to extend beyond 10:00 p.m. Office buildings may contain printing plants, communications operations, broadcasting studios, and computing centers, which could operate 24 h per day. Therefore, for economical air-conditioning design, the intended uses of an office building must be very well established before design development.

Occupancy varies considerably. In accounting or other sections where clerical work is done, the maximum density is approximately one person per 7 m² of floor area. Where there are private offices, the density may be as little as one person per 19 m². The most serious cases, however, are the occasional waiting rooms, conference rooms, or directors' rooms where occupancy may be as high as one person per 1.9 m².

The lighting load in an office building constitutes a significant part of the total heat load. Lighting and normal equipment electrical loads average from 20 to 50 W/m², but may be considerably higher, depending on the type of lighting and the amount of equipment. Buildings with computer systems and other electronic equipment can have electrical loads as high as 50 to 110 W/m². An accurate appraisal should be made of the amount, size, and type of computer equipment anticipated for the life of the building to size the air-handling equipment properly and provide for future installation of air-conditioning apparatus.

Where electrical loading is 65 W/m² or more, heat should be withdrawn from the source by exhaust air or water tubing. About 30% of the total lighting heat output from recessed fixtures can be withdrawn by exhaust or return air and, therefore, will not enter into space conditioning supply air requirements. By connecting a duct to each fixture, the most balanced air system can be provided. However, this method is expensive, so the suspended ceiling is often used as a return air plenum with the air drawn from the space to above the suspended ceiling through the lights.

Miscellaneous allowances (for fan heat, duct heat pickup, duct leakage, and safety factors) should not exceed 12% of the total load.

Building shape and orientation are often determined by the building site, but certain variations in these factors can produce increases of 10 to 15% in the refrigeration load. Shape and orientation should, therefore, be carefully analyzed in the early design stages.

Design Concepts

The variety of functions and range of design criteria applicable to office buildings have allowed the use of almost every available air-conditioning system. While multistory structures are discussed here, the principles and criteria are similar for all sizes and shapes of office buildings.

Attention to detail is extremely important, especially in modular buildings. Each piece of equipment, duct and pipe connections, and the like may be duplicated hundreds of times. Thus, seemingly minor design variations may substantially affect construction and operating costs. In initial design, each component must be analyzed not only as an entity, but also as part of an integrated system. This systems design approach is essential to achieve optimum results.

There are several classes of office buildings, determined by the type of financing required and the tenants who will occupy the building. Design evaluation may vary considerably based on specific tenant requirements; it is not enough to consider typical floor patterns only. Included in many larger office buildings are stores, restaurants, recreation facilities, data centers, telecommunication centers, radio and television studios, and observation decks.

Built-in system flexibility is essential for office building design. Business office procedures are constantly being revised, and basic building services should be able to meet changing tenant needs.

The type of occupancy may have an important bearing on the selection of the air distribution system. For buildings with one owner or lessee, operations may be defined clearly enough so that a system can be designed without the degree of flexibility needed for a less well-defined operation. However, owner-occupied buildings may require considerable design flexibility, because the owner will pay for all alterations. The speculative builder can generally charge alterations to tenants. When different tenants occupy different floors, or even parts of the same floor, the degree of design and operation complexity increases to ensure proper environmental comfort conditions to any tenant, group of tenants, or all tenants at once. This problem is more acute if tenants have seasonal and variable overtime schedules.

Stores, banks, restaurants, and entertainment facilities may have hours of occupancy or design criteria that differ substantially from those of office buildings; therefore, they should have their own air distribution systems and, in some cases, their own refrigeration equipment.

Main entrances and lobbies are sometimes served by a separate system because they buffer the outside atmosphere and the building interior. Some engineers prefer to have a lobby summer temperature 2.3 to 3.4 K above office temperature to reduce thermal shock to people entering or leaving the building. This also reduces operating costs.

The unique temperature and humidity requirements of data processing system installations, and the fact that they often run 24 h per day for extended periods, generally warrant separate refrigeration and air distribution systems. Separate backup systems may be required for data processing areas in case the main building HVAC system fails. Chapter 16 has further information.

The degree of air filtration required should be determined. The service cost and the effect air resistance has on energy costs should be analyzed for various types of filters. Initial filter cost and air pollution characteristics also need to be considered. Activated charcoal filters for odor control and reduction of outdoor air requirements are another option to consider.

There is seldom justification for continuous 100% outdoor air systems for office buildings; therefore, most office buildings are designed to minimize outdoor air usage, except during economizer operation. However, recent attention to indoor air quality may dictate higher levels of ventilation air. In addition, the minimum volume of outdoor air should be maintained in variable volume air-handling systems. Dry-bulb or enthalpy controlled economizer cycles should be considered for reducing energy costs.

When an economizer cycle is used, systems should be zoned so that energy waste will not occur by heating outside air. This is often accomplished by a separate air distribution system for the interior and each major exterior zone.

High-rise office buildings have traditionally used perimeter dual-duct, induction, or fan-coil systems. Where fan-coil or induction systems have been installed at the perimeter, separate all-air systems have been generally used for the interior. More recently, variable air volume systems and deluxe self-contained perimeter unit systems have also been used. With variable air volume systems serving the interior, perimeter variable volume dual-duct or baseboard radiation or infrared systems are being used at the perimeter. Baseboard and infrared panels may be hydronic or electric.

Many office buildings without an economizer cycle have a bypass multizone unit installed on each floor, with a heating coil in each exterior zone duct. Variable air volume variations of the bypass multizone and other floor-by-floor, all-air systems are also being used. These systems are popular because of low fan power, low initial cost, and energy savings, resulting from independent operating schedules, which are possible between floors occupied by tenants with different operating hours.

It may be more economical for smaller office buildings to have perimeter radiation or infrared systems with conventional, single-duct, low-velocity air-conditioning systems furnishing air from packaged air-conditioning units or multizone units. The need for a perimeter system should be carefully analyzed, as this system is a function of exterior glass percentage, external wall thermal value, and climate severity.

A perimeter heating system separate from the cooling system is preferable, since then air distribution devices can then be selected for a specific duty, rather than a compromise between heating and cooling performance. The higher cost of additional air-handling or fan-coil units and ductwork may lead the designer to a less expensive option, such as fan-powered terminal units with heating coils serving perimeter zones in lieu of a separate heating system. Radiant ceiling panels for the perimeter zones are another option.

Interior space usage usually requires that interior air-conditioning systems permit modification to handle all load situations. Variable air volume systems are often used. When using these systems, a careful evaluation of low load conditions should be made to determine if adequate air movement and outdoor air can be provided at the proposed supply air temperature without overcooling. Increases in supply air temperature tend to nullify energy savings in fan power, which are characteristic of variable air volume systems. Low temperature air distribution for additional savings in transport energy is seeing increased use. This is especially true when coupled with an ice storage system.

In small to medium-sized office buildings, air source heat pumps may be chosen. In larger buildings, internal source heat pump systems (water-to-water) are feasible with most types of air-conditioning systems. Heat removed from core areas is either rejected to a cooling tower or perimeter circuits. The internal source heat pump can be supplemented by a boiler on extremely cold days or over extended periods of limited occupancy. Removed excess heat may also be stored in hot water tanks.

Many heat recovery or internal source heat pump systems exhaust air from conditioned spaces through lighting fixtures. Approximately 30% of lighting heat can be removed in this manner. One design advantage is a reduction in required air quantities. In addition, lamp life is extended by operation in a much lower ambient temperature.

Suspended ceiling return air plenums eliminate sheet metal return air ductwork to reduce floor-to-floor height requirements. However, suspended ceiling plenums may increase the difficulty of proper air balancing throughout the building. Problems often connected with suspended ceiling return plenums are as follows:

* Air leakage through cracks, with resulting smudges
* Tendency of return air openings nearest to a shaft opening or collector duct to pull too much air, thus creating uneven air motion and possible noise
* Noise transmission between office spaces

Air leakage can be minimized by proper workmanship. To overcome drawing too much air, return air ducts can be run in the suspended ceiling pathway from the shaft, often in a simple radial pattern. The ends of the ducts can be left open or dampered. Generous sizing of return air grilles and passages will lower the percentage of circuit resistance attributable to the return air path. This bolsters effectiveness of supply air-balancing devices and reduces the significance of air leakage and drawing too much air. Structural blockage can be solved by locating openings in beams or partitions with fire dampers, where required.

Total office building electromechanical space requirements are approximately 8 to 10% of the gross area. Clear height required for fan rooms varies from approximately 3 to 5.5 m, depending on the distribution system and equipment complexity. On office floors, perimeter fan-coil or induction units require approximately 1 to 3% of the floor area. Interior shafts require 3 to 5% of the floor area. Therefore, ducts, pipes, and equipment require approximately 3 to 5% of each floor's gross area. Electrical and plumbing space requirements per floor average an additional 1 to 3% of the gross area.

Where large central units supply multiple floors, shaft space requirements depend on the number of fan rooms. In such cases, one mechanical equipment room usually furnishes air requirements for 8 to 20 floors (above and below for intermediate levels), with an average of 12 floors. The more floors served, the larger the duct shafts and equipment required. This results in higher fan room heights, and greater equipment size and mass.

The fewer floors served by an equipment room, the more equipment rooms will be required to serve the building. This axiom allows greater flexibility in serving changing floor or tenant requirements. Often, one mechanical equipment room per floor and complete elimination of vertical shafts requires no more total floor area than a few larger mechanical equipment rooms, especially when there are many small rooms and they are the same height as typical floors. Equipment can also be smaller, although maintenance costs will be higher. Energy costs may be reduced, with more equipment rooms serving fewer areas, because the equipment can be shut off in unoccupied areas, and high-pressure ductwork will not be required.

Equipment rooms on upper levels generally cost more to install because of rigging and transportation logistics, but this cost can be balanced against increased revenues for desirable lower level spaces.

In all cases, mechanical equipment rooms must be thermally and acoustically isolated from office areas.

Cooling towers are the largest single piece of equipment required for air-conditioning systems. Cooling towers require approximately 1 m^2 of floor area per 400 m^2 of total building area and are from 4 to 12 m high. When towers are located on the roof, the building structure must be capable of supporting the cooling tower and dunnage, full water load (approximately 590 to 730 kg/m^2), and wind load stresses.

Where cooling tower noise may affect neighboring buildings, towers should be designed to include sound traps or other suitable noise baffles. This may affect tower space, mass of the units, and motor power. Slightly oversizing cooling towers can also reduce noise and power consumption due to lower speeds, but this may increase the initial cost.

Cooling towers are sometimes enclosed in a decorative screen for aesthetic reasons; therefore, calculations should ascertain that the screen has sufficient free area for the tower to obtain its required air quantity and to prevent recirculation.

If the tower is placed in a rooftop well, near a wall, or split into several towers at various locations, design becomes more complicated and initial and operating costs increase substantially. Also, towers should not be split and placed on different levels because hydraulic problems increase. Finally, the cooling tower should be built high enough above the roof so that the bottom of the tower and the roof can be maintained properly.

Special Considerations

Office building areas with special ventilation and cooling requirements include elevator machine rooms, electrical and telephone closets, electrical switchgear, plumbing rooms, refrigeration rooms, and mechanical equipment rooms. The high heat loads in some of these rooms may require air-conditioning units for spot cooling.

In larger buildings having intermediate elevator, mechanical, and electrical machine rooms, it is desirable to have these rooms on the same level or possibly on two levels. This may simplify the horizontal ductwork, piping, and conduit distribution systems and permit more effective ventilation and maintenance of these equipment rooms.

An air-conditioning system cannot prevent occupants at the perimeter from feeling direct sunlight. Venetian blinds and drapes are often provided but seldom used. External shading devices (screens, overhangs, etc.) or reflective glass are preferable.

Tall buildings in cold climates experience severe stack effect. No amount of additional heat provided by the air-conditioning system can overcome this problem. The following features help combat infiltration due to the stack effect:

- Revolving doors or vestibules at exterior entrances
- Pressurized lobbies
- Tight gaskets on stairwell doors leading to the roof
- Automatic dampers on elevator shaft vents
- Tight construction of the exterior skin
- Tight closure and seals on all dampers opening to the exterior

LIBRARIES AND MUSEUMS

In general, libraries have stack areas, working and office areas, a main circulation desk, reading rooms, rare book vaults, and small study rooms. Many libraries also contain seminar and conference rooms, audiovisual rooms, rooms for listening to audio recordings, special exhibit areas, computer rooms, and perhaps an auditorium. This wide diversity of functions requires careful analysis to provide proper environmental conditions.

Museums fall into several categories:

- Art museums and galleries
- Natural and social history
- Scientific
- Specialized topics

In general, museums have exhibit areas, work areas, offices, and storage areas. Some of the larger museums may have shops, a restaurant, etc., but these areas are not basic to this type of building and are discussed in other sections.

Specialized topic museums, such as reconstructed or preserved residences, or industrial museums showing product development and growth, usually have less complex air-conditioning requirements. In some scientific museums, the necessity for reproducing the results of various exhibits or experiments may require close environmental control.

Most art museums and galleries, and some natural history museums, have their exhibits exposed within the viewing area. However, some exhibits are kept in enclosed cases, cubicles, or rooms. These exhibits may require special conditions that differ markedly from human comfort requirements. In this case, separate systems may be set up for maintaining the proper temperature and humidity.

Work areas in art museums consist of rooms for restoration and touch-up, picture framing, sculpture mounting, and repair. Paints, chemicals, plaster of paris, and other materials requiring special temperature, humidity, and air circulating conditions may be used. Noise level is not critical but should not be objectionable to the occupants.

A greater variety of functions, such as animal stuffing and reconstruction of fossils or cultural exhibits, may be performed in the work areas of natural and social history museums. Some museums have research facilities and laboratories, and odors and chemicals in these areas may require larger exhaust air quantities. Individual room or area zone control will generally be necessary.

Storage areas in most museums contain articles that must be repaired or large numbers of articles for which exhibit space is not available. These areas may require closely controlled environmental conditions.

Load Characteristics

Many libraries, especially college libraries, operate up to 16 h per day and may run the air-conditioning equipment about 5000 h per year. Such constant usage requires the selection of heavy-duty, long-life equipment, which requires little maintenance. Museums are generally open about 8 to 10 h per day, 5 to 7 days per week. Many people who visit museums do not remove their outer clothing.

The ambient conditions should not vary in temperature or relative humidity. The conditions should remain constant 24 h per day, year-round. Cold or hot walls and windows, and hot steam or water pipes should be avoided. Object humidity may be destructive, even if the ambient relative humidity is under control. If the ambient dry-bulb temperature varies or if the collection is subjected to radiant effects, the temperature of objects will vary, always lagging behind the atmospheric changes.

Some of the specific factors of particular importance in determining the heating and cooling loads follow.

Sun Gain. Libraries and museums usually have windows, sometimes of stained glass, and skylights—more in traffic areas, than in book stacks or storage areas. Care must be taken to minimize the effects of the sun; shortwave (actinic) rays are particularly injurious. Heat gain from skylights, often over artificially lighted frosted glass ceilings, can be reduced by a separate forced ventilation system.

Transmission. In winter, effects on objects located close to outside walls and possible condensation of moisture on the objects and the surface of outside walls must be evaluated. In summer, possible radiant effects from exposure should be considered.

People. Some areas may have concentrations as high as 0.9 m² per person, while office space will have closer to 9 or 14 m² per person, and book stack areas up to 90 m² per person.

When smoking is permitted, return air should be contained, and the recirculated part of the air should be deodorized with activated charcoal and similar odor-removal devices or exhausted.

Lights. A detailed examination should be made of the wattage provided in various rooms and the length of operation of the lights. In book stacks, various storage rooms, and vaults, lights may be discounted completely because of occasional use.

Stratification. Main reading rooms, large entrance halls, and large art galleries often have high ceilings that may allow the air temperature to stratify.

After individual room loads are evaluated at their optimum values to determine air requirements, the instantaneous refrigeration load should be calculated using proper diversity and storage factors.

Design Concepts

All-air systems are preferred in library areas where steam or water may ruin rare books, manuscripts, tapes, and so forth. This is also true for museums, because exhibit items are generally irreplaceable. However, there are many libraries that have used air-water systems with satisfactory results.

In some libraries with auditoriums, it is possible to use the spaces under the seats and behind the backs of the seats for handling supply and return air distribution. In picture galleries, the wall space below the rail height may be made available. Under some circumstances, the space above and below the wall cases may be used. In book stacks, each tier or deck should have individual air supply and return. Newly constructed book stacks would probably be interior spaces, with air ductwork and other services worked into the steel shelf-supporting structure. The most important consideration in designing air distribution should be to avoid stagnant spaces in all exhibits, especially in storage rooms and vaults. Steam or water piping should not run through exhibit or storage rooms, to avoid possible damage by accidental leaks and radiation.

In museums, patron traffic may follow a planned or random pattern, depending on the size of the museum, the number of exhibits and people, or the organization of the exhibits. The pattern may affect the type of air-conditioning system. People loads vary, depending on whether there is a new exhibit, the time of day, weather, and other factors. Thus, individually controlled zones are required to maintain optimal environmental conditions.

The most difficult problem encountered in designing an air-conditioning system for a museum is that partitioned areas may be radically changed from one exhibit to another. Attempts to establish a modular system for partitions have been only partially successful because of the wide range of sizes of items in the exhibits. Air distribution and lighting systems must be set up in the most flexible manner possible to minimize problems.

In art museums, particularly, partitions may create local pockets with hot air supply or exhaust; transfer grilles may be placed in the partitions to obtain some air movement.

Another problem is the location of room thermostats and humidistats. Sometimes it is not practicable to locate them either in the room or in the common return air duct because conditions may be typical of only a small area. One solution is to have the basic floor set up with small, individual zones. This, of course, is one of the most costly solutions. Other potential solutions are to locate the thermostats in return air ducts or on aspirating diffusers.

Special Considerations

Many old manuscripts, books, museum exhibits, and works of art have been damaged or destroyed because they were not kept in a properly air-conditioned environment. The need for better preservation of such valuable materials, together with a rising popular interest in the use of libraries and museums, requires that most of them, whether new or existing, be air conditioned.

Air-conditioning problems for museums and libraries are generally similar, but differ in design concept and application. The temperature and humidity ranges that are best for books, museum exhibits, and works of art do not usually fall within the human comfort range. Thus, compensations must be made to balance the value of preserving contents against human comfort, as well as initial and operating costs of air conditioning.

Design Criteria. In an average library or museum, less stringent design criteria are usually provided than for archives, because the value of the books and collections does not justify the higher initial and operating costs. Low-efficiency air filters are often provided. Relative humidity is held below 55%. Room temperatures are held within the 20 to 22°C range.

Archival libraries and museums should have 85% or better air filtration, a relative humidity of 35% for books, and temperatures of 13 to 18°C in book stacks and 20°C in reading rooms. Canister-type filters or spray washers should be installed if chemical pollutants are present in the outdoor air.

Art storage areas are often maintained at 15 to 22°C or lower, and 50% rh (±2%). Stuffed, fur-bearing animals should be stored at about 4 to 10°C and 50% rh for maximum preservation; fossils and old bones are better preserved at higher humidities. Museum authorities should be consulted to ensure optimal conditions for specific collections.

Building Contents. Because preservation of the collections housed within libraries and museums is so important, the reaction of each of the materials in the collections to room conditions should be carefully considered.

Paper used in books and manuscripts before the eighteenth century is very stable and is not significantly affected by the room environment. Produced by a cottage industry, paper was made in small lots by breaking down the wood fibers by stamping, using naturally alkaline water, and applying a gelatin sizing. Industrialized production, in which wood fibers were cut with steel knives, ordinary water was used, and rosin sizing was substituted for gelatin, made a paper susceptible to deterioration because of its acid content. For archival preservation, this paper should be stored at very low temperatures. It is estimated that for each 5.5 K dry bulb the room temperature is lowered, the life of the paper will double, and that any humidity reduction will also lengthen the life of paper.

Libraries, however, house more than books; they also store and use films and tapes. The desiccation point for microfilm and magnetic tape is below 35% rh. This, then, is the lower limit for relative humidity in libraries, with the optimum humidity just above this point to minimize paper deterioration. The upper limit for humidity when the room dry-bulb temperature exceeds 18°C is 67% rh, because mold forms above this point.

Many materials housed in museums are organically based and also benefit from lower room temperatures. Museums that display only part of their collection at one time and keep the rest in storage rooms should consider reducing the storage temperature to lengthen the life of organic materials.

Chemical pollutants and dirt are other factors that affect the preservation of books and organic materials. Chemical pollutants causing the greatest concern include sulfur dioxide, the oxides of nitrogen, and ozone. Electrostatic filtration of the air is not recommended because it can generate ozone. Sulfur dioxide combines with water to form sulfuric acid. In the past, sulfur dioxide was removed from outdoor air by using a washer in which the spray water was kept at a pH value between 8.5 and 9.0. Some museum and library system designs use special canister-type filters to remove sulfur dioxide.

Effect of Ambient Atmosphere. The temperature and, particularly, the relative humidity (not humidity ratio) of the air have a marked influence on the appearance, behavior, and general quality of hygroscopic materials such as paper, textiles, wood, and leather, because the moisture content of these substances comes into equilibrium with the moisture content of the surrounding air.

This process is of particular importance because it can multiply the destructive effect of changes in the ambient atmosphere. If any object in a collection (such as a book, painting, tapestry, or other article on exhibit) is at a temperature higher or lower than that of the air in the museum or library, the relative humidity of the air close to the object will differ considerably from that of the ambient room air.

The *object humidity* is the relative humidity of the thin film of air in close contact with the surface of an object and at a temperature cooler or warmer than the ambient dry bulb (Banks 1974). Object humidity differs from that of the ambient air because the dry-bulb temperature of various layers of the air film approaches that of the object, while the dew-point temperature remains constant.

If objects in a museum are permitted to cool overnight, the next day they will be enveloped by layers of air having progressively higher relative humidities. These may range from the ambient of 45 to 60% to 97% immediately next to the object surface, thus effecting a change in material regain or even condensation. This, combined with the hygroscopic or salty dust often found on objects recovered from excavations, can be destructive. If the particular material is warmed, however, the object's humidity will be lower than the humidity of the surrounding space. This warming may be caused by spotlights or any hot, radiating surface.

Sound and Vibration. Air-conditioning equipment should be treated with sound and vibration isolation to ensure quiet comfort for visitors and staff. Acoustical isolation is also necessary to avoid transmitting or setting up resonant (sympathetic) vibration within objects on exhibit, which may be damaged by such motion. Sound level should be low, but not so low as to produce an environment where normal sounds will be objectionable. It should also be noted that exhibit spaces tend to be acoustically reverberant (see Chapter 43).

Case Breathing. Many objects and exhibits are housed in cases, and unless they are sealed tightly, the cases tend to breathe (Banks 1974). For example, the Declaration of Independence and the U.S. Constitution are inscribed on sheepskin parchment and enclosed within sealed receptacles filled with helium. The air in a case expands and contracts as the air temperature is changed by variation in ambient temperature, lights, and atmospheric pressure. Consequently, the air within the case will change more or less frequently. The worst offender in this instance may be a spotlight thought to be installed far enough from the case so as to have no effect.

Special Rooms. When a library or museum has seminar and conference rooms, audiovisual, rooms for listening to audio recordings, and special exhibit areas, individual environmental room control is needed. Seminar and conference rooms may be exposed to heavy smoking, so auxiliary 100% exhaust should be provided. The other rooms listed may require a slightly quieter environment. Separate temperature and humidity controls should be provided for record and tape storage rooms.

Location of mechanical equipment rooms and air-handling equipment should be as remote as possible from the reading and exhibit areas to minimize the need for expensive sound and vibration isolation measures.

BOWLING CENTERS

Bowling centers may also contain a bar, restaurant, children's play area, offices, locker rooms, and other types of facilities. Such auxiliary areas are not discussed in this section, except as they may affect design for the bowling center, which consists of alleys and a spectator area.

Load Characteristics

Bowling alleys usually have their greatest period of use in the evenings, but weekend daytime use may also be heavy. Thus, when designing for the peak air-conditioning load on the building, it is necessary to compare the day load and its high outside solar and off-peak people loads with the evening peak people load and no solar loads. Because bowling areas generally have little fenestration, the solar load may not be important.

If the building contains auxiliary areas, these areas may be included in the refrigeration and air distribution systems for the bowling alleys, with suitable provisions for zoning the different areas as dictated by load analysis. Alternatively, separate systems may be established for each area having different load operation characteristics.

Heat buildup due to lights, external transmission load, and pin-setting machinery in front of the foul line can be reduced by exhausting some air above the alleys or from the area of the pin-setting machines; however, this gain should be compared against the cost of conditioning additional makeup air. In the calculation of the air-conditioning load, a portion of the unoccupied alley space load is included. Because this consists mainly of lights and some transmission load, about 15 to 30% of this heat load may have to be taken into account. The higher figure may apply when the roof is poorly insulated, no exhaust air is taken from this area, or no vertical baffle is used at the foul line. One estimate is 16 to 32 W per square metre of vertical surface at the foul line, depending mostly on the type and intensity of the lighting.

The heat load from bowlers and spectators may be found in Chapter 26 of the 1993 *ASHRAE Handbook—Fundamentals*. The proper heat gain should be applied for each person to avoid too large a design heat load.

Design Concepts

As with other building types having high occupancy loads, heavy smoke and odor concentration, and low sensible heat factors, all-air systems are generally the most suitable for bowling alley areas. Since most bowling alleys are almost windowless structures except for such areas as entrances, exterior restaurants, and bars, it is uneconomical to use terminal unit systems because of the small number required. Where required, radiation in the form of baseboard or radiant ceiling panels is generally placed at perimeter walls and entrances.

It is not necessary to maintain normal indoor temperatures down the length of the alleys; temperatures may be graded down to the pin area. Unit heaters are often used at this location.

Air Pressurization. Spectator and bowling areas must be well shielded from entrances so that no cold drafts are created in these areas. To minimize infiltration of outdoor air into the alleys, the exhaust and return air system should handle only 85 to 90% of the total supply air, thus maintaining a positive pressure within the space.

Air Distribution. Packaged units without ductwork produce uneven space temperatures, and unless they are carefully located and installed, the units may cause objectionable drafts. Central ductwork systems are recommended for all but the smallest buildings, even where packaged refrigeration units are used. Because only the areas behind the foul line are air conditioned, the ductwork should provide comfortable conditions within this area.

The return and exhaust air systems should have a large number of small registers uniformly located at high points, or pockets, to draw off the hot, smoky, and odorous air. In some parts of the country and for larger bowling alleys, it may be desirable to use all outdoor air to cool during intermediate seasons.

Special Considerations

People in sports and amusement centers engage in a high degree of physical activity, which makes them feel warmer and increases their rate of evaporation. In these places, odor and smoke control are important environmental considerations.

Bowling centers are characterized by the following:

- A large number of people concentrated in a relatively small area of a very large room. A major portion of the floor area is unoccupied.
- Heavy smoking, high physical activity, and high latent heat load.
- Greatest use from about 6:00 to 12:00 p.m.

The first two items make it mandatory that large amounts of outdoor air be furnished to minimize odors and smoke in the atmosphere.

The area between the foul line and the bowling pins need not be air conditioned or ventilated. Transparent or opaque vertical partitions are sometimes installed to separate the upper portions of the occupied and unoccupied areas so that air distribution is better contained within the occupied area.

COMMUNICATION CENTERS

Communication centers include telephone terminal buildings, teletype centers, radio stations, television studios, and transmitter and receiver stations.

Most telephone terminal rooms are air conditioned because constant temperature and relative humidity help prevent breakdowns and increase equipment life. In addition, air conditioning permits the use of a lower number of air changes, which, for a given filter efficiency, decreases the chances of damage to relay contacts and other delicate equipment.

Teletype centers are similar to telephone terminal rooms except that people operate the teletype machines, so more care is required in the design of the air distribution systems.

Radio and television studios require critical analysis for the elimination of heat buildup and the control of noise. Television studios have the added problem of air movement, lighting, and occupancy load variations. This section deals with television studios because they encompass most of the problems also found in radio studios.

Load Characteristics

Human occupancy is limited, so the air-conditioning load for telephone terminal rooms is primarily equipment heat load. Teletype centers are similar, except for the load from the people who operate the teletype machines.

Television studios have very high lighting capacities, which may fluctuate considerably in intensity over short periods. The operating hours may vary every day. In addition, there may be from one to several dozen people onstage for short times. The air-conditioning system must be extremely flexible and capable of handling wide load variations quickly, accurately, and efficiently, similar to the conditions of a theater stage. The studio may also have an assembly area with a large number of spectator seats. Generally, studios are located so that they are completely shielded from external noise and thermal environments.

Design Concepts

The critical areas of a television studio consist of the performance studio and control rooms. The audience area may be handled in a manner similar to that for a place of assembly. Each area should have its own air distribution system or at least its own zone control separate from the studio system. The heat generated in the studio area should not be allowed to permeate the audience atmosphere.

The air distribution system selected must have the capabilities of a dual-duct, single-duct with cooling and heating booster coils, variable air volume, or multizone system to satisfy design criteria. The air distribution system should be designed so that simultaneous heating and cooling cannot occur, unless such heating is achieved solely by heat recovery methods.

Studio loads seldom exceed 350 kW of refrigeration. Even if the studio is part of a large communications center or building, it is desirable for the studio to have its own refrigeration system in case of emergencies. In this size range, the refrigeration machine may be of the reciprocating type, which requires a remote location so that machine noise is isolated from the studio.

Special Considerations

On-Camera Studios. This is the stage of the television studio and requires the same general considerations as a concert hall stage. Air movement must be uniform and, because scenery, cameras, and equipment may be moved during the performance, ductwork must be planned carefully to avoid interference with proper studio operation.

Control Rooms. Each studio may have one or more control rooms serving different functions. The video control room, which is occupied by the program and technical directors, contains monitoring sets and picture-effect controls. The room may require up to 30 air changes per hour to maintain proper conditions. The large number of air changes required and the low sound level that must be maintained require the air distribution system to be specially analyzed.

If a separate control room is furnished for the announcer, the heat load and air distribution problems will not be as critical as those for control rooms for the program, technical, and audio directors.

Thermostatic control should be furnished in each control room, and provisions should be made to enable occupants to turn the air conditioning on and off.

Noise Control. Studio microphones are moved throughout the studio during a performance, and they may be moved past or set near air outlets or returns. These microphones are considerably more sensitive than the human ear; therefore, air outlets or returns should be located away from areas where microphones are likely to be used. Even a leaky pneumatic thermostat can be a problem.

Air Movement. It is essential that air movement within the stage area, which often contains scenery and people, be kept below 0.13 m/s within 3.7 m of the floor. The scenery is often fragile and will move in air velocities above 0.13 m/s; also, actors' hair and clothing may be disturbed.

Air Distribution. Ductwork must be fabricated and installed so that there are no rough edges, poor turns, or improperly installed dampers to cause turbulence and eddy currents within the ducts. Ductwork should contain no holes or openings that might create whistles. Air outlet locations and the distribution pattern must be carefully analyzed to eliminate turbulence and eddy currents within the studio, which might cause noise that could be picked up by studio microphones.

At least some portions of supply, return, and exhaust ductwork will require acoustical material to maintain noise criteria (NC) levels from 20 to 25. Any duct serving more than one room should acoustically separate each room by means of a sound trap. All ductwork should be suspended by means of neoprene or rubber in shear-type vibration mountings. Where ductwork goes through wall or floor slabs, the openings should be sealed with acoustically deadening material. The supply fan discharge and the return and exhaust fan inlets should have sound traps; all ductwork connections to fans should be made with nonmetallic, flexible material. Air outlet locations should be coordinated with ceiling-mounted tracks and equipment. Air distribution for control rooms may require a perforated ceiling outlet or return air plenum system.

Piping Distribution. All piping within the studio, as well as in adjacent areas that might transmit noise to the studio, should be supported by suitable vibration isolation hangers. To prevent transmission of vibration, piping should be supported from rigid structural elements to maximize absorption.

Mechanical Equipment Rooms. These rooms should be located as remotely from the studio as possible. All equipment should be selected for very quiet operation and should be mounted on suitable vibration-eliminating supports. Structural separation of these rooms from the studio is generally required.

Offices and Dressing Rooms. The functions of these rooms are quite different from each other and from the studio areas. It is recommended that such rooms be treated as separate zones, with their own controls.

Air Return. Whenever practicable, the largest portion of studio air should be returned over the banks of lights. This is similar to theater stage practice. Sufficient air should also be removed from studio high points to prevent heat buildup.

TRANSPORTATION CENTERS

The major transportation facilities are airports, ship docks, bus terminals, and passenger car garages. Airplane hangars and freight and mail buildings are also among the types of buildings to be considered. Freight and mail buildings are usually handled as standard warehouses.

Load Characteristics

Airports, ship docks, and bus terminals operate on a 24-h basis, although on a reduced schedule during late evening and early morning hours.

Airports. Terminal buildings consist of large, open circulating areas, one or more floors high, often with high ceilings, ticketing counters, and various types of stores, concessions, and convenience facilities. Lighting and equipment loads are generally average, but occupancy varies substantially. Exterior loads are, of course, a function of architectural design. The largest single problem often results from thermal drafts created by large entranceways, high ceilings, and long passageways, which have many openings to the outdoors.

Ship Docks. Freight and passenger docks consist of large, high-ceilinged structures with separate areas for administration, visitors, passengers, cargo storage, and work. The floor of the dock is usually exposed to the outdoors just above the water level. Portions of the side walls are often open while ships are in port. In addition, the large portion of ceiling (roof) area presents a large heating and cooling load. Load characteristics of passenger dock terminals generally require the roof and floors to be well insulated. Occasional heavy occupancy loads in visitor and passenger areas must be considered.

Bus Terminals. This building type consists of two general areas: the terminal building, which contains passenger circulation, ticket booths, and stores or concessions, and the bus loading area. Waiting rooms and passenger concourse areas are subject to a highly variable people load. Occupancy density may reach 1 m^2 per person, and, at extreme periods, 0.3 to 0.5 m^2 per person.

Design Concepts

Since heating and cooling plants may be centralized or provided for each building or group in a complex, they will not be discussed here. In large, open circulation areas of transportation centers, any all-air system with zone control can be used. Where ceilings are high, air distribution is often along the side wall to concentrate the air conditioning where desired and avoid disturbing stratified air. Perimeter areas may require heating by radiation, a fan-coil system, or hot air blown up from the sill or floor grilles, particularly in colder climates. Hydronic perimeter radiant ceiling panels may be especially suited to these high load areas.

Airports. Airports generally consist of one or more central terminal buildings connected by long passageways, or trains, to rotundas containing departure lounges for airplane loading. Most terminals have portable telescoping-type loading bridges connecting departure lounges to the airplanes. These passageways eliminate the heating and cooling problems associated with traditional permanent structure passenger loading.

Because of difficulties in controlling the air balance resulting from the many outside openings, high ceilings, and long, low passageways (which often are not air conditioned), the terminal building (usually air conditioned) should be designed to maintain a substantial positive pressure. Zoning will generally be required in passenger waiting areas, departure lounges, and at ticket counters to take care of the widely variable occupancy loads.

Main entrances may be designed with vestibules and windbreaker partitions to minimize undesirable air currents within the building.

Hangars must be heated in cold weather, and ventilation may be required to eliminate possible fumes (although fueling is seldom permitted in hangars). Gas-fired, electric, and low- and high-intensity heaters are used extensively in hangars because they provide comfort for employees at relatively low operating costs.

Hangars may also be heated by large air blast heaters or floor-buried heated liquid coils. Local exhaust air systems may be used to evacuate fumes and odors that result in smaller ducted systems. Under some conditions, exhaust systems may be portable and may possibly include odor-absorbing devices.

Ship Docks. In severe climates, occupied floor areas may contain heated floor panels. The roof should be well insulated and, in appropriate climates, evaporative spray cooling substantially reduces the summer load. Freight docks are usually heated and well ventilated but seldom cooled.

High ceilings and openings to the outdoors may present serious draft problems unless the systems are designed properly. Vestibule entrances or air curtains help minimize cross drafts. Air door blast heaters at cargo opening areas may be quite effective.

Ventilation of the dock terminal should prevent noxious fumes and odors from reaching occupied areas. Therefore, occupied areas should be under a positive pressure and the cargo and storage areas exhausted to maintain negative air pressure. Occupied areas should be enclosed to simplify the possibility of local air conditioning.

In many respects, these are among the most difficult buildings to heat and cool because of their large open areas. If each function is properly enclosed, any commonly used all-air or large fan-coil system is suitable. If areas are left largely open, the best approach is to concentrate on proper building design and the heating and cooling of the openings. High-intensity infrared spot heating can often be advantageous (see Chapter 15 of the 1992 *ASHRAE Handbook—Systems and Equipment*). Exhaust ventilation from tow truck and cargo areas should be exhausted through the roof of the dock terminal.

Bus Terminals. Conditions are similar to those for airport terminals, except that all-air systems are more practical because ceiling heights are often lower, and perimeters are usually flanked by stores or office areas. The same types of systems are applicable as for airport terminals, but ceiling air distribution will generally be feasible.

Properly designed radiant hydronic or electric ceiling systems may be used if high occupancy latent loads are fully considered. This may result in smaller duct sizes than are required for all-air systems and may be advantageous where bus loading areas are above the terminal and require structural beams. This heating and cooling system reduces the volume of the building that must be conditioned. In areas where latent load is a concern, heating-only panels may be used at the perimeter, with a cooling-only interior system.

The terminal area air-supply system should be under a high positive pressure to ensure that no fumes and odors infiltrate from bus areas. Positive exhaust from bus loading areas is essential for a properly operating total system (see Chapter 12).

Special Considerations

Airports. Filtering outdoor air with activated charcoal filters should be considered for areas subject to excessive noxious fumes from jet engine exhausts. However, locating outside air intakes as remotely as possible from airplanes is a less expensive and more positive approach.

Where ionization filtration enhancers are used, outdoor air quantities are sometimes reduced due to cleaner air. However, care must be taken to maintain sufficient amounts of outside air for space pressurization.

Ship Docks. Ventilation design must ensure that fumes and odors from forklifts and cargo in work areas do not penetrate occupied and administrative areas.

Bus Terminals. The primary concerns with enclosed bus loading areas are health and safety problems, which must be handled by proper ventilation (see Chapter 12).

Although diesel engine fumes are generally not as noxious as gasoline fumes, bus terminals often have many buses loading and unloading at the same time, and the total amount of fumes and odors may be quite disturbing.

Enclosed Garages. From a health and safety viewpoint, enclosed bus loading areas and car garages present the most serious problems in these buildings (see Chapter 12).

Three major problems are encountered. The first and most serious is carbon monoxide (CO) gas emission by cars and oxides of nitrogen by buses, which can cause serious illness and, possibly, death.

The second problem is oil and gasoline fumes, which may cause nausea and headaches and can also create a fire hazard. The third involves lack of air movement and the resulting stale atmosphere that develops because of the increased carbon dioxide (CO_2) content in the air. This condition may cause headaches or grogginess.

Most codes require a minimum of four to six air changes per hour. This is predicated on maintenance of a maximum safe CO concentration in the air, assuming short periods of occupancy in the garage.

All underground garages should have facilities for testing the CO concentration or should have the garage checked periodically. Clogged duct systems, improperly operating fans, motors or dampers, clogged air intake or exhaust louvers, etc., may not allow proper air circulation. Proper maintenance is required to minimize any operational defects.

Carbon Monoxide Criteria. Minimum ventilation requirements, as set up by the National Institute of Standards and Technology (NIST) and ASHRAE are primarily concerned with preventing buildup of noxious concentrations of carbon monoxide.

However, keeping the CO level within safe limits is no guarantee that patrons or employees in the garage will not experience discomfort. The air furnished will probably be sufficient to eliminate any atmospheric staleness, but the greatest discomfort can be caused by oil and gasoline fumes, which are particularly noticeable in areas with poor air circulation and at poorly ventilated ramps. Therefore, a properly designed air distribution system is essential to a comfortable and safe human environment.

WAREHOUSES

Warehouses are used to store merchandise and may be open to the public at times. They are also used to store equipment and material inventories as part of an industrial facility. The buildings are generally not air conditioned, but often have heat and ventilation sufficient to provide a tolerable working environment. Facilities such as shipping, receiving, and inventory control offices, associated with warehouses and occupied by office workers, are generally air conditioned.

Load Characteristics

Internal loads from lighting, people, and miscellaneous sources are generally low. Most of the load is thermal transmission and infiltration. An air-conditioning load profile tends to flatten where materials stored are massive enough to cause the peak load to lag.

Design Concepts

Most warehouses are only heated and ventilated. Forced flow unit heaters may, in many instances, be located near heat entrances and work areas. Even though comfort for warehouse workers may not be desired, it may be necessary to keep the temperature above 4°C to protect sprinkler piping or stored materials from freezing.

Thermal qualities of the building, which would be included if the building might later be air conditioned, also minimize required heating and aid in comfort. For maximum summer comfort without air conditioning, excellent ventilation with noticeable air movement in work areas is necessary. Even greater comfort can be achieved in appropriate climates by adding roof spray cooling. This can reduce the roof's surface temperature by 20 to 30 K, thereby reducing ceiling radiation inside. Low- and high-intensity radiant heaters can be used to maintain the minimum ambient temperature throughout a facility above freezing. Radiant heat may also be used for occupant comfort in areas permanently or frequently open to the outside.

Special Considerations

Powered forklifts and trucks using gasoline, propane, and other fuels are often used inside warehouses. Proper ventilation is necessary to alleviate the buildup of CO and other noxious fumes. Proper ventilation of battery-charging rooms for electrically powered forklifts and trucks is also required.

REFERENCES

ASHRAE. 1976. Method of testing air-cleaning devices used in general ventilation for removing particulate matter. *Standard* 52-1976.

ASHRAE. 1989. Ventilation for acceptable indoor air quality. *Standard* 62-1989.

Banks, P.N. 1974. Environmental standards for storage of books and manuscripts. *The Library Journal* 99(3).

HUD *Bulletin* LR-11. Housing and Urban Development Agency, Washington, D.C.

Library of Congress. 1975. Environmental protection of books and related material. Library of Congress Preservation Leaflet No. 2 (February). Washington, D.C.

Smith, R. 1969. Paper impermanence as a consequence of pH and storage conditions. *Library Quarterly* 39(2).

PLACES OF ASSEMBLY

THIS chapter covers the design considerations associated with enclosed assembly buildings. (Chapter 3 covers general criteria for commercial and public buildings that also apply to public assembly buildings.)

This chapter is organized into six sections. The first, Common Characteristics, applies to all places of assembly and includes information on general criteria and system considerations. Design criteria include ventilation, lighting loads, indoor air conditions, filtration, and noise and vibration control.

The remaining sections apply to the following specific buildings: houses of worship, auditoriums, arenas and stadiums, convention and exhibition centers, natatoriums, fairs and other temporary exhibits, and atriums. These sections include information on load characteristics, design concepts, design criteria, and systems applicability, as appropriate.

COMMON CHARACTERISTICS

Assembly rooms are generally large, have relatively high ceilings, and are few in number for any given facility. They usually have a periodically high density of occupancy per unit floor area, as compared to other buildings, and thus have a relatively low design sensible heat ratio. Space volume per person is usually high, thus allowing fewer air changes than for many other building types.

GENERAL CRITERIA

Energy conservation codes and standards have a major impact on system design as well as performance and require consideration.

Assembly buildings have relatively few actual hours of use per week and are seldom in full use when maximum outdoor temperatures or solar effects occur. The designer must obtain as much information as possible regarding the anticipated hours of use, particularly the times of full seating, so that simultaneous loads may be considered to obtain optimum air-conditioning loads and operating economy. These buildings often are fully occupied for as little as 1 to 2 h, and the load may be materially reduced by precooling. Latent cooling requirements should be considered before reducing equipment size. The intermittent or infrequent nature of the cooling loads may allow these buildings to benefit from thermal storage systems.

The occupants usually generate the major room cooling and ventilation load. The number of occupants is best determined from the seat count, but when this is not available, it can be estimated at 0.7 to 0.9 m² per person for the entire seating area, including exit aisles but not the stage, performance areas, or entrance lobbies.

Ventilation

Ventilation is a major contributor to total load. ASHRAE *Standard* 62 lists outdoor air requirements for various occupancies. Typi-

cal minimum ventilation rates range from 8 to 30 L/s per person. These rates may be reduced somewhat if the facility is used only for short periods and if it can be flushed out between performances. Some building codes require higher ventilation rates.

Assembly buildings lend themselves to automatic recirculation and outdoor air control so that preoccupancy warm-up or cooling can be accomplished with low ventilation loads, and light occupancy use can be handled with reduced outdoor air loads.

The evaluation of ventilation loads is important in considering the effect of infiltration in any structure. Generally, sufficient outside air should be introduced into the air-handling system to offset the effect of infiltration and keep the structure under positive pressure, provided adjacent areas, such as ancillary facilities, are not adversely affected.

Lighting Loads

Lighting loads are one of the few major loads that vary from one type of assembly building to another. Lighting may be at the level of 1600 lux in convention halls where color television cameras are expected to be used, or lighting may be virtually absent, as during a presentation in a motion picture theater. In many assembly buildings, lights are controlled by dimmers or other means to present a suitably low level of light during performances, with much higher lighting levels during cleanup, when the house is nearly empty. The designer should ascertain what light levels will be associated with maximum occupancies, not only in the interest of economy but also to determine the proper room sensible heat ratio.

Indoor Air Conditions

Indoor air temperature and humidity should parallel the ASHRAE comfort recommendations (see Chapter 8 of the 1993 *ASHRAE Handbook—Fundamentals*). In addition, the following should be considered:

1. In arenas and stadiums, gymnasiums, and some motion picture theaters, people generally dress informally in summer. The summer indoor conditions may favor the warmer end of the thermal comfort scale with no major ill effect. By the same reasoning, the winter indoor temperature may favor the cooler end of the scale.

2. In churches, concert halls, and legitimate theaters, people are usually fairly well dressed, with most men wearing jackets and ties and women often wearing suits. The temperature should favor the middle range of design, and there should be little summer-to-winter variation.

3. In convention and exhibition centers, the visiting public is continually walking. Here the indoor temperature should favor the lower range of comfort conditions both in summer and in winter.

4. For spaces of high population density or sensible heat factors of 0.75 or less, a lower dry-bulb temperature results in less latent heat from people, thus reducing the need for reheat and saving energy. Therefore, the optimum space dry-bulb temperatures should be the result of detailed design analysis.

5. Restrictions of energy conservation codes must be considered in system design and operation.

The preparation of this chapter is assigned to TC 9.8, Large Building Air-Conditioning Applications.

Because of their low room sensible heat ratio, assembly areas generally require some form of reheat to maintain the relative humidity at a suitably low level during periods of maximum occupancy. Refrigerant hot gas or condenser water reject heat is frequently used for this purpose. Face and bypass control of low-temperature cooling coils is also effective. In colder climates, it may also be desirable to provide humidification. High rates of internal gain may make evaporative humidification attractive during economizer cooling.

Filtration

Most places of assembly are minimally filtered with filters rated at 30 to 35% efficiency, as tested in accordance with ASHRAE *Standard* 52. Where smoking is permitted, however, filters with a minimum rating of 80% are required before any effective amount of tobacco smoke is removed. Filters with 80% or higher efficiency are also recommended for those facilities having particularly expensive interior decor. Because of the few operating hours of these facilities, the added expense of higher efficiency filters can be justified by their longer life. Low-efficiency prefilters are generally used with high-efficiency filters to extend their useful life. Ionization and chemically reactive filters should be considered where high concentrations of smoke or odors are present.

Noise and Vibration Control

The desired noise criteria (NC) vary with the type and quality of the facility. The need for noise control is minimal in a gymnasium or natatorium, but it is important in a concert hall. Facilities that are used for varied functions require noise control evaluation over the entire spectrum of use.

In most cases, sound and vibration control is required for both equipment and duct systems, as well as in the selection of diffusers and grilles. When designing a project like a theater or concert hall, consultation with an experienced acoustics engineer is recommended. In these projects, the quantity and quality or characteristic of the noise is important.

Transmission of vibration and noise can be decreased by mounting pipes, ducts, and equipment on a separate structure independent of the music hall. If the mechanical equipment space is close to the music hall, it may be necessary to float the entire mechanical equipment room on isolators, including the floor slab, structural floor members, and other structural elements, supporting pipes, or similar materials that can carry vibrations. Properly designed inertia pads are often used under each piece of equipment. The equipment is then mounted on vibration isolators.

Manufacturers of vibration isolating equipment have devised methods to float large rooms and entire buildings on isolators. Where subway and street noise may be carried into the structure of a music hall, it is necessary to float the entire music hall on isolators. If the music hall is isolated from outside noise and vibration, it also must be isolated from mechanical equipment and other internal noise and vibrations.

External noise from mechanical equipment such as cooling towers should not be allowed to enter the building. Care should be taken to avoid air-conditioning designs that permit noises to enter the space through air intakes or reliefs and carelessly designed duct systems.

SYSTEM CONSIDERATIONS

Ancillary Facilities

Ancillary facilities are generally a part of any assembly building; almost all have some office space. Convention centers and many auditoriums, arenas, and stadiums have restaurants and cocktail lounges. Churches may have apartments for the clergy or a school. Many facilities have parking structures. These varied ancillary facilities are discussed in other chapters of this volume. However, for reasonable operating economy, these facilities should be served by separate systems when their hours of use differ from those of the main assembly areas.

Air-Conditioning Systems

Because of their characteristic large size and need for considerable ventilation air, assembly buildings are frequently served by all-air systems, usually of the single-zone or variable-volume type. Separate air-handling units usually serve each zone, although multizone, dual-duct, or reheat types can also be applied with lower operating efficiency. In larger facilities, separate zones are generally provided for the entrance lobbies and arterial corridors that surround the seating space. Low intensity radiant heating is often an efficient alternative. In some assembly rooms, folding or rolling partitions divide the space for different functions, so that a separate zone of control for each resultant space is best. In extremely large facilities, several air-handling systems may serve a single space, due to the limits of equipment size and also for energy and demand considerations.

Precooling

Cooling the building mass several degrees below the desired indoor temperature several hours before it is occupied allows it to absorb a portion of the peak heat load. This precooling reduces the equipment size needed to meet short-term loads. The effect can be used if precooling time (at least 1 h) is available prior to occupancy, and then only when the period of peak load is relatively short (2 h or less).

The designer must advise the owner that the space temperature will be cold to most people as occupancy begins, but that it will warm up as the performance progresses. This may be satisfactory, but it should be understood by all concerned before proceeding with a precooling design concept. Precooling is best applied when the space is used only occasionally during the hotter part of the day and when provision of full capacity for an occasional purpose is not economically justifiable.

Stratification

Because most applications involve relatively high ceilings, some heat may stratify above the occupied zone, thereby reducing the equipment load. Heat gain from lights can be stratified, except for the radiant effect (about 50% for fluorescent and 65% for incandescent or mercury-vapor fixtures). Similarly, only the radiant effect of the upper wall and roof load (about 33%) reaches the occupied space. Stratification can be achieved only when air is admitted and returned at a sufficiently low elevation so that it does not mix with the upper air.

Conversely, stratification may increase heating loads during periods of minimal occupancy in winter months. In these cases, reduction of stratification by using ceiling fans, air-handling systems, or high/low air distribution may be desirable. Balconies may also be affected by stratification and should therefore be well ventilated.

Air Distribution

In assembly buildings, people generally remain in one place throughout a performance, so they cannot avoid drafts. Good air distribution is essential to a successful installation.

Heating is seldom a major problem, except at entrances or during warm-up prior to occupancy. Generally, the seating area is isolated from the exterior by lobbies, corridors, and other ancillary spaces. For cooling, air can be supplied from the overhead space, where it mixes with heat from the lights and occupants. Return air openings can also aid air distribution. Air returns located below seating or at a low level around the seating can effectively distribute air with minimum drafts. For returns located below the seats, register velocities in excess of 1.4 m/s may cause objectionable drafts and noise.

Because of the configuration of these spaces, it is sometimes necessary to install jet-type nozzles with long throw requirements of 15 to 45 m for sidewall supplies. For ceiling distribution, downward throw is not critical provided returns are low. This approach has been successful in applications that are not particularly noise-sensitive, but the designer needs to carefully select air distribution nozzles. The application data should ensure proper performance for specific projects.

The air-conditioning systems must be quiet. This is difficult to achieve if the air supply is expected to travel 9 m or more from sidewall outlets to condition the center of the seating area. Due to the large size of most houses of worship, theaters, and halls, high air discharge velocities from the wall outlets are required. These high velocities can produce objectionable noise levels for people sitting near the outlets. This can be avoided if the return air system does some of the work. The supply air must be discharged from the air outlet (preferably at the ceiling) at the highest velocity consistent with an acceptable noise level. Although this velocity does not allow the conditioned air to reach all seats, the return air registers, which are located near seats not reached by the conditioned air, pull the air to cool or heat the audience, as required. In this way, the supply air blankets the seating area and is pulled down uniformly by the return air registers under or beside the seats.

A certain amount of exhaust air should be taken from the ceiling of the seating area, preferably over the balcony (if there is one) to prevent the formation of pockets of hot air, which can produce a radiant effect and cause discomfort, as well as increase the cost of air conditioning. Where the ceiling is close to the audience (e.g., below balconies and mezzanines), specially designed plaques or air-distributing ceilings should be provided to absorb noise.

Regular ceiling diffusers placed more than 9 m apart normally give acceptable results if careful engineering is applied in the selection of the diffusers. Because large air quantities are generally involved and because the building is large, it is common to select fairly large capacity diffusers, which tend to be noisy. Linear diffusers are more acceptable architecturally and perform well if selected properly. Integral dampers in diffusers should not be used as the sole means of balancing the system because they generate intolerable amounts of noise, particularly in larger diffusers.

Mechanical Equipment Rooms

The location of mechanical and electrical equipment rooms affects the degree of sound attenuation treatment required. Mechanical equipment rooms located near the seating area are more critical because of the normal attenuation of sound through space. Mechanical equipment rooms located near the stage area are critical because the stage is designed to project sound to the audience. If possible, mechanical equipment rooms should be located in an area separated from the main seating or stage area by buffers such as lobbies or service areas. The economies of the structure, attenuation, equipment logistics, and site must be considered in the selection of locations for mechanical equipment rooms.

At least one mechanical equipment room is placed near the roof to house the toilet exhaust, general exhaust, cooling tower, kitchen, and emergency stage exhaust fans, if any. Individual roof-mounted exhaust fans may be used, thus eliminating the need for a mechanical equipment room. However, to reduce sound problems, mechanical equipment should not be located on the roof over the music hall or stage but rather over offices, storerooms, or auxiliary areas.

HOUSES OF WORSHIP

Houses of worship seldom have full or near-full occupancy more than once a week, but they have considerable use for smaller functions (meetings, weddings, funerals, christenings, or daycare) throughout the balance of the week. It is important to determine how and when the building will be used. When thermal storage systems are used, longer operation of equipment prior to occupancy may be required due to the high thermal mass of the structure. The seating capacity of houses of worship is usually well defined. Some houses of worship have a movable partition to form a single large auditorium for special holiday services. It is important to know how often this maximum use is expected.

The design of many houses of worship is inspired by the classic Gothic cathedral, and a high vaulted ceiling creates thermal stratification. Where stained glass is used, a shade coefficient is assumed to be approximately equal to solar glass (S.C. = 0.70).

Houses of worship test a designer's ingenuity to secure an architecturally acceptable solution to the problem of locating equipment and air-diffusion devices. Because occupants are often seated, drafts and cold floors should be avoided.

Houses of worship may also have auxiliary rooms that should be air conditioned. To ensure privacy, sound transmission between adjacent areas should be considered in the air distribution scheme. Diversity in the total air-conditioning load requirements should be evaluated to take full advantage of the characteristics of each area.

In houses of worship, it is desirable to provide some degree of individual control for the platform, sacristy, and bema or choir area.

AUDITORIUMS

The types of auditoriums considered are motion picture theaters, playhouses, and concert halls. Other types of auditoriums in elementary schools and the large auditoriums in some convention centers follow the same principles, with varying degrees of complexity.

Motion Picture Theaters

Motion picture theaters are the simplest of the auditorium structures mentioned here. They run continuously for periods of 4 to 8 h and, thus, are not a good choice for precooling techniques, except for the first matinee peak. They operate frequently at low occupancy levels, and low-load performance must be considered.

Motion picture sound systems make noise control less important than it is in other kinds of theaters. The lobby and exit passageways in a motion picture theater are seldom densely occupied, although some light to moderate congestion can be expected for short times in the lobby area. A reasonable design for the lobby space is one person per 1.8 to 2.8 m^2.

The lights are usually dimmed when the house is occupied; full lighting intensity is used only during cleaning. A reasonable value for lamps above the seating area during a performance is 5 to 10% of the installed wattage. Designated smoking areas should be handled with separate exhaust or air-handling systems to avoid contamination of the entire facility.

Projection Booths

The projection booth represents the major problem in motion picture theater design. For large theaters using high-intensity lamps, projection room design must follow applicable building codes. If no building code applies, the projection equipment manufacturer usually has specific requirements. The projection room may be air conditioned, but it is normally exhausted or operated at negative pressure. Exhaust is normally taken through the housing of the projectors. Additional exhaust is required for the projectionist's sanitary facilities. Other heat sources include the sound and dimming equipment, which require a continuously controlled environment and necessitate a separate system.

Smaller theaters using 16-mm safety film have fewer requirements for projection booths. It is a good idea to condition the projection room with filtered supply air to avoid soiling lenses. In addition to the projector light, heat sources in the projection room include the sound equipment, as well as the dimming equipment.

Legitimate Theaters

The legitimate theater differs from the motion picture theater in the following ways:

1. Performances are seldom continuous. Where more than one performance occurs in a day, the performances are usually separated by a period of 2 to 4 h. Accordingly, precooling techniques are applicable, particularly for afternoon performances.
2. Legitimate theaters seldom play to houses that are not full or near full.
3. Legitimate theaters usually have intermissions, and the lobby areas are used for drinking and socializing. The intermissions are usually relatively short, seldom exceeding 15 to 20 min.; however, the load may be as dense as one person per 0.5 m^2.
4. Because sound amplification is less used than that in a motion picture theater, background noise control is more important.
5. Stage lighting contributes considerably to the total cooling load in the legitimate theater. Lighting loads can vary from performance to performance.

Stages

The stage presents the most complex problem. It consists of the following loads:

1. A heavy, mobile lighting load
2. Intricate or delicate stage scenery, which varies from scene to scene and presents difficult air distribution requirements
3. Actors may perform tasks that require exertion

Approximately 40 to 60% of the lighting heat load can be negated by exhausting air around the lights. This procedure works for lights around the proscenium. However, for the light strips over the stage, it is more difficult to locate and arrange exhaust air ducts directly over the lights because of the scenery and light drops. Careful coordination is required to achieve an effective and flexible design layout.

Conditioned air should be introduced from the low side and back stages and returned or exhausted around the lights. Some exhaust air must be taken from the top of the tower directly over the stage containing lights and equipment (i.e., the fly). The air distribution design is further complicated because pieces of scenery may consist of lightweight materials that flutter in the slightest air current. Even the vertical stack effect created by the heat from lights may cause this motion. Therefore, low air velocities are essential and air must be distributed over a wide area with numerous supply and return registers.

With multiple scenery changes, low supply or return registers from the floor of the stage are almost impossible to provide. However, some return air at the footlights and for the prompter should be considered. Air conditioning should also be provided for the stage manager and the control board areas.

One phenomenon encountered in many theaters with overhead flies is the billowing of the stage curtain when it is down. This situation is primarily due to the stack effect created by the height of the main stage tower, the heat from the lights, and the temperature difference between the stage and seating areas. Proper air distribution and balancing can minimize this phenomenon. Bypass damper arrangements with suitable fire protection devices may be feasible.

Loading docks adjacent to stages located in cold climates should be heated. The doors to these areas may be open for long periods, for example, while scenery is being loaded or unloaded for a performance.

On the stage, local code requirements must be followed for emergency exhaust ductwork or skylight (or blow-out hatch) require-

ments. These openings are often sizable and should be incorporated in the early design concepts.

Concert Halls

Concert halls and music halls are similar to the legitimate theater in many ways. They normally have a full stage, complete with fly gallery, for the presentation of operas, ballet, and musicals. There are dressing areas for performers. Generally, the only differences between the two are size and decor, with the concert hall usually being larger and more elaborately decorated.

Air-conditioning design must consider that the concert hall is used frequently for special charity and civic events, which may be preceded by or followed by parties (and may include dancing) in the lobby area. Concert halls often have cocktail lounge areas that become very crowded, with heavy smoking during intermissions. These areas should be equipped with flexible exhaust-recirculation systems. Concert halls may also have full restaurant facilities.

As in theatres, noise control is important in concert halls. The design must avoid characterized or narrow-band noises in the level of audibility. Much of this noise is structure-borne, resulting from inadequate equipment and piping vibration isolation. An experienced acoustical engineer is essential for help in the design of these applications.

ARENAS AND STADIUMS

Functions at arenas and stadiums may be quite varied, so the air-conditioning loads will vary. Arenas and stadiums are not only used for sporting events such as basketball, ice hockey, boxing, and track meets but may also house circuses; rodeos; convocations; social affairs; meetings; rock concerts; car, cycle, and truck events; and special exhibitions such as home, industrial, animal, or sports shows. For multipurpose operations, the designer must provide mechanical systems having a high degree of flexibility. High-volume ventilation systems may be satisfactory in many instances, depending on load characteristics and outside air conditions.

Load Characteristics

Depending on the range of use, the load may vary from a very low sensible heat ratio for events such as boxing to a relatively high sensible heat ratio for industrial exhibitions. Multispeed fans often improve the performance at these two extremes and can aid in sound control for special events such as concerts or convocations. When using multispeed fans, the designer should consider the performance or air distribution devices and cooling coils when the fan is operating at lower speeds.

Because total comfort cannot be ensured in an all-purpose facility, the designer must determine the level of discomfort that can be tolerated, or at least the type of performances for which the facility is primarily intended.

As with other assembly buildings, seating and lighting combinations are the most important load considerations. Boxing events, for example, may have the most seating, because the arena area is very small. For the same reason, however, the area that needs to be intensely illuminated is also small. Thus, boxing matches may represent the worst latent load situation. Other events that present latent load problems are rock concerts and large-scale dinner dances, although the audience at a rock concert is generally less concerned with thermal comfort. A good exhaust ventilation system is a must, however, since the audience at rock concerts may be eating and/or smoking during the performance to a much greater degree than at pop concerts. Good exhaust ventilation is also essential in the removal of fumes at car, cycle, and truck events. Circuses, basketball, and hockey have a much larger arena area and less seating. The sensible load from lighting the arena area does improve the sensible heat ratio. The large expanse of ice in hockey games represents a considerable reduction in both latent and sensible loads. High latent loads caused by occupancy or ventilation will create severe problems in ice arenas such as condensation on interior surfaces and fog. Special attention should be paid to the ventilation system, air distribution, and construction materials.

Enclosed Stadiums

An enclosed stadium may have either a retractable or fixed roof. When the roof is open, mechanical ventilation is not required. However, when it is closed, ventilation is needed. Ductwork must be run in the permanent sections of the stadium. The large air volumes required and the long air throws make proper air distribution difficult to achieve; thus, the duct distribution systems must be capable of substantial flexibility and adjustment.

Some open stadiums have radiant heating coils in the floor slabs of the seating areas for use during cold weather. Another means of providing comfort is the use of gas-fired or electric high- or low-intensity radiant heating systems located above the occupants.

Open racetrack stadiums may present a ventilation problem if the grandstand is enclosed. The grandstand area may have multiple levels and may be in the range of 400 m long and 60 m deep. The interior (ancillary) areas must be ventilated to control odors from toilet facilities, concessions, and the high population density. General practice provides about four air changes per hour for the stand seating area and exhausts the air through the rear of the service areas. More efficient ventilation systems may be selected if architectural considerations permit. Fogging of windows is a winter concern with glass-enclosed grandstands. This can be minimized by double glazing, humidity control, moving dry air across the glass, or a radiant heating system for perimeter glass areas.

Air-supported structures require the continuous operation of a fan to maintain a properly inflated condition. The possibility of condensation on the underside of the air bubble should be considered. The U-value of the roof should be sufficient to prevent condensation at the lowest expected ambient temperature. Heating and air-conditioning functions can either be incorporated into the inflating system or they can be separately furnished. Solar and radiation control is also possible through the structure's skin. Applications, though increasing rapidly, still require working closely with the enclosure manufacturer to achieve proper and integrated results.

Ancillary Spaces

The concourse areas of arenas and stadiums contain concession stands that are heavily populated during entrance, exit, and intermission periods. Considerable odor is generated in these areas by food, drink, and smoke, requiring considerable ventilation. If energy conservation is an important factor, carbon filters and controllable recirculation rates should be considered.

Ticket offices, restaurants, and similar facilities are often expected to be operative during hours that the main arena is closed, and, therefore, separate systems should be considered for these areas.

Locker rooms require little treatment other than excellent ventilation, usually not less than 10 to 15 L/s per square metre. To reduce the outdoor air load, excess air from the main arena or stadium may be transferred into the locker room areas. However, reheat or recooling by water or primary air should be considered to maintain the locker room temperature. To maintain proper air balance under all conditions, locker room areas should have separate supply and exhaust systems.

Concourse area air systems should be considered for their flexibility of returning or exhausting air because these areas are subject to heavy smoking between periods of sports events. The economics of this type of flexibility should be evaluated with regard to the associated problem of air balance and freeze-up in cold climates.

When an ice skating rink is designed into the facility, the concerns of groundwater conditions, site drainage, structural foundations, insulation, and waterproofing become even more important,

with the potential of freezing of soil or fill under the floor and subsequent expansion. The rink floor may have to be strong enough to support heavy trucks. The floor insulation also must be strong enough to take this load. Ice-melting pits of sufficient size with steam pipes may have to be furnished. If the arena is to be air conditioned, the possibility of combining the air-conditioning system with the ice rink system may be analyzed. The designer should be aware that both systems operate at vastly different temperatures and may operate at different capacity levels at any given time. The radiant effects of the ice on the people and of the heat from the roof and lights on the ice must be considered in the design and operation of the system. Also, low air velocity at the floor is related to minimizing the refrigeration load. High air velocities will cause moisture to be drawn from the air by the ice sheet. Fog is caused by the uncontrolled introduction of airborne moisture through ventilation with warm, moist outside air and generally develops within the boarded area (playing area). Fog can be controlled by reducing outdoor air ventilation rates, using appropriate air velocities to bring the air in contact with the ice, and using a dehumidification system. Air-conditioning systems have limited impact on reducing the dew-point temperature sufficiently to prevent fog.

The type of lighting used over ice rinks must be carefully considered when precooling is used prior to hockey games and between periods. Main lights should be capable of being turned off, if feasible. Incandescent lights require no warm-up time and are more applicable than types requiring warm-up. Low emissivity ceilings with reflective characteristics successfully reduce condensation on roof structures; they also reduce lighting requirements.

Gymnasiums

Smaller gymnasiums, such as those found in school buildings, are miniature versions of arenas and often incorporate multipurpose features. For further information, see Chapter 6, Educational Facilities.

Many school gymnasiums are not air conditioned. Application of low-intensity perimeter radiant systems with a central ventilation system having four to six air changes are effective and energy efficient. An alternative method of heating with unit heaters located in the ceiling area is also effective. Ventilation must be provided due to the high activity levels and the resulting odors.

Most gymnasiums are located in schools. However, public and private organizations and health centers may also have gymnasiums. During the day, gymnasiums are usually used for physical activities, but in the evening and on weekends, they may be used for sports events, social affairs, or meetings. Thus, their activities fall within the scope of those of a civic center. More gymnasiums are being considered for air conditioning to make them more suitable for civic center activities.

The design criteria are similar to arenas and civic centers when used for nonstudent training activities. However, for schooltime use, space temperatures are often kept between 18 and 20°C, when weather permits. Occupancy and the degree of activity during daytime use does not usually require high quantities of outdoor air, but if used for other functions, system flexibility is required.

CONVENTION AND EXHIBITION CENTERS

Convention-exhibition centers schedule diverse functions similar to those at arenas and stadiums and present a unique challenge to the designer. The center generally is a high-bay, long-span space. These centers can be changed weekly, for example, from an enormous computer room into a gigantic kitchen, large machine shop, department store, automobile showroom, or miniature zoo. They can also be the site of gala banquets or used as major convention meeting rooms.

The income earned by these facilities is directly affected by the time it takes to change from one activity to the next, so highly flexible utility distribution and air-conditioning systems are needed.

Ancillary facilities include restaurants, bars, concession stands, parking garages, offices, television broadcasting rooms, and multiple meeting rooms varying in capacity from small (10 to 20 people) to large (hundreds or thousands of people). Often, an appropriately sized full-scale auditorium or arena will also be incorporated.

By their nature, these facilities are much too large and diverse in usage to be served by a single air-handling system. Multiple air-handling systems with several chillers can be economical.

Load Characteristics

The main exhibition room undergoes a variety of loads, depending on the type of activity in progress. Industrial shows provide the highest sensible loads, which may have a connected capacity of 215 W/m^2 along with one person per 3.7 to 4.6 m^2. Loads of this magnitude are seldom considered because large power-consuming equipment is seldom in continuous operation at full load. An adequate design accommodates (in addition to lighting load) about 108 W/m^2 and one person per 3.7 to 4.6 m^2 as a maximum continuous load.

Alternative loads that are very different in character may be encountered. When the main hall is used as a meeting room, the load will be much more latent in character. Thus, multispeed fans or variable volume systems may provide a better balance of load during these high latent, low sensible periods of use. The determination of accurate occupancy and usage information is critical in any plan to design and operate such a facility efficiently and effectively.

System Applicability

The main exhibition hall is normally handled by one or more all-air systems. These systems should be capable of operating on all outdoor air, because during set-up time, the hall may contain a number of highway-size trucks bringing in or removing exhibit materials. There are also occasions when the space is used for equipment that produces an unusual amount of fumes or odors, such as restaurant or printing industry displays. It is helpful to build some flues into the structure to duct noxious fumes directly to the outside. Perimeter radiant ceiling systems have been successfully applied to exhibition halls with large expanses of glass.

The groups of smaller meeting rooms are best handled with either separate individual room air-handling systems, or with variable volume central systems, because these rooms have high individual peak loads but are are not used frequently. Constant volume systems of the dual- or single-duct reheat type waste considerable energy when serving empty rooms, unless special design features are incorporated.

The offices and restaurant spaces often operate for many more hours than the meeting areas or exhibition areas and should be served from separate systems. Storage areas can generally be conditioned by exhausting excess air from the main exhibit hall through these spaces.

NATATORIUMS

The materials that are to be used in the construction of the walls, floors, and roof, and their method of application, must be carefully analyzed and selected to ensure that the building will not be damaged by its humid, corrosive environment or by condensation. The careful selection of materials for the heating, ventilating, and/or air-conditioning systems and their controls is important for the same reason. Ventilation of the pool enclosure helps control corrosion and condensation. However, ventilation is the dominant source of heat loss, thus heat recovery should be considered.

The design of the heating and air-conditioning or ventilation systems must be planned to provide comfort for spectators and swimmers, in or out of the pool. Excessive air motion or drafts in the pool area must be avoided.

Pools located in the interior of a building may provide less of a design problem than pools with wall and roof exposures. The use of glass on pool exposures complicates the design problem and adds cost, especially in cold climates. Cold draft conditions and condensation are difficult and expensive to eliminate if glass walls or exposures are used. The pool area should be isolated from adjacent building areas, if possible, by providing a negative pressure at the pool.

Load Characteristics

Natatoriums are characterized by high latent loads that should be controlled to minimize corrosion and condensation on the building construction. Outdoor air of proper moisture content may be used for this purpose. During periods when outdoor humidity is high, it may be necessary to provide some form of reheat for humidity control. Mechanical dehumidifiers with energy recycling features are increasingly used to control humidity as well as to heat pool water.

Pools used for open or free swimming have a large number of people engaged in fairly strenuous physical activity. Air must be introduced to the space without causing discomfort to occupants, in and out of the pool.

Design Concepts

The following three basic types of pools are discussed in this section:

1. Natatoriums with no spectator facilities
2. Natatoriums with spectator facilities
3. Therapeutic pools

Pools require humidity control to maintain comfort conditions. Pool air-handling systems are generally designed to use up to 100% outdoor air for cooling and/or dehumidification. On a winter cycle, when outdoor temperature and humidity are below pool design conditions, the amount of outdoor air can be controlled by a humidistat to maintain the desired humidity level. On a summer cycle, when outdoor temperature and humidity are above pool design conditions, minimum outdoor air is used. Mechanical dehumidifier systems that provide year-round humidity control make air conditioning possible for year-round temperature control and may be considered for year-round comfort control.

Spectator areas should have their own separate air supply; pool air should not return or exhaust through the spectator area because of its high moisture content and chloramine odor.

System Applicability

All-air systems are required to remove large quantities of moisture in the air. Warm wall and floor surfaces increase occupant comfort and are provided by application of insulation in walls, heating of the pool piping tunnels where available, the use of warm air curtains on perimeter walls and glass, and the use of radiation and radiant panels.

Close attention should be paid to latent chloramine and moisture levels. High levels of humidity and corrosive elements are more likely to occur during periods of high activity. When systems have marginal moisture control and when pool water treatment is unbalanced, humidity levels and corrosive elements are again likely to increase. Building material selection is extremely critical. Selection criteria should include corrosion resistance and maintaining inside surface temperatures above dew point under design outside winter conditions.

Design Criteria

Design conditions for pools include the following:

Indoor air

Pleasure swimming	24 to 29°C, 50 to 60% rh
Therapeutic	27 to 29°C, 50 to 60% rh

Pool water

Pleasure swimming	24 to 29°C
Therapeutic	29 to 35°C
Competitive swimming	22 to 24°C
Whirlpool/spa	36 to 39°C

Relative humidities should not be maintained below recommended levels because of the evaporative cooling effect on a person emerging from the pool and because of the increased rate of evaporation from the pool, which increases pool heating requirements. Humidities higher than recommended encourage corrosion and condensation problems as well as occupant discomfort.

Air velocity at any point 2.4 m above the walking deck of the pool should not exceed 0.13 m/s. In a diving area, air velocity around the divers should be below this level.

In spectator seating areas, air velocity may be increased to 0.2 to 0.25 m/s, unless seats are located in an area also occupied by swimmers.

Calculation for Minimum Air Requirements. When using outside air for moisture removal, sufficient air must be supplied to the pool to remove the water that evaporates from its surface. When using mechanical dehumidification only, the rate of evaporation needs to be determined to size the dehumidification capacity. The rate of evaporation (Carrier 1918) can be found from empirical Equation (1). This equation is for pools at normal activity, allowing for splashing and a limited area of wetted deck. Other pool uses may have more or less evaporation (Smith et al. 1993).

$$w_p = A (p_w - p_a) (0.089 + 0.0782V) / Y \qquad (1)$$

where

w_p = evaporation of water, kg/s
A = area of pool surface, m^2
V = air velocity over water surface, m/s
Y = latent heat required to change water to vapor at surface water temperature, kJ/kg
p_a = saturation pressure at room air dew point, kPa
p_w = saturation vapor pressure taken at the surface water temperature, kPa

The units for the constant, 0.089, are $W/(m^2 \cdot Pa)$. The units for the constant, 0.0782, are $W \cdot s/(m^3 \cdot Pa)$.

For Y values of about 2330 kJ/kg and V values ranging from 0.05 to 0.15 m/s, Equation (1) can be reduced to

$$w_p = 4.1 \times 10^{-5} \times A (p_w - p_a) \qquad (2)$$

The minimum air quantity required to remove this evaporated water can be determined from the following equation:

$$Q = w_p / \rho (W_i - W_o) \qquad (3)$$

where

Q = quantity of air, m^3/s
ρ = standard air density, 1.204 kg/m^3
W_i = humidity ratio of pool air at design criteria, kg/kg
W_o = humidity ratio of outdoor air at design criteria, kg/kg

The values of W can be obtained from the ASHRAE psychrometric charts.

When using makeup air for humidity control, design outdoor conditions should be reviewed carefully in cold climates, because this optimum condition for moisture removal occurs infrequently. Several outdoor conditions should be reviewed and calculated before establishing the minimum air volumes.

This calculated quantity of air often produces only one or two air changes per hour—a low rate of air movement. The following airflow rates are recommended, assuming the minimum air requirement from the previous calculation falls below the minimum

recommendations. Most codes require six air changes per hour, except where air conditioning is furnished.

Pools with no spectator facilities	4 to 6 air changes
Spectator facilities	6 to 8 air changes
Therapeutic pools	4 to 6 air changes

Air volume above the minimum calculated value is recirculated, but the fan system incorporates return-exhaust fans and outdoor air, exhaust air, and return air controls to enable larger amounts (up to 100%) of outdoor air to be introduced during milder weather for temperature and humidity control. The return-exhaust fan facilitates the introduction of outdoor air and the maintenance of recommended pressure. Heating elements should be large enough to handle more than design amounts of outdoor air to control temperature during periods of heavy pool use. With mechanical dehumidification being used for moisture removal, greater control of both temperature and humidity can be achieved.

Air Distribution and Filtering

The type of air distribution system influences the amount of air introduced into the pool area. Air volume above the minimum calculated value may be recirculated, provided the recirculated air is dehumidified and filtered to reduce air contaminants to safe levels. Care should be taken in the choice of filtration used, especially where filter media may react with chloramines in the air. The air-handling system should have filters of 45 to 60% (based on ASHRAE *Standard* 52) for occupant comfort and protection and to minimize streaking of walls and floors from dirt contacting moist surfaces.

Noise Level

Pool air system noise levels may be designed for NC 45 to 50 without causing discomfort or annoyance. However, wall, floor, and ceiling surfaces should be evaluated for their contribution to increasing noise levels.

Special Considerations

Condensation and corrosion from the humid, corrosive atmosphere of the pool can cause damage and even failure of materials and equipment within or serving a natatorium. Ferrous metals should be eliminated from all areas of pool construction. Roofs and walls must be protected by a vapor barrier. Suspended ceilings should be discouraged because they provide a high humidity enclosure that requires separate ventilation. Despite these precautions, the ceiling and appurtenances, such as lights and supports, are subject to hidden corrosion.

All components of the pool-heating, air-handling, and air distribution systems that are exposed directly or indirectly to the moist, corrosive pool atmosphere should be noncorrosive. If this is economically infeasible, they should be protected with a high-quality corrosion-resistant coating.

All ductwork serving pool areas should be constructed of aluminum, galvanized coated steel, or other rust-resistant material. The high moisture content of the pool air being conveyed in ductwork requires extra design considerations. Return or exhaust ductwork located above ceilings should be insulated on the outside and sealed against moisture penetration.

The pool should be maintained at a slightly negative pressure of 12 to 37 Pa to prevent moisture and chloramine odors from migrating to other areas of the building. Pool air may be used as makeup for showers and toilet rooms, but a separate system is more desirable. Locker rooms and offices should have separate supplies and a positive pressure relationship with respect to the pool. Openings from the pool to other areas should be minimized, and passageways should have a vestibule (air lock) or some other arrangement to discourage the passage of air and moisture.

A properly designed air distribution system is essential for uniform, draft-free conditions. Although it is more difficult to design, side wall distribution is generally less costly than overhead supply. Perimeter air distribution with air distribution upward against outside windows provides the best draft-free supply and is very effective in preventing condensation buildup. Any air outlet over the pool is difficult to adjust manually and may increase air motion over the water surface, thus increasing the rate of evaporation. It is not necessary for return and/or exhaust outlets to be located low to pick up the moisture, because the water vapor tends to rise.

In cold climates, the combination of high humidities and the use of glass requires careful design and engineering to avoid condensation problems. Double or triple glazing combined with radiation is relatively ineffective where ceiling air is directed toward high glass areas. This results in a forced downdraft condition, which will quickly overwhelm the natural convection caused by radiation. It is far more noticeable in pool areas due to evaporative cooling of bodies.

Where radiation from cold glass is not desirable, a radiant panel, perimeter floor radiant, or infrared system may be used. Because radiant floors alone may not be sufficiently effective against large glass areas, a combination of supplemental supply air to eliminate downdrafts may be used. Locating the return slots at the bottom of the glass reduces draft effects away from the wall. The use of skylights is discouraged in cold winter climates due to the difficulty of preventing condensation buildup.

Pool water-filtering equipment and chemical gases require special ventilation rates for occupational safety. This requirement is the decisive factor in the design of the pool equipment room air-handling system.

FAIRS AND OTHER TEMPORARY EXHIBITS

At frequent intervals, large-scale exhibits are constructed throughout various parts of the world to stimulate business, present new ideas, and provide cultural exchanges. Fairs of this type take years to construct, are open from several months to several years, and are sometimes designed considering future use of some of the buildings. Fairs, carnivals, or exhibits, which may consist of prefabricated shelters and tents that are moved from place to place and remain in a given location for only a few days or weeks, are not covered here because they seldom require the involvement of architects and engineers.

Design Concepts

One consultant or agency should be responsible for setting uniform utility service regulations and practices to ensure proper organization and operation of all exhibits. Exhibits that are open only during the intermediate spring or fall months require a much smaller heating or cooling plant than those designed for for peak summer or winter design conditions. This information is required in the earliest planning stages so that system and space requirements can be properly analyzed.

Occupancy

Fair buildings have heavy occupancy during visiting hours, but patrons seldom stay in any one building for long periods. The approximate length of time that patrons stay in a building determines the design of the air-conditioning system. The shorter the anticipated stay, the greater the leeway in designing for less-than-optimum operating design conditions, equipment, and duct layout. If patrons wear outer garments while in the building, this will affect operating design conditions.

Equipment and Maintenance

Heating and cooling equipment used solely for maintaining proper comfort conditions and not for exhibit purposes may be secondhand or leased equipment, if available and of the proper capacities. Another possibility is to rent all air-conditioning equipment to reduce the exhibitors' capital investment and eliminate disposal problems when the fair is over.

Depending on the size of the fair, the length of time it will operate, the types of exhibitors, and the policies of the fair sponsors, it may be desirable to analyze the potential for a centralized heating and cooling plant versus individual plants for each exhibit. The proportionate cost of a central plant to each exhibitor, including utility and maintenance costs, may be considerably less than having to furnish space and plant utility and maintenance costs. The larger the fair, the more savings may result. It may be practical to consider making the plant a showcase, suitable for exhibit and possibly added revenue. A central plant may also form the nucleus for the commercial or industrial development of the area after the fair is over.

If exhibitors furnish their own air-conditioning plants, it is advisable to analyze shortcuts that may be taken to reduce equipment space and maintenance aids. For a 6-month to 2-year maximum operating period, for example, tube pull or equipment removal space is not needed or may be drastically reduced. Higher fan and pump motor power and smaller equipment is permissible to save on initial costs. Ductwork and piping costs should be kept as low as possible because these are usually the most difficult items to salvage; cheaper materials may be substituted wherever possible. The job must be thoroughly analyzed to eliminate all unnecessary items and reduce all others to bare essentials.

The central plant may be designed for short-term use as well. However, if the plant is to be used after the fair closes, the central plant should be designed in accordance with the best practice for long-life plants. It is difficult to determine how much of the piping distribution system can be used effectively for permanent installations. For that reason, piping should be simply designed initially, preferably in a grid, loop, or modular layout, so that future additions can be made easily and economically.

Air Cleanliness

The efficiency of the filters needed for each exhibit is determined by the nature of the area served. Because the life of an exhibit is very short, it is desirable to furnish the least expensive filtering system. If possible, one set of filters should be selected to last for the life of the exhibit. In general, the filtering efficiencies do not have to exceed 30%. (See ASHRAE *Standard* 52.)

System Applicability

If a central air-conditioning plant is not built, the systems installed in each building should be the least costly to install and operate for the life of the exhibit. These units and systems should be designed and installed to occupy the minimum usable space.

Whenever feasible, heating and cooling should be performed by one medium, preferably air, to avoid running a separate piping and radiation system for heating and a duct system for cooling. Air curtains used on an extensive scale may, on analysis, simplify the building structure and lower total costs.

Another possibility when both heating and cooling are required is a heat pump system, which may be less costly than separate heating and cooling plants. Economical operation may be possible, depending on the building characteristics, lighting load, and occupant load. If well or other water is available, it may produce a more economical installation than an air-source heat pump.

Exhibits and buildings can serve many varied functions, so that when specific problems or applications arise, reference should be made to the building type most closely resembling the exhibit building.

ATRIUMS

Atriums have diverse functions and occupancies. An atrium may (1) connect buildings; (2) serve as an architectural feature, leisure space, greenhouse, and/or smoke reservoir; and (3) afford energy and lighting conversation. The designer must be fully aware of the intended uses of an atrium.

The temperature, humidity, and hours of usage of an atrium are directly related to those of the adjacent buildings. Glass window wall systems and skylight systems are common. Atriums are generally large in volume with relatively small floor areas. The temperature and humidity conditions, air distribution, impact from adjacent buildings, and fenestration loads to the space must be considered in the development of design concepts for an atrium.

Perimeter radiant heating systems (e.g., overhead radiant type, wall finned-tube or radiant type, floor radiant type, or combinations thereof) are commonly used for the expansive glass window and skylight systems. Air-conditioning systems can heat, cool, and control smoke. The distribution of air across window and skylight systems can also control heat transfer and condensation. Low supply and high return air distribution can control heat stratification, as well as wind and stack effects. Some atrium designs include a combination of high/low supply and high/low return air distribution to control heat transfer, condensation, stratification, and wind/stack effects.

The energy usage of an atrium can be reduced by installing double- and triple-panel glass and mullions with thermal breaks, as well as shading devices such as external, internal and interior screens, shades, and louvers.

Extensive landscaping is common in atriums. Humidity levels are generally maintained between 10 and 35%. Hot and cold air should not be distributed directly onto plants and trees.

REFERENCES

ASHRAE. 1976. Method of testing air-cleaning devices used in general ventilation for removing particulate matter. *Standard* 52-1976.

ASHRAE. 1989. Ventilation for acceptable indoor air quality. *Standard* 62-1989.

Carrier, W.H. 1918. The temperature of evaporation. *ASHVE Transactions* 24:25-50.

Smith, C.C., R.W. Jones, G.O.G. Lof. 1993. Energy requirements and potential savings for heated indoor swimming pools. *ASHRAE Transactions* 99(2):864-74.

DOMICILIARY FACILITIES

DOMICILIARY facilities may be single-room or multiroom, long- or short-term dwelling (or residence) units; they may be stacked sideways and/or vertically. Domiciliary facilities include apartment houses (high- and low-rise rental, cooperative, and condominium, either single-purpose structures or portions of multipurpose buildings), dormitories, hotels, motels, nursing homes, and other similar buildings. Apartment houses are discussed in Chapter 1, and nursing homes in Chapter 7.

Ideally, each room of each unit should be able to be ventilated and cooled, heated, or dehumidified independently of any other room. If this ideal is not available, optimum air conditioning for each room will be compromised.

LOAD CHARACTERISTICS

1. The space is not necessarily occupied at all times. Adequate flexibility should allow each unit's cooling and ventilation to be shut off (except when humidity control is required) and heating to be shut off or turned down.
2. Concentrations of lighting and occupancy are low; activity is generally sedentary or light. Occupancy is transient in nature, with greater night use of bedrooms. There is occasional heavy occupancy, smoking, and physical activity in dining rooms, living rooms, and other areas used for entertaining.
3. There is a potential for high appliance loads, odor generation, and large exhaust requirements in kitchens, whether integrated with or separate from residential quarters.
4. Rooms are generally exterior, except sometimes for kitchens, toilets, and dressing rooms; the building as a whole usually has multiple exposure, as may many individual dwelling units.
5. Toilet, washing, and bathing facilities are almost always incorporated within dwelling units in hostelries and nursing homes, but only occasionally in dormitories. Toilet areas requiring exhaust air are usually incorporated in each domicile unit.
6. The building has relatively high domestic hot water demands, generally for periods of an hour or two, several times a day. This can vary from a fairly moderate and level daily load profile in senior citizens' buildings to sharp, unusually high peaks at about 6:00 p.m. in dormitories. Chapter 45 includes details on service hot water systems.
7. Load characteristics of rooms, dwelling units, and buildings can be well defined without anticipation of any substantial future design load changes other than the addition of a service such as cooling that may not have been incorporated originally.

The prevalence of shifting, transient interior loads and exterior exposures with glass results in high diversity factors; the long-hour usage results in fairly high load factors.

DESIGN CONCEPTS AND CRITERIA

Wide load swings and diversity within and between rooms require the design of a flexible system for 24-h comfort. Besides

The preparation of this chapter is assigned to TC 9.8, Large Building Air-Conditioning Applications.

opening windows, the only method of providing flexible temperature control is having individual room components under individual room control that can cool, heat, and ventilate independently of the equipment in other rooms.

In some climates, summer humidity levels become objectionable because of the low internal sensible loads that result when cooling is on-off controlled. Modulated cooling and/or reheat may be required to achieve comfort. Reheat should be avoided unless some sort of heat recovery is involved.

Dehumidification can be achieved by lowering cooling coil temperatures and reducing airflow. Another means of dehumidification is desiccant dehumidifiers.

Odor control within any single dwelling unit is impractical. Most designers strive simply to confine odors to the dwelling unit. Systems in which makeup air is drawn into the dwelling unit from pressurized public corridors are the most successful. Due to the high operating cost of providing makeup air, high local odor concentrations, and codes that prohibit centralized air return from dwelling units to common supply air fans, centralized all-air systems are not in common use.

Some people have a noise threshold low enough that certain types of equipment disturb their sleep. Higher noise levels may be acceptable in areas where there is little need for air conditioning. Medium and better quality equipment is available with noise criteria (NC) 35 levels at 3 to 4 m in medium to soft rooms and little sound change when the compressor cycles.

Perimeter fan-coil systems are usually quieter than unitary systems, but unitary systems provide more redundancy in case of failure.

SYSTEMS

Energy-Efficient Systems

The systems used for this type of occupancy should be energy-efficient. The most efficient systems generally include water-source and air-source heat pumps. In areas with ample solar radiation, water-source heat pumps may be solar assisted. Energy-efficient equipment generally has the lowest operating cost and is relatively simple, an important factor in multiple-dwelling and domiciliary facilities, where skilled operating personnel are unlikely to be available. Most systems allow individual operation and thermostatic control. The typical system allows individual metering so that most, if not all, of the cooling and heating costs can be metered directly to the occupant (McClelland 1983). Existing buildings can be retrofitted with heat flow meters and timers on fan motors for individual metering.

The water-loop heat pump system has lower operating costs than air-cooled unitary equipment, especially where electric heat is used. The lower installed cost encourages the use of this system in mid- and high-rise buildings where individual dwelling units have floor areas of 75 m² or larger. Some systems incorporate sprinkler system piping as the water loop.

Except for the central circulating pump, heat rejector fans, and supplementary heater, the water-loop heat pump system is predominantly decentralized; individual metering allows most of the operating costs to be paid by the occupants. System life should be longer than for other unitary systems because most of the mechanical equipment is within the building and not exposed to outdoor conditions; also, the load on the refrigeration circuit is not as severe because water temperatures are controlled for optimum system operation. Low operating costs are due to the energy conservation inherent in the system. Excess heat may be stored during the daytime for the following night, and heat may be transferred from one part of the building to another.

Simultaneous heating and cooling occurs during cool weather because while heating is required in many areas, cooling may be necessary in rooms having high solar or internal loads. On a mild day, surplus heat throughout the building is frequently transferred into the hot water loop by water-cooled condensers on cooling cycle so that water temperature rises. The heat remains stored in water from which it can be extracted at night; a water heater is not needed. This heat storage is improved by the presence of a greater mass of water in the pipe loop; some systems include a storage tank for this reason. Because the system is designed to operate during the heating season with water supplied at temperatures as low as 15°C, the water-loop heat pump lends itself to solar assist; relatively high solar collector efficiencies result from the low water temperature.

The installed cost of the water-loop heat pump system is higher in very small buildings. In severe cold climates with prolonged heating seasons, even where natural gas or fossil fuels are available at reasonable cost, the operating cost advantages of this system may diminish unless heat can be recovered from some other source, such as solar collectors, geothermal systems, or internal heat loads from a commercial area served by the same system.

Energy-Neutral Systems

Energy-neutral systems do not allow simultaneous cooling and heating. Some examples are (1) packaged terminal air conditioners (PTACs) (through-the-wall units), (2) window units or radiant ceiling panels for cooling combined with finned or baseboard radiation for heating, (3) unitary air conditioners with an integrated heating system, (4) fan coils with remote condensing units, and (5) variable air volume (VAV) systems with either perimeter radiant panel heating or baseboard heating. To qualify as energy-neutral, systems must have controls that prevent simultaneous operation of the cooling and heating cycles. For unitary equipment, control may be as simple as a heat-cool switch. For other types, dead-band thermostatic control may be required.

PTACs are frequently installed in buildings where most of the individual units are relatively small, serving one or two rooms. In common arrangement for two rooms, a supply plenum is added to the discharge of the PTAC unit so that some portion of the conditioned air serving one room can be diverted into the second, usually smaller, room. Multiple PTAC units allow additional zoning in dwellings having more rooms. Additional radiation heat is sometimes needed in cold climates to distribute heat around the perimeter.

Heating in a PTAC system may be supplied either by electric resistance heaters or by hot water or steam heating coils. Initial costs are lower for a decentralized system using electric resistance heat. Operating costs are lower for a system using heating coils heated by combustion fuels. Despite relatively inefficient refrigeration circuits, operating costs of PTAC systems are quite reasonable, mostly because of the individual thermostatic control of each machine, which eliminates the use of reheat while preventing the space from being overheated or overcooled. Also, because the equipment is located in the space being served, little power is devoted to circulating the room air. Servicing is simple—a defective machine is replaced by a spare chassis and is then forwarded to a service organization for repair. Thus, building maintenance requires relatively unskilled personnel.

Noise levels are generally no higher than NC40, but some units are noisier than others. Installations near a seacoast should be specially constructed (usually with stainless steel or special coatings) to prevent the accelerated corrosion to aluminum and steel components caused by salt. In high-rise buildings of more than 12 stories, special care is

required, both in the design and construction of outside partitions and in the installation of air conditioners, to avoid the operating problems associated with leakage (caused by stack effect) around and through the machines.

Frequently, the least expensive system to install is finned or baseboard radiation for heating and window-type room air conditioners for cooling. The window units are often purchased individually by the building occupants. This system offers reasonable operating costs and is relatively simple to maintain. However, window units have the shortest equipment life, the highest operating noise level, and the poorest distribution of conditioned air of any of the systems evaluated in this section.

Fan coils with remote condensing units are used in smaller buildings. Fan coil units are located in closets, and ductwork distributes air to the rooms in the dwelling. Condensing units may be located on roofs, at ground level, or on balconies.

Low capacity residential warm air furnaces may be used for heating, but with gas- or oil-fired units, the products of combustion must be vented. In a one- or two-story structure, it is possible to use individual chimneys or flue pipes, but in a high-rise structure, a multiple-vent chimney or a manifold vent system is necessary. Local codes should be consulted.

Sealed combustion furnaces have been developed for domiciliary use. They draw all the combustion air from outside and discharge the flue products through a windproof vent to the outdoors. The unit must be located near an outside wall, and exhaust gases must be directed away from windows and intakes. In one- or two-story structures, outdoor units mounted on the roof or on a pad at ground level may also be used. All of these heating units can be obtained with cooling coils, either built-in or add-on. Evaporative-type cooling units are popular in motels, low-rise apartments, and residences in mild climates.

Desiccant dehumidification should be considered when independent control of temperature and humidity is required to avoid reheat.

Energy-Inefficient Systems

Energy-inefficient systems allow simultaneous cooling and heating. Examples of these systems are two-, three-, and four-pipe fan coil units, terminal reheat systems, and induction systems. Some systems, such as the four-pipe fan coil, can be controlled so that they are energy-neutral. Their primary use is for humidity control.

Four-pipe systems and two-pipe systems with electric heaters can be designed for complete temperature and humidity flexibility during summer and intermediate season weather, although none provides winter humidity control. Both of these systems provide full dehumidification and cooling with chilled water, reserving the other two pipes or the electric coil for space heating or reheat. The systems and necessary controls are expensive, and only the four-pipe system, if equipped with an internal-source heat-recovery design for the warm coil energy, can operate at low cost. When year-round comfort is essential, four-pipe systems or two-pipe systems with electric heat should be considered.

Total Energy Systems

A total energy system is an option for any multiple or large housing facility with large year-round domestic hot water requirements. Total energy systems are a form of cogeneration in which all or most electrical and thermal energy needs are met by on-site systems, as described in Chapter 7 of the 1992 *ASHRAE Handbook—Systems and Equipment*. A detailed load profile must be analyzed to determine the merits of using a total energy system. The reliability and safety of the heat-recovery system must also be considered.

Any of the previously described systems can perform the HVAC function of a total energy system. The major considerations as they apply to total energy in choosing an HVAC system are as follows:

1. Optimum use must be made of the thermal energy recoverable from the prime mover during all or most operating modes, not just during conditions of peak HVAC demand.

2. Heat recoverable via the heat pump may become less useful because the heat required during many of its potential operating hours will be recovered from the prime mover. The additional investment for heat pump or heat recovery cycles may be more difficult to justify because operating savings are lower.
3. The best application for recovered waste heat is for those services that use only heat (i.e., domestic hot water, laundry facilities, and space heating).

Special Considerations

Local building codes govern ventilation air quantities for most domiciliary buildings. Where they do not, ASHRAE *Standard* 62 should be followed. Table 2.3 of the standard lists outdoor air requirements for single and multiple private dwellings. It requires a minimum of 25 L/s intermittent or 10 L/s continuous or openable windows in bath and toilet areas and a minimum of 50 L/s intermittent or 12 L/s continuous or openable windows in kitchens. The quantity of outdoor air introduced into the corridors is usually slightly in excess of the exhaust quantities to pressurize the building. To avoid adding any load to the individual systems, outdoor air should be treated to conform to indoor air temperature and humidity conditions. In humid climates, special attention must be given to controlling the humidity introduced through the outdoor air system. Otherwise, the outdoor air may reach corridor temperature while still retaining a significant amount of humidity.

In buildings having centrally controlled exhaust and supply systems, the systems are regulated by a time clock or a central management system for certain periods of the day. In other cases, the outside air may be reduced or shut off during extremely cold periods, although this practice is not recommended and may be prohibited by local codes. These factors should be considered when estimating heating load.

For buildings using exhaust and supply air systems on a 24-h basis, consideration of air-to-air heat recovery devices may be merited (see Chapter 44 of the 1992 *ASHRAE Handbook—Systems and Equipment*). Such recovery devices can reduce energy consumption by capturing 60 to 80% of the sensible and latent heat extracted from the air source.

Infiltration loads in high-rise buildings without ventilation openings for perimeter units are not controllable on a year-round basis by general building pressurization. When outer walls are pierced to supply outdoor air to unitary or fan-coil equipment, combined wind and thermal stack-effect forces create problems. These factors must be considered for high-rise buildings (see Chapter 23 of the 1993 *ASHRAE Handbook—Fundamentals*).

Interior public corridors should have tempered supply air with transfer into individual area units, if necessary, to provide kitchen and toilet makeup air requirements through acoustically lined transfer louvers. Corridors, stairwells, and elevators should be pressurized for fire and smoke control (see Chapter 48).

Kitchen air can be recirculated through hoods with activated charcoal filters rather than exhausted. Toilet exhaust can be VAV with a damper operated by the light switch. A controlled source of supplementary heat in each bathroom is recommended to ensure comfort while bathing.

Air-conditioning equipment must be isolated to reduce noise generation or transmission. The cooling tower or condensing unit must be designed and located to avoid disturbing occupants of the building or of adjacent buildings.

An important but frequently overlooked load is the heat gain from piping for hot water services. Insulation thickness should conform to the latest local energy codes and standards at a minimum. In large, luxury-type buildings, the installation of a central energy or building management system allows supervision of individual air-conditioning units for operation and maintenance.

Some domiciliary facilities achieve energy conservation by reducing indoor temperatures during the heating season. Such a strategy should be pursued with caution, however, if there are aged occupants because they are susceptible to hypothermia.

DORMITORIES

Dormitory buildings frequently have large commercial dining and kitchen facilities, laundering facilities, and common areas for indoor recreation and bathing. Such ancillary loads may make heat pump or total energy systems valid, economical alternatives, especially on campuses with year-round activity.

When dormitories are shut down during cold weather, the heating system must supply sufficient heat to prevent freezeup. If the dormitory contains nondwelling areas such as administrative offices or eating facilities, these facilities should be designed as a separate zone or with a separate system for flexibility, economy, and odor control.

Subsidiary facilities should be controlled separately for flexibility and shutoff capability, but they may share common refrigeration and heating plants. With internal-source heat pumps, this interdependence of unitary systems permits the reclamation of all internal heat usable for building heating, domestic water preheating, and snow melting. It is easier and less expensive to place heat reclaim coils in the building's exhaust airstream than to use air-to-air heat recovery devices. Heat reclaim can easily be sequence controlled to add heat to the building's chilled water systems when required.

HOTELS AND MOTELS

Hotel and motel accommodations are usually single rooms with toilet and bath adjacent to a corridor, flanked on both sides by other guest rooms. The building may be single-story, low-rise, or high-rise. Multipurpose subsidiary facilities range from stores and offices to ballrooms, dining rooms, kitchens, lounges, auditoriums, and meeting halls. Luxury motels may be built with similar facilities. Occasional variations are seen, such as kitchenettes, multiroom suites, and, more frequently than for apartments, outside doors to patios and balconies.

Load Characteristics

The diversity of use of all hostelry areas makes load profile studies essential for avoiding unnecessary oversizing and duplication and for generating low operating cost designs. Loads in hostelries without guest room cooking facilities usually peak an hour or two earlier in the morning than those in apartment houses and have lower noon and evening peaks. Electrical and cooling peaks for hostelries are highest between 6:00 and 8:00 p.m.

Because of the lower guest room peaks, the substantial use of subsidiary facilities with noncoincident peaks, and the more concentrated living areas, load factors for hostelries are higher than for apartment buildings.

Design Concepts and Criteria

Air conditioning in hotel rooms should be quiet, easily adjustable, and draft-free. It must also provide ample outside air. The hotel business is competitive, and space is at a premium, so systems that require little space and have low total owning and operating costs should be selected.

Common central plant and air distribution systems for each auxiliary area can lead to major investment savings. When odor generation is common to the facilities on one air system, or when diverse odors are generated but controlled in a common apparatus, a VAV system with zone reheat sequencing can offer many advantages over the unitary system approach. Chapters 3 and 4 provide information on various public and assembly areas.

Systems must be trouble-free and easy to maintain because system shutdown can often cause loss of revenue. Multiple heating and refrigeration plant units ensure continuity of services. Emergency generators may be required for protection against electrical power outages, although many utilities provide dual service and transfer devices to prevent long power outages.

The ballroom is the showplace of a hotel; special attention must be paid to incorporating a satisfactory system into the interior design of the room. A ballroom should be served by its own system. Many ballrooms do not provide proper comfort for the functions they serve. When used for banquets, ballrooms generally have more odors and smoke than other places of assembly. Most ballrooms are designed as multipurpose rooms, and the peak loads often occur when they are used as meeting rooms.

Initial and operating costs can be lowered by proper transfer of air from air-conditioned areas to service areas such as kitchens, workshops, and laundry and storage rooms. Smaller air quantities may achieve satisfactory results and eliminate the need for separate supply systems.

In hotel guest wing towers, space is at a premium, so the placement of chases to the top of the building for kitchen and boiler flues and emergency generator exhaust stacks (which take away potential guest room space) should be analyzed and determined early in the design review. Such discharges should be located at least 30 m away from the nearest outside air intake.

Hotels require good instrumentation for all major systems and services to provide accurate operating data for analysis to ensure economy of operation. A good maintenance staff and well-instructed cleaning staff can regulate or shut off energy services in unoccupied areas. Many hotels and motels have some type of guest room energy management system that allows units in unoccupied rooms to be shut off, temperature to be reset, or all functions to be controlled from the front desk and overridden by the occupant.

Applicability of Systems

For guest rooms, most hotels use fan coil systems or deluxe self-contained (PTAC) systems. Depending on the architectural design and the climate, the units may be located under the exterior windows, in the hung ceiling over the bathroom or vestibule, or in the wall between the bathroom and the bedroom.

Hotels are suited to VAV systems because they have simple room layouts and no guest cooking facilities. VAV systems are particularly suitable in tropical or high-humidity climates and if doors open to the exterior, creating condensation problems on the exposed casings of fan-coil units.

ASHRAE *Standard* 62 requires at least 18 L/s of outdoor air in bathrooms. This air usually comes from the ventilation air delivered to each room or through the room's PTAC unit, if provided. Some systems supply the corridors with outdoor air, which is transferred into each room by the bathroom exhaust; however, this arrangement provides no odor and smoke control. In tropical climates and where outdoor air supplied to the corridors has not been previously treated, this indirect method also provides poor humidity control; in many instances, makeup air treatment systems are added. In large systems, desiccant dehumidification used in conjunction with direct or indirect evaporative cooling or heat pipes can be very energy-efficient in providing the required outdoor air for the facility.

There may be great latitude in the selection of systems for other hotel areas, as long as the systems satisfy the space and hotel operating requirements. Systems are predominantly all-air because of high ventilation requirements.

Because of the substantial public area cooling required during the heating season, as well as the large heat sources in kitchens and laundries, internal-source heat pumps with storage should be considered for hostelries. These heat pumps are good candidates for use with total energy systems because of their constant thermal demand.

Special Considerations

In some climates, supplementary heat should be available in guest bathrooms. Guest rooms opening to the outdoors should have self-closing doors to prevent excess condensation from forming on the cooling coils within room units, as well as to save energy.

Design information for administration, eating, entertainment, assembly, and other areas can be found in Chapters 3 and 4. In tropical climates, special design techniques for proper insulation, ventilation, and vapor retardation are necessary to prevent the moisture buildup in closets and storage areas that results in mildew and sometimes costly damage to clothing and other materials.

These problems can be minimized by ensuring that outdoor air is adequately dried before it is allowed into the building. Infiltration of humid outdoor air into wall cavities is reduced if the building is kept under slightly positive pressure with dehumidified air. This practice can help compensate for small leaks in air/vapor barriers and for minor building water leaks.

The likelihood of excess humidity in the wall can be reduced by drying ventilation air to approximately a 12°C dew point before it enters the building. This ensures that air will not exceed 60% relative humidity, even near surfaces that are at 20°C (i.e., when guest rooms are overcooled). By keeping the air in the building structure below 60% rh, the HVAC system can prevent moisture absorption by building materials and thus prevent the growth of mold and mildew behind walls and in building furnishings.

Before a building is accepted as complete by the owner, a certified air balance contractor should be engaged to demonstrate that the volume of dry makeup air exceeds the volume of exhaust air. As the building ages, it is important that this slight positive pressure is maintained inside the building. Otherwise, the infiltration of humid air into building cavities will allow absorption regardless of how dry the room air is maintained (Banks 1992).

A desiccant dehumidifier or refrigeration type with heat pipes can be added to an existing makeup air system for humidity control.

MULTIPLE-USE COMPLEXES

These complexes combine retail facilities, office space, hotel space, residential space, and/or other commercial space into a single site. The peak HVAC demands of the various facilities may occur at different times of the day and year. Loads should be determined independently for each occupancy. Where a central plant is considered, a block load should also be determined.

Separate air handling and distribution should serve the separate facilities. However, heating and cooling units can be combined economically into a central plant. A central plant provides good opportunities for heat recovery, thermal storage, and other similarly sophisticated techniques that may not be economical in a single-use facility. A multiple-use complex is a good candidate for a centralized system of fire and smoke control, security, remote monitoring, billing for central facility use, maintenance control, building operations control, and energy management.

REFERENCES

Banks, N.J. 1992. Field test of a desiccant-based HVAC system for hotels. *ASHRAE Transactions* 98(1):1303-10.

McClelland, L. 1983. Tenant paid energy costs in multi-family rental housing. DOE, University of Colorado, Campus Box 108, Boulder, CO 80309.

BIBLIOGRAPHY

Kimbrough, J. 1990. The essential requirements for mold and mildew. Plant Pathology Department, University of Florida, Gainesville, FL.

Peart, V. 1989. Mildew and moisture problems in hotels and motels in Florida. Institute of Food and Agricultural Sciences, University of Florida, Gainesville, FL.

Wong, S.P. and S.K. Wang. 1990. Fundamentals of simultaneous heat and mass transfer between the building envelope and the conditioned space. *ASHRAE Transactions* 96(2).

EDUCATIONAL FACILITIES

THIS chapter contains technical and environmental factors and considerations that will assist the design engineer in the proper application of heating, ventilating, and air-conditioning systems and equipment for educational facilities. For further information on the HVAC systems mentioned in this chapter, see the 1992 *ASHRAE Handbook—Systems and Equipment*.

GENERAL CONSIDERATIONS

All forms of educational facilities require an efficiently controlled atmosphere to help provide a proper learning environment. This involves the selection of heating, ventilating, and air-conditioning systems, equipment, and controls to provide adequate ventilation, comfort, and a quiet atmosphere. The system must also be easily maintained by the facility's maintenance staff.

The selection of the HVAC equipment and systems depends upon whether the facility is new or existing and whether it is to be totally or partially renovated. For minor renovations, existing HVAC systems are often expanded in compliance with current codes and standards with equipment that matches the existing types. For major renovations or new construction, new HVAC systems and equipment can be installed. In some cases, old and outdated systems and equipment in existing facilities can be replaced during new construction if the budget allows. The remaining useful life of existing equipment and distribution systems should be considered.

Energy use and its associated life-cycle costs should be considered in evaluating HVAC systems and equipment. Energy analysis may justify new HVAC equipment and systems when a good return on investments can be shown. The engineer must take care to review all assumptions in the energy analysis with the school administration. Assumptions, especially as they relate to hard-to-measure items such as infiltration and part-load factors, can have a significant influence on the energy use calculated.

Other considerations for existing facilities are (1) whether the central plant is of adequate capacity to handle the additional loads of new or renovated facilities, (2) the age and condition of the existing equipment, pipes, and controls, and (3) the capital and operating costs of new equipment. Educational facilities usually have very limited budgets.

When new facilities are built or major renovations to existing facilities are made, seismic bracing of the HVAC equipment should be considered. Refer to Chapter 50, Seismic Restraint Design, for further information. Seismic codes may also apply in areas where tornados and hurricanes necessitate additional bracing. This consideration is especially important if there is an agreement with local officials to let the facility be used as a disaster relief shelter.

The type of HVAC equipment selected also depends upon the climate and the months of operation. In hot, dry climates, for instance, evaporative cooling may be the primary type of cooling; some school districts may choose not to provide air conditioning. In hot,

humid climates, it is recommended that air conditioning or dehumidification be operated year-round to prevent the growth of mold and mildew.

The final general consideration in the selection of HVAC systems and equipment is the project budget. As mentioned previously, most school districts have limited funds, so the costs of equipment and systems and the energy they use must be justified. Any savings in capital expenditures and energy costs may be available for the maintenance and upkeep of the HVAC systems and equipment and for other facility needs.

DESIGN CONSIDERATIONS

Climate control systems for educational facilities may be similar to systems applied to other types of buildings. To suit the mechanical systems to an educational facility, the designer must understand not only the operation of the various systems but also the functions of the school. Year-round school programs, adult education, night classes, and community functions put ever-increasing demands on these facilities; the environmental control system must be designed carefully to meet these needs. When occupied, classrooms, lecture rooms, and assembly rooms have uniformly dense occupancies. Gymnasiums, cafeterias, laboratories, shops, and similar functional spaces have variable occupancies and varying ventilation and temperature requirements. Many schools use gymnasiums as auditoriums and community meeting rooms. Wide variation in use and occupancy makes system flexibility important in these multiuse rooms.

Design considerations for the indoor environment of educational facilities include heat loss, heat gain, humidity, air movement, ventilation methods, ventilation rates, noise, and vibration. Winter and summer design dry-bulb temperatures for the various spaces in schools are given in Table 1. These values should be evaluated for each application to avoid energy waste while ensuring comfort.

Table 1 Recommended Winter and Summer Design Dry-Bulb Temperatures for Various Spaces Common in Schools[a]

Space	Winter Design, °C	Summer Design, °C
Laboratories	22	26[d]
Classrooms, auditoriums, libraries, administrative areas, etc.	22	26
Shops	22	26[b]
Locker, shower rooms	24[d]	c,d
Toilets	22	c
Storage	18	c,d
Mechanical rooms	16	c
Corridors	20	27[b]

[a]For spaces of high population density and where sensible heat factors are 0.75 or less, lower dry-bulb temperatures will result in generation of less latent heat, which may reduce the need for reheat and thus save energy. Therefore, optimum dry-bulb temperatures should be the subject of detailed design analysis.
[b]Frequently not air conditioned.
[c]Usually not air conditioned.
[d]Provide ventilation for odor control.

The preparation of this chapter is assigned to TC 9.8, Large Building Air-Conditioning Applications.

Table 2 Typical Ranges of Sound Design Levels

Space	A-Sound Levels, dB	Desired NC (Noise Criteria)
Libraries, classrooms	35-45	30-40
Laboratories, shops	40-50	35-45
Gyms, multipurpose corridors	40-55	35-50
Kitchens	45-55	40-55

Sound levels are also an important consideration in the design of educational facilities. Typical ranges of sound design levels are given in Table 2. Whenever possible and as the budget allows, the design sound level should be at the lower end of these ranges. For additional information, see Chapter 43, Sound and Vibration Control.

Proper ventilation can control odors and inhibit the spread of respiratory diseases. Refer to ASHRAE *Standard* 62 for the ventilation rates required for various spaces within educational facilities.

Equipment for school facilities should be reliable and require minimal maintenance. Equipment location is also an important consideration. The easier equipment is to access and maintain, the more promptly it will be serviced. For example, roof hatches, ladders, and walkways should be provided for access to rooftop equipment.

Controls. Controls for most educational facilities should be kept as simple as possible because a wide variety of people need to use, understand, and maintain them. The different demands of the various spaces indicate the need for individual controls within each space. However, unless each space is served by its own self-contained unit, this approach is usually not practical. Many times, a group of similar spaces will be served by a single thermostat controlling one heating or cooling unit. Rooms grouped together should have similar occupancy schedules and the same outside wall orientation, so that their cooling loads peak at approximately the same time.

Because most facilities are unoccupied at night and on weekends, night setback, boiler optimization, and hot water temperature reset controls are usually incorporated. When setbacks are used, morning warm-up and cool-down controls should be considered. After morning warm-up, automatic controls should change systems from heating to cooling as spaces heat up due to light and occupancy loads.

A manual or programmable clock is desirable for the control of most systems. These clock systems should have a manually wound spring mechanism or a battery backup for power outages. More sophisticated systems may include computerized control. Computerized systems can do much more than start and stop equipment; capabilities include resetting water temperatures, optimizing chillers and boilers, and keeping maintenance logs. See Chapter 36, Computer Applications, for more information about computerized management systems. For further information on controls, see Chapter 42, Automatic Control.

Preschools

Commercially operated preschools are generally provided with a standard architectural layout based on owner-furnished designs. This may include the HVAC systems and equipment. All preschool facilities require quiet and economical systems. The equipment should be easy to operate and maintain. The design should provide for warm floors and no drafts. Consideration should be given to the fact that the children are smaller and will play on the floor, while the teacher is taller and will be walking around. The teacher also requires a place for desk work; the designer should consider treating this area as a separate zone.

Preschool facilities generally operate on weekdays from early in the morning to 6:00 or 7:00 p.m. This schedule usually coincides with the normal working hours of the children's parents. The HVAC systems will therefore operate 12 to 14 hours per work day and be either off or on night setback at night and on weekends.

Supply air outlets should be located so that the floor area is maintained at about 24°C without the introduction of drafts. Both supply and return air outlets should be placed where they will not be blocked by furniture that may be positioned along the walls. Coordination with the architect about the location of these outlets is essential. Proper ventilation is essential for controlling odors and helping prevent the spread of diseases between the children.

The use of floor-mounted heating equipment, such as electric baseboard heaters, within the space should be carefully evaluated, as children must be prevented from coming in contact with hot surfaces or electrical devices. Also, because floor space may be at a premium, placing equipment in the area may not be allowed.

Elementary Schools

Elementary schools are similar to preschools except that they have more facilities, such as gymnasiums and cafeterias. Such facilities are generally occupied from about 7:00 a.m to about 3:00 p.m. Peak cooling loads usually occur at the end of the school day. Peak heating is usually early in the day, when classrooms begin to be occupied and outdoor air is introduced into the facility. The following are facilities found in elementary schools:

Classrooms. Each classroom should at a minimum be heated and ventilated. Air conditioning should be considered for facilities in warm, humid climates, for school districts that have year-round classes, and as the construction budget allows. In humid climates, consideration should be given to providing dehumidification during the summer months, when the school is unoccupied, to prevent the growth of mold and mildew. Economizer cooling during the winter months, to provide cooling and ventilation after morning warm-up cycles, should also be evaluated.

Gymnasiums. Gyms may be used after regular school hours for evening classes, meetings, and other functions. There may also be weekend use for group activities. The loads for these occasional uses should be considered when selecting and sizing the systems and equipment for the gym, so that small part loads can be met using as little energy as possible. Gymnasiums should have independent HVAC control.

Administrative Areas. The school's office area should be set up for individual control because it is usually occupied after school hours. As the offices are also occupied before school starts in the fall, air conditioning for the area should be considered.

Science Classrooms. Science rooms are now being provided for elementary schools. Although the children do not usually perform experiments, odors may be generated if the teacher demonstrates an experiment or if animals are kept in the classroom. Under these conditions, adequate ventilation for science classrooms is essential; an exhaust fan with a local on-off switch may be needed for the occasional removal of excessive odors.

Libraries. If possible, libraries should be air conditioned to preserve the books and materials stored in them. See Chapter 3, Commercial and Public Buildings, for additional information.

Middle and Secondary Schools

Middle and secondary schools are usually occupied for longer periods of time than elementary schools, and they have additional special facilities. The following are typical special facilities for these schools:

Auditoriums. These facilities require a quiet atmosphere as well as heating, ventilating, and in some cases air conditioning. Auditoriums are not often used except for special assemblies, practice for special programs, and special events. For other considerations see Chapter 4, Places of Assembly.

Computer Classrooms. These facilities have a high sensible heat load due to the computer equipment. Computer rooms may require additional cooling equipment such as small spot-cooling units to offset the additional load. Humidification may also be required. See Chapter 16, Data Processing System Areas, for additional information.

Science Classrooms. Science facilities in middle and secondary schools may require fume hoods with special exhaust systems. A makeup air system may be required if there are several fume hoods within a room. If there are no fume hoods, a room exhaust system may be required for odor removal, depending on the type of experiments conducted in the room and whether animals are kept within the room. Any associated storage and preparation rooms are generally exhausted continuously to remove odors and vapors emanating from stored materials. The amount of exhaust and the location of exhaust grilles may be dictated by NFPA or local codes. See Chapter 13, Laboratory Systems, for further information.

Auto Repair Shops. These facilities require outdoor air ventilation to remove odors and fumes and to provide makeup air for exhaust systems. The shop is usually heated and ventilated but not air conditioned. To contain odors and fumes, return air should not be supplied to other spaces, and the shop should be kept at a negative pressure relative to the surrounding spaces. Special exhaust systems such as welding exhaust or direct-connected carbon monoxide exhaust systems may be required. See Chapter 26, Industrial Exhaust Systems, for further information.

Ice Rinks. These facilities require special air-conditioning and dehumidification systems to keep the spectators comfortable, maintain the ice surface at freezing conditions, and prevent the formation of fog at the surface. See Chapter 33 of the 1994 *ASHRAE Handbook—Refrigeration* and Chapter 4, Places of Assembly, for further discussion of these systems.

School Stores. These facilities contain school supplies and paraphernalia and are usually open for short periods of time. The heating and air-conditioning systems serving these areas should be able to be shut off when the store is closed to save energy.

Natatoriums. These facilities, like ice rinks, require special humidity control systems. In addition, special materials of construction are required. See Chapter 4, Places of Assembly, for further information.

Industrial Shops. These facilities are similar to auto repair shops and have special exhaust requirements for welding, soldering, and paint booths. In addition, a dust collection system is sometimes provided and the collected air is returned to the space. Industrial shops have a high sensible load due to the operation of the shop equipment. When calculating loads, the design engineer should consult the shop teacher about the shop operation, and, where possible, diversity factors should be applied. See Chapter 26, Industrial Exhaust Systems, for more information.

Locker Rooms. It is required by code that these facilities be maintained at a negative pressure with respect to the surrounding spaces. They are usually heated and ventilated only. Makeup air is required; the exhaust and makeup air systems should be coordinated and should operate only when required.

Home Economics Rooms. These rooms usually have a high sensible heat load due to appliances such as washing machines, dryers, stoves, ovens, and sewing machines. Different options should be considered for the exhaust of the stoves and dryers. If local codes allow, residential-style range hoods may be installed over the stoves. A central exhaust system could be applied to the dryers as well as the stoves. If enough appliances are located within the room, a makeup air system may be required. See Chapter 28, Kitchen Ventilation, for more information.

Colleges and Universities

College and university facilities are similar to those of middle and secondary schools, except that there are more of them, and they may be located in several buildings on the campus. Some colleges and universities have satellite campuses scattered throughout a city or a state. The design criteria for each building are established by the requirements of the users. The following is a list of major facilities commonly found on college and university campuses but not in middle and secondary schools:

Administrative Buildings. These buildings should be treated as office buildings. They usually have a constant interior load and additional part-time loads associated with such spaces as conference rooms.

Animal Facilities. These spaces are commonly associated with laboratories, but usually have their own separate areas. Animal facilities need close temperature control and require a significant amount of outdoor air ventilation to control odors and prevent the spread of disease among the animals. Discussion of animal facilities is found in Chapter 13, Laboratory Systems.

Dormitory Buildings. Discussion of dormitories can be found in Chapter 5, Domiciliary Facilities.

Storage Buildings. These buildings are usually heated and ventilated only; however, a portion of the facility may require temperature and humidity control, depending upon the materials stored there.

Student and Faculty Apartment Buildings. Discussion of these facilities can be found in Chapter 5, Domiciliary Facilities.

Large Gymnasiums. These facilities are discussed in Chapter 4, Places of Assembly.

Laboratory Buildings. These buildings house research facilities and may contain fume hoods, machinery, lasers, animal facilities, areas with controlled environments, and departmental offices. The HVAC systems and controls must be able to accommodate the diverse functions of the facility, which has the potential for 24-hour, year-round operation, and yet be easy to service and quick to repair. Variable air volume (VAV) systems can be used in laboratory buildings, but a proper control system should be applied to introduce the required quantities of outdoor air. Significant energy savings can be achieved by recovering the energy from the exhaust air and tempering the outdoor makeup air. Some other energy-saving systems employed for laboratory buildings are (1) ice storage, (2) economizer cycles, (3) heat reclaim chillers to produce domestic hot water or hot water for booster coils in the summer, and (4) cooling-tower free cooling.

The design engineer should discuss expected contaminants and concentrations with the owner to determine which materials of construction should be used for fume hoods and fume exhaust systems. Backup or standby systems for emergency use should be considered, as should alarms on critical systems. The maintenance staff should be thoroughly trained in the upkeep and repair of all systems, components, and controls. See Chapter 13, Laboratory Systems, for additional information.

Museums. These facilities are discussed in Chapter 4, Places of Assembly.

Central Plants. These facilities contain the generating equipment and other supporting machinery for campus heating and air conditioning. Central plants are generally heated and ventilated only. Central plants are covered in Chapter 11 of the 1992 *ASHRAE Handbook—Systems and Equipment.*

Television and Radio Studios. Discussion of these facilities can be found in Chapter 3, Commercial and Public Buildings.

Hospitals. Hospitals are discussed in Chapter 7, Health Care Facilities.

Chapels. Discussion of these facilities can be found in Chapter 4, Places of Assembly.

Student Unions. These facilities may contain a dining room, a kitchen, meeting rooms, lounges, and game rooms. Student unions are generally open every day from early in the morning to late at night. An HVAC system such as VAV cooling with perimeter heating may be suitable. Extra ventilation air is required for makeup air for the kitchen systems and due to the high occupancy of the building. See Chapter 3, Commercial and Public Buildings, and Chapter 28, Kitchen Ventilation, for additional information.

Lecture Halls. These spaces house large numbers of students for a few hours several times a day and are vacant the rest of the time, so heat loads vary throughout the day. The systems that serve lecture halls must respond to these load variations. In addition, because the rooms are configured such that students at the rear of the room are seated near the ceiling, the systems must be quiet and must not produce any drafts. Coordination with the design team is essential for determining the optimal location of diffusers and return grilles and registers. Placement of the supply diffusers at the front of the room and the return air grilles and registers near the front and rear of the room or under the seats should be considered. An analysis of the system noise levels expected in the hall is recommended. See Chapter 43, Sound and Vibration Control, for further information.

ENERGY CONSIDERATIONS

Most new buildings and renovated portions of existing buildings must comply with the energy codes and standards enforced by local code officials. Use of energy-efficient equipment, systems, and building envelope materials should always be considered and evaluated. Methods of reducing energy usage should be discussed by the design team.

Outdoor air economizer cycles use cool outdoor air to accommodate high internal loads for many hours of the year, as dry- and wet-bulb temperatures permit. A relief air system sized to remove up to 100% of the outdoor air that could be supplied is required. Some packaged rooftop air-conditioning units do not have adequate relief air outlets, which could lead to overpressurization of the space. Additional relief vents or an exhaust system may be necessary.

Night setback, activated by a time clock controller and controlled by night setback thermostats, conserves energy by resetting the heating and cooling space temperatures during unoccupied hours. Intermittent use of educational facilities sometimes requires that individual systems have their own night setback control times and settings.

Coupled with night setback is *morning warm-up and cool-down*. This allows the space temperatures to be reset for normal occupancy at a predetermined time. This can be achieved either by setting a time clock for approximately one hour before occupancy or through the optimization program of a central energy management system. Such a program determines the warm-up or cool-down time based on the mass of the building, the outdoor air temperature, and a historical log of building occupancy.

Hot water temperature reset can save energy by resetting the building hot water temperature based upon outdoor air temperature. Care must be taken to not lower the water temperature below that recommended by the boiler manufacturer.

During cool weather, *cooling-tower free cooling* uses the cooling towers and either chiller refrigerant migration or a heat exchanger to cool the chilled water system without using the chillers. This system is useful for facilities with high internal heat loads.

Heat recovery systems can save energy when coupled to systems that generate a substantial amount of exhaust air. A study must be performed to determine whether a heat recovery system is cost-effective. Various heat recovery systems include runaround loops, heat wheels, and heat exchangers.

Boiler economizers can save energy by recovering the heat lost in the stacks of large boilers or in central plants. The recovered heat can be used to preheat the boiler makeup water or combustion air. Care should be exercised not to reduce the stack temperature enough to cause condensation. Special stack materials are required for systems in which condensation may occur.

Variable-speed drives for large fans and pumps can save energy, especially if coupled with energy-efficient motors. The HVAC systems should be evaluated to determine whether there is enough variation in the airflows or water flows to justify the application of these devices. A cost analysis should also be performed.

Passive solar techniques, such as the use of external shading devices at fenestrations, can be incorporated into the building design. Coordination with the owner and design team is essential for this strategy. A life-cycle cost analysis may be necessary to evaluate the cost-effectiveness of this method.

Solar energy systems can either supplement other energy sources or be the sole energy source for an HVAC system. A life-cycle cost analysis should be performed to evaluate the cost-effectiveness of such a system.

For additional information on energy conservation, see Chapter 31, Energy Resources.

EQUIPMENT AND SYSTEM SELECTION

The architectural configuration of an educational building has a considerable influence on the building's heating and cooling loads and therefore on the selection of its HVAC systems. Other influences on system selection are the budget and whether the project is to expand an existing system or to integrate an HVAC system into the design of a new building. ASHRAE *Standard* 100.5 has recommendations for energy conservation in existing institutional buildings. ASHRAE *Standard* 90.1 should be followed for the energy-efficient design of new buildings. ASHRAE *Standard* 62 addresses the criteria necessary to meet indoor air quality requirements.

The only trend evident in HVAC design for educational facilities is that air-conditioning systems are being installed whenever possible. In smaller single buildings, systems such as unit ventilators, heat pumps, small rooftop units, and VAV systems are being applied. In larger facilities, VAV systems, large rooftop units with perimeter heating systems, and heat pump systems coupled with central systems are utilized. When rooftop units are used, stairs to the roof or penthouses make maintenance easier.

Single-Duct with Reheat

Reheat systems are restricted from use by ASHRAE *Standard* 90.1 unless recovered energy is used as the reheat medium. Booster heating coils should be considered if the heat is to be applied after mechanical cooling has been turned off or locked out (e.g., during an economizer cycle). The booster heating coil system can use recovered energy to further reduce energy costs.

Multizone

The original multizone design, which simply heated or cooled by individual zones, is seldom used because mixing hot and cold airstreams is not energy-efficient. Multizone system units can be equipped with reset controllers that receive signals from zone thermostats, allowing them to maintain optimum hot and cold temperatures and minimize the mixing of airstreams.

Another energy-conserving technique for multizone systems is the incorporation of a neutral zone that mixes heated or cooled air with return air, but does not mix heated and cooled airstreams. This technique minimizes the energy-wasting practice of requiring the simultaneous production of both heating and cooling.

Multizone systems are also available that have individual heating and cooling within each zone and therefore avoid simultaneous heating and cooling. The zone thermostats energize either the heating system or the cooling system as required.

Replacement multizone units with the energy-saving measures mentioned above are also available for existing systems. The replacement of multizone units with VAV systems described in the next section has been successful.

Variable Air Volume

Types of variable air volume systems include single-duct, single-duct with perimeter heating, and dual-duct systems. The supply air volumes from the primary air system are adjusted according to the space load requirements. VAV systems are usually zoned by exposure and occupancy schedules. The primary system usually provides supply air to the zones at a constant temperature either for cooling only or, in the case of dual-duct systems, for heating or cooling. To conserve energy, heating and cooling should not be provided simultaneously to the same zone. Heating should be provided only during economizer cooling or when mechanical cooling has been turned off or locked out.

Perimeter heating can be provided by duct-mounted heating coils, fan terminal units with duct- or unit-mounted heating coils, perimeter fin tubes, baseboard heaters, cabinet unit heaters, or convectors. Radiant ceiling or floor heating systems can also be applied around the perimeter. Each zone or exposure should be controlled to prevent overheating of the space. Perimeter air-distribution systems should be installed so that drafts are not created.

Fan terminal units are utilized when constant air motion is required. These units operate either constantly or only during heating periods. Noisy units should be avoided. A system of fan terminal units with heating coils allows the primary heating system to be shut off during night setback or unoccupied periods.

Packaged Units

Unit ventilators or through-the-wall packaged units with either split or integral direct-expansion cooling may serve one or two rooms with a common exposure. Heat pumps can also be used. When a single packaged unit covers more than one space, zoning is critical. As is the case for VAV systems, spaces with similar occupancy schedules are grouped together. If exterior spaces are to be served, exterior wall orientation must also match.

For water-source heat pumps, a constant flow of water is circulated to the units. This water is tempered by boilers for heating in the winter or by a fluid cooler provided for heat rejection (cooling). Heat rejection may be necessary for much of the school year, depending upon the internal loads. Certain operating conditions balance the water loop for water-source heat pumps, which could result in periods in which neither primary heating nor cooling is required. There may be simultaneous heating and cooling by different units throughout the facility, depending upon exposure and room loads. A water storage tank may be incorporated into the water loop to store heated water during the day for use at night.

Air-to-air heat pumps are also used in schools, as are small rooftop and self-contained unit ventilators featuring heat pump cycles.

REFERENCES

ASHRAE. 1989. Energy conservation design of new buildings except low-rise residential buildings. *Standard* 90.1-1989.

ASHRAE. 1989. Ventilation for acceptable indoor air quality. *Standard* 62-1989.

ASHRAE. 1991. Energy conservation in existing buildings—Institutional. *Standard* 100.5

HEALTH CARE FACILITIES

CONTINUAL advances in medicine and technology necessitate constant reevaluation of the air-conditioning needs of hospitals and medical facilities. While medical evidence has shown that proper air conditioning is helpful in the prevention and treatment of many conditions, the relatively high cost of air conditioning demands efficient design and operation to ensure economical energy management (Woods et al. 1986, Murray et al. 1988, Demling and Maly 1989, and Fitzgerald 1989).

Health care occupancy classification, based on the latest occupancy guidelines from the National Fire Protection Association (NFPA), should be considered early in the design phase of the project. Health care occupancy is important for fire protection (smoke zones, smoke control) and for the future adaptability of the HVAC system for a more restrictive occupancy.

Health facilities are becoming increasingly diversified in response to a trend toward outpatient services. The term clinic may refer to any building from the ubiquitous residential doctor's office to a specialized cancer treatment center. Prepaid health maintenance provided by integrated regional health care organizations is becoming the model for medical care delivery. These organizations, as well as long established hospitals, are constructing buildings that look less like hospitals and more like office buildings.

For the purpose of this chapter, health care facilities are divided into the following categories:

1. Hospital facilities
2. Outpatient health care facilities
3. Nursing home facilities

The specific environmental conditions required by a particular medical facility may vary from those in this chapter, depending on the agency responsible for the environmental standard. Among the agencies that may have standards for medical facilities are state and local health agencies, the Department of Health and Human Services, Indian Health Service, Public Health Service, Medicare/Medicaid, the Department of Defense (Army, Navy, Air Force), the Department of Veterans Affairs, and the Joint Commission on Accreditation of Healthcare Organizations (JCAHO). It is advisable to discuss infection control objectives with the hospital's infection control committee.

The general hospital was selected as the basis for the fundamentals outlined in the first section of this chapter because of the variety of services it provides. Environmental conditions and design criteria apply to comparable areas in other health facilities.

The general acute care hospital has a core of critical care spaces, including operating rooms, labor rooms, delivery rooms, and a nursery. Usually the functions of radiology, laboratory, central sterile, and the pharmacy are located close by the critical care space. Inpatient nursing, including intensive care nursing, is in the complex.

The facility also incorporates an emergency room, a kitchen, dining and food service, a morgue, and central housekeeping support.

Criteria for outpatient facilities are given in the second section of this chapter. Outpatient surgery is performed with the anticipation that the patient will not stay overnight. An outpatient facility may be part of an acute care facility, a freestanding unit, or part of another medical facility such as a medical office building.

Nursing homes are addressed separately in the third section of the chapter because their fundamental requirements differ greatly from those of other medical facilities.

AIR CONDITIONING IN THE PREVENTION AND TREATMENT OF DISEASE

Hospital air conditioning assumes a more important role than just the promotion of comfort. In many cases, proper air conditioning is a factor in patient therapy; in some instances, it is the major treatment.

Studies show that patients in controlled environments generally have more rapid physical improvement than do those in uncontrolled environments. Patients with thyrotoxicosis do not tolerate hot, humid conditions or heat waves very well. A cool, dry environment favors the loss of heat by radiation and evaporation from the skin and may save the life of the patient.

Cardiac patients may be unable to maintain the circulation necessary to ensure normal heat loss. Therefore, air conditioning cardiac wards and rooms of cardiac patients, particularly those with congestive heart failure, is necessary and considered therapeutic (Burch and Depasquale 1962). Individuals with head injuries, those subjected to brain operations, and those with barbiturate poisoning may have hyperthermia, especially in a hot environment, due to a disturbance in the heat regulatory center of the brain. Obviously, an important factor in recovery is an environment in which the patient can lose heat by radiation and evaporation—namely, a cool room with dehumidified air.

A hot, dry environment of 32°C dry bulb and 35% rh has been successfully used in treating patients with rheumatoid arthritis.

Dry conditions may constitute a hazard to the ill and debilitated by contributing to secondary infection or infection totally unrelated to the clinical condition causing hospitalization. Clinical areas devoted to upper respiratory disease treatment and acute care, as well as the general clinical areas of the entire hospital, should be maintained at a relative humidity of 30 to 60%.

Patients with chronic pulmonary disease often have viscous respiratory tract secretions. As these secretions accumulate and increase in viscosity, the patient's exchange of heat and water dwindles. Under these circumstances, the inspiration of warm, humidified air is essential to prevent dehydration (Walker and Wells 1961).

Patients needing oxygen therapy and those with a tracheotomy require special attention to ensure a warm, humid supply of inspired air. Cold, dry oxygen or the bypassing of the nasopharyngeal mucosa presents an extreme situation. Rebreathing techniques for

---

The preparation of this chapter is assigned to TC 9.8, Large Building Air-Conditioning Applications.

anesthesia and enclosure in an incubator are special means of addressing impaired heat loss in therapeutic environments.

Burn patients need a hot environment and high relative humidity. A ward for severe burn victims should have temperature controls that permit adjusting the room temperature up to 32°C dry bulb and the relative humidity up to 95%.

HOSPITAL FACILITIES

Although proper air conditioning is helpful in the prevention and treatment of disease, the application of air conditioning to health facilities presents many problems not encountered in the usual comfort conditioning system.

The basic differences between air conditioning for hospitals (and related health facilities) and that for other building types stem from (1) the need to restrict air movement in and between the various departments; (2) the specific requirements for ventilation and filtration to dilute and remove contamination in the form of odor, airborne microorganisms and viruses, and hazardous chemical and radioactive substances; (3) the different temperature and humidity requirements for various areas; and (4) the design sophistication needed to permit accurate control of environmental conditions.

Infection Sources and Control Measures

Bacterial Infection. Examples of bacteria that are highly infectious and transported within air or air and water mixtures are *Mycobacterium tuberculosis* and *Legionella pneumophila* (Legionnaire's disease). Wells (1934) showed that droplets or infectious agents of 5 μm or less in size can remain airborne indefinitely. Isoard et al. (1980) and Luciano (1984) have shown that 99.9% of all bacteria present in a hospital are removed by 90 to 95% efficient filters (ASHRAE *Standard* 52.1). This is because bacteria are typically present in colony-forming units that are larger than 1 μm. Some authorities recommend the use of high efficiency particulate air (HEPA) filters having dioctyl phthalate (DOP) test filtering efficiencies of 99.97% in certain areas.

Viral Infection. Examples of viruses that are transported by and virulent within air are *Varicella* (chicken pox/shingles), *Rubella* (German measles), and *Rubeola* (regular measles). Epidemiological evidence and other studies indicate that many of the airborne viruses that transmit infection are submicron in size; thus, there is no known method to effectively eliminate 100% of the viable particles. HEPA and/or ultra low penetration (ULPA) filters provide the greatest efficiency currently available. Attempts to deactivate viruses with ultraviolet light and chemical sprays have not proven reliable or effective enough to be recommended by most codes as a primary infection control measure. Therefore, isolation rooms and isolation anterooms with appropriate ventilation-pressure relationships are the primary means used to prevent the spread of airborne viruses in the hospital environment.

Molds. Evidence indicates that some molds such as *Aspergillis* can be fatal to advanced leukemia, bone marrow transplant, and other immunocompromised patients.

Outdoor Air Ventilation. If outdoor air intakes are properly located, and areas adjacent to outdoor air intakes are properly maintained, outdoor air, in comparison to room air, is virtually free of bacteria and viruses. Infection control problems frequently involve a bacterial or viral source within the hospital. Ventilation air dilutes the viral and bacterial contamination within a hospital. If ventilation systems are properly designed, constructed, and maintained to preserve the correct pressure relations between functional areas, they remove airborne infectious agents from the hospital environment.

Temperature and Humidity. These conditions can inhibit or promote the growth of bacteria and activate or deactivate viruses. Some bacteria such as *Legionella pneumophila* are basically waterborne and survive more readily in a humid environment. Codes and guidelines specify temperature and humidity range criteria in some hospital areas as a measure for infection control as well as comfort.

AIR QUALITY

Systems must also provide air virtually free of dust, dirt, odor, and chemical and radioactive pollutants. In some cases, outside air is hazardous to patients suffering from cardiopulmonary, respiratory, or pulmonary conditions. In such instances, systems that intermittently provide maximum allowable recirculated air should be considered.

Outdoor Intakes. These intakes should be located as far as practical (on directionally different exposures whenever possible), but not less than 9 m, from combustion equipment stack exhaust outlets, ventilation exhaust outlets from the hospital or adjoining buildings, medical-surgical vacuum systems, cooling towers, plumbing vent stacks, and areas that may collect vehicular exhaust and other noxious fumes. Outdoor intakes may be close to outlets that exhaust air suitable for recirculation; however, exhaust air must not short-circuit into the intakes of outdoor air units or fan systems used for smoke control. The bottom of outdoor air intakes serving central systems should be located as high as practical but not less than 1.8 m above ground level or, if installed above the roof, 0.9 m above the roof level (AIA 1992).

Exhaust Outlets. These exhausts should be located a minimum of 3 m above ground level and away from doors, occupied areas, and operable windows. Preferred location for exhaust outlets is at roof level projecting upward or horizontally away from outdoor intakes. Care must be taken in locating highly contaminated exhausts (e.g., from engines, fume hoods, biological safety cabinets, kitchen hoods, and paint booths). Prevailing winds, adjacent buildings, and discharge velocities must be taken into account (see Chapter 31 of the 1992 *ASHRAE Handbook—Systems and Equipment*). In critical applications, wind tunnel studies or computer modeling may be appropriate.

Air Filters. A number of methods are available for determining the efficiency of filters in removing particulates from an airstream (see Chapter 25 of the 1992 *ASHRAE Handbook—Systems and Equipment*). All central ventilation or air-conditioning systems should be equipped with filters having efficiencies no lower than those indicated in Table 1. Where two filter beds are indicated, Filter Bed No. 1 should be located upstream of the air-conditioning equipment, and Filter Bed No. 2 should be downstream of the supply fan, any recirculating spray water systems, and water-reservoir type humidifiers. Appropriate precautions should be observed to prevent wetting of the filter media by free moisture from humidifiers. Where only one filter bed is indicated, it should be located upstream of the air-conditioning equipment. All filter efficiencies are based on ASHRAE *Standard* 52.1.

The following are guidelines for filter installations:

1. HEPA filters having DOP test efficiencies of 99.97% should be used on air supply systems serving rooms for clinical treatment of patients with a high susceptibility to infection due to leukemia, burns, bone marrow transplant, organ transplant, or acquired immunodeficiency syndrome (AIDS). HEPA filters should also be used on the discharge air from fume hoods or safety cabinets in which infectious or highly radioactive materials are processed. The filter system should be designed and equipped to permit safe removal, disposal, and replacement of contaminated filters.

2. All filters should be installed to prevent leakage between the filter segments and between the filter bed and its supporting frame. A small leak that permits any contaminated air to escape through the filter can destroy the usefulness of the best air cleaner.

3. A manometer should be installed in the filter system to measure the pressure drop across each filter bank. This precaution furnishes a more accurate indication of when filters should be replaced than does visual observation.

Table 1 Filter Efficiencies for Central Ventilation and Air-Conditioning Systems in General Hospitals

Minimum Number of Filter Beds	Area Designation	Filter Efficiencies, %		
		Filter Bed		
		No. 1[a]	No. 2[a]	No. 3[b]
3	Orthopedic operating room Bone marrow transplant operating room Organ transplant operating room	25	90	99.97[c]
2	General procedure operating rooms Delivery rooms Nurseries Intensive care units Patient care rooms Treatment rooms Diagnostic and related areas	25	90	
1	Laboratories Sterile storage	80		
1	Food preparation areas Laundries Administrative areas Bulk storage Soiled holding areas	25		

[a]Based on ASHRAE *Standard* 52.1-1992.
[b]Based on DOP test.
[c]HEPA filters at air outlets.

Table 2 Influence of Bedmaking on Airborne Bacterial Count in Hospitals

Item	Count per Cubic Metre	
	Inside Patient Room	Hallway near Patient Room
Background	1200	1060
During bedmaking	4940	2260
10 min after	2120	1470
30 min after	1270	950
Background	560	
Normal bedmaking	3520	
Vigorous bedmaking	6070	

Source: Greene et al. (1960).

4. High efficiency filters should be installed in the system, with adequate facilities provided for maintenance without introducing contamination into the delivery system or the area served.

5. Because high efficiency filters are expensive, the hospital should project the filter bed life and replacement costs and incorporate these into their operating budget.

6. During construction, openings in ductwork and diffusers should be sealed to prevent intrusion of dust, dirt, and hazardous materials. Such contamination is often permanent and provides a medium for the growth of infectious agents. Existing or new filters may rapidly become contaminated by construction dust.

Air Movement

The data given in Table 2 illustrate the degree to which contamination can be dispersed into the air of the hospital environment by one of the many routine activities for normal patient care. The bacterial counts in the hallway clearly indicate the spread of this contamination.

Because of the dispersal of bacteria resulting from such necessary activities, air-handling systems should provide air movement patterns that minimize the spread of contamination. Undesirable airflow between rooms and floors is often difficult to control because of open doors, movement of staff and patients, temperature differentials, and

stack effect, which is accentuated by the vertical openings such as chutes, elevator shafts, stairwells, and mechanical shafts common to hospitals. While some of these factors are beyond practical control, the effect of others may be minimized by terminating shaft openings in enclosed rooms and by designing and balancing air systems to create positive or negative air pressure within certain rooms and areas.

Systems serving highly contaminated areas, such as autopsy rooms and isolation rooms for contagious or immunocompromised patients, should maintain a positive or negative air pressure within these rooms relative to adjoining rooms or the corridor (Murray et al. 1988). The pressure is obtained by supplying less air to the area than is exhausted from it. This induces a flow of air into the area around the perimeters of doors and prevents an outward airflow. The operating room offers an example of an opposite condition. This room, which requires air that is free of contamination, must be positively pressurized relative to adjoining rooms or corridors to prevent any airflow from these relatively highly contaminated areas.

A differential in air pressure can be maintained only in an entirely closed room. Therefore it is important to obtain a reasonably close fit of all doors or other barriers between pressurized areas. This is best accomplished by using weather stripping and drop bottoms on doors. The opening of a door or closure between two areas instantaneously reduces any existing pressure differential between them to such a degree that its effectiveness is nullified. When such openings occur, a natural interchange of air takes place because of the thermal currents resulting from temperature differences between the two areas.

For critical areas requiring both the maintenance of pressure differentials to adjacent spaces and personnel movement between the critical area and adjacent spaces, the use of appropriate air locks or anterooms is indicated.

Figure 1 shows the bacterial count in a surgery room and its adjoining rooms during a normal surgical procedure. These bacterial counts were taken simultaneously. The relatively low bacterial counts in the surgery room, compared with those of the adjoining rooms, are attributable to the lower level of activity and higher air pressure within operating rooms.

In general, outlets suppling air to sensitive ultraclean areas and highly contaminated areas should be located on the ceiling, with perimeter or several exhaust inlets near the floor. This arrangement provides a downward movement of clean air through the breathing and working zones to the contaminated floor area for exhaust. The bottoms of return or exhaust openings should be at least 75 mm above the floor.

The laminar airflow concept developed for industrial clean room use has attracted the interest of some medical authorities. There are advocates of both vertical and horizontal laminar airflow systems, with and without fixed or movable walls around the surgical team (Pfost 1981). Some medical authorities do not advocate laminar

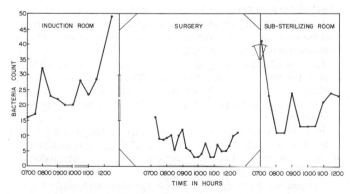

Fig. 1 Typical Airborne Contamination in Surgery and Adjacent Areas

airflow for surgeries but encourage air systems similar to those described in this chapter.

Laminar airflow in surgical operating rooms is defined as airflow that is predominantly unidirectional when not obstructed. The unidirectional laminar airflow pattern is commonly attained at a velocity of 0.46 ± 0.10 m/s.

Laminar airflow systems have shown promise for rooms used for the treatment of patients who are highly susceptible to infection (Michaelson et al. 1966). Among such patients would be the badly burned and those undergoing radiation therapy, concentrated chemotherapy, organ transplants, amputations, and joint replacement.

Temperature and Humidity

Specific recommendations for design temperatures and humidities are given in the next section, Specific Design Criteria. Temperature and humidity requirements for other inpatient areas not covered should be 22°C or less and 30% to 60% rh.

Pressure Relationships and Ventilation

Table 3 covers ventilation standards for comfort, asepsis, and odor control in areas of acute care hospitals that directly affect patient care. Table 3 does not necessarily reflect the criteria of the American Institute of Architects (AIA) or any other group. If specific organizational criteria must be met, refer to that organization's literature. Ventilation in accordance with ASHRAE *Standard* 62, Ventilation for Acceptable Indoor Air Quality, should be used for areas where specific standards are not given. Where a higher outdoor air requirement is called for in ASHRAE *Standard* 62 than in Table 3, the higher value should be used. Specialized patient care areas, including organ transplant and burn units, should have additional ventilation provisions for air quality control as may be appropriate.

Design of the ventilation system must as much as possible provide air movement from clean to less clean areas. In critical care areas, constant volume systems should be employed to assure proper pressure relationships and ventilation, except in unoccupied rooms. In noncritical patient care areas and staff rooms, variable air volume (VAV) systems may be considered for energy conservation. When using VAV systems within the hospital, special care should be taken to ensure that minimum ventilation rates (as required by codes) are maintained and that pressure relationships between various departments are maintained. With VAV systems, a method such as air volume tracking between supply, return, and exhaust could be used to control pressure relationships (Lewis 1988). In Table 3, those areas that require continuous control are labeled P, N, or E for positive, negative, or equal pressure. Where ± is used, there is no requirement for continuous directional control.

The number of air changes may be reduced to 25% of the indicated value when the room is unoccupied if provisions are made to ensure that (1) the number of air changes indicated is reestablished whenever the space is occupied and (2) the pressure relationship with the surrounding rooms is maintained when the air changes are reduced.

In areas requiring no continuous directional control (±), ventilation systems may be shut down when the space is unoccupied and ventilation is not otherwise needed.

Because of the cleaning difficulty and potential for buildup of contamination, recirculating room units must not be used in areas marked "No." Note that the standard recirculating room unit may also be impractical for primary control where exhaust to the outside is required.

In rooms having hoods, extra air must be supplied for hood exhaust so that the designated pressure relationship is maintained. Refer to Chapter 13, Laboratory Systems, for further discussion of laboratory ventilation.

For maximum energy conservation, use of recirculated air is preferred. If an all-outdoor air system is used, an efficient heat recovery method should be considered.

Smoke Control

As the ventilation design is developed, a proper smoke control strategy must be considered. Passive systems rely on fan shutdown, smoke and fire partitions, and operable windows. Proper treatment of duct penetrations must be observed.

Active smoke control systems use the ventilation system to create areas of positive and negative pressures that, along with fire and smoke partitions, limit the spread of smoke. The ventilation system may be used in a smoke removal mode in which the products of combustion are exhausted by mechanical means. As design of active smoke control systems continues to evolve, the engineer and code authority should carefully plan system operation and configuration. Refer to Chapter 48 and NFPA *Standards* 90A, 92A, 99, and 101.

SPECIFIC DESIGN CRITERIA

There are seven principal divisions of an acute care general hospital: (1) surgery and critical care, (2) nursing, (3) ancillary, (4) administration, (5) diagnostic and treatment, (6) sterilizing and supply, and (7) service. The environmental requirements of each of the departments/spaces within these divisions differ to some degree according to their function and the procedures carried out in them. This section describes the functions of these departments/spaces and covers details of design requirements. Close coordination with health care planners and medical equipment specialists in the mechanical design and construction of health facilities is essential to achieve the desired conditions.

Surgery and Critical Care

No area of the hospital requires more careful control of the aseptic condition of the environment than does the surgical suite. The systems serving the operating rooms, including cystoscopic and fracture rooms, require careful design to reduce to a minimum the concentration of airborne organisms.

The greatest amount of the bacteria found in the operating room comes from the surgical team and is a result of their activities during surgery. During an operation, most members of the surgical team are in the vicinity of the operating table, creating the undesirable situation of concentrating contamination in this highly sensitive area.

Operating Rooms. Studies of operating room air distribution systems and observation of installations in industrial clean rooms indicate that delivery of the air from the ceiling, with a downward movement to several exhaust inlets located on opposite walls, is probably the most effective air movement pattern for maintaining the concentration of contamination at an acceptable level. Completely perforated ceilings, partially perforated ceilings, and ceiling-mounted diffusers have been applied successfully (Pfost 1981).

In the average hospital, operating rooms are in use no more than 8 to 12 h per day (excepting emergencies). For this reason and for energy conservation, the air-conditioning system should allow a reduction in the air supplied to some or all of the operating rooms. However, positive space pressure must be maintained at reduced air volumes to ensure sterile conditions. Consultation with the hospital surgical staff will determine the feasibility of providing this feature.

A separate air exhaust system or special vacuum system should be provided for the removal of anesthetic trace gases (NIOSH 1975). Medical vacuum systems have been used for removal of nonflammable anesthetic gases (NFPA *Standard* 99). One or more outlets may be located in each operating room to permit connection of the anesthetic machine scavenger hose.

Although good results have been reported from air disinfection of operating rooms by irradiation, this method is seldom used. The reluctance to use irradiation may be attributed to the need for special designs for installation, protective measures for patients and personnel, constant monitoring of lamp efficiency, and maintenance.

Table 3 General Pressure Relationships and Ventilation of Certain Hospital Areas

Function Space	Pressure Relationship to Adjacent Areas	Minimum Air Changes of Outdoor Air per Hour[a]	Minimum Total Air Changes per Hour[b]	All Air Exhausted Directly to Outdoors	Air Recirculated Within Room Units
SURGERY AND CRITICAL CARE					
Operating room (all outdoor air system)	P	15[c]	15	Yes	No
(recirculating air system)	P	5	25	Optional	No
Delivery room (all outdoor air system)	P	15	15	Optional	No
(recirculating air system)	P	5	25	Optional	No
Recovery room	E	2	6	Optional	No
Nursery suite	P	5	12	Optional	No
Trauma room[d]	P	5	12	Optional	No
Anesthesia storage (see code requirements)	±	Optional	8	Yes	No
NURSING					
Patient room[e]	±	2	4	Optional	Optional
Toilet room[f]	N	Optional	10	Yes	No
Intensive care	P	2	6	Optional	No
Protective isolation[g]	P	2	15	Yes	Optional[h]
Infectious Isolation[g]	±	2	6	Yes	No
Isolation alcove or anteroom	±	2	10	Yes	No
Labor/delivery/recovery/postpartum (LDRP)	E	2	4	Optional	Optional
Patient corridor[e]	E	2	4	Optional	Optional
ANCILLARY					
Radiology X-ray (surgery and critical care)	P	3	15	Optional	No
X-ray (diagnostic and treatment)	±	2	6	Optional	Optional
Darkroom	N	2	10	Yes[i]	No
Laboratory, general	N	2	6	Yes	No
Laboratory, bacteriology	N	2	6	Yes	No
Laboratory, biochemistry	P	2	6	Optional	No
Laboratory, cytology	N	2	6	Yes	No
Laboratory, glasswashing	N	Optional	10	Yes	Optional
Laboratory, histology	N	2	6	Yes	No
Laboratory, nuclear medicine	N	2	6	Yes	No
Laboratory, pathology	N	2	6	Yes	No
Laboratory, serology	P	2	6	Optional	No
Laboratory, sterilizing	N	Optional	10	Yes	No
Laboratory, media transfer	P	2	4	Optional	No
Autopsy	N	2	12	Yes	No
Nonrefrigerated body-holding room[j]	N	Optional	10	Yes	No
Pharmacy	P	2	4	Optional	Optional
ADMINISTRATION					
Admitting and Waiting Rooms	N	2	6	Yes	Optional[h]
DIAGNOSTIC AND TREATMENT					
Bronchoscopy, sputum collection, and pentamidine administration	N	2	10	Yes	Optional[h]
Examination room[e]	±	2	6	Optional	Optional
Medication room	P	2	4	Optional	Optional
Treatment room[e]	±	2	6	Optional	Optional
Physical therapy and hydrotherapy	N	2	6	Optional	Optional
Soiled workroom or soiled holding	N	2	10	Yes	No
Clean workroom or clean holding	P	2	4	Optional	Optional
STERILIZING AND SUPPLY					
Sterilizer equipment room	N	Optional	10	Yes	No
Soiled or decontamination room	N	2	6	Yes	No
Clean workroom and sterile storage	P	2	4	Optional	Optional
Equipment storage	±	2 (Optional)	2	Optional	Optional
SERVICE					
Food preparation centers[k]	±	2	10	Yes	No
Warewashing	N	Optional	10	Yes	No
Dietary day storage	±	Optional	2	Optional	No
Laundry, general	N	2	10	Yes	No
Soiled linen sorting and storage	N	Optional	10	Yes	No
Clean linen storage	P	2 (Optional)	2	Optional	Optional
Linen and trash chute room	N	Optional	10	Yes	No
Bedpan room	N	Optional	10	Yes	No
Bathroom	N	Optional	10	Optional[f]	No
Janitor's closet	N	Optional	10	Optional	No

P = Positive   N = Negative   ± = Continuous directional control not required[e]

[a]Ventilation in accordance with ASHRAE *Standard* 62-1989, Ventilation for Acceptable Indoor Air Quality, should be used for areas for which specific ventilation rates are not given. Where a higher outdoor air requirement is called for in *Standard* 62 than in Table 3, the higher value should be used.

[b]Total air changes indicated should be either supplied or, where required, exhausted.

[c]For operating rooms, 100% outside air should be used only when codes require it and only if heat recovery devices are used.

[d]The term trauma room as used here is the first aid room and/or emergency room used for general initial treatment of accident victims. The operating room within the trauma center that is routinely used for emergency surgery should be treated as an operating room.

[e]Although continuous directional control is not required, variations should be minimized, and in no case should a lack of directional control allow the spread of infection from one area to another. Boundaries between functional areas (wards or departments) should have directional control. Lewis (1988) describes methods for maintaining directional control by applying air tracking controls.

[f]For a discussion of design considerations for central toilet exhaust systems, see the section on Patient Rooms.

[g]The infectious isolation rooms described in this table are those that might be used for infectious patients in the average community hospital. The rooms are negatively pressurized. Some isolation rooms may have a separate anteroom. Refer to the discussion in the chapter for more detailed information. Where highly infectious respirable diseases such as tuberculosis are to be isolated, increased air change rates should be considered.

Protective isolation rooms are those used for immunosuppressed patients. The room is positively pressurized to protect the patient. Anterooms are generally required and should be negatively pressurized with respect to the patient room.

[h]Recirculation is allowed in rooms with possible respiratory isolation patients if supply air is HEPA filtered.

[i]All air need not be exhausted if darkroom equipment has scavenging exhaust duct attached and meets ventilation standards of NIOSH, OSHA, and local employee exposure limits.

[j]The nonrefrigerated body-holding room exists only in facilities that do not perform autopsies on-site and use the space for short periods while waiting for the body to be transferred.

[k]Food preparation centers should have an excess of air supply for positive pressure when hoods are not in operation. The number of air changes may be reduced or varied for odor control when the space is not in use. Minimum total air changes per hour should be that required to provide proper makeup air to kitchen exhaust systems. See Chapter 28, Kitchen Ventilation.

The following conditions are recommended for operating, catheterization, cystoscopic, and fracture rooms:

1. There should be a variable range temperature capability of 20 to 24°C.
2. Relative humidity should be kept between 50 and 60%.
3. Air pressure should be maintained positive with respect to any adjoining rooms by supplying 15% excess air.
4. Differential pressure indicating device should be installed to permit air pressure readings in the rooms. Thorough sealing of all wall, ceiling, and floor penetrations and tight-fitting doors are essential to maintaining readable pressure.
5. Humidity indicator and thermometers should be located for easy observation.
6. Filter efficiencies should be in accordance with Table 1.
7. Entire installation should conform to the requirements of NFPA *Standard* 99, Health Care Facilities.
8. All air should be supplied at the ceiling and exhausted or returned from at least two locations near the floor (see Table 3 for minimum ventilating rates). Bottom of exhaust outlets should be at least 75 mm above the floor. Supply diffusers should be of the unidirectional type. High-induction ceiling or side wall diffusers should be avoided.
9. Acoustical materials should not be used as duct linings unless 90% efficient minimum terminal filters are installed downstream of the linings. Internal insulation of terminal units may be encapsulated with approved materials. Duct-mounted sound traps should be of the packless type or have polyester film linings over acoustical fill.
10. Any spray-applied insulation and fireproofing should be treated with fungi growth inhibitor.
11. Sufficient lengths of watertight, drained stainless steel duct should be installed downstream of humidification equipment to assure complete evaporation of water vapor before air is discharged into the room.

Control centers that monitor and permit adjustment of temperature, humidity, and air pressure may be located at the surgical supervisor's desk.

Obstetrical. The pressure in the obstetrical department should be positive or equal to that in other areas.

Delivery Rooms. The design for the delivery room should conform to the requirements of operating rooms.

Recovery Rooms. Postoperative recovery rooms used in conjunction with the operating rooms should be maintained at a temperature of 24°C and a relative humidity between 50 and 60%. Because residual anesthesia odor sometimes creates an odor problem in recovery rooms, ventilation is important, and a balanced air pressure relative to the air pressure of adjoining areas should be provided.

Nursery Suite. Air conditioning in nurseries provides the constant temperature and humidity conditions essential to care of the newborn in a hospital environment. Air movement patterns in nurseries should be carefully designed to reduce the possibility of drafts.

All air supplied to nurseries should enter at or near the ceiling and be removed near the floor with the bottom of exhaust openings located at least 75 mm above the floor. Air system filter efficiencies should conform to Table 1. Finned tube radiation and other forms of convection heating should not be used in nurseries.

Full-Term Nursery. A temperature of 24°C and relative humidity from 30 to 60% are recommended for the full-term nursery, examination room, and work space. The maternity nursing section should be controlled similarly to protect the infant during visits with the mother. The nursery should have a positive air pressure relative to the work space and examination room, and any rooms interposed between the nurseries and the corridor should be similarly pressurized relative to the corridor. This prevents the infiltration of contaminated air from outside areas.

Special Care Nursery. The design conditions for this nursery require a variable range temperature capability of 24 to 27°C and relative humidity from 30 to 60%. This nursery is usually equipped with individual incubators to regulate temperature and humidity. It is desirable to maintain these same conditions within the nursery proper to accommodate both infants removed from the incubators and those not placed in incubators. The pressurization of this nursery should correspond to that of the regular nurseries.

Observation Nursery. Temperature and humidity requirements for this nursery are similar to those for the full-term nursery. Because infants in these nurseries have unusual clinical symptoms, the air from this area should not enter other nurseries. A negative air pressure relative to the air pressure of the workroom should be maintained in the nursery. The workroom, usually interposed between the nursery and the corridor, should be pressurized relative to the corridor.

Emergency. This department, in most instances, is the most highly contaminated area in the hospital as a result of the soiled condition of many arriving patients and the relatively large number of persons accompanying them. Temperatures and humidities of offices and waiting spaces should be within the comfort range.

Trauma Rooms. Trauma rooms should be ventilated in accordance with requirements in Table 3. Emergency operating rooms located near the emergency department should have the same temperature, humidity, and ventilation requirements as those of an operating room.

Anesthesia Storage Room. The anesthesia storage room must be ventilated in conformance with NFPA *Standard* 99. However, mechanical ventilation only is recommended.

Nursing

Patient Rooms. When central systems are used to air condition patients' rooms, the recommendations in Tables 1 and 3 for air filtration and air change rates should be followed to reduce cross-infection and to control odor. Rooms used for isolation of infected patients should have all air exhausted directly outdoors. A winter design temperature of 24°C with 30% rh is recommended; 24°C with 50% rh is recommended for summer. Each patient room should have individual temperature control. Air pressure in patient suites should be neutral in relation to other areas.

Most governmental agency design criteria and codes require that all air from toilet rooms be exhausted directly outdoors. The requirement appears to be based on odor control. Chaddock (1986) analyzed odor from central (patient) toilet exhaust systems of a hospital and found that large central exhaust systems generally have sufficient dilution to render the toilet exhaust practically odorless.

Where room unit systems are used, it is common practice to exhaust through the adjoining toilet room an amount of air equal to the amount of outdoor air brought into the room for ventilation. The ventilation of toilets, bedpan closets, bathrooms, and all interior rooms should conform to applicable codes.

Intensive Care Unit. This unit serves seriously ill patients, from the postoperative to the coronary patient. A variable range temperature capability of 0 to 0°C, a relative humidity of 30% minimum and 60% maximum, and positive air pressure are recommended.

Protective Isolation Units. Immunosuppressed patients (including bone marrow or organ transplant, leukemia, burn, and AIDS patients) are highly susceptible to diseases. Some physicians prefer an isolated laminar airflow unit to protect the patient. Others are of the opinion that the conditions of the laminar cell have a psychologically harmful effect on the patient and prefer flushing out the room and reducing spores in the air. An air distribution of 15 air changes per hour supplied through a nonaspirating diffuser is often recommended. The sterile air is drawn across the patient and returned near the floor, at or near the door to the room.

In cases where the patient is immunosuppressed but not contagious, a positive pressure should be maintained between the patient room and adjacent area. Some jurisdictions may require an anteroom, which maintains a negative pressure relationship with respect to the adjacent isolation room and an equal pressure relationship with respect to the corridor, nurses' station, or common area. Exam and treatment rooms should be controlled in the same manner. A positive pressure should also be maintained between the entire unit and the adjacent areas to preserve sterile conditions.

When a patient is both immunosuppressed and contagious, isolation rooms within the unit may be designed and balanced to provide a permanent equal or negative pressure relationship with respect to the adjacent area or anteroom. Alternatively, when it is permitted by the jurisdictional authority, such isolation rooms may be equipped with controls that enable the room to be positive, equal, or negative in relation to the adjacent area. However, in such instances, controls in the adjacent area or anteroom must maintain the correct pressure relationship with respect to the other adjacent room(s).

A separate, dedicated air-handling system to serve the protective isolation unit simplifies pressure control and quality (Murray 1988).

Infectious Isolation Unit. The infectious isolation room is used to protect the remainder of the hospital from the patients' infectious diseases. Recent multidrug-resistant strains of tuberculosis have increased the importance of pressurization, air change rates, filtration, and air distribution design in these rooms (Rousseau 1993). Temperatures and humidities should correspond to those specified for patient rooms.

The designer should work closely with health care planners and the code authority to determine the appropriate isolation room design. It may be desirable to provide more complete control, with a separate anteroom used as an air lock to minimize the potential that airborne particulates from the patients' area reach adjacent areas.

Some designers have provided isolation rooms that allow maximum space flexibility by using an approach that reverses the airflow direction by varying exhaust flow rate. This approach is useful only if appropriate adjustments can be ensured for different types of isolation procedures.

Floor Pantry. Ventilation requirements for this area depend on the type of food service adopted by the hospital. Where bulk food is dispensed and dishwashing facilities are provided in the pantry, the use of hoods over equipment, with exhaust to the outdoors, is recommended. Small pantries used for between-meal feedings require no special ventilation. The air pressure of the pantry should be in balance with that of adjoining areas to reduce the movement of air into or out of it.

Labor/Delivery/Recovery/ Postpartum (LDRP). The procedures for normal childbirth are considered noninvasive, and rooms are controlled similarly to patient rooms. Some jurisdictions may require higher air change rates than in a typical patient room. It is expected that invasive procedures such as cesarean section are performed in a nearby delivery or operating room.

Ancillary

Radiology Department. Among the factors that affect the design of ventilation systems in these areas are the odorous characteristics of certain clinical treatments and the special construction designed to prevent radiation leakage. The fluoroscopic, radiographic, therapy, and darkroom areas require special attention.

Fluoroscopic, Radiographic, and Deep Therapy Rooms. These rooms require temperatures of 24 to 27°C and humidity of 40 to 50%. Depending on the location of air supply outlets and exhaust intakes, lead lining may be required in supply and return ducts at the points of entry to the various clinical areas to prevent radiation leakage to other occupied areas.

The darkroom is normally in use for longer periods than the X-ray rooms and should have an independent system to exhaust the air to the outdoors. The exhaust from the film processor may be connected into the darkroom exhaust.

Laboratories. Air conditioning is necessary in laboratories for the comfort and safety of the technicians (Degenhardt and Pfost

1983). Chemical fumes, odors, vapors, heat from the equipment, and the undesirability of open windows all contribute to this need.

Particular attention should be given to the size and type of equipment heat gain used in the various laboratories, as equipment heat gain usually constitutes the major portion of the cooling load.

The general air distribution and exhaust systems should be constructed of conventional materials following standard designs for the type of systems used. Exhaust systems serving hoods in which radioactive materials, volatile solvents, and strong oxidizing agents such as perchloric acid are used should be fabricated of stainless steel. Washdown facilities should be provided for hoods and ducts handling perchloric acid. Perchloric acid hoods should have dedicated exhaust fans.

Hood use may dictate other duct materials. Hoods in which radioactive or infectious materials are to be used must be equipped with ultrahigh efficiency filters at the exhaust outlet and have a procedure and equipment for the safe removal and replacement of contaminated filters. Exhaust duct routing should be as short as possible with a minimum of horizontal offsets. This is especially true for perchloric acid hoods because of the extremely hazardous, explosive nature of this material.

Determining the most effective, economical, and safe system of laboratory ventilation requires considerable study. Where the laboratory space ventilation air quantities approximate the air quantities required for ventilation of the hoods, the hood exhaust system may be used to exhaust all ventilation air from the laboratory areas. In situations where hood exhaust exceeds air supplied, a supplementary air supply may be used for hood makeup. The use of VAV supply/ exhaust systems in the laboratory has gained acceptance but requires special care in design and installation.

The supplementary air supply, which need not be completely conditioned, should be provided by a system that is independent of the normal ventilating system. The individual hood exhaust system should be interlocked with the supplementary air system. However, the hood exhaust system should not shut off if the supplementary air system fails. Chemical storage rooms must have a constantly operating exhaust air system with terminal fan.

Exhaust fans serving hoods should be located at the discharge end of the duct system to prevent any possibility of exhaust products entering the building. For further information on laboratory air-conditioning and hood exhaust systems, see Chapter 13, NFPA *Standard* 99, and Control of Hazardous Gases and Vapors in Selected Hospital Laboratories (Hagopian and Doyle 1984).

The exhaust air from the hoods in the units for biochemistry, histology, cytology, pathology, glass washing/sterilizing, and serology-bacteriology should be discharged to the outdoors with no recirculation. Typically, exhaust fans discharge vertically at a minimum of 2.1 m above the roof at velocities up to 20 m/s. The serology-bacteriology unit should be pressurized relative to the adjoining areas to reduce the possibility of infiltration of aerosols that might contaminate the specimens being processed. The entire laboratory area should be under slight negative pressure to reduce the spread of odors or contamination to other hospital areas. Temperatures and humidities should be within the comfort range.

Bacteriology Laboratories. These units should not have undue air movement, so care should be exercised to limit air velocities to a minimum. The sterile transfer room, which may be within or adjoining the bacteriology laboratory, is a room where sterile media are distributed and where specimens are transferred to culture media. To maintain a sterile environment, an ultrahigh efficiency HEPA filter should be installed in the supply air duct near the point of entry to the room. The media room, essentially a kitchen, should be ventilated to remove odors and steam.

Infectious Disease and Virus Laboratories. These laboratories, found only in large hospitals, require special treatment. A minimum ventilation rate of six air changes per hour or makeup equal to hood exhaust volume is recommended for these laboratories, which should have a negative air pressure relative to any other area in the vicinity to prevent the exfiltration of any airborne contaminants. The exhaust air from fume hoods or safety cabinets in these laboratories must be sterilized before being exhausted to the outdoors. This may be accomplished by the use of electric or gas-fired heaters placed in series in the exhaust systems and designed to heat the exhaust air to 315°C. A more common and less expensive method of sterilizing the exhaust is to use ultrahigh efficiency filters in the system.

Nuclear Medicine Laboratories. Such laboratories administer radioisotopes to patients orally, intravenously, or by inhalation to facilitate diagnosis and treatment of disease. There is little opportunity in most cases for airborne contamination of the internal environment, but exceptions warrant special consideration.

One important exception involves the use of iodine 131 solution in capsules or vials to diagnose disorders of the thyroid gland. Another involves use of xenon 133 gas via inhalation to study patients with reduced lung function. Capsules of iodine 131 occasionally leak part of their contents prior to use. Vials emit airborne contaminants when opened for preparation of a dose.

It is common practice for vials to be opened and handled in a standard laboratory fume hood. A minimum face velocity of 0.5 m/s should be adequate for this purpose. This recommendation applies only where small quantities are handled in simple operations. Other circumstances may warrant provision of a glove box or similar confinement.

Use of xenon 133 for patient study involves a special instrument that permits the patient to inhale the gas and to exhale back into the instrument. The exhaled gas is passed through a charcoal trap mounted in lead and often (but not always) vented outdoors. The process suggests some potential for escape of the gas into the internal environment.

Due to the uniqueness of this operation and the specialized equipment involved, it is recommended that system designers determine the specific instrument to be used and contact the manufacturer for guidance. Other guidance is available in U.S. Nuclear Regulatory Commission Regulatory Guide 10.8 (NRC 1980). In particular, emergency procedures to be followed in case of accidental release of xenon 133 should include temporary evacuation of the area and/or increasing the ventilation rate of the area.

Recommendations concerning pressure relationships, supply air filtration, supply air volume, recirculation, and other attributes of supply and discharge systems for histology, pathology, and cytology laboratories are also relevant to nuclear medicine laboratories. There are, however, some special ventilation system requirements imposed by the NRC where radioactive materials are used. For example, NRC (1980) provides a computational procedure to estimate the airflow necessary to maintain xenon 133 gas concentration at or below specified levels. It also contains specific requirements as to the amount of radioactivity that may be vented to the atmosphere; the disposal method of choice is adsorption onto charcoal traps.

Autopsy Rooms. The autopsy room is an area of the pathology department that requires special attention. These rooms are subject to heavy bacterial contamination and odor. Exhaust intakes should be located both at the ceiling and in the low sidewall. The exhaust system should discharge the air above the roof of the hospital. A negative air pressure relative to adjoining areas should be provided in the autopsy room to prevent the spread of contamination. Where large quantities of formaldehyde are used, special exhaust hoods may be needed to keep concentration levels below legal requirements.

For smaller hospitals where the autopsy room is used infrequently, local control of the ventilation system and an odor control system with either activated charcoal or potassium permanganate-impregnated activated alumina may be desirable.

Animal Quarters. This area is found only in larger hospitals. Principally because of odor, animal quarters require a mechanical exhaust system that discharges the contaminated air above the hospital roof. To prevent the spread of odor or other contaminants from the animal quarters to other areas, a negative air pressure of at least 25 Pa relative to adjoining areas must be maintained. Chapter 13 has further information on animal room air conditioning.

Pharmacy. The pharmacy should be treated for comfort and requires no special ventilation. Room air distribution must be coordinated with any laminar airflow benches that may be needed. See Chapter 13, Laboratory Systems.

Administration

This department includes the main lobby, admitting and business offices, and medical records. Admissions and waiting rooms are areas where there is a potential risk of the transmission of undiagnosed airborne infectious disease. The use of local exhaust systems that move air toward the admitting patient should be considered. A separate air-handling system is considered desirable to segregate this area from the hospital proper because it is usually unoccupied at night.

Diagnostic and Treatment

Bronchoscopy, Sputum Collection, and Pentamidine Administration. These spaces are remarkable because of the high potential for large discharges of possibly infectious water droplet nuclei into the room air. Although the procedures performed may indicate the use of a patient hood, the general room ventilation should be increased under the assumption that higher than normal levels of airborne infectious contaminants will be generated.

Magnetic Resonance Imaging (MRI). These rooms should be treated as exam rooms in terms of temperature, humidity, and ventilation. However, special attention is required in the control room due to the high heat release of computer equipment and in the exam room due to the cryogens used to cool the magnet. The manufacturer of the magnet should be consulted.

Treatment Rooms. Patients are brought to these rooms for special treatments that cannot be conveniently performed in the patient rooms. To accommodate the patient, who may be brought from bed, the rooms should have individual temperature and humidity control. Temperatures and humidities should correspond to those specified for patients' rooms.

Physical Therapy Department. The cooling load of the electrotherapy section is affected by the shortwave diathermy, infrared, and ultraviolet equipment used in this area.

Hydrotherapy Section. This section, with its various water treatment baths, is generally maintained at temperatures up to 27°C. The potential latent heat buildup in this area should not be overlooked. The exercise section requires no special treatment, and temperatures and humidities should be within the comfort zone. The air may be recirculated within the areas, and an odor control system is suggested.

Occupational Therapy Department. In this department, spaces for activities such as weaving, braiding, artwork, and sewing require no special ventilation treatment. Recirculation of the air in these areas using medium-grade filters in the system is permissible.

Larger hospitals and those specializing in rehabilitation have a greater diversity of skills and crafts, including carpentry, metalwork, plastics, photography, ceramics, and painting. The air-conditioning and ventilation requirements of the various sections should conform to normal practice for such areas and to the codes relating to them. Temperatures and humidities should be maintained within the comfort zone.

Inhalation Therapy Department. Inhalation therapy is for treatment of pulmonary and other respiratory disorders. The air must be very clean, and the area should have a positive air pressure relative to adjacent areas.

Workrooms. The clean workroom serves as a storage and distribution center for clean supplies and should be maintained at a positive air pressure relative to the corridor.

The soiled workroom serves primarily as a collection point for soiled utensils and materials. It is considered a contaminated room and should have a negative air pressure relative to adjoining areas. Temperatures and humidities should be within the comfort range.

Sterilizing and Supply

Used and contaminated utensils, instruments, and equipment are brought to this unit for cleaning and sterilization prior to reuse. The unit usually consists of a cleaning area, a sterilizing area, and a storage area where supplies are kept until requisitioned. If these areas are in one large room, air should flow from the clean storage and sterilizing areas toward the contaminated cleaning area. The air pressure relationships should conform to those indicated in Table 3. Temperature and humidity should be within the comfort range.

The following guidelines are important in the central sterilizing and supply unit:

1. Insulate sterilizers to reduce heat load.
2. Amply ventilate sterilizer equipment closets to remove excess heat.
3. Where ethylene oxide (ETO) gas sterilizers are used, provide a separate exhaust system with terminal fan (Samuals and Eastin 1980). Provide adequate exhaust capture velocity in the vicinity of sources of ETO leakage. Install an exhaust at sterilizer doors and over the sterilizer drain. Exhaust aerator and service rooms. ETO concentration sensors, exhaust flow sensors, and alarms should also be provided. ETO sterilizers should be located in dedicated unoccupied rooms that have a highly negative pressure relationship to adjacent spaces and 10 air changes per hour. Many jurisdictions require that ETO exhaust systems have equipment to remove ETO from exhaust air. See OSHA 29 CFR, Part 1910.
4. Maintain storage areas for sterile supplies at a relative humidity of no more than 50%.

Service

Service areas include dietary, housekeeping, mechanical, and employee facilities. Whether these areas are air conditioned or not, adequate ventilation is important to provide sanitation and a wholesome environment. Ventilation of these areas cannot be limited to exhaust systems only; provision for supply air must be incorporated in the design. Such air must be filtered and delivered at controlled temperatures. The best designed exhaust system may prove ineffective without an adequate air supply. Experience has shown that reliance on open windows results only in dissatisfaction, particularly during the heating season. The incorporation of air-to-air heat exchangers in the general ventilation system offers possibilities for economical operation in these areas.

Dietary Facility. This area usually includes the main kitchen, bakery, dietitian's office, dishwashing room, and dining space. Because of the various conditions encountered (i.e., high heat and moisture production and cooking odors), special attention in design is needed to provide an acceptable environment. Refer to Chapter 28 for information on kitchen facilities.

The dietitian's office is often located within the main kitchen or immediately adjacent to it. It is usually completely enclosed to ensure privacy and noise reduction. Air conditioning is recommended for the maintenance of normal comfort conditions.

The dishwashing room should be enclosed and ventilated at a minimum rate to equal the dishwasher hood exhaust. It is not uncommon for the dishwashing area to be divided into a soiled area and a clean area. When this is done, the soiled area should be kept at a negative pressure relative to the clean area.

Kitchen Compressor/Condenser Space. Ventilation of this space should conform to all codes, with the following additional considerations: (1) 220 L/s of ventilating air per compressor kilowatt should be used for units located within the kitchen; (2) condensing units should operate optimally at 32.2°C maximum ambient temperature; and (3) where air temperature or air circulation is marginal, combination air- and water-cooled condensing units should be specified. It is often worthwhile to use condenser water coolers or remote condensers. Heat recovery from water-cooled condensers should be considered.

Dining Space. Ventilation of this space should conform to local codes. The reuse of dining space air for ventilation and cooling of food preparation areas in the hospital is suggested, provided the reused air is passed through 80% efficient filters. Where cafeteria service is provided, serving areas and steam tables are usually hooded. The air-handling capacities of these hoods should be at least 380 L/s per square metre of perimeter area.

Laundry and Linen. Of these facilities, only the soiled linen storage room, the soiled linen sorting room, the soiled utility room, and the laundry processing area require special attention.

The room provided for storage of soiled linen prior to pickup by commercial laundry is odorous and contaminated and should be well ventilated and maintained at a negative air pressure.

The soiled utility room is provided for inpatient services and is normally contaminated with noxious odors. This room should be exhausted directly outside by mechanical means.

In the laundry processing area, the washers, flatwork ironers, tumblers, and so forth, should have direct overhead exhaust to reduce humidity. Such equipment should be insulated or shielded whenever possible to reduce the high radiant heat effects. A canopy over the flatwork ironer and exhaust air outlets near other heat-producing equipment capture and remove heat best. The air supply inlets should be located to move air through the processing area toward the heat-producing equipment. The exhaust system from flatwork ironers and tumblers should be independent of the general exhaust system and equipped with lint filters. Air should exhaust above the roof or where it will not be obnoxious to occupants of other areas. Heat reclamation from the laundry exhaust air may be desirable and practicable.

Where air conditioning is contemplated, a separate supplementary air supply, similar to that recommended for kitchen hoods, may be located in the vicinity of the exhaust canopy over the ironer. Alternatively, spot cooling for the relief of personnel confined to specific areas may be considered.

Mechanical Facilities. The air supply to boiler rooms should provide both comfortable working conditions and the air quantities required for maximum rates of combustion of the particular fuel used. Boiler and burner ratings establish maximum combustion rates, so the air quantities can be computed according to the type of fuel. Sufficient air must be supplied to the boiler room to supply the exhaust fans as well as the boilers.

At workstations, the ventilation system should limit temperatures to 32°C effective temperature. When ambient outside air temperature is higher, indoor temperature may be that of the outside air up to a maximum of 36°C to protect motors from excessive heat.

Maintenance Shops. Carpentry, machine, electrical, and plumbing shops present no unusual ventilation requirements. Proper ventilation of paint shops and paint storage areas is important because of fire hazard and should conform to all applicable codes. Maintenance shops where welding occurs should have exhaust ventilation.

CONTINUITY OF SERVICE AND ENERGY CONCEPTS

Zoning

Zoning—using separate air systems for different departments—may be indicated to (1) compensate for exposures due to orientation or for other conditions imposed by a particular building configuration,

(2) minimize recirculation between departments, (3) provide flexibility of operation, (4) simplify provisions for operation on emergency power, and (5) conserve energy.

By ducting the air supply from several air-handling units into a manifold, central systems can achieve a measure of standby capacity. When one unit is shut down, air is diverted from noncritical or intermittently operated areas to accommodate critical areas, which must operate continuously. This or other means of standby protection is essential if the air supply is not to be interrupted by routine maintenance or component failure.

Separation of supply, return, and exhaust systems by department is often desirable, particularly for surgical, obstetrical, pathological, and laboratory departments. The desired relative balance within critical areas should be maintained by interlocking the supply and exhaust fans (i.e., exhaust should cease when the supply airflow is stopped).

Heating and Hot Water Standby Service

The number and arrangement of boilers should be such that when one boiler breaks down or is temporarily taken out of service for routine maintenance, the capacity of the remaining boilers is sufficient to provide hot water service for clinical, dietary, and patient use; steam for sterilization and dietary purposes; and heating for operating, delivery, birthing, labor, recovery, intensive care, nursery, and general patient rooms. However, reserve capacity is not required in climates where a design dry-bulb temperature of –4°C is equaled or exceeded for 99% of the total hours in any one heating period as noted in the tables in Chapter 24 of the 1993 *ASHRAE Handbook—Fundamentals*.

Boiler feed pumps, heat circulation pumps, condensate return pumps, and fuel oil pumps should be connected and installed to provide normal and standby service. Supply and return mains and risers for cooling, heating, and process steam systems should be valved to isolate the various sections. Each piece of equipment should be valved at the supply and return ends.

Some supply and exhaust systems for delivery and operating room suites should be designed to be independent of other fan systems and to operate from the hospital emergency power system in the event of power failure. The operating and delivery room suites should be ventilated such that the hospital facility retains some surgical and delivery capability in cases of ventilating system failure.

Boiler steam is often treated with chemicals that cannot be released in the air-handling units serving critical areas. In this case, a clean steam system should be considered for humidification.

Mechanical Cooling

The source of mechanical cooling for clinical and patient areas in a hospital should be carefully considered. The preferred method is an indirect refrigerating system using chilled water or antifreeze solutions. When using direct refrigerating systems, consult codes for specific limitations and prohibitions. Refer to ASHRAE *Standard* 15, Safety Code for Mechanical Refrigeration.

Insulation

All exposed hot piping, ducts, and equipment should be insulated to maintain the energy efficiency of all systems and protect building occupants. To prevent condensation, ducts, casings, piping, and equipment with outside surface temperature below ambient dew point should be covered with insulation having an external vapor barrier. Insulation, including finishes and adhesives on the exterior surfaces of ducts, pipes, and equipment, should have a flame spread rating of 25 or less and a smoke developed rating of 50 or less, as determined by an independent testing laboratory in accordance with NFPA *Standard* 255, as required by NFPA 90A. The smoke developed rating for pipe insulation should not exceed 150 (DHHS 1984a).

Linings in air ducts and equipment should meet the Erosion Test Method described in Underwriters Laboratories *Standard* 181. These linings, including coatings, adhesives, and insulation on exterior surfaces of pipes and ducts in building spaces used as air supply plenums, should have a flame spread rating of 25 or less and a smoke developed rating of 50 or less, as determined by an independent testing laboratory in accordance with ASTM *Standard* E84.

Duct linings should not be used in systems supplying operating rooms, delivery rooms, recovery rooms, nurseries, burn care units, or intensive care units, unless terminal filters of at least 90% efficiency are installed downstream of linings. Duct lining should be used only for acoustical improvement; for thermal purposes, external insulation should be used.

When existing systems are modified, asbestos materials should be handled and disposed of in accordance with applicable regulations.

Energy

Health care is an energy-intensive, energy-dependent enterprise. Hospital facilities are different from other structures in that they operate 24 h a day year-round, require sophisticated backup systems in case of utility shutdowns, use large quantities of outside air to combat odors and dilute microorganisms, and must deal with problems of infection and solid waste disposal. Similarly, large quantities of energy are required to power diagnostic, therapeutic, and monitoring equipment and support services such as food storage, preparation, and service and laundry facilities.

Hospitals conserve energy in various ways, such as using larger energy storage tanks and employing energy conversion devices that transfer energy from hot or cold building exhaust air to heat or cool incoming air. Heat pipes, runaround loops, and other forms of heat recovery are receiving increased attention. Solid waste incinerators, which generate exhaust heat to develop steam for laundries and hot water for patient care, are becoming increasingly common. Large health care campuses use central plant systems, which may include thermal storage, hydronic economizers, primary/secondary pumping, cogeneration, heat recovery boilers, and heat recovery incinerators.

The construction design of new facilities, including alterations of and additions to existing buildings, has a major influence on the amount of energy required to provide such services as heating, cooling, and lighting. The selection of building and system components for effective energy use requires careful planning and design. Integration of building waste heat into systems and use of renewable energy sources (e.g., solar under some climatic conditions) will provide substantial savings (Setty 1976).

OUTPATIENT HEALTH CARE FACILITIES

An outpatient health care facility may be a free-standing unit, part of an acute care facility, or part of a medical facility such as a medical office building (clinic). Any surgery is performed without anticipation of overnight stay by patients (i.e., the facility operates 8 to 10 h per day.)

If physically connected to a hospital facility and served by the hospital's HVAC systems, spaces within the outpatient health care facility should conform to requirements in the section on Hospital Facilities. Outpatient health care facilities that are totally detached and have their own HVAC systems may be categorized as diagnostic clinics or treatment clinics.

DIAGNOSTIC CLINICS

A diagnostic clinic is a facility where patients are regularly seen on an ambulatory basis for diagnostic services or minor treatment, but where major treatment requiring general anesthesia or surgery is not performed. Diagnostic clinic facilities have design criteria as shown in Tables 4 and 5 (see the section on Nursing Home Facilities).

TREATMENT CLINICS

A treatment clinic is a facility that provides, on an outpatient basis, major or minor treatment for patients that would render them incapable of taking action for self-preservation under emergency conditions without assistance from others. (See NFPA *Standard* 99.)

Design Criteria

The system designer should refer to the following paragraphs from the section on Hospital Facilities:

- Infection Sources and Control Measures
- Air Quality
- Air Movement
- Temperature and Humidity
- Pressure Relationships and Ventilation
- Smoke Control

Air-cleaning requirements correspond to those in Table 1 for operating rooms. A recovery area need not be considered a sensitive

Table 4 Filter Efficiencies for Central Ventilation and Air-Conditioning Systems in Nursing Homes[a]

Area Designation	Minimum Number of Filter Beds	Filter Efficiency of Main Filter Bed, %
Patient care, treatment, diagnostic, and related areas	1	80
Food preparation areas and laundries	1	80
Administrative, bulk storage, and soiled holding areas	1	30

[a]Ratings based on ASHRAE *Standard* 52.1-92.

area. The bacteria concern is the same as in an acute care hospital. The minimum ventilation rates, desired pressure relationships, desired relative humidity, and design temperature ranges are similar to the requirements for hospitals shown in Table 3 except for operating rooms, which may meet the criteria for trauma rooms.

The following functional areas in a treatment clinic facility have design criteria similar to those in hospitals:

- Surgical—operating rooms, recovery rooms, and anesthesia storage rooms
- Ancillary
- Diagnostic and Treatment—generally a small radiology area
- Sterilizing and Supply
- Service—soiled workrooms, mechanical facilities, and locker rooms

Continuity of Service and Energy Concepts

Some owners may desire that the heating, air-conditioning, and service hot water systems have standby or emergency service capability and that these systems be able to function after a natural disaster.

To reduce utility costs, facilities should include energy-conserving measures such as recovery devices, variable air volume, load shedding, or systems to shut down or reduce the ventilation of certain areas when unoccupied. Mechanical ventilation should take advantage of outside air by using an economizer cycle, when appropriate, to reduce heating and cooling loads.

The subsection on Continuity of Service and Energy Concepts in the section on Hospital Facilities includes information on zoning and insulation that applies to treatment clinics as well.

NURSING HOME FACILITIES

Nursing homes may be classified as follows:

Extended care facilities are for the recuperation of hospital patients who no longer require hospital facilities but do require the

Table 5 Pressure Relationships and Ventilation of Certain Areas of Nursing Homes

Function Area	Pressure Relationship to Adjacent Areas	Minimum Air Changes of Outdoor Air per Hour Supplied to Room	Minimum Total Air Changes per Hour Supplied to Room	All Air Exhausted Directly to Outdoors	Air Recirculated Within Room Units
PATIENT CARE					
Patient room	±	2	2	Optional	Optional
Patient area corridor	±	Optional	2	Optional	Optional
Toilet room	N	Optional	10	Yes	No
DIAGNOSTIC AND TREATMENT					
Examination room	±	2	6	Optional	Optional
Physical therapy	N	2	6	Optional	Optional
Occupational therapy	N	2	6	Optional	Optional
Soiled workroom or soiled holding	N	2	10	Yes	No
Clean workroom or clean holding	P	2	4	Optional	Optional
STERILIZING AND SUPPLY					
Sterilizer exhaust room	N	Optional	10	Yes	No
Linen and trash chute room	N	Optional	10	Yes	No
Laundry, general	±	2	10	Yes	No
Soiled linen sorting and storage	N	Optional	10	Yes	No
Clean linen storage	P	Optional	2	Yes	No
SERVICE					
Food preparation center	±	2	10	Yes	Yes
Warewashing room	N	Optional	10	Yes	Yes
Dietary day storage	±	Optional	2	Yes	No
Janitor closet	N	Optional	10	Yes	No
Bathroom	N	Optional	10	Yes	No

P = Positive   N = Negative   ± = Continuous directional control not required therapeutic and rehabilitative services of skilled nurses. This type of facility is either a direct hospital adjunct or a separate facility having close ties with the hospital. Clientele may be of any age, usually stay from 35 to 40 days, and usually have only one diagnostic problem.

Skilled nursing homes are for the care of people who require assistance in daily activities; many of them are incontinent and non-ambulatory, and some are disoriented. Clientele come directly from home or residential care homes, are generally elderly (average age of 80), stay an average of 47 months, and frequently have multiple diagnostic problems.

Residential care homes are generally for elderly people who are unable to cope with regular housekeeping chores but have no acute ailments and are able to care for all their personal needs, lead normal lives, and move freely in and out of the home and the community. These homes may or may not offer skilled nursing care. The average length of stay is four years or more.

Functionally, these buildings have five types of areas that are of concern to the HVAC designer: (1) administrative and supportive areas, inhabited by the staff; (2) patient areas that provide direct normal daily services; (3) treatment areas that provide special medical services; (4) clean workrooms for storage and distribution of clean supplies; and (5) soiled workrooms for collection of soiled and contaminated supplies and for sanitization of nonlaundry items.

DESIGN CONCEPTS AND CRITERIA

Controlling bacteria levels in nursing homes is not as critical as it is in the acute care hospital. Nevertheless, the designer should be aware of the necessity for odor control, filtration, and airflow control between certain areas.

Table 4 lists recommended filter efficiencies for air systems serving specific nursing home areas. Table 5 lists recommended minimum ventilation rates and desired pressure relationships for certain areas in nursing homes.

Recommended interior winter design temperature is 24°C for areas occupied by patients and 21°C for nonpatient areas. Provisions for maintenance of minimum humidity levels in winter depend on the severity of the climate and are best left to the judgment of the designer. Where air conditioning is provided, the recommended interior summer design temperature and humidity is 24°C and 50% rh.

The general design criteria in the sections on Heating and Hot Water Standby Service, Insulation, and Energy for hospital facilities apply to nursing home facilities as well.

APPLICABILITY OF SYSTEMS

The occupants of nursing homes are usually frail, and many are incontinent. They may be ambulatory, but some are bedridden, with illnesses in advanced stages. The selected HVAC system must dilute and control odors and should not cause drafts. Local climatic conditions, costs, and designer judgment determine the extent and degree of air conditioning and humidification. Odor may be controlled with large volumes of outside air and some form of heat recovery. To conserve energy, odor may be controlled with activated carbon or potassium permanganate-impregnated activated alumina filters instead.

Temperature control should be on an individual room basis. In geographical areas with severe climates, patients' rooms should have supplementary heat along exposed walls. In moderate climates, i.e., where outside winter design conditions are −1°C or above, heating from overhead may be used.

REFERENCES

AIA. 1992. Guidelines for construction and equipment of hospital and medical facilities. The American Institute of Architects, Washington, DC.

ASHRAE. 1989. Ventilation for acceptable indoor air quality. *Standard* 62-1989.

ASHRAE. 1992. Gravimetric and dust-spot procedures for testing air-cleaning devices used in general ventilation for removing particulate matter. *Standard* 52.1-1992.

ASHRAE. 1992. Safety code for mechanical refrigeration. *Standard* 15-1992.

ASTM. 1991. Standard test method for surface burning characteristics of building materials. *Standard* E84 REV A-91. American Society for Testing and Materials, Philadelphia, PA.

Burch, G.E. and N.P. Pasquale. 1962. *Hot climates, man and his heart.* C.C. Thomas, Springfield, IL.

Chaddock, J.B. 1986. Ventilation and exhaust requirements for hospitals. *ASHRAE Transactions* 92(2A):350-95.

Degenhardt, R.A. and J.F. Pfost. 1983. Fume hood design and application for medical facilities. *ASHRAE Transactions* 89(2B):558-70.

Demling, R.H. and J. Maly. 1989. The treatment of burn patients in a laminar flow environment. Annals of the New York Academy of Sciences 353:294-259.

DHHS. 1984. Guidelines for construction and equipment of hospital and medical facilities. *Publication* No. HRS-M-HF, 84-1. United States Department of Health and Human Services.

Fitzgerald, R.H. 1989. Reduction of deep sepsis following total hip arthroplasty. Annals of the New York Academy of Sciences 353:262-269.

Greene, V.W., R.G. Bond, and M.S. Michaelsen. 1960. Air handling systems must be planned to reduce the spread of infection. *Modern Hospital* (August).

Hagopian, J.H. and E.R. Hoyle. 1984. Control of hazardous gases and vapors in selected hospital laboratories. *ASHRAE Transactions* 90(2A):341-53.

Isoard, P., L. Giacomoni, and M. Payronnet. 1980. Proceedings of the 5th International Symposium on Contamination Control, Munich (September).

Lewis, J.R. 1988. Application of VAV, DDC, and smoke management to hospital nursing wards. *ASHRAE Transactions* 94(1):1193-1208.

Luciano, J.R. 1984. New concept in French hospital operating room HVAC systems. *ASHRAE Journal* 26(2):30-34.

Michaelson, G.S., D. Vesley, and M.M. Halbert. 1966. The laminar air flow concept for the care of low resistance hospital patients. Paper presented at the annual meeting of American Public Health Association, San Francisco (November).

Murray, W.A., A.J. Streifel, T.J. O'Dea, and F.S. Rhame. 1988. Ventilation protection of immune compromised patients. *ASHRAE Transactions* 94(1):1185-92.

NFPA. 1990. Standard method of test of surface burning characteristics of building materials. *Standard* 255-90. National Fire Protection Agency, Quincy, MA.

NFPA. 1993. Standard for health care facilities. *Standard* 99-93. National Fire Protection Agency, Quincy, MA.

NFPA. 1993. Standard for the installation of air conditioning and ventilation systems. *Standard* 90A-93. National Fire Protection Agency, Quincy, MA.

NFPA. 1993. Recommended practice for smoke-control systems. *Standard* 92A-93. National Fire Protection Agency, Quincy, MA.

NFPA. 1994. Code for safety to life from fire in buildings and structures. *Code* 101-94. National Fire Protection Agency, Quincy, MA.

NIOSH. 1975. Elimination of waste anesthetic gases and vapors in hospitals, *Publication* No. NIOSH 75-137 (May). United States Department of Health, Education, and Welfare.

NRC. 1980. *Regulatory Guide* 10.8. Nuclear Regulatory Commission.

OSHA. Occupational exposure to ethylene oxide. OSHA 29 CFR, Part 1910. United States Department of Labor.

Pfost, J.F. 1981. A re-evaluation of laminar air flow in hospital operating rooms. *ASHRAE Transactions* 87(2):729-39.

Rousseau, C.P. and W.W. Rhodes. 1993. HVAC system provisions to minimize the spread of tuberculosis bacteria. *ASHRAE Transactions* 99(2):1201-04.

Samuals, T.M. and M. Eastin. 1980. ETO exposure can be reduced by air systems. *Hospitals* (July).

Setty, B.V.G. 1976. Solar heat pump integrated heat recovery. *Heating, Piping and Air Conditioning* (July).

UL. 1990. Factory-made air ducts and air connectors. *Standard* 181-90. Underwriters Laboratories Inc., Northbrook, IL.

Walker, J.E.C. and R.E. Wells. 1961. Heat and water exchange in the respiratory tract. *American Journal of Medicine* (February):259.

Wells, W.F. 1934. On airborne infection. Study II: Droplets and droplet nuclei. *American Journal of Hygiene* 20:611.

Woods, J.E., D.T. Braymen, R.W. Rasussen, G.L. Reynolds, and G.M. Montag. 1986. Ventilation requirement in hospital operating rooms—Part I: Control of airborne particles. *ASHRAE Transactions* 92(2A):396-426.

BIBLIOGRAPHY

DHHS. 1984. Energy considerations for hospital construction and equipment. *Publication* No. HRS-M-HF, 84-1A. United States Department of Health and Human Services.

Gustofson, T.L. et al. 1982. An outbreak of airborne nosocomial Varicella. *Pediatrics* 70(4):550-56.

Rhodes, W.W. 1988. Control of microbioaerosol contamination in critical areas in the hospital environment. *ASHRAE Transactions* 94(1):1171-84.

CHAPTER 8

SURFACE TRANSPORTATION

AUTOMOBILE AIR CONDITIONING

WITH increased ease of operation, improved road isolation, lower noise levels, and increasing urban travel time, modern automobile users have become conscious of the in-vehicle environment. All passenger cars sold in the United States must meet federal defroster requirements, so ventilation systems and heaters are included in the basic vehicle design. Even though trucks are excluded from federal requirements, all manufacturers include heater-defrosters as standard equipment. Air conditioning remains an extra-cost option on nearly all vehicles.

ENVIRONMENTAL CONTROL

Environmental control in modern automobiles consists of one or more of the following systems: (1) heater-defroster, (2) ventilation, and (3) cooling and dehumidifying (air-conditioning). The integration of the heater-defroster and ventilation systems is common.

Heating

Outdoor air passes through a heater core, using engine coolant as a heat source. To avoid visibility-reducing condensation on the glass due to raised air dew point from occupant respiration and interior moisture gains, interior air should not recirculate through the heater.

Temperature control is achieved by either water flow regulation or heater air bypass and subsequent mixing. A combination of ram effect from forward movement of the car and the electrically driven blower provides the airflow.

Heater air is generally distributed into the lower forward compartment, under the front seat, and up into the rear compartment. Heater air exhausts through body leakage points. At higher vehicle speeds, the increased heater air quantity (ram assist through the ventilation system) partly compensates for the infiltration increase. Air exhausters are sometimes installed to increase airflow and reduce the noise of air escaping from the car.

The heater air distribution system is usually adjustable between the diffusers along the floor and on the dashboard. Supplementary ducts are sometimes required when consoles, panel-mounted air conditioners, or rear seat heaters are installed. Supplementary heaters are frequently available for third-seat passengers in station wagons and for the rear seats in limousines and luxury sedans.

Defrosting

Some heated outdoor air is ducted from the heater core to defroster outlets at the base of the windshield. This air absorbs moisture from the interior surface of the windshield and raises the glass temperature above the interior dew point. Induced outdoor air has a lower dew point than the air inside the vehicle, which absorbs moisture from the occupants and car interior. Heated air provides the energy necessary to melt or sublime ice and snow from the glass exterior. The defroster air distribution pattern on the windshield is developed by test for conformity with federal standards, satisfactory distribution, and rapid defrost.

Some systems operate the air-conditioning compressor to dry the induced outdoor air and/or to prevent a wet evaporator from increasing the dew point when the compressor is disengaged. Some vehicles are equipped with side window demisters that direct a small amount of heated air and/or air with lowered dew point to the front side windows. Rear windows are defrosted primarily by heating wires embedded in the glass.

Ventilation

Fresh air ventilation is achieved by one of two systems: (1) ram air or (2) forced air. In both systems, air enters the vehicle through a screened opening in the cowl just forward of the base of the windshield. The cowl plenum is usually an integral part of the vehicle structure. Air entering this plenum can also supply the heater and evaporator cores.

In the ram air system, ventilation air flows aft and up toward the front seat occupants' laps and then over the remainder of their bodies. Additional ventilation occurs by turbulence and air exchange through open windows. Directional control of ventilation air is frequently unavailable. Airflow rate varies with relative wind-vehicle velocity but may be adjusted with windows or vents.

Forced air ventilation is available in many automobiles. The cowl inlet plenum and heater/air-conditioning blower are used together with instrument panel outlets for directional control. Positive air pressure from the ventilation fan or blower helps reduce the amount of exterior pollutants entering the passenger compartment. In air-conditioned vehicles, the forced air ventilation system uses the air-conditioning outlets. Body air exhausts and vent windows exhaust air from the vehicle. With the increased popularity of air conditioning and forced ventilation, most late model vehicles are not equipped with vent windows.

Air Conditioning

There are two basic types of systems: combination evaporator-heater and dealer-installed add-on systems.

The combination evaporator-heater system in conjunction with the ventilation system is the prevalent type of factory-installed air conditioning. This system is popular because (1) it permits dual use of components such as blower motors, outdoor air ducts, and structure; (2) it permits compromise standards where space considerations dictate (ventilation reduction on air-conditioned cars); (3) it generally reduces the number and complexity of driver controls; and (4) it typically features capacity control innovations such as automatic reheat.

Outlets in the instrument panel distribute air to the car interior. These are individually adjustable, and some have individual shutoffs. The dashboard end outlets are for the driver and front seat passenger; center outlets are primarily for rear seat passengers.

The preparation of this chapter is assigned to TC 9.3, Transportation Air Conditioning.

The dealer-installed system is normally available only as a service or after-market installation. In recent designs, the existing air outlets, blower, and controls are used. Evaporator cases are styled to look like factory-installed units. These units are integrated with the heater as much as possible to provide outdoor air and to take advantage of existing air-mixing dampers. Where it is not possible to use existing air ducts, custom ducts distribute the air in a manner similar to those in factory-installed units. On occasion, rather than installing a second duct, the dealer will create a slot in the existing duct to accept an evaporator.

As most of the air for the rear seat occupants flows through the center outlets of the front evaporator unit, a passenger in the center of the front seat impairs rear seat cooling. Supplemental trunk units for luxury sedans and roof units for station wagons improve the cooling of rear seat passengers.

The trunk unit is a blower-evaporator, complete with expansion valve, installed in the trunk of limousines and premium line vehicles. The unit uses the same high-side components as the front evaporator unit. It cools and recirculates air drawn from the base of the rear window (back light) of the car.

A roof-mounted unit is similar to the trunk unit, except that it is intended for station wagon use and is attached to the inside of the roof, toward the rear of the wagon.

GENERAL CONSIDERATIONS

General considerations include ambient temperatures and contaminants, vehicle and engine concessions, flexibility, physical parameters, durability, electrical power, refrigeration capacity, occupants, infiltration, insulation, solar effect, and noise.

Ambient and Vehicle Criteria

Ambient Temperature. Heaters are evaluated for performance at temperatures from −40 to 21°C. Air-conditioning systems with reheat are evaluated from 4 to 43°C; add-on units are evaluated from 10 to 38°C, although ambient temperatures above 52°C are occasionally encountered. Because the system is an integral part of the vehicle detail, the effects of vehicle heat and local heating must be considered.

Ambient Contaminants. Airborne bacteria, pollutants, and corrosion agents must also be considered when selecting materials for seals and heat exchangers. Electronic air cleaners have been installed in premium automobiles.

Vehicle Performance. Proper engine coolant temperature, freedom from gasoline vapor lock, adequate electrical charging system, acceptable vehicle ride, minimum surge due to compressor clutch cycling, and good handling must be maintained.

Flexibility. Engine coolant pressure at the heater core inlet ranges up to 280 kPa (gage) in cars and 380 kPa on trucks. The engine coolant thermostat remains closed until coolant temperature reaches 71 to 96°C. Coolant flow is a function of pressure differential and system restriction but ranges from 40 mL/s at idle to 630 mL/s at 100 km/h (lower for water valve regulated systems because of the added restriction).

Present-day antifreeze coolant solutions have specific heats from 2.7 to 4.2 kJ/(kg·K) and boiling points from 121 to 133°C (depending on concentration) when a 100 kPa radiator pressure cap is used.

Multiple-speed (usually four-speed) blowers supplement the ram air effect through the ventilation system and produce the necessary velocities for distribution. Heater air quantities range from 60 to 90 L/s. Defroster air quantities range from 42 to 68 L/s. Other considerations are (1) compressor/engine speed, from 8.3 to 92 rev/s; (2) drive ratio, from 0.89:1 to 1.41:1; (3) condenser air, from 10 to 52°C and from 150 to 1400 L/s (corrected for restriction and distribution factors); and (4) evaporator air, from 50 to 190 L/s (limits established by design but selective at operator's discretion) and from 2 to 66°C.

Physical Parameters

Parameters include engine rock, proximity to adverse environments, and durability.

Engine Rock. Relative to the rest of the car, the engine moves both fore and aft because of inertia and in rotation because of torque. Fore and aft movement may be as much as 6 mm; rotational movements at the compressor may be as much as 20 mm due to acceleration and 13 mm due to deceleration.

Proximity to Adverse Environments. Wiring, refrigerant lines, hoses, vacuum lines, and so forth, must be protected from exhaust manifold heat and sharp edges of sheet metal. Accessibility to the normal service items such as oil filler caps, power steering filler caps, and transmission dipsticks cannot be impaired. Removal of air-conditioning components should not be necessary for servicing other components.

Durability. Hours of operation are short compared to commercial systems (260 000 km ÷ 65 km/h = 4000 h), but the shock and vibration the vehicle receives or produces must not cause a malfunction or failure.

Systems are designed to meet the recommendations of SAE *Standard* J 639, which states that the burst strength of those components subjected to high-side refrigerant pressure must be at least 2.5 times the venting pressure (or pressure equivalent to venting temperature) of the relief device. *Standard* J 639 now also requires electrical cut-out of the clutch coil prior to pressure relief to prevent unnecessary refrigerant discharge. Components for the low-pressure side frequently have burst strengths in excess of 2.1 MPa. The relief device should be located as close as possible to the discharge gas side of the compressor, preferably in the compressor itself.

Power and Capacity

Fan size is kept to a minimum, not only for space and mass effect, but because of power consumption. If a standard vehicle has a heater that draws 10 A, and the air-conditioning system requires 20 A, an alternator and wiring system must be redesigned to supply this additional 10 A, with obvious cost and mass penalties.

The refrigeration capacity of a system must be adequate to reduce the vehicle interior temperature to the comfort temperature quickly and then to maintain the selected temperature at reasonable humidity during all operating conditions and environments. A design may be established by mathematical modelling or empirical evaluation of all the known and predicted factors. A design trade-off in capacity is sought relative to the vehicle's mass, component size, and fuel economy needs.

Occupancy per unit volume is high in automotive applications. The system (and auxiliary evaporators and systems) is matched to the intended vehicle occupancy.

Other Considerations

Infiltration varies with relative wind-vehicle velocity. It also varies with assembly quality. Body sealing is part of air-conditioning design for automobiles. Occasionally, sealing beyond that required for dust, noise, and draft control is required.

Due to cost, insulation is seldom added to reduce thermal load. Insulation for sound control is generally considered adequate. Roof insulation is of questionable benefit, as it retards heat loss during the nonoperating, or soak, periods. Additional dashboard and floor insulation helps reduce cooling load. Typical maximum ambient temperatures are 90°C above mufflers and catalytic converters, 50°C for other floor areas, 63°C for dash and toe board, and 43°C for sides and top. The following three solar effects add to the cooling load:

Vertical. Maximum intensity occurs at or near noon. Solar heat gain through all glass surface area normal to the incident light is a substantial fraction of the cooling load.

Horizontal and Reflected Radiation. The intensity is significantly less, but the glass area is large enough to merit consideration.

Surface Heating. The temperature of the surface is a function of the solar energy absorbed, the interior and ambient temperatures, and the automobile's velocity.

The temperature control system should not produce objectionable sounds. During maximum heating or cooling operation, a slightly higher noise level is acceptable. Thereafter, it should be possible to maintain comfort at a lower blower speed with acceptable noise level. In air-conditioning systems, compressor-induced vibrations, gas pulsations, and noise must be kept to a minimum. Suction and discharge mufflers are often used to reduce noise. Belt-induced noises, engine torsional vibration, and compressor mounting all require particular attention.

COMPONENTS

Compressors

Much research and development has been conducted in the past several years to make compressors suitable for car cooling use. Scroll compressors are now also used in mobile applications.

Displacement. Fixed displacement compressors have displacements of 100 to 206 mL/rev. Variable displacement compressors have a minimum displacement of about 16 mL/rev, about 10% of their maximum displacement. A typical variable capacity scroll compressor has a maximum displacement of 120 mL/rev and a minimum displacement of just 3% of the maximum.

Physical Size. Fuel economy, lower hood lines, and more engine accessories all decrease compressor installation space. These features, along with the fact that smaller engines have less accessory power available, promote the use of smaller compressors.

Speed Range. Because compressors are belt driven directly from the engine, they must withstand speeds of over 100 rev/s and remain smooth and quiet down to 8.3 rev/s. In the absence of a variable drive ratio, there are instances when the maximum compressor speed may need to be higher in order to achieve sufficient pumping capacity at idle.

Torque Requirements. Because torque pulsations cause or aggravate vibration problems, it is best to minimize them. Minimizing peak torque benefits the compressor drive and mount systems. Multicylinder reciprocating and rotary compressors aid in reducing vibration. An economical single-cylinder compressor reduces cost; however, any design must reduce peak torques and belt loads, which would normally be at a maximum in a single-cylinder design.

Compressor Drives. A magnetic clutch, energized by power from the car engine electrical system, drives the compressor. The clutch is always disengaged when air conditioning is not required. The clutch can also be used to control evaporator temperature (see the section on Controls).

Variable Displacement Compressors. Wobble plate-type compressors are used for automobile air conditioning. The angle of the wobble plate changes in response to the evaporator and discharge pressure to achieve a constant suction pressure just above freezing, regardless of load. A bellows valve or electronic sensor-controlled valve routes internal gas flow to control the wobble plate.

Refrigerants and Lubricants

New Vehicles. The phaseout of Refrigerant 12 production has led to extensive use of non-CFC alternatives in new vehicles. The popular choice is R-134a. Its physical and thermodynamic properties closely match those of R-12.

R-134a requires the use of synthetic lubricants such as polyalkylene glycol (PAG) or polyol ester (POE) because its miscibility with mineral oil is poor. PAGs are currently preferred. These lubricants—especially PAGs—are more hygroscopic than mineral oil; production and service procedures must emphasize the need for system dehydration and limited exposure of the lubricants to atmospheric moisture.

Older Vehicles. Vehicles originally equipped with R-12 do not require retrofit to another refrigerant as long as an affordable supply of new or reclaimed R-12 exists. When circumstances require it, most systems can be modified to use R-134a and POE. When PAG oil is used in a retrofit, the remaining mineral oil should be removed to levels established by the original equipment manufacturer to prevent residual chlorides from breaking down the PAG oils.

R-22 is not suitable for topping off or retrofit in R-12 systems, as it will produce unacceptable discharge pressures (up to 4.5 MPa at idle) and temperatures. R-12 idle discharge pressures can reach 3.1 MPa; in addition, it may be incompatible with the elastomeric seals and hose materials.

There are several blends of two or more refrigerants being sold as "drop-in" replacements for R-12. While some may have merit, they should not be used without thorough investigation into their effect on system performance and reliability under all operating conditions. Other issues requiring attention include (1) flammability, (2) fractionation, (3) evaporator temperature glide, (4) increased probability of adding the wrong refrigerant or oil in future service, and (5) compounded complexity of refrigerant reclaim.

Condensers

Condensers must be sized adequately. High discharge pressures reduce the compressor capacity and increase power requirements. When the condenser is in series with the radiator, condenser air restriction must be compatible with the engine cooling fan and engine cooling requirements. Generally, the most critical condition occurs at engine idle under high load conditions. An undersized condenser can raise head pressures sufficiently to stall small displacement engines. Condensers may be of the following designs: (1) tube-and-fin with mechanically bonded fins, (2) serpentine tube with brazed, multilouvered fins, and (3) header extruded tube brazed to multilouvered fins. Aluminum is popular for its lower cost and mass.

An oversized condenser may produce condensing temperatures significantly below the engine compartment temperature. This can result in evaporation of refrigerant in the liquid line where the liquid line passes through the engine compartment (the condenser is ahead of the engine and the evaporator is behind it). Engine compartment air has not only been heated by the condenser but by the engine and radiator as well. Typically, this establishes a minimum condensing temperature of 17 K above ambient.

Liquid flashing occurs more often at reduced load, when the liquid line velocity decreases, allowing the liquid to be heated above saturation temperature before reaching the expansion valve. This is more apparent on cycling systems than on systems that have a continuous liquid flow. Liquid flashing is audibly detected as gas enters the expansion valve. This problem can be reduced by adding a subcooler or additional fan power to the condenser.

Condensers generally cover the entire radiator surface to prevent air bypass. Accessory systems designed to fit several different cars are occasionally over-designed. Internal pressure drop should be minimized to reduce power requirements. Condenser-to-radiator clearances as low as 6 mm have been used, but 13 mm is preferable. Primary-to-secondary surface area ratio varies from 8:1 to 16:1. Condensers are normally painted black so that they are not visible through the vehicle's grille.

A condenser ahead of the engine-cooling radiator not only restricts air but also heats the air entering the radiator. The addition of air conditioning requires supplementing the engine cooling system. Radiator capacity is increased by adding fins, depth, or face area or by raising pump speed to increase coolant flow. Pump cavitation at high speeds is the limiting factor. It should be noted that increased coolant velocity may contribute to excessive tube erosion. In the case of direct fan drive, increasing the speed of the water pump increases the engine cooling fan speed, which not only supplements the engine cooling system but also provides more air for the condenser. When air conditioning is installed in an automobile, fan size, number of blades, and blade width and pitch are frequently increased, and fan shrouds are often added. Increases in fan speed, diameter, and pitch raise the noise level and power consumption.

Temperature- and torque-sensitive drives (viscous drives or couplings) or flex-fans reduce these increases in noise and power. They rely on ram air produced by the forward motion of the car to reduce the amount of air the radiator fan must move to maintain adequate coolant temperatures. As vehicle speed increases, fan requirements drop.

Front-wheel-drive vehicles typically have electric motor-driven cooling fans. Some vehicles also have a side-by-side condenser and radiator, each with its own motor-driven fan.

Automotive condensers must satisfy several key durability requirements, including internal and external corrosion, pressure cycle, burst, and vibration requirements.

Evaporator Systems

Current materials and construction include (1) copper or aluminum tube and aluminum fin, (2) brazed aluminum plate and fin, and (3) brazed serpentine tube and fin. Design parameters include air pressure drop, capacity, and condensate carryover. Fin spacing must permit adequate condensate drainage to the drain pan below the evaporator.

Condensate must drain outside the vehicle. At road speeds, the vehicle exterior is generally at a higher pressure than the interior by 250 to 500 Pa. Drains are usually on the high-pressure side of the blower; they sometimes incorporate a trap and are as small as possible. Drains can become plugged not only by contaminants but also by road splash. Vehicle attitude (slope of the road and inclines), acceleration, and deceleration must be considered when designing condensate systems.

High refrigerant pressure loss in the evaporator requires externally equalized expansion valves. A bulbless expansion valve, which provides external pressure equalization without the added expense of an external equalizer, is available. The evaporator must provide stable refrigerant flow under all operating conditions and have sufficient capacity to ensure rapid cool-down of the vehicle after it has been standing in the sun.

The conditions affecting evaporator size and design are different from those in residential and commercial installations in that the average operating time, from a hot-soaked condition, is less than 20 min. Inlet air temperature at the start of the operation can be as high as 65°C, and it decreases as the duct system is ventilated. In a recirculating system, the temperature of inlet air decreases as the car interior temperature decreases; in a system using outdoor air, inlet air temperature decreases to a few degrees above ambient (perpetual heating by the duct system). During longer periods of operation, the system is expected to cool the entire vehicle interior rather than just produce a flow of cool air.

During sustained operation, vehicle occupants want less air noise and velocity, so the air quantity must be reduced; however, sufficient capacity must be preserved to maintain satisfactory interior temperatures. Ducts must be kept as short as possible and should be insulated from engine compartment and solar-ambient heat loads. Thermal lag resulting from the added heat sink of ducts and housings increases cool-down time.

Filters and Hoses

Air filters are not common. Coarse screening prevents such objects as facial tissues, insects, and leaves from entering fresh-air ducts. Studies show that wet evaporator surfaces reduce the pollen count appreciably. In one test, an ambient of 23 to 96 mg/mm^3 showed 53 mg/mm^3 in a non-air-conditioned car and less than 3 mg/mm^3 in an air-conditioned car. Rubber hose assemblies are used where flexible refrigerant transmission connections are needed

due to relative motion between components or because stiffer connections cause installation difficulties and noise transmission problems. Refrigerant effusion through the hose wall is a design consideration. Effusion occurs at a reasonably slow and predictable rate that increases as pressure and temperature increase. Hose with a nylon core is less flexible (pulsation dampening), has a smaller OD, is generally cleaner, and allows practically no effusion. It is recommended for Refrigerant 134a.

Heater Cores

The heat transfer surface in an automotive heater is generally either copper/brass cellular, aluminum tube and fin, or aluminum brazed tube and center. Each of these designs can currently be found in production in either straight-through or U-flow designs. The basics of each of the designs are outlined below.

The copper/brass cellular design (Figure 1) uses brass tube assemblies (0.15 to 0.41 mm) as the water course and convoluted copper fins (0.08 to 0.20 mm) held together with a lead-tin solder. The tanks and connecting pipes are usually brass (0.66 to 0.86 mm) and again are attached to the core by a lead-tin solder. Capacity is adjusted by varying the face area of the core to increase or decrease the heat transfer surface area.

The aluminum tube and fin design generally uses round copper or aluminum tubes mechanically joined to aluminum fins. U-tubes can take the place of a conventional return tank. The inlet/outlet tank and connecting pipes are generally plastic and clinched onto the core with a rubber gasket. Capacity can be adjusted by varying face area, adding coolant-side turbulators, or varying air-side surface geometry for turbulence and air restriction.

The aluminum brazed tube and center design uses flat aluminum tubes and convoluted fins or centers as the heat transfer surface. Tanks can be either plastic and clinched onto the core or aluminum and brazed to the core. Connecting pipes can be constructed of various materials and attached to the tanks a number of ways, including brazing, clinching with an o-ring, fastening with a gasket, and so forth. Capacity can be adjusted by varying face area, core depth, or air-side surface geometry.

Receiver-Dryer Assembly

The receiver-dryer assembly accommodates charge fluctuations from changes in system load (refrigerant flow and density). It accommodates an overcharge of refrigerant (0.25 to 0.5 kg) to compensate for system leaks and hose effusion. The assembly houses the high-side filter and desiccant. Several types of desiccant are used,

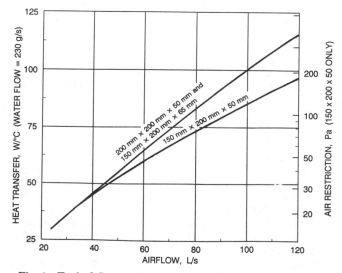

Fig. 1 Typical Copper-Brass Cellular Heater Core Capacity

the most common of which is spherical molecular sieves; silica gel is occasionally used. Mechanical integrity (freedom from powdering) is important because of the vibration to which the assembly is exposed. For this reason, molded desiccants have not obtained wide acceptance.

Moisture retention at elevated temperatures is also important. The rate of release with temperature increase and the reaction while accumulating high concentration should be considered. Design temperatures of at least 60°C should be used.

The receiver-dryer often houses a sight glass that allows visual inspection of the system charge level. It houses safety devices such as fusible plugs, rupture discs, or high-pressure relief valves. High-pressure relief valves are gaining increasing acceptance because they do not vent the entire charge. Location of the relief devices is important. Vented refrigerant should be directed so as not to endanger personnel.

Receivers are usually (though not always) mounted on or near the condenser. They should be located so that they are ventilated by ambient air. Pressure drops should be minimal.

Expansion Valves

Thermostatic expansion valves (TXVs) control the flow of refrigerant through the evaporator. These are applied as shown in Figures 4, 5, and 6. Both liquid- and gas-charged power elements are used. Internally and externally equalized valves are used as dictated by system design. Externally equalized valves are necessary where high evaporator pressure drops exist. A bulbless expansion valve, usually block-style, that senses evaporator outlet pressure without the need for an external equalizer, is now widely used. There is a trend toward a variable compressor pumping rate and electrically controlled expansion valves.

Orifice Tubes

The use of an orifice tube instead of an expansion valve to control refrigerant flow through the evaporator has come into widespread use in original equipment installations, primarily due to the lower component costs. The components of the system must be carefully matched to obtain proper performance. Even so, under some conditions there is flood-back of liquid refrigerant to the compressor with this system.

Suction Line Accumulators

A suction line accumulator is required with an orifice tube system to ensure uniform return of refrigerant and oil to the compressor to prevent slugging and to cool the compressor. It also stores excess refrigerant. A typical suction line accumulator is shown in Figure 2. A bleed hole at the bottom of the standpipe meters oil and liquid refrigerant back to the compressor. The filter and desiccant are contained in the accumulator because no receiver-dryer is used with this system. The amount of refrigerant charge is more critical when a suction line accumulator is used than it is with a receiver-dryer.

Refrigerant Flow Control

The cycling clutch refrigerant systems shown in Figures 3 and 4 are common in late model cars for both factory- and dealer-installed units. The clutch is cycled by a thermostat that senses evaporator temperature or by a pressure switch that senses evaporator pressure. Some dealer-installed units use an adjustable thermostat, which controls car temperature by controlling evaporator temperature. The thermostat also prevents evaporator icing. Most units use a fixed thermostat or pressure switch set to prevent evaporator icing. Temperature is then controlled by blending the air with warm air coming through the heater core.

Cycling the clutch sometimes causes noticeable surges as the engine is loaded and unloaded by the compressor. This is more evident in cars with smaller engines. Reevaporation of condensate from the evaporator during the off-cycle may cause objectionable temperature fluctuation or odor. This system cools faster and at lower cost than a continuously running system.

In orifice tube-accumulator systems, the clutch cycling switch disengages at about 170 kPa and cuts in at about 310 kPa (gage). Thus, the evaporator defrosts on each off-cycle. The flooded evaporator has enough thermal inertia to prevent rapid clutch cycling. It is

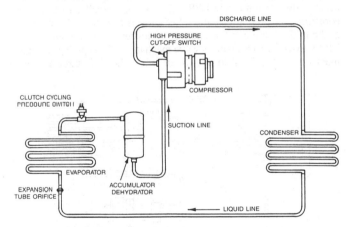

Fig. 3   Clutch Cycling Orifice Tube Air-Conditioning System Schematic

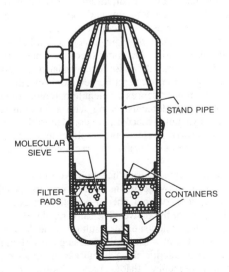

Fig. 2   Typical Suction Line Accumulator

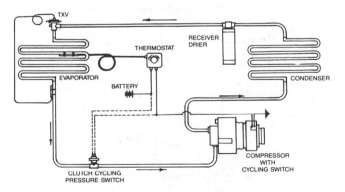

Fig. 4   Clutch Cycling System with Thermostatic Expansion Valve (TXV)

desirable to limit clutch cycling to a maximum of 4 cycles per minute because heat is generated by the clutch at engagement.

The pressure switch can be used with a TXV in a dry evaporator if the pressure switch is damped to prevent rapid cycling of the clutch.

Continuously running systems, as shown in Figures 5 and 6, have also been widely used. An evaporator pressure regulator (EPR) keeps the evaporator pressure above the condensate freezing level. Temperature is controlled by reheat or by blending the air with warm air from the heater core. The valve, which is located on the downstream side of the evaporator, may be self contained in its own housing and installed in the compressor suction line, or it may be placed in the suction cavity of the compressor. The device is pressure-sensitive and operates to maintain a minimum evaporator pressure (saturation pressure). The setting depends on the minimum airflow across the evaporator, the condensate-draining ability of the evaporator coil, and the lowest ambient temperature at which it is anticipated that the system will be operated.

The continuously running system possesses neither of the previously mentioned disadvantages of the cycling clutch system, but it does increase the suction line pressure drop, hence causing a slight reduction in system performance at maximum load. A solenoid version of this valve that is controlled by an antifreeze switch that senses evaporator fin temperature has also been used. The switch closes the valve electrically when the fin temperature drops to the control point.

Two basic refrigeration circuits, shown in Figures 5 and 6, use the EPR. Figure 5 shows the conventional dry-type system. Figure 6 shows a flooded evaporator system, which uses a plate-and-separator type of evaporator with a tank at the top and bottom. It is also a unique piping arrangement. The TXV external equalizer line is connected downstream of the EPR valve. Also, a small oil return line containing an internal pressure relief valve is connected downstream of the EPR. These two lines may be connected to the housing of the EPR valve downstream of its valve mechanism. This piping arrangement causes the TXV to open wide when the EPR throttles, thus

flooding the evaporator and causing the EPR to act as an expansion valve. The system allows the air conditioner to run at ambients as low as 2°C to defog windows while maintaining adequate refrigerant flow and ensuring oil return to the compressor.

CONTROLS

Manual

The fundamental control mechanism is a flexible control cable that transmits linear motion with reasonable efficiency and little hysteresis. Rotary motion is obtained by crank mechanisms. Common applications are (1) movement of the damper doors that control discharge air temperature in blend air systems, (2) regulation of the amount of defrost bleed, and (3) regulation of valves to control the flow of engine coolant through the heater core.

Vacuum

Vacuum provides a silent, powerful method of manual control, requiring only the movement of a lever or a switch by the operator. Vacuum is obtained from the engine intake manifold. A vacuum reservoir (0.4 to 4 L) may be used to ensure an adequate source. Most systems are designed to function at minimum vacuum of −17 to −20 kPa, even though as much as −88 kPa may be available at times.

Linear or rotary slide valves select functions. Vacuum-modulating, temperature-compensating (bimetal) valves control thermostatic coolant valves. Occasionally, solenoid valves are used, but they are generally avoided because of their cost and associated wiring.

Electrical

Electrical controls regulate blower motors. Blowers have three or four speeds. The electrically operated compressor clutch frequently sees service as a secondary system control operated by function (mode) selection integration or evaporator temperature or pressure sensors.

Temperature Control

Air-conditioning capacity control to regulate car temperature is achieved in one of two ways—the clutch can be cycled in response to an adjustable thermostat sensing evaporator discharge air, or the evaporator discharge air can be blended with or reheated by air flowing through the heater core. The amount of reheat or blend is usually controlled by driver-adjusted damper doors.

Solid-state logic interprets system requirements and automatically adjusts to heating or air conditioning, depending on the operator's selection of temperature and on ambient temperature. Manual override enables the occupant to select the defrost function. The system not only regulates this function but also controls capacity and air quantity. An in-car thermistor measures the temperature of the air within the passenger compartment and compares it to the setting at the temperature selector. An ambient sensor, sometimes a thermistor, senses ambient temperature to prevent offset or droop. These elements, along with a vacuum supply line from an engine vacuum reservoir, are coupled to the control package, which consists of an electronic amplifier and a transducer. The output is a vacuum or electrical signal regulated by the temperature inputs.

This regulated signal is supplied to servo-units that control the quantity of coolant flowing in the heater core, the position of the heater blending door, or both; the air-conditioning evaporator pressure; and the speed of the blower. The same regulated signal controls sequencing of damper doors, either resulting in the discharge of air at the occupant's feet or at waist level or causing the system to work on the recirculation of interior air. A number of interlocking assurance devices are usually required that (1) prevent blower operation before engine coolant is up to temperature, (2) prevent the air-conditioning

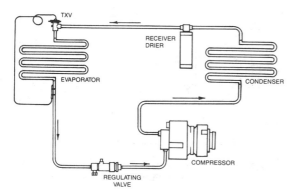

Fig. 5   Suction Line Regulation System

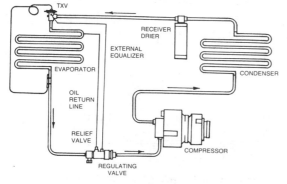

Fig. 6   Flooded Evaporator Suction Line Regulation System

compressor clutch from being energized at low ambient temperatures, and (3) provide other features for passenger comfort.

Other Controls

A pressure switch is located in the suction line at either the block valve or the accumulator to cycle the clutch off at pressures below, and on at pressures above, that at which water would freeze onto the evaporator surface. A cycling switch may be included to start an electric fan when insufficient ram air flows over the condenser. Other sophistications include a charge loss/low ambient switch, transducer evaporator pressure control, and thermistor control.

BUS AIR CONDITIONING

Bus air-conditioning design must consider highly variable passenger and climatic loads. It is usually not cost-effective to design for a specific climate. Therefore, the design should consider all likely climates, from the high ambient temperatures of the dry southwestern United States to the high humidity of the cooler East. Units should operate satisfactorily at ambient temperatures up to 50°C. A unit designed for both extremes has a greater sensible cooling capacity in hot, dry climates than in humid climates. The quality of the ambient air must also be considered. Frequently, intakes are subjected to thermal contamination from road surfaces, condenser air recirculation, or radiator air discharge. Vehicle motion also introduces pressure variables that affect condenser fan performance. In addition, engine speed will affect compressor capacity. Bus air-conditioning units are generally tested separately in climate-controlled test cells. The large vehicle test chambers now available make realistic as-installed testing possible.

INTERURBAN BUSES

The following conditions have been adopted as standard for interurban vehicles:

1. Capacity of 47 passengers
2. Insulation thickness of 25 to 38 mm
3. Double-pane tinted windows
4. Outdoor air intake of 190 L/s
5. Road speed of 100 km/h
6. Inside design temperatures of 27°C dry-bulb and 19.4°C wet-bulb, or 11 K lower than ambient

Outside conditions typical of the United States give loads from 12 to 35 kW. Unless a specific area requires a custom design, the designer should consider extreme conditions. The design is sized for

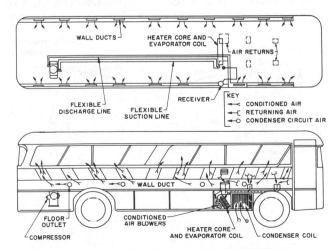

Fig. 7 Typical Airflow Pattern and Air-Conditioning Equipment Location in an Interurban Bus

standby conditions. Using the engine as the compressor drive for standby results in excessive capacity over the road; appropriate unloading devices are necessary. Higher engine idle speed lessens the amount of compressor oversizing. Some interurban buses have a separate engine-driven air conditioner to give more constant performance; however, the space required, complexity, cost, and maintenance requirement are greater for these than for other units. Figure 7 shows a typical arrangement of equipment. Winter features, including warm sidewalls and window diffusers, are important to offset downdrafts. Return air openings near the floor help reduce stratification. These features, while desirable, are not as important on urban buses because passengers are exposed for only a short time and wear appropriate clothing. Some additional thermal losses occur with sidewall distribution and underside ducting; therefore, effective insulation and avoidance of thermal short circuits are necessary. Placing the condenser behind the front axle has been satisfactory; it should not be placed low and near the rear axle on rear engine-driven buses because hot air (up to 60°C) from the radiator reduces the condenser capacity.

Rooftop-mounted units are increasingly installed in new and retrofitted buses. They consist of the entire air-conditioning unit except the compressor, which is shaft- or belt-driven from the bus engine, as with other types of units, and connected by refrigerant piping extending up through the bus structure to the unit overhead. The components are arranged in a low enclosure that contains compartments for the condenser, its associated fans, and the evaporator blowers and controls. Often, a third compartment houses an outdoor air fan for supply of ventilation air and for emergency ventilation in case of a cooling system failure. Conditioned air is directed from the unit through openings in the top of the bus to an interior duct/diffuser distribution system, which in new buses is usually two linear diffusers extending the length of the bus. In retrofitted systems, the conditioned air is often directed through a duct that is fitted with a linear diffuser of narrow grilles and extends the length of one side of the bus above the windows.

Due to the need to maintain a low profile, the roof-mounted systems usually have multiple small supply-air blowers and distribute air to the in-bus system from several different points; sometimes, on larger buses, up to eight blowers are used. Return air is taken from a central grille in the bus ceiling directly up into the unit above, where it mixes with outside ventilation air before passing across the cooling coils. Multiple condenser air fans are also used.

Rooftop systems are serviced by removing the upper cover of the unit, exposing all parts of the condenser and evaporator sections. Filters are serviced by removal of the return air grille within the bus. Due to the location of the unit, servicing is difficult and requires ladders or special scaffolding for safe access to the bus roof.

Shock and Vibration

Most transport air-conditioning manufacturers design components for a shock loading between 29 and 50 m/s². Vibration eliminators, flexible lines, and other shock-cushioning devices interconnect the various air-conditioning components. The vibration characteristics of each component are different. On a direct, engine-connected compressor, whether the connection is made by belts, flexible drives, gearing, or clutches, there is relative motion between the main engine and the compressor. This motion may be taken care of by shock arms, vibration dampening devices, or automatic belt-tensioning devices. It is good practice to tie the compressor onto the engine by bracket extensions or other mounting devices to eliminate the relative motion between the engine and the compressor.

The relative motion between the compressor and other air-conditioning components is usually taken care of by flexible refrigerant hoses. The permeability of flexible lines has presented some difficulty with R-22; thus R-134a, because it operates at lower pressures, is usually chosen. The use of R-22 is then limited to higher capacity systems such as double-decker and articulated interurban buses. The use of metallic bellow hoses is preferred.

Controls

Most interurban coaches have a simple driver control to select air conditioning or heating. In both modes, a thermal sensing element (thermistor, mercury tube, etc.) controls these systems with on/off circuitry and actuators. When dehumidification but not cooling is needed, the air-conditioning system is engaged and the air is reheated for comfort. Higher than needed engine power draw (13 kW) occurs with on/off control. More sophisticated controls and actuators that limit the need to reheat and lower the power draw are becoming popular. They improve comfort and fuel economy.

URBAN BUSES

Urban bus heating and cooling loads are greater than those of the interurban bus. A city bus may seat up to 50 people and carry a crush load of standees. The fresh air load is greater because of constant door opening and infiltration around doors. Table 1 shows the results of a test that recorded door openings. An urban bus stops frequently and may open both front and rear doors to take on or discharge passengers.

Table 1   Door Operation of a City Bus

	Front Entrance Door	Center Exit Door
Door open		
Times per kilometre	4.7	3.4
Times per hour	70	44
% of operating time	35	15
Longest time open	55 s	34 s
Shortest time open	3 s	3 s
Average time open	13.5 s	12.5 s

Note: Observed during the rush period on a 35-passenger bus in San Antonio, Texas.

The cooling capacity required for the typical 50-seat urban bus is from 20 to 35 kW of refrigeration. The equipment must be flexible enough to range from low to maximum capacity promptly because the passenger load varies greatly. Buses with engine-driven compressors suffer a loss of compressor capacity at idle. This loss is compensated for by sizing other parts of the system somewhat larger than indicated by the load analysis to meet a time average capacity equal to the comfort goal.

At maximum engine speed, the compressor may have a capacity in excess of the cooling load or the rating of the remainder of the system—a condition that must also be considered. In general, the amount of power available is a limiting factor; more recently, fuel consumption has influenced specifications. Equipment must be

designed for maximum efficiency, and the vehicle, for minimum thermal losses.

System Types

Air-conditioning systems for urban buses generally fall into three categories. The newer, advanced-design buses are usually equipped with a roof- or rear-mounted package unit similar to those shown in Figures 8 and 9 that include all system components except the compressor. The compressor is usually belt or shaft driven from the main traction engine.

The heater is located just downstream of the evaporator. Hot coolant from the engine cooling system provides sufficient heat for most operations; however, additional sources may be required in colder climates for long durations of idle. Additional floor heaters may also be required to reduce the effects of stratification. Air distribution for these systems is either through linear diffuser(s) in the ceiling fed from a duct concealed in the space between the roof and ceiling panels or through longitudinal ducts concealed in the sidewalls behind the lighting fixtures above the windows.

Many older urban buses similar to the one shown in Figure 9 have an evaporator/blower unit mid-mounted below the bus floor that feeds conditioned air through sidewall voids to slotted or perforated diffusers at the windowsills. These systems have either an upper, rear-mounted condenser or a condenser mounted in tandem with the engine radiator. The tandem installations are susceptible to recirculation of radiator discharge air, which can severely reduce capacity, especially at idle. Compressors are either shaft driven from the transmission auxiliary power take-off or belt driven directly from the engine. In either case, a clutch connects or disconnects the drive for system control. The wide separation of components in this design requires considerable lengths of interconnecting piping and many fittings and joints that are susceptible to leakage from vibration and physical damage. In addition, the air distribution systems located in the lower sidewalls of the coach are easily damaged during a traffic accident and are most difficult to repair and keep airtight.

Retrofit Systems

As the public has demanded a more comfortable riding environment, there has been an upsurge in retrofitting older buses with air-conditioning systems and in replacing older systems with newer, more efficient designs. Available systems follow the configuration of systems supplied for new buses in that roof- and rear-mounted units can be fitted into old coaches. Some manufacturers also offer a simple system of two to four interior evaporator/blower units, depending on the bus size and the operating environment. These units are installed over the windows along one side of the bus. They sometimes replace the lighting fixtures formerly at that location,

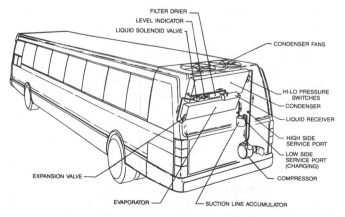

Fig. 8   Typical Mounting Location of Urban Bus Air-Conditioning Equipment

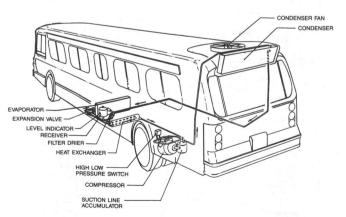

Fig. 9   Mounting Location of Air-Conditioning Equipment in Older Urban Buses

providing the necessary cooling without the need for duct installation. Single evaporator/blower units can be mounted in the rear of small buses, along with a skirt-mounted condenser (Figure 7), to supplement the factory-installed dash unit. Refrigerant piping in these systems is somewhat more extensive than in packaged units, but the overall installation is faster and less expensive. In addition, renovations to the bus interior are kept to a minimum, and the structural concern of cutting an opening in the roof is eliminated.

Compressors and Drives

Several different types and styles of compressors and drives are seen on urban buses. The four- or six-cylinder reciprocating compressor, in which some cylinders are equipped with unloaders, is popular. The compressor is mounted so that it can be belt driven directly from the engine. It is fitted with a clutch that is either pneumatically or electrically actuated by the temperature control system and refrigerant pressure. Several designs have one or more high-speed, axial piston or rotary compressors that are belt driven and electric-clutch controlled. Helical screw compressors have also been successfully demonstrated on urban transit buses.

Articulated Coaches

Several urban transit authorities operate articulated buses in which each of the two sections is equipped with a separate roof- or rear-mounted package unit. Each section has its own belt-driven compressor; both compressors are usually located in the rear-section engine compartment. The package for the rear section has a dual condenser, while the package for the forward section has no condenser. Long runs of flexible refrigerant piping connect the front package to its compressor/condenser at the extreme rear of the rear section.

System Maintenance

Due to the extremely adverse operating environment of the urban transit bus, the cost of air-conditioning system inspection and repairs is far greater than the initial cost. For this reason, it is generally cost-effective to arrange the system for ease of access to the repairable parts and to provide convenient points for checking the critical pressures, fluid levels, and temperatures. Ease of access to the air filters for replacement or cleaning is essential because this is the most frequently performed maintenance task. Refrigerant piping joints should be kept to a minimum, and all controls, safety devices, and accessory items should be heavy-duty and able to withstand the extreme conditions and environments to which the bus will be subjected. Brushless dc motors have replaced the brush type because the maintenance requirement is much lower.

Controls

The typical urban transit coach is relatively simply controlled. Cooling thermostats for full and part capacity and a heating thermostat to operate the pump or valve serving the heating core are usually included. Many current systems use solid-state control modules to interpret interior and ambient temperatures and to generate signals to operate the full or part cooling or heating functions. Thermistor temperature sensors, which are usually more stable and reliable than electromechanical controls, are used in these systems.

RAILROAD AIR CONDITIONING

Railroad air-conditioning systems are electromechanical with direct-expansion evaporators usually using R-22. R-134a is being considered as CFCs are eliminated. Electronic solid-state automatic controls are common, with a trend toward microprocessor control with fault monitoring and feedback to a central car information system. Electric heating elements installed in the air-conditioning unit

or supply duct temper outdoor air brought in for ventilation and are often used as reheat for humidity control.

Passenger Car Construction

Passenger car design has emphasized lighter car construction to lower building, operating, and maintenance costs. The drive to reduce car mass and cost has caused a reduction in the size and mass of air-conditioning equipment and other auxiliaries.

Vehicle Types

Mainline long-haul railroads generally operate single and bilevel cars hauled by a locomotive. Locomotive-driven alternators or static inverters distribute power via an intercar cable power bus to the air-conditioning systems. A typical rail car has a control package and two self-contained heating/cooling units mounted in equipment rooms. Split systems and underfloor and roof-mounted package systems are less common.

Commuter cars operating around large cities are similar in size to mainline cars and generally have two evaporator-heater fan units mounted above the ceiling with a common or two separate underfloor compressor-condenser unit(s) and a control package. These cars may be locomotive hauled but are often self propelled by high-voltage direct current (dc) or alternating current (ac) power supplied from an overhead catenary or from a dc-supplied third rail system. On such cars, the air conditioning may operate on ac or dc power. Diesel-driven vehicles still operate in a few areas.

Subways usually operate on a third rail dc power supply. The car air conditioning operates on the normal dc supply voltage or on three-phase ac supply from an alternator or inverter mounted under the car. Split air-conditioning systems are common, with evaporators at roof level and underfloor-mounted condensing sections.

Streetcars, light rail vehicles, and downtown people movers usually run on the city ac or dc power supply and have air-conditioning equipment similar to subway cars. During rush hours, interior space is at a premium on these cars, so roof-mounted packages are used more often than are split systems.

Equipment Selection

The source and type of power dictate the type of air-conditioning equipment installed on a railroad car; mass is also a major consideration. Thus, ac-powered semihermetic compressors, which are lighter than open machines with dc motor drives, are a common choice. However, each car design must be carefully examined in this respect because dc/ac inverters may increase not only the total system mass, but also the total system power draw, due to conversion losses.

Other considerations in equipment selection include space requirement, location, accessibility, reliability, and maintainability. Interior and exterior equipment noise levels must be considered both during the early stages of design and later, when the equipment is coordinated with the car-builder's ductwork and grilles.

Design Limitations

Space underneath and inside a railroad car is at a premium, especially on self-propelled light rail vehicles and subway and commuter cars; this generally rules out unitary interior or underfloor-mounted systems. System components are usually built to fit the configuration of the available space. Overall car height, roof profile, ceiling cavity, and undercar clearance restrictions determine the shape and size of equipment.

Because a mainline railroad car must operate in various parts of the country, the air-conditioning system must be able to handle the national seasonal extreme design days. Commuter cars and subway cars operate in a small geographical area, so only the local design temperatures and humidities need be considered.

Dirt and corrosion constitute an important design factor, especially if the equipment is beneath the car floor, where it is subject to all types of weather, as well as severe dirt conditions. For this reason, corrosion-resistant materials and coatings must be selected wherever possible. Aluminum has not proved durable enough; the sandblasting effect destroys any surface treatment on it. Because dirt pickup cannot be avoided, the equipment must be designed for quick and easy cleaning; access doors are provided and evaporator and condenser fin spacing is usually limited to 2.5 to 3 mm. Closer spacing causes more rapid dirt buildup and higher cleaning costs. Dirt and severe environmental conditions must be considered in selecting motors and controls.

Railroad HVAC equipment requires more maintenance and servicing than stationary units. A modern railroad car, having sealed windows and a well-insulated structure, becomes almost unusable if the air conditioning fails. The equipment therefore has many additional components that permit quick diagnosis and correction of the failure. Motors, compressors, and valves must be easily accessible for inspection or repair. The liquid receiver should have sight glasses that allow the amount of refrigerant to be checked quickly. Likewise, a readily accessible liquid charging valve should be available. Pressure gages and test switches allow a fast check of the system while the train is stopped at intermediate stations.

Safety must be considered, especially on equipment located beneath the car. Vibration isolators and supports should be designed to prevent bolt shears and loose nuts. A piece of equipment that dangles or drops off could cause a train derailment. All belt drives and other revolving items must be safety guarded. High-voltage controls and equipment must be labeled by approved warning signs. Pressure vessels and coils must meet ASME test specifications for protection of the passengers and maintenance personnel. Materials selection criteria include low flammability, toxicity, and smoke emission.

For undermount condensers, consideration must be given to air circulation that occurs at passenger loading platforms or in tunnels. To prevent a total system shutdown due to high discharge pressure, the system is forced into a reduced cooling mode to limit load on the condenser, thus reducing discharge pressure.

Interior Comfort

Air-conditioning and heating comfort conditions may be selected in accordance with ASHRAE *Standard* 55. However, the selected temperature and humidity levels must consider the passenger's metabolic rate on entry, clothing insulation, and journey time. Chapter 8 of the 1993 *ASHRAE Handbook—Fundamentals* has more details.

Vibration and dusty conditions preclude the use of commercial humidity controllers. Evaporator sectional staging and compressor cylinder unloading, coupled with electric reheat, are used to provide part-load humidity control. Studies are focusing on the use of variable compressor and fan speed control to make inherently wasteful reheat control more efficient. A maximum relative humidity for optimum comfort is usually specified. In winter, humidity control is usually not provided.

The dominant summer cooling load is due to passengers, followed by ventilation, internal heat, car body transmission, and solar gain. Heating loads are due to car body losses and ventilation. The heating load calculation does not credit heat from passengers and internal sources. Comfortable internal conditions in ventilated non-air-conditioned cars can be maintained only when ambient conditions permit.

Due to the continuous variation in passenger and solar loads in mass transit cars, the interior conditions are difficult to maintain. Air-conditioning systems in North American cars are selected to maintain temperatures of 23 to 24°C, with a maximum relative humidity of 55 to 60%. In Europe and elsewhere, the conditions are usually set at a dry-bulb temperature from 0 to 5 K below ambient, with a coincidental relative humidity of 50 to 66%. In the heating mode, the car interior is kept in the 18 to 21°C range. Outdoor conditions are given in Chapter 24 of the 1993 *ASHRAE Handbook—Fundamentals*.

System Requirements

Most cars are equipped with both overhead and floor heat. The overhead heat raises the temperature of the recirculated and ventilation air mixture to slightly above the car design temperature. The floor heat offsets heat loss through the car body and reduces temperature stratification.

Cooling and heating loads are calculated in accordance with Chapters 25 and 26 of the 1993 *ASHRAE Handbook—Fundamentals*. The times of maximum occupancy, outdoor ambient, and solar gain must be ascertained. The peak cooling load on urban transit cars usually coincides with the evening rush hour; on intercity railroads the peak load occurs in the midafternoon.

Heating capacity for the car depends on envelope construction, size, and the design area-averaged relative wind-vehicle velocity. In some instances, minimum car warmup time may be the governing factor. In long-distance trains, the toilets, the galley, and the lounges often have exhaust fans. Ventilation airflow must exceed forced exhaust air rates sufficiently to maintain positive car pressure.

Minimum per occupant fresh air ventilation rates of 2.5 to 3.5 L/s in nonsmoking cars, and 5 to 7 L/s in smoking-permitted cars are desirable. Ventilation air pressurizes the car and reduces infiltration.

Air Distribution

The most common type of air distribution system is a centerline supply duct running the length of the car in the space between the ceiling and the roof. The air outlets are usually ceiling-mounted linear slot air diffusers. Egg crate recirculation grilles are positioned in the ceiling beneath the evaporator units. The main supply duct must be insulated from the ceiling cavity to prevent summer thermal gain/loss and condensation. Taking ventilation air from both sides of the roofline overcomes the effect of wind. Adequate snow and rain louvers must be installed on the outdoor air intakes. Separate outdoor air filters are usually paired with a return air filter. Disposable media or permanent, cleanable air filters are used and are usually serviced every month.

Some long-haul cars, such as a sleeper, require a network of delivered-air and return ducts. Duct design should consider noise and static pressure losses. The latter can impact negatively on HVAC system performance.

Piping Design

Standard refrigerant piping practice is followed. Pipe joints should be accessible for inspection and not concealed in car walls. Evacuation, leak testing, and dehydration must be completed successfully after system installation and prior to charging. Piping should be supported adequately and installed without traps that could retard the flow of oil back to the compressor. Pipe sizing and arrangement should be in accordance with Chapter 2 of the 1994 *ASHRAE Handbook—Refrigeration*. Evacuation, dehydration, charging, and testing should be performed as described in Chapter 45 of the 1992 *ASHRAE Handbook—Systems and Equipment*. Piping on packaged units should also conform to these standards.

Control Requirements

Car HVAC systems are automatically controlled for year-round comfort. In the cooling mode, load variations are handled by two-stage direct-expansion coils and compressors equipped with suction pressure unloaders, electric unloaders, or speed control. Under low-load conditions, cooling is provided by outdoor air supplied through the ventilation system. During part-load cooling, electric reheat maintains humidity control. Under low loads not requiring humidity control, the system assumes the ventilation mode.

A pump-down cycle and low ambient lockout protect the compressor from damage caused by liquid slugging. In addition, the compressor may be fitted with a crankcase heater that is energized during the compressor off-cycle. During the heating mode, floor and overhead heaters are staged to maintain the car interior temperature.

Today's control systems use thermistors and solid-state auxiliary electronics instead of electromechanical devices. The control circuits are normally powered by low-voltage dc or, occasionally, by single-phase ac.

Future Trends

The demand for lighter, more efficient railway cars remains strong. The use of rooftop packaged air-conditioning units has reduced mass and improved reliability by eliminating long piping runs. Most manufacturers offer hermetic compressors, which are isolated to withstand the shock and vibration normally encountered. Some manufacturers market heat pump units. Neither hermetic compressors nor heat pumps have been fully field tested in North America. However, some operating authorities are considering using them to reduce costs. Specifications stipulate high reliability combined with low maintenance. The use of microprocessors for maintenance purposes and diagnostics allows a variety of possibilities for data retrieval, storage, and transmission.

FIXED GUIDEWAY VEHICLE AIR CONDITIONING

Fixed guideway systems, commonly referred to as people movers, are increasing in popularity. They can be monorails or rubber-tired cars running on an elevated or grade level guideway, as seen at many airports and in urban areas such as Miami. The guideway directs and steers the vehicle and provides the electrical power necessary to operate the car's traction motors, lighting, electronics, and air-conditioning and heating systems.

People movers are usually unmanned and computer controlled from some central point. Operations control determines vehicle speed and headway and the length of time doors stay open based on telemetry from the individual cars or trains. Therefore, a reliable and effective environmental control system is essential.

People movers are smaller than most other mass transit vehicles, generally having spaces for 20 to 40 seated passengers and generous floor space for standing passengers. It has been found that under some conditions of passenger loading, a 12-m car can accommodate 100 passengers. The wide range of possible passenger loads and the continual movement of the car from full sunlight to deep shade make it essential that the car's air-conditioning system be especially responsive to the amount of cooling required at a given moment.

System Types

The HVAC system for a people mover is usually one of three types:

1. Conventional undercar condensing unit connected with refrigerant piping to an evaporator/blower unit mounted above the car ceiling
2. Packaged, roof-mounted unit having all components within one enclosure and mated to an air distribution system built into the car ceiling
3. Packaged, undercar-mounted unit mated to supply and return air ducts built into the car body

Two systems are usually installed in each car, one at each end; each system provides one-half of the maximum cooling requirement. The systems, whether unitary or split, operate on the guideway's power supply, which is usually 60 Hz, 460 to 600 V (ac).

Refrigeration Components

Due to the availability of commercial electrical power, standard semihermetic motor-compressors and commercially available fan motors and other components can be used. Compressors generally have one or two stages of unloaders, and hot gas bypass is used to maintain cooling at low loads. Condenser and cooling coils are copper tube, copper, or aluminum fin units. Generally, flat fins are preferred for undercar systems to make it simpler to clean the coils. Evaporator/blower sections must often be designed for the specific vehicle and fitted to its ceiling contours. The condensing units must also be arranged to fit within the limited space available and still ensure good airflow across the condenser coil. R-22 is employed in these systems almost universally.

Heating

Where heating must be provided, electric resistance heaters are introduced into the system at the evaporator unit discharge. These heaters operate on the guideway power supply. One or two stages of heat control are used, depending on the size of the heaters.

Controls

A solid-state control system is usually used to maintain interior conditions. The cooling set point is usually between 23 and 24°C. For heating, the set point is 20°C or lower. Control systems are available that provide some humidity control by using the electric heat. Between the cooling and heating set points, blowers continue to operate on a ventilation cycle. Often, two-speed blower motors are used, switching to low speed for the heating cycle. Some control systems have internal diagnostic capability; they are able to signal the operations center when a cooling or heating malfunction occurs.

Fresh Air

With overhead air-handling equipment, fresh air is introduced into the return airstream at the evaporator entrance. Fresh air is usually taken from a grilled or louvered opening in the end or side of the car and, depending on the configuration of components, filtered separately or directed so that the return air filter can handle both airstreams. For undercar systems, a similar procedure is used, except the air is introduced into the system through an intake in the undercar enclosure. In some cases, a separate fan is used to induce fresh air into the system. The amount of outdoor air ventilation is difficult to estimate due to the widely varying passenger loads to be provided for. Older vehicles provided an amount equivalent to 20 to 25% of the total air circulated. Newer designs attempt to relate this to the passenger load by providing 5 to 7 L/s per seated and comfortably standing passenger. In some cases, the outdoor air proportion will reach 30% or more of the total airflow.

Air Distribution

With overhead systems, air is distributed through linear ceiling diffusers that are often constructed as a part of the overhead lighting fixtures. Undercar systems usually make use of the void spaces in the sidewalls and below fixed seating. In all cases, the spaces used for air supply must be adequately insulated to prevent condensation on surfaces and, in the case of voids below seating, to avoid cold seating surfaces. Supply air discharge from undercar systems is typically through a windowsill diffuser. Recirculation air from overhead systems flows through ceiling-mounted grilles. For undercar systems, return air grilles are usually found in the door wells or beneath seats.

Because of the small size of the vehicle and its low ceilings, extreme care must be taken to design the air supply system to avoid direct impingement of supply air on passengers' heads or shoulders. High rates of diffusion are needed, and diffuser placement and arrangement should permit the discharge airstreams to hug the ceiling and walls of the car. Total air quantity and discharge temperature must be carefully balanced to minimize cold drafts and air currents.

REFERENCES

ASHRAE. 1992. Thermal environmental conditions for human occupancy. *Standard* 55-1992.

SAE. 1991. Safety and containment of refrigerant for mechanical vapor compression systems used for mobile air-conditioning systems, recommended practice. *Standard* J 639-91. Society of Automotive Engineers, Warrendale, PA.

BIBLIOGRAPHY

Kuffe K.W. 1978. Air conditioning and heating systems for trucks. SP-425. Society of Automotive Engineers, Warrendale, PA.

Towers, J.A. and R.H. Krueger. 1985. Refrigerant additive and method for reducing corrosion in refrigeration systems. U.S. Patent 4,599,185.1.

Wojtkowski, E.F. 1964. System contamination and cleanup. *ASHRAE Journal* 6(6):49-52.

AIRCRAFT

THE type of air-conditioning system used in an aircraft depends on the function and performance of the aircraft. The information in this chapter applies to subsonic commercial transport aircraft. Aircraft environmental control systems consist of air-conditioning, cabin pressurization, oxygen, ice protection, and pneumatic systems. Safety requirements for these systems are included in the Federal Aviation Regulations (FARs), which are legal constraints on system design and operation. Requirements for aircraft air-conditioning systems are different from requirements for residential or commercial air-conditioning systems. In addition, aircraft systems must be lightweight, accessible for quick inspection and servicing, and highly reliable and must tolerate high-altitude environments and accommodate system failures.

Requirements for aircraft air-conditioning systems can be divided into two categories: (1) those related to safety and health and (2) those related to occupant comfort. The safety requirements should not be violated under any circumstances.

DESIGN CONDITIONS

Aircraft air-conditioning systems operate under unique conditions. Outside air at altitude is relatively free from pollution compared to ground air; however, the aircraft spend little time on the ground relative to the time in flight. The ambient conditions change quickly from ground to flight. Conditions in the aircraft and the location of the occupants are controlled, and emissions from furnishings and point sources are well defined. Aircraft air-conditioning systems must be able to operate under conditions caused by component failures.

Ambient Temperature, Humidity, and Pressure

Figure 1 shows typical design ambient temperature profiles for hot, standard, and cold days. The ambient temperatures used for the design of a particular aircraft may be higher or lower than those shown in Figure 1, depending on the regions in which the airplane is to be operated. The design ambient moisture content at various attitudes that is recommended for commercial aircraft is shown in Figure 2. The moisture content may be as high as 29.2 g per kilogram of dry air at sea level. The variation in ambient pressure with altitude is shown in Figure 3.

Air-Conditioning Performance

Heating and cooling capabilities vary from aircraft to aircraft, based on the demands of the operator and manufacturer. Generally, the performance of an air-conditioning system is related to ground-level heat up and cool down times and is a function of the aircraft's auxiliary power unit. A system should be able to bring the cabin temperature to between 21 and 27°C within 30 min from a soaked ground ambient temperature in the range −32 to 46°C. Flight air-conditioning requirements are typically defined as the ability to control cabin temperature to within 1 K of a selected set point in the range 18 to 29°C.

The preparation of this chapter is assigned to TC 9.3, Transportation Air Conditioning.

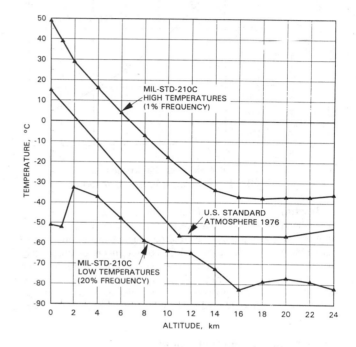

Fig. 1 Typical Ambient Temperature Profiles

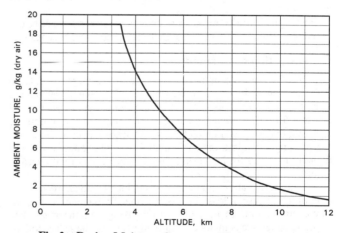

Fig. 2 Design Moisture Content at Various Altitudes

The system should be able to maintain temperature uniformity within 2.8 K.

Cooling. During ground operations, the air-conditioning system should be capable of cooling the cabin to an average temperature of 27°C within 30 min and maintaining that temperature with a full passenger load and all external doors closed. During cruise, the system should maintain an average cabin temperature of 24°C with a full passenger load.

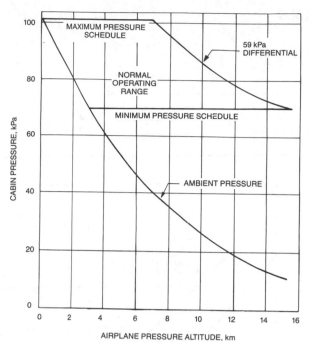

Fig. 3   Typical Cabin Pressure Schedule

Heating. During ground operations, the system should be capable of heating the cabin to an average temperature of 21°C within 30 min, starting with a cold soaked airplane at a temperature of −32°C, an ambient temperature of −40°C, no passengers or other internal heat loads, and all external doors closed. During cruise, the system should maintain an average cabin temperature of 24°C with a 20% passenger load, a cargo compartment bulk air temperature above 4°C, and cargo floor temperatures above 0°C to prevent the freezing of cargo.

Operational Recommendations

Preconditioning. Air-conditioning systems should be operated for about one-half hour prior to passenger boarding and during all ground operation.

Postflight Conditioning. Air-conditioning systems should be operated for approximately 15 min after the last passenger has deplaned.

Pressurization Performance

Figure 3 shows typical operating ranges for cabin pressure. Generally, aircraft designed for airline service are capable of maintaining a cabin altitude of 1.5 to 2.1 km (i.e., a cabin pressure equivalent to that at 1.5 to 2.1 km) at typical operating altitudes and of limiting cabin altitude to a maximum of 2.4 km at maximum cruise altitudes. For passenger comfort, the cabin altitude rate of change is typically limited to 2.5 m/s for increasing altitude and 1.5 m/s for decreasing altitude.

Aircraft flying above 3 km must be equipped with an oxygen system for use in case of loss of pressurization. To maintain cabin pressure control, air inflow must exceed air leaking from the pressurized cabin. Leakage areas include controlled vents for the galley, toilets, and electronic equipment as well as uncontrolled leakage through door seals and structural joints.

Ventilation

Crew and passenger compartments should be ventilated whenever the aircraft is in operation, except for brief periods when all engine power is needed to produce thrust. Outside air ventilation rates as low as 2.4 L/s per person are adequate. Actual total ventilation rates (recirculating plus outside air) in commercial aircraft range

from 6 to 12 L/s per person. In current practice, 3.3 to 7.1 L/s per person of outside air are provided for comfort considerations. However, temperature control and air distribution requirements may establish higher ventilation rates. These ventilation rates are calculated assuming a 100% passenger load, no failures in the air-conditioning system, and control of harmful contaminants. Generally, much higher ventilation rates are provided in the cockpit, primarily to cool all the electrical and electronic equipment, compensate for increased solar loads, and provide positive pressurization of the cockpit relative to the cabin. Cabin air is often recirculated. Recirculated air is generally filtered with high-efficiency filters before it is reintroduced into occupied areas.

Emergency ventilation is sometimes necessary and may consist of (1) a ram air circuit supplying outside air through the normal distribution system or (2) provisions for manipulating cabin pressure control valves to draw ambient air through the cabin and flight deck, door seals, negative relief valves, hatches, and so forth. These methods require that the cabin is unpressurized.

Air Quality

Cabin air quality is a complex function of many parameters, including ambient air quality, cabin volume, design of ventilation and pressurization systems, system operation and maintenance procedures, and contaminant concentrations. Part 25 of the Federal Aviation Regulations (FARs), stipulates that crew and passenger compartment air be free of harmful or hazardous concentrations of gases or vapors.

An important aspect of air quality for international flights is the dilution of tobacco smoke to acceptable levels. A number of tests have been run to determine smoker and nonsmoker response to various dilution indices. The dilution index (DI) is defined as the number of litres of fresh air per milligram of tobacco burned. Figure 4 shows how irritation varies with the DI, and Figure 5 shows smoker and nonsmoker satisfaction with different DIs.

Smoking Zones. The data in Figure 5 and an estimated smoking rate of 280 µg/s per smoker may be used to calculate the ventilation required to obtain any desired acceptance rate in a smoking zone.

Ventilation should be at a level so that at least 80% acceptance by smokers is achieved in smoking zones. For an 80% acceptance level, the DI is 24 L/mg for smokers in a smoking zone. At a smoking rate of 280 µg/s, a ventilation rate of 6.8 L/s of outside air per smoker is indicated for acceptance by 80% of smokers. Filtered recirculated air may be used if it provides the equivalent dilution of contaminants, but the ventilation rate may need to be modified.

Ozone. In certain areas above 40° latitude, during given times of the year, and under certain atmospheric conditions, aircraft operating at altitudes above approximately 9 km may encounter atmospheric

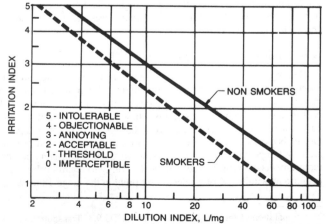

Fig. 4   Irritation Versus Dilution of Smoke

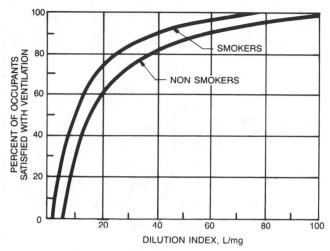

Fig. 5   Effect of Dilution Index on Occupant Satisfaction

ozone concentrations of sufficient magnitude to affect cabin air quality adversely.

Physiologically, ozone affects the soft tissues of the lung, causing pulmonary edema, dyspnea, and reduced lung capacity.

The effects of ozone are a function of concentration and exposure time. Currently, Part 25 of the Federal Aviation Regulations (FARs) sets the following concentration and time of exposure limits for ozone in cabins of transport aircraft:

- 0.25 mL/m^3, sea level equivalent, at any time above 9.8 km

- 0.1 mL/m^3, sea level equivalent, time-weighted average during any 3-h interval above 8.2 km

Ozone concentration in an airstream can be reduced by an absorption process, a chemical reaction with a filter surface, or a catalytic decomposition process.

Ozone decomposes at a rate determined by temperature. Thermal decomposition can be accelerated by allowing the air-ozone mixture to come in contact with metals (e.g., magnesium dioxide or stainless steel) or other materials (e.g., activated charcoal) that act as catalysts.

Catalytic converters similar to that shown in Figure 6 are installed in the pneumatic supply air of aircraft to decompose atmospheric ozone.

Load Determination

The cooling and heating loads for a particular airplane can be determined from a heat transfer study and from an analysis of the solar and internal heat from occupants and electrical equipment. The study should consider all possible flow paths through the complex aircraft structure. Air film coefficients vary with altitude and should also be considered. For high-speed aircraft, the increase in air temperature and pressure due to ram effects is appreciable and may be calculated from the following equations:

$$\Delta t_r = 0.2 M^2 T_a F_r \qquad (1)$$

$$\Delta p_r = (1 + 0.2 M^2)^{3.5} p_a - p_a \qquad (2)$$

where

F_r = recovery factor, dimensionless
Δt_r = increase in temperature due to ram effect, K
M = Mach number, dimensionless
T_a = absolute static temperature of ambient air, K
Δp_r = increase in pressure due to ram effect, kPa
p_a = absolute static pressure of ambient air, kPa

The average increase in airplane skin temperature for subsonic flight is generally based on a recovery factor F_r of 0.9.

Ground and flight requirements may be quite different. For an aircraft sitting on the ground in bright sunlight, local skin temperatures are considerably higher than the ambient for surfaces that have high absorptivity and are perpendicular to the sun's rays. This temperature increase is reduced considerably if a breeze blows across the airplane.

Other considerations for ground operations include cool-down or warm-up requirements, time that doors are open for loading, and whether ground heating and cooling are provided using equipment on-board or from an external source.

DESIGN APPROACHES

State-of-the-art design approaches should be examined when establishing the air-conditioning and pressurization system design for a particular aircraft. The design selected should meet the performance requirements and rate the best overall from the standpoint of aircraft penalty, life-cycle cost, and development risk.

Air Distribution

The design should ensure air supply to, and exhaust from, the occupied compartments. The location of inlets is usually based on previous experience confirmed by quantitative and subjective testing. Generally, passenger compartment air is introduced at a high level and exhausted at floor level. The design of air exhausts should preclude the possibility of blockage by luggage, clothing, or litter. The overall flow pattern should keep contaminated air that is generated by failures occurring under the floor, behind furnishings, or in electronic equipment from entering occupied compartments. Air from toilets and galleys should not be exhausted into occupied areas.

The flight crew should have means to direct and vary the airflow. Such adjustments should not significantly affect the overall balance of air distribution and should not allow complete shutoff of the air supply to the flight compartment.

With individual air supplies closed and the air-conditioning system operating normally, the air velocity in the vicinity of seated occupants should be between 0.1 and 0.2 m/s for optimum comfort and should not exceed 0.3 m/s. To avoid the sensation of no airflow, air

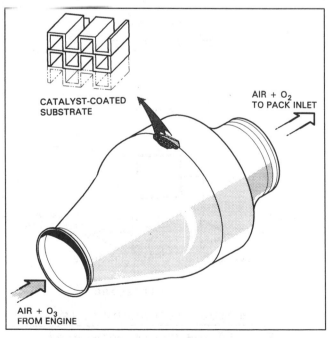

Fig. 6   Ozone Converter
(Reprinted with permission from SAE Paper #932057.
©1993 Society of Automotive Engineers, Inc.)

velocities should not be less than 0.05 m/s at seated head level. When individual air supplies are provided, the flow should be adjustable, and the jet velocity at seated head level should be at least 1 m/s. Cabin distribution ducts and air inlets should be sized to limit air velocities so that air noise levels are not objectionable to occupants. Longitudinal movement of air in the passenger cabin should be minimized.

In stabilized temperature control system phases, compartment air distribution should be such that the temperature variation measured in a vertical plane from 50 mm above the floor to seated head level should not exceed 2.8 K.

The temperature of air entering occupied compartments should not be less than 2°C or more than 71°C during normal operation. Where cabin air supplies combine, or if cold or hot air is added for temperature control, the various supplies should be effectively mixed to achieve a uniform temperature prior to distribution in occupied compartments.

Free water present in the air supply system should be removed to prevent it from being discharged onto passengers or equipment. Any drainage facilities should ensure that the water presents no hazards to equipment or the airplane structure.

Air Source

Engine compressor bleed air is the source of cabin pressurization and ventilation air on most current aircraft. Normally, bleed air contamination problems do not exist with current turbine engines. The available compressor bleed air usually more than adequately meets the air-conditioning and pressurization requirements for ground and flight operations. Under many operating conditions, the bleed air temperature and pressure are higher than required and must be regulated to lower values. Use of engine compressor bleed air can be a significant penalty to the aircraft, and this penalty should be considered when conducting studies to select the system design approach.

Airflow from auxiliary compressors can be used in lieu of engine bleed air. These compressors can be shaft driven off the engine gearbox or a remote gearbox. Auxiliary compressors may also be driven by pneumatic, hydraulic, or electrical power.

Engine bleed is a convenient means of obtaining pressurized air. However, as a power source, bleed air is generally not as efficient as other alternatives, partly because power is often dissipated by regulation to acceptable pressure levels. Electrical and hydraulic power can be provided at relatively high efficiencies, but they are limited by the available sources and their equipment adds considerable mass. Direct mechanical power use is restricted by component location, but it is most efficient on high bypass ratio engines.

REFRIGERATION SYSTEMS

Both air-cycle and vapor-cycle refrigeration systems are used on aircraft. Air-cycle systems are more commonly used. Their lightweight, compact equipment and readily available bleed air usually offset the inherent low efficiency.

Air Cycle

Air-cycle refrigeration systems may be designed and operated as open or closed. In the closed air-cycle system, the air refrigerant remains in the piping of component parts at all times, usually at a pressure above atmospheric. The closed air-cycle system is also known as the dense-air system, a term derived from the higher pressures maintained in comparison with the open system. Closed air-cycle systems have not been used on aircraft.

Open air-cycle systems are widely used in military and commercial aircraft. In the open air-cycle system, air is the refrigerant that directly cools and ventilates the space requiring conditioning. Unlike the refrigerant of a vapor-cycle system, which continually changes phase from liquid to gas and back to liquid again, the refrigerant of an air-cycle system remains in the gaseous phase throughout the cycle.

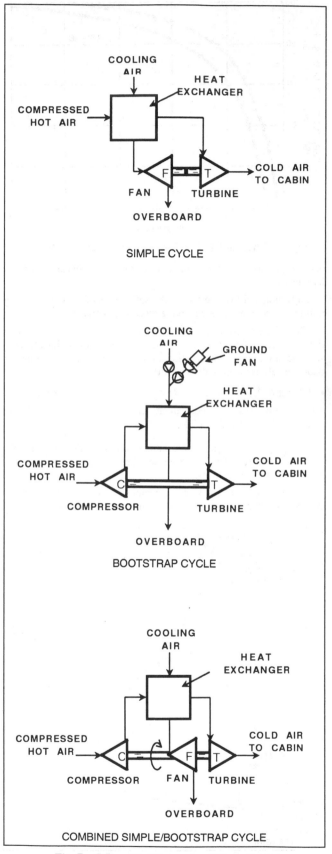

Fig. 7   Schematic Diagrams of Three Basic Air-Cycle Configurations
(Reprinted with permission from SAE Paper #932057. ©1993 Society of Automotive Engineers, Inc.)

Generally, the open air-cycle refrigeration process begins with the compression of ambient air, followed by cooling in a heat exchanger to remove the heat generated by compression. The cooled high-pressure air then expands through a turbine, where its temperature falls well below the ambient temperature. Energy produced by the turbine can be used either to (1) assist in compression, in which case the arrangement is called a bootstrap cycle; (2) drive a fan to pump cooling air through the heat exchanger, in what is termed a simple cycle; or (3) assist in compression and drive a fan, in what is known as a combined simple-bootstrap arrangement. Simplified schematics of these three cycles are shown in Figure 7. Dozens of other air-cycle arrangements are possible, but these three are used the most in aircraft applications.

The bootstrap cycle is the most efficient of the three cycles because it requires the lowest inlet pressure to produce a given amount of refrigeration. During flight, the forward motion of the aircraft produces sufficient heat exchanger cooling airflow; however, for ground operation, separate cooling air fans are needed.

The simple cycle is the least efficient, requiring higher supply pressure because no turbine energy is used to assist compression. Due to the higher temperatures associated with this higher supply pressure, larger heat exchangers are required. Because the simple cycle drives its own cooling air fan, it can operate on the ground or in helicopters, which do not always travel forward at high speed.

The combined simple/bootstrap cycle performs with an intermediate efficiency and includes a cooling air fan. Each of the three arrangements has advantages and disadvantages, but all continue to be widely applied.

The advantages of an air-cycle refrigeration system include its light weight, compact size, high reliability, and the availability of bleed air. The advantages usually offset the disadvantages of low efficiency and poor ground cooling. Ground cooling can be provided by an external air-conditioning cart. Cooling with the installed equipment can be obtained with an onboard auxiliary power unit, usually a gas turbine engine, that supplies large quantities of high-pressure air. It is not practical to run the jet engines of an airplane on the ground for cooling because of the high power required, the high rate of fuel consumption, and the associated noise levels.

The newest open air-cycle systems use high pressure water removal, cabin air recirculation, and air-bearing cooling turbines.

Water Separation. Under humid ambient conditions, condensation occurs in the expansion of air discharged from an air-cycle refrigeration unit. The condensation can be in the form of mist, fog, or even snow. The most severe fogging occurs during low-altitude flight, such as the landing approach, and on the ground, where good visibility is essential. In addition, moisture in the aircraft cabin causes deterioration of the equipment and reduces its reliability, accelerates structural corrosion, and promotes the growth of bacteria. Therefore, it is important to install provisions in an air-cycle refrigeration unit for the removal of entrained moisture. Low- or high-pressure water separators may be used.

The heart of a conventional low-pressure water separator in an air-cycle system is a cloth bag interposed in the airstream at the turbine discharge. This bag coalesces the water formed in the turbine as the air is cooled. The resulting large water droplets are separated from the airstream by centrifuge baffles and traps and either drained overboard or reevaporated in the ram air heat sink supply to enhance the cycle efficiency.

Temperature sensors measure the temperature of the air entering the water separator, and an anti-ice control system limits the dry-bulb temperature of this air to a minimum nominal value of 1.7°C. The control system positions the anti-ice valve so that the proper amount of turbine bypass hot air is mixed with the cold turbine discharge air. This prevents freezing of the water condensed in the expansion process.

The high-pressure water removal system condenses water at the turbine inlet. A large amount of the moisture from high-humidity

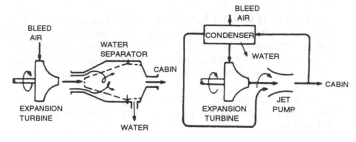

Fig. 8   Low-Pressure and High-Pressure Water Separators

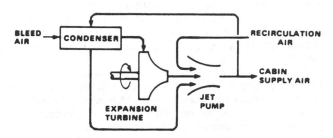

Fig. 9   Recirculation Technique as Implemented with High-Pressure Separation

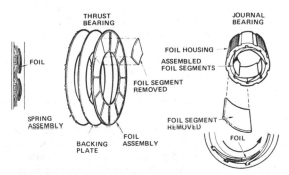

Fig. 10   Self-Acting Compliant Foil Air Bearing

conditions may be condensed and removed at this point, primarily because of the higher dew-point temperature associated with the high pressures. Condensation is enhanced by cooling the turbine inlet temperature below the dew-point temperature in a small heat exchanger using turbine discharge air. The velocity is low, and the water can be induced to separate from the airstream and drain away. This high-pressure water separator is maintenance-free, whereas the cloth bag of a low-pressure separator must be periodically cleaned. Figure 8 shows low- and high-pressure water separators.

Recirculation. Air-cycle systems with low-pressure water separators are handicapped by their inability to use turbine discharge temperatures that are below freezing because the frozen water particles would quickly block either a low-pressure water separator or the cabin supply duct system. When moisture is removed in a high-pressure water separator, air can be delivered at a temperature below −18°C, which nearly doubles the cooling capacity of each kilogram of air obtained from the engine compressor stage. This air is mixed with filtered return air and, for temperature control, is mixed with bypass hot bleed air (trim air) to provide sufficient cabin air supply at a temperature that is comfortable to passengers near the outlets.

Figure 9 shows recirculation in conjunction with high-pressure water separation. The recirculation could be achieved with a jet pump or a motor-driven fan.

Air-Bearing Cooling Turbine. In the air-bearing cooling turbine, the oil-lubricated ball bearings have been replaced with a set of journal and thrust bearings lubricated by air. Figure 10 shows the bearings for an air-bearing turbine. Air lubrication eliminates the maintenance associated with the oil lubrication system and improves reliability.

Air bearings and high-pressure water separation eliminate the need for periodic maintenance of an air-cycle environmental control system.

Vapor Cycle

In most air-cycle systems, the fresh air is refrigerated (often to subfreezing temperatures) as needed to control cabin temperatures and is then mixed with recirculated air for distribution in the cabin. The recirculated air does not pass through the refrigeration system. However, in vapor-cycle (direct-expansion) systems, it is more convenient to mix the fresh and recirculated airflows first and then to cool them in the vapor-cycle evaporator at a temperature above freezing.

In addition to lower power consumption, several developments have made vapor-cycle systems attractive for use in commercial aircraft. Substantial improvements in electronics and compressors have led to the development of the variable-speed refrigerant compressor. It incorporates a highly efficient, permanent magnet motor and operates on process gas bearings that use Refrigerant-134a as the lubricant. This eliminates oil/refrigerant incompatibility.

Several of the early jet transports had vapor-cycle refrigeration systems with a cooling capacity of approximately 70 kW. Smaller vapor-cycle units are used for galley refrigerators. Vapor-cycle systems are widely used on general aviation aircraft and are used as supplementary systems for cooling electronic equipment on commercial and military aircraft.

As discussed in the section on Load Determination, aircraft operating at high mach numbers and low altitudes experience very high skin and ram air temperatures. Heat rejection to ram air may not be practical. Heat rejection to fuel is currently used in military applications.

Figure 11 shows a typical aircraft vapor-cycle system. Cabin air is normally recirculated by an electric fan through the evaporator when the airplane is on the ground. Capacity is modulated by an evaporator pressure regulator that raises or lowers the temperature level at which the refrigerant is evaporated. The expansion valve is thermostatic and must control to superheat the refrigerant sufficiently to ensure that no liquid enters the compressor. As flow is throttled at the evaporator pressure regulator or expansion valve, the surge control valve must open to bypass refrigerant and keep a minimum flow through the compressor.

Other methods of capacity modulation include bypassing air around the evaporator, throttling refrigerant at the evaporator inlet, varying compressor speed, or unloading the compressor. Compressor drive speed can be varied through use of a permanent magnet motor in conjunction with a solid state variable-frequency inverter power source. For evaporator inlet throttling with a motorized expansion valve and a centrifugal compressor, the function of the valve may be combined with surge control on the same shaft. As the expansion valve closes, the bypass circuit opens.

Turbocompressor air from outside or engine compressor bleed air may be manually selected after engines are started. It then switches on automatically at takeoff for cabin pressurization. The hot turbocompressor or engine bleed air is first passed through an air-to-air heat exchanger, where it is partially cooled before entering the evaporator. Precooling is used to match the required in-flight capacity to that required on the ground and to improve system reliability.

The condenser fan provides cooling air on the ground and can also be turned on in flight to supplement ram air at low speeds and low altitude. In this system, the fan windmills in flight when turned off, but a separate circuit is provided for ram air on some airplanes. The cooling air is automatically modulated to maintain a minimum condensing pressure, both to reduce drag and to maintain pressure across the expansion valve at the low ambient encountered at high altitude.

Temperature Control System

Independent, automatic temperature controls should be provided for the flight crew and passenger compartments. In large airplanes, additional zonal temperature controls should be provided to balance uneven thermal loads caused by conditions such as mixed cargo/passenger configurations or nonuniform occupancy. Flight crew compartment temperature control should not be affected significantly by these other compartment controls. Each compartment or zone air temperature should be selectable within the range 18 to 29°C. When the initial or reselected temperature stabilizes, it should be maintained within about 1 K at the compartment sensor. Means should be provided for temperature sensors to sample compartment air at a rate compatible with control sensitivity requirements. The air sample should be taken from a location representative of average compartment temperature. Following reselection, compartment temperature should not overshoot the newly selected value by more than 1.5 K. Compartment supply temperature should be controlled within limits appropriate for required heating and cooling rates and with consideration for occupant safety.

Figure 12 shows a basic temperature control system for a source of cold air from an air-cycle refrigeration unit. Hot bleed air is added to the cold air source in response to a control system. A flow control valve holds the total flow delivered to the cabin relatively constant regardless of the modulation valve position. Basically, the crew selects a knob position that represents a desired cabin temperature. A sensor located at some point in the cabin measures the actual cabin temperature. The associated control equipment then compares these inputs and moves the valve to the position that will bring the cabin temperature to the selected value.

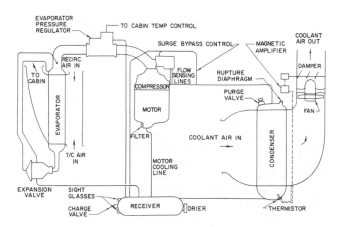

Fig. 11   Vapor-Cycle Refrigeration Unit

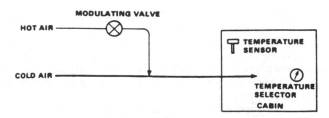

Fig. 12   Basic Temperature Control System

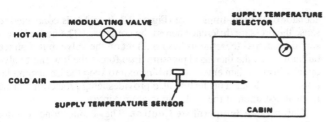

Fig. 13 Duct Control System Regulating Air Temperature Before It Enters the Cabin

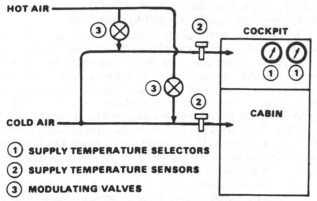

Fig. 14 Duct Control System Regulating Air Temperature to Cabin and Cockpit Separately

Figures 13 and 14 show variations of the basic concept and show a duct control system (which, instead of controlling the temperature in the cabin, controls the temperature of the air entering the cabin) and a two-compartment system that controls the temperature of the cockpit and the cabin independently.

Cabin Pressurization Control System

The cabin pressurization control system must meter the exhaust and ventilating air to maintain the selected low-altitude cabin as the airplane altitude and cabin inflow air vary. Both pneumatic and electronic systems have been used.

Cabin pressure control systems (1) maintain pressure in the cabin at levels acceptable to humans, (2) change pressure during climb and descent at acceptable rates, and (3) limit transient pressure changes, commonly called bumps, to magnitudes that are not annoying.

Pressure changes result in stretching of the human ear drum, which causes unpleasant sensations. Contraction of the eustachian tube dilator muscles, as during swallowing or yawning, helps open the tube and equalize the pressure. A positive pressure within the middle ear is easier to neutralize because it helps force the air through the tube. A negative pressure tends to collapse and seal the tube, hindering relief. Therefore, a lower rate of sustained pressure change is used for increasing ambient pressure than for decreasing it. Rates of 2.5 m/s ascending and 1.5 m/s descending have been the industry-accepted maximums for some time.

Positive pressure relief at some maximum pressure must be provided to protect the airplane in the event of a pressure control system failure. A negative (vacuum) pressure relief mechanism to let air in when outdoor pressure exceeds cabin pressure must also be provided.

Other desirable features are a barometric correction selector to help select the proper landing field altitude so that the pressure differential at landing approaches zero, and a limit control to maintain a maximum cabin altitude if other control components fail. An indicating system should be provided that includes a rate of climb indicator, an altitude warning horn, an altimeter, and a differential pressure indicator.

TYPICAL SYSTEM

Figure 15 shows the air-cycle air-conditioning system for a commercial transport aircraft. It operates from a source of preconditioned engine bleed air, creating a supply of conditioned air controlled to maintain the selected temperatures and ventilation rates within two passenger zones and the flight station. The various automatic temperature control functions are regulated by electronic controllers and electric valve actuation. The two refrigeration packs are installed in the unpressurized area beneath the wing center section and are supplied with preconditioned bleed air from the two main engines or from the tail-mounted auxiliary power unit (APU). An underfloor distribution bay mixes conditioned air from the two packs with filtered recirculated cabin air and distributes it throughout the pressurized areas. It should be noted that more than two air-conditioning packs may be used.

Preconditioned bleed air enters each of the two parallel air-conditioning packs through a flow control valve (Figure 15, Item 1), which also functions as a pack shutoff valve. Most of the basic refrigeration equipment is assembled into a package ready for installation into the aircraft. Each pack includes a three-wheel air-cycle machine (ACM), four heat exchangers, and temperature control valves, as well as protective devices and all necessary ducting and hardware.

The ACM (Figure 15, Item 2) has three rotating wheels—a compressor, a turbine, and a fan—which are mounted on a common shaft supported by air bearings. After the bleed air passes through the primary heat exchanger (Item 3), where it is cooled by ram air, it enters the compressor section of the ACM, where it is compressed to a higher pressure and temperature. The bleed air is cooled again by ram air in the secondary heat exchanger (Item 4) and, after passing through the reheater (Item 5), enters the condenser (Item 6). The air discharging from the condenser contains free moisture, which is removed in the extractor (Item 7) before the air enters the other side of the reheater. The function of the reheater is to cool the air on the first pass, thereby reducing the amount of cooling required of the condenser, and to reheat the air before it enters the turbine. This reheating process evaporates small amounts of entrained moisture that may still be present and creates a higher temperature at the turbine inlet, with an attendant increase in turbine power. The energy removed from the turbine airflow causes a substantial temperature reduction, permitting a turbine discharge temperature well below the ram temperature.

Ram cooling air for the heat exchangers enters through a variable-area ram inlet scoop, passes through the secondary and primary heat exchangers in series, and exhausts through a variable-area exhaust door. The ram inlet and exhaust door actuators (both shown in Figure 15, Item 8) are operated under the control of the pack temperature controller.

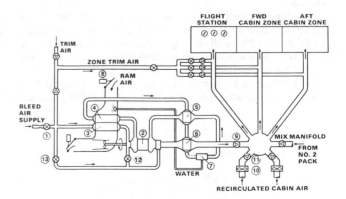

Fig. 15 Air-Conditioning Schematic for Commercial Aircraft

Conditioned air from the two packs is delivered through bulkhead check valves (Figure 15, Item 9) to a manifold, where it mixes with recirculated cabin air for delivery to the flight deck zone and the two passenger zones. Both the pack and cabin zone temperature controls limit the total cabin supply flow to 2°C minimum at all altitudes. Conditioned air for the flight deck is removed upstream of the mix manifold, and fresh air is used for flight deck cooling.

Cabin recirculation air is supplied to the manifold by electric fans (Figure 15, Item 10) through check valves (Item 11) that prevent reverse flow when the fans are not operating. Recirculation maintains the desired level of cabin ventilation while minimizing the use of bleed air.

Air-Conditioning Controls

Flow Control Operation. Maximum pack airflow is limited by flow control valves at the pack inlets. A dual flow schedule capability allows a pack flow to be increased over normal flow. With the higher supply flow, one pack will maintain approximately 80% of normal cabin airflow. The high-flow schedule is automatically selected through aircraft wiring circuits whenever a cooling pack or cabin recirculating air fan is shut off.

Pack Temperature Control. When the air-conditioning system is operating at maximum refrigeration capacity, the pack outlet temperature is determined either by the capabilities of the system or by the action of the low-limit controls. The maximum cooling capacity is necessary only for extremely hot conditions or to cool a heat-soaked aircraft. For most operating conditions, a warmer air supply is needed to satisfy the actual cabin cooling or heating demands, and the pack refrigeration capacity must be modulated.

The actual pack supply temperature requirement is determined by the zone controller and satisfies the cabin zone requiring the most cooling. Both packs are controlled to produce the required supply temperature. Each pack is operated independently by its own controller and temperature sensor, but both operate to produce the same outlet temperature due to the single temperature demanded by the zone system. This results in the temperature control valve and ram doors operating at approximately the same relative position for both packs.

To adjust pack outlet temperature, the controller modulates the temperature control valve (Figure 15, Item 12) and the ram inlet and outlet actuators (both Item 8). Although each of these devices is independently actuated, the controller maintains a definite positional relationship between them. For maximum refrigeration, the ram doors are positioned fully open and the bypass valve is closed. To increase pack discharge temperature, the ram doors are partially closed, thus reducing cooling airflow while the bypass valve simultaneously begins opening to divert warm air around the air-cycle machine. Proper scheduling of this pack control scheme minimizes ram air drag without exceeding equipment maximum temperature limitations.

The cabin air supply temperature leaving the mix manifold is measured by a duct air temperature sensor at the outlet and is limited to a minimum of 2°C at all conditions. The pack controller overrides the normal control schedule and modulates the temperature control valve and the ram air door actuators, as required.

An all-pneumatic backup control protects against icing that might occur during automatic control system failures. The pneumatic actuator of the low-limit valve (Figure 15, Item 13) is connected to sense the pressure difference across the condenser. The normal pressure drop across these items has no effect on the valve, but if an ice buildup starts, the increased pressure drop forces the low-limit valve open. Warm air then enters the turbine exit to keep the ice accumulation at a low level. This feature also provides icing protection during manual operation of the pack temperature control.

Cabin Zone Temperature Control. The cabin zone control subsystem incorporates a programmed duct air temperature control for accurate regulation during aircraft transient conditions. An inherent feature of the control is the limitation of zone inlet duct temperature that results from electronic damping of the inlet temperature demand reference signal within predetermined limits.

Temperature is independently controlled in the flight station zone and in the fore and aft passenger zones by automatic feedback controls. The flight crew selects the desired temperature for each zone on selector units mounted in the flight station. Signals from each selector and corresponding zone temperature sensors are processed in the cabin zone controller, which produces a zone supply air temperature demand signal for each zone according to a predetermined schedule. The demand signal for each zone is appropriately supplied to each trim air control loop. Trim air control allows a portion of hot air to mix with the air-cycle discharge air to provide the demanded temperature at each zone inlet.

The cabin zone controller also includes a discriminator function that selects the lowest of these demand signals and passes it to the pack controllers at the pack temperature command. Both packs are then modulated to provide this temperature. Thus, both packs are operated at the same supply duct temperature, which, when mixed with the cabin recirculation air, satisfies the zone requiring the lowest inlet temperature. When the temperature sensed at the mix manifold outlet differs from the temperature demand, the temperature command signal to the packs is automatically changed until the sensed outlet temperature equals the temperature demand from the zone requiring the greatest cooling. This guarantees that the demand of the zone requiring the greatest cooling is satisfied by the pack controls and that the trim air valve for that zone is closed.

BIBLIOGRAPHY

Crabtree, R.E, M.P. Saba, and J.E. Strang. 1980. The cabin air conditioning and temperature control system for the Boeing 767 and 757 airplanes. ASME-80-ENAS-55. American Society of Mechanical Engineers, New York.

Dieckmann, R.R., O.R. Kosfeld, and I.C. Jenkins. 1986. Increased avionics cooling capacity for F-15 aircraft. SAE 860910. Society of Automotive Engineers, Warrendale, PA.

Linnet, K. and R. Crabtree. 1993. What's next in commercial aircraft systems? SAE 93057. Society of Automotive Engineers, Warrendale, PA.

Payne, G. 1980. Environmental control systems for executive jet aircraft. SAE 800607. Society of Automotive Engineers, Warrendale, PA.

SAE. 1969. Aerospace applied thermodynamics manual. AIR 1168. Society of Automotive Engineers, Warrendale, PA.

SAE. 1976. Aircraft cabin pressurization control criteria. *Standard* ARP 1270. Society of Automotive Engineers, Warrendale, PA.

Thayer, W.W. 1982. Tobacco smoke dilution recommendations for comfortable ventilation. Paper 7092. Douglas Aircraft Company, Long Beach, CA.

CHAPTER 10

SHIPS

THIS chapter covers air conditioning for oceangoing surface vessels, including luxury liners, tramp steamers, and naval vessels. Although the general principles of air conditioning that apply to land installations also apply to marine installations, some systems are not suitable for ships due to their inability to meet shock and vibration requirements. The chapter focuses on load calculations and air distribution systems for these ships.

GENERAL CRITERIA

Air conditioning in ships provides an environment in which personnel can live and work without heat stress, increases crew efficiency, improves the reliability of electronic and similar critical equipment, and prevents rapid deterioration of special weapons equipment aboard naval ships.

The following factors should be considered in the design of an air-conditioning system for shipboard use:

1. The system should function properly under conditions of roll and pitch.
2. The construction materials should withstand the corrosive effects of salt air and seawater.
3. The system should be designed for uninterrupted operation during the voyage and continuous year-round operation. Because ships en route cannot be easily serviced, some standby capacity, spare parts for all essential items, and extra refrigerant charges should be carried.
4. The system should have no objectionable noise or vibration, and must meet the noise criteria required by the shipbuilding specifications.
5. The system should meet the special requirements for operation given in section 4 of ASHRAE *Standard* 26.
6. The equipment should occupy a minimum of space commensurate with its cost and reliability. The weight should be kept to a minimum.
7. Because a ship may pass through one or more complete cycles of seasons on a single voyage and may experience a change from winter to summer operation in a matter of hours, the system should be flexible enough to compensate for climatic changes with minimal attention of the ship's operating personnel.
8. Infiltration through weather doors is generally disregarded. However, specifications for merchant ships occasionally require an assumed infiltration load for heating steering gear rooms and the pilothouse.
9. Sun load must be considered on all exposed surfaces above the waterline. If a compartment has more than one exposed surface, the surface with the greatest sun load is used, and the other exposed boundary is calculated at outside ambient temperature.

10. Cooling load inside design conditions are given as a dry-bulb temperature with a maximum relative humidity. For merchant ships, the cooling coil leaving air temperature is assumed to be 9°C dry bulb. For naval ships, it is assumed to be 10.8°C dry bulb. For both naval and merchant ships, the wet bulb is consistent with 95% rh. This off-coil air temperature is changed only when humidity control is required in the cooling season.
11. When calculating winter heating loads, heat transmission through boundaries of machinery spaces in either direction is not considered.

Calculations for Merchant Ship Heating, Ventilation, and Air Conditioning Design, which is a bulletin available from the Society of Naval Architects and Marine Engineers (SNAME), gives sample calculation methods and estimated values.

MERCHANT SHIPS

DESIGN CRITERIA

Outdoor Ambient Temperatures

The service and type of vessel determines the proper outdoor design temperature. Some luxury liners make frequent off-season cruises where more severe heating and cooling loads may be encountered. The selection of the ambient design should be based on the temperatures prevalent during the voyage. In general, for the cooling cycle, outdoor design conditions for North Atlantic runs are 35°C dry bulb and 25.5°C wet bulb; for semitropical runs, 35°C dry bulb and 26.5°C wet bulb; and for tropical runs, 35°C dry bulb and 28°C wet bulb. For the heating cycle, −18°C is usually selected as the design temperature, unless the vessel will always operate in higher temperature climates. The design temperatures for seawater are 30°C in summer and −2°C in winter.

Indoor Temperatures

Effective temperatures (ETs) from 21.5 to 23°C are generally selected as inside design conditions for commercial oceangoing surface ships.

Inside design temperatures range from 24.5 to 26.5°C dry bulb and approximately 50% rh for summer and from 18 to 24°C dry bulb for winter.

Comfortable room conditions during intermediate outside ambient conditions should be considered in the design. Quality systems are designed to (1) provide optimum comfort when outdoor ambient conditions of 18 to 24°C dry bulb and 90 to 100% rh exist and (2) ensure that proper humidity and temperature levels are maintained during periods when sensible loads are light.

Ventilation Requirements

Ventilation must meet the requirements given in ASHRAE *Standard* 62, except when special consideration is required for ships.

The preparation of this chapter is assigned to TC 9.3, Transportation Air Conditioning.

Air-Conditioned Spaces. In public spaces (such as mess rooms, dining rooms, and lounges), 10 to 15 L/s of outdoor air per person or three air changes per hour (ACH) are required. In all other spaces, a minimum of 8 L/s of outside air per person or 2 ACH must be provided. However, the maximum outside air for any air-conditioned space is 25 L/s per person.

Ventilated Spaces. The fresh air to be supplied to a space is determined by the required rate of change or the limiting temperature rise. The minimum quantity of air to a space is 15 L/s per occupant or 18 L/s per terminal. In addition to these requirements, exhaust requirements must be balanced.

Load Determination. The cooling load estimate for air conditioning considers factors discussed in Chapter 26 of the 1993 *ASHRAE Handbook—Fundamentals*, including the following:

- Solar radiation
- Heat transmission through hull, decks, and bulkheads
- Heat (latent and sensible) dissipation from occupants
- Heat gain due to lights
- Heat (latent and sensible) gain due to ventilation air
- Heat gain due to motors or other electrical equipment
- Heat gain from piping, machinery, and equipment

The cooling effect of adjacent spaces is not considered unless temperatures are maintained with refrigeration or air-conditioning equipment. The latent heat from the scullery, galley, laundry, washrooms, and similar ventilated spaces is assumed to be fully exhausted overboard.

The heating load estimate for air conditioning should consist of the following:

- Heat losses through decks and bulkheads
- Ventilation air
- Infiltration (when specified)

No allowances are made for heat gain from warmer adjacent spaces.

Heat Transmission Coefficients

The overall heat transmission coefficient *U* for the composite structures common to shipboard construction do not lend themselves to theoretical derivation; they are most commonly obtained from full-scale panel tests. The Society of Naval Architects and Marine Engineers *Bulletin* 4-7 gives a method for determining these coefficients when tested data are unavailable.

Heat Dissipation from People

The rate at which heat and moisture dissipate from people depends on their activity levels and the ambient dry-bulb temperature. Values that can be used at 27°C room dry bulb are listed in Table 1.

Table 1 Heat Gain from Occupants

Activity at 27°C	Heat Rate, W		
	Sensible	Latent	Total
Dancing	72	177	249
Eating (mess rooms and dining rooms)	64	97	161
Waiters	88	205	293
Moderate activity (lounge, ship's office, chart rooms)	59	73	132
Light activity (staterooms, crew's berthing)	57	60	117
Workshops	73	149	222

Heat Gain from Sources Within a Space

The heat gain from motors, appliances, lights, and other equipment should be obtained from the manufacturers. Data in Chapter 26 of the 1993 *ASHRAE Handbook—Fundamentals* may be used when manufacturer's data is unavailable.

EQUIPMENT SELECTION

General

The principal equipment required for an air-conditioning system can be divided into four categories:

1. Central station air-handling, consisting of fans, filters, central heating and cooling coils, and sound treatment
2. Distribution network, including air ductwork, water piping, and steam piping
3. Required terminal treatment, consisting of heating and cooling coils, terminal mixing units, and diffusing outlets
4. Refrigeration equipment

These factors should be considered in the selection of air-conditioning equipment:

- Installed initial cost
- Space available in fan rooms, passageways, machinery rooms, and staterooms
- Operation costs, including system maintenance
- Noise levels
- Weight

High-velocity air distribution offers many advantages. The use of unitary (factory-assembled) central air-handling equipment and prefabricated piping, clamps, and fittings facilitates installation for both new construction and conversions. Substantial space saving is possible as compared to the conventional low-velocity sheet metal duct system. Maintenance is also reduced.

Fans must be selected for stable performance over their full range of system operation and should have adequate isolation to prevent transmission of vibration to the deck. Because fan rooms are often located adjacent to or near living quarters, effective sound treatment is essential.

In general, equipment used for ships is considerably more rugged than equipment for land applications, and it must withstand the corrosive environment of the salt air and sea. Materials such as stainless steel, nickel-copper, copper-nickel, bronze alloys, and hot-dipped galvanized steel are used extensively. Sections 6 through 10 of ASHRAE *Standard* 26 list equipment requirements.

Fans

The U.S. Maritime Administration specifies a family of standard vaneaxial, tubeaxial, and centrifugal fans. Selection curves found on the standard drawings are used in selecting fans (Standard Plans S38-1-101, S38-1-102, S38-1-103). Belt-driven centrifugal fans must conform to these requirements, except for speed and drive.

Cooling Coils

Cooling coils are based on an inlet water temperature of 5.6°C with a temperature rise of about 5.5 K, and must meet the following requirements:

- Maximum face velocity of 2.5 m/s
- At least six rows of tubes and 25% more rows than required by the manufacturer's published ratings
- Construction and materials as specified in USMA (1965)

Heating Coils

Heating coils must meet the following requirements:

- Maximum face velocity of 5 m/s

- Supply system preheaters with a final design temperature between 13°C and 16°C with outside air at −18°C
- Preheaters for air-conditioning systems suit design requirements but have a discharge temperature not less than 7°C with 100% outside air at −18°C
- Tempering heaters for supply systems have a final design temperature between 10°C and 21°C with outside air at −18°C
- Pressure drop for preheater and reheater in series, with full fan volume, does not exceed 125 Pa
- Capacity of steam heaters is based on a steam pressure of 240 kPa (gage), less an allowed 35 kPa (gage) line pressure drop and the design pressure drop through the control valve

Filters

Filters must meet the following requirements:

- All supply systems fitted with cooling and/or heating coils have manual roll, renewable marine-type air filters
- Maximum face velocity of 2.5 m/s
- Weather protection
- Located so that they are not bypassed when the fan room door is left open
- Clean, medium rolls are either fully enclosed or arranged so that the outside surface of the clean roll becomes the air-entering side as the medium passes through the airstream
- Dirty medium is wound with the dirty side inward

Medium rolls are 19.8 m long and are readily available as standard factory-stocked items in nominal widths of 0.6, 0.9, 1.2, or 1.5 m. A permanently installed, dry-type filter gage, graduated to read from 0 to 250 Pa, is installed on each air filter unit.

Air-Mixing Boxes

If available, air-mixing boxes should fit between the deck beams. Other requirements include

- Volume regulation over the complete mixing range within +5% of the design volume, with static pressure in the hot and cold ducts equal to at least the fan design static pressure
- Leakage rate of the hot and cold air valve(s) less than 2% when closed
- Calculations for space air quantities allow for leakage through the hot air damper

Box sizes are based on manufacturer's data, as modified by the above requirements.

Air Diffusers

Air diffusers ventilate the space without creating drafts. In addition, diffusers must meet the following requirements:

- Diffusers serving air-conditioned spaces should be a high induction type and constructed so that moisture will not form on the cones when a temperature difference of 17 K is used with the space dew point at least 5 K above that corresponding to summer inside design conditions. Compliance with this requirement should be demonstrated in a mock-up test, unless previously tested and approved.
- Volume handled by each diffuser should not, in general, exceed 240 L/s.
- Diffusers used in supply ventilation systems are of the same design, except have adjustable blast skirts.
- Manufacturer's published data is used in selecting diffusers.

Air-Conditioning Compressors

Compressors should be the same types used for ship's service and cargo refrigeration, except for the number of cylinders or the speed (see Chapter 29, Marine Refrigeration, in the 1994 *ASHRAE Handbook—Refrigeration*).

TYPICAL SYSTEMS

General

Comfort air-conditioning systems installed on merchant ships are classified as those serving (1) passenger staterooms, (2) crew's quarters and similar small spaces, and (3) public spaces.

Single-Zone Central System

Public spaces are treated as one zone and are effectively handled by a single-zone central system. Exceptionally large spaces may require two systems. Generally, either built-up or factory-assembled fan coil central station systems are selected for large public spaces. A schematic of a typical system, which is also known as the *Type A* system, is shown in Figure 1. Under certain conditions, it is desirable to avoid the use of return ducts and to use all outdoor air, which requires more refrigeration. Outdoor air and return air are both filtered before being preheated, cooled, and reheated as required at the central station unit. A room thermostat maintains the desired temperature by modulating the valve regulating the flow of steam to the reheat coil. Humidity control is often provided. During mild weather, the dampers frequently are opened to admit 100% outdoor air automatically.

Multizone Central System

A multizone central system, also known as the *Type C* system, is usually confined to the crew's and officers' quarters (Figure 2). Spaces are divided into zones in accordance with similarity of loads and

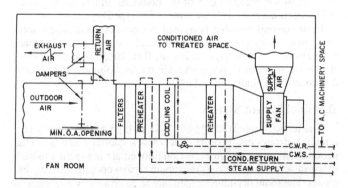

Fig. 1 Single-Zone Central (Type A) System

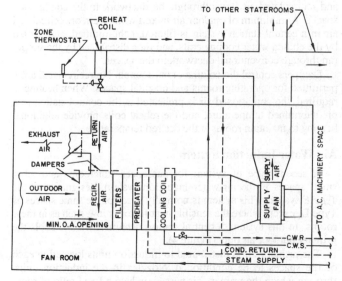

Fig. 2 Multizone Central (Type C) System

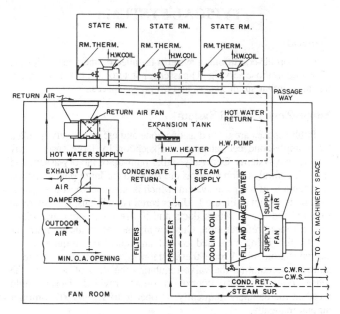

Fig. 3   Terminal Reheat (Type D) System

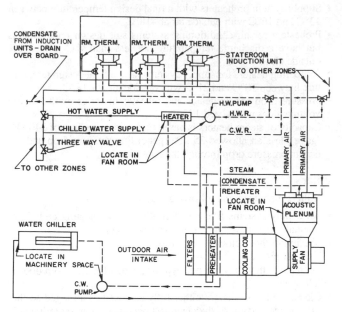

Fig. 4   Air-Water Induction (Type E) System

exposures. Each zone has a reheat coil to supply air at a temperature adequate for all the spaces served. These coils, usually steam-heated, are controlled thermostatically. Manual control of the air volume is the only means for occupant control of conditions. Filters, cooling coils, and dampers are essentially the same as for Type D systems.

Each internal sensible heat load component, particularly solar and lights, varies greatly. The thermostatic control methods cannot compensate for these large variations; therefore, this system cannot always satisfy individual space requirements. Volume control is conducive to noise, drafts, and odors.

Terminal Reheat System

The terminal reheat, or *Type D,* system is generally used for passengers' staterooms, officers' and crew's quarters, and miscellaneous small spaces (Figure 3). Conditioned air is supplied to each space in accordance with its maximum design cooling load requirements. The room dry-bulb temperature is controlled by a reheater. A room thermostat automatically controls the volume of hot water passing through the reheat coil in each space. A mixture of outdoor and recirculated air flows through the ductwork to the conditioned spaces. A minimum of outdoor air mixed with return or recirculated air in a central station system is filtered, dehumidified, and cooled by the chilled water cooling coils, and then distributed by the supply fan through conventional ductwork to the spaces.

Dampers control the volume of outdoor air. No recirculated air is permitted for operating rooms and hospital spaces. When heating is required, the conditioned air is preheated at the control station to a predetermined temperature, and the reheat coils provide additional heating to maintain rooms at the desired temperature.

Air-Water Induction System

A second type of system for passenger staterooms and other small spaces is the air-water induction system, designated as the *Type E* system. This system is normally used for the same spaces as Type D, except where the sensible heat factor is low, such as in mess rooms. In this system, a central station dehumidifies and cools the primary outdoor air only (Figure 4).

The primary air is distributed to induction units located in each of the spaces to be conditioned. Nozzles in the induction units, through which the primary air passes, induce a fixed ratio of room (secondary) air to flow through a water coil and mix with the pri-

mary air. The mixture of treated air is then discharged to the room through the supply grille. The room air is either heated or cooled by the water coil. Water flow to the coil (chilled or hot) can be controlled either manually or automatically to maintain the desired room conditions.

This system requires no return or recirculated air ducts because only a fixed amount of outdoor (primary) air needs to be conditioned at the central station equipment. This relatively small amount of conditioned air must be cooled to a sufficiently low dew point to take care of the entire latent load (outdoor air plus room air). It is distributed at high velocity and pressure and thus requires relatively little space for air distribution ducts. However, this saving of space is offset by the additional space required for water piping, secondary water pumps, induction cabinets in staterooms, and drain piping.

Space design temperatures are maintained during intermediate conditions (i.e., the outdoor temperature is above the changeover point and chilled water is at the induction units), with primary air heated at the central station unit according to a predetermined temperature schedule. Unit capacity in spaces requiring cooling must be sufficient to satisfy the room sensible heat load plus the load of the primary air.

When outdoor temperatures are below the changeover point, the chiller is secured, and space design temperatures are maintained by circulating hot water to the induction units. Preheated primary air provides cooling for spaces that have a cooling load. Unit capacity in spaces requiring heating must be sufficient to satisfy the room heat load plus the load of the primary air. Detailed analysis is required to determine the changeover point and temperature schedules for water and air.

High-Velocity Dual-Duct System

The high-velocity dual-duct system, also known as the *Type G* system, is normally used for the same kinds of spaces as Types D and E. In the Type G system, all air is filtered, cooled, and dehumidified in the central units (Figure 5). Blow-through coil arrangements are essential to ensure efficient design. A high-pressure fan circulates air at high velocities approaching 30 m/s through two ducts or pipes, one carrying cold air and the other warm air. A steam reheater in the central unit heats the air as required.

In each space served, the warm and cold air flow to an air-mixing unit that has a control valve to proportion hot and cold air to obtain

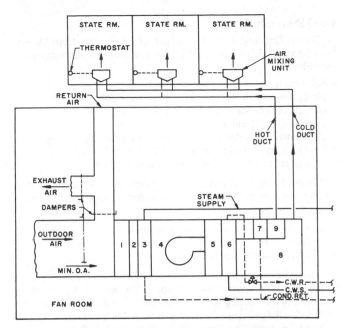

Fig. 5 Dual-Duct (Type G) System

Table 2 Minimum Thickness of Steel Ducts

All vertical exposed ducts	16 USSG	1.52 mm
Horizontal or concealed vertical ducts less than 150 mm	24 USSG	0.61 mm
Horizontal or concealed vertical ducts 160 to 300 mm	22 USSG	0.76 mm
Horizontal or concealed vertical ducts ducts 310 to 460 mm	20 USSG	0.91 mm
Horizontal or concealed vertical ducts 470 to 760 mm	18 USSG	1.21 mm
Horizontal or concealed vertical ducts over 760 mm	16 USSG	1.52 mm

The increased application of high-velocity, high-pressure systems has resulted in a greater use of prefabricated round pipe and fittings, including spiral formed sheet metal ducts. It is important that the field fabrication of ducts and fittings ensure that all are airtight. The use of factory-fabricated fittings, clamps, and joints effectively minimizes air leakage for these high-pressure systems.

In addition to the space advantage, small ductwork saves weight, another important consideration for this application.

the desired temperature. Within the capacity limits of the equipment, any temperature can quickly be obtained and maintained, regardless of load variations in adjacent spaces. The air-mixing unit also incorporates self-contained regulators that maintain a constant volume of total air delivery to the various spaces, regardless of adjustments in the air supplied to rooms down the line.

The following are some advantages of this system:

- All conditioning equipment is centrally located, simplifying maintenance and operation
- The system can heat and cool adjacent spaces simultaneously without cycle changeover and with a minimum of automatic controls
- Because only air is distributed from fan rooms, no water or steam piping, electrical equipment, or wiring are in the conditioned spaces
- With all conditioning equipment centrally located, direct-expansion cooling using halocarbon refrigerants is possible, eliminating all intermediary water-chilling equipment

AIR DISTRIBUTION METHODS

Good air distribution in staterooms and public spaces is difficult because of low ceiling heights and compact space arrangements. The design should consider room dimensions, ceiling height, volume of air handled, air temperature difference between supply and room air, location of berths, and allowable noise level. For major installations, mock-up tests are often used to establish the exacting design criteria required for satisfactory performance.

Air usually returns from individual small spaces either by a sight-tight louver mounted in the door or by an undercut in the door leading to the passageway. An undercut door can only be used with small air quantities of 35 L/s or less. Louvers are most commonly sized for a velocity of 2 m/s based on net area.

Ductwork

Ductwork on merchant ships is constructed of steel. Ducts, other than those requiring heavier construction because of susceptibility to damage or corrosion, are usually made with riveted seams sealed with hot solder or fire-resistant duct sealer, welded seams, or hooked seams and laps. They are fabricated of hot-dipped, galvanized, copper-bearing sheet steel, suitably stiffened externally. The minimum thickness of material is determined by the diameter of round ducts or by the largest dimension of rectangular ducts, as listed in Table 2.

CONTROLS

The conditioning load, even on a single voyage, varies over a wide range in a short period. Not only must the refrigeration plant meet these variations in load, but the controls must readily adjust the system to sudden climatic changes. Accordingly, it is general practice to equip the plant with automatic controls. Since comfort is a matter of individual taste, adjustable room thermostats are placed in living spaces.

Manual volume control has also been used to regulate temperatures in cabins and staterooms. For a low-velocity air distribution system, however, manual volume control tends to disturb the air balance of the remainder of the spaces served. Manual controls are also applied on high-velocity single-duct systems if a constant volume regulator is installed in the terminal box.

Design conditions in staterooms and public spaces can be controlled by one or a combination of the following:

Volume Control. (Used in Type C Systems.) This is the least expensive and simplest control. Its basic disadvantages are an inability to meet simultaneous heating and cooling demands in adjacent spaces, unsatisfactory air distribution, objectionable noise, and inadequate ventilation because of reduction in air delivery.

Reheater Control. Zone reheaters of Type C systems are controlled by regulating the amount of steam to the zone coils by one of two methods. The first method uses a room thermostat in a *representative* space with other spaces in the zone using manual dampers. Spaces served by this type of control do not always develop the desired comfort conditions. The second method uses a master-submaster control, which adjusts the reheater discharge temperature according to a predetermined schedule, so that full heat is applied at the design outside heating temperature and less heat is applied as the temperature rises. This control does not adjust to meet variations in individual room loads; however, it is superior to the first method because it is more dependable.

Preheater Control. A duct thermostat, which modulates steam through the valve, controls the preheaters. To prevent bucking, the set point usually is a few degrees below the design cooling off-coil setting.

Coil Control. Except for Type A systems with humidity control and Type E system primary air coils, cooling coils (water) are controlled by a dew-point thermostat to give a constant off-coil temperature during the entire cooling cycle. No control is provided for coils of Type E systems because they are in series with the flow-through induction units, and maximum dehumidification must be accomplished by the primary coil to maintain dry coil operation in the room units.

Damper Control. Outdoor, return, and exhaust dampers are either manually (as a group) or automatically controlled. In automatic control, one of two methods is used: (1) controlling the damper settings by thermostats exposed to weather air or (2) using a duct thermostat that restricts the outdoor airflow only when the design temperature leaving the cooling coil cannot be achieved with full flow through the coil.

REGULATORY AGENCIES

Merchant vessels that operate under the United States flag come under the jurisdiction of the Coast Guard. Accordingly, the installation and components must conform to the Marine Engineering Rules and Marine Standards of the Coast Guard. Equipment design and installation must also comply with the requirements of the U.S. Public Health Service and Department of Agriculture. Principally, this involves ratproofing.

Comfort air-conditioning installations do not primarily come under the American Bureau of Shipping. However, equipment should be manufactured, wherever possible, to comply with the American Bureau of Shipping Rules and Regulations. This is important when the vessels are equipped for carrying cargo refrigeration, because the air-conditioning compressors may serve as standby units in the event of a cargo compressor failure. This compliance eliminates the necessity of a separate spare cargo compressor.

NAVAL SURFACE SHIPS

DESIGN CRITERIA

Outdoor Ambient Temperatures

Design conditions for naval vessels have been established as a compromise, considering the large cooling plants required for internal heat loads generated by machinery, weapons systems, electronics, and personnel. Temperatures of 32°C dry bulb and 27°C wet bulb are used as design requirements for worldwide applications, together with 29.5°C seawater temperatures. Heating season temperatures are assumed to be −12°C for outdoor air and −2°C for seawater.

Indoor Temperatures

Naval ships are generally designed for space temperatures of 26.5°C dry bulb with a maximum of 55% rh for most areas requiring air conditioning. The *Air Conditioning, Ventilation and Heating Design Criteria Manual for Surface Ships of the United States Navy* (USN 1969) gives design conditions established for specific areas. *Standard Specification for Cargo Ship Construction* (USMA 1965) gives temperatures for ventilated spaces.

Ventilation Requirements

Ventilation must meet the requirements given in ASHRAE *Standard* 62, except when special consideration is required for ships.

Air-Conditioned Spaces. Naval ship design requires that air-conditioning systems serving living and berthing areas on surface ships replenish air in accordance with damage control classifications, as specified in USN (1969):

1. Class Z systems: 2.4 L/s per person.
2. Class W systems for troop berthing areas: 2.4 L/s per person.
3. All other Class W systems: 4.7 L/s per person. The flow rate is increased only to meet either a 35.4 L/s minimum branch requirement or to balance exhaust requirements. Outdoor air should be kept at a minimum to prevent the air-conditioning plant from becoming excessively large.

Load Determination

The cooling load estimate consists of coefficients from Design Data Sheet DDS511-2 or USN (1969) and includes allowances for the following:

- Solar radiation
- Heat transmission through hull, decks, and bulkheads
- Heat (latent and sensible) dissipation of occupants
- Heat gain due to lights
- Heat (latent and sensible) gain due to ventilation air
- Heat gain due to motors or other electrical equipment
- Heat gain from piping, machinery, and equipment

Loads should be derived from requirements indicated in USN (1969). The heating load estimate for air conditioning should consist of the following:

- Heat losses through hull, decks, and bulkheads
- Ventilation air
- Infiltration (when specified)

Some electronic spaces listed in USN (1969) require adding 15% to the calculated cooling load for future growth and require using one-third of the cooling season equipment heat dissipation (less the 15% added for growth) as heat gain in the heating season.

Heat Transmission Coefficients. The overall heat transmission coefficient U between the conditioned space and the adjacent boundary should be estimated from Design Data Sheet DDS511-2. Where new materials or constructions are used, new coefficients may be used from SNAME or calculated using methods found in DDS511-2 and SNAME.

Heat Dissipation from People. USN (1969) gives heat dissipation values for people in various activities and room conditions.

Heat Gain from Sources Within the Space. USN (1969) gives heat gain from lights and motors driving ventilation equipment. Heat gain and use factors for other motors and electrical and electronic equipment may be obtained from the manufacturer or from Chapter 26 of the 1993 *ASHRAE Handbook—Fundamentals*.

EQUIPMENT SELECTION

The equipment described for merchant ships also applies for naval vessels, except as follows.

Fans

The navy has a family of standard vaneaxial, tubeaxial, and centrifugal fans. Selection curves used for system design are found on NAVSEA Standard Drawings 810-921984, 810-925368, and 803-5001058. Manufacturers are required to furnish fans dimensionally identical to the standard plan and within 5% of the delivery. No belt-driven fans are included in the fan standards.

Cooling Coils

The navy uses eight standard sizes of direct-expansion and chilled water cooling coils. All coils have eight rows in the direction of airflow, with a range in face area of 0.06 to 0.93 m².

The coils are selected for a face velocity of 2.5 m/s maximum; however, sizes 54 DW to 58 DW may have a face velocity up to 3.2 m/s if the bottom of the duct on the discharge is sloped up at 15° for a distance equal to the height of the coil.

Chilled water coils are most commonly used and are selected based on 7.2°C inlet water with approximately a 3.7 K rise in water temperature through the coil. This is equivalent to 65 mL/s per kilowatt of cooling.

Construction and materials are specified in MIL-C-2939.

Heating Coils

The navy has standard steam and electric duct heaters with specifications as follows:

Table 3 Minimum Thickness of Materials for Ducts

Sheet for Fabricated Ductwork

Diameter or Longer Side	Nonwatertight		Watertight	
	Galvanized Steel	Aluminum	Galvanized Steel	Aluminum
Up to 150	0.46	0.64	1.90	2.69
160 to 300	0.76	1.02	2.54	3.56
310 to 460	0.91	1.27	3.00	4.06
470 to 760	1.22	1.52	3.00	4.06
Above 760	1.52	2.24	3.00	4.06

Welded or Seamless Tubing

Tubing Size	Nonwatertight Aluminum	Watertight Aluminum
50 to 150	0.89	2.69
160 to 300	1.27	3.56

Spirally Wound Duct (Nonwatertight)

Diameter	Steel	Aluminum
Up to 200	0.46	0.64
Over 200	0.76	0.81

Note: All dimensions in millimetres.

Steam Duct Heaters

- Maximum face velocity is 9.1 m/s
- Preheater leaving air temperature is 5.5 to 10°C
- Steam heaters are served from a 350 kPa (gage) steam system

Electric Duct Heaters

- Maximum face velocity is 7.1 m/s.
- Temperature rise through the heater is per MIL-H-22594A, but in no case is more than 26.5 K.
- Power supply for the smallest heaters is 120 V, 3 phase, 60 Hz. All remaining power supplies are 440 V, 3 phase, 60 Hz.
- Pressure drop through the heater must not exceed 85 Pa at 5 m/s. Manufacturers tested data should be used in system design.

Filters

The navy uses seven standard filter sizes with the following characteristics:

- Filters are available in steel or aluminum
- Filter face velocity is between 1.9 and 4.6 m/s.
- A filter-cleaning station on board ship includes facilities to wash, oil, and drain filters

Air Diffusers

Although the navy has standard diffusers for air-conditioning systems, a commercial type similar to those for merchant ships is generally used.

Air-Conditioning Compressors

The navy uses reciprocal compressors up to approximately 530 kW. For larger capacities, open, direct-drive centrifugal compressors are used. Seawater is used for condenser cooling at the rate of 90 mL/s per kilowatt for reciprocal compressors and 72 mL/s per kilowatt for centrifugal compressors.

Typical Systems

On naval ships, zone reheat systems are used for most applications. Some ships with sufficient electric power have used low-velocity terminal reheat systems with electric heaters in the space. Some newer ships have used a fan coil unit with fan, chilled water cooling coil, and electric heating coil in spaces with low to medium sensible heat per unit area of space requirements. The unit is supplemented by conventional systems serving spaces with high sensible or high latent loads.

Air Distribution Methods

Methods used on navy ships are similar to those discussed in the section for merchant ships. The minimum thickness of materials for ducts is listed in Table 3.

Controls

The navy's principal air-conditioning control uses a two-position dual thermostat that controls a cooling coil and an electric or steam reheater. This thermostat can be set for summer operation and does not require resetting for winter operation.

Steam preheaters use a regulating valve with (1) a weather bulb controlling approximately 25% of the valve's capacity to prevent freeze-up and (2) a line bulb in the duct downstream of the heater to control the temperature between 5.5 and 10°C.

Other controls are used to suit special systems. For example, pneumatic/electric controls can be used when close tolerances in temperature and humidity control are required, as in operating rooms. Thyristor controls are sometimes used on electric reheaters in ventilation systems.

REFERENCES

ASHRAE. 1985. Mechanical refrigeration installations on shipboard. *Standard 26-1985*.

ASHRAE. 1989. Ventilation for acceptable indoor air quality. *Standard 62-1989*.

SNAME. Calculations for merchant ship heating, ventilation and air conditioning design. *Technical and Research Bulletin No. 4-16*. Society of Naval Architects and Marine Engineers, Jersey City, NJ.

SNAME. Thermal insulation report. *Technical and Research Bulletin No. 4-7*. Society of Naval Architects and Marine Engineers, Jersey City, NJ.

SNAME. 1971. *Marine engineering*. Chapter 19 by J. Markert. Society of Naval Architects and Marine Engineers, Jersey City, NJ.

USMA. 1965. Standard specification for cargo ship construction. U.S. Maritime Administration, Washington, D.C.

USMA. *Standard Plan S38-1-101, Standard Plan S38-1-102, and Standard Plan S38-1-103*. U.S. Maritime Administration, Washington, D.C.

USN. 1969. The air conditioning, ventilation and heating design criteria manual for surface ships of the United States Navy. Washington, D.C.

USN. NAVSEA Drawing No. 810-921984, NAVSEA Drawing No. 810-925368, and NAVSEA Drawing No. 803-5001058. Naval Sea Systems Command, Dept. of the Navy, Washington, D.C.

USN. *General specifications for building naval ships*. Naval Sea Systems Command, Dept. of the Navy, Washington, D.C.

Note: MIL specifications are available from Commanding Officer, Naval Publications and Forms Center, ATTN: NPFC 105, 5801 Tabor Ave., Philadelphia, PA 19120.

INDUSTRIAL AIR CONDITIONING

INDUSTRIAL plants, warehouses, laboratories, nuclear power plants and facilities, and data processing rooms are designed for specific processes and environmental conditions that include proper temperature, humidity, air motion, air quality, and cleanliness. Airborne contaminants generated must be collected and treated before being discharged from the building or returned to the area.

Many industrial buildings require large quantities of energy, both in manufacturing and in the maintenance of building environmental conditions. Energy can be saved by the proper use of insulation, ventilation, and solar energy and by the recovery of waste heat and cooling.

For worker efficiency, the environment should be comfortable, minimize fatigue, facilitate communication, and not be harmful to health. Equipment should (1) control temperature and humidity or provide spot cooling to prevent heat stress, (2) have low noise levels, and (3) control health-threatening fumes.

GENERAL REQUIREMENTS

Typical temperatures, relative humidities, and specific filtration requirements for the storage, manufacture, and processing of various commodities are listed in Table 1. Requirements for a specific application may differ from those in the table. Improvements in processes and increased knowledge may cause further variation; thus, systems should be flexible to meet future requirements.

Inside temperature, humidity, filtration levels, and allowable variations should be established by agreement with the owner. A compromise between the requirements for product or process conditions and those for comfort may optimize quality and production costs.

A work environment that allows a worker to perform assigned duties without fatigue caused by temperatures that are too high or too low and without exposure to harmful airborne contaminants results in better, continued performance. It may also improve worker morale and reduce absenteeism.

PROCESS AND PRODUCT REQUIREMENTS

A process or product may require control of one or more of the following: (1) moisture regain; (2) rates of chemical reactions; (3) rates of biochemical reactions; (4) rate of crystallization; (5) product accuracy and uniformity; (6) corrosion, rust, and abrasion; (7) static electricity; (8) air cleanliness; and (9) product formability. Discussion of each of these factors follows.

Moisture Regain

In the manufacture or processing of hygroscopic materials such as textiles, paper, wood, leather, and tobacco, the air temperature and relative humidity have a marked influence on the production rate, product mass, strength, appearance, and quality.

The preparation of this chapter is assigned to TC 9.2, Industrial Air Conditioning.

Moisture in vegetable or animal materials (and some minerals) reaches equilibrium with the moisture of the surrounding air by *regain*. Regain is defined as the percentage of absorbed moisture in a material compared to its bone-dry mass. If a material sample with a mass of 110 g has a mass of 100 g after a thorough drying under standard conditions of 104 to 110°C, the mass of absorbed moisture is 10 g—10% of the sample's bone-dry mass. The regain, therefore, is 10%.

Table 2 lists typical values of regain for materials at 24°C in equilibrium at various relative humidities. Temperature change affects the rate of absorption or drying, which generally varies with the nature of the material, its thickness, and its density. Sudden temperature changes cause a slight regain change, even with fixed relative humidity; but temperature has little effect on regain compared to the effect of relative humidity.

Hygroscopic Materials. In absorbing moisture from the air, hygroscopic materials deliver sensible heat to the air in an amount equal to that of the latent heat of the absorbed moisture. Moisture gains or losses by materials in processes are usually quite small, but if significant, the amount of heat liberated should be included in the load estimate. Actual values of regain should be obtained for a particular application. Manufacturing economy requires regain to be maintained at a level suitable for rapid and satisfactory manipulation. Uniform humidity allows high-speed machinery to operate efficiently.

Conditioning and Drying. Materials may be exposed to the required humidity during manufacturing or processing, or they may be treated separately after conditioning and drying. Conditioning removes or adds hygroscopic moisture. Drying removes both hygroscopic moisture and free moisture in excess of that in equilibrium. Free moisture may be removed by evaporation, physically blowing it off, or other means.

Drying and conditioning may be combined to remove moisture and accurately regulate final moisture content in, for example, tobacco and some textile products. Conditioning or drying is frequently a continuous process, in which the material is conveyed through a tunnel and subjected to controlled atmospheric conditions. Chapter 22 of the 1992 *ASHRAE Handbook—Systems and Equipment* describes dehumidification and pressure-drying equipment.

Rates of Chemical Reactions

Some processes require temperature and humidity control to regulate chemical reactions. For example, in rayon manufacture, pulp sheets are conditioned, cut to size, and passed through a mercerizing process. The temperature directly controls the rate of the reaction, while the relative humidity maintains a solution of constant strength and a constant rate of surface evaporation.

The oxidizing process in drying varnish depends on temperature. Desirable temperatures vary with the type of varnish. High relative humidity retards surface oxidation and allows internal gases to escape as chemical oxidizers cure the varnish from within. Thus, a bubble-free surface is maintained, with a homogeneous film throughout.

Table 1 Temperatures and Humidities for Industrial Air Conditioning

Process	Dry Bulb (°C)	rh (%)
ABRASIVE		
Manufacture	26	50
CERAMICS		
Refractory	43 to 66	50 to 90
Molding room	27	60 to 70
Clay storage	16 to 27	35 to 65
Decalcomania production	24 to 27	48
Decorating room	24 to 27	48

Use high-efficiency filtration in decorating room. To minimize the danger of silicosis in other areas, a dust-collecting system or medium-efficiency particulate air filtration may be required.

Process	Dry Bulb (°C)	rh (%)
DISTILLING		
General manufacturing	16 to 24	45 to 60
Aging	18 to 22	50 to 60

Low humidity and dust control are important where grains are ground. Use high-efficiency filtration for all areas to prevent mold spore and bacteria growth. Use ultrahigh efficiency filtration where bulk flash pasteurization is performed.

Process	Dry Bulb (°C)	rh (%)
ELECTRICAL PRODUCTS		
Electronics and X-ray		
Coil and transformer winding	22	15
Semiconductor assembly	20	40 to 50
Electrical instruments		
Manufacture and laboratory	21	50 to 55
Thermostat assembly and calibration	24	50 to 55
Humidistat assembly and calibration	24	50 to 55
Small mechanisms		
Close tolerance assembly	22*	40 to 45
Meter assembly and test	24	60 to 63
Switchgear		
Fuse and cutout assembly	23	50
Capacitor winding	23	50
Paper storage	23	50
Conductor wrapping with yarn	24	65 to 70
Lightning arrester assembly	20	20 to 40
Thermal circuit breakers assembly and test	24	30 to 60
High-voltage transformer repair	26	5
Water wheel generators		
Thrust runner lapping	21	30 to 50
Rectifiers		
Processing selenium and copper oxide plates	23	30 to 40

*Temperature to be held constant.

Dust control is essential in these processes. Minimum control requires medium-efficiency filters. Degree of filtration depends on the type of function in the area. Smaller tolerances and miniature components suggest high-efficiency particulate air filters.

Process	Dry Bulb (°C)	rh (%)
FLOOR COVERING		
Linoleum		
Mechanical oxidizing of linseed oil*	32 to 38	
Printing	27	
Stoving process	70 to 120	

*Precise temperature control required.

Medium-efficiency particulate air filtration is recommended for the stoving process.

Process	Dry Bulb (°C)	rh (%)
FOUNDRIES*		
Core making	16 to 21	
Mold making		
Bench work	16 to 21	
Floor work	13 to 18	
Pouring	4	
Shakeout	4 to 10	
Cleaning room	13 to 18	

*Winter dressing room temperatures. Spot coolers are sometimes used in larger installations.

In mold making, provide exhaust hoods at transfer points with wet-collector dust removal system. Use 280 to 380 L/s per hood.

In shakeout room, provide exhaust hoods with wet-collector dust removal system. Exhaust 190 to 240 L/s in grate area. Room ventilators are generally not effective.

In cleaning room, provide exhaust hoods for grinders and cleaning equipment with dry cyclones or bag-type collectors. In core making, oven and adjacent cooling areas require fume exhaust hoods. Pouring rooms require two-speed powered roof ventilators. Design for minimum of 10 L/s per square metre of floor area at low speed. Shielding is required to control radiation from hot surfaces. Proper introduction of air minimizes preheat requirements.

Process	Dry Bulb (°C)	rh (%)
FUR		
Drying	43	
Shock treatment	−8 to −7	
Storage	4 to 10	55 to 65

Shock treatment or eradication of any insect infestations requires lowering the temperature to −8 to −7°C for 3 to 4 days, then raising it to 16 to 21°C for 2 days, then lowering it again for 2 days and raising it to the storage temperature.

Furs remain pliable, oxidation is reduced, and color and luster are preserved when stored at 4 to 10°C.

Humidity control is required to prevent mold growth (which is prevalent with humidities above 80%) and hair splitting (which is common with humidities lower than 55%).

Process	Dry Bulb (°C)	rh (%)
GUM		
Manufacturing	25	33
Rolling	20	63
Stripping	22	53
Breaking	23	47
Wrapping	23	58
LEATHER		
Drying	20 to 52	75
Storage, winter room temperature	10 to 16	40 to 60

After leather is moistened in preparation for rolling and stretching, it is placed in an atmosphere held at room temperature with a relative humidity of 95%.

Leather is usually stored in warehouses without temperature and humidity control. However, it is necessary to keep humidity sufficiently low to prevent mildew. Medium-efficiency particulate air filtration is recommended for fine finish.

Process	Dry Bulb (°C)	rh (%)
LENSES (OPTICAL)		
Fusing	24	45
Grinding	27	80

Table 1 Temperatures and Humidities for Industrial Air Conditioning (*Concluded*)

Process	Dry Bulb (°C)	rh (%)
MATCHES		
Manufacture	22 to 23	50
Drying	21 to 24	60
Storage	16 to 17	50

Water evaporates with the setting of the glue. The amount of water evaporated is 8 to 9 kg per million matches. The match machine turns out about 750,000 matches per hour.

Process	Dry Bulb (°C)	rh (%)
PAINT APPLICATION		
Lacquers: Baking	150 to 180	
Oils paints: Paint spraying	16 to 32	80

The required air filtration efficiency depends on the painting process. On fine finishes, such as car bodies, high-efficiency particulate air filters are required for the outdoor air supply. Other products may require only low- or medium-efficiency filters.

Makeup air must be preheated. Spray booths must have 0.5 m/s face velocity if spraying is performed by humans; lower air quantities can be used if robots perform spraying. Ovens must have air exhausted to maintain fumes below explosive concentration. Equipment must be explosion-proof. Exhaust must be cleaned by filtration and solvents reclaimed or scrubbed.

Process	Dry Bulb (°C)	rh (%)
PHOTO STUDIO		
Dressing room	22 to 23	40 to 50
Studio (camera room)	22 to 23	40 to 50
Film darkroom	21 to 22	45 to 55
Print darkroom	21 to 22	45 to 55
Drying room	32 to 38	35 to 45
Finishing room	22 to 24	40 to 55
Storage room (b/w film and paper)	22 to 24	40 to 60
Storage room (color film and paper)	4 to 10	40 to 50
Motion picture studio	22	40 to 55

The above data pertain to average conditions. In some color processes, elevated temperatures as high as 40°C are used, and a higher room temperature is required.

Conversely, ideal storage conditions for color materials necessitate refrigerated or deep-freeze temperatures to ensure quality and color balance when long storage times are anticipated.

Heat liberated during printing, enlarging, and drying processes is removed through an independent exhaust system, which also serves the lamp houses and dryer hoods. All areas except finished film storage require a minimum of medium-efficiency particulate air filters.

Process	Dry Bulb (°C)	rh (%)
PLASTICS		
Manufacturing areas		
Thermosetting molding compounds	27	25 to 30
Cellophane wrapping	24 to 27	45 to 65

In manufacturing areas where plastic is exposed in the liquid state or molded, high-efficiency particulate air filters may be required. Dust collection and fume control are essential.

Process	Dry Bulb (°C)	rh (%)
PLYWOOD		
Hot pressing (resin)	32	60
Cold pressing	32	15 to 25
RUBBER-DIPPED GOODS		
Manufacture	32	
Cementing	27	25 to 30*
Dipping surgical articles	24 to 27	25 to 30*
Storage prior to manufacture	16 to 24	40 to 50*
Laboratory (ASTM Standard)	23.0	50*

*Dew point of air must be below evaporation temperature of solvent.

Solvents used in manufacturing processes are often explosive and toxic, requiring positive ventilation. Volume manufacturers usually install a solvent-recovery system for area exhaust systems.

Process	Dry Bulb (°C)	rh (%)
TEA		
Packaging	18	65

Ideal moisture content is 5 to 6% for quality and mass. Low-limit moisture content for quality is 4%.

Process	Dry Bulb (°C)	rh (%)
TOBACCO		
Cigar and cigarette making	21 to 24	55 to 65*
Softening	32	85 to 88
Stemming and stripping	24 to 29	70 to 75
Packing and shipping	23 to 24	65
Filler tobacco casing and conditioning	24	75
Filter tobacco storage and preparation	25	70
Wrapper tobacco storage and conditioning	24	75

*Relative humidity fairly constant with range as set by cigarette machine.

Before stripping, tobacco undergoes a softening operation.

Rates of Biochemical Reactions

Fermentation requires temperature and humidity control to regulate the rate of biochemical reactions.

Rate of Crystallization

The cooling rate determines the size of crystals formed from a saturated solution. Both temperature and relative humidity affect the cooling rate and change the solution density by evaporation.

In the coating pans for pills, a heavy sugar solution is added to the tumbling mass. As the water evaporates, sugar crystals cover each pill. Blowing the proper quantity of air at the correct dry- and wet-bulb temperatures forms a smooth opaque coating. If cooling and drying are too slow, the coating is rough, translucent, and unsatisfactory in appearance; if cooling and drying are too fast, the coating chips through to the interior.

Product Accuracy and Uniformity

In the manufacture of precision instruments, tools, and lenses, air temperature and cleanliness affect the quality of work. If manufac-

turing tolerances are within 5 μm, close temperature control prevents expansion and contraction of the material. Constant temperature is more important than the temperature level; thus conditions are usually selected for personnel comfort and to prevent a film of moisture on the surface. High- or ultrahigh-efficiency particulate air filtration may be required.

Corrosion, Rust, and Abrasion

In the manufacture of metal articles, the temperature and relative humidity are kept sufficiently low to prevent hands from sweating to protect the finished article from fingerprints, tarnish, and/or etching. The salt and acid in body perspiration can cause corrosion and rust within a few hours. The manufacture of polished surfaces usually requires medium- to high-efficiency particulate air filtering to prevent surface abrasion. This is also true for steel-belted radial tire manufacturing.

Static Electricity

In processing light materials such as textile fibers and paper, and where explosive atmospheres or materials are present, humidity can

Table 2 Regain of Hygroscopic Materials
Moisture Content Expressed in Percent of Dry Mass of the Substance at Various Relative Humidities—Temperature 24°C

| Classification | Material | Description | Relative Humidity | | | | | | | | |
			10	20	30	40	50	60	70	80	90
Natural textile fibers	Cotton	Sea island—roving	2.5	3.7	4.6	5.5	6.6	7.9	9.5	11.5	14.1
	Cotton	American—cloth	2.6	3.7	4.4	5.2	5.9	6.8	8.1	10.0	14.3
	Cotton	Absorbent	4.8	9.0	12.5	15.7	18.5	20.8	22.8	24.3	25.8
	Wool	Australian merino—skein	4.7	7.0	8.9	10.8	12.8	14.9	17.2	19.9	23.4
	Silk	Raw chevennes—skein	3.2	5.5	6.9	8.0	8.9	10.2	11.9	14.3	18.3
	Linen	Table cloth	1.9	2.9	3.6	4.3	5.1	6.1	7.0	8.4	10.2
	Linen	Dry spun—yarn	3.6	5.4	6.5	7.3	8.1	8.9	9.8	11.2	13.8
	Jute	Average of several grades	3.1	5.2	6.9	8.5	10.2	12.2	14.4	17.1	20.2
	Hemp	Manila and sisal rope	2.7	4.7	6.0	7.2	8.5	9.9	11.6	13.6	15.7
Rayons	Viscose nitrocellulose	Average skein	4.0	5.7	6.8	7.9	9.2	10.8	12.4	14.2	16.0
	Cuprammonium cellulose acetate		0.8	1.1	1.4	1.9	2.4	3.0	3.6	4.3	5.3
Paper	M.F. newsprint	Wood pulp—24% ash	2.1	3.2	4.0	4.7	5.3	6.1	7.2	8.7	10.6
	H.M.F. writing	Wood pulp—3% ash	3.0	4.2	5.2	6.2	7.2	8.3	9.9	11.9	14.2
	White bond	Rag—1% ash	2.4	3.7	4.7	5.5	6.5	7.5	8.8	10.8	13.2
	Comm. ledger	75% rag—1% ash	3.2	4.2	5.0	5.6	6.2	6.9	8.1	10.3	13.9
	Kraft wrapping	Coniferous	3.2	4.6	5.7	6.6	7.6	8.9	10.5	12.6	14.9
Miscellaneous organic materials	Leather	Sole oak—tanned	5.0	8.5	11.2	13.6	16.0	18.3	20.6	24.0	29.2
	Catgut	Racquet strings	4.6	7.2	8.6	10.2	12.0	14.3	17.3	19.8	21.7
	Glue	Hide	3.4	4.8	5.8	6.6	7.6	9.0	10.7	11.8	12.5
	Rubber	Solid tires	0.11	0.21	0.32	0.44	0.54	0.66	0.76	0.88	0.99
	Wood	Timber (average)	3.0	4.4	5.9	7.6	9.3	11.3	14.0	17.5	22.0
	Soap	White	1.9	3.8	5.7	7.6	10.0	12.9	16.1	19.8	23.8
	Tobacco	Cigarette	5.4	8.6	11.0	13.3	16.0	19.5	25.0	33.5	50.0
Miscellaneous inorganic materials	Asbestos fiber	Finely divided	0.16	0.24	0.26	0.32	0.41	0.51	0.62	0.73	0.84
	Silica gel		5.7	9.8	12.7	15.2	17.2	18.8	20.2	21.5	22.6
	Domestic coke		0.20	0.40	0.61	0.81	1.03	1.24	1.46	1.67	1.89
	Activated charcoal	Steam activated	7.1	14.3	22.8	26.2	28.3	29.2	30.0	31.1	32.7
	Sulfuric acid		33.0	41.0	47.5	52.5	57.0	61.5	67.0	73.5	82.5

be used to reduce static electricity, which is often detrimental to processing and extremely dangerous in explosive atmospheres. Static electricity charges are minimized when the air has a relative humidity of 35% or higher. The power driving the processing machines is converted into heat that raises the temperature of the machines above that of the adjacent air, where humidity is normally measured. Room relative humidity may need to be 65% or higher to maintain the high humidity required in the machines.

Air Cleanliness

Each application must be evaluated to determine the filtration needed to counter the adverse effects of (1) minute dust particles, (2) airborne bacteria, and (3) other air contaminants such as smoke, radioactive particles, spores, and pollen on the product or process. These effects include chemically altering production material, spoiling perishable goods, or clogging small openings in precision machinery. See Chapter 25 in the 1992 *ASHRAE Handbook—Systems and Equipment*.

Product Formability

The manufacture of pharmaceutical tablets requires close control of humidity for optimum tablet formation.

EMPLOYEE REQUIREMENTS

Space conditions required by health and safety standards to avoid excess exposure to high temperatures and airborne contaminants are often established by the American Conference of Governmental Industrial Hygienists (ACGIH). In the United States, the National

Institute of Occupational Safety and Health (NIOSH) does research and recommends guidelines for workspace environmental control. The Occupational Safety and Health Administration (OSHA) sets standards from these environmental control guidelines and enforces them through compliance inspection at industrial facilities. Enforcement may be delegated to a corresponding state agency.

Standards for safe levels of contaminants in the work environment or in air exhausted from facilities do not cover all contaminants encountered. Minimum safety standards and facility design criteria are available, however, from various U.S. Department of Health, Education, and Welfare (DHEW) agencies such as the National Cancer Institute, the National Institutes of Health, and the Public Health Service (Centers for Disease Control and Prevention). For radioactive substances, standards established by the U.S. Nuclear Regulatory Commission (NRC) should be followed.

Thermal Control Levels

In industrial plants that need no specific control for products or processes, conditions range from 20 to 38°C and 25 to 60% rh. For a more detailed analysis, the work rate, air velocity, quantity of rest, and effects of radiant heat must be considered (see Chapter 8 of the 1993 *ASHRAE Handbook—Fundamentals*). To avoid stress to workers exposed to high work rates and hot temperatures, the ACGIH established guidelines to evaluate high-temperature air velocity and humidity levels in terms of heat stress (Dukes-Dobos and Henschel 1971).

If a comfortable environment is of concern rather than the avoidance of heat stress, the thermal control range becomes more specific (McNall et al. 1967). In still air, nearly sedentary workers (120-W metabolism) prefer 22 to 24°C dry-bulb temperature with 20 to 60%

rh, and they can detect a 1 K change per hour. Workers at a high rate of activity (300-W metabolism) prefer 17 to 19°C dry bulb with 20 to 50% rh, but they are less sensitive to temperature change (ASHRAE *Standard* 55). Workers at a high activity level can also be cooled by increasing the air velocity (ASHRAE *Standard* 55).

Contaminant Control Levels

Toxic materials are present in many industrial plants and laboratories. In such plants, the air-conditioning and ventilation systems must minimize human exposure to toxic materials. When these materials become airborne, their range expands greatly, thus exposing more people. Chapter 11 of the 1993 *ASHRAE Handbook—Fundamentals*, current OSHA regulations, and *Threshold Limit Values of Airborne Contaminants* (ACGIH 1992) give guides to evaluating the health impact of contaminants.

In addition to being a health concern, gaseous flammable substances must also be kept below explosive concentrations. Acceptable concentration limits are 20 to 25% of the lower explosive limit of the substance. Chapter 11 of the 1993 *ASHRAE Handbook—Fundamentals* includes information on flammable limits and means of control.

Instruments are available to measure concentrations of common gases and vapors. For less common gases or vapors, the air is sampled by drawing it through an impinger bottle or by inertial impaction-type air samplers, which support biological growth on a nutrient gel for subsequent identification after incubation.

Gases and vapors are found near acid baths and tanks holding process chemicals. Machine processes, plating operations, spraying, mixing, and abrasive cleaning operations generate dusts, fumes, and mists. Many laboratory procedures, including grinding, blending, homogenizing, sonication, weighing, dumping of animal bedding, and animal inoculation or intubation, generate aerosols.

DESIGN CONSIDERATIONS

To select equipment, the required environmental conditions, both for the product and personnel comfort, must be known. Consultation with the owner establishes design criteria such as temperature and humidity levels, energy availability and opportunities to recover it, cleanliness, process exhaust details, location and size of heat-producing equipment, lighting levels, frequency of equipment use, load factors, frequency of truck or car loadings, and sound levels. Consideration must be given to separating dirty processes from manufacturing areas that require relatively clean air. If not controlled by physical barriers, hard-to-control contaminants—such as mists from presses and machining, or fumes and gases from welding—can migrate to areas of final assembly, metal cleaning, or printing and can cause serious problems.

Because of changing mean radiant temperature, insulation should be evaluated for initial and operating cost, saving of heating and cooling, elimination of condensation (on roofs in particular), and maintenance of comfort. When high levels of moisture are required within buildings, the structure and air-conditioning system must prevent condensation damage to the structure and ensure a quality product. Condensation can be prevented by (1) proper selection of insulation type and thickness, (2) proper placing of vapor retarders, and (3) proper selection and assembly of construction components to avoid thermal short circuits. Chapters 20 and 21 in the 1993 *ASHRAE Handbook—Fundamentals* have further details.

Personnel engaged in some industrial processes may be subject to a wide range of activity levels for which a broad range of indoor temperature and humidity levels are desirable. Chapter 8 of the 1993 *ASHRAE Handbook—Fundamentals* addresses recommended indoor conditions at various activity levels.

If layout and construction drawings are not available, a complete survey of existing premises and a checklist for proposed facilities is necessary (Table 3).

Table 3 Facilities Checklist

Construction
1. Single or multistory
2. Type and location of doors, windows, crack lengths
3. Structural design live loads
4. Floor construction
5. Exposed wall materials
6. Roof materials and color
7. Insulation type and thicknesses
8. Location of existing exhaust equipment
9. Building orientation

Use of Building
1. Product needs
2. Surface cleanliness; acceptable airborne contamination level
3. Process equipment: type, location, and exhaust requirements
4. Personnel needs, temperature levels, required activity levels, and special workplace requirements
5. Floor area occupied by machines and materials
6. Clearance above floor required for material-handling equipment, piping, lights, or air distribution systems
7. Unusual occurrences and their frequency, such as large cold or hot masses of material moved inside
8. Frequency and length of time doors open for loading or unloading
9. Lighting, location, type, and capacity
10. Acoustical levels
11. Machinery loads, such as electric motors (size, diversity), large latent loads, or radiant loads from furnaces and ovens
12. Potential for temperature stratification

Design Conditions
1. Design temperatures—indoor and outdoor dry and wet bulb
2. Altitude
3. Wind velocity
4. Makeup air required
5. Indoor temperature and allowable variance
6. Indoor relative humidity and allowable variance
7. Indoor air quality definition and allowable variance
8. Outdoor temperature occurrence frequencies
9. Operational periods: one, two, or three
10. Waste heat availability and energy conservation
11. Pressurization required
12. Mass loads from the energy release of productive materials

Code and Insurance Requirements
1. State and local code requirements for ventilation rates, etc.
2. Occupational health and safety requirements
3. Insuring agency requirements

Utilities Available and Required
1. Gas, oil, compressed air (pressure), electricity (characteristics), steam (pressure), water (pressure), wastewater, interior and site drainage
2. Rate structures for each utility
3. Potable water

New industrial buildings are commonly single-story with flat roofs and ample height to distribute utilities without interfering with process operation. Fluorescent fixtures are commonly mounted at heights up to 3.5 m, high-output fluorescent fixtures up to 6 m, and high-pressure sodium or metal halide fixtures above 6 m.

Lighting design considers light quality, degree of diffusion and direction, room size, mounting height, and economics. Illumination levels should conform to recommended levels of the Illuminating Engineering Society of North America. Air-conditioning systems can be located in the top of the building. However, the designs require coordination because air-handling equipment and ductwork compete for space with sprinkler systems, piping, structural elements, cranes, material-handling systems, electric wiring, and lights.

Operations within the building must also be considered. Production materials may be moved through outside doors and allow large amounts of outdoor air to enter. Some operations require close control of temperature, humidity, and contaminants. A time schedule of operation helps in estimating the heating and cooling load.

LOAD CALCULATIONS

Table 1 and specific product chapters discuss product requirements. Chapter 26 of the 1993 *ASHRAE Handbook—Fundamentals* covers load calculations for heating and cooling.

Solar and Transmission

The solar load on the roof is generally the largest perimeter load and is usually a significant part of the overall load. Roofs should be light colored to minimize solar heat gain. Wall loads are often insignificant, and most new plants have no windows in the manufacturing area, so a solar load through glass is not present. Large windows in old plants may be closed in.

Internal Heat Generation

The process, product, facility utilities, and employees generate internal heat. People, power or process loads, and lights are internal sensible loads. Of these, production machinery often creates the largest sensible load. The design should consider anticipated brake power, rather than connected motor loads. The lighting load is generally significant. Heat gain from people is usually negligible in process areas.

Heat from operating equipment is difficult to estimate. Approximate values can be determined by studying the load readings of the electrical substations that serve the area.

In most industrial facilities, the latent heat load is minimal, with people and outdoor air being the major contributors. In these areas, the sensible heat factor approaches 1.0. Some processes, such as papermaking, release large amounts of moisture. This moisture and its condensation on cold surfaces must be managed.

Stratification Effect

The cooling load may be dramatically reduced in a work space that takes advantage of temperature stratification, which establishes a stagnant blanket of air directly under the roof by keeping air circulation to a minimum. The convective component of high energy sources such as the roof, upper walls, and high-level lights has little impact on the cooling load in the lower occupied zones. A portion of the heat generated by high-energy sources (such as process equipment) in the occupied zone is not part of the cooling load. Due to radiation and a buoyancy effect, 20-60% of the energy rises to the upper strata. The amount that rises depends on the building construction, air movement, and temperatures of the source, other surfaces, and air.

Supply- and return-air ducts should be as low as possible to avoid mixing the warm boundary layers. The location of supply-air diffusers generally establishes the boundary of the warmer stratified air. For areas with supply air quantities greater than 10 L/s per square metre, the return air temperature is approximately that of the air entrained by the supply airstream and only slightly higher than that at the end of the throw. With lower air quantities, the return air is much warmer than the supply air at the end of the throw. The amount depends on the placement of return inlets relative to internal heat sources. The average temperature of the space is higher, thus reducing the effect of outside conditions on heat gain. Spaces with a high area-to-employee ratio adapt well to low quantities of supply air and to spot cooling. For design specifics, refer to Chapter 24, the section on General Comfort and Dilution Ventilation.

Makeup Air

Makeup air that has been filtered, heated, cooled, humidified, and/or dehumidified is introduced to replace exhaust air, provide ventilation, and pressurize the building. For exhaust systems to function, air must enter the building by infiltration or the air-conditioning equipment. The space air-conditioning system must be large enough to heat or cool the outside air required to replace the exhaust. Cooling or heat recovery from exhaust air to makeup air can substantially reduce the outdoor air load.

Exhaust from air-conditioned buildings should be kept to a minimum by proper hooding or by relocating exhausted processes to areas not requiring air conditioning. Frequently, excess makeup air is provided to pressurize the building slightly, thus reducing infiltration, flue downdrafts, and ineffective exhaust under certain wind conditions. Makeup air and exhaust systems can be interlocked so that outdoor makeup air can be reduced as needed, or makeup air units may be changed from outdoor air heating to recirculated air heating when process exhaust is off.

In some facilities, outdoor air is required for ventilation because of the function of the space. Recirculation of air is avoided to reduce concentrations of health-threatening fumes, airborne bacteria, and radioactive substances. Ventilation rates for human occupancy should be determined from ASHRAE *Standard* 62 and applicable codes.

Outdoor air dampers of industrial air-conditioning units should handle 100% supply air, so modulating the dampers satisfies the room temperature under certain weather conditions. For 100% outside air, a modulating exhaust damper or fan must be used. The infiltration of outdoor air often creates considerable load in industrial buildings due to poorly sealed walls and roofs. Infiltration should be minimized to conserve energy.

Fan Heat

The heat from air-conditioning return-air fans or supply fans goes into the refrigeration load. This energy is not part of the room sensible heat, except for supply fans downstream of the conditioning apparatus.

SYSTEM AND EQUIPMENT SELECTION

Industrial air-conditioning equipment includes heating and cooling sources, air-handling and air-conditioning apparatus, filters, and an air distribution system. To provide low life-cycle cost, components should be selected and the system designed for long life with low maintenance and operating costs.

Systems may consist of (1) heating only in cool climates, where ventilation air provides comfort for workers; (2) air washer systems, where high humidities are desired and where the climate requires cooling; or (3) heating and mechanical cooling, where temperature and/or humidity control are required by the process and where the activity level is too high to be satisfied by other means of cooling. All systems include air filtration appropriate to the contaminant control required.

A careful evaluation will determine the zones that require control, especially in large, high-bay areas where the occupied zone is a small portion of the space volume. ASHRAE *Standard* 55 defines the occupied zone as 80 to 1830 mm high and more than 600 mm from the walls.

Air-Handling Units

Air-handling units heat, cool, humidify, and dehumidify air that is distributed to the workspace. They can supply all or any portion of outdoor air so that in-plant contaminants do not become too concentrated. Units may be factory assembled or field constructed.

HEATING SYSTEMS

Floor Heating

Floor heating is often desirable in industrial buildings, particularly in large high-bay buildings, garages, and assembly areas where workers must be near the floor, or where large or fluctuating outdoor air loads make maintenance of ambient temperature difficult.

As an auxiliary to the main heating system, floors may be tempered to 18 or 21°C by embedded hydronic systems, electrical resistance cables, or warm-air ducts. The heating elements may be buried deep (150 to 450 mm) in the floor to permit slab warm-up at off-peak times, thus using the floor mass for heat storage to save energy during periods of high use.

Floor heating may be the primary or sole heating means, but floor temperatures above 29°C are uncomfortable, so such use is limited to small, well-insulated spaces.

Unit Heaters

Gas, oil, electricity, hot water, or steam unit heaters with propeller fans or blowers are used for spot heating areas or are arranged in multiples for heating an entire building. Temperatures can be varied by individual thermostatic control. Unit heaters are located so that the discharge (or throw) will reach the floor and flow adjacent to and parallel with the outside wall. They are spaced so that the discharge of one heater is just short of the next heater, thus producing a ring of warm air moving peripherally around the building. In industrial buildings with heat-producing processes, much heat stratifies in high-bay areas. In large buildings, additional heaters should be placed in the interior so that their discharge reaches the floor to reduce stratification. Downblow unit heaters in high bays and large areas may have a revolving discharge.

Gas- and oil-fired unit heaters should not be used where corrosive vapors are present. Unit heaters function well with regular maintenance and periodic cleaning in dusty or dirty conditions. Propeller fans generally require less maintenance than centrifugal fans. Centrifugal fans or blowers usually are required if heat is to be distributed to several areas. Gas- and oil-fired unit heaters require proper venting.

Ducted Heaters

Ducted heaters include large direct- or indirect-fired heaters, door heaters, and heating and ventilating units. They generally have centrifugal fans.

Code changes and improved burners have led to increased use of direct-fired gas heaters (the gas burns in the air supplied to the space) for makeup air heating. With correct interlock safety precautions of supply air and exhaust, no harmful effect from direct firing occurs. The high efficiency, high turndown ratio, and simplicity of maintenance make these units suitable for makeup air heating.

For industrial applications of ducted heaters, the following is a list of problems commonly encountered and their solutions:

• *Steam Coil Freeze-up.* Use steam-distributing-type (sometimes called nonfreeze) coils, face-and-bypass control with the steam valve wide open for entering air below 2°C, and a thermostat in the exit to stop the airflow when it falls below 5°C. Ensure free condensate drainage.

• *Hot Water Coil Freeze-up.* Ensure adequate circulation through drainable coil at all times; a thermostat in the air and in the water leaving the coil opens the water valve wide and stops the airflow under freezing conditions.

• *Temperature Override.* Because of wiping action on coil with face-and-bypass control, zoning control is poor. Locate face damper carefully and use room thermostat to reset the discharge air temperature controller.

• *Bearing Failure.* Follow bearing manufacturer's recommendations for application and lubrication.

• *Insufficient Air Quantity.* Require capacity data based on testing; keep forward-curved blade fans clean or use backward-inclined blade fans; inspect and clean coils; install low- to medium-efficiency particulate air filters and change when required.

Door Heating

Unit heaters and makeup air heaters commonly temper outdoor air that enters the building at open doors. These door heaters may have directional outlets to offset the incoming draft or may resemble a vestibule where air is recirculated.

Unit heaters successfully heat air at small doors open for short periods. They temper the incoming outdoor air through mixing and quickly bring the space to the desired temperature after the door is closed. The makeup air heater should be applied as a door heater in buildings where the doors are large (those that allow railroad cars or large trucks to enter) and open for long periods. Unit heaters are also needed in facilities with a sizable negative pressure or that are not tightly constructed. These units help pressurize the door area, mix the incoming cold air and temper it, and bring the area quickly back to the normal temperature after the door is closed.

Often, door heater nozzles direct heated air at the top or down the sides of a door. Doors that create large, cold drafts when open can best be handled by introducing air in a trench at the bottom of the door. When not tempering cold drafts through open doors, some door heaters direct heated air to nearby spots. When the door is closed, they can switch to a lower output temperature.

The door heating units that resemble a vestibule operate with air flowing down across the opening and recirculating from the bottom, which helps reduce cold drafts across the floor. This type of unit is effective on doors routinely open and no higher than 3 m.

Infrared

High-intensity infrared heaters (gas, oil, or electric) transmit heat energy directly to warm the occupants, floor, machines, or other building contents, without appreciably warming the air. Some air heating occurs by convection from objects warmed by the infrared. These units are classed as near- or far-infrared heaters, depending on the closeness of the wavelengths to visible light. Near-infrared heaters emit a substantial amount of visible light.

Both vented and unvented gas-fired infrared heaters are available as individual radiant panels, or as a continuous radiant pipe with burners 4.5 to 9 m apart and an exhaust vent fan at the end of the pipe. Unvented heaters require exhaust ventilation to remove flue products from the building, or moisture will condense on the roof and walls. Insulation reduces the exhaust required. Additional information on both electric and gas infrared is given in Chapter 15 of the 1992 *ASHRAE Handbook—Systems and Equipment.*

Infrared heaters are used in the following:

1. High-bay buildings, where the heaters are usually mounted 3 to 9 m above the floor, along outside walls, and tilted to direct maximum radiation to the floor. If the building is poorly insulated, the controlling thermostat should be shielded to avoid influence from the radiant effect of the heaters and the cold walls.
2. Semiopen and outdoor areas, where people can comfortably be heated directly and objects can be heated to avoid condensation.
3. Loading docks, where snow and ice can be controlled by strategic placement of near-infrared heaters.

COOLING SYSTEMS

Common cooling systems include refrigeration equipment, evaporative coolers, and high-velocity ventilating air.

For manufacturing operations, particularly heavy industry where mechanical cooling cannot be economically justified, evaporative cooling systems often provide good working conditions. If the operation requires heavy physical work, spot cooling by ventilation, evaporative cooling, or refrigerated air can be used. To minimize summer discomfort, high outdoor air ventilation rates may be adequate in some hot process areas. In all these operations, a mechanical air supply with good distribution is needed.

Refrigerated Cooling Systems

The refrigeration cooling source may be located in a central equipment area, or smaller packaged cooling systems may be placed near or combined with each air-handling unit. Central mechanical equipment uses positive-displacement, centrifugal, or absorption refrigeration to chill water. Pumped through unit cooling coils, the chilled water absorbs heat, then returns to the cooling equipment.

Central system condenser water rejects heat through a cooling tower. The heat may be transferred to other sections of the building where heating or reheating is required, and the cooling unit becomes a heat recovery heat pump. Refrigerated heat recovery is particularly advantageous in buildings with a simultaneous need for heating exterior sections and cooling interior sections.

When interior spaces are cooled with a combination of outdoor air and chilled water obtained from reciprocating or centrifugal chillers with heat recovery condensers, hot water at temperatures up to 45°C is readily available. Large quantities of air at room temperature, which must be exhausted because of contaminants, can first be passed through a chilled water coil to recover heat. Heating or reheating obtained by refrigerated heat recovery occurs at a COP approaching 4, regardless of outdoor temperature, and can save considerable energy.

Mechanical cooling equipment should be selected in multiple units. This enables the equipment to match its response to fluctuations in the load and allows maintenance during nonpeak operation. Small packaged refrigeration equipment commonly uses positive-displacement (reciprocating or screw) compressors with air-cooled condensers. These units usually provide up to 700 kW of cooling. Because equipment is often on the roof, the condensing temperature may be affected by warm ambient air, which is often 5 to 10 K higher than the design outdoor air temperature. In this type of system, the cooling coil may receive refrigerant directly.

ASHRAE *Standard* 15, *Safety Code for Mechanical Refrigeration*, limits the type and quantity of refrigerant in direct air-to-refrigerant exchangers. Fins on the air side improve heat transfer, but they increase the pressure drop through the coil, particularly as they get dirty.

Evaporative Cooling Systems

Evaporative cooling systems may be evaporative coolers or air washers. Evaporative coolers have water sprayed directly on a wetted surface through which air moves. An air washer recirculates water, and the air flows through a heavily misted area. Water atomized in the airstream evaporates, and the water cools the air. Refrigerated water simultaneously cools and dehumidifies the air.

Evaporative cooling offers energy conservation opportunities, particularly in intermediate seasons. In many industrial facilities, evaporative cooling controls both temperature and humidity. In these systems, the sprayed water is normally refrigerated, and a reheat coil is often used. Temperature and humidity of the exit airstream may be controlled by varying the temperature of the chilled water and the reheat coil and by varying the quantity of air passing through the reheat coil with a dewpoint thermostat.

Care must be taken that accumulation of dust or lint does not clog the nozzles or evaporating pads of the evaporative cooling systems. It may be necessary to filter the air before the evaporative cooler. Fan heat, air leakage through closed dampers, and the entrainment of room air into the supply airstream all affect design. Chemical treatment of the water may be necessary to prevent mineral buildup or biological growth on the pads or in the pans.

AIR FILTRATION SYSTEMS

Air filtration systems remove contaminants from air supplied to, or exhausted from, building spaces. Supply air filtration (most frequently on the intake side of the air-conditioning apparatus) removes particulate contamination that may foul heat transfer surfaces, contaminate products, or present a health hazard to people, animals, or plants. Gaseous contaminants must sometimes be removed to prevent exposing personnel to health-threatening fumes or odors. The supply airstream may consist of air recirculated from building spaces and/or outdoor air for ventilation or exhaust air makeup. Return air with a significant potential for carrying contaminants should be recirculated only if it can be filtered enough to minimize personnel exposure. If monitoring and contaminant control cannot be ensured, the return air should be exhausted.

The supply air filtration system usually includes a collection medium or filter, a medium-retaining device or filter frame, and a filter house or plenum. The filter medium is the most important component of the system; a mat of randomly distributed small-diameter fibers is commonly used.

Depending on fiber material, size, density, and arrangement, fibrous filters have a wide range of performance. Low-density filter media with relatively large-diameter fibers remove large particles, such as lint. These roughing filters collect a large percentage by mass of the particulates, but are ineffective in reducing the total particle concentration. The fibers are sometimes coated with an adhesive to reduce particle re-entrainment.

Small-fiber, high-density filter media effectively collect essentially all particulates. Ultrahigh-efficiency particulate air filters reduce total particle concentration by more than 99.9%. As filter efficiency increases, so does the resistance to airflow, with typical pressure drops ranging from 12 to 250 Pa. Conversely, dust-holding capacity decreases with increasing filter efficiency, and fibrous filters should not be used for dust loading greater than 4 mg/m^3 of air. For more discussion of particulate filtration systems, refer to Chapter 25 of the 1992 *ASHRAE Handbook—Systems and Equipment*.

Exhaust Air Filtration Systems

Exhaust air systems are either (1) general systems that remove air from large spaces or (2) local systems that capture aerosols, heat, or gases at specific locations within a room and transport them so they can be collected (filtered), inactivated, or safely discharged to the atmosphere. The air in a general system usually requires minimal treatment before discharging to the atmosphere. The air in local exhaust systems can sometimes be safely dispersed into the atmosphere, but sometimes contaminants must be removed so that the emitted air meets air quality standards.

Many types of contamination collection or inactivation systems are applied in exhaust air emission control. Fabric bag filters, glass-fiber filters, venturi scrubbers, and electrostatic precipitators all collect particulates. Packed bed or sieve towers can absorb toxic gases. Activated carbon columns or beds, often with oxidizing agents, are frequently used to absorb toxic or odorous organics and radioactive gases.

Outdoor air intakes should be carefully located to avoid recirculation of contaminated exhaust air. Because wind direction, building shape, and the location of the effluent source strongly influence concentration patterns, exact patterns are not predictable.

Air patterns resulting from wind flow over buildings are discussed in Chapter 14 of the 1993 *ASHRAE Handbook—Fundamentals*. The leading edge of a roof interrupts smooth airflow, resulting in reduced air pressure at the roof and on the lee side. To prevent fume damage to the roof and roof-mounted equipment, and to keep fumes from the building air intakes, fumes must be discharged through either (1) vertical stacks terminating above the turbulent air boundary or (2) short stacks with a velocity high enough to project the effluent through the boundary into the undisturbed air passing over the building. A high vertical stack is the safest and simplest solution to fume dispersal.

Contaminant Control

In addition to maintaining thermal conditions, air-conditioning systems should control contaminant levels to provide (1) a safe and healthy environment, (2) good housekeeping, and (3) quality control for the processes. Contaminants may be gases, fumes, mists, or airborne particulate matter. They may be produced by a process within the building or contained in the outside air.

Contamination can be controlled by (1) preventing the release of aerosols or gases into the room environment and (2) diluting room air contaminants. If the process cannot be enclosed, it is best to capture aerosols or gases near their source of generation with local exhaust systems that include an enclosure or hood, ductwork, a fan, a motor, and an exhaust stack.

Dilution controls contamination in many applications but may not provide uniform safety for personnel within a space (West 1977). High local concentrations of contaminants can exist within a room, even though the overall dilution rate is quite high. Further, if tempering of outdoor air is required, high energy costs can result from the increased airflow required for dilution.

Exhaust Systems

An exhaust system draws the contaminant away from its source and removes it from the space. An exhaust hood that surrounds the point of generation contains the contaminant as much as is practical. The contaminants are transported through ductwork from the space, cleaned as required, and exhausted to the atmosphere. The suction air quantity in the hood is established by the velocities required to contain the contaminant.

Design values for average and minimum face velocities are a function of the characteristics of the most dangerous material that the hood is expected to handle. Minimum values may be prescribed in codes for exhaust systems. Contaminants with greater mass may require higher face velocities for their control.

Properly sized ductwork keeps the contaminant flowing. This requires very high velocities for heavy materials. The selection of materials and the construction of exhaust ductwork and fans depend on the nature of the contaminant, the ambient temperature, the lengths and arrangement of duct runs, the method of hood fan operation, and the flame and smoke spread rating.

Exhaust systems remove chemical gases, vapors, or smokes from acids, alkalis, solvents, and oils. Care must be taken to minimize the following:

1. *Corrosion*, which is the destruction of metal by chemical or electrochemical action; commonly used reagents in laboratories are hydrochloric, sulfuric, and nitric acids, singly or in combination, and ammonium hydroxide. Common organic chemicals include acetone, benzene, ether, petroleum, chloroform, carbon tetrachloride, and acetic acid.
2. *Dissolution*, which is a dissolving action. Coatings and plastics are subject to this action, particularly by solvent and oil fumes.
3. *Melting*, which can occur in certain plastics and coatings at elevated hood operating temperatures.

Low temperatures that cause condensation in ferrous metal ducts increase chemical destruction. Ductwork is less subject to attack when the runs are short and direct to the terminal discharge point. The longer the runs, the longer the period of exposure to fumes and the greater the degree of condensation. Horizontal runs allow moisture to remain longer than it can on vertical surfaces. Intermittent fan operation can contribute to longer periods of wetness (because of condensation) than continuous operation. High loading of condensables in exhaust systems should be avoided by installing condensers or scrubbers as close to the source as possible.

Operation and Maintenance of Components

All designs should allow ample room to clean, service, and replace any component quickly so that design conditions are affected as little as possible. Maintenance of refrigeration and heat-rejection equipment is essential for proper performance without energy waste.

For system dependability, water treatment is essential. Air washers and cooling towers should not be operated unless the water is properly treated.

Maintenance of heating and cooling systems includes changing or cleaning system filters on a regular basis. Industrial applications are usually dirty, so proper selection of filters, careful installation to avoid bypassing, and prudent changing of filters to prevent overloading and blowout are required. Dirt that lodges in ductwork and on forward-curved fan blades reduces air-handling capacity appreciably.

Fan and motor bearings require lubrication, and fan belts need periodic inspection. Infrared and panel systems usually require less maintenance than equipment with filters and fans, although gas-fired units with many burners require more attention than electric heaters.

The direct-fired makeup heater has a relatively simple burner that requires less maintenance than a comparable indirect-fired heater. The indirect oil-fired heater requires more maintenance than a comparable indirect gas-fired heater. With either type, the many safety devices and controls require periodic maintenance to ensure constant operation without nuisance cutout. Direct- and indirect-fired heaters should be inspected at least once per year.

Steam and hot water heaters have fewer maintenance requirements than comparable equipment having gas and oil burners. When used for makeup air in below-freezing conditions, however, the heaters must be correctly applied and controlled to prevent frozen coils.

REFERENCES

ASHRAE. 1989. Ventilation for acceptable indoor air quality. *Standard* 62-1989.

ASHRAE. 1992. Safety code for mechanical refrigeration. *Standard* 15-1992.

ASHRAE. 1992. Thermal environmental conditions for human occupancy. *Standard* 55-1992.

Dukes-Dobos, F. and A. Henschel. 1971. *The modification of the WNGT Index for establishing permissible heat exposure limits in occupational work.* U.S. HEW, USPHS, ROSH joint publication TR-69.

McNall, P.E., J. Juax, F.H. Rohles, R.G. Nevins, and W. Springer. 1967. Thermal comfort (thermally neutral) conditions for three levels of activity. *ASHRAE Transactions* 73(1):I.3.1-14.

West, D.L. 1977. Contaminant dispersion and dilution in a ventilated space. *ASHRAE Transactions* 83(1):125-40.

BIBLIOGRAPHY

ACGIH. 1992. *Industrial ventilation: A manual of recommended practice*, 21st ed. American Conference of Governmental Industrial Hygienists, Cincinnati, OH.

AIHA. 1975. *Heating and cooling for men in industry*, 2nd ed. American Industrial Hygienist Association, Akron, Ohio.

Azer, N.Z. 1982a. Design guidelines for spot cooling systems. Part 1—Assessing the acceptability of the environment. *ASHRAE Transactions* 88(2):81-95.

Azer, N.Z. 1982b. Design guidelines for spot cooling systems. Part 2—Cooling jet model and design procedure. *ASHRAE Transactions* 88(2):97-116.

Gorton, R.L. and Bagheri. 1987a. Verification of stratified air-conditioning design. *ASHRAE Transactions* 93(2):211-27.

Gorton, R.L. and Bagheri. 1987b. Performance characteristics of a system designed for stratified cooling operating during the heating season. *ASHRAE Transactions* 93(2):367-81.

Harstad, J., et al. 1967. Air filtration of submicron virus aerosols. *American Journal of Public Health* 57:2186-93.

IES. 1982. *Lighting handbook*, 6th ed. Illuminating Engineering Society of North America, New York.

O'Connell, W.L. 1976. How to attack air-pollution control problems. *Chemical Engineering* (October), desktop issue.

Whitby, K.T. and D.A. Lundgren. 1965. Mechanics of air cleaning. *ASAE Transactions* 8:3, 342. American Society of Agricultural Engineers, St. Joseph, MI.

Yamazaki, K. 1982. Factorial analysis on conditions affecting the sense of comfort of workers in the air-conditioned working environment. *ASHRAE Transactions* 88(1):241-54.

CHAPTER 12

ENCLOSED VEHICULAR FACILITIES

THIS chapter deals with the ventilation requirements for cooling, pollution control, and emergency smoke and temperature control for road tunnels, rapid transit tunnels and stations, enclosed parking structures, bus garages, and bus terminals. Also included are the design approach and type of equipment applied to these ventilation systems.

ROAD TUNNELS

All internal combustion engines produce exhaust gases that contain toxic compounds and smoke. To dilute the concentrations of obnoxious or dangerous contaminants to acceptable levels, road tunnels require ventilation, which may be provided by natural means, traffic-induced piston effect, or mechanical equipment. The selected ventilation system should be the most economical in terms construction and operating costs. Natural and traffic-induced ventilation systems are considered adequate for tunnels of relatively short length and low traffic volume (or density). Long and heavily traveled tunnels should have mechanical ventilation systems.

The exhaust constituent of greatest concern is carbon monoxide (CO), because of its notorious asphyxiant nature. The ventilation system dilutes the CO content of the tunnel atmosphere to a safe and comfortable level. Tests and operating experience indicate that when CO is properly diluted, the other dangerous and objectionable exhaust by-products are also diluted to acceptable levels. The section on Bus Terminals includes further information on diesel engine operation.

The ventilation systems discussed here are for incorporation as a permanent part of the finished tunnel and are intended primarily to serve the needs of the traveling public passing through the tunnel. Ventilation needed by workers during construction of the facility or while working in the finished tunnel is not covered. These ventilation requirements are specified in detail by state or local mining laws, industrial codes, or in the standards set by the U.S. Occupational Safety and Health Administration (OSHA).

Allowable Carbon Monoxide Concentrations

In 1975, the U.S. Environmental Protection Agency (EPA) issued a supplement to its *Guidelines for Review of Environmental Statements for Highway Projects* that evolved into a design approach based on keeping a CO concentration of 143 mg/m^3 (125 ppm) or below for a maximum of one hour exposure time for tunnels located at or below an altitude of 1000 m (see the section on Minimum Ven-

tilation Rate). In 1988, the EPA revised its recommendations for maximum CO levels in tunnels located at or below an altitude of 1500 m to the following:

Max. 137 mg/m^3 (120 ppm) for 15 min exposure time
Max. 74 mg/m^3 (65 ppm)   for 30 min exposure time
Max. 52 mg/m^3 (45 ppm)   for 45 min exposure time
Max. 40 mg/m^3 (35 ppm)   for 60 min exposure time

The new guidelines do not apply to existing tunnels.

Above an elevation of 1000 m, the CO emission of vehicles is greatly increased, and human tolerance to CO exposure is reduced. For tunnels above 1000 m, the designer should consult with medical authorities to establish a proper design value for CO concentrations. Unless specified otherwise, the material in this chapter refers to tunnels at or below an altitude of 1000 m.

Carbon Monoxide Emission

The CO content in exhaust gases of individual vehicles varies greatly because of such factors as the age of the vehicle, carburetor adjustment, quality of fuel, engine power, level of vehicle maintenance, and different driving habits of motorists. The vehicle CO emission for any given calendar year can be predicted using the Mobile Source Emissions Factor Model (EPA 1993).

All suggested ventilation rates given in this chapter are thus derived using uncontrolled emission rates, but the results can be corrected as indicated to reflect the impact of emission control devices.

VENTILATION AIR REQUIRED

The ventilation system must have the capacity to protect the traveling public during the most adverse and dangerous conditions, as well as during normal conditions. Establishing air requirements is complicated by many uncontrollable variables, such as the extensive number of possible vehicle combinations and traffic situations that could occur during the lifetime of the facility.

Ventilation rates that meet the general criteria are computed in two parts (shown in Example 1). The first considers a lane of tunnel traffic composed entirely of passenger cars. The following assumptions are necessary:

1. A major traffic stoppage has occurred outside the tunnel, and traffic is blocked. This stoppage causes tunnel traffic to slow down from normal operating speeds and finally come to a stop with engines idling. It is assumed that traffic reverses the process (i.e., idle first, then proceed at 10 km/h, then up to 40 km/h) when the stoppage is cleared.

The preparation of this chapter is assigned to TC 5.9, Enclosed Vehicular Facilities.

Table 1   Space Between Cars at Various Speeds

Speed, km/h	Space, m
40	22.0
35	18.6
25	11.6
15	4.6
10	3.0
Idle	1.2

Table 2   Emission Factors for Highway Vehicles

Calendar Year	Average Emission Factor (F)
1991	0.159
1992 and after	0.151

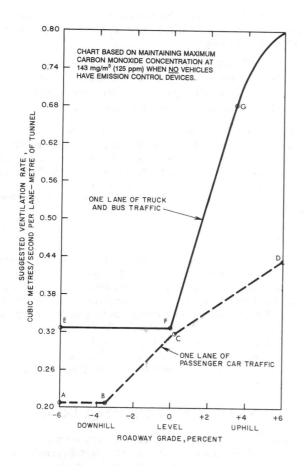

CHART BASED ON MAINTAINING MAXIMUM CARBON MONOXIDE CONCENTRATION AT 143 mg/m³ (125 ppm) WHEN NO VEHICLES HAVE EMISSION CONTROL DEVICES.

Key Values

Point	Grade	Ventilation Rate, m³/s per lane-metre of Tunnel
A	−6.0	0.206
B	−3.5	0.206
C	zero	0.313
D	+6.0	0.433
E	−6.0	0.325
F	zero	0.325
G	+3.5	0.682

Fig. 1   Tunnel Ventilation Rates for Different Roadway Gradients

2. Although CO emission rates during acceleration and deceleration are higher than at constant speed, the effect of changing speed is neglected. (The error introduced by this assumption will be offset by a 10% safety factor included in the computations.)
3. Traffic is assumed to move as a unit, with space between vehicles remaining constant regardless of roadway grade.
4. Passenger cars are assumed to be 5.8 m long (the usual length of a highway design vehicle), and spaces between cars at various speeds are shown in Table 1.

From this, the governing ventilation rates for different roadway gradients, from 6% downhill to 6% uphill, are determined using the data plotted in Figure 1. In most cases, the critical traffic situation from a ventilation standpoint occurs when the lane is congested with vehicles moving at a speed of about 15 km/h. On the steeper downgrades, the governing condition occurs when traffic is stopped and vehicles are bumper-to-bumper with engines idling. Bumper-to-bumper traffic calls for a constant ventilation rate regardless of roadway grade.

Figure 1 shows the results of a second set of computations made for a lane of tunnel traffic composed entirely of trucks and buses. About 40% of the vehicles are assumed to be diesel powered and averaging 15.2 m in length, and 60% are taken as light- to medium-duty gasoline-powered trucks averaging 9.1 m in length. The governing traffic conditions are similar to the passenger car lane except that on steeper upgrades, trucks traveling at a crawl speed of about 10 km/h become the critical traffic condition.

In both cases, the ventilation rates given in Figure 1 are based on emission rates from vehicles not equipped with control devices and on a maximum allowable CO concentration of 143 mg/m³ (125 ppm).

Conversion of Different Traffic Mix

To correct the ventilation rate for emission control, the average emission factors for highway vehicles given in Table 2 can be used; they are derived from EPA estimates for CO emission rate for 80 000 travel kilometres. These emission factors are for the vehicle population mix for the calendar year shown, not for vehicles of that model year only. The values are projected emission factors based on (1) actual test results of existing sources and control systems and (2) projected values for future years based on required emission reductions, as stipulated in the present law (the EPA emission standard for CO remains at 2.1 g/km after the 1983 model year). The designer should be aware of any possible future revisions of the predicted factors and of emission control laws; actual emission rates versus predicted emission rates must also be evaluated.

Converting to any desired calendar year may be accomplished with the following equation:

$$Q = VF \tag{1}$$

where

Q = converted ventilation rate
V = unconverted ventilation rate
F = emission factor for desired calendar year from Table 2

Conversion of Carbon Monoxide Concentration

Converting ventilation rates to a CO level other than 143 mg/m³,

$$Q = V(143/C) \tag{2}$$

where C = desired CO concentration, mg/m³.

Adjustment of Ventilation Rate for Ambient CO Level

Ventilation rates shown in Figure 1 were calculated based on the assumption that the ventilation air contains little or no CO when it is introduced. If this is not the case, the ventilation rates may be adjusted as follows:

$$Q = 143V / (143 - E) \tag{3}$$

where E = ambient CO level, mg/m^3.

Conversion of Truck Traffic Mix

The conversion constants in the following equation consider the size of the truck and the percentage of the truck type in the traffic stream.

$$Q = V (1.22\, P_G + 0.85\, P_D) \tag{4}$$

where

P_G = fraction of gasoline-powered vehicles
P_D = fraction of diesel-powered vehicles

Minimum Ventilation Rate

In addition to diluting vehicle emission contaminants, the road tunnel ventilation system should be capable of effecting smoke control during a fire emergency. Smoke control practice has been to utilize an exhaust rate of at least 0.15 m^3/s per lane-metre of tunnel. However, an analysis of life safety and smoke control requirements and ventilation system configuration could permit a reduction in this value. At the time of publication of this *Handbook*, ASHRAE was participating in the Memorial Tunnel Fire Ventilation Test Program to evaluate the effectiveness of specific types and configurations of road tunnel ventilation systems. These tests should help to determine the minimum ventilation needed to control smoke in the tunnel to maintain a safe environment for the evacuation of motorists and the entry of fire fighters.

Example 1. Calculate the volume flow rate of a single-bore tunnel with two lanes of unidirectional traffic. The tunnel has the grades and dimensions shown below. It has a design level of 143 mg/m^3 and an ambient CO level of 6 mg/m^3.

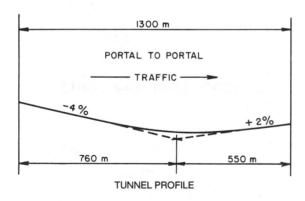

TUNNEL PROFILE

Solution: From Figure 1,

−4% grade	0.206	m^3/s per lane-metre for cars
	0.325	m^3/s per lane-metre for trucks
	0.531	m^3/s per lane-metre
+2% grade	0.355	m^3/s per lane-metre for cars
	0.535	m^3/s per lane-metre for trucks
	0.890	m^3/s per lane-metre

$$0.531 \times 760 = 404$$
$$0.890 \times 550 = 490$$
$$894\ m^3/s$$

Adjusting for ambient CO level of 6 mg/m^3 [Equation (3)] and converting traffic mix for calendar year 1992 [Equation (1)],

$$Q = VF\,[143/ (143 - 6)]$$

From Table 2, $F = 0.151$.

$$Q = 894 \times 0.151\,[143/ (143 - 6)]$$
$$= 141\,m^3/s$$

Check for minimum ventilation rate:

For −4% grade: $0.531 \times 0.151\,[143/(143 - 6)] = 0.08$ m^3/s

For +2% grade: $0.890 \times 0.151\,[143/(143 - 6)] = 0.14$ m^3/s

Both grade requirements fall short of 0.15 m^3/s per lane-metre; therefore, adequacy under fire conditions should be analyzed.

Computer Computation Program

A computer computation program called TUNVEN that solves the coupled one-dimensional steady-state tunnel aerodynamic and advection equations is available (NTIS). This program can predict the quasi-steady-state longitudinal air velocities and the concentrations of CO, oxides of nitrogen (NO_x), and total hydrocarbons (THC) along a highway tunnel for a wide range of tunnel designs, traffic loads, and external ambient conditions. The program will model all common ventilation system types: natural, longitudinal, semitransverse, and transverse.

VENTILATION SYSTEM TYPES

Natural Ventilation

Naturally ventilated tunnels rely primarily on meteorological conditions to maintain a satisfactory environment within the tunnel. The piston effect of traffic provides additional airflow when the traffic is moving. The chief meteorological condition affecting the tunnel environment is the pressure differential created by differences in elevation, ambient temperature, or wind between two portals. Unfortunately, none of these factors can be relied on for continued, consistent results. A sudden change in wind direction or velocity can negate all these natural effects, including the piston effect. The total of all pressures must be of sufficient magnitude to overcome the tunnel resistance, which is influenced by tunnel length, coefficient of friction, hydraulic radius, and air density.

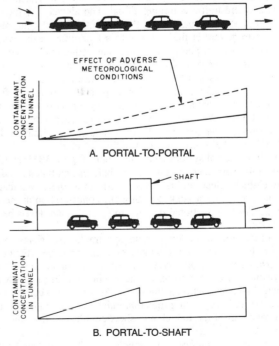

Fig. 2   Natural Ventilation

Airflow through a naturally ventilated tunnel can be portal-to-portal (Figure 2A) or portal-to-shaft (Figure 2B). Portal-to-portal flow functions best with unidirectional traffic, which produces a consistent, positive airflow. The air velocity within the roadway is uniform, and the contaminant concentration increases to a maximum at the exit portal. Under adverse meteorological conditions, the velocity is reduced and the CO concentration increases, as shown by the dashed line on Figure 2A. The introduction of bidirectional traffic into such a tunnel further reduces the airflow.

The naturally ventilated tunnel with an intermediate shaft (Figure 2B) is best suited for bidirectional traffic. However, airflow through such a shafted tunnel is also at the mercy of the elements. The benefit of the stack effect of the shaft depends on air and rock temperatures, wind, and shaft height. The addition of more than one shaft to a naturally ventilated tunnel is more of a disadvantage than an advantage because a pocket of contaminated air can be trapped between the shafts.

Most naturally ventilated urban tunnels over 150 m long require an emergency mechanical ventilation system to purge smoke and hot gases generated during an emergency and to remove stagnated, polluted gases during an severe adverse meteorological conditions. Because of the aforementioned uncertainties, reliance on natural ventilation for all tunnels over 150 m long should be thoroughly evaluated, especially the effect of adverse meteorological and operating conditions. This is particularly true for a tunnel with an anticipated heavy or congested traffic flow. If the natural ventilation is inadequate, the installation of a mechanical system with fans should be considered.

Mechanical Ventilation

The most appropriate mechanical ventilation systems for tunnels are longitudinal ventilation, semitransverse ventilation, and full transverse ventilation.

Longitudinal Ventilation. This applies to any system that introduces or removes air from the tunnel at a limited number of points, thus creating a longitudinal flow of air within the roadway.

The injection-type longitudinal system has frequently been used in rail tunnels and has also found application in road tunnels. Air injected at one end of the tunnel mixes with air brought in by the piston effect of the incoming traffic (Figure 3A). This system is most effective where traffic is unidirectional. The air velocity stays uniform throughout the tunnel, and the concentration of contaminants increases from zero at the entrance to a maximum at the exit. Adverse external atmospheric conditions can reduce the effectiveness of this system. The contaminant level at the exit portal increases as the airflow decreases or the tunnel length increases.

The longitudinal form of ventilation is recognized as an effective method of smoke control within a highway tunnel bore with unidirectional traffic. The air velocity necessary to prevent the backlayering of smoke over the stalled vehicles is the minimum velocity needed for smoke control in a longitudinal system.

The longitudinal system with a fan shaft (Figure 3B) is similar to the naturally ventilated system with a shaft, except that it provides a positive stack effect. Bidirectional traffic in a tunnel ventilated in this manner causes a peak contaminant concentration at the shaft location. For unidirectional tunnels, the contaminant levels become unbalanced.

Another form of the longitudinal system has two shafts near the center of the tunnel: one for exhaust and one for supply (Figure 3C). This arrangement reduces contaminant concentration in the second half of the tunnel. A portion of the air flowing in the roadway is replaced in the interaction at the shafts. Adverse wind conditions can reduce the airflow, causing the contaminant concentration to rise in the second half of the tunnel, and short-circuit the flow from fan to exhaust.

In a growing number of tunnels, longitudinal ventilation is achieved with fans mounted at the tunnel ceiling (Figure 3D). Such a

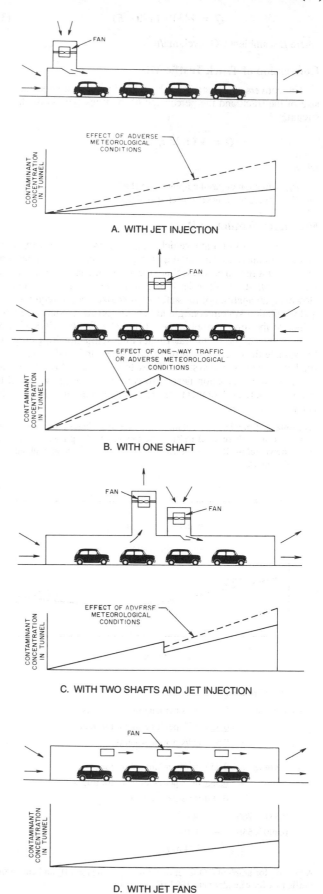

Fig. 3   Longitudinal Ventilation

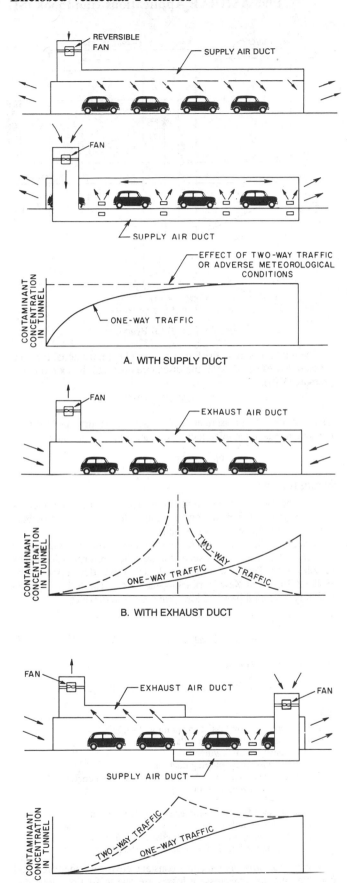

A. WITH SUPPLY DUCT

B. WITH EXHAUST DUCT

C. WITH SUPPLY AND EXHAUST DUCT

Fig. 4   Semitransverse Ventilation

system eliminates the space needed to house ventilation fans in a separate structure or ventilation building; however, it may require a tunnel of greater height or width to accommodate the jet fans.

Standard longitudinal ventilation systems (excluding the jet fan system) with either supply or exhaust at a limited number of locations within the tunnel are the most economical because they require the least number of fans, place the least operating burden on these fans, and require no distribution air ducts. As the length of the tunnel increases, however, the disadvantages of these systems, such as excessive air velocities in the roadway and smoke being drawn the entire length of the roadway during an emergency, become apparent. Uniform air distribution would alleviate these problems.

Semitransverse Ventilation. This system distributes or collects air uniformly throughout the length of a tunnel. The supply air version of the system applied to a bidirectional tunnel (Figure 4A) produces a uniform level of contaminants throughout the tunnel because the air and the vehicle exhaust gases enter the roadway area at the same rate. In a tunnel with unidirectional traffic, additional airflow is generated within the roadway area, thus reducing the contaminant level in portions of the tunnel, as shown in Figure 4A.

Because of the fan-generated flow, this system is not adversely affected by atmospheric conditions. The air flows the length of the tunnel in a duct fitted with periodic supply outlets. Fresh air is best introduced at vehicle exhaust pipe level to dilute the exhaust gases immediately. An adequate pressure differential between the duct and the roadway must be generated to counteract piston effect and atmospheric winds.

If a fire occurs within the tunnel, the supply air will initially dilute the smoke. The ventilation system should be operated for the emergency so that fresh air enters the tunnel through the portal to create a respirable environment for fire-fighting efforts and emergency egress. Therefore, the ventilation configuration for a supply semitransverse system should have a ceiling supply and reversible fans so that the smoke will be drawn upward toward the ceiling in an emergency.

An exhaust semitransverse system (Figure 4B) in a unidirectional tunnel produces a maximum contaminant concentration at the exit portal. In a bidirectional tunnel, the maximum level of contaminants occurs near the center of the tunnel. A combination supply and exhaust system (Figure 4C) applies only in a unidirectional tunnel where the air entering with the traffic stream is exhausted in the first half, and air supplied in the second half is exhausted through the exit portal.

The supply semitransverse system is the only ventilation system not affected by adverse meteorological conditions or opposing traffic. Semitransverse systems are used in tunnels up to about 1000 m,

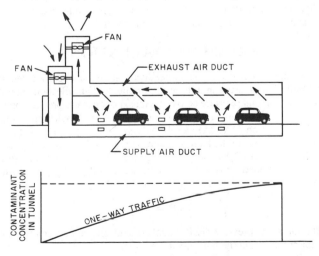

Fig. 5   Full Transverse Ventilation

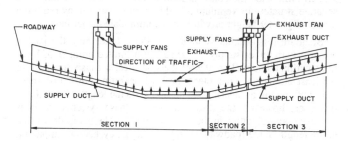

Fig. 6 Combined Ventilation System

at which point the tunnel air velocities near the portals become excessive.

Full Transverse Ventilation. This is used in large tunnels. A full exhaust duct added to a supply-type semitransverse system achieves uniform distribution of supply air and uniform collection of vitiated air (Figure 5). With this arrangement, pressure throughout the roadway is uniform, and there is no longitudinal airflow except that generated by traffic piston effect, which tends to reduce contaminant levels. An adequate pressure differential between the ducts and the roadway is required to assure proper air distribution under all ventilation conditions.

Full-scale tests conducted by the U.S. Bureau of Mines (Fieldner et al. 1921) showed that for rapid dilution of exhaust gases, supply air inlets should be at exhaust pipe level, and exhaust outlets should be in the ceiling. The air distribution can be one- or two-sided.

Other Ventilation. Many variations and combinations of the described systems exist. Figure 6 shows a combined system for a unidirectional tunnel approximately 425 m long. Section 3 uses a full transverse system because of the upgrade roadway; Section 2 uses a semitransverse supply with a longitudinal exhaust; and the remainder of the tunnel (Section 1) uses a semitransverse supply system. This type of system is not recommended for long tunnels.

Emergency Conditions

In any road tunnel, an emergency condition, particularly one generating smoke and heat, such as a fire, can lead to a disaster if the tunnel ventilation system is designed or operated improperly. The primary objective in such a situation is the rapid removal of smoke and heat from the tunnel to enable motorists to exit safely and fire fighters to reach the fire. The criteria for emergency conditions, including the minimum level of ventilation required to create a safe environment, must be established prior to the design of a tunnel.

The Memorial Tunnel Fire Ventilation Test Program should provide the data needed to determine both the type and level of ventilation suitable to control smoke in a road tunnel.

PRESSURE EVALUATIONS

Air pressure losses in the tunnel duct system must be evaluated to compute fan pressure and drive requirements. Fan selection should be based on total pressure across the fans, not on static pressure only.

Fan total pressure (FTP) is defined by the Air Moving and Control Association (AMCA) (ASHRAE *Standard* 51) as the algebraic difference between the total pressures at the fan discharge (TP_2) and at the fan inlet (TP_1), as shown in Figure 7. The fan velocity pressure (FVP) is defined by AMCA as the pressure (VP_2) corresponding to the air velocity and air density at the fan discharge.

$$FVP = VP_2$$

The fan static pressure (FSP) equals the difference between the fan total pressure and the fan velocity pressure.

$$FSP = FTP - FVP$$

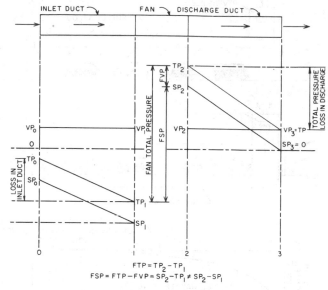

Fig. 7 Fan Total Pressure

The total pressure at the fan discharge (TP_2) must equal the total pressure losses (ΔTP_{2-3}) in the discharge duct and the exit velocity pressure (VP_3).

$$TP_2 = \Delta TP_{2-3} - VP_3$$

Likewise, the total pressure at the fan inlet (TP_1) must equal the total pressure losses in the inlet duct system and the inlet pressure.

$$TP_1 = TP_0 + \Delta TP_{0-1}$$

Straight Ducts

Straight ducts in tunnel ventilation systems can be classified as (1) those that transport air and (2) those that uniformly distribute (supply) or uniformly collect (exhaust) air.

Several methods have been developed to predict pressure losses in a duct of constant cross-sectional area that uniformly distributes or collects air. The most widely used method was developed for the Holland Tunnel in New York (Singstad 1929). The following relationships give total pressure loss at any point in the duct.

For a *supply duct*,

$$p = p_1 + \rho V_0^2/2 \, [aLZ^3/3H - (1-K) \, Z^2/2] \qquad (5)$$

For an *exhaust duct*,

$$p = p_1 + \rho V_0^2/2 \, [aLZ^3/(3+c) \, H + 3Z^2/(2+c)] \qquad (6)$$

where

p = total pressure loss at any point in duct, kPa
p_1 = pressure at last outlet, kPa
ρ = density of air, kg/m³
V_0 = velocity of air entering duct, m/s
a = constant related to coefficient of friction for concrete = 0.0035
L = total length of duct, m
Z = $(L - X)/L$
X = distance from duct entrance to any location, m
H = hydraulic radius, m
K = constant accounting for turbulence = 0.615
c = constant relating to turbulence of exhaust port = 0.25

For a *transport duct* having constant cross-sectional area and constant velocity, the pressure losses are due to friction alone and can be computed using the standard expressions for losses in ducts and fittings (see Chapter 32 of the 1993 *ASHRAE Handbook—Fundamentals*).

CARBON MONOXIDE ANALYZERS AND RECORDERS

The air quality in a tunnel should be monitored constantly at several key points. Carbon monoxide is the impurity usually selected as the prime indicator of tunnel air quality. CO-analyzing instruments base their measurements on one of three processes: catalytic oxidation, infrared absorption, and electrochemical oxidation.

The *catalytic oxidation (metal oxide) instrument*, the most widely used in road tunnels, offers reliability and stability at a moderate initial cost. Maintenance requirements are low, and the instruments can be calibrated and serviced by maintenance personnel after only brief instruction.

The *infrared analyzer* has the advantage of sensitivity and response but the disadvantage of high initial cost. A precise and complex instrument, it requires a highly trained technician for maintenance and servicing.

The *electrochemical analyzer* is precise, compact, and lightweight. The units are of moderate cost and are easily maintained.

No matter which type of CO analyzer is selected, each air sampling point must be located where significant readings can be obtained. For example, the area near the entrance of a unidirectional tunnel usually has very low CO concentrations, and sampling points there yield little information for ventilation control. The length of piping between the sampling point and the CO analyzer should be as short as possible to maintain a reasonable air sample transport time.

Where intermittent analyzers are used, provisions should be made to prevent the loss of more than one sampling point during an air pump outage. Each analyzer should be provided with a strip chart recorder to keep a permanent record of tunnel air conditions. Usually, recorders are mounted on the central control board.

Haze or smoke detectors have been used on a limited scale, but most of these instruments are optical devices and require frequent or constant cleaning with a compressed air jet. If traffic is predominantly diesel powered, smoke haze and oxides of nitrogen require monitoring in addition to CO.

CONTROL SYSTEMS

To reduce the number of operating personnel at a tunnel, all ventilating equipment should be controlled at a central location. At many older tunnel facilities, fan operation is manually controlled by an operator at the central control board. Many new tunnels, however, have partial or total automatic control for fan operation.

CO Analyzer Control System

In this system, adjustable contacts within CO analyzers turn on additional fans or increase fan speeds as CO readings increase. The reverse occurs as CO levels decrease. The system requires a fairly complex wiring arrangement because fan operation must normally respond to the single highest level being recorded at several analyzers. A manual override and delay devices are required to prevent the ventilation system from responding to short-lived high or low levels.

Time Clock Control Systems

This type of automatic fan control is best suited for those installations that experience heavy rush-hour traffic. A time clock is set to increase the ventilation level in preset increments before the anticipated traffic increase. The system is simple and easily revised to suit changing traffic patterns. Because it anticipates an increased air requirement, the ventilation system can be made to respond slowly and thus avoid expensive demand charges by the power company. As with the CO system, a manual override is needed to cope with unanticipated conditions.

Traffic-Actuated Systems

Several automatic fan control systems based on recorded traffic flow information have been devised. Most of them require the installation of computers and other electronic equipment needing maintenance expertise.

Local Fan Control

In addition to a central control board, each fan unit should have local control within sight of the unit. It should be interlocked to permit positive isolation of the fan from remote or automatic operation during maintenance and servicing.

RAPID TRANSIT SYSTEMS

Most rapid transit systems run at least part of their lines underground, particularly in business districts of urban centers. Older subway systems relied heavily on natural ventilation, the primary air mover being the piston effect of the vehicles themselves. In recent years, the piston effect has been supplemented by forced mechanical ventilation. Most new systems use air-conditioned vehicles, but it is still important to provide a reasonable environment within subway tunnels and stations. The *Subway Environmental Design Handbook* (DOT 1976), based on unproven operating experience but validated by field and model tests, provides comprehensive and authoritative design aids.

DESIGN CONCEPTS

The factors to be considered, although interrelated, may be divided into four categories: natural ventilation, mechanical ventilation, emergency ventilation, and station air conditioning.

Natural Ventilation

Natural ventilation in subway systems (infiltration and exfiltration) is primarily the result of train operation in tightly fitting trainways, where air generally moves in the direction of train travel. The positive pressure in front of a train expels air from the system through portals and station entrances; the negative pressure in the wake of the train induces airflow into the system through these same openings.

Considerable short-circuiting occurs in subway structures when two trains traveling in opposite directions pass each other, especially in stations or in tunnels with perforated or no dividing walls. Such short-circuiting reduces the net ventilation rate and causes increased air velocities on station platforms and in station entrances. During the time of peak operation and peak ambient temperatures, it can cause an undesirable heat buildup.

To help counter these negative effects, ventilation shafts are customarily placed near the interfaces between tunnels and stations. Shafts in the approach tunnel are often called *blast shafts*; part of the air pushed ahead of the train is expelled from the system through them. Shafts in the departure tunnel are called *relief shafts* because they relieve the negative pressure created during the departure of the train and induce outside air through the shaft rather than through station entrances. Additional ventilation shafts may be provided between stations (or between portals, for underwater crossings), as dictated by tunnel length. The high cost of such ventilation structures necessitates a design for optimum effectiveness. Internal resistance due to offsets and bends should be kept to a minimum, and shaft cross-sectional areas should be approximately equal to the cross-sectional area of a single-track tunnel (DOT 1976).

Mechanical Ventilation

Mechanical ventilation in subway systems (1) supplements the ventilation effect created by train piston action, (2) expels heated

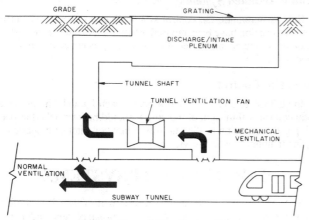

Fig. 8   Tunnel Ventilation Shaft

air from the system, (3) introduces cool outside air, (4) supplies makeup air for exhaust, (5) restores the cooling potential of the heat sink through extraction of heat stored during off-hours or system shutdown, (6) reduces the flow of air between the tunnel and the station, (7) provides outside air for passengers in stations or tunnels in an emergency or during other unscheduled interruptions of traffic, and (8) purges smoke from the system in case of fire.

The most cost-effective design for mechanical ventilation is one that serves two or more purposes. For example, a vent shaft provided for natural ventilation may also be used for emergencies if a fan is installed in a parallel with the bypass (Figure 8).

A system of several vent shafts working together may be capable of meeting many, if not all, of the eight objectives. Depending on shaft location and the given train situation, a shaft with the bypass

damper open and the fan damper closed may serve as a blast or relief shaft. With the fan in operation and the bypass damper closed, air can be supplied or exhausted by mechanical ventilation, depending on the direction of the fan rotation.

Except for emergencies, fan rotation is usually predetermined based on the overall ventilation concept. If subway stations are not air conditioned, the heated system air should be exchanged at a maximum rate with cooler outside air. If stations are air conditioned below ambient temperatures, the inflow of warmer outside air should be limited and controlled.

Figure 9 illustrates a typical tunnel ventilation system between two subway stations. Here the flow of heated tunnel air into a cooler station is kept to a minimum. The dividing wall separating Tracks No. 1 and 2 is discontinued in the vicinity of the emergency fans. As shown in Figure 9A, air pushed ahead of the train on Track No. 2 diverts partially to the emergency fan bypasses and partially into the wake of a train on Track No. 1 as a result of pressure differences. Figure 9B shows an alternative operation with the same ventilation system. When outdoor temperatures are favorable, the midtunnel fans operate as exhaust fans, with makeup air introduced through the emergency fan bypasses. This system can also provide or supplement station ventilation. To achieve this goal, emergency fan bypasses would be closed, and the makeup air for midtunnel exhaust fans would enter through station entrances.

A more direct ventilation system removes station heat at its primary source, the underside of the train. Figure 10 illustrates a trackway ventilation system. Tests have shown that such systems not only reduce the upwelling of heated air onto platform areas, but also remove significant portions of the heat generated by dynamic braking resistor grids and air-conditioning condensers underneath the train (DOT 1976). Ideally, makeup air for the exhaust should be introduced at track level to provide a positive control over the direction of airflow (Figure 10A).

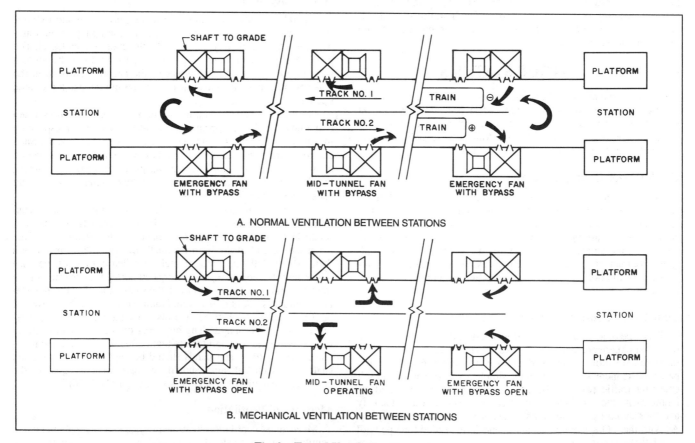

Fig. 9   Tunnel Ventilation Concept

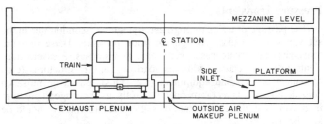

A. OUTSIDE AIR MAKEUP AT TRACK LEVEL

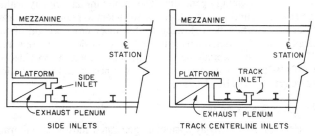

B. NO MAKEUP AIR SUPPLY

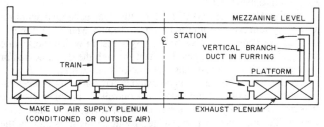

C. MAKEUP AIR SUPPLY AT CEILING LEVEL

Fig. 10   Trackway Ventilation Concepts (Cross-Sections)

Underplatform exhaust systems without makeup supply air, as illustrated in Figure 10B, are the least effective and, under certain conditions, could cause heated tunnel air to flow into the station. Figure 10C shows a cost-effective compromise in which makeup air is introduced at the ceiling above the platform. Although the heat removal effectiveness may not be as good as that of the system illustrated in Figure 10A, the inflow of hot tunnel air that might occur without supply air makeup is negated.

Emergency Ventilation

During a subway tunnel fire, mechanical ventilation is a major control strategy. An increase in air supply over stoichiometric requirements will reduce the progression of the fire by lowering the flame temperature. Further, ventilation can control the direction of smoke migration to permit safe evacuation of passengers and facilitate access by fire fighters.

Emergency ventilation must allow for the unpredictable location of a disabled train or source of fire and smoke. Therefore, emergency ventilation fans should have nearly full reverse flow capability so that fans on either side of a stalled train can operate together to control the direction of airflow and counteract the migration of smoke. When a train is stalled between two stations and smoke is present, outside air is supplied from the nearest station and contaminated air is exhausted at the opposite end of the train (unless the location of the fire dictates otherwise). Passengers can then be evacuated along walkways in the tunnel via the shortest route (Figure 11).

It is essential that provisions be made to (1) quickly assess any emergency situation; (2) communicate the situation to central control; (3) establish the location of the train; and (4) start, stop, and reverse emergency ventilation fans from the central console as quickly as possible to establish smoke control.

Midtunnel and station trackway ventilation fans may be used to enhance the emergency ventilation system; these fans must withstand elevated temperatures for a prolonged period and have reverse flow capacity.

Station Air Conditioning

Higher approach speeds and closer headways, made possible by computerized train control systems, have increased the amount of heat gains. The net internal sensible heat gain in a typical double-track subway station, with 40 trains per hour per track traveling at top speeds of 80 km/h, may reach 1.5 MW even after heat is removed by the heat sink, by station underplatform exhaust systems, and by tunnel ventilation. To remove such a quantity of heat by station ventilation with outside air at a temperature rise of 1.5 K, for example, requires roughly 830 m³/s.

Not only would such a system be costly, but the air velocities on station platforms would be objectionable to passengers. The same amount of sensible heat gain, plus latent heat, plus outside air load with a station design temperature 4 K lower than ambient, could be handled by about 2.2 MW of refrigeration. Even if station air conditioning is initially more expensive, long-term benefits include (1) reduced design airflow rates, (2) improved environment for passengers, (3) increased equipment life, (4) reduced maintenance of equipment and structures, and (5) increased acceptance of the subway as a viable means of public transportation.

Air conditioning should be considered for other ancillary station areas such as concourses, concession areas, and transfer levels. However, unless these walk-through areas are designed to attract patronage to concessions, the cost of air conditioning is not usually warranted.

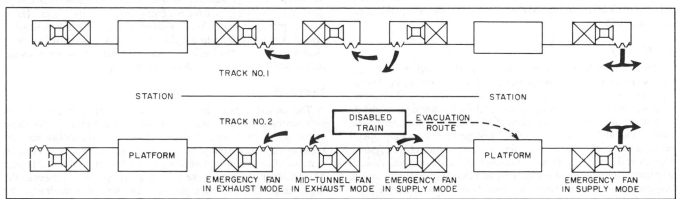

Fig. 11   Emergency Ventilation Concept

The physical configuration of the platform level usually determines the cooling distribution pattern. Areas with high ceilings, local hot spots due to train location, high-density passenger accumulation, or high-level lighting may need spot cooling. Conversely, where train length equals platform length and ceiling height above the platform is limited to 3 to 3.5 m, isolation of heat sources and application of spot cooling is not usually feasible.

The use of available space in the station structure for air distribution systems should be of prime concern because of the high cost of underground construction. Overhead distribution ductwork, which adds to building height in commercial construction, could add to the depth of excavation in subway construction. The space beneath a subway platform is normally an excellent area for low-cost distribution of supply, return, and/or exhaust air.

DESIGN APPROACH

A subway system may have to satisfy two separate sets of environmental criteria: one for normal operations and one for emergencies. Criteria for normal operations generally include limits on temperature and humidity for various times of the year, a minimum ventilation rate to dilute contaminants generated within the subway, and limits on air velocity and the rate of air pressure change to which commuters may be exposed. Some of these criteria are subjective and may vary based on demographics. Criteria for emergencies generally include a minimum purge time for sections of the subway in which smoke or fire may occur and minimum and maximum fan-induced tunnel air velocities.

Given a set of criteria, a set of outdoor design conditions, and appropriate tools for estimating interior heat loads, earth heat sink, ventilation, air velocity, and air pressure changes, design engineers select the elements of the environmental control system. Air temperature control, air velocity control, air quality control, and air pressure control should be considered. The system selected generally includes a combination of unpowered ventilation shafts, powered ventilation shafts, underplatform exhaust, and air conditioning.

The train propulsion/braking systems and the configurations of tunnels and stations greatly affect the subway environment, which must often be considered during the early stages of design. The factors affecting a subway environmental control system are discussed in this section. The *Subway Environmental Design Handbook* (DOT 1976) and NFPA *Standard* 130, Standard for Fixed Guideway Transit Systems, have additional information.

Comfort Criteria

Because of the transient nature of a person's exposure to the subway environment, comfort criteria are not as strict as for continuous occupancy. As a general principle, the environment within a subway station should provide a smooth transition between outside conditions and those within the transit vehicles. Based on nuisance considerations, it is recommended that peak air velocities in public areas be limited to 5 m/s.

Air Quality

Air quality within a subway system is influenced by many factors, some of which are not under the direct control of the HVAC engineer. Some particulates, gaseous contaminants, and odorants existing on the outside can be prevented from entering the subway system by the judicious location of ventilation shafts. Particulate matter, including iron and graphite dust generated by train operations, is best controlled by a regular cleaning of the subway system. However, the only viable way to control gaseous contaminants such as ozone (from electrical equipment) and CO_2 (generated by human respiration) is through adequate ventilation from outside. A minimum of 3.5 L/s of filtered outside air per person should be introduced into tunnels and stations to dilute gaseous contaminants.

Pressure Transients

The passage of trains through aerodynamic discontinuities in the subway system causes changes in static pressure. These pressure transients can irritate passengers' ears and sinuses. Pressure transients may also cause additional load on various structures (e.g., acoustical panels) and equipment (e.g., fans). Based on nuisance considerations, it is recommended that if the total change in pressure is greater than 700 Pa, the rate of static pressure change should be kept below 400 Pa/s.

During emergencies, it is essential to provide ventilation to control smoke and reduce air temperatures to permit passenger evacuation and fire-fighting operations. The minimum air velocity within the affected tunnel section should be sufficient to prevent the smoke from backlayering (i.e., flowing in the upper cross-section of the tunnel in the direction opposite to the forced outside ventilation air). The method for ascertaining this minimum velocity is provided in DOT (1976). The maximum air velocity experienced by evacuating passengers should be 11 m/s.

Interior Heat Loads

Heat within a subway system is generated mostly by (1) train braking; (2) train acceleration; (3) car air conditioning and miscellaneous equipment; (4) station lights, people, and equipment; and (5) ventilation (outside) air.

Deceleration. From 40 to 60% of the heat generated within a subway system arises from the braking of trains. Many rapid transit vehicles use nonregenerative braking systems, in which the kinetic energy of the train is dissipated as heat through the dynamic and/or friction brakes, rolling resistance, and aerodynamic drag.

Acceleration. Heat is also generated as a result of train acceleration. Many operational trains use cam-controlled variable resistance elements to regulate voltage across dc traction motors during acceleration. Electrical power is dissipated by these resistors and by the third rail as heat within the subway system. Heat released during acceleration also includes that due to traction motor losses, rolling resistance, and aerodynamic drag and generally amounts to 10 to 20% of total heat released within a subway system.

For closely spaced stations, more heat is generated because trains frequently undergo only acceleration or braking, with little operation at constant speed.

Car Air-Conditioning Systems. Most new cars are fully climate controlled. Air-conditioning equipment removes patron and lighting heat from the cars and deposits it, along with the condenser fan heat and compressor heat, into the subway system. Air-conditioning capacities generally range from 35 kW per car for the shorter cars (about 15 m) up to about 70 kW for the longer cars (about 21 m). Heat from car air conditioners and other accessories is generally 25 to 30% of total heat generated within a subway system.

Other Sources. Within a subway system, heat is also released by people, lighting, induced outside air, and miscellaneous equipment (fare vending machines, escalators, and the like). These sources provide 10 to 30% of total heat generated within a subway system. For analysis of heat balance within a subway, it is convenient to define a control volume about each station that includes the station and its approach and departure tunnels.

Heat Sink

The amount of heat flow from subway walls to subway air varies on a seasonal basis, as well as for morning and evening rush-hour operations. Short periods of abnormally high or low outside air temperature may cause a temporary departure from the heat sink effect in portions of a subway system that are not heated or air conditioned. This results in a change in the subway air temperature. However, the departure of subway air temperature from normal is diminished by the thermal inertia of the subway structure. During abnormally hot periods, heat flow to the subway structure increases. Similarly, during abnormally cold periods, heat flow from the subway structure to the air increases.

For subways where daily station air temperatures are held constant by heating and cooling, heat flux from station walls is negligible. Depending on the infiltration of station air into the adjoining tunnels, heat flux from tunnel sections may also be reduced. Other factors affecting the heat sink component are type of soil (dense rock or light dry soil), migrating groundwater, and the surface configuration of the tunnel walls (ribbed or flat).

Measures to Limit Heat Loads

Various measures have been proposed to limit the interior heat load in subway systems. Among these are regenerative braking, thyristor (chopper) motor controls, track profile optimization, and underplatform exhaust systems.

Electrical regenerative braking converts kinetic energy to electrical energy for use by other trains. Flywheel energy storage, an alternative form of regenerative braking, stores part of the braking energy of a train in high-speed flywheels for subsequent use in vehicle acceleration. Using these methods, the reduction in heat generated by braking is limited by present technology to approximately 25%.

Conventional cam-controlled propulsion systems apply a set of resistance elements to regulate traction motor current during acceleration. Electrical energy dissipated by these resistors appears as waste heat within a subway system. *Thyristor motor controls* replace the acceleration resistors with solid-state controls, which reduce acceleration heat loss by about 10% on high-speed subways and about 25% on low-speed subways.

Track profile optimization refers to a depressed trackway between stations that reduces vehicle heat emissions. In this way, less power is used for acceleration, since some of the potential energy of the standing train is converted to kinetic energy as the train accelerates toward the tunnel low point. Conversely, some of the kinetic energy of the train at maximum speed is converted to potential energy as it approaches the next station. Using this method, the maximum reduction in total vehicle heat loss from acceleration and braking is approximately 10%.

An *underplatform exhaust system* make use of a hooding technique to prevent some of the heat generated by vehicle underfloor equipment (e.g., resistors and air-conditioning condensers) from entering the station environment. Exhaust ports beneath the edge of the station platform withdraw heated undercar air.

For preliminary calculations, it may be assumed that (1) train heat release (due to braking and train air conditioning) within the station box is about two-thirds of the control volume heat load and (2) the underplatform exhaust system is about 50% effective.

A quantity of air equal to that withdrawn by the underplatform exhaust system enters the control volume from the outside. Thus, when station design temperature is below outside ambient, an underplatform exhaust system reduces the subway heat load by drawing off undercar heat but increases the heat load by drawing in outside air. A proposed technique to reduce the uncontrolled infiltration of outside air is to provide a complementary supply of outside air on the opposite side of the trackway underplatform exhaust ports. While in principle the underplatform exhaust system with complementary supply tends to reduce the mixing of outside air with air in public areas of the subway, test results on such systems are unavailable.

PARKING GARAGES

Automobile parking garages are either fully enclosed or partially open. Fully enclosed parking areas are usually underground and require mechanical ventilation. Partially open parking levels are generally above-grade structural decks having open sides (except for barricades), with a complete deck above. Natural ventilation, mechanical ventilation, or a combination of both can be used for partially open garages.

The operation of automobiles presents two concerns. The most serious is the emission of carbon monoxide, with its known risks. The second concern is the presence of oil and gasoline fumes, which may cause nausea and headaches as well as present a fire hazard. Additional concerns regarding oxides of nitrogen and smoke haze from diesel engines may also require consideration. However, the ventilation required to dilute carbon monoxide to acceptable levels will also control the other contaminants satisfactorily.

Two factors are required to determine the ventilation quantity: the number of cars in operation and the emission quantities. Most codes simplify this determination by requiring four to six air changes per hour, or 4 L/s per square metre, for fully enclosed parking garages. For partially open parking garages, 2.5 to 5% of the floor area is required as a free opening to permit natural ventilation. Applicable codes and National Fire Protection Association (NFPA) standards should be consulted for specific requirements.

Number of Cars in Operation

The number of cars in operation depends on the type of facility served by the parking garage. For a distributed, continuous use such as an apartment house or shopping area, the variation is generally from 3 to 5% of the total vehicle capacity. It could reach 15 to 20% for peak use, such as in a sports stadium or short-haul airport.

The length of time that a car remains in operation within a parking garage is a function of the size and layout of the garage and the number of cars attempting to enter or exit at a given time. This time could vary from 60 to 600 s, but on the average it ranges from 60 to 180 s.

A survey should be conducted of existing parking garages serving facilities having use and physical characteristics similar to those of the proposed design. Data on car entry and exiting rate by hour and time in operation within the parking garage can be recorded.

Contaminant Level Criteria

It is recommended that the ventilation rate be designed to maintain a CO level of 29 mg/m^3, with peak levels not to exceed 137 mg/m^3. The American Conference of Governmental Industrial Hygienists (ACGIH 1994) recommends a threshold limit of 29 mg/m^3 for an 8-h exposure, and the EPA has determined that at or near sea level, exposure to a CO concentration of 40 mg/m^3 for up to 1 h would be safe. For installations above 1000 m, far more stringent limits would be required.

Design Approach. The operation of a car engine within a parking garage differs considerably from normal vehicle operation, including normal operation within a road tunnel. On entry, the car travels slowly. As the car proceeds from the garage, the engine is cold and thus at full choke; it operates with a rich mixture and in low gear. Emissions for a cold start are considerably higher, so the distinction between hot and cold emission plays a critical role in determining the ventilation rate. Motor vehicle emission factors for hot and cold start operation are presented in Table 3. An accurate analysis requires correlation of CO readings with the survey data on car movements (Hama et al. 1974). Table 4 lists approximate data for vehicle movements. These data should be adjusted to suit the specific physical configuration of the facility.

Table 3 Predicted CO Emissions Within Parking Garages

Season	Hot Emissions (Stabilized), mg/s		Cold Emissions, mg/s	
	1991	1996	1991	1996
Summer (32°C)	42.3	31.5	71.2	61.0
Winter (0°C)	60.2	56.3	345.7	316.0

Results from EPA Mobile 3, version NYC-2.2; sea level location.
Note: Assumed vehicle speed is 8 km/h.

Table 4 Average Entrance and Exit Times for Vehicles

Level	Average Entrance Time, s	Average Exit Time, s
1	35	45
3[a]	40	50
5	70	100

Source: Stankunas et al. (1980).
[a]Average pass-through time = 30 s.

Access tunnels or fully enclosed ramps should be designed in accordance with the recommendations for road tunnels. When natural ventilation is used, the wall opening free area should be as large as possible. A portion of the free area should be at floor level. For parking levels with large interior floor areas, a central emergency smoke exhaust system should be considered for smoke removal or fume removal during calm weather.

The ventilation system, whether mechanical, natural, or both, should be designed to meet applicable codes and maintain an acceptable contaminant level. To conserve energy, fan systems should be controlled by CO meters to vary the amount of air supplied, if permitted by local codes. For example, fan systems could consist of multiple fans with single- or variable-speed motors or variable-pitch blades. In multilevel parking garages or single-level structures of extensive area, independent fan systems, each under individual control, are preferred.

Systems can be classified as supply-only, exhaust-only, or combined. Regardless of which system is chosen, the following should be considered: (1) the contaminant level of the outside air drawn in for ventilation; (2) avoiding short circuiting of supply air; (3) avoiding long flow fields that permit the contaminant levels to build up above an acceptable level at the end of the flow field; (4) providing short flow fields in areas of high pollutant emission, thereby limiting the extent of mixing; (5) providing an efficient, adequate flow throughout the parking structure; and (6) stratification of engine exhaust gases.

Noise. In general, parking garage ventilation systems move large quantities of air through large openings without extensive ductwork. These conditions, in addition to the highly reverberant nature of the space, contribute to high noise levels. For this reason, sound attenuation should be considered. This is a safety concern as well, since high noise levels may mask the sound of an approaching car.

Ambient Standards and Pollution Control. Some state and municipal authorities have developed ambient air quality standards. The exhaust system discharge should meet these requirements.

BUS GARAGES

Bus garages generally include a maintenance and repair area; a service lane where buses are fueled and cleaned; a storage area where buses are parked; and support areas such as offices, a stock room, a lunch room, and locker rooms. The location and layout of these spaces can depend on such factors as climate, bus fleet size, and type of fuel used by the buses. Servicing and storage functions may be located outdoors in temperate regions but are often found indoors in colder climates. While large fleets cannot always be stored indoors, maintenance areas may double as storage space for small fleets. Local building and/or fire codes may prohibit the dispensing of certain types of fuel indoors.

In general, areas where buses are maintained, serviced, or stored should be kept under slightly negative pressure to aid in the removal of fumes and contaminants. Ventilation should be accomplished using 100% outside air with no recirculation. Therefore, the use of heat recovery devices should be considered in colder climates. Direct exhaust of tailpipe emissions is recommended in maintenance and repair areas. Offices and similar support areas should be kept under positive pressure to prevent the infiltration of vehicle emissions.

Maintenance and Repair Areas

ASHRAE *Standard* 62 and most model codes require a minimum outdoor air ventilation rate of 7.5 L/s per square metre in vehicle repair garages, with no recirculation recommended. However, because the interior ceiling height may vary greatly from garage to garage, the designer should consider making a volumetric analysis of contaminant generation and air exchange rates rather than using the above value as a blanket standard. See the section on Bus Terminals for information on diesel engine emissions and the ventilation rates necessary to control their concentrations in areas where buses are operated.

Maintenance and repair areas often include below-grade inspection and repair pits for working underneath buses. Because the vapors produced by conventional bus fuels are heavier than air, they tend to settle in these pit areas, so a separate exhaust system should be provided to prevent their accumulation. NFPA *Standard* 88B recommends a minimum of 12 complete air changes per hour in pit areas and the location of exhaust registers near the pit floor.

Fixed repair stations, such as inspection and repair pits or hydraulic lifts, should include a direct exhaust system for tailpipe emissions. Such systems have a flexible hose and coupling that is attached to the bus tailpipe; emissions are discharged to the outdoors through an exhaust fan. The system may be of the overhead reel type, overhead tube type, or underfloor duct type, depending on the location of the tailpipe. For heavy diesel engines, a minimum exhaust rate of 285 L/s per station is recommended to capture emissions without creating excessive back pressure in the vehicle. Fans, ductwork, and hoses should be able to receive vehicle exhaust at temperatures exceeding 260°C without degradation.

Garages often include areas for battery charging, a process that can produce potentially explosive concentrations of corrosive, toxic gases. There are no published code requirements for the ventilation of battery-charging areas, but DuCharme (1991) has suggested using a combination of floor and ceiling exhaust registers to remove gaseous by-products. Recommended exhaust rates are 11.5 L/s per square metre of room area at floor level to remove acid vapors and 3.75 L/s per square metre of room area at ceiling level to remove hydrogen gases. Supply air volume should be 10 to 20% less than exhaust air volume and directed to provide 0.5 m/s terminal velocity at floor level. If the battery-charging space is located in the general maintenance area rather than in a dedicated room, an exhaust hood should be provided to capture gaseous by-products. Chapter 26 contains specific information on exhaust hood design. Makeup air should be provided to replace that removed by the exhaust hood.

Garages may also contain spray booths or rooms for painting buses. Most model codes reference NFPA *Standard* 33 for spray booth construction requirements, and that document should be consulted when designing heating and ventilating systems for such areas.

Servicing Areas

For indoor service lanes, ASHRAE *Standard* 62 and several model codes specify a minimum of 7.5 L/s per square metre of ventilation air, while NFPA *Standard* 30A requires only 5.0 L/s per square metre. The designer should determine which minimum requirement is applicable for the project location; additional ventilation may be necessary due to the nature of servicing operations. Buses often queue up in the service lane with their engines running during the service cycle. Depending on the length of the queue and the time to service each bus, the contaminant levels may exceed maximum allowable concentrations. A volumetric analysis examining contaminant generation should thus be considered.

Because of the increased potential for concentrations of flammable or combustible vapor, service lane HVAC systems should not be interconnected with systems serving other parts of the garage. Service lane systems should be interlocked with the dispensing equipment to prevent operation of the latter if the former is shut off or fails. Exhaust inlets should be located at ceiling level and between 80 and 300 mm above the finished floor, with supply and exhaust diffusers/registers arranged to provide air movement across all planes of the dispensing area.

Another feature common to modern service lanes is the cyclone cleaning system, which is used to vacuum out the interior of a bus. These devices have a dynamic connection to the front door(s) of the bus, through which a large-volume fan vacuums dirt and debris from inside the bus. A large cyclone assembly then removes the dirt and debris from the airstream and deposits it into a large hopper for disposal. Because of the large volume of air involved in the cyclone cleaning process, the designer should be consulted in the selection of the discharge and makeup air systems required to complete the cycle. Consideration should be given to recirculation or energy recovery, especially in the winter months. To aid in contaminant and heat removal during the summer months, some systems currently in use discharge the cyclone system air to the outdoors and provide untempered makeup air through relief hoods above the service lane.

Storage Areas

Where buses are stored indoors, the minimum ventilation standard of 7.5 L/s per square metre should be provided, subject to volumetric considerations. The designer should also consider the increased contaminant levels present during peak traffic periods. An example is the morning pull-out, when the majority of the bus fleet is dispatched for rush-hour commute. It is common practice to start and idle a large number of buses during this period to warm up the engines and check for defects. As a result, the concentration of emissions in the storage area rises, and additional ventilation may be required to maintain contaminant levels within acceptable limits. The use of supplemental purge fans is a common solution to this problem. These fans can be (1) interlocked with a timing device to operate during peak traffic periods, (2) started manually on an as-needed basis, or (3) connected to an air quality monitoring system that activates them when contaminant levels exceed some preset limit.

Design Considerations and Equipment Selection

Most model codes require that open flame heating equipment such as unit heaters be located at least 2.5 m above the finished floor or, where located in active trafficways, 0.6 m above the tallest vehicle. Fuel-burning equipment located outside of the garage area, such as boilers in a mechanical room, should be installed with the combustion chamber at least 45 cm above the floor. Combustion air should be taken from outside the building. Exhaust fans should be of the nonsparking type, with the motor located outside the airstream.

Infrared heating systems and air curtains are often considered for repair garages because of the size of the building and the high amount of infiltration through the large overhead doors needed to move buses in and out of the garage. However, caution must be exercised in applying infrared heating in areas where buses are parked or stored for extended periods of time; the buses may absorb the majority of the heat, which is then lost when they leave the garage. This is especially true during the morning pull-out. Infrared heating systems are more successfully applied in the service lane or at fixed repair positions. Air curtains should be considered for high-traffic doorways to limit heat loss and infiltration of cold air.

Where air quality monitoring systems are considered for controlling ventilation equipment, maintainability is a key factor in determining the success of the application. The high concentration of particulate matter in bus emissions can adversely affect the performance of the monitoring equipment, which often has filtering media at sampling ports to protect the sensors and instrumentation. Location of sampling ports, effects of emissions fouling, and calibration requirements should be considered when selecting the monitoring equipment to control the ventilation systems and air quality of the garage. Nitrogen dioxide (NO_2) and CO exposure limits published by OSHA and the EPA should be consulted to determine the contaminant levels at which exhaust fans should be activated.

Effects of Alternative Fuel Use

Because of recent federal legislation limiting contaminant concentrations in diesel bus engine emissions, the transit industry has begun testing buses operating on so-called alternative fuels. Such fuels include methanol, ethanol, compressed natural gas (CNG), and liquefied petroleum gas (LPG). Because these fuels have different flammability, emissions, and vapor dispersion characteristics from the conventional fuels for which current code requirements and design standards were developed, the requirements may not be valid for garage facilities in which alternative-fuel vehicles are maintained, serviced, and stored. The designer should consult the latest available literature regarding the design of HVAC systems for these facilities rather than relying on conventional practice. Information can be obtained from the Alternative Fuels Data Center at the U.S. Department of Energy in Washington, DC.

BUS TERMINALS

Bus terminals vary considerably in physical configuration. Most terminals consist of a fully enclosed space containing passenger waiting areas, ticket counters, and some food vending service. Buses load and unload outside the building, generally under a canopy for weather protection. In larger cities, where space is at a premium and bus service is extensive, comprehensive customer services and multiple levels with attendant busway tunnels and/or ramps may be required.

Waiting rooms and consumer spaces should have a controlled environment in accordance with normal practice for public terminal occupancy. The space should be pressurized against intrusion of the busway environment. Waiting rooms and passenger concourse areas are subject to a highly variable people load. The occupant density may reach 1 m² per person; during periods of extreme congestion, 0.3 to 0.5 m² per person.

Basically, there are two types of bus service—urban-suburban and long distance. Urban-suburban service is characterized by frequent bus movements and the requirement of rapid loading and

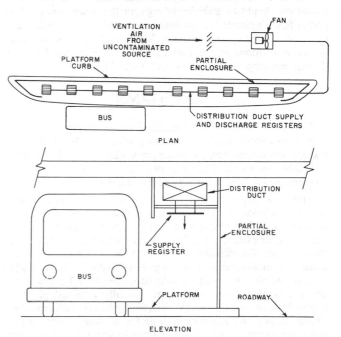

Fig. 12 Partially Enclosed Platform, Drive-Through Type

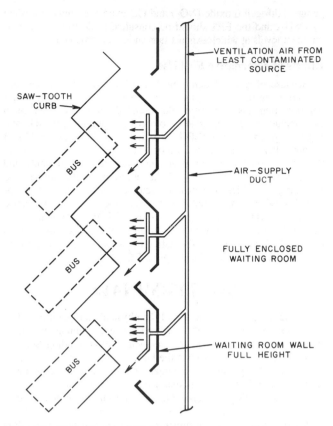

Fig. 13   Fully Enclosed Waiting Room with Sawtooth Gates

unloading. Therefore, ideal passenger platforms are long, narrow, and the drive-through type, which requires no bus backup movements on departure. Long-distance operations and those with greater headways generally use sawtooth gate configurations.

Ventilation systems to serve bus operating levels can be either natural or forced ventilation. When natural ventilation is selected, the bus levels should be open on all sides, and the slab-to-ceiling dimension should be sufficiently high, or the space contoured, to permit free air circulation. Jet fans improve the natural airflow at a relatively low energy requirement. Mechanical systems that ventilate open platforms or gate positions should be configured to serve the bus operating areas, as shown in Figures 12 and 13.

PLATFORMS

Naturally ventilated drive-through platforms may expose passengers to inclement weather and strong winds. Enclosed platforms (except for an open front) with the appropriate mechanical systems should be considered. Even partially enclosed platforms may trap contaminants and require mechanical ventilation.

Multilevel bus terminals have limited headroom, thus restricting natural ventilation. For such terminals, mechanical ventilation should be selected, and all platforms should be partially or fully enclosed. Ventilation supplied to partially enclosed platforms should minimize the induction of contaminated air from the busway. Figure 12 shows a partially enclosed drive-through platform with an air distribution system. Supply air velocity should be limited to 1.3 m/s to avoid drafty conditions on the platform. Partially enclosed platforms require large amounts of outside air to hinder fume penetration; experience indicates that a minimum of 85 L/s per square metre of platform area is required during rush hours and about half of this quantity during the remaining time.

Platform air quality remains essentially the same as that of the ventilation air introduced. Because of the piston effect, however, some momentary higher concentrations of pollutants will occur on the platform. Separate ventilation systems with two-speed fans for each platform permit operational flexibility. Fans should be controlled automatically to conform to bus operating schedules. In northern areas, it may be possible to reduce mechanical ventilation during extreme winter weather.

Fully enclosed platforms are strongly recommended for large terminals with heavy bus traffic. They can be pressurized adequately and ventilated with the normal heating and cooling air quantities, depending on the tightness of construction and the number of boarding doors and other openings. Conventional air distribution can be used; air should not be recirculated. Openings around doors and in the enclosure walls are usually adequate to relieve air unless platform construction is extraordinarily tight.

BUS OPERATION AREAS

Most buses are powered by diesel engines. Certain models have small auxiliary gasoline engines to drive the air-conditioning system. Tests performed on the volume and composition of exhaust gases emitted from diesel engines in various traffic conditions indicate large variations depending on temperature and humidity; manufacturer, size, and adjustment of the engine; and the fuel burned.

Contaminants

The components of diesel exhaust gases that affect the ventilation system design are oxides of nitrogen, hydrocarbons, formaldehyde, odor constituents, aldehydes, smoke particulates, and a relatively small amount of carbon monoxide. Operation of diesel engines in enclosed spaces causes visibility obstruction and odors, as well as contaminants. Table 5 provides approximate data on the major health-threatening contaminants found in diesel engine exhaust gas. The nature of bus engines should be determined for each project, however.

The U.S. Code of Federal Regulations (CFR) Occupational Safety and Health Standards, Subpart Z, sets the contaminant levels for an 8-h time-weighted average (TWA) exposure as follows: carbon monoxide (CO), 55 mg/m^3; nitric oxide (NO), 30 mg/m^3; and formaldehyde (HCHO), 0.92 mg/m^3. Subpart Z also sets the short-term exposure ceiling as 9 mg/m^3 for nitrogen dioxide (NO$_2$) and 2.5 mg/m^3 (ppm) for formaldehyde.

ACGIH (1994) recommends threshold limit values (TLVs) for an 8-h TWA exposure of 29 mg/m^3 for CO, 39 mg/m^3 for NO, and 5.6 mg/m^3 for NO$_2$. It also recommends short-term exposure limits (STELs) of 9.4 mg/m^3 for NO$_2$ and 0.37 mg/m^3 for formaldehyde.

Oxides of nitrogen occur in two basic forms: NO$_2$ and NO. Nitrogen dioxide is the major contaminant to be considered in the design of a bus terminal ventilation system. Exposure to concentrations of 19 mg/m^3 and higher causes health problems. Furthermore, NO$_2$ affects light transmission, causing visibility reduction. It is intensely colored and absorbs light over the entire visible spectrum, especially at shorter wavelengths. Odor perception is immediate at 0.79 mg/m^3 and can be perceived by some at levels as low as 0.23 mg/m^3.

Terminal operation also affects the quality of surrounding ambient air. The dilution rate and the location and design of the intakes and discharges control the impact of the terminal on ambient air quality. State and local regulations, which require consideration of local atmospheric conditions and ambient contaminant levels, must be followed.

Calculation of Ventilation Rate

To calculate the ventilation rate, the total amount of engine exhaust gases should be determined using the bus operating schedule, the amount of time that buses are in different modes of operation (i.e., cruising, decelerating, idling, and accelerating) and Table 5.

Table 5 Approximate Diesel Bus Engine Emissions (ppm)

	Idling 55 L/s	Accelerating 225 L/s	Cruising 163 L/s	Decelerating 143 L/s
Carbon monoxide	215	500	230	130
Hydrocarbons	390	210	90	330
Oxides of nitrogen (NO_x)	60	850	235	30
Formaldehydes (HCHO)	9	17	11	30

Note: For additional information on diesel bus and truck engine emissions, see Watson et al. (1988).

Further use of Table 5 permits computation of the contaminant level. The design engineer must ascertain the grade (if any) within the terminal and whether the platforms are drive-through, drive-through with bypass lanes, or sawtooth. Bus headways, operating speeds, and operating modes must also be evaluated.

For instance, with sawtooth platforms, the departing bus must accelerate backward, brake, and then accelerate forward. The drive-through platform requires a different mode of operation. Certain codes prescribe a maximum idling time for engines, usually 180 to 300 s. However, it should be recognized that 60 to 120 s of engine operation are required to build up brake air pressure.

The discharged contaminant quantities should be diluted by natural and/or forced ventilation to acceptable, legally prescribed levels. To maintain odor control and visibility, the exhaust gas contaminants should be diluted with outside air in the proportion of 75 to 1. Where urban-suburban operations are involved, the ventilation rate will vary considerably throughout the day and between weekdays and weekends. Control of fan speed or blade pitch should be used to conserve energy.

Source of Ventilation Air

Because dilution is the primary means of contaminant level control, the source of ventilation air is extremely important. The cleanest available ambient air, which in an urban area is generally above the roof, should be used. Surveys of ambient air contaminant levels should be conducted and the most favorable source located. The possibility of short circuiting of exhaust air due to prevailing winds and building airflow patterns should be evaluated.

Control by Contaminant Level Monitoring

Time clocks or tapes coordinated with bus movement schedules and smoke monitors (obscurity meters) provide the most practical means of controlling the ventilation system.

Instrumentation is available for monitoring various contaminants. Control by instrumentation can be simplified by monitoring carbon dioxide (CO_2) levels, as studies have shown a relationship between the levels of various diesel engine pollutants and CO_2. However, the mix and quantity of pollutants varies with the rate of operation and the condition of the bus engines. Therefore, if CO_2 monitoring is employed, actual conditions under specific bus traffic conditions should be determined in order to verify the selected CO_2 settings.

Dispatcher's Booth

The dispatcher's booth should be kept under positive pressure to prevent the intrusion of engine fumes. Because the booth is occupied for sustained periods, normal interior comfort conditions and OSHA contaminant levels must be maintained.

EQUIPMENT

The ability of an enclosed vehicular facility to function depends mostly on the effectiveness and reliability of its ventilation system. The system must be completely effective under the most adverse environmental and traffic conditions and during periods when not all equipment is operational. For a tunnel, it is necessary to provide more than one dependable source of power to prevent an interruption of service.

FANS

The fan manufacturer should be prequalified and should be responsible under one contract for furnishing and installing the fans, bearings, drives (including variable-speed components), motors, vibration devices, sound attenuators, discharge and inlet dampers, and limit switches.

The prime considerations in selecting the type and number of fans include the total theoretical ventilating air capacity required and a reasonable safety factor. Selection is also influenced by the manner in which reserve ventilation capacity is provided when a fan is inoperative and during repair of equipment or of the power supply.

Selection of fans (number and size) to meet normal and reserve capacity requirements is based on the principles of parallel fan operation. Actual capacities can be determined by plotting fan performance and system curves on the same pressure-volume diagram.

It is important that fans selected for parallel operation operate in the region of their performance curves where transferral of capacity back and forth between fans does not occur. This is accomplished by selecting a fan size and speed such that the duty point, no matter how many fans are operating, falls well below the unstable range of fan performance. Fans operating in parallel on the same system should be of equal size and have identical performance curves. For axial flow fans, blades should be at the same pitch or stagger angle; if flow is regulated through dampers, dampers should be at the same angle or setting; if flow is regulated through rotational speed control, all fans should operate at the same speed (i.e., if one fan is operating at low speed, any other operating fan must be at the same low speed).

Number and Size of Fans

At locations where no space limitations or other restrictions are placed on the structures that house the ventilation equipment, the number and size of fans should be selected by comparing several alternative fan arrangements. Comparison should be based on the practicality and overall economy of the layouts, including an estimate of (1) annual power cost for operation; (2) annual capital cost of the equipment (usually capitalized over an assumed equipment life of 30 years for rapid transit tunnel fans and 50 years for road tunnel fans); and (3) annual capital cost of the structure required to house equipment (usually capitalized over an arbitrary structure life of 50 years).

There are two opposing views on the proper number and size of fans. The first advocates a few large-capacity fans; the second prefers numerous small units. In most cases, a compromise arrangement produces the greatest operating efficiency. Regardless of the design philosophy, the number and size of fans should be selected to build sufficient flexibility into the system to meet the varying air demands created by daily and seasonal traffic fluctuations.

Type of Fan

Normally, the ventilation system of a vehicular facility requires large air volumes at relatively low pressure. Some fan designs have low efficiencies under these conditions, so the choice of suitable fan type is often limited to either centrifugal or vaneaxial.

Special Considerations. For rapid transit systems, any fan operating in the presence of the flow and pressure transients caused by train passage needs special consideration. If the transient tends to increase flow to the fan (i.e., the positive flow in front of a train to an exhaust fan or negative flow behind the train to a supply fan), it is important that blade loading does not become high enough to produce long-term fatigue failures.

If the disturbance tends to decrease flow to the fan (i.e., the negative flow behind the train to an exhaust fan or the positive flow in front of the train to a supply fan), the fan performance characteristic must have adequate margin to prevent aerodynamic stall.

The ability to rapidly reverse the rotation of tunnel supply, exhaust, and emergency fans is usually important in an emergency. This requirement must be considered in the selection and design of the fan and drive system.

Fan Design and Operation

Fans and components (e.g., blade-positioning mechanisms, drives, bearings, motors, controls, etc.) that are required to operate in the exhaust airstream during a fire or smoke emergency should be capable of operating at maximum speed under the following conditions:

- For rapid transit tunnels, at temperatures as specified in NFPA *Standard* 130
- For road tunnels, at maximum expected temperatures (based on computer simulations or other calculations)

Fans and dampers that are operated infrequently or for emergency service only should be operated at least once a month for 30 min to ensure that all rotating elements are working and lubricated. The fans do not have to be operated at high speed.

Inlet boxes can be used to protect the bearings and drives of centrifugal fans from the high temperatures and corrosive gases and particles present in the exhaust air during the emergency operating mode.

Axial flow fans that must be reversible should be designed so that they can, when operating in one direction at maximum design speed and capacity, reverse direction and reach maximum design speed and capacity in the opposite direction within a 35-s interval.

All components of reversible fans should be designed to withstand under actual installed conditions at least four reversible cycles per year for 50 years without damage or overstressing of the components.

Housing for variable-pitch axial flow fans should be furnished with flow-measuring instruments designed to measure the airflow in both directions. Capped connections should be provided for measuring the pressure developed across the fan. The fan should be protected from operating in a stall region.

To minimize blade failure in axial flow fans, the following is recommended:

- Blades should be secured to the hub with positive locking devices.
- The fan inlet and discharge should be protected to prevent objects that could damage the blades from coming in contact with the rotor assembly.
- Fan blades should be designed and tested to withstand 4 000 000 cycles of cyclical reverse bending due to flow reversal.
- To ensure that the fan blades will not be overstressed when the fan is placed in service, strain at the maximum stress point in the hub area of each blade and at the natural frequency point of each blade should be measured under simulated flow conditions.

Fan Shafts. Fan shafts should be designed so that the maximum deflection of the assembled fan components, including the forces associated with the fan drive, does not exceed 0.4 mm per metre of shaft length between centers of bearings. For centrifugal fans where the shaft overhangs the bearing, the maximum deflection at the centerline of the fan drive pulley should not exceed 0.4 mm per metre of shaft length between the center of the bearing and the center of the fan drive pulley.

It is good practice to design fan shafts so that the fundamental bending mode frequency of the assembled shaft, wheel, or rotor is more than 50% higher than the highest fan operating speed. The first

resonant speed of all rotational components should be at least 125% above the maximum operating speed. The fan assembly should be designed to withstand for at least 3 min. all stresses and loads resulting from an overspeed test at 110% of the maximum design fan operating speed.

Bearings. Fan and motor bearings should have a minimum equivalent L_{10} rated life, as defined by the Anti-Friction Bearing Manufacturers Association, of 20 000 h. Improper tension (overtension) on the belts can drastically reduce the life of the bearings that support the drive and possibly the life of the belts and shafts as well.

Axial flow fans. Each fan motor bearing and fan bearing should be equipped with a monitoring system that senses individual bearing vibrations and temperatures and provides a warning alarm if either rises above the normal range.

Centrifugal fans. Due to their low speed (generally less than 450 rpm), centrifugal fans are not always provided with bearing vibration sensors, but they do require temperature sensors with warning alarm and automatic fan shutdown. Bearing pedestals for centrifugal fans should provide rigid support for the bearing with negligible impediment to the airflow. The static and dynamic loading of the shaft and impeller and the maximum force due to the tension in the belts should be taken into account.

Corrosion-Resistant Materials. Choosing a particular material or coating to protect a fan from corrosive gas is a matter of economics. The selection of the material and/or coating should be based on the site environment, the fan duty, and an expected service life of 50 years.

Sound. For the normal ventilation mode, the construction documents should specify the following:

- The speed and direction of airflow and number of fans operating
- The maximum DBA rating or NC curve or curves acceptable under installed conditions and the locations at the fan inlet and exhaust stack outlet where measurements are to be taken
- That the DBA rating measured at a height of 1.5 m anywhere in the fan room may not exceed OSHA or local requirements
- That the DBA rating measured at locations such as intake louvers, discharge louvers, or discharge stacks may not exceed OSHA or local requirements
- That if the measured sound values exceed the specified maximum values, the fan manufacturer must furnish and install the acoustic treatment needed to bring the sound level to an acceptable value

DAMPERS

Shutoff dampers can be installed in multiple-fan systems (1) to isolate any parallel nonoperating fan from those operating to prevent short circuiting and consequent pressure and flow loss through the inoperative fan; (2) to prevent serious windmilling of an inoperative fan; and (3) to provide a safe environment for maintenance and repair work on each fan. Single-fan installations could have an isolating damper to prevent serious windmilling due to natural or piston-effect drafts and to facilitate fan maintenance.

Two types of dampers have generally been used in ventilation systems: (1) the trapdoor type, which is installed in a vertical duct so that the door lies flat when closed; and (2) the multiple-blade louver type with parallel operating blades. Both types are usually driven by a reversible gear drive motor, which is operated by the fan controller, which closes the damper when the fan is off and reopens it when the fan is on.

The trapdoor damper is simple and works satisfactorily where a vertical duct enters a plenum-type fan room through an opening in the floor. The damper, usually constructed of steel plate, reinforced with welded angle iron, and hinged on one side, closes by gravity against the embedded angle frame of the opening. The opening mechanism is usually a shaft sprocket-and-chain device. The drive motor and gear drive must develop sufficient force to open the damper door against the maximum static pressure difference the fan can develop. This pressure can be obtained from fan performance curves. Limit switches start and stop the gear-motor drive at the proper position.

Dampers placed in other than vertical ducts should be of the multiple-blade louver type. These dampers usually consist of a rugged channel frame, the flanges of which are bolted to the flanges of the fan, duct, or duct opening. Louver blades are mounted on shafts that turn in bearings mounted on the outside of the channel frame. This arrangement requires access space on the outside of the duct for bearing and shaft lubrication and maintenance and space for operating linkages. Louver dampers should have edge and end seals to make them airtight.

The trapdoor-type damper, if properly fabricated, is inherently airtight due to its weight and the overlap at its edges. However, louver dampers must be carefully constructed to ensure tightness on closing. The pressure drop across a fully opened damper and the leakage rate across a fully closed louver damper should be verified by the appropriate test procedure in AMCA *Standard* 500. A damper that leaks under pressure will cause the fan to rotate counter to its power rotation, thus making restarting dangerous and possibly damaging the drive motor.

Louver dampers and damper operators should be selected carefully because their function is independent of the performance of their respective fans after installation. Damper motors located in the exhaust airstream should be selected to operate under the same maximum temperature and time conditions as their respective fans and may have to be explosionproof. Damper bearings should be selected on the same criteria as those of their respective fans. The corrosion-resistant characteristics of a damper bearing should be determined by the environment in which it will operate.

SUPPLY AIR INTAKE

Supply air intakes require careful design to ensure that the quality of air drawn into the system is the best available. Such factors as recirculation of exhaust air or intake of contaminants from nearby sources should be considered.

Louvers or grillework are usually installed over air intakes for aesthetic, security, or safety reasons. Bird screens are also important if openings between louver blades or grillework are large enough to allow birds to enter.

Due to the large quantities of air required in some ventilation systems, it may not be possible for intake louvers to have sufficiently low face velocities to be weatherproof. Therefore, intake plenums, shafts, fan rooms, and fan housings need water drains. Blowing snow can also fill the fan room or plenum with snowdrifts, but this usually does not stop the ventilation system from operating satisfactorily if additional floor drains are located in the area of the louvers.

Noise elimination devices may have to be installed in fresh air intakes to keep fan and air noise from disturbing the outside environment. If sound reduction is required, the total system—fan, fan plenum, building, fan housing, and air intake (location and size)—should be investigated. Fan selection should also be based on a total system, including the pressure drop that results from sound attenuation devices.

EXHAUST OUTLETS

The discharge of exhaust air should be remote from street level and from areas with human occupancy. Contaminant concentrations in the exhaust air are not of concern if the system is working effectively. However, odors and entrained particulate matter make this air undesirable in occupied areas. Exhaust stack discharge velocity should be high enough to disperse contaminants into the atmosphere. A minimum of 10 m/s is usually necessary.

In the past, evasé outlets were used to regain some static pressure and thereby reduce the energy consumption of the exhaust fan.

Unless the fan discharge velocity is in excess of 2 m/s, however, the energy savings may not offset the cost of the evasé.

In a vertical or near-vertical exhaust fan discharge connection to an exhaust duct or shaft, rainwater will run down the inside of the stack into the fan. This water will dissolve material deposited from vehicle exhausts on the inner surface of the stack and become extremely corrosive. Therefore, fan housing should be made of a corrosion-resistant material or specially coated to protect the metal from corrosion.

Discharge louvers and gratings should be sized and located so that their discharge will not be objectionable to pedestrians or contaminate intake air louvers; at the same time, the air resistance across the louver or grating should be minimized. Discharge velocities through sidewalk gratings are usually limited to 2.5 m/s. Bird screens should be provided if the exhaust air is not continuous (24 h, 7 days a week) and the openings between the louver blades are large enough to allow birds to enter.

The corrosion-resistant characteristics of the louver or grating should be determined by the corrosiveness of the exhaust air and the installation environment. The pressure drop across the louvers should be verified by the appropriate test procedure in AMCA *Standard* 500.

REFERENCES

ACGIH. 1992. Industrial ventilation: A manual of recommended practice, 21st ed. American Conference of Governmental Industrial Hygienists, Cincinnati, OH.

ACGIH. 1994. Threshold limit values for chemical substances and physical agents and biological exposure indices 1994-1995. American Conference of Governmental Industrial Hygienists, Cincinnati, OH.

AMCA. 1989. Test methods for louvers, dampers and shutters. *Standard* 500. Air Movement and Control Association, Arlington Heights, IL.

ASHRAE. 1985. Laboratory methods of testing fans for rating. *Standard* 51-1985 (AMCA *Standard* 210-85).

ASHRAE. 1989. Ventilation requirements for acceptable indoor air quality. *Standard* 62-1989.

Code of Federal Regulations. 29 CFR 1910. Occupational Safety and Health Standards, Subpart Z.

DOT. 1976. *Subway environmental design handbook.* Urban Mass Transportation Administration, U.S. Government Printing Office.

DuCharme, G.N. 1991. Ventilation for battery charging. *Heating/Piping/Air-Conditioning* (February).

EPA. Average emission factors for highway vehicles for selected calendar years. *Supplement No. 7 for compilation of air pollutant emission factors,* 3rd ed., Table I-3. Environmental Protection Agency, Washington, D.C.

EPA. 1993. User's guide to Mobile 5A (Mobile source emissions factor model). Ann Arbor, MI.

Fieldner, A.C. et al. 1921. Ventilation of vehicular tunnels. Report of the U.S. Bureau of Mines to New York State Bridge and Tunnel Commission and New Jersey Interstate Bridge and Tunnel Commission. American Society of Heating and Ventilating Engineers (ASHVE).

Hama, G.M., W.G. Frederick, and H.G. Monteith. 1974. How to design ventilation systems for underground garages. *Air Engineering.* Study by the Detroit Bureau of Industrial Hygiene, Detroit (April).

NFPA. 1989. Standard for spray application using flammable and combustible materials. *Standard* 33-89. National Fire Protection Association, Quincy, MA.

NFPA. 1991. Standard for repair garages. *Standard* 88B-91. National Fire Protection Association, Quincy, MA.

NFPA. 1992. Standard for compressed natural gas (CNG) vehicular fuel systems. *Standard* 52-92. National Fire Protection Association, Quincy, MA.

NFPA. 1993. Automotive and marine service station code. *Standard* 30A-93. National Fire Protection Association, Quincy, MA.

NFPA. 1993. Standard for fixed guideway transit systems. *Standard* 130-93. National Fire Protection Association, Quincy, MA.

NTIS. TUNVEN user's guide. *Publication* PB80141575. National Technical Information Service, Springfield, VA.

Singstad, O. 1929. *Ventilation of vehicular tunnels.* World Engineering Congress, Tokyo, Japan.

Stankunas, A.R., P.T. Bartlett, and K.C. Tower. 1980. Contaminant level control in parking garages. *ASHRAE Transactions* 86(2):584-605.

Watson, A.Y., R.R. Bates, and D. Kennedy. 1988. Air pollution, the automobile, and public health. Sponsored by the Health Effects Institute. National Academy Press, Washington, D.C.

BIBLIOGRAPHY

Ball, D. and J. Campbell. 1973. Lighting, heating and ventilation in multi-story and underground car park. Paper presented at the Institution for Structural Engineers and The Institution of Highway Engineers' Joint Conference on Multi-story and Underground Car Parks (May).

Federal Register. 1974. 39(125), June.

Ricker, E.R. 1948. *The traffic design of parking garages.* ENO Foundation for Highway Traffic Control, Saugatuck, CT.

Round, F.G. and H.W. Pearall. Diesel exhaust odor: Its evaluation and relation to exhaust gas composition. Research Laboratories, General Motors Corporation, *Society of Automotive Engineers Technical Progress Series* 6.

Turk, A. 1963. Measurements of odorous vapors in test chambers: Theoretical. *ASHRAE Journal* 5(10):55-58.

Wendell, R.E., J.E. Norco, and K.G. Croke. 1973. Emission prediction and control strategy: Evaluation and pollution from transportation systems. *Air Pollution Control Association Journal* (February).

LABORATORY SYSTEMS

MODERN laboratories require the regulation of temperature, humidity, relative static pressure, air motion, air cleanliness, sound, and exhaust. This chapter addresses biological, chemical, animal, and physical laboratories. Within these generic descriptions, some laboratories have their own unique requirements. This chapter provides an overview of the heating, ventilating, and air conditioning characteristics and design criteria applicable to laboratories and provides a brief overview of architectural and utility concerns. This chapter does not cover scale-up process laboratories, commonly called pilot plants, which are small manufacturing units.

The function of the laboratory is important in deciding the heating, ventilating, and air-conditioning system selection and design. Air-handling, hydronic, control, life safety, and heating and cooling systems must function as a unit and not as independent systems. HVAC systems must conform to applicable safety and environmental regulations.

Providing a safe environment for all personnel is a primary objective in the design of HVAC systems for laboratories. A vast amount of information relating to heating, ventilating, and air conditioning of laboratories is available, and HVAC engineers must study the subject thoroughly to understand all the factors that relate to proper and optimum design. This chapter serves only as an introduction to the topic of laboratory HVAC design.

HVAC systems must integrate with architectural planning and design, electrical systems, structural systems, other utility systems, and functional requirements of the laboratory. The HVAC engineer, then, is a member of a team that includes other designers of the facility, users, industrial hygienists, safety officers, operators, and the maintenance staff. Decisions or recommendations by the HVAC engineer may have significant impact on costs of construction, operations, and maintenance.

Laboratories frequently use 100% outside air, which broadens the operating range of conditions to which the systems must respond. They seldom operate at maximum design conditions, so the HVAC engineer must pay particular attention to partial load operations that are continually changing due to internal space load requirements, external conditions, and day-night variances.

Most laboratories will be modified at some time. As a consequence, the HVAC engineer must consider to what extent laboratory systems are adaptable for other needs. Both economics and integration of the systems with the rest of the facility must be considered.

The preparation of this chapter is assigned to TC 9.10, Laboratory Systems.

LABORATORY TYPES

Laboratories can be divided into the following four generic types:

- *Biological laboratories* include any kind of laboratory in which biologically active materials or the chemical manipulation of these materials are found. This would include laboratories that support scientific disciplines such as biochemistry, microbiology, cell biology, biotechnology, immunology, botany, pharmacology, and toxicology. Both chemical fume hoods and biological safety cabinets are commonly installed in these areas.
- *Chemical laboratories* include both organic and inorganic synthesis and analytical functions. They may also include laboratories in the material and electronic sciences. Chemical laboratories commonly include the installation of a number of fume hoods.
- *Animal laboratories* include areas for manipulation, surgical modification, and pharmacological observation of laboratory animals. They also includes animal holding rooms, which are similar to laboratories in many of the performance requirements, but they have an additional subset of requirements.
- *Physical laboratories* include spaces associated with physics and commonly consist of facilities associated with lasers, optics, nuclear materials, high and low temperature materials, electronics, and analytical instruments.

HAZARD ASSESSMENT

Laboratory research potentially involves some hazard. Nearly all laboratories contain some type of hazardous materials. The assessment of this hazard must be completed before the laboratory can be designed. A comprehensive hazard assessment for the laboratory should be performed by the owner's designated *safety officers*. They include but are not limited to the chemical hygiene officer, radiation safety officer, biological safety officer, and fire and loss prevention official. This hazard assessment should be incorporated into the *chemical hygiene plan, radiation safety plan*, and *biological safety protocols*.

The facility should be designed in accordance with the appropriate safety plans. The nature of the contaminant, quantities present, types of operations, and degree of hazard will dictate the types of containment and local exhaust devices. For functional convenience, operations posing less hazard are conducted in devices that use directional air flows for personnel protection, e.g., laboratory fume hoods and biological safety cabinets; however, these devices do not provide absolute containment. Operations that have a significant hazard potential are conducted in devices that provide greater protection but are more restrictive, e.g., sealed glove boxes.

The design team should visit similar laboratories to assess successful design approaches and safe operating practices. Each laboratory is somewhat different. It's design must be evaluated using appropriate standards and practices rather than duplicating existing and possibly outmoded facilities.

DESIGN PARAMETERS

The following design parameters must be established for a laboratory space:

- Temperature and humidity, both indoor and outdoor
- Air quality from both process and safety perspectives, including air filtration and special treatment, which may include charcoal, HEPA, or other filtration of supply or exhaust air
- Equipment and process heat gains, both sensible and latent
- Minimum ventilation rates
- Equipment and process exhaust quantities
- Exhaust and air intake locations
- Style of the exhaust device, capture velocities, and usage factors
- Need for standby equipment and emergency power
- Alarm requirements
- Future increase in the size and number of fume hoods
- Anticipated increase in the electrical power consumption
- Room pressurization requirements

It is important to (1) review design parameters with the safety officers and scientific staff, (2) establish limitations that cannot be exceeded, and (3) establish the desirable operating conditions. For areas requiring variable temperature or humidity, these parameters must be carefully reviewed to establish a clear understanding of the expected operating conditions and system performance.

Because laboratory HVAC systems often incorporate 100% outside air systems, the selection of design parameters will have a substantial impact on capacity, first cost, and operating costs. The selection of proper and prudent design conditions is very important.

Internal Thermal Considerations

In addition to the heat gain from people and lighting, laboratories frequently have significant heat gain from equipment and processes that introduce both sensible and latent loads. Table 1 has a partial list of equipment that may be found in laboratories. Often data for equipment used in laboratories is not available or the equipment has been custom built by the scientific staff. Heat release from animals that may be housed in the space can be found in Chapter 9 of the 1993 *ASHRAE Handbook—Fundamentals* and in Alereza and Breen (1984).

Due to the variable use of equipment in laboratories, the heat gain to the space varies substantially; the simultaneous use of all equipment may seldom (if ever) occur. Due to highly variable equipment heat gains, individual laboratories should have dedicated temperature controls.

Careful review, detailed understanding of how the laboratory will be used, and prudent judgment are required to obtain good estimates of the heat gains in a laboratory. The HVAC system designer needs to evaluate equipment nameplate ratings, applicable load and use factors, and overall diversity. Heat released from equipment located within exhaust devices can be discounted. Heat from equipment that is directly vented or heat from water-cooled equipment should not be considered part of the heat released to the room. Any unconditioned makeup air that is not directly captured by an exhaust device must be included in the load calculation for the room. In many cases, additional equipment will be obtained by the time a laboratory facility has been designed and constructed. The design should allow for this additional equipment.

Table 1   Recommended Rate of Heat Gain from Hospital Equipment Located in the Air-Conditioned Area

Appliance Type	Size	Maximum Input Rating, Watts	Recommended Rate of Heat Gain, Watts[a]
Autoclave (bench)	0.02 m^3	1250	140
Bath, hot or cold circulating, small	3.7 to 36.7 litres, −30 to 100°C	750 to 1800	130 to 310 (sensible) 250 to 590 (latent)
Blood analyzer	120 samples/h	735	735
Blood analyzer with CRT screen	115 samples/h	1500	1500
Centrifuge (large)	8 to 24 places	1100	1050
Centrifuge (small)	4 to 12 places	150	140
Chromatograph	—	2000	2000
Cytometer (cell sorter/analyzer)	1000 cells/s	21 460	21 460
Electrophoresis power supply	—	400	250
Freezer, blood plasma, medium	0.37 m^3, down to −40°C	3530[b]	1410[b]
Hot plate, concentric ring	4 holes, 100°C	1100	870
Incubator, CO_2	0.14 to 0.28 m^3, up to 55°C	2830	1410
Incubator, forced draft	0.28 m^3, 27 to 60°C	720	360
Incubator, general application	0.04 to 0.31 m^3, up to 70°C	1660 to 2260[b]	850 to 1130[b]
Magnetic stirrer	—	600	600
Microcomputer	16 to 256 kbytes[c]	100 to 600	88 to 528
Minicomputer	—	2200 to 6000	2200 to 6600
Oven, general purpose, small	0.04 to 0.08 m^3, 240°C	21 900[b]	2970[b]
Refrigerator, laboratory	0.63 to 3.0 m^3, 4°C	880[d]	350[d]
Refrigerator, blood, small	0.20 to 0.56 m^3, 4°C	2680[b]	1060[b]
Spectrophotometer	—	500	500
Sterilizer, freestanding	0.11 m^3, 100 to 132°C	20 900	2370
Ultrasonic cleaner, small	0.04 m^3	120	120
Washer, glassware	0.22 m^3 load area	4460	2930
Water still	19 to 57 litres	1120[e]	25[e]

Source: Alereza and Breen (1984).
[a]For hospital equipment installed under a hood, the heat gain is assumed to be zero.
[b]Heat gain per cubic metre of interior space.
[c]Input is not proportional to memory size.
[d]Heat gain per 10 m^3 of interior space.
[e]Heat gain per litre of capacity.

Two cases encountered frequently are building programs based on generic laboratory modules and laboratory spaces that are to be highly flexible and adaptive. Both situations require the design team to establish heat gain on an area basis. The values for area-based heat gain vary substantially for different types of laboratories. Heat gains of 50 to 270 W/m^2 or more are common for laboratories with high concentrations of equipment.

Architectural Considerations

The integration of utility systems into the architectural planning, design, and detailing is key to providing highly successful research facilities. Both the architect and the HVAC system engineers must seek an early understanding of each other's requirements and develop integrated solutions. The following play key roles in the design of research facilities:

- Modular planning
- Development of laboratory units
- Provisions for adaptability and flexibility
- Early understanding of space requirements for shafts and ceiling spaces
- High quality envelope integrity
- Planned locations of air intakes and exhaust

Most laboratory programming and planning is based on developing a module that becomes the base building block for the floor plan. Laboratory planning modules are generally 3 to 3.6 m wide and 6 to 9 m deep. The laboratory modules may be developed as single work areas or combined to form multiple-station work areas. Utility systems should be arranged to reflect the architectural planning module, with services provided for each module or pair of modules, as appropriate.

NFPA *Standard* 45 requires that laboratory units be designated. Similarly, both the Uniform and BOCA model codes require the development of control areas. The development of laboratory units or control areas and the determination of the appropriate hazard levels should occur early in the design process. The HVAC designer should review the requirements for maintaining separations between laboratories and note requirements for exhaust ductwork to serve only a single laboratory unit.

Additionally, NFPA *Standard* 45 requires that no fire dampers be installed in laboratory exhaust ductwork. Building codes offer no relief from maintaining required floor to floor fire separations. The resulting combination of these requirements commonly require that dedicated fire-rated shafts be constructed from each occupied floor to the penthouse or building roof. These criteria and the proposed solutions should be reviewed early in the design process with the appropriate building code officials.

Research objectives frequently require change in laboratory operations and programs. To accommodate these changes, laboratories must be designed to be flexible and adaptable, without requiring significant modifications to the laboratory's infrastructure. For example, the utility system design can be flexible enough to supply ample cooling to support the addition of heat-producing analytical equipment, without requiring modifications to the HVAC system. Adaptable designs should allow for programmatic research changes that require modifications to the laboratory's infrastructure within the limits of the individual laboratory area and/or interstitial and utility corridors. For example, an adaptable design would allow the addition of a fume hood without requiring work outside of that laboratory work area. The degree of flexibility and adaptability that the laboratory HVAC system is designed for is determined from discussion with the researchers, laboratory programmer, and laboratory planner. The HVAC designer should have a clear understanding of these requirements.

The amount of utility space and its location is significantly more important in the design of research facilities than in the design of most other buildings. The available ceiling space and the frequency of vertical distribution shafts are interdependent and can have a significant impact on the architectural planning. The HVAC designer must take an early lead in establishing these parameters, and the resultant design needs to reflect these constraints. The designer should review alternate utility distribution schemes, including their advantages and disadvantages.

The building envelope may need to be designed to accommodate relatively high levels of humidification and slightly negative building pressures without moisture condensation in the winter or excessive infiltration. Additionally, mechanical equipment rooms and the associated air intakes and exhaust stacks must be located to avoid intake of fumes into the building. Air intake locations must be chosen to avoid fumes from loading docks, cooling tower discharge, vehicular traffic, etc.

LABORATORY EXHAUST AND CONTAINMENT DEVICES

FUME HOODS

The Scientific Equipment and Furniture Association defines (SEFA 1992) a laboratory fume hood as "a ventilated enclosed work space intended to capture, contain, and exhaust fumes, vapors, and particulate matter generated inside the enclosure. It consists basically of side, back and top enclosure panels, a floor or counter top, an access opening called the face, a sash(es), and an exhaust plenum equipped with a baffle system for air flow distribution." Figure 1 illustrates the basic elements of a general purpose benchtop fume hood.

Fume hoods may be equipped with a variety of accessories, including internal lights, service outlets, sinks, air bypass openings, airfoil entry devices, flow alarms, special linings, ventilated base storage units, and exhaust filters. Undercounter cabinets for storage of flammable materials require special attention to ensure safe installation. NFPA *Standard* 30, Flammable and Combustible Liquids Code, does not recommend venting of these cabinets; however, ventilation is often required to avoid accumulation of toxic or hazardous vapors. Ventilation of these cabinets by a separately ducted

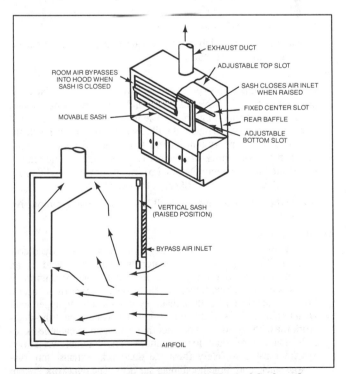

Fig. 1   Bypass Fume Hood with Vertical Sash and Bypass Air Inlet

supply and exhaust that will maintain the temperature rise of the cabinet interior within the limits defined by NFPA *Standard* 30 should be considered.

Types of Fume Hoods

The primary types of fume hoods and their application are as follows:

Standard (constant volume airflow with variable face velocity)
Hood that meets basic SEFA definition. Sash may be vertical, horizontal, or combination type.

Application: Research laboratories—frequent or continuous use. Moderate to highly hazardous processes; varying procedures.

Bypass (constant volume airflow with constant face velocity)
A standard vertical sash hood modified with openings above and below the sash. The openings are sized to minimize the change in the face velocity, which is generally to 3 or 4 times the full-open velocity, as the sash is lowered.

Application: Research laboratories—frequent or continuous use. Moderate to highly hazardous processes; varying procedures.

Variable Volume (constant face velocity)
Hood has an opening or bypass designed to provide a prescribed minimum air intake when the sash is closed and an exhaust system designed to vary the air flow in accordance with sash opening. Sash may be vertical, horizontal, or a combination of both.

Application: Research laboratories—frequent or continuous use. Moderate to highly hazardous processes; varying procedures.

Auxiliary Air (constant volume airflow with constant face velocity)
A plenum above the face receives air from a secondary air supply that provides partially conditioned or unconditioned outside air.

Application: Research laboratories—frequent or continuous use. Moderate to highly hazardous processes; varying procedures.

Note: Many organizations restrict the use of this type of hood.

Process (constant volume airflow with constant face velocity)
Standard hood without a sash. By some definitions, this is not a fume hood. Considered a ventilated enclosure.

Application: Process laboratories—intermittent use. Low hazard processes; known procedures.

Radioisotope

A standard hood with special integral work surface, linings impermeable to radioactive materials, and structure strong enough to support lead shielding bricks. The interior must be constructed to prevent radioactive material buildup and allow complete cleaning. The ductwork system should have flanged neoprene gasketed joints with quick disconnect fasteners that can be readily dismantled for decontamination. Provisions may need to be made for HEPA and or charcoal filters in the exhaust duct.

Application: Process and research laboratories using radioactive isotopes.

Perchloric Acid

A standard hood with special integral work surfaces, coved corners, and nonorganic lining materials. Perchloric acid is an extremely active oxidizing agent. Its vapors can form unstable deposits in the ductwork that present a potential explosion hazard. To alleviate this hazard, the exhaust system must be equipped with an internal water washdown and drainage system, and the ductwork must be constructed of smooth, impervious, cleanable materials that are resistant to acid attack. The internal washdown system must completely flush the ductwork, exhaust fan, discharge stack, and fume hood inner surfaces. The ductwork system should be kept as short as possible with minimum elbows. Perchloric acid exhaust systems with longer ductwork runs may need

the washdown system zoned to avoid water flow rates in excess of the ability to drain the water from the hood. Because perchloric acid is an extremely active oxidizing agent, organic materials should not be used in the exhaust system in places such as joints and gaskets. Ductwork should be constructed of a stainless steel material, with a chromium and nickel content not less than that of 316 stainless steel, or of a suitable nonmetallic material. Joints should be welded and ground smooth. A perchloric acid exhaust system should only be used for work involving perchloric acid.

Application: Process and research laboratories using perchloric acid. Mandatory use because of explosion hazard.

California

A special hood with sash openings on multiple sides (usually horizontal).

Application: For enclosing large and complex research apparatus that require access from two or more sides.

Walk-In

A standard hood with sash openings to the floor. Sash can be either horizontal or vertical.

Application: For enclosing large or complex research apparatus. Not for personnel to enter while chemical operations are in progress.

Distillation

A standard fume hood with extra depth and 1/3- to 1/2-height benches.

Application: Research laboratory. For enclosing tall distillation apparatus.

Canopy

An open hood with an overhead capture structure.

Application: Not a fume hood. Useful for heat or water vapor removal from some work areas. Not to be substituted for a fume hood. Not recommended when workers must bend over the source of heat or water vapor.

Fume Hood Sash Configurations

The work opening has operable glass sash(es) for observation and shielding. A sash may be vertically operable, horizontally operable, or a combination of vertically and horizontally operable. A vertically operable sash can incorporate single or multiple vertical panels. A horizontal operable sash incorporates multiple panels that slide in multiple tracks, allowing the open area to be positioned across the face of the hood. The combination of horizontally operable sash mounted within a single vertically operable sash section allows the entire hood face to be opened for setup. Then the opening area can be limited by closing the vertical panel, with only the horizontally sliding sash sections used during experimentation. Either multiple vertical sash sections or the combination sash arrangement allow the use of larger fume hoods with limited opening areas, resulting in reduced exhaust airflow requirements. Fume hoods with vertically rising sash sections should include closures around the sash to prevent the bypass of ceiling plenum air into the fume hood.

Fume Hood Performance

Containment of hazards in a fume hood is based on the principle that a flow of air entering at the face of the fume hood, passing through the enclosure, and exiting at the exhaust port prevents the escape of airborne contaminants from the hood into the room.

The following affect the performance of the fume hood:

- Face velocity
- Size of the face opening
- Shape and configuration of the opening surfaces
- Inside dimensions and location of work area relative to the face area
- Size and number of exhaust ports

- Back baffle and exhaust plenum arrangement
- Bypass arrangement, if applicable
- Auxiliary air supplies, if applicable

The following also affect the effectiveness of the fume hood:

- Arrangement and type of replacement supply air outlets
- Air velocities near the hood
- Openings to spaces outside the laboratory, including distance from doors into the laboratory
- Pedestrian traffic past the hood face
- Movements of the researcher within the hood opening
- Location, size, and type of research apparatus placed in the hood
- Distance from the research apparatus to the worker's breathing zone

Air currents external to the fume hood can jeopardize the fume hood's effectiveness and expose the researcher to materials used in the hood. Detrimental air currents can be produced by the following:

- Air supply distribution patterns in the laboratory
- Movements of the researcher
- People walking past the fume hood
- Thermal convection
- Opening of doors and windows

Caplan and Knutson (1977, 1978) conducted tests to determine the interactions between room air motion and fume hood capture velocities with respect to the spillage of contaminants into the room. Test conclusions indicated that the effect of room air currents is significant and of the same order of magnitude as the effect of the hood face velocity. Consequently, improper design and/or installation of the replacement air supply negatively affects the performance of the fume hood.

Terminal air supply at the face of the hood should be no more than one half and, preferably, one fifth the face velocity of the hood. This becomes an especially critical factor in designs that use low face velocities. As an example, a fume hood with a face velocity of 0.5 m/s could tolerate a maximum disturbance velocity of 0.25 m/s. If the design face velocity is 0.3 m/s, the maximum disturbance velocity would be reduced to 0.15 m/s.

Air currents are also generated by traffic past the fume hood. To the extent possible, fume hoods should be located so that traffic flow past the hood is minimal. Additionally, the fume hood should be placed to avoid potential air currents generated from the opening of windows and doors. To assure the optimum placement of the fume hoods, the HVAC system designer must take an active role early in the design process.

ANSI/AIHA *Standard* Z 9.5 discourages the use of auxiliary air fume hoods. These hoods incorporate an air supply at the fume hood to reduce the amount of room air required to be exhausted. The following difficulties and installation criteria are associated with auxiliary air fume hoods:

- The auxiliary air supply must be introduced outside the fume hood to maintain appropriate velocities past the researcher
- The flow pattern of the auxiliary air must not degrade the containment performance of the fume hood
- Auxiliary air must be conditioned to avoid blowing cold air on the researcher; often the air must be cooled to maintain the required temperature and humidity within the hood
- Auxiliary air may introduce additional heating and cooling loads in the laboratory
- Only vertical sash may be used in the hood
- Control system for the exhaust, auxiliary, and supply airstreams must be coordinated
- Additional coordination of utilities during installation is required to avoid spatial conflicts caused by the additional duct system
- Humidity cannot be controlled when auxiliary air is used

Components of the fume hood that affect its performance include entrance conditions, corner and intermediate posts, deep deck lip projections, baffle and slot adjustments, and service fittings near the face of the fume hood. Plain entrance edges, on either of the sides, the bottom, or on intermediate posts can produce a vena contracta within 25 mm of the surface and to a depth of 150 mm. Fumes generated in this area are disturbed and may escape the hood enclosure. Airfoil shapes at the entry edges correct this condition. Correcting this defect on existing hoods may make satisfactory hoods from previously unacceptable units. Deep deck lip depressions also produce similar vena contracta.

Flow from most fume hoods is controlled by horizontal slots in the back baffle. A slot at the bottom of the back baffle draws air across the working surface, another slot at the top exhausts the canopy, and a third slot is frequently located midway on the baffle. These adjustable openings regulate the flow distribution for specific applications. The openings should be set and locked by the commissioning engineer.

Sinks and service fittings should be located at least 150 mm behind the face of the fume hood to avoid detrimental disturbances of the airflow pattern.

Fume Hood Performance Criteria. ASHRAE *Standard* 110, Method of Testing Performance of Laboratory Fume Hoods, describes a quantitative method of determining the containment performance of a fume hood. The method requires the use of a tracer gas and instrumentation to measure the amount of tracer gas that enters the breathing zone of a mannequin, which simulates the containment capability of the fume hood as a researcher conducts operations in the hood.

The following tests are commonly used to judge the performance of the fume hood: (1) face velocity test, (2) flow visualization test, (3) smoke tests, and (4) tracer gas tests. These tests should be performed under the following conditions:

- Fume hood sash in multiple positions
- Usual amount or research equipment in the hood; the room air balance set
- Doors and windows in their normal positions
- Fume hood sash set in varying positions to simulate both static and dynamic performance

The following descriptions partially summarize the test procedures. ASHRAE *Standard* 110 provides specific requirements and procedures.

Face Velocity Test

The desired face velocity should be determined by the facility safety officer and researcher. The velocity is a balance between safe operation of the fume hood, airflow needed for the hood operation, and energy costs. Face velocity measurements are taken on a vertical/horizontal grid, with each measurement point representing not more than 0.1 m^2. The measurements should be taken with a device that is accurate in the intended operating range, and an instrument holder should be used to improve accuracy. Computerized multipoint grid measurement devices provide the greatest accuracy.

Flow Visualization

1. Swab a strip of titanium tetrachloride along both walls and the hood deck in a line parallel to the hood face and 150 mm back into the hood. *Caution:* Titanium tetrachloride is corrosive to the skin and extremely irritating to the eyes and respiratory system.
2. Swab a 200 mm circle on the back of the hood. Define air movement toward the face of the hood as reverse airflow and define the lack of movement as dead airspace.
3. Swab the work surface of the hood, being sure to swab lines around all equipment in the hood. All smoke should be carried to the back of the hood and out.
4. Test the operation of the deck airfoil bypass by running the cotton swab under the airfoil.
5. Before going to the next test, move the cotton swab around the face of the hood; if there is any outfall, the exhaust capacity test should not be made.

Large Volume Flow Visualization

1. Ignite and place a smoke generator near the center of the work surface 150 mm inside the rear of the sash. Some smoke sources generate a jet of smoke that produces an unacceptably high challenge to the hood. Care is required to ensure that the generator does not disrupt the hood performance, leading to erroneous conclusions.
2. After the smoke bomb is ignited, pick it up with tongs and move it around the hood. The smoke should not be seen or smelled outside the hood. Appropriate measures should be taken prior to undertaking a smoke test to avoid accidental activation of the building's smoke detection system.

Tracer Gas Test

1. Place the sulfur hexafluoride gas ejector in the required test locations (i.e., the center and near each side). Similarly position a mannequin with a detector in its breathing zone in the corresponding location at the hood.
2. Release the tracer gas and record measurements over a 5 min time span.
3. After testing with the mannequin is complete, remove it, traverse the hood opening with the detector probe, and record the highest measurement.

All fume hoods should be tested annually and their performance certified.

BIOLOGICAL SAFETY CABINETS

A biological safety cabinet protects the researcher and, in some configurations, protects the research materials as well. Biological safety cabinets are sometimes called safety cabinets, ventilated safety cabinets, laminar flow cabinets, and glove boxes. Biological safety cabinets are categorized into six groups (several are shown in Figure 2):

Class I   Similar to chemical fume hood, no research material protection, 100% exhaust through a HEPA filter

Class II

   Type A   70% recirculation within the cabinet, 30% exhaust through a HEPA filter, common plenum configuration, can be recirculated into the laboratory

   Type B1   30% recirculation within the cabinet, 70% exhaust through a HEPA filter, separate plenum configuration, must be exhausted to the outside

   Type B2   100% exhaust through a HEPA filter to the outside

   Type B3   70% recirculation within the cabinet, 30% exhaust through a HEPA filter, common plenum configuration, must be exhausted to the outside

Class III   Special applications, 100% exhaust through a HEPA filter to the outside, researcher manipulates material within cabinet through physical barriers (gloves)

Several key decisions must be made by the researcher prior to the selection of a biological safety cabinet (Eagleston 1984). An important difference in biological safety cabinets is their ability to handle chemical vapors properly (Stuart et al. 1983). Of special concern to the HVAC engineer are the proper placement of the biological safety cabinet in the laboratory and the room's air distribution. Rake (1978) concluded the following:

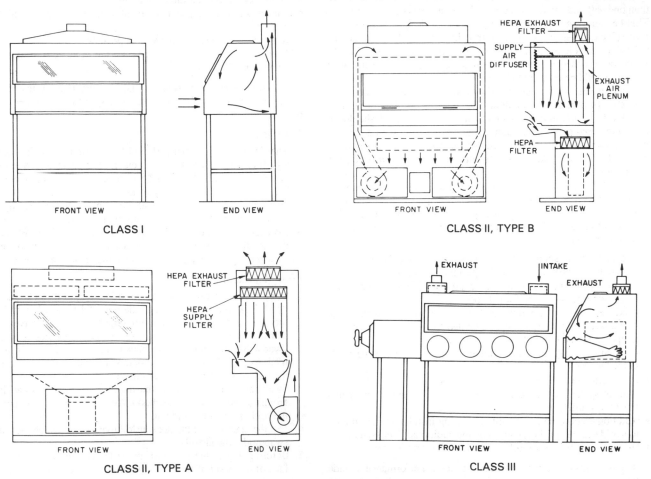

Fig. 2 Types of Biological Safety Cabinets

"A general rule of thumb should be that, if the crossdraft or other disruptive room airflow exceeds the velocity of the air curtain at the unit's face, then problems do exist. Unfortunately, in most laboratories such disruptive room airflows are present to various extent. Drafts from open windows and doors are the most hazardous sources because they can be far in excess of 1 m/s and accompanied by substantial turbulence. Heating and air-conditioning vents perhaps pose the greatest threat to the safety cabinet because they are much less obvious and therefore seldom considered.... It is imperative then that all room airflow sources and patterns be considered before laboratory installation of a safety cabinet."

Class II biological safety cabinets should only be placed in the laboratory in compliance with the National Sanitation Foundation (NSF) *Standard* 49, Class II (Laminar Flow) Biohazard Cabinetry. Assistance in procuring, testing, and evaluating performance parameters of Class II biological safety cabinets is available from the NSF as part of the standard. The cabinets should be located away from drafts, active walkways, and doors. The air distribution system should be designed to avoid air patterns that impinge on the biological safety cabinet.

The different biological safety cabinets have varying static pressure resistance requirements. Generally, Class II Type A cabinets have pressure drops ranging between 1 and 25 Pa. Class II Type B1 cabinets have pressure drops in the range of 150 to 300 Pa, and Class II Type B1 cabinets have pressure drops ranging from 380 to 580 Pa. The manufacturer must be consulted to verify specific requirements.

Additionally, the pressure requirements will vary based on filter loadings and the intermittent operation of individual biological safety cabinets. Exhaust systems for biological safety cabinets must be designed with these considerations in mind. Also, care must be exercised when manifolding biological safety cabinet exhausts to ensure that the varying pressure requirements are met.

The manufacturer of the biological safety cabinet may be able to supply the transition from the biological safety cabinet to the duct system. The transition should include an access port for testing and balancing and an air tight damper for decontamination. As with any containment ductwork, high integrity duct fabrication and joining systems are necessary. The responsible safety officer should be consulted to determine the need for and placement of isolation dampers to facilitate decontamination operations.

Class I Cabinets

The Class I cabinet is a partial containment device designed for research operations with low and moderate risk etiologic agents. It does not provide protection for the materials used in the cabinet. Room air flows through a fixed opening and prevents aerosols, which may be generated within the cabinet enclosure, from escaping into the room. The air exhausted through the cabinet may be HEPA filtered prior to being discharged into the exhaust system. The use of the optional HEPA filter depends on the cabinet usage. The fixed opening through which the researcher works is usually 200 mm high. To provide adequate personnel protection, the air velocity through the fixed opening is usually at least 0.38 m/s.

The Class I cabinet can be modified to contain chemical carcinogens by adding an appropriate exhaust air treatment system and increasing the velocity through the opening to 0.5 m/s. Specific OSHA criteria should be consulted. Large pieces of research equipment can be placed in the cabinet, if adequate shielding is provided.

The Class I cabinet is not appropriate for containing systems that are vulnerable to airborne contamination, because the air flowing into the cabinet is untreated. Also, the Class I cabinet is not recommended for use with highly infectious agents, because an interruption of the inward airflow may allow aerosolized particles to escape.

Class II Cabinets

The Class II cabinet provides protection to personnel, product, and the environment. The cabinet features an open front with inward airflow and HEPA-filtered recirculated and exhaust air.

The Class II Type A cabinet has a fixed opening with a minimum inward airflow velocity of 0.38 m/s. The average minimum downward velocity of the internal airflow is 0.38 m/s. The Class II Type A cabinet is suitable for use with agents meeting Biosafety Level 2 criteria (National Safety Council 1979), and, if properly certified, can meet Biosafety Level 3. However, because approximately 70% of the airflow is recirculated, the cabinet is not suitable for use with flammable, toxic or radioactive agents.

The Class II Type B1 cabinet has a vertical sliding sash and maintains an inward airflow of 0.5 m/s at a sash opening of 200 mm. The average downward velocity of the internal airflow is 0.5 m/s. The Class II Type B cabinet is suitable for use with agents meeting Biosafety Level 3. Approximately 70% of the internal airflow is exhausted through HEPA filters and thus allows the use of biological agents treated with limited quantities of toxic chemicals and trace amounts of radionuclides—provided the work is performed in the direct exhaust area of the cabinet.

The Class II Type B2 cabinet maintains an inward airflow velocity of 0.5 m/s through the work opening. The cabinet is 100% exhausted through HEPA filters to the outdoors; therefore, all downward velocity air is drawn from the laboratory or other supply source and is HEPA filtered prior to being introduced into the work space. The Class II Type B2 cabinet may be used for the same level of work as the Class II Type B1 cabinet. In addition, the design permits use of toxic chemicals and radionuclides in microbiological studies.

The Class II Type B3 cabinet maintains an inward airflow velocity of 0.5 m/s and is similar in performance to the Class II Type A cabinet.

In Class II Type A and Type B3 cabinets, exhaust air delivered to the outlet of the cabinet by internal blowers must be handled by the laboratory exhaust system. This arrangement requires a delicate balance between the cabinet and the laboratory's exhaust system, and it may incorporate a thimble-type connection between the cabinet and the laboratory exhaust ductwork. Thimble (or canopy) connections incorporate an air gap between the biosafety cabinet and the exhaust duct. The exhaust system must pull more air than exhausted by the biological safety cabinet to make air flow in through the gap. The designer should confirm the amount of air to be drawn through the air gap. A minimum flow is required to provide the specified level of containment and a maximum flow cannot be exceeded without causing an imbalance through aspiration.

The Class II Type B1 and Type B2 cabinets require the building exhaust system to pull the air from the cabinet's work space and through the exhaust HEPA filters. The pressure resistance that must be overcome by the building exhaust system can be obtained from the cabinet manufacturer. Because containment in this type cabinet depends on the building's exhaust system, the exhaust fan(s) should have redundant backups.

Class III Cabinets

The Class III cabinet is a gastight, negative pressure containment system that physically separates the agent from the worker. These cabinets provide the highest degree of personnel protection. Work is performed through arm length rubber gloves attached to a sealed front panel. Room air is drawn into the cabinet through HEPA filters. Particulate materials entrained in the exhaust air are removed by a HEPA filtration or by incineration before discharge to the atmosphere. A Class III system may be designed to enclose and isolate incubators, refrigerators, freezers, centrifuges, and other research equipment. Double-door autoclaves, liquid disinfectant dunk tanks, and pass boxes are used to transfer materials into and out of the cabinet.

Class III systems contain highly infectious materials and radioactive contaminants. Although there are operational inconveniences with these cabinets, they are the equipment of choice when a high degree of personnel protection is required. It should be noted that explosions have occurred in Class III cabinets used for research involving volatile substances.

MISCELLANEOUS EXHAUST DEVICES

Snorkels are used in laboratories to remove heat or nontoxic particulates that may be generated from benchtop research equipment or apparatus. Snorkels usually have funnel-shaped inlet cones connected to 75 to 150 mm diameter flexible or semiflexible ductwork extending from the ceiling to above the benchtop level.

Typically, *canopy hoods* are used to remove heat or moisture that may be generated by a specific piece of research apparatus (e.g., steam sterilizer) or process. Canopy hoods cannot contain hazardous fumes adequately to protect the researcher.

The *laboratory*, if maintained at negative relative static pressure, provides a second level of containment, protecting occupied spaces outside of the laboratory from operations and processes undertaken therein.

LAMINAR FLOW CLEAN BENCHES

Laminar flow clean benches are available in two configurations—horizontal (crossflow) and vertical (downflow). Both configurations filter the supply air and usually discharge the air out the front opening into the room. Clean benches protect the experiment or product but do not protect the researcher; therefore, they should not be used with any potentially hazardous or allergenic substances. Clean benches are not recommended for any work involving hazardous biological, chemical, or radionuclide materials.

LABORATORY VENTILATION SYSTEMS

The total airflow rate for a laboratory is dictated by either the total amount of exhaust from containment and exhaust devices, the cooling required to offset internal heat gains, or the amount to provide minimum ventilation rates. The fume hood requirements (including evaluation of alternate fume hood sash configurations as described in the section on Fume Hoods) must be determined in consultation with the safety officers. The HVAC engineer must determine the expected heat gains from the research equipment after consulting with the research staff. The section on Internal Thermal Considerations describes the criteria for determining the amount of internal heat gain from research equipment.

Minimum ventilation rates are generally in the range of 6 to 10 air changes per hour when occupied; however, some spaces (e.g., animal holding areas) have minimum ventilation rates established by specific standards or may have ventilation rates established by internal facility policies. The maximum ventilation rate for the laboratories should be reviewed to ensure that appropriate supply air delivery methods are chosen so supply airflows do not impede the performance of the exhaust devices.

Laboratory ventilation systems can be arranged for either constant or variable volume airflow. The specific type should be selected with the research staff and maintenance personnel. Special attention should be given to unique areas such as glass washing areas, hot and cold environmental rooms and labs, fermentation rooms, and cage washing rooms. Emergency power systems to operate the laboratory ventilation equipment should be provided based on hazard assessment. Additional criteria for selection are described in the sections on Hazard Assessment and Operation and Maintenance.

In many laboratories, all hoods and safety cabinets are seldom needed at the same time. A system usage factor represents the maximum number of exhaust devices in operation at one time. The system usage factor depends on several factors including the type and size of the facility, the total number of fume hoods and the number of fume hoods per researcher, the type of fume hood controls, the type of laboratory ventilation systems, the quantity of devices required to operate continuously due to chemical storage requirements or contamination prevention, and the number of current plus projected research programs.

Usage factors should be applied carefully when sizing equipment. As mentioned previously, the usage factor will vary based on the size of the facility and specific research operations. For example, teaching laboratories may have a usage factor of 100%. If too low a usage factor is selected, design airflow and containment performance cannot be maintained. It is usually expensive and disruptive to add capacity to an operating laboratory's supply or exhaust system. Detailed discussions with research staff are required to ascertain maximum usage rates of exhaust devices.

Noise in the laboratory should be considered at the beginning of the design. By considering noise criteria as a requirement of the design, it is possible to achieve reasonable levels suitable for scientific work. For example, at the National Institute of Health (NIH), sound levels of 45 NC (including fume hoods) are required in regularly occupied laboratories. The requirement is relaxed to 55 NC for instrument rooms. Noise criteria not addressed as part of the design can result in NC levels of 65 or greater, which are unacceptable to most occupants. Additionally, sound generated by the building HVAC equipment should be evaluated to assure that excessive levels do not escape to the outdoors. Remedial correction of excessive sound levels can be difficult and expensive.

SUPPLY SYSTEMS

Supply air systems for laboratories provide the following:

- Thermal comfort for the occupants
- Minimum ventilation rates
- Replacement for air exhausted through fume hoods, biological safety cabinets, or other exhaust devices
- Space pressurization control
- Environmental control to meet process or experimentation criteria

The design parameters must be well defined to select, lay out, and size the supply air system. Installation and setup should be verified as part of a commissioning process. Design parameters are covered in the section on Design Parameters, and commissioning is covered in the section on Commissioning.

Laboratories in which chemicals and compressed gases are used generally require nonrecirculating air supply systems or 100% outside air supplies. The selection of 100% outside air supply systems versus return air systems should be made as part of the hazard assessment process, which is discussed in the section on Hazard Assessment. The use of 100% outside air requires a system with a very wide range of heating and cooling capacity, which requires special design and control.

Supply air systems for laboratories include both constant and variable volume systems that incorporate either single-duct reheat or dual-duct configurations, with distribution through low, medium, or high pressure ductwork.

Filtration

The filtration for the air supply depends on the requirements for the laboratory. Conventional chemistry and physics laboratories commonly use 85% dust spot efficient filters (ASHRAE *Standard* 52.1). Biological and biomedical laboratories usually require 85 to 95% dust spot efficient filtration. High-efficiency particulate air (HEPA) filters should be provided for special spaces where research materials or animals are particularly susceptible to contamination from external sources. HEPA filtration of the supply air is necessary in such applications as environmental studies, specific pathogen-free research animals, nude mice, dust-sensitive work, and electronic assemblies. In many instances, biological safety cabinets or laminar flow clean benches (which are HEPA filtered), rather than HEPA filtration for the entire room, may be used.

Air Distribution

Air supplied to a laboratory must keep temperature gradients and air currents to minimum. Air outlets (preferably nonaspirating diffusers) must not discharge into the face of a fume hood. Acceptable room air velocities are covered in the sections on Fume Hoods and Biological Safety Cabinets.

EXHAUST SYSTEMS

Laboratory exhaust systems remove air from containment devices and from the laboratory itself. The exhaust system must be controlled and coordinated with the supply air system to maintain the correct pressurization of laboratories.

Design parameters must be well defined to select, lay out, and size the exhaust air system. Additional information on the control of exhaust systems is included in the section on Controls. Also, installation and setup should be verified as part of a commissioning process. See the sections on Design Parameters and Commissioning.

Laboratory exhaust systems should be designed for high reliability and for ease of maintenance. One way to achieve this design is to provide redundant exhaust fans and a means of sectionalizing equipment so that an individual exhaust fan may be maintained while the system is operating. To the extent possible, components of exhaust systems should allow maintenance without exposing maintenance personnel to the exhaust airstream. Access to filters and the need for bag-in, bag-out filter housings need to be considered during the design process.

Types of Exhaust Systems

Both constant and variable volume (or high-low volume) systems that incorporate low, medium, or high pressure ductwork are used for laboratory exhaust systems. Either individual exhaust fans for each fume hood or manifolded fume hoods connected to central exhaust fans are installed. Maintenance, functional requirements, and safety must be considered when selecting an exhaust system. Part of the hazard assessment analysis determines the appropriate use of variable volume systems and the need for individually ducted exhaust systems. Laboratories with a high degree of hazard may not be good candidates for variable volume systems. In addition, fume hoods or other devices in which extremely hazardous or radioactive materials are used should receive special review to determine if they should be connected to a manifolded exhaust system.

All exhaust devices installed in a laboratory are seldom run simultaneously at full capacity. This allows the HVAC engineer to conserve energy and, potentially, to reduce equipment capacities by installing variable volume systems that include an overall system usage factor. The selection of an appropriate usage factor is discussed in the section on Laboratory Ventilation Systems.

Manifolded exhaust systems can be categorized as pressure-dependent and pressure-independent. Pressure-dependent systems are constant volume only and incorporate manually adjusted balancing dampers for each exhaust device. If an additional fume hood is added to a pressure-dependent exhaust system, the entire system must be rebalanced, and the speed of the exhaust fans may need to be readjusted. Due to the greater flexibility of pressure-independent systems, pressure-dependent systems are not common in current designs.

A pressure-independent system can be constant, variable, or a mix of constant and variable volume. It incorporates pressure-independent volume regulators with each exhaust device. The system offers two advantages: (1) the flexibility of being able to add exhaust devices without having to rebalance the entire system and (2) variable volume control.

The volume regulators can incorporate either direct measurement of the exhaust airflow rate or positioning of a calibrated pressure-independent air valve. The input to the volume regulator can be either (1) a manual or timed switch to index the fume hood airflow from minimum to operational airflow, (2) sash position sensors, or (3) fume hood cabinet pressure sensors. The section on Controls covers this topic in greater detail. Running many exhaust devices into the manifold of a common exhaust system offers the following benefits:

- Lower ductwork costs
- Fewer pieces of equipment to operate and maintain
- Fewer roof penetrations and exhaust stacks
- Opportunity for energy recovery
- Centralized locations for exhaust discharge
- Ability to take advantage of exhaust system diversity
- Ability to provide a redundant exhaust system by adding one spare fan per manifold

Individually ducted exhaust systems comprise a separate duct, exhaust fan, and discharge stack for each exhaust device or laboratory. The exhaust fan can be either single-speed, multiple-speed, or variable-speed and can be configured for constant, variable, or a combination of constant and variable volume. An individually ducted exhaust system has the following potential benefits:

- Provision for installation of special exhaust filtration or treatment systems
- Customized ductwork and exhaust fan corrosion control for specific applications
- Provision for selected emergency power backup
- Simpler initial balancing

Maintaining correct flow at each exhaust fan requires (1) periodic maintenance and balancing and (2) consideration of the flow rates with the fume hood sash in different positions. One problem encountered with individually ducted exhaust systems occurs when an exhaust fan is shut down. In this case, air is drawn, in reverse flow, through the exhaust ductwork into the laboratory because the laboratory is maintained at a negative pressure.

One challenge in designing independently ducted exhaust systems for multistory buildings is to provide extra vertical ductwork, extra space, and other provisions for the future installation of additional exhaust devices. In multiple story buildings, dedicated fire-rated shafts are required from each floor to the penthouse or roof level. As a result, individually ducted exhaust systems (or vertically manifolded systems) consume greater floor space than horizontally manifolded systems. However, individual systems may require less height between floors.

Ductwork Leakage

Ductwork systems should have low leakage rates and should be tested to confirm that the specified leakage rates have been obtained. Leaks from positive pressure exhaust ductwork can contaminate the building, so they must kept to a minimum. Designs that minimize the amount of positive pressure ductwork are desirable. All positive pressure ductwork should be of the highest possible integrity. The fan discharge should connect directly to the vertical discharge stack. Careful selection and proper installation of airtight flexible connectors at the exhaust fans are essential, including providing flexible connectors on the exhaust fan inlet only. Flexible connectors are not used on the discharge side of the exhaust fan because a connector failure could result in the leakage of hazardous fumes into the equipment room.

The leakage of the containment devices themselves must also be considered. For example, in vertical sash fume hoods, the clearance to allow sash movement creates an opening from the top of the fume hood into the ceiling space or area above. The air introduced through this leakage path also contributes to the exhaust airstream. The amount that these leakage sources contribute to the exhaust airflow depends on the fume hood design. Edge seals can be placed around sash tracks to minimize leaks. While the volume of air exhausted through a fume hood is based on the actual face opening, appropriate allowances for air introduced through paths other than the face opening must be included.

Materials and Construction

The selection of materials and the construction of exhaust ductwork and fans depend on the following:

- Nature of the effluents
- Ambient temperature
- Effluent temperature
- Length and arrangement of duct runs
- Constant or intermittent flow
- Flame spread and smoke developed ratings
- Duct velocities and pressures

Effluents may be classified generically as organic or inorganic chemical gases, vapors, fumes, or smoke; and qualitatively as acids, alkalis, solvents, or oils. Exhaust system ducts, fans, dampers, flow sensors, and coatings are subject to (1) corrosion, which destroys metal by chemical or electrochemical action; (2) dissolution, which dissolves materials such as coatings and plastics; and (3) melting, which can occur in certain plastics and coatings at elevated temperatures.

Common reagents used in laboratories include acids and bases. Common organic chemicals include acetone, ether, petroleum ether, chloroform, and acetic acid. Discussions should be held with the safety officer and scientists because the specific research being conducted defines the chemicals used and, therefore, the necessary ductwork material and construction.

The ambient temperature in the space housing the ductwork and fans affects the condensation of vapors in the exhaust system. Condensation contributes to the corrosion of metals, and the chemicals used in the laboratory may further accelerate corrosion.

Ductwork is less subject to corrosion when runs are short and direct, the flow is maintained at reasonable velocities, and condensation is avoided. Horizontal ductwork may be more susceptible to corrosion if condensate accumulates in the bottom of the duct. Applications with moist airstreams (cage washers, sterilizers, etc.) may require condensate drains that are connected to process sewers.

If flow through the ductwork is intermittent, condensate may remain for longer periods as it will not be able to reevaporate into the airstream. Moisture can also condense on the outside of ductwork exhausting cold environmental rooms.

Flame spread and smoke development ratings, which are specified by codes or insurance underwriters, must also be considered when selecting ductwork materials.

In determining the appropriate ductwork material and construction the HVAC engineer should

- Determine the types of effluents (and possibly combinations) handled by the exhaust system
- Classify effluents as either organic or inorganic and determine whether they occur in the gaseous, vapor, or liquid state
- Classify decontamination materials
- Determine the concentration of the reagents used and the temperature of the effluents at the hood exhaust port, which may be impossible in research laboratories
- Estimate the highest possible dew point of the effluent
- Determine the ambient temperature of the spaces where the exhaust system is installed
- Estimate the degree to which condensation may occur
- Determine if flow will be constant or intermittent (intermittent flow conditions may be improved by adding time delays to run the

exhaust system sufficiently long enough to dry the duct interior prior to shutdown)
- Determine whether insulation, watertight construction, or sloped and drained ductwork are required
- Select materials and construction most suited for the application

Considerations in selecting materials include resistance to chemical attack and corrosion, condensation, flame and smoke ratings, ease of installation, ease of repair or replacement, and maintenance costs.

Appropriate materials can be selected from standard references and by consulting with manufacturers of specific materials. Materials for chemical fume exhaust systems and their characteristics include the following:

Galvanized steel. Subject to acid and alkali attack, particularly at cut edges and under wet conditions; cannot be field welded without destroying galvanization; easily formed; low in cost.

Stainless steel. Subject to acid and chloride compound attack depending on the nickel and chromium content of the alloy; relatively high in cost. The most common stainless steel alloys used for laboratory exhaust systems are 304 and 316. Costs increase with the increasing chromium and nickel content.

Asphaltum-coated steel. Resistant to acids; subject to solvent and oil attack; high flame and smoke rating; base metal vulnerable when exposed by coating imperfections and cut edges; cannot be field welded without destroying galvanization; moderate cost.

Epoxy-coated steel. Epoxy phenolic resin coatings on mild black steel can be selected for particular characteristics and applications; these have been successfully applied for both specific and general use, but no one compound is inert or resistive to all effluents. Requires sand blasting to prepare the surface for a shop-applied coating, which should be specified as pinhole-free, and field touch-up of coating imperfections or damage caused by shipment and installation; cannot be field welded without destroying coating; cost is moderate.

Polyvinyl-coated galvanized steel. Subject to corrosion at cut edges; cannot be field welded; easily formed; moderate in cost.

Fiberglass. When additional glaze coats are used, this is particularly good for acid applications including hydrofluoric. May require special fire suppression provisions. Special attention to hanger types and spacing is needed to prevent damage.

Plastic materials. Have particular resistance to specific corrosive effluents; their physical strength is limited; flame spread and smoke development rating; heat distortion; and high cost of fabrication. Special attention to hanger types and spacing is needed to prevent damage.

Borosilicate glass. For specialized systems with high exposures to certain chemicals such as chlorine.

FIRE SAFETY FOR VENTILATION SYSTEMS

Most local authorities have laws that incorporate National Fire Protection Association (NFPA) *Standard* 45, Fire Protection for Laboratories Using Chemicals. Laboratories located in patient care buildings require fire standards based on the NFPA *Standard* 99, Health Care Facilities Handbook. Selected NFPA *Standard* 45 design criteria include the following:

Air balance. "Laboratory...shall be maintained at an air pressure that is negative to the corridors or adjacent nonlaboratory areas." (para. 6-4.2)

Controls. "Controls and dampers...shall be of a type that, in the event of failure, will fail in an open position to assure a positive draft." (para. 6-6.7)

Diffuser locations. "Care shall be exercised in the selection and placement of air supply diffusion devices to avoid air currents that would adversely affect performance of laboratory hoods..." (para. 6-4.3)

Exhaust stacks. "Exhaust stacks should extend at least 7 ft [2.1 m] above the roof." (para A-6-8.7)

Fire dampers. "Automatic fire dampers shall not be used in laboratory hood exhaust systems. Fire detection and alarm systems shall not be interlocked to automatically shut down laboratory hood exhaust fans..." (para. 6-11.3)

Hood alarms. "Airflow indicators shall be installed on new laboratory hoods or on existing laboratory hoods, when modified." (para. 6-9.7)

Hood placement. "A second means or access to an exit shall be provided from a laboratory work area if...a hood...is located adjacent to the primary means of exit access." (para. 3-3.2(d))

"For new installations, laboratory hoods shall not be located adjacent to a single means of access or high traffic areas." (para. 6-10.2)

Recirculation. "Air exhausted from laboratory hoods or other special local exhaust systems shall not be recirculated." (para. 6-5.1)

"Air exhausted from laboratory work areas shall not pass unducted through other areas." (para. 6-5.3)

The designer should review the entire NFPA *Standard* 45 and local building codes to determine applicable requirements. Then the designer should inform the other members of the design team of their responsibilities (such as proper fume hood placement). The incorrect placement of exhaust devices is a frequent design error and a common cause of costly redesign work.

CONTROLS

Laboratory control systems must suitably regulate the environment. Controls regulate temperature and humidity, control and monitor laboratory safety devices that protect personnel, and control and monitor secondary safety barriers used to protect the environment outside the laboratory from laboratory operations (West 1978). Reliability, redundancy, accuracy, and monitoring are important factors in controlling the lab environment. Many laboratories require the precise control of temperature, humidity, and airflows. In addition, components of the control system must provide the necessary accuracy and corrosion resistance if they are exposed to corrosive environments.

Controls for the laboratory should provide fail-safe operation, which should be defined jointly with the safety officer. A fault tree can be developed to evaluate the impact of any failure of a control system component and to ensure that safe conditions are maintained.

Thermal Control

The temperature in laboratories with a constant volume air supply is generally regulated with a thermostat that controls the position of a control valve on a reheat coil, thereby regulating the supply air temperature. Laboratories with a variable volume ventilation system, in addition to regulating the supply air temperature and volume, generally regulate room exhaust device(s) as well. The general room exhaust device(s) are modulated to provide greater airflow in the laboratory when additional cooling is needed. The exhaust device(s) potentially determine the total supply air quantity for that laboratory.

Most microprocessor-based laboratory control systems are able to use proportional-integral-derivative (PID) algorithms to eliminate the error between the measured temperature and the temperature set point. The use of anticipatory control strategies (Marsh 1988) increases accuracy in the regulation of laboratory temperatures. Anticipatory control strategies recognize the increased reheat requirements associated with changes in the ventilation flow rates and make adjustments in the position of reheat control valves prior to space temperature changes measured by the thermostat.

Room Pressure Control

In order for the laboratory to act as a secondary confinement barrier (see Chapter 25 for example of secondary confinement), it must be maintained at a slightly negative pressure with respect to adjoining areas to contain odors and fumes. Exceptions are sterile facilities or clean spaces that may need to be maintained at a positive pressure with respect to adjoining spaces.

The common methods of room pressure control include manual balancing, direct pressure, flow tracking, and cascade control. All methods modulate the same control variable—supply airflow rate; however, each method measures a different variable.

Direct Pressure Control. This method measures the pressure differential across the room envelope and adjusts the amount of supply air into the laboratory to maintain the required differential pressure. Challenges encountered with the direct pressure control arrangement include (1) maintaining the pressure differential when the laboratory door is open, (2) finding suitable locations for the sensor, (3) maintaining a well-sealed laboratory envelope, and (4) obtaining and maintaining accurate pressure sensing devices. The direct pressure control arrangement requires tightly constructed and compartmentalized facilities and may require a vestibule on entry/exit doors. Engineering parameters pertinent to envelope integrity and associated flow rates are difficult to predict.

Because direct pressure control works to maintain the pressure differential, the control system automatically reacts to transient disturbances. Doors to rooms may need a switch to disable the control system when they are open. Pressure controls recognize and compensate for unquantified disturbances such as stack effects, infiltration, and the influences of other systems in the building. Expensive, complex controls are not required, but the controls must be sensitive and reliable. In noncorrosive environments, controls can support a combination of exhaust applications, and they are insensitive to minimum duct velocity conditions. Successful pressure control provides the desired directional airflow but cannot guarantee a specific volumetric flow differential.

Volumetric Flow Tracking Control. This method measures both the exhaust and supply airflows and controls the amount of the supply air to maintain the desired differential pressure. Volumetric control requires that the air at each supply and exhaust point be controlled. Volumetric controls do not recognize or compensate for unquantified disturbances such as stack effects, infiltration, and influences of other systems in the building. Flow tracking is essentially independent of room door operation. Engineering parameters are easy to predict, and extremely tight construction is not required. Balancing is critical and must be addressed across the full operating range.

Controls may be located in corrosive and contaminated environments; however, the controls may be subject to fouling, corrosive attack, and/or loss of calibration. Flow measurement controls are sensitive to minimum duct velocity conditions. Volumetric control may not guarantee directional airflow.

Cascade Control. This method measures the pressure differential across the room envelope to reset the flow tracking differential set point. Cascade control includes the merits and problems of both direct pressure control and flow tracking control; however, first cost is greater and the control system is more complex to operate and maintain.

Fume Hood Control

The control criteria for fume hood control differs depending on the type of fume hood. The exhaust flow volume is kept constant for standard, auxiliary air, and air-bypass fume hoods. In variable volume fume hoods, the exhaust flow is varied to maintain a constant face velocity. Selection of the fume hood control method should be made in consultation with the safety officer.

The constant volume fume hood can further be split into either pressure-dependent or pressure-independent systems. Although simple in configuration, the pressure-dependent system is unable to adjust the damper position in response to any fluctuation in system pressure across the exhaust damper.

Variable volume fume hood control strategies can be grouped into two categories. The first measures the air velocity entering a small sensor in the wall of the fume hood. The measured velocity is used to infer the average face velocity based on an initial calibration. This calculated face velocity is then used to modulate the exhaust flow rate to maintain the desired set point of the face velocity.

The second category of variable volume fume hood control measures the fume hood sash opening and computes the exhaust flow requirement by multiplying the sash opening by the face velocity set point. The controller then adjusts the fume hood exhaust device (which may be a variable-frequency drive on the exhaust fan or a damper) to maintain the desired exhaust flow rate. The control system measures the exhaust flow for closed loop control, or it may use linear-calibrated, flow control dampers.

STACK HEIGHTS AND AIR INTAKES

Laboratory exhaust stacks should release effluent to the atmosphere without producing undesirable high concentrations at fresh air intakes, operable doors and windows, and locations on or near the building where access is uncontrolled.

Three primary factors that influence the proper disposal of effluent gases are the effects of stack/intake separation, stack height, and stack height plus momentum. Chapter 14 of the 1993 *ASHRAE Handbook—Fundamentals* covers the criteria and formulas to calculate the effects of these physical relationships. More extensive treatment of wind flow and diffusion can be found in Hosker (1984). For complex buildings or buildings with unique terrain or other obstacles to the airflow around the building, either detailed computer simulation or scale model wind tunnel testing should be considered.

Stack/Air Intake Separation

Separation of the stack discharge and air intake locations allows the atmosphere to dilute the effluent. Separation with the use of short to medium height stacks is simple to apply; however, to achieve adequate atmospheric dilution of the effluent, greater distances than are physically possible may be required and the building roof near the stack will be exposed to higher concentrations of the effluent.

Stack Height

Chapter 14 of the 1993 *ASHRAE Handbook—Fundamentals* describes a technique to determine the discharge height of a stack sufficiently above the turbulent zone around the building so that little or no effluent gas impinges on air intakes of the emitting building. The technique is conservative and generally requires tall stacks that may be visually unacceptable, or may not meet building code or zoning requirements. Also, the technique does not ensure acceptable concentrations of effluents at air intakes (e.g., if there are large releases of hazardous materials or elevated intake locations on nearby buildings). *Note*: NFPA *Standard* 45 requires a minimum 2.1 m stack height and ANSI/AIHA *Standard* Z 9.5 requires a minimum 3 m stack height.

To increase the effective height of the exhaust stacks, both the volumetric flow and discharge velocity can be increased to increase the discharge momentum (Momentum Flow = Density × Volume Flow × Velocity). The momentum of the large vertical flow in the emergent jet lifts the plume a substantial distance above the stack top, thereby reducing the physical stack height and making it easier to screen from view. This technique is particularly suitable when (1) many small exhaust streams can be clustered together or manifolded prior to the exhaust fan to provide the large volume flow and

(2) atmospheric air can be added through automatically controlled dampers to provide constant exhaust velocity under variable load. The drawbacks to this arrangement are the amount of energy consumed to achieve the constant high velocities and the added complexity of the controls to maintain the constant flow rates. Dilution equations presented in Chapter 14 of the 1993 *ASHRAE Handbook—Fundamentals* or mathematical plume analysis (e.g., Halitsky 1989) can be used to predict the performance of this arrangement, or performance can be validated through wind tunnel testing. Current mathematical procedures tend to have a high degree of uncertainty, and the results should be judged accordingly.

Criteria for Suitable Dilution

An example criterion might be that a release of 7.5 L/s of pure gas through any stack in a moderate wind (with speeds of 5 to 30 km/h) from any direction with a near-neutral atmospheric stability (Pasquill Gifford Class C or D) must not produce concentrations at any air intake that exceed

> 3 ppm simulating an accidental release such as would occur in a spill of an evaporating liquid or after the fracture of the neck of a small lecture bottle of gas in a fume hood.

The intent of this criterion is to limit the concentration of exhausted gases at the air intake locations to levels below the odor thresholds of gases released in fume hoods, excluding highly odorous gases such as mercaptans. Laboratories that use extremely hazardous substances should conduct a chemical specific analysis based on published health limits. A more lenient limit may be justified for laboratories with restricted chemical usage. Project-specific requirements need to be developed in consultation with the safety officer.

The equations presented in Chapter 14 of the 1993 *ASHRAE Handbook—Fundamentals* are presented in terms of dilution and the ratio of stack exit concentration to receptor location. For the criterion with the emission of 7.5 L/s of a pure gas, the exit concentration will vary with the total volume flow rate of the exhaust stack. Therefore, the dilution required to meet the criterion will vary with the volume flow rate of the stack. For example, a small stack with a 500 L/s flow rate will have an exit concentration of 7.5/500 or 15,000 ppm, using the 7.5 L/s emission of pure gas. A dilution of 1:5000 is needed to achieve an intake concentration of 3 ppm. A larger stack with a volume flow rate of 5000 L/s will have a lower exit concentration of 7.5/5000 or 1500 ppm and would need a lower dilution of 1:500 to achieve the 3 ppm intake concentration.

The above criterion is preferred over a pure dilution standard because a defined release scenario (7.5 L/s) is related to a defined intake concentration (3 ppm) based on odor thresholds or health limits. A dilution requirement may yield safe intake concentrations for a large flow rate stack but not for a small flow rate stack.

APPLICATIONS

LABORATORY ANIMAL FACILITIES

Laboratory animals must be housed in comfortable, clean, temperature and humidity controlled rooms. Animal welfare must be considered in the design. Also, the air-conditioning system must provide the desired microenvironment (animal cage environment) as specified by the facility's veterinarian (Woods 1980, Besch 1975, ILAR 1985).

Early detailed discussions with the laboratory animal veterinarian concerning airflow patterns, cage layout, and risk assessment help ensure a successful animal room HVAC design. The elimination of research variables (fluctuating temperatures and humidities, drafts, and spreading of airborne diseases) is another reason for a high quality air-conditioning system.

Table 2 gives minimum ventilation rates recommended by the Institute of Laboratory Animal Resource (ILAR). The ability to control odor in animal facilities depends primarily on the number and species of animals housed, the amount of cleanable surfaces (including exposed ducts and pipes), the sanitation practices, and the proper design and operation of the ventilation system. The required ventilation rate may be greater than that required by thermal loads.

Table 2 Recommended Ventilation for Laboratory Animal Rooms

Species	Minimum Room Air Changes per Hour (Using 100% Outside Air)	Reference
Mouse	15	ILAR (1977)
Hamster	15	ILAR (1977)
Rat	15	ILAR (1977)
Guinea pig	15	ILAR (1977)
Rabbit	10	—
Cat	10	ILAR (1978)
Dog	10	—
Nonhuman primate	10-15	ILAR (1980)

The air-conditioning flow rate for an animal room should be determined by the following factors:

- Desired animal microenvironment (Besch 1975, 1980; ILAR 1985)
- Species of animal(s)
- Animal population
- Recommended minimum ventilation rate (Table 2)
- Recommended ambient temperature and humidity (Table 3)
- Heat produced by motors on specialized animal housing units (such as laminar flow racks or HEPA-filtered air supply units for ventilated racks)
- Heat produced by the animals (Table 4)

Additional design factors include method of animal cage ventilation; operational use of a fume hood or a biological safety cabinet during procedures such as animal cage cleaning and animal examination; airborne contaminants (generated by animals, bedding, cage cleaning, and room cleaning); and institutional animal care standards (Besch 1980, ILAR 1985).

Table 3 Recommended Ambient Temperatures and Humidity Ranges for Animal Rooms

Species	Temperature, °C	Rel. Humidity, %	Reference
Mouse	18-26	40-70	ILAR (1977)
Hamster	18-26	40-70	ILAR (1977)
Rat	18-26	40-70	ILAR (1977)
Guinea pig	18-26	40-70	ILAR (1977)
Rabbit	16-21	40-60	—
Cat	18-29	30-70	ILAR (1978)
Dog	18-29	30-70	—
Nonhuman primate	18-29	30-70	ILAR (1980)

Note: The above ranges permit the scientific personnel who will use the facility to select optimum conditions (set points). The ranges do not represent acceptable fluctuation ranges.

Table 4 Heat Generated by Laboratory Animals

Species	Mass, kg	Heat Generation, Normally Active Watts per Animal		
		Sensible	Latent	Total
Mouse	0.021	0.33	0.16	0.49
Hamster	0.118	1.18	0.58	1.76
Rat	0.281	2.28	1.12	3.40
Guinea pig	0.41	2.99	1.47	4.46
Rabbit	2.46	11.49	5.66	17.15
Cat	3.00	13.35	6.58	19.93
Nonhuman primate	5.45	20.9	10.3	31.2
Dog	10.31	30.7	16.5	47.2
Dog	22.70	67.6	36.4	104.0

Due to the nature of research programs, air-conditioning design temperature and humidity control points may be required. Research animal facilities require more precise control of the environment than farm animal or production facilities because variations affect the experimental results. A totally flexible system permits control of the temperature of individual rooms to within ±1 K for any temperature set point in a range of 18 to 29°C. This requires significant capital expenditure, which can be mitigated by designing the facility for selected species and their specific requirements. Recommended ranges of room temperatures and relative humidities from which set points are selected are listed in Table 3. The relative humidity should be maintained between 30 to 70% throughout the year, according to the needs of the species.

If the entire animal facility or extensive portions of the facility are permanently planned for species with similar requirements, the range of individual adjustments should be reduced. Each animal room or group of rooms serving a common purpose should have temperature and humidity controls.

Control of air pressure in animal housing and service areas is important to ensure directional airflow. For example, quarantine, isolation, soiled equipment, and biohazard areas should be kept under negative pressure, whereas clean equipment and pathogen-free animal housing areas and research animal laboratories should be kept under positive pressure (ILAR 1985).

The animal facility and human occupancy areas should be conditioned separately. The human areas may use a return air HVAC system and may be shut down on weekends for energy conservation. Separation also prevents exposure of personnel to biological agents, allergens, and odors present in animal rooms.

Supply air outlets should not cause drafts on research animals. An efficient air distribution system for animal rooms is essential; this may be accomplished effectively by supplying air through ceiling outlets and exhausting air at floor level (Hessler and Moreland 1984). Supply and exhaust systems should be sized to minimize noise.

A study by Neil and Larsen (1982) showed that the predesign evaluation of a full-size mock-up of the animal room and its HVAC system was a cost-effective way to select a system that distributes air to all areas of the animal-holding room effectively. Wier (1983) describes many typical design problems and their resolutions. Room air distribution should be evaluated using ASHRAE *Standard* 113 procedures to evaluate drafts and temperature gradients.

Air-conditioning systems must remove the sensible and latent heat produced by laboratory animals. The literature concerning the metabolic heat production appears to be divergent, but new data is consistent. Current recommended values are given in Table 4. These values are based on experimental results and the following formula:

$$ATHG = 2.5 M$$

where

ATHG = average total heat gain, watts per animal
M = metabolic rate of animal, watts per animal = $3.5 W^{0.75}$
W = mass of animal, kg

Conditions in animal rooms must be maintained at constant values. This requires year-round availability of refrigeration and, in some cases, dual/standby chillers and emergency electrical power for motors and control instrumentation. The storage of critical spare parts is one alternative to the installation of a standby refrigeration system.

HVAC ductwork and utility penetrations must present few cracks in animal rooms so that all wall and ceiling surfaces can be easily cleaned. Exposed ductwork is not recommended; however, if constructed of 316 stainless steel in a fashion to facilitate removal for cleaning, it can provide a cost-effective alternative. Joints around diffusers, grilles, and the like should be sealed. Exhaust air grilles with 25 mm washable or disposable filters are normally used to prevent animal hair and dander from entering the ductwork.

Multiple cubicles within animal rooms enhance the operational flexibility of the animal room (i.e., housing multiple species in the same room, quarantine, and isolation). Each cubicle should be treated as if it were a separate animal room with the same air exchange/balance, temperature, and humidity control.

CONTAINMENT LABORATORIES

With the initiation of biomedical research involving recombinant DNA technology, federal guidelines about laboratory safety were published that influence design teams, researchers, and others. Containment describes safe methods for managing hazardous chemicals and infectious agents in laboratories. The three elements of containment are laboratory operational practices and procedures, safety equipment, and facility design. Thus, the HVAC design engineer helps decide two of the three containment elements during the design phase.

In the United States, the U.S. Department of Health and Human Services (DHHS), Center for Disease Control and Prevention (CDC), and the National Institutes of Health (NIH) classify biological laboratories into four levels—Biosafety Level 1 through Biosafety Level 4—listed in DHHS Publication No. (CDC) 93-8395.

Biosafety Level 1

Biosafety Level 1 is suitable for work involving agents of no known or of minimal potential hazard to laboratory personnel and the environment. The laboratory is not required to be separated from the general traffic patterns in the building. Work may be conducted either on an open bench top or in a chemical fume hood. Special containment equipment is neither required nor generally used. The laboratory can be cleaned easily and contains a sink for washing hands. The federal guidelines for these laboratories contain no specific HVAC requirements, and typical college laboratories are usually acceptable. Many colleges and research institutions require directional airflow from the corridor into the laboratory, chemical fume hoods, and approximately 3 to 4 air changes per hour of outside air. These requirements protect the research materials from contamination, thereby improving research efficiency. Directional airflow from the corridor into the laboratory helps to control odors.

Biosafety Level 2

Biosafety Level 2 is suitable for work involving agents of moderate potential hazard to personnel and the environment. The DHHS guideline (1993) contains lists that explain the levels of containment needed for various hazardous agents. Laboratory access is limited when certain work is in progress. The laboratory can be cleaned easily and contains a sink for washing hands. Biological safety cabinets (Class I or II) are used whenever

- Procedures with a high potential for creating infectious aerosols are conducted. The procedures include centrifuging, grinding, blending, vigorous shaking or mixing, sonic disruption, opening containers of infectious materials, inoculating animals intranasally, and harvesting infected tissues or fluids from animals or eggs.
- High concentrations or large volumes of infectious agents are used. Federal guidelines for these laboratories contain minimum facility standards, as described above.

At this level of biohazard, most research institutions have a full-time safety officer (or safety committee) who establishes facility standards. The federal guidelines for Biosafety Level 2 contain no specific HVAC requirements; however, typical HVAC design criteria can include the following:

- 100% outside air systems
- 6 to 15 air changes per hour
- Directional airflow into the laboratory rooms

- Site-specified hood face velocity at fume hoods (many institutions specify 0.4 to 0.5 m/s)
- An assessment of research equipment heat load in a room
- Inclusion of biological safety cabinets

Most biomedical research laboratories are designed for Biosafety Level 2. However, the laboratory director must evaluate the risks and determine the correct containment level before design begins.

Biosafety Level 3

Biosafety Level 3 applies to facilities in which work is done with indigenous or exotic agents that may cause serious or potentially lethal disease as a result of exposure by inhalation. All procedures involving the manipulation of infectious materials are conducted within biological safety cabinets. The laboratory should have special engineering (HVAC) features such as air locks and a nonrecirculating ventilation system.

Biosafety Level 4

Biosafety Level 4 is required for work with dangerous and exotic agents that pose a high risk of aerosol-transmitted laboratory infections and life-threatening disease. The design of HVAC systems for these areas will have stringent requirements that must be determined by the biological safety officer.

TEACHING LABORATORIES

Laboratories in academic settings can generally be classified as those used for instruction or those used for research. Research laboratories vary significantly depending on the discipline of work being performed in the laboratory and generally fit in one of the categories of laboratories described previously.

As with research laboratories, the design requirements for teaching laboratories vary based on their function. The designer should become familiar with the specific teaching program, so that a suitable hazard assessment can be made. For example, the requirements for the number and size of fume hoods vary significantly for undergraduate inorganic and graduate organic chemistry teaching laboratories.

Unique aspects of teaching laboratories include the need for the instructor to be visually in contact with the students at their work stations and to have ready access to the controls for the fume hood operations and any safety shut-off devices and alarms. Frequently, students have not received extensive safety instruction, thus easily understood controls and labeling are necessary. Because the teaching environment depends on verbal communication, sound from the building ventilation system becomes an important concern.

CLINICAL LABORATORIES

Clinical laboratories are found in hospitals and as stand-alone operations. The work in these laboratories generally consists of handling human specimens (blood, urine, etc.) and using chemical reagents for analysis. Some samples may be potentially infectious and because it is impossible to know which samples may be contaminated, good work practices require that all be handled as biohazardous materials. The primary protection of the staff at clinical laboratories depends on the techniques and laboratory equipment used (e.g., biological safety cabinets) to control aerosols, spills, or other inadvertent releases of samples and reagents. People outside the laboratory must also be protected from infectious or chemical agents.

Additional protection can be provided by the building HVAC system with suitable exhaust, ventilation, and filtration. The HVAC engineer is responsible for providing an HVAC system that meets the biological and chemical safety requirements. The engineer should consult with appropriate senior staff and safety professionals to ascertain what potentially hazardous chemical or biohazardous conditions will be in the facility and then provide suitable engineering controls to minimize risks to staff and the community. The use of biological safety cabinets, chemical fume hoods, and other specific exhaust systems should be considered by appropriate staff of the laboratory and the design engineer.

RADIOCHEMISTRY LABORATORIES

In the United States, laboratories located on Department of Energy (DOE) facilities are governed by DOE regulations. All other laboratories using radioactive materials are governed by the Nuclear Regulatory Commission (NRC), state, and local regulations. Other agencies may be responsible for the regulation of other toxic and carcinogenic materials present in the facility.

Laboratory containment equipment for nuclear processing facilities are treated as primary, secondary, or tertiary containment zones depending on the level of radioactivity anticipated for the area and the materials to be handled. Chapter 23 has additional information on nuclear laboratories.

OPERATION AND MAINTENANCE

Laboratories may need to maintain design performance conditions continuously with no interruptions for long periods due to long-term research studies. Even when research needs are not so demanding, systems that maintain air balance, temperature, and humidity in laboratories must be highly reliable, with a minimal amount of downtime. Operations and maintenance personnel, as well as users, should work with the designer early in the design of systems to gain their input and agreement.

System components must be of adequate quality to achieve reliable HVAC operation, and they should be easily accessible for maintenance. Laboratory work surfaces should be protected from possible leakage of coils, pipes, and humidifiers. Changeout of supply and exhaust filters should require minimum downtime.

Centralized monitoring of laboratory variables (e.g., pressure differentials and face velocity of fume hoods, supply flows, and exhaust flows) are useful in predictive maintenance of equipment and in assuring safe conditions.

For their safety, laboratory users should be instructed in the proper use of laboratory fume hoods, safety cabinets, ventilated enclosures, and local ventilation devices. They should be trained to understand the operation of the devices and the indicators and alarms that show whether they are safe to operate. Users should request periodic testing of the devices to be sure that they and the connected ventilation systems are operating properly.

Personnel that know the nature of the contaminants in a particular laboratory should be responsible for decontamination of equipment and ductwork before they are turned over to maintenance personnel for work.

Maintenance personnel should be trained to keep laboratory systems in good operating order and should understand the critical safety requirements of those systems. Preventative maintenance of the equipment and periodic checks of air balance should be scheduled. High maintenance items should be placed outside the actual laboratory (in service corridors or interstitial space) to reduce disruption of laboratory operations and exposure of the maintenance staff to laboratory hazards.

Maintenance personnel need to be aware of and trained in the procedures for maintaining good indoor air quality (IAQ) in laboratories. Many IAQ problems have been traced to poor maintenance due to poor accessibility (Woods et al. 1987).

ENERGY

Due to the nature of the functions they support, HVAC systems serving laboratories consume large amounts of energy. Any efforts to reduce energy use must not compromise the safety standards established by safety officers. Typically HVAC systems supporting laboratories and animal areas use 100% outside air and operate continuously. All HVAC systems serving laboratories can benefit from energy reduction techniques that are either an integral part of the original design or added later. Energy reduction techniques should be analyzed in terms of appropriateness to the facility and in economic terms considering payback.

The HVAC engineer must understand and respond to the scientific requirements of the facility. Research requirements typically include continual control of temperature, humidity, relative static pressure, and air quality. Energy reduction systems must maintain the required environmental conditions during both occupied and unoccupied modes.

Energy Conservation

Energy can be conserved in laboratories by reducing the exhaust air requirements. For example, the exhaust air requirements for fume hoods can be reduced by closing part of the hood opening during operation, thereby reducing the airflow needed to obtain the desired capture velocities. The sash styles that may be adjusted are described in the section on Laboratory Exhaust and Containment Devices.

Another way to reduce exhaust airflow is to use variable volume control of exhaust air through the fume hoods in accordance with the fume hood sash opening. This control method reduces exhaust airflow when the fume hood sash is not fully open. A variation of this arrangement incorporates a user-initiated selection of the fume hood airflow from a minimum flow rate to a maximum flow rate when the hood is in use. Any airflow control must be integrated with the laboratory control system, which is described in the section on Controls, and must not jeopardize the safety and function of the laboratory.

A third energy conservation method uses night setback controls when the laboratory is unoccupied to reduce the air volume being exhausted from one quarter to one half the minimum volume required when the laboratory is occupied. Timing devices, sensors, manual override, or a combination of these methods can be used to set back the controls at night. If this strategy is considered, the safety and function of the laboratory must be considered and appropriate safety officers should be consulted.

Energy Recovery

Energy from the exhaust airstream can often be recovered economically in laboratory buildings with large quantities of exhaust air. Many energy recovery systems are available including rotary air-to-air energy exchangers or heat wheels, coil energy recovery loops (run-around cycle), twin tower enthalpy recovery loops, heat pipe heat exchangers, fixed-plate heat exchangers, thermosiphon heat exchangers, and direct evaporative cooling. Some of these technologies can be combined with indirect evaporative cooling for further energy recovery. These systems are described in Chapter 44 of the 1992 *ASHRAE Handbook—Systems and Equipment*.

Specific concerns for the use of any energy recovery device in laboratory HVAC systems include the potential for cross contamination of chemical and biological materials from the exhaust air to the intake airstream and for corrosion and fouling of the devices located in the exhaust airstream. NFPA *Standard* 45 specifically prohibits the use of latent heat recovery devices in fume hood exhaust systems.

Energy recovery possibilities also exist for hydronic systems associated with the HVAC systems. Rejected heat from centrifugal chillers can be used to produce low temperature reheat water. Potential also exists in plumbing systems where hot waste from washing operations can be recovered to heat makeup water.

COMMISSIONING

In addition to HVAC systems, electrical systems and chemical handling and storage areas must be commissioned. Training of technicians, scientists, and maintenance personnel is a critical aspect of the commissioning process. Users must understand the systems and their operation.

Commissioning is defined in Chapter 39 and the process for commissioning is outlined in ASHRAE *Guideline* 1. Laboratory commissioning is more demanding than described in ASHRAE *Guideline* 1 because areas that are not associated with the normal office complex must be considered.

Laboratory commissioning starts with the intended use of the laboratory and should include development of a commissioning plan, as outlined in ASHRAE *Guideline* 1. Commissioning should include the validation of individual components initially; after successful individual component validation, the entire system should be evaluated. This requires verification and documentation that the design meets applicable codes and standards and that it has been constructed in accordance with the design intent. Before commissioning begins the following must be obtained:

- Complete set of the laboratory utility drawings
- Definition of the use of the laboratory and an understanding of the work being performed
- Equipment requirements
- All test results
- Understanding of the intent of the system operation

With respect to the laboratory's HVAC system, the issues that should be addressed as part of this process include verification and documentation of the following:

- Fume hood design face velocities have been met.
- Manufacturer's requirements for airflows for biological safety cabinets and laminar flow clean benches have been met.
- Exhaust system configuration, damper locations, and performance characteristics, including any required emission equipment, are correct.
- Control system operates as specified. Controls include fume hood alarm; miscellaneous safety alarm systems; fume hood and other exhaust volume regulation; laboratory pressurization control system; laboratory temperature control system; and main ventilation unit controls for supply, exhaust, and heat recovery systems. Control system performance verification should include speed of response, accuracy, repeatability, turndown, and stability.
- Desired laboratory pressurization relationships are maintained throughout the laboratory including at entrances, adjoining areas, air locks, interior rooms, and hallways. *Note:* To accomplish desired air balance relationships, balancing of terminal devices within 10% of design requirements will not provide adequate results. Additionally, internal pressure relationships can be affected by airflow around the building itself. Refer to Chapter 14 of the 1993 *ASHRAE Handbook—Fundamentals* for additional information.
- Fume hood containment performance is within specification. ASHRAE *Standard* 110 provides criteria for this evaluation.
- Dynamic response of the laboratory's control system is satisfactory. One method of testing the control system's performance is to open and shut the doors of the laboratory during the fume hood performance testing.
- System failure modes are acceptable.
- Standby electrical power systems function properly.
- Design NC levels of occupied spaces have been met.

Early in the design process, it should be determined whether any laboratory systems must comply with Food and Drug Administration (FDA) regulations because these systems have additional design and commissioning requirements.

ECONOMICS

In laboratories, HVAC systems make up a significant part (often 30 to 50%) of the overall construction budget. The design criteria and system requirements must be reconciled with the budget allotment for HVAC early in the planning stages and continually throughout the design stages to ensure that the project remains within budget.

Every project must be evaluated on both its technical features and business aspects. The following common economic terms are discussed in Chapter 33 and defined here as follows:

Initial cost. Cost to design, install, and test an HVAC system such that it is fully operational and suitable for use.

Operating cost. Cost to operate a system (including energy, maintenance, and component replacements) such that the total system can reach the end of it's normal useful life.

Life-cycle costs. Cost related to the total cost over the life of the HVAC system, including capital cost, considering the time value of money. Mechanical and electrical costs related to HVAC systems are commonly assigned a depreciation life based on current tax policies. This depreciation life may be different than the projected functional life of the equipment, which is influenced by the quality of the components of the system and the maintenance they receive. Some portions of the system, such as ductwork, could last the full life of the building. Other components, such as air-handling units, may have a useful life of 15 to 30 years depending on the decisions related to original quality and ongoing maintenance efforts. Estimated service life of equipment is listed in Chapter 33.

Engineering economics can be used to evaluate life-cycle costs of configuration (utility corridor versus interstitial space), systems, and major equipment. The user or owner makes a business decision concerning quality of the system and its ongoing operating costs. The HVAC engineer may be asked to provide an objective analysis of energy, maintenance, and construction costs, so that an appropriate life-cycle cost analysis can be made. Other considerations that may be appropriate include economic influences related to the long-term use of energy and governmental laws and regulations.

Many technical considerations and the great variety of available equipment influence the design of HVAC systems. Factors affecting design must be well understood to ensure appropriate comparisons between various systems and to determine the impact on either first or operating costs.

LABORATORY SAFETY RESOURCE MATERIALS

Occupational Exposure to Chemicals in Laboratories. Appendix VII, 29 CFR 1910.1450. Available from U.S. Government Printing Office, Washington, DC.

Uniform Building, Mechanical, and Fire Prevention Model Codes. International Conference of Building Officials, Whittier, CA.

BOCA Building, Mechanical, and Fire Prevention Model Codes. Building Officials and Code Administrators International, Country Club Hills, IL.

Prudent Practices for Handling Hazardous Chemicals in Laboratories. National Research Council, National Academy Press, Washington, DC. (1981).

Biosafety in Microbiological and Biomedical Laboratories, 3rd ed. U.S. Department of Health and Human Services *Publication* (CDC) 93-8395.

Guidelines for Construction and Equipment of Hospital and Medical Facilities. American Institute of Architects, Washington, DC.

Medical Laboratory Planning and Design. College of American Pathologists, Skokie, IL.

Occupational Safety and Health Administration (OSHA) Safety and Health requirements.

Fire Protection for Laboratories Using Chemicals. NFPA *Standard* 45. National Fire Protection Association, Quincy, MA.

Health Care Facilities Handbook. NFPA *Standard* 99. National Fire Protection Association, Quincy, MA.

Laboratory Fume Hoods Recommended Practices. SEFA 1-1992. Scientific Furniture and Equipment Association.

Industrial Ventilation: A Manual of Recommended Practice. American Conference of Governmental Industrial Hygienists, Cincinnati, OH.

Laboratory Ventilation. ANSI/AIHA *Standard* Z 9.5-92. American Industrial Hygiene Association, Fairfax, VA.

Americans with Disabilities Act (ADA).

Other regulations and guidelines may apply to laboratory design. All applicable institutional, local, state, and federal requirements should be identified prior to the start of design.

REFERENCES

Alereza, T. and J. Breen, III. 1984. Estimates of recommended heat gains due to commercial appliances and equipment. *ASHRAE Transactions* 90(2A):25-58.

AIA. 1993. *Guidelines for construction and equipment of hospital and medical facilities.* Committee on Architecture for Health, American Institute of Architects, Washington, D.C.

AIHA. 1992. Laboratory ventilation. *Standard* Z9.5-92. American Industrial Hygiene Association, Fairfax, VA.

ASHRAE. 1985. Method of testing performance of laboratory fume hoods. *Standard* 110-1985.

ASHRAE. 1989. Guideline for commissioning of HVAC systems. *Guideline* 1-1989.

ASHRAE. 1990. Method of testing for room air diffusion. *Standard* 113-1990.

ASHRAE. 1992. Gravimetric and dust-spot procedures for testing air-cleaning devices used in general ventilation for removing particulate matter. *Standard* 52.1-1992.

Besch, E. 1975. Animal cage room dry bulb and dew point temperature differentials. *ASHRAE Transactions* 81(2):459-58.

Besch, E. 1980. Environmental quality within animal facilities. *Laboratory Animal Science* 30(2II):385-406.

Caplan, K. and G. Knutson. 1977. The effect of room air challenge on the efficiency of laboratory fume hoods. *ASHRAE Transactions* 83(1):141-56.

Caplan, K. and G. Knutson. 1978. Laboratory fume hoods: Influence of room air supply. *ASHRAE Transactions* 84(1):511-37.

DHHS. 1993. Biosafety in microbiological and biomedical laboratories, 3rd ed. Publication No. (CDC) 93-8395. U.S. Department of Health and Human Services, NIH, Bethesda, MD.

Eagleson, J., Jr. 1984. Aerosol contamination at work. In *The international hospital federation yearbook.* Sabrecrown Publishing, London.

Halitsky, J. 1989. A jet plume model for short stacks. *APCA Journal* 39(6).

Hessler, J. and A. Moreland. 1984. Design and management of animal facilities. In *Laboratory animal medicine,* J. Fox, B. Cohen, F. Loew, eds. Academic Press.

ILAR. 1977. Laboratory animal management—Rodents. *ILAR News* 20(3). Institute of Laboratory Animal Resources, National Institutes of Health, Bethesda, MD.

ILAR. 1978. Laboratory animal management—Cats. *ILAR News* 21(3). Institute of Laboratory Animal Resources.

ILAR. 1980. Laboratory animal management—Nonhuman primates. *ILAR News* 23(2-3). Institute of Laboratory Animal Resources.

ILAR. 1985. Guide for the care and use of laboratory animals. *NIH Publication* No. 85-23. Institute of Laboratory Animal Resources.

Marsh, C. 1988. DDC systems for pressurization, fume hood face velocity, and temperature control in variable air volume laboratories. *ASHRAE Transactions* 94(2B).

NFPA. 1991. Hazardous chemical data. *Standard* 49. National Fire Protection Association, Quincy, MA.

NFPA. 1991. *Standard* on fire protection for laboratories using chemicals. *Standard* 45. National Fire Protection Association, Quincy, MA.

NFPA. 1993. Flammable and combustible liquids code. *Standard* 30. National Fire Protection Association, Quincy, MA.

NFPA. 1993. Health care facilities. *Standard* 99. National Fire Protection Association, Quincy, MA.

NRC. 1981. *Prudent practices of handling hazardous chemicals in laboratories.* National Research Council. National Academy Press, Washington, D.C.

NSF. 1992. Class II (laminar flow) biohazard cabinetry. NSF *Standard* 49-92. National Sanitation Foundation, Ann Arbor, MI.

Neil, D. and R. Larsen. 1982. How to develop cost-effective animal room ventilation: Build a mock-up. *Laboratory Animal Science* (Jan-Feb): 32-37.

Rake, B. 1978. Influence of crossdrafts on the performance of a biological safety cabinet. *Applied and Environmental Microbiology* (August): 278-83.

SEFA. 1992. Laboratory fume hoods recommended practices. SEFA 1-1992. Scientific Equipment and Furniture Association, McLean, VA.

Stuart, D., M. First, R. Rones, and J. Eagleston. 1983. Comparison of chemical vapor handling by three types of Class II biological safety cabinets. *Particulate & Microbial Control* (March/April).

Weir, R. 1983. Toxicology and animal facilities for research and development. *ASHRAE Transactions* 89(2B).

West, D. 1978. Assessment of risk in the research laboratory: A basis for facility design. *ASHRAE Transactions* 84(1B).

Woods, J. 1980. The animal enclosure—A microenvironment. *Laboratory Animal Science* 30(2II):407-13.

Woods, J., J. Janssen, P. Morey, and D. Rask. 1987. Resolution of the "sick" building syndrome. *Proceedings* of ASHRAE Conference: Practical Control of Indoor Air Problems, pp. 338-48.

BIBLIOGRAPHY

Abramson, B. and T. Tucker. 1988. Recapturing lost energy. *ASHRAE Journal* 30(6):50-52.

Adams, J.B., Jr. 1989. Safety in the chemical laboratory: Synthesis—Laboratory fume hoods. *Journal of Chemical Education* 66(12).

Ahmed, O., J.W. Mitchell, and S.A. Klein. 1993. Dynamics of laboratory pressurization. *ASHRAE Transactions* 99(2):223-29.

Ahmed, O. and S.A Bradley. 1990. An approach to determining the required response time for a VAV fume hood control system. *ASHRAE Transactions* 96(2):337-42.

Albern, W., F. Darling, and L. Farmer. 1988. Laboratory fume hood operation. *ASHRAE Journal* 30(3):26-30.

Anderson, S. 1987. Control techniques for zoned pressurization. *ASHRAE Transactions* 93(2B):1123-39.

Anderson, C.P. and K.M. Cunningham. 1988. HVAC controls in laboratories—A systems approach. *ASHRAE Transactions* 94(1):1514-20.

ASHRAE. 1989. Ventilation for acceptable indoor air quality. *Standard* 62-1989.

Barker, K.A., O. Ahmed, and J.A. Parker. 1993. A methodology to determine laboratory energy consumption and conservation characteristics using an integrated building automation system. *ASHRAE Transactions* 99(2):1155-67.

Baylie, C.L. and S.H. Schultz. 1994. Manage change: Planning for the validation of HVAC Systems for a clinical trials production facility. *ASHRAE Transactions* 100(1):1660-68.

Bertoni, M. 1987. Risk management considerations in design of laboratory exhaust stacks. *ASHRAE Transactions* 93(2B):2149-64.

Bossert, K.A. and S.M. McGinley. 1994. Design characteristics of clinical supply laboratories relating to HVAC systems. *ASHRAE Transactions* 94(100):1655-59.

Brow, K. 1989. AIDS research laboratories—HVAC criteria. In *Building Systems: Room Air and Air Contaminant Distribution,* L.L. Christianson, ed. ASHRAE cosponsored *Symposium,* pp. 223-25.

Brown, W.K. 1993. An integrated approach to laboratory energy efficiency. *ASHRAE Transactions* 99(2):1143-54.

Carnes, L. 1984. Air-to-air heat recovery systems for research laboratories. *ASHRAE Transactions* 90(2).

Code of Federal Regulations. (Latest Edition). Good laboratory practices. CFR 21 Part 58. U.S. Government Printing Office.

Coogan, J.J. 1994. Experience with commissioning VAV laboratories. *ASHRAE Transactions* 100(1):1635-40.

Crane, J. 1994. Biological laboratory ventilation and architectural and mechanical implications of biological safety cabinet selection, location, and venting. *ASHRAE Transactions* 100(1):1257-65.

CRC. 1989. *CRC handbook of laboratory safety,* 3rd ed. CRC Press, Boca Raton, FL.

Dahan, F. 1986. HVAC systems for chemical and biochemical laboratories. *Heating, Piping and Air Conditioning* (May):125-30.

Davis, S. and R. Benjamin. 1987. VAV with fume hood exhaust systems. *Heating, Piping and Air Conditioning* (August):75-78.

Degenhardt, R. and J. Pfost. 1983. Fume hood system design and application for medical facilities. *ASHRAE Transactions* 89(2B):558-70.

DHHS. 1981. NIH guidelines for the laboratory use of chemical carcinogens. NIH *Publication No.* 81-2385. Department of Health and Human Services, National Institutes of Health, Bethesda, MD.

DiBeradinis, L., J. Baum, M. First, G. Gatwood, E. Groden, and A. Seth. 1992. *Guidelines for laboratory design: Health and safety considerations.* John Wiley & Sons, Boston.

Doyle, D.L., R.D. Benzuly, and J.M. O'Brien. 1993. Variable volume retrofit of an industrial research laboratory. *ASHRAE Transactions* 99(2): 1168-80.

Flanherty, R.J. and R. Gracilieri. 1994. Documentation required for the validation of HVAC systems. *ASHRAE Transactions* 100(1):1629-34.

Ghidoni, D.A. and R.L. Jones, Jr. 1994. Methods of exhausting a BSC to an exhaust system containing a VAV component. *ASHRAE Transactions* 100(1):1275-81.

Hayter, R. and R. Gorton. 1988. Radiant cooling in laboratory animal caging. *ASHRAE Transactions* 94(1B).

Hitchings, D.T. and R.S. Shull 1993. Measuring and calculating laboratory exhaust diversity—Three case studies. *ASHRAE Transactions* 99(2): 1059-71.

Knutson, G. 1984. Effect of slot position on laboratory fume hood performance. *Heating, Piping and Air Conditioning* (February):93-96.

Knutson, G. 1987. Testing containment laboratory hoods: A field study. *ASHRAE Transactions* 93(2B):1801-12.

Koenigsberg, J. and H. Schaal. 1987. Upgrading existing fume hood installations. *Heating, Piping and Air Conditioning* (October):77-82.

Koenigsberg, J. and E. Seipp. 1988. Laboratory fume hood—An analysis of this special exhaust system in the post "Knutson-Caplan" era. *ASHRAE Journal* 30(2):43-46 (February).

Laboratory & health facilities programming & planning. Planning *Monograph* 1.0. Planning Collaborative Ltd., Reston, VA.

Lacey, D.R. 1994. HVAC for a low temperature biohazard facility. *ASHRAE Transactions* 100(1):1282-86.

Lentz, M. and A. Seth. 1989. A procedure for modeling diversity in laboratory VAV systems. *ASHRAE Transactions* 95(1A).

Maghirang, R.G., G.L. Riskowski, P.C. Harrison, H.W. Gonyou, L. Sebek, and J. McKee. 1994. An individually ventilated caging system for laboratory rats. *ASHRAE Transactions* 100(1):913-20.

Maust, J. and R. Rundquist. 1987. Laboratory fume hood systems—Their use and energy conservation. *ASHRAE Transactions* 93(2B):1813-21.

McDiarmid, M. 1988. A quantitative evaluation of air distribution in full scale mock-up of animal holding rooms. *ASHRAE Transactions* 94(1B).

Mikell, W. and F. Fuller. 1988. Safety in the chemical laboratory: Good hood practices for safe hood operation. *Journal of Chemical Education* 65(2).

Moyer, R. 1983. Fume hood diversity for reduced energy conservation. *ASHRAE Transactions* 89(2B).

Moyer, R. and J. Dungan. 1987. Turning fume hood diversity into energy savings. *ASHRAE Transactions* 93(2B):1822-34.

Murray, W., A. Strifel, T. O'Dea, and F. Rhame. 1988. Ventilation for protection of immune compromised patients. *ASHRAE Transactions* 94(1B).

Neuman, V. 1989. Disadvantage of auxiliary air fume hoods. *ASHRAE Transactions* 95(1A).

Neuman, V. 1989. Health and safety in laboratory plumbing. *Plumbing Engineering* (March):21-24.

Neuman, V. and H. Guven. 1988. Laboratory building HVAC systems optimization. *ASHRAE Transactions* 94(2A).

Neuman, V. and W. Rousseau. 1986. VAV for laboratory hoods—Design and costs. *ASHRAE Transactions* 92(1A).

Neuman, V., F. Sajed, and H. Guven. 1988. A comparison of cooling thermal storage and gas air conditioning for lab building. *ASHRAE Transactions* 94(2A).

Neuman, V. 1989. Design considerations for laboratory HVAC system dynamics. *ASHRAE Transactions* 95(1A).

Parker, J.A., O. Ahmed, and K.A. Barker. 1993. Application of building automation system (BAS) in evaluating diversity and other characteristics of a VAV laboratory. *ASHRAE Transactions* 99(2):1081-89.

Peterson, R., E. Schofer, and D. Martin. 1983. Laboratory air systems—Further testing. *ASHRAE Transactions* 89(2B).

Peterson, R. 1987. Designing building exhausts to achieve acceptable concentrations of toxic effluents. *ASHRAE Transactions* 93(2):2165-85.

Pike, R. 1976. Laboratory-associated infections: Summary and analysis of 3921 cases. *Health Laboratory Science* 13(2):105-14.

Rabiah, T.M. and J.W. Wellenbach. 1993. Determining fume hood diversity factors. *ASHRAE Transactions* 99(2):1090-96.

Rhodes, W. 1988. Control of microbiological contamination in critical areas of the hospital environment. *ASHRAE Transactions* 94(IA).

Richardson, G. 1994. Commissioning of VAV laboratories and the problems encountered. *ASHRAE Transactions* 100(1):1641-45.

Rizzo, S. 1994. Commissioning of laboratories: A case study. *ASHRAE Transactions* 100(1):1646-52.

Schuyler, G. and W. Waechter. 1987. Performance of fume hoods in simulated laboratory conditions. Report No. 487-1605 by Rowan Williams Davies & Irwin, Inc., under contract for Health and Welfare Canada.

Sessler, S. and R. Hoover. 1983. Laboratory fume hood noise. *Heating, Piping, and Air Conditioning* (September):124-37.

Simons, C.G. 1991. Specifying the correct biological safety cabinet. *ASHRAE Journal* 33(8).

Simons, C.G. and R. Davoodpour. 1994. Design considerations for laboratory facilities using molecular biology techniques. *ASHRAE Transactions* 100(1):1266-74.

Smith, W. 1994. Validating the direct digital control (DDC) system in a clinical supply laboratory. *ASHRAE Transactions* 100(1):1669-75.

Streets, R. and B. Setty. 1983. Energy conservation in institutional laboratory and fume hood systems. *ASHRAE Transactions* 89(2B).

Stuart, D., R. Greenier, R. Rumery, and J. Eagleson. 1982. Survey, use, and performance of biological safety cabinets. *American Industrial Hygiene Association Journal* 43:265-70.

Varley, J.O. 1993. The measurement of fume hood use diversity in an industrial laboratory. *ASHRAE Transactions* 99(2):1072-80.

Wilson, D. 1983. A design procedure for estimating air intake contamination from nearby exhaust vents. *ASHRAE Transactions* 89(2A).

Yoshida, K., H. Hachisu, J.A. Yoshida, and S. Shumiya. 1994. Evaluation of the environmental conditions in a filter-capped cage using a one-way airflow system. *ASHRAE Transactions* 100(1):901-905.

CHAPTER 14

ENGINE TEST FACILITIES

FOR industrial testing of internal combustion engines, test spaces are enclosed to control noise and isolate the test for safety or security, and ventilated to control environmental conditions and fumes. Isolated engines are tested in test cells; engines situated inside automobiles are tested in chassis dynamometers. The ventilation and safety principles for test cells also apply when large open areas in the plant are used for production testing and emissions measurements.

Enclosed test cells are normally found in research or emission test facilities. Test cells may require instruments to measure cooling water flow and temperature; exhaust gas flow, temperature, and emission concentrations; fuel flow; power output; and combustion air volume and temperature. These measurements are affected by changes in the temperature and humidity of the test cell. Accurate control of the testing environment is becoming more critical. For example, the U.S. Environmental Protection Agency, under the direction of the 1970 Clean Air Act and its 1990 amendment, has developed certification procedures for automobiles sold in the United States. The recent amendments require that tests demonstrate control of automobile pollutants in both hot and cold environments.

VENTILATION SYSTEMS

Ventilation and air conditioning of test cells must (1) supply and exhaust proper quantities of air to remove heat and control temperature; (2) exhaust sufficient quantities of air at proper locations to prevent buildup of combustible vapors; (3) modulate large quantities of air to meet changing conditions; (4) remove engine exhaust fumes; (5) supply combustion air; (6) prevent noise transmission through the system; and (7) provide for human comfort and safety during setup, testing, and tear-down procedures. Supply and exhaust systems for test cells are unitary, central, or a combination of the two. Mechanical exhaust is necessary in all cases, and ventilation is generally controlled on the exhaust side (Figure 1).

Constant-volume systems with variable supply temperatures can be used; however, variable-volume, variable-temperature systems are usually selected. Unitary variable-volume systems (Figure 1A) use an individual exhaust fan and makeup air supply for each cell. Supply and exhaust fans are interlocked, and their operation is coordinated with that of the engine, usually by sensing the temperature of the cell. Some systems have exhaust only, with supply induced directly from outside (Figure 1B). Volume variation can be accomplished by changing fan speed or damper position.

Ventilation systems with central supply fans, central exhaust fans, or both (Figure 1C) regulate air quantities by test cell temperature control of individual dampers or by two-position switches actuated by dynamometer operations. Air balance is maintained by static pressure regulation within the cell. Constant pressure in the supply duct is obtained by controlling inlet vanes, modulating dampers, or varying fan speed.

In systems with individual exhaust fans and central supply, the exhaust system is controlled by cell temperature or a two-position switch actuated by dynamometer operation, while the central supply system is controlled by a static pressure device located within the cell to maintain room pressure (Figure 1D). Variable volume exhaust airflow should not drop below minimum requirements. Exhaust requirements should override cell temperature requirements; thus, reheat may be needed.

Ventilation should be interlocked with the fire protection system to shut down supply to and exhaust from the cell in case of fire. Exhaust fans should be nonsparking, and makeup air should be heated by steam or hot water coils.

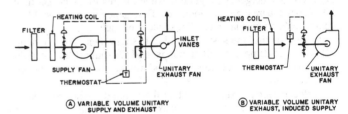

(A) VARIABLE VOLUME UNITARY SUPPLY AND EXHAUST

(B) VARIABLE VOLUME UNITARY EXHAUST, INDUCED SUPPLY

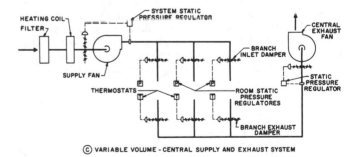

(C) VARIABLE VOLUME - CENTRAL SUPPLY AND EXHAUST SYSTEM

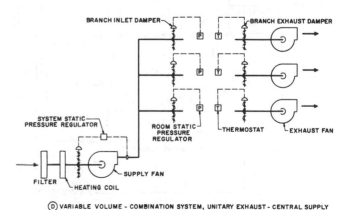

(D) VARIABLE VOLUME - COMBINATION SYSTEM, UNITARY EXHAUST - CENTRAL SUPPLY

Fig. 1   Heat Removal Ventilation Systems

The preparation of this chapter was assigned to TC 9.2, Industrial Air Conditioning.

TEST CELL EXHAUST

Ventilation for test cells is based on exhaust requirements for (1) removal of heat generated by the engine, (2) emergency purging (removal of fumes after fuel spills), and (3) cell scavenging, i.e., continuous exhaust during nonoperating periods.

During engine operation, heat convects to the air in the cell and also radiates to surrounding surfaces. The flow of air needed to remove convected heat from the cell is determined by

$$Q = 830H/(t_e - t_s) \qquad (1)$$

where

Q = airflow, L/s
H = engine heat release, W
t_e = temperature of exhaust air, °C
t_s = temperature of supply air, °C

Heat radiated from the engine, dynamometer, and exhaust piping warms the surrounding surfaces, which release the heat to the air by convection. The rate of release is governed by the temperature differences, film coefficients, and other factors discussed in Chapter 5 of the 1993 *ASHRAE Handbook—Fundamentals*. The value for $(t_e - t_s)$ in Equation (1) cannot be arbitrarily set when a portion of H is radiated heat. Determination of H is discussed in the section on Engine Heat Release.

Vapor removal exhaust should remove vapors as quickly as possible. The air required for engine heat removal is generally the quantity used for emergency purging; however, it should not be less than 50 L/s per square metre of floor area. Emergency purging should be controlled by a manual overriding switch for each cell.

In case of fire, the system should (1) shut down all equipment, (2) close fire dampers at all openings, and (3) shut off the solenoid valves that control fuel flow. Fire and smoke control is discussed in Chapter 48.

Cell scavenging exhaust is either the minimum required to keep combustible vapors from fuel leaks from accumulating or the amount required for heating and cooling, whichever is greater. In general, 10 L/s per square metre of floor area is sufficient.

Because gasoline vapors are heavier than air, exhaust grilles should be low even when ceiling exhaust is used. Removing exhaust close to the engine minimizes the convective heat that escapes into the cell. In some installations, all air is exhausted through a floor grating surrounding the engine bedplate into a cubicle or duct beneath. In this arrangement, supply slots are located in the ceiling over the bedplate to cover the engine with a curtain of air to remove the heat. This scheme has worked quite successfully in a number of installations and is particularly suitable for central exhaust systems (Figure 2). Water sprays in the cubicle or exhaust duct lessen the danger of fire or explosion in case of fuel spills.

Trenches or pits should be avoided in test cells. If they exist, as is the case in most chassis dynamometer rooms, they should be mechanically exhausted at all times. Excessively long trenches may require multiple takeoff exhaust points. The exhaust should sweep the entire area, leaving no dead spaces. Because of fuel spills and vapor accumulation, suspended ceilings or basements should not be located directly below test cells. If such spaces do exist, they should be ventilated continuously and have no fuel lines run through them.

Although the exhaust quantities should be calculated for each individual test cell on the basis of heat to be removed, evaporation of possible fuel spills, and minimum downtime ventilation, a general idea of current practice can be obtained from Table 1.

Table 1   Exhaust Quantities for Test Cells

	Minimum Exhaust Rates per Unit Floor Area	
	L/s per m²	Air changes per hᵃ
Engine testing—cell operating	50	60ᵇ
Cell idle	10	12
Trenchesᶜ and pits	50	—
Accessory testing	20	24
Control rooms and corridors	5	6

ᵃBased on a cell height of 3 m.
ᵇFor chassis dynamometer rooms, this quantity is usually determined by test requirements.
ᶜFor large trenches, use 0.5 m/s across the cross-sectional area of the trench.

TEST CELL SUPPLY

The quantity of air supplied to a test cell should be slightly less than that exhausted; this yields a slightly negative pressure. Recirculation of test cell air during engine testing is not recommended. Air from other nontest areas can be used when available, provided good ventilation practice is followed.

The method for introducing the large quantities of ventilation air is important, especially if the cell is occupied. Ventilation should keep the heat released from the engine away from cell occupants. Slot-type outlets with automatic dampers that maintain constant discharge velocity have been used with variable-volume systems.

A variation of Systems C and D in Figure 1 includes a separate air supply sized for the minimum, downtime ventilation rate; a room thermostat regulates its heated and chilled water coils to control the temperature within the cell. This system is useful in installations where much time is devoted to the setup and preparation of tests, or where constant temperature is required for complicated or sensitive instrumentation. Except for production and endurance testing, the actual engine operating time in test cells may be surprisingly low. Industry-wide, test cells are used approximately 15 to 20% of the time.

Filtration of all air to the cell to remove atmospheric particulates and insects is ordinarily required. The degree of filtration is determined by the nature of the test. Facilities located in clean environments sometimes use unfiltered outdoor air.

Heating coils are needed to temper supply air if there is danger of freezing equipment or if low temperatures adversely affect test performance.

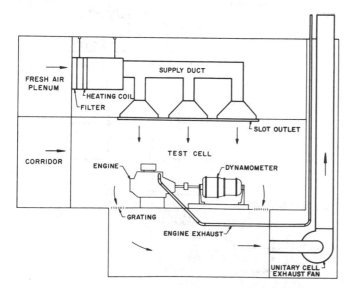

Fig. 2   Engine Test Cell Showing Direct Engine Exhaust—Unitary Ventilation System

Table 2 Heat Balance for Internal Combustion Engine at Wide-Open Throttle

Engine Type and Size	Engine Heat Balance, %			
	Work Output	To Exhaust	To Coolant	Engine Convection, Radiation, and Oil
Air-cooled I.C.				
To 11 kW	15	45	—	40
11 to 110 kW	20	43	—	37
Water-cooled I.C.				
110 to 370 kW	23.5	36	26	14.5
Diesel	32	36	22	10
G.T. regenerator				
To 300 kW	23	66	—	10

ENGINE HEAT RELEASE

Engine heat released to the space is the total energy input to the engine from the fuel less the energy transmitted to the dynamometer as work, the heat removed by the jacket cooling water, and the heat discharged in the exhaust gas. These quantities vary with the engine, its operating speed and load, the fuel used, the type of dynamometer, and the piping configurations external to the engine. The heat balance of an internal combustion engine at full load is approximately one-third to the shaft for useful work, one-third to the exhaust, and one-third to the jacket cooling water; for a gas turbine at full load, the heat balance is roughly one-third to the shaft and two-thirds to the exhaust. In both cases, there are losses by convection and radiation from the engine itself and from the exhaust piping. Table 2 shows a representative breakdown of heat release by engine type and size.

For an L-head engine, heat rejected to the jacket water in kilowatts is RPM divided by 1110 per litre displacement, which is about 1 kW per kilowatt of output. For an overhead valve engine, it is RPM divided by 1210 per litre of displacement, which is 0.9 kW per kilowatt output.

The energy absorbed by the dynamometer as work is removed from the test cell, except for that portion transmitted to the space from the dynamometer if it is located in the test cell. Absorbed energy is returned to the electrical system, dissipated in remote resistance grids, or, when dynamometers are used, removed by cooling water.

Engine coolant system heat is normally removed by recirculated tower water through heat exchangers. In some cases, once-through city water cooling may be used; however, there may be regulatory limitations, and the water-use cost may be high. Heat exchangers for engine oil are required under many test conditions. Radiators within the cell dissipate this heat, so the exhaust air quantity must be increased.

The amount of heat lost by convection from the engine block, head, and oil pan depends largely on engine coolant temperature, air temperature, and the quantity of air passing over the engine. A surface temperature of 80 to 150°C may be expected for internal combustion engines, and 200 to 260°C for gas turbines.

Because exhaust gas temperatures are between 650 and 1000°C, engine exhaust pipes release considerable heat through radiation and convection. A minimum amount of piping should be located within the cell. Commonly, engine exhaust pipes are run into a vault below the cell. Catalytic converters and other emission control devices are usually installed within the cell, adding to the heat load. Cooling the engine exhaust system by jacketing or injecting water directly into the pipe is common. Insulating the exhaust pipes inside the test cell also reduces radiant heat.

Methods for calculating convection and radiation losses from exhaust pipe surfaces are found in Chapter 3 of the 1993 *ASHRAE Handbook—Fundamentals*. Because exhaust piping, mufflers, and add-on devices are subject to change from test to test, this heat loss is usually estimated. Maximum heat loss is obtained by using the largest engine the dynamometer can handle at wide-open throttle. For a diesel engine, the heat loss to the test cell ambient from the exhaust can be estimated to be 3% of the rating of the dynamometer per metre of exposed exhaust pipe, including the exhaust manifold. For a gasoline-powered engine, the heat loss can be estimated to be 4.5% per metre of pipe. Due to several compensating factors, the actual diameter of the pipe has only a secondary effect.

In summary, the total engine heat release is

$$P(L_d + L_e + L_c + C)$$

where

P = engine brake power, kW
L_d = dynamometer loss per kW
L_e = exhaust pipe losses per kW
L_c = engine connection losses per kW
C = cooling system losses per kW

GAS TURBINE TEST CELLS

Large gas turbine test cells must handle the large quantities of air required by the turbine itself, attenuate the noise generated, and operate safely with respect to the large flow of fuel. Such cells are unitary and use the turbine to draw in the untreated air and exhaust it through mufflers.

Small gas turbines can generally be tested in a conventional engine test cell with relatively minor modifications. Test cell ventilation supply and exhaust are sized for the heat generated by the turbine as for a conventional engine. Combustion air supply for the turbine is considerable; it may be drawn from the cell, from the outdoors, or through separate conditioning units that handle only combustion air.

Exhaust quantities are higher than from internal combustion engines and are usually ducted directly to the outdoors through muffling devices that provide little restriction to flow. The exhaust air may be water-cooled, as temperature may exceed 530°C.

CHASSIS DYNAMOMETER ROOMS

A chassis dynamometer (Figure 3) simulates road driving and acceleration conditions. The drive wheels of the vehicle rest on a large roll, which drives a dynamometer. Air quantities calibrated to correspond to air velocities at a particular road speed are blown across the front of the vehicle for radiator cooling and to approximate the effect on the body of the vehicle. Additional refinements may vary the air temperature within prescribed limits from ambient to 55°C. Relative humidity is often controlled to simulate outside conditions. Air is usually introduced through an area approximating the frontal area of the vehicle. A duct with a return grille may be lowered to a position at the rear of the vehicle, so that the air remains near the floor rather than short-cycling through a ceiling grille. The air is recirculated to the air-handling equipment above the ceiling.

Chassis dynamometers are also installed (1) in cold rooms, where temperatures may be as low as −70°C; (2) in altitude chambers, where heights up to 3700 m can be simulated; (3) in noise, vibration, and harshness chambers for sound evaluations; (4) in electromagnetic cells for evaluation of electrical components; and (5) in full-sized wind tunnels with throat areas many times the cross-sectional area of the vehicle. Combustion air is drawn directly from the room, but a mechanical engine exhaust must be installed in a way that will preserve the low temperatures and humidities.

For chassis dynamometer rooms having a controlled temperature, there is often a soak space nearby. This space is used to precool or preheat those automobiles scheduled to enter the room. It generally

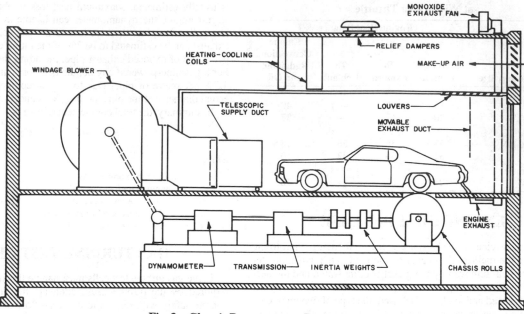

Fig. 3   Chassis Dynamometer Room

takes 18 to 24 h for the vehicle's temperature to stabilize at the temperature of the room. The soak space and the temperature-controlled room are often isolated from the rest of the facility, with entry and egress through an airlock-type vestibule.

ENGINE EXHAUST

Engine exhaust systems remove combustion products, unburned fuel vapors, and water vapor resulting from the injection of water for gas cooling. Design criteria are flow loads and the system operating pressure.

Flow loads are calculated based on the number of engines, engine size, and load and use factors.

System operating pressure is determined by engine discharge pressures and the range of allowable suction or back pressure at the point of connection to the exhaust system. Systems may operate at positive pressure, using available engine tail pipe pressure to force the flow of gas, or at negative pressure, with mechanically induced flow.

The simplest method for inducing engine exhaust from a test cell is to size the exhaust pipe to minimize variations in pressure on the engine and to connect it directly to the outside (Figure 4A).

The length of run of such direct systems is limited by the cost of piping of a size adequate to keep back pressure to a minimum. Direct exhaust to the outdoors is subject to wind currents and air pressures, may be hazardous due to positive pressures within the system, and cannot be as closely regulated as mechanically ventilated systems. Allowances for explosion relief should be made in the ductwork, especially in close-coupled engine exhaust systems.

Mechanical engine exhaust systems are unitary or central. A *unitary system* (Figure 4B) serves an individual test cell with a fan and conductors and can be closely regulated to match operation to the engine's requirements. A *central system* (Figure 4D) serves multiple cells with single or multiple fans, main ducts, and branch connections to individual cells.

Pressures in engine exhaust systems fluctuate with changes in engine capacity. The exhaust system should be designed so that load variations in individual cells have a minimum effect on the system. Dampers and pressure regulators, if needed, keep the pressures within test tolerances. Engine characteristics and diversity of operation determine the maximum exhaust to be handled by the

system. Allowable back pressure and tolerance for its variation are the basis for system size and regulation.

An indirect connection between the engine exhaust pipe and a mechanical removal system (Figure 4C) simplifies back pressure regulation and provides augmentation of exhaust gas flow rates by inducing room air into the exhaust stream. The exhaust pipe terminates 50 to 75 mm inside of an inlet to the augmented exhaust system, which removes a mixture of exhaust fumes and room air from the test cell.

If it is not within the test cell, an indirect connection requires noise control. Also, this connection can self-ignite if the engine is motored by the dynamometer, which has a fuel-rich mixture exhausting from the pipe at temperatures above 350°C.

Because the exhaust piping is open to the room, the indirect-connected system is essentially self-regulating. Room air mixes with the exhaust and cools the mixture. Materials are not required to withstand the higher temperature conditions of a direct-connected system, although corrosion may be more severe.

To reduce exhaust gas temperatures (usually about 800°C but sometimes as high as 1000°C), water may be injected directly into

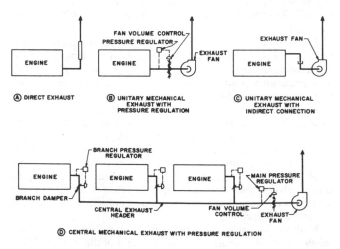

Fig. 4   Engine Exhaust Systems

the exhaust pipe, the pipe may be water-jacketed, or a heat shield or shroud may be placed around the exhaust pipe, with air mechanically exhausted through the shrouded enclosure. Exhaust piping and fans are often made of high-strength, high-temperature alloys that withstand corrosion, thermal stresses, and pressure pulsations resulting from rapid flow fluctuations. The equipment must be adequately supported and anchored to relieve thermal expansion stresses.

Exhaust systems for chassis dynamometer installations must capture the high-velocity exhaust from the tail pipe to prevent fume buildup in the test area. Exhaust airflow of 330 L/s has been used effectively for automobiles at 105 km/h.

Engine exhaust should be discharged through a stack extending above the roof at an elevation and velocity sufficient to allow the fumes to clear the building windstream wake. If high stacks are not practical, shorter stacks with increased ejection velocity may propel the exhaust gases above the building wake. Chapter 14 of the 1993 *ASHRAE Handbook—Fundamentals* has further details. Ejection systems must adjust to keep the discharge velocity relatively constant as the gas volume varies with load changes. Local or federal laws may require that exhaust gases be cleaned before they enter the atmosphere.

COOLING WATER SYSTEMS

Dynamometers absorb and measure the useful output of an engine or its components. Two basic classes of dynamometers are the water-cooled induction type and the electrical type. In the *water-cooled* dynamometer, engine work is converted to heat, which is absorbed by a circulating water system. *Electrical* dynamometers convert engine work to electrical energy, which can be used or dissipated as heat in resistance grids. Grids should be outdoors or adequately ventilated.

Heat loss from electric dynamometers is approximately 8% of the measured output, plus a constant load of about 5 kW for auxiliaries within the cell. Recirculating water systems absorb heat from engine jackets, oil coolers, and water-cooled dynamometers through a system of circulating pumps, cooling towers, or atmospheric coolers and hot and cold well-collecting tanks (Figure 5).

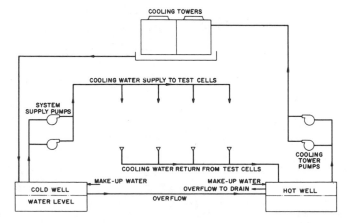

Fig. 5   Cooling Water System Using Cooling Towers

COMBUSTION AIR SUPPLY SYSTEMS

Combustion air is usually drawn from the test cell or introduced directly from the outdoors. If combustion air conditions must be closely regulated, and conditioning of the entire test cell is impractical, separate units that condition only the combustion air can be used. These units filter, heat, and cool the supply air and regulate its humidity and barometric pressure; they usually provide air directly to the air intake system. Combustion air systems may be central units or portable, packaged units.

NOISE

The characteristics of the noise generated by internal combustion engines and gas turbines must be considered in the design of a test cell air-handling system. Part of the engine noise is discharged through the tail pipe. If possible, internal mufflers should be installed to attenuate this noise at its source. Any ventilation ducts or pipe trenches that penetrate the cells must be insulated against sound transmission to other building areas or to the outdoors. Attenuation equivalent to that provided by the structure of the cell should be applied to the duct penetrations. Table 3 lists typical noise levels in test cells during engine operation.

Table 3   Typical Noise Levels in Test Cells

Type and Size of Engine	Decibel Reading 0.9 m from Engine			
	Dba	124 Hz	500 Hz	2000 Hz
Diesel				
Full load	105	107	98	99
Part load	70	84	56	49
Gasoline engine, 7.2 L at 5000 rpm				
Full load	107	108	104	104
Part load	75	—	—	—
Rotary engine, 75 kW				
Full load	90	90	83.5	86
Part load	79	78	75	72

BIBLIOGRAPHY

Bannasch, L.T. and G.W. Walker. 1993. Design factors for air-conditioning systems serving climatic automobile emission test facilities. *ASHRAE Transactions* 99(2):614-23.

Computer controls engine test cells. *Control Engineering* 16(75):69.

Hazardous gases need ventilation for safety. 1968. *Power,* 112(May):92.

Heldt, P.M. 1956. *High speed combustion engines.* Chilton Co., Philadelphia.

NFPA. 1993. Flammable and combustible liquids code. NFPA *Code* 30-93. National Fire Protection Association, Quincy, MA.

Paulsell, C.D. 1990. Description and specification for a cold weather emissions testing facility. U.S. Environmental Protection Agency, Washington, D.C.

Ricardo and Hempson. 1968. *The high speed internal combustion engine.* Blacke and Son Limited, London.

Schuett, J.A. and T.J. Peckham. 1986. Advancements in test cell design. *SAE Transactions*, paper no. 861215. Society of Automotive Engineers, Warrendale, PA.

CLEAN SPACES

DESIGN of clean spaces or clean rooms encompasses much more than the traditional controlling of temperature and humidity. In clean space design, other factors include particulate and microbial contamination control, airflow pattern control, sound and vibration control, industrial engineering aspects, and manufacturing equipment layout.

The performance of a clean room is judged by the quality of control of particulate concentration and dispersion, temperature, humidity, vibration, noise, airflow pattern, and construction. The objective of good clean room design is to control these parameters while maintaining reasonable installation and operating costs.

TERMINOLOGY

Acceptance criteria. The upper and lower limits of the room environment (critical parameters); if these limits are exceeded, the product may be considered adulterated.

As-built clean room. A clean room that is complete and ready for operation, with all services connected and functional, but without production equipment or personnel in the room.

Aseptic space. A space controlled such that bacterial growth is contained within acceptable limits; not a "sterile" space.

At-rest clean room. A clean room that is complete with the production equipment installed and operating, but without personnel in the room.

CFR. Code of Federal Regulations

CFU (colony forming unit). A visible growth of microorganisms arising from a single cell or multiple cells.

Challenge. A dispersion of known particle size and concentration used to test filter integrity and efficiency.

Class M1.5. Particle count not to exceed 35 particles/m^3 of a size 0.5 μm (or appropriate other size, in accordance with Figure 1 and Table 1) and larger. This criterion should be based on a large sampling of counts.

Class M2.5. Particle count not to exceed 350 particles/m^3 of a size 0.5 μm and larger, with no particle exceeding 5.0 mm.

Class M3.5. Particle count not to exceed 3500 particles/m^3 of a size 0.5 μm and larger.

Class M5.5. Particle count not to exceed 353 000 particles/m^3 of a size 0.5 μm and larger or 2300 particles/m^3 of a size 5.0 μm and larger.

Class M6.5. Particle count not to exceed 3 530 000 particles/m^3 of a size 0.5 μm and larger or 24 700 particles/m^3 of a size 5.0 μm and larger.

Clean room. A specially constructed enclosed area environmentally controlled with respect to airborne particulates, temperature, humidity, air pressure, air pressure flow patterns, air motion, vibration, noise, viable organisms, and lighting.

Clean space. A defined area in which the concentration of airborne particles is controlled at or below specified limits.

Conventional flow clean room. A clean room with nonunidirectional or mixed airflow patterns and velocities.

Critical parameter. A room variable (such as temperature, humidity, air changes, room pressure, particulates, viables, etc.) that, by law or by determination from product development data, affects product strength, identity, safety, purity, or quality (SISPQ).

Critical surface. The surface of the work part to be protected from particulate contamination.

Design conditions. The environmental conditions for which the clean space is designed.

Exfiltration. Leakage of air out of a room through cracks in doors and pass-throughs, through material transfer openings, etc., due to a difference in space pressures

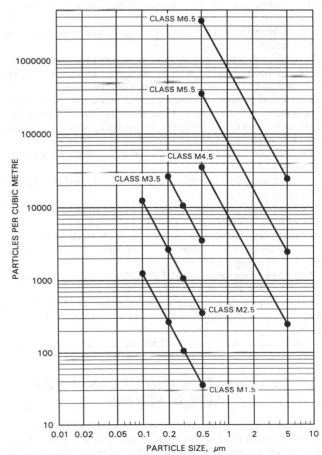

Fig. 1 Air Cleanliness Class Limits

The preparation of this chapter is assigned to TC 9.2, Industrial Air Conditioning.

Table 1 SI Classes for Clean Rooms

Class Name[b]		Class Limits[a]				
SI	Inch-Pound	0.1 µm Particles per m³	0.2 µm Particles per m³	0.3 µm Particles per m³	0.5 µm Particles per m³	5 µm Particles per m³
M1		350	75.0	30.9	10.0	—
M1.5	1	1 240	265	106	35.3	—
M2		3 500	757	309	100	—
M2.5	10	12 400	2 650	1 060	353	—
M3		35 000	7 570	3 090	1 000	—
M3.5	100		26 500	10 600	3 530	—
M4			75 700	30 900	10 000	—
M4.5	1 000		—	—	35 300	247
M5			—	—	100 000	618
M5.5	10 000		—	—	353 000	2 470
M6			—	—	1 000 000	6 180
M6.5	100 000		—	—	3 530 000	24 700
M7			—	—	10 000 000	61 800

Source: Federal *Standard* 209E, Airborne Particulate Cleanliness Classes in Clean Rooms and Clean Zones, September 11, 1992.

[a]The class limits shown are defined for classification purposes only and do not necessarily represent the size distribution to be found in any particular situation.

[b]Concentration limits for intermediate classes can be calculated, approximately, from the following equation:

$$\text{Particles}/m^3 = 10^M (0.5/d)^{2.2}$$ where M is the numerical designation of the class based on SI units and d is the particle size in micrometres.

First air. The air that issues directly from the HEPA filter before it passes over any work location.

GMP. Good manufacturing practice, as defined by CFR 21 (also, cGMP = current GMP)

High efficiency particulate air (HEPA) filter. A filter with an efficiency in excess of 99.97% of 0.3 µm particles.

Makeup air. Air introduced to the air system for ventilation, pressurization, and replacement of exhaust air.

Monodispersed particles. A replacement for DOP in determining the cleanliness level of a room using a concentration of particles approximately 0.2 µm in size.

Nonunidirectional flow work station. A work station without uniform airflow patterns and velocities.

Operational clean room. A clean room in normal operation with all services functioning and with production equipment and personnel present and performing their normal work functions.

Parenteral product. A pharmaceutical product normally meant to be injected into the patient. Parenterals are manufactured under aseptic conditions or are terminally sterilized to destroy bacteria and meet aseptic requirements.

Particle concentration. The number of individual particles per unit volume of air.

Particle size. The apparent maximum linear dimension of a particle in the plane of observation.

Polydispersed particles. A filter challenge utilizing particles of different sizes to determine the cleanliness of a space.

Primary air. Air that recirculates through the work space.

Secondary air. That portion of the primary air circulated through the air-conditioning equipment.

Ultra low penetration air (ULPA) filter. A filter with a minimum of 99.999% efficiency on 0.12 µm particles.

Unidirectional flow. Formerly called laminar flow. Flow of air in generally parallel streamlines with uniform velocity and in the same direction.

Work station. An open or enclosed work surface with direct air supply.

CLEAN SPACES AND CLEAN ROOM APPLICATIONS

The use of clean space environments in manufacturing, packaging, and research continues to grow as technological advances and the need for cleaner work environments increase. The following major industries use clean spaces for their products.

Semiconductors. The advances in semiconductor microelectronics continues to drive the state of the art in clean room design. Semiconductor facilities account for a significant percentage of all clean rooms in operation in the United States with most of the newer semiconductor clean rooms being Class M3.5 or cleaner.

Pharmaceutical. Preparation of pharmaceutical, biological, and medical products, and genetic engineering research are prime examples of this industry. Clean spaces control viable (living) particles that would produce undesirable bacteria growth.

Aerospace. Clean rooms were first developed for aerospace applications to manufacture and assemble satellites, missiles, and aerospace electronics. Most applications involve clean spaces of large volumes with cleanliness levels of Class M5.5 or higher.

Miscellaneous Applications. Clean rooms are also used in aseptic food processing and packaging, manufacture of artificial limbs and joints, automotive paint booths, laser/optic industries, and advanced materials research.

Hospital operating rooms may be classified as clean rooms, yet their primary function is to limit particular types of contamination rather than the quantities of particles present. Clean rooms are used in patient isolation and surgery where risks of infection exist.

AIRBORNE PARTICLES AND PARTICLE CONTROL

Airborne particles occur in nature as pollen, bacteria, miscellaneous living and dead organisms, and windblown dust and sea spray. Industry generates particles from combustion processes, chemical vapors, and friction in manufacturing equipment. People in the work space are a prime source of particles in the form of skin flakes, lint, cosmetics, and respiratory emissions. These airborne particles vary in size from 0.001 µm to several hundred micrometres. Particles larger than 5 µm tend to settle quickly. With many manufacturing processes these airborne particles are viewed as a source of contamination, where contact between the particulate and the product will cause product failure.

Particle Sources in Clean Spaces

In general, particle sources with respect to the clean space are grouped into two categories—external and internal.

External Sources. External sources are those particles that enter the clean space from outside sources, normally via infiltration through doors, windows, and wall penetrations for pipes, ducts, etc. But the largest external source is outside makeup air entering through the air-conditioning system.

In an operating clean room, external particle sources normally present little impact on overall clean room particle concentration. However research has shown a direct correlation between ambient particle concentrations and indoor particle concentrations for clean spaces "at rest." Control of external particulate sources is primarily through air filtration, room pressurization, and sealing of process related space penetrations.

Internal Sources. Particulate contamination generated within the clean space is the result of people, clean room surface shedding, process equipment, and the manufacturing process itself. Clean room personnel are potentially the largest source of internal particles. Personnel may generate several thousand to several million particles per minute in a clean room. These particles are usually skin flakes, moisture droplets, cosmetics, hair, etc. Personnel generated particles are controlled with airflow designed to continually "wash" the personnel with clean air, new clean room garments, and proper gowning procedures. As personnel work in the clean room, their body movements may re-entrain airborne particles from other sources. Other activities, such as writing, may also cause higher particle concentrations.

Particle concentrations within the clean room may be used to define clean room class, but actual particle deposition on the product is of greater concern. The science of aerosols, filter theory, and fluid motion are the primary sources of understanding in the areas of contamination control. Clean room designers may not be able to control or prevent internal particle generation completely, but they may anticipate internal sources and design control mechanisms to limit their impact on the product.

Fibrous Air Filters

Most externally generated particles are prevented from entering the clean room with proper air filtration. Present technology for high efficiency air filters centers around two types: high efficiency particulate air (HEPA) filters and ultra low penetration air (ULPA) filters.

Both HEPA and ULPA filters use glass fiber paper technology. HEPA and ULPA filters are usually constructed in a deep pleated form with either aluminum, coated string, or filter paper as pleating separators. Filters may vary from 50 to 300 mm in depth; correspondingly higher media area is available with deeper filters and more concentrated pleat spacing.

Theories and models describing the different fibrous filter capture mechanisms have been developed and verified by empirical data. The consensus today is that interception and diffusion are the dominant capture mechanisms for HEPA filters. Fibrous filters have their lowest removal efficiency at the most penetrating particle size (MPPS), which is determined by filter fiber diameter, volume fraction or packing density, and air velocity. For most HEPA filters the MPPS is between 0.1 and 0.3 μm. Thus HEPA and ULPA filters have rated efficiencies based on 0.3 μm and 0.12 μm particle sizes, respectively.

Air Pattern Control

Air turbulence within the clean space is strongly influenced by air supply and return configurations, people traffic, and process equipment layout. The selection of the air pattern configurations is the first step for good clean room design. User requirements for cleanliness level, process equipment layout, available space for installation of air pattern control equipment (i.e., air handlers, clean work stations, environmental control components, etc.), and project financial considerations all influence the final air pattern design selection. Project financial aspects govern air pattern control concepts where operating and capital costs may limit the type and size of air handling equipment to be used.

Numerous air pattern configurations are in use, but in general they comprise two categories, unidirectional airflow (commonly, but incorrectly, called laminar flow) and nonunidirectional airflow (commonly called turbulent airflow).

Unidirectional airflow, although not truly laminar airflow, is characterized as air flowing in a single pass in a single direction through a clean room or clean zone with generally parallel streamlines. Ideally, the flow streamlines would be uninterrupted, and although personnel and equipment in the airstream do distort the streamlines, a state of constant velocity is approximated. Most particles that encounter an obstruction in a laminar airflow strike the obstruction and continue around it as the laminar airstream reestablishes itself downstream of the obstruction.

Nonunidirectional airflow does not meet the definition of unidirectional airflow by having either multiple pass circulating characteristics or a nonparallel flow direction.

Nonunidirectional Airflow

Variations of nonunidirectional airflow are primarily based on the location of supply air inlets and outlets and air filter locations. Some examples of different nonunidirectional airflow systems can be seen in Figures 2 and 3. Airflow is typically supplied to the space through supply diffusers with HEPA filters (Figure 2) or through supply diffusers with HEPA filters in the ductwork or air handler (Figure 3). Air can also be prefiltered in the supply system components, with HEPA filtered work stations located in the clean space.

Nonunidirectional airflow may provide satisfactory contamination control results for cleanliness levels of Class M4.5 through Class M6.5. Attainment of desired cleanliness classes with designs similar to Figures 2 and 3 presupposes that the major space contamination is from external sources (i.e., makeup air) and that contamination is removed in air handler or ductwork filter housings, or through HEPA filter supply devices. When internally generated particles are of primary concern, clean work stations are provided in the clean space.

Air turbulence is harmful in flow control systems, but air turbulence is needed in dilution control to enhance the mixing of low and high particle concentrations and produce a homogeneous particle concentration level acceptable to the process.

Unidirectional Airflow

Air patterns and air turbulence reduction are optimized in unidirectional airflow systems. In a vertical laminar flow (VLF) room, air is typically introduced through the ceiling and returned through a raised floor or at the base of sidewalls. This configuration produces nominally parallel airflow. In a horizontal flow clean room, air enters one wall and returns on the opposite wall.

A *downflow clean room* has a ceiling with HEPA filters. In a clean room with a low class number, the greater part of the ceiling requires HEPA filters. For a Class M3.5 room, the entire ceiling will usually require HEPA filtration. Ideally, a grated or perforated floor serves as the air exhaust. This type of floor is inappropriate in pharmaceutical clean rooms, which typically have solid floors and low level returns.

In a downflow clean room a uniform shower of air bathes the entire room in a downward flow of ultraclean air. Contamination generated in the space will not move laterally against the downward flow of air (it is swept down and out through the floor) and will not contribute to a buildup of contamination in the room.

Care must be taken in the design, selection, and installation of the system to seal the HEPA ceiling. Assuming that the HEPA filters installed in the ceiling have been properly sealed, this design can provide the cleanest working environment presently available.

In a *horizontal flow clean room* the supply wall consists entirely of HEPA filters supplying air at a velocity of approximately 0.46 m/s across the entire section of the room. The air then exits through the

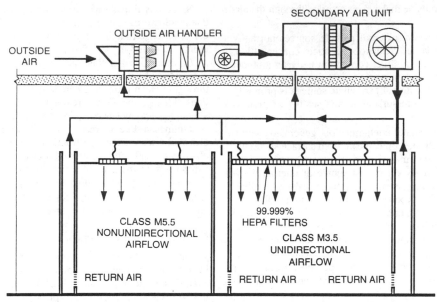

Fig. 2 Class M5.5 Nonunidirectional Clean Room with Ducted HEPA Filter Supply Elements and Class M3.5 Unidirectional Clean Room with Ducted HEPA Filter Ceiling

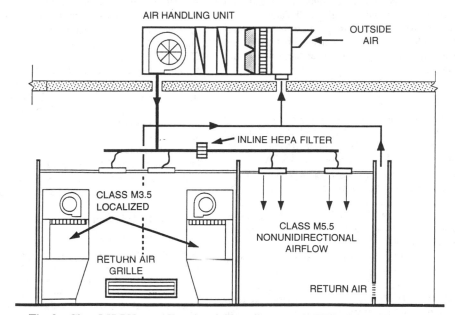

Fig. 3 Class M5.5 Nonunidirectional Clean Room with HEPA Filters Located in Supply Duct and Class M3.5 Local Work Stations

return wall at the opposite end of the room and recirculates within the system. As with the downflow room, this design removes contamination generated in the space at a rate equal to the air velocity and does not allow cross-contamination perpendicular to the airflow. However, a major limitation to this design is that downstream air becomes contaminated. In this design, the air first coming from the filter wall is as clean as air in a downflow room. The process activities can be oriented to have the cleanest, most critical operations in the first air, or at the clean end of the room, with progressively less critical operations located toward the return air, or dirty end of the room.

U.S. Federal *Standard* 209E, Airborne Particulate Cleanliness Classes in Clean Rooms and Clean Zones (1992), does not specify velocity standards, so the actual velocity is as specified by the owner or his agent. The 0.46 m/s velocity specified in a previous standard

(FS 209B) is still widely accepted in the clean room industry. Current research suggests lower velocities may be possible. Care should be taken to assure that required cleanliness levels are maintained. Other reduced air volume designs use a mixture of high and low pressure drop HEPA filters, reduced coverage in traffic areas, or lower velocities in personnel corridor areas.

Unidirectional airflow systems have a predictable airflow path that airborne particulates tend to follow. Without good filtration practices, unidirectional airflow will only ensure a predictable path for particulates. However, superior clean room performance may be obtained with a good understanding of unidirectional airflow. This airflow remains parallel (or within 18° of parallel) to below the normal work surface height of 760 to 910 mm. But this flow deteriorates when the air encounters obstacles such as process equipment and work benches. Personnel movement also degrades the flow. The result is a clean room with areas of good unidirectional airflow and areas of turbulent airflow.

Turbulent zones have countercurrents of air with high velocities, reverse flow, or even no flow (stagnancy). The effect of countercurrents is to produce stagnant zones where small particles may cluster and finally settle onto the product; or countercurrents may lift particles from contaminated surfaces. Once lifted, these particles may deposit on product surfaces.

Clean room mockups may help the designer avoid turbulent airflow zones and countercurrents. Smoke, neutral buoyancy helium filled soap bubbles, and nitrogen vapor fogs can make air streamlines visible within the clean room mockup.

Computer Aided Flow Modeling

Computer models of particle trajectories, transport mechanisms, and contamination propagation are commercially available. Flow analysis with computer models may compare flow fields associated with different process equipment, workbenches, robots, and building structural design. Airflow analysis of flow patterns and air streamlines is performed by solving equations of fluid mechanics for laminar and turbulent flow where incompressibility and uniform thermophysical properties are assumed. Design parameters may be modified to determine the effect of airflow on particle transport and flow streamlines, thus avoiding the cost of mockups.

Major features and benefits associated with most computer flow models are

1. Two- or three-dimensional modeling of simple clean room configurations.
2. Modeling of both laminar and turbulent airflows.
3. Multiple air inlets and outlets of varying sizes and velocities.
4. Allowances for varying boundary conditions associated with walls, floors, and ceilings.
5. Aerodynamic effects of process equipment, workbenches, and people.
6. Prediction of specific airflow patterns of all or part of a clean room.
7. Reduced costs associated with new clean room design verification.
8. Graphical representation of flow streamlines and velocity vectors to assist in flow analysis (Figures 4 and 5).
9. Graphical representation of simulated particle trajectories and propagation (Figure 6).

Research has shown good correlation between flow modeling by computer and that done in simple mockups. However, the computer flow modelling software should not be considered a panacea for clean room design, as this is a new technique and possible inaccuracies may be discovered in the future.

TESTING CLEAN AIR AND CLEAN SPACES

Since early clean rooms were largely for the government, the testing procedures have been dictated by government standards. U.S. Federal *Standard* 209 is widely accepted, as it defines air cleanliness levels for clean spaces around the world. Aspects of clean room performance other than air cleanliness are no longer covered by FS 209E. Standardized testing methods and practices have been developed and published by the Institute of Environmental Sciences (IES), the American Society for Testing and Materials (ASTM), and other groups.

Three basic test modes for clean room systems are used to evaluate a facility properly: (1) as built, (2) at rest, and (3) operational. A clean room facility cannot be fully evaluated until it has performed under full occupancy, and the process to be performed in it is operational. Thus, the techniques for conducting initial performance tests and operational monitoring must be similar.

Sources of contamination, as previously described, are both external and internal. For both laminar and nonlaminar flow clean

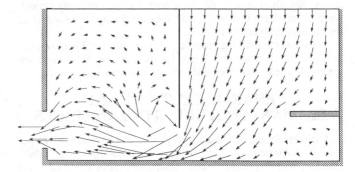

Note: Velocity vectors with length proportional to magnitude

Fig. 4   Clean Room Airflow Velocity Vectors Generated by Computer Simulation

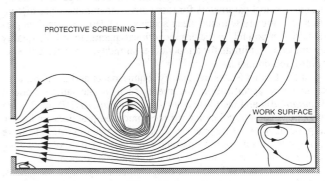

Fig. 5   Computer Modelling of Clean Room Airflow Streamlines

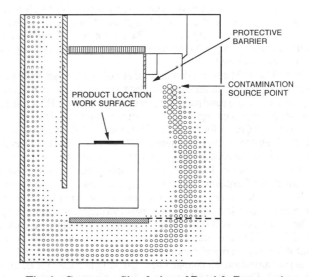

Fig. 6   Computer Simulation of Particle Propagation Within a Clean Room

rooms, the primary air loop is the major source for external contamination. Discrete particle counters using laser or light scattering principles may be used to detect particles of 0.01 to 5 μm. For particles 5.0 μm and larger, microscopic counting can be used, with the particles collected on a membrane filter through which a sample of air has been drawn.

HEPA filters should be tested for pinhole leaks at the following places: the filter media, the sealant between the media and the filter frame, the filter frame gasket, and the filter bank supporting frames. The area between the wall or ceiling and the frame should also be tested. A pinhole leak at the filter bank can be extremely critical, since the concentration of the leak varies inversely as the square of the pressure drop across the hole.

The clean room testing procedure of the Institute of Environmental Sciences (1993) describes 12 tests for clean rooms. Which tests are applicable to each specific clean room project must be determined.

PHARMACEUTICAL AND BIOMANUFACTURING CLEAN SPACES

Facilities for the manufacture of pharmaceutical products require careful assessment of many integrated entities including HVAC, controls, room finishes, process equipment, room operations, and utilities. Flow of equipment, personnel, and product must also be considered. It is important to involve the designers, operators, commissioning staff, quality control, maintenance, constructors, and the production representative during the conceptual stage of a facility design. Critical parameters for room environment and types of controls vary greatly with the clean space's intended purpose. It is particularly important to determine critical parameters with quality assurance to set limits for temperature, humidity, pressure, and other control requirements.

In the United States, regulatory requirements and a multitude of clean space specification documents such as the current Federal *Standard* 209 and the *Code of Federal Regulations* 210 and 211 are available. These regulations and recommended practices are known as good manufacturing practice (GMP). The goal of GMP is to serve as a guide to achieve a proper and repeatable method of producing sterile products free from microbial and particle contaminants.

Various countries have their own regulatory agencies and guides, which a pharmaceutical manufacturer must meet if it is to dispense product to that country. The Commission of the European Communities (CEC) adopted the European Union (EU) Guide to Good Manufacturing Practice for Medicinal Products in 1992. Much of the regulation is based on the United Kingdom's Guide to Good Pharmaceutical Manufacturing Practice. Except for the EU guide, other documents are strictly advisory and may not be construed as legally binding specifications.

In the United States, the one factor that makes pharmaceutical processing suites most different from clean spaces for other purposes (e.g., electronic and aerospace) is the requirement to pass FDA inspections for product licensing. It is important to include the FDA's regulatory arms, the Center for Biologics Evaluation and Research (CBER) and the Center for Drug Evaluation and Research (CDER), for facility design early in the concept design process to test for objection.

Also, early in the design process, a qualification plan (QP) must be considered. Functional requirement specifications (FRSs), critical parameters and acceptance criteria, installation qualification (IQ), operational qualification (OQ), and performance qualification (PQ) in the clean room suites are all requirements for product validation. IQ, OQ, and PQ protocols, in part, set the acceptance criteria and control limits for critical parameters such as temperature, humidity, room pressurization, air change rates, and particle counts (or air classifications). These protocols must receive defined approvals in compliance with the owner's guidelines. The qualification plan must also address standard operating procedures (SOPs), preventive maintenance (PM), and operator and maintenance personnel training.

Biomanufacturing and pharmaceutical aseptic clean spaces are typically arranged in suites, with clearly defined operations in each suite. For example, common convention positions an "aseptic core" (Class M3.5) filling area in the innermost room, which is at the highest pressure, surrounded by areas of descending pressure and increasing particulate classes (see Figure 7).

In aseptic processing facilities, the highest quality air is intentionally placed within the lesser quality areas and separated by room pressure differences via air locks. A commonly used pressure difference is 15 Pa between rooms, with the higher quality room having

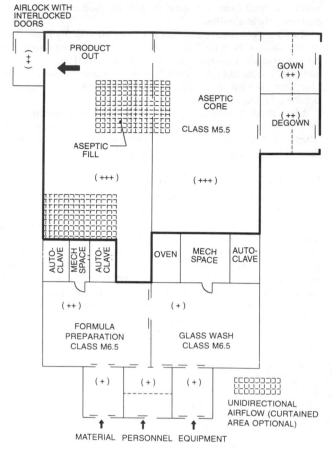

(+) = ROOM PRESSURIZATION LEVEL ABOVE ZERO REFERENCE

Fig. 7   Typical Aseptic Suite

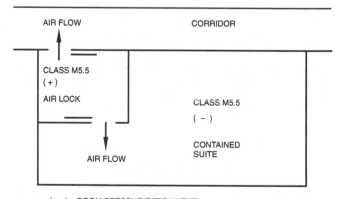

(–) = ROOM PRESSURIZATION LEVEL
ABOVE ZERO REFERENCE

Fig. 8   Contained Suite Arrangement

the higher pressure. Other pressure differences are acceptable if they are proven effective. This pressure differential is generally accepted as good management practice to inhibit particles from entering a clean suite.

Where product containment is an issue (i.e., pathogens), the suite requires a lower pressure than the adjacent rooms, but it may still require a higher air quality. In this case the air lock may be designed to maintain one pressure level higher than the adjacent room and also higher than the contained suite (Figure 8). Biological containment requirements are addressed by the U.S. National Institutes of Health where potentially hazardous organisms are grouped in biosafety levels BL-1 to BL-4.

START-UP AND QUALIFICATION OF PHARMACEUTICAL CLEAN ROOMS

Qualification of HVAC for Aseptic Pharmaceutical Manufacturing

Qualification of pharmaceutical clean room HVAC systems is part of the overall commissioning of the building and its systems, except that documentation is more rigorous. The qualification covers the equipment affecting critical parameters and their control. Other groups within the manufacturing company, such as the safety or environmental groups, may also require similar commissioning documentation for systems in their areas of concern. The most important objective in meeting the requirements of the approving agency is to (1) state what you are going to do and then verify that it was done and (2) show that you are protecting the product and meeting your acceptance criteria.

Qualification Plan and Acceptance Criteria

Early in the design, the owner and designer should discuss who will be responsible for as-built drawings, setting up maintenance files, and training. They should create a qualification plan for the HVAC systems, which includes (1) a functional description of what the systems will do, (2) maps of room pressures and cleanliness zones served by each air handler, (3) a list of critical components to be qualified, (4) a list of owner's procedures that must be followed for the qualification of equipment and systems that affect critical parameters, (5) a list of qualification procedures (IQ/OQ/PQ protocols) that must be written especially for the project, and (6) a list of needed commissioning equipment.

The approval should also be defined in the QP. It is important to be able to measure and document the critical parameters of a system (such as room pressure), but it is also important to document the performance of system components that affect that critical parameter (i.e., fan performance, motor brake power, duct pressure, airflow volume, etc.) for GMP as well as business records. Documentation helps ensure that replacement parts (such as a motor) can be specified, purchased, and installed to satisfy the critical parameter.

It is important to determine what the critical components are (including instruments and alarms), the performance of which could affect critical parameters, and the undetected failure of which could lead to adulteration of the product. If performance data are in the qualification records, replacement parts of different manufacture can be installed without major change control approvals—as long as they meet performance requirements. Owner approvals for the qualification plan should be obtained while proceeding with detailed design.

Qualification is the successful completion of the following activities for critical components and systems. The designer should understand the requirements for owner's approval of each protocol (usually, the owner approves the blank protocol form and the subsequently executed protocol):

The *installation qualification (IQ) protocol* is a record of inspection of construction to verify compliance with contract documents, including completion of punchlist work, for critical components. It may include material test reports, receipt verification forms, shop inspection reports, motor rotation tests, and contractor-furnished testing and balancing. This record also includes calibration records for instrumentation used in commissioning and for installed instrumentation (such as sensors, transmitters, controllers, and actuators) traceable to National Institute of Standards and Technology (NIST) instruments.

Control software should be bench tested, and preliminary (starting) tuning parameters should be entered. Control loops should be dry-loop checked to verify that the installation is correct. Equipment and instruments should be tagged and wiring labelled. Documentation must attest the completion of these activities. This includes as-built drawings and installation-operation-maintenance (IOM) manuals from the contractors and vendors.

The *operational qualification (OQ) protocol* documents the start-up of the systems that includes the critical components. This includes individual wet-loop testing of control loops under full system operating pressure performed in a logical order (i.e., fan control before room pressure control). The commissioning agent must produce verification that operating parameters are within acceptance criteria.

The system should be challenged under extremes of design load (where possible) to verify operation of alarms and recorders, to determine (and correct, if significant) weak points in the system, and to verify control and door interlocks. Based on observations of the system, informal alert values of critical parameters, which might signify abnormal system operation, may be considered. Although product would not be adulterated at these parameter values, the owner's staff could assess an alarm and react to it before further excursions from normal operation were experienced.

Documented smoke tests verify space pressure and airflow in critical rooms or inside containment hoods, and show airflow patterns and directions around critical parts of production equipment. Many smoke tests have been documented with videotape, especially when room pressure differentials are lower than acceptance criteria require and cannot be corrected.

Files should include an updated functional description of the system, which describes how the system operates, and schematics and room pressure maps that accompany it. Copies should be readily accessible and properly filed.

Other Documentation. Documentation should also include test reports for HEPA filters (efficiency or pinhole-scan integrity tests) at final operating velocities. If the tests were performed by the filter installer, these data should have been part of the IQ package.

Documentation should verify that instrumentation displays, tracks, and stores critical parameters and action alarms. (Consider recording data by exception and routine logging of data at minimal frequency.)

Systems and equipment should be entered into the owner's maintenance program, with rough drafts of new maintenance procedures (final drafts should reflect commissioning experience).

Records should attest the completion of these activities, including final as-builts and air and water balance reports.

Performance qualification (PQ) is proof that the entire system performs as intended under actual production conditions. PQ is the beginning of the ongoing verification that the system meets the acceptance criteria (often called validation). This includes documentation of

- Maintenance record keeping and final operating and maintenance procedures in place, with recommended frequency of maintenance. The owner may also want a procedure for periodic challenge of the control and alarm systems.
- Logs of critical parameters that prove the system maintains acceptance criteria over a prescribed time period.
- Records of training of operators and maintenance personnel.
- Final loop tuning parameters.

After acceptance of PQ, the owner's change control procedure should limit further modifications to critical components (as shown on IQ and OQ forms) that affect the product. Although much of the building's HVAC system equipment should not fall under the need for qualification, it is good business practice to keep records for the entire facility's system and to correct problems before they become significant. Records of the corrections should also be maintained.

Once the system is operational, pharmaceutical product trial lots are run in the facility (process validation), and the owner should monitor viable (microbial) and nonviable particulates in the room.

Nonaseptic Pharmaceutical Manufacturing

Although the previous discussion centers on aseptic manufacturing clean spaces, the design and qualification of pharmaceutical manufacturing for nonaseptic products (like topical and oral products) generally follows the same approach, but with fewer critical parameters and fewer components to be qualified. However, in many facilities that manufacture powdered materials, humidity may become a more important, and more tightly controlled, critical parameter. In that case, HEPA filters perform more of a dust capture role than bacteria control and may be efficiency tested rather than pinhole scanned. To avoid overdesign and overqualification costs, the owner must determine the function of the facility and the qualification documentation needed.

SEMICONDUCTOR CLEAN ROOMS

Most microelectronic facilities manufacturing semiconductors require a clean room providing Class M3.5 and cleaner. State of the art clean room technology has been driven by the reducing size of microelectronics circuitry. A deposited particle having a diameter of 10% of the circuit width may result in a circuit failure. With circuit line widths approaching 0.25 μm, particles of 0.025 μm are of concern. Many of today's facilities are designed to meet as-built air cleanliness of less than 35 particles 0.1 μm and larger per cubic metre of air.

Semiconductor Clean Room Configuration Classes

Semiconductor clean rooms today are of two major configurations—the *clean tunnel* or the *open bay* design. The clean tunnel is composed of narrow modular clean rooms that may be completely isolated from each other. Fully HEPA filtered pressurized plenums, ducted HEPA filters, or individual fan modules are used in clean tunnel installations. Production equipment may be located within the tunnel or installed through the wall where a lower cleanliness level (nominally Class M5.5) service chase is adjacent to the clean tunnel. The service chase is used in conjunction with either sidewall or a raised floor. A basement return system is also used with a raised floor.

The primary advantage of the tunnel design is reduced HEPA filter coverage and ease of expanding additional tunnel modules into unfacilitated areas. The tunnel is typically between 3.4 and 4.3 m. If the tunnel is narrower, production equipment cannot be placed on both sides. If it is wider, the flow becomes too turbulent and tends to break toward the walls before it leaves the work plane.

The tunnel design also has the drawback of restricting new equipment layouts. Clean room flexibility is very valuable to semiconductor manufacturing logistics. As processes change and new equipment is installed, the clean tunnel may restrict equipment location to the point that a new module must be added. The tunnel approach does not allow easy movement of product from one type of equipment to another.

The open bay design involves large (up to 5000 m²) open construction clean room layouts. Interior walls may be placed wherever manufacturing logistics dictate, thus providing maximum equipment layout flexibility. Replacement of process equipment with newer equipment is an ongoing process for most wafer fabrication facilities where support services must be designed to handle different process equipment layouts and even changes in process function. Many times a manufacturer may completely redo the equipment layout if a new product is being made.

In the open bay design, pressurized plenum or ducted filter modules are used, but pressurized plenum systems are becoming more common. When the pressurized plenum design is used, either one large plenum with multiple supply fans or small adjacent plenums may be configured. Small plenums provide the ability to shut down areas of the clean room without disturbing other clean areas. Small plenums may also include one or more supply fans.

Major semiconductor facilities, with total manufacturing areas of 3000 m² and larger, may incorporate both open bay and tunnel design configurations. Flexibility to allow equipment layout revisions warrants the open bay design. Process equipment suitable for through-the-wall installation, such as diffusion furnaces, may use the tunnel design or open bay; whereas process equipment such as lithographic steppers and coaters must be located entirely under laminar flow conditions—thus, open bay designs are more suitable. The decisions on which method to use, tunnel or open bay, should be discussed among the clean room designer, production personnel, and contamination control specialist.

Many semiconductor facilities contain separate clean rooms for process equipment ingress into the main factory. These ingress areas are staged levels of cleanliness. For instance, the equipment receiving area may be Class M6.5 for equipment uncrating, while the next stage may be Class M5.5 for preliminary equipment setup and inspection. The final stage, where equipment is cleaned and final installation preparations made prior to the fabrication entrance, is Class M4.5. In some cases, these staged clean rooms must have adequate clear heights to allow for lifting of equipment subassemblies.

Vertical Unidirectional Airflow Systems

Current semiconductor industry clean rooms use the vertical unidirectional airflow system, which produces a uniform shower of clean air throughout the entire clean room. Particles are swept from personnel and process equipment, with contaminated air leaving at the floor level; this produces clean air for all space above the work surface. In contrast, horizontal airflow systems produce decreasing cleanliness downstream from the air inlet.

In vertical unidirectional airflow systems with cleanliness Classes M1.5, M2.5, and M3.5, the clean room ceiling area consists of HEPA filters set in a nominal grid size of 600 mm by 1200 mm. HEPA filters are set into a T-bar grid with gasket or caulked seals for many Class M3.5 systems; Class M1.5 and M2.5 often use either low vapor pressure petrolatum fluid or silicone dielectric gel to seal the HEPA filters into a channel shaped ceiling grid. Whether T-bar or channel shaped grids are used, the HEPA filters normally cover 85 to 95% of the ceiling area, with the remainder of the ceiling area composed of gridwork and fire protection sprinkler panels.

HEPA filters in vertical unidirectional airflow designs are installed (1) with a pressurized plenum above the filters, (2) through individually ducted filters, or (3) with individually fan powered filter modules. A system with a plenum must provide even pressurization to maintain uniform airflow through each filter. Ducted systems typically have higher static pressure losses from the ducting and balance dampers. Higher maintenance costs may also be incurred due to the balance method involved with ducted systems.

Individual fan powered filter modules use fractional kilowatt fans (usually forward-curved fans) that provide airflow through one filter assembly. This method allows airflow to be varied throughout the clean room and requires less space for mechanical components. The disadvantages of this method are the large number of fans involved, low fan and motor efficiencies due to the small size, potentially higher fan noises, and higher maintenance costs.

Vertical unidirectional airflow is normally returned through the floor with perforated raised floor panels or through floor grates in a waffle-type floor structure. When raised floor panels are used, vibration may be a problem due to the lack of rigidity. Insufficient raised floor height may also propagate air turbulence below the floor and lift up particles; or turbulence may increase system static pressure requirements. When through-the-floor return grating is used, a basement return is normally included to provide a more uniform return as well as floor space for "dirty" production support equipment.

Sidewall returns are an alternative to through-the-floor returns; however, airflow may not be uniform throughout the work area. As previously stated, these returns are most applicable for clean rooms with double sidewall returns and widths less than 4.3 m.

With any HEPA filtered air system, adequate prefiltration is an economical way to increase HEPA filter life. Prefilters are located in the recirculation airflow, either in the return basement or in the air handler. If the prefilter installation requires that dirty prefilters be removed through the clean room, the potential for contaminating the clean space exists. Another disadvantage is that in the event of a chemical spill, prefilters will become contaminated with sometimes very hazardous chemicals, which complicates the cleanup procedure. Spill control may be handled more effectively with the basement-type return system.

Air Ionization Systems. In addition to clean room particle control with fibrous filters, air ionization technology is sometimes used to control particle deposition on product surfaces. However, these devices may deposit particulates on the product in question.

ENVIRONMENTAL SYSTEMS

Cooling Loads and Cooling Methods

Two major internal heat load components in semiconductor facilities are process equipment and fan energy. Because most clean rooms are located entirely within conditioned space, traditional heat sources of infiltration, fenestration, and heat conductance from adjoining spaces are typically less than 2 to 3% of the total load. Some clean rooms have been built with windows to the outside, usually for daylight awareness, and a corridor separating the clean room window from the exterior window.

The major cooling sources designed to remove clean room heat and maintain environmental conditions are makeup air units, primary and secondary air units, and the process equipment cooling water system. Some process heat, typically heat from electronic sources in process equipment computers and controllers, may be removed by the process exhaust system.

Fan energy is a very large heat source in Class M3.5 or better clean rooms. Recirculated airflow rates of 460 L/s per square metre, which equals 600 air changes per hour, are usual for Class M3.5 or better clean rooms.

Latent loads are primarily associated with makeup unit dehumidification. The low dry-bulb leaving air temperature (2 to 7°C) associated with dehumidified makeup air supplements sensible cooling. Supplemental cooling by makeup air may account for as much as 950 W per square metre of clean room.

Process cooling water (PCW) is used in process equipment heat exchangers, performing either simple heat transfer to cool internal heat sources or process specific heat transfer, in which the PCW contributes to the process reaction.

The diversity of the manufacturing heat load sources, that is, the portion of the total heat load that is transferred to each cooling medium, should be well understood. When bulkhead or through-the-wall equipment is used, the equipment heat loss to support chases versus to the production area will impact the cooling system design when the support chase is served by a different cooling system than the production area.

Makeup Air Systems

The control of makeup air and clean room exhaust affects clean room pressurization, humidity, and room cleanliness level. The flow requirements of makeup air are dictated by the amount required to replace process exhaust and by air volumes for pressurization. Makeup air volumes can be much greater than the total process exhaust volume in order to provide adequate pressurization and safe ventilation standards.

Makeup volumes are adjusted with zone dampers and makeup fan controls using speed controllers, inlet vanes, etc. Opposed blade dampers should have low leak characteristics and minimum hysteresis.

Makeup air should be filtered prior to injection into the clean room because it is the primary source of external particulates in the clean room. If the makeup air is injected upstream of the clean room HEPA filters, minimum 95% (ASHRAE Atmospheric Dust Test) efficient filters should be used to avoid high dust loading on the HEPA filters. In addition, prefilters of 30% and then 85% should be used to prolong the life of the 95% filter. When makeup air is injected downstream of the main HEPA filter system, HEPA filtering of the makeup air should be of the same removal efficiency (minimum 95%).

In addition to particle filtering, many makeup air handler systems require filters to remove chemical contaminants present in the outside air. These contaminants include salts and pollutants from industries and automobiles. Chemical filtration may be accomplished with chemical absorbers such as activated carbon.

Makeup air is frequently introduced into the primary air path on the suction side of the primary fan(s). Makeup air may also be introduced into the pressurized HEPA plenum, but additional energy is required to offset the HEPA plenum static pressure.

Process Exhaust Systems

Process exhaust systems for semiconductor facilities handle acid, solvent, toxic, pyrophoric (self-igniting) fumes, and process heat exhaust. Process exhaust systems should be dedicated for each fume category, dedicated by process area, or based on the chemical nature of the fume and its compatibility with exhaust duct material. Typically, process exhausts are segregated into corrosive fumes, which are ducted through plastic or fiberglass reinforced plastic (FRP) ducts, and flammable (normally from solvents) gases and heat exhaust, which are ducted in metal duct systems. Care must be taken to ensure that gases cannot combine into hazardous compounds that can ignite or explode within the ductwork. Segregated heat exhaust systems are sometimes installed to recover heat, or heat may be exhausted into the suction side of the primary air path.

Required process exhaust volumes vary from 5 L/s per square metre of clean room for photolithographic process areas to 50 L/s per square metre for wet etch, diffusion, and implant process areas. When specific process layouts have not been designated prior to exhaust system design, an average of 25 L/s per square metre is normally acceptable for fan and abatement equipment sizing. Fume exhaust ductwork should be sized at low velocities (5 m/s) to allow for future requirements.

For many airborne substances, the American Conference of Governmental Industrial Hygienists (ACGIH) has established requirements to avoid excessive worker exposure. Specific standards for the allowable concentration of airborne substances are set by the U.S. Occupational Safety and Health Administration (OSHA). These limits are based on working experience, laboratory research, and medical data; and they are subject to constant revision. Reference to the latest standards available should be made when evaluating a clean room exposure. The ACGIH publishes *Industrial Ventilation: A Manual of Recommended Practices,* which may be referred to when limits are to be determined.

Fire Safety for the Exhaust System. The Uniform Building Code (UBC) designates semiconductor fabrication facilities as Group H, Division 6, occupancies. The H-6 occupancy requirements impact architectural, mechanical, and electrical designs significantly when compared to the previous B-2 class in the UBC. The H-6 occupancy class should be reviewed even if the local jurisdiction does not use the UBC because it is currently the only major code specially written for the semiconductor industry and, hence, can be considered "usual practice." This review is particularly helpful if the local jurisdiction has few semiconductor facilities.

Uniform Fire Code (UFC) Article 51, Semiconductor Fabrication Facilities Using Hazardous Production Materials, addresses the specific requirements for process exhaust systems relating to fire safety and minimum exhaust standards. UFC Article 80, Hazardous Materials, is relevant to many semiconductor clean room projects due to the large quantities of hazardous materials stored in these areas. Areas covered include ventilation and exhaust standards for production and storage areas, control requirements, use of gas detectors, redundancy and emergency power, and duct fire protection.

Temperature and Humidity

Precise temperature control is required in most semiconductor clean rooms. Specific chemical processes may change under different temperatures, or masking alignment errors may occur due to product dimensional changes as a result of the coefficient of expansion. Temperature tolerances of ±0.5 K are common, and precision of 0.05 to 0.3 K is likely in wafer or mask writing process areas. Wafer reticle writing by electron beam technology requires ±0.05 K while photolithographic projection printers require ±0.3 K tolerance. Specific process temperature control zones must be small enough to counteract the large air volume inertia in vertical laminar flow clean rooms. Internal environmental controls, which allow for room tolerances of ±0.5 K and larger temperature control zones, are used in many process areas.

Within temperature zones of the typical semiconductor factory, latent heat loads are normally small enough to be offset by incoming makeup air. Sensible temperature is controlled with either (1) cooling coils in the primary air stream or (2) unitary sensible cooling units that bypass primary air through the sensible air handler and blend conditioned air with unconditioned primary air.

In most clean rooms of Class M4.5 or better, production personnel wear full coverage protective smocks that require clean room temperatures of 20°C or less. Process temperature set points may be higher as long as tolerances are maintained. If full coverage smocks are not used, lower temperature set points are recommended for comfort.

Semiconductor humidity levels vary from 30 to 50% rh. Humidity control and precision are functions of process requirements, prevention of condensation on cold surfaces within the clean room, and control of static electric forces. Humidity tolerances vary from 0.5 to 5% rh, primarily dictated by process requirements. Photolithographic areas have the more precise standards and lower set points. Photoresists are chemicals used in photolithography, and their exposure timing can be affected with varying relative humidities. Negative resists typically require low (35 to 45%) relative humidity. Positive resists tend to be more stable, so the relative humidity can go up to 50% rh where there is less of a static electricity problem.

Independent makeup units should control dew point in places where direct-expansion systems, chilled water/glycol cooling coils, or chemical dehumidification are used. Chemical dehumidification is rarely used in semiconductor facilities due to the high maintenance costs and the potential for chemical contamination in the clean room. While an operating clean room generally does not require reheat, systems are typically designed to provide heat to the space when new clean rooms are being built and no production equipment has been installed.

Makeup air is humidified by steam humidifiers, water spray nozzles, or evaporative coolers. Steam humidifiers are most commonly used. Care should be taken to avoid the potential for the release of water treatment chemicals. Stainless steel unitary packaged boilers with high purity water and stainless steel piping have also been used. Water sprayers, located in the clean room return, use air-operated, water jet sprayers. Evaporative coolers have been used, taking advantage of the sensible cooling effect in dry climates.

Pressurization

Pressurization of semiconductor clean rooms is an additional method of contamination control, providing resistance to infiltration of external sources of contaminants. Outside particulate contaminates enter the clean room by infiltration through doors, cracks, pass-throughs, and other process related penetrations for pipes, ducts, etc. Positive pressure in the clean room (as referenced to any less clean space) ensures that air flows from the cleanest space to the less clean space. Positive differential pressure in the clean room inhibits the entrance of unfiltered external particulate contamination.

A differential pressure (DP) of 12.5 Pa is a widely used standard. The cleanest clean room should maintain the highest pressure, with pressure levels decreasing corresponding to decreasing levels of cleanliness.

Pressure levels within the clean room are principally established by the balance between process exhaust and makeup air volumes and the supply and return air volumes. Process exhaust requirements are dictated by process equipment vendors and industrial hygienists, and they cannot be changed without risk to safety. Clean room supply air volumes are also set by contamination control specialists. Control of makeup air and return air volumes is the primary means for pressure control. Pressure differences should be kept as low as possible while still creating the proper flow direction. Large pressure differences can create eddy currents at wall openings and cause vibration problems.

Static or active control methods are normally used in clean room pressure control. One method is used in lieu of the other based upon pressure control tolerances. Pressure level control precision is typically ±2.5 to ±7.5 Pa; the owner's contamination control specialist specifies the degree of precision required. Many semiconductor processes where clean room pressure affects the process itself (e.g., glass deposition with silane gas) require process chamber pressure precision of ±0.6 Pa.

Static pressure control methods are suited for static or unchanging clean room environments, where the primary pressure control parameters, process exhaust and supply air volumes, either do not change or change slowly over weeks or months at a time. Static control systems provide initial room pressure levels, and monthly or quarterly maintenance adjusts makeup and return volumes if the pressure level has changed. Static systems may include DP gages for visual monitoring by maintenance personnel.

Active system designs provide closed loop control where pressure control is critical. Standard control systems normally do not have a quick enough response to maintain the DP when doors are opened. Active systems should be carefully evaluated as to their need, however.

Air locks are also used to segregate pressure levels within the factory, but typically air locks are used only between uncontrolled personnel corridors and entrance foyers and the protective clothing (smocks) gowning area. Air locks may also be used between the gowning room and the main wafer fabrication area and for process equipment staging areas prior to ingress into the wafer fabrication area. Within the main portion of the factory, air locks are rarely used because they restrict personnel access, evacuation routes, and traffic control.

System Sizing and Redundancy

The design of environmental systems must consider future requirements of the factory. Semiconductor products can become obsolete in as little as two years, and process equipment may be replaced as new product designs dictate. As new processes are added or old ones deleted (e.g., wet etch versus dry etch) the function of one clean room may change from high humidity requirements to low humidity, or the heat load may increase or decrease. Thus the clean room designer must design for flexibility and growth. Unless specific process equipment layouts are available, maximum cooling capability should be provided in all process areas at the time of installation, or space should be provided for future installations.

Because clean room space relative humidity must be held to close tolerances, and humidity excursions cannot be tolerated, the latent load removal air systems should be based on high ambient dew points and not on the high mean coincident dry-bulb/wet-bulb data.

In addition to proper equipment sizing, system redundancy is also desirable when economics dictate it. Many semiconductor wafer facilities operate 24 hours per day, seven days per week and shut down only during holidays and scheduled nonwork times. Mechanical and electrical system redundancy is required if the loss of such equipment would shut down critical manufacturing processes. For example, process exhaust fans must operate continuously for safety reasons, while particularly hazardous exhaust should have two fans, both running. The majority of the process equipment is computer controlled with interlocks to provide safety for personnel and products. Electrical redundancy or uninterrupted power supplies may be necessary to prevent costly downtime during power outages.

Energy Conservation in Clean Rooms

The major operating costs associated with a clean room include conditioning of makeup air, air movement within the clean room, and process exhaust. Environmental control, contamination control, and process equipment electrical loads can be as much as 3000 W/m^2. Besides process equipment electrical loads, the major energy users are cooling systems, air movement, and process liquid transport (i.e., deionized water and process cooling water pumping). It is important to provide a system with low static pressure duct systems and efficient fans.

Fan Energy. Because flow rates in typical semiconductor facilities are 90 to 100 times greater than in conventional HVAC systems, the fan system should be closely examined for areas of energy conservation. System static pressures and total airflow requirements should be designed to reduce system operating costs. The fan energy required to move recirculation air may be reduced by reducing the air volume or system static pressure.

Air volumes may be reduced by reducing HEPA filter coverage or by reducing the clean room average velocity. When air volumes are reduced, each square metre of reduced HEPA coverage saves 250 to 500 W/m^2 in fan energy and the same amount in cooling load. Reducing room average velocity from 0.45 m/s to 0.40 m/s saves 50 W/m^2 in fan energy and the same amount in cooling energy. If the amount of air supplied to the clean room cannot be lowered, reductions in system static pressure can produce significant savings. With good fan selection and transport system design, up to 150 W/m^2 can be saved per 250 Pa reduction in static pressure. Static pressure may be reduced by installing low pressure drop HEPA filters, pressurized plenums in lieu of ducted filters, and proper fan inlets and outlets. Operations in many clean rooms only run for one shift. Air volumes may be reduced during nonworking hours by using two-speed motors, inverters, inlet vanes, and variable pitch fans; or, in multifan systems, by using only a portion of the fans.

Additional fan energy savings may be achieved by installing high-efficiency motors instead of standard-efficiency motors. Good fan selection also influences system energy costs. The choice of forward-curved centrifugal fans versus backward-inclined, airfoil, or vaneaxial fans affects total system efficiency. The number of fans used in a pressurized plenum design influences system redundancy as well as total energy usage. The fan size specified changes power requirements as well. Different options for a prescribed configuration should be investigated.

Makeup and Exhaust System Energy. As stated previously, process exhaust requirements in the typical semiconductor facility vary from 5 to 50 L/s per square metre. Makeup air requirements vary correspondingly, with an added amount for leakage and pressurization. The energy required to supply the conditioned makeup air can be quite large. The quantity of exhaust in a given facility is normally determined by the type of equipment installed by the user. Heat recovery has been used effectively in process exhaust systems. When heat recovery is used, the selection of the heat exchanger material is very important due to the potentially corrosive atmosphere.

Makeup air quantities cannot normally be reduced without reducing the process exhaust. Due to safety and contamination control requirements, process exhaust may be difficult to reduce. Therefore the costs of conditioning the makeup should be investigated. Conventional HVAC methods of high efficiency chillers, good equipment selection, and precise control system design can also save energy. One example is the use of multiple-temperature chillers to bring outdoor air temperatures to a desired dew point in steps.

NOISE AND VIBRATION

Noise is one of the most difficult variables to control. Particular attention must be given to the noise generated by contamination control equipment. Prior to the start of the design, the noise and vibration criteria should be established. Chapter 43 provides more complete information on sound control.

In normal applications of contamination control equipment, vibration displacement levels need not be dampened below 0.5 µm in the 1 to 50 Hz range. However, electron microscopes and other ultrasensitive instrumentation may require smaller deflections in different frequency ranges. Photolithography areas may prohibit floor deflections greater than 0.076 µm. As a general rule, the displacement should not exceed one-tenth of the line width.

For highly critical areas, vaneaxial fans should be considered. These fans generate less noise in the lower frequencies, and they may be dynamically balanced to displacements of less than 4 µm, which will decrease the likelihood of vibration transmittal to sensitive areas.

ROOM CONSTRUCTION AND OPERATION

Controlling particulate contamination from other than supply air will depend upon the classification of the space and the type of system and operation involved. Typical items, which may vary with the room class, include the following:

Construction Finishes

- *General.* Smooth, monolithic, cleanable, and chip resistant, with minimum seams, joints, and no crevices or moldings.

- *Floors.* Sheet vinyl, epoxy, or polyester coating with carried up wall base, or raised floor with and without perforations using the above materials.

- *Walls.* Plastic epoxy-coated drywall, baked enamel, polyester, or porcelain with minimum projections.

- *Ceilings.* Plaster covered with plastic, epoxy, or polyester coating or with plastic-finished acoustical tiles when entire ceiling is not fully HEPA filtered.

- *Lights.* Teardrop shaped single lamp fixtures mounted between filters or flush mounted and sealed.

- *Service penetrations.* All penetrations for pipes, ducts, conduit runs, etc., should be fully sealed or gasketed.

- *Appurtenances.* All doors, vision panels, switches, clocks, etc., should have either flush mounted or sloped tops.

Personnel and Garments

- Hands and face are cleaned before entering area.
- Lotions and soap containing lanolin are used to lessen the emission of skin particles.
- Wearing cosmetics and skin medications is not permitted.
- Smoking and eating are not permitted.
- Lint-free smocks, coveralls, gloves, and head and shoe covers are worn.

Materials and Equipment

- Equipment and materials are cleaned before entry.
- Nonshedding paper and ballpoint pens are used. Pencils and erasers are not permitted.

- Work parts are handled with gloved hands, finger cots, tweezers, and other methods to avoid transfer of skin oils and particles.

Particulate Producing Operations

- Grinding, welding, and soldering operations are shielded and exhausted.
- Containers are used for transfer and storage of materials.

Entries

- Air locks and pass-throughs are used to maintain pressure differentials and reduce contamination.

CHAPTER 16

DATA PROCESSING SYSTEM AREAS

DATA processing system areas contain the computer equipment and ancillary equipment needed to perform a particular data processing function. Computers generate heat and contain components sensitive to extremes of temperature and humidity and to the presence of dust. Exposure to conditions outside prescribed limits can cause improper operation or complete shutdown of the equipment.

Ancillary spaces for computer-related activities or for the storage of computer components and materials (including magnetic tape, disk packs and cartridges, data cells, paper, and punch cards) require environmental conditions comparable to those where the computers are housed, although tolerances are generally wider and the degree of criticality is usually much lower. If components and supplies are exposed to temperature and humidity levels outside the limits established by the manufacturer, they must be conditioned to the operating environment in accordance with the manufacturer's recommendations.

Other areas in the data processing complex may house such auxiliary equipment as engine generators, motor generators, uninterruptible power supplies (UPSs), and transformers. The air-conditioning and ventilating quality requirements are less stringent than those for computers, but the continuing satisfactory operation of this equipment is vital to the proper functioning of the computer system.

DESIGN CRITERIA

The data processing system spaces that house the computers, computer personnel, and associated equipment require air conditioning to maintain proper environmental conditions for both the equipment and the personnel.

Computer room air conditioning does not require quick response to changes in set points for the environmental conditions, but *maintaining* conditions within established limits is essential to the operation. System reliability is so vital that the potential cost of system failure often justifies redundant capacity and/or components.

The environmental conditions required by computer equipment vary widely, depending on the manufacturer. Table 1 lists general recommendations on conditions for the computer room. Most manufacturers recommend that the computer equipment draw conditioned air from the room; some, however, permit or recommend direct cooling by supply air. Criteria for air introduced directly to computers differ from those for usual room air; this air must remove computer equipment heat adequately and preclude the possibility of condensation within the equipment (see Table 2).

Because of the large amount of energy needed to maintain the proper environment in data processing areas, the design should hold *net* energy use as low as possible. Heat energy recovery is quite feasible in many installations.

The preparation of this chapter is assigned to TC 9.2, Industrial Air Conditioning.

Table 1 Typical Computer Room Design Conditions[a]

Temperature set point and offset	22 ± 1°C
Relative humidity set point and offset	50 ± 5%
Filtration quality[b]	45%, minimum 20%

[a]These conditions are typical of those recommended by most computer equipment manufacturers.
[b]From ASHRAE *Standard* 52.1-1992, dust-spot efficiency test.

Table 2 Design Conditions for Air Supply Direct to Computer Equipment[a]

Conditions	Recommended Levels
Temperature	As required for heat removal but no lower than 16°C
Relative humidity	Maximum 65% (some manufacturers permit up to 80%)
Filtration quality[b]	45%

[a]These conditions are typical of those recommended by most computer equipment manufacturers, but under no circumstances should conditioned air be supplied directly to a computer unless it is controlled within the manufacturer's limits for the particular piece of equipment.
[b]From ASHRAE *Standard* 52.1-1992, dust-spot efficiency test.

Computer Room Environment

Computer rooms should be kept at the lower end of the temperature tolerance of 22 ± 1°C for two reasons. First, the air distribution and/or the response of the controls may not match the nonuniform heat distribution of the equipment. Setting the controls no higher than 22°C generally ensures that all equipment will remain at a temperature within the established range for satisfactory operation. Secondly, the lower control temperature provides a cushion for short-term peak load temperature rise without adversely affecting computer operation.

High relative humidity levels may cause improper feeding of cards and paper and, in extreme cases, condensation on machine surfaces. Low relative humidity in combination with other factors may result in static discharge, which can adversely affect the operation of data processing and other electronic equipment.

To maintain proper relative humidity in a computer room, vapor transmission retarders sufficient to restrain moisture migration during the maximum expected vapor pressure differences between the computer room and surrounding areas should be installed around the entire envelope. Cable and pipe entrances should be sealed and caulked with a vaporproof material. Door jambs should fit tightly. Windows in colder climates should be double- or triple-glazed.

In localities where the outdoor air contains unusually high quantities of dust, dirt, salt, or corrosive gases, it may be necessary to pass the air through higher efficiency filters or adsorption chemicals before it is introduced into the computer room.

The presence of dust can affect the operation of data processing equipment, so good quality filtration and proper maintenance of filters are very important in the computer room. Dirty filters can reduce airflow, thereby decreasing the sensible heat ratio of computer room air-conditioning equipment. This in turn loads the computer room with energy-intensive humidification, needlessly increasing the operating cost.

Computer room systems should provide only enough outdoor air for personnel requirements and to maintain the room under a positive pressure relative to surrounding spaces. Since most computer rooms have few occupants, and the quantity of conditioned air circulated is higher than that required for comfort, the need to maintain positive pressure is usually the controlling design criterion. In most computer rooms, an outdoor air quantity of less than 5% of total supply air will satisfy ventilation requirements and prevent inward leakage. Outdoor air in excess of the required minimum increases the cooling and heating loads and makes control of atmospheric contaminants and winter humidity more difficult.

The air-conditioning system should operate within the noise tolerance of the computer equipment. This is ordinarily not a difficult criterion to satisfy. Equipment should be served by isolated power sources, and vibration should be isolated to prevent transmission to the computer equipment. The computer manufacturer should be consulted regarding computer equipment sound tolerance and specific special requirements for vibration isolation.

Personnel comfort is also important. Temperature, humidity, and filtration requirements for the equipment are within the comfort range for room occupants, but drafts and cold surfaces must be kept to a minimum in occupied areas.

Some manufacturers have established criteria for allowable rates of environmental change to prevent shock to the computer equipment. These can usually be satisfied by high-quality commercially available controls responding to changes of ±0.5 K and ±5% rh. The manufacturer's requirements should be reviewed and fulfilled to ensure that the system will function properly during normal operation and during periods of start-up and shutdown.

Computer equipment will tolerate a somewhat wider range of environmental conditions when not in operation; but to keep the room within those limits and to minimize thermal shock, it may be desirable to operate the air conditioning.

Computer technology is continually changing; during the life of a system, the computer equipment will almost certainly be changed and/or rearranged. The air-conditioning system must be sufficiently flexible to facilitate this rearrangement of components and permit expansion without requiring the system to be rebuilt. In the usual applications, it should be possible to make modifications without extensive air-conditioning shutdowns, and, in highly critical applications, with no shutdown at all.

Ancillary Spaces Environment

Spaces for storing products such as paper, cards, and tape generally require the same environmental conditions as the computer room itself.

Electrical power supply and conditioning equipment can tolerate more variation in temperature and humidity than can computer equipment. Equipment in this category includes motor generators, uninterruptible power supplies (UPSs), batteries, voltage regulators, and transformers. Ventilation to remove heat from the equipment is normally sufficient. Manufacturers' data should be consulted to determine the amount of heat release and the design conditions for satisfactory operation.

Battery rooms require ventilation to remove hydrogen and to control the space temperature (optimum 25°C). Hydrogen accumulation may be no greater than 3% by volume, and the ventilation system should be designed to prevent pockets of concentration, particularly at the ceilings.

Many computer installations include an engine generator for emergency power that requires a large amount of ventilation when running. It is easier to start if low ambient temperatures are avoided.

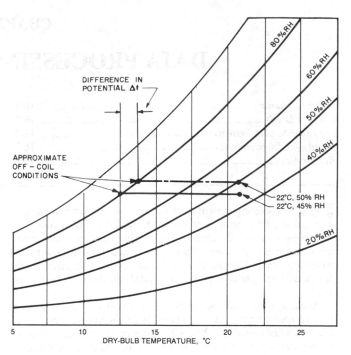

Fig. 1 Potential Effect of Room Design Conditions on Design Supply Air Quantity

COOLING LOADS

The major heat source in a computer room is the equipment; this heat is highly concentrated and distributed nonuniformly. Heat gain from lights should be no greater than that in good quality office space; occupancy loads and outdoor air requirements will be low to moderate. Heat gains through the structure depend on the location and construction of the room. Transmission heat gain to supply spaces should be carefully evaluated and provided for in the design.

Information on computer equipment heat release should be obtained from the computer manufacturer. In general, the data are used without a reduction for diversity, unless the computer manufacturer or experience with a similar installation recommends a reduction.

Due to the relatively low occupancy and proportion of outdoor air, computer room heat gains are almost entirely sensible. For this reason and because of the low room design temperatures, air supply quantity per unit of cooling load will be greater than that for most comfort applications. Figure 1 shows that choosing room design at 22°C, 45% rh can reduce air supply quantity by approximately 10% compared to a slightly more humid room design of 22°C, 50% rh.

A sensible heat ratio of approximately 0.9 to 1.0 is common for computer room applications. This relatively small amount of latent cooling is adequate to handle the minimum moisture loads incurred and will not cause needless dehumidification.

AIR-CONDITIONING SYSTEMS

The air-handling apparatus for the computer room air-conditioning system should be independent of other systems in the building, although cross-connection with systems inside or outside the data processing area may be desirable for backup. Redundant air-handling equipment is frequently used, normally with automatic operation. The air-handling facilities should provide air filtration, cooling and dehumidification, humidification, and heating.

The refrigeration systems should be independent of other systems and capable of year-round operation. It may be desirable to cross-connect refrigeration equipment for backup, as suggested for air-handling equipment. Redundant refrigeration may be required; the extent of the redundancy will depend on the importance of the computer installation. In many cases, standby power is justified for the computer room air-conditioning system.

Computer rooms are being successfully conditioned with a wide variety of systems, including (1) complete self-contained packaged units consisting of air-handling and refrigeration apparatus close-coupled within a common housing and installed inside the computer room; (2) chilled water packaged units located within the computer room and served by remotely located refrigeration equipment; and (3) central station air-handling units with both air-handling and refrigeration equipment located outside the computer room.

Self-Contained Packaged Units

Self-contained units should be specifically designed for computer room applications. These units are built to higher overall standards of performance and reliability than conventional packaged air conditioners intended for comfort, although some major components are identical.

Packaged units are available with (1) multiple reciprocating compressors and separate multiple refrigeration circuits through the cooling coil and condensers; (2) air filters to meet computer room criteria; (3) humidifiers; (4) a reheat coil; (5) corrosion-resistant construction for coils, humidifiers, and other components; (6) controls; (7) instrumentation such as indicator lights to show which equipment components are in operation and alarms to signal dirty filters and component failure; and (8) filter gages to indicate the status of filter loading. Status and/or alarm devices may be connected to remote monitoring panels.

Although component placement varies with the manufacturer, a typical unit arrangement is shown in Figure 2. The units can supply air either downward to the floor cavity or upward to overhead ducts and/or a ceiling plenum.

The refrigeration cycles for computer room units require a means for condensing, and this is provided by water-cooled condensers connected to a remote cooling tower, water- or glycol-cooled condensers connected to a remote radiator or remote air-cooled or evaporative condensers.

Self-contained air conditioners are usually located within the computer room, but may also be remotely located and ducted to the conditioned space. If they are remotely located, their temperature

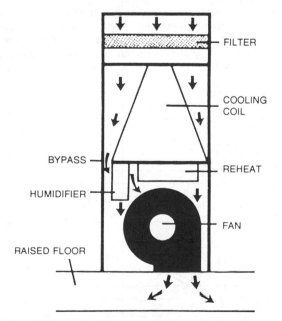

Fig. 2 Self-Contained Computer Room Air Conditioner

and humidity controls should be located in the conditioned space. A major benefit of locating the air conditioner close to the load is the flexibility of such an arrangement, which accommodates the ever-changing load pattern in many computer rooms. The space occupied by packaged units in the computer room may be expensive, but security considerations alone make it practical to place them there because a central system serving a computer room normally has no security protection beyond standard building maintenance.

In systems with multiple unitary conditioners, it may be advantageous to introduce outdoor air through one conditioner that serves all of the spaces in the data processing area. Self-contained systems achieve redundancy by providing multiple units so that the loss of one or more units will have a minimum effect on system performance. Expansion of the data processing facility is generally much easier to handle with self-contained units.

Chilled Water Packaged Units

These units are similar to completely self-contained packaged units, except that they contain no refrigeration equipment because they are served by remotely located refrigeration units, usually through chilled water connections. Since computer components in some systems require a source of chilled water for cooling, the use of chilled water packaged units may be preferable in this application. Chilled water supply temperatures suitable for water-cooled computer equipment can range from 5 to 15°C. These temperatures are normally compatible with those required for computer room air-conditioning units.

When chilled water packaged units are used, the reliability of the remote refrigeration system must be considered. This system generally must be capable of operating 24 h a day, year-round. In severe winter locations, operation at low ambient temperatures must be guaranteed. Packaged units with chilled water coils and direct-expansion coils are available and afford an alternate refrigeration source when the chilled water plant shuts down.

Chilled water packaged units occupy space within the computer room, but because they do not contain refrigeration equipment, they require less servicing within the room than self-contained equipment.

Central Station Air-Handling Units

Central station supply systems have a larger capacity than self-contained equipment. Because the equipment is not located within the computer room, a greater variety of choices in air-conditioning system design and arrangement is available.

Central station air-handling equipment must satisfy computer room performance criteria; it should be arranged to facilitate servicing and maintenance. Redundancy can be achieved by cross-connecting systems, by providing standby equipment, or by combining these approaches. No floor space in the computer room is required, and virtually all servicing and maintenance operations are performed in areas devoted specifically to air-conditioning equipment. However, security is lessened.

Central station supply systems must be designed with the capability to accommodate expanding loads in the computer areas. These systems must include humidification, reheat, dehumidification, and controls. Ductwork penetration of the computer space must be sealed to prevent moisture migration to the area. Central systems for computer rooms should not be tied together with building air-handling systems unless there are special controls that provide for year-round cooling and high sensible heat ratio design factors.

SUPPLY AIR DISTRIBUTION

Computer components that generate large quantities of heat are normally constructed with internal fans and passages to convey cooling air through the machine. The inlet usually draws the air from the computer room, but some manufacturers recommend that certain components of their systems take cooling air directly from the conditioned air supply.

Computer room heat gains are often highly concentrated. For minimum room temperature gradients, supply air distribution should closely match load distribution, and the control thermostat must be located where it will sense the average conditions in the area it serves. Because of the constant change in the arrangement of computer equipment, return-air thermostats have generally met this requirement the best. The distribution system should be sufficiently flexible to accommodate changes in the location and magnitude of the heat gains with minimum change in the basic distribution system.

Supply air systems usually require approximately 75 L/s per kilowatt of cooling to satisfy computer room requirements. This airflow rate cools hot spots and maintains an even temperature distribution.

The construction materials and methods chosen for air distribution should ensure a clean air supply. Duct or plenum material that may erode must be avoided. Access for cleaning is desirable.

Zoning

Computer rooms should be adequately zoned to maintain temperatures within the design criteria. Zoning is not expensive because of the relatively small number of rooms and the generally open character of the spaces involved. Except in the smallest computer rooms, some zoning is usually required to minimize temperature variations due to load fluctuations. At a minimum, individual control for each major space is desirable. In larger areas, there may be potential for temperature variations within a single room.

Packaged air-conditioning systems using an underfloor air supply plenum (Figure 3) self-zone a large computer room area adequately. These systems have automatic return-air temperature and humidity control of the air being circulated in the zone. The size of the zoned area is controlled by various floor registers and perforated floor panels. These systems have the flexibility to accommodate relocation of computer equipment and future additional heat loads.

Underfloor Plenum Supply

To facilitate the interconnection of equipment components by electric cables, data processing equipment is usually set above a false floor, which affords a flat walking surface over the space where the connecting cables are installed. This space can be used as an air distribution channel, either as a plenum or, less often, to accommodate ducts. Underfloor air is distributed to the room through perforated panels or registers set or built into floor panels around the room, especially in the vicinity of computer equipment with high heat release.

Some manufacturers of computer room floors produce perforated floor panels similar in appearance to, and interchangeable with, conventional computer room floor panels. These panels allow the supply air to accommodate shifting equipment loads. The free area of various manufacturers' panels varies significantly. One configuration has slide-type dampers that allow airflow balance comparable to that provided by a floor-mounted register (Figures 3 and 4).

Floor-mounted registers allow volume adjustment and are capable of longer throws and better directional control than perforated floor outlets. However, they are usually located outside traffic areas because some types are not completely flush mounted, and almost all tend to be drafty for nearby personnel. Perforated floor panels are especially suitable for installation in normal traffic aisles because they are completely level with the floor. In addition, when operated with moderate airflow, they produce a high degree of mixing near the point of discharge. For this reason, they cause less direct injection of unmixed conditioned air to the computer when located fairly close to equipment air inlets than do registers that have lower induction ratios.

Air flows through all openings in the floor cavity; if direct flow to a computer unit is not desired, all openings between the unit and the floor cavity should be sealed. Most openings between the underfloor space and the equipment accommodate cables; collars are available that fit the cable and seal the opening.

It is important that there be sufficient clearance area within the raised floor cavity to permit airflow. At least 300 mm of clearance is desirable and 250 mm is the usual minimum; in applications where cabling is extensive and/or air quantities are especially high, additional clearance may be required.

The supply connection from unitary equipment to an underfloor cavity should allow minimum turbulence; turning vanes at the unit discharge sometimes helps accomplish this. Where possible, the supply to the cavity should be central to the area served, and abrupt changes in direction should be avoided. Piping and conduit to unitary equipment should not interfere with airflow from the unit.

Where multiple zones are served from the underfloor plenum, dividing baffles may generally be omitted, as it can be predicted that supply and return air will follow the course of least resistance. Dividing baffles installed to meet zone configurations or fire codes can impede the modification of computer systems having cabling interconnections between zones; every cabling change entails penetration of the zone baffles.

Surfaces of underfloor plenums may require insulation or vapor barriers if supply air temperatures drop low enough to cause condensation on either side of plenum surfaces or if temperatures on the floor above a ceiling plenum become low enough to cause discomfort to occupants. Insulation may also be required to reduce heat

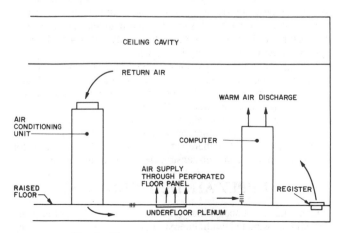

Fig. 3 Typical Underfloor Distribution

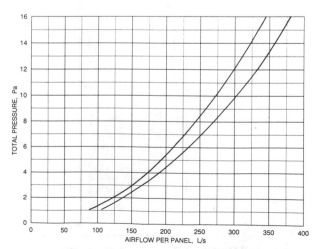

Fig. 4 Floor Panel Air Performance

transmission through plenum surfaces. Plenums should be of airtight construction and be thoroughly cleaned and smoothly finished to prevent entrainment of foreign materials in the airstream. The use and method of construction of plenums may be limited by local codes or fire underwriter regulations.

The potential adverse effects of direct air supply on the computer equipment (condensation within the machine) are serious enough to discourage its use unless the manufacturer insists on it. If this is the case, sufficient controls and safety devices must be installed to ensure operation within the specified limits.

Ceiling Plenum Supply

Overhead supply through perforated ceiling panels used as diffusers may be suited to computer rooms; this arrangement is compatible with either central station or packaged unitary equipment. Ceiling plenum supply systems can satisfy equipment and personnel comfort requirements, but they are generally not as flexible as underfloor plenum supply systems.

Distribution of air can be regulated by selective placement of acoustical pads on the perforated panels of a metal panel ceiling or by the location of *active* perforated sections in a lay-in acoustical tile ceiling. Where precise distribution is essential, active ceiling panels or air supply zones may be equipped with air valves. When required for zoning or to meet codes, baffles may be installed in overhead plenums to avoid interfering with underfloor computer system cabling. Ceiling plenums, if properly constructed and cleaned, are more likely to remain clean than are underfloor plenums.

Plenums must be sufficiently deep to permit airflow without turbulence; the required depth will depend on air quantities. Best conditions may be achieved by the use of distribution ductwork, with the air discharged into the plenum through adjustable outlets above the ceiling (Figure 5). The sealing, surfaces, and insulation of ceiling plenums must be constructed as described in the section on Underfloor Plenum Supply.

Overhead Ducted Supply

Overhead ducted supply systems should be limited to applications where air supply concentrations are low or the need for flexibility is small. These considerations rule out the use of overhead ducted supply systems in most computer room applications. Where loads are high, overhead diffusers may cause drafts, especially if the ceilings are low. It can be difficult to relocate outlets in a facility that must remain in operation.

RETURN AIR

It is common to install the minimum number of openings for return air. With packaged units, the use of a free return from supply

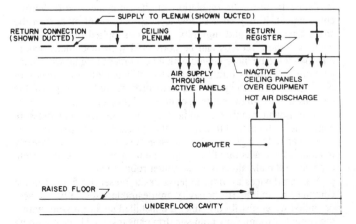

Fig. 5 Typical Ceiling Plenum Distribution

outlets back to the return on the unit is typical (Figure 3). Inlets should be located near high heat loads, and the effect of future modifications of the computer installation should be considered.

Ceiling plenum returns successfully capture a portion of the heat from the computers and the lights directly in the return airstream. This allows a reduction in the supply air circulation rate. A duct collar is usually provided to the top of the self-contained unit; the unit draws the air from the return-plenum area, treats it, and discharges it into the underfloor supply plenum.

WATER-COOLED COMPUTER EQUIPMENT

Some computer equipment requires water cooling to maintain the equipment environment within the limits established by the manufacturer. Generally, a closed system circulates distilled water through passages in the computer to cool it. The manufacturer supplies this cooling system as part of the computer equipment, and a water-to-water heat exchanger attached to a chilled water connection to the air-conditioning system does the cooling.

In systems with water-cooled components, the majority of the cooling is still accomplished by air. The overall heat release to the computer room from water-cooled computers is usually equal to or greater than that for most air-cooled computer systems.

Chilled water may be provided by either a small separate chiller matched in capacity to the computer equipment or a branch of the chilled water system serving the air-handling units. Construction and insulation of chilled water piping and selection of the operating temperatures should minimize the possibility of leaks and condensation, especially within the computer room, while satisfying the requirements of the systems served.

Chilled water systems for water-cooled computer equipment must be designed to provide water at a temperature within the manufacturers' tolerances and to be capable of operating year-round, 24 h a day.

AIR-CONDITIONING COMPONENTS

Controls

A well planned control system must coordinate the performance of the temperature and humidity equipment.

Controls of self-contained, packaged computer room air-conditioning systems have been tested, coordinated, and most likely improved upon over several installations. Controls for systems air-conditioned by other than packaged equipment may be included in a project-specific control system that may be computer based.

Space thermostats and humidistats should be carefully located to sample room conditions accurately.

Where backup power is provided for the computer installation and some portion of the HVAC equipment, it is imperative that the necessary control equipment also have power backup. In underfloor plenum systems, liquid detectors are useful in warning of accidental water spillage that might otherwise go undetected for some time.

Refrigeration

A separate refrigeration facility for the data processing area is desirable because performance requirements often differ drastically from those for comfort systems. Refrigeration systems should generally satisfy the following criteria: (1) match the cooling load; (2) be capable of expansion; (3) have a degree of flexibility comparable to that of other components; (4) be capable of year-round and continuous operation; (5) provide the degree of reliability and redundancy required by the particular operation; (6) be capable of being operated, serviced, and maintained without interfering with normal operation; and (7) be operable with emergency power if this is a system requirement. Fulfillment of these requirements results in the use of multiple units or cross-connections with other reliable systems.

Self-contained systems invariably use reciprocating compressors and direct-expansion cooling. Central systems and systems using decentralized air handling are more flexible; either reciprocating, centrifugal, or absorption refrigeration equipment can be selected.

If the installation is especially critical, it may be necessary to install up to 100% standby refrigeration capacity to provide the minimum requirements in the event of equipment failure.

Condensing Methods

Heat rejection equipment must be designed and selected for *worst case* operating conditions to prevent computer shutdown during weather extremes.

In some geographical areas, water-cooled equipment using cooling towers may be satisfactory; in other areas, the need for winterization and water treatment may preclude the use of systems vulnerable to freezing or open to the atmosphere. Air-cooled condensers are widely used and are equipped with pressure controls capable of operation at greatly varying outdoor ambient temperatures. They ordinarily should not have separate condenser circuits, hot gas lines, and liquid lines for each refrigeration circuit in the system.

Glycol-cooled radiators, or water-cooled radiators in climates not subject to freezing, permit the use of a closed system on the condensing side and facilitate head pressure control by allowing control of the condensing medium. In systems where multiple refrigeration systems or units are remote from the point of heat rejection to the atmosphere, radiator-cooled systems may significantly simplify piping and capacity control (Figure 6).

Glycol- or water-cooled systems require water piping in the computer areas, so precautions must be taken to prevent and detect possible leaks.

Humidification

Many types of humidifiers are used to serve data processing system areas. Humidifiers used include (1) steam with outside source of steam, (2) steam-generating, (3) pan with immersion element, (4) pan with infrared (quartz lamp), and (5) wetted-pad. Where a clean source is continuously available, steam humidification should be considered. The humidification method chosen must be responsive to control, low-maintenance, and free of moisture carryover.

Chilled Water

Chilled water distribution systems should be designed to the same standards of quality, reliability, and flexibility as the rest of the computer room system. Multiple units should be provided for those components that must be shut down for servicing or routine maintenance. Where any likelihood of growth exists, the chilled water system should be designed for expansion or the addition of new conditioners without extensive shutdown.

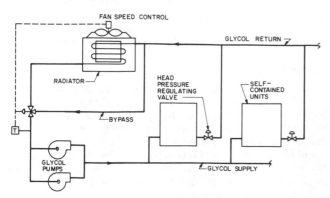

Fig. 6   Glycol System for Condensing

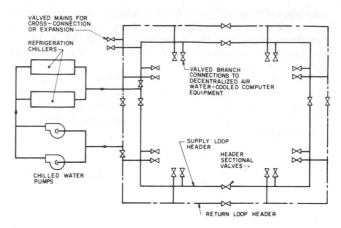

Fig. 7   Chilled Water Loop Distribution

Figure 7 illustrates a looped chilled water system with sectional valves and multiple valved branch connections. The branches could serve air handlers or water-cooled computer equipment. The valves permit modifications or repairs without complete shutdown.

Chilled water temperature should match the load and minimize the possibility of condensation, especially if portions of the system are installed within the computer room. Since computer room loads are primarily sensible, relatively high-temperature chilled water should be circulated. Although this may result in some reduction in unit capacity or require a deeper cooling coil, the net effect on system operation and efficiency will be positive.

Water temperatures as high as 9°C are still slightly below the dew point of a 22°C, 45% rh room and several degrees below that of a 22°C, 50% rh room. For this reason and because the system controls may not regulate chilled water temperature precisely, the chilled water piping must be fully insulated.

The piping system should be pressure tested. If a leak could interfere with the operation of the computer system, the test pressure should be applied in increments in occupied areas. Drip pans should be placed below any valves or other components within the computer room that cannot be satisfactorily insulated. A good quality strainer should be used to prevent heat exchanger passages from clogging.

If cross-connections with other systems are made, the possible effects on the computer room system of the introduction of dirt, scale, or other impurities must be evaluated and dealt with.

INSTRUMENTATION

Because computer equipment malfunctions may be caused by or attributed to improper regulation of the computer room thermal environment, it may be desirable to keep permanent records of the space temperature and humidity. If air is supplied directly to the machines, these records can be correlated with machine function. Alarms should be incorporated with the recorder to signal when temperature or humidity limits are violated. Where chilled water is supplied directly to computer units, records should be kept of the entering and exiting water temperature and pressure.

A sufficient number of indicating thermometers and pressure gages should be placed throughout the system so that operating personnel can tell at a glance when unusual conditions prevail. Properly maintained and accurate filter gages can help prevent loss of system capacity and maintain correct computer room conditions.

Sensing devices to indicate leaks or the presence of water in the computer room underfloor cavity are desirable, especially if glycol or chilled water distribution lines are installed under pressure within the computer room. Low points in drip pans installed beneath piping should also be monitored.

All monitoring and alarm devices should give local indication; if the system is in a building with a remote monitoring point, indications of system malfunctions should be transmitted to activate alarms in the remote location.

FIRE PROTECTION

Fire protection for the air-conditioning system should be fully integrated with fire protection for the computer room and the building as a whole. Applicable codes must be complied with, and the owner's insurers must be consulted. Automatic extinguishing systems afford the highest degree of protection. Fire underwriters often recommend an automatic sprinkler system (Jacobson 1967). However, most computer system owners are reluctant to install such a system because it uses water, so most computer rooms are not so protected. Nonwater extinguishing systems are available and should be considered.

At a minimum, sensing devices should be placed in both the room and the air-conditioned airstream to warn of fire. Sensing devices should be located in supply- and return-air passages and in the underfloor cavity if it contains electric cables, whether it is used for air supply or not. These devices should provide early warning of combustion products, even if no smoke is visible and temperatures are at or near normal levels.

Any fire protection system should include either manual or automatic shutdown of computer power, depending on the importance of the system and on the potential effect of unwarranted shutdown (NFPA 1993).

HEAT RECOVERY AND ENERGY CONSERVATION

Computer room systems are good candidates for heat recovery because of their large, relatively steady year-round loads. If the heat removed by the conditioning system can be efficiently transferred and applied elsewhere in the building, some cost saving may be realized. However, reliability is a prime requirement in most computer installations, and any complication to the basic conditioning system that might impair its reliability or otherwise adversely affect performance should be carefully considered before incorporation into the design. Potential monetary loss from system malfunction or unscheduled shutdown will often outweigh savings from increased operating efficiency.

Heat rejected during condensing can be used for space heating, domestic water heating, or other process heat. Packaged, self-contained units containing such heat recovery components are available. When an air-side economizer is used in the cooling of computer areas, outdoor air humidification requirements should be considered. Use of a liquid heat exchanger between cool outdoor air and interior air is economically feasible in some climates. Packaged, self-contained units are available with an additional cooling coil for chilled glycol solution.

Energy conservation can be achieved by effecting optimum operating conditions. For instance, coil surface temperatures should be maintained as high as possible to cool the space adequately. Coil temperatures lower than necessary waste energy, both through higher refrigeration energy consumption and over-dehumidification. Annual energy-use calculations or simulations to evaluate the economics of various designs are often desirable.

REFERENCES

ASHRAE. 1992. Gravimetric and dust-spot procedures for testing air-cleaning devices used in general ventilation for removing particulate matter. *Standard* 52.1-1992.

Jacobson, D.W. 1967. Automatic sprinkler protection for essential electric and electronic equipment. *NFPA Fire Journal* (January):48.

NFPA. 1993. National electrical code. NFPA 70-93. National Fire Protection Association, Quincy, MA.

PRINTING PLANTS

THIS chapter outlines air-conditioning requirements for key printing operations. Air conditioning of printing plants can provide controlled, uniform air moisture content and temperature in working spaces. Paper, the principal material used in printing, is hygroscopic and very sensitive to variations in the humidity of the surrounding air. Printing problems caused by paper expansion and contraction can be avoided by controlling the moisture content throughout the manufacture and printing of the paper.

GENERAL DESIGN CRITERIA

The following are three basic methods of printing:
1. Relief printing (letterpress): ink is applied to a raised surface
2. Lithography: the inked surface is neither in relief nor recessed
3. Gravure (intaglio printing): the inked areas are recessed below the surface

Figure 1 shows the general work flow through a printing plant. The operation begins at the publisher and ends with the finished printed product and paper waste. Paper waste, which may be as high as 20% of the total paper used, affects the profitability of a printing operation. Proper air conditioning can help reduce the amount of paper wasted.

In sheetfed printing, individual sheets are fed through a press from a stack or load of sheets and collected after printing. In webfed rotary printing, a continuous web of paper is fed through the press from a roll. The printed material is cut, folded, and delivered from the press as signatures, which form the sections of a book.

Sheetfed printing is a slow process in which the ink is essentially dry as the sheets are delivered from the press. *Offsetting,* the transference of an image from one sheet to another, is prevented by applying a powder or starch to separate each sheet as it is delivered from the press. Starches present a housekeeping problem: the particles (30 to 40 μm in size) tend to fly off, eventually settling on any horizontal surface.

If both temperature and relative humidity are maintained within normal human comfort limits, they have little to do with web breaks or the runnability of paper in a webfed press. At extremely low humidity, static electricity causes the paper to cling to the rollers, creating undue stress on the web, particularly with high-speed presses. Static electricity is also a hazard when flammable solvent inks are used.

Special Considerations

Various areas in printing plants require special attention to processing and heat loads.

An engraving department must have very clean air not as clean as that for an industrial clean room, but cleaner than that for an office. Engraving and photographic areas have special ventilation needs because of the chemicals used. Nitric acid fumes from powderless etching require stainless steel or aluminum ducts. Composing rooms, which contain computer equipment, can be treated the same as similar office areas. The excessive dust from cutting in the press folders must be controlled. Stereotype departments have very high heat loads.

The preparation of this chapter is assigned to TC 9.2, Industrial Air Conditioning. This chapter received a major revision in 1974.

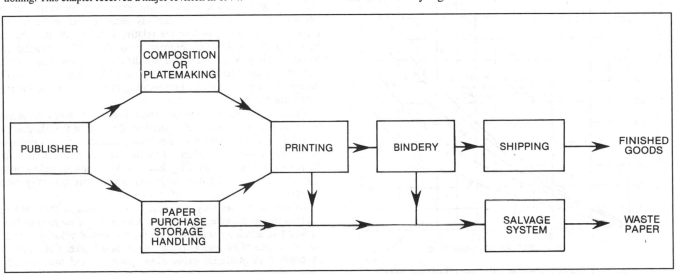

Fig. 1   Work Flow Through a Printing Plant

The air distribution in pressrooms must not cause the web to flutter and must not force contaminants or heat (which normally would be removed by roof vents) down to the occupied level. Pressroom exhaust air must be treated to eliminate pollutants (pigments, oils, resins, solvents, and dust). Suppression systems in the presses reduce the ink mist carried into the exhaust filtration system.

Conventional air-conditioning and air-handling equipment is used in printing plants. Ventilation of storage areas should be about one-half air change per hour; bindery ventilation should be about one air change per hour. Storage areas with materials piled high may need roof-mounted smoke- and heat-venting devices.

In the bindery, loads of loose signatures are stacked near the equipment, which makes it difficult to supply air to occupants without scattering the signatures. One approach is to run the main ducts at the ceiling with many supply branches dropped to within 2.5 to 3 m of the floor. Conventional adjustable blow diffusers, often the linear type, are used.

CONTROL OF PAPER MOISTURE CONTENT

Controlling the moisture content and temperature of paper is important in all printing, particularly multicolor lithography. Paper should be received at the printing plant in moistureproof wrappers, which are not broken or removed until the paper is brought to the pressroom temperature. When exposed to room temperature, paper at temperatures substantially below the room temperature rapidly absorbs moisture from the air, causing distortion. Figure 2 shows the time required to temperature-condition wrapped paper. Printers usually order paper with a moisture content approximately in equilibrium with the relative humidity maintained in their pressrooms. Papermakers find it difficult to supply paper in equilibrium with a relative humidity higher than 50%.

The sword hygrometer (or paper hygroscope) can be used to check the hygroscopic condition of paper relative to the surrounding air. The blade contains a moisture-sensitive element that expands or

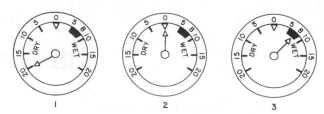

Fig. 3 Sword Hygroscope Readings Taken from Three Lots of Paper

contracts to actuate the pointer in a dial on the handle. The hygrometer is waved in the air until the pointer comes to rest, and the dial is set at zero. The sword is then inserted into the paper; the pointer movement from zero indicates the paper's moisture content relative to that of the surrounding air.

Figure 3 illustrates sword readings for three lots of paper in a pressroom. The first reading indicates paper that is too dry to be exposed to room air without developing wavy edges. The second reading indicates paper that is in equilibrium with the air, which is the correct condition for all sheetfed printing, except multicolor lithography. The third reading indicates paper that has a greater moisture content than the air in the pressroom. This paper was conditioned to balance with the air in the paper stockroom, with a relative humidity 7% above that in the pressroom, which is optimal for multicolor lithographic printing.

Intact mill wrappings and the tightness of the roll normally protect a paper roll for about six months. If the wrapper is damaged, moisture will usually penetrate no more than 3 mm.

PLATEMAKING

Humidity and temperature control are important in making lithographic and collotype plates, photoengravings, and gravure plates and cylinders. If the moisture content and temperature of the plates increase, the coatings increase in light sensitivity, which necessitates adjustments in the light intensity or the length of exposure to give uniformity.

If platemaking rooms are maintained at constant dry-bulb temperature and relative humidity, plates can be produced at known control conditions. As soon as it is dry, a bichromated colloid coating starts to age and harden at a rate that varies with the atmospheric conditions, so exposures made a few hours apart may be quite different. The rate of aging and hardening can be estimated more accurately when the space is air conditioned. Exposure then can be reduced progressively to maintain uniformity. An optimum relative humidity of 45% or less substantially increases the useful life of bichromated colloid coatings; the relative humidity control should be within 2%. A dry-bulb temperature of 24 to 27°C maintained within 1°C is good practice. The ventilation air requirements of the plate room should be investigated. A plant with a large production of deep-etch plates should consider locating this operation outside the conditioned area.

Exhausts for platemaking operations consist primarily of lateral or downdraft systems at each operation. Because of their bulkiness or weight, plates or cylinders are generally conveyed by an overhead rail system to the workstation, where they are lowered into the tank for plating, etching, or grinding. Exhaust ducts must be below or to one side of the working area, so lateral exhausts are generally used for open surface tanks.

Exhaust quantities vary, depending on the nature of the solution, but they should provide 500 L/s per square metre of solution surface (at standard conditions) and/or a minimum control velocity of 0.25 m/s at the side of the tank opposite the exhaust intake. Tanks should be covered to minimize exhaust air quantities and increase efficiency. Excessive supply air ventilation across open tanks should be

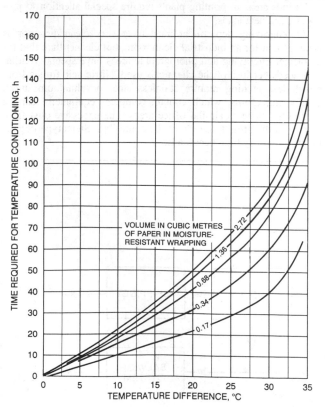

Fig. 2 Temperature Conditioning Chart for Paper

(Graph: x-axis "TEMPERATURE DIFFERENCE, °C" from 0 to 35; y-axis "TIME REQUIRED FOR TEMPERATURE CONDITIONING, h" from 0 to 170. Curves labeled 0.17, 0.34, 0.68, 1.36, 2.72. Text: "VOLUME IN CUBIC METRES OF PAPER IN MOISTURE-RESISTANT WRAPPING")

avoided. Because of the nature of the exhaust, ducts should be acid-proof and liquidtight to prevent moisture condensation.

Webfed offset operations and related departments are similar to webfed letterpress operations, without the heat loads created in the composing room and stereotype departments. Special attention should be given to air cleanliness and ventilation in platemaking to avoid flaws in the plates from chemical fumes and dust.

A rotogravure plant can be hazardous because highly volatile solvents are used. Equipment must be explosionproof, and air-handling equipment must be sparkproof. Clean air must be supplied at controlled temperature and relative humidity. Reclamation or destruction systems are used to prevent photosensitive hydrocarbons from being exhausted into the atmosphere. Some systems use activated carbon for continuous processing or rapid oxidation to eliminate pollutants. The amount of solvents reclaimed may exceed that added to the ink.

RELIEF PRINTING

In relief printing (letterpress), rollers apply ink only to the raised surface of a printing plate. Then pressure is applied to transfer the ink from the raised surface directly to the paper. Only the raised surface touches the paper to transfer the desired image.

Air conditioning in newspaper pressrooms and other webfed letterpress printing areas minimizes problems caused by static electricity, ink mist, and expansion or contraction of the paper during printing. A wide range of operating conditions is satisfactory. The temperature should be selected for operator comfort.

At web speeds of 5 to 10 m/s it is not essential to control the relative humidity because inks are dried with heat. In some types of printing, moisture is applied to the web, and the web is passed over chill rolls to further set the ink.

Webfed letterpress ink is heat-set, made with high boiling, slow evaporating synthetic resins and petroleum oils dissolved or dispersed in a hydrocarbon solvent. The solvent must have a narrow boiling range with a low volatility at room temperatures and a fast evaporating rate at elevated temperatures. The solvent is vaporized in the printing press dryers at temperatures of 120 to 200°C, leaving the resins and oils on the paper. Webfed letterpress inks are dried after all colors are applied to the web.

The inks are dried by passing the web through dryers at speeds of 5 to 10 m/s. There are several types of dryers: open-flame gas cup, flame impingement, high-velocity hot air, or steam drum.

Exhaust quantities through a press dryer system vary from about 3300 to 7000 L/s at standard conditions, depending to the type of dryer used and the speed of the press. Exhaust temperatures range between 120 to 200°C.

Solvent-containing exhaust is heated to 700°C in an air pollution control device to incinerate the effluent. A catalyst can be used to reduce the temperature required for combustion to 540°C but it requires periodic inspection and rejuvenation. Heat recovery reduces the fuel required for incineration and can be used to heat pressroom makeup air.

LITHOGRAPHY

Lithography uses a grease-treated printing image receptive to ink, on a surface that is neither raised nor depressed. Both grease and ink repel water. Water is applied to all areas of the plate, except the printing image. Ink is then applied only to the printing image and transferred to the paper in the printing process.

Offset printing transfers the image first to a rubber blanket and then to the paper. Sheetfed and web offset printing are similar to letterpress printing. The inks are similar but contain water-resistant vehicles and pigment. In web offset and gravure printing, the relative humidity in the pressroom should be maintained at a low level, and the temperature should be selected for comfort or, at least, to avoid heat stress. It is important to maintain steady conditions.

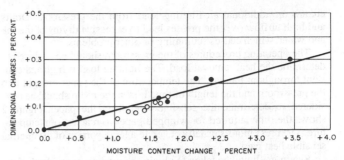

Fig. 4 Effects of Variation in Moisture Content on Dimensions of Printing Papers
(Weber and Snyder 1934)

The pressroom for sheet multicolor offset printing has more exacting humidity requirements than other printing processes. The paper must remain flat with constant dimensions during multicolor printing in which the paper may make six or more passes through the press over a period of a week or more. If the paper does not have the right moisture content at the start, or if there are significant changes in atmospheric humidity during the process, the paper will not retain its dimensions and flatness, and misregistering will result. In many cases of color printing, a register accuracy of 100 μm is required. Figure 4 shows the close control of the air relative humidity that is necessary to achieve this register accuracy. The data shown in this figure are for composite lithographic paper.

Maintaining constant moisture content of the paper is complicated because paper picks up moisture from the moist offset blanket during printing—0.1 to 0.3% for each impression. When two or more printings are made in close register work, the paper at the start of the printing process should have a moisture content in equilibrium with air at 5 to 8% rh above the pressroom air. At this condition, the moisture evaporated from the paper into the air nearly balances the moisture added by the press. In obtaining register, it is important to keep the sheet flat and free from wavy or tight edges. To do this, the relative humidity balance of the paper should be slightly above that of the pressroom atmosphere. This balance is not as critical in four-color presses because the press moisture does not penetrate the paper fast enough between colors to affect sheet dimensions or sheet distortion.

Recommended Environment

The Graphic Arts Technical Foundation recommends ideal conditions in a lithographic pressroom of 24 to 27°C dry-bulb temperature and 43 to 47% rh, controlled to ±1°C dry-bulb temperature and ±2% rh (Reed 1970). Comfort and economy of operation influence the choice of temperature. The effect of relative humidity variations on register can be estimated for offset paper from Figure 4. Closer relative humidity control of the pressroom air is required for multicolor printing of 1930-mm sheets than for 560-mm sheets with the same register accuracy. Closer control is needed for multicolor printing, where the sheet makes two or more trips through the press, than for one-color printing.

Ink drying is affected by temperature and humidity, so uniform results are difficult to obtain without controlling the atmospheric conditions. Printing inks must dry rapidly to prevent offsetting and smearing. High relative humidity and high moisture content in paper tend to prevent ink penetration, so more ink remains on the surface than can be quickly oxidized. This affects drying time, intensity of color, and uniformity of ink on the surface. Relative humidity below 60% is favorable for drying at a comfortable temperature.

The air-conditioning system for the pressroom of a lithographic plant should control air temperature and relative humidity, filter the air, supply ventilation air, and distribute the air without pronounced drafts around the presses. The use of anti-offset sprays to set the ink creates an additional air-filtering load from the pressroom. Drafts and high airflow over the presses lead to excessive drying of the ink and water, which causes scumming or other problems.

The operating procedures of the pressroom should be analyzed to determine the heat removal load. The lighting load is high and constant throughout the day. The temperature of the paper brought into the pressroom and the length of time it is in the room should be considered to determine the sensible load from the paper. Figure 2 shows the time required for wrapped paper to reach room temperature. The press motors usually generate a large portion of the internal sensible heat gain.

Readings should be taken to obtain the running power load of the larger multicolor presses. The moisture content of the paper fed to the press and the relative humidity of the air must be considered when computing the internal latent heat gain. Paper is used that is in equilibrium with air at a relative humidity somewhat higher than that of the pressroom, so the paper gives up moisture to the space as it absorbs moisture during printing. If the moisture transfer is in balance, the water used in the printing process should be included in the internal moisture load. It is preferable to determine the water evaporation from the presses by test.

Air-Conditioning Systems

Precise multicolor offset lithography printing requires refrigeration with provision for separate humidity control or sorption dehumidifying equipment for independent humidity control and provision for cooling. The need for humidity control in the pressroom may be determined by calculating the dimensional change of the paper for each percent change in relative humidity and checking this with the required register for the printing process.

Air conditioning of the photographic department is usually considered next in importance to that of the pressroom. Most of the work in offset lithography is done on film. Air conditioning controls cleanliness and comfort and maintains the size of the film for register work.

Air conditioning is important in the stripping department, both for comfort and for maintaining size and register. Curling of the film and flats, as well as shrinkage or stretch of materials, can be minimized by maintaining constant relative humidity. This is particularly important for close register color work. The photographic area, stripping room, and platemaking area usually are maintained at the same conditions as the pressroom. Dryers used for web offset printing are the same type as for webfed letterpress. Drying is less complex because less ink is applied and presses run at lower speeds—4 to 9 m/s.

ROTOGRAVURE

Rotogravure printing uses a cylinder with minute ink wells etched in the surface to form the printing image. Ink is applied to the cylinder, filling the wells. Excess ink is then removed from the cylinder surface by doctor blades, leaving only the ink in the wells. The image is then transferred to the paper as it passes between the printing cylinder and an impression cylinder.

In sheetfed gravure printing, as in offset printing, expansion, contraction, and distortion should be prevented to obtain correct register. The paper need not be in equilibrium with air at a relative humidity higher than that of the pressroom, because no moisture is added to the paper in the printing process. Humidity and temperature control should be exacting, as in offset printing. The relative humidity should be 45 to 50%, controlled to within ±2%, with a comfort temperature controlled to within ±1°C.

Gravure printing ink dries principally by evaporating the solvent in the ink, leaving a solid film of pigment and resin. The solvent is a low-boiling hydrocarbon, and evaporation takes place rapidly, even without the use of heat. The solvents have closed-cup flash points from −5 to 27°C and are classified as Group I or special hazard liquids by local code and insurance company standards. As a result,

in areas adjacent to gravure press equipment and solvent and ink storage areas, electrical equipment must be Class I, Division 1 or 2, as described by the National Electrical Code, and ventilation requirements (both supply and exhaust) are stringent. Ventilating systems should be designed for high reliability, with sensors to detect unsafe pollutant concentrations and then to initiate alarm or safety shutdown.

Rotogravure printing units operate in tandem, each superimposing print over that from the preceding unit. Press speeds range from 6 to 12 m/s. Each unit is equipped with its own dryer to prevent subsequent smearing or smudging.

A typical drying system consists of four dryers connected to an exhaust fan. Each dryer is equipped with fans to recirculate 2500 to 4000 L/s (at standard conditions) through a steam or hot water coil and then through jet nozzles. The hot air (55°C) impinges on the web and drives off the solvent-laden vapors from the ink. It is normal to exhaust half of this air. The system should be designed and adjusted to prevent solvent vapor concentration from exceeding 25% of its lower flammable limit (Marsailes 1970). If this is not possible, constant lower-flammable-limit (LFL) monitoring, concentration control, and safety shutdown capability should be included.

In exhaust system design for a particular process, solvent vapor should be captured from the printing unit where paper enters and leaves the dryer, from the fountain and sump area, and from the printed paper, which continues to release solvent vapor as it passes from printing unit to unit. Details of the process, such as ink and paper characteristics and rate of use, are required to determine exhaust quantities.

When dilution-type ventilation is used, exhaust of 500 to 700 L/s (at standard conditions) at the floor is often provided between each unit. The makeup air units are adjusted to supply slightly less air to the pressroom than that exhausted in order to keep the pressroom negative with respect to the surrounding areas.

OTHER PLANT FUNCTIONS

Flexography

Flexography uses rubber raised printing plates and functions much like a letterpress. Flexography is used principally in the packaging industry to print labels and also to print on smooth surfaces, such as plastics and glass.

Collotype Printing

Collotype or photogelatin printing is a sheetfed printing process related to lithography. The printing surface is bichromated gelatin with varying affinity for ink and moisture, depending on the degree of light exposure received. There is no mechanical dampening as in lithography, and the necessary moisture in the gelatin printing surface is maintained by operating the press in an atmosphere of high relative humidity, usually about 85%. Since the tone values printed are very sensitive to changes in the moisture content of the gelatin, the relative humidity should be maintained within ±2%.

Because tone values are very sensitive to changes in ink viscosity, temperature must be closely maintained; 27 ± 2°C is recommended. Collotype presses are usually partitioned off from the main plant, which is kept at a lower relative humidity, and the paper is exposed to high relative humidity only while it is being printed.

Salvage Systems

Salvage systems remove paper trim and shredded paper waste from production areas and carry airborne shavings to a cyclone collector, where they are baled for removal. Air quantities required are 2.5 to 2.8 m³ per kilogram of paper trim, and the transport velocity in the ductwork is 23 to 25 m/s (Marsailes 1970). Humidification may be provided to prevent the buildup of a static charge and the consequent system blockage.

Air Filtration

Ventilation and air-conditioning systems for printing plants commonly use automatic moving-curtain dry-media filters with renewable media having a weight arrestance of 80 to 90% (ASHRAE *Standard* 52.1).

In sheetfed pressrooms, a high-performance bag-type after-filter is used to filter starch particles, which require about 85% ASHRAE dust spot efficiency. In film processing areas, which require relatively dust-free conditions, high-efficiency air filters are installed, with 90 to 95% ASHRAE dust spot efficiency.

A different type of filtration problem in printing is ink mist or ink fly, which is common in newspaper pressrooms and in heatset letterpress or offset pressrooms. Minute droplets of ink (5 to 10 μm) are dispersed by ink rollers rotating in opposite directions. The cloud of ink droplets is electrostatically charged. Suppressors, charged to repel the ink back to the ink roller, are used to control ink mist. Additional control is provided by automatic moving curtain filters.

Binding and Shipping

Some printed materials must be bound. Two methods of binding are perfect binding and stitching. In perfect binding, sections of a book (signatures) are gathered, ruffed, glued, and trimmed. The glued edge is flat. Large books are easily bound by this type of binding. A low pressure compressed air system and a vacuum system are usually required to operate a perfect binder, and paper shavings are removed by a trimmer. The use of heated glue requires an exhaust system if the fumes are toxic.

In stitching, sections of a book are collected and stitched (stapled) together. Each signature is opened individually and laid over a moving chain. Careful handling of the paper is important. This system has the same basic air requirements as perfect binding.

Mailing areas of a printing plant wrap, label, and ship the manufactured goods. Operation of the wrapper machine can be affected by low humidity. In winter, humidification of the bindery and mailing area to about 40 to 50% rh may be necessary to prevent static buildup.

REFERENCES

ASHRAE. 1992. Gravimetric and dust-spot procedures for testing air-cleaning devices used in general ventilation for removing particulate matter. *Standard* 52.1-1992.

Marsailes, T.P. 1970. Ventilation, filtration and exhaust techniques applied to printing plant operation. *ASHRAE Journal* (December):27.

Reed, R.F. 1970. What the printer should know about paper. Graphic Arts Technical Foundation, Pittsburgh, PA.

Weber, C.G. and L.W. Snyder. 1934. Reactions of lithographic papers to variations in humidity and temperature. *Journal of Research*, National Bureau of Standards 12 (January).

CHAPTER 18

TEXTILE PROCESSING

THIS chapter covers (1) basic processes for making fiber, yarn, and fabric; (2) various types of air-conditioning systems used in textile manufacturing plants; (3) relevant health considerations; and (4) energy conservation procedures.

Most textile manufacturing processes may be put into one of three general classifications: (1) synthetic fiber making, (2) yarn making, and (3) fabric making. Synthetic fiber manufacturing is divided into staple processing, tow-to-top conversion, and continuous fiber processing; yarn making is divided into spinning and twisting; and fabric making is divided into weaving and knitting. Although these processes vary, their descriptions reveal the principles on which air-conditioning design is based.

FIBER MAKING

Processes preceding fiber extrusion have diverse ventilating and air-conditioning requirements, based on principles similar to those that apply to chemical plants.

Synthetic fibers are extruded from metallic spinnerets and solidified as continuous parallel filaments. This process, called *continuous spinning*, differs from the mechanical spinning of fibers or tow into yarn, which is generally referred to as *spinning*.

Synthetic fibers may be formed by melt-spinning, dry-spinning, or wet-spinning. Melt-spun fibers are solidified by cooling the molten polymer; dry-spun fibers by evaporating a solvent, leaving the polymer in fiber form; and wet-spun fibers by hardening the extruded filaments in a liquid bath. The selection of a spinning method is affected by economic and chemical considerations. Generally, nylons, polyesters, and glass fibers are melt-spun, acetates dry-spun, rayons and aramids wet-spun, and acrylics dry- or wet-spun.

For melt-spun and dry-spun fibers, the filaments of each spinneret are usually drawn through a long vertical tube called a *chimney* or *quench stack*, within which solidification occurs. For wet-spun fibers, the spinneret is suspended in a chemical bath where coagulation of the fibers takes place. Wet-spinning is followed by washings, applying a finish, and drying.

Synthetic continuous fibers are extruded as a heavy denier tow for cutting into short lengths called staple or somewhat longer lengths for tow-to-top conversion, or they are extruded as light denier filaments for processing as continuous fibers. An oil is then applied to lubricate, give antistatic properties, and control fiber cohesion. The extruded filaments are usually drawn (stretched) both to align the molecules along the axis of the fiber and to improve the crystalline structure of the molecules, thereby increasing the fiber's strength and resistance to stretching.

Heat applied to the fiber when drawing heavy denier or high-strength synthetics releases a troublesome oil mist. In addition, the mechanical work of drawing generates a high localized heat load. If the draw is accompanied by twist, it is called *draw-twist;* if not, it is

The preparation of this chapter is assigned to TC 9.2, Industrial Air Conditioning.

called *draw-wind*. After draw-twisting, continuous fibers may be given additional twist or may be sent directly to warping.

When tow is cut to make staple, the short fibers are allowed to assumed random orientation. The staple, alone or in a blend, is then usually processed as described in the Cotton System section. However, tow-to-top conversion, a more efficient process, has become more popular. The longer tow is broken or cut to maintain parallel orientation. Most of the steps of the cotton system are bypassed; the parallel fibers are ready for blending and mechanical spinning into yarn.

In the manufacture of glass fiber yarn, light denier multifilaments are formed by attenuating molten glass through platinum bushings at high temperatures and speeds. The filaments are then drawn together while being cooled with a water spray, and a chemical size is applied to protect the fiber. This is all accomplished in a single process prior to winding the fiber for further processing.

YARN MAKING

The fiber length determines whether spinning or twisting must be used. Spun yarns are produced by loosely gathering synthetic staple, natural fibers, or blends into rope-like form; drawing them out to increase fiber parallelism, if required; and then applying twist. Twisted (continuous filament) yarns are made by twisting mile-long monofilaments or multifilaments. Ply yarns are made in a similar manner from spun or twisted yarns.

The principles of mechanical spinning are applied in three different systems: cotton, woolen, and worsted. The cotton system is used for all cotton, most synthetic staple, and many blends. Woolen and worsted systems are used to spin most wool yarns, some wool blends, and synthetic fibers such as acrylics.

Cotton System

The cotton system was originally developed for spinning cotton yarn, but now its basic machinery is used to spin all varieties of staple, including wool, polyester, and blends. Most of the steps from raw materials to fabrics, along with the ranges of frequently used humidities, are outlined in Figure 1.

Opening, Blending, and Picking. The compressed tufts are partly opened, most foreign matter and some short fibers are removed, and the mass is put in an organized form. Some blending is desired to average the irregularities between bales or to mix different kinds of fiber. Synthetic staple, which is cleaner and more uniform, usually requires less preparation. The product of the picker is pneumatically conveyed to the feed rolls of the card.

Carding. This process lengthens the lap into a thin web, which is gathered into a rope-like form called a *sliver*. Further opening and fiber separation follows, as well as partial removal of short fiber and trash. The sliver is laid in an ascending spiral in cans of various diameters.

For heavy, low-count (length per unit of mass) yarns of average or lower quality, the card sliver goes directly to drawing. For lighter,

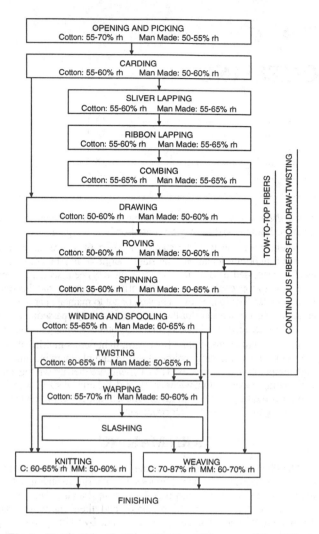

Fig. 1   Textile Process Flow Chart and Ranges of Humidity

high-count yarns requiring fineness, smoothness, and strength, the card sliver must first be combed.

Lapping. In lapping, several slivers are placed side by side and drafted. In ribbon lapping, the resulting ribbons are laid one on another and drafted again. The doubling and redoubling averages out sliver irregularities; drafting improves fiber parallelism. Some recent processes lap only once before combing.

Combing. After lapping, the fibers are combed with fine metal teeth to substantially remove all fibers below a predetermined length, to remove any remaining foreign matter, and to improve fiber arrangement. The combed lap is then attenuated by drawing rolls and again condensed into a single sliver.

Drawing. Drawing follows either carding or combing and improves uniformity and fiber parallelism by doubling and drafting several individual slivers into a single composite strand. Doubling averages the thick and thin portions; drafting further attenuates the mass and improves parallelism.

Roving. Roving continues the processes of drafting and paralleling until the strand is a size suitable for spinning. A slight twist is inserted, and the strand is wound on large bobbins used for the next roving step or for spinning.

Spinning. Mechanical spinning simultaneously applies draft and twist. The packages (any form into or on which one or more ends can be wound) of roving are creeled at the top of the frame. The unwinding strand passes progressively through gear-driven drafting rolls, a yarn guide, the C-shaped traveler, and then to the bobbin.

The vertical traverse of the ring causes the yarn to be placed in predetermined layers.

The difference in peripheral speed between the back and front rolls determines the draft. Twist is determined by the rate of front roll feed, spindle speed, and drag, which is related to the traveler weight.

The space between the nip or bite of the rolls is adjustable and must be slightly greater than the longest fiber. The speeds of front and back rolls are independently adjustable. Cotton spindles normally run at 8000 to 9000 rpm, but may exceed 14 000 rpm. In ring twisting, drawing rolls are omitted, and a few spindles run as high as 18 000 rpm.

Open-end or turbine spinning, combines drawing, roving, lapping, and spinning. Staple fibers are fragmented as they are drawn from a sliver and fed into a fast-spinning, small centrifugal device. In this device, the fibers are oriented and discharged as yarn; twist is imparted by the rotating turbine. This system is faster, quieter, and less dusty than ring spinning.

Spinning is the final step in the cotton system, and the feature that distinguishes it from twisting is the application of draft. The amount and point of draft application accounts for many of the subtle differences that require different humidities for apparently identical processes.

Atmospheric Conditions. From carding to roving, the loosely bound fibers are vulnerable to static electricity. In most instances, static can be adequately suppressed with humidity, which should not be so high as to cause other problems. In other instances, it is necessary to suppress electrostatic properties with antistatic agents. Wherever draft is applied, constant humidity is needed to maintain optimum frictional uniformity between adjacent fibers and, hence, cross-sectional uniformity

Woolen and Worsted Systems

The woolen system generally makes coarser yarns, while the worsted system makes finer ones of a somewhat harder twist. Both may be used for lighter blends of wool, as well as for synthetic fibers with the characteristics of wool. The machinery used in both systems applies the same principles of draft and twist but differs greatly in detail and is more complex than that used for cotton.

Compared to cotton, wool fibers are dirtier, greasier, and more irregular. They are scoured to remove grease and are then usually reimpregnated with controlled amounts of oil to make them less hydrophilic and to provide better interfiber behavior. Wool fibers are scaly and curly, so they are more cohesive and require different treatment. Wool, in contrast to cotton and synthetic fibers, requires higher humidities in the processes prior to and including spinning than it does in the processes that follow. Approximate humidities are presented in Table 1.

Table 1   Recommended Humidities for Wool Processing at 24 to 27°C

Departments	Humidity, %
Raw wool storage	50 to 55
Mixing and blending	65 to 70
Carding —worsted	60 to 70
—woolen	60 to 75
Combing, worsted	65 to 75
Drawing, worsted —Bradford system	50 to 60
—French system	65 to 70
Spinning—Bradford worsted	50 to 55
—French (mule)	75 to 85
—woolen (mule)	65 to 75
Winding and spooling	55 to 60
Warping, worsted	50 to 55
Weaving, woolen, and worsted	50 to 60
Perching or clothroom	55 to 60

Reprinted by permission of copyright owner, Interscience Division of John Wiley and Sons, Inc., New York.

Twisting Filaments and Yarns

Twisting was originally applied to silk filaments; several filaments were doubled and then twisted to improve strength, uniformity, and elasticity. Essentially, the same process is used today, but it is now extended to spun yarns, as well as to single or multiple filaments of synthetic fibers. Twisting is widely used in the manufacture of sewing thread, twine, tire cord, tufting yarn, rug yarn, ply yarn, some knitting yarns, etc.

Twisting and doubling is done on a *down-* or *ring-twister*, which draws in two or more ends from packages on an elevated creel, twists them together, and winds them into a package. Except for the omission of drafting, down-twisters are similar to conventional ring-spinning frames.

When yarns are to be twisted without doubling, an *up-twister* is used. Up-twisters are used primarily for throwing synthetic monofilaments and multifilaments to add to or vary elasticity, light reflection, and abrasion resistance. As with spinning, yarn characteristics are controlled by making the twist hard or soft, right (S) or left (Z). Quality is determined largely by the uniformity of twist, which, in turn, depends primarily on the tension and stability of the atmospheric conditions (Figure 1 and Table 1). As the frame may be double- or triple-decked, twisting requires concentrations of power. The frames are otherwise similar to those used in spinning and present the same problems in air distribution. In twisting, lint is not a serious problem.

FABRIC MAKING

Preparatory Processes

When spinning or twisting is complete, the yarn may be prepared for weaving or knitting by processes including winding, spooling, creeling, beaming, slashing, sizing, and dyeing. These processes have two purposes: (1) to transfer the yarn from the type of package dictated by the preceding process to a type suitable for the next and (2) to impregnate some of the yarn with sizes, gums, or other chemicals that may not be left in the final product.

Filling Yarn. Filling yarn is wound on quills for use in a loom shuttle. It is sometimes predyed and must be put into a form suitable for package or skein dyeing before it is quilled. If the filling is of relatively hard twist, it may be put through a twist-setting or conditioning operation in which internal stresses are relieved by applying heat, moisture, or both.

Warp Yarn. Warp yarn is impregnated with a transient coating of size or starch that strengthens the yarns' resistance to chafing it will receive in the loom. The yarn is first rewound onto a cone or other large package from which it will unwind speedily and smoothly. The second step is warping, which rewinds a multiplicity of ends in parallel arrangement on large spools, called warp or section beams. In the third step, slashing, the threads pass progressively through the sizing solution, through squeeze rolls, and then around cans, around steam-heated drying cylinders, or through an air-drying chamber. A thousand kilograms or more may be wound on a single loom beam.

Knitting Yarn. If hard spun, knitting yarn must be twist-set to minimize kinking. Filament yarns must be sized to reduce strip-backs and improve other running qualities. Both must be put in the form of cones or other suitable packages.

Uniform tension is of great importance in maintaining uniform package density. Yarns tend to hang up when unwound from a hard package or slough off from a soft one, and both tendencies are aggravated by spottiness. The processes that require air conditioning, along with recommended relative humidities, are presented in Figure 1 and Table 1.

Weaving

In the simplest form of weaving, harnesses raise or depress alternate warp threads to form an opening called a *shed*. A shuttle containing a quill is kicked through the opening, trailing a thread of filling behind it. The lay and the reed then beat the thread firmly into one apex of the shed and up to the fell of the previously woven cloth. Each shuttle passage forms a pick. These actions are repeated at frequencies up to five per second.

Each warp thread usually passes through a drop-wire that is released by a thread break and automatically stops the loom. Another automatic mechanism inserts a new quill in the shuttle as the previous one is emptied, without stopping the loom. Other mechanisms are actuated by filling breaks, improper shuttle boxing, and the like, which stop the loom until it is manually restarted. Each cycle may leave a stopmark sufficient to cause an imperfection that may not be apparent until the fabric is dyed.

Beyond this basic machine and pattern are many complex variations in harness and shuttle control, which result in intricate and novel weaving effects. The most complex is the jacquard, with which individual warp threads may be separately controlled. Other variations appear in looms for such products as narrow fabrics, carpets, and pile fabrics. In the Sulzer weaving machine, a special filling carrier replaces the conventional shuttle. In the rapier, a flat, spring-like tape uncoils from each side and meets in the middle to transfer the grasp on the filling. In the water jet loom, a tiny jet of high-pressure water carries the filling through the shed of the warp. Other looms transport the filling with compressed air.

High humidity increases the abrasion resistance of the warp. Weave rooms require 80 to 85% humidity or higher for cotton and up to 70% humidity for synthetic fibers. Many looms run faster when room humidity and temperature are precisely controlled.

In the weave room, power distribution is uniform, with an average concentration somewhat lower than in spinning. The rough treatment of fibers liberates many minute particles of both fiber and size, thereby creating considerable amounts of airborne dust. Due to high humidity, air changes average from four to eight per hour. Special provisions must be made for maintaining conditions during production shutdown periods, usually at a lower relative humidity.

Knitting

Typical products are seamless articles knitted on circular machines (e.g., undershirts, socks, and hosiery) and those knitted flat (e.g., full-fashioned hosiery, tricot, milanese, and warp fabrics).

Knitted fabric is generated by forming millions of interlocking loops. In its simplest form, a single end is fed to needles that are actuated in sequence. In more complex constructions, hundreds of ends may be fed to groups of elements that function more or less in parallel.

Knitting yarns may be either single strand or multi-ply and must be of uniform high quality and free from neps or knots. These yarns, particularly the multifilament type, are usually treated with special sizes to provide lubrication and to keep broken filaments from stripping back.

The necessity for precise control of yarn tension, through controlled temperature and relative humidity, increases with the fineness of the product. For example, in finer gages of full-fashioned hosiery, a 1 K change in temperature is the limit, and a 10% change in humidity may add or subtract 75 mm in the length of a stocking. For knitting, desirable room conditions are approximately 24°C dry-bulb and 45 to 65% rh.

Dyeing and Finishing

Finishing, which is the final readying of a mill product for its particular market, ranges from cleaning to imparting special characteristics. The specific operations involved vary considerably, depending on the type of fiber, yarn, or fabric, and the end product usage. Operations are usually done in separate plants.

Inspection is the only finishing operation to which air conditioning is regularly applied, although most of the others require ventilation. Finishing operations that use wet processes usually keep their solutions at high temperatures and require special ventilation to prevent destructive condensation and fog. Spot cooling of workers may be necessary for large releases of sensible, latent, or radiant heat.

AIR-CONDITIONING DESIGN

Air washers are especially important in textile manufacturing and may be either conventional low-velocity or high-velocity units in built-up systems. Unitary high-velocity equipment using rotating eliminators, although no longer common, is still be found in some plants.

Contamination of air washer by airborne oils often dictates separation of air washers and process chillers by heat exchangers, usually of the plate or frame type.

Integrated Systems

Many mills use a refined air washer system that combines the air-conditioning system and the collector system (see section on Collector Systems) into an integrated unit. The air handled by the collector system fans and any air required to make up total return air are delivered back to the air-conditioning apparatus through a central duct. The quantity of air returned by individual yarn-processing machine cleaning systems must not exceed the air-conditioning supply air quantity. The air discharged by these individual suction systems is carried by return air ducts directly to the air-conditioning system. Before entering the duct, some of the cleaning system air passes over the yarn-processing machine drive motor and through a special enclosure to capture heat losses from the motor.

When integrated systems occasionally exceed the supply air requirements of the area served, the surplus air must be reintroduced after filtering.

Individual suction cleaning systems that can be integrated with air conditioning are available for cards, drawing frames, lap winders, combers, roving frames, spinning frames, spoolers, and warpers. The following advantages result from this integration:

1. With a constant air supply, the best uniform air distribution can be maintained year-round.
2. Downward airflow can be controlled; crosscurrents in the room are minimized or eliminated; drift or fly from one process to another is minimized or eliminated. Room partitioning between systems serving different types of manufacturing processes further enhances the value of this integration by controlling room air pattern year-round.
3. Heat losses of the yarn-processing frame motor and any portion of the processing frame heat captured in the duct, as well as the heat of the collector system equipment, cannot affect room conditions; hot spots in motor alleys are eliminated, and although this heat goes into the refrigeration load, it does not enter the room, and, as a result, the supply air quantity can be reduced.
4. Uniform conditions in the room improve production; conditioned air is drawn directly to the work areas on the machines, minimizing or eliminating wet or dry spots.
5. Maximum cleaning use is made of the air being moved. A guide for cleaning air requirements follows:

Pickers	1200 to 1900 L/s per picker
Cards	300 to 700 L/s per card
Spinning	2 to 4 L/s per spindle
Spooling	19 L/s per spool

Collector Systems

A collector system is a waste-capturing device that uses many orifices operating at high suction pressures. Each piece of production machinery is equipped with suction orifices at all points of major lint generation. The captured waste is generally collected in a fan and filter unit located either on each machine or centrally to accept waste from a group of machines.

A collector in the production area may discharge waste-filtered air either back into the production area or into a return duct to the air-conditioning system. It then enters the air washer or is relieved through dampers to the outdoors.

Figure 2 shows a mechanical spinning room with air-conditioning and collector systems combined into an integrated unit. In this case, the collector system returns all of its air to the air-conditioning system. If supply air from the air-conditioning system exceeds the maximum that can be handled by the collector system, additional air should be returned by other means.

Figure 2 also shows return air entering the air-conditioning system through damper T, passing through air washer H, and being delivered by fan J to the supply duct, which distributes it to maintain conditions within the spinning room. At the other end of each spinning frame are unitary-type filter-collectors consisting of enclosure N, collector unit screen O, and collector unit fan P.

Collector fan P draws air through the intake orifices spaced along the spinning frame. This air passes through the duct that runs lengthwise to the spinning frame, through the screen O, and is then discharged into the enclosure base (beneath the fan and screen). The air quantity is not constant, but drops slightly as material builds up on the filter screen.

Because the return air quantity must remain constant and the air quantity discharged by fan P is slightly reduced at times, relief openings are necessary. Relief openings also may be required when the return air volume is greater than the amount of air the collector suction system can handle.

The discharge of fan P is split, so part of the air cools the spinning frame drive motor before rejoining the rest of the air in the return air tunnel. Regardless of whether the total return air quantity enters the return air tunnel through collector units, or through a combination of collector units and floor openings beneath spinning frames, return-air fan R delivers it into the apparatus, ahead of return-air damper T. Consideration should be given to filtering the return air prior to its delivery into the air-conditioning apparatus.

Mild-season operation causes more outdoor air to be introduced through damper U. This air is relieved through motorized damper S, which opens gradually as outdoor damper U opens, while return damper T closes in proportion. All other components perform as typical central station air-washer systems.

A system having the general configuration shown in Figure 2 may also be used for carding; the collector system portion of this arrangement is shown in Figure 3. A central collector filters the lint-laden air taken from multiple points on each card. This air is discharged to return-air duct A and is then either returned to the air-conditioning system, exhausted outside, or returned directly to the room. A central collector filter may also be used with the spinning room system of Figure 2.

Air Distribution

Textile plants served by *generally uniform air distribution* may still require special handling for areas of load concentration.

Continuous Spinning Area. Methods of distribution are diverse and generally not critical. However, spot cooling or localized heat removal may be required. This area may be cooled by air conditioning, evaporative cooling, or ventilation.

Chimney (Quench Stack). Carefully controlled and filtered air or other gas is delivered to the chimneys; it is returned for conditioning and recovery of any valuable solvents present. Distribution of the air is of the utmost importance. Nonuniform temperature, humidity, or airflow disturbs the yarn, causing variations in fiber diameter, crystalline structure, and orientation. A fabric made of such fibers streaks when dyed.

In melt spinning, the solvent concentration in the chimney air must be maintained below its explosive limit. Care still must be

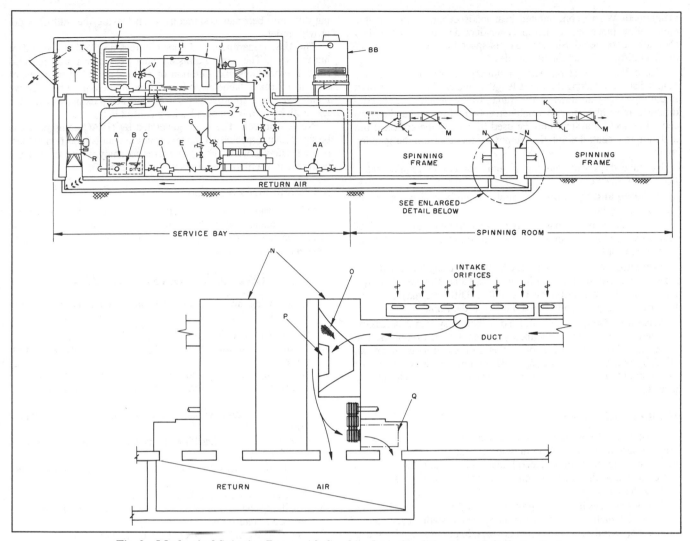

Fig. 2   Mechanical Spinning Room with Combined Air-Conditioning and Collector System

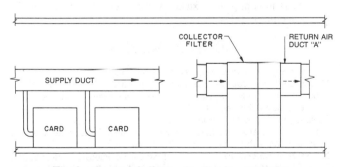

Fig. 3   Central Collector for Carding Machine

exercised that the vapors are not ignited by a spark or flame. The air-conditioning system must be reliable, because an interruption of the spinning causes the solution to solidify in the spinnerets.

Wind-Up or Take-Up Areas of Continuous Spinning. A heavy air-conditioning load is developed. Air is often delivered through branch ducts alongside each spinning machine. Diffusers are of the low-velocity, low-aspiration type, sized so as not to agitate delicate fibers.

Draw-Twist or Draw-Wind Areas of Fiber Manufacture. A heavy air-conditioning load is developed. Distribution, diffusion, and return systems are similar to those for the continuous spinning take-up area.

Opening and Picking. Usually opening and picking require only a uniform distribution system. The area is subject to shutdown of machinery during portions of the day. Generally, an all-air system with independent zoning is installed.

Carding. A uniform distribution system is generally installed. There should be little air movement around the web in cotton carding. Central lint collecting systems are available, but must be incorporated into the system design. An all-air system is often selected for cotton carding.

In wool carding, there should be less air movement than in cotton carding, not only to avoid disturbing the web, but also to reduce cross-contamination between adjacent cards. This is because different colors of predyed wool may be run side by side on adjacent cards. A split system may be considered for wool carding to reduce air movement. The method of returning air is also critical for achieving uniform conditions.

Drawing and Roving. Generally, a uniform distribution all-air system works well.

Mechanical Spinning Areas. A heavy air-conditioning load is generated, consisting of spinning frame power uniformly distributed along the frame length and frame driver motor losses concentrated in the motor alley at one end of the frame.

Supply air ducts should run across the frames at right angles. Sidewall outlets between each of the two adjacent frames then direct the supply air down between the frames, where conditions must be maintained. Where concentrated heat loads occur, as in a double motor alley, placement of a supply air duct directly over the alley should be considered. Sidewall outlets spaced along the bottom of the duct diffuse air into the motor alley.

The collecting system, whether unitary or central, with intake points distributed along the frame length at the working level, assists in pulling supply air down to the frame, where maintenance of conditions is most important. A small percentage of the air handled by a central collecting system may be used to convey the collected lint and yarn to a central point, thus removing that air from the spinning room.

Machine design in spinning systems sometimes requires interfloor air pressure control.

Winding and Spooling. Generally, a uniform distribution, all-air system is used.

Twisting. This area has a heavy air-conditioning load. Distribution considerations are similar to those in spinning. Either all-air or split systems are installed.

Warping. This area has a very light load. Long lengths of yarn may be exposed unsupported in this area. Generally, an all-air system with uniform distribution is installed. Diffusers may be of the low-aspiration type. Return air is often near the floor.

Weaving. Generally, a uniform distribution system is necessary. Synthetic fibers are more commonly woven than natural fibers. The lower humidity requirements of synthetic fibers allow the use of an all-air system rather than the previously common split system. When lower humidity is coupled with the water jet loom, a high latent load results.

Health Considerations

Control of Oil Mist. Whenever textiles coated with lubricating oils are heated above 93°C in drawing operations in ovens, heated rolls, tenterframes, or dryers, an oil mist is liberated. If the oil mist is not collected at the source of emission and disposed of, a slightly odorous haze results.

Various devices have been proposed to separate oil mist from the exhaust air, such as fume incinerators, electrostatic precipitators, high-energy scrubbers, absorption devices, high-velocity filters, and condensers.

Spinning operations that generate oil mist must be provided with a high percentage (30 to 75%) of outside air. In high-speed spinning, 100% outside air is commonly used.

Operations such as drum cooling and air texturizing, which could contaminate the air with oil, require local exhausts.

Control of Monomer Fumes. Separate exhaust systems for monomers are required, with either wet- or dry-type collectors, depending on the fiber being spun. For example, caprolactam nylon spinning requires wet exhaust scrubbers.

Control of Hazardous Solvents. Provisions must be made for the containment, capture, and disposal of hazardous solvents.

Control of Cotton Dust. Byssinosis, also known as brown or white lung disease, is believed to be caused by a histamine-releasing substance in cotton, flax, and hemp dust. A cotton worker returning to work after a weekend experiences difficulty in breathing that is not relieved until later in the week. After 10 to 20 years, the breath-

ing difficulty becomes continuous; even leaving the mill does not provide relief.

The U.S. Department of Labor enforces an OSHA standard of lint-free dust. The most promising means of control are improved exhaust procedures and filtration of recirculated air. Lint particles are 1 to 15 µm in diameter, so filtration equipment must be effective in this size range. Improvements in carding and picking, which leave less trash in the raw cotton, also help control lint.

Noise Control. The noise generated by HVAC equipment can be significant, especially if the textile equipment is modified to meet present safety criteria. For procedures to analyze and correct the noise due to ventilating equipment, see Chapter 43.

Safety and Fire Protection

Oil mist can accumulate in ductwork and create a fire hazard. Periodic cleaning reduces the hazard, but provisions should be made to contain a fire with suppression devices such as fire-activated dampers and interior duct sprinklers.

ENERGY CONSERVATION

The following are some steps that can be taken to reduce energy consumption:

1. Heat recovery applied to water and air.
2. Automation of high-pressure dryers to save heat and compressed air.
3. Decreasing the hot water temperatures and increasing the chilled water temperatures for rinsing and washing in dyeing operations.
4. Replacing running washes with recirculating washes, where possible.
5. Charging double-bleaching procedures to single-bleaching, where possible.
6. Eliminating rinses and final wash in dye operations, where possible.
7. Drying by means of the "bump and run" process.
8. Modifying the drying or curing oven air-circulation systems to provide counterflow.
9. Using energy-efficient electric motors and textile machinery.
10. For drying operations, using discharge air humidity measurements to control the exhaust versus recirculation rates in full economizer cycles.

BIBLIOGRAPHY

Carrol-Porezynski. *Inorganic fibers.* Academic Press, New York.
Hearle and Peters. *Moisture in textiles.* Textile Book Publishers, New York.
Kirk and Othmer, eds. *Encyclopedia of chemical technology*, 2nd ed. Vol. 9. Interscience Publishers, New York.
Labarthe, J. *Textiles: Origins to usage.* Macmillan, New York.
Mark, J.F., S.M. Atlas, and E. Cernin, eds. *Man-made fibers: Science & technology*, Vol. 1. Interscience Publishers, New York.
Nissan. *Textile engineering processes.* Textile Book Publishers, New York.
Press, J.J., ed. *Man made textile encyclopedia.* Textile Book Publishers, New York.
Sachs, A. 1987. Role of process zone air conditioning. *Textile Month* (October):42.
Schicht, H.H. 1987. Trends in textile air engineering. *Textile Month* (May): 41.

PHOTOGRAPHIC MATERIALS

THE processing and storage of sensitized photographic products requires temperature, humidity, and air quality control.

Manufacturers of photographic products and processing equipment provide specific recommendations for facility design that should always be consulted. This chapter contains general information that can be used in conjunction with these recommendations.

STORING UNPROCESSED PHOTOGRAPHIC MATERIALS

Virtually all photosensitive materials deteriorate with age; the rate of photosensitivity deterioration depends largely on the storage conditions. Photosensitivity deterioration increases both at high temperature and at high relative humidity and usually decreases at lower temperature and humidity.

High humidity can accelerate loss of sensitivity and contrast, increase shrinkage, produce mottle, cause softening of the emulsion (which can lead to scratches), and promote fungal growth. Low relative humidity can increase the susceptibility of film or paper to static markings, abrasions, brittleness, and curl.

Because different photographic products require different handling, product manufacturers should be consulted about the proper temperature and humidity conditions for storage. Refrigerated storage may be necessary for some products in some climates.

Products not packaged in sealed vaportight containers are vulnerable to contaminants. They must be kept away from formaldehyde vapor (emitted by particleboard, some insulation, some plastics, and some glues), industrial gases, engine exhaust, and vapors of solvents and cleansers. In hospitals, industrial plants, and laboratories, all photosensitive products, regardless of their packaging, must be protected from x-rays, radium, and other sources of radioactivity. For example, films stored 8 m away from 100 mg of radium require the protection of 90 mm of lead.

PROCESSING AND PRINTING PHOTOGRAPHIC MATERIALS

Ventilation with clean, fresh air maintains a comfortable working environment and prevents fume-related health problems. It is also necessary for the high-quality processing, safe handling, and safe storage of photographic materials.

Processing produces odors, fumes, high humidity, and heat (from lamps, electric motors, dryers, mounting presses, and high-temperature processing solutions). It is thus important to supply plentiful clean, fresh air at the optimum temperature and relative humidity to all processing rooms. Modern ventilation systems filter, supply, and distribute air, keep it moving, control its temperature and humidity, and adjust its pressure.

The preparation of this chapter is assigned to TC 9.2, Industrial Air Conditioning.

Air Conditioning for Preparatory Operations

During receiving operations, exposed film is removed from its protective packaging for presplicing and processing. Presplicing combines many individual rolls of film into a long roll to be processed. At high relative humidity, photographic emulsions become soft and can be mechanically damaged (scratched). At excessively low relative humidity, the film base is prone to static, sparking, and curl deformation. The presplice work area should be maintained at 50 to 55% rh and 21 to 24°C dry bulb.

Air Conditioning for Processing Operations

Processing exposed films or paper involves the use of a series of tempered chemical and wash tanks that emit heat, humidity, and fumes. A room exhaust system must be provided, along with local exhaust at noxious tanks. To conserve energy, air from the pressurized presplice rooms can be used as makeup for processing room exhaust. Further supply air should maintain the processing space at a maximum of 27°C dry bulb and 50% rh.

The processed film or paper proceeds from the final wash to the dryer, which controls the moisture remaining in the product. Too little drying will cause the film to stick when wound, while too much drying will cause undesirable curl. Drying can be regulated by controlling drying time, humidity, and temperature.

The volume of incoming air should be sufficient to change the air in a processing room completely in about 8 min. The flow of air should be diffused or distributed to avoid objectionable drafts. Apart from causing personnel discomfort, drafts can cause dust problems and disturb the surface temperature uniformity of drying drums and other heated equipment. For automated processing equipment, tempered fresh air should be supplied from the ceiling above the feed or head end of the machine at a minimum rate of 70 L/s per machine (see Figure 1). If the machine extends through a wall into another room, both rooms should be ventilated.

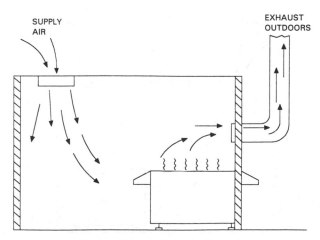

Fig. 1   Open Machine Ventilation

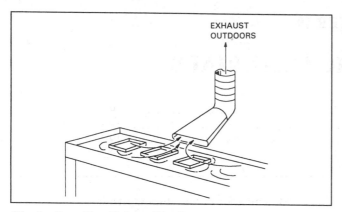

Fig. 2 Open Tray Exhaust Ventilation from a Processing Sink

An exhaust system should be installed to remove humid or heated air and chemical vapors directly to the outside of the building. The room air from an open machine or tank area should be exhausted to the outdoors at a minimum rate of 80 L/s per machine. An exhaust rate higher than the supply rate produces a negative pressure and makes less likely the escape of vapors or gases to adjoining rooms. The exhaust opening should be positioned so that the flow of exhausted air is away from the operator, as illustrated in Figure 2. This air should not be recirculated. The exhaust opening should always be placed as close as possible to the source of the contaminant for efficient removal. For a processing tank, an exhaust hood having a narrow opening at the back of and level with the top edge of the tank should be used.

If the processing tanks are enclosed and equipped with an exhaust connection, the minimum room air supply rate can be reduced to 42 L/s and the air removal rate to 47 L/s per machine (see Figure 3).

Air distribution to the drying area must provide a tolerable environment for operating personnel. The exposed sides of the dryer should be insulated as much as is practical to reduce the large radiant and convected heat losses to the space. Return or exhaust grilles above the dryer can directly remove much of its rejected heat and moisture. The supply air should be directed to offset the remaining radiant losses.

The use of processor dryer heat to preheat cold incoming air during winter conditions can save energy. An economic evaluation is necessary to determine whether the energy savings justify the additional cost of the heat recovery equipment.

A canopy exhaust hood over the drying drum of continuous paper processors extracts heat and moisture. The dryers should be exhausted to the outside of the building to avoid undesirable humidity buildup. A similar hood over a sulfide-toning sink can be used to

vent hydrogen sulfide. However, the exhaust duct must be placed on the side opposite the operator so that the vapor is not drawn toward the operator's face.

Exhaust should draw off the vapor from the solvent and wax mixture that is normally applied to lubricate the motion picture film as it leaves the dryer.

Air Conditioning for the Printing/Finishing Operation

In printing, where a second sensitized product is exposed through the processed original, the amount of environmental control needed depends on the size and type of operation. For small-scale printing, close control of the environment is not necessary, except to minimize dust. In photo-finishing plants, printers for colored products emit substantial heat. The effect on the room can be reduced by removing the lamphouse heat directly. Computer-controlled electronic printers transport the original film and raw film or paper at high speed. Proper temperature and humidity are especially important because in some cases two or three images from many separate films may be superimposed in register onto one film. For best results, the printing room should be maintained at between 21 and 24°C and at 50 to 60% rh to prevent curl, deformation, and static. Curl and film deformation affect the register and sharpness of the images produced. Static charge should be eliminated because it leaves static marks and may also attract dust to the final product.

Mounting of reversal film into slides is a critical finishing operation requiring a 21 to 24°C dry-bulb temperature with 50 to 55% rh.

Particulates in Air

Air-conditioning systems for most photographic operations require 85% efficient disposable bag-type filters with lower efficiency prefilters to extend filter life. In critical applications (such as high-altitude aerial films) and for microminiature images, filtering of foreign matter is extremely important. These products are handled in a laminar airflow room or workbench with HEPA filters plus prefilters.

Other Exhaust Requirements

A well ventilated room should be provided for mixing the chemicals used in color processing and high-volume black-and-white work. The room should contain one or more movable exhaust hoods that provide a capture velocity of 0.5 m/s. Figure 4 shows a suitable mixing tank arrangement.

If prints are lacquered regularly, a spray booth is needed. A concentration of lacquer spray is both hazardous and very objectionable to personnel; spray booth exhaust must be discharged outside.

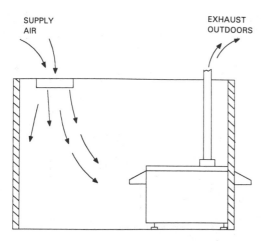

Fig. 3 Enclosed Machine Ventilation

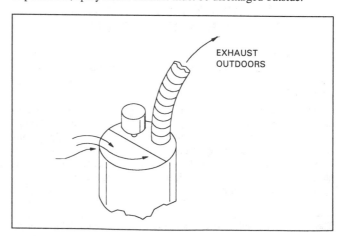

Fig. 4 Chemical Mixing Tank Exhaust

Processing Temperature Control

The density of a developed image on photographic material depends on characteristics of the emulsion, the exposure it has received, and the degree of development. With a particular emulsion, the degree of development depends on the time of development, the developer temperature, the degree of agitation, and the developer activity. Therefore, developer solution temperature control is critical. Once a developer temperature is determined, it should be maintained within 0.5 K for black-and-white products and within 0.3 K or better for color materials. Other solutions in the processor have a greater temperature tolerance.

Low processing volumes are typically handled with minilabs, which are often installed in retail locations. Minilabs are usually self-contained and equipped with temperature controls, heaters, pumps, and so forth. Typically, the owner has only to hook the lab up to water, electricity, exhaust (thimble connection), and a drain.

Higher volume processing is handled with processors that come from the manufacturer complete with controls, heat exchangers, pumps, and control valves designed for the process that the owner has specified. Electricity, hot water, cold water, drainage, and steam may be required, depending on the manufacturer, who typically provides the specifications for these utilities.

STORING PROCESSED FILM AND PAPER

Storage of developed film and paper differs from storage of the raw stock because the developed materials are no longer photosensitive, are seldom sealed against moisture, and are generally stored for much longer periods. Required storage conditions depend on (1) the value of the records, (2) the length of storage time, (3) whether the films are on nitrate or safety base, (4) whether the paper base is resin coated, and (5) the type of photographic image.

Photographic materials must be protected against fire, water, mold, chemical or physical damage, extreme relative humidity, and high temperature. Relative humidity is much more important than temperature. High relative humidity can cause films to stick together —particularly roll films, but also sheet films. High humidity also damages gelatin, encourages the growth of mold, increases dimensional changes, accelerates the decomposition of nitrate support, and accelerates the deterioration of both black-and-white and color images. Low relative humidity causes a temporary increase in curl and decrease in flexibility, but these conditions are usually reversed when the humidity rises again. An exception occurs when motion picture film is stored for a long time in loosely wound rolls at very low humidities. The curl causes the film roll to resemble a polygon rather than a circle when viewed from the side. This *spokiness* occurs because a highly curled roll of film resists being bent in the length direction when it is already bent in the width direction. When a spoky roll is stored for a long time, the film flows permanently into the spoky condition, resulting in film distortion. Very low relative humidity in storage may also cause the film or paper to crack or break if handled carelessly.

Low temperature is desirable for film and paper storage provided that (1) the relative humidity of the cold air is controlled and (2) the material is sufficiently warmed before opening to prevent moisture condensation. High temperature can accelerate film shrinkage, which may produce physical distortions, and the fading of dye images. High temperature is also detrimental to the stability of any nitrate film still in storage.

PHOTOGRAPHIC FILMS AND PAPER PRINTS

ANSI (1991a) defines longevities of films with a life expectancy (LE) rating. The LE rating is the minimum number of years information can be retrieved if the subject film is stored under long-term stor-

age conditions. In order to achieve the maximum LE rating, a product must be stored under long-term storage conditions. Polyester black-and-white silver gelatin films have an LE rating of 500, while acetate black-and-white silver gelatin films have an LE rating of 100. No LE ratings have been assigned to color films or black-and-white silver papers. Medium-term storage conditions have been defined for materials that are to retain their information for a minimum of 10 years.

Medium-Term Storage

Rooms for the medium-term storage of safety base film should be protected from accidental water damage by rain, flood, or pipe leaks. Air conditioning with controlled relative humidity is desirable but not always essential in moderate climates. Extremes of relative humidity are detrimental to film.

The most satisfactory storage relative humidity for processed film is about 50%, although the range from 30 to 60% is satisfactory. Air conditioning is required where the relative humidity of the storage area exceeds 60% for any appreciable period. For a small room, a dehumidifier may be used if air conditioning cannot be installed. The walls should be coated with a vapor retarder and the controlling humidistat should be set at about 40% rh. If the prevailing relative humidity is under 25% for long periods, and problems from curl or brittleness are encountered, humidity should be controlled by a mechanical humidifier with a controlling humidistat set at 40%.

For medium-term storage, a room temperature between 20 and 25°C is recommended. Higher temperatures may cause shrinkage, distortion, and dye fading. Occasional peak temperatures of 32°C should not have a serious effect. Color films should be stored at below 10°C to reduce dye fading. Films stored below the ambient dew point should be allowed to warm up before being opened to prevent moisture condensation.

An oxidizing or reducing atmosphere may deteriorate the film base and gradually fade the photographic image. Oxidizing agents may also cause microscopically small colored spots on fine grain film such as microfilm (Adelstein et al. 1970). Typical gaseous contaminants include hydrogen sulfide, sulfur dioxide, peroxides, ozone, nitrogen oxides, and paint fumes. If such fumes are present in the intended storage space, they must be eliminated, or the film must be protected from contact with the atmosphere. Chapters 26 and 41 have further information on this subject.

Long-Term Storage

For films or records that are to be preserved indefinitely, long-term storage conditions should be maintained. The recommended space relative humidity ranges from 20 to 50% rh, depending on the film type. When several film types are stored within the same area, 30% rh is a good compromise. The recommended storage temperature is below 21°C. Low temperature aids preservation, but if the storage temperature is below the dew point of the outdoor air, the records must be allowed to warm up in a closed container before they are used, to prevent moisture condensation. Temperature and humidity conditions must be maintained year-round and should be continuously monitored.

Requirements of a particular storage application can be met by any one of several air-conditioning equipment/system combinations. Standby equipment should be considered. Sufficient outdoor air should be provided to keep the room under a slight positive pressure for ventilation and to retard the entrance of untreated air. The air-conditioning unit should be located outside the vault for ease of maintenance, with precautions taken to prevent water leakage into the vault. The conditioner casing and all ductwork must be well insulated. Room conditions should be controlled by a dry-bulb thermostat and either a wet-bulb thermostat, a humidistat, or a dew-point controller (see Chapter 42).

Air-conditioning installations and fire dampers in ducts carrying air to or from the storage vault should be constructed and maintained according to National Fire Protection Association recommendations for air conditioning (NFPA 1993) and for fire-resistant file rooms (NFPA 1991).

All supply air should be filtered with noncombustible HEPA filters to remove dust, which may abrade the film or react with the photographic image. As is the case with medium-term storage, gaseous contaminants such as paint fumes, hydrogen sulfide, sulfur dioxide, peroxides, ozone, and nitrogen oxides may cause slow deterioration of the film base and gradual fading of the photographic image. When these substances cannot be avoided, an air scrubber, activated carbon adsorber, or other purification method is required.

Films should be stored in metal cabinets with adjustable shelves or drawers and with louvers or openings located to facilitate circulation of conditioned air through them. The cabinets should be arranged in the room to permit free circulation of air around them.

All films should be protected from water damage due to leaks, fire sprinkler discharge, or flooding. Drains should have sufficient capacity to keep the water from sprinkler discharge from reaching a depth of 75 mm. The lowest cabinet, shelf, or drawer should be at least 150 mm off the floor and constructed so that water cannot splash through the ventilating louvers onto the records.

When fire-protected storage is required, the film should be kept in either fire-resistant vaults or insulated record containers (Class 150). Fire-resistant vaults should be constructed in accordance with NFPA *Standard* 232 (NFPA 1991). Although the NFPA advises against air conditioning in valuable-paper record rooms because of the possible fire hazard from outside, properly controlled air conditioning is essential for long-term preservation of archival films. The fire hazard introduced by the openings in the room for air-conditioning ducts may be reduced by fire and smoke dampers activated by smoke detectors in the supply and return ducts.

Storage of Nitrate Base Film

Although photographic film has not been manufactured on cellulose nitrate film base for several decades, many archives, libraries, and museums still have valuable records on this material. The preservation of the nitrate film will be of considerable importance until the records have been printed on safety base.

Cellulose nitrate film base is chemically unstable and highly flammable. It decomposes slowly but continuously even under normal room conditions. The decomposition produces small amounts of nitric oxide, nitrogen dioxide, and other gases. Unless the nitrogen dioxide can escape readily, it reacts with the film base, accelerating the decomposition (Carrol and Calhoun 1955). The rate of decomposition is further accelerated by moisture and is approximately doubled with every 5 K increase in temperature.

All nitrate film must be stored in an approved vented cabinet or vault. Nitrate films should never be stored in the same vault with safety base films because any decomposition of the nitrate film will cause decomposition of the safety film. Cans in which nitrate film is stored should never be sealed, since this traps the nitrogen dioxide gas. Standards for the storage of nitrate film have been established (NFPA 1988). The National Archives and the National Institute of Standards and Technology have also investigated the effect of a number of factors on fires in nitrate film vaults (Ryan et al. 1956).

The storage temperature should be kept as low as economically possible. The film should be kept at below 50% rh. The temperature and humidity recommendations for the cold storage of color film in the following section also apply to nitrate film.

Storage of Color Film and Prints

All dyes fade in time. ANSI (1991a) does not define an LE for color films or black-and-white images on paper. However, many valuable color films and prints exist, and it is important to preserve them for as long as possible.

Light, heat, moisture, and atmospheric pollution contribute to the fading of color photographic images. Storage temperature should be

as low as possible for the preservation of dyes. For maximum permanence of images, the materials should be stored in lighttight sealed containers or in moistureproof wrapping materials at a temperature below freezing and at a relative humidity of 20 to 50%. The containers should be warmed to room temperature prior to opening to avoid moisture condensation on the surface. Photographic films can be brought to the recommended humidity by passing them through a conditioning cabinet with circulating air at about 20% rh for about 15 min.

An alternative is the use of a storage room or cabinet controlled at a steady (noncycled) low temperature and maintained at the recommended relative humidity. This eliminates the necessity of sealed containers but involves an expensive installation. The dye-fading rate decreases rapidly with decreasing storage temperature.

Storage of Black-and-White Prints

The recommended storage conditions for processed black-and-white paper prints are given by ANSI (1982). The optimum limits for relative humidity of the ambient air are 30 to 50%, but daily cycling between these limits should be avoided.

A variation in temperature can drive relative humidity beyond the acceptable range. A temperature between 15 and 25°C is acceptable, but daily variations of more than 4 K should be avoided. Prolonged exposure to temperatures above 30°C should also be avoided. The degradative processes in black-and-white prints can be slowed considerably by low storage temperature. Exposure to airborne particles and oxidizing or reducing atmospheres should also be avoided, as mentioned for films.

REFERENCES

Adelstein, P.Z., C.L. Graham, and L.E. West. 1970. Preservation of motion picture color films having permanent value. *Journal of the Society of Motion Picture and Television Engineers* 79(November):1011.

ANSI. 1982. Photography (film and slides)—Practice for storage of black-and-white photographic paper prints. *Standard* PH1.48-82. American National Standards Institute, New York.

ANSI. 1991a. Imaging media—Processed safety photographic film—Storage. *Standard* IT9.11-91. American National Standards Institute, New York.

ANSI. 1991b. Photography (photographic film)—Specifications for safety film. *Standard* IT9.6-91. American National Standards Institute, New York.

Carrol, J.F. and J.M. Calhoun. 1955. Effect of nitrogen oxide gases on processed acetate film. *Journal of the Society of Motion Picture and Television Engineers* 64(September):601.

NFPA. 1988. Standard for the storage and handling of cellulose nitrate motion picture film. *Standard* 40-88. National Fire Protection Association, Quincy, MA.

NFPA. 1991. Standard for the protection of records. *Standard* 232-91. National Fire Protection Association, Quincy, MA.

NFPA. 1993. Standard for the installation of air conditioning and ventilating systems. *Standard* 90A-93. National Fire Protection Association.

Ryan, J.V., J.W. Cummings, and A.C. Hutton. 1956. Fire effects and fire control in nitro-cellulose photographic-film storage. Building Materials and Structures Report, No. 145. U.S. Department of Commerce, Washington, D.C. (April).

BIBLIOGRAPHY

ACGIH. 1992. *Industrial ventilation: A manual of recommended practice*, 21st ed. American Conference of Governmental Industrial Hygienists, Cincinnati, OH.

ANSI. 1992. Imaging media (film)—specifications for stability of ammonia-processed diazo films. *Standard* IT9.5-92. American National Standards Institute, New York.

Carver, E.K., R.H. Talbot, and H.A. Loomis. 1943. Film distortions and their effect upon projection quality. *Journal of the Society of Motion Picture and Television Engineers* 41(July):88.

Kodak. 1987. Current information summary—General guidelines for ventilating photographic process areas. CIS-58. Eastman Kodak Company, Rochester, NY.

Kodak. 1990. Photolab design for professionals. *Publication* K-13. Eastman Kodak Company, Rochester, NY.

UL. 1983. Tests for fire resistance of record protection equipment, 12th ed. *Standard* 72-83. Underwriters Laboratories, Northbrook, IL.

CHAPTER 20

ENVIRONMENTAL CONTROL FOR ANIMALS AND PLANTS

THE design of plant and animal housing is complicated by the many environmental factors affecting the growth and production of living organisms and the financial constraint that equipment must repay costs through improved economic productivity. The engineer must balance the economic costs of modifying the environment against the economic losses of a plant or animal in a less-than-ideal environment.

Thus, the design of plant and animal housing is affected by (1) economics, (2) concern for both workers and the care and welfare of animals, and (3) regulations on pollution, sanitation, and health assurance.

DESIGN FOR ANIMAL ENVIRONMENTS

Typical animal production systems modify the environment, to some degree, by housing or sheltering animals year-round or for parts of a year. The amount of modification is generally based on the expected increase in production. Animal sensible heat and moisture production data, combined with information on the effects of environment on growth, productivity, and reproduction, help designers select optimal equipment (Chapter 9 of the 1993 ASHRAE Handbook—Fundamentals). Detailed information is available in a series of handbooks published by the MidWest Plan Service. These include Mechanical Ventilating Systems for Livestock Housing (1990), Natural Ventilating Systems for Livestock Housing and Heating (1989), and Cooling and Tempering Air for Livestock Housing (1990). ASAE Monograph No. 6, Ventilation of Agricultural Structures (1983), also gives more detailed information.

Design Approach

The environmental control system is typically designed to maintain thermal and air quality conditions within an acceptable range and as near the ideal for optimal animal performance as is practicable. Equipment is usually sized assuming steady-state energy and mass conservation equations. Experimental measurements confirm that heat and moisture production by animals is not constant and that there may be important thermal capacitance effects in livestock buildings. Nevertheless, for most design situations, the steady-state equations are acceptable.

Achieving the appropriate fresh air exchange rate and establishing the proper distribution within the room are generally the two most important design considerations. The optimal ventilation rate is selected according to the ventilation rate logic curve (Figure 1).

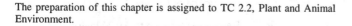
The preparation of this chapter is assigned to TC 2.2, Plant and Animal Environment.

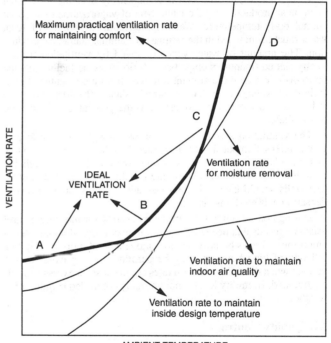

Fig. 1 Logic for Selecting the Appropriate Ventilation Rate in Livestock Buildings
(adapted from Christianson and Fehr 1983)

During the coldest weather, the ideal ventilation rate is that required to maintain indoor humidity at or below the maximum desired and air contaminants within acceptable ranges (rates A and B in Figure 1). Supplemental heating is usually required to prevent the temperature from dropping below optimal levels.

In milder weather, the ventilation rate required for maintaining optimal room air temperatures is greater than that required for moisture and air quality control (rates C and D in Figure 1). In hot weather, the ventilation rate is chosen to minimize the temperature rise above ambient and to provide optimal air movement over animals. Cooling is sometimes used in hot weather. The maximum rate (D) is often set at 60 air changes per hour as a practical maximum.

Temperature Control

The temperature within an animal structure is computed from the sensible heat balance of the system, usually disregarding transient effects. Nonstandard systems with low airflow rates and/or large thermal mass may require transient analysis. Steady-state heat transfer through walls, ceiling or roof, and ground is calculated as presented in Chapter 25 of the 1993 *ASHRAE Handbook—Fundamentals*.

Mature animals typically produce more heat per unit floor area than do young stock. Chapter 9 of the 1993 *ASHRAE Handbook—Fundamentals* presents estimates of animal heat loads. Lighting and equipment heat loads are estimated from power ratings and operating times. Typically, the designer selects indoor and outdoor design temperatures and calculates the ventilation rate to maintain the temperature difference. Outdoor design temperatures are given in Chapter 24 of the 1993 *ASHRAE Handbook—Fundamentals*. The section on Recommended Practices by Species in this chapter presents indoor design temperature values for various livestock.

Moisture Control

Moisture loads produced in an animal building may be calculated from data in the 1993 *ASHRAE Handbook—Fundamentals*. The mass of water vapor produced is estimated by dividing the animal latent heat production by the latent heat of vaporization of water at animal body temperature. Water spilled and evaporation of fecal water must be included in the estimates of animal latent heat production. The amount of water vapor removed by ventilation from a totally slatted (manure storage beneath floor) swine facility may be up to 40% less than the amount removed from a conventional concrete floor (solid). If the floor is partially slatted, the 40% maximum reduction is decreased in proportion to the percentage of the floor that is slatted.

The ventilation system should remove enough moisture to prevent condensation but should not reduce the relative humidity so low (less than 50%) as to create dusty conditions. Design indoor relative humidities for winter ventilation are usually between 70 and 80%. The walls should have sufficient insulation to prevent surface condensation at 80% rh inside.

During cold weather, the ventilation needed for moisture control usually exceeds that needed to control temperature. Minimum ventilation must always be provided to remove animal moisture. Up to a full day of high humidity may be permitted during extreme cold periods when normal ventilation rates could cause an excessive heating demand. Humidity level is not normally controlled in mild or hot weather.

Air Quality Control

The amount of dust varies with animal density, size, and degree of activity, type of litter or bedding, type of feed, and relative humidity of the air. A moisture content of 25 to 30%, wet basis, in the litter or bedding keeps dust to a minimum.

Gaseous contaminants within a building are often controlled by ventilation for moisture or temperature control. Ammonia, which results from the decomposition of manure, is the most important chronically present contaminant gas. Its production can be minimized by removing wastes from the room and keeping floor surfaces or bedding dry. Covering manure solids in gutters and pits with water also reduces ammonia, which is highly soluble in water. Ammonia should be maintained below 20 mg/m³ and, ideally, below 7 mg/m³.

Hydrogen sulfide, a by-product of the microbial decomposition of stored manure, is the most important acute gas contaminant. During normal operation, hydrogen sulfide concentration is usually insignificant. During agitation of stored manure, hydrogen sulfide levels sometimes exceed 1100 mg/m³.

Since adverse effects on production begin to occur at 28 mg/m³, ventilation systems should be designed to maintain hydrogen sulfide levels below 28 mg/m³ during agitation. When manure is agitated and removed from the storage, the building should be well ventilated

and all animals and occupants evacuated because of potential deadly concentrations of gases.

Air quality control based on carbon dioxide concentrations was suggested by Donham et al. (1989). They suggested a carbon dioxide concentration of 2770 mg/m³ as a threshold level above which symptoms of respiratory disorders occurred in a population of swine building workers. For other industries, a carbon dioxide concentration of 9000 mg/m³ is suggested as the time-weighted threshold limit value for 8 h of exposure (ACGIH 1992).

Barber et al. (1993), reporting on 173 pig buildings, found that carbon dioxide concentrations were below 5400 mg/m³ in nearly all instances when the external temperature was above 0°C. Carbon dioxide concentrations were rarely below 5400 mg/m³ when the outside temperature was below 0°C. They indicated that there was a very high penalty in heating costs in cold climates if the maximum allowed carbon dioxide concentration was less than 9000 mg/m³.

Other gas contaminants can also be important. Carbon monoxide from improperly operating unvented space heaters sometimes reaches problem levels. Methane is another occasional concern.

Disease Control

Airborne microbes can transfer disease-causing organisms among animals. For some situations, typically with young animals where there are low-level infections, it is important to minimize air mixing among pens. It is especially important to minimize air exchange between different animal rooms.

Air Distribution

Air speeds should be maintained below 0.25 m/s for most animal species in cold and mild weather. Animal sensitivities to draft are comparable to those of humans, although some animals are more sensitive at different stages. Riskowski and Bundy (1988) document that air velocities for optimal rates of gain and feed efficiencies can be below 0.13 m/s for young pigs at thermoneutral conditions.

Increased air movement during hot weather increases growth rates and improves heat tolerance. There are conflicting and limited data defining optimal air velocities in hot weather. Bond et al. (1965) and Riskowski and Bundy (1988) determined that both young and mature swine perform best when air speeds are less than 1 m/s (Figure 2). Mount et al. (1980) did not observe performance penalties at air speeds increased to a maximum of 0.76 m/s.

Degree of Shelter

Livestock, especially young animals, need some protection from adverse climate. On the range, mature cattle and sheep need

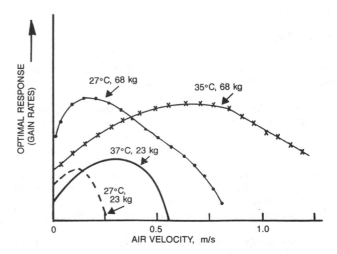

Fig. 2 Response of Swine to Air Velocity

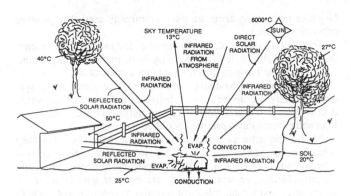

Fig. 3 Energy Exchange Between a Farm Animal and Its Surroundings in a Hot Environment

protection during severe winter conditions. In winter, dairy cattle and swine may be protected from precipitation, wind, and drafts with a three-sided, roofed shelter open on the leeward side. The windward side should also have approximately 10% of the wall surface area open to prevent a negative pressure inside the shelter; this pressure could cause rain and snow to be drawn into the building on the leeward side. Such shelters do not protect against extreme temperatures or high humidity.

In warmer climates, shades often provide adequate shelter, especially for large, mature animals such as dairy cows. Shades are commonly used in Arizona; research in Florida has shown an approximate 10% increase in milk production and a 75% increase in conception efficiency for shaded versus unshaded cows. The benefit of shades has not been documented for areas with less severe summer temperatures. Although shades for beef are also common practice in the southwest, beef cattle are somewhat less susceptible to heat stress, and extensive comparisons of various shade types in Florida have detected little or no differences in daily weight gain or feed conversion.

The energy exchange between an animal and various areas of the environment is illustrated in Figure 3. A well-designed shade makes maximum use of radiant heat sinks, such as the cold sky, and gives maximum protection from direct solar radiation and high surface temperatures under the shade. Good design considers geometric orientation and material selection, including roof surface treatment and insulation materials on the lower surface.

An ideal shade has a top surface that is highly reflective to solar energy and a lower surface that is highly absorptive to solar radiation reflected from the ground. A white-painted upper surface reflects solar radiation, yet emits infrared energy better than aluminum. The undersurface should be painted a dark color to prevent multiple reflection of shortwave energy onto animals under the shade.

COOLING AND HEATING

Evaporative Cooling

Supplemental cooling of animals in intensive housing conditions may be necessary during heat waves to prevent heat prostration, mortality, or serious losses in production and reproduction. Evaporative cooling systems, which may reduce ventilation air to 27°C or lower in most of the United States, are popular for poultry houses and are sometimes used for swine and dairy housing.

Evaporative cooling is well suited to animal housing because the high air exchange rates effectively remove odors and ammonia and increase air movement for convective heat relief. Initial cost, operating expense, and maintenance problems are all relatively low compared with those of other types of heating and ventilation. Evaporative cooling works best in areas with low relative humidity,

but significant benefits can be obtained even in the humid southeastern United States.

System Design. The pad area should be sized to maintain air velocities between 1.0 and 1.4 m/s through the pads. For most pad systems, these velocities produce evaporative efficiencies between 75 and 85%; they also increase pressures against the ventilating fans from 10 to 30 Pa.

The building must be tight because air leaks due to the negative pressure ventilation will reduce the airflow through the pads, and hence reduce the cooling effectiveness.

The most serious problem encountered with evaporative pad systems for agricultural applications is clogging by dust and other airborne particles. Whenever possible, fans should exhaust away from pads on adjacent buildings. Regular preventive maintenance is essential. Water bleed-off and the addition of algicides to the water are recommended. When pads are not used in cool weather, they should be sealed to prevent dusty inside air from exhausting through them.

High-pressure fogging systems with water pressures of 3.5 MPa are preferred to pad coolers for cooling the air in broiler houses with built-up litter. The high pressure creates a fine aerosol, causing minimal litter wetting. Timers and/or thermostats control the system. Evaporative efficiencies and installation costs are about one-half those of a well-designed evaporative pad system. Foggers can also be used with naturally ventilated, open-sided housing. Low-pressure systems are not recommended for poultry, but may be used during emergencies.

Nozzles that produce water mist or spray droplets to wet animals directly are used extensively during hot weather in swine confinement facilities with solid concrete or slatted floors. Currently, misting or sprinkling systems that directly wet the skin surface of the animals (not merely the outer portion of the hair coat) are preferred. Timers that operate a system periodically, (e.g., 30 s on a 5-min cycle) help to conserve water.

Mechanical Refrigeration

Mechanical refrigeration systems can be designed for effective animal cooling, but they are considered uneconomical for most production animals. Air-conditioning loads for dairy housing may require 2.5 kW or more per cow. Recirculation of refrigeration air is usually not feasible due to high contaminant loads in the air in the animal housing. Sometimes zone cooling of individual animals is used instead of whole-room cooling, particularly in swine farrowing houses where a lower air temperature is needed for sows than for unweaned piglets. Refrigerated air, 10 to 20K below ambient, is supplied through insulated ducts directly to the head and face of the animal. Air delivery rates are typically 5 to 15 L/s per sow for snout cooling, and 30 to 40 L/s per sow for zone cooling.

Earth Tubes

Some livestock facilities obtain cooling in summer and heating in winter by drawing ventilation air through tubing buried 2 to 4 m below grade. These systems are most practical in the north central region for animals that benefit from both cooling in summer and heating in winter.

Goetsch and Muehling (1983) detail design procedures for these systems. A typical design uses 15 to 50 m of 200-mm diameter pipe to provide 150 L/s of tempered air. Soil type and moisture, pipe depth, airflow, climate, and other factors affect the efficiency of buried pipe heat exchangers. The pipes must slope to drain condensation and must not have dips that could plug with condensation.

Heat Exchangers

Ventilation accounts for 70 to 90% of the heat losses in typical livestock facilities during winter. Heat exchangers can reclaim some of the heat lost with the exhaust ventilating air. However, predicting fuel savings based on savings obtained during the coldest periods will overestimate yearly savings from a heat exchanger. Estimates of energy savings based on air enthalpy can improve the accuracy of the predictions.

Heat exchanger design must address the problems of condensate freezing and/or dust accumulation on the heat exchanging surfaces. These problems result in either reduced efficiency and/or the inconvenience of frequent cleaning.

Supplemental Heating

For poultry weighing 1.5 kg or more, for pigs heavier than 23 kg, and for other large animals such as dairy cows, the body heat of the animals at recommended space allocations is usually sufficient to maintain moderate temperatures (i.e., above 10°C) in a well-insulated structure. Combustion-type heaters are used to supplement heat for baby chicks and pigs. Supplemental heating also increases the moisture-holding capacity of the air, which reduces the quantity of air required for removal of moisture. Various types of heating equipment may be included in ventilation systems, but they need to perform well in dusty and corrosive atmospheres.

Insulation Requirements

The amount of building insulation required depends on climate, animal space allocations, and animal heat and moisture production. Usually, structures with an overall heat transmission coefficient of 0.45 to 0.68 $W/(m^2 \cdot K)$ are adequate for northern climates. In moderate climates, values ranging from 0.85 to 1.4 $W/(m^2 \cdot K)$ are adequate for adult animals, but may be decreased in cold areas to conserve fuel when heating for young animals. In warm weather, ventilation between the roof and insulation helps reduce the radiant heat load from the ceiling. Insulation in warm climates can be more important for reducing radiant heat loads in summer than reducing building heat loss in winter.

VENTILATION SYSTEMS

Mechanical Ventilation

Mechanical ventilation uses fans to create a static pressure difference between the inside and outside of a building. Farm buildings use either positive pressure, with fans forcing air into a building, or negative pressure, with exhaust fans. Some ventilation systems use a combination of positive pressure to introduce air into a building and separate fans to remove air. These zero-pressure systems are particularly appropriate for heat exchangers.

Positive Pressure Systems. Fans blow outside air into the ventilated space, forcing humid air out through any planned outlets and through leaks in walls and ceilings. If vapor barriers are not complete, moisture condensation will occur within the walls and ceiling during cold weather. Condensation causes deterioration of building materials and reduces insulation effectiveness. The energy used by fan motors and rejected as heat is added to the building—an advantage in winter but a disadvantage in summer.

Negative Pressure Systems. Fans exhaust air from the ventilated space while drawing outside air in through planned inlets and leaks in walls, in ceilings, and around doors and windows. Air distribution in a negative pressure system is often less complex and costly. Simple openings and baffled slots in walls control and distribute air in the building. However, at low airflow rates, negative pressure systems may not distribute air uniformly because of air leaks and wind pressure effects. Supplemental air mixing may be necessary.

Allowances should be made for reduced fan performance due to dust, guards, and corrosion of louver joints (Person et al. 1979). Totally enclosed fan motors are protected from exhaust air contaminants and humidity. Periodic cleaning helps prevent overheating.

Negative pressure systems are more commonly used than positive pressure systems.

Ventilation systems should always be designed so that manure gases are not drawn into the building from manure storages connected to the building by underground pipes or channels.

Neutral Pressure Systems. Neutral pressure (push-pull) systems typically use intake fans to distribute air down a distribution duct to room inlets and exhaust fans to remove air from the room. Inlet and exhaust fan capacities should be matched.

Neutral pressure systems are often more expensive, but they achieve better control of the air. They are less susceptible to wind effects and to building leakage than are positive or negative pressure systems. Neutral pressure systems are most frequently used for young stock and for animals most sensitive to environmental conditions, primarily where cold weather is a concern.

Natural Ventilation

Either natural or mechanical ventilation systems are used to modify environments in livestock shelters. Natural ventilation is most common for mature animal housing, such as free-stall dairy, poultry growing, and swine finishing houses. Natural ventilation depends on pressure differences caused by wind and temperature differences. A well-designed natural ventilation system keeps temperatures reasonably stable, if automatic controls regulate ventilation openings. Usually, a design includes an open ridge (with or without a rain cover) and openable sidewalls, which should cover at least 50% of the wall for summer operation. Ridge openings are about 17 mm wide for each metre of house width, with a minimum ridge width of 150 mm.

Openings can be adjusted automatically, with control based on air temperature. Some designs, referred to as flex housing, include a combination of mechanical and natural ventilation usually dictated by outside air temperature and/or the amount of ventilation required.

VENTILATION MANAGEMENT

Air Distribution

Pressure differences across walls and inlet or fan openings are usually maintained between 10 and 15 Pa. (The exhaust fans are usually sized to provide proper ventilation at 30 Pa to compensate for wind effects.) This pressure difference creates inlet velocities of 3 to 5 m/s, sufficient for effective air mixing, but low enough to cause only a small reduction in fan capacity. A properly planned inlet system distributes fresh air equally throughout the building. Negative pressure systems that rely on cracks around doors and windows do not distribute fresh air effectively. Inlets require adjustment, since winter airflow rates are typically less than 10% of summer rates. Automatic controllers are available to regulate inlet area.

Positive pressure systems, with fans connected directly to perforated polyethylene air distribution tubes, may combine heating, circulation, and ventilation in one system. Air distribution tubes or ducts connected to circulating fans are sometimes used to mix the air in negative pressure systems. However, dust and dirtiness are of concern when air is recirculated, particularly when cold incoming air condenses moisture in the tubes.

Inlet Design. Inlet location and size most critically affect air distribution within a building. Continuous or intermittent inlets can be placed along the entire length of one or both outside walls. Building widths less than 6 m may need only a single inlet along one wall. The total inlet area may be calculated by the system characteristic technique, which follows. Because the distribution of the inlet area is based on the geometry and size of the building, which makes specific recommendations are difficult.

System characteristic technique. This technique determines the operating points for the ventilation rate and pressure difference across inlets. Fan airflow rate as a function of pressure difference across the

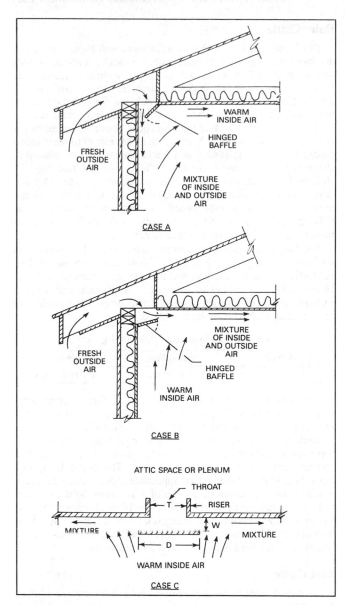

Fig. 4   Typical Livestock Building Inlet Configurations

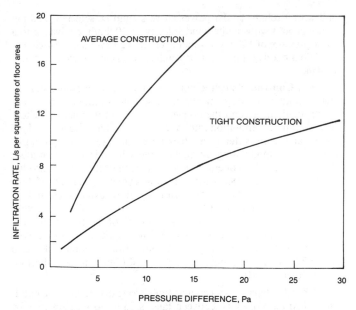

Fig. 5   Infiltration Due to Cracks for Average and Tight Animal Building Construction
(Adapted from ASAE Data: EP270.5)

fan should be available from the manufacturer. Allowances should be made for additional pressure losses from fan shutters or other devices such as light restriction systems or cooling pads.

Inlet flow characteristics are available for hinged baffle and center-ceiling flat baffle slotted inlets (Figure 4). Airflow rates can be calculated for the baffles in Figure 4 by the following:

For Case A:

$$Q = 1.1 Wp^{0.5} \qquad (1)$$

For Case B:

$$Q = 0.71 Wp^{0.5} \qquad (2)$$

For Case C:

$$Q = 1.3 Wp^{0.5} (D/T)^{0.08} e^{(-0.867 \, W/T)} \qquad (3)$$

where

Q = airflow rate, L/s per metre length of slot opening
W = slot width, mm
p = pressure difference across the inlet, Pa
D = baffle width, mm
T = width of slot in ceiling, mm

Air Infiltration. In addition to airflow through the inlet, infiltration airflow should be included. Figure 5 illustrates infiltration rates for two types of dairy barn construction (appropriate for other animal buildings also). The infiltration rates can be described as follows:

For average construction:

$$I = 3.4p^{0.67} \qquad (4)$$

For tight construction:

$$I = 1.2p^{0.67} \qquad (5)$$

where I = infiltration rate, L/s per square metre of floor area.

The above equation can be graphed as a system ventilation rate curve for various slot widths. Then, together with fan data, a ventilation schedule for expected weather and seasonal conditions can be developed.

Example. A dairy barn with a total floor area of 600 m^2 is to be ventilated with hinged baffle, wall flow, and slotted inlets. A total inlet length of 140 m is available. The barn is considered to have tight construction. The total fresh air ventilation rate Q_{tot} is given by the following:

$$Q_{tot} = Q + I = 1.1 \ Wp^{0.5} (140/1000) + 600 \times 1.2p^{0.67}$$

$$Q_{tot} = 0.154 Wp^{0.5} + 720p^{0.67}$$

Room Air Velocity. The average air velocity inside a slot ventilated structure relates to the inlet air velocity, inlet slot width (or equivalent continuous length for boxed inlets), building width, and ceiling height. Estimates of air velocity within a barn, based on air exchange rates, may be very low due to the effects of jet velocity and recirculation. Conditions are usually partially turbulent, and there is no reliable way to predict room air velocity at animal level. General design guidelines keep the throw distance less than 6 m from slots and less than 3 m from polyethylene tubes with holes.

Fans

Fans should not exhaust against prevailing winds. If structural or other factors require installing fans on the windward side, fans rated to deliver the required capacity against at least 30 Pa static pressure and with a relatively flat power curve should be selected. The fan motor should withstand a wind velocity of 50 km/h, equivalent to a static pressure of 100 Pa, without overloading beyond its service factor. Wind hoods on the fans or windbreak fences reduce the effects of wind.

Flow Control. Since the numbers and size of livestock and climatic conditions vary, means to modulate ventilation rates are often required beyond the conventional off/on thermostat switch. The minimum ventilation rate to remove moisture, reduce noxious gases, and keep water from freezing should always be provided. Methods of modulating ventilation rates include (1) intermittent fan operation—fans operate for a percentage of the time controlled by a percentage timer with a 10-min cycle; (2) staging of fans using multiple units or fans with high/low exhaust capability; (3) the use of multispeed fans—larger fans (400 W and up) with two flow rates, the lower being about 60% of the maximum rate; and (4) the use of variable-speed fans—split capacitor motors designed to modulate fan speed smoothly from maximum down to 10 to 20% of the maximum rate (the controller is usually thermostatically adjusted).

Generally, fans are spaced uniformly along the lee side of a building. Maximum distance between fans is 35 to 40 m. Fans may be grouped in a bank if this range is not exceeded. In housing with side curtains, exhaust fans that can be reversed or removed and placed inside the building in the summer are sometimes installed to increase air movement in combination with doors, walls, or windows being opened for natural ventilation.

Thermostats

Control thermostats should be placed where they respond to a representative temperature as sensed by the animals. Thermostats need protection and should be placed to prevent potential physical damage, i.e., away from animals, ventilation inlets, water pipes, lights, heater exhausts, outside walls, or any other objects that will unduly affect performance. Control thermostats also require periodic adjustment based on accurate thermometer readings taken in the immediate area of the animal.

Emergency Warning System

Animals housed in a high-density, mechanically controlled environment are subject to considerable risk of heat prostration if a failure of power or ventilation equipment occurs. To reduce this danger, an alarm system and an automatic standby electric generator are highly recommended. Many alarm systems will detect failure of the ventilation system. These range from inexpensive "power off" alarms to systems that sense temperature extremes and certain gases. Automatic telephone dialing systems are effective as alarms and are relatively inexpensive. Building designs that allow some side wall panels (e.g., 25% of wall area) to be removed for emergency situations are also recommended.

RECOMMENDED PRACTICES
BY SPECIES

Mature animals readily adapt to a broad range of temperatures, but efficiency of production varies. Younger animals are more temperature sensitive. Figure 6 illustrates animal production response to temperature.

Relative humidity has not been shown to influence animal performance, except when accompanied by thermal stress. Relative humidity consistently below 50% may contribute to excessive dustiness; above 80%, it may increase building and equipment deterioration. Disease pathogens also appear to be more viable at either low or high humidities.

Dairy Cattle

Dairy cattle shelters include confinement stall barns, free stalls, and loose housing. In a stall barn, cattle are usually confined to stalls approximately 1.2 m wide, and all chores, including milking and feeding, are conducted there. Such a structure requires environmental modification, primarily through ventilation. Total space requirements are 5 to 7 m^2 per cow. In free-stall housing, cattle are not confined to stalls but are free to move about. Space requirements per cow are 7 to 9 m^2. In loose housing, cattle are free to move within a fenced lot containing resting and feeding areas. Space required in sheltered loose housing is similar to that in free-stall housing. The shelters for resting and feeding areas are generally open-sided and require no air conditioning or mechanical ventilation, but supplemental air mixing is often beneficial during warm weather. The milking area is in a separate area or facility and may be fully or partially enclosed, thus requiring some ventilation.

For dairy cattle, climate requirements for minimal economic loss are broad and range from 2 to 24°C with 40 to 80% rh. Below 2°C, production efficiency declines and management problems increase. However, the effect of low temperature on milk production is not as extreme as are high temperatures, where evaporative coolers or other cooling methods may be warranted.

Ventilation Rates for Each 500-kg Cow

Winter	Spring/Fall	Summer
17 to 22 L/s	67 to 90 L/s	110 to 220 L/s

Required ventilation rates depend on specific thermal characteristics of individual buildings and internal heating load. The relative humidity should be maintained between 50 and 80%.

Both loose housing and stall barns require an additional milk room to cool and hold the milk. Sanitation codes for milk production contain minimum ventilation requirements. The market being supplied should be consulted for all applicable codes. Some state codes require positive pressure ventilation of milk rooms. Milk rooms are usually ventilated with fans at rates of 4 to 10 air changes per hour to satisfy requirements of local milk codes and to remove heat from milk coolers. Most milk codes require ventilation in the passageway (if any) between the milking area and the milk room.

Beef Cattle

Beef cattle ventilation requirements are similar to those of dairy cattle on a unit weight basis. Beef production facilities often provide only shade and wind breaks.

Swine

Swine housing can be grouped into four general classifications:

1. Farrowing pigs, from birth to 7 kg
2. Nursery pigs, from 7 to 23 kg
3. Growing pigs, from 23 kg to market size
4. Breeding and gestation

In farrowing barns, two environments must be provided: one for sows and one for piglets. Because each requires a different temperature, zone heating and/or cooling is used. The environment within the nursery is similar to that within the farrowing barn for piglets. The requirements for growing barns and breeding stock housing are similar.

Currently recommended practices for **farrowing houses**:

- Temperature: 10 to 20°C, with small areas for piglets warmed to 28 to 32°C by means of brooders, heat lamps, or floor heat. Avoid cold drafts and extreme temperatures. Hovers are sometimes used. Provide supplemental cooling (usually sprinklers or evaporative cooling systems) in extreme heat.
- Relative humidity: Up to 75% maximum

- Ventilation rate: 18 to 250 L/s per sow and litter (about 180 kg total mass). The low rate is for winter; the high rate is for summer temperature control.
- Space: 3.2 m² per sow and litter (stall); 6.0 m² per sow and litter (pens)

  Recommendations for **nursery barns**:

- Temperature: 26°C for first week after weaning. Lower room temperature 1.5 K per week to 21°C. Provide warm, draft-free floors. Provide supplemental cooling for extreme heat (temperatures 28°C and above).
- Ventilation rate: 1.5 to 18 L/s per pig, 6 to 36 kg each
- Space: 0.16 to 0.36 m² per pig, 6 to 14 kg each

  Recommendations for **growing and gestation barns**:

- Temperature: 13 to 22°C preferred. Provide supplemental cooling (sprinklers or evaporative coolers) for extreme heat.

- Relative humidity: 75% maximum in winter; no established limit in summer
- Ventilation rate:
  Growing pig (34 to 68 kg)—3 to 40 L/s
  Finishing pig (68 to 100 kg)—5 to 60 L/s
  Gestating sow (110 to 230 kg)—0.5 L/s per kilogram
- Space:
  0.54 m² per pig, 34 to 68 kg each
  0.72 m² per pig, 68 to 100 kg each
  1.3 to 2.2 m² per pig, 110 to 130 kg each

Poultry

In broiler and brooder houses, growing chicks require changing environmental conditions, and heat and moisture dissipation rates increase as the chicks grow older. Supplemental heat, usually from brooders, is used until the sensible heat produced by the birds is adequate to maintain an acceptable air temperature. At early stages of

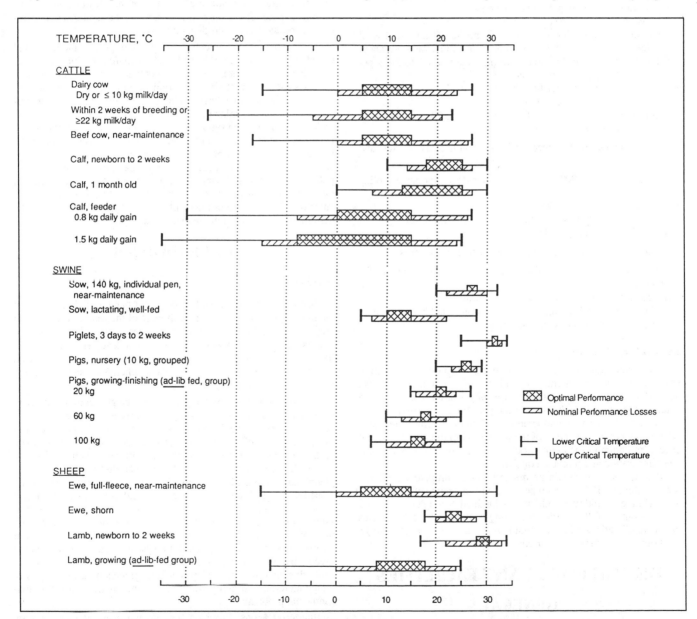

Fig. 6 Critical Ambient Temperatures and Temperature Zones for Optimum Performance and Nominal Performance Losses in Farm Animals
(Hahn 1985)

growth, moisture dissipation per bird is low. Consequently, low ventilation rates are recommended to prevent excessive heat loss. Litter is allowed to accumulate over 3 to 5 flock placements. Lack of low-cost litter material may justify the use of concrete floors. After each flock, caked litter is removed and fresh litter is added.

Housing for poultry may be open, curtain-sided (dominant style in the southern United States), or totally enclosed (dominant style in northern climates). Mechanical ventilation depends on the type of housing used. For open-sided housing, ventilation is generally natural airflow in warm weather, supplemented with stirring fans, and by fans with closed curtains in cold weather or during the brooding period. Mechanical ventilation is used in totally enclosed housing. Newer houses have smaller curtains and well-insulated construction to accommodate both natural and mechanical ventilation operation.

Recommendations for **broiler houses**:

- Room temperature: 15 to 27°C
- Temperature under brooder hover: 30 to 33°C, reducing 3 K per week until room temperature is reached
- Relative humidity: 50 to 80%
- Ventilation rate: Sufficient to maintain house within 1 to 2 K of outside air conditions during summer. Generally, rates are about 0.1 L/s per kilogram live mass during winter and 1 to 2 L/s per kilogram for summer conditions.
- Space: 0.06 to 0.1 m² per bird (for the first 21 days of brooding, only 50% of floor space is used)
- Light: Minimum of 10 lux to 28 days of age; 1 to 20 lux for growout (in enclosed housing).

Recommendations for **breeder houses** with birds on litter and slatted floors:

- Temperature: 10 to 30°C maximum; consider evaporative cooling if higher temperatures are expected.
- Relative humidity: 50 to 75%
- Ventilation rate: Same as for broilers on live mass basis.
- Space: 0.2 to 0.3 m² per bird

Recommendations for **laying houses** with birds in cages:

- Temperature, relative humidity, and ventilation rate: Same as for breeders.
- Space: 0.032 to 0.042 m² per hen minimum
- Light: Controlled day length using light-controlled housing is generally practiced (January through June).

Laboratory Animals

Proper management of laboratory animals includes any system of housing and care that permits animals to grow, mature, reproduce, behave normally, and be maintained in physical comfort and good health. Most recommendations for temperature and relative humidity are based on room temperature and humidity measurements, which may not be indicative of the microenvironment of the animal cage. Ventilation of the animal room (or cage) is necessary to regulate temperature and promote comfort.

Ideally, a system should permit individual compartment adjustments within ±1 K for any temperature within a range of 18 to 30°C. The relative humidity should be maintained throughout the year within a range of 40 to 70%, according to the needs of the species and local climatic conditions. Room air changes at a rate of 10 to 15 per hour are recommended for odor control.

DESIGN FOR PLANT FACILITIES

GENERAL

Greenhouses, plant growth chambers, and other facilities for indoor crop production overcome adverse outdoor environments and provide conditions conducive to economical crop production.

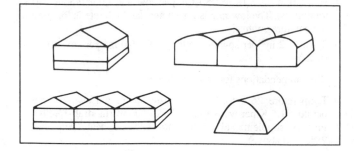

Fig. 7   Structural Shapes of Commercial Greenhouses

The basic requirements of indoor crop production are (1) adequate light; (2) favorable temperatures; (3) favorable air or gas content; (4) protection from insects and disease; and (5) suitable growing media, substrate, and moisture. Because of their lower cost per unit of usable space, greenhouses are preferred over plant growth chambers for protected crop production.

This section covers greenhouses and plant growth facilities, and Chapter 9 of the 1993 *ASHRAE Handbook—Fundamentals* describes the environmental requirements in these facilities. Figure 7 shows the structural shapes of typical commercial greenhouses. Other greenhouses may have Gothic arches, curved glazing, or simple lean-to shapes. Glazing, in addition to traditional glass, now includes both film and rigid plastics. High light transmission by the glazing is usually important; good location and orientation of the house are important in providing desired light conditions. Location also affects heating and labor costs, exposure to plant disease and air pollution, and material handling requirements. As a general rule in the northern hemisphere, a greenhouse should be placed at a distance of at least 2.5 times the height of the object closest to it in the eastern, western, and southern directions.

GREENHOUSES

Site Selection

Sunlight. Sunlight provides energy for plant growth and is often the limiting growth factor in greenhouses of the central and northern areas of North America during the winter. When planning greenhouses that are to be operated year-round, a designer should design for the greatest sunlight exposure during the short days of midwinter. The building site should have an open southern exposure, and if the land slopes, it should slope to the south.

Soil and Drainage. When plants are to be grown in the soil covered by the greenhouse, a growing site with deep, well-drained, fertile soil, preferably sandy loam or silt loam, should be chosen. Even though organic soil amendments can be added to poor soil, fewer problems occur with good natural soil. However, when good soil is not available, growing in artificial media should be considered. The greenhouse should be level, but the site can and often should be sloped and well-drained to reduce salt buildup and insufficient soil aeration. A high water table or a hardpan may produce water-saturated soil, increase greenhouse humidity, promote diseases, and prevent effective use of the greenhouse. These problems can, if necessary, be alleviated by tile drain systems under and around the greenhouse. Ground beds should be level to prevent water from concentrating in low areas. Slopes within greenhouses also increase temperature and humidity stratification and create additional environmental problems.

Sheltered Areas. Provided they do not shade the greenhouse, surrounding trees act as wind barriers and help prevent winter heat loss. Deciduous trees are less effective than coniferous ones in midwinter, when the heat loss potential is greatest. In areas where

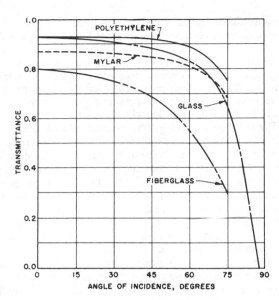

Fig. 8 Transmittance of Solar Radiation Through Glazing Materials for Various Angles of Incidence

snowdrifts occur, windbreaks and snowbreaks should be 30 m or more from the greenhouse to prevent damage.

Accessibility

Accessibility to utilities and transportation facilities are extremely important. Greenhouse operation is a year-round function requiring continuous service for all utilities and access to roads or other means of transportation for moving supplies into the facility and the crop to market.

Orientation

Generally in the northern hemisphere, for single-span greenhouses located north of 35° latitude, maximum transmission during winter is attained by an east-west orientation. South of 35° latitude, orientation is not important, provided headhouse structures do not shade the greenhouse. North-south orientation provides more light on an annual basis.

Gutter-connected or ridge and furrow greenhouses preferably are oriented with the ridge line north-south regardless of latitude. This orientation permits the shadow pattern caused by the gutter superstructure to move from the west to the east side of the gutter during the day. With an east-west orientation, the shadow pattern would remain north of the gutter, and the shadow would be widest and create the most shade during winter when light levels are already low. Also, the north-south orientation allows rows of tall crops, such as roses and staked tomatoes, to align with the long dimension of the house—an alignment that is generally more suitable to long rows and the plant support methods preferred by many growers.

The slope of the greenhouse roof is a critical part of greenhouse design. If the slope is too flat, a greater percentage of sunlight is reflected from the roof surface (Figure 8). A slope with a 1:2 rise-to-run ratio is the usual inclination for a gable roof.

HEATING

Structural Heat Loss

Estimates for heating and cooling a greenhouse consider conduction, infiltration, and ventilation energy exchange. In addition, the calculations must consider solar energy load and electrical input, such as light sources, which are usually much greater for greenhouses than for conventional buildings. Generally, conduction q_c

Table 1 Suggested Heat Transmission Coefficients

		U, W/(m²·K)
Glass		
	Single-glazing	6.4
	Double-glazing	4.0
	Insulating	Manufacturer's data
Plastic film		
	Single film[a]	6.8
	Double film, inflated	4.0
	Single film over glass	4.8
	Double film over glass	3.4
Corrugated glass fiber		
	Reinforced panels	6.8
Plastic structured sheet[b]		
	16 mm thick	3.3
	8 mm thick	3.7
	6 mm thick	4.1

[a]Infrared barrier polyethylene films reduce heat loss; however, use this coefficient when designing heating systems because the structure could occasionally be covered with non-IR materials.
[b]Plastic structured sheets are double-walled, rigid plastic panels.

Table 2 Construction U-Factor Multipliers

Metal frame and glazing system, 400 to 600 mm spacing	1.08
Metal frame and glazing system, 1200 mm spacing	1.05
Fiberglass on metal frame	1.03
Film plastic on metal frame	1.02
Film or fiberglass on wood	1.00

Table 3 Suggested Design Air Changes (N)

New Construction	
Single glass lapped, unsealed	1.25
Single glass lapped, laps unsealed	1.0
Plastic film covered	0.6 to 1.0
Structured sheet	1.0
Film plastic over glass	0.9

Old Construction	
Good maintenance	1.5
Poor maintenance	2 to 4

plus infiltration q_i are used to determine the peak requirements q_t for heating.

$$q_t = q_c + q_i \tag{6}$$

$$q_c = UA\,(t_i - t_o) \tag{7}$$

$$q_i = 0.5VN\,(t_i - t_o) \tag{8}$$

where

U = overall heat loss coefficient, W/(m²·K) (Table 1)
A = exposed surface area, m²
t_i = inside temperature, °C
t_o = outside temperature, °C
V = greenhouse internal volume, m³
N = number of air exchanges per hour (Table 3)

Type of Framing

The type of framing should be considered in determining overall heat loss. Aluminum framing and glazing systems may have the metal exposed to the exterior to a greater or lesser degree, and the heat transmission of this metal is higher than that of the glazing material. To allow for such a condition, the U-factor of the glazing material should be multiplied by the factors shown in Table 2.

Infiltration

Equation (8) may be used to calculate heat loss by infiltration. Table 3 suggests values for air changes N.

Radiation Energy Exchange. Solar gain can be estimated using the procedures outlined in Chapter 26 of the 1993 *ASHRAE Handbook—Fundamentals*. As a guide, when a greenhouse is filled with a mature crop of plants, one-half the incoming solar energy is converted to latent heat, and one-quarter to one-third, to sensible heat. The rest is either reflected out of the greenhouse or absorbed by the plants and used in photosynthesis.

Radiation from a greenhouse to a cold sky is more complex. Glass admits a large portion of solar radiation but does not transmit long-wave thermal radiation in excess of approximately 5000 nm. Plastic films transmit more of the thermal radiation, but, in general, the total heat gains and losses are similar to those of glass. Newer plastic films containing infrared (IR) inhibitors reduce the thermal radiation loss. Plastic films and glass with improved radiation reflection are available at a somewhat higher cost. Normally, radiation energy exchange is not considered in calculating the design heat load.

Heating Systems

Greenhouses may have a variety of heating systems. One is a convection system that circulates hot water or steam through plain or finned pipe. The pipe is most commonly placed along walls and occasionally beneath plant benches to create desirable convection currents. A typical temperature distribution pattern created by perimeter heating is shown in Figure 9. More uniform temperatures can be achieved when about one-third the total heat comes from pipes spaced uniformly across the house. These pipes can be placed above or below the crop, but temperature stratification and shading are avoided when they are placed below. Outdoor weather conditions affect temperature distribution, especially on windy days in loosely constructed greenhouses. Manual or automatic overhead pipes are also used for supplemental heating to prevent snow buildup on the roof. In a gutter-connected greenhouse in a cold climate, a heat pipe should be placed under each gutter to prevent snow accumulation.

Overhead tube systems consist of a unit heater that discharges into 300- to 750-mm diameter plastic film tubing perforated to provide uniform air distribution. The tube is suspended at 2 to 3 m intervals and extends the length of the greenhouse. Variations include a tube and fan receiving the discharge of several unit heaters. The fan and tube system is used without heat to recirculate the air and, during cold weather, to introduce ventilation air. However, tubes sized for heat distribution may not be large enough for effective ventilation during warm weather.

Perforated tubing, 150 to 250 mm in diameter, placed at ground level (underbench) can also improve heat distribution. Ideally, the ground-level tubing should draw air from the top of the greenhouse

for recirculation or heating. Tubes on or near the floor have the disadvantage of being obstacles to workers and reducing usable floor space.

Underfloor heating can supply up to 25% or more of the peak heating requirements of northern greenhouses. A typical underfloor system uses 20-mm plastic pipe with nylon fittings in the floor, 100 mm below the surface, spaced 300 to 400 mm on centers, and covered 0.13 to 0.48 L/s per loop with regular gravel or porous concrete. Hot water, not exceeding 40°C, circulates at a rate of 0.5 to 1.0 L/s. The pipe loops should generally not exceed 130 m in length. This system can provide 50 to 65 W/m² from a bare floor, and about 75% as much when potted plants or seedling flats cover most of the floor.

Similar systems can heat soil directly, but root temperature must not exceed 25°C. When used with water from solar collectors or other heat sources, the underfloor area can store heat. This storage consists of a vinyl swimming pool liner placed on top of insulation and a moisture barrier at a depth of 200 to 300 mm below grade, and filled with 50% void gravel. Hot water from solar collectors or other clean sources enters and is pumped out on demand. Some heat sources, such as cooling water from power plants, cannot be used directly but require a closed-loop heat transfer system to avoid fouling the storage system and the power plant cooling water.

Greenhouses can also be bottom heated with 6-mm diameter EPDM tubing (or variations of that method) in a closed-loop system. The tubes can be placed directly in the growing medium of ground beds or under plant containers on raised benches. The best temperature uniformity is obtained by flow in alternate tubes in opposite directions. This method can supply all the greenhouse heat needed in mild climates.

Bottom heat, underfloor heating, and underbench heating are, because of the location of the heat source, more effective than overhead or peripheral heating and can reduce energy loss by 20 to 30%.

Unless properly located and aimed, overhead unit heaters, whether hydronic or direct fired, do not give uniform temperature at the plant level and throughout the greenhouse. Horizontal blow heaters positioned so that they establish a horizontal airflow around the outside of the greenhouse offer the best distribution. The airflow pattern can be supplemented with the use of horizontal blow fans or circulators.

When direct combustion heaters are used in the greenhouse, combustion gases must be adequately vented to the outside to minimize danger to plants and humans from products of combustion. One manufacturer recommends that combustion air must have access to the space through a minimum of two permanent openings in the enclosure, one near the bottom. A minimum of 2200 mm² of free area per kilowatt input rating of the unit, with a minimum of 0.65 m² for each opening, whichever is greater, is recommended. Unvented direct combustion units should not be used inside the greenhouse.

In many greenhouse heating systems, a combination of overhead and perimeter heating is used. Regardless of the type of heating, it is common practice first to calculate the overall heat loss, and then to calculate the individual elements such as the roof, sidewalls, and gables. It is then simple to allocate the overhead portion to the roof loss and the perimeter portions to the sides and gables, respectively.

The annual heat loss can be approximated by calculating the design heat loss and then, in combination with the annual degree-day tables using the 18.3°C base, estimating an annual heat loss and computing fuel usage on the basis of the rating of the particular fuel used. If a 10°C base is used, it can be prorated.

Heat curtains for energy conservation are becoming more important in greenhouse construction. Although this energy saving may be considered in the annual energy use, it should not be used when calculating design heat load; the practice is to open the heat curtains during snowstorms to facilitate the melting of snow, thereby nullifying its contribution to the design heat loss value.

Air-to-air and water-to-air heat pumps have been used experimentally on small-scale installations. Their usefulness is especially sensitive to the availability of a low-cost heat source.

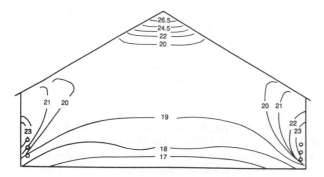

Fig. 9 Temperature Profiles in a Greenhouse Heated with Radiation Piping Along the Sidewalls

Radiant (Infrared) Heating

Radiant heating is used in some limited applications for greenhouse heating. Steel pipes spaced at intervals and heated to a relatively high temperature by special gas heaters serve as the source of radiation. Because the energy is transmitted by radiation from a source of limited size, proper spacing is important to completely cover the heated area. Further, heavy foliage crops can shade the lower parts of the plants and the soil, thus restricting the radiation from warming the root zone, which is important to plant growth.

Cogenerated Sources of Heat

Greenhouses have been built near or adjacent to power plants to use the heat and electricity generated by the facility. While this energy may cost very little, an adequate standby energy source must be provided, unless the power supplier can assure that it will supply a reliable, continuous source of energy.

COOLING

Solar radiation is a considerable source of sensible heat gain; even though some of this energy is reflected from the greenhouse, some of it is converted into latent heat as the plants transpire moisture, and some is converted to plant material by photosynthesis. Natural ventilation, mechanical ventilation, shading, and evaporative cooling are common methods used to remove this heat. Mechanical refrigeration is seldom used to air condition greenhouses because the cooling load and resulting cost is so high.

Natural Ventilation

Most older greenhouses and many new ones rely on natural ventilation with continuous roof sashes on each side of the ridge and continuous sashes in the sidewalls. The roof sashes are hinged at the ridge, and the wall sashes are hinged at the top of the sash. During much of the year, vents admit enough ventilating air for cooling without the added cost of running fans.

The principles of natural ventilation are explained in Chapter 23 of the 1993 *ASHRAE Handbook—Fundamentals*. Ventilation air is driven by wind and thermal buoyancy forces. Proper vent openings

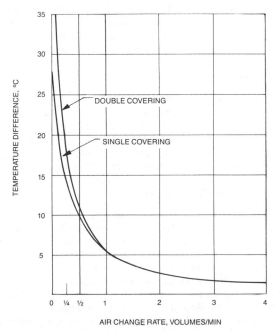

Fig. 10 Influence of Air-Exchange Rate on Temperature Rise in Single- and Double-Covered Greenhouses

take advantage of pressure differences created by wind. Thermal buoyancy caused by the temperature difference between the inside and the outside of the greenhouse is enhanced by the area of the vent opening and the stack height (vertical distance between the center of the lower and upper opening). Within the limits of typical construction, the larger the vents, the greater the ventilating air exchanged. For a single greenhouse, the combined area of the sidewall vents should equal that of the roof vents. In ranges of several greenhouses connected at the gutters, the sidewall area cannot equal the roof vent area.

Mechanical (Forced) Ventilation

Exhaust fans provide positive ventilation without depending on wind or thermal buoyancy forces. The fans are installed in the side or end walls of the greenhouse and draw air through vents on the opposite side or end walls. The air velocity through the inlets should not exceed 2 m/s.

Air exchange rates between 0.75 and 1 change per minute effectively control the temperature rise in a greenhouse. As shown in Figure 10, the temperature inside the greenhouse rises rapidly at lower airflow rates. At higher airflow rates the reduction of the temperature rise is small, fan power requirements are increased, and plants may be damaged by the high air speed.

Shading

Shading compounds can be applied in varying amounts to the exterior of the roof of the greenhouse to achieve up to 50% shading. The durability of these compounds varies—ideally the compound will wear away during the summer and leave the glazing clean in the fall when shading is no longer needed. In practice, some physical cleaning is needed. Compounds used formerly usually contained lime, which corroded aluminum and attacked some types of caulking. Most compounds used currently are formulated to avoid this problem.

Mechanically operated shade cloth systems with a wide range of shade levels are also available. They are mounted inside the greenhouse to protect them from the weather. Not all shading compounds or shade cloths are compatible with all plastic glazings, so the manufacturer's instructions and precautions should be followed.

Evaporative Cooling

Fan-and-Pad Systems. Fans for the fan-and-pad evaporative cooling system are installed in the same manner as fans used for mechanical ventilation. Pads of cellulose material in a honeycomb form are installed on the inlet side. The pads are kept wet continuously when evaporative cooling is needed. As air is drawn through the pads, the water evaporates and cools the air. New pads cool the air by about 80% of the difference between the outdoor dry-bulb and wet-bulb temperature, or to 1.5 to 2 K above the wet-bulb temperature.

The empirical base rate of airflow is 40 L/s per square metre of floor area. This flow rate is modified by multiplying it by factors for elevation (F_e), maximum interior light intensity (F_l), and the allowable temperature rise between the pad and the fans (F_t). These factors are listed in Table 4. The overall factor for the house is given by the following equation:

$$F_h = F_e \times F_l \times F_t \qquad (9)$$

The maximum fan-to-pad distance should be kept to 53 m, although some greenhouses with distances of 68 m have shown no serious reduction in effectiveness. With short distances, the air velocity becomes so low that the air feels clammy and stuffy, even though the airflow is sufficient for cooling. Therefore, a velocity factor F_v listed in Table 5 is used for distances less than 30 m. For distance less than 30 m, F_v is compared to F_h. The factor that gives the

Table 4 Multipliers for Calculating Airflow for Fan-and-Pad Cooling

Elevation (Above Sea Level)		Max. Interior Light Intensity		Fan-to-Pad Temp. Difference	
m	F_e	klx	F_l	K	F_t
<300	1.00	40	0.74	5.5	0.71
300	1.03	45	0.84	5.0	0.78
600	1.08	50	0.93	4.5	0.87
900	1.12	55	1.02	4.0	0.98
1200	1.16	60	1.12	3.5	1.12
1500	1.20	65	1.21	3.0	1.31
1800	1.25	70	1.30	2.5	1.58
2100	1.29	75	1.39		
2400	1.33	80	1.49		
2700	1.37	85	1.58		

Table 5 Velocity Factors for Calculating Airflow for Fan-to-Pad Cooling

Fan-to-Pad Distance, m	F_v	Fan-to-Pad Distance, m	F_v
6	2.26	20	1.23
8	1.96	22	1.17
10	1.75	24	1.13
12	1.60	26	1.08
14	1.48	28	1.04
16	1.38	30	1.00
18	1.30		

Table 6 Recommended Air Velocity Through Various Pad Materials

Pad Type and Thickness	Air Face Velocity Through Pad[a], m/s
Corrugated cellulose, 100 mm thick	1.25
Corrugated cellulose, 150 mm thick	1.75

[a]Speed may be increased by 25% where construction is limiting.

Table 7 Recommended Water Flow and Sump Capacity for Vertically Mounted Cooling Pad Materials

Pad Type and Thickness	Minimum Water Rate per Linear Metre of Pad, L/s	Minimum Sump Capacity per Unit Pad Area, L/m²
Corrugated cellulose, 100 mm thick	0.10	30
Corrugated cellulose, 150 mm thick	0.16	40

greatest air flow is used to modify the empirical base rate. For fan-to-pad distances greater than 30 m, F_v can be ignored.

For best performance, pads should be installed on the windward side and fans should be spaced within 7.5 m of each other. Fans should not blow toward pads of an adjacent house unless it is at least 15 m away. Fans in adjacent houses should be offset if they blow toward each other and are within 4.5 m of each other.

Recommended air velocities through commonly used pads are listed in Table 6. Water flow and sump capacities are shown in Table 7. The system should also include a small, continuous bleed-off of water to reduce the buildup of dirt and other impurities.

Unit Evaporative Coolers. This equipment contains the pads, water pump, sump, and fan in one unit. Unit coolers are primarily used for small compartments. They are mounted 4.5 to 6 m apart on the sidewall and blow directly into the greenhouse. They cool up to a distance of 15 m from the unit. A side sash on the outside opposite wall is the best outlet, but roof vents may also work. The roof vent on the same side as the unit should be slightly open for better air distribution. If the roof vent on the opposite side is opened instead, air

may flow directly out the vent and not cool the opposite side of the greenhouse.

Fog Systems. In a direct-pressure atomizing system, a high-pressure pump forces water at 5.5 to 7 MPa through a special fog nozzle. Fog is considered to be a water droplet smaller than 40 µm in diameter. The direct-pressure atomizing system generates droplets of 35 µm or less. This system requires a very good filter system to minimize clogging of the very small nozzle orifices.

A line of nozzles placed along the top of the vent opening can cool the entering air nearly to its wet-bulb temperature. Additional lines in the greenhouse continue to cool the air as it absorbs heat in the space.

Fogging systems will cool satisfactorily with less airflow than fan-and-pad systems, but the fan capacity must still be based on one air change per minute to ventilate the greenhouse when the system will be used without fog.

OTHER ENVIRONMENTAL CONTROLS

Humidity Control

At various times during the year, humidity may need to be controlled in the greenhouse. When the humidity is too high at night, it can be reduced by adding heat and ventilating at the same time. When the humidity is too low during the day, it can be increased by turning on a fog or mist nozzle system.

Winter Ventilation

During the winter, houses are normally closed tight to conserve heat, but photosynthesis by the plants may lower the carbon dioxide level to such a point that it slows plant growth. Some ventilation helps maintain carbon dioxide levels inside. A normal rate of airflow for winter ventilation is 10 to 15 L/s per square metre of floor area.

Air Circulation

Continuous air circulation within the greenhouse reduces still-air conditions that favor plant diseases. Recirculating fans, heaters that blow air horizontally, and fans attached to polyethylene tubes are used to circulate air. The amount of recirculation has not been well defined, except that some studies have shown high air velocities (greater than 1.0 m/s) can harm plants or reduce growth.

Insect Screening

Insect screening is being used to cover vent inlets and outlets. These fine-mesh screens increase the resistance to airflow, which must be considered when selecting ventilation fans. The screen manufacturer should provide static pressure data for its screens. The pressure drop through the screen can be reduced by framing out from the vent opening to increase the area of the screen.

Carbon Dioxide Enrichment

Carbon dioxide is added in some greenhouse operations to increase growth and enhance yields. However, CO_2 enrichment is practical only when little or no ventilation is required for temperature control. Carbon dioxide can be generated from solid CO_2 (dry ice), bottled CO_2, and misting carbonated water. Bulk or bottled CO_2 gas is usually distributed through perforated tubing placed near the plant canopy. Carbon dioxide from dry ice is distributed by passing greenhouse air through an enclosure containing dry ice. Air movement around the plant leaf increases the efficiency with which the plant absorbs whatever CO_2 is available. One study found an air speed of 0.5 m/s to be equivalent to a 50% enrichment in CO_2 without forced air movement.

Table 8   Constants to Convert to W/m²

Light Source	klx	μmol/s²·m²
400-700 nm		
Incandescent (INC)	3.99	0.20
Fluorescent cool white (FCW)	2.93	0.22
Fluorescent warm white (FWW)	2.81	0.21
Discharge clear mercury (HG)	2.62	0.22
Metal halide (MH)	3.05	0.22
High-pressure sodium (HPS)	2.45	0.20
Low-pressure sodium (LPS)	1.92	0.20
Daylight	4.02	0.22

Note: 1 μmol/(s·m²) = 1 einstein/(s·m²)

Table 9   Suggested Radiant Energy, Duration, and Time of Day for Supplemental Lighting in Greenhouses

Plant and Stage of Growth	W/m²	Duration Hours	Duration Time
African violets early-flowering	12 to 24	12 to 16	0600-1800 0600-2200
Ageratum early-flowering	12 to 48	24	
Begonias—fibrous rooted branching and early-flowering	12 to 24	24	
Carnation branching and early-flowering	12 to 24	16	0800-2400
Chrysanthemums vegetable growth branching and multiflowering	12 to 24 12 to 24	16 8	0800-2400 0800-1600
Cineraria seedling growth (four weeks)	6 to 12	24	
Cucumber rapid growth and early-flowering	12 to 24	24	
Eggplant early-fruiting	12 to 48	24	
Foliage plants (Philodendron, Schefflera) rapid growth	6 to 12	24	
Geranium branching and early-flowering	12 to 48	24	
Gloxinia early-flowering	12 to 48 6 to 12	16 24	0800-2400
Lettuce rapid growth	12 to 48	24	
Marigold early-flowering	12 to 48	24	
Impatiens—New Guinea branching and early-flowering	12	16	0800-2400
Impatiens—Sultana branching and early-flowering	12 to 24	24	
Juniper vegetative growth	12 to 48	24	
Pepper early-fruiting, compact growth	12 to 24	24	
Petunia branching and early-flowering	12 to 48	24	
Poinsettia—vegetative growth branching and multiflowering	12 12 to 24	24 8	0800-1600
Rhododendron vegetative growth (shearing tips)	12	16	0800-2400
Roses (hybrid teas, miniatures) early-flowering and rapid regrowth	12 to 48	24	
Salvia early-flowering	12 to 48	24	
Snapdragon early-flowering	12 to 48	24	
Streptocarpus early-flowering	12	16	0800-2400
Tomato rapid growth and early-flowering	12 to 24	16	0800-2400
Trees (deciduous) vegetative growth	6	16	1600-0800
Zinnia early-flowering	12 to 48	24	

Radiant Energy

Light is normally the limiting factor in greenhouse crop production during the winter. North of the 35th parallel, light levels are especially inadequate or marginal in fall, winter, and early spring. Artificial light sources, usually high-intensity discharge (HID) lamps, may be added to greenhouses to supplement low natural light levels. High-pressure sodium (HPS), metal halide (MH), low-pressure sodium (LPS), and occasionally mercury lamps coated with a color-improving phosphor are currently used. Since differing irradiance or illuminance ratios are emitted by the various lamp types, the incident radiation is best described as radiant flux density (W/m²) between 400 and 850 nm, or as photon flux density between 400 and 700 nm, rather than in photometric terms of lux.

To assist in relating irradiance to more familiar illuminance values, Table 8 shows constants for converting the irradiance (W/m²) of HPS, MH, LPS, and other lamps to illuminance (lux) and photon flux density.

Table 9 gives values of suggested irradiance at the top of the plant canopy, duration, and time of day for supplementing natural light levels for specific plants.

Luminaires have been developed specifically for greenhouse use of HID lamps. The lamps in these special greenhouse luminaires are often placed in a horizontal position, which may decrease both the light output and the life of the lamp. These drawbacks may be balanced by improved horizontal and vertical uniformity as compared to industrial parabolic reflectors.

Photoperiod Control

Artificial light sources are also used to lengthen the photoperiod during the short days of winter. Photoperiod control requires much lower light levels than those needed for photosynthesis and growth. Photoperiod illuminance needs to be only 6 to 12 W/m². The incandescent lamp is the most effective light source for this purpose due to its higher far-red component. Lamps such as 150 W (PS-30) silverneck lamps spaced 3 to 4 m on centers and 4 m above the plants provide a cost-effective system. Where a 4-m height is not practical, 60-W extended service lamps on 2-m centers are satisfactory. One method of photoperiod control is to interrupt the dark period by turning the lamps on at 2200 and off at 0200. The 4-h interruption, initially based on chrysanthemum response, induces a satisfactory long-day response in all photoperiodically sensitive species. Many species, however, respond to interruptions of 1 h or less. Demand charges can be reduced in large installations by operating some sections from 2000 to 2400 and others from 2400 to 0400. The biological response to these schedules, however, is much weaker than with the 2200 to 0200 schedule, so some varieties may flower prematurely. If the 4-h interruption period is used, it is not necessary to keep the light on throughout the interruption period. Photoperiod control of most plants can be accomplished by operating the lamps on light and dark cycles with 20% "on" times; for example, 12 s/min. The length of the dark period in the cycle is critical, and the system may fail if the dark period exceeds about 30 min. Demand charges can be reduced by alternate scheduling of the "on" times between houses or benches without reducing the biological effectiveness of the interruption.

Plant displays in places such as showrooms or shopping malls require enough light for plant maintenance and a spectral distribution that best shows the plants. Metal halide lamps, with or without incandescent highlighting, are often used for this purpose. Fluorescent lamps, frequently of the special phosphor plant-growth type, enhance color rendition but are more difficult to install in aesthetically pleasing designs.

Design Conditions

Plant requirements vary from season to season and during different stages of growth. Even different varieties of the same species may vary in their requirements. State and local cooperative extension offices are a good source of specific information on design conditions affecting plants. These offices also provide current, area-specific information on greenhouse operations.

Alternate Energy Sources and Energy Conservation

Limited progress has been achieved in heating commercial greenhouses with solar energy. Collecting and storing the heat requires a volume at least one-half the volume of the greenhouse. Passive solar units work at certain times of the year and, in a few localities, year-round.

If available, reject heat is a possible source of winter heat. Winter energy and solar (photovoltaic) sources are possible future energy sources for greenhouses, but the development of such systems is still in the research stage.

Energy Conservation. A number of energy-saving measures (e.g., thermal curtains, double glazing, and perimeter insulation) have been retrofitted to existing greenhouses and incorporated into new construction. Sound maintenance is necessary to keep heating system efficiency at a maximum level.

Automatic controls, such as thermostats, should be calibrated and cleaned at regular intervals, and heating-ventilation controls should interlock to avoid simultaneous operation. Boilers that can burn more than one type of fuel permit use of the lowest cost fuel available.

Modifications to Reduce Heat Loss

Film covers that reduce heat loss are widely used in commercial greenhouses, particularly for growing foliage plants and other species that grow under low light levels. Irradiance (intensity) is reduced 10 to 15% per layer of plastic film.

One or two layers of transparent 0.10 or 0.15 mm continuous-sheet plastic is stretched over the entire greenhouse (leaving some vents uncovered), or from the ridge to the sidewall ventilation opening. When two layers are used, (outdoor) air at a pressure of 50 to 60 Pa is introduced continuously between the layers of film to maintain the air space between them. When a single layer is used, an air space can be established by stretching the plastic over the glazing bars and fastening it around the edges, or a length of polyethylene tubing can be placed between the glass and the plastic and inflated (using outside air) to stretch the plastic sheet.

Double-Glazing Rigid Plastic. Double-wall panels are manufactured from acrylic and polycarbonate plastics, with walls separated by about 10 mm. Panels are usually 1.2 m wide and 2.4 m or more long. Nearly all types of plastic panels have a high thermal expansion coefficient and require about 1% expansion space (10 mm/m). When a panel is new, light reduction is roughly 10 to 20%. Moisture accumulation between the walls of the panels must be avoided.

Double-Glazing Glass. The framing of most older greenhouses must be modified or replaced to accept double glazing with glass.

Light reduction is 10% more than with single glazing. Moisture and dust accumulation between glazings increases light loss. As with all types of double glazing, snow on the roof melts slowly and increases light loss. Snow may even accumulate sufficiently to cause structural damage, especially in gutter-connected greenhouses.

Silicone Sealants. Transparent silicone sealant in the glass overlaps of conventional greenhouses reduces infiltration and may produce heat savings of 5 to 10% in older structures. There is little change in light transmission.

Precautions. The various methods described above reduce heat loss by reducing conduction and infiltration. They may also cause more condensation, higher relative humidity, lower carbon dioxide concentrations, and an increase in ethylene and other pollutants. Combined with the reduced light levels, these factors may cause delayed crop production, elongated plants, soft plants, and various deformities and diseases, all of which reduce the marketable crop.

Thermal blankets are any flexible material that is pulled from gutter to gutter and end to end in a greenhouse, or around and over each bench, at night. Materials ranging from plastic film to heavy cloth, or laminated combinations, have successfully reduced heat losses by 25 to 35% overall. Tightness of fit around edges and other obstructions is more important than the kind of material used. Some films are vaportight and retain moisture and gases. Others are porous and permit some gas exchange between the plants and the air outside the blanket. Opaque materials can control crop day length when short days are part of the requirement for that crop. Condensation may drip onto and collect on the upper sides of some blanket materials to such an extent that they collapse.

Multiple-layer blankets, with two or more layers separated by air spaces, have been developed. One such system combines a porous-material blanket and a transparent film blanket; the latter is used for summer shading. Another system has four layers of porous, aluminum foil-covered cloths, with the layers separated by air.

Thermal blankets may be opened and closed manually as well as automatically. The decision to open or close should be based on the irradiance level and whether it is snowing, rather than the time of day. Two difficulties with thermal blankets are the physical problems of installation and use in greenhouses with interior supporting columns, and the loss of space due to shading by the blanket when it is not in use during the day.

Other Recommendations. While the foundation can be insulated, the insulating materials must be protected from moisture, and the foundation wall should be protected from freezing. All or most of the north wall can be insulated with opaque or reflective-surface materials. The insulation reduces the amount of diffuse light entering the greenhouse, and, in cloudy climates, causes reduced crop growth near the north wall.

Ventilation fan cabinets should be insulated, and fans not needed during the winter should be sealed against air leaks. Efficient management and operation of existing facilities are the most cost-effective ways of reducing energy use.

PLANT GROWTH ENVIRONMENTAL FACILITIES

Controlled-environment rooms (CERs), also called plant growth chambers, include all controlled or partially controlled environmental facilities for growing plants, except greenhouses. CERs are indoor facilities. Units with floor areas less than 5 m^2 may be moveable with self-contained or attached refrigeration units. CERs usually have artificial light sources, provide control of temperature, and, in some cases, control relative humidity and CO_2 level.

CERs are used to study all aspects of plant science. Some growers use growing rooms to increase seedling growth rate, produce more uniform seedlings, and grow specialized, high-value crops. The main components of the CER are (1) an insulated room or an insulated box with an access door; (2) a heating and cooling mechanism with associated air-moving devices and controls; and (3) a lamp module at the top of the insulated box or room. CERs are similar to walk-in cold storage rooms, except for the lighting and larger refrigeration system needed to handle heat produced by the lighting.

Location

The location for a CER must have space for the outside dimensions of the chamber, refrigeration equipment, ballast rack, and control panels. Additional space around the unit is necessary for servicing the various components of the system, and, in some cases, for substrate, pots, nutrient solutions, and other paraphernalia associated with plant research. The location also requires an electrical supply (on the order of 300 W/m^2 of controlled environment space), a water supply, and a compressed air supply.

Construction and Materials

Wall insulation should have a unit thermal conductance value of less than 0.15 W/(m²·K). Materials should resist corrosion and moisture. The interior wall covering should be metal, with a high-reflectance white paint or specular aluminum with a reflectivity of not less than 80%. Reflective films or similar materials can be used but will require periodic replacement.

Floors and Drains

Floors that are part of the CER should be corrosion-resistant. Tar or asphalt waterproofing materials and volatile caulking compounds should not be used because they will likely release phytotoxic gases into the chamber atmosphere. The floor must have a drain to remove spilled water and nutrient solutions. The drains should be trapped and equipped with screens to catch plant and substrate debris.

Plant Benches

Three bench styles for supporting the pots and other plant containers are normally encountered in plant growth chambers: (1) stationary benches; (2) benches or shelves built in sections that are adjustable in height; and (3) plant trucks, carts, or dollies on casters, which are used to move plants between chambers, greenhouses, and darkrooms. The bench supports containers filled with moist sand, soil, or other substrate and is usually designed for a minimum loading of 240 kg/m². The bench or truck top should be constructed of nonferrous, perforated metal or metal mesh to allow free passage of air around the plants and to let excess water drain from the containers to the floor and subsequently to the floor drain.

Normally, benches, shelves, or truck tops are adjustable in height so that small plants can be placed close to the lamps and thus receive a greater amount of light. As the plants grow, the shelf or bench is lowered so that the tops of the plants continue to receive the original radiant flux density.

Controls

Environmental chambers require complex controls to provide the following:

1. Automatic transfer from heating to cooling with 1 K or less dead zone and adjustable time delay.
2. Automatic daily switching of the temperature set point for different day and night temperatures (setback may be as much as 5 K).
3. Protection of sensors from radiation. Ideally the sensors are located in a shielded, aspirated housing, but satisfactory performance can be attained by placing them in the return air duct.
4. Control of the daily duration of light and dark periods. Ideally, this control should be programmable to change the light period each day to simulate the natural progression of day length. Photoperiod control, however, is normally accomplished with mechanical time clocks, which must have a control interval of 5 min or less for satisfactory timing.
5. Protective controls to prevent the chamber temperature from going more than a few degrees above or below the set point. Controls should also prevent short cycling of the refrigeration system, especially when the condensers are remotely located.
6. Audible and visual alarms to alert personnel of malfunctions.
7. Maintenance of relative humidity to prescribed limits.

Data loggers, recorders, or recording controllers are recommended to aid in monitoring daily operation of the system. Solid-state, microprocessor-based control is not yet widely used. However, programming flexibility and control performance are expected to improve as microprocessor control is developed for CER use.

Heating, Air Conditioning, and Airflow

When the lights are on, cooling will normally be required, and the heating system will rarely be called on to operate. When the lights are off, however, both heating and cooling may be needed. Conventional refrigeration systems are generally used with some modification. Direct expansion units usually operate with a hot-gas bypass to prevent numerous on-off cycles, and secondary coolant systems may use aqueous ethylene glycol rather than chilled water. Heat is usually provided by electric heaters, but other energy sources can be used, including hot gas from the refrigeration system.

The plant compartment is the heart of the growth chamber. The primary design objective, therefore, is to provide the most uniform, consistent, and regulated environmental conditions possible. Thus, airflow must be adequate to meet specified psychrometric conditions, but it is limited by the effects of high air speeds on plant growth. As a rule, the average air speed in CERs is restricted to about 0.5 m/s.

To meet the uniform conditions required by a CER, conditioned air is normally moved through the space from bottom to top, although an increasing number of CERs use top-to-bottom airflow. There is no apparent difference in plant growth between horizontal, upward, or downward airflow when the speed is less than 0.9 m/s. Regardless of the method, a temperature gradient is certain to exist, and the design should keep the gradient as small as possible. Uniform airflow is more important than the direction of flow; thus selection of properly designed diffusers or plenums with perforations is essential for achieving it.

The ducts or false sidewalls that direct air from the evaporator to the growing area should be small, but not so small that the noise level increases appreciably more than acceptable building air duct noise. CER design should include some provision for cleaning the interior of the air ducts.

Air-conditioning equipment for relatively standard chambers provides temperatures that range from 7 to 35°C. Specialized CERs that require temperatures as low as −20°C need low-temperature refrigeration equipment and devices to defrost the evaporator without increasing the growing area temperature. Other chambers that require temperatures as high as 45°C need high-temperature components. The air temperature in the growing area must be controlled with the least possible variation about the set point. Temperature variation about the set point can be held to 0.3 K using solid-state controls, but in most existing facilities, the variation is 0.5 to 1 K.

The relative humidity in many CERs is simply an indicator of the existing psychrometric conditions and is usually between 50 and 80%, depending on the temperature. Relative humidity in the chamber can be increased by steam injection, misting, hot water evaporators, and other conventional humidification methods. Steam injection causes the least temperature disturbance, and sprays or misting causes the greatest disturbance. Complete control of relative humidity requires dehumidification as well as humidification.

A typical humidity control system includes a cold evaporator or steam injection to adjust the chamber air dew point. The air is then conditioned to the desired dry-bulb temperature by electric heaters, a hot-gas bypass evaporator, or a temperature-controlled evaporator. A dew point lower than about 5°C cannot be obtained with a cold plate dehumidifier because of icing. Dew points lower than 5°C usually require a chemical dehumidifier in addition to the cold evaporator.

Lighting Environmental Chambers

The type of light source and the number of lamps used in CERs are determined by the desired plant response. Traditionally, cool-white fluorescent plus incandescent lamps that produce 10% of the fluorescent illuminance are used. Nearly all illumination data are based on either cool-white or warm-white fluorescent, plus incandescent. A number of fluorescent lamps have special phosphors hypothesized to be the spectral requirements of the plant. Some of

Table 10 Input Power Conversion of Light Sources

Lamp Identification		Total Input Power, W	Radiation (400-700 nm), %	Radiation (400-850 nm), %	Other Radiation, %	Conduction and Convection, %	Ballast Loss, %
Incandescent	INC, 100A	100	7	15	75	10	0
Fluorescent							
Cool white	FCW	46	21	21	32	34	13
Cool white	FCW	225	19	19	34	35	12
Warm white	FWW	46	20	20	32	35	13
Plant growth A	PGA	46	13	13	35	39	13
Plant growth B	PGB	46	15	16	34	37	13
Infrared	FIR	46	2	9	39	39	13
Discharge							
Clear mercury	HG	440	12	13	61	17	9
Mercury deluxe	HG/DX	440	13	14	59	18	9
Metal halide	MH	460	27	30	42	15	13
High-pressure sodium	HPS	470	26	36	36	15	15
Low-pressure sodium	LPS	230	27	31	25	22	22

Note: Conversion efficiency is for lamps without luminaire. Values compiled from manufacturers' data, published information, and unpublished test data by R.W. Thimijan.

these lamps are used in CERs, but there is little data to suggest that they are superior to cool-white and warm-white lamps. In recent years, high-intensity discharge lamps have been installed in CERs, either to obtain very high radiant flux densities, or to reduce the electrical load while maintaining a light level equal to that produced by the less efficient fluorescent-incandescent systems.

One approach to lighting design for biological environments is to base light source output recommendations on photon flux density $\mu mol/(s \cdot m^2)$ between 400 and 700 nm, or, less frequently, as radiant flux density between 400 and 700 nm, or 400 and 850 nm. Rather than basing illuminance measurements on human vision, this enables comparisons between light sources as a function of plant photosynthetic potential. Table 8 shows constants for converting various measurement units to W/m^2. However, instruments that measure the 400 to 850 nm spectral range are generally not available, and some controversy exists about the effectiveness of 400 to 850 nm as compared to the 400 to 700 nm range in photosynthesis. The power conversion of various light sources are listed in Table 10.

The design requirements for plant growth lighting differ greatly from those for vision lighting. Plant growth lighting requires a greater degree of horizontal uniformity and, usually, higher light levels than vision lighting. In addition, plant growth lighting should have as much vertical uniformity as possible—a factor rarely important in vision lighting. Horizontal and vertical uniformity are much easier to attain with linear or broad sources, such as fluorescent lamps, than with point sources, such as HID lamps. Tables 11 and 12 show the type and number of lamps, mounting height, and spacing required to obtain several levels of incident energy. Since the data were taken directly under lamps with no reflecting wall surfaces nearby, the incident energy is perhaps one-half of what the plants would receive if the lamps had been placed in a small chamber with highly reflective walls.

Extended-life incandescents or traffic signal lamps, which have a much longer life, will lower lamp replacement requirements. These lamps have lower lumen output, but are nearly equivalent in the red portion of the spectrum. For safety, porcelain lamp holders and heat-resistant lamp wiring should be used. Lamps used for CER lighting include fluorescent lamps (usually 1500 mA), 250-, 400-, and occasionally 1000-W HPS and MH lamps, 180-W LPS lamps, and various sizes of incandescent lamps. In many installations, the abnormally short life of incandescent lamps is due to vibration from the lamp loft ventilation or from cooling fans. Increased incandescent lamp life under these conditions can be attained by using lamps constructed with a C9 filament.

Energy-saving lamps have approximately equal or slightly lower irradiance per input watt. Since the irradiance per lamp is lower, there is no advantage to using these lamps, except in tasks that can be accomplished with low light levels. Light output of all lamps declines with use, except perhaps for low-pressure sodium (LPS) lamps, which appear to maintain approximately constant output but require an increase in input watts during use.

Fluorescent and metal halide designs should be based on 80% of the initial lumens. Most CER lighting systems have difficulty maintaining a relatively constant light level over considerable periods of time. Combinations of MH and HPS lamps compound the problem, because the lumen depreciation of the two light sources is significantly different. Thus, over time, the spectral energy distribution at plant level will shift toward the HPS. Lumen output can be maintained in two ways: (1) individual lamps, or a combination of lamps, can be switched off initially and activated as the lumen output decreases, and (2) the oldest 25 to 33% of the lamps can be replaced periodically. Solid-state dimmer systems are commercially available only for low-wattage fluorescent lamps and for mercury lamps; in the future, it may be possible to apply the technology to metal halide lamps.

Large rooms, especially those constructed as an integral part of the building and retrofitted as CERs, rarely separate the lamps from the growing area with a transparent barrier. Rooms designed as CERs at the time a building is constructed and freestanding rooms or chambers usually separate the lamp from the growing area with a barrier of glass or rigid plastic. Light output from fluorescent lamps is a function of the temperature of the lamp. Thus, the barrier serves a two-fold purpose: (1) to maintain optimum lamp temperature when the growing area temperature is higher or lower than optimum, and (2) to reduce the thermal radiation entering the growing area. Fluorescent lamps should operate in an ambient temperature and airflow environment that will maintain the tube wall temperature at 40°C. Under most conditions, the light output of HID lamps is not affected by ambient temperature. The heat must be removed, however, to prevent high thermal radiation from causing adverse biological effects (see Figure 11).

Transparent glass barriers remove nearly all radiation from about 350 to 2500 nm. Rigid plastic is less effective than glass; however, the lighter weight and lower breakage risk of plastic makes it a popular barrier material. Ultraviolet is also screened by both glass and plastic (more by plastic). Special UV-transmitting plastic (which degrades rapidly) can be obtained if the biological process requires

Table 11 Approximate Mounting Height and Spacing of Luminaires in Greenhouses

Lamp and Wattage	Irradiation, W/m²			
	6	12	24	48
	Height and Spacing, m			
HPS (400 W)	3.0	2.3	1.6	1.0
LPS (180 W)	2.4	1.7	1.2	0.8
MH (400 W)	2.7	2.0	1.4	0.9

Table 12   Height and Spacing of Luminaires

Light Source	Radiant Flux Density, W/m^2						
	0.3	0.9	3	9	18	24	50
Fluorescent—Cool White							
40 W single 1.2 m lamp, 3.2 klm							
Radiant power, W/m^2, 400-700 nm	0.3	0.9	2.9	8.8			
Illumination, klx	0.10	0.30	1.0	3.0			
Lamps per 10 m^2	1.1	3.3	11	33			
Distance from plants, m	2.9	1.7	0.92	0.53			
40 W 2-lamp fixtures (1.2 m), 6.4 klm							
Radiant power, W/m^2, 400-700 nm	0.3	0.9	2.9	8.8			
Illumination, klx	0.10	0.30	1.0	3.0			
Fixtures per 10 m^2	0.6	1.7	5.5	16.7			
Distance from plants, m	4.1	2.4	1.3	0.75			
215 W, 2-2.4 m lamps, 31.4 klm							
Radiant power, W/m^2, 400-700 nm	0.3	0.9	2.9	8.8	17.6	23.5	49.0
Illumination, klx	0.10	0.30	1.0	3.0	6.0	8.0	16.7
Lamps per 10 m^2	0.1+	0.4	1.2	3.6	7.1	9.3	20
Distance from plants, m	8.8	5.1	2.8	1.6	1.1	1.0	0.7
High-Intensity Discharge							
Mercury-1 400 W parabolic reflector							
Radiant power, W/m^2, 400-700 nm	0.28	0.84	2.80	8.39	16.8	22.4	46.6
Illumination, klx	0.1	0.32	1.1	3.2	6.4	8.6	18.0
Lamps per 10 m^2	0.2	0.5	1.6	4.8	9.3	13.0	27
Distance from plants, m	7.6	4.4	2.4	1.4	1.0	0.8	0.6
Metal halide-1 400 W							
Radiant power, W/m^2, 400-700 nm	0.77	0.80	2.68	8.03	16.1	21.4	44.6
Illumination, klx	0.09	0.26	0.88	2.6	5.3	7.0	15.0
Lamps per 10 m^2	0.09	0.2	0.7	2.2	4.4	5.8	12.0
Distance from plants, m	11.3	6.5	3.6	2.1	1.5	1.3	0.87
High-pressure sodium 400 W							
Radiant power, W/m^2, 400-700 nm	0.22	0.65	2.18	6.52	13.0	17.4	36.2
Illumination, klx	0.09	0.27	0.89	2.7	5.3	7.1	15.0
Lamps per 10 m^2	0.05	0.14	0.5	1.4	2.8	3.6	7.6
Distance from plants, m	14.2	8.2	4.5	2.6	1.8	1.6	1.1
Low-pressure sodium 180 W							
Radiant power, W/m^2, 400-700 nm	0.26	0.79	2.64	7.93	15.9	21.1	44.0
Illumination, klx	0.14	0.41	1.4	4.1	8.3	11.0	23.0
Lamps per 10 m^2	0.08	0.24	0.8	2.4	4.9	6.5	13.6
Distance from plants, m	10.7	6.2	3.4	2.0	1.4	1.2	0.83
Incandescent							
Incandescent 100 W							
Radiant power, W/m^2, 400-700 nm	0.14	0.41	1.38	4.14	8.28	11.0	23.0
Illumination, klx	0.033	0.10	0.33	1.0	2.0	2.7	5.6
Lamps per 10 m^2	0.5	1.6	5.2	15.8	32	42	87
Distance from plants, m	4.2	4.2	1.3	0.77	0.54	0.47	0.33
Incandescent 150 W flood							
Radiant power, W/m^2, 400-700 nm	0.14	0.41	1.38	4.14	8.28	11.0	23.0
Illumination, klx	0.033	0.098	0.33	1.0	2.0	2.6	5.5
Lamps per 10 m^2	0.3	0.9	3.3	9.3	19.5	26	54
Distance from plants, m	5.4	3.1	1.7	1.0	0.7	0.6	0.4
Incandescent-Hg 160 W							
Radiant power, W/m^2, 400-700 nm	0.14	0.41	1.38	4.14	8.28	11.0	23.0
Illumination, klx	0.050	0.15	0.50	1.5	3.0	4.0	8.3
Lamps per 10 m^2	0.7	2.0	6.9	20.4	42	56	111
Distance from plants, m	3.7	2.1	1.2	0.67	0.47	0.41	0.28
Sunlight							
Radiant power, W/10 m^2	0.22	0.66	2.21	6.65	13.3	17.7	76.9
Illumination, klx	0.054	0.16	0.54	1.6	3.2	4.3	8.9

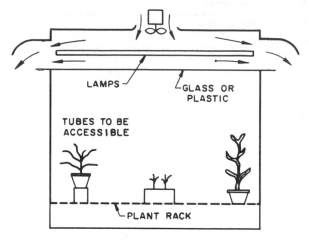

GROWTH CABINET - LIGHTS
AIR-COOLED

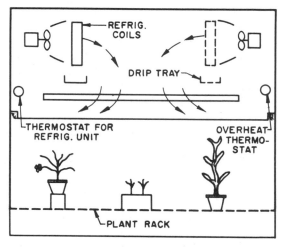

GROWTH CHAMBER - LIGHTS COOLED
BY REFRIGERATION

Fig. 11 Cooling Lamps in Growth Chambers

Table 13 Mounting Height for Luminaires in Storage Areas

	Survival = 3 W/m²		Maintenance = 9 W/m²	
	Distance, m	lux	Distance, m	lux
Fluorescent (F)				
FCW 2-40 W	0.9	1000	0.75	3000
FWW	0.9	1000	0.75	3000
FCW 2-215 W	2.8	1000	1.6	3000
Discharge (HID)				
MH 400 W	3.3	800	2.0	2400
HPS 400 W	4.5	800	2.5	2400
LPS 180 W	3.4	1300	1.2	4000
Incandescent (INC)				
INC 160 W	1.3	350	0.3	1000
INC-HG 160 W	1.2	500	1.6	1500
DL	—	500	—	1500

insulated building. Ventilation and cooling may be required. Illumination levels depend on plant requirements. Table 13 shows approximate mounting heights for two levels of illumination. Luminaires mounted on chains permit lamp height to be adjusted to compensate for varying plant heights.

The main concerns for interior landscape lighting are how it renders the color of plants, people, and furnishings, as well as how it meets the minimum irradiation requirements of plants. The temperature required for human occupancy is normally acceptable for plants. Light level and duration determine the types of plants that can be grown or maintained. Plants grow when exposed to higher levels, but do not survive below the suggested minimum levels. Plants may be grouped into three levels based on the following levels of irradiance:

Low (survival): A minimum light level of 0.75 W/m² and a preferred level of 3 W/m² irradiance for 8 to 12 h daily.

Medium (maintenance): A minimum of 3 W/m² and a preferred level of 9 W/m² irradiance for 8 to 12 h daily.

High (propagation): A minimum of 9 W/m² and a preferred level of 24 W/m² irradiance for 8 to 12 h daily.

Fluorescent (warm-white), metal halide, or incandescent lighting is usually chosen for public places. Table 12 lists irradiance levels of various light sources.

UV light. When irradiance is very high, especially from HID lamps or large numbers of incandescent lamps or both, rigid plastic can soften from the heat and fall from the supports. Furthermore, very high irradiance and the resulting high temperatures can cause plastic to darken, which can increase the absorptivity and temperature enough to destroy it. Under these conditions, heat-resistant glass may be necessary. The lamp compartment and barrier absolutely require positive ventilation regardless of the light source, and the lamp loft should have limit switches that will shut down the lamps if the temperature rises to a critical level.

OTHER PLANT ENVIRONMENTAL FACILITIES

Plants may be held or processed in warehouse-type structures prior to sale or use in interior landscaping. Required temperatures range from slightly above freezing for cold storage of root stock and cut flowers, to 20 to 25°C for maintaining growing plants, usually in pots or containers. Provision must be made for venting fresh air to avoid CO_2 depletion.

Light duration must be controlled by a time clock. When they are in use, lamps and ballasts produce almost all the heat required in an

BIBLIOGRAPHY

ANIMALS

Handbooks and Proceedings

Albright, L.D. 1990. *Environment control for animals and plants, with computer applications.* American Society of Agricultural Engineers, St. Joseph, MI.

ASAE. 1982. *Dairy housing* II. Second National Dairy Housing Conference Proceedings. American Society of Agricultural Engineers, St. Joseph, MI.

ASAE. 1982. *Livestock environment* II. Second International Livestock Environment Symposium. American Society of Agricultural Engineers, St. Joseph, MI.

ASAE. 1988. *Livestock environment* III. Proceedings of the Third International Livestock Environment Symposium. American Society of Agricultural Engineers, St. Joseph, MI.

ASAE. 1993. *Livestock environment* IV. Proceedings of the Fourth International Livestock Environment Symposium. American Society of Agricultural Engineers, St. Joseph, MI.

ASAE. 1993. Design of ventilation systems for livestock and poultry shelters. *Standard* EP270.5. American Society of Agricultural Engineers, St. Joseph, MI.

Christianson, L.L., ed. 1989. *Building systems: Room air and air contaminant distribution.* ASHRAE, Atlanta.

Christianson, L.L. and R.L. Fehr. 1983. Ventilation—Energy and economics. In Ventilation of agricultural structures, pp. 335-49. American Society of Agricultural Engineers, St. Joseph, MI.

Curtis, S.E. 1983. *Environmental management in animal agriculture.* Iowa State University Press, Ames, IA.

Curtis, S.E., ed. 1988. Guide for the care and use of agricultural animals in agricultural research and teaching. Consortium for Developing a Guide for the Care and Use of Agricultural Animals in Agricultural Research and Teaching, 309 W. Clark Street, Champaign, IL 61820.

Hahn, G.L. 1985. Management and housing of farm animals in hot environments. In *Stress physiology in livestock*, Vol. II. pp. 151-76. M.K. Yousef, ed. CRC Press, Boca Raton, FL.

Hellickson, M.A. and J.N. Walker, eds. 1983. Ventilation of agricultural structures. ASAE *Monograph* No. 6. American Society of Agricultural Engineers, St. Joseph, MI.

HEW. 1978. Guide for the care and use of laboratory animals. Publication No. (NIH)78-23. U.S. Department of Health, Education and Welfare.

MWPS. 1989. Natural ventilating systems for livestock housing and heating. MidWest Plan Service, Ames, IA.

MWPS. 1990. Mechanical ventilating systems for livestock housing. Mid-West Plan Service, Ames, IA.

Rechcigl, M., Jr., ed. 1982. *Handbook of agricultural productivity.* Vol. II, *Animal productivity.* CRC Press, Boca Raton, FL.

Air Cooling

Canton, G.H., D.E. Buffington, and R.J. Collier. 1982. Inspired-air cooling for dairy cows. *Transactions of ASAE* 25(3):730-34.

Hahn, G.L. and D.D. Osburn. 1969. Feasibility of summer environmental control for dairy cattle based on expected production losses. *Transactions of ASAE* 12(4):448-51.

Hahn, G.L. and D.D. Osburn. 1970. Feasibility of evaporative cooling for dairy cattle based on expected production losses. *Transactions of ASAE* 12(3):289-91.

Heard, L., D. Froelich, L. Christianson, R. Woerman, and R. Witmer. 1986. Snout cooling effects on sows and litters. *Transactions of ASAE* 29(4): 1097-1101.

Kimball, B.A., D.S. Benham, and F. Wiersma. 1977. Heat and mass transfer coefficients for water and air in aspen excelsior pads. *Transactions of ASAE* 20(3):509.

Morrison, S.R., H. Heitman, Jr., and R.L. Givens. 1979. Effect of air movement and type of slotted floor on sprinkled pigs. *Tropical Agriculture* 56(3):257.

Morrison, S.R., M. Prokop, and G.P. Lofgreen. 1981. Sprinkling cattle for heat stress relief: Activation, temperature, duration of sprinkling, and pen area sprinkled. *Transactions of ASAE* 24(5):1299-1300.

Stewart, R.E., et al. 1966. Field tests of summer air conditioning for dairy cattle in Ohio. *ASHRAE Transactions* 72(1):271.

Timmons, M.B. and G.R. Baughman. 1983. Experimental evaluation of poultry mist-fog systems. *Transactions of ASAE* 26(1):207-10.

Timmons, M.B. and G.R. Baughman. 1984. A plenum concept applied to evaporative pad cooling for broiler housing. *Transactions of ASAE* 27(6):1877-81.

Wilson, J.L., H.A. Hughes, and W.D. Weaver, Jr. 1983. Evaporative cooling with fogging nozzles in broiler houses. *Transactions of ASAE* 26(2): 557-61.

Air Pollution in Buildings

ACGIH. 1992. *1992-1993 Threshold limit values for chemical substances and physical agents and biological exposure indices.* American Conference of Governmental Industrial Hygienists, Cincinnati, OH.

Avery, G.L., G.E. Merva, and J.B. Gerrish. 1975. Hydrogen sulphide production in swine confinement units. *Transactions of ASAE* 18(1):149.

Barber, E.M., J.A. Dosman, C.S. Rhodes, G.I. Christison, and T.S. Hurst. 1993. Carbon dioxide as an indicator of air quality in swine buildings. Proceedings of Third International Livestock Environment Symposium. American Society of Agricultural Engineers, St. Joseph, MI.

Bundy, D.S. and T.E. Hazen. 1975. Dust levels in swine confinement systems associated with different feeding methods. *Transactions of ASAE* 18(1): 137.

Deboer, S. and W.D. Morrison. 1988. *The effects of the quality of the environment in livestock buildings on the productivity of swine and safety of humans—A literature review.* Department of Animal and Poultry Science, University of Guelph, Ontario, Canada.

Donham, J.K., P. Haglind, Y. Peterson, R. Rylander, and L. Belin, 1989. Environmental and health studies of workers in Swedish swine confinement buildings. *British Journal of Industrial Medicine* 40:31-37.

Grub, W., C.A. Rollo, and J.R. Howes. 1965. Dust problems in poultry environment. *Transactions of ASAE* 8(3):338.

Logsdon, R.F. 1965. Methods of air filtration and protection of air-tempering equipment. *Transactions of ASAE* 8(3):345.

Van Wicklen, G. and L.D. Albright. 1982. An empirical model of respirable aerosol concentration in an enclosed calf barn. Proceedings of the Second International Livestock Environment Symposium, pp. 534-39. American Society of Agricultural Engineers, St. Joseph, MI.

Effects of Environment on Production and Growth of Animals

Cattle

Anderson, J.F., D.W. Bates, and K.A. Jordan. 1978. Medical and engineering factors relating to calf health as influenced by the environment. *Transactions of ASAE* 21(6):1169.

Berry, I.L., M.C. Shanklin, and H.D. Johnson. 1964. Dairy shelter design based on milk production decline as affected by temperature and humidity. *Transactions of ASAE* 7:329-33.

Garrett, W.N. 1980. Factors influencing energetic efficiency of beef production. *Journal of Animal Science* 51(6):1434.

Gebremedhin, K.G., C.O. Cramer, and W.P. Porter. 1981. Predictions and measurements of heat production and food and water requirements of Holstein calves in different environments. *Transactions of ASAE* 24 (3): 715.

Gebremedhin, K.G., W.P. Porter, and C.O. Cramer. 1983. Quantitative analysis of heat exchange through the fur layer of Holstein calves. *Transactions of ASAE* 28(1):188-93.

Holmes, C.W. and N.A. McLean. 1975. Effects of air temperature and air movement on the heat produced by young Friesian and Jersey calves, with some measurements of the effects of artificial rain. *New Zealand Journal of Agricultural Research* 18(3):277.

Morrison, S.R., G.P. Lofgreen, and R.L. Givens. 1976. Effect of ventilation rate on beef cattle performance. *Transactions of ASAE* 19(3):530.

Morrison, S.R. and M. Prokop. 1983. Beef cattle performance on slotted floors: Effect of animal weight on space allotment. *Transactions of ASAE* 26(2):525-28.

Webster, A.J.F., J.G. Gordon, and J.S. Smith. 1976. Energy exchanges of veal calves in relation to body weight, food intake and air temperature. *Animal Production* 23(1):35.

Phillips, P.A. and F.V. MacHardy. 1983. Predicting heat production in young calves housed at low temperature. *Transactions of ASAE* 26(1):175-78.

General

Hahn, G.L. 1982. Compensatory performance in livestock: Influences on environmental criteria. Proceedings of the Second International Livestock Environment Symposium, pp. 285-94. American Society of Agricultural Engineers, St. Joseph, MI.

Hahn, G.L. 1981. Housing and management to reduce climatic impacts on livestock. *Journal of Animal Science* 52(1):175-86.

Pigs

Boon, C.R. 1982. The effect of air speed changes on the group postural behaviour of pigs. *Journal of Agricultural Engineering Research* 27(1):71-79.

Bruce, J.M. and J.J. Clark. 1971. Models of heat production and critical temperature for growing pigs. *Animal Production* 13(2):285.

Christianson, L.L., D.P. Bane, S.E. Curtis, W.F. Hall, A.J. Muehling, and G.L. Riskowski. 1989. *Swine care guidelines for pork producers using environmentally controlled housing.* National Pork Producers Council, Des Moines, IA.

Close, W.H., L.E. Mount, and I.B. Start. 1971. The influence of environmental temperature and plane of nutrition on heat losses from groups of growing pigs. *Animal Production* 13(2):285.

Driggers, L.B., C.M. Stanislaw, and C.R. Weathers. 1976. Breeding facility design to eliminate effects of high environmental temperatures. *Transactions of ASAE* 19(5):903.

Holmes, C.W. and N.A. McLean. 1977. The heat production of groups of young pigs exposed to reflective and non-reflective surfaces on walls and ceilings. *Transactions of ASAE* 20(3):527.

McCracken, K.J., B.J. Caldwell, and N. Walker. 1979. A note on the performance of early-weaned pigs. *Animal Production* 29(3):423.

McCracken, K.J. and R. Gray. 1984. Further studies on the heat production and affective lower critical temperature of early-weaned pigs under commercial conditions of feeding and management. *Animal Production* 39:283-90.

Morrison, S.R., H. Heitman, Jr., and R.L. Givens. 1979. Effect of air movement and type of slotted floor on sprinkled pigs. *Tropical Agriculture* 56(3):257.

Mount, L.E. and I.B. Start. 1980. A note on the effects of forced air movement and environmental temperature on weight gain in the pig after weaning. *Animal Production* 30(2):295.

Nienaber, J.A. and G.L. Hahn. 1988. Environmental temperature influences on heat production of ad-lib-fed nursery and growing-finishing swine. *Livestock environment* III. Proceedings of the Third International Livestock Environment Symposium. American Society of Agricultural Engineers, St. Joseph, MI, 73-78.

Phillips, P.A., B.A. Young, and J.B. McQuitty. 1982. Liveweight, protein deposition and digestibility responses in growing pigs exposed to low temperature. *Canadian Journal of Animal Science* 62:95-108.

Phillips, P.A. and F.V. MacHardy. 1982. Modelling protein and lipid gains in growing pigs exposed to low temperature. *Canadian Journal of Animal Science* 62:109-21.

Riskowski, G.L. and D.S. Bundy. 1988. Effects of air velocity and temperature on weanling pigs. *Livestock environment* III. Proceedings of the Third International Livestock Environment Symposium, pp. 117-24. American Society of Agricultural Engineers, St. Joseph, MI.

Poultry

Buffington, D.E., K.A. Jordan, W.A. Junnila, and L.L. Boyd. 1974. Heat production of active, growing turkeys. *Transactions of ASAE* 17(3):542.

Carr, L.E., T.A. Carter, and K.E. Felton. 1976. Low temperature brooding of broilers. *Transactions of ASAE* 19(3):553.

Riskowski, G.L., J.A. DeShazer, and F.B. Mather. 1977. Heat losses of white leghorn laying hens as affected by intermittent lighting schedules. *Transactions of ASAE* 20(4):727-31.

Siopes, T.D., M.B. Timmons, G.R. Baughman, and C.R. Parkhurst. 1983. The effect of light intensity on the growth performance of male turkeys. *Poultry Science* 62:2336-42.

Sheep

Schanbacher, B.D., G.L. Hahn, and J.A. Nienaber. 1982. Photoperiodic influences on performance of market lambs. Proceedings of the Second International Livestock Environment Symposium, pp. 400-05. American Society of Agricultural Engineers, St. Joseph, MI.

Vesely, J.A. 1978. Application of light control to shorten the production cycle in two breeds of sheep. *Animal Production* 26(2):169.

Modeling and Analysis

Albright, L.D. and N.R. Scott. 1974a. An analysis of steady periodic building temperature variations in warm weather—Part I: A mathematical model. *Transactions of ASAE* 17(1):88-92, 98.

Albright, L.D. and N.R. Scott. 1974b. An analysis of steady periodic building temperature variations in warm weather—Part II: Experimental verification and simulation. *Transactions of ASAE* 17(1):93-98.

Albright, L.D. and N.R. Scott. 1977. Diurnal temperature fluctuations in multi-air spaced buildings. *Transactions of ASAE* 20(2):319-26.

Bruce, J.M. and J.J. Clark. 1979. Models of heat production and critical temperature for growing pigs. *Animal Production* 28:353-69.

Buffington, D.E. 1978. Simulation models of time-varying energy requirements for heating and cooling buildings. *Transactions of ASAE* 21(4):786.

Christianson, L.L. and H.A. Hellickson. 1977. Simulation and optimization of energy requirements for livestock housing. *Transactions of ASAE* 20(2):327-35.

Ewan, R.C. and J.A. DeShazer. 1988. Mathematical modeling the growth of swine. *Livestock environment* III. Proceedings of the Third International Livestock Environment Symposium, pp. 211-18. American Society of Agricultural Engineers, St. Joseph, MI.

Hellickson, M.L., K.A. Jordan, and R.D. Goodrich. 1978. Predicting beef animal performance with a mathematical model. *Transactions of ASAE* 21(5):938-43.

Teter, N.C., J.A. DeShazer, and T.L. Thompson. 1973. Operational characteristics of meat animals—Part I: Swine; Part II: Beef; Part III: Broilers. *Transactions of ASAE* 16:157-59; 740-42; 1165-67.

Timmons, M.B. 1984. Use of physical models to predict the fluid motion in slot-ventilated livestock structures. *Transactions of ASAE* 27(2):502-07.

Timmons, M.B., L.D. Albright, and R.B. Furry. 1978. Similitude aspects of predicting building thermal behavior. *Transactions of ASAE* 21(5):957.

Timmons, M.B., L.D. Albright, R.B. Furry, and K.E. Torrance. 1980. Experimental and numerical study of air movement in slot-ventilated enclosures. *ASHRAE Transactions* 86(1):221-40.

Shades for Livestock

Bedwell, R.L. and M.D. Shanklin. 1962. Influence of radiant heat sink on thermally-induced stress in dairy cattle. Missouri Agricultural Experiment Station *Research Bulletin* No. 808.

Bond, T.E., L.W. Neubauer, and R.L. Givens. 1976. The influence of slope and orientation of effectiveness of livestock shades. *Transactions of ASAE* 19(1):134-37.

Roman-Ponce, H., W.W. Thatcher, D.E. Buffington, C.J. Wilcox, and H.H. VanHorn. 1977. Physiological and production responses of dairy cattle to a shade structure in a subtropical environment. *Journal of Dairy Science* 60(3):424.

Transport of Animals

Ashby, B.H., D.G. Stevens, W.A. Bailey, K.E. Hoke, and W.G. Kindya. 1979. *Environmental conditions on air shipment of livestock.* USDA, SEA, Advances in Agricultural Technology, Northeastern Series No. 5.

Ashby, B.H., A.J. Sharp, T.H. Friend, W.A. Bailey, and M.R. Irwin. 1981. Experimental railcar for cattle transport. *Transactions of ASAE* 24(2):452.

Ashby, B.H., H. Ota, W.A. Bailey, J.A. Whitehead, and W.G. Kindya. 1980. Heat and weight loss of rabbits during simulated air transport. *Transactions of ASAE* 23(1):162.

Grandin, R. 1988. *Livestock trucking guide.* Livestock Conservation Institute, Madison, WI.

Jackson, W.T. 1974. Air transport of Hereford cattle to the People's Republic of China. *Veterinary Records* 9(1):209.

Scher, S. 1980. Lab animal transportation receiving and quarantine. *Lab Animal* 9(3):53.

Stermer, R.A., T.H. Camp, and D.G. Stevens. 1982. Feeder cattle stress during handling and transportation. *Transactions of ASAE* 25(1):246-48.

Stevens, D.G., G.L. Hahn, T.E. Bond, and J.H. Langridge. 1974. *Environmental considerations for shipment of livestock by air freight.* USDA, APHIS (May).

Stevens, D.G. and G.L. Hahn. 1981. Minimum ventilation requirement for the air transportation of sheep. *Transactions of ASAE* 24(1):180.

Laboratory Animals

NIH. 1978. Laboratory animal housing. Proceedings of a symposium (p. 220) held at Hunt Valley, MD, September 1976. National Academy of Sciences, Washington, D.C.

McSheehy, T. 1976. *Laboratory animal handbook* 7—Control of the animal house environment. Laboratory Animals, Ltd.

Soave, O., W. Hoag, et al. 1980. The laboratory animal data bank. *Lab Animal* 9(5):46.

Ventilation Systems

Albright, L.D. 1976. Air flows through hinged-baffle, slotted inlets. *Transactions of ASAE* 19(4):728, 732, 735.

Albright, L.D. 1978. Air flow through baffled, center-ceiling, slotted inlets. *Transactions of ASAE* 21(5):944-47, 952.

Albright, L.D. 1979. Designing slotted inlet ventilation by the systems characteristic technique. *Transactions of ASAE* 22(1):158.

Goetsch, W.D., D.P. Stombaugh, and A.T. Muehling. 1984. *Earth-tube heat exchange systems.* AED-25, Midwest Plan Service, Ames, IA.

McGinnis, D.S., J.R. Ogilvie, D.R. Pattie, K.W. Blenkhorn, and J.E. Turnbull. 1983. Shell-and-tube heat exchanger for swine buildings. *Canadian Agricultural Engineering* 25(1):69-74.

Person, H.L., L.D. Jacobson, and K.A. Jordan. 1979. Effect of dirt, louvers and other attachments on fan performance. *Transactions of ASAE* 22(3):612-16.

Pohl, S.H. and M.A. Hellickson. 1978. Model study of five types of manure pit ventilation systems. *Transactions of ASAE* 21(3):542.

Randall, J.M. 1975. The prediction of air flow patterns in livestock buildings. *Journal of Agricultural Engineering Research* 20(2):199-215.

Randall, J.M. 1980. Selection of piggery ventilation systems and penning layouts based on the cooling effects of air speed and temperature. *Journal of Agricultural Engineering Research* 25(2):169-87.

Randall, J.M. and V.A. Battams. 1979. Stability criteria for air flow patterns in livestock buildings. *Journal of Agricultural Engineering Research* 24(4):361-74.

Sokhansanj, S., K.A. Jordan, L.A. Jacobson, and G.L. Messers. 1980. Economic feasibility of using heat exchangers in ventilation of animal buildings. *Transactions of ASAE* 23(6):1525-28.

Spengler, R.W. and D.P. Stombaugh. 1983. Optimization of earth-tube exchangers for winter ventilation of swine housing. *Transactions of ASAE* 26(4):1186-93.

Timmons, M.B. 1984. Internal air velocities as affected by the size and location of continuous inlet slots. *Transactions of ASAE* 27(5):1514-17.

Timmons, M.B. and G.R. Baughman. 1983. The FLEX house: A new concept in poultry housing. *Transactions of ASAE* 26(2):529-32.

Witz, R.L., G.L. Pratt, and M.L. Buchanan. 1976. Livestock ventilation with heat exchanger. *Transactions of ASAE* 19(6):1187.

Alarm Systems

Clark, W.D. and G.L. Hahn. 1971. Automatic telephone warning systems for animal and plant laboratories or production systems. *Journal of Dairy Science* 54(k6):932-35.

Natural Ventilation

Bruce, J.M. 1982. Ventilation of a model livestock building by thermal buoyancy. *Transactions of ASAE* 25(6):1724-26.

Jedele, D.G. 1979. Cold weather natural ventilation of buildings for swine finishing and gestation. *Transactions of ASAE* 22(3):598-601.

Timmons, M.B. and G.R. Baughman. 1981. Similitude analysis of ventilation by the stack effect from an open ridge livestock structure. *Transactions of ASAE* 24(4):1030-34.

Timmons, M.B., R.W. Bottcher, and G.R. Baughman. 1984. Nomographs for predicting ventilation by thermal buoyancy. *Transactions of ASAE* 27(6):1891-93.

PLANTS

Greenhouse and Plant Environment

Albright, L.D. 1990. Environment control for animals and plants, with computer applications. American Society of Agricultural Engineers, St. Joseph, MI.

Aldrich, R.A. and J.W. Bartok. 1984. *Greenhouse engineering.* Department of Agricultural Engineering, University of Connecticut, Storrs, CT.

ASAE. 1992. Guidelines for measuring and reporting environmental parameters for plant experiments in growth chambers. *Engineering Practice* EP411.2-92. American Society of Agricultural Engineers, St. Joseph, MI.

Clegg, P. and D. Watkins. 1978. *The complete greenhouse book.* Garden Way Publishing, Charlotte, VT.

Downs, R.J. 1975. *Controlled environments for plant research.* Columbia University Press, New York.

Hellmers, H. and R.J. Downs. 1967. Controlled environments for plant-life research. *ASHRAE Journal* (February):37.

Hellickson, M. and J. Walker. 1983. Ventilation of agricultural structures. *Monograph* No. 6, American Society of Agricultural Engineers, St. Joseph, MI.

Langhans, R.W. 1985. *Greenhouse management.* Halcyon Press, Ithaca, NY.

Mastalerz, J.W. 1977. *The greenhouse environment.* John Wiley and Sons, New York.

Nelson, P.V. 1978. *Greenhouse operation and management.* Reston Publishing Co., Reston, VA.

Pierce, J.H. 1977. *Greenhouse grow how.* Plants Alive Books, Seattle, WA.

Riekels, J.W. 1977. *Hydroponics.* Ontario Ministry of Agriculture and Food, Fact Sheet No. 200-24, Toronto, Ontario, Canada.

Riekels, J.W. 1975. *Nutrient solutions for hydroponics.* Ontario Ministry of Agriculture and Food, Fact Sheet No. 200-532, Toronto, Ontario, Canada.

Sheldrake, R., Jr. and J.W. Boodley. *Commercial production of vegetable and flower plants.* Research Park, 1B-82, Cornell University, Ithaca, NY.

Tibbitts, T.W. and T.T. Kozlowski, eds. 1979. *Controlled environment guidelines for plant research.* Academic Press, New York.

Light and Radiation

Armitage, A.M. and M.J. Tsugita. 1974. The effect of supplemental lights source, illumination, and quantum flux density on the flowering of seed propagated geraniums. *Journal of the American Society for Horticultural Science* 54:195.

Bickford, E.D. and S. Dunn. 1972. *Lighting for plant growth.* Kent State University Press, Kent, OH.

Boodley, J.W. 1970. Artificial light sources for gloxinia, African violet, and tuberous begonia. *Plants & Gardens* 26:38.

Carpenter, G.C. and L.J. Mousley. 1960. The artificial illumination of environmental control chambers for plant growth. *Journal of Agricultural Engineering Research* [England] 5:283.

Carpenter, G.A., L.J. Mousley, and P.A. Cottrell. 1964. Maintenance of constant light intensity in plant growth chambers by group replacement of lamps. *Journal of Agricultural Engineering Research* 9.

Campbell, L.E., R.W. Thimijan, and H.M. Cathey. 1975. Special radiant power of lamps used in horticulture. *Transactions of ASAE* 18(5):952.

Cathey, H.M. and L.E. Campbell. 1974. Lamps and lighting: A horticultural view. *Lighting Design & Application* 4:41.

Cathey, H.M. and L.E. Campbell. 1975. Effectiveness of five vision lighting sources on photo-regulation of 22 species of ornamental plants. *Journal of the American Society for Horticultural Science* 100(1):65.

Cathey, H.M. and L.E. Campbell. 1979. Relative efficiency of high- and low-pressure sodium and incandescent filament lamps used to supplement natural winter light in greenhouses. *Journal of the American Society for Horticultural Science* 104(6):812.

Cathey, H.M. and L.E. Campbell. 1980. Light and lighting systems for horticultural plants. *Horticultural Reviews* 11:491. AVI Publishing Co., Westport, CT.

Cathey, H.M., L.E. Campbell, and R.W. Thimijan. 1978. Comparative development of 11 plants grown under various fluorescent lamps and different duration of irradiation with and without additional incandescent lighting. *Journal of the American Society for Horticultural Science* 103:781.

Fonteno, W.C. and E.L. McWilliams. 1978. Light compensation points and acclimatization of four tropical foliage plants. *Journal of the American Society for Horticultural Science* 103:52.

Hughes, J., M.J. Tsujita, and D.P. Ormrod. 1979. *Commercial applications of supplementary lighting in greenhouses.* Ontario Ministry of Agriculture and Food, Fact Sheet No. 290-717, Toronto, Ontario, Canada.

Kaufman, J.E., ed. 1981. IES *Lighting handbook*, Application Volume. IES, New York.

Kaufman, J.E., ed. 1981. IES *Lighting handbook*, Reference Volume. IES, New York.

Poole, R.T. and C.A. Conover. Influence of shade and nutrition during production and dark storage stimulating shipment on subsequent quality and chlorophyll content of foliage plants. *HortScience* 14:617.

Robbins, F.V. and C.K. Spillman. 1980. Solar energy transmission through two transparent covers. *Transactions of ASAE* 23(5).

Sager, J.C., J.L. Edwards, and W.H. Klein. 1982. Light energy utilization efficiency for photosynthesis. *Transactions of ASAE* 25(6):1737-46.

Photoperiod

Cathey, H.M. and H.A. Borthwick. 1961. Cyclic lighting for controlling flowering of chrysanthemums. *Proceedings of ASAE* 78:545.

Heins, R.D., W.H. Healy, and H.F. Wilkens. 1980. Influence of night lighting with red, far red, and incandescent light on rooting of chrysanthemum cuttings. *HortScience* 15:84.

Piringer, A.A. and H.M. Cathey. 1960. Effect of photoperiod, kind of supplemental light, and temperature on the growth and flowering of petunia plants. *Proceedings of the American Society for Horticultural Science* 76:649.

Carbon Dioxide

Bailey, W.A., et al. 1970. CO_2 systems for growing plants. *Transactions of ASAE* 13(2):63.

Gates, D.M. 1968. Transpiration and leaf temperature. *Annual Review of Plant Physiology* 19:211.

Holley, W.D. 1970. CO_2 enrichment for flower production. *Transactions of ASAE* 13(3):257.

Kretchman, J. and F.S. Howlett. 1970. Enrichment for vegetable production. *Transactions of ASAE* 13(2):252.

Pettibone, C.A., et al. 1970. The control and effects of supplemental carbon dioxide in air-supported plastic greenhouses. *Transactions of ASAE* 13(2):259.

Tibbitts, T.W., J.C. McFarlane, D.T. Krizek, W.L. Berry, P.A. Hammer, R.H. Hodgsen, and R.W. Langhans. 1977. Contaminants in plant growth chambers. *Horticulture Science* 12:310.

Watt, A.D. 1971. Placing atmospheric CO_2 in perspective. IEEE *Spectrum* 8(11): 59.

Wittwer, S.H. 1970. Aspects of CO_2 enrichment for crop production. *Transactions of ASAE* 13(2):249.

Heating, Cooling, and Ventilation

ASAE. 1990. Heating, ventilating and cooling greenhouses. *Engineering Practice* EP406.1-90. American Society of Agricultural Engineers, St. Joseph, MI.

Albright, L.D., I. Seginer, L.S. Marsh, and A. Oko. 1985. In situ thermal calibration of unventilated greenhouses. *Journal of Agricultural Engineering Research* 31(3):265-81.

Buffington, D.E. and T.C. Skinner. 1979. Maintenance guide for greenhouse ventilation, evaporative cooling, and heating systems. Publication No. AE-17. Department of Agricultural Engineering, University of Florida, Gainesville, FL.

Duncan, G.A. and J.N. Walker. 1979. Poly-tube heating ventilation systems and equipment. Publication No. AEN-7. Agricultural Engineering Department, University of Kentucky, Lexington, KY.

Elwell, D.L., M.Y. Hamdy, W.L. Roller, A.E. Ahmed, H.N. Shapiro, J.J. Parker, and S.E. Johnson. 1985. *Soil heating using subsurface pipes.* Department of Agricultural Engineering, Ohio State University, Columbus, OH.

Gray, H.E. 1948. A study of the problems of heating, ventilating greenhouses. PhD Thesis, Cornell University, Ithaca, NY.

Gray, H.E. 1954. Greenhouse heating. Cornell University Experiment Station *Bulletin* No. 906. Ithaca, NY.

Hanan, J.J. Ozone and ethylene effects on some ornamental plant species. *General Series Bulletin* No. 974. Colorado State University, Fort Collins, CO.

Heins, R. and A. Rotz. 1980. Plant growth and energy savings with infrared heating. *Florists' Review* (October):20.

Kimball, B.A. 1983. *A modular energy balance program including subroutines for greenhouses and other latent heat devices.* Agricultural Research Service, 4331 East Broadway, Phoenix, AZ 85040.

NGMA. 1989. Greenhouse heat loss. National Greenhouse Manufacturers' Association, Taylors, SC.

NGMA. 1989. Standards for ventilating and cooling greenhouses. National Greenhouse Manufacturers' Association, Taylors, SC.

NGMA. 1993. Recommendation for using insect screens in greenhouse structures. National Greenhouse Manufacturers' Association, Taylors, SC.

Roberts, W.J. and D. Mears. 1984. *Floor heating and bench heating extension bulletin for greenhouses.* Department of Agricultural and Biological Engineering, Cook College, Rutgers University, New Brunswick, NJ.

Roberts, W.J. and D. Mears. 1984. *Heating and ventilating greenhouses.* Department of Agricultural and Biological Engineering, Cook College, Rutgers University, New Brunswick, NJ.

Roberts, W.J. and D.R. Mears. 1979. *Floor heating of greenhouses.* Miscellaneous Publication, Rutgers University, New Brunswick, NJ.

Rogers, B.T. 1979. Wood and the curious case of the rock salt greenhouse. *ASHRAE Journal* (November):74 (see also Silverstein).

Rotz, C.A. and R.D. Heins. 1980. An economic comparison of greenhouse heating systems. *Florists' Review* (October):24.

Silverstein, S.D. 1976. Effect of infrared transparency on heat transfer through windows: A clarification of the greenhouse effect. *Science* 193:229 (see also Rogers).

Skinner, T.C. and D.E. Buffington. 1977. Evaporative cooling of greenhouses in Florida. Publication No. AE-14. Department of Agricultural Engineering, University of Florida, Gainesville, FL.

Walker, J.N. 1965. Predicting temperatures in ventilated greenhouses. *Transactions of ASAE* 8(3):445.

Walker, J.N. and G.A. Duncan. 1975. Greenhouse heating systems. Publication No. AEN-31. Agricultural Engineering Department, University of Kentucky, Lexington, KY.

Walker, J.N. and G.A. Duncan. 1979. Greenhouse ventilation systems. Publication No. AEN-30. Agricultural Engineering Department, University of Kentucky, Lexington, KY.

Walker, P.N. 1979. Greenhouse surface heating with power plant cooling water: Heat transfer characteristics. *Transactions of ASAE* 22(6):1370, 1380.

Energy Conservation

Badger, P.C. and H.A. Poole. 1979. *Conserving energy in Ohio greenhouses.* OARDC *Special Circular* No. 102. Ohio Agricultural Research and Development Center, Ohio State University, Wooster, OH.

Blom, T., J. Hughes, and F. Ingratta. 1978. Energy conservation in Ontario greenhouses. Publication No. 65. Ontario Ministry of Agriculture and Food, Toronto, Ontario, Canada.

Roberts, W.J., J.W. Bartok, Jr., E.E. Fabian, and J. Simpkins. 1985. *Energy conservation for commercial greenhouses.* NRAES-3. Department of Agricultural Engineering, Cornell University, Ithaca, NY.

Solar Energy Use

Albright, L.D., R.W. Langhans, G.B. White, and A.J. Donohoe. 1980. Enhancing passive solar heating of commercial greenhouses. Proceedings of the ASAE National Energy Symposium, St. Joseph, MI.

Albright, L.D., et al. 1980. Passive solar heating applied to commercial greenhouses. Publication No. 115, Energy in Protected Civilization. Acta Horticultura.

Cathey, H.M. 1980. Energy-efficient crop production in greenhouses. *ASHRAE Transactions* 86(2):455.

Duncan, G.A., J.N. Walker, and L.W. Turner. 1979. *Energy for greenhouses,* Part I: Energy Conservation. Publication No. AEES-16. College of Agriculture, University of Kentucky, Lexington, KY.

Duncan, G.A., J.N. Walker, and L.W. Turner. 1980. *Energy for greenhouses,* Part II: Alternative Sources of Energy. College of Agriculture, University of Kentucky, Lexington, KY.

Gray, H.E. 1980. Energy management and conservation in greenhouses: A manufacturer's view. *ASHRAE Transactions* 86(2):443.

Roberts, W.J. and D.R. Mears. 1980. Research conservation and solar energy utilization in greenhouses. *ASHRAE Transactions* 86(2):433.

Short, T.H., M.F. Brugger, and W.L. Bauerle. 1980. Energy conservation ideas for new and existing commercial greenhouses. *ASHRAE Transactions* 86(2):448.

Interior Plantscaping

Gaines, R.L. 1980. *Interior plantscaping.* Architectural Record, McGraw-Hill, New York.

Cathey, H.M. and L.E. Campbell. 1978. Indoor gardening artificial lighting, terrariums, hanging baskets and plant selection. USDA *Home and Garden Bulletin* No. 220.

CHAPTER 21

DRYING AND STORING FARM CROPS

CONTROL of moisture content and temperature during storage is critical to preserving the quality of farm crops as they move from the field to the market. Relative humidity and temperature affect mold growth, which is reduced to a minimum if the crop is kept cooler than 10°C and if the relative humidity of the air in equilibrium with the stored crop is less than 60% (Figure 1).

Mold growth and spoilage are a function of elapsed storage time, temperature, and moisture content above critical values. The approximate allowable storage life for cereal grains is shown in Table 1. For example, corn at 16°C and 20% moisture has a storage life of about 25 days. If the corn is dried to 18% after 12 days, one-half of its storage life has elapsed. Thus, the remaining storage life at 16°C and 18% moisture content is 25 days, not 50 days.

Insects thrive in stored grain if the moisture content and temperature are too high. At low moisture contents and temperatures less than 10°C, insects remain dormant or die.

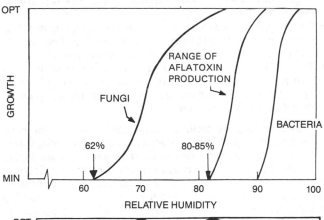

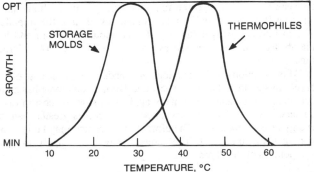

Fig. 1   Microbial Growth as Affected by Relative Humidity and Temperature (University of Kentucky 1973)

The preparation of this chapter is assigned to TC 2.2, Plant and Animal Environment.

Table 1   Approximate Allowable Storage Time (Days) for Cereal Grains

Moisture Content, % w.b.[1]	Temperature, °C					
	−1	4	10	16	22	27
14	*	*	*	*	200	140
15	*	*	*	240	125	70
16	*	*	230	120	70	40
17	*	280	130	75	45	20
18	*	200	90	50	30	15
19	*	140	70	35	20	10
20	*	90	50	25	14	7
22	190	60	30	15	8	3
24	130	40	15	10	6	2
26	90	35	12	8	5	2
28	70	30	10	7	4	2
30	60	25	5	5	3	1

Based on composite of 0.5% maximum dry matter loss calculated on the basis of USDA research; *Transactions of ASAE* 333-337, 1972; and "Unheated Air Drying," Manitoba Agriculture Agdex 732-1, rev. 1986.

[1]Grain moisture content calculated as percent wet basis: (Mass of water in a given amount of wet grain ÷ Mass of the wet grain) × 100.
*Approximate allowable storage time exceeds 300 days.

Most farm crops must be dried to, and maintained at, a moisture content of 12 to 13%, depending on the specific crop, storage temperature, and length of storage. Oil seeds such as peanuts, sunflowers, and flaxseed must be dried to a moisture content of 8 to 9%. Grain stored for more than a year and seed stock should be dried to a lower moisture content. Moisture levels above these critical values lead to the growth of fungi, which may produce toxic compounds.

The maximum yield of dry matter can be obtained by harvesting when the corn has dried in the field to an average moisture content of 26%. Wheat can be harvested when it has dried to 20%. However, harvesting at these moisture contents requires expensive mechanical drying. Although field drying requires less expense than operating drying equipment, it may be more costly in the end because field losses generally increase as the moisture content decreases.

The price of grain to be sold through commercial market channels is based on a specified moisture content, with price discounts for moisture levels above the specified amount. These discounts compensate for the mass of excess water, cover the cost of water removal, and control the supply of wet grain delivered to market. Grain dried to below the base moisture content set by the market (15.0% for corn, 13.0% for soybeans, and 13.5% for wheat) is not generally sold at a premium; thus, the seller loses the opportunity to sell water for the price of grain.

Grain Quantity

The bushel is the common measure used for marketing grain in the United States, while the tonne (Mg) is the more common international measure. A bushel is a volume measure equal to 0.03524 m³.

Table 2 Calculated Densities of Grains and Seeds Based on Weights and Measures Used by the U.S. Department of Agriculture

	Bulk Density, kg/m³
Alfalfa	768
Barley	614
Beans, dry	768
Bluegrass	180 to 384
Clover	768
Corn	
Ear, husked	448
Shelled	717
Cottonseed	410
Oats	410
Peanuts, unshelled	
Virginia type	218
Runner, Southeastern	269
Spanish	317
Rice, rough	576
Rye	717
Sorghum	640
Soybeans	768
Sudan grass	512
Sunflower	
Non-oil	307
Oil seed	410
Wheat	768

The legal mass for the bushel in the United States is 25.40 kg for corn and 27.22 kg for wheat. The densities of some crops are listed in Table 2.

The percent of initial mass lost due to water removed may be calculated by the following equation:

$$\text{Moisture shrink, \%} = \frac{M_o - M_f}{100 - M_f} \times 100$$

where

M_o = original or initial moisture content, wet basis
M_f = final moisture content, wet basis

Applying the formula to drying a crop from 25% to 15%,

$$\text{Moisture shrink} = \frac{25 - 15}{100 - 15} \times 100 = 11.76\%$$

In this case, the moisture shrink is 11.76%, or an average 1.176% mass reduction for each percentage point of moisture reduction. The moisture shrink varies depending on the final moisture content. For example, the average shrink per point of moisture when drying from 20% to 10% is 1.111.

Economics

Producers generally have the choice of drying their grain on the farm before delivering it to market or delivering wet grain at a price discount for excess moisture. The expense of drying on the farm includes both fixed and variable costs. Once a dryer is purchased, the costs of depreciation, interest, taxes, and repairs are fixed and minimally affected by volume. The costs of labor, fuel, and electricity vary directly with the volume dried. Total drying costs vary widely,

Table 3 Estimated Corn Drying Energy Requirement

Dryer Type	kJ/kg of Water Removed
Unheated air	2300-2800
Low temperature	2800-3500
Batch-in-bin	3500-4700
High temperature	
Air recirculating	4200-5100
Without air recirculating	4700-7000

Note: Includes all energy requirements for fans and heat.

depending on the volume dried, the drying equipment, and fuel and equipment prices. Energy consumption depends primarily on dryer type. Generally, the faster the drying speed, the greater the energy consumption (Table 3).

DRYING EQUIPMENT AND PRACTICES

Contemporary crop-drying equipment depends largely on mass and energy transfer between the drying air and the product to be dried. The drying rate depends on the initial temperature and moisture content of the crop, the air-circulation rate, the entering condition of the circulated air, the length of flow path through the products, and the time elapsed since the beginning of the drying operation. Outdoor air is frequently heated before it is circulated through the product. Heating increases the rate of heat transfer to the product, raises its temperature, and increases the vapor pressure of the product moisture.

Most crop-drying equipment consists of (1) a fan to move the air through the product, (2) a controlled heater to increase the ambient air temperature to the desired level, and (3) a container to distribute the drying air uniformly through the product. The exhaust air is vented to the atmosphere. Where climate and other factors are favorable, unheated air is used for drying, and the heater is omitted.

Fans

The fan selected for a given drying application should meet the same requirements important in any air-moving application. It must deliver the desired amount of air against the static resistance of the product in the bin or column, the resistance of the delivery system, and the resistance of the air inlet and outlet.

Foreign material in the grain can significantly change the required air pressure in the following ways:

- Foreign particles larger than the grain (straw, plant parts, and larger seeds) reduce airflow resistance. The airflow rate may be increased by 60% or more.
- Foreign particles smaller than the grain (broken grain, dust, and small seeds) increase the airflow resistance. The effect may be dramatic, decreasing the airflow rate by 50% or more.
- The method used to fill the dryer or the agitation of the grain after it is placed in the dryer can increase pressure requirements up to 100%. In some grain high moisture causes less pressure drop than does low moisture.

Vaneaxial fans are normally recommended when static pressures are less than 0.75 kPa. Backward-curved centrifugal fans are commonly recommended when static pressures are higher than 1.0 kPa. Low-speed centrifugal fans operating at 1750 rpm perform well up to about 1.75 kPa, and high-speed centrifugal fans operating at about 3500 rpm have the ability to develop static pressure up to about 2.5 kPa. The in-line centrifugal fan consists of a centrifugal fan impeller mounted in the housing of an axial flow fan. A bell intake funnels the air into the impeller. The in-line centrifugal fan operates at about 3450 rpm and has the ability to develop pressures up to 2.5 kPa on 6 kW or larger fans.

After functional considerations are made, the initial cost of the dryer fan should be taken into account. Drying equipment has a low percentage of annual use in many applications, so the cost of dryer ownership per unit of material dried is sometimes greater than the energy cost of operation. Therefore, fan efficiency may be less important than initial cost. The same considerations apply to other components of the dryer.

Heaters

Crop dryer heaters are fueled by natural gas, liquefied petroleum gas, or fuel oil; some electric heaters are also used. Dryers using coal, biomass (such as corn cobs, stubble, or wood), and solar energy have been built.

Fuel combustion in crop dryers is similar to combustion in domestic and industrial furnaces. Heat is transferred to the drying air either indirectly, by means of a heat exchanger, or directly, by combining the combustion gases with the drying air. Most grain dryers use direct combustion. Indirect heating is used in drying products such as hay and sunflowers because of their greater fire hazard.

Controls

In addition to the usual temperature controls for drying air, all heated air units must have safety controls similar to those found on space-heating equipment. These safety controls shut off the fuel in case of flame failure and stop the burner in case of overheating or excessive drying air temperatures. All controls should be set up to operate the machinery safely in case of power failure.

SHALLOW LAYER DRYING

Batch Dryers

The batch dryer cycles through the loading, drying, cooling, and unloading of the grain. Fans force hot air through grain columns, which are generally about 300 mm thick. Drying time depends on the type of grain and the amount of moisture to be removed. A general rule is 10 min per percentage point (percent) of moisture to be removed. Many dryers circulate and mix the grain to prevent significant moisture content gradients from forming across the column. A circulation rate that is too fast or a poor selection of handling equipment may cause undue damage and loss of market quality. Batch dryers are suitable for farm operations and are often portable.

Continuous Flow Dryers

This type of self-contained dryer passes a continuous stream of grain through the drying chamber. In a second chamber the hot, dry grain is cooled prior to storage. Handling and storage equipment must be available at all times to move grain to and from the dryers. These dryers have crossflow, concurrent flow, or counterflow designs.

Crossflow Dryers. A crossflow dryer is a column dryer that moves air perpendicular to the grain movement. These dryers commonly consist of two or more vertical columns surrounding the drying and cooling air plenums. The columns range in thickness from 200 to 400 mm. Airflow rates range from 0.7 to 2.7 m^3/s per cubic metre of grain. The thermal efficiency of the drying process increases as column width increases and decreases as airflow rate increases. However, moisture uniformity and drying capacity increase as airflow rate increases and as column width decreases. Dryers are designed to obtain a desirable balance of airflow rate and column width for the expected moisture content levels and drying air temperatures. Performance is evaluated in terms of drying capacity, thermal efficiency, and dried product uniformity.

As with the batch dryer, a moisture gradient forms across the column because the grain nearest the inside of the column is exposed to the driest air during the complete cycle. Several methods minimize the problem of uneven drying. One method is to place in the columns turnflow devices that split the grain stream and move the inside half of the column to the outside and the outside half to the inside. Although effective, turnflow devices tend to plug if the grain is trashy. Under these conditions, a scalper/cleaner should be used to clean the grain before it enters the dryer. Another method is to divide the drying chamber into sections and duct the hot air so that its direction through the grain is reversed in alternate sections. This method produces about the same effect as the turnflow method. A third method is to divide the drying chamber into sections and reduce the drying air temperature in each section consecutively. This method is the least effective.

Rack-Type Dryers. In this special type of crossflow dryer, grain flows over alternating rows of heated air supply ducts and air exhaust ducts (Figure 2). This action mixes the grain and alternates exposure to relatively hot drying air and air cooled by previous contact with the grain, promoting moisture uniformity and equal exposure of the product to the drying air.

Concurrent-Flow Dryers. In the concurrent-flow dryer, grain and drying air move in the same direction in the drying chamber. The drying chamber is coupled to a counterflow cooling section. Thus, the hottest air is in contact with the wettest grain, allowing the use of higher drying air temperatures (up to 230°C). Rapid evaporative cooling in the wettest grain prevents the grain temperature from reaching excessive levels. Because higher drying air temperatures are used, the energy efficiency is better than that obtained with a conventional crossflow dryer. In the cooling section, the coolest air initially contacts the coolest grain. The combination of drying and cooling chambers results in lower thermal stresses in the grain kernels during drying and cooling, and thus a higher quality product.

Counterflow Dryers. The grain and drying air move in opposite directions in the drying chamber of this dryer. Counterflow is common for in-bin dryers. Drying air enters from the bottom of the bin and exits from the top. The wet grain is loaded from overhead, and floor sweep augers can be used to bring the hot, dry grain to a center sump, where it is removed by another auger. The travel of the sweep is normally controlled by temperature-sensing elements.

A drying zone exists only in the lower layers of the grain mass and is truncated at its lower edge so that the grain being removed is not overdried. As a part of the counterflow process, the warm, saturated or near-saturated air leaving the drying zone passes through the cool incoming grain. Some energy is used to heat the cool grain, but

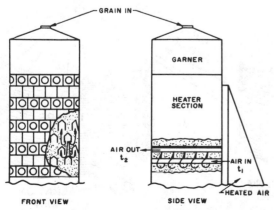

Fig. 2   Rack-Type Continuous-Flow Grain Dryer with Alternate Rows of Air Inlet and Outlet Ducts

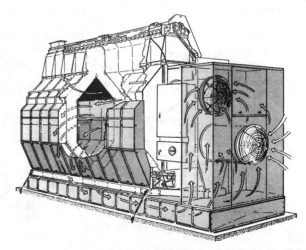

Fig. 3   Crop Dryer Recirculation Unit
(Courtesy Farm Fans, Inc.)

some moisture may condense on the cool grain if the bed is deep and the initial grain temperature is low.

Reducing Energy Costs

Recirculation. In most commercially available dryers there are optional ducting systems to recycle some of the exhaust air from the drying and cooling chambers back to the inlet of the drying chamber (Figure 3). Systems vary, but most make it possible to recirculate all of the air from the cooling chamber and the air from the lower two-thirds of the drying chamber. The relative humidity of this recirculated air for most crossflow dryers is less than 50%. Energy savings of up to 30% can be obtained in a well-designed system.

Dryeration. This is another means of reducing energy consumption and improving grain quality. In this process, hot grain with a moisture content one or two percentage points above that desired for storage is removed from the dryer (Figure 4). The hot grain is placed in a dryeration bin, where it tempers without airflow for at least 4 to 6 h. After the first grain delivered to the bin has tempered, the cooling fan is turned on as additional hot grain is delivered to the bin. The air cools the grain and removes 1 to 2% of its moisture before the grain is moved to final storage. If the cooling rate equals the filling rate, cooling is normally completed about 6 h after the last hot grain is added. The crop cooling rate should equal the filling rate of the dryeration bin. A faster cooling rate cools the grain before it has tempered. A slower rate may result in spoilage, since the allowable storage time for hot damp grain may be only a few days. The required airflow rate is based on dryer capacity and crop density. An airflow rate of 0.2 m³/s for each cubic metre per hour (m³/h) of dryer capacity provides the cooling capacity to keep up with the dryer when it is drying corn that has a mass of 900 kg/m³. Recommended airflow rates for some crops are listed in Table 4.

Combination Drying. This method was developed to improve drying thermal efficiency and corn quality. First, a continuous-flow dryer dries the corn to 18 to 20% moisture content. Then it is transferred to a bin, where the in-bin drying system brings the moisture down to a safe storage level. For energy savings, operating temperatures of batch and continuous-flow dryers are usually set at the highest level that will not damage the product for its particular end use. The energy requirements of a conventional crossflow dryer as a function of drying air temperature and airflow rate are shown in Figure 5.

DEEP BED DRYING

A deep bed drying system can be installed in any structure that holds grain. Most grain storage structures can be designed or adapted

Table 4   Recommended Airflow Rates for Dryeration

Crop	Density, kg/m³	Recommended Dryeration Airflow Rate, m³/s per m³/h of Dryer Capacity
Barley	768	0.170
Corn	896	0.200
Durum	960	0.210
Edible beans	960	0.210
Flaxseed	896	0.200
Millet	800	0.180
Oats	512	0.110
Rye	896	0.200
Sorghum	896	0.200
Soybean	960	0.210
Non-oil sunflower	384	0.090
Oil sunflower	512	0.110
HRS Wheat	960	0.210

Note: Basic air volume is 0.80 m³/kg.

for drying if a means of distributing the drying air uniformly through the grain is provided. A perforated floor (Figure 6) and duct systems placed on the floor of the bin (Figure 7) are the two most common means.

Perforations in the floor should have a total area of at least 10% of the floor area; 15% is better. A perforated floor distributes air more uniformly and offers less resistance to airflow than do ducts, but a duct system is less expensive for larger floor area systems. Ducts can be removed after the grain is removed, and the structure can be cleaned and used for other purposes. Ducts should not be spaced farther apart than one-half times the depth of the grain. The amount of perforated area or the duct length will affect airflow distribution uniformity.

Air ducts and tunnels that disperse air into the grain should be large enough to prevent the air velocity from exceeding 10 m/s; slower speeds are desirable. Sharp turns, obstructions, or abrupt changes in duct size should be eliminated as they cause pressure loss. Operating methods for drying grain in storage bins are full-bin drying, layer drying, and batch drying.

Full-Bin Drying

Full-bin drying is generally performed with unheated air or air heated up to 11 K above ambient. A humidistat is frequently used to sense the humidity of the drying air and turn off the heater if the weather conditions are such that heated air would cause overdrying. A humidistat setting of 55% stops drying at approximately the 12% moisture level for most farm grains, assuming that the ambient relative humidity does not go below this point.

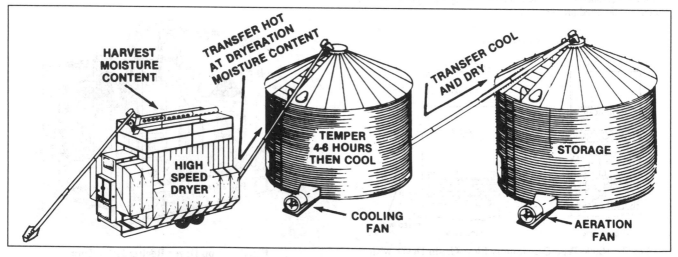

Fig. 4   Dryeration System Schematic

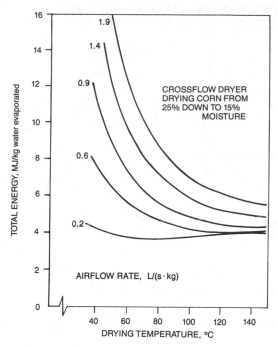

Fig. 5   Energy Requirements of a Conventional Crossflow Dryer as a Function of Drying Air Temperature and Airflow Rate
(University of Nebraska)

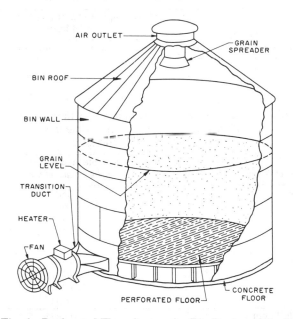

Fig. 6   Perforated Floor System for Bin Drying of Grain

Airflow rate requirements for full-bin drying are generally calculated on the basis of m^3/s of air required per cubic metre of grain. The airflow rate recommendations depend on the weather conditions and on the type of grain and its moisture content. Airflow rate is important for successful drying. Because faster drying results from higher airflow rates, the highest economical airflow rate should be used. However, the cost of full-bin drying at high airflow rates may exceed the cost of using column dryers.

Recommendations for full-bin drying with unheated air are shown in Tables 5, 6, and 7. These recommendations apply to the principal production areas of the continental United States and are based on experience under average conditions; they may not be

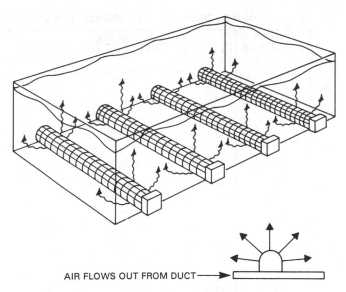

AIR FLOWS OUT FROM DUCT ➔

Fig. 7   Tunnel or Duct Air Distribution System

applicable under unusual weather conditions or even usual weather conditions in the case of late maturing crops. Much of the weather hazard can be eliminated by the use of supplemental heat. Full-bin drying may not be feasible in some geographical areas.

The maximum practical depth of grain to be dried (distance of air travel) is limited by the cost of the fan, motor, air distribution system, and power required. This depth seems to be 6 m for corn and beans, and about 4.5 m for wheat.

To ensure satisfactory drying, heated air may be used during periods of prolonged fog or rain. Burners should be sized to raise the temperature of the drying air by no more than 11 K. The temperature should not exceed about 27°C after heating. Overheating the drying air causes the grain to overdry and dry nonuniformly; heat is recommended only to counteract adverse weather conditions. Electric controllers are available to assist in controlling fan and heater operation to achieve the final desired grain moisture content.

Drying takes place in a drying zone, which advances upward through the grain (Figure 8). Grain above this drying zone remains at or slightly above the initial moisture content, while grain below the drying zone is at a moisture content in equilibrium with the drying air. The equilibrium moisture content of a material is the moisture content it approaches when exposed to air at a specified relative humidity.

As the direction of air movement does not affect the rate of drying, other factors must be considered in choosing the direction. A pressure system moves the moisture-laden air up through the grain, and it is discharged under the roof. If there are insufficient roof outlets, moisture may condense on the underside of metal roofs. During pressure system ventilation, the wettest grain is near the top surface and is easy to monitor. Fan and motor waste heat enter into the airstream and contribute to drying.

A suction system moves the air down through the grain. The moisture-laden air discharges from the fan into the outside atmosphere; thus, roof condensation is not a problem. However, the wettest grain is near the bottom of the mass and is difficult to sample. Of the two systems, the pressure system is recommended because it is easier to manage.

The following management practices must be observed to ensure the best performance of the dryer:

1. Minimize foreign material. A scalper-cleaner is recommended for cleaning the grain to reduce air pressure and energy requirements and to help provide uniform airflow for the elimination of wet spots.
2. Distribute the remaining foreign material uniformly by installing a grain distributor on the end of the storage filling chute.

Table 5   Maximum Corn Moisture Contents, Wet Mass Basis, for Single-Fill Unheated Air Drying

Zone	Full-Bin Airflow Rate, m³/s per m³ of Grain	9-1	9-15	10-1	10-15	11-1	11-15	12-1
		Harvest Date — Initial Moisture Content, %						
A	0.013	18	19.5	21	22	24	20	18
	0.017	20	20.5	21.5	23	24.5	20.5	18
	0.020	20	20.5	22.5	23	25	21	18
	0.027	20.5	21	23	24	25.5	21.5	18
	0.040	22	22.5	24	25.5	27	22	18
B	0.013	19	20	20	21	23	20	18
	0.017	19	20	20.5	21.5	24	20.5	18
	0.020	19.5	20.5	21	22.5	24	21	18
	0.027	20	21	22.5	23.5	25	21.5	18
	0.040	21	22.5	23.5	24.5	26	22	18
C	0.013	19	19.5	20	21	22	20	18
	0.017	19	20	20.5	21.5	22.5	20.5	18
	0.020	19.5	20	21	22	23.5	21.5	18
	0.027	20	21	22	23	24.5	21.5	18
	0.040	21	22	23.5	24.5	25.5	22	18
D	0.013	19	19.5	20	21	22	20	18
	0.017	19	19.5	20.5	21	22.5	20.5	18
	0.020	19	19.5	21	22	23	21	18
	0.027	19.5	21	21.5	23	24	21.5	18
	0.040	20.5	21.5	23	24	25	22	18

Developed by T.L. Thompson, University of Nebraska.

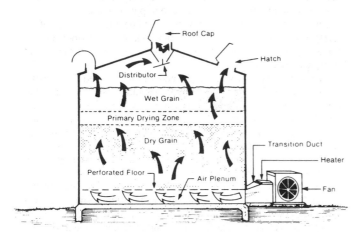

Fig. 8   Three Zones Within Grain During Full-Bin Drying

3. Place the grain in layers and keep it leveled.

4. Start the fan as soon as the floor or ducts are covered with grain.

5. Operate the fan continuously unless it is raining heavily or there is a dense ground fog. Once all the grain is within 1% of storage moisture content, run the fans only when the relative humidity is below 70%.

Layer Drying

In layer drying, successive layers of wet grain are placed on top of dry grain. When the top 150 mm has dried to within 1% of the desired moisture content, another layer is added (Figure 9). In comparison to full-bin drying, layering reduces the time that the top layers of grain are left wet. Because the effective airflow rate is greater for lower layers, allowable harvest moisture content of grain in these levels can be greater than that in the upper levels. Either unheated air or air heated 6 to 11 K above the ambient may be used, but the use of heated air controlled with a humidistat to prevent overdrying is most common. The first layer may be about 2 m in depth, with successive layers of about 1 m.

Batch-in-Bin. A storage bin adapted for drying may be used to dry several batches of grain during a harvest season, if the grain is

Table 6   Minimum Airflow Rates for Unheated Air Low-Temperature Drying of Small Grains and Sunflower in the Northern Plains of the United States

Airflow Rate, m³/s per m³ of Grain	Maximum Initial Moisture Content, % Wet Basis	
	Small Grains	Sunflower
0.007	16	15
0.013	18	17
0.027	20	21

Table 7   Recommended Unheated Air Airflow Rates for Different Grains and Moisture Contents in the Southern United States

Type of Grain	Grain Moisture Content, %	Recommended Airflow Rate, m³/s per m³ of Grain
Wheat	25	0.0800
	22	0.0667
	20	0.0400
	18	0.0267
	16	0.0133
Oats	25	0.0400
	20	0.0267
	18	0.0200
	16	0.0133
Shelled Corn	25	0.0667
	20	0.0400
	18	0.0267
	16	0.0133
Ear Corn	25	0.1067
	18	0.0533
Grain Sorghum	25	0.0800
	22	0.0667
	18	0.0400
	15	0.0267
Soybeans	25	0.0800
	22	0.0667
	18	0.0400
	15	0.0267

Compiled from USDA Leaflet 332 (1952) and Univ. of Georgia *Bulletin* NS 33 (1958).

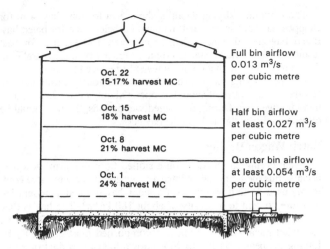

Fig. 9   Example of Layer Filling of Corn

kept to a shallow layer so that higher airflow rates and temperatures can be used. After the batch is dry, the bin is emptied, usually by a system of augers, and the cycle is repeated. The drying capacity of the batch system is greater than that of other in-storage drying systems. In a typical operation, batches of corn in 1-m depths are dried from an initial moisture content of 25% with 55°C air at the rate of about 0.33 m³/s per cubic metre. Considerable nonuniformity of moisture content may be present in the batch after drying is stopped; therefore, the grain should be well mixed as it is placed into storage. If the mixing is done well, grain that is too wet equalizes in moisture with grain that is too dry before spoilage can occur.

Grain may be cooled in the dryer to ambient temperature before it is stored. Cooling is accomplished by operating the fan without the heater for about 1 h. Some additional drying occurs during the cooling process, particularly in the wetter portions of the batch.

Grain stirring devices are used with both full-bin and batch-in-bin drying systems. Typically, these devices consist of one or more open, 50-mm diameter, standard pitch augers suspended from the

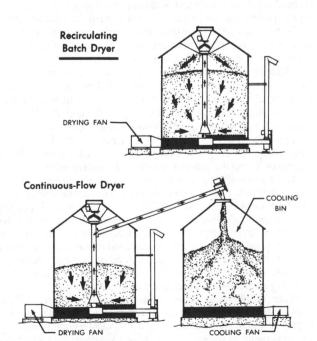

Fig. 10   Grain Recirculators Convert Bin Dryer to High-Speed Continuous Flow Dryer

bin roof and extending to near the bin floor. The augers rotate and simultaneously travel horizontally around the bin, mixing the drying grain to reduce moisture gradients and prevent overdrying of the bottom grain. The augers also loosen the grain, allowing a higher airflow rate for a given fan. Stirring equipment reduces bin capacity by about 10% and may cost as much as, and do no more than, a larger dryer fan. Furthermore, commercial stirring devices are available only for round storage enclosures. The improper use of stirring devices can result in increased grain breakage.

Recirculating/Continuous Flow Bin Dryer. This type of dryer incorporates a tapered sweep auger that removes grain from the bottom of the bin as it dries (Figure 10). The dry grain is then redistributed on top of the pile of grain or moved to a second bin for cooling. The sweep auger may be controlled by temperature or moisture sensors. When the desired condition is reached, the sensor starts the sweep auger, which removes a layer of grain. After a complete circuit of the bin, the sweep auger stops until the sensor determines that another layer is dry. Some drying takes place in the cooling bin. Up to two percentage points of moisture may be removed, depending on the management of the cooling bin.

DRYING SPECIFIC CROPS

SOYBEANS

Soybeans usually need drying only when there is inclement weather during the harvest season. Mature soybeans left exposed to rain or damp weather develop a dark brown color and a mealy or chalky texture. Seed quality deteriorates rapidly. Oil from weather-damaged beans costs more to refine and is often not of edible grade. In addition to preventing deterioration, the artificial drying of soybeans offers the advantage of early harvest, which reduces the chance of loss from bad weather and reduces natural and combine shatter loss. Soybeans harvested with a wet basis moisture content greater than 13.5% exhibit less damage.

Drying Soybeans for Commercial Use

Conventional corn-drying equipment can be used for soybeans, with some limitations on heat input. Soybeans for commercial use can be dried at 54 to 60°C; drying temperatures of 88°C reduce the oil yield. If the relative humidity of the drying air is below 40%, excessive seedcoat cracking occurs, causing many split beans in subsequent handling. Physical damage can cause fungal growth on the beans, storage problems, and a slight reduction in oil yield and quality. Flow-retarding devices should be used, and beans should not be dropped more than 6 m onto concrete floors.

Drying Soybeans for Seed and Food

The relative humidity of the drying air should be kept above 40%, regardless of the amount of heat used. The maximum drying temperature to avoid germination loss is 43°C. Natural air drying at a flow rate of 0.0267 m³/s per cubic metre is adequate for drying seed with an initial moisture content of up to 16%.

Low-temperature drying in bins is another method used in the midwestern United States. The ambient temperature is raised no more than 3 K. This drying method is slow, but it results in excellent quality and avoids overdrying. However, drying must be completed before spoilage occurs. At an airflow rate of 0.0267 m³/s per cubic metre, soybeans usually take about five days to dry from 16% to 13%. At higher moisture contents, good results have been obtained using an airflow rate of 0.0534 m³/s per cubic metre with humidity control. Data on the allowable drying time for soybeans are unavailable. In the absence of better information, an estimate of storage life for oil crops can be made based on the values for corn, using an adjusted moisture content calculated by the following equation:

$$\text{Comparable moisture content} = \frac{\text{Oil seed moisture content}}{100 - \text{Seed oil content}} \times 100$$

A corn moisture content 2% greater than that of the soybeans should generally be used to estimate allowable drying time; i.e., 12% soybeans are comparable to 14% corn. Soybeans are dried from a lower initial moisture content than corn.

Soybean seed technologists suggest drying high-moisture soybeans in a bin with the air temperature controlled to keep the relative humidity at 40% or higher. Airflow rates of 0.13 m³/s per cubic metre are recommended, with the depth of the beans not to exceed 1.2 m.

HAY

Hay normally contains 65 to 80% wet basis moisture at cutting. Field drying to 20% may result in a large loss of leaves. Alfalfa hay leaves average about 50% of the crop by mass, but they contain 70% of the protein and 90% of the carotene. The quality of hay can be increased and the risk of loss due to bad weather reduced if the hay is put under shelter when partially sun dried (35% moisture content) and then artificially dried to a safe storage moisture content (about 20%). In good drying weather, hay conditioned by mechanical means can be dried sufficiently in one day and placed in the dryer. Hay may be long, chopped, or baled for this operation; unheated or heated air can be used.

In-Storage Drying

Unheated air is normally used for in-storage or mow drying. The hay is dried in the field to 30 to 40% moisture content before being placed in the dryer. For unheated air drying, the airflow should be at least 0.10 m³/s per tonne. The fan should be capable of delivering the required airflow against a static pressure of 250 to 500 Pa.

Slotted floors, with at least 50% of the area open, are generally used for drying baled hay. For long or chopped hay in mows less than 11 m wide, the center duct system is the most popular. A slotted floor should be placed on each side of the duct to within 1.5 m of its ends and the outside walls (Figure 11). If the width is greater than 11 m, the mow should be divided crosswise into units of 8.5 m or less. These should then be treated as individual dryers. If the storage depth exceeds about 4 m, vertical flues and/or additional levels of ducts may be used. If tiered ducts are used, a vertical air chamber, about 75% of the probable hay depth, should be used. The supply ducts are then connected at 2- to 3-m vertical intervals as the mow is filled. With either of these methods, hay in total depths up to 9 m can be dried. The duct size should be such that the air velocity is less than 5 m/s. The maximum depth of wet hay that should be placed on a hay-drying system at any time depends on hay moisture content, weather conditions, the physical form of the hay, and the airflow rate.

The maximum drying depth is about 5 m for long hay, 4 m for chopped hay, and seven small rectangular bales deep for baled hay. Baled hay should have a density of about 130 kg/m³. For best results, bales should be stacked tightly together on edge (parallel to the stems) to ensure that no openings exist between them.

For mow drying, the fan should run continuously during the first few days. Afterwards, it should be operated only during low relative humidity weather. During prolonged wet periods, the fan should be operated only enough to keep the hay cool.

Batch Wagon Drying

Batch drying can be done on a slotted floor platform; however, because this method is labor-intensive, wagon dryers are more commonly used. With a wagon dryer system, hay is baled at about 45% moisture content to a density of about 180 kg/m³. The hay is then stacked onto a wagon with tight, high sides and a slotted or expanded metal floor. Drying is accomplished most efficiently by forcing the heated air (up to 70°C) down the canvas duct of a plenum chamber secured to the top of the wagon. After 4 or 5 h of drying, the exhaust air is no longer saturated with moisture, and about 75% of it may be recirculated or passed through a second wagon of wet hay for greater drying efficiency.

With this method, the amount of hay harvested each day is limited by the capacity of the drying wagons. In this 24-h process, the hay cut one day is stored the following day; only enough hay to load the drying wagons should be harvested each day.

The airflow rate with this method is normally much higher than when unheated air is used. About 0.2 m³/s per square metre of wagon floor space is required. As with mow drying, the duct size should be such that the air velocity is less than 5 m/s.

COTTON

Producers normally allow cotton to dry naturally in the field to 12% moisture content or less before harvest. Cotton harvested under these conditions can be stored in trailers, baskets, or compacted stacks for extended periods with little loss in fiber or seed quality. Thus, cotton is not normally aerated or artificially dried prior to ginning. Cotton harvested during inclement weather and stored cotton exposed to rain or snow must be dried at the cotton gin within a few days to prevent self-heating and deterioration of the fiber and seed.

Even though cotton may be safely stored at moisture contents as high as 12%, moisture levels near the upper limit are too high for efficient ginning and for obtaining optimum fiber grade. The cleaning efficiency of cotton is inversely proportional to its moisture content, with the most efficient level being 5% fiber moisture content. However, fiber quality is best preserved when the fiber is separated from the seed at moisture contents between 6.5 and 8%. Therefore, if cotton comes into the system below this level, it can be cleaned, but moisture should be added prior to separating the fiber from the seed to improve the ginning quality. Dryers in the cotton gins are capable of drying the cotton to the desired moisture level.

Although several types of dryers are available commercially, the tower dryer is the most commonly used. This device operates on a parallel flow principle: 0.015 to 0.025 m³/s of drying air per kilogram of cotton also serves as the conveying medium. As it moves through the dryer's serpentine passages, cotton impacts on the walls. This action agitates the cotton for improved drying and lengthens its exposure time. The drying time depends on many variables, but total exposure seldom exceeds 15 s. For extremely wet cotton, two stages of drying are necessary for adequate moisture control.

Wide variations in initial moisture content dictate different amounts of drying for each load of cotton. Rapid changes in drying requirements are accommodated by automatically controlling drying air temperature in response to moisture measurements taken before or after drying. These control systems prevent overdrying and reduce energy requirements. For safety and to preserve fiber quality,

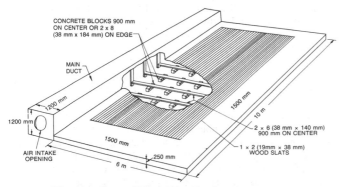

Fig. 11 Central Duct Hay-Drying System with Lateral Slotted Floor for Wide Mows

drying air temperature should not exceed 177°C in any portion of the drying system.

If the internal cottonseed temperature does not exceed 60°C, germination is unimpaired by drying. This temperature is not exceeded in a tower dryer; however, the moisture content of the seed after drying may be above the 12% level recommended for safe long-term storage. Wet cottonseed is normally processed immediately at a cottonseed oil mill. Cottonseed under the 12% level is frequently stored for several months prior to milling or prior to delinting and treatment at a planting seed processing plant. The aeration that cools deep beds of stored cottonseed effectively maintains viability and prevents an increase in free fatty acid content. For aeration, ambient air is normally drawn downward through the bed at a minimum rate of 0.0004 m³/s per cubic metre of oil mill seed and 0.0021 m³/s per cubic metre of planting seed.

PEANUTS

Peanuts normally have a moisture content of about 50% at the time of digging. Allowing the peanuts to dry on the vines in the windrow for a few days removes much of this water. However, peanuts usually contain 20 to 30% moisture when removed from the vines, and some artificial drying is necessary. Drying should begin within 6 h after harvesting to keep the peanuts from self-heating. Both the maximum temperature and the rate of drying must be carefully controlled to maintain quality.

High temperatures result in an off flavor or bitterness. Drying too rapidly without high temperatures results in blandness or nuts that do not develop flavor on roasting. High temperatures, rapid drying, or excessive drying cause the skin to slip easily and the kernels to become brittle. These conditions result in high damage rates in the shelling operation but can be avoided if the moisture removal rate does not exceed 0.5% per hour. Because of these limitations, continuous flow drying is not usually recommended for peanuts.

Peanuts can be dried in bulk bins using unheated air or air with supplemental heat. Under poor drying conditions, unheated air may cause spoilage, so supplemental heat is preferred. Air should be heated no more than 7 or 8 K to a maximum temperature of 35°C. An airflow rate of 0.050 to 0.130 m³/s per cubic metre of peanuts should be used, depending on the initial moisture content.

The most common method of drying peanuts is bulk wagon drying. Peanuts are dried in depths of 1.5 to 1.8 m, using airflow rates of 0.050 to 0.075 m³/s per cubic metre of peanuts and air heated 6 to 8 K above ambient. This method retains quality and usually dries the peanuts in three to four days. Wagon drying reduces handling labor but may require additional investment in equipment.

RICE

Of all grains, rice is probably the most difficult to process without quality loss. Rice containing more than 13.5% moisture cannot be safely stored for long periods, yet the recommended harvest moisture content for best milling and germination ranges from 20 to 26%. When rice is harvested at this moisture content, drying must be started promptly to prevent souring. Normally, heated air is used in continuous flow dryers, where large volumes of air are forced through 100- to 250-mm layers of rice. Temperatures as high as 55°C may be used, if (1) the temperature drop across the rice does not exceed 11 to 17 K, (2) the moisture reduction does not exceed two percentage points in a 0.5-h exposure, and (3) the rice temperature does not exceed 38°C. During the tempering period following drying, the rice should be aerated to ambient temperature before the next pass through the dryer. This removes additional moisture and eliminates one to two dryer passes. It is estimated that full use of aeration following dryer passes could increase the maximum daily drying capacity by about 14%.

Unheated air or air with a small amount of added heat (7 K above ambient, but not exceeding 35°C) should be used for deep-bed rice drying. Too much heat overdries the bottom, resulting in checking (cracking), reduced milling qualities, and possible spoilage in the top. Because unheated air drying requires less investment and attention than supplemental heat drying, it is preferred when conditions permit. In the more humid rice-growing areas, supplemental heat is desirable to ensure that the rice dries. The time required for drying varies with weather conditions, moisture content, and airflow rate. In California, the recommended airflow rate is 0.001 to 0.012 m³/s per cubic metre. Because of less favorable drying conditions in Arkansas, Louisiana, and Texas, greater airflow rates are recommended: e.g., a minimum of 0.010 m³/s per cubic metre is recommended in Texas. Whether unheated air or supplemental heat is used, the fan should be turned on as soon as rice uniformly covers the air distribution system. The fan should then run continuously until the moisture content in the top 300 mm of rice is reduced to about 15%. At this point, the supplemental heat should be turned off. The rice can then be dried to a safe storage level by operating the fan only when the relative humidity is below 75%.

STORAGE PROBLEMS AND PRACTICES

MOISTURE MIGRATION

Redistribution of moisture generally occurs in stored grain when grain temperature is not controlled (Figure 12). Localized spoilage can occur even when the grain is stored at a safe moisture level. Grain placed in storage in the fall at relatively high temperatures cools nonuniformly through contact with the outside surfaces of the storage bin as winter approaches. Thus, the grain near the outside walls and roof may be at cool outdoor temperatures while the grain nearer the center is still nearly the same temperature it was at harvest. These temperature differentials induce air convection currents that flow downward along the outside boundaries of the porous grain mass and upward through the center. When the cool air from the outer regions contacts the warm grain in the interior, the air is heated and its relative humidity is lowered, increasing its capacity to absorb moisture from the grain. When the warm, humid air reaches the cool grain near the top of the bin, it cools again and transfers vapor to the grain. Under extreme conditions, water condenses on the grain. The moisture concentration near the center of the grain surface causes significant spoilage if moisture migration is uncontrolled. During spring and summer, the temperature gradients are reversed. The grain moisture content increases most at depths of 0.6 to 1.2 m below the surface. Daily variations in temperature do not cause significant moisture migration. Aside from seasonal temperature variations, the size of the grain mass is the most important factor in fall and winter moisture migration. In storages containing less than 35 m³, there is less trouble with moisture migration. The problem becomes critical in large storages and is aggravated by incomplete

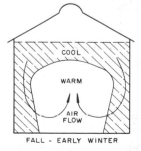
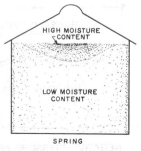

Fig. 12   Grain Storage Conditions Associated with Moisture Migration During Fall and Early Winter

cooling of artificially dried grain. Artificially dried grain should be cooled to near ambient temperature when dried.

GRAIN AERATION

Aeration by mechanically circulating ambient air through the grain mass is the best way to control moisture migration. Aeration systems are also used to cool grain after harvest, particularly in warmer climates where grain may be placed in storage at temperatures exceeding 38°C. After the harvest heat is removed, aeration may be continued in cooler weather to bring the grain to a temperature within 11 K of the coldest average monthly temperature. The temperature must be maintained at less than 10°C.

Aeration systems are not a means of drying because airflow rates are too low. However, in areas where the climate is favorable, carefully controlled aeration may be used to remove small amounts of moisture. Commercial storages may have pockets of higher moisture grain if, for example, some batches of grain are delivered after a rain shower or early in the morning. Aeration can control heating damage in the higher moisture pockets.

Aeration Systems Design

Aeration systems include fans capable of delivering the required amount of air at the required static pressure, suitable ducts to distribute the air into the grain, and controls to regulate the operation of the fan. The airflow rate determines how many hours are required to cool the crop (Table 8). Most aeration systems are designed with airflow rates between 0.0007 to 0.0027 m^3/s per cubic metre of grain.

Stored grain is aerated by forcing air up or down through the grain. Moving air up through the grain is more common because it is easier to observe when the cooling front has moved through the entire grain mass. Also, with upward airflow there is little hazard of vent screens freezing over and fan suction collapsing the bin roof. In large, flat storages with long ducts, upward airflow results in more uniform air distribution than downdraft systems.

During aeration, a warming or cooling front moves through the crop (Figure 13); it is important to run the fan long enough to move the front completely through the crop.

In tall tower storages, airflow rates should be limited to avoid excessive power requirements caused by the long flow path through the grain. Crossflow aeration reduces the length of the flow path, resulting in reduced system static pressure for a given aeration rate (Figure 14). Crossflow systems, however, are less easily adapted to aeration of partially filled bins than up- or downflow systems, and installation costs are higher.

Static pressure for an aeration system can be determined using the airflow resistance information in Chapter 10 of the 1993 *ASHRAE Handbook—Fundamentals*. All common types of fans are used in

aeration systems. Attention should be given to noise levels with fans that are operated near residential areas or where people work for extended periods. The supply ducts connecting the fan to the distribution ducts in the grain should be designed and constructed according to the standards of good practice for any air-moving application. A maximum air velocity of 13 m/s may be used, but 8 to 10 m/s is preferred. In large systems, one large fan may be attached to a manifold duct leading to several distribution ducts in one or more storages, or smaller individual fans may serve individual distribution ducts.

Table 8 Airflow Rates Corresponding to Approximate Grain Cooling Time

Airflow Rate, m^3/s per m^3 of Grain	Cooling Time, h
0.0007	240
0.0013	120
0.0027	60
0.0040	40
0.0054	30
0.0067	24
0.0080	20
0.0107	15
0.0134	12

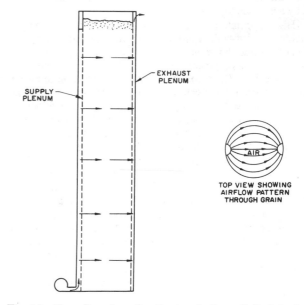

Fig. 14 Crossflow Aeration System in Deep Cylindrical

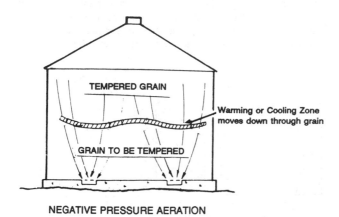

NEGATIVE PRESSURE AERATION

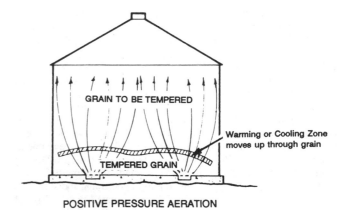

POSITIVE PRESSURE AERATION

Fig. 13 Aerating to Change Grain Temperature

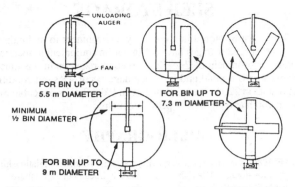

Fig. 15 Common Duct Patterns for Round Grain Bins

Where a manifold is used, valves or dampers should be installed at each takeoff to allow adjustment or closure of airflow when part of the system is not needed.

Distribution ducts are usually perforated sheet metal with a circular or inverted U-shaped cross section, although many functional arrangements are possible. The area of the perforations should be at least 10% of the total duct surface. The holes should be uniformly spaced and small enough to prevent the passage of the grain into the duct. For example, 2.5-mm holes or 2-mm wide slots do not pass wheat.

Since most problems develop in the center of the storage, and the crop cools naturally near the wall, the aeration system must provide good airflow in the center. If ducts placed directly on the floor are to be held in place by the crop, the crop flow should be directly on top of the ducts to prevent movement and damage. Flush floor systems work well in storages with sweep augers and unloading equipment. Ducts should be easily removable for cleaning. Duct spacing should not exceed the depth of the crop; the distance between the duct and storage structure wall should not exceed one-half the depth of the crop for bins and flat storages. Common duct patterns for round bins are shown in Figure 15. Duct spacing for flat storages is shown in Figure 16.

When designing the distribution duct system for any type of storage, the following should be considered: (1) the cross-sectional area and length of the duct, which influences both the air velocity within the duct and the uniformity of air distribution; (2) the duct surface area, which affects the static pressure losses in the grain surrounding the duct; and (3) the distance between ducts, which influences the uniformity of airflow.

For upright storages where the distribution ducts are relatively short, distribution duct velocities up to 10 m/s are permissible. Maximum recommended air velocities in ducts for flat storages are shown in Table 9. Furthermore, these velocities should not be exceeded in the air outlets from the storage; therefore, an air outlet area at least equal to the duct cross-sectional area should be provided.

Table 9 Maximum Recommended Air Velocities Within Ducts for Flat Storages

Grain	Airflow Rate, m³/s per cubic metre of Grain	Air Velocity (m/s) Within Ducts for Grain Depths of:				
		3 m	6 m	9 m	12 m	15 m
Corn, soybeans, and other large grains	0.00067	—	3.8	5.0	6.3	6.3
	0.00134	3.8	5.0	6.3	7.6	8.8
	0.00268	5.0	6.3	—	—	—
Wheat, grain sorghum, and other small grains	0.00067	—	5.0	7.6	8.8	10.0
	0.00134	3.8	7.6	10.0	—	—
	0.00268	5.0	10.0	—	—	—

BUILDINGS UP TO 18 m TO 36 m WIDE

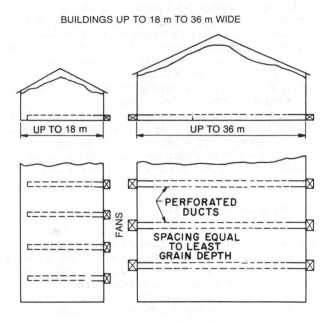

BUILDINGS UP TO 12 m WIDE AND 30 m LONG

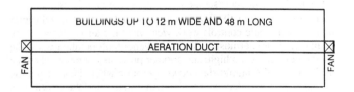

GRAIN PEAKED WITH SHALLOW DEPTH NEAR WALL

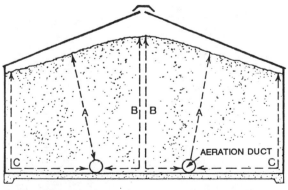

(A) IS SHORTEST AIR PATH
MAKE (B) OR (C) NO MORE THAN 1.5 TIMES (A)

Fig. 16 Duct Arrangements for Large Flat Storages

The duct surface area that is perforated or otherwise open for air distribution must be great enough that the air velocity through the grain surrounding the duct is low enough to avoid excessive pressure loss. When a semicircular perforated duct is used, the entire surface area is effective; only 80% of the area of a circular duct resting on the floor is effective. For upright storages, the air velocity through the grain near the duct (duct face velocity) should be limited to 0.15 m/s or less; in flat storages, to 0.10 m/s or less.

Duct strength and anchoring are important. Distribution ducts buried in the grain must be strong enough to withstand the pressure the grain exerts on them. In tall, upright storages, static grain pressures may reach 70 kPa. When ducts are located in the path of the grain flow, as in a hopper, they may be subjected to many times this pressure during grain unloading.

Operating Aeration Systems

The operation of aeration systems depends largely on the objectives to be attained and the locality. In general, cooling should be carried out any time the outdoor air temperature is about 8 K cooler than the grain. Stored grain should not be aerated when the air humidity is much above the equilibrium humidity of the grain because moisture will be added. The fan should be operated long enough to cool the crop completely, but it should then be shut off and covered, thus limiting the amount of grain that is rewetted.

Aeration to cool the grain should be started as soon as the storage is filled. Aeration to prevent moisture migration should be started whenever the average air temperature is 6 to 8 K below the highest grain temperature. Aeration is usually continued as weather permits until the grain is uniformly cooled to within 11 K of the average temperature of the coldest month or to 0 to 4°C.

Grain temperatures of about 0 to 10°C are desirable. In the northern corn belt, aeration may be resumed in the spring to equalize the grain temperature and raise it to between 5 to 10°C. This reduces the risk of localized heating from moisture migration. Storage problems are the only reason to aerate when air temperatures are above 15°C. Aeration fans and ducts should be covered when not in use.

In storages where the fans are operated daily in the fall and winter months, automatic controls work well when the air is not too warm or humid. One thermostat usually prevents fan operation when the air temperature is too high, and another prevents operation when the air is too cold. A humidistat allows operation when the air is not too humid. Fan controllers are available that determine the equilibrium moisture content of the crop based on existing air conditions and regulate the fan based on entered information.

SEED STORAGE

Seed must be stored in a cool, dry environment to maintain viability. Most seed storages have refrigeration equipment to maintain a storage environment of 7 to 12°C. Seed storage conditions must be achieved before mold and insect damage occur. Desired conditions can be met in 220 h at an airflow rate of 0.0007 m³/s per cubic metre and in 140 h at 0.0013 m³/s per cubic metre.

BIBLIOGRAPHY

ASAE. 1989a. Density, specific gravity, and weight-moisture relationships of grain for storage. ASAE D241.2. American Society of Agricultural Engineers, St. Joseph, MI.

ASAE. 1989b. Moisture relationships of grain. ASAE D245.4. American Society of Agricultural Engineers, St. Joseph, MI.

ASAE. 1989c. Resistance of airflow through grains, seeds, and perforated metal sheets. ASAE D272.1. American Society of Agricultural Engineers, St. Joseph, MI.

Brooker, D.B., F. Bakker-Arkema, and C.A. Hall. 1974. *Drying cereal grains.* AVI Publishing, Westport, CT.

Cotton ginners handbook. 1977. Agricultural handbook No. 503, Agricultural Research Service, U.S. Department of Agriculture, Washington, D.C.

Midwest Plan Service. 1988. *Grain drying, handling and storage handbook.* MWPS-13. Iowa State University, Ames, IA.

Midwest Plan Service. 1980. *Low temperature and solar grain drying handbook.* MWPS-22. Iowa State University, Ames, IA.

Midwest Plan Service. 1980. *Managing dry grain in storage.* AED-20. Iowa State University, Ames, IA.

Foster, G.H. 1982. *Storage of cereal grains and their products*, 3rd ed. American Association of Cereal Chemists, St. Paul, MN.

Hall, C.A. 1980. *Drying and storage of agricultural crops.* AVI Publishing, Westport, CT.

Hellevang, K.J. 1989. *Crop storage management.* AE-791. NDSU Extension Service, North Dakota State University, Fargo, ND.

Hellevang, K.J. 1987. *Grain drying.* AE-701. NDSU Extension Service, North Dakota State University, Fargo, ND.

Hellevang, K.J. 1983. *Natural air/Low temperature crop drying.* EB-35. NDSU Extension Service, North Dakota State University, Fargo, ND.

Henderson, S.M. and R.L. Perry. 1976. *Agricultural process engineering.* AVI Publishing, Westport, CT.

Schuler, R.T., B.J. Holmes, R.J. Straub, and D.A. Rohweder. 1986. *Hay drying.* A3380. University of Wisconsin-Extension, Madison, WI.

CHAPTER 22

AIR CONDITIONING OF WOOD AND PAPER PRODUCTS FACILITIES

THIS chapter covers some of the standard requirements for air conditioning facilities that manufacture finished wood products such as pulp and paper.

GENERAL WOOD PRODUCT OPERATIONS

Finished lumber products to be used in heated buildings should be stored in areas that are heated 6 to 11°C above ambient. This provides sufficient protection for furniture stock, interior trim, cabinet material, and stock for products such as ax handles and glue-laminated beams. Air should be circulated within the storage areas. Lumber that is kiln-dried to a moisture content of 12% or less can be kept within a given moisture content range through storage in a heated shed. The moisture content can be regulated either manually or automatically by altering the dry-bulb temperature (Figure 1).

Some special materials require close control of moisture content. For example, musical instrument stock must be dried to a given moisture level and maintained there because the moisture content of the wood affects the harmonics of most stringed wooden instru-

The preparation of this chapter is assigned to TC 9.2, Industrial Air Conditioning.

ments. This degree of control requires an air-conditioning system with reheat and a heating system with humidification.

Process Area Air Conditioning

Temperature and humidity requirements within wood product process areas vary according to product, manufacturer, and governing code. For example, in match manufacturing, the match head must be cured (i.e., dried) after dipping. This requires careful control of the humidity and temperature to avoid temperatures near the ignition point. Any process involving the application of flammable substances should follow the ventilation recommendations of the National Fire Protection Association, the National Fire Code, and the U.S. Occupational Safety and Health Act.

Finished Product Storage

Finished lumber products manufactured from predried stock (moisture content of 10% or less) regain moisture if exposed to high relative humidities for an extended period. All storage areas housing finished products such as furniture and musical instruments should be conditioned to avoid moisture regain.

Designers should be familiar with the client's entire operation and should be aware of the potential for moisture regain.

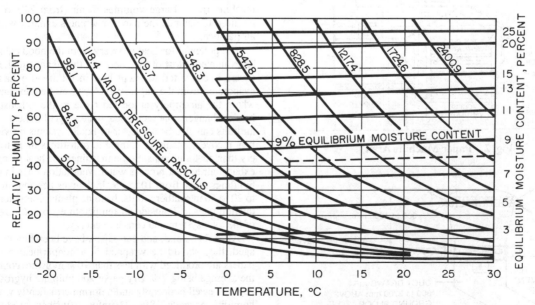

Fig. 1   Relationship Between the Temperature, Relative Humidity, and Vapor Pressure of Air and the Equilibrium Moisture Content of Wood

PULP AND PAPER OPERATIONS

The papermaking process comprises two basic steps: (1) wood is reduced to pulp, that is, wood fibers; and (2) the pulp is converted to paper. Wood can be pulped by either mechanical action (e.g., grinding in a groundwood mill), chemical action (e.g., kraft pulping), or a combination of both.

Many different types of paper can be produced from pulp, ranging from the finest glossy finish to newsprint to bleached board to fluff pulp for disposable diapers. To make newsprint, a mixture of mechanical and chemical pulps is fed into the paper machine. To make kraft paper (e.g., grocery bags and corrugated containers), however, only unbleached chemical pulp is used. Disposable diaper material and photographic paper require bleached chemical pulp with a very low moisture content of 6 to 9%.

Paper Machine Area

In papermaking, extensive air systems are required to support and enhance the process (e.g., by preventing condensation) and to provide reasonable comfort for machine personnel. Radiant heat from steam and hot water sources and mechanical energy dissipated as heat can result in summer temperatures in the machine room as high as 40 to 50°C. In addition, high paper machine operating speeds of 600 to 1200 m/min and stock temperatures near 50°C produce warm vapor in the machine room.

Outside air makeup units absorb and remove water vapor released from the paper as it is dried (Figures 2 and 3). The makeup air is distributed to the working areas above and below the operating floor. Part of the air delivered to the basement migrates to the operating floor through hatches and stairwells. Motor cooling systems distribute cooler basement air to the paper machine dc drive motors. The intake to the personnel coolers should be located in the basement below the warm stratified air that accumulates under the

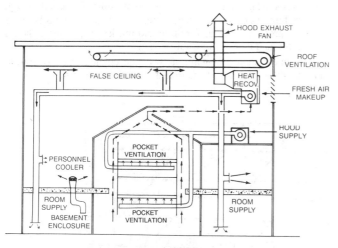

Fig. 2 Paper Machine Area

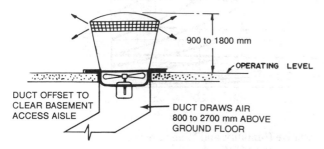

Fig. 3 Personnel Cooler

operating floor. A basement exhaust system for the wet end should vent this warm air during warmer months.

The most severe ventilation demand occurs in the area between the wet end forming section and press section and the dryer section. In the forming section, the pulp slurry, which contains about 90% water, is deposited on a traveling screen. Gravity, rolls, foils, vacuum, steam boxes, and three or more press roll nips are sequentially used to remove up to 50% of the water content in the forming section and press section. The wet end is very humid due to the evaporation of moisture and the mechanical generation of vapor by turning rolls and cleaning showers. Baffles and custom-designed exhaust systems in the forming section help control the vapor. A drive-side exhaust system in the wet end keeps heat from the dc drive motor vent air and keeps the vapor generated to a minimum. To prevent condensation or accumulated fiber from falling on the traveling web, a false ceiling is used with duct connections to roof exhausters that remove humid air not captured at a lower elevation. The wet end usually has a heated inside air circulation system that scrubs the underside of the roof to prevent condensation in cold weather. Additional roof exhaust may also remove accumulated heat from the dryer section and the dry end during warmer periods. Ventilation in the wet end should be predominantly by roof exhaust.

In the dryer section, the paper web is dried as it travels in a serpentine path around rotating steam-heated drums. Exhaust hoods control heat from the dryers and moisture evaporated from the paper web. Most modern machines have enclosed hoods, which reduce the airflow mass required to less than 50% of that required for an open hood exhaust. Temperatures inside an enclosed hood range from 50 to 60°C at the operating floor level to 80 to 95°C in the hood exhaust plenum at 70 to 90% rh, with an exhaust rate generally ranging from 140 to 190 m^3/s.

Where possible, pocket ventilation air and hood supply air are drawn from the upper level of the machine room to take advantage of the preheating of makeup air by process heat as it rises. The basement of the dryer section is also enclosed to control infiltration of machine room air to the enclosed hood. The hood supply and pocket ventilation air typically operate at 95°C; however, some systems run at temperatures as high as 120°C. Enclosed hood exhaust is typically 0.16 L/s per kilogram of machine capacity. The pocket ventilation and hood supply are designed for 75 to 80% of the exhaust, with the balance infiltrated from the basement and machine room. Large volumes of air (from 240 to 380 m^3/s) are required to balance the paper machine's exhaust with the building air balance.

The potential for heat recovery from the hood exhaust air should be evaluated. Most of the energy in the steam supplied to the paper dryers is converted to latent heat in the hood exhaust as water is evaporated from the paper web. Air-to-air heat exchangers are used where the air supply is located close to the exhaust. Air-to-liquid heat exchangers that recirculate water-glycol to heat remote makeup air units can also be used. Air-to-liquid systems provide more latent heat recovery, resulting in three to four times more total heat recovery than air-to-air units. Some machines use heat recovered from the exhaust air to heat process water. Ventilation rates in paper machine buildings range from 10 to 25 air changes per hour in northern mills to 20 to 50 in southern mills. In some plants, computers monitor production rates and outside air temperatures to optimize the operation of the total system and conserve energy.

After fine, bond, and cut papers have been bundled and/or packaged, they should be wrapped in a nonpermeable material. Most papers are produced with less than 10% moisture content by weight, the average being 7%. Dry paper and pulp are hygroscopic and will begin to swell noticeably and deform permanently when the relative humidity exceeds 38%. Therefore, finished products should be stored under controlled conditions to maintain their uniform moisture content.

Finishing Area

To produce a precisely cut paper that will stabilize at a desirable equilibrium moisture content, the finishing areas require temperature and humidity controls. Further converting operations such as printing and die cutting require optimum sheet moisture content for efficient processing. Finishing room conditions range from 21 to 24°C db and from 40 to 45% rh. The system should maintain the selected conditions within reasonably close limits. Without precise environmental control, the paper equilibrium moisture content will vary, influencing dimensional stability, the tendency to curl, and further processing.

PROCESS AND MOTOR CONTROL ROOMS

In most pulp and paper applications, process control, motor control, and switchgear rooms are separate from the process environment. Air conditioning removes heat generated by equipment, lights, etc., and reduces the air-cleaning requirements. (See Chapter 16 for air conditioning in control rooms that include a computer, a computer terminal, or data processing equipment.) Ceiling grilles or diffusers should be located above access aisles to avoid the risk of condensation on control consoles or electrical equipment during start-up and recovery after an air-conditioning system shutdown. Electrical rooms are usually maintained in the range of 24 to 27°C, with control rooms at 23°C; the humidity is maintained in the range of 45 to 55% in process control rooms and is not normally controlled in electrical equipment rooms.

Electrical control rooms for distributed control and process control contain electronic equipment that is susceptible to corrosion. The typical pulp and paper mill environment contains both particulate and vapor-phase contaminants with sulfur- and chloride-based compounds. To protect the equipment, multistage particulate and adsorbent filter systems should be used that have treated activated charcoal and potassium permanganate-impregnated alumina sections for vapor-phase contaminants, as well as fiberglass and cloth media for particulates. A minimum amount of outside air should be used. Air conditioning and filtration of the outside air and a portion of the recirculated air controls contaminants if the filters are carefully maintained.

Switchgear and motor control centers are not as heat-sensitive as control rooms, but the moisture-laden air carries chemical residues onto the contact surfaces. Arcing, corrosion, and general deterioration can result. A minimum amount of outside air and air conditioning are used to protect these areas.

Paper Testing Laboratories

Design conditions within paper mill laboratories are critical and must be followed rigidly. The most recognized standard for testing environments for paper and paper products (paperboard, fiberboard, and containers) is TAPPI (the Technical Association of the Pulp and Paper Industry) *Standard* T402, Standard Conditioning and Testing Atmospheres for Paper, Board, Pulp Handsheets, and Related Products. Other standards include ASTM E171, Standard Specification for Standard Atmospheres for Conditioning and Testing Materials, and ISO/TC 125, Enclosures and Conditions for Testing. TAPPI *Standard* T402 is discussed in this section.

Standard pulp and paper testing laboratories have three environments: preconditioning, conditioning, and testing. The physical properties of a sample are different if it is brought to the testing humidity from a high humidity than if it is brought from a lower humidity. Preconditioning at lower relative humidity tends to eliminate hysteresis. For a preconditioning atmosphere, TAPPI T402 recommends 10 to 35% rh and 22 to 40°C db. This is usually accomplished in a controlled, conditioned cabinet.

Conditioning and testing atmospheres should be maintained at 50 ± 2.0% rh and 23 ± 1°C db. The designer should realize, however, that a change of 1°C db at 23°C makes the relative humidity fluctu-

ate as much as 3%. A dry-bulb temperature tolerance of ±0.5°C must be held to maintain a ±2.0% rh. These low humidity variations suggest reheat or chemical dryers of some type. Humidistatic and temperature instrumentation as well as graphical wet- and dry-bulb recorders should be provided for the laboratories.

Miscellaneous Areas

The pulp digester area contains many components that release heat and contribute to dusty conditions. For batch digesters, the chip feeders are a source of dust and need hooded exhaust and makeup air. The wash and screen areas have numerous components with hooded exhausts that require considerable makeup air. Good ventilation controls fumes and humidity. The lime kiln feed-end releases extremely large amounts of heat and requires high ventilation rates or air conditioning.

Recovery-boiler and power-boiler buildings have conditions similar to those of power plants; the ventilation rates are also similar. The control rooms are generally air conditioned. The grinding motor room, in which groundwood is made, contains many large motors that require ventilation to keep the humidity low.

SYSTEM SELECTION

The system and equipment selection for air conditioning a pulp and paper mill depends on a great number of variables, including the plant layout and atmosphere, geographic location, roof and ceiling heights (which can exceed 30 m), and degree of control desired.

Chilled water systems are economical and practical for most pulp and paper operations, because they have both the large cooling capacity needed by mills and the precision of control to maintain the temperature and humidity requirements of laboratories and finishing areas. In the bleach plant, the manufacture of chlorine dioxide is enhanced by the use of water with a temperature of 7°C or lower; this water is often supplied by the chilled water system. If clean plant or process water is available, water-cooled chillers are satisfactory if they are supplemented by water-cooled direct-expansion package units for remote, small areas. However, if plant water is not clean enough, a separate cooling tower and condenser water system must be installed for the air-conditioning systems.

Most manufacturers prefer water-cooled over air-cooled systems because of gases and particulates present in most paper mill atmospheres. The most prevalent contaminants are chlorine gas, caustic soda, borax, phosphates, and sulfur compounds. With efficient air cleaning, the air quality in and about most mills is adequate for properly placed air-cooled chillers or condensing units that have properly applied coil and housing coatings. Phosphor-free brazed coil joints are recommended in areas where sulfur compounds are present.

Heat is readily available from processing operations and should be recovered whenever possible. Most plants have quality hot water and steam, which can be used for unit heater, central station, or reheat quite easily. Evaporative cooling should not be ignored. Newer plant air-conditioning methods using energy conservation techniques, such as temperature destratification and stratified air conditioning, have application in large structures. Absorption systems should be considered for pulp and paper mills because they allow some degree of energy recovery from high-temperature steam processes.

BIBLIOGRAPHY

ACGIH. 1992. *Industrial ventilation. A manual of recommended practice,* 21st ed. American Conference of Governmental Industrial Hygienists, Cincinnati, OH.

ASTM. 1987. Standard specification for standard atmospheres for conditioning and testing materials. ASTM E171-87. American Society for Testing and Materials, Philadelphia, PA.

Britt, K.W. 1970. *Handbook of pulp and paper technology,* 2nd ed. Van Nostrand Reinhold, New York.

Casey, J.T. 1960. *Pulp and paper chemistry and chemical technology,* Vol. 3. Interscience Publishers.

ISO/TC 125. Enclosures and conditions for testing. International Organization for Standardization, Geneva, Switzerland.

Rasmussen, E.F. 1961. *Dry kiln operator's manual.* Agriculture Handbook No. 188, USDA.

Southland Paper Company. *Houston Mill Guide.* Houston, TX.

Stephenson, J.N. 1950. *Preparation and treatment of wood pulp,* Vol. 1. McGraw-Hill, New York.

TAPPI. 1988. Standard conditioning and testing atmospheres for paper, board, pulp handsheets, and related products. Test Method T402 OM-88. Technical Association of the Pulp and Paper Industry, Atlanta, GA.

NUCLEAR FACILITIES

THE HVAC requirements for facilities using radioactive materials are discussed in this chapter. Such facilities include nuclear power plants, fuel fabrication and processing plants, plutonium processing plants, hospitals, corporate and academic research facilities, and other facilities housing nuclear operations or materials. The information presented here should serve as a guide for dealing with radioactive materials; however, careful and individual analysis of each system is required.

BASIC TECHNOLOGY

Criticality, radiation fields, and regulation are three considerations that are more important in the design of nuclear-related HVAC systems than in that of other special HVAC systems.

Criticality. Criticality considerations are unique to nuclear facilities. Criticality is the condition reached when the chain reaction of fissionable material, which produces extreme radiation and heat, becomes self-sustaining. Unexpected or uncontrolled conditions of criticality must be prevented at all costs. Only a limited number of facilities—including fuel plants, weapons facilities, and some national laboratories—handle special nuclear material (SNM) subject to criticality concerns.

Radiation Fields. Radiation fields are found in all facilities using nuclear materials. They pose problems of material degradation and personnel exposure. Although material degradation is usually addressed by regulation, it must be considered in all designs. The personnel exposure hazard is more difficult to measure than the amount of material degradation because a radiation field cannot be detected without special instrumentation. It is the responsibility of the designer and of the end user to monitor radiation fields and limit personnel exposure.

Regulation. In the United States, the Department of Energy (DOE) regulates weapons-related facilities and national laboratories; the Nuclear Regulatory Commission (NRC) controls civil, industrial, and power facilities. Further complicating the issue are other government regulations at the local, state, and federal levels. For example, meeting an NRC requirement does not relieve the designer or operator of the responsibility of meeting Occupational Safety and Health Administration requirements. The design of an HVAC system to be used near radioactive materials must follow all guidelines set by these agencies and by the state, local, and federal governments.

As Low as Reasonably Achievable

As low as reasonably achievable (ALARA) means that all aspects of a nuclear facility are designed so that worker exposure is limited to the minimum amount of radiation that is reasonably achievable. This refers not to meeting legal requirements, but rather to attaining the lowest cost-effective below-legal levels.

Design

HVAC requirements for a facility using radioactive materials depend on the type of facility and the specific service required. The following are design considerations:

1. Physical layout of the HVAC system that minimizes the accumulation of material within piping and ductwork
2. Control of the system so that portions can be safely shut down as needed to facilitate maintenance and testing or in the case of any event, accident, or natural catastrophe
3. Modular design for facilities that change operations regularly
4. Preservation of confinement system integrity to limit the spread of radioactive contamination and exposure in the physical plant and surrounding areas

The design basis in nuclear facilities requires that safety-class systems and their components have active control during and after any event, accident, or natural catastrophe

Normal or Power Design Basis

The *normal* or *power design basis* for nuclear power plants covers normal plant operation, including normal operation mode and normal shutdown mode. This design basis imposes no requirements more stringent than those specified for standard indoor conditions.

Safety Design Basis

The *safety design basis* establishes the special requirements necessary for a safe work environment and public protection from radiation exposure. Any system designated essential or safety-class must mitigate the effect of any event, accident, or natural catastrophe that may cause the release of radioactivity to the surroundings or to the plant atmosphere. These safety systems must be operative at all times. Safety analysis reports (SARs) determine which components must function during and after a design basis accident (DBA) or the simultaneous occurrence of such events as a safe shutdown earthquake (SSE), a tornado, a loss of coolant accident (LOCA), and loss of off-site electrical power (LOEP). Non-safety-related equipment should not adversely affect safety-related equipment.

System Redundancy. Systems important to safety must be redundant so that their function is performed even if a component of one system fails. Such a failure should not cause a failure in the backup system. For additional redundancy requirements, refer to the section on Nuclear Regulatory Commission Facilities.

Seismic Qualification. All safety-class components, including equipment, pipe, duct, and conduit, must be seismically qualified by testing or calculation to document their ability to withstand and

---

The preparation of this chapter is assigned to TC 9.2, Industrial Air conditioning.

perform under the shock and vibration caused by an SSE or an operating basis earthquake (the largest earthquake postulated for the region in which the plant is located). This qualification also covers any amplification by the building structure. In addition, any HVAC component, the failure of which could jeopardize the essential function of a safety-related component, must be seismically qualified or restrained to prevent such failure.

Environmental Qualification. Safety-class components must be environmentally qualified; i.e., the useful life of the component in the environment in which it is operating must be determined through a program of accelerated aging. Environmental factors such as temperature, humidity, pressure, acidity, and accumulated radioactivity must be considered.

Quality Assurance. All designs and components of safety-class systems must comply with the requirements of a quality assurance (QA) program for design control, inspection, documentation, and traceability of material. Refer to Appendix B of Title 10 of the Code of Federal Regulations, Part 50 (10 CFR 50) or ANSI/ASME NQA-1 for quality assurance program outdoor conditions.

Emergency Power. All safety-class systems must be powered by a backup source such as an emergency diesel generator.

Outdoor Conditions

The 1993 *ASHRAE Handbook—Fundamentals*, the National Weather Service, or site meteorology can provide information on outdoor conditions, temperature, humidity, solar load, altitude, and wind. DOE 6430.1A may specify outdoor conditions.

Nuclear facilities generally consist of heavy structures having high thermal inertia. Time lag should be considered in determining solar loads. For some applications, such as diesel generator buildings or safety-related pump houses in nuclear power plants, 24-h averages suffice.

Indoor Conditions

Indoor temperatures are dictated by occupancy, equipment or process requirements, and personnel activities. HVAC system temperatures are dictated by the environmental qualification of the safety-class equipment located in the space and by ambient conditions during the different operating modes of the equipment.

Indoor Pressures

Where control of air pattern is required, a specific pressure relative to the outside atmosphere or to adjacent areas must be maintained. For process facilities having pressure zones, the pressure relationships are specified in the section on Confinement Systems.

In facilities where zoning is different from that in process facilities, and in cases where any airborne radioactivity must not spread to rooms within the same zone, this airborne radioactivity must be controlled by airflow.

Airborne Radioactivity

The level of airborne radioactivity within a facility and the amount released to the surroundings must be controlled to meet the requirements of DOE *Order* 5400.5, DOE *Order* 5480.11, 10 CFR 20, 10 CFR 50, and 10 CFR 100.

Tornado/Wind Protection

Protection from tornados and the objects or missiles generated by tornados or design basis wind is normally required to prevent the release of radioactive material to the atmosphere. A tornado passing over a facility causes a sharp drop in ambient pressure. If exposed to this transient pressure, ducts and filter housings could collapse because the pressure inside the structure would still be that of the environment prior to the pressure drop. Protection is usually provided by tornado dampers and missile barriers in all appropriate

openings to the outside. Tornado dampers are heavy-duty, low leakage dampers designed for pressure differences in excess of 2 kPa. They are normally considered safety-class and are environmentally and seismically qualified.

Fire Protection

Fire protection for the HVAC and filtration systems must comply with the applicable requirements of Appendix R of 10 CFR 50 and NFPA, UL, and ANSI standards. Design criteria should be developed for all building fire protection systems, including secondary sources, filter plenum protection, fire dampers, and detection/suppression systems. Fire protection systems may consist of a combination of building sprays, hoses and standpipes, and gaseous or foam suppression. The type of fire postulated in the SAR determines which kind of system is used.

Secondary sources may include one or more of the following: (1) a water tower with a diesel-powered or electric water pump, (2) a pressurized water tank, or (3) a second water main. These systems are not unique to nuclear facilities.

A requirement specific to nuclear facilities is protection of the filter plenums and the ventilation ductwork. Water sprays (window nozzles, fog nozzles, or standard dry pipe/wet pipe system spray heads) are usually used to protect the filters in filter plenums. In the case of metal filters, fire protection equipment may also be used for cleansing. Sprinkler location, system balancing, and spray patterns should be designed for maximum effect. Flow requirements should be consistent with the design criteria for the facility. Performance of the room ventilation system in response to fire scenarios must be analyzed to determine the operational requirements and optimum location of fire dampers. Various codes may limit alternatives.

Heat detectors and fire suppression systems should be considered for special equipment such as glove boxes. Application of the two systems in combination allows the shutdown of one system at a time for repairs, modifications, or maintenance. Smoke control criteria can be found in NFPA 801 and NFPA 901.

Control Room Habitability Zone

The HVAC system in a control room is a safety-related system that must fulfill the following requirements during all normal and postulated accident conditions to ensure that occupancy conditions are maintained.

1. Maintain conditions comfortable to personnel and ensure the continuous functioning of control room equipment.
2. Protect personnel from exposure to potential airborne radioactivity or toxic chemicals present in the outside atmosphere or surrounding plant areas.
3. Protect personnel from the effects of breaks in high-energy lines in the surrounding plant areas.
4. Protect personnel from combustion products emitted from on-site fires.

Filtration

HVAC filtration systems can be designed to remove either radioactive particles or radioactive gaseous iodine from the airstream. They filter potentially contaminated exhaust air prior to discharge to the environment and may also filter potentially contaminated makeup air for power plant control rooms and technical support centers.

The composition of the filter train is dictated by the type and concentration of the contaminant, the process air conditions, and the filtration levels required by the applicable regulations (e.g., RG 1.52, RG 1.140, ANSI/ASME AG-1, ANSI/ASME N509, 10 CFR 20, and 10 CFR 100). Filter trains may consist of one or more of the following components: prefilters, high efficiency particulate air (HEPA) filters, charcoal filters (adsorbers), sand filters, heaters, and demisters.

Dust Filters/Prefilters. Dust filters are selected for the efficiency required by the particular application. High efficiency dust filters are often used as prefilters for the special filters listed below to prevent them from being loaded with atmospheric dust and to minimize replacement costs.

HEPA Filters. Nuclear HEPA filters are used where there is a risk of particulate airborne radioactivity. The construction and pre-use quality assurance testing of HEPA filters is specified in DOE *Standards* F 3-43 and F 3-45. Filter performance requirements are based on penetration at a specified airflow and static pressure. For a 0.3 μm particle, the penetration at rated airflow must not exceed 0.03%. The construction and QA testing of HEPA filters for use in nuclear power plants is specified in ANSI/ASME N509, ANSI/ASME N510, and ANSI/ASME AG-1. HEPA filters are in-place tested and inspected when first installed, and then tested every 12 to 18 months thereafter. One or both of the following two methods may be used for in-place testing of a stage of filtration: (1) mass flow testing of the stage as a whole or (2) testing of the individual filters and frame that make up the stage. In both pre-use and in-place tests, an approved challenge agent must be used; two agents in current use are the aerosols dioctyl phthalate (DOP) and dioctyl sebacate (DOS).

Sand Filters. Sand filters consists of multiple beds of sand and gravel through which air is drawn. The air enters an inlet tunnel that runs the entire length of the filter. Smaller cross-sectional laterals running perpendicular to the inlet tunnel distribute the air across the base of the sand. The air rises through the several layers of various sizes of sand and gravel, typically at a rate of 25 mm/s. It is then collected in the outlet tunnel for discharge to the atmosphere.

Charcoal Filters. Activated charcoal adsorbers are used mainly to remove radioactive iodine, which is a vapor or gas. Bed depths are typically 25 or 50 mm. These filters have an efficiency of 99.9% for elemental iodine and 95 to 99% for organic iodine. Charcoal filters lose efficiency rapidly as the relative humidity increases. They are preceded by a heating element to keep the relative humidity of the entering air below 70%.

To control the argon content in the primary coolant, argon is adsorbed on charcoal at extremely low (cryogenic) temperatures. In a high-temperature gas-cooled reactor (HTGR) or in the off-gas system of a boiling water reactor (BWR), helium is similarly adsorbed.

Heaters. Electric heating coils and/or demisters may be used to meet the relative humidity conditions requisite for charcoal filters. For safety-class systems, electric heating coils should be connected to the emergency power supply. Interlocks should be provided to prevent heater operation when the exhaust fan is de-energized.

Demisters (Mist Eliminators). Demisters are required to protect HEPA and charcoal filters if entrained moisture droplets are expected in the airstream. They should be fire resistant.

DEPARTMENT OF ENERGY FACILITIES

Nonreactor nuclear HVAC systems must be designed in accordance with DOE *Order* 6430.1A and DOE *Order* 5480.6. Critical items and systems in plutonium processing facilities are designed to confine radioactive materials under both normal operation and DBA conditions, as required by 10 CFR 100.

CONFINEMENT SYSTEMS

Zoning

Typical process facility confinement systems are shown in Figure 1. Process facilities comprise several zones.

Primary Confinement Zone. This zone includes the interior of the hot cell, canyon, glove box, or other means of containing radioactive material. Containment must prevent the spread of radioactivity

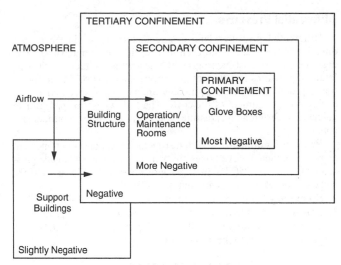

Fig. 1 Typical Process Facility Confinement Categories

within or from the building under both normal conditions and upset conditions up to and including a facility DBA. Complete isolation from neighboring facilities is necessary. Multistage HEPA and/or sand filtration of the exhaust is required

Secondary Confinement Zone. This zone is bounded by the walls, floors, roofs, and associated ventilation exhaust systems of the cell or enclosure surrounding the primary confinement zone. Except in the case of glove box operations, this zone is usually unoccupied.

Tertiary Confinement Zone. This zone is bounded by the walls, floors, roofs, and associated ventilation exhaust systems of the facility. They provide a final barrier against the release of hazardous material to the environment. Radiation monitoring may be required at exit points.

Uncontaminated Zone. This zone includes offices and cold shop areas.

Air Locks

Air locks in nuclear facilities are used as safety devices to maintain a negative differential pressure when a confinement zone is accessed. They are used for placing items in primary confinement areas and for personnel entry into secondary and tertiary confinement areas. Administrative controls ensure proper operation of the air lock doors.

There are three methods of ventilating personnel air locks (ventilated vestibules):

1. The clean conditioned supply air (CCS) method, where the air lock is at positive pressure with respect to the adjacent zones. In order for this method to be effective, the air lock must remain uncontaminated at all times.
2. The flow-through ventilation air (FTV) method, where no conditioned air is supplied to the air lock, and the air lock stays at negative pressure with respect to the less contaminated zone.
3. The combined ventilation air (CV) method, which is a combination of the CCS and FTV methods. Testing has shown that this is the most effective method, when properly designed.

Zone Pressure Control

Negative static pressure increases (becomes more negative) from the uncontaminated zone to the primary confinement zone, causing any air leakage to be inward, toward areas of higher potential contamination. All zones should be maintained at negative pressure with respect to the atmospheric pressure. Zone pressure control cannot be achieved through the ventilation system alone. Special consideration must be given to ensure that confinement barrier construction meets all applicable specifications.

Differential Pressures

Differential pressures help ensure that air flows in the proper direction in the event of a breach in a confinement zone barrier. The design engineer must incorporate the desired magnitudes of the differential pressures into the design process early to avoid later operational problems. These magnitudes are normally specified in the design basis document of the SAR. The following are approximate values for differential pressures between the three confinement zones.

Primary Confinement. With respect to the secondary confinement area, air-ventilated glove boxes are typically maintained at gage pressures of −175 to −250 Pa, inert gas glove boxes at −175 to −370 Pa, and canyons and cells at a minimum of −250 Pa.

Secondary Confinement. Differential pressures of −7 to −37 Pa (gage) with respect to the tertiary confinement area are typical.

Tertiary Confinement. Differential pressures of −3 to −37 Pa (gage) with respect to the atmosphere are typical.

VENTILATION

Ventilation systems are designed to confine radioactive materials under normal and DBA conditions and to limit radioactive discharges to the required minimum. They ensure that airflows are, under all normal conditions, toward areas (zones) of progressively higher potential radioactive contamination. Air-handling equipment should be sized conservatively so that upsets in the airflow balance do not cause the direction of airflow to reverse. Examples of upsets include improper use of an air lock, occurrence of a credible breach in the confinement barrier, or excessive loading of HEPA filters.

HEPA filters at the ventilation inlets in all primary confinement zone barriers prevent movement of contamination toward zones of lower potential contamination in the event of an airflow reversal. Ventilation system balancing helps ensure that the building air pressure is always negative with respect to the outside atmosphere.

Recirculating refers to the reuse of air in a particular zone or area. Room air recirculated from a space or zone may be returned to the primary air-handling unit for reconditioning and then, with the approval of health personnel, be returned to the same space (zone) or to a zone of greater potential contamination. All air recirculated from secondary and tertiary zones must be HEPA-filtered prior to reintroduction to the same space. Recirculating air is not permitted in primary confinement areas, except those with inert atmospheres.

A safety analysis is necessary to establish minimum acceptable response requirements for the ventilation system and its components, instruments, and controls under normal, abnormal, and accident conditions.

Analysis determines the number of exhaust filtration stages required in different areas of the facility to limit (in conformance with the applicable standards, policies, and guidelines) the amount of radioactive or toxic material released to the environment during normal and accident conditions. Consult DOE *Order* 6430.1A and DOE *Order* 5480 for air-cleaning system criteria.

Ventilation Requirements

A partial recirculating ventilation system may be considered for economic reasons. However, such systems must be designed to prevent contaminated exhaust from entering the room air-recirculating systems.

The exhaust system is designed to (1) clean radioactive contamination from the discharge air, (2) safely handle combustion products, and (3) maintain the building under negative pressure relative to the outside.

Provisions may be made for independent shutdown of ventilation systems or isolation of portions of the systems to facilitate operations, filter change, maintenance, or emergency procedures such as fire fighting. All possible effects of partial shutdown on the airflows in interfacing ventilation systems should be considered. Positive means must be provided to control the backflow of air that might transport contamination. A HEPA filter installed at the interface between the enclosure and ventilation system minimizes contamination in the ductwork; a prefilter reduces HEPA filter loading. These filters should not to be considered the first stage of an airborne contamination cleaning system.

Ventilation Systems

The following is a partial list of elements that may be included in the overall air filtration and air-conditioning system of a nuclear facility:

- Air-sampling devices
- Carbon bed adsorbers
- Prefilters and absorption, HEPA, sand, and glass fiber filters
- Scrubbers
- Demisters
- Process vessel vent systems
- Condensers
- Distribution baffles
- Fire suppression systems
- Fire and smoke dampers
- Exhaust stacks
- Fans
- Coils
- Heat removal systems
- Pressure- and flow-measuring devices
- Radiation-measuring devices
- Critically safe drain systems
- Tornado dampers

The ventilation system and associated fire suppression system are designed for fail-safe operation. The ventilation system is equipped with alarms and instruments that report and record its behavior through readouts in control areas and utility services areas.

Control Systems

Control systems for HVAC systems in nuclear facilities have some unique safety-related features. Since the exhaust system is to remain in operation during both normal and accident-related conditions, redundancy in the form of standby fans is often provided. These standby fans and their associated isolation dampers energize automatically upon a set reduction in either airflow rate or specific location pressure, as applicable. For DOE facilities, maintaining exhaust airflow is important, so fire dampers are excluded from all potentially contaminated exhaust ducts.

Pressure control in the facility interior maintains zones of increasing negative pressure in areas of increasing contamination potential. Care must be taken to prevent windy conditions from unduly effecting the atmospheric control reference. Pulsations can cause the pressure control system to oscillate strongly, resulting in potential reversal of relative pressures. One alternative is to use a variety of balancing and barometric dampers to establish an air balance at the desired differential pressures, lock the dampers in place, and then control the exhaust air to a constant flow rate.

Air and Gaseous Effluents Containing Radioactivity

Air and all other gaseous effluents are exhausted through a ventilation system designed to remove radioactive particulates. Exhaust ducts or stacks located downstream of final filtration that may contain radioactive contaminants should have two monitoring systems, one a continuous air monitor (CAM) and the other a fixed sampler. These systems may be a combination unit. Exhaust stacks from nuclear facilities are usually equipped with an isokinetic sampling system that relies on a relatively constant airflow rate. The sensing probe is a symmetrically arranged series of pickup tubes connected through sweeping bends to a common tube, usually stainless steel, that leads to a nearby CAM. Typically, an exhaust system flow controller modulates the exhaust fan inlet dampers to hold the exhaust airflow rate steady while the HEPA filters load.

Continuous air monitors can also be located in specific ducts where a potential for radiological contamination has been detected. These CAMs are generally placed beyond the final stage of HEPA filtration, as specified in ANSI N13.1. Each monitoring system is connected to an emergency power supply.

NUCLEAR REGULATORY COMMISSION FACILITIES

NUCLEAR POWER PLANTS

The two kinds of commercial light-water power reactors used in the United States today are the pressurized water reactor (PWR) and the boiling water reactor (BWR). For both types of reactors, the main objective of the HVAC systems, in addition to ensuring personnel comfort and reliable equipment operation, is protecting operating personnel and the general public from airborne radioactive contamination during all normal and emergency modes of plant operation. 10 CFR 20 sets forth the requirements for maintaining radiation exposure as low as reasonably achievable (ALARA). The ALARA concept is the design objective of the HVAC system. In no case is the radiological dose allowed to exceed the limits as defined in 10 CFR 50 and 10 CFR 100.

The NRC has developed regulatory guides (RGs) for meeting the design criteria that delineate techniques of evaluating specific problems and provide guidance to licensed applicants concerning information needed by the NRC for its review of the facility. Four regulatory guides that relate directly to HVAC system design are RG 1.52, RG 1.78, RG 1.95, and RG 1.140. Deviations from RG criteria must be justified by the owner and approved by the NRC.

The design of the HVAC systems for a nuclear power generating station must ultimately be approved by the NRC staff in accordance with Appendix A of 10 CFR 50. The NRC developed standard review plans (SRPs) as part of *Regulatory Report* NUREG-0800 to provide an orderly and thorough review. As such, the SRP provides a good basis or checklist for the preparation of a safety analysis report (SAR). The safety review plan is based primarily on the information provided by an applicant in an SAR as required by Section 50.34 of 10 CFR 50. Technical specifications for nuclear power plant systems are developed by the owner and approved by the NRC as outlined in Section 50.36 of 10 CFR 50. Technical specifications define safety limits, limiting conditions for operation, and surveillance requirements for all systems important to plant safety.

Minimum requirements for the performance, design, construction, acceptance testing, and quality assurance of equipment used in safety-related air and gas treatment systems in nuclear facilities are found in ANSI/ASME AG-1.

PRESSURIZED WATER REACTOR HVAC SYSTEMS

Reactor Containment Building

The containment building houses the reactor in a nuclear power plant. The temperature and humidity conditions are dictated by the nuclear steam supply system (NSSS). These conditions are generally specified for three modes of operation: normal operation, refueling operation, and loss of coolant accident (LOCA) condition. General design requirements are contained in ANSI/ANS 56.6.

Normal Operating Conditions. Nuclear steam supply system temperature and humidity requirements are specified by the NSSS

supplier. Some power plants require recirculation filtration trains in the containment building to control the level of airborne radioactivity. Cooling is provided by a reactor containment cooling system.

Refueling Condition. The maximum allowable temperature during refueling is determined by the refueling personnel. Because they work in protective clothing, their activities are slowed by discomfort, and the refueling outage is prolonged. Cooling can be accomplished by normal cooling units since the cooling load is low when the reactor is shut down. Ventilation with outdoor air is necessary.

Loss of Coolant Accident Condition. In the event of a LOCA (breakage of the primary cooling loop), circulating water at high pressure and temperature flashes and fills the containment building with radioactive steam. The major source of radioactivity is iodine in the water.

Should such an accident occur, the primary measures taken are directed at reducing the pressure in the containment building and lowering the amount of radioactive products in the containment atmosphere. Pressure is reduced by the reactor containment cooling units and/or sprays, which cool the atmosphere and condense the steam.

Containment Cooling. The following systems are typical for containment cooling:

Reactor containment coolers. These units remove most of the heat load. Distribution of the air supply depends on the containment layout and the location of the major heat sources.

Reactor cavity air-handling units or fans. These units are usually transfer fans without coils that provide cool air to the reactor cavity.

Control rod or control element drive mechanism (CRDM or CEDM) air-handling units. The CRDM and CEDM are usually cooled by an induced draft system using exhaust fans. Because the flow rates, pressure drops, and heat loads are generally high, it is desirable to cool the air before it is returned to the containment atmosphere.

Essential reactor containment cooling units. The containment air-cooling system, or a part of it, is normally designed to provide cooling after a postulated accident. The system must be capable of performing at high temperature, high pressure, high humidity, and a high level of radioactivity. Cooling coils are provided with essential service water.

System design must accommodate both normal and accident conditions. The ductwork must be able to endure the rapid pressure buildup associated with accident conditions. Fan motors must be sized to handle the high-density air associated with accident conditions.

Radioactivity Control. Airborne radioactivity is controlled by the following means:

Essential containment air filtration units. Some older power plants rely on redundant filter units powered by two Class IE buses to reduce the amount of post-LOCA airborne radioactivity. The system consists of a demister, a heater, a HEPA filter, and a charcoal adsorber followed by a second HEPA filter. The electric heater is designed to reduce the relative humidity from 100% to less than 70% at the design inlet air temperature. All the components must be designed and manufactured to meet the requirements of a LOCA environment.

In the case of LOCA and the subsequent operation of the filter train, the charcoal becomes loaded with radioactive iodine such that the decay heat could cause the charcoal to self-ignite if the airflow stops. If the primary fan stops, a secondary fan maintains a minimum airflow through the charcoal bed to remove the heat generated by the radioactive decay. The decay heat fan is powered Class IE power supply. The filtration units are located inside the containment.

Containment power access purge or minipurge. It is necessary to ventilate during normal operation, when the reactor is under pressure, to control containment pressure or the level of airborne radioactivity within the containment. The maximum opening size allowed in the containment boundary during normal operation is 200 mm.

The system consists of a supply fan, double containment isolation valves in each of the containment wall penetrations (supply and exhaust), and an exhaust filtration unit with a fan. The filtration unit contains a HEPA filter and a charcoal adsorber followed by a second HEPA filter.

This system should not be connected to any duct system inside the containment. It should include a debris screen within the containment over the inlet and outlet ducts, so the containment isolation valves can close even if blocked by debris or collapsed ducts.

Containment refueling purge. Ventilation is required to control the level of airborne radioactivity during refueling. Since the reactor is not under pressure during refueling, there are no restrictions on the size of the penetrations through the containment boundary. Large openings of 1 to 1.2 m, each protected by double containment isolation valves, may be provided. The required ventilation rate is typically based on one air change per hour.

The system consists of a supply air-handling unit, double containment isolation valves at each supply and exhaust containment penetration, and an exhaust fan. Filters are recommended.

Containment combustible gas control. In case of LOCA, when a strong solution of sodium hydroxide or boric acid is sprayed into the containment, various metals react and produce hydrogen. Also, if some of the fuel rods are not covered with water, the fuel rod cladding can react with steam at elevated temperatures to release hydrogen into the containment. Therefore, redundant hydrogen recombiners are needed to remove the air from the containment atmosphere, recombine the hydrogen with the oxygen, and return the air to the containment. The recombiners may be backed up by special exhaust filtration trains.

BOILING WATER REACTOR
HVAC SYSTEMS

Primary Containment

The boiling water reactor (BWR) primary containment is a low leakage, pressure-retaining structure that surrounds the reactor pressure vessel and related piping. Sometimes referred to as the *drywell*, it is designed to withstand, with minimum leakage, the high temperature and pressure caused by a major reactor coolant line break. General design requirements are found in ANSI/ANS 56.7.

The primary containment HVAC system consists of recirculating cooler units. It normally recirculates and cools the primary containment air to maintain the environmental conditions specified by the NSSS supplier. In an accident, the system performs the safety-related function of recirculating the air to prevent stratification of any hydrogen that may be generated. The cooling function may or may not be safety related, depending on the specific plant design.

Temperature problems have been experienced in many BWR primary containments due to temperature stratification effects and underestimation of heat loads. The ductwork should adequately mix the air to prevent stratification. Heat load calculations should include a sufficient safety factor to allow for deficiencies in insulation installation. In addition, a temperature monitoring system should be installed in the primary containment to ensure that bulk average temperature limits are not exceeded.

Reactor Building

The reactor building completely encloses the primary containment, auxiliary equipment, and refueling area. Under normal conditions, the reactor building HVAC system maintains the design space conditions and minimizes the release of radioactivity to the environment. The HVAC system consists of a 100% outside air cooling system. Outside air is filtered, heated, or cooled as required prior to being distributed throughout the various building areas. The exhaust air flows from areas with the least potential of contamination to the areas of most potential contamination. Prior to exhausting to the

environment, potentially contaminated air is filtered with HEPA filters and charcoal adsorbers, and all exhaust air is monitored for radioactivity. To ensure that no unmonitored exfiltration occurs during normal operations, the ventilation systems maintain the reactor building at a negative pressure relative to the atmosphere.

Upon detection of abnormal plant conditions, such as a line break, high radiation in the ventilation exhaust, or loss of negative pressure, the HVAC system's safety-related function is to isolate the reactor building. Once isolated via fast-closing, gastight isolation valves, the reactor building serves as a secondary containment boundary. This boundary is designed to contain any leakage from the primary containment or refueling area following an accident.

Once the secondary containment is isolated, pressure rises due to the loss of the normal ventilation system and the thermal expansion of the confined air. A safety-related exhaust system, the standby gas treatment system (SGTS), is started to reduce and maintain the building's negative pressure. The SGTS exhausts air from the secondary containment to the environment through HEPA filters and charcoal adsorbers. The capacity of the SGTS is based on the amount of exhaust air needed to reduce the pressure in the secondary containment and maintain it at the design level, given the containment leakage rates and required drawdown times.

In addition to the SGTS, some designs include safety-related recirculating air systems within the secondary containment to mix, cool, and/or treat the air during accident conditions. These recirculation systems use portions of the normal ventilation system ductwork; therefore, the ductwork must be classified as safety related.

If the isolated secondary containment area is not to be cooled during accident conditions, it is necessary to determine the maximum temperatures that could be reached during an accident. All safety-related components in the secondary containment must be environmentally qualified to operate at these temperatures. In most plant designs, safety-related unit coolers handle the high heat release with emergency core cooling system (ECCS) pumps.

Turbine Building

Only a BWR supplies radioactive steam directly to the turbine, which could cause a release of airborne radioactivity to the surroundings. Therefore, areas of the BWR turbine building where release of airborne radioactivity is a possibility should be enclosed. These areas must be ventilated and the exhaust filtered to ensure that no radioactivity is released to the surrounding atmosphere. Filtration trains consist of a prefilter, a HEPA filter, and a charcoal adsorber followed by a second HEPA filter. Filtration requirements are based on the plant and site configuration.

AREAS COMMON TO BOTH
PWRS AND BWRS

All areas located outside the primary containment are designed to the general requirements contained in ANSI/ANS 59.2.

Auxiliary Building

The auxiliary building contains a large amount of support equipment, much of which handles potentially radioactive material. The building is air conditioned for equipment protection, and the exhaust is filtered to prevent the release of potential airborne radioactivity. The filtration trains consist of a prefilter, HEPA filters, and a charcoal adsorber followed by a second HEPA filter.

The HVAC system is a once-through system, as needed for general cooling. Ventilation is augmented by local recirculation air-handling units located in the individual equipment rooms requiring additional cooling due to localized heat loads. The building is maintained at negative pressure relative to the outside.

If the equipment in these rooms is not safety related, the area is cooled by normal air-conditioning units. If it is safety related, the area is cooled by safety-related or essential air-handling units powered from the same Class IE (according to ANSI/IEEE 323) power as the equipment in the room.

The normal and essential functions may be performed by one unit having both a normal and an essential cooling coil and a safety-related fan served from a Class IE bus. The normal coil is served with chilled water from a normal chilled water system and the essential coil operates with chilled water from a safety-related chilled water system.

Control Room

The control room HVAC system serves the control room habitability zone—those spaces that must be habitable following a postulated accident to allow the orderly shutdown of the reactor—and performs the following functions:

- Control indoor environmental conditions
- Provide pressurization to prevent infiltration
- Reduce the radioactivity of the influent
- Protect the zone from hazardous chemical fume intrusion
- Protect the zone from fire
- Remove noxious fumes, such as smoke

The design requirements are described in detail in SRP 6.4, SRP 9.4.1, and a paper by Murphy and Campe (1974). Regulatory guides that directly affect control room design are RG 1.52, RG 1.78, and RG 1.95. NUREG-CR-3786 provides a summary of the documents affecting control room system design. ANSI/ASME N509 also provides guidance for the design of control room habitability systems and methods of analyzing pressure boundary leakage effects.

Control Cable Spreading Rooms

These rooms, which contain many cables, are located directly above and below the control room. They are usually served by the same air-handling units that serve the electric switchgear room or the control room.

Diesel Generator Building

Nuclear power plants have auxiliary power plants to generate electric power for all essential and safety-related equipment in the event of loss of off-site electrical power. The auxiliary power plant consists of at least two independent diesel generators, each sized to meet the emergency power load. The heat released by the diesel generator and associated auxiliary systems is normally removed through outside air ventilation.

Emergency Electrical Switchgear Rooms

These rooms house the electrical switchgear that controls essential or safety-related equipment. The switchgear located in these rooms must be protected from excessive temperatures to (1) ensure that its useful life, as determined by environmental qualification, is not cut short and (2) prevent the loss of power circuits required for proper operation of the plant, especially its safety-related equipment.

Battery Rooms

Battery rooms should be maintained at 25°C with a temperature gradient of not more than 2.8 K, according to IEEE 484. The minimum room design temperature should be taken into account in determining battery size. Because batteries produce hydrogen gas during charging periods, the HVAC system must be designed to limit the hydrogen concentration to the lowest of the levels specified by IEEE 484, OSHA, and the lower explosive limit (LEL). The minimum number of room air changes per hour is five. As hydrogen is lighter than air, the system exhaust duct inlet openings should be located on the top side of the duct to prevent hydrogen pockets from forming at the ceiling. If the ceiling is supported by structural beams, there should be an exhaust air opening in each beam pocket.

Fuel-Handling Building

New and spent fuel is stored in the fuel-handling building. The building is air conditioned for equipment protection and ventilated with a once-through air system to control potential airborne radioactivity. Normally, the level of airborne radioactivity is so low that the exhaust need not be filtered, although it should be monitored. If significant airborne radioactivity is detected, the building is sealed and kept under negative pressure by exhaust through filtration trains powered by Class IE buses.

Personnel Facilities

For nuclear power plants, this area usually includes decontamination facilities, laboratories, and medical treatment rooms.

Pump Houses

Cooling water pumps are protected by houses that are often ventilated by fans to remove the heat from the pump motors. If the pumps are essential or safety related, the ventilation equipment must also be considered safety related.

Radwaste Building

Radioactive waste other than spent fuel is stored, shredded, baled, or packaged for disposal in this building. The building is air conditioned for equipment protection and ventilated to control potential airborne radioactivity. The air may require filtration through HEPA filters and/or charcoal adsorbers prior to release to the atmosphere.

Technical Support Center

The technical support center (TSC) is an outside facility located close to the control room and is used by plant management and technical support personnel to provide assistance to control room operators under accident conditions.

In case of an accident, the TSC HVAC system must provide the same comfort and radiological habitability conditions maintained in the control room. The system is generally designed to commercial HVAC standards. An outside air filtration system (HEPA-charcoal-HEPA) pressurizes the facility with filtered outside air during emergency conditions. The TSC HVAC system does have to be designed to safety-related standards.

NUCLEAR NONPOWER MEDICAL AND RESEARCH REACTORS

The requirements for HVAC and filtration systems for nuclear nonpower medical and research reactors are set by the NRC. The criteria depend on the type of reactor (ranging from a nonpressurized swimming pool type to a 10 MW or more pressurized reactor), the type of fuel, the degree of enrichment, and the type of facility and environment. Many of the requirements discussed in the sections on various nuclear power plants apply to a certain degree to these reactors. It is therefore imperative for the designer to be familiar with the NRC requirements for the reactor under design.

LABORATORIES

The requirements for HVAC and filtration systems for laboratories involved with the use of radioactive materials are determined by the DOE and/or the NRC. Laboratories located at DOE facilities are governed by DOE regulations. All other laboratories using radioactive materials are regulated by the NRC. Other agencies may be responsible for the regulation of other toxic and carcinogenic materials present in the facility.

Laboratory containment equipment for nuclear processing facilities is treated as a primary, secondary, or tertiary containment zone, depending on the level of radioactivity anticipated for the area and on the materials to be handled. For additional information see Chapter 13, Laboratories.

Glove Boxes

Glove boxes are windowed enclosures equipped with one or more flexible gloves for handling material inside the enclosure from the outside. The gloves are attached to a porthole in the enclosure and seal the enclosure from the surrounding environment. Glove boxes permit hazardous materials to be manipulated while preventing the release of material to the environment.

Since the glove box is usually used to handle hazardous materials, the exhaust is HEPA filtered before leaving the box and prior to entering the main exhaust duct. In nuclear processing facilities, a glove box is considered Zone 1, or primary confinement (Figure 1), and is therefore subject to the regulations governing those areas. For nonnuclear processing facilities, the designer should be aware of the designated application of the glove box and design the system according to the regulations governing that particular application.

Radiobenches

A radiobench has the same geometric shape as a glove box except that in lieu of the panel for the gloves, there is an open area. Air velocity across the opening is generally the same as for laboratory hoods. The level of radioactive contamination handled in a radiobench is much lower than that handled in a glove box.

Laboratory Fume Hoods

Nuclear laboratory fume hoods are similar to those used in nonnuclear applications. Air velocity across the hood opening must be sufficient to capture and contain all contaminants in the hood. Excessive air velocities should be avoided. High hood face velocities cause contaminants to escape when an obstruction (e.g., an operator) is positioned at the hood face. For information on fume hood testing, refer to ASHRAE *Standard* 110.

CODES AND STANDARDS

ANSI N13.1	Guide for Sampling Airborne Radioactive Materials in Nuclear Facilities
ANSI/ANS 51.1	Nuclear Safety Criteria for the Design of Stationary Pressurized Water Reactor Plants
ANSI/ANS 52.1	Nuclear Safety Criteria for the Design of Stationary Boiling Water Reactor Plants
ANSI/ANS 56.6	Pressurized Water Reactor Containment Ventilation Systems
ANSI/ANS 56.7	Boiling Water Reactor Containment Ventilation Systems
ANSI/ANS 59.2	Safety Criteria for HVAC Systems Located Outside Primary Containment
ANSI/ASME AG-1	Code on Nuclear Air and Gas Treatment
ANSI/ASME N509	Nuclear Power Plant Air-Cleaning Units and Components
ANSI/ASME N510	Testing of Nuclear Air Treatment Systems
ANSI/ASME NQA-1	Quality Assurance Program Requirements for Nuclear Facilities
ANSI/IEEE 323	Standard for Qualifying Class IE Equipment for Nuclear Power Generating Stations
ASHRAE 110	Method of Testing Performance of Laboratory Fume Hoods

10 CFR	Title 10 of the Code of Federal Regulations
Part 20	Standards for Protection Against Radiation (10 CFR 20)
Part 50	Domestic Licensing of Production and Utilization Facilities (10 CFR 50)
	Appendix A: General Design Criteria for Nuclear Power Plants
Part 100	Reactor Site Criteria (10 CFR 100)
DOE Order 6430.1A	Department of Energy General Design Criteria
ERDA 76-21	Nuclear Air Cleaning Handbook
IEEE 484	Recommended Practice for Design and Installation of Large Lead Storage Batteries for Generating Stations and Substations.
NFPA 801	Recommended Fire Protection Practice for Facilities Handling Radioactive Materials
NFPA 901	Uniform Coding for Fire Protection
NUREG-0696	Functional Criteria for Emergency Response Facilities
NUREG-0800	Standard Review Plans
SRP 6.2.3	Secondary Containment Functional Design
SRP 6.2.5	Combustible Gas Control in Containment
SRP 6.4	Control Room Habitability Systems
SRP 6.5.1	ESF Atmosphere Cleanup Systems
SRP 6.5.3	Fission Product Control Systems and Structures
SRP 9.4.1	Control Room Area Ventilation System
SRP 9.4.2	Spent Fuel Pool Area Ventilation System
SRP 9.4.3	Auxiliary and Radwaste Building Ventilation Systems
SRP 9.4.4	Turbine Area Ventilation System
SRP 9.4.5	Engineered Safety Feature Ventilation System
NUREG-CR-3786	A Review of Regulatory Requirements Governing Control Room Habitability
Regulatory Guides	Nuclear Regulatory Commission
RG 1.52	Design, Testing, and Maintenance Criteria for Engineered Safety Feature Atmospheric Cleanup System Air Filtration and Adsorption Units of LWR Nuclear Power Plants
RG 1.78	Assumptions for Evaluating the Habitability of Nuclear Power Plant Control Room During a Postulated Hazardous Chemical Release
RG 1.95	Protection of Nuclear Power Plant Control Room Operators Against Accidental Chlorine Release
RG 1.140	Design, Testing, and Maintenance Criteria for Normal Ventilation Exhaust System Air Filtration and Adsorption Units of LWR Nuclear Power Plants

REFERENCE

Murphy, K.G. and K.M. Campe. 1974. Nuclear Power Plant Control Room Ventilation System Design for Meeting General Design Criterion 19. Paper 6-6, 13th AEC Air Cleaning Conference (August).

CHAPTER 24

VENTILATION OF THE INDUSTRIAL ENVIRONMENT

GENERAL ventilation controls heat, odors, and hazardous chemical contaminants that could affect the health and safety of industrial workers. For better control, heat and contaminants should be exhausted at their sources by local exhaust systems, which require lower airflows than general (dilution) ventilation (Goldfield 1980). Chapter 26 supplements this chapter.

General ventilation can be provided by natural draft, by mechanical systems, or by their combination. Examples of combination systems include mechanical supply with air relief through louvers and/or other types of vents and mechanical exhaust with air replacement inlet louvers and/or doors.

As a rule, mechanical supply systems provide the best control and the most comfortable environment. They consist of an inlet section, a filter, heating and/or cooling equipment, a fan, ducts, and air diffusers for distributing air within the building. When toxic gases and particles are not present, air that is cleaned in the general exhaust system or in free-hanging filter units can be recirculated via a return duct. Air recirculation can reduce heating costs in winter.

A general exhaust system, which removes air contaminated by gases or particles not captured by local exhausts, usually consists of inlets, ducts, an air cleaner, and a fan. After passing through filters, cleaned air is discharged outside or part is returned to the building. The cleaning efficiency of an air filter should conform to environmental regulations and depends on factors such as building location, background concentrations in the atmosphere, nature of the contaminants, and height and velocity of the discharge. In some cases, for example when the industrial zone is located away from residential areas, a general exhaust system may not have an air cleaner.

Many industrial ventilation systems must handle simultaneous exposures to heat and hazardous substances. In these cases, ventilation can be provided by a combination of local exhaust, general supply, and exhaust systems. The ventilation engineer must carefully analyze supply and exhaust air requirements to determine the optimum balance between them. For example, air supply makeup for hood exhaust may be insufficient for control of heat exposure. It is also important to consider the effect of seasonal conditions on the performance of ventilation systems.

The specification of acceptable toxic chemical and heat exposure levels for design by the industrial hygienist or industrial hygiene engineer requires the use of the appropriate government standards and guidelines given either in this chapter or in reference materials. The standard levels for most chemical and heat exposures are time-weighted averages that allow excursions above the limit as long as they are compensated by equivalent excursions below the limit during the workday. However, exposure level standards for heat and contaminants are not fine lines of demarcation between safe and unsafe exposures. Rather, they represent conditions to which, it is believed, nearly all workers may be exposed day after day without adverse effects (ACGIH 1993). Because a small percentage of workers may be overly stressed at exposure values below the standards, it is prudent that the ventilation engineer design for exposure levels below the limits.

In the case of exposures to toxic chemicals, the number of contaminant sources, their generation rates, and the effectiveness of exhaust hoods are rarely known. Consequently, the ventilation engineer must rely on common ventilation/industrial hygiene practice when designing toxic chemical controls. Close cooperation among the industrial hygienist, the process engineer, and the ventilation engineer is required (Schroy 1986).

This chapter describes principles of good ventilation practice and includes other information to help the engineer appreciate the hygiene concerns involved in the industrial environment. Various publications from NIOSH (1986), the British Occupational Hygiene Society (1987), the National Safety Council (1988), and the U.S. Department of Health and Human Services (1986) provide in-depth coverage of industrial hygiene principles and their application.

Ventilation control measures alone are frequently inadequate for meeting heat stress standards. Optimum solutions may involve additional controls, such as spot air conditioning, changes in work-rest patterns, and radiation shielding. Goodfellow and Smith (1982) summarized the technical progress being made in the industrial ventilation field by different investigators throughout the world. Proceedings from international symposiums (e.g., Ventilation '85, Ventilation '88, and Ventilation '91) are also valuable sources of information on ventilation technology.

Supplemental information can be found in Chapters 16 through 21, 24, 25, and 26 of the 1992 *ASHRAE Handbook—Systems and Equipment*. Chapters 12 through 28 of this volume include ventilation requirements for specific applications, and Chapter 41 covers control of gaseous contaminants. Fundamentals of heating, cooling, and ventilation are covered in Chapter 31 of the 1993 *ASHRAE Handbook—Fundamentals*.

HEAT CONTROL IN INDUSTRIAL WORK AREAS

Ventilation for Heat Relief

Many industrial work situations involve processes that release large amounts of heat and moisture to the environment. In such environments, maintaining comfort conditions (ASHRAE *Standard* 55), particularly during the hot summer months, may not be economically feasible. Comfortable conditions are not physiologically necessary; the body must be in thermal balance with the environment, but this can occur at temperature and humidity conditions well above the comfort zone. In areas where heat and moisture gains from a process are low to moderate, comfort conditions may not be provided simply because personnel exposures are infrequent and of short duration. In such cases, ventilation is one of many controls that may be necessary to prevent excessive physiological strain from heat stress.

The engineer must distinguish between the control needs for *hot-dry* industrial areas and *warm-moist* conditions. In hot-dry areas, a process only gives off sensible and radiant heat, without adding

The preparation of this chapter is assigned to TC 5.8, Industrial Ventilation.

moisture to the air. This increases the heat load on exposed workers, but the rate of cooling by evaporation of sweat is not reduced. Heat balance may be maintained, but it may be at the expense of excessive sweating. In warm-moist conditions, a wet process mainly gives off latent heat. The rise in the heat load on workers may be insignificant, but the increased moisture content of the air seriously reduces heat loss by the evaporation of sweat. The warm-moist condition is potentially more hazardous than the hot-dry condition.

Hot-dry work situations occur around furnaces, forges, metal-extruding and rolling mills, glass-forming machines, and so forth. Typical warm-moist operations are found in textile mills, laundries, dye houses, and deep mines where water is used extensively for dust control.

The industrial heat problem is affected by the local climate. Solar heat gain and elevated outdoor temperatures increase the heat load at the workplace, but these contributions may be insignificant when compared with the process heat generated locally. The moisture content of the outdoor air is an important factor that can affect hot-dry work situations by seriously restricting an individual's evaporative cooling. For warm-moist conditions, solar heat gain and elevated outdoor temperatures are more important because, compared with the moisture release on the job, the moisture contributed by the outdoor air is of little significance.

Both ASHRAE and the International Standards Organization (ISO) have revised standards for thermal comfort conditions for humans (ASHRAE *Standard* 55 and ISO *Standard* 7730). The research these standards are based on was performed mainly under environmental conditions similar to those in commercial and residential buildings, with relatively low activity levels (mainly sedentary, metabolic rate of 1.2 met), normal indoor clothing (with an insulation value of 0.5 to 1.0 clo), and a limited range of environmental parameters. One met is defined as 58.2 W/m^2; one clo is equivalent to 0.155 $m^2 \cdot K/W$. There are other standards and guidelines for evaluating more severe and stressful thermal environments.

Analyses by Zhivov and Olesen (1993) and Olesen and Zhivov (1994) show that existing thermal comfort standards can be extended to workplaces with higher levels of activity.

Methods for evaluating the general thermal state of the body both in comfort conditions and under heat and cold stress are based on an analysis of the heat balance for the human body, which is discussed in Chapter 8 of the 1993 *ASHRAE Handbook—Fundamentals*. A person may find the thermal environment unacceptable or intolerable due to local effects on the body caused by asymmetric radiation, the air velocity, vertical air temperature differences, or contact with hot or cold surfaces (floors, machinery, tools, etc.).

Moderate Thermal Environments

ISO *Standard* 7730 defines the Predicted Mean Vote-Predicted Percent Dissatisfied (PMV-PPD) index (Fanger 1982) as a method for evaluating moderate thermal environments. To quantify comfort, the PMV index gives a value on the 7-point ASHRAE thermal sensation scale: +3 hot, +2 warm, +1 slightly warm, 0 neutral, −1 slightly cool, −2 cool, −3 cold. An equation in the standard for calculating the PMV index is based on six factors: clothing, activity, air and mean radiant temperatures, air speed, and humidity. Even if the PMV is zero, at least 5% of the occupants will be dissatisfied with the thermal environment. This method is also discussed in Chapter 8 of the 1993 *ASHRAE Handbook—Fundamentals*.

The PMV index is determined assuming that all evaporation from the skin is transported through the clothing to the environment; therefore, the PMV index is applicable only within the range −2 < PMV < +2, that is, for thermal environments where sweating is minimal. The PMV index is not applicable for hot environments.

Another method used to estimate combined effects in moderate environments is the Effective Temperature ET*, which is described in Chapter 8 of the 1993 *ASHRAE Handbook—Fundamentals*. In the comfort range, it gives similar results to the PMV index.

ASHRAE *Standard* 55 specifies ranges for operative temperatures that will be acceptable to at least 90% of the occupants. For example, for a sedentary activity level (1.2 met) and typical indoor clothing, the standard recommends the following operative temperature ranges: in winter (heating period, 0.9 to 1.0 clo), 20 to 24°C; in summer (cooling period, 0.5 clo), 23 to 26°C.

Operative temperatures t_o^{act} for activities higher than 1.2 met (but less than 3 met) can be found from ISO *Standard* 7730 or can be calculated from the operative temperatures t_o^{sed} at sedentary conditions using the following equation (ASHRAE *Standard* 55):

$$t_o^{act} = t_o^{sed} - 3(1 + I_{cl})(met - 1.2) \qquad (1)$$

where I_{cl} = insulation value for garment ensemble, clo.

Heat Stress—Thermal Standards

Heat stress is the thermal condition of the environment that, in combination with metabolic heat generation of the body, causes the deep body temperature to exceed 38°C. The recommended heat stress index for evaluating an environment's heat stress potential is the wet-bulb globe temperature (WBGT), which is defined as follows:

Outdoors with solar load:

$$WBGT = 0.7t_{nw} + 0.2t_g + 0.1t_{db} \qquad (2)$$

Indoors or outdoors with no solar load:

$$WBGT = 0.7t_{nw} + 0.3t_g \qquad (3)$$

where

t_{nw} = natural wet-bulb temperature (no defined range of air velocity; not the same as adiabatic saturation temperature or psychrometric wet bulb)

t_{db} = dry-bulb temperature (shielded thermometer)

t_g = globe temperature (Vernon bulb thermometer, 150-mm diameter)

The threshold limit value (TLV) for heat stress is set for different levels of physical stress, as shown in Figure 1 (ACGIH 1992). This graph depicts the allowable work regime, in terms of rest periods

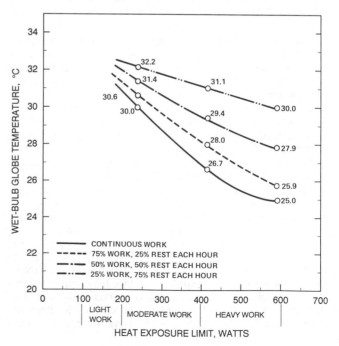

Fig. 1 Heat Stress by WBGT Method

and work periods each hour, for different levels of work, over a range of WBGT. For applying Figure 1, it is assumed that the rest area has the same WBGT as the work area. If the rest area is at or below 24°C WBGT, the resting time is reduced by 25%. The curves are valid for people acclimatized to heat. Refer to criteria of the National Institute for Occupational Safety and Health (NIOSH 1986) for recommended WBGT ceiling values and time-weighted average exposure limits for both acclimatized and unacclimatized workers.

The WBGT index is an international standard (ISO *Standard* 7243) for the evaluation of hot environments. The WBGT index and activity levels should be evaluated on 1-h mean values, that is, WBGT and activity are measured and estimated as time-weighted averages on a 1-h basis for continuous work, or on a 2-h basis when the exposure is intermittent. Although recommended by NIOSH, the WBGT has not been accepted as a legal standard by the Occupational Safety and Health Administration (OSHA). It is generally used in conjunction with other methods to determine heat stress.

Although Figure 1 is useful for evaluating heat stress, it is of limited use for control purposes or for the evaluation of comfort. Air velocity and psychrometric wet-bulb measurements are usually needed in order to specify proper controls, and neither is measured in WBGT determinations. However, Harris (1988) uses the adiabatic wet-bulb temperature line on a psychrometric chart to represent natural wet bulb as a conservative substitute in heat stress situations. More useful tools, including the Heat Stress Index (HSI), may be found in Chapter 8 of the 1993 *ASHRAE Handbook—Fundamentals* and in ISO *Standards* 7730 and 7933.

The thermal relationship between humans and their environment depends on four independent variables: air temperature, radiant temperature, moisture content of the air, and air velocity. Together with the rate of internal heat production (the metabolic rate), these factors may combine in various ways to create different degrees of heat stress. Heat stress index formulas include calculations of the relative contributions to stress resulting from metabolism, radiant heat gain (or loss), convective heat gain (or loss), and evaporative (sweat) heat gain (or loss) (AIHA 1975). For supplemental information on the evaluation and control of heat stress using such methods as reduction of radiation, changes in work-rest pattern, spot cooling, and cooling vests and suits, refer to NIOSH (1986), ACGIH (1992), Constance (1983), Caplan (1980), AIHA (1975), and Brief et al. (1983).

Local Discomfort and Individual Parameters

Even if the combined effect of comfort parameters provides a heat balance for the body as a whole, parts of the body may experience discomfort from drafts, radiant asymmetry, and/or vertical temperature differences.

Air Speed. Air at increased speeds is often used in industry to provide cooling in warm environments. Depending on the conditions, an air speed that is too high may cause a draft, which is defined as unwanted convective cooling of one or more parts of the body. Drafts are probably the largest source of complaints from occupants at low activity levels (sedentary, standing) in air-conditioned spaces.

The maximum value of the mean air speed in an occupied zone depends on the air temperature t_a and the turbulence intensity Tu, which is the ratio of the standard deviation of the air speed to the mean air speed.

ASHRAE *Standard* 55 provides information on air speed and temperature limits for sedentary (less than 1.2 met), light (1.6 met), medium (2 met), and high (greater than 3 met) levels of human activity; however, precise relationships are not available from this standard. Air speeds as high as 1.5 m/s are acceptable, unless they cause problems that are not related to thermal comfort, for example, the blowing of papers or the blowing of shield gas in semiautomatic welding.

Based on the data presented in ISO *Standard* 7730, combinations of air temperature and speed related to comfort conditions in winter and summer periods for different levels of human activity (assuming typical indoor clothing) are presented in Figure 2. Other research and normative data on acceptable and optimal thermal comfort environments are discussed by Olesen and Zhivov (1994).

The effect of turbulence intensity on the sensation of draft has been investigated by Fanger et al. (1988) for different methods of air distribution. Turbulence intensity is the ratio of the magnitude of the velocity fluctuation to the mean velocity:

$$Tu = \sqrt{\frac{\overline{V'^2}}{\overline{V}}} \qquad (4)$$

where

Tu = turbulence intensity

V' = velocity fluctuation

$\overline{V}$ = mean velocity = $\dfrac{1}{\tau} \int_{\tau_1}^{\tau_2} V d\tau$

Based on studies of different air distribution methods, Fanger et al. (1988) found that turbulence intensity in ventilated spaces depends on the type of ventilation system and varies from 10 to 70%. For air velocities ranging from 0.05 to 0.4 m/s and air temperatures ranging from 20 to 26°C, tests were conducted at three levels of turbulence intensity: low (*Tu* less than 12%), medium (*Tu* between 20% and 35%), and high (*Tu* greater than 55%). For displacement systems having the same mean air velocity and temperature, they also found that periodically fluctuating airflow is less comfortable than airflow that does not fluctuate. Very little data on turbulence intensity of different air distribution methods are available, and the effect of turbulence on draft sensation at higher activity levels has not been studied.

Radiant Temperature Asymmetry. Limits for the radiant asymmetry from cold walls and warm ceilings of 10 K and 5 K, respectively, are recommended in both ASHRAE *Standard* 55 and ISO *Standard* 7730. These data are based on testing with sedentary persons in 0.6 clo clothing (Fanger et al. 1985). Similar data are available for warm walls and cold ceilings. In industrial workplaces, the radiant asymmetry from overhead radiant heaters or a hot roof is the main cause of problems. The values established for sedentary persons are too conservative for the higher activity levels and higher ceilings encountered in industry.

Langkilde et al. (1985) showed that significantly higher radiant asymmetry is acceptable. Based on criteria similar to that just described, and requiring that no more than 5% of the occupants are dissatisfied, they recommend an asymmetry limit of 10 to 14 K.

Vertical and Horizontal Air Temperature Differences. Another comfort parameter described in ASHRAE *Standard* 55 and ISO *Standard* 7730 is the temperature difference between the head and feet. The recommended limit of 3 K is based on studies with sedentary persons (Olesen et al. 1979) and cannot be used directly with higher activity levels. It is not the air temperature difference alone, but the combined effect of differences in air temperature, air velocity, and radiant temperature between the head and feet that cause discomfort. Data for industrial environments are available only in *Building Norms and Regulations* (GOSSTROY 1992), in which the vertical temperature difference in occupied zones is limited to 4 K, and the horizontal temperature difference cannot exceed 4 K at light activity, 5 K at medium activity, and 6 K at high activity.

Spot Cooling. If the workplace is located near a source of radiant heat that cannot be entirely controlled by radiation shielding, spot cooling is recommended (Olesen and Nielsen 1980, 1981, 1983; Azer 1982a, b, 1984). The air temperature of the workplace is limited to 20 to 30°C, depending on the heat load and the air speed shown in Table 1.

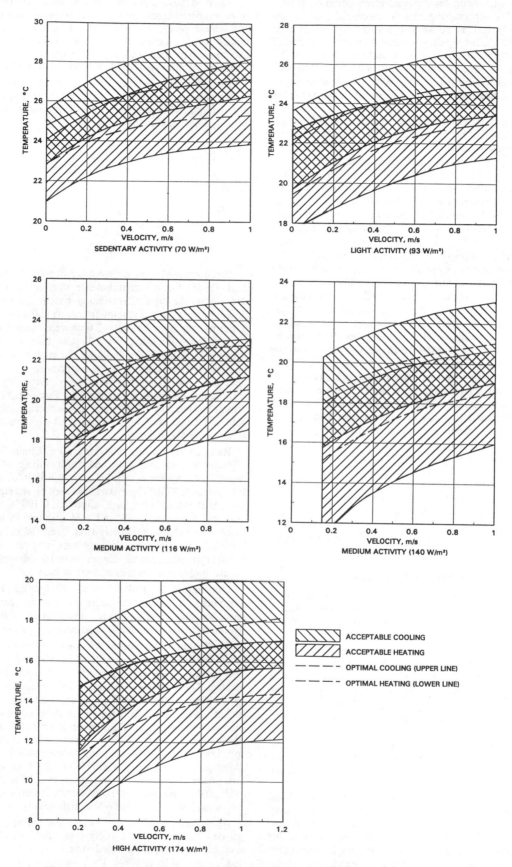

Fig. 2 Optimal and Acceptable Ranges (ISO 7730) of Air Temperature and Air Speed in Occupied Zone for Different Levels of Human Activity

Table 1 Acceptable Air Speed in Workplace

Activity Level	Air Speed, m/s
Continuous exposure	
Air-conditioned space	0.25 to 0.4
Fixed workstation, general ventilation or spot cooling	
Sitting	0.4 to 0.6
Standing	0.5 to 1.0
Intermittent exposure, spot cooling or relief stations	
Light heat loads and activity	5 to 10
Moderate heat loads and activity	10 to 15
High heat loads and activity	15 to 20

Table 2 Recommended Spot Cooling Air Speed and Temperature

Activity Level	Air Speed in Jet, m/s, Averaged on 0.1 m^2 of Workplace	Average Air Temperature, °C, in Jet Cross Section				
		Heat Flux Density, W/m^2				
		140-350	700	1400	2100	2800
Light—I	1	28	24	21	16	—
	2	—	28	26	24	20
	3	—	—	28	26	24
	3.5	—	—	—	27	25
Moderate—II	1	27	22	—	—	—
	2	28	24	21	16	—
	3	—	27	24	21	18
	3.5	—	28	25	22	19
Heavy—III	2	25	19	16	—	—
	3	26	22	20	18	17
	3.5	—	23	22	20	19

Information about combinations of air temperature and speed in jets used for spot cooling is presented in Table 2. More detailed information is provided in SNiP 2.04.05-92 (GOSSTROY 1992). If the air temperature in the occupied zone is less than or greater than the temperatures listed in Table 2, the air temperature in the jet should be increased or decreased, respectively, by 0.4 K for each degree of the temperature difference. The average temperature in the jet may not be lower than 16°C.

The temperature of the radiating surface should be averaged over the period of radiation. If the radiation lasts less than 15 min or more than 30 min, the jet air temperature can be 2 K greater than or less than, respectively, the values listed in Table 2.

HEAT EXPOSURE CONTROL

Control at Source

The heat exposure can be reduced by insulating hot equipment, locating such equipment in zones with good general ventilation within buildings or outdoors, covering steaming water tanks, providing covered drains for direct removal of hot water, and maintaining tight joints and valves where steam may escape.

Local Exhaust Ventilation

Local exhaust ventilation removes heated air generated by a hot process and/or nonbuoyant gases emitted by process equipment. This is done with the removal of a minimum of air from the surrounding space.

The following criterion can be used to evaluate the economics of local exhaust versus general (dilution) ventilation:

$$\frac{C_{le} - C_o}{K_G(C_{oz} - C_o)} > (1.2 \text{ to } 1.5) \qquad (5)$$

where

C_o = concentration of gas, vapor, or particles in air supplied, kg/kg
C_{oz} = concentration of gas, vapor, or particles in occupied zone, kg/kg
C_{le} = concentration of gas, vapor, or particles evacuated by local exhausts, kg/kg
K_G = coefficient of air exchange efficiency for removing gas contaminants from occupied zone

If the value calculated is greater than 1.2 to 1.5, it is more economical to use local exhaust. Chapter 26 covers the design of exhaust hoods and duct systems for local exhaust.

Radiation Shielding

In some industries, the major environmental heat load is radiant heat from hot objects and surfaces, such as, furnaces, ovens, furnace flues and stacks, boilers, molten metal, hot ingots, castings, and forgings. Because air temperature has no significant effect on radiant heat flow, ventilation is of little help in controlling such exposure. The only effective control is reducing the amount of radiant heat impinging on the workers. Radiant heat exposures can be reduced by insulating or placing radiation shields around the source (Chapter 3 of the 1993 *ASHRAE Handbook—Fundamentals*).

Radiation shields are effective in the following forms (AIHA 1975):

- *Reflective shielding.* Sheets of reflective material or insulating board, semipermanently attached to the hot equipment or arranged in a semiportable floor stand.
- *Absorptive shielding* (*water-cooled*). These shields absorb and remove heat from hot equipment.
- *Transparent shields.* Heat reflective tempered plate glass, reflective metal chain curtains, and close mesh wire screens moderate radiation without obstructing view of the hot equipment.
- *Flexible shielding.* Aluminum-treated fabrics give a high degree of radiation shielding.
- *Protective clothing.* Reflective garments such as aprons, gauntlet gloves, and face shields provide moderate radiation shielding. For extreme radiation exposures, complete suits with vortex tube cooling may be required.

If the shield is a good reflector, it will remain relatively cool in severe radiant heat. Bright or highly polished tinplate, stainless steel, and ordinary flat or corrugated aluminum sheets are efficient and durable. Foil-faced plasterboard, though less durable, gives good reflectivity on one side. To be efficient, however, the reflective shield must remain bright.

The best radiation shields are infrared reflectors, which should reflect the radiant heat back to the primary source, where it can be removed by local exhaust. However, unless the shield completely surrounds the primary source, some of the infrared energy will be reflected into the cooler surroundings and possibly into an occupied area. To avoid this, the direction of the reflected heat should be studied to ensure proper installation of the shielding.

VENTILATION DESIGN PRINCIPLES

General Ventilation

General ventilation supplies and/or exhausts air to provide heat relief, dilute contaminants to an acceptable level, and replace (makeup) exhaust air. It can be provided by either natural or mechanical supply and/or exhaust systems. Outdoor air is unacceptable for ventilation if it is known to contain any contaminant at a concentration above that given in ASHRAE *Standard* 62. If air is thought to contain any contaminant not listed in the standard, guidance on acceptable exposure levels should be obtained from OSHA standards. General ventilation rates must be high enough to dilute the carbon dioxide produced by the occupants.

Ventilation design for the industrial environment can be established using the following techniques (Goodfellow 1985, 1986, 1987):

- Field testing methodology
- Fluid dynamic modeling
- Computer modeling

For complex industrial ventilation problems, fluid dynamic modeling or computer modeling are often used in addition to field testing.

Field Testing Methodology

A field testing program must be organized and developed in detail prior to its trial in the field. Ventilation flow in large process buildings is usually complex, so an inexperienced sampling team could collect irrelevant or insufficient data for the subsequent analysis. Therefore, the field testing program should be reviewed and approved by plant operating personnel before the actual field testing begins. A good approach is to (1) perform preliminary calculations for all sources, (2) identify information gaps, and (3) determine what data must be collected.

Although each processing or manufacturing operation has its unique ventilation flows, design parameters, and field testing protocol, many elements of a field testing program can be applied to all ventilation problems. The major engineering activities in a ventilation field testing program are listed in order in Table 3, which includes a brief descriptions of the tasks and scope of the work required for each activity.

Step 1 in Table 3, gathering information, includes obtaining and studying all reports and drawings pertinent to the plant. Specific information about current and future operating practices is required, as well as information on the nature of dust, chemical contaminants, and heat or cold stresses. A visit to the site should include a walk-through ventilation survey using a questionnaire or data sheet developed by the ventilation engineer and experienced process and operating personnel for the specific industry. The questionnaire is used to identify the operating practices (present and future) that will be used as the design basis for the ventilation system. An industrial hygienist should work with the ventilation engineer to establish the scope and extent of the sampling program.

Step 4 in Table 3, developing details of the field testing program, includes preparation of the field data log sheets prior to the actual field testing.

Step 5 in Table 3, carrying out the field testing program, includes the following:

- Measurement of air velocities through all openings. Sufficient time must be allowed to determine representative velocities
- Measurement of the temperature for each velocity measured. For air entering the building, the ambient temperature should be recorded hourly. Of course, the thermometer or temperature probe should not be exposed to sunlight or radiant heat from hot objects. Good temperature readings are important for heat balance calculations, and there must be sufficient temperature data to evaluate the air density distribution.
- Measurement of the mean surface temperatures of hot surfaces to determine the subsequent heat release and air movement.
- Recording weather data. Weather data also should be obtained from the nearest airport or meteorological station. The data, which should be recorded hourly, include ambient temperature, relative humidity, and wind speed and direction.
- Recording plant activities during the testing program. This includes the operational status of all major process and environmental equipment, as well as production levels. Plant records and charts from process operations should be obtained. It is helpful if the ventilation field testing team can work with plant personnel.

Step 8 in Table 3, the preparation of a report on the ventilation field testing program, includes reporting all the field data, calculations, and test results. Using this report, an experienced ventilation engineer can recommend cost-effective solutions for any plant ventilation problem.

Fluid Dynamic Modeling

Fluid dynamic modeling is a valuable tool for modeling problems and developing alternative cost-effective solutions. It has been used extensively for a wide variety of industrial ventilation applications (Baturin 1972; Goodfellow 1985, 1987). Applications of scale modeling include the following:

- Finalizing building ventilation flow rates and schemes
- Examining internal flow patterns and contaminant concentrations at any location
- Examining external flow patterns, including quantitative measurements of downwash and transport of contaminants to other buildings
- Establishing the effectiveness of source hoods

Fluid dynamic modeling can be used in the design of ventilation systems for greenfield plants and for solving ventilation problems in existing plants. For an existing plant, fluid dynamic modeling can supplement the field testing program.

For the design of ventilation systems for new process buildings, the following steps are recommended: (1) the overall ventilation concepts and architectural constraints are developed using computer models; (2) the project design team designs the structural steel while the small-scale model is constructed and tested; and (3) the results of the small-scale modeling are used to refine the ventilation design and to finalize all requirements.

Fluid dynamic modeling includes the following activities:

- Define contaminant source characteristics
- Define environment flows
- Develop details of the scope of the work
- Design the model system
- Construct the model system
- Test the program
- Produce the report

Table 3 Engineering Activities for Ventilation Field Testing Program

Activity	Specific Tasks
1. Gather information	Obtain drawings, reports, operating procedures Review existing data and studies Visit the plant Define problem (summer, winter, heat/cold stress, chemical supply, exhaust)
2. Collect data on ventilation openings	Develop isometric drawings: plans, sections Develop schedule of openings (type, size, location)
3. Develop plant questionnaire	Develop process flow sheet and general layout Identify significant heat sources in building Identify typical operating practice Identify gaps in data to be filled in by field testing program
4. Develop details of field testing program	Determine building ventilation flow rates Determine in-plant flows Develop field data log sheets for ventilation measurements, weather conditions, plant operating records, etc.
5. Carry out field testing program	Perform field measurements (velocity, temperature, pressure, etc.)
6. Analyze data and perform calculations	Determine plant ventilation flow balance Determine in-plant flow patterns Perform heat balance calculations (total plant/ventilation) Calculate air set-in-motion volumes
7. Perform computer simulation of ventilation flows (natural ventilation)	Calibrate using field test data Run program for different conditions (summer, winter, etc.)
8. Produce field testing report	Summarize test conditions and test results Make recommendations

In defining the contaminant source characteristics, the size of the industrial building and details about the source flux (e.g., heat and contaminant release rates) must be determined. Information about the major sources of heat is used to calculate heat balances and volumes of air set in motion.

In defining environment flows, data are required on the external and internal flow conditions for the prototype. Information may be needed about site conditions such as wind speed and direction.

Design of the model system requires an examination of the scaling parameters and possible fluid media to select the best model for the specific ventilation problem. Although the most common media used in models are air and water, a variety of working or buoyancy-driven fluids are available. Fluid systems used include air and heated air, water and saltwater, water and carbon tetrachloride, and mercury and carbon tetrachloride. Usually, air models are the simplest and least expensive models to build and test, but they must be large to ensure fully turbulent flow. Because water has a smaller kinematic viscosity than air, a small model is required to ensure a high Reynolds number and turbulent flow. Flow visualization is easier with water-based models because velocities are lower than with air. Data measured in the model flow can be related quantitatively to the full-scale prototype flow by establishing dynamic similarity (geometric and kinematic) between the model and the prototype. If the model and prototype are to have similar ventilation, their Archimedes numbers must be equal (Baturin 1972):

$$Ar_m = Ar_p$$

$$\left(\frac{gL_o(t_o - t_s)}{V_o^2 T_s}\right)_m = \left(\frac{gL_o(t_o - t_s)}{V_o^2 T_s}\right)_p$$

where

g = gravitational acceleration rate, m/s^2
L_o = length scale, m
t_o = initial temperature of jet, °C
t_s = temperature of surrounding air, °C
V_o = initial air velocity of jet, m/s
T_s = room air temperature (absolute), K
p = subscript identifying prototype
m = subscript identifying model

An adequate simulation of the convective flows requires that the value of the Grashof-Prandtl number exceed 2×10^7:

$$Gr \times Pr = \frac{gL^3 \Delta t}{\nu^2 T_s} \times \frac{c_p \nu \rho}{k} = \frac{gL^3 \Delta t\, c_p \rho}{\nu T_s k} \geq 2 \times 10^7$$

where

ν = kinematic viscosity, m^2/s
Δt = temperature difference between two points, K
c_p = specific heat at constant pressure, kJ/(kg·K)
k = thermal conductivity, W/(m·K)
ρ = density, kg/m^3

The technique used to measure a contaminant must be evaluated for its impact on the cost of the testing program and on the degree of accuracy of the measurements.

For any model testing program, the use of photography and video cameras to record results is suggested. The photographs and videos are invaluable for analyzing test results and for presenting the proposed solutions to management.

Computer Modeling

Figure 3 shows the flowchart for a computer ventilation model based on design equations for flow and heat balances.

Once the observed ventilation flow rates are correctly modeled, the computer program can be used to study the effects of different weather conditions (e.g., summer/winter, wind direction and speed) on the ventilation. The computer program is useful also for evaluating and comparing proposed schemes to find the most cost-effective method of improving the ventilation.

Computer ventilation models can reliably predict the gross ventilation rates for complex process buildings. High-speed computers enable the designer to examine the impact of architectural changes, wind conditions, or process changes on the performance of the proposed ventilation scheme. Problems such as contamination resulting from cross drafts or high temperatures in the work environment can be identified quickly and corrected. Other ventilation questions, such as internal flow patterns, intermittent flows, 2-D or 3-D flow, the location of fresh air, and contaminant concentrations in the breathing zone can be answered using computer models.

Computer programs available for microcomputers can solve complex 3-D ventilation problems (Cawkwell and Goodfellow 1990). Proceedings for conferences such as Ventilation '85, Ventilation '88,

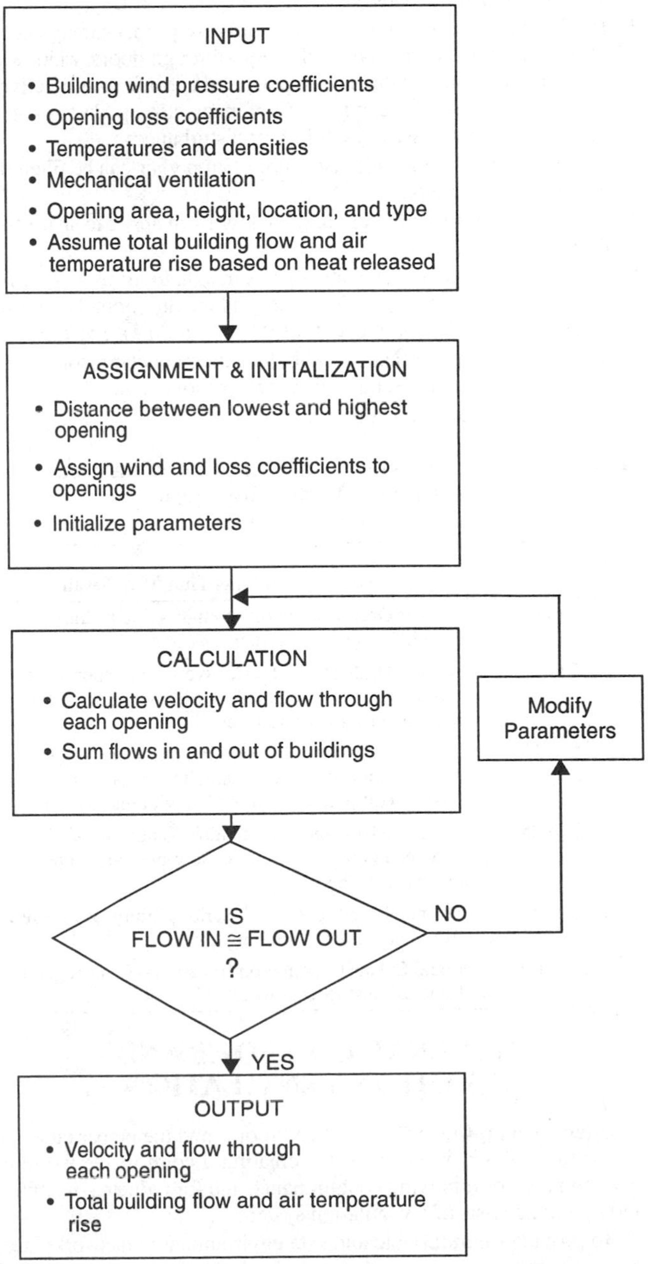

Fig. 3 Computer Ventilation Model Flow Diagram
(Goodfellow 1985)

Ventilation '91, RoomVent '90, RoomVent '92, and RoomVent '94 include many technical papers on new computer programs for calculating contaminant levels as a function of the location and time spent in a workplace environment. These computer programs are now being used in the routine engineering design of ventilation systems.

Need for Makeup Air

For safe, effective operation, most industrial plants require makeup air to replace the large volumes of air exhausted to provide conditions for personnel comfort, safety, and process operations. Makeup air consistently provided by good air distribution results in more effective cooling in the summer and more efficient and effective heating in the winter. The use of windows or other inlets that cannot be used in stormy weather is discouraged. The most important requirements for makeup air can be summarized as follows:

1. To replace air being exhausted through combustion processes and local and general exhaust systems (see Chapter 26).
2. To eliminate uncomfortable cross drafts by proper arrangement of supply air and to prevent infiltration (through doors, windows, and similar openings) that may make hoods unsafe or ineffective, defeat environmental control, bring in or stir up dust, or adversely affect processes by cooling or disturbances.
3. To obtain air from the cleanest source. Supply air can be filtered; infiltration air cannot.
4. To control building pressure and airflow from space to space for three reasons:
   a) To avoid positive or negative pressures that will make it difficult or unsafe to open doors, to replace air (Item 1), and to eliminate drafts and prevent infiltration (Item 2). See Table 4.
   b) To confine contaminants and reduce their concentration and to control temperature, humidity, and air movement.
   c) To recover heat and conserve energy.

Table 4   Negative Pressures That May Cause Unsatisfactory Conditions Within Buildings
(ACGIH 1992)

Negative Pressure, Pa	Adverse Conditions That May Result
2.5 to 5.0	Worker Draft Complaints—High-velocity drafts through doors and windows
2.5 to 12	Natural Draft Stacks Ineffective—Ventilation through roof exhaust ventilators, flow through stacks with natural draft greatly reduced
5 to 12	Carbon Monoxide Hazard—Back drafting will take place in hot water heaters, unit heaters, and other combustion equipment not provided with induced draft
7 to 25	General Mechanical Ventilation Reduced—Airflows reduced in propeller fans and low-pressure supply and exhaust systems
12 to 25	Doors Difficult to Open—Serious injury may result from unchecked, slamming doors
25 to 60	Local Exhaust Ventilation Impaired—Centrifugal fan fume exhaust flow reduced

GENERAL COMFORT AND DILUTION VENTILATION

Effective air diffusion in ventilated rooms and the proper quantity of conditioned air are essential for creating a comfortable working environment, for removing contaminants, and for reducing the initial and operating costs of a ventilation system.

To provide a comfortable and safe environment in the workplace, general ventilation systems must supply air that has the proper speed and temperature, as well as contaminant concentrations that are within permissible limits. In most cases, the objective is to provide tolerable (acceptable) working conditions rather than total comfort (optimal) conditions.

The design of a general ventilation system is based on the assumption that local exhaust ventilation, radiation shielding, and equipment insulation and encapsulation have been selected to minimize both the heat load and the contamination level in the workplace. In cold climates, infiltration and heat loss through the building shell also must be minimized.

Quantity of Supplied Air

Sufficient air must be supplied to replace air exhausted by process ventilation and local exhausts, to provide dilution of contaminants (gases, vapors, or airborne particles) not captured by local exhausts, and to provide the required thermal environment. The amount of supplied air should be the maximum of that needed for temperature control, dilution, or replacement.

The airflow rate Q_o to be supplied through diffusers for temperature control can be estimated from the following equation:

$$Q_o = Q_{exh} + \frac{W - \rho c_p Q_{exh} (t_{oz} - t_o)}{\rho c_p K_t (t_{oz} - t_o)} \qquad (6)$$

where

Q_o = required airflow rate, m³/s
Q_{exh} = flow rate of air evacuated from the occupied zone by general and/or local exhaust and process equipment, m³/s
W = surplus heat gain or heat supplied to warm the air, W
c_p = specific heat at constant pressure, kJ/(kg·K)
ρ = density, kg/m³
t_{oz} = air temperature in the occupied zone, °C
t_o = temperature of supplied air, °C
t_{exh} = temperature of exhausted air, °C
K_t = coefficient of air exchange efficiency for removing heat from the occupied zone
$\quad = \dfrac{t_{exh} - t_o}{t_{oz} - t_o}$

The airflow rate Q_o required to remove water vapor released into the space at a rate M_w (kg/s) is given by the following equation:

$$Q_o = Q_{exh} + \frac{M_w - \rho Q_{exh} (W_{oz} - W_o)}{\rho K_w (W_{oz} - W_o)} \qquad (7)$$

where

W_o = humidity ratio, kg of water per kg of dry air supplied
W_{oz} = humidity ratio, kg of water per kg of dry air in the occupied zone
W_{exh} = humidity ratio, kg of water per kg of dry air in exhausted air
K_w = coefficient of air exchange efficiency for removing water vapor from the occupied zone
$\quad = \dfrac{W_{exh} - W_o}{W_{oz} - W_o}$

The airflow rate Q_o required for dilution ventilation can be calculated as follows:

$$Q_o = Q_{exh} + \frac{G - \rho Q_{exh} (C_{oz} - C_o)}{\rho K_G} \qquad (8)$$

where

G = rate of contaminant release into the space, kg/s
C_o = concentration of gas, vapor, or particles in air supplied, kg/kg
C_{oz} = concentration of gas, vapor, or particles in occupied zone, kg/kg
C_{exh} = concentration of gas, vapor, or particles in exhaust, kg/kg
K_G = coefficient of air exchange efficiency for removing gas contaminant from the occupied zone
$\quad = \dfrac{C_{exh} - C_o}{C_{oz} - C_o}$

The values of the air exchange coefficients (K_t, K_w, and K_G) depend on the method of air distribution, characteristics of the ventilated space, and activities within the space. They can be determined by laboratory and field tests.

Table 5 Coefficients of Ventilation Efficiency for Mechanically Ventilated Spaces with Insignificant (Less Than 23 W/m³) Heat Load

Air Supply Method	Ventilation Efficiency Coefficients K_t/K_w			
	Air Exchange Rates, ACH			
	3	5	10	> 15
Concentrated air jets	0.95/1.1	1.0/1.05	1.0/1.0	1.0/1.0
Concentrated air jets and vertical and/or horizontal directing jets	1.0/1.0	1.0/1.0	1.0/1.0	1.0/1.0
Inclined air jets from a height				
greater than 4 m	1.15/1.4	1.1/1.2	1.0/1.1	1.0/1.0
less than 4 m	1.0/1.2	1.0/1.1	1.0/1.05	1.0/1.0
Trough ceiling-mounted air diffusers with				
radial attached jets	0.95/1.1	1.0/1.05	1.0/1.0	1.0/1.0
conical (compact) jets	1.05/1.1	1.0/1.05	1.0/1.0	1.0/1.0
linear attached jets	1.1/1.2	1.05/1.1	1.0/1.05	1.0/1.0

In some cases, the value of K_t also can be obtained analytically (Shilkrot 1993; Pozin 1993). For the mixing-type air distribution, the values of K_t, K_w, and K_G are 1 ± 0.1. Displacement ventilation and air supply with inclined jets lower than 4 m take advantage of thermal stratification. When the sources of heat and gaseous emissions are close to each other, both K_t and K_G can be greater than 1.2 and, in some cases, can reach 2.5 or more. Values of K_t and K_w for typical mixing-type air distribution methods are listed in Table 5. Values of K_t and K_w for displacement ventilation are close to those for natural ventilation and are listed in Table 9. When the sources of heat and gaseous emissions are separated or are distributed uniformly in the occupied zone, the value of K_G for displacement ventilation does not differ much from 1, although K_t may be greater than 1.2 (Shilkrot and Zhivov 1992).

Local air supply does not provide the desired air quality for the entire occupied zone, but only for certain areas that are permanently occupied. This dramatically reduces the amount of supplied air.

The design value of the concentration C_{oz} of a contaminant in the occupied zone should not exceed an acceptable level of exposure such as a permissible exposure limit (PEL), a threshold limit value (TLV), or an in-house TLV. It is desirable to set design objectives below statutory PELs and TLVs because of variations in sensitivity to contaminants and because acceptable limits can be lowered.

Determination of the generation rate of the contaminant G (in Equation 8) can be based on production records, material balance, similar operations, and experience. However they are obtained, the acceptable exposure level and the contaminant generation rate are necessary to design a dilution ventilation system properly. Design based on the number of air changes per hour, or other estimates, are inadequate and could lead to unacceptably high exposure levels or to unnecessarily high installation costs and/or energy consumption.

Air Distribution

Methods of room air distribution can be classified as mixing-type, displacement, and local.

Mixing-type air distribution. This creates a more or less uniform distribution of air, temperature, humidity, velocity, and concentration of contaminants in the occupied zone, as well as along the room height.

Shepelev (1978), Grimitlyn and Pozin (1993), Tarnopolsky (1992), and Zhivov (1992, 1993) describe typical methods of mixing-type air distribution:

A. Air is supplied by concentrated air jets, and ventilation of the occupied zone is provided by reverse flow (Figure 4a).

B. Air is supplied by concentrated air jets attached to the ceiling, and ventilation of the occupied zone is provided by reverse flow (Figure 4b).

C. Air is supplied by concentrated air jets and vertical and/or horizontal directing jets, and ventilation of the occupied zone is provided by reverse flow and vertical directing jets (Figure 4c).

D. Air is supplied by inclined air jets (Figure 5a,b).

E. Air is supplied through ceiling-mounted diffusers (Figure 5c,d,e).

F. Air is supplied through wall-mounted grilles, and ventilation of the occupied zone is provided by the jet directly (Figure 5f).

G. Air is supplied through the wall-mounted grille, and ventilation of the occupied zone is provided by the jet and reverse flow (Figure 5g).

Displacement ventilation. Cold fresh air is discharged through panels installed in the occupied zone, spreads along the floor under the influence of gravitational forces, and floods the lower zone of the room (Jackman 1991, Skaret 1985, Shilkrot and Zhivov 1992). See Figure 6. The air close to the heat source is heated and rises upward as a convective airstream. In the upper zone, this stream spreads along the ceiling. The lower part of the convective stream induces the cold air of the lower zone of the room, and the upper part of the convective airstream induces the heated air of the upper zone of the room.

The height of the lower zone depends on the air volume discharged through the panels into the occupied zone and on the amounts of convective heat discharged by the sources. In some applications of displacement ventilation (in computer rooms or in hot industrial buildings), air can be supplied into the occupied zone through a false floor.

Air supplied locally. Air is supplied toward the breathing zone of the occupants to create comfortable conditions and/or to reduce the concentration of pollutants.

Guidelines for Selection of an Air Distribution Method

The economical and hygienic efficiencies of general ventilation systems depend mainly on the proper choice of the method of air distribution. Zhivov (1992) recommends the following steps for selecting a method of air distribution:

1. Select several possible methods of air distribution that are suitable for the given room, the demands for comfort conditions in the occupied zone, the size of the equipment, and the internal loads and operation modes of the HVAC or air-conditioning systems in a year-round cycle (variable/constant air volume system).

2. Determine the air volume requirements.

3. Determine the dimensions (length L and width B) of the area ventilated through one air diffuser, the type and size of air diffuser, the height of its installation, and the position of the controls for the design mode of system operation (e.g., highest heat loads in summer).

4. Check air flow pattern and air parameters in the occupied zone for other characteristic modes of system operation (e.g., heating regime). This is extremely important for VAV systems.

5. Compare the possible methods, and choose the most economical one.

Guidelines for the selection of an air distribution method for typical situations (GOSSTROY 1992) are presented in Table 6.

If there are significant surplus heat gains in a room, the methods shown in Figures 5a, 5e, or 6 should be used with air diffusers installed lower than 4 m to supply conditioned air closer to the occupants and to take advantage of thermal stratification.

In variable-air-volume (VAV) HVAC systems, variations in the air volume supplied through diffusers and in the initial temperature difference during a year-round cycle of the system operation cause changes in the ratio between gravitational and inertial forces in the air jets. This affects the trajectory and throw of a jet, as well as the distance to the point where the jet separates from the ceiling. Undesirable airflow patterns or larger zones that are poorly ventilated may result, particularly in spaces with many obstructions. Operation

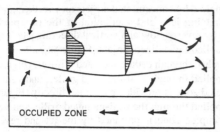

a. Air is supplied by nonattached horizontally projected jet, and occupied zone is ventilated by reverse flow.

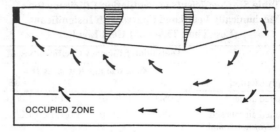

b. Air is supplied by horizontally projected jet attached to the ceiling, and occupied zone is ventilated by reverse flow.

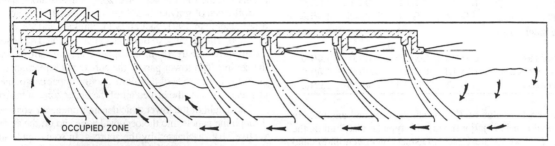

c. Air is supplied by horizontally projected concentrated air jets and vertical and/or horizontal directing jets, and occupied zone is ventilated by reverse flow and vertical directing jets.

Fig. 4   Concentrated Air Supply Methods

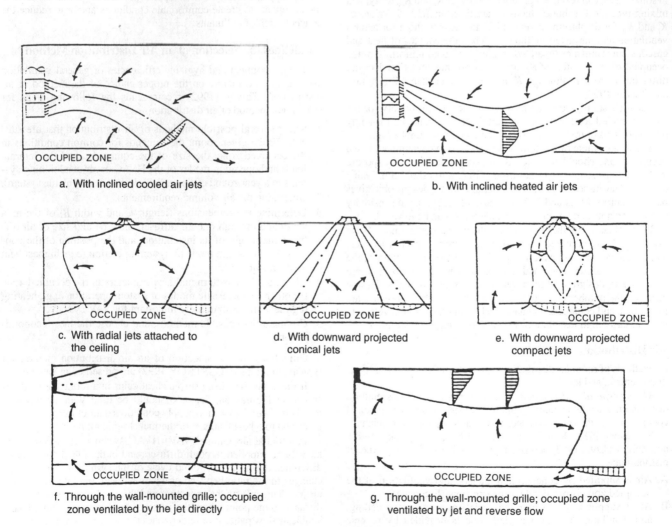

a. With inclined cooled air jets

b. With inclined heated air jets

c. With radial jets attached to the ceiling

d. With downward projected conical jets

e. With downward projected compact jets

f. Through the wall-mounted grille; occupied zone ventilated by the jet directly

g. Through the wall-mounted grille; occupied zone ventilated by jet and reverse flow

Fig. 5   Nonconcentrated Air Supply Methods

Table 6 Guidelines for Selection of Air Distribution Method

Characteristics of the Ventilated Space						
Height, m	Air Change Rate, ACH	Activity Type	Demand for Parameter Uniformity in Occupied Zone	Obstructions	HVAC System Type*	Recommended Air Supply Method
> 6-8	5-7	Moderate and heavy	No special demands	Insignificant	CAV VAV	See Fig. 4 (a,b,c); Fig. 5 (a,b,e); Fig. 6 See Fig. 4c; Fig. 5 (a,b,e)
				> 3 m	CAV VAV	See Fig. 4c; Fig. 5 (a,b); Fig. 6 See Fig. 4c; Fig. 5 (a,b)
< 6-8	> 5	Sedentary and light	High degree of uniformity	Insignificant	CAV VAV	See Fig. 5 (c,d,e,f,g); Fig. 6 See Fig. 5 (c,e,f,g)
				> 2 m	CAV VAV	See Fig. 5 (d,e); Fig. 6 See Fig. 5e

*CAV = constant air volume; VAV = variable air volume.

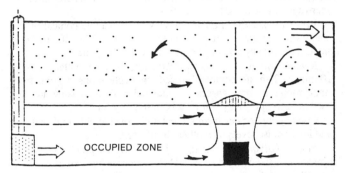

Fig. 6 Displacement Ventilation

modes and air distribution design with VAV HVAC systems are discussed by Zhivov (1990).

If the initial temperature differential ($t_o - t_{oz}$) of the supplied air is increased, the airflow supplied for temperature control can be reduced. Mixing-type air distribution methods provide high entrainment of the ambient air into the diffuser air jet, making it possible to increase the supply air temperature when in the heating mode or to supply low temperature air when in the cooling mode.

Displacement ventilation cannot be used to supply heated air and can only be used to supply chilled air if the temperature differential is less than 5 to 6 K.

When local exhaust ventilation is required for contaminant control, the exhaust rates are frequently greater than the required general ventilation for comfort control. Under these conditions, the exhaust rate determines makeup air rates. A supply system with air distribution arranged so that the makeup air is provided without disturbance at the hoods and process is mandatory.

Caplan and Knutson (1978) and Peterson et al. (1983) established that the manner of air supply to a room with a hood has a major impact on the hood performance. The hood performance factor (or hood index) is defined as the logarithm of the ratio of contamination concentration within the hood to that just outside the hood. Adequate hood performance factors range from 4 to 6, with the higher number indicating a more effective hood installation (Fuller and Etchells 1979).

Local Area or Spot-Cooling Ventilation

In hot workplaces that have only a few work areas, it is impractical and wasteful to maintain a comfortable environment in the entire building. However, working conditions in areas occupied by workers can be improved by air-conditioned cabins, individual cooling, and spot cooling.

Air-conditioned cabins effectively provide thermal comfort, but they are expensive. Control rooms for monitoring production and manufacturing processes can use this technology.

Individual cooling can be provided by air- or water-cooled suits, by vests, or by helmets. Air-cooled suits are appropriate for moder-

ately high temperatures and activity levels. The supply air can be cooled by a conventional heat exchanger or by a vortex tube. The suits are simple and self-controlled (provided there is sufficient airflow). When the supply air is at skin temperature, heat is removed by increasing the evaporation of sweat. The cooling capacity can be increased if the convection heat loss is improved by lowering the supply air temperature. Water-cooled suits have almost unlimited cooling capacity and lower pumping requirements, making them superior to air-cooled suits. A water-cooled garment needs a control system to protect the wearer from undercooling or overcooling. Various physiological measurements, including oxygen consumption and skin temperature measured at selected sites, have been used in the control of water-cooled garments.

Individual cooling with vests or helmets is appropriate if the thermal conditions are not extreme. Vests and helmets remove less heat than cooled suits, but they cost less, are easier to use, and allow increased mobility.

Spot cooling is probably the most popular method of improving the thermal environment. Spot cooling can be provided by radiation (decreasing the mean radiant temperature), by convection (increasing the air velocity), or by a combination of the two methods. Spot cooling equipment is fixed at the workplace, whereas individual cooling has the worker wearing the cooling equipment.

Radiant spot cooling is provided by cooling panels installed near each workstation. Cooling panels decrease the mean radiant temperature, which allows greater radiant heat loss from the workers. The cooling power of radiant spot cooling can be changed either by altering the surface temperature of the cooling panels or by altering the angle factor between the cooling panels and the workers. (The angle factor is determined by the distance from the cooling panel to the worker and by the position of the cooling panel relative to the worker.) Radiant spot cooling is not very efficient. It may improve the thermal comfort of the workers but also may create local discomfort due to radiant asymmetry. In the design of radiant spot cooling, water condensation on the cooling panels should be considered. Condensation occurs when the surface temperature of the panel is equal to or less than the dew point temperature of the room air. Below 0°C, the panel will be covered with ice. In practice, the poor efficiency, water condensation, and positioning of the panels may limit the application of radiant spot cooling.

Convective spot cooling by air jets is an efficient way of providing acceptable thermal comfort conditions. Convective spot cooling exposes workers to air with an increased velocity; the jet air may be at the same temperature as the air temperature in the workplace or cooler. Different combinations of jet outlet velocity and temperature can produce the same cooling effect. The optimal combination of jet velocity and temperature depends on the type of work performed, the clothing worn, the surface area of the exposed body parts, the size of the target area, and the direction of the jet. A design procedure for spot cooling by air jets has been developed and optimized (Azer 1984, Hwang et al. 1984). The procedure is based on a semi-empirical model of the fully developed region of a cold jet projected downward in a hot environment. Skin wettedness, which is defined as the fraction of the subject's body surface area covered by evaporative moisture, is used as a physiological index. The cooling air jet is supposed to provide a skin wettedness of 0.5 as an acceptable physiological strain. For maximum cooling, studies (Robertson and Downie 1978; Melikov et al. 1994) show that the initial (potential core) region and the transition region of the jet should be used for spot cooling workers. The fully developed region of the jet does not cool as well because intensive mixing of cold jet air with warm room air increases the jet target temperature, which decreases the cooling capacity of the jet. Subjective studies (Olesen and Nielsen 1983, Melikov et al. 1991) found significant differences in the target velocities preferred by individuals; therefore, individual control of the jets and jet temperature are recommended. Spot cooling systems that also allow workers to adjust the distance to the jet outlet and the direction of the jet are preferable. The use of convective spot cooling, especially at high temperatures, reduces heat stress but may cause local discomfort due to draft. In hot environments, such as those in foundries and steel mills, velocities as high as 15 to 20 m/s are common (Table 1). When high-velocity air is used, it is important to avoid hot air convection, dust entrainment (which can be hazardous to eyes), and disturbing local exhaust systems.

Air Distribution Design in Industrial Spaces

Chapter 31 of the 1993 *ASHRAE Handbook—Fundamentals* should be reviewed for basic principles of air distribution design and air diffuser selection. Design methods for air distribution in large industrial spaces (Figures 4, 5, and 6) are discussed by Zhivov (1990, 1993).

Air Distribution for Local Relief. When designing air distribution for local relief, the following factors should be considered:

Location. For general low-level ventilation, the outlets should be at about 3 m, although 2.5 to 3.5 m is acceptable. For spot cooling, the outlets should be kept close to the worker to minimize mixing with warmer air in the space. In most spot cooling installations, the outlets should be brought down to the 2-m level.

Discharge Velocity, Temperature and Air Volume. Discharge velocity, temperature, and air volume should be designed to provide thermal comfort parameters for workers as recommended in Tables 1 and 2. Guidelines are discussed in Chapter 31 of the 1993 *ASHRAE Handbook—Fundamentals*.

Air Diffusers and Their Performance. A wide variety of air diffusion devices can be used with different air distribution methods.

Grilles are one of the most universal types of air diffusers. They can have one or two rows of vertical or horizontal vanes and different aspect and vane ratios. Vanes affect grille performance if their depth is at least equal to the distance between the vanes. A grille discharging air uniformly forward (with the vanes in a straight position) has a spread of 14 to 24°, depending on the duct approach, the discharge velocity, and the type of diffuser. Turning the vanes affects the direction and throw of the discharged airstream. Parallel, horizontal vanes direct the airstream vertically within 45°. If the vane ratio is less than two, the jet inclination will be smaller than the angle of the vanes.

Vertical vanes spread the air horizontally, and horizontal vanes spread the air vertically. A grille with diverging vanes (i.e., vertical vanes with uniformly increasing angular deflection from the centerline to a maximum at each end of 45°) has a spread of about 60° and reduces the throw considerably. With increasing divergence, the quantity of air discharged by the grille decreases for a given total upstream pressure.

A grille with converging vanes (i.e., vertical vanes with uniformly decreasing angular deflection from the centerline) has a slightly higher throw than a grille with straight vanes, but the spread is approximately the same for both. Compared to the airstream dis-

charged from a grille with straight vanes, the airstream from a grille with converging vanes converges slightly for a short distance in front of the outlet and then spreads more rapidly.

Ceiling-mounted air diffusers can be round, rectangular, or linear and have outlets covered with grilles, perforations, flat plaques, or vanes forming a slot. Ceiling-mounted air diffusers can be regulated or nonregulated. Depending on their design, they can form attached radial, concentrated, or linear jets as well as nonattached conical or concentrated air jets. Rectangular air diffusers with triangular or four-sided grilles form nonuniform circular flow and can be considered as three or four separate jets.

For a variable-air-volume (VAV) application with a considerable air volume and initial temperature differential, air diffusers with a regulated outlet area and/or a regulated direction of air supply, as well as those with induction of room air, perform better within the year-round cycle of system operation.

Wall-mounted round air diffusers with adjustable conical or flat plaques can be used to supply small air volumes (50 to 400 m^3/h) to small rooms. They also can be used for evacuating air.

Round, square, and rectangular nozzles with outlet sizes of 10 mm to 2 m are commonly used for applications ranging from small rooms to large spaces in industrial buildings. They are also used to form directing jets for a mixing-type system.

Converging nozzles form air jets with considerably higher throw and lower noise levels than other air diffusers.

Round, half/quarter round, or flat perforated panels supply air directly into the occupied zone. They discharge air with low velocities (0.2 to 0.5 m/s) and low turbulence. These diffusers can be installed either near walls and columns or inside walls or other interior structures. Flat perforated plenums can be integrated into the ceiling construction to supply air by jets projected downward. Typical applications for air supply diffusers are summarized in Table 7.

Outlet dampers for volume and directional control should always be provided. Figure 7 shows some of the directional outlets that have

Table 7 Typical Air Diffusers and Their Applications

Type of Air Diffuser	Constants of Air Diffuser Performance		Method of Air Distribution	Application
	K_1*	K_2*		
Large grilles	2-6	1.8-5.1	Fig. 4 (a,b,c); Fig. 5 (a,b)	Large shops
Sidewall grilles	2-6	1.8-5.1	Fig. 5 (f,g)	Small rooms
Ceiling-mounted grilles	2-4	1.8-2.8	Fig. 5 (c,e)	Small rooms
Circular ceiling-mounted air diffusers	1-3.0	0.9-3.2	Fig. 5(c,d,e)	Low (< 6 m) rooms
Square ceiling-mounted air diffusers	1-2.8	1.2-3.2	Fig. 5 c,d	Low (< 6 m) rooms
Linear ceiling-mounted air diffusers	2.5	2	Fig. 5 (c,f)	Low (< 6 m) rooms
Perforated panels (round, half/quarter round, flat, mounted on the floor)			Fig. 6	
Perforated ceiling-mounted panels or perforated ceilings	2.1	1.7	Fig. 5e	Low industrial rooms with high air exchange rate
Nozzles	6-6.8	4.2-4.8	Fig. 4 (a,b,c); Fig. 5 (f,g)	

* Approximate ranges are given for air diffuser performance constants:
  K_1 = coefficient of velocity decay along the jet
  K_2 = coefficient of temperature decay along the jet
For actual values of these constants, consult manufacturers' guides.

been used for low-level general ventilation and spot cooling. Outlet A (Navy Type E) in various forms has been used for many years in ship machinery spaces. Outlets B and C are excellent for local area or spot cooling. The adjustable louvers of D applied to directional outlets such as E or F provide excellent control; commercial directional grilles serve the same purpose. The two-way damper arrangement in E is used in local area or aisleway ventilation; it directs the supply air upward in winter to mix discharge air with warm or hot air rising from internal sources. Outlet G is a commercial directional diffuser that can be adjusted to provide a variable downward airflow pattern that ranges from flat to vertical. The outlet is shown on a roof supply fan, which is a common application, but the outlet drop must extend through the layer of hot ceiling air to provide effective relief ventilation.

NATURAL VENTILATION

Natural ventilation is a controlled flow of air caused by thermal and wind pressures. It is commonly used in Canada and in European countries. Numerous studies of natural ventilation have been conducted, and methods of design have been developed (Baturin 1972; Shepelev 1978; Goodfellow 1985; Shilkrot 1993). Shilkrot in Stroiizdat (1992) summarizes these design methods.

Natural ventilation can be used in spaces with a significant heat release as long as contamination of the incoming air by gases and particles does not exceed 30% of the design exposure limits. Natural ventilation is not recommended when air filtration and treatment are required by process or when the incoming outdoor air causes mist or condensation.

In the summer, air inlets for natural ventilation are located in exterior walls; the lower level of the openings is 0.3 to 1.8 m above the floor. Air inlets can be arranged in one, two, or more rows in the longitudinal exterior walls. Windows, doorways, other types of openings in the exterior walls, or apertures in floors over basements (with air transportation along special channels) also can be used as air inlets.

During periods of the year other than summer, air inlets must be located higher than 3 m above the floor level in rooms with ceilings lower than 6 m. If the room height exceeds 6 m, low-level inlets must be located higher than 4 m. These inlets must be supplied with baffles to direct air at an upward angle. Schematics of and pressure loss coefficients for inlets are given in Table 8.

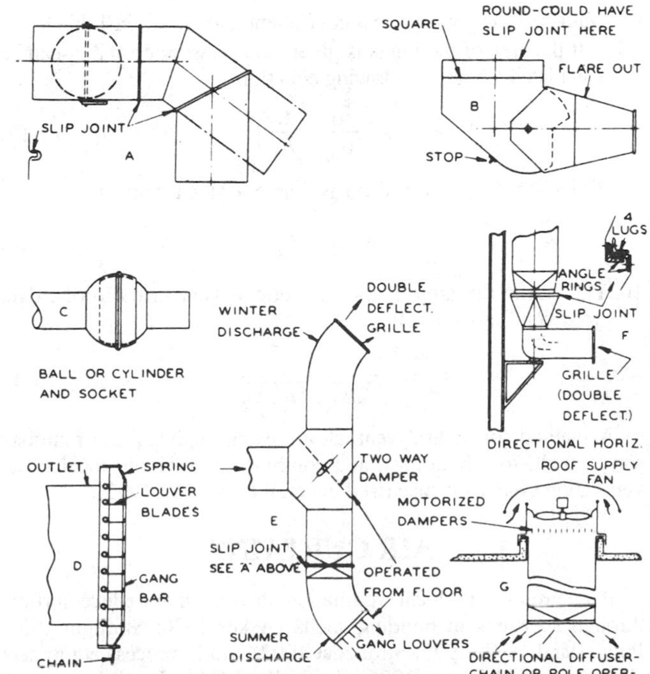

Fig. 7 Directional Outlets for Spot Cooling

Table 8 Pressure Loss Coefficients for Inlet Apertures

Schematic	Baffle Type	h/b	Pressure Loss Coefficient ξ at Different Baffle Angles α: 15°	30°	45°	60°	90°
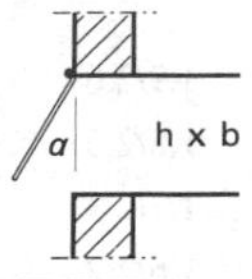	Single, top-hinged baffle	0	30.8	9.2	5.2	3.5	2.6
		0.5	20.6	6.9	4.0	3.2	2.6
		1	16.0	5.7	3.7	3.1	2.6
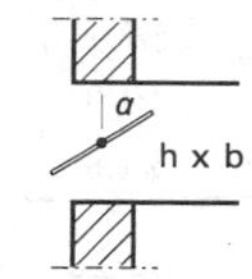	Single, center-hinged baffle	0	59	13.6	6.6	3.2	2.7
		1	45.3	11.1	5.2	3.2	2.4
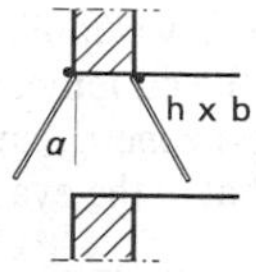	Double, top-hinged baffle	0.5	30.8	9.8	5.2	3.5	2.4
		1.0	14.8	4.9	3.8	3.0	2.4
	Gate	—	—	—	—	—	2.4

Source: Stroiizdat (1992).

Interior ventilated halls can obtain outside air through ridge vents in adjacent "cold" halls. "Cold" halls must be separated from the "hot" ones by screens dropped from the roof to create an airway above the floor that is 2 to 4 m deep. Air supply via ridge vents on the roof is not recommended but can be used if necessary.

Air is evacuated from naturally ventilated spaces through wind-protected continuous ridge vents, skylights, or round roof ventilators. Many types of ridge vents and roof ventilators have been designed; their sizes and pressure loss coefficients are available from the manufacturers. All inlets and outlets for natural ventilation must be supplied with controls for easy opening and closing.

The air exchange rate G_o required for temperature control in the occupied zone can be calculated from the following room heat balance equation:

$$G_o = \frac{W}{c_p K_t (t_{uz} - t_{oz})} \qquad (9)$$

where

G_o = air exchange rate, kg/s
W = surplus heat releases in the space, kW
c_p = specific heat of air, kJ/(kg·K)
K_t = coefficient of natural ventilation efficiency calculated from the air temperatures of the occupied zone t_{oz}, the air removed from the upper zone t_{exh}, and the outdoor air t_o

$$= \frac{t_{exh} - t_o}{t_{oz} - t_o}$$

The method of zone-by-zone heat balances (Shilkrot 1993) can be used to define K_t values. Computed values of K_t for different industrial applications are given in Table 9.

Equation (9) is similar to Equations (6) through (8).

Both the method for natural ventilation design and the method for displacement ventilation design assume temperature stratification throughout the room height. Air close to the heat source is heated and rises as a convective stream. Part of this heated air is evacuated through outlets in the upper zone, and part of it remains in the upper zone, in the so-called heat cushion. Separation level Z (Figure 8), which is the lower level of the upper zone, is defined in

Table 9 Coefficients K_t and K_w of Natural Ventilation Effectiveness for Industrial Spaces with Significant (Higher Than 23 W/m³) Heat Load

Room Category	K_t/K_w
Blacksmith and press shops, oven bays at foundries, rail rolling mills, bloomings, cooling bays, etc.	2.0/2.7
Heat treatment shops	1.9/2.6
Drying shops	1.8/2.5
Casting shops	1.7/2.3
Blast-furnace shops	1.6/2.2
Rolling mills	1.5/2.1
Electrolysis shops, machine and compressor rooms	1.4/1.9
Shops for vulcanization and plastic production	1.3/1.8

terms of the equality of G_{conv} and G_o, which are the airflow rate in the convective flows above the heat source and the airflow supplied to the occupied zone, respectively. It is assumed that the air temperature in the lower zone is equal to that in the occupied zone t_{oz} and that the air temperature in the upper zone is equal to that of the evacuated air t_{exh}.

The incoming air jet can destroy the air temperature stratification when the rarefaction Δp_{jet} created by this jet on the level of separation Z is strong enough to move the heated air from the upper zone to the lower zone:

$$\Delta p_{jet} \geq g (\rho_{oz} - \rho_{uz}) (Z - h_o) \qquad (10)$$

where

Δp_{jet} = rarefaction, Pa
g = gravitational acceleration, m/s²
ρ_{oz} = density of air in the occupied zone, kg/m³
ρ_{uz} = density of air in the upper zone, kg/m³
h_o = height of air supply opening, m
Z = separation level, m

Calculations by Shilkrot (1993) show that when the supply air velocity V_o in the inlet is 2 to 3 m/s and the separation level Z is more than 1 m higher than the air supply opening h_o (i.e., $Z - h_o > 1$ m), a temperature difference of 2 to 3 K between the air in the upper and lower zones ensures the stability of the stratification. Shilkrot has also shown that when the Richardson number is greater than 5, the turbulent exchange between the upper and the lower zones can be neglected.

For single bay shops, the following steps are recommended for natural ventilation design:

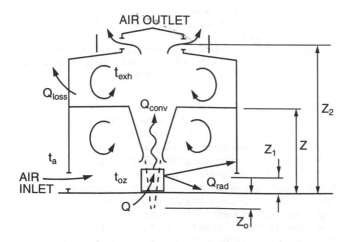

Fig. 8 Natural Ventilation of Single Bay Building

1. Determine the surplus heat load (i.e., heat gains minus heat losses) W (kW) in the space.
2. Select the K_t coefficient from Table 9.
3. Compute the air change rate G_o (kg/s) for temperature control in the occupied zone:

$$G_o = \frac{W}{c_p K_t (t_{oz} - t_o)} \qquad (11)$$

4. Compute the outgoing air temperature, °C:

$$t_{exh} = t_o + \frac{W}{c_p \rho_o Q} \qquad (12)$$

5. Select the separation level Z:

$$Z = 0.4H \text{ to } 0.7H \qquad (13)$$

where

H = ventilated space height, m ($6 \leq H \leq 24$)
$Z = 0.7H$ when $H = 6$ m; $Z = 0.4H$ when $H = 24$ m

6. Determine the pressure difference available to transport air through the inlet and outlet apertures:

$$\Delta p = g (Z - Z_1) (\rho_o - \rho_{oz}) + g (Z_2 - Z) (\rho_o - \rho_{exh}) \qquad (14)$$

where $\rho = \rho_{20°C} \times T_{20°C}/T = 1.2 \times 293/T = 353/T$ in kg/m³.

7. Calculate the pressure loss in the air inlets, Pa:

$$\Delta p_1 = \beta \Delta p \qquad (15)$$

β is part of the pressure available for air transportation through inlets. To reduce the velocity of the incoming air, the inlet area can be increased to keep β in the range 0.1 to 0.4.

8. Calculate the area (m²) of the inlet apertures:

$$F_1 = \frac{G}{\sqrt{2\rho_{out}\Delta p_1 / \xi_1}} \qquad (16)$$

Values of the pressure loss coefficient ξ are listed in Table 8.

If the area of the inlets is given, the pressure drop Δp_1 can be calculated from the following equation:

$$\Delta p_1 = \frac{\xi_1}{2\rho_{out}} \left(\frac{G}{F_1}\right)^2 \qquad (17)$$

9. Calculate the pressure drop available for the air outlets:

$$\Delta p_2 = \Delta p - \Delta p_1 \qquad (18)$$

10. Determine the area of the air outlets (vent in the roof ridge, roof ventilators, etc.):

$$F_2 = \frac{G}{\sqrt{2\rho_{exh}\Delta p_2 / \xi_2}} \qquad (19)$$

A method for natural ventilation design applicable to multibay shops, multistory buildings, and combined natural and mechanical ventilation systems is discussed in detail (Stroiizdat 1992).

AIR CURTAINS

Air curtains are local ventilation devices that reduce airflow through apertures in building shells (Asker 1970; Strongin 1993; Powlesland 1971, 1973; Stroiizdat 1992) and in process equipment (Bintzer and Malehom 1976, Goodfellow 1985, Ivanitskaya et al. 1986, Strongin and Nikulin 1991). They are also used to localize gaseous and particulate emissions near their sources and to convey them toward local exhausts (Stoler and Savelyev 1977, Posokhin and Broida 1980, Posokhin 1985). Different applications of air curtains for jet assisted hoods are discussed in Chapter 26.

Air curtains provide better thermal environments for workstations located near doorways. They can also reduce the energy consumption of HVAC systems.

Shutter-type air curtains create a resistance to airflow through a door and thus reduce the incoming airflow (Figure 9). They direct air toward the incoming outside air at an angle ranging from 30 to 40°. Shutter-type air curtains may be *single-sided* or *double-sided* and projected upward or downward. Shutter-type air curtains are often double-sided because they deter the airflow better. Upward projected air curtains are recommended when the gate width is greater than its height. They also provide better coverage of the lower area of the door opening.

Air curtains can supply heated air, air at room temperature, or air with the outdoor temperature. Air curtains with heated air are recommended for doors smaller than 3.6 m × 3.6 m and for process apertures that are frequently opened (e.g., more than five times or for longer than 40 min during an 8-hour shift) and that are located in regions with design outdoor winter temperatures of −15°C and lower. Air curtains provide desired air temperatures for workstations near apertures; however, the heat consumption is relatively high. Air curtains that supply unheated indoor air have application in spaces with (1) a heat surplus, (2) temperature stratification along the room height, (3) low air temperatures (less than 8°C) near the aperture area, and (4) in regions with a mild climate. Shutter-type air curtains also are recommended for use in the cooled spaces.

Air curtains with a lobby are shown in Figure 10. Performance is based on transition of the supply air jet impulse into counter pressure, which prevents outdoor airflow into the room. Air is supplied in a direction opposite to that of the outdoor airflow or at a small angle to it. A curtain jet propagates along the channel walls, slows down, and makes a U-turn, reversing along its axis. The lobby length is chosen so that it exceeds 2.5 times its width (for double-sided air curtains) and so that air is not forced outside. To shorten the lobby, air is usually supplied by a jet with a coerced angle of divergence. Air curtains with a lobby for supplying outdoor air are shown in Figure 11.

Combined air curtains are used in regions having a very cold climate (with winter temperatures as low as −65°C), for doors larger than 3.6 m × 3.6 m, and for spaces with several doors. A combined air curtain with a specially designed lobby is shown in Figure 12 (Strongin and Nikulin 1987). The lobby has corrugated iron walls that reduce the wind pressure on the gate aperture. Air curtains in the lobby supply untreated outdoor air, while air curtains inside the building supply heated air. Combined air curtains without a lobby are shown in Figure 9c.

Air curtain controls should be designed so that the curtains are turned on when the door is opened and turned off when the door is closed or when the air temperature near the door reaches the target value.

Principles of Air Curtain Design

The velocity V_o of the supplied air can be calculated from the following equation:

$$V_o = \left(\frac{\Delta p f}{2\beta_o \rho E} \right)^{0.5} \qquad (20)$$

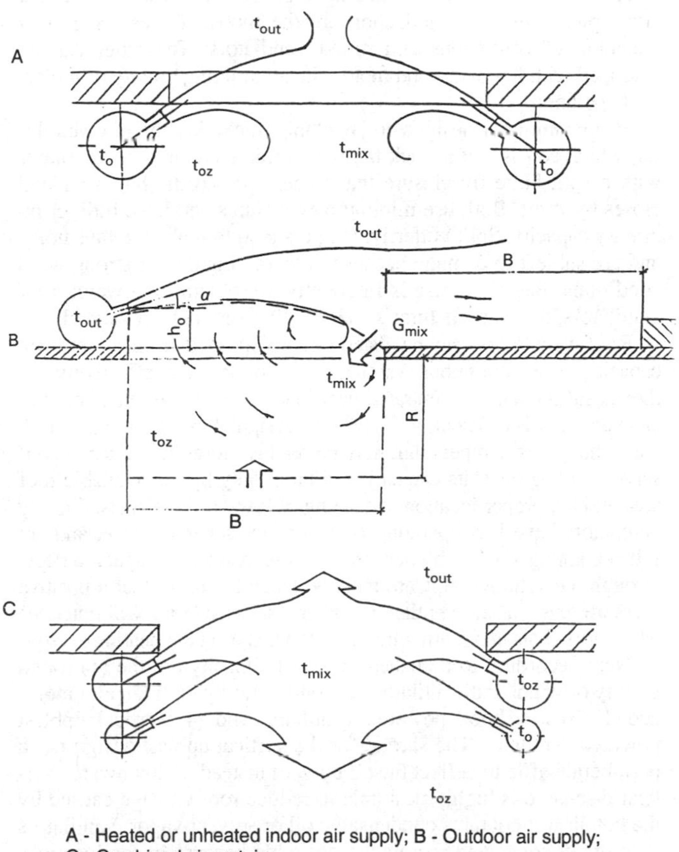

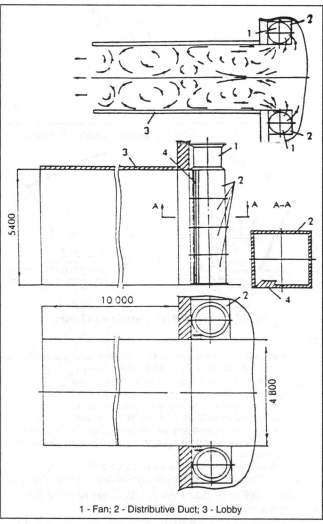

A - Heated or unheated indoor air supply; B - Outdoor air supply;
C - Combined air curtain

Fig. 9   Shutter-Type Air Curtains

1 - Fan; 2 - Distributive Duct; 3 - Lobby

Fig. 10   Air Curtain for Medium-Sized Gate with Lobby

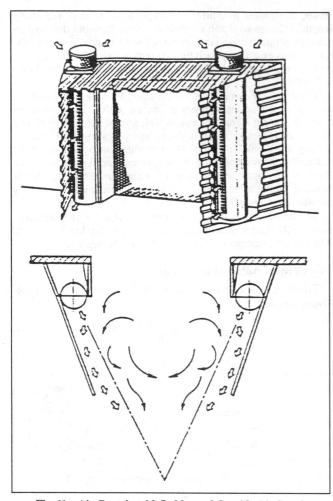

Fig. 11   Air Curtain with Lobby and Outside Air Supply

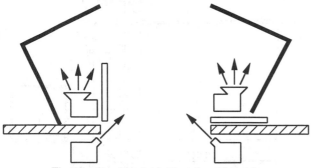

Fig. 12   Combined Air Curtain with Lobby

where

Δp = average pressure difference (Pa) between inside and outside air near the aperture with the air curtain turned on

f = ratio of area of air supply slots A_{ap} to door area A_o

= A_{ap}/A_o

For air curtains with heated air, unheated indoor air, and for combined air curtains, f = 10 to 20. For air curtains supplying outdoor air or protecting air-conditioned spaces, f = 20 to 40.

β_o = Bussinesque coefficient; for air curtain supply nozzle, ranges from 1.05 to 1.1

ρ = density of air supplied by air curtain, kg/m³

E = coefficient of air curtain dynamic efficiency (given in Table 10)

Airflow supplied by the air curtain (kg/s) can be calculated from the following equation:

$$G_o = \rho V_o A_o \qquad (21)$$

Table 10   Air Curtain Dynamic Efficiency Coefficient E

	Air Curtain Type	
$\sin \alpha$	Air Curtains Supplying Heated Air, Unheated Air, and Combinations	Air Curtains Supplying Outdoor Air or Protecting Air-Conditioned Spaces
0.1	0.10	0.15
0.2	0.15	0.20
0.3	0.20	0.25
0.4	0.25	0.30
0.5	0.30	0.40
0.6	0.35	—

Table 11   Coefficients m_1 and m_2

f	m_1	m_2
10	2.0	−0.6
15	2.3	−1.0
20	2.5	−1.3

The temperature t_o of the supplied air can be estimated from the heat balance equation for the aperture under consideration:

$$t_o = t_{out} + m_1 (t_{mix} - t_{out}) + m_2 (t_{oz} - t_{out}) \qquad (22)$$

where

t_{mix} = normative air mixture temperature

m_1 and m_2 = coefficients for the air curtain in Figure 9a; values are given in Table 11

ROOF VENTILATORS

Roof ventilators are heat escape ports located high in a building and properly enclosed for weathertightness (Goodfellow 1985). Stack effect plus some wind induction are the motive forces for gravity operation of continuous and round ventilators. The latter can be equipped with fan barrel and motor, thus permitting gravity or motorized operation.

Many ventilator designs are available, including the low ventilator, which consists of a stack fan with a rainhood, and the ventilator with a split butterfly closure that floats open to discharge air and closes by itself. Both use minimum enclosures and have little or no gravity capacity. Split butterfly dampers tend to make the fans noisy and are subject to damage because of slamming during strong wind conditions. Because noise is frequently a problem in powered roof ventilators, the manufacturer's sound rating should be reviewed.

Roof ventilators can be listed according to their heat removal capacity. The continuous ventilation monitor most effectively removes substantial, concentrated heat loads. The streamlined continuous ventilator is efficient, weathertight, designed to prevent backdraft, and usually has dampers that may be readily closed in winter to conserve building heat. Its capacity is limited only by the available roof area and the proper location and sizing of low-level air inlets. Gravity ventilators have low operating costs, do not generate noise, and are self-regulating (i.e., a higher heat release results in higher airflow through the ventilators). Care must be taken to ensure that a positive pressure exists at the ventilators, otherwise outside air will enter the ventilators. This is of particular importance during the heating season.

Next according to their heat removal capacity are the (1) round gravity or windband ventilator, (2) round gravity with fan and motor added, (3) low hood powered ventilator, and (4) vertical upblast powered ventilator. The shroud for the vertical upblast design has a peripheral baffle to deflect the air upward instead of downward. Vertical discharge is highly desirable to reduce roof damage caused by the hot air if it contains condensable oil or solvent vapor. Ventilators with direct-connected motors are desirable because of the locations of the units and the belt maintenance required for units having short shaft centerline distances. Round gravity ventilators are applicable to warehouses with light heat loads and to manufacturing areas having high roofs and light loads.

Streamlined continuous ventilators must operate effectively without mechanical power. Efficient ventilator operation is generally obtained when the difference in elevation between the average air inlet level and the roof ventilation is at least 10 m and the exit temperature is 14 K above the prevailing outdoor temperature. Chapter 23 of the 1993 *ASHRAE Handbook—Fundamentals* has further details. Under these conditions and with a wind velocity of 8 km/h, the ventilator throat velocity will be about 1.9 m/s and it will remove $1.0 \text{ kJ/(kg·K)} \times 1.2 \text{ kg/m}^3 \times 14 \text{ K} \times 1.9 \text{ m/s} = 32 \text{ kW/m}^2$.

To ensure this level of performance, sufficient low level openings must be provided for the incoming air. Manufacturers recommend an inlet velocity of 1.3 to 2.5 m/s. Insufficient inlet area and significant air currents are the most common reasons gravity roof ventilators malfunction. A positive supply of air around the hot equipment may be necessary in large buildings where the external wall inlets are remote from the equipment.

The cost of electrical power for mechanical ventilation is offset by the advantage of constant airflow. Mechanical ventilation also can create the pressure differential necessary for good airflow, even with small inlets. Inlets should be sized correctly to avoid infiltration and other problems caused by high negative pressure in the building. Often, a mechanical system is justified to supply enough makeup air to maintain the work area under positive pressure.

Careful study of airflow around buildings is necessary to avoid reintroducing contaminants from the exhaust into the ventilation system. Even discharging the exhaust at the roof level or from an area opposite the outside intake may still allow it to reenter the building even when intakes are inside walls opposite the exhaust discharge point. Chapter 14 of the 1993 *ASHRAE Handbook—Fundaments* describes the nature of airflow around buildings.

Fusible link dampers, which close in case of fire, may be required by building codes.

HEAT CONSERVATION AND RECOVERY

Because of the large volumes required for the ventilation of industrial plants, heat conservation and recovery should be practiced and will provide substantial savings. Therefore, heat conservation and recovery should be incorporated in preliminary planning for an industrial plant.

In some cases, it is possible to provide unheated or partially heated makeup air to the building. Rotary, regenerative heat exchangers recover up to 80% of the heat available from the difference between the exhaust and the outdoor air temperatures. Reductions of 5 to 10 K in the heating requirement for most industrial systems should be routine. Although most of the heat conservation and recovery methods in this section apply to heating, the savings possible with air-conditioning systems are equally impressive. The following are some methods of heat conservation and recovery:

1. In the original design of the building, process, and equipment, provide insulation and heat shields to minimize heat loads. Vaporproofing and reduction of glass area may be required. Changes in process design may be required to keep the building heat loads within reasonable bounds. Review the exhaust needs for hoods and process and keep those to a practical, safe minimum.

2. Design the supply and exhaust general ventilation systems for optimal operation modes throughout the year. Use variable-air volume (VAV) HVAC systems and efficient air distribution methods that provide minimal restrictions on the systems operation. The air should be supplied as close to the occupied zone as possible (Zhivov 1990). Recirculated air should be used in the winter makeup, and unheated or partially heated air should

be brought to hoods or process (ACGIH 1992, Holcomb and Radia 1986).

3. Design the system to achieve the highest efficiency and lowest residence time for contaminants. In the design, consider that it is good practice to permit workers to adjust the air patterns to which they are exposed; people want direct personal control over their working environment.

4. When possible, recycle exhaust air. For example, office exhaust can be directed first to work areas, then to locker rooms or process areas, and finally, to the outside. Air heated from cooling equipment in motor or generator rooms can be recycled. The cooling systems for many large motors and generators can be arranged to discharge into the building in the winter to provide heat and to the outside in the summer to avoid heat loads.

5. Supply air can be passed through air-to-air, liquid-to-air, or hot-gas-to-air heat exchangers to recover building or process heat. Rotary, regenerative, and air-to-air heat exchangers are discussed in Chapter 44 of the 1992 *ASHRAE Handbook—Systems and Equipment*.

6. Operate the system for economy. Shut the systems down at night or on weekends whenever possible, and operate the makeup air in balance with the needs of operating the process equipment and hoods. Keep supply air temperatures at the minimum for heating and the maximum for cooling, consistent with the needs of process and employee comfort. Keep the building in balance so that uncomfortable drafts do not require excessive heating.

REFERENCES

ACGIH. 1992. *Industrial ventilation: A manual of recommended practice.* 21st ed. American Conference of Governmental Industrial Hygienists, Cincinnati, OH.

ACGIH. 1993. Threshold limit values and biological exposure indices for 1993-94. American Conference of Governmental Industrial Hygienists, Cincinnati, OH.

AIHA. 1975. *Heating and cooling for man in industry.* 2nd ed. American Industrial Hygiene Association, Akron, OH.

ASHRAE. 1989. Ventilation for acceptable indoor air quality. *Standard* 62-1989.

ASHRAE. 1992. Thermal environmental conditions for human occupancy. *Standard* 55-1992.

Asker, G.S.F. 1970. What, where and how of air curtain systems. *Heating, Piping and Air Conditioning* (June).

Azer, N.Z. 1982a. Design guidelines for spot cooling systems: Part 1—Assessing the acceptability of the environment. *ASHRAE Transactions* 88(1):81-95.

Azer, N.Z. 1982b. Design guidelines for spot cooling systems: Part 2—Cooling jet model and design procedure. *ASHRAE Transactions* 88(1):97-116.

Azer, N.Z. 1984. Design of spot cooling systems for hot industrial environments. *ASHRAE Transactions* 90(1B):460-75.

Baturin, V.V. 1972. *Fundamentals of industrial ventilation.* 3rd ed. Pergamon Press, Oxford, U.K.

Bintzer, W. and F.A. Malehom. 1976. Air curtains on electric arc furnaces at Lukens Steel Co. *Iron and Steel Engineer* (July):53-55.

Brief, R.S., S. Lipton, S. Amarmani, and R.W. Powell. 1983. Development of exposure control strategy for process equipment. *Annals American Conference of Governmental Industrial Hygienists* (5).

British Occupational Hygiene Society (BOHS). 1987. Controlling airborne contaminants in the workplace. Tech. Guide No. 7. Science Review Ltd. and H&H Sci. Consult., Leeds, U.K.

Caplan, K.J. 1980. Heat stress measurements. *Heating, Piping and Air Conditioning* (February):55-62.

Caplan, K.J. and G.W. Knutson. 1978. Laboratory fume hoods: Part 2—Influence of room air supply. *ASHRAE Transactions* 84(1):522-37.

Cawkwell, G.C. and H.D. Goodfellow. 1990. Multiple cell ventilation model with time-dependent emission sources. Proceedings of the 2nd International Conference on Engineering Aero- and Thermodynamics of Ventilated Rooms, A1-9 (Oslo, Norway, June 13-15).

Constance, J.D. 1983. *Controlling in-plant airborne contaminants.* Marcel Dekker, Inc., New York.

Fanger, P.O. 1982. *Thermal comfort.* Robert E. Krieger Publishing Company, Malabar, FL.

Fanger, P.O., A.K. Melikov, H. Hanzawa, and J. Ring. 1988. Air turbulence and sensation of draft. *Energy and Buildings* 12(1).

Fanger, P.O., B.M. Ipsen, G. Langkilde, B.W. Olesen, M.K. Christiansen, and S. Tanabe. 1985. Comfort limits for asymmetric thermal radiation. *Energy and Buildings* 8:225-26.

Fuller, F.H. and A.W. Etchells. 1979. The rating of laboratory hood performance. *ASHRAE Journal* (October):49-53.

Goldfield, J. 1980. Contaminant concentration reduction general ventilation versus local exhaust ventilation. *American Industrial Hygienists Association Journal* 41 (November).

Goodfellow, H.D. 1985. *Advanced design of ventilation systems for contaminant control.* Elsevier Science Publishers, B.V., Amsterdam.

Goodfellow, H.D. 1987. Ventilation, industrial. *Encyclopedia of physical science and technology* (14). Academic Press, Inc., San Diego, CA.

Goodfellow, H.P. and J. W. Smith. 1982. Industrial ventilation—A review and update. *American Industrial Hygiene Association Journal* 43 (March): 175-84.

GOSSTROY. 1992. SNiP 2.04.05-92. Building norms and regulations. Heating, ventilation and air conditioning. Moscow (in Russian).

Grimitlyn M.I. and G.M. Pozin. 1993. Fundamentals of optimizing air distribution in ventilated spaces. *ASHRAE Transactions* 99(1):1128-38.

Harris, R.L. 1988. Design of dilution ventilation for sensible and latent heat. *Applied Industrial Hygiene* 3(1).

Holcomb, M.L. and J.T. Radia. 1986. An engineering approach to feasibility assessment and design of recirculating exhaust systems. Proceedings of Ventilation '85. Elsevier Science Publishers, B.V., Amsterdam.

Hwang, C.L., F.A. Tillman, and M.J. Lin. 1984. Optimal design of an air jet for spot cooling. *ASHRAE Transactions* 90(1B):476-98.

ISO. 1982. Hot environments—Estimation of the heat stress on a working man, based on the WBGT-index (wet bulb globe temperature). *Standard 7243.* International Standards Organization, Geneva.

ISO. 1989. Hot environments—An analytical determination and interpretation of thermal stress using calculation of required sweat rate. *Standard 7933.* International Standards Organization, Geneva.

ISO. 1993. Moderate thermal environments—Determination of the PMV and PPD indices and specifications of the conditions for thermal comfort. *Standard 7730.* International Standards Organization, Geneva.

Ivanitskaya, M.Yu., A.S. Strongin, and E.A. Visotskaya. 1986. Studies of the air curtain application for the channel of the localizing ventilation. Proceedings of Heating and Ventilation. TsNIIpromzdanii (Moscow).

Jackman, P. 1991. Displacement ventilation. CIBSE National Conference (April). University of Kent, Canterbury.

Langkilde, G., L. Gunnarsen, and N. Mortensen. 1985. Comfort limits during infrared radiant heating of industrial spaces. CLIMA 2000, Copenhagen.

Melikov, A.K., L. Halkjaer, R.S. Arakelian, and P.O. Fanger. 1991. Human response to cooling with air jets. ASHRAE Research Project 518-R.

Melikov, A.K., L. Halkjaer, R.S. Arakelian, and P.O. Fanger. 1994. Spot cooling—Parts 1 and 2. *ASHRAE Transactions* 100(2).

National Safety Council. 1988. *Fundamentals of industrial hygiene,* 3rd ed. Chicago, IL.

NIOSH. 1986. Occupational exposure to hot environments. Revised criteria. National Institute for Occupational Safety and Health, Washington, D.C.

Olesen, B.W. and R. Nielsen. 1980. Spot cooling of workplaces in hot industries. Final Report to the European Coal and Steel Union. Research Programme Ergonomics—Rehabilitation III, No. 7245-35-004 (October):88.

Olesen, B.W. and R. Nielsen. 1981. Radiant spot cooling of hot working places. *ASHRAE Transactions* 87(1):593-608.

Olesen, B.W. and R. Nielsen. 1983. Convective spot-cooling of hot working environments. Proceedings of the XVIth International Congress of Refrigeration (September), Paris.

Olesen, B.W. and A.M. Zhivov. 1994. Evaluation of thermal environment in industrial work spaces. *ASHRAE Transactions* 100(2).

Olesen, B.W., M. Scholer, and P.O. Fanger. 1979. Discomfort caused by vertical air temperature differences. Indoor climate. Danish Research Institute, Copenhagen.

Peterson, R.L., E.L. Schofer, and D.W. Martin. 1983. Laboratory air systems—Further testing. *ASHRAE Transactions* 89(2B):571-98.

Posokhin, V.N. and V.A. Broida. 1980. Local exhausts incorporated with air curtain. Hydromechanics and heat transfer in sanitary technique equipment. KNTI, Kazan (in Russian).

Posokhin, V.N. 1985. *Design of local ventilation systems for the process equipment with heat and gas release.* Mashinostroyeniye, Moscow (in Russian).

Powlesland, J.W. 1971. Air curtains. *Canadian Mining Journal* (October): 84-93.

Powlesland, J.W. 1973. Air curtains in controlled energy flows. American Conference of Industrial Hygienists (February).

Pozin, G.M. 1993. Determination of the ventilating effectiveness in mechanically ventilated spaces. Proceedings of the 6th International Conference on Indoor Air Quality (IAQ '93), Helsinki.

Robertson, K.S. and Downie, R.H. 1978. Personal cooling in industry. Technical Report No. TR-18. Highett, Victoria.

RoomVent '90. 1990. Proceedings of the 2nd International Conference on Engineering Aero- and Thermodynamics of Ventilated Rooms (Oslo, Norway, June 13-15). ScanVac, Denmark.

RoomVent '92. 1992. Proceedings of the 3rd International Conference on Engineering Aero- and Thermodynamics of Ventilated Rooms (Aalborg, Denmark). ScanVac, Denmark.

RoomVent '94. 1994. Proceedings of the 4th International Conference on Engineering Aero- and Thermodynamics of Ventilated Rooms (Krakow, Poland, June 15-17). ScanVac, Denmark.

Schroy, J.M. 1986. A philosophy on engineering controls for workplace protection. *Annals of Occupational Hygiene* 30(2):231-36.

Shepelev, I.A. 1978. *Aerodynamics of air flows in rooms.* Stroiizdat, Moscow (in Russian).

Shilkrot E.O. 1993. Determination of design loads on room heating and ventilation systems using the method of zone-by-zone balances. *ASHRAE Transactions* 99(1):987-91.

Shilkrot, E.O. and A.M. Zhivov. 1992. Room ventilation with designed temperature stratification. Roomvent '92. Proceedings of the 3rd International Conference on Engineering Aero- and Thermodynamics of Ventilated Rooms (Aalborg, Denmark).

Skaret, E. 1985. *Ventilation by displacement—Characterization and design applications.* Elsevier Science Publishers, B.V., Amsterdam.

Stoler, V.D. and Yu.L. Savelyev. 1977. Push-pull systems design for the etching tanks. *Heating, Ventilation, Water Supply, and Sewage Systems Design* 8(124). TsINIS, Moscow (in Russian).

Stroiizdat. 1992. *Design manual: Ventilation and air conditioning.* 4th ed. Part 3(1). Moscow (in Russian).

Strongin, A.S. 1993. Aerodynamic protection of hangars against cold ingress through apertures. *Building Services Engineering Research and Technology* 14(1). CIBSE Ser. A, London, UK.

Strongin, A.S. and M.V. Nikulin. 1987. Air curtains with outdoors air supply. Heating and ventilation of industrial buildings. TsNIIpromzdanii, Moscow (in Russian).

Strongin, A.S. and M.V. Nikulin. 1991. A new approach to the air curtain design. *Construction and Architecture— Izvestiya VUZOV* (January):84-87 (in Russian).

Tarnopolsky M. 1992. Design of air distribution systems in air conditioned spaces. Technion Research and Development Foundation, Haifa, Israel.

U.S. Department of Health and Human Services. 1986. Advanced industrial hygiene engineering. PB87-229621. Cincinnati, OH.

Ventilation '85. 1986. Proceedings of the 1st International Symposium on Ventilation for Contaminant Control. Elsevier Science Publishers, B.V., Amsterdam.

Ventilation '88. 1989. Proceedings of the 2nd International Symposium on Ventilation for Contaminant Control. Elsevier Science Publishers, B.V., Amsterdam.

Ventilation '91. 1993. Proceedings of the 3rd International Symposium on Ventilation for Contaminant Control. American Conference of Governmental Industrial Hygienists (ACGIH), Cincinnati, OH.

Zhivov, A.M. 1990. Variable-air-volume ventilation systems for industrial buildings. *ASHRAE Transactions* 96(2):367-72.

Zhivov, A.M. 1992. Selection of general ventilation method for industrial spaces. Presented at 1992 ASHRAE Annual Meeting ("Supply Air Systems for Industrial Facilities" Seminar).

Zhivov, A.M. 1993. Theory and practice of air distribution with inclined jets. *ASHRAE Transactions* 99(1):1152-59.

Zhivov, A.M. and B.W. Olesen. 1993. Extending existing thermal comfort standards to work spaces. Proceedings of the 6th International Conference on Indoor Air Quality (IAQ'93), Helsinki.

BIBLIOGRAPHY

Akinchev N. 1984. *General ventilation of hot shops.* Stroiizdat, Moscow (in Russian).

Alden, J.L. and J.M. Kane. 1982. *Design of industrial ventilation systems.* 5th ed. Industrial Press, New York.

Anderson, R. and M. Mehos. 1988. Evaluation of indoor air pollutant control techniques using scale experiments. ASHRAE Indoor Air Quality Conference.

Anufriev, L. et al. 1974. *Thermophysical calculations for agricultural industrial (production) buildings.* Stroiizdat, Moscow (in Russian).

Astelford, W. 1975. Engineering Control of Welding Fumes. DHEW (NIOSH) Pub. No. 75-115. NIOSH, Cincinnati, OH.

Balchin, N.C., ed. 1991. *Health and safety in welding and allied processes.* 4th ed. Abington Publishing, Cambridge.

Bartnecht, W. 1989. *Dust explosions: Course, prevention, protection.* Springer-Verlag, Berlin.

Baturin, V. and V. Elterman. 1963. *Aeration of industrial buildings.* Stroygiz, Moscow (in Russian).

BOCA. 1990. BOCA National building code. 11th ed. Building Officials and Code Administrators International, Country Club Hills, IL.

Boshnyakov, E. 1985. *Ventilation in the main productions of non-ferrous metallurgy.* Mettallurgiya, Moscow (in Russian).

Burgess, W.A., M. J. Ellenbecker, and R.D. Treitman. 1989. *Ventilation for control of the work environment.* John Wiley and Sons, New York.

Chamberlin, L.A. 1988. Use of controlled low velocity air patterns to improve operator environment at industrial work stations. Masters thesis, University of Massachusetts, September.

Cralley, L.V. and L.J. Cralley, eds. 1986. Industrial hygiene aspects of plant operations, Vol. 3 of *Patty's industrial hygiene and toxicology.* John Wiley and Sons, New York.

Crawford, M. 1976. *Air pollution control theory.* McGraw-Hill, New York.

Croome-Gale, D.J. and B.M. Roberts. 1981. *Air conditioning and ventilation of buildings.* 2nd ed. Pergamon Press, Oxford, U.K.

Curd, E.F. 1981. Possible applications of wall jets in controlling air contaminants. *Annals of Occupational Hygiene* 24(1):133-46.

Elterman V. 1985. *Pollution control at chemical and oil-chemical industry.* Chemia, Moscow (in Russian).

Eto, J.H. and C. Meyer. 1988. The HVAC costs of fresh air ventilation. *ASHRAE Journal* (September):31-35.

Flagan, R.C. and J.H. Seinfeld. 1988. *Fundamentals of air pollution engineering.* Prentice-Hall, Englewood Cliffs, NJ.

Godish, T. 1989. *Indoor air pollution control.* Lewis Publishers, Chelsea, MI.

Goodfellow, H.D. 1986. Proceedings of Ventilation '85. Elsevier Science Publishers, B.V., Amsterdam.

Goodier, J.L., E. Boudreau, G. Coletta, and R. Lucas. 1975. Industrial health and safety criteria for abrasive blast cleaning operations. DHEW (NIOSH) Pub. No. 75-122. NIOSH, Cincinnati, OH.

Graedel, T.W. 1978. *Chemical compounds in the atmosphere.* Academic Press, New York.

Grimitlyn, M. 1983. *Ventilation and heating of plastics processing shops.* Chemia, Leningrad (in Russian).

Grimitlyn, M. 1993. *Ventilation and heating of shops of machinery building plants.* 2nd ed. Mashinostroyeniye, Moscow (in Russian).

Hagopian, J.H. and E.K. Bastress. 1976. Recommended industrial ventilation guidelines. DHEW (NIOSH) Pub. No. 76-162. NIOSH, Cincinnati, OH.

Hayashi, T., R.H. Howell, M. Shibata, and K. Tsuji. 1987. Industrial ventilation and air conditioning. CRC Press, Boca Raton, FL.

Heinsohn, R.J. 1991. *Industrial ventilation: Engineering principles.* John Wiley and Sons, New York.

Hughes, D. 1980. A literature survey and design study of fume cupboards and fume-dispersal systems. (Occup. Hyg. Manual 4). Science Review, Leeds, U.K.

Licht, W. 1988. *Air pollution control engineering.* 2nd ed. Marcel Dekker, Inc., New York.

McDermott, H.J. 1985. *Handbook of ventilation for contaminant control.* 2nd ed. Ann Arbor Science Publishers, Ann Arbor, MI.

Mehta, M.P., H.E. Ayer, B.E. Saltzman, and R. Ronk. 1988. Predicting concentration for indoor chemical spills. ASHRAE Indoor Air Quality Conference.

Murmann, H. 1980. *Lufttechnische Anl. fuer gewerbl. Betriebe. 2nd ed. (Ventilation systems for industrial buildings).* C. Marhold, Berlin.

NFPA. 1973. National Fire Codes, Vol. 4 (NFPA 90-1973, NFPA 91-1973, and NFPA 96-1973). National Fire Protection Association, Quincy, MA.

NIOSH. 1973. The industrial environment—Its evaluation and control. National Institute for Occupational Safety and Health, Washington, D.C.

NIOSH. 1978. Engineering control technology assessment for the plastics and resins industry. DHEW (NIOSH) Pub. No. 78-159. NIOSH, Cincinnati, OH.

NIOSH. 1978. Proceedings of the Recirculation of Industrial Exhaust Air. DHEW (NIOSH) Pub. No.78-141. NIOSH, Cincinnati, OH.

NIOSH. 1989. Guidance for indoor air quality investigations. Cincinnati, OH.

O'Brien, D.M. and D.E. Hurley. 1981. An evaluation of engineering control technology for spray painting. DHEW (NIOSH) Pub. No. 81-121. NIOSH, Cincinnati, OH.

Partridge, L.J., Nayak, R.S. Stricoff, and J.H. Hagopian. 1978. A recommended approach to recirculation of exhaust air. DHEW (NIOSH) Pub. No. 78-124. NIOSH, Cincinnati, OH.

Perry, R. and P.W. Kirk, eds. 1988. Indoor and ambient air quality. Publication Division, Selper Ltd., London.

Pisarenko, V. and M. Roginsky. 1981. *Ventilation of working places in welding production.* Mashinostroyeniye, Moscow (in Russian).

RAPRA. 1979. *Ventilation handbook for the rubber and plastics industries.* Rubber and Plastics Research Association of Great Britain, Shrewsbury.

Rymkevich, A.A. 1990. *Optimization of general ventilation and air conditioning systems operation.* Stroiizdat, Moscow (in Russian).

Sandberg, M. 1983. Ventilation efficiency as a guide to design. *ASHRAE Transactions* 89(2B):455.

Sandberg, M. and C. Blomqvist. 1989. Displacement ventilation systems in office rooms. *ASHRAE Transactions* 95(2).

Skaret, E. Ventilation efficiency. *ASHRAE Transactions* 89(2B):480.

Speight, F. and H. Campbell, eds. 1979. Fumes and gases in the welding environment. American Welding Society, Miami, FL.

Stephanov, S.P. 1986. Investigation and optimization of air exchange in industrial halls ventilation. Proceedings of Ventilation '85. Elsevier Science Publishers, B.V., Amsterdam.

Teterevnikov, V.N., T.V. Kuksinskaya, and L.V. Pavlukhin. 1974. Industrial microclimate and air conditioning. TsNIIOT, Moscow (in Russian).

Theodore, L. and A. Buonicore, eds. 1982. *Air pollution control equipment: Selection, design, operation, and Maintenance.* Prentice-Hall, Englewood Cliffs, NJ.

Van Wagenen, H.D. 1979. Assessment of Selected Control Technology for Welding Fumes. DHEW (NIOSH) Pub. No. 79-125. NIOSH, Cincinnati, OH.

Vincent, J.H. 1989. *Aerosol sampling, science and practice.* John Wiley and Sons, Oxford, U.K.

Volkavein, J.C., M.R. Engle, and T.D. Raether. 1988. Dust control with clean air from an overhead air supply island (oasis). *Applied Industrial Hygiene* 3(August):8.

Wadden, R.A. and P.A. Scheff. 1982. *Indoor air pollution: Characterization, prediction, and control.* John Wiley and Sons, New York.

Wilson, D.J. 1982. A design procedure for estimating air intake contamination from nearby exhaust vents. *ASHRAE Transactions* 89(2A):136.

MINE AIR CONDITIONING AND VENTILATION

IN underground mines, excess humidity, high temperatures, and inadequate oxygen supplies have always been points of concern because they lower worker efficiency and productivity and can cause illness and death. Air cooling and ventilation are needed in deep underground mines to minimize heat stress. As mines have become deeper, heat removal and ventilation problems have become more difficult to solve.

WORKER HEAT STRESS

Mine air must be conditioned to maintain the temperature and humidity at levels that ensure the health and comfort of miners so they can work safely and productively. Chapter 8 of the 1993 *ASHRAE Handbook—Fundamentals* addresses human response to heat and humidity. The upper temperature limit for humans at rest in still, saturated air is about 32°C. If the air is moving at 1 m/s the upper limit is 35°C. In a hot mine environment, a relative humidity of less than 80% is desirable.

Hot, humid environments are improved by providing air movement of 0.8 to 2.5 m/s. Although a greater air volume lowers the mine temperature, air velocity has a limited range in which it improves worker comfort.

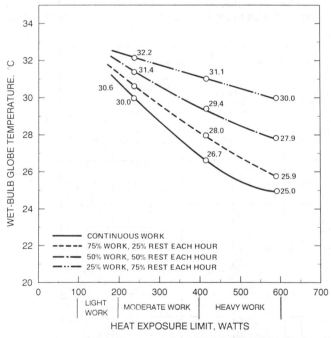

Fig. 1 Permissible Heat Exposure Threshold Limit Value

The preparation of this chapter is assigned to TC 9.2, Industrial Air Conditioning.

Indices for defining acceptable temperature limits include the following:

- *Effective temperature scale.* An effective temperature of 26.7°C is the upper limit for ensuring worker comfort and productivity.
- *Wet-bulb globe temperature (WBGT) index.* A WBGT of 26.7°C is the permissible temperature exposure limit for moderate continuous work; a WBGT of 25°C is the limit for heavy continuous work.

Figure 1 shows acceptable heat exposure limits for different activity levels.

DESIGN CONSIDERATIONS: SOURCES OF HEAT ENTERING MINE AIR

Adiabatic Compression

Air descending a shaft increases in pressure (due to the mass of air above it) and temperature. As air flows down a shaft, it is heated as if compressed in a compressor, even if there is no heat interchange with the shaft and no evaporation of moisture.

One kilojoule is added to each kilogram of air for every 102-m decrease in elevation or is removed for the same elevation increase. For dry air, the specific heat is 1.006 kJ/kg·K, and the dry-bulb temperature change is $1/(1.006 \times 102 \times 1) = 0.00975$ K per metre or 1 K per 102 m of elevation. For constant air-vapor mixtures, the change in dry-bulb temperatures is $(1 + W)/(1.006 + 1.84 W)$ per 102 m of elevation, where W is the humidity ratio in kilograms of water per kilogram of dry air.

Theoretically, when 50 m³/s of standard air (density = 1.204 kg/m³) is delivered underground via an inlet airway, the heat of autocompression for every 100 m of depth is calculated as follows:

$$50 \text{ m}^3 \times 1.204 \frac{\text{kg}}{\text{m}^3} \times \frac{1 \text{ kJ/kg}}{102 \text{ m}} \times 100 \text{ m} = 59 \text{ kW of refrigeration}$$

Autocompression of air may be masked by the presence of other heating or cooling sources, such as shaft wall rock, groundwater, air and water lines, or electrical facilities. The actual temperature increase for air descending a shaft does not usually match the theoretical adiabatic temperature increase, due to the following:

- Effect of night cool air temperature on the rock or lining of the shaft
- Temperature gradient of ground rock related to depth
- Evaporation of moisture within the shaft, which decreases the temperature while increasing the moisture content of the air

The seasonal variation in surface air temperature has a major effect on the temperature of air descending a shaft. When the surface air temperature is high, much of its heat is absorbed by the shaft walls; thus, the temperature rise for the decensding air may not reach the adiabatic prediction. When the surface temperature is low, heat is absorbed from the shaft walls, and the temperature increases more than predicted adiabatically. Similar diurnal variations may occur. As air flows down a shaft and increases in temperature and density, its cooling ability and volume decrease. Additionally, the mine ventilation requirements increase with depth. Fan static pressures up to 2.5 kPa (gage) are common in mine ventilation and raise the temperature of the air about 1 K per kPa.

Electromechanical Equipment

Power-operated equipment transfers heat to the air. In mines, systems are commonly powered by electricity, diesel fuel, and compressed air.

For underground diesel equipment, about 90% of the heat value of the fuel consumed, or 35 MJ/L, is dissipated to the air as heat. If exhaust gases are bubbled through a wet scrubber, the gases are cooled by adiabatic saturation, and both the sensible heat and the moisture content of the air are increased.

Vehicles with electric drives or electric-hydraulic systems release one-third to one-half the heat released by diesel equipment.

All energy used in a horizontal plane appears as heat added to the mine air. Energy required to elevate a load gives potential energy to the material and will not appear as heat.

Groundwater

Transport of heat by groundwater is the largest variable in mine ventilation. Groundwater usually has the same temperature as the virgin rock. If there is an uncovered ditch containing hot water, ventilation cooling air can pick up more heat from the ditch water than from the hot wall rock. Thus, hot drainage water should be contained in pipelines or in a covered ditch.

Heat release from open ditches becomes more significant as airways get older and the flow of heat from the surrounding rock decreases. In one Montana mine, water in open ditches was 22 K cooler than when it issued from the wall rock; the heat was transferred to the air. Evaporation of water from wall rock surfaces lowers the surface temperature of the rock, which increases the temperature gradient of the rock, depresses the dry-bulb temperature of the air, and allows more heat to flow from the rock. Most of this extra heat is expended in evaporation.

WALL ROCK HEAT FLOW

Wall rock is a major heat source in mines. The heat flow from wall rock with constant thermal conductivity is considerably higher when a mine is first excavated and transient heat flow applies than it is several years later, when steady-state conditions have developed. Heat flow from wall rock to air can be calculated using the following empirical equation:

$$Q = UA\Delta t \qquad (1)$$

where

Q = heat flow, kW
U = overall coefficient of heat transfer, W/m$^2\cdot$K (from Figure 2 or 3)
A = area, m^2
Δt = temperature difference between the virgin rock and the dry-bulb air temperature, K

Equation (1) can be used with mine ventilation rates, virgin rock temperatures, and the age of airways to calculate rock heat flow throughout a mine.

In young mining areas, heat inflow must be computed by transient heat flow techniques. Figures 2 and 3 show heat flow from wall rock to air over 1000 days in two types of mines: a salt envelope mine and a granite envelope mine. The temperature difference between wall rock and air is not constant along the entire length of an airway, so the airway can be treated as a series of consecutive lengths. Figures 2 and 3 are based on the following:

- The cross-sectional areas of the mine openings range from 21 to 42 m^2.
- The heat transfer coefficient decreases from the time the heading is excavated and ventilated.
- Steady-state heat transfer coefficients are approximately 0.57 to 0.85 W/(m$^2\cdot$K) for salt or granite envelopes.

The Starfield and Goch-Patterson methods of calculating heat flow rates from underground exposed rock surfaces are often used. The Starfield method, which is theoretically more accurate, is frequently difficult to apply because it depends on air velocity criteria.

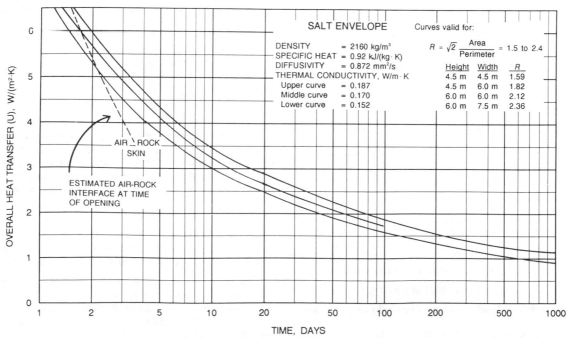

Fig. 2   Heat Flow for a Salt Envelope Mine

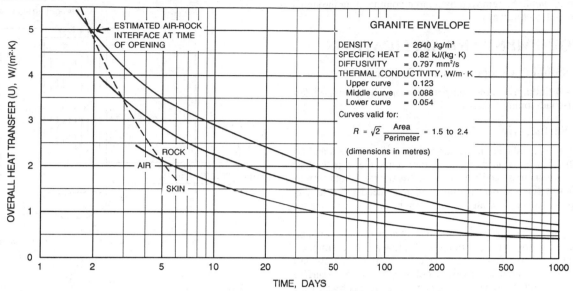

Fig. 3   Heat Flow for a Granite Envelope Mine

Table 1   Maximum Virgin Rock Temperatures
(Fenton 1972)

Mining District	Depth, m	Temperature, °C
Kolar Gold Field, India	3353	67
South Africa	3048	52 to 54
Morro Velho, Brazil	2438	54
N. Broken Hill, Australia	1076	44
Great Britain	1219	46
Braloroe BC Canada	1250	45
Kirkland Lake, Ontario	1219 to 1829	19 to 27
Falconbridge Mine, Ontario	1219 to 1829	21 to 29
Lockerby Mine, Ontario	914 to 1219	19 to 36
Levac Borehole (Inco) Ontario	2134 to 3048	37 to 53
Garson Mine, Ontario	610 to 1524	12 to 26
Lake Shore Mine, Ontario	1829	23
Hollinger Mine, Ontario	1219	14
Creighton Mine, Ontario	610 to 3048	16 to 59
Superior, AZ	1219	60
San Manuel, AZ	1372	48
Butte, MT	1585	63 to 66
Ambrosia Lake, NM	1219	60
Brunswick, No. 12, New Brunswick, CA	1128	23
Belle Island Salt Mine, LA	427	31

The errors in measuring or estimating air velocities and the length and cross-sectional dimensions of the rock surfaces are often greater than the error introduced by the Starfield method.

The Goch-Patterson method, which assumes that the rock face temperature is the same as that of the air, is widely used. Heat load calculated for airways is overestimated by 10 to 20%, so a contingency is normally not added to the heat load. When using the Goch-Patterson tables, the correct value of instantaneous versus average heat load data should be used for shafts and older connecting airways. Time-average heat load data should be used for excavations that are continuous and active.

The rate of heat release from wall rock can be accurately estimated horizontally and vertically. The virgin rock temperature (VRT) should be measured at various elevations throughout the mine.

Before constructing a gradient to estimate the temperatures at various elevations, a base point for measurement must be selected. The VRT 15 m below the earth's surface is often used; it is usually equal to the mean of surface air temperatures (dry-bulb) recorded over a number of years.

Table 1 lists maximum VRTs at the bottom of various mines.

Heat from Blasting

The heat produced by blasting can be appreciable. Virtually all the explosive energy is converted to heat. The typical heat potential in various types of explosives is similar to that of 60% dynamite, about 4.2 MJ/kg.

AIR COOLING AND DEHUMIDIFICATION

Sources of heat in mines include wall rock, hot water issuing from underground sources, adiabatic compression of the inlet air down deep shafts, electromechanical equipment, hot compressed air lines and lines draining hot mine water, oxidation of timber and sulfide minerals, blasting, friction and shock losses from moving air, body metabolism, friction from rock movement, and internal combustion engines (Fenton 1972).

Geothermic gradients of the world's major mining districts vary from approximately 0.9 K to over 7.3 K per 100 m of elevation.

When air is cooled underground, the heat extracted must be removed. If enough cool, uncontaminated air is available, the warmed air can be mixed with it and removed from the mine. In a water heat exchanger, the water is warmed while the air is cooled. The warm water is either returned in a closed circuit to the cooling plant (cooling tower, underground mechanical refrigeration unit, or heat exchanger), discharged to the mine drainage system, or sprayed into the mine exhaust air system.

The high heat capacity of water makes it a better heat transfer medium than air. For water, a relatively small pipeline can remove as much heat as would be removed by a large volume of air through a large shaft and inlet airways.

EQUIPMENT AND APPLICATIONS

Air-Conditioning Components and Practices

Figures 4 through 7 show components normally used in underground air-conditioning systems.

Cooling Surface Air

If cold air is available on the surface, as in far northern and far southern hemisphere winters, it can be forced into the mine. In deep mines with small cross-sectional airways, this supply of cold air may be insufficient for adequate cooling and may need to be supplemented by cooled air. In winter, it may even be necessary to heat the intake air.

Cooling air on the surface is the least expensive method of cooling. Factors favoring surface cooling are (1) low adiabatic compression (which applies to shafts that are not too deep) and (2) low heat gain in the shaft and intake airways. Cooling inlet air on the surface rather than in an underground cooling plant saves energy because the cooling water does not need to be returned to the surface.

In a surface plant installation, air is chilled using cold water or brine cooled by mechanical refrigeration units. The chilled air is then delivered to the intake shaft to be taken underground. Condenser water is usually cooled in towers or spray ponds. Such installations have been used in Morro Velho, Brazil; Robinson Deep Mine, Rand, South Africa; Kolar, India; and Rieu du Coeur, Belgium.

There are serious disadvantages to surface air cooling. By the time the chilled air reaches the working headings, it has been warmed considerably by autocompression and heat from the wall rock. Furthermore, a plant of sufficient size for peak load has an annual load factor as low as 60%. For these reasons, surface plant systems are no longer being installed. The trend is to install air-cooling plants underground, with the cooling coils as close to the working areas of the mine as possible. If a large supply of cold water is available, it can be piped underground to the air-conditioning plants.

Cooling Surface Water

Surface-made ice or ice slurry in 0°C water can be delivered down a pipeline and used for activities such as drilling and wetting down muckpiles. The use of chilled water at 0°C reduces the required circulated flow rate to 50 to 70% of that in chilled water air-conditioning plants using warmer water. However, ice-making machines have a lower coefficient of performance and higher capital costs than do normal mechanical refrigeration units used to chill mine water.

In one Canadian mine, large, cold, mined-out areas are sprayed with water during the winter months, allowing ice to form in the openings. In summer, intake air is cooled by drawing it over this ice.

Evaporative Cooling of Mine Chilled Water

Geographic areas with both low average winter temperatures and low relative humidity during the summer have a natural cooling capacity that is adaptable to mine air-cooling methods. A system in such an area cools water or brine on the surface, carries it through closed-circuit piping to an underground air-cooling plant near the working zone, and returns it to the cooling tower through a second pipeline. At the mine underground air-cooling plant, heat is absorbed by the water circulating through the system and is dissipated to the surface atmosphere in the surface cooling tower. The closed-circuit piping balances the hydrostatic pressure, so pumping power must only overcome frictional resistance.

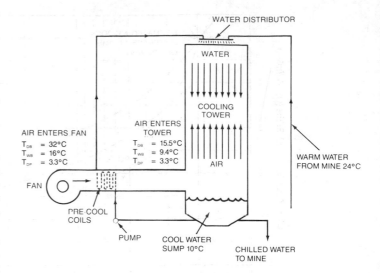

Fig. 4   Evaporative Cooling Tower System
(Richardson 1950)

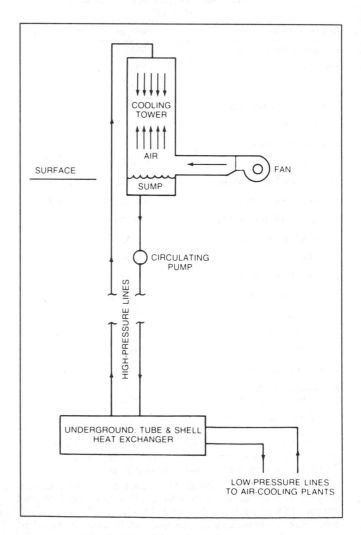

**Fig. 5   Underground Heat Exchanger,
Pressure Reduction System**

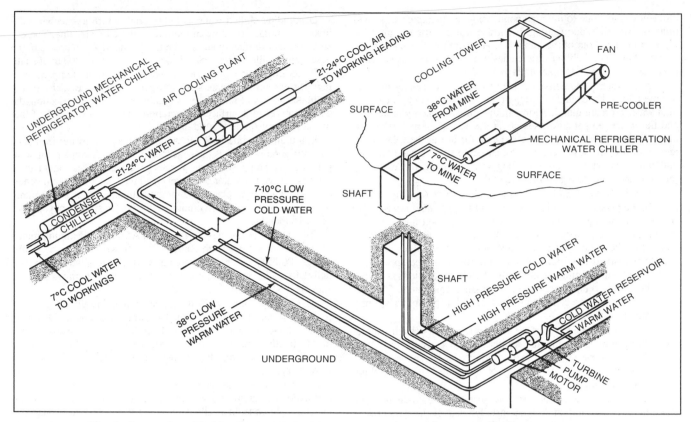

Fig. 6 Layout for a Turbine-Pump-Motor Unit with Air-Cooling Plants and Mechanical Refrigeration

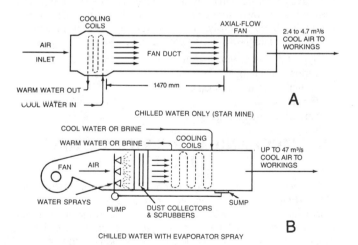

Fig. 7 Underground Air-Cooling Plants

An evaporative cooling tower was installed at a mine in the northwest United States. This "dew-point" cooling system reduces the temperature of the cooling medium to below the wet-bulb temperature of the surface atmosphere (Figure 4). Precooling coils were installed between the fan and the cooling tower. Some cool water from the sump at the base of the cooling tower is pumped through the coils to the top of the cooling tower, where this heated flow joins the warm return water from the airflow; the moisture content in the airstream passing over the coils is unchanged, so that the dew-point temperature remains constant. The heat content of the air is reduced, and the equivalent heat is added to the water circulating in coil adsorbers. The dry- and wet-bulb temperature of the air entering the bottom of the cooling tower is less than the temperature of the air entering the fan. The temperature of the water leaving the tower approaches the wet-bulb temperature of the surface atmosphere.

Evaporative Cooling Plus Mechanical Refrigeration

On humid summer days, the wet-bulb temperature may increase over extended periods, severely hampering the effectiveness of evaporative cooling. This factor, plus warming of the air entering the mine, may necessitate the series installation of a mechanical refrigeration unit to chill the water delivered underground.

Performance characteristics at one northern United States mine on a spring day were as follows:

Evaporative cooling tower	4305 kW
Mechanical refrigeration unit (in series)	2870 kW
Water volume circulated	110 L/s
Water temperature entering mine	4.4°C
Water temperature leaving mine	20°C

Combination Systems

Components may be arranged in various ways for the greatest efficiency. For example, air-cooling towers may be used to cool water during the cool months of the year and to supplement a mechanical refrigerator during the warm months of the year.

Surface-installed mechanical refrigeration units provide the bulk of cooling in summer. In winter, much of the cooling comes from the precooling tower when the ambient wet-bulb temperature is usually lower than the temperature of water entering the tower. The precooling tower is normally located above the return water storage reservoir. An evaporative cooling tower is more cost-effective (capital and operating costs) than mechanical refrigeration with comparable capacity.

Reducing Water Pressure

The use of underground refrigerated water chillers is increasing because the systems are efficient and can be located close to the work. Transfer of heat from the condensers is the major problem with these systems. If hot mine water is used to cool the condenser, efficiency is lost due to the high condensing temperature, the possibility of corrosion, and fouling. If surface water is used, it must be piped both in and out of the mine. If water is noncorrosive and nonfouling, fairly good chiller efficiencies can be obtained for entering condenser water with temperatures up to 52°C.

Surface water being delivered in a vertical pipe is usually allowed to flow into tanks located at different levels in the shaft to break the high water pressure that develops. In this process, energy is wasted, and the water temperature rises about 1.8 K for every 1000 m of drop. The water pressure can be reduced for use at the mine level and, after use, the water can be discharged to the drainage system. Although low-pressure mine cooling is convenient, the costs for pumping the water to the surface are high.

If a pipe 1000 m high were filled with water (density = 998 kg/m³), the pressure would be $998 \times 1000 \times 9.807$ m/s² = 9.8 MPa. In an open piping system, the pressure at the bottom is further increased by the pressure necessary to raise the water up the pipe and out of the mine. Water pipes and coils in deep mines must be able to withstand this high pressure. Fittings and pipe specialties for high-pressure equipment are costly. Safety precautions and care must be taken when operating high-pressure equipment. Closed-circuit piping has the same static pressures, but pumps must only overcome pipe friction.

Frequent movement of surface cooling towers, a desired feature for shifting mining operations, results in high construction costs. Closed-circuit systems have been used in various mines in the United States to overcome the cost of pumping brine or cooling water out of the mine.

To take advantage of both low-pressure and closed-circuit systems, the Magma mine in Arizona installed heat exchangers underground at the mining horizon. Shell-and-tube heat exchangers convert surface chilled water in a high-pressure closed circuit to a low-pressure chilled water system on mine production levels. Air-cooling plants and chilled water lines can be constructed of standard materials, permitting frequent relocation (Figure 5). Although desirable, this system has not been widely used.

Energy-Recovery Systems

Pumping costs can be reduced by combining a water turbine with the pump. The energy of high-pressure water flowing to a lower pressure drives the pump needed for the low-pressure water circuit. Rotary-type water pumps have been developed to pump against 15 MPa, and water turbines are also available to operate under pressure. Figure 6 shows a turbine pump-motor combination. Only the shaft and the pipe and fittings to the unit on the working level need strong pipes. The system connects to underground refrigeration water-chilling units; the return chilled water is used for condensing before being pumped out.

Two types of turbines are suitable for mine use—the Pelton wheel and a pump in reverse. The Pelton wheel has a high-duty efficiency of about 80%, is simply constructed, and is readily controlled. A pump in reverse is only 10 to 15% efficient, but mine maintenance and operating personnel are familiar with this equipment. Turbine energy recovery may encounter difficulties when operating on chilled mine service water because mine demands fluctuate widely, often outside the operating range of the turbine. Operating experience shows that coupling a Pelton wheel to an electric generator is the best approach.

South African mines have mechanical refrigeration units, surface heat-recovery systems, and turbine pumps incorporated into their air-cooling plants in a closed circuit.

A surface-sited plant has two disadvantages: (1) chilled water delivered underground at low operating pressure heats up at a rate of 1.8 K per 1000 m of shaft, and (2) pumping this water back to the surface is expensive. Energy-recovery turbines underground and a heat-recovery system on the surface help offset pumping costs.

Descending chilled water is fed through a turbine mechanically linked to pumps operating in the return chilled water line. The energy recovery turbine reduces the rate of temperature increase in the descending chilled water column to about 0.5 K per 1000 m of shaft.

Precooling towers on the surface reduce the water temperature a few degrees before it enters the refrigeration plant. Because of the unlimited supply of relatively cool ambient air for heat rejection, the operating cost of a surface refrigeration plant is about one-half that of a comparable underground plant.

In a uranium mine in South Africa, condensing water from surface refrigeration units is the heat source for a high condensing temperature heat pump, which discharges 55°C water. This water can be used as service hot water or as preheated feedwater for steam generation in the uranium plant. The total additional cost of a heat pump over a conventional refrigeration plant has a simple payback of about 18 months.

Pelton turbines are used by the South African gold-mining industry to recover energy from chilled water flowing down the shafts. About 1000 kW can be recovered at a typical installation; this partially offsets the power requirements for pumping return water to the surface.

One mining company in the United States installed an energy-recovery system that includes two separate units, one at the 750-m level and one at the 1500-m level. Each installation consists of turbines directly connected to 150 kW, 3600 rpm induction motors operating as generators. Rated output is 144 kW at 34.7 L/s water flow at approximately 7.5 MPa.

A surface-level installation chills the service water used throughout the mining operation. A 150-mm line feeds chilled water to the turbine through a pneumatically controlled valve. The water level in an adjacent discharge reservoir is monitored to modulate the position of this valve proportional to the water demand in the reservoir. When the system is shut down, spring pressure slowly closes the inlet valve to the turbine. Simultaneously, the bypass valve opens slowly and water is discharged into the sump. Typical operating data are as follows:

Bypass discharge temperature	6.5°C
Turbine inlet temperature	4.5°C
Turbine discharge temperature	4.9°C
Turbine output	144 kW
Volume through turbine	34.7 L/s

MECHANICAL REFRIGERATION PLANTS

In most underground applications, mechanical refrigeration plants (Figure 6) provide chilled water for delivery underground, where some or all may be supplied condensed to direct-expansion air-to-air cooling systems. Air that passes over direct-expansion evaporator coils is cooled and delivered to working headings.

Underground mechanical refrigeration plants avoid the additional heat due to the autocompression of air coming down the shafts. The main disadvantage of these units is the disposal of heat from the condensers. The condenser discharge water can either be run into the mine drainage system, returned to the surface for cooling, or sprayed into the mine discharge air system.

Hot water discharged from the condensers of underground refrigeration units may be ejected into sumps of the mine water-pumping system. This procedure requires a constant resupply of condenser water from freshwater lines or mine drainage sources and increases mine pumping costs. Underground cooling towers are seldom used because they encounter air shortages, which result in higher condensing temperatures than those on the surface. It is difficult to predict the performance of underground cooling towers and the condensing temperatures of equipment over the life of a mine. High condenser temperatures cause excessive power consumption. In addition, water is often contaminated with dust and fumes, which cause fouling, scaling, and corrosion of the piping system and condenser tubes. These problems make placing mechanical refrigeration units on the surface more favorable.

Spot Cooling

Spot cooling permits the driving of long headings for exploration or extended development prior to the installation of primary ventilation equipment. Using mechanical refrigeration systems with direct-expansion air coolers, spot cooling services a single heading or localized mining area. Spot coolers allow development headings to be advanced more rapidly and under more comfortable conditions. Development headings where rock temperatures exceed 38°C may also use spot cooling.

UNDERGROUND HEAT EXCHANGERS

Two types of heat exchangers are used underground: air-to-water and water-to-water. Air-to-water exchangers include those with (1) water sprays in the airstream and (2) finned tubes.

Air Cooling Versus Working Place Cooling

Figure 7 shows a typical underground central plant that provides a large area in a mine with air cooled by chilled water. Air flows through ventilation ducts or headings from the chiller to the working headings. When working headings are distant, the air warms and, in some cases, picks up moisture in transit. This limits the cooled air in the production area. The performance characteristics of one plant are as follows:

Air volume cooled	47 m³/s
Entering air temperature	
Dry-bulb	26.9°C
Wet-bulb	23.5°C
Discharge air temperature	13.6°C
Entering water temperature	5.6°C
Discharge water temperature	19.4°C
Heat extracted from air	1790 kW

If one of the air-conditioning plants from the workings is not operating, it is preferable to pump chilled water in closed circuit to cooling plants close to or serving individual productive headings.

Another underground cooling plant arrangement (Figure 7A) at a mine in Wallace, Idaho, comprises a light, portable unit that attaches to the normal ventilation system and requires no extra excavation. No separate spray or dust-collection system is used, and the condensed water drips from the duct. The fan gives good service downstream from the cooler. A mechanical refrigeration unit located on the same, or a nearby, level chills water to cool the air; potable water is used in the condenser and discharged to the mine sumping system.

This underground system avoids the central plant problem of cooled air being heated as it travels down a shaft. The cooling coils are normally located within 90 m of the working place and reduce overall air-cooling plant requirements. Table 2 shows typical performance characteristics.

Table 2   Typical Performance of Portable, Underground Cooling Units

Size Rating	760 by 1220 mm (141 kW)		610 by 910 mm (70 kW)	
Location	Drift	Shaft	Stope	Stope
Entering air temperature	27°C sat.	27°C sat.	27°C sat.	27°C sat.
Discharge air temperature	21°C sat.	18°C sat.	20°C sat.	27°C sat.
Volume, m³/s	5.7	5.7	2.8	2.8
Calculated kW	158	211	79	91

Cooling Coils and Fan Position

The fan may be upstream or downstream of the cooling coils in underground cooling plants. An upstream fan provides a discharge temperature that is a few degrees lower, but a downstream fan distributes air more efficiently over the coils.

WATER SPRAYS AND EVAPORATIVE COOLING SYSTEMS

Finned-tube heat exchangers need periodic cleaning, especially when they are upstream from the fan. Downstream exchangers have spray water, which helps to wash some of the dirt away.

Another type of bulk-spraying cooling plant in South Africa consists of a spray chamber serving a section of isolated drift up to 120 m long. Chilled water is introduced through a manifold of spray nozzles. Warm, mild air flows countercurrent to the various stages of water sprays. Air cooled by this direct air-to-water cooling system is delivered to active mine workings by the primary and secondary ventilation system. These bulk-spray coolers are efficient and economical.

At one uranium mine, a portable bulk-spray cooling plant was developed that can be advanced with the working faces to overcome the high heat load between a stationary spray chamber and the production heading.

The 1060-kW capacity plant has a stainless steel chamber 2060 mm long, 2210 mm wide, and 2590 mm high that contains two stages of spray nozzles and a demister baffle. The skid-mounted unit weighs about 1000 kg and is divided into four components (spray chamber, demister assembly, and two sump halves).

Portable bulk-spray coolers have a wide range of cooling capacity and are cost-effective. A smaller portable spray cooling plant has been developed to cool mine air adjacent to the workplace. It cools and cleans the air through direct air-to-water contact. The cooler is tube-shaped and is normally mounted in a remote location. The mine inlet and discharge air ventilation ducts are connected to duct transitions from the unit. Chilled water is piped to an exposed manifold, and warm water is discharged from the unit into a sump drain.

Hot, humid air enters the cooler at the bottom; it then slows down and flows through egg crate flow straighteners. Initial heat exchange occurs as the air passes through plastic mesh and contacts suspended water droplets. Vertically sprayed water in a spray chamber then directly contacts the ascending warm air. The air passes through the mist eliminator, which removes suspended water droplets. Cool, dehumidified air exits from the cooler through the top outlet transition. The warmed spray water drops to the sump and is discharged through a drainpipe.

BIBLIOGRAPHY

Anonymous. 1980. Surface refrigeration proves energy efficient at Anglo mine. *Mine Engineering* (May).

Bell, A.R. 1970. Ventilation and refrigeration as practiced at Rhokana Corporation Ltd., Zambia. *Journal of the Mine Ventilation Society of South Africa* 23(3):29-35.

Beskine, J.M. 1949. Priorities in deep mine cooling. *Mine and Quarry Engineering* (December):379-84.

Bossard, F.C. 1983. *A manual of mine ventilation design practices.*

Bossard, F.C. and K.S. Stout. Underground mine air-cooling practices. USBM Sponsored Research Contract G0122137.

Bromilow, J.G. 1955. Ventilation of deep coal mines. *Iron and Coal Trades Review.* Part I, February 11:303-08; Part II, February 18:376; Part III, February 25:427-34.

Brown, U.E. 1945. Spot coolers increase comfort of mine workers. *Engineering and Mining Journal* 146(1):49-58.

Caw, J.M. 1953. Some problems raised by underground air cooling on the Kolar Gold Field. *Journal of the Mine Ventilation Society of South Africa* 2(2):83-137.

Caw, J.M. 1957. Air refrigeration. *Mine and Quarry Engineering* (March): 111-17; (April):148-56.

Caw, J.M. 1958. Current ventilation practice in hot deep mines in India. *Journal of the Mine Ventilation Society of South Africa* 11(8):145-61.

Caw, J.M. 1959. Observations at an underground air conditioning plant. *Journal of the Mine Ventilation Society of South Africa* 12(11):270-74.

Cleland, R. 1933. Rock temperatures and some ventilation conditions in mines of Northern Ontario. C.I.M.M. *Bulletin Transactions Section* (August):370-407.

Fenton, J.L. 1972. Survey of underground mine heat sources. Masters Thesis, Montana College of Mineral Science and Technology.

Field, W.E. 1963. Combatting excessive heat underground at Bralorne. *Mining Engineering* (December):76-77.

Goch, D.C. and H.S. Patterson. 1940. The heat flow into tunnels. *Journal of the Chemical Metallurgical and Mining Society of South Africa* 41(3): 117-28.

Hartman, H.L. 1961. *Mine ventilation and air conditioning.* The Ronald Press Company, New York.

Hill, M. 1961. Refrigeration applied to longwall stopes and longwall stope ventilation. *Journal of the Mine Ventilation Society of South Africa* 14(5):65-73.

Kock, H. 1967. Refrigeration in industry. *The South African Mechanical Engineer* (November):188-96.

Le Roux, W.L. 1959. Heat exchange between water and air at underground cooling plants. *Journal of the Mine Ventilation Society of South Africa* 12(5):106-19.

Marks, John. 1969. Design of air cooler—Star Mine. Hecla Mining Company, Wallace, ID.

Minich, G.S. 1962. The pressure recuperator and its application to mine cooling. *The South African Mechanical Engineer* (October):57-78.

Muller, F.T. and M. Hill. 1966. Ventilation and cooling as practiced on E.R.P.M. Ltd., South Africa. *Journal of the South African Institute of Mining and Metallurgy.*

Richardson, A.S. 1950. A review of progress in the ventilation of the mines of the Butte, Montana District. *Quarterly of the Colorado School of Mines* (April), Golden, CO.

Sandys, M.P.J. 1961. The use of underground refrigeration in stope ventilation. *Journal of the Mine Ventilation Society of South Africa* 14(6):93-95.

Schlosser, R.B. 1967. The Crescent Mine cooling system. Northwest Mining Association Convention (December).

Short, B. 1957. Ventilation and air conditioning at the Magma Mine. *Mining Engineering* (March):344-48.

Starfield, A.M. 1966. Tables for the flow of heat into a rock tunnel with different surface heat transfer coefficients. *Journal of the South African Institute of Mining and Metallurgy* 66(12):692-94.

Thimons, E., R. Vinson, and F. Kissel. 1980. Water spray vent tube cooler for hot stopes. USBM TPR 107.

Thompson, J.J. 1967. Recent developments at the Bralorne Mine. *Canadian Mining and Metallurgy Bulletin* (November):1301-05.

Torrance, B. and G.S. Minish. 1962. Heat exchanger data. *Journal of the Mine Ventilation Society of South Africa* 15(7):129-38.

Van der Walt, J., E. de Kock, and L. Smith. Analyzing ventilation and cooling requirements for mines. Engineering Management Services, Ltd., Johannesburg, Republic of South Africa.

Warren, J.W. 1958. The science of mine ventilation. Presented at the American Mining Congress, San Francisco (September).

Warren, J.W. 1965. Supplemental cooling for deep-level ventilation. *Mining Congress Journal* (April):34-37.

Whillier, A. 1972. Heat—A challenge in deep-level mining. *Journal of the Mine Ventilation Society of South Africa* 25(11):205-13.

CHAPTER 26

INDUSTRIAL EXHAUST SYSTEMS

INDUSTRIAL exhaust ventilation systems collect and remove airborne contaminants consisting of particulates (dust, fumes, smokes, fibers), vapors, and gases that can create an unsafe, unhealthy, or undesirable atmosphere. Exhaust systems also salvage usable material and improve plant housekeeping. Chapter 11 of the 1993 *ASHRAE Handbook—Fundamentals* covers definitions, particle sizes, allowable concentrations, and upper and lower explosive limits of various air contaminants.

There are two types of exhaust systems: (1) *general exhaust*, in which an entire work space is exhausted without considering specific operations; and (2) *local exhaust*, in which a contaminant is controlled at its source. Local exhaust is preferable because it offers better contaminant control with minimum air volumes, thereby lowering the cost of air cleaning and replacement air equipment. Chapter 24 of this volume and Chapter 2 of *Industrial Ventilation—A Manual of Recommended Practice* (ACGIH 1992) detail steps to determine the air volumes necessary to dilute the contaminant concentration for general exhaust.

Replacement air, which is usually conditioned, provides air to the work space to replace exhausted air. Therefore, the systems are not isolated from each other. Refer to Chapter 24 for further information on replacement and makeup air. A complete industrial ventilation program includes replacement air systems that provide a total volumetric flow rate equal to the total exhaust rate. If insufficient replacement air is provided, the pressure of the building will be negative relative to local atmospheric pressure. Negative pressure allows air to infiltrate through open doors, window cracks, and combustion equipment vents. As little as 12 Pa of negative pressure can cause drafts and might cause downdrafting of combustion vents, thereby creating a potential health hazard. Negative plant pressures can also cause excessive energy use. If workers near the plant perimeter complain about cold drafts, unit heaters are often installed. Heat from these units is usually drawn into the plant interior because of the velocity of the infiltration air, leading to overheating. Too often, the solution is to exhaust more air from the interior, causing increased negative pressure and more infiltration. Negative plant pressures reduce the exhaust volumetric flow rate because of increased system resistance. Balanced plants that have equal exhaust and replacement air rates use less energy. Wind effects on building balance are discussed in Chapter 14 of the 1993 *ASHRAE Handbook—Fundamentals*.

FLUID MECHANICS

Chapters 2 and 32 of the 1993 *ASHRAE Handbook—Fundamentals* describe basic fluid mechanics and its applications to duct systems. A thorough understanding of fluid mechanics is essential to an economical local exhaust system design.

The preparation of this chapter is assigned to TC 5.8, Industrial Ventilation.

The equation for *volumetric flow rate* is

$$Q = VA \tag{1}$$

where

Q = volumetric flow rate, m³/s
V = average flow velocity, m/s
A = flow cross-sectional area, m²

The equation that relates velocity to velocity pressure is

$$V = (2p_v/\rho)^{0.5} \tag{2}$$

where

p_v = velocity pressure, Pa
ρ = density, kg/m³

If the air temperature is 20°C ± 15°C, the ambient pressure is the standard 101 kPa, the duct pressure is no more than 5 kPa different from the ambient pressure, the dust loading is low (0.2 to 2 g/m³), and moisture is not a consideration, then the density in Equation (2) is standard without significant error. For a standard air density of 1.2 kg/m³, Equation (2) simplifies to

$$V = 1.29\sqrt{p_v} \tag{3}$$

LOCAL EXHAUST SYSTEM COMPONENTS

Local exhaust systems have five basic components: (1) the *hood*, or entry point, of the system; (2) the *duct system*, which transports air; (3) the *air-cleaning device*, which removes contaminants from the airstream; (4) the *air-moving device*, which provides motive power for overcoming system resistance; and (5) the *exhaust stack*, which discharges system air to the atmosphere.

Hoods

The most effective hood uses the minimum exhaust volumetric flow rate to provide maximum contaminant control. The capturing effectiveness should be high, but it would be difficult and costly to develop a hood that is 100% efficient. Makeup air supplied by general ventilation to replace locally exhausted volumes can dilute contaminants that are not captured by the hood (Posokhin 1985). Knowledge of the process or operation is essential before a hood can be designed.

Hoods are either *enclosing* or *nonenclosing* (Figure 1). Enclosing hoods provide better and more economical contaminant control because their exhaust rates and the effects of room air currents are minimal compared to those for nonenclosing hoods. Hood access openings for inspection and maintenance should be as small as possible and placed out of the natural path of the contaminant. Hood

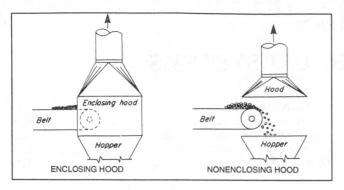

Fig. 1   Enclosing and Nonenclosing Hoods

performance (i.e., how well it controls the contaminant) should be checked by an industrial hygienist.

A nonenclosing hood can be used if access requirements make it necessary to leave all or part of the process open. Careful attention must be paid to airflow patterns around the process and hood and to characteristics of the process to make nonenclosing hoods functional.

Nonenclosing hoods can be classified according to their location relative to the contaminant source (Posokhin 1985) as either updraft coaxial, sidedraft, or downdraft (Figure 2).

Local exhaust systems with nonenclosing hoods can be stationary (i.e., having a fixed hood position), moveable, portable, or built into the process equipment. Moveable (turnable) hoods are used when process equipment must be accessed for repairing, loading, and unloading, (e.g., in electric ovens for melting steel). Hoods attached to flexible fume extraction arms (Figure 3) are used for small, medium-sized, and large workplaces when the source of contamination is not fixed, as in arc welding (Zhivov 1993). Flexible fume extraction arms usually have a hood connected to a duct 140 to 160 mm in diameter and have higher efficiencies for lower air volumes compared to stationary hoods. When the source of contamination is confined to a small, poorly ventilated space, such as a tank, an additional flexible hose extension with a hood and a magnetic foot can be hooked on the fume extraction arm instead of the standard hood.

The portable fume extractor (Figure 4) with a built-in fan and filter and a linear or round nozzle attached to a flexible hose about 45 mm in diameter is commonly used for the temporary extraction of fumes and solvents in confined spaces or during maintenance.

Built-in local exhausts, such as gun-mounted exhaust hoods and fume extractors built into stationary or turnover welding tables, are commonly used to evacuate welding fumes. Lateral exhaust hoods, which exhaust air through slots on the periphery of open vessels

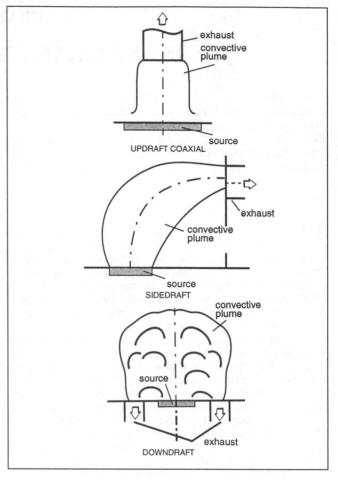

Fig. 2   Nonenclosing Hoods

such as those used for galvanizing metals, are another example of built-in local exhausts.

Nonenclosing hoods should be located so that the contaminant is drawn away from the operator's breathing zone. Canopy hoods should not be used where the operator must bend over a tank or process.

Capture Velocities

Capture velocities are air velocities at the point of contaminant generation upstream of a hood. The contaminant enters the moving

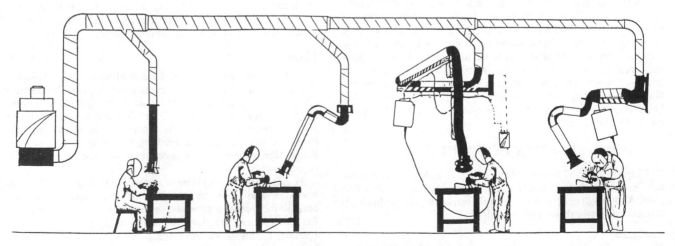

Fig. 3   Hoods Attached to Flexible Fume Extraction Arms

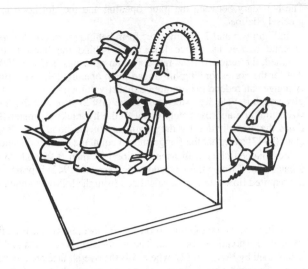

Fig. 4 Portable Fume Extractor with Built-in Fan and Filter

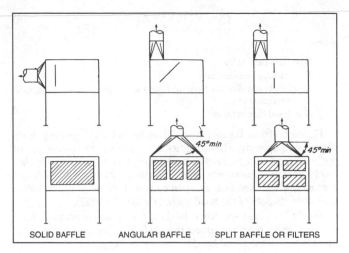

SOLID BAFFLE   ANGULAR BAFFLE   SPLIT BAFFLE OR FILTERS

Fig. 5 Use of Interior Baffles to Ensure Good Air Distribution

airstream at the point of generation and is conducted along with the air into the hood. Designers use capture velocities to select a volumetric flow rate Q to withdraw air through a hood. Table 1 shows ranges of capture velocities for several industrial operations. These capture velocities are based on successful experience under ideal conditions. If velocities anywhere upstream of a hood are known [$v = f(Q, x, y, z)$], the capture velocity is set equal to v at points (x, y, z) where contaminants are to be captured and Q is found. To ensure that contaminants enter an inlet, the transport equations between the source and the hood have to be solved.

Hood Volumetric Flow Rate

After the hood configuration and capture velocity are determined, the exhaust volumetric flow rate can be calculated.

For *enclosing hoods*, the exhaust volumetric flow rate is the product of the hood face area and the velocity required to prevent outflow of the contaminant [Equation (1)]. The inflow velocity is

typically 0.5 m/s. However, research with laboratory hoods indicates that lower velocities can reduce the vortex downstream of the human body, thus lessening the reentrainment of contaminant into the operator's breathing zone (Caplan and Knutson 1977, Fuller and Etchells 1979). These lower face velocities require that the replacement air supply be distributed to minimize the effects of room air currents. This is one reason replacement air systems must be designed with exhaust systems in mind. Because air must enter the hood uniformly, interior baffles are sometimes necessary (Figure 5).

For *nonenclosing hoods*, the air velocity at the point of contaminant release must equal the capture velocity and the air must be directed so that the contaminant enters the hood. The simplest form of nonenclosing hood is the *plain opening* shown in Figure 6 (Alden and Kane 1982). For plain round and rectangular openings, the required flow rate can be approximated by the following equation:

Table 1 Range of Capture (Control) Velocities

Condition of Contaminant Dispersion	Examples	Capture Velocity, m/s
Released with essentially no velocity into still air	Evaporation from tanks, degreasing, plating	0.25 to 0.5
Released at low velocity into moderately still air	Container filling, low-speed conveyor transfers, welding	0.5 to 1.0
Active generation into zone of rapid air motion	Barrel filling, chute loading of conveyors, crushing, cool shakeout	1.0 to 2.5
Released at high velocity into zone of very rapid air motion	Grinding, abrasive blasting, tumbling, hot shakeout	2.5 to 10

In each category above, a range of capture velocities is shown. The proper choice of values depends on several factors (Alden and Kane 1982):

Lower End of Range	Upper End of Range
1. Room air currents are favorable to capture.	1. Distributing room air currents.
2. Contaminants of low toxicity or of nuisance value only.	2. Contaminants of high toxicity.
3. Intermittent, low production.	3. High production, heavy use.
4. Large hood; large air mass in motion.	4. Small hood; local control only.

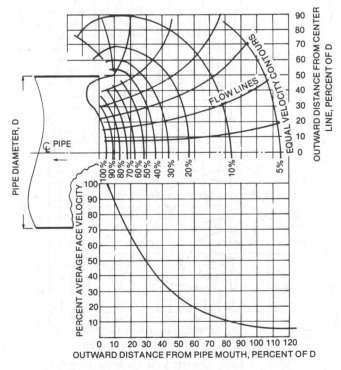

Fig. 6 Velocity Contours for Plain Round Opening

$$Q = V(10X^2 + A) \qquad (4)$$

where

Q = flow rate, m³/s
V = capture velocity, m/s
X = center line distance from the hood face to the point of contaminant generation, m
A = hood face area, m²

Figure 6 shows lines of equal velocity (velocity contours) for a plain round opening. The velocities are expressed as percentages of the hood face velocity. Studies have established the *principle of similarity of contours*, which states that the positions of the velocity contours, when expressed as a percentage of the hood face velocity, are functions only of the hood shape (DallaValle 1952).

Figure 7 (Alden and Kane 1972) shows velocity contours for a rectangular hood with an aspect ratio of 0.333. The profiles are similar to those for the round hood but are more elongated.

If the aspect ratio is lower than about 0.2 (0.15 for flanged openings), the shape of the flow pattern in front of the hood changes from approximately spherical to approximately cylindrical, and Equation (4) is no longer valid. For this type of nonenclosing hood, commonly called a *slot hood*, the required flow rate is predicted by the following equation for openings of 13 to 50 mm in width (Silverman 1942):

$$Q = 3.7LVX \qquad (5)$$

where L = long dimension of the slot, m.

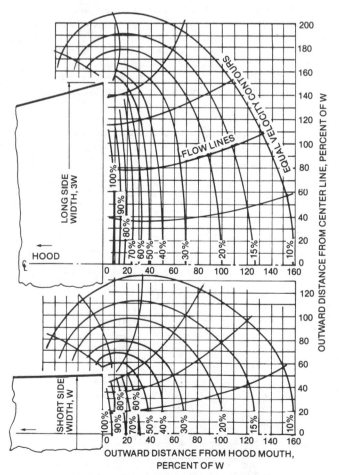

Fig. 7 Velocity Contours for Plain Rectangular Opening with Sides in a 1:3 Ratio

The boundary between the plain opening and the slot hood is not precisely defined.

In Figures 6 and 7, air migrates from behind the hood. If a *flange* (a solid barrier to reduce flow from behind the hood face) is installed, the required flow rate for plain openings is about 75% of that for the corresponding unflanged plain opening (Hemeon 1963). A flange can reduce the required exhaust volumetric flow rate for a given capture velocity. Alden and Kane (1982) state that the flange size should be approximately equal to the hydraulic diameter (four times the area divided by the perimeter) of the hood face. Brandt et al. (1947) state that the flange should equal the capture distance.

For flanged slots with aspect ratios less than 0.15 and flanges greater than 3 times the slot width, Silverman (1942) reported that the required flow rate can be calculated from the following equation:

$$Q = 2.6LVX \qquad (6)$$

Multiple slots are often used on a hood face; however, Equations (5) and (6) should *not* be used. The volumetric flow rate should be determined by Equation (4), where A is the overall face area, because the slots on such a hood are for air distribution over the hood face only.

Another device that helps increase the effectiveness of a nonenclosing hood is a *baffle*, which is a solid barrier that prevents airflow from unwanted areas in front of the hood. For example, if a rectangular hood is placed on a bench surface, flow cannot come from below the hood and the hood is more effective. Equation (4) can be modified for this situation by assuming that a mirror image of the flow exists on the underside of the bench:

$$Q = V(5X^2 + A) \qquad (7)$$

This is known as the DallaValle half-hood equation (DallaValle 1952). For better collection, the hood should be flanged. Flanging and baffling a nonenclosing hood create a situation more similar to an enclosure.

Nonenclosing hoods should be placed as near as possible to the source of contamination to prevent excessive exhaust rates. Figures 6 and 7 show that the velocity decreases extremely quickly as the distance in front of the hood increases, reaching 10% of the face velocity at approximately one hydraulic diameter in front of the hood. Because the volumetric flow rate is a direct function of the square of the distance (except for single slots), the required flow rate increases rapidly with capture distance. Typically, nonenclosing hoods are ineffective if the capture distance is greater than about 1 m. Large capture distances can decrease contaminant control. Room air currents caused by thermal currents, supply air grilles, mechanical action of the process, or personnel-cooling fans can disturb the flow toward the hood enough to deflect the contaminant-laden airstream away from the hood. If capture distances greater than 1 m are required, the designer should consider a push-pull ventilation system, which is discussed later.

Example 1. A nonenclosing hood is to be designed to capture a contaminant that is liberated with a very low velocity 0.6 m in front of the face of the hood. The nature of the operation requires hood face dimensions of 0.45 by 1.2 m. The hood rests on a bench, and a flange is placed on the sides and top of the face (Figure 8). Determine the volumetric flow rate required to capture the contaminant. Assume that (1) the room air currents are variable in direction but less than 0.25 m/s, (2) the contaminant has low toxicity, and (3) the hood is used continuously.

Solution: From Table 1, a capture velocity of 0.25 to 0.5 m/s is required. The selected capture velocity must be higher than the room air currents; assume that 0.4 m/s is sufficient. The flow rate required can be predicted by modifying Equation (7) to account for the flanges. Thus,

$$Q = 0.75V(5X^2 + A)$$

For this example, the required flow rate is

$$Q = (0.75)(0.4)[(5)(0.6)^2 + (0.45)(1.2)](1000) = 702 \text{ L/s}$$

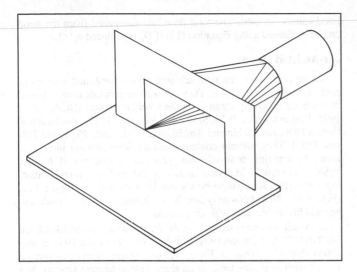

Fig. 8   Hood on a Bench

Hot Process Hoods. The exhaust from hot processes requires special consideration because of the buoyant effect of heated air near the hot process. The minimum exhaust rate is obtained by completely enclosing the process and placing the exhaust connection at the top of the enclosure. Determining the exhaust rate for a hot process requires an understanding of the convectional heat transfer rate (see Chapter 3 of the 1993 *ASHRAE Handbook—Fundamentals*) and the physical size of the process.

If the process cannot be completely enclosed, the canopy hood should be placed above the process so that the contaminant moves toward the hood. Canopy hoods should be designed with caution to avoid drawing contaminants across the operator's breathing zone. The hood's height above the process should be kept to a minimum to reduce the total exhaust air rate.

A *low canopy hood* is within 1 m of a process and requires the least volumetric flow rate of all nonenclosing hoods. A *high canopy hood* is more than 3 m above a process and requires a greater volumetric flow rate because room air is entrained in the column of hot, contaminated air rising from the process; this situation should be avoided. Lateral exhaust requires the highest exhaust volumetric flow rate because the airflow toward the hood must overcome the tendency of the air to rise.

Hemeon (1963) lists Equations (8) through (12) for determining the volumetric flow rate of hot gases for low canopy hoods. Note that canopy hoods located 1 to 3 m above the process cannot be analyzed using these equations.

$$Q_o = (2gR/pc_p)^{1/3} (qLA_p^2)^{1/3} \qquad (8)$$

where

   Q_o = volumetric flow rate, m³/s
   g = gravitational acceleration, 9.8 m/s²
   R = air gas constant, 287 J/(kg·K)
   p = local atmospheric pressure, Pa
   c_p = constant pressure specific heat, 1004 J/(kg·K)
   q = convection heat transfer rate, W
   L = vertical height of hot object, m
   A_p = cross-sectional area of airstream at upper limit of hot body, m

For a standard atmospheric pressure of 101.325 kPa, Equation (8) can be written as

$$Q_o = 0.038 (qLA_p^2)^{1/3} \qquad (9)$$

For three-dimensional bodies, the area A_p in Equation (8) is approximated by the plan view area of the hot body (Figure 9a). For horizontal cylinders, A_p is the product of the length and the diameter of the rod.

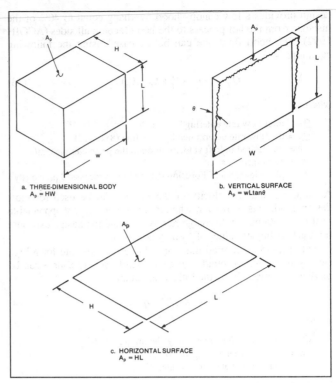

a. THREE-DIMENSIONAL BODY
   A_p = HW

b. VERTICAL SURFACE
   A_p = wLtanθ

c. HORIZONTAL SURFACE
   A_p = HL

Fig. 9 A_p for Various Situations

For vertical surfaces, the area A_p in Equation (8) is the area of the airstream (viewed from above) as the flow leaves the vertical surface (Figure 9b). As the airstream moves upward on a vertical surface, it appears to expand at an angle of approximately 4 to 5°. Thus, A_p is given by

$$A_p = wL\tan\theta \qquad (10)$$

where

   w = width of vertical surface, m
   L = height of vertical surface, m
   θ = angle at which the airstream expands

For horizontal heated surfaces, A_p is the surface area of the heated surface and L is the longest length (conservative) of the horizontal surface, or its diameter if it is round (Figure 9c).

If the heat transfer is caused by steam from a hot water tank,

$$q = h_{fg}GA_p \qquad (11)$$

where

   q = heat transferred, kW
   h_{fg} = latent heat of vaporization, kJ/kg
   G = steam generation rate, kg/(s·m²)
   A_p = surface area of the tank, m²

At 100°C, the latent heat of vaporization is 2257 kJ/kg. Using this value and Equation (11), Equation (8) simplifies to

$$Q_o = 5A_p (GL)^{1/3} \qquad (12)$$

The exhaust volumetric flow rate determined by Equation (8) or (12) is the required exhaust flow rate when (1) a low canopy hood of the same dimensions as the hot object or surface is used and (2) side and back baffles are used to prevent room air currents from disturbing the rising air column. If side and back baffles cannot be used, the canopy hood size and the exhaust flow rate should be increased to reduce the possibility of spillage from the hood. A good design provides a low canopy hood overhang equal to 40% of the distance from the hot process to the hood face on all sides (ACGIH 1992). The hood flow rate can be increased using the following equation:

$$Q_T = Q_o + V_f (A_f - A_p) \qquad (13)$$

where

Q_T = total flow rate entering hood, m³/s
Q_o = the flow rate determined by Equation (8) or (12)
V_f = the desired indraft velocity through the perimeter area, m/s
A_f = hood face area, m²
A_p = plan view area of Equation (8) or (12), as discussed previously

A minimum indraft velocity of 0.5 m/s should be used for most design conditions. However, when room air currents are appreciable or if the contaminant discharge rate is high and the design exposure limit is low, higher values of V_f may be required.

Sutton (1950) reported that the volumetric flow rate for a high canopy hood over a round, square, or nearly square source can be predicted empirically by the following equation:

$$Q_z = 0.008Z^{3/2}q^{1/3} \qquad (14)$$

where

Q_z = volumetric flow rate at any elevation Z, m³/s
Z = vertical distance (Figure 10), m
q = convection heat transfer rate, W

Z can be obtained from

$$Z = Y + 2B \qquad (15)$$

where

Y = distance from the hot process to the hood face (Figure 10), m
B = maximum dimension of hot process plan view (Figure 10), m

At Z, the rising airstream will be nearly round, with a diameter determined by

$$D_z = 0.5Z^{0.88} \qquad (16)$$

where D_z = flow diameter at any elevation Z above the apparent point source, m.

High canopy hoods are extremely susceptible to room air currents. Therefore, they are typically much larger (often twice as large) than indicated by Equation (16) and are used only if a low canopy hood cannot be used. The total flow rate exhausted from the hood can be evaluated using Equation (13) if Q_o is replaced by Q_z.

Jet-Assisted Hoods

Jet-assisted hoods are nonenclosing hoods combined with compact, linear or radial air jets. They are used to separate contaminated zones from relatively clean zones in working spaces (Boshnyakov 1975, Romeyko et al. 1976, Stoler and Savelyev 1977, Posokhin and Broida 1980, Anichkhin and Anichkhina 1984, Cesta 1988, and Zhivov 1993). They prevent contaminated air from moving into clean zones by creating positive static pressure (Strongin and Marder 1988), for example, in drying chambers and cooling tunnels of casting conveyors. Jet-assisted hoods can be used over welding robots (Figure 11). Near laboratory hoods and industrial ovens, hoods can be used in combination with air curtains.

Contaminated air is moved toward the hood by a jet-assisted hood (ACGIH 1992, Heinsohn 1991, Elinskii 1989, Stroiizdat 1992, Sciola 1993). Push-pull exhausts (Figure 12) can be used over open surface vessels such as plating tanks as an alternative to lateral exhausts. In a push-pull system, air is blown through one slot on the periphery of a vessel and exhausted from the opposite slot. The inlet to the hood must be large to accommodate the jet and the room air entrained by

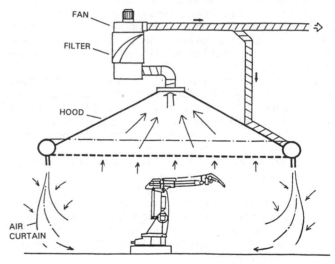

Fig. 11   Jet-Assisted Hood over Welding Robot

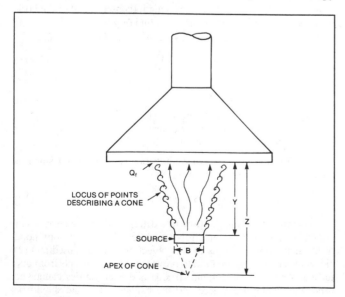

Fig. 10   High Canopy Hood Parameters for Y > 3 m

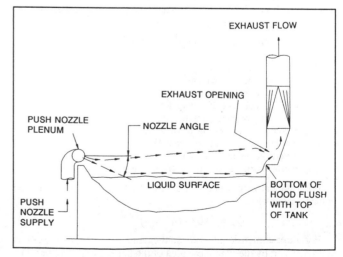

Fig. 12   Push-Pull Tank Ventilation
(Courtesy American Conference of Governmental Industrial Hygienists, Inc.)

the jet. Push-pull systems can be used on wider vessels than lateral exhaust hoods. Chapter 10 of the *Industrial Ventilation Manual* (ACGIH 1992) provides a design procedure for these systems.

The jets cause swirling airflows near the exhaust hood, increasing its capturing efficiency (Ljungqist and Waering 1898). The capturing efficiency of jet-assisted hoods is 15 to 20% higher than that of conventional hoods for the same operating costs (Strongin and Marder 1988). For the same capturing efficiency, they are 25 to 30% more cost-effective due to lower airflow. When exhausted air must be cleaned, reduced cleaning costs make jet-assisted hoods 100 to 150% more cost-effective. If compensating-type jet-assisted hoods are used, operating costs for heating and cooling in general ventilation systems are also reduced.

Principles of Hood Design

Numerous studies of local exhausts and common practices have led to the development of the following list of hood design principles:

- The hood should be located as close as possible to the source of contamination.
- The hood opening should be positioned so that it causes the contaminant to deviate the least from its natural path.
- The hood should be located so that the contaminant is drawn away from the operator's breathing zone.
- The hood must be the same size as or larger than the flow entering the hood (Posokhin 1985). If the hood is smaller than the flow, a higher volumetric flow rate will be required.

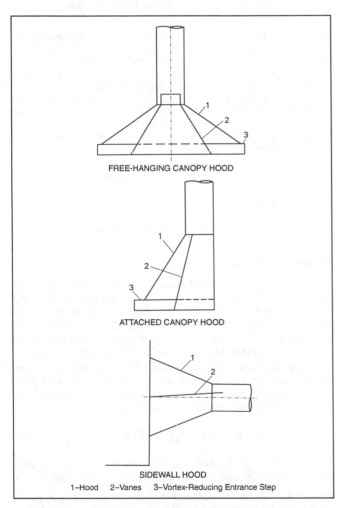

FREE-HANGING CANOPY HOOD

ATTACHED CANOPY HOOD

SIDEWALL HOOD

1–Hood 2–Vanes 3–Vortex-Reducing Entrance Step

Fig. 13 Hoods with Nonuniform Velocities in Opening Cross Sections

- The velocity distribution in the hood opening cross section should be nonuniform (Figure 13), following the velocity profile of the incoming flow. This can be achieved by incorporating vanes in the hood opening. In the case of a stationary hood and a contaminant source that is not fixed (e.g., welding or soldering), the air velocity along the hood must be uniform; this can be achieved using vanes or perforations.

Special Situations

Some operations may require exhaust flow rates different from those calculated using the previous equations. Typical reasons for different flow rates include the following:

1. Induced air currents are created whenever anything is projected into an airspace. For example, high-speed rotating machines such as pulverizers, high-speed belt material transfer systems, falling granular materials, and escaping compressed air from pneumatic tools all produce air currents. The size and direction of the airflow should be considered in hood design.
2. Exhaust flow rates may dilute combustible vapor-air mixtures insufficiently. They should be diluted to less than about 25% of the lower explosive limit of the vapor (NFPA 1991).
3. Room air currents caused by cross drafts, compensating air, spot cooling, or motion of the machinery are especially significant when designing high canopy hoods for hot process exhaust.

Duct Considerations

The second component of a local exhaust ventilation system is the duct through which contaminated air is transported from the hood(s). Round ducts are preferred because they (1) offer a more uniform air velocity to resist settling of material and (2) can withstand the higher static pressures normally found in exhaust systems. When design limitations require rectangular ducts, the aspect ratio (height to width ratio) should be as close to unity as possible.

Table 2 Contaminant Transport Velocities
Adapted from *Industrial Ventilation—A Manual of Recommended Practices* (ACGIH 1992)

Nature of Contaminant	Examples	Minimum Transport Velocity, m/s
Vapors, gases, smoke	All vapors, gases, smokes	Usually 5 to 10
Fumes	Welding	10 to 13
Very fine light dust	Cotton lint, wood flour, litho powder	13 to 15
Dry dusts and powders	Fine rubber dust, Bakelite molding powder dust, jute lint, cotton dust, shavings (light), soap dust, leather shavings	15 to 20
Average industrial dust	Grinding dust, buffing lint (dry), wool jute dust (shaker waste), coffee beans, shoe dust, granite dust, silica flour, general material handling, brick cutting, clay dust, foundry (general), limestone dust, asbestos dust in textile industries	18 to 20
Heavy dusts	Sawdust (heavy and wet), metal turnings, foundry tumbling barrels and shakeout, sandblast dust, wood blocks, hog waste, brass turnings, cast-iron boring dust, lead dust	20 to 23
Heavy or moist dusts	Lead dust with small chips, moist cement dust, asbestos chunks from transite pipe cutting machines, buffing lint (sticky), quicklime dust	23 and up

Minimum transport velocity is the velocity required to transport particulates without settling. Table 2 lists some generally accepted transport velocities as a function of the nature of the contaminants (ACGIH 1992). The values listed are typically higher than theoretical and experimental values to account for (1) damages to ducts, which would increase system resistance and reduce volume and duct velocity; (2) duct leakage, which tends to decrease velocity in the duct system upstream of the leak; (3) fan wheel corrosion or erosion and/or belt slippage, which could reduce fan volume; and (4) reentrainment of settled particulate caused by improper operation of the exhaust system. Design velocities can be higher than the minimum transport velocities but should never be significantly lower. When particulate concentrations are low, the effect on fan power is negligible.

Standard duct sizes and fittings should be used to cut cost and delivery time. Information on available sizes and the cost impact of nonstandard sizes can be obtained from the contractor(s).

Duct Size Determination

The size of the round duct attached to the hood can be calculated using Equation (1), the volumetric flow rate, and the minimum transport velocity.

Example 2. Suppose the contaminant captured by the hood in Example 1 requires a minimum transport velocity of 15 m/s. What diameter round duct should be specified?

Solution: Using Equation (1), the duct area required is

$$A = 702 / [(15)(1000)] = 0.047 \text{ m}^2$$

Generally the area calculated will not correspond to a standard duct size. The area of the standard size chosen should be less than that calculated. For this example, a 225-mm diameter with an area of 0.0398 m² should be chosen. The actual duct velocity is then

$$V = 702 / [(0.0398)(1000)] = 17.6 \text{ m/s}$$

Hood Entry Loss

When air enters a hood, a loss of total pressure occurs because of dynamic losses. This is called the *hood entry loss* and may have several components, each given by

$$h_e = C_o p_v \qquad (17)$$

where

h_e = hood entry loss, Pa
C_o = loss factor, dimensionless
p_v = appropriate velocity pressure, Pa

Loss factors for various hood shapes are given in Figure 14. The graph shows an optimum hood entry angle to minimize the entry losses. However, this total included angle of 45° is impractical in many situations because of the required transition length. A 90° angle, with a corresponding loss factor of 0.25 (for rectangular openings), is standard for most tapered hoods.

Total pressure is difficult to measure in a duct system because it varies from point to point across a duct, depending on the local velocity. On the other hand, static pressure remains constant across a straight duct. Therefore, a single measurement of static pressure in a straight duct downstream of the hood can monitor the volumetric flow rate. The absolute value of this static pressure, *hood suction*, is given by

$$p_{hs} = p_v + h_e \qquad (18)$$

where p_{hs} = hood suction, Pa.

Simple Hoods

A simple hood has only one dynamic loss. The hood suction becomes

$$p_{hs} = (1 + C_o) p_v \qquad (19)$$

where p_v is the duct velocity pressure.

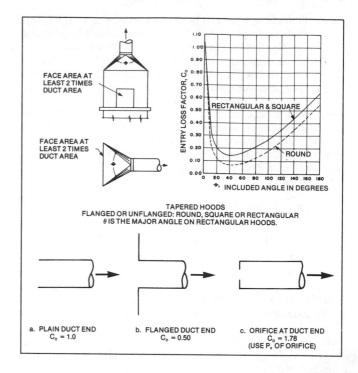

Fig. 14   Entry Losses for Typical Hoods

Example 3. Suppose the hood in Example 1 was designed such that the largest angle of transition between the hood face and the duct equals 90°. What is the suction for this hood? Assume standard air density.

Solution: The two transition angles cannot be equal. Whenever this is true, the largest angle is used to determine the loss factor from Figure 14. Because the transition piece originates from a rectangular opening, the curve marked "rectangular" must be used. This corresponds to a loss factor of 0.25. The duct diameter and the velocity required were determined in Example 2. Equation (3), which assumes standard air density, can be used to determine the duct velocity pressure:

$$p_v = (17.6 / 1.29)^2 = 186 \text{ Pa}$$

From Equation (19),

$$p_{hs} = (1 + 0.25)(186) = 232 \text{ Pa}$$

Compound Hoods

The losses for multislotted hoods (see Figure 15) or single-slot hoods with a plenum (called *compound hoods*) must be analyzed somewhat differently. The slots distribute air over the hood face and do not influence capture efficiency. The slot velocity should be approximately 10 m/s to provide the required distribution at the minimum energy cost. Higher velocities dissipate more energy.

Losses occur when air passes through the slot and when air enters the duct. Because the velocities, and, therefore, the velocity pressures, can be different at the slot and at the duct entry locations, the hood suction must reflect both losses and is given by

$$p_{hs} = p_v + (C_o p_v)_s + (C_o p_v)_d \qquad (20)$$

where the first p_v is generally the higher of the two velocity pressures, *s* refers to the slot, and *d* refers to the duct entry location.

Example 4. A multislotted hood has 3 slots, each 25 mm by 1 m. At the top of the plenum is a 90° transition into the 250-mm duct. The volumetric flow rate required for this hood is 0.78 m³/s. Determine the hood suction. Assume standard air.

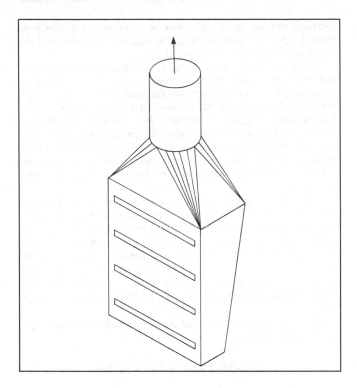

Fig. 15   Multislotted Nonenclosing Hood

Solution: The slot velocity V_s from Equation (1) is

$$V_s = 0.78 / (3 \times 0.025 \times 1) = 10.4 \text{ m/s}$$

Substituting this velocity in Equation (3),

$$p_v = (10.4 / 1.29)^2 = 65 \text{ Pa}$$

The duct area is 0.0491 m^2. Therefore, the duct velocity determined from Equation (1) is

$$V_d = 0.78 / 0.0491 = 15.9 \text{ m/s}$$

Substituting this velocity in Equation (3),

$$p_v = (15.9 / 1.29)^2 = 152 \text{ Pa}$$

For a 90° transition into the duct, the loss factor is 0.25. For the slots, the loss factor is 1.78 (Figure 14). The duct velocity pressure is added to the sum of the two losses because it is larger than the slot velocity pressure, therefore, using Equation (20),

$$p_{hs} = 152 + (1.78)(65) + (0.25)(152) = 306 \text{ Pa}$$

Hood suction is the negative static pressure measured about three duct diameters downstream of the hood. A larger distance is required for included angles 180° or larger.

Exhaust volume requirements, minimum duct velocities, and entry loss factors for many specific operations are given in Chapter 10 of the *Industrial Ventilation Manual* (ACGIH 1992).

DUCT LOSSES

Chapter 32 of the 1993 *ASHRAE Handbook—Fundamentals* covers basics of duct design and the design of metalworking exhaust systems. The design method in Chapter 32 is based on total pressure loss, including the fitting coefficients, while *Industrial Ventilation* (AGCIH 1992) calculates static pressure loss. Loss coefficients are in the *ASHRAE Duct Fitting Database* (ASHRAE 1994), which runs on a personal computer.

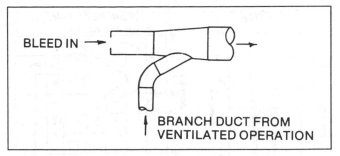

Fig. 16   Air Bleed-In

For systems conveying particulates, elbows with a centerline radius-to-diameter ratio (r/D) greater than 1.5 are most suitable. If r/D is 1.5 or less, abrasion in dust-handling systems can reduce the life of elbows. Elbows are often made of seven or more gores, especially in large diameters. For converging flow fittings, a 30° entry angle is recommended to minimize energy losses and abrasion in dust-handling systems (Fitting ED5-1 of ASHRAE 1994).

Where exhaust systems handling particulates must allow for a substantial increase in future capacity, required transport velocities can be maintained by providing open-end stub branches in the main. Air is admitted into the systems through these stub branches at the proper pressure and volumetric flow rate until the future connection is installed. Figure 16 shows such an air bleed-in. The use of outside air minimizes replacement air requirements. The size of the opening can be calculated by first determining the pressure drop required across the orifice from the duct calculations. Then the orifice velocity pressure can be determined from one of the following equations:

$$p_{v,o} = \Delta p_{t,o} / C_o \tag{21a}$$

or

$$p_{v,o} = \Delta p_{s,o} / (C_o + 1) \tag{21b}$$

where

$p_{v,o}$ = orifice velocity pressure, Pa
$\Delta p_{t,o}$ = total pressure to be dissipated across the orifice, Pa
$\Delta p_{s,o}$ = static pressure to be dissipated across the orifice, Pa
C_o = orifice loss coefficient referenced to the velocity at the orifice cross-sectional area, dimensionless (Figure 14)

Equation (21a) should be used if the total pressure is calculated through the system; Equation (21b) should be used if the static pressure is calculated through the system. Once the velocity pressure is known, Equation (2) or (3) can be used to determine the orifice velocity. Equation (1) can then be used to determine the orifice size.

EXHAUST STACKS

The exhaust stack must be designed and located to prevent the reentrainment of discharged air into supply system inlets. The building's shape and surroundings determine the atmospheric airflow over it. Chapter 14 of the 1993 *ASHRAE Handbook—Fundamentals*, has more details on exhaust stack design. The following guidelines can be used:

1. For many one- or two-story industrial or laboratory buildings, 5-m stacks may be adequate for discharging above the roof cavity.
2. The minimum stack height should be at least 2.5 m above the roof for safety purposes.
3. The discharge should be above the point where wind flow is unaffected by the building.

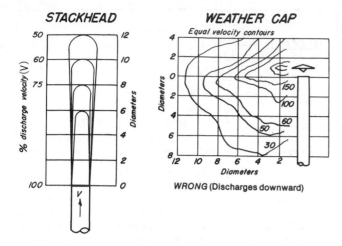

Fig. 17 Comparison of Flow Patterns for Stackheads and Weathercaps

If shorter stacks must be used, the stack discharge velocity should be high enough to project the contaminant-laden air above the roof recirculation zone. However, this usually requires wastefully high velocities (up to 40 m/s). Wind velocities of 16 to 24 km/h reduce exhaust plume rise by 85 to 90% compared to no-wind conditions.

If rain protection is important, stackhead design is preferable to weathercaps. Weathercaps, which are not recommended, have three disadvantages:

1. They deflect air downward, increasing the chance that contaminants will recirculate into air inlets.
2. They have high friction losses.
3. They provide less rain protection than a properly designed stackhead.

Figure 17 contrasts the flow patterns of weathercaps and stackheads. Loss data for weathercaps are presented in Chapter 32 of the 1993 *ASHRAE Handbook—Fundamentals*. Losses in the straight duct form of stackheads (Figure 15 F or G, Chapter 14 of the 1993 *ASHRAE Handbook—Fundamentals*) are balanced by the pressure regain at the expansion to the larger diameter stackhead.

INTEGRATING DUCT SEGMENTS

Most systems have more than one hood. If the pressures are not designed to be the same for merging parallel airstreams, the system adjusts to equalize pressure at the common point; however, the flow rates of the two merging airstreams will not necessarily be the same as designed. As a result, the hoods can fail to control the contaminant adequately, exposing workers to potentially hazardous contaminant concentrations.

Two design methods ensure that the two pressures will be equal. The preferred design self-balances without external aids. The second design, which uses adjustable balance devices such as blast gates or dampers, is not recommended, especially when conveying abrasive material.

AIR CLEANERS

Air-cleaning equipment is usually selected to (1) conform to federal, state, or local emission standards and regulations; (2) prevent reentrainment of contaminants to work areas, where they may become a health or safety hazard; (3) reclaim usable materials; (4) permit cleaned air to recirculate to work spaces and/or processes; (5) prevent physical damage to adjacent properties; and (6) protect neighbors from contaminants.

Factors to consider when selecting air-cleaning equipment include the type of contaminant (number of components, particulate versus gaseous, and concentration), the contaminant removal efficiency required, the disposal method, and the air or gas stream characteristics. Chapter 26 of the 1992 *ASHRAE Handbook—Systems and Equipment* covers industrial gas-cleaning and air pollution control equipment such as dry centrifugal collectors, fabric collectors, electrostatic precipitators, and wet collectors. A qualified applications engineer should be consulted when selecting equipment.

The collector's pressure loss must be added to overall system pressure calculations. In some collectors, specifically some fabric filters, the loss varies as the operation time increases. The system design should incorporate the maximum pressure drop of the collector, or hood flow rates will be lower than designed during most of the duty cycle. Also, fabric collector losses are usually given only to the clean air plenum. A reacceleration to the duct velocity, with the associated entry losses, must be calculated in the design phase.

Most other collectors are rated flange-to-flange with reacceleration included in the loss.

AIR-MOVING DEVICES

The type of air-moving device used depends on the type and concentration of contaminant, the pressure rise required, and the allowable noise levels. Fans are usually selected. Chapter 18 of the 1992 *ASHRAE Handbook—Systems and Equipment* describes available fans and refers the reader to the Air Movement and Control Association (AMCA) *Publication 201, Fans and Systems*, for proper connection of the fan(s) to the system. The fan should be located downstream of the air cleaner whenever possible to (1) reduce possible abrasion of the fan wheel blades and (2) create a negative pressure in the air cleaner so that air will leak into it and maintain positive control of the contaminant. In some instances, however, the fan is located upstream from the cleaner to help remove dust. This is especially true when using cyclone collectors, such as in the woodworking industry.

If explosive, corrosive, flammable, or sticky materials are handled, an *injector* can transport the material to the air-cleaning equipment. Injectors create a shear layer that induces airflow into the duct. Injectors should be the last choice because their efficiencies seldom exceed 10%.

ENERGY RECOVERY

The transfer of energy from exhausted air to replacement air may be economically feasible, depending on (1) the location of the exhaust and replacement air ducts, (2) the temperature of the exhausted gas, and (3) the nature of the contaminants being exhausted.

DUCT CONSTRUCTION

Elbows and converging flow fittings should be made of thicker material than the straight duct, especially if abrasives are conveyed. Some cases require elbows constructed with a special wear strip in the heel.

When corrosive material is present, alternatives such as special coatings or different duct materials (fiberglass, stainless steel, or special coatings) can be used. Industrial duct construction is described in Chapter 16 of the 1992 *ASHRAE Handbook—Systems and Equipment*. For further construction details, refer to SMACNA (1990) for industrial duct construction standards.

SYSTEM TESTING

After installation, an exhaust system should be tested to ensure that it operates properly with the required flow rates through each hood. If the actual installed flow rates are different from the design values, they should be corrected before the system is used. Testing is also necessary to obtain baseline data to determine (1) compliance with federal, state, and local codes; (2) by periodic inspections, whether maintenance on the system is needed to ensure design operation; (3) whether a system has sufficient capacity for additional airflow; and (4) whether system leakage is acceptable. Chapter 34 of AMCA *Bulletin* 203 and Chapter 9 of the *Industrial Ventilation Manual* (ACGIH 1992) contain detailed information on the preferred methods for testing systems.

OPERATION AND MAINTENANCE

Periodic inspection and maintenance is required for the proper operation of exhaust systems. Systems are often changed or damaged after installation, resulting in low duct velocities and/or incorrect volumetric flow rates. Low duct velocities can cause the contaminant to settle and plug the duct, reducing flow rates at the affected hoods. Installing additional hoods in an existing system can change volumetric flow at the original hoods. In both cases, changed hood volumes can increase worker exposure and health risks.

The maintenance program should include (1) inspecting ductwork inspection for particulate accumulation and damage by erosion or physical abuse, (2) checking exhaust hoods for proper volumetric flow rates and physical condition, (3) checking fan drives, and (4) maintaining air-cleaning equipment according to manufacturers' guidelines.

REFERENCES

ACGIH. 1992. *Industrial ventilation—A manual of recommended practice,* 21st ed. Committee on Industrial Ventilation, American Conference of Governmental Industrial Hygienists.

Alden, J.L. and J.M. Kane. 1982. *Design of industrial ventilation systems,* 5th ed. Industrial Press, New York.

Baturin, V.V. 1972. *Fundamentals of Industrial Ventilation,* 3rd English ed. Pergamon Press, New York.

Boshnyakov, E.N. 1975. Local exhaust with air curtains. *Water supply and sanitary techniques,* #3. Moscow (in Russian).

Brandt, A.D., R.J. Steffy, and R.G. Huebscher. 1947. Nature of air flow at suction openings. *ASHRAE Transactions* 53:55.

Caplan, K.J. and G.W. Knutson. 1977. The effect of room air challenge on the efficiency of laboratory fume hoods. *ASHRAE Transactions* 83(1): 141.

Cesta, T. 1988. Capture of pollutants from a buoyant point source using a lateral exhaust hood with and without assistance from air curtains. Proceedings of the 2nd International Symposium on Ventilation for Contamination Control, Ventilation '88. Pergamon Press, UK.

DallaValle, J.M. 1952. *Exhaust hoods,* p.22. Industrial Press, New York.

Elinskii, I.I. 1989. Ventilation and heating of galvanic shops of machine-building plants. Mashinostroyeniye, Moscow (in Russian).

Flynn, M.R. and M.J. Ellenbecker. 1985. The potential flow solution for air flow into a flanged circular hood. *American Industrial Hygiene Journal* 46(6):318-22.

Fuller, F.H. and A.W. Etchells. 1979. The rating of laboratory hood performance. *ASHRAE Journal* 21(10):49-53.

Heinsohn, R.J. 1991. *Industrial Ventilation: Engineering Principles.* John Wiley & Sons, New York.

Heinsohn, R.J., K.C. Hsieh, and C.L. Merkle. 1985. Lateral ventilation systems for open vessels. *ASHRAE Transactions* 91(1B):361-82.

Hemeon, W.C.L. 1963. *Plant and process ventilation,* p. 77. Industrial Press, New York.

Huebener, D.J. and R.T. Hughes. 1985. Development of push-pull ventilation. *American Industrial Hygiene Association Journal* 46(5):262-67.

Ljungqist, B. and C. Waering. 1988. Some observations on "modern" design of fume cupboards. Proceedings of the 2nd International Symposium on Ventilation for Contaminant Control, Ventilation '88. Pergamon Press, United Kingdom.

NFPA. 1991. Standard for ovens and furnaces—Design, location and equipment. National Fire Protection Association, *Standard* 86A, Item 4-2.1.

Posokhin, V.N. 1985. Design of local ventilation systems for the process equipment with heat and gas release. Mashinostroyeniye, Moscow (in Russian).

Posokhin, V.N. and V.A. Broida. 1980. Local exhausts incorporated with air curtains. Hydromechanics and heat transfer in sanitary technique equipment. KHTI, Kazan (in Russian).

Romeyko, N.F., N.E. Siromyatnikova, and E.V. Schebraev. 1976. Design of air curtains near an oven opening supplied with a hood. Heating and Ventilation, Proceedings of the A.I. Mikoyan Institute of Civil Engineers (in Russian).

Sciola, V. 1993. The practical application of reduced flow push-pull plating tank exhaust systems. 3rd International Symposium on Ventilation for Contaminant Control, Ventilation '91 (Cincinnati, Ohio).

Sepsy, C.F. and D.B. Pies. 1973. An experimental study of the pressure losses in converging flow fittings used in exhaust systems. *Document* PB 221 130. Prepared by Ohio State University for National Institute for Occupational Health.

Shibata, M., R.H. Howell, and T. Hayashi. 1982. Characteristics and design method for push-pull hoods: Part 1—Cooperation theory of air flow; Part 2—Streamline analysis of push-pull flow. *ASHRAE Transactions* 88.

Silverman, L. 1942. Velocity characteristics of narrow exhaust slots. *Journal of Industrial Hygiene and Toxicology* 24 (November):267.

SMACNA. 1990. *Round industrial duct construction standards.* Sheet Metal and Air Conditioning Contractors' National Association, Vienna, VA.

SMACNA. 1990. *Rectangular industrial duct construction standards.* Sheet Metal and Air Conditioning Contractors' National Association, Vienna, VA.

Stoler, V.D. and Yu. L. Savelyev. 1977. Push-pull systems design for etching tanks. *Heating, Ventilation, Water Supply, and Sewage Systems Design* 8(124). TsINIS, Moscow (in Russian).

Strongin, A.S. and M.L. Marder. 1988. Complex solution of painting shops ventilation. Proceedings of Utilization of Natural Resources and New Ventilation and Dust Transportation Systems Design. Penza (in Russian).

Sutton, O.G. 1950. The dispersion of hot gases in the atmosphere. *Journal of Meteorology* 7(5):307.

Zarouri, M.D., R.J. Heinsohn, and C.L. Merkle. 1983. Computer-aided design of a grinding booth for large castings. *ASHRAE Transactions* 89 (2A):95-118.

Zarouri, M.D., R.J. Heinsohn, and C.L. Merkle. 1983. Numerical computation of trajectories and concentrations of particles in a grinding booth. *ASHRAE Transactions* 89 (2A):119-35.

Zhivov, A.M. 1993. Principles of source capturing and general ventilation design for welding premises. *ASHRAE Transactions* 99(1):979-86.

BIBLIOGRAPHY

Braconnier, R. 1988. Bibliographic review of velocity field in the vicinity of local exhaust hood openings. *American Industrial Hygienist Association Journal* 49(4):185-98.

Glinski, M. 1978. Influence of disturbing streams on efficiency of suction pipes in local ventilation installations. *Transactions of Central Institute for Labor Protection* 28:45-60. Warsaw.

Pozin, G.M. 1977. Calculation of the effect of limitation planes on suction flows. *Transactions of the Central Institute for Labor Protection of the VCSPS* 105:8-13. Profizdat, Moscow (in Russian).

Qiang, Y.L. 1984. The effectiveness of hoods in windy conditions. Kungl Tekriska Hoggskolan. Stockholm, Sweden.

Safemazandarani, P. and H.D. Goodfellow. 1989. Analysis of remote receptor hoods under the influence of cross-drafts. *ASHRAE Transactions* 95(1):465-71.

Strongin, A.S., M.Yu. Ivanitskaya, and E.A. Visotskaya. 1986. Studies of the application of air curtains in tunnels for local ventilation. *Heating and Ventilation, Transactions of TsNIIpromzdanii* (in Russian).

INDUSTRIAL DRYING SYSTEMS

DRYING removes water and other liquids from gases, liquids, and solids. Drying is most commonly used, however, to describe the removal of water or solvent from solids by thermal means. *Dehumidification* refers to the drying of a gas, usually by condensation or by absorption with a drying agent (see Chapter 19 of the 1993 *ASHRAE Handbook—Fundamentals*). *Distillation*, particularly *fractional distillation*, is used to dry liquids.

It is cost-effective to separate as much water as possible from a solid using mechanical methods *before* drying using thermal methods. Mechanical methods such as filtration, screening, pressing, centrifuging, or settling require less power and less capital outlay per unit mass of water removed.

This chapter describes systems used for industrial drying and their advantages, disadvantages, relative energy consumption, and applications.

MECHANISM OF DRYING

When a solid dries, two processes occur simultaneously: (1) the transfer of heat to evaporate the liquid and (2) the transfer of mass as vapor and internal liquid. Factors governing the rate of each process determine the drying rate.

The principal objective in commercial drying is to supply the required heat efficiently. Heat transfer can occur by convection, conduction, radiation, or by a combination of these. Types of industrial dryers differ in their methods of transferring heat to the solid. In general, heat must flow first to the outer surface of the solid and then into the interior. An exception is drying with high-frequency electrical currents, where heat is generated within the solid, producing a higher temperature at the interior than at the surface and causing heat to flow from inside the solid to the outer surfaces.

APPLYING HYGROMETRY TO DRYING

In many applications, recirculating the drying medium improves thermal efficiency. The optimum proportion of recycled air balances the lower heat loss associated with more recirculation against the higher drying rate associated with less recirculation.

Because the humidity of drying air is affected by the recycle ratio, the air humidity throughout the dryer must be analyzed to determine whether the predicted moisture pickup of the air is physically attainable. The maximum ability of air to absorb moisture corresponds to the difference between saturation moisture content at wet-bulb (or adiabatic cooling) temperature and moisture content at supply air dew point. The actual moisture pickup of air is determined by heat and mass transfer rates and is always less than the maximum attainable.

ASHRAE psychrometric charts for normal and high temperatures (No. 1 and No. 3) can be used for most drying calculations. The process will not exactly follow the adiabatic cooling lines because some

heat is transferred to the material by direct radiation or by conduction from the metal tray or conveyor.

Example 1. Assume a dryer with a capacity of 41 kg of bone-dry gelatin per hour. Initial moisture content is 228% bone-dry basis, and final moisture content is 32% bone-dry basis. For optimum drying, the supply air is at 50°C dry bulb and 30°C wet bulb in sufficient quantity so that the condition of exhaust air is 40°C dry bulb and 29.5°C wet bulb. Makeup air is available at 27°C dry bulb and 18.6°C wet bulb.

Find (1) the required amount of makeup and exhaust air and (2) the percentage of recirculated air.

Solution: In this example, the humidity in each of the three airstreams is fixed; hence, the recycle ratio is also determined. Refer to ASHRAE Psychrometric Chart No. 1 to obtain the humidity ratio of makeup air and exhaust air. To maintain a steady-state condition in the dryer, water evaporated from the material must be carried away by exhaust air. Therefore, the pickup (which is the difference between the humidity ratio of exhaust air and that of makeup air) is equal to the rate at which water is evaporated from the material divided by the mass of dry air exhausted per unit of time.

Step 1. From ASHRAE Psychrometric Chart No. 1, the humidity ratios can be found as follows:

	Dry bulb, °C	Wet bulb, °C	Humidity ratio, g/kg dry air
Supply air	50	30	18.7
Exhaust air	40	29.5	22
Makeup air	27	18.6	10

Moisture pickup is $22 - 10 = 12$ kg per kilogram of dry air. The rate at which water is evaporated in the dryer is

$$[(41 \times 1000 \text{ g/kg})(3600 \text{ s/h})][(228 - 32)/100] = 22.3 \text{ g/s}$$

The dry air required to remove the evaporated water is $22.3/12 = 1.86$ kg/s.

Step 2. Assume x = percentage of recirculated air and $(100 - x)$ = percentage of makeup air. Then

Humidity ratio of supply air =
  (Humidity ratio of exhaust and recirculated air) $(x/100)$
  + (Humidity ratio of makeup air)$(100 - x)/100$

Hence,

$$18.7 = 22(x/100) + 10(100 - x)/100$$

$x = 72.5\%$ recirculated air
$100 - x = 27.5\%$ makeup air

DETERMINING DRYING TIME

The following are three methods of finding drying time, listed in order of preference:

1. Conduct tests in a laboratory dryer simulating conditions for the commercial machine, or obtain performance data using the commercial machine.
2. If the specific material is not available, obtain drying data on similar material by either of the above methods. This is subject to the investigator's experience and judgment.

The preparation of this chapter is assigned to TC 9.2, Industrial Air Conditioning.

3. Estimate drying time from theoretical equations (see Bibliography). Care should be taken in using the approximate values obtained by this method.

When designing commercial equipment, tests are conducted in a laboratory dryer that simulates commercial operating conditions. Sample materials used in the laboratory tests should be identical to the material found in the commercial operation. Results from several tested samples should be compared for consistency. Otherwise, the test results may not reflect the drying characteristics of the commercial material accurately.

When laboratory testing is impractical, commercial drying data can be based on the equipment manufacturer's experience.

Commercial Drying Time

When selecting a commercial dryer, the estimated drying time determines what size machine is needed for a given capacity. If the drying time has been derived from laboratory tests, the following should be considered:

- In a laboratory dryer, considerable drying may be the result of radiation and heat conduction. In a commercial dryer, these factors are usually negligible.
- In a commercial dryer, humidity conditions may be higher than in a laboratory dryer. In drying operations with controlled humidity, this factor can be eliminated by duplicating the commercial humidity condition in the laboratory dryer.
- Operating conditions are not as uniform in a commercial dryer as in a laboratory dryer.
- Because of the small sample used, the test material may not be representative of the commercial material.

Thus, the designer must use experience and judgment to modify the test drying time to suit the commercial conditions.

Dryer Calculations

To estimate the preliminary cost estimates for a commercial dryer, the circulating airflow rate, the makeup and exhaust airflow rate, and the heat balance must be determined.

Circulating Air. The required circulating or supply airflow rate is established by the optimum air velocity relative to the material. This can be obtained from laboratory tests or previous experience, keeping in mind that the air also has an optimum moisture pickup. (See the section on Applying Hygrometry to Drying.)

Makeup and Exhaust Air. The makeup and exhaust airflow rate required for steady-state conditions within the dryer is also discussed in the section on Applying Hygrometry to Drying. In a *continuously operating* dryer, the relation between the moisture content of the material and the quantity of makeup air is given by

$$G_T(W_2 - W_1) = M(w_1 - w_2) \tag{1}$$

where

- G_T = dry air supplied as makeup air to the dryer, kg/s
- M = stock dried in a continuous dryer, kg/s
- W_1 = humidity ratio of entering air, kg of water vapor per kg of dry air
- W_2 = humidity ratio of leaving air, kg of water vapor per kg of dry air (In a continuously operating dryer, W_2 is constant; in a batch dryer, W_2 varies during a portion of the cycle.)
- w_1 = moisture content of entering material dry basis, kg of water per kg
- w_2 = moisture content of leaving material dry basis, kg of water per kg

In *batch* dryers, the drying operation is given as

$$G_T(W_2 - W_1) = M_1(dw/d\theta) \tag{2}$$

where

- M_1 = mass of material charged in a discontinuous dryer, kg per batch
- $dw/d\theta$ = instantaneous rate of evaporation corresponding to w

The makeup air quantity is constant and is based on the average evaporation rate. Equation (2) then becomes identical to Equation (1), where $M = M_1/\theta$. Under this condition, the humidity in the *batch dryer* varies from a maximum to a minimum during the drying cycle, whereas in the *continuous dryer*, the humidity is constant with constant load.

Heat Balance. To estimate the fuel requirements of a dryer, a heat balance consisting of the following is needed:

- Radiation and convection losses from the dryer
- Heating of the commercial dry material to the leaving temperature (usually estimated)
- Vaporization of the water being removed from the material (usually considered to take place at the wet-bulb temperature)
- Heating of the vapor from the wet-bulb temperature in the dryer to the exhaust temperature
- Heating of the total water in the material from the entering temperature to the wet-bulb temperature in the dryer
- Heating of the makeup air from its initial temperature to the exhaust temperature

The energy absorbed must be supplied by the fuel. The selection and design of the heating equipment is an essential part of the overall design of the dryer.

Example 2. Magnesium hydroxide is to dried from 82% to 4% moisture content (wet basis) in a continuous conveyor dryer with a fin-drum feed (see Figure 7). The desired production rate is 0.4 kg/s. The optimum circulating air temperature for drying is 71°C, which is not limited by the existing steam pressure of the dryer.

Step 1. Laboratory tests indicate the following:

Specific heats

air (c_a)	=	1.00 kJ/(kg·K)
material (c_m)	=	1.25 kJ/(kg·K)
water (c_w)	=	4.18 kJ/(kg·K)
water vapor (c_v)	=	1.84 kJ/(kg·K)

Temperature of material entering dryer	=	15°C
Temperature of makeup air		
dry bulb	=	21°C
wet bulb	=	15.5°C
Temperature of circulating air		
dry bulb	=	71°C
wet bulb	=	38°C
Air velocity through drying bed	=	1.3 m/s
Dryer bed loading	=	33.3 kg/m²
Test drying time	=	25 min

Step 2. Previous experience indicates that the commercial drying time is 70% greater than the time obtained in the laboratory test. Therefore, the commercial drying time is estimated to be 1.7 × 25 = 42.5 min.

Step 3. The holding capacity of the dryer bed can be calculated as follows:

$$0.4 \, (42.5 \times 60) = 1020 \text{ kg at 4\% (wet basis)}$$

The required conveyor area is 1020/33.3 = 30.6 m². Assuming the conveyor is 2.4 m wide, the length of the drying zone is 30.6/2.4 = 12.8 m.

Step 4. The amount of water in the material entering the dryer is

$$0.4 \, [82/(100 + 4)] = 0.315 \text{ kg/s}$$

The amount of water in the material leaving is

$$0.4 \, [4/(100 + 4)] = 0.015 \text{ kg/s}$$

Thus, the moisture removal rate is 0.315 − 0.015 = 0.300 kg/s.

Step 5. The air circulates perpendicular to the perforated plate conveyor, so the air volume is the face velocity times the conveyor area:

$$\text{Air volume} = 1.3 \times 30.6 = 39.8 \text{ m}^3/\text{s}$$

ASHRAE Pyschrometric Charts 1 and 3 show the following air properties:

Supply air (71°C db, 38°C wb)

Humidity ratio	=	29.0 g per kg of dry air
Specific volume	=	1.02 m³ per kg of dry air

Makeup air (21°C db, 15.5°C wb)
Humidity ratio W_1 = 8.7 g per kg of dry air

The mass flow rate of dry air is

$$39.8/1.02 = 39.0 \text{ kg/s}$$

Step 6. The amount of moisture pickup is

$$0.300/39.0 = 7.7 \text{ g per kg of dry air}$$

The humidity ratio of the exhaust air is

$$W_2 = 29.0 + 7.7 = 36.7 \text{ g per kg of dry air}$$

Substitute in Equation (1) and calculate G_T as follows:

$$G_T(36.7 - 8.7)(1 \text{ kg}/1000 \text{ g}) = (0.4/1.04)(82 - 4)/100$$

$$G_T = 10.7 \text{ kg dry air per second}$$

Therefore,

Makeup air	=	$100 \times 10.7/39.0 = 27.4\%$
Recirculated air	=	72.6%

Step 7. Heat Balance

Sensible heat of material	=	$M(t_{m2} - t_{m1})c_m$
	=	$(0.4/1.04)(38 - 15)\,1.25$
	=	11.1 kW
Sensible heat of water	=	$M_{w1}(t_w - t_{m1})c_w$
	=	$0.315\,(38 - 15)\,4.18$
	=	30.2 kW
Latent heat of evaporation	=	$M(w_1 - w_2)\,H$
	=	$0.300 \text{ kg/s} \times 2411 \text{ kJ/(kg·K)}$
	=	723.3 kW
Sensible heat of vapor	=	$M(t_2 - t_w)c_v$
	=	$0.300\,(71 - 38)\,1.84$
	=	18.2 kW
Required heat for material	=	782.8 kW

The temperature drop $(t_2 - t_3)$ through the bed is

$$\frac{\text{Required heat}}{\text{Supplied air, kg/s} \times c_a} = \frac{782.8}{39.0 \times 1.00} = 20 \text{ K}$$

Therefore, the exhaust air temperature is $71 - 20 = 51°C$.

Required heat for makeup air	=	$G_T(t_3 - t_1)\,c_a$
	=	$10.7\,(51 - 21)\,1.00$
	=	321 kW

The total heat required for material and makeup air is

$$782.8 + 321 = 1100 \text{ kW}$$

Additional heat that must be provided to compensate for radiation and convection losses can be calculated from the known construction of the dryer surfaces.

DRYING SYSTEM SELECTION

A general procedure for selecting a drying system consists of the following:

1. Survey of suitable dryers.
2. Preliminary cost estimates of various types.
   a) Initial investment
   b) Operating cost
3. Drying tests conducted in prototype or laboratory units, preferably using the most promising equipment available. Sometimes a pilot plant is justified.
4. Summary of tests evaluating quality of samples of the dried products.

Factors that can overshadow the operating or investment cost include the following:

- Product quality, which should not be sacrificed
- Dusting, solvent, or other product losses
- Space limitation
- Bulk density of the product, which can affect packaging cost

Friedman (1951) and Parker (1963) discuss additional aids to dryer selection.

TYPES OF DRYING SYSTEMS

Radiant Infrared Drying

Thermal radiation may be applied by infrared lamps, gas-heated incandescent refractories, steam-heated sources, and, most often, electrically heated surfaces. Infrared heats only near the surface of a material, so it is best used to dry thin sheets.

The use of infrared heating to dry webs such as uncoated materials has been relatively unsuccessful because of process control problems. Thermal efficiency can be low; heat transfer depends on the emitter's characteristics and configuration, and on the properties of the material to be dried.

Radiant heating is used for drying ink and other coatings on paper, textile fabrics, paint films, and lacquers. Inks have been specifically formulated for curing with tuned or narrow wavelength infrared radiation.

Ultraviolet Radiation Drying

Ultraviolet (UV) drying uses electromagnetic radiation. Inks and other coatings based on monomers are cure dried when exposed to UV radiation. Ultraviolet drying of inks has been justified because of superior properties (Chatterjee and Ramaswamy 1975): the print resists scuff, scratch, acid, alkali, and some solvents. Printing can also be done at higher speeds without damage to the web.

Major barriers to wider acceptance of UV drying include the high capital installation cost and the increased cost of inks. The cost and frequency of replacing UV lamps are greater than those for infrared ovens.

Overexposure to radiation and ozone, which is formed by UV radiation's effect on atmospheric oxygen, can cause severe sunburn and possibly blood and eye damage. Safety measures include fitting the lamp housings with screens, shutters, and exhausts.

Conduction Drying

Drying rolls or drums (Figure 1), flat surfaces, open kettles, and immersion heaters are examples of direct-contact drying. The heating surface *must* have close contact with the material, and agitation may increase uniform heating or prevent overheating.

Conduction drying is used to manufacture and dry paper products. It (1) does not provide a high drying rate, (2) does not furnish uniform heat and mass transfer conditions, (3) usually results in a poor moisture profile across the web, (4) lacks proper control, (5) is costly to operate and install, and (6) usually creates undesirable working conditions in areas surrounding the machine. Despite these disadvantages, replacing existing systems with other forms of drying is expensive. For example, Joas and Chance (1975) report that RF (dielectric) drying of paper requires approximately four times the capital cost, six times the operating (heat) cost, and five times the maintenance cost of steam cylinder conduction drying. However,

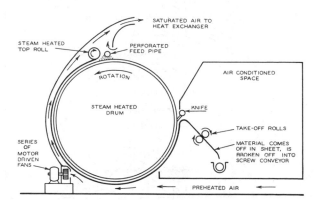

Fig. 1   Drum Dryer

augmenting conduction drying with dielectric drying sections offsets the high cost of RF drying and may produce savings and increased profits from greater production and higher final moisture content.

Further use of large conduction drying systems depends on reducing heat losses from the dryer, improving heat recovery, and incorporating other drying techniques to improve final product quality.

Dielectric Drying

When wet material is placed in a strong, high-frequency (2 to 100 MHz) electrostatic field, heat is generated within the material. More heat is developed in the wetter areas than in the drier areas, resulting in automatic moisture profile correction. Water is evaporated without unduly heating the substrate. Therefore, in addition to its leveling properties, dielectric drying provides uniform heating throughout the web thickness.

Dielectric drying is controlled by varying field or frequency strength; varying field strength is easier and more effective. Response to this variation is quick, with neither time nor thermal lag in heating. The dielectric heater is a sensitive moisture meter.

There are several electrode configurations. The platen type (Figure 2) is used for drying and baking foundry cores, heating plastic preforms, and drying glue lines in furniture. The rod or stray field types (Figure 3) are used for thin web materials such as paper and textile products. The double-rod types (over and under material) are used for thicker webs or flat stock, such as stereo-type matrix board and plywood.

Dielectric drying is popular in the textile industry. Because air is entrained between fibers, convection drying is slow and uneven. This can be overcome by dielectric drying after yarn drying.

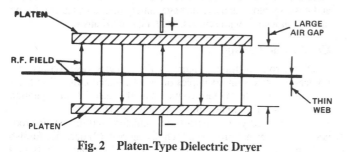

Fig. 2   Platen-Type Dielectric Dryer

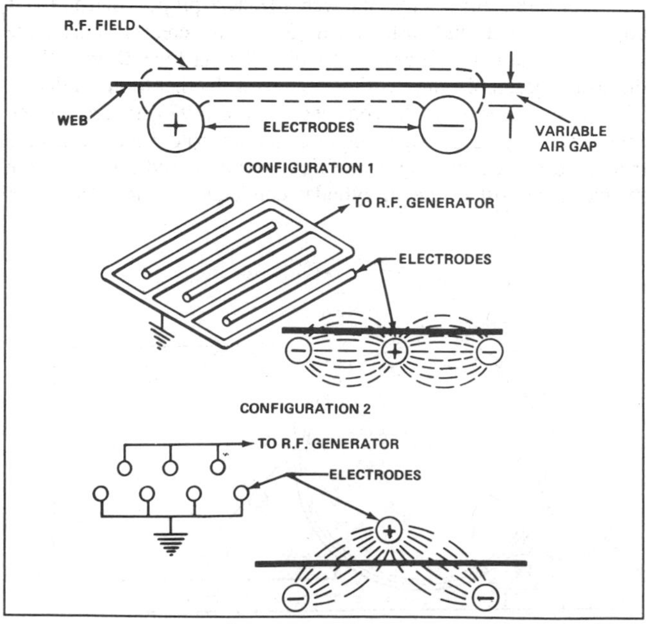

Fig. 3   Rod-Type Dielectric Dryers

Because the yarn is usually transferred to large packages immediately after drying, even and correct moisture content can be obtained by dielectric drying. Knitting wool seems to benefit from internal steaming in hanks.

Warping caused by nonuniform drying is a serious problem for plywood and linerboard. Dielectric drying yields warp-free products.

Dielectric drying is not cost-effective for overall paper drying, but has advantages when used at the dry end of a conventional steam drum dryer. It corrects moisture profile problems in the web without overdrying. This combination of conventional and dielectric drying is synergistic; the drying effect of the combination is greater than the sum of the two types of drying. This is more pronounced in thicker web materials, accounting for as much as a 16% line speed increase and a corresponding 2% energy input increase.

Microwave Drying

Microwave drying or heating uses ultrahigh frequency (900 to 5000 MHz) radiation. It is a form of dielectric heating and is applied to heating nonconductors. Because of its high frequency, microwave equipment is capable of generating extreme power densities.

Microwave drying is applied to thin materials in strip form by passing the strip through the gap of a split waveguide. Entry and exit shielding requirements make continuous process applications difficult. Its many safety concerns make microwave drying more expensive than dielectric drying. Control is also difficult because microwave drying lacks the self-compensating properties of dielectrics.

Convection Drying (Direct Dryers)

Some convection drying occurs in almost all dryers. True convection dryers, however, use circulated hot air or other gases as the principal heat source. Each means of mechanically circulating air or gases has its advantages.

Rotary Dryers. These cylindrical drums cascade the material being dried through the airstream (Figure 4). The dryers are heated directly or indirectly, and air circulation is parallel or counterflow. A variation is the rotating-louver dryer, which introduces air beneath the flights to provide close contact.

Cabinet and Compartment Dryers. These batch dryers range from the heated loft (with only natural convection and usually poor and nonuniform drying) to self-contained units with forced draft and properly designed baffles. Several systems may be evacuated to dry delicate or hygroscopic materials at low temperatures. These dryers are usually loaded with material spread in trays to increase the exposed surface. Figure 5 shows a dryer that can dry water-saturated products.

When designing dryers to process products saturated with solvents, special features must be included to prevent explosive gases from forming. Safe operation requires exhausting 100% of the air circulated during the initial drying period or during any part of the drying cycle when the solvent is evaporating at a high rate. At the end of the purge cycle, the air is recirculated and heat is gradually applied. To prevent explosions, laboratory dryers can be used to determine the amount of air circulated, the cycle lengths, and the

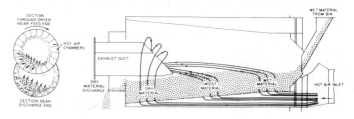

**Fig. 4   Cross Section and Longitudinal Section
of Rotary Dryer**

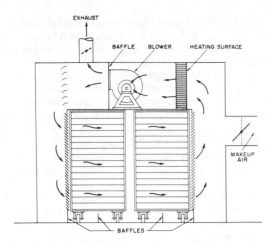

Fig. 5   Compartment Dryer Showing Trucks with Air Circulation

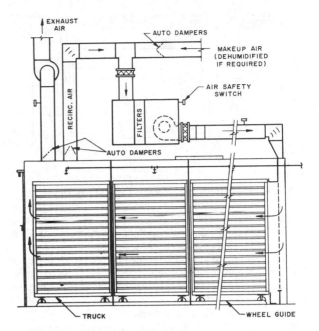

Fig. 6   Explosionproof Truck Dryer Showing Air Circulation and Safety Features

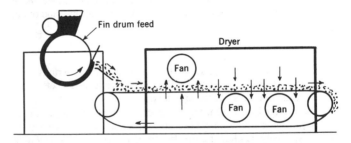

Fig. 7   Section of Blow-Through-Type Continuous Dryer

rate that heat is applied for each product. In the drying cycle, dehumidified air, which is costly, should be recirculated as soon possible. The air *must not* be recirculated when cross-contamination of products is prohibited.

Dryers must have special safety features in case any part of the drying cycle fails. The following are some of the safety design features described in *Industrial Ovens and Driers* (FMEA 1990):

1. Each compartment must have separate supply and exhaust fans and an explosion-relief panel.
2. The exhaust fan blade tip speed should be 25 m/s for a forward-inclined blade, 35 m/s for a radial-tip blade, and 38 m/s for a backward-inclined blade. These speeds produce high static pressures at the fan, ensuring constant air exhaust volumes under conditions such as negative pressures in the building or downdrafts in the exhaust stacks.
3. An airflow failure switch in the exhaust duct must shut off fans and the heating coil and must sound an alarm.
4. An airflow failure switch in the air supply system must shut off fans and the heating coil and must sound an alarm.
5. A high-temperature limit controller in the supply duct must shut off the heat to the heating coil and must sound an alarm.
6. An electric interlock on the dryer door must interrupt the drying cycle if the door is opened beyond a set point, such as that wide enough for a person to enter for product inspection.

Tunnel Dryers. Tunnel dryers are modified compartment dryers that operate continuously or semicontinuously. Heated air or combustion gas is circulated by fans. The material is handled on trays or racks on trucks and moves through the dryer either intermittently or continuously. The airflow may be parallel, counterflow, or a combination obtained by center exhaust (Figure 6). Air may also flow across the tray surface, vertically through the bed, or in any combination of directions. By reheating the air in the dryer or recirculating it, a high degree of saturation is reached before the air is exhausted, thus reducing the sensible heat loss.

The following problems with tunnel dryers have been experienced and should be considered in future designs:

• Operators may overload product trays to increase output, but this can overtax the system and increase the drying time.
• Sometimes air from the drying tunnel is discharged into the production area, increasing the humidity. Air from the drying tunnel should be discharged to the drying system return or outside.
• Overloaded product trays add pressure drop, which decreases flow through the tunnel dryer. The control panel should indicate validated flow through the tunnel. High and low flow and high moisture levels should trigger alarms.

• Cycle times can be reduced by designing dryers for cross-flow rather than end-to-end flow.

A variation of the tunnel dryer is the strictly continuous dryer, which has one or more mesh belts that carry the product through it, as shown in Figure 7. Many combinations of temperature, humidity, air direction, and velocity are possible. Hot air leaks at the entrance and exit can be minimized by baffles or inclined ends, with the material entering and leaving from the bottom.

High-Velocity Dryers. High-velocity hoods or dryers have been tried as supplements to conventional cylinder dryers for drying paper. When used with conventional cylinder dryers, web instability and lack of process control result. Applications such as thin permeable webs, where internal diffusion is not the controlling factor in the drying rate, offer more promise.

Spray Dryers. Spray dryers have been used in the production of dried milk, coffee, soaps, and detergents. Because the dried product (in the form of small beads) is uniform and the drying time is short (5 to 15 s), this drying method has become more important. When a liquid or slurry is dried, the spray dryer has high production rates.

Spray drying involves the atomization of a liquid feed in a hot-gas drying medium. The spray can be produced by a two-fluid nozzle, a high-pressure nozzle, or a rotating disc. Inlet gas temperatures range from 93 to 760°C, with the high temperatures requiring special construction materials. Because thermal efficiency increases with the inlet gas temperature, high inlet temperatures are desirable. Even heat-sensitive products can be dried at higher temperatures because of the short drying time. Hot-gas flow may be either concurrent or countercurrent

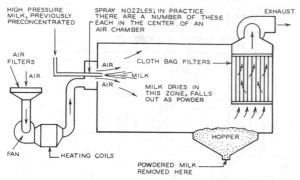

Fig. 8   Pressure-Spray Rotary-Type Spray Dryer

to the falling droplets. Dried particles settle out by gravity. Fine material in the exhaust air is collected in cyclone separators or bag filters. Figure 8 shows a typical spray drying system.

The bulk physical properties of the dried product (such as particle size, bulk density, and dustiness) are affected by atomization characteristics and the temperature and direction of flow of the drying gas. The product's final moisture content is controlled by the humidity and temperature of the exhaust gas stream.

Currently, pilot-plant or full-scale production operating data are required for design purposes. The drying chamber design is determined by the nozzle's spray characteristics and heat and mass transfer rates. There are empirical expressions that approximate mean particle diameter, drying time, chamber volume, and inlet and outlet gas temperatures.

Freeze Drying

Freeze drying has been applied to pharmaceuticals, serums, bacterial and viral cultures, vaccines, fruit juices, vegetables, coffee and tea extracts, seafoods, meats, and milk.

The material is frozen, then placed in a high-vacuum chamber connected to a low-temperature condenser or chemical desiccant. Heat is slowly applied to the frozen material by conduction or infrared radiation, allowing the volatile constituent, usually water, to sublime and condense or be absorbed by the desiccant. Most freeze-drying operations occur between −10 and −40°C under minimal pressure. While this process is expensive and slow, it has advantages for heat-sensitive materials (see Chapter 8 of the 1994 *ASHRAE Handbook—Refrigeration* and Perry and Chilton 1978).

Vacuum Drying

Vacuum drying takes advantage of the decrease in the boiling point of water that occurs as as the pressure is lowered. Vacuum drying of paper has been partially investigated. Serious complications arise if the paper breaks and massive sections must be removed. Vacuum drying is used successfully for pulp drying, where lower speeds and higher weights make breakage relatively infrequent.

Fluidized-Bed Drying

A fluidized-bed system contains solid particles through which a gas flows with a velocity higher than the incipient fluidizing velocity but lower than the entrainment velocity. Heat transfer between the individual particles and the drying air is efficient because there is close contact between powdery or granular material and the fluidizing gas. This contact makes it possible to dry sensitive materials without danger of large temperature differences.

The dried material is free-flowing and, unlike that from convection-type dryers, is not encrusted on trays or other heat-exchanging surfaces. Automatic charging and discharging are possible, but the greatest advantage is reduced process time. Only simple controls are important, that is, control over fluidizing air or gas temperatures and the drying time of the material.

All fluid-bed dryers should have explosion-relief flaps. Both the pressure and flames of an explosion are dangerous. Also, when toxic materials are used, uncontrolled venting to the atmosphere is prohibited. Explosion suppression systems, such as pressure-actuated ammonium-phosphate extinguishers, have been used instead of relief venting. An inert dryer atmosphere is preferable to suppression systems because it prevents explosive mixtures from forming.

When organic and inflammable solvents are used in the fluid-bed system, the closed system offers advantages other than explosion protection. A portion of the fluidizing gas is continuously run through a condenser, which strips the solvent vapors and greatly reduces air pollution problems, thus making solvent recovery convenient.

Materials dried in fluidized-bed installations include coal, limestone, cement rock, shales, foundry sand, phosphate rock, plastics, medicinal tablets, and foodstuffs. Leva (1959) and Othmer (1956) discuss the theory and methods of fluidization of solids. Clark (1967) and Vanecek et al. (1966) developed design equations and cost estimates.

Agitated-Bed Drying

Uniform drying is ensured by periodically or continually agitating a bed of preformed solids with a vibrating tray or conveyor, a mechanically operated rake, or, in some cases, by partial fluidization of the bed on a perforated tray or conveyor through which recycled drying air is directed. Drying and toasting cereals is an important application.

Drying in Superheated Vapor Atmospheres

When drying solids with air or another gas, the vaporized solvent (water or organic liquid) must diffuse through a stagnant gas film to reach the bulk gas stream. Because this film is the main resistance to mass transfer, the drying rate depends on the solvent vapor diffusion rate. If the gas is replaced by solvent vapor, resistance to mass transfer in the vapor phase is eliminated, and the drying rate depends only on the heat transfer rate. Drying rates in solvent vapor, such as superheated steam, are greater than those in air for equal temperatures and mass flow rates of the drying media (Chu et al. 1953).

This method also has higher thermal efficiency, easier solvent recovery, a lower tendency to overdry, and eliminates oxidation or other chemical reactions that occur when air is present. In drying cloth, superheated steam reduces the migration tendency of resins and dyes. Superheated vapor drying cannot be applied to heat-sensitive materials because of the high temperatures.

Commercial drying equipment with recycled solvent vapor as the drying medium is available. Installations have been built to dry textile sheeting and organic chemicals.

Flash Drying

Finely divided solid particles that are dispersed in a hot-gas stream can be dried by flash drying, which is rapid and uniform. Commercial applications include drying pigments, synthetic resins, food products, hydrated compounds, gypsum, clays, and wood pulp.

REFERENCES

Chatterjee, P.C. and R. Ramaswamy. 1975. Ultraviolet radiation drying of inks. *British Ink Maker* 17(2):76.

Chu, J.C., A.M. Lane, and D. Conklin. 1953. Evaporation of liquids into their superheated vapors. *Industrial and Engineering Chemistry* 45:1586.

Clark, W.E. 1967. Fluid bed drying. *Chemical Engineering* 74(March 13): 177.

FMEA. 1990. Industrial ovens and driers. Data Sheet No. 6-9. Factory Mutual Engineering Association, Norcross, GA.

Friedman, S.J. 1951. Steps in the selection of drying equipment. *Heating and Ventilating* (February):95.

Joas, J.G. and J.L. Chance. 1975. Moisture leveling with dielectric, air impingement and steam drying—A comparison. *Tappi* 58(3):112.

Leva, M. 1959. *Fluidization*. McGraw-Hill, New York.

Othmer, D.F. 1956. *Fluidization*. Reinhold Publishing, New York.

Parker, N.H. 1963. Aids to drier selection. *Chemical Engineering* 70(June 24):115

Perry, R.H. and C.H. Chilton, ed. 1978. *Chemical engineers' handbook*, 5th ed. Section 17, Sublimation, and Section 20, Gas-Solid Systems. McGraw-Hill, New York.

Vanecek, Markvart, and Drbohlav. 1966. *Fluidized bed drying*. Chemical Rubber Company, Cleveland, OH.

BIBLIOGRAPHY

[*NOTE: ABIPC stands for Abstract Bulletin of the Institute of Paper Chemistry. Currently, the publication is ABIPST and is published by the Institute of Paper Science and Technology, Atlanta, GA.*]

Alt, C. 1964. A comparison between infrared and other ink drying methods. *Polygraph* 17(4):200; ABIPC 34:1455.

Appel, D.W. and S.H. Hong. 1969. Condensate distribution and its effect on heat transfer in steam heated driers. *Pulp and Paper Canada* 70(4):66, T51; ABIPC 39:943.

Balls, B.W. 1970. The control of drying cylinders. *Paper Technology* 1(5):483; ABIPC 31:1119.

Bell, J.R. and P. Grosberg. 1962. The movement of vapor and moisture during the falling rate period of drying of thick textile materials. *Journal of the Textile Institute*, Transactions 53(5):T250; ABIPC 33:72.

Booth, G.L. 1970. Factors in selecting an air heating system for drying coatings. *Paper Trade Journal* 154(23); Graphic Arts Abstracts 24(7):71.

Booth, G.L. 1970. General principles in the drying of paper coatings. *Paper Trade Journal* 154(17):48; Graphic Arts Abstract 24(7):72.

Brown, G.G. and Associates. 1950. *Unit operations*, p. 564. John Wiley & Sons, New York.

Chu, J.C., S. Finelt, W. Hoerrner, and M.S. Lin. 1959. Drying with superheated steam-air mixtures. *Industrial and Engineering Chemistry* 51:275.

Church, F. 1968. How dielectric heating helps to control moisture content. *Pulp and Paper International* 10(2):50; ABIPC 39:202.

Daane, R.A. and S.T. Han. 1961. An analysis of air-impingement drying. *Tappi* 44(1):73; ABIPC 31:1120.

Dooley, J.A. and R.D. Vieth. 1965. High velocity drying gives new impetus to solution coatings. *Paper, Film, Foil Converter* 39(4):53; ABIPC 36.

Dyck, A.W.J. 1969. Focus on paper drying. *American Paper Industry* 51(6):49; ABIPC 40:458.

Foust, A.S. et al. 1962. *Simultaneous heat and mass transfer 11: Drying. Principles of unit operations*. John Wiley & Sons, New York.

Friedman, S.J., R.A. Gluckert, and W.R. Marshall, Jr. 1952. Centrifugal disk atomization. *Chemical Engineering Progress* 48:181.

Gardner, T.A. 1964. Air systems and Yankee drying. *Tappi* 47(4):210; ABIPC 34:1787.

Gardner, T.A. 1968. Pocket ventilation and applied fundamentals spell uniform drying. *Paper Trade Journal* 152(4):46, 48, 51-2; ABIPC 39:115.

Gavelin, G. 1964. Paper and paperboard drying—Theory and practice (monograph). Lockwood Trade Journal Co., Inc., New York, 85 pp.; ABIPC 35:480.

Gavelin, G. 1970. New heat recovery system for paper machine hood exhaust. *Paper Trade Journal* 154(8):38; ABIPC 41:1225.

Haley, N.A. 1976. High frequency heating on a linerboard machine. Tappi Papermakers Conference (April 26-29, Atlanta) Preprints, 217.

Heating and Ventilating. 1942. What the air conditioning engineer should know about drying (December).

Hoyle, R. 1963. Thermal conditions in a steam drying cylinder. *Paper Technology* 4(3):259; ABIPC 34:341.

Janett, L.G., A.J. Schregenberger, and J.C. Urbas. 1965. Forced convection drying of paper coatings. *Pulp and Paper Canada* 66(January):T20; ABIPC 35:1433.

Larsson, T. 1962. Comparing high velocity driers—Aspects of theory and design. *Paper Trade Journal* 146(38):36; ABIPC 33:547.

Marshall, W.R., Jr. *Drying section in Encyclopedia of Chemical Technology*, 2nd ed., Vol. 7., p. 326. Interscience Publishers, New York.

Metcalf, W.K. 1970. What pocket ventilation systems can do and what they cannot. *Pulp and Paper* 44(2):95-7; ABIPC 41:3231.

Mill, D.N. 1961. High velocity air drying. *Paper Technology* 2(4), August; ABIPC 32:588.

Nissan, A.H. 1968. Drying of sheet materials. *Textile Research Journal* 38:447.

Nissan, A.H. and D. Hansen. 1962. Fundamentals of drying of porous materials. Errata. *Tappi* 45(7):608; ABIPC 33:395.

Olmedo, E.B. 1966. *Steam control in paper machine driers*. ATCP 6(2):142.

Priestly, R.J. 1962. Where fluidized solids stand today. *Chemical Engineering*(July 9):125.

Scheuter, K.R. 1968. Drier theory and dryer systems. *Druckprint* 105(12):939; ABIPC 40:24.

Simon, E. Containment of hazards in fluid bed technology. *Manufacturers of Chemical Aerosol News* 49(1):23.

Sloan, C.E., T.D. Wheelock, and G.T. Tsao. 1967. Drying. *Chemical Engineering* 74:167.

Spraker, W.A., G.B. Wallis, and B.R. Yaros. 1969. Analysis of heat and mass transfer in the Yankee drier. *Pulp and Paper Canada* 70(1):55, T1-5, ABIPC 39:947.

Stangl, K. 1966. Progress in flash drying. *Pulp and Paper National* 8(6):65; ABIPC 37:305.

Streaker, W.A. 1968. Drying pigmented coatings with infrared heat. *Tappi* 51(10):105; ABIPC 39:659.

Tarnawski, Z. 1962. Drying of paper and calculation of the drying surface of the paper machine. *Przeglad Papier* 18(7):218; ABIPC 34:639.

Wen, C.Y. and W.E. Loos. 1969. Rate of veneer drying an a fluidized bed. *Wood Science and Technology* 3.

Wilhoit, D.L. 1968. Theory and practice of drying aqueous coatings with a high velocity air drier. *Tappi* 51(1); ABIPC 38:803.

Yoshida, T., and T. Hyodo. 1963. Superheated vapor as a drying agent in spinning fiber. *Industrial and Engineering Chemistry, Process Design and Development*(January):52.

KITCHEN VENTILATION

KITCHEN ventilation is a complex application of HVAC systems. The design of a kitchen ventilation system includes aspects of air conditioning, fire safety, ventilation, building pressurization, refrigeration, air distribution, food service equipment, and systems commissioning. Kitchens are found in many buildings, including restaurants, hotels, hospitals, retail malls, single and multifamily dwellings, and correctional facilities. Each of these building types has special requirements for its kitchens, but many of the basic needs are common to all.

Kitchen ventilation systems are provided in buildings for at least two reasons: (1) to provide a comfortable environment in the kitchen and (2) to enhance the safety of personnel working in the kitchen and of other building occupants. Obviously, the kitchen ventilation system can affect the temperature and humidity conditions in the kitchen. "Comfortable" in this context has different meanings because, depending on the local climate, some kitchens are not air conditioned. The ventilation system can also affect the acoustical conditions of a kitchen.

The centerpiece of almost any kitchen ventilation system is an exhaust hood, which is used primarily to remove effluent from kitchens. Effluent includes the gaseous, liquid, and solid contaminants produced by the cooking process. These contaminants must be removed for both comfort and safety. The effluent can range from simply annoying to potentially life threatening and, under certain conditions, flammable. Not only is the exhaust hood itself very important to the efficient operation of the kitchen, but the way the food service equipment is arranged and coordinated with the hood(s) greatly affects the operating costs of the kitchen.

HVAC system designers are most frequently involved in commercial kitchen applications, those in which large amounts of grease or water vapor are produced in the cooking effluent. Residential kitchens typically use a totally different type of hood. The amount of grease produced in residential applications is significantly less than in commercial applications, so the health and fire hazard is much lower.

COOKING EQUIPMENT AND PROCESSES

Effluent Generation

Cooking is the process of creating chemical and physical changes in food by applying heat to the raw or precooked food. Cooking improves edibility, taste, or appearance or delays decay processes. As heat is applied to the food, effluent is released into the surrounding atmosphere. This effluent includes heat that does not transfer to the food, water vapor, and organic material released from the food. The heat source, especially if it involves a combustion process, may release other contaminants, which must be removed.

All cooking methods release some heat, some of which radiates from all hot surfaces; but most is dissipated by natural convection

via a rising plume of heated air. Most of the effluent released from the food and the heat source is entrained in this plume, so primary contaminant control should be based on capturing and removing the air and effluent that compose the plume. A quantitative analysis, or even a relative determination, of volumetric plume flow rates and combustion product flow rates is not available at present, although research is in progress.

Hot Process Versus Cold Process

The most common method of contaminant control is to install an air inlet device (a hood) so that the plume can enter it and be conveyed away by an exhaust system. The hood is generally located above or behind the heated surface to intercept the normal upward flow path. Understanding the behavior of the plume is central to designing effective ventilation systems.

Effluent released from a noncooking cold process, such as metal grinding, is captured and removed by locating air inlets so they catch forcibly ejected material, or by creating air streams with sufficient velocity to induce the flow of effluent into an inlet. This technique has led to an empirical concept of capture velocity that is often misapplied to heated processes. Effluent released from a hot process and contained in a plume may be captured by locating an inlet hood so that the plume flows into it by buoyant forces. The exhaust rate from the hood must equal or slightly exceed the plume flow rate, but the hood need not actively capture the effluent if the hood is large enough at its height above the cooking operation to encompass the plume as it expands during its rise. Additional exhaust airflow may be needed to resist crossdrafts that might carry the plume away from the hood.

A plume, in the absence of crossdrafts or other interference, rises vertically. As it rises additional air is entrained, which causes the plume to enlarge and the average velocity and temperature to decrease. In most cooking processes, the distance between the heated surface and the hood is so short that entrainment is negligible and the plume loses very little of its velocity or temperature before it reaches the hood. If a surface parallel to the plume centerline (e.g., a back wall) is located nearby, the plume will attach to the surface by the Coanda effect. This tendency also directs the plume into hoods.

Appliance Types (Steam, Electricity, Solid Fuel, Gas)

The heat source has an effect on the type and quantity of effluent released. Steam is contained in a closed vessel, so no contaminants are released from the steam itself. When electricity is used as a heating source, no significant contaminants are released.

Solid (wood, charcoal) or gaseous (natural or LP gas) fuels are common sources of heat for cooking. Their combustion generates water vapor and carbon dioxide, and it may also generate carbon monoxide and other potentially harmful gases. Hence, these effluents must be controlled along with those released from the food. In some cases, the food or its container is directly exposed to the flame; as a result the combustion effluent and the food effluent are mixed, and a single plume is generated. In other cases, such as ovens, the

The preparation of this chapter is assigned to Task Group for Kitchen Ventilation.

combustion products are ducted to an outlet adjacent to the plume, but the effluents still mix.

HOODS

The kitchen exhaust hood captures, contains, and evacuates heat, smoke, odor, steam, grease, vapor, and other contaminants generated from cooking in order to provide a safe, healthy, comfortable, and productive work environment for kitchen personnel. This section discusses all aspects of kitchen hood design; however, it is based primarily on model codes and standards in the United States.

The design, engineering, construction, installation, and maintenance of commercial kitchen exhaust hoods are controlled by the major nationally recognized standards (e.g., NFPA *Standard* 96) and model codes. In some cases, local codes may prevail. Prior to designing a kitchen ventilation system, governing codes and the authority having jurisdiction should be identified and consulted. Local authorities having jurisdiction may have amendments to these standards and codes, or additional requirements.

Hood Types

Many types, categories, and styles of hoods are available, and hood selection depends on many factors. Hood types are classified based on whether or not they are designed for grease-handling applications. The model codes distinguish between grease-handling and non-grease-handling hoods, although not all model codes use Type I/Type II terminology. Type I refers to hoods designed for removal of grease and smoke, and Type II refers to all other hoods. A Type I hood may be used where a Type II hood is required, but not vice versa. However, the characteristics of the cooking equipment under the hood, and not the hood type, determine the requirements for the entire exhaust system, including the hood.

A Type I hood is used for collection and removal of grease and smoke. It includes both listed grease filters, baffles, or extractors for removal of the grease and fire suppression equipment. This type of hood is required over restaurant equipment, such as ranges, fryers, griddles, broilers, ovens, and steam kettles, that produces smoke or grease-laden vapors.

A Type II hood is a general hood for collection and removal of steam, vapor, heat, and odors where grease is not present. The Type II hood may or may not have grease filters or baffles and typically does not have a fire suppression system. It is typically used over dishwashers, steam tables, and so forth. The Type II hood is sometimes used over ovens, steamers, or kettles if they do not produce smoke or grease-laden vapors and if the authority having jurisdiction allows it.

Type I Hood Categories

Type I hoods fall into two categories. One is the standard (non-listed) category, which meets the design, construction, and performance criteria of the applicable national and local codes. The second is the listed category of design, construction, and performance. The listed category may differ from applicable code criteria, but, by meeting special certification criteria established by nationally recognized standards, it seeks to show equivalency with the criteria of the applicable codes.

Among Type I Listed hoods, there are two basic subcategories, as defined and listed by UL *Standard* 710. These subcategories are (1) exhaust hoods without exhaust dampers and (2) exhaust hoods with exhaust dampers. In 1991, UL changed the categories for exhaust hoods. The previous categories of listed grease extractors (water-wash hoods with exhaust dampers) and listed automatic damper and hood assemblies (dry hoods with exhaust dampers) were combined into the current category of exhaust hoods with exhaust dampers. The UL listings do not distinguish between water-wash and dry hoods, except that water-wash hoods with fire-actuated water systems are identified in UL's product directory. The old category of classified exhaust hoods (hoods with or without water-wash and without exhaust dampers) has been changed to a listed category.

Standard hoods must meet all the construction requirements of the applicable model codes. Standard hoods may not have fire-actuated dampers.

Listed hoods are not generally designed, constructed, or operated in accordance with requirements of the model codes, but are constructed in accordance with the terms of the hood manufacturer's listing. This is allowed because the model codes include exceptions if the hood is listed to show equivalency with the safety criteria of the model code requirements.

All listed hoods are subjected to electrical, temperature, and fire and cooking smoke capture tests. The listed exhaust hood with exhaust damper includes a fire-actuated damper, typically located at the exhaust duct collar (and the makeup air duct collar, depending on the hood configuration). In the event of a fire, the damper closes to prevent fire from entering the duct.

Listed exhaust hoods with fire-actuated water systems are typically water-wash hoods whose wash system also operates as a fire extinguishing system. In addition to meeting the requirements of UL *Standard* 710, they are also tested under UL *Standard* 300 and may be listed for plenum extinguishment, duct extinguishment, or both.

Type I Hoods—Grease Removal

Most grease removal devices in Type I hoods operate on the same general principle—the exhaust air passes through a series of baffles in which a centrifugal force is created as the exhaust air passes around the baffles to extract the grease. Mesh filters cannot meet the requirements of UL *Standard* 1046 and therefore cannot be used as primary grease filters. Grease removal devices generally fall into the following categories:

- *Baffle Filter.* The baffle filter is a series of vertical baffles designed to capture grease and drain it away to a container. The filters are arranged in a channel or bracket for easy insertion into, and removal from, the hood for cleaning. Each hood usually has two or more baffle filters. The filters are usually constructed of aluminum, steel, or stainless steel, and they come in various standard sizes. Filters are cleaned by running them through a dishwasher or by soaking and rinsing. NFPA *Standard* 96 requires grease filters to be listed. Listed grease filters are tested and certified by a nationally recognized test laboratory under UL *Standard* 1046.

- *Removable Extractor.* Removable extractors are an integral component of listed exhaust hoods designed to use them. They are typically constructed of stainless steel and contain a series of horizontal baffles designed to remove grease and drain it away to a container. Removable extractors come in various sizes. They are cleaned by running them through a dishwasher or by soaking and rinsing.

- *Stationary Extractor.* The stationary extractor (also called a water-wash hood) is an integral component of listed exhaust hoods that use them. They are typically constructed of stainless steel and contain a series of horizontal baffles that run the full length of the hood. The baffles are not removable for cleaning. The stationary extractor includes one or more water manifolds with spray nozzles that, upon activation, wash the grease extractor with hot, detergent-injected water, removing the accumulation of grease from the extractor. The wash cycle is typically activated at the end of the day, after the cooking equipment and fans have been turned off (although it can be activated at more frequent schedules); it runs for 5 to 10 minutes, depending upon the hood manufacturer, the type of cooking, the number of hours of operation, and the water temperature and pressure. Most water-wash hood manufacturers recommend a water temperature of 55 to 82°C and water pressure of 200 to 550 kPa. The average water consumption varies between 0.1 and 0.3 L/s per linear metre of hood, depending upon the hood manufacturer.

Some manufacturers of water-wash hoods provide continuous cold water mist as an option. The mist runs continuously during cooking operation and may or may not be recirculated, depending on the manufacturer. Typical cold water usage is 3.5 mL/s per linear metre of hood. The advantage of continuous cold water mist is that it improves grease extraction and removal, thereby improving the cleaning process. Many hood manufacturers recommend continuous cold water mist in hoods that are located over solid fuel-burning equipment, as the mist will also extinguish hot embers that may be drawn up into the hood and help cool the exhaust stream.

UL *Standards* 1046 and 710 do not include grease extraction tests because, at present, no industry-accepted tests are available in the United States. Grease extraction rates published by filter and hood manufacturers are usually derived from tests conducted by independent test laboratories retained by the manufacturer. Because of this, the test methods and results vary greatly.

Type I Hoods—Styles

Figure 1 shows the six basic styles of hoods for both Type I and Type II applications. These style names are not universally used in all standards and codes, but are well accepted in the industry. The styles are as follows:

1. *Wall-mounted canopy.* Used for all types of cooking equipment located against a wall.
2. *Single island canopy.* Used for all types of cooking equipment in a single line island configuration.
3. *Double island canopy.* Used for all types of cooking equipment mounted back to back in an island configuration.
4. *Backshelf.* Used for counter-height equipment typically located against a wall, but could be freestanding.
5. *Eyebrow.* Used for direct mounting to ovens and some dishwashers.
6. *Pass-over style.* Used over counter-height equipment when pass-over configuration (from the cooking side to the serving side) is required.

Type I Hoods—Sizing

The size of the exhaust hood in relation to cooking appliances is an important aspect of hood performance. Usually the hood must extend beyond the cooking appliances (overhang)—on all open sides of canopy-style hoods and over the ends on backshelf and pass-over hoods—to capture the expanding thermal currents rising from the appliances. This overhang varies with the style of the hood, the distance between the hood and the cooking appliance, and the characteristics of the cooking equipment. With backshelf and pass-over hoods, the front of the hood must be kept behind the front of the cooking equipment (set back) to allow head clearance for the cooks. These hoods may require a higher front inlet velocity to catch and contain the expanding thermal currents. All hoods may have full or partial side panels to close the area between the appliances and the hood. This may eliminate the overhang requirement and generally reduces the exhaust flow rate requirements.

For standard hoods, the hood size is dictated by the prevailing model code, and for listed hoods, by the terms of the manufacturer's listing. Typically, the overhang requirements applied to listed hoods are similar to those for standard hoods. General overhang requirements are shown in Table 1.

Type I Hoods—Exhaust Flow Rates

Exhaust flow rate requirements to capture, contain, and remove the effluent vary considerably depending on the hood style, the amount of overhang, the distance from the hood to the cooking appliances, the presence and size of end panels, and the cooking equipment and product involved. The hot cooking surfaces and product vapors create thermal air currents that are "received" or captured by the hood and then exhausted. The velocity of these currents depends largely upon the surface temperature and tends to vary from 76 mm/s over steam equipment to 0.76 m/s over charcoal broilers. The actual required flow rate is determined by these thermal currents, a safety allowance to absorb crossdrafts and flare-ups, and a safety factor for the style of hood.

Overhangs, distance from the cooking surface to the hood, and the presence or absence of side panels all play an important part in the safety factor of different hood styles. Use of gas-fired cooking equipment may require an additional allowance for the exhaust of inherent combustion products and combustion air. Because it is not practical to place a separate hood over each piece of equipment, general practice is to categorize the equipment into four groups. While published lists vary, and accurate documentation does not yet exist, the following is a consensus opinion list (great variance in product or volume could shift an appliance into another category):

1. Light duty, such as ovens, steamers, and small kettles (up to 200°C)
2. Medium duty, such as large kettles, ranges, griddles, and fryers (up to 200°C)
3. Heavy duty, such as upright broilers, charbroilers, and woks (up to 320°C)
4. Extra heavy duty, such as solid fuel-burning equipment (up to 370°C)

The exhaust flow rate requirement is based on the group of equipment under the hood. If there is more than one group, the flow rate is based on the heaviest duty group, unless the hood design permits different volumes over different sections of the hood.

For areas where the model codes or other regional codes have been adopted, the exhaust flow rate requirement for standard hoods is dictated by the codes; therefore, the manufacturer's method may not be used without consultation with the authority having jurisdiction. The model code required exhaust flow rates for standard canopy hoods are typically calculated by determining the area A of the hood opening and multiplying the area by a given air velocity. Table 2 indicates typical formulas, taken from the model codes, for determining the exhaust flow rate Q for standard canopy hoods. Some jurisdictions may use the length of the open perimeter of the hood times the vertical height between the hood and the appliance instead of the horizontal hood area.

The Uniform Mechanical Code (UMC) and some state codes have alternate formulas that allow lower flow rates for equipment

Table 1 Typical Overhang Requirements for Both Listed and Standard Hoods

Type of Hood	End Overhang	Front Overhang	Rear Overhang
Wall-mounted canopy	150 mm	300 mm	—
Single island canopy	300 mm	300 mm	300 mm
Double island canopy	150 mm	300 mm	300 mm
Eyebrow	0 mm	300 mm	—
Backshelf/Pass-over	150 mm	Front set back 1/2 over equipment	

Note: The model codes typically require a 150-mm minimum overhang, but most manufacturers design for a 300-mm overhang.

Table 2 Typical Model Code Exhaust Flow Rates for Standard Hoods

Wall-mounted canopy	$Q = 0.5A$
Single island canopy	$Q = 0.75A$
Double island canopy	$Q = 0.5A$
Eyebrow	$Q = 0.5A$
Backshelf/Pass-over	$Q = 0.47 \times$ Length of hood

Note: Q = exhaust flow rate, m^3/s; A = area of hood, m^2

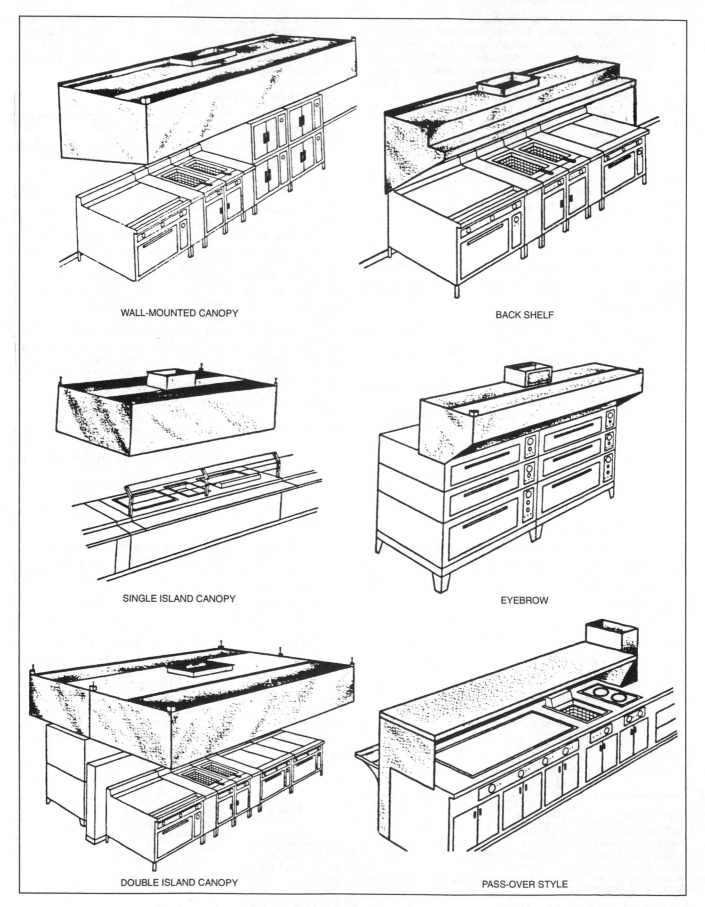

WALL-MOUNTED CANOPY

BACK SHELF

SINGLE ISLAND CANOPY

EYEBROW

DOUBLE ISLAND CANOPY

PASS-OVER STYLE

Fig. 1   Styles of Kitchen Exhaust Hoods

Table 3 Typical Minimum Exhaust Flow Rates for Listed Hoods by Cooking Equipment Type

	Minimum Exhaust Flow Rate, m³/s per Linear Metre of Hood			
Type of Hood	Light Duty Equipment	Medium Duty Equipment	Heavy Duty Equipment	Extra Heavy Duty Equipment
Wall-mounted canopy	0.23 to 0.31	0.31 to 0.46	0.31 to 0.62	0.54+
Single island	0.39 to 0.46	0.46 to 0.62	0.46 to 0.93	0.85+
Double island (per side)	0.23 to 0.31	0.31 to 0.46	0.39 to 0.62	0.77+
Eyebrow	0.23 to 0.39	0.23 to 0.39	—	—
Backshelf/ Pass-over	0.15 to 0.31	0.31 to 0.46	0.46 to 0.62	Not recommended

that produces less heat and smoke; however, the UMC does require 1 m³/s per square metre of hood area for hoods covering charbroilers. Backshelf and pass-over style nonlisted hoods are usually calculated at 0.46 m³/s per linear metre of exhaust hood.

Listed hoods are allowed to operate at their listed exhaust flow rates by exceptions in the model codes. Most manufacturers of listed hoods verify their listed flow rates by conducting the tests per UL *Standard* 710. Typically the average flow rates are much lower than those dictated by the model codes. It should be noted that these listed values are established under draft-free laboratory conditions. The four categories of equipment groups mentioned are tested and marked according to cooking surface temperatures: light and medium duty up to 200°C, heavy duty up to 320°C, and extra heavy duty up to 370°C. Each of these groups would have an air quantity factor assigned for each style of hood, with the total exhaust flow rate typically calculated by multiplying this factor times the length of the hood.

Minimum exhaust flow rates for listed hoods serving single categories of equipment vary from manufacturer to manufacturer, but are generally as shown in Table 3.

Actual exhaust flow rates for hoods with internal "short circuit" makeup air are typically higher than those in Table 3, though the net exhaust (actual exhaust less makeup air quantity) may be similar. The specific hood manufacturer should be contacted for exact exhaust and makeup flow rates.

Type I Hoods—Makeup Air Options

Air exhausted from the kitchen space must be replaced. This makeup air can be brought in through the traditional method of ceiling registers; however, they must be located so that the discharged air does not disrupt the air pattern entering the hood. Air should be supplied either (1) as far from the hood as possible or (2) close to the hood and directed away from the hood or straight down at very low velocity.

Another way of distributing makeup air is through systems built as an integral part of the hood. Figure 2 shows three designs of internal makeup air that are available. Combinations of these designs are also available. Because the actual flows and percentages vary with all hoods, the manufacturer should be consulted about specific applications. The following are typical descriptions:

Front Face Discharge. This method of introducing makeup air into the kitchen is flexible and has many advantages. Typical supply volume is 70 to 80% of the exhaust, depending on the air balance desired. Supply air temperatures should range from 15 to 18°C but may be as low as 10°C, depending on flow rate, distribution, and internal heat load. This air should be directed away from the hood, but the closer the air outlet's lower edge is to the bottom of the hood, the lower the velocity must be to avoid drawing effluent out of the hood.

Down Discharge. This method of introducing makeup air to the kitchen area is typically used when spot cooling of the cooking staff

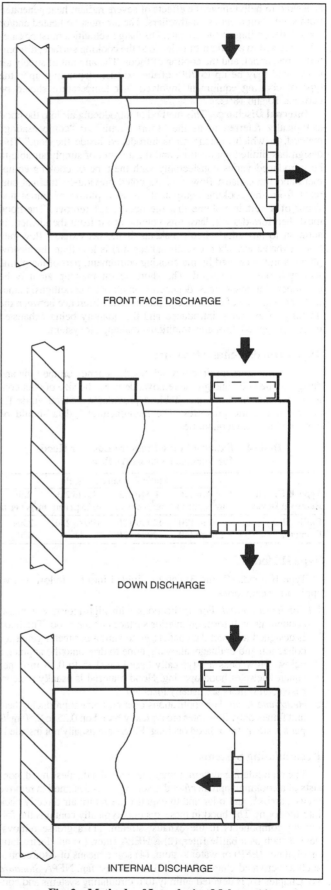

FRONT FACE DISCHARGE

DOWN DISCHARGE

INTERNAL DISCHARGE

Fig. 2 Methods of Introducing Makeup Air

is desired to help relieve the effects of severe radiant heat generated from such equipment as charbroilers. The air must be heated and/or cooled, depending on the climate. Discharge velocities must be carefully selected to avoid air turbulence at the cooking surface, discomfort to personnel, and the cooling of foods. The amount of supply air introduced may be up to 70% of the exhaust, depending upon the type of cooking equipment involved. Air temperature should be between 10 and 18°C.

Internal Discharge. This method of introducing air into the hood is typically referred to as the "short circuit" or "compensating" method, in which makeup air is introduced inside the hood. This design has limited application, and the amount of supply air able to be introduced varies considerably with the type of cooking equipment and the exhaust flow rate. As noted previously, thermal currents from the cooking equipment create a plume of fumes and vapor of a certain volume that the hood must remove. The hood must therefore draw at least this volume of air from the kitchen, in addition to any short circuit makeup. If the net exhaust flow rate (total exhaust less short circuit makeup air) is less than the airflow plume volume created by the cooking equipment, part of the plume will spill out of the hood. The short circuit makeup air may be untempered in most areas, depending on climatic conditions, manufacturer, and type of cooking equipment. The difference between the quantity of air being introduced and the quantity being exhausted must be supplied through a traditional makeup air system.

Type I Hoods—Static Pressure

The static pressure drop through hoods depends on the type and design of the hood and grease removal devices, the size of duct connections, and the flow rate. Table 4 provides a general guide for determining static pressures. The manufacturers' data should be consulted for actual quantities.

Table 4   Exhaust Static Pressure Loss for Hoods for Various Exhaust Airflows

Type of Grease Removal Device	Static Pressure Loss, Pa			
	0.8 to 1.3 m³/s per m	1.3 to 1.8 m³/s per m	1.8 to 2.3 m³/s per m	2.5+ m³/s per m
Baffle filter	60 to 120	120 to 190	190 to 250	250+
Extractor	250 to 340	320 to 420	420	420

Type II Hoods

Type II hoods (Figure 3) can be divided into the following two application categories:

1. *Condensate hood.* For applications with high-moisture exhaust, condensate will form on interior surfaces of the hood. The hood is designed to direct the condensate toward a perimeter gutter for collection and drainage, allowing none to drip onto the appliance below. Flow rates are typically based on 0.25 to 0.38 m³/s per square metre of hood opening. Hood material is usually noncorrosive, and filters are usually installed.
2. *Heat/fume hood.* For applications over equipment producing heat and fumes only, flow rates are typically based on 0.25 to 0.50 m³/s per square metre of hood opening. Filters are usually not installed.

Recirculating Systems

The recirculating system, sometimes called a ductless hood, consists of a cooking appliance/hood assembly; it is designed to remove grease, smoke, and odor and to exhaust the return air directly back into the room. The hood in these systems typically contains the following components in the exhaust stream: (1) a grease removal device such as a baffle filter, (2) a HEPA filter, (3) an electrostatic precipitator (ESP) or water system, (4) some means of odor control such as activated charcoal, and (5) an exhaust fan. NFPA *Standard* 96, Chapter 10, is devoted entirely to recirculating systems and contains specific requirements such as (1) design, including various

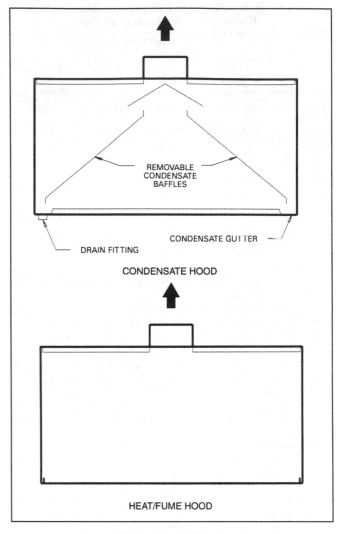

Fig. 3   Type II Hoods

interlocks of all critical components to prevent operation of the cooking appliance if any of the components are not operating; (2) fire extinguishment, including specific nozzle locations; (3) maintenance, which includes a specific schedule for cleaning filters, ESPs, hoods, and blowers; and (4) inspection and testing of the total operation and interlocks. In addition, NFPA 96 requires that all recirculating systems be listed with a testing laboratory. The recognized test standard for a recirculating system is UL *Standard* 197, Supplement SB, Commercial Electric Cooking Appliances with Recirculating Systems. Ductless systems should not be used over gas-fired or solid fuel-fired equipment, which needs to be exhausted directly outdoors to remove products of combustion.

Designers should thoroughly review NFPA *Standard* 96 requirements and contact a manufacturer of recirculating systems to obtain specific information prior to incorporating this type of system in a food service design. It should be remembered that with recirculating systems, the total heat and moisture of the cooking operation will be discharged into the same space, which adds to the air-conditioning load.

EXHAUST SYSTEMS

Exhaust systems remove effluent from the appliances and cooking processes for reasons of fire and health safety, comfort, and aesthetics. Typical exhaust systems simultaneously incorporate fire prevention designs and fire suppression equipment. In most cases, these functions complement each other, but in other cases they may seem to conflict. Designers should recognize that designs must balance the various functions. For example, fire-actuated dampers may be installed to minimize the spread of fires to ducts, yet maintaining an open duct might be better for removing the smoke of an appliance fire from the kitchen.

Effluent Control

Over 90% of kitchen ventilation systems rely entirely on the grease removal device in the hood for effluent control. Effluent controls today are primarily located on the roofs of food service facilities. Designers have typically considered it prudent to carry effluent out of hoods into ducts and then treat the effluent components after transport through the ducts. Minimum velocities in the duct have been established by NFPA *Standard* 96 to minimize grease buildup. Recently, efforts have been made to develop more effective effluent controls integral to, or immediately downstream of, the hood to limit the contaminant spread and therefore reduce the fire hazard. This approach can dramatically reduce duct cleaning costs as well, especially in multistory buildings and in multiple-hood systems. Additionally, it could reduce the strict construction requirements for ductwork and fans, once the fire safety of the components has been proven. However, exhaust system designers should pay special attention to effluent control equipment that traps grease in the removal device, rather than draining it away to a remote, fire-safe container.

Complete effluent control requires control of grease (in the solid, liquid, and vapor state), other solid particulates, and volatile organic compounds (VOCs or low carbon aromatics, commonly referred to as odors). Grease should be removed in all forms first, and then the VOCs. Failure to remove grease first will result in fouling of the odor control hardware.

Grease removal typically starts in the hood with baffle filters or grease removal devices. The more effective these devices are, the less it is necessary to install devices in the field. The next device is almost always another grease removal device, which removes any remaining grease. The VOC removal device follows.

The term grease extraction filters may be a misnomer. These filters are tested and listed for their ability to limit, but not totally prevent, flame penetration into the hood plenum and duct, and not for their grease extraction ability. Additionally, research is beginning to indicate that grease particles are generally so small (less than 20 μm in diameter) that they are aerodynamic and not easily removed by the centrifugal impingement principle usually used in grease extraction devices.

The following technologies are available today and applied to varying degrees for control of cooking effluent. These technologies are listed in the common order of use, with particulate control upstream of the VOC controls. Following the description of each technology are some of the qualifications and concerns about its use.

Electrostatic precipitators (ESPs). Particulate removal is by high-voltage ionization, then by collection on flat plates.

- In a cool environment, collected grease can block airflow.
- As the ionizer section gets dirty, efficiency drops because the effective plate surface area is reduced.

Water mist, waterfall, and water bath. Passage of the effluent stream through water mechanically entraps particulates.

- Airflow separates the bath and the waterfall, so they are less effective.
- Bath types have a very high static pressure loss.
- Spray nozzles need much attention. Water may need to be softened to minimize clogging.
- Drains, no matter how large, tend to become blocked.

Pleated or bag filters of finely sized natural and synthetic fibers. Very fine particulate removal is by mechanical filtration. Some types have an activated carbon face coating for odor control.

- These filters become blocked quickly if too much grease enters.
- Static loss builds quickly with extraction, and airflow drops.
- Almost all filters are disposable and very expensive.

Active carbon filters. VOC control is through adsorption by fine activated charcoal particles.

- Require a large volume and thick bed to be effective.
- Heavy and can be difficult to clean.
- Expensive to change and recharge. Many now are disposable.
- Ruined quickly if they become grease coated or subjected to water.
- Some concern about increased fuel available in fire.

Oxidizing pellet bed filters. VOC and odor control is by oxidation of gaseous effluent into solid compounds.

- Require a large volume and long bed to be effective.
- Heavy to handle and can be difficult to clean.
- Expensive to change out.
- Some concern about increased oxygen available in fire.

Incineration. Particulate, VOC, and odor control is by high-temperature oxidation (burning) into solid compounds.

- Must be at system terminus and clear of combustibles.
- Expensive to install with adequate clearances.
- Can be difficult to access for service.
- Very expensive to operate.

Catalytic conversion. A catalytic or "assisting" material, when exposed to relatively high-temperature air, provides additional temperature that is adequate to decompose (oxidize) most particulates and VOCs.

- Requires high temperature (260°C minimum).
- Expensive to operate due to high temperature requirement.

In addition, for fire safety, grease should be eliminated entirely or drained away from the duct to a safe container.

Duct Systems

The exhaust ductwork conveys the exhaust air from the hood to outdoors, along with any grease, smoke, VOCs, and odors that are not extracted from the airstream along the way. In addition, this ductwork may be used to exhaust smoke from a fire. To be effective, the ductwork must be greasetight; it must be clear of combustibles, or the combustible material must be protected so it cannot be ignited by a duct with a fire in it; and ducts must be sized to convey the volume of air necessary to remove the effluent.

The ductwork should not have any traps that can hold grease, which would be an extra fuel source in the event of a fire, and ducts should pitch toward the hood for constant drainage of liquefied grease or condensates. On long duct runs, allowance must be made for possible thermal expansion due to a fire.

Single-duct systems carry effluent from a single hood or section of a large hood to a single exhaust termination. In multiple-hood systems, several branch ducts carry effluent from several hoods to a single master duct that has a single termination.

For correct flow through the branch duct in multiple-hood systems, the static pressure loss of the branch must match the static pressure loss of the common duct upstream from the point of connection. Any exhaust points subsequently added or removed must be closely designed to maintain the balance of the remaining system.

Ducts may be constructed of field-fabricated round or rectangular sections. The standards and model codes contain minimum specifications for duct materials, including gage, joining methods, and minimum clearances to combustible materials. UL Listed prefabricated duct systems may also be used. These

Fig. 4   Upblast Exhaust Fan

systems typically allow reductions in the clearance to combustible materials requirements.

Types of Exhaust Fans

Exhaust fans for kitchen ventilation applications must be capable of handling hot, grease-laden air. The fan should be designed to keep the motor out of the airstream and effectively cooled to prevent premature failure. To prevent roof damage, the fan should contain and properly drain all grease removed from the airstream.

The following types of exhaust fans are in common use (all have centrifugal wheels with backward-inclined blades):

* *Upblast* (Figure 4). These fans are designed for roof mounting directly on top of the exhaust stack, with an upward discharge. These are generally aluminum and must be listed for the service. They typically can only provide static pressures of up to 250 Pa, but are available with higher pressures. They may have an integral grease drainage path and collection container. These fans allow easy access for duct cleaning because they generally hinge back from the duct.

* *Utility set.* These fans are usually steel, and they are usually roof mounted. The inlet and outlet are at 90° to each other (single width, single inlet), and the outlet can usually be rotated to discharge at many different angles around a vertical circle. They can operate at medium to high and (with special provisions) very high static pressures. Care must be taken to drain the low part of the fan to a safe remote container.

* *Inline.* These fans are typically located in the duct run inside a building where exterior fan mounting is not practical for wall or roof exhaust. They are always constructed of steel. The gasketed flange mounting must be greasetight and still be removable for service. A catch pan must be placed under the entire assembly in event of a grease leak at the flanges. If the fan housing is lower than the duct, there must be a drain to a separate collection container.

Exhaust System Terminations

Rooftop. Rooftop terminations are preferred because the discharge can be directed away from the building, the fan is at the end of the system, and the fan is in an accessible location. The common concerns with rooftop terminations are as follows:

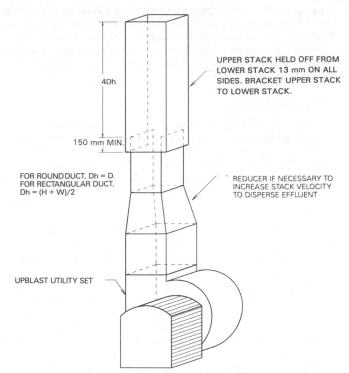

UPPER STACK HELD OFF FROM LOWER STACK 13 mm ON ALL SIDES. BRACKET UPPER STACK TO LOWER STACK.

4Dh

150 mm MIN.

FOR ROUND DUCT, Dh = D.
FOR RECTANGULAR DUCT, Dh = (H + W)/2

REDUCER IF NECESSARY TO INCREASE STACK VELOCITY TO DISPERSE EFFLUENT

UPBLAST UTILITY SET

Fig. 5   Rooftop Upblast Utility Set

* The discharge of the exhaust system should be arranged to minimize reentry of effluent into any fresh air intake or other opening to any building. This requires not only a separation of the exhaust from intakes, but also knowledge of the direction of the prevailing winds. Some codes specify a minimum distance to air intakes.

* In the event of a fire, neither the flames, nor the radiant heat, nor the dripping grease should be able to ignite the roof or other nearby structures.

* All grease from the fan or the duct termination should be collected and drained to a remote closed container to preclude ignition.

* Rain water should be kept out of the exhaust system and especially the grease container. If this is not possible, then the grease container should have a design that separates the water and grease and drains the water back onto the roof. Figure 5 shows a rooftop upblast utility set with a stackhead fitting, which directs the exhaust away from the roof and minimizes rain penetration. Discharge caps should not be used because they direct the exhaust back towards the roof and can become grease fouled.

Outside Wall. Wall terminations are less common today but still exist. The fan may or may not be the terminus of the system, located on the outside of the wall. The common concerns with wall terminations are as follows:

* Discharge from the exhaust system should not be able to enter any fresh air intake or other opening to any building.

* Adequate clearance from combustibles must be maintained.

* Duct sections should pitch back to the hood inside, or a grease drain should be provided to drain the grease back into a safe container inside the building in order to not have grease draining down the side of the building.

* The discharge must not be directed downward or toward any pedestrian areas.

* Louvers should be designed to minimize grease extraction effects and to prevent staining the building facade.

Recirculating Systems. With these units, it is extremely critical to keep the components in good working order to maintain optimal performance. If not, excessive grease and odor will accumulate in the premises.

As with other terminations, containing and removing the grease and keeping the discharge as far as possible from combustibles are the main concerns. Some of these units are fairly portable and can readily be set in an unsafe location. The operator should be made aware of the importance of safely locating the unit.

A major concern is the quality of maintenance of the system. If the hood is not properly maintained, it will allow grease, heat, and odors to accumulate in the space. These units are best for large, unconfined areas that have a separate outside exhaust to keep the environment comfortable.

MAKEUP AIR SYSTEMS

The air in the kitchen that is exhausted to remove the cooking effluent must be replaced with air from outside the building to avoid excess negative pressure in the space, which may degrade the exhaust system performance. The standards and model codes require 100% makeup air. The effectiveness of short circuit hoods is debatable—some local authorities approve them, and some do not. The Uniform Mechanical Code (ICBO) requires that at least 20% of the makeup air come from the kitchen area. In any case, all applicable codes must be consulted to assure that proper criteria are being followed.

Makeup Air Through HVAC System

In many cases, the ideal means of providing makeup air is through the HVAC system, the reason being that the air is comfort conditioned, enhancing the kitchen environment. In smaller stand-alone restaurants, rooftop units are typical, and these can be obtained with outdoor air dampers that are controlled to automatically open for the proper amount of outside air when the hoods are operational and close when they are turned off. An even better approach is the use of an enthalpy or temperature control not only to provide the make-up air, but to provide a degree of free cooling, by keeping the compressors off, when outdoor conditions warrant by adjusting the outside and return air dampers. Designers must exercise extreme caution when applying outdoor air economizer cycles to kitchen systems in multiple-occupancy applications because the economizer can upset the air balance between occupancies and cause odor migration problems.

The amount of air brought in through the HVAC units should be limited so that comfort conditions are maintained. Typically, dining room units should be limited to 25 to 30% outside air to avoid unacceptable variations in discharge temperature. Kitchen units, on the other hand, may be set to bring in up to 50% outside air because the temperature swings are less critical. For commercial kitchens where there is no on-site dining, such as take-out stores, commissaries, and so forth, the entire area can be treated under the kitchen design criteria. Rooftop units are available to deliver 100% outside air with discharge air temperature control on cooling and heating.

Indoor Air Quality

ASHRAE *Standard* 62, Ventilation for Acceptable Indoor Air Quality, requires that 10 L/s outdoor air per person based on a maximum occupancy of 70 persons per 100 square metres (100 persons per 100 square metres for cafeterias and fast-food restaurants) be brought into the dining area. The amounts are 15 L/s per person in bars/cocktail lounges and 8 L/s per person in kitchens based on 20 persons per 100 square metres These requirements may be increased or decreased in certain areas by local code and sometimes dictate the size of an HVAC system even more than thermal loading or necessitate the use of another means of introducing outside air. A further requirement of *Standard* 62, that outside air be sufficient to provide

for an exhaust rate of not less than 7.5 L/s per square metre of kitchen space, is generally easily met.

Makeup Air for Large Buildings

In large buildings with commercial kitchens (e.g., malls, hospitals, supermarkets, schools, etc.), adequate outside air from other parts of the building is generally more available to provide makeup air for the hoods. However, in designing such a facility, the kitchen needs should be factored into the outside air provisions even though the amount may be a small fraction of the overall requirement. Also the source of the makeup air must be considered so that air contaminated with odors, germs, dust, etc., is not drawn to the kitchen area, where cleanliness is required. Most malls and other multiple-occupancy buildings require a minimum amount of air to be taken from their space to keep cooking odors in the kitchen, but they also specify a maximum amount that may be taken to hold down the cost of tempering the outside air.

Non-HVAC System Makeup Air

In situations where the outside air from the HVAC system is insufficient to replace all the air being drawn from the kitchen, air must be brought directly to the kitchen with limited or no conditioning. In some climates outdoor conditions may be temperate enough that outside air does not require heating or cooling, particularly if it is mixed with the conditioned air such that comfort is not unduly compromised. On the other hand, in colder climates it is essential to heat the air at least to a minimal comfort level. This may be done using electric resistance heat or direct or indirect gas-fired equipment (local codes must be consulted before direct gas-fired equipment is selected). In more arid areas, evaporative cooling can be effective, and direct-expansion HVAC equipment may not be needed in either the kitchen or dining areas.

Again, makeup air is used primarily for proper operation of the exhaust hoods and not for comfort of the kitchen personnel. Comfort can be achieved through adequate flow of air around the workers, and higher velocity air movement can compensate for higher temperatures and humidities, up to a point. Still, a restaurant owner may want greater comfort for the employees, and then comfort becomes an important design concern.

Some attempts have been made to temper outside air using heat from the exhaust system itself. However, effective heat transfer from the exhaust stream to the incoming air stream is made difficult by the strict requirements of the exhaust duct design, (i.e., cleanability, minimal grease entrapment, etc.). Careful selection of the exhaust hood system to minimize the amount of air exhausted will help minimize the cost of conditioning the makeup air and the cost of operating the exhaust and makeup air fans.

Non-Code Areas

In areas where commercial kitchens are not governed strictly by various codes, air exhausted through the ventilation hoods is often replaced through open windows, doors, and gravity vents. Doors should not be used to provide makeup air, as adequate filters cannot be provided to prevent the intake of dirt and insects. Windows and gravity vents can be used if they are adequately filtered and are sized to admit enough air that hood performance is maintained and all cooking effluent is exhausted.

Air Distribution

The outside makeup air should be distributed in the general vicinity of the hoods. It should be delivered so that high velocities, eddies, swirls, or stray currents do not degrade the performance of the hood. Makeup air from any source should be distributed throughout the kitchen with low-velocity ceiling diffusers. Some hood manufacturers place diffusers in the face or lower front edge of the canopy to attempt to have makeup air enter the hood in an effective but nonintrusive manner. The important point is to have makeup air delivered to the hoods (1) with proper velocity and (2) uniformly from all directions to which the hood is open in order to minimize excessive crosscurrents that could cause spillage from the capture area.

SYSTEM INTEGRATION AND BALANCING

System integration and balancing brings the many ventilation components together in a way that provides the most comfortable, efficient, and economical performance of each component and of the entire system. In commercial kitchen ventilation, the supply air system (typically referred to as the HVAC system) must integrate and balance with the exhaust system. Optimal performance is achieved more by effective use of components, through controls and airflow adjustments, than by the selection or design of the system and components. The following fundamentals relative to restaurants, and kitchens in particular, should be considered and applied within the constraints of any particular location and its equipment and systems.

Principles. The principles of integrating and balancing restaurant systems are as follows:

1. The building is at all times pressurized compared to the outdoors. A slight negative pressure is desirable in multiple-occupancy buildings to minimize odor migration from the restaurant to other occupancies.
2. The kitchen is negative at all times compared to adjacent zones.
3. Airflow between zones is minimized during temperate weather.
4. There are minimal drafts and temperature differences within and between zones.

These principles are viewed as fundamental, logical concepts in both comfort control and economical operation in a restaurant. While there are exceptions that will be addressed, the following are the rationales for the above statements.

1. In a freestanding restaurant, the overall building should be slightly pressurized compared to outside to minimize the infiltration of untempered air, as well as dirt, dust, and insects, when doors are opened.
2. Every kitchen should be slightly negative compared to the rooms or areas immediately surrounding it for the following reasons:
   - To better contain the unavoidable grease vapors in the kitchen area and limit the extent of cleanup necessary
   - To keep the cooking odors in the kitchen area and not adversely affect the dining experience
   - To prevent the generally hotter and more humid kitchen air from upsetting the comfort level of the adjacent spaces, especially the dining areas
3. Cross-zoning of airflow should be minimal, especially in the temperate seasons when adjacent zones might be in different modes (e.g., economizer versus air-conditioning or heating). Two situations to consider are as follows:
   - In the transitions from winter to spring and from summer to fall, the kitchen zone could be in the economizer mode, or even the mechanical cooling mode, while the dining areas are in the heating mode. To bring heated dining area supply air into a kitchen that is in the cooling mode only adds to the cooling load. In some areas, these conditions are present every day.
   - When all zones are in the same mode, it is more acceptable, and even more economical, to bring dining area air into the kitchen. However, the controls to automatically effect this method of operation are more complex and costly.
4. If there are noticeable drafts or temperature differences, customers and restaurant personnel will be distracted and will not enjoy dining or working. Typically, no drafts should be noticeable and temperatures should vary no more than 0.6 K in dining areas or 1.6 K in

kitchen areas. These conditions can be achieved with even distribution and thorough circulation of air in each zone by an adequate number of properly sized registers to preclude high air velocities.

Both design concepts and operating principles for proper integration and balance are involved in bringing about the desired results under the varying conditions. The same principles are important in almost every aspect of ventilation in a restaurant. The more variable the exhaust, the more complex the design. The more numerous and smaller the zones involved, the more complex the design. But the overall pressure relationship principles must be maintained to provide optimum comfort, efficiency, and economy.

In designing restaurant ventilation, all the exhaust is assumed to be in operation at one time, and the makeup air is a maximum requirement because it is for these maximum design conditions that the heating and cooling equipment are sized.

In restaurants with a single large exhaust hood, balancing should be set for this one operation only. In restaurants with multiple exhaust hoods, some are operated only during heavy business hours or for special menu items. In this case, makeup air must be controlled to maintain minimum building positive pressure and to maintain the kitchen at a negative pressure.

A different application is a kitchen having one side exposed to a larger building with common or remote dining. Such a case would be a food court in a mall or a small restaurant in a hospital, airport, or similar building. Pressurizing the kitchen at the front of its space might cause some of the cooking grease vapor and odors to spread into the common building space, which would be undesirable. In such cases, the kitchen area is held at a negative pressure relative to other common building areas, as well as to its own back room storage or office space.

Air Balancing

Balancing is best performed when the manufacturers of all the various equipment are able to provide a certified reference method of measuring the airflows, rather than depending on generic measurements of duct flows or other forms of measurement in the field. These field measurements can be in error by 20% or more. The manufacturer of the equipment should be able to develop a reference method of measuring airflow in a portion of the equipment that is dynamically stable in the laboratory as well as in the field. This method should relate directly to airflow by graph or formula.

The general steps for air balancing in restaurants are as follows:

1. The exhaust hoods should first be set to their proper flow rates. This should be done with the supply and exhaust fans on.
2. Next, the air supply flow rate, whether part of combined HVAC units or separate makeup air units, should be set to the design values through the coils and the design supply flows from each outlet, with the approximate correct settings on the outside air flow rate. Then the correct amounts of outside and return air should be set proportionately for each unit, as applicable.

   Where the outside and return airflows of a particular unit are expected to modulate, there should ideally be similar static losses through both airflow paths to preclude large changes in total supply air from the unit. Such changes, if large enough, could affect the efficiency of heat exchange and could also change the airflows within and between zones, thereby upsetting the air distribution and balance.

   These settings should also be made with the exhaust on, to ensure adequate relief for the outside air.
3. Next, outside air should be set with all fans (exhaust and supply) operating.

   The pressure between inside and outside should be checked to see that (1) the nonkitchen zones of the building are under pressure compared to outside and (2) the kitchen zone pressure is negative compared to the surrounding zones and negative or neutral compared to outside.

For applications with modulating exhaust, every step of exhaust and makeup should be shut off, one step at a time. Each combination of operation should be rechecked to be sure that the design pressures and flows are maintained within each zone and between zones. This requires that the makeup airflow rate compensate automatically with each increment of exhaust. It may require some adjustments in controls or in damper linkage settings to get the correct proportional response.

4. When the above steps are complete, the system is properly integrated and balanced. At this time, all fan speeds and damper settings (at all modes of operation) should be recorded on the equipment with permanent markings and in the test and balance report. The complete air balance records of exhaust, supply, return air, fresh air, and individual register airflows must also be completed. These records should be kept by the food service facility for future reference.

5. For new facilities, after two or three days in operation (no longer than a week and usually before the facility opens), all belts in the system should be checked and readjusted because new belts wear in quickly and could begin slipping.

6. Once the facility is operational, the performance of the ventilation system should be checked to see if the design is adequate for the actual operation, particularly at maximum cooking and at outdoor environmental extremes. If any changes are required, they should be made, and all the records should be updated to show the changes.

Rechecking of the air balance should not be necessary more than once every two years, unless basic changes are made in the facility operation. If there are any changes, such as a new type of cooking equipment, or added or deleted exhaust connections, the system should be modified accordingly.

Multiple-Hood Systems

Kitchen exhaust systems serving more than a single hood face several design challenges that are not encountered with single-hood systems. One of the main challenges to multiple-hood exhaust systems is air balancing. Because balancing dampers are not permitted in the exhaust ductwork, the system must be balanced by design. Some hoods and grease filters are available that have adjustable baffles that allow airflow to be adjusted at the hood. These may be helpful for relatively fine balancing, but the system must provide most of the balancing. Adjustable filters should not be used if they can be interchanged between hoods or within the same hood because an interchange could disrupt the previously achieved balance.

If one designer selects and lays out the entire exhaust system, such as for a large kitchen with several hoods, the ductwork can be designed for controlled pressure loss in all branches. In cases such as master kitchen exhaust systems, which are sometimes used in shopping center food courts, no single group is responsible for the entire design. The base building designer typically lays out the ductwork to (or through) each tenant space, and each tenant selects the hood and lays out the connecting ductwork. Often the base building designer has incomplete information regarding the tenants' exhaust requirements. Therefore, one engineer must be responsible for defining criteria for each tenant's design and for evaluating proposed tenant work to ensure that tenants' designs match the system's capacity as much as possible. The engineer should also evaluate any proposed changes to the system, such as changing tenancy. Rudimentary computer modeling of the exhaust system may be helpful (Elovitz 1992). Given the unpredictability and volatility of tenant requirements, it may not be possible to balance the entire system perfectly. However, without adequate supervision, chances are very good that at least parts of the system will be badly out of balance.

For greatest success with multiple-hood exhaust systems, pressure losses in the ductwork should be minimized by keeping velocities low, minimizing sharp transitions, and using hoods with relatively high pressure drops. When the pressure loss in the ductwork is low compared to the loss through the hood, changes in pressure loss in the ductwork due to field conditions or changes in design airflow will have a smaller impact on total pressure loss, and thus on actual airflow.

The minimum, code-required air velocity must be maintained in all portions of the exhaust ductwork at all times. If fewer or smaller hoods are installed than the design anticipated (resulting in low velocities in portions of the ductwork), the velocities must be brought up to the minimum. One way is to introduce air, preferably untempered makeup air, directly into the exhaust duct where it is required (see Figure 6). The bypass duct should connect to the top or sides (at least 50 mm from the bottom) of the exhaust duct, to prevent backflow of water or grease through the bypass duct when the fans are off. This arrangement is shown in NFPA *Standard* 96 and should be discussed with the authority having jurisdiction.

A fire damper should be provided in the bypass duct, located close to the exhaust duct. The bypass duct construction should be the same as the exhaust duct construction, including enclosure and clearance requirements, for about a metre beyond the fire damper. Means to adjust the bypass airflow must be provided upstream of the fire damper. All dampers must be in the clean bypass air duct so that they will not be exposed to any grease-laden exhaust air. The difference in pressure between the makeup air and the exhaust air duct may be great, and the balancing device must be able to make a fine airflow adjustment against this pressure difference. It is best to provide two balancing devices in series, such as an orifice plate or blast gate for coarse adjustment followed by an opposed-blade damper for fine adjustment.

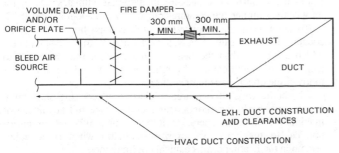

Fig. 6   Method of Introducing Makeup Air Directly into Exhaust Duct

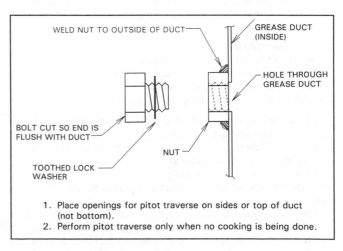

Fig. 7   Metal-to-Metal Seal for Pitot Tube Access Port in Grease Duct

Direct measurement of air velocities in the exhaust ductwork to assess exhaust system performance may be desirable. Velocity (pitot tube) traverses may be performed in kitchen exhaust systems, but the holes drilled for the pitot tube must be liquidtight to maintain the fire-safe integrity of the ductwork. Figure 7 shows a means of providing a metal-to-metal seal for these holes. Holes should never be drilled in the bottom of a duct, where they may collect grease. Velocity traverses should not be performed when cooking is in progress, since grease will collect on the instrumentation.

ENERGY CONSIDERATIONS

Many people believe that kitchen ventilation rates are too high and that existing kitchen ventilation requirements could be reduced by 50%. Restaurants are by far the largest consumer of energy on a unit area basis, considering all the building types. This section briefly discusses ventilation conservation issues with respect to climate, type of restaurant, type of hood, and equipment characteristics, as energy conservation in restaurants depends upon these variables. Means of conserving energy in kitchen ventilation systems using existing codes and standards are also covered.

Climate. Average outdoor temperature is the principal predictor of restaurant energy use. Because different climate zones vary dramatically in temperature and humidity, kitchen ventilation conservation retrofits will have widely varying economic recovery periods based on existing HVAC system design practices and geographical location. In new facilities the designer can select conservation measures that consider the climate zone and the HVAC system to maximize the economic benefits.

Restaurant Type. One study defined five restaurant types: fast food, full service, coffee shop, pizza, and cafeteria. Restaurant associations identify many additional restaurant types. Energy use varies significantly among the types. As a result, some energy conservation measures work in all restaurants while others may work only in specific types. For simplicity, this discussion classifies restaurants as cafeteria, fast food, full menu, and pizza.

Hood Type and Equipment Characteristics. The type of exhaust hood selected depends on such factors as restaurant type, restaurant menu, and food service equipment used. Exhaust flow rates will largely be determined by the food service equipment and hoods installed. The effectiveness of conservation measures, whether in the new design or as a retrofit, will also be affected by the type of hood and food service equipment installed.

Energy Conservation Measures

Energy costs in restaurants vary from 2 to 6% of gross sales. The following major energy conservation measures can reduce commercial kitchen ventilation costs:

- Custom-designed hood
- Dining room air for kitchen ventilation
- Heat recovery from exhaust hood ventilation air
- Extended use of economizer operation
- Reduced exhaust and makeup air volumes

Custom-Designed Hood. A typical cooking lineup combines food service equipment with different thermal updraft characteristics as well as different exhaust flow rate requirements depending on cooking load, cooking temperature, product cooked, fuel source, and so forth. Exhaust hoods are usually sized to handle worst-case cooking requirements, so kitchen exhaust requirements are often overstated.

Custom-designed hoods for each piece of equipment can lower ventilation requirements and, consequently, reduce fan sizes, energy use, and energy costs for all types of restaurants. Either thermal current charts or empirical tests developed by hood manufacturers and their recommended exhaust flow rates are the basis for hood sizing. The total hood exhaust flow rate and size are determined by summing the individual equipment airflow rate and size requirements for the entire cooking lineup.

Dining Room Air for Kitchen Ventilation. Most fast-food and pizza restaurants do not physically segregate the kitchen from the dining room, so conditioned air moves from the dining area to the kitchen. Dining room air helps cool the kitchen to some extent and provides some makeup air for the kitchen exhaust. Full menu and cafeteria food service facilities are frequently designed so that no air transfers between the dining room and the kitchen. This segregation means that kitchen exhaust must be made up entirely within the kitchen by conditioned or unconditioned air.

In restaurants in which the kitchen and dining room are physically segregated, ducts between the two areas permit conditioned air to flow to the kitchen. This design can reduce makeup air requirements and enhance employee comfort, especially if the kitchen is not air conditioned. Of course, the dining room must have makeup air to replace the air transferred to the kitchen. Also, care must be taken to ensure that enough makeup air is introduced into the kitchen to meet the requirements of applicable codes and standards. In new construction, this conservation measure can pay back the initial investment in less than one year.

Heat Recovery from Exhaust Hood Ventilation Air. High-temperature effluent, often in excess of 200°C, heats the exhaust air to 40 to 90°C as it travels over the cooking surfaces and areas. It is frequently assumed that this heated exhaust air is suitable for heat recovery; however, smoke and grease in the exhaust air will, with time, cover any heat transfer surface. Under these conditions the heat exchangers require constant maintenance, such as automatic wash-down, to maintain acceptable heat recovery.

Because heat recovery systems are very expensive, only those food service facilities with large amounts of cooking equipment and large cooking loads are good candidates for this equipment. Heat recovery may work in full menu and cafeteria restaurants and in institutional food service facilities such as prisons and colleges. An exhaust hood equipped with a heat recovery system is more likely to be cost-effective where the climate is extreme, such as the northern tier states or the southern states. A mild climate, such as found in California, is not conducive to use of this conservation measure.

Extended Use of Economizer Operation. To keep the staff and customers cool in most restaurants, the mechanical cooling is called on as soon as someone is too warm. By operating an economizer system instead, similar cooling can be achieved at less cost because only fans are operated and not compressors. Economizers can be used in lieu of mechanical refrigeration when the outside temperature (and sometimes humidity) is lower than that of the return air. Economizer systems may be designed to increase ventilation by 25% or less over the normal ventilation, or about 50 to 60% of capacity. Yet economizers have the capability of increasing the outside air to 100%. This operation can dramatically extend the time that the mechanical cooling is kept off.

Use of full capacity economizer operation requires variable rather than one-step control of the exhaust and makeup air. The savings can justify the initial cost, especially in moderate climates where, for most of the summer, temperatures range between about 25 and 28°C.

To allow economizer operation with canopy hoods, it may be adequate to increase the exhaust at each hood. With backshelf hoods, which are lower than the heads of the crew, it may be better to increase the flow of exhaust from the ceiling of the kitchen in order to lower the upper space temperature.

The entire food service area, including all cooking and dining zones, must be in the economizer mode to take full advantage of maximum economizer operation. It may be uncomfortable or uneconomical to maximize economizer operation when some zones remain in the heating or mechanical cooling mode.

Reduced Exhaust and Makeup Airflow Rates. The tempering of outdoor makeup air can account for a large part of a food service facility's heating and cooling costs. By reducing the exhaust flow rates (and the corresponding makeup air quantity) when no product is being cooked, the cost of energy can be significantly reduced. Field evaluations by one large restaurant chain have suggested that cooking appliances may be at no-load for 75% or more of an average business day. During these periods when no smoke or grease-laden vapors are being produced, NFPA *Standard* 96 allows for the reduction of exhaust quantities. The only restriction on the reduced exhaust quantity is that it be "... sufficient to capture and remove flue gases and residual vapors... ."

FIRE PROTECTION

The combination of flammable grease and particulates carried by kitchen ventilation systems and the potential for the cooking equipment to be an ignition source results in a higher level of hazard than normally found in HVAC systems. Where an exhaust system serves cooking equipment that may produce grease-laden vapors, the design must provide, at a minimum, a reasonable level of protection for the safety of building occupants and fire fighters. The design can also be enhanced to provide extra protection for property.

Exhaust systems serving only cooking equipment that produces no grease-laden vapor, makeup air systems, and air-conditioning systems serving a kitchen have no specific fire protection requirements beyond those applicable to similar systems not located in kitchens. However, an exhaust system that serves any grease-producing cooking equipment must be treated entirely as a grease exhaust system, even if it also serves non-grease-producing equipment.

It is important to prevent a fire from starting in the first place through proper operation and maintenance of the cooking equipment and the exhaust system. After that, the two primary aspects of fire protection in a grease exhaust system are (1) to extinguish a fire quickly once it has started and (2) to prevent the spread of fire from or to the grease exhaust system.

Fire Suppression

NFPA *Standard* 96 requires that exhaust systems serving grease-producing equipment, except where certain listed grease removal devices or systems are installed, must include provision for fire suppression. The fire suppression system must protect the cooking surfaces, the hood interior, the hood filters or grease extractors, the ductwork, and any additional grease removal devices in the system. The most common fire suppression systems are dry chemical, wet chemical, and water systems. Carbon dioxide and deluge foam-water systems are an alternative, but they are rarely used. Exhaust systems may be protected by one or a combination of these systems.

Operation. Actuation of any fire suppression system should not depend on normal building electricity. If actuation relies on electric power, it should be supplied with standby power.

All sources of fuel and heat under the hood must shut off automatically in the event of actuation of the fire suppression system. This includes gas supply to all cooking equipment under the hood and power to all cooking equipment with electric heating elements, whether grease producing or not.

If the hood is in a building that has a fire alarm system, actuation of the hood fire suppression system should send a signal to the fire alarm system.

Dry and Wet Chemical Systems. Dry and wet chemical fire suppression systems are the most common systems in use today for protection of grease exhaust systems. Dry chemical systems are covered in NFPA *Standard* 17, Dry Chemical Extinguishing Systems; wet chemical systems are covered in NFPA *Standard* 17A, Wet Chemical Extinguishing Systems. Both standards provide detailed application information. These systems are tested for their ability to suppress fires occurring in cooking operations in accordance with UL *Standard* 300, Fire Testing of Fire Extinguishing Systems for Protection of Restaurant Cooking Areas.

Both dry and wet chemicals suppress a fire by reacting with fats and grease to saponify, or form a soapy layer of foam that prevents oxygen from reaching the hot surface. This suppresses the fire and prevents reignition. Saponification is particularly important with deep fat fryers, where the frying medium may be above the autoignition point for some time after the fire is extinguished. Should the foam layer disappear or be disturbed before the frying medium has cooled below its autoignition temperature, it could reignite.

Frying media commonly used today have autoignition points of about 363 to 375°C when they are new. Contamination through normal use reduces the autoignition point. While suppressing the fire, the chemical agent will also contaminate the frying medium, which can further reduce the autoignition point by 28 to 33 K. One advantage of wet chemical systems over dry chemical systems is that the wet chemical provides extra cooling to the frying medium, so it falls below the autoignition point more quickly.

For a chemical system protecting the entire exhaust system, fire suppression nozzles are located over the cooking equipment being protected, in the hood to protect the grease removal devices and hood plenum, and at the duct collar (downstream from any fire dampers) to protect the ductwork. The duct nozzle is rated to protect an unlimited length of ductwork, so additional nozzles are not required further downstream in the ductwork. Fire detection is required at the entrance to each duct (or ducts, in hoods with multiple duct takeoffs) and over each piece of cooking equipment that requires protection. The detector at the duct entrance may also cover the piece of cooking equipment directly below it.

Chemical fire suppression systems are available as listed, pre-engineered (packaged) systems. Wet chemical systems were developed more recently than dry chemical systems and are the more common of the two. Both systems typically consist of one or more tanks of chemical agent (dry or wet), a propellant gas, piping to the suppression nozzles, fire detectors, and auxiliary equipment. The fire detectors are typically fusible links that melt at a set temperature associated with a fire, although electronic devices are also available. Auxiliary equipment may include manual pull stations, gas shutoff valves (spring-loaded or solenoid-actuated), and auxiliary electric contacts.

Actuation of the dry and wet chemical suppression systems is typically completely mechanical, requiring no electric power. The fire detectors are typically interconnected with the system actuator by steel cable in tension, such that melting any of the fusible links releases the tension on the steel cable, causing the actuator to release the propellant and suppressant. The total length of the steel cable and the number of pulley elbows permitted are limited. A manual pull station is typically connected to the system actuator by steel cable. If a mechanical gas valve is used, it is also connected to the system actuator by steel cable. Actuation of the system also switches auxiliary dry electrical contacts, which can be used to shut off electrical cooking equipment, operate an electric gas valve, shut off a makeup air fan, and/or send an alarm signal to the building fire alarm system.

Manual pull stations are generally required to be at least 3 m from the cooking appliance and in a path of egress. Some authorities may prefer that the pull station be installed closer to the cooking equipment, for faster response. However, if the pull station is too close to the cooking equipment, it may not be possible to approach it once a fire has started. Refer to the applicable code requirements for each jurisdiction to determine specific requirements for location and mounting heights of pull stations.

Water Systems. Water can be used in several ways for protection of grease exhaust systems. Standard fire sprinklers may be used throughout the system, except over deep fat fryers, where special automatic spray nozzles specifically listed for the application must be used. These nozzles must be aimed properly and be supplied with the correct water pressure. These sprinkler systems are applicable only to canopy hood systems. A water system used for cleaning or grease removal in a hood can also protect the hood or duct in the case of fire, if the system is listed for this purpose.

Application of standard fire sprinklers for protection of grease exhaust systems is covered by NFPA *Standard* 13, Installation of Sprinkler Systems. NFPA *Standard* 96 covers maintenance of sprinkler systems serving an exhaust system. Where water sprinklers are used in a grease exhaust system, they must connect to a wet-pipe building sprinkler system installed in compliance with NFPA *Standard* 13.

Contrary to the common wisdom that water is inappropriate for fighting grease fires, water systems can be used to protect cooking equipment if the water spray is a fine mist. The water spray works in two ways to suppress the fire: the mist spray first absorbs heat from the fire, and then it becomes steam, which displaces air and suffocates the fire.

Although standard sprinklers may be used to protect cooking equipment other than deep fat fryers, care must be taken that the sprinklers be properly selected for a fine mist discharge; if the pressure is too high or the spray too narrow, the water spray could push the flames off of the cooking equipment. Most canopy hood manufacturers that use water to protect the cooking equipment use the sprinklers listed for deep fryers over all of the cooking equipment.

One advantage of a sprinkler system is that it has virtually unlimited capacity, whereas chemical systems have limited chemical supplies. While this can be an advantage in suppressing a serious fire, all of that water must be safely removed from the space. Where sprinklers are used in ductwork, the ductwork should be pitched to drain safely. Improper design or installation could cause water in the ductwork to flood back onto the cooking equipment.

When sprinklers are used to protect ductwork, the current requirement is that they must be installed every 3 m on center in horizontal ductwork, at the top of every vertical riser, and in the middle of any vertical offset. Care must be taken to protect any sprinklers exposed to freezing temperatures.

Hoods that use water either for periodic cleaning (water-wash) or for grease removal (cold water mist) can use this feature as a fire extinguishing system to protect the hood and grease removal device in the event of a fire, if the system has been tested and listed for the purpose. The water supply for these systems may be from the kitchen water supply if flow and pressure requirements are met. These hoods can also act as a fire stop because of their multipass configuration and the fact that the grease extractors are not removable.

CO$_2$ Systems. Design and application of carbon dioxide (CO$_2$) fire extinguishing systems are covered by NFPA *Standard* 12, Carbon Dioxide Extinguishing Systems. Although more common many years ago, CO$_2$ systems are rarely used in grease exhaust systems today. They are more expensive to install and maintain than other systems; the exhaust airflow must be shut off in order to maintain a sufficient concentration to suppress a fire; the CO$_2$ will not cool a vat of frying medium to prevent reignition, as other systems will; and CO$_2$ can be hazardous to health at the concentrations required to suppress a fire. Evacuation of the kitchen is generally necessary if the system discharges. Since CO$_2$ is colorless and odorless, an accidental discharge may go unnoticed.

Deluge Foam-Water Systems. Design and application of deluge foam-water fire extinguishing systems are covered by NFPA *Standard* 16, Deluge Foam-Water Sprinkler and Foam-Water Spray Systems. Such systems are rarely applied in grease exhaust systems.

Combination Systems. Different system types may protect different parts of the grease exhaust system, as long as the entire system is protected. Examples include (1) an approved water-wash or water mist system to protect the hood in combination with a dry or wet chemical system to protect the ductwork and the cooking surface or (2) a chemical system in the hood backed up by water sprinklers in the ductwork. Combination systems are more common in multiple-hood systems, where a separate extinguishing system may be used to protect a common duct. Where combination systems will discharge their suppressing agents together in the same place, the agents must be compatible.

Steam Systems. Steam suppression systems were commonly used into the 1970s, but they are unsafe and have limited effectiveness.

Multiple-Hood Systems. All hoods connected to a multiple-hood exhaust system must be on the same floor of the building to prevent fire spread through the duct from floor to floor. Preferably, there should be no walls requiring greater than a 1-h fire resistance rating between hoods.

The multiple-hood exhaust system must be designed (1) to prevent a fire in one hood or in the duct from spreading through the ductwork to another hood and (2) to protect against a fire starting in the common ductwork. Of course, the first line of protection for the ductwork is to keep it clean. Especially in a multiple-tenant system, a single entity must assume responsibility for cleaning the common ductwork frequently.

Each hood must have its own fire suppression system to protect the hood and cooking surface. A single system could serve more than one hood, but in the event of fire under one hood, the system would discharge its suppressant under all of the hoods served, resulting in unnecessary cleanup expense and inconvenience. A water mist system could serve multiple hoods if the sprinkler heads were allowed to operate independently.

Because of the possibility of a fire spreading through the ductwork from one hood to another, the common ductwork must have its own fire protection system. The appendices of NFPA *Standards* 17 and 17A present detailed examples of how the common ductwork can be protected either by one separate system or by a combination of separate systems serving individual hoods. Different types of fire suppression systems may be used to protect different portions of the exhaust system; however, in any case where two different types of system can discharge into the common duct at the same time, the agents must be compatible.

As always, actuation of the fire suppression system protecting any hood must shut off fuel or power to all cooking equipment under that hood. When a common duct, or portion thereof, is protected by a chemical fire suppression system, and it activates from a fire in a single hood, NFPA *Standards* 17 and 17A require shutoff of fuel or power to the cooking equipment under every hood served by that common duct, or portion of it protected by the activated system, even if there was no fire in the other hoods served by that duct.

From an operational standpoint, it is usually most sensible to provide one or more fire suppression systems to detect and protect against fire in the common ductwork and a separate system to protect each hood and its connecting ductwork. This would (1) prevent a fire in the common duct from causing discharge of fire suppressant under an unaffected hood and (2) allow unaffected hoods to continue operations in the event of a fire under one hood unless the fire spread to the common duct.

Preventing Fire Spread

The exhaust system must be designed and installed both to prevent a fire starting in the grease exhaust system from spreading to other areas of the building or damaging the building and to prevent a fire in one area of the building from spreading to other parts of the building via the grease exhaust system. This protection has two main aspects: (1) maintaining a clearance from the duct to other portions of the building and (2) enclosing the duct in a fire resistance rated enclosure. Both aspects are sometimes addressed by a single action.

Clearance to Combustibles. A grease exhaust duct fire can generate gas temperatures of 1100°C or greater in the duct. In such a grease fire, heat radiating from the hot duct surface can ignite combustible materials in the vicinity of the duct. Most codes require a minimum clearance of 460 mm from the grease exhaust duct to any combustible material. However, even 460 mm may not be sufficient clearance to prevent ignition of combustibles in the case of a major grease fire, especially in larger ducts.

Several methods to protect combustible materials from the radiant heat of a grease duct fire and permit reduced clearance to combustibles are described in NFPA *Standard* 96. Previous editions required that these protections be applied to the combustible material rather than to the duct, on the theory that if the ductwork is covered with an insulating material, the duct itself will get hotter than it would if it were free to radiate its heat to the surroundings. The hotter temperatures may result in structural damage to the duct. The 1994 edition of NFPA *Standard* 96 allows materials to be applied to the actual ductwork. However, because this edition may not have been accepted by some jurisdictions, the authority having jurisdiction should be consulted before clearances to combustible materials are reduced.

Listed grease ducts, typically double-wall ducts with or without insulation between the walls, may be installed with reduced clearance to combustibles in accordance with manufacturers' installation instructions, which should include specific information regarding the listing. Listed grease ducts are tested under UL *Standard* 1978.

NFPA *Standard* 96 also requires a minimum clearance of 75 mm to "limited combustible" materials (e.g., gypsum wallboard). At present, none of the standards and model codes requires any clearance to noncombustible materials except for enclosures.

Enclosures. Normally, when a duct penetrates a fire resistance rated wall or floor, a fire damper is used to maintain the integrity of the wall or floor. Since fire dampers may not be installed in a grease exhaust duct unless specifically approved for such use, there must be an alternate means of maintaining the integrity of rated wall or floor assemblies. Therefore, grease exhaust ducts that penetrate a fire resistance rated wall or floor-ceiling assembly must be continuously enclosed in a fire rated enclosure from the point the duct penetrates the first fire barrier until the duct leaves the building. Listed grease ducts are also subject to these enclosure requirements. The requirements are similar to those for a vertical shaft (typically 1-h rating if the shaft penetrates fewer than three floors, 2-h rating if the shaft penetrates three or more floors), except that the shaft can be both vertical and horizontal. In essence, the enclosure extends the room containing the hood through all of the other compartments of the building without creating any unprotected openings to those compartments.

Where a duct is enclosed in a rated enclosure, whether vertical or horizontal, clearance must be maintained between the duct and the shaft. The model codes vary considerably in this area: NFPA *Standard* 96 and BOCA require minimum 150-mm clearance. UMC requires 75-mm minimum and 300-mm maximum clearance. NFPA *Standard* 96 and UMC require the shaft to be vented to the outdoors. UMC requires that each exhaust duct have its own dedicated enclosure.

Some materials are available that are listed to serve as a fire resistance rated enclosure for a grease duct when used to cover a duct directly with minimal clear space between the duct and the material. These listed materials must be applied in strict compliance with the manufacturer's installation instructions, which may limit the size of duct to be covered, and which may specify required clearances for duct expansion and other installation details. When a duct is directly covered with an insulating material there is a greater chance of structural damage to the duct due to the heat of a severe fire. The structural integrity of any exhaust duct should be assessed after any serious duct fire.

Insulation materials that have not been specifically tested and approved for use as fire protection for grease exhaust ducts should not be used in lieu of rated enclosures or to reduce clearance to combustibles. Even insulation that is approved for other fire protection applications, such as to protect structural steel, may not be appropri-

ate for grease exhaust ducts because of the high temperatures that may be encountered in a grease fire.

Exhaust and Supply Fire-Actuated Dampers. Because of the risk that the damper may become coated with grease and become a source of fuel in a fire, balancing and fire-actuated dampers are not permitted at any point in a grease exhaust system except where specifically listed for each use or required as part of a listed device or system. Typically, fire dampers are found only at the hood collar and only if provided by the manufacturer as part of a listed hood.

Opinions differ regarding whether any fire-actuated dampers should be provided in the exhaust hood. On one hand, a fire-actuated damper at the exhaust collar may prevent a fire under the hood from spreading to the exhaust duct. However, like anything in the exhaust airstream, the fire-actuated damper and linkage may become coated with grease if not properly maintained, which may impede damper operation. On the other hand, without fire-actuated dampers the exhaust fan will draw smoke and fire away from the hood. While this cannot be expected to remove all of the smoke from the kitchen in the event of a fire, it can help to contain the smoke in the kitchen and minimize migration of smoke to other areas of the building.

A hood fire-actuated damper will generally close only in the event of a severe fire; most kitchen fires are extinguished before enough heat is released to trigger the fire-actuated damper. Thus the hood fire-actuated damper remains open during relatively small fires, allowing the hood to remove smoke, but can close in the event of a severe fire, helping to contain the fire in the kitchen area.

Fan Operations. If it is over 1 m^3/s, the makeup air supply to the kitchen is generally required to be shut down in case of fire to avoid feeding air to the fire. However, if the exhaust system is intended to operate during a fire to remove smoke from the kitchen (as opposed to just containing it in the kitchen), the makeup air system must operate as well. If the hood has an integral makeup air plenum, a fire-actuated damper must be installed in the makeup air collar to prevent a fire in the hood from entering the makeup air ductwork. NFPA *Standard* 96 details the instances where fire-actuated dampers are required in a hood makeup air assembly.

Regardless of whether fire-actuated dampers are installed in the exhaust system, NFPA *Standard* 96 calls for the exhaust fan to continue to run in the event of a fire unless fan shutdown is required by a listed component of the ventilating system or of the fire extinguishing system. Dry and wet chemical fire extinguishing systems protecting ductwork are tested both with and without airflow. Exhaust airflow is not necessary for proper operation of these systems.

Special Filtration Systems

Concerns about air quality have emphasized the need for higher efficiency grease extraction from the exhaust airstream than can be provided by filters or grease extractors in the exhaust hoods. Cleaner exhaust discharge to outdoors may be required by increasingly stringent air quality regulations or where the exhaust discharge configuration is such that grease or odors in the discharge would create a nuisance. In some cases, the exhaust air is cleaned so that it can be discharged indoors (e.g., through recirculating hoods). Several systems have been developed to clean the exhaust airstream, each of which presents special fire protection risks. These systems are as follows:

- Systems that use one or more stages of filtration to remove the grease from the airstream capture and collect the grease on the filters. In the event of a fire, grease-coated filters may become a source of fuel. Each bank of filters must be protected by an extinguishing system. Where filters with deep pleats, such as HEPA filters, are used, the extinguishing agent must be able to penetrate deep into the pleats to extinguish the fire.

- Systems that use electrostatic precipitators (ESPs) to clean the exhaust stream must also have appropriate fire protection. ESPs use electricity, which could be a source of ignition. ESPs in grease ducts should use self-limiting (cold spark) transformers to minimize the possibility of electrical fire. ESP plates must be cleaned frequently to prevent grease buildup and to maintain operating efficiency. ESPs are available with automatic washdown features that simplify the task of keeping the plates clean.

Where odor control is required in addition to grease removal, activated charcoal or other oxidizing bed filters are typically used downstream of the grease filters. Because much of the cooking odor is gaseous, and therefore not removed by air filtration, the filtration upstream of the charcoal filters must remove virtually all of the grease in the airstream in order to prevent grease buildup on the charcoal filters. However, charcoal filters themselves are flammable and must be protected from fire.

OPERATION AND MAINTENANCE

Operation and maintenance are closely related; operation creates the need for maintenance, and maintenance permits continued operation.

Operation

All components of the ventilation system are designed to operate in balance with each other, even under variable loads, to properly capture, contain, and remove the cooking effluent and heat and maintain proper temperature control in the spaces in the most efficient and economical manner. Deterioration in any of these components will unbalance the system, affecting one or more of its design concepts. The system design intent should be fully understood by the operator so that any deviations in the operation can be noted and corrected. In addition to creating health and fire hazards, normal cooking effluent deposits can also unbalance the system, so they must be regularly removed.

All the various components of the exhaust and makeup air systems affect proper capture, containment, and removal of the cooking effluent. In the exhaust system this includes the cooking equipment itself, the exhaust hood, all filtration devices, the ductwork, the exhaust fan, and any dampers and louvers that might be in the exhaust system. In the makeup air system this includes the air-handling unit(s), with the intake louvers, dampers, filters, fan wheels, heating and cooling coils, ductwork, and supply registers. In systems that obtain their makeup air from the general HVAC system for the space, this also includes the return air registers and ductwork.

When the system is first set up and balanced in new condition, these components are set to optimum efficiency. In time, all of the components become dirty; filtration devices, dampers, louvers, heating and cooling coils, and ducts become restricted; fan blades change shape as they accumulate dirt and grease; and fan belts become loose. In addition, dampers can come loose and change position, even closing, and ductwork can develop leaks or be shut off when internal insulation sheets fall down.

All of these changes deteriorate the performance of the system. For this reason the operator should know how the system performed when it was new, in order to better recognize when it is no longer performing the same way. Because of this knowledge, problems can be found and corrected sooner, and the peak efficiency and safety of the system operation can better be maintained.

Maintenance

Maintenance may be classified as preventive or emergency (breakdown). Preventive maintenance keeps the system operating in the most narrow range of variation (deviation from optimal performance). Preventive maintenance provides for continued optimal performance of the system, including the maximum production and least shutdown. It is the most effective maintenance and is preferred.

Preventive maintenance is able to prevent most emergency shutdown and maintenance of the system. Preventive maintenance has a modest ongoing cost and fewer unexpected costs. It is clearly the lowest cost maintenance in the long run because it keeps the system components in peak condition, which extends the operating life of all components.

Emergency maintenance must be applied when a breakdown occurs. Adequate staffing and money must be applied to the situation to bring the system back on line in the shortest possible time. Such emergencies can be of almost any nature. They are impossible to predict or address in advance, except to presume the type of component failures that could shut the system down and keep spares of these components on hand, or readily accessible, so they can be quickly replaced. Preventive maintenance, which includes regular inspection of critical system components, is the most effective way to avoid emergency maintenance.

Following are brief descriptions of typical operations of various components of kitchen ventilation systems and the type of maintenance and cleaning that would be applied to the abnormally operating system to bring it back to normal. Many nontypical operations are not listed here.

Cooking Equipment:

Normal Operation. Produces properly cooked product, of correct temperature, within expected time period. Minimum smoke during cooking cycles.

Abnormal Operation. Produces undercooked product, of lower temperature, with longer cooking times. Increased amount of smoke during cooking cycles.

Cleaning/Maintenance. Clean solid cooking surfaces between each cycle if possible, or at least once a day. (Baked-on product insulates and retards heat transfer). Filter frying medium daily and change it on schedule recommended by supplier. Check that (1) fuel source is at correct rating, (2) thermostats are correctly calibrated, and (3) conditioned air is not blowing on cooking surface.

Exhaust Systems:

Normal Operation. All cooking vapors are readily drawn into the exhaust hood, where they are captured and removed from the space. The ambient environment immediately around the cooking operation is clear and fresh.

Abnormal Operation. Many cooking vapors do not enter the exhaust hood at all, and some that enter subsequently escape. The ambient environment around the cooking operation, and likely in the entire kitchen, is contaminated with cooking vapors.

Cleaning/Maintenance. Clean all grease removal devices in the exhaust system. Hood filters should be cleaned at least daily. For other devices, follow the minimum recommendations of the manufacturer; even these may not be adequate in very high volumes or with products producing large amounts of effluent. Check that (1) all dampers are in their original position, (2) fan belts are properly tensioned, (3) the exhaust fan is operating at the proper speed and turning in the proper direction, (4) the exhaust duct is not restricted, and (5) the fan blades are clear.

NFPA *Standard* 96 design requirements for access to the system should be followed to facilitate cleaning of the exhaust hood, ductwork, and fan. If the cleaner cannot get to parts of the system, these parts of the system will not be cleaned. Cleaning should be done before the grease has built up to 6 mm in any part of the system. Cleaning should be by a method that cleans to bare metal, and the metal should be left clean. Cleaning agents should be thoroughly rinsed off, and all loose grease particles should be removed, as they can ignite more readily. Agents should not be added to the surface after cleaning, as their textured surfaces merely collect more grease more quickly. Fire suppression systems may need to be disarmed prior to cleaning to prevent accidental discharge and then reset by authorized personnel after cleaning. All access panels removed must be reinstalled after cleaning, making sure that proper gasketing is in place to prevent grease leaks and escape of fire.

Supply, Makeup, and Return Air Systems:

Normal Operation. The ambient environment in the kitchen area is clear, fresh, comfortable, and free of drafts and excessive air noise.

Abnormal Operation. The ambient environment in the kitchen is smoky, choking, hot, and humid, and it can be very drafty with excessive air noise.

Cleaning/Maintenance. Check that the makeup air system is operating and that it is providing the correct amount of air to the space. If it is not, the exhaust system cannot operate properly. Check that the dampers are set correctly, the filters and exchangers are clean, the belts are tight, the fan is turning in the correct direction, and the supply and return ductwork and registers are open with the supply air discharging in the correct direction and pattern. If drafts persist, the system may need to be rebalanced. If noise persists in a balanced system, system changes may be required.

Filter cleaning or changing frequency varies widely depending on the quantity of airflow and the contamination of the local air. Once determined, the cleaning schedule must be maintained regularly.

With makeup air systems, the air-handling unit, coils, and fan are usually cleaned in spring and fall, at the beginning of the seasonal change. More frequent cleaning or better quality filtering may be required in some contaminated environments. Duct cleaning for the system is on a much longer cycle, but local codes should be checked as stricter requirements are invoked. Ventilation systems should be cleaned by professionals to ensure that none of the expensive system components are damaged. Cleaning companies should be required to carry adequate liability insurance. The International Kitchen Exhaust Cleaning Association (IKECA) and the National Air Duct Cleaners Association (NADCA), both located in Washington, D.C., can provide both descriptions of proper cleaning and inspection techniques and lists of their members.

Control Systems (Operation and Safety):

Normal Operation. Control systems do not permit the cooking equipment to operate unless both the exhaust and makeup air systems are operating and the fire suppression system is armed. With multiple exhaust and makeup air systems, the controls maintain the proper balance as cooking equipment is turned on and shut off. In the event that a fire suppression system operates, the energy source to the cooking equipment it serves is shut off. On ducted systems, the exhaust fan usually keeps running to take any fire and smoke out of the building. On ductless systems, the fan may or may not keep running, but a discharge damper closes to keep the flames away from the ceiling. The makeup air system may continue to run, or it may be shut off by a separate local area fire and smoke sensor. If the control system does not operate in this way, changing to this operation should be considered.

Abnormal Operation. The cooking equipment can operate when the exhaust and supply are turned off, perhaps because the fire suppression system is unarmed or has been bypassed. When extra cooking systems are turned on or off, the operator must remember to manually turn the exhaust and makeup air fans on or off as well.

When the exhaust and makeup air systems are not interlocked, the system can be out of balance. This can cause many of the abnormal operations described with the previously mentioned systems. With gas-fired cooking equipment, the fire suppression system may have a false discharge if the exhaust system is not operating. Whenever cooking is permitted with the fire suppression system inoperable, there is a great chance of a serious fire, and the operator is liable because insurance usually does not cover this situation.

Cleaning/Maintenance. Cleaning is usually restricted to the mechanical operators and electrical sensors that are exposed to the grease environment. These would be in the fire suppression system and within the hood. If they become excessively coated with grease, mechanical operators cannot move and sensors cannot sense. The result is a decrease in control and safety. The mechanical operators should be cleaned as often as required to maintain free movement. Fire suppression system operators may need to be changed annually rather than cleaned. Check with a local fire suppression system dealer before attempting to clean any of these components.

Maintenance of control systems (presuming they were properly designed) is mostly a matter of checking the performance of the entire system on a regular basis (minimum every three months) to be sure it is still performing as designed. Mechanical linkages on dampers and cleanliness of sensors should be checked regularly (minimum once a month). All electrical screw terminals in the components should be checked for tightness, relay contacts should be checked for cleanliness, and all exposed conducting surfaces should be checked for corrosion on an annual basis.

RESIDENTIAL HOODS AND KITCHEN VENTILATION FANS

Even though commercial and residential cooking processes are similar, ventilation requirements and procedures for the two are different. These differences include exhaust airflow rate and installation height. In addition, residential kitchen ventilation has no makeup air requirement and energy consumption is insignificant.

Residential Cooking Equipment and Processes

Although the physics of the cooking process and the resulting effluent are about the same, residential cooking is done more conservatively. Cooking equipment such as upright broilers and solid fuel-burning equipment, described in the section on Hoods as Group 3 (heavy duty) and Group 4 (extra heavy duty), is not used. Therefore, the high rates of ventilation and the ventilation equipment associated with them are not used in residential kitchens. On the other hand, cooking effluent and by-products of open flame combustion must be more closely controlled in a residence than in a commercial kitchen because of the greater sensitivity to effluent and the lower background ventilation rate. This situation presents a different challenge because it is not possible to overcome problems by simply increasing the ventilation rate at the cooking process.

Residential cooking always produces a convective plume, which is driven by the heat rising from the cooking surface. The plume carries upward with it both the cooking effluent and any by-products of combustion. Residential kitchen hoods depend almost entirely on buoyancy capture. Sometimes there is spatter as well, but those particles are so large that they are not removed by ventilation.

Hoods and Other Residential Kitchen Ventilation Equipment

Most residential kitchens are ventilated by wall-mounted, conventional range hoods. Overall, they do the best job at the lowest installed cost. A variation is the deep decorative canopy hood, which is usually chosen for aesthetic reasons. The canopy hood is significantly more costly, but somewhat more effective. Downdraft range-top ventilators are currently popular in higher priced homes. These are an exception because they are designed to capture the contaminants by producing velocities greater than those of the convective plume over the cooking surface. With sufficient velocity, their operation can be satisfactory, but with gas cooking velocity is limited to prevent an adverse effect on the flame.

Kitchen exhaust fans were more widely used in the past, but they are still used and probably will be for several years. Mounted in the wall or ceiling, they ventilate the kitchen generally rather than at the source of contaminant. For general ventilation in all possible kitchen arrangements, 15 air changes per hour are typical; for ceiling-mounted fans, this is usually sufficient, but for wall-mounted fans, it may be marginal.

Continuous low-level ventilation throughout the house is increasingly used as houses are built tighter and the public awareness of the importance of good indoor air quality grows. This continuous ventilation is for a distinctly different purpose than kitchen ventilation. ASHRAE *Standard* 62 recommends 0.35 air changes per hour but not less than 8 L/s per person. Some installations with equipment for continuous ventilation use that same equipment for kitchen ventilation. The systems work by intermittently boosting the flow to a considerably higher level or by simply running longer to achieve the needed reduction in concentration. These dual-purpose systems require good design for satisfactory operation and seldom achieve the results of a conventional range hood.

Differences Between Commercial and Residential Equipment

Residential hoods usually meet UL *Standard* 507 requirements. Fire-actuated dampers are never part of the hood and are almost never used. Grease filters in residential hoods are much simpler; grease collection through channels is undesirable because inadequate maintenance could result in a pool of grease that would be a fire and health hazard.

Conventional residential wall hoods are usually produced in standard height (from 150 to 230 mm or 300 mm) and standard depth (430 to 560 mm). Width is usually the same as the cooking surface; 760 mm is almost a standard in the United States. The current HUD Manufactured Housing Standard calls for 75 mm overhang per side.

Mounting height is usually 460 to 610 mm, with some mounted even higher at a sacrifice in collection efficiency. The lower the hood, the more effective the capture; the velocity of the convective plume is not great, and the plume is easily disturbed by air currents and disturbances. For suitable access to the cooking surface, studies show that 460 mm is a minimum height. Some codes call for a minimum of 760 mm from the cooking surface to combustible cabinets. In that case, the bottom of a 150-mm hood can be 610 mm above the cooking surface.

A minimum flow rate (exhaust capacity) of 60 L/s per linear metre of hood width is recommended by the Home Ventilating Institute (HVI). Informal tests were conducted in which a user of a hood was told to keep adjusting the variable speed control on the hood until it was running slow enough to minimize the sound level while still maintaining good performance. After a trial period, the chosen setting was evaluated; most users selected a speed providing flow within 15% of the recommended minimum 60 L/s per linear metre. Although these tests were too few to be statistically valid and were quite informal, the outcome was predicted. Additional capacity by means of speed control is desirable to support unusually vigorous cooking and cooking mistakes. The airflow can be increased for a brief period to clear the air.

Residential Exhaust Systems

Residential hoods offer little opportunity for custom design of an exhaust system. A duct connector is built into the hood, and the installation should use the same size duct. A hood includes either an axial or a centrifugal fan. The centrifugal fan has higher pressure capabilities, but for low-volume hoods, the axial fan is usually adequate. Virtually all residential hoods on the market in the United States have airflow performance certified by HVI.

Makeup Air

The exhaust rate of residential hoods is generally low enough to avoid the need for makeup air systems, so they rely on natural infiltration. This may result in slight negative pressurization of the residence, but it is usually less than that caused by other equipment such as the clothes dryer. Still, the penalties for backdrafts through the flue of a combustion appliance are high. Unless a gas furnace and

water heater are of the sealed combustion type, the flue should be checked for adequate flow with all exhaust devices turned on.

Sometimes commercial-style cooking equipment with associated higher ventilation requirements is installed in residences. In such cases, designers should refer to the earlier sections of this chapter, possibly including the section on Makeup Air Systems.

Energy Conservation

Energy costs of residential hoods are quite low because of the few hours of running time and the low rate of exhaust. For example, it typically costs less than $10 per heating season in Chicago to run a hood and heat the makeup air, based on running at 70 L/s for an hour a day and using gas heat.

Fire Protection and Codes for Residential Hoods

Residential hoods must be installed with metal (preferably steel) duct, and the duct should be positioned to avoid pooling of grease. Residential hood exhaust ducts are almost never cleaned, and there is no evidence of fires caused by that. Several attempts have been made to make fire extinguishers available in residential hoods, but none has met with broad acceptance. In contrast, residential grease fires on the cooking surface are still a problem. Although the fires always ignite at the cooking surface when cooking grease is left unattended and overheats, the hood is sometimes blamed. That problem is not solved with fire extinguishing equipment.

Maintenance of Residential Kitchen Ventilation Equipment

All UL Listed hoods and kitchen exhaust fans have provisions for cleaning, which should be done at intervals consistent with the cooking practices of the user. Although cleaning is often thought to be for fire prevention, it also benefits health by removing nutrients available for the growth of organisms.

BIBLIOGRAPHY

Abrams, S. 1989. Kitchen exhaust systems. *The Consultant* (FCSI) 22(4).

Alereza, T. and J.P. Breen. 1984. Estimates of recommended heat gains due to commercial appliances and equipment. *ASHRAE Transactions* 89(2A): 25-58.

Marn, W.L. 1962. Commercial gas kitchen ventilation studies. *Research Bulletin* 90. American Gas Association Laboratories, Cleveland, OH.

Annis, J.C., and P.J. Annis. 1989. Size distributions and mass concentrations of naturally generated cooking aerosols. *ASHRAE Transactions* 95(1): 735-43.

ASHRAE. 1989. Ventilation for acceptable indoor air quality. *Standard* 62-1989.

ASTM. 1990. Standard specification for food service equipment hoods for cooking appliances. *Standard* F1299-90. American Society for Testing and Materials, Philadelphia.

Balogh, E.A. 1989. Tandem wet/dry chemical cooking area fire extinguishing concept. *The Consultant* (FCSI) 22(4).

Bevirt, D.W. 1994. What engineers need to know about testing and balancing. *ASHRAE Transactions* 100(1).

Black, D.K. 1989. Commercial kitchen ventilation—Efficient exhaust and heat recovery. *ASHRAE Transactions* 95(2):780-86.

Capalbo, F.J. 1991. Efficient kitchen ventilation: Grease filters are the key. *The Consultant*.

Chih-Shan, L., L. Wen-Hai, and J. Fu-Tien. 1993. Removal efficiency of particulate matter by a range exhaust fan. *Environmental International* 19:371-80.

Chih-Shan, L., L. Wen-Hai, and J. Fu-Tien. 1993. Size distributions of submicrometer aerosols from cooking. *Environmental International*.

Cini, J.C. 1989. Innovative kitchen exhaust systems. *The Consultant* (FCSI) 22(4).

Claar, C.N., R.P. Mazzucchi, and J.A. Heidell. 1985. The project on restaurant energy performance (PREP)—End use monitoring and analysis. Office of Building Energy Research and Development. U.S. Department of Energy, Washington, DC.

Conover, D.R. 1992. Building construction regulation impacts on commercial kitchen ventilation and exhaust systems. *ASHRAE Transactions* 98(1):1227-35.

Crabtree, S. 1989. Ventilation: Where less is better. *The Consultant* (FCSI) 22(4).

Dallavalle, J. 1953. Design of kitchen range hoods. *Heating and Ventilating* (August).

de Albuquerue, A.J. 1972. Equipment loads in laboratories. *ASHRAE Journal* 14(9):59-62.

Department of Mechanical Engineering, University of Manitoba-BETT Program. 1984. Analysis of commercial kitchen exhaust systems design and energy conservation. Report prepared for Energy, Mines and Resources Canada.

Deppisch, J.R. and L.J. Irwin. 1974. Energy efficiencies of gas and electric range top sections. *Research Report* 1504. American Gas Association Laboratories, Cleveland, OH.

Donnelly, F. 1966. Preventing fires in kitchen ventilating ducts. *Air Conditioning, Heating and Ventilating* (December).

Drees, K.H., J.D. Wenger, and G. Janu. 1992. Ventilation airflow measurement for ASHRAE *Standard* 62-1989. *ASHRAE Journal* 34(10):40.

Electrification Council. 1975. New electric kitchen ventilation design criteria—Leader's guide. The Electrification Council, New York.

Elovitz, G. 1992. Design considerations to master kitchen exhaust systems. *ASHRAE Transactions* 98(1):1199-1213.

EPRI. 1989. Assessment of building codes, standards and regulations impacting commercial kitchen design. *Final Report* for National Conference of States on Building Codes and Standards. Electric Power Research Institute, Palo Alto, CA.

EPRI. 1989. Analysis of building codes for commercial kitchen ventilation systems. *Final Report* CU-6321 for National Conference of States on Building Codes and Standards. Electric Power Research Institute, Palo Alto, CA.

EPRI. 1991. Proposed testing protocols for commercial kitchen ventilation research. *Final Report* CU-7210 for Underwriters Laboratories. Electric Power Research Institute, Palo Alto, CA.

Esmen, N.A., D.A. Wegel, and F.P. McGuigan. 1986. Aerodynamic properties of exhaust hoods. *American Industrial Hygiene Association Journal* 47(8).

Farnsworth, C., A. Walters, R.M. Kelso, and D. Fritzsche. 1989. Development of a fully vented gas range. *ASHRAE Transactions* 95(1).

Farnsworth, C.A., and D.E. Fritzsche. Improving the ventilation of residential gas ranges. American Gas Association Laboratories, Cleveland, OH.

Fitzgerald, H. 1969. Code reduces kitchen exhaust fires. *Air Conditioning, Heating and Ventilating* (May).

Frey, D.J., C.N. Claar, and P.A. Oatman. 1989. The model electric restaurant. *Final Report* CU-6702. Electric Power Research Institute, Palo Alto, CA.

Frey, D.J., K.F. Johnson, and V.A. Smith. 1993. Computer modeling analysis of commercial kitchen equipment and engineered ventilation. *ASHRAE Transactions* 99(2).

Fritz, R.L. 1989. A realistic evaluation of kitchen ventilation hood designs. *ASHRAE Transactions* 95(1):769-79.

Fugler, D. 1989. Canadian research into the installed performance of kitchen exhaust fans. *ASHRAE Transactions* 95(1):753-58.

Garrett, R.T. 1984. Commercial/Institutional kitchen and make-up air analysis: Design and selection application guide. Muckler Industries, St. Louis, MO.

Garrett, R.T. 1989. Are you willing to save money on heating costs just to blow your air conditioning out the exhaust? *The Consultant* (FCSI) 22(4).

Gordon, E.B. and N.D. Burk. 1993. A two-dimensional finite-element analysis of a simple commercial kitchen ventilation system. *ASHRAE Transactions* 99(2).

Greenheck Fan Corporation. 1981. Cooking equipment ventilation—Application and design. CEV-5-82. Greenheck Fan Corporation, Schofield, WI.

Guffey, S.E. 1992. A computerized data acquisition and reduction system for velocity traverses in a ventilation laboratory. *ASHRAE Transactions* 98(1):98-106.

Hane, A.E. 1989. Discussion of ventilation and exhaust hoods. *The Consultant* (FCSI) 22(4).

Hicks, B. 1989. Who wants ventilation anyway? And why? *The Consultant* (FCSI) 22(4).

Himmel, R.L. 1983. Effects of ventilation on equipment performance. *Commercial Cooking Equipment Improvement*, Volume IV. GRI-80/0079.4 Final Report. American Gas Association Laboratories, Cleveland, OH.

Horton, D.J., J.N. Knapp, and E.J. Ladewski. 1993. Combined impact of ventilation rates and internal heat gains on HVAC operating costs in commercial kitchens. *ASHRAE Transactions* 99(2).

Hunt, C.M., D.R. Showalter, and S.J. Treado. 1974. Tests of a grease interceptor similar to those used in galleys. Center for Building Technology, NBS, Washington, DC.

Jamboretz, J.S. 1982. Before you specify a "no-heat" hood. *The Consultant* (FCSI). 15(2):40.

Jin, Y. and J.R. Ogilvie. 1992. Isothermal airflow characteristics in a ventilated room with a slot inlet opening. *ASHRAE Transactions* 98(2):296-306.

Kelso, R.M., A.J. Baker, and S. Roy. An efficient CFD algorithm for prediction of contaminant dispersion in room air motion. University of Tennessee, Knoxville.

Kelso, R.M., L.E. Wilkening, E.G. Schaub, and A.J. Baker. 1992. Computational simulation for kitchen airflows with commercial hoods. *ASHRAE Transactions* 98(1):1219-26.

Kim, I.G. and H. Homma. 1992. Possibility for increasing ventilation efficiency with upward ventilation. *ASHRAE Transactions* 98(1):723-29.

Knapp, J. 1989. Ventilation in the food service industry. *The Consultant* (FCSI) 22(4).

Knapp, J.N., and W.A. Cheney. 1993. Development of high-efficiency air cleaners for grilling and deep-frying operations. *ASHRAE Transactions* 99(2).

Kuehen, T.H., J. Ramsey, H. Han, M. Perkovich, and S. Youssef. 1989. Study of kitchen range exhaust systems. *ASHRAE Transactions* 95(1):744-52.

Laderoute, M. 1989. Under the hood (don't let fire put you out of business). *The Consultant* (FCSI) 22(4).

Leipman, M.A. 1989. Who do you trust? *The Consultant* (FCSI) 22(4).

Levenback, G. 1989. The kitchen ventilation troika: Underwriters Laboratories, National Fire Protection Association, and Council of American Building Officials. *The Consultant* (FCSI) 22(4).

Lockhart, C. 1989. Restaurant ventilation: Our environmental concerns. *The Consultant* (FCSI) 22(4).

Locklin, D.W., R.D. Giammer, and S.G. Talbert. 1971. Preliminary study of ventilation requirements for commercial kitchens. *ASHRAE Journal* 13(11):51.

Loving, W.H. 1964. Make-up air for commercial kitchens. *Air Engineering* (December).

McGuire, A.B. 1993. Commercial and institutional kitchen exhaust systems. *Heating, Piping and Air Conditioning* (May)

Murakami, S., S. Kato, T. Tanaka, D.-H. Choi, and T. Kitazawa. 1992. The influence of supply and exhaust openings on ventilation efficiency in an air-conditioned room with a raised floor. *ASHRAE Transactions* 98(1):738-55.

NFPA. 1980. Standard for ventilation control and fire protection of commercial cooking operations. *Standard* 96-94. National Fire Protection Association, Quincy, MA.

Niemeier, R.E. 1989. Acoustical considerations for commercial kitchen ventilators. *The Consultant* (FCSI) 22(4).

Otenbaker, J. 1989. Parting the curtain on ventilation codes: A challenge for specifiers and manufacturers. *The Consultant* (FCSI) 22(4).

Parikh, J.S. 1992. Testing and certification of fire and smoke dampers. *ASHRAE Journal* 34(11).

Parikh, J.S. 1989. UL's certification program for restaurant ventilation equipment. *The Consultant* (FCSI) 22(4).

Pekkinen, J. and T.H. Takki-Halttunen. 1992. Ventilation efficiency and thermal comfort in commercial kitchens. *ASHRAE Transactions* 98(1):1214-18.

Riemenschneider, J. 1989. Ventilation and heat reclaim systems. *The Consultant* (FCSI) 22(4).

Romano, S. 1989. Kitchen cooking equipment ventilation is a lot of hot air. *The Consultant* (FCSI) 22(4).

Schaelin, D., J. van der Maas, and A. Moser. 1992. Simulation of airflow through large openings in buildings. *ASHRAE Transactions* 98(2):319-28.

Seller, J. and B. Ward. 1991. The environment and you, Part 1. *Foodservice Equipment & Supplies Specialist.*

Shibata, M., R.H. Howell, and T. Hayashi. 1982. Characteristics and design method for push-pull hoods: Part 2—Streamline analyses of push-pull flows. *ASHRAE Transactions* 88(1):557-570.

Shute, R.W. 1992. Integrating access floor plenums for HVAC air distribution. *ASHRAE Journal* 34(10).

Snyder, O.P., D.R. Thompson, and J.F. Norwig. 1983. Comparative gas/electric food service equipment energy consumption ratio study. University of Minnesota, St. Paul.

Soling, S.P. and J. Knapp. 1985. Laboratory design of energy efficient exhaust hoods. *ASHRAE Transactions* 91(1B):383-92.

Talbert, S.G., L.J. Flanigan, and J.A. Eibling. 1973. An experimental study of ventilation requirements of commercial electric kitchens. *ASHRAE Transactions* 79(1):34-47.

UL. 1990. Standard for safety exhaust hoods for commercial cooking equipment, 4th ed. *Standard* 710-90. Underwriters Laboratories, Northbrook, IL.

UL. 1991. Commercial electric cooking appliances with recirculative systems. *Supplement* 197. Underwriters Laboratories, Northbrook, IL.

UL. 1991. Outline of investigation for power ventilation for restaurant exhaust appliances. *Subject* 762-89, Issue No. 1. Underwriters Laboratories, Northbrook, IL.

UL. 1992. Standard for safety fire extinguishing systems for protection of restaurant cooking areas. *Standard* 300-92f. Underwriters Laboratories, Northbrook, IL.

VDI Verlag GmbH. 1984. Raumlufttechnische Anlagen für Küchen (Ventilation equipment for kitchens). VDI 2052.

Wolbrink, D.W. and J.R. Sarnosky. 1992. Residential kitchen ventilation—A guide for the specifying engineer. *ASHRAE Transactions* 91(1):1187-98.

Wood, C.A. 1989. Hot air. *The Consultant* (FCSI) 22(4).

Zhang, J. S., G.J. Wu, and L.L. Christianson. 1992. Full-scale experimental results on the mean and turbulent behavior of room ventilation flows. *ASHRAE Transactions* 98(2):307-18.

Zimmerman, B.A. 1989. Fire dampers…facts and fallacies. *The Consultant* (FCSI) 22(4).

GEOTHERMAL ENERGY

THE use of geothermal resources can be broken down into three general categories: high-temperature (>150°C) electric power production, intermediate- and low-temperature direct-use applications (<150°C), and ground-source heat pump applications (generally <32°C). This chapter covers only the direct-use and ground-source heat pump categories. After an overview of resources, the chapter is divided into two sections. The section on Direct-Use Applications contains information on wells, equipment, and applications. The section on Ground-Source Heat Pump Systems includes information only on the ground-source portion of systems. Information on system design within the building may be found in Chapter 8 of the 1992 *ASHRAE Handbook—Systems and Equipment.*

RESOURCES

Geothermal energy is the thermal energy within the earth's crust—the thermal energy in rock and the fluid (water, steam, or water containing large amounts of dissolved solids) that fills the pores and fractures within the rock, sand, and gravel. Calculations show that the earth, originating from a completely molten state, would have cooled and become completely solid many thousands of years ago without an energy input in addition to that of the sun. It is believed that the ultimate source of geothermal energy is radioactive decay within the earth (Dullard 1973).

Through plate motion and vulcanism, some of this energy is concentrated at high temperature near the surface of the earth. Energy is also transferred from the deeper parts of the crust to the earth's surface by conduction and by convection in regions where geological conditions and the presence of water permit.

Because of variation in volcanic -activity, radioactive decay, rock conductivities, and fluid circulation, different regions have different heat flows (through the crust to the surface), as well as different temperatures at a particular depth. The normal increase of temperature with depth (i.e., the normal geothermal gradient) is about 24 K per kilometer of depth, with gradients of about 9 to 48 K/km being common. The areas that have higher temperature gradients and/or higher-than-average heat flow rates constitute the most interesting and viable economic resources. However, areas with normal gradients may be valuable resources if certain geological features are present.

Geothermal resources of the United States are categorized into the following types:

- Igneous point sources
- Deep convective circulation in areas of high regional heat flow
- Geopressured resources
- Concentrated radiogenic heat sources
- Deep regional aquifers in areas of near-normal gradient

Igneous point sources are associated with magma bodies, which result from volcanic activity. These bodies heat the surrounding and overlying rock by conduction and convection as permitted by the rock permeability and the fluid content of the rock pores.

Deep circulation of water in areas of high regional heat flow can result in hot fluids near the surface of the earth. Known as *hydrothermal convection systems*, these geothermal resources are widely used. The fluids near the surface have risen from natural convection circulation between the hotter, deeper formation and the cooler formations near the surface. The passageway that provides for this deep circulation must consist of adequately permeable fractures and faults.

The geopressured resource, widely present in the Gulf Coast area, consists of regional occurrences of confined hot water in deep sedimentary strata, where pressures of greater than 75 MPa are common. This resource also contains methane, which is dissolved in the geothermal fluid.

Radiogenic heat sources exist in various regions as granitic plutonic rocks that are relatively rich in uranium and thorium. These plutons have a higher heat flow than the surrounding rock; if the plutons are blanketed by sediments of low thermal conductivity, elevated temperatures at the base of the sedimentary section can result. This resource has been identified in the eastern United States.

Deep regional aquifers of commercial value can occur in deep sedimentary basins, even in areas of only normal temperature gradient. For deep aquifers to be of commercial value, (1) the basins must be deep enough to provide usable temperature levels at the prevailing gradient, and (2) the permeabilities within the aquifer must be adequate for flow.

The thermal energy in geothermal resource systems exists primarily in the rocks and only secondarily in the fluids that fill the pores and fractures within them. Thermal energy is usually extracted by bringing to the surface the hot water or steam that occurs naturally in the open spaces in the rock. Where rock permeability is low, the energy extraction rate is low. To extract the thermal energy from the rock itself, water must be recharged into the system as the initial water is extracted. In permeable aquifers, or where there are natural fluid conductors, the produced fluid may be injected back into the aquifer at some distance from the production well to pass through the aquifer again and recover some of the energy in the rock.

Temperatures

The temperature of fluids produced in the earth's crust and used for their thermal energy content varies from below 4°C to 360°C. As indicated in Figure 1, local gradients also vary with geologic conditions. The lower value represents the fluids used as the low-temperature energy source for heat pumps, and the higher temperature represents an approximate value for the HGP-A well at Hilo, Hawaii.

The following system of classifying resources by temperature level is used in the geothermal industry:

The preparation of this chapter is assigned to TC 6.8, Geothermal Energy Utilization.

High temperature	$t > 150°C$
Intermediate temperature	$90°C < t < 150°C$
Low temperature	$t < 90°C$

Electricity generation is generally not economically feasible for resources with temperatures below about 150°C, which is the reason for the division between high- and intermediate-temperature systems. However, binary power plants, with the proper set of circumstances, have demonstrated that it is possible to generate electricity economically above 110°C. In 1988, there were 86 binary plants worldwide, generating a total of 126.3 MW (Di Pippo 1988).

The 90°C division between intermediate and low temperatures is common in resource inventories, but it is somewhat arbitrary. At 90°C and above, applications such as district heating can be readily implemented with equipment used in conventional applications of the same type; at lower temperatures, these applications require redesign to take the greatest advantage of the geothermal resource.

Geothermal systems at lower temperature levels are more common. The frequency by reservoir temperature of identified convective systems at temperatures above 90°C is shown in Figure 2.

Geothermal Fluids

Geothermal energy is currently extracted from the earth through the naturally occurring fluids in rock pores and fractures; in the future, however, an additional fluid may be introduced into the geothermal system and circulated through it to recover the energy. The fluids being produced are steam, hot water, or a two-phase mixture of both. These may contain various amounts of impurities, notably dissolved gases and dissolved solids.

Geothermal systems that produce essentially dry steam are *vapor dominated*. While these systems are valuable resources, they are rare. Hot water (*fluid-dominated*) systems are much more common than vapor-dominated systems and can be produced either as hot water or as a two-phase mixture of steam and hot water, depending on the pressure maintained on the production system. If the pressure in the production casing or in the formation around the casing is reduced below the saturation pressure at that temperature, some of the fluid will flash, and a two-phase fluid will result. If the pressure is maintained above the saturation pressure, the fluid will remain single-phase. In fluid-dominated systems, both dissolved gases and dissolved solids are significant. The quality of the fluid varies from site to site, from a few hundred ppm to over 300 000 ppm dissolved solids.

Table 1 presents the composition of fluids from a number of geothermal wells in the United States. The list illustrates the types of substances and the range of concentrations that can be expected in the fluids. The harshness of the fluid is not only site-dependent but also temperature-dependent; harshness increases with temperature. This chapter concentrates on systems produced as a single-phase hot liquid.

Present Use

Discoveries of concentrated radiogenic heat sources and deep regional aquifers in areas of near-normal temperature gradient indicate

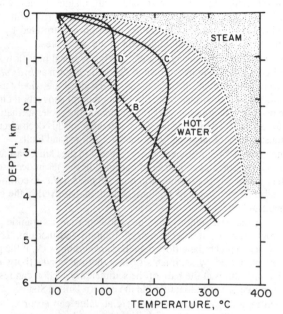

A Near-normal temperature gradient
B High conductive gradient
C and D Temperature resulting from convective flow

Fig. 1   Representative Temperature-Depth Relations in the Earth's Crust
(Combs et al. 1980)

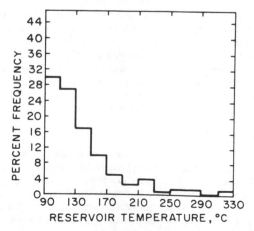

Fig. 2   Frequency of Identified Hydrothermal Convection Systems by Reservoir Temperature
(Muffler et al. 1980)

Table 1   Representative Fluid Compositions from Geothermal Wells in Various Resource Areas of the United States

Material	Concentrations, ppm		
	Boise, ID[a]	Klamath Falls, OR[b]	Salton Sea, CA
Total dissolved solids	290	795	220 000
SiO_2	160	48	350
Na^+	90	205	5 100
K^+	1.6	4.3	12 500
Ca^{2+}	1.7	26	23 000
Mg^{2+}	0.05	—	150
Cl^-	10	51	133 000
F^-	14	1.5	13
SO_4^{2-}	23	330	5
NO_3^-	—	4.9	—
NH_4^+	—	1.3	—
H_2S	trace	1.5	—
HCO_3^-	70	20	7 025
CO_3^{2-}	4	15	—
CO_2	0.2	—	—
B^{3+}	0.14	—	350
Fe^{2+}, Fe^{3+}	0.13	0.3	1 300
O_2	0.0029	0.2	—
Temperature, °C	80	89	250

Source: Boise and Salton Sea data from Cosner and Apps (1978); Klamath Falls data from Ellis and Conover (1981).
[a]Well name unknown, but near old penitentiary.   [b]Wendling Well (Lund et al. 1976).

that 37 states in the United States have economically exploitable direct-use geothermal resources (Interagency Geothermal Coordinating Council 1980).

The geysers resource area in northern California is the largest single geothermal development in the world. The total electricity generated by geothermal development in the world was 5175 MW in 1988 (Di Pippo 1988). The direct application of geothermal energy for space heating and cooling, water heating, agricultural growth-related heating, and industrial processing represented about 8600 MW worldwide in 1988 (Lienau et al. 1988 and Gudmundsson 1985). In the United States in 1988, direct-use installed capacity amounted to 1670 MW, providing 4.98×10^6 MWh (Lienau et al. 1988).

The major uses of geothermal energy in agricultural growth applications are for heating greenhouse and aquaculture facilities. The principal industrial uses of geothermal energy in the United States are for food processing (dehydration) and gold processing. Worldwide, the main applications include space and water heating, space cooling, agricultural growth, and food processing. Exceptions are diatomaceous earth processing in Iceland and pulp and paper processing in New Zealand.

DIRECT-USE SYSTEMS

Geothermal energy it is a good choice of energy for many applications because many processes require thermal energy at a temperature compatible with geothermal energy, and geothermal energy can be exploited in most states. However, an evaluation of using geothermal energy, as well as the proper design for its use in specific applications, requires consideration of the characteristics of geothermal systems and their interaction with specific equipment. This section explains geothermal system characteristics and the direct use of geothermal energy in residential, commercial, and industrial applications.

Figure 3 is a schematic illustration of a typical direct-use system. Such a system may consist of five subsystems: (1) the production system, including the producing wellbore and associated wellhead equipment; (2) the transmission and distribution system that transports the geothermal energy from the resource site to the user site

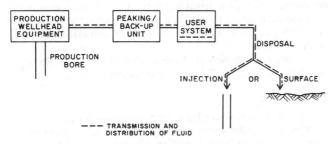

Fig. 3   Basic Geothermal Direct-Use System

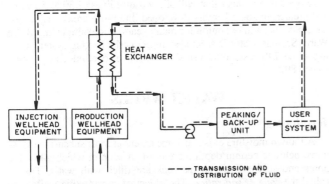

Fig. 4   Geothermal Direct-Use System with Wellhead Heat Exchanger and Injection Disposal

and then distributes it to the individual user loads; (3) the user system; (4) the disposal system, which can be either surface disposal or injection back into a formation; and (5) a peaking/backup system. None of the 14 major geothermal district heating systems in the United States includes a peaking/backup component as part of the main distribution system. Backup is most commonly included in the end-user systems, and peaking is rarely employed.

In a typical direct-use system, the geothermal fluid is produced from the production borehole by a lineshaft multistage centrifugal pump. (For wells that free-flow adequate quantities of fluid, a pump may not be required. However, the more common commercial-sized operation is expected to require pumping to provide the necessary flow rate.) When the geothermal fluid reaches the surface, it is delivered to the application site through the transmission and distribution system.

In the system shown in Figure 4, the geothermal production and disposal systems are closely coupled, and they are both separated from the remainder of the system by a heat exchanger, which limits the contact of the geothermal fluid with the user equipment. This secondary loop is desirable, especially when the geothermal fluid is particularly harsh in terms of corrosion and/or scaling. This type of system is environmentally advantageous because the geothermal fluid is pumped directly back into the ground without loss to the surrounding surface environment.

CHARACTERISTICS

Several characteristics of geothermal energy systems greatly influence their applicability and the design for their use. These characteristics arise from (1) the resource, (2) the application, and (3) the interaction between the two. The characteristics take the form of either constraints or design variables. Although the constraints are fixed for a particular resource and application, they influence the economic feasibility of using the geothermal resource in the particular application. The design variables are those parameters that can be varied to improve the feasibility of geothermal energy use.

The following are characteristics that influence the cost of energy delivered from geothermal systems:

- Depth of resource
- Distance between resource location and application site
- Well flow rate
- Resource temperature
- Temperature drop
- Load size
- Load factor
- Composition of fluid
- Ease of disposal
- Resource life

Many of these characteristics have a major influence because the costs of geothermal systems are primarily front-end capital costs; annual operating costs are relatively low.

Depth of the Resource

The well cost is usually one of the larger items in the overall cost of a geothermal system, and the cost of the overall system increases with the depth of the resource. Compared to many other geothermal areas worldwide, the western United States is fortunate in that well depth requirements are relatively shallow; most larger geothermal systems operate with production wells of less than 600 m, and many at less than 300 m.

Distance Between Resource Location and Application Site

The direct use of geothermal energy must occur near the resource location. The reason is primarily economic; although the geothermal fluid (or a secondary fluid) could be transmitted over moderately long distances (greater than 100 km) without a great temperature loss, such transmission would not generally be economically feasible. Most existing geothermal projects are characterized by transmission distances of less than 600 m.

Well Flow Rate

The energy output from a production well varies directly with the fluid flow rate. The energy cost at the wellhead varies inversely with the well flow rate. Typical good resources have production rates of about 25 to 50 L/s per production well; however, geothermal direct-use wells have been designed that produce up to 126 L/s.

Resource Temperature

In geothermal systems, the available temperature is associated with the prevailing resource. This temperature is an approximately fixed value for a given resource. The temperature may or may not increase with deeper drilling. Natural convection in fluid-dominated systems keeps the temperature relatively uniform throughout the depth of the resource (see Figure 1), but if there are deeper separate aquifers (producing zones) in the area, deeper drilling can result in recovery of energy at a higher temperature.

The temperature limitation can restrict potential applications. It often requires a reevaluation of the accepted application temperatures, as these have been developed in systems served by conventional fuels for which the application temperature could be selected at any value within a relatively broad range. When geothermal energy is used directly, the application temperature must be lower than the temperature of the produced fluid; however, when heat pumps are used, the application temperature may be somewhat higher than the produced fluid temperature.

Temperature Drop

Because the well flow rate is limited, the power output from the geothermal well is directly proportional to the temperature drop of the geothermal fluid effected by the user system. Consequently, a larger temperature drop results in lower energy cost at the wellhead. This is in great contrast to many conventional and solar systems that circulate a heating fluid with a small temperature drop, so a different design philosophy and different equipment are required.

Cascading the geothermal fluid to uses with lower temperature requirements can be advantageous in achieving a large temperature difference Δt. Most geothermal systems have been designed for a Δt of between 17 and 28 K, although one was designed for a Δt of 56 K with an 88°C resource temperature.

Load Size

Large-scale applications benefit from economy of scale, particularly in regard to reduced resource development and transmission system costs. For smaller developments, matching the size of the application with the production rate from the geothermal system is important because the total output varies in increments of one well's output.

Load Factor

Defined as the ratio of the average load to the design capacity of the system, the load factor effectively reflects the fraction of time that the initial investment in the system is working. Again, because geothermal system costs are primarily initial costs rather than operating costs, this factor significantly affects the viability of a geothermal system. As this factor increases, so does the economy of using geothermal energy. The two main ways of increasing the load factor are (1) to select applications where it is naturally high and (2) to use peaking equipment so that the design load is not the application peak load, but rather a reduced load that occurs over a longer period.

Composition of Fluid

The quality of the produced fluid is site-specific and may vary from less than 1000 ppm total dissolved solids (TDS) to heavily brined. The quality of the fluid influences two aspects of the system design: (1) material selection to avoid corrosion and scaling effects and (2) disposal or ultimate end use of the fluid.

Many direct-use geothermal systems operate with fluids containing less than 1000 ppm TDS. Despite these low levels, such fluids can create substantial corrosion and scaling problems. It is thus advisable to isolate the geothermal fluid from the balance of the system. While it is more expensive than using the fluid directly in the process or heating equipment, isolation is preferred for minimizing long-term system maintenance requirements.

Ease of Disposal

Most systems dispose of the geothermal effluent on the surface, including discharge to irrigation systems, rivers, and lakes. This method of disposal is considerably less expensive than the construction of injection wells. However, the magnitude of geothermal development in certain areas where surface disposal has historically been employed (e.g., Klamath Falls, Oregon, and Boise, Idaho) has caused aquifer water levels to decline. As a result, regulatory authorities in these and many other areas favor the use of injection in order to maintain reservoir fluid levels.

In addition, geothermal fluids sometimes contain chemical constituents that cause surface disposal to become a problem. Some of these constituents are listed in Table 2.

Table 2   Selected Chemical Species Affecting Fluid Disposal

Species	Reason for Control
Hydrogen sulfide (H_2S)	Odor
Boron (B^{3+})	Damage to agricultural crops
Fluoride (F^-)	Level limited in drinking water sources
Radioactive species	Levels limited in air, water, and soil

Source: Lunis (1989).

If injection is required, the depth at which the fluid can be injected affects well costs substantially. Some jurisdictions allow considerable latitude in terms of injection level; others require that the fluid be returned to the same or similar aquifers. In the latter case, it may be necessary to bore the injection well to the same depth as the production well.

The costs associated with disposal, particularly when injection is involved, can substantially affect development costs.

Resource Life

The life of the resource has a direct bearing on the economic viability of a particular geothermal application. There is little experience on which to base projections of resource life for heavily developed geothermal resources. However, resources can readily be developed in a manner that will allow useful lives of 30 to 50 years and greater. In some heavily developed direct-use areas, major systems have been in operation for many years. For example, the Boise Warm Springs Water District system (a district heating system serving some 240 residential users) has been in continuous operation since 1892.

WATER WELLS

Terminology

Although moisture exists to some extent at most subsurface locations, below a certain depth there exists a zone in which all of the pores and spaces between the rock are filled with water. This is called the *zone of saturation*. The object of constructing wells is to gain access to *groundwater*. Groundwater is the water that exists within the zone of saturation. An *aquifer* is a geologic unit that is capable of yielding groundwater to a well in sufficient quantities to be of practical use (UOP 1975).

In most projects, the construction of the well (or wells) is handled through a separate contract between the owner and the driller. As a result, the engineer is not responsible for its design. However, because the design of the building system is dependent upon the performance of the well, it is critical that the engineer be familiar with well terminology and test data. The most important consideration with regard to wells is that they be completed and tested (for flow volume and water quality) prior to final system design.

Figure 5 presents a summary of the more important terms relating to wells. Several references (Anderson 1984, EPA 1975, Roscoe Moss Company 1985, and Campbell and Lehr 1973) are available that cover well drilling and well construction in detail; these aspects are beyond the scope of this chapter.

Static water level (SWL) is the level in the well under static (non-pumping) conditions. In some cases, this level is much closer to the surface than that at which the driller first encounters water during drilling. *Pumping water level* (PWL) is different for different pumping rates (higher pumping rates yield lower pumping levels). The difference between the SWL and the PWL is the *drawdown*. The *specific capacity* of a well is frequently quoted in units of L/s per metre of drawdown. For example, for a well with a static level of 15 m that produces 30 L/s at a pumping level of 60 m, drawdown = 60 − 15 = 45 m; specific capacity = 30/45 = 0.67 L/s per metre.

For groundwater having carbonate scaling potential, *water entrance velocity* (through the screen or perforated casing) is an important design concern. Velocity should be limited to 0.03 m/s to avoid incrustation of the entrance area. A common cause of high entrance velocity is *overpumping* (installation of a pump that is too large relative to the well's capacity and produces excessive drawdown).

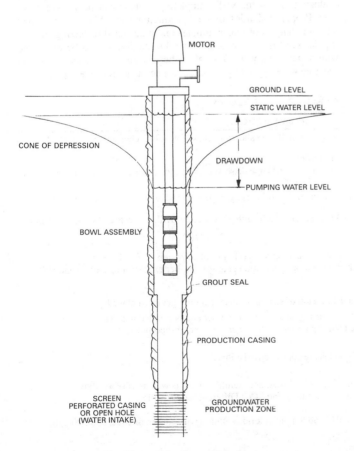

Fig. 5 Water Well Terminology

Table 3 Surface Casing Size Requirements

Production Rate, L/s	Nominal Pump Diameter, mm	Nominal Surface Casing Diameter, mm
< 6	102	152
6 to 11	127	203
11 to 22	152	254
22 to 44	203	305
44 to 63	254	356
63 to 100	305	406
100 to 189	356	457

Note: The table assumes the use of a vertical lineshaft 1750-rpm pump.

The *pump bowl* assembly (impeller housings and impellers) is always placed sufficiently below the expected pumping level to prevent cavitation at the peak production rate. For the previous example, the pump should be placed at least 67 m below the casing top (pump setting depth = 67 m) to allow for adequate submergence at peak flow. The specific net positive suction pressure required for a pump varies with each application and should be carefully considered in the design process.

For the well pump, *total pump pressure* is composed of three primary factors: lift, column friction, and wellhead pressure. *Lift* is the vertical distance that water must be pumped to reach the surface. In the example, lift would be 60 m. The additional 6 m of submergence imposes no static pump pressure. *Column friction* is calculated from pump manufacturer data in a similar manner to other pipe friction calculations (see Chapter 33 of the 1992 *ASHRAE Handbook—Systems and Equipment*). Surface pressure requirements account for friction losses through piping, heat exchangers, controls, and injection pressure (if any).

The well pump also influences the diameter of the casing required in the upper portion of the well. The casing diameter depends on the diameter of the pump (bowl assembly) necessary to produce the required flow rate. Table 3 presents nominal casing sizes for a range of water flow rates. The table assumes the use of vertical lineshaft 1750-rpm pumps. Submersible pumps and 3500-rpm lineshaft pumps are generally characterized by smaller casing diameter requirements than those indicated.

In addition to the production well, most systems should include an injection well to dispose of the fluid after it has passed through the system. Injection stabilizes the aquifer from which the fluid is withdrawn and helps assure long-term productivity. A brief discussion of injection wells is presented in Chapter 40, Thermal Storage.

Flow Testing

Well testing should be completed prior to mechanical system design. Only with final flow test data and water chemistry analysis information can proper design proceed.

Flow testing can be divided into three different types of tests: rig, short-term, and long-term (Stiger et al. 1991). Rig tests are generally shorter than 24 h and are accomplished while the drilling rig is on site. The primary purpose of this type of test is to purge the well of remaining drilling fluids and cuttings and to get a preliminary indication of yield. The length of the test is generally governed by the time required for the water to run clean. The rate is determined by the available pumping equipment. Frequently the well is blown, or pumped using air lift, with the drilling rig's air compressor. As a result, little can be learned about the production characteristics of the well from a rig test. If the well is air lifted, it may not be useful to collect water samples for chemical analysis because certain chemical constituents may be oxidized by the compressed air.

If properly conducted, short-term single-well tests of 1- to 7-day duration yield information about the well flow rate, temperature, pressure, drawdown, and recovery. These data can provide initial estimates of reservoir parameters (Stiger et al. 1989). The test is generally run with a temporary electric submersible pump or lineshaft turbine pump driven by an internal combustion engine. The work may be performed by the drilling contractor or by a well pump contractor.

The test should involve a minimum of three different production rates, the largest being equal to the design flow rate for the system served. The three points are the minimum required for the determination of a productivity curve for the well that relates production to drawdown (Stiger et al. 1989). Water level and pumping rate should be stabilized at each point before the flow is increased. In most cases, water level is monitored with a "bubbler" or an electric sounder, and flow rate is measured using an orifice meter. This level of testing is generally used for small projects or in areas where reservoir parameters are already established.

Long-term tests of up to 30 days provide information on the reservoir. Normally these tests involve monitoring nearby wells to evaluate interference effects. The data are useful in calculating average permeability thickness, storativity, reservoir boundaries, and recharge areas (Stiger et al. 1989).

It is also important to collect background information prior to the test and water level recovery data after pumping has ceased. Recovery data in particular can be used to evaluate skin effect, which is a type of well flow resistance caused by residual drilling fluids, insufficient screen or slotted liner area, or improper filter pack.

Water Quality Testing

It is necessary to evaluate the chemical nature of the fluid, the presence of suspended solids, and the possibility of biological contamination.

Geothermal fluids commonly contain seven key chemical species that produce a significant corrosive effect (Ellis 1989). The key species include

- Oxygen (generally from aeration)
- Hydrogen ion (pH)
- Chloride ion
- Sulfide species
- Carbon dioxide species
- Ammonia species
- Sulfate ion

The principal effects of these species are summarized in Table 4. Except as noted, the described effects are for carbon steel. Kindle and Woodruff (1981) present recommended procedures for complete chemical analysis of geothermal well waters.

Two of the species are not reliably detected by standard water chemistry tests and deserve special mention. Dissolved oxygen does not occur naturally in low-temperature (50 to 100°C) geothermal fluids that contain traces of hydrogen sulfide. However, because of slow reaction kinetics, oxygen from air in-leakage may persist for some minutes. Once the geothermal fluid is produced, it is extremely difficult to prevent contamination, especially if pumps other than downhole submersible or lineshaft turbine pumps are used to move the fluid. Even if the fluid systems are maintained at positive pressure, air in-leakage at the pump seals is likely, particularly with the low level of maintenance in many installations.

Hydrogen sulfide is ubiquitous in extremely low concentrations in geothermal fluids above 50°C. This corrosive species also occurs naturally in many cooler groundwaters. For alloys such as cupronickels, which are strongly affected by it, hydrogen sulfide concentrations in the low ppb range may have a serious detrimental effect, especially if oxygen is also present. At these levels, the characteristic rotten egg odor of hydrogen sulfide may be absent, so field methods may be required for detection. Hydrogen sulfide levels down to 50 ppb can be detected using a simple field kit; however, absence of hydrogen sulfide at this low level may not preclude damage by this species.

Two other key characteristics that should be measured in the field are pH and carbon dioxide concentration. This is necessary because most geothermal fluids release carbon dioxide rapidly, causing a rise in pH.

Production of suspended solids (sand) from a well should be evaluated during the well completion with gravel pack, screen, or both. Proper evaluation of sand production should be accomplished through analysis of water samples taken during flow testing. If substantial sand is produced, the size distribution should be evaluated with a sieve analysis. Only after these results are available can an accurate screen/gravel pack selection be made. Surface separation is

Table 4 Principal Effects of the Key Corrosive Species

Species	Principle Effects
Oxygen	• Extremely corrosive to carbon and low alloy steels; 30 ppb shown to cause fourfold increase in carbon steel corrosion rate. • Concentrations above 50 ppb cause serious pitting. • In conjunction with chloride and high temperature, <100 ppb dissolved oxygen can cause chloride-stress corrosion cracking (chloride-SCC) of some austenitic stainless steels.
Hydrogen ion (pH)	• Primary cathodic reaction of steel corrosion in air-free brine is hydrogen ion reduction. Corrosion rate decreases sharply above pH 8. • Low pH (5) promotes sulfide stress cracking (SSC) of high strength low alloy (HSLA) steels and some other alloys coupled to steel. • Acid attack on cements.
Carbon dioxide species (dissolved carbon dioxide, bicarbonate ion, carbonate ion)	• Dissolved carbon dioxide lowers pH, increasing carbon and HSLA steel corrosion. • Dissolved carbon dioxide provides alternative proton reduction pathway, further exacerbating carbon and HSLA steel corrosion. • May exacerbate SSC. • Strong link between total alkalinity and corrosion of steel in low-temperature geothermal wells.
Hydrogen sulfide species (hydrogen sulfide, bisulfide ion, sulfide ion)	• Potent cathodic poison, promoting SSC of HSLA steels and some other alloys coupled to steel. • Highly corrosive to alloys containing both copper and nickel or silver in any proportions.
Ammonia species (ammonia, ammonium ion)	• Causes stress corrosion cracking (SCC) of some copper-based alloys.
Chloride ion	• Strong promoter of localized corrosion of carbon, HSLA, and stainless steel, as well as of other alloys. • Chloride-dependent threshold temperature for pitting and SCC. Different for each alloy. • Little if any effect on SSC. • Steel passivates at high temperature in pH 5, 6070 ppm chloride solution with carbon dioxide. 133 500 ppm chloride destroys passivity above 150°C.
Sulfate ion	• Primary effect is corrosion of cements.

Source: Ellis (1989).   *Note*: Except as indicated, the described effects are for carbon steel.

less desirable because a surface separator requires the sand to pass first through the pump, reducing its useful life.

Biological fouling is largely a phenomenon of low-temperature (< 32°C) wells. The most prominent organisms are various strains (*Galionella, Crenothrix*) of what are commonly referred to as iron bacteria. These organisms typically inhabit water characterized by a pH range of 6.0 to 8.0, dissolved oxygen content of less than 5 ppm, ferrous iron content of less than 0.2 ppm, and a temperature of 8 to 16°C (Hackett and Lehr 1985). Iron bacteria can be identified microscopically.

The most common treatment for iron bacteria infestation is a combination of chlorination, surging, and flashing. Successful use of this treatment is dependent upon the maintenance of proper pH (less than 8.5), dosage, free residual chlorine content (13 to 32 L/s), contact time (24 h minimum), and agitation or surging. Hackett and Lehr (1985) provide additional detail on treatment methods.

EQUIPMENT AND MATERIALS

The primary equipment used in geothermal systems includes pumps, heat exchangers, and piping. While some aspects of these components are unique to geothermal applications, many of them are of routine design. However, the great variability and general aggressiveness of the geothermal fluid necessitate limiting corrosion and scale buildup rather than relying on system cleanup. Corrosion and scaling can be limited through (1) proper system and equipment design or (2) treatment of the geothermal fluid, which is generally precluded by cost and environmental regulations relating to disposal.

Performance of Materials

Carbon Steel. The Ryznar Index has traditionally been used to estimate the corrosivity and scaling tendencies of potable water supplies. However, one study found no significant correlation (at the 95% confidence level) between carbon steel corrosion and the Ryznar Index (Ellis and Smith 1983). Therefore, the Ryznar and other indices based on calcium carbonate saturation should not be used to predict corrosion in geothermal systems.

In Class Va geothermal fluids as described by Ellis (1989) [< 5000 ppm total key species (TKS), total alkalinity 207 to 1329 ppm as $CaCO_3$, pH 6.7 to 7.6], corrosion rates of about 100 to 500 µm per year can be expected, often with severe pitting.

In Class Vb geothermal fluids as described by Ellis (1989) [< 5000 ppm TKS, total alkalinity <210 ppm as $CaCO_3$, pH 7.8 to 9.85], carbon steel piping has given good service in a number of systems, provided the system design rigorously excluded oxygen. However, introduction of 30 ppb oxygen under turbulent flow conditions causes a fourfold increase in uniform corrosion. Saturation with air often increases the corrosion rate by at least 15 times. Oxygen contamination at the 50 ppb level often causes severe pitting. Chronic oxygen contamination causes rapid failure.

In the case of buried steel pipe, the external surfaces must be protected from contact with groundwater. Groundwater is aerated and has caused pipe failures by external corrosion. Required external protection can be obtained by use of coatings, pipe-wrap, or preinsulated piping, provided the selected material will resist the system operating temperatures and thermal stresses.

At temperatures above 57°C, galvanizing (zinc coating) will not reliably protect steel from either geothermal fluid or groundwater. Hydrogen blistering can be prevented by using void-free (killed) steels.

Low-alloy steels (steels containing no more than 4% alloying elements) have corrosion resistance similar, in most respects, to carbon steels. As in the case of carbon steels, sulfide promotes entry of atomic hydrogen into the metal lattice. If the steel exceeds a hardness of Rockwell C22, sulfide stress cracking may occur.

Copper and Copper Alloys. Copper fan-coil units and copper-tubed heat exchangers have a consistently poor performance due to traces of sulfide species found in geothermal fluids in the United States. Copper tubing rapidly becomes fouled with cuprous sulfide films more than 1 mm thick. Serious crevice corrosion occurs at cracks in the film, and uniform corrosion rates of 50 to 150 µm per year appear typical, based on failure analyses.

Experience in Iceland also indicates that copper is unsatisfactory for heat exchange service and that most brasses (Cu-Zn) and bronzes (Cu-Sn) are even less suitable. Cupronickels often perform more poorly than does copper in low-temperature geothermal service because of trace sulfide.

Much less information is available regarding copper and copper alloys in non-heat-transfer service. Copper pipe shows corrosion behavior similar to that of copper heat exchange tubes under conditions of moderate turbulence (Reynolds numbers of 40 000 to 70 000). The internals of the few yellow brass valves analyzed showed no significant corrosion. However, silicon bronze CA 875 (12-16Cr, 3-5Si, <0.05Pb, <0.05P), an alloy normally resistant to dealloying, failed in less than three years when used for a pump impeller. Leaded red brass (CA 836 or 838) and leaded red bronze (SAE 67) appear viable for pump internals. Based on a few tests at Class Va sites, aluminum bronzes have shown potential for corrosion in heavy-walled components.

Solder is yet another problem area for copper equipment. Lead-tin solder (50Pb, 50Sn) was observed to fail by dealloying after a few years of exposure. Silver solder (1Ag, 7P, Cu) was completely removed from joints in under two years. If the designer elects to accept this risk, solders containing at least 70% tin should be used.

Stainless Steel. Unlike copper and cupronickels, stainless steels are not affected by traces of hydrogen sulfide. Their most likely application is heat exchange surfaces. For economic reasons, most heat exchangers are probably of the plate-and-frame type, most of which are fabricated with one of two standard alloys, Type 304 and Type 316 austenitic stainless steel. Some pump and valve trim also is fabricated from these or other stainless steels.

These alloys are subject to pitting and crevice corrosion above a threshold chloride level that depends on the chromium and molybdenum content of the alloy and on the temperature of the geothermal fluid. Above this threshold, the passivation film, which gives the stainless steel its corrosion resistance, is ruptured, and local pitting and crevice corrosion occur. Figure 6 shows the relationship between temperature, chloride level, and occurrence of localized corrosion for Type 304 and Type 316 stainless steel. This figure indicates, for example, that localized corrosion of Type 304 may occur in 27°C geothermal fluid if the chloride level exceeds approximately

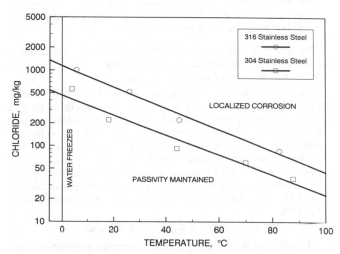

Fig. 6 Chloride Concentration Required to Produce Localized Corrosion of Stainless Steel as a Function of Temperature
(Efrid and Moeller 1978)

210 ppm, while Type 316 is resistant at that temperature until the chloride level reaches approximately 510 ppm. Due to its 2 to 3% molybdenum content, Type 316 is always more resistant to chlorides than Type 304.

Aluminum. Aluminum alloys are not acceptable in most cases because of catastrophic pitting.

Titanium. This material has extremely good corrosion resistance and could be used for heat exchanger plates in any low-temperature geothermal fluid, regardless of dissolved oxygen content. Great care is required if acid cleaning is to be performed. The vendor's instructions must be followed. The titanium should not be scratched with iron or steel tools, as this can cause pitting.

Chlorinated Polyvinyl Chloride (CPVC) and Fiber Reinforced Plastic (FRP). These materials are easily fabricated and are not adversely affected by oxygen intrusion. External protection against groundwater is not required. The mechanical properties of these materials at higher temperatures may vary greatly from those at ambient temperature, and the mechanical limits of the materials should not be exceeded. The usual mode of failure is creep rupture; the strength decays with time. Manufacturer's directions for joining should be followed to avoid premature failure of joints.

Elastomeric Seals. Tests on o-ring materials in a low-temperature system in Texas indicated that fluoroelastomer is the best material for piping of this nature; Buna-N is also acceptable. Neoprene, which developed extreme compression set, was a failure. Natural rubber and Buna-S should also be avoided. Ethylene-propylene terpolymer (EPDM) has been used successfully for gaskets, o-rings and valve seats in many systems.

Corrosion Engineering and Design

The design of the geothermal system is as critical to controlling corrosion as the selection of suitable materials (Ellis 1989). Furthermore, the design and material selection processes are interactive in that certain design decisions force the use of certain materials, while selection of material may dictate design. In all cases, the objective should be the same—to produce an adequately reliable system with the lowest possible lifetime cost.

There are three different basic corrosion engineering design philosophies for geothermal systems:

1. Use corrosion-resistant materials throughout the system.
2. Exclude or remove oxygen, and use carbon steel throughout the system.
3. Transfer the heat via an isolation heat exchanger to a noncorrosive working medium so that the kind and number of components contacting the geothermal fluid are minimized, and make those components of corrosion-resistant materials.

The first philosophy would produce a reliable, low-maintenance system. However, the cost would be high, and many of the desired components are not available in the required alloys.

The second philosophy may be considered for district-sized heating projects with attendant surface storage and potential for oxygen intrusion; for such projects it may be economical to inhibit the geothermal fluid by continuous addition of excess sulfite as an oxygen scavenger. When this is done, carbon steel can be used for heat exchange equipment, provided fluid pH is greater than 8. This approach is widely and successfully used for municipal heating systems in Iceland. System design should minimize the introduction of oxygen to reduce sulfite costs. Use of vented tanks should also be minimized.

The second philosophy has three major drawbacks:

1. The sulfite addition plant is relatively complex and requires careful maintenance and operation.
2. Failure of the sulfite addition plant or insufficient treatment is likely to cause rapid failure of the carbon steel heat exchangers.

3. Sulfite addition will probably not be economical for smaller systems because of the complexity and excessive maintenance and operation requirements of the plants. Without oxygen scavenging, even with careful design, some oxygen contamination will occur, and carbon steel heat exchangers will probably not be satisfactory. In addition, environmental considerations regarding disposal of the treated fluid may be a problem.

The third philosophy—transferring the heat via isolation heat exchangers to a noncorrosive secondary heat transfer medium—has several advantages for systems of all sizes. The system is much simpler to operate and therefore more reliable than those mentioned above. Only a small number of corrosion-resistant components are required, and it is feasible to design systems in which oxygen exclusion is not critical. Even with careful material selection and design, geothermal heating systems require more maintenance than conventional systems. Minimizing the number and kinds of components in contact with geothermal fluid will further reduce maintenance costs.

Pumps

Pumps are used for production, circulation, and disposal. For circulation and disposal, whether surface disposal or injection, standard hot water circulating pumps, almost exclusively of centrifugal design, are used. These are routine engineering design selections, the only special consideration being the selection of appropriate materials. Small circulating in-line pumps of all-iron construction have shown acceptable performance in many applications. For larger pumps (base-mounted, vertical in-line, double suction), cast-iron impellers and volute with stainless steel shaft, screws, keys, and washers are typically employed (Rafferty 1989a). In addition, mechanical seals are preferred over packing.

Production well pumps are among the most critical components in a geothermal system and have in the past been the source of much system downtime. Therefore, proper selection and design of the production well pump system is extremely important. Well pumps are available for larger systems in two general configurations—lineshaft and submersible. The lineshaft type is the most often used for direct-use systems (Rafferty 1989a).

Lineshaft Pumps. Lineshaft pumps are similar to those typically used in irrigation applications. An aboveground driver, typically an electric motor, rotates a vertical shaft extending down the well to the pump. The shaft rotates the pump impellers within the pump bowl assembly, which is positioned at such a depth in the wellbore that adequate net positive suction pressure is available when the unit is operating. Two designs for the shaft/bearing portion of the system are available—open and enclosed.

In the *open* lineshaft design, the shaft bearings are supported in "spiders," which are anchored to the pump column pipe at 1.5- to 3-m intervals. The shaft and bearings are lubricated by the fluid flowing up the pump column. In geothermal applications, bearing materials for open lineshaft designs have consisted of both bronze and various elastomer compounds. The shaft material is typically stainless steel. Experience with this type of design in geothermal applications has been mixed. Two large district heating systems have successfully operated such pumps for approximately 10 years; many more systems, however, initiated operation with open lineshaft pumps and subsequently changed to enclosed lineshaft design after numerous failures. Open lineshaft pumps are generally less expensive than enclosed lineshaft pumps for the same application. In addition, the open lineshaft type is easier to apply to artesian wells.

In an *enclosed* lineshaft pump, an enclosing tube protects the shaft and bearings from exposure to the pumped fluid. A lubricating fluid is admitted to the enclosed tube at the wellhead. It flows down the tube, lubricating the bearings, and exits at the base of the pump column where the column attaches to the bowl assembly. The bowl shaft and bearings are lubricated by the pumped fluid. Oil-lubricated enclosed lineshaft pumps have the longest service life in low-temperature direct-use applications.

These pumps typically include carbon or stainless steel shafts and bronze bearings in the lineshaft assembly, and stainless steel shafts and leaded red bronze bearings in the bowl assembly. Keyed-type impeller connections (to the pump shaft) are superior to collet-type connections (Rafferty 1989a).

Because of the lineshaft bearings, the reliability of lineshaft pumps decreases as the pump-setting depth increases. Nichols (1978) indicates that at depths greater than about 240 m, reliability is questionable even under good pumping conditions.

Submersible Pumps. The electrical submersible pump system consists of three primary components located downhole—the pump, the drive motor, and the motor protector. The pump is a vertical multistage centrifugal type. The motor is usually a three-phase induction type that is oil filled for cooling and lubrication; it is cooled by heat transfer to the pumped fluid moving up the well. The motor protector is located between the pump and the motor and isolates the motor from the well fluid while allowing pressure equalization between the pump intake and the motor cavity.

The electrical submersible pump has several advantages over lineshaft pumps, particularly for wells requiring greater pump bowl setting depths. The deeper the well, the more economical is the submersible to install. Moreover, it is more versatile in that it adapts more easily to different depths. The break-over point is at a pump depth of about 240 m; the submersible pump is desirable at greater pump depths, and the lineshaft is preferred at lesser depths.

Submersible pumps have not demonstrated acceptable lifetimes in most geothermal applications. Although they are commonly used in high-temperature downhole applications in the oil and gas industry, the acceptable overhaul interval in that industry is much shorter than in a geothermal application. In addition, most submersibles are 3600-rpm machines, which results in greater susceptibility to erosion in aquifers that produce moderate amounts of sand. None of the 14 largest geothermal district heating systems in the United States employs submersibles. They have, however, been applied in geothermal projects in which it is necessary to use an existing well that is of relatively small diameter. Because the submersibles operate at 3600 rpm, they are capable of providing greater flow capacity for a given bowl size than an equivalent 1750-rpm lineshaft pump.

Variable-Speed Drives

In many geothermal applications, particularly space heating, well flow requirements vary over a substantial range. To avoid inefficient throttling of the pump to meet those requirements, variable-speed drives are frequently applied. In most cases, one of two varieties of controls has been employed—fluid coupling or variable-frequency drive. One major geothermal system employs a two-speed induction motor on the production pump.

The fluid coupling was most often used prior to the general availability of electronic frequency controls. These units are typically installed on lineshaft pumps between the electric motor and the pump shaft. They cannot be installed on submersible pumps. Advantages of the fluid coupling include relatively low cost and mechanical simplicity.

The efficiency of the fluid coupling is a function of the ratio of the output speed to the input speed. As a result, in applications that require a large turndown (to less than about 70% of full speed), operating economy declines. In most geothermal applications, the minimum pump speed is controlled by the necessity to generate sufficient pressure to raise the fluid from the pumping level to the wellhead, resulting in only moderate turndown requirements.

Variable-frequency drives are commonly used on geothermal production well pumps. They provide a more efficient overall drive than the fluid coupling approach.

The savings from the use of a variable-speed drive in geothermal applications are subject to some unique pumping considerations. Most

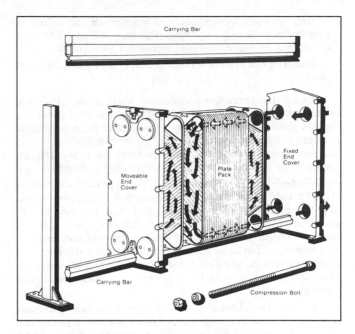

Fig. 7 Plate Heat Exchanger

well pumping applications involve a considerable static pressure requirement for lifting the fluid out of the well. This static requirement is not subject to the usual cubic relationship between flow and power requirement experienced in a 100% friction pressure situation. As a result, for an equal reduction in flow rate, pump energy savings are less than those for a circulating pump (friction pressure) system.

Heat Exchangers

In the system shown in Figure 3, heat exchangers are located at the individual use or process. For the system in Figure 4, one or more large heat exchangers are located near the wellhead. In both cases, the principle is to isolate the geothermal fluid from complicated systems or those which cannot be readily designed to be compatible with the geothermal fluid.

The principal types of heat exchangers used in transferring energy from the geothermal fluid are (1) plate, (2) shell-and-tube, and (3) downhole.

Plate Heat Exchangers. In a plate heat exchanger (Figure 7), the geothermal fluid is routed along one side of each plate, and the heated fluid is routed along the other side. The plates can readily be manifolded for various combinations of series and parallel flow. Plate heat exchangers have been widely used for many years in the food-processing industry and in marine applications. They have two main characteristics that make them desirable for many geothermal applications:

1. They are readily cleaned. By loosening the main bolts, the header plate and the individual heat exchanger plates can be removed and cleaned.
2. The stamped plates are very thin and may be made of a wide variety of materials. When expensive materials are required, the thinness of the plates keeps this type of heat exchanger much less expensive than other types.

Plate heat exchangers have additional characteristics that influence their selection in specific applications:

- Approach temperature differences are usually smaller than those for shell-and-tube heat exchangers; this is particularly important in low-temperature geothermal applications. Many geothermal applications have temperatures differences of 5 K, and some as low as 2 K.

- In water-to-water applications, heat transfer per unit volume is usually larger than for shell-and-tube heat exchangers because overall heat transfer coefficients approach 5.7 kW/(m$^2 \cdot$K).
- Increased capacity can be accommodated easily by the addition of plates.

In most low-temperature direct-use applications, plate heat exchangers are constructed of Type 316 stainless steel plates and Buna-N gaskets (Rafferty 1989a). Stainless steel plates have failed in at least two applications in which the geothermal fluid was used to heat swimming pool water. This was probably due to high chlorine content in the pool water, which promoted pitting of the plates.

Shell-and-Tube Heat Exchangers. This type of heat exchanger is used in only a limited number of geothermal applications because plate heat exchangers are more economical when specialized materials are required to minimize corrosion. However, when mild steel shells and copper or silicon bronze tubes can be used, shell-and-tube heat exchangers can be more economical. With the geothermal fluid passing through the tube side of a shell-and-tube heat exchanger, the tubes should be in a straight configuration to facilitate mechanical cleaning.

Downhole Heat Exchangers. The downhole heat exchanger (DHE) consists of an arrangement of pipes or tubes suspended in a wellbore (Culver and Reistad 1978). A secondary fluid circulates from the user system through the exchanger and back to the system in a closed loop. The primary advantage of a DHE is that only heat is extracted from the earth, eliminating the need for disposal of spent fluids. Other advantages are the elimination of (1) pumps, with their initial, operating, and maintenance costs; (2) the potential for depletion of groundwater; and (3) environmental and institutional restrictions on surface disposal.

One disadvantage of DHEs is the limited amount of heat that can be extracted from or rejected to the well. This depends on the hydraulic conductivity of the aquifer and on the well design.

DHEs in low- to moderate-temperature geothermal wells are installed in a casing, as shown in Figure 8. The U-bend design is the most common, but the concentric tube design is also used. Pipes are usually black iron or steel; however, epoxy-fiberglass DHEs have also been installed.

Downhole heat exchangers with higher outputs rely on water circulation within the well, whereas lower output DHEs rely on earth conduction. Circulation within the well can be accomplished by one of two methods: (1) undersized casing and (2) convection tube. Both methods rely on the difference in density between the water surrounding the DHE and that in the aquifer.

Circulation provides the following advantages:

- Water circulates around the DHE at velocities that, at optimum conditions, can approach those in the shell of a shell-and-tube exchanger.
- Hot water moving up the annulus heats the upper rocks, and the well becomes nearly isothermal.
- Some of the cool water, being more dense than the water in the aquifer, sinks into the aquifer and is replaced by hotter water, which flows up the annulus.

Figure 8 shows well construction in competent formation, i.e., where the wellbore will stand open without casing. An undersized casing having perforations at the lowest producing zone (usually near the bottom) and just below the static water level is installed. A packer near the top of the competent formation permits the installation of cement between it and the surface. When the DHE is installed and heat is extracted, thermosyphoning causes cooler water inside the casing to move to the bottom, and hotter water moves up the annulus outside the casing.

Because most DHEs are used for space heating (an intermittent operation), the heated rocks in the upper portion of the well provide heat storage for the next cycle.

In areas where the well will not stand open without casing, the convection tube can be used. This is pipe one-half the diameter of the casing, either hung with its lower end above the well bottom and its upper end below the surface or set on the bottom with perforations at the bottom and below the static water level. If a U-bend DHE is use, it can either be outside the convection tube or have one leg in the convection tube. A concentric tube DHE should be outside the convection tube.

DHEs operate best in high hydraulic conductivity aquifers, which provide the water movement for heat and mass transfer.

Valves

Resilient lined butterfly valves have been the design of choice for geothermal applications. The lining material protects the valve body from exposure to the geothermal fluid. The rotary rather than reciprocating motion of the stem makes the valve less susceptible to leakage and the buildup of scale deposits. For many direct-use applications, these valves are composed of Buna-N or EPDM seats, stainless steel shafts, and bronze or stainless steel disks.

Gate valves have been employed in some larger geothermal systems but have been subject to stem leakage and seizure. After several years of use, they are no longer capable of 100% shutoff.

Piping

Due to the aggressive nature of most geothermal fluids, nonmetallic piping has been widely applied to direct-use projects. Although steel has occasionally been used for distribution and transmission piping, such materials as fiberglass and polybutylene are much more common. Asbestos cement (AC) piping have been widely applied, but it is now being phased out.

Fiberglass reinforced piping (FRP)—available with a wide variety of joining methods—has been used in a number of projects. The epoxy adhesive bell and spigot joint have seen the widest use.

In comparison to metallic piping, fiberglass piping offers lighter weight and smoother interior surfaces for lower pressure drop. Although it can be formulated for service at temperatures as high as 150°C, most cost-competitive products are limited to approximately

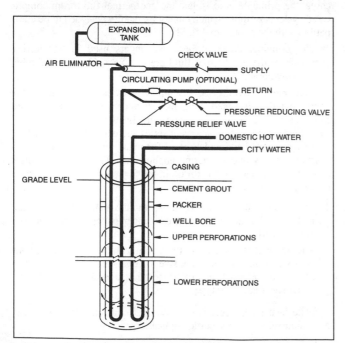

Fig. 8   Typical Connection of a Downhole Heat Exchanger for Space and Domestic Hot Water Heating
(Reistad et al. 1979)

95 to 120°C. The rate of expansion for FRP is about twice that of steel. Due to the very low modulus, however, the forces exerted are only about 3 to 5% of those of steel. As a result, in buried application, expansion loops and joints are generally not required. Fitting costs, particularly in large sizes, can constitute a significant portion of the overall material costs for a system. Fiberglass piping is generally available in sizes larger than 50 mm.

Polybutylene (PB) piping has been used widely in small-diameter (<50 mm) sizes for service lines on geothermal district heating systems. A larger size (50 to 150 mm) was used successfully in one geothermal distribution system. Polybutylene piping can be used up to a temperature of approximately 82°C.

Polybutylene is a member of the polyolefin family and, as such, cannot be joined using solvent welding. The preferred method of joining for small sizes and the only method for larger sizes is thermal fusion. Small-diameter piping can be socket fused, and large piping can be butt fused.

Polybutylene is a flexible material and, in small sizes, is available in rolls. As with many other plastics, its pressure rating is dependent on temperature and is influenced by wall thickness. The pipe is available in various wall thicknesses as designated by the standard dimension ratio (SDR), the ratio of the nominal diameter to the wall thickness. For example, an SDR 13.5 pipe carries a pressure rating of 720 kPa at 65°C, while the rating of an SDR 11 pipe is 900 kPa at 65°C.

Polyethylene (PE), another polyolefin, has seen only limited use in geothermal direct-use applications. It shares many of the same considerations discussed for polybutylene, except for a lower maximum service temperature (60 to 65°C). It is much lower in cost than polybutylene. Cross-linked polyethylene (PEX) has been used extensively in European district heating applications. It carries higher temperature and pressure ratings (690 kPa at 82°C) than standard polyethylene, along with much higher costs.

Polyethylene products can be employed in direct buried applications without the use of expansion joints or loops.

Because steel piping is susceptible to corrosion from both soil moisture (external) and geothermal fluid (internal), it has not been used widely in direct burial applications for transporting geothermal fluid. Two large projects in which steel piping was used for geothermal transmission lines involved tunnel or in-building type installation, so the piping was not exposed to soil moisture.

Steel piping is employed in many cases for mechanical room and in-building piping. It is critical that oxygen be carefully excluded from the system when steel pipe is used to transport geothermal fluid.

Preinsulated Pipe. Most large transmission and distribution piping systems for transporting geothermal fluids employ preinsulated piping products. These products consist of a carrier pipe through which the geothermal fluid flows, a layer (nominal 25 to 50 mm) of polyurethane insulation and a jacket for protecting the insulation from physical damage and moisture penetration. The most common carrier/jacket material combinations are AC/AC, FRP/FRP, FRP/PVC, steel/FRP, steel/PE, and PE/PE.

In most existing direct buried applications, the connections between lengths of pipe are left uninsulated unless steel piping is used.

RESIDENTIAL AND COMMERCIAL APPLICATIONS

The primary applications for the direct use of geothermal energy in residential and commercial areas are space heating, sanitary water heating, and space cooling (using the absorption process). While geothermal space and sanitary water heating are widespread, space cooling is rare. Where space heating is accomplished, sanitary water heating, or at least preheating, is almost universally accomplished.

SPACE AND SANITARY WATER HEATING

Figure 9 illustrates a system that uses geothermal fluid at 77°C (Austin 1978). The geothermal fluid is used in two main equipment components for heating of the buildings: (1) a plate heat exchanger that supplies energy to a closed heating loop previously heated by a natural gas boiler (the boiler remains as a standby unit) and (2) a water-to-air coil used for preheating ventilation air. In this system, proper control is crucial for economical operation.

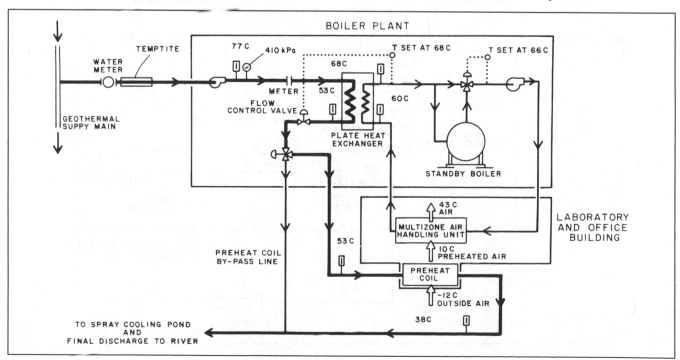

Fig. 9   Heating System Schematic

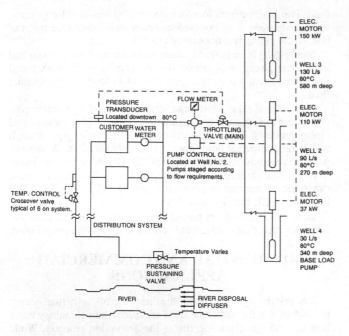

Fig. 10   Open-Type Geothermal District Heating System
(Rafferty 1989)

The average temperature of the discharged fluid is 49 to 54°C. The geothermal fluid is used directly in the preheat terminal equipment within the buildings (this would probably not be the case if the system were being designed today). Several corrosion problems have arisen in the direct use, mainly because of the action of hydrogen sulfide on copper-based equipment parts (Mitchell 1980). Even with these difficulties, the geothermal system appears to be highly cost-effective (Lienau 1979).

Figure 10 illustrates a geothermal district heating system. The geothermal fluid is produced from three wells. Depending on the load, one, two, or three wells may be in operation. Each well is equipped with a single-speed electric motor and is capable of producing a constant flow rate. The main system flowmeter controls which pumps are in operation. System flow rate is also controlled by a large throttling valve located in the main production line. This valve responds to a signal from a pressure transducer located in the downtown area on the supply side of the distribution system.

The open-type distribution system is composed of preinsulated pipe for supply fluid and uninsulated pipe for disposal fluid. Temperature-sensitive, self-actuated valves are located at six strategic points around the distribution system. These valves open when necessary to maintain acceptable supply water temperatures in the system. Water meters are located at each valve to allow system operators to monitor the amount of fluid required for temperature maintenance.

Figure 11 shows a geothermal district heating system that is unique in terms of its design based on a peak load Δt of 56 K using an 88°C resource. It has a closed-loop design with central heat exchangers. The production well has an artesian shut-in pressure of

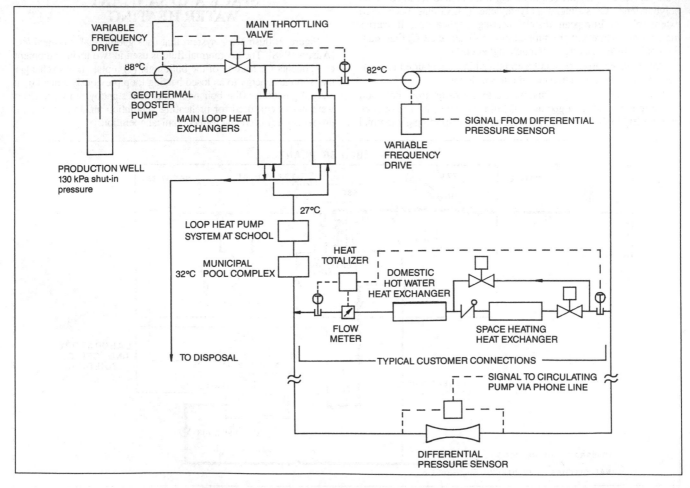

Fig. 11   Closed-Type Geothermal District Heating System
(Rafferty 1989)

175 kPa, so the system operates with no production pump for most of the year. During colder weather, a surface centrifugal pump located at the wellhead boosts the pressure.

Geothermal flow from the production well is initially controlled by a throttling valve on the supply line to the main heat exchanger, which responds to a temperature signal from the supply water on the closed-loop side of the heat exchanger. When the throttling valve has reached the full open position, the production booster pump is enabled. The pump is controlled through a variable-frequency drive that responds to the same supply water signal as the throttling valve. The booster pump is designed for a peak flow rate of 19 L/s of 88°C water.

Types of Terminal Heating Equipment. The terminal equipment used in geothermal systems is the same as that used in nongeothermal heating systems. However, certain types of equipment are better suited to geothermal design than others.

In many cases, buildings that derive their heat from geothermal fluids operate their heating systems at less than conventional supply water temperatures because of low resource temperatures and the use of heat exchangers to isolate the fluids from the building loop. It is frequently advisable to design for a larger than normal Δt in geothermal systems. The selection of equipment to accommodate these considerations can enhance the feasibility of using the geothermal source.

Finned coil, forced-air systems are generally the most capable of functioning under the low-temperature/high Δt conditions described above. One or two additional rows of coil depth compensate for lower supply water temperatures. While increased Δt affects coil circuiting, it improves controllability. This type of system should be capable of using supply water temperatures as low as 38 to 49°C.

Radiant floor panel systems are able to use very low water temperatures, particularly in industrial applications with little or no floor covering. The availability of new nonmetallic piping has renewed the popularity of this type of system. In industrial settings, with a bare floor and relatively low space temperature requirements, average water temperatures could conceivably be as low as 35°C. For higher space temperatures and/or thick floor coverings, higher water temperatures may be required.

Baseboard convectors and similar equipment are the least capable of operating at low supply water temperature. At 65°C average water temperatures, derating factors for this type of equipment are on the order of 0.43 to 0.45. As a result, the quantity of equipment required to meet the design load is generally uneconomical. This type of equipment can be operated at low temperatures from the geothermal source to provide base-load heating capacity. Peak load can be supplied by a conventional boiler.

Heat pump systems take advantage of the lowest temperature geothermal resources. Loop heat pump systems operate with water temperatures in the 15 to 32°C range, and central station heat pump plants (supplying four-pipe systems) in the 7°C range.

Domestic Water Heating

Domestic water heating in a district space heating system is beneficial because it increases the overall size of the energy load, the energy demand density, and the load factor. For those resources that cannot heat water to the required temperature, preheating is usually desirable. Whenever possible, the domestic hot water load should be placed in series with the space heating load in order to reduce system flow rates.

Space Cooling

Geothermal energy has seldom been used for cooling, although emphasis on solar energy and waste heat has created interest in systems that cool using thermal energy. Although there are a number of methods for producing cooling effect using a heat source, the absorption cycle is the most often used. Lithium bromide/

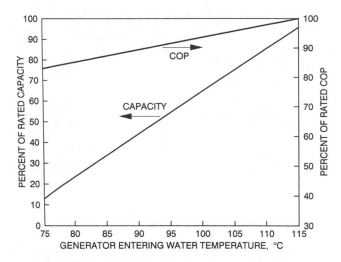

Fig. 12 Typical Lithium Bromide Absorption Chiller Performance Versus Temperature
(Christen 1977)

water absorption machines are commercially available in a wide range of capacities. At least two geothermal/absorption space cooling applications are in successful operation in the United States.

Temperature and flow requirements for absorption chillers run counter to the general design philosophy for geothermal systems. They require high supply water temperatures and a small Δt on the hot water side. Figure 12 illustrates the effect of reduced supply water temperature on machine performance. The machines are rated at a 115°C input temperature, so derating factors must be applied if the machine is operated below this temperature. For example, operation with 93°C supply water would result in a 50% decrease in capacity, which seriously affects the economy of absorption cooling at low resource temperatures.

Coefficient of performance (COP) is less seriously affected by reduction in supply water temperatures. The nominal COP of a single-stage machine at 115°C is 0.65 to 0.70; that is, for each kilowatt of cooling output, a heat input of 1 kW divided by 0.65, or 1.54 kW, would be required.

Most absorption equipment is designed for steam input (an isothermal process) to the generator section. As a result, when this equipment is operated from a hot water resource, a relatively small Δt must be employed. This creates a mismatch between a building's flow requirements for space heating and cooling. For example, assume an 18 600 m² building is to use a geothermal resource for heating and cooling. At 79 W/m² and a design Δt of 22.2 K, the flow requirement for heating would amount to 15.8 L/s. At 95 W/m², a Δt of 8.3 K, and a COP of 0.65, the flow requirement for cooling would amount to 77.6 L/s.

Some small-capacity (10- to 90-kW) absorption equipment has been optimized for low-temperature operation in conjunction with solar systems. While this equipment could also be applied to geothermal resources, the prospects are questionable. Small absorption equipment would generally compete with packaged direct-expansion units in this range; the absorption equipment requires a great deal more mechanical auxiliary equipment for a given capacity. The cost of the chilled water piping, pump, and coil; cooling water piping, pump, and tower; and hot water piping raises the capital cost of the absorption system substantially. Only in larger sizes (>35 kW) and in areas with high electric rates and high cooling requirements (>2000 full load hours) would this type of equipment offer an attractive investment to the owner (Rafferty 1989b).

Table 5 Geothermal Industrial Uses in the United States

Application	Number of Sites	Resource Temperature, °C
Sewage digester heating	2	54; 77
Laundry	4	40 to 83[a]
Vegetable dehydration	1	132
Mushroom growing	1	113
Greenhouse heating	37	35 to 99[a]
Aquaculture	8	16 to 96[a]
Grain drying	1	93
Highway deicing	3	8; 88
Gold mine heap leaching	2	114; 86

Source: Lienau et al. (1988).
[a]Varies with site.

INDUSTRIAL APPLICATIONS

The use of geothermal energy in industrial applications, including agricultural facilities, requires design philosophies similar to those for space conditioning. However, these applications have the potential for much more economical use of the geothermal resource, primarily because (1) they operate year-round, which gives them greater load factors than possible with space-conditioning applications; (2) they do not require an extensive (and expensive) distribution to dispersed energy consumers, as is common in district heating; and (3) they often require various temperature levels and, consequently, may be able to make greater use of a particular resource than space conditioning, which is restricted to a specific temperature level. Lienau et al. (1988) summarize the industrial applications of geothermal energy in the United States (Table 5).

GROUND-SOURCE HEAT PUMP SYSTEMS

Ground-source heat pumps were originally developed in the residential arena and are now being applied in the commercial sector. Many of the installation recommendations, design guides, and rules of thumb appropriate to residential design must be amended for large buildings. Consult Kavanaugh (1994) for a more complete overview of ground-source heat pumps. OSU (1988) and Kavanaugh (1991) provide a more detailed treatment of the design and installation of ground-source heat pumps, but the focus of these two documents is primarily on residential and light commercial applications. The emphasis of this section is on larger commercial, institutional, and industrial applications. Comprehensive coverage of commercial and institutional design and construction of ground-source heat pump systems is provided in CSA (1993).

TERMINOLOGY

The term ground-source heat pump (GSHP) is applied to a variety of systems that use the ground, groundwater, and surface water as a heat source and sink. Included under the general term are ground-coupled (GCHP), groundwater (GWHP), and surface water (SWHP) heat pump systems. Many parallel terms exist [e.g., geothermal heat pumps (GHPs), earth energy systems, and ground-source (GS) systems] and are used to meet a variety of marketing or institutional needs (Kavanaugh 1992).

Ground-Coupled Heat Pumps

GCHPs are a subset of GSHPs and are often referred to as closed-loop ground-source heat pumps. A GCHP is a system that consists of a reversible vapor compression cycle linked to a closed ground heat exchanger buried in soil (Figure 13). The most widely used unit is the water-to-air heat pump, which circulates water or a water-antifreeze solution through a liquid-to-refrigerant heat exchanger and a

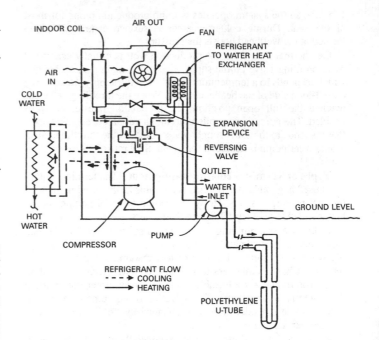

Fig. 13 Vertical Closed-Loop Ground-Coupled Heat Pump System
(Kavanaugh 1985)

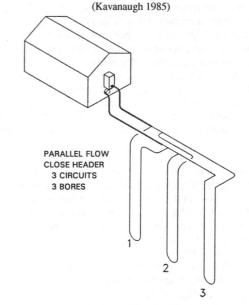

Fig. 14 Vertical Ground-Coupled Heat Pump Piping Arrangements

buried thermoplastic piping network. A second type of GCHP is the direct-expansion (DX) GCHP, which uses a buried copper piping network through which refrigerant is circulated. To distinguish them from DX GCHPs, systems using water-to-air and water-to-water heat pumps are often referred to as GCHPs with secondary solution loops.

GCHPs are further subdivided according to ground heat exchanger design: vertical and horizontal. Vertical GCHPs (Figure 14) generally consist of two small-diameter high-density PE tubes in a vertical borehole. The tubes are thermally fused at the bottom of the bore to a close return U-bend. The nominal diameter of vertical tubes ranges from 20 to 40 mm. Bore depths range from 15 to 185 m depending on local drilling conditions and available equipment.

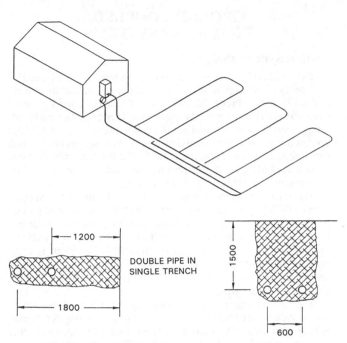

**Fig. 15   Horizontal Ground-Coupled Heat Pump
Piping Arrangements**

The advantages of vertical GCHPs are that they (1) require relatively small plots of ground, (2) are in contact with soil that varies very little in temperature and thermal properties, (3) require the smallest amount of pipe and pumping energy, and (4) can yield the most efficient GCHP system performance. The disadvantage is that they are typically higher in cost because of the limited availability of appropriate equipment and installation personnel.

Horizontal GCHPs (Figure 15) can be divided into three subgroups: single-pipe, multiple-pipe, and spiral. Single-pipe horizontal GCHPs were initially placed in narrow trenches at least four deep. These designs require the greatest amount of ground area. Multiple pipes (usually two or four) placed in a single trench can reduce the required amount of ground area. With multiple-pipe GCHPs, if trench length is reduced, total pipe length must be increased in order to overcome thermal interference with adjacent pipes. The spiral coil is reported to further reduce the required ground area. These horizontal ground heat exchangers are made by stretching small-diameter PE tubing from the tight coil in which it is shipped into an extended coil that can be placed vertically in a narrow trench or laid flat at the bottom of a wide trench. Required trench lengths are only 20 to 30% of those for single-pipe horizontal GCHPs, but pipe lengths may be double for equivalent thermal performance.

The advantages of horizontal GCHPs are that (1) they are typically less expensive than vertical GCHPs because appropriate installation equipment is widely available, (2) many residential applications have adequate ground area, and (3) trained equipment operators are more widely available. Disadvantages include, in addition to the larger ground area requirement, (1) greater adverse variations in performance because ground temperatures and thermal properties fluctuate with season, rainfall, and burial depth; (2) slightly higher pumping energy requirements; and (3) lower system efficiencies. OSU (1988) and Svec (1990) cover the design and installation of horizontal GCHPs.

Groundwater Heat Pumps

The second subset of GSHPs is groundwater heat pumps (Figure 16). Until the recent development of GCHPs, they were the most widely used type of GSHP. In the commercial sector, GWHPs can be an attractive alternative because large quantities of water

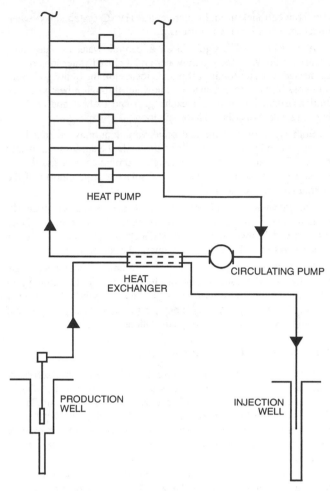

Fig. 16   Unitary Groundwater Heat Pump System

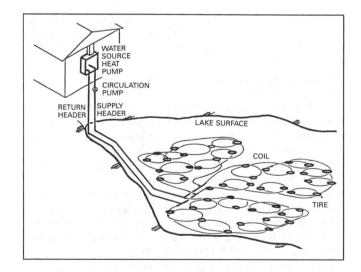

Fig. 17   Spread Coil Loop

can be delivered from and returned to relatively inexpensive wells that require very little ground area. While the cost per kilowatt of the ground heat exchanger is relatively constant for GCHP systems, the cost per kilowatt of a well water system is much lower for a larger GWHP system. A single pair of high-volume wells can serve an entire building. Properly designed groundwater loops with well-developed water wells require no more maintenance than conventional air and water central HVAC systems, and when injection is used, net water use is zero.

A widely used design places a central water-to-water heat exchanger between the groundwater and a closed water loop that is connected to water-to-air heat pumps located in the building. A second possibility is to circulate groundwater through a heat recovery chiller (isolated with a heat exchanger) and to heat and cool the building with a distributed hydronic loop.

Both types of systems and other variations may be suited for direct preconditioning in much of the United States. Groundwater below 15°C can be circulated through hydronic coils in series or in parallel with heat pumps. This can displace a large amount of the energy required for cooling.

The advantages of GWHPs are that (1) they cost less than GCHP systems, (2) the water well is very compact, (3) water well contractors are widely available, and (4) the technology has been used for decades in some of the largest commercial systems.

Disadvantages are that (1) local environmental regulations may be restrictive, (2) water availability may be limited, (3) fouling precautions may be necessary if the wells are not properly developed or if water quality is poor, and (4) pumping energy may be excessive if the pump is oversized or poorly controlled.

Surface Water Heat Pumps

Surface water heat pumps have been included as a subset of GSHPs because of the similarities in applications and installation methods. SWHPs can be either closed-loop systems similar to GCHPs or open-loop systems similar to GWHPs. However, thermal characteristics of surface water bodies are quite different from those of the ground. Some unique applications are possible, and special precautions are warranted.

Closed-loop SWHPs (Figure 17) consist of water-to-air or water-to-water heat pumps connected to a piping network placed in a lake, river, or other open body of water. A pump circulates water or a water-antifreeze solution through the heat pump water-to-refrigerant heat exchanger and the submerged piping loop, which transfers heat to or from the lake. The recommended piping material is thermally fused high-density PE tubing with some type of ultraviolet (UV) radiation protection.

The advantages of closed-loop SWHPs are (1) relatively low cost (compared to GCHPs), (2) low pumping energy requirements, (3) high reliability, (4) low maintenance requirements, and (5) low operating costs. Disadvantages are (1) the possibility of coil damage in public lakes and (2) wide temperature variations with outdoor conditions if lakes are small and/or shallow. This would result in some undesirable variations in efficiency and capacity, though not as severe as those of air-source heat pumps.

Open-loop SWHPs can use surface water bodies the way cooling towers are used, but without the need for fan energy or frequent maintenance. In warm climates, lakes can also serve as heat sources during the winter heating mode, but in colder climates, closed-loop systems are the only viable option for heating.

Lake water can be pumped directly to water-to-air or water-to-water heat pumps or through an intermediate heat exchanger that is connected to the units with a closed piping loop. Direct systems tend to be smaller, having only a few heat pumps. In deep lakes (12 m or more), there is often enough thermal stratification throughout the year that direct cooling or precooling is possible. Water can be pumped from the bottom of deep lakes through a coil in the return air duct. Total cooling is a possibility if the water is 10°C or below. Precooling is possible with warmer water, which can then be circulated through the heat pump units.

GROUND-COUPLED HEAT PUMP SYSTEMS

Vertical Systems Design

The design of vertical ground heat exchangers is complicated by the variety of geological formations and properties that affect thermal performance. Proper identification of materials, moisture content, and water movement is important. However, the necessary information for complex analysis is usually unavailable. One design approach is to apply empirical data to a simplified system of heated (or cooled) pipes placed in the ground. The thermal properties can be estimated using values for soils in a particular group and a moisture content characteristic of local conditions.

This method has proven successful for residential and light commercial GCHP systems. Some permanent change in the local ground temperature may be expected for systems having large annual differences between the amount of heat extracted in the heating mode and the amount rejected in the cooling mode. This problem is compounded in commercial systems because the earth heat exchangers are more likely to be installed in close proximity due to limited ground area availability.

The method described here makes use of a limited amount of information from commercial systems. A major missing component is long-term, field-monitored data. These data are needed to further validate the design method so that the effects of water movement and long-term heat storage are more fully addressed. Currently, the conservative designer can assume no benefit from water movement, while the gambler can assume maximum benefit by ignoring annual imbalances in heat rejection and absorption.

One design method is based upon the solution of the equation for heat transfer from a cylinder buried in the earth. This equation was developed and evaluated by Carslaw and Jaeger (1947) and was suggested by Ingersoll and Zobel (1954) as an appropriate method of sizing ground heat exchangers. Kavanaugh (1994) has adjusted the method of Ingersoll and Zobel to account for the U-bend arrangement and hourly heat rate variations.

It has been demonstrated that the thermal performance of a ground heat exchanger is a strong function of the amount of heat that has been extracted from or rejected to the ground (Eskilson 1987). Minimum and maximum temperatures may take several years to occur, especially if multiple vertical bores are located in close proximity. However, short-term variations in heat transfer are damped by the large thermal mass of the ground surrounding the coils. Therefore, an estimate of the annual net amount is sufficient for design purposes. This concept can also be extended to monthly variations with good accuracy. However, greater care must be exercised in handling daily or hourly variations in heat transfer to the ground.

The method of Ingersoll and Zobel (1954) can be used to handle these shorter term variations. This method uses a steady-state heat transfer equation:

$$q = L\,(t_g - t_w)\,/R \tag{1}$$

where

q = heat transfer rate, kW
L = required bore length, m
t_g = ground temperature, °C
t_w = liquid temperature, °C
R = effective thermal resistance of the ground, (m·K)/kW

The equation is rearranged to solve for the required bore length L. The steady-state equation can be transformed to represent the variable heat rate of a ground heat exchanger by using a series of constant heat rate "pulses." The thermal resistance of the ground per unit length is calculated as a function of time corresponding to the timespan over which a particular heat pulse occurs. A term is also included to account for the thermal resistance of the pipe wall and interfaces between the pipe and the fluid and between the pipe and the ground. The resulting equation takes this form for cooling:

$$L_c = \frac{q_a R_{ga} + (q_{lc} - W_c)(R_p + \text{PLF}_m R_{gm} + R_{gd} F_{sc})}{t_g - \dfrac{t_{wi} + t_{wo}}{2} - t_p} \tag{2}$$

The required length for heating is

$$L_h = \frac{q_a R_{ga} + (q_{lh} - W_h)(R_p + \text{PLF}_m R_{gm} + R_{gd} F_{sc})}{t_g - \dfrac{t_{wi} + t_{wo}}{2} - t_p} \tag{3}$$

where

F_{sc} = short circuit heat loss factor
L_c = required bore length for cooling, m
L_h = required bore length for heating, m
PLF_m = part load factor during design month
q_a = net annual average heat transfer to the ground, kW
q_{lc} = building design cooling block load, kW
q_{lh} = building design heating block load, kW
R_{ga} = effective thermal resistance of the ground—annual pulse, (m·K)/kW
R_{gd} = effective thermal resistance of the ground—daily pulse, (m·K)/kW
R_{gm} = effective thermal resistance of the ground—monthly pulse, (m·K)/kW
R_p = thermal resistance of pipe, (m·K)/kW
t_g = undisturbed ground temperature, °C
t_p = temperature penalty for interference of adjacent bores, K
t_{wi} = liquid temperature at heat pump inlet, °C
t_{wo} = liquid temperature at heat pump outlet, °C
W_c = power input at design cooling load, kW
W_h = power input at design heating load, kW

Note: Heat transfer rate, building loads, and temperature penalties are positive for heating and negative for cooling.

Equations (2) and (3) consider three different pulses of heat to account for long-term heat imbalances (q_a), average monthly heat rates during the design month, and maximum heat rates for a short-term period during a design day. This period could be as short as 1 h, but a 4-h block is recommended.

The required bore is the larger of the two lengths L_c and L_h found from Equations (2) and (3). If L_c is larger than L_h, the benefits of an oversized coil could be enjoyed during the heating season. A second option is to install the smaller heating length along with a cooling tower to compensate for the undersized coil. If L_h is larger, the designer should install this length, and during the cooling mode the efficiency benefits of an oversized ground coil could be used to compensate for the higher first cost.

The thermal resistance of the ground is calculated from ground properties, pipe dimensions, and operating periods of the representative heat rate pulses. Table 6 lists typical thermal properties for soils, and Table 7 gives equivalent thermal resistances and diameters for vertical U-bend heat exchangers.

The most difficult parameters in Equations (2) and (3) to evaluate are the equivalent thermal resistance of the ground. The solutions of Carslaw and Jaeger (1947) require that the time of operation, the outside pipe diameter, and the thermal diffusivity of the ground be related in the dimensionless Fourier Number (Fo):

$$\text{Fo} = \frac{4\alpha_g \tau}{d^2} \tag{4}$$

where

α_g = thermal diffusivity of the ground
τ = time of operation
d = outside pipe diameter

Table 6 Typical Thermal Properties of Soils

	Conductivity, W/(m·K)	Specific Heat, kJ/(kg·K)	Density, kg/m³	Diffusivity, m²/day
Granite	2.1 to 4.5	0.84	2640	0.078 to 0.18
Limestone	1.4 to 5.2	0.88	2480	0.056 to 0.20
Marble	2.1 to 5.5	0.80	2560	0.084 to 0.23
Sandstone				
Dry	1.4 to 5.2	0.71	2240	0.074 to 0.28
Wet	2.1 to 5.2			0.11 to 0.28
Clay				
Damp	1.4 to 1.7	1.3 to 1.7		0.046 to 0.056
Wet	1.7 to 2.4	1.7 to 1.9	1440 to 1920	0.056 to 0.074
Sand				
Damp		1.3 to 1.7		0.037 to 0.046
Wet[a]	2.1 to 2.6	1.7 to 1.9	1440 to 1920	0.065 to 0.084

[a]Water movement will substantially improve thermal properties.

Table 7 Equivalent Diameters and Thermal Resistances of Polyethylene U-Bend Ground Couplings

Nominal Diameter, mm	SDR or Schedule	Equivalent Diameter, m	Thermal Resistance, (m·K)/W Turbulent Flow, Wet Soil	Laminar Flow, Wet Soil	Turbulent or Laminar Flow, Dry Soil
20	SDR 11	0.046	0.058	0.13	0.29+
	SDR 9	0.046	0.069	0.14	0.29+
	Sch 40	0.046	0.064	0.14	0.29+
25	SDR 11	0.055	0.058	0.13	0.29+
	SDR 9	0.055	0.069	0.14	0.29+
	Sch 40	0.055	0.069	0.13	0.29+
38	SDR 11	0.076	0.058	0.13	0.29+
	SDR 9	0.076	0.069	0.14	0.29+
	Sch 40	0.076	0.046	0.12	0.29+

Source: Kavanaugh (1984).

The method of Carslaw and Jaeger may be modified to permit calculation of equivalent thermal resistances for varying heat pulses. A system can be modeled by three heat pulses, a 10-year (3650-day) pulse q_a, a one-month (30-day) pulse q_m, and a 6-h (0.25-day) pulse q_d. Three times are defined as

$$\tau_1 = 3650 \text{ days}$$
$$\tau_2 = 3650 + 30 = 3680 \text{ days}$$
$$\tau_f = 3650 + 30 + 0.25 = 3680.25 \text{ days}$$

The Fourier number is then computed using the following values:

$$\text{Fo}_f = 4\alpha\tau_f/d^2$$
$$\text{Fo}_1 = 4\alpha(\tau_f - \tau_1)/d^2$$
$$\text{Fo}_2 = 4\alpha(\tau_f - \tau_2)/d^2$$

An intermediate step in the computation of the thermal resistance of the ground using the methods of Ingersoll and Zobel (1954) is the determination of a "G-factor." A G-factor for each of the Fourier values is then determined from Figure 18. The three equivalent thermal resistances during each heat pulse are found from

$$R_{ga} = \frac{G_f - G_1}{k_g} \tag{5a}$$

$$R_{gm} = \frac{G_1 - G_2}{k_g} \tag{5b}$$

$$R_{gd} = \frac{G_2}{k_g} \tag{5c}$$

The thermal conductivity of the ground k_g can be found from Table 6.

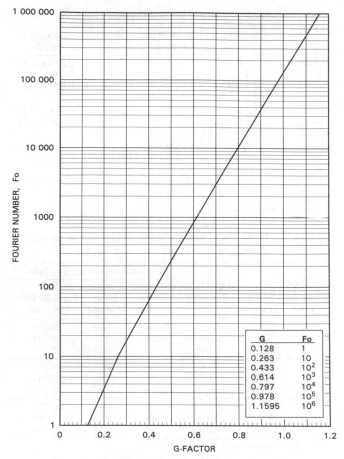

Fig. 18 Fourier/G-Factor Graph for Ground Thermal Resistance

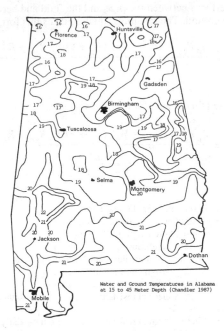

Fig. 19 Water and Ground Temperatures in Alabama at Depths of 15 to 45 m (Chandler 1987)

There is some degradation of performance due to short-circuit heat losses between the upward- and downward-flowing legs of a conventional U-bend loop. When liquid flow rates are 0.054 L/s per kilowatt, the degradation is approximately 4% (Kavanaugh 1984). Losses can be accounted for by multiplying the equivalent thermal resistance for the daily pulse R_{gd} by 1.04 (Kavanaugh 1984). The loss is reduced considerably if there are multiple boreholes on a single parallel loop (see following table).

	F_{sc}	
Bores per Loop	0.054 L/s per kW	0.036 L/s per kW
1	1.06	1.04
2	1.03	1.02
3	1.02	1.01

System Temperatures. The remaining terms in Equations (2) and (3) are temperatures. The local deep ground temperature t_g can best be obtained from local water well logs and geological surveys. A second, less accurate source is temperature contour maps similar to Figure 19 prepared by state geological surveys. A third source, which can yield ground temperatures within 2 K, is a U.S. map with contours, such as Figure 20. Comparison of Figures 19 and 20 indicates the complex variations that would not be accounted for without detailed contour maps.

Selecting the temperature t_{wi} of the water entering the unit is a critical choice in the design process. Choosing a value close to the ground temperature results in higher system efficiency, but the required ground coil length will be very long and thus unreasonably expensive. Choosing a value far from t_g may allow the selection of a small, inexpensive ground coil, but the system's heat pumps will have both greatly reduced capacity during heating and high demand

in cooling. Selecting t_{wi} to be 11 to 14 K higher than t_g in cooling and 8 to 11 K lower than t_g in heating is a good compromise between first cost and efficiency in most regions of the United States.

A final temperature to consider is the temperature penalty t_p resulting from thermal interference from adjacent bores. Again the designer must select a reasonable separation distance in order to minimize required land area without causing large increases in the required bore length (L_c, L_h). The suggested approach is to assume some reasonable value (± 0.5 to 3 K over a 10-year period) when Equations (2) or (3) are applied, calculate the actual penalty, and either increase the separation distance or add length to the bore based upon conditions at the building site. Kavanaugh (1994) gives details on calculating the temperature penalty.

Groundwater movement has a large impact upon the long-term temperature change in a densely packed ground coil. Because the amount of impact has not been thoroughly studied, the design engineer must establish a range of design lengths between one based on minimal groundwater movement, as in very tight clay soils with poor percolation rates, and a second based on the higher rates characteristic of porous aquifers.

Horizontal Systems

The buried pipe of a closed-loop GSHP may theoretically produce a change in temperature in the ground as far as 5 m away. For all practical purposes, however, the ground temperature is essentially unchanged beyond about 1 m from the pipe loop. For that reason, the pipe can be buried relatively near the ground surface and still benefit from the moderating temperatures that the earth provides. Because the ground temperature may fluctuate by as much as 5 K at a depth of 2 m, an antifreeze solution must be used in most heating-dominated regions. The critical design aspect of horizontal applications is to have enough buried pipe loop within the available land area to serve the equipment. The design guidelines for residential horizontal loop installations can be found in NRECA/OSU (1988).

A horizontal loop design has several advantages over a vertical loop design for a closed-loop ground source system:

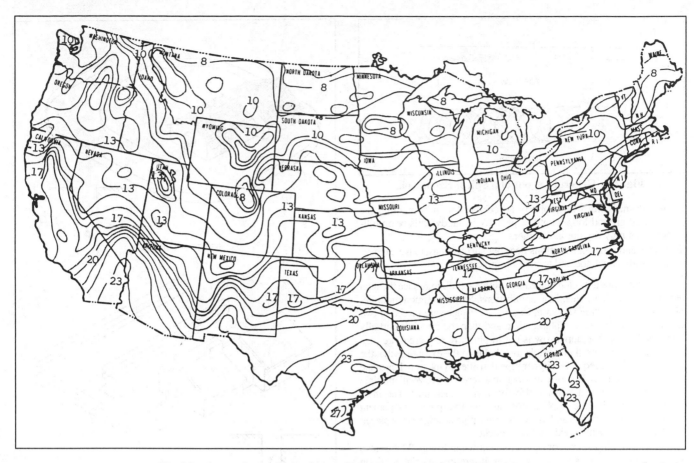

Fig. 20   Approximate Groundwater Temperatures (°C) in the United States
(Courtesy National Ground Water Association)

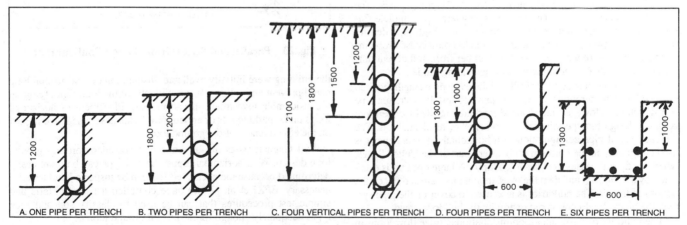

Fig. 1   Horizontal Ground Loop Configurations

1. Installation of the ground loop is usually less expensive than for vertical well designs because the capital cost of a backhoe or trencher is only a fraction of the cost of most drilling rigs.
2. Most ground source contractors must have a backhoe anyway to dig out the header pits and trenches to the building, so they can perform the entire job without having to schedule a special contractor.
3. Large drilling rigs may not be able to get to all locations due to their size and weight.
4. There is usually no potential for aquifer contamination due to the shallow depth of the trench.
5. There is minimal residual temperature effect from unbalanced annual loads on the ground loop because the heat transfer to or from the ground loop is usually small compared to the normal heat transfer at the ground surface.

There are also some limitations on selecting a horizontal loop design:

1. The minimum land area needed for most nonspiral horizontal loop designs for an average house is on the order of 2000 m², so horizontal systems are not feasible for most urban houses, which are commonly built on 1000- to 1200-m² lots.
2. The larger length of pipe buried relatively near the surface risks being cut during excavations for other utilities.

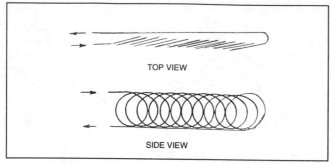

Fig. 22 General Layout of a Spiral Earth Coil

3. Soil dryness must be properly accounted for in computing the required ground loop length, especially in sandy soils or on hilltops that may dry out in summer.

4. Rocks and other obstructions near the surface may make excavation with a backhoe and trencher impractical.

Multiple pipes are often placed in a single trench to reduce the land area needed for horizontal loop applications. Some common multiple-pipe arrangements are shown in Figure 21. If pipes are placed at two depths, the bottom row is placed first, and then the trench is partially backfilled before the upper row is put in place. Rarely are more than two layers of pipe used in a single trench due to the extra time that the partial backfilling requires. Higher pipe densities in the trench provide diminishing returns because thermal interference between multiple pipes reduces the heat transfer effectiveness of each pipe. The most common multiple-pipe applications are the two-pipe arrangement used with chain trenches and the four- or six-pipe arrangements placed in a trench made with a wide backhoe bucket.

An overlapping spiral configuration, shown in Figure 22, has also been used with some success. However, it requires special attention during the backfilling process to ensure that soil fills all the pockets formed by the overlapping pipe. Large quantities of water often must be added to compact the soil around the overlapping pipes. The backfilling must be performed in stages to guarantee complete filling around the pipes and good soil contact. The high pipe density (up to 10 m of pipe per linear metre of trench) could cause problems in prolonged severe weather conditions, either from soil dryout during cooling mode or from freezing during heating mode.

The extra time needed to backfill and the extra pipe length required make spiral configurations just as expensive to install as straight pipe configurations. However, the reduced land area needed for the more compact design may permit their use on smaller residential lots. The flattened spiral pipe configuration in a horizontal pit arrangement is used commonly in the northern midwestern part of the United States, where sandy soil causes trenches to collapse. A large open pit is excavated by a bulldozer, and then the overlapping pipes are laid flat on the bottom of the pit. The bulldozer is also used to cover up the pipe system; the pipes should not be run over with the bulldozer tread.

The majority of horizontal loop installations utilize flow loops in a parallel rather than a single (series) loop to reduce pumping power, as shown in Figure 23. Parallel loops may require slightly more pipe, but they have smaller internal volumes and so need less antifreeze. Also, the smaller pipe is typically much cheaper per metre, so total pipe costs will be less for parallel loops. An added benefit is that parallel loops can be flushed out with a smaller pump than would be required for a larger single-pipe loop.

The time required to install a horizontal loop system is not much different from that for a vertical system. For the arrangements described above, a two-person crew can typically install the ground loops for an average house in a single day.

While not restricted to single-family residential applications, horizontal loop installations are rarely used in larger commercial buildings due to the land area that is required. Even if the land adjacent to

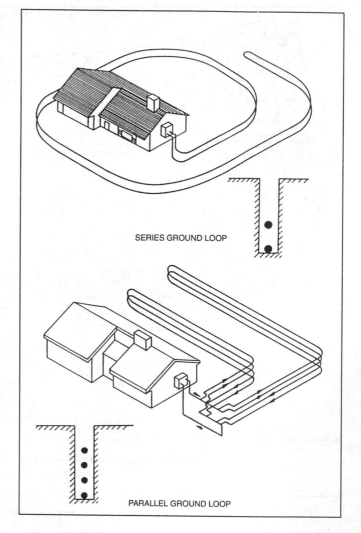

Fig. 23 Parallel and Series Ground Loop Configurations

the building were initially available, installation of a horizontal loop could prevent any future construction above the loop field, tying up a considerable investment in vacant land. Placement of horizontal loops under parking lots or agricultural fields has no negative impact on the effectiveness of the ground loop.

Soil characteristics are an important consideration any ground loop design. With horizontal loops, the soil type can be more easily determined because the excavated soil can be inspected and tested if necessary. EPRI et al. (1989) have compiled a list of criteria and simple test procedures that can be used to classify soil and rock types adequately enough for horizontal ground loop design.

Leaks in the heat-fused plastic pipe are quite rare when proper attention is paid to pipe cleanliness and fusion technique. Should a leak occur, it is usually best to try to isolate the leaking parallel loop and abandon it in place. The time and effort required to find the source of the leak usually far outweigh the cost of replacing the defective loop. Because the loss of as little as 1 L of water from the system will cause it to shut down, leaks cannot be located by looking for wet soil, as is common with water lines.

GROUNDWATER HEAT PUMP SYSTEMS

Groundwater heat pump systems remove groundwater from a well and deliver it to a heat pump (or an intermediate heat exchanger) for the purpose of serving as a heat source or sink. The section of the GWHP system inside the building can be of the unitary or central plant type. In the unitary approach, a large number (up to 20) of small water-to-air heat pumps are distributed throughout the building. The central plant design employs one or a small number of large-capacity chillers supplying hot and chilled water to a two- or four-pipe distribution system.

Regardless of the type of building system, the groundwater-specific components are similar. The primary items include (1) the wells (supply and, if required, injection), (2) the well pump, and (3) the groundwater heat exchanger. The specifics of these items are discussed in the section on Direct-Use Systems. In addition to those comments, the following considerations apply.

Well Pumps

Submersible pumps have not performed well in higher temperature direct-use projects. However, for normal groundwater temperatures such as those encountered in heat pump applications, the submersible pump may be a cost-effective option, particularly for very small projects. The low temperature eliminates the necessity of specifying an industrial design for the motor/protector, thereby greatly reducing the first cost relative to direct-use projects. Caution should still be exercised for wells that are expected to produce moderate amounts of sand. The high speed (3500 rpm) of most submersibles makes them susceptible to erosion damage under these conditions.

Small groundwater systems have frequently been identified with excessive well pump energy consumption. In many cases this is true; however, the reasons for excessive pump energy consumption (high water flow rates, coupling to the domestic pressure tank, and low efficiency of small submersible pumps) are generally not present in large commercial groundwater systems. In large systems, the groundwater flow rate per kilowatt is frequently less than half that of residential systems. The pressure at the wellhead is not the 200 to 350 kPa typical of domestic systems but rather a function only of the losses through the groundwater loop. Finally, large lineshaft well pumps are characterized by bowl efficiencies of up to 83% compared to the 35 to 40% range for small submersible pumps. Together these factors result in a substantial reduction in unit pumping power for large commercial systems compared to small residential systems.

Heat Exchangers

It may be possible to slightly reduce heat exchanger costs in groundwater applications by employing Type 304 stainless steel plates rather than the Type 316 or titanium plates common in direct-use projects. The low temperatures and generally low chloride content of heat pump fluids frequently make the less expensive Type 304 material acceptable.

Direct Systems

Direct systems (those in which the groundwater is used directly in the heat pump without an intermediate heat exchanger) are not recommended except on the very smallest installations. Although some systems with open designs have been successfully implemented, others have had serious difficulty even with groundwater of apparently benign chemistry. As a result, it is considered prudent design for commercial/industrial-scale projects to isolate the groundwater from the building system with an intermediate heat exchanger. The increased capital cost arising from the installation of the heat exchanger amounts to a small percentage of the total cost. In view of the greatly reduced maintenance requirements of closed systems, the increase in capital cost is quickly recovered.

Indirect Systems

Unitary Systems. In unitary systems, the groundwater heat exchanger is placed at the point occupied by the tower/boiler in a conventional water-loop system. If a tower/boiler is to be used for backup purposes, the groundwater exchanger should be placed upstream of it. For most larger building applications, the design of the heat exchanger will be governed by the cooling duty. The exception to this would be a building having little or no central core area in a northern climate with very low temperature groundwater.

The designer should be careful to evaluate the need for insulation on the water-loop piping in the building. If a significant heating load exists, and groundwater temperature is low, it is likely that loop temperatures will be such that insulation will be required in order to avoid sweating.

Groundwater flow requirements in the cooling mode are generally much lower than loop flow rates. For a building with a 280-kW peak cooling load employing an 18°C groundwater resource with a maximum loop temperature of 32°C, the groundwater flow requirement would amount to only 7.6 L/s with a 2.8 K heat exchanger approach temperature. This amounts to only 0.027 L/s per kilowatt. Heating season flow requirements are a function of building load characteristics and groundwater temperatures. The use of extended-range heat pumps, however, preserves the ability of the building to operate at low groundwater flow rates even in a northern climate.

Control of the groundwater flow has been accomplished by a number of methods, including cycling, multiple wells, two-speed well pumps, and well pumps equipped with variable-speed drives. The last approach is the most commonly used today.

A control valve responding to temperature is generally located at the heat exchanger. The valve throttles groundwater flow in response to loop temperature, causing a pressure change in the groundwater line. The variable-speed drive adjusts pump speed in response to pressure in the line. This strategy is especially useful if the pump is remote from the building. The variable-speed drive can also respond directly to loop temperature if it is convenient to run control communications to the wellhead. In either case, it is important that the controls do not allow the pump to operate below some minimum flow rate. It is possible with a vertical well pump for the drive to operate the pump at a speed that does not generate enough pressure to lift water to the surface; operation under these conditions can be damaging to the pump. Therefore, the control scheme should limit the turndown to avoid these conditions.

Central Plant Systems. For central plant groundwater systems, two heat exchangers are normally employed, one in the chilled water loop and one in the condenser water loop (Figure 24A). The evaporator-loop exchanger provides a heat source for heating-dominated operation, and the condenser-loop exchanger provides a heat sink for cooling-dominated operation. A variation of the design involves a heat recovery chiller with an auxiliary condenser to which the groundwater heat exchanger is connected (Figure 24B).

Sizing the condenser loop exchanger is simply a function of providing sufficient capacity to reject the condenser load in the absence of any building heating requirement.

Sizing of the chilled water loop exchanger depends on two loads. The primary criterion is the load required during heating-dominated operation. The exchanger must transfer sufficient heat (when combined with compressor heat) from the groundwater to the chilled water loop to meet the space heating requirement of the building. Depending on the relative groundwater and chilled water temperatures and on the design rise, the exchangers may also provide some free cooling during cooling-dominated operation. If the groundwater temperature is lower than the temperature of the chilled water returning to the exchanger, some of the chilled water load can be met by the exchanger. This mode is most likely to be available in northern climates with groundwater temperatures below 15°C.

For the unitary indirect system, controls unique to the groundwater approach are largely related to the well pump. For the central plant design, chiller controls must also allow for the unique operation with a groundwater source. Well pump control options are as discussed in the section on Unitary Systems. Controls for the chiller can be similar to those on a heat recovery chiller system with a

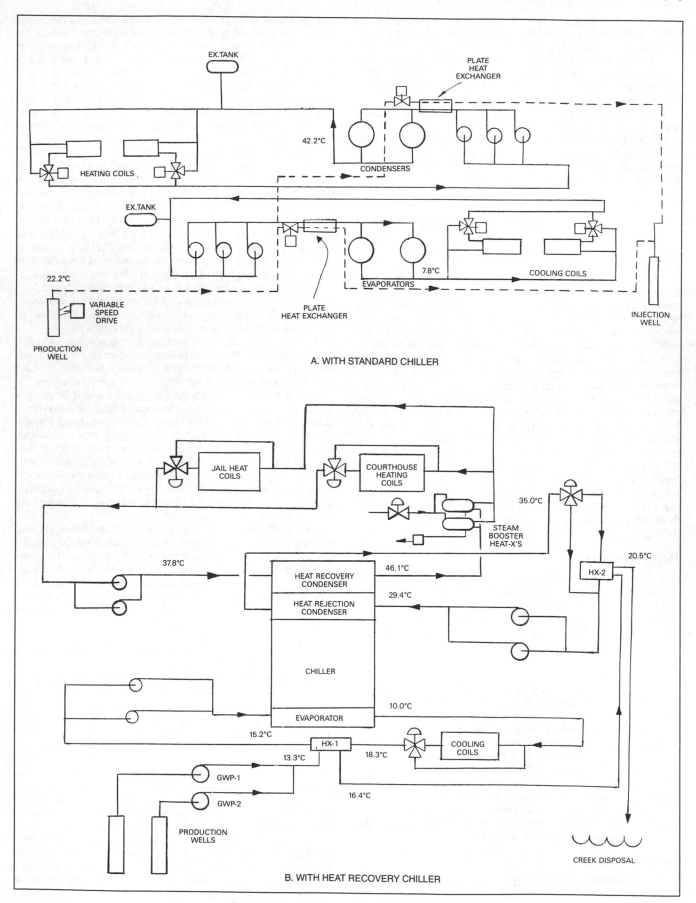

A. WITH STANDARD CHILLER

B. WITH HEAT RECOVERY CHILLER

Fig. 24   Central Plant Groundwater Systems

tower, with one important difference. In a conventional heat recovery chiller system, waste heat is available only when there is a building chilled water (or conditioning) load. In a groundwater system, a heat source (the groundwater) is available year-round. To take advantage of this source, it is necessary during the heating season to load the chiller in response to the heating load instead of the chilled water load. That is, the control scheme must include a heating-dominated mode and a cooling-dominated mode. Two general approaches are available for this: (1) the chiller capacity remains controlled by chilled water (supply or return) temperature, and the groundwater flow through the chilled water exchanger is varied in response to heating load, or (2) the chiller capacity is controlled by the heating water (condenser) loop temperature, and the groundwater flow through the chilled water exchanger is controlled by chilled water temperature. For buildings with a significant heating load, the former may be more attractive, while the latter may be appropriate for conventional buildings in moderate-to-warm climates.

SURFACE WATER HEAT PUMP SYSTEMS

Surface water bodies can be very good heat sources and sinks if properly utilized. In some cases, lakes can be the best water systems for cooling operation. A variety of water circulation systems are possible; several of the more common are presented.

In a closed-loop system, water-to-air heat pumps are linked to a submerged coil. Heat is exchanged to (cooling mode) or from (heating mode) the lake by the fluid (usually a water-antifreeze mixture) circulating inside the coil. The heat pumps are used to transfer heat to or from the air in the building.

In an open-loop system, water is pumped from the lake through a heat exchanger and returned to the lake some distance from the point at which it was removed. The pump can be located either slightly above or submerged below the lake water level. For heat pump operation in the heating mode, this type of system is restricted to warmer climates; water temperatures must remain above 5.5°C.

Thermal stratification of water often causes large quantities of cold water to remain undisturbed near the bottom of deep lakes. This water is cold enough to adequately cool buildings by simply being circulated through heat exchangers. A heat pump is not needed for cooling, and energy use is substantially reduced. Closed-loop coils may also be used in colder lakes. Heating can be provided by a separate source or with heat pumps in the heating mode. Precooling or supplemental total cooling are also permitted when water temperatures are between 10 and 15°C.

Heat Transfer in Lakes

Heat is transferred to lakes by three primary modes: radiation from the sun, convective heat transfer from the surrounding air (when the air temperature is greater than the water temperature), and conduction from the ground. Solar radiation, which can exceed 950 W per square metre of lake area, is the dominant heating mechanism, but it occurs primarily in the upper portion of the lake unless the lake is very clear. About 40% of the solar radiation is absorbed at the surface (Pezent and Kavanaugh 1990). Approximately 93% of the remaining energy is absorbed at depths visible to the human eye.

Convective heat transfer to the lake occurs when the lake surface temperature is lower than the air temperature. Wind speed increases the rate at which heat is transferred to the lake, but maximum heat gain by convection is usually only 10 to 20% of maximum solar heat gain. The conduction gain from the ground is even less than convection gain (Pezent and Kavanaugh 1990).

Cooling of lakes is accomplished primarily by evaporative heat transfer at the surface. Convective cooling or heating in warmer months contributes only a small percentage of the total because of the relatively small temperature differences between the air and the lake surface temperature. Back radiation, which typically occurs at night when the sky is cool, can account for a significant amount of cooling. The relatively warm water surface radiates heat to the cooler sky. For example, on a clear night, a cooling rate of up to 160 W/m^2 is possible for a lake 14 K warmer than the sky. The last major mode of heat transfer, conduction to the ground, does not play a major role in lake cooling (Pezent and Kavanaugh 1990).

To put these heat transfer rates in perspective, consider a 4000 m^2 lake that is used in connection with a 35-kW heat pump. In the cooling mode, the unit will reject approximately 44 kW to the lake. This is 11 W/m^2, or approximately 1% of the maximum heat gain from solar radiation in the summer. In the winter, a 35-kW heat pump would absorb only about 26 kW, or 6.5 kW/m^2, from the lake.

Thermal Patterns in Lakes

The maximum density of water occurs at 4.0°C, not at the freezing point of 0°C. This phenomenon, in combination with the normal modes of heat transfer to and from lakes, produces temperature profiles advantageous to efficient heat pump operation.

In the winter, the coldest water is at the surface. It tends to remain at the surface and freeze. The bottom of a deep lake stays 3 to 5 K warmer than the surface. This condition is referred to as winter stagnation. The warmer water is a better heat source than the colder water at the surface.

As spring approaches, surface water warms until the temperature approaches the maximum density point of 4.0°C. The winter lake stratification becomes unstable, and circulation loops begin to develop from top to bottom. This condition is called spring overturn (Peirce 1964). The lake temperature becomes fairly uniform.

Later in the spring as the water temperatures rise above 7.2°C, the circulation loops are in the upper portion of the lake. This pattern continues throughout the summer. The upper portion of the lake remains relatively warm, with evaporation cooling the lake and solar radiation warming it. The lower portion (hypolimnion) of the lake remains cold because most radiation is absorbed in the upper zone, circulation loops do not penetrate to the lower zone, and conduction to the ground is quite small. The result is that in deeper lakes with small or medium inflows, the upper zone is 21 to 32°C, the lower zone is 4 to 13°C, and the intermediate zone (thermocline) has a sharp change in temperature within a small change in depth. This condition is referred to as summer stagnation.

As fall begins, the water surface begins to cool through back radiation and evaporation. With the approach of winter, the upper portion begins to cool toward the freezing point, and the lower levels approach the maximum density temperature of 4.0°C. An ideal temperature versus depth chart is shown in Figure 25 for each of the four seasons (Peirce 1964).

Many lakes do exhibit near-ideal temperature profiles. However, a variety of circumstances can disrupt the profile. These circumstances include (1) high inflow/outflow rates, (2) insufficient depth for stratification, (3) level fluctuations, (4) wind, and (5) lack of enough cold weather to establish sufficient amounts of cold water for summer stratification. Therefore, it is suggested that a thermal survey of the lake be conducted or that existing surveys of similar lakes in similar geographic locations be consulted.

Closed-Loop Lake Water Heat Pump Systems

The closed-loop lake water heat pump system shown in Figure 17 has several advantages over open-loop systems. The most obvious advantage is the reduced fouling resulting from the circulation of clean water (or water-antifreeze solution) through the heat pump. A second advantage is the reduced pumping power requirement. This results from the absence of an elevation pressure from the lake surface to the heat pumps. A third advantage of closed-loop systems is that they are the only type recommended if winter lake temperatures below 4°C are possible. The outlet temperature of the fluid will be

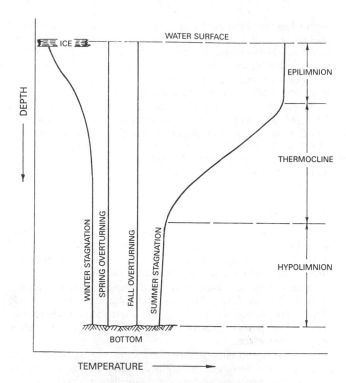

Fig. 25   Idealized Diagram of Annual of Thermal Stratification Cycle

about 3 K below the inlet for a flow of 0.054 L/s per kilowatt. Frosting will occur on the heat exchanger surfaces when the bulk water temperature is in the 1 to 3°C range.

There are disadvantages to the closed-loop system. The performance of the heat pump is slightly reduced because the circulation fluid temperature degrades 2 to 7 K compared to the lake temperature. A second disadvantage of the closed-loop system is the possibility of damage to coils located in public lakes. Thermally fused polyethylene or polybutylene loops are much more resistant to damage than copper, glued plastic (PVC), or tubing with band-clamped joints. A third possible disadvantage is fouling on the outside of the lake coil. This is a possibility in murky lakes or for installations in which coils are located on or near the lake bottom.

Currently acceptable options for in-lake piping are high-density polyethylene (PE 3408) and polybutylene. All connections must be either thermally socket fused or butt fused. These plastic pipes should also have protection from UV radiation, especially when near the surface. Polyvinyl chloride (PVC) pipe and plastic pipe with band-clamped joints are not acceptable or recommended.

The piping networks of closed-loop systems resemble those used in ground-coupled heat pump systems. Both a large-diameter header between the heat pump and lake coil and several parallel loops of piping in the lake are required. The loops are spread out to limit thermal interference, hot spots, and cold pockets. While this layout is preferred in terms of performance, installation is more time consuming. Many contractors simply unbind plastic pipe coils and submerged them in a loose bundle. Some compensation for thermal interference is obtained by making the bundled coils longer than the spread coils. A diagram of this type of installation is shown in Figure 26.

Copper coils have also been used successfully. Copper tubes have a very high thermal conductivity, so coils only one-fourth to one-third the length of plastic coils are required. However, copper pipe does not have the durability of PE 3408 or polybutylene, and if the possibility of fouling exists, coils must be significantly longer. The percent degradation will be much greater because copper coils are so much shorter than plastic.

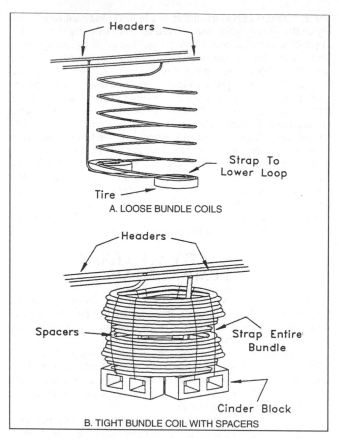

Fig. 20   Closed-Loop Lake Coil in Bundles

Three antifreeze solutions are commonly used. Propylene glycol has a very low toxicity without inhibitors. It is the most viscous of the three and has the poorest performance. Ethylene glycol has similar properties. Its viscosity is slightly lower, but it is toxic and some authorities may limit its use in public lakes. Alcohol-water mixtures are a good compromise in terms of freeze protection and performance.

REFERENCES

Anderson, K.E. 1984. *Water well handbook.* Missouri Water Well and Pump Contractors Association, Belle, MD.

Austin, J.C. 1978. A low temperature geothermal space heating demonstration project. *Geothermal Resources Council Transactions* 2(2).

Bullard, E. 1973. *Basic theories* (Geothermal Energy; Review of Research and Development), UNESCO, Paris, France.

CSA. 1993. Design and construction of earth energy heat pump systems for commercial and institutional buildings. C447-93. Canadian Standards Association, Rexdale, ON.

Campbell, M.D. and J.H. Lehr. 1973. *Water well technology.* McGraw-Hill, New York.

Carslaw, H.S. and J.C. Jaeger. 1947. *Heat conduction in solids.* Claremore Press, Oxford.

Chandler, R.V. 1987. Alabama streams, lakes, springs and ground waters for use in heating and cooling. *Bulletin* 129. Geological Survey of Alabama, Tuscaloosa, AL.

Christen, J.E. 1977. *Central cooling—Absorption chillers.* Oak Ridge National Laboratories, Oak Ridge, TN.

Combs, J., J.K. Applegate, R.O. Fournier, C.A. Swanberg, and D. Nielson. 1980. Exploration, confirmation and evaluation of the resource. In Geothermal Resources Council *Special Report* No. 7, Direct utilization of geothermal energy: Technical handbook.

Cosner, S.R. and J.A. Apps. 1978. A compilation of data on fluids from geothermal resources in the United States. DOE *Report* LBL-5936. Lawrence Berkeley Laboratory, Berkeley, CA.

Culver, G.G. and G.M. Reistad. 1978. Evaluation and design of downhole heat exchangers for direct applications. DOE *Report* RLO-2429-7.

Di Pippo, R. 1988. Industrial developments in geothermal power production. *Geothermal Resources Council Bulletin* 17(5).

Efrid, K.D. and G.E. Moeller. 1978. Electrochemical characteristics of 304 and 316 stainless steels in fresh water as functions of chloride concentration and temperature. *Paper* 87. Corrosion/78, Houston, TX (March).

Electric Power Research Institute, National Rural Electric Cooperative Association, and Oklahoma State University. 1989. Soil and rock classification for the design of ground-coupled heat pump systems. International Ground Source Heat Pump Association, Stillwater, OK.

Ellis, P. 1989. Materials selection guidelines. *Geothermal direct use engineering and design guidebook*, Ch. 8. Oregon Institute of Technology, Geo-Heat Center, Klamath Falls, OR.

Ellis, P. and C. Smith. 1983. Addendum to material selection guidelines for geothermal energy utilization systems. Radian Corporation, Austin, TX.

Ellis, P.F. and M.F. Conover. 1981. Material selection guidelines for geothermal energy utilization systems. DOE *Report* RA/27026-1. Radian Corporation, Austin, TX.

EPA. 1975. Manual of water well construction practices. EPA-570/9-75-001. U.S. Environmental Protection Agency, Washington, D.C.

Eskilson, P. 1987. *Thermal analysis of heat extraction boreholes*. University of Lund, Sweden.

Gudmundsson, J.S. 1985. Direct uses of geothermal energy in 1984. Geothermal Resources Council Proceedings, 1985 International Symposium on Geothermal Energy, International Volume, Davis, CA.

Hackett, G. and J.H. Lehr. 1985. *Iron bacteria occurrence problems and control methods in water wells*. National Water Well Association, Worthington, OH.

Ingersoll, L.R. and A.C. Zobel. 1954. *Heat conduction with engineering and geological applications*, 2nd ed. McGraw-Hill, New York.

Interagency Geothermal Coordinating Council. Geothermal energy, research, development and demonstration program. DOE *Report* RA-0050, IGCC-5. U.S. Department of Energy, Washington, D.C.

Kavanaugh, S.P. 1985. Simulation and experimental verification of a vertical ground-coupled heat pump system. Ph.D. diss., Oklahoma State University, Stillwater, OK.

Kavanaugh, S.P. 1991. *Ground and water source heat pumps*. Oklahoma State University, Stillwater, OK.

Kavanaugh, S.P. 1992. Ground-coupled heat pumps for commercial buildings. *ASHRAE Journal* 34(9):30-37.

Kavanaugh, S.P. 1994. *Ground source heat pumps for commercial buildings*. Energy Information Services, Tuscaloosa, AI.

Kavanaugh, S.P. and M.C. Pezent, 1990. Lake water applications of water-to-air heat pumps. *ASHRAE Transactions* 96(1):813-20.

Kindle, C.H. and E.M. Woodruff. 1981. Techniques for geothermal liquid sampling and analysis. Battelle Pacific Northwest Laboratory, Richland, WA.

Lienau, P.J. 1979. Materials performance study of the OIT geothermal heating system. Geo-Heat Utilization Center Quarterly Bulletin, Oregon Institute of Technology, Klamath Falls, OR.

Lienau, P.J., G.G. Culver, and J.W. Lund. 1988. Geothermal direct use developments in the United States. Oregon Institute of Technology, Geo-Heat Center, Klamath Falls, OR.

Lund, J.W., P.J. Lienau, G.G. Culver, and C.V. Higbee. 1979. Klamath Falls geothermal heating district. *Geothermal Resources Council Transactions* 3.

Lunis, B. 1989. Environmental considerations. *Geothermal direct use engineering and design guidebook*, Ch. 20. Oregon Institute of Technology, Geo-Heat Center, Klamath Falls, OR.

Mitchell, D.A. 1980. Performance of typical HVAC materials in two geothermal heating systems. *ASHRAE Transactions* 86(1):763-68.

Muffler, L.J.P., ed. 1979. Assessment of geothermal resources of the United States—1978. U.S. Geological Survey *Circular* No. 790.

Nichols, C.R. 1978. Direct utilization of geothermal energy: DOE's resource assessment program. Direct Utilization of Geothermal Energy: A Symposium. Geothermal Resources Council.

NRECA/OSU. 1988. *Closed-loop/ground-source heat pump systems installation guide*. International Ground Source Heat Pump Association, National Rural Electric Cooperative Association, Oklahoma State University, Stillwater, OK.

OSU. 1988. *Closed loop ground source heat pump systems*. Oklahoma State University, Stillwater, OK.

Peirce, L.B. 1964. Reservoir temperatures in north central Alabama. Geological Survey of Alabama, *Bulletin* 82, Tuscaloosa, AL.

Pezent, M.C. and S.P. Kavanaugh. 1990. Development and verification of a thermal model of lakes used with water-source heat pumps. *ASHRAE Transactions* 96(1):574-82.

Rafferty, K. 1989a. A materials and equipment review of selected U.S. geothermal district heating systems. Oregon Institute of Technology, Geo-Heat Center, Klamath Falls, OR.

Rafferty, K. 1989b. Absorption refrigeration. *Geothermal direct use engineering and design guidebook*, Ch. 14. Oregon Institute of Technology, Geo-Heat Center, Klamath Falls, OR.

Reistad, G.M., G.G. Culver, and M. Fukuda. 1979. Downhole heat exchangers for geothermal systems: Performance, economics and applicability. *ASHRAE Transactions* 85(1):929-39.

Roscoe Moss Company. 1985. *The engineers manual for water well design*. Roscoe Moss Company, Los Angeles, CA.

Stiger, S., J. Renner, and G. Culver. 1989. Well testing and reservoir evaluation. *Geothermal and direct use engineering and design guidebook*, Ch. 7. Oregon Institute of Technology, Geo-Heat Center, Klamath Falls, OR.

Svec, O.J. 1990. Spiral ground heat exchangers for heat pump applications. Proceedings of 3rd IEA Heat Pump Conference, Pergamon Press, Tokyo.

UOP. 1975. *Ground water and wells*. Johnson Division, UOP Inc., St. Paul, MN.

BIBLIOGRAPHY

Allen, E. 1980. Preliminary inventory of western U.S. cities with proximate hydrothermal potential. Eliot Allen and Associates, Salem, OR.

Anderson, D.A. and J.W. Lund, eds. 1980. Direct utilization of geothermal energy: Technical handbook. Geothermal Resource Council *Special Report* No. 7.

SOLAR ENERGY UTILIZATION

SOLAR water heaters were first used in the United States in the early 1900s. However, the introduction of low-cost water heaters using fossil fuels and electricity ended the use of solar water heaters except for a few applications where conventional fuels were unavailable or cost-prohibitive. In the 1920s and 1930s, interest in the use of solar radiation for space heating increased. Early solar houses (located north of the equator) used large expanses of south-facing glass. While the extra glass admitted large amounts of solar radiation, it also lost so much heat at night and on cloudy days that the actual energy savings were minimal.

Some passive solar homes use movable insulation and efficient glazing systems to minimize heat loss when the sun is not shining. The installation of solar energy collectors on roofs or walls, and the use of water or air as a heat transfer medium with rock beds or heat-of-fusion materials for energy storage have received equal interest.

The major obstacles encountered in solar heating and cooling are economic, resulting from the high cost of the equipment needed to collect and store solar energy. In some cases, the cost of the solar equipment is greater than the resulting savings in fuel costs. Some problems that must be overcome are inherent in the nature of solar radiation, for example,

- It is relatively low in intensity, rarely exceeding 950 W/m². Consequently, when large amounts of energy are needed, large collectors must be used.
- It is intermittent because of the inevitable variation in solar radiation intensity from zero at sunrise to a maximum at noon and back to zero at sunset. Some means of energy storage must be provided at night and during periods of low solar irradiation.
- It is subject to unpredictable interruptions because of clouds, rain, snow, hail, or dust.
- Systems should be chosen that make maximum use of the solar energy input by effectively using the energy at the lowest temperatures possible.

QUALITY AND QUANTITY OF SOLAR ENERGY

Solar Constant, I_{sc}

Solar energy approaches the earth as electromagnetic radiation, with wavelengths ranging from 0.1 μm (X rays) to 100 m (radio waves). The earth maintains a thermal equilibrium between the annual input of shortwave radiation (0.3 to 2.0 μm) from the sun and the outward flux of longwave radiation (3.0 to 30 μm). This equilibrium is global rather than local, because at any given time, some places are too cold for human existence while others are too hot.

Only a limited band need be considered in terrestrial applications, because 99% of the sun's radiant energy has wavelengths

The preparation of this chapter is assigned to TC 6.7, Solar Energy Utilization.

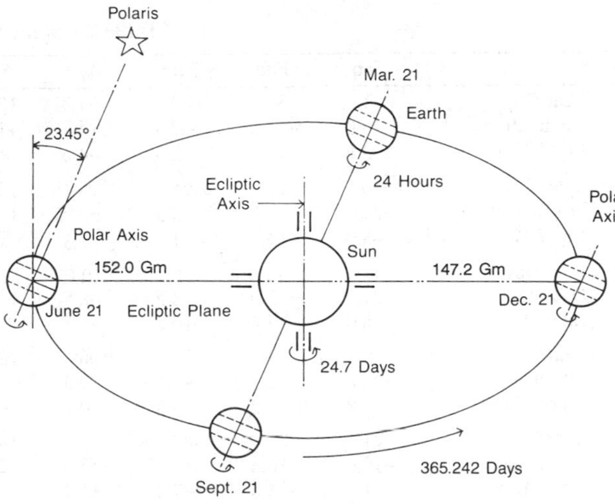

Fig. 1   Annual Motion of the Earth about the Sun

between 0.28 and 4.96 μm. The most probable value of the solar constant (which is defined as the intensity of solar radiation on a surface normal to the sun's rays, just beyond the earth's atmosphere at the average earth-sun distance) is 1355 W/m² ± 1.5%. The radiation scale in use since 1956 is considered to be 2% too low and varies slightly because of changes in the sun's output of ultraviolet radiation.

The major variations in solar radiation intensity and air temperature are the result of the slightly elliptical nature of the earth's orbit around the sun and, with respect to the orbital plane, the tilt of the axis about which the earth rotates (see Figure 1). The sun is located at one focus of the earth's orbit and is 147 Gm away in late December and early January, while the earth-sun distance on July 1 is about 152 Gm. Because radiation follows the inverse square law, the normal incidence intensity on an extraterrestrial surface varies from 1401 W/m² on January 1 to 1311 W/m² on July 5.

Solar Angles

The axis about which the earth rotates is tilted at an angle of 23.45° to the plane of the ecliptic that contains the earth's orbital plane and the sun's equator. The earth's tilted axis results in a day-by-day variation of the angle between the earth-sun line and the earth's equatorial plane, called the *solar declination* δ. This angle varies with the date, as shown in Table 1 for the year 1964 and in Table 2 for 1977. For other dates, the declination may be estimated by the following equation:

$$\delta = 23.45 \sin [360° (284 + N)/365] \qquad (1)$$

Table 1   Date, Declination, and Equation of Time for the 21st Day of Each Month of 1964, with Data (A, B, C) Used to Calculate Direct Normal Radiation Intensity at the Earth's Surface

	Jan	Feb	Mar	Apr	May	June	July	Aug	Sept	Oct	Nov	Dec
Year Day	21	52	80	111	141	172	202	233	264	294	325	355
Declination δ, degrees	−19.9	−10.6	0.0	+11.9	+20.3	+23.45	+20.5	+12.1	0.0	−10.7	−19.9	−23.45
Equation of time, minutes	−11.2	−13.9	−7.5	+1.1	+3.3	−1.4	−6.2	−2.4	+7.5	+15.4	+13.8	+1.6
Solar noon		late			early			late			early	
A, W/m²	1230	1215	1186	1136	1104	1088	1085	1107	1152	1193	1221	1234
B, dimensionless	0.142	0.144	0.156	0.180	0.196	0.205	0.207	0.201	0.177	0.160	0.149	0.142
C, dimensionless	0.058	0.060	0.071	0.097	0.121	0.134	0.136	0.122	0.092	0.073	0.063	0.057

A = apparent solar irradiation at air mass zero for each month.
B = atmospheric extinction coefficient.
C = ratio of the diffuse radiation on a horizontal surface to the direct normal irradiation.

Table 2   Solar Position Data for 1977

Date		Jan	Feb	Mar	Apr	May	June	July	Aug	Sept	Oct	Nov	Dec
1	Year Day	1	32	60	91	121	152	182	213	244	274	305	335
	Declination δ	−23.0	−17.0	−7.4	+4.7	+15.2	+22.1	+23.1	+17.9	+8.2	−3.3	−14.6	−21.9
	Eq of Time	−3.6	−13.7	−12.5	−4.0	+2.9	+2.4	−3.6	−6.2	+0.0	+10.2	+16.3	+11.0
6	Year Day	6	37	65	96	126	157	187	218	249	279	310	340
	Declination δ	−22.4	−15.5	−5.5	+6.6	+16.6	+22.7	+22.7	+16.6	+6.7	−5.3	−16.1	−22.5
	Eq of Time	−5.9	−14.2	−11.4	−2.5	+3.5	+1.6	−4.5	−5.8	+1.6	+11.8	+16.3	+9.0
11	Year Day	11	42	70	101	131	162	192	223	254	284	315	345
	Declination δ	−21.7	−13.9	−3.5	+8.5	+17.9	+23.1	+22.1	+15.2	+4.4	−7.2	−17.5	−23.0
	Eq of Time	−8.0	−14.4	−10.2	−1.1	+3.7	+0.6	−5.3	−5.1	+3.3	+13.1	+15.9	+6.8
16	Year Day	16	47	75	106	136	167	197	228	259	289	320	350
	Declination δ	−20.8	−12.2	−1.6	+10.3	+19.2	+23.3	+21.3	+13.6	+2.5	−8.7	−18.8	−23.3
	Eq of Time	−9.8	−14.2	−8.8	+0.1	+3.8	−0.4	−5.9	−4.3	+5.0	+14.3	+15.2	+4.4
21	Year Day	21	52	80	111	141	172	202	233	264	294	325	355
	Declination δ	−19.6	−10.4	+0.4	+12.0	+20.3	+23.4	+20.6	+12.0	+0.5	−10.8	−20.0	−23.4
	Eq of Time	−11.4	−13.8	−7.4	+1.2	+3.6	−1.5	−6.2	−3.1	+6.8	+15.3	+14.1	+2.0
26	Year Day	26	57	85	116	146	177	207	238	269	299	330	360
	Declination δ	−18.6	−8.6	+2.4	+13.6	+21.2	+23.3	+19.3	+10.3	−1.4	−12.6	−21.0	−23.4
	Eq of Time	−12.6	−13.1	−5.8	+2.2	+3.2	−2.6	−6.4	−1.8	+8.6	+15.9	+12.7	−0.5

Source: ASHRAE *Standard* 93-1991.
Notes: Units for declination are angular degrees; units for equation of time are minutes. Values of declination and equation of time vary slightly for specific dates in other years.

where N = year day, with January 1 = 1. For values of N, see Tables 1 and 2.

The relationship between δ and the date varies to an insignificant degree. The daily change in the declination is the primary reason for the changing seasons, with their variation in the distribution of solar radiation over the earth's surface and the varying number of hours of daylight and darkness.

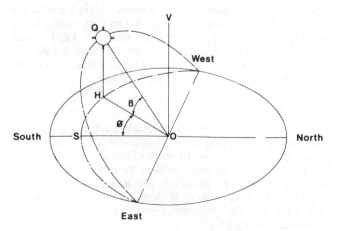

Fig. 2   Apparent Daily Path of the Sun Showing Solar Altitude (β) and Solar Azimuth (φ)

The earth's rotation causes the sun's apparent motion (Figure 2). The position of the sun can be defined in terms of its altitude β above the horizon (angle HOQ) and its azimuth φ, measured as angle HOS in the horizontal plane.

At solar noon, the sun is, by definition, exactly on the meridian, which contains the south-north line and, consequently, the solar azimuth φ is 0.0°. The *noon altitude* β_N is given by the following equation:

$$\beta_N = 90° - LAT + \delta \tag{2}$$

where LAT = latitude.

Because the earth's daily rotation and its annual orbit around the sun are regular and predictable, the solar altitude and azimuth may be readily calculated for any desired time of day when the latitude, longitude, and date (declination) are specified. Apparent solar time (AST) must be used, expressed in terms of the hour angle H, where

$$H = \text{(number of hours from solar noon) } 15°$$
$$= \text{(number of minutes from solar noon) } /4 \tag{3}$$

Solar Time

Apparent solar time (AST) generally differs from local standard time (LST) or daylight saving time (DST), and the difference can be significant, particularly when DST is in effect. Because the sun appears to move at the rate of 360° in 24 h, its apparent rate of motion is 4 min per degree of longitude. The AST can be determined from the following equation:

$$AST = LST + \text{Equation of Time}$$
$$+ (4 \text{ min}) \text{ (LST Meridian} - \text{Local Longitude)} \qquad (4)$$

The longitudes of the seven standard time meridians that affect North America are Atlantic ST, 60°; Eastern ST, 75°; Central ST, 90°; Mountain ST, 105°; Pacific ST, 120°; Yukon ST, 135°; and Alaska-Hawaii ST, 150°.

The *equation of time* is the measure, in minutes, of the extent by which solar time, as determined by a sundial, runs faster or slower than local standard time (LST), as determined by a clock that runs at a uniform rate. Table 1 gives values of the declination of the sun and the equation of time for the 21st day of each month for the year 1964 (when the ASHRAE solar radiation tables were first calculated), while Table 2 gives values of δ and the equation of time for six days each month for the year 1977.

Example 1. Find AST at noon DST on July 21 for Washington, D.C., longitude = 77°, and for Chicago, longitude = 87.6°.

Solution: Noon DST is actually 11:00 A.M. LST. Washington is in the eastern time zone, and the LST meridian is 75°. From Table 1, the equation of time for July 21 is −6.2 min. Thus, from Equation (4), noon DST for Washington is actually

$$AST = 11:00 - 6.2 + 4 (75 - 77) = 10:45.8 \text{ AST} = 10.76 \text{ h}$$

Chicago is in the central time zone, and the LST meridian is 90°. Thus, from Equation (4), noon central DST is

$$AST = 11:00 - 6.2 + 4 (90 - 87.6) = 11:03.4 \text{ AST} = 11.06 \text{ h}$$

The hour angles H, for these two examples (see Figure 2) are

for Washington, $H = (12.00 - 10.76) \, 15° = 18.6°$ east
for Chicago, $H = (12.00 - 11.06) \, 15° = 14.10°$ east

To find the solar altitude β and the azimuth ϕ when the hour angle H, the latitude LAT, and the declination δ are known, the following equations may be used:

$$\sin \beta = \cos (\text{LAT}) \cos \delta \cos H + \sin (\text{LAT}) \sin \delta \qquad (5)$$

$$\sin \phi = \cos \delta \sin H / \cos \beta \qquad (6)$$

or

$$\cos \phi = [\sin \beta \sin (\text{LAT}) - \sin \delta] / \cos \beta \cos (\text{LAT}) \qquad (7)$$

Tables 12 through 18 in Chapter 27 of the 1993 *ASHRAE Handbook—Fundamentals* give values for latitudes from 16 to 64° north. For any other date or latitude, interpolation between the tabulated values will give sufficiently accurate results. More precise values, with azimuths measured from the north, are given in the U.S. Hydrographic Office Bulletin No. 214 (1958).

Incident Angle

The angle between the line normal to the irradiated surface (OP′ in Figure 3) and the earth-sun line OQ is called the incident angle. It is important in solar technology because it affects the intensity of the direct component of the solar radiation striking the surface and the ability of the surface to absorb, transmit, or reflect the sun's rays.

To determine θ, the surface azimuth ψ and the surface-solar azimuth γ must be known. The surface azimuth (angle POS in Figure 3) is the angle between the south-north line SO and the normal PO to the intersection of the irradiated surface with the horizontal plane, shown as line OM. The surface-solar azimuth, angle HOP, is designated by γ and is the angular difference between the solar

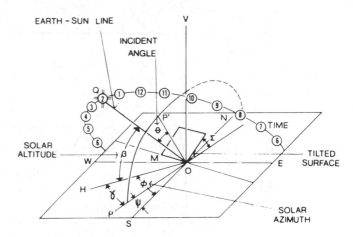

Fig. 3 Solar Angles with Respect to a Tilted Surface

azimuth ϕ and the surface azimuth ψ. For surfaces facing *east* of south, $\gamma = \phi - \psi$ in the morning and $\gamma = \phi + \psi$ in the afternoon. For surfaces facing *west* of south, $\gamma = \phi + \psi$ in the morning and $\gamma = \phi - \psi$ in the afternoon. For south-facing surfaces, $\psi = 0°$, so $\gamma = \phi$ for all conditions. The angles δ, β, and ϕ are always positive.

For a surface with a tilt angle Σ (measured from the horizontal), the angle of incidence θ between the direct solar beam and the normal to the surface (angle QOP′ in Figure 3) is given by:

$$\cos \theta = \cos \beta \cos \gamma \sin \Sigma + \sin \beta \cos \Sigma \qquad (8)$$

For vertical surfaces, $\Sigma = 90°$, $\cos \Sigma = 0$, and $\sin \Sigma = 1.0$, so Equation (8) becomes

$$\cos \theta = \cos \beta \cos \gamma \qquad (9)$$

For horizontal surfaces, $\Sigma = 0°$, $\sin \Sigma = 0$, and $\cos \Sigma = 1.0$, so Equation (8) leads to

$$\theta_H = 90° - \beta \qquad (10)$$

Example 2. Find θ for a south-facing surface tilted upward 30° from the horizontal at 40° north latitude at 4:00 P.M., AST, on August 21.

Solution: From Equation (3), at 4:00 P.M. on August 21,

$$H = 4 \times 15° = 60°$$

From Table 1,

$$\delta = 12.1°$$

From Equation (5),

$$\sin \beta = \cos 40° \cos 12.1° \cos 60° + \sin 40° \sin 12.1°$$
$$\beta = 30.6°$$

From Equation (6),

$$\sin \phi = \cos 12.1° \sin 60° / \cos 30.6°$$
$$\phi = 79.7°$$

The surface faces south, so $\phi = \gamma$. From Equation (8),

$$\cos \theta = \cos 30.6° \cos 79.7° \sin 30° + \sin 30.6° \cos 30°$$
$$\theta = 58.8°$$

Tabulated values of θ are given in ASHRAE *Standard* 93, Methods of Testing to Determine the Performance of Solar Collectors, for horizontal and vertical surfaces and for south-facing surfaces tilted upward at angles equal to the latitude minus 10°, the latitude, the latitude plus 10°, and the latitude plus 20°. These tables cover the latitudes from 24° to 64° north, in 8° intervals.

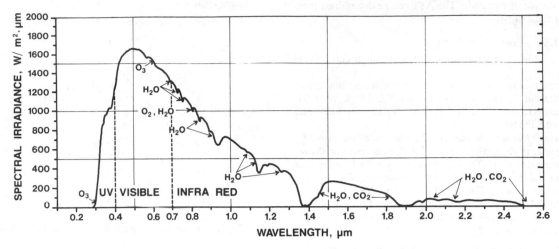

Fig. 4   Spectral Solar Irradiation at Sea Level for Air Mass = 1.0

Solar Spectrum

Beyond the earth's atmosphere, the effective black body temperature of the sun is 5760 K. The maximum spectral intensity occurs at 0.48 µm in the green portion of the visible spectrum (Figure 4). Thekaekara (1973) presents tables and charts of the sun's extraterrestrial spectral irradiance from 0.120 to 100 µm, the range in which most of the sun's radiant energy is contained. The ultraviolet portion of the spectrum below 0.40 µm contains 8.73% of the total, another 38.15% is contained in the visible region between 0.40 and 0.70 µm, and the infrared region contains the remaining 53.12%.

Solar Radiation at the Earth's Surface

In passing through the earth's atmosphere, some of the sun's direct radiation I_D is scattered by nitrogen, oxygen, and other molecules, which are small compared to the wavelengths of the radiation; and by aerosols, water droplets, dust, and other particles with diameters comparable to the wavelengths (Gates 1966). This scattered radiation causes the sky to be blue on clear days, and some of it reaches the earth as diffuse radiation I_d.

Attenuation of the solar rays is also caused by absorption, first by the ozone in the outer atmosphere, which causes a sharp cutoff at 0.29 µm of the ultraviolet radiation reaching the earth's surface. In the longer wavelengths, there are a series of absorption bands caused by water vapor, carbon dioxide, and ozone. The total amount of attenuation at any given location is determined by (1) the length of the atmospheric path through which the rays traverse and (2) the composition of the atmosphere. The path length is expressed in terms of the air mass m, which is the ratio of the mass of atmosphere in the actual earth-sun path to the mass that would exist if the sun were directly overhead at sea level ($m = 1.0$). For all practical purposes, at sea level, $m = 1.0/\sin \beta$. Beyond the earth's atmosphere, $m = 0$.

Prior to 1967, solar radiation data was based on an assumed solar constant of 1324 W/m² and on a standard sea level atmosphere containing the equivalent depth of 2.8 mm of ozone, 20 mm of precipitable moisture, and 300 dust particles per cm³. Threlkeld and Jordan (1958) considered the wide variation of water vapor in the atmosphere above the United States at any given time, and particularly the seasonal variation, which finds three times as much moisture in the atmosphere in midsummer as in December, January, and February. The basic atmosphere was assumed to be at sea level barometric pressure, with 2.5 mm of ozone, 200 dust particles per cm³, and an actual precipitable moisture content that varied throughout the year from 8 mm in midwinter to 28 mm in mid-July. Figure 5 shows the variation of the direct normal irradiation with

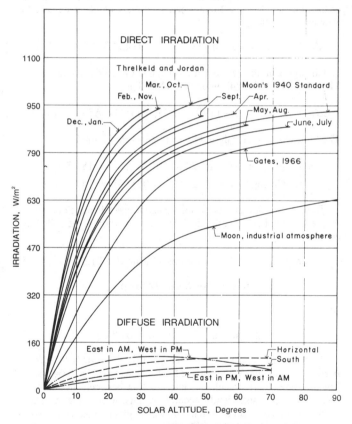

Fig. 5   Variation with Solar Altitude and Time of Year for Direct Normal Irradiation

solar altitude, as estimated for clear atmospheres and for an atmosphere with variable moisture content.

Stephenson (1967) showed that the intensity of the direct normal irradiation I_{DN} at the earth's surface on a clear day can be estimated by the following equation:

$$I_{DN} = A/\exp(B/\sin \beta) \qquad (11)$$

where A, the apparent extraterrestrial irradiation at $m = 0$, and B, the atmospheric extinction coefficient, are functions of the date and take into account the seasonal variation of the earth-sun distance and the air's water vapor content.

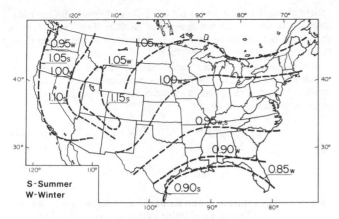

Fig. 6 Clearness Numbers for the United States for Summer (S) and Winter (W)

The values of the parameters A and B given in Table 1 were selected so that the resulting value of I_{DN} would be in close agreement with the Threlkeld and Jordan (1958) values on average cloudless days. The values of I_{DN} given in Tables 12 through 18 in Chapter 27 of the 1993 *ASHRAE Handbook—Fundamentals*, were obtained by using Equation (11) and data from Table 1. The values of the solar altitude β and the solar azimuth φ may be obtained from Equations (5) and (6).

Because local values of atmospheric water content and elevation can vary markedly from the sea level average, the concept of *clearness number* was introduced to express the ratio between the actual clear-day direct radiation intensity at a specific location and the intensity calculated for the standard atmosphere for the same location and date.

Figure 6 shows the Threlkeld-Jordan map of winter and summer clearness numbers for the continental United States. Irradiation values should be adjusted by the clearness numbers applicable to each particular location.

Design Values of Total Solar Irradiation

The total solar irradiation $I_{t\theta}$ of a terrestrial surface of any orientation and tilt with an incident angle θ is the sum of the direct component $I_{DN} \cos \theta$ plus the diffuse component $I_{d\theta}$ coming from the sky plus whatever amount of reflected shortwave radiation I_r may reach the surface from the earth or from adjacent surfaces:

$$I_{t\theta} = I_{DN} \cos \theta + I_{d\theta} + I_r \tag{12}$$

The diffuse component is difficult to estimate because of its nondirectional nature and its wide variations. Figure 5 shows typical values of diffuse irradiation of horizontal and vertical surfaces. For clear days, Threlkeld (1963) has derived a dimensionless parameter (designated as C in Table 1), which depends on the dust and moisture content of the atmosphere and thus varies throughout the year:

$$C = I_{dH}/I_{DN} \tag{13}$$

where I_{dH} is the diffuse radiation falling on a horizontal surface under a cloudless sky.

The following equation may be used to estimate the amount of diffuse radiation $I_{d\theta}$ that reaches a tilted or vertical surface:

$$I_{d\theta} = C I_{DN} F_{ss} \tag{14}$$

where

$$F_{ss} = (1 + \cos \Sigma)/2 \tag{15}$$
$$= \text{angle factor between the surface and the sky}$$

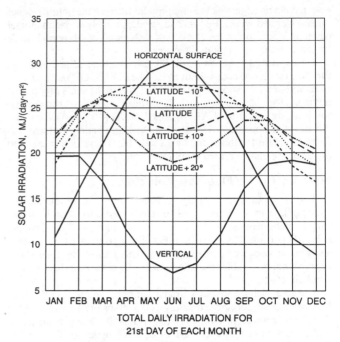

Fig. 7 Total Daily Irradiation for Horizontal, Tilted, and Vertical Surfaces at 40° North Latitude

The reflected radiation I_r from the foreground is given by the following equation:

$$I_r = I_{tH} \rho_g F_{sg} \tag{16}$$

where

ρ_g = reflectance of the foreground
I_{tH} = total horizontal irradiation

$$F_{sg} = (1 - \cos \Sigma)/2 \tag{17}$$
$$= \text{angle factor between the surface and the earth}$$

The intensity of the reflected radiation that reaches any surface depends on the nature of the reflecting surface and on the incident angle between the sun's direct beam and the reflecting surface. Many measurements made of the reflection (albedo) of the earth under varying conditions show that clean, fresh snow has the highest reflectance (0.87) of any natural surface.

Threlkeld (1963) gives values of reflectance for commonly encountered surfaces at solar incident angles from 0 to 70°. Bituminous paving generally reflects less than 10% of the total incident solar irradiation; bituminous and gravel roofs reflect from 12 to 15%; concrete, depending on its age, reflects from 21 to 33%. Bright green grass reflects 20% at θ = 30° and 30% at θ = 65°.

The maximum daily amount of solar irradiation that can be received at any given location is that which falls on a flat plate with its surface kept normal to the sun's rays so it receives both direct and diffuse radiation. For fixed flat-plate collectors, the total amount of clear day irradiation depends on the orientation and slope. As shown by Figure 7 for 40° north latitude, the total irradiation of horizontal surfaces reaches its maximum in midsummer, while vertical south-facing surfaces experience their maximum irradiation during the winter. These curves show the combined effects of the varying length of days and changing solar altitudes.

In general, flat-plate collectors are mounted at a fixed tilt angle Σ (above the horizontal) to give the optimum amount of irradiation for each purpose. Collectors intended for winter heating benefit from higher tilt angles than those used to operate cooling systems in summer. Solar water heaters, which should operate satisfactorily throughout the year, require an angle that is a compromise between the optimal values for summer and winter. Figure 7 shows the monthly variation of total day-long irradiation on the 21st day of each month at 40° north latitude for flat surfaces with various tilt angles.

Tables in ASHRAE *Standard* 93 give the total solar irradiation for the 21st day of each month at latitudes 24° to 64° north on surfaces with the following orientations: normal to the sun's rays (direct normal, DN, data *do not* include diffuse irradiation); horizontal; south-facing, tilted at (LAT−10), LAT, (LAT+10), (LAT+20), and 90° from the horizontal. The day-long total irradiation for fixed surfaces is highest for those that face south, but a deviation in azimuth of 15° to 20° causes only a small reduction.

Solar Energy for Flat-Plate Collectors

The preceeding data apply to clear days. The irradiation for average days may be estimated for any specific location by referring to publications of the U.S. Weather Service. The *Climatic Atlas of the United States* (U.S. GPO 1968) gives maps of monthly and annual values of percentage of possible sunshine, total hours of sunshine, mean solar radiation, mean sky cover, wind speed, and wind direction.

The total daily horizontal irradiation data reported by the U.S. Weather Bureau for approximately 100 stations prior to 1964 show that the percentage of total clear-day irradiation is approximately a linear function of the percentage of possible sunshine. The irradiation is not zero for days when the percentage of possible sunshine is reported as zero, because substantial amounts of energy reach the earth in the form of diffuse radiation. Instead, the following relationship exists:

$$\frac{\text{Daylong actual } I_{tH}}{\text{Clear day } I_{tH}} 100 = a + b \text{ (possible sunshine percent)} \quad (18)$$

where a and b are constants for any specified month at any given location. See also Jordan and Liu (1977) and Duffie and Beckman (1974).

Longwave Atmospheric Radiation

In addition to the shortwave (0.3 to 2.0 μm) radiation it receives from the sun, the earth receives longwave radiation (4 to 100 μm, with maximum intensity near 10 μm) from the atmosphere. In turn, a surface on the earth emits longwave radiation q_{Rs} in accordance with the Stefan-Boltzmann law:

$$q_{Rs} = e_s \sigma (T_s / 100)^4 \quad (19)$$

where

e_s = surface emittance
σ = constant, 5.67 W/(m²·K⁴)
T_s = absolute temperature of the surface, K

For most nonmetallic surfaces, the longwave hemispheric emittance is high, ranging from 0.84 for glass and dry sand to 0.95 for black built-up roofing. For highly polished metals and certain selective surfaces, e_s may be as low as 0.05 to 0.20.

Atmospheric radiation comes primarily from water vapor, carbon dioxide, and ozone (Bliss 1961); very little comes from oxygen and nitrogen, although they make up 99% of the air.

Approximately 90% of the incoming atmospheric radiation comes from the lowest 90 m. Thus, the air conditions at ground level largely determine the magnitude of the incoming radiation. The downward radiation from the atmosphere q_{Rat} may be expressed as

$$q_{Rat} = e_{at} \sigma (T_{at} / 100)^4 \quad (20)$$

The emittance of the atmosphere is a complex function of air temperature and moisture content. The dew point of the atmosphere near the ground determines the total amount of moisture in the atmo-

sphere above the place where the dry-bulb and dew-point temperatures of the atmosphere are determined (Reitan 1963). Bliss (1961) found that the emittance of the atmosphere is related to the dew-point temperature, as shown by Table 3.

The apparent sky temperature is defined as the temperature at which the sky (as a blackbody) emits radiation at the rate actually emitted by the atmosphere at ground level temperature with its actual emittance e_{at}. Then,

$$\sigma (T_{sky} / 100)^4 = e_{at} \sigma (T_{at} / 100)^4 \quad (21)$$

or

$$T_{sky}^4 = e_{at} T_{at}^4 \quad (22)$$

Table 3 Sky Emittance and Amount of Precipitable Moisture Versus Dew-Point Temperature

Dew Point, °C	Sky Emittance, e_a	Precipitable Water, mm
−30	0.68	3.3
−25	0.70	3.7
−20	0.72	4.3
−15	0.74	5.0
−10	0.76	6.2
−5	0.78	8.2
0	0.80	11.2
5	0.82	15.0
10	0.84	20.3
15	0.86	28.0
20	0.88	38.6

Example 3. Consider a summer night condition when the ground level temperatures are 17.5°C dew point and 30°C dry bulb. By interpolation from Table 3, e_{at} at 17.5°C dew point is 0.87, and the apparent sky temperature is

$$T_{sky} = 0.87^{0.25} (30 + 273.15) = 292.8 \text{ K}$$

Thus, T_{sky} = 292.8 − 273.15 = 19.6°C, which is 10.4 K below the ground level dry-bulb temperature.

For a winter night in Arizona, when the temperatures at ground level are 15°C dry bulb and −5°C dew point, the emittance of the atmosphere is 0.78 from Table 3, and the apparent sky temperature is 270.8 K or −2.4°C.

A simple relationship, which ignores vapor pressure of the atmosphere, may also be used to estimate the apparent sky temperature:

$$T_{sky} = 0.0552 (T_{at})^{1.5} \quad (23)$$

where T is in kelvins.

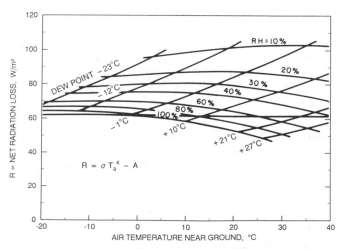

Fig. 8 Radiation Heat Loss to Sky from Horizontal Blackbody

If the temperature of the radiating surface is assumed to equal the atmospheric temperature, the loss of heat from a black surface ($e_s = 1.00$) may be found from Figure 8.

Example 4. For the conditions in the previous example for summer, 30°C dry bulb and 17.5°C dew point, the rate of radiative heat loss is about 72 W/m^2. For winter, 15°C dry bulb and −5°C dew point, the heat loss is about 85 W/m^2.

Where a bare, blackened roof is used as a heat dissipater, the rate of heat loss rises rapidly as the surface temperature goes up. For the summer example, a black-painted metallic roof, $e_s = 0.96$, at 38°C (311 K) will have a heat loss rate of

$$q_{RAD} = 0.96 \times 5.67 \, [\, (311/100)^4 - (292.8/100)^4]$$

$$= 109 \text{ W/m}^2$$

This analysis shows that radiation alone is not an effective means of dissipating heat under summer conditions of high dew-point and high ambient temperature. In spring and fall, when both the dew-point and dry-bulb temperatures are relatively low, radiation becomes much more effective.

On overcast nights, when the cloud cover is low, the clouds act much like blackbodies at ground level temperature, and virtually no heat can be lost by radiation. The exchange of longwave radiation between the sky and terrestrial surfaces occurs in the daytime as well as at night, but the much greater magnitude of the solar irradiation masks the longwave effects.

SOLAR ENERGY COLLECTION

Solar energy can be used by (1) heliochemical, (2) helioelectrical, and (3) heliothermal processes. Photosynthesis is a heliochemical process that produces food and converts CO_2 to O_2. Photovoltaic converters employ a helioelectrical process and can be used to power spacecraft and in many terrestrial applications. The heliothermal process, the primary subject of this chapter, provides thermal energy for space heating and cooling, domestic water heating, power generation, distillation, and process heating.

Solar Heat Collection by Flat-Plate Collectors

The solar irradiation data presented in the foregoing sections may be used to estimate how much energy is likely to be available at any specific location, date, and time of day for collection by either a concentrating device, which uses only the direct rays of the sun, or by a flat-plate collector, which can use both direct and diffuse irradiation. The temperatures needed for space heating and cooling do not exceed 90°C, even for absorption refrigeration, and they can be attained with carefully designed flat-plate collectors. Depending on the load and ambient temperatures, single-effect absorption systems can use energizing temperatures of 43 to 110°C.

A flat-plate collector generally consists of the following components (see Figure 9):

- *Glazing.* One or more sheets of glass or other diathermanous (radiation-transmitting) material.
- *Tubes, fins,* or *passages.* To conduct or direct the heat transfer fluid from the inlet to the outlet.
- *Absorber plates.* Flat, corrugated, or grooved plates, to which the tubes, fins, or passages are attached. The plate may be integral with the tubes.
- *Headers* or *manifolds.* To admit and discharge the fluid.
- *Insulation.* To minimize heat loss from the back and sides of the collector.
- *Container* or *casing.* To surround the aforementioned components and keep them free from dust, moisture, etc.

Flat-plate collectors have been built in a wide variety of designs from many different materials (Figure 10). They have been used to heat fluids such as water, water plus an antifreeze additive, or air. Their major purpose is to collect as much solar energy as possible at

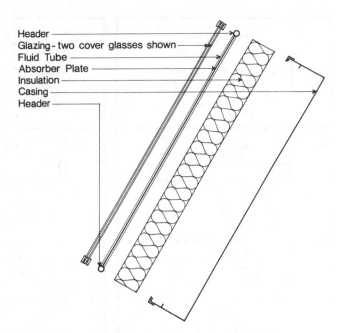

Fig. 9 Exploded Cross Section Through Double-Glazed Solar Water Heater

the lowest possible total cost. The collector should also have a long effective life, despite the adverse effects of the sun's ultraviolet radiation; corrosion or clogging because of acidity, alkalinity, or hardness of the heat transfer fluid; freezing or air-binding in the case of water, or deposition of dust or moisture in the case of air; and breakage of the glazing because of thermal expansion, hail, vandalism, or other causes. These problems can be minimized by the use of tempered glass.

Glazing Materials

Glass has been widely used to glaze solar collectors because it can transmit as much as 90% of the incoming shortwave solar irradiation while transmitting virtually none of the longwave radiation emitted outward by the absorber plate. Glass with low iron content has a relatively high transmittance for solar radiation (approximately 0.85 to 0.90 at normal incidence), but its transmittance is essentially zero for the longwave thermal radiation (5.0 to 50 μm) emitted by sun-heated surfaces.

Plastic films and sheets also possess high shortwave transmittance, but because most usable varieties also have transmission bands in the middle of the thermal radiation spectrum, they may have longwave transmittances as high as 0.40.

Plastics are also generally limited in the temperatures they can sustain without deteriorating or undergoing dimensional changes. Only a few finds of plastics can withstand the sun's ultraviolet radiation for long periods. However, they are not broken by hail and other stones, and, in the form of thin films, they are completely flexible and have low mass.

The glass generally used in solar collectors may be either single-strength (2.2 to 2.5 mm thick) or double-strength (2.92 to 3.38 mm thick). The commercially available grades of window and greenhouse glass have normal incidence transmittances of about 0.87 and 0.85, respectively. For direct radiation, the transmittance varies markedly with the angle of incidence, as shown in Table 4, which gives transmittances for single- and double-glazing using double-strength clear window glass.

The 4% reflectance from each glass-air interface is the most important factor in reducing transmission, although a gain of about 3% in transmittance can be obtained by using water-white glass.

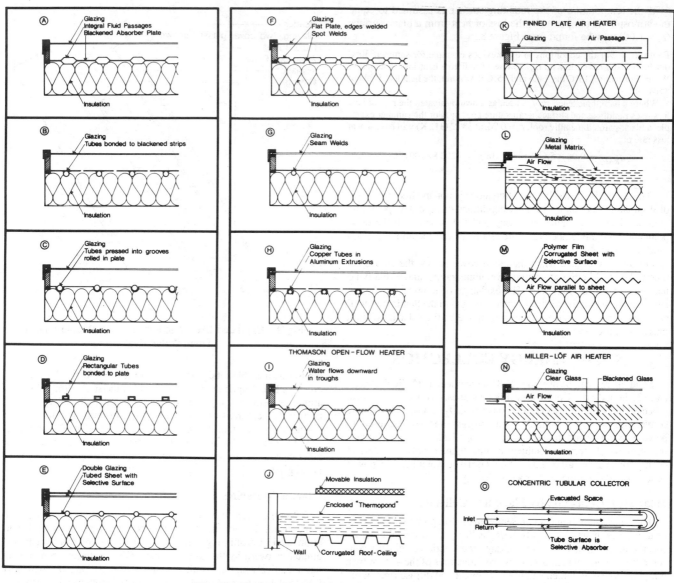

Fig. 10   Various Types of Solar Water and Air Heaters

Antireflective coatings and surface texture can also improve transmission significantly. The effect of dirt and dust on collector glazing may be quite small, and the cleansing effect of an occasional rainfall is usually adequate to maintain the transmittance within 2 to 4% of its maximum value.

The glazing should admit as much solar irradiation as possible and reduce the upward loss of heat as much as possible. Although glass is virtually opaque to the longwave radiation emitted by collector plates, absorption of that radiation causes an increase in the glass temperature and a loss of heat to the surrounding atmosphere by radiation and convection. This type of heat loss can be reduced by using an infrared reflective coating on the underside of the glass; however, such coatings are expensive and reduce the effective solar transmittance of the glass by as much as 10%.

In addition to serving as a heat trap by admitting shortwave solar radiation and retaining longwave thermal radiation, the glazing also reduces heat loss by convection. The insulating effect of the glazing is enhanced by the use of several sheets of glass, or glass plus plastic. The upward heat loss may be expressed by the following equation:

$$Q_{up} = A_p U_L (t_p - t_{at})$$ (24)

where

Q_{up} = heat loss upward from the absorber, W
A_p = absorber plate area, m^2
U_L = upward heat loss coefficient, W/(m^2·K)
t_p = absorber plate temperature, °C
t_{at} = ambient air temperature, °C

The loss from the back of the plate rarely exceeds 10% of the upward loss.

Collector Plates

The collector plate absorbs as much of the irradiation as possible through the glazing, while losing as little heat as possible upward to the atmosphere and downward through the back of the casing. The collector plates transfer the retained heat to the transport fluid. The absorptance of the collector surface for shortwave solar radiation depends on the nature and color of the coating and on the incident angle, as shown in Table 4 for a typical flat black paint.

By suitable electrolytic or chemical treatments, surfaces can be produced with high values of solar radiation absorptance α and low values of longwave emittance e_s. Essentially, typical selective surfaces consist of a thin upper layer, which is highly absorbent to shortwave solar radiation but relatively transparent to longwave thermal radiation, deposited on a substrate that has a high reflectance

**Table 4 Variation with Incident Angle
of Transmittance for Single and Double Glazing and
Absorptance for Flat Black Paint**

Incident Angle, Deg	Transmittance		Absorptance for Flat Black Paint
	Single Glazing	Double Glazing	
0	0.87	0.77	0.96
10	0.87	0.77	0.96
20	0.87	0.77	0.96
30	0.87	0.76	0.95
40	0.86	0.75	0.94
50	0.84	0.73	0.92
60	0.79	0.67	0.88
70	0.68	0.53	0.82
80	0.42	0.25	0.67
90	0.00	0.00	0.00

and a low emittance for longwave radiation. Selective surfaces are particularly important when the collector surface temperature is much higher than the ambient air temperature.

For fluid-heating collectors, passages must be integral with or firmly bonded to the absorber plate. A major problem is obtaining a good thermal bond between tubes and absorber plates without incurring excessive costs for labor or materials. Materials most frequently used for collector plates are copper, aluminum, and steel. UV-resistant plastic extrusions are used for low-temperature application. If the entire collector area is in contact with the heat transfer fluid, the thermal conductance of the material is not important.

Whillier (1964) concluded that steel tubes are as effective as copper if the bond conductance between tube and plate is good. Potential corrosion problems should be considered for any metals. Bond conductance can range from a high of 5700 W/(m^2·K) for a securely soldered or brazed tube to a low of 17 W/(m^2·K) for a poorly clamped or badly soldered tube. Plates of copper, aluminum, or stainless steel with integral tubes are among the most effective types available.

Figure 10, adapted from Van Straaten (1961) and other sources, shows a few of the solar water and air heaters that have been used with varying degrees of success. Figure 10A shows a bonded sheet design, in which the fluid passages are integral with the plate to ensure good thermal contact between the metal and the fluid. Figures 10B and 10C show fluid heaters with tubes soldered, brazed, or otherwise fastened to upper or lower surfaces of sheets or strips of copper, steel, or aluminum. Copper tubes are used most often because of their superior resistance to corrosion.

Thermal cement, clips, clamps, or twisted wires have been tried in the search for low-cost bonding methods. Figure 10D shows the use of extruded rectangular tubing to obtain a larger heat transfer area between tube and plate. Mechanical pressure, thermal cement, or brazing may be used to make the assembly. Soft solder must be avoided because of the high plate temperatures encountered at stagnation conditions.

Figure 10E shows a double-glazed collector with a tubed copper sheet, and Figure 10F shows the use of thin parallel sheets of malleable metal (copper, aluminum, or stainless steel) that are seam-welded along their edges and spot-welded at intervals to provide fluid passages that are developed by expansion. Nontubular types are limited in the internal pressure they can sustain; they are more adapted to space heating than to domestic hot water heating due to the relatively high water pressures used in cities in the United States.

Figure 10G shows a seam-welded stainless steel absorber with integral tubes. Figure 10H shows copper tubes pressed into appropriately shaped aluminum extrusions. Differential thermal expansion may break the thermal bond.

Figure 10I shows a proprietary open flow collector, which uses black-painted corrugated aluminum sheets, generally mounted on steeply pitched south-facing roofs. Water is distributed to the channels by perforated copper or plastic piping running along the peak of the roof. The sun-warmed water is collected at the bottom of the channels by a galvanized or plastic trough and conducted through plastic pipe to the basement storage tank.

Figure 10J shows another proprietary collector that consists of horizontally arranged transparent plastic bags filled with treated water. A corrugated metal roof ceiling with a black waterproof liner supports these water-filled bags. The water is thus in thermal contact with the ceiling. Movable horizontal insulating panels cover the ponds on winter nights; during winter days, they are opened to admit solar radiation. In summer, the operation is reversed: the panels are opened at night to dissipate heat but are closed during the day to keep out unwanted solar radiation.

Air or other gases can be heated with flat-plate collectors, particularly if some type of extended surface (Figure 10K) is used to counteract the low heat transfer coefficients between metal and air. Metal or fabric matrices (Figure 10L), or thin corrugated metal sheets (Figure 10M) may be used, with selective surfaces applied to the latter when a high level of performance is required. The principal requirement is a large contact area between the absorbing surface and the air. The Miller-Löf air heater, shown in Figure 10N, provides this area by overlapping glass plates, the upper being clear and the lower blackened to absorb the incoming solar radiation.

Reduction of heat loss from the absorber can be accomplished either by a selective surface to reduce radiative heat transfer or by suppressing convection. Francia (1961) showed that a honeycomb made of transparent material, placed in the airspace between the glazing and the absorber, was beneficial.

Tubular collectors (Figure 10O) with evacuated jackets have demonstrated that the combination of a selective surface and an effective convection suppressor can result in good performance at temperatures higher than a flat-plate collector can attain.

Concentrating Collectors

Temperatures far above those attainable by flat-plate collectors can be reached if a large amount of solar radiation is concentrated on a relatively small collection area. Simple flat reflectors can markedly increase the amount of direct radiation reaching a flat-plate collector, as shown in Figure 11A.

Because of the apparent movement of the sun across the sky, conventional concentrating collectors must follow the sun's daily motion. There are two methods by which the sun's motion can be readily tracked. The altazimuth method requires the tracking device to turn in both altitude and azimuth; when performed properly, this method enables the concentrator to follow the sun exactly. Paraboloidal solar furnaces, Figure 11B, generally use this system. The polar, or equatorial, mounting points the axis of rotation at the North Star, tilted upward at the angle of the local latitude. By rotating the collector 15° per hour, it follows the sun perfectly (on March 21 and September 21). If the collector surface or aperture must be kept normal to the solar rays, a second motion is needed to correct for the change in the solar declination. This motion is not essential for most solar collectors.

The maximum variation in the angle of incidence for a collector on a polar mount will be ±23.45° on June 21 and December 21; the incident angle correction would then be cos 23.45° = 0.917.

Horizontal reflective parabolic troughs, oriented east and west, as shown in Figure 11C, require continuous adjustment to compensate for the changes in the sun's declination. There is inevitably some morning and afternoon shading of the reflecting surface if the concentrator has opaque end panels. The necessity of moving the concentrator to accommodate the changing solar declination can be reduced by moving the absorber or by using a trough with two sections of a parabola facing each other, as shown in Figure 11D. Known as a *compound parabolic concentrator* (CPC), this design can accept incoming radiation over a relatively wide range of angles. By using multiple internal reflections, any radiation that is accepted finds its way to the absorber surface located at the bottom of the

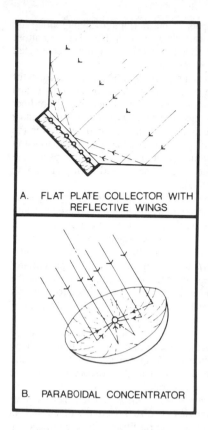

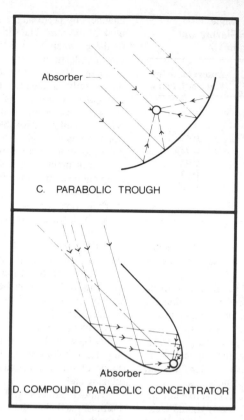

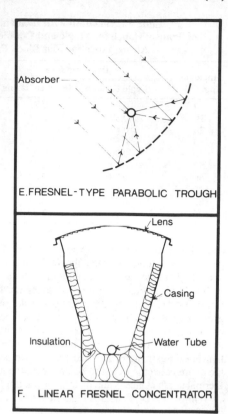

Fig. 11   Types of Concentrating Collectors

apparatus. By filling the collector shape with a highly transparent material having an index of refraction greater than 1.4, the acceptance angle can be increased. By shaping the surfaces of the array properly, total internal reflection is made to occur at the medium-air interfaces, which results in a high concentration efficiency. Known as a *dielectric compound parabolic concentrator* (DCPC), this device has been applied to the photovoltaic generation of electricity (Cole et al. 1977).

The parabolic trough of Figure 11C can be simulated by many flat strips, each adjusted at the proper angle so that all reflect onto a common target. By supporting the strips on ribs with parabolic contours, a relatively efficient concentrator can be produced with less tooling than the complete reflective trough.

Another concept applied this segmental idea to flat and cylindrical lenses. A modification is shown in Figure 11F, in which a linear Fresnel lens, curved to shorten its focal distance, can concentrate a relatively large area of radiation onto an elongated receiver. Using the equatorial sun-following mounting, this type of concentrator has been used as a means of attaining temperatures well above those that can be reached with flat-plate collectors.

One disadvantage of concentrating collectors is that, except at low concentration ratios, they can use only the direct component of solar radiation, because the diffuse component cannot be concentrated by most types. However, an advantage of concentrating collectors is that, in summer, when the sun rises and sets well to the north of the east-west line, the sun-follower, with its axis oriented north-south, can begin to accept radiation directly from the sun long before a fixed, south-facing flat plate can receive anything other than diffuse radiation from the portion of the sky that it faces. Thus, at 40° north latitude, for example, the cumulative *direct* radiation available to a sun-follower on a clear day is 36.1 MJ/m², while the *total* radiation falling on the flat plate tilted upward at an angle equal to the latitude is only 25.2 MJ/m² each day. Thus, in relatively cloudless areas, the concentrating collector may capture more radiation per unit of aperture area than a flat-plate collector.

For extremely high inputs of radiant energy, a multiplicity of flat mirrors, or *heliostats,* using altazimuth mounts, can be used to reflect their incident direct solar radiation onto a common target. Using slightly concave mirror segments on the heliostats, large amounts of thermal energy can be directed into the cavity of a steam generator to produce steam at high temperature and pressure.

Collector Performance

The performance of collectors may be analyzed by a procedure originated by Hottel and Woertz (1942) and extended by Whillier (ASHRAE 1977). The basic equation is

$$q_u = I_{t\theta}\,(\tau\alpha)_\theta - U_L\,(t_p - t_{at})$$
$$= \dot{m}\,c_p\,(t_{fe} - t_{fi})\,/\,A_{ap} \qquad (25)$$

Equation (25) also may be adapted for use with concentrating collectors:

$$q_u = I_{DN}\,(\tau\alpha)_\theta\,(\rho\Gamma) - U_L\,(A_{abs}/A_{ap})\,(t_{abs} - t_a) \qquad (26)$$

where

q_u = useful heat gained by collector per unit of aperture area, W/m²
$I_{t\theta}$ = total irradiation of collector, W/m²
I_{DN} = direct normal irradiation, W/m²
$(\tau\alpha)_\theta$ = transmittance τ of cover times absorptance α of plate at prevailing incident angle θ
U_L = upward heat loss coefficient, W/(m²·K)
t_p = temperature of the absorber plate, °C
t_a = temperature of the atmosphere, °C
t_{abs} = temperature of the absorber, °C
$\dot{m}$ = fluid flow rate, kg/s
c_p = specific heat of fluid, kJ/(kg·K)
t_{fe}, t_{fi} = temperatures of the fluid leaving and entering the collector, °C
$\rho\Gamma$ = reflectance of the concentrator surface times fraction of reflected or refracted radiation that reaches the absorber
A_{abs}, A_{ap} = areas of absorber surface and of aperture that admit or receive radiation, m²

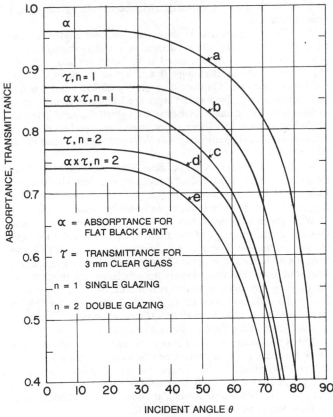

Fig. 12   Variation of Absorptance and Transmittance with Incident Angle

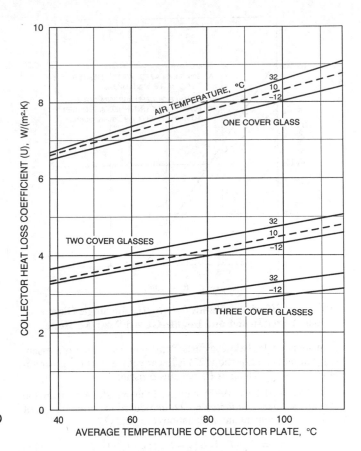

Fig. 13   Variation of Upward Heat Loss Coefficient U_L with Collector Plate Temperature and Ambient Air Temperatures for Single-, Double-, and Triple-Glazed Collectors

The total irradiation and the direct normal irradiation for clear days may be found in ASHRAE *Standard* 93. The transmittance for single and double glazing and the absorptance for flat black paint may be found in Table 4 for incident angles from 0 to 90°. These values, and the products of τ and α, are also shown in Figure 12. Little change occurs in the solar-optical properties of the glazing and absorber plate until θ exceeds 30°, but, since all values reach zero when $\theta = 90°$, they drop off rapidly for values of θ beyond 40°.

For nonselective absorber plates, U_L varies with the temperature of the plate and the ambient air, as shown in Figure 13. For selective surfaces, which effect major reductions in the emittance of the absorber plate, U_L will be much lower than the values shown in Figure 13. Manufacturers of such surfaces should be asked for values applicable to their products, or test results that give the necessary information should be consulted.

Example 5. A flat-plate collector is operating in Denver, latitude = 40° north, on July 21 at noon solar time. The atmospheric temperature is assumed to be 30°C, and the average temperature of the absorber plate is 60°C. The collector is single-glazed with flat black paint on the absorber. The collector faces south, and the tilt angle is 30° from the horizontal. Find the rate of heat collection and the collector efficiency. Neglect the losses from the back and sides of the collector.

Solution: From Table 2, $\delta = 20.6°$.
From Equation (2),

$$\beta_N = 90° - 40° + 20.6° = 70.6°$$

From Equation (3), $H = 0$; therefore from Equation (6), $\sin \phi = 0$ and thus, $\phi = 0°$. Because the collector faces south, $\psi = 0°$, and $\gamma = \phi$. Thus $\gamma = 0°$.
Then Equation (8) gives

$$\cos \theta = \cos 70.6° \cos 0° \sin 30° + \sin 70.6° \cos 30°$$
$$= (0.332) (1) (0.5) + (0.943) (0.866)$$
$$= 0.983$$
$$\theta = 10.6°$$

From Table 1, $A = 1085$ W/m², $B = 0.207$, and $C = 0.136$. Using Equation (11),

$$I_{DN} = 1085/\exp \ (0.207/\sin 70.6°)$$
$$= 871 \ \text{W/m}^2$$

Combining Equations (14) and (15) gives

$$I_{d\theta} = 0.136 \times 871 \ (1 + \cos 30°) \ /2$$
$$= 111 \ \text{W/m}^2$$

Assuming $I_r = 0$, Equation (12) gives a total solar irradiation on the collector of

$$I_{t\theta} = 871 \cos 10.6° + 111 + 0 = 967 \ \text{W/m}^2$$

From Figure 12, for $n = 1$, $\tau = 0.87$ and $\alpha = 0.96$.

From Figure 13, for an absorber plate temperature of 60°C and an air temperature of 30°C, $U_L = 7.3$ W/(m²·K).

Then from Equation (25),

$$q_u = 967 \ (0.87 \times 0.96) - 7.3 \ (60 - 30)$$
$$= 589 \ \text{W/m}^2$$

The collector efficiency η is

$$589/967 = 0.60$$

The general expression for collector efficiency is

$$\eta = (\tau\alpha)_\theta - U_L \ (t_p - t_{at})/I_{t\theta} \qquad (27)$$

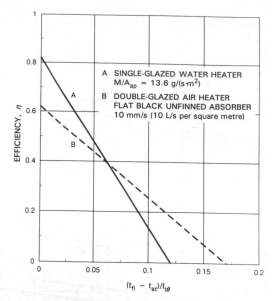

Fig. 14 Efficiency Versus $(t_{fi} - t_{at})/I_{t\theta}$ for Single-Glazed Solar Water Heater and Double-Glazed Solar Air Heater

For incident angles below about 35°, the product τ times α is essentially constant and Equation (27) is linear with respect to the parameter $(t_p - t_{at})/I_{t\theta}$, as long as U_L remains constant.

Whillier (ASHRAE 1977) suggested that an additional term, the *collector heat removal factor* F_R, be introduced to permit the use of the fluid inlet temperature in Equations (25) and (27):

$$q_u = F_R \left[I_{t\theta} (\tau\alpha)_\theta - U_L (t_{fi} - t_{at}) \right] \qquad (28)$$

$$\eta = F_R (\tau\alpha)_\theta - F_R U_L (t_{fi} - t_{at}) / I_{t\theta} \qquad (29)$$

where F_R equals the ratio of the heat actually delivered by the collector to the heat that would be delivered if the absorber were at t_{fi}. F_R is found from the results of a test performed in accordance with ASHRAE *Standard* 93.

The results of such a test are plotted in Figure 14. When the parameter is zero, because there is no temperature difference between the fluid entering the collector and the atmosphere, the value of the y-intercept equals $F_R(\tau\alpha)$. The slope of the efficiency line equals the heat loss factor U_L multiplied by F_R. For the single-glazed, nonselective collector with the test results shown in Figure 14, the y-intercept is 0.82, and the x-intercept is 0.12 $m^2 \cdot K/W$. This collector used high transmittance single glazing, $\tau = 0.91$, and black paint with an absorptance of 0.97, so Equation (29) gives $F_R = 0.82/(0.91 \times 0.97) = 0.93$.

Assuming that the relationship between η and the parameter is actually linear, as shown, then the slope is $-0.82/0.12 = -6.78$; thus $U_L = 6.78/F_R = 6.78/0.93 = 7.29$ $W/(m^2 \cdot K)$. The tests on which Figure 14 is based were run indoors. Factors that affect the measured efficiency are wind speed and fluid velocity.

Figure 14 also shows the efficiency of a double-glazed air heater with an unfinned absorber coated with flat black paint. The y-intercept for the air heater *B* is considerably less than it is for water heater *A* because (1) transmittance of the double glazing used in *B* is lower than the transmittance of the single glazing used in *A* and (2) F_R is lower for *B* than for *A* because of the lower heat transfer coefficient between air and the unfinned metal absorber.

The x-intercept for air heater *B* is greater than it is for the water heater *A* because the upward loss coefficient U_L is much lower for the double-glazed air heater than for the single-glazed water heater. The data for both *A* and *B* were taken at near-normal incidence with high values of $I_{t\theta}$. For Example 5, using a single-glazed water

heater, the value of the parameter would be close to $(60 - 30)/967 = 0.031$ $m^2 \cdot K/W$, and the expected efficiency, 0.60, agrees closely with the test results.

As ASHRAE *Standard* 93 shows, the incident angles encountered with south-facing tilted collectors vary widely throughout the year. Considering a surface located at 40° north latitude with a tilt angle $\Sigma = 40°$, the incident angle θ will depend on the time of day and the declination δ. On December 21, $\delta = 23.456°$; at 4 h before and after solar noon, the incident angle is 62.7°, and it remains close to this value for the same solar time throughout the year. The total irradiation at these conditions varies from a low of 140 W/m^2 on December 21 to approximately 440 W/m^2 throughout most of the other months.

When the irradiation is below about 315 W/m^2, the losses from the collector may exceed the heat that can be absorbed. This situation varies with the temperature difference between the collector inlet temperature and the ambient air, as suggested by Equation (28).

When the incident angle rises above 30°, the product of the transmittance of the glazing and the absorptance of the collector plate begins to diminish; thus, the heat absorbed also drops. The losses from the collector are generally higher as the time moves farther from solar noon, and consequently the efficiency also drops. Thus, the daylong efficiency is lower than the near-noon performance. During the early afternoon hours, the efficiency is slightly higher than at the comparable morning time, because ambient air temperatures are lower in the morning than in the afternoon.

ASHRAE *Standard* 93 describes the *incident angle modifier*, which may be found by tests run when the incident angle is set at 30°, 45°, and 60°. Simon (1976) showed that for many flat-plate collectors, the incident angle modifier is a linear function of the quantity $(1/\cos\theta - 1)$. For evacuated tubular collectors, the incident angle modifier may grow with rising values of θ.

ASHRAE *Standard* 93 specifies that the efficiency be reported in terms of the gross collector area A_g rather than the aperture area A_{ap}. The reported efficiency will be lower than the efficiency given by Equation (29), but the total energy collected is not changed by this simplification:

$$\eta_g = \eta_{ap} A_{ap} / A_g \qquad (30)$$

HEAT STORAGE SYSTEMS

Storage may be part of solar heating, cooling, and power generating systems. For approximately one-half of the 8760 h in a year, any location is in darkness, and heat storage is necessary if the system must operate continuously. For some applications, such as swimming pool heating, daytime air heating, and irrigation pumping, intermittent operation is acceptable, but most other uses of solar energy require operating at night and when the sun is obscured by clouds. Chapter 40 provides further information on general thermal storage technologies.

Packed Rock Beds

Packed bed, pebble bed, or rock pile storage uses the heat capacity of a bed of loosely packed particles through which a fluid, usually air, is circulated to add heat to, or remove heat from, the bed. Although a variety of solids may be used, rock from 20 to 50 mm is the most common. Well-designed packed rock beds have several desirable characteristics for energy storage: (1) the cost of the storage material is low, (2) the conductivity of the bed is low when there is no airflow, (3) the heat transfer coefficient between the air and the solid is high, and (4) a large heat transfer area is achieved at low cost by using small storage particles. Figure 15 shows a schematic of a packed bed storage unit. Essential features include a container, a porous structure (such as a wire screen) to support the bed, and air plenums at both inlet and outlet. The container can be made of wood, concrete blocks, or poured concrete. Insulation requirements

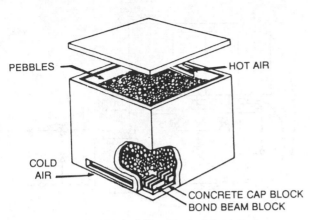

Fig. 15   Vertical Flow Packed Rock Bed

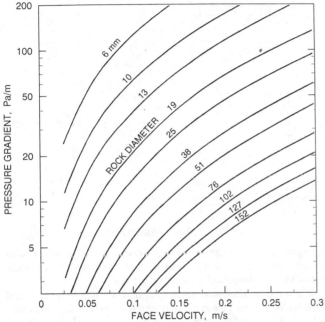

Fig. 16   Pressure Drop Through a Packed Rock Bed
(Cole et al. 1980)

outside the bed are small for short-term storage, because the lateral thermal conductivity of the bed is low. For a rock bed used for daily cycling, the 24-h heat loss should not exceed 5% and should preferably be only 2%.

Fluid flows through the bed in one direction during the addition of heat and in the opposite direction during the removal of heat. Therefore, heat cannot be added to and removed from these storage devices simultaneously. Packed bed storage devices operated in this manner thermally stratify naturally. For efficiency, the bed temperature should increase during downward heat flow and decrease during upward heat flow. During standby periods, this flow (hot top, cool bottom) inhibits convection currents which decrease storage efficiency. To prevent channeling (in which a narrow jet of heat transfer fluid travels from inlet to outlet without sweeping through the entire bed), both the inlet and the outlet should have plenum chambers. To further equalize pressure drops along all flow paths, it is best to locate the inlet and outlet on opposite sides of the storage bed (see Figure 15).

The plenum chambers distribute the flow over the entire cross section of the bed. Tests by Jones and Loss (1982) on a packed bed

like the one in Figure 15 showed that the inlet jet shoots across the rock top, hits the far wall, and forms two counter-rotating flow cells along the side walls. The flow distribution is improved by placing a block slightly larger than the inlet opening inside the top plenum at a distance of 1.5 times the vertical inlet dimension.

Because of the large surface area of the pebbles exposed to the air passing through the bed, pebble beds exhibit good heat transfer between the air and the storage medium. An empirical relation for the volumetric heat transfer coefficient h_v is given by the following equation (Lof and Hawley 1948):

$$h_v = 0.79 \, (G/D)^{0.7} \tag{31}$$

where

h_v = volumetric heat transfer coefficient, W/(m^3·K)
G = superficial mass velocity, kg/(s·m^2)
D = equivalent spherical diameter of the particles, m
   = [(6/π)(net volume of particles)/(number of particles)]$^{1/3}$ (32)

The superficial mass velocity is the mass flow per unit time through the rock bed per unit area of bed face.

The particle size in a packed bed should be uniform to obtain a large void fraction and thus minimize the pressure drop through the bed. Figure 16 (Cole et al. 1980) shows the pressure drop as a function of face velocity and particle size diameter. Analytical or numerical methods can be used to evaluate the performance of packed bed storage devices, but the calculations for an arbitrary time-dependent inlet air temperature variation are laborious. Hughes et al. (1976) and Mumma and Marvin (1976) give methods of solution.

SOLAR WATER HEATING SYSTEMS

A solar water heater includes a solar collector array that absorbs solar radiation and converts it to heat, which is then absorbed by a heat transfer fluid (water, a nonfreezing liquid, or air) that passes through the collector. The heat transfer fluid's heat can be stored or used directly.

Portions of the solar energy system are exposed to the weather, so they must be protected from freezing. The systems must also be protected from overheating caused by high insolation levels during periods of low energy demand.

In solar water heating systems, potable water that is heated directly in the collector or indirectly by a heat transfer fluid that is heated in the collector, passes through a heat exchanger and transfers its heat to the domestic or service water. The heat transfer fluid is transported by either natural or forced circulation. Natural circulation occurs by natural convection (thermosiphoning), whereas forced circulation uses pumps or fans. Except for thermosiphon systems, which need no control, solar domestic and service hot water systems are controlled using differential thermostats.

Six types of solar energy systems are used to heat domestic and service hot water: thermosiphon, direct circulation, drain-back, indirect, integral collector storage, and air. The term recirculation has been used to describe a freeze protection method for direct solar water heating systems. Similarly, the term drain-down has also been used to describe a freeze protection method for direct circulation systems.

Thermosiphon Systems

Thermosiphon systems (Figure 17) heat potable water or a heat transfer fluid and use natural convection to transport it from the collector to storage. For direct systems, pressure-reducing valves are required when the city water pressure is greater than the working pressure of the collectors. In a thermosiphon system, the storage tank must be elevated above the collectors, which sometimes requires designing the upper level floor and ceiling joists to bear this additional load. Extremely hard or acidic water can cause scale deposits that clog or corrode the absorber fluid passages.

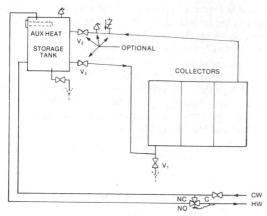

Fig. 17   Thermosiphon System

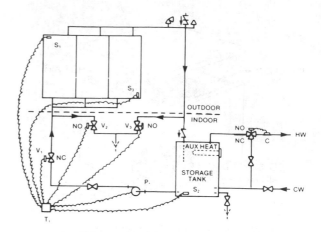

Fig. 19   Drain-Down System

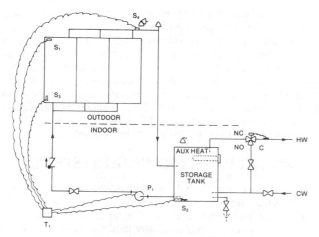

Fig. 18   Direct Circulation System

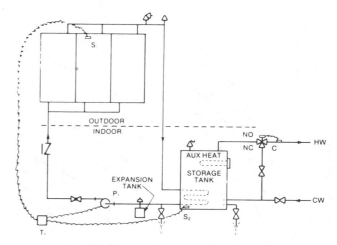

Fig. 20   Indirect Water Heating

Thermosiphon flow is induced whenever there is sufficient sunshine, so these systems do not need pumps.

Direct Circulation Systems

Direct circulation systems (Figure 18) are direct water heating systems that pump potable water from storage to the collectors when there is enough solar energy available to warm it, and then return the heated water to the storage tank until it is needed. A pump circulates the water, so the collectors can be mounted either above or below the storage tank. Direct circulation systems are feasible only in areas where freezing is infrequent. Freeze protection for extreme weather conditions is provided either by recirculating warm water from the storage tank or by flushing the collectors with cold water. Direct water heating systems should not be used in areas where the water is extremely hard or acidic because scale deposits may clog or corrode the absorber fluid passages, rendering the system inoperable.

Direct circulation systems are exposed to city water line pressures and must withstand test pressures, as required by local codes. Pressure-reducing valves and pressure relief valves are required when the city water pressure is greater than the working pressure of the collectors. Direct circulation systems often use a single storage tank for both solar energy storage and the auxiliary water heater, but two-tank storage systems can be used.

Drain-Down Systems. Drain-down systems (Figure 19) are direct circulation, water heating systems in which potable water is pumped from storage to the collector array where it is heated. Circulation continues until usable solar heat is no longer available. When a freezing condition is anticipated or a power outage occurs, the system drains automatically by isolating the collector array and exterior

piping from the city water pressure and draining it using one or more valves. The solar collectors and associated piping *must* be carefully sloped to drain the collector's exterior piping.

Drain-down systems are also exposed to city water pressures and must withstand test pressures, as required by local codes. Pressure-reducing valves and pressure relief valves are required when the city water pressure is greater than the working pressure of the collectors. One- or two-tank storage systems can be used. Scale deposits and corrosion can occur in the collectors if the water is hard or acidic.

Indirect Water Heating Systems

Indirect water heating systems (Figure 20) circulate a freeze-protected heat transfer fluid through the closed collector loop to a heat exchanger, where its heat is transferred to the potable water. The most commonly used heat transfer fluids are water/ethylene glycol and water/propylene glycol solutions, although other heat transfer fluids such as silicone oils, hydrocarbons, and refrigerants can also be used (ASHRAE 1983). These fluids are nonpotable, sometimes toxic, and normally require double-wall heat exchangers. The double-wall heat exchanger can be located inside the storage tank, or an external heat exchanger can be used. The collector loop is closed and therefore requires an expansion tank and a pressure relief valve. A one- or two-tank storage system can be used. Additional over-temperature protection may be needed to prevent the collector fluid from decomposing or becoming corrosive.

Designers should avoid automatic water makeup in systems using water/antifreeze solutions because a significant leak may raise the freezing temperature of the solution above the ambient temperature, causing the collector array and exterior piping to freeze. Also,

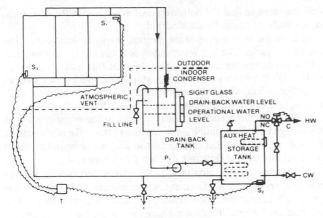

Fig. 21 Drain-Back System

antifreeze systems with large collector arrays and long pipe runs may need a time-delayed bypass loop around the heat exchanger to avoid freezing the heat exchanger on start-up.

Drain-Back Systems. Drain-back systems are generally indirect water heating systems that circulate treated or untreated water through the closed collector loop to a heat exchanger, where its heat is transferred to the potable water. Circulation continues until usable energy is no longer available. When the pump stops, the collector fluid drains by gravity to a storage tank or a drain-back tank. In a pressurized system, the tank also serves as an expansion tank when the system is operating and must be protected from excessive pressure with a temperature and pressure relief valve. In an unpressurized system (Figure 21), the tank is open and vented to the atmosphere.

The collector loop is isolated from the potable water, so valves are not needed to actuate draining, and scaling is not a problem. The collector array and exterior piping must be sloped to drain completely, and the pumping pressure must be sufficient to lift water to the top of the collector array.

Integral Collector Storage Systems

Integral collector storage (ICS) systems use hot water storage as part of the collector. Some types use the surface of a single tank as the absorber, and others use multiple, long, thin tanks placed side-by-side horizontally to form the absorber surface. In this type of ICS, hot water is drawn from the top tank and cold replacement water enters the bottom tank. Because of the greater nighttime heat loss from ICS systems, they are typically less efficient than pumped systems, and selective surfaces are recommended. ICS systems are normally installed as a solar preheater without pumps or controllers. Flow through the ICS system occurs on demand, as hot water flows from the collector to a hot water auxiliary tank in the structure.

Air Systems

Air systems (Figure 22) are indirect water heating systems that circulate air through the collectors via ductwork to an air-to-liquid heat exchanger. There, heat is transferred to the potable water, which is pumped through the tubes of the exchanger and returned to the storage tank. Circulation continues as long as usable heat is available. Air systems can use single or double storage tank configurations. The two-tank storage system is used most often, because air systems are generally used for preheating domestic hot water and may not be capable of reaching the 50 to 70°C delivery temperatures.

Air does not need to be protected from freezing or boiling, is non-corrosive, and is free. However, air ducts and air-handling equipment need more space than piping and pumps. Ductwork is laborious to seal, and air leaks are difficult to detect. Power consumption is generally higher than that of a liquid system because of high collector and

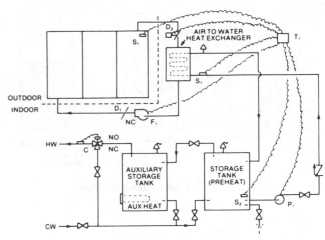

Fig. 22 Air System

heat exchanger static pressure loss. All dampers installed in air systems must fit tightly to prevent leakage and heat loss.

In areas with freezing temperatures, tight dampers are needed in the collector ducts to prevent reverse thermosiphoning at night, which could freeze the water in the heat exchanger coil. No special precautions are needed to control overheating conditions in air systems.

Pool Heating Systems

Solar pool heating systems require no separate storage tank, because the pool itself serves as storage. In most cases, the pool's filtration pump forces the water through the solar panels or plastic pipes. In some retrofit applications, a larger pump may be required to handle the needs of the solar system, or a small pump may be added to boost the pool water to the solar collectors.

Automatic controls may be used to direct the flow of filtered water to the collectors when solar heat is available; this may also be accomplished manually. Normally, solar systems are designed to drain down into the pool when the pump is turned off; this provides the collectors with freeze protection.

Four primary types of collector designs are used for swimming pool heat: (1) rigid black plastic panels (polypropylene), usually 1.2 by 3 m or 1.2 by 2.4 m; (2) tube-on-sheet panels, which usually have a metal deck (copper or aluminum) with copper water tubes; (3) EPDM rubber mat, extruded with the water passages running its length; and (4) arrays of black plastic pipe, usually 38-mm diameter ABS plastic (Root et al. 1985).

HOT WATER
RECIRCULATION SYSTEMS

Domestic hot water (DHW) recirculation systems (Figures 23 and 24), which continuously circulate domestic hot water throughout a building, are common in motels, hotels, hospitals, dormitories, office buildings, and other commercial buildings. The recirculation heat losses in these systems are usually a significant part of the total water heating load. A properly integrated solar energy system can make up much of this heat loss.

System Components

A solar domestic hot water system has a variety of components, which are selected based on function, component compatibility, climatic conditions, required performance, site characteristics, and architectural requirements. For optimum system performance, the components of a solar energy system must fit and function together satisfactorily. The system must also be properly integrated into the conventional domestic or service hot water system. The major com-

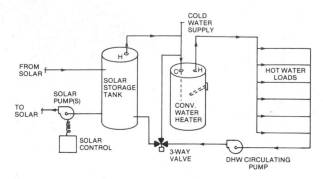

Fig. 23   DHW Recirculation System

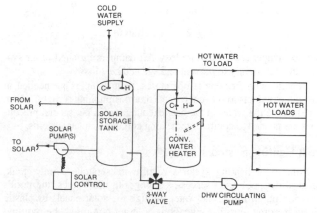

Fig. 24   DHW Recirculation System with Makeup Preheat

ponents involved in the collection, storage, transportation, control, and distribution of solar energy are discussed in this section.

Collectors. Flat-plate collectors are most commonly used in water heating applications because of the year-round load requiring temperatures of 30 to 80°C. For discussions of other collectors and applications, see ASHRAE *Standard* 93 and previous sections of this chapter. Collectors must withstand extreme weather conditions (such as freezing, stagnation, and high winds), as well as system pressures.

Heat Transfer Fluids. Heat transfer fluids transport heat from the solar collectors to the domestic water. The potential safety problems that exist in this transfer are both chemical and mechanical in nature and apply primarily to liquid transfer and storage systems in which a heat exchanger interface exists with the potable water supply. Both the chemical compositions of the heat transfer fluids (pH, toxicity, and chemical durability), as well as their mechanical properties (specific heat and viscosity) must be considered.

Except in unusual cases or when potable water is being circulated, the energy transport fluid is nonpotable and has the potential for contaminating potable water during the heat transfer process. Even potable or nontoxic fluids in closed systems are likely to become nonpotable because of contamination from metal piping, solder joints, and packing, or by the inadvertent installation of a toxic fluid at a later date.

Trade-offs between the thermal efficiency, cost effectiveness, and risk of both heat transfer fluids and heat exchangers may be necessary to ensure acceptable safety. Existing codes regulate the need for and design of heat exchangers.

Thermal Energy Storage. Thermal energy (heat) storage in hydronic solar domestic and service heating systems is virtually always liquid stored in tanks. All storage tanks and bins should be well insulated so that the collected heat is not lost to the surroundings, which are normally unoccupied.

In domestic hot water systems, thermal energy is usually stored in one or two tanks. The hot water outlet is at the top of the tank, and cold water enters the tank through a dip tube that extends down to within 100 to 150 mm of the tank bottom. The outlet on the tank to the collector loop should be approximately 100 mm above the tank bottom to prevent scale deposits from being drawn into the collectors. Water from the collector array returns to the upper portion of the storage tank. This plumbing arrangement may take advantage of thermal stratification, depending on the delivery temperature from the collectors and the flow rate through the storage tank.

Single-tank electric auxiliary systems often incorporate storage and auxiliary heating in the same vessel. Conventional electric water heaters commonly have two heating elements: one near the top and one near the bottom. If a dual element tank is used in a solar energy system, the bottom element should be disconnected and the top left functional to take advantage of fluid stratification. Standard gas- and oil-fired water heaters should not be used in single-tank arrangements. In gas and oil water heaters, heat is added to the bottom of the tanks, which reduces both stratification and collection efficiency in single-tank systems.

Dual-tank systems often use the solar domestic hot water storage tank as a preheat tank. The second tank is normally a conventional domestic hot water system tank and contains the auxiliary heat source. Multiple tank systems are sometimes used in large institutions, where they operate similarly to dual-tank systems. Although the use of a two-tank system may increase collector efficiency and the solar fraction, which is the total heating load supplied by the solar energy, it increases tank heat losses. The water inlet to these tanks is usually a dip tube that extends close to the bottom of the tank.

Estimates for sizing storage tanks usually range from 40 to 100 L per square metre of solar collector area. The estimate used most often is 75 L per square metre of collector area, which usually provides enough heat for a sunless period of about a day. Storage volume should be analyzed and sized according to the project water requirements and draw schedule; however, solar applications typically require larger tanks than would normally be used.

Some electric utilities offer reduced electric rates to customers willing to charge their tanks during off-peak hours only. The charge time is controlled by a time clock or by a signal sent by the utility either over its power lines (ripple control) or by radio. Depending on the nature of the hot water draw by the individual user, a tank larger than the one normally used is recommended. Thus, solar heating is an ideal adjunct to this off-peak water heating application.

Heat Exchangers. Heat exchangers transfer thermal energy between two fluids. All solar energy systems that use indirect water heating require one or more heat exchangers. The potential exists for contamination problems in the transfer of heat energy from solar collectors to potable hot water supplies.

Heat exchangers influence the effectiveness with which collected energy is made available to heat domestic water. They also separate and protect the potable water supply from contamination when nonpotable heat transfer fluids are used. As in transport fluid selection, heat exchanger selection should consider thermal performance, cost effectiveness, reliability, safety, and the following:

• Heat exchange effectiveness
• Pressure drop, operating power, and flow rate
• Physical design, design pressure, configuration, size, materials, and location in the system
• Cost and availability
• Reliable protection of the potable water supply from contamination by the heat transfer fluids
• Leak detection, inspection, and maintainability
• Material compatibility with other system elements such as metals and fluids

• Thermal compatibility with system design parameters such as operating temperatures, flow rate, and fluid thermal properties

Heat exchanger selection depends on the characteristics of the fluids that pass through the heat exchanger and the properties of the exchanger itself. Fluid characteristics to consider are fluid type, specific heat, mass flow rate, and hot and cold fluid inlet and outlet temperatures. Physical properties of the heat exchanger to consider are the overall heat transfer coefficient of the heat exchanger and the heat transfer surface area. When these variables are known, the heat transfer rate can be determined.

For most solar domestic hot water system designs, only the hot and cold inlet temperatures are known; the other temperatures must be calculated using the physical properties of the heat exchanger.

This information can be used to evaluate two quantities that are useful in determining the heat transfer in a heat exchanger and the performance characteristics of a collector when it is combined with a given heat exchanger. These quantities are (1) the fluid capacitance rate, which for a given fluid is the product of the mass flow rate and the specific heat of the fluid passing through the heat exchanger and (2) the heat exchanger effectiveness, which relates the capacitance rate of the two fluids to the fluid inlet and outlet temperatures. The effectiveness is equal to the ratio of the actual heat transfer rate to the maximum heat transfer rate theoretically possible. Generally, a heat exchanger effectiveness of 0.4 or greater is recommended.

Expansion Tanks. Indirect solar water heating systems operating in a closed collector loop require an expansion tank to control excessive pressure increases. Fluid in solar collectors under stagnation conditions can easily boil, resulting in excessive pressure increases in the collector loop, and solar system expansion tanks must be sized to account for this. The ASHRAE expansion tank sizing formulas for closed loop hydronic systems found in Chapter 12 of the 1992 *ASHRAE Handbook—Systems and Equipment*, may be used for solar system expansion tank sizing, but the expression for volume change due to temperature increase should be replaced with the total volume of fluid in the solar collectors and of any piping located above the collectors, if significant. This sizing method provides a passive means for eliminating fluid loss due to over temperature or stagnation, common problems in closed loop solar systems. This results in larger expansion tanks than those typically found in hydronic systems, but the increase in cost is small compared to the savings in fluid replacement and maintenance costs (Lister and Newell 1989).

Pumps. Pumps circulate heat transfer liquid through collectors and heat exchangers. The two most commonly used types are centrifugal pumps and positive displacement pumps. In solar domestic hot water systems, the pump is usually a centrifugal circulator driven by a small fractional horsepower motor. The flow rate for collectors generally ranges from 0.010 to 0.027 L/(s·m²). Pumps used in drainback systems must provide pressure to overcome friction and to lift the fluid to the collectors.

System Piping. Piping can be plastic, copper, galvanized steel, or stainless steel. The most widely used is sweat-soldered L-type copper tubing. M-type copper is also acceptable if permitted by local building codes. If water/glycol is the heat transfer fluid, galvanized pipes or tanks must not be used because unfavorable chemical reactions will occur; copper piping is recommended instead. Also, if glycol solutions or silicone fluids are used, they may leak through joints where water would not. System piping should be compatible with the collector fluid passage material, for example, copper or plastic piping should be used with collectors having copper fluid passages.

Piping that carries potable water can be plastic, copper, galvanized steel, or stainless steel. In indirect systems, corrosion inhibitors must be checked and adjusted routinely, preferably every three months. Inhibitors should also be checked if the system overheats during stagnation conditions. If dissimilar metals are joined, dielectric or nonmetallic couplings should be used. The best protection is

sacrificial anodes or getters in the fluid stream. Their location depends on the material to be protected, the anode material, and the electrical conductivity of the heat transfer fluid. Sacrificial anodes consisting of magnesium, zinc, or aluminum are often used to reduce corrosion in storage tanks. Because there are so many possibilities, each combination must be evaluated on its own merits. A copper-aluminum or copper-galvanized steel joint is unacceptable because of severe galvanic corrosion. Systems containing aluminum, copper, and iron metals have a greatly increased potential for corrosion.

Insufficient consideration of air elimination requirements, pipe expansion control, and piping slope can cause serious system failures. Collector pipes (particularly manifolds) should be designed to allow expansion from stagnation temperatures to extreme cold weather temperatures. Expansion control can be achieved with offset elbows in piping, hoses, or expansion couplings. Expansion loops should be avoided unless they are installed horizontally, particularly in systems that must drain for freeze protection. The collector array piping should slope 5 mm per metre for drainage (DOE 1978b).

Air can be eliminated by placing air vents at all piping high points and by air purging during filling. Flow control, isolation, and other valves in the collector piping must be chosen carefully so that these components do not restrict drainage significantly or back up water behind them. The collectors must drain completely.

Valves and Gages. Valves in solar domestic hot water systems must be located to ensure system efficiency, satisfactory performance, and the safety of equipment and personnel. Drain valves must be ball-type; gate valves may be used if the stem is installed horizontally.

Check valves or other valves used for freeze protection or for reverse thermosiphoning must be reliable to avoid significant damage.

Auxiliary Heat Sources. On sunny days, a typical solar energy system should supply water at a predetermined temperature, and the solar storage tank should be large enough to hold sufficient water for a day or two. Because of the intermittent nature of solar radiation, an auxiliary heater must be installed to handle hot water requirements. If a utility is the source of auxiliary energy, operation of the auxiliary heater can be timed to take advantage of off-peak utility rates. The auxiliary heater should be carefully integrated with the solar energy system to obtain maximum solar energy use. For example, the auxiliary heater should not destroy any stratification that may exist in the solar-heated storage tank, which would reduce collector efficiency.

Fans. In air systems, fans circulate air by forcing air that has been heated in the solar collectors through ductwork to an air-to-water heat exchanger. The heat from the air is transferred to water that is pumped through the coil section of the heat exchanger. For detailed information on fan performance and selection, refer to Chapter 18 of the 1992 *ASHRAE Handbook—Systems and Equipment*.

Ductwork, particularly in systems with air-type collectors, must be sealed carefully to avoid leakage in duct seams, damper shafts, collectors, and heat exchangers. Ducts should be sized using conventional air duct design methods.

Control Systems. Control systems regulate solar energy collection by controlling fluid circulation, activate system protection against freezing and overheating, and initiate auxiliary heating when it is required.

The three major control components are sensors, controllers, and actuated devices. Sensors detect conditions or measure quantities, such as temperatures. Controllers receive output from the sensors, select a course of action, and signal a system component to adjust the condition. Actuated devices are components, such as pumps, valves, dampers, and fans, that execute controller commands and regulate the system.

Temperature sensors measure the temperature of the absorber plate near the collector outlet and near the bottom of the storage tank. The sensors send signals to a controller, such as a differential temperature thermostat, for interpretation.

The differential thermostat compares the signals from the sensors with adjustable set points for high and low temperature differentials. The controller performs different functions, depending on which set points are met. In liquid systems, when the temperature difference between the collector and storage reaches a high set point, usually 10 K, the pump starts, automatic valves are activated, and circulation begins. When the temperature difference reaches a low set point, usually 2 K, the pump is shut off and the valves are deenergized and returned to their normal positions. To restart the system, the high-temperature set point must again be met. If the system has either freeze or overtemperature protection, the controller opens or closes valves or dampers and starts or stops pumps or fans to protect the system when its sensors detect that either a freezing or an overheating condition is about to occur.

Sensors must be selected to withstand high temperatures, such as those that may occur during collector stagnation. Collector loop sensors can be located on the absorber plate, in a pipe above the collector array, on a pipe near the collector, or in the collector outlet passage. Although any of these locations may be acceptable, attaching the sensor on the collector absorber plate is recommended. When attached properly, the sensor gives accurate readings, can be installed easily, and is basically unaffected by ambient temperature, as are sensors mounted on exterior piping.

A sensor installed on an absorber plate reads temperatures 2 K higher than the temperature of the fluid leaving the collector. However, such temperature discrepancies can be compensated for in the differential thermostat settings.

The sensor must be attached to the absorber plate with good thermal contact. If a sensor access cover is provided on the enclosure, it must be gasketed for a watertight fit. Adhesives and adhesive tapes should not be used to attach the sensor to the absorber plate.

The storage temperature sensor should be near the bottom of the storage tank to detect the temperature of the fluid before it is pumped to the collector or heat exchanger. The storage fluid is usually coldest at that location because of thermal stratification and the location of the makeup water supply. The sensor should be either securely attached to the tank and well insulated, or immersed inside the tank near the collector supply.

The freeze protection sensor, if required, should be located so that it will detect the coldest liquid temperature when the collector is shut down. Common locations are the back of the absorber plate at the bottom of the collector, the collector intake or return manifolds, or the center of the absorber plate. The center absorber plate location is recommended because reradiation to the night sky will freeze the collector heat transfer fluid, even though the ambient temperature is above freezing. Some systems, such as the recirculation system, have two sensors for freeze protection, while others, such as the drain-down, use only one.

Control on-off temperature differentials affect system efficiency. The turn-on temperature differential must be selected properly because, if the differential is too high, the system starts later than it should, and if it is too low, the system starts too soon. The turn-on differential for liquid systems usually ranges from 8 to 17 K and is most commonly 10 K. For air systems, the range is usually 14 to 25 K.

The turn-off temperature differential is more difficult to estimate. Selection depends on a comparison between the value of the energy collected and the cost of collecting it. It varies with individual systems, but a value of 2 K is typical.

Water temperature in the collector loop depends on ambient temperature, solar radiation, radiation from the collector to the night sky, and collector loop insulation. Freeze protection sensors should be set to detect temperatures of 4°C.

Sensors are important but often overlooked control system components. They must be selected and installed properly because no control system can produce accurate outputs from unreliable sensor inputs. Sensors are used in conjunction with a differential temperature controller and are usually supplied by the controller manufacturer. Sensors must survive the anticipated operating conditions without physical damage or loss of accuracy. Low-voltage sensor circuits must be located away from 120/240 V (ac) lines to avoid electromagnetic interference. Sensors attached to collectors should be able to withstand stagnation temperatures.

Sensor calibration, which is often overlooked by installers and maintenance personnel, is critical to system performance; a routine calibration maintenance schedule is essential.

Another control system is the photovoltaic (PV) pumped system. A PV panel converts sunlight into direct current (dc) electricity and provides it directly to a small circulating pump. No additional sensing is required because the PV panel and pump output increase with sunlight intensity and stop when no sunlight (collector energy) is available. Cromer (1984) has shown that with proper matching of pump and PV electrical characteristics, PV panel sizes as low as 1.35 W per m^2 of thermal panel may be used successfully. Difficulty with late starting and running too long in the afternoon can be alleviated by tilting the PV panel slightly to the east.

System Performance Evaluation Methods

The performance of any solar energy system is directly related to the (1) heating load requirements, (2) amount of solar radiation available, and (3) solar energy system characteristics. Various calculation methods use different procedures and data when considering the available solar radiation. Some simplified methods consider only average annual incident solar radiation, while complex methods may use hourly data.

Solar energy system characteristics, as well as individual component characteristics, are required to evaluate performance. The degree of complexity with which these systems and components are described varies from system to system.

The cost effectiveness of a solar domestic and service hot water heating system depends on the initial cost and energy cost savings. A major task is to determine how much energy is saved. The *annual solar fraction*—the annual solar contribution to the water heating load divided by the total water heating load—can be used to estimate these savings. It is expressed as a decimal fraction or percentage and generally ranges from 0.3 to 0.8 (30 to 80%), although more extreme values are possible.

Water Heating Load Requirements

The amount of water required must be estimated as accurately as possible because it affects system component selection. Chapter 45 gives methods to determine the load.

Oversized storage may result in low-temperature water that requires auxiliary heating to reach the desired supply temperature. Undersizing can prevent the collection and use of available solar energy.

Integration of the solar energy system into a recirculating hot water system, if one is used, must also be considered carefully to maximize solar energy use (see Figures 23 and 24).

COOLING BY SOLAR ENERGY

A review of solar-powered refrigeration by Swartman (1974) emphasizes various absorption systems. Newton (Jordan and Liu 1977) discusses commercially available water vapor/lithium bromide absorption refrigeration systems. Standard absorption chillers are generally designed to give rated capacity for activating fluid temperatures well above 90°C at full load and design condenser water temperature. Few flat-plate collectors can operate efficiently in this range; therefore, lower hot fluid temperatures are used when solar energy provides the heat. Both the temperature of the condenser water and the percentage of design load are determinants of the optimum energizing temperature, which can be quite low, sometimes below 50°C. Proper control can raise the coefficient of performance (COP) at these part-load conditions.

Many large commercial or institutional cooling installations must operate year-round, and Newton (Jordon and Liu 1977) showed that the low-temperature cooling water available in winter enables the LiBr/H$_2$O to function well with hot fluid inlet temperatures below 88°C. Residential chillers in sizes as low as 5.3 kW, with inlet temperatures in the range of 80°C, have been developed.

COOLING BY NOCTURNAL RADIATION AND EVAPORATION

Radiative cooling is a natural heat loss mechanism that causes the formation of dew, frost, and ground fog. Because its effects are the most obvious at night, it is sometimes termed *nocturnal radiation*, although the process continues throughout the day. Thermal infrared radiation, which affects the surface temperature of a building wall or roof, has been treated in an approximate manner using the sol-air temperature concept. Radiative cooling of window and skylight surfaces can be significant, especially under winter conditions when the dew-point temperature is low.

The most useful parameter for characterizing the radiative heat transfer between horizontal nonspectral emitting surfaces and the sky is the *sky temperature* T_{sky}. If S designates the total down-coming radiant heat flux emitted by the atmosphere, then T_{sky} is defined as

$$T_{sky}^4 = S/\sigma \tag{33}$$

where $\sigma = 5.67 \times 10^{-8}$ W/(m^2·K^4).

The sky radiance is treated as if it originates from a blackbody emitter of temperature T_{sky}. The *net radiative cooling rate* R_{net} of a horizontal surface with absolute temperature T_{rad} and a nonspectral emittance ε is then

$$R_{net} = \varepsilon\sigma \left(T_{rad}^4 - T_{sky}^4 \right) \tag{34}$$

Values of ε for most nonmetallic construction materials are usually about 0.9.

Radiative building cooling has not been fully developed. Design methods and performance data compiled by Hay and Yellott (1969) and Marlatt et al. (1984) are available for residential roof pond systems that use a sealed volume of water covered by sliding insulation panels as the combined rooftop radiator and thermal storage element. Other conceptual radiative cooling designs have been proposed, but more developmental work is required (see Givoni 1981 and Mitchell and Biggs 1979).

The sky temperature is a function of atmospheric water vapor, the amount of cloud cover, and air temperature; the lowest sky temperatures occur under an arid, cloudless sky. The monthly average sky temperature depression, which is the average of the difference

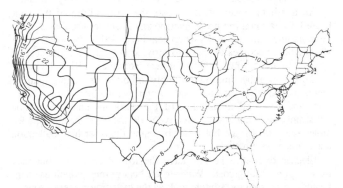

Fig. 25   Average Monthly Sky Temperature Depression
$(T_{air} - T_{sky})$ **for July, °C**
(Adapted from Martin and Berdahl 1984)

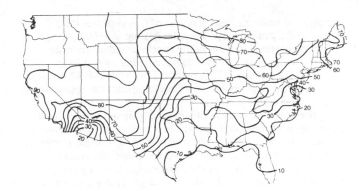

Fig. 26   Percentage of Monthly Hours when Sky Temperature Falls below 16°C
(Adapted from Martin and Berdahl 1984)

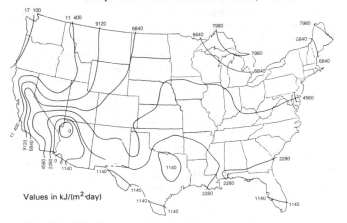

Values in kJ/(m^2·day)

Fig. 27   July Nocturnal Net Radiative Cooling Rate from Horizontal Dry Surface at 25°C
(Adapted from Clark 1981)

between the ambient air temperature and the sky temperature, typically lies between 5 and 24 K throughout the continental United States. Martin and Berdahl (1984) have calculated this quantity using hourly weather data from 193 sites, as shown in the contour map for the month of July (Figure 25).

The sky temperature may be too high at night to effectively cool the structure. Martin and Berdahl (1984) suggest that the sky temperature should be less than 16°C to achieve reasonable cooling in July (Figure 26). In regions where sky temperatures fall below 16°C 40% or more of the month, all nighttime hours are effectively available for radiative cooling systems.

Clark (1981) modeled a horizontal radiator at various surface temperatures in convective contact with outdoor air for 77 U.S. locations. The average monthly cooling rates for a surface temperature of 25°C are plotted in Figure 27. If effective steps are taken to reduce the surface convection coefficient by modifying the radiator geometry or using an infrared-transparent glazing, it may be possible to improve performance beyond these values.

SOLAR HEATING AND COOLING SYSTEMS

The components and subsystems discussed earlier may be combined to create a wide variety of solar heating and cooling systems. There are two principal categories of such systems: passive and active.

Passive solar systems require little, if any, nonrenewable energy to make them function (Yellott 1977, Yellott et al. 1976). Every building is passive in the sense that the sun tends to warm it by day and it loses heat at night. Passive systems incorporate solar collection, storage, and distribution into the architectural design of the building and make minimal or no use of fans to deliver the collected energy to the structure. Passive solar heating, cooling, and lighting design must consider the building envelope and its orientation, the thermal storage mass, and window configuration and design. DOE (1980, 1982), LBL (1981), Mazria (1979), and ASHRAE (1984b) give estimates of energy savings resulting from the application of passive solar design concepts.

Active solar systems use either liquid or air as the collector fluid. Active systems must have a continuous availability of nonrenewable energy, generally in the form of electricity, to operate pumps and fans. A complete system includes solar collectors, energy storage devices, and pumps or fans for transferring energy to storage or to the load. The load can be space cooling, heating, or hot water. Although it is technically possible to construct a solar heating and cooling system to supply 100% of the design load, such a system would be uneconomical and oversized for seasonal operation. The size of the solar system, and thus its ability to meet the load, is determined by life-cycle cost analysis that weighs the cost of energy saved against the amortized solar system cost.

Active solar energy systems have been combined with heat pumps for water and/or space heating. The most economical arrangement in residential heating is a solar system in parallel with a heat pump, which supplies auxiliary energy when the solar source is not available. For domestic water systems requiring high water temperatures, a heat pump placed in series with the solar storage tank may be advantageous. Freeman et al. (1979) and Morehouse and Hughes (1979) present information on performance and estimated energy savings for solar-heat pump systems.

Hybrid systems combine elements of both active and passive systems. Hybrid systems require some nonrenewable energy, but the amount is so small that they can maintain a coefficient of performance of about 50.

Passive Systems

Passive systems may be divided into several categories. The first residence to which the name *solar house* was applied used a large expanse of south-facing glass to admit solar radiation; this is known as a *direct gain* passive system.

Indirect gain solar houses use the south-facing wall surface or the roof of the structure to absorb solar radiation, which causes a rise in temperature that, in turn, conveys heat into the building in several ways. This principle was applied to the pueblos and cliff dwellings of the southwestern United States. Glass has led to modern adaptations of the indirect gain principle (Trombe et al. 1977, Balcomb et al. 1977).

By glazing a large south-facing, massive masonry wall, solar energy can be absorbed during the day, and conduction of heat to the inner surface provides radiant heating at night. The mass of the wall and its relatively low thermal diffusivity delays the arrival of the heat at the indoor surface until it is needed. The glazing reduces the loss of heat from the wall back to the atmosphere and increases the collection efficiency of the system.

Openings in the wall, near the floor, and near the ceiling allow convection to transfer heat to the room. The air in the space between the glass and the wall warms as soon as the sun heats the outer surface of the wall. The heated air rises and enters the building through the upper openings. Cool air flows through the lower openings, and convective heat gain can be established as long as the sun is shining.

In another indirect gain passive system, a metal roof-ceiling supports transparent plastic bags filled with water (Hay and Yellott 1969). Movable insulation above these water-filled bags is rolled away during the winter day to allow the sun to warm the stored water. The water then transmits heat indoors by convection and radiation. The insulation remains over the water bags at night or during overcast days. During the summer, the water bags are exposed at night for cooling by (1) convection, (2) radiation, and (3) evapora-

tion of water on the water bags. The insulation covers the water bags during the day to protect them from unwanted irradiation.

Pittenger et al. (1978) tested a building for which water rather than insulation was moved to provide summer cooling and winter heating.

Add-on greenhouses (sunspaces) can be used as solar attachments when the orientation and other local conditions are suitable. The greenhouse can provide a buffer between the exterior wall of the building and the outdoor conditions. During daylight hours, warm air from the greenhouse can be introduced into the house by natural convection or a small fan.

In most passive systems, control is accomplished by moving a component that regulates the amount of solar radiation admitted into the structure. Manually operated window shades or venetian blinds are the most widely used and simplest controls.

Passive heating and cooling systems have been shown to be effective in field demonstrations (Howard and Pollock 1982; Howard and Sanders 1989).

Active Systems

Active systems absorb solar radiation with collectors and convey it to storage using a suitable fluid. As heat is needed, it is obtained from storage via heated air or water. Control is exercised by several types of thermostats, the first being a differential device that starts the flow of fluid through the collectors when they have been sufficiently warmed by the sun. It also stops the fluid flow when the collectors no longer gain heat. In locations where freezing conditions occur only rarely, a low-temperature sensor on the collector controls a circulating pump when freezing is impending. This process wastes some stored heat, but it prevents costly damage to the collector panels. This system is not suitable for regions where subfreezing temperatures persist for long periods.

The space heating thermostat is generally the conventional double-contact type that calls for heat when the temperature in the controlled space falls to a predetermined level. If the temperature in storage is adequate to meet the heating requirement, a pump or fan is started to circulate the warm fluid. If the temperature in the storage subsystem is inadequate, the thermostat calls on the auxiliary or standby heat source.

Space Heating and Service Hot Water

Figure 28 shows one of the many systems for service hot water and space heating. In this case, a large, atmospheric pressure storage tank is used, from which water is pumped to the collectors by pump P_1 in response to the differential thermostat T_1. The drain-back system is used to prevent freezing, since the amount of antifreeze required in such a system would be prohibitively expensive. Service hot water is obtained by placing a heat exchanger coil in the tank near the top, where, if stratification is encouraged, the hottest water will be found.

An auxiliary water heater boosts the temperature of the sun-heated water when required. Thermostat T_2 senses the indoor temperature and starts pump P_2 when heat is needed. If the water in the storage tank becomes too cool to provide enough heat, the second contact on the thermostat calls for heat from the auxiliary heater.

Standby heat becomes increasingly important as heating requirements increase. The heating load, winter availability of solar radiation, and cost and availability of the auxiliary energy must be determined. It is rarely cost effective to do the entire heating job for either space or service hot water by using the solar heat collection and storage system alone.

Electric resistance heaters have the lowest first cost, but may have high operating costs. Water-to-air heat pumps, which use sun-heated water from the storage tank as the evaporator energy source, are an alternative auxiliary heat source. The heat pump's COP is high enough to yield 11 to 15 MJ of heat for each kilowatt-hour of energy supplied to the compressor. When summer cooling as well as

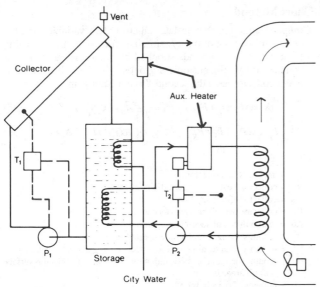

Fig. 28   Solar Collection, Storage, and Distribution System for Domestic Hot Water and Space Heating

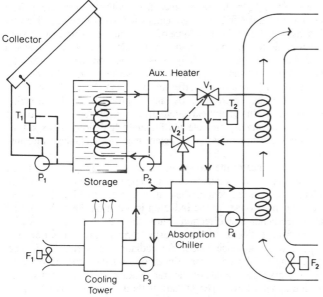

Fig. 29   Space Heating and Cooling System Using Lithium Bromide-Water Absorption Chiller

winter heating are needed, the heat pump becomes a logical solution, particularly in large systems where a cooling tower is used to dissipate the heat withdrawn from the system.

The system shown in Figure 28 may be retrofitted into a warm air furnace. In such systems, the primary heater is deleted from the space heating circuit, and the coil is located in the return duct of the existing furnace. Full backup is thus obtained, and the auxiliary heater provides only the heat not available at the storage temperature of the solar system.

Solar Cooling with Absorption Refrigeration

When solar energy is used for cooling as well as for heating, the absorption system shown in Figure 29, or one of its many modifications, may be used. The collector and storage subsystems must operate at temperatures approaching 90°C on hot summer days when the water from the cooling tower exceeds 27°C, but considerably lower operating water temperatures may be used when cooler water is

available from the tower. The controls for the collection, cooling, and distribution subsystems are generally separated, with the circulating pump P_1 operating in response to the collector thermostat T_1, which is located within the air-conditioned space. When T_2 calls for heating, valves V_1 and V_2 direct the water flow from the storage tank through the unactivated auxiliary heater to the fan coil in the air distribution system. The fan F_1 in this unit may respond to the thermostat also, or it may have its own control circuit so that it can bring in outdoor air when suitable temperature conditions are present.

When thermostat T_2 calls for cooling, the valves direct the hot water into the absorption unit's generator, and pumps P_3 and P_4 are activated to pump the cooling tower water through the absorber and condenser circuits and the chilled water through the cooling coil in the air distribution system. A relatively large hot water storage tank allows the unit to operate when no sunshine is available. A chilled water storage tank (not shown) may be added so that the absorption unit can operate during the day whenever water is available at a sufficiently high temperature to make the unit function properly. The COP of a typical lithium bromide-water absorption unit may be as high as 0.75 under favorable conditions, but frequent on-off cycling of the unit to meet a high variable cooling load may cause significant loss in performance because the unit must be heated to operating temperature after each shutdown. Modulating systems are analyzed differently than on-off systems.

Water-cooled condensers are required with the absorption cycles, because the lithium bromide-water cycle operates with a relatively delicate balance among the temperatures of the three fluid circuits—cooling tower water, chilled water, and activating water. The steam-operated absorption systems, from which solar cooling systems are derived, customarily operate at energizing temperatures of 110 to 116°C, but these are above the capability of most flat-plate collectors. The solar cooling units are designed to operate at considerably lower temperatures, but unit ratings are also lowered.

Smaller domestic units may operate with natural circulation, or *percolation*, which carries the lithium bromide-water solution from the generator (to which the activating heat is supplied) to the separator and condenser; there, the reconcentrated LiBr is returned to the absorber while the water vapor goes to the condenser before being returned to the evaporator where cooling takes place. Larger units use a centrifugal pump to transfer the fluid.

SIZING SOLAR HEATING AND COOLING SYSTEMS—ENERGY REQUIREMENTS

Methods used to determine solar heating and/or cooling energy requirements for both active and passive/hybrid systems are described by Feldman and Merriam (1979) and Hunn et al. (1987). Descriptions of public and private domain methods are included. An overview of the simulation techniques suitable for active heating and cooling systems analysis, and for passive/hybrid heating, cooling, and lighting analysis follows.

Simplified Analysis Methods

Simplified analysis methods have advantages of computational speed, low cost, rapid turnaround (especially important during iterative design phases), and ease of use by persons with little technical experience. Disadvantages include limited flexibility for design optimization, lack of control over assumptions, and a limited selection of systems that can be analyzed. Thus, if the system application, configuration, or load characteristics under consideration are significantly nonstandard, a detailed computer simulation may be required to achieve accurate results. This section describes the *f*-Chart method for active solar heating and the Solar Load Ratio method for passive solar heating (Dickinson and Cheremisinoff 1980, Lunde 1980, and Klein and Beckman 1979).

Active Heating/Cooling Systems

Beckman et al. (1977) developed the *f*-Chart method using an hourly simulation program (Klein et al. 1976) to evaluate space heating and service water heating systems in many climates and conditions. The results of these analyses correlate the fraction *f* of the heat load met by the solar energy system. The correlations give the fraction *f* of the monthly heating load (for space heating and hot water) supplied by solar energy as a function of collector characteristics, heating loads, and weather. The standard error of the differences between detailed simulations in 14 locations in the United States and the *f*-Chart predictions was about 2.5%. Correlations also agree within the accuracy of measurements of long-term system performance data. Beckman et al. (1977, 1981), and Duffie and Beckman (1980) discuss the method in detail.

The *f*-Chart method requires the following data:

- Monthly average daily radiation on a horizontal surface
- Monthly average ambient temperatures
- Collector thermal performance curve slope and intercept from standard collector tests, i.e., $F_R U_L$ and $F_R(\tau\alpha)_n$ (see ASHRAE *Standard* 93 and Chapter 34, 1992 *ASHRAE Handbook—Systems and Equipment*)
- Monthly space and water heating loads

Standard Systems

The *f*-Chart assumes several standard systems and applies only to these liquid configurations. The standard *liquid system* uses water, an antifreeze solution, or air as the heat transfer fluid in the collector loop and water as the storage medium (Figure 30). Energy is stored in the form of sensible heat in a water tank. A water-to-air heat exchanger transfers heat from the storage tank to the building. A liquid-to-liquid heat exchanger transfers energy from the main storage tank to a domestic hot water preheat tank, which in turn supplies solar heated water to a conventional water heater. A conventional furnace or heat pump is used to meet the space heating load when the energy in the storage tank is depleted.

Figure 31 shows the assumed configuration for a solar *air heating system* with a pebble-bed storage unit. Energy for domestic hot water is provided by heat exchange from the air leaving the collector to a domestic water preheat tank as in the liquid system. The hot water is further heated, if necessary, by a conventional water heater. During summer operation, a seasonal, manually operated storage bypass damper is used to avoid heat loss from the hot bed into the building.

The standard solar *domestic water heating system* collector heats either air or liquid. Collected energy is transferred by a heat exchanger to a domestic water preheat tank that supplies solar-heated water to a convectional water heater. The water is further heated to the desired temperature by conventional fuel if necessary.

f-Chart Method

Computer simulations correlate dimensionless variables and the long-term performance of the systems. The fraction *f* of the monthly space and water heating loads supplied by solar energy is empirically related to two dimensionless groups. The first dimensionless group *X* is collector loss; the second *Y* is collector gain:

$$X = F_R U_L (F_r / F_R) (t_{ref} - \bar{t}_a) \Delta\theta A_c / L \tag{35}$$

$$Y = F_R (\tau\alpha)_n (F_r / F_R) [(\overline{\tau\alpha}) / (\tau\alpha)_n] \bar{H}_T N A_c / L \tag{36}$$

where

A_c = area of solar collector, m²
F_r = collector-heat exchanger efficiency factor
F_R = collector efficiency factor
U_L = collector overall energy loss coefficient, W/(m²·K)
$\Delta\theta$ = total number of seconds in month
$\bar{t}_a$ = monthly average ambient temperature, °C
L = monthly total heating load for space heating and hot water, J
$\bar{H}_T$ = monthly averaged, daily radiation incident on collector surface per unit area, J/d·m²
N = number of days in month
$(\overline{\tau\alpha})$ = monthly average transmittance-absorptance product
$(\tau\alpha)_n$ = normal transmittance-absorptance product
t_{ref} = reference temperature, 100°C

$F_R U_L$ and $F_R(\tau\alpha)_n$ are obtained from collector test results. The ratios F_r / F_R and $(\overline{\tau\alpha})/(\tau\alpha)_n$ are calculated using methods given by Beckman et al. (1977). The value of $\bar{t}_a$ is obtained from meteorological records for the month and location desired. $\bar{H}_T$ is calculated from the monthly average, daily radiation on a horizontal surface by the methods in Chapter 34 of the 1992 *ASHRAE Handbook—Systems and Equipment* or in Duffie and Beckman (1980). The monthly load *L* can be determined by any appropriate load estimating method, including analytical techniques or measurements. Values of the collector area A_c are selected for the calculations. Thus, all the terms in these equations can be determined from available information.

Transmittance of the transparent collector cover system τ and the absorptance of the collector plate α depend on the angle at which solar radiation is incident on the collector surface. Collector tests are usually carried out with the radiation incident on the collector in a nearly perpendicular direction. Thus, the value of $F_R(\tau\alpha)_n$ determined from collector tests ordinarily corresponds to the transmittance and absorptance values for radiation at normal incidence. Depending on the collector orientation and the time of year, the monthly averaged values of the transmittance and absorptance can be significantly lower. The *f*-Chart method requires a knowledge of

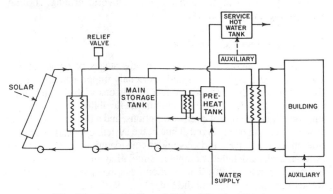

Fig. 30 Liquid-Based Solar Heating System
(Adapted from Beckman et al. 1977)

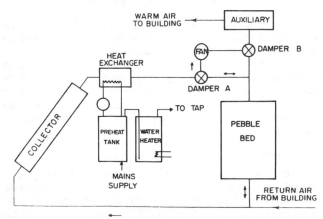

Fig. 31 Solar Air Heating System
(Adapted from Beckman et al. 1977)

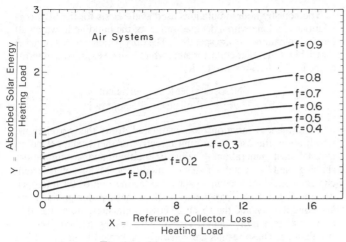

Fig. 32 Chart for Air System
(Adapted from Beckman et al. 1977)

the ratio of the monthly average to normal incidence transmittance-absorptance.

The f-Chart method for liquid systems is similar to that for air systems. The fraction of the monthly total heating load supplied by the solar air heating system is correlated with the dimensionless groups X and Y, as shown in Figure 32. To determine the fraction of the heating load supplied by solar energy for a month, values of X and Y are calculated for the collector and heating load in question. The value of f is determined at the intersection of X and Y on the f-Chart, or from the following equivalent equations.

Air system: $f = 1.04\ Y - 0.065\ X - 0.159\ Y^2$
$$+ 0.00187\ X^2 - 0.0095\ Y^3 \qquad (37)$$

Liquid system: $f = 1.029\ Y - 0.065\ X - 0.245\ Y^2$
$$+ 0.0018\ X^2 + 0.025\ Y^3 \qquad (38)$$

This is done for each month of the year. The solar energy contribution for the month is the product of f and the total heating load L for the month. Finally, the fraction F of the annual heating load supplied by solar energy is the sum of the monthly solar energy contributions divided by the annual load:

$$F = \Sigma f L / \Sigma L$$

Example 6. Calculating the heating performance of a residence, assume that a solar heating system is to be designed for use in Madison, WI, with two-cover collectors facing south, inclined 58° with respect to the horizontal. The air heating collectors have the characteristics $F_R U_L = 2.84$ W/(m²·K) and $F_R (\tau\alpha)_n = 0.49$. The $\overline{t}_a$ is −7°C, the total space and water heating load for January is calculated to be 36 GJ, and the solar radiation incident on the plane of the collector is calculated to be 13 MJ/(d·m²). Determine the fraction of the load supplied by solar energy with a system having a collector area of 50 m².

Solution: For air systems, there is no heat exchanger penalty factor and $F_r / F_R = 1$. The value of $(\overline{\tau\alpha})/(\tau\alpha)_n$ is 0.94 for a two-cover collector in January. Therefore, the values of X and Y are

$$X = (2.84\ \text{W/m}^2 \cdot \text{K})\ [100°\text{C} - (-7°\text{C})]\ (31\text{d})\ (86\ 400\ \text{s/d})$$
$$(50\ \text{m}^2 / 36\ \text{GJ}) = 1.13$$

$$Y = (0.49)\ (1)\ (0.94)\ (13\ \text{MJ/d} \cdot \text{m}^2)$$
$$(31\ \text{d})\ (50\ \text{m}^2 / 36\ \text{GJ}) = 0.26$$

Then the fraction f of the energy supplied for January from Figure 16 is 0.19. The total solar energy supplied by this system in January is

$$fL = 0.19 \times 36\ \text{GJ} = 6.84\ \text{GJ}$$

The annual system performance is obtained by summing the energy quantities for all months. The result is that 37% of the annual load is supplied by solar energy.

The collector heat removal factor F_R that appears in X and Y is a function of the collector fluid flow rate. Because of the higher cost of power for flowing fluid through air collectors than through liquid collectors, the capacitance rate used in air heaters is ordinarily much lower than that in liquid heaters. As a result, air heaters generally have a lower value of F_R. Values of F_R corresponding to the expected airflow rate in the collector must be used in calculating X and Y.

An increase in airflow rate tends to improve collector performance by increasing F_R, but it tends to decrease system performance by reducing the degree of thermal stratification in the pebble bed (or water storage tank). The f-Chart for air systems is based on a collector airflow rate of 10 L/s per square metre of collector area. The performance of systems with different collector airflow rates can be estimated by using the appropriate values of F_R in both X and Y. A further modification to the value of X is required to account for the change in degree of stratification in the pebble bed.

The performance of air systems is less sensitive to storage capacity than that of liquid systems for two reasons: (1) air systems can operate with air delivered directly to the building in which the storage component is not used, and (2) pebble beds are highly stratified and additional capacity is effectively added to the cold end of the bed, which is seldom heated and cooled to the same extent as the hot end. The f-Chart for air systems is for a nominal storage capacity. The performance of systems with other storage capacities can be determined by modifying the dimensionless group X as described in Beckman et al. (1977).

With modification, f-Charts can be used to estimate the performance of a solar water heating system operating in the range of 50 to 70°C. The main water supply temperature and the minimum acceptable hot water temperature (i.e., the desired delivery temperature) both affect the performance of solar water heating systems. The dimensionless group X, which is related to collector energy losses, can be redefined to include these effects. If monthly values of X are multiplied by a correction factor, the f-Chart for liquid-based solar space and water heating systems can be used to estimate monthly values of f for water heating systems. Experiments and analysis show that the load profile for a well-designed system has little effect on long-term performance. Although the f-Chart was originally developed for two-tank systems, it may be applied to single- and double-tank domestic hot water systems with and without collector tank heat exchangers.

For industrial process heating, absorption air conditioning, or other processes for which the delivery temperature is outside the normal f-Chart range, modified f-Charts are applicable (Klein et al. 1976). The concept underlying these charts is that of solar usability, which is the fraction of the total solar energy that is useful in the given process. This fraction depends on the required delivery temperature as well as collector characteristics and solar radiation. The procedure allows the energy delivered to be calculated in a manner similar to that for f-Charts. An example of the application of this method to solar-assisted heat pumps is presented in Svard et al. (1981).

Other Active System Methods

The *relative areas method*, based on correlations of the f-Chart method, predicts annual rather than monthly active heating system performance (Barley and Winn 1978). An hourly simulation program has been used to develop the *monthly solar-load ratio* (SLR) *method*, another simplified procedure for residential systems (Dickinson and Cheremisinoff 1980). Based on hour-by-hour simulations, a method was devised to estimate system performance based on monthly values of horizontal solar radiation and heating degree-days. This SLR method has also been extended to nonresidential buildings for a range of design water temperatures (Dickinson and Cheremisinoff 1980, Schnurr et al. 1981).

Passive Heating Systems

A widely accepted simplified passive space heating design tool is the solar-load ratio method (DOE 1980, 1982; ASHRAE 1984b). It can be applied manually, although like the *f*-Chart, it is available on microcomputer software. The SLR method for passive systems is based on correlating results of multiple hour-by-hour computer simulations, the algorithms of which have been validated against test cell data for generic passive heating system types: direct gain, thermal storage wall, and attached sunspace. Monthly and annual performance, as expressed by the auxiliary heating requirement, is predicted by this method. The method applies to single-zone, envelope-dominated buildings. A simplified, annual-basis distillation of SLR results, the *load collector ratio* (LCR) *method*, and several simple-to-use rules have grown out of the SLR method. Several hand-held calculator and microcomputer programs have been written using the methodology (Nordham 1981).

The SLR method uses a single dimensionless correlating parameter (SLR), which Balcomb et al. (1982) define as a particular ratio of solar energy gains to building heating load:

$$SLR = \frac{\text{Solar energy absorbed}}{\text{Building heating load}} \qquad (39)$$

A correlation period of 1 month is used; thus the quantities in the SLR are calculated for a 1-month period.

The parameter that is correlated to the SLR, the *solar savings fraction* (SSF), is defined as

$$SSF = 1 - \frac{\text{Auxiliary heat}}{\text{Net reference load}} \qquad (40)$$

The SSF measures energy savings expected from the passive solar building, relative to a reference nonpassive solar building.

In Equation (40), the net reference load is equal to the kelvin-day load DD of the nonsolar elements of the building:

$$\text{Net reference load} = (NLC)(DD) \qquad (41)$$

where NLC is the net load coefficient, which is a modified *UA* coefficient computed by leaving out the solar elements of the building. The nominal units are kJ/K·day. The term DD is the kelvin-days computed for an appropriate base temperature. A building energy analysis based on the SLR correlations begins with a calculation of the monthly SSF values. The monthly auxiliary heating requirements are then calculated by

$$\text{Auxiliary heat} = (NLC)(DD)(1 - SSF) \qquad (42)$$

The annual auxiliary heat is calculated by summing the monthly values.

By definition, SSF is the fraction of the kelvin-day load of the nonsolar portions of the building met by the solar element. If the solar elements of the building (south-facing walls and window) were replaced by other elements so that the net annual flow of heat through these elements was zero, the annual heat consumption of the building would be the net reference load. The savings achieved by the solar elements would therefore be the net reference load in Equation (41) minus the auxiliary heat in Equation (42), which gives

$$\text{Solar savings} = (NLC)(DD)(SSF) \qquad (43)$$

Although simple, in many situations and climates, Equation (43) is only approximately true because a normal solar-facing wall, with a normal complement of opaque walls and windows, has a near-zero effect over the entire heating season. In any case, the auxiliary heat estimate is the primary result and does not depend on this assumption.

The hour-by-hour simulations used as the basis for the SLR correlations are done with a detailed model of the building in which all the design parameters are specified. The only parameter that remains a variable is the solar collector area, which can be expressed in terms of the load collector ratio (LCR):

$$LCR = \frac{NLC}{A_p} = \frac{\text{Net load coefficient}}{\text{Projected collector area}} \qquad (44)$$

Performance variations are estimated from the correlations, which allow the user to account directly for thermostat set point, internal heat generation, glazing orientation, and configuration, shading, and other solar radiation modifiers. Major solar system characteristics are accounted for by selecting one of 94 reference designs. Other design parameters, such as thermal storage thickness and conductivity, and the spacing between glazings, are included in a series of sensitivity calculations obtained using the hour-by-hour simulations. These results are generally presented in graphic form so that the designer can see the effect of changing a particular design parameter.

Solar radiation correlations for the collector area have been determined using hour-by-hour simulations and typical meteorological year (TMY) weather data. These correlations are expressed as ratios of incident-to-horizontal radiation, transmitted-to-incident radiation, and absorbed-to-transmitted radiation as a function of the latitude minus mid-month solar declination and the atmospheric clearness index K_T.

The performance predictions of the SLR method have been compared to predictions made by the detailed hour-by-hour simulations for a variety of climates in the United States. The standard error in the prediction of the annual SSF, compared to the hour-by-hour simulation, is typically 2 to 4%.

An annual solar savings fraction calculation involves summing the results of 12 monthly calculations. For a particular city, the resulting SSF depends only on the LCR of Equation (44), the system type, and the temperature base used in calculating degree-days. Thus, tables that relate SSF to LCR for the various systems and for various degree-day base temperatures may be generated for a particular city. Such tables are easier for hand analysis than are the SLR correlations.

Annual SSF versus LCR tables have been developed for 209 locations in the United States and 14 cities in southern Canada for 94 reference designs and 12 base temperatures (ASHRAE 1984b).

Example 7. Consider a small office building located in Denver, CO, with 279 m^2 of usable space and a sunspace entry foyer that faces due south; the projected collector area A_p is 39 m^2. A sketch and preliminary plan are shown in Figure 33. Distribution of solar heat to the offices is primarily by convection through the doorways from the sunspace. The principle thermal mass is in the common wall that separates the sunspace from the offices and in the sunspace floor. This example is abstracted from the detailed version given in Balcomb et al. (1982). Even though lighting and cooling are likely to have the greatest energy costs for this building, heating is a significant energy item and should be addressed by a design that integrates passive solar heating, cooling, and lighting strategies.

Solution: From calculations of the net load coefficient shown in Table 5,

$$NLC = 86\,400 \times 277.5 = 24.0 \text{ MJ}/(\text{K} \cdot \text{d})$$

and the total load coefficient includes the solar aperture:

$$TLC = 86\,400 \times 357 = 30.8 \text{ MJ}/(\text{K} \cdot \text{d})$$

From Equation (44), the load collector ratio is

$$LCR = 24.0/39 = 0.615 \text{ MJ}/(\text{m}^2 \cdot \text{K} \cdot \text{d})$$

Suppose the daily internal heat is 135 MJ/day, and the average thermostat setting is 20°C. Then,

$$T_{base} = 20 - 135/30.8 = 15.6°C$$

(Note that solar gains to the space are not included in the internal gain term as they customarily are for nonsolar buildings.)

Table 5 Estimated Heat Load Coefficients for Example Building

	Area A, m²	U-Factor, W/(m²·K)	UA, W/K
Opaque wall	186	0.23	42.8
Ceiling	279	0.17	47.4
Floor (over crawl space)	279	0.23	64.2
Windows (E, W, N)	9	3.12	28.1
		Subtotal	18.25
Infiltration			95.0
		Subtotal	277.5
Sunspace (treated as unheated space)			79.5
		Total	357

ASHRAE procedures are approximated with V = volume, m³; c = heat capacity of Denver air, kJ/(m³·K); ACH = air changes/h; equivalent UA for infiltration = VcACH. In this case, V = 680 m³, c = 1.006 kJ/(m³·K), and ACH = 0.5.

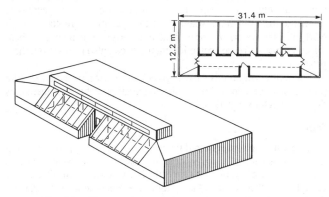

Fig. 33 Floor Plan and Perspective Drawing for Commercial Building in Example 7

In this example, the solar system is type SSD1, defined in ASHRAE (1984b) and Balcomb et al. (1982). It is a semiclosed sunspace with a 300-mm masonry common wall between it and the heated space (offices). The aperture is double-glazed, with a 50° tilt and no night insulation. To achieve a projected area of 39 m², a sloped glazed area of 39/sin 50° = 51 m² is required.

The SLR correlation for solar system type SSD1 is shown in Figure 34. Values of the absorbed solar energy S and heating kelvin-days DD to base temperature 15.6°C are determined monthly. For S, the solar radiation correlation presented in ASHRAE (1984b) for tabulated Denver, CO, weather data are used. For January, the horizontal surface incident radiation is 9.54 MJ/(m²·d). The 15.6°C base degree-days are 518. Using the tabulated incident-to-absorbed coefficients found in ASHRAE (1984b), S = 495.4 MJ/m². Thus S/DD = 0.9564 MJ/(m²·K·d). From Figure 34, at an LCR = 0.615 MJ/(m²·K·d), the SSF = 0.51. Therefore, for January

$$\text{Net reference load} = 518\,(24.0) = 12.43 \text{ GJ}$$
$$\text{Solar savings} = 12.43\,(0.51) = 6.34 \text{ GJ}$$
$$\text{Auxiliary heat} = 12.43 - 6.34 = 6.09 \text{ GJ}$$

Repeating this calculation for each month and adding the results for the year yields an annual auxiliary heat of 21.0 GJ.

Other Passive Heating System Methods

The concept of usability has been applied to passive buildings. In this approach, the energy requirements of zero- and infinite-capacity buildings are calculated. The amount of solar energy that enters the building and exceeds the instantaneous load of the zero-capacity building is then calculated. This excess energy must be dumped in the zero-capacity building, but it can be stored to offset heating loads in a finite-capacity building. Methods are provided to interpolate between the zero- and infinite-capacity limits for finite-capacity buildings. Equations and graphs for direct gain and collector-storage wall systems are given in Monsen et al. (1981, 1982).

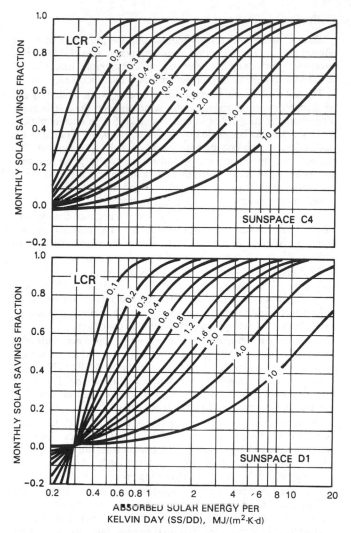

Fig. 34 Monthly SSF Versus Monthly S/DD for Various LCR Values

INSTALLATION GUIDELINES

Most solar system components are the same as those in HVAC and hot water systems (pumps, piping, valves, and controls), and their installation is not much different from conventional system installation.

Solar collectors are the most unfamiliar component used in a solar energy system. They are located outdoors, which necessitates penetration of the building envelope. They also require a structural element to support them at the proper tilt and orientation toward the sun.

Site considerations must be taken into account. The collectors should be (1) located so that shading is minimized and (2) installed so that they are attractive both on and off site. They should also be located to minimize vandalism and avoid a safety hazard.

Collectors should be placed as near to the storage tank as possible to reduce piping costs and heat losses. The collector and piping must be installed so that they can be drained without trapping fluid in the system.

For best annual performance, collectors should be installed at a tilt angle above the horizontal that is appropriate for the local latitude. They should be oriented toward true south, not magnetic south. Small variations in tilt (±10°) and orientation (±20°) are acceptable without significant performance degradation.

Collector Mounting

Solar collectors are usually mounted on the ground or on flat or pitched roofs. A roof location necessitates penetration of the building envelope by mounting hardware, piping, and control wiring. Ground or flat roof-mounted collectors are generally rack-mounted.

Pitched roof mounting can be done several ways. Collectors can be mounted on structural *standoffs,* which support them at an angle other than that of the roof to optimize solar tilt. In another pitched roof-mounting technique known as *direct mounting,* collectors are placed on a waterproof membrane on top of the roof sheeting. The finished roof surface together with the necessary collector structural attachments and flashing are then built up around the collector. A weatherproof seal between the collector and the roof must be maintained to prevent leakage, mildew, and rotting.

Integral mounting can be done for new pitched roof construction. The collector is attached to and supported by the structural framing members. The top of the collector then serves as the finished roof surface. Weather tightness is crucial to avoid damage and mildew.

Collectors should support snow loads that occur on the roof area they cover. The collector tilt usually expedites snow sliding with only a small loss in efficiency. The roof structure should be free of objects that could impede snow sliding, and the collectors should be raised high enough to prevent snow buildup over them.

The mounting structure should be built to withstand winds of at least 160 km/h, which impose a wind load of 1.9 kPa on a vertical surface or an average of 1.2 kPa on a tilted roof (HUD 1977). Wind load requirements may be higher, depending on local building codes. Flat-plate collectors mounted flush with the roof surface should be constructed to withstand the same wind loads.

When mounted on racks, the collector array becomes more vulnerable to wind gusts as the angle of the mount increases. Collectors can be uplifted by wind striking the undersides. This wind load, in addition to the equivalent roof area wind loads, should be determined according to accepted engineering procedures.

Expansion and contraction of system components, material compatibility, and the use of dissimilar metals must be considered. Collector arrays and mounting hardware (bolts, screws, washers, and angles) must be well protected from corrosion. Steel-mounting hardware in contact with aluminum, and copper piping in contact with aluminum hardware are both examples of metal combinations that have a high potential for corrosion.

Dissimilar metals can be separated by washers made of fluorocarbon polymer, phenolic, or neoprene rubber.

Freeze Protection

Freeze protection is important and is often the determining factor when selecting a system. Freezing can occur at ambient temperatures as high as 6°C because of radiation to the night sky.

One simple way of protecting against freezing is to drain the fluid from the collector array and interior piping when potential freezing conditions exist. The drainage may be automatic, as in drain-down and drain-back systems, or manual, as in direct thermosiphon systems. Automatic systems should be capable of fail-safe drainage operation—even in the event of pump failure or power outage. Special pump considerations may also be required to permit water to drain back through the pump and to permit system refilling without causing cavitation.

In areas where freezing is infrequent, recirculating water from storage to the collector array can be used as freeze protection.

Freeze protection can be provided by using fluids that resist freezing. Fluids such as water/glycol solutions, silicone oils, and hydrocarbon oils are circulated by pumps through the collector array and double wall heat exchanger. Draining the collector fluid is not required, because these fluids have freezing points well below the coldest anticipated outdoor temperature.

In mild climates where recirculation freeze protection is used, a second level of freeze protection can be provided by flushing the collector with cold supply water when the collector approaches near-freezing temperatures. This can be accomplished with a temperature-controlled valve that is set to open a small port at a near-freezing temperature of about 4.5°C and then close at a slightly higher temperature.

Over-Temperature Protection

During periods of high insolation and low hot water demand, overheating can occur in the collectors or storage tanks. Protection against overheating must be considered for all portions of the solar hot water system. Liquid expansion or excessive pressure can burst piping or storage tanks. Steam or other gases within a system can restrict liquid flow, making the system inoperable.

The most common methods of overheat protection stop circulation in the collection loop until the storage temperature decreases, discharge the overheated water from the system and replace it with cold makeup water, or use a heat exchanger as a means of heat rejection. Some freeze protection methods can also provide overheat protection.

For nonfreezing fluids such as glycol antifreezes, over-temperature protection is needed to limit fluid degradation at high temperatures during collector stagnation.

Safety

Safety precautions required for installing, operating, and servicing a solar domestic hot water system are essentially the same as those required for a conventional domestic hot water system. One major exception is that some solar systems use nonpotable heat transfer fluids. Local codes may require a double wall heat exchanger for potable water installations.

Pressure relief must be provided in all parts of the collector array that can be isolated by valves. The outlet of these relief valves should be piped to a container or drain, and not where workers could be affected.

Start-Up Procedure

After completing the installation, certain tests must be performed before charging or filling the system. The system must be checked for leakage, and pumps, fans, valves, and sensors must be checked to see that they are functional. Testing procedures vary with system type.

Closed-loop systems should be hydrostatically tested. The system is filled and pressurized to 1.5 times the operating pressure for one hour and inspected for leaks and any appreciable pressure drop.

Drain-down systems should be tested to be sure that all water drains from the collectors and piping located outdoors. All lines should be checked for proper pitch so that gravity drains them completely. All valves should be verified to be in working order.

Drain-back systems should be tested to ensure that the collector fluid is draining back to the reservoir tank when circulation stops and that the system refills properly.

Air systems should be tested for leaks before insulation is applied by starting the fans and checking the ductwork for leaks.

Pumps and sensors should be inspected to verify that they are in proper working order. Proper cycling of the system pumps can be checked by a running time meter. A sensor that is suspected of being faulty can be dipped alternately in hot and cold water to see if the pump starts or stops.

Following system testing and before filling or charging it with heat transfer fluid, the system should be flushed to remove debris.

System Maintenance

All systems should be checked at least once a year in addition to any periodic maintenance that may be required for specific components. A log of all maintenance performed should be kept, along with an owner's manual that describes system operational characteristics and maintenance requirements.

The collectors' outer glazing should be hosed down periodically. Leaves, seeds, dirt, and other debris should be carefully swept from the collectors. Care should be taken not to damage plastic covers.

Without opening up a sealed collector panel, the absorber plate should be checked for surface coating damage caused by peeling, crazing, or scratching. Also, the collector tubing should be inspected to ensure that it contacts the absorber. If the tubing is loose, the manufacturer should be consulted for repair instructions.

Heat transfer fluids should be tested and replaced at intervals suggested by the manufacturer. Also, the solar energy storage tank should be drained about every six months to remove sediment.

Performance Monitoring / Minimum Instrumentation

Temperature sensors and temperature differential controllers are required to operate most solar systems. However, additional instruments should be installed for system monitoring, checking, and troubleshooting.

Thermometers should be located on the collector supply and return lines so that the temperature difference in the lines can be determined visually.

A pressure gage should be inserted on the discharge side of the pump. The gage can be used to monitor the pressure that the pump must work against and to indicate if the flow passages are blocked.

Running time meters on pumps and fans may be installed to determine if the system is cycling properly.

DESIGN, INSTALLATION, AND OPERATION CHECKLIST

The following checklist is for designers of solar heating and cooling systems. Specific values have not been included because these vary for each application. The designer must decide whether design figures are within acceptable limits for any particular project (see DOE 1978a for further information). The review order listed does not reflect their precedence or importance during design.

Collectors

- Check flow rate for compliance with manufacturer's recommendations.
- Check that collector area matches application and claimed solar participation.
- Review collector instantaneous efficiency curve and check match between collector and system requirements.
- Relate collector construction to end use; two cover plates are not required for low-temperature collection in warm climates and may, in fact, be detrimental. Two cover plates are more efficient when the temperature difference between the absorber plate and outdoor air is high, such as in severe winter climates or when collecting at high temperatures for cooling. Radiation losses only become significant at relatively high absorber plate temperatures. Selective surfaces should be used in these cases. Flat black surfaces are acceptable and sometimes more desirable for low collection temperatures.
- Check match between collector tilt angle, latitude, and collector end use.
- Check collector azimuth.
- Check collector location for potential shading and exposure to vandalism or accidental damage.
- Review provisions made for high stagnation temperatures. If not used, are liquid collectors drained or left filled in the summer?
- Check for snow hang-up and ice formation. Will casing vents become blocked?

- Review precautions, if any, against outgassing.
- Check access for cleaning covers.
- Check mounting for stability in high winds.
- Check for architectural integration. Do collectors on roof present rainwater drainage or condensation problems? Do roof penetrations present potential leak problems?
- Check collector construction for structural integrity and durability. Will materials deteriorate under operating conditions? Will any pieces fall off?
- Are liquid collector passages organized in such a way as to allow natural fill and drain? Does mounting configuration affect this?
- Does air collector duct connection configuration promote a balanced airflow and an even heat transfer? Are connections potentially leaky?

Hydraulics

- Check that the flow rate through the collector array matches system parameters.
- If antifreeze is used, check that the flow rate has been modified to allow for the viscosity and specific heat.
- Review properties of proposed antifreeze. Some fluids are highly flammable. Check toxicity, vapor pressure, flash point, and boiling and freezing temperatures at atmospheric pressure.
- Check means of makeup into antifreeze system. An automatic water makeup system can result in freezing.
- Check that provisions are made for draining and filling the system. (Air vents at high points, drains at low points, pipes correctly graded in-between, drain-back systems vented to storage or expansion tank.)
- If system uses drain-back freeze protection, check that
  1. Provision is made for drain-back volume and back venting
  2. Pipes are graded for drain back
  3. Solar primary pump is sized for lift head
  4. Pump is self-priming if tank is below pump
- Check that collector pressure drop for drain-back system is slightly higher than static pressure between the supply and return headers.
- Optimum pipe arrangement is reverse return with collectors in parallel. Series collectors reduce flow rate and increase head. A combination of parallel/series can sometimes be beneficial, but check that equipment has been sized and selected properly.
- Cross-connections under different operating modes sometimes result in pumps operating in opposition or tandem, causing severe hydraulic problems.
- If heat exchangers are used, check that approach temperature differential has been recognized in the calculations.
- Check that adequate provisions are made for water expansion and contraction. Use specific volume/temperature tables for calculation. Each unique circuit must have its own provision for expansion and contraction.
- Three-port valves tend to leak through the closed port. This, together with reversed flows in some modes, can cause potential hydraulic problems. As a general rule, simple circuits and controls are better.

Airflow

- Check that the flow rate through the collector array matches the system design parameters.
- Check temperature rise across collectors using air mass flow and specific heat.
- Check that duct velocities are within the system parameters.
- Check that cold air or water cannot flow from collectors by gravity under "no-sun" conditions.
- Verify duct material and construction methods. Ductwork must be sealed to reduce losses.

- Check duct configuration for balanced flow through collector array.
- More than two collectors in series can reduce collection efficiency.

Thermal Storage

- Check that thermal storage capacity matches parameters of collector area, collection temperature, utilization temperature, and system load.
- Verify that thermal inertia does not impede effective operation.
- Check provisions for promoting temperature stratification during both collection and use.
- Check that pipe and duct connections to storage are compatible with the control philosophy.
- If liquid storage is used for high temperatures (above 90°C), check that tank material and construction can withstand the temperature and pressure.
- Check that storage location does not promote unwanted heat loss or gain and that adequate insulation is provided.
- Verify that liquid storage tanks are treated to resist corrosion. This is particularly important in tanks that are partially filled.
- Check that provision is made to protect liquid tanks from exposure to either an overpressure or vacuum.

Uses

Domestic Hot Water

- Characteristics of domestic hot water loads include short periods of high draw interspersed with long dormant periods. Check that domestic hot water storage matches solar heat input.
- Check that provisions have been made to prevent reverse heating of the solar thermal storage by the domestic hot water backup heater.
- Check that system allows cold makeup water preheating on days of low solar input.
- Verify that tempering valve limits domestic hot water supply to a safe temperature during periods of high solar input.
- Depending on total dissolved solids, city water heated above 65°C may precipitate a calcium carbonate scale. If collectors are used to heat water directly, check provisions for preventing scale formation in absorber plate waterways.
- Check whether the system is required to have a double wall heat exchanger and that it conforms to appropriate codes if the collector uses nonpotable fluids.

Heating

- Warm air heating systems have the potential of using solar energy directly at moderate temperatures. Check that air volume is sufficient to meet the heating load at low supply temperatures and that the limit thermostat has been reset.
- At times of low solar input, solar heat can still be used to meet part of the load by preheating return air. Check location of solar heating coil in system.
- Baseboard heaters require relatively high supply temperatures for satisfactory operation. Their output varies as the 1.5 power of the log mean temperature difference and falls off drastically at low temperatures. If solar is combined with baseboard heating, check that supply temperature is compatible with heating load.
- Heat exchangers imply an approach temperature difference that must be added to the system operating temperature to derive the minimum collection temperature. Verify calculations.
- Water-to-air heat pumps rely on a constant solar water heat source for operation. When the heat source is depleted, the backup system must be used. Check that storage is adequate.

Cooling

- Solar activated absorption cooling with fossil fuel backup is currently the only commercially available active cooling. Be assured of all design criteria and a large amount of solar participation. Verify calculations.
- Storing both hot water and chilled water may make better use of available storage capacity.

Controls

- Check that control philosophy matches the desired modes of operation.
- Verify that collector loop controls recognize solar input, collector temperature, and storage temperature.
- Verify that controls allow both the collector loop and the utilization loop to operate independently.
- Check that control sequences are reversible and will always revert to the most economical mode.
- Check that controls are as simple as possible within the system requirements. Complex controls increase the frequency and possibility of breakdowns.
- Check that all controls are fail-safe.

Performance

- Check building heating, cooling, and domestic hot water loads as applicable. Verify that building thermal characteristics are acceptable.
- Check solar energy collected on a monthly basis. Compare with loads and verify solar participation.

REFERENCES

ASHRAE. 1977. *Applications of solar energy for heating and cooling of buildings.*

ASHRAE. 1983. *Solar domestic and service hot water manual.*

ASHRAE. 1984a. Bin Weather Data, ASHRAE Research Project 385. Larry Degelman, Principal Investigator; bin data are available on computer diskette for 58 sites, based on Weather Year for Energy Calculations (WYEC) data.

ASHRAE. 1984b. *Passive solar heating analysis: A design manual.*

ASHRAE. 1991. Methods of testing to determine the thermal performance of solar collectors. *Standard* 93-1986 (Reaffirmed 1991).

Balcomb, D. et al. 1977. Thermal storage walls in New Mexico. *Solar Age* 2(8):20.

Balcomb, J.D., R.W. Jones, R.D. McFarland, and W.O. Wray. 1982. Expanding the SLR method. *Passive Solar Journal* 1:2.

Barley, C.D. and C.B. Winn. 1978. Optimal sizing of solar collectors by the method of relative areas. *Solar Energy* 21:4.

Beckman, W.A., S.A. Klein, and J.A. Duffie. 1977. *Solar heating design by the f-Chart method.* John Wiley, New York.

Beckman, W.A., S.A. Klein, and J.A. Duffie. 1981. Performance predictions for solar heating systems. *Solar energy handbook*, J.F. Kreider and F. Kreith, eds. McGraw Hill, New York.

Bliss, R.W. 1961. Atmospheric radiation near the surface of the earth. *Solar Energy* 59(3):103.

Clark, G. 1981. Passive/hybrid comfort cooling by thermal radiation. *Proceedings of the International Passive and Hybrid Cooling Conference.* American Section of the International Solar Energy Society, Miami Beach, FL.

Cole, R.L. et al. 1977. Applications of compound parabolic concentrators to solar energy conversion. *Report* No. AMLw42. Argonne National Laboratory, Chicago.

Cromer, C.J. 1984. Design of a DC-pump, photovoltaic-powered circulation system for a solar domestic hot water system. Florida Solar Energy Center (June).

Dickinson, W.C. and P.N. Cheremisinoff, eds. 1980. *Solar energy technology handbook*, Part B: Application, systems design and economics. Marcel Dekker, Inc., New York.

DOE. 1978a. DOE facilities solar design handbook. DOE/AD-0006/1. U.S. Department of Energy.

DOE. 1978b. SOLCOST—Solar hot water handbook; A simplified design method for sizing and costing residential and commercial solar service hot water systems, 3rd ed. DOE/CS-0042/2. U.S. Department of Energy.

DOE. 1980 and 1982. Passive Solar Design Handbooks. Vols. 2 and 3, Passive solar design analysis. DOE Reports/CS-0127/2 and CS-0127/3. January, July. U.S. Department of Energy.

Duffie, J.A. and W.A. Beckman. 1974. *Solar energy thermal processes.* John Wiley and Sons, New York.

Duffie, J.A. and W.A. Beckman. 1980a. *Solar engineering of thermal processes.* John Wiley and Sons, New York.

Duffie, J.A. and W.A. Beckman. 1980b. *Solar thermal energy processes.* Wiley Interscience, New York.

Edwards, D.K. et al. 1962. Spectral and directional thermal radiation characteristics of selective surfaces. *Solar Energy* 6(1):1.

Feldman, S.J. and R.L. Merriam. 1979. Building energy analysis computer programs with solar heating and cooling system capabilities. Arthur D. Little, Inc. *Report* No. EPRIER-1146 (August) to the Electric Power Research Institute.

Francia, G. 1961. A new collector of solar radiant energy. U.N. Conference on New Sources of Energy (Rome) 4:572.

Freeman, T.L., J.W. Mitchell, and T.E. Audit. 1979. Performance of combined solar-heat pump systems. *Solar Energy* 22:2.

Gates, D.M. 1966. Spectral distribution of solar radiation at the earth's surface. *Science* 151(3710):523.

Givoni, B. 1981. Experimental studies on radiant and evaporative cooling of roofs. *Proceedings of the International Passive and Hybrid Cooling Conference.* American Section of the International Solar Energy Society, Miami Beach, FL.

Hay, H.R. and J.I. Yellott. 1969. Natural air conditioning with roof ponds and movable insulation. *ASHRAE Transactions* 75(1):165-77.

Hottel, H.C. and B.B. Woertz. 1942. The performance of flat-plate solar collectors. *Transactions of ASME* 64:91.

Howard, B.D. and E.O. Pollock. 1982. Comparative report—Performance of passive solar heating systems. Vitro Corp. U.S. DOE National Solar Data Program. U.S. TIC Box 62, Oak Ridge, TN 37829.

Howard, B.D. and D.H. Saunders. 1989. Building thermal performance monitoring. *Thermal Performance of the Exterior Envelopes of Buildings II.* ASHRAE.

HUD. 1977. Intermediate minimum property standards supplement for solar heating and domestic hot water systems. SD Cat. No. 0-236-648. U.S. Department of Housing and Urban Development.

Hunn, B.D., N. Carlisle, G. Franta, and W. Kolar. 1987. Engineering principles and concepts for active solar systems. SERI/SP-271-2892. Solar Energy Research Institute, Golden, CO.

Jordan, R.C. and B.Y.H. Liu, eds. 1977. Applications of solar energy for heating and cooling of buildings. ASHRAE Publication GRP 170.

Klein, S.A. and W.A. Beckman. 1979. A general design method for closed-loop solar energy systems. *Solar Energy* 22(3):269-82.

Klein, S.A., W.A. Beckman, J.A. Duffie. 1976. TRNSYS—A transient simulation program. *ASHRAE Transactions* 82(1):623-33.

LBL. 1981. DOE-2 Reference Manual Version 2.1A. Los Alamos Scientific Laboratory, Report LA-7689-M, Version 2.1A. Report LBL-8706 Rev. 2, Lawrence Berkeley Laboratory, May.

Lister, L. and T. Newell. 1989. Expansion tank characteristics of closed loop, active solar collection systems; Solar engineering—1989. *American Society of Mechanical Engineers*, New York.

Lunde, P.J. 1980. *Thermal engineering.* John Wiley and Sons, New York.

Macriss, R.A. and R.H. Elkins. 1976. Standing pilot gas consumption. *ASHRAE Journal* 18(6):54-57.

Marlatt, W., C. Murray, and S. Squire. 1984. Roofpond systems energy technology engineering center. Rockwell International, *Report* No. ETEC6, April.

Martin, M. and P. Berdahl. 1984. Characteristics of infrared sky radiation in the United States. *Solar Energy* 33(3/4):321-36.

Mazria, E. 1979. *The passive solar energy book.* Rodale Press, Emmaus, PA.

Mitchell, D. and K.L. Biggs. 1979. Radiative cooling of buildings at night. *Applied Energy* 5:263-75.

Monsen, W.A., S.A. Klein, and W.A. Beckman. 1981. Prediction of direct gain solar heating system performance. *Solar Energy* 27(2):143-47.

Monsen, W.A., S.A. Klein, and W.A. Beckman. 1982. The un-utilizability design method for collector-storage walls. *Solar Energy* 29(5):421-29.

Morehouse, J.H. and P.J. Hughes. 1979. Residential solar-heat pump systems: Thermal and economic performance. Paper 79-WA/SOL-25, ASME Winter Annual Meeting, New York, December.

Nordham, D. 1981. Microcomputer methods for solar design and analysis. Solar Energy Research Institute, SERI-SP-722-1127, February.

Pittenger, A.L., W.R. White, and J.I. Yellott. 1978. A new method of passive solar heating and cooling. *Proceedings of the Second National Passive Systems Conference*, Philadelphia, ISES and DOE.

Reitan, C.H. 1963. Surface dew point and water vapor aloft. *Journal of Applied Meteorology* 2(6):776.

Root, D.E., S. Chandra, C. Cromer, J. Harrison, D. LaHart, T. Merrigan, and J.G. Ventre. 1985. *Solar water and pool heating course manual,* 2 vols. Florida Solar Energy Center, Cape Canaveral, FL.

Schnurr, N.M., B.D. Hunn, and K.D. Williamson. 1981. The solar load ratio method applied to commercial buildings active solar system sizing. *Proceedings of the ASME Solar Energy Division Third Annual Conference on System Simulation, Economic Analysis/Solar Heating and Cooling Operational Results*, Reno, NV, May.

Simon, F.F. 1976. Flat-plate solar collector performance evaluation. *Solar Energy* 18(5):451.

Stephenson, D.G. 1967. Tables of solar altitude and azimuth; Intensity and solar heat gain tables. Technical Paper No. 243, Division of Building Research, National Research Council of Canada, Ottawa.

Svard, C.D., J.W. Mitchell, and W.A. Beckman. 1981. Design procedure and applications of solar-assisted series heat pump systems. *Journal of Solar Energy Engineering* 103(5):135.

Swartman, R.K., Vinh Ha, and A.J. Newton. 1974. Review of solar-powered refrigeration. *Paper* No. 73-WA/SOL-6. American Society of Mechanical Engineers, New York.

Thekaekara, M.P. 1973. Solar energy outside the earth's atmosphere. *Solar Energy* 14(2):109 (January).

Threlkeld, J.L. 1963. Solar irradiation of surfaces on clear days. *ASHRAE Transactions* 69:24.

Threlkeld, J.L. and R.C. Jordan. 1958. Direct radiation available on clear days. *ASHRAE Transactions* 64:45.

Trombe, F. et al. 1977. Concrete walls for heat. *Solar Age* 2(8):13.

U.S. GPO. 1968. *Climatic atlas of the U.S.* U.S. Government Printing Office, Washington, D.C.

U.S. Hydrographic Office. 1958. Tables of computed altitude and azimuth. Hydrographic Office Bulletin No. 214, Vols. 2 and 3. U.S. Superintendent of Documents, Washington, D.C.

Van Straaten, J.F. 1961. Hot water from the sun. Ref. No. D-9, National Building Research Institute of South Africa, Council for Industrial and Scientific Research, Pretoria, South Africa.

Whillier, A. 1964. Thermal resistance of the tube-plate bond in solar heat collectors. *Solar Energy* 8(3):95.

Yellott, J.I. 1977. Passive solar heating and cooling systems. *ASHRAE Transactions* 83(2):429.

Yellott, J.I., D. Aiello, G. Rand, and M.Y. Kung. 1976. Solar-oriented architecture. Arizona State University Architecture Foundation, Tempe, AZ.

BIBLIOGRAPHY

Angstrom, A. 1915. A study of the radiation of the atmosphere. *Smithsonian Miscellaneous Collection* 65(3).

ASHRAE. 1986. Methods of testing to determine the thermal performance of flat-plate solar collectors containing a boiling liquid. *Standard* 109-1986.

ASHRAE. 1987. Methods of testing to determine the thermal performance of solar domestic water heating systems. *Standard* 95-1981 (Reaffirmed 1987).

ASHRAE. 1988. *Active solar heating systems design manual.*

ASHRAE. 1989. Methods of testing to determine the thermal performance of unglazed flat-plate liquid-type solar collectors. *Standard* 96-1980 (Reaffirmed 1989).

Bennett, I. 1965. Monthly maps of daily insolation in the U.S. *Solar Energy* 9(3):145.

Bennett, I. 1967. Frequency of daily insolation in Anglo North America during June and December. *Solar Energy* 11(1):41.

Butler, C.P. et al. 1964. Surfaces for solar spacecraft power. *Solar Energy* 8(1):2.

Colorado State University. 1980. Solar heating and cooling of residential buildings: Design of systems. Superintendent of Documents, U.S. Government Printing Office, Washington, D.C.

Colorado State University. 1980. Solar heating and cooling of residential buildings: Sizing, installation and operation of systems. Superintendent of Documents, U.S. Government Printing Office, Washington, D.C.

Cook, J., ed. 1989. *Passive cooling.* MIT Press, Cambridge, MA.

Daniels, F. 1964. *Direct use of the sun's energy.* Yale University Press, New Haven, CT.

Diamond, S.C. and J.G. Avery. 1986. Active solar energy system design, installation and maintenance: Technical applications manual. LA-UR-86-4175.

Durlak, E.R. 1986. Evaluation of installed solar systems at navy, army and air force bases. U.S. Naval Civil Engineering Laboratory. NCEL TN-1750.

Foresti, F.G. 1981. Corrosion and scaling in solar heating systems. Solar/0909-81/70; DE82006139. U.S. Department of Energy, Vitro Laboratories Division, Automation Industries, Inc.

Gier, J.T. and R.V. Dunkle. 1958. Selective spectral characteristics as an important factor in the efficiency of solar energy collectors. Transactions of the Conference on Scientific Uses of Solar Energy (1955) 2(1-A):41.

Hall, I.J., R.R. Prairie, H.E. Anderson, and E.C. Boes. 1979. Generation of typical meteorological years: 426 Solmet stations. *ASHRAE Transactions* 85(2):507.

HUD. 1979. Solar domestic hot water: HUD 000-1230. U.S. Department of Housing and Urban Development.

HUD. 1980. Installation guidelines for solar DHW systems in one- and two-family dwellings. U.S. Department of Housing and Urban Development, 2nd ed. (May).

HUD. 1980. Solar terminology HUD-PDR-465(2) (March). U.S. Department of Housing and Urban Development.

ITT. 1976. *Solar heating systems design manual.* International Telephone and Telegraph Corporation.

Knapp, C.L., T.L. Stoffel, and S.D. Whitaker. 1980. Insolation data manual. SERI/SP-755-789. Solar Energy Research Institute, Golden, CO.

Kreider, J.F. 1989. Solar design: Components, systems, economics. Hemisphere Publication, New York.

Lameiro, G.F. and P. Bendt. 1978. The GFL method for designing solar energy space heating and domestic hot water systems, 2.1. *Proceedings at the 1978 Annual Meeting of the American Section of the International Solar Energy Society,* Denver, CO.

Lane, G.A. 1986. *Solar heat storage: Latent heat materials,* 2 vols. CRC Press, Boca Raton, FL.

Lof, G.O., J.A. Duffie, and C.D. Smith. 1966. World distribution of solar radiation. *Report* No. 21. Solar Energy Laboratory, University of Wisconsin, Madison, WI.

Morrison, C.A. and E.A. Farber. 1974. Development and use of solar insolation data for south facing surfaces in northern latitudes. *ASHRAE Transactions* 80(2):350.

Mueller Associates, Inc. 1980. Economic analysis of commercial solar combined space-heating and hot-water systems. MAI *Report* No. 204. Argonne National Laboratory.

Mueller Associates, Inc. 1980. Economic analysis of residential and commercial solar heating and hot water systems; Summary report. MAI *Report* No. 206. Argonne National Laboratory.

Mueller Associates, Inc. 1985. Active solar thermal design manual. U.S. Department of Energy, Solar Energy Research Institute and ASHRAE.

Mumma, S.A. 1985. Solar collector tilt and azimuth charts for rotated collectors on sloping roofs. *Proceedings Joint ASME-ASES Solar Energy Conference,* Knoxville, TN.

Mumma, S., L. Milnarist, and J. Rodriquez-Anza. 1973. Innovative double walled heat exchanger for use in solar water heating. EWxG-03.

Newton, A.B. and S.F. Gilman. 1982. Solar collector performance manual. ASHRAE Publication SP 32.

Newton, A.B. U.S. Patents 2,343,211 and 2,396,338.

Parmalee, G.V. and W.W. Aubele. 1952. Radiant energy transmission of the atmosphere. *ASHVE Transactions* 58:85.

Ruegg, R.T., G.T. Sav, J.W. Powell, and E.T. Pierce. 1982. Economic evaluation of solar energy systems in commercial buildings; Methodology and case studies. NBSIR 82-2540. U.S. National Bureau of Standards.

Solar Energy Research Institute. 1981. Solar design workbook—Solar federal buildings program. SERI/SP-62-308. U.S. Department of Energy and Los Alamos Scientific Laboratory.

Solar Energy Research Institute. 1981. Solar radiation energy resource atlas of the United States. SERI/SP642-1037. Golden, CO.

Solar Environmental Engineering Co., Inc. 1981. Solar domestic hot water system inspection and performance evaluation handbook SERI/SP-98189-1B. Solar Energy Research Institute, Golden, CO.

Tables of radiation powers. Paper No. 105, Division of Building Research, National Research Council of Canada, Ottawa.

Tabor, H. 1958. Selective radiation. I. Wavelength discrimination. *Transactions of the Conference on Scientific Uses of Solar Energy* (1955) 2(1-A):1, University of Arizona Press.

Telkes, M. 1949. A review of solar house heating. *Heating and Ventilating,* 68.

U.S. National Bureau of Standards. 1982. Performance criteria for solar heating and cooling systems in residential buildings. NBS Building Science Series 147.

U.S. National Bureau of Standards. 1984. Performance criteria for solar heating and cooling systems in commercial buildings. NBS Technical Note 1187.

Ward, D.S. and H.S. Oberoi. 1980. Handbook of experience in the design and installation of solar heating and cooling systems. ASHRAE (July).

Zarem, A.M. and D.D. Erway. 1963. *Introduction to the utilization of solar energy.* McGraw-Hill, New York.

ENERGY RESOURCES

BUILDINGS and facilities of various types may be heated, ventilated, air conditioned, and refrigerated—using systems and equipment designed for that purpose and using the site energy forms commonly available—without concern for the original energy resources from whence those energy forms came. Since the energy used in buildings and facilities comprises a significant amount of the total energy used for all purposes, and since the use of this energy has an impact on energy resources, ASHRAE recognizes the "effect of its technology on the environment and natural resources to protect the welfare of posterity" (ASHRAE 1990).

Many governmental agencies regulate energy conservation legislation for obtaining building permits (Conover 1984). The application of specific values to building energy use situations has a considerable effect on the selection of HVAC&R systems and equipment and how they are applied.

CHARACTERISTICS OF ENERGY AND ENERGY RESOURCE FORMS

The HVAC&R industry deals with energy forms as they occur on or arrive at a building site. Generally, these energy forms are fossil fuels (natural gas, oil, and coal) and electricity. Solar energy and wind energy are also available at most sites, as is low-level geothermal energy (energy source for heat pumps). Direct-use (high-temperature) geothermal energy is available at some. These are the prime forms of energy used to power or heat the improvements on a site.

Forms of On-Site Energy

Fossil fuels and electricity are commodities that are usually metered or measured for payment by the facility owner or operator. On the other hand, solar or wind, each of which might be considered a dispersed energy form in its natural state (i.e., requiring neither central processing nor a distribution network) costs nothing for the commodity itself, but does incur cost for the means to make use of it. High-temperature geothermal energy, which is not universally available, may or may not be a sold commodity, depending on the particular locale and local regulations (Chapter 29).

Some prime on-site energy forms require further processing or conversion into other forms more directly suited for the particular systems and equipment needed in a building or facility. For instance, natural gas or oil is burned in a boiler to produce steam or hot water, a form of thermal energy which is then distributed to various use points (such as heating coils in air-handling systems, unit heaters, convectors, fin-tube elements, steam-powered cooling units, humidifiers, and kitchen equipment) throughout the building. Although electricity is not converted in form on-site, it is nevertheless used in a variety of ways, including lighting, running motors for fans and pumps, powering electronic equipment and office

The preparation of this chapter is assigned to TC 1.10, Energy Resources.

machinery, and space heating. While the methods and efficiencies with which these processes take place fall within the scope of the HVAC&R designer, the process by which a prime energy source arrives at a given facility site is not under direct control of the professional. On-site energy choices, if available, may be controlled by the designer based in part on the present and future availability of the associated resource commodities.

The basic energy source for heating may be natural gas, oil, coal, or electricity. Cooling may be produced by electricity, thermal energy, or natural gas. If electricity is generated on-site, the generator may be turned by an engine using natural gas or oil, or by a turbine using steam or gas directly.

The term *energy source* refers to on-site energy in the form in which it arrives at or occurs on a site (e.g., electricity, gas, oil, or coal). *Energy resource* refers to the raw energy, which (1) is extracted from the earth (wellhead or mine-mouth), (2) is used in the generation of the energy source delivered to a building site (coal used to generate electricity), or (3) occurs naturally and is available at a site (solar, wind, or geothermal energy).

Nonrenewable and Renewable Energy Resources

From the standpoint of energy conservation, energy resources may be classified in two broad categories: (1) nonrenewable (or discontinuous) resources, which have definite, although sometimes unknown, limitations; and (2) renewable (or continuous) resources, which can generally be freely used without depletion or have the potential to renew in a reasonable period. Resources used most in industrialized countries, both now and in the past, are nonrenewable (Gleeson 1951).

Nonrenewable resources of energy include
- Coal
- Crude Oil
- Natural gas
- Uranium 235 (atomic energy)

Renewable resources of energy include
- Hydropower
- Solar
- Wind
- Earth heat (geothermal)
- Biomass (wood, wood wastes and municipal solid waste)
- Tidal power
- Ocean thermal
- Atmosphere or large body of water (as used by the heat pump)
- Crops (for alcohol production)

Characteristics of Fossil Fuels and Electricity

Most on-site energy for buildings in developed countries involves electricity and fossil fuels as the prime on-site energy sources. Both fossil fuels and electricity can be described in terms of their energy content. This implies that the two energy forms are comparable and that an equivalence can be established. In reality, however, fossil fuels and electricity are only comparable in energy terms when they are used to generate heat. Fossil fuels, for example, cannot directly drive motors or energize light bulbs. Conversely, electricity gives off heat as a by-product regardless of whether it is used for running a motor or lighting a light bulb, and regardless of whether that heat is needed. Thus, electricity and fossil fuels have different characteristics, uses, and capabilities aside from any differences relating to their derivation.

Beyond the building site, further differences between these energy forms may be observed, such as methods of extraction, transformation, transportation, and delivery, and the characteristics of the resource itself. Natural gas arrives at the site in virtually the same form in which it was extracted from the earth. Oil is processed (distilled) before arriving at the site; having been extracted as crude oil, it arrives at a given site as, for example, No. 2 oil or diesel fuel. Electricity is created (converted) from a different energy form, often a fossil fuel, which itself may first be converted to a thermal form. The total electricity conversion, or generation, process includes energy losses governed largely by the laws of thermodynamics.

Fuel cells, which are used only on a small scale, convert a fossil fuel to electricity by chemical means.

Fossil fuels undergo a conversion process by combustion (oxidation) and heat transfer to thermal energy in the form of steam or hot water. The conversion equipment used is a boiler or a furnace in lieu of a generator, and it usually occurs on a project site rather than off-site. (District heating is an exception.) Inefficiencies of the fossil fuel conversion occur on-site, while the inefficiencies of most electricity generation occur off-site, before the electricity arrives at the building site. (Cogeneration is an exception.)

WORLD ENERGY RESOURCES

Production

Energy production trends for the world, leading producers, and world areas from 1982 to 1991 are shown on Figure 1. World primary energy production in 1991 was up 24% since 1982, 41% since 1973. The countries with the largest total energy production in 1991 were the United States (19%), the U.S.S.R. (19%), and

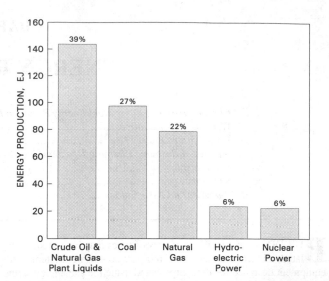

Fig. 2   **World Primary Energy Production by Resource Type: 1991**

China (9%). Together these countries produced just under half of the world's energy. Saudi Arabia (6%) was the world's fourth largest producer. (Note that the U.S.S.R. officially dissolved on January 1, 1992.) Total world energy production by resource type is shown in Figure 2.

Crude Oil. World crude oil production was $9.5 \times 10^6 \text{ m}^3$ per day in 1992—up 8% since 1973. The biggest crude oil producers in 1992 were the eight nations comprising the Organization of Petroleum Exporting Countries (OPEC) at 42%, the former U.S.S.R. (14%), Saudi Arabia (14%), and the United States (12%). Oil production declined markedly in the former U.S.S.R. in the 1990s, and that region was then the second largest producer after Saudi Arabia. The other primary non-OPEC producers were China, Mexico, the United Kingdom, and Canada.

Natural Gas. World production reached $2.1 \times 10^{12} \text{ m}^3$ in 1991—up 39% from the 1982 level. The biggest producers in 1991 were the U.S.S.R. (38%) and the United States (24%).

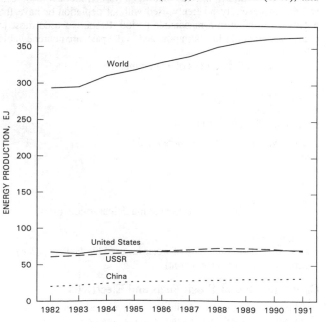

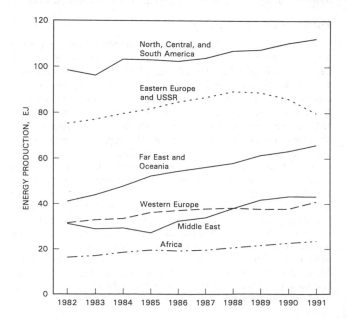

Fig. 1   **World Primary Energy Production Trends**

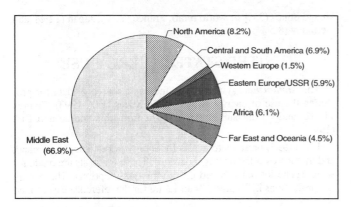

Fig. 3 World Crude Oil Reserves: January 1, 1992
(Basis: *Oil and Gas Journal*)

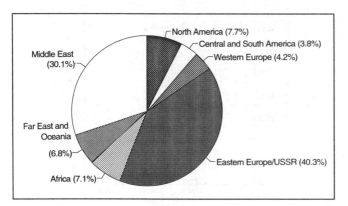

Fig. 4 World Natural Gas Reserves: January 1, 1992
(Basis: *Oil and Gas Journal*)

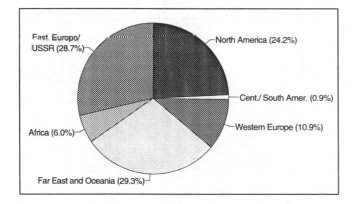

Fig. 5 World Recoverable Coal Reserves: January 1, 1992
(Basis: World Energy Council)

Coal. At 4.6 Tg in 1991, coal production had risen 19% since 1982 and comprised 27% of the world's energy production. The leading producers of coal were China (24%), the United States (20%), and the U.S.S.R. (15%).

Reserves

On January 1, 1992 the estimated world reserves of crude oil and gas were distributed by world region according to Figures 3 and 4. Saudi Arabia was estimated to have 39% of the Middle Eastern crude oil reserves. Iraq, the United Arab Emirates, Kuwait, and Iran were each estimated to have more crude oil reserves than any world region outside the Middle East. Outside of the Middle East and the

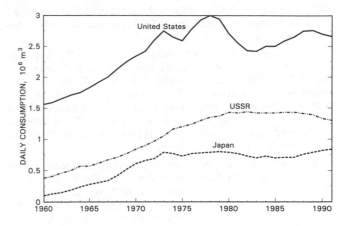

Fig. 6 Petroleum Consumption Trends of Leading Consumers: 1960 - 1990

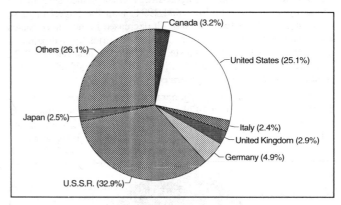

Fig. 7 World Natural Consumption: 1991

United States, the countries with the biggest estimated reserves were Venezuela, the U.S.S.R., and Mexico. The single country with the largest gas reserves by far was the U.S.S.R.

World coal reserves as of January 1, 1992 are shown by world area in Figure 5. The countries with the most plentiful reserves, as a percent of total, were the United States (23%), the U.S.S.R. (23%), China (11%), and Australia (9%).

Consumption

Data on world energy consumption are available only by type of resource rather than by total energy consumed.

Petroleum. The consumption trends of the leading consumers from 1960 to 1991 are depicted in Figure 6. In 1991, the United States consumed far more petroleum than any other country—26% of world consumption and 44% of the consumption of Organization for Economic Cooperation and Development (OECD) countries. By contrast, Japan, another OECD country, consumed just 8% of the world total and 14% of that of the OECD countries. Of the non-OECD countries, the U.S.S.R. was the biggest consumer (12% of world consumption).

Natural Gas. This energy resource tends to be consumed close to the site of production, and indeed, in 1991 the two biggest natural gas producers were also the two biggest consumers. Figure 7 depicts natural gas consumption by the leading consumer countries as a percentage of world consumption. Of the major consumers, the United States consumed more than it produced (107%), and the U.S.S.R. consumed less (88%). Germany, the third largest consumer, produced very little. Canadian consumption was 62% of its production. World consumption of natural gas increased 43% between 1980 and

1991, with the U.S.S.R. up 88% and the United States down 4%. After the United States and the U.S.S.R., no single country consumed more than 5% of the world total.

Coal. Here, the three largest producers were also the three largest consumers. Figure 8 depicts the percentage of world consumption of the leading consumers during 1991. Since 1980, world coal consumption had increased 23% (a slight drop-off of 4% since its peak in 1989). In the same period, China's consumption increased 77%, the United States' 26%, and the U.S.S.R.'s 4%. In 1990 the leading coal exporters by a wide margin were the United States and Australia, while the leading importer was Japan.

Electricity. Figure 9 shows the world's electricity generation by energy resource in 1990. Figure 10 shows installed capacity at the end of 1990 for the same resources. Both net generation and installed capacity were dominated by the United States (26% for both), the U.S.S.R. (15% and 13%, respectively), and Japan (7% for both). China was close to Japan's net electricity generation capacity, with 5% of the world total.

Electricity generated by hydroelectric means increased in the world by 18% between 1982 and 1991, with the largest increase occurring in the Far East/Oceania. (It decreased in the United States over that same period, however.) The top countries for hydroelectric generation in 1991 were Canada, the United States, the U.S.S.R., and Brazil—collectively accounting for close to half of the world total quantity of electricity generated by that means.

Total world electricity generation from nuclear resources increased 325% between 1979 and 1992, with higher-than-average increases occurring in Western Europe (425%), and the Far East and Africa (432%). The top-generating countries in 1992 were the

United States (35% of world total), France (18%), Japan (12%), and Germany (8.5%).

UNITED STATES ENERGY USE

The *Annual Energy Outlook* is the basic source of data for projecting the use of energy in the United States (EIA 1993). Figures 11, 12, and 13 are summaries of data from this source and EIA (1992).

EIA (1993) presents forecasts for energy prices, supply, demand, and imports over the next two decades. These forecasts are made for seven scenarios, each based on different assumptions. The energy use projections in Figures 11 and 12 are for the reference case, called

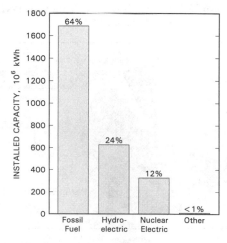

Fig. 10   World Installed Electricity Generation Capacity by Resource: January 1, 1991

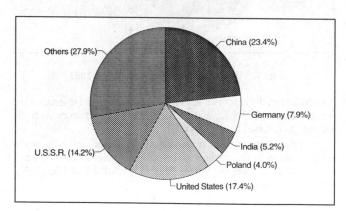

Fig. 8   World Coal Consumption: 1990

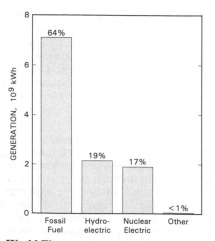

Fig. 9   World Electricity Generation by Resource: 1990

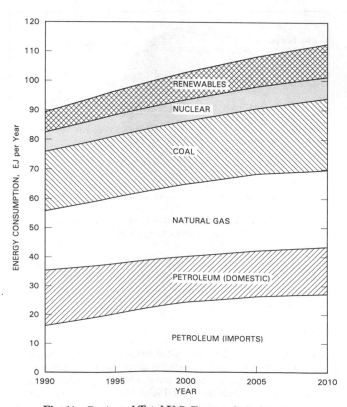

Fig. 11   Projected Total U.S. Energy Consumption by Resource

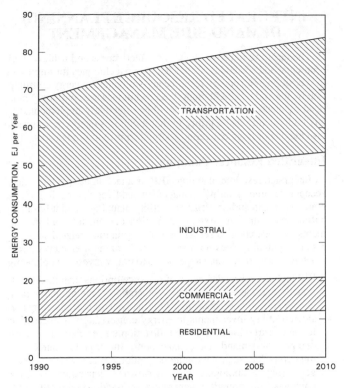

**Fig. 12   Projected Total U.S. Energy Consumption
by End-Use Sector**

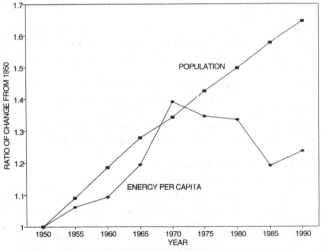

**Fig. 13   Per Capita End-Use Energy Consumption
in the United States**

the baseline case by EIA. The reference case combines the assumption of an annual economic growth rate of 2.0% with a mid-level path for the world oil price. The relatively slow world oil price rise assumes that demands for oil continue to grow parallel with economic growth, particularly in developing and newly industrialized countries. Total demand for energy grows at an annual rate of 1.2% per year under these assumptions. The reference case also assumes that economically recoverable oil and natural gas resources are 14.9×10^9 m^3 and 25.3×10^{12} m^3, respectively.

Figure 13, which is based on EIA (1992), presents a capsule overview of past U.S. energy use intensity by relating per capita energy use since 1950 to population growth.

Projected Overall Energy Consumption

The following observations apply to the overall picture of projected energy use in the United States over the next two decades (Figure 11):

- The Energy Policy Act of 1992, which has a goal of more efficient use of energy in the United States, among other things promotes use of alternative fuels, revises efficiency standards for buildings, simplifies licensing for nuclear power plants, promotes advanced coal use technologies, and allows utilities to operate independent generating plants outside their service territories.
- The introduction, adoption, and use of more efficient energy-using technologies in all sectors (because of the Energy Policy Act of 1992) and general improvements in technology should limit energy demand to less than half that of overall economic growth.
- A renewed focus on conserving energy through improved efficiency in energy end use will develop. The trend of the past two decades toward improved (lowered) U.S. energy intensity will continue.
- Despite expected improvements, the need for energy in the United States will continue to increase.
- More than 50% of the increase of energy use is expected to be for electricity, but because of the impact of the Energy Policy Act of 1992, the projections show lower growth in electricity than the projections of previous years.
- Stable electricity prices, the Energy Policy Act of 1992 and continuing consumer preference for electricity will cause growth in the demand for electricity to increase only slightly more slowly than economic growth.
- Nonutility generators are expected to provide a large share of new electric capacity.
- Environmental quality and the link between the use of energy and environmental problems will remain key issues for policymakers.
- No agreement exists in the scientific community about human-caused global warming. Many factors that affect climate or provide potential offsets to global warming are neither understood nor accurately represented in the forecast models.

Projected Energy Consumption by Resource

- Fossil fuels (petroleum, natural gas, and coal) comprise the preponderance of projected total energy consumption.
- Although petroleum does show an increase in the relative proportion of imports compared to domestic, the wide range of possible world prices leads to a high degree of uncertainty as to the level of imports. Several factors beyond the control of the United States affect these prices and thus the U.S. oil market outlook.
- Cost of transportation and storage precludes the availability of natural gas on international markets to the same extent as petroleum.
- Though natural gas wellhead prices are projected to increase two to three times in the next two decades, these projections (which affect gas use proper) are sensitive to assumptions about improvements in exploration technology and uncertainty about the U.S. resource base.
- Imports (mainly from Canada) will also have an impact on the domestic gas market.
- Over the next 15 years, the majority of new nonutility electric generators will be natural gas-fired.
- Although consumption for nonelectric uses will remain relatively stagnant over the next two decades, coal should continue to be the mainstay of baseload electricity generation.
- Growth in coal-fired generation, along with doubling of coal exports, is expected to stimulate strong growth in coal production.
- The United States is expected to regain its position as the world's leading coal exporter by the end of the forecast period.
- The Clean Air Act Amendments of 1990 primarily affect electricity generation and petroleum refining. To comply, electric utilities will increase their use of low-sulfur coal, low-sulfur residual fuel oil,

and natural gas; will purchase sulfur dioxide allowances; and will add scrubbers to older units. The Act, along with increased exports, will also increase demand for higher-cost coal from central Appalachia, which in turn will cause increased production of less expensive, Western coal.

- Carbon emissions from fossil fuel combustion will continue to grow and may contribute to global climate change, increasing the "greenhouse gas effect."
- Trends toward conservation will be complemented by expanding use of renewable energy in the form of alcohol fuels, biomass (wood and municipal solid waste), solar, geothermal, and wind.
- Use of renewables in generating electricity will also grow because of incentives of the Energy Policy Act and, to a lesser extent, the Clean Air Act Amendments of 1990.

Projected Energy Consumption by Sector

Figure 12 shows energy use by end-use sector, with the major end-use sectors being residential, commercial, industrial, and transportation. HVAC&R engineers are primarily concerned with the first three sectors. Figure 12 shows less total energy consumption than Figure 11 primarily because it excludes the thermodynamic losses of electricity generation and the processing and delivery burdens of various energy forms. Some observations include the following:

- For the transportation sector, the dominant fuel will remain petroleum, but increases in vehicle efficiency will continue as the large stock of older vehicles is replaced by more efficient ones.
- The alternative fuels provision of the Energy Policy Act of 1992 may further accelerate the trend toward less use of petroleum for transportation. (Two alternatives could be increased use of compressed natural gas as a fuel and further development and use of electric vehicles.)
- The state of industrial energy intensity has become an issue because of its direct relationship to global competitiveness and the emission of "greenhouse" gases.
- Significant improvements can be made in the most energy-intensive industries through the use of more efficient motors and demand side management (DSM) programs. DSM programs by utilities are projected to reduce electricity demand by almost 10% within the decade.
- Projections reflect increased efficiency for virtually every sector, with significant conservation potential in commercial buildings due primarily to improved building shells, more efficient appliances, greater use of energy management and control systems, improved lighting and better insulation.

Figure 13 superimposes per capita end use over a comparable graph of population growth. Per capita energy use has varied significantly. The 1960s experienced a sharp increase in per capita energy use, which leveled off during the 1970s due to higher energy prices and the emphasis on energy conservation. In the early 1980s, however, a significant drop in per capita energy use occurred as industrial output decreased, efficiency of use improved, and global economic pressures mounted. The last half of the 1980s indicates an increase in per capita energy use at a rate paralleling that of the population—but at a considerably lower level.

Outlook Summary

In general, notwithstanding enactment of the Energy Policy Act of 1992, the following major issues will continue to dominate energy matters in the next two decades:

- Rising dependency of the United States on imported oil
- Concerns about global climate changes and other environmental issues
- Potential of energy conservation and energy efficiency as alternatives to energy production
- Growth of population

INTEGRATED RESOURCE PLANNING/ DEMAND-SIDE MANAGEMENT

The desire to improve economic effectiveness and achieve environmental and societal goals has led to two techniques for improving the selection of an energy resource and influencing energy consumption by the end user. Although these techniques are commonly practiced by electric utilities, they could be effectively implemented by any entity acting *like* a utility in that it chooses energy resources to provide one or more converted energy forms to multiple users.

Integrated Resource Planning

Integrated resource planning (IRP) is a technique that has gained acceptance among utility management and regulators. It goes by many names, including "least cost utility planning," and it has many definitions. Although a single definition that includes all the elements considered in discussions and regulations related to utility resource planning does not exist, several common elements distinguish IRP from traditional approaches to utility investment planning.

- IRP includes a broader range of resource options than those traditionally considered in electricity generation planning, particularly energy conservation measures. Originally, least cost planning considered utility investments in energy conservation as alternatives to the construction of new central-station power plants (similar to one goal of demand-side management). Integrated resource planning considers such nontraditional options as renewable resource generating technologies, customer-owned generation, purchase contracts from nonutility generators, transactions with other utilities, and many others.
- IRP considers the costs and benefits of these options beyond traditional measures of revenue requirements. Regulators typically require determining costs and benefits from a "total resource" or societal perspective.
- IRP attempts to optimize resource decisions by determining the costs and benefits of all options on the same basis and selecting the options that provide the most favorable balance of benefits against costs.
- IRP addresses a wider population of stakeholders than traditional utility planning. Many state regulatory commissions involve the public in the review of utility resource plans—and sometimes even in their formation. Customers, environmentalists, and other public interest groups are often prominent in these proceedings.

Demand-Side Management

Broadly defined, demand-side management (DSM) includes any actions taken by an electric utility or industrial plant that provide the customer (or end user) with the same level of energy services, but at lower overall costs, and the energy provider with economic or other desirable benefits. Some public utility commissions have requested that gas utilities apply these techniques as well. The fundamental objectives of DSM are to reduce demand, conserve energy, and reduce pollutants. DSM may be integrated into an energy provider's integrated resource plan. Many (but not all) DSM plans offer some form of economic incentive in exchange for a modified energy use pattern.

Supporters of DSM may be primarily interested in one of its benefits. For example, industry may use DSM to reduce operating expenses. A public service commission may require a utility to incorporate DSM as part of its IRP to minimize the need for new electric generating capacity and subsequent rate increases. Environmental advocates may encourage DSM to reduce energy consumption. However, supporters may also share reasons they support DSM. For example, industry and environmental advocates may both be interested in delaying new electric generation and in reducing contaminants. Commissions and environmental advocates are concerned with the economic health of industry (Pritchett et al. 1993).

RELATIONSHIPS

In designing the systems required for a facility, an HVAC&R designer sooner or later must consider the use of one or more forms of prime energy. Most likely, these would be nonrenewable energy sources (fossil fuels and electricity), although installations are sometimes designed using a single energy source (e.g., only a fossil fuel or only electricity).

Solar energy normally impinges on the site (and on the facilities to be put there), so it will have an impact on the energy consumption of the facility. The designer must account for this impact as well as decide whether to make active use of solar energy. When solar energy is used beneficially, it can reduce the requirements for nonrenewable energy forms. Other naturally occurring and distributed renewable forms such as wind power and earth heat (if available) might also be considered.

If they are to understand and be concerned with the earth's energy resources, designers must be aware of the relationship between on-site energy sources and raw energy resources—including how these resources are used and what they are used for. The relationship between energy sources and energy resources involves two parts: (1) quantifying the energy resource units expended and (2) considering the societal impact of the depletion of one energy resource (caused by on-site energy use) with respect to others. The following two sections describe those parts in more specific terms.

Quantifiable Relationships

As on-site energy sources are consumed, a corresponding amount of resources are consumed to produce that on-site energy. For instance, for every 1000 L of No. 2 oil consumed by a boiler at a building site, some greater number of litres of crude oil was extracted from the earth. On leaving the well, the crude oil is transported and processed into its final form, perhaps stored, and then transported to the site where it will be used.

Even though gas requires no significant processing, it is transported, often over long distances, to reach its final destination, which causes some energy loss. Electricity may have as its raw energy resource a fossil fuel, uranium, or an elevated body of water (hydroelectric generating plant).

Data to assist in determining the amount of resource use per delivered on-site energy source unit are available. In the United States, data are available from entities within the U.S. Department of Energy and from the agencies and associations listed at the end of this chapter.

A resource utilization factor (RUF) is the ratio of resources consumed to energy delivered (for each form of energy) to a building site. Specific RUFs may be determined for various energy sources normally consumed on-site, including nonrenewable sources such as coal, gas, oil, and electricity, and renewable sources such as solar, geothermal, waste, and wood energy. With electricity, which may derive from several resources depending on the particular fuel mix of the generating stations in the region served, the overall RUF is the weighted combination of individual factors applicable to electricity and a particular energy resource. Grumman (1984) gives specific formulas for calculating RUFs.

While a designer does not need to determine the amount of energy resources attributable to a given building or building site for its design or operation, this information may be helpful when planning the long-range availability of energy for a building or the building's impact on energy resources. Currently, factors or fuel-quantity-to-energy resource ratios are used, which suggests that energy resources are of concern to the HVAC&R industry.

Intangible Relationships

Energy resources should not simply be converted into common energy units [e.g., gigajoule] because the commonality thus established gives a misleading picture of the equivalence of these resources.

Other differences and limitations of each of the resources defy easy quantification but are nonetheless real.

For instance, consider electricity that arrives and is used on a site and the resources from which it is derived. Electricity can be generated from coal, oil, natural gas, uranium, or elevated bodies of water (hydropower). The end result is the same: electricity at X kilovolts, Y hertz. However, is a megajoule of electricity generated by hydropower equal in societal impact to that same megajoule generated by coal? by uranium? by domestic oil? or by imported oil? In other words, electricity generated by hydropower, though identical in quantity with the electricity generated by imported oil, might be considered more desirable from a societal impact standpoint.

Intangible factors such as safety, environmental acceptability, availability, and national interest also are affected in different ways by the consumption of each resource. Heiman (1984) proposes a procedure for weighting the following intangible factors:

National/Global Considerations

- Balance of trade
- Environmental impacts
- International policy
- Employment
- Minority employment
- Availability of supply
- Alternative uses
- National defense
- Domestic policy
- Effect on capital markets

Local Considerations

- Exterior environmental impact—air
- Exterior environmental impact—solid waste
- Exterior environmental impact—water resources
- Local employment
- Local balance of trade
- Use of distribution infrastructure
- Local energy independence
- Land use
- Exterior safety

Site Considerations

- Reliability of supply
- Indoor air quality
- Aesthetics
- Interior safety
- Anticipated changes in energy resource prices

SUMMARY

In designing HVAC&R systems, the need to address immediate issues such as economics, performance, and space constraints often prevents designers from fully considering the energy resources affected. Today's energy resources are less certain because of issues such as availability, safety, national interest, environmental concerns, and the world political situation. As a result, the reliability, economics, and continuity of many common energy resources over the potential life of a building being designed are unclear. For this reason, the designer of building energy systems must consider the energy resources on which the long-term operation of the building will depend. If the continued viability of those resources is reason for concern, the design should provide for, account for, or address such an eventuality.

AGENCIES AND ASSOCIATIONS

American Gas Association (AGA), Arlington, VA

American Petroleum Institute (API), Washington, D.C.

Bureau of Mines, Department of Interior, Washington, D.C.

Council on Environmental Quality (CEQ), Washington, D.C.

Edison Electric Institute (EEI), Washington, D.C.

Electric Power Research Institute (EPRI), Palo Alto, CA

Energy Information Administration (EIA), Washington, D.C.

Gas Research Institute (GRI), Chicago, IL

National Coal Association (NCA), Washington, D.C.

North American Electric Reliability Council (NAERC), Princeton, NJ

Organization of Petroleum Exporting Countries (OPEC), Vienna, Austria

REFERENCES

ASHRAE. 1990. Energy position statement (June).

Bartlett, A.A. 1983. *Forgotten fundamentals of the energy crisis.* Boulder, CO.

Conover, D.R. 1984. Accounting for energy resource use in building regulations. *ASHRAE Transactions* 90(1B):547-63.

EIA. 1992. *Annual energy review.* DOE/EIA—0384(92). Energy Information Administration, Office of Energy Markets and End Use, U.S. Department of Energy, Washington, D.C.

EIA. 1993. *Annual energy outlook 1993.* DOE/EIA—0383(93). Energy Information Administration, U.S. Department of Energy, Washington, D.C.

Gleeson, G.W. 1951. Energy—Choose it wisely today for safety tomorrow. *ASHVE Transactions* 57:523-40.

Grumman, D.L. 1984. Energy resource accounting: ASHRAE *Standard* 90C-1977R. *ASHRAE Transactions* 90(1B):531-46.

Heiman, J.L. 1984. Proposal for a simple method for determining resource impact factors. *ASHRAE Transactions* 90(1B):564-70.

Pritchett, T., L. Moody, and M. Brubaker. 1993. Why industry demand-side management programs should be self-directed. *Electricity Journal.* Seattle, WA.

BIBLIOGRAPHY

Anderson, R.J. 1984. The energy resource picture in 1984 in the U.S. and abroad. *ASHRAE Transactions* 90(1B):521-30.

DOE. 1979. *Impact assessment of a mandatory source-energy approach to energy conservation in new construction.* U.S. Department of Energy, Washington, D.C.

Pacific Northwest Laboratory. 1987. *Development of whole-building energy design targets for commercial buildings phase 1 planning.* PNL-5854, Vol. 2. U.S. Department of Energy, Washington, D.C.

ENERGY MANAGEMENT

ENERGY conservation can be defined as more efficient or effective use of energy. As fuel costs rise and environmental concerns grow, more efficient energy conversion and utilization technologies become cost-effective. However, technology alone cannot produce sufficient results without a continuing management effort. Energy management begins with the commitment and support of an organization's top management.

A suggested flowchart for developing an energy management program is shown in Figure 1. First, a team with the right blend of skills is selected to manage and execute the program. This team establishes objectives, priorities, and a time frame, sometimes with the help of outside consultants. The assembly of a historic database can serve as a basis for evaluating future energy conservation opportunities (ECOs). A monitoring system can be set up to obtain detailed use data (see Chapter 37, Building Energy Monitoring). Then a detailed energy audit, which is discussed later in this chapter, should be performed. After the ECOs and the estimated savings have been determined, the team issues a report and adjusts or fine-tunes the objectives and priorities. A few lower cost conservation measures that result in substantial savings can be done first to show progress. At this stage it is also critical to monitor and collect data, because energy management programs that can show progress tend to succeed. Metering can be installed to monitor energy consumption for each major piece of equipment and for each consumption area. Large charts showing the progress of everyone in the facility help increase awareness of saving energy. The program will require regular reports and readjustments of the objectives and priorities, and the implemented ECOs will need to be maintained.

ORGANIZATION

Because energy management is performed in existing facilities, most of this chapter is devoted to these facilities. Information on energy conservation in new design can be found in all volumes of the *ASHRAE Handbook* and in the ASHRAE 90 Series Standards. The area most likely to be overlooked in new design is the ability to measure and monitor energy consumption and trends for each energy use category given in Chapter 37.

To be effective, energy management must be given the same emphasis as management of any other cost/profit center. In this regard, the functions of top management are as follows:

- Establish the energy cost/profit center
- Assign management responsibility for the program
- Hire or assign an energy manager
- Allocate resources
- Ensure that the energy management program is clearly communicated to all departments to provide necessary support for achieving effective results
- Monitor the cost-effectiveness of the program
- Clearly set the program goals
- Encourage ownership of the program at the lowest possible level in the organization
- Set up an ongoing reporting and analysis procedure to monitor the energy management program

An effective energy management program requires that the manager (supported by a suitable budget) act and be held accountable for those actions. It is common for a facility to allocate 3 to 10% of the annual energy cost for the administration of an energy management program. The budget should include funds for additional personnel as needed and for continuing education for the energy manager and staff.

If it is not possible to add a full-time, first-line manager to the staff, an existing employee, preferably with a technical background, should be considered for either a full- or part-time position. This person must be trained to organize an energy management program. Energy management should not be an alternate or collateral duty of a person who is already fully occupied.

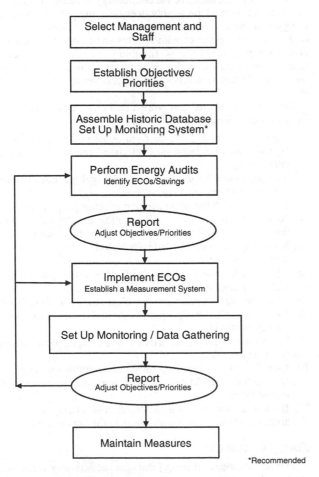

Fig. 1   Energy Management Program

The preparation of this chapter is assigned to TC 9.6, System Energy Utilization.

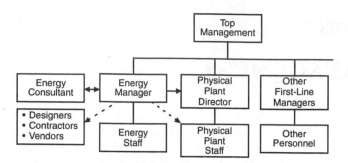

Fig. 2 Organizational Relationships

Another option is to hire a professional energy management consultant to design, implement, and maintain energy efficiency improvements. Some energy services companies (ESCOs) and other firms provide energy management services as part of a contract, with payments based on realized savings.

Figure 2 shows the organizational functions for effective energy management. The solid lines indicate normal reporting relationships within the organization; the dotted lines indicate new relationships created by energy management. The arrows indicate the primary directions initiatives related to energy conservation normally flow. This chart shows the ideal organization. In actuality, the functions shown may overlap, especially in smaller organizations. For example, the assistant principal of a high school may also serve as the energy manager and physical plant director. However, it is important that all functions be covered.

Energy Manager

The functions of an energy manager fall into four broad categories: technical, policy-related, planning and purchasing, and public relations. The energy manager will need assistance with some of the tasks in the following lists. A list of the specific tasks and a plan for their implementation must be clearly documented.

Technical functions include the following:

1. Conducting energy audits and identifying energy conservation opportunities (ECOs)
2. Establishing a baseline from which energy-saving improvements can be measured
3. Acting as an in-house technical consultant on new energy technologies, alternative fuel sources, and energy-efficient practices
4. Evaluating the energy efficiency of proposed new construction, building expansion, remodeling, and new equipment purchases
5. Setting performance standards for efficient operation and maintenance of machinery and facilities
6. Reviewing state-of-the-art energy management hardware
7. Selecting the most appropriate technology
8. Reviewing operation and maintenance
9. Implementing energy conservation measures (ECMs)
10. Establishing an energy accounting program for continuing analysis of energy usage and the results of ECMs
11. Maintaining the effectiveness of ECMs
12. Measuring energy use in the field whenever possible to verify design and operation conditions

Policy-related functions include the following:

1. Fulfilling the energy policy established by top management
2. Monitoring federal and state legislation and regulatory activities, and recommending policy/response on such issues
3. Representing the organization in energy associations
4. Administering government-mandated reporting programs

Planning and purchasing functions include the following:

1. Monitoring energy supplies and costs to take advantage of fuel-switching and load management opportunities

2. Ensuring that systems and equipment are purchased based on economics, their energy requirements, and their ability to perform the required functions, not simply on the lowest initial cost
3. Maintaining a current understanding of energy conservation, grant programs, and demand side management (DSM) programs offered by utilities and agencies
4. Negotiating or advising on major utility contracts
5. Developing contingency plans for supply interruptions or shortages
6. Forecasting the organization's short- and long-term energy requirements and costs
7. Developing short- and long-term energy conservation plans and budgets
8. Reporting periodically to top management

Public relations functions include the following:

1. Making fellow employees aware of the benefits of efficient energy use
2. Establishing a mechanism to elicit and evaluate energy conservation suggestions from employees
3. Recognizing successful energy conservation projects through awards to plants or employees
4. Setting up a formal mechanism for reporting to top management
5. Establishing an energy communications network within the organization, including bulletins, manuals, and conferences
6. Making the community aware of the energy conservation achievements of the organization with press releases and appearances at civic group meetings

General qualifications of the staff energy manager include (1) a technical background, preferably in engineering, with experience in energy-efficient design of building systems and processes; (2) practical, hands-on experience with systems and equipment; (3) a goal-oriented management style; (4) the ability to work with people at all levels, from operations and maintenance personnel to top management; and (5) technical report writing and verbal communications skills.

Desirable educational and professional qualifications of the staff energy manager include the following:

1. Bachelor of Science degree from an accredited four-year college, preferably in mechanical, electrical, industrial, or chemical engineering
2. Thorough knowledge of the principles and practices of energy resource planning and conservation
3. Familiarity with the administrative governing organization
4. Ability to analyze and compile technical and statistical information and reports with particular regard to energy usage
5. Knowledge of resources and information relating to energy conservation and planning
6. Ability to develop and establish effective working relationships with other employees and to motivate people
7. Ability to interpret plans and specifications for building facilities
8. Knowledge of the basic types of automatic controls and systems instrumentation
9. Knowledge of energy-related metering equipment and practices (see Chapter 37)
10. Knowledge of the organization's manufacturing processes
11. Knowledge of building systems design and operation and/or maintenance
12. Interest in and enthusiasm for efficient energy use and the ability to present ideas to all levels in the organization

Energy Consultant

In many instances, an energy manager needs outside assistance to conduct the energy management program. An energy consultant may be called on to assist in any of the energy manager's functions and, in addition, may be responsible for training the energy management, operations, and maintenance staffs. In some cases, the energy consultant may be responsible for implementing and monitoring the energy management program.

The basic qualifications of an energy consultant should be similar to those of the energy manager. The consultant should be an objective party with no connections to the sale of equipment or systems.

Energy service companies (ESCOs) and other contracting firms provide energy management services. Often these companies are compensated by a percentage of realized savings based on measured energy usage before and after the improvements. It is vital that the compensation methods be clearly specified and include provisions to adjust for changes in building use, production levels, and other factors.

Optimum energy conservation in a comprehensive and integrated energy management program can be accomplished only if system interaction is thoroughly understood and accounted for by the energy consultant and manager. For example, a lighting retrofit could reduce lighting energy while simultaneously causing other systems that interact with the lighting system to increase their energy consumption.

Motivation

The success of an energy management program depends on the interests and motivation of the people implementing it (Turner 1993). Participation and communication are key points. Employees can be stimulated to support an energy management program through awareness of the following:

1. Amount of energy they use
2. Cost of the energy
3. Critical role of energy in the continued viability of their jobs
4. Meaning of energy saving in their operations
5. Relationship between production rate and energy consumption
6. Benefits of participation, such as greater comfort and improved air quality

Energy management activities can be made a part of each supervisor's performance or job standards. If the supervisor knows that top management is solidly behind the energy management program and the overall performance rating depends, to some extent, on the energy savings the department or group achieves, the supervisor will motivate employee interest and cooperation.

FINANCING A PROGRAM

Several financing options exist for energy management programs.

Internal financing. The building owner, manager, or occupant makes a decision to finance the modifications internally, taking advantage of all the savings and assuming all the risk.

Shared savings contracts. Typically, independent contractors install energy-related modifications at their expense. The building owner then pays a prearranged fee, which is usually related to the savings achieved. The length of these contracts runs from several years to as many as 20 years. The terms and conditions of the contract must be carefully documented in advance. In particular, methods of determining energy savings and of accounting for other influences on energy consumption must be carefully and precisely described.

Guaranteed savings contracts. These are similar to shared savings contracts, except that the costs of the modifications are specified up front. The owner can pay for the retrofit at the beginning of the contract, finance it over a period of years, or have the contractor finance it with repayment out of the savings. The owner should enter into these agreements carefully, with an understanding of the financial liability if the savings do not materialize and the contractor is unable to make good on the guarantee. Energy savings are guaranteed, usually on an annual basis. If greater savings are achieved, some contracts require that a bonus be paid to the contractor. If less

than the guaranteed savings are realized, the owner is reimbursed for the difference. Insurance policies should be examined carefully. The performance record of the contractor and insurance company with this type of contract and policy should be checked.

Demand side management. Demand side management (DSM) programs are offered by utilities and most often subsidize the installation of energy modifications. These programs usually take the form of rebates to contractors or building owners for the installation of specified equipment and are often carried out by subcontractors or unregulated subsidiaries of utilities. The purpose of these programs is to reduce demand or consumption of the form of energy supplied by the utility.

Utility consultants. Utility consultants review utility bills for building owners, checking for billing errors and proper rates and riders. Usually, no consideration is given to the quantity of energy consumed. Contracts are on a contingent fee basis, frequently with payment being half the savings for five years.

Government programs. There are some federal, state and local government programs that provide subsidies or tax credits for certain energy modifications.

Tariff analysis. Most utilities offer a variety of rates. The customer selects the rates and riders, provided they meet their criteria. Because rates change frequently, it is often advantageous to examine the applicability of other utility rates that may lower costs.

Tax exemption. Many states exempt certain uses of energy from sales tax. As a general rule, utilities charge sales tax to all customers, unless they are advised that the customer is exempt. Typical exemptions include those for nonprofit organizations, research and development, and some types of manufacturing.

IMPLEMENTING A PROGRAM

There are six basic stages in implementing an energy management program:

1. Develop a thorough understanding of how energy is used.
2. Conduct a planned, comprehensive search to identify all potential opportunities for energy conservation activities. Contact utilities to find out about applicable rates and to learn about any DSM program offerings.
3. Determine and clearly specify the energy and cost savings analysis methods to be used. See Chapter 33, Owning and Operating Costs; MacDonald and Wasserman (1989); and PNL (1990).
4. Identify, acquire, allocate, and prioritize the resources necessary to implement energy conservation opportunities.
5. Implement the energy conservation measures in a rational order. This is usually a series of independent activities that take place over a period of years.
6. Monitor and maintain the energy conservation measures that have been taken. Reevaluate them as building functions change over time.

For existing buildings, the ASHRAE 100 Series Standards have been developed to provide criteria that will result in conservation of nonrenewable energy resources. These standards are directed toward upgrading existing building thermal performance, increasing system energy efficiency, and providing procedures and programs essential to the operation, maintenance, and monitoring of energy-conserving measures.

Energy efficiency and/or energy conservation efforts should not be equated with discomfort, nor should they interfere with the primary function of the organization or facility. Energy conservation activities that disrupt or impede the normal functions of workers and/or processes and adversely affect productivity constitute false economies.

Database

The compilation of a base of past energy usage and cost is important in developing an energy management program. Any reliable utility data that is applicable should be examined. Utilities can usually provide demand metered data on computer disks, often with measurement intervals as short as 15 minutes. This is preferable to monthly data because anomalies are more apparent. High consumption at certain times may reveal opportunities for conservation (Haberl and Komar 1990). If monthly data are used, they should be analyzed over several years. A base year should be established as a reference point for future energy conservation and energy cost avoidance activities. In tabulating such data, the actual dates of meter readings should be recorded so that energy use can be normalized for differences in the number of days in a period. Any periods during which consumption was estimated rather than measured should be noted.

If energy is available for more than one building and/or department under the authority of the energy manager, each of these should be tabulated separately. Initial tabulations should include both energy and cost per unit area. (In an industrial facility, this may be energy and cost per unit of goods produced.) Available information on variables that may have affected past energy use should also be tabulated. These might include heating or cooling degree-days, percent occupancy for a hotel, quantity of goods produced in a production facility, or average daily weather conditions, which can be obtained from the National Oceanic and Atmospheric Administration (NOAA), Environmental Data and Information Service, National Climatic Center Federal Building, Asheville, NC 28801. Because such variables may not be directly proportional to energy use, it is best to plot information separately or to superimpose one plot over another, rather than developing such units as kilojoules per square meter per degree day. As such data are tabulated, ongoing energy accounting procedures should be developed for regular data collection and use in the future.

Comparing a building's energy use with that of many different buildings is a valuable way to check its relative efficiency. The Energy Information Administration (EIA) within the U.S. Department of Energy collects data on buildings in all sectors. This information is summarized in DOE/EIA-0246 and DOE/EIA-0318 for nonresidential buildings and in DOE/EIA-0321/1 for households.

Tables 1 and 2 list physical characteristics of the buildings surveyed. Tables 3 and 4 list measured demand and energy consumption by building type. Table 5 lists energy sources for various end uses. The EIA also collects data on household energy consumption, which is summarized in Table 6.

When an energy management program for a new building is established, the energy use database may consist solely of typical energy use data for similar buildings, as illustrated in Tables 3 and 4. This may be supplemented by energy simulation data for the specific building if such data were developed during the design of the building. In addition, a new building and its systems should be properly commissioned upon completion of construction to ensure proper operation of all systems, including any energy conservation features. Refer to ASHRAE *Guideline* 1, Guideline for Commissioning of HVAC Systems.

All the data presented in these tables are derived from detailed reports of consumption patterns in buildings. Before using the data, however, it is important to understand how it was derived. For example, all the household energy consumption data presented in Table 6 are average data, and they may not reflect variations in appliances or fuel situations for different buildings. Therefore, when using the data, verify the correct use of it with the original EIA documents. Because these surveys are performed regularly, there may be newer data.

ANSI/ASHRAE *Standard* 105, Standard Methods of Measuring and Expressing Building Energy Performance, contains information that allows uniform, consistent expressions of energy consumption in both proposed and existing buildings. Its use is recommended. However, the data presented here are not in accordance with this standard.

ASHRAE Special Project Report 56 (ASHRAE 1989) has a list of data elements useful for normalizing and comparing utility billing information. Metered energy consumption and cost data are also gathered and published by trade associations that represent building owners. Typical trade associations include the Building Owners and Managers Association and the American Hotel and Motel Association.

The quality of published energy consumption data for buildings varies because the data are collected for different purposes by people with different levels of technical knowledge of buildings. The data presented here are primarily national data. In some cases, local energy consumption data may be available from local utility companies or state or provincial energy offices.

At this point in the development of an energy management program, it is useful to compile a list of previously accomplished energy conservation measures and the actual energy and/or cost savings of such measures. Energy and cost savings analysis methods (MacDonald and Wasserman 1989, section 2) can be used to calculate savings. These measures and savings should be studied during subsequent energy audits to determine their present effectiveness and the effort(s) necessary to maintain and/or improve them.

Table 1   1992 Commercial Building Characteristics in the United States

	Size, m²				Hours Per Week		% of Buildings Less Than 51%	
	Per Building		Per Worker					
Building Type	Mean	Median	Mean	Median	Mean	Median	Heated	Cooled
Education	2620	840	114	90	49	45	0	34
Food Sales	540	240	84	93	108	105	0	0
Food Service	530	320	62	57	91	95	8	16
Health Care	2590	400	48	46	69	50	0	0
Lodging	1750	740	133	158	158	168	0	30
Mercantile/Service	900	370	72	100	62	54	22	62
Office	1520	460	42	46	53	45	6	11
Parking Garage	6490	560	712	279	83	52	58	88
Public Assembly	1520	550	154	174	50	45	13	45
Public Order/Safety	1270	460	95	98	123	168	0	50
Religious Worship	950	410	151	189	22	10	5	33
Warehouse and Storage	1400	460	240	179	50	48	75	94
Other	1520	370	85	118	75	56	0	65
Vacant	1280	390	416	186	20	0	74	84
All Buildings	1310	420	89	94	58	50	28	52
Source Table	A1	A1	A1	A1	A1	A1	A44	A45

Source: DOE/EIA (1994). Based on 6600 buildings.

Table 2 1986, 1989, and 1992 Commercial Building Characteristics in the United States

	% of Buildings		
	'86	'89	'92
Predominant lighting			
T13 Fluorescent	64	87	91
T13 Efficient fluorescent	27		
TA55 Incandescent	41	53	56
Efficient incandescent	10		
High intensity discharge	6	10	8
High efficiency ballast		24	
Other	1	1	6
District heating or cooling			
T6 Any source	2		
Steam	2		
Hot water			
Chilled water			
Weekly schedules			
T10 Closed	8		
Less than 24 h			
Mon to Fri	28	26	
Mon to Sat	27	23	
Mon to Sun	17	17	
Open 24 h	8	9	
Other	11	24	
Weekly operating hours			
T11 39 or fewer	19	19	22
T11 40 to 48	27	25	27
TA18 49 to 60	23	22	21
61 to 84	14	14	13
85 to 167	9	11	10
Continuous	8	9	8
Building size, m²			
T11 93 to 465	53	56	56
T11 466 to 929	23	20	20
TA8 930 to 2322	14	14	13
2323 to 4645	6	5	6
4646 to 9290	3	3	2
9291 to 18 580	1	1	1
18 581 to 46 450	1	1	1
Over 46 450			
Occupancy, nongovernment building			
T13 Owner-occupied			76
T11 Single	88	87	90
TA22 Multiple	12	13	10
Nonowner-occupied			19
Single	72	59	61
Multiple	24	23	39
Vacant	4	18	5
Year constructed			
T11 1900 or before	4	4	4
T11 1901 to 1920	6	5	5
TA10 1921 to 1945	15	15	15
1946 to 1960	21	19	18
1961 to 1970	18	18	16
1971 to 1979	20	20	20
1980 to 1986	16	14	14
1987 to 1989		5	4
1990 to 1992			3
Number of floors			
T13 One	64	64	63
T13 Two	24	23	24
TA12 Three	8	9	9
More than three	4		
Four to nine		4	4
Ten or more			

	% of Buildings		
	'86	'89	'92
Energy sources used			
T11 Electricity	99	95	100
T11 Natural gas	55	53	58
TA29 Fuel oil	13	13	12
District heat	2	2	2
District cool		1	1
Propane	9	8	7
Other	4	3	2
Heating sources used*			
T11 Electricity	29	28	36
T11 Natural gas	51	48	58
TA36 Fuel oil	13	12	11
District heat	2	2	2
Propane	6	5	6
Other	4	2	2
Not heated	9		
Cooling sources used*			
T11 Electricity	68	68	97
TA40 Natural gas	3	2	3
District cool		1	1
Other	4		
Not cooled	28		
Water heating sources*			
T11 Electricity	35	34	48
T11 Natural gas	33	31	47
TA41 Fuel oil	3	3	4
District heat	1	1	1
Propane	3	2	2
Other			
None	28		
Cooking Energy Sources*			
T11 Electricity	6	9	49
T11 Natural gas	8	10	59
TA43 Propane	2	2	10
None	28		
Percent heated			
T11 Not heated	8	15	14
T11 1 to 50% m²	15	14	14
TA44 51 to 99% m²	11	11	13
100% m²	65	60	59
Percent cooled			
T11 Not cooled	28	30	27
T11 1 to 50% m²	24	23	24
TA45 51 to 99% m²	13	13	14
100% m²	36	34	35
Percent lit open hours			
T13 Not lit	2	7	9
T13 1 to 50% m²	16	22	18
TA46 51 to 99% m²	16	21	17
100% m²	66	50	56
Heating equipment			
T13 Furnaces	45	36	40
T13 Boilers	16	16	15
TA47 In space	27	31	35
Packaged	13	19	21
Air heat pump	8	10	11
District heat	2		2
Air ducts		44	
Heat/reheat coil		5	
Fan coil units		4	
Baseboard/radiators		11	
Other		1	1

	% of Buildings		
	'86	'89	'92
Cooling equipment			
T13 Central	28	4	4
T13 Individual	23	24	29
TA49 Packaged	18	44	42
Air heat pumps	8	10	13
District cooling			1
Air ducts		38	
Residential central air			23
Other		2	5
HVAC distribution			
T13 Forced air (Ducts)			
TA1 Heating only	15		12
Cooling only	4		7
Heat and cool	44		49
VAV system			5
Baseboard/radiators			10
Steam	6		
Hot water	7		
Fan coil units			
Heating only	5		2
Cooling only	1		1
Heat and cool	4		
Heating panels	5		
Shell conservation			
T13 Any	82		88
T13 Roof/ceiling insulation	68	67	70
TA58 Wall insulation	50	45	48
Multiglazing	31	32	35
Glass film/shade	22	21	22
Awnings/shading	32	33	39
Weatherstripping	63	61	
Windows that open			44
None		17	
HVAC conservation			
T13 Any	54		
TA1 Preventive maintenance	52		52
Heat recovery	4		
Energy management system	5		
Time clock/thermostat	2		
Economizer cycle			9
VAV system			5
Other	2		
Lighting conservation			
T13 Any	36		
TA1 High efficiency ballasts	25		
Delamping	8		
Natural light sensors	4		2
Other controls	10		17
Reflectors			12
Other	2		2
In response to audit			
T59 Any	5		11
TA1 HVAC	2		
Building shell	2		
Lighting	3		
HVAC control			
T13 Heating only	16	53	
T13 Cooling only	2	44	
Both	50		
Reduction in off hours			
T13 Heating only	19	17	
T13 Cooling only	3	6	
Both	58	53	

Sources: DOE/EIA (1988, 1989, 1992, 1994). T# designates table number therein.
Notes: There are some differences in sample sizes between references. Differences in characteristics from 1986 to 1989 to 1992 are due to questionnaire design.
* = Solely or in combination

Table 3   1989 Commercial Building Median Electric Demand and Median Load Factor Characteristics in the United States

	Demand, W/m²	Load Factor, %		Demand, W/m²	Load Factor, %		Demand, W/m²	Load Factor, %
Building use			**Ownership and occupancy**			**Cooling source**		
Assembly	51.3	13.7	Owner-occupied	51.3	25.0	Electricity	56.2	25.3
Education	44.7	20.2	Single	53.0	24.7	Others	54.7	29.1
Food sales	104.1	50.3	Multiple	44.9	25.9	No cooling	25.8	18.6
Food service	122.0	35.1	Nonowner-occupied	47.3	22.5			
Health care	69.2	23.5	Single	50.8	21.6	**Water heating sources**		
Lodging	41.1	34.1	Multiple	49.7	26.2	Electricity	57.3	25.0
Mercantile/Service	47.1	23.8	Vacant	29.4	14.5	Others	50.8	26.4
Office	65.3	26.4	Government	47.4	23.4	No water heating	34.1	18.6
Parking garage	36.6	23.5						
Public order	43.1	21.0	**Percent lit open hours**			**Cooking energy sources**		
Warehouse	20.0	20.8	Not lit	13.5	9.5	Electricity	66.8	32.0
Other	94.2	26.6	1 to 50% m²	32.3	19.8	Others	68.0	31.3
Vacant	27.4	15.0	51 to 99% m²	53.8	25.5	No cooking	44.6	22.4
			100% m²	58.3	25.3			
						Percent heated		
			Reduction in off hours			Not heated	26.9	21.0
Weekly operating hours			Heating only	33.2	20.3	1 to 50% m²	28.0	22.0
39 or fewer	40.4	12.5	Cooling only	51.3	29.5	51 to 99% m²	59.7	27.0
40 to 48	44.5	20.5	Both	52.2	23.5	100% m²	57.4	24.3
49 to 60	40.4	22.7						
61 to 84	54.7	29.5	**Energy sources used**			**Percent cooled**		
85 to 167	68.4	34.2	Electricity	50.0	24.1	Not cooled	25.8	18.6
Continuous	65.9	41.3	Natural gas	48.4	24.4	1 to 50% m²	32.9	23.5
			Fuel oil	38.6	25.1	51 to 99% m²	63.0	28.1
			District heat	47.3	43.2	100% m²	73.2	25.6
			District cool	45.1	37.2			
			Propane	51.1	25.7	**Year constructed**		
Building size, m²			Other	26.3	21.0	1899 or before	26.0	22.8
93 to 465	68.0	21.1				1900 to 1919	33.0	18.6
466 to 929	40.9	24.0	**Main heating source**			1920 to 1945	40.4	22.8
930 to 2322	32.3	25.4	Electricity	80.7	25.3	1946 to 1959	43.1	21.7
2323 to 4645	30.2	28.2	Natural gas	46.4	23.5	1960 to 1969	51.9	24.3
4646 to 9290	31.5	33.8	Fuel oil	39.5	24.0	1970 to 1979	57.4	25.9
9291 to 18 580	34.0	37.2	District heat	47.0	43.2	1980 to 1983	62.8	25.2
18 581 to 46 450	41.9	45.8	Propane	51.5	20.1	1984 to 1986	58.7	26.2
Over 46 450	35.5	50.6	Other	16.9	17.9	1987 to 1989	54.8	29.9

Source: DOE/EIA (1992), Tables 35 and 36.

Table 4   1989 Commercial Building Energy Use in the United States

Building Type	Percent Bldg.	Percent m²	GJ/m² All Buildings Using Any Fuel Elec.	Gas	Total	GJ/m² Major Fuel Energy Intensity by Region NE	MW	SO	WE	kWh per m² Electric Energy Intensity by Region NE	MW	SO	WE	m³/m² Natural Gas Intensity by Region NE	MW	SO	WE	kWh per m² Total Electric Use Heat	HVAC	AC	m³/m² Total Gas Use Heat
Assembly	14	11	307	443	727	704	875	636	772		76	102	79		17.5	7.9	14.0	113	129	98	13.0
Education	6	13	307	534	988	897	1147	738	1250	79	80	94	85	8.6	18.6	10.2	20.2	141	138	87	15.3
Food sales	2	1	1511	545	1999		2011					426						520	524	433	14.8
Food service	5	2	1102	1727	2476	2283	2079	2636	3283	234	206	424	357	59.2	34.3	46.2	61.7	302	301	301	45.9
Health care	2	3	852	1284	2488	2295	3113	1749	1977	216	243	252	216	22.6	47.7	25.0	21.1	206	222	243	37.5
Lodging	3	6	454	818	1386	1999	1659	943	1329	115	128	133	115	16.3	26.4	16.2	26.4	165	159	126	26.1
Mercantile/Service	28	20	500	523	966	1318	1147	704	829	163	135	130	144	19.2	18.9	9.0	9.5	165	167	143	15.8
Office	15	19	750	364	1181	1113	1363	1091	1238	175	180	225	241	6.3	17.0	7.2	8.2	243	245	215	13.0
Parking garage	1	2	204		488		329			64	46							38		66	
Public order/Safety	1	1	534	613	1443															144	18.8
Warehouse	14	15	307	443	659	863	784	579	341	142	93	58	71	4.8	14.4	16.8	5.1	81	85	98	13.2
Other	1	2	1488	1204	2556	2204	1818					323						375	376	442	37.1
Vacant	7	7	148		273	386	170		170	75	18	39	50		3.4			47	43	46	
All buildings	100	100	511	557	1045	1136	1181	852	1102	139	123	145	168	12.3	19.2	12.7	14.2	175	184	156	16.6
Source table	11	11	21	38	13	16	16	16	16	23	23	23	23	40	40	40	40	29	28	28	45

Source: DOE/EIA (1992). Based on 4528 buildings.
Notes: GJ/m² are for buildings that are supplied with those fuels.

The standard error for the nationwide total electricity or natural gas consumption is 6%.
Blank spaces indicate insufficient data.

Table 5 Energy Sources for End Uses—Percent of All Buildings
(More Than One May Apply)

Sources	All Buildings Using Energy	Primary Space Heating	Secondary Space Heating	Water Heating	Cooling	Cooking	Manu-facturing	Electric Generation
All Buildings	100	90	15	76	76	16	3	3
Electricity	100	24	9	37	74	8	2	
Natural Gas	58	49	3	36	2	9	0	1
Fuel Oil	12	9	2	3				2
District Heat	2	2	0	1		0		
Chilled Water	1				1			
Propane	7	5	1	2		2	0	0
Other	4	1	1					1

Source: DOE/EIA (1994), Table A69.   *Note*: Reported percentages do not exclude buildings where end use is not performed.

Since most energy management activities are dictated by economics, the energy manager must understand the utility rates that apply to each facility. Special rates are commonly applied for such variables as time of day, interruptible service, on peak/off peak, summer/winter, and peak demand. There are more than 1000 electric rate variations in the United States. The energy manager should work with local utilities to develop the most cost-effective methods of metering and billing and enable energy cost avoidance to be calculated effectively.

For example, it is common for electric utilities to meter both electric consumption and demand. Demand is the peak rate of consumption, typically integrated over a 15- or 30-min. period. Electric utilities may also establish a ratchet billing procedure for demand. In a simplified version of ratchet demand billing, the billing demand is established as either the *actual demand* for the month in question or a percentage of the *highest demand* during the previous 11 months, whichever is greatest. Figure 3 illustrates a gas-heated, electrically cooled building with the highest electric demands occurring in the summer, and shows actual demand versus billing demand under an 85% ratchet. The winter demand is approximately 700 kW each month, with chiller operation causing the August demand to peak at 1000 kW. Since an 85% ratchet applies, all months following August with actual demand lower than 850 kW are billed at 850 kW. Therefore, the following can be concluded about this building:

1. ECMs that reduce winter demand would not reduce the billed cost of demand, unless they also reduced summer demand; and if such measures were implemented in September, they would not produce savings in demand billings for the first 11 months.

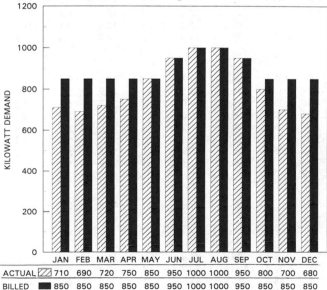

Fig. 3 Actual Demand Versus Billing Demand
(85%, 11-Month Ratchet)

	JAN	FEB	MAR	APR	MAY	JUN	JUL	AUG	SEP	OCT	NOV	DEC
ACTUAL	710	690	720	750	850	950	1000	1000	950	800	700	680
BILLED	850	850	850	850	850	950	1000	1000	950	850	850	850

2. ECMs that reduce peak demand in each of the summer months (for example, 50 kW reduction in chiller peaks) may produce savings in demand billings throughout the year.

A billing method called Real Time Pricing (RTP) is beginning to be used by utilities. With this method, the utility calculates the marginal cost of power per hour for the next day, determines the customer price, and sends this hourly price (for the next day) to customers. The customer can then determine the amount of power to be consumed at different times of the day.

There are many variations of the above billing methods, and it is important to understand the applicable rates. Caution is advised in designing or installing energy management and energy retrofit systems that take advantage of utility rate provisions, because these provisions can change. The structure or provisions of utility rates cannot be guaranteed for the life of the ECMs. Some of the provisions that change include on-peak times, declining block rates, and demand ratchets.

Priorities

Having established a database, the energy manager should assign priorities to future work efforts. If there is more than one building or department under the energy manager's care, their energy use and cost databases should be compared overall and on the basis of energy use and cost per unit area, of cost per unit of production, or of some other index that demonstrates an acceptable level of accuracy. Comparisons should also be made with realistic energy targets, if they are known. Using such comparisons, it is often possible to set priorities that allocate the available resources most effectively.

One approach to comparing that is used in England is called "Monitoring and Targeting." The monitoring portion involves collecting short-term metered data and analyzing it with respect to occupancy, weather prediction, and other variables. This permits a better understanding of the factors that influence energy consumption so that improvements can be made. Targeting involves comparisons with other similar buildings. When the building being monitored uses more energy than the norm for similar buildings, it is likely that reductions can be achieved.

At this point, a report should be prepared for top management outlining the data collected, the priorities assigned, plans for continued development of the energy management program, and projected budget needs. This should be the beginning of a regular monthly, quarterly, or semiannual reporting procedure.

Energy Audits

Three levels of energy audits or analysis have been defined (ASHRAE 1989). Depending on the physical and energy use characteristics of a building and the needs and resources of the owner, these steps require different levels of effort. Following a preliminary energy use evaluation, an energy analysis can generally be classified into the following three categories:

Table 6   1990 Household Energy Consumption

Table Number	**18**	**21**	**22**	**23**	**24**	**25**	**26**	**29**	**29**	**29**	**29**	**29**	**29**
								Consumption by End Uses*					
Characteristics	**All Major Sources**	**Electricity**	**Gas**	**Fuel Oil**	**Kerosene**	**LPG**	**Wood Main Fuel**	**All Uses**	**Space Heat**	**Air Cond.**	**Water Heat**	**Refrigerator**	**Appliances**
Total U.S. households	103.5	34.1	88.8	88.0	13.0	36.0	81.4	103.5	54.8	8.5	18.8	5.6	19.8
Urban status													
Urban	104.7	32.6	88.5	89.9	10.3	31.0	78.0	104.7	54.4	8.8	19.6	5.6	19.7
Central city	103.1	29.3	87.7	75.9	14.0	20.2	63.7	103.1	54.5	8.9	19.9	5.4	18.3
Suburban	105.7	34.9	89.3	96.7	8.7	32.7	80.0	105.7	54.4	8.8	19.4	5.8	20.9
Rural	99.4	38.9	90.6	82.3	16.7	40.3	83.7	99.4	55.8	7.7	15.9	5.7	20.2
Climate zone													
<1100 CDD, >3900 HDD	116.7	30.6	109.7	93.8	29.9	50.6	118.8	116.7	75.9	3.2	18.4	4.5	18.1
3056 to 3900 HDD	130.1	29.6	116.8	103.3	8.2	35.4	81.9	130.1	80.8	4.1	22.7	5.0	20.3
2200 to 3055 HDD	107.2	33.5	88.2	77.9	12.1	39.5	68.7	107.2	58.6	6.5	19.7	5.6	20.0
Under 2200 HDD	76.9	32.2	61.0	34.8	12.6	33.2	56.1	76.9	30.8	7.9	16.9	5.5	19.3
>1100 CDD, <2200 HDD	79.3	45.8	59.4			21.7		79.3	23.2	17.7	14.0	7.5	20.7
Type of housing unit													
Single family	117.0	39.1	99.6	97.5	9.7	38.3	83.0	117.0	62.9	9.6	19.5	6.3	23.3
Detached	119.5	39.8	100.6	98.7	9.9	38.6	83.0	119.5	64.4	9.8	19.8	6.4	23.8
Attached	92.1	33.1	87.7	84.1	7.1			92.1	49.0	7.3	16.6	5.0	17.4
Mobile home	82.3	32.7	80.6	52.2	27.6	31.0	78.3	82.3	42.7	7.9	13.7	4.2	17.4
Multifamily	72.3	20.9	63.8	67.1		20.2		72.3	36.0	5.8	17.9	4.0	11.3
2 to 4 units	99.7	21.2	88.9	94.5		24.9		99.7	59.1	5.3	20.8	4.3	13.6
5 or more units	53.3	20.8	43.0	48.1				53.3	20.0	6.1	15.9	3.8	9.8
Heated floorspace, m²													
Fewer than 93	67.1	23.8	59.4	56.7	22.6	27.1	57.7	67.1	32.3	6.4	15.5	4.3	12.6
93 to 185	103.9	36.2	88.5	79.4	11.0	39.8	81.8	103.9	52.8	8.9	18.9	5.8	21.0
186 to 278	132.1	42.4	106.6	95.3	4.0	42.1	98.0	132.1	72.0	9.9	21.6	6.8	25.6
279 or more	186.1	45.6	152.7	156.6	0.8	53.3	99.9	186.1	116.9	10.8	25.3	7.2	31.2
Number of rooms													
1 to 3	52.6	19.6	46.0	46.1	21.6	26.9	48.3	52.6	23.5	6.3	13.4	3.7	9.1
4 to 6	96.5	32.4	83.1	77.4	14.6	34.9	81.6	96.5	50.3	7.9	18.1	5.4	18.7
7 or more	144.4	44.9	119.3	120.4	7.0	44.7	90.0	144.4	80.1	10.7	23.0	7.1	27.9
Ownership of unit													
Owned	116.6	38.5	99.7	95.4	11.7	37.6	85.3	116.6	63.2	9.1	19.2	6.2	23.0
Rented	77.8	25.2	68.9	66.8	17.0	29.0	60.6	77.8	38.1	7.3	18.1	4.4	13.7
Public housing	60.6	20.6	50.1	51.6				60.6	25.0	4.7	18.3	3.8	11.5
Not public housing	79.2	25.6	70.4	69.5	17.0	29.2	60.6	79.2	39.2	7.5	18.0	4.4	13.8
Year of construction													
1939 or before	126.3	25.1	103.5	96.1	15.0	43.0	101.6	126.3	81.3	4.2	20.8	5.1	18.7
1940 to 1949	111.0	28.8	91.8	73.6	8.8	37.1	87.4	111.0	65.5	6.3	18.9	5.6	18.9
1950 to 1959	115.5	32.4	90.7	89.8	9.5	34.6	75.1	115.5	64.0	7.7	20.7	6.1	21.7
1960 to 1969	100.6	34.2	80.5	79.3		23.8	80.3	100.6	49.6	9.2	18.6	5.9	20.8
1970 to 1979	89.8	40.5	76.1	85.5	18.9	32.9	65.1	89.8	39.4	10.2	17.9	5.7	19.9
1980 to 1984	75.9	42.3	66.1	58.8		43.3	70.6	75.9	27.6	10.4	15.7	5.8	19.5
1985 to 1987	71.3	39.2	89.3			30.2	65.9	71.3	26.9	10.8	13.8	5.1	17.2
1988 to 1990	108.8	40.2	112.0			39.2		108.8	53.5	11.1	20.8	5.8	22.9
All utilities paid by household													
Yes	107.5	34.9	94.4	95.8		35.6		107.5	57.0	9.1	18.5	5.9	21.3
No	81.6	21.2	65.3	61.9		44.4		81.6	42.7	5.0	20.6	3.9	12.2
Household size													
1 person	75.0	22.2	70.5	72.0		27.5	67.3	75.0	45.9	5.7	11.1	4.7	10.6
2 persons	103.0	33.0	88.2	91.6		35.1	77.4	103.0	56.7	8.5	17.2	5.8	18.6
3 to 5 persons	119.8	41.6	100.0	92.8		41.1	88.4	119.8	58.7	10.2	23.9	6.0	25.8
6 or more persons	126.3	43.0	99.8	87.0		35.7	73.2	126.3	56.7	10.1	30.2	5.9	30.6

Source: DOE/EIA (1993).   CDD = cooling degree-days.   HDD = heating degree-days.

* = Averages are distributed over all households, not just those with that end use; households had an average of 1.2 refrigerators.

Table 6 1990 Household Energy Consumption (*Concluded*)

	Gigajoules per Household per Year										Wh/m² per CDD	
Table Number	33	34	35	36	37	38	41	41	42	42	37	38
	Main Space Heating				Type of Air Conditioning		Water Heating by Fuel				Type of Air Conditioning	
Characteristics	Electric	Gas	Fuel Oil	LPG	Central	Room	Electric	Gas	Fuel Oil	LPG	Central	Room
Total U.S. households	13.0	68.0	83.7	43.4	11.6	3.8	10.0	24.4	22.7	19.2	17.6	17.2
Urban status												
Urban	11.1	67.3	83.1	34.9	11.7	3.6	9.5	24.6	22.7	18.6	17.4	16.7
Central city	9.8	68.4	68.4		12.4	3.8	9.2	24.4	20.4		20.3	17.1
Suburban	11.9	66.4	90.4	36.4	11.3	3.6	9.6	24.8	24.2	17.9	16.3	16.5
Rural	21.0	72.1	85.5	50.6	11.3	4.4	11.3	23.1	22.6	19.6	20.0	18.4
Climate zone												
<1100 CDD, >3900 HDD	23.9	88.0	103.1	68.3	4.6	1.5	11.3	25.1	21.4	21.8	16.5	13.6
3056 to 3900 HDD	23.2	93.1	93.6	59.0	6.5	1.9	11.8	27.4	22.3	21.3	17.4	14.3
2200 to 3055 HDD	17.1	73.3	71.1	52.4	9.1	2.7	11.1	24.9	23.4	18.9	17.1	14.5
Under 2200 HDD	10.2	38.6	35.8	32.7	9.4	4.5	9.3	21.9		16.7	17.2	18.2
>1100 CDD, <2200 HDD	6.1	37.7		17.1	20.3	10.7	7.7	20.5		16.8	21.1	21.7
Type of housing unit												
Single family	16.5	75.5	94.1	47.3	12.8	4.4	11.0	24.9	24.2	19.5	16.3	16.1
Detached	17.6	75.6	96.0	47.3	13.3	4.4	11.3	25.0	24.8	19.5	16.3	15.9
Attached	11.9	73.4	75.7		9.0	3.7	8.3	23.7	19.4		17.4	19.0
Mobile home	18.3	58.4	53.4	32.7	10.3	5.3	9.8	22.4		15.7	40.1	35.8
Multifamily	7.1	48.7	58.1		8.2	2.3	7.7	23.3	20.8		26.0	16.9
2 to 4 units	10.1	69.6	86.0		9.2	2.2	8.3	25.4	26.2		23.1	16.5
5 or more units	6.0	28.6	35.6		7.9	2.3	7.4	21.5	18.8		27.3	17.4
Heated floorspace, m²												
Fewer than 93	8.7	42.7	51.2	35.2	9.0	3.6	8.4	21.5	18.5	15.5	30.4	24.8
93 to 185	14.8	65.4	75.4	42.7	11.7	4.6	10.4	24.2	24.6	19.7	20.0	19.0
186 to 278	18.6	85.0	90.1	59.2	13.1	3.2	12.1	26.8	22.8	22.7	16.5	11.6
279 or more	25.6	126.0	147.1	65.9	14.3	2.0	13.2	29.6	27.1	22.5	14.0	7.8
Number of rooms												
1 to 3	6.9	32.6	38.5	49.3	8.5	3.1	7.0	19.3	15.5	12.1	30.8	21.1
4 to 6	13.1	63.1	74.1	38.2	10.9	4.2	10.0	23.3	23.1	19.2	19.6	18.2
7 or more	19.4	91.4	113.2	57.7	13.7	3.1	12.3	28.7	25.8	23.7	15.7	13.2
Ownership of unit												
Owned	16.1	77.1	90.6	44.1	12.1	4.1	10.7	24.7	23.4	19.0	16.9	15.9
Rented	9.1	50.2	62.1	39.5	10.3	3.4	9.0	23.8	21.2	20.4	25.2	21.5
Public housing	8.2	32.5	45.7		6.1	3.7	9.0	22.8	22.9		23.1	19.4
Not public housing	9.2	51.6	64.4		10.6	3.4	9.0	23.9	20.8	20.4	25.2	21.9
Year of construction												
1939 or before	21.9	88.0	90.3	61.3	8.9	3.1	10.6	24.5	20.8	21.3	17.1	14.5
1940 to 1949	19.4	72.6	75.5	52.4	10.3	4.3	10.6	22.8	18.1	22.9	17.8	17.8
1950 to 1959	18.8	68.2	83.0	43.1	11.0	4.2	10.0	24.5	27.3	13.0	17.4	16.1
1960 to 1969	13.3	56.4	77.9	30.5	13.0	3.8	10.2	22.5	22.2	16.8	17.8	18.4
1970 to 1979	13.1	54.9	80.4	39.0	11.9	4.6	10.3	25.3	21.8	21.9	19.8	23.1
1980 to 1984	10.9	44.3		38.9	12.2	3.7	10.1	23.8		17.3	17.6	16.1
1985 to 1987	8.3	63.0		30.1	10.4	6.2	8.0	28.1		19.2	15.9	20.2
1988 to 1990	10.9	82.1		41.4	11.9	1.9	9.7	33.7		18.5	14.5	18.2
All utilities paid by household												
Yes	13.1	71.2	92.5	43.0	11.7	3.9	10.2	24.8	24.1		17.6	17.1
No	11.2	50.3	52.4	49.9	8.9	3.0	7.6	23.0	20.6		26.9	23.3
Household size												
1 person	11.0	61.7	69.7	36.5	7.6	3.2	5.8	14.7	13.9	10.0	17.2	15.9
2 persons	13.1	69.5	87.2	44.1	11.4	3.7	9.2	22.4	21.4	18.3	17.2	16.3
3 to 5 persons	14.3	71.2	87.0	48.6	13.7	4.4	13.1	30.8	27.9	23.9	18.0	19.0
6 or more persons	18.0	64.2	99.5	29.6	15.3	5.1	18.3	36.9		24.1	17.4	18.8

Source: DOE/EIA (1993).   CDD = cooling degree-days.   HDD = heating degree-days.
* = Averages are distributed over all households, not just those with that end use; households had an average of 1.2 refrigerators.

Level I—Walk-Through Assessment. This energy audit level involves the assessment of a building's energy cost and efficiency through the analysis of energy bills and a brief survey of the building. A Level I energy analysis will identify and provide a savings and cost analysis of low-cost/no-cost measures. It will also provide a listing of potential capital improvements that merit further consideration, along with an initial judgment of potential costs and savings. The level of detail depends on the experience of the person performing the audit or on the specifications of the client paying for the audit.

Level II—Energy Survey and Analysis. This level includes a more detailed building survey and energy analysis. A breakdown of energy use within the building is provided. A Level II energy analysis identifies and provides the savings and cost analysis of all practical measures that meet the owner's constraints and economic criteria, along with a discussion of any effect on operation and maintenance procedures. It also provides a listing of potential capital-intensive improvements that require more thorough data collection and analysis, along with an initial judgment of potential costs and savings. This level of analysis will be adequate for most buildings and measures.

Level III—Detailed Analysis of Capital-Intensive Modifications. This level of analysis focuses on potential capital-intensive projects identified during Level II and involves more detailed field data gathering and engineering analysis. It provides detailed project cost and savings information with a high level of confidence sufficient for major capital investment decisions.

The levels of energy audits do not have sharp boundaries between them. They are general categories for identifying the type of information that can be expected and an indication of the level of confidence in the results; that is, various measures may be subjected to different levels of analysis during an energy analysis of a particular building.

In the complete development of an energy management program, Level II audits should be performed on all facilities, although Level I audits are useful in establishing the program. Figure 4 illustrates Level II energy audit input procedures in which the following data are collected:

- General building data
- Historic energy consumption data
- Energy systems data

The collected data is used to calculate an energy use profile that includes all end-use categories. From the energy use profiles, it is possible to develop and evaluate energy conservation opportunities.

In conducting an energy audit, a thorough systems approach produces the best results. This approach has been described as starting at the end rather than at the beginning. As an example of this approach, consider a factory with steam boilers in constant operation. An expedient (and often cost-effective) approach would be to measure the combustion efficiency of each boiler and to improve boiler efficiency. Beginning at the end would require observing all or most of the end uses of steam in the plant. It is possible that this would result in the discovery of considerable quantities of steam being wasted by venting to the atmosphere, venting through defective steam traps, uninsulated lines, and passing through unused heat exchangers. Elimination of such end-use waste could produce greater savings than those easily and quickly developed by improving boiler efficiency. When using this approach, care must be taken to make cost-effective use of the energy auditor's time. It may not be cost-effective to track down every end use.

When conducting an energy audit, it is important to become familiar with operating and maintenance procedures and personnel. The energy manager can then recommend, through the appropriate departmental channels, energy-saving operating and maintenance procedures. The energy manager should determine, through continued personal observation, the effectiveness of the recommendations.

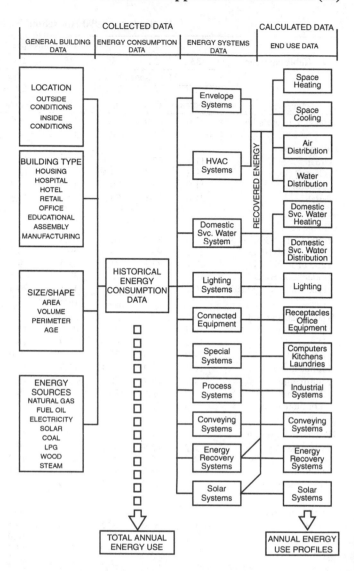

Fig. 4   Energy Audit Input Procedures

Stewart et al. (1984) tabulated 139 different energy audit input procedures and forms for 10 different building types, each using 62 factors. They discuss features of selected audit forms that can help in developing or obtaining an audit procedure.

To calculate the energy cost avoidance of various energy conservation opportunities, it is helpful to develop an energy cost distribution chart similar to that shown for a hospital in Figure 5. Preliminary information of this nature can be developed from monthly utility data by calculating end-use energy profiles (Shehadi et al. 1984).

Analysis of electrical operating costs starts with the recording of data from the bills on a form similar to that in Table 7. By dividing the consumption by the days between readings, the average daily consumption can be calculated. This consumption should be plotted to detect errors in meter readings or reading dates and to detect consumption variances (see Figure 6). For this example, 312 kWh/day is chosen as the "base electrical consumption" to cover year-round electrical needs such as lighting, business machines, domestic hot water, terminal reheat, security, and safety lighting. At this point, consumptions or spans that appear to be in error should be reexamined and corrected as necessary. If the reading date for the 10Nov bill in Table 7 was 05Nov, the curve in Figure 6 would be more continuous. On the basis of a 05Nov reading, the minimum daily consumption of 312 kWh on the continuous curve (Figure 6) occurred in the February billing.

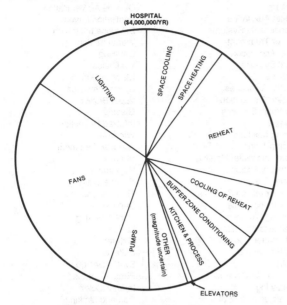

Fig. 5 Energy Cost Distribution

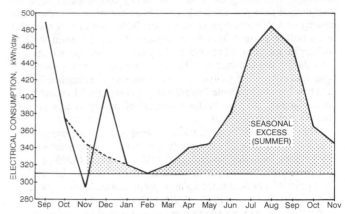

Fig. 6 Average Daily Electrical Consumption

$$ELF = \frac{\text{Base consumption}}{\text{Base demand} \times 24} = \frac{312}{792} = 0.394 \text{ or } 39.4\%$$

In this example, the electrical load factor (39.4%) is higher than the occupancy factor (29.8%).

$$\frac{\text{Occupancy}}{\text{factor}} = \frac{\text{Occupied hours}}{24 \text{ h} \times 7 \text{ days}} = \frac{50}{168} = 0.298 = 29.8\%$$

One reason for this difference may be that the lights are left energized beyond the occupied hours.

Because of the air-conditioning demand, summer demand is 46.8 kW. If this additional demand of 13.8 kW had operated 24 h each day, the summer extra would equal 13.8 × 24 or 331.2 kWh/day. The ratio of each summer month's excess as a percent of 331.2 kWh yields the summer electrical load factor. Summer ELFs higher than the occupancy factor indicate that air conditioning is not shut off as early as possible in the evening. Winter excess demand and consumption are analyzed in the same way to yield winter monthly ELFs.

Base load energy use is the amount of energy consumed independent of weather. When a building has electric cooling and no electric heating, the base load energy use is normally the energy consumed during the winter months. The opposite is true for heating. The annual estimated base load energy consumption can be obtained by establishing the average monthly consumption during the non-heating or noncooling months and multiplying by 12. For many

To start the analysis, the monthly base consumption is calculated (base daily consumption times billing days) and is subtracted from each monthly total to obtain the difference. These differences fall under either summer excess or winter excess, depending on the season. Excess consumption in the summer is primarily due to the air-conditioning load. A similar analysis is made of the actual monthly demand. In Table 7, the base demand is 33.0 kW, and it is usually found in the same or adjacent months as the month with the base consumption. Substantial errors may arise if missing bills are not accounted for.

The base consumption can be further analyzed by calculating the electrical load factor (ELF) associated with this consumption. If the base demand had operated 24 h per day, then base consumption would be

$$33.0 \text{ kW} \times 24 \text{ h/day} = 792 \text{ kWh/day}$$

But if the daily base consumption is 312 kWh/day, then the electrical load factor is

Table 7 Example of Analytical Method for Analyzing Electrical Operating Costs

| Billing Date | Billing Days | Consumption, kWh | | | Air Conditioning, kWh | | | Demand, kW | | |
		Total Actual	Actual per Day	Base[b]	Difference	ELF, %	Excess	Actual	Winter Excess	Summer Excess
12Aug								46.2		0.6
11Sep	30	14 700	490	9 360	5 340	53.7	1 859	46.8		1.2
10Oct	29	10 860	374.5	9 048	1 812	18.9		46.8		1.2
10Nov	31	9 120	294.2	9 672	1 008[a]	11.7		45.6		0.0
06Dec	26	10 680	410.8	8 112	1 008[a]			33.0[c]	0.0	
09Jan	34	10 860	319.4	10 608	252			33.0	0.0	
13Feb	35	10 920	312[b]	10 920	0			33.6	0.6	
11Mar	27	8 700	322.2	8 424	276			33.0	0.0	
10Apr	30	10 140	338	9 360	780			33.6	0.6	
12May	32	11 020	344.4	9 984	1 036	9.8		45.6[d]		0.0
12Jun	31	11 760	379.4	9 672	2 088	20.3		46.8		1.2
13Jul	31	14 160	456.8	9 672	4 488	43.7	893	46.8		1.2
11Aug	29	14 340	494.5	9 048	5 292	55.2	1 937	46.8		1.2
10Sep	30	13 740	458	9 360	4 380	44.1	904	46.2		0.6
14Oct	34	12 120	356.5	10 608	1 512	13.4		45.6		0.0
10Nov	27	9 360	346.7	8 424	936			33.6	0.6	

[a]Estimated from corrected monthly consumption from Figure 6.
[b]Base electrical consumption = 312 kWh/day.
[c]Base winter demand = 33 kW.
[d]Base summer demand = 45.6 kW. Base summer excess demand = (45.6 − 33) = 12.6.

buildings, subtracting the base load energy consumption from the total annual energy consumption yields an accurate estimate of the heating or cooling energy consumption. This approach is not valid when building use differs from summer to winter; when there is cooling in operation 12 months of the year; or when space heating is used during the summer months, as for reheat. In many cases, the base load energy use analysis can be improved by using hourly load data that may be available from the utility company. ELFs and occupancy factors can also be used instead of hourly energy profiles (Haberl and Komor 1990).

Although it is difficult to relate heating and cooling energy used in commercial buildings directly to the severity of the weather, several authors, including Shehadi et al. (1984) and Fels (1986), suggest that this is possible using a curve-fitting method to calculate the balance point of a building (the balance point of a building is discussed in Chapter 28 of the 1993 *ASHRAE Handbook—Fundamentals*). The pitfalls of such an analysis are (1) estimated rather than actual utility usage data are used, (2) the actual dates of the metered information must be used, together with average billing period weather data, and (3) building use and/or operation are not regular.

A more detailed breakdown of energy usage requires that some metered data be collected on a daily basis (winter days versus summer days, weekdays versus weekends) and that some hourly information be collected to develop profiles for night (unoccupied), morning warmup, day (occupied), and the shutdown period. This discussion presumes that submetering is not installed within the building. Individual metering of the various energy end uses provides the energy manager with the information to apply energy management principles optimally. Ideally, each energy end use in Figure 4 would be metered separately. See also Chapter 37, Building Energy Monitoring.

Energy Conservation Opportunities

It is possible to quantitatively evaluate various energy conservation opportunities (ECOs) from end-use energy profiles. Important considerations in this process are as follows:

- System interaction
- Utility rate structure
- Payback
- Installation requirements
- Life of the measure
- Maintainability
- Impact on building operation and appearance

Accurate energy savings calculations can be made only if system interaction is fully understood and allowed for. Annual simulation models may be necessary to accurately estimate the interactions between various ECOs. The calculated remaining energy use should be verified against a separately calculated zero-based energy target.

Further, the actual energy cost avoidance may not be proportional to the energy saved, depending on the method of billing for energy used. Using average costs per unit of energy in calculating the energy cost avoidance of a particular measure is likely to result in incorrect values.

Figure 7 is an example of a list of potential ECOs. The references also contain such lists. A discussion of 118 ECOs can be found in PNL (1990).

In addition, previously implemented energy conservation measures should be evaluated—first, to ensure that they have remained effective, and second, to consider revising them to reflect changes in technology, building use, and/or energy cost.

Prioritize Resources

Once a list of ECOs is established, it should be evaluated, prioritized, and implemented. In establishing priorities, the capital cost, cost-effectiveness, and resources available must be considered.

Boilers	Outside Air Ventilation
Boiler Auxiliaries	Ventilation Layout
Condensate Systems	Envelope Infiltration
Water Treatment	Weatherstripping
Fuel Acquisition	Caulking
Fuel Systems	Vestibules
Chillers	Elevator Shafts
Chiller Auxiliaries	Space Insulation
Steam Distribution	Vapor Barrier
Hydronic Systems	Glazing
Pumps	Infrared Reflection
Piping Insulation	Windows
Steam Traps	Window Treatment
Domestic Water Heating	Shading
Lavatory Fixtures	Vegetation
Water Coolers	Trombe Walls
Fire Protection Systems	Thermal Shutters
Swimming Pools	Surface Color
Cooling Towers	Roof Covering
Condensing Units	Lamps
City Water Cooling	Fixtures
Air Handling Units	Ballasts
Coils	Switch Design
Outside Air Control	Photo Controls
Balancing	Interior Color
Air Volume Control	Demand Limiting
Shutdown	Current Leakage
Air Purging	Power Factor
Minimizing Reheat	Transformers
Air Heat Recovery	Power Distribution
Filters	Cooking Practices
Dampers	Hoods
Humidification	Refrigeration
Duct Resistance	Dishwashing
System Air Leakage	Laundry
Diffusers	Vending Machines
System Interaction	Chiller Heat Recovery
System Reconfiguration	Heat Storage
Space Segregation	Time-of-Day Rates
Equipment Relocation	Computer Controls
Fan-Coil Units	Cogeneration
Heat Pumps	Active Solar Systems
Radiators	Staff Training
System Infiltration	Occupant Indoctrination
Relief Air	Documentation
Space Heaters	Management Structure
Controls	Financial Practices
Thermostats	Building Geometry
Setback	Space Planning
Instrumentation	

Fig. 7 Potential Energy Conservation Opportunities

Factors involved in evaluating the desirability of a particular energy conservation retrofit measure are as follows:

- Rate of return (simple payback, life-cycle cost)
- Total savings (energy, cost avoidance)
- Initial cost (required investment)
- Other benefits (safety, comfort, improved system reliability, and improved productivity)
- Liabilities (increased maintenance costs and potential obsolescence)
- Risk of failure (confidence in predicted savings, rate of increase in energy costs, maintenance complications, and success of others with the same measures)

To reduce the risk of failure, documented performance of ECOs in similar situations should be obtained and evaluated. One common problem is that energy consumption for individual end uses is overestimated, and the predicted savings are not achieved. When doubt exists about the energy consumption, temporary measurements should be made and evaluated. Also, some owners are reluctant to implement ECOs because of past experiences with energy projects.

The reasons for past failures should be analyzed carefully to minimize the possibility of their reoccurrence.

The resources available to accomplish an energy conservation retrofit measure should include the following:

- Management attention, commitment, and follow-through
- Skills
- Manpower
- Investment capital

EMCs may be financed with the following:

- Profit/investment
- Loans
- Rearranging budget priorities
- Energy savings
- Shared savings plans with outside firms and investors
- Utility incentives
- Grant programs
- Tax avoidance
- Donations

When all of these considerations have been weighed and a prioritized list of recommendations is developed, a report should be prepared for management. Each recommendation should include the following:

- Present condition of the system or equipment to be modified
- Recommended action
- Who should accomplish the action
- Necessary documentation or follow-up required
- Potential interferences to successful completion of the recommendation
- Staff effort required
- Risk of failure
- Interactions with other end uses and ECOs
- Economic analysis (including payback, investment cost, and estimated savings figures) using corporate economic evaluation criteria
- Schedule for implementation

The energy manager must be prepared to sell the plans to upper management. Energy conservation measures must generally be financially justified if they are to be adopted. Every organization has limited funds available and must use these funds in the most effective way. The energy manager is competing with others in the organization for the same funds. A successful plan must be presented in a form that is easily understood by the decision makers. Finally, the energy manager must present benefits other than the financial ones, such as improved quality of product or the possibility of postponing other expenditures.

Accomplish Measures

Following approval by management, the energy manager directs the completion of energy conservation retrofit measures selected from the above prioritized list. Certain measures require that an architect or engineer prepare plans and specifications for the retrofit work. The package of services required usually includes drawings, specifications, assistance in obtaining competitive bids, evaluation of the bids, selection of the best bid, construction observation, final check-out, and assistance in training personnel in the proper application of the revisions.

Maintain Measures

Once energy conservation measures have begun, it is necessary to establish procedures to record, on a frequent and regular basis, energy consumption and costs for each building and/or end-use category in a manner consistent with functional cost accountability. Additional metering may be needed to monitor the energy consumption accurately. Metering can be in the form of devices that automat-

ically read and transmit data to a central location or in the form of less expensive metering devices that require regular readings by building maintenance and/or security personnel. Many energy managers find it beneficial to collect energy consumption information hourly. Data may also be obtained from the utility. However, if the energy manager is not able to evaluate data as frequently as it is collected, it may be more practical to collect data less frequently. The energy manager should review data while it is current and take immediate action if profiles indicate a trend in the wrong direction. Such trends could be caused by uncalibrated control systems, changes in operating practices, or failures of mechanical systems, all problems that should be isolated and corrected as soon as possible.

Energy Accounting

The energy manager continues the meter reading, monitoring, and tabulation of facility energy use and profiles. These tabulations indicate the cost of energy management efforts and the resulting energy cost avoidance. In conjunction with this effort, the energy manager periodically reviews pertinent utility rates, rate structures, and their trends as they affect the facility. Many utilities maintain free mailing lists for changes in their rate tariffs. The energy manager provides periodic reports of the energy management efforts to top management, summarizing the work accomplished, the cost-effectiveness of the work, the plans and suggested budget for future work, and projections of future utility costs. If energy conservation measures are to retain their cost-effectiveness, continued monitoring and periodic reauditing are necessary, because many energy conservation measures become less effective if they are not carefully monitored and maintained.

BUILDING EMERGENCY ENERGY USE REDUCTION

The need for occasional reductions in energy use during specific periods has become more common due to rising energy costs and supply reductions (which may be voluntary or mandatory) or equipment failures. Emergency periods include short-term shortages of a particular energy source(s) brought about by natural disasters, extreme weather conditions, utility system equipment disruptions, labor strikes, failures in building systems or equipment, self-imposed cut-backs in energy use, or world political activities or other forces beyond the control of the building owner and operator. This section provides information to help building owners and operators maintain near normal operation of facilities during energy emergencies.

The following terms are applicable to such programs:

Energy Emergency. A period in which energy supply reductions and/or climatic and natural forces or equipment failures preclude the normal operation of a building and necessitate a reduction in building energy use.

Level of Energy Emergency. A measure of the severity of the emergency that calls for the implementation of various energy use reduction measures. Typical levels include the following:

1. Green (Normal)—All occupant or building functions maintained and systems operating under the normal operating circumstances for the building
2. Blue—All or most of the occupant or building functions maintained, with reductions in building system output that may result in borderline occupancy comfort
3. Yellow—A minor reduction of occupant or building functions with reductions in output
4. Red—A major reduction of occupant or building functions and systems that barely sustains building occupancy capability
5. Black—The orderly shutdown of systems and occupancy that maintains the minimum building conditions needed to protect the building and its systems

Implementation of Energy Reductions

Each building owner, lessor, and operator should use the energy team approach and identify an individual with the necessary authority to review and fit recommendations into a plan for the particular building. For each class of emergency, the responsible party recommends a specific plan to reduce building energy use that still maintains the best building environment under the given circumstances. Implementation of the particular recommendations should then be coordinated through the building operator with assistance from the responsible party and the building occupants, as necessary. The plan should be tested occasionally.

Depending on the type of building, its use, the energy source(s) for each function, and local conditions such as climate and availability of other similar buildings, the following steps should be taken in developing a building energy plan:

1. Develop a list of measures that are applicable to the building.
2. Estimate the amount and type of energy savings for each measure and appropriate combination of measures (e.g., account for air-conditioning savings resulting from reduced lighting and other internal loads). Tabulate demand and usage savings separately for response to different types of emergencies.
3. For the various levels of energy emergency, develop a plan that would maintain the best building environment under the circumstances. Include both short- and long-term measures in the plan. Operational changes may be implemented quickly and prove adequate for short-term emergencies.
4. Experiment with the plan developed, record energy consumption and demand reduction data, and revise the plan as necessary. Much of the experimentation may be done on weekends to minimize disruptive effects.
5. Meet with the local utility company(s) to review the plan.

In the emergency planning process, some measures can be implemented permanently. Depending on the level of energy emergency and the building priority, the following actions may be considered in developing the plan for emergency energy reduction in the building:

- Change operating hours
- Move personnel into other building areas (consolidation)
- Shut off nonessential equipment

Thermal Envelope

- Use all existing blinds, draperies, etc., during summer
- Install interior window insulation
- Caulk and seal around unused exterior doors and windows
- Install solar shading devices in summer
- Seal all unused vents and ducts to outside

Heating, Ventilating, and Air-Conditioning Systems and Equipment

- Modify controls or control set points to raise and lower temperature and humidity as necessary
- Shut off or isolate all nonessential equipment
- Tune up equipment
- Lower thermostat set points in winter
- Raise chilled water temperature
- Lower hot water temperature (*Note*: Keep hot water temperature higher than 63°C if a gas boiler is used.)
- Reduce the level of reheat or eliminate it in winter
- Reduce or eliminate ventilation and exhaust airflow
- Raise thermostat set points in summer
- Reduce the amount of recooling in summer

Lighting Systems

- Remove lamps or reduce lamp wattage
- Use task lighting, where appropriate
- Move building functions to exterior or daylight areas
- Turn off electric lights in areas with adequate natural light
- Lower luminaire height, where appropriate

- Wash all lamps and luminaires
- Replace fluorescent ballasts with high-efficiency or multilevel ballasts
- Revise building cleaning and security procedures to minimize lighting periods
- Consolidate parking and turn off unused parking security lighting

Special Equipment

- Take transformers off-line during periods of nonuse
- Shut off or regulate the use of vertical transportation systems
- Shut off unused or unnecessary equipment, such as photocopying equipment, music systems, typewriters, and computers
- Reduce or turn off hot water supply

Building Operation Demand Reduction

- Sequence or interlock heating or air-conditioning systems
- Disconnect or turn off all nonessential loads
- Turn off some lights
- Preheat or precool prior to the emergency period

REFERENCES

ASHRAE. 1984. Standard methods of measuring and expressing building energy performance. *Standard* 105-1984 (RA 90).

ASHRAE. 1989. Guideline for commissioning of HVAC systems. *Guideline* 1-1989.

ASHRAE. 1989. Development of a guide for analyzing and reporting building characteristics and energy use in commercial buildings. *Special Project Report* 56.

DOE/EIA. 1988. Nonresidential buildings energy consumption survey: Characteristics of commercial buildings 1986. DOE/EIA-0246(86).

DOE/EIA. 1989. Nonresidential buildings energy consumption survey: Commercial buildings consumption and expenditures 1986. DOE/EIA-0318 (86).

DOE/EIA. 1991. Nonresidential buildings energy consumption survey: Characteristics of commercial buildings 1989. DOE/EIA-0246(89).

DOE/EIA. 1992. Nonresidential buildings energy consumption survey: Commercial buildings consumption and expenditures 1989. DOE/EIA-0318(89).

DOE/EIA. 1993. Household energy consumption and expenditures 1990. Part 1: National Data. DOE/EIA-0321/1(90).

DOE/EIA. 1994. Nonresidential buildings energy consumption survey: Characteristics of commercial buildings 1992. DOE/EIA-0246(92).

Fels. M. 1986. *Energy and buildings*, Vol. 9, Nos. 1 and 2, Special issue devoted to the Princeton Scorekeeping Method (PRISM).

Haberl, J. and P. Komor. 1990. Improving energy audits: How annual, monthly, daily and hourly data can help. *ASHRAE Journal* (August and September).

MacDonald, J.M. and D.M. Wasserman. 1989. Investigation of metered data analysis methods for commercial and related buildings. ORNL/CON-279. Oak Ridge National Laboratories, Oak Ridge, TN.

PNL (Pacific Northwest Laboratories). 1990. Architect's and engineer's guide to energy conservation in existing buildings. Vol. 2, Chapter 1. DOE/RL/01830 P-H4.

Shehadi, M.T., J.D. Cowan, I.A. Jarvis, L.G. Spielvogel, and D.R. Wul-finghoff. 1984. Energy use evaluation. Symposium AT84-8. *ASHRAE Transactions* 90(1B):401-50.

Stewart, R., S. Stewart, and R. Joy. 1984. Energy audit input procedures and forms. *ASHRAE Transactions* 90(1A):350-62.

Turner, W.C. 1993. *Energy management handbook* p.11. John C. Wiley and Sons, New York.

BIBLIOGRAPHY

Energy for the year 2000. 1993. National Electrical Contractors Association, Bethesda, MD.

Freund, J.K. 1980. Selling energy management to an owner's engineer's management. *Heating, Piping and Air-Conditioning* (September).

Guide to a successful project energy conservation & management, model competitive procurement procedure and professional selection of professional engineers. NSPE-PEPP, Alexandria, VA.

Landsberg, D. and R. Stewart. 1980. *Improving energy efficiency in buildings*. State University of New York Press, Albany, NY.

CHAPTER 33

OWNING AND OPERATING COSTS

OWNING and operating cost information for the HVAC system should be part of the investment plan of a facility. This information can be used for preparing annual budgets, managing assets, and selecting design options. Table 1 shows a representative form that summarizes these costs.

A properly engineered system must also be economical. Economics are difficult to assess because of the complexities surrounding the effective management of money and the inherent difficulty of predicting future operating and maintenance expenses. Complex tax structures and the time value of money can affect the final engineering decision. This does not imply the use of either the cheapest or the most expensive system; instead, it demands an intelligent analysis of the financial objectives and requirements of the owner.

Certain tangible and intangible costs or benefits must also be considered when assessing owning and operating costs. Local codes may require highly skilled or certified operators for specific types of equipment. This could be a significant cost over the life of the system. Similarly, such intangible items as aesthetics, acoustics, comfort, safety, security, flexibility, and environmental impact may be important to a particular building or facility.

OWNING COSTS

The following elements must be established to calculate annual owning costs: (1) initial cost, (2) analysis period (study period), (3) interest or discount rate, and (4) other periodic costs such as insurance, property taxes, refurbishment, or disposal fees. Once established, these elements are coupled with operating costs to develop an economic analysis, which may be a simple payback evaluation or an in-depth analysis such as outlined in the section on Economic Analysis Techniques.

Initial Cost of System

Major decisions affecting annual owning and operating costs for the life of the building must generally be made prior to the completion of contract drawings and specifications. To achieve the best performance and economics, comparisons between alternate methods of solving the engineering problems peculiar to each project must be made in the early stages of architectural design. Oversimplified estimates can lead to substantial errors in evaluating the system.

A thorough understanding of the installation costs and accessory requirements must be established. Detailed lists of materials, controls, space and structural requirements, services, installation labor, and so forth can be prepared to increase the accuracy in preliminary cost estimates. A reasonable estimate of the cost of components may be derived from cost records of recent installations of comparable design or from quotations submitted by manufacturers and contractors. Table 2 shows a representative checklist for initial cost.

Analysis Period

The time frame (analysis period or study period) over which an economic analysis is performed has a major impact on results of the analysis. The analysis period is usually determined by specific analysis objectives, such as length of planned ownership or loan repayment period. The chosen analysis period is often unrelated to depreciation period or equipment service life, although these factors may be important in the analysis.

Table 3 lists representative estimates of various system component service lives. Service life as used here is the time during which a particular system or component remains in its original service application. Replacement may be for any reason, including, but not limited to, failure, general obsolescence, reduced reliability, excessive maintenance cost, and changed system requirements due to such influences as building characteristics or energy prices.

Depreciation periods are usually set by federal, state, or local tax law, which change periodically. Applicable tax laws should be consulted for more information on depreciation.

Interest or Discount Rate

Most major economic analyses consider the cost of borrowing money (or the value of alternative uses of money), inflation, and the time value of money. *Opportunity cost* of money reflects the earnings that investing (or loaning) the money can produce. *Time value* of money reflects the fact that money received today is more useful than the same amount received a year from now, even with zero inflation, because the money is available for reinvestment earlier. *Inflation* (price escalation) decreases the purchasing or investing power (value) of future money because it can buy less (in the future).

The cost or value of money must also be considered. When borrowing money, a percentage fee or interest rate must normally be paid. However, the interest rate may not necessarily be the correct cost of money to use in an economic analysis. Another factor, called the discount rate, is more commonly used to reflect the true cost of money. Discount rates used for analyses vary depending on individual investment, profit, and other opportunities. Interest rates, in contrast, tend to be more centrally fixed by lending institutions.

The vague definition and variable nature of discount rate often causes confusion. To avoid this confusion, the U.S. government has specified particular discount rates that can be used in economic analyses relating to federal expenditures. These discount rates are updated annually (Lippiatt 1994) but may not be appropriate for private sector economic analyses.

Other Periodic Costs

Regularly or periodically recurring costs apply to varying situations. Examples are insurance, property taxes, income taxes, refurbishment expenses, disposal fees (e.g., refrigerant recycling costs), occasional major repair costs, and decommissioning expenses.

Insurance. Insurance reimburses a property owner for a financial loss so that equipment can be repaired or replaced. Insurance

The preparation of this chapter is assigned to TC 1.8, Owning and Operating Costs.

Table 1   Owning and Operating Cost Data and Summary

OWNING COSTS

I.   Initial Cost of System
- A.   Equipment (see Table 2 for items included)   _____
- B.   Control systems—Complete   _____
- C.   Wiring and piping costs attributable to system   _____
- D.   Any increase in building construction cost attributable to system   _____
- E.   Any decrease in building construction cost attributable to system   _____
- F.   Installation costs   _____

   TOTAL INITIAL COST   _____

II.   Annual Fixed Charges
- A.   Equivalent uniform annual cost   _____
- B.   Income taxes   _____
- C.   Property taxes   _____
- D.   Insurance   _____
- E.   Rent   _____

   TOTAL ANNUAL FIXED CHARGES   _____

OPERATING COSTS

III.   Annual Maintenance Allowances
- A.   Replacement or servicing of oil, air, or water filters   _____
- B.   Contracted maintenance service   _____
- C.   Lubricating oil and grease   _____
- D.   General housekeeping cost   _____
- E.   Replacement of worn parts (labor and material)   _____
- F.   Refrigerant   _____

   TOTAL ANNUAL MAINTENANCE ALLOWANCE   _____

IV.   Annual Energy, Fuel, and Water Costs
- A.   Electric Energy Costs
  - 1.   Chiller or compressor   _____
  - 2.   Pumps
    - a)   Chilled water   _____
    - b)   Heating water   _____
    - c)   Condenser or tower water   _____
    - d)   Potable hot water recirculation   _____
    - e)   Well water   _____
    - f)   Boiler auxiliaries (including fuel oil heaters)   _____
  - 3.   Fans
    - a)   Condenser or tower   _____
    - b)   Inside air handling   _____
    - c)   Exhaust   _____
    - d)   Makeup air   _____
    - e)   Boiler auxiliaries and equipment room ventilation   _____

IV.   Annual Energy, Fuel and Water Costs (*continued*)
-    4.   Resistance heaters (primary or supplementary)   _____
-    5.   Heat pump   _____
-    6.   Domestic water heating   _____
-    7.   Lighting   _____
-    8.   Cooking and food service equipment   _____
-    9.   Miscellaneous (e.g., elevators, escalators, and computers)   _____
- B.   Gas, Oil, Coal, or Purchased Steam Costs
  - 1.   On-site generation of the electrical power requirements under A of this section   _____
  - 2.   Heating
    - a)   Direct heating   _____
    - b)   Ventilation
      - (1)   Preheaters   _____
      - (2)   Reheaters   _____
    - c)   Supplementary heating (i.e., oil preheating)   _____
    - d)   Other   _____
  - 3.   Domestic water heating   _____
  - 4.   Cooking and food service equipment   _____
  - 5.   Air conditioning
    - a)   Absorption   _____
    - b)   Chiller or compressor
      - (1)   Gas/Diesel engine driven   _____
      - (2)   Gas turbine driven   _____
      - (3)   Steam turbine driven   _____
  - 6.   Miscellaneous   _____
- C.   Water
  - 1.   Condenser makeup water   _____
  - 2.   Sewer charges   _____
  - 3.   Chemicals   _____
  - 4.   Miscellaneous   _____

   TOTAL ANNUAL FUEL, ENERGY, AND WATER COSTS

V.   **ANNUAL WAGES OF ENGINEERS AND OPERATORS**   _____

SUMMARY

II.   Total Annual Fixed Charges   _____
III.   Total Annual Maintenance Allowance   _____
IV.   Total Annual Energy, Fuel, and Water Costs   _____
V.   Annual Wages of Engineers and Operators   _____

   TOTAL ANNUAL OWNING AND OPERATING COSTS   _____

often indemnifies the owner from liability as well. Financial recovery may also include replacing the loss of income, rents, or profits resulting from property damage.

Some of the principal factors that influence the total annual insurance premium are building size, construction materials, amount and size of mechanical equipment, geographic location, and policy deductibles. Some regulations set minimum required insurance coverages and premiums that may be charged for various forms of insurable property.

Property Taxes. Property taxes differ widely and may be collected by one or more agencies, such as state, county, or local governments or special assessment districts. Furthermore, property taxes may apply to both real (land, buildings) and personal (everything else) property. Property taxes are most often calculated as a percentage of assessed value but are also determined in other ways,

such as fixed fees, license fees, registration fees, etc. Moreover, definitions of assessed value vary widely in different geographic areas. Tax experts should be consulted for applicable practices in a given area.

Income Taxes. Taxes are generally imposed in proportion to net income, after allowance for expenses, depreciation, and numerous other factors. Special tax treatment is often granted to encourage certain investments. Income tax experts can provide up-to-date information on income tax treatments.

Additional Periodic Costs. Examples of additional costs include changes in regulations that require unscheduled equipment refurbishment to eliminate use of hazardous substances, and disposal costs for such substances. Moreover, at the end of the equipment's useful life there may be negative salvage value (i.e., removal, disposal, or decommissioning costs).

Table 2 Initial Cost Checklist

Energy and Fuel Service Costs

Fuel service, storage, handling, piping, and distribution costs

Electrical service entrance and distribution equipment costs

Total energy plant

Heat-Producing Equipment

Boilers and furnaces

Steam-water converters

Heat pumps or resistance heaters

Make-up air heaters

Heat-producing equipment auxiliaries

Refrigeration Equipment

Compressors, chillers, or absorption units

Cooling towers, condensers, well water supplies

Refrigeration equipment auxiliaries

Heat Distribution Equipment

Pumps, reducing valves, piping, piping insulation, etc.

Terminal units or devices

Cooling Distribution Equipment

Pumps, piping, piping insulation, condensate drains, etc.

Terminal units, mixing boxes, diffusers, grilles, etc.

Air Treatment and Distribution Equipment

Air heaters, humidifiers, dehumidifiers, filters, etc.

Fans, ducts, duct insulation, dampers, etc.

Exhaust and return systems

System and Controls Automation

Terminal or zone controls

System program control

Alarms and indicator system

Building Construction and Alteration

Mechanical and electric space

Chimneys and flues

Building insulation

Solar radiation controls

Acoustical and vibration treatment

Distribution shafts, machinery foundations, furring

OPERATING COSTS

Operating costs result from the actual operation of the system. They include fuel and electrical costs, wages, supplies, water, material, and maintenance parts and services. Chapter 28 of the 1993 *ASHRAE Handbook—Fundamentals* outlines how fuel and electrical requirements are estimated. Note that total energy consumption cannot generally be multiplied by a per unit energy cost to arrive at annual utility cost.

Electrical Energy

The total cost of electrical energy is usually a combination of several components: energy consumption charges, fuel adjustment charges, special allowances or other adjustments, and demand charges.

Energy Consumption Charges. Most utility rates have step rate schedules for consumption, and the cost of the last unit of energy consumed may be substantially different from that of the first. The last unit may be cheaper than the first because the fixed costs to the utility may already have been recovered from earlier consumption costs. Alternatively, the last unit of energy may be sold at a higher rate to encourage conservation.

To reflect time-varying operating costs, some utilities charge different rates for consumption according to the time of use and season; typically, costs rise toward the peak period of use. This may justify the cost of shifting the load to off-peak periods.

Fuel Adjustment Charge. Due to substantial variations in fuel prices, electric utilities may apply a fuel adjustment charge to recover costs. This adjustment may not be reflected in the rate schedule. The fuel adjustment is usually a charge per unit of energy and may be positive or negative depending on how much of the actual fuel cost is recovered in the energy consumption rate.

Power plants with multiple generating units that use different fuels typically have the greatest effect on this charge (especially during peak periods, when more expensive units must be brought on-line). Although this fuel adjustment charge can vary monthly, the utility should be able to estimate an average annual or seasonal fuel adjustment for calculations.

Allowances or Adjustments. Special allowances may be available for customers who can receive power at higher voltages or for those who own transformers or similar equipment. Special rates may be available for specific interruptible loads such as domestic water heaters.

Certain facility electrical systems may produce a low power factor, which means that the utility must supply more current on an intermittent basis, thus increasing their costs. These costs may be passed on as an adjustment to the utility bill if the power factor is below a level established by the utility. The power factor is the ratio of active (real) kilowatt power to apparent (reactive) kVA power.

When calculating power bills, utilities should be asked to provide detailed cost estimates for various consumption levels. The final calculation should include any applicable special rates, allowances, taxes, and fuel adjustment charges.

Demand Charges. Electric rates may also have demand charges based on the customer's peak kilowatt demand. While consumption charges typically cover the operating costs of the utility, demand charges typically cover the owning costs.

Demand charges may be formulated in a variety of ways:

1. Straight charge—cost per kilowatt per month, charged for the peak demand of the month.
2. Excess charge—cost per kilowatt above a base demand (e.g., 50 kW), which may be established each month.
3. Maximum demand (ratchet)—cost per kilowatt for the maximum annual demand, which may be reset only once a year. This established demand may either benefit or penalize the owner.
4. Combination demand—cost per hour of operation of the demand. In addition to a basic demand charge, utilities may include further demand charges as demand-related consumption charges.

The actual level of demand represents the peak energy use averaged over a specific period, usually 15, 30, or 60 min. Accordingly, high electrical loads of only a few minutes' duration may never be recorded at the full instantaneous value. Alternatively, peak demand is recorded as the average of several consecutive short periods (i.e., 5 min out of each hour).

The particular method of demand metering and billing is important when load shedding or shifting devices are considered. The portion of the total bill attributed to demand may vary widely, from 0% to as high as 70%.

Natural Gas

Rates. Conventional natural gas rates are usually a combination of two main components: (1) utility rate for gas consumption and (2) purchased gas adjustment (PGA) charges.

Although gas is usually metered by volume, it is often sold by energy content (therm). The utility rate is the amount the local

Table 3 Estimates of Service Lives of Various System Components[a]

Equipment Item	Median Years	Equipment Item	Median Years	Equipment Item	Median Years
Air conditioners		Air terminals		Air-cooled condensers	20
Window unit	10	Diffusers, grilles, and registers	27	Evaporative condensers	20
Residential single or split package	15	Induction and fan-coil units	20	Insulation	
Commercial through-the-wall	15	VAV and double-duct boxes	20	Molded	20
Water-cooled package	15	Air washers	17	Blanket	24
Heat pumps		Ductwork	30	Pumps	
Residential air-to-air	15[b]	Dampers	20	Base-mounted	20
Commercial air-to-air	15	Fans		Pipe-mounted	10
Commercial water-to-air	19	Centrifugal	25	Sump and well	10
Roof-top air conditioners		Axial	20	Condensate	15
Single-zone	15	Propeller	15	Reciprocating engines	20
Multizone	15	Ventilating roof-mounted	20	Steam turbines	30
Boilers, hot water (steam)		Coils		Electric motors	18
Steel water-tube	24 (30)	DX, water, or steam	20	Motor starters	17
Steel fire-tube	25 (25)	Electric	15	Electric transformers	30
Cast iron	35 (30)	Heat Exchangers		Controls	
Electric	15	Shell-and-tube	24	Pneumatic	20
Burners	21	Reciprocating compressors	20	Electric	16
Furnaces		Package chillers		Electronic	15
Gas- or oil-fired	18	Reciprocating	20	Valve actuators	
Unit heaters		Centrifugal	23	Hydraulic	15
Gas or electric	13	Absorption	23	Pneumatic	20
Hot water or steam	20	Cooling towers		Self-contained	10
Radiant heaters		Galvanized metal	20		
Electric	10	Wood	20		
Hot water or steam	25	Ceramic	34		

Source: Data obtained from a survey of the United States by ASHRAE Technical Committee TC 1.8 (Akalin 1978).
[a]See Lovvorn and Hiller (1985) and Easton Consultants (1986) for further information.
[b]Data updated by TC 1.8 in 1986.

distribution company charges per unit of energy to deliver the gas to a particular location. This rate may be graduated in steps, so that the first 10 GJ of gas consumed may not be the same price as the last 10 GJ. The PGA is an adjustment for the cost of the gas per unit of energy to the local utility. It is similar to the electric fuel adjustment charge. The total cost per therm is then the sum of the appropriate utility rate and the PGA, plus taxes and other adjustments.

Interruptible Gas Rates and Contract / Transport Gas. Large industrial plants usually have the ability to burn alternate fuels at their facility and can qualify for special interruptible gas rates. During peak periods of severe cold weather, these customers may be curtailed by the gas utility and may have to switch to propane, fuel oil, or some other backup fuel. The utility rate and PGA are usually considerably cheaper for these interruptible customers than they are for firm rate (noninterruptible) customers.

Deregulation of the natural gas industry allows end users to negotiate for gas supplies on the open market. The customer actually contracts with a producer or gas broker and pays for the gas at the source. Then, transport fees must be negotiated with the pipeline companies carrying the gas to the customer's local gas utility. This can be a very complicated administrative process and is usually economically feasible for large gas users only. Some local utilities have special rates for delivering contract gas volumes through their system; others simply charge a standard utility fee (PGA is not applied since the customer has already negotiated with the supplier for the cost of the fuel itself).

When calculating natural gas bills, be sure to determine which utility rate and PGA and/or contract gas price is appropriate for the particular interruptible or firm customer. As with electric bills, the final calculation should include any taxes, prompt payment discounts, or other applicable adjustments.

Other Fuels

Propane, fuel oil, and diesel are examples of other fuels in widespread use today. Calculating the cost of these fuels is usually much easier than calculating typical utility rates. When these fuels are used, other items that can affect owning or operating costs must be considered.

The cost of the fuel itself is usually a simple per unit volume or per unit mass charge. The customer is free to negotiate for the best price. However, trucking or delivery fees must also be included in final calculations. Some customers may have their own transport trucks, while most shop around for the best delivered price. Rental fees for storage tanks must be considered if they are not customer-owned. Periodic replacement of diesel-type fuels may be necessary due to storage or shelf-life limitations and must also be considered. The final fuel cost calculation should include any of these costs that are applicable, as well as the appropriates taxes.

MAINTENANCE COSTS

The quality of maintenance and maintenance supervision can be a major factor in the energy cost of a building. Chapter 35 covers the maintenance, maintainability, and reliability of systems. Dohrmann and Alereza (1986) obtained maintenance costs and HVAC system information from 342 buildings located in 35 states in the United States. In 1983 U.S. dollars, data collected showed a mean HVAC system maintenance cost of $3.40 per square metre per year, with a median cost of $2.60/m² per year. The age of the building has a statistically significant but minor effect on HVAC maintenance costs. When analyzed by geographic location, the data revealed that location does not significantly affect maintenance costs. Analysis also indicated that building size is not statistically significant in explaining cost variation.

The type of maintenance program or service agency that the building management contracts for can also have a significant effect on total HVAC maintenance costs. While extensive or thorough routine and preventive maintenance programs cost more to administer, they usually produce benefits such as extended equipment life, improved reliability, and less system downtime.

Estimating Maintenance Costs

Using data from Table 4, the following method may be used for estimating or comparing the total building HVAC maintenance costs for various equipment combinations. (Manufacturers and other industry sources should be consulted for current information on new types of equipment). Several important limitations of the data presented here should be noted:

- Only selected data from Table 12 of the report by Dohrmann and Alereza (1986) are presented here.
- Data were collected for office buildings only.
- Data measure total HVAC building maintenance costs and do not measure the costs associated with maintaining particular items or individual pieces of equipment.
- The cost data are not intended for use in selecting new HVAC equipment or systems. This information is for equipment and systems already in place and may not be representative of the maintenance costs expected with newer equipment.

This method assumes that the base HVAC system in the building consists of fire-tube boilers for heating equipment, centrifugal chillers for cooling equipment, and VAV distribution systems. The total annual building HVAC maintenance cost for this system is $3.59/m^2$. Adjustment factors from Table 4 are then applied to this base cost to account for building age and variations in type of HVAC equipment as follows:

C = Total annual building HVAC maintenance cost ($/m^2$)
   = Base system maintenance costs
   + (Age adjustment factor) × (age in years n)
   + Heating system adjustment factor h
   + Cooling system adjustment factor c
   + Distribution system adjustment factor d

$C = 3.59 + 0.019 n + h + c + d$

Table 4   Annual HVAC Maintenance Cost Adjustment Factors (in $ per square metre, 1983 U.S. dollars)

Age Adjustment	$0.019 n$
Heating Equipment h	
Water tube boiler	+0.083
Cast iron boiler	+0.101
Electric boiler	−0.287
Heat pump	−1.043
Electric resistance	−1.432
Cooling Equipment c	
Reciprocating chiller	−0.431
Absorption chiller (single stage)[a]	+2.072
Water-source heat pump	−0.508
Distribution System d	
Single zone	+0.892
Multizone	−0.502
Dual duct	−0.031
Constant volume	+0.948
Two-pipe fan coil	−0.298
Four-pipe fan coil	+0.624
Induction	+0.734

[a]These results pertain to buildings with older, single-stage absorption chillers. The data from the survey are not sufficient to draw inferences about the costs of HVAC maintenance in buildings equipped with new absorption chillers.

Example 1. Estimate the total annual building HVAC maintenance costs per square metre for a building that is 10 years old and has an electric boiler, a reciprocating chiller, and a constant volume distribution system.

$$C = 3.59 + 0.019 (10) - 0.287 - 0.431 + 0.948$$

$$C = \$4.01/m^2 \text{ in 1983 dollars}$$

This estimate can be adjusted to current dollars by multiplying the maintenance cost estimate by the current Consumer Price Index (CPI) divided by the CPI in July 1983. In July 1983, the CPI was 100.1. Monthly CPI statistics are recorded in *Survey of Current Business*, a U.S. Department of Commerce publication. This estimating method is limited to one equipment variable per situation. That is, the method can estimate maintenance costs for a building having either a centrifugal chiller or a reciprocating chiller, but not both. Assessing the effects of combining two or more types of equipment within a single category requires a more complex statistical analysis.

Impact of Refrigerant Phaseouts

Due to the phaseout of some refrigerants, many building owners and operators are making decisions regarding the replacement of existing chillers or the conversion to alternative refrigerants. Several factors must be considered when evaluating chiller replacement versus conversion. These include

- Additional maintenance costs to recycle refrigerants.
- Cost to upgrade mechanical equipment rooms to meet ASHRAE *Standard* 15.
- Remaining useful life and the impact of the level of maintenance performed over the life of the chiller.
- Impact of taxes. In many cases, the cost of replacement is considered a capital improvement and is depreciated, while a conversion may be considered a repair or operating expense.
- Utility incentives for improved energy efficiency.

Although Table 4 identifies maintenance cost variations by type of system and age of basic equipment, the impact of conversions or upgrades requires further analysis. The resulting maintenance costs will vary depending on the extent of the work performed on the chiller.

Another impact on the total cost of annual maintenance is the material cost of the refrigerants used. Refrigerant leakage rates for new chillers are generally much less than those of older designs. Note that the impact of chiller conversion versus replacement is most significant in the area of annual energy costs. That is, converting existing systems to operate on alternative refrigerants generally reduces both refrigeration capacity and system efficiency. Original manufacturers can recommend changes to offset some of these losses by re-engineering the machine to operate on the new refrigerant.

Reassessing the original equipment's capacity offers another opportunity to improve system efficiency. Existing buildings may have excess chiller capacity due to changes in the use of the building or from conservative estimates of old design requirements.

New technologies and load management strategies such as thermal storage or absorption chillers should also be considered when evaluating refrigerant phaseout options. This information should be used as a guide for completing Table 1 for each option. The economic analysis techniques described at the end of the chapter may then be applied to compare the relative values of each option.

INNOVATIVE FINANCING ALTERNATIVES

Many financing approaches have been developed since the 1970s to make the economics of advanced energy-conserving equipment and productivity-enhancement devices more attractive. These alternatives generally reduce the initial capital investment or level of risk in return for sharing in the operating cost or productivity-enhancement benefits. Examples include shared savings programs, low interest financing, cost sharing, and leasing programs. All of these innovative owning and operating cost reduction approaches have important tax consequences that should be investigated on a case-by-case basis.

Shared Savings

In this arrangement, the equipment supplier and the purchaser contractually agree to share in the savings or productivity enhancements provided by the equipment. In return the supplier offers the equipment at reduced cost. This approach has been particularly effective at overcoming reservations or uncertainties about future benefits.

Low Interest Financing

In this approach, the supplier offers the equipment with special financing arrangements at below-market interest rates, thereby reducing the annual ownership cost compared to alternatives.

Cost Sharing

Several variants of cost sharing programs exist. In some examples, two or more groups jointly purchase and share new equipment or facilities, thereby increasing the utilization of the equipment and improving the economic benefits for both parties. In other examples, equipment suppliers or independent third parties (such as utilities) who receive an indirect benefit may share part of the equipment cost to establish a market foothold for the product.

Leasing Programs

In this program, equipment suppliers or independent third parties may retain ownership of new equipment and lease it to the user. Leasing results in almost no initial or annual capital and maintenance costs, but somewhat higher annual operating costs than if the equipment was purchased. Leasing also reduces risk in the event of uncertain future benefits.

ECONOMIC ANALYSIS TECHNIQUES

Analysis of overall owning and operating costs and comparisons of alternatives requires an understanding of the cost of lost opportunities, inflation, and the time value of money. This process of economic analysis, which considers all cost factors, is sometimes called life-cycle cost analysis.

Simplified Economic Analysis

Various levels of sophistication can be applied to economic analysis, depending on the desired accuracy of the results. One commonly used technique for evaluating systems relative to their projected savings is the *simple years to payback* approach. This approach is particularly useful for quick screening of alternatives. More detailed analysis, such as *present value* (also termed current value, current worth, or present worth) or *uniform annualized cost* analysis is usually warranted for the most attractive alternatives or when more complex time-dependent variations in maintenance requirements and other effects must be considered.

Simple Years to Payback. In the simple payback technique, a projection of the revenue stream, cost savings, and so forth, is estimated and compared to the magnitude of the initial capital outlay. This simple technique ignores the cost of borrowing money (interest) and lost opportunity costs. It also ignores inflation and the time value of money.

Example 2. Equipment item 1 costs $10 000 and will save $2000 per year in operating costs, while equipment item 2 costs $12 000 and saves $3000 per year. Which item has the best simple payback?

Item 1 — $10 000/($2000/yr) = 5 year simple payback

Item 2 — $12 000/($3000/yr) = 4 year simple payback

Because the analysis of equipment for the duration of its realistic life can produce a much different result, the simple years to payback technique should be used with caution.

Present Value (Present Worth) Analysis

All sophisticated economic analysis methods use the basic principles of present value analysis to account for the time value of money. Therefore, a good understanding of these principles is important.

Single Payment Present Value Analysis. The cost or value of money is a function of the available interest rate and inflation rate. The future value F of a present sum of money P over n periods with compound interest rate i per period is

$$F = P(1+i)^n \tag{1}$$

Conversely, the present value or present worth P of a future sum of money F is given by

$$P = F/(1+i)^n \tag{2}$$

or

$$P = F \times \text{PWF}(i,n)_{sgl} \tag{3}$$

where the single payment present worth factor $\text{PWF}(i,n)_{sgl}$ is defined as

$$\text{PWF}(i,n)_{sgl} = 1/(1+i)^n \tag{4}$$

Example 3. Calculate the future value of a system presently valued at $10 000 in 10 years at 10% per year interest.

$$F = P(1+i)^n = \$10\,000\,(1+0.1)^{10} = \$25\,937.42$$

Example 4. Using the present worth factor for 10% per year interest and an analysis period of 10 years, calculate the present value of a future sum of money valued at $10 000. (Another way of stating this problem is to determine what sum of money must be invested today at 10% per year interest to yield an amount of $10 000 ten years from now.)

$$P = F \times \text{PWF}(i,n)_{sgl}$$

$$P = \$10\,000 \times 1/(1+0.1)^{10}$$

$$= \$3855.43$$

Series of Equal Payments. The present worth factor for a series of future equal payments (e.g., operating costs) is given by

$$\text{PWF}(i,n)_{ser} = \frac{(1+i)^n - 1}{i(1+i)^n} \tag{5}$$

The present value P of those future equal payments (PMT) is then the product of the present worth factor and the payment (i.e., $P = \text{PWF}(i,n)_{sgl} \times \text{PMT}$).

The future equal payments to repay a present value of money is determined by the capital recovery factor (CRF), which is the reciprocal of the present worth factor for a series of equal payments:

$$\text{CRF} = \text{PMT}/P \tag{6}$$

$$\text{CRF}(i,n) = \frac{i(1+i)^n}{(1+i)^n - 1} = \frac{i}{1-(1+i)^{-n}} \tag{7}$$

The capital recovery factor is often used to describe periodic uniform mortgage or loan payments.

Table 5 gives abbreviated annual CRF values for several values of analysis period n and annual interest rate i. Some of the values of Table 5 are plotted versus time in Figure 1.

Note that when payment periods other than annual are to be studied, the interest rate must be expressed per appropriate period. For example, if monthly payments or return on investment are being analyzed, then interest must be expressed per month, not per year, and n must be expressed in months, not years.

Example 5. Determine the present value of an annual operating cost of $1000/year over a 10-year period, assuming 10% per year interest rate.

$$\text{PWF}(i,n)_{ser} = [(1+0.1)^{10}-1] / [0.1(1+0.1)^{10}] = 6.14$$

Present Value $= \$1000 (6.14) = \$6140$

Example 6. Determine the uniform monthly mortgage payments for a loan of $100 000 to be repaid over 30 years at 10% per year interest. Since we wish monthly payment periods, the payback duration is 30(12) = 360 monthly periods, and the interest rate per period is 0.1/12 = 0.00833 per month.

$$\text{CRF}(i,n) = 0.00833(1+0.00833)^{360} / [(1+0.00833)^{360}-1]$$

$\quad$ CRF $= 0.008773$

$\quad$ PMT $= P\,(\text{CRF})$

$\quad$ PMT $= \$100\,000\,(0.008773)$

$\quad$ PMT $= \$877.30$ per month

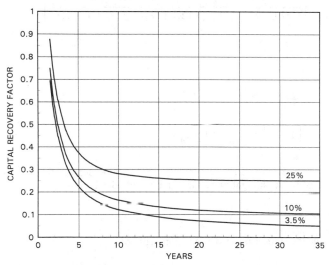

Fig. 1   Capital Recovery Factor Versus Time

Improved Years to Payback Analysis. This somewhat more sophisticated years to payback approach is similar to the simple years to payback method, except that the cost of money (interest rate, discount rate, etc.) is considered. Solving Equation (7) for n yields the following:

$$n = \frac{\ln\,[\,\text{CRF}/\,(\text{CRF}-i)\,]}{\ln\,(1+i)} \qquad (8)$$

Given known investment amounts and earnings, CRFs can be calculated for the alternative investments. Subsequently, the number of periods until payback has been achieved can be calculated using Equation (8). Alternatively, a period-by-period (e.g., month-by-month or year-by-year) tabular cash flow analysis may be performed, or the necessary period to yield the calculated capital recovery factor may be obtained from a plot of CRFs, such as shown in Figure 1.

Example 7. Compare the years to payback of the same items described in Example 2 if the value of money is 10% per year.

Item 1

$\quad$ cost $\quad = \$10\,000$

$\quad$ savings $= \$2000/\text{year}$

$\quad$ CRF $\quad = \$2000/\$10\,000 = 0.2$

$\quad$ $n \quad = \ln[0.2/(0.2-0.1)]/\ln(1+0.1) = 7.3$ years

Item 2

$\quad$ cost $\quad = \$12\,000$

$\quad$ savings $= \$3000/\text{year}$

$\quad$ CRF $\quad = \$3000/\$12\,000 = 0.25$

$\quad$ $n \quad = \ln[0.25/(0.25-0.1)]/\ln(1+0.1) = 5.4$ years

Accounting For Inflation

Different economic goods inflate at different rates. Inflation reflects the rise in the real cost of a commodity over time and is a separate issue from the time value of money. Inflation must often be accounted for in an economic evaluation. One way to account for inflation is to substitute effective interest rates that account for inflation into the equations given in this chapter.

The effective interest rate i', sometimes called the real rate, accounts for inflation rate j and interest rate i or discount rate i_d; it can be expressed as follows (Kreith and Kreider 1978, Kreider and Kreith 1982):

$$i' = \frac{1+i}{1+j} - 1 = \frac{i-j}{1+j} \qquad (9)$$

Table 5   Annual Capital Recovery Factors

Years	Rate of Return or Interest Rate, % per Year								
	3.5	**4.5**	**6**	**8**	**10**	**12**	**15**	**20**	**25**
2	0.52640	0.53400	0.54544	0.56077	0.57619	0.59170	0.61512	0.65455	0.69444
4	0.27225	0.27874	0.28859	0.30192	0.31547	0.32923	0.35027	0.38629	0.42344
6	0.18767	0.19388	0.20336	0.21632	0.22961	0.24323	0.26424	0.30071	0.33882
8	0.14548	0.15161	0.16104	0.17401	0.18744	0.20130	0.22285	0.26061	0.30040
10	0.12024	0.12638	0.13587	0.14903	0.16275	0.17698	0.19925	0.23852	0.28007
12	0.10348	0.10967	0.11928	0.13270	0.14676	0.16144	0.18448	0.22526	0.26845
14	0.09157	0.09782	0.10758	0.12130	0.13575	0.15087	0.17469	0.21689	0.26150
16	0.08268	0.08902	0.09895	0.11298	0.12782	0.14339	0.16795	0.21144	0.25724
18	0.07582	0.08224	0.09236	0.10670	0.12193	0.13794	0.16319	0.20781	0.25459
20	0.07036	0.07688	0.08718	0.10185	0.11746	0.13388	0.15976	0.20536	0.25292
25	0.06067	0.06744	0.07823	0.09368	0.11017	0.12750	0.15470	0.20212	0.25095
30	0.05437	0.06139	0.07265	0.08883	0.10608	0.12414	0.15230	0.20085	0.25031
35	0.05000	0.05727	0.06897	0.08580	0.10369	0.12232	0.15113	0.20034	0.25010
40	0.04683	0.05434	0.06646	0.08386	0.10226	0.12130	0.15056	0.20014	0.25006

Different effective interest rates can be applied to individual components of cost. Projections for future fuel and energy prices are available in the *Annual Supplement to NIST Handbook* 135 (Lippiatt 1994).

Example 8. Determine the present worth P of an annual operating cost of $1000 over 10 years, given a discount rate of 10% per year and an inflation rate of 5% per year.

$$i' = (0.1 - 0.05) / (1 + 0.05) = 0.0476$$

$$\text{PWF}(i',n)_{ser} = \frac{(1 + 0.0476)^{10} - 1}{0.0476 (1 + 0.0476)^{10}} = 7.813$$

$$P = \$1000 (7.813) = \$7813$$

More Sophisticated Economic Analysis Methods

The most frequently used economic analysis techniques arc to examine all costs and incomes to be incurred over the analysis period (1) in terms of their present value (i.e., today's or initial year's value, also called constant value); (2) in terms of equal periodic costs or payments (uniform annualized costs); or (3) in terms of periodic cash flows (e.g., monthly or annual cash flows). Each method provides a slightly different insight. The present value method allows easy comparison of alternatives over the analysis period chosen. The uniform annualized costs method allows comparison of average annual costs of different options. The cash flow method allows comparison of actual cash flows rather than average cash flows; it can identify periods of overall positive and negative cash flow, which is helpful for cash management purposes.

Computer analysis software such as the *NIST Building Life Cycle Computer Program* (NIST 1993) eases calculations and performs parametric or sensitivity analyses. Some commercial spreadsheet programs include economic analysis functions.

The examples presented up to this point have been fairly simple. However, any of the analyses should consider more details of both positive and negative costs over the analysis period, such as varying inflation rates, capital and interest costs, salvage costs, replacement costs, interest deductions, depreciation allowances, taxes, tax credits, mortgage payments, and all other costs associated with a particular system.

Present Value Method. The total present value (present worth) for any analysis is determined by summing the present worths of all individual items under consideration, both future single payment items and series of equal future payments, as described earlier. The scenario with the highest present value is the preferred alternative.

Uniform Annualized Costs Method. It is sometimes useful to project a uniform periodic (e.g., average annual) cost over the analysis period. The basic procedure for determining uniform annualized costs is to first determine the present worth of all costs and then apply the capital recovery factor to determine equal payments over the analysis period.

Uniform annualized mechanical system owning, operating, and maintenance costs can be expressed, for example, as

$$C_y = \quad \begin{aligned} &- \text{capital and interest} + \text{salvage value} - \text{replacements} \\ &- \text{disposals} - \text{operating energy} - \text{property tax} \\ &- \text{maintenance} - \text{insurance} \\ &+ \text{interest tax deduction} + \text{depreciation} \end{aligned}$$

where

capital and interest	$=$	$(C_{s,init} - \text{ITC}) \, \text{CRF}(i',n)$
salvage value	$=$	$C_{s,salv} \text{PWF}(i',n) \text{CRF}(i',n) (1 - T_{salv})$
replacements or disposals	$=$	$\sum_{k=1}^{n} [R_k \text{PWF}(i',k)] \, \text{CRF}(i',n) (1 - T_{inc})$
operating energy	$=$	$C_e [\text{CRF}(i',n)/\text{CRF}(i'',n)] (1 - T_{inc})$
property tax	$=$	$C_{s,assess} T_{prop} (1 - T_{inc})$
maintenance	$=$	$M (1 - T_{inc})$

insurance	$=$	$I (1 - T_{inc})$
interest tax deduction	$=$	$T_{inc} \sum_{k=1}^{n} [i_m P_{k-1,i} \text{PWF}(i_d,k)] \, \text{CRF}(i',n)$
depreciation (for commercial systems)	$=$	$T_{inc} \sum_{k=1}^{n} [D_k \text{PWF}(i_d,k)] \, \text{CRF}(i',n)$

The outstanding principle P_k during year k at market mortgage rate i_m is given by

$$P_k = (C_{s,init} - \text{ITC}) \left[(1 + i_m)^{k-1} + \frac{(1 + i_m)^{k-1} - 1}{(1 + i_m)^{-n} - 1} \right] \quad (10)$$

Note: P_k is in current dollars and must, therefore, be discounted by the discount rate i_d, not i'.

Likewise, the summation term for interest deduction can be expressed as

$$\sum_{k=1}^{n} [i_m P_k / (1 + i_d)^k] = (C_{s,init} - \text{ITC}) \times$$

$$\left[\frac{\text{CRF}(i_m,n)}{\text{CRF}(i_d,n)} + \frac{1}{(1 + i_m)} \frac{i_m - \text{CRF}(i_m,n)}{\text{CRF} [(i_d - i_m) / (1 + i_m) ,n]} \right] \quad (11)$$

If $i_d = i_m$,

$$\sum_{k=1}^{n} [i_m P_k / (1 + i_d)^k] = (C_{s,init} - \text{ITC}) \times$$

$$\left[1 + \frac{n}{1 + i_m} [i_m - \text{CRF}(i_m,n)] \right] \quad (12)$$

Depreciation terms commonly used include depreciation calculated by the straight line depreciation method, which is

$$D_{k,SL} = (C_{s,init} - C_{s,salv}) / n \quad (13)$$

and the sum-of-digits depreciation method:

$$D_{k,SD} = (C_{s,init} - C_{s,salv}) [2 (n - k + 1)] / n (n + 1) \quad (14)$$

Riggs (1977) and Grant et al. (1982) present further information on advanced depreciation methods. Certified accountants may also be consulted for information regarding accelerated methods allowed for tax purposes. The following example illustrates the use of the uniform annualized cost method. Additional examples are presented by Haberl (1993).

Example 9. Calculate the annualized system cost using constant dollars for a $10 000 system considering the following factors: a 5-year life, a salvage value of $1000 at the end of the 5 years, no investment tax credits, a $500 replacement in year 3, a discount rate i_d of 10%, a general inflation rate j of 5%, a fuel inflation rate j_e of 8%, a market mortgage rate i_m of 10%, an annual operating cost for energy of $500, a $100 annual maintenance cost, a $50 annual insurance cost, straight line depreciation, an income tax rate of 50%, a property tax rate of 1% of assessed value, an assessed system value equal to 40% of the initial system value, and a salvage tax rate of 50%.

Effective interest rate i' for all but fuel

$$i' = (i_d - j) / (1 + j) = (0.10 - 0.05) / (1 + 0.05) = 0.047619$$

Effective interest rate i'' for fuel

$$i'' = (i_d - j_e) / (1 + j_e) = (0.10 - 0.08) / (1 + 0.08) = 0.018519$$

Capital recovery factor CRF(i',n) for items other than fuel

$$CRF(i',n) = i'/[1 - (1+i')^{-n}]$$
$$= 0.047619/[1 - (1.047619)^{-5}] = 0.229457$$

Capital recovery factor CRF(i'',n) for fuel

$$CRF(i'',n) = i''/[1 - (1+i'')^{-n}]$$
$$= 0.018519/[1 - (0.018519)^{-5}] = 0.211247$$

Capital recovery factor CRF(i_m,n) for loan or mortgage

$$CRF(i_m,n) = i_m/[1 - (1+i_m)^{-n}]$$
$$= 0.10/[1 - (1.10)^{-5}] = 0.263797$$

Loan payment = $10 000 (0.263797) = $ 2637.97

Present worth factor PWF(i_d, years 1 to 5)

$$PWF(i_d,1) = 1/(1.10)^1 = 0.909091$$
$$PWF(i_d,2) = 1/(1.10)^2 = 0.826446$$
$$PWF(i_d,3) = 1/(1.10)^3 = 0.751315$$
$$PWF(i_d,4) = 1/(1.10)^4 = 0.683013$$
$$PWF(i_d,5) = 1/(1.10)^5 = 0.620921$$

Present worth factor PWF(i', years 1 to 5)

$$PWF(i',1) = 1/(1.047619)^1 = 0.954545$$
$$PWF(i',2) = 1/(1.047619)^2 = 0.911157$$
$$PWF(i',3) = 1/(1.047619)^3 = 0.869741$$
$$PWF(i',4) = 1/(1.047619)^4 = 0.830207$$
$$PWF(i',5) = 1/(1.047619)^5 = 0.792471$$

Capital and interest

$$(C_{s,init} - ITC)\, CRF(i',n) = (\$10\,000 - \$0)\, 0.229457 = \$2294.57$$

Salvage value

$$C_{s,salv} PWF(i',n) CRF(i',n)(1 - T_{salv})$$
$$= \$1000 \times 0.792471 \times 0.229457 \times 0.5 = \$90.92$$

Replacements

$$\sum_{k=1}^{n} [R_k PWF(i',k)]\, CRF(i',n)(1 - T_{inc})$$
$$= \$500 \times 0.869741 \times 0.229457 \times 0.5 = \$49.89$$

Operating energy

$$C_e [CRF(i',n)/CRF(i'',n)](1 - T_{inc})$$
$$= 500 [0.229457/0.211247] 0.5 = \$271.55$$

Property tax

$$C_{s,assess} T_{prop}(1 - T_{inc}) = \$10\,000 \times 0.40 \times 0.01 \times 0.5 = \$20.00$$

Maintenance

$$M(1 - T_{inc}) = \$100(1 - 0.5) = \$50.00$$

Insurance

$$I(1 - T_{inc}) = \$50(1 - 0.5) = \$25.00$$

Interest deduction

$$T_{inc} \sum_{k=1}^{n} [i_m P_{k-1} PWF(i_d,k)]\, CRF(i',n) = \dots \text{ see Table 6}$$

Table 6 summarizes the interest and principle payments for this example. Annual payments are the product of the initial system cost $C_{s,init}$ and the capital recovery factor $CRF(i_m,5)$. Also, Equation (10) can be used to calculate total discounted interest deduction directly.

Next, apply the capital recovery factor $CRF(i',5)$ and tax rate T_{inc} to the total of the discounted interest sum.

$$\$2554.66\, CRF(i',5) T_{inc} = \$2554.66 \times 0.229457 \times 0.5 = \$293.09$$

Depreciation

$$T_{inc} \sum_{k=1}^{n} [D_{k,SL} PWF(i_d,k)]\, CRF(i',n) \dots$$

Use the straight line depreciation method to calculate depreciation:

$$D_{k,SL} = (C_{s,init} - C_{s,salv})/n = (\$10\,000 - \$1000)/5 = \$1800.00$$

Next, discount the depreciation.

Year	$D_{k,SL}$	PWF(i_d,k)	Discounted Depreciation
1	$1800.00	0.909091	$1636.36
2	$1800.00	0.826446	$1487.60
3	$1800.00	0.751315	$1352.37
4	$1800.00	0.683013	$1229.42
5	$1800.00	0.620921	$1117.66
		Total	$6823.42

Finally, the capital recovery factor and tax are applied.

$$\$6823.42\, CRF(i',n) T_{inc} = \$6823.42 \times 0.229457 \times 0.5 = \$782.84$$

U.S. tax code recommends estimating the salvage value prior to depreciating. Then depreciation is claimed as the difference between the initial and salvage value, which is the way depreciation is treated in this example. The more common practice is to initially claim zero salvage value, and at the end of ownership of the item, treat any salvage value as a capital gain.

Table 6 Interest Deduction Summary (for Example 9)

Year	Payment Amount, Current $	Interest Payment, Current $	Principal Payment, Current $	Outstanding Principal, Current $	PWF(i_d,k)	Discounted Interest, Discounted $	Discounted Payment, Discounted $
0	—	—	—	10 000.00	—	—	—
1	2 637.97	1 000.00	1 637.97	8 362.02	0.909091	909.09	2 398.17
2	2 637.97	836.20	1 801.77	6 560.26	0.826446	691.07	2 180.14
3	2 637.97	656.03	1 981.95	4 578.31	0.751315	492.89	1 981.95
4	2 637.97	457.83	2 180.14	2 398.17	0.683013	312.70	1 801.77
5	2 637.97	239.82	2 398.17	0	0.620921	148.91	1 637.97
Total	—	3 189.88	10 000.00	—	—	2 554.66	10 000.00

Table 7 Summary of Cash Flow (for Example 10)

	1	2	3	4	5	6	7	8	9	10	11
								Present Worth of Net Cash Flow			
		Cash Outlay, $	Net Income Before Taxes, $	Depreciation, $	Net Taxable Income,[a] $	Income Taxes @50%, $	Net Cash Flow,[b] $	**10% Rate**		**15% Rate**	**20% Rate**
Year								PWF	P, $	P, $	P, $
0	120 000	0	0	0	0	−120 000	1.000	−120 000	−120 000	−120 000
1	0	20 000	15 000	5 000	2 500	17 500	0.909	15 900	15 200	14 600
2	0	30 000	15 000	15 000	7 500	22 500	0.826	18 600	17 000	15 600
3	0	40 000	15 000	25 000	12 500	27 500	0.751	20 600	18 100	15 900
4	0	50 000	15 000	35 000	17 500	32 500	0.683	22 200	18 600	15 700
5	0	50 000	15 000	35 000	17 500	32 500	0.621	20 200	16 200	13 100
6	0	50 000	15 000	35 000	17 500	32 500	0.564	18 300	14 100	10 900
7	0	50 000	15 000	35 000	17 500	32 500	0.513	16 700	12 200	9 100
8	0	50 000	15 000	35 000	17 500	32 500	0.467	15 200	10 600	7 600
						Total cash flow		27 700	2 000	−17 500
						Investment value		147 500	122 000	102 500

[a]Net taxable income = net income − depreciation.

[b]Net cash flow = net income − taxes.

Summary of terms

Capital and interest	−$ 2294.57
Salvage value	+$ 90.92
Replacements	−$ 49.89
Operating costs	−$ 271.55
Property tax	−$ 20.00
Maintenance	−$ 50.00
Insurance	−$ 25.00
Interest deduction	+$ 293.09
Depreciation deduction	+$ 782.84
Total annualized cost	−$ 1544.16

Cash Flow Analysis Method

The cash flow analysis method accounts for costs and revenues on a period-by-period (e.g., year-by-year) basis, both actual and discounted to present value. This method is especially useful for identifying periods when net cash flow will be negative due to intermittent large expenses.

Example 10. An eight-year study for a $120 000 investment with depreciation spread equally over the assigned period. The benefits or incomes are variable. The marginal tax rate is 50%. The rate of return on the investment is required. Table 7 has columns showing year, cash outlays, income, depreciation, net taxable income, taxes and net cash flow.

Solution. To evaluate the effect of interest and time, the net cash flow must be multiplied by the single payment present worth factor. An arbitrary interest rate of 10% has been selected and the PWF_{sgl} is obtained by using Equation (4). Its value is listed in Table 7, column 8. Present worth of the net cash flow is obtained by multiplying columns 7 and 8. Column 9 is then added to obtain the total cash flow. If year 0 is ignored, an investment value is obtained for a 10% required rate of return.

The same procedure is used for 15% interest (column 10, but the PWF is not shown) and for 20% interest (column 11).

Discussion. The interest at which the summation of present worth of net cash flow is zero gives the rate of return. In this example, the investment has a rate of return by interpolation of about 15.4%. If this rate offers an acceptable rate of return to the investor, the proposal should be approved; otherwise, it should be rejected.

Another approach would be to obtain an investment value at a given rate of return. This is accomplished by adding the present worth of the net cash flows, but not including the investment cost. In the example, under the 10% given rate of return, $147 700 is obtained as an investment value. This amount, when using money that costs 10%, would be the acceptable value of the investment.

REFERENCES

Akalin, M.T. 1978. Equipment life and maintenance cost survey. *ASHRAE Transactions* 84(2):94-106.

Dohrmann, D.R. and T. Alereza. 1986. Analysis of survey data on HVAC maintenance costs. *ASHRAE Transactions* 92(2A):550-65.

Easton Consultants. 1986. Survey of residential heat pump service life and maintenance issues. Available from American Gas Association, Arlington, VA (Catalog No. S-77126).

Grant, E., W. Ireson, and R. Leavenworth. 1982. *Principles of engineering economy.* John Wiley and Sons, New York.

Haberl, J. 1993. Economic calculations for ASHRAE Handbook. Energy Systems Laboratory *Report* No. ESL-TR-93/04-07. Texas A&M University, College Station, TX.

Kreider, J. and F. Kreith. 1982. *Solar heating and cooling.* Hemisphere Publishing, Washington, D.C.

Kreith, F. and J. Kreider. 1978. *Principles of solar engineering.* Hemisphere Publishing, Washington, D.C.

Lippiatt, B.L. 1994. Energy prices and discount factors for life-cycle cost analysis 1993. *Annual Supplement to NIST Handbook* 135 and *NBS Special Publication* 709. NISTIR 85-3273.7. National Institute of Standards and Technology, Gaithersburg, MD.

Lovvorn, N.C. and C.C. Hiller. 1985. A study of heat pump service life. *ASHRAE Transactions* 91(2B):573-88.

NIST. 1993. Building life cycle cost (BLCC) computer program, version 4.1. NISTIR 4481. National Institute of Standards and Technology, Gaithersburg, MD.

Riggs, J.L. 1977. *Engineering economics.* McGraw-Hill, New York.

Ruegg, R.T. Life-cycle costing manual for the federal energy management program. *NIST Handbook* 135. National Institute of Standards and Technology, Gaithersburg, MD.

U.S. Department of Commerce, Bureau of Economic Analysis. *Survey of current business.* U.S. Government Printing Office, Washington, D.C.

BIBLIOGRAPHY

ASTM. 1992. Standard terminology of building economics. *Standard* E833 Rev A-92. American Society for Testing and Materials, Philadelphia.

Kurtz, M. 1984. *Handbook of engineering economics: A guide for engineers, technicians, scientists, and managers.* McGraw-Hill, New York.

Quirin, D.G. 1967. *The capital expenditure decision.* Richard D. Win, Inc., Homewood, IL.

Van Horne, J.C. 1980. *Financial management and policy.* Prentice Hall, Englewood Cliffs, NJ.

TESTING, ADJUSTING, AND BALANCING

THE system that controls the environment within a building is a dynamic entity that changes with time and use, and it must be rebalanced accordingly. The designer must consider initial and supplementary testing and balancing requirements for commissioning. Complete and accurate operating and maintenance instructions and manuals that include intent of design and how to test, adjust, and balance the building systems are essential. Building operating personnel must be well trained, or qualified operating service organizations must be employed, to ensure optimum comfort, proper process operations, and economy of operation.

This chapter does not suggest which groups or individuals should perform the functions of a complete testing, adjusting, and balancing procedure. However, the procedure must produce repeatable results that meet the intent of the designer and accurately reflect the requirements of the owner. Overall, one source must be responsible for testing, adjusting, and balancing all systems. As part of this responsibility, the testing organization should check the performance of all equipment under field conditions to ensure compliance.

Testing and balancing should be repeated as the systems are renovated and changed. The testing of boilers and other pressure vessels for compliance with safety codes is not the primary function of the testing and balancing firm; rather it is to verify and adjust operating conditions in relation to design conditions for flow, temperature, pressure drop, noise, and vibration. ASHRAE *Standard* 111 provides detailed procedures not covered in this chapter.

DEFINITIONS

System testing, adjusting, and balancing is the process of checking and adjusting all the environmental systems in a building to produce the design objectives. This process includes (1) balancing air and water distribution systems, (2) adjusting the total system to provide design quantities, (3) electrical measurement, (4) establishing quantitative performance of all equipment, (5) verifying automatic controls, and (6) sound and vibration measurement. These procedures are accomplished by (1) checking installations for conformity to design, (2) measuring and establishing the fluid quantities of the system, as required to meet design specifications, and (3) recording and reporting the results.

The following definitions are used in this chapter. Refer to ASHRAE *Terminology of Heating, Ventilation, Air Conditioning, and Refrigeration* (1991) for additional definitions.

Test. To determine quantitative performance of equipment.

Balance. To proportion flows within the distribution system (submains, branches, and terminals) according to specified design quantities.

Adjust. To regulate the specified fluid flow rate and air patterns at the terminal equipment (e.g., reduce fan speed, adjust a damper).

Procedure. An approach to and execution of a sequence of work operations to yield repeatable results.

Report forms. Test data sheets arranged for collecting test data in logical order for submission and review. The data sheets should also form the permanent record to be used as the basis for any future testing, adjusting, and balancing.

Terminal. A point where the controlled medium (fluid or energy) enters or leaves the distribution system. In air systems, these may be variable air or constant volume boxes, registers, grilles, diffusers, louvers, and hoods. In water systems, these may be heat transfer coils, fan coil units, convectors, or finned-tube radiation or radiant panels.

GENERAL CRITERIA

Effective and efficient testing, adjusting, and balancing require a systematic, thoroughly planned procedure implemented by experienced and qualified staff. All activities, including organization, calibrated instrumentation, and execution of the actual work, should be scheduled. Because many systems function differently on a seasonal basis, and because temperature performance is significant, it is important to coordinate air-side with water-side work. Preparatory work includes planning and scheduling all procedures, collecting necessary data (including all change orders), reviewing data, studying the system to be worked on, preparing forms, and making preliminary field inspections.

Duct systems must be designed, constructed, and installed to minimize and control air leakage. During construction, all duct systems should be sealed and tested for air leakage; and water, steam, and pneumatic piping should be tested for leakage. Any leakage can have a marked effect on system performance.

Design Considerations

Testing, adjusting, and balancing begin as design functions, with most of the devices required for adjustments being integral parts of the design and installation. To ensure that proper balance can be achieved, the engineer should show and specify a sufficient number of dampers, valves, flow measuring locations, and flow balancing devices; these must be properly located in required straight lengths of pipe or duct for accurate measurement. The testing procedure depends on the system's characteristics and layout. The interaction between individual terminals varies with the system pressures, flow requirements, and control devices.

The design engineer should specify balancing tolerances. Suggested tolerances are ±10% for individual terminals and branches in noncritical applications and ±5% for main ducts. For critical

---

The preparation of this chapter is assigned to TC 9.7, Testing and Balancing.

applications where differential pressures must be maintained, the following tolerances are suggested:

Positive zones
Supply air	0 to +10%
Exhaust and return air	0 to −10%

Negative zones
Supply air	0 to −10%
Exhaust and return air	0 to +10%

AIR VOLUMETRIC MEASUREMENT METHODS

General

The pitot-tube traverse is the generally accepted method of measuring airflow in duct systems. Other methods of measuring airflow at individual terminals are described by the terminal manufacturers. The primary objective is to establish repeatable measurement procedures that correlate with the pitot-tube traverse.

Laboratory tests, data, and techniques prescribed by equipment and air terminal manufacturers must be reviewed and checked for accuracy, applicability, and repeatability of the results. Conversion factors that correlate field data with laboratory results must be developed to predict the equipment's actual field performance.

Air Devices

Generally, K factors given by air diffuser manufacturers should be checked for accuracy by field measurement, comparing actual flow measured by pitot-tube traverse to actual measured velocity. Air diffuser manufacturers usually base their volumetric test measurements on a deflecting vane anemometer. The velocity is multiplied by an empirical effective area to obtain the air diffuser's delivery. Accurate results are obtained by measuring at the vena contracta with the probe of the deflecting vane anemometer.

The methods advocated for measuring the airflow of troffer-type terminals are similar to the methods described for air diffusers. The capture hood is frequently used to measure device airflows, primarily of diffusers and slots. K factors should be established for hood measurements with varying flow rates and deflection settings. If the air does not fill the measurement grid, the readings will require a correction factor (similar to the K factor).

Rotating vane anemometers are commonly used to measure airflow from sidewall grilles. Effective areas (K factors) should be established with the face dampers fully open and deflection set uniformly on all grilles. Correction factors are required when measuring airflow in open ducts, i.e., damper openings and fume hoods (Sauer and Howell 1990).

All flow measuring instruments should be field verified by running pitot-tube traverses to establish correction and/or density factors.

Duct Flow

Most procedures for testing, adjusting, and balancing air-handling systems rely on measuring volumes in the ducts rather than at the terminals. These measurements are more reliable than those obtained at the terminals, which are based on manufacturer's data. In such procedures, terminal measurements are relied on only for proportionally balancing the distribution within a space or zone.

The preferred method of duct volumetric flow measurement is the pitot-tube traverse average. Care should be taken to obtain the maximum straight run before and after the traverse station. Measuring points should be located as shown in Chapter 13 of the 1993 ASHRAE *Handbook—Fundamentals* and ASHRAE *Standard* 111 to obtain the best duct velocity profile. Where factory-fabricated volume measuring stations are used, the measurements should be checked against a pitot-tube traverse.

The power input to a fan's driver should be used only as a guide to indicate its delivery. It may be used to verify performance determined by a reliable method (e.g., pitot-tube traverse of system's main) considering system effects that may be present. The flow rate from some fans is not proportional to the power needed to drive them. In some cases, as with forward-curved blade fans, the same power is required for two or more flow rates. The backward-curved blade centrifugal fan is the only type with suitable characteristics, i.e., flow rate that varies directly with the power input. If an installation has an inadequate straight length of ductwork or no ductwork to allow a pitot-tube traverse, a procedure prescribed by Sauer and Howell (1990) can be followed. In this procedure a vane anemometer is used to read air velocities at multiple points across the face of a coil to determine a K factor.

Mixture Plenums

Approach conditions are often so unfavorable that the air quantities comprising a mixture (e.g., outdoor air and return air) cannot be determined accurately by volumetric measurements. In such cases, the temperature of the mixture indicates the balance (proportions) between the component airstreams. Temperatures must be measured carefully to account for stratification of the air, and the difference between the outside and return air temperatures must be greater than 10 K. The temperature of the mixture can be calculated from Equation (1) as follows:

$$Q_t t_m = Q_o t_o + Q_r t_r \tag{1}$$

where

Q_t = total measured air quantity, %
Q_o = outside air quantity, %
Q_r = return air quantity, %
t_m = temperature of outside and return mixture, °C
t_o = outdoor temperature, °C
t_r = return temperature, °C

Pressure Measurements

The air pressures measured include barometric pressure, static pressure, velocity pressure, total pressure, and differential pressure. Pressure measurement for field evaluation of air-handling system performance should be measured as recommended in ASHRAE *Standard* 111 and analyzed together with the manufacturers' fan curves and system effect as predicted from application of methods in AMCA *Standard* 210. When measured in the field, pressure readings, air quantity, and power input often do not correlate with the manufacturers' certified performance curves unless proper correction is made.

Pressure drops through equipment such as coils, dampers, or filters should not be used to measure airflow. Pressure is an acceptable means of establishing flow volumes only where it is required by, and performed in accordance with, the manufacturer certifying the equipment.

Stratification

Normal design minimizes conditions causing air turbulence in order to produce the least friction, resistance, and consequent pressure losses in the system. However, under certain conditions, air turbulence is desirable and necessary. For example, two airstreams of different temperatures can stratify in smooth, uninterrupted flow conditions. In this situation, mixing should be promoted in the design. The return and outside airstreams at the inlet side of the air-handling unit tend to stratify where enlargement of the inlet plenum or casing size decreases the air velocity. Without a deliberate effort to mix the two airstreams (i.e., in cold climates, placing the outdoor air entry at the top of the plenum and the return air at the bottom of the plenum to allow natural mixing), stratification can exist and be carried throughout the system (e.g., filter, coils, eliminators, fans, and ducts). Stratification can cause damage by freezing coils and rupturing tubes. It can also affect the temperature control in plenums, spaces, or both.

Stratification can also be reduced by adding vanes to break up and mix the two airstreams. No solution to stratification problems is guaranteed; each condition must be evaluated by field measurements and experimentation.

BALANCING PROCEDURES FOR AIR DISTRIBUTION SYSTEMS

General procedures for testing and balancing are described here, although no one established procedure is applicable to all systems. The bibliography lists sources of additional information.

Instrumentation for Testing and Balancing

The minimum instruments necessary for air balance are

- Manometer calibrated in 1 Pa divisions
- Combination inclined and vertical manometer (0 to 2.5 kPa)
- Pitot tubes in various lengths, as required
- Tachometer (direct contact, self-timing type) or strobe light
- Clamp-on ammeter with voltage scales (rms type)
- Deflecting vane anemometer
- Rotating vane anemometer
- Flow hood
- Dial thermometers (50-mm diameter minimum and 0.5 K graduations minimum) and glass stem thermometers (0.5 K graduations minimum)
- Sound level meter with octave filter set, calibrator, accelerometer, and integrator
- Vibration analyzer capable of measuring displacement velocity and acceleration
- Combustion analyzer
- Water flowmeters (0 to 12 kPa and 0 to 100 kPa ranges)
- Compound gage
- Test gages (700 kPa and 2000 kPa)
- Sling psychrometer
- Etched stem thermometer (0 to 50°C in 0.05 K increments)
- Hygrometers
- Digital thermometers

The instrumentation must be checked periodically to verify its accuracy and repeatability prior to use in the field.

Preliminary Procedure for Air Balancing

Before operating the system, the following steps should be performed:

1. Obtain as-built design drawings and specifications, and become thoroughly acquainted with the design intent.
2. Obtain copies of approved shop drawings of all air-handling equipment, outlets (supply, return, and exhaust), and temperature control diagrams including performance curves. Compare design requirements with shop drawing capacities.
3. Compare design to installed equipment and field installation.
4. Walk the system from the air-handling equipment to terminal units to determine variations of installation from design.
5. Check dampers (both volume and fire) for correct and locked position and temperature control for completeness of installation before starting fans.
6. Prepare report test sheets for both fans and outlets. Obtain manufacturer's outlet factors and recommended testing procedure. A summation of required outlet volumes permits a cross-checking with required fan volumes.
7. Determine best locations in main and branch ductwork for most accurate duct traverses.
8. Place all outlet dampers in the full open position.
9. Prepare schematic diagrams of system as-built ductwork and piping layouts to facilitate reporting.

10. Check filters for cleanliness and proper installation (no air bypass). If specifications require, establish procedure to simulate dirty filters.
11. For variable volume air systems, develop a plan to simulate diversity.

Equipment and System Check

1. Place all fans (supply, return, and exhaust) in operation and immediately check the following items:
   a) Motor amperage and voltage to guard against overload.
   b) Fan rotation.
   c) Operability of static pressure limit switch.
   d) Automatic dampers for proper position.
   e) Air and water resets operating to deliver required temperatures.
   f) Air leaks in the casing and in the scarfing around the coils and filter frames should be checked by moving a bright light along the outside of the duct joints while observing the darkened interior of the casing. Any leaks should be caulked. Note points where piping enters the casing to ensure that escutcheons are right. Do not rely on pipe insulation to seal these openings because the insulation may shrink. In prefabricated units, check that all panel-fastening holes are filled to prevent whistling.
2. Traverse the main supply ductwork whenever possible. All main branches should also be traversed where duct arrangement permits. Selection of traverse points and method of traverse should be as follows:

   a) Traverse each main or branch after the longest possible straight run for the duct involved.
   b) For test hole spacing, refer to Chapter 13 of the 1993 ASHRAE *Handbook—Fundamentals*.
   c) Traverse using a pitot tube and manometer where velocities are over 3 m/s. Below this velocity, use either a micromanometer and pitot tube or a recently calibrated thermal anemometer.
   d) Note temperature and barometric pressure to determine if they need to be corrected for standard air quantity. Corrections are normally insignificant below 600-m elevation; however, where accurate results are desirable, corrections are justified.
   e) After establishing the total air being delivered, adjust the fan speed to obtain the design airflow, if necessary. Check power and speed to see that motor power and/or critical fan speed have not been exceeded.
   f) Proportionally adjust branch dampers until each has the proper air volume.
   g) With all the dampers and registers in the system open and with the supply, return, and exhaust blowers operating at or near design speed, set the minimum outdoor and return air ratio. If duct traverse locations are not available, this can be done by measuring the mixture temperature with thermometers in the return air, outdoor air louver, and filter section. As an approximation, the temperature of the mixture may be calculated from Equation (1).

   The greater the temperature difference between hot and cold air, the easier it is to get accurate damper settings. Take the temperature at many points in a uniform traverse to be sure there is no stratification.

   After the minimum outdoor air damper has been set for the proper percentage of outdoor air, take another traverse of mixture temperatures and install baffling if the variation from the average is more than 5%. Remember that stratified mixed air temperatures vary greatly with the outdoor temperature in cold weather, while return air temperature has only a minor effect.

3. Carefully set the system for balance using the following procedures:

   a) Adjust the system with mixing dampers positioned for minimum outdoor air.

   b) When adjusting multizone or double-duct constant volume systems, establish the ratio of the design volume through the cooling coil to total fan volume to achieve the desired diversity factor. Keep the proportion of cold to total air constant during the balance. However, check each zone or branch with this component on full cooling. If the design calls for full flow through the cooling coil, the entire system should be set to full flow through the cooling side while making tests. Perform the same procedure for the hot air side.

4. Balance the terminal outlets in each control zone in proportion to each other. The following steps may be followed to balance the terminals.

   a) Once the preliminary fan quantity is set, proportion the terminal outlet balance from the outlets into the branches to the fan. Concentrate on proportioning the flow rather than the absolute quantity. As changes are made to the fan settings and branch dampers, the outlet terminal quantities remain proportional. Branch dampers should be used for major adjusting and terminal dampers for trim or minor adjustment only. It may be necessary to install additional subbranch dampers to decrease the use of terminal dampers that create objectionable noise.

   b) Normally, several passes through the entire system are necessary to obtain proper outlet values.

   c) The total tested outlet air quantity compared to duct traverse air quantities may be an indicator of duct leakage.

   d) With total design air established in the branches and at the outlets, perform the following: (1) take new fan motor amperage readings, (2) find static pressure across the fan, (3) read and record static pressure across each component (intake, filters, coils, and mixing dampers), and (4) take a final duct traverse.

Dual-Duct Systems

Most constant volume dual-duct systems are designed to handle a portion of the total system's supply through the cold duct and smaller air quantities through the hot duct. Balancing should be accomplished as follows:

1. Check the leaving air temperature at the nearest terminal to verify that the hot and cold damper inlet leakage is not greater than the maximum allowable leakage established.

2. Check apparatus and main trunks, as outlined in the section on Equipment and System Check.

3. Determine whether the static pressure at the end of the system (the longest duct run) is at or above the minimum required for mixing box operation. Proceed to the extreme end of the system and check the inlet static pressure with an inclined manometer. The inlet static pressure should exceed the minimum static pressure recommended by the mixing box manufacturer. Additional static pressure is required for the low-pressure distribution system downstream of the box.

4. Proportionately balance the diffusers or grilles on the low-pressure side of the box, as described for low-pressure systems in the previous section.

5. Change the control settings to full heating, and make certain that the controls and dual-duct boxes function properly. Spot-check the airflow at several diffusers. Check for stratification.

6. If the engineer has included a diversity factor in selecting the main apparatus, it will not be possible to get full flow from all boxes simultaneously, as outlined in item 3b in the previous section. Mixing boxes closest to the fan should be set to the opposite hot or cold deck to the more critical season air flow to force the air to the end of the system.

VARIABLE VOLUME SYSTEMS

Many types of variable volume systems have been developed to conserve energy. These systems can be categorized as pressure dependent or pressure independent.

Pressure-dependent systems incorporate air terminal boxes that have a thermostat signal controlling a damper actuator. The air volume to the space varies to maintain the space temperature, while the air temperature supplied to the terminal boxes remains constant. The balance of this system constantly changes with loading changes; therefore, any balancing procedure will not produce repeatable data unless changes in system load are simulated by using the same configuration of thermostat settings each time the system is tested—that is, the same terminal boxes are fixed in the minimum and maximum positions for the test.

Pressure-independent systems incorporate air terminal boxes that have a thermostat signal used as a master control to open or close the damper actuator and a velocity controller used as a submaster control to maintain the maximum and minimum amounts of air to be supplied to the space. The air volume to the space varies to maintain the space temperature, while the air temperature supplied to the terminal remains constant. Care should be taken to verify the operating range of the damper actuator as it responds to the velocity controller to prevent dead bands or overlap of control in response to other system components (e.g., double-duct VAV, fan-powered boxes, and retrofit systems). Care should also be taken to verify the action of the thermostat with regard to the damper position, as the velocity controller can change the control signal ratio or reverse the control signal.

In a pressure-dependent system, the setting of minimum airflows to the space, other than at no flow, is not suggested, unless the terminal box has a normally closed damper and the manufacturer of the damper actuator provides adjustable mechanical stops. The pressure-independent system requires verification that the velocity controller is operating properly. Inlet duct configuration can adversely affect the operation of the velocity controller (Griggs et al. 1990). The primary difference between the two systems is that the pressure-dependent system supplies a different amount of air to the space as the pressure upstream of the terminal box changes. If the thermostats are not calibrated properly to meet the space load, several zones may overcool or overheat. When the zones overcool and receive greater amounts of supply air than required, they decrease the amount of air that can be supplied to overheated zones. The pressure-independent system is not affected by improper thermostat calibration in the same way that a pressure-dependent system is because the minimum and maximum airflow limits may be set for each zone.

System Static Control

System static control is important to save energy and to prevent overpressurizing the duct system. The following procedures and equipment are some of the means used to control static pressure.

No Fan Volumetric Control. This is sometimes referred to as "riding the fan curve." This type of system should be limited to systems with minimum airflows of 50% of peak design and flat forward-curved fans. Pressure and noise are potential problems, and this control is not energy-efficient.

System Bypass Control. As the system pressure increases due to terminal boxes closing, a relief damper bypasses the system air back to the fan inlet. With this type of control, the economy of varied fan output is nonexistent, and the relief damper is usually a major source of duct leakage and noise. The relief damper should be modulated to maintain a minimum duct static pressure.

Discharge Damper. System losses and noise should be considered with this system.

Vortex Damper. System losses due to inlet air conditions are a problem, and the vortex damper does not completely close. The minimum expected airflow should be evaluated.

Variable Inlet Cones. System loss can be a problem because typically the cone does not close completely. The minimum expected airflow should be evaluated.

Varying Fan Speed Mechanically. Slippage, loss of belts, cost of belt replacement, and the initial cost of the components are of concern.

Variable Pitch-in-Motion Fans. Maintenance and the means to prevent the fan from running in the stall condition must be evaluated.

Varying Fan Speed Electrically. This system is usually the most efficient and is accomplished by varying the voltage or the frequency to the fan motor. Some versions of motor drives may cause electrical noise and affect other devices.

In controlling VAV fan systems, the location of the static pressure sensors is critical and should be field verified to give the most representative point of operation. After the terminal boxes have been proportioned, the static pressure control can be verified by observing static pressure changes at the fan discharge and the static pressure sensor as the load is simulated from maximum airflow to minimum airflow (i.e., set all terminal boxes to balanced airflow conditions and determine whether any changes in static pressure occur by placing one terminal box at a time to minimum airflow, until all terminals are placed at the minimal airflow setting). Care should be taken to verify that the maximum to minimum air volume changes are within the fan curve performance (speed or total pressure).

Diversity

Diversity may be used on a VAV system, assuming that the total system airflow volume is lower by design and that all of the system terminal boxes will never open fully at the same time. Care should be taken to avoid any duct leakage. All ductwork upstream of the terminal box should be considered as medium-pressure ductwork, whether in a low- or medium-pressure system.

A procedure to test the total air on the system should be established by setting terminal boxes to the zero or minimum position nearest to the fan. During peak load conditions, care should be taken to verify that an adequate pressure is available upstream of all terminal boxes to achieve design airflow to the spaces.

Outside Air Requirements

Maintaining the space under a slight positive or neutral pressure to atmosphere is difficult with all variable volume systems. In most systems, the exhaust requirement for the space is constant; hence, the outside air used to equal the exhaust air and meet the minimum outside air requirements for the building codes must also remain constant. Due to the location of the outside air intake and the changes in pressure, this does not usually happen. The outside air should enter the fan at a point of constant pressure (i.e., supply fan volume can be controlled by proportional static pressure control, which can control the volume of the return air fan).

Return Air Fans

If return air fans are required in series with a supply fan, the type of control and sizing of the fans is most important. Serious over- and underpressurization can occur, especially during the economizer cycle.

Types of VAV Systems

Single-Duct VAV. This system incorporates a pressure-dependent or -independent terminal and usually has reheat at some predetermined minimal setting on the terminal unit or separate heating system.

Bypass System. This system incorporates a pressure-dependent damper, which, on demand for heating, closes the damper to the space and opens to the return air plenum.

This system sometimes incorporates a constant bypass airflow or a reduced amount of airflow bypassed to the return plenum in relation to the amount supplied to the space. No economical value can be obtained by varying the fan speed with this system. A control problem can exist if any return air sensing is done to control a warm-up or cool-down cycle.

VAV System Using Single-Duct VAV and Fan-Powered, Pressure-Dependent Terminals. This system has a primary source of air from the fan to the terminal and a secondary powered fan source that pulls air from the return air plenum before the additional heat source. This system places additional maintenance of terminal filters, motors, and capacitors on the building owner. In certain fan-powered boxes, backdraft dampers are a source of system duct leakage when the system calls for the damper to be fully closed. Typical applications include geographic areas where the ratio of heating hours to cooling hours is low.

Double-Duct VAV. This type of terminal incorporates two single-duct variable terminals. It is controlled by velocity controllers that operate in sequence so that both hot and cold ducts can be opened or closed. Some control systems have a downstream flow sensor in the terminal unit to maintain either the heating or the cooling. The other flow sensor is in the inlet controlled by the thermostat. As this inlet damper closes, the downstream controller opens the other damper to maintain the set airflow. Often, low system pressure in the decks controlled by the thermostat causes unwanted mixing of air, which results in excess energy use or discomfort in the space. On most direct digital control (DDC) systems, inlet control on both ducts is favored in lieu of the downstream controller.

Balancing the VAV System

The general procedure for balancing a VAV system is

1. Determine the required maximum air volume to be delivered by the supply and return air fans. Diversity of load usually means that the volume will be somewhat less than the outlet total.
2. Obtain fan curves on these units, and request information on surge characteristics from the fan manufacturer.
3. If an inlet vortex damper control is to be used, obtain the fan manufacturer's data pertaining to the deaeration of the fan when used with the damper. If speed control is used, find the maximum and minimum speed that can be used on the project.
4. Obtain from the manufacturer the minimum and maximum operating pressures for terminal or variable volume boxes to be used on the project.
5. Construct a theoretical system curve, including an approximate surge area. The system curve starts at the minimum inlet static pressure of the boxes, plus system loss at minimum flow, and terminates at the design maximum flow. The operating range using an inlet vane damper is between the surge line intersection with the system curve and the maximum design flow. When variable speed control is used, the operating range is between (a) the minimum speed that can produce the necessary minimum box static at minimum flow still in the fan's stable range and (b) the maximum speed necessary to obtain maximum design flow.
6. Position the terminal boxes to the proportion of maximum fan air volume to total installed terminal maximum volume.
7. Set the fan to operate at approximate design speed (increase about 5% for a full open inlet vane damper).
8. Check a representative number of terminal boxes. If a wide variation in static pressure is encountered, or if the airflow at a number of boxes is below minimum at maximum flow, check every box.
9. Run a total air traverse with a pitot tube.
10. Increase the speed if static pressure and/or volume are low. If the volume is correct, but the static is high, reduce the speed. If the static is high or correct, but the volume is low, check for system effect at the fan. If there is no system effect, go over all terminals and adjust them to the proper volume.

11. Run steps (7) through (10) with the return or exhaust fan set at design and measured by a pitot-tube traverse and with the system set on minimum outdoor air.
12. Proportion the outlets, and verify the design volume with the VAV box on the maximum flow setting. Verify the minimum flow setting.
13. Set the terminals to minimum, and adjust the inlet vane or speed controller until minimum static pressure and airflow are obtained.
14. The temperature control personnel, the balancing personnel, and the design engineer should agree on the final placement of the sensor for the static pressure controller. This sensor must be placed in a representative location in the supply duct to sense average maximum and minimum static pressures in the system.
15. Check the return air fan speed or its inlet vane damper that tracks or adjusts to the supply fan airflow to ensure proper outside air volume.
16. Operate the system on 100% outside air (weather permitting), and check supply and return fans for proper power and static pressure.

Induction Systems

Most induction systems use high-velocity air distribution. Balancing should be accomplished as follows:

1. Perform steps outlined under the basic procedures common to all systems for apparatus and main trunk capacities.
2. Determine the primary airflow at each terminal unit by reading the unit plenum pressure with a manometer and locating the point on the charts (or curves) of air quantity versus static pressure supplied by the unit manufacturer.
3. Normally, about three complete passes around the entire system are required for proper adjustment. Make a final pass without adjustments to record the end result.
4. To provide the quietest possible operation, adjust the fan to run at the slowest speed that provides sufficient nozzle pressure to all units with minimum throttling of all unit and riser dampers.
5. After balancing each induction system with minimum outdoor air, reposition to allow maximum outdoor air and check power and static pressure readings.

Report Information

To be of value to the consulting engineer and owner's maintenance department, the air-handling report should consist of at least the following items:

1. *Design*
   a) Air quantity to be delivered
   b) Fan static pressure
   c) Motor power
   d) Percent of outside air under minimum conditions
   e) Speed of the fan
   f) Power required to obtain this air quantity at design static pressure

2. *Installation*
   a) Equipment manufacturer (indicate model number and serial number)
   b) Size of unit installed
   c) Arrangement of the air-handling unit
   d) Fan class
   e) Nameplate power, nameplate voltage, phase, cycles, and full-load amperes of the motor installed

3. *Field tests*
   a) Fan speed
   b) Power readings (voltage, amperes of all phases at motor terminals)
   c) Total pressure differential across unit components

   d) Fan suction and fan discharge static pressure (equals fan total pressure)
   e) Plot of actual readings on manufacturer's fan performance curve to show the installed fan operating point
   f) Measured airflow rate

It is important to establish the initial static pressures accurately for the air treatment equipment and the duct system so that the variation in air quantity due to filter loading can be calculated. It enables the designer to ensure that the total fan quantity will never be less than the minimum requirements. It also serves as a check of dirt loading in coils, since the design air quantity for peak loading of the filters has already been calculated.

4. *Terminal Outlets*
   a) Outlet by room designation and position
   b) Outlet manufacture and type
   c) Outlet size (using manufacturer's designation to ensure proper factor)
   d) Manufacturer's outlet factor (Where no factors are available, or field tests indicate the listed factors are incorrect, a factor must be determined in the field by traverse of a duct leading to a single outlet.)
   e) Design air quantity and the required velocity to obtain it
   f) Test velocities and resulting air quantity
   g) Adjustment pattern for every air terminal

5. *Additional Information (if applicable)*
   a) Air-handling units
      (1) Belt number and size
      (2) Drive and driven sheave size
      (3) Belt position on adjusted drive sheaves (bottom, middle, and top)
      (4) Motor speed under full load
      (5) Motor heater size
      (6) Filter type and static pressure at initial use and full load; time to replace
      (7) Variations of velocity at various points across the face of the coil
      (8) Existence of vortex or discharge dampers, or both
   b) Distribution system
      (1) Unusual duct arrangements
      (2) Branch duct static readings in double-duct and induction system
      (3) Ceiling pressure readings where plenum ceiling distribution is being used; tightness of ceiling
      (4) Relationship of building to outdoor pressure under both minimum and maximum outdoor air
      (5) Induction unit manufacturer and size (including required air quantity and plenum pressures for each unit) and a test plenum pressure and resulting primary air delivery from the manufacturer's listed curves
   c) All equipment nameplates visible and easily readable

Many independent firms have developed detailed procedures suitable to their own operations and the area in which they function. These procedures are often available for information and evaluation on request (see Bibliography).

PRINCIPLES AND PROCEDURES FOR BALANCING HYDRONIC SYSTEMS

Both air- and water-side balance techniques must be performed with sufficient accuracy to ensure that the system operates economically, with minimum energy, and with proper distribution. Air-side balance requires a precise flow measuring technique because air, which is usually the prime heating or cooling transport medium, is more difficult to measure in the field. Reducing the airflow to less than the design requirement directly reduces the heat transfer. In contrast, the rate of heat transfer for the water side of the terminal does not vary linearly with the water flow rate. Therefore, the proper balancing valve with the correct control characteristics must be provided at each terminal to proportionally balance the water flow rate according to the design conditions.

Heat Transfer at Reduced Flow Rate

The typical heating-only hydronic terminal gradually reduces its heat output as flow is reduced (Figure 1). Decreasing water flow to 50% of design reduces the heat transfer to 90% of that at full design flow. The control valve must reduce the water flow to 10% to reduce the heat output to 50%. The reason for the relative insensitivity to changing flow rates is that the governing coefficient for heat transfer is the air-side coefficient. A change in internal or water-side coefficient with flow rate does not materially affect the overall heat transfer coefficient. This means that (1) heat transfer for water-to-air terminals is established by the mean air-to-water temperature difference, (2) the heat transfer is measurably changed, and (3) a change in the mean water temperature requires a greater change in the water flow rate.

A secondary safety factor also applies to heating terminals. Unlike chilled water, hot water can be supplied at a wide range of temperatures. So, in some cases, an inadequate terminal heating capacity

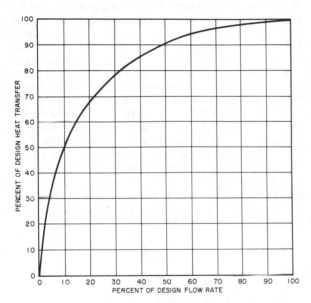

Fig. 1   Effects of Flow Variation on Heat Transfer from a Hydronic Terminal
(Design Δt = 10 K and supply temperature = 93°C)

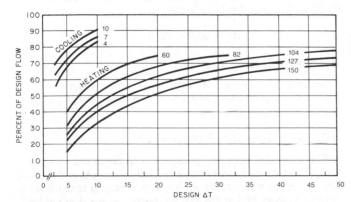

Fig. 2   Variation of Design Flow Versus Design Δt to Maintain 90% Terminal Heat Transfer

caused by insufficient flow can be overcome by raising the system supply water temperature. Design below the temperature limit of 120°C of the ASME low pressure code must be considered.

The previous comments apply to heating terminals selected for a 10 K temperature drop (Δt) and with a supply water temperature of about 93°C. Figure 2 shows the flow variation when 90% terminal capacity is acceptable. Note that heating system tolerance decreases with temperature and flow rates and that chilled water terminals are much less tolerant to flow variation than hot water terminals.

Dual-temperature heating/cooling hydronic systems are sometimes completed and started during the heating season. Adequate heating ability in the terminals may suggest that the system is balanced. Figure 2 shows that 40% of design flow through the terminal provides 90% of design heating with 60°C supply water and a 5 K temperature drop. Increased supply water temperature establishes the same heat transfer at terminal flow rates of less than 40% design.

In some cases, dual-temperature water systems may experience a decreased flow during the cooling season because of the chiller pressure drop; this could cause a flow reduction of 25%. For example, during the cooling season, a terminal that originally heated satisfactorily would only receive 30% of the design flow rate.

While the example of reduced flow rate at Δt = 10 K only affects the heat transfer by 10%, this reduced heat transfer rate may have the following negative effects:

1. The object of the system is to deliver (or remove) heat where required. When the flow is reduced from the design rate, the system must supply heating or cooling for a longer period to maintain room temperature.
2. As the load reaches design conditions, the reduced flow rate is unable to maintain room design conditions.

Terminals with lower water temperature drops have a greater tolerance for unbalanced conditions. However, larger water flows are necessary, requiring larger pipes, pumps, and pumping costs. Also, automatic valve control is more difficult.

System balance becomes more important in terminals with a large temperature difference. Less water flow is required, which reduces the size of pipes, valves, and pumps as well as pumping costs. A more linear emission curve gives better system control.

Generalized Chilled Water Terminal— Flow Versus Heat Transfer

The heat transfer for a typical chilled water coil in an air duct versus water flow rate is shown in Figure 3. The curves shown are based on ARI rating points: 7.2°C inlet water at a 5.6 K rise with entering air at 26.7°C dry bulb and 19.4°C wet bulb.

The basic curve applies to catalog ratings for lower dry-bulb temperatures providing a consistent entering air moisture content (e.g., 23.9°C dry bulb, 18.3°C wet bulb). Changes in inlet water temperature, temperature rise, air velocity, and dry-bulb and wet-bulb temperatures will cause terminal performance to deviate from the curves. Figure 3 is only a general representation of the total heat transfer change versus flow for a hydronic cooling coil and does not apply to all chilled water terminals. Comparing Figure 3 with Figure 1 indicates the similarity of the nonlinear heat transfer in comparison with flow for both the heating and the cooling terminal.

Table 1   Load Flow Variations

Load Type	% Design Flow at 90% Load	Other Load, Order of %		
		Sensible	Total	Latent
Sensible	65	90	84	58
Total	75	95	90	65
Latent	90	98	95	90

Note: Dual-temperature systems are designed to chilled flow requirements and often operate on a 5 K temperature drop at full-load heating.

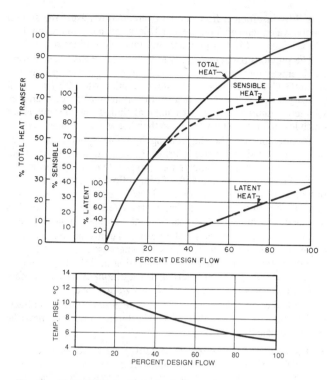

Fig. 3 Chilled Water Terminal Flow Versus Heat Transfer

Table 1 shows that if the coil is selected for the load, and the flow is reduced to 90% of the load, three flow variations can be interpreted to satisfy the reduced load at various sensible and latent combinations.

Flow Tolerance and Balance Procedure

The design procedure rests on a design flow rate and an allowable flow tolerance. The designer must define both the terminal's flow rates and feasible flow tolerance, bearing in mind that the cost of balancing rises with tightened flow tolerance. Also, the designer must remember that any overflow increases pumping cost, and any decrease of flow reduces the maximum heating or cooling at design conditions.

WATER-SIDE BALANCING

The water side should be balanced by direct flow measurement. This approach is accurate because it avoids the compounding errors introduced by temperature difference procedures. Measuring the flow at each terminal enables proportional balancing and ultimate matching of the pump to the actual system requirements (by trimming the pump impeller or reducing the pump motor power, for example). In many cases, reduction in pump operating cost will pay for the cost of water-side balancing.

Equipment

Proper equipment selection and preplanning is needed to successfully balance hydronic systems. Circumstances sometimes dictate that flow, temperature, and pressure be measured. The designer should specify the water flow balancing devices for installation during construction and testing during the balancing of the hydronic system. The devices may consist of all or a combination of the following:

- Flowmeters (ultrasonic stations, turbines, venturi, orifice plate, multiported pitot tubes, and flow indicators)
- Manometers, ultrasonic digital meters, and differential pressure gages (either analog or digital)

- Portable digital meter to measure flow and pressure drop
- Portable pyrometers to measure temperature differentials when test wells are not provided
- Test pressure taps, pressure gages, thermometers, and wells
- System components used as flowmeters (terminal coils, chillers, heat exchangers, or control valves if using the manufacturer's factory-certified flow versus drop curves)
- Flow-limiting or regulator devices (to add a variable load to the pump)
- Pumps with factory-certified pump curves
- Balancing valve with a factory-rated flow coefficient, a flow versus handle position and pressure drop table, or a slide rule flow calculator.

Record Keeping

Balancing requires keeping accurate records while making field measurements. The actual dated and signed field test report helps the designer or customer in approval of the work, and the owner will have a valuable reference when documenting future changes.

Sizing Balancing Valves

A balancing valve is placed in the system to adjust water flow to a terminal, branch, zone, riser, or main. A common valve-sizing approach is to select for the line size; this may be unwise, however, as too much pressure drop may have to be added, which may damage the valve seat. Many balancing valves and measuring meters can give an accuracy of ±5% of range down to a pressure drop of 3 kPa with the balancing valve in the wide open position. Some designers select a balancing valve such that the pressure drop is between 5 and 10% of the total loop pressure drop at design flow when the valve is wide open. Too large a balancing valve pressure drop will affect the performance and flow characteristic of the control valve. Equation (2) may be used to determine the flow coefficient A_v for a balancing valve. The equation may also be used to size a control valve.

The pressure drop across a control valve is a function of a flow coefficient A_v, which represents the flow rate in cubic metres per second through a fully open valve at a pressure drop of 1 Pa. The flow coefficient is defined as follows:

$$A_v = Q\,(\rho/\Delta p)^{0.5} \qquad (2)$$

where

Q = design flow for terminal or valve, m³/s
ρ = density of fluid, kg/m³
Δp = pressure drop across valve, Pa

The traditional inch-pound flow coefficient C_v may be converted to the SI coefficient A_v by the following factor:

$$A_v = 24.0 \times 10^{-6} C_v \qquad (3)$$

Hydronic Balancing Methods

A variety of techniques for balancing hydronic systems is used. Balance by temperature difference and water balance by proportional method are the most common.

Balance by Temperature Difference

This often-used balancing procedure is based on water temperature difference measurement between supply and return at the terminal. The designer selects the cooling and/or heating terminal for a calculated design load at full-load conditions. At less than full load, which is true for most operating hours, the temperature drop is proportionately less. Figure 4 demonstrates this relationship for a heating

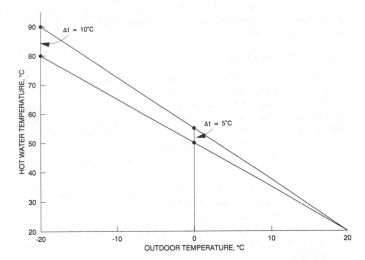

Fig. 4 Water Temperature Versus Outdoor Temperature Showing Approximate Temperature Difference

system at a design Δt of 10 K for outside design of –20°C and room design of 20°C.

At any outside temperature, other than design, the balancing technician should construct a similar chart and read off the Δt for balancing. For example, at 50% load, or –1°C outdoor air, the Δt required is 5 K, or 50% of the design drop.

The temperature balance method is a rough approximation and should not be used where great accuracy is required. It is not accurate enough for a cooling system or for high temperature drop heating systems.

Water Balance by Proportional Method

The proportional method of water-side balance uses as-built conditions and adapts well to design diversity factors. Circuits in a system are proportionately balanced to each other by a flow quotient, which is the actual flow rate divided by the design flow rate through a circuit:

$$\text{Flow quotient} = \frac{\text{Actual flow rate}}{\text{Design flow rate}} \qquad (4)$$

To balance a branch system proportionally, first open the balancing valves and the control valves in that branch to their wide-open position, and calculate each balancing valve's quotient based on actual measurements at each terminal. Record these values on the test form, and note the circuit with the lowest flow quotient.

The circuit with the lowest quotient is the circuit with the greatest pressure loss and is the reference circuit. Adjust all the other balancing valves in that branch until they have the same quotient as the reference circuit (at least one valve in the branch should be fully open).

Note that when a second valve is adjusted, the flow quotient in the reference valve will also change, and continued adjustment is required to make their flow quotients equal. Once they are equal, they will remain equal or in proportional balance to each other while other valves in the branch are adjusted or there is a change in pressure or flow.

When all the balancing valves are adjusted to their branch's respective flow quotient, the total system water flow is adjusted to the design by setting the balancing valve at the pump discharge to a flow quotient of one.

The pressure drop across the balancing valve at the pump discharge is pressure produced by the pump that is not required to provide the design flow rate to the system. This excess pressure can be removed by trimming the pump impeller or reducing the pump speed. Once the trimming is done, the pump discharge balancing valve must be reopened to its wide-open position to provide the design flow.

As in variable-speed pumping, diversity and flow changes are well accommodated by a system that has been proportionately balanced. Since the balancing valves have been balanced to each other at a particular flow (design), any changes in that flow are proportionately distributed.

Balancing of the water side in a system that uses diversity must be done at full flow. The system components are selected based on heat transfer at full flow, so they must be balanced to this point. To accomplish full-flow proportional balance, shut off part of the system while balancing the remaining sections. When a section has been balanced, shut it off and open the section that was open originally to complete full balance of the system. (When balancing, proper care should be taken if the building is occupied or if near full load conditions exist).

Variable-Speed Pumping. To achieve hydronic balance, full flow through the system is required during the balancing procedure. Once balance is achieved, the system can be placed on automatic control, and the pump speed allowed to change.

After the full-flow condition is balanced and the system differential pressure set point is established, to control the variable-speed pumps, observe the flow on the circuit with the greatest resistance as the other circuits are closed one at a time. The flow in the observed circuit should remain equal to or more than the previously set flow. Note that water flow may become laminar at less than 0.6 m/s, which alters the heat transfer characteristics of the system.

Other Balancing Techniques

Flow Balancing by Rated Differential Procedure. This procedure depends on deriving a performance curve for the test coil that compares water temperature difference Δt_w to entering water temperature t_{ew} minus entering air temperature t_{ea}. One point of the desired curve can be determined from the manufacturer's ratings since these are published as $(t_{ew} - t_{ea})$. A second point is established by observing that the heat transfer from air to water is zero (and consequently $\Delta t_w = 0$) when $(t_{ew} - t_{ea})$ is zero. With these two points, an approximate performance curve can be drawn (see Figure 5). Then, for any other $(t_{ew} - t_{ea})$, this curve is used to determine the appropriate Δt_w.

Example 1. From manufacturer's data:

$$\begin{aligned} \text{Capacity} &= 3 \text{ kW} \\ t_{ew} &= 95°\text{C} \\ t_{ea} &= 15°\text{C} \\ \text{Water flow} &= 0.1 \text{ L/s} \\ c_p &= 4.18 \text{ kJ/(kg·K)} \\ \rho &= 1.0 \text{ kg/L} \end{aligned}$$

Solution:

1. Calculate rated Δt_w.

$$\Delta t_w = \frac{3}{4.18 \times 0.1 \times 1} = 7.18 \text{ K}$$

2. Construct a performance curve as illustrated in Figure 5.
3. From test data:

$$\begin{aligned} t_{ew} &= 80°\text{C} \\ t_{ea} &= 20°\text{C} \\ t_{ew} - t_{ea} &= 60°\text{C} \end{aligned}$$

4. From Figure 5 read $\Delta t_w = 5.4$ K, which is required to balance water flow at 0.1 L/s. The water temperature difference may also be calculated as proportion of the rate value as follows:

$$\frac{(t_{ew} - t_{ea})_{\text{test}}}{(t_{ew} - t_{ea})_{\text{rated}}} (\Delta t_w)_{\text{rated}} = (\Delta t_w)_{\text{required}}$$

$$\frac{80 - 20}{95 - 15} \times 7.18 = 5.4 \text{ K}$$

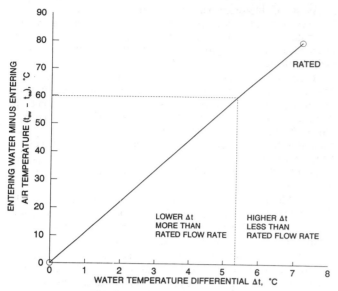

Fig. 5　Coil Performance Curve
(Derived from manufacturers' data)

This procedure is useful for balancing terminal devices such as finned tube convectors, where flow measuring devices do not exist and where airflow measurements cannot be made. It may also be used for cooling coils for sensible transfer (dry coil).

Flow Balancing by Total Heat Transfer. This procedure determines water flow by running an energy balance around the coil. From field measurements of airflow, wet- and dry-bulb temperatures both upstream and downstream of the coil, and the difference Δt_w between the entering and leaving water temperatures, water flow can be determined by the following equations:

$$Q_w = q/4180 \,\Delta t_w \tag{5}$$

$$q_{cooling} = 1.20 \, Q_a (h_1 - h_2) \tag{6}$$

$$q_{heating} = 1.23 \, Q_a (t_1 - t_2) \tag{7}$$

where

Q_w = water flow rate, L/s
q = load, W
$q_{cooling}$ = cooling load, W
$q_{heating}$ = heating load, W
Q_a = airflow rate, L/s
h = enthalpy, kJ/kg
t = temperature, °C

Example 2. Find the water flow for a cooling system having the following characteristics:

Test data

t_{ewb} = entering wet-bulb temperature = 20.3°C
t_{lwb} = leaving wet-bulb temperature = 11.9°C
Q_a = airflow rate = 10 000 L/s
t_{lw} = leaving water temperature = 15.0°C
t_{ew} = entering water temperature = 8.6°C

From psychrometric chart

h_1 = 76.52 kJ/kg
h_2 = 52.01 kJ/kg

Solution: From Equations (5) and (6),

$$Q_w = \frac{1.20 \times 10\,000\,(76.52 - 52.01)}{4180\,(15.0 - 8.6)} = 11.0 \text{ L/s}$$

The desired water flow is achieved by successive manual adjustments and recalculations. Note that these temperatures can be

greatly influenced by the heat of compression, stratification, bypassing, and duct leakage.

General Balance Procedures

All the variations of balancing hydronic systems cannot be listed; however, the general approach should balance the system while minimizing operating cost. Excess pump pressure (excess operating power) can be eliminated by trimming the pump impeller. Allowing excess pressure to be absorbed by throttle valves adds a lifelong operating cost penalty to system operation.

The following is a general procedure based on setting the balance valves on the site:

1. Develop a flow diagram if one is not included in the design drawings. Illustrate all balance instrumentation, and include any additional instrumentation requirements.
2. Compare pumps, primary heat exchangers, and terminal units specified and determine whether a design diversity factor can be achieved.
3. Examine the control diagram and determine the control adjustments needed to obtain design flow conditions.

Balance Procedure—Primary and Secondary Circuits

1. Inspect the system completely to ensure that (a) it has been flushed out, it is clean, and all air is out of the system; (b) all manual valves are open or in operating position; (c) all automatic valves are in their proper positions and operative; and (d) the expansion tank is properly charged.
2. Place the controls in position for design flow.
3. Examine the flow diagram and piping for obvious short circuits; check flow and adjust the balance valve.
4. Take pump suction, discharge, and differential pressure readings at both full flow and no flow. For larger pumps, a no flow condition may not be safe. In any event, valves should be closed slowly.
5. Read pump motor amperage and voltage, and determine approximate power.
6. Establish a pump curve, and determine the approximate flow rate.
7. If a total flow station exists determine the flow, and compare this flow with the pump curve flow.
8. If possible, set the total flow about 10% high using the total flow station first and the pump differential pressure second; then maintain pumped flow to a constant value as balance proceeds by adjusting the pump throttle valve.
9. If branch main flow stations exist, these should be tested and set, starting by setting the shortest runs low as balancing proceeds to the longer branch runs.
10. If the system incorporates primary and secondary pumping circuits, a reasonable balance must be obtained in the primary loop before the secondary loop can be considered. The secondary pumps must be running and terminal units must be open to flow when the primary loop is being balanced, unless the secondary loop is decoupled.

SYSTEM COMPONENTS AS FLOWMETERS

Flow Measurement Based on Manufacturer's Data

Any system component (terminal, control valve, or chiller) that has an accurate, factory-certified flow/pressure drop relationship can be used as a flow-indicating device. The flow and pressure drop may be used to establish an equivalent flow coefficient as shown in Equation (3).

According to the Bernoulli equation, pressure drop varies as the square of the velocity or flow rate, assuming density is constant:

$$Q_1^2 / Q_2^2 = \Delta h_1 / \Delta h_2 \qquad (8)$$

For example, a chiller has a certified pressure drop of 80 kPa at 10 L/s. The calculated flow with a field-measured pressure drop of 90 kPa is

$$Q_2 = Q_1 (\Delta h_2 / \Delta h_1)^{0.5} = 10 (90/80)^{0.5} = 10.6 \text{ L/s}$$

However, flow results calculated in this manner have limited accuracy. Accuracy of system components used as flow indicators depends on (1) the accuracy of cataloged information concerning flow/pressure drop relationships and (2) the accuracy of the pressure differential readings. As a rule, the component should be factory-certified flow tested if it is to be used as a flow indicator.

Pressure Differential Readout by Gage

Either gages or manometers can be used to read differential pressures. Gages are usually used for high differential pressures and mercury manometers for lower differentials. Accurate gage readout is diminished when two gages are used, especially when the gages are permanently mounted and, as such, subject to malfunction.

A single high-quality gage should be used for differential readout (Figure 6). This gage should be alternately valved to the high- and low-pressure side to establish the differential. A single gage needs no static height correction, and errors caused by gage calibration are eliminated.

Differential pressure can also be read from differential gages, thus eliminating the need to subtract outlet from inlet pressures to establish differential pressure. Differential pressure gages are usually dual gages mechanically linked to read differential pressure.

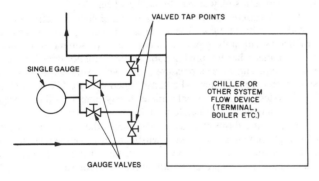

Fig. 6   Single Gage for Reading Differential Pressure

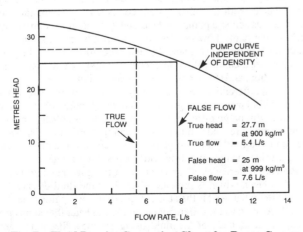

Fig. 7   Fluid Density Correction Chart for Pump Curves

Conversion of Differential Pressure to Head

Gages can be calibrated to metres of fluid head, which, historically, refers to the height of water above a point in the fluid circuit. Head is a function of density and is related to pressure by the hydrostatic law as follows:

$$h = p / \rho g \qquad (9)$$

where

    h = fluid head, m
    p = pressure, kPa
    ρ = density, kg/m^3
    g = standard force of gravity = 9.806 65 N/kg

But because the calibration only applies at one density (typically at a standard temperature of 15°C), the reading of head may require correction when the gage is applied to high-temperature water with a significantly lower density.

For example, a manufacturer may test a boiler control valve with 40°C water. Differential pressures from another test made in the field at 120°C may be correlated with the manufacturer's data by using Equation (8) to account for the density differences of the two tests.

When differential heads are used to estimate flow, a density correction must be made because of the shape of the pump curve. For example, in Figure 7 the uncorrected differential reading for pumped water with a density of 900 kg/m^3 is 25 m; the gage conversion was assumed to be for water with a density of 999 kg/m^3. The uncorrected or false reading gives a 40% error in flow estimation.

Differential Head Readout with Mercury Manometers

Mercury manometers are used for differential pressure readout, especially when very low differentials, great precision, or both, are required. Mercury manometers should not be used for field testing because mercury could blow out into the water system and rapidly deteriorate the components. Mercury manometers must be handled with care. A proposed manometer arrangement is shown in Figure 8.

Reference to Figure 8 and the following instructions provide accurate manometer readings with minimum risk of mercury blowout.

1. Make sure that both legs of the manometer are filled with water.
2. Open the purge bypass valve.
3. Open valved connections to high and low pressure.
4. Open the bypass vent valve slowly and purge air here.

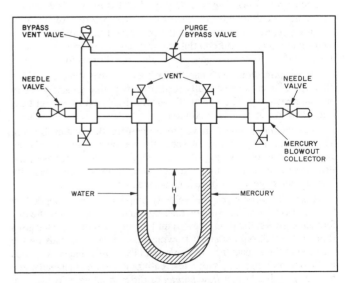

**Fig. 8   Mercury Manometer Arrangement for
Accurate Reading and Blowout Protection**

Table 2   Conversion Table

Water Temperature, °C	Multiplier to Convert mm Hg to Kilopascals
15	0.1232
60	0.1251
90	0.1275
120	0.1304
150	0.1343
170	0.1371

5. Open manometer block vents and purge air at each point.

6. Close the needle valves. The mercury columns should zero in if the manometer is free of air. If not, vent again.

7. Open the needle valves and begin throttling the purge bypass valve slowly, watching the mercury columns. If the manometer has an adequate available mercury column, the valve can be closed and the differential reading taken. However, if the mercury column reaches the top of the manometer before the valve is completely closed, this indicates insufficient manometer height, and further throttling will blow mercury into the blowout collector. A longer manometer or the single gage readout method should then be used.

An error is often introduced when converting millimetres of mercury to kilopascals because many conversion factors correct for a mercury column exposed to air. The conversion factor for air does not consider that the mercury column is counterbalanced by an equal height of liquid. When a mercury manometer is used to determine pressure in a water flow test, the conversion factor is 1 mm of mercury equals 0.123 kPa when the water is about 15°C. The conversion factor changes with test fluid density, which is a function of temperature as shown in Table 2.

Other Flow Information Devices, Orifice Plates, and Flow Indicators

Manufacturers provide flow information for several devices used in hydronic system balance. In general, the devices can be classified as (1) orifice flowmeters, (2) venturi flowmeters, (3) velocity impact meters, (4) pitot-tube flowmeters, (5) bypass spring impact flowmeters, (6) calibrated balance valves, (7) turbine flowmeters, and (8) ultrasonic flowmeters.

The *orifice flowmeter* is widely used and is extremely accurate. The meter is calibrated and shows differential pressure versus flow. Accuracy generally increases as the pressure differential across the meter increases. The differential pressure readout instrument may be a manometer, differential gage, or single gage (Figure 6).

The *venturi flowmeter* has lower pressure loss than the orifice plate meter because a carefully formed flow path increases velocity pressure recovery. The venturi flowmeter is placed in a main flow line where it can be read continuously.

Velocity impact meters have precise construction and calibration. The meters are generally of specially contoured glass or plastic, which permits observation of a flow float. As flow increases, the flow float rises in the calibrated tube to indicate flow rate. Velocity impact meters generally have high accuracy.

A special version of the velocity impact meter is applied to hydronic systems. This version operates on the velocity pressure difference between the pipe sidewall and the pipe center, which causes fluid to flow through a small flowmeter. Accuracy depends on the location of the impact tube and on a velocity profile that corresponds to theory and the laboratory test calibration base. Generally, the accuracy of this *bypass flow impact* or differential velocity pressure flowmeter is less than a flow-through meter, which can operate without creating a pressure loss in the hydronic system.

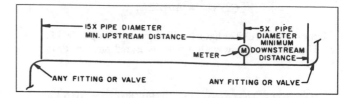

Fig. 9   Minimum Installation Dimensions for Flowmeter

The *pitot-tube flowmeter* is also used for pipe flow measurement. Manometers are generally used to measure velocity pressure differences because these differences are low.

The *bypass spring impact flowmeter* uses a defined piping pressure drop to cause a correlated bypass side branch flow. The side branch flow pushes against a spring that increases in length with increased side branch flow. Each individual flowmeter is calibrated to relate extended spring length position to main flow. The bypass spring impact flowmeter has, as its principal merit, a direct readout. However, dirt on the spring reduces accuracy. The bypass is opened only when a reading is made. Flow readings can be taken at any time.

The *calibrated balance valve* is an adjustable orifice flowmeter. Balance valves can be calibrated so that a flow/pressure drop relationship can be obtained for each incremental setting of the valve. A ball, rotating plug, or butterfly valve may have its setting expressed in percent open or degree open; a globe valve, in percent open or number of turns. The calibrated balance valve must be manufactured with precision and care to ensure that each valve of a particular size has the same calibration characteristics.

The *turbine flowmeter* is a mechanical device. The velocity of the liquid spins a wheel in the meter, which generates a 4 to 20 mA output that may be calibrated in units of flow. The meter must be well maintained, as wear or water impurities on the bearing may slow the wheel, and debris may clog or break the wheel.

The *ultrasonic flowmeter* senses sound signals, which are calibrated in units of flow. The ultrasonic metering station may be installed as part of the piping, or it may be a strap-on meter. In either case, the meter has no moving parts to maintain, nor does it intrude into the pipe and cause a pressure drop. Two distinct types of ultrasonic meter are available—the transit time meter for HVAC or clear water systems and the Doppler meter for systems handling sewage or large amounts of particulate matter.

If any of the above meters are to be useful, the minimum distance of straight pipe upstream and downstream as recommended by the meter manufacturer and flow measurement handbooks must be adhered to. Figure 9 presents minimum installation suggestions.

Using a Pump as an Indicator

Although the pump is not a meter, it can be used as an indicator of flow together with the other system components. Differential pressure readings across a pump can be correlated with the pump curve to establish the pump flow rate. Accuracy depends on (1) accuracy of readout, (2) pump curve shape, (3) actual conformance of the pump to its published curve, (4) pump operation without cavitation, (5) air-free operation, and (6) velocity pressure correction.

When a differential pressure reading must be taken, a single gage with manifold provides the greatest accuracy (Figure 10). The pump suction to discharge differential can be used to establish pump differential pressure and, consequently, pump flow rate. The single gage and manifold may also be used to check for strainer clogging by measuring the pressure differential across the strainer.

Pressure differential, as obtained from the gage reading, is converted to head, which is pressure divided by the fluid density. The pump differential head is then used to determine pump flow rate (Figure 11). As long as the differential head used to enter the pump curve is expressed as head of the fluid being pumped, the pump curve shown by the manufacturer should be used as described. The pump curve may state that the curve was defined by test with 30°C water. This is unimportant, since the same curve applies unchanged from 15 to 120°C water, or to any fluid within a broad viscosity range.

Generally, pump-derived flow information, as established by the performance curve, is questionable unless the following precautions are observed:

1. The installed pump should be factory calibrated by a test to establish the actual flow-pressure relationship for that particular pump. Production pumps can vary from the cataloged curve because of minor changes in impeller diameter, interior casting tolerances, and machine fits.
2. When a calibration curve is not available for a centrifugal pump being tested, the discharge valve can be closed briefly to establish the no-flow shutoff pressure, which can be compared to the published curve. If the shutoff pressure differs from that published, draw a new curve parallel to the published curve. While not exact, the new curve will usually fit the actual pumping circumstance more accurately. Clearance between the impeller and casing minimize the danger of damage to the pump during a no-flow test, but verify with the manufacturer to be sure.
3. Differential pressure should be determined as accurately as possible, especially for pumps with flat flow curves.
4. The pump should be operating air-free and without cavitation. A cavitating pump will not operate to its curve, and differential readings will provide false results.
5. Ensure that the pump is operating above the minimum net positive suction pressure.

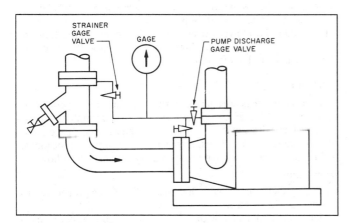

Fig. 10 Single Gage for Differential Readout Across Pump and Strainer

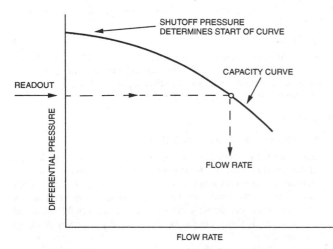

Fig. 11 Differential Pressure Used to Determine Pump Flow

6. Power readings can be used as a check for the operating point when the pump curve is flat or as a reference check when the pump is suspected to be cavitating or providing false readings because of air.
7. The flow determined by the pump curve should be compared to the flow measured at the flowmeters, flow measured by pressure drops through circuits, and flow measured by pressure drops through exchangers.

The power draw should be measured in watts. Ampere readings cannot be trusted because of voltage and power factor problems. If motor efficiency is known, the wattage drawn can be related to pump brake power, as described on the pump curve, and the operating point determined.

Central Plant Chilled Water Systems

In existing installations establishing thermal load profiles accurately is of prime importance in establishing proper primary chilled water supply temperature and flow. In new installations, actual load profiles can be compared with design load profiles to obtain valid operating data.

To perform proper testing and balancing, all interconnecting points between the primary and secondary systems must be designed with sufficient temperature, pressure, and flow connections so that adequate data may be indicated and/or recorded.

Water Flow Instrumentation

As indicated previously, the proper location and use of system instrumentation is vital to the accuracy of the system balance. Instruments for testing temperature and pressure at various points of information are listed in Table 3. With a table of this nature, instrumentation can be tailored to a specific design with ease and accuracy. Flow-indicating devices should be placed in water systems as follows:

- At each major heating coil bank (0.6 L/s or more)
- At each major cooling coil bank (0.6 L/s or more)
- At each bridge in primary-secondary systems
- At each main pumping station

Table 3   Instruments for Monitoring a Water System

Point of Information	Manifold Gage	Single Gage	Thermometer	Test Well	Pressure Tap
Pump—Suction, discharge	x				
Strainer—In, out					x
Cooler—In, out		x	x		
Condensers—In, out		x	x		
Concentrator—In, out		x	x		
Absorber—In, out		x	x		
Tower cell—In, out				x	x
Heat exchanger—In, out	x		x		x
Coil—In, out				x	x
Coil bank—In, out		x	x		
Booster coil—In, out					x
Cool panel—In, out					x
Heat panel—In, out				x	x
Unit heater—In, out					x
Induction—In, out					x
Fan coil—In, out					x
Water boiler—In, out			x		
Three-way valve—All ports					x
Zone return main			x		
Bridge—In, out			x		
Water makeup		x			
Expansion tank		x			
Strainer pump					x
Strainer main	x				
Zone three-way—All ports				x	x

- At each water chiller evaporator
- At each water chiller condenser
- At each water boiler outlet
- At each floor takeoff to booster reheat coils, fan coil units, induction units, ceiling panels, and radiation (Do not exceed 25 terminals off of any one zone meter probe.)
- At each vertical riser to fan coil units, induction units, and radiation
- At the point of tie-in to existing systems

STEAM DISTRIBUTION SYSTEMS

Procedures for Steam Balancing Variable Flow Systems

A steam distribution system cannot be balanced by adjustable flow-regulating devices. Flow regulation is accomplished by fixed restrictions built into the piping system in accordance with carefully designed pipe and orifice sizes.

It is important to have a balanced distribution of steam to all portions of the steam piping system at all load levels. This is best accomplished by the proper design of the steam distribution piping system by carefully considering steam pressure, steam quantities required by each branch circuit, pressure drops, steam velocities, and pipe sizes. Just as other flow systems are balanced, steam distribution systems are balanced by ensuring that the pressure drops are equalized at design flow rates for all portions of the piping system. Only marginal balancing can be done by pipe sizing. Therefore, additional steps must be taken to achieve a balanced performance.

Steam flow balance can be improved by using spring-type packless supply valves equipped with precalibrated orifices. The valves should have a tight shutoff between 80 to 400 kPa (gage). These valves have a nonrising stem, are available with a lockshield, and have a replaceable disk. Orifice flanges can also be used to regulate and measure steam flow at appropriate locations throughout the system. The orifice sizes are determined by the pressure drop required to provide a given flow rate at a given location in the system. A schedule should be prepared showing (1) orifice sizes, (2) valve or pipe sizes, (3) required flow rates, and (4) corresponding pressure differentials for each flow rate. It may be useful to calculate pressure differentials for several flow rates for each orifice size. Such a schedule should be maintained for future reference.

After the appropriate regulating orifices are installed in the proper locations, the system should be tested for tightness by sealing all openings in the system and applying a vacuum of 35 kPa (absolute), held for 2 hours. Next, the system should be readied for warmup and pressurizing with steam following the procedures outlined in Section VI of the ASME *Boiler and Pressure Vessel Code*. After the initial warm-up and system pressurization, evaluate system steam flow, and compare it to system requirements. The orifice schedule calculated earlier will now be of value should any of the orifices need to be changed.

Steam Flow Measuring Devices

Many devices are available for measuring flow in steam piping systems: (1) steam meters, (2) condensate meters, (3) orifice plates, (4) venturi fittings, (5) steam recorders, and (6) manometers for reading differential pressures across orifice plates and venturi fittings. Some of these devices are permanently affixed to the piping system to facilitate taking instantaneous readings that may be necessary for proper system operation and control. A surface pyrometer used in conjunction with a pressure gage is a convenient way to determine steam saturation temperature and the degree of superheat at various locations in the system. Such information can be used to evaluate performance characteristics of the system.

COOLING TOWERS

Field testing cooling towers is a demanding and difficult task. ASME *Standard* PTC 23 (1986) and CTI *Standard Specification*

ATC-105 (1990) establish procedures for these tests. Certain general guidelines for testing cooling towers are as follows:

Conditions at Time of Test
- Water flow within 15% of design flow
- Heat load within 30% of design heat load and stabilized
- Entering wet bulb within 6.7 K of entering design wet bulb

Using the above limitations and field readings that are as accurate as possible, a projection to design conditions produces an accuracy of ±5% of tower performance.

Conditions for Performing Test
- Water-circulating system serving the tower should be thoroughly cleaned of all dirt and foreign matter. Samples of water should be clear and indicate clear passage of water through pumps, piping, screens, and strainers.
- Fans serving the cooling tower should operate in proper rotation. Foreign obstructions should be removed. Permanent obstruction should be noted.
- Interior filling of cooling tower should be clean and free of foreign materials such as scale, algae, or tar.
- Water level in the tower basin should be maintained at the proper level. Visually check the basin sump during full flow to determine that the centrifugal action of the water is not causing entrainment of air, which could cause pump cavitation.
- Water-circulating pumps should be tested with full flow through the tower. If flow exceeds design, it should be valved down until design is reached. The flow finally set should be maintained throughout the test period. All valves, except necessary balancing valves, should be in the full-open position.
- If makeup and blowdown have facilities to determine flow, set them to design flow at full flow through the tower. If flow cannot be determined, shut off both.

Instruments

Testing and balancing agencies provide instruments to perform the required tests. Mechanical contractors provide and install all components such as orifice plates, venturis, thermometer wells, gage cocks, and corporation cocks. Designers specify measuring point locations.

The instruments used should be recently calibrated as follows:

Temperature
- Mercury thermometer with divisions of 0.1 K in a proper well for water should be used.
- Mercury thermometer with solar shield and 0.1 K divisions or thermocouple with 0.1 K readings having mechanical aspiration for wet-bulb readings should be used.
- Sling psychrometer may be used for rough checks.
- Mercury thermometer with 0.1 K divisions should be used for dry-bulb readings.

Water Flow
- Orifice or venturi drops can be read using a water-over-mercury manometer or a recently calibrated differential pressure gage.
- Where corporation cocks are installed, a pitot tube and manometer traverse can be made by trained technicians.

Test Method

The actual test consists of the following steps:

1. Conduct water flow tests to determine volume of water on the tower, and volume of makeup and blowdown water.
2. Conduct water temperature tests, if possible, in suitable wells as close to the tower as possible. Temperature readings at pumps or the condensing element are not acceptable in tower evaluation. If there are no wells, surface pyrometer readings are acceptable.
3. Take makeup water volume and temperature readings at the point of entry to the system.

4. Take blowdown volume and temperature readings at the point of discharge from system.

5. Take inlet and outlet dry- and wet-bulb temperature readings using the prescribed instruments.

   a) Use wet-bulb entering and leaving temperatures to determine tower actual performance as against design.

   b) Use wet-bulb and dry-bulb entering and leaving temperatures to determine evaporation involved.

6. If the tower has a ducted inlet or outlet where a reasonable duct traverse can be made, use this air volume as a cross-check of tower performance.

7. Take wet- and dry-bulb temperature readings between 1 and 1.5 m from the tower on all inlet sides. These readings shall be taken halfway between the base and the top of the inlet louvers at no more than 1.5-m spacing horizontally; then they should be averaged. Note any unusual inlet conditions.

8. Note wind velocity and direction at the time of test.

9. Take test readings continually with a minimum time lapse between readings.

10. If the first test indicates a tower deficiency, perform two additional tests to verify the original readings.

TEMPERATURE CONTROL VERIFICATION

The test and balance technician should work closely with the temperature control installer to ensure that the project is completed correctly. The balancing technician needs to verify proper operation of the control system and communicate findings back to the agency responsible for ensuring that the system has been installed correctly. This is usually the HVAC system designer, although others may be involved. Generally, the balancing technician does not adjust, relocate, or calibrate the controls. However, this is not always the case, and differences do occur with VAV terminal unit controllers. The balancing technician should be familiar with the specifications and design intent of the project so that all responsibilities are understood.

During the design and specification phase of the project, the designer should specify verification procedures for the control systems and the responsibilities of the contractor who installs the temperature controls. It is important that the designer specify (1) the degree of coordination between the installer of the control system and the balancing technician and (2) the testing responsibilities of each.

Verification of the control system operation starts with a review by the balancing technician of the submitted documents and the shop drawings of the control system. In some cases the controls technician should instruct the balancing technician in the operation of certain control elements, such as digital terminal unit controllers. This is followed by schedule coordination between the control and balancing technicians. In addition, the balancing technician and the controls technician need to work together when reviewing the operation of some sections of the HVAC system (not all). This is particularly important with regard to VAV systems and the setting of the parameters for flow measurement in digital terminal unit controllers.

Major mechanical systems should be verified after the testing, adjusting, and balancing is completed. The control system should be operated in stages to prove its capability of matching system capacity to varying load conditions. Mechanical subsystem controllers should be verified when balancing data is collected, considering that the entire system may not be completely functional at the time of verification. Testing and verification should account for seasonal variations; tests should be performed under varying outdoor loads to assure operational performance. A retest of a random sampling of terminal units may be desirable to verify the work of the controls technician.

Suggested Procedures

Although both electrical and pneumatic control system styles are used, the following verification procedures may be used with either.

1. Obtain design drawings and documentation, and become thoroughly acquainted with the design intent and specified responsibilities.

2. Obtain copies of approved control system shop drawings.

3. Compare design to installed field equipment.

4. Obtain recommended operating and test procedures from the manufacturers.

5. Verify with the controls contractor that all controllers are calibrated and commissioned.

6. Check the location of transmitters and controllers. Note adverse conditions that would affect control, and suggest relocation as necessary.

7. Note settings on controllers. Note discrepancies between set point for controller and actual measured variable.

8. Verify operation of all limiting controllers, positioners and relays (e.g., high- and low-temperature thermostats, high and low differential pressure switches, etc.).

9. Activate controlled devices, checking for free travel and proper operation of stroke for both dampers and valves. Verify normally open (NO) or normally closed (NC) operation.

10. Verify sequence of operation of controlled devices. Note line pressures and controlled device positions. Correlate to air or water flow measurements. Note speed of response to step change.

11. Confirm interaction of electrically operated switch transducers.

12. Confirm interaction of interlock and lockout systems.

13. Coordinate balancing and control system technicians' schedules to avoid duplication of work and prevent testing errors.

Pneumatic System Modifications

1. Verify main control supply air pressure and observe compressor and dryer operation.

2. For hybrid systems using electronic transducers for pneumatic actuation, modify procedures accordingly.

Electronic Systems

1. Monitor voltages of power supply and controller output. Determine whether the system operates on a grounded or nongrounded power supply, and check condition. Although electronic control systems are being designed with more robust electronic circuits, improper grounding can cause functional variation in controller and actuator performance from system to system.

2. Note operation of electric actuators using spring return. Generally, actuators should be under control and use springs only upon power failure to return to a failsafe position.

Direct Digital Controllers

Direct digital control (DDC) offers nontraditional challenges to the balancing technician. Many of the control devices, such as sensors and actuators, are the same as used in electronic and pneumatic systems. Currently DDC is dominated by two types of controllers—one is fully programmable and one is application-specific. Fully programmable controllers offer a group of functions that are linked together in an applications program to control a system such as an air-handling unit. Application-specific controllers are functionally defined with the programming necessary to carry out the functions required for a system, but not all of the adjustments and settings are defined. Both types of controllers and their functions have some variations. One of the functions is adaptive control, which includes control algorithms that automatically adjust settings of various controller functions.

The balancing technician must understand the controller functions so that they do not interfere with the test and balance functions. The balancing technician does not need to be literate in computer programming, although it does help. When testing the DDC, the following steps should be included:

1. Obtain controller application program. Discuss application of the designer's sequence with the control system programmer.
2. Coordinate testing and adjustment of controlled systems with mechanical systems testing. Avoid duplication of efforts between technicians.
3. Coordinate storage (e.g., saving to central DDC database and controller memory) of all required system adjustments with controls technician.

In cases where the balancing agency is required to test discrete points in the control system, the following steps are necessary:

1. Establish criteria for test with the designer.
2. Use reference standards that test the end device through the entire controller chain (e.g., device, wiring, controller, communications, and operator monitoring device). An example of this would be using a dry block temperature calibrator (a testing device that allows a temperature to be set, monitored, and maintained in a small chamber) to test a space temperature sensor. The sensor is installed with extra wire so that it may be removed from the wall and placed in the calibrator chamber. After the system is thermally stabilized, the temperature is read at the controller and the central monitor, if installed.
3. Report findings of reference and all points of reading.

FIELD SURVEY FOR ENERGY AUDIT

An energy audit is an organized survey of a specific building to identify and measure all energy uses, determine probable sources of energy losses, and list energy conservation opportunities. This is usually performed as a team effort under the direction of a qualified energy engineer. The field data can be gathered by firms employing technicians trained in testing, adjusting, and balancing.

Instruments

To determine a building's energy use characteristics, an accurate measurement of existing conditions must be made with proper instruments. Accurate measurements point out opportunities to reduce waste and provide a record of the actual conditions in the building before energy conservation measures were taken. They provide a compilation of installed equipment data and a record of equipment performance prior to changes. Judgments will be made based on the information gathered during the field survey; that which is not accurately measured cannot be properly evaluated.

Generally, the instruments required for testing, adjusting, and balancing are sufficient for energy conservation surveying. Possible additional instruments include a power factor meter, a light meter, combustion testing equipment, refrigeration gages, and equipment for recording temperatures, fluid flow rates, and energy use over time. Only high-quality instruments should be used.

Observation of system operation and any information the technician can obtain from the operating personnel pertaining to the operation should be included in the report.

Data Recording

Organized record keeping is extremely important. A camera is also helpful. Photographs of building components and mechanical and electrical equipment can be reviewed later when the data is analyzed.

Data sheets needed for energy conservation field surveys contain different and, in some cases, more comprehensive information than those used for testing, adjusting, and balancing. Generally, the energy engineer determines the degree of fieldwork to be performed; data sheets should be compatible with the instructions received.

Building Systems

The most effective way to reduce building energy waste is to identify and define the energy load by building system. This provides an orderly procedure for tabulating the load. Also, the most effective energy conservation opportunities can be achieved more quickly because high priorities can be assigned to systems that consume the most energy.

For this purpose, load is defined as the quantity of energy used in a building, or by one of its subsystems, for a given period.

A building can be divided into nonenergized systems and energized systems. Nonenergized systems do not require outside energy sources such as electricity and fuel. Energized systems (e.g., mechanical and electrical systems) require outside energy sources. Energized and nonenergized systems can be divided into subsystems defined by function. Nonenergized subsystems are (1) building site, envelope, and interior; (2) building utilization; and (3) building operation.

Building Site, Envelope, and Interior. The site, envelope, and interior are surveyed to determine how they can be changed to reduce the building load that the mechanical and electrical systems must meet without adversely affecting the building's appearance. This requires uncovering energy-wasting items and recording any existing conditions affecting the practicability of making changes to eliminate waste.

It is important to compare actual conditions with conditions assumed by the designer, so that the mechanical and electrical systems can be adjusted to balance their capacities to satisfy the real needs.

Building Use. The functioning of people within the building envelope subsystem is most important when considering the building load; it must be observed because the action of people affects the energy usage of all other building subsystems.

Building-use loads can be classified as (1) people occupancy loads and (2) people operation loads. People occupancy loads are related to schedule, density, and mixing of occupancy types (e.g., process and office). People operation loads are varied, such as (1) operation of manual window shading devices; (2) setting of room thermostats; and (3) conservation-related habits such as turning off lights, closing doors and windows, turning off energized equipment when not in use, and not wasting domestic hot or chilled water.

Building Operation. This subsystem consists of the operation and maintenance of all the building subsystems. The load on the building operation subsystem is affected by factors such as (1) the time at which janitorial services are performed, (2) janitorial crew size and time required to clean, (3) amount of lighting used to perform janitorial functions, (4) quality of the equipment maintenance program, (5) system operational practices, and (6) equipment efficiencies.

Building Energized Systems

The energized subsystems of the building are generally plumbing, heating, ventilating, cooling, space conditioning, control, electrical, and food service. Although these systems are interrelated and often use common components, it is important to evaluate the energy use of each subsystem as independently as possible, for a logical organization of data. In this way, proper energy conservation measures for each subsystem can be developed.

Process Loads

In addition to building subsystem loads, the process load in most buildings must be evaluated. The energy field auditor must be able to determine its impact.

Most tasks not only require energy for performing a service but also affect the energy consumption of other building subsystems. For example, if a process releases large amounts of heat to the space, the process consumes energy and also imposes a large load on the cooling system.

Guidelines for Developing a Field Study Form

A brief checklist follows that outlines requirements for a field study form needed to conduct an energy audit.

Inspection and Observation of All Systems. Record physical and mechanical condition of the following:

- Fan blades, fan scroll, drives, belt tightness, and alignment
- Filters, coils, and housing tightness
- Ductwork (equipment room and space, where possible)
- Strainers
- Insulation ducts and piping
- Makeup water treatment and cooling tower

Interview of Physical Plant Supervisor. Record answers to the following survey:

- Is the system operating as designed? If not, what changes have been made to ensure its performance?
- Have there been modifications or additions to the system?
- If the system has been a problem, list problems by frequency of occurrence.
- Are any systems cycled? If so, which systems and when, and would building load permit cycling systems?

Recording System Information. Record the following system/equipment identification:

- Type of system—single-zone, multizone, double-duct, low- or high-velocity, reheat, variable volume, or other
- System arrangement—fixed minimum outside air, no relief, gravity or power relief, economizer gravity relief, exhaust return, or other
- Air-handling equipment—fans (supply, return, and exhaust): manufacturer, model and size, type, and class; dampers (vortex, scroll, or discharge); motors: manufacturer, power requirement, full load amperes, voltage, phase, and service factor
- Chilled and hot water coils—area, tubes on face, fin spacing, and number of rows (coil data necessary when shop drawings are not available)
- Terminals—high-pressure mixing box: manufacturer, model, and type (reheat, constant volume, variable volume, induction); grilles, registers, and diffusers: manufacturer, model, style, and K factor to convert field-measured velocity to flow rate
- Main heating and cooling pumps, over 3.5 kW—manufacturer, pump service and identification, model and size, impeller diameter, speed, flow rate, pressure at full flow, and pressure at no flow; motor data: power, speed, voltage, amperes, and service factor
- Refrigeration equipment—chiller manufacturer, type, model, serial number, nominal cooling capacity, input power, total heat rejection, motor (kilowatts, amperes, volts), chiller pressure drop, entering and leaving chilled water temperatures, condenser pressure drop, condenser entering and leaving water temperatures, running amperes and volts, no-load running amperes and volts
- Cooling tower—manufacturer, size, type, nominal cooling capacity, range, flow rate, and entering wet-bulb temperature
- Heating equipment—boiler (small through medium) manufacturer, fuel, energy input (rated), and heat output (rated)

Recording Test Data. Record the following test data:

- Systems in normal mode of operation (if possible)—fan motor: running amperes and volts and power factor (over 3.5 kW); fan: speed, total air (pitot-tube traverse where possible), and static pressure (discharge static minus inlet total); static profile drawing (static pressure across filters, heating coil, cooling coil, and dampers); static pressure at ends of runs of the system (identifying locations)

- Cooling coils—entering and leaving dry- and wet-bulb temperatures, entering and leaving water temperatures, coil pressure drop (where pressure taps permit and manufacturer's ratings can be obtained), flow rate of coil (when other than fan), outdoor wet and dry bulb, time of day, and conditions (sunny or cloudy)

- Heating coils—entering and leaving dry-bulb temperatures, entering and leaving water temperatures, coil pressure drop (where pressure taps permit and manufacturer's ratings can be obtained), and flow rate through the coil (when other than fan)

- Pumps—no-flow pressure, full-flow discharge pressure, full-flow suction pressure, full-flow differential pressure, motor running amperes and volts, and power factor (over 3.5 kW)

- Chiller (under cooling load conditions)—chiller pressure drop, entering and leaving chilled water temperatures, condenser pressure drop, entering and leaving condenser water temperatures, running amperes and volts, no-load running amperes and volts, chilled water on and off, and condenser water on and off

- Cooling tower—water flow rate on tower, entering and leaving water temperatures, entering and leaving wet bulb, fan motor [amperes, volts, power factor (over 3.5 kW), and ambient wet bulb]

- Boiler (full fire)—input energy (if possible), percent CO_2, stack temperature, efficiency, and complete Orsat test on large boilers

- Boiler controls—description of the operation

- Temperature controls—operating and set point temperatures for mixed air controller, leaving air controller, hot deck controller, cold deck controller, outdoor reset, interlock controls, and damper controls; description of complete control system and any malfunctions

- Outside air intake versus exhaust air—total airflow measured by pitot-tube traverses of both outside air intake and exhaust air systems, obtained where possible. Determine whether there is an imbalance in the exhaust system that causes infiltration. Observe exterior walls to determine whether outside air can infiltrate into the return air system (record outside air, return air, and return air plenum dry- and wet-bulb temperatures). The greater the differential between outside and return air, the more evident the problem will be.

TESTING FOR SOUND AND VIBRATION

Testing for sound and vibration ensures that equipment is operating satisfactorily and that no objectionable noise and vibration are transmitted to the building structure and occupied space. Although sound and vibration are specialized fields that require expertise not normally developed by the HVAC engineer, the procedures to test HVAC systems are relatively simple and can be performed with a minimum of equipment by following the steps outlined in this section. Although this section provides useful information for resolving common noise and vibration problems, Chapter 43 should be consulted for information on problem solving or the design of HVAC systems.

TESTING FOR SOUND

Present technology does not permit tests to determine whether equipment is operating with desired sound levels. Field tests can determine only sound pressure levels, and equipment ratings are almost always in terms of sound power levels. Until new techniques are developed, the testing engineer can determine only (1) whether sound pressure levels are within desired limits and (2) which equipment, systems, or components are the source of excessive or disturbing transmission.

Sound-Measuring Instruments

Although an experienced listener can often determine whether systems are operating in an acceptably quiet manner, sound-measuring instruments are necessary to determine whether system noise levels are in compliance with specified criteria and, if not, to obtain and report detailed information to evaluate the cause of noncompliance. Instruments normally used in field testing are as follows:

The *precision sound level meter* is used to measure sound pressure level. The most basic sound level meters measure overall sound pressure level and have up to three weighted scales that provide limited filtering capability. The instrument is useful in assessing outdoor noise levels in certain situations and can provide limited information on the low-frequency content of overall noise levels, but it provides insufficient information for problem diagnosis and solution. Its usefulness in evaluating indoor HVAC sound sources is thus limited.

Proper evaluation of HVAC sound sources requires a sound level meter capable of filtering overall sound levels into frequency increments of one octave or less.

Sound analyzers provide detailed information about the sound pressure levels at various frequencies through filtering networks. The most popular sound analyzers are the octave band and center frequency types, which break the sound into the eight octave bands of audible sound. Instruments are also available for 0.33, 0.1, and narrower spectrum analysis; however, these are primarily for laboratory and research applications. Sound analyzers (octave band or center frequency) are required where specifications are based on noise criteria (NC) curves or similar frequency criteria, and for problem jobs where a knowledge of frequency is necessary to determine proper corrective action.

Personal computers are a versatile sound-measuring tool. Software used in conjunction with portable computers has all the functional capabilities described above, plus many that previously required a fully equipped acoustical laboratory. This type of sound-measuring system is many times faster and much more versatile than conventional sound level meters. With suitable accessories, it can also be used to evaluate vibration levels.

A *stethoscope* is an invaluable instrument in measuring sound levels and tracking down problems, as it enables the listener to determine the direction of the sound source.

Regardless of which sound-measuring system is used, it should be calibrated prior to each use. Some systems have built-in calibration; others use external calibrators. Much information is available on the proper application and use of sound-measuring instruments.

Air noise caused by air flowing at a velocity of over 5 m/s or by winds over 19 km/h can cause substantial error in sound measurements due to wind effect on the microphone. For outdoor measurements or in places where air movement is prevalent, either a wind screen for the microphone or a special microphone is required.

Sound Level Criteria

In the absence of specified values, the testing engineer must determine whether sound levels are within acceptable limits (Chapter 43). Note that complete absence of noise is seldom a design criterion, except for certain critical locations such as sound and recording studios. In most locations, a certain amount of noise is desirable to mask other noises and provide speech privacy; it also provides an acoustically pleasing environment, since few people can function effectively in extreme quiet. Table 2 of Chapter 7 of the 1993 *ASHRAE Handbook—Fundamentals* lists typical sound pressure levels. In determining allowable HVAC equipment noise, it is as inappropriate to demand 30 dB for a factory where the normal noise level is 75 dB as it is to specify 60 dB for a private office where normal noise level might be 35 dB.

Most field sound-measuring instruments and techniques yield an accuracy of ±3 dB, the smallest difference in sound pressure level that the average person can discern. A reasonable tolerance for

sound criteria is 5 dB; if 35 dBA is considered the maximum allowable noise, the design engineer should specify 30 dBA.

The measured sound level of any location is a combination of all sound sources present, including sound generated by HVAC equipment as well as sound from other sources such as plumbing systems and fixtures, elevators, light ballasts, and outside noises. In testing for sound, all sources from other than HVAC equipment are considered background or ambient noise.

Background sound measurements generally have to be made (1) when the specification requires that the sound levels from HVAC equipment only, as opposed to the sound level in a space, not exceed a certain specified level; (2) when the sound level in the space exceeds a desirable level, in which case the noise contributed by the HVAC system must be determined; and (3) in residential locations where little significant background noise is generated during the evening hours and where generally low allowable noise levels are specified or desired. Because background noise from outside sources such as vehicular traffic can fluctuate widely, sound measurements for residential locations are best made in the normally quiet evening hours.

Sound-Testing Procedures

Ideally, a building should be completed and ready for occupancy before sound level tests are taken. All spaces in which readings will be taken should be furnished with drapes, carpeting, and furniture, as these affect the room absorption and the subjective quality of the sound. In actual practice, since most tests have to be conducted before the space is completely finished and furnished for final occupancy, the testing engineer must make some allowances. Because furnishings increase the absorption coefficient and reduce to 4 dB the sound pressure level that can be expected between most live and dead spaces, the following guidelines should suffice for measurements made in unfurnished spaces. If the sound pressure level is 5 dB or more over specified or desired criterion, it can be assumed that the criterion will not be met, even with the increased absorption provided by furnishings. If the sound pressure level is 0 to 4 dB greater than specified or desired criterion, recheck when the room is furnished to determine compliance.

Follow this general procedure:

1. Obtain a complete set of accurate, as-built drawings and specifications, including duct and piping details. Review specifications to determine sound and vibration criteria and any special instructions for testing.

2. Visually check systems for noncompliance with plans and specifications, obvious errors, and poor workmanship. Turn system on for aural check. Listen for noise and vibration, especially duct leaks and loose fittings.

3. Adjust and balance equipment, as described in other sections, so that final acoustical tests are made with the system as it will be operating. It is desirable to perform acoustical tests for both summer and winter operation, but where this is not practical, make tests for the summer operating mode, as it usually has the potential for higher sound levels. Tests must be made for all mechanical equipment and systems, including standby.

4. Check calibration of instruments.

5. Measure sound levels in all areas as required, combining measurements as indicated in item 3 if equipment or systems must be operated separately. Before final measurements are made in any particular area, survey the area using an A-weighted scale reading (dBA) to determine the location of the highest sound pressure level. Indicate this location on a testing form, and use it for test measurements. Restrict the preliminary survey to determine location of test measurements to areas that can be occupied by standing or sitting personnel. For example, measurements would not be made directly in front of a diffuser located in the ceiling, but they would be made as close to the diffuser as standing or sitting

personnel might be situated. In the absence of specified sound criteria, the testing engineer should measure sound pressure levels in all occupied spaces to determine compliance with criteria indicated in Chapter 43 and to locate any sources of excessive or disturbing noise.

6. Determine whether background noise measurements must be made.

   a) If specification requires determination of sound level from HVAC equipment only, it will be necessary to take background noise readings by turning HVAC equipment off.

   b) If specification requires compliance with a specific noise level or criterion (e.g., sound levels in office areas shall not exceed 35 dBA), ambient noise measurements must be made only if the noise level in any area exceeds the specified value.

   c) For residential locations and areas requiring very low noise levels, such as sound recording studios and locations that are used during the normally quieter evening hours, it is usually desirable to take sound measurements in the evening and/or take ambient noise measurements.

7. For outdoor noise measurements to determine noise radiated by outdoor or roof-mounted equipment such as cooling towers and condensing units, the section on Sound Control for Outdoor Equipment in Chapter 43, which presents proper procedure and necessary calculations, should be consulted.

Noise Transmission Problems

Regardless of the precautions taken by the specifying engineer and the installing contractors, situations can occur where the sound level exceeds specified or desired levels, and there will be occasional complaints of noise in completed installations. A thorough understanding of Chapter 43 and the section on Vibration Testing in this chapter is desirable before attempting to resolve any noise and vibration transmission problems. The following is intended as an overall guide rather than a detailed problem-solving procedure.

All noise transmission problems can be evaluated in terms of the source-path-receiver concept. Objectionable transmission can be resolved by (1) reducing the noise at the source by replacing defective equipment, repairing improper operation, proper balancing and adjusting, and replacing with quieter equipment; (2) attenuating the paths of transmission with silencers, vibration isolators, and wall treatment to increase transmission loss; and (3) reducing or masking objectionable noise at the receiver by increasing room absorption or introducing a nonobjectionable masking sound. The following discussion includes (1) ways to identify actual noise sources using simple instruments or no instruments and (2) possible corrections.

When troubleshooting in the field, the engineer should listen to the offending sound. The best instruments are no substitute for careful listening, as the human ear has the remarkable ability to identify certain familiar sounds such as bearing squeak or duct leaks and is able to discern small changes in frequency or sound character that might not be apparent from meter readings only. The ear is also a good direction and range finder; because noise generally gets louder as one approaches the source, direction can often be determined by turning the head. Hands can also identify noise sources. Air jets from duct leaks can often be felt, and the sound of rattling or vibrating panels or parts often changes or stops when these parts are touched.

In trying to locate noise sources and transmission paths, the engineer should consider the location of the affected area. In areas that are remote from equipment rooms containing significant noise producers but adjacent to shafts, noise is usually the result of structure-borne transmission through pipe and duct supports and anchors. In areas adjoining, above, or below equipment rooms, noise is usually caused by openings (acoustical leaks) in the separating floor or wall or by improper, ineffective, or maladjusted vibration isolation systems.

Unless the noise source or path of transmission is quite obvious, the best way to identify it is by eliminating all sources systematically as follows:

1. Turn off all equipment to make sure that the objectionable noise is caused by the HVAC system. If the noise stops, the HVAC system components (compressors, fans, and pumps) must be operated separately to determine which are contributing to the objectionable noise. Where one source of disturbing noise predominates, the test can be performed starting with all equipment in operation and turning off components or systems until the disturbing noise is eliminated. Tests can also be performed starting with all equipment turned off and operating various component equipment singularly, which permits evaluation of noise from each individual component.

   Any system component can be termed a predominant noise source if, when the equipment is shut off, the sound level drops 3 dBA or if, when measurements are taken with equipment operating individually, the sound level is within 3 dBA of the overall objectionable measurement.

   When a sound level meter is not used, it is best to start with all equipment operating and shut off components one at a time because the ear can reliably detect differences and changes in noise but not absolute levels.

2. When some part of the HVAC system is established as the source of the objectionable noise, try to further isolate the source. By walking around the room, determine whether the noise is coming through the air outlets or returns, the hung ceiling, or the floors or walls.

3. If the noise is coming through the hung ceiling, check that ducts and pipes are isolated properly and not touching the hung ceiling supports or electrical fixtures, which would provide large noise radiating surfaces. If ducts and pipes are the source of noise and are isolated properly, the possible remedies to reduce the noise are changing flow conditions, installing silencers, and/or wrapping the duct or pipe with an acoustical barrier such as a lead blanket.

4. If noise is coming through the walls, ceiling, or floor, check for any openings to adjoining shafts or equipment rooms, and check vibration isolation systems to ensure that there is no structure-borne transmission from nearby equipment rooms or shafts.

5. Noise traced to air outlets or returns usually requires careful evaluation by an engineer or acoustical consultant to determine the source and proper corrective action (see Chapter 43). In general, air outlets can be selected to meet any acoustical design goal by keeping the velocity sufficiently low. For any given outlet, the sound level increases about 2 dB for each 10% increase in airflow velocity over the vanes, and doubling the velocity increases the sound level by about 16 dB. Also, the sound approach conditions caused by improperly located control dampers or improperly sized diffuser necks can easily increase sound levels by 10 to 20 dB.

   A simple yet effective instrument that aids in locating noise sources is a microphone mounted on a pole. It can be used to localize noises in hard-to-reach places, such as hung ceilings and behind heavy furniture.

6. If the noise is traced to an air outlet, measure the A-sound level close to it but with no air blowing against the microphone. Then, remove the inner assembly or core of the air outlet and repeat the reading with the meter and the observer in exactly the same position as before. If the second reading is more than 3 dB below the first, a significant amount of noise is caused by airflow over the vanes of the diffuser or grille. In this case, check whether the system is balanced properly. As little as 10% too much air will increase the sound generated by an air outlet by 2.5 dB. As a last resort, a larger air outlet could be substituted to obtain lower air

velocities and hence less turbulence for the same air quality. Before this is considered, however, the air approach to the outlet should be checked.

Noise far exceeding the normal rating of a diffuser or grille is generated when a throttled damper is installed close to it. Air jets impinge on the vanes or cones of the outlet and produce *edge tones* similar to the hiss heard when blowing against the edge of a ruler. The material of the vanes has no effect on this noise, although loose vanes may cause additional noise from vibration.

When balancing air outlets with integral volume dampers, consider the static pressure drop across the damper, as well as the air quantity. Separate volume dampers should be installed sufficiently upstream from the outlet that there is no jet impingement. Plenum inlets should be brought in from the side, so that the jets do not impinge on the outlet vanes.

7. If the air outlets are eliminated as sources of excessive noise, inspect the fan room. If possible, change the fan speed by about 10%. If resonance is involved, this small change can make a significant difference.

8. Sometimes fans are poorly matched to the system. If a belt-driven fan delivers air at a higher static pressure than is needed to move the design air quantity through the system, reduce the fan speed by changing sheaves. If the fan does not deliver enough air, consider increasing the fan speed only after checking the duct system for unavoidable losses. Turbulence in the air approach to the fan inlet not only increases the fan sound generation, but decreases its air capacity. Other parts that may cause excessive turbulence are dampers, duct bends, and sudden enlargements or contractions of the duct.

When investigating fan noise, seek assistance from the supplier or manufacturer of the fan.

9. If additional acoustical treatment is to be installed in the ductwork, obtain a frequency analysis. This involves the use of an octave-band analyzer and should generally be left to a trained acoustician.

TESTING FOR VIBRATION

Vibration testing is necessary to ensure that (1) equipment is operating within satisfactory vibration levels and (2) objectionable vibration and noise are not transmitted to the building structure. Although these two factors are interrelated, they are not necessarily interdependent. A different solution is required for each, and it is essential to test both the isolation system and the vibration levels of the equipment.

General Procedure

The general order of steps in vibration testing are as follows:

1. Make a visual check of all equipment for obvious errors that must be corrected immediately.

2. Make sure all isolation systems are *free floating* and not short-circuited by any obstruction between equipment or equipment base and building structure.

3. Turn on the system for an aural check of any obviously rough operation. Check bearings with a stethoscope. Bearing check is especially important because bearings can become defective in transit and/or if equipment was not properly stored, installed, or maintained. Defective bearings should be replaced immediately to avoid damage to the shaft and other components.

4. Adjust and balance equipment and systems so that final vibration tests are made on equipment as it will actually be operating.

5. Test equipment vibration.

Instrumentation

Although instruments are not required to test vibration isolation systems, they are essential to test equipment vibration properly.

Sound level meters and *computer-driven sound-measuring systems* are the most useful instruments for measuring and evaluating vibration. Usually, they are fitted with accelerometers or vibration pickups for a full range of vibration measurement and analysis. Other instruments used for testing vibration in the field are described below.

Reed vibrometers are relatively inexpensive instruments often used for testing vibration, but relative inaccuracy limits their usefulness.

Vibrometers are moderately priced instruments that measure vibration amplitude by means of a light beam projected on a graduated scale.

Vibration meters are moderately priced electronic instruments that measure vibration amplitude on a meter scale and are simple to use.

Vibrographs are moderately priced mechanical instruments that measure both amplitude and frequency. They are useful for analysis and testing because they provide a chart recording showing amplitude, frequency, and actual wave form of vibration. They can be used for simple yet accurate determination of the natural frequency of shafts, components, and systems by a *bump test*.

Vibration analyzers are relatively expensive electronic instruments that measure amplitude and frequency, usually incorporating a variable filter.

Strobe lights are often used with many of the aforementioned instruments for analyzing and balancing rotating equipment.

Stethoscopes that amplify sound are available as inexpensive mechanic's type (basically, a standard stethoscope with a probe attachment); relatively inexpensive types incorporating a tuneable filter; and moderately priced powered types that electronically amplify sound and provide some type of meter and/or chart recording. Stethoscopes are often used to determine whether bearings are bad.

The choice of instrumentation depends on the test. A stethoscope should be part of every tester's kit as it is one of the most practical, yet least expensive, instruments and one of the best means of checking bearings. Vibrometers and vibration meters can be used to measure vibration amplitude as an acceptance check. Since they cannot measure frequency, they cannot be used for analysis and primarily function as a go-no-go instrument. The best acceptance criteria consider both amplitude and frequency. However, because vibrometers and vibration meters are moderately priced and easy to use, they are widely used. Anyone seriously concerned with vibration testing should use an instrument that can determine frequency as well as amplitude, such as a vibrograph or vibration analyzer.

Testing Vibration Isolation Systems

The following steps should be taken to ensure that vibration isolation systems are functioning properly:

1. Ensure that the system is *free floating* by applying an unbalanced load, which should cause the system to move freely and easily. On floor-mounted equipment, check that there are no obstructions between the base or foundation and the building structure that would cause transmission while still permitting equipment to rock relatively free because of the application of an unbalanced force (Figure 12). On suspended equipment, check that hanger rods are not touching the hanger. Rigid connections such as pipes and ducts can prohibit mounts from functioning properly and from providing a transmission path. Note that the fact that the system is free floating does not mean that the isolators are functioning properly. For example, a 500 rpm fan installed on isolators having a natural frequency of 500 cycles per minute (8.33 Hz) could be free floating but

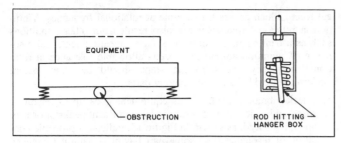

Fig. 12   Obstructed Isolation Systems

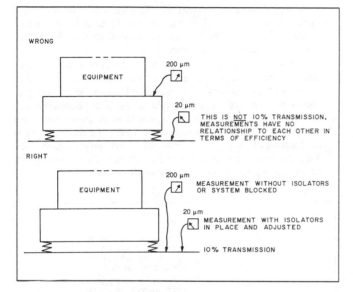

Fig. 13   Testing Isolation Efficiency

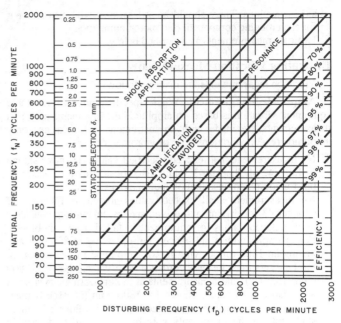

Fig. 14   Isolator Natural Frequencies and Efficiencies

If isolators are in the 90% efficiency range, and there is transmission to the building structure, either the equipment is operating roughly or there is a flanking path of transmission, such as connecting piping or obstruction, under the base.

Testing Equipment Vibration

Testing equipment vibration is necessary as an acceptance check to determine whether equipment is functioning properly and to ensure that objectionable vibration and noise are not transmitted. Although a person familiar with equipment can determine when it is operating roughly, instrumentation is usually required to determine accurately whether vibration levels are satisfactory.

Vibration Tolerances

Vibration tolerance criteria are provided in Table 41 of Chapter 43. These criteria are based on equipment installed on vibration isolators and can be met by any reasonably smoothly running equipment.

Procedure for Testing Equipment Vibration

The following steps should be taken to ensure that equipment vibration is properly tested:

1. Determine operating speeds of equipment from nameplates, drawings, or a speed-measuring device such as a tachometer or strobe, and indicate them on the test form. For any equipment where the driving speed (motor) is different from the driven speed (fan wheel, rotor, impeller) because of belt drive or gear reducers, indicate both driving and driven speeds.

2. Determine acceptance criteria from specifications, and indicate them on the test form. If specifications do not provide criteria, use those shown in Chapter 43.

3. Ensure that the vibration isolation system is functioning properly (see the section on Testing Vibration Isolation Systems).

4. Operate equipment and make visual and aural checks for any apparent rough operation. Check all bearings with a stethoscope. Any defective bearings, misalignment, or obvious rough operation should be corrected before proceeding further. If not corrected, equipment should be considered unacceptable.

5. Measure and record on the test form vibration at bearings of driving and driven components in horizontal, vertical, and, if possible,

would actually be in resonance, resulting in transmission to the building and excessive movement.

2. Determine whether isolators are adjusted properly and providing desired isolation efficiency. All isolators supporting a piece of equipment should have approximately the same deflection (i.e., they should be compressed the same under the equipment). If not, they have been improperly adjusted, installed, or selected; this should be corrected immediately. Note that isolation efficiency cannot be checked by comparing vibration amplitude on equipment to amplitude on the structure (Figure 13).

The only accurate check of isolation efficiencies is to compare vibration measurements of equipment operating with isolators to measurements of equipment operating without isolators. Since these tests are usually impractical, it is best to check isolator deflection to determine whether deflection is as specified and whether specified or desired isolation efficiency is being provided. Figure 14 shows natural frequency of isolators as a function of deflection and indicates the theoretical isolation efficiencies for various frequencies at which the equipment operates.

While it is easy to determine the deflection of spring mounts by measuring the difference between the free heights with a ruler (information as shown on submittal drawings or available from a manufacturer), such measurements are difficult with most pad or rubber mounts. Further, most pad and rubber mounts do not lend themselves to accurate determination of natural frequency as a function of deflection. For such mounts, the most practical approach is to check that there is no excessive vibration of the base and no noticeable or objectionable vibration transmission to the building structure.

axial directions. There should be at least one axial measurement for each rotating component (fan motor, pump motor).

6. Evaluate measurements as described below.

Evaluating Vibration Measurements

Vibration measurements are made as follows:

Amplitude Measurement. When specification for acceptable equipment vibration is based on amplitude measurements only, and measurements are made with an instrument that measures only amplitude (e.g., a vibration meter or vibrometer):

1. No measurement shall exceed specified values or values shown in Table 42 of Chapter 43, taking into consideration reduced values for equipment installed on inertia blocks

2. No measurement shall exceed values shown in Table 42 of Chapter 43 for driving and driven speeds, taking into consideration reduced values for equipment installed on inertia blocks. For example, with a belt-driven fan operating at 800 rpm and having an 1800 rpm driving motor, amplitude measurements at fan bearings must be in accordance with values shown for 800 cpm (13.3 Hz), and measurements at motor bearings must be in accordance with values shown for 1800 cpm (30 Hz). If measurements at motor bearings exceed specified values, take measurements of the motor only with belts removed to determine whether there is feedback vibration from the fan.

3. No axial vibration measurement should exceed maximum radial (vertical or horizontal) vibration at the same location.

Amplitude and Frequency Measurements. When specification for acceptable equipment vibration is based on both amplitude and frequency measurements, and measurements are made with instruments that measure both amplitude and frequency (e.g., a vibrograph or vibration analyzer):

1. No amplitude measurements at driving and driven speeds shall exceed specified values or values shown in Table 42 of Chapter 43, taking into consideration reduced values for equipment installed on inertia blocks. Measurements that exceed acceptable amounts may be evaluated as explained in the section on Vibration Analysis.

2. No axial vibration measurement shall exceed maximum radial (vertical or horizontal) vibration at the same location.

3. The presence of any vibration at frequencies other than driving or driven speeds is generally reason to rate operation unacceptable, and such vibration should be analyzed as explained in the section on Vibration Analysis.

Vibration Analysis

The following guide covers most vibration problems that may be encountered.

Axial Vibration Exceeds Radial Vibration. When the amplitude of axial vibration (parallel with shaft) at any bearing exceeds radial vibration (perpendicular to shaft—vertical or horizontal), it usually indicates misalignment. This is most common on direct-driven equipment because flexible couplings accommodate parallel and angular misalignment of shafts. Such misalignment can generate forces that cause axial vibration. Axial vibration can cause premature bearing failure, so misalignment should be checked carefully and corrected promptly. Other possible causes of large-amplitude axial vibration are resonance, defective bearings, insufficient rigidity of bearing supports or equipment, and loose hold-down bolts.

Vibration Amplitude Exceeds Allowable Tolerance at Rotational Speed. The allowable vibration limits established by Table 41 of Chapter 43 are based on vibration caused by rotor imbalance, which results in vibration at rotational frequency. While vibration caused by imbalance must be at the frequency at which the part is rotating, a vibration at rotational frequency does not have to be caused by imbalance. An unbalanced rotating part develops centrifu-

gal force, which causes it to vibrate at rotational frequency. Vibration at rotational frequency can also result from other conditions such as a bent shaft, an eccentric sheave, misalignment, and resonance. If vibration amplitude exceeds allowable tolerance at rotational frequency, the following steps should be taken before performing field balancing of rotating parts.

1. Check vibration amplitude as equipment goes up to operating speed and as it coasts to a stop. Any significant peaks at or near operating speed, as shown in Figure 15, indicate a probable condition of resonance, i.e., some part having a natural frequency close to the operating speed, resulting in greatly amplified levels of vibration.

A bent shaft or eccentricity usually causes imbalance that results in significantly higher vibration amplitude at lower speeds, as shown in Figure 16, whereas vibration caused by imbalance generally increases as speed increases.

If a bent shaft or eccentricity is suspected, check the dial indicator. A bent shaft or eccentricity between bearings as shown in Figure 17A can usually be compensated for by field balancing, although some axial vibration might remain. Field balancing cannot correct vibration caused by a bent shaft on direct-connected equipment, on belt-driven equipment where the shaft is bent at the location of sheave, or if the sheave is eccentric (Figure 17B). This is because the center-to-center distance of the sheaves will fluctuate, each revolution resulting in vibration.

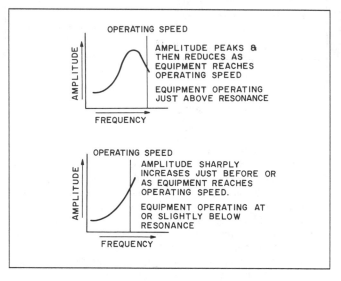

Fig. 15   Vibration from Resonant Condition

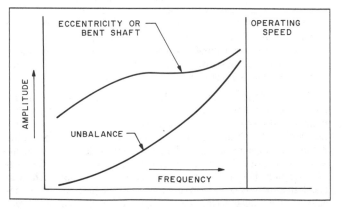

Fig. 16   Vibration Caused by Eccentricity

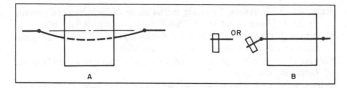

Fig. 17 Bent Shafts

2. For belt- or gear-driven equipment where vibration is at motor driving frequency rather than driven speed, it is best to disconnect the drive to perform tests. If the vibration amplitude of the motor operating by itself does not exceed specified or allowable values, excessive vibration (when the drive is connected) is probably a function of bent shaft, misalignment, eccentricity, resonance, or loose hold-down bolts.

3. Vibration caused by imbalance can be corrected in the field by firms specializing in this service or by testing personnel if they have appropriate equipment and experience.

Vibration at Other than Rotational Frequency. Vibration at frequencies other than driving and driven speeds is generally considered unacceptable. Table 4 shows some common conditions that can cause vibration at other than rotational frequency.

Resonance. If resonance is suspected, determine which part of the system is in resonance.

Isolation Mounts. The natural frequency of the most commonly used spring mounts is a function of spring deflection, as shown in Figure 11 of Chapter 7 of the 1993 *ASHRAE Handbook—Fundamentals*, and it is relatively easy to calculate by determining the difference between the free and operating height of the mount, as explained in the section on Testing Vibration Isolation Systems. This technique cannot be applied to rubber, pad, or fiberglass mounts, which have a natural frequency in the 5 to 50 Hz range. Natural frequency for such mounts is determined by a bump test. Any resonance with isolators should be immediately corrected as it results in excessive movement of equipment and more transmission to the building structure than if equipment were attached solidly to the building (installed without isolators).

System Component. Resonance can occur with any system component shaft, structural base, casing, and connected piping. The easiest way to determine natural frequency is to perform a bump test with a vibrograph. This test consists of bumping the part and measuring with an instrument; the part will vibrate at its natural frequency, which is recorded on instrument chart paper. Similar tests, though not as convenient or accurate, can be made with a reed vibrometer or a vibration analyzer. However, most of these instruments are restricted to frequencies above 8.3 Hz. They therefore cannot be used to determine natural frequencies of most isolation systems, which usually have natural frequencies lower than 8.3 Hz.

Checking for Vibration Transmission. The source of vibration transmission can be checked by determining frequency with a vibration analyzer and tracing back to equipment operating at this speed. However, the easiest and usually the best method (even if test equipment is being used) is to shut off components one at a time until the source of transmission is located. Most transmission problems cause disturbing noise; listening is the most practical approach to determine a noise source because the ear is usually better than sound-measuring instruments at distinguishing small differences and changes in character and amount of noise. Where disturbing transmission consists solely of vibration, a measuring instrument will probably be helpful, unless vibration is significantly above the sensory level of perception. Vibration below the sensory level of perception is generally not objectionable.

If equipment is located near the affected area, check isolation mounts and equipment vibration. If vibration is not being transmitted through the base, or if the affected area is remote from equipment, the probable cause is transmission through connected piping and/or ducts. Ducts can usually be isolated by isolation hangers. However, transmission through connected piping is very common and presents numerous problems that should be understood before attempting to correct them as discussed in the following section.

Vibration and Noise Transmission in Piping Systems

Vibration and noise in connected piping can be generated by either equipment (e.g., pump or compressor) or flow (velocity). Mechanical vibration due to equipment can be transmitted through the walls of pipes or by a water column. Flexible pipe connectors, which provide system flexibility to permit isolators to function properly and protect equipment from stress caused by misalignment and thermal expansion, can be useful in attenuating mechanical vibration transmitted through a pipe wall. However, they rarely suppress flow vibration and noise and only slightly attenuate mechanical vibration as transmitted through a water column.

Tie rods are often used with flexible rubber hose and rubber expansion joints (Figure 18). While they accommodate thermal movements, they hinder the isolation of vibration and noise. This is because pressure in the system causes the hose or joint to expand until resilient washers under tie rods are virtually rigid. To isolate noise adequately with a flexible rubber connector, tie rods and anchor piping should not be used. However, this technique generally cannot be used with pumps that are on spring mounts because they would still permit the hose to elongate. Flexible metal hose can be used with spring-isolated pumps since wire braid serves as tie rods; metal hose controls vibration but not noise.

Problems of transmission through connected piping are best resolved by changes in the system to reduce noise (improve flow characteristics, turn down impeller) or by completely isolating piping from the building structure. Note, however, that it is almost

Table 4 Common Causes of Vibration Other than Unbalance at Rotation Frequency

Frequency	Source
0.5 × rpm	Vibration at approximately 0.5 rpm can result from improperly loaded sleeve bearings. This vibration will usually disappear suddenly as equipment coasts down from operating speed.
2 × rpm	Equipment is not tightly secured or bolted down.
2 × rpm	Misalignment of couplings or shafts usually results in vibration at twice rotational frequency and generally a relatively high axial vibration.
Many × rpm	Defective antifriction (ball, roller) bearings usually result in low-amplitude, high-frequency, erratic vibration. Because defective bearings usually produce noise rather than any significantly measurable vibration, it is best to check all bearings with a stethoscope or similar listening device.

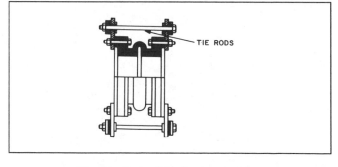

Fig. 18 Typical Tie Rod Assembly

impossible to isolate piping completely from the structure, as required resiliency is inconsistent with rigidity requirements of pipe anchors and guides. Chapter 43 contains information on flexible pipe connectors and resilient pipe supports, anchors, and guides, which should help resolve any piping noise transmission problems.

REFERENCES

AMCA. 1985. Laboratory methods of testing fans for rating. *Standard* 210-85. Also ASHRAE *Standard* 51-1985. Air Movement and Control Association, Arlington Heights, IL.

ASHRAE. 1988. Practices for measurement, testing, adjusting and balancing of building heating, ventilation, air conditioning and refrigeration systems. *Standard* 111-1988.

ASHRAE. 1989. Guideline for commissioning of HVAC systems. *Guideline* 1-1989.

ASME. 1986. Atmospheric water cooling equipment. *Standard* PTC 23-86. American Society of Mechanical Engineers, New York.

CTI. 1990. Standard specifications for thermal testing of wet/dry cooling towers. *Standard Specification* ATC-105. Cooling Tower Institute, P.O. Box 73383, Houston, TX.

Griggs, E.I., W.B. Swim, and H.G. Yoon. 1990. Placement of air control sensors. *ASHRAE Transactions* 96(1).

Sauer, H.J. and R.H. Howell. 1990. Airflow measurements at coil faces with vane anemometers: Statistical correction and recommended field measurement procedure. *ASHRAE Transactions* 96(1):502-11.

BIBLIOGRAPHY

AABC. 1967. Field balancing depends on dampers. Associated Air Balance Council, Washington, D.C. (January).

AABC. 1989. National standards for total system balance, 5th ed. Associated Air Balance Council, Washington, D.C.

AMCA. 1987. Fan application manual. Air Movement and Control Association, Arlington Heights, IL.

Armstrong Pump. 1986. Technology of balancing hydronic heating and cooling systems. Armstrong Pump, North Tonawanda, NY.

ASA. 1983. Specification for sound level meters. *Standard* 1.4-83. Acoustical Society of America, New York.

ASHRAE. 1974. Energy conservation pumping systems. *ASHRAE Journal* 16(6).

Carlson, G.F. 1968-1969. Hydronic systems analysis and evaluation. *ASHRAE Journal* (October through March).

Carlson, G.F. 1972. Central plants chilled water systems. *ASHRAE Journal* (February through April).

Carlson, G.F. 1974. Liquid viscosity effects on pumping systems, Parts I and II. *ASHRAE Journal* 16(7, 9).

Carrier Corporation. *Carrier air conditioning manual*. Carrier Corporation, Syracuse, NY.

Choat, E.E. 1976. An evaluation of the temperature difference method for balancing hydronic coils. *ASHRAE Transactions* 82(1).

Coad, W.J. 1985. Variable flow in hydronic systems for improved stability, simplicity and energy economics. *ASHRAE Transactions* 91(1B):224-37.

Eads, W.G. 1983. Testing, balancing and adjusting of environmental systems. In *Fan Engineering*, 8th ed. Buffalo Forge Company, Buffalo, NY.

Gladstone, J. 1981. *Air conditioning—Testing and balancing: A field practice manual*. Van Nostrand Reinhold, New York.

Gupton, G. 1989. *HVAC controls, operation and maintenance*. Van Nostrand Reinhold, New York.

Haines, R.W. 1987. *Control systems for heating, ventilating and air conditioning*, 4th ed. Van Nostrand Reinhold, New York.

Hansen, E.G. 1985. *Hydronic system design and operation*. McGraw-Hill, New York.

Hayes, F.C. and W.F. Stoecker. 1966. The effect of inlet conditions on flow measurements at ceiling diffusers. *ASHRAE Transactions* 72(2).

Hayes, F.C. and W.F. Stoecker. 1966. Tables of application factors for flow measurement at return intakes. *ASHRAE Transactions* 72(2).

Hightower, G.B. 1971. Testing, balancing and adjusting of HVAC induction systems. *ASHRAE Journal* 13(6).

ITT Bell and Gossett. Balance procedure manual. TEB-985. ITT Bell and Gossett, Morton Grove, IL.

McQuiston, F.C. and J.D. Parker. 1988. *Heating, ventilating and air conditioning: Analysis and design*, 3rd ed. John Wiley and Sons, New York.

Miller, R.W. 1983. *Flow measurement engineering handbook*. McGraw-Hill, New York.

National Joint Steamfitter-Pipefitter Apprenticeship Committee. 1976. *Start, test and balance manual*. Washington, DC.

NEBB. 1986. Testing, adjusting, balancing manual for technicians, 1st ed. (April).

NEBB. 1991. Procedural standards for testing, balancing and adjusting of environmental systems, 5th ed. National Environmental Balancing Bureau, Vienna, VA.

Nevins, R.G. and E.D. Ward. 1968. Room air distribution with an air distribution ceiling. *ASHRAE Transactions* 74(1).

SMACNA. 1977. Round industrial duct construction standards, 1st ed. Sheet Metal and Air Conditioning Contractors' National Association, Merrifield, VA.

SMACNA. 1980. Rectangular industrial duct construction standards, 1st ed.

SMACNA. 1985. HVAC air duct leakage test manual, 1st ed.

SMACNA. 1985. HVAC duct construction standards—Metal and flexible, 1st ed.

SMACNA. 1993. HVAC systems—Testing, adjusting and balancing, 2nd ed.

Tamura, G.T. and A.G. Wilson. 1966. Pressure differences for a nine-story building as a result of chimney effect and ventilating system operation. *ASHRAE Transactions* 72(1).

Tamura, G.T. and A.G. Wilson. 1967. Pressure differences caused by chimney action and mechanical ventilation. *ASHRAE Transactions* 73(2).

Tamura, G.T. and A.G. Wilson. 1967. Building pressures caused by chimney action and mechanical ventilation. *ASHRAE Transactions* 73(2).

Tamura, G.T. and A.G. Wilson. 1968. Pressure differences caused by wind on two tall buildings. *ASHRAE Transactions* 74(2).

Trane Company. 1988. *Trane air conditioning manual*. The Trane Company, LaCrosse, WI.

Yerges, L.F. 1978. *Sound, noise and vibration control*, 2nd ed. Van Nostrand Reinhold, New York.

CHAPTER 35

OPERATION AND MAINTENANCE MANAGEMENT

WHILE mechanical maintenance was once the responsibility of trained technical personnel, increasingly sophisticated systems and equipment require overall management programs to handle organization, staffing, planning, and control. These programs should meet present and future energy management requirements, upgrade management skills, and increase communication among those who benefit from cost-effective operation and maintenance.

Good maintenance management planning includes a proper life-cycle cost analysis and a process to ensure that occupant comfort, energy planning, and safety and security systems are optimal for all facilities. Appropriate technical expertise—in-house or contracted—is also important. The following issues are addressed in this chapter:

- Life-cycle costs
- Variations in approaches according to the number and size of buildings or systems within a program
- Organizational frameworks according to the number and size of buildings or systems in a program
- Documentation and record keeping
- Responsibilities of designers, contractors, manufacturers/suppliers, and owners in relation to operation and maintenance

Operation and maintenance of all HVAC&R systems should be considered and documented in the original design of a building. Newly installed systems must then be commissioned to ensure that they are functioning as designed. It is the responsibility of operation and maintenance staff to ensure that the systems continue to function throughout the life of the building. Existing systems may need to be recommissioned to accommodate changes.

TERMINOLOGY

System operation describes what the building or systems operator adjusts to satisfy tenant comfort requirements, process requirements, and the strategy for optimum energy use and minimum maintenance.

The *maintenance program* documents objectives, establishes evaluation criteria, and commits the maintenance department to basic areas of performance, such as prompt response to mechanical failure and attention to planned functions, which protect a capital investment and minimize downtime or failure response.

Failure response classifies maintenance department resources expended or reserved to handle interruptions in the operation or the function of equipment covered by the maintenance program. This classification includes two types of response—repair and service.

Repair is to make good, or to restore to good or sound condition with the following constraints: (1) operation is fully restored without embellishment and (2) the response is triggered by failure.

Service provides what is necessary, short of repair, to effect a maintenance program. It is usually based on procedures recommended by manufacturers.

Planned maintenance classifies maintenance department resources that are invested in prudently selected functions at specified intervals. All functions and resources within this classification must be planned, budgeted, and scheduled. Planned maintenance embodies two concepts—preventive and corrective maintenance.

Preventive maintenance classifies resources allotted to ensure the proper operation of a system or equipment under the maintenance program. Durability, reliability, efficiency, and safety are the principal objectives.

Corrective maintenance classifies resources expended or reserved for predicting and correcting conditions of impending failure. Corrective action is strictly remedial and always performed before failure occurs. An identical procedure performed in response to failure is classified as a repair. Action taken during a shutdown caused by failure may be termed corrective if it is optional and unrelated.

Predictive maintenance is a function of corrective maintenance. Statistically supported objective judgment is implied. Nondestructive testing, chemical analysis, vibration and noise monitoring, as well as routine visual inspection and logging are classified as predictive, providing that the item tested or inspected is part of the planned maintenance program.

Durability is the average expected service life of a system or facility. Table 3 in Chapter 33 lists median years of service life of various equipment. More specifically, individual manufacturers quantify durability as design life, which is the average number of hours of operation before failure, extrapolated from accelerated life tests and stressing components critical to economic destruction.

Reliability implies that a system or facility will perform its intended function for a specified period without failure.

QUANTITATIVE MANAGEMENT CONCEPT

The following life-cycle management concept works well for operation and maintenance programs. It can be used for planning during the construction cycle as well as for day-to-day operation and maintenance management. Derived from value engineering, this life-cycle concept of management involves three interdependent dimensions: effectiveness (value), durability (time), and life-cycle cost (money). See Figure 1. Their numerical values are affected by the type and extent of the operation and maintenance programs.

The *effectiveness* of a system is the probability of successfully providing required services over a given period under specified conditions. To be perfectly effective, a facility must provide the required

---

The preparation of this chapter is assigned to TC 9.2, Industrial Air Conditioning.

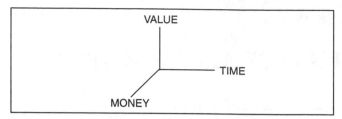

Fig. 1   Life-Cycle Concept of Management

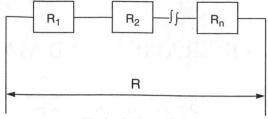

Fig. 2   Series System

services satisfactorily and operate dependably without failure for the given period. Mathematically, effectiveness E is the product of capability C and dependability D.

$$E = CD \qquad (1)$$

Effectiveness may be assigned a numerical value according to the type of service required to meet a tenant's functional need. Table 1 is a sample guide for assigning effectiveness values. The values given in this table are examples only; values must be developed for each project to create proper relationships between the different systems in each facility.

Table 1   Examples of Effectiveness Values

Range of Effectiveness	Effectiveness Category	Example of Service
0.50 to 0.75	Regular	Environmental control
0.76 to 0.85	Essential	Freeze protection
0.86 to 1.00	Critical	Smoke control

Capability is the measure of a system's ability to satisfactorily provide required service. It is the probability of meeting functional requirements, providing the system is operated under previously designated conditions. An example of capability is the ability of a heating system to cope with the heating load at the design winter temperature. Capability must be verified when the system is first commissioned and whenever the functional requirements change.

Capability can be calculated from the time N the system cannot meet the requirements and the time Y it is expected to operate in one year:

$$C = 1 - (N/Y) \qquad (2)$$

Dependability is the measure of a system's condition. Assuming that the system operated at the beginning of its service life, dependability is the probability of its operating at any given time until the end of its life.

For systems that cannot be repaired during use, dependability is the probability that there will be no failure during use. For systems that can be repaired, dependability is governed by the ease and rapidity with which repairs can be made. This ease is the system maintainability.

Dependability D can be expressed as the product of reliability R and maintainability M:

$$D = RM \qquad (3)$$

Reliability implies that a system will perform its function without failure for a portion of a specified period. Reliability of two or more systems is calculated as follows:

1. If systems are arranged in *series*, where the output of one is the input of the next (Figure 2), the combined reliability is the product of their individual reliabilities:

$$R = R_1 R_2 ... R_N \qquad (4)$$

2. If systems are arranged in *parallel*, such as dual pumping systems with automatic changeover on failure of one of the two pumps (Figure 3), their combined reliability is calculated from the following equation:

$$R = 1 - (1 - R_1)(1 - R_2) ... (1 - R_n) \qquad (5)$$

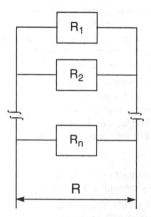

Fig. 3   Parallel System

Maintainability is the ease and rapidity with which a system can be repaired. It complements reliability by defining the specific length of time a system can operate in a fully restored condition.

Maintainability is calculated from the time B the system is being repaired in regular occupancy and the time Y the system is expected to operate in one year.

$$M = 1 - (B/Y) \qquad (6)$$

The previous terms can be interrelated by an algorithm (shown in Figure 4) that can be manipulated either manually or by computer.

DOCUMENTATION

Operation and maintenance documentation should be prepared during the delivery cycle of any project. Information should be documented as soon as it becomes available to support design, construction activities, and training of operation and maintenance staff in preparation for system commissioning.

A complete operation and maintenance documentation package consists of the following:

- Complete set of design criteria and summary results
- Complete set of specifications, including all addenda and all approved and applied changes
- Complete set of approved shop drawings, including all subsequent modifications
- Complete set of as-built drawings
- Operating manual, which includes control sequences of operation and, if there is direct digital control, a flowchart and hard copy of the software code/database

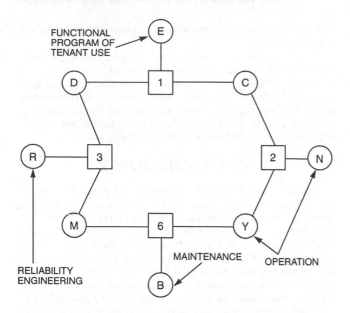

Fig. 4 Maintainability Algorithm

- Maintenance manual, including proposed equipment maintenance programs to facilitate staffing
- Electrical power coordination report
- Air and water balancing report
- Performance verification report
- Copies of all certificates by the inspectors representing authorities having jurisdiction
- Copy of the commissioning report

All of these documents should be available for large, complex facilities. However, small projects require only the documentation appropriate for their size and complexity.

The most important documents are operation and maintenance manuals. The following suggested organization of these manuals reflects the needs of their various users.

Operation Manual

This manual should consist of two parts. Part I, Operation Instruction, contains information a qualified operator needs (1) to start and stop equipment, (2) to control and monitor the performance of the equipment in normal modes of operation, (3) to change from one mode of operation to another, and (4) to operate equipment in emergency situations. Operation procedures with proper flowcharts for all integrated systems are also required. The system function should be represented pictorially and in writing.

Part II, Performance Verification Procedures, contains all the information a qualified operator needs to verify equipment and overall system performance at the design load as well as at part loads, where applicable. The design calculations are needed for performance verification and should be included in this manual.

Maintenance Manual

This manual should also consist of two parts. Part I, Inventory, contains a listing of all systems and pieces of equipment to be maintained as well as all the technical information needed to order spare parts. Manufacturers' catalogs are considered useful adjuncts only. Part II, Maintenance Program, contains the information necessary to perform breakdown, preventive, and predictive maintenance. These programs include written information regarding when or how often to perform maintenance in the most efficient and economical fashion to satisfy tenant needs.

KNOWLEDGE AND SKILLS

For effective and economical operation and maintenance, a system requires personnel with the right combination of technical and managerial skills. Relevant technical skills include not only the skills of operation and maintenance mechanics, but also the engineering skills of the physical plant engineer. Managerial skills involve managing the facility on both a life-cycle and a day-to-day basis. This management administers both the accommodation contract with the tenant and the service contracts, including maintenance service for all equipment. Specialized contractual maintenance companies require yet another level of management.

Physical plant engineers must coordinate the equipment selected by the designer, the operating and maintenance personnel, and the requirements of the investment plan. Good physical plant engineering solutions start when the investment plan is being formulated and continue throughout the life of the facility.

TYPES OF BUILDINGS

The size and complexity of the system to be operated and maintained need to be considered, because buildings range from houses through commercial offices, institutional buildings and complexes, processing plants, refrigerated storage facilities, and large central plants. All installed systems, no matter how simple, should have a commissioning plan. This gives the building owner(s) or system operator(s) a means of operating the system in the most economical manner—minimizing energy consumption and maintenance costs while meeting user requirements. Because these goals may conflict, the commissioning plan should identify the trade-offs in operating a system that provides comfort while affecting the frequency of various types of maintenance.

Small Buildings

Small systems have relatively simple operating procedures, such as adjusting a single zone thermostat from heating to cooling if it is not automatic, and resetting temperatures as desired by the occupants. Maintenance procedures for this type of system are limited to those recommended by most manufacturers.

Most building managers call on-site maintenance staff to change filters, belts, and motors. However, cleaning condenser and evaporator coils and assessing refrigeration and control systems require a qualified technician. In most small facilities, maintenance contractors provide this service on a maintenance program basis. The frequency of maintenance depends on hours of system use, proximity to major transportation routes (dirt accumulation), and type of operation in the facility.

Smaller facilities may also have complex control systems, especially when zoned for a variety of load conditions. Whenever the operator does not have the knowledge to service and repair a system or component, the owner should hire qualified contractors.

Medium-Sized Buildings

The systems in medium-sized buildings are generally more complex, with several pieces of mechanical equipment operating together through a control system to provide a variety of comfort zones. Commissioning is particularly important with this type of system to ensure optimum comfort at minimal energy cost. Without detailed documentation, the operation staff cannot consider energy budgets while they are addressing building occupants' comfort demands.

Maintenance programs in medium-sized facilities must be detailed for the maintenance personnel and should be implemented on a sequenced basis to reduce the failure response required. Over time, the maintenance programs can be adjusted for the particular characteristics of the system. Computerized maintenance programs can assist management in overseeing the effectiveness of the program.

The building owner or leaseholder may employ operation and maintenance personnel, who should have the technical knowledge to operate these systems if they observed a full system commissioning process. The operating personnel may also be responsible for system maintenance. The maintenance procedures for the system should be detailed in the commissioning documentation (with individual equipment maintenance frequency detailed in the manufacturers' literature). The maintenance program should reflect the unique nature of each building. Again, certain maintenance programs for highly technical pieces of equipment may be contracted out if the staff does not have the technical knowledge.

Large Buildings

Large buildings, including central plants, require a management structure to hire, direct, and control two staffs—one for operation personnel and the other for maintenance personnel.

The operations budget should be large enough to support computerized maintenance programs, which detail proper timing of system maintenance procedures so that all systems operate at maximum efficiency while ensuring occupant comfort. Annual and life-cycle cost planning are essential to ensure the most cost-effective operation and maintenance of these systems.

To facilitate proper commissioning, management should be involved in the design and construction of the facility. Because facilities are long-term investments, any first-cost compromises must take into account both life-cycle cost and occupant comfort while maintaining reasonable energy budgets.

Logged information can be used with proper database management systems as predictive maintenance to reduce failure response requirements. These systems may be used to indicate the weaknesses in the systems so that management can make appropriate decisions or system changes. As with small and medium-sized systems, large systems may require the support of outside contractors or manufacturers for specific equipment.

RESPONSIBILITIES

To allow self-management, the building owner should be apprised of the design intent of the system. The system design must include proper operation and maintenance information; flowcharts and instrumentation requirements must be indicated. The installing contractor must provide the director of operations and maintenance with an organized and comprehensive turnover of the systems that have been installed.

An effective director should be able to organize, staff, train, plan, and control the operation and maintenance of a facility with the cooperation of senior management and all departments. A manager's responsibilities include administering the operation and maintenance budget and protecting the life-cycle objectives. Before selecting a least-cost alternative, a manager should determine the consequential effect on durability and loss prevention (Loveley 1973).

NEW TECHNOLOGY

Operation and maintenance programs are based on the technology available at the time of their preparation. The programs should be adhered to throughout the required service life of the facility or system. In the course of the service life, a new technology may become available that would affect the operation and/or maintenance program. When this occurs, the switch from the existing to the new technology must be assessed in life-cycle terms. The existing technology must be assessed for the degree of loss due to shorter return on investment; the new technology must be assessed for (1) all initial and operation and maintenance costs, (2) the correlation between its service life and the remaining service life of the facility, and (3) the cost of conversion, including the revenue losses due to the associated downtime.

REFERENCE

Loveley, J.D. 1973. Durability, reliability, and serviceability. *ASHRAE Journal* 15(1):67.

BIBLIOGRAPHY

Fuchs, S.J. 1982. *Complete building equipment maintenance desk book.* Prentice Hall, Englewood Cliffs, NJ.

Lawson, C.N. 1989. Commissioning—The construction phase. *ASHRAE Transactions* 95(1):887-94.

Trueman, C.J. 1989. Commissioning: An owner's approach for effective operation. *ASHRAE Transactions* 95(1):895-99.

Petrocelly, K.L. 1989. *Physical plant operations handbook.* Fairmont Press, Englewood Cliffs, NJ.

Criswell, J.W. 1987. *Planned maintenance for productivity and energy conservation,* 2nd ed. Fairmont Press, Englewood Cliffs, NJ.

CHAPTER 36

COMPUTER APPLICATIONS

Computer System Architecture...36.1
Hardware Options..36.1
Software Options...36.3
Design Calculations...36.4
Administrative Uses of Computers....................................36.10
Productivity Tools...36.11
Computer-Aided Design 36.12
Computer Graphics and Modeling...................................36.13
Applications of Artificial Intelligence 36.14
Communications...36.15
Data Acquisition..36.16
Monitoring and Control...36.16

THE use of digital computers in the heating, refrigeration, and air-conditioning industry has come about because of the wide variety of easily used engineering analysis programs for the HVAC industry, an even larger number and range of programs for business use, and the low cost of powerful computers on which to run them. The ordinary calculations required in the HVAC industry, such as heating and cooling loads, can be performed easily and inexpensively on a computer. In addition, computers sometimes allow the solution of more complicated problems that would otherwise be impractical to solve. The operation and maintenance of buildings has also benefited from more powerful computers that monitor, control, and in certain cases diagnose problems in HVAC equipment. This chapter covers (1) alternatives for obtaining computing capability, (2) available computer hardware, and (3) computer programs or software for HVAC applications. ASHRAE (1990) lists many commercially available programs.

In the following discussion of computers, the terms hardware and software appear. They refer to the distinction between the physical equipment (hardware) and the programs or sets of instructions (software) that direct the equipment to perform the desired tasks.

COMPUTER SYSTEM ARCHITECTURE

Current technology in the computer industry is changing rapidly. By the time a computer system is purchased and implemented, new technology is already on the market that renders it obsolete. This situation, along with the need to share information and applications among various computer systems, makes a system design that can readily adapt to change a necessity. An information system has an architecture in much the same way that a building does. If laid out properly, this architecture can accommodate the introduction of new technologies while allowing the continuing use of existing software.

Before choosing computer products, it is important to understand the business organization that the information system will serve. End users can best specify the principles most important to a particular business. The system and applications can then be selected to match the needs of the business, and a technical framework that supports a consistent computing environment can be established. When the computer architecture is in place, the current products that match its standards are chosen.

Standards take several forms. There are industry standards, such as ISO, ANSI/IEEE and SQL, which are defined and recognized by industry groups. Proprietary standards are often formed where other standards are not applicable, such as within a company. There are also de facto standards, such as the operating system software for personal computers.

The preparation of this chapter is assigned to TC 1.5, Computer Applications.

HARDWARE OPTIONS

Computer hardware options may be examined by considering the classes of equipment that can be used: large (mainframe computers), intermediate (minicomputers, laptops, notebooks, and personal computers), and small (personal digital assistants and programmable calculators). These categories indicate rough relative rankings in computing speed, number of simultaneous users, and cost.

Mainframe Computers

Large computers require a substantial personnel commitment. Any organization with plans to use such equipment needs more expertise and guidance than is given here. The following three approaches to obtaining the use of a large computer may be considered:

1. Own a large mainframe computer.
2. Contract the work to a computer service bureau.
3. Work with a time-sharing service company.

The second method, contracting the work out-of-house, may be the easiest and fastest to implement. The responsibilities and problems are placed on the computer bureau; there is no need for the business to purchase expensive hardware and software packages, establish training programs for in-house personnel, or acquire experienced computer staff. In most cases, costs are high, but they can be included in the total project cost.

This method also has some disadvantages. In-house personnel gain no experience for future projects. The money spent is not recoverable for the next job, and the amount of time required for a job is not readily controllable. A firm that does little work requiring computers and plans to perform only short-term projects finds this approach reasonable, but a firm performing medium to large projects and requiring skilled personnel to handle a large variety of tasks in a relatively short time finds this approach limiting.

The third method, working with a time-sharing service, requires the purchase or lease of a minimum amount of hardware (a terminal and a telephone coupler or modem) and the use of a telephone-linked time-sharing network. As with the second approach, there is little or no first-cost expense. However, the responsibilities and problems are not transferred as with the computer bureau; the firm using the service does the greater part of the work. Most computer service firms supply some degree of technical support; a total service provides consulting, programming, training, and documentation. In addition, interactive and batch processing are available, providing quick response and versatility, as well as economy and computing power.

Probably the greatest advantages of time-sharing are the accessibility of a large software library and the enormous computing power of a large mainframe computer at relatively low cost. However, there are some disadvantages to this method. Training for in-house personnel is essential. The staff must become familiar with the services as well as the specific program(s) before they can begin to use them.

36.1

In addition, it is necessary to keep track of operating and storage charges to reduce hidden costs.

The cost of using a time-sharing service is related to the computer time used, plus any royalty fees for third-party programs and fees for access to enhanced programs. The computer time is often taken as a function of time spent on the computer, type of computing done, size of memory space used, and operating expense of personnel and equipment. Other surcharges encountered are telephone connect time, charges for printed output, and storage charges for retaining information on disks or tapes. Because most time-sharing services charge separately for each resource, careful oversight can reduce costs substantially. Unnecessary connect time, on-line storage of unneeded files, interactive processing of large programs, and output to slow terminals can all escalate costs dramatically.

Support services such as consulting, training, and programming are often considered part of the total service included in the charge for computer time (usually referred to as time and maintenance) as long as they are not used too extensively. However, use of support services can result in additional cost, depending on the type or level of support. In-depth training sessions, specific modifications to software, and technical consulting on a specific program application or project usually incur an extra charge. Before those additional charges are assessed, an overall cost or stated rate should be presented to the user.

One benefit of time-sharing for which there is no substitute is access to specialty databases. The amount and diversity of information available on these systems is vast: databases that contain corporate, financial, legal, medical, scientific, cultural, bibliographic, and many other types of data are readily available, although the cost of using the more specialized databases is significant. The user connects to the database through a terminal or a microcomputer with a modem and terminal program, initiates a search for the information desired, and receives the result of the query. In many instances, the considerable cost of accessing a specialized database system (to search for legal precedents, for example) is offset by the hours of expensive professional effort that would otherwise be required.

Minicomputers, Microcomputers, Laptops, and Notebooks

The availability of inexpensive workstations and personal computers at prices affordable to any company has made in-house computing an attractive option for even the smallest firm. The obvious advantages of rapid turnaround and unlimited computing for a fixed investment are compelling. Time-sharing and service bureau costs may be sufficiently high that a purchased system paid for out of savings is the economical choice.

The cost of a large computer system can be a burden on a small firm, and is thus impossible to write off on the first few projects. In addition to hardware and installation costs, software purchases and maintenance add substantially to the total operating cost of a computer system. Replacing and updating hardware and software is also included in the overall costs. To justify owning a large computer system, a firm must use it frequently. However, with the increasing power and networking capabilities of personal computers (PCs), many of the functions previously performed only on a large mainframe can now be performed on PCs.

A PC is often the most promising computing option for a first-time user. The experience and knowledge gained from using these computers can later be applied to a larger computer system. A PC can also interface with the large mainframe computer of the time-sharing service. A personal computer system operates at lower costs to provide interactive usage and storage for the small tasks a first-time user generally takes on. Data can be transferred for processing by a specific program available on a time-shared computer. This combination can enable a firm to handle a larger number and variety of projects than would be possible with time-sharing or a purchased system alone.

There are additional expenses incurred by owning a computer, including maintenance, software, and personnel costs. Software must be either bought or developed, and although it may seem expensive to buy, it is almost always more expensive to develop. A firm should employ at least one person knowledgeable about the system and the software, or have contractual access to such a person.

Micro- or personal computers offer the small office enormous power at low cost. These machines should be considered in the class of business machines, in that they offer possibilities far beyond technical analysis (see sections on Administrative Uses of Computers and Productivity Tools).

Laptop computers are basically battery-powered personal computers that have been miniaturized to the extent that they can be used on one's lap, in the field, hence the name. These computers have also been called palmtops, portables, or notebooks by some manufacturers. In almost all cases, laptop computers are fully functional personal computers that are more or less portable.

Table 1 summarizes the various benefits of four approaches to computing. Any of the hardware options listed in the table may be connected through a local-area network (LAN), further discussed in the section on Communications. When machines are thus connected, one or more of the computers may be used as a file server. Any computer, once connected locally or via telephone to the server, has access to common data and programs through the network. This method allows data stored in one logical database to be accessed from many separate computers without the necessity of copying the data onto numerous storage devices.

An emerging hardware technology is parallel processing. In this paradigm, the processors in one or more networked computers are used in parallel, each solving a portion of a problem. A control program determines what processing power is needed and available for a particular computational task. The task is then assigned to a processor. When the computation is complete, the results are passed back. Depending on the system and the control program, a large number of software pieces may be completed at the same time. This method shows great promise in the area of simulations, which tend to be computationally intensive and which may be easily divided into smaller pieces.

Personal Digital Assistants

Recently, the combination of ultraminiaturization, digital telecommunications, and advanced programming has made possible a new form of computing device called the computerized assistant or personal digital assistant (PDA). Such devices offer the user unique capabilities that were previously obtainable only through a combination of equipment (phone, facsimile machine, laptop, electronic mail, and more). PDAs could have as profound an impact on the business world as did each of the individual components they combine.

Programmable Calculators and Personal Organizers

Calculators exist that can be programmed to carry out operations of hundreds of steps and can be outfitted with printers and/or connected to personal computers to exchange data. A range of commercial programs is available for these machines. Program listings may be found in engineering magazines, and several manufacturers are also making programs available. Top-of-the-line programmable calculators have magnetic card readers or memory modules. These machines can store user-created programs on cards or modules and can read in programs created in machine-readable form by the user or by a vendor. Some manufacturers make program modules that plug into the calculator; others have optical bar code readers that can read printed programs.

The primary advantages of the programmable calculator are low cost and portability, although with a printer attached, the calculators are not nearly as portable. For field calculations, they are unexcelled,

Table 1   Ways of Using Computers

	Mainframe Computer	Computer Bureau	Time-Sharing Service	Micro- or Personal Computer
Equipment required	Computer, terminal, printer, hard copier, storage devices	—None—	Terminal, modem	Computer, monitor, printer, storage (floppy or hard disk)
Personnel	Computer specialists, systems analysts, trained users	—None—	Trained users	Trained users
Program types	Dependent on availability of public domain programs, leasing arrangements with private companies, or internally developed programs	Usually specializing in a few programs	Public domain programs, third-party programs with royalty fees, company program with access fees, internally developed programs, very large databases	Wide variety for technical and business applications from software vendors, manufacturers, and users' groups
Charges and costs	High operating costs, high first cost, cost of maintenance and update to avoid obsolescence	Large costs for entire work, personnel costs, computer charges, etc.	Computer time costs, printing charges, telephone connect costs, storage charges	Low operating costs, modest first cost and maintenance cost, cost of initial training, modest updating costs
Type of user	Frequent users, large to medium-sized firms with several uses for computer system besides outside projects	Small firms seldom having projects with large amounts of data to be analyzed	Frequent to nonfrequent users requiring access to large computing power of a mainframe or to extensive library of software programs	Small to large firms
Advantages (Disadvantages)	Speed and data storage capacity	Easiest approach for avoiding involvement with computer technology	Accessibility, support services (Becoming less popular)	Low cost, fast turnaround time, can be used for many functions other than technical, becoming the norm in most applications

but are being challenged by battery-powered, notebook-sized laptop computers and even hand-held computers.

Many of these calculators have additional features, such as programmable calendars, alarm clocks, and even language interpreters, which allow them to serve also as personal organizers. In many cases programmable calculators perform the same functions as laptop computers. Likewise, some laptops are becoming nearly as small as programmable calculators.

SOFTWARE OPTIONS

A computer must have both hardware and software to perform useful tasks. Osborne (1980), Norton (1986), and Goodman (1988) provide further information. Software can be divided into four major categories: system software, languages, utilities, and application programs.

System software, otherwise known as the operating system, is the environment in which other programs run. It handles input and output (keyboard, video display, and printer) and file transfer between disks and memory; it also supports the operation of other programs. Operating system software can be obtained from the computer manufacturer or from software companies. The operating system is specific to a particular type of computer.

Languages are used to write computer programs. They range from assembly language, which involves coding at the machine instruction level, to high-level languages such as FORTRAN, BASIC, Pascal, C, or C++. Many high-level languages exist to satisfy various programming requirements. FORTRAN is useful for scientific or mathematical applications. BASIC established microcomputers (personal computers) as viable business machines. New dialects of BASIC are overcoming many of the language's previous limitations. Pascal is a structured language originally designed for teaching programming. C and C++ are emerging as preferred standards for professional programming in many industries, including the HVAC industry. It compiles to a very efficient and fast code, and the C source code from one computer can be recompiled with other C compilers to run on many types of machines. The major disadvantage of C is the high level of programming skill required to create programs.

Utility software programs perform standard organizing and data-handling tasks for a specific computer, such as copying files from one disk to another, printing file directories, printing files, and merging files (splicing two or more files together in a specific order). Utility software generally performs one or two specific functions, while applications software has a particular application that can require many functions and utilities.

HVAC application programs are designed to use the computer's power to calculate items such as loads, energy, and piping design. General-purpose applications software, such as accounting and word processing programs, is discussed in the section on Productivity Tools. Another specialized area of applications software is artificial intelligence, discussed in the section on Applications of Artificial Intelligence.

Purchased Software

While manufacturers of computer equipment also offer software, other distribution channels exist. PC software can be purchased from the manufacturer, software companies, distributors, and discount houses. Software price, level of support, return policies, and distribution method vary from vendor to vendor. Most software companies support their product in some way. Some companies have a poor reputation for supporting customers, so potential purchasers should determine whether the level of support from a particular vendor is acceptable. Some vendors allow customers to try software and return it if it is not acceptable; others make demonstration or limited function versions available for a minimal charge.

Another area of concern is how the software is distributed. It must be compatible with the computer it will run on. Some vendors still use copy protection schemes, which prevent the user from running the software on more than one computer. This copy protection inconveniences the user, and the trend is to move away from these protection methods. Upon purchase of PC software, the user is given a license with specific restrictions on how that software may be used, typically that it is to be used on only one computer. Often the user implicitly agrees to the terms of the license by opening the package. A separate signed license is not as common but is occasionally used. Because the distribution of unauthorized copies has been so costly to software companies, many are actively prosecuting illegal software use.

Before purchasing general-purpose software (e.g., spreadsheet, database, or word processor programs), the user should look into how the software runs and how easy it is to use. A full-service computer store may offer advice and demonstrations unavailable from mail-order or discount software sources. The user interface is important; some programs are very difficult to operate and understand, while others are very easy to use. Computer magazines often publish comparisons of general-purpose software that can be used for a first appraisal. The final determination should be made by the user. Once a package is chosen, the user may have difficulty changing to another.

Occasionally suppliers offer the program's source code, which is a set of human-readable instructions in a computer language. Skilled programmers can modify a source code to change the operation of the program. This is not generally recommended, because once the program is changed, support must be supplied internally.

Public Domain Software

Public domain software is available to the public either without charge or for a minimal charge (usually for maintenance and support). These programs are developed through government-supported projects, at universities, and by individuals. The source code for many public domain programs is available, though usually poorly supported. However, some programs are well documented and supported. Some programs offer only executable code to prevent unauthorized use of the source code in proprietary programs.

Public domain software can be obtained from computer bulletin boards, other individuals, and companies that distribute it for a small duplication charge. Care should be used when obtaining public domain software because its origin is difficult to trace. Some public domain programs have computer viruses; programs on bulletin boards are particularly suspect. A virus is a small program inserted into another program that can destroy stored data, lock up the computer, and otherwise cause problems (McAfee and Haynes 1989).

Royalty Software in Time-Sharing

The fact that unused computers cost almost as much to own as heavily used computers created a large industry in the 1960s—computer time-sharing, which allows more than one user to access a computer or family of computers simultaneously. Most major time-sharing service organizations have networks of mainframe computers and communications equipment capable of serving thousands of customers on an international basis. Customers may access a time-sharing computer across the country simply by dialing a telephone number.

Time-share organizations sell computer resources. Purchased software can be installed on the networks for the private use of individual customers. Under this arrangement, the cost of computer time varies with the activity level. The software is developed by either the time-share company or third party authors who sublicense to the time-share company. The development costs are billed to the user through the time-share company.

The advent of the PC has reduced the number of time-share users because the cost of a PC is low compared to the cost of a mainframe computer. More and more of the programs that formerly ran only on mainframe computers are now being run on PCs.

Custom Programming

There are three major strategies for obtaining custom programs: contracting to an outside firm, using an outside firm to provide consultation and help, or developing the programs internally.

Contracting to an outside firm for the entire programming effort should be considered if the host organization does not have the personnel or desire to support the software on an ongoing basis. Funds should be budgeted for the outside organization to support any modifications or enhancements that become necessary. Contracting outside

is a good approach for an organization that does not want to get involved with programming. The main drawbacks are the expense and the lack of control over the program. Licensing and ownership issues should be carefully spelled out in the contract.

Using an outside firm to provide consulting is practical if internal skill is insufficient for programming, but long-term support and maintenance of the software are to be done internally. Outside firms can provide the expertise to get a project going quickly. The design specification is critical to the success of the software.

Developing the program internally is viable if the skill and resources are available. Internal projects are easier to control because the people involved are usually under one roof. Most of the major vendors of software-based HVAC systems develop the software internally with occasional consultation from outside firms.

No matter which approach is chosen, the user must provide a detailed functional design specification. The calculations, human interface, reports, user documents, and testing procedures should be carefully detailed and agreed on by all parties before the development begins. To create a useful software program, a thorough understanding of the subject matter and a solid knowledge of computer programming are required. Software testing should also be specified at the beginning of the project to avoid the common problem of low quality due to hasty and inadequate testing. Design testing should address the human interface, a wide range of input values (including improper inputs), the algorithm, and any outputs (to paper, disk, or other media). Field testing should be done under field conditions with the final users of the software.

With any of the development approaches, good, understandable documentation is required. If for any reason the software cannot be adequately supported, the program will have to be either abandoned, replaced, or redone, causing a substantial drain on the organization.

Custom programming should be used to create only programs or features unavailable with existing software: no matter how high the cost of existing software, custom software that produces comparable results will cost more in the long run.

Sources for HVAC Programs

ASHRAE Journal's HVAC&R Software Directory (1990) is a comprehensive listing of HVAC-related software. It contains a brief description of each program, along with information on availability.

DESIGN CALCULATIONS

Although computers are now widely used in the design process, most programs perform not design, but simulation. That is, the engineer proposes a design and the computer program calculates the consequences of that design. When alternatives are easily cataloged, a program may design by simulating a range of alternatives and then selecting the best according to predetermined criteria. Thus, a program to calculate the annual energy usage of a building requires specifications for the building and its systems; it then simulates the performance of that building under certain conditions of weather, occupancy, and scheduling. A duct-design program may actually size ductwork, but usually an engineer must still decide air quantities, duct routing, and so forth.

Because computers do repetitive calculations rapidly, accurately, and tirelessly, it is possible for the designer to explore a wider range of alternatives and to use selection criteria based on annual energy costs or life-cycle costs, which would be much too tedious without a computer.

Heating and Cooling Loads

The calculation of design thermal loads in a building is a necessary step in the selection of HVAC equipment for virtually any building project. To ensure that heating equipment can maintain satisfactory building temperature under all conditions, peak heating loads are usually calculated for steady-state conditions without solar or internal heat gains. This relatively simple calculation can be performed with or without a computer.

Peak cooling loads are more transient than heating loads. Radiative heat transfer within a space and thermal storage cause thermal loads to lag behind instantaneous heat gains and losses. Especially with cooling loads this lag can be important, as the peak is both reduced in magnitude and delayed in time compared to the heat gains that cause it. Early methods of calculating peak cooling loads tended to overestimate loads; this resulted in oversized cooling equipment with penalties of a high first cost and part-load operating inefficiencies. Today various calculation methods are used to account for the transient nature of cooling loads. Some widely used methods of calculating the design loads for building elements include the following:

- Transfer Function Method (TFM)
- Cooling Load Temperature Difference/Cooling Load Factor (CLTD/CLF) Method
- Total Equivalent Temperature Differential/Time Averaging (TETD/TA) Method
- Steady-State Heat Transfer Method
- Response Factor Method
- Finite Difference Method

Chapters 25 and 26 of the 1993 *ASHRAE Handbook—Fundamentals* describe some of these methods in detail.

Both the TETD/TA and the TFM methods require a history of thermal gains and loads. Since histories are not initially known, they are assumed zero and the building under analysis is taken through a number of daily weather and occupancy cycles to establish a proper 24-hour load profile. Thus, procedures required for the TFM and TETD/TA methods involve so many individual calculations that noncomputerized calculations are almost ruled out. The CLTD/CLF method, on the other hand, is meant to be a manual calculation method; it can be implemented by a spreadsheet program. The CLTD and CLF tables presented in the 1993 *ASHRAE Handbook— Fundamentals* are, in fact, based on application of the TFM to certain geometries and building constructions. An automated version of the TFM is the preferred cooling load calculation method. The automated TETD/TA method, however, can provide a good approximation with significantly less computational effort than the TFM.

Characteristics of a Loads Program. In general, a loads program requires user input for most or all of the following:

- Full building description, including the construction of the walls, roof, windows, etc., and the geometry of the rooms, zones, and building. Shading geometries may also be included.
- Sensible and latent internal loads due to lights and equipment, and their corresponding operating schedules.
- Sensible and latent internal loads due to people.
- Indoor and outdoor design conditions.
- Geographic data such as latitude and elevation.
- Ventilation requirements and amount of infiltration.
- Number of zones per system and number of systems.

With this input, loads programs will calculate both the heating and cooling loads as well as perform a psychrometric analysis. Output typically includes peak room and zone loads, supply air quantities, and total system (coil) loads.

Selecting a Loads Program. In addition to general characteristics, such as hardware and software requirements, type of interface (icon-based, menu-driven versus command-driven), availability of manuals and support, and cost, some loads program-specific characteristics should be considered when selecting a loads program. The following are among the items to be considered:

- Type of building to be analyzed—residential versus commercial. Residential-only loads programs tend to be simpler to use than the more general-purpose programs meant for commercial and industrial use. However, residential-only programs have limited abilities.

- Method of calculation for the cooling load, as discussed previously in this section.
- Program limits on such items as number of systems, zones, rooms, and surfaces per room.
- Sophistication of modeling techniques, for example, the capability of handling exterior or interior shading devices, tilted walls, daylighting, and skylights.
- Units of input and output.
- Complexity of program. In general, the more sophisticated and flexible programs require more input and are somewhat more difficult to use than the simpler programs.
- Capability of handling the system types under investigation.
- Ability to share data with other programs, such as computer-aided design (CAD) and energy analysis.

Energy and System Simulation

Building energy simulation programs differ from peak load calculation programs; in the former, loads are integrated over time (usually a year), the systems serving the loads are considered, and the energy required by the equipment to support the system is usually the calculated output. Most energy programs simulate the performance of already designed systems, although programs are now available that make selections formerly left to the designer, such as equipment sizes, system air volume, and fan power. Energy programs are fundamental in making decisions regarding building energy use and, along with life-cycle costing routines, quantify the impact of proposed energy conservation measures during the design phase. In new building design, energy programs help determine the appropriate type and size of building systems and components; they also explore the effects of design tradeoffs and can be used to evaluate the benefits of innovative control strategies and the efficiency of new equipment.

Energy programs that track building energy use accurately can help in determining whether a building is operating efficiently or wastefully. They have also been used to allocate costs from a central heating/cooling plant among customers of the plant. However, one must be certain that such programs have been adequately calibrated to measured data from the building under consideration.

Characteristics of Building Energy Simulation Programs. Most programs simulate a wide range of buildings, mechanical equipment, and control options. While the importance of approximating solar loads on sunny days is widely recognized, computational results differ substantially from program to program. The shading effect from overhangs, side projections, and adjacent buildings is frequently a factor in the energy consumption of a building; however, the diversity of approaches to the load calculation results in a wide range of answers. Depending on the requirements of each program, various weather data are used. These can be broken down into five groups:

1. Typical hourly data for one year only, from averaged weather data
2. Typical hourly data for one year, as well as design conditions for typical design days
3. Reduced data, commonly a typical day or days per month for the year
4. Typical reduced data, nonserial or bin format
5. Actual hourly data, recorded on-site or nearby, for analysis where the simulation is being compared to actual utility billing data or measured hourly data

The calculation of heating and cooling loads is dependent on the choice of weather data.

Heat extraction is the rate at which the HVAC system removes heat from a conditioned space. When the space temperature is kept constant, this rate equals the cooling load; because this rarely happens, heat extraction is generally either smaller or larger than the cooling load. This concept is important for the analysis of intermittently operated HVAC systems; it provides information on the relationship between load and space temperature and leads to the calculation of the preheat/cool load for various combinations of equipment capacity and preheat/cool periods.

Both air-side and energy conversion simulations are required to handle the wide variations among central heating, ventilation, and air-conditioning systems. For proper estimation of energy use, simulations must be performed for each combination of system design, operating scheme, and control sequence.

System Simulation Techniques. Two basic approaches are currently used in computer simulation of energy systems: the fixed schematic technique and the component relation technique.

The fixed-schematic-with-options technique is the most prevalent program organization. Used in the development of the first generally available energy analysis programs, this technique involves writing a calculation procedure that defines a given set of systems. The system schematic is then fixed, with the user's options usually limited to equipment performance characteristics, fuel types, and the choice of certain components.

Component Relation Technique. Advances in system simulation and increased interest in special and innovative systems have prompted interest in the component approach to system simulation. The component relation technique differs from the fixed schematic in that it is organized around components rather than systems. Each of the components is described mathematically and placed in a library for use in constructing the system. The user input includes the definition of the system schematic, as well as equipment characteristics and capacities. Once all the components of a system have been identified and a mathematical model for each has been formulated, the components may be connected together in the desired manner, and information may be transferred between them. Although certain inefficiencies are built into this approach because of its more general organization, the component relation technique does offer versatility in defining system configurations.

Selecting an Energy Program. In selecting an energy analysis program, factors such as cost, availability, ease of use, technical support, and accuracy are important. There is, however, another fundamental consideration: whether the program will do what is required of it. It should be sensitive to the parameters of concern, and its output should include the necessary data. For other considerations, see Chapter 28 of the 1993 *ASHRAE Handbook—Fundamentals*.

In the past, computer simulation technology concentrated on thermal load calculation techniques; system simulation was limited to a number of common systems. Simulation development in the future will likely focus on system assimilation techniques, i.e., the means by which a system can be configured, given the mathematical models of its components.

Time can be saved if the initial input file for an energy program can also be used for load calculations. Some programs interface directly with computer-aided design (CAD) files, greatly reducing the time needed to create an energy program input file.

Comparisons of Energy Programs. Because energy analysis computer programs employ different methods of calculation, their results vary significantly. Many comparisons, verifications, and validations of simulation programs have been made and reported (see Bibliography). Conclusions from these reports can be summarized as follows:

- The results obtained by using several computer programs on the same building range from good agreement to no agreement at all. The degree of agreement depends on the interpretations of the program user and the ability of the computer programs to model the building.
- Several people using several programs on the same building will probably not agree on the results of an energy analysis.
- The same person using different programs on the same building may or may not find good agreement, depending on the complexity

of the building and its systems and on the ability of the computer programs to model the specific conditions in that building.

- "Forward" computer simulation programs, which calculate the performance of a building given a set of descriptive inputs, weather conditions, and occupancy conditions, are best used for design purposes.
- Calibration of hourly forward computer simulation programs is possible, but can require considerable effort for a moderately complex commercial office building. Detailed information concerning scheduled use, equipment set points, and even certain on-site measurements may be necessary for a closely calibrated model. Special-purpose graphic plots are useful in calibrating a simulation program with data from monthly, daily, or hourly measurements.
- "Inverse," empirical, or system parameter identification models may also be useful in determining the characteristics of building energy usage. Such models can determine the relevant building parameters from a given set of actual performance data. This is the inverse of the traditional building modeling approach, hence the name.

Energy Programs to Model Existing Buildings. Computer energy analysis of existing buildings can accommodate complex situations, evaluate the energy impacts of many alternatives, such as changes in control settings, occupancy, and equipment performance, and predict the relative magnitude of energy use. There are many programs available, varying widely in cost, degree of complexity, and ease of use (ASHRAE 1990).

A general procedure should be followed in computer energy analysis of existing buildings. First, energy consumption data must be obtained for a one- to two-year period. For electricity, these data usually consist of metered electrical energy consumption and demand on at least a month-by-month basis. For natural gas, the data are in a similar form and are almost always on a monthly basis. For both electricity and natural gas, it is helpful to record the dates when the meters were read. These dates are important in weather normalization for determining average billing period temperatures. For other types of fuel, such as oil and coal, the only data available may be the delivery amounts and dates.

Unless fuel use is metered or measured daily or monthly, consumption for any specific period shorter than one season or year is difficult to determine. The data should be converted to a per-day usage or adjusted to account for differences in the length of metering periods. The data tell how much energy went into the building on a gross basis. Unless extensive submetering is used, it is nearly impossible to determine when and how that energy was used and what it was used for. It may be necessary to install such meters to determine energy use.

The thermal and electricity usage characteristics of a building and its energy-consuming systems as a time-varying function of ambient conditions and occupancy must also be determined. Most computer programs can use as much detailed information about the building and its mechanical and electrical system as is available. Where the energy implications of these details are significant, it is worth the effort of obtaining them. Testing fan systems for air quantities, pressures, control set points, and actions can provide valuable information on deviations from design conditions. Test information on pumps can also be useful.

Data on building occupancy are among the most difficult to obtain. Since most energy analysis computer programs simulate the building on an hourly basis for a one-year period, it is necessary to know how the building is used for each of those hours. Frequent observation of the building during days, nights, and weekends shows which energy-consuming systems are being used and to what degree. Measured, submetered hourly data for at least one week is necessary for a beginning of an understanding of weekday/weekend schedule-dependent loads.

Weather data, usually one year of hour-by-hour weather data, are necessary for simulation (see Chapter 24 of the 1993 *ASHRAE Handbook—Fundamentals*). The actual weather data for the year in which energy consumption data were recorded will significantly improve the simulation of an existing building. Where the energy-consuming nature of the building is related more to internal than to external loads, the selection of weather data is less important; however, with residential buildings or buildings with large outside air loads, the selection of weather data can affect results significantly. The purpose of the simulation should also be considered when choosing weather data; i.e., either specific year data, data representative of long-term averages, or data showing temperature extremes may be needed, depending on the goal of the simulation.

It is likely that the results of the first computer runs will not agree with the actual metered energy consumption data. The following are possible reasons for this discrepancy:

- Insufficient understanding of the energy-consuming systems that create the greatest use
- Inaccurate information on occupancy and time of building use
- Inappropriate design information on air quantities, set points, and control sequences

The input building description must be adjusted and trial runs continued until the results approximate the actual energy use. It is usually difficult to match the metered energy consumption precisely. In any month, results within 10% are considered adequate.

The following techniques for calibrating a simulation program to measured data from a building should be considered:

- Matching submetered loads or 24-hour day profiles of simulated whole-building electric loads to measured data
- Matching *x-y* scatter plots of simulated daily whole-building thermal loads versus average daily temperature to measured data
- Matching simulated monthly energy use and demand profiles to utility billing data

Having a simulation of the building as it is being used permits subsequent computer runs to evaluate the energy impact of various alternatives or modifications. The evaluation may be accomplished simply by changing the input parameters and running the program again. The impacts of the various alternatives may then be compared and an appropriate one selected.

Duct Design

Two major needs exist in duct design: sizing and flow distribution. Duct sizing and equipment selection are part of any new duct system design. Flow distribution is the calculation of flows through the duct sections and terminals for an existing system with known cross sections and fan characteristics.

Duct Sizing. There are two major approaches to computerized duct sizing: (1) application of manual procedures, which, although computerized, are still limited in capability, and (2) optimization.

Selecting and Using a Program. Duct design involves laying out the ductwork, selecting the fittings, and sizing the ducts. Computer programs comply with many duct system constraints that require recomputation of the duct size. Computer printouts provide detailed documentation. Any calculation requires the preparation of accurate estimates of pressure losses in duct sections and the definition of the interrelations of velocity heads, static pressures, total pressures, and fitting losses.

The general computer procedure is to designate nodes (the beginning and end of duct sections) by number. Details about each node (e.g., divided flow fitting and terminal) and each section of duct between nodes (maximum velocity, flow rate, length, fitting codes, size limitation, insulation, and acoustic liner) are used as input data (Figure 1).

Characteristics of a duct design program include the following:

- Calculations for supply, return, and exhaust systems
- Sizing by constant friction, velocity reduction, static regain, and constant velocity methods
- Analysis of existing duct systems

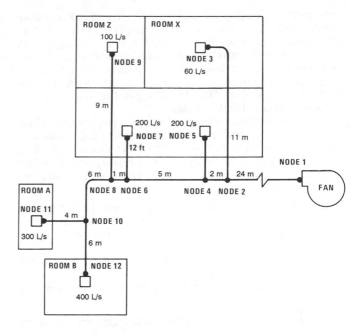

Fig. 1 Example of Duct System Node Designation

- Inclusion of fitting codes for a variety of common fittings
- Selection of the duct run with the highest pressure loss, and tabulation of all individual losses in each run
- Printout of all input data for verification
- Provision for error messages
- Calculation and printout of airflow for each duct section
- Printout of velocity, fitting pressure loss, duct pressure loss, and total static pressure change for each duct section
- Graphic showing of a schematic or line diagram indicating duct size and shape and flow rate and temperature in the duct system
- Calculation of heat gain/loss in the system and correction of temperatures and flow rates, including possible resizing of the system
- Specification of maximum velocities, size constraints, and insulation thicknesses
- Consideration of insulated or acoustically lined duct
- Bill of materials for sheet metal, insulation, and acoustic liner
- Acoustic calculations for each section of the system
- File-sharing with other programs, such as spreadsheet and CAD

Since many duct design programs are available, the following factors should be considered in program selection:

- Maximum number of branches that can be calculated
- Maximum number of terminals that can be calculated
- Types of fittings that can be selected
- Number of different types of fittings that can be accommodated in each branch
- Ability of the program to balance pressure losses in branches
- Ability to handle two- and three-dimensional layouts
- Ability to size a double-duct system
- Ability to handle draw-through and blow-through systems
- Ability to prepare cost estimates
- Ability to calculate fan motor power
- Provision for determining acoustical requirements at each terminal
- Ability to update the fitting library

Optimization Techniques for Duct Sizing. For an optimized duct design, fan pressure and duct cross sections are selected by minimizing the life-cycle cost of the system, an objective function that includes the initial cost and energy cost. A large number of constraints, including constant pressure balancing, acoustic restrictions, and size limitations, must be satisfied. Duct optimization is a mathematical programming problem with a nonlinear objective function and many nonlinear constraints. The solution must be taken from a set of standard diameters and standard equipment. Several numerical methods for duct optimization exist, such as the T-Method (Tsal et al. 1988), Coordinate Descent (Tsal and Chechik 1968), Lagrange Multipliers (Stoecker et al. 1971, Kovarik 1971), Dynamic Programming (Tsal and Chechik 1968), and Reduced Gradient (Arkin and Shitzer 1979).

Flow Distribution. Another problem is the prediction of airflows in each section of a presized system with known fan characteristics. This is called the flow distribution or air duct simulation problem. Whenever a retrofit to an existing duct system is considered, the need to calculate flow distribution occurs. An HVAC engineer may then ask the following questions:

- How will the retrofit influence the flow at existing terminals?
- Is it possible to change only the motor and leave the same fan?
- What is the new working point on the fan characteristic?
- Which duct sizes should be changed?
- What are the new duct sizes and what are the flows in the system with fully opened dampers?
- What is the best way to connect additional diffusers to an existing system?

A simulation program can help answer these questions. It can also analyze the efficiency of a control system effectively, check the performance of a number of parallel fans if one is not running, and predict the flows during field air balancing. The T-Method (Tsal et al. 1988) and the Gradient Steepest Descent Method (Tsal and Chechik 1968) have been used for simulating a duct system.

Piping Design

A large number of computer programs exist to size or calculate the flexibility of piping systems. Sizing programs normally size piping and estimate pump head for systems based on velocity and pressure drop limits, and some consider heat gain or loss from piping sections. Several programs produce bill of materials or cost estimates for the piping system. Piping flexibility programs assist in stress and deflection analysis. Many of the piping design programs can account for thermal effects in pipe sizing, as well as deflections, stresses, and moments.

The general technique for computerizing piping design problems is similar to that for duct design. A typical piping problem in its nodal representation is shown in Figure 2.

Useful piping programs do the following:

- Provide sufficient design information
- Perform calculations for both open and closed systems
- Calculate the flow, pipe size, and pressure drop in each section of the system
- Handle three-dimensional piping systems

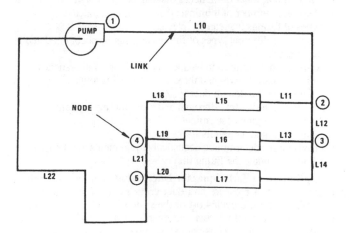

Fig. 2   Nodal System for Piping System

- Cover a wide selection of commonly used valves and fittings, including such specialized types as solenoid and pressure-regulating valves
- Consider different piping materials such as steel, copper, and plastic by including generalized friction factor routines
- Accommodate liquids, gases, and steam by providing property information for a multiplicity of fluids
- Calculate pump capacity and head required for liquids
- Calculate the available terminal pressure for nonreturn pipe systems
- Calculate the required expansion tank size
- Estimate heat gain/loss for each portion of the system
- Prepare a system cost estimate, including costs of pipe and insulation materials and associated labor
- Print out a bill of materials for the complete system
- Calculate balance valve requirements
- Perform a pipe flexibility analysis
- Perform a stress analysis for the pipe system
- Print out a graphic display of the system
- Allow customization of specific design parameters and conditions, such as maximum and minimum velocities, maximum pressure drops, condensing temperature, superheat temperature, and subcooling temperature
- Allow evaluation of piping systems for off-design conditions
- Provide links for calling by other programs, such as equipment simulation programs

Limiting factors to consider in piping program selection include the following:

- Maximum number of terminals the program can accommodate
- Maximum number of circuits the program can handle
- Maximum number of nodes each circuit can have
- Maximum number of nodes the program can handle
- Compressibility effects for gases and steam
- Provision for two-phase fluids

Acoustic Calculations

Chapter 43 presents a concise summary of sound generation and attenuation in HVAC systems. Application of these data and this methodology to the analysis of noise in HVAC systems is straightforward, but difficult in the amount of computation involved. All sound generation mechanisms and sound transmission paths are potential candidates for analysis. Adding to the computational work load is the necessity of extending the analysis over, at a minimum, octave bands 1 through 8 (63 Hz through 8 kHz). A computer can save a great amount of time and difficulty in the analysis of any noise situation, but the HVAC system designer should be wary of using unfamiliar software.

Caution and critical acceptance of analytical results are mandatory at all frequencies, but particularly at low frequencies. Not all manufacturers of equipment and sound control devices provide data below 125 Hz. Thus, the HVAC system designer conducting the analysis and the programmer developing the software must make assumptions based on experience for these critical low-frequency ranges.

The designer/analyst should be well satisfied if predictions are within 5 dB of field-measured results. In the low-frequency "rumble" regions, results within 10 dB are often as accurate as can be expected, particularly in areas of fan discharge. Conservative analysis and application of the results is necessary, especially if the acoustic environment of the space being served is critical.

Noise computations for HVAC duct systems should account not only for obvious noise generators such as fans, but also the potential for flow generation in duct elements (e.g., tees and branches) and downstream elements (e.g., VAV boxes). Whenever flow velocities exceed 7.5 m/s, flow-generated noise is a significant possibility.

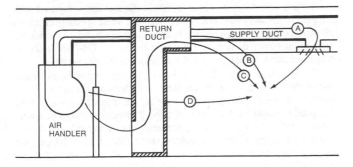

Ⓐ Discharge Airborne Sound
Ⓑ Discharge Breakout
Ⓒ Inlet or Return
Ⓓ Radiated

Fig. 3   Significant Indoor Air Handler Sound Paths

Sound in ducts propagates upstream as well as downstream, and sound from the fan of a single system travels down at least four paths (Figure 3):

• Fan discharge sound travels down the supply duct through the diffusers into the space (path A).
• Fan discharge sound can break out of the supply duct. This path is significant for occupants within 15 m of the fan discharge (path B).
• Fan inlet sound travels down the return duct, through the return grilles or ceiling and into the occupied space (path C).
• Radiated fan sound can penetrate the unit casing and mechanical room wall and enter the occupied space (path D).

The sound pressure level in a space is the composite of all sound paths to the space. For example, one diffuser in a room may meet noise criteria curve NC-30, but the combined sound pressure level of ten diffusers in a room may be NC-40 or greater.

Several currently available acoustics programs are generally easy to use but are often less detailed than the custom programs developed by acoustic consultants for their own use. Acoustics programs are designed for comparative sound studies and allow the design of a comparatively quiet system. Acoustic analysis should address the following key areas of the HVAC system:

• Sound generation by HVAC equipment
• Sound attenuation and regeneration in duct elements
• Wall and floor sound attenuation
• Ceiling sound attenuation
• Sound break-out or break-in in ducts or casings
• Room absorption effect (relation of sound power criteria to sound pressure experienced)

Algorithm-based programs are preferred because they cover more situations (see Chapter 43). However, assumptions are an essential ingredient of algorithms. These basic algorithms, along with sound data from the acoustics laboratories of equipment manufacturers, are incorporated to various degrees in acoustics programs. The HVAC equipment sound levels in acoustics programs should come from the manufacturer and be based on measured data, because there is a wide variation in the sound generated by similar pieces of equipment. Some generic equipment sound generation data, which may be used as a last resort in the absence of specific measured data, are found in Chapter 43. Whenever possible, equipment sound power data by octave band (including 32 Hz and 63 Hz) should be obtained for the path under study. A good sound prediction program relates all performance data.

Many other more specialized acoustics programs are available. Various manufacturers provide equipment selection programs that not only select the optimum equipment for a specific application, but also provide associated sound power data by octave bands. These programs can help in the design of a specific aspect of a job. Data from these programs should be incorporated in the general acoustic analysis. For example, duct design programs may contain sound predictions for discharge airborne sound based on the discharge sound power of the fans, noise generation/attenuation of duct fittings, attenuation and end reflections of variable air volume (VAV) terminals, attenuation of ceiling tile, and room effect. VAV terminal selection programs generally contain subprograms that estimate the space NC level near the VAV unit in the occupied space. However, projected space NC levels alone are not acceptable substitutes for octave-band data. The designer/analyst should be aware of assumptions, such as room effect, made by the manufacturer in the presentation of acoustical data.

Predictive acoustic software allows system designers to look at HVAC-generated sound in a realistic, affordable time frame. HVAC-oriented acoustic consultants generally assist designers by providing cost-effective sound control ideas for sound-critical applications. A well-executed analysis of the various components and sound paths enables the designer to assess the relative importance of each and to direct corrective measures, where necessary, to the most critical areas. Computer-generated results should supplement the designer's skills, not replace them.

Equipment Selection and Simulation

Three types of equipment-related computer programs are equipment selection, equipment optimization, and equipment simulation programs.

Equipment selection programs are basically computerized catalogs. The program locates an existing equipment model that satisfies the entered criteria. The output is a model number, performance data, and sometimes alternative selections.

Equipment optimization programs display all possible equipment alternatives and let the user establish ranges of performance data or first cost to narrow the selection. The user continues to narrow the performance ranges until the best selection is found. The performance data used for optimizing selections vary by product family.

Equipment simulation programs calculate the full- and part-load performance of specific equipment over time, generally one year. The calculated performance is matched against an equipment load profile to determine energy requirements per hour. Utility rate structures and related economic data are then used to project equipment operating cost, life-cycle cost, and comparative payback.

Some advantages of equipment programs include the following:

1. High speed and accuracy of the selection procedure
2. Pertinent data presented in an orderly fashion
3. More consistent selections than with manual procedures
4. More extensive selection capability
5. Multiple or alternate solutions
6. Small changes in specifications or operating parameters easily and quickly evaluated
7. Data-sharing with other programs, such as spreadsheets and CAD

Simulation programs have the advantage of (1) projecting part-load performance quickly and accurately, (2) establishing minimum part-load performance, and (3) projecting operating costs and payback-associated higher performance product options.

There are programs for nearly every type of HVAC equipment, and industry standards apply to the selection of many types. The more common programs and their optimization parameters include the following:

Air distribution units	Pressure drop, first cost, sound, throw
Air-handling units, rooftop units	Power, first cost, sound, filtration, heating and cooling capacity
Boilers	First cost, efficiency, stack losses
Cooling towers	First cost, design capacity, power, flow rate, air temperatures
Chillers	Condenser head, evaporator head, capacity, power input, first cost, compressor size, evaporator size, condenser size, refrigerant type
Coils	Capacity, first cost, water pressure drop, air pressure drop, rows, fin spacing
Fan coils	Capacity, first cost, sound, power
Fans	Volume flow, power, sound, first cost, minimum volume flow
Heat recovery equipment	Capacity, first cost, air pressure drop, water pressure drop (if used), effectiveness
Pumps	Capacity, head, impeller size, first cost, power
Air terminal units (variable and constant volume flow)	Volume flow rate, air pressure drop, sound, first cost

Some extensive selection programs have evolved. For example, coil selection programs can select steam, hot water, chilled water, and refrigerant (direct expansion) coils. Generally, they select coils according to procedures in ARI *Standards* 410 and 430 (ARI 1991, 1989).

Chiller and refrigeration equipment selection programs can choose optimal equipment based on such factors as lowest first cost, highest efficiency, best load factor, and best life-cycle performance. In addition, some manufacturers offer modular equipment for customization of their product. This type of equipment is ideal for computer selection. Recently, chiller selection programs have begun to emphasize the type of refrigerant and its ozone depletion potential.

However, equipment selection programs have limitations. The logic of most manufacturers' programs is proprietary and not available to the user. All programs incorporate built-in approximations or assumptions, some of which may not be known to the user. Equipment selection programs should be qualified before use.

ADMINISTRATIVE USES OF COMPUTERS

Word Processing

Word processing creates documents electronically rather than physically, as in typing. Editing text is easier and less expensive, as the entire document does not have to be retyped with each draft. Documents are usually stored on magnetic disk, although optical disk storage systems with vast capacities are now becoming available.

Text can be entered, corrected, reformatted, moved, copied, and deleted in blocks. Heading and footing text can be specified for all pages in a report. Global search and replace functions locate and replace words or phrases wherever desired in a document, reformatting the paragraphs where changes were made.

Many word processing programs can do arithmetic with columns of figures, renumber paragraphs to match an outline, and organize notes, either as footnotes on the page of the reference or as end notes at the end of the document. It is also possible to incorporate images from graph or drawing programs into the documents.

The following are commonly found additional functions:

Mail-merging, in which text, such as names and addresses, is inserted from one file into a form letter; everyone on the list receives an individualized copy of the letter.

Spelling checking, in which words in the text are compared with a master dictionary and errors are flagged. Most spelling checkers allow the addition of specialty words to the main standard dictionary or to an auxiliary dictionary for future reference. On-line thesaurus programs allow the user to ask for a list of equivalent words.

Grammar and style checking, in which long sentences, excessive use of the passive voice, incomplete sentences, incorrect choice of homonyms (e.g., there, their, they're), improper capitalization, unpaired punctuation (i.e., lack of closing quotation marks or parentheses), lack of subject-verb agreement, and many other potential flaws are flagged. If these programs have a fault, it is in flagging too many things; they can bog the user down with excessive citations. Nonetheless, they give valuable guidance.

Indexing, in which indexes or tables of contents are created from specially marked section and paragraph headings in the text.

Specification Writing

The efficiency and flexibility of a computerized master specification in conjunction with a word processor make it a valuable tool for preparing project specifications. Those who use the master specification should standardize language to avoid repeated editing of text. A master specification is most beneficial to firms that do not specialize in unique, one-of-a-kind designs. If the master can be used with only minor revisions for the majority of specifications, the time spent on editing may be justifiable. In general, the larger the size and the greater the number of documents produced, the greater the savings in time and money compared to traditional preparation and typing.

Even without a master specification, however, it is possible for the writer and typist to save considerable time by "cutting and pasting" from previous specifications with minor revisions.

Desktop Publishing

Desktop publishing is the creation of documents that integrate text with graphics. Many high-end word processors have the ability to import graphic images and to print in decorative typefaces, but the full capability to arrange page layouts on a graphics screen is found only in desktop publishing programs. However, word processing programs are incorporating more and more desktop publishing features as time progresses.

Most desktop publishing programs are limited in their ability to create text and drawings; they normally import text from a word processing program and images from separate drawing programs. High-resolution graphic displays for personal computers and laser printers have allowed the creation and printing of high-quality graphic images.

Management Planning and Decision Making

Computers are ideal for such repetitive operations as inventory control, accounting, purchasing, and project control, providing managers with up-to-date information. The ability to store and manipulate large amounts of data is a powerful basis for forecasting. With sufficient historical data, statistical methods can be employed to forecast sales, inventory needs, production capability, manpower needs, and capital requirements. For example, econometric modeling may be included to help forecast the effects of varying inflation rates.

Much of the data required for such techniques may not be in the company's files and must be found elsewhere. Technical papers, market trends, industry reports, financial forecasts, codes, standards, regulatory data, and demographic or geographic data are examples of information that can be searched for by computer. In most cases, only a local phone call must be made, and the company is charged a time-sharing fee. Management information systems should also have the capability to provide special or supporting information in case it becomes necessary to delve deeper into an operation.

Employee Records and Accounting

All corporate accounting functions, from payroll to aging accounts receivable to taxes, can be automated with a wide choice of engineering-specific accounting software on even the smallest PCs. A firm with unique accounting requirements may often meet them by developing specific applications, called templates or overlays, for common spreadsheet programs. Continuously updated financial reports needed for project evaluation and government use can be generated more readily by computer than by manual methods or off-site accounting services. Labor expenses for various projects or departments can easily be separated and monitored on a regular basis. Personnel records for each employee can also be maintained on a computer.

Project Scheduling and Job Costing

The success of any project, no matter what its size, depends on good scheduling and control to ensure that manpower and materials are where they are supposed to be at the proper time. Whenever portions of a project can be done simultaneously, are dependent on the completion of one or more prior parts, or must share a limited amount of manpower and equipment, a project scheduling technique such as CPM (Critical Path Method) or PERT (Program Evaluation and Review Technique) can establish priorities. These techniques are well implemented on microcomputers. Current microcomputer programs can also perform manpower leveling to make the best use of available resources, and some may provide graphics suitable for client presentations as well.

Commercially available computerized scheduling techniques quickly provide updated schedules and priorities, even if the critical path changes due to a delay in one part of the job. This capability is valuable for determining whether scheduled completion dates and costs can be met and for finding possible alternatives. Once a computerized system of personnel, payroll, and accounting functions has been established, a history of project costs, manpower requirements, and material requirements can be maintained. The system can categorize and tabulate them for future reference, permitting more accurate estimates for future projects.

Security and Integrity

While computers are generally reliable, provision must be made for potential loss of important data in the event of equipment failure, theft, sabotage, natural catastrophe, or a previously undiscovered problem in the software. The most common method of protection is periodic duplication of master tape or disk files. Duplication of the most important data files, such as the firm's accounts, should take place with every posting. Duplicate files should be kept in a secure area, such as a fireproof vault on the premises or totally off-site.

Another security concern is unauthorized access to particular data. The implementation of a formal, written data security policy, no matter how minimal it is initially, is a critical step in the computerization of any firm. It requires careful planning to make certain that only authorized personnel have access to confidential files. Most major accounting packages, as well as many popular spreadsheet and database management programs, contain built-in safeguards, such as multilevel password protection, to avoid this problem. Some packages also contain a transactions log to track who has accessed a particular file and what functions were performed.

PRODUCTIVITY TOOLS

Spreadsheet Software

One important type of microcomputer program used routinely by managers and engineers is the spreadsheet. Spreadsheets are two- or three-dimensional electronic tables containing thousands of cells into which the user may enter labeling text, numbers, formulas, or even macro-commands. Macro-commands can transfer control to another program, automatically move the user around in the spreadsheet, jump to other spreadsheets or databases, or execute other programs and return with a value. Automatic recalculation of all cells defined by formulas following data entry keeps all areas of the spreadsheet current. Advantages of spreadsheets include the following:

- Rapid assembly and testing of a chain of equations
- Preformatted output/input and "template" structures
- Prepackaged or "canned" graphics
- Easy file transfer between spreadsheets or between different proprietary spreadsheet programs and ASCII or database files
- Ability to sort lists of information
- Matrix and statistical computations
- Automatic recording of multiple keystrokes into a macro-command
- Additional programs with special features that "add on" to the existing features of a spreadsheet

Spreadsheets also allow numeric and alphanumeric data to be imported or exported from other nonspreadsheet programs. Graphs from spreadsheets can be imported into sophisticated graphics presentation programs for additional editing and preparation.

Graphics and Imaging Software

Microcomputer software graphics packages readily produce x-y and bar graphs, pie charts, and CAD drawings, and often have a library of graphic elements that make possible the rapid creation of custom presentations. Graphics files can be imported and incorporated into desktop publishing documents. Many specialty programs allow graphics files to be transferred between different proprietary application programs and images digitized from photographs to be imported into an existing graphics drawing. The new technology of multimedia tools allows video and sound to be combined with traditional text and graphics.

Communications Software

Another evolving technology is electronic communication. Communication between computers takes place through telephone lines or other networks of various configurations. Coaxial cables, twisted pair wiring, superimposed ac carrier frequencies, fiber optics, radio transmissions, and infrared light are among the media used. Features available on such networks include bulletin boards, on-line retrieval services, shared services, electronic mail, and file transfer.

Electronic communication using existing telephone lines is rapidly becoming possible anywhere there is telephone service, even through overseas exchanges or with cellular phones. Telephone systems are being upgraded to transfer digital signals, which will vastly expand capabilities.

Electronic communication continues to require both appropriate software and a modem, which performs the translation of the digital computer signals into analog signals suitable for current telephone lines. The modems most commonly used at present transfer data at 1200, 2400, 4800, 9600, or 14,400 bits per second (bps) using one of several transmission standards.

With the simplest of communications programs, the user can type messages at the keyboard to be sent to a remote computer, specify a file in a local storage to be sent to the remote computer, or receive a file from the remote computer and save it locally. Other capabilities include executing error-checking protocols with the remote computer to eliminate transmission errors and allowing the remote computer to take control of the local one, which is called terminal emulation.

Database Software

Microcomputer database systems allow the user to record, organize, and store information, and can be viewed as computerized record-keeping systems (Date 1988). Most database software includes a query-based command language, as well as a programming language that can be used to create applications. Some programs allow for query-by-example, windowing, or an icon-driven user interface.

One example of an application could be a client list containing a record of the following information for each client: firm name, street address, city, state, mail code, telephone number, contact person, dates of contact, and links to others in the list. Database programs allow very large numbers of records to be alphabetized, selected by specific index (e.g., city, state), and printed onto mailing labels or incorporated into a report. With object-oriented design, the query structure is different, and data can be encapsulated and reused in other databases more easily.

Special-Purpose Software

New types of special-purpose software are created almost daily, some of them specific to the HVAC industry. Some of the programs, which coexist with other programs in the computer memory, provide accessories such as notepads; appointment calendars; battery charge (in laptop PCs); name and address files; alarm clocks; binary, octal, and hexadecimal calculators; telephone dialers and on-line dictionaries; spell checkers; thesaurus; and "last command" recall. Specialty programs also include file organizers, routine backup programs, file restoration programs, and hardware diagnostics. Many software packages integrate word publishing programs, spreadsheets, communication programs, and graphics programs.

Many special-purpose HVAC application programs exist, including proprietary equipment selection packages and microcomputer versions of public domain programs such as DOE-2, BLAST, TRNSYS, and other powerful simulation programs previously available only on large mainframe computers. Many non-HVAC advanced analysis programs are also available, including statistical and mathematical packages, specialty artificial intelligence or knowledge-based system development shells, and neural network development programs.

Advanced Input and Output Options

With the rapid expansion of input and output options have come advanced input capabilities, including digitizing of images and text, scanners and bar code input, pen-based input, and downloading of data directly from portable field instruments and facsimile (fax) machines.

Traditionally, only text and numbers have been stored and manipulated in computers. Each character of text can be represented internally by a single byte of storage. Graphic images are more difficult to store and manipulate; they can be represented in bitmap or in vector format. In the bitmap or raster representation, the image is decomposed into points in a rectangular array (pixels). Storage requires a minimum of one bit per pixel for monochrome images and several bits per pixel for color. In vector representation, only end points of line segments are stored (as numbers). While the output of bitmap images is simple on raster devices such as graphic printers, scaling of images does not always produce satisfactory output. Vector representations plot easily on pen plotters and scale to a high degree of accuracy, but must be converted to raster format for output on graphic printers.

Computer-aided design and other drawing systems use vector representation, which allows for easy manipulation. Paint programs use bitmap images. Scanners are now available that scan an image on paper and produce a raster representation (text and images are both in bitmap format). Paint programs can take scanned images directly and facilitate their manipulation by the user. It is possible to scan drawings into raster format, and then use conversion software to change the raster representation into vector form for use in CAD systems. Similarly, optical character representation (OCR) software can take a raster image of text and convert it to individual text characters in the computer for input into word processing programs.

Fax equipment scans input images into raster format and sends the images bit-by-bit over telephone lines to a receiver, which prints them out as raster images. Computers can capture incoming fax images, allowing either conversion to internal text form or manipulation of the images.

Just as input capabilities have improved, so have output options, including the porting of graphic routines to VCR tapes for assembly into training tapes, direct 35-mm slide production, and direct transmission of output via fax machine. Newly available hardware and specialty software for laptops allow the user to preassemble and display an animated electronic slide show, digitized images, or even CAD drawings using a special translucent output screen (LCD) that is placed on top of a traditional overhead projector. Certain hardware even allows the user to view a television image in a separate window while working on a document or spreadsheet. Such an application might be the precursor of a video-based energy management and control system (EMCS).

COMPUTER-AIDED DESIGN

Capabilities

Computer-aided design (CAD) systems give designers computerized tools for the basic drafting of HVAC contract and shop drawings, and simplify HVAC analytical and design tasks. Computer graphics of buildings and their systems help coordinate interdisciplinary design and simplify modifications. A distinction is sometimes drawn between CAD (computer-aided drafting) and CADD (computer-aided design and drafting) in that the former is a subset of the latter: CADD encompasses both the creation of graphic images and the storing and use of attributes of the graphic elements, such as automated area takeoff or the storage of design characteristics in a database. In this section, CAD refers to all aspects of design and drafting systems.

Scale drawings express not only physical size, but also how different elements are connected and where they are located within an area. It is easy to extract lengths, areas, and volumes, as well as the orientation and connectedness (topology) of building parts from computerized drawings of buildings. Drawings can be changed easily and do not require redrawing, as in manual drafting. Building walls or ductwork, for example, can be added, removed, enlarged, reduced, and moved to a new location on a drawing. The user can rotate an entire building on the site plan without redrawing the building or destroying the previous drawing.

Some computer-aided design systems also help in (1) material and cost estimating, (2) design and analysis of HVAC systems, (3) visualization and interference checking, and even (4) manufacturing of HVAC components such as ductwork. These systems are now customized to HVAC applications; for instance, a two-line duct layout can be generated automatically from a one-line drawing and section size information.

Computer-generated drawings of building parts can be linked to nongraphic characteristics, which can be extracted for reports, schedules, and specifications. For example, an architectural designer can draw wall partitions and windows, and then have the computer automatically tabulate the number and size of windows, the areas and lengths of walls, and the areas and volumes of rooms and zones. Similarly, an operator can store information about a drawing for later use in reports, schedules, design procedures, or drawing notes. For example, the airflow, voltage, weight, manufacturer, model number, cost, and other data about a fan displayed on a drawing can be stored in a file or database. The linkage between the graphics and attributes makes it possible to review the characteristics of any item or to generate schedules of items located in a certain area on a drawing. Conversely, the graphics-to-data links allow the designer to enhance graphic items having a particular characteristic by searching for that characteristic in the data files and then making the associated graphics brighter, flashing, bolder, or different-colored in the display.

Computer-generated drawings, especially those created in layers, can also help the designer visualize the building and its systems and check for interferences, such as the structure and ductwork occupying the same space. Some software packages now perform interference checking themselves. This layering feature also helps an HVAC designer or drafter coordinate designs with other disciplines.

HVAC system CAD drawings and their associated data can be used by building owners and maintenance personnel for ongoing facilities management, strategic planning, and maintenance. They are useful for computer-aided manufacturing such as duct construction. Two-dimensional CAD drawings can be used to create a three-dimensional model of the building and its components, which is useful in the visualization of part or all of the building. The three-dimensional building model can also aid in the development of sections and details for building drawings. CAD programs can also automate the cross-referencing of drawings, drawing notations, and building documents.

Hardware Considerations

Computer-aided design can be implemented on a wide range of computers, terminals, and display screens using many output and storage devices. The nature of the user's work, such as the size and type of buildings to be designed, may influence CAD hardware selection. Small PC-based systems have basic CAD capabilities but are somewhat limited in the size and number of drawings they can work with and in their ability to coordinate drawings. Unless they are networked, microcomputer systems are inherently single-user systems, which means that other designers generally cannot share the drawings another user is working on. On the other hand, minicomputer and mainframe computer-based CAD systems are multiuser, multitasking systems designed for sharing drawings and data among several users, as well as managing a large inventory of drawings of almost unlimited size and complexity. These differences are important for both small, single-discipline design firms and larger, integrated, multidisciplinary design firms that may require that CAD drawings be available to all in-house designers working on a common project. Other hardware considerations include a system's speed, cost, memory and storage capability, and ease of use. Turnkey CAD systems are available for which the hardware is developed, installed, maintained, and upgraded by the CAD supplier, allowing the operator to use the system without worrying about start-up and support.

Software Considerations

Ease of use is perhaps the most critical consideration in selecting a CAD system. CAD software tailored to HVAC design and drafting functions is preferable to software that has only basic graphics and data management capabilities. CAD software with flexible, user-definable features is advantageous to users who may need to customize or enhance the CAD HVAC task. Built-in error checking and error reporting features are important in keeping the design process moving ahead when problems occur. The availability and quality of software support, check-out, training, and enhancement are also important and can determine whether an in-house computer programmer is necessary or not. Turnkey CAD systems are generally customized to particular applications, such as HVAC design, and offer more substantial support.

Another software consideration is transferability of drawings and data among different CAD systems, because the building owner, architect, and other engineers may have different CAD systems. Several graphic exchange standards make transfers possible. The Initial Graphics Exchange Specification (IGES), the Standard Interchange Format (SIF), and the Data Exchange Format (DXF) are the three most prominent formats. Software interfaces to these formats are important.

COMPUTER GRAPHICS AND MODELING

Computer graphics programs may be used to create and manipulate pictorial information between the computer and the user (Rankin 1989). Information can be assimilated more easily when presented in the form of graphical displays, diagrams, and models. Historically, the building of scale models has improved the design engineer's understanding and analysis of early prototype designs. The psychrometric chart has also been invaluable to HVAC engineers for many years (Li et al. 1986).

Combining alphanumerics with computer graphics enables the design engineer to quickly evaluate design alternatives. Computers can pictorially simulate design problems with three-dimensional color displays that can predict the performance of mechanical systems before they are constructed. Simulation graphics software is also used to evaluate design conditions that cannot ordinarily be tested with scale models due to high costs or time constraints. Product manufacturability and economic feasibility may thus be determined without the construction of a working prototype. Where historically only a few dozen scale model tests could be performed prior to full-scale manufacturing, combined computer and scale model testing allows hundreds or thousands of tests to be performed, improving reliability in product design.

Integrating computer graphics software with expert system software can further enhance design capability. Integrated CAD and expert systems can help the building designer and planner with construction design drawings and with construction simulation for planning and scheduling complex building construction scenarios (Potter 1987).

A simple yet helpful tool available to the HVAC engineer is the graphic representation of thermodynamic properties and thermodynamic cycle analysis. These programs may be as simple as computer-generated psychrometric charts with cross-hair cursor retrieval of properties and computer display zooming or magnification of chart areas for easier data retrieval. More complex software is also available. Using graphic-assisted numeric analysis, these programs provide graphic representation of thermodynamic cyclic paths overlaid on two- or three-dimensional thermodynamic property graphs such as P-v, P-h-T, or T-h-s diagrams. With a simultaneous graphic display of the calculated results of Carnot efficiency and COP, the user can understand more readily the cycle fundamentals and the practicality of the cycle synthesis (Abtahi et al. 1986).

HVAC engineers use complex simulation software to assist in the design of large district heating and cooling systems. Typical piping and duct system software produces large amounts of data to be analyzed. The introduction of computer graphics into the simulation process simplifies the task. Pictorially enhanced simulation output of piping system curves, pump curves, and load curves speeds the sizing and selection process for the engineer. What-if scenarios with numeric/graphic output increase the designer's understanding of the system and help avoid design problems that may otherwise come to light only after installation (Chen 1988). Graphic-assisted fan and duct system design and analysis programs are also available to system designers (Mills 1989).

As the numerical analysis of system design becomes more complex, involving three-dimensional simulations with highly complex coupling between the various system variables, massive amounts of data are generated. When the volume of data becomes so large that analysis and understanding are inhibited, graphic simulation improves visualization of the process and allows for such complex studies as fluid flow field analysis. Graphic simulation and analysis may replace (or at least complement) the time-consuming and costly scale model testing in wind and water tunnels normally associated with these problems. Graphic modeling and simulation of complex fluid flow fields are also used to simulate airflow fields in unidirectional clean rooms. This simulation technology may make clean room mock-ups obsolete (Busnaina et al. 1988). Computer models of particle trajectories, transport mechanisms, and contamination propagation are also commercially available (Busnaina 1987).

Airflow analysis of flow patterns and air streamlines is done by solving the fundamental equations of fluid mechanics for laminar and turbulent flows, where incompressibility and uniform thermophysical properties are assumed (Kuehn 1988). For example, the Navier-Stokes equations for conservation of momentum and the continuity equation for conservation of mass can be used. Finite element and finite volume modeling techniques are used to produce two- or three-dimensional pictorial-assisted displays, with the velocity vectors and velocity pressures at individual nodes solved and numerically displayed.

An example of pictorial output is seen in Figure 5 of Chapter 15, where calculated airflow streamlines have been overlaid on the computer against a two-dimensional graphic of the clean room. With this display, the clean room designer can see potential problems in the configuration, such as the circular flow pattern in the lower left and right.

Major features and benefits associated with most computer flow models include the following:

- Two- or three-dimensional modeling of simple clean room configurations
- Modeling of both laminar and turbulent airflows
- CAD of air inlets and outlets of varying sizes and of room construction features
- Allowance for varying boundary conditions associated with walls, floors, and ceilings
- Pictorial display of aerodynamic effects of process equipment, workbenches, and people
- Prediction of specific airflow patterns in all or part of a clean room
- Reduced costs of design verification for new clean rooms
- Graphical representation of flow streamlines and velocity vectors to assist in flow analysis

A graphical representation of simulated particle trajectories and propagation is shown in Figure 6 of Chapter 15. For this kind of graphic output, the user inputs a concentrated particle contamination and the program simulates the propagation of particle populations. In this example, the circles represent particle sources, where the diameter is a function of the concentration present. Such simulations assist the clean room designer in placing protective barriers.

While research has shown excellent correlation between flow modeling by computer and flow modeling done in simple mock-ups, the modeling software should not be considered a panacea for clean room design; simulation of flow around complex shapes is still being developed, as are improved simulations of low Reynolds number flow.

The acceptance and success of computer graphic-assisted programs depend on the accuracy of graphic output and on the method used to convey the display. Whether the simulations are two-dimensional and black and white, or three-dimensional, color and fully animated, the choice of display type will influence user acceptance. Many three-dimensional color outputs are no longer used only for attractive presentations, but are necessary design tools that increase the productivity of the user (Mills 1989).

APPLICATIONS OF ARTIFICIAL INTELLIGENCE

There are many different applications of artificial intelligence (AI), ranging from systems that learn, reason, and recognize speech like human beings to those that can help make decisions based on incomplete, uncertain, and/or complicated information. Artificial intelligence techniques are generally used in three areas—design, control, and diagnosis. The idea of intelligent building design tools has attracted much attention (Brambley et al. 1988).

Some of the technology of interest to engineers is discussed below. Although each technology is discussed separately here, there are software packages available that combine several AI techniques into one development environment. An engineer who would like to try out some of these packages can usually start with a small application to get a feel for the technology. Some software packages require no programming experience and are relatively easy to learn, while others are best learned at a short course.

Knowledge-Based Systems

Knowledge-based systems (KBSs), often called expert systems, can make decisions similar to those made by human experts. These systems are used to enhance the productivity of engineers. KBSs typically use facts, if-then rules, and/or models to make decisions. Most KBS tools can also link with other programs, access databases, and import graphics. Expert systems shells are usually complete development packages that include rule and database development tools, debugging facilities, and good user interfaces. Some shells take advantage of windowing environments and incorporate object-oriented features, hypertext, and graphics. Considerations for selecting a shell are discussed by Shams et al. (1994a), and the development of expert systems is discussed by Shams et al. (1994b) and Nielsen and Walters (1988).

Knowledge-based systems have been used to make decisions, diagnose system faults, monitor and control system operation, and develop tutorials. Some HVAC applications include KBSs that select HVAC equipment for small office buildings (Shams et al. 1994a, 1994b), analyze building energy consumption (Haberl and Claridge 1987, Haberl et al. 1988), recommend energy conservation options (Meadows and Brothers 1989), help with the conceptual design of building energy systems (Doheny and Monaghan 1987, Mayer et al. 1991), incorporate building regulations into design (Cornick et al. 1990), monitor HVAC equipment (Kaler 1990), manage cooling loads (Potter et al. 1991), and diagnose HVAC equipment problems.

Artificial Neural Networks

Artificial neural networks (ANNs) are collections of small individual interconnected processing units. Information is passed between these units along interconnections. An incoming connection has two values associated with it: an input value and a weight. The output of the unit is a function of the summed value. ANNs, while implemented on computers, are not programmed to perform specific tasks. Instead, they are trained with repeated data sets until they learn the patterns presented to them. Once the net is trained, new patterns may be presented to it for classification.

There are numerous techniques and architectures for implementing neural nets, but these details are best covered in other texts, such as Wasserman (1989). Specialized neural computers are currently being developed, but most neural nets can be implemented by software. Some of the areas in which ANNs have been successful include speech recognition, image enhancement, pattern association, and global optimization problems.

Fuzzy Logic

Fuzzy logic is a form of mathematics that allows precise computations using inexact and uncertain input. It operates by assigning a membership value in a fuzzy set to the values of each input. For example, the set of (membership value, fuzzy value) for a temperature of 22°C might be [(0.0, frigid), (0.1, cold), (0.7, comfortable), (0.3, warm), (0.0, hot)]. The membership values do not have to add up to 1.0. Mathematics and rules are then applied to the fuzzy values (frigid, cold, comfortable, warm, hot) to arrive at the conclusion, which is the output fuzzy set. The output fuzzy set is defuzzyfied, using a centroid method, to produce a real value if needed. Applications of fuzzy logic include chiller controls, antilock braking systems, autofocus video systems, and washing machine control. For further information on fuzzy logic, see Kosko (1992).

Case-Based Reasoning

Case-based reasoning (CBR) techniques employ search methodologies to find a match between entered information and a previous case on file in a case library. This technique is beginning to gain some popularity because it requires less programming maintenance than a rule-based system. There are a variety of search algorithms, but generally the user is required only to answer a few questions which seek to match parameters to the cases on file. While not as generic in providing solutions as rule-based systems, CBR tools increase in general applicability each time a new case is added to the file. When a match is found, the solution for the previous case is presented to the user.

The most obvious application of CBR is at help desks, where many of the cases handled are similar. If they use a CBR tool to settle routine matters, help desk experts have more time to spend on unusual calls.

Genetic Algorithms

Genetic algorithms (GAs) are search methods based on the mechanics of natural selection. Each element of a string represents a parameter in the process being optimized. Each set of strings represents a possible solution for the process. Genetic algorithms differ from traditional search and optimization techniques in that they manipulate the coding of parameter sets, not the parameters themselves. They search for the optimal solution for a number of points, rather than for a single point, using probabilistic rather than deterministic rules. There are a limited number of applications today that use GAs; most of these are in design and in the aerospace industry.

Natural Language Processing

Natural language processing (NLP) is concerned with using a computer to understand the meaning of everyday language. There are three steps in the process. First, the language must be made available to the NLP program. This may be accomplished through digitized voice input, keyboard input, or reading a file. Second, the sentences must be parsed to extract significant words from phrases, and to arrange the words in an unambiguous manner. Third, the meaning of the words must be determined based on context, and the meaning of the entire phrase or sentence must be constructed. These steps can be performed through neural networks or pattern matching, semantic, or knowledge-based techniques, among others. The main applications of NLPs to date are in accessing databases. Some commercially available NLPs are beginning to appear in telephone applications, including hands-free dialing in automobiles. NLPs are most effective when the knowledge domain is somewhat narrow.

Optical Character Recognition

Optical character recognition (OCR) is the technique used to turn bit-mapped images into alphanumeric codes, usually ASCII characters. An image must first be scanned into the computer and processed. This involves filtering, locating individual characters, format conversion, image orientation, and feature extraction. After this process is completed, the character can be classified and stored. Most of the current applications for OCR have been developed by scanner manufacturers. The technique is essential for the conversion of paper documentation to a computerized format.

Virtual Reality

Virtual realities are computer-generated environments in which users interact visually, acoustically, and tactilely with three-dimensional objects or information. The paradigm has numerous useful applications in design, hazardous operations, and space. Virtual reality is often used to enable an operator to control a remotely situated robot. It can also allow one to experience data other than visually. Besides a computer, the equipment required includes a head-mounted display, a data glove, an audio device, and possibly a motion-producing device.

COMMUNICATIONS

Computer communications usually involves the transmittal of digital information. Traditional computer communications encompasses the connection of computer terminals to a time-sharing system, of word processing workstations to a central time-shared processor, of one computer to another, and of a computer to various peripherals such as plotters, printers, and mass storage units. In the building services industry, this definition is expanded to include connecting controllers to supervisory computers or other controllers, and connecting HVAC systems to computers or fire or security alarm systems. Information passed includes sensor and status information needed for control or coordination, data logging, system tuning or maintenance, and set point or schedule information for performance modification or remote programming.

The following are some of the benefits of computer or digital communications:

- Several computers may share peripheral resources, such as printers, plotters, and mass storage devices, that would be underused by one computer.
- Small computers may access a central time-sharing system when heavy computing is needed.
- Computers may access a central database containing software, product, customer, inventory, or vendor information.
- Computers may be networked so that all computers on the network share storage and peripherals, as well as memory and computational capabilities.
- Electronic mail through networks and bulletin boards via telephone links allows efficient distribution of information among individuals.
- Building controls share occupancy, performance, and set point information for control coordination, energy-saving strategies, remote monitoring, and programming capabilities.

Bytes are groups of eight binary digits, or bits, which are the fundamental units of information. In most computers each character is represented by a byte. The two most common character sets are ASCII and EBCDIC. These codes assign the specific bit patterns within a byte to digits, alphabetic characters, and punctuation.

Generally, computers and peripherals communicate to each other through input/output cable connections called ports. A parallel port uses eight signal lines plus control lines for sending complete characters at one time. While this mode of transmission is quick, it is restricted to short (typically 5-m) cables and cannot be used through telephone lines.

A serial port sends characters one bit at a time, or serially, requiring as few as two or three wires. For serial communication, the receiver and transmitter must coordinate the beginning and end of each character, and the receiver must accommodate the bit rate (speed) at which the transmitter is sending information. To this end, communication can be achieved synchronously or asynchronously. Synchronous communications generally begin with a string of bits in a predetermined pattern. This pattern allows the receiver to synchronize with the transmitter. With asynchronous communication, each character has a specific length that includes a start bit and one or two stop bits. The receiver resynchronizes on every character sent. Although there is a time penalty of about 25% for sending start and stop bits for each character with asynchronous serial communication, most smaller computers do it because of the greater simplicity of both the hardware and the software required. For communications over telephone lines, both asynchronous and synchronous systems can support dial-up connections, where a new telephone call is made for each connection. Synchronous systems can also support connections over dedicated leased telephone lines, where the connection is permanent.

When communication errors are a concern and/or multiple devices are using a network to exchange information, a protocol is needed to check the data for errors, to correct the data if errors are found, to provide equitable access for all devices on the network, and to ensure that only one device transmits at a time. A communication network limited to an office, a building, or several buildings close together is called a local area network (LAN). LANs are typically used to share data (such as CAD drawings or a common database with an order entry system) or peripherals (such as printers, plotters, or large disk drives). Workstations on a LAN can even share processing power. LANs vary in the type of medium (twisted pair, coaxial, or fiber optic cabling), interconnection scheme or topology (bus, ring, or star configurations), access method (contention, token passing, or time slice), and operating system they have. The LAN operating system can be added to the normal computer operating system, which handles the communication functions on the network. With twisted pair wiring and coaxial cables, information can be transported throughout a LAN at speeds of 10 megabits per second (Mbps). With fiber optic cabling, speeds can reach 100 Mbps.

Gateways are devices that interconnect similar or dissimilar LANs, dissimilar large computers, LANs and large computers, and so forth. A wide-area network (WAN) connects disparate elements, which can be LANs, over a large geographic area.

The voltage levels used for short-distance communications cannot be used with switched telephone lines or fiber optic cables. A modem (modulator/demodulator) interfaces the two media by converting bits between voltage levels and tones or light. The quality of the medium determines the communication speed. Current technology provides up to 38,400 bits per second (bps) over normal telephone lines, at 9600 bps with 4:1 data compression. Speeds of up to 1 Mbps can be supported over dedicated T1 telephone lines. As more information is passed around buildings, voice, computer data, building services data, and CATV networks may be called on to integrate in order to reduce installation and reconfiguration costs (Tanenbaum 1981, Bellamy 1982).

Access to a central database containing data, product information, and software is quickly becoming a major communications feature. A mainframe database serves as a repository for information available to numerous users. Each of these users may have different needs and require different interface options. Gateways to the mainframe repositories may be via modem, a network connection, or a direct connection. These gateways have numerous protocols depending on the class of computer being connected.

Recently, use of the public domain, worldwide Internet has increased dramatically for such functions as e-mail and file transfer. The Internet was originally developed to allow information exchange between government and academic researchers. Reasons for this expanded use include the opening of the Internet to commercial access and the development of Mosaic software, which provides a user-friendly text and graphics display package.

DATA ACQUISITION

Data Acquisition System Hardware

Data from the field is rapidly replacing estimates derived from simulations, nomograms, and tables. Most microcomputer and data logger combinations can be classified as (1) analog data loggers (strip charts); (2) digital data loggers (analog and digital data are captured in electronic format, stored internally, and transferred to a microcomputer via serial communications lines); or (3) internal add-on data acquisition cards for a microcomputer. Data acquisition systems (DASs) vary in number of channels, types of channels (e.g., digital, analog, and pulse counting), scanning rate, and error checking capability. The

sensor signal can be delivered to the data acquisition system by direct wiring, RF carrier systems, power-line carrier systems, and, in special instances, infrared communications. Because most commercially available DASs represent devices that are less than 10 years old, government standards for accuracy testing have yet to be developed. It is therefore recommended that the user obtain a copy of the manufacturer's certified tests and/or perform a set of tests before blindly accepting the data from the DAS (Sparks et al. 1992).

Data Acquisition System Software

Software for a data acquisition system can be purchased from DAS hardware vendors or specialty software vendors; it can also be adapted from public domain software (Ryan 1986, Feuermann and Kempton 1987, Sparks et al. 1994a, 1994b) or developed in-house.

Software developed in-house requires the establishment of an experiment plan, the development of an interface to the data-gathering hardware, and the selection and calibration of analog sensors and digital inputs. Coding DAS software includes (1) selecting the programming language, (2) defining the data channels, (3) selecting the time intervals, (4) designing the archive database, (5) developing error detection and correction algorithms, (6) analyzing the data, (7) graphic considerations, (8) transferring the data (e.g., modem, RF, floppy), and (9) final storage of the data (e.g., tape, optical disk, floppy).

MONITORING AND CONTROL

Direct digital control (DDC) of HVAC components is more accurate and flexible than the proportional control commonly used in pneumatics. Such flexibility permits controllers to be tuned and allows control algorithms to be replaced or extended (Nesler 1986). For example, a DDC microprocessor can control temperature in a variable air volume terminal box, with control based only on a dry-bulb temperature sensor. The same microprocessor can be modified to monitor relative humidity and mean radiant temperature and to change the dry-bulb set point to better maintain comfort conditions (Int-Hout 1986). It can adjust airflows and temperatures based on occupancy indicators in an individual office. By transferring data back to a computer controlling the central fans, the microprocessor also facilitates minimization of the fan power required to deliver a given flow of air (Englander and Norford 1988). Finally, it can control flow on the basis of CO_2 measurements to provide a supply of outdoor air matched to the number of occupants in the space.

Microprocessor-based devices are currently used to turn on chillers and boilers at the optimal time for a building to recover from an unconditioned period. The required programs monitor the building's thermal behavior and adjust the equipment start time; such programs are examples of parameter estimation routines and give the controller an adaptive capability. This kind of optimization has been performed off-line (Hackner et al. 1985) and, more effectively, as part of an on-line control system (Cumali 1988). The control can be extended to include weather forecasting algorithms that can improve control of thermal storage or to schedule precooling by night ventilation (Shapiro et al. 1988).

Linking local microprocessor controllers with a central, supervisory computer makes it possible to integrate HVAC control with such services as security, life safety monitoring, and lighting. The network can be extended to the electric utility, which provides spot pricing information as input to load-shedding programs, and even to dropping specified equipment in the event of a power shortage. This technology has been demonstrated in pilot projects (Peddie and Butteit 1985).

Central computers are already used to collect data from individual controllers or meters, but typically perform a minimal amount of data analysis. Analysis performed off-site by consultants identifies long-term trends in energy consumption, isolates beneficial or harmful changes in equipment operation, and normalizes energy consumption for changes in weather. These analysis programs can be incorporated into on-site computers, providing operators and management with up-to-date, readily available information (Anderson et al. 1989).

Hardware is available in a wide range of sizes and capabilities, from specialized packaged units, such as individual air handler controls, to intelligent field panels or programmable controllers that gather information from several inputs and control multiple outputs, to stand-alone computers, which not only monitor and control but are also suitable for program development.

REFERENCES

Abtahi, H., T.L. Wong, and J. Villanueva, III. 1986. Computer aided analysis in thermodynamic cycles. Proceedings of the 1986 ASME International Computers in Engineering Conference 2.

Anderson, D., L. Graves, W. Reinert, J.F. Kreider, J. Dow, and H. Wubbena. 1989. A quasi-real-time expert system for commercial building HVAC diagnostics. *ASHRAE Transactions* 95(2):954-60.

ARI. 1991. Forced-circulation air-cooling and air-heating coils. ARI *Standard* 410-91. Air-Conditioning and Refrigeration Institute, Arlington, VA.

ARI. 1989. Central station air-handling units. ARI *Standard* 430-89. Air-Conditioning and Refrigeration Institute, Arlington, VA.

Arkin, H. and A. Shitzer. 1979. Computer aided optimal life-cycle design of rectangular air supply duct systems. *ASHRAE Transactions* 85(1):197-213.

ASHRAE. 1990. *ASHRAE Journal's HVAC&R Software Directory.*

Bellamy, J. 1982. *Digital telephony.* John Wiley and Sons, New York.

Brambley, M.R., D.B. Crawley, D.D. Hostetler, R.C. Stratton, M.S. Addison, J.J. Deringer, J.D. Hall, and S.E. Selkowitz. 1988. Advanced energy design and operation technologies research. Pacific Northwest Laboratory *Report* PNL-6255.

Busnaina, A.A. 1987. Modeling of clean rooms on the IBM personal computer. *Proceedings of the Institute of Environmental Sciences,* 292-97.

Busnaina, A.A., S. Abuzeid, and M.A.R. Sharif. 1988. Three-dimensional numerical simulation of fluid flow and particle transport in a clean room. *Proceedings of the Institute of Environmental Sciences,* 326-30.

Chen, T.Y.W. 1988. Optimization of pumping system design. Proceedings of the 1988 ASME International Computers in Engineering Conference.

Cornick, S.M., D.A. Leishman, and J.R. Thomas. 1990. Incorporating building regulations into design systems: An object-oriented approach. *ASHRAE Transactions* 96(2):542-49.

Cumali, Z. 1988. Global optimization of HVAC system operations in real time. *ASHRAE Transactions* 94(1):1729-44.

Date, C. 1988. *Database—A primer.* Addison Wesley Publishing Company, Reading, MA.

Doheny, J.G., and P.F. Monaghan. 1987. IDABES: An expert system for the preliminary stages of conceptual design of building energy systems. *Artificial Intelligence in Engineering* 2(2).

Englander, S.L. and L.K. Norford. 1988. Fan energy savings: Analysis of a variable speed drive retrofit. Proceedings of the American Council for an Energy-Efficient Economy, 1988 summer study on energy efficiency in buildings, Asilomar, CA.

Feuermann, D. and W. Kempton. 1987. ARCHIVE: Software for management of field data. Center for Energy and Environmental Studies, *Report* No. 216. Princeton University, Princeton, NJ.

Goodman, D. 1988. *The complete hypercard handbook.* Bantam Books, New York.

Haberl, J. and D. Claridge. 1987. An expert system for building energy consumption analysis: Prototype results. *ASHRAE Transactions* 93(1):979-98.

Haberl, J.S., L.K. Smith, K.P. Cooney, and F.D. Stern. 1988. An expert system for building energy consumption analysis: Applications at a university campus. *ASHRAE Transactions* 94(1):1037-62.

Hackner, R.J., J.W. Mitchell, and W.A. Beckman. 1985. System dynamics and energy use. *ASHRAE Journal* 27(6).

Int-Hout, D., III. 1986. Microprocessor control of zone comfort. *ASHRAE Transactions* 92(1B):528-38.

Kaler, G.M., Jr. 1990. Embedded expert system development for monitoring packaged HVAC equipment. *ASHRAE Transactions* 96(2):733-42.

Kosko, B. 1992. *Neural networks and fuzzy systems.* Prentice Hall, Englewood Cliffs, NJ.

Kovarik, M. 1971. Automatic design of optimal duct systems. Use of computers for environmental engineering related to buildings. National Bureau of Standards, Building Science Series 39 (October).

Kuehn, T.H. 1988. Computer simulation of airflow and particle transport in cleanrooms. *The Journal of Environmental Sciences,* 31.

Li, K.W., W.K. Lee, and J. Stanislo. 1986. Three-dimensional graphical representation of thermodynamic properties. Proceedings of the 1986 ASME International Computers in Engineering Conference 1.

Mayer, R., L.O. Degelman, C.J. Su, A. Keen, P. Griffith, J. Huang, D. Brown, and Y.S. Kim. 1991. A knowledge-aided design system for energy-efficient buildings. *ASHRAE Transactions* (97)2:479-99.

McAfee, J. and C. Haynes. 1989. Computer viruses, worms, data diddlers, killer programs and other threats to your system. St. Martin's Press, New York.

Meadows, K.L. and P.W. Brothers. 1989. Decision analysis for prioritizing recommended energy conservation options. *ASHRAE Transactions* 95 (2): 934-53.

Mills, R.B. 1989. Why 3D graphics? *Computer Aided Engineering.* Penton Publishing, Cleveland, OH.

Neilson, N.R. and J.R. Walters. 1988. *Crafting knowledge-based systems: Expert systems made realistic.* John Wiley and Sons, New York.

Nesler, C.G. 1986. Automated controller tuning for HVAC applications. *ASHRAE Transactions* 92(2B):189-201.

Norton, P. 1986. *Inside the IBM PC.* Brady Books, Prentice Hall, New York.

Osborne, A. 1980. *An introduction to microcomputers,* Vols. 1, 2, and 3. Osborne-McGraw Hill, Berkeley, CA.

Peddie, R.A. and D.A. Butteit. 1985. Managing electricity demand through dynamic pricing. In Energy sources: Conservation and renewables. Conference Proceedings 135. American Institute of Physics, New York.

Potter, C.D. 1987. CAD in construction. Penton Publishing, Cleveland, OH.

Potter, R.A., J.F. Kreider, M.J. Brandemuehl, and L.M. Windingland. 1991. Development of a knowledge-based system for cooling load demand management at large installations. *ASHRAE Transactions* 97(2):669-75.

Rankin, J.R. 1989. *Computer graphics software construction.* Prentice Hall, New York, 1-4.

Ryan, L. 1986. Documentation of DAS: A data acquisition software package. Center for Energy and Environmental Studies—*Working Paper* No. 87. Princeton University, Princeton, NJ.

Shams, H., R.M. Nelson, G.M. Maxwell, and C. Leonard. 1994a. Development of a knowledge based system for the selection of HVAC system types for small buildings: Part I—Knowledge acquisition. *ASHRAE Transactions* 100(1):203-10.

Shams, H., R.M. Nelson, and G.M. Maxwell. 1994b. Development of a knowledge-based system for the selection of HVAC system types for small buildings: Part II—Expert system shell. *ASHRAE Transactions* 100(1):211-17.

Shapiro, M.M., A.J. Yager, and T.H. Ngan. 1988. Test hut validation of a microcomputer predictive HVAC control. *ASHRAE Transactions* 94(1): 644-63.

Sparks, R., J. Haberl, S. Bhattacharrya, M. Rayaprolu, J. Wang, and S. Vadlamani. 1992. Testing data acquisition systems for HVAC system monitoring. Proceedings of the ASME Solar Energy Conference, New York, pp. 325-28.

Sparks, R., A. Baranowski, K. Weber, and J. Haberl. 1994a. POLLC180 Software. Energy Systems Laboratory, Texas A&M University, College Station, TX.

Sparks, R., C. Sims, and J. Haberl. 1994b. Monitor Software. Energy Systems Laboratory, Texas A&M University, College Station, TX.

Stoecker, W.F., R.C. Winn, and C.O. Pedersen. 1971. Optimization of an air-supply duct system. Use of computers for environmental engineering related to buildings. National Bureau of Standards, Building Science Series 39 (October).

Tanenbaum, A.S. 1981. *Computer networks.* Prentice Hall, New York.

Tsal, R.J. and E.I. Chechik. 1968. *Use of computers in HVAC systems.* Budivelnick Publishing House, Kiev. Available from the Library of Congress, Service -TD153.T77 (Russian).

Tsal, R.J., H.F. Behls, and R. Mangel. 1988. T-method duct design: Part I, optimization theory; Part II, calculation procedure and economic analysis. *ASHRAE Technical Data Bulletin* (June).

Wasserman, P.D. 1989. *Neural computing: Theory and practice.* Van Nostrand Reinhold, New York.

BIBLIOGRAPHY

Barr, A. and E. Feigenbaum. 1982. *Handbook of artificial intelligence*, Vols. I and II. William Kaufman, Los Altos, CA.

Brothers, P.W. 1988. Knowledge engineering for HVAC expert systems. *ASHRAE Transactions* 94(1):1063-73.

Brothers, P.W. and K.P. Cooney. 1989. A knowledge-based system for comfort diagnostics. *ASHRAE Journal* 31(9).

Christensen, C. 1984. Digital and color energy maps for graphic display of hourly data. Proceedings of the 9th Annual Passive Solar Conference, Columbus, OH (September).

Christensen, C. and K. Ketner. 1986. Computer graphic analysis of class B hourly data. Proceedings of the 11th Annual Passive Solar Conference. Boulder, CO (June).

Culp, C.H. 1989. Expert systems in preventive maintenance and diagnosis. *ASHRAE Journal* 31(8).

Date, C. and C. White. 1987. SQL/DS—A user's guide to the IBM product structured query language/data system. Addison Wesley Publishing Company, Reading, MA.

Diamond, S.C., C.C. Cappiello, and B.D. Hunn. 1985. User-effect validation tests of the DOE-2 building energy analysis computer program. *ASHRAE Transactions* 91(2B):712-24.

Diamond, S.C. and B.D. Hunn. 1981. Comparison of DOE-2 computer program simulations to metered data for seven commercial buildings. *ASHRAE Transactions* 87(1):1222-31.

Feigenbaum, E.A., P. McCorduck, and H.P. Nii. 1988. *The rise of the expert company*. The York Times Books, New York.

Haberl, J.S. and D. Claridge. 1985. Retrofit energy studies of a recreation center. *ASHRAE Transactions* 91(2B):1421-33.

Haberl, J.S., M. MacDonald, and A. Eden. 1988. An overview of 3-D graphical analysis using DOE-2 hourly simulation data. *ASHRAE Transactions* 94(1):212-27.

Haberl, J.S., L.K. Norford, and J.S. Spadero. 1989. Expert systems for diagnosing operation problems in HVAC systems. *ASHRAE Journal* 31(6).

Hall, J.D. and J.J. Deringer. 1989. Overview of knowledge-based systems research and development activities in the HVAC industry. *ASHRAE Journal* 31(7).

Harmon, P. and D. King. 1985. *Expert systems: Artificial intelligence in business*. John Wiley and Sons, New York.

Harmon, P., R. Maus, and W. Morrisey. 1988. *Expert systems: Tools and applications*. John Wiley and Sons, New York.

Hsieh, E. 1988. Calibrated computer models of commercial buildings and their role in building design and operation. *Report* No. 230, Center for Energy and Environmental Studies, Princeton University, Princeton, NJ.

Jones, L. 1979. The analyst as a factor in the prediction of energy consumption. Proceedings of the Second International CIB Symposium on Energy Conservation in the Building Environment, Session 4, Copenhagen, Denmark.

Judkoff, R. 1988. International energy agency design tool evaluation procedure. Solar Energy Research Institute *Report* No. SERI/ TP-254-3371 (July).

Judkoff, R., D. Wortman, and B. O'Doherty. 1981. A comparative study of four building energy simulations, phase II: DOE-2.1, BLAST-3.0, SUN-CAT-2.4 and DEROB. Solar Energy Research Institute *Report* No. SERI/TP-721-1326 (July).

Kaplan, M.B., J. McFerran, J. Jansen, and R. Pratt. 1990. Reconciliation of a DOE2.1C model with monitored end-use data for a small office building. *ASHRAE Transactions* 96(1):981-93.

Kusuda, T. and J. Bean. 1981. Comparison of calculated hourly cooling load and indoor temperature with measured data for a high mass building tested in an environmental chamber. *ASHRAE Transactions* 87(1):1232-40.

Kusuda, T., T. Pierce, and J. Bean. 1981. Comparison of calculated hourly cooling load and attic temperature with measured data for a Houston test house. *ASHRAE Transactions* 87(1):1185-99.

McQuiston, F.C. and J.D. Spitler. 1992. *Cooling and heating load calculation manual*. ASHRAE.

Milne, M. and S. Yoshikawa. 1978. Solar-5 an interactive computer-aided passive solar building design system. Proceedings of the Third National Passive Solar Conference (Newark, DE).

Olgyay, V. 1963. *Design with climate: Bioclimatic approach to architectural regionalism*. Princeton University Press, Princeton, NJ.

Rabl, A. 1988. Parameter estimation in buildings. Methods for dynamic analysis of measured energy use. ASME *Journal of Solar Energy Engineering*, 110.

Reddy, A. 1989. Application of dynamic building inverse models to three occupied residences monitored non-intrusively. Proceedings of the Thermal Performance of the Exterior Envelopes of Buildings IV. ASHRAE/DOE/BTECC/CIBSE (December).

Reiter, P.D. 1986. Early results from commercial ELCAP buildings: Schedules as a primary determinant of load shapes in the commercial sector. *ASHRAE Transactions* 92(2B):297-309.

Robertson, D.K. and J. Christian. 1985. Comparison of four computer models with experimental data from test buildings in New Mexico. *ASHRAE Transactions* 91(2B):591-607.

Shanmugavelu, I., T.H. Kuehn, and B.Y.H. Liu. 1987. Numerical simulation of flow fields in clean rooms. *Proceedings of the Institute of Environmental Sciences*, 298-303.

Sorrell, F., T. Luckenback, and T. Phelps. 1985. Validation of hourly building energy models for residential buildings. *ASHRAE Transactions* 91(2).

Spielvogel, L.G. 1975. Computer energy analysis for existing buildings. *ASHRAE Journal* 7(August):40.

Spielvogel, L.G. 1977. Comparisons of energy analysis computer programs. *ASHRAE Transactions* 83(2).

Sturgess, G.J., W.P.C. Inko-Tariah, and R.H. James. 1986. Postprocessing computational fluid dynamic simulations of gas turbine combustor. Proceedings of the 1986 ASME International Computers in Engineering Conference 3.

Subbarao, K. 1988. PSTAR Primary and secondary terms analysis and renormalization, a unified approach to building energy simulations and short-term monitoring. Solar Energy Research Institute *Report* No. SERI/TR-254-3175.

Wortman, D. and B. O'Doherty. 1981. The implementation of an analytical verification technique on three building energy analysis codes: SUNCAT-2.4, DOE-2.1 and DEROB III. Solar Energy Research Institute *Report* No. SERI/TP-721-1008 (January).

Yuill, G. 1985. Verification of the BLAST computer program for two houses. *ASHRAE Transactions* 91(2B):687-700.

Yuill, G. and E. Phillips. 1981. Comparison of BLAST program predictions with the energy consumptions of two buildings. *ASHRAE Transactions* 87(1):1200-06.

Zadeh, L.A. 1965. Fuzzy sets. *Information and Control* 8.

CHAPTER 37

BUILDING ENERGY MONITORING

C OLLECTING empirical data to determine building energy performance is an important but complex and costly task. Careful and thorough planning facilitates timely and cost-effective data collection and analysis. Although they can have different project objectives and scopes, building energy monitoring projects have several issues in common, which allow methodologies and procedures (monitoring protocols) to be standardized.

This chapter provides general guidelines for developing building monitoring protocols, which, if applied, will yield quality empirical data. The protocols are designed to ensure that data products are planned to meet project objectives. The intended audience comprises building energy monitoring practitioners and data end users, such as energy suppliers, energy end users, building system designers, public and private research organizations, utility program managers and evaluators, equipment manufacturers, and officials who regulate residential and commercial building energy systems.

TYPES OF PROJECTS

Building energy monitoring projects develop empirical information from field data to give a better understanding of building energy performance. Monitoring projects can be noninstrumented (performance tracking using utility billing data) or instrumented (billing data supplemented by additional sources of data, such as an installed instrumentation package or a building energy management system). Noninstrumented approaches are generally simpler and less costly than instrumented approaches, but they are subject to more uncertainty in interpretation. The analyst must decide whether the extra cost of an instrumented approach is justified by the greater detail and accuracy obtained.

Instrumented field monitoring projects generally involve data acquisition systems, which are typically comprised of various sensors, and data recording devices such as data loggers. Field monitoring projects may involve a single building or hundreds of buildings, and may be carried out over periods ranging from weeks to years. Most monitoring projects involve the following activities:

- Project planning
- Site installation of data acquisition equipment
- Ongoing data collection and verification
- Data analysis and reporting

These activities often require support by various professional disciplines (e.g., engineering, data analysis, and management) and construction trades (electrical installers or plumbers).

Useful building energy performance data extend over a wide range of time and space. Space dimensions range from individual end use (lighting) or components (HVAC equipment, walls) in a building, to the building itself (meter readings) and to the utility network (utility load factors, excess capacity). Time dimensions of the data range from

seconds (controller actuation) to many years (building/component lifetimes). This wide range of building data corresponds to the wide range of data uses and analysis methods.

Current monitoring practices vary considerably. For example, a utility load research project tends to characterize the average performance of buildings with relatively few data points per building, whereas an elaborate and complex monitoring project may involve a test of new technology performance. Practitioners should use accepted standards of monitoring practices to communicate results. Key elements in this process are classifying the types of project monitoring and building consensus on the purposes, approaches, and problems associated with each type (Misuriello 1987).

Monitoring projects can be broadly categorized by the goals, objectives, experimental approach, level of monitoring detail, and uses. (Table 1). Other factors such as data analysis procedures, duration and frequency of data collection, and instrumentation are common to most, if not all, projects.

Whole-Building Energy Use

Whole-building energy use measures the total consumption for a building. Often these measurements are used to provide the average energy use of a representative sample of buildings or building systems. Typically, these projects have a small number of data points in each building. The data from a large number of buildings can be statistically analyzed to provide results with the desired level of accuracy.

Examples of this type of project include utility load research projects that use short-interval whole-building data. Other examples are evaluations of government or utility energy conservation programs using billing data analysis.

Energy End Use

Energy end-use projects focus on individual energy systems in particular buildings, typically for large samples. Monitoring usually requires separate meters or collection channels for each end use, and analysts must account for all factors that may affect energy use. Examples of this approach include detailed utility load research efforts and projects to evaluate or simulate building energy systems. Depending on the project objectives, the frequency of data collection may range from one-time measurements of full-load operation to continuous time-series measurements (Mazzucchi 1987).

Specific Technology Assessment

Specific technology assessment projects monitor the field performance of specific equipment or technologies that affect building energy use, including the impact of envelope retrofit measures, major end uses (such as lighting), or mechanical equipment.

The typical goal of retrofit performance monitoring projects is to estimate savings resulting from the retrofit under a variety of indoor/outdoor conditions and occupant behavior. The frequency and complexity of data collection depend on project objectives and site-specific conditions. Projects in this category assess perfor-

The preparation of this chapter is assigned to TC 9.6, Systems Energy Utilization.

Table 1   Characteristics of Major Monitoring Project Types

| | | —Project Type— | | |
Characteristic	Aggregate Energy Use	Energy End Use	Specific Technology Assessment	Diagnostics
Goals and objectives	Infer average performance of large sample of buildings or of technology applications.	Determine characteristics of specific energy end uses within building.	Measure field performance of building system technology or retrofit measure in individual buildings.	Solve problems. Measure physical or operating parameters that affect energy use or that are needed to model building or system performance.
General approach	Large, statistically designed sample with small number of data points per building. Often uses whole-building utility meters.	Often uses large statistically designed sample. Monitor energy demand or use profile of each end use of interest within building.	Characterize individual building or technology, occupant, and operation. Account and correct for variations.	Typically uses one-time and/or short-term measurement with special methods, such as infrared imaging, flue gas analysis, blower door, or coheating.
Level of detail	Average performance of sample, not individual building. Cannot explain variations between buildings. Often uses seasonal or annual data.	Detailed data on end uses metered. Collect building and operating data that affect end use.	Uses detailed audit, submetering, indoor temperature, onsite weather, and occupant surveys. May use weekly, hourly, or short-term data.	Focused on specific building component or system. Amount and frequency of data varies widely between projects.
Uses	Load forecasting. Conservation or demand-side management program evaluation. Rate design.	Load forecasting. Identify and confirm energy conservation or demand-side management opportunities. Simulation calculations. Rate design.	Technology evaluation. Retrofit performance. Validate models and predictions.	Energy audit. Identify and solve operation and maintenance, indoor air qualities, or system problems. Provide input for models. Building commissioning.

mance variations between different buildings or for the same buildings, before and after retrofit.

Field tests of end-use equipment are characterized by detailed monitoring of all critical performance parameters and operational modes. Data are often needed to develop or verify algorithms predicting technology performance. The project scope may include reliability, maintenance, design, energy efficiency, sizing, and environmental effects (Hughes and Clark 1986).

Building Diagnostics

Diagnostic projects are designed to measure physical and operating parameters that determine the energy use of buildings and systems. Usually, the project goal is to determine the cause of problems, provide parameters to model energy performance, or isolate the effects of components. Diagnostic tests frequently involve one-time measurements or short-term monitoring. The frequency of measurement varies from several seconds to about an hour; it must be at least three times faster than the time constant of the effect being monitored. Sometimes diagnostic tests require intermittent or long-term continuous data collection.

The residential sector, particularly single-family dwellings, has a large number of diagnostic measurement procedures. Typical measurements for single-family residences include (1) flue gas analysis to determine steady-state furnace efficiency, (2) fan pressurization tests to measure and locate building envelope air leakage, and (3) infrared thermography to locate thermal defects in the building envelope.

For single-family residences, diagnostics include techniques to measure or determine (1) the efficiency of major end-use appliances, such as furnaces, air conditioners, water heaters, and refrigerators; (2) the airtightness of air distribution systems (Modera 1989, Robison and Lambert 1989, Modera 1993); (3) overall energy loss *UA*, i.e., electric coheating or the dynamic signature of a building envelope (Subbarao et al. 1986, Subbarao 1988); and (4) the thermal signature of building envelope components (Duffy et al. 1988).

In multifamily residential buildings, diagnostic measurements are mostly in the developmental stage (Porterfield 1988). Some techniques are designed to determine the operating efficiency of steam and hot water boilers and to measure the air leakage between apartments (Modera et al. 1986, Bohac et al. 1989).

In research, diagnostic measurement techniques have been designed to measure the overall airtightness of building envelopes and the thermal performance of walls (Persily and Grot 1988). Practicing engineers also employ a host of diagnostic techniques to analyze energy performance. For example, short-term chart-recording measurements can be used to diagnose poor commercial building chiller performance (Misuriello 1988). Although chart recorders can be used, electronic data monitors are preferred, because data can be read and analyzed with a variety of microcomputers.

Diagnostic measurements are also well suited to support the development and implementation of building energy management programs (Chapter 32). In this role, diagnostic measurements can be used in conjunction with energy audits to identify existing energy usage patterns, to determine energy performance parameters of building systems and equipment, and to identify and quantify areas of inefficient energy usage (Misuriello 1988, Haberl and Kornor 1990a, b). Diagnostic measurement projects can generally be designed using procedures adapted to specific project requirements as appropriate (see the section on Design Methodology).

Equipment for diagnostic measurement may be installed temporarily or permanently to aid energy management efforts. Designers should consider providing permanent or portable check metering of major electrical loads in new building designs (ASHRAE 1989). The same concept can be extended to fuel and thermal energy use.

COMMON MONITORING ISSUES

Field monitoring projects are costly and involve unique challenges. A strong, lasting project commitment is necessary, because failing to anticipate or deal effectively with problems at any stage of implementation may jeopardize the data. The key issues in monitoring are the accuracy and reliability of the data. Projects have been compromised by inaccurate or missing data. Periodic monitoring projects should have sensor calibration and ongoing data verification to ensure the reliability and quality of the collected data.

Field monitoring projects also require effective management of various professional skills. Project staff must understand the building systems being examined, data management, data acquisition, and sensor technology. The logistics of field monitoring projects pose additional challenges, such as coordinating equipment procurement, delivery, and installation, as well as data collection, processing, and analysis.

Effective Planning

Many common problems in monitoring projects can be avoided by effective and comprehensive planning.

Project goals. Project goals and data requirements should be established before selecting hardware. Unfortunately, projects are often driven by the selection of hardware rather than by the project objectives because monitoring hardware must be ordered several months before data collection begins or because project initiation procedures are lacking. As a result, the hardware may be inappropriate for the particular monitoring task, or critical data points may be overlooked.

Data products. It is important to establish what data products are needed before selecting data points. Failure to plan these products first may lead to failure to answer critical research questions.

Data management. Failure to anticipate the (typically) large amounts of data collected can lead to major project difficulties. The computer and personnel resources needed to verify, retrieve, analyze, and archive data can be estimated based on experience with previous projects.

Commitment. A long-term commitment regarding personnel and resources is necessary. Project success depends on daily, long-term attention to detail and staff continuity.

Accuracy requirements. The accuracy requirements of the final data and the experimental design needed to meet these requirements should be determined early on. After the required accuracy is specified, the iterative process of choosing the sample size (number of buildings, control buildings, or pieces of equipment) and of determining the required measurement precision (including error propagation into the final data products) must be performed. This process is often complicated by a large number of independent variables (occupants, operating modes) and the stochastic nature of many variables (weather).

Implementation and Verification

The following steps can facilitate smooth project implementation and data verification when detailed field managements are made:

- Calibrate sensors before installation. Spot-check the calibration on-site. During long-term monitoring projects, recalibrate sensors periodically.
- Track sensor performance on a regular basis. Quick detection of sensor failure or calibration problems is essential. Ideally, this should be an automated or a daily task. Moreover, the value of lost data is high, as reconstructing it may be difficult or impossible.
- Generate data on a timely, periodic basis. Problems often occur in developing final data products. These problems include missing data from failed sensors, data points not installed due to planning oversights, and anomalous data for which there are no explanatory records. If data products are specified as part of general project planning and produced periodically, production problems can be identified and resolved as they occur. Automating the process of checking for data reliability and data accuracy can be invaluable in keeping the project on track and in preventing sensor failure and data loss.

Follow-Through on Data Analysis and Reporting

For most projects, after the data are obtained, they must be analyzed and put into reports. To ensure that the data collected and analyzed are the data needed, particular emphasis is placed on the analytical selection of data points.

Close attention must be paid to resource allocation to ensure that adequate resources are dedicated to verification, management, and analysis of the data. As a quality control procedure and to make data analysis more manageable, these activities should be ongoing. The data analysis should be defined carefully before the project begins.

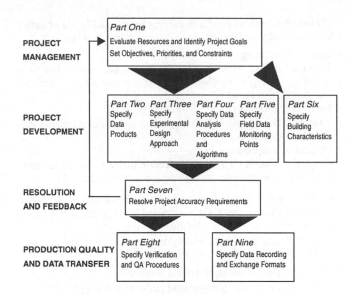

Fig. 1 Methodology for Designing Field Monitoring Projects

DESIGN METHODOLOGY

A methodology for designing effective field monitoring projects is described in this section. The task components and relationships among the nine activities constituting this methodology are identified in Figure 1. These activities fall into four categories: project management, project development, resolution and feedback, and production quality and data transfer. Field monitoring projects (resources, goals and objectives, data product requirements, or others) vary, which affects how the methodology should be applied. Nonetheless, the methodology provides a proper framework for advance planning, which minimizes or helps avoid project implementation problems.

An iterative approach to planning activities is best. The scope, accuracy, and techniques can be adjusted based on cost estimates and resource assessments. The initial design should be performed simply and quickly to estimate cost and evaluate resources. Recent monitoring projects have compiled useful cost data and equipment choices (Katipamula and Claridge 1992, O'Neal et al. 1992, Parker and Stokes 1985). Formulation and reassessment are also suggested in specifying data products and analytical methods—a crucial step in obtaining high-quality results. After a reasonable design is developed, pretesting of the overall monitoring process (from data collection to final data production) is beneficial for identifying and eliminating potential problems before full-scale monitoring begins.

An iterative approach to project planning is also necessary when the desired levels of instrumentation exceed the resources available for the project. The planning process should identify and resolve various trade-offs necessary to execute the project within a given budget. In many instances, it is preferable to reduce the scope of the project rather than relax instrumentation specifications or accuracy requirements.

Part One:
Identify Project Goals, Objectives, and Research Questions

First formulate a statement of project outcome(s). The goals and objectives statement determines the overall direction and scope of the data collection and analysis effort. The statement should also list the research questions to be answered by the empirical data, noting the error or uncertainty associated with the desired result. A realistic assessment of error is needed, since too small an uncertainty will lead to an overly complex and expensive project.

Research questions can have varying scopes and levels of detail, addressing entire systems or specific components as follows:

- Classes of buildings
  Example: To an accuracy of 20%, how much energy has been saved by using a mandated building construction/performance standard in this state?
- Particular buildings
  Example: Has a lapse in building maintenance caused energy performance to degrade?
- Particular components
  Example: What is the average reduction in demand charges due to the installation of an ice storage system in this building?

In general, more detailed and precise goal statements are better. They ensure that the project is constrained in scope and developed to meet specific accuracy and reliability requirements.

Research questions vary widely in technical complexity, generally taking one of the following two forms:

- How does the building/component perform?
- Why does the building/component perform as it does?

The first type of question can sometimes be answered generically for a class of typical buildings without detailed monitoring and analysis. The second type of question usually requires more detailed monitoring and analysis, and thus more detailed planning.

Part Two:
Specify Data Products and Project Output

This activity identifies the specific data products needed to answer critical research questions and provides a clear specification of the minimum acceptable data requirements for the project.

Evaluation results can be presented in many forms, often as interim and final reports (by heating and/or cooling season), technical notes, or technical papers. These documents must convey specific results of the field monitoring clearly and concisely. They should also contain estimates of the accuracy of the results.

The composition of data presentations and analysis summaries should be determined early to ensure that no critical parameters are overlooked (Hough et al. 1987). For instance, mock-ups of data tables, charts, and graphs can be used to identify requirements. Results reported previously can be used to provide examples of useful output. Data products should also be prioritized to accommodate possible cost trade-offs or revisions resulting from error analysis (see Part 7). A worksheet can be used to determine information needs and to facilitate organization of specific research questions and technology issues by project objectives. Next, specific data presentation formats should be determined.

It is important to identify the type and amount of comparative and evaluation analysis needed. For example, any need for data normalization for weather or usage profiles, along with the method to be employed, should be determined up front.

Although requirements for the acceptable minimum data results can often be specified during planning, data analysis will also reveal further requirements. Thus, budget plans should include allowances and data product specifications to adequately deal with additional or unique project output requirements uncovered during analysis of the results.

In specifying project data products, future information needs should be anticipated to provide project personnel with a more comprehensive understanding of testing requirements. For example, a project may have short- and long-term data needs (e.g., demonstrating peak electrical demand reductions versus demonstrating cost-effectiveness or reliability to a target audience). The initial results on demand reductions may not be the ultimate goal, but rather a step towards later presentations on energy consumption and initial cost. Thus, it is prudent to consider long-term and potential future data needs so that additional qualitative data needs, such as photographs or testimonials, may be identified.

Part Three:
Specify Experimental Design Approach

A general experiment design that defines two interacting factors—the number of buildings admitted to the study and the experimental approach—must be developed. For example, a less detailed or precise approach can be considered if the number of buildings is increased, and vice versa. Note that if the goal is related to a specific product, the experiment design must isolate the effects of that product. Haberl et al. (1990) discusses an experimental design.

Specifying the experimental design approach is particularly important, because the total building performance is a complex function of several variables, all of which are subject to changes that are difficult to monitor and to translate into performance. Unless care is taken with measurement organization and accuracy, uncertainties and errors (noise sources) can make it difficult to detect performance changes of less than 20 to 30% (Fracastoro and Lyberg 1983).

In some cases, judgment should be used in selecting the number of buildings involved in the project. If an owner seeks information about a particular building, the choice is simple—the number of buildings in the experiment is fixed at one. However, for other monitoring applications, such as drawing conclusions regarding effects in a sample population of buildings, some degree of choice is involved. Generally, the error in the derived conclusions decreases as the square root of the number of buildings (Box 1978).

Accuracy requirements depend on the effect to be measured and on the specified experimental design and analysis methods. Without special measurements, energy savings of less than 20% of the total consumption are difficult to detect from an analysis of monthly utility bills. A specific project may be directed at one of the following:

- Fewer buildings or systems with more detailed measurements
- Many buildings or systems with less detailed measurements
- Many buildings or systems with more detailed measurements

For projects of the first type, accuracy requirements are usually resolved initially by determining the expected variations of measured quantities (dependent variables) about their average values in response to the expected variations of the independent variables. For buildings, a typical concern is the response of heating and cooling loads (dependent variables) to temperature or solar input (independent variables). The response of building lighting energy to daylighting systems using natural light is also an example of the relationship between dependent and independent variables. The fluctuations in response are caused by (1) outside influences not quantified by the measured energy use data and (2) limitations and uncertainties associated with measurement equipment and procedures. Thus, accuracy must often be determined using statistical methods to describe mean tendencies of dependent data variables.

For projects of the second and third types, the increased number of buildings facilitates improved confidence in the mean tendencies of the dependent response(s) of interest. Larger sample sizes are also needed to use experimental designs with control groups, which are used to adjust for some outside influences. For further information, see Box (1978), Fracastoro and Lyberg (1983), Palmer (1989), and Hirst and Reed (1991).

Most monitoring procedures use one or more of the following general experimental approaches:

On-Off. If the retrofit or product can be activated or deactivated at will, energy consumption can be measured in a number of repeated on-off cycles. The on-period consumption is then compared to the off-period consumption (Woller 1989, Cohen et al. 1987).

Before-After. Building energy consumption is monitored before and after a new component or retrofit is installed. Changes in the

Table 2 Advantages and Disadvantages of Common Experimental Approaches

Mode	Advantages	Disadvantages
Before-After	No reference building required. Same occupants implies smaller occupant variations. Modeling processes will be mostly identical before/after.	Weather different before/after. More than one heating/cooling season may be needed. Model is required to account for weather and other changes.
Test reference	One season of data may be adequate. Small climate difference between buildings.	Reference building required. Calibration phase required (may extend testing to two seasons). Occupants in either or both buildings can change behavior.
On-Off	No reference building required. One season may be adequate. Modeling processes will be mostly identical before/after. Most occupancy changes will be small.	Requires reversible product. Cycle may be too long if time constants are large. Model is required to account for weather differences in cycles. Dynamic model accounting for transients may be needed.
Simulated occupancy	Noise from occupancy effects is eliminated. A variety of standard schedules can be studied.	Not "real" occupants. Expensive apparatus required. Building must be unoccupied with attendant extra cost.
Nonexperimental reference	Cost of actual reference building eliminated. With simulation, weather variation is eliminated.	Data base may be lacking in strata entries. Simulation errors and definition of reference problematic. With data base, weather changes usually not possible.
Engineering field test	Information focused on the product of interest. Minimal number of buildings required. Same occupants during the test.	Extensive instrumentation of product processes required. Models required to extrapolate to other buildings and climates. Occupancy effects not determined.

weather and building operation during the two periods must be accounted for, often requiring a model-based analysis (Hirst et al. 1983, Fels 1986, Robison and Lambert 1989, Sharp and MacDonald 1990, Kissock et al. 1992).

Test-Reference. The building energy consumption data of two identical buildings, one with the product or retrofit, are compared. Because buildings cannot be identical (e.g., different air leakage distributions, insulation effectiveness, temperature settings, and solar exposure), measurements should be taken prior to installation as well, to allow calibration. Once the product or retrofit is installed, any deviation from the calibration relationship can be attributed to the product or retrofit (Levins and Karnitz 1986, Fracastoro and Lyberg 1983).

Simulated Occupancy. In some cases, the desire to reduce noise can lead the experimenter to postulate certain standard profiles for temperature set points, internal gains, moisture release, or window manipulation and to introduce this profile into the building by computer-controlled devices. The reference is often given by the test-reference design. In this case, both occupant and weather variations are nearly eliminated in the comparison (Levins and Karnitz 1986).

Nonexperimental Reference. A reference for assessing the performance of a building can be derived nonexperimentally by using (1) a normalized, stratified performance database—energy use per unit area by some building-type classification (MacDonald and Wasserman 1989) or (2) a reasonable standard building, simulated by a calculated hourly or bin-method building energy performance model subject to the same weather and occupancy as the monitored building.

Engineering Field Test. When the experiment is focused on the performance of a test on a particular piece of equipment, the total building performance is not of primary interest. The building provides a realistic environment for testing the equipment for reliability, maintenance requirements, and comfort and noise levels, as well as energy usage. The whole-building energy advantages of the product are then derived using one of the preceding approaches. The equipment may be instrumented extensively.

Some of the general advantages and disadvantages of these approaches are listed in Table 2 (Fracastoro and Lyberg 1983). Note that experimental design choices have been successfully combined, for example, the before-after and test-reference approaches.

These are some of the questions to be considered in choosing an experimental approach:

- Can the product be turned on and off at will? The on-off design offers considerable advantages.
- Are occupancy and occupant behavior critical? Changing building tenants, use schedules, internal gains, temperature set points, and natural or forced ventilation practices should be considered, since any one of these can ruin an experiment if it is not constant or accounted for.
- Are actual baseline energy performance data critical? In before-after designs, time must be allotted to characterize the before case as precisely as the after case. For instances in which heating and cooling systems are being evaluated, data may be required for a complete range of anticipated ambient conditions.
- Is it a test of an individual technology, or are multiple technologies, installed as a package, being tested? If individual impacts are desired, detailed component data and careful model-based analyses are required.
- Does the technology have a single or multiple mode(s) of operation? Can the modes be controlled to suit the experiment? If many modes are involved, testing over a variety of conditions and conducting model-based analysis will be necessary.

Part Four:
Specify Data Analysis Procedures and Algorithms

Data are useless unless they are distilled into meaningful products that allow conclusions to be drawn. Too often, data are collected and never analyzed. This planning step focuses on specifying the minimum acceptable data analysis procedures and algorithms and detailing how collected data will be processed to produce desired data products. This step determines which parameters will be assumed as constants and which will be calculated by actual continuous data or be monitored continuously as a field data point. Based on this information, monitoring practitioners should do the following:

- Prepare a list of sensors and analysis constants to be determined in the field (e.g., fan power, lights and receptacles power, indoor air temperature).
- Develop engineering calculations and equations (algorithms) necessary to convert field data to end products. This includes the use of statistical methods and simulation modeling.
- Specify detailed items, such as the frequency of data collection and the reasons certain data must be obtained at different intervals. For example, 15-minute interval demand data is assembled into hourly data streams to match utility billing data.

Data analysis procedures and algorithms should take into account established testing and rating standards. For details see Doebelin (1990), EEI (1981), Huang (1991), Hurley and Schooley (1984), Hurley (1985), Hyland and Hurley (1983), ISA (1976), Kulwicki (1991), Miller (1989), and Ramboz and McAuliff (1983). However, it is usually not practical to implement standards in the field. For example, maintaining the length of straight ductwork required for an airflow sensor is usually difficult in the field, necessitating compromise.

Algorithm inputs can be assumed values (such as the energy value of a unit volume of natural gas), one-time measurements (the leakage area of a house), or time-series measurements (fuel consumption and outside and inside temperatures at the site). The spatial level of the algorithms may pertain to (1) utility level aggregates of buildings, (2) particular whole-building performance, or (3) performance of instrumented components.

Analysis methods can generally be classified as empirical or model-based.

Empirical Methods. While empirical methods are the simplest, they have large uncertainty and generate little or no cause-effect information for small sample sizes. The simplest empirical methods are based on annual consumption values, tracking the annual numbers and looking for degradation. Questions about building performance relative to other buildings are based on comparing certain performance indices between the building and an appropriate reference, for example, ANSI/ASHRAE/IES *Standard 105*.

For commercial buildings, the most common index is the energy use intensity (EUI), which is the annual consumption by fuel type or summed over all fuel types, divided by the floor area. Comparison is often made only on the basis of general building type, implying large variations in climate, building occupancy, and HVAC systems. Such variations can be accommodated to some degree by stratifying the database from which the reference EUI is chosen. Although no empirical database is currently large enough, generating such a database is one of the motivations for standardizing monitoring practices. A common alternative is to use computer simulations to set reasonable standards. Generally, however, simple empirical methods that are applied to retrofit applications should include, as a minimum, data on daily energy use and average daily temperature (recorded locally) to account for variations in occupancy and building schedules.

Monthly EUI or billing data provide more information for empirical analysis and can be used to detect billing errors, improper equipment operation during unoccupied hours, and seasonal space condition problems (Haberl and Komor 1990a, b). Daily data are often used in these analyses, and raw hourly total building consumption data, when available, provide more detailed information on occupied versus unoccupied performance. Hourly, daily, monthly, and annual EUI across buildings can be directly compared when reduced to average power per unit area (power density). To avoid false correlations, the method of analysis should have statistical significance that can be traced to realistic parameters.

Model-Based Methods. These techniques allow a wide range of additional data normalization to improve the accuracy of comparisons and provide cause-effect relations. To generate a model-based analysis, the analyst must carefully define the system and postulate a useful form of the governing energy balance equation. Explicit terms are retained for equipment or processes of particular interest. As part of the data analysis, whole-building data (driving forces and thermal or energy response) are used to determine the model's thermally descriptive parameters. The parameters themselves can provide insight for diagnosing the building, although parameter interpretation can be difficult, particularly with time-integrated billing data methods. The model can then be used for a number of normalization processes as well as future diagnostic and control applications. Two general classes of models are used in analysis methods: time-integrated methods and dynamic techniques (Burch 1986).

Time-integrated methods. Based on algebraic calculation of the building energy balance, time-integrated methods are often used prior to data comparison to correct annual consumption for variations in outdoor temperature, internal gains, and internal temperature (Fels 1986, Busch et al. 1984, Haberl and Claridge 1987, Claridge et al. 1991). This type of correction is essential for most retrofit applications.

Time-integrated methods can be used with whole-building energy consumption data (billing data) or with submetered end use data. For example, standard time-integration methods are often used to separately integrate end-use consumption data (heating, cooling, domestic water heating) for comparison and analysis. Time-integrated methods are generally reliable, as long as three conditions are accounted for:

- *Appropriate time step.* Generally, the time step should be as long, or longer, than the response time of the building or building system for which energy use is being integrated using a model-based method. For example, the response of daylighting controls to natural illumination levels (which result in changes to lighting energy use) can be rapid, allowing short time steps for data integration. By contrast, the response of cooling system energy use to changes in cooling load (from internal gains, ambient temperature, and solar gains) can be comparatively slow. In this instance, either a sufficiently long time step that will average over these slow variations or a dynamic model should be used. In general, an appropriate time step should account for the physical behavior of the energy system(s) and the expression of this behavior in model parameters.

- *Linearity of model results.* Generally, time-integrated models should not be applied to model parameters used to estimate nonlinear effects. Air infiltration, for example, is nonlinear when estimated using wind speed and indoor/outdoor temperature difference parameters in certain models. Estimation errors would result if these parameters were independently time-integrated and then used to calculate air infiltration. This problem can be avoided by modeling such nonlinear effects at each time step (1 hour).

- *End-use uniformity within data set.* Analysts who apply time-integrated models to end-use data sets should ensure that the particular data set is uniform, i.e., that it does not inadvertently contain observations with measurements of end uses other than those intended. During mild weather, for example, HVAC systems may provide both heating and cooling over the course of a day, creating data observations of both heating and cooling measurements. In a time-integrated model of heating energy usage, these cooling energy observations will lead to error. In such cases, observations should be identified or otherwise flagged by their true end use.

For whole-building energy consumption data (billing data), reasonable results can be expected from heating analysis models when the building is dominantly responsive to inside-outside temperature differences. Note that the billing data analysis method yields little of interest when internal gains are large compared to skin loads, as in large commercial buildings and industrial applications. Daily, weekly, and monthly whole-building heating season consumption integration steps have been employed (Fels 1986, Ternes 1986, Sharp and MacDonald 1990, Claridge et al. 1991). As might be expected, cooling analysis results have been less reliable because cooling load is not strictly proportional to cooling degree-days (Fels 1986). Problems also arise when solar gains are dominant and vary by season.

Dynamic techniques. Dynamic models — both macrodynamic (whole-building model) or microdynamic (component-specific) — offer great promise for reducing monitoring duration and increasing conclusion accuracy. Furthermore, individual effects from multiple measures and system interactions can be examined explicitly. Dynamic whole-building analysis will generally accompany detailed instrumentation of specific technologies.

Table 3 Whole Building Analysis Guidelines

| Project Goal | Class of Method | | |
	Empirical (Billing Data)	Time-Integrated Model[a]	Dynamic Model
Building evaluation	Yes, but expect fluctuations in 20 to 30% range.	Yes, extra care needed beyond 15% uncertainty.	Yes, extra care needed beyond 10% uncertainty.
Building retrofit evaluation	Not generally applicable using monthly data, unless large samples are used. Requires daily data and various normalization techniques for reasonable accuracy.	Yes, but difficult beyond 15% uncertainty. Method cannot distinguish multiple retrofit effects.	Yes, can resolve 5% change with short-term tests. Can estimate multiple retrofit effects.
Component evaluation	Not applicable.	Not applicable unless submetering is done to supplement.	Yes, about 5% accuracy, but best with submetering.

Note: Error figures are approximate for total energy use in a single building. All methods improve with selection of more buildings.

[a]Accuracy can be improved by decreasing time step to weekly or daily. These methods are of little use when the *outdoor* temperature approaches the *balance* temperature.

Dynamic techniques pose a dynamic physical model for the building, adjusting the parameters of the model to fit the experimental data (Subbarao 1988, Duffy et al. 1988). In residential applications, computer-controlled electric heaters can be used to maintain a steady interior temperature overnight, extracting from this data an experimental value for the building steady-state load coefficient. These methods can also use a cool down period to extract information on building internal thermal mass storage. Daytime data can be used to renormalize the response (computed from a microdynamic model) of the building to solar radiation, which is particularly appropriate for buildings with glazing areas over 10% of the building floor area (Subbarao et al. 1986). Once the data with electric heaters has been taken, the building can be used as a dynamic calorimeter to assess the performance of auxiliary heating and cooling systems.

Similar techniques have been applied to commercial buildings (Norford et al. 1985, Burch et al. 1990). In these cases, delivered energy from the HVAC must be directly monitored in lieu of using electric heaters. Because ventilation is a major variable term in the building energy balance, the outside airflow rate should also be directly monitored. Simultaneous heating and cooling, common in large buildings, requires a multizone treatment which has not been adequately tested in any of the dynamic techniques.

Table 3 provides a guide to analysis method selection. The error quotations are intended as rough estimates for a single building scenario.

Part Five: Specify Field Data Monitoring Points

Careful specification of field monitoring points is critical in planning and organizing a monitoring project. Its purpose is to identify the variables that need to be monitored or measured in the field to produce the required data.

Because metering projects are dynamic, special consideration must be given to identifying and controlling significant changes in climate, systems, and operation during the monitoring period. Additional monitoring points may be required to measure variables that are assumed to be constant, insignificant, or related to other measured variables in order to draw sound conclusions from the measurements. Since the necessary data may be obtained in several ways, it is recommended that data analysts, equipment installers, and data acquisition system engineers work together to develop tactics that best suit the project requirements.

The costs of data collection are nonlinear functions of the number, accuracy, and duration of measurements that must be considered while planning within budget constraints. If the extent of data applications is unknown, such as in research projects, consideration should be given to the value of other concurrent measurements because the incremental cost of alternative analyses may be small.

For any project involving large amounts of data, data quality should be verified in an automated fashion (Lopez and Haberl 1992). While this approach may require adding monitoring points to facilitate energy balances or redundancy checks, the added costs are likely to be offset by savings in data verification for large projects.

If multiple sites are to be monitored, common protocols for the selection and description of all field monitoring points should be established so that data can be more readily verified, normalized, compared, and averaged. The protocol also adds consistency in selecting monitoring points. Pilot installations should be conducted to provide data for a test of the system and to ensure that the necessary data points have been properly specified and described.

General considerations in selecting monitoring equipment include the following:

- Evaluating the equipment thoroughly under actual test conditions before committing to large-scale procurement. Particular attention should be paid to any sensitivity to power outages and lightning.
- Avoiding state-of-the-art systems and systems that have not been proven. If the ability or reliability of the data acquisition equipment is questionable, the equipment should be avoided (Sparks et al. 1992).
- Considering the costs and benefits of remote data interrogation and programming.
- Verifying vendor claims by calling references or obtaining performance guarantees.

An examination of the analysis method will determine the data to be measured in the field. The simplest methods require no on-site instrumentation. As the methods become more complex, data channels increase. For engineering field tests conducted with dynamic building methods, up to 100 data channels may be required.

Particular attention must be paid to sensor location. For example, if the method requires an average indoor temperature, examine the potential for internal temperature variations, which often require that data from several temperature sensors be averaged. Alternatively, temperature sensors adjacent to HVAC thermostats will detect the temperature to which the HVAC equipment reacts.

After the channel list is compiled, sensor accuracy and scan rates should be assigned. Some sensors, such as indoor and outdoor temperature sensors, may require low scan rates (once every 5 min). Others, such as total electric or solar radiation sensors, may contain high-frequency transients that require rapid sampling (every 10 s). The maximum sampling rate is usually programmed into the logger, and averages are stored at a specified time step (hourly). Some loggers can scan different channels at different rates. The logger's interrupt capability can also be used for rapid, infrequent transients. Interrupt channels signal the data logger to start monitoring an event only once it begins. In some cases, the on-line computation of derived quantities must be considered. For example, if heat flow in an air duct is required, it might be computed from a differential tem-

Table 4 General Characteristics of Data Acquisition Systems

Types of Data Acquisition Systems (DAS)	Typical Use	Typical Data Retrieval	Comments
Manual readings	Total energy use	Monthly or daily written logs	Human factors may affect accuracy and reading period. Data must be manually entered for computer analysis.
Pulse counter, cassette tapes (1 to 4 channels)	Total energy use (some end use)	Monthly pickup of cassette tapes	Data loss due to cassette is a common problem. Pulse data must be read and converted before it can be analyzed.
Pulse counter, solid state (1, 4, or 8 channels)	Total energy use (some end use)	Telephone protocols to mainframe or minicomputer	Computer hardware and software is needed for transfer and conversion of pulse data. Can be expensive. Can handle large numbers of sites. User-friendly.
Plug-in A/D boards for PCs	Diagnostics, technology assessment, and control	On-site real-time collection and storage	Usually small quantity, unique applications. PC programming capability needed to set up data software and configure boards.
Simple field DAS (usually 16 to 32 channels)	Technology assessment, residential end use (some diagnostics)	Phone retrieval to host computer for primary storage (usually daily to weekly)	Can use PCs as hosts for data retrieval. Good A/D conversion available. Low cost per channel. Requires programming skills to set up field unit and configure communications for data transfer.
Advanced field DAS (usually > 40 channels/units)	Diagnostics, energy control systems, commercial end use	On-site real-time collection and data storage, or phone retrieval	Usually designed for single buildings. Can be PC-based or stand-alone unit. Can run applications/diagnostic programs. User-friendly.

perature measurement multiplied by an air mass flow rate that was determined from a one-time measurement. However, it should only be computed and only totaled when the fan is operating.

Once the required field data monitoring points are specified, these requirements should be clearly communicated to all members of the project team to ensure that the actual monitoring points are accurately described. This can be accomplished by publishing handbooks for measurement plan development and equipment installation and by outlining procedures for diagnostic tests and technology assessments.

Scanning and recording intervals. The frequency of data measurements and data storage can have an impact on the accuracy of project output. Scanning differs from storage in that data channels may be read (scanned) once per second, for example, while data may be recorded and stored every 15 min. Most data loggers maintain temporary storage registers, accumulating an integrated average of channel readings from each scan. The average is then recorded at the specified storage interval.

The data scan rate must be sufficiently fast to ensure that all significant effects are monitored. The frequency of data recording required can vary by data channel. As mentioned previously, transient and/or random events are often best monitored using the interrupt capabilities of modern data loggers.

Sensors and data acquisition systems. Sensors should be selected to obtain each measurement on the field data list. Next, conversion and proportion constants should be specified for each sensor type on the field data list, and the accuracy, resolution, and repeatability of each sensor should be noted. Sensors should be calibrated before they are installed in the field, preferably with a NIST-traceable calibration procedure. They should be checked periodically for drift and recalibrated, and postcalibrated at the conclusion of the experiment (Turner et al. 1992, Haberl et al. 1992).

Because hardware needs vary considerably by project, specific selection guidelines are not provided here. However, general characteristics of generic data acquisition hardware components are shown in Table 4. Some typical concerns for selecting data acquisition hardware are outlined in Table 5.

Part Six: Specify Building Characteristics

The measured energy data will not be meaningful at a later date to people who were not involved in the project unless the characteristics of the building being monitored and its use (activities) have been documented. To meet this need, a data structure (e.g., a characteristic database) can be developed to describe the buildings.

Table 5 Practical Concerns for Selecting and Using Data Acquisition Hardware

Components	Field Application Concerns
Data logger unit and peripherals	• Select equipment for field application. • Equipment should store data in electronic form such as on floppy disks, magnetic tape, or in memory for easy transfer to the computer that will perform the analysis. • Remote programming capability should be available to minimize on-site software modifications. • Avoid equipment with cooling fans. • Use high-quality, proven modems. • Make sure logger and modem reset after power outage.
Cabling and interconnection hardware	• Use only signal-grade cable-shielded, twisted-pair with drain wire for analog signals. • Mitigate sources of common mode and normal mode signal noise.
Sensors	• Use rugged, reliable sensors that are rated for field application. • Use a signal splitter if sharing existing sensors or signals with other recorders or energy management control system (EMCS). • Operate sensors at 50 to 75% of full scale. • Choose sensors that do not require special signal conditioning. • Precalibrate sensors and recalibrate periodically.

Building characteristics can be collected at many levels of detail, but it is important to provide at least enough detail to understand the following:

• General building configuration and envelope (particularly energy-related aspects)
• Building occupant information (number, occupancy schedule, activities)
• Internal loads
• Type and quantity of energy-using systems
• Any building changes that occur during the monitoring project

The minimum level of detail is known as summary characteristics data (Ternes 1986). Simulation level characteristics—detailed information collected for hourly simulation model input—may be desirable for some buildings. Regardless of the level of detail, the data should provide a context for analysts, who may not be familiar with the project, to understand the building(s) and their energy use.

For occupied existing buildings, characteristics information should be collected in four areas:

- Building descriptive information, summarizing key building envelope and internal heat gain parameters to be collected
- HVAC system descriptive information, characterizing key parameters affecting HVAC system performance
- Entrance interview information, focusing on the energy-related behavior of building occupants prior to monitoring
- Exit interview information, documenting physical or lifestyle changes at the test site that may affect data analysis

Part Seven: Resolve Data Product Accuracies

System accuracy requirements and equipment selection can be determined by addressing the following factors, typically in an iterative manner.

Determine Measurement System Uncertainties. To determine the overall accuracy and precision of the final data products, all sources of bias and uncertainty should be included. The precision of the final data products can usually be determined directly from (1) the precision of the sensor/data acquisition systems, (2) the relationship (due to stochastic variables such as weather, occupant behavior, operational variations) between the variables being measured and the length of the measurement period, and (3) the number of cases studied.

On the other hand, the accuracy of the final data cannot be improved by monitoring more buildings or including more heating seasons and is usually not dominated by the accuracies associated with the sensors. Rather, the accuracy of the data is usually controlled by limitations on the number and type of measurements made (single-point temperature measurements, one-time flow measurements) and on the simplifications required to analyze and interpret the data. Many of these simplifications are associated with the analyses or physical models used (see Part 4) to account for the stochastic or random nature of the parameters being measured and for the indirect nature of most of the measurements.

Once a specific algorithm or equation for obtaining final data from physical measurements has been established, standard techniques that incorporate measurement uncertainties into the final data product are available. The general procedure for determining the uncertainty in a data product Y as a function of the uncertainties in the measured parameters x_i is to assume that the uncertainties of the measured parameters are small compared to the mean parameter values, which is not always true. In cases when this is not true, the equation for the final data products can be expanded in a Taylor series, and the higher order terms can be neglected. Thus, the overall uncertainty (or standard deviation) in the data product Y can be expressed in terms of the uncertainties in the measured parameters σ_{x_i} as

$$\sigma_Y = \sqrt{\left(\frac{\delta Y}{\delta x_1}\right)^2 \sigma_{x_1}^2 + \left(\frac{\delta Y}{\delta x_2}\right)^2 \sigma_{x_2}^2 + \left(\frac{\delta Y}{\delta x_1}\right)\left(\frac{\delta Y}{\delta x_2}\right)\sigma_{x_1 x_2} + \dots} \quad (1)$$

As the errors in a given measurement do not correlate with (i.e., are independent of) the errors in another measurement, the covariance terms, $\sigma_{x_i x_j}$ are usually zero. For example, if the errors in x_1 and x_2 are uncorrelated, the percentage uncertainty in $Y = x_1 x_2$ reduces to:

$$\frac{\sigma_Y}{\overline{Y}} = \sqrt{\left(\frac{\sigma_{x_1}}{\overline{x}_1}\right)^2 + \left(\frac{\sigma_{x_2}}{\overline{x}_2}\right)^2} \quad (2)$$

On the other hand, if the errors are still uncorrelated, but $Y = x_1 + x_2$, the percentage uncertainty in Y becomes

$$\frac{\sigma_Y}{\overline{Y}} = \sqrt{\left(\frac{\sigma_{x_1}}{\overline{Y}_1}\right)^2 + \left(\frac{\sigma_{x_2}}{\overline{Y}_2}\right)^2} \quad (3)$$

Estimating the accuracy or potential bias of the data products based on a given experimental plan is more complicated. The usual technique for quantifying the accuracy or bias in final data products is to make a statistically significant set of Monte Carlo simulations of building/system operation and performance. The bias of the final data products is determined by comparing the mean values of the experimentally determined data products with those based on the known (assumed) performance of the building and system. Such a procedure also provides a direct measure of the overall uncertainty in the final data products based on the observed scatter of the measurement-based results. For a more complete discussion of these issues see, Coleman and Steele (1986).

Errors can also be estimated based on historical experience, for example, using results from previous similar projects. Alternatively, a pilot study can be launched to obtain an estimate of potential errors in a proposed analysis. Some estimate of potential error must be available to determine if the project goals and objectives are reasonable.

This step of estimating data uncertainty represents one part of the iterative procedure associated with proper experimental design. If the final data product uncertainty that was determined using the preceding evaluation procedure is unacceptable, the uncertainty can be reduced by (1) reducing overall measurement uncertainty (improving sensor precision), (2) increasing the duration of the monitoring period (to average out stochastic variations), and/or (3) increasing the number of buildings tested. On the other hand, if simulations indicate that the expected bias in the final data products is unacceptable, the bias may be reduced by (1) employing additional sensors to get an unbiased measurement of the desired quantity, (2) using more detailed models and analysis procedures, and/or (3) increasing the data acquisition frequency (in combination with a more detailed model) to eliminate biases due to sensor or system nonlinearities.

Review System Accuracy and Cost Requirements. The process of specifying system accuracies is iterative. After the first round of analysis and decision-making, cost constraints and accuracy levels should be reviewed. Adjustments should be made based on accuracy needs and costs. Several iterations of determining the overall system accuracy and specific equipment options may be necessary.

Part Eight: Specify Verification and Quality Assurance Procedures

Establishing and using data quality assurance (QA) procedures can be very important to the successful implementation of a field test. The amount and importance of the data to be collected help determine the extent and formality of the QA procedures. For most projects, the entire data path, from sensor installation to procedures that generate results for the final report, should be considered for verification tests. In addition, the data flow path should be checked routinely for failure of sensors or test equipment, as well as unexpected or unauthorized modifications to equipment.

Quality assurance often requires complex data handling. Building energy monitoring projects collect data from sensor(s) and manipulate that data into results. Data handling in a project with only a few sensors and required readings can be handled with a relatively simple data flow on paper. Computers, which are generally used in one or more stages of the process, require a different level of process documentation because much of what occurs has no direct paper trail.

Computers also facilitate the collection of large data sets and increase project complexity. To achieve maximum automation of the process, computers require the development of specific software. Often, separate computers are involved in each step, so passing information from one computer system to another must be automated in large projects. To move the data as smoothly as possible, an automated data pipeline should be developed. This pipeline minimizes the time delay from data collection to the production of results and maximizes the cost-effectiveness of the entire project.

An automated data verification procedure should be used when possible. Verification procedures should be performed at frequent intervals (daily or weekly), depending on the importance of missing data. This minimizes data loss due to equipment failure and/or

Table 6   Table of Quality Assurance Elements

Time Frame	Hardware	Engineering Data	Characteristics Data
Initial start-up	Bench calibration (1)	Installation verification (1)	Field verification (1)
	Field calibration (1)	Collection verification (1,2)	Completeness check (1)
	Installation verification (1)	Processing verification (1,2)	Reasonableness check (1,2)
		Result production (1,2)	Result production (1,2)
Ongoing	Functional testing (1)	Quality checking (2)	Problem diagnosis (3)
	Failure mode diagnosis (3)	Reasonableness checking (2)	Data reconstruction (4)
	Repair/maintenance (4)	Failure mode diagnosis (3)	Change control (1)
	Change control (1)	Data reconstruction (4)	
		Change control (1)	
Periodic	Preventative maintenance (1)	Summary report preparation	Scheduled updates/resurveys (1)
	Calibration (1)	and review (2)	Summary report preparation and review (2)

(1) Actions to ensure good data.   (2) Actions to check data quality.   (3) Actions to diagnose problems.   (4) Actions to repair problems.

changes at a building site. It also allows the processed information to be applied quickly.

The following QA actions should take place:

1. Calibrate hardware and establish a good control procedure to facilitate the collection of good data.
2. Verify data, check for reasonableness, and prepare a summary report to assure the quality of the data after it is collected.
3. Perform initial analysis of the data. Significant findings may lead to changes in procedures for checking data quality.
4. Thoroughly document and control procedures applied to remedy problems. These procedures may entail changes in hardware or collected data (such as data reconstruction), which can have a fundamental impact on the results reported.

Three aspects of a monitoring project that require quality assurance are shown in Table 6: hardware, engineering data, and characteristics data. Three QA reviews are necessary for each of these aspects: (1) initial QA confirms that the project starts correctly; (2) ongoing QA confirms that the information collected by the project continues to satisfy quality requirements; and (3) periodic QA involves additional checks, established at the beginning of the project to ensure continued performance at acceptable levels of quality.

Information about the data quality and the quality assurance process should be readily available to data users. Otherwise, significant

Table 7   Documentation Included with Computer Data to be Transferred

1. Title and/or acronym
2. Contact person (name, address, phone number)
3. Description of file (number of records, geographic coverage, spatial resolution, time period covered, temporal resolution, sampling methods, uncertainty/reliability)
4. Definition of data values (variable names, units, codes, missing value representation, location, method of measurement, variable derivation)
5. Original uses of file
6. Size of file (number of records, bytes)
7. Original source (person, agency, citation)
8. Pertinent references (complete citation) on materials providing additional information
9. Appropriate reference citation for the file
10. Credit line (for use in acknowledgments)
11. Restrictions on use of data/program
12. Disclaimer (examples follow):
    (a) Unverified/unvalidated data; use at your own risk.
    (b) Draft data; use with caution.
    (c) Clean data to the best of our knowledge. Please let us know of any possible errors or questionable values.
    (d) Program under development.
    (e) Program tested under the following conditions...(conditions specified by author).

analytical resources may be expended to determine data quality, or the analyses may never be performed due to uncertainties.

Part Nine:
Specify Recording and Data Exchange Formats

This step involves specifying the formats in which the data will be supplied to the end user or other data analysts. Both raw and processed (adjusted for missing data or anomalous readings) data formats should be specified. In addition, if any supplemental analyses are planned, the media and format to be used (magnetic tape, disk, spreadsheet, ASCII) should be specified. These requirements can be determined by analyzing the software data format specifications. Common formats for raw data are comma delimited and blank delimited ASCII, which do not require data conversion.

Data documentation is essential for all monitoring projects, especially when several organizations are involved. Data usability is facilitated by specifying and adhering to data recording and exchange formats. Most data transfer problems are related to inadequate documentation. Other problems include hardware or software incompatibilities, errors in the tape or disk, errors or inconsistencies in the data, and transmittal of the wrong data set. Some of these problems can be avoided by the following precautions:

- Provide documentation to accompany the data exchange (Table 7). Because these guidelines apply to general data, models, programs, and other types of information, all items listed in Table 7 may not apply to every case.
- Provide documentation of transfer media, including the computer operating system, software used to create the files, media format (ASCII, EBCDIC, binary), tape or disk characteristics (tracks, density, record length, block, size).
- Provide procedures to check the accuracy and completeness of the data transfer, including statistics or frequency counts for variables and hard copy versions of the file. Test input data and corresponding output results for models on other programs.
- Keep all raw data, including erroneous records.
- Convert and correct data; save routines for later use (FSF 1989).
- Limit access to equipment to authorized individuals.
- Check incoming data soon after they are collected, using simple time-series and *x-y* inspection plots.
- Automate as many routines as possible to avoid operator error.

PROTOCOLS FOR RETROFIT PERFORMANCE MONITORING

Residential Retrofit Monitoring

Protocols for residential building retrofit performance can answer specific research questions associated with the actual measured performance (Ternes 1986). Discrepancies between predicted and actual performance, as measured by the energy bill, are common. Protocols improve on previous methods in two ways: (1) internal temperature is monitored, which eliminates a major source of noise or an unknown variable in data interpretation; and (2) data are taken more frequently than monthly, which potentially shortens the monitoring duration. (Utility bill analysis generally requires a full season of pre- and postaction data. The above protocols may require only a single season.)

For example, Ternes (1986) developed a single-family retrofit monitoring protocol, which consists of a data specification guideline that identifies important parameters to be measured. Both one-time and time-sequential data parameters are covered, and the parameters are defined carefully to ensure consistency and comparability between experiments. The guideline identifies a minimum set of data, which must be collected in all field studies that use the guideline. The guideline also describes optional extensions to the minimum data set that can be used to study additional issues, such as those related to occupant behavior, effects of microclimate, and effects of the distribution system on energy performance. Szydlowski and Diamond (1989) have developed a similar method for multifamily buildings.

The single-family retrofit monitoring protocol recommends a before-after experimental design approach, and the minimum data set allows performance to be measured on a normalized basis with weekly time-series data. The minimum time-series data for heating and/or cooling consumption are total consumption by fuel type and indoor temperature. The protocol also allows hourly recording intervals for time-integrated parameters—an extension of the basic data requirements in the minimum data set. The minimum one-time data measurements cover descriptive information on house- and space-conditioning systems, occupant interviews and infiltration measurements at the beginning and end of the field study period, performance measurement of space-conditioning systems, and an assessment of retrofit installation quality.

The minimum data set may also be extended through optional data parameter sets for users seeking more analytical information. This protocol has standardized the experimental design and data collection specifications, enabling independent researchers to compare project results more readily. Moreover, the approach of including both minimum and optional data sets and two recording intervals accommodates research projects of varying financial resources.

The data parameters in this protocol have been grouped into four data sets: basic, occupant behavior, microclimate, and distribution system (Table 8). The minimum data set consists of a weekly option of the basic data parameter set. Time-sequential measurements are monitored continuously throughout the field study period. These are all time-integrated parameters, i.e., the appropriate average value of a parameter over the recording period, rather than the instantaneous values.

This protocol also addresses instrumentation installation, accuracy, and measurement frequency and expected ranges for all time-sequential parameters (Table 9). The minimum data set (weekly option of the basic data) must always be collected. At the user's discretion, hourly data may be collected, which allows two optional parameters to be monitored. Parameters from the optional data sets may be chosen, or other data not described in the protocol added, to arrive at the final data set.

Commercial Retrofit Monitoring

A protocol has been developed for use in field monitoring studies of energy improvements (retrofits) for commercial buildings (MacDonald et al. 1989). Similar to the residential protocol, it addresses data requirements for monitoring studies. Commercial buildings are more complex, with a diverse array of potential efficiency improvements. Consequently, the approach to specifying

Table 8 Data Parameters for Residential Retrofit Monitoring

	Recording Period	
	Option 1	Option 2
Basic Parameters		
House description	once	
Space-conditioning system description	once	
Entrance interview information	once	
Exit interview information	once	
Pre- and post-retrofit infiltration rates	once	
Metered space-conditioning system performance	once	
Retrofit installation quality verification	once	
Heating and cooling equipment energy consumption	weekly	hourly
Weather station climatic information	weekly	hourly
Indoor temperature	weekly	hourly
Indoor humidity	—	hourly
House gas or oil consumption	weekly	hourly
House electricity consumption	weekly	hourly
Wood heating use	—	hourly
Domestic hot water energy consumption	weekly	hourly
Optional Parameters		
Occupant behavior		
Additional indoor temperatures	weekly	hourly
Heating thermostat set point	—	hourly
Cooling thermostat set point	—	hourly
Indoor humidity	weekly	—
Microclimate		
Outdoor temperature	weekly	hourly
Solar radiation	weekly	hourly
Outdoor humidity	weekly	hourly
Wind speed	weekly	hourly
Wind direction	weekly	hourly
Shading	once	
Shielding	once	
Distribution system		
Evaluation of ductwork infiltration	once	

measurement procedures, describing buildings, and determining the range of analysis must differ.

The strategy used for this protocol is to specify data requirements for projects, analysis, performance data with optional extensions, and a building core data set that describes the field performance of efficiency improvements. The key advance made in developing this protocol is the specification of those building characteristics deemed important in determining the building energy consumption. Since data are often collected by different investigators, this protocol helps to ensure that data from different projects are reasonably consistent and comparable.

With this protocol, data should be reported in the following categories:

- Project or program description—general information, including identification of the project or program, the reasons it was conducted, and what improvements were made to the buildings or systems studied.
- Analysis methods and results—summary of analysis (evaluation) methods, experimental design, and project results.
- Performance data—summary of monthly (billing) data, submetered or detailed energy consumption, inclusion of any demand data, and temperature and weather data.
- Building description data—survey data describing each building and associated building systems, functional use areas, tenants, schedules, base energy data, and energy improvements. Core data must be provided to describe each building before and after any energy improvements are made (two distinct time periods).

This protocol requires that the approach used for analyzing building energy performance be described. The description must cover the data used in the performance calculations and the methods

Table 9   Time-Sequential Parameters for Residential Retrofit Monitoring

Data Parameter	Accuracy[a]	Range	Stored Value per Recording Period	Scan Rate[b] Option 1	Scan Rate[b] Option 2
Basic Parameters					
Heating and cooling equipment energy consumption	3%		Total consumption	15 s	15 s
Indoor temperature	0.6°C	10 to 35°C	Average temperature	1 h	1 min
Indoor humidity	5% rh	10 to 95% rh	Average humidity		1 min
House gas or oil consumption	3%		Total consumption	15 s	15 s
House electricity consumption	3%		Total consumption	15 s	15 s
Wood heating use	0.6°C	10 to 450°C	Average surface temperature or total use time		1 min
Domestic hot water	3%		Total consumption	15 s	15 s
Optional data parameter sets					
Optional Data Parameter Sets					
Occupant Behavior					
Additional indoor temperatures	0.6°C	10 to 35°C	Average temperature	1 h	1 min
Heating thermostat set point	0.6°C	10 to 35°C	Average set point		1 min
Cooling thermostat set point	0.6°C	10 to 35°C	Average set point		1 min
Indoor humidity	5% rh	10 to 95% rh	Average humidity	1 h	
Microclimate					
Outdoor temperature	0.6°C	−40 to 50°C	Average temperature	1 h	1 min
Solar radiation	30 W/m²	0 to 1100 W/m²	Total horizontal radiation	1 min	1 min
Outdoor humidity	5% rh	10 to 95% rh	Average humidity	1 h	1 min
Wind speed	0.2 m/s	0 to 10 m/s	Average speed	1 min	1 min
Wind direction	5°	0 to 360°	Average direction	1 min	1 min

[a]All accuracies are stated values.
[b]Applicable scan rates if nonintegrating instrumentation is employed.

Table 10   Performance Data Requirements of the Commercial Retrofit Protocol

Projects with Submetering			
	Before Retrofit	**After Retrofit**	
Utility billing data (for each fuel)	12 month minimum	3 month minimum (optional update to 12 months)	
Submetered data (for all recording intervals)	All data for each major end use up to 12 months	All data for each major end use up to 12 months	
	Type	**Recording Interval**	**Period Length**
Temperature data (daily maximum and minimum must be provided for any periods without integrated averages)	Maximum and minimum —or— Integrated averages	Daily —or— Same as for submetered data but not longer than daily	Same as billing data length —or— Length of submetering
Projects Without Submetering			
	Before Retrofit	**After Retrofit**	
Utility billing data (for each fuel)	12 month minimum	12 month minimum	
	Type	**Recording Interval**	**Period Length**
Temperature data	Maximum and minimum —or— Integrated averages	Daily	Same as billing data length

used to account for any performance variations caused by changes in building characteristics. The model or equations used to describe building or system performance must also be specified.

In addition, the general experimental design must be indicated. Common designs such as side-by-side (test-reference) testing, before-after testing, on-off testing, or control-treatment group testing, can simply be listed on the form. Other designs should be fully described.

The analysis of retrofit performance must be provided. Results must include both cost and energy savings (including demand savings for electricity, if applicable). Savings may be reported as a percentage of some cost or energy quantity, but the absolute savings (cost and energy use per year) must be reported. Savings reports must indicate the breakdown between energy cost savings, electric kilowatt demand cost savings, and operation and maintenance (O&M) cost savings. The results should indicate any useful data extractions, aggregations, combinations, or normalization methods.

The cost of implementing energy improvement and any additional costs for increased demand or changes in O&M should be reported and used to report measures of cost-effectiveness.

The level of confidence in the results should also be reported. Confidence can be either a statistical determinant or a qualitative description and should include a discussion of the expected longevity of the results. For example, if an initial study on building performance included some short-term testing (assume for 4 months after the improvement was made), the report should indicate the 4-month validity of the results. If a later report covered a 2-year period after the change was made, this length of time would be indicated in a follow-up report. Availability of results data and contact information should also be reported.

The objectives and resources of different projects usually influence the amount and detail of monitored data at two levels: (1) monthly (billing) and (2) more detailed field recording intervals. Monthly data are required as a minimum, but submetered data or more detailed whole-building data are often necessary to provide more meaningful results.

Because monitoring at short time intervals can produce a large volume of data, the maximum amount of submetered or detailed monitored data that can be accepted varies, depending on the recording interval. The necessary performance data, including identification of a minimum data set, are outlined in Table 10.

The parameters in the minimum data set were selected to obtain as much information about a building and its performance as possible, while maximizing the number of projects that could participate. Because of the selection of readily available billing and outdoor temperature data as minimum requirements, data from buildings where energy improvements have been made without submetering are sufficient to complete the set.

Pre- and postretrofit measuring periods should be consecutive, ensuring that outside operational conditions remain as constant as possible (e.g., minimize changes in occupancy, operating schedules, operating modes, building use, and other variables). If these variables can all be held constant, any change in energy use can be attributed to the retrofit.

The protocol is intended to (1) promote more consistent results for better understanding of the field performance of energy improvements in these buildings, (2) facilitate comparisons of energy use and savings between buildings, (3) support continued advances in the reporting of energy improvement results, and (4) provide data for use in estimating regional benefits.

EXISTING TEST PROCEDURES AND POTENTIAL PROTOCOL APPLICATIONS

In addition to the specialized protocols for particular monitoring applications, a number of specific laboratory and field measurement standards exist (see Chapter 51), and many monitoring source books are in circulation (e.g., Fracastoro and Lyberg 1983).

REFERENCES

ASHRAE. 1984. Standard methods of measuring and expressing building energy performance. *Standard* 105-1984.

ASHRAE/IESNA. 1989. Energy efficient design of new buildings except low-rise residential buildings. *Standard* 90.1-1989.

Bohac, D.L., R. Landry, R. Voegtline, T. Dunsworth, and M. Hewett. 1989. Field measurement of multifamily boiler losses. MEO/TR89-7-MF. Center for Energy and Environment, Minneapolis, MN.

Box, G.E.P. 1978. *Statistics for experimenters: An introduction to design, data analysis and model-building.* John Wiley and Sons, New York.

Burch, J.D. 1986. Thermal performance monitoring methods. *Passive Solar Journal* 3.

Burch, J.D., K. Subbarao, A. Lekov, M. Warren, L. Norford, and M. Krarti. 1990. Short-term energy monitoring in a large commercial building. *ASHRAE Transactions* 96(1):1459-77.

Busch, J.F., A.K. Meier, and T.S. Nagpal. 1984. Measured heating performance of new, low-energy homes: Updated results from the BECA-A database. LBL-17883, Lawrence Berkeley Laboratory, Berkeley, CA.

Claridge, D.E., J.S. Haberl, W.D. Turner, D.L. O'Neal, W.M. Heffington, C. Tombari, M. Roberts, and S. Jaeger. 1991. Improving energy conservation retrofits with measured savings. *ASHRAE Journal* 33(10):14-22.

Cohen, R.R., P.W. O'Callaghan, S.D. Probert, N.M. Gibson, D.J. Nevrala, and G.F. Wright. 1987. Energy storage in a central heating system: Spa school field trial. *Building Service Engineering Research Technology* (Great Britain) 8:79-84.

Coleman, H.W. and W.G. Steel. 1986. *Experimentation and uncertainty analysis for engineers.* John Wiley and Sons, New York.

Doebelin, E. 1990. *Measurement systems.* McGraw-Hill, New York.

Duffy, J.J., D. Saunders, and J. Spears. 1988. Low-cost method for evaluation of space heating efficiency of existing homes. Proceedings of the 12th Passive Solar Conference, ASES, Boulder, CO.

EEI. 1981. *Handbook for electricity metering.* Edison Electric Institute, Washington, DC.

Fels, M.F., ed. 1986. Measuring energy savings: The scorekeeping approach. *Energy and Buildings* 9.

Fracastoro, G.V. and M.D. Lyberg. 1983. Guiding principles concerning design of experiments, instruments, instrumentation, and measuring techniques. Swedish Council for Building Research, Stockholm, Sweden (December).

FSF. 1989. PC version of the Unix-based AWK toolkit. Free Software Foundation, 675 Massachusetts Ave., Cambridge, MA 02139.

Haberl, J.S. and D.E. Claridge. 1987. An expert system for building energy consumption analysis: Prototype results. *ASHRAE Transactions* 93(1):979-98.

Haberl, J.S. and P.S. Komor. 1990a. Improving [commercial building] energy audits: How annual and monthly consumption data can help. *ASHRAE Journal* 32(8):26-33.

Haberl, J.S. and P.S. Komor. 1990b. Improving [commercial building] energy audits: How daily and hourly data can help. *ASHRAE Journal* 32(9):26-36.

Haberl, J., D.E. Claridge, and D. Harrje. 1990. The design of field experiments and demonstration. Proceedings of the IEA Filed Monitoring Workshop (Gothenburg, Sweden, April), pp. 33-58.

Haberl, J., W.D. Turner, C. Finstad, F. Scott, J. Bryant, and D. Coonrod. 1992. Calibration of flowmeters for use in HVAC systems monitoring. Solar Engineering 1992—Proceedings of the ASME-JSES-KSES Solar Engineering Conference (Maui, Hawaii, April 5-8).

Hirst, E. and J. Reed (ed.). 1991. Handbook of evaluation of utility DSM programs. ORNL/CON-336, Oak Ridge National Laboratory, Oak Ridge, TN.

Hirst, E., D. White, and R. Goeltz. 1983. Comparison of actual electricity savings with audit predictions in the BPA residential weatherization pilot program. ORNL/CON-142. Oak Ridge National Laboratory, Oak Ridge, TN. (See also ORNL/CON-174.)

Hough, R.E., P.J. Hughes, R.J. Hackner, and W.E. Clark. 1987. Results-oriented methodology for monitoring HVAC equipment in the field. *ASHRAE Transactions* 93(1):1569-79.

Huang, P. 1991. Humidity measurements and calibration standards. *ASHRAE Transactions* 97(2):278-304.

Hughes, P. and W. Clark. 1986. Planning and design of field data acquisition and analysis projects: A case study. Proceedings of the National Workshop on Field Data Acquisition for Building and Equipment Energy Use Monitoring, Oak Ridge National Laboratory, Oak Ridge, TN (March).

Hurley, C.W. 1985. Measurement of temperature, humidity and fluid flow. Field data acquisition for building and equipment energy-use monitoring. ORNL Report No. CONF-8501218 (March).

Hurley, C.W. and J.F. Schooley. 1984. Calibration of temperature measurement systems installed in buildings. NBS Building Science Series 153 (January).

Hyland, R.W. and C.W. Hurley. 1983. General guidelines for the on-site calibration of humidity and moisture control systems in buildings. NBS Building Science Series 157 (September).

ISA. 1976. Recommended environments for standards laboratories. Recommended Practice—Instrument Society of America, Research Triangle Park, NC.

Katipamula, S. and D.E. Claridge. 1992. Use of simplified system models to measure retrofit energy savings. Solar Engineering 1992—Proceedings of the 1992 ASME-JSES-KSES International Solar Engineering Conference (Maui, Hawaii, April 5-8), pp. 349-360.

Kissock, J.K., D.E. Claridge, J.S. Haberl, and T.A. Reddy. 1992. Measuring retrofit savings for the Texas LoanSTAR program: Preliminary methodology and results. Solar Engineering 1992—Proceedings of the 1992 ASME-JSES-KSES International Solar Engineering Conference (Maui, Hawaii, April 5-8), pp. 299-308.

Kulwicki, B. 1991. Humidity sensors. *Journal of the American Ceramic Society* 74:697-707.

Levins, W.P. and M.A. Karnitz. 1986. Cooling-energy measurements of unoccupied single-family houses with attics containing radiant barriers. ORNL/CON-200. Oak Ridge National Laboratory, Oak Ridge, TN. (See also ORNL/CON-2136.)

Lopez, R. and J.S. Haberl. 1992. Data processing routines for monitored building energy data. Solar Engineering 1992—Proceedings of the 1992 ASME-JSES-KSES International Solar Engineering Conference (Maui, Hawaii, April 5-8), pp. 329-36.

MacDonald, J.M. and D.M. Wasserman. 1989. Investigation of metered data analysis methods for commercial and related buildings. ORNL/CON-279. Oak Ridge National Laboratory, Oak Ridge, TN.

MacDonald, J.M., T.R. Sharp, and M.B. Gettings. 1989. A protocol for monitoring energy efficiency improvements in commercial and related buildings. ORNL/CON-291. Oak Ridge National Laboratory, Oak Ridge, TN (September).

Mazzucchi, R.P. 1987. Commercial building energy use monitoring for utility load research. *ASHRAE Transactions* 93(1).

Miller, R. 1989. *Flow measurements handbook.* McGraw-Hill, New York.

Misuriello, H. 1987. A uniform procedure for the development and dissemination of monitoring protocols. *ASHRAE Transactions* 93(1).

Misuriello, H. 1988. Instrumentation applications for commercial building energy audits. *ASHRAE Transactions* 95(2).

Modera, M. 1993. Field comparison of alternative techniques for measuring air distribution system leakage. Lawrence Berkeley Laboratory Report, LBL-33818.

Modera, M.P. 1989. Residential duct system leakage: Magnitude, impacts, and potential for reduction. *ASHRAE Transactions* 95(2).

Modera, M.P., R.C. Diamond, and J.T. Brunsell. 1986. Improving diagnostics and energy analysis for multifamily buildings: A case study. LBL-20247, Lawrence Berkeley Laboratory, Berkeley, CA.

Norford, L.K, A. Rabl, and R.H. Socolow. 1985. Measurement of thermal characteristics of office buildings. ASHRAE Conference on the Thermal Performance of Building Envelopes (Clearwater Beach, FL).

O'Neal, D.L., J. Bryant, and K. Boles. 1992. Building energy instrumentation for determining retrofit savings: Lessons learned. Solar Engineering 1992—Proceedings of the 1992 ASME-JSES-KSES International Solar Engineering Conference (Maui, Hawaii, April 5-8), pp. 1263-1268.

Palmer, J. 1989. Energy performance assessment: A guide to procedures, Vols. I and II. Energy Technology Support Unit, Harwell, United Kingdom.

Parker, G.B. and R.A. Stokes. 1985. An overview of ELCAP. Proceedings of Conservation in Buildings: Northwest Perspective (Butte, MT, May 19-22). PNL-SA-13179. Pacific Northwest Laboratory, Richland, WA.

Persily, A.K. and R. Grot. 1988. Diagnostic techniques for evaluating office building envelopes. *ASHRAE Transactions* 94(1).

Porterfield, J.M. 1988. Alternatives for monitoring multidwelling energy measures. *ASHRAE Transactions* 94(1).

Ramboz, J.D. and R.C. McAuliff. 1983. A calibration service for wattmeters and watt-hour meters. NBS Technical Note 1179. National Bureau of Standards.

Robison, D.H. and L.A. Lambert. 1989. Field investigation of residential infiltration and heating duct leakage. *ASHRAE Transactions* 95(2): 542-50.

Sharp, T.R. and J.M. MacDonald. 1990. Effective, low-cost HVAC controls upgrade in a small bank building. *ASHRAE Transactions* 96 (1):1011-17.

Sparks, R., J.S. Haberl, S. Bhattacharyya, M. Rayaprolu, J. Wang, and S.Vadlamani. 1992. Use of simplified system models to measure retrofit energy savings. Solar Engineering 1992—Proceedings of the 1992 ASME-JSES-KSES International Solar Engineering Conference (Maui, Hawaii, April 5-8), pp. 325-28.

Subbarao, K. 1988. PSTAR—A unified approach to building energy simulations and short-term monitoring. SERI/TR-254-3175. Solar Energy Research Institute, Golden, CO.

Subbarao, K., J.D. Burch, and H. Jeon. 1986. Building as a dynamic calorimeter: Determination of heating system efficiency. SERI/TR-254-2947. Solar Energy Research Institute, Golden, CO.

Szydlowski, R.F. and R.C. Diamond. 1989. Data specification protocol for multifamily buildings. LBL-27206, Lawrence Berkeley Laboratory, Berkeley, CA.

Ternes, M.P. 1986. Single-family building retrofit performance monitoring protocol: Data specification guideline. ORNL/CON-196. Oak Ridge National Laboratory, Oak Ridge, TN.

Turner, D., J. Haberl, J. Bryant, C. Finstad, and J. Robinson. 1992. Calibration facility for the LoanSTAR program. Solar Engineering 1992—Proceedings of the ASME-JSES-KSES Solar Engineering Conference (Maui, Hawaii, April 5-8).

Woller, B.E. 1989. Data acquisition and analysis of residential HVAC alternatives. *ASHRAE Transactions* 95(1):679-86.

BIBLIOGRAPHY

Alereza, T., D.R. Dohrmann, M. Martinez, and D. Mort. 1990. End-use metered data for commercial buildings. *ASHRAE Transactions* 96(1): 1004-10.

ASHRAE. 1986. Energy use in commercial buildings: Measurements and models. *Technical Data Bulletin* 2(4).

ASHRAE. 1987. Energy measurement applications. *Technical Data Bulletin* 3(3).

Baird, G., M.R. Donn, F. Pool, W.D.S. Brander, and C.S. Aun. 1984. Energy performance of buildings. CRC Press, Boca Raton, FL.

Burch, J., K. Subbarao, A. Lekov, M. Warren, and L. Norford. 1990. Short-term energy monitoring in a large commercial building. *ASHRAE Transactions* 96(1):1459-77.

Cowan, J.D. and I.A. Jarvis. 1984. Component analysis of utility bills: A tool for the energy auditor. *ASHRAE Transactions* 90(1B):411-23.

Douglass, J.C. 1989. Assessing in situ performance of advanced residential heat recovery ventilator systems: A case study of monitoring protocol. *ASHRAE Transactions* 95 (1):697-705.

EPRI. 1983. Monitoring methodology handbook for residential HVAC systems. EPRI Em-3003, Electric Power Research Institute (May).

EPRI. 1985. Survey of utility commercial sector activities. EPRI Em-4142, Electric Power Research Institute (July).

EPRI. 1985. Survey of residential end-use projects. EPRI Em-4578, Electric Power Research Institute (May).

Heidel, J. 1985. Commercial building end-use metering inventory. PNL-5027. Pacific Northwest Laboratory, Richland, WA (March).

Jones, J.R. and S. Boonyatikarn. 1990. Factors influencing overall building efficiency. *ASHRAE Transactions* 96(1):1449-58.

Kaplan, M.B., J.B. McFerran, J. Jansen, and R.G. Pratt. 1990. Reconciliation of a DOE2.1C model with monitored end-use data for a small office building. *ASHRAE Transactions* 96(1):981-93.

Piette, M.A. 1986. A comparison of measured end-use consumption for 12 energy-efficient new commercial buildings. Proceedings of ACEEE 3:176.

Pratt, R.G. 1990. Errors in audit predictions of commercial lighting and equipment loads and their impacts on heating and cooling estimates. *ASHRAE Transactions* 96(1):994-1003.

Shehadi, M.T. 1984. Automated utility management system. *ASHRAE Transactions* 90(1B):401-10.

Spielvogel, L.G. 1984. One approach to energy use evaluation. *ASHRAE Transactions* 90(1B):424-36.

Subbarao, K., J.D. Balcomb, J.D. Burch, C.E. Hancock, and A. Lekov. 1990. Short-term energy monitoring: Summary of results from four houses. *ASHRAE Transactions* 96(1):1478-83.

Ternes, M.P. 1987. A data specification guideline for DOE's single-family building energy retrofit research program. *ASHRAE Transactions* 93(1): 1607-18.

Van Hove, J. and L. van Loon. 1990. Long-term cold energy storage in aquifer for air conditioning in buildings. *ASHRAE Transactions* 96(1):1484-88.

Wulfinghoff, D.R. 1984. Common sense about building energy consumption analysis. *ASHRAE Transactions* 90(1B):437-47.

CHAPTER 38

BUILDING OPERATING DYNAMICS AND STRATEGIES

COMPUTERIZED energy management and control systems were developed during the 1970s to reduce utility costs associated with maintaining environmental conditions in commercial buildings. Within such systems, advanced control strategies that respond to the dynamics of weather and building conditions can be implemented to minimize operating costs.

HVAC systems are typically controlled using a two-level hierarchical control structure. The lower level provides local-loop control of a single set point through manipulation of an actuator. For example, the supply air temperature from a cooling coil is typically controlled by adjusting the opening of a valve that provides chilled water to the coil. The second control level, or supervisor, specifies the set points and other modes of operation that are time-dependent.

The performance of large commercial HVAC systems can be optimized by improving local-loop and supervisory control. Proper tuning of local-loop controllers can enhance comfort and reduce energy use. Set points and operating modes for cooling plant equipment can be adjusted by the supervisor to maximize the overall operating efficiency. Dynamic control strategies for ice or chilled water storage systems can significantly reduce on-peak electrical energy and demand costs to minimize total utility costs. Similarly, the thermal storage inherent in a building's structure can be dynamically controlled to minimize utility costs. In general, strategies that take advantage of thermal storage work best when forecasts of future energy requirements are available.

This chapter focuses on control strategies associated with computerized control systems. The chapter also presents guidelines, methods, and cost savings estimates for the effective control of air-handling, heating plant, cooling plant, and thermal storage systems. General methods are described, so that designers and building operators may apply the strategies to other facilities.

AIR-HANDLING SYSTEMS

Air-handling systems in commercial buildings provide conditioned air for human comfort or for special-purpose equipment, such as computers. Typical air-handling systems include equipment such as fans, dampers, valves, actuators, humidifiers, and coils. The performance of these systems and equipment has significant impact on the overall energy consumption of a building. In discussing the operation of these systems, both local-loop control and control strategy selection are addressed. The manner in which air-handling equipment is controlled can directly affect comfort conditions in a building, the amount of energy consumed, and the maintenance requirements and useful life of the mechanical equipment. From an optimal control standpoint, the factors of most concern include (1) the choice of set points, (2) the ability of individual loop controllers to maintain these set points, (3) the response of the system being controlled, and (4) the quality of control achieved.

This section complements and supplements Chapter 42, Automatic Control.

Choice of Set Points

The set points for some air-handling equipment are fixed values selected by the designer based on a design analysis with the intent of maintaining fixed operating conditions. For example, the discharge air temperature set point of a preheat coil controller is typically fixed at a value above freezing to ensure that equipment, such as coils, downstream of the preheat coil does not freeze.

In some applications, setting and resetting local-loop set points by a central energy management system or by local controllers may be desired to maintain comfort and to help minimize energy costs. For example, space temperature set points may be reset to (1) maintain the highest space temperature during the cooling cycle and the lowest space temperature during the heating cycle that are compatible with occupant comfort and equipment requirements, (2) minimize simultaneous heating and cooling, and (3) provide minimum or no conditioning to unoccupied spaces, whenever possible.

Constant volume systems may benefit from resetting the heating and/or cooling coil discharge air temperature set points. In a multizone system with simultaneous heating and cooling, resetting the discharge air temperature set points so that the temperatures are closer together reduces energy waste.

Caution should be exercised in resetting the set point of a cooling coil in a variable air volume (VAV) system. For example, increasing the set point may increase the airflow, but the extra fan power required to move the additional air may negate any potential savings in chiller COP. In addition, a discharge temperature that is set too high may cause comfort problems. It may be desirable to reset the cooling coil set point to accommodate low-load conditions for improved comfort control and to avoid the use of reheat. The extent to which the set point can be beneficially reset in a VAV system is much less than in a constant volume system. The section on System Energy Effects discusses the effects of resetting supply air temperature.

When reset, set points should be adjusted according to designer-defined and calculated conditions (i.e., a schedule). For example, in a reset schedule for a coil discharge air application, the designer may select the design air temperature set point to be 60°C when the coldest zone is 13°C. However, at reduced load, the designer may determine that the discharge air temperature set point should be 38°C when the coldest zone is 21°C.

Airflow Control

Variable air volume (VAV) systems are popular because they save large amounts of heating, cooling, and fan energy in comparison with other HVAC systems. In a VAV system, the flow may be modulated by (1) system dampers on the outlet side of the fan, (2) inlet vanes on the fan, or (3) variable-speed control of the fan motor. Flow modulation for vaneaxial fans may be accomplished by controllable-pitch fan blades.

The preparation of this chapter is assigned to TC 4.6, Building Operation Dynamics.

VAV systems typically incorporate a duct static pressure control loop to control the flow as described in Chapter 42, Automatic Control. In a single-duct VAV system, the duct static pressure set point is typically selected by the designer. The sensor should be located at a point in the duct work such that the established set point ensures proper operation of the zone VAV boxes under varying load (supply airflow) conditions. A shortcoming of this approach is that static pressure control is based on the readings of a single sensor intended to represent the pressure available to all VAV boxes. If the sensor malfunctions or is placed in a location that is not representative, operating problems will result.

An alternative approach to supply fan control in a VAV system uses flow readings from the direct digital control (DDC) zone terminal boxes to integrate zone VAV requirements with supply fan operation (Englander and Norford 1992, Hartman 1993, Warren and Norford 1993). Englander and Norford suggest that duct static pressure and fan energy can be reduced without sacrificing occupant comfort or adequate ventilation. They compared modified proportional-plus-integral (PI) and heuristic control algorithms via simulation and demonstrated that either static pressure or fan speed can be regulated directly using a flow error signal from one or more zones. They are careful to note that system component modeling limitations constrain their results primarily to a comparison of the control algorithms. The results show that both PI and heuristic control schemes work, but the authors suggest that a hybrid of the two might be ideal.

VAV systems that use return or relief fans require control of airflow through the return or relief air duct systems. Return fans are commonly used in VAV systems to help ensure adequate air distribution and acceptable zone pressurization. The control techniques for these systems are described in Chapter 42.

In a return fan VAV system, there is significant potential for control system instability due to the interaction of control variables (Avery 1986). In a typical system, these variables might include supply fan speed, supply duct static pressure, return fan speed, mixed air temperature, outside and return air damper flow characteristics, and wind pressure effect on the relief louver. The interaction of these variables and the selection of control schemes to minimize or eliminate interaction must be considered carefully (Avery 1986, Delp et al. 1993). Mixed air damper sizing and selection is particularly important (Dickson 1987a, Avery 1989). Set point selection for a VAV system return fan flow controller is described by Dickson 1987b. Zone pressurization, building construction, and outdoor wind velocity must be considered. The resultant design helps ensure proper air distribution, especially through the return air duct. Using the technique described by Dickson, the designer may be able to eliminate the return fan altogether.

The fan energy consumed by VAV systems is strongly influenced by the device used to vary the airflow rate. Brothers and Warren (1986) compared the fan energy consumption for three typical flow modulation devices: (1) system dampers on the outlet side of the fan, (2) inlet vanes on the fan, and (3) variable-speed control of the fan motor. They used the building simulation program DOE-2.1 to simulate the annual energy consumption of a 930 m^2 commercial building employing a VAV system with the three different flow control methods. Both centrifugal and vaneaxial fans were considered, even though system dampers are not usually used with vaneaxial fans because they may overload the fan motor. Although variable-pitch blades for a vaneaxial fan are a viable option, they were not included in the study. The VAV system used an enthalpy economy cycle and provided cooling only during working hours, 7 a.m. to 5 p.m., five days per week. The analysis was performed for five different cities: Fresno, CA; Forth Worth, TX; Miami, FL; Phoenix, AZ; and Washington, D.C.

Values for the annual energy consumption are presented in Table 1. In all locations, the centrifugal fan uses less energy than the vaneaxial fan. The vaneaxial fan has a higher efficiency at the

Table 1 Annual Energy Use with Different Flow Throttling Techniques

| Type of Fan | Flow Control | Fan Energy Use, MWh | | | | |
		Fresno	Fort Worth	Miami	Phoenix	Washington, D.C.
Backward-inclined centrifugal	System damper	12.6	12.9	13.9	17.9	11.5
	Inlet vane	10.2	10.4	10.1	14.4	9.5
	Variable speed	4.1	4.5	5.9	5.8	3.3
Vaneaxial	System damper[a]	16.0	16.2	15.5	22.8	15.0
	Inlet vane	10.8	11.0	10.9	15.4	10.0
	Variable speed	6.8	7.0	7.7	9.6	5.9

[a]Not normally used with vaneaxial fans due to possible overload of fan motor.

full-load design point, but the centrifugal fan has better off-design characteristics that lead to lower annual energy consumption.

For a centrifugal fan, inlet vane control uses about 20% less energy (for the five locations) than a system damper control. Compared to inlet vane control, variable-speed control savings range from 42% for Miami to 65% for Washington D.C., with an average savings of 57% for the five locations.

Control Response

The ability of a local-loop controller to maintain set point is dependent on the type and quality of the control hardware, the nature of the process being controlled, and the tuning of the controller.

Significant advances in control hardware quality have been made in recent years, especially in digital controls, which are proving to be very accurate and reliable. Proportional-only (P) control hardware and algorithms are being used much less in favor of proportional-plus-integral (PI) and proportional-plus-integral-plus-derivative (PID) controls.

P, PI, and PID controllers are well suited for linear or near-linear processes with short and constant or near-constant time delays (transport delays). Many HVAC processes fit this category, but others do not, and they require special consideration in the design, selection, and subsequent tuning of the system controls. This is also true when considering the system as a whole. There may be interaction between control loops, for example, in coupled linear or nonlinear systems and in systems where a single controller is used to control multiple processes in sequence, such as a single-zone system with a single controller modulating the heating coil, cooling coil, and return/outside air dampers.

Tuning of PI controllers and digital controls is discussed as part of the control system commissioning process in Chapter 42, Automatic Control.

Industry bench mark tuning techniques were introduced by Ziegler and Nichols (1942) and Nesler and Stoecker (1984). Ziegler and Nichols introduced a quantitative technique based on the process reaction curve, whereas Nesler and Stoecker introduced an effective trial-and-error technique. More recently, Borresen and Grindal (1990) and Bekker et. al. (1991) introduced controller tuning techniques. Borresen and Grindal extend the Ziegler and Nichols technique by introducing the concept of relative control difficulty to classify the control system and choose the type of controller (P, PI, or PID). Bekker et. al. followed the lead of MacArthur et al. (1989) in defining a systematic, although less mathematically rigorous, technique to tune a PI controller using pole-zero cancellation and root locus concepts on first-order processes. The resultant response, unlike that of the Ziegler and Nichols technique, is intended to be critically damped.

Self-tuning, adaptive, and other forms of advanced and intelligent control are not new in concept but are in their infancy in application to the HVAC industry. Advanced local-loop control concepts, issues, and techniques have been addressed by Nesler (1986), Dexter and Haves (1989), MacArthur et al. (1989), Huang and Nelson (1991), Wallenborg (1991), and Ho (1993). With the maturing of microprocessor-based HVAC controls, advanced forms of control are seeing greater application in the HVAC industry.

Self-tuning PID controllers can alleviate at least some of the time, training, and experience required to manually tune a controller. They have been demonstrated to be effective in many HVAC applications, including those with significant time delays, which, in practice, are difficult to tune manually (Wallenborg 1991).

Intelligent controls, such as those that use adaptive (Curtiss et al. 1993) or fuzzy logic algorithms, have significant potential for use in HVAC applications as an alternative to PID control. Processes that are highly nonlinear or have varying transport delays are especially good candidates for fuzzy logic control (Huang and Nelson 1991). However, there are currently no formal design or tuning methods for

fuzzy logic controllers; the rules and scaling parameters must be selected by trial and error to achieve satisfactory control for each application (Huang and Nelson 1994).

HVAC system design often gives primary consideration to steady-state performance. To achieve good control response, consideration should also be given to component sizing and selection.

Compared to an actuator without a positioner, a pneumatic valve or damper actuator mounted with a positive positioner responds more rapidly and accurately to the control signal because the actuator is provided up to full main air pressure. Shavit and Brandt (1982) conducted a simulation that showed the effect of integral gain (I) on the performance of a system employing an equal percentage valve with a positioner (no hysteresis). Although a lack of integral gain causes droop (offset from the set point), the optimum integral gain forces the response of the system to operate uniformly around the set point (Figure 1). As the integral gain increases slightly from

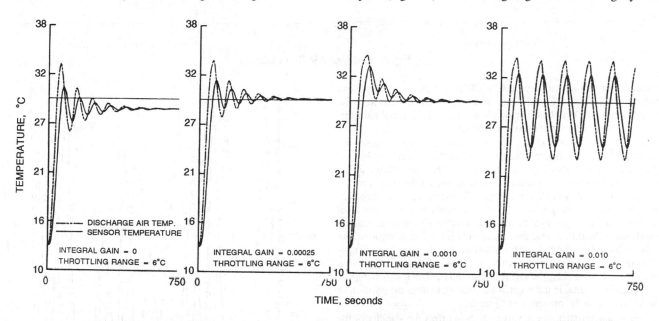

Fig. 1   Effect of Integral Gain on PI Control with a Positive Positioner (Using an Equal Percent Valve)

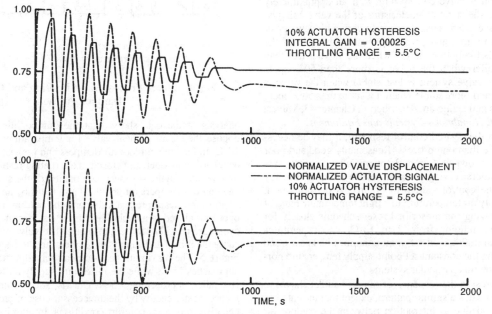

Fig. 2   P and PI Control Without a Positive Positioner (Using an Equal Percent Valve)

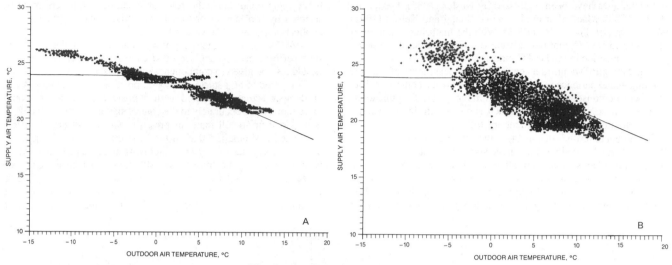

Fig. 3   Outdoor Air Reset History

optimum for this specific system (0.00025 units/K·sec) to less than optimum (0.001 units/K·sec), the response initially deviates from the set point but still matches it at steady state. As the integral gain further increases, the system becomes unstable.

As part of the same study, Shavit and Brandt (1982) compared the impact of removing the positioner from this system (by imposing a hysteresis of 10%). At identical controller gain settings, as the first two cases shown in Figure 1, the response of the system is less stable, as illustrated in Figure 2. Figure 2 shows the valve displacement and actuator output signal of the controllers. For proportional control, once the system reaches equilibrium, all variables reach a steady-state condition, and the system freezes in this position unless it is disturbed (Figure 2, upper illustration). In the proportional-plus-integral (PI) case, as the valve reaches equilibrium, deviation from the set point (i.e., the error signal) may occur if the change in the controller signal is not significant enough to alter the position of the valve due to the presence of hysteresis.

The characteristics of the valve can also affect the stability of the control system. For many heat exchangers with an equal percentage valve, the energy output of the coil as a function of valve position is linear. This results in a valve-coil system with an approximately fixed gain. However, the static characteristics of the valve-coil system with a linear valve often cause the gain to vary with valve position. In comparison to an equal percentage valve, the gain of the linear valve system is likely to be larger when the valve is less than 50% open, and smaller when the valve is more than 50% open. Therefore, the linear valve system is less stable when the valve is less than 50% open, and more stable when it is more than 50% open. Valve characteristics and sizing are described in Chapter 43, Valves, of the 1992 *ASHRAE Handbook—Systems and Equipment*.

In order to obtain the desired control response, it may be necessary to consider more than one process. Reasonably good and poor set point control are shown in Figure 3 for pneumatic control systems on two air handlers serving an actual office building (Bushby and Kelly 1988). The control system in Figure 3A performs best, maintaining the supply air temperature in a tight temperature band of about 1 K and following the prescribed reset schedule closely for outdoor temperatures ranging from 2 to 13°C. At temperatures below 2°C, the actual maintained set point continues its linear trend instead of maintaining the constant set point supply temperature normally implemented on this particular system.

Of particular interest is the response of the air handler shown in Figure 3B, which indicates a similar pattern, except that the data has a much greater spread due to interaction between the outdoor air dampers and the steam preheat valve in this air-handling unit. An

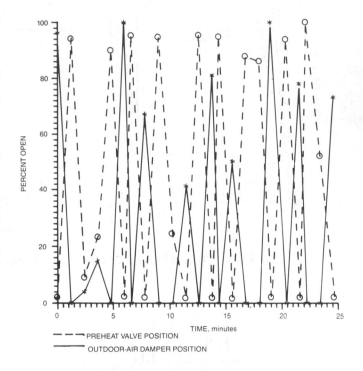

Fig. 4   Steam Preheat Valve Position and Outdoor Air Damper Position

analysis of the data shows that the outdoor air dampers tend to oppose the steam preheat valve for temperatures between −5 and 10°C. In fact, the outdoor air dampers and preheat valve have a tendency to swing open and closed in alternating fashion in a very short time period. Figure 4 shows the position of the steam preheat valve and outdoor dampers for about 20 min (Bushby and Kelly 1988).

Both the steam preheat valve and the outdoor air dampers swing open and closed in a cyclic fashion in just less than 2 min. The two cycles are roughly 180° out of phase. The valve and damper cycling cause the mixed and supply air temperatures to oscillate, as shown in Figure 5. The maximum peak-to-peak mixed air temperature swing approaches 15 K, while the maximum supply air temperature swing is smaller, approaching 2.5 K. Clearly, such control system performance wastes energy by the unnecessary use of preheating and has a negative impact on comfort conditions. In addition, it causes needless wear and tear on valves, dampers, and actuators. These problems

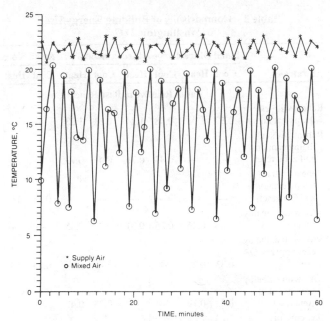

Fig. 5   Time History of Mixed Air and Supply Air Temperatures

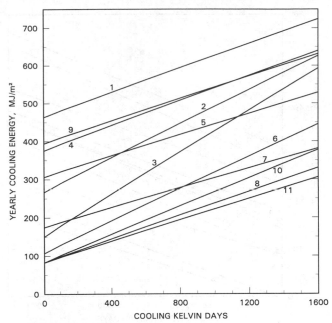

1   Reheat, base case
2   Reheat, dry-bulb economizer
3   Reheat, enthalpy economizer
4   Reheat, supply air reset based on outside air
5   Reheat, supply air reset based on zone
6   Reheat, supply air reset based on zone, enthalpy economizer
7   VAV entire building
8   VAV entire building, enthalpy economizer
9   Int. zones: VAV
    Ext. zones: Dual duct
10 Int. zones: VAV, enthalpy economizer
    Ext. zones: Dual duct, enthalpy economizer, zone reset
11 Int. zones: VAV, enthalpy economizer
    Ext. zones: Dual duct, enthalpy economizer, zone reset, 4-pipe zone fan coils

Fig. 6   Cooling Energy Consumption of Large Office Buildings

may be corrected by proper tuning of the controls and/or resizing or adjusting the system components.

System Energy Effects

The energy use of a building is influenced by its shell construction, its use, the type of HVAC system employed, the operation of equipment and systems, and the heating and cooling equipment efficiencies. Control strategies applied to air-handling systems play a key role in determining the effectiveness of these systems.

Kao (1985a) investigated the energy effect of the most commonly used control strategies on different air-handling systems in four different building types. The results were generated using the building energy program BLAST (Hittle 1979) and are useful as general guidelines for the design and operation of HVAC systems; however, a more detailed analysis is required to incorporate the unique aspects of a particular building and its intended use.

The control strategies studied by Kao (1985a) included two different techniques for resetting the supply air temperature by (1) either outdoor air temperature or space demand, or (2) a combination of these approaches. The heating, cooling, and fan energy consumption was computed at the air-handling system level and thus does not include the effects of plant and distribution efficiencies. Typical Meteorological Year (TMY) hourly weather data were used in simulations for six cities in the United States: Lake Charles, LA; Madison, WI; Nashville, TN; Santa Maria, CA; Seattle, WA; and Washington, D.C. In all cases, the air-handling system was operational only during occupied periods in the cooling season, and the conditioned zones used temperature setback at night during the heating season.

Figures 6 and 7 (Kao 1985a) show the yearly cooling and heating energy per square area of floor in a large office building. Cases 1 through 6 simulate a terminal reheat system for the entire building. Adding a dry-bulb economy cycle (case 2) allows supply air temperature to be maintained by using cooler outside air, consequently saving a large amount of cooling energy while using a substantial amount of heating. Changing to an enthalpy economizer (case 3) results in more cooling savings and greater heat use. The increase in heating energy for both economizer cycles results mainly from the lowering of the supply air temperature by the proportional controls during the period that the economy cycles are in operation. Both the

absolute amount and the percentage of cooling savings are greater in low cooling degree-day areas than in high degree-day areas.

Resetting the supply air temperature benefits both heating and cooling because reheat is not needed if the extra cooling capacity is reduced. Resetting by sensing zone temperature (case 5) reduces cooling energy approximately twice as much as resetting by using outside air temperature (case 4), and heating energy reduction is even more dramatic. By adding an enthalpy economy cycle to the supply air temperature reset strategy (case 6), cooling energy is further reduced substantially, while heating energy increases for the reason cited previously.

Cases 7 and 8 use variable air volume (VAV) systems for the entire building. The perimeter zones have reheat coils, which operate in sequence with the zone dampers. The dampers allow supply air to be reduced to 20% of design air volume. The interior zones have only damper controls. The base VAV system (case 7), which has no special strategy, has cooling energy consumption comparable to the best reheat case, case 6, which uses an enthalpy economy cycle and supply air temperature reset by zone demand. When an enthalpy economy cycle is added to the VAV system, 15% (Lake Charles) to 59% (Seattle) of cooling energy is saved. Contrary to results obtained with the reheat system, adding an enthalpy cycle to the VAV system (case 8) does not significantly increase the heating energy requirement in most cities. The exceptions are Santa Maria (20% increase) and Lake Charles (7% increase).

Case 9 simulates a base dual-duct system for the perimeter zones and a simple VAV system for interior zones. Case 10 adds an enthalpy

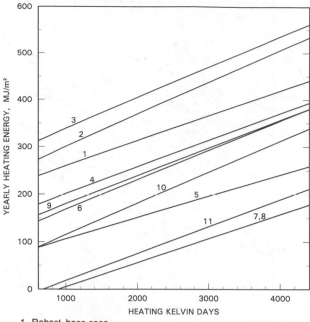

1 Reheat, base case
2 Reheat, dry-bulb economizer
3 Reheat, enthalpy economizer
4 Reheat, supply air reset based on outside air
5 Reheat, supply air reset based on zone
6 Reheat, supply air reset based on zone, enthalpy economizer
7 VAV entire building
8 VAV entire building, enthalpy economizer
9 Int. zones: VAV
   Ext. zones: Dual duct
10 Int. zones: VAV, enthalpy economizer
   Ext. zones: Dual duct, enthalpy economizer, zone reset
11 Int. zones: VAV, enthalpy economizer
   Ext. zones: Dual duct, enthalpy economizer, zone reset, 4-pipe zone fan coils

Fig. 7 Heating Energy Consumption of Large Office Buildings

economy cycle to the entire building, with both the hot and the cold air of the dual-duct system reset by sensing the space temperature. It is difficult to pinpoint the effects of the individual strategies and systems for these two cases. The overall results show that much cooling and heating energy is saved as compared to the base VAV and dual-duct systems (case 9). The perimeter zones are also simulated with a four-pipe fan-coil system having both the cold and hot water available year-round (case 11). A fixed amount of outside air is introduced directly to the fan-coil units during operating hours. The interior system remains a VAV system with an economy cycle. The energy consumption for both heating and cooling is similar to that in case 8.

Kao (1985a) also analyzed the cooling and heating energy consumption of the combined classroom and office areas of a school building. Reheat, dual-duct, VAV, and cooling-type unit ventilator systems were simulated for these areas. Roughly the same relative pattern as for the large office building was observed for the cooling consumption of the first three systems. The cooling results of the various VAV system strategies were close together, the best being the one applying the enthalpy economy cycle and resetting the supply air temperature by zone sensing. The unit ventilator system performed slightly poorer on cooling than the VAV systems. Unlike the large office building discussed previously, applying economy cycles to the reheat systems of the school building did not increase the heating energy much.

Kao (1985a) suggests that most of the conclusions reached for the previous two buildings will also apply to retail and small office buildings. For a retail store, among the best strategies and systems is

Table 2   Comparisons of Building Energy Use in Washington, D.C.

Strategies	Small Office		Large Office		School[a]		Retail Store	
	Cool	Heat	Cool	Heat	Cool	Heat	Cool	Heat
	Ratios Relative to Base Reheat Cases							
Reheat with enthalpy economy	0.57	1.08	0.58	1.24	0.61	1.03	—	—
Reheat with enthalpy economy and zone reset	—	—	0.43	0.76	0.43	0.73	—	—
Reheat with enthalpy economy and OA reset	—	—	—	—	—	—	0.54	0.67
VAV with enthalpy economy	0.36	0.26	0.25	0.31	0.37	0.45	—	—
VAV with enthalpy economy and OA reset	0.33	0.20	—	—	—	—	0.43	0.17
VAV with enthalpy economy and zone reset	0.33	0.19	—	—	0.33	0.43	—	—
Fan-coil for perimeter	—	—	0.32	0.29	—	—	—	—
Unit ventilator	—	—	—	—	0.42	0.10	—	—

[a]Classroom, office, and library only.

a VAV system with the supply air temperature reset by the outside air temperature and a packaged direct-expansion (DX) system, both with enthalpy economy cycles. For a small office building, a VAV system with enthalpy economy cycle and supply air temperature reset by zone demand sensing may be the best choice.

Table 2 (Kao 1985a) illustrates the energy consumption ratios of selected cases for Washington, D.C. (1415 cooling degree-days and 4211 heating degree-days). The ratios are defined as the building energy consumption obtained using a particular control strategy divided by the energy consumption of the reheat base case (case 1 in Figures 6 & 7). These data indicate that substantial improvements in a building's energy use may be obtained using different control strategies and air-handling systems. For example, when the enthalpy economy cycle is employed on the VAV system of the large office building, it consumes only 25% of the cooling energy and less than one-third of the heating energy consumed by a reheat system using a fixed amount of outside air year-round.

CONTROL OF SYSTEMS WITHOUT STORAGE

In the optimal control of an HVAC system, the cost of operation is minimized. In general, the cost of operation includes energy and maintenance costs and other costs associated with running the equipment. When only the operation of a certain HVAC component, such as the air distribution system, or a particular piece of equipment, such as a chiller, is optimized, the process is termed local optimization. Simultaneous optimization of the entire system and the building is termed global optimization. In design optimization, the structure, type, or components of an HVAC system are altered to obtain the best system. Design optimization includes optimal operation in its scope. System design must be integrated with operation in order to achieve global optimization. Because of the strong interactions among the many components comprising an HVAC system, the distinction between local and global optimization is important. The benefits and complexity of optimal operation are increased as the process is broadened to include global optimization. If a given system with a prescribed operating mode is evaluated, it is generally found that the component selection is not optimal.

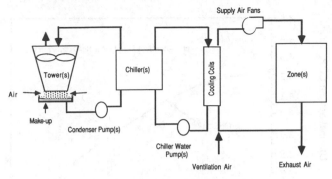

Fig. 8   Typical Chilled Water System

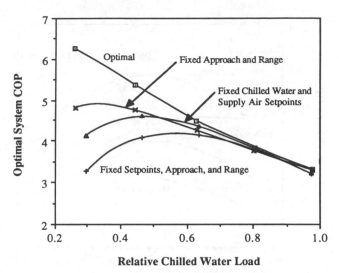

Fig. 9   Comparisons of Optimal Control with Conventional Control Strategies

Optimal control guidelines can be determined for systems that have no storage components (e.g., chilled water or ice tanks). Because there is no significant storage of energy that shifts loads in time, optimal control minimizes the instantaneous total power used by the system for every operating condition. Figure 8 illustrates a central cooling facility with a variable air volume (VAV) system consisting of multiple centrifugal chillers, cooling towers, and pumps. Chilled water is sent to a number of air-handling units in order to cool the air that is supplied to the conditioned space. For any given condition of loads and weather, it is possible to meet the same cooling demand with many different modes of operation and set points; however, only one set of modes and set points will be optimal.

The global optimization procedure for an HVAC system is complex and difficult to implement. However, for many systems, guidelines for some of the control variables may be developed that yield near-optimal performance. These guidelines simplify the selection of the control process required to achieve near-optimal performance and allow focus on the remaining control variables. The guidelines presented here may be readily implemented by supervisory controllers. The emphasis in this section is on optimization of cooling systems; similar studies have been made on heating systems by House et al. (1991) and Nuorkivi (1990).

Optimal Versus Alternative Control Strategies

Optimization of plant operation is most important when loads vary and when the system operates far from design conditions for a significant amount of time. A variety of strategies are used in chilled water systems at off-design conditions. Commonly, the chilled water and supply air set point temperatures are changed only according to the ambient dry-bulb temperature. In some systems, cooling tower airflow and condenser water flow are not varied in response to changes in the load and ambient wet bulb. In other systems, these flow rates are controlled to maintain constant temperature differences between the cooling tower outlet and the ambient wet bulb (approach) and between the cooling tower inlet and outlet (range), regardless of the load and wet bulb. Sometimes both humidity and temperature are controlled within the zones. Although these strategies seem reasonable, they generally do not minimize operating costs.

No one fixed set of chilled water temperature, supply air set point temperature, and tower approach and range is optimal over the range of load and weather experienced by a cooling plant. Fixed values that result in optimal performance at design conditions do not produce optimal performance over the entire range of operation. To illustrate this, the performance of the system shown in Figure 8 was determined for optimal control and three alternative strategies (Braun et al. 1989a). Figure 9 shows the system coefficient of performance (COP) as a function of load for a fixed wet-bulb temperature for these three conventional strategies:

1. Fixed chilled water and supply air temperature set points 4 and 11°C, respectively, with optimal condenser loop control

2. Fixed tower approach and range 2.5 and 6.5 K, respectively, with optimal chilled water-loop control

3. Fixed set points, approach, and range

Because the fixed values are chosen to be optimal at design conditions, the differences in performance for all the strategies are minimal at high loads. However, at part-load conditions, Figure 9 shows that the savings associated with the use of optimal control become significant. For part-load ratios less than about 50%, optimal control of the chilled water loop results in greater savings than that of the condenser loop. The overall savings over a cooling season depend on the time variation of the load. If the cooling load is relatively constant and near the design load, fixed values of temperature set points, approach, and range may be chosen to give near-optimal performance. However, for typical building loads with significant daily and seasonal variations, the penalty for using a fixed set point control strategy ranges from 5 to 20% of the HVAC system energy used.

Near-Optimal Control Guidelines

Near-optimal control guidelines are relatively simple to implement and may reduce energy use significantly. Such guidelines are presented for a number of situations commonly found in electrically driven central chilled water systems.

Multiple Chillers. Normally, multiple chillers operate in parallel and are controlled to supply chilled water at identical temperatures. For parallel chiller combinations, this control is either optimal or near-optimal (when not constrained by equipment). Besides the chilled water set point, additional variables that may be controlled are the relative chilled and condenser water flow rates.

In general, the condenser water flow to each chiller should be controlled to give identical leaving condenser water temperatures. This condition approximately corresponds to relative condenser flow rates equal to the relative loads on the chillers, even if the chillers are loaded unevenly. Figure 10 shows results for four sets of two chillers operated in parallel based on data from the Dallas-Fort Worth (D/FW) airport (Braun 1988, Braun et al. 1989a, and Hackner et al. 1984). The curves in Figures 10 and 11 represent data from chillers at three different installations: (1) the 19.3 MW variable-speed chiller at the Dallas-Fort Worth Airport, in Dallas, Texas; (2) the 1.9 MW fixed-speed chiller at the IBM Building in Atlanta, Georgia (Hackner et al. 1984, 1985); and (3) the 4.4 MW fixed-speed chill at the IBM Building in Charlotte, North Carolina (Lau

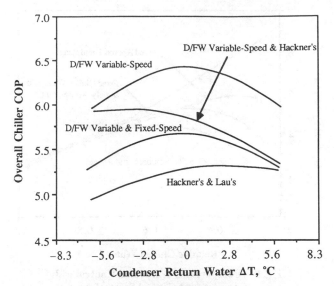

Fig. 10 Effect of Condenser Water Flow Distribution for Two Chillers in Parallel

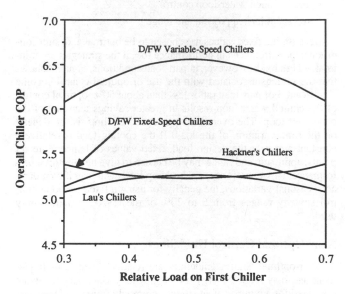

Fig. 11 Effect of Relative Loading for Two Identical Parallel Chillers

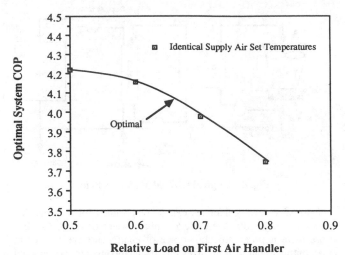

Fig. 12 Comparison of Performance for Optimal Supply Air Set Temperature with that for Identical Values

$$f_{L,i} = q_{cap,i} / \Sigma q_{cap,i} \text{ for } i = 1 \text{ to } n \qquad (1)$$

where

$q_{cap,i}$ = cooling capacity of the ith chiller
n = total number of chillers

The relative loadings determined with Equation (1) could result in either minimum or maximum power consumption. However, this solution gives a minimum when the chillers are operating at loads greater than the point at which the maximum COP occurs. Typically, the maximum COP occurs at loads that are about 40 to 60% of a chiller's cooling capacity. With loads greater than about 50% of cooling capacity, this control strategy results in minimum power consumption. When loads are less than 50% it may be best to operate with fewer chillers. This is discussed in the section on Chiller Sequencing.

Figure 11 shows the effect of the relative loading on chiller COP for different sets of identical chillers loaded at approximately 70% of their overall capacities. Three of the chillers have maximum COPs when evenly loaded, while the fourth (D/FW fixed-speed) obtains a minimum at that point. The part-load characteristic of the D/FW fixed-speed chiller is unusual in that the maximum COP occurs at its maximum capacity. This chiller was retrofit with a different refrigerant and drive motor, which derated its capacity from 30.6 to 19.3 MW. As a result, the evaporators and condensers are oversized for its current capacity. This illustrates the necessity of knowing the part-load characteristics of the equipment for optimization.

Chiller Sequencing. One of the important issues in the control of multiple chillers is chiller sequencing, which specifies the conditions at which specific chillers are brought on-line or off-line. The optimal sequencing of chillers depends on the part-load characteristics of the chiller and on the power associated with the various pumps. Chillers should be brought on-line only if the total power to operate with the additional chiller is less than without it. For systems with chilled water and condenser pumps that are dedicated to each chiller, it is optimal to operate each chiller to its full capacity before bringing on another one. For systems with nondedicated pumps, the optimal sequencing of chillers should be coupled to the optimization of the rest of the system. The characteristics of the system change when a chiller is brought on-line or off-line due to changes in the system pressure distribution and overall part-load performance. The optimal point for switching chiller operation may differ significantly from the switch points determined if only chiller performance was considered. It is important to evaluate all power-consuming devices.

et al. 1985). The performance of the chillers in the IBM buildings were scaled up for comparison with the D/FW chiller.

The overall chiller coefficient of performance (COP) is plotted versus the difference between the condenser water return temperatures for equal loadings on the chillers. For identical chillers, and for either variable-speed or fixed-speed, the optimal temperature difference between the chillers is almost zero. For situations in which chillers do not have identical performance, equal leaving condenser water temperatures result in chiller performance that is close to the optimum. Even for the variable- and fixed-speed chiller combinations that have very different performance characteristics, the penalty associated with the use of identical condenser leaving water temperatures is small.

For chillers with different cooling capacities but identical part-load characteristics, each chiller should be loaded at the same load fractions (i.e., according to the ratio of its capacity to the sum total capacity of all operating chillers). For a given chiller i, the part-load fraction $f_{L,i}$ is given by the following equation:

Multiple Air Handlers. A large central chilled water facility may provide cooling to several buildings, each of which may have a number of air-handling units in parallel. It is nearly optimal to set identical supply air temperatures for each air handler. The cost over the optimal strategy with different temperatures is small, even when the loading on the various cooling coils differs significantly.

Figure 12 shows a comparison of the system COP between optimal and identical supply air set temperatures as a function of the relative loading on one of the air handlers for a system with two identical air handlers. The difference between optimal and identical set temperatures is small over the practical range of relative loadings. Identical set temperatures should be used for air handlers operating in parallel even if they do not have identical characteristics. However, if any of the air handlers are operating at a required minimum level of flow, the set temperature should be set to meet the load. For the purpose of determining optimal control of the entire system, it is sufficiently accurate to combine all air handlers into a single effective air handler and to assume that all zones are maintained at the same air temperature and that the heat transfer characteristics of the coils are similar.

Multiple Cooling Tower Cells. The power consumption of a chiller is sensitive to the condensing water temperature, which, in turn, is affected by both the condenser water and cooling tower airflow rates. Increasing either of these flows reduces the chiller power requirement, but at the expense of an increase in the pump or fan power consumption.

Braun et al. (1987) and Nugent et al. (1988) showed that using all the cooling tower cells at an appropriate speed consumes the minimum amount of power for a system with variable-speed fans under most conditions. Because the power consumption of the fans is proportional to the cube of the fan speed, operating all cells in parallel with the same airflow lowers the overall fan power consumption compared to operating some at higher flow while some are off. An additional benefit associated with using all cells is a lower pressure drop across the water spray nozzles, which also results in lower pumping power requirements. However, at very low pressure drops, inadequate spray distribution may adversely affect the thermal performance of the cooling tower. Water loss from drift may be reduced when cells are not operated at high airflows.

Many cooling towers use multiple-speed fans rather than continuously adjustable variable-speed fans. In this case, it is not optimal to operate all tower cells under all conditions. The optimal number of cells operating and their individual fan speeds will depend on the system characteristics and on ambient conditions and needs to be determined for each operating condition. In almost all cases, when additional tower capacity is required, the best sequencing for cooling tower fans requires that fans operating at the lowest speed (including those that are off) should be increased in speed first. Similarly, when removing tower capacity, the fans with the highest speeds should be reduced first.

These guidelines are derived from evaluating the trade-off between the incremental power increase associated with increasing the fan currently on or switching on another fan. For two-speed fans, the incremental power increase associated with turning on a low-speed fan is less than that for increasing one to high speed if the low speed is less than 79% of the high fan speed. Most commonly, the low speed of a two-speed cooling tower fan is between one-half and three-quarters of full speed, so that the incremental increase in airflow is greater and thermal performance is better if the low-speed fan is added. Similarly, for three-speed fans, the best strategy is to increment the lowest fan speeds first when adding tower capacity and to decrement the highest fan speeds when removing capacity. Typical three-speed combinations that allow optimization in this manner are (1) one-third, two-thirds, and full speed or (2) one-half, three-quarters, and full speed. Braun and Diderrich (1990) present a method for determining the optimal control for multispeed fans.

Another issue related to the control of multiple cooling tower cells having multiple-speed fans is the distribution of water flow to the individual cells. The best overall thermal performance of the cooling tower occurs when the water flow is divided so that the ratio of water flow rate to airflow rate, rather than the ratio of water flow to the total, is identical for all tower cells. However, a comparison of performance with equal flow rate ratios to that for equal flow rates showed only a maximum 5% increase in the heat transfer effectiveness with one cell operating at one-half speed and the other cell at full speed; this resulted in less than a 1% decrease in the chiller power. There is also lower water pressure drop across the spray nozzles (lower pumping power) associated with equally divided flow. Finally, the performance differences are smaller when more than two cells are operating and when a majority of cells are operating at the same speed. Overall then, equal water flow distribution between cooling tower cells is near-optimal.

Multiple Pumps. In a common control strategy for sequencing condenser and chilled water pumps, the pumps are brought on-line or off-line as the chillers are turned on or off. For fixed-speed pumps, this strategy is not optimal. When a chiller is brought on-line in parallel with the other units, the flow increases through both the condenser and the chilled water loops if the pump control is not changed. The increased flow rates tend to improve the overall chiller performance. However, if the pumps are operating near their peak efficiency before the additional chiller is brought on-line, the pump efficiency drops when the additional chiller is added. Most often, the improvements in chiller performance offset the degradation in pump performance, so that no additional pump is needed at the chiller switch point.

Figure 13 shows the optimal system performance for different combinations of chillers and fixed-speed pumps in parallel, as a function of load for a given wet bulb. The optimal switch point for a second pump occurs at a much higher relative chilled load (0.62) than the switch point for adding or removing a chiller (0.38). If the second pump were controlled to be turned on with the second chiller, the optimal switch point would be at the maximum chiller capacity, with a penalty in performance approximately 10% over that of the optimal pump strategy. The optimal control for sequencing fixed-speed pumps depends on the load and the ambient wet-bulb temperature and should not be directly coupled to chiller sequencing.

Because variable-speed pumps have relatively constant efficiencies over their range of operation, their sequencing is more straightforward than that for fixed-speed pumps. As a result, in the best sequencing strategy, pumps that operate near their peak efficiencies for each possible combination of chillers are selected. Because the system pressure drop characteristics change when chillers are added or removed, the sequencing of variable-speed pumps should be

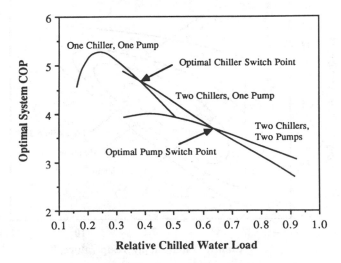

Fig. 13 Effect of Chiller and Pump Sequencing on Optimal System Performance

directly coupled to the sequencing of chillers. For identical variable-speed pumps operated in parallel, the best overall efficiency is obtained by operating them at identical speeds. For nonidentical pumps, near-optimal efficiency is realized by operating them at equal fractions of their maximum speed.

Effects of Load and Ambient Conditions on Optimal Control Settings

For a system in which the relative loads in each zone are basically the same with time, the optimal control variables are functions of (1) the total sensible and latent gains to the zones and (2) the ambient dry- and wet-bulb temperatures. The effect of the ambient dry-bulb temperature alone is small because air enthalpies depend primarily on wet-bulb temperatures and wet surface heat exchangers are driven primarily by enthalpy differences. Typically, the zone latent gains are on the order of 15 to 25% of the total zone gains, and the effect of changes in latent gains have a relatively small effect on system performance for a given total load. Consequently, results for overall system performance and optimal control may be correlated in terms of only the ambient wet-bulb temperature and the total chilled water load. If the load distribution between zones changes significantly over time, the load distribution must also be included in a correlation.

Chilled Water Loop. Both the optimal chilled water and supply air temperatures decrease with increasing load for a fixed wet-bulb temperature because the rate of change in air handler fan power with respect to load changes is larger than that for the chiller. As the wet-bulb temperature increases for a fixed load, the optimal set point temperatures also increase.

Figure 14 shows the sensitivity of system power consumption to chilled water and supply air set point temperatures for a given load and wet-bulb temperature. Within about 1.5 K of the optimum values, the power consumption is within 1% of the minimum. Outside this range, the sensitivity to the set points increases significantly. The penalty associated with operation away from the optimum is greater in the direction of smaller differences between the supply air and chilled water set points. As this temperature difference is reduced, the required flow of chilled water to the coil increases, and the chilled water pumping power is greater. For a given chilled water or supply air temperature, the temperature difference is limited by the heat transfer characteristics of the coil. As this limit is approached, the required water flow and pumping power would

become infinite if the pump speed were not constrained. It is generally better to have too large rather than too small a temperature difference between the supply air and chilled water set points.

Condenser Water Loop. The primary controllable variables associated with heat rejection to the environment are the condenser water and tower airflow rates. Both optimal airflow and water flow increase with load and wet-bulb temperature. Higher condensing temperatures and reduced chiller performance result from either increasing loads or wet-bulb temperatures for a given control. Increasing the airflow and water flow under these circumstances reduces the chiller power consumption at a faster rate than the increases in fan and pump power.

Figure 15 shows the sensitivity of the total power consumption to tower fan and condenser pump speed. Near the optimum, power consumption is not sensitive to either of these control variables, but increases significantly away from the optimum. The rate of increase in power consumption is particularly large at low condenser pump speeds. A minimum pump speed is necessary to overcome the static pressure associated with the height of the water discharge in the cooling tower above the sump. As the pump speed approaches this value, the condenser flow approaches zero and the chiller power increases dramatically. A pump speed that is too high is generally better than a pump speed that is too low. The broad area near the optimum indicates that it is not necessary to accurately determine the optimal setting.

Zone Humidity Control. In a variable air volume (VAV) system, it is generally possible to control the chilled water temperature, supply air temperature, and supply airflow. However, for a given chilled water temperature, only one combination of the supply air temperature, airflow rate, and water flow rate will maintain both the zone temperature and humidity at their set points. Acceptable bounds on the room temperature and humidity for human comfort are defined in Chapter 8 of the 1993 *ASHRAE Handbook—Fundamentals*. For a zone being cooled, the equipment operating costs are minimized when the zone temperature is at the upper bound of the comfort region. However, operating simultaneously at the upper limit of humidity does not minimize operating costs.

Figure 16 shows a comparison of system COP and zone humidity associated with fixed and free-floating zone humidity as a function of the relative load. Over the range of loads for this system, allowing the humidity to float within the comfort zone produces a lower cost and a lower zone humidity than setting the humidity at the highest

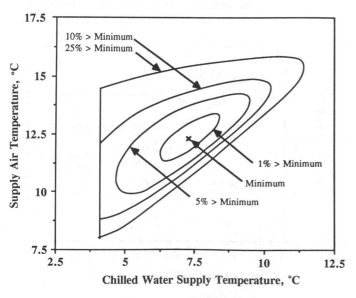

Fig. 14 Power Contours for Chilled Water and Supply Air Temperatures

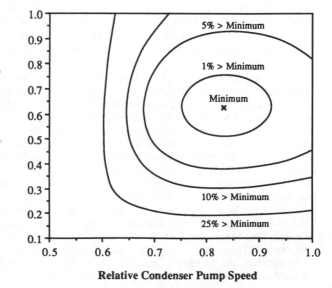

Fig. 15 Power Contours for Condenser Loop Control Variables

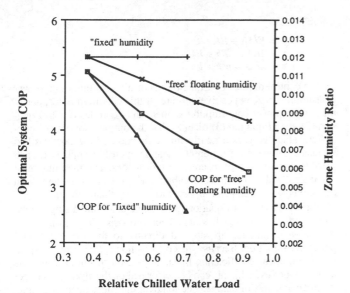

Fig. 16 Comparison of Free-Floating and Fixed Humidity Control

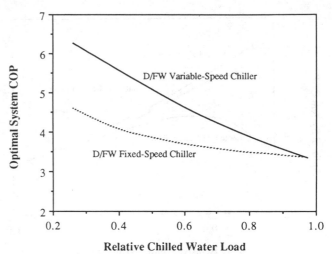

Fig. 17 Optimal System Performance for Variable and Fixed-Speed Chillers

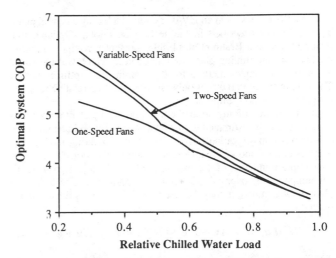

Fig. 18 Comparison of One-Speed, Two-Speed, and Variable-Speed Cooling Tower Fans (Four Cells)

acceptable value. The largest differences occur at the highest loads. Operation with the zone at the upper humidity bound results in lower latent loads than with a free-floating humidity, but this humidity control constraint requires higher supply air temperatures that, in turn, result in greater air handler power consumption. When determining optimal control points, the humidity should be allowed to float freely unless it falls outside the bounds of human comfort.

Variable-Speed Equipment

Variable-speed equipment (e.g., chillers, fans, and pumps) facilitates control and reduces operating costs. The overall savings associated with the use of variable-speed equipment over a cooling season depends on the time variation of the load. Overall, the use of variable-speed drives results in operating costs that are 20 to 50% lower than costs for equipment with fixed-speed drives.

The part-load performance of a centrifugal chiller is generally better for variable-speed control than it is with vane control. Figure 17 shows a comparison between the overall optimal system performance for variable-speed and variable-vane control of a large chilled water facility (Braun 1988, Braun et al. 1989a). At part-load conditions, the performance associated with the use of variable-speed equipment is improved as much as 25%. However, as expected, the power requirements are similar at conditions associated with peak loads because, at full load, the vanes are wide open, and the speed is the same under variable-speed control and fixed-speed operation.

The most common design for cooling towers places multiple tower cells in parallel with a common sump. Each tower cell has a fan with one, two, or possibly three operating speeds. Although multiple cells with multiple fan settings offer wide flexibility in control, the use of variable-speed tower fans can provide additional improvements in overall system performance.

Figure 18 shows a comparison of optimal system performance for single-speed, two-speed, and variable-speed tower fans as a function of load for a given wet bulb. The variable-speed option results in higher system COP under all conditions. In contrast, for discrete fan control, the tower cells are isolated when their fans are off, and the performance is poorer. Below about 70% of full-load conditions, there is a 15% difference in total energy consumption between single-speed and variable-speed fans. Between two-speed and variable-speed fans, the differences are much smaller, on the order of 3 to 5% over the entire range.

Fixed-speed pumps sized to give proper flow to a chiller at design conditions are oversized for part-load conditions. As a result, the

system will have higher operating costs than with a variable-speed pump of the same design capacity. Control through multiple pumps with different capacities offers increased flexibility, and the use of a smaller fixed-speed pump for low-load conditions can reduce overall power consumption. The optimal system performance for variable-speed and fixed-speed pumps applied to both the condenser and chilled water flow loops is shown in Figure 19. Large fixed-speed pumps were sized for the design conditions, and the small pumps were sized to have one-half the flow capacity of the large pumps. Below about 60% of full-load conditions, variable-speed pumps show a significant improvement over single fixed-speed pumps. With the addition of small fixed-speed pumps, the improvements with variable-speed pumps become significant at about 40% of the maximum load.

Simplified Optimal Control Method

Optimal supervisory control of central cooling systems without thermal storage minimizes total plant power consumption at each instant of time (or over each decision interval). In practice, optimization techniques are applied to a steady-state model for the plant based on physical measurements. The plant model may take any number of different forms. Cumali (1988) describes a global optimization method based on fundamental thermodynamic models that

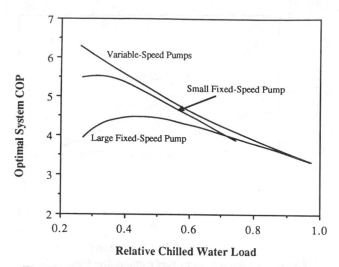

Fig. 19 Comparison of Variable and Fixed-Speed Pumps

simulate the processes in an HVAC system, including a central plant. This is further described in the section on Global Optimization. Braun (1988) and Braun et al. (1989b) present a component-based method for determining plant optimal control, which was then used to develop a simpler method for determining the optimal control. The simpler method requires only one empirical correlating function for plant power consumption.

In the vicinity of any optimal control point, the plant power consumption may be estimated with a quadratic function in terms of the continuous control variables for each of the operating modes (i.e., discrete control mode). A quadratic function also correlates power consumption in terms of the uncontrolled variables (i.e., load, ambient temperature) over a wide range of conditions. The following plant power function may be used to determine optimal control points:

$$J(f,M,u) = u^T A u + b^T u + f^T C f + d^T f + f^T E u + g \qquad (2)$$

where

$$
\begin{aligned}
J &= \text{total plant power} \\
u &= \text{vector of the continuous control variables} \\
f &= \text{vector of the uncontrolled variables} \\
M &= \text{vector of the discrete control variables} \\
T &= \text{superscript designating the transpose vector} \\
A, C, E &= \text{coefficient matrices} \\
b, d &= \text{coefficient vectors} \\
g &= \text{a scalar}
\end{aligned}
$$

Because the empirical coefficients of the above plant power function depend on the operating modes, these constants must be determined for each feasible combination of discrete control modes.

A solution for the optimal control vector that minimizes the power may be determined analytically by applying the first-order condition for a minimum. Equating the Jacobian of Equation (2) with respect to the control vector to zero and solving for the optimal control set points gives the following equation:

$$u^* = k + Kf \qquad (3)$$

where

$$
\begin{aligned}
k &= -(1/2)\,A^{-1}b \\
K &= -(1/2)\,A^{-1}E
\end{aligned}
$$

The cost associated with the unconstrained control defined by Equation (3) is given by the following equation:

$$J^* = f^T \theta f + \sigma^T f + \tau \qquad (4)$$

where

$$
\begin{aligned}
\theta &= K^T A K + E K + C \\
\sigma &= 2 K A k + K b + E k + d \\
\tau &= k^T A k + b^T k + g
\end{aligned}
$$

The control defined by Equation (3) results in a minimum power consumption if A is positive definite. If this condition holds, and if the system power consumption is adequately correlated with Equation (2), then Equation (3) dictates that the optimal continuous control variables vary as a nearly linear function of the uncontrolled variables. However, a different linear relationship applies to each feasible combination of discrete control modes. The minimum cost associated with each mode combination must be computed from Equation (4) and compared in order to identify the minimum.

The coefficients of Equation (3) must be determined empirically, and a variety of approaches have been proposed. In one approach, regression techniques are applied directly to the measured total power consumption. Because the cost function is linear with respect to the empirical coefficients, linear regression techniques may be used. A set of experiments can be performed on the system to collect the large amounts of data needed to cover the entire operating range and to account for measurement uncertainty.

The regression can be performed on-line using least-squares recursive parameter updating (Ljung and Soderstron 1983). Rather than fitting empirical coefficients of the system cost function of Equation (2), the coefficients of the optimal control Equation (3) and the minimum cost function of Equation (4) can be estimated directly. For a limited set of conditions, optimal values of the continuous control variables can be estimated through trial-and-error variations in the system. If the load and wet bulb are the only uncontrolled variables, only three independent conditions are needed to determine the coefficients of the linear control law given by Equation (3). The coefficients of the minimum cost function can then be determined from system measurements with the linear control law in effect. A disadvantage of this approach is that there is no direct way to handle constraints on the controls.

Employing the guidelines for optimal control described previously, the important independent control variables are (1) supply air set temperature, (2) chilled water set temperature, (3) relative tower airflow, (4) relative condenser water flow, and (5) the number of operating chillers. The supply air and chilled water set points are continuously adjustable control variables. However, because the chilled water flow requirements depend on these controls, discrete changes in power consumption may be associated with these controls if the pump operation undergoes discrete control changes. For the same total flow rate, the overall pumping efficiency changes with the number of operating pumps. However, this has a relatively small effect on the overall power consumption and may be neglected in fitting the overall cost function to changes in the control variables.

For variable-speed cooling tower fans and condenser water pumps, the relative tower air and condenser water flows are continuous control variables. Analogous to the chilled water flow, the overall condenser pumping efficiency changes with the number of operating pumps, so that the power consumption associated with continuous changes in the overall relative condenser water flow may have a discontinuity. This discontinuity may be neglected in fitting the overall cost function to changes in this control variable.

With variable-speed pumps and fans, the only significant discrete control variable is the number of operating chillers. The chiller mode defines which of the available chillers are to be on-line. The optimization involves determining optimal values of only four continuous control variables for each of the feasible chiller modes. The chiller mode giving the minimum overall power consumption represents the optimum. For a chiller mode to be feasible, the specified chillers must operate safely within their capacity and surge limits. In practice, abrupt changes in the chiller modes should be avoided. Large chillers should not be cycled on or off except when the savings associated with the change is significant.

For fixed-speed cooling tower fans and condenser water pumps, the relative flows are discrete values. One method of handling these variables is to consider each of the discrete combinations as separate modes. However, for multiple cooling tower cells with multiple fan speeds, the number of possible combinations may be large. In a simpler approach that works satisfactorily, the relative flows are treated as continuous control variables during the optimization, and the discrete relative flow that is closest to the optimal value is selected. At least three relative flows (discrete flow modes) are necessary for each chiller mode in order to fit the quadratic cost function. The number of possible sequencing modes for fixed-speed pumps is generally much more limited than it is for cooling tower fans, with two or three possibilities (at most) for each chiller mode. In fact, with many current designs, individual pumps are physically coupled with chillers, so that it is impossible to operate more or fewer pumps than the number of operating chillers. Thus, it is generally better to treat the control of fixed-speed condenser water pumps with a set of discrete control possibilities rather than to use a continuous control approximation.

The methodology for determining the near-optimal control of a chilled water system may be summarized as follows:

1. Change the chiller operating mode if the system operation is at the limits of chiller operation (near surge or maximum capacity).
2. For the current set of conditions (load and wet bulb), estimate the feasible modes of operation that would avoid operating the chiller and condenser pump at their limits.
3. For the current operating mode, determine optimal values of the continuous controls using Equation (3).
4. Determine a constrained optimum if controls exceed their bounds.
5. Repeat steps 3 and 4 for each feasible operating mode.
6. Change the operating mode if the optimal cost associated with the new mode is significantly less than that associated with the current mode.
7. Change the values of the continuous control variables. When treating multiple-speed fan control as a continuous variable, use the discrete control closest to the optimal continuous value.

If the linear optimal control Equation (3) is directly determined from optimal control results, then the constraints on controls may be handled directly. A simple solution is to constrain the individual control variables as necessary and neglect the effects of the constraints on the optimal values of the other controls and the minimum cost function. The variables of primary concern with regard to constraints are the chilled water and supply air set temperatures. These controls must be bounded for proper comfort and safe operation of the equipment. On the other hand, the cooling tower fans and condenser water pumps should be sized so that the system performs efficiently at design loads, and constraints on control of this equipment should only occur under extreme conditions.

The optimal value of the chilled water supply temperature is coupled to the optimal value of the supply air temperature, so that decoupling these variables in evaluating constraints is generally not justified. However, optimization studies indicate that when either control is operated at a bound, the optimal value of the other "free" control is approximately bounded at a value that depends only on the ambient wet-bulb temperature. The optimal value of this free control (either chilled water or supply air set point) may be estimated at the load at which the other control reaches its limit. Coupling between optimal values of the chilled water and condenser water-loop controls is not as strong, so that interactions between constraints on these variables may be neglected.

Braun et al. (1987) correlated the power consumption of the Dallas/Fort Worth airport chiller, condenser pumps, and cooling tower fans with the quadratic cost function given by Equation (2) and showed good agreement with the data. Because the chilled water-loop control was not considered, the chilled water set point was treated as a known

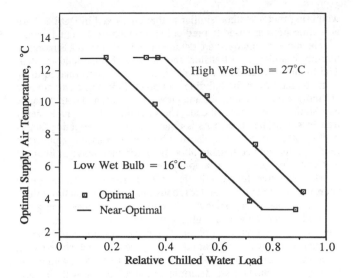

Fig. 20 Comparisons of Optimal Chilled Water Temperature

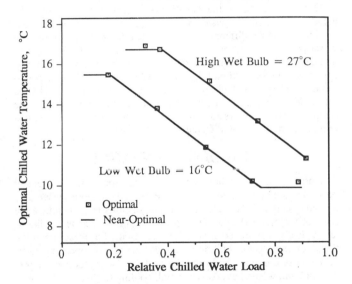

Fig. 21 Comparisons of Optimal Supply Air Temperature

uncontrolled variable. The discrete control variables associated with the four tower cells with two-speed fans and the three condenser pumps were treated as continuous control variables. The optimal control determined by the near-optimal Equation (3) also agreed well with that determined using a nonlinear optimization applied to a detailed simulation of the system. Figures 20 and 21 show values of the chilled water and supply air temperatures obtained from the quadratic approach and the true optimal values. The chilled water temperature was constrained between 3 and 13°C, while the supply air set point was allowed to float freely. Also, for conditions under which the chilled water temperature is constrained, the optimal supply air temperature is also bounded at a value that depends on the ambient wet bulb.

Global Optimization

Cumali (1988, 1993) presents a method for real-time global optimization of HVAC systems, including the central plant and associated piping and duct networks. The method utilizes a building thermal load prediction model for the zones based on a coupled weighting factor method similar to that employed in DOE-2. Variable time steps are used to predict loads on a 5- to 15-min period.

The models employed are detailed mechanistic models based on thermodynamics, heat transfer, and fluid mechanics fundamentals. The models are calibrated to match the actual performance using data obtained from the building and plant. Pipe and duct networks are represented as incidence and circuit matrices, and both dynamic and static losses are included. The coupling of the fluid energy transfers with the zone loads is done via custom weighting factors calibrated for each zone. The resulting equations are grouped to represent feasible equipment allocations for each range of building loads and are solved using a nonlinear solver.

The objective function is the cost of delivering or removing energy to meet the loads, and it is constrained by the comfort criteria for each zone. The objective function is minimized using the reduced gradient method, subject to constraints on comfort and equipment operation. The optimization starts with the feasible points as determined by a nonlinear equation solver for each combination of equipment allocation. The values of the set points that minimize the objective function are determined; the allocation with the least cost is the desired operation mode.

The results obtained with this approach have been applied to San Francisco high-rise office buildings with central plants, VAV, dual duct, and induction systems. Electrical demand reductions of 8 to 12% and energy savings of 18 to 23% were achieved.

Optimal operation is the yardstick of performance and defines the achievable limit of the system operation. It may be used as the basis of comparison for different operating scenarios and algorithms. The worst possible operation of a building system is difficult to define and determine, but optimal operation defines the minimum cost case and is unique for each building and HVAC system. Thus the penalties associated with other operational modes can be measured by the difference in energy or demand from that of the optimal case. The optimal approach offers an unbiased basis for comparing HVAC system performance under different control options.

BUILDING DYNAMICS

Optimal start algorithms determine times for turning equipment on so that the building zones reach the desired conditions when they become occupied. The goal of these algorithms is to minimize the precool (or preheat) time. During occupied hours, zone conditions are typically maintained at specified set points. For these strategies, the assumption is that building mass acts to increase operating costs. A massless building would require no time for precooling (or preheating) and would have lower overall cooling (or heating) loads.

In some situations, the thermal mass of a building represents a storage medium that may be used for reducing operating costs. Dynamic building control strategies attempt to manage the thermal mass of the building within acceptable comfort limits to (1) limit peak electrical demands and (2) minimize the daily operating costs in response to favorable utility rates and outside air cooling possibilities.

Recovery from Night Setback

During times that a building is not occupied, a significant savings in operating costs may be realized by raising the building set point temperature for cooling and by lowering the set point for heating. Bloomfield and Fisk (1977) showed energy savings of 12% for a heavyweight building and 34% for a lightweight building.

An optimal controller for return from night setback returns zone temperatures to the comfort range precisely when the building becomes occupied. Seem et al. (1989) compared seven different algorithms for determining the optimal time for return from night setback. Each algorithm requires the estimation of parameters from measurements of the actual time for return from night setback.

Seem et al. (1989) showed that the optimal return time for cooling was not strongly affected by the outdoor temperature. The following

quadratic function of the initial zone temperature was found to be adequate for estimating the return time:

$$\tau = a_0 + a_1 t_{z,i} + a_2 t_{z,i}^2 \qquad (5)$$

where

τ = estimate of the optimal return time
$t_{z,i}$ = initial zone temperature at the beginning of the return period
a_0, a_1, a_2 = empirical parameters

The parameters of Equation (5) may be estimated by applying linear least-squares techniques to the difference between the actual return time and the estimates. These parameters may be continuously corrected using recursive updating schemes as outlined by Ljung and Soderstron (1983).

For heating, ambient temperature has a significant effect on the return time. The following is one relationship for estimating the return time for heating:

$$\tau = a_0 + (1 - w)(a_1 t_{z,i} + a_2 t_{z,i}^2) + w a_3 t_a \qquad (6)$$

where

t_a = ambient temperature
a_0, a_1, a_2, a_3 = empirical parameters

and

$$w = 1000^{-(t_{z,i} - t_{unoccupied}) / (t_{occupied} - t_{unoccupied})} \qquad (7)$$

where w is a weighting function and $t_{unoccupied}$ and $t_{occupied}$ are the zone set points for unoccupied and occupied periods. Within the context of Equation (6), this function weights the outdoor temperature more heavily when the initial zone temperature is close to the set point temperature during the unoccupied time. The parameters of Equation (6) may be estimated by applying linear least-squares techniques to the difference between the actual return time and the estimates.

Ideally, separate equations should be used for zones that have significantly different return times. Equipment operation is initiated for the zone with the earliest return time. In a building with a central cooling system, the equipment should be operated above some minimum load limit. With this constraint, some zones need to be returned to their set points earlier than the optimum time.

Optimal start algorithms often use a measure of the building mass temperature rather than the space temperature for determining return time. Although the use of space temperature results in lower energy costs (i.e., shorter return time), the mass temperature may result in better comfort conditions at the time of occupancy.

Building Precooling Studies

For certain building types, it is possible to take advantage of the storage capabilities of the building structure by shifting a significant portion of a building's on-peak cooling requirements to off-peak periods, thereby reducing both energy and demand costs. This can be a significant demand management tool that does not require a major capital retrofit in a building.

Braun (1990) demonstrated the maximum potential savings associated with optimal control of building thermal mass as compared with conventional night setback control. He used optimization routines applied to computer simulations of buildings and their associated cooling systems to show that there is a significant potential for reducing operating costs. The savings are primarily due to the following four effects: (1) use of low cost off-peak electrical energy, (2) reduced demand charges due to lower peak power, (3) reduced mechanical cooling due to the use of cool nighttime air for ventilative precooling, and (4) improved mechanical cooling efficiency due to increased operation at more favorable part-load and ambient conditions.

In order to have an effective control method, these benefits must be balanced against the increase in the total cooling requirement due to precooling of the thermal mass. As a result of these trade-offs, Braun (1990) found that the cost savings are very sensitive to both the control method and several design and operating characteristics. Even without time-of-day utility prices, this study showed a large benefit associated with effective utilization of building thermal mass for many situations. For a design day, energy cost savings for cooling associated with optimal control of building thermal mass ranged from 0 to 35%, depending on the system. These results were obtained without adjustment of the temperature set points during the occupied period. Through optimal set point adjustment, it was possible to reduce the total building peak electrical demand by 15% to 35% on the design day. The primary factors that affected the comparisons between optimal and conventional night setback control were utility rates, cooling system part-load characteristics, weather, occupancy schedule, and building construction. Other simulation studies that have produced results consistent with those of Braun include those of Snyder and Newell (1990) and Andresen and Brandemuehl (1992).

One of the most important conclusions that can be drawn from previous simulation studies is that appropriate methods for controlling building thermal storage and the associated cost savings as compared with conventional control are very sensitive to the application and operating conditions. Some previously documented experimental studies in buildings did not include sufficient preliminary analysis to identify (1) whether the system was a good candidate for effective utilization of building thermal mass and (2) an effective method for control. As a result, the conclusions from these studies were not as positive as might have been expected.

Ruud et al. (1990) evaluated the effect of precooling on the on-peak cooling requirements for an existing building. The results showed only a 10% reduction in the cooling energy required during the occupied period with a substantial increase in the total cooling required and no reduction in the peak cooling requirement. The poor results were primarily attributed to the fact that the building was not a good candidate for effective utilization of building thermal storage. The building construction was such that there was a relatively small amount of thermal mass, a weak heat transfer coupling between the thermal mass and the internal air space, and a strong heat transfer coupling to the external ambient conditions. Furthermore, the control was not optimized for this application. However, the optimal control for this application may well have been conventional night setback control.

Conniff (1991) performed experiments using a laboratory test facility in which the structure was designed to represent a thermal zone within a multistory building. External boundary conditions for the floor, ceiling, and walls could be controlled independently. The structure was of relatively lightweight construction, having a 50-mm concrete floor, suspended ceiling, and gypsum walls. The floor was carpeted and there were no special provisions for coupling to the thermal mass. The purpose of their study was to investigate the effect of alternative control procedures on the peak air-conditioning load. The strategies investigated resulted in less than a 3% reduction in the peak cooling load. However, the strategies were not optimized for the facility.

Recently, Morris et al. (1994) devised and performed a set of experiments at the same facility used by Conniff (1991) in order to demonstrate the potential for load shifting and load leveling when control is optimized. Prior to performing experiments, a simulation of the test facility was developed that included a model of the structure, a model for human comfort, a specific profile of internal gains, and a model of a hypothetical cooling system that would serve the structure. Results of the zone simulation were in good agreement with the measurements. In order to determine the control strategy to use in the experiments, optimization routines were applied to the simulation with the constraint that thermal comfort must be maintained during the occupied period. Two separate objectives resulting in two different control strategies were considered: (1) minimum cooling system energy use and (2) minimum peak cooling system electrical demand. The two strategies were implemented in the facility and compared with night setback control.

Morris et al. (1994) demonstrated that for a representative day, approximately 40% of the daytime cooling requirement could be shifted from the occupied period to the unoccupied period. Depending on the cooling system and utility rates, this could result in energy cost savings of up to 30%. In addition, Morris et al. showed that the peak cooling demand could be reduced by about 40% through optimal control of the building thermal mass. Comfort conditions were also monitored for the tests, and, overall, were well within acceptable limits during the occupied period.

The contradictory results of Morris et al. (1994) and Conniff (1991) for the same test facility emphasize the importance of developing a control strategy that is specific to the application. Conniff's results were not nearly as encouraging as those of Morris et al. for the same test facility because the control was not optimized. There is a need to develop control methods and application guidelines for effective use of building thermal storage. Clearly, there is a great potential associated with effective use of the thermal mass in buildings. However, there is also potential for misuse of thermal mass that could lead to increased operating costs as compared with conventional night setback control.

One of the more effective methods for precooling a building involves the use of "free" cool night air to reduce cooling requirements for the next day. In reality, the use of outside air is not free, because energy is required to operate the air-handling fans. Precooling with outside air should only be considered if (1) heating is not required during the occupancy period, (2) the humidity of the ambient air is lower than an acceptable comfort limit, and (3) the cost of operating air-handling fans is less than the reduction in operating costs for mechanical cooling during the occupied period. The task of evaluating the cost savings of precooling with outside air is not straightforward. Depending on the cooling requirements and equipment design, precooling with outside air is usually advantageous when the temperature difference between the zones and the ambient air is greater than about 5 K. In a simple precooling control strategy, the zone temperature is decreased to, and maintained at, the lower comfort limit whenever ambient cooling is available. After precooling, the space temperature set point is adjusted upward (within the comfort zone) during the occupied period to take advantage of the precooling. In a simple strategy, the set point is immediately adjusted to the upper comfort limit at the onset of occupancy.

CONTROL OF THERMAL STORAGE SYSTEMS

As described in Chapter 40, thermal energy storage can significantly reduce the costs of providing cooling in commercial buildings by shifting a portion of the daytime cooling requirements to the nighttime hours. The savings result from reductions in both demand and energy charges. Most of the thermal storage systems being installed use ice as the storage medium. Although the material in this section emphasizes ice storage applications, much of it is relevant to chilled water storage as well.

Figure 22 shows the schematic of a typical partial ice storage system. The system consists of one or more chillers, cooling tower cells, condenser water pumps, chilled water/glycol distribution pumps, ice storage tanks, and valves for controlling charging and discharging modes of operation. Ice is made at night and used during the day to provide a portion of a building's cooling requirements. The partial storage terminology comes about because the storage is not sized to handle the on-peak load requirement on the design day. Typically, the storage and chiller capacity are sized such that the chiller operates at full capacity during the on-peak period on the design day.

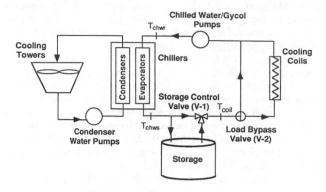

Fig. 22   Ice-Storage System

Typical modes of operation for the system in Figure 22 are as follows:

1. *Storage Charging Mode.* Typically, charging of storage (i.e., ice making) only occurs when the building is unoccupied and off-peak electric rates are in effect. In this mode, the load bypass valve (V-2) is fully closed to the building cooling coils, the storage control valve (V-1) is fully open to the ice storage tank (i.e., the total chilled water/glycol flow is through the tank), and the chiller produces temperatures low enough (e.g., $-7°C$) to make ice in the tank.

2. *Storage Discharging Mode.* The discharging of storage (i.e., ice melting) only occurs when the building is occupied. In this mode, valve V-2 is open to the building cooling coils, and valve V-1 modulates the mixture of flows from the storage tank and chiller in order to maintain a constant supply temperature to the building cooling coils (e.g., $3°C$). Individual valves at the cooling coils modulate their chilled water/glycol flow to maintain supply air temperatures to the zones.

3. *Direct Chiller Mode.* The chiller may operate to meet the load directly without the use of storage during the occupied mode (typically when off-peak electric rates are in effect). In this mode, valve V-1 is fully closed with respect to the storage tank.

For a typical partial storage system, the storage meets only a portion of the on-peak cooling loads on the design day, and the chiller operates at capacity during the on-peak period. As a result, the peak power is limited by the capacity of the chiller. For off-design days, many different control strategies meet the building's cooling requirements, and each method has a different overall operating cost.

In chiller-priority control, which is the most commonly used method for controlling the use of ice storage, the chiller provides as much cooling as possible to meet the on-peak building requirements, and the ice storage medium provides the rest. The principal advantages of this strategy are that it is extremely easy to implement and it ensures that sufficient cooling capacity will be available throughout the day. The disadvantage is that both the energy and demand costs are greater than those of a better control method.

Other control strategies for ice storage systems have been presented in the literature, including those of Rawlings (1985) and Braun (1992). In contrast to chiller-priority control, these storage-priority control methods attempt to maximize the use of storage to provide cooling during the on-peak period. As a result, they require the use of forecasts of cooling requirements. Although these control strategies are improvements over chiller-priority control, they are not optimal for all systems and conditions.

Chiller-Priority Control

With chiller-priority control, the ice storage tank is fully charged each day during the off-peak, unoccupied period. During the storage discharge mode, the chiller operates at full cooling capacity (or less

if it is sufficient to meet the load) and storage matches the difference between the building requirement and chiller capacity. The operation of equipment for chiller-priority control during different parts of the day can be described as follows:

- *Unoccupied, Off-Peak Period.* The charging mode (as defined previously) begins after the building becomes unoccupied and at the onset of low electric rates. Ice making continues until maximum ice formation or until the building is occupied. During the charging mode, the chillers operate at maximum cooling capacity. Typically, this is accomplished by establishing a very low chilled water supply set point (T_{chws}) for the chiller controller.

- *Occupied Period.* T_{chws} is set equal to the desired supply temperature for the coils (T_{coil}). If the capacity of the chiller is sufficient to maintain this set point at any time, then storage is not utilized, and the system operates in the direct chiller mode. Otherwise, the storage control valve modulates the flow through storage to maintain the supply set point, providing a cooling rate that matches the difference between the building load and the maximum cooling capacity of the chillers.

This strategy is easy to implement and does not require the use of a load forecast. It works well for design conditions, but can result in relatively high demand and energy costs for off-design conditions because the chiller operates at full capacity during the on-peak period.

Load-Limiting, Storage-Priority Control

In contrast to chiller-priority control, storage-priority control attempts to maximize the use of storage to provide cooling during the on-peak period. The chiller operates to meet the difference between the building requirement and the rate at which the storage is providing cooling. The rate at which cooling is provided by storage must be modulated so that its cooling capacity is not fully depleted (i.e., there is no ice remaining) before the end of the on-peak period. As a result, storage-priority control methods require forecasting building cooling requirements.

Braun (1992) presents a storage-priority control strategy, termed load-limiting control, that tends to minimize the peak cooling plant power demand. It also gives nearly optimal performance in terms of energy costs for the particular system studied. The operation of equipment for load-limiting control during different parts of the day can be described as follows:

1. *Off-Peak, Unoccupied Period.* Ice making is initiated at the time when the building is both unoccupied and off-peak electrical rates are in effect. During the ice making period, the chiller operates at full capacity (e.g., all cylinders loaded). With feedback control of the chilled water/glycol supply temperature, this is accomplished by establishing a set point (e.g., $-7°C$) that is low enough. Ice making continues until full ice formation or until the building is occupied.

   On days the storage capacity exceeds the total building cooling requirements, it is not necessary to reach maximum ice formation. In fact, there is an energy penalty associated with over-charging storage due to an increase in the heat transfer resistance (especially for area-constrained ice tanks). In this case, the amount to charge storage can be estimated using a forecast of the building cooling requirement for the next day.

2. *Off-Peak, Occupied Period.* During this period, the goal is to minimize the use of storage (i.e., chiller priority). This is accomplished by establishing a T_{chws} equal to the T_{coil}. If, at any time, the chiller has sufficient capacity to maintain this set point, then there is no flow through the storage tank and the storage remains fully charged. If the chiller cannot maintain the set point, it operates at full capacity and flow through the storage tank necessary to maintain the coil supply set point provides the extra cooling to meet the building load.

3. *On-Peak, Occupied Period.* During this period, the goal is to operate the chillers at a constant load while discharging the ice storage such that the ice is completely melted when the off-peak period begins. This requires the use of a building cooling load forecaster. At each decision interval (e.g., 15 min), the following steps are applied:

- Forecast the total integrated building cooling requirement until the end of the discharging period.
- Estimate the total remaining integrated cooling capacity of the ice storage tank.
- Estimate the chiller load by dividing the difference between the integrated cooling requirement and the storage capacity by the time until the end of the discharge period.
- Determine the chiller set point temperature necessary to achieve the desired loading from the following equation:

$$T_{chws} = T_{chwr} - \dot{Q}_{ch}/C_{chw}$$

where

T_{chwr} = temperature of water/glycol returned to the chiller
$\dot{Q}_{ch}$ = chiller load
C_{chw} = capacitance rate (mass flow times specific heat) of the flow stream

Hourly forecasts of cooling loads can be determined with the methods outlined in the section on Forecasting Diurnal Energy Requirements. These hourly forecasts are integrated to forecast the total cooling requirement (either for charging or discharging strategies). To ensure sufficient cooling capacity, a worst-case forecast of cooling requirements is estimated to be the sum of the best forecast and two or three times the standard deviation of the errors of previous forecasts.

Comparison with Optimal Control

Neither the chiller-priority nor storage-priority strategies are optimal in terms of both energy and demand costs. By design, the chiller-priority strategy only gives a demand reduction for the month containing the design day. In addition, chiller-priority control will result in relatively high energy costs when the ratio of on-peak to off-peak rates is significantly greater than one. Although the load-limiting strategy tends to maximize demand reduction for all months, it is also not optimal in terms of energy costs.

The optimal energy cost strategy depends on the ratio of on-peak to off-peak electric rates and the performance of the cooling system in the ice-making and chiller-only modes. The use of ice storage generally results in an energy penalty due to lower chiller efficiencies associated with making ice. However, with respect to energy costs, it is usually more economical to maximize the use of storage (i.e., fully charging and discharging each day) as a result of low off-peak electrical rates. Optimal control involves charging and discharging storage such that the total cost of operating a system over the day is minimized.

Braun (1992) compared the energy costs of (1) chiller-priority, (2) load-limiting, and (3) optimal control using simulations for a single ice storage system. The system analyzed was an unconstrained ice-on-pipe partial storage system similar in configuration to that shown in Figure 22. Figure 23 shows relative comparisons of daily cooling system energy costs for the three methods as a function of the ratio of on-peak to off-peak energy charges for three different types of days. For the design day, there is practically no difference between the costs associated with each control, regardless of the on and off-peak electrical rates. For a properly sized partial storage system operating on the design day, the chillers must operate at maximum capacity throughout both the unoccupied, off-peak period and the occupied period in order to satisfy the building load. The optimal, load-limiting, and chiller-priority control strategies all approach this condition for the design day.

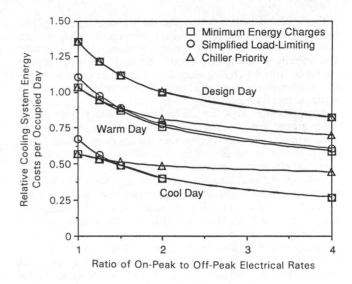

Fig. 23 Comparison of Energy Charges for Three Control Strategies

For off-design days, Figure 23 shows a difference between the daily energy charges for the three control strategies that depends on the electric rates. At low ratios of on-to-off peak electric rates, the chiller-priority strategy gives energy costs that are close to optimal, while the load-limiting strategy results in somewhat greater energy costs. There is an energy penalty associated with making ice as compared with using the chillers directly to meet the load. In the absence of time-of-day electric rates (i.e., rate ratio of 1 in Figure 23), the optimal strategy minimizes the use of storage as is accomplished with the chiller-priority control strategy. For high on-peak rates, the best strategy involves maximizing the use of storage. For ratios greater than about 1.3, Figure 23 shows that the load-limiting control strategy gives energy costs that are nearly optimal. At on to off peak ratios greater than about 2, the chiller-priority control results in a significant energy cost penalty for off-design days.

Braun (1992) also compared the demand costs associated with (1) chiller-priority, (2) load-limiting, and (3) optimal control. Table 3 compares the total building demand charges and the portion of the demand charges associated with the cooling system (in parentheses) for the three control strategies and the three day types. The demand costs were determined for each day type assuming that it would result in the peak power demand for a month during the year. For the design day, the difference between the demand charges associated with each type of control is relatively small. Again, the chillers must operate at maximum capacity throughout both the unoccupied, off-peak period and the occupied period in order to satisfy the building load on the design day. For off-design conditions, there can be a significant demand charge penalty (e.g., 25%) associated with using the chiller-priority control strategy. However, the load-limiting strategy utilizes storage to reduce the demand charges to within about 5% of the minimum demand costs.

Table 3 Example of Monthly Building and Cooling Plant Demand Charges for Three Control Strategies

Peak Day	Chiller-Priority	Load-Limiting	Minimum Demand
Design	$4880 ($2200)	$4832 ($2151)	$4673 ($1993)
Warm	$4676 ($1996)	$4140 ($1459)	$3919 ($1238)
Cool	$4111 ($1430)	$3124 ($444)	$3124 ($444)

Notes: $7.50 per peak kilowatt.
Cooling plant portions of demand charges are in parentheses.

Figure 23 and Table 3 indicate that the load-limiting strategy provides near-optimal control in terms of both energy and demand costs over a range of environmental conditions. In studying these results, it is interesting to note that monthly demand and energy charges associated with the cooling plant are on the same order of magnitude, each between about $500 and $2000 per month, assuming 20 occupied days for Figure 23.

For off-design days, the chiller-priority strategy gave significantly greater demand costs and somewhat greater energy costs in the presence of time-of-day rates than did the optimal control strategy. However, in the absence of time-of-day energy charges, the chiller-priority strategy gave lower energy costs.

Although the load-limiting control strategy worked well for the system considered, further work is necessary to evaluate its performance for other systems. For example, the optimal strategy for charging storage depends on the part-load performance of the cooling system during the charging cycle. In this study, the optimal charging strategy involved operating the chillers at maximum capacity or equivalently for minimum charging time. For systems with good part-load performance (e.g., centrifugal chillers, variable-speed drives), a better charging strategy might involve operation of the chillers at part-load over the entire off-peak, unoccupied period (i.e., maximum charging time). In addition, the optimal amount to charge storage depends on the energy penalty associated with making ice. Compared with the unconstrained system considered in this study, an area-constrained ice-on-pipe storage would have a greater penalty as it approaches full charge.

For discharging during the on-peak period, the load-limiting strategy resulted in daily peak power rates that were close to the minimum. However, this result is sensitive to the noncooling electrical use profile. If the noncooling electric use varies significantly (i.e., is not level) during the on-peak period, then the load-limiting strategy does not perform nearly as well as with true demand-limiting control.

FORECASTING DIURNAL ENERGY REQUIREMENTS

As discussed previously, forecasts of cooling requirements and electrical use in buildings are often necessary for the control of systems that use thermal storage to shift electrical usage from on-peak to off-peak periods. In addition, forecasts can help plant operators anticipate major changes in system operating modes, such as bringing additional chillers on-line.

Overview of Forecasting Methods

Several forecasting methods for thermal storage applications have been presented in the literature. In most of the methods, predictions are estimated as a function of time-varying input variables that affect the cooling requirements and electrical use. Examples of inputs that affect building energy use include (1) ambient dry-bulb temperature, (2) ambient wet-bulb temperature, (3) solar radiation, and (4) building occupancy. Forecasting methods that include time-varying measured input variables are often termed deterministic methods.

It is not easy to measure all inputs that affect cooling requirements and electrical use. For example, building occupancy is difficult to determine, solar radiation measurements are expensive, and wet-bulb temperatures are often unreliable. In addition, the accuracy of forecast models depends on the accuracy of predictions of the inputs. As a result, most of the inputs that affect building energy use are typically not utilized. When ambient dry-bulb temperature is included as an input, National Weather Service forecasts of ambient conditions should be used.

Much of the time-dependent variation in cooling loads and electrical use for a building can be captured with time as a deterministic input. For example, building occupancy follows a regular schedule that depends on the time of day and the time of year. In addition, variations in ambient conditions follow a regular daily and seasonal pattern due to the rotation of the earth about its axis and around the sun. Many forecasting methods use time in place of unmeasured deterministic inputs in a functional form that captures the average time dependence of the variation in energy use.

The use of a deterministic model has a limited accuracy for forecasts due to both unmeasured and unpredictable (random) input variables. Short-term forecasts can be improved significantly by adding previous values of deterministic inputs and previous output measurements (e.g., cooling requirements or electrical use) as inputs to the forecasting model. The time history of these inputs provides valuable information about recent trends in the time variation of the forecasted variable and the unmeasured input variables that affect it. Most forecasting methods use variables that reflect past history to predict the future.

Any forecasting method requires that a functional form is defined and that parameters of the model are learned based on measured data. Either off-line or on-line methods can be used to estimate parameters. Off-line methods estimate parameters from a batch of data that has been collected. The parameters of the process are assumed to be constant over time. Typically, the parameters are determined by minimizing the sum of the squares of the forecast errors. On-line methods allow the parameters of the forecasting model to vary slowly with time. Again, the sum of the squares of the forecast errors are minimized, but this is accomplished in a sequential or recursive manner. Often, a forgetting factor is utilized in order to give additional weight to the recent data. Because of the effect of seasonal variations in weather, the ability to track time-varying systems can be important when forecasting cooling requirements or electrical usage in buildings.

Forrester and Wepfer (1984) presented an algorithm for forecasting cooling loads and electrical use in commercial buildings. The algorithm uses current and previous ambient temperatures and previous loads to predict future requirements. Hourly trends are determined using measured inputs for a few hours preceding the current time. Day-to-day trends are determined using the load that occurred 24 hours earlier as an input. One of the major limitations of this model is its inability to accurately predict loads when an occupied day (e.g., Monday) follows an unoccupied day (e.g., Sunday) or when an unoccupied day follows an occupied day (e.g., Saturday). The cooling load for a particular hour of the day on a Monday cannot be accurately predicted from the requirement 24 hours earlier on Sunday. Forrester and Wepfer (1984) described a number of methods for eliminating this 24 hour indicator. MacArthur et al. (1989) also presented a load profile prediction algorithm that uses a 24 hour regressor.

A very simple algorithm for forecasting either cooling or electrical requirements that does not use the 24 hour regressor was presented by Armstrong et al. (1989) and then further developed and validated by Seem and Braun (1991). The average time-of-day and time-of-week trends are modeled using a lookup table with time of day and type of day (e.g., occupied vs. unoccupied) as the deterministic input variables. Entries in the table are updated using an exponentially-weighted moving-average (EWMA) model. Short-term trends are modeled using previous hourly measurements of cooling requirements in an autoregressive (AR) model. Model parameters adapt to slow changes in the system characteristics. The combination of updating the table and modifying model parameters works well in adapting the forecasting algorithm to changes in season and occupancy schedule. The algorithm is discussed in the section on Simple Forecasting Algorithm.

Kreider and Wang (1991) used artificial neural networks (ANNs) to predict energy consumption of various HVAC equipment in a commercial building. Data inputs to the ANN included (1) the previous hour's electrical power consumption, (2) building occupancy, (3) wind speed, (4) ambient relative humidity, (5) ambient dry-bulb temperature, (6) the previous hour's ambient dry-bulb temperature, (7) two hours' previous ambient temperature, and (8) sine and cosine of the hour number to roughly represent the diurnal change of temperature and insolation. The models were developed primarily to detect changes in equipment and system performance for monitoring purposes. However, the authors suggest that an ANN-based predictor may be valuable for predicting energy consumption based on recent historical data. To apply this method, forecasts of all deterministic input variables are necessary.

Gibson and Kraft (1993) used an ANN to predict building electrical consumption as part of the operation and control of a thermal energy storage (TES) cooling system. The ANN used the following inputs: (1) electric demand of occupants (lighting and other loads), (2) electric demand of TES cooling tower fans, (3) outside ambient temperature, (4) outside ambient temperature minus inside target temperature, (5) outside ambient relative humidity, (6) on-off status for building cooling, (7) cooling system on-off status, (8) chiller #1 direct-cooling mode on-off status, (9) chiller #2 direct-cooling mode on-off status, (10) ice storage discharging mode on-off status, (11) ice storage charging mode on-off status, (12) chiller #1 charging mode on-off status, and (13) chiller #2 charging mode on-off status. In order to use this forecaster, the values of each of these deterministic inputs must be predicted. Although the authors suggest the use of average occupancy demand profile as an input, they do not clearly state how the other input variables should be forecast.

Simple Forecasting Algorithm

The hourly cooling requirements or electrical usage in buildings can be forecast using the following simple algorithm, which is based on the method developed by Seem and Braun (1991). At a given hour, n, the forecast value is

$$\hat{E}(n) = \hat{X}(n) + \hat{D}(h,d) \tag{8}$$

where

$\hat{E}(n)$ = forecast cooling load or electrical usage for hour n
$\hat{X}(n)$ = stochastic part of the forecast for hour n
$\hat{D}(h,d)$ = deterministic part of the forecast at hour n associated with the hth hour of the day and the current day type d

The deterministic part of the forecast is simply a lookup table for the forecasted variable in terms of hour of the day h and the type of day d. Seem and Braun recommend the use of three distinct day types: unoccupied days, occupied days following unoccupied days, and occupied days following occupied days. The three day types account for differences between the building response associated with return from night setback and return from weekend setback. The building operator or control engineer must specify the number of day types and a calendar of day types.

Given the hour of the day and the day type, the deterministic part of the forecast is simply the value stored within that location in the table. Table entries are updated when a new measurement becomes available for that hour and day type. Updates are made using an exponentially-weighted moving-average (EWMA) model as given in the following equation:

$$\hat{D}(h,d) = \hat{D}(h,d)_{old} + \lambda[E(n) - \hat{D}(h,d)_{old}] \tag{9}$$

where

$E(n)$ = the measured value of cooling load or electrical usage for the current hour n
λ = the exponential smoothing constant; $0 < \lambda < 1$
$\hat{D}(h,d)_{old}$ = previous table entry for $\hat{D}(h,d)$

As λ increases, the more recent observations have more effect on the average. As λ approaches zero, the table entry approaches the average of all the data for that hour and day type. When λ equals one, the table entry is updated with the most recent measured value.

Seem and Braun recommend using a value of 0.30 for λ in conjunction with three day types and 0.18 with two day types.

The stochastic portion of the forecast is estimated with a third-order autoregressive model, AR(3), of the forecasting errors associated with the deterministic model. With this model, an estimate of the next hour's error in the deterministic model forecast is given by the following equation:

$$\hat{X}(n+1) = \phi_1 X(n) + \phi_2 X(n-1) + \phi_3 X(n-2) \tag{10}$$

where

$X(n)$ = difference between the measurement and the forecast of the cooling load or electrical use at any hour n
ϕ_1, ϕ_2, ϕ_3 = parameters of the AR(3) model that must be learned

The error in the deterministic forecast at any hour is simply

$$X(n) = E(n) - \hat{D}(h,d) \tag{11}$$

For forecasting more than one hour ahead, conditional expectation is used to estimate the deterministic model forecast errors using the AR(3) model as follows:

$$\hat{X}(n+2) = \phi_1 \hat{X}(n+1) + \phi_2 X(n) + \phi_3 X(n-1) \tag{12}$$

$$\hat{X}(n+3) = \phi_1 \hat{X}(n+2) + \phi_2 \hat{X}(n+1) + \phi_3 X(n) \tag{13}$$

$$\hat{X}(n+k) = \phi_1 \hat{X}(n+k-1) + \phi_2 \hat{X}(n+k-2)$$
$$+ \phi_3 \hat{X}(n+k-3) \text{ for } k > 3 \tag{14}$$

For on-line estimation of the AR(3) model parameters, the following time-dependent cost function is minimized:

$$J(\phi) = \sum_{k=1}^{n} \alpha^{n-k} [X(k) - \hat{X}(k)]^2 \tag{15}$$

where the constant α is called the forgetting factor and has a value between 0 and 1. With this formulation, the residual for the current time step has a weight of one and the residual for k time steps back has a weight of α^k. By choosing a value of α that is positive and less than one, recent data has greater influence on the parameter estimates. In this manner, the model can track changes due to seasonal or other effects. Seem and Braun (1991) recommend using a forgetting factor of 0.99. Parameters of the AR(3) model should be updated each hour a new measurement becomes available.

Ljung and Soderstron (1983) describe on-line estimation methods for determining coefficients of an AR model. The parameter estimates should be evaluated for stability. If an AR model is not stable, the forecasts will grow without bound as the time over which the forecasts are made increases. Ljung and Soderstron discuss methods for checking stability.

Seem and Braun (1991) compared forecasts of electrical usage with both simulated and measured data. Figure 24 shows the standard deviation of the 1- through 24-hour errors in electrical use forecasts for annual simulation results. Results are given for the deterministic model alone, deterministic plus AR (2), and deterministic plus AR(3). For the combined models, the standard deviation of the residuals increases as the forecast length increases. For short time steps (i.e., less than 6 hours), the combined deterministic and stochastic models provide much better forecasts than the purely deterministic model (i.e., lookup table).

Seem and Braun (1991) also investigated a method for adjusting the deterministic forecast based on the use of the maximum daily ambient temperature as an input. For short time periods (i.e., less than 4 hours), the forecasts for the temperature-dependent model were nearly identical to the forecasts for the temperature-independent

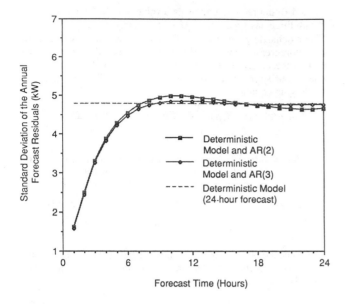

Fig. 24　Standard Deviation of Annual Errors for Forecasts 1 to 24 Hours Ahead

model. For longer time periods, the temperature-dependent model provided better forecasts than the temperature-independent model.

REFERENCES

Andresen, I. and M.J. Brandemuehl. 1992. Heat storage in building thermal mass: A parametric study. *ASHRAE Transactions* 98(1):910-18.

Armstrong, P.R., T.N. Bechtel, C.E. Hancock, S.E. Jarvis, J.E. Seem, and T.E. Vere. 1989. Environment for structured implementation of general and advanced HVAC controls. Phase II—Final Report. Small Business Innovative Research Program, Chapter 7, January. DOE Contract DE-AC02-85ER 80290.

Avery, G. 1986. VAV—Designing and controlling an outside air economizer cycle. *ASHRAE Journal* 28(12):26-30.

Avery, G. 1989. Updating the VAV outside air economizer cycle. *ASHRAE Journal* 31(4):14-16.

Bekker, J.E., P.H. Meckl, and D.C. Hittle. 1991. A tuning method for first-order processes with PI controllers. *ASHRAE Transactions* 97(2):19-23.

Bloomfield, D.P. and D.J. Fisk. 1977. The optimization of intermittent heating. *Buildings and Environment* 12:43-55.

Borresen, B.A. and A. Grindal. 1990. Controllability—Back to basics. *ASHRAE Transactions* 96(2):817-19.

Braun, J.E. 1988. Methodologies for the design and control of central cooling plants. Ph.D. dissertation, University of Wisconsin-Madison.

Braun, J.E. 1990. Reducing energy costs and peak electrical demands through optimal control of building thermal storage. *ASHRAE Transactions* 96(2):876-88.

Braun, J.E. 1992. A comparison of chiller-priority, storage-priority, and optimal control of an ice-storage system. *ASHRAE Transactions* 98(1):893-902.

Braun, J.E. and G.T. Diderrich. 1990. Near-optimal control of cooling towers for chilled-water systems. *ASHRAE Transactions* 96(2):806-13.

Braun, J.E., J.W. Mitchell, S.A. Klein, and W.A. Beckman. 1987. Performance and control characteristics of a large central cooling system. *ASHRAE Transactions* 93(1):1830-52.

Braun, J.E., S.A. Klein, J.W. Mitchell, and W.A. Beckman. 1989a. Applications of optimal control to chilled water systems without storage. *ASHRAE Transactions* 95(1):663-75.

Braun, J.E., S.A. Klein, J.W. Mitchell, and W.A. Beckman. 1989b. Methodologies for optimal control of chilled water systems without storage. *ASHRAE Transactions* 95(1):652-62.

Brothers, P.W. and M.L. Warren. 1986. Fan energy use in variable air volume systems. *ASHRAE Transactions* 92(2B):19-29.

Bushby, T.B. and G.E. Kelly. 1988. Comparison of digital control and pneumatic control systems in a large office building. NBSIR 88-3739. National Bureau of Standards.

Conniff, J.P. 1991. Strategies for reducing peak air-conditioning loads by using heat storage in the building structure. *ASHRAE Transactions* 97(1): 704-709.

Cumali, Z. 1988. Global optimization of HVAC system operations in real time. *ASHRAE Transactions* 94(1):1729-44.

Cumali, Z. 1988. Optimization of HVAC operation. In *Proceedings of the International Building Program Simulation Association (IBPSA)*, Vancouver, Canada.

Cumali, Z. 1993. Application of real-time optimization to building systems. Presented at the ASHRAE Summer Meeting, Denver.

Curtiss, P.S., J.F. Kreider, and M.J. Brandemuehl. 1993. Adaptive control of HVAC processes using predictive neural networks. *ASHRAE Transactions* 99(1):496-504.

Delp, W.W., R.H. Howell, H.J. Sauer, and B. Subbarao. 1993. Control of outside air and building pressurization in VAV systems. *ASHRAE Transactions* 99(1):565-89.

Dexter, A.L. and P. Haves. 1989. A robust self-tuning predictive controller for HVAC applications. *ASHRAE Transactions* 95(2):431-38.

Dickson, D.K. 1987a. Control dampers. *ASHRAE Journal* 29(6):40-42.

Dickson, D.K. 1987b. Return fans: Are they necessary? *ASHRAE Journal* 29(9):30-33.

Englander, S.L. and L.K. Norford. 1992. Saving fan energy in VAV systems—Part 2: Supply fan control for static pressure minimization using DDC zone feedback. *ASHRAE Transactions* 98(1):19-32.

Forrester, J.R. and W.J. Wepfer. 1984. Formulation of a load prediction algorithm for a large commercial building. *ASHRAE Transactions* 90(2B): 536-51.

Gibson, G.L. and T.T. Kraft. 1993. Electric demand prediction using artificial neural network technology. *ASHRAE Journal* 35(3):60-68.

Hackner, R.J., J.W. Mitchell, and W.A. Beckman. 1984. HVAC system dynamics and energy use in buildings—Part I. *ASHRAE Transactions* 90(2B):523-35.

Hackner, R.J., J.W. Mitchell, and W.A. Beckman. 1985. HVAC system dynamics and energy use in buildings—Part II. *ASHRAE Transactions* 91(1B):781-95.

Hartman, T. 1993. Terminal regulated air volume (TRAV) systems. *ASHRAE Transactions* 99(1):791-800.

Hittle, D.C. 1979. The building loads analysis and systems thermodynamics (BLAST) program, Version 2.0. U.S. Army Construction Engineering Research Laboratory, Users Manual, Vol. 1. Available from NTIS, Springfield, VA 22151.

Ho, W.F. 1993. Development and evaluation of a software package for self-tuning of three-term DDC controllers. *ASHRAE Transactions* 99(1):529-34.

House, J.M., T.F. Smith, and J.S. Arora. 1991. Optimal control of a thermal system. *ASHRAE Transactions* 97(2):991-1001.

Huang, S. and R.M. Nelson. 1991. A PID-law-combining fuzzy controller for HVAC applications. *ASHRAE Transactions* 97(2):768-74.

Huang, S. and R.M. Nelson. 1994. Rule development and adjustment strategies of a fuzzy logic controller for an HVAC system: Part 1—Analysis. *ASHRAE Transactions* 100(1):841-50.

Kao, J.Y. 1985a. Control strategies and building energy consumption. *ASHRAE Transactions* 91(2B):810-17.

Kreider, J.F. and X.A. Wang. 1991. Artificial neural networks demonstration for automated generation of energy use predictors for commercial buildings. *ASHRAE Transactions* 97(2):775-79.

Lau, A.S., W.A. Beckman, and J.W. Mitchell. 1985. Development of computerized strategies for a large chilled water plant. *ASHRAE Transactions* 91(1B):766-80.

Ljung, L. and T. Soderstron. 1983. *Theory and practice of recursive identification*. MIT Press, Cambridge, MA.

MacArthur, J.W., A. Mathur, and J. Zhao. 1989. On-line recursive estimation for load profile prediction. *ASHRAE Transactions* 95(1):621-28.

Morris, F.B., J.E. Braun, and S.J. Treado. 1994. Experimental and simulated performance of optimal control of building thermal storage. *ASHRAE Transactions* 100(1):402-14.

Nesler, C.G. 1986. Automated controller tuning for HVAC applications. *ASHRAE Transactions* 92(2B):189-201.

Nesler, C.G. and W.F. Stoecker. 1984. Selecting the proportional and integral constants in the direct digital control of discharge air temperature. *ASHRAE Transactions* 90(2B):834-44.

Nugent, D.R., S.A. Klein, and W.A. Beckman. 1988. Investigation of control alternatives for a steam turbine driven chiller. *ASHRAE Transactions* 94(1):627-43.

Nuorkivi, A. 1990. Real-time optimization system of a district heat network operation. *ASHRAE Transactions* 96(1):946-48.

Rawlings, L. 1985. Strategies to optimize ice storage. *ASHRAE Journal* 27(5):39-44.

Ruud, M.D., J.W. Mitchell, and S.A. Klein. 1990. Use of building thermal mass to offset cooling loads. *ASHRAE Transactions* 96(2):820-29.

Seem, J.E. and J.E. Braun. 1991. Adaptive methods for real-time forecasting of building electrical demand. *ASHRAE Transactions* 97(1):710-21.

Seem, J.E., P.R. Armstrong, and C.E. Hancock. 1989. Comparison of seven methods for forecasting the time to return from night setback. *ASHRAE Transactions* 95(2):439-46.

Shavit, G. and S.G. Brandt. 1982. The dynamic performance of a discharge air-temperature system with a P-I controller. *ASHRAE Transactions* 88(2):826-38.

Snyder, M.E. and T.A. Newell. 1990. Cooling cost minimization using building mass for thermal storage. *ASHRAE Transactions* 96(2):830-38.

Wallenborg, A.O. 1991. A new self-tuning controller for HVAC systems. *ASHRAE Transactions* 97(1):19-25.

Warren, M. and L.K. Norford. 1993. Integrating VAV zone requirements with supply fan operation. *ASHRAE Journal* 35(4):43-46.

Ziegler, J.G. and N.B. Nichols. 1942. Optimum settings for automatic controllers. *Trans. ASME* 64:759-68.

BIBLIOGRAPHY

ASHRAE. 1985. Thermal storage. *ASHRAE Technical Data Bulletin* (January).

Hartman, P.E. 1988. Dynamic control: A new approach. *Heating/Piping/Air Conditioning* (April):97-100.

Johnson, G.A. 1985. Optimization techniques for a centrifugal chiller plant using a programmable controller. *ASHRAE Transactions* 91(2B):835-47.

Kao, J.Y. 1985b. Sensor errors in VAV systems. *ASHRAE Journal* 27(1):100-104.

Ljung, L. 1987. *System identification theory for the user.* Prentice-Hall, Englewood Cliffs, NJ.

Marcev, C.L. 1980. Steady-state modeling and simulation as applied to energy optimization of a large office building integrated HVAC system. MS thesis, University of Tennessee.

Marcev, C.L., C.R. Smith, and D.D. Bruns. 1984. Supervisory control and system optimization of chiller-cooling tower combinations via simulation using primary equipment component models—Part II, Simulation and optimization. In *Proceedings of the Workshop on HVAC Controls Modeling and Simulation* (February). Georgia Institute of Technology.

Marseille, T.J. and J.S. Schliesing. 1991. Integration of water loop heat pumps and building structural thermal energy storage, PNL-7850. Pacific Northwest Laboratories, Richland, WA.

Miller, D.E. 1980. The impact of HVAC process dynamics on energy use. *ASHRAE Transactions* 86(2):535-53.

Nizet, J.L., J. Lecomte, and F.X. Litt. 1984. Optimal control applied to air conditioning in buildings. *ASHRAE Transactions* 90(1B):587-600.

Seem, J.E. 1987. Modeling of heat transfer in buildings. Ph.D. dissertation, University of Wisconsin.

Shapiro, M.M., A.J. Yager, and T.H. Ngan. 1988. Test hut validation of a microcomputer predictive HVAC control. *ASHRAE Transactions* 94(1):644-63.

Sud, I. 1984. Control strategies for minimum energy usage. *ASHRAE Transactions* 90(2A):247-77.

Treichler, W.W. 1985. Variable speed pumps for water chillers, water coils, and other heat transfer equipment. *ASHRAE Transactions* 91(1B):202-13.

CHAPTER 39

BUILDING COMMISSIONING

THIS chapter addresses the generic commissioning process, which is a function of management. Other *ASHRAE Handbook* chapters set forth methods of commissioning specific system components (e.g., boilers, chillers, pumps, fans, etc.).

Basics of Commissioning

The commissioning of an *installation* is the process for achieving, verifying, and documenting the performance of *buildings* to meet the operational needs of the building within the capabilities of the design and to meet the design documentation and the owner's functional criteria, including preparation of operator personnel. This definition refers to the building as a total system, including structural systems, building envelope, life safety, security systems, elevators, escalators, plumbing systems, electrical systems, and HVAC systems. This implies the need to consider the interface requirements of the commissioning plans of the individual building systems.

The commissioning of a *building system* is the process for achieving, verifying, and documenting the performance of that *system* to meet the operational needs of the building within the capabilities of the design and to meet the design documentation and the owner's functional criteria, including preparation of operator personnel. The result should be a fully functional, fine-tuned system that can be recommissioned throughout the useful life of the building. Individual systems can be selected by reference to the standard Construction Specifications Institute (CSI) specification format and the U.S. DOE General Design Criteria (GPO #643.1A4698), which serve as a guide to the nature and scope of building systems requiring commissioning.

ASHRAE *Guideline* 1 provides procedures for documenting and verifying the performance of HVAC systems so that the systems operate in conformity with the design intent. *Guideline* 1 includes procedures for

- Documentation of occupancy requirements and HVAC design assumptions
- Documentation of the design intent for use by contractors, owners, and operators
- Functional performance testing and documentation necessary for evaluating the HVAC systems for acceptance and for adjusting the capability of the design intent

The *Guideline* is intended for use by all members of the design, construction, and operation team: owner, designer, contractor, supplier, operator, and others working on the given project.

Testing, Adjusting, and Balancing. The commissioning of building systems is not simply the testing, adjusting, and balancing (TAB) process as presently applied in the construction industry. Although testing, adjusting, and balancing is an important part of the construction and operation phases of the commissioning process, the commissioning process extends through all phases of the project, from concept through occupancy, and requires participation of all parties involved (i.e., the commissioning team).

Advantages of Commissioning

The advantages of the commissioning process vary from one building project to another. It is important to consider and evaluate

The preparation of this chapter is assigned to TC 9.9, Building Commissioning.

the advantages applicable to each project for two reasons. First, an evaluation helps in setting a commissioning scope that is consistent with the complexity of the site and the criticality (e.g., with respect to life safety and/or environmental impact) of the project. Second, it helps in quantifying both the costs and benefits of commissioning. Advantages of commissioning often include the following:

- Presets channels of communication
- Predelegates responsibility
- Representation on the commissioning team
- Improves planning for verification and acceptance
- Formats inspection/testing procedures
- Coordinates responsibilities
- Establishes continuous monitoring of priorities and schedules
- Includes operation and maintenance program
- Prepares and trains operator personnel
- Helps keep fully functional building systems on schedule
- Documents indoor air quality (IAQ) and comfort control
- Helps ensure on-schedule occupancy
- Includes flexibility to change with occupancy requirements

Commissioning Team

The size and makeup of the commissioning team depends both on the size and complexity of the project and on the commissioning parameters specified in the contract documents. The responsibility of each member of the commissioning team is documented during the predesign phase and included in the contract documents.

An ideal commissioning team would include all participants in the construction project: the commissioning authority, owner, design professionals, construction manager, general contractor, subcontractors, operation and maintenance manager, suppliers, and equipment manufacturers. It is important that the operation and maintenance manager be brought into the commissioning process early, preferably during the predesign phase. Knowledge of occupancy, special lighting requirements, anticipated equipment loads, and other factors may influence the design. Of equal importance is the early participation of the TAB contractor, whose experience in testing, adjusting, and balancing systems should provide recommendations about space requirements, probe and control locations, and access.

Commissioning Authority

The role of the commissioning authority is often exaggerated and confused with that of the construction manager. The appointment of a commissioning authority does not alter other professional or contractual obligations. The owner may select the commissioning authority from among the owner's staff, the operation and maintenance manager, the construction manager, the commissioning specialist, the design professional, the general contractor, and the TAB contractor.

The commissioning authority is the qualified person, company, or agency that implements the overall commissioning process in cooperation with the commissioning team as specified in the contract documents. as specified by the design professional. The commission authority may or may not participate in the design of the commissioning plan, which is primarily the responsibility of the design professional. Responsibilities of the commission authority include, but are not limited to, the following:

1. Participating in all project meetings, predesign through final acceptance

2. Participating in the selection and qualification of commissioning team members
3. Scheduling and conducting all meetings required by the commissioning plan
4. Recording and distributing minutes and other pertinent information
5. Providing or approving the format for documentation necessary to the commissioning team for evaluating inspections, systems testing, and functional performance testing
6. Providing the communication format; directing and recording the transfer of information between all commissioning plan participants
7. Coordinating inspection and testing of all interacting building systems
8. Scheduling and witnessing all training of operation and maintenance personnel required by the contract documents
9. Recording and reporting compliance and noncompliance with the requirements of the contract documents

Cost Factors

No reliable data are available to determine the cost or cost-contributing factors of building commissioning. The number of cost-contributing factors depends on the size and complexity of the project and the willingness of the owners to invest in the appropriate commissioning process.

A wider range of professional services, comprehensive documentation, system and subsystems testing, and operator training are among the cost-contributing factors that may be offset or discounted by the advantages they bring. One intent of commissioning is to fully recover the capital investment over the life of the system through management, efficiency, and user satisfaction.

Phases of Commissioning

Predesign Phase. During the predesign phase, the design professional (1) sets forth building commissioning parameters, responsibilities, and documentation for all phases of construction; (2) ensures that project team members are aware of these requirements; and (3) sets a framework for commissioning during all phases. The predesign phase documents the basis used to develop the design and benchmark information that is used to evaluate final performance including: occupancy requirements, design assumptions, building construction, building loads and zoning, building use, cost factors, and design compromises.

Design Phase. During the design phase, the commissioning requirements for all building systems to have a comprehensive commissioning process are developed. Requirements include

- Design criteria and assumptions
- Description of each individual building system, including the intended operation and performance of each system
- Commissioning plan
- Documentation requirements
- Verification requirements
- Maintenance requirements

Construction Phase. Prior to the construction and/or installation of each building system or subsystem, shop drawings and operation and maintenance manuals are submitted, reviewed, and documented. During the construction phase, the building systems are installed in conformance with the contract documents and the commissioning plan. Upon completion of the construction of each system, pre-start-up inspections are performed. The TAB work commences upon acceptance of the systems for start-up. After the TAB work is completed, functional performance testing of equipment, subsystems, interacting systems, and components commences.

Acceptance procedures include (1) equipment and subsystem functional performance tests, (2) verification and documentation, (3) deferred functional performance tests, (4) corrective measures, (5) intersystem functional performance tests, (6) acceptance documentation, (7) operator training, and (8) final acceptance.

During the functional performance testing procedures, the results are documented, system performance is evaluated, and modifications to the systems are made. Functional performance tests are completed to satisfy the design intent. As-built drawings, as-installed specifications, operation and maintenance manuals, and all documentation are turned over to the owner's staff, which is trained in the operations and maintenance of the building systems.

Training Operating Personnel. The length of time required for the instruction of operating personnel is dictated by the complexity of the building systems. Training schedules, prepared during the design phase, are implemented by the commissioning authority. Operator participation in, or witnessing of, initial equipment testing and adjusting procedures is an important part of the training strategy. Instruction is provided by several different sources:

1. The design professional instructs the operators on the proper operation and limitations of the system during full- and part-load modes of operation; IAQ control; energy conservation; and emergency procedures. Review and use of operations and maintenance manuals is part of this training.
2. Equipment manufacturers provide instruction on the maintenance of installed equipment, the recommended parts inventory, supply sources, and the availability of parts and service.
3. The automatic temperature control contractor provides training in the proper operation, adjustment, and maintenance of the automatic temperature control system and the energy management control system.
4. The TAB contractor provides training in the testing, adjusting, and balancing procedures necessary to maintain and verify system and subsystem performance.
5. Other participants in the operating personnel training program may include specialty contractors such as fire and smoke control, lighting, and building security systems professionals, as required by the contract documents.

Post-Acceptance Phase. Post-acceptance commissioning is a critical step for the effective, ongoing functioning of a building's systems. It includes operation and maintenance, retesting of systems and subsystems, alterations and changes, and recommissioning.

During the operation of a facility, the owner's staff (1) regularly maintains and services the building systems and equipment in accordance with procedures in the maintenance manuals and (2) keeps accurate records of work performed. A standard method of recording complaints regarding system operation should be developed and maintained. Consistent complaints may indicate the necessity of recommissioning the system or reviewing the commissioning plan. Subsystems and systems should be retested periodically to measure actual performance. The system functional performance test checklist used in the construction phase should be a guide for retesting. Discrepancies between predicted performance and actual performance should be analyzed.

As the occupancy and use requirements of a facility changes, building systems need to be adapted. It is appropriate to maintain a history of the facility, recording changes and revising as-built drawings and other documents to reflect modifications made to any part of the facility or building systems. Any change in usage, installed equipment, loads, or occupancy must be carefully monitored and documented. Evaluation of the impact on operations of planned alterations to the building systems must be provided; as-built documentation, including commissioning reports, must be updated to reflect all alterations.

CHAPTER 40

THERMAL STORAGE

THERMAL storage systems remove heat from or add heat to a storage medium for use at another time. Thermal storage for HVAC applications can involve storage at various temperatures associated with heating or cooling processes. High-temperature storage is typically associated with solar energy or high-temperature heating processes, and cool storage with air-conditioning, refrigeration, or cryogenic-temperature processes. Energy may be charged, stored, and discharged daily, weekly, annually, or in seasonal or rapid batch process cycles. The *Design Guide for Cool Thermal Storage* (ASHRAE 1992) covers cool storage issues and design parameters in more detail.

Thermal storage may be an economically attractive approach to meeting heating or cooling loads if one or more of the following conditions apply:

- loads are of short duration
- loads occur infrequently
- loads are cyclical in nature
- loads are not well matched to the availability of the energy source
- energy costs are time-dependent (e.g., time-of-use energy rates or demand charges for peak energy consumption)
- utility rebates, tax credits, or other economic incentives are provided for the use of load-shifting equipment
- energy supply from the utility is limited, thus preventing the use of full-size nonstorage systems

Definitions

Heat storage. As used in this chapter, the storage of thermal energy at temperatures above the nominal temperature required by the space or process.

Cool storage. As used in this chapter, the storage of thermal energy at temperatures below the nominal temperature required by the space or process.

Sensible energy storage (sensible heat storage). Heat storage or cool storage systems in which all of the energy stored is in the form of sensible heat associated with a temperature change in the storage medium.

Latent energy storage (latent heat storage). Heat storage or cool storage systems in which the energy stored is predominantly in the form of the latent heat (usually of fusion) associated with a phase change (usually between solid and liquid states) in the storage medium.

Off-peak air conditioning. An air-conditioning system that uses cool storage.

Off-peak heating. A heating system that uses heat storage.

The preparation of this chapter is assigned to TC 6.9, Thermal Storage.

Storage Media

A wide range of materials can be used as the storage medium. Desirable characteristics include the following:

- commonly available
- low cost
- environmentally benign
- nonflammable
- nonexplosive
- nontoxic
- compatible with common HVAC materials
- noncorrosive
- inert
- well-documented physical properties
- high density
- high specific heat (for sensible heat storage)
- high heat of fusion (for latent heat storage)
- high heat transfer characteristics
- storage at ambient pressure
- characteristics unchanged over long use

Common storage media for sensible heat storage include water, soil, rock, brick, ceramics, concrete, and various portions of the building structure (or process fluid) being heated or cooled. In HVAC applications such as air conditioning, space heating, and water heating, water is often the chosen thermal storage medium; it provides virtually all of the desirable characteristics when kept between its freezing and boiling points. In lower temperature applications, aqueous secondary coolants (typically glycol solutions) are often used as the heat transfer medium, enabling certain storage media to be used below their freezing or phase-change points. For high-temperature heat storage, the storage medium is often rock, brick, or ceramic materials for residential or small commercial applications and oil, oil-rock combinations, or molten salt for large industrial or solar energy power plant applications. Use of the building structure itself as passive thermal storage offers advantages under some circumstances (Treado et al. 1994).

Common storage media for latent heat storage include water-ice, aqueous brine-ice solutions, and other phase-change materials (PCMs) such as hydrated salts and polymers. Clathrates, carbon dioxide, and paraffin waxes are among the alternative storage media used in latent storage systems at various temperatures. For air-conditioning applications, water-ice is the most common storage medium; it provides virtually all of the previously listed desirable characteristics.

A challenge common to all latent heat storage systems is to find an efficient and economical means of achieving the heat transfer necessary to alternately freeze and thaw the storage medium. Various methods have been developed to limit or deal with the heat transfer approach temperatures associated with freezing and melting; however, leaving fluid temperatures (from storage during melting) must be higher than the freezing point, while entering fluid temperatures (to storage during freezing) must be lower than the freezing point. Ice storage can provide leaving temperatures well below those normally used for comfort and nonstorage air-conditioning applications. However, entering temperatures are also much lower than normal. Certain PCM storage systems can be charged using temperatures near those for comfort cooling, but they produce warmer leaving temperatures.

Benefits of Thermal Storage

The primary reasons for the use of thermal storage are economic. The following are some of the key benefits of storage systems:

Reduced Equipment Size. If thermal storage is used to meet all or a portion of peak heating or cooling loads, equipment sized to meet peak loads can be downsized to meet an average load.

Capital Cost Savings. Capital savings can result both from equipment downsizing and from certain utility cash incentive programs. Even in the absence of utility cash incentives, the savings from downsizing cooling equipment can offset the cost of the storage system. Storage systems integrated with low-temperature air and water distribution systems can also provide an initial cost savings due to the use of smaller chillers, pumps, piping, ducts, and fans. Storage has the potential to provide capital savings for systems having heating or cooling peak loads of extremely short duration.

Energy Cost Savings. The significant reduction of time-dependent energy costs such as electric demand charges and on-peak time-of-use energy charges is a major economic incentive for the use of thermal storage.

Energy Savings. Although thermal storage is generally designed primarily to *shift* energy use rather than to *conserve* energy, storage often reduces energy consumption. Cool storage systems permit chillers to operate more at night when lower condensing temperatures improve equipment efficiency; and storage permits the operation of equipment at full-load conditions, avoiding inefficient part-load performance. Documented examples include chilled water storage installations that reduce annual energy consumption for air conditioning by up to 12% (Fiorino 1994, ITSAC 1992).

Improved System Operation. The use of storage adds an element of thermal capacitance to a heating or cooling system, allowing the decoupling of the thermal load profile from the operation of the equipment. This decoupling can be used to provide increased flexibility, reliability, or backup capacity for the control and operation of the system.

Other Benefits. Storage can bring about other beneficial system synergies. As already noted, cool storage can be integrated with cold air distribution systems. Thermal storage can be configured to serve a secondary function as fire protection, as is often the case with chilled water storage installations. Some cool storage systems can be configured for recharge via free cooling (Holness 1992, Hussain and Peters 1992, Meckler 1992).

ECONOMICS

Thermal storage is installed for two major reasons: (1) to lower initial costs and (2) to lower operating costs.

Lower Initial Costs. Applications such as in churches and sports facilities, where the load is short in duration and there is a long time between load occurrences, generally have the lowest initial costs for thermal storage. These buildings have a relatively large space-conditioning load for fewer than 6 h per day and only a few days per week. The relatively small refrigeration plant for these applications would operate continuously for up to 100 h or more to recharge the thermal storage. The initial cost of such a design could

be less than that of a standard cooling plant designed to meet the highest instantaneous cooling load.

Secondary capital costs may also be lower for thermal storage. For example, the electrical supply (kVa) can sometimes be reduced because peak energy demand is lower. For cold air distribution systems, the volume of air is reduced, and smaller air-handling equipment and ducts may generally be used. The smaller air-handling unit sizes may reduce the mechanical room space required, providing more usable space. The smaller duct sizes may allow the floor-to-floor height of the building to be reduced; in some cases, an additional floor may be added without an increase in building height.

An example of reduction in equipment size is shown in Figure 1. The load profiles shown are for a 9000 m commercial building. If the 6120 kWh (22 GJ) load is met by a nonstorage air-conditioning system, as shown in Figure 1A, a 660 kW chiller is required to meet the peak cooling demand.

If a load-leveling partial storage system is used, as shown in Figure 1B, an 255 kW chiller meets the demand. The design-day cooling load in excess of the chiller output (3060 kWh) is supplied by the storage. The cost of the storage system approximates the amount saved by downsizing the chiller, cooling tower, electrical service, etc., so load-leveling partial storage systems are often competitive with nonstorage systems on an initial cost basis.

If a full storage system is installed, as shown in Figure 1C, the entire peak load is shifted to the storage, and a 360 kW chiller is required. The size of the chiller equipment may be reduced, but the total equipment cost including the storage is usually higher for the full storage system than for nonstorage systems. Although the initial cost is higher than for the load-leveling system (Figure 1B), a full storage system offers substantially reduced operating costs because the entire chiller demand is shifted to the off-peak period.

In order to prepare a complete cost analysis, the initial costs must be determined. Equipment costs should be obtained from each of the manufacturers under consideration, and an estimate of installation cost should be made.

Lower Operating Costs. The other reason for installing a thermal storage system is to reduce operating costs. Most electric utilities charge less during the night or weekend off-peak hours than during the times of highest electrical demand, which often occur on hot summer afternoons due to air-conditioning use. Electric rates are normally divided into demand charges and consumption charges.

Monthly demand charges are based on the building's highest recorded demand for electricity during the month and are measured over a brief period (usually 1 to 15 min). Ratchet billing, another form of demand charge, is based on the highest annual monthly demand. This charge is assessed each month, even if the maximum demand in the particular month is less than the annual peak.

Consumption charges are based on the total measured use of electricity in kilowatt-hours (kWh) over a longer period and are generally representative of the utility's cost of fuel to operate its generation facilities. In some cases, consumption charges are lower during off-peak hours because a higher proportion of the electricity is generated by baseload plants that are less expensive to operate. Rates that reflect this difference are known as time-of-use billing structures.

To compare the costs of different systems, the annual operating costs of each system being considered must be estimated, including both electrical demand and consumption costs. To determine demand costs, the monthly peak demand for each system is multiplied by the demand charge and totaled for the year; any necessary adjustments for ratchet billing must be made. Electrical consumption costs are determined by totaling the annual energy use for each system in kilowatt-hours and multiplying it by the cost per kilowatt-hour. For time-of-use billing, energy use must be classified by time-of-use period and multiplied by the corresponding rate.

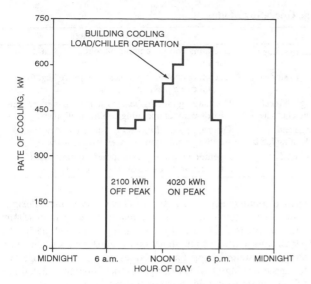

A. CONVENTIONAL COOLING

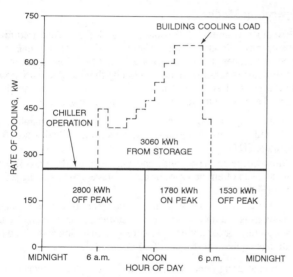

B. LOAD-LEVELING PARTIAL STORAGE

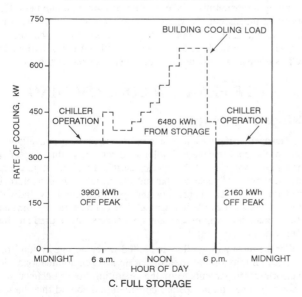

C. FULL STORAGE

Fig. 1   Hourly Cooling Load Profiles

The first step is to estimate the annual utility cost for a nonstorage system. For example, the peak demand in Figure 1A, which occurs at 3 p.m. (660 kW), is multiplied by the demand charge and, if necessary, adjusted for ratchet billing. The energy used during the off-peak period for the nonstorage system (corresponding to 2100 kWh of cooling) would be multiplied by the off-peak electrical rate. The energy used during the on-peak period (4020 kWh) is multiplied by the on-peak electrical rate. This procedure is repeated (1) for each day of each month to determine the energy charges and (2) for the peak day in each month to determine the demand charges.

Annual costs for a load-leveling partial thermal storage system are calculated the same way. The peak demand in Figure 1B (255 kW) is multiplied by the demand charge and adjusted for any ratchet billing. This process is repeated for each month. The demand charges for the load-leveling system are lower than those for a nonstorage system; the difference represents the demand savings. If time-of-use billing is used, the energy used during the off-peak period (2800 + 1530 = 4330 kWh) is multiplied by the off-peak electrical rate, and the energy consumed during the on-peak period (1780 kWh) is multiplied by the on-peak electrical rate. The total energy costs are lower because less energy is used during the on-peak hours.

The full storage system in Figure 1C eliminates all energy use during the on-peak hours, except energy used by the auxiliary equipment that transfers stored energy into the system. The energy costs for the off-peak periods (3960 + 2160 = 6120 kWh) are multiplied by the off-peak rate. This procedure should be repeated for each month to determine the annual energy costs.

The operating costs and the net capital costs should be analyzed using the life-cycle cost method or other suitable method to determine which system is best for the project. Other items to be considered are space requirements, reliability, and ease of interface to the planned delivery system. An optimal application of thermal energy storage balances the savings in utility charges against the initial cost of the installation needed to achieve the savings.

APPLICATIONS

Thermal storage can take many forms to suit a variety of applications. This section addresses several groups of thermal storage applications: off-peak air conditioning, industrial/process cooling, off-peak heating, and other applications.

General Applicability and Control Strategies

Although applications and technologies vary significantly, certain characteristics and design options are common to all thermal storage systems. Whether for heat storage or cool storage, and whether for storing sensible or latent heat, storage designs follow one of two control strategies: full storage or partial storage.

Full storage (load-shifting) designs are those that use storage to fully decouple the operation of the heating or cooling generating equipment from the peak heating or cooling load. The peak heating or cooling load is met through the use (i.e., discharging) of storage while the heating or cooling generating equipment is idle. Full storage systems are likely to be economically advantageous only under one or more of the following conditions:

- spikes in the peak load curve are of short duration
- time-of-use energy rates are based on short-duration on-peak periods
- there are short overlaps between peak loads and peak energy periods
- high cash incentives are offered for using thermal storage
- high peak demand charges apply

For example, a school or business in which electrical demand drops dramatically after 5 p.m. in an electric utility territory where on-peak energy and demand charges apply between 1 p.m. and 9 p.m. can economically apply a full cool storage system. The brief 4-h

Table 1 Thermal Storage Operating Modes

Operating Mode	For Heat Storage	For Cool Storage
Charging storage	Operating heating equipment to add heat to storage	Operating cooling equipment to remove heat from storage
Charging storage while meeting loads	Operating heating equipment to add heat to storage *and* meet heating loads	Operating cooling equipment to remove heat from storage *and* meet cooling loads
Discharging storage (full load shift)	Discharging (removing heat from) storage to meet heating loads without operating heating equipment	Discharging (adding heat to) storage to meet cooling loads without operating cooling equipment
Discharging storage (partial load shift)	Discharging (removing heat from) storage *and* operating heating equipment to meet heating loads	Discharging (adding heat to) storage *and* operating cooling equipment to meet cooling loads
Meeting loads (real-time)	Operating nonstorage equipment to meet loads (no net heat flow to or from storage)	Operating nonstorage equipment to meet loads (no net heat flow to or from storage)

overlap between 5 and 9 p.m. allows the full load to be shifted with a relatively small and cost-effective storage system and without oversizing the chiller equipment.

Partial storage more often provides the best economics, and, therefore, represents the majority of thermal storage installations. Although they do not shift as much load (on a design day) as full storage systems, partial storage systems can have lower initial costs, particularly if the design incorporates smaller equipment by using low-temperature water (greater Δt) and cold air distribution systems. For many applications, a form of partial storage known as load leveling can be used with minimum capital cost. A load-leveling system is designed with the heating or cooling equipment sized to

operate continuously at or near its full capacity to meet design-day loads; thus, equipment of the minimum capacity (and minimum cost) can be used. During operation at less than peak design loads, partial storage designs may function as full storage systems. For example, a system designed as a load-leveling partial storage system for space heating at winter design temperatures may function as a full shift system on mild spring or autumn days.

Operating Modes

Another feature of all thermal storage designs is the need for various operating modes. Table 1 shows the five basic thermal storage operating modes. Figure 2 provides simplified flow schematics for the five basic operating modes, including three distinct variations for the partial load shift discharge mode. Partial storage can be discharged according to any of the following arrangements:

- with storage and process equipment operating in parallel, as is often the case for cool storage systems using chilled water storage (Figure 2D1)
- with storage in series with and upstream of the process equipment, as is often the case for cool storage systems using ice storage (Figure 2D2)
- with storage in series with and downstream of the process equipment, as is often the case for cool storage systems using hydrated salt storage or for low-temperature systems using some kind of ice storage equipment (Figure 2D3)

Optimal design choices are related to the storage technology, to the discharge temperatures needed to serve the loads, and to the operating temperatures of the heating or cooling equipment. Other operating modes not shown here are of course possible and sometimes desirable. For example, recharge of cool storage via free cooling, which is particularly adaptable for chilled water or hydrated salt PCM systems, can enhance the economy of either technology.

OFF-PEAK AIR CONDITIONING

Refrigeration Design

Packaged compression or absorption equipment is available for most of the cool storage technologies discussed in this chapter. The maximum cooling load that can be satisfied with a single package and the number of packages that must be installed in parallel to yield the required capacity depend on the specific technology. Individual components such as ice builders, chiller bundles, falling-film chillers, ice makers, compressors, and condensers can be used and interconnected on-site.

For the most part, chillers in chilled water storage systems operate at conditions similar to those for nonstorage applications. However, a greater percentage of the operating hours occur at lower ambient temperatures; special consideration should thus be given to providing condensing temperatures that maintain compressor differential.

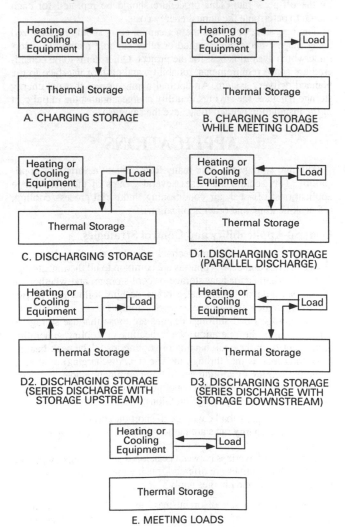

Fig. 2 Thermal Storage Operating Modes

The lower suction temperatures necessary for making ice impose a higher compression ratio on the refrigeration equipment. Positive displacement compressors (e.g., reciprocating, screw, and scroll compressors) are usually better suited to these higher compression ratios than centrifugal compressors.

Ice storage systems must operate over a wider range of conditions than nonstorage systems. Care should be taken in the design and installation of these systems to ensure proper operation at all operating points.

A detailed discussion of refrigeration equipment is provided in Chapter 2 of the 1994 *ASHRAE Handbook—Refrigeration*. The following is a general categorization of the various types of refrigeration equipment used for ice storage systems:

Direct-expansion systems feed refrigerant to the heat exchanger through expansion valves. The wide range of operating loads requires careful selection and control of expansion valves.

Flooded or overfeed systems circulate liquid refrigerant from a large low-pressure receiver to the ice-making heat exchangers. Chapter 1 of the 1994 *ASHRAE Handbook—Refrigeration* covers liquid overfeed systems in detail.

Secondary coolant systems use chillers to chill and circulate low-temperature secondary coolant to the ice-producing and storage equipment. The corrosion properties of the coolant should be considered; a solution containing corrosion inhibitors is normally added. It is also important that the chiller be properly derated both for the lower heat transfer properties of the secondary coolant and for the lower than normal operating temperatures. Secondary coolant systems may use direct-expansion, flooded, or liquid overfeed refrigeration systems.

The following steps should be considered for all ice thermal storage refrigeration systems:

Design for part-load operation. Refrigerant flow rates, pressure drops, and velocities are reduced during part-load operation. Components and piping must be designed so that, at all load conditions, control of the system can be maintained and oil can be returned to the compressors.

Design for pull down load. Because ice-making equipment is designed to operate at water temperatures approaching 0°C, a higher load is imposed on the refrigeration system during the initial start-up, when the inlet water is warmest. The components must be sized to handle this higher load.

Plan for chilling versus ice making. Most ice-making equipment has a much higher instantaneous chilling capacity than ice-making capacity. This higher chilling capacity can be used advantageously if the refrigeration equipment and interconnecting piping are properly sized to handle it.

Protect compressors from liquid slugging. Ice builders and ice harvesters tend to contain more refrigerant than chillers of similar capacity used for nonstorage systems. This provides more opportunity for compressor liquid slugging. Care should be taken to oversize suction accumulators and equip them with high-level compressor cutouts and suction heat exchangers to evaporate any remaining liquid.

Oversize the receivers. Every opportunity should be taken to make the system easy to maintain and service. Maintenance flexibility may be provided in a liquid overfeed system by oversizing the low-pressure receiver and in a direct-expansion system by oversizing the high-pressure receiver.

Prevent oil trapping. Refrigerant lines should be arranged to prevent the trapping of large amounts of oil anywhere in the system and to ensure its return to compressors under all operating conditions, especially during periods of low compressor loads. All suction lines should slope toward the suction line accumulators, and all discharge lines should pitch toward the oil separators. Oil tends to collect in the evaporator because that is the location at the lowest temperature and pressure. Because refrigerant accumulators trap oil as well as liquid refrigerant, the larger accumulators needed for ice storage sys-

tems require special provisions to ensure adequate oil return to the compressors.

It is very important to fully analyze design trade-offs; *equivalent comparisons* should be made when comparing cool storage to other available cost- and energy-saving opportunities. For example, if energy-efficient motors are assumed for a cool storage design, they should be assumed for the nonstorage system being compared. Current designs, whether based on HCFC, HFC, or ammonia refrigerants, or on any of the proven refrigeration techniques (e.g., direct-expansion, flooded, or liquid overfeed), should be used.

Cold Air Distribution

Reducing the temperature of the distribution air is attractive because smaller air-handling units, ducts, pumps, and piping can be used, resulting in a lower initial cost. In addition, the reduced ceiling space required for ductwork can significantly reduce building height, particularly in high-rise construction. These cost reductions can make thermal storage systems competitive with nonstorage systems on an initial cost basis (Landry and Noble 1991).

The optimum supply air temperature should be determined through an analysis of initial and operating costs for the various design options. Depending on the load, the additional latent energy removed at the lower discharge air temperature may be offset by the reduction in fan energy associated with the lower air flow.

The minimum achievable supply air temperature is determined by the chilled water temperature and the temperature rise between the cooling plant and the terminal units. With some ice storage systems, the fluid temperature may rise during discharge; the supply temperatures normally achievable with various types of ice storage plants must therefore be carefully investigated with the equipment supplier.

A heat exchanger, which is sometimes required with storage tanks that operate at atmospheric pressure or between a secondary coolant and chilled water system, adds at least 1 K to the final chilled water supply temperature. The rise in chilled water temperature between the cooling plant discharge and the chilled water coil depends on the length of piping and the amount of insulation; it could be up to 0.3 K over long exposed runs.

The difference between the temperature of the chilled water entering the cooling coil and the temperature of the air leaving the coil is generally between 3 and 6 K. A closer approach (smaller temperature differential) can be achieved with more rows or a larger face area on the cooling coil; but extra heat transfer surface to provide a closer approach is often uneconomical. A 1.5 K temperature rise due to heat gain in the duct between the cooling coil and the terminal units can be assumed for preliminary design analysis. With careful design and adequate insulation, this rise can be reduced to as little as 0.5 K.

A blow-through configuration provides the lowest supply air temperature and the minimum supply air volume. The lowest temperature rise achievable with a draw-through configuration is about 1.5 K because heat from the fan is added to the air. A draw-through configuration should be used if space for flow straightening between the fan and coil is limited. A blow-through unit should not be used with a lined duct because air with high relative humidity enters the duct.

Face velocity determines the size of the coil for a given supply air volume. The coil size determines the size of the air-handling unit. A lower face velocity generates a lower supply air temperature, whereas a higher face velocity results in smaller equipment and lower first costs. The face velocity is limited by moisture carryover from the coil. The face velocity for cold air distribution systems should be 1.8 to 2.3 m/s, with an upper limit of 2.8 m/s.

Cold primary air can be tempered with room air or plenum return air by using fan-powered mixing boxes or induction boxes. The primary air should be tempered before it is supplied to the space. The energy use of fan-powered mixing boxes is significant and negates the savings from downsizing central supply fans (Hittle and Smith 1994, Elleson 1993). Diffusers designed for cold air distribution can provide supply air directly to the space without causing drafts, thereby eliminating the need for fan-powered boxes.

If the supply airflow rate to occupied spaces is expected to be below 2 L/s per square metre, fan-powered or induction boxes should be used to boost the air circulation rate. At supply air rates of 2 to 3 L/s per square metre, a diffuser with a high ratio of induced room air to supply air should be used to ensure adequate dispersion of ventilation air throughout the space. A diffuser that relies on turbulent mixing rather than induction to temper the primary air may not be effective at these flow rates.

Cold air distribution systems normally maintain space humidity between 30 and 45% rh, as opposed to the 50 to 60% rh generally maintained by other systems. At this lower humidity level, equivalent comfort conditions are provided at higher dry-bulb temperatures. The increased dry-bulb set point generally results in decreased energy consumption.

The surfaces of any equipment that may be cooled below the ambient dew point, including air-handling units, ducts, and terminal boxes, should be insulated. All vapor barrier penetrations should be sealed to prevent migration of moisture into the insulation. Prefabricated, insulated round ducts should be insulated externally at joints where internal insulation is not continuous. If ducts are internally insulated, access doors should also be insulated.

Duct leakage is undesirable because it represents cooling capacity that is not delivered to the conditioned space. In cold air distribution systems, leaking air can cool nearby surfaces to the point that condensation forms. Designers should specify acceptable methods of sealing ducts and air-handling units and establish allowable leakage rates and test procedures. These specifications must be followed up with on-site supervision and inspection during construction.

A thorough commissioning process is important for the optimal operation of any large space-conditioning system, particularly thermal storage and cold air distribution systems. Reductions in initial and operating costs are major selling points for cold air distribution, but the commitment to provide a successful system must not be compromised by the desire to reduce first costs. While a commissioning procedure may appear to involve additional expense, it actually decreases costs by reducing future system malfunctions and troubleshooting expenses and provides increased value by ensuring optimal system operation. The commissioning process is discussed in greater detail in the section on Commissioning.

Storage of Heat in Cool Storage Units

Some cool storage installations may be used to provide storage for heating duty and/or heat reclaim. Many commercial buildings have areas that require cooling even during the winter, and the refrigeration plant may be in operation year-round. When cool storage is charged during the off-peak period, the rejected heat may be directed to the heating system or to storage rather than to a cooling tower. Depending on the building loads, the chilled water or ice could be considered a byproduct of heating. One potential use of the heat may be to provide morning warm-up on the following day.

Heating with a cool storage system can be achieved using heat pumps or heat-reclaim equipment such as double bundle condensers, two condensers in series on the refrigeration side, or a heat exchanger in the condenser water circuit. Cooling is withdrawn from storage as needed; when heat is required, the cool storage is recharged and used as the heat source for the heat pump. If needed, additional heat must be obtained from another source because this type of system can supply usable heating energy only when the heat pump or heat-recovery equipment is running. Possible secondary heat sources include a heat storage system, solar collectors, or waste heat recovery from exhaust air. When more cooling than heating is required, excess energy can be rejected through a cooling tower.

Individual storage units may be alternated between heat storage and cool storage service. The stored heat can meet the building's heating requirements, which may include morning warm-up and/or evening heating, and the stored cooling can provide midday cooling. If both the chiller and storage are large enough, all compressor and boiler operation during the on-peak period can be avoided, and the heating or cooling requirements can be satisfied from storage. This method of operation replaces the air-side economizer cycle that involves the use of outside air to cool the building when the enthalpy of the outside air is lower than that of the return air.

The extra cost of equipping a chilled water storage facility for heat storage is small. The necessary plant additions include a partition in the storage tank to convert a portion to warm water storage, some additional controls, and a chiller equipped for heat reclaim. The additional cost may be offset by savings in heating energy. In fact, adding heat storage capability may increase the economic advantage of cool storage.

STORAGE FOR RETROFIT APPLICATIONS

Because latent heat storage (ice and other phase-change materials) has a smaller volume, it is preferable to water storage for retrofit installations where space is limited. Latent heat storage is also preferred for building sites where rock or a high water table make deep excavation for an underground tank expensive. The relatively small size of ice-on-coil units is an advantage in cases where access to the equipment room is limited or where the storage units can be distributed throughout the building and located near the cooling coils.

Existing installations that already contain chilled water piping and chillers are more easily retrofitted to accept chilled water storage than latent heat storage. The requirement for secondary coolant in some ice storage systems may complicate pumping and heat exchange for existing equipment. Secondary coolant temperatures colder than the chilled water temperatures in the original design typically offset the reduction in heat transfer.

INDUSTRIAL/PROCESS COOLING

Refrigeration and process cooling typically require lower temperatures than air conditioning. Applications such as vegetable hydrocooling, milk cooling, carcass spray-cooling, and storage room dehumidification can use cold water from an ice storage system. Lower temperatures for freezing food, storing frozen food, and so forth can be obtained by using either a low-temperature hydrated salt phase-change material (PCM) charged and discharged by a secondary coolant or a nonaqueous PCM such as CO_2.

Refrigeration applications have traditionally used heavy-duty industrial refrigeration equipment. The size of the loads has often required custom-engineered, field-erected systems, while the occupancy has often allowed the use of ammonia. Cool storage systems for these applications can be optimized by the choice of refrigerant (halocarbon or ammonia) and the choice of refrigerant feed control (direct-expansion, flooded, mechanically pumped, or gas-pumped).

Some of the advantages of cool storage may be particularly important for refrigeration and process cooling applications. Cool storage systems can provide steady supply temperatures regardless of large, rapid changes in cooling load or return temperature. Charging equipment and discharging pumps can be placed on a separate electrical service to economically provide cooling during a power blackout or to take advantage of low, interruptible electrical rates. Storage tanks can be oversized to provide water for emergency cooling or fire fighting as well. Cooling can be extracted very quickly from harvested ice storage or chilled water storage to satisfy a very large load and then be recharged slowly using a small charging system. System design may be simplified when the load profile is determined by production scheduling rather than occupancy and outdoor temperatures.

An industrial application for cool storage is the precooling of inlet air for combustion turbines. Oil- or gas-fired combustion turbines typically generate full rated output at an inlet air temperature of 15°C. At higher inlet air temperatures, the mass flow of air is reduced, as are the shaft power and fuel efficiency. The capacity and efficiency can be increased by precooling the inlet combustion air with chilled water coils or direct contact cooling. The optimum inlet air temperature is typically in the range of 5 to 10°C, depending on the turbine model.

On a 38°C summer day, a combustion turbine driving a generator may lose up to 25% of its rated output. This drop in generator output occurs at the most inopportune time, when an electrical utility or industrial user is most in need of the additional electrical output available from combined cycle or peaking turbines. This loss in capacity can be regained if the inlet air is artificially cooled (MacCracken 1994, Mackie 1994, Ebeling et al. 1994, Andrepont 1994). Storage systems using chilled water or ice are well suited to this application because the cooling is generated and stored during off-peak periods, and maximum net increase in generator capacity is obtained by turning off the refrigeration compressor during peak demand periods. While instantaneous cooling systems are also used for inlet air precooling, they often use 25% to 50% of the increase in power output to drive the refrigeration compressors. Evaporative coolers are a simple and well established technology for inlet air precooling, but they cannot deliver the low air temperatures available from refrigeration systems.

OFF-PEAK HEATING

Service Water Heating

The tank-equipped service water heater, which is the standard water heater in North America, is a thermal storage device; some electric utilities provide incentives for off-peak water heating. The heater is equipped with a control system that is activated by a clock or by the electric utility to curtail power use during peak demand. An off-peak water heater generally requires a larger tank than a conventional heater.

An alternate system of off-peak water heating consists of two tanks connected in series, with the hot water outlet of the first tank supplying water to the second tank (ORNL 1985). This arrangement minimizes the mixing of hot and cold water in the second tank. Tests performed on this configuration show that it can supply 80 to 85% of rated capacity at suitable temperatures, compared to 70% for single-tank configurations. The wiring for the heating elements in the two tanks must be modified to accommodate the dual-tank configuration.

Solar Space and Water Heating

Rock beds, water tanks, or PCMs can be used for solar space and water heating, depending on the type of solar system employed. Various applications are described in Chapter 30.

Space Heating

Thermal storage for space heating can be carried out by radiant floor heating, brick storage heaters, water storage heaters, or PCMs. Radiant floor heating is generally applied to single-story buildings. The choice between the other storage methods depends on the type of heating system. Air systems use brick, while water systems may use water or a PCM. Water systems can be charged either electrically or thermally.

Insulation is more important for heat storage than for cool storage because the difference between storage and ambient temperatures is greater. Methods of determining the cost-effective thickness of insulation are given in Chapter 20 of the 1993 *ASHRAE Handbook—Fundamentals*.

A rational design procedure requires an hourly simulation of the design heat load and discharge capacity of the storage device. The

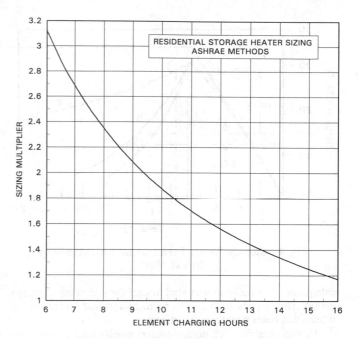

Fig. 3   Representative Sizing Factor Selection Graph

first criterion for satisfactory performance is that the discharge rate be no less than the design heating load at any hour. During the off-peak period, energy is added to storage at the rate determined by the connected load of the resistance elements, while the design heating load is subtracted. The second criterion for satisfactory performance, the daily energy balance, is checked at the end of a simulated 24-h period. Assuming that the design day is preceded and followed by similar days, the storage energy level at the end of the simulated day should equal the starting level.

A simpler procedure is recommended by Hersh et al. (1982) for typical residential heating system designs. For each zone of the building, the design heat loss is calculated in the usual manner and multiplied by the selected sizing factor. The resulting value (rounded to the next kilowatt) is the required storage heater capacity. Sizing factors in the United States range from 2.0 to 2.5 for an 8-h charge period and from 1.6 to 2.0 for a 10-h charge period. The lower end of the range is marginal for the northeastern United States. The designer should consult the manufacturer for specific sizing information. Figure 3 shows a typical sizing factor selection graph.

Brick Storage Heaters. In this storage, heat is transferred from circulating air to bricks inside a flexible cover. The bed cover is a potentially serious source of leakage. For example, in an air system pressurized to 250 Pa, the force on a 1.8-m by 2.4-m cover is approximately 1.1 N. Covers should be sealed with silicone sealants or a soft rubber gasket and should be screwed and bolted to the container sides.

Figure 4 shows the operating characteristics of electrically charged room storage heaters. Curve 1 represents theoretical performance. In reality, heat is continually transferred to the room during charging by radiation and convection from the exterior surface of the device. Curve 2 shows this static discharge. When the thermostat calls for heat, the internal fan starts. The resulting faster dynamic discharge corresponds to Curves 3 and 4.

The operating characteristics of electrically charged room storage heaters differ from those of electric heaters in a nonstorage application in two important respects. First, because the heating elements are energized only during the off-peak period, they must store the total daily heating requirement during that period, typically 8 or 10 h. Second, the rate at which heat can be delivered to the heated space decreases as the charge level decreases. Thus, the heaters are less

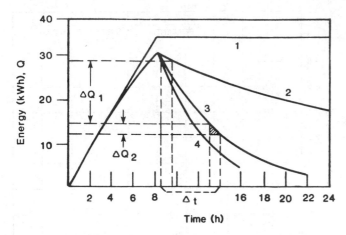

Fig. 4   Storage Heater Performance Characteristics
(Hersch 1981)

able to meet the building heat load toward the end of the discharge period. This applies primarily to room heaters and less to central heaters that contain tempering dampers.

Central ceramic brick storage heaters usually have a night heating section that supplies heat to the occupied space during the off-peak period while storage is being charged. Thus, the amount of heat that must be put into storage is equal to the space heating needs during the peak period only. For best efficiency, the unit should not be charged to a higher temperature than required to satisfy the heating needs of the following day. An outdoor anticipatory temperature sensor is thus recommended.

Water Storage Heaters. Electrically charged pressurized water storage tanks, in most applications, must be able to recharge the full daily heating requirement during the off-peak period, while simultaneously supplying heat during the off-peak period. The heat exchanger design allows a constant discharge and does not decrease near the end of the cycle.

Thermally charged hot water storage tanks are similar in design to the cool storage tanks described in the section on Off-Peak Air Conditioning. Many are also used for cooling. Unlike off-peak cooling, off-peak heating seldom permits a reduction in the size of the heating plant. The size of a storage tank used for both heating and cooling is based more on cooling than on heating for lowest life-cycle cost, except in some residential applications.

Care should be taken to avoid thermal shock in cast-in-place concrete tanks. Tamblyn (1985) showed that the seasonal change from heat storage to cool storage caused sizable leaks to develop if the cool-down period was fewer than five days. Raising the temperature of a concrete tank causes no problems because it generates compressive stresses, which concrete can sustain. Cool-down, on the other hand, causes tensile stresses, under which concrete has low strength.

OTHER APPLICATIONS

Storage in Aquifers

Aquifers can be used to store large quantities of thermal energy. Aquifers are underground, water-yielding geological formations, either unconsolidated (gravel and sand) or consolidated (rocks). In general, the natural aquifer water temperature is slightly warmer than the local mean annual air temperature. Aquifer thermal energy storage has been used for process cooling, space cooling, space heating, and ventilation air preheating (Jenne 1992). Aquifers can be used as heat pump sinks or sources and to store energy from ambient winter air, waste heat, and renewable sources.

The thickness and porosity of the aquifer determine the storage volume. Two separate wells are normally used to charge and dis-

charge aquifer storage. A well pair may be pumped either (1) constantly in one direction or (2) alternatively from one well to the other, especially when both heating and cooling are provided. Back-flushing is recommended to maintain well efficiency.

The length of the storage period depends on the local climate and on the type of building or process being supplied with cooling or heating. Aquifer thermal energy storage may be used on a short-term or long-term basis, as the sole source of energy or as partial storage, and at a temperature useful for direct application or needing upgrade. It may also be used in combination with a dehumidification system such as desiccant cooling. The cost-effectiveness of aquifer thermal energy storage is based on the avoidance of equipment capital cost and on lower operating costs.

The incorporation of aquifer thermal energy storage into a building system may be accomplished in a variety of ways, depending on the other components present and the intentions of the designer (Hall 1993b, Canadian Standards Association 1993). Control is simplified by separate hot and cold wells that operate on the basis that the last water in is the first water out. This principle ensures that the hottest or coldest water is always available when needed.

Seasonal storage has a large saving potential and began as an environmentally sensitive improvement on the large-scale mining of groundwater (Hall 1993a). The current approach is to reinject all pumped water and to attempt annual thermal balancing (Morofsky 1994; Public Works Canada 1991, 1992; Snijders 1992). Recent research directed at community-based use of aquifer thermal energy storage, perhaps integrated with other community services, promises large-scale storage implementation.

Building Mass Effects

The thermal storage capabilities inherent in building mass can have a significant effect on the temperature within the space as well as on the performance and operation of HVAC systems. Effective use of structural mass for thermal storage has been shown to reduce building energy consumption, reduce and delay peak heating and cooling loads (Braun 1990), and, in some cases, improve comfort (Simmonds 1991b, Treado et al. 1994). Perhaps the best-known use of thermal mass to reduce energy consumption is in buildings that include passive solar techniques (Balcomb 1983).

Cooling energy can be reduced by precooling the structure at night using ventilation air. Braun (1990), Ruud et al. (1990), and Andresen and Brandemuehl (1992) suggested that mechanical precooling of a building can reduce and delay peak cooling demand; Simmonds (1991b) suggested that the correct building configuration may even eliminate the need for a cooling plant. Mechanical precooling may require more energy use; however, the reduction in electrical demand costs may give lower overall energy costs. Moreover, the installed capacity of air-conditioning equipment may also be reduced, providing lower installation costs.

The effective use of thermal mass can be considered incidental and be allowed for in the heating or cooling design, or it may be considered intentional and form an integral part of the system design. The effective use of building structural mass for thermal energy storage depends on such factors as (1) the physical characteristics of the structure, (2) the dynamic nature of the building loads, (3) the coupling between the mass and zone air (Akbari et al. 1986), and (4) the strategies for charging and discharging the stored thermal energy. Some buildings, such as frame buildings with no interior mass, are inappropriate for thermal storage. Many other physical characteristics of a building or an individual zone, such as carpeting, ceiling plenums, interior partitions, and furnishings, affect thermal storage and the coupling of the building with zone air.

Incidental Thermal Mass Effects. The principal thermal mass effect on heating and cooling systems serving spaces in heavyweight buildings is that a greater amount of thermal energy must be removed or added to bring the room to a suitable condition before occupancy than for a similar lightweight building. Therefore, the system must either start conditioning the spaces earlier or operate at a greater output. During the occupied period, a heavyweight building requires a lower output, as a higher proportion of heat gains or losses are absorbed by the thermal mass.

Advantage can be taken of these effects if low-cost electrical energy is available during the night; the air-conditioning system can be operated during this period to precool the building. This can reduce both the peak and total energy required during the following day (Braun 1990, Andresen and Brandemuehl 1992) but may not always be energy-efficient.

Intentional Thermal Mass Effects. To make best use of thermal mass, the building should be designed with this objective in mind. Intentional use of the thermal mass can be either passive or active. Passive solar heating is a common application that applies the thermal mass of the building to provide warmth outside the sunlit period. This effect is discussed in further detail in Chapter 30. Passive cooling applies the same principles to limit the temperature rise during the day. The spaces can be naturally ventilated overnight to absorb surplus heat from the building mass. This technique works well in moderate climates with a wide diurnal temperature swing and low relative humidities, but it is limited by the lack of control over the cooling rate.

Active systems overcome some of the disadvantages of passive systems by using (1) mechanical power to help heat and cool the building and (2) appropriate controls to limit the output during the release or discharge period.

STORAGE TECHNOLOGIES

WATER STORAGE

Water is commonly used in both heat storage and cool storage applications. Water has the highest specific heat of all common materials [4.18 kJ/(kg·K)] and is well suited for thermal storage. ASHRAE *Standard* 94.3 gives procedures for measuring the thermal performance of sensible heat storage systems. Tanks are available in many shapes; however, vertical cylinders are the most common. Tanks can be located above ground, partially buried, or completely buried. They are usually atmospheric, unpressurized tanks and may have clear-span spherical dome roofs or column-supported flat roofs.

Water storage is based on maintaining a state of thermal separation between cool water and warm water. This discussion focuses on chilled water storage because it is the most common system in use. The same techniques apply to hot water.

Temperature Range and Storage Size

The amount of energy stored in a chilled water storage is directly related to the volume of water in the storage times the temperature differential Δt between the entering and leaving water. For economical storage, the cooling coils should provide as large a Δt as practical. For example, chillers that cool water in storage to 5°C and coils that return water to storage at 15°C provide a Δt of 10 K. The cost of the extra coil surface required to provide this range can be offset by savings in pipe size, insulation, and pumping energy. Further, fan energy costs are reduced if the extra coil surface is added as face area rather than as extra rows of coils.

The initial cost of chilled water storage benefits from a dramatic economy of scale; that is, large installations can compare favorably with ice storage as well as with equivalent nonstorage chilled water plants (Andrepont 1992).

Techniques for Separating Cool and
Warm Water Volumes

The following four methods have all been applied successfully in chilled water storage. However, thermal stratification has become

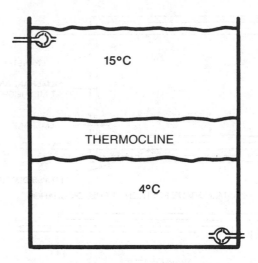

Fig. 5 Thermally Stratified Storage with Concentric Pipe Diffusers

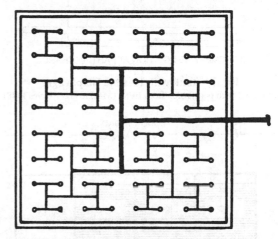

Fig. 6 Plan View of Distributed Nozzle Inlet/Outlet for Thermally Stratified Storage

the dominant method due to its simplicity, reliability, efficiency, and low cost.

Thermal Stratification. In stratified storage, the warmer, less dense returning water floats on top of the stored chilled water. The water from storage is supplied and withdrawn in low velocity, horizontal flow so that buoyancy forces dominate inertial effects. Pure water is densest at 4°C; therefore, it can not be stratified below this temperature.

When the stratified storage tank is charged, chilled supply water, typically between 4 and 6.7°C, enters through the diffuser at the bottom of the tank (Figures 5 and 6), and return water exits to the chiller through the diffuser at the top of the tank. Typically, the incoming water mixes with the water in the tank to form a 0.3- to 0.9-m thick thermocline (region with vertical temperature and density gradients). The thermocline precludes further mixing of the water above it with that below it. The thermocline rises as recharging continues and subsequently falls during the discharging process. It thickens somewhat during charging and discharging due to heat conduction through the water and heat transfer to and from the walls of the tank.

The storage tank may have any cross-section, but the walls are usually vertical. Horizontal cylindrical tanks are generally not good candidates for stratified storage.

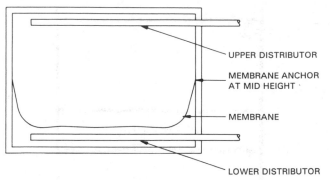

A. MEMBRANE POSITION AT FULL DISCHARGE

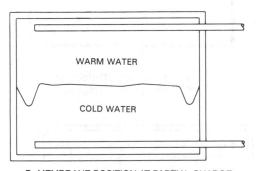

B. MEMBRANE POSITION AT PARTIAL CHARGE

Fig. 7 Tank with Flexible Diaphragm
Reprinted with permission from EPRI (Mackie and Reeves 1988)

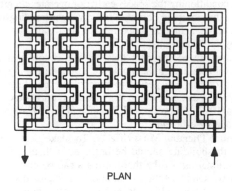

PLAN

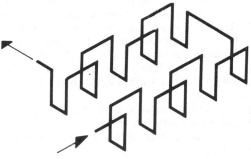

SCHEMATIC OF WATER PATH

Fig. 9 Labyrinth Tank

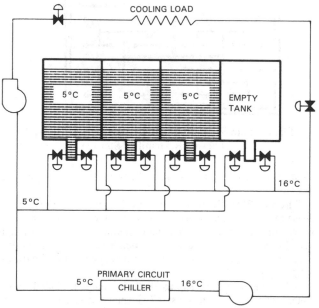

Fig. 8 Empty-Tank System
Reprinted with permission from EPRI (Mackie and Reeves 1988)

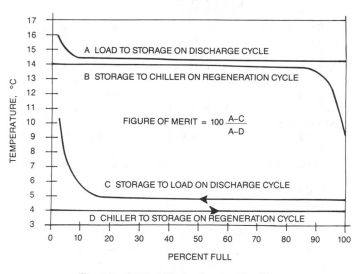

$$\text{FIGURE OF MERIT} = 100 \,\frac{A-C}{A-D}$$

Fig. 10 Chilled Water Storage Profiles

Flexible Diaphragm. A flexible diaphragm (normally horizontal) is used in a tank to separate the cold water from the warm water (Figure 7).

Multiple Compartments. Multiple compartments in a single tank or a series of two or more tanks can also be used (Figure 8). Pumping is scheduled so that one compartment is always at least partially empty; water returning from the system during the occupied period and from the chiller during storage regeneration is then received into the empty tank. Water at the different temperatures is thus stored in separate compartments, minimizing blending.

Labyrinth Tank. A labyrinth tank has both horizontal and vertical traverses (Figure 9). This design commonly takes the form of successive cubicles with high and low ports. Strings of cylindrical tanks and tanks with successive vertical weirs may also be used.

Performance of Chilled Water Storage

A perfect storage tank would deliver water at the same temperature at which it was stored. It would also require that the water returning to storage neither mix nor exchange heat with the stored water or the tank. In practice, however, both types of heat exchange occur.

Typical temperature profiles of water entering and leaving a storage tank are shown in Figure 10. Tran et al. (1989) tested several large chilled water storage systems and developed the figure of merit, which is used as a measure of the amount of cooling available from the tank.

$$\text{Figure of Merit } (\%) = \frac{\text{Area between A and C}}{\text{Area between A and D}} \times 100 \qquad (1)$$

Well-designed storage tanks have figures of merit of 90% or higher for daily complete charge/discharge cycles and between 80 and 90% for partial charge/discharge cycles.

Design of Stratification Diffusers

Stratification diffusers must be designed and constructed to produce and maintain stratification at the maximum flow rate through storage. A further discussion on design parameters is in the *Design Guide for Cool Thermal Storage* (Dorgan and Ellison 1993). Wildin (1990) gives more information concerning the design of stratified storage hardware.

Inlet and outlet streams must be kept at sufficiently low velocity that buoyancy forces predominate over inertia forces to produce a gravity current (density current) across the bottom or the top of the tank. For this purpose, the diffuser dimensions should be selected to create an inlet Froude number of 2.0 or less (Yoo et al. 1986). The inlet Froude number Fr is defined in Equation (2).

$$\text{Fr} = \frac{Q}{\sqrt{gh^3 (\Delta\rho/\rho)}} \qquad (2)$$

where

Q = volume flow rate per unit length of diffuser, m³/(s·m)
g = gravitational acceleration, m/s²
h = inlet opening height, m
ρ = inlet water density, kg/m³
$\Delta\rho$ = difference in density between stored water and incoming or outflowing water, kg/m³

The density difference $\Delta\rho$ can be obtained from Table 2. The inlet Reynolds number is defined as follows:

$$\text{Re} = Q/\nu \qquad (3)$$

where ν = kinematic viscosity, m²/s.

Table 2 Chilled Water Density

°C	kg/m³	°C	kg/m³	°C	kg/m³
0	999.87	7	999.93	14	999.27
1	999.93	8	999.88	15	999.13
2	999.97	9	999.81	16	998.97
3	999.99	10	999.73	17	998.80
4	1000.00	11	999.63	18	998.62
5	999.99	12	999.52	19	998.43
6	999.97	13	999.40	20	998.23

The inlet Reynolds number affects mixing on the inlet side of the thermocline, and it can significantly affect tank performance under some circumstances. Wildin and Truman (1989), observing results from a 1.5-m deep, 6.1-m diameter vertical cylindrical tank, found that reduction of the inlet Reynolds number from 850 (using a radial disk diffuser) to 240 (using a diffuser comprised of pipes in an octagonal array) reduced mixing to negligible proportions. This is consistent with subsequent results obtained by Wildin (1991) in a 0.9 m deep scale model tank, which indicated negligible mixing at Reynolds numbers below approximately 450. For tall tanks with water depths of 12 m or more, the inlet Reynolds number is less important. Data indicate that when the thermocline is more than about 3 m above or below the diffuser, mixing is negligible. Successful operation of tanks with 20- to 21-m water depths and inlet Reynolds numbers above 4000 has been reported (Bahnfleth and Joyce 1994).

Storage Tank Insulation

Exposed tank surfaces should be insulated to help maintain the temperature differential in the tank. Insulation is especially important for smaller storage tanks because the ratio of surface area to stored volume is relatively high. Heat transfer between the stored water and the tank contact surfaces (including divider walls) is a primary source of capacity loss. Not only does the stored fluid lose heat to (or gain heat from) the ambient by conduction through the floor and wall, but heat flows vertically along the tank walls from the warmer to the cooler region. Exterior insulation of the tank walls does not inhibit this heat transfer.

Other Factors

The cost of chemicals for water treatment may be significant, especially if the tank is filled more than once during its life. A filter system helps keep the stored water clean. Exposure of the stored water to the atmosphere may require the occasional addition of biocides. While tanks should be designed to prohibit leakage, the designer should understand the potential impact of leakage on the selection of chemical water treatment.

The storage circulating pumps should be installed below the minimum operating water level to assure flooded suction. The required net positive suction pressure (NPSH) must be maintained to avoid subatmospheric conditions at the pumps.

ICE STORAGE AND OTHER PHASE-CHANGE MATERIALS

Thermal energy can be stored in the latent heat of fusion of water (ice) or other materials. Water has the highest latent heat of fusion of all common materials—334 kJ/kg at the melting or freezing point of 0°C. The PCM other than water most commonly used for thermal storage is a hydrated salt with a latent heat of fusion of 95.4 kJ/kg and a melting or freezing point of 8.3°C. The volume required to store energy in the form of latent heat is considerably less than that stored as sensible heat, which contributes to lower capital cost.

External Melt Ice-on-Coil Storage Systems

The oldest type of ice storage is the refrigerant-fed ice builder, which consists of refrigerant coils inside a storage tank filled with water. A compressor and evaporative condenser freeze the tank water on the outside of the coils to a thickness of up to 65 mm. Ice is melted from the outside of the formation (hence the term external melt) by circulating the return water through the tank, whereby it again becomes chilled. Air bubbled through the tank agitates the water to promote uniform ice buildup and melting.

Instead of refrigerant, a secondary coolant (e.g., 25% ethylene glycol and 75% water) can be pumped through the coils inside the storage tank. The coolant has the advantage of greatly decreasing the refrigerant inventory. However, a refrigerant-to-coolant heat exchanger between the refrigerant and the storage tank is required.

Major concerns specific to the control of ice-on-coil systems are (1) to limit ice thickness (and thus excess compressor energy) during the build cycle and (2) to minimize the bridging of ice between individual tubes in the ice bank. Bridging must be avoided because it restricts the free circulation of water during the discharge cycle. While not physically damaging to the tank, this blockage reduces performance, allowing a higher leaving water temperature due to the reduced heat transfer surface.

Regardless of the refrigeration method (direct-expansion, pumped liquid overfeed, or secondary coolant), the compressor is controlled by (1) a time clock, which restricts operation to the periods dictated by the utility rate structure, and (2) an ice-thickness override control, which stops the compressor(s) at a predetermined ice thickness. At least one ice-thickness device should be installed per ice bank; the devices should be connected in series. If there are multiple refrigeration circuits per ice bank, one ice-thickness control per circuit should be installed. Placement of the ice-thickness device(s) should be determined by the ice-bank manufacturer, based on circuit geometry and flow pressure drop, to minimize bridging.

Ice-thickness controls are either mechanically or electrically operated. Mechanical controls typically consist of a fluid-filled probe positioned at the desired distance from the coil. As ice builds, it encapsulates the probe, causing the fluid to freeze and apply pressure. The pressure signal controls the refrigeration system via a pneumatic-electric (P/E) switch. Electric controls sense the three times greater electrical conductivity of ice as compared to water. Multiple probes are installed at the desired thickness, and the change in current flow between probes provides a control signal. Consistent water treatment is essential to maintaining constant conductivity and thus accurate control.

Because energy use is related to ice thickness on the coil, a partial-load ice inventory management system should be considered. This system maintains the ice inventory at the minimum level needed to supply immediate future cooling needs, rather than topping off the inventory after each discharge cycle. This method also helps prevent bridging by ensuring that the tank is completely discharged at regular intervals, thereby allowing ice to build evenly.

The most common method of measuring ice inventory is based on the fact that ice has a greater volume than water; thus, a sensed change in water level can be taken to indicate a change in the amount of stored ice. Water level can be measured by either an electrical probe or a pressure-sensing transducer.

Internal Melt Ice-on-Coil Storage Systems

In this type of storage device, coils or tubing is placed inside a water tank (Figure 11). The coils occupy approximately 10% of the tank volume; another 10% of the volume is left empty to allow for the expansion of the water upon freezing; and the rest is filled with water. A secondary coolant solution (e.g., 25% ethylene glycol and 75% water) is cooled by a liquid chiller and circulated through the coils to freeze the water in the tank. The thickness of ice on the coils and the percentage of the water in the tank that is frozen depend on the coil configuration and on the type of system. During discharge, the secondary coolant circulates to the system load and returns to the tank to be cooled again by the coils submerged in ice.

A standard chiller can provide the refrigeration for these systems. During the charging cycle for a typical system, the chilled secondary coolant exits the chiller at a constant −3.5°C and returns at −0.5°C. When the tanks are 90% charged, the chiller inlet and outlet temperatures fall rapidly because there is little water left to freeze. When the chiller exit temperature reaches approximately −5.5°C, the chiller is shut down and locked off for the night so that it does not

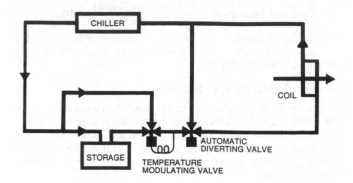

Fig. 12 Partial Storage—Cooling Mode with Water Tempered to Standard Coil Temperature

short-cycle or recirculate due to convection flow through the pipes. The chiller remains fully loaded through the entire cycle and keeps the system running at its maximum efficiency because the exit-temperature thermostat is set at −5.5°C. The temperatures for a given system may vary from this example.

Because the same heat transfer surface freezes and melts the water, the coolant may freeze the water completely every night, minimizing loss in efficiency. A temperature-modulating valve at the outlet of the tanks keeps a constant flow of liquid to the load (see Figure 12). Under a full storage control strategy, the chiller is kept off during discharge, and the modulating valve allows some fluid to bypass the tanks to supply the load as needed. If a partial storage control strategy is followed, the chiller thermostat is reset from the −5.5°C setting up to the design temperature of the cooling coils (e.g., 6.7°C) during the discharge cycle. If the load on the building is low, the chiller operates to meet the 6.7°C setting without depleting the storage. If the load is greater than the chiller's capacity, the exiting temperature of the secondary coolant rises, and the temperature-modulating valve automatically opens to maintain the design temperature to the coils (Kirshenbaum 1991).

Because water increases 9% in volume when it turns to ice, the water level varies directly with the amount of ice in the tank as long as all of the ice remains submerged. This water displaced by the ice must not be frozen, or it will trap ice above the original water level. Therefore, no heat exchange surface area can be above the original water level.

The change in water level in the tank due to freezing and thawing is typically 150 mm. This change can be measured with either a pressure gage or a standard electrical transducer. On projects with multiple tanks, a reverse-return piping system ensures uniform flow through all the tanks, so measuring the level on one tank is sufficient to determine the proportion of ice remaining.

Encapsulated Ice

This type of thermal storage system relies on plastic containers filled with deionized water and an ice-nucleating agent. Commercially available systems have either spherical containers of approximately 100-mm diameter (Figure 13) or rectangular containers approximately 35 mm by 300 mm by 760 mm (Figure 14). These primary containers are placed in storage tanks, which may be either steel pressure vessels, open concrete tanks, or suitable fiberglass or polyethylene tanks. In tanks with spherical containers, water flows vertically through the tank, and in tanks with rectangular containers, water flows horizontally. The type, size, and shape of the storage tank is limited only by its ability to achieve even flow of heat transfer fluid between the containers. The refrigeration system may be any type of liquid chiller rated for the lower temperatures required.

A secondary coolant (e.g., 25% ethylene glycol or propylene glycol and 75% water) is cooled to about −4°C by a liquid chiller and circulates through the tank and over the outside surface of the plastic

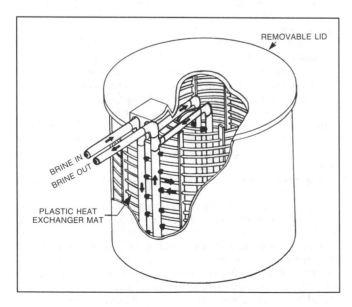

Fig. 11 Internal Melt Ice-on-Coil System

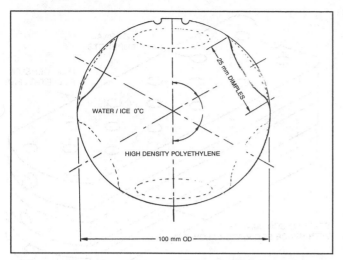

Fig. 13 Encapsulated Ice—Spherical Container

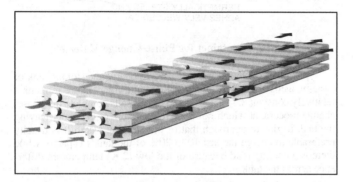

Fig. 14 Encapsulated Ice—Rectangular Container

containers, causing ice to form inside the containers. As for the internal melt ice-on-coil storage system, the temperature at the end of the charge cycle is lower (e.g., −5.5°C), and the chiller must be capable of operating at this reduced temperature. The plastic containers must be flexible to allow for change of shape during ice formation; the spherical type has preformed dimples in the surface, and the rectangular type is designed for direct flexure of the walls. During discharge, coolant flows either directly to the system load or to a heat exchanger, thereby removing heat from the load and melting the ice within the plastic containers. As the ice melts, the plastic containers return to their original shape.

Ice inventory is measured and controlled by an inventory/expansion tank normally located at the high point in the system and connected directly to the main storage tank. As ice forms, the flexing plastic containers force the surrounding secondary coolant into the inventory tank. The liquid level in the inventory tank may be monitored to account for the ice available at any point during the charge or discharge cycle.

Ice-Harvesting Systems

Ice-harvesting systems separate ice formation from ice storage. Ice is typically formed on both sides of a hollow, flat plate or on the outside or inside (or both) of a cylindrical evaporator surface. The evaporators are arranged in vertical banks located above the storage tank. Ice is formed to thicknesses between 6 and 10 mm. This ice is then harvested, often through the introduction of hot refrigerant gas into the evaporator. The gas warms the evaporator, which breaks the bond between the ice and the evaporator surface and allows the ice to drop into the storage tank below. Other types of ice harvesters use a mechanical means of separating the ice from the evaporator surface. Figure 15 shows a schematic of an ice-harvesting system.

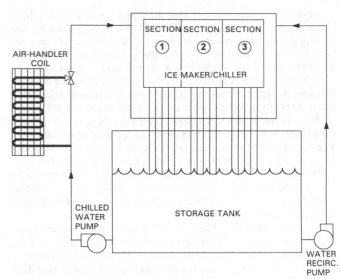

Fig. 15 Ice-Harvesting Schematic

Ice is generated by circulating 0°C water from the storage tank over the evaporators for a 10- to 30-min. build cycle. The defrost time is a function of the amount of energy required to warm the system and break the bond between the ice and the evaporator surface. Depending on the control methods, the evaporator configuration, and the discharge conditions of the compressor, defrost can be accomplished in 20 to 90 s. Typically, the evaporators are grouped in sections that are defrosted individually such that the heat of rejection from the active sections provides the energy for defrost. Knebel (1991) showed that the net available capacity from an ice harvester is the compressor gross capacity minus the fraction of time spent in defrost times the total heat of rejection from the compressor.

Tests done by Stovall (1991) indicate that for a four-section plate ice harvester, all of the total heat of rejection is introduced into the evaporator during the defrost cycle. To maximize ice production, harvest time must be kept to a minimum.

In Figure 15 chilled water is pumped from the storage tank to the load and returned to the ice generator. A low-pressure recirculation pump is used to provide minimum flow for wetting the evaporator in the ice-making mode. The system may be applied to load-leveling or load-shifting applications.

In load-leveling applications, ice is generated and the storage tank charged when there is no building load. When a building load is present, the return chilled water flows directly over the evaporator surface, and the ice generator functions either as a chiller or as both an ice generator and a chiller. Cooling capacity as a chiller is a function of the water velocity on the evaporator surface and the entering water temperature. The defrost cycle must be energized any time the exit water from the evaporator is within a few degrees of freezing. In chiller operation, maximum performance is obtained with minimum system water flows and highest entering water temperatures. In load-shifting applications, the compressors are turned off during the electric utility on-peak period.

Positive displacement compressors are usually used with ice harvesters, and saturated suction temperatures are usually between −7.8 and −5.5°C. Condensing temperatures should be kept as low as possible to reduce energy consumption. The minimum allowable condensing temperature depends on the type of refrigeration system used and the defrost characteristics of the system. Several systems operating with evaporative cooled condensers have operated with a compressor specific power consumption of 0.26 to 0.28 kW per kilowatt of refrigeration (Knebel 1986, 1988a; Knebel and Houston 1989).

Ice-harvesting systems can melt the stored ice very quickly. Individual ice fragments are characteristically less than 150 mm by 150 mm by 6 mm and provide a minimum of 4 m^2 of surface area per kilowatt-hour of ice stored. When properly wetted, a 24-h charge of ice can be melted in less than 30 min for emergency cooling demands.

During the ice generation mode, the system is energized if the ice is below the high ice level. A partial storage system is energized only when the entering water is at or above a temperature that will permit chilling during the discharge mode; otherwise, the system is off, and the ice tank is discharged during the on-peak period to meet the load. The high ice level sensor can be mechanical, optical, or electronic. The entering water temperature thermostat is usually electronic.

When ice is floating in the tank, the water level will always be constant, so it is impossible to measure ice inventory by measuring water level. The following methods are generally used to determine the ice inventory:

Water Conductivity Method. As water freezes, dissolved solids are forced out of the ice into the liquid water, thus increasing their concentration in the water. Accurate ice inventory information can be maintained by measuring conductivity and recalibrating daily from the high ice level indicator.

Heat Balance Method. The cooling effect obtained from a system may be determined by measuring the power input to, and heat of rejection from, the compressor. The cooling load on the system is determined by measuring coolant flow and temperature. This may be accomplished by measuring the ice inventory and knowing the time to recharge the storage. The ice inventory is then determined by integrating cooling input minus load and calibrating daily from the high ice level sensor. A variant of the heat balance method is determined by running a heat balance on the compressor only, using performance data from the compressor manufacturer and measured load data (Knebel 1988b).

Optimal performance of ice-harvesting systems may be achieved by recharging the ice storage tank over a maximum amount of time with minimum compressor capacity. Ice inventory measurement and the known recharge time are used to accomplish this goal. Knebel and Houston (1989) demonstrated that efficiency can be dramatically increased with proper selection of multiple compressors and unloading controls.

Design of the storage tank is important to the operation of the system. The amount of ice stored in a storage tank depends on the shape of the storage, the location of the ice entrance to the tank, the angle of repose of the ice (between 15 and 30°, depending on the shape of the ice fragments), and the water level in the tank. If the water level is high, voids occur under the water due to the buoyancy of the ice. Ice-water slurries have been reported to have a porosity of 0.50 and typical storage densities of 0.0235 m^3 per kWh. Gute et al. (1995), and Stewart et al. (1995a, 1995b) describe models for determining the amount of ice that can be stored in rectangular tanks and the discharge characteristics of various tank configurations. Dorgan and Elleson (1993) provide further information.

Other Phase-Change Materials

Like ice and chilled water storage systems, hydrated salts have been in use for many decades. PCMs with various phase-change points have been developed. To date, the hydrated salt most commonly used for cool storage applications changes phase at 8.3°C; it is often encapsulated in plastic containers, as described in the section on Encapsulated Ice. This material is a mixture of inorganic salts, water, and nucleating and stabilizing agents. It has a latent heat of fusion of 95.4 kJ/kg and a density of 1490 kg/m^3. The PCM's latent heat of fusion requires a capacity of about 0.044 m^3/kWh for the entire tank assembly, including piping headers and water in the tank.

Another option is an internal melt ice-on-coil system in which the water is replaced by salts or polymers that change phase at −11°C or −2°C. The hydrated salts typically have a latent heat of fusion of 267 to 314 kJ/kg and a density of 1040 to 1130 kg/m^3.

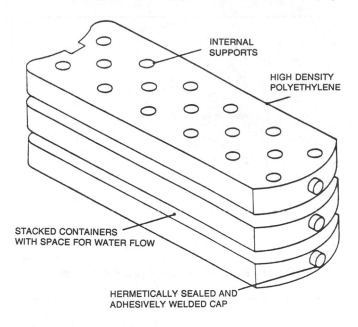

Fig. 16 Container for Phase-Change Material

Typically, 5°C chilled water is used to charge the storage tank of systems using the 8.3°C PCM. The leaving temperature remains a relatively constant 8.3°C until the hydrated salts complete the phase-change process, at which point the temperature of the water leaving the tank begins to approach that of the entering water. It is usually preferable to charge the last 10 to 20% of the tank's capacity while there is a cooling load because of the low (2 K) temperature difference across the tank.

During discharge, exit temperatures begin at 5.5°C and rise to 9 to 10°C. The usable storage capacity is determined by the highest discharge temperature that can be used by the system.

Any new or existing centrifugal, screw, or reciprocating chiller can be used to charge the storage system because conditions are comparable to those for standard air conditioning. As a result, this technology is particularly appropriate for retrofit applications.

As shown in Figure 16, the PCM is sealed inside self-stacking plastic containers measuring approximately 610 mm by 200 mm by 45 mm. The containers are designed to provide space for water to pass between them in a meandering flow pattern. The containers are placed in an open tank, typically of below-grade concrete or sprayed-on concrete. The containers displace about two-thirds of the volume of the tank, so that one-third of the tank is occupied by the water used as the heat transfer medium. The containers do not float or expand when the PCM changes phase, so the container spacing arrangement is maintained throughout the phase-change cycle. The top of the tank can be landscaped or designed for traffic or use as a parking lot.

Precoolers with partial storage provide some storage and colder water temperatures to the load. The 8.3°C PCM can be used with the chiller downstream of the storage to supply water at 6.7°C or lower to the load.

Circuitry for Ice Storage Systems

Piping for ice storage systems can be configured in a variety of ways. The optimum configuration depends on the type of system; the operational, performance, and temperature requirements; and the configuration of the building loop. The chiller may be located upstream or downstream of the building load, and the cooling plant may also be decoupled from the building load. An example of a basic piping schematic for an ice storage system providing partial storage is shown in Figure 17.

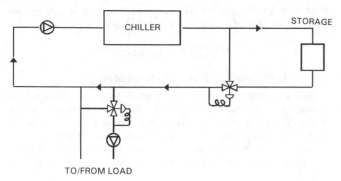

A. CHARGING STORAGE

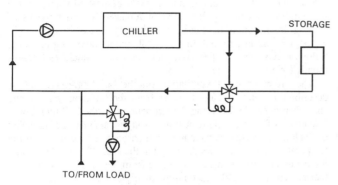

B. DISCHARGING STORAGE

Fig. 17   Thermal Storage System with Chiller Upstream

During off-peak periods, the loop supplying the building load is bypassed, and the chiller charges the ice storage unit (Figure 17A). When cooling is required during the on-peak period, the building load may be met by the chiller, the storage, or a combination of both (Figure 17B). A downstream modulating valve maintains the chilled water supply to the building loop at the desired temperature. If demand limiting is desired, the chiller electrical demand must also be controlled. The remainder of the load is met by the storage.

Numerous studies have been conducted to determine optimum configurations and control schemes for a variety of designs (Braun 1992, Simmonds 1994).

Storage Tank Insulation

Because of the low system temperatures associated with ice storage, insulation is a high priority. All ice storage tanks located above ground should be insulated to limit standby losses. For external melt ice-on-coil systems and some internal melt ice-on-coil systems, the insulation and vapor barrier are part of the factory-supplied containers; most other storage tanks require field-applied insulation and vapor barrier. Below-ground tanks used with ice harvesters may not need insulation below the first metre. Because the tank temperature does not drop below 0°C at any time, there is no danger of freezing and thawing groundwater. All below-ground tanks using fluids below 0°C during the charge cycle should have an insulation and vapor barrier system that is well designed and installed, generally on the exterior. Interior insulation is susceptible to damage from the ice and should be avoided.

Because the 8.3°C hydrated salt operates at chilled water temperatures, most of the considerations in this section apply to it as well.

ELECTRICALLY CHARGED HEAT STORAGE DEVICES

Thermal energy can also be stored in electrically charged, thermally discharged storage devices. For devices that use a solid mass

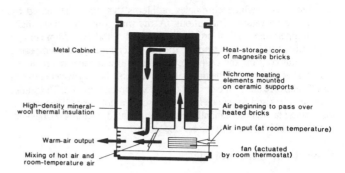

Fig. 18   Room Storage Heater

as the storage medium, equipment size is typically specified by the nominal power rating (to the nearest kilowatt) of the internal heating elements, and the nominal storage capacity is taken as the amount of energy supplied (to the nearest kilowatt-hour) during an 8-h charge period. For example, a 5-kW heater would have a nominal storage capacity of 40 kWh. ASHRAE *Standard* 94.2 describes methods for testing these devices.

Room Storage Heaters

Room storage heaters have olivine or magnetite brick cores encased in shallow metal cabinets that can fit under windows (Figure 18). The core is heated to 760°C during off-peak hours by resistance heating elements located throughout the cabinet. Storage heaters are discharged by natural convection, radiation, and conduction (static heaters) or by a fan. The air flowing through the core is mixed with room air to limit the outlet air temperature to a comfortable range. Room storage heaters tend to overheat on mild winter days unless equipped with a controller that adjusts the thermal charge depending on the outdoor temperature. Room heaters are used extensively in European homes, and many units of European manufacture have been installed in the United States. These heaters have storage capacities of 14 to 50 kWh and a power draw of 2 to 6 kW, respectively.

Central Storage Air Heaters

Central storage heaters operate on the same principle as room storage heaters; they contain a brick core that is heated to 760°C during off-peak hours. They are suitable for buildings with forced-air systems because they are discharged by a fan through the building's warm air ducts. A damper section mixes return air with the heated air to maintain a comfortable outlet air temperature during discharge. Central storage heaters also contain a night heating section, which provides thermal comfort to occupants during the off-peak period when the core is being charged.

A charge controller senses outdoor air temperature as well as core temperature to determine the required charge and to prevent overheating. The core heating elements are activated in steps to prevent current surges. A manual override permits activation of the night section during the on-peak period in case the storage is exhausted. In addition to the power required for night heating, the units store from 112 to 240 kWh and have a charging draw of 14 to 30 kW, respectively.

Pressurized Water Storage Heaters

This storage device consists of an insulated cylindrical steel tank containing immersion electrical resistance elements near the bottom of the tank and a water-to-water heat exchanger near the top (Figure 19). During off-peak periods, the resistance elements are sequentially energized until the storage water reaches a maximum temperature of 138°C, corresponding to 240 kPa (gage). The *ASME Boiler Code* considers such vessels unfired pressure vessels, so they are not required to meet the provisions for fired vessels. The heaters are controlled by a pressure sensor, which eliminates problems that could be caused by unequal temperature distributions. A thermal controller gives high-limit temperature control protection. Heat is withdrawn from storage by running service water through the heat exchangers and a tempering device that controls the output temperature to a predetermined level. The storage capacity of the device is the sensible heat of water between 138°C and 5 K above the desired output water temperature. The output water can be used for space heating or service hot water. The water in the storage tank is permanently treated and sealed, requires no makeup, and does not interact with the service water.

Underfloor Heat Storage

This storage method typically uses electric resistance cables buried in a bed of sand 300 to 900 mm below the floor of a building. It is suitable for single-story commercial buildings, such as factories and warehouses. An underfloor storage heater acts as a flywheel; while it is charged only during the nightly off-peak, it maintains the top of the floor slab at a constant temperature slightly higher than the desired space temperature. Because the cables spread heat in all directions, they do not have to cover the entire slab area. For most buildings, a cable location of 450 mm below the floor elevation is optimum. The sand bed should be insulated along its perimeter with 50 mm of rigid, closed-cell foam insulation to a depth of 1200 mm (see Figure 20).

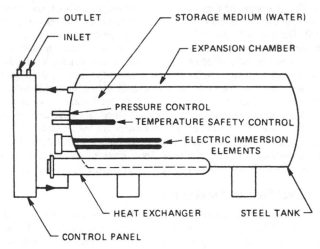

Fig. 19   Pressurized Water Heater

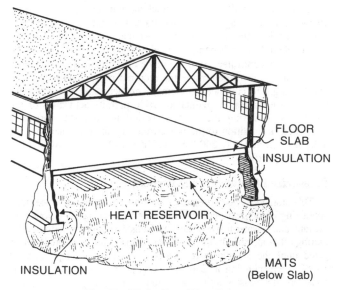

Fig. 20   Underfloor Heat Storage

Even with a well-designed and -constructed underfloor storage system, 10% or more of the input heat may be lost to the ground.

BUILDING MASS

Principle of Building Mass Thermal Storage

Building elements such as floor slabs and internal walls form the storage, and the materials (concrete, brick, etc.) are the storage media. The objective for building mass thermal storage is not to maintain a constant temperature in the conditioned space, but rather to limit the rise in temperature over a normal working day. Energy savings are usually the greatest when it is used without mechanical cooling.

The method usually takes advantage of "free cooling" with outside air, which is available during the night in most climates; but the free cooling can be supplemented or replaced by other means of cooling. In air-conditioned buildings, running the systems overnight to cool the building mass limits the peak and total cooling loads the following day (Braun 1990, Andresen and Brandemuehl 1992, Simmonds 1991a).

The effect of night ventilation on the building mass can be described as follows. Heat transfer from the building mass to the space is reduced by lowering the temperature of the building fabric. The effect of time on the heat flow through a slab can be modelled by including thermal capacitance, which adds a time delay to the temperature and heat flows predicted in the slab (Kreith 1973). During occupied periods, heat is supplied from the outside, loading the thermal capacitance. If no cooling or night ventilation is used to reject the heat, the heat will dissipate, particularly following occupied periods when the heat energy is transferred back into the room.

With night ventilation, both sides of the building fabric are subjected to lower temperatures, and heat is extracted from the building mass. The building fabric is generally designed for the winter, when heat flow is from inside to outside. When the building fabric is exposed on both sides to a temperature that is lower than its own temperature, the thermal capacitance and the thermal resistance (thus the time constant) are split into two parts. This results in heat transfer in two directions and therefore a quicker thermal response.

Systems

Both night ventilation and precooling have limitations. The amount of heat stored in a slab equals the product of mass, specific heat and temperature rise. The amount of heat available to the space depends on the rate at which heat can be extracted from the slab, which in simple terms is

$$q_s = \rho c_p V \frac{dt_s}{d\theta} = h_o A (t_s - t) \qquad (4)$$

where

q_s = rate of heat flow from slab, kW
ρ = density, kg/m³
c_p = specific heat, kJ/(kg·K)
V = slab volume, m³
θ = time, s
h_o = heat transfer coefficient, W/(m²·K)
A = area of slab, m²
t_s = temperature of slab, °C
t = temperature of space, °C

Equation (4) also applies to transferring heat to the storage medium; while the potential is equivalent to $c_p V(t_s - t)$, the heat released during the daytime period is related to the transfer coefficients. Building transfer coefficients are quite low; for example, a typical value for room surfaces is 8 W/(m²·K), which is the maximum amount of energy that can be released.

Effective Storage Capacity. The total heat capacity (THC) (Ruud et al. 1990) is the maximum amount of thermal energy stored or released due to a uniform change in temperature Δt of the material and is given by

$$THC = \rho c_p V \Delta t \qquad (5)$$

The diurnal heat capacity (DHC) is a measure of the thermal capacity of a building component exposed to periodically varying temperatures.

Many factors must be considered when an energy source is time-dependent. Minimum temperatures occur around dawn, which may be at the end of the off-peak tariff; the optimum charge period may run into the working day. Beginning the charge earlier may be less expensive but also less energy-efficient. In addition, the energy stored in the building mass is neither isolated nor insulated, so some energy is lost during charging; and the amount of available free energy varies and must be balanced against the energy cost of mechanical power. As a result, there is a trade-off that varies with time between the amount of free energy that can be stored and the power necessary for charging.

As the cooling capacity is, in effect, embedded in the building thermal mass, conventional techniques of assessing the peak load cannot be used. Detailed weather records that show peaks over 3- to 5-day periods, as well as data on either side of the peaks, should be examined to ensure that (1) the temperature at which the building fabric is assumed to be before the peak period is realistic and (2) the consequences of running the system with an exhausted storage after peak are considered. This level of analysis can only be carried out effectively using a dynamic simulation program. Experience has shown that these programs should be used with a degree of caution, and the results should be compared with both experience and intuition.

Storage Charging and Discharging

The building mass can be charged (cooled or warmed) either indirectly or directly. Indirect charging is usually accomplished by heating or cooling either the bounded space or an adjacent void. Almost all passive and some active cooling systems are charged by cooling the space overnight (Arnold 1978). Most indirect active systems charge the store by ventilating the void beneath a raised floor (Herman 1980, Crane 1991). Where this is an intermediate floor, cooling can be radiated into the space below and convected from the floor void the following day. By varying the rate of ventilation through the floor void, the rate of discharge can be controlled. Proprietary floor slabs are commonly of the hollow-core type (Anderson et al. 1979, Willis and Wilkins 1993). The cores are continuous, but when used for thermal storage, they are plugged at each end, and holes are drilled to provide the proper airflow. Charging is carried out by circulating cool or warm air through the hollow cores and exhausting it to the room. Discharge can be controlled by a ducted switching unit that directs air through the slab or straight into the space.

A directly charged slab, used commonly for heating and occasionally for cooling, can be constructed with an embedded hydronic coil. The temperature of the slab is only cycled 2 to 3 K to either side of the daily mean temperature of the slab. Consequently the technique can use very low grade free cooling (approximately 19°C) (Meierhans 1993) or low-grade heat rejected from condensers (approximately 28°C). In cooling applications the slab is used as a cool radiant ceiling, and for warming it is usually a heated floor. Little control is necessary due to the small temperature differences and the high heat capacity of the slab.

INSTALLATION, OPERATION, AND MAINTENANCE

The design professional must consider that almost all thermal storage systems require more space than nonstorage systems. Having selected a system, the designer must decide on the physical location; the piping interface to the air-conditioning equipment; and the water treatment, control, and optimization strategies to transfer theoretical benefits into realized benefits. The design must also be documented, the operators trained, and the performance verified (i.e., the system must be properly commissioned). Finally, the system must be properly maintained over its projected service life. For further information on operation and maintenance management, see Chapter 35.

SPECIAL REQUIREMENTS

The location and space required by a thermal storage system are functions of the type of storage and the architecture of the building and site. Building or site constraints often shift the selection from one option to another.

Chilled Water Systems

Chilled water systems are associated with large volume. As a result, many stratified chilled water storage systems are located outdoors (such as in industrial plants or suburban campus locations). A tall tank is desirable for stratification, but a buried tank may be required for architectural or zoning reasons. Tanks are traditionally constructed of steel or prestressed concrete. A systems supplier who assumes full responsibility for the complete system performance often constructs the tank at the site and installs the entire distribution system.

Ice-on-Coil Systems (External and Internal Melt)

Ice-on-coil systems are available in many configurations with differing space and installation requirements. Because of the wide variety available, these systems often best meet the unique requirements of many types of buildings.

Bare coils are available for installation in concrete cells, which are a part of the building structure. The bare steel coil concept can be used with direct cooling, in which the refrigerant is circulated through the coils, and the water is circulated over the coils to be chilled or frozen. This external melt system has very stringent installation requirements. Coil manufacturers do not normally design or furnish the tank, but they do provide design assistance, which covers distribution and air agitation design as well as side and end clearance requirements. These recommendations must be followed exactly to ensure success.

The bare coil concept can also be used with a secondary coolant circulation system to provide the cooling necessary to build the ice. In an internal melt configuration, the ice and water, which remains in the tank and is not circulated to the cooling system, cools the secondary coolant during discharge. This indirect chilling system can also be used with an external melt discharge if it is not desirable to circulate the secondary coolant to the cooling load. The indirect system can use either steel or plastic tubes in the ice builder.

Coils with factory-furnished containers come in a variety of sizes and shapes. A suitable style can usually be found to fit the available space. Round plastic containers with plastic coils are available in several sizes. These are offered only in an internal melt configuration and can be above ground or partially or completely buried.

Rectangular steel tanks are available with both steel and plastic pipe in a wide variety of sizes and capacities. Steel coil modules have the option of either internal or external melt. These steel tank systems are not normally buried. Each system comes prepackaged; installation requires only placement of the tank and proper piping connections. Any special support or insulation requirements of the manufacturer must be strictly followed.

Encapsulated Ice

Cylindrical steel containers with encapsulated water modules are also available. These offer yet another shape to fit available space.

With proper precautions, these containers can be installed below grade. Standards and recommendations for corrosion protection published by the Steel Pipe Institute and the National Association of Corrosion Engineers should be followed, as should the manufacturer's instructions. These systems are not shipped assembled. The containers must be placed in the shell at the job site in a way that channels the secondary coolant through passages where the desired heat transfer will be achieved.

Ice-Harvesting Systems

These systems are generally used with field-built concrete ice tanks. The ice harvester manufacturer may furnish assistance in tank design and piping distribution within the tank. The tank may be completely or partially buried or installed above ground. Where the ground is dry and free of moving water, tanks have been buried without insulation. In this situation the ground temperature eventually stabilizes, and the heat loss becomes minimal. However, a minimum of 50 mm of closed cell insulation should be applied to the external surface. Because the shifting ice creates strong dynamic forces, internal insulation should not be used except on the underside of the tank cover. In fact, only very rugged components should be placed in the tank; exit water distribution headers should be of stainless steel or rugged plastic suitable for the cold temperatures encountered. PVC is not an acceptable material due to its extreme brittleness at ice-water temperatures. An underfloor system that is a part of the concrete structure is preferred.

As with the chilled water and hydrated salt PCM tanks, close attention to the design and construction is critical to prevent leakage. Unlike a system where the manufacturer builds the tank and assumes responsibility for its integrity, an ice-harvesting system needs an on-site engineer familiar with concrete construction requirements to monitor each pour and to check all water stops and pipe seals. Unlined tanks that do not leak can be built. If liners are used, the ice equipment suppliers will provide assistance in determining a suitable type; the liner should be installed only by a qualified installer trained in the proper methods of installation by the liner maker.

The sizing and location of the ice openings is critical; the tank design engineer should check all framed openings against the certified drawings before the concrete is poured.

An ice harvester is generally installed by setting in place a prepackaged unit that includes the ice-making surface, the refrigerant piping, the refrigeration equipment, and, in some cases, the heat rejection equipment and the prewired control system. To ensure proper ice harvest, the unit must be properly positioned with respect to the drop opening. As the internal piping is not normally insulated, the drop opening should extend under the piping so that condensate drops into the tank. A grating below the piping is desirable. To prevent air or water leakage, gasketing between the unit frame and caulking must be installed in accordance with the manufacturer's instructions. External piping and power and control wiring complete the installation.

Other PCM Systems

Coolant normally flows horizontally in salt and polymeric systems, so the tanks tend to be shallower than the ideal chilled water storage tank. As in chilled water systems, the chilled water supply to and return from the tank must be designed to distribute water uniformly through the tank without channeling. Tanks are traditionally of concrete. The system supplier normally designs the tank and its distribution system, builds the tank, and installs the salt solution containers.

SYSTEM INTERFACE

Open Systems

Chilled water; salt and polymeric PCMs; external melt ice-on-coil; and ice-harvesting systems are all open chilled water piping systems. Drain-down must be prevented by isolation valves, pressure-sustaining valves, or heat exchangers. Due to the potential for drain-down, the open nature of the system, and the fact that the water being pumped may be saturated with air, the construction contractor must follow the piping details carefully to prevent pumping or piping problems.

Closed Systems

Closed systems normally circulate an aqueous secondary coolant (25 to 30% glycol solution) either directly to the cooling coils or to a heat exchanger interface to the chilled water system. A domestic water makeup system should not be the automatic makeup to the secondary coolant system. An automatic makeup system that pumps a premixed solution into the system is recommended, along with an alarm signal to the building automation system to indicate makeup operation. The secondary coolant must be an industrial solution (not automotive antifreeze) with inhibitors to protect the steel and copper found in the piping system. The water should be deionized; as portable deionizers can be rented, the solution can be mixed on-site. A calculation, backed up by metering the water as it is charged into the piping system for flushing, is needed to determine the specified concentration. Premixed coolant made with deionized water is also available, and tank truck delivery with direct pumping into the system is recommended on large systems. An accurate estimate of volume is required.

INSULATION

Because the chilled water, secondary coolant, or refrigerant temperatures are generally 6 to 11 K below those found in nonstorage systems, special care must be taken to prevent damage. Although fiberglass or other open-cell insulation is theoretically suitable when supplied with an adequate vapor barrier, experience has shown that its success is highly dependent upon workmanship. Therefore, a two-layer closed-cell material with staggered joints and carefully sealed joints is recommended. A thickness of 40 to 50 mm is normally adequate to prevent condensation in a normal room. Provisions must be made to ensure that the relative humidity in the equipment room is less than 80%. This can be done with heating or with cooling and dehumidification.

Special attention must be paid to pump and heat exchanger insulation covers. Valve stem, gage, and thermometer penetrations and extensions should be carefully sealed and insulated to prevent condensation. PVC covers over all insulation in the mechanical room provide an excellent appearance and limited protection. Damaged sections are easily replaced. Insulation located outdoors should be protected by an aluminum jacket.

REFRIGERATION EQUIPMENT

The refrigeration system may be packaged chillers, field-built refrigeration, or a refrigeration system furnished as a part of a package. The refrigeration system must be installed in accordance with the manufacturer's recommendations. Due to the high cost of refrigerant, refrigerant vapor detectors are suggested even for Class A1 refrigerants. Equipment rooms must be designed and installed to meet ASHRAE *Standard* 15. Relief valve lines should be monitored to detect valve weeping; any condensate that collects in the relief lines must be diverted and trapped so that it does not flow to the relief valve and eventually damage the seat.

WATER TREATMENT

Open Systems

Water treatment must be given close scrutiny in open systems. While the evaporation and concentration of solids associated with cooling towers does not occur, the water may be saturated with air, so the corrosion potential is greater than in a closed system. Treatment against algae, scale, and corrosion must be provided. No matter what type of treatment is chosen (i.e., traditional chemical or nontraditional treatment), some type of filtration that is effective at least down to 24 μm should be provided. To prevent damage to the system, the water treatment system must be operational immediately following the completion of the cleaning procedure. Corrosion coupon assemblies should be included to monitor the effectiveness of the treatment. Water testing and service should be performed at least once a month by the water treatment supplier.

Closed Systems

The secondary coolant should be pretreated by the supplier. A complete analysis should be done annually. Monthly checks on the solution concentration should be made using a refractive indicator. Automotive-type testers are not suitable. For normal use, the solution should be good for many years without needing new inhibitors. However, provision should be made for the injection of new inhibitors through a shot feeder if recommended by the manufacturer. The need for filtering, whether it be the inclusion of a filtering system or filtering the water or solution before it enters the system, should be carefully considered. Combination filter feeders and corrosion coupon assemblies may be needed for monitoring the effect of the solution on copper and steel.

CONTROLS

A direct digital control system to monitor and control all of the equipment associated with the central plant is preferred. Monitoring of electrical use by all primary plant components, individually if possible but at least as a group, is strongly recommended. Monitoring of the refrigeration capacity produced by the refrigeration equipment, by direct measurement where possible or by manufacturer's capacity ratings related to suction and condensing pressures, should be incorporated. This ensures that a performance rating can be calculated for use in the commissioning process and reevaluated on an ongoing basis as a management tool to gage system performance.

Optimization Software

Optimization software should be installed to obtain the best performance from the system. This software must be able to predict, monitor, and adjust to meet load requirements, as well as adapt to daily or weekly storage, full or partial storage, chiller or ice priority, and a wide variety of rate schedules.

COMMISSIONING AND MAINTENANCE

Chapters 35 and 39 and ASHRAE *Guidelines* 1 and 4 provide information regarding design documentation and operator training.

Performance Verification

The commissioning authority should verify performance and document all operating parameters. This information should be used to establish a database for future reference to normal conditions based on a constant design condensing temperature. Some of the performance data for various systems are as follows:

External Melt Ice-on-Coil Storage System

- evaporator and suction temperatures at start of ice build
- evaporator and suction temperatures at end of ice build
- ice thickness at end of ice build
- time to build ice
- efficiency at start versus theoretical efficiency
- efficiency at end versus theoretical efficiency
- refrigeration capacity based on published ratings (deviation can indicate refrigerant loss or surface fouling)

Internal Melt Ice-on-Coil Storage System

- secondary coolant temperature and suction temperature at start
- secondary coolant temperature and suction temperature at end
- secondary coolant flow
- tank water level at start
- tank water level at end
- time to build ice
- efficiency at start versus theoretical efficiency
- efficiency at end versus theoretical efficiency
- capacity based on measured flow, heat balance, and published rating

Ice-Harvesting System

- suction temperature at start
- suction temperature at harvest
- harvest time/condensing temperature
- time from start to bin full signal
- efficiency
- tank water level at start
- tank water level at bin full signal
- capacity based on published rating

While tank water level cannot be used as an indicator of the amount of ice in storage in a dynamic system, the water level at the end of the discharge cycle is a good indicator of conditions in the system. In systems with no gain or loss of water, the shutdown level should be consistent, and it can be used as a backup to determine when the bin is full for shutdown requirements. Conversely, a change in level at shutdown can indicate a water gain or loss.

Maintenance Requirements

Following the manufacturer's maintenance recommendations is essential to satisfactory long-term operation. These recommendations vary, but their objective is to maintain the refrigeration equipment, the refrigeration charge, the coolant circulating system, the ice builder surface, the water distribution system, water treatment, and controls so that they continue to perform at the same level as when the system was commissioned. Monitoring ongoing performance against kilowatt-hours per megagram of ice built gives a continuing report of system performance.

REFERENCES

Akbari, H. et al. 1986. The effect of variations in convection coefficients on thermal energy storage in buildings: Part 1—Interior partition walls. Lawrence Berkeley Laboratory, Berkeley, CA.

Anderson, L.O., K.G. Bernander, E. Isfalt, and A.H. Rosenfeld. 1979. Storage of heat and cooling in hollow-core concrete slabs. Swedish Experience and Application to Large, American Style Building. 2nd International Conference on Energy Use Management, Los Angeles.

Andrepont, J.S. 1992. Central chilled water plant expansions and the CFC refrigerant issue—Case studies of chilled water storage. Proceedings of the Association of Higher Education Facilities Officers 79th Annual Meeting, Indianapolis, IN.

Andrepont, J.S. 1994. Performance and economics of CT inlet air cooling using chilled water storage. *ASHRAE Transactions* 100(1):587.

Andresen, I. and M.J. Brandemuehl. 1992. Heat storage in building thermal mass: A parametric study. *ASHRAE Transactions* 98(1):910-18.

Arnold, D. 1978. Comfort air conditioning and the need for refrigeration. *ASHRAE Transactions* 84(2):293-303.

ASHRAE. 1981. Methods of testing thermal storage devices with electrical input and thermal output based on thermal performance. *Standard* 94.2-1981 (RA 89).

ASHRAE. 1986. Methods of testing active sensible thermal energy storage devices based on thermal performance. *Standard* 94.3-1986 (RA 90).

ASHRAE. 1989. Commissioning of HVAC systems. *Guideline* 1-1989.

ASHRAE. 1992. Safety code for mechanical refrigeration. *Standard* 15-1992.

ASHRAE. 1993. Preparation of operating and maintenance documentation for building systems. *Guideline* 4-1993.

ASME. 1992. Boiler and pressure vessel codes. American Society of Mechanical Engineers, New York.

Balcomb, J.D. 1983. Heat storage and distribution inside passive solar buildings. Los Alamos National Laboratory, Los Alamos, NM.

Bahnfleth, W.P. and W.S. Joyce. 1994. Energy use in a district cooling system with stratified chilled water storage. *ASHRAE Transactions* 100(1):1767-78.

Braun, J.E. 1990. Reducing energy costs and peak electrical demand through optimal control of building thermal storage. *ASHRAE Transactions* 96(2):876-88.

Braun, J.E. 1992. A comparison of chiller-priority, storage-priority, and optimal control of an ice-storage system. *ASHRAE Transactions* 98(1):893-902.

Canadian Standards Association. 1993. Design and construction of earth energy heat pump systems for commercial and institutional buildings. *Standard* C447-93. Canadian Standards Association, Rexdale, Ontario.

Crane, J.M. 1991. The Consumer Research Association new office/laboratory complex, Milton Keynes: Strategy for environmental services. CIBSE National Conference (pp. 2-29), Canterbury, England.

Dorgan, C.E. and J.S. Elleson. 1993. *Design guide for cool thermal storage.* ASHRAE.

Ebeling, J.A., L. Beaty, and S.K. Blanchard. 1994. Combustion turbine inlet air cooling using ammonia-based refrigeration for capacity enhancement. *ASHRAE Transactions* 100(1):583-6.

Elleson, J.S. 1993. Energy use of fan-powered mixing boxes with cold air distribution. *ASHRAE Transactions* 99(1):1349-58.

Fiorino, D.P. 1994. Energy conservation with thermally stratified chilled-water storage. *ASHRAE Transactions* 100(1):1754-66.

Gute, G.D., W.E. Stewart, Jr., and J. Chandrasekharan. 1995. Modelling the ice-filling process of rectangular thermal energy storage tanks with multiple ice makers. *ASHRAE Transactions* 101(1).

Hall, S.H. 1993a. Environmental risk assessment for aquifer thermal energy storage. PNL-8365. Pacific Northwest Laboratory, Richland, WA.

Hall, S.H. 1993b. Feasibility studies for aquifer thermal energy storage. PNL-8364. Pacific Northwest Laboratory, Richland, WA.

Herman, A.F.E. 1980. Underfloor, structural storage air-conditioning systems. Proc. Paper 7. FRIGAIR 80, Pretoria, South Africa.

Hersh, H., G. Mirchandani, and R. Rowe. 1982. Evaluation and assessment of thermal energy storage for residential heating. ANL SPG-23. Argonne National Laboratory, Argonne, IL.

Hittle, D.C. and T.R. Smith. 1994. Control strategies and energy consumption for ice storage systems using heat recovery and cold air distribution. *ASHRAE Transactions* 100(1):1221-29.

Holness, G.V.R. 1992. Case study of combined chilled-water thermal energy storage and fire protection storage. *ASHRAE Transactions* 98(1):1119-22.

Hussain, M.A. and D.C.J. Peters. 1992. Retrofit integration of fire protection storage as chilled-water storage—A case study. *ASHRAE Transactions* 98(1):1123-32.

ITSAC. 1992. *Advisory Newsletter* (March). International Thermal Storage Advisory Council, San Diego.

Jenne, E.A. 1992. Aquifer thermal energy (heat and chill) storage. Papers presented at the 1992 Intersociety Energy Conversion Engineering Conference, PNL-8381. Pacific Northwest Laboratory, Richland, WA.

Kirshenbaum, M.S. 1991. Chilled-water production in ice-based thermal storage systems. *ASHRAE Transactions* 97(2):422-27.

Knebel, D.E. 1986. Thermal storage—A showcase on cost savings. *ASHRAE Journal* 28(5):28-31.

Knebel, D.E. 1988a. Economics of harvesting thermal storage systems: A case study of a merchandise distribution center. *ASHRAE Transactions* 94(1):1894-1904. Reprinted in ASHRAE Technical Data Bulletin 5(3):35-39, 1989.

Knebel, D.E. 1988b. Optimal control of harvesting ice thermal storage systems. AICE Proceedings, 1990, pp. 209-214.

Knebel, D.E. and S. Houston. 1989. Case study on thermal energy storage—The Worthington Hotel. *ASHRAE Journal* 31(5):34-42.

Knebel, D.E. 1991. Optimal design and control of ice harvesting thermal energy storage systems. ASME 91-HT-28. American Society of Mechanical Engineers, New York.

Kreith, F. 1973. *Principles of heat transfer,* 3rd ed. Harper and Row, Singapore.

Landry, C.M. and C.D. Noble. 1991. Case study of cost-effective low-temperature air distribution, ice thermal storage. *ASHRAE Transactions* 97(1):854-62.

MacCracken, C.D. 1994. An overview of the progress and the potential of thermal storage in off-peak turbine inlet cooling. *ASHRAE Transactions* 100(1):569-71.

Mackie, E.I. 1994. Inlet air cooling for a combustion turbine using thermal storage. *ASHRAE Transactions* 100(1):572-82.

Mackie, E.I. and G. Reeves. 1988. *Stratified chilled-water storage design guide.* EPRI EM-4852s. Electric Power Research Institute, Palo Alto, CA.

Meckler, M. 1992. Design of integrated fire sprinkler piping and thermal storage systems: Benefits and challenges. *ASHRAE Transactions* 98(1):1140-48.

Meierhans, R.A. 1993. Slab cooling and earth coupling. *ASHRAE Transactions* 99(2):511-18.

Morofsky, E. 1994. Procedures for the environmental impact assessment of aquifer thermal energy storage. PWC/RDD/106E. Environment Canada and Public Works and Government Services Canada, Ottawa K1A OM2.

Morris, F.B., J.E. Braun, and S.J. Treado. 1994. Experimental and simulated performance of optimal control of building thermal storage. *ASHRAE Transactions* 100(1):402-14.

Oak Ridge National Laboratories. 1985. Field performance of residential thermal storage systems. EPRI EM 4041 (May).

Public Works Canada. 1991. Workshop on generic configurations of seasonal cold storage applications. International Energy Agency, Energy Conservation Through Energy Storage Implementing Agreement (Annex 7). PWC/RDD/89E (September). Public Works Canada, Ottawa K1A OM2.

Public Works Canada. 1992. Innovative and cost-effective seasonal cold storage applications: Summary of national state-of-the-art reviews. International Energy Agency, Energy Conservation Through Energy Storage Implementing Agreement (Annex 7). PWC/RDD/96E (June). Public Works Canada, Ottawa K1A OM2.

Ruud, M.D., J.W. Mitchell, and S.A. Klein. 1990. Use of building thermal mass to offset cooling loads. *ASHRAE Transactions* 96(2):820-30.

Simmonds, P. 1991a. A building's thermal inertia. CIBSE National Conference, Canterbury, England.

Simmonds, P. 1991b. The utilization and optimization of a building's thermal inertia in minimizing the overall energy use. *ASHRAE Transactions* 97(2):1031-42.

Simmonds, P. 1994. A comparison of energy consumption for storage priority and chiller priority for ice-based thermal storage systems. *ASHRAE Transactions* 100(1):1746-53.

Snijders, A.L. 1992. Aquifer seasonal cold storage for space conditioning: Some cost-effective applications. *ASHRAE Transactions* 98(1):1015-22.

Stewart, W.E., G.D. Gute, J. Chandrasekharan, and C.K. Saunders. 1995a. Modelling of the melting process of ice stores in rectangular thermal energy storage tanks with multiple ice openings. *ASHRAE Transactions* 101(1).

Stewart, W.E., G.D. Gute, and C.K. Saunders. 1995b. Ice melting and melt water discharge temperature characteristics of packed ice beds for rectangular storage tanks. *ASHRAE Transactions* 101(1).

Stovall, T.K. 1991. Turbo Refrigerating Company ice storage test report. ORNL/TM-11657. Oak Ridge National Laboratory, Oak Ridge, TN.

Tamblyn, R.T. 1985. College Park thermal storage experience. *ASHRAE Transactions* 91(1B):947-51.

Tran, N., J.F. Kreider, and P. Brothers. 1989. Field measurement of chilled water storage thermal performance. *ASHRAE Transactions* 95(1):1106-12.

Wildin, M.W. 1990. Diffuser design for naturally stratified thermal storage. *ASHRAE Transactions* 96(1):1094-1102.

Wildin, M.W. 1991. Flow near the inlet and design parameters for stratified chilled water storage. ASME 91-HT-27. American Society of Mechanical Engineers, New York.

Wildin, M.W. and C.R. Truman. 1989. Performance of stratified vertical cylindrical thermal storage tanks—Part I: Scale model tank. *ASHRAE Transactions* 95(1):1086-95.

Willis, S. and J. Wilkins. 1993. Mass appeal. *Building Services Journal* 15(1):25-27.

Yoo, J., M. Wildin, and C.W. Truman. 1986. Initial formation of a thermocline in stratified thermal storage tanks. *ASHRAE Transactions* 92(2A):280-90.

Sources of certain figures are as follows: Figure 11 courtesy Calmac Manufacturing Corp., Figure 13 courtesy Cryogel, Figure 14 courtesy Reaction Thermal Systems, Figure 16 courtesy Transphase Systems, and Figure 18 courtesy Control Electric Corp.

BIBLIOGRAPHY

ASHRAE. 1989. *Cool Storage Applications.* ASHRAE Technical Data Bulletin 5(3).

ASHRAE. 1989. *Cool Storage Modeling and Design.* ASHRAE Technical Data Bulletin 5(4).

AWWA. 1984. Standard for welded steel tanks for water storage. *Standard* D100-84. American Water Works Association, Denver, CO.

AWWA. 1984. Standard for wire-wound circular prestressed-concrete water tanks. *Standard* D110-86. American Water Works Association, Denver, CO.

Ames, D.A. 1986. Design and control of eutectic salt thermal storage systems. Transphase System Inc. Brochure, August 1986.

Ames, D.A. 1990. Thermal storage forum: Eutectic cool storage—Current developments. *ASHRAE Journal* 32(4):46-53.

Anderson, E.D., N.W. Hanson, and D.M. Schultz. 1987. Concrete storage tank construction techniques and quality control. EPRI Seminar Proceedings: Commercial Cool Storage, State of the Art, EM-5454-SR (October).

Andersson, O. 1993. Scaling and corrosion, International Energy Agency, Energy Conservation Through Energy Storage Implementing Agreement (Annex 6—Environmental and chemical aspects of thermal energy storage in aquifers and research and development of water treatment methods), Report 38-E.

Arnold, D. 1991. Laboratory performance of an encapsulated-ice store. *ASHRAE Transactions* 97(2):1170-78.

Arnold, D. 1994. Dynamic simulation of encapsulated ice stores—Part II: Model development and validation. *ASHRAE Transactions* 100(1):1245-54.

Athienitis, A.K. and T.Y. Chen. 1993. Experimental and theoretical investigation of floor heating with thermal storage. *ASHRAE Transactions* 99(1):1049-60.

Bellecci, C. and M. Conti. 1993. Transient behaviour analysis of a latent heat thermal storage module. *International Journal of Heat and Mass Transfer* 36(15):3851.

Berlund, L.G. 1991. Comfort benefits for summer air conditioning with ice storage. *ASHRAE Transactions* 97(1):843-47.

Bhansali, A. and D.C. Hittle. 1990. Estimated energy consumption and operating cost for ice storage systems with cold air distribution. *ASHRAE Transactions* 96(1):418-27.

Brady, T.W. 1994. Achieving energy conservation with ice-based thermal storage. *ASHRAE Transactions* 100(1):1735-45.

Cai, L., W.E. Stewart, Jr., and C.W. Sohn. 1993. Turbulent buoyant flows into a two-dimensional storage tank. *International Journal of Heat and Mass Transfer* 36(17):4247-56.

Cao, Y. and A. Faghri. 1992. A study of thermal energy storage systems with conjugate turbulent forced convection. *Journal of Heat Transfer* 114(4):1019.

CBI. 1991. CBI Strata-Therm: Thermally stratified water storage systems. CBI *Brochure* ST 0191. Chicago Bridge and Iron Company, Chicago.

Cole, R.T. et al. 1980. Design and installation manual for thermal energy storage, 2nd ed. ANL-79-15. Argonne National Laboratory, Argonne, IL.

Conniff, J.P. 1991. Strategies for reducing peak air-conditioning loads by using heat storage in the building structure. *ASHRAE Transactions* 97(1):704-9.

Crane, R.J. and M.J.M. Krane. 1992. The optimum design of stratified thermal energy storage systems—Part II: Completion of the analytical model, presentation and interpretation of the results. *Journal of Energy Resources Technology* 114(3):204.

Darkwa, K. and P.W. O'Callaghan. 1993. Thermal energy storage in the transport sector. *Energy World* 209(June):15.

Dorgan, C.E. and J.S. Elleson. 1987. Low temperature air distribution: Economics, field evaluation, design. EM-5454-SR (October). EPRI Seminar Proceedings: Commercial Cool Storage, State of the Art.

Dorgan, C.E. and J.S. Elleson. 1988. Cold air distribution design guide. EM-5730 (March). Electric Power Research Institute, Palo Alto, CA.

Dorgan, C.E. and J.S. Elleson. 1989. Design of cold air distribution systems with ice storage. *ASHRAE Transactions* 95(1):1317-22. Reprinted in ASHRAE Technical Data Bulletin 5(4):23-28.

Dorgan, C.E. and J.S. Elleson. 1994. ASHRAE design guide for cool thermal storage. *ASHRAE Transactions* 100(1):33-48.

Ebeling, J.A. and R. R. Balsbaugh. 1993. Combustion turbine inlet air cooling with thermal energy storage. *Turbomachinery International* 34(1):30.

Elleson, J.S., S.S. Dingle, and S.P. Leight. 1993. Field evaluation of a eutectic salt thermal storage system. *Draft Report.* Electric Power Research Institute, Palo Alto, CA. (Final report in press).

Etaion, Y. and E. Erell. 1991. Thermal storage mass in radiative cooling systems. *Building and Environment* 26(4):389.

Fiorino, D.P. 1990. Thermal energy storage retrofit project at a large manufacturing facility. *Energy and environmental strategies for the 1990's*, Ch. 82, pp. 485-509.

Fiorino, D.P. 1991. Case study of a large, naturally stratified, chilled water thermal energy storage system. *ASHRAE Transactions* 97(2):1161-69.

Fiorino, D.P. 1992. Thermal energy storage program for the 1990's. *Energy Engineering* 89(4):23-33.

GPC-1 199(4)R. The HVAC Commissioning Process. (Public Review Draft expected to be available 2nd qtr 1994).

Gallagher, M.W. 1991. Integrated thermal storage/Life safety systems in tall buildings. *ASHRAE Transactions* 97(1):833-38.

Gatley, D.P. 1992. Cool storage ethylene glycol design guide. EPRI TR-100945. Electric Power Research Institute, Palo Alto, CA.

Goncalves, L.C.C. and S.D. Probert. 1993. Thermal energy storage: Dynamic performance characteristics of cans each containing a phase-change material, assembled as a packed bed. *Applied Energy* 45(2):117.

Gretarsson, S.P., C.O. Pedersen, and R.K. Strand. 1994. Development of a fundamentally based stratified thermal storage tank model for energy analysis calculations. *ASHRAE Transactions* 100(1):1213-20.

Guven, H. and J. Flynn. 1992. Commissioning TES systems. *Heating/Piping/Air Conditioning* (January):82-84.

Guven, H. and S. Spaeth. 1993. TES commissioning guidelines. California Institute of Energy Efficiency, Berkeley, CA.

Harmon, J.J. and H.C. Yu. 1989. Design considerations for low-temperature air distribution systems. *ASHRAE Transactions* 95(1):1295-99.

Harmon, J.J. and H.C. Yu. 1991. Centrifugal chillers and glycol ice thermal storage units. *ASHRAE Journal* 33(12):25-31.

Hensel, E.C. Jr., N.L. Robinson, J. Buntain, J.W. Glover, B.D. Birdsell, and C.W. Sohn. 1991. Chilled-water thermal storage system performance monitoring. *ASHRAE Transactions* 97(2):1151-60.

Hensen, J.L.M. 1990. Literature review on thermal comfort in transient conditions. *Building and Environment* 25(4):309-16.

Holness, G.V.R. 1987. Thermal storage forum: Industrial plants offer opportunities. *ASHRAE Journal* 29(5):24-25.

Holness, G.V.R. 1988. Thermal storage retrofit restores dual-temperature system. *ASHRAE Transactions* 94(1):1866-78. Reprinted in ASHRAE Technical Data Bulletin 5(3):51-59.

Humphreys, M.A. and J. Nicol. 1991. An investigation in the thermal comfort of office workers. Current Paper CP14/71. Building Research Establishment, Garston, UK.

Hussain, M.A. and M.W. Wildin. 1991. Studies of mixing on the inlet side of the thermocline in diurnal stratified storage. Proceedings THERMASTOCK 1991, Fifth International Conference on Thermal Energy Storage, Scheveningen, Netherlands, May 13-16.

Jekel, T.B., J.W. Mitchell, and S.A. Klein. 1993. Modeling of ice-storage tanks. *ASHRAE Transactions* 99(1):1016-24.

Joyce, W.S. and W.P. Bahnfleth. 1992. Cornell thermal storage project saves money and electricity. *District Heating and Cooling* (2):22-29.

Kamel, A.A., M.V. Swami, S. Chandra, and C.W. Sohn. 1991. An experimental study of building-integrated off-peak cooling using thermal and moisture ("enthalpy") storage systems. *ASHRAE Transactions* 97(2):240-44.

Kleinbach, E., W. Beckman, and S. Klein. 1993. Performance study of one-dimensional models for stratified thermal storage tanks. *Solar Energy* 50(2):155.

Landry, C.M. and C.D. Noble. 1991. Making ice thermal storage first-cost competitive. *ASHRAE Journal* 33(5):19-22.

Laybourn, D.R. 1988. Thermal energy storage with encapsulated ice. *ASHRAE Transactions* 94(1):1971-88. Reprinted in ASHRAE Technical Data Bulletin 5(4):95-102.

Laybourn, D.R. 1990. Encapsulated ice thermal energy storage. AICE Proceedings, pp. 215-20.

Leight, S.P. and J.S. Elleson. 1993. Case study of an ice storage system with cold air distribution and heat recovery, draft report. Electric Power Research Institute, Palo Alto, CA. (Final report in press).

MacCracken, C.D. 1985. Control of brine-type ice storage systems. *ASHRAE Transactions* 91(1B):32-43. Reprinted in ASHRAE Technical Data Bulletin: *Thermal Storage* 1985(January):26-36.

Mackie, E.I. and W.V. Richards. 1992. Design of off-peak cooling systems. ASHRAE Professional Development Seminar, 1992.

McQuiston, F.C. and J.D. Spitler. 1992. *Cooling and heating load calculation manual*, 2nd ed. ASHRAE.

Midkiff, K.C., Y.K. Song, and C.E. Brett. 1991. Thermal performance and challenges for a seasonal chill energy storage based air-conditioning system. ASME 91-HT-29. American Society of Mechanical Engineers, New York.

Mirth, D.R., S. Ramadhyani, and D.C. Hittle. 1993. Thermal performance of chilled-water cooling coils operating at low water velocities. *ASHRAE Transactions* 99(1):43-53.

Mirza, C. 1992. Proceedings of Workshop on Design and Construction of Water Wells for ATES, International Energy Agency, Energy Conservation Through Energy Storage Implementing Agreement (Annex 6—Environmental and chemical aspects of thermal energy storage in aquifers and research and development of water treatment methods), Report 41-E (August). The Netherlands.

Molson, J.W., E.O. Frind, and C.D. Palmer. 1992. Thermal energy storage in an unconfined aquifer—II: Model development, validation, and application. *Water Resources Research* 28(10):2857.

Najafi, M. and W.J. Schaetzle. 1991. Cooling and heating with clathrate thermal energy storage system. *ASHRAE Transactions* 97(1):177-83.

Nussbaum, O.J. 1990. Using glycol in a closed circuit system. *Heating/Piping/Air Conditioning* (January):75-85.

Ott, V.J. and D. Limaye. 1986. Thermal storage air conditioning with clathrates and direct contact heat transfer. EPRI Proceedings: International Load Management Conference, June 1986, Section 47.

Paris, J., M. Falareau, and C. Villeneuve. 1993. Thermal storage by latent heat: A viable option for energy conservation in buildings. *Energy Sources* 15(1):85.

Peters, D., W. Chadwick, and J. Esformes. 1986. Equipment sizing concepts for ice storage systems. EPRI Proceedings: International Load Management Conference, June 1986, Section 46.

Prusa, J., G.M. Maxwell, and K.J. Timmer. 1991. A Mathematical model for a phase-change, thermal energy storage system utilizing rectangular containers. *ASHRAE Transactions* 97(2):245-61.

Rabl, V.A. 1987. Load management: Issues and opportunities. EM-5454-SR (October). EPRI Seminar Proceedings: Commercial Cool Storage, State of the Art. Electric Power Research Institute, Palo Alto, CA.

Rogers, E.C. and B.A. Stefl. 1993. Ethylene glycol: Its use in thermal storage and its impact on the environment. *ASHRAE Transactions* 99(1):941-49.

Rudd, A.F. 1993. Phase-change material wallboard for distributed thermal storage in buildings. *ASHRAE Transactions* 99(2):339-46.

Ryu, H.W., S.A. Hong, and B.C. Shin. 1991. Heat transfer characteristics of cool-thermal storage systems. *Energy* 16(4):727.

Slabodkin, A.L. 1992. Integrating an off-peak cooling system with a fire suppression system in an existing high-rise office building. *ASHRAE Transactions* 98(1):1133-39.

Sodha, M.S., J. Kaur, and R.L. Sawhney. 1992. Effect of storage on thermal performance of a building. *International Journal of Energy Research* 16(8):697.

Sohn, C.W. and J.J. Tomlinson. 1988. Design and construction of an ice-in-tank diurnal ice storage for the PX building at Fort Stewart, GA.

CERL-TR-E-88/07 *Final Report*. Construction Engineering Research Lab.

Sohn, C.W., G.L. Cler, and R.J. Kedl. 1990a. Performance of an ice-in-tank diurnal ice storage cooling system at Fort Stewart, GA. U.S.A. CERL *Technical Report* E-90/10. Construction Engineering Research Lab.

Sohn, C.W., G.L. Cler, and R.J. Kedl. 1990b. Ice-on-coil diurnal ice storage cooling system for a barracks/office/dining hall facility at Yuma Proving Ground, AZ. CERL *Technical Report* E-90/13. Construction Engineering Research Lab.

Sohn, C.W. 1991a. Thermal performance of an ice storage cooling system. ASME 91-HT-26. American Society of Mechanical Engineers, New York.

Sohn, C.W. 1991b. Field performance of an ice harvester storage cooling system. *ASHRAE Transactions* 97(2):1187-96.

Somasundaram, S., D. Brown, and K. Drost. 1992. Cost evaluation of diurnal thermal energy storage for cogeneration applications. *Energy Engineering* 89(4):8.

Sozen, M., K. Vafai, and L.A. Kennedy. 1992. Thermal charging and discharging of sensible and latent heat storage packed beds. *Journal of Thermophysics and Heat Transfer* 5(4):623.

Spethmann, D.H. 1989. Optimal control for cool storage. *ASHRAE Transactions* 95(1):1189-93. Reprinted in ASHRAE Technical Data Bulletin 5(4):57-61.

Spethmann, D.H. 1993. Application considerations in optimal control of cool storage. *ASHRAE Transactions* 99(1).

Stewart, R.E. 1990. Ice formation rate for a thermal storage system. *ASHRAE Transactions* 96(1):400-5.

Stewart, W.E., R.L. Kaupang, C.G. Tharp, R.D. Wendland, and L.A. Stickler. 1993. An approximate numerical model of falling-film iced crystal growth for cool thermal storage. *ASHRAE Transactions* 99(2):347-55.

Stovall, T.K. 1991a. Baltimore Aircoil Company (BAC) ice storage test report. ORNL/TM-1 1342. Oak Ridge National Laboratory, Oak Ridge, TN.

Stovall, T.K. 1991b. CALMAC ice storage test report. ORNL/TM-11582. Oak Ridge National Laboratory, Oak Ridge, TN.

Stovall, T.K. and J.J. Tomlinson. 1991. Laboratory performance of a dynamic ice storage system. *ASHRAE Transactions* 97(2):1179-86.

Strand, R.K., C.O. Pedersen, and G.N. Coleman. 1994. Development of direct and indirect ice-storage models for energy analysis calculations. *ASHRAE Transactions* 100(1):1230-44.

Tamblyn, R.T. 1985c. Control concepts for thermal storage. *ASHRAE Transactions* 91(1B):5-11. Reprinted in ASHRAE Technical Data Bulletin: *Thermal Storage* 1985(January):1-6. Also reprinted in *ASHRAE Journal* 27(5):31-34.

Tamblyn, R.T. 1990. Optimizing storage savings. *Heating/Piping/Air Conditioning* (August):43-46.

Wildin, M.W., E.I. Mackie, and W.E. Harrison. 1990. Thermal storage forum—Stratified thermal storage: A new/old technology. *ASHRAE Journal* 32(4):29-39.

Wildin, M.W. and C.R. Truman. 1985a. A summary of experience with stratified chilled water tanks. *ASHRAE Transactions* 91(1B):956-76. Reprinted in ASHRAE Technical Data Bulletin: *Thermal Storage* 1985 (January):104-123.

Wildin, M. W. and C.R. Truman. 1985b. Evaluation of stratified chilled-water storage techniques. EPRI EM-4352. Electric Power Research Institute, Palo Alto, CA.

Yoo, H. and E.-T. Pak. 1993. A theoretical model of the charging process for stratified thermal storage tanks. *Solar Energy* 51(6):513.

CHAPTER 41

CONTROL OF GASEOUS INDOOR AIR CONTAMINANTS

AIR throughout the world contains nearly constant amounts of nitrogen (78% by volume), oxygen (21%), and argon (0.9%), with varying amounts of carbon dioxide (about 0.03%) and water vapor (up to 3.5%). In addition, trace quantities of inert gases (neon, xenon, krypton, helium, etc.) are always present. Gases other than those listed are usually considered contaminants or pollutants. Their concentrations are almost always small, but they may have serious effects on building occupants, construction materials, or contents. Removal of these gaseous pollutants is often desirable or necessary.

Traditionally, indoor gaseous contaminants are controlled with ventilation air drawn from outdoors, but minimizing outdoor airflow by using a high recirculation rate and filtration is an attractive means of energy conservation. However, recirculated air cannot be made equivalent to fresh outdoor air by removing only particulate contaminants. Noxious, odorous, and toxic gaseous contaminants must also be removed by gaseous contaminant control equipment, which is different from particulate filtration equipment. In addition, available outdoor air may contain undesirable gaseous contaminants at unacceptable concentrations. If so, it too will require treatment by gaseous contaminant removal equipment.

This chapter deals with design procedures for gaseous pollutant control systems for occupied spaces only. The control of gaseous pollutants from industrial processes and stack gases is covered in Chapter 11 of the 1992 *ASHRAE Handbook—Systems and Equipment.*

GASEOUS CONTAMINANT CHARACTERISTICS

Before a gaseous contaminant control system can be designed, at least the following information must be determined:

1. Exact chemical identity of the contaminants present in significant concentrations
2. Rates at which the contaminants are generated in the space,
3. Rates at which contaminants are brought into the space with outdoor air.

This information may be difficult to obtain. Designers must often make do with a chemical family name (e.g., aldehydes) and a qualitative description of generation (e.g., from plating tank) or perceived concentration (e.g., at odorous levels). System design under these conditions is uncertain. When the exact chemical identity of a contaminant is known, the chemical and physical properties influencing its collection by control devices can usually be obtained from handbooks and technical publications. Factors of special importance are

---

The preparation of this chapter is assigned to TC 2.3, Gaseous Air Contaminants and Gas Contaminant Removal Equipment.

- Molecular weight
- Normal boiling point (i.e., at standard pressure)
- Heat of vaporization
- Polarity
- Chemical reactivity and chemisorption velocity

With this information, the effectiveness of control devices on contaminants for which no specific tests have been made may be estimated. The following sections on control methods discuss the usefulness of such estimations. Some gaseous contaminants, including ozone, radon, and sulfur trioxide, have unique properties that must be considered in system design.

Ozone will reach an equilibrium concentration in a ventilated space without a filtration device. It does so partly because ozone molecules join to form normal oxygen, but also because it reacts with people, plants, and materials in the space. This oxidation is harmful to all three, and therefore natural ozone decay is not a satisfactory way to control ozone except at low concentrations (< 0.2 mg/m³). Fortunately, activated carbon adsorbs ozone readily, both reacting with it and catalyzing its conversion to oxygen.

Radon is a radioactive gas that decays by alpha-particle emission, eventually yielding individual atoms of polonium, bismuth, and lead. These atoms form extremely fine aerosol particles, called radon daughters or radon progeny, which are also radioactive; they are especially toxic in that they lodge deep in the human lung, where they emit cancer-producing alpha particles. Radon progeny, both attached to larger aerosol particles and unattached, can be captured by particulate air filters. Radon gas itself may be removed with activated carbon (Thomas 1974), but in HVAC systems this method costs too much for the benefit derived. Control of radon emission at the source and ventilation are the preferred methods of radon control.

Another gaseous contaminant that often appears in particulate form is sulfur trioxide (SO_3), which should not to be confused with sulfur dioxide (SO_2). At ambient temperatures, it reacts rapidly with water vapor to form a fine mist of sulfuric acid. If this mist collects on a particulate filter, and no means is provided to remove it, the acid will vaporize and reenter the protected space. Because many such conversions and desorptions are possible, the designer must understand the problems they cause.

Major chemical families of gaseous pollutants and examples of specific compounds are given in Table 1. The *Merck Index* (Windholtz 1983), the *Toxic Substances Control Act Chemical Substance Inventory* (EPA 1979), and *Dangerous Properties of Industrial Materials* (Sax and Lewis 1988) are all useful in identifying contaminants, including some known by trade names only. Note that a single chemical compound, especially if it is organic, may have several scientific names.

41.1

Table 1 Major Chemical Families of Gaseous Air Pollutants (with Examples)

Inorganic Pollutants

1. Single-element atoms and molecules
   - chlorine
   - radon
   - mercury
2. Oxidants
   - ozone
   - nitrogen dioxide
   - nitrous oxide
   - nitric oxide
3. Reducing agents
   - carbon monoxide
4. Acid gases
   - sulfur dioxide
   - sulfuric acid
   - hydrochloric acid
   - hydrogen sulfide
   - nitric acid
5. Nitrogen compounds
   - ammonia
6. Miscellaneous
   - arsine

Organic Pollutants

7. n-Alkanes
   - methane
   - n-butane
   - n-hexane
   - n-octane
   - n-hexadecane
8. Branched alkanes
   - 2-methyl pentane
   - 2-methyl hexane
9. Alkenes and cyclohexanes
   - 1-octene
   - 1-decene
   - cyclohexane
10. Chlorofluorocarbons
    - R-11 (trichlorofluoromethane)
    - R-114 (dichlorotetrafluoroethane)
11. Chlorinated hydrocarbons
    - 1,1,1-trichloroethane
    - carbon tetrachloride
    - chloroform

- perchloroethane
- tetrachloroethylene
12. Halide compounds
    - methyl bromide
    - methyl iodide
13. Alcohols
    - methanol
    - ethanol
    - 2-propanol (isopropanol)
    - phenol
    - cresol
    - diethylene glycol
14. Ethers
    - vinyl ether
    - methoxyvinyl ether
    - n-butoxyethanol
15. Aldehydes
    - formaldehyde
    - acetaldehyde
    - acrolein
    - benzaldehyde
16. Ketones
    - 2-butanone (MEK)
    - 2-propanone (acetone)
    - methyl isobutyl ketone (MIBK)
    - chloroacetophenone
17. Esters
    - ethyl acetate
    - n-butyl acetate
    - diethylhexyl phthalate
    - dioctyl phthalate (DOP)
    - di-n-butyl phthalate
    - butyl formate
    - methyl formate
18. Nitrogen compounds other than amines
    - nitromethane
    - acetonitrile
    - acrylonitrile
    - urea
    - uric acid
    - skatole
    - putrescine
    - hydrogen cyanide
    - peroxyacetal nitrate

19. Aromatic hydrocarbons
    - benzene
    - toluene
    - ethyl benzene
    - naphthalene
    - p-xylene
    - benz-α-pyrene
20. Terpenes
    - 2-pinene
    - limonene
21. Heterocylics
    - ethylene oxide
    - furan
    - tetrahydrofuran
    - pyrrole
    - pyridine
    - methyl furfural
    - nicotine
    - 1,4-dioxane
    - caffeine
22. Organophosphates
    - malathion
    - tabun
    - sarin
    - soman
23. Amines
    - methylamine
    - diethylamine
    - n-nitrosodimethylamine
24. Monomers
    - vinyl chloride
    - methyl formate
    - ethylene
25. Mercaptans and other sulfur compounds
    - bis-2-chloroethyl sulfide (mustard gas)
    - ethyl mercaptan
    - methyl mercaptan
    - carbon disulfide
    - carbonyl sulfide
26. Organic acids
    - formic acid
    - acetic acid
    - butyric acid
27. Miscellaneous
    - phosgene

HARMFUL EFFECTS OF GASEOUS CONTAMINANTS

The only reason to remove a gaseous contaminant from an airstream is that it has harmful or annoying effects on the ventilated space or its occupants. These effects are noticeable at different concentration levels. In most cases, contaminants become annoying through their odors before they reach levels toxic to humans, but this is not always true. For example, the highly toxic (even deadly) contaminant carbon monoxide has no odor.

There are four categories of harmful effects: toxicity, odor, irritation, and material damage. These effects depend on the contaminant concentrations, which are usually expressed in the following units:

ppm = parts of contaminant by volume per million parts of air by volume

ppb = parts of contaminant by volume per billion parts of air by volume

mg/m^3 = milligrams of contaminant per cubic metre of air

μg/m^3 = micrograms of contaminant per cubic metre of air

The conversions between ppm and mg/m^3 (based on the general gas law) are

$$\text{ppm} = 8.314 \, (\text{mg}/\text{m}^3)(273.15 + t)/(Mp) \qquad (1)$$

$$\text{mg}/\text{m}^3 = (\text{ppm})(Mp)/8.314\,(273.15 + t) \qquad (2)$$

where

> M = relative molecular mass of contaminant
> p = mixture pressure, kPa
> t = mixture temperature, °C

Concentration data are often reduced to a standard temperature and pressure (i.e., 25°C and 101.325 kPa), in which case

$$\text{ppm} = 24.46\,(\text{mg/m}^3)/M \qquad (3)$$

Toxicity

The harmful effects of gaseous pollutants on a person depend on both short-term peak concentrations and the time-integrated exposure received by the person. The allowable concentration for short exposures is higher than that for long exposures. In the United States, the Occupational Safety and Health Administration (OSHA) has defined three concentration-averaging periods for workplaces and has assigned allowable average concentrations for the three periods for over 490 compounds, mostly gaseous contaminants. The abbreviations for concentrations for the three averaging periods are

AMP = acceptable maximum peak for a short exposure
ACC = acceptable ceiling concentration, not to be exceeded during an 8-h shift, except for periods where AMP applies
TWA8 = time-weighted average, not to be exceeded in any 8-h shift of a 40-h week

In non-OSHA literature, ACC is sometimes called STEL (short-term exposure limit), and TWA8 is sometimes called TLV (threshold limit value). The medical community disagrees on what values should be assigned to AMP, ACC, and TWA8 for different contaminants. OSHA values, which change periodically, are published yearly in the *Code of Federal Regulations* (29 CFR 1900, 1000 ff) and intermittently in the *Federal Register*. A similar list is available from the American Conference of Governmental Industrial Hygienists (ACGIH 1993). The National Institute for Occupational Safety and Health (NIOSH) is charged with researching toxicity problems, and it greatly influences the legally required levels. NIOSH annually publishes the *Registry of Toxic Effects of Chemical Substances* as well as numerous *Criteria for Recommended Standard for Occupational Exposure to (compound)*. Some compounds not in the OSHA list are covered by NIOSH literature, and recommended levels are sometimes lower than the legal requirements set by OSHA. The *NIOSH/OSHA Pocket Guide to Chemical Hazards* (NIOSH 1990) is a condensation of these references and is convenient for engineering purposes. This publication also lists values for the following toxic limit:

IDLH = immediately dangerous to life and health

Although this toxicity limit is rarely a factor in HVAC design, HVAC engineers should consider it when deciding how much recirculation is safe in a given system. Ventilation airflow must never be so low that the concentration of any gaseous contaminant could rise to the IDLH level. The levels set by OSHA and NIOSH define acceptable occupational exposures, but cannot be used by themselves as acceptable standards for residential or commercial spaces. They do, however, suggest some upper limits for contaminant concentrations for design purposes.

Personnel exposure to radioactive gases is addressed not by OSHA regulations, but by rules promulgated by the Nuclear Regulatory Commission. Allowable concentrations, in terms of radioactivity, are listed annually in the *Code of Federal Regulations* (10 CFR 50). Radioactive gases are controlled in much the same way as nonradioactive gases, but their high toxicity demands more careful system design. Chapter 23 introduces the procedures unique to nuclear systems.

Another toxic effect that may influence design is the loss of sensory acuity due to gaseous contaminant exposure. Carbon monoxide, for example, affects psychomotor responses and could be a problem in areas such as air traffic control towers. Clearly, waste anesthetic gases in operating suites should not be allowed to reach levels that affect the alertness of any of the personnel. NIOSH recommendations are frequently based on such subtle effects.

Odors

Concentrations as high as those set by OSHA regulations are rarely encountered in commercial or residential spaces. In such spaces, gaseous contaminant problems usually appear as complaints about odors or stuffiness that are the result of concentrations considerably below TWA8 values. Each individual has different sensitivities to odors, and this sensitivity decreases even during a relatively brief exposure. One pollutant may enhance or mask the odor of another. Odors are usually stronger when relative humidity is high. In addition, the human olfactory response S is nonlinear; perceived intensities are approximate power functions of the contaminant concentration C:

$$S = kC^n \qquad (4)$$

where n is typically about 0.6.

This nonlinear response means that the concentration of some odorants must be reduced substantially before the odor level is perceived to change. For these reasons, determining acceptable concentrations of pollutants on the basis of their odors is as imprecise as for their toxicity. Far fewer data are available on odors, since odors are more of an annoyance than a hazard. In fact, it is desirable for a toxic or explosive material to have an odor threshold well below toxic levels, for this can warn building occupants that the pollutant is present.

At some low concentration of an odorant, an individual ceases to be aware of its presence. Odor studies seek to establish the level at which a percentage of the general population—usually 50%—is no longer aware of the odor of the compound studied. This concentration is the *odor threshold* for that compound. Fazzalari (1978) compiled odor thresholds for a large number of gaseous compounds in commercial and industrial situations. Additional values, reflecting newer measurement techniques but largely limited to compounds found only in industrial workplaces, are listed in AIHA (1989), Moore and Houtala (1983), and Van Germert and Mettenbreijer (1977). Table 2 shows the wide range of odor threshold values for different contaminants. Chapter 12 of the 1993 *ASHRAE Handbook—Fundamentals* discusses odor perception.

Irritation

Although some gaseous pollutants may have no discernible continuing health effects, exposure to such elements may irritate building occupants. Coughing; sneezing; eye, throat, and skin irritation; nausea; breathlessness; drowsiness; headaches; and depression have all been attributed to air pollutants. Rask (1988) suggests that when 20% of a single building's occupants suffer such irritations, the structure is said to suffer from "sick building syndrome" or "tight building syndrome." For the most part, case studies of such occurrences have consisted of analyses of questionnaires submitted to building occupants. Some attempts to relate irritations to gaseous contaminant concentrations are reported (Lamm 1986, Cain et al. 1986, Berglund et al. 1986, Molhave et al. 1982). The correlation of reported complaints with gaseous pollutant concentrations is not strong; many factors affect these less serious responses to pollution. In general, physical irritation does not occur at odor threshold concentrations.

Damage to Materials

Material damage from gaseous pollutants may take such forms as corrosion, embrittlement, and discoloration. Because such effects usually involve chemical reactions that need water, material damage from air pollutants is less in the relatively dry indoor environment

Table 2   Characteristics of Selected Gaseous Air Pollutants

Pollutant	Allowable Concentration, mg/m³				Odor Threshold, mg/m³	Chemical and Physical Properties			
	IDLH[a]	AMP[a]	ACC[a]	TWA8[a]		Family[b]	BP,[c] °C	M[d]	Retentivity,[e,f] %
Acetaldehyde	18 000			360	1.2	15	21	44	8*
Acetone	4 800		3 200	2 400	47	16	56	58	16
Acetonitrile	7 000	105		70	>0	18	82	41	1
Acrolein	13		0.75	0.25	0.35	15	52	56	
Acrylonitrile	10			45	50	18	77	53	3
Allyl chloride	810		9	3	1.4	12	44	77	
Ammonia	350		35	38	33	5	−33	17	0
Benzene	10 000	25		5	15	19	80	78	12
Benzyl chloride	50			5	0.2	12	179	127	0.3
2-Butanone (MEK)	8 850			590	30	16	79	72	12
Carbon dioxide	90 000		54 000	9 000	0	4	−78	44	
Carbon monoxide	1 650		220	55	0	3	−192	28	0
Carbon disulfide	1 500	300	90	60	0.6	25	46	76	4*
Carbon tetrachloride	1 800	1 200	150	60	130	11	77	154	8
Chlorine	75		1.5	3	0.007	1	−34	71	2*
Chloroform	4 800		9.6	240	1.5	11	124	119	11
Chloroprene	1 440		3.6	90		12	120	89	
p-Cresol	1 100			22	0.056	13	305	108	5*
Dichlorodifluoromethane	250 000			4 950	5 400	10	−30	121	
Dioxane	720			360	304	21	100	68	13
Ethylene dibromide	3 110	271	233	155		12	131	188	11
Ethylene dichloride	4 100	818	410	205	25	12	84	99	12
Ethylene oxide	1 400		135	90	196	21	10	44	0
Formaldehyde	124	12	6	4	1.2	15	97	30	0.4*
n-Heptane	17 000			2 000	2.4	7	98	100	7*
Hydrogen chloride	140		7	7	12	4	−121	37	0.7*
Hydrogen cyanide	55			11	1	18	26	27	0.4
Hydrogen fluoride	13		5	2	2.7	4	19	20	1*
Hydrogen sulfide	420	70	28	30	0.007	4	−60	34	0.8*
Mercury	28			0.1	0	1	357	201	13
Methane	ASPHY[g]					7	−164	16	0
Methanol	32 500			260	130	13	64	32	6*
Methyl chloride	59 500		1 783	1 189	595	12	74	133	
Methylene chloride	7 500		3 480	1 740	750	12	40	85	9
Nitric acid	250			5		4	84	63	3*
Nitric oxide	120	45		30	>0	2	−152	30	
Nitrogen dioxide	90		1.8	9	51	2	21	46	2*
Ozone	20			2	0.2	2	−112	48	h
Phenol	380		60	19	0.18	13	182	94	5*
Phosgene	8		0.8	0.4	4	27	8	90	1
Propane	36 000				1 800	7	−42	44	5*
Sulfur dioxide	260			13	1.2	4	−10	64	1.7*
Sulfuric acid	80			1	1	4	270	98	1.9*
Tetrachloroethane	1 050			35	24	11	146	108	
Tetrachloroethylene	3 430	2 060	1 372	686	140	11	121	166	
o-Toluidine	440			22	24	23	199	107	
Toluene	7 600	1 900	1 140	760	8	19	111	92	17*
Toluene diisocyanate	70		0.14	0.14	15	18	251	174	
1,1,1-Trichloroethane	2 250			45	1.1	11	113	133	5
Trichloroethylene	5 410	1 620	1 080	541	120	11	87	131	
Vinyl chloride monomer			0.014	0.003	1 400	24	−14	63	0.13
Xylene	43 500		870	435	2	19	137	106	16*

Sources: ASTM *Standard* D 1605 (1988), Balieu et al. (1977), Dole and Klotz (1946), Freedman et al. (1973), Gully et al. (1969), Miller and Reist (1977), Revoir and Jones (1972), Turk (1954).

[a]IDLH, AMP, ACC, and TWA8 are defined in the section on Toxicity.

[b]Chemical family numbers are as given in Table 1.

[c]BP = boiling point at 101.325 kPa (1 atmosphere) pressure

[d]M = molecular mass

[e]Retentivities are for typical commercial-grade activated carbon, either measured at TWA8 levels or corrected to TWA8 inlet concentrations using the expression for breakthrough time t_b given in Nelson and Correia (1976):

$$t_{b2} = t_{b1}\,(C_2/C_1)^{-2/3}$$

Multiplying breakthrough time by inlet concentration gives retentivity, so

$$R_2 = R_1\,(C_2/C_1)^{1/3}$$

Both concentrations must be in the same units, here mg/m³.

[f]Retentivities marked (*) were calculated from values given in ASTM *Standard* D 1605 (1988) and Turk (1954) assuming that the listed retentivities were measured at 1000 ppm.

[g]ASPHY = Simple asphyxiant; causes breathing problems when concentration reaches about 1/3 atmospheric pressure.

[h]Ozone life is extremely long; activated carbon assists the essentially complete conversion of ozone to normal oxygen by both chemisorption and catalysis.

than outdoors, even at similar gaseous contaminant concentrations. To maintain this advantage, indoor condensation should be avoided. However, damage to some materials can be significant, especially from ozone, hydrogen peroxide and other oxidants, sulfur dioxide, and hydrogen sulfide.

These effects are most serious in museums, as any loss of color or texture changes the essence of the object. Libraries and archives are also vulnerable, as are pipe organs and textiles. Ventilation is often a poor method of protecting collections of rare objects; facilities are usually located in the centers of cities, which have relatively polluted ambient outdoor air.

Filtration systems for rare-object protection must be applied with great care. Pollution-control methods that claim to modify rather than remove damaging contaminants offer no protection for fragile materials. Likewise, delaying the passage of a contaminant—capturing it at a relatively high concentration and releasing it slowly at low concentration—may reduce its odor or health effects but has almost no effect on material damage. One control device involves a chemical reaction that converts a pollutant into another compound, which may reduce or increase material damage, depending on the reaction product. The performance of the pollution-control device must be demonstrated by actual tests at the concentrations and environmental conditions expected for the ventilation air.

Caution must be taken with systems incorporating electrostatic air cleaners, which can generate ozone. In rare-object protection systems, electrostatic air cleaners must be followed by an effective ozone adsorber; the system must ensure that ozone levels in the ventilation air downstream of the purification equipment are always below an acceptable level. Various concerns on material damage by indoor air pollutants are discussed by Walsh et al. (1977), Chiarenzelli and Joba (1966), NTIS (1982, 1984), Braun and Wilson (1970), Jaffe (1967), Grosjean et al. (1987), American Guild of Organists (1966), Mathey et al. (1983), Haynie (1978), Graminski et al. (1978), and Thomson (1986).

CONTAMINANT SOURCES AND GENERATION RATES

Tobacco Smoke

Tobacco smoke is a prevalent and potent source of indoor air pollutants. Cigar and pipe smoking produce somewhat different compounds than cigarette smoking, but almost all tobacco pollution arises from cigarette smoking. Table 3 lists some of the important gaseous compounds found in cigarette smoke and gives average concentrations for the two types of smoke produced—mainstream and sidestream. Mainstream smoke is that which goes directly from the cigarette to the respiratory tract of the smoker. The part that is not trapped returns to the room air when exhaled by the smoker and joins the sidestream smoke that has not been directly inhaled by the smoker (including the smoke produced between puffs). Tobacco types and cigarette configurations vary widely; the data represent averages of values from several sources. The effect of cigarette tip filters on these values is not striking.

Cleaning Agents and Other Consumer Products

Commonly used liquid detergents, waxes, polishes, spot removers, and cosmetics all contain organic solvents that volatilize either slowly or quickly. Mothballs and other pest control agents also emit organic volatiles. Knoeppel and Schauenburg (1989), Black and Bayer (1986), and Tichenor (1989) report data on the release of these volatile organic compounds (VOCs). Field studies have shown that such products contribute significantly to indoor pollution; however, a large variety of compounds are in use, and few studies have been made that allow calculation of typical emission rates. Pesticides, both those applied indoors and those applied outdoors to control termites, also pollute building interiors.

Table 3 Major Gaseous Compounds in Typical Cigarette Smoke

Pollutant	Mainstream Smoke Generation Rate,[a] mg/cigarette	Sidestream Smoke Generation Rate,[b] mg/cigarette	Sources[c]
Acetaldehyde	0.7	4.4*	2,3
Acetone	0.4	—	1
Acetonitrile	0.1	0.4	1
Acrolein	0.1	0.6*	2,3
Ammonia	0.1	5.9	1
2-Butanone	0.1	0.3	1
Carbon dioxide	45	360	1
Carbon monoxide	17	43	1,4
Ethane	0.4	1.2*	2
Ethene	—	0.8*	2
Formaldehyde	0.05	2.3	2,3
Hydrogen cyanide	0.4	0.1	1
Isoprene	0.3	3.1	2
Methane	1.0	0.4	1
Methyl chloride	0.4	0.8	1
Nitrogen dioxide	—	0.2	1
Nitric oxide	0.2	1.8	1
Propane	0.2	0.7*	2
Propene	0.2	0.8	1

[a]Averages of values reported in listed sources.
[b]Entries without asterisks were obtained by multiplying mainstream values by SS/MS ratios given in Surgeon General (1979), pp. 11-6, 14-3, and 14-37. Entries with asterisks were obtained by using average chamber concentrations in Loefroth et al. (1989), using a chamber volume of 24 m[3].
[c]Sources:
1. Surgeon General (1979)   3. Newsome et al. (1965)
2. Loefroth et al. (1989)   4. Cohen et al. (1971)

Building Materials and Furnishings

Particleboard, which is usually made from wood chips bonded with a phenol-formaldehyde or other resin, is widely used in current construction, especially for mobile homes, carpet underlay, and case goods. These materials, along with ceiling tiles, carpeting, wall coverings, office partitions, adhesives, and paint finishes, emit formaldehyde and other VOCs. Latex paints containing mercury have been shown to emit mercury vapor. While the emission rates for these materials decline steadily with age, the half-life of emissions is surprisingly long. Black and Bayer (1989), Nelms et al. (1986), and Molhave et al. (1982) report on these sources. Measured emission rates are listed in Table 4.

Equipment

Commercial and residential spaces have internal sources of gaseous contaminants, though generation rates are substantially lower than in the industrial environment. As equipment is rarely hooded, emissions go directly to the occupants. In commercial spaces, the chief sources of gaseous contaminants are office equipment, including electrostatic copiers (ozone) and diazo printers (ammonia and related compounds), and office supplies, including carbonless copy paper (formaldehyde), correction fluids, inks, and adhesives (various VOCs). Medical and dental activities generate pollutants from the escape of anesthetic gases (nitrous oxide and halomethanes) and sterilizers (ethylene oxide). The potential for asphyxiation is always a concern when compressed gases are present, even if the gas is nitrogen.

In residences, the main sources of equipment-derived pollutants are gas ranges, wood stoves, and kerosene heaters. Venting is helpful, but some pollutants escape into the occupied area. The pollutant contribution by gas ranges is somewhat mitigated by the fact that they operate for shorter periods than heaters. The same is true of showers, which can contribute to radon and halocarbon concentrations indoors.

Table 4 Generation of Gaseous Pollutants by Building Materials

Contaminant	Average Generation Rate, $\mu g/(h \cdot m^2)$						
	Caulk	Adhe-sive	Lino-leum	Carpet	Paint	Varnish	Lac-quer
C-10 Alkane	1200						
n-Butanol	7300						760
n-Decane	6800						
Formaldehyde			44	150			
Limonene		190					
Nonane	250						
Toluene	20	750	110	160	150		310
Ethyl benzene	7300						
Trimethyl benzene		120					
Undecane						280	
Xylene	28						310

Contaminant	Average Generation Rate, $\mu g/(h \cdot m^2)$						
	GF Insula-tion	GF Duct Liner	GF Duct Board	UF Insula-tion	Particle Under-board	lay	Printed Ply-wood
Acetone					40		
Benzene					6		
Benzaldehyde					14		
2-Butanone					2.5		
Formaldehyde	7	2	4	340	250	600	300
Hexanal					21		
2-Propanol					6		

GF = glass fiber; UF = ureaformaldehyde foam
Sources: Matthews et al. (1983, 1985), Nelms et al. (1986), and White et al. (1988).

Traynor et al. (1985a, 1985b) report on emission rates from indoor combustion devices (Table 5). Emission rates for equipment depend greatly on equipment type and manner of use. Typical values are difficult to obtain (Traynor 1987). Equipment suspected to be a significant source of pollutants should be hooded. Filtration is not likely to be an economical or safe solution to such problems.

Occupants

Humans and animals emit a wide array of pollutants by breath, sweat, and flatus. Some of these emissions are conversions from solids or liquids within the body. Many volatile organics emitted are, however, reemissions of pollutants inhaled earlier, with the tracheobronchial system acting like a gas chromatograph or a saturated physical adsorber. For such pollutants, the occupant may be considered a filter, and, in that sense, a pollutant-control device. Some data on lung filter efficiency are listed in Table 6. Several studies (mostly in relation to spacecraft habitability) have measured pollutant generation by humans (Table 7).

Outdoor Air

The model (Meckler and Janssen 1988) in the section on Influence of Ventilation Patterns allows calculation of the effect of outdoor pollution on indoor air quality. Outdoor concentrations of gaseous contaminants must be included as inputs to this model. Outdoor concentrations of a few pollutants are available for cities in the United States from Environmental Protection Agency (EPA) summaries of the National Air Monitoring Stations network data and in similar reports for other countries. A summary of this data is available (EPA 1990). In many cases, the concentrations listed in the National Ambient Air Quality Standards, which are the levels the

Table 5 Generation of Gaseous Pollutants by Indoor Combustion Equipment

	Generation Rates, $\mu g/kJ$					Typical Heating kW	Typical Use, hour/day	Vented or Unvented	Fuel
	CO_2	NO	NO_2	NO	HCHO				
Convective heater	51 000	83	12	17	1.4	9.1	4	U	Natural gas
Controlled combustion wood stove		13	0.04	0.07		3.8	10	V	Oak, pine
Range oven		20	10	22		9.4	1.0[a]	U	Natural gas
Range-top burner		65	10	17	1.0	2.8/burner	1.7	U	Natural gas

[a]Sterling and Kobayashi (1981) found that gas ranges are used for supplemental heating by about 25% of users in older apartments. This increases the time of use per day to that of unvented convective heaters.

Sources: Wade et al. (1975), Sterling and Kobayashi (1981), Cole (1983), Traynor et al. (1985b), Leaderer et al. (1987), and Moschandreas and Relwani 1989).

Table 6 Efficiency of Lung/Tracheobronchial System in Sorption of Some Volatile Organic Pollutants

Pollutant	Average Concentrations, mg/m^3		Lung and Tracheobronchial Efficiency, %	Sources[a]
	Ambient	Breath		
Benzene	5	1.5	70	1
Chloroform	4	1.8	80	1
Vinylidene chloride	7	1.5	79	1
1,1,1-Trichloroethyene	60	5	92	1
Trichloroethylene	5.5	1	82	1
	5.5	1.7	70	7
n-Dichlorobenzene	3	2	33	1
Formaldehyde	< 0.005		98	3
Acetaldehyde	< 0.0004		60	2
Acrolein	< 0.002		82	3
Carbon monoxide			55	4,5
Sulfur dioxide	14.3	4.3	70	6
Mercury	0.8		74	8
Furan	310		80	9

[a]*Sources:*
1. Wallace et al. (1983)
2. Egle (1971)
3. Egle (1972)
4. Cohen et al. (1971)
5. Surgeon General (1979)
6. Wolff et al. (1975)
7. Stewart et al. (1974)
8. Hursh et al. (1976)
9. Egle (1979)

Table 7 Total-Body Emission of Some Gaseous Pollutants by Humans

Contaminant	Typical Emission, $\mu g/h$	Contaminant	Typical Emission, $\mu g/h$
Acetaldehyde	35	Methane	1710
Acetone	475	Methanol	6
Ammonia	15 600	Methylene chloride	88
Benzene	16	Propane	1.3
2-Butanone (MEK)	9700	Tetrachloroethane	1.4
Carbon dioxide	32×10^6	Tetrachloroethylene	1
Carbon monoxide	10 000	Toluene	23
Chloroform	3	1,1,1-Trichloroethane	42
Dioxane	0.4	Vinyl chloride monomer	0.4
Hydrogen sulfide	15	Xylene	0.003

Sources: Anthony and Thibodeau (1980), Brugnone et al. (1989); Cohen et al. (1971), Conkle et al. (1975), Gorban et al. (1964), Hunt and Williams (1977), and Nefedov et al. (1972).

Table 8 Primary Ambient Air Quality Standards for the United States

Contaminant	Long Term Concentration, µg/m³	Long Term Averaging Period	Short Term Concentration, µg/m³	Short Term Averaging Period, h
Sulfur dioxide	80	1 year	365	24
Carbon monoxide			10 000	1
			40 000	8
Nitrogen dioxide	100	1 year		
Ozone[a]			235	1
Hydrocarbons				
Total particulate (PM10)[b]	75	1 year	260	24
Lead particulate	1.5	3 months		

[a]Standard is met when the number of days per year with maximum hour-period concentration above 235 µg/m³ is less than one.
[b]PM10 = Particulates below 10µm diameter.

Table 9 Typical Outdoor Concentrations of Selected Gaseous Air Pollutants

Pollutant	Typical Concentration, µg/m³	Pollutant	Typical Concentration, µg/m³
Acetaldehyde	20	Methylene chloride	2.4
Acetone	3	Nitric acid	6
Ammonia	1.2	Nitric oxide	10
Benzene	8	Nitrogen dioxide	51
2-Butanone (MEK)	0.3	Ozone	40
Carbon dioxide	612 000[a]	Phenol	20
Carbon monoxide	3 000	Propane	18
Carbon disulfide	310	Sulfur dioxide	240
Carbon tetrachloride	2	Sulfuric acid	6
Chloroform	1	Tetrachloroethylene	2.5
Ethylene dichloride	10	Toluene	20
Formaldehyde	20	1,1,1-Trichloroethane	4
n-Heptane	29	Trichloroethylene	15
Mercury (vapor)	0.005	Vinyl chloride monomer	0.8
Methane	1 100	Xylene	10
Methyl chloride	9		

[a]Normal concentration of carbon dioxide in air. The concentration in occupied spaces should be maintained at no greater than three times this level (1000 ppm).
Sources: Braman and Shelley (1980), Casserly and O'Hara (1987), Chan et al. (1990), Cohen et al. (1989), Coy (1987), Fung and Wright (1987), Hakov et al. (1987), Hartwell et al. (1985), Hollowell et al. (1982), Lonnemann et al. (1974), McGrath and Stele (1987), Nelson et al. (1987), Sandalls and Penkett (1977), Shah and Singh (1988), Singh et al. (1981), Wallace et al. (1983), and Weschler and Shields (1989).

EPA considers acceptable for six pollutants (Table 8), are satisfactory. This is the model recommended by ASHRAE *Standard* 62.

Table 9 gives outdoor concentrations for gaseous pollutants at urban sites. These values are typical; however, they may be exceeded if the building under consideration is located near a fossil fuel power plant, refinery, chemical production facility, sewage treatment plant, municipal refuse dump or incinerator, animal feed lot, or other major source of gaseous contaminants. If such sources will have a significant influence on the intake air, a field survey or dispersion model must be run. Many computer programs have been developed to expedite such calculations.

INFLUENCE OF VENTILATION PATTERNS

A recirculating air-handling system is shown schematically in Figure 1. In this case, mixing is not perfect; the horizontal dashed line represents the boundary of the region close to the ceiling through which air passes directly from the inlet diffuser to the return-air intake. Ventilation effectiveness E_v is the fraction of the

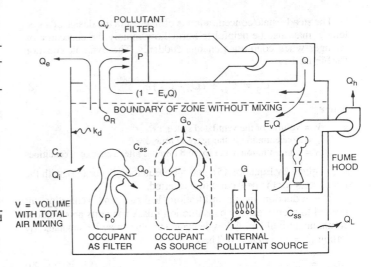

Fig. 1 Recirculatory Air-Handling and Gaseous Pollutant System

total air supplied to the space that mixes with the room air and does not bypass the room along the ceiling. Meckler and Janssen (1988) suggest a value of 0.8 for E_v. Any people in the space are additional sources and sinks for gaseous contaminants. In the ventilated space, the steady-state concentration C_{ss} for a single gaseous contaminant is

$$C_{ss} = a/b \qquad (5)$$

where

$$a = C_x(Q_i + 0.01 P E_v Q_v/f) + (G_i + NG_o) \qquad (6)$$

$$b = Q_e + Q_h + Q_L + k_d A + NQ_o(1 - 0.01P_o)$$
$$+ (E_v Q - Q_v)(1 - 0.01P)/f \qquad (7)$$

and

Q = system total flow, m³/s
Q_e = exhaust airflow, m³/s
Q_v = ventilation (makeup) airflow, m³/s
Q_h = hood flow, m³/s
Q_i = infiltration flow, m³/s
Q_L = leakage (exfiltration) flow, m³/s
Q_o = average respiratory flow for a single occupant, m³/s
P = filter penetration for pollutant, %
P_o = penetration of pollutant through human lung, %
A = surface area inside the ventilated space on which pollutant can be adsorbed, m²
k_d = deposition velocity on A for pollutant, m/s
C_x = outdoor concentration of pollutant, mg/m³
C_{ss} = steady-state indoor concentration of pollutant, mg/m³
G_o = generation rate for pollutant by an occupant, mg/s
G_i = generation rate for pollutant by nonoccupant sources, mg/s
N = number of occupants
E_v = ventilation effectiveness, fraction
$f = 1 - 0.01P(1 - E_v)$

Flow continuity allows the expression for b to be simplified to the following alternate form, which may make it easier to determine flows:

$$b = Q_i + Q_v + k_d A + NQ_o(1 - 0.01P_o) \qquad (7a)$$

The parameters for this model must be determined carefully so that nothing significant is ignored. The leakage flow Q_L, for example, may include flow up chimneys or toilet vents.

The steady-state concentration is of interest in the design of a system. It may also be helpful to know how rapidly the concentration changes when conditions change suddenly. The dynamic equation for Figure 1 is

$$C_\theta = C_{ss} + (C_0 - C_{ss})\, e^{-b\theta/V} \tag{8}$$

where

V = volume of the ventilated space, m^3
C_0 = concentration in the space at time $\theta = 0$
C_θ = concentration in the space θ minutes after a change of conditions

C_{ss} is given by Equation (5), and b by Equation (7) or (7a), with the parameters for the new condition inserted.

The reduction in air infiltration, leakage, and ventilation air needed to reduce energy consumption raises concerns about indoor contaminant buildup. A low-leakage structure may be simulated by letting $Q_i = Q_L = Q_h = 0$. Then

$$C_{ss} = \frac{0.01\, P E_v Q_v C_x / f + 2119\,(G_i + N G_o)}{Q_e + k_d A + N Q_o\,(1 - 0.01 P_o) + (E_v Q - Q_v)\,(1 - 0.01 P)/f} \tag{9}$$

Even if ventilation airflow Q_v is reduced to zero, a low-penetration (high-efficiency) gaseous contaminant filter and a high recirculation rate help lower the internal contaminant concentration. In commercial structures, infiltration and exfiltration are never zero. The only inhabited spaces operating on 100% recirculated air are space capsules, undersea structures, and other structures where life-support systems eliminate carbon dioxide and carbon monoxide and supply oxygen.

Real systems may have many rooms, with multiple sources of gaseous contaminants and complex room-to-room air changes. In addition, there may be mechanisms other than adsorption that eliminate gaseous contaminants on building interior surfaces. Nazaroff and Cass (1986) provide estimates for pollutant deposition velocity k_d in Equations (5) through (9) that range from 0.003 to 0.6 mm/s for surface adsorption only. A worst-case analysis, yielding the highest estimate of indoor concentration, is obtained by setting $k_d = 0$. Sparks (1988) and Nazaroff and Cass (1986) describe computer programs to handle these calculations.

The assumption of bypass and mixing used in the models presented here approximates the multiple-room case, since gaseous contaminants are readily dispersed by airflows. In addition, a contaminant diffuses from a zone of high concentration to a zone of low concentration even with low rates of turbulent mixing.

Quantities appropriate for the various flows in Equations (5) through (9) are discussed in the sections on Hooding and Local Exhaust and General Ventilation. Infiltration flow can be determined approximately by the techniques described in Chapter 23 of the 1993 *ASHRAE Handbook—Fundamentals* or, for existing buildings, by tracer or blower-door measurements. ASTM *Standard* E 741 defines procedures for tracer-decay measurements. ASTM (1980) covers tracer and blower-door techniques; DeFrees and Amberger (1987) describe a useful variation on the blower-door technique applicable to large structures.

CONTROL TECHNIQUES

Elimination of Sources

This approach to gaseous contaminant control is one of the most effective and least expensive. Control of radon gas begins with the installation of traps in sewage drains and the sealing and venting of leaky foundations and crawl spaces to prevent entry of the gas into the structure (EPA 1986, 1987). Prohibition of smoking in a building, or its isolation to limited areas, greatly reduces indoor pollution, even when rules are poorly enforced (Elliott and Rowe 1975, Lee

et al. 1986). The use of waterborne materials instead of those requiring organic solvents may reduce VOCs, although Girman et al. (1984) show that the reverse is sometimes true. The substitution of compressed carbon dioxide for halocarbons in spray-can propellants is an example of the use of a relatively innocuous pollutant instead of a more troublesome one. The growth of mildew and other organisms that emit odorous pollutants can be restrained by controlling condensation and applying fungicides and bactericides.

Hooding and Local Exhaust

When sources within a building generate substantial amounts of gaseous pollutants, direct exhaust through hoods is more effective than control by general ventilation. If these pollutants are toxic, irritating, or strongly odorous, hooding is essential. The minimum transport velocity required for the capture of large particles differs from that required for gaseous contaminants; otherwise, the problems of capture are the same for both.

Hoods are normally provided with exhaust fans and stacks that vent to the outdoors. Hoods are great consumers of ventilation air, requiring large fan power for exhaust and makeup and wasting heating and cooling energy. Makeup for air exhausted by a hood should be supplied, so that the general ventilation system balance is not upset when a hood exhaust fan is turned on. Back diffusion from an open hood to the general work space can be eliminated by surrounding the work space near the hood with an isolation enclosure (Figure 2). Such a space not only isolates the pollutants, but also tends to keep unnecessary personnel out of the area. Glass walls for the enclosure decrease the claustrophobic effect of working in a small space.

Increasingly, filtration of hood exhausts is required to prevent toxic releases to the outdoors. Hoods should be equipped with controls that decrease their flow when maximum protection is not needed. Hoods are sometimes arranged to exhaust air back into the occupied space, thus saving heating and cooling of that air. This practice must be limited to hoods exhausting the most innocuous pollutants because of the risk of filtration system failure. The design and operation of effective hoods are described in Chapters 14, 25, and 28 and in *Industrial Ventilation: A Manual of Recommended Practice* (ACGIH 1992).

General Ventilation

In residential and commercial buildings, the chief use of hoods and local exhaust occurs in kitchens, bathrooms, and occasionally around distinctive point sources such as diazo printers. Where there is no local exhaust of contaminants, the general ventilation distribution system provides contaminant control. Such systems must meet both thermal load requirements and pollutant control standards. Complete

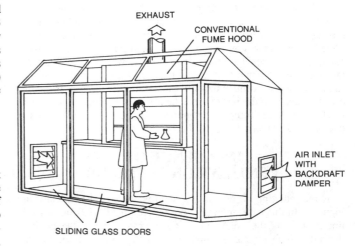

Fig. 2 Glass-Walled Enclosure for Pollutant Isolation

mixing and a relatively uniform air supply per occupant are desirable for both purposes. The guidelines for air distribution system design given in Chapters 31 and 32 of the 1993 *ASHRAE Handbook— Fundamentals* are appropriate for pollution control by general ventilation. In addition, the standards set by ASHRAE *Standard* 62 must be met.

When local exhaust is combined with general ventilation, a proper supply of makeup air must equal the exhaust flow for any hoods present. Supply fans may be needed to provide enough pressure to maintain flow balance. Systems such as cleanrooms are sometimes designed so that static pressure forces the air to flow from clean to less clean spaces. With these systems, the effects of door openings, wind pressures, etc. may require backdraft dampers. Chapter 15 covers clean rooms in detail.

Absorbers

Air washers, which are used primarily to control the temperature and humidity of the air, also remove some water-soluble gaseous pollutants by absorption (Swanton et al. 1971). Absorbers are used in life-support systems in spacecraft and in undersea craft. There, both solid and liquid reactive absorbers reduce carbon dioxide and carbon monoxide to carbon and return the oxygen to the system. Liquid absorption devices (scrubbers) are also used to remove gaseous pollutants from stack gases.

During absorption, the gaseous pollutant is dissolved in or reacts with the body of the absorbing medium, which can be either a porous solid or a liquid. Pollutant gases are absorbed in liquids when the partial pressure of the pollutant in the bulk gas flow is greater than the vapor pressure of the pollutant in the solution. The rate of transfer of pollutant to liquid is a complex function of (1) the physical properties of the pollutant, carrier gas, and liquid; (2) the geometry of the gas/liquid contactor device; and (3) the temperature and velocity of the gas flowing through the contactor. The effectiveness of liquid absorbers can be improved by adding reagents to the scrubbing liquid. Chapter 26 of the 1992 *ASHRAE Handbook—Systems and Equipment* introduces liquid absorber design, and many chemical engineering texts also cover the subject.

The use of liquid absorbers in occupied spaces has several disadvantages. The airstream tends to be saturated with the vapor of the scrubbing liquid, which then becomes a pollutant itself; or, if the liquid is water, relative humidity may be raised to undesirable levels. To avoid excess humidification, some systems have been installed that use chilled water and subsequent thermal recovery (Pedersen and Fisk 1984). Systems with desiccants, which are regenerated, and heat exchanger wheels, which recover the energy expended, have also been tried (Novosel et al. 1987). Unfortunately, desiccants that have enough affinity for the pollutants present are not readily available.

Physical Adsorbers

Adsorption, in contrast to absorption, is a surface phenomenon, similar in many ways to condensation. Pollutant gas molecules that strike a surface and remain bound to it for an appreciable time are said to be adsorbed by that surface. This effect takes place to some degree on all surfaces, but the available surfaces of gaseous contaminant control adsorption media are expanded in two ways. First, these adsorbers are provided in granular or fibrous form to increase the gross surface exposed to an airstream. Second, the surface of the adsorber is treated to develop pores of microscopic dimensions, which greatly increases the area available for molecular contact. Typical treated (activated) aluminas have surface areas of 200 m^2 per gram of adsorber; typical activated carbons have areas from 1000 to 1500 m^2 per gram of adsorber. Pores of various sizes and shapes provide minute traps that can fill with condensed vapors (multiple layers of pollutant molecules).

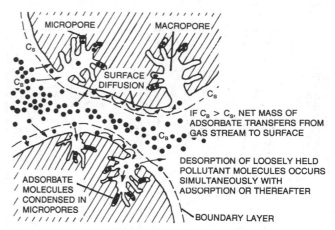

Fig. 3   Steps in Pollutant Adsorption

In physical adsorption, the forces binding the pollutant molecules to the adsorber surface are the same forces that are important to heat and mass transfer. Several steps must occur in the adsorption of a molecule (Figure 3):

1. The molecule must be transported from the carrier gas stream across the boundary layer surrounding the adsorber granule. This process is random, with molecular movement both to and from the surface; the net flow of molecules cannot be toward the surface unless the concentration of the pollutant in the gas flow is greater than at the granule surface. For this reason, adsorption action decreases as pollutant load on the adsorber surface increases. Very low concentrations in the gas flow also result in low adsorption rates.

2. The molecules of the pollutant must diffuse into the pores to occupy that portion of the surface.

3. The pollutant molecules must be bound to the surface.

Any of these steps may determine the rate at which adsorption occurs. In general, step (3) is very fast for physical adsorption, but reversible. This means that an adsorbed molecule can be desorbed at a later time, either when cleaner air passes through the adsorber bed, or when another pollutant arrives that binds more tightly to the adsorber surface. Coupled with the above effects are thermal effects; a physically adsorbed molecule gives up energy when it is bound to the adsorber surface. This energy raises the temperature of the adsorber and of the carrier gas stream.

When a pollutant is fed at constant concentration and constant gas flow rate to an adsorber of bed depth L, the gas stream concentration varies with time and bed depth, as shown in Figure 4A. When bed loading begins ($\theta = 0$), the pollutant concentration decreases logarithmically with bed depth; deeper into the bed, the slope of the concentration-versus-bed depth curve flattens. At later times, the entrance portion of the adsorber bed becomes loaded with pollutant, so pollutant concentrations in the gas stream are higher at each bed depth.

The pattern of downstream concentration versus time for an adsorber of bed depth L is shown in Figure 4B. Usually the downstream concentration is very low until time θ_{b1}, when the concentration rises rapidly until the downstream concentration is the same as the upstream. This penetration is called *breakthrough* because it tends to occur suddenly. Not all adsorber-pollutant combinations show as sharp a breakthrough as Figure 4B indicates; therefore, to evaluate a given breakthrough time, the entering pollutant concentration, downstream concentration at breakthrough, bed depth, and the velocity, temperature, and relative humidity of the airstream used in the test must also be specified.

In spite of the complexity of physical adsorption, comparatively simple expressions have been developed to describe the behavior of

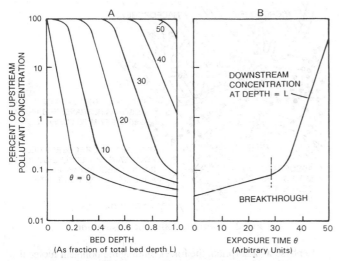

Fig. 4 Dependence of Adsorbate Concentration on Bed Depth and Exposure Time

adsorbers. Expressions fitting available test data over a wide range of operating conditions for many pollutants are given in articles by Nelson and Correia (1976) and Yoon and Nelson (1984a, 1984b, 1988). The expressions apply to the case of constant flow and constant pollutant inlet concentration. Nelson and Correia developed a semi-empirical expression:

$$\theta_{10} = K_{a1} C^{-2/3} \qquad (10)$$

where

θ_{10} = breakthrough time for a downstream concentration equal to 10% of upstream concentration

K_{a1} = constant for a given adsorber/pollutant combination; factors within this constant include adsorption media mass per unit airflow and pollutant molecular mass and boiling point

C = upstream pollutant concentration, ppm

Yoon and Nelson's articles present a more rigorous expression, giving the breakthrough time θ_b at any desired downstream/upstream concentration ratio:

$$\theta_b = \theta_{50} + K_{a2} \ln[C_b / (C - C_b)] \qquad (11)$$

where

θ_{50} = breakthrough time for a downstream concentration equal to 50% of upstream concentration

C_b = downstream concentration, ppm

K_{a2} = a constant determined experimentally for the adsorber/pollutant combination by tests run at two inlet concentrations

Most studies were run at concentrations of interest for short-term exposures (~ 1000 ppm), which may be misleading for the low concentrations met in HVAC applications. Low-concentration studies are reported by Ensor et al. (1988), Miller and Reist (1977), Ostajic (1985), Nelson and Correia (1976), Nunez et al. (1989), Stampfer (1982), Stankavich (1969), and Jonas and Rehrmann (1972). All showed that breakthrough time depends on inlet concentration. Yoon and Nelson (1984a) showed the equivalence between their expressions and those of Wheeler and Robell (1969) and Mecklenberg (1925) for low bed loadings. These theories, assisted by the Dubinin-Raduskevich isotherm theory (Dubinin 1975), relate dynamic adsorption breakthrough times to isotherm data obtained under static equilibrium conditions. (An isotherm is a plot of equilibrium vapor mass adsorption as a function of vapor pressure.) Isotherms are, however, normally plotted over the complete pressure range from 0 to 101.325 kPa (0 to 1 atmosphere). The range of

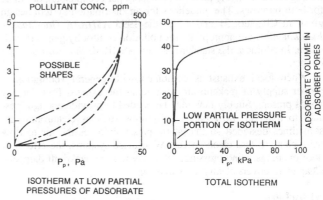

Fig. 5 Uncertainty in Isotherm Shapes at Low Partial Pressures of Adsorbate

interest for indoor pollutants is from about 0.0001 to 1000 Pa (10^{-9} to 10^{-2} atmospheres or 1 ppb to 10 000 ppm), which is barely visible on such plots. The behavior at very low pressures cannot be reliably extrapolated from higher pressures (Figure 5).

Yoon and Nelson (1988) and Underhill et al. (1988) discuss the effect of relative humidity on physical adsorption. In essence, water vapor acts as a second pollutant, altering the adsorption parameters by reducing the amount of the first pollutant that can be held by the adsorber and shortening the breakthrough times. The simultaneous adsorption of several pollutants shortens breakthrough times in similar fashion. Jonas et al. (1983) calculate mixture breakthrough times at low bed loadings by determining the volumes of the individual components that the adsorber will hold separately, then summing the actual partial volumes (based on the partial pressures of each) to obtain an effective volume held for the mixture.

Several sources list maximum possible retentivity (mass of adsorbate held per mass of adsorber) for various pollutants on typical activated charcoals. These values were normally obtained by tests with concentrations on the order of 1000 ppm and therefore grossly overestimate the amount of pollutants trapped at low indoor concentrations. Calculations based on such loadings should be viewed skeptically in determining the economic advantages of using adsorbers versus using ventilation. The maximum breakthrough time for an adsorber, assuming no concentration effect on efficiency (100% collection of pollutant), is

$$\theta_{max} = 1000 f_a W_a / Q_a C \qquad (12)$$

where

θ_{max} = maximum breakthrough time, s

W_a = mass of adsorber, g

f_a = ratio of maximum pollutant mass adsorbed to adsorber mass, g/g

Q_a = flow through the adsorber, m³/s

C = pollutant concentration, mg/m³

ASTM *Standard* D 3686 lists maximum adsorptivities for about 130 vapors in high-quality activated carbons. In the standard, the column headed "Recommended Maximum Tube Loading, mg" represents the maximum percentage retentivity for the listed pollutants at TWA8 levels. The percentage retentivities in ASTM D 3686 should be corrected to TWA8 levels by the following equation, derived from Equation (10):

$$R_{TWA8} = R_{1000} (TWA8/1000)^{1/3} \qquad (13)$$

where

R_{TWA8} = percent retentivity at TWA8 pollutant concentration in ppm

R_{1000} = percent retentivity at pollutant concentration of 1000 ppm (test level used in ASTM D 3686)

Table 10 Low-Temperature Absorbers, Chemisorbers, and Catalysts

Material	Impregnant	Typical Vapors or Gases Captured
Physical Adsorbers		
Activated carbon	none	organic vapors, ozone, acid gases
Activated alumina	none	polar organic compounds[a]
Activated bauxite	none	polar organic compounds[a]
Silica gel	none	water, polar organic compounds[a]
Molecular sieves (zeolites)	none	carbon dioxide, iodine
Porous polymers	none	various organic vapors
Chemisorbers		
Activated alumina	$KMnO_4$	hydrogen sulfide, sulfur dioxide
Activated carbon	I_2, Ag, S	mercury vapor
Activated carbon[b]	I_2, KI_3, amines	radioactive iodine and organic iodides
Activated carbon	$NaHCO_3$	nitrogen dioxide
LiO_3, NaO_3, KO_3	none	carbon dioxide
LiO_2, NaO_2, KO_2, $Ca(O_2)_2$	none	carbon dioxide
Li_2O_2, Na_2, O_2	none	carbon dioxide
LiOH	none	carbon dioxide
$NaOH + Ca(OH)_2$	none	acid gases
Activated carbon	KI, I_2	mercury vapor
Catalysts		
Activated carbon	none	ozone
Activated[g] carbon[c]	$Cu + Cr + Ag + NH_4$	acid gases, chemical warfare agents
Activated alumina	$CuCl_2 + PdCl_2$	formaldehyde
Activated alumina	Pt, Rh oxides	carbon monoxide

[a]Polar organics = alcohols, phenols, aliphatic and aromatic amines, etc.
[b]Mechanism may be isotopic exchange as well as chemisorption.
[c]"ASC Whetlerite"

Table 10 lists some physical adsorbers used for indoor air applications. Activated carbon is by far the most popular because of the wide range of pollutants it can adsorb.

Chemisorbers

Chemisorption is similar in many ways to physical adsorption. The three steps shown in Figure 3 also apply to chemisorption. However, the third step in chemisorption is different from physical adsorption, for surface binding in chemisorption is by chemical reaction with electron exchange between the pollutant molecule and the chemisorber. This action differs in the following ways from physical adsorption:

1. It is highly specific; only certain pollutant compounds will react with a given chemisorber.
2. Chemisorption improves as temperature increases; physical adsorption improves as temperature decreases.
3. Chemisorption does not generate heat, but instead may require heat input.
4. Chemisorption is not generally reversible. Once the adsorbed pollutant has reacted, it is not readily desorbed. However, one or more reaction products, different from the original pollutant, may be formed in the process. These products may themselves have undesirable effects.
5. Water vapor often helps chemisorption or is necessary for it; it usually hinders physical adsorption.
6. Chemisorption per se is a monomolecular layer phenomenon; the pore-filling effect that takes place in physical adsorption does not occur, except where adsorbed water condensed in the pores forms a reactive liquid.

Although the overall pattern of the concentration/time/bed depth relations in chemisorption is the same as described for physical adsorption, and the same equations apply, the rate constant K in Equations (10) and (11) depends on the kinetics of the chemical reaction between the pollutant and the chemisorber. Test data on individual chemisorber/pollutant combinations under operating conditions provide the only way to determine chemisorber performance in typical applications. However, the upper limit of capacity for a chemisorber/pollutant combination may be determined if both the chemical reaction at the surface and the amount of reactant available are known. A chemical reaction cannot involve more than stoichiometric amounts of the reactants unless catalytic effects are present.

Equation (12) may be used for chemisorbers, with the following substitutions:

W_a = mass of chemisorber reactant available for reaction, g
f_a = number of grams of pollutant that combines stoichiometrically with each gram of reactant

Solid chemisorbers may be porous, chemically homogeneous materials, but this is usually wasteful and ineffective because only the exposed (generally small) surface reacts with the pollutant. For this reason, most chemisorptive media are formed from a highly porous support (e.g., activated alumina or activated carbon) coated or impregnated with a chemical reactant. The mass W_a is the mass of the reactant, not of the entire support. The support may have physical adsorption ability that comes into play when chemisorptive action ceases. Table 10 lists a few chemisorbers used in indoor air pollution control.

Catalysis

A catalytic air-purifying medium is one that operates at ambient temperatures or at least generates its own activation energy by stimulating an exothermic chemical reaction. Catalytic combustion, which enables a pollutant to burn (oxidize) at lower temperatures than is possible with unassisted combustion, is described later; it requires substantial heat input and elevation of the temperature of the whole gas stream. Only a few catalysts are known to be effective at ambient temperatures for the removal of gaseous pollutants; some of these are listed in Table 10.

Catalytic air purification is closely related to chemisorption, for chemical reactions occur at the surface of a catalyst; the significant difference is that the gaseous pollutant does not react stoichiometrically with the catalyst itself. The catalyst assists whatever reaction takes place but is not used up in the reaction. Because of this feature, catalytic purifiers have the potential for much longer lives than adsorbers or chemisorbers. Of course the reaction that occurs must convert the gaseous pollutant to something innocuous. This conversion may divide the gaseous pollutant into smaller molecules or join the pollutant to a supplied reactant. The most satisfactory supplied reactant is oxygen, which is present in unlimited quantity in the carrier airstream.

Some catalysts convert carbon monoxide to carbon dioxide, which is not controlled by physical adsorbers such as activated carbon (Collins 1986). Thus, in 100% recirculatory systems, sequential beds can be used to remove carbon monoxide by catalysis, carbon dioxide by absorption, and odors by adsorption. This type of filtration is found in spacecraft, but the systems are complex and impractical for more conventional ventilation applications.

Combustion

Combustion is widely used to remove toxic gaseous pollutants from stack gases. In many cases, volatile organics can be converted to innocuous concentrations of steam and carbon dioxide. The entire carrier gas stream must, however, be raised to a temperature at which substantial oxidation of the pollutant occurs (from 650 to 820°C for most organic pollutants). In most applications, the required heat is produced by direct firing of natural gas. Combustion is not a suitable method of gaseous pollution control for indoor air. Even in stack gas combustors, regenerators are frequently installed to reduce the amount of heating gas required. Still more cooling and heat recovery would be needed in order to apply direct combustion to indoor environmental control; and the unavoidable combustion products would defeat the purpose of a gaseous pollution control system.

Catalytic Combustion

Gaseous pollutants can sometimes be oxidized at temperatures somewhere between ambient and the temperature of ignition of the pollutant by passing the gas stream through a heated bed of catalyst. Nuclear submarines incorporate a heat recovery unit and return the airstream temperature to ambient following a catalytic combustor (Piatt and White 1962). These systems oxidize volatile organic contaminants as well as the carbon monoxide and hydrogen in the submarine atmosphere. In general, such complexity is unlikely to be economical in commercial HVAC systems. Catalytic combustors are used in millions of automobile exhaust pipes and in wood stoves to oxidize unburned hydrocarbons that would otherwise reduce the quality of urban air.

Nwanko and Turk (1975) describe a system in which styrene, toluene, and methyl ethyl ketone (MEK) are adsorbed at 25°C on activated carbon impregnated with palladium. The carrier stream is then heated slowly to 310 to 420°C, almost completely oxidizing the organics without destroying the activated carbon. Such a cyclic system can possibly be adapted to HVAC applications.

Cryogenic Condensation

Physical adsorption is far more effective at low temperatures than at ambient temperatures. For this reason, some adsorbers in nuclear safety systems are operated at cryogenic temperatures for capture of noble gases. It is also possible to condense a gas or vapor out of an airstream by cooling the airstream below the boiling point of the gas. However, the gas partial pressures are so low at normal odor thresholds and TWA8 concentration levels that temperatures far below boiling points are necessary to obtain sufficient mass transfer to the cooling surface. The resulting air desiccation, surface icing, and cost make this scheme impractical for HVAC systems.

Plants

Plants can accomplish indoors some of the same pollution-control activity that they perform outdoors. Wolverton et al. (1989) describe a series of experiments in which benzene, formaldehyde, and trichloroethylene were removed from a sealed chamber atmosphere by various decorative plants (see results in Table 11). The specific combination of plant and pollutant is important. These authors also describe a scheme in which air is drawn through the potting soil of several plants, and pollutants are adsorbed by activated carbon in the soil. They postulate that plant root bacteria are able to regenerate the activated carbon and maintain its effectiveness far longer than is normal.

Occupants

The model shown in Figure 1 includes the inhalation and exhalation of gas by a single occupant. A typical person engaged in office work has an average breathing flow of about 0.23 L/s. About 70% of this flow reaches the lung alveoli, where volatile organics as well as water-soluble inorganic pollutant gases are absorbed to a considerable degree. Wallace et al. (1983) show the relation between ambient concentrations and exhaled air concentrations for seven organic pollutants. The high removal efficiency (> 70%) for these subjects shows distinct absorption in the tracheobronchial system as well as in the alveoli (see Table 6).

The body is not a perfect trap for such organics but behaves somewhat like a physical adsorber and desorbs some of the captured, but non-metabolized, pollutants when cleaner air is subsequently breathed. The time span for this process can be several days. Inorganics (acid gases and so forth) are captured more efficiently than volatile organics by the breathing system and are more readily taken up by the body.

Odor Counteractants and Odor Masking

An odor counteractant is a vapor supplied to a space to reduce the sensitivity of the nose to a known odor present in the space. An odor mask is a vapor supplied in sufficient quantity to overshadow the existing odor. These solutions apply only to specific odors and may be effective for a limited percentage of the occupants of a given space. They offer no improvement of toxic effects or material damage and are pollutants themselves.

CONTROL SYSTEM CONFIGURATIONS

Adsorbers, chemisorbers, and catalysts are supplied as powders and granular or pelletized media. They must be held in a retaining structure that allows the treated air to pass through the media with an acceptable pressure drop at the operating airflow. Typical configurations of units where granular or pelletized media are held between

Table 11 Adsorption of Pollutants from a Sealed Chamber by Houseplants

	Formaldehyde			Benzene			Trichloroethylene		
	Total Leaf Area, m²	Concentration, mg/m³	Deposition Velocity, mm/s	Total Leaf Area, m²	Concentration, mg/m³	Deposition Velocity, mm/s	Total Leaf Area, m²	Concentration, mg/m³	Deposition Velocity, mm/s
Plant		θ = 0 θ = 24 h			θ = 0 θ = 24 h			θ = 0 θ = 24 h	
None (control)	0.0	23 22	0.039	0.0	65 61	0.056	0.0	109 98	0.083
Mass cane *(Dracaena massangeana)*		25 7.5			45 36			87 76	
Pot chrysanthemum *(Chrysanthemum morifolium)*		22 8.7	0.42		188 88	5.30	0.72	109 71	5.14
Gerbera daisy *(Gerbera jamesonii)*		20 10			210 68		0.46	109 71	3.00
Wareckei *(Dracaena deremensis W.)*		10 5.0		0.72	87 42	8.72	0.72	109 98	1.28
Ficus *(Ficus benjamina)*		24 12			65 45			103 92	

Source: Wolverton et al. (1989).

Deposition rates are calculated by use of the equation $dC/dt = -(kA/V)C$, which gives k (deposition velocity, mm/s) $= -1000 (V_\theta/A)\ln(C_\theta/C_0)$, where V is the chamber volume, A the area of the deposition surface, θ the time for the final concentration measurement, and C_0 and C_θ the concentrations at time 0 and time θ, respectively. The volume V of the chamber used in calculation was 0.44 m³, and area A was 3.47 m². Only the leaf areas were used in the calculation of deposition velocities when plants were in the chamber.

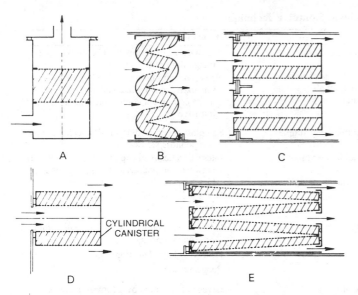

Fig. 6 Configurations of Gaseous Pollutant Filters Using Dry Granular Adsorbers, Chemisorbers, and Catalysts

perforated metal sheets or wire screens are shown in Figure 6. The perforated metal or screens must have holes smaller than the smallest particle of the active media. A margin without perforations must be left around the edges of the retaining sheets or screens to minimize the amount of air that can slip around the active media. Media must be tightly packed within the structure so that open passages through the beds do not develop. Aluminum; stainless, painted, plated, or conversion-coated steel; and plastics are all used for retainers. Adsorptive media are also retained in fibrous filter media, which can then be pleated into large filter structures as shown in Figure 6B. The active media must be bound to the fibers in such a way that media micropores are not damaged, and adequate overall adsorptive capacity is maintained. These factors must be evaluated by tests on the complete structure, not the base adsorptive media alone.

The configuration shown in Figure 6A is used for such applications as the collection of noble gases under cryogenic conditions in nuclear installations. Velocities are low and contact times long (typically > 1 s). Similar configurations are used for portable, temporary filter systems during cleanup of chemical spills. Figures 6B, 6C, and 6D are typical fixed-bed units used in nuclear safety systems and highly toxic situations, such as the disposal of chemical warfare agents. Contact times here are on the order of 0.25 s, and in order to maximize the effective mass transfer, the active media particles are finer than those used in commercial HVAC systems. In addition, great attention is paid to keeping the active media tightly packed and to maintaining highly reliable perimeter gaskets.

Figure 6E is typical of commercial HVAC equipment. The retaining elements are flat panels or trays that slide into a holding rack that has gaskets to seal all joints. This construction allows the trays to be removed easily for replacement or regeneration of the active media. Contact times are usually from 0.05 to 0.25 s; efficiency is higher for longer contact times.

Large systems can be supplied with banks of cells of the types shown in Figures 6B through 6E, in both face-access and side-access arrangements. Clamping and gasketing between cells prevents the bypass of untreated air.

ENVIRONMENTAL INFLUENCES ON CONTROL SYSTEMS

High relative humidity in the treated airstream lowers the gaseous-pollutant removal efficiency of physical adsorbers. Very low relative humidities may make some chemisorption activity impossible. Therefore, the performance of these media must be tested over the expected range of operation, and the relative humidity of an operating system must be held within limits acceptable for the sorption media. Temperatures must also be held within acceptable limits.

All adsorption and catalytic media have a modest ability to capture dust particles and lint, which eventually plug the openings between the media granules and cause a rapid rise in the pressure drop across the media or a decrease in system airflow. All granular gaseous adsorption beds need to be protected against particulate buildup by installing appropriate particulate filters upstream. Vibration breaks up the granules to some degree, depending on the granule hardness. ASTM *Standard* D 3802 is a test procedure for measuring the resistance of activated carbon to abrasion, and ASTM *Standard* D 4058 specifies the procedure for measuring the same property of somewhat harder catalysts and catalyst carriers. Critical systems using activated carbon require hardness above 92%, as described by ASTM D 3802.

Adsorption media, chemisorbers, and catalysts sometimes accelerate the corrosion of metals that they touch. For this reason, media holding cells, trays, modules, and so forth, should not be constructed of uncoated aluminum. Painted steel or ABS plastic are in common use and exhibit good material service lives in most applications. Organically coated or stainless steel components may be required in more aggressive environments. Liquid-scrubbing absorption devices must be made of corrosion-resistant materials, especially when reagents are present in water solutions.

CONTAMINANT MEASUREMENT AND MONITORING

The concentration of various pollutants in the air must be measured to determine whether the gaseous contaminant quality of air is acceptable. Such testing may show that air quality is acceptable, even when occupants complain. It may also show that gaseous contaminant control through hood exhaust, filtration, or added ventilation is needed to maintain safe and pleasant conditions.

Acceptable carbon dioxide content and odor conditions are defined by ASHRAE *Standard* 62. The alternative Indoor Air Quality Procedure specified in that standard, which applies where outdoor air is reduced to low levels by filtering recirculated air, sets limits for indoor concentrations for several contaminants (see the section on Energy Consumption). The standard also cautions that any unusual contaminants must be controlled to one-tenth the TLV levels specified by ACGIH (1993). When outdoor air levels are reduced, the actual gaseous contaminant concentrations must be measured to ensure that ASHRAE *Standard* 62 is met. The measurement of gaseous contaminant concentrations near levels acceptable for indoor air is not as straightforward as measurement of temperature or humidity. Relatively costly analytical equipment is needed that must be calibrated and operated by experienced personnel.

Measurement of gaseous contaminant concentrations has two aspects: sampling and analysis. Currently available sampling techniques are listed in Table 12. Some analytical or detection techniques are specific to a single pollutant compound; others are capable of presenting a concentration spectrum for many compounds simultaneously (Table 13). Pollutant concentration measurement instruments should be able to detect contaminants of interest at about one-tenth TWA8 levels; if odors are of concern, detection sensitivity must be at odor threshold levels. As indicated in Table 12, pollutants may be accumulated or concentrated over time so that very low average concentrations can be measured. Procedures for evaluating odor levels are given in Chapter 12 of the 1993 *ASHRAE Handbook—Fundamentals*.

When sampling and analytical procedures appropriate to the application have been selected, a pattern of sampling locations and times must be carefully planned. The building and air-handling system layout

Table 12 Gaseous Contaminant Sampling Techniques

Technique[a]	Advantages	Disadvantages
1. Direct flow to detectors	Real-time readout, continuous monitoring possible Several pollutants possible with one sample (when coupled with chromatograph, spectroscope, or multiple detectors)	Average concentration must be determined by integration No preconcentration possible before detector; sensitivity may be inadequate On-site equipment complicated, expensive, intrusive
2. Capture by pumped flow through solid adsorbent; subsequent desorption for concentration measurement	On-site sampling equipment relatively simple and inexpensive Preconcentration and integration over time inherent in method Several pollutants possible with one sample	Sampling media and desorption techniques are compound-specific Interaction between captured compounds and between compounds and sampling media; bias may result Gives only average values over sampling period, no peaks
3. Colorimetric detector tubes	Very simple, relatively inexpensive equipment and materials Immediate readout Integration over time	Rather long sampling period normally required One pollutant per sample Relatively high detection limit Poor precision
4. Collection in evacuated containers	Very simple on-site equipment No pump (silent) Several pollutants possible with one sample	Gives only average over sampling period, no peaks
5. Collection in nonrigid containers (plastic bags) held in an evacuated box	Simple, inexpensive on-site equipment (pumps required) Several pollutants possible with one sample	Cannot hold some pollutants
6. Cryogenic condensation	Wide variety of organic pollutants can be captured Minimal problems with interferences and media interaction Several pollutants possible with one sample	Water vapor interference
7. Passive diffusional samplers	Simple, unobtrusive, inexpensive No pumps; mobile; may be worn by occupants to determine average exposure	Gives only average over sampling period, no peaks
8. Liquid impingers (bubblers)	Integration over time Several pollutants possible with one sample	May be noisy

Sources: NIOSH (1977, 1984), Lodge (1988), Taylor et al. (1977), and ATC (1990).
[a]All techniques except 1 and 3 require laboratory work after completion of field sampling. Only technique 1 is adaptable to continuous monitoring and able to detect short-term excursions.

and the space occupancy and use patterns must be considered so that representative concentrations will be measured. Traynor (1987) and Nagda and Rector (1983) offer guidance in planning such surveys.

DEVICE TESTING

No standard procedure has been developed to test gaseous contaminant filters for general ventilation applications. Many articles have, however, described procedures where a steady concentration of a single contaminant is fed to a filter and the downstream concentration is determined as a function of the total contaminant captured by the filter (Nelson and Correia 1976; Mahajan 1987, 1989; and ASTM 1989a). Rivers (1988) summarizes the desirable elements of a standardized test and some of the problems involved in formulating one. One problem with gaseous contaminant filter testing is that tests need to be run at the low concentrations comparable to those found in occupied spaces (TWA8 or odor threshold).

Testing at high concentrations does not necessarily predict performance at low concentrations; the corrections described in the section on Physical Adsorbers must be applied to the test data. If attempts are made to speed the test process by high-concentration loading, pollutant desorption from the filter may confuse the results (Ostajic 1985). An adsorber cannot be tested for every pollutant, and there is no general agreement on which contaminants should be considered typi-

cal. Nevertheless, tests run according to the previously mentioned references do give useful measures of filter performance on single contaminants, and they do give a basis for estimates of filter penetrations and filter lives. Filter penetration data thus obtained can be used as P in Equation (9) to estimate steady-state indoor concentrations.

Since physical adsorbers, chemisorbers, and catalysts are affected by the temperature and relative humidity of the carrier gas and the moisture content of the filter bed, they should be tested over the range of conditions expected in the application. The capture of contaminants and the generation of reaction products need to be evaluated by sweeping the test filter with unpolluted air and measuring downstream concentrations. Reaction products may be as toxic, odorous, or corrosive as captured pollutants.

Gaseous contaminant control devices have been tested in sealed chambers by recirculating contaminated air through them and measuring the decay of an initial contaminant concentration over time. This method introduces factors extraneous to the device itself, such as possible errors introduced by adsorption on the chamber surfaces, leaks in the facility, and the drawing of test samples. Daisey and Hodgson (1988) compared the pollutant decay rate with and without the control device to overcome these uncertainties.

In critical applications, such as chemical warfare protective devices and nuclear safety applications, sorption media are evaluated in a small canister, using the same carrier-gas velocity as in the full-scale

Table 13 Gaseous Contaminant Concentration Measurement Methods

Method	Description	Typical Application
1. Gas chromatography (Using the following detectors)	Separation of gas mixtures by time of passage down an absorption column	
Flame ionization	Change in flame electrical resistance due to ions of pollutant gas	Volatile, nonpolar organics
Flame photometry	Measures light produced when pollutant is ionized by a flame	Sulfur, phosphorous compounds
Photoionization	Measures ion current for ions created by ultraviolet light	Most organics (except methane)
Electron capture	Radioactively generated electrons attach to pollutant atoms; current measured	Halogenated organics Nitrogenated organics
Mass spectroscopy	Pollutant atoms are charged, passed through electrostatic magnetic fields in a vacuum; path curvature depends on mass of atom, allowing separation and counting of each type	Volatile organics
2. Infrared spectroscopy and Fourier transform infrared spectroscopy (FTIR)	Absorption of infrared light by pollutant gas in a transmission cell; a range of wavelengths is used, allowing identification and measurement of individual pollants	Acid gases Many organics
3. High-performance liquid chromatography (HPLC)	Pollutant is captured in a liquid, which is then passed through a liquid chromatograph (analogous to a gas chromatograph)	Aldehydes, ketones Phosgene Nitrosamines Cresol, phenol
4. Colorimetry	Chemical reaction with pollutant in solution yields a colored product whose light absorption is measured	Ozone Nitrogen oxides Formaldehyde
5. Fluorescence and Pulsed fluorescence	Pollutant atoms are stimulated by a monochromatic light beam, often ultraviolet; they emit light at characteristic fluorescent wavelengths, whose intensity is measured	Sulfur dioxide Carbon monoxide
6. Chemiluminescence	Reaction (usually with a specific injected gas) results in photon emission proportional to concentration	Ozone Nitrogen compounds Several organics
7. Electrochemical	Pollutant is bubbled through reagent/water solution, changing its conductivity or generating a voltage	Ozone Hydrogen sulfide Acid gases
8. Titration	Pollutant is absorbed into water	Acid gases
9. Ultraviolet absorption	Absorption of UV light by a cell through which the polluted air passes	Ozone Aromatics Sulfur dioxide Oxides of nitrogen Carbon monoxide
10. Atomic absorption	Contaminant is burned in a hydrogen flame; a light beam with a spectral line specific to the pollutant is passed through the flame; optical absorption of the beam is measured	Mercury vapor

Sources: NIOSH (1977, 1984), Lodge (1988), Taylor et al. (1977), and ATC (1990).

unit. The full-scale unit is then checked for leakage through gaskets, structural member joints, and thin spots or gross open passages in the sorption media by feeding a readily adsorbed contaminant to the filter and probing for its presence downstream (ASME 1989).

The test used to evaluate fundamental properties of sorption media is a static test, the measurement of the adsorption isotherm. In this test, a small sample of the adsorber is exposed to the pollutant vapor at sucessively increasing pressures, and the mass of pollutant adsorbed at each pressure is measured. The low-pressure section of an isotherm can be used to predict kinetic behavior, although the calculation is not simple. For many years, the test outlined in ASTM *Standard* D 3467 was widely used for specifying the performance activated carbons. This test measured a single point on the isotherm of an activated carbon using carbon tetrachloride (CCl_4) vapor. The test has been replaced by one described in ASTM *Standard* D 5228 due to the toxicity of CCl_4. The method in ASTM D 5228 uses

butane as a test contaminant. A correlation has been developed between the results of the two procedures so that users accustomed to CCl_4 numbers can recognize the performance levels given by ASTM D 5228. It is a qualitative measure of performance at other conditions, and is a useful quality control procedure. Another qualitative measure of performance is the Brunauer-Emmett-Teller (BET) method (ASTM *Standard* D 4567), in which the surface area is determined by measuring the mass of an adsorbed monolayer of nitrogen. The results of this test are reported in square metres per gram of sorbent or catalyst. This number is often used as an index of adsorber quality, with high numbers indicating high quality.

ENERGY CONSUMPTION

The choice between using outside air only and outside air plus filtration may be made on the basis of technical or maintenance factors,

Table 14 Additional Indoor Air Quality Criteria Specified by ASHRAE *Standard* 62 when Ventilation Air is Recirculated

Pollutant	Maximum Allowed Concentration	Sampling Period
Carbon dioxide	1.8 g/m³	Continuous
Chlordane	5 µg/m³	Continuous
Ozone	100 µg/m³	Continuous
Radon	0.027 WL[a]	Annual average

[a]WL = Working Level, a unit of measure of human exposure to radon gas and radon progeny.

convenience, economics, or a combination of these. An energy consumption calculation must be made before an economic analysis is possible. Presumably, a building load calculation will be available that will include the various flows in Figure 1 and the total heating and cooling energy requirements for various seasons of the year.

Enough ventilation air must be supplied to every building space to satisfy the metabolic requirements of the occupants, including the dilution of carbon dioxide to nontoxic levels. This dilution requires only 0.05 L/s per person. ASHRAE *Standard* 62, however, requires a minimum of 8 L/s of ventilation air per person; higher flows are required for certain spaces, such as where smoking is expected. The minimum quality of the outdoor air to be used is specified in Table 8. If the amount of outdoor air used is less than that specified in *Standard* 62, a different set of criteria must be met. The quality of the air in the occupied space must be no worse than that given in Table 8, and it must also have lower concentrations than those given in Table 14 for the four additional pollutants. Systems may use outdoor air or filtered recirculated air in any ratio, provided the air quality level is maintained. The logic of these requirements is discussed by Janssen (1989).

Where building habitability can be maintained with ventilation alone, economizer cycles are recommended under appropriate outdoor conditions. When a system is in the economizer mode, some energy is required to move air through the system. For fixed-flow systems, this energy requirement is almost constant. For variable air volume (VAV) systems, power consumption varies, depending on the flow rate and on the design of the control, fan, and motor system.

Data on the resistance of the filtration device as a function of airflow and on the resistance of the heating/cooling coils must be provided by the manufacturer. In addition, the resistance of any particulate filters required to protect the gaseous contaminant filter must be included in the energy analysis.

VAV systems involve considerably more calculation than fixed-flow systems; fortunately, computer programs can help in the calculations. However, treating a VAV system as operating at two or three conditions representative of heating, cooling, and economizer modes should show which systems are economically sound.

ECONOMIC CONSIDERATIONS

Most meaningful comparisons are made using a discounted cash flow analysis in which the capital and operating costs of the competing systems are estimated separately. If the system with lower capital cost also costs less to operate, it is the obvious choice. However, if the more expensive system costs less to operate, some time will be required to recoup the added investment. The difference in the cash flows generated by each system must be estimated for each year of operation. Usually the cash flow is not constant because inflation increases the costs of energy, material, and labor. For example, the analysis may be run for 20 years, which is a conservative estimate of the life of the equipment. Then a constant compound interest rate is calculated for this period to discount each year's cash flow difference, from the last year back to the first. The sum of the discounted cash flow differences equals the investment difference. If this rate of

return on the added investment is sufficient, the building owner should buy the more expensive system.

All capital and operating costs for each competing system should be identified. Table 15 is a checklist of these items; the actual values to be used vary by location and by year. The life of any filtration media used is an important factor in this economic analysis, as is the cost of utilities for heating and cooling. Often the cost of replacing the filtration media can be reduced by sending spent media to a reactivation service. However, some fresh media must always be added to maintain performance, and reactivation is not permissible for media containing highly toxic pollutants.

Table 15 Checklist of Items Included in Economic Comparisons Between Competing Gaseous Contaminant Control Systems

Ventilation	
Capital Costs	**Operating Costs**
Fans	Utilities
Motors	Water treatment
Coil	Maintenance labor
Plenum	
Chiller	
Boiler	
Piping	
Pumps	
Controls	
Installation	
Floor space	

Filtration	
Capital Costs	**Operating Costs**
Added filtration equipment	Replacement or reactivation of
Fan	gaseous pollutant filter media
Motor	Disposal of spent gaseous
Controls	pollutant filter media
Plenum	Added electric power
Spare media holding units	Maintenance labor
Floor space	

REFERENCES

ACGIH. 1992. *Industrial ventilation: A manual of recommended practice*, 21st ed. American Conference of Governmental Industrial Hygienists, Cincinnati, OH.

ACGIH. 1993. *Threshold limit values and biological exposure indices for 1993-94.* American Conference of Governmental Industrial Hygienists, Cincinnati, OH.

AIHA. 1989. *Odor thresholds for chemicals with established occupational health standards.* American Industrial Hygiene Association, Akron, OH.

American Guild of Organists. 1966. Air pollution and organ leathers: Panel discussion. *AGO Quarterly* 11(2):62-73.

Anthony, C.P. and G.A. Thibodeau. 1980. *Textbook of anatomy and physiology.* C.V. Mosby Co., St. Louis, MO.

ASHRAE. 1989. Ventilation for acceptable indoor air quality. *Standard* 62-1989.

ASME. 1989. Testing of nuclear air treatment systems. *Standard* N 510-89. American Society of Mechanical Engineers, New York.

ASTM. 1980. Building air change rate and infiltration measurements. STP 719. American Society for Testing and Materials, Philadelphia.

ASTM. 1986. Standard test method for single-point determination of specific surface area of catalysts using nitrogen adsorption by continuous flow method. *Standard* D 4567-86. American Society for Testing and Materials, Philadelphia.

ASTM. 1988. Recommended practices for sampling atmospheres for analysis of gases and vapors. *Standard* D 1605-60. American Society for Testing and Materials, Philadelphia. [Standard withdrawn 1992.]

ASTM. 1989a. Standard practice for sampling atmospheres to collect organic compound vapors (activated charcoal tube adsorption method). *Standard* D 3686-89. American Society for Testing and Materials, Philadelphia.

ASTM. 1989b. Standard test method for nuclear-grade activated carbon. *Standard* D 3803-89. American Society for Testing and Materials, Philadelphia.

ASTM. 1990. Standard test method for ball-pan hardness of activated carbon. *Standard* D 3802-79. American Society for Testing and Materials, Philadelphia.

ASTM. 1992a. Standard test method for attrition and abrasion of catalysts and catalyst carriers. *Standard* D 4058-92. American Society for Testing and Materials, Philadelphia.

ASTM. 1992b. Standard test method for the determination of the butane working capacity of activated carbon. *Standard* D 5288-92. American Society for Testing and Materials, Philadelphia.

ASTM. 1993a. Standard test method for carbon tetrachloride activity of activated carbon. *Standard* D 3467-93. American Society for Testing and Materials, Philadelphia.

ASTM. 1993b. Standard test method for determining air change in a single zone by tracer dilution. *Standard* E 741-93. American Society for Testing and Materials, Philadelphia.

ATC. 1990. Technical assistance document for sampling and analysis of toxic organic compounds in ambient air. EPA/600/8-90-005. Environmental Protection Agency, Research Triangle Park, NC.

Balieu, E., T.R. Christiansen, and L. Spindler. 1977. Efficiency of b-filters against hydrogen cyanide. Staub-Reinhalt. *Luft* 37(10):387-90.

Berglund, B., U. Berglund, and T. Lindvall. 1986. Assessment of discomfort and irritation from the indoor air. In *IAQ '86: Managing Indoor Air for Health and Energy Conservation*, 138-49. ASHRAE, Atlanta.

Black, M.S. and C.W. Bayer. 1986. Formaldehyde and other VOC exposures from consumer products. In *IAQ '86: Managing Indoor Air for Health and Energy Conservation*. ASHRAE, Atlanta.

Braman, R.S. and T.J. Shelley. 1980. Gaseous and particulate ammonia and nitric acid concentrations: Columbus, Ohio area—Summer 1980. PB 81-125007. National Technical Information Service, Springfield, VA.

Braun, R.C. and M.J.G. Wilson. 1970. The removal of atmospheric sulphur by building stones. *Atmospheric Environment* 4:371-78.

Brugnone, F., L. Perbellini, F.B. Faccini, G. Pasini, G. Maranelli, L. Romeo, M. Gobbi, and A. Zedde. 1989. Breath and blood levels of benzene, toluene, cumene and styrene in non-occupational exposure. *International Archives of Environmental Health* 61:303-11.

Cain, W.S., L.C. See, and T. Tosun. 1986. Irritation and odor from formaldehyde chamber studies, 1986. In *IAQ '86: Managing Indoor Air for Health and Energy Conservation*. ASHRAE, Atlanta.

Casserly, D.M. and K.K. O'Hara. 1987. Ambient exposures to benzene and toluene in southwest Louisiana. *Paper* 87-98.1. Air and Waste Management Association, Pittsburgh.

Chan, C.C., L. Vanier, J.W. Martin, and D.T. Williams. 1990. Determination of organic contaminants in residential indoor air using an adsorption-thermal desorption technique. *Journal of Air and Waste Management Association* 40(1):62-67.

Chiarenzelli, R.V. and E.L. Joba. 1966. The effects of air pollution on electrical contact materials: A field study. *Journal of the Air Pollution Control Association* 16(3):123-27.

Code of Federal Regulations. 29 CFR 1900. U.S. Government Printing Office, Washington, D.C. Revised annually.

Code of Federal Regulations. National primary and secondary ambient air quality standards. 40 CFR 50. U.S. Government Printing Office, Washington, D.C. Revised annually.

Code of Federal Regulations. 10 CFR 50. U.S. Government Printing Office, Washington, D.C. Revised annually.

Cohen, M.A., P.B. Ryan, Y. Yanagisawa, J.D. Spengler, H. Ozkaynak, and P.S. Epstein. 1989. Indoor/outdoor measurements of volatile organic compounds in the Kanawha Valley of West Virginia. *Journal of the Air Pollution Control Association* 39(8):1086-93.

Cohen, S.I., N.M. Perkins, H.K. Ury, and J.R. Goldsmith. 1971. Carbon monoxide uptake in cigarette smoking. *Archives of Environmental Health* 22(1):55-60.

Cole, J.T. 1983. Constituent source emission rate characterization of three gas-fired domestic ranges. APCA *Paper* 83-64.3. Air and Waste Management Association, Pittsburgh.

Collins, M.F. 1986. Room temperature catalyst for improved indoor air quality. In *Indoor Air Quality in Cold Climates*, D.S. Walkinshaw, ed., 448-60. Air and Waste Management Association, Pittsburgh.

Conkle, J.P., B.J. Camp, and B.E. Welch. 1975. Trace composition of human respiratory gas. *Archives of Environmental Health* 30(6):290-95.

Coy, C.A. 1987. Regulation and control of air contaminants during hazardous waste site remediation. APCA *Paper* 87-18.1. Air and Waste Management Association, Pittsburgh.

Daisey, J.M. and A.P. Hodgson. 1988. Efficiencies of portable air cleaners for removal of nitrogen dioxide and volatile organic compounds. *Report* LBL-24964. Lawrence Berkeley Laboratory, Berkeley, CA.

DeFrees, J.A. and R.F. Amberger. 1987. Natural infiltration analysis of a residential high-rise building. In *IAQ '87: Practical Control of Indoor Air Problems*, 195-210. ASHRAE, Atlanta.

Dole, M. and I.M. Klotz. 1946. Sorption of chloropicrin and phosgene on charcoal from a flowing gas stream. *Industrial Engineering Chemistry* 38(12):1289-97.

Dubinin, M.M. 1975. Physical adsorption of gases and vapors in micropores. *Progress in Surface and Membrane Science* 9:1-70.

Egle, J.L. 1971. Single-breath retention of acetaldehyde in man. *Archives of Environmental Health* 23(12):427-33.

Egle, J.L. 1972. Retention of inhaled formaldehyde, proprionaldehyde, and acrolein in the dog. *Archives of Environmental Health* 25(8):119-24.

Egle, J.L. 1979. Respiratory retention and acute toxicity of furan. *Archives of Environmental Health* 40(4):310-14.

Elliott, L.P. and D.R. Rowe. 1975. Air quality during public gatherings. *Journal of the Air Pollution Control Association* 25(6):635-36.

Ensor, D.S., A.S. Viner, J.T. Hanley, P.A. Lawless, K. Ramanathan, M.K. Owen, T. Yamamoto, and L.E. Sparks. 1988. Air cleaner technologies for indoor air pollution. In *IAQ '88: Engineering Solutions to Indoor Air Problems*, 111-29. ASHRAE, Atlanta.

EPA. 1979. Toxic substances control act chemical substance inventory, Volumes I-IV. Environmental Protection Agency, Office of Toxic Substances, Washington, D.C.

EPA. 1986. Radon reduction techniques for detached houses: Technical guidance. EPA/625/5-86/O19. Environmental Protection Agency, Center for Environmental Research Information, Cincinnati.

EPA. 1987. *Radon reduction methods: A homeowner's guide*, 2nd ed. EPA-87-010. Environmental Protection Agency, Center for Environmental Research Information, Cincinnati.

EPA. 1990. National air quality and emission trends report, 1988. Environmental Protection Agency, Research Triangle Park, NC.

Fazzalari, F.A., ed. 1978. Compilation of odor and taste threshold values data. DS 48A. American Society for Testing and Materials, Philadelphia.

Freedman, R.W., D.I. Ferber, and A.M. Hartstein. 1973. Service lives of respirator cartridges versus several classes of organic vapors. *American Industrial Hygiene Association Journal* (2):55-60.

Fung, K. and B. Wright. 1990. Measurement of formaldehyde and acetaldehyde using 2,4-dinitrophenylhydrazine-impregnated cartridges. *Aerosol Science and Technology* 12(1):44-48.

Girman, J.R., A.P. Hodgson, A.F. Newton, and A.P. Winks. 1984. Emissions of volatile organic compounds from adhesives for indoor application. *Report* LBL-17594. Lawrence Berkeley Laboratory, Berkeley, CA.

Gorban, C.M., I.I. Kondrat'yeva, and L.Z. Poddubnaya. 1964. Gaseous activity products excreted by man in an airtight chamber. In *Problems of Space Biology*. JPRS/NASA. National Technical Information Service, Springfield, VA.

Graminski, E.L., E.J. Parks, and E.E. Toth. 1978. The effects of temperature and moisture on the accelerated aging of paper. NBSIR 78-1443. National Institute of Standards and Technology, Gaithersburg, MD.

Grosjean, D., P.M. Whitmore, C. Pamela de Moor, G.R. Cass, and J.R. Druzik. 1987. Fading of alizarinad-related artist's pigments by atmospheric ozone. *Environmental Science and Technology* 21:635-43.

Gully, A.J., R.M. Bethea, R.R. Graham, and M.C. Meador. 1969. Removal of acid gases and oxides of nitrogen from spacecraft cabin atmospheres. NASA-CR-1388. National Aeronautics and Space Administration. National Technical Information Service, Springfield, VA.

Hakov, R., J. Jenks, and C. Ruggeri. 1987. Volatile organic compounds in the air near a regional sewage treatment plant in New Jersey. *Paper* 87-95.1. Air and Waste Management Association, Pittsburgh.

Hartwell, T.D., J.H. Crowder, L.S. Sheldon, and E.D. Pellizzari. 1985. Levels of volatile organics in indoor air. *Paper* 85-30B.3. Air and Waste Management Association, Pittsburgh.

Haynie, F.H. 1978. Theoretical air pollution and climate effects on materials confirmed by zinc corrosion data. In *Durability of building materials and components* (ASTM STP 691). American Society for Testing and Materials, Philadelphia.

Hollowell, C.D., R.A. Young, J.V. Berk, and S.R. Brown. 1982. Energy-conserving retrofits and indoor air quality in residential housing. *ASHRAE Transactions* 88(1):875-93.

Hunt, R.D. and D.T. Williams. 1977. Spectrometric measurement of ammonia in normal human breath. *American Laboratory* (June):10-23.

Hursh, J.B., T.W. Clarkson, M.G. Cherian, J.J. Vostal, and R. Vandermallie. 1976. Clearance of mercury (Hg-197, Hg-203) vapor inhaled by human subjects. *Archives of Environmental Health* 31(6):302-309.

Jaffe, L.S. 1967. The effects of photochemical oxidants on materials. *Journal of the Air Pollution Control Association* 17(6):375-78.

Janssen, J.H. 1989. Ventilation for acceptable indoor air quality. *ASHRAE Journal* 31(10):40-45.

Jonas, L.A. and J.A. Rehrmann. 1972. The kinetics of adsorption of organophosphorous vapors from air mixtures by activated carbons. *Carbon* 10:657-63.

Jonas, L.A., E.B. Sansone, and T.S. Farris. 1983. Prediction of activated carbon performance for binary vapor mixtures. *American Industrial Hygiene Association Journal* 44:716-19.

Knoeppel, H. and H. Schauenburg. 1989. Screening of household products for the emission of volatile organic compounds. *Environment International* 15:413-18.

Lamm, S.H. 1986. Irritancy levels and formaldehyde exposures in U.S. mobile homes. In *Indoor Air Quality in Cold Climates*, 137-47. Air and Waste Management Association, Pittsburgh.

Leaderer, B.P., R.T. Zagranski, M. Berwick, and J.A.J. Stolwijk. 1987. Predicting NO_2 levels in residences based upon sources and source uses: A multi-variate model. *Atmospheric Environment* 21(2):361-68.

Lee, H.K., T.A. McKenna, L.N. Renton, and J. Kirkbride. 1986. Impact of a new smoking policy on office air quality. In *Indoor Air Quality in Cold Climates*, 307-22. Air and Waste Management Association, Pittsburgh.

Lodge, J.E., ed. 1988. *Methods of air sampling and analysis*, 3rd ed. Lewis Publishers, Chelsea, MI.

Loefroth, G., R.M. Burton, L. Forehand, S.K. Hammond, R.L. Seila, R.B. Zweidlinger, and J. Lewtas. 1989. Characterization of environmental tobacco smoke. *Environmental Science and Technology* 23(5):610-14.

Lonnemann, W.A., S.L. Kopczynski, P.E. Darley, and F.D. Sutterfield. 1974. Hydrocarbon composition of urban air pollution. *Environmental Science and Technology* 8(3):229-35.

Mahajan, B.M. 1987. A method of measuring the effectiveness of gaseous contaminant removal devices. NBSIR 87-3666. National Institute of Standards and Technology, Gaithersburg, MD.

Mahajan, B.M. 1989. A method of measuring the effectiveness of gaseous contaminant removal filters. NBSIR 89-4119. National Institute of Standards and Technology, Gaithersburg, MD.

Mathey, R.G., T.K. Faison, and S. Silberstein. 1983. Air quality criteria for storage of paper-based archival records. NBSIR 83-2795. National Institute of Standards and Technology, Gaithersburg, MD.

Matthews, T.G., T.J. Reed, B.J. Tromberg, C.R. Daffron, and A.R. Hawthorne. 1983. Formaldehyde emissions from combustion sources and solid formaldehyde resin-containing products. In *Proceedings of Engineering Foundation Conference on Management of Atmospheres in Tightly Enclosed Spaces*. ASHRAE, Atlanta.

Matthews, T.G., W.G. Dreibelbis, C.V. Thompson, and A.R. Hawthorne. 1985. Preliminary evaluation of formaldehyde mitigation studies in unoccupied research homes. In *Indoor Air Quality in Cold Climates*, 137-47. Air and Waste Management Association, Pittsburgh.

McGrath, T.R. and D.B. Stele. 1987. Characterization of phenolic odors in a residential neighborhood. *Paper* 87-75A.5. Air and Waste Management Association, Pittsburgh.

Mecklenberg, W. 1925. *Z. Elektrochem.* 31:488.

Meckler, M. and J.E. Janssen. 1988. Use of air cleaners to reduce outdoor air requirements. In *IAQ '88: Engineering Solutions to Indoor Air Problems*, 130-47. ASHRAE, Atlanta.

Miller, G.C. and P.C. Reist. 1977. Respirator cartridge service lives for exposure to vinyl chloride. *American Industrial Hygiene Association Journal* 38:498-502.

Molhave, L., I. Anderson, G.R. Lundquist, and O. Nielson. 1982. Gas emission from building materials. *Report* 137. Danish Building Research Institute, Copenhagen.

Moore, J.E. and E. Houtala. 1983. Odor as an aid to chemical safety: Odor thresholds compared with threshold limit values and volatility for 214 industrial chemicals in air and water dilution. *Journal of Applied Toxicology* 3:272.

Moschandreas, D.J. and S.M. Relwani. 1989. Field measurement of NO_2 gas-top burner emission rates. *Environment International* 15:489-92.

Nagda, N.L. and H.E. Rector. 1983. Guidelines for monitoring indoor-air quality. EPA 600/4-83-046. Environmental Protection Agency, Research Triangle Park, NC.

Nazaroff, W.W. and G.R. Cass. 1986. Mathematical modeling of chemically reactive pollutants in indoor air. *Environmental Science and Technology* 20:924-34.

Nefedov, I.G., V.P. Savina, and N.L. Sokolov. 1972. Expired air as a source of spacecraft carbon monoxide. *Paper*, 23rd International Aeronautical Congress, International Astronautical Federation, Paris.

Nelms, L.H., M.A. Mason, and B.A. Tichenor. 1986. The effects of ventilation rates and product loading on organic emission rates from particleboard. In *IAQ '86: Managing Indoor Air for Health and Energy Conservation*, 469-85. ASHRAE, Atlanta.

Nelson, E.D.P., D. Shikiya, and C.S. Liu. 1987. Multiple air toxics exposure and risk assessment in the south coast air basin. *Paper* 87-97.4. Air and Waste Management Association, Pittsburgh, PA.

Nelson, G.O. and A.N. Correia. 1976. Respirator cartridge efficiency studies: VIII. Summary and conclusions. *American Industrial Hygiene Association Journal* 37:514-25.

Newsome, J.R., V. Norman, and C.H. Keith. 1965. Vapor phase analysis of tobacco smoke. *Tobacco Science* 9:102-10.

NIOSH. 1977. *NIOSH manual of sampling data sheets*. U.S. Department of Health and Human Services, National Institute for Occupational Safety and Health, Washington, D.C.

NIOSH. 1984. *NIOSH manual of analytical methods*, 3rd ed. 2 vols. U.S. Department of Health and Human Services, National Institute for Occupational Safety and Health, Cincinnati.

NIOSH. 1990. *NIOSH/OSHA pocket guide to chemical hazards*. DHHS (NIOSH) *Publication* 90-117. U.S. Department of Labor, OSHA, Washington, DC. Users should read Cohen (1983) before using this reference.

NIOSH. Annual registry of toxic effects of chemical substances. U.S. Department of Health and Human Services, National Institute for Occupational Safety and Health, Washington, D.C.

NIOSH. (intermittent). Criteria for recommended standard for occupational exposure to (compound). U.S. Department of Health and Human Services, National Institute for Occupational Safety and Health, Washington, D.C.

Novosel, D., S.M. Relwani, and D.J. Moschandreas. 1987. Development of a desiccant-based environmental control unit. In *IAQ '87: Practical Control of Indoor Air Problems*, 195-210. ASHRAE, Atlanta.

NTIS. 1982. Air pollution effects on materials. *Published Search* PB 82-809427. National Technical Information Service, Springfield, VA.

NTIS. 1984. Air pollution effects on materials. *Published Search* PB 84-800267. National Technical Information Service, Springfield, VA.

Nunez, C., M. Kosusko, and B. Daniel. 1989. Effect of humidity on carbon adsorption performance in removing organics from contaminated air streams. *Paper* 89-29.2. Air and Waste Management Association, Pittsburgh.

Nwanko, J.N. and A. Turk. 1975. Catalytic oxidation of vapors adsorbed on activated carbon. *Environmental Science and Technology* 9(9):846-49.

Ostajic, N. 1985. Test method for gaseous contaminant removal devices. *ASHRAE Transactions* 91(2):11-31.

Pedersen, B.S. and W.J. Fisk. 1984. Air washing for the control of formaldehyde in indoor air. In *Proceedings of the 3rd International Conference on Indoor Air and Climate*. Swedish Council for Building Research, Stockholm.

Piatt, V.R. and J.C. White. 1962. The present status of chemical research in atmosphere purification and control on nuclear-powered submarines. NRL *Report* 5814, U.S. Naval Research Laboratory, Washington, D.C.

Rask, D. 1988. Indoor air quality and the bottom line. *Heating, Piping and Air Conditioning* 60(10).

Revoir, W.H. and J.A. Jones. 1972. Superior adsorbents for removal of mercury vapor from air. AIHA conference paper. American Industrial Hygiene Association, Akron, OH.

Rivers, R.D. 1988. Practical test method for gaseous gontaminant removal devices. In *Proceedings of the Symposium on Gaseous and Vaporous Removal Equipment Test Methods* (NBSIR 88-3716). National Institute of Standards and Technology, Gaithersburg, MD.

Sandalls, F.J. and S.A. Penkett. 1977. Measurements of carbonyl sulfide and carbon disulphide in the atmosphere. *Atmospheric Environment* 11:197-99.

Sax, N.I. and R.J. Lewis, Sr. 1988. *Dangerous properties of industrial materials*, 6th ed. 3 vols. Van Nostrand Reinhold, New York.

Shah, J.J. and H.B. Singh. 1988. Distribution of volatile organic chemicals in outdoor and indoor air. *Environmental Science and Technology* 22(12): 1381-88.

Singh, H.B., L.J. Salas, A. Smith, R. Stiles, and H. Shigeishi. 1981. Atmospheric measurements of selected hazardous organic chemicals. PB 81-200628, National Technical Information Service, Springfield, VA.

Sparks, L.E. 1988. Indoor air quality model, version 1.0. EPA-600/8-88-097a. Environmental Protection Agency, Research Triangle Park, NC.

Stampfer, J.F. 1982. Respirator canister evaluation for nine selected organic vapors. *American Industrial Hygiene Association Journal* 43(5):319-28.

Stankavich, A.J. 1969. The capacity of activated charcoal under dynamic conditions for selected atmospheric contaminants in the low parts-per-million range. In ASHRAE Symposium *Odors and Odorants: The Engineering View.* ASHRAE, Atlanta.

Sterling, T.D. and D. Kobayashi. 1981. Use of gas ranges for cooking and heating in urban dwellings. *Journal of the Air Pollution Control Association* 31(2):162-65.

Stewart, R.D., C.L. Hake, and J.E. Peterson. 1974. Use of breath analysis to monitor trichloroethylene exposures. *Archives of Environmental Health* 29(1):6-13.

Surgeon General. 1979. *Smoking and health.* U.S. Department of Health and Human Services, National Institute for Occupational Safety and Health, Washington, D.C.

Swanton, J.R., Jr. 1971. Field study of air quality in air-conditioned spaces. *ASHRAE Transactions* 77(1):124-39.

Taylor, D.G., R.E. Kupel, and J.M. Bryant. 1977. *Documentation of the NIOSH validation tests.* U.S. Department of Health and Human Services, National Institute for Occupational Safety and Health, Washington, D.C.

Thomas, J.W. 1974. Evaluation of activated carbon canisters for radon protection in uranium mines. HASL-280. National Technical Information Service, Springfield, VA.

Thomson, G. 1986. *The museum environment*, 2nd ed. Butterworths, London.

Tichenor, B.A. 1989. Measurement of organic compound emissions using small test chambers. *Environment International* 15:389-96.

Traynor, G.W. 1987. Field monitoring design considerations for assessing indoor exposures to combustion pollutants. *Atmospheric Environment* 21(2):377-83.

Traynor, G.W., J.R. Girman, M.G. Apte, J.F. Dillworth, and P.D. White. 1985a. Indoor air pollution due to emissions from unvented gas-fired space heaters. *Journal of the Air Pollution Control Association* 35:231-37.

Traynor, G.W., I.A. Nitschke, W.A. Clarke, G.P. Adams, and J.E. Rizzuto. 1985b. A detailed study of thirty houses with indoor combustion sources. *Paper* 85-30A.3. Air and Waste Management Association, Pittsburgh.

Turk, A. 1954. Odorous atmospheric gases and vapors: Properties, collection, and analysis. *Annals of the New York Academy of Sciences* 58:193-214.

Underhill, D.W., G. Micarelli, and M. Javorsky. 1988. Effects of relative humidity on adsorption of contaminants on activated carbon. In *Proceedings of the Symposium on Gaseous and Vaporous Removal Equipment Test Methods* (NBSIR 88-3716). National Institute of Standards and Technology, Gaithersburg, MD.

Van Gemert, L.J. and A.H. Mettenbreijer. 1977. *Compilation of odor threshold values in air and water.* Central Institute for Nutrition and Food Research, TNO Delft (Netherlands) or National Institute for Water Supply, Zeist/Voorburg (Netherlands).

Wade, W.A., W.A. Cote, and J.E. Yocum. 1975. A study of indoor air quality. *Journal of the Air Pollution Control Association* 25(9):933-39.

Wallace, L.A., E.D. Pellizzari, T.D. Hartwell, C. Sparacino, and H. Zelon. 1983. Personal exposure to volatile organics and other compounds indoors and outdoors—The TEAM-study, 1983. APCA *Paper*, Air and Waste Management Association, Pittsburgh.

Walsh, M., A. Black, A. Morgan, and G.H. Crawshaw. 1977. Sorption of SO_2 by typical indoor surfaces including wool carpets, wallpaper and paint. *Atmospheric Environment* 11:1107-11.

Weschler, C.J. and H.C. Shields. 1989. The effects of ventilation, filtration, and outdoor air on the composition of indoor air at a telephone office building. *Environment International* 15:593-604.

Wheeler, A. and A.J. Robell. 1969. Performance of fixed bed catalytic reactors with poison in the feed. *Journal of Catalysis* 13:299-305.

White, J.B., J.C. Reaves, P.C. Reist, and L.S. Mann. 1988. A data base on the sources of indoor air pollution emissions. In *IAQ '88: Engineering Solutions to Indoor Air Problems.* ASHRAE, Atlanta.

Windholz, M., ed. 1983. *The Merck Index*, 10th ed. Merck and Company, Rahway, NJ.

Wolff, R.K., M. Dolovich, C.M. Rossman, and M.T. Newhouse. 1975. Sulfur dioxide and tracheobronchial clearance in man. *AMA Archives of Environmental Health* 30(11):521-27.

Wolverton, B.C., A. Johnson, and K. Bounds. 1989. Interior landscape plants for indoor air pollution abatement. NASA, Stennis Space Center, Bay St. Louis, MS.

Yoon, Y.H. and J.H. Nelson. 1984a. Application of gas adsorption kinetics: I. A theoretical model for respirator cartridge service life. *American Industrial Hygiene Association Journal* 45(8):509-16.

Yoon, Y.H. and J.H. Nelson. 1984b. Application of gas adsorption kinetics: II. A theoretical model for respirator cartridge service life and its practical applications. *American Industrial Hygiene Association Journal* 45(8): 517-24.

Yoon, Y.H. and J.H. Nelson. 1988. A theoretical study of the effect of humidity on respirator cartridge service life. *American Industrial Hygiene Association Journal* 49(7):325-32.

BIBLIOGRAPHY

Cain, W.S., C.R. Shoaf, S.F. Velasquez, S. Selevan, and W. Vickery. 1992. Reference guide to odor thresholds for hazardous air pollutants listed in the Clean Air Act Amendments of 1990. EPA/600/R-92/047. Environmental Protection Agency.

Chung, T.-W., T.K. Ghosh, A.L. Hines, and D. Novosel. 1993. Removal of selected pollutants from air during dehumidification by lithium chloride and triethylene glycol solutions. *ASHRAE Transactions* 99(1):834-41.

Cohen, H.J., ed. 1983. AIHA's respiratory protection committee's critique of NIOSH/OSHA pocket guide to chemical hazards. *American Industrial Hygiene Association Journal* 44(9):A6-A7.

Foresti, R., Jr. and G. Dennison. 1986. Formaldehyde originating from foam insulation. In *IAQ '86: Managing Indoor Air for Health and Energy Conservation*, 523-37. ASHRAE, Atlanta.

Kelly, T.J. and D.H. Kinkead. 1993. Testing of chemically treated adsorbent air purifiers. *ASHRAE Journal* 35(7):14-23.

Liu, R.-T. 1991. Modeling activated carbon adsorbers for the control of volatile organic compounds in indoor air. *Far East Conference on Environmental Quality*, 119-26. ASHRAE, Atlanta.

Mehta, M.P., H.E. Ayer, D.E. Saltzman, and R. Ronk. 1988. Predicting concentrations for indoor chemical spills. In *IAQ '88: Engineering Solutions to Indoor Air Problems*, 231-50. ASHRAE, Atlanta.

Nirmalakhandan, N.N. and R.E. Speece. 1993. Prediction of activated carbon adsorption capacities for organic vapors using quantitative structure-activity relationship methods. *Environmental Science and Technology* 27:1512-16.

NBS. 1959. Maximum permissible body burdens and maximum permissible concentrations of radionuclides in air and water for occupational exposure. NBS *Handbook* 69. U.S. Government Printing Office, Washington, D.C.

Riggen, R.M., W.T. Wimberly, and N.T. Murphy. 1990. Compendium of methods for determination of toxic organic compounds in ambient air. EPA/600/D-89/186. Environmental Protection Agency, Research Triangle Park, NC. Also second supplement, 1990, EPA/600/ 4-89/017.

Spicer, C.W., R.W. Coutant, G.F. Ward, A.J. Gaynor, and I.H. Billick. 1986. Removal of nitrogen dioxide from air by residential materials. In *IAQ '86: Managing Indoor Air for Health and Energy Conservation*, 584-90. ASHRAE, Atlanta.

Sterling, T.D., E. Sterling, and H.D. Dimich-Ward. 1983. Air quality in public buildings with health related complaints. *ASHRAE Transactions* 89(2A):198-212.

Webb, P. 1964. *Bioastronautics data book*. NASA SP-3006. National Technical Information Service, Springfield, VA.

Winberry, W.J. et al. 1990. *EPA compendium of methods for the determination of air pollutants in indoor air.* EPA/600/S4-90/010. Environmental Protection Agency. (NTIS-PB 90200288AS).

Woods, J.E., J.E. Janssen, and B.C. Krafthefer. 1986. Rationalization of equivalence between the ventilation rate and air quality procedures in ASHRAE *Standard* 62. In *IAQ '86: Managing Indoor Air for Health and Energy Conservation.* ASHRAE, Atlanta.

Yu, H.H.S. and R.R. Raber. 1992. Air-cleaning strategies for equivalent indoor air quality. *ASHRAE Transactions* 98(1):173-81.

AUTOMATIC CONTROL

A UTOMATIC control of HVAC systems and equipment usually includes control of temperature, humidity, pressure, and flow rate. Automatic control sequences equipment operation to meet load requirements and provides safe operation of the equipment, using pneumatic, mechanical, electrical, and electronic control devices.

This chapter focuses on the design of automatic control systems. It covers (1) control fundamentals, including terminology; (2) the different types of control components; (3) the methods of connecting these components together to form various types of individual control loops or subsystems; (4) the synthesis of these subsystems into a complete control system; (5) design considerations; and (6) start-up and testing.

CONTROL FUNDAMENTALS

CONTROL

A *closed loop* or *feedback* control system measures changes in the controlled variable and actuates the control device to take corrective action, which continues until the variable is brought to a desired value within the design limitations of the controller. This system of transmitting the value of the controlled variable back to the controller is known as *feedback*.

In an *open loop* or *feedforward* control system there is no direct link between the value of the controlled variable and the controller. Feedforward control anticipates the effect an external variable will have on the system. An example is an outdoor thermostat that controls heat flow to a building in proportion to the calculated load of changes in outdoor temperature. In essence, the designer of this system presumes a fixed relationship between outdoor air temperature and the heat requirement of the building and takes control action based on the outdoor air temperature. The actual space temperature has no effect on this controller. Because there is no feedback of the controlled variable (space temperature), the control is known as open loop.

Feedforward control can also be used effectively on a feedback controller to modify the control based on an external variable affecting the control loop. The controller is then a feedback controller with feedforward compensation.

The preparation of this chapter is assigned to TC 1.4, Control Theory and Application.

TERMINOLOGY

Figure 1 illustrates the components of a typical *control loop*.

The *sensor* measures the controlled variable and conveys values to the controller. The *controller* compares those values with the set point and sends a signal for corrective action to the controlled device. Thermostats, humidistats, flow controllers, and pressure controllers are examples of controllers.

The controller seeks to maintain the *set point*, which is the desired value of the controlled variable.

The *controlled device* reacts to signals received from the controller to vary the flow of the control agent. It may be a valve, a damper, an electric relay, or a motor driving a pump or a fan.

The *control agent* is the medium manipulated by the controlled device. It may be air flowing through a damper or gas, steam, or water flowing through a valve.

The *process plant* is the apparatus being controlled. It reacts to the control agent and effects the change in the controlled variable.

The *controlled variable* is the temperature, humidity, pressure, or flow being controlled.

A control loop can be represented in the form of a block diagram, in which each component is modeled and represented in its own block (Figure 2). The flow of information from one component to the next is shown by lines between the blocks. The figure shows the set point being compared to the feedback of the controlled variable. The difference, or error, is fed into the controller, which sends a control signal to the controlled device. In this case, the controlled device is a valve that can change the amount of steam flow through the coil (Figure 1). The amount of steam flow is the input to the next block,

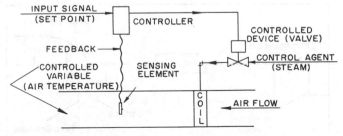

Fig. 1 Discharge Air Temperature Control—Example of Feedback Control System

which represents the output. The process plant block affects the controlled variable, which is temperature. The output is sensed by the sensing element and fed to the controller as feedback, completing the loop.

Each component of Figure 2 can be represented by a transfer function, which is an idealized mathematical representation of the relationship between the input and output variables of the component. The transfer function must be sufficiently detailed to cover both the dynamic and static characteristics of the device. The dynamics of the component are represented in the time domain by a differential equation. In environmental control systems, the transfer function of many of the components can be adequately described by a first order differential equation, implying that the dynamic behavior is dominated by a single capacitance factor. The differential equation is converted to its *LaPlace transform* or *z-transform*.

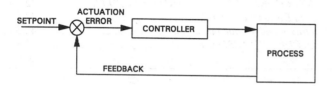

Fig. 2   Block Diagram of Discharge Air Temperature Control System

The time constant is defined as the time it takes for the output to reach 63.2% of its final value when a step change in the input is effected. If the time constant of the component is small, its output will react rapidly to reflect changes in the input; conversely, components with a large time constant will be sluggish in responding to changes in the input.

Dead time is a nonlinearity that can cause control and modeling problems. It is the time between a change in the process input and when that change affects the output of the process. In the control loop of Figure 1, dead time can occur due to the transportation time of the air. The temperature the sensor records represents the conditions at the coil some time in the past because the air it is sensing went through the coil at some time in the past. Dead time can result from a slow sensor, a time lag in the signal from the controller, or the transportation time of the control agent. If dead time is small, it can be ignored in the model of the control system; if it is significant, it must be considered.

The *gain* of a transfer function is the amount by which the output of the component will change for a given change of input under steady-state conditions. If the element is linear, its gain remains constant. However, many control system components are nonlinear and have varying gains, depending on the operating conditions.

Figure 3 shows the response of a first order plus dead time process to a step change of the input signal. Notice that the process shows no reaction during the dead time, followed by a response that resembles a first order exponential.

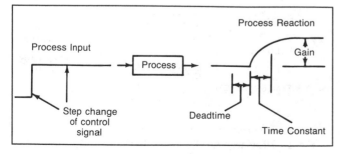

Fig. 3   Process Subject to a Step Input

The principle behind controllers is that the sensor sends a signal (pneumatic, electric, or electronic); the pressure, voltage, or current of this signal is proportional to the value of the variable being measured. The controller contains circuitry that compares the signal from the sensor to the desired value and outputs a control signal based on this comparison. The controller is an analog device that receives and acts on data continuously.

TYPES OF CONTROL ACTION

Closed loop control systems are commonly classified by the type of corrective action the controller is programmed to take when it senses a deviation of the controlled variable from the set point. The following are the most common types of control action.

Two-Position Action.   The controlled device shown in Figure 4 can be positioned only to a maximum or minimum state, or can be either on or off. A typical home thermostat that starts and stops a furnace is a good example of two-position action.

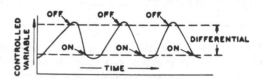

Fig. 4   Two-Position Control

Controller differential, as it applies to two-position control action, is the difference between a setting at which the controller operates to one position and a setting at which it operates to the other. There are two kinds of controller differential. Thermostat ratings usually refer to the differential that becomes apparent by raising and lowering the dial setting. This differential is known as the *manual differential* of the thermostat. The *operating differential* is the total change in temperature that occurs between a call for more heat and a call for less heat; it is usually greater than the manual differential caused by thermostat lag. For example, a thermostat with a 1 K manual differential may produce temperature variations of 1.5 K between the "system on" and "system off" states.

Timed Two-Position Action.   This is a common variation of straight two-position action that is often used on room thermostats to reduce operating differential. In heating thermostats, a heater element is energized during "on" periods, shortening the on-time as the heater warms the thermostat prematurely. This is known as heat anticipation. The same anticipating action can be obtained in cooling thermostats by energizing a heater during thermostat "off" periods. In both cases, the percentage of on-time is varied in proportion to the system load, while the total cycle time remains relatively constant.

Floating Action.   In floating action, the controller can perform only two operations—moving the controlled device either toward its open position or toward its closed position, usually at a constant rate (see Figure 5). There is generally a neutral zone between the two positions in which the controlled device may stop at any position when the controlled variable is within the differential of the controller. When the controlled variable moves outside the differential, the controller moves the controlled device in the proper direction.

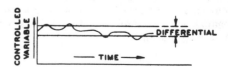

Fig. 5   Floating Control Showing Variations in Controlled Variable as Load Changes

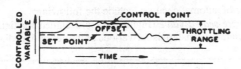

Fig. 6 Proportional Control Showing Variations in Controlled Variable as Load Changes

Proportional Action. In proportional action, the controlled device is positioned proportionally in response to changes in the controlled variable (Figure 6). A proportional controller can be described by

$$V_p = K_p e + V_o \qquad (1)$$

where

V_p = output of the proportional controller
K_p = proportional gain (proportional to 1/throttling range)
e = error signal, or offset
V_o = offset adjustment parameter

The output of the controller is proportional to the difference between the sensed value and its set point. The controlled device is normally adjusted by offset adjustment to stay in the middle of its control range at set point. This control is similar to that shown in Figure 1.

Throttling range is the amount of change in the controlled variable that causes the controller to move the controlled device from one extreme to the other. It can be adjusted to meet job requirements. Throttling range is inversely proportional to proportional gain.

Control point is the actual value of the controlled variable at which the instrument is controlling. It varies within the throttling range of the controller and changes with changing load on the system and other variables.

Offset, or *error signal*, is the difference under steady-state conditions between the set point and the control point. It is sometimes called drift, deviation, droop, or steady-state error.

Proportional-plus-Integral. Proportional-plus-integral (PI) control improves on simple proportional control with the addition of a component that eliminates the offset typical of proportional control. The reset action is most easily shown by the following equation:

$$V_p = K_p e + K_i \int e \, d\theta + V_o \qquad (2)$$

where

K_i = integral gain
$\int e \, d\theta$ = summation of error over difference in time

The second term in Equation (2) implies that the longer the period is during which the error e exists, the more the controller output will change in attempting to eliminate the error.

Selecting proportional and integral gain constants is critical to system stability. Proper tuning eliminates offset, obtaining greater control accuracy.

Proportional-Integral-Derivative (PID). This type of control is PI control with a term added to the controller. It varies with the value of the derivative of the error. The equation for PID control is

$$V_p = K_p e + K_i \int e \, d\theta + K_d \, de/d\theta + V_o \qquad (3)$$

where

K_d = derivative gain of controller
$de/d\theta$ = time derivative of error

Adding the derivative term allows the controller to take some anticipatory action, which can reduce overshoot on controlled variables that respond quickly. However, the addition of the derivative term also makes the controller more sensitive to noisy signals and harder to tune than a PI controller. Most HVAC control loops perform satisfactorily with PI control without the added

derivative term. Adaptive control, or self-tuning, is an enhancement of PID control in which the gain factors (K_p, K_i, and K_d) are continuously or periodically modified to compensate for the control loop offset.

PERFORMANCE REQUIREMENTS OF CONTROL SYSTEMS

The control system performance in an air-conditioning application is evaluated in terms of speed of response and stability. A stable control loop keeps the controlled variable near set point while avoiding long-term oscillations. A control loop with a fast response time reacts quickly to process disturbances. The requirements of accuracy, speed, and stability often conflict; that is, a change made in one of these parameters could adversely affect the others. Control systems must be selected to suit the application and must be evaluated in terms of control, comfort, and energy conservation.

CLASSIFICATIONS OF CONTROL SYSTEMS

Control components can be classified into the following three categories, according to primary source of energy:

- *Pneumatic* components use compressed air, usually at a pressure of 100 to 250 kPa (gage), as an energy source. The air is supplied to the controller, which regulates the pressure supplied to the controlled device.

- *Electric* components use electrical energy, either low or line voltage, as an energy source. The electric energy supplied to the controlled device is regulated by the controller. Controlled devices in this category include motors; relays; contactors; and electromechanical, electromagnetic, and solid-state regulating devices. The components that include signal conditioning, modulation, and amplification in their operation are classified as electronic.

  A direct digital controller (DDC) receives electronic signals from the sensors, converts the electronic signals to numbers, and performs mathematical operations on these numbers inside the microprocessor. The output from the microprocessor takes the form of a number, which can be converted to an electric or pneumatic signal to operate the actuator. The digital controller must sample its data because the microprocessor must have time for other operations besides reading data. If the sampling interval is properly chosen, no significant degradation in control performance will be seen due to sampling.

- *Self-powered* components apply the power of the measured system to induce the necessary corrective action, with no auxiliary source of energy. Temperature changes at the sensor result in pressure or volume changes in the enclosed media that are transmitted directly to the operating device of the valve or damper. A component using a thermopile in a pilot flame to generate electrical energy is also self-powered.

This method of classification can be extended to individual control loops and to complete control systems. For example, the room temperature control for a particular room that includes a pneumatic room thermostat and a pneumatically actuated reheat coil would be referred to as a pneumatic control loop. Some control systems use a combination of different energy sources and are more accurately called *hybrid systems*. For example, the control system for an air handler could include electric components for on/off control of the fan, pneumatic components for control of the heating and cooling coils, and self-powered safety controls (e.g., a freezestat).

COMPUTERS FOR AUTOMATIC CONTROL

Computers can perform the control schemes described in this chapter. Chapter 36 covers computer components and some of the ways computers are being used in the HVAC industry. Other technical publications are available that describe the application of computers in the HVAC control industry.

CONTROL COMPONENTS

Although control components may be classified in several ways, this section groups them by their function within a complete control system. The first subsection considers controlled devices, or final control elements, including (1) relays; (2) valves; (3) dampers; and (4) variable air volume (VAV) boxes, which contain a damper or damper-like mechanism. Operators, which are used to drive the valve or damper assembly, are also covered.

The second subsection considers the sensing element that measures changes in the controlled variable. Examples of sensor types included are temperature, humidity, water and air pressure, water and air differential pressure, and water and airflow rate. While many other kinds of special sensors are available, these represent the majority of those found in the HVAC control systems and subsystems described in the section on Control System Design and Application.

In the third subsection, various types of controllers are reviewed. Controllers are classified according to the control action they initiate to maintain the desired condition (set point). They may use two-position, floating, proportional, PI, or PID control. This section also describes the various devices, such as pneumatic, electronic, and digital controllers, available for making the control decision in a modulating control system. Thermostats (devices that combine a temperature sensor and controller into a single unit) are also covered.

Many control systems can be constructed using only the types of components described in the first three subsections. In practice, however, a fourth type is sometimes necessary. These components are neither sensing elements nor controlled devices nor controllers, but are referred to as auxiliary control components and include transducers, relays, switches, power supplies, and air compressors.

CONTROLLED DEVICES

The controlled device most often regulates or varies the flow of steam, water, or air within an HVAC system. Water and steam flow regulators are known as *valves*, and airflow control devices are called *dampers*; they perform essentially the same function and must be properly sized and selected for the particular application. The control system's link to the valve or damper is a component called an *operator*, or *actuator*. This device uses electricity, compressed air, or hydraulic fluid to power the motion of the valve stem or damper linkage through its operating range.

Valves

An *automatic valve* is designed to control the flow of steam, water, gas, and other fluids and may be thought of as a variable orifice positioned by an electric or pneumatic operator in response to impulses or signals from the controller. It may be equipped with a throttling plug or V-port specially designed to provide the desired flow characteristics.

Renewable composition disks are common. They are made of the materials best suited to the media handled by the valve, the operating temperature, and the pressure. For high pressures or for superheated steam, metal disks are often used. Internal parts of valves, such as the seat ring, throttling plug or V-port skirt, disk holder, and stem, are sometimes made of stainless steel or other hard and corrosion-resistant metals.

Types of automatic valves include the following:

A *single-seated valve* (Figure 7A) is designed for tight shutoff. Appropriate disk materials for various pressures and media are used.

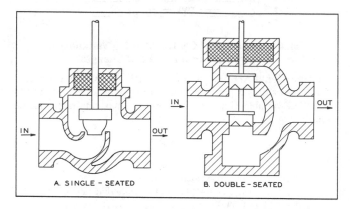

Fig. 7   Typical Single- and Double-Seated Two-Way Valves

A *double-seated* or *balanced valve* (Figure 7B) is designed so that the pressure acting against the valve disk is essentially balanced, reducing the operator force required. It is widely used where fluid pressure is too high to permit a single-seated valve to close. It cannot be used where tight shutoff is required.

A *three-way mixing valve* (Figure 8A) has two inlet connections, one outlet connection, and a double-faced disk operating between two seats. It is used to mix two fluids entering through the inlet connections and leaving through the common outlet, according to the position of the valve stem and disk.

A *three-way diverting valve* (Figure 8B) has one inlet connection, two outlet connections, and two separate disks and seats. It is used either to divert the flow to one of the outlets or to proportion the flow to both outlets.

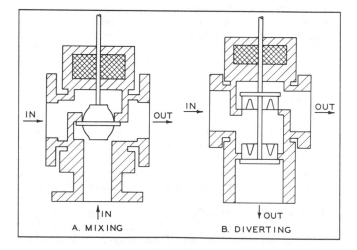

Fig. 8   Typical Three-Way Mixing and Diverting Valves

A *butterfly* consists of a heavy ring enclosing a disk that rotates on an axis at or near its center; it is similar in principle to a round single-blade damper. The disk seats against a ring machined within the body or a resilient liner in the body. Two butterfly valves used together can reproduce the mixing or diverting function of the three-way valve.

Characteristics. The performance of a valve is expressed in terms of its flow characteristics as it operates through its stroke, based on a *constant pressure drop*. Three common characteristics are shown in Figure 9 and are defined as follows:

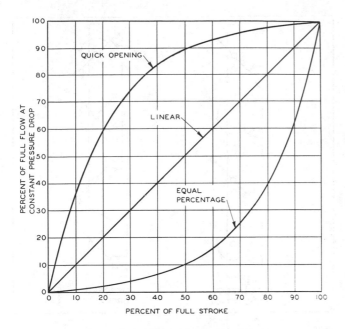

Fig. 9   Typical Flow Characteristics

Quick-opening. Maximum flow is approached rapidly as the device begins to open.

Linear. Opening and flow are related in direct proportion.

Equal percentage. Each equal increment of opening increases the flow by an equal percentage over the previous value.

Because the pressure drop across a valve seldom remains constant as its opening changes, actual performance usually deviates from the published characteristic curve. The magnitude of the deviation is determined by the overall system design. For example, in a system arranged so that control valves or dampers can shut off all flow, the pressure drop across a controlled device increases from a minimum at design conditions to the total system pressure drop at no flow. Figure 10 shows the extent of the resulting deviations for a valve or damper designed with a linear characteristic. To approximate the

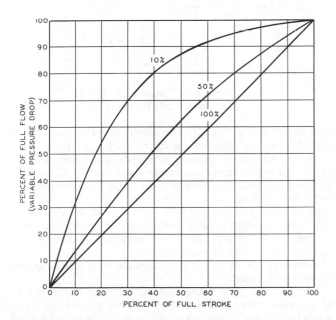

Fig. 10   Typical Performance Curves for Linear Devices at Various Percentages of Total System Pressure Drop

designed characteristic of the valve or damper, the design pressure drop should be a reasonably large percentage of the total system pressure drop, or the system should be designed and controlled so that this pressure drop remains relatively constant.

Selection and Sizing. Higher pressure drops for controlled devices are obtained by decreasing their sizes and possibly increasing the size of other equipment in the system. Since sizing techniques are different for steam, water, and air, each is discussed separately.

Steam Valves. Steam-to-water and steam-to-air heat exchangers are typically controlled through regulation of steam flow rate with a two-way throttling valve. One-pipe steam systems require a line-size two-position valve for proper condensate drainage and steam flow, while two-pipe steam systems can be controlled by two-position or modulating (throttling) valves.

Water Valves. Valves used for water service may be two- or three-way and two-position or proportional. Proportional valves are used most often, but two-position valves are not unusual and are sometimes essential (e.g., on steam preheat coils). Two-position valves are normally of the quick-opening type, while proportioning valves are normally of the linear or equal percentage type.

The *flow coefficient* K_v is generally used to compare valve capacities in I-P units only. K_v is defined as the volume of water flow in L/s at 16°C through a control valve in the fully open position with a 6.9 kPa differential across the valve.

While it is possible to design a water system in which the pressure differential from supply to return is kept constant, this is seldom done. It is safer to assume that the pressure drop across the valve will increase as it modulates from fully open to fully closed. Figure 11 shows the effect in a simple system with one pump and one two-way control valve with a heat exchanger. The *system curve* represents the pressure loss in the piping system and heat exchanger at various flow rates. The *pump curve* is the typical curve for a centrifugal pump. The valve is selected for a specific pressure drop A-A′ at design flow rates. At part load, the valve must close partially to provide a higher pressure drop B-B′. The ratio of the design pressure drop A-A′ to the zero-flow pressure drop C-C′ affects the control capability of the valve.

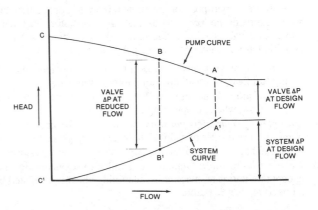

Fig. 11   Pump and System Curves with Valve Control

Better control at part load is obtained by using equal percentage valves, particularly in hot water coils where the heat output of the coil is not linearly related to flow. As flow is reduced, a greater amount of heat is transferred from each unit volume of water, counteracting the reduction in flow. The use of equal percentage valves linearizes the heat transfer from the coil with respect to the control signal.

Two-way control valves should be sized to provide from 20 to 60% of the total system pressure drop. The valve operator should be sized to ensure complete shut-off during no-flow condition. For additional information on control valve sizing and selection, see

Chapters 12 and 43 of the 1992 *ASHRAE Handbook—Systems and Equipment.*

Operators. Valve operators are of five general types:

1. A *solenoid* consists of a magnetic coil operating a movable plunger. Most are for two-position operation, but modulating solenoid valves are available with a pressure equalization bellows or piston to achieve modulation. Solenoid valves are generally limited to relatively small sizes (100 mm or less).

2. An *electric motor* operates the valve stem through a gear train and linkage. Electric motor operators are of the following three types:

   a) *Unidirectional*—for two-position operation. The valve opens during one half revolution of the output shaft and closes during the other half revolution. Once started, it continues until the half revolution is completed, regardless of subsequent action by the controller. Limit switches built into the operator stop the motor at the end of each stroke. If the controller has been satisfied during this interval, the operator continues to the other position.

   b) *Spring-return*—for two-position operation. Electric energy drives the valve to one position, and a spring returns it to its normal position.

   c) *Reversible*—for floating and proportional operation. The motor can run in either direction and can stop in any position. It is sometimes equipped with a return spring. In proportional control applications, the motor also drives a feedback potentiometer for rebalancing the control circuit.

3. A *pneumatic operator* consists of a spring-opposed, flexible diaphragm or bellows attached to the valve stem. An increase in air pressure above the minimum point of the spring range compresses the spring and simultaneously moves the valve stem. Springs of various constants can sequence the operation of two or more devices, if properly selected or adjusted. For example, a chilled water valve operator may modulate the valve from fully closed to fully open over a range of 20 to 60 kPa (gage), while a sequenced steam valve may operate from 60 to 90 kPa (gage).

   Pneumatic operators are used primarily for proportional control. Two-position control is accomplished using a two-position controller or a two-position pneumatic relay to apply either full air pressure or no pressure to the valve operator. Pneumatic valves and valves with spring-return electric operators can be classified as normally open or normally closed.

   a) A *normally open* valve will assume an open position, providing full flow, when all operating force is removed.

   b) A *normally closed* valve will assume a closed position, stopping flow, when all operating force is removed.

4. *Springless pneumatic operators*, which use two opposed diaphragms or two sides of a single diaphragm, are generally limited to special applications involving large valves or high pressures.

5. An *electric-hydraulic actuator* is similar to a pneumatic one, except that it uses an incompressible fluid, which is circulated by an internal electric pump.

Dampers

Types and Characteristics. Automatic dampers are used in air-conditioning and ventilation systems to control airflow. They may be used (1) for modulating control, to maintain a controlled variable such as mixed air temperature or supply air duct static pressure; or (2) for two-position control, to initiate system operation (e.g., by opening outdoor air dampers when a fan system is started).

Two damper arrangements are used for air-handling system flow control—parallel-blade and opposed-blade (see Figure 12). Parallel-blade dampers are adequate for two-position control and can be used for modulating control when they are the primary source of system pressure drop. However, opposed-blade dampers are preferable because they normally provide better control (Figures 13 and 14). In

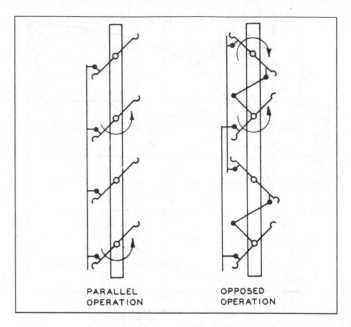

Fig. 12   Typical Multiblade Dampers

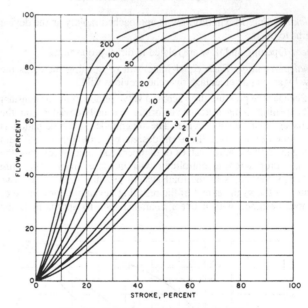

**Fig. 13   Installed Characteristic Curves
of Parallel Blade Dampers**

these figures, the parameter α is the ratio of the system pressure drop to the drop across the damper at maximum (fully open) flow.

Damper leakage is important, particularly where tight shutoff is required to reduce energy consumption significantly. For another example, outdoor air dampers must close tightly to prevent coils and pipes from freezing. Low-leakage dampers are more costly and require larger operators because of the friction of the seals in the closed position; therefore, they should be used only when necessary.

Operators. Damper operators are available using either electricity or compressed air as a power source.

Electric damper operators (actuators) can be reversible, spring-return or unidirectional.

The reversible type is frequently used for accurate control in modulating damper applications. There are two sets of motor windings within the housing of a reversible electric actuator. Energizing one set of windings causes the actuator output shaft to turn clockwise, and energizing the other causes it to turn counterclockwise.

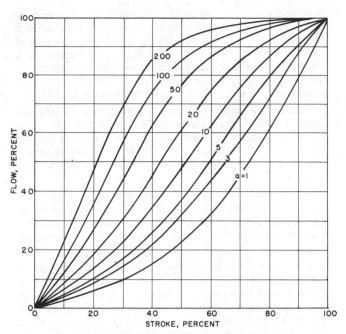

**Fig. 14 Installed Characteristic Curves
of Opposed Blade Dampers**

When neither set of windings is energized, the shaft remains in its last position. The simplest form of control for this actuator is a floating point controller, which causes a contact closure to drive the motor clockwise and counterclockwise. The reversible electric actuator is available with a wide range of options for rotational shaft travel (expressed in degrees of rotation) and timing (expressed in the number of seconds it takes to move through the rotational range). In addition, a variety of standard electronic signals from electronic controllers (such as 4-20 mA dc or 0-10 V dc) can be used to drive this type of modulating actuator.

A spring-return actuator moves in one direction whenever voltage is applied to its internal windings; when no power is present, the actuator is returned via spring force to its normal position. Depending on how the actuator is connected to the dampers, this will open or close the dampers.

Pneumatic damper operators are similar to pneumatic valve operators, except that they have a longer stroke, or the stroke is increased by a multiplying lever. Increasing the air pressure produces a linear motion of the shaft, which, through a linkage, moves the crank arm to open or close the dampers. Normally open or normally closed operation refers to the position of the dampers when no air pressure is applied at the operator. This position depends on the mounting of the operator and on the linkage connection.

Mounting. Damper operators are mounted in several different ways, depending on damper size and accessibility and on the power required to move the dampers. They can be mounted in the airflow on the damper frame and connected by a linkage directly to a damper blade, or they can be mounted outside the duct and connected to a crank arm attached to a shaft extension of one of the blades. On large dampers, two or more operators may be needed. In this case, they are usually mounted at separate points on the damper. An alternative is to install the damper in two or more sections, each section being controlled by a single damper operator; however, proper airflow control is easier with a single modulating damper. Positive positioners may be required for proper sequencing. A small damper with a two-position operator for minimum outside flow may be used with a large, independently controlled damper for economy cycle cooling.

Positive Positioners

A pneumatic operator may not respond quickly or accurately enough to small changes in control pressure due to friction in the actuator or load or to changing load conditions, such as wind acting on a damper blade. Where accurate positioning of a modulating damper or valve in response to load is required, positive positioners should be used.

A positive positioner provides up to full main control air pressure to the actuator for any change in position required by the controller. This is achieved by the sample arrangement shown schematically in Figure 15. An increase in branch pressure from the controller (A) moves the relay lever (B), opening the supply valve (C). This allows main air to flow to the relay chamber and the actuator cylinder, moving the piston (not shown). The piston movement is transmitted through a linkage and spring (D) to the other end of the lever (B), and when the force due to movement balances out the control force, the supply valve closes, leaving the actuator in the new position. A decrease in control pressure allows the exhaust valve (E) to open and remain open until a new balance is obtained. Thus, full main air pressure is available, if needed, even though the control pressure may have changed only a fraction of that amount. The movement feedback linkage may be mounted internally or externally. Positioners may be connected for direct or reverse action.

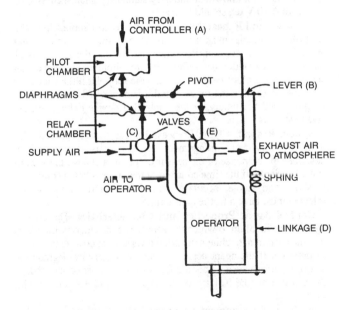

Fig. 15 Positive Positioner

A positive positioner provides finite and repeatable changes in positioning and permits adjustment of the control range (spring range of the actuator) to provide proper sequencing control of two or more controlled devices.

SENSORS

The sensor is the component in the control system that measures the value of the controlled variable. A change in the controlled variable (such as the temperature of water flowing in a pipe) produces a change in some physical or electrical property of the primary sensing element that is then translated or amplified by mechanical or electrical signal. A device that converts one form of energy (mechanical or thermal) to another (electrical) is known as a *transducer*. In some cases, the sensing element is a transducer, such as a thermistor, in which a change in temperature causes a change in electrical resistance.

With the trend toward electronic miniaturization and solid-state sensing elements, sensor selection has become a specialty. New measurement technologies and manufacturers are emerging regularly, expanding the options available to the control system designer and outdating older texts on sensor application. This section does not describe all the technologies for measurement and transmission of even the common variables of temperature, humidity, pressure, and flow rate. Chapter 13 of the 1993 *ASHRAE Handbook—Fundamentals*, manufacturers' catalogs and tutorials, and the bibliography give specific applications information.

In selecting a particular sensor product for a specific application, the following elements should be considered:

Operating Range of the Controlled Variable. The sensor must be capable of varying its output signal detectably and significantly over the expected operating range. In the case of an office room temperature measurement, for example, the output of a pneumatic transmitter may change from 20 to 100 kPa (gage) over a temperature range of 13 to 20°C.

Compatibility of the Controller Input. Pneumatic receiver controllers for HVAC systems will typically accept input signals of 20 to 100 kPa (gage). Electronic and digital controllers accept various ranges and types of electronic signals. In the selection of an electronic sensor, the specific controller to be used must be considered; if this is not known, an industry standard signal, such as 4-20 mA dc or 0-10 V dc, should be used.

Accuracy and Repeatability. For some control applications, the controlled variable must be maintained within a narrow band around a desired set point. Both the accuracy and the sensitivity of the sensor selected must reflect this requirement. However, an accurate sensor alone cannot maintain the set point if the controller is unable to resolve the input signal, the controlled device cannot be positioned accurately, or the controlled device exhibits excessive hysteresis.

System Response Time (Process Dynamics). Associated with a sensor/transducer arrangement is a response curve, which describes the response of the sensor output to a change in the controlled variable. If the time constant of the process being controlled is short, and stable, accurate control is important, the sensor selected must have a fast response time.

Control Agent Properties and Characteristics. The control agent is the medium to which the sensor is exposed, or with which it comes in contact, while measuring a variable such as temperature or pressure. If the agent acts to corrode or otherwise degrade the sensor's performance, either a different sensor must be selected, or the sensor must be isolated or protected from direct contact with the agent.

Ambient Environment Characteristics. Even when the sensor's components are isolated from direct contact with the control agent, the ambient environment in which they are located must be considered. The temperature and humidity range of the ambient environment must not adversely affect the sensor or its accuracy. Likewise, the presence of certain gases, chemicals, and electromagnetic interference (EMI) can cause component degradation. In such cases, a special sensor or transducer housing can be used to protect the element while ensuring true measurement of the control variable.

Temperature Sensors

Temperature-sensing elements depend on one of three physical phenomena: (1) a change in relative dimension due to differences in thermal expansion, (2) a change in state in a vapor- or liquid-filled bellows, or (3) a change in some electrical property. Within each category, there are a variety of sensing element configurations to measure room, duct, water, and surface temperatures.

The specific types of temperature-sensing technologies commonly used in HVAC applications are as follows:

- A *bimetal* element is composed of two thin strips of dissimilar metals fused together. Because the two metals have different thermal expansion coefficients, the element bends as the temperature varies and produces a change in position. Depending on the space available and the movement required, the strip may be straight, U-shaped, or wound into a spiral. This element is commonly used in room, insertion, and immersion thermostats.

- A *rod-and-tube* element consists of a high-expansion metal tube containing a low-expansion rod with one end attached to the rear of the tube. The tube changes length with changes in temperature, causing the free end of the rod to move. This element is commonly used in certain types of insertion and immersion thermostats.

- A *sealed bellows* element is filled with vapor, gas, or liquid after being evacuated of air. Temperature changes cause changes in the pressure or volume of the gas or liquid, resulting in a change in force or movement. This element is often used in room thermostats.

- A *remote bulb* element is a sealed bellows or diaphragm to which a bulb or capsule is attached by means of a capillary tube; the entire system is filled with vapor, gas, or liquid. Temperature changes at the bulb result in volume or pressure changes, which are communicated to the bellows or diaphragm through the capillary tube. The remote bulb element is useful where the temperature measuring point is remote from the desired thermostat location. It usually is provided with fittings suitable for insertion into a duct, pipe, or tank. *Averaging bulbs* are used in large ducts where a single sensing point may not be representative.

- A *thermistor* makes use of the change in electrical resistance of a semiconductor material due to a representative change in temperature. The characteristic curve of a thermistor is nonlinear over a wide range. It has a negative temperature coefficient; that is, the resistance decreases as the temperature increases. For electronic control systems, a variety of techniques is available to ensure a linear change over a particular temperature range. In a digital control system, one technique for converting the nonlinear response is to store a computer "look-up table" that maps the temperatures corresponding to measured resistances. Thermistors are used because of their relatively low cost and the large change in resistance possible for a small change in temperature.

- A *resistance temperature detector* (RTD) is another sensor that makes use of the temperature dependence of electrical resistance. Most metallic materials increase in resistance with increasing temperature. Over limited ranges, this variation is linear for such metals as platinum, copper, tungsten, and some nickel/iron alloys. Platinum, for example, is linear within 0.3% from −20 to 150°C. The RTD sensing element is available in several different forms for surface or immersion mounting. Flat grid windings are used for measurements of surface temperatures. For direct measurement of fluid temperatures, the winding of resistance wire is encased in a stainless-steel bulb to protect it from corrosion. The RTD measurement circuit typically consists of three wires, to correct for line resistance. In many cases, the three-wire RTD is mated to an electronic circuit (transmitter) to produce a 4 to 20 mA current signal over a finite temperature range.

- A *thermocouple* is formed by the junction of two wires of dissimilar metals. An electromotive force dependent on the wire materials and the junction temperature exists between the wires. When the wires are joined at two points, a thermocouple circuit is formed. As long as one junction is kept at a constant temperature different from that of the other junction, an electric current flows through the circuit as a result of the difference in voltage potential developed between the two junctions. The constant temperature junction is called the cold junction. Various systems are used to provide cold junction compensation. Advances in solid-state circuitry have produced thermocouple transmitters with built-in cold junction and linearization circuits. When attached to the proper thermocouple, this circuit can provide a linearized signal (such as 4 to 20 mA dc) to a controller or an indicating digital meter.

Humidity Sensors

Humidity sensors, or *hygrometers,* can be used to measure the relative humidity or dew point of ambient or moving air. Hygroscopic (water-absorbent) materials respond directly to atmospheric moisture. Two types of hygrometers are available for use in central systems—mechanical and electronic.

A mechanical hygrometer makes use of the fact that a hygroscopic material, when exposed to water vapor, retains moisture and expands. The change in size or form is detected by a mechanical linkage and converted to a pneumatic or electronic signal. The most direct indicator of relative humidity is the change in length of an untampered human hair. Human hair exhibits a total change of about 2.5% of its original length as the relative humidity changes from 0 to 100%. A human hair hygrometer can measure relative humidity from 5 to 100% to within 5% at temperatures above 0°C. Other materials used in mechanical humidity sensors include organic materials (e.g., wood fibers, paper, cotton) and manufactured materials (e.g., nylon).

Electronic hygrometers can be of the resistance or capacitance type. The resistance type uses a conductive grid coated with a hygroscopic substance. The conductivity of the grid varies with the water retained; thus, the resistance varies according to the relative humidity. The conductive element is arranged in an alternating current-excited wheatstone bridge and responds quickly to humidity changes. The capacitance type is a stretched membrane of nonconductive film with metal electrodes on both sides that is mounted within a perforated plastic capsule. The change in the sensor's capacitance is nonlinear with respect to rising relative humidity. The signal is linearized and compensated for temperature in the amplifier circuit to provide an output signal as the relative humidity changes from 0 to 100%.

Pressure Transmitters, Controllers, and Transducers

A pneumatic *pressure transmitter* converts a change in absolute, gage, or differential pressure to a mechanical motion using a bellows, diaphragm, or Bourdon tube mechanism. When corrected through appropriate linkage, this mechanical motion produces a change in the air pressure to a controller. In some instances, the sensing and control functions are combined in a single component, a *pressure controller.*

An electronic *pressure transducer* moves a potentiometer or differential transformer by mechanical linkage of a diaphragm or Bourdon tube. Another type of transducer uses a strain gage bonded to a diaphragm. The strain gage detects the displacement resulting from the force applied to the diaphragm. Electronic circuitry for temperature compensation and amplification produces a standard output signal.

Flow Rate Sensors

The following basic sensing principles and devices are used to sense water or fluid flow: orifice plate, pitot tube, venturi, turbine meter, magnetic flowmeter, vortex shedding meter, and Doppler effect meter. These vary in range, accuracy, cost, and suitability for use with clean or dirty fluids, making them appropriate for different applications. In general, the pressure differential devices (orifice plates, venturi tubes, and pitot tubes) are less expensive and simpler to use but have limited range; thus, their accuracy depends on how they are applied and where in a system they are located.

More sophisticated flow devices, such as turbine, magnetic flow, and vortex shedding meters, usually have a better range and are more accurate over a wide range. When retrofitting, however, the expense of shutting down a system and cutting into a pipe must be considered. In this case, a noninvasive meter, such as a Doppler effect meter, can be more cost-effective than a flowmeter.

CONTROLLERS

Controllers take the sensor input, compare it with the desired control condition (set point), and regulate an output signal to cause a control action at the controlled device. The controller and sensor can be combined in a single instrument, such as a room thermostat, or they may be two separate devices. When separate pneumatic units are used, the pneumatic controller is usually referred to as a *receiver-controller.*

Electric/Electronic Controllers

For two-position control, the controller output may be a simple electrical contact that starts a burner or pump or actuates a spring-return valve or damper operator. Single-pole, double-throw (SPDT) switching circuits are used to control a three-wire unidirectional motor operator; they are also used for heating-cooling applications. Either single-pole, single-throw (SPST), or SPDT circuits can be modified for timed two-position action.

For floating control, the controller output is an SPDT switching circuit with a neutral zone where neither contact is made. This control is used with reversible motor operators.

Proportional control provides a continuously or incrementally changing output signal to position an electrical actuator or controlled device.

Nonindicating, Indicating, and Recording Controllers

The *nonindicating* controller is most common in HVAC work and includes all varieties in which the sensing element provides no visual indication of the value of the controlled variable. A separate thermometer, relative humidity indicator, or pressure gage is required for an indication. For example, a separate thermometer is often attached to the cover of room thermostats.

An *indicating* controller has a pointer added to the sensing element or attached to it by a linkage, so that the value of the controlled variable is indicated on a suitable scale.

A *recording* controller is similar to an indicating controller, except that the indicating pointer is replaced by a recording pen that provides a permanent record on special chart paper.

Pneumatic Receiver-Controllers

Pneumatic receiver-controllers are normally combined with sensing elements having a force or position output in order to obtain a variable air pressure output. The control mode is usually proportional, but other modes, such as proportional-plus-integral, can be used. These controllers are generally classified as nonrelay, relay direct, or relay indirect.

A *nonrelay* pneumatic controller uses a restrictor in its air supply and a bleed nozzle. The sensing element positions an air exhaust flapper, which varies the nozzle opening, causing a variable air pressure output to be applied to the controlled device, usually a pneumatic operator. The response time is relatively long, since all air must flow through the small orifice to position the actuator.

A *relay-type* pneumatic controller, either directly or indirectly through a restrictor, nozzle, and flapper, actuates a relay device that amplifies the air volume available for control. This arrangement provides quicker response to a measured variable change.

Controllers are further classified by construction as direct- or reverse-acting.

Direct-acting controllers increase the output signal as the controlled variable increases. For example, a direct-acting pneumatic thermostat increases output pressure when the sensing element detects a temperature rise.

Reverse-acting controllers increase the output signal as the controlled variable decreases. A reverse-acting pneumatic thermostat increases output pressure when the temperature drops.

Direct Digital Controllers

A direct digital controller uses a digital computer to implement control algorithms on one or multiple control loops. It differs fundamentally from pneumatic or electronic controllers in that the control algorithm is stored as a set of program instructions in memory (software or firmware). The controller itself calculates the proper control signals digitally rather than using an analog circuit or mechanical change.

A digital controller can be either single- or multiloop. Interface hardware allows the digital computer to process signals from various input devices, such as the electronic temperature, humidity, and pressure sensors described in the section on Sensors. Based on the digitized equivalents of the voltage or current signals produced by the inputs, the control software calculates the required state of the output devices, such as valve and damper actuators and fan starters. The output devices are then moved to the calculated position via interface hardware, which converts the digital signal from the computer to the analog voltage or current required to position the actuator or energize a relay.

The user enters parameters such as set points, proportional or integral gains, minimum on- and off-times, or high and low limits. The control algorithms stored in the computer's memory, in conjunction with actual input values, make the control decisions. The computer scans the input devices, executes the control algorithms, and then positions the output device(s) in a time-multiplex scheme. Digital controllers can be classified with regard to the way they store control algorithms in memory.

Preprogrammed control routines are typically stored in permanent memory, such as programmable read-only memory (PROM). The user can modify parameters such as set points, limits, and minimum off-times within the control routines, but the program logic cannot be changed without replacing the memory chips. This prevents unauthorized alteration.

User-programmable controllers allow the algorithms to be changed by the user. The programming language provided with the controller can vary from a derivation of a standard language to a custom language developed by the controller's manufacturer. Preprogrammed routines for proportional control, PI control, Boolean logic, timers, and so forth, are typically included in the language. Standard energy management routines may also be preprogrammed and may interact with other control loops where appropriate.

A terminal allows the user to communicate with and, where applicable, modify the program in the controller. These terminals can range from hand-held units with an LCD and several buttons to a full-size console with a cathode-ray tube (CRT) and typewriter-style keyboard. The terminal can be limited in function to allow only the display of sensor and parameter values or powerful enough to allow changing or reprogramming the control strategies. In some instances, the terminal can communicate remotely with one or more controllers, allowing centralized system displays, alarms, and commands.

Thermostats

Thermostats combine the control and sensing functions into a single device. Because thermostats are so prevalent, this section describes the various types and their operating characteristics.

- The *occupied-unoccupied* or *dual-temperature room thermostat* reduces temperature at night. It may be indexed (changed from occupied to unoccupied operation or vice versa) individually from a remote point or in a group by a manual or time switch. Some electric types have an individual clock and switch built in.

  The *pneumatic day-night thermostat* uses a two-pressure air supply system, the two pressures often being 90 and 120 kPa (gage) or 100 and 140 kPa (gage). Changing the pressure at a central point from one value to the other actuates switching devices in the thermostat and indexes it. Supply air mains are often divided into two or more circuits, so that switching can be done in separate areas of the building at different times. For example, a school building may have separate circuits for classrooms, offices and administrative areas, the auditorium, the gymnasium, and locker rooms.

- The *heating-cooling* or *summer-winter thermostat* can have its action reversed and its set point changed by indexing. It is used to actuate controlled devices, such as valves or dampers, that regulate a heating source at one time and a cooling source at another. The indexing is often done manually by switch or automatically by another thermostat that senses the temperature of the control agent (outdoor temperature or another suitable variable).

  The *pneumatic heating-cooling thermostat* uses a two-pressure air supply similar to that described for occupied-unoccupied thermostats.

- *Multistage thermostats* are arranged to operate two or more successive steps in sequence.

- A *submaster thermostat* has its set point raised or lowered over a predetermined range, in accordance with variations in output from a master controller. The master controller can be a thermostat, manual switch, pressure controller, or similar device. For example, a master thermostat measuring outdoor air temperature can be used to readjust the set point of a submaster thermostat that controls the water temperature in a heating system. Master-submaster combinations are sometimes said to have single-cascade action. When reset is accomplished by a single thermostat having more than one measuring element, it is known as *compensated control.*

- A *wet-bulb thermostat* is often used (in combination with a dry-bulb thermostat) for humidity control. Both a wick or other means of keeping the bulb wet with pure (distilled) water and rapid air motion, to ensure a true wet-bulb measurement, are essential. Because of maintenance problems, wet-bulb thermostats are seldom used.

- A *dew-point thermostat* is designed to control dew-point temperatures. Dew point is measured in several ways; the most accurate is to measure the temperature of an electrically heated chemical layer.

- A *dead-band thermostat* has a wide differential over which the thermostat remains neutral, requiring neither heating nor cooling. This differential may be adjustable up to 5 K. The thermostat controls to maximum or minimum output over a small differential at each end of the dead band, as shown in Figure 16.

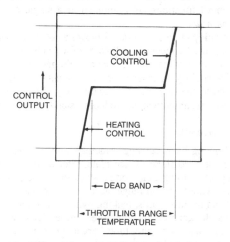

Fig. 16   Dead-Band Thermostat

AUXILIARY CONTROL DEVICES

In addition to conventional controllers and controlled devices, many electric control systems require one or more of the following auxiliary devices:

- *Transformers* to provide current at the required voltage.
- *Electric relays* to control electric heaters or to start and stop oil burners, refrigeration compressors, fans, pumps, or other apparatus for which the electrical load is too large to be handled directly by the controller. Other uses include time-delay and circuit-interlocking safety applications.
- *Potentiometers* for manual positioning of proportional control devices, for remote set point adjustment of electronic controllers, and for feedback.
- *Manual switches* for performing several operations. These can be two-position or multiple-position, with single or multiple poles.
- *Auxiliary switches* on valve and damper operators for providing a selected sequence of operation.

Auxiliary control equipment for pneumatic systems includes the following:

- *Air compressors* and accessories, including dryers and filters, to provide clean, dry air at the required pressure.
- *Electropneumatic relays*, which are electrically actuated air valves that operate pneumatic equipment in accordance with variations in electrical input.
- *Pneumatic-electric switches*, actuated by air pressure to make or break an electrical circuit.
- *Pneumatic relays*, actuated by the pressure from a controller to perform several functions. They may be divided into two groups:
  1. *Two-position relays*, which permit a controller actuating a proportional device to actuate one or more two-position devices as well.
  2. *Proportional relays*, which are used to reverse the action of a proportional controller. They can select the highest or lowest of two or more pressures, average two or more pressures, respond to the difference between two pressures, add or subtract pressures, and amplify or retard pressure changes.
- *Positive positioning relays*, which are devices for ensuring the accurate positioning of a valve or damper operator in response to changes in pressure from a controller. They are affected by the position of the operator and the pressure from the controller. Whenever the two are out of balance, the relays use full control air pressure to change the pressure applied to the operator until balance is restored.
- *Switching relays*, which are pneumatically operated air valves for diverting air from one circuit to another or opening and closing air circuits.
- *Pneumatic switches*, which are manually operated devices for diverting air from one circuit to another or opening and closing air circuits. They can be two-position or multiple-position.

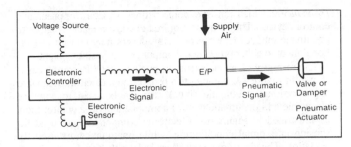

Fig. 18   Example of Electronic and Pneumatic Control Components Combined with Electronic-to-Pneumatic (E/P) Transducer

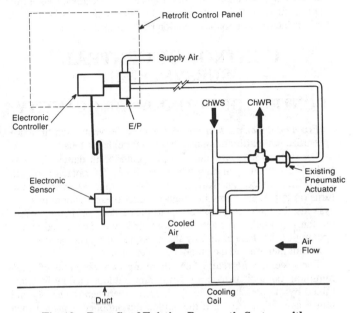

Fig. 19   Retrofit of Existing Pneumatic System with Electronic Sensors and Controllers

- *Gradual switches*, which are proportional devices that manually vary the air pressure in a circuit.
- *Logic networks* and *square root extractors*.

Auxiliary control devices common to both electric and pneumatic systems include the following:

- *Step controllers* for operating a number of electric switches in sequence by means of a proportional electric or pneumatic operator. They are commonly used for controlling several steps of refrigeration and may be arranged to prevent the simultaneous starting of compressors and to alternate the sequence to equalize wear. They may also be used for sequence operation of electric heating elements and other equipment.
- *Power controllers* for controlling electrical power input to resistance-type electric heating elements. The final controlled device may be a variable autotransformer, a saturable-core reactor, or a solid-state power controller. They are available with various ratings for single- or three-phase heater loads and are usually arranged to regulate power input to the heater in response to the demands of proportional electronic or pneumatic controllers. Solid-state controllers may also be used in two-position control modes.
- *Clocks* or *timers* for turning apparatus on and off at predetermined times, for switching control systems from day to night operation, and for other time sequence functions.

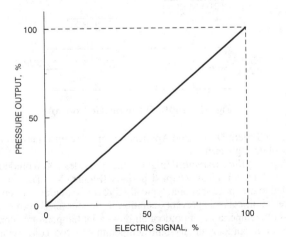

Fig. 17   Response of Electronic-to-Pneumatic (E/P) Transducer

• *Transducers*, which consist of combinations of electric or pneumatic control devices. They may be required to convert electric signals to pneumatic output or vice versa. Transducers may convert proportional input to either proportional output or two-position output.

The response of an electronic-to-pneumatic (E/P) transducer is shown in Figure 17. It converts a proportional electronic signal to a proportional pneumatic signal and can be used to combine electronic and pneumatic control components into a control loop, as illustrated in Figure 18. Electronic components are used for sensing and signal conditioning, while pneumatics are used for actuation. The electronic controller can be either analog or digital.

The E/P transducer presents a special option for retrofit applications. An existing HVAC system with pneumatic controls can be retrofitted with electronic sensors and controllers while retaining its existing pneumatic actuators (Figure 19). One advantage of this approach is that the retrofit can be accomplished with only a minor interruption to the operation of the control system.

CONTROL OF CENTRAL SUBSYSTEMS

CONTROL OF OUTDOOR AIR QUANTITY

Fixed minimum outdoor air control provides ventilation air, space pressurization (exfiltration), and makeup for exhaust fans.

For systems *without* return fans, the outdoor air damper is interlocked to remain open only when the supply fan operates (Figure 20A). The outdoor air damper should open quickly when the fan turns on to prevent excessive negative duct pressurization. In some systems, the fan's on-off switch opens the outdoor air damper before the fan is started. The rate of outdoor airflow is determined by the opening of the damper and by the pressure difference between the mixed air plenum and the outdoor air plenums.

For systems *with* return fans, there are two variations of fixed minimum outdoor air control. Minimum outdoor airflow is determined by the pressure (airflow) difference between the supply and return fans (Figure 20B). If the outdoor air supplied is greater than the difference between the supply and return fan airflows, a variation of economizer cycle control is used (Figure 20C).

In systems using *100% outdoor air*, all air goes to the fan and no air is returned (Figure 21). The outdoor air damper is interlocked and usually opens before the fan starts.

Economizer cycle control reduces cooling costs when outside conditions are suitable, that is, when the outdoor air is cool enough to be used as a cooling medium. If the outdoor air is below a high-temperature limit, typically 18°C, the return, exhaust, and outdoor air dampers modulate to maintain a ventilation cooling set point, typically 13 to 16°C (Figure 22). The relief dampers are interlocked to close, and the return air dampers to open, when the supply fan is not operating. When the outdoor air temperature exceeds the high-temperature limit set point, the outdoor air damper is closed to a fixed minimum and the exhaust and return air dampers close and open, respectively.

In *enthalpy economizer control*, the high-temperature limit interlock system of the economizer cycle is replaced in order to further reduce energy costs when latent loads are significant. The interlock function (Figure 22) can be based instead on (1) a fixed enthalpy upper limit, (2) a comparison with return air so as not to exceed return air enthalpy, or (3) a combination of enthalpy and high-temperature limits.

VAV warm-up control during unoccupied periods requires no outdoor air; typically, outdoor and exhaust dampers remain closed. However, in systems with a return fan (Figure 23), the outdoor air damper should be positioned at its minimum position, and supply airflow (volume) should be limited to return air airflow (volume) to minimize positive or negative duct pressurization. See the section on

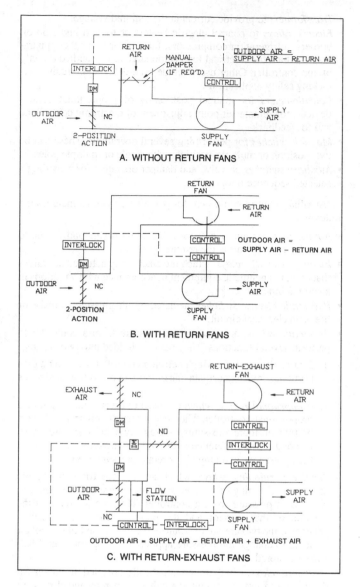

Fig. 20 Fixed Minimum Outdoor Air for Various Systems

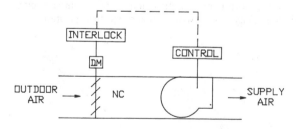

Fig. 21 100% Outdoor Air Control

Control System Design and Application for more information on fan control during warm-up.

Night cool-down control (night purge) provides 100% outdoor air for cooling during unoccupied periods (Figure 24). The space is cooled to the space set point, typically 5 K above outdoor air temperature. Limit controls prevent operation if outdoor air is above space dry-bulb temperature, if outdoor air dew-point temperature is excessive, or if outdoor air dry-bulb temperature is too cold, typically 10°C or below. The night cool-down cycle is initiated before sunrise, when overnight outside temperatures are usually the coolest. When

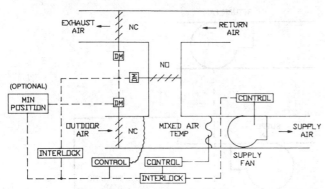

Fig. 22   Economizer Cycle Control

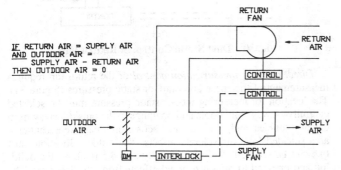

IF RETURN AIR = SUPPLY AIR
AND OUTDOOR AIR =
     SUPPLY AIR - RETURN AIR
THEN OUTDOOR AIR = 0

Fig. 23   Warm-Up Control

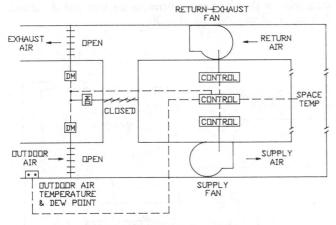

Fig. 24   Night Cool-Down Control

outside air conditions are acceptable and the space requires cooling, the cool-down cycle is the first phase of the optimum start sequence.

FAN CONTROL

Fans may be modulated by

- Varying speed
- Fan inlet or discharge dampers
- Fan scroll dampers
- Fan runaround or scroll bypass
- Controllable pitch
- Fan inlet vanes

Constant Volume Control

Constant volume control fixes the airflow rate when duct resistances vary (Figure 25).

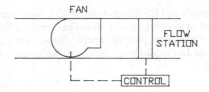

Fig. 25   Constant Volume Control

Duct Static Control

Duct static control for VAV and other terminal-type systems maintains a static pressure at a measurement location. The sensor is typically placed at 75 to 100% of the distance between the first and last air terminals (Figure 26). Care must be taken in selecting the reference static sensor location. Controller upset due to opening and closing doors, elevator shafts, and other sources of air turbulence should also be prevented. The pressure selected provides minimum static pressure to all air terminal units during all supply fan design conditions.

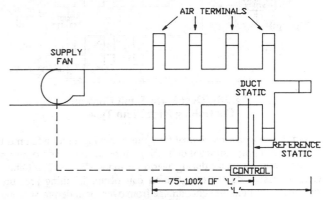

Fig. 26   Duct Static Control

Multiple static sensors (Figure 27) are required when more than one duct runs from the supply fan. The sensor with the lowest static requirement controls the fan. Since duct run-outs may vary, a control that uses individual set points for each measurement is recommended.

Duct static limit control prevents excessive duct pressures—usually at the discharge of the supply fan. Two variations are used: (1) the *fan shutdown* type, which is a safety high limit control that

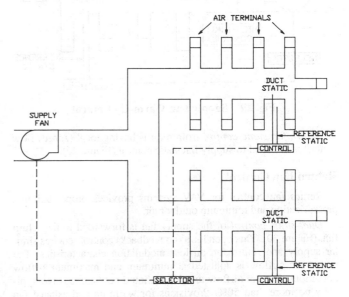

Fig. 27   Multiple Static Sensors

turns the fans off, and (2) the *controlling high limit* type (Figure 28), which is used in systems having zone fire dampers. When the zone fire damper closes, duct pressure drops, causing the duct static control to increase fan modulation; however, the controlling high limit will override.

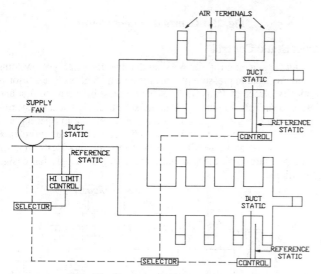

**Fig. 28   Duct Static Limit Control:
Controlling High Limit Type**

Supply fan warm-up control for systems having a return fan must prevent the supply fan from delivering more airflow than the return fan maximum capacity during warm-up mode (Figure 29). Otherwise, excessive duct pressurization can occur. Limiting pressure controllers can help prevent damage from over- or underpressurizing the ductwork.

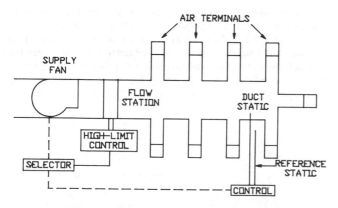

Fig. 29   Supply Fan Warm-Up Control

Return fan static control from returns having local (zoned) flow control is identical to supply fan static control (Figure 26).

Return Fan Control

Return fan control for VAV systems provides proper building pressurization and minimum outdoor air.

Duct static control of the supply fan is forwarded to the return fan. (Figure 30). This open loop (no feedback) control requires similar supply and return fan airflow modulation characteristics. The return fan airflow is adjusted at minimum and maximum airflow conditions. System airflow turndown should not be excessive, typically no more than 50%. Provisions for warm-up and exhaust fan switching are impractical.

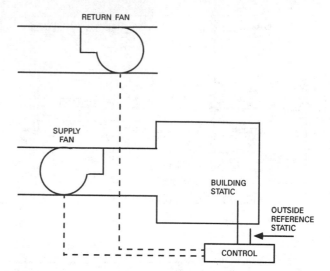

Fig. 30   Duct Static Control of Return Fan

Direct building pressurization control of the return fan requires measurement of the space and outdoor static pressures (Figure 31). The location for measuring indoor static pressure must be selected carefully—away from doors and openings to the outside, away from elevator lobbies, and, when using a sensor, in a large representative area shielded from air velocity effects. The outdoor location must likewise be selected carefully, typically 3 to 4.5 m above the building and oriented to minimize wind effects from all directions. The amount of minimum outdoor air varies with building permeability and exhaust fan switching. During the warm-up mode, the building static pressure is reset to zero differential pressure, and all exhaust fans are turned off.

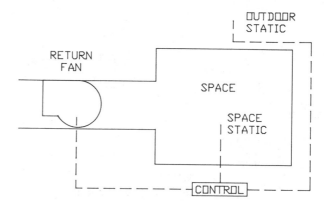

Fig. 31   Direct Building Pressurization Control

Airflow tracking uses duct airflow measurements to control the return air fans (Figure 32). Sensors, called flow stations, are typically multiple-point, pitot tube, and averaging. Provisions must be made for exhaust fan switching to maintain pressurization of the building. Warm-up is accomplished by setting the return airflow equal to the supply fan airflow—usually with exhaust fans turned off—and limiting supply fan volume to return fan capability.

During night cool-down, the return fan operates in the normal mode.

Fan Sequencing

Sequencing of fans for VAV systems gives greater airflow reduction, resulting in greater operating economy and stable fan operation if airflow reductions are significant. Alternation of fans usually provides greater system reliability.

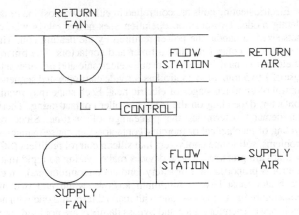

Fig. 32 Airflow Tracking Control

Centrifugal fans are controlled to keep system disturbances to a minimum when additional fans are started. The added fan is started and slowly brought to capacity while the capacity of the operating fans is *simultaneously* reduced. The combined output of all fans then equals the output before fan addition.

Vaneaxial fans usually cannot be sequenced in the same manner as centrifugal fans. To avoid stall, the operating fans must be reduced to some minimum level of airflow. Then, additional fans may be started and all fans modulated to achieve equilibrium.

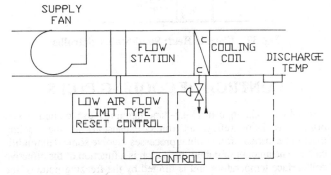

Fig. 33 Coil Reset Control to Prevent Supply Fan Instability

Unstable Fan Operation

Fan instability in VAV systems can usually be avoided by proper fan sizing. However, if airflow reduction is large (typically over 60%), a technique such as fan sequencing is usually required for maintaining airflow within the fan's stable range.

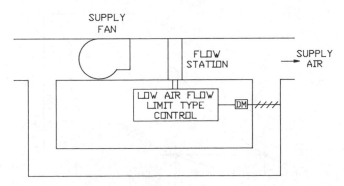

Fig. 34 Fan Bypass Control to Prevent Supply Fan Instability

Coil reset avoids fan instability by resetting the cooling coil discharge temperature higher (Figure 33) so that the building cooling loads require greater airflow. Because of the time lag between temperature reset and demand for more airflow, the value at which reset starts should be selected on the safe side of the fan instability point. When this technique is used, it must be ascertained that dehumidification requirements can be met.

Fan bypass allows airflow to short circuit around the fan so that a minimum airflow can be maintained (Figure 34). This technique uses constant volume control to limit the low airflow.

CONTROL OF HEATING COILS

Heating coils in central air-handling units preheat, reheat, or heat, depending on the climate and the amount of minimum outdoor air needed.

Preheating Coils

Control of steam or hot water preheating coils must include protection against freezing, unless the minimum outdoor air quantity is small enough to keep the mixed air temperature above freezing, and there is enough mixing to prevent stratification. Even though the average mixed air temperature is above freezing, inadequate mixing may allow a freezing air stratum to impinge on the coil.

Steam preheat coils should have two-position valves and vacuum breakers to prevent a buildup of condensate in the coil. The valve should be fully open when outdoor air (or mixed air) temperature is below freezing. This causes unacceptably high coil discharge temperatures at times, necessitating face and bypass dampers for final temperature control (Figure 35). The bypass damper should be sized to provide the same pressure drop at full bypass airflow as the combination of face damper and coil does at full airflow.

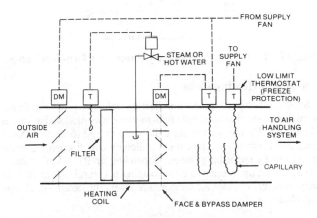

Fig. 35 Preheat with Face and Bypass Dampers

Hot water preheat coils must maintain a minimum water velocity in the tubes of 0.9 m/s to prevent freezing. A two-position valve combined with face and bypass dampers can usually be used to control the water velocity. More commonly, a secondary pump control in one of two configurations (Figures 36 and 37) is used. The control valve modulates to maintain the desired coil air discharge temperature, while the pump operates to maintain the minimum tube water velocity when outdoor air is below freezing. The system in Figure 37 uses less pump power, allows variable flow in the hot water supply main, and is preferred for energy conservation. The system in Figure 36 may be required on small systems with only one or two air handlers, or where constant main water flow is needed.

Some systems may use a glycol solution in combination with any of these methods.

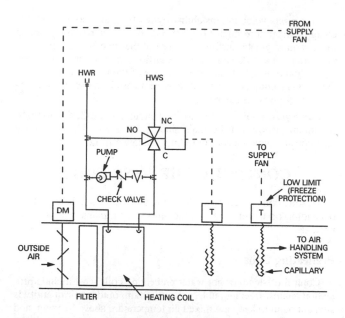

Fig. 36   Preheat with Secondary Pump and Three-Way Valve

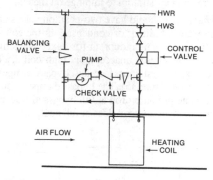

Fig. 37   Preheat with Secondary Pump and Two-Way Valve

Reheat and Heating Coils

Steam or hot water reheat and heating coils not subject to freezing can be controlled by simple two- or three-way modulating valves (Figure 38). Steam distributing coils are required to ensure proper steam coil control. The valve is controlled by coil discharge air temperature or by space temperature, depending on the HVAC system. Valves are set up to open to allow heating if control power fails. In many systems, outdoor air temperature resets the heating discharge controller.

Electric heating coils are controlled in either two-position or modulating mode. Two-position operation uses power relays with contacts sized to handle the power required by the heating coil. Timed two-position control requires a timer and contactors. The timer can be electromechanical, but it is usually electronic and provides a time base of 1 to 5 min. Step controllers provide cam-operated sequencing control of up to ten stages of electric heat. Each stage may require a contactor, depending on the step controller contact rating. Thermostat demand determines the percentage of on-time. Since rapid cycling of mechanical or mercury contactors can cause maintenance problems, solid-state controllers like silicon control rectifiers (SCRs) or triacs are preferred. These devices make cycling so rapid that the control is proportional. For safety (and code requirements), an electric heater must have a minimum airflow switch and two high-temperature limit sensors—one with manual reset and one with automatic reset. Therefore, face and bypass dampers are not used. A control system with a solid-state controller and safety controls is shown in Figure 39.

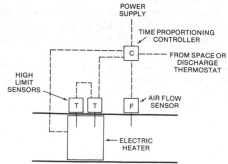

Fig. 39   Electric Heat: Solid-State Controller

CONTROL OF COOLING COILS

Cooling coils in central air-handling units use chilled water, brine, glycol, or refrigerant (direct-expansion) as the cooling medium. Most comfort cooling processes involve some dehumidification. The amount of dehumidification is a function of the effective coil surface temperature and is limited by the freezing point of the coolant. If water condensing out of the airstream freezes on the coil surface, airflow is restricted and, in severe cases, may be shut off. The practical limit is about 4°C dew point on the coil surface. As indicated in Figure 40, this results in a relative humidity of about 30% at a space temperature of 24°C, which is adequate for most commercial applications. When lower humidities are necessary, chemical dehumidifiers are required (see the section on Humidity Control).

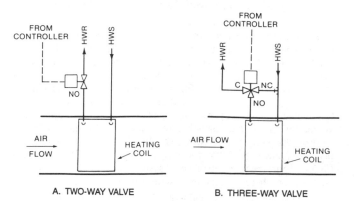

Fig. 38   Heating Control

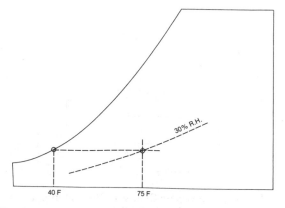

Fig. 40   Cooling and Dehumidifying—Practical Low Limit

Chilled water or brine cooling coils are controlled by two- or three-way valves (Figure 41). These valves are similar to those used for heating control, but are usually closed to prevent cooling when the fan is off. The valve typically modulates in response to coil air discharge temperature or space temperature.

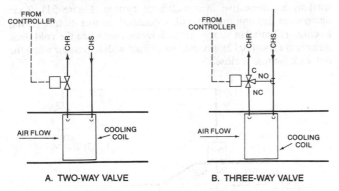

Fig. 41   Chilled Water Control

When maximum relative humidity control is required, a space or return air humidistat is provided in addition to the space thermostat. To limit maximum humidity, a control function selects the higher of the output signals from the two devices and controls the cooling coil valve accordingly. A reheat coil is needed to maintain space temperature (Figure 42). If humidification is also provided, this cycle is sometimes referred to as a constant temperature, constant humidity cycle.

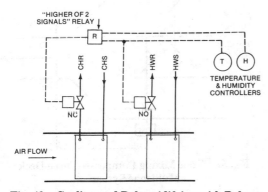

Fig. 42   Cooling and Dehumidifying with Reheat

Direct-expansion (DX) cooling coils are usually controlled by solenoid valves in the refrigerant liquid line (Figure 43). Face and bypass dampers are not recommended because they permit the formation of ice on the coil when airflow is reduced. Control can be improved by using two or more stages; the solenoid valves are controlled in sequence, and there is a differential of 0.5 or 1 K between stages (Figure 44). The first stage should be the first coil row on the

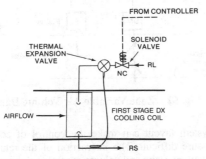

Fig. 43   Direct Expansion—Two-Position Control

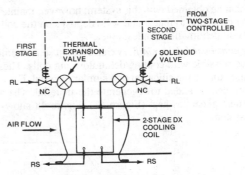

Fig. 44   Two-Stage Direct-Expansion Cooling

entering air side; the following rows form the second and succeeding stages. Side-by-side stages tend to generate icing on the stage in use, which causes reduction of airflow and loss of control. Modulating control is achieved using a variable suction pressure controller (Figure 45). This type of control is uncommon but necessary if accurate control of discharge or space temperature is required. A refrigerant compressor capacity reduction control should be supplied with modulation control.

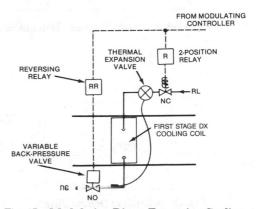

Fig. 45   Modulating Direct-Expansion Cooling

HUMIDITY CONTROL

Although simple cooling by refrigeration maintains an upper limit to space humidity, without additional equipment it does not control humidity.

Dehumidification

The *sprayed coil dehumidifier* (Figure 46) was formerly used for dehumidification. Space relative humidities ranging from 35 to 55%

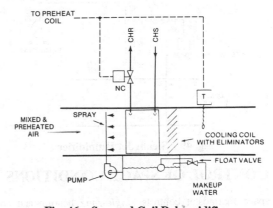

Fig. 46   Sprayed Coil Dehumidifier

at 24°C can be obtained with this system; however, maintenance and operating costs due to reheat and solid deposits on the coil make the sprayed coil dehumidifier undesirable.

Desiccant-based dehumidifiers can develop space humidities below those possible with cooling/dehumidifying coils. These devices adsorb moisture using silica gel or a similar material. For continuous operation, heat is added to regenerate the material. The adsorption process also generates heat (Figure 47). Figure 48 shows a typical control system.

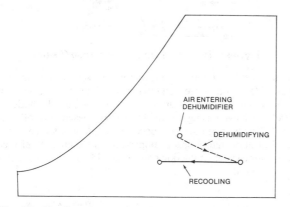

Fig. 47   Psychrometric Chart: Chemical Dehumidification

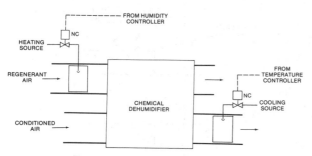

Fig. 48   Chemical Dehumidifier

Humidification

Evaporative pans (usually heated), *steam jets*, and *atomizing spray tubes* are all used for space humidification. A space or return air humidistat is used for control. A high limit duct humidistat should also be used to minimize moisture carryover or condensation in the duct (Figure 49). With proper use and control, humidifiers can achieve high space humidity, although they more often maintain design minimum humidity during the heating season.

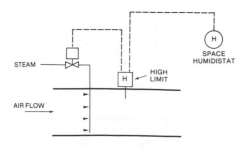

Fig. 49   Steam Jet Humidifier

CONTROL OF SPACE CONDITIONS

A space thermostat controls *single-zone* heating and cooling directly. If used, a humidifier is controlled by the space humidistat.

Multizone and *dual-zone* units include mixing dampers that are controlled by each zone's space thermostat (Figure 50). If used, humidifiers are usually controlled by a return air humidistat. Some jurisdictions no longer permit the mixing of hot and cold air to provide simultaneous heating and cooling. A three-deck unit may be used as an alternative in a multizone system (Figure 51). Zone dampers in this unit operate with sequenced damper motors (DMs) to either (1) mix hot supply air with bypass air when the cold deck damper is closed or (2) mix cold supply air with bypass air when the hot deck damper is closed.

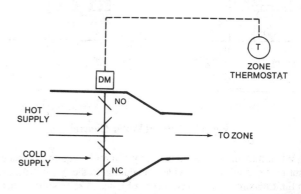

Fig. 50   Zone Mixing Dampers—Multizone System

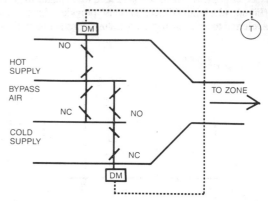

Fig. 51   Zone Mixing Dampers—Three-Deck Multizone System

Variable air volume (VAV) units have motorized dampers in each zone supply duct. A related zone space thermostat controls each damper by a flow sensor/controller that is reset by the thermostat (Figure 52). If used, humidifiers are controlled by a return air humidistat or a humidistat in a representative critical zone.

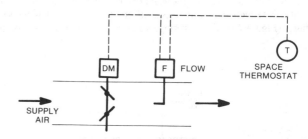

Fig. 52   Zone Variable Air Volume Damper

The system layout can make the control of economizers and static pressure difficult. The interaction of the return exhaust fan with the static pressure and volume control is a particular concern in the control system layout.

Pressure Control

The most common application for static pressure controls is fan capacity control in VAV systems. Static pressure controls can also be used to pressurize a building or space relative to adjacent spaces or the outdoors. Typical applications include clean rooms (positive pressure to prevent infiltration), laboratories (positive or negative, depending on use), and various manufacturing processes, such as spray-painting rooms. The pressure controller usually modulates dampers in the supply ducts to maintain the desired pressure as exhaust volumes change.

CONTROL SYSTEM DESIGN AND APPLICATION

CENTRAL AIR-HANDLING SYSTEMS

Variable Air Volume

Variable air volume systems vary the amount of air supplied by terminal units to individual zones as the loads in those zones vary. Hybrid systems that use bypass terminal units to vary air volume to a space while handling a constant air volume from the central fan are treated as constant volume (CV) systems.

From a control standpoint, VAV systems can be classified further as follows:

1. Single-duct cooling-only (Figure 53)
2. Single-duct heating/cooling
3. Dual-duct (see the section on Dual-Duct systems)
   a) Dual-duct single supply fan
   b) Dual-duct dual supply fan

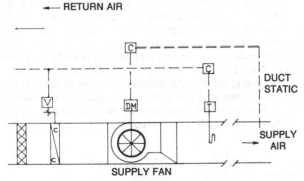

Fig. 53 VAV Single-Duct Cooling-Only

Single-Duct Cooling-Only Systems. *Fan Control.* In a VAV system having a supply fan with no means of modulation, supply fan volume controls should be used. Otherwise, as terminal units reduce total airflow, the duct static pressure increases as the fan moves up its operating curve. If uncontrolled, this pressure can damage ductwork. Even in strong ducts, terminal unit dampers must work against a higher pressure, which results in poor control, increased noise, and increased fan energy usage.

Fan volume control is based on supply duct static pressure. To conserve fan energy, the static pressure controller should be set at the lowest point permitting proper air distribution at design conditions. The controller requires PI control because it eliminates offset while maintaining stability. In proportional-only control, the low proportional gain required to stabilize fan control loops allows static pressure to offset upward as the load decreases, which causes the fan in the terminal unit to consume more energy. The pressure sensor must be properly placed to maintain optimum pressure throughout the supply duct.

Experience indicates that most systems perform satisfactorily with the sensor located at 75 to 100% of the distance from the first to the most remote terminal. If the sensor is located at less than 100% of the distance, the control set point should be adjusted to account for the pressure loss between the sensor and the remote terminal.

In addition to the remote static pressure controllers, a high limit static pressure controller should be placed at the fan discharge to turn the fan off or limit discharge static pressure in the event of excessive duct pressure (e.g., when a fire or smoke damper closes between the fan and the remote sensor). Supply fan static pressure control devices such as inlet guide vanes and variable-speed drives should be interlocked to move to the minimum flow or closed position when the fan is not running; this precaution prevents fan overload or damage to ductwork upon start-up.

Temperature and Ventilation Control. VAV systems are typically designed to supply constant temperature air at all times. To conserve central plant energy, the temperature of the supply air can be raised in response to demand from the zone with the greatest load (load analyzer control). However, because more cool air must then be supplied to match a given load, the mechanical cooling energy saved may be offset by an increase in fan energy. Equipment operating efficiency should be studied closely before the implementation of temperature reset in cooling-only VAV systems.

Outdoor air (OA), return air (RA), and exhaust air (EA) ventilation dampers are controlled by the discharge air temperature controller to provide free cooling as the first stage in the cooling sequence. When outdoor air temperature rises to the point that it can no longer be used for cooling, an outdoor air limit (economizer) control overrides the discharge controller and moves ventilation dampers to the minimum ventilation position. An enthalpy control system can replace outdoor air limit control in some climatic areas.

Single-Duct Heating/Cooling Systems. Single-duct VAV systems, which supply warm air to all zones when heating is required and cool air to all zones when cooling is required, have limited application and are used where heating is required only for morning warm-up. They are not recommended if some zones require heating at the same time that others require cooling. These systems, like single-duct cooling-only systems, are generally controlled during occupancy.

During warm-up periods, as determined by a time clock or manual switch, a constant heating supply air temperature is maintained. Because the terminal unit may be fully open, uncontrolled overheating can occur. It is preferable to allow unit thermostats to maintain complete control of their terminal units by reversing their action to the unit. During warm-up and unoccupied cycles, outdoor air dampers should be closed.

Constant Volume

Constant volume systems supply a constant amount of variable temperature air to individual zones. In all single-duct CV systems, fans and ductwork must be sized to accommodate design conditions in all zones simultaneously.

Humidity control is commonly provided by a representative zone humidistat or a return air humidity sensor-controller and is interlocked with the humidifying system to operate only during the heating season. A high-limit humidity sensor-controller in the supply air duct prevents excessive duct moisture.

Ventilation damper control (OA, RA, and EA) is very similar to that for VAV single-duct cooling-only systems.

Constant volume systems can be classified as follows:

1. Single-duct single-zone (see the section on Single-Zone systems)
2. Single-duct with zone reheat
3. Single-duct bypass
4. Dual-duct multiple-zone (see the section on Dual-Duct systems)

Single-Duct with Zone Reheat Systems. These systems (Figure 54) use a single central CV fan system to serve multiple zones.

All delivered air is cooled to satisfy the greatest zone cooling load. Air delivered to other zones is then reheated with heating coils in individual zone ducts. Because these systems consume more energy than VAV systems, they are generally limited to those applications with larger ventilation needs, such as hospitals and special process or laboratory applications. Some jurisdictions no longer permit simultaneous heating and cooling.

To provide unoccupied heating or preoccupancy warm-up, the central fan system can include a heating coil. During warm-up or unoccupied periods, a constant supply duct heating temperature is maintained, with all bypass terminal units closed to the bypass and the cooling coil valve closed. To prevent overheating, an unoccupied mode zone thermostat can cycle the fan, or the terminal unit thermostat action can be reversed.

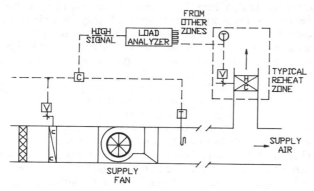

Fig. 54   Single-Duct with Zone Reheat

No *fan control* is required because the design, selection, and adjustment of fan system components determine system air volume and duct static pressure.

These systems generally supply constant temperature air at all times. *Load analyzer control* permits the supply temperature to rise in response to a demand from the greatest cooling load that is less than the design load, thus conserving energy.

Single-Duct with Zone Bypass Systems. These systems (Figure 55) are a compromise between single-duct VAV systems and CV reheat systems. The primary systems and all ducts supply a constant volume of air. During space partial load conditions, however, the terminal unit diverts some air directly back to the return system instead of reheating it, thus bypassing the space. These terminals are often added to single-zone CV systems to provide zoning without the energy penalty for reheat.

Control is similar to that for single-duct zone reheat systems. Supply temperature should be reset from load analyzer control that monitors individual space temperatures or from a return air temperature sensor to prevent a high percentage of the air from bypassing the space and thus reducing air circulation and comfort.

Dual-Duct

Dual-duct systems may be either VAV or CV, depending on whether terminal mixing box units operate their dampers individually, allowing zone air volume to vary with variations in load conditions, or together.

VAV Dual-Duct Single Supply Fan Systems. These systems (Figure 56) use a single fan to supply separate heating and cooling ducts. Terminal mixing box units in which the heating and cooling dampers operate in sequence are used to satisfy space load requirements. Frequently, the space thermostat provides for zero energy band operation for greater energy savings.

Fan Control. Static control is similar to that in VAV single-duct systems, except that static pressure sensors are needed in each supply duct. A comparator control allows the sensor sensing the lowest pressure to govern the fan volume control system, thus ensuring that there is adequate static pressure under all load conditions to supply the necessary air for all zones, whether they are supplied from the hot or the cold duct.

If the system includes a return air fan, its volume control considerations are similar to those already described in the section on Return Fan Control. Flow stations are usually located in each supply duct, and a signal corresponding to the sum of the two airflows is transmitted to the RA fan volume controller to establish the set point of the return fan controller.

Temperature Control. The hot deck has its own heating coil, and the cold deck has its own cooling coil. Each coil is controlled by its own discharge air temperature controller. The controller set point may be reset from the greatest representative demand zone, based on zone temperature—the hot deck may be reset from the zone with the greatest heating demand, and the cold deck from the zone with the greatest cooling demand. An alternative to greatest-demand reset is reset from a parameter that predicts load, such as OA temperature, duct air volume, or RA temperature. Positive control is required to prevent simultaneous heating and cooling of the air (a code mandate in many areas).

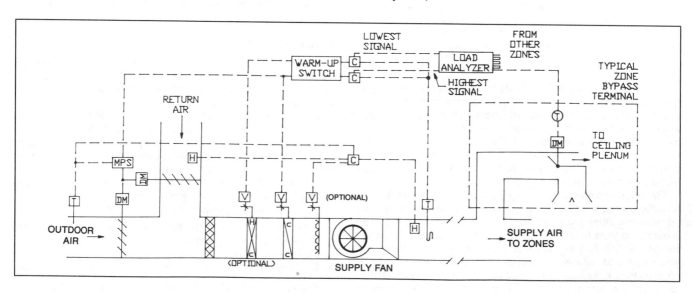

Fig. 55   Single-Duct with Zone Bypass

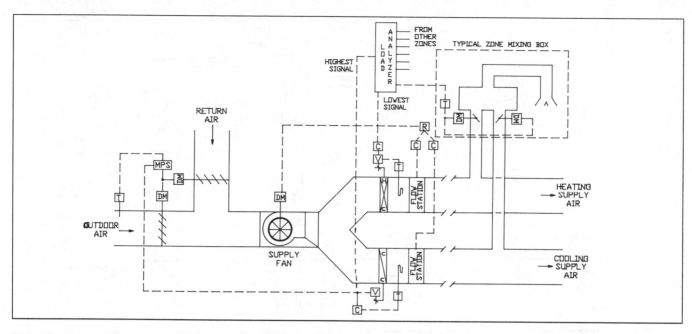

Fig. 56 VAV Dual-Duct Single Supply Fan System

Control based on the zone requiring the most heating or cooling increases operation economy because it reduces the energy delivered at less-than-maximum load conditions. Figure 56 shows this reset control of duct temperatures. However, the expected economy is lost if (1) air quantity to a zone is too low, (2) thermostats in some spaces are reset to an extreme value by occupants, (3) a thermostat is placed so that it senses spot loads (due to coffee pots, the sun, copiers, etc.), or (4) a thermostat malfunctions. In these cases, a weighted average of zone signals can recover the benefit at the expense of some comfort in specific zones.

In some applications, the highest demand heating thermostat should not dictate supply air temperature because the highest demand may come from a temporarily unoccupied zone.

Ventilation Control. Ventilation dampers (OA, RA, and EA) are controlled for cooling, with outdoor air as the first stage of cooling in sequence with the cooling coil from the cold deck discharge temperature controller. Control is similar to that in single-duct systems. A more accurate OA flow-measuring system can replace the minimum positioning switch.

Humidity Control. Humidity is commonly controlled based on RA humidity. It is coordinated with the OA temperature sensor so the humidifying system operates only during the heating season. A high-limit humidistat in the supply air duct prevents excessive duct moisture.

VAV Dual-Duct Dual Supply Fan Systems. These systems (Figure 57) use separate supply fans for the heating and cooling ducts. Static pressure control is similar to that for VAV dual-duct single supply fan systems, except that each supply fan has its own static pressure sensor and control system. If the system has a return air fan, volume control considerations are similar to those described in the section on Return Fan Control. Temperature, ventilation, and

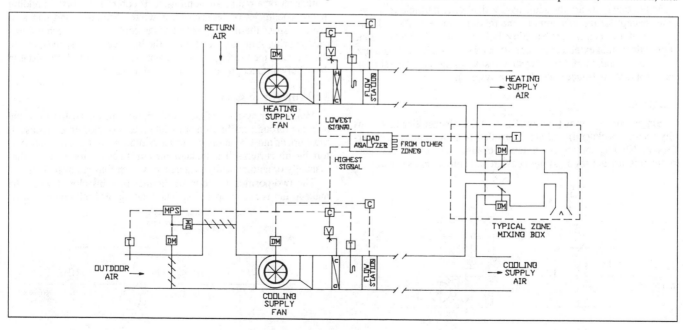

Fig. 57 VAV Dual-Duct Dual Supply Fan System

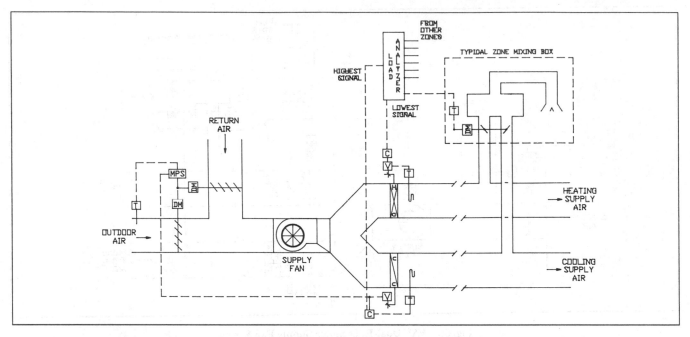

Fig. 58   Constant Volume Dual-Duct or Multizone Single-Fan System

humidity control considerations are similar to those for VAV dual-duct single supply fan systems.

Constant Volume Dual-Duct Multiple-Zone Systems. These systems (Figure 58) use a single CV supply fan and multiple zone mixing dampers in terminal units that supply a constant volume of air to individually controlled zones. These systems require automatic control of neither duct static pressure nor fan volume, as these operating characteristics are set by system design and component selection and adjustment. Temperature, ventilation, and humidity control considerations are similar to those for VAV dual-duct single supply fan systems.

Multizone Units

Multizone units are the same as CV dual-duct multiple-zone systems, except for the location of zone mixing dampers. These dampers are at the central air-handling unit and feed individual zone supply ducts that extend to the conditioned spaces. Fan, temperature, ventilation, and humidity control considerations are the same as those for CV dual-duct multiple-zone systems.

Single Zone

Single-zone systems (Figure 59) use a CV air-handling unit (usually factory-packaged) and are controlled from a single space thermostat. No fan control is required because fan volume and duct static pressure are set by system design and component selection.

Temperature and Ventilation Control. A single space controller or thermostat controls the heating coil, ventilation dampers, and cooling coil in sequence as thermal load varies in the conditioned space. Ventilation dampers (OA, RA, and EA) are controlled for outdoor air cooling as first-stage cooling. When outdoor air temperature rises to the point that it can no longer be used for cooling, an outdoor air limit control overrides the signal to the ventilation dampers and moves them to the minimum ventilation position, as determined by the minimum positioning switch. Where applicable, an enthalpy control system can replace the OA limit control for appropriate climatic areas. A zero energy band thermostat can separate the heating and cooling control ranges, thus saving energy.

Humidity Control. Humidity is controlled by either a space humidistat or a return air humidistat. It controls a moisture-adding device (steam jet or pan-type) for winter operation and can also override space thermostat control of the cooling coil for dehumidification during summer operation. In the latter case, the space thermostat activates the heating coil for reheat. The high-limit humidistat in the supply air duct prevents excessive duct moisture.

Makeup Air Systems

Makeup air systems (Figure 60) replace air exhausted from the building through exfiltration or by laboratory or industrial processes. Makeup air must be at or near space conditions to minimize uncomfortable air currents. The makeup air fan is usually turned on, either manually or automatically, whenever exhaust fans are turned on.

The two-position outdoor air damper remains closed when the makeup fan is not in operation. The outdoor air limit control opens

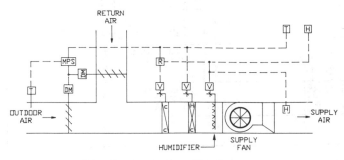

Fig. 59   Single-Zone Fan System

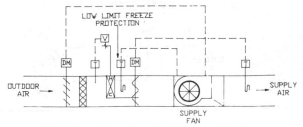

Fig. 60   Throttling VAV Terminal Unit

the preheat coil valve when outdoor air temperature drops to the point where the air requires heating to raise it to the desired supply air temperature. The discharge temperature controller then positions the face and bypass dampers to maintain that temperature. A capillary element thermostat located adjacent to the coil shuts the fan down for freeze protection if air temperature approaches freezing at any spot along the sensing element.

ZONE CONTROL SYSTEMS

VAV Terminal Units

Variable air volume terminal units are used in conjunction with VAV central air-handling fan systems to vary the volume of air into individual zones as required by the thermal load on the area. Because they are available in several configurations, control considerations vary. VAV terminal unit controls can be classified as follows:

Single-Duct VAV	Dual-Duct VAV
Throttling	Variable constant volume
Variable constant volume	Variable constant volume
Bypass	(zero energy band)
Induction	
Fan-powered	
Bypass fan induction	

Throttling VAV terminal units (Figure 61) are sometimes pressure-dependent; that is, the volume of air entering the conditioned space at any given space temperature varies as static pressure in the supply duct varies. The space thermostat controls the damper directly.

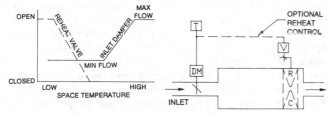

Fig. 61   Throttling VAV Terminal Unit

A reheat coil can be added. The space thermostat also controls the reheat coil valve in sequenced mode to open the valve after the damper has closed to its minimum flow position.

For perimeter areas, convectors or radiation with automatic control valves can provide reheat. Functionally, these are controlled in a similar way to reheat coil valves.

Variable constant volume terminal units are sometimes pressure-independent; that is, with varying supply duct static pressure, the unit continues to deliver the same volume of air to the conditioned space at the given space temperature. Airflow is controlled by either a mechanical controller or a receiver-controller (Figure 62).

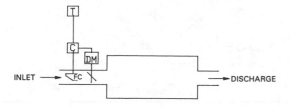

Fig. 62   Pressure-Independent VAV Terminal Unit

A flow sensor in the box airflow stream controls air volume. The room sensor resets the airflow controller set point as thermal load on the area changes. The airflow controller can be set with a minimum flow condition to ensure comfortable distribution of supply air into the space under light loading. Maximum flow can be set to limit flow to that required for design conditions.

A reheat coil, separate convection, or radiation can provide reheat for this unit. Reheat control is similar to that for throttling VAV terminal units.

Bypass VAV terminal units (Figure 63) have a space thermostat-controlled diverting damper that proportions the amount of entering supply air between the discharge duct and the bypass opening into the return plenum. A manual balancing damper in the bypass is adjusted to match the resistance in the discharge duct. In this way, the supply of air from the primary system remains at a constant volume.

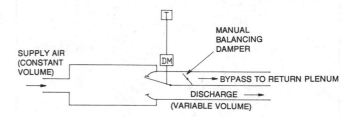

Fig. 63   Bypass VAV Terminal Unit

Induction VAV terminal units (Figure 64) provide return air reheat by routing air through the unit to induce air from the return air plenum into the supply airstream. In addition to the primary inlet damper, there is also a damper on the return air inlet. Both dampers are controlled simultaneously so that as the primary air opening decreases, the return air opening increases.

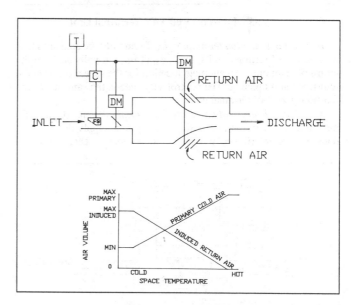

Fig. 64   Induction VAV Terminal Unit

An airflow controller in the unit controls the volume of air coming through the primary air damper. The space thermostat resets the set point of this controller as required by the thermal load on the conditioned space.

Fan-powered VAV terminal units (Figure 65) are similar to throttling VAV terminal units, except that they include an integral fan that recirculates space air at constant volume. In addition to enhancing air distribution in the space, they also provide a reheat coil and a means of maintaining a lowered unoccupied temperature in the space when the primary system is off.

Figure 65 shows a typical control sequence. A space thermostat resets an integral airflow controller as it senses space load changes. As primary air decreases, the fan activates to ensure adequate air circulation. The units serving the perimeter area of a building usually include a reheat coil that is sequenced with the primary air damper to supply heat when required. When the primary air system is not operating (nighttime or unoccupied control mode), the night operating mode of the thermostat cycles the fan with the reheat coil valve open to maintain the lowered temperature in the space.

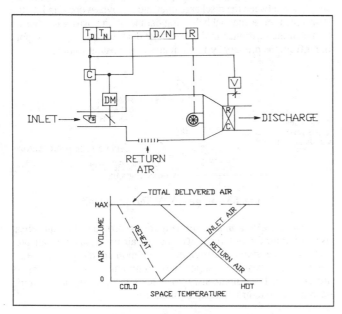

Fig. 65   Fan-Powered VAV Terminal Unit

Bypass fan induction terminal units (Figure 66) are similar to fan-powered VAV terminal units, except that the fan pulls air from the return plenum only. An alternate location for the reheat coil is in the return plenum opening. The control sequence is the same as for the fan-powered VAV terminal unit.

Variable constant volume dual-duct terminal units (Figure 67) have inlet dampers on the heating and cooling supply ducts. These dampers are interlinked to operate in reverse of each other and

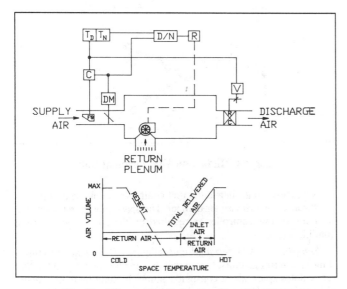

Fig. 66   Bypass Fan Induction Terminal Unit

require a single control actuator. There is also a total airflow volume damper with its own actuator.

The space thermostat controls inlet mixing dampers directly, and the airflow controller controls the volume damper. The space thermostat resets the airflow controller from maximum to minimum flow as the thermal load on the conditioned area changes. Figure 67 shows the control schematic and damper operation. Note that in a portion of the control range, heating and cooling supply air mix.

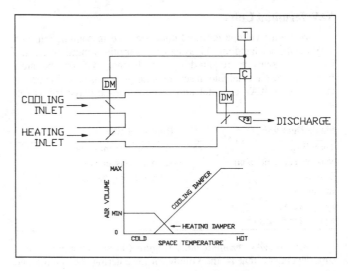

Fig. 67   Pressure-Independent Dual-Duct VAV Terminal Unit

Variable constant volume (zero energy band) dual-duct terminal units (Figure 68) have inlet dampers (with individual damper actuators and airflow controllers) on the cooling and heating supply ducts and no total airflow volume damper. The zero energy band (ZEB) space thermostat resets airflow controller set points in sequence as space load changes. The airflow controllers maintain adjustable minimum flows for ventilation, with no overlap of damper operations, during the ZEB when neither heating nor cooling is required. Figure 68 shows the control schematic and damper operation. Energy consumption of variable constant volume dual-duct terminal units with and without ZEB is essentially the same.

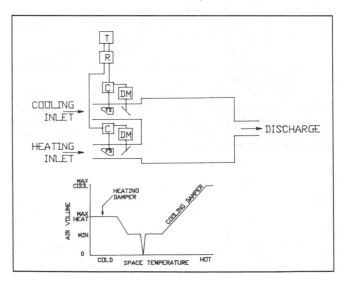

Fig. 68   Variable Constant Volume (ZEB) Dual-Duct Terminal Unit

Constant Volume Terminal Units

Multiple-zone systems using CV air distribution can be classified as follows:

Single-Duct	Dual-Duct
Zone reheat	Mixing box
Positive constant volume	Constant volume mixing box

Single-duct zone reheat systems have a heating coil (hot water, steam, or electric) in the branch supply duct to each zone. The central air-handling unit supplies constant temperature air. The space thermostat positions the reheat coil valve (or electric heating elements) as required to maintain space condition (Figure 69).

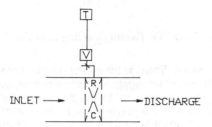

Fig. 69   Constant Volume Single-Duct Zone Reheat

Single-duct positive constant volume terminal units supply a constant volume of distribution air to the space, even though static pressure varies in the supply duct system. This is accomplished by an integral mechanical constant volume regulator or an airflow constant volume control furnished by the terminal unit manufacturer. If a reheat coil comes with the unit, a space thermostat controls the reheat coil valve as required to maintain space condition (Figure 70).

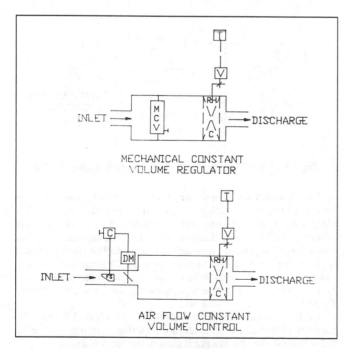

MECHANICAL CONSTANT
VOLUME REGULATOR

AIR FLOW CONSTANT
VOLUME CONTROL

**Fig. 70   Single-Duct Pressure-Independent Constant
Volume Terminal Unit**

Dual-duct mixing box terminal units generally apply to low-static-pressure systems that require large amounts of ventilation. The warm duct damper and the cool duct damper are linked to operate in reverse of each other. A space thermostat positions the mixing

dampers through a damper actuator to mix warm and cool supply air to maintain space condition. Discharge air quantity depends on the static pressure in each supply duct at that location. Static pressures in the supply ducts vary because of the varying airflow in each duct (Figure 71).

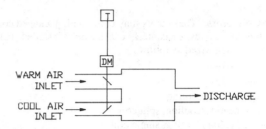

Fig. 71   Dual-Duct Mixing Box Terminal Unit

Dual-duct constant volume mixing box terminal units are typically used on high-static-pressure systems where the airflow quantity to each space is critical. The units are the same as those described in the previous paragraph, except that they include either an integral mechanical constant volume regulator or an airflow constant volume control furnished by the unit manufacturer (Figure 72).

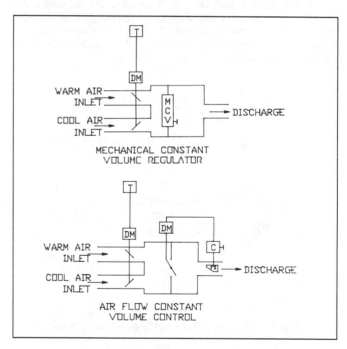

MECHANICAL CONSTANT
VOLUME REGULATOR

AIR FLOW CONSTANT
VOLUME CONTROL

**Fig. 72   Dual-Duct Pressure-Independent Constant
Volume Mixing Box Terminal Units**

Perimeter Radiation and Convection

Radiators or convectors can provide either total room heat or supplemental heat at the perimeter to offset building transmission losses. The control strategy depends on which function the radiation performs.

For a total room heating application, rooms are usually controlled individually; each radiator and convector is equipped with an automatic control valve. Depending on room size, one thermostat may control one valve or several valves in unison. The thermostat can be located in the return air to the unit or on a wall at occupant level. Return air control is generally less accurate and results in wider space temperature fluctuations. When the space is controlled for the comfort of seated occupants, wall-mounted thermostats give the best results.

For supplemental heating applications, where perimeter radiation is used only to offset perimeter heat losses (the zone or space load is handled separately by a zone air system), outdoor reset of the water temperature to the radiation should be considered. Radiation can be zoned by exposure, and the compensating outdoor sensor can be located to sense compensated indoor (outdoor) temperature, solar load, or both.

Fan-Coil Units. These units may come with packaged controls for the fan and valves, or controls can be field installed. Fan-coil units can be categorized as follows:

- Two-pipe heating
- Two-pipe cooling
- Two-pipe heating/cooling
- Four-pipe heating/cooling, split coil
- Four-pipe heating/cooling, single coil

Chapters 11, 12, 13, and 14 of the 1992 *ASHRAE Handbook—Systems and Equipment* describe design features and applications for the central plant and distribution systems associated with these fan-coil units.

Figure 73 shows typical control of a *two-pipe heating/cooling fan-coil unit*. Because the single coil may be used for either heating or cooling, depending on the season, hot water temperatures can be much lower than in standard heating coils. The unit requires an integral return-air or a wall-mounted heating/cooling thermostat that reverses its action at a remote changeover signal. Alternately, a pipe-mounted aquastat sensing a change in supply water temperature can initiate the heating/cooling changeover. Frequently, a local fan switch or a central time clock operates the fan.

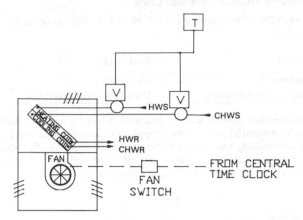

Fig. 74   Four-Pipe Heating/Cooling Split Fan-Coil Unit

space temperature. When the thermostat senses a need for cooling, valves to the hot water supply and return circuits remain closed, and the supply water valve cold water port is throttled to maintain space temperature. The supply water valve is adjusted so that both supply ports are closed in the dead band between heating and cooling operations. In this dead band, the return water valve is positively positioned to connect the correct return water circuit to the coil before either supply port opens. A local fan switch or a local time clock can control the fan.

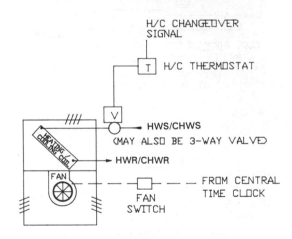

Fig. 73   Two-Pipe Heating/Cooling Fan-Coil Unit

In *four-pipe heating/cooling split coil systems,* the heating and cooling systems are hydraulically separated; the control precautions for two-pipe systems do not apply. The four-pipe fan-coil unit has a split coil, typically with one heating and two cooling rows. Hot water or steam can be used to heat, while direct expansion or chilled water can be used to cool. Figure 74 shows a room thermostat connected to the hot water and chilled water valves. The hot water valve closes when room temperature increases. With further increases, the chilled water valve begins to open. Valve ranges should be adjusted to provide a dead band within which both valves are closed. A local fan switch or a central time clock can control the fan.

In *four-pipe heating/cooling single coil systems* (Figure 75), special three-way control valves are used on both the supply water side and the return water side of the coil. The space thermostat controls both valves. When the thermostat senses a need for heat, valves to the chilled water supply and return circuits remain closed while the supply water valve hot water port is throttled to maintain

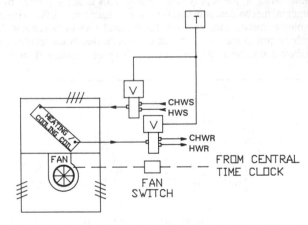

Fig. 75   Four-Pipe Heating/Cooling Single Fan-Coil Unit

Unit Ventilators. These are designed to heat, ventilate, and cool a space by introducing outdoor air in quantities up to 100%. Optionally, they can cool and dehumidify with a cooling coil (either chilled water or direct-expansion). Heating can be by hot water, steam, or electric resistance. The control of these coils can be by valves or face and bypass dampers. Consequently, control systems applied to unit ventilators are many and varied. This section describes the three most commonly used control schemes: Cycle I, Cycle II, and Cycle III.

Cycle I. Except during the warm-up stage, Cycle I (Figure 76) supplies 100% outdoor air at all times. During warm-up, the heating valve is open, the OA damper is closed, and the RA damper is open. As temperature rises into the operating range of the space thermostat, the OA damper opens fully, and the RA damper closes. The heating valve is positioned as required to maintain space temperature. The airstream thermostat can override space thermostat action on the heating valve to prevent discharge air from dropping below a minimum temperature. Figure 78 shows the positions of the heating valve and ventilation dampers in relation to space temperature.

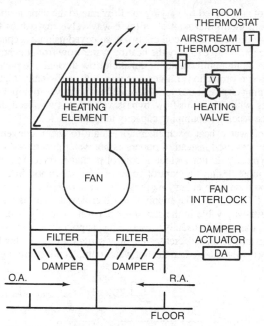

Fig. 76 Cycles I and II Control Arrangements

Cycle II. During the heating stage, Cycle II (Figure 76) supplies a set minimum quantity of outdoor air. Outdoor air is gradually increased as required for cooling. During warm-up, the heating valve is open, the OA damper is closed, and the RA damper is open. As the space temperature rises into the operating range of the space thermostat, ventilation dampers move to their set minimum ventilation positions. The heating valve and ventilation dampers are operated in sequence as required to maintain space temperature. The airstream thermostat can override space thermostat action on the heating valve and ventilation dampers to prevent discharge air from dropping below a minimum temperature. Figure 78 shows the relative positions of the heating valve and ventilation dampers with respect to space temperature.

Cycle III. During the heating, ventilating, and cooling stages, Cycle III (Figure 77) supplies a variable amount of outdoor air as required to maintain at a fixed temperature (typically 13°C) the air

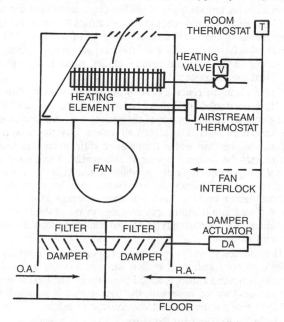

Fig. 77 Cycle III Control Arrangement

entering the heating coil. When heat is not required, this air is used for cooling. During warm-up, the heating valve is open, the OA air damper is closed, and the RA damper is open. As the space temperature rises into the operating range of the space thermostat, ventilation dampers control the air entering the heating coil at the set temperature. Space temperature is controlled by positioning the heating valve as required. Figure 78 shows the relative positions of the heating valve and ventilation dampers with respect to space temperature.

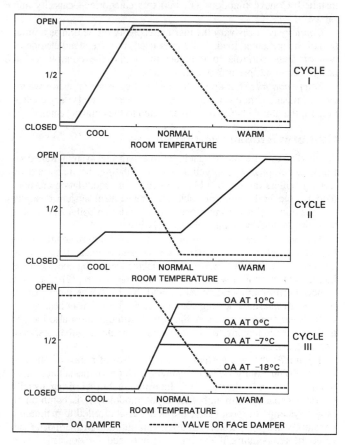

Fig. 78 Valve and Damper Positions with Respect to Room Temperature

Day/night thermostats are frequently used with any of these control schemes to maintain a lower space temperature during unoccupied periods by cycling the fan with the outdoor air damper closed. Another common option is a freezestat placed next to the heating coil that shuts the unit off when near-freezing temperatures are sensed.

Radiant Panels. These combine controlled-temperature room surfaces with central air conditioning and ventilation. The radiant panel can be in the floor, walls, or ceiling. Panel temperature is maintained by circulating water or air or by electric resistance. The central air system can be a basic one-zone, constant-temperature, CV system, with the radiant panel operated by individual room control thermostats; or it can include some or all the features of dual-duct, reheat, multizone, or VAV systems, with the radiant panel operated as a one-zone, constant-temperature system. The one-zone radiant heating panel system is often operated from an outdoor temperature reset system to vary panel temperature as outdoor temperature varies.

Radiant panels for both heating and cooling applications require controls similar to those described for the four-pipe heating/cooling single-coil fan coil. To prevent condensation, ventilation air supplied to the space during the cooling cycle should have a dew-point temperature below that of the radiant panel surface.

CONTROL OF THE CENTRAL PLANT

The term central plant, as used here, refers not only to a plant supplying heating and/or cooling media to multiple buildings, but also to a single building having central chillers and boilers.

The manufacturer almost always supplies boilers (both steam and hot water) and chillers (all types) with an automatic control package installed. Control functions fall into two categories—capacity and safety.

Capacity controls vary the thermal capacity of the unit as a function of the presented load. A designer needs to understand the operation of these controls to integrate them into the control system design for a multiple-unit plant.

Safety controls shut down the unit and generate an alarm whenever an unsafe condition is detected. If there is a supervisory control system, the alarm should be retransmitted to the control center.

Control of Hydronic Heating Systems

Load affects the rate of heat input to a hydronic system. Rate control is accomplished by cycling and modulating the flame and by turning boilers on and off. Flame cycling and modulation are handled by the boiler control package. The control system designer decides under what circumstances to add or drop a boiler and at what temperature to control the boiler supply water.

Hot water distribution control includes temperature control at the hot water boilers or converter, reset of heating water temperature, and control for multiple zones. Other factors requiring consideration include (1) minimum water flow through the boilers, (2) protection of boilers from temperature shock, and (3) coil freeze protection. If multiple or alternate heating sources (such as condenser heat recovery or solar storage) are used, the control strategy must also include a means of sequencing hot water sources or selecting the most economical source.

Figure 79 shows a system for load control of a gas- or oil-fired boiler. Boiler safety controls usually include flame-failure, high-temperature, and other cutouts. Intermittent burner firing usually controls capacity, although fuel input modulation is common in larger systems. In most cases, the boiler is controlled to maintain a constant water temperature, although an outdoor air thermostat can reset the temperature if the boiler is not used for domestic water heating. In Figure 79, a master-submaster arrangement is used in which the outdoor thermostat resets supply water temperature according to the reset schedule shown. To minimize condensation of flue gases and boiler damage, water temperature should not be reset below that recommended by the manufacturer, typically 60°C.

In the system shown in Figure 79, three-way control valves at the heating coils ensure minimum water flow through the boiler. Zone control is achieved by varying the flow rate to the zone heating coil. A room thermostat controls a three-way valve that varies the coil output. Proper preheat coil control is achieved using a second three-way control valve to vary the heating water flow from the boiler; the recirculating pump maintains constant flow through the preheat coil for freeze protection. Larger systems with sufficiently high pump operating costs can use variable-speed pump drives, pump discharge valves with minimum flow bypass valves, or two-speed drives to reduce secondary pumping capacity to match the load.

Hot water heat exchangers or steam-to-water converters are sometimes used instead of boilers as hot water generators. Converters typically do not include a control package; therefore, the engineer must design the control scheme. The schematic in Figure 80 can be used with either low-pressure steam or boiler water ranging from 93 to 127°C. The supply water thermostat controls a modulating two-way valve in the steam (or hot water) supply line. An outdoor thermostat usually resets the supply water temperature downward as load decreases. A flow switch interlock should close the two-way valve when the hot water pump is not operating.

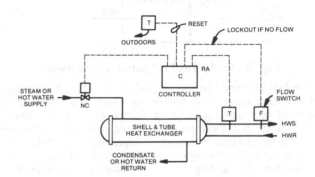

Fig. 80　Steam-to-Hot-Water Heat Exchanger Control

Control of Chiller Plants

Because of the wide variety of chiller types, sizes, drives, manufacturers, piping configurations, pumps, cooling towers, distribution systems, and loads, it is almost inevitable that each central chiller plant, including its controls, is designed on a custom basis.

Chapter 42 of the 1994 *ASHRAE Handbook—Refrigeration* gives information on various types of chillers (e.g., centrifugal and reciprocating). Each type has specific characteristics that must match the requirements of the installation. Chapter 11 of the 1992 *ASHRAE Handbook—Systems and Equipment* covers variations in piping configurations (e.g., series and parallel chilled water flow) and some associated control concepts.

Chiller plants are generally one of two types—*variable flow* (Figure 81) or *constant flow* (Figure 82). Both figures show parallel flow piping configuration. Control of the remote load determines which type should be used. Throttling coil valves vary the flow rate in response to the load and a temperature differential that tends to remain near the design temperature differential. The chilled water supply temperature typically establishes the base flow rate. To improve energy efficiency, the set point is reset based on the zone with the greatest load (load reset) or other variances. Different methods of piping and pumping can accomplish the desired task while maintaining the manufacturer's desired minimum flow through an operating chiller.

The constant flow system (Figure 82) is actually constant flow under each combination of chillers on line; a major upset always occurs whenever a chiller is added or dropped. The load reset function ensures that the zone with the largest load is satisfied, while supply or return water control treats average zone load.

Optimize Refrigerant Pressure. Chiller efficiency is a function of both the percent of full load on the chiller and the difference

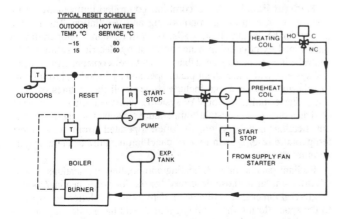

Fig. 79　Load and Zone Control in a Simple Hydronic System

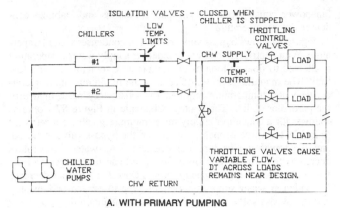

A. WITH PRIMARY PUMPING

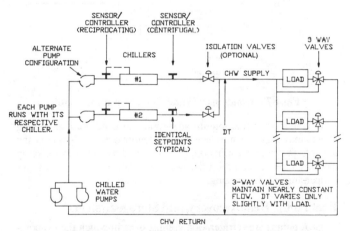

B. WITH PRIMARY-SECONDARY PUMPING

Fig. 81   Variable Flow Chilled Water Systems

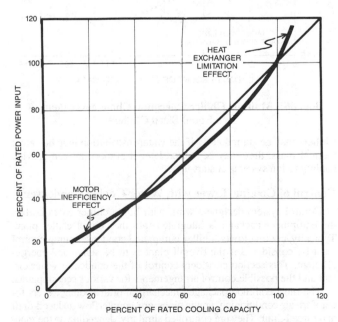

Fig. 82   Constant Flow Chilled Water System

in refrigerant pressure between the condenser and the evaporator. In practice, this pressure difference is represented by condenser water exit temperature minus chilled water supply temperature. To reduce the pressure difference, the chilled water supply temperature must be increased and/or the condenser water temperature decreased. There is a 2 to 3% energy saving for each Kelvin reduction.

There are two effective methods for reducing refrigerant pressure difference:

1. Use *chilled water load reset* to raise the supply setpoint as load decreases. Figure 83 shows the basic function of this method. Varying degrees of sophistication are available, including computer control.

2. *Lower condenser temperature* to the lowest safe temperature (use manufacturer's recommendations) by keeping the cooling tower

bypass valve closed, operating at full condenser water pump capacity, and maintaining full airflow in all cells of the cooling tower until water temperature is within about 3 K of outdoor air wet-bulb temperature. However, pumps and fans consume power. Consider this and the fan power of the VAV air handlers in calculating net energy savings.

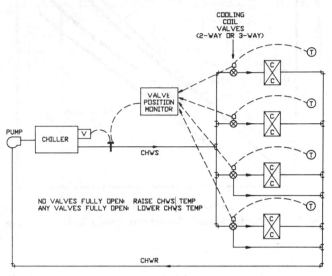

Fig. 83   Chilled Water Load Reset

Optimize Operation. Multiple chiller plants should be operated at the most efficient point on the part-load curve. Figure 84 shows a typical part-load curve for a centrifugal chiller operated at design conditions. Figure 85 shows similar curves at different pressure-limiting conditions. Figure 86 indicates the point at which a chiller should be added or dropped in a two-unit plant. In general, the part-load curves are plotted for all combinations of chillers; then, the break-even point between n and $n + 1$ chillers can be determined.

Minimize Run Time. Daily start-up of the chiller plant should be optimized based on start-up time of the air-handling units. Chillers are generally started at the same time as the first fan system.

Fig. 84   Chiller Part-Load Characteristics at Design Refrigerant Pressure Difference

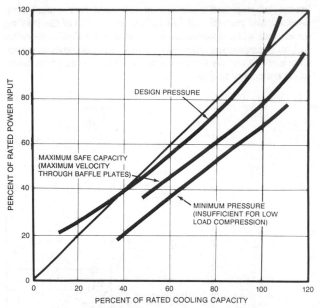

Fig. 85   Chiller Part-Load Characteristics with Variable Pressure

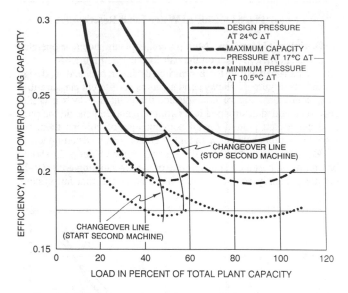

**Fig. 86   Multiple Chiller Operation Changeover Point—
Two Equal-Sized Chillers**

Chillers may be started early if the water distribution loop has great thermal mass; they may be started later if outdoor air can provide cooling to fan systems at start-up.

Control of Cooling Tower with Water-Cooled Condenser

Control system designers work with liquid chiller control when the equipment package is integrated into the central chiller plant. Typically, cooling tower, chiller pump, and condenser pump control must be considered if the overall plant is to be stable and energy-efficient. This section considers control of the condenser water circuit and the possible control arrangements for various central plants.

The most common packaged mechanical-draft cooling towers for comfort air-conditioning applications are counterflow induced-draft and forced-draft. They are controlled similarly, depending on the manufacturer's recommendations. On larger towers, two-speed or variable-speed fans (and associated motor control circuitry) can reduce

fan power consumption at part-load conditions and stabilize condenser water temperature.

Figure 87 shows bypass valve control of condenser water temperature. With centrifugal chillers, condenser supply water temperature is allowed to float as long as the temperature remains above a low limit. The manufacturer should specify the minimum entering condenser water temperature required for satisfactory performance of the particular chiller. The control schematic in Figure 87 works as follows: for a condenser supply temperature (e.g., above a set point of 24°C), the valve is open to the tower, the bypass valve is closed, and the tower fan or fans are operating. As water temperature decreases (e.g., to 18°C), tower fan speed can be reduced to low-speed operation if a two-speed motor is used. On a further decrease in condenser water supply temperature, the tower fan or fans stop and the bypass valve begins to modulate to maintain the acceptable minimum water temperature.

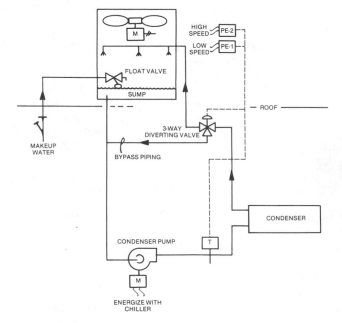

Fig. 87   Condenser Water Temperature Control

In colder areas that require year-round air conditioning, cooling towers may require sump heating and continuous full flow over the tower to prevent ice formation. In that case, the cooling tower sump thermostat would control a hot water or steam valve to keep water temperature above freezing.

Heat Pump, Heat Recovery, and Storage Systems

Heat pumps are refrigeration compressors in which the evaporator is used for cooling and the condenser is used for heating. Many conventional means can control heating and cooling cycles. Chapters 8 and 46 of the 1992 *ASHRAE Handbook—Systems and Equipment* have details.

Heat recovery and storage systems almost always require customized control systems to provide the desired sequence. Chapter 8 of the 1992 *ASHRAE Handbook—Systems and Equipment* has details on heat recovery systems; Chapter 40 of this volume has information on thermal storage.

DESIGN CONSIDERATIONS AND PRINCIPLES

In total building HVAC system selection and design, the type, size, use, and operation of the structure must be considered. Subsystems such as fan and water supply are normally controlled by localized automatic control or local loop control. A *local loop control system* includes the sensors, controllers, and controlled devices used with a single HVAC system and excludes any supervisory or remote functions such as reset and start-stop. However, extension of local control to a central control point is frequently included in the design when justified by the need to diagnose system malfunction when it occurs, thus reducing any damage that might result from delay, and by reduced labor and energy costs.

The growing popularity of distributed processing using microprocessors has augmented computer use at many locations other than the central control point. The local loop controller can be a direct digital controller (DDC) instead of a pneumatic or electric thermostat, and some energy management functions may be performed by a DDC.

Because HVAC systems are designed to meet maximum design conditions, they nearly always function at partial capacity. It is important to be able to control the system at all times; because the system must be adjusted and operated for many years, the simplest system that produces the necessary results is usually the best.

Coordination Between Mechanical and Electrical Systems

Even when a system is basically pneumatic, the electrical engineer must design wiring, conduit, switchgear, and electrical distribution for many electrical devices.

The mechanical designer must inform the electrical designer of the total electrical requirements of the control system if the controls are to be wired by the electrical contractor. Requirements include (1) the devices to be furnished and/or connected, (2) loads in watts, (3) location of electrical items, and (4) a description of each control function. Proper coordination should produce a schematic control diagram that interfaces properly with other control elements to form a complete and usable system. As an option, the control engineer may develop a complete performance specification and require the control system contractor to install all wiring related to the specified sequence.

Coordination is essential. The control system designer must make the final checks of drawings and specifications. Both mechanical and electrical specifications must be checked for compatibility and uniformity.

Building and System Subdivision

The following must be considered in building and mechanical system subdivision:

- Heating and cooling loads as they vary—the ability to heat or cool interior or exterior areas of a building at any time may be required.
- Occupancy schedules and the flexibility to meet needs without undue initial and/or operating costs.
- Fire and smoke control and possible compartmentation that matches the air-handling system layout and operation.

Control Principles for Energy Conservation

After the general needs of a building have been established, and building and system subdivision has been made, the mechanical system and its control approach can be considered. Designing systems that conserve energy requires knowledge of (1) the building, (2) its operating schedule, (3) the systems to be installed, and (4) ASHRAE *Standard* 90.1. The principles or approaches that conserve energy are as follows:

1. *Run equipment only when needed.* Schedule HVAC unit operation for occupied periods. Run heat at night only to maintain internal temperature between 10 and 13°C to prevent freezing. Start morning warm-up as late as possible to achieve design internal temperature by occupancy time, considering residual space temperature, outdoor temperature, and equipment capacity (optimum start control). Under most conditions, equipment can be shut down

some time before the end of occupancy, depending on internal and external load and space temperature (optimum stop control). Calculate shutdown time so that space temperature does not drift out of the selected comfort zone before the end of occupancy.
2. *Sequence heating and cooling.* Do not supply heating and cooling simultaneously. Central fan systems should use cool outdoor air in sequence between heating and cooling. Zoning and system selection should eliminate, or at least minimize, simultaneous heating and cooling. Also, humidification and dehumidification should not take place concurrently.
3. *Provide only the heating or cooling actually needed.* Reset the supply temperature of hot and cold air (or water) according to actual need. This is especially important on systems or zones that allow simultaneous heating and cooling.
4. *Supply heating and cooling from the most efficient source.* Use free or low-cost energy sources first, then higher cost sources as necessary.
5. *Apply outdoor air control.* When on minimum outdoor air, use no less than that recommended by ASHRAE *Standard* 62. In areas where it is cost-effective, use enthalpy rather than dry-bulb temperature to determine whether outdoor or return air is the most energy-efficient air source for the cooling mode.

System Selection

The mechanical system significantly affects the control of zones and subsystems. The type of system and the number and location of zones influence the amount of simultaneous heating and cooling that occurs. For exterior building sections, heating and cooling should be controlled in sequence to minimize simultaneous heating and cooling. In general, this sequencing must be accomplished by the control system because only a few mechanical systems (e.g., two-pipe systems and single-coil systems) have the ability to minimize simultaneous heating and cooling. Systems that require engineered control systems to minimize simultaneous heating and cooling include the following:

- *VAV cooling with zone reheat.* Reduce cooling energy and/or air volume to a minimum before applying reheat.
- *Four-pipe heating and cooling for unitary equipment.* Sequence heating and cooling.
- *Dual-duct systems.* Condition only one duct (either hot or cold) at a time. The other duct should supply a mixture of outdoor and return air.
- *Single-zone heating/cooling.* Sequence heating and cooling.

Some exceptions exist, as in the case of dehumidification with reheat.

Control zones are determined by the location of the thermostat or temperature sensor that sets the requirements for heating and cooling supplied to the space. Typically, control zones are for a room or an open area portion of a floor.

Many jurisdictions in the United States no longer permit CV systems that reheat cold air or that mix heated and cooled air. Such systems should be avoided. If selected, they should be designed for minimal use of the reheat function through zoning to match actual dynamic loads and resetting cold and warm air temperatures based on the zone(s) with the greatest demand. Heating and cooling supply zones should be structured to cover areas of similar load. Areas with different exterior exposures should have different supply zones.

Systems that provide changeover switching between heating and cooling prevent simultaneous heating and cooling. Some examples are hot or cold secondary water for fan coils or single-zone fan systems. They usually require small operational zones, which have low load diversity, to permit changeover from warm to cold water without occupant dissatisfaction.

Systems for building interiors usually require year-round cooling and are somewhat simpler to control than exterior systems. These interior areas normally use all-air systems with constant supply air temperature, with or without VAV control. Proper control techniques and operational understanding can reduce the energy used to treat these areas. Reheat should be avoided.

General load characteristics of different parts of a building may lead to selecting different types of systems for each.

Load Matching

With individual room control, it is possible to control a space more accurately and conserve energy if the whole system can be controlled in response to the major factor influencing the system load. Thus, water temperature in a hot water heating system, steam temperature or pressure in a steam heating system, or delivered air temperature in a central fan system can be varied as building load varies. This puts a reasonable control on the whole system, relieves individual space controls of part of their burden, and allows more accurate space control. Also, modifying the basic rate of heating or cooling input to the system in accordance with system load reduces losses in the distribution system.

The system must always satisfy the area or room with the greatest demand. Individual controls handle demand variations within the area the system serves. The more accurate the system zoning, the greater the control by the overall system, the smaller the system distribution losses, and the more effectively space conditions are maintained by individual controls.

Buildings or zones with a modular arrangement can be designed for subdivision to meet occupant needs. Before subdivision, operating inefficiencies can occur if a zone has more than one thermostat. If a system allows one thermostat to activate heating should be controlled while another activates cooling, the two zones or terminals should be controlled from a single thermostat until the area is properly subdivided.

Size of Controlled Area

No individually controlled area should exceed about because the difficulty of obtaining good distribution and of finding a representative location for the space controls increases with zone area. Each individually controlled area must have similar load characteristics throughout. Equitable distribution, provided through competent engineering design, careful equipment sizing, and proper system balancing, is necessary for maintaining uniform conditions throughout an area. The control can measure conditions only at its location; it cannot compensate for nonuniform conditions caused by improper distribution or inadequate design. Areas or rooms having dissimilar load characteristics or different conditions to be maintained should be controlled individually. The smaller the controlled area, the better the control and the better the system performance and flexibility.

Location of Space Sensors

Space sensors and controllers must be located where they accurately sense the variables they control and where the condition is representative of the area (zone) they serve. In large open areas having more than one zone, thermostats should be located in the middle of their zones to prevent them from sensing conditions in surrounding zones. There are three common locations for space temperature controllers or sensors.

Wall-mounted thermostats or *sensors* are usually placed on inside walls or columns in the space they serve. Avoid outside wall locations. Mount thermostats where they will not be affected by heat from sources such as direct sun rays; wall pipes or ducts; convectors; or direct air currents from diffusers or equipment (e.g., copy machines, coffee makers, or refrigeration cases). Air circulation should be ample and unimpeded by furniture or other obstructions, and the thermostat should be protected against mechanical injury. Thermostats located in spaces such as corridors, lobbies, or foyers should be used to control only those areas.

Return air thermostats can control floor-mounted unitary conditioners such as induction or fan-coil units and unit ventilators. On induction and fan-coil units, the sensing element is behind the return air grille. On classroom unit ventilators that use up to 100% outdoor air for natural cooling, however, a forced flow sampling chamber should be provided for the sensing element. The sensing element should be located carefully to avoid radiant effect and to ensure adequate air velocity across the element.

If return air sensing is used with central fan systems, locate the sensing element as near as possible to the space being controlled to eliminate influence from other spaces and the effect of any heat gain or loss in the duct. Where supply/return light fixtures are used to return air to a ceiling plenum, the return air sensing element can be located in the return air opening. (Be sure to offset the set point to compensate for the heat from the light fixtures.)

Diffuser-mounted thermostats usually have sensing elements mounted on circular or square ceiling supply diffusers and depend on aspiration of room air into the supply airstream. They should be used only on high-aspiration diffusers adjusted for a horizontal air pattern. The diffuser on which the element is mounted should be in the center of the occupied area of the controlled zone.

Lowered Night Temperature

When temperatures during unoccupied periods are lower than those normally maintained during occupied periods, an automatic timer often establishes the proper day and night temperature time cycle. Allow sufficient time in the morning to pick up the conditioning load well before there is any heavy increase. Night setback temperatures are often monitored and controlled more closely with supervisory control systems (see the section on Control of Central Subsystems). These supervisory systems take into account variables such as outdoor temperature, system capacity, and building mass to determine optimal start-up and shutdown times.

Cost Analysis

Due to the rising cost of energy, it has been popular to promote various types of control equipment in terms of "months to payback" based on projected energy savings. While this is one appropriate measure of the value of a purchase, many other factors can affect the true value. For example, if the estimated payback of an energy management system is 24 months, it does not necessarily deliver an annual rate of return of 50%. For a detailed explanation of life-cycle costing, see Chapter 33.

SPECIAL DESIGN CONSIDERATIONS

Controls for Mobile Units

The operating point of any control that relies on pressure to operate a switch or valve varies with atmospheric pressure. Normal variations in atmospheric pressure do not noticeably change the operating point, but a change in altitude affects the control point to an extent governed by the change in absolute pressure. This is especially important when controls are selected for use in land and aerospace vehicles that in normal use are subject to wide variations in altitude. The effect can be substantial; for example, barometric pressure decreases by nearly one-third as altitude increases from sea level to 3000 m.

In mobile applications, three detrimental factors are always present in varying degrees—vibration, shock, and G-forces. Controls selected for service in mobile units must qualify for the specific conditions expected in the installation. In general, devices containing mercury switches, slow-moving or low-force contacts, or mechanically balanced components are unsuitable for mobile applications; electronic solid-state devices are generally less susceptible to the three factors.

Explosive Atmospheres

Sealed-in-glass contacts are not considered explosionproof; therefore, other means must be provided to eliminate the possibility of a spark within an explosive atmosphere.

When using electric control systems, the designer can surround the control case and contacts with an explosionproof case, permitting only the capsule and the capillary tubing to extend into the conditioned space. It is often possible to use a long capillary tube and mount the instrument case in a nonexplosive atmosphere. The latter method can be duplicated with an electronic control system by placing an electronic sensor in the conditioned space and feeding its signal to an electronic transducer located in the nonexplosive atmosphere.

Because a pneumatic control system uses compressed air as its energy source, it is safe in otherwise hazardous locations. However, many pneumatic systems include interfaces to electrical components. All electrical components require appropriate explosionproof protection.

Sections 500 to 503 of the *National Electrical Code* (NFPA 1993) include detailed information on electrical installation protection requirements for various types of hazardous atmospheres.

Safety and Limit Controls

When selecting automatic controls, it is important to consider the type of system needed to control high or low limit (safety). This control may be inoperative for days, months, or years, and then operate immediately to prevent serious damage to equipment or property. Separate operating and limit controls are always recommended, even for the same functions.

Steam or hot water exchangers tend to be self-regulating and, in that respect, differ from electrical resistance heat transfer devices. For example, if airflow through a steam or hot water coil stops, coil surfaces approach the temperature of the entering steam or hot water, but cannot exceed it. Convection or radiation losses from the steam or hot water to the surrounding area take place, so the coil is not usually damaged. Electric coils and heaters, on the other hand, can be damaged when air stops flowing around them. Therefore, control and power circuits must interlock with heat transfer devices (pumps and fans) to shut off electrical energy when the device shuts down. Flow or differential pressure switches may be used for this purpose; however, they should be calibrated to energize only when there is airflow. This precaution shuts off power in case a fire damper closes or some duct lining blocks the air passage. Limit thermostats should also be installed to deenergize the heaters when temperatures exceed safe operating levels.

Safety for Duct Heaters

The current in individual elements of electric duct heaters is normally limited to a maximum safe value established by the *National Electric Code* or local codes. Two safety devices in addition to the airflow interlock device are usually applied to duct heaters (Figure 88). The automatic reset high limit thermostat normally deenergizes the control circuit; if the control circuit has an inherent time delay or uses solid-state switching devices, a separate safety contactor may be desirable. The manual reset backup high limit safety device is generally set independently to interrupt all current to the heater in case other control devices fail.

OPERATION AND MAINTENANCE

The benefit gained from HVAC equipment ultimately depends on properly designed, installed, operated, and maintained automatic control systems. Control systems duplicate and, in certain situations, eliminate some of the functions that plant operators have historically performed. The trade-off is greater complexity and an increase in the number of items to be maintained.

Current construction practices complicate design because responsibilities are divided between different groups of specialists. Often, each specialist has different objectives. Direct communication is essential to integrate all the related skills and thus ensure that the HVAC system will provide the healthy and comfortable space conditions desired. The designer who accurately evaluates the design from the perspective of the installers, operators, and maintenance personnel has the best chance of assuring a successful control system.

The following is a list of questions that should be addressed by the system designer to clarify some operation and maintenance objectives.

Cost

1. Will trained personnel be employed to operate the equipment?
2. If not, will the automatic control system safely and cost-effectively operate all equipment as required to satisfy all comfort needs without trained operating personnel?
3. What is the relative cost of these alternative approaches? Are these costs in documented form? Are the owners aware of and in agreement with the analysis?
4. What is the estimated cost for maintaining the automatic control system? Has it been verified with the maintenance sources available to the owner? Are these costs in documented form? Are the owners aware of and in agreement with the estimate?
5. If the owners are going to hire their own operating and/or maintenance personnel, what has the designer provided (in the form of documentation and training recommendations) to ensure that the design intent will be available throughout the life of the facility? Has this been discussed with the owners, and do they concur with the plan?

General comment: Control system maintenance costs can easily be underestimated. They may range from 5 to 20% of the purchase or replacement cost annually. It is not unusual for the personnel requirement for the maintenance of automatic controls—furnished as stand-alone systems or integral components of HVAC equipment—to exceed that for the maintenance of the remainder of the HVAC components and their related electrical distribution system. The most common reason for inadequate control system maintenance is the management's lack of understanding of and emphasis on this cost.

Flexibility

1. How will standby, spare, or redundant system components (fans, pumps, switchgear) be interfaced with the control system to provide backup service when their paired components are taken out of service due to failure or scheduled maintenance requirements? If the procedure is not automatic, are operational instructions conveniently provided and easy to understand?
2. Has the designer or supplier evaluated the impact of each component of the control system on system performance during failure

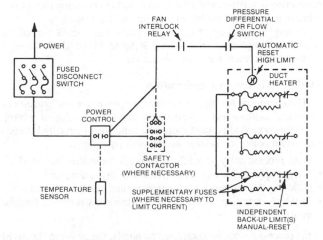

Fig. 88   Duct Heater Control

modes? What is the appropriate operator function at each such occurrence? Is it documented?

3. Can the actuation devices (valves, dampers, relays) be manipulated by operators to facilitate control system component maintenance, replacement, or emergency operation? Are proper instructions for their emergency use provided?

General comment: Many designers provide two operator interface stations for each actuation device. These stations provide an automatic/manual selection and a positioning or mode selection capability for use during manual operation. This may be a desirable feature despite the added cost.

Maintainability

1. What provisions has the designer made to ensure that each control component is easily accessible for operation and maintenance? Are the number of components requiring ladders, supplemental lighting, special access panels, and so forth, reduced to the lowest practical level? At the conclusion of the installation, will components be difficult to access because of conduit, piping, insulation, or proximity to other system components?

2. Are the system components that are most subject to failure of the quick-replacement, plug-in type?

3. Are piping and wiring practices specified? Are the materials subject to damage from normal maintenance functions or operation of other HVAC system components? Do they interfere with any other building maintenance function?

4. Are adequate gages, meters, information displays, and other diagnostic tools provided for convenient, detailed troubleshooting of system components?

5. Can components be replaced without draining hydronic systems, losing refrigerant charge, or shutting down major system components?

6. Will repair parts or substitute components be readily available at a reasonable price during the economic life of the facility?

7. Is equipment maintenance and/or operation based on the installation-acquired knowledge of one or more key individuals rather than on documentation available for new people?

Reliability

1. Have provisions been made to ensure a clean, dry control air supply during all weather conditions?

2. Are power supplies protected from lightning and other electrical transients where applicable?

3. Are electronic components located in a clean, dry enclosure with adequate temperature regulation to prevent product damage?

4. Are the specified and supplied components adequately field proven? If not, are both the manufacturer and the owner willing to assume the risk of added warranty and maintenance costs? If the components need replacement, what are the alternatives? Who pays? Who demonstrates that the components are not reliably functional?

5. What are the quality control procedures used by the manufacturer? What type of functional tests are made at the factory? In the field? Is every component checked, or is random sampling the accepted practice?

6. Is the control system, as designed, dependent on the accurate function and calibration of a large number of system components (e.g., high- or low-temperature signal feedback, valve position feedback, occupancy feedback, etc.)? If so, what provisions have been made for detecting any component malfunction or HVAC equipment application or installation problem to ensure that the fault does not cause the system to operate in maximum energy consumption mode by default?

Documentation

1. Who is responsible for checking the completeness and accuracy of the project's documentation deliverables?
   a) "As-built" control diagrams
   b) Laminated or framed sequences of operation and maintenance interface instructions
   c) Component and subcomponent parts lists
   d) Calibration and/or maintenance procedures for each component
   e) Operating software in field-loadable form
   f) Operating software in engineering logic form, at least for all operational sequences
   g) User's manual, including software generation manual

2. What is the schedule for documentation deliverables? Does it coincide with training schedules?

3. What provisions have been made for updating documentation to accommodate future corrections or changes?

4. Who is responsible for furnishing the deliverables listed in item 1? Are systems and components that must be an integral part of major items of equipment included?

Expansion and Modification

1. What are the alternatives for upgrading the control system to accommodate expansion or system modification?

2. Will the owner be required to accept installation of side-by-side systems to facilitate competitive bidding for system add-on components or services?

3. Have provisions been made to identify and control the future cost of software and hardware additions and modifications?

START-UP AND TESTING

A successful control system receives a proper start-up and testing, not merely the adjustment of a few parameters (set points and throttling ranges) and a few quick system checks. However, the controls used for many efficient HVAC systems present a formidable challenge for proper start-up and testing and require the services of experienced professionals. The increased use of VAV systems and digital controls has generally increased the importance and complexity of start-up and testing.

Design and construction specifications should include specific commissioning procedures. Start-up should be coordinated with testing and balancing (see Chapter 34) because each affects the other. The start-up and testing procedure begins with the checking of each control device to ensure that it is installed and connected according to approved shop drawings. Each electrical and pneumatic connection is verified, and all interlocks to fan and pump motors and primary heating and cooling equipment are checked.

Testing and start-up of a control system are a small part of the procedure known as commissioning. Refer to ASHRAE *Guideline* 1 for further information on commissioning.

Tuning—General Considerations

The systematic tuning of controllers improves the performance of all control systems and is particularly important for digital control. First, the controlled process should be controlled manually between various set points to evaluate the following questions:

- Is the process noisy (rapid fluctuations in controlled variable)?
- Is there appreciable hysteresis (backlash) in the actuator?
- How easy (or difficult) is it to maintain and change set point?
- In which operating region is the process most sensitive (highest gain)?

If the process cannot be controlled manually, the problem should be identified and corrected before the controller is tuned.

Tuning selects control parameters that determine the steady-state and transient characteristics of the control system. HVAC processes are nonlinear, and characteristics change on a seasonal basis. Controllers tuned under one operating condition may become unstable as conditions change. A well-tuned controller (1) minimizes the steady-state error from set point, (2) responds quickly to disturbances, and (3) remains stable under all operating conditions. Tuning of proportional controllers is a compromise between minimizing steady-state error and maintaining margins of stability. Proportional-plus-integral control minimizes this compromise because the integral action reduces steady-state error while the proportional term determines the controller's response to disturbances.

Control loops should be tuned under conditions of highest process gain to ensure stability over a wide range of process conditions.

Tuning PI Controllers

Popular methods of determining PI controller tuning parameters include the closed- and open-loop process identification methods and the trial-and-error method (Stoecker and Stoecker 1989). The closed-loop method involves increasing the gain of the controller in a proportional-only mode until the system cycles continuously after a set point change (Figure 89, where $K_p = 40$). Proportional and integral terms are then computed from the loop's period of oscillation and the proportional gain value that caused cycling. The open-loop method introduces a step change in input into the opened control loop. A graphical technique is used to estimate the process transfer function parameters. Proportional and integral terms are calculated from the estimated process parameters using a series of equations (Ziegler and Nichols 1942).

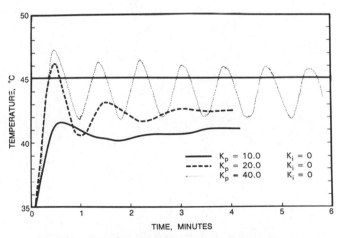

Fig. 89 Response of Discharge Air Temperature to a Step Change in Set Point at Various Proportional Constants and No Integral Action

The trial-and-error method involves adjusting the gain of the controller in proportional-only mode until the desired response to a set point change is observed. Conservative tuning dictates that this response should be a small initial overshoot that quickly damps to steady-state conditions (Figure 89, $K_p = 10$). Set point changes should be made in the range where controller saturation, or output limits, is avoided. The integral term is then increased until changes in set point produce the same dynamic response as the controller under proportional-only control, but with the response now centered about the set point (Figure 90).

Tuning Digital Controllers

In tuning digital controllers, it may be necessary to specify additional parameters. The selection of a digital controller sampling

interval is a compromise between computational resources and control requirements. A sampling interval of about one-half of the time constant of the controlled process usually provides adequate control. Many digital control algorithms include an error dead band to eliminate unnecessary control actions when the process is near set point. Hysteresis compensation is possible with digital controllers, but it must be carefully applied because overcompensation can cause continuous cycling of the control loop.

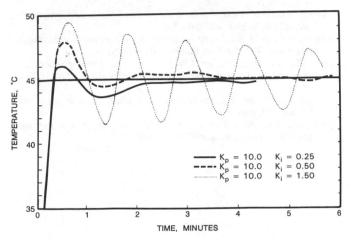

Fig. 90 Response of Discharge Air Temperature to a Step Change in Set Point at Various Integral Constants with a Fixed Proportional Constant

REFERENCES

ASHRAE. 1989. Energy efficient design of new buildings except low-rise residential buildings. *Standard* 90.1-1989.

ASHRAE. 1989. Ventilation for acceptable indoor air quality. *Standard* 62-1989.

ASHRAE. 1989. Guideline for commissioning of HVAC systems. *Guideline* 1-1989.

NFPA. 1993. National electrical code. *Code* 70-1993. National Fire Protection Association, Quincy, MA.

Stoecker, W.F. and D.A. Stoecker. 1989. Microcomputer control of thermal and mechanical systems. Van Nostrand Reinhold, New York.

Ziegler, J.G. and N.B. Nichols. 1942. Optimal settings for automatic controllers. ASME *Transactions* (64):759-65.

BIBLIOGRAHPY

Baker, D.W. and C.W. Hurley. 1984. On-site calibration of flow metering systems installed in buildings. NBS Science Series *Report* 159. National Bureau of Standards, Washington, D.C.

Carlson, R.A. and R.A. Di Giandomenico. 1991. *Understanding building automation systems*. R.S. Means, Kingston, MA.

Coad, W.J. 1985. Variable flow in hydronic systems for improved stability, simplicity and energy economics. *ASHRAE Transactions* 91(1B).

Coggan, D.A. 1986. Control fundamentals apply more than ever to DDC. *ASHRAE Transactions* 92(1).

Deshpande, P.B. and R.H. Ash. 1981. Elements of computer process control. Instrument Society of America, Research Triangle Park, NC.

Dressel, L.J. 1982. Improved operator interface techniques. *ASHRAE Transactions* 88(1).

Haines, R.W. 1990. *Roger Haines on HVAC controls*. TAB Books, Blue Ridge Summit, PA.

Hartman, T.B. 1993. *Direct digital controls for HVAC systems*. McGraw-Hill, New York.

Hurley, C.W. and J.F. Schooley. 1984. Calibration of temperature measurement systems installed in buildings. NBS Science Series *Report* 153. National Bureau of Standards, Washington, D.C.

Hyland, R.W. and C.W. Hurley. 1983. General guidelines for the on-site calibration of humidity and moisture control systems in buildings. NBS Science Series *Report* 157. National Bureau of Standards, Washington, D.C.

Johnson, G.A. 1985. Retrofit of a constant volume air system for variable speed fan control. *ASHRAE Transactions* 91(1).

Kao, J.Y. and W.J. Snyder. 1982. Application information on typical hygrometers used in heating, ventilating, and air conditioning (HVAC) systems. *Report* NBSIR 81-2460. National Bureau of Standards, Washington, D.C.

Kuo, B.C. 1975. *Automatic control systems*, 3rd ed. Prentice Hall, Englewood Cliffs, NJ.

Langley, B.C. 1985. *Control systems for air conditioning and refrigeration.* Prentice Hall, Englewood Cliffs, NJ.

Langley, B.C. 1989. *Solid state electronic controls for air conditioning and refrigeration.* Prentice Hall, Englewood Cliffs, NJ.

Lau, A.S., W.A. Beckman, and J.W. Mitchell. 1985. Development of computerized control strategies for large chilled water plants. *ASHRAE Transactions* 91(1B).

Levenhagen, J.I. and D.H. Spethmann. 1993. *HVAC controls and systems.* McGraw-Hill, New York.

McMillan, G.K. 1983. Tuning and control loop performance. Instrument Society of America, Research Triangle Park, NC.

Murrill, P.W. 1981. Fundamentals of process control theory. Instrument Society of America, Research Triangle Park, NC.

Nesler, C.G. and W.F. Stoecker. 1984. Selecting the proportional and integral constants in the direct digital control of discharge air temperature. *ASHRAE Transactions* 90(2).

Ogata, K. 1978. *System dynamics.* Prentice-Hall, Englewood Cliffs, NJ.

Treichler, W.W. 1985. Variable speed pumps for water chillers, water coils, and other heat transfer equipment. *ASHRAE Transactions* 91(1).

Walker, C.A. 1984. Application of direct digital control to a variable air volume system. *ASHRAE Transactions* 90(2).

Westphal, B. 1982. Human engineering: The man/system interface. *ASHRAE Transactions* 88(1):1435-39.

Wichman, P.E. 1984. Improved local loop control systems. *ASHRAE Transactions* 90(2).

Williams, T.J. 1984. The use of digital computers in process control. Instrument Society of America, Research Triangle Park, NC.

Yaeger, G.A. 1986. Flow charting and custom programming. *ASHRAE Transactions* 92(1).

Zell, B.P. 1985. Design and evaluation of variable speed pumping systems. *ASHRAE Transactions* 91(1).

CHAPTER 43

SOUND AND VIBRATION CONTROL

MECHANICAL equipment is one of the major sources of sound in a building. The primary considerations given to the selection and use of mechanical equipment in buildings have generally been only those directly related to the intended use of the equipment. However, with low mass building structures and variable-volume air distribution systems, the sound generated by mechanical equipment and its effects on the overall acoustical environment in a building must be considered. Thus, the selection of mechanical equipment and the design of equipment spaces should be undertaken not only with an emphasis on the intended uses of the equipment, but also with a desire to provide acceptable sound and vibration levels in occupied spaces of the building.

Air distribution duct systems, unless adequately designed, act as transmission paths for unwanted sound throughout a building. The direction of the airflow has little to do with the transmission of sound. Sound travels just as effectively upstream in a return air duct as it does downstream in a supply air duct.

Adequate noise and vibration control in a heating, ventilating, and air-conditioning (HVAC) system is not difficult to achieve during the design phase of the system, if noise and vibration control principles are understood. This chapter discusses these principles and provides data needed by HVAC system designers. The chapter is organized differently than previous ASHRAE Handbook chapters on sound and vibration control in that more information is provided on acoustic design guidelines and system design requirements. Most of the equations associated with sound and vibration control design in HVAC systems have been replaced with related tables and simpler design procedures. The equations can be found in Chapter 42 of the 1991 *ASHRAE Handbook—Applications*, in the Additions and Corrections section of the 1992 *ASHRAE Handbook—Systems and Equipment*, and in Chapter 11 of *HVAC Systems Duct Design* (SMACNA 1990). Complete technical discussions, along with detailed HVAC component and system design examples based on all the equations used to generate the tables and design procedures covered in this chapter, are contained in *Algorithms for HVAC Acoustics* (Reynolds and Bledsoe 1991) and *Sound and Vibration Design and Analysis* (Reynolds and Bevirt 1994a).

Chapter 7 of the 1993 *ASHRAE Handbook—Fundamentals* covers the fundamentals of sound and vibration in HVAC systems. *A Practical Guide to Noise and Vibration Control for HVAC Systems* (Schaffer 1991) provides specific guidelines (1) for acoustic design and related construction associated with HVAC systems, (2) for

trouble shooting sound and vibration problems in HVAC systems, and (3) for HVAC sound and vibration specifications. *Procedural Standards for Measurement and Assessment of Sound and Vibration* (Reynolds and Bevirt 1994b) discusses instrument requirements, instrument and measurement calibration procedures, measurement procedures, and specification and construction installation review procedures associated with sound and vibration measurements related to HVAC systems in buildings.

ACOUSTICAL DESIGN OF HVAC SYSTEMS

The solution to nearly every HVAC system noise and vibration control problem involves examining the sound sources, transmission paths, and receivers. For most HVAC systems, the sound sources are associated with the building mechanical and electrical equipment. As indicated in Figure 1, there are many possible sound and/or

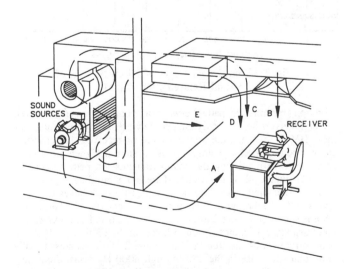

Path A: Structure-borne path through floor
Path B: Airborne path through supply air system
Path C: Duct breakout from supply air duct
Path D: Airborne path through return air system
Path E: Airborne path through mechanical equipment room wall

Fig. 1 Typical Paths in HVAC Systems

The preparation of this chapter is assigned to TC 2.6, Sound and Vibration Control.

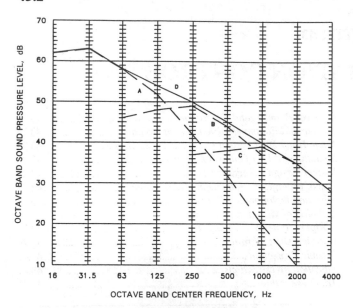

Curve A: Fan sound pressure levels
Curve B: VAV valve sound pressure levels
Curve C: Diffuser sound pressure levels
Curve D: Total sound pressure levels

**Fig. 2   Illustration of Well-Balanced HVAC Sound Spectrum
for Occupied Spaces**

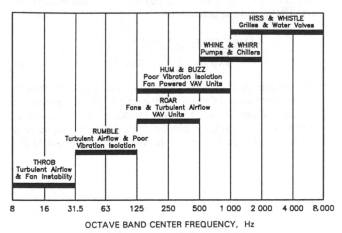

**Fig. 3   Frequency Ranges of Likely Sources of
Sound-Related Complaints**

vibration transmission paths between a sound source and receiver. Sound receivers are generally the people who occupy a building. For most HVAC systems, the sound source or receiver characteristics can not be modified or changed. Thus, system designers are usually constrained to modifying the sound transmission paths to achieve desired sound levels in the occupied areas of a building.

In an occupied space, fan noise generally contributes to sound levels in the 16 through 250 Hz octave frequency bands. This sound level is shown in Figure 2 as curve A. Valve noise from VAV equipment usually contributes to sound levels in the 63 through 1000 Hz octave frequency bands (curve B in Figure 2). Diffuser noise usually contributes to noise in the 250 through 8000 Hz octave frequency bands (curve C in Figure 2). The overall sound pressure levels L_p associated with all of these sound sources is shown as curve D in Figure 2. Curve D represents acceptable and desirable octave band sound pressure levels in many occupied spaces.

When an acceptable sound spectrum is not achieved in an occupied space, occupants may complain. Figure 3 shows the frequency

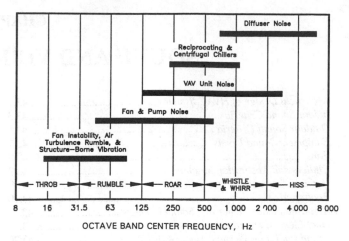

**Fig. 4   Frequencies at Which Different Types of Mechanical
Equipment Generally Control Sound Spectra**

ranges of the most likely causes of complaints associated with HVAC noise, and Figure 4 shows the frequencies at which different types of mechanical equipment generally control the sound spectra in a room.

The following factors should be considered when selecting fans and related mechanical equipment and when designing air distribution systems to minimize the sound transmitted from different components of the system to the occupied spaces that they serve:

1. Design the air distribution system to minimize flow resistance and turbulence. High flow resistance increases the required fan pressure, which results in more noise generated by the fan. Turbulence increases the flow noise generated by duct fittings and dampers in the air distribution system, especially at low frequencies.

2. Select a fan to operate as near as possible to its rated peak efficiency. Also, select a fan that generates the lowest possible noise, but still meets the required design conditions for which it is selected. Oversized or undersized fans that do not operate at or near rated peak efficiencies can cause substantially higher noise levels.

3. Design duct connections at both the fan inlet and outlet for uniform and straight airflow. Failure to do this can result in severe turbulence at the fan inlet and outlet and in flow separation at the fan blades. Both of these factors can significantly increase the noise generated by the fan.

4. Select duct silencers that do not significantly increase the required fan total static pressure. Duct silencers can significantly increase the required fan static pressure if improperly selected. Silencer airflow regenerated noise can be minimized by selecting silencers with static pressure losses of 87 Pa or less.

5. Do not place fan-powered mixing boxes associated with variable-volume air distribution systems over or near noise-sensitive areas.

6. When possible, locate elbows and duct branch takeoffs at least four to five duct diameters from each other to minimize flow-generated noise. For high-velocity systems, this distance may need to be increased to as much as 10 duct diameters in critical noise areas. The use of "egg crate" grids in the necks of short takeoffs that lead directly to grilles, registers, and diffusers is preferred to the use of volume extractors that protrude into the main duct airflow.

7. Near critical noise areas, it be may desirable to expand the duct cross-section area to keep the flow velocity in the duct as low as possible (7.6 m/s or less). However, do not exceed an included expansion angle of 15°. Expansion angles greater than 15° may result in flow separation, which produces rumble noise.

8. Use turning vanes in large 90° rectangular elbows and branch takeoffs. Turning vanes provide a smoother transition in which the air can change direction, thus reducing turbulence.

9. Place grilles, diffusers, and registers in occupied spaces as far as possible from elbows and branch takeoffs.

10. Minimize the use of volume dampers near grilles, diffusers, and registers in acoustically critical situations.

11. If mechanical equipment is located on upper floors or is roof mounted, all vibrating reciprocating and rotating equipment must be vibration isolated. Usually mechanical equipment located in the basement must be vibration isolated as well. When tenant space is located directly above, piping supported from the ceiling slab of a basement must also be vibration isolated.

12. Flexible connectors between rotating or reciprocating equipment and pipes and ducts that are connected to the equipment may be necessary.

13. Use spring and/or neoprene hangers to vibration isolate ducts and pipes within a minimum of the first 15 m of the vibration isolated equipment.

14. Use flexible conduit between rigid electrical conduit and reciprocating and rotating equipment.

15. Use barriers when noise associated with outdoor equipment will disturb adjacent properties if barriers are not used. In normal practice, barriers typically produce no more than 15 dB of sound attenuation in the mid-frequency range.

Table 1 lists several common sound sources associated with mechanical equipment noise. Anticipated sound transmission paths and recommended noise reduction methods are also listed in the table. Airborne and/or structure-borne sound can follow any or all of the transmission paths associated with a specified sound source. Schaffer (1991) has more information.

INDOOR SOUND CRITERIA

Room Criterion Curves

HVAC-related sound is often the major source of background sound in many indoor areas. The primary method used to determine the acceptability of background HVAC-related sound in unoccupied indoor areas uses the room criterion (RC) curves shown in Figure 5 (Blazier 1981). RC curves have the following advantages: (1) they account for the influence of both spectrum shape and level on the subjective assessment process, (2) they include data in the 16 Hz and 31.5 Hz octave frequency bands, and (3) they account for the ability of low-frequency acoustic energy to induce perceptible vibration in light building construction.

The RC curves in Figure 5 extend from the 16 Hz octave band through the 4000 Hz octave band. These are the general frequency limits for noise produced by HVAC systems. Technical data, based on recognized standards, for the relevant properties of mechanical and architectural building components are not available for frequencies as low as the 16 Hz octave band. In some cases technical data may extend to the 63 Hz octave band; however, in most cases, the data only extend to the 125 Hz octave band. When determining the RC level of a room based on calculated octave band sound pressure levels, the lower frequency limit of the analysis can only extend to the frequency for which valid data exist. When determining the RC level based on measured octave band sound pressure levels, the lower frequency limit of the analysis is determined by the capability of the instrument(s) used to make the sound measurements. Four factors should be considered when assessing the acceptability of HVAC system background noise: level, spectrum shape or balance, tonal content, and temporal fluctuations.

There are two parts to determining the RC noise rating associated with HVAC background noise. The first is the calculation of a number

Table 1   Sound Sources, Transmission Paths, and Recommended Noise Reduction Methods

Sound Source	Path No.
Circulating fans; grilles; registers; diffusers; unitary equipment in room	1
Induction coil and fan-powered VAV mixing units	1, 2
Unitary equipment located outside of room served; remotely located air-handling equipment, such as fans, blowers, dampers, duct fittings, and air washers	2, 3
Compressors, pumps, and other reciprocating and rotating equipment (excluding air-handling equipment)	4, 5, 6
Cooling towers; air-cooled condensers	4, 5, 6, 7
Exhaust fans; window air conditioners	7, 8
Sound transmission between rooms	9, 10

No.	Transmission Paths	Noise Reduction Methods
1	Direct sound radiated from sound source to ear Reflected sound from walls, ceiling, and floor	Direct sound can be controlled only by selecting quiet equipment. Reflected sound is controlled by adding sound absorption to the room and to equipment location.
2	Air- and structure-borne sound radiated from casings and through walls of ducts and plenums is transmitted through walls and ceiling into room	Design duct and fittings for low turbulence; locate high velocity ducts in noncritical areas; isolate ducts and sound plenums from structure with neoprene or spring hangers.
3	Airborne sound radiated through supply and return air ducts to diffusers in room and then to listener by Path 1	Select fans for minimum sound power; use ducts lined with sound-absorbing material; use duct silencers or sound plenums in supply and return air ducts.
4	Noise transmitted through equipment room walls and floors to adjacent rooms	Locate equipment rooms away from critical areas; use masonry blocks or concrete for equipment room walls and floor.
5	Vibration transmitted via building structure to adjacent walls and ceilings, from which it radiates as noise into room by Path 1	Mount all machines on properly designed vibration isolators; design mechanical equipment room for dynamic loads; balance rotating and reciprocating equipment.
6	Vibration transmission along pipes and duct walls	Isolate pipe and ducts from structure with neoprene or spring hangers; install flexible connectors between pipes, ducts, and vibrating machines.
7	Noise radiated to outside enters room windows	Locate equipment away from critical areas; use barriers and covers to interrupt noise paths; select quiet equipment.
8	Inside noise follows Path 1	Select quiet equipment.
9	Noise transmitted to an air diffuser in a room, into a duct, and out through an air diffuser in another room	Design and install duct attenuation to match transmission loss of wall between rooms.
10	Sound transmission through, over, and around room partition	Extend partition to ceiling slab and tightly seal all around; seal all pipe, conduit, duct, and other partition penetrations.

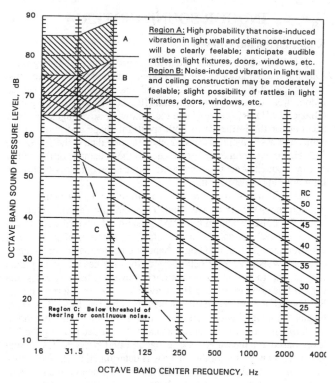

Fig. 5   Room Criterion Curves

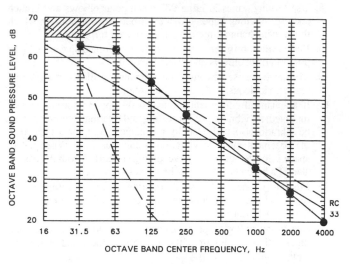

Fig. 6   RC Level for Example 1

that corresponds to the speech communication or masking properties of the noise. The second is designating the quality or character of the background noise. The procedure for determining the RC rating is

1. Calculate the arithmetic average of the octave band sound pressure levels in the 500, 1000, and 2000 Hz octave frequency bands. Round to the nearest integer. This is the RC level associated with the background noise.

2. Draw a line with a −5 dB/octave slope so that it passes through the calculated RC level at 1000 Hz. For example, if the calculated RC level is RC 32, the line will pass through a value of 32 dB at the 1000 Hz octave band.

3. Classify the subjective quality or character of the background noise as follows:

   Neutral. Noise that is classified as neutral has no particular identity with frequency. It is usually bland and unobtrusive. Background noise that is neutral usually has an octave band spectrum shape similar to the RC curves in Figure 5. If the octave band data do not exceed the RC curve by more than 5 dB at frequencies of 500 Hz and below and by more than 3 dB at frequencies of 1000 Hz and above, the background sound is neutral, and (N) is placed after the RC level.

   Rumble. Noise that has a rumble has an excess of low-frequency sound energy. If any of the octave band sound pressure levels below the 500 Hz octave band are more than 5 dB above the RC curve associated with the background noise in the room, the noise is judged to have a "rumbly" quality or character, and (R) is placed after the RC level.

   Hiss. Noise that has an excess of high-frequency sound energy will have a "hissy" quality. If any of the octave band sound pressure levels above the 500 Hz octave band are more than 3 dB above the RC curve, the noise is judged to have a hissy quality, and (H) is placed after the RC level.

   Acoustically Induced Perceptible Vibration. The cross-hatched region of the RC curves in Figure 5 indicates the sound pressure levels in the 16 to 63 Hz octave frequency bands at which perceptible vibration in the walls and ceiling of a room can occur. These sound levels are associated with rattles in cabinet doors, pictures,

ceiling fixtures, and other furnishings in contact with walls or ceilings. If the background sound levels fall in this region, (RV) is placed after the RC level.

The octave band spectrum of background sound should have a neutral character or quality. If the noise spectrum has a rumble, hiss, or tonal character, it will generally be objectionable.

Room criterion procedures can be used to determine the acceptability of sound in an area based on sound pressure levels measured in the area. However, care must be exercised in using room criterion procedures for determining the acceptability of sound in an area based on calculated sound pressure levels. A balanced sound spectrum in a room is comprised of sound from the fan and duct system and sound from airflow through the air diffusers in the room. Sound from both sources must be included in the calculations when using room criterion procedures to determine the acceptability of sound.

Example 1. The octave band sound pressure levels L_p of background noise in an office area are as follows:

	Octave Band Center Frequency, Hz							
	31.5	**63**	**125**	**250**	**500**	**1000**	**2000**	**4000**
L_p, dB	63	62	54	46	40	33	27	20

Determine the RC level and the corresponding character of the noise.
Solution:

The RC level is determined by obtaining the arithmetic average of the octave band sound pressure levels in the 500 Hz, 1000 Hz, and 2000 Hz octave bands.

$$RC = \left[\frac{40 + 33 + 27}{3}\right] = 33$$

The octave band sound pressure levels for the background noise are plotted in Figure 6. The RC 33 curve (level in 1000 Hz octave band is 33 dB) is shown. A dashed line 5 dB above the RC 33 curve for frequencies below 500 Hz and a dashed line 3 dB above the RC 33 curve for frequencies above 500 Hz are also shown in the figure. An examination of the figure indicates that at frequencies below the 250 Hz octave band, the octave band sound pressure levels of the background noise are 5 dB or more above the RC 33 curve. Thus, the background noise has a rumble character. The octave band sound pressure levels above 500 Hz are equal to or below the RC 33 curve, so there is no problem at these frequencies. The RC rating of the background noise is RC 33(R).

Criteria for Acceptable HVAC Sound Levels in Rooms

Sound associated with HVAC systems is usually considered part of the background sound in a building. Therefore, to be judged acceptable, it must neither noticeably mask sounds people want to hear nor be otherwise intrusive or annoying in character. In an office environment, the acceptable level of background sound is generally established by speech requirements, which vary widely depending on the space use. In contrast, the acceptable level of background sound for unamplified performance spaces, such as recital and concert halls, is governed by the need to avoid masking the faintest of sounds that are likely to occur in a typical performance.

Table 2 lists normally accepted HVAC background sound levels for a variety of space uses. The acceptable sound levels in Table 2 are specified in terms of room criterion, which should be used whenever the quality of space use dictates the need for a neutral, unobtrusive background sound. When applying the levels in Table 2 as a basis for design, sound from non-HVAC sources such as traffic and office equipment may be the lower limit for sound levels in a space.

Table 2   Design Guidelines for HVAC System Noise in Unoccupied Spaces

Space	RC(N) Level[a,b]
Private residences, apartments, condominiums	25-35
Hotels/Motels	
Individual rooms or suites	25-35
Meeting/banquet rooms	25-35
Halls, corridors, lobbies	35-45
Service/support areas	35-45
Office buildings	
Executive and private offices	25-35
Conference rooms	25-35
Teleconference rooms	25 (max)
Open plan offices	30-40
Circulation and public lobbies	40-45
Hospitals and clinics	
Private rooms	25-35
Wards	30-40
Operating rooms	25-35
Corridors	30-40
Public areas	30-40
Performing arts spaces	
Drama theaters	25 (max)
Concert and recital halls	c
Music teaching studios	25 (max)
Music practice rooms	35 (max)
Laboratories (with fume hoods)	
Testing/research, minimal speech communication	45-55
Research, extensive telephone use, speech communication	40-50
Group teaching	35-45
Churches, mosques, synagogues	25-35
With critical music programs	c
Schools	
Classrooms up to 70 m²	40 (max)
Classrooms over 70 m²	35 (max)
Lecture rooms for more than 50 (unamplified speech)	35 (max)
Libraries	30-40
Courtrooms	
Unamplified speech	25-35
Amplified speech	30-40
Indoor stadiums and gymnasiums	
School and college gymnasiums and natatoriums	40-50[d]
Large seating capacity spaces (with amplified speech)	45-55[d]

[a]The values and ranges are based on judgment and experience, not on quantitative evaluations of human reactions. They represent general limits of acceptability for typical building occupancies. Higher or lower values may be appropriate and should be based on a careful analysis of economics, space usage, and user needs. They are not intended to serve by themselves as a basis for a contractual requirement.

[b]When the quality of sound in the space is important, specify criteria in terms of RC(N). If the quality of the sound in the space is of secondary concern, the criteria may be specified in terms of NC levels.

[c]An experienced acoustical consultant should be retained for guidance on acoustically critical spaces (below RC 30) and for all performing arts spaces.

[d]Spectrum levels and sound quality are of lesser importance in these spaces than overall sound levels.

Noise criterion (NC) levels (see Chapter 7 of the 1993 *ASHRAE Handbook—Fundamentals*) may be substituted when the quality of space use is not as demanding and rumbly, hissy, or tonal characteristics in the background sound can be tolerated as long as it is not too loud. The RC levels specified in Table 2 address the following factors: human comfort, speech communication, and acoustically critical spaces. If these levels are to be used as a basis for contractual requirements, the following questions must be considered:

1. What sound levels are to be measured? (Specify required sound pressure levels in each octave frequency band.)
2. Where and how are the sound levels to be measured? (Specify space average over a defined area or specific points.)
3. What type(s) of instruments are to be used to make the sound measurements? (Specify ANSI Type 1 or Type 2 sound level meters with octave band filters.)
4. How are the instruments used for the sound measurements to be calibrated?
5. How are the results of the sound measurements to be interpreted?

Unless these five points are clearly stipulated, the specified sound criteria may be unenforceable.

OUTDOOR SOUND CRITERIA

Acceptable outdoor sound levels are generally specified by local noise ordinances. In the absence of a local noise ordinance, composite noise rating (CNR) procedures can be used to evaluate the acceptability of HVAC and other types of mechanical and electrical equipment noise (Stevens et al. 1955). The non-normalized composite noise rating is determined by plotting the octave band sound pressure levels associated with an intruding noise on the set of non-normalized CNR curves shown in Figure 7. The octave band sound pressure levels should be measured at several representative times at each location of interest in the community. The measurements should span a period long enough to give confidence that the average octave band sound pressure levels of the noise are truly representative. If the daytime and nighttime noise

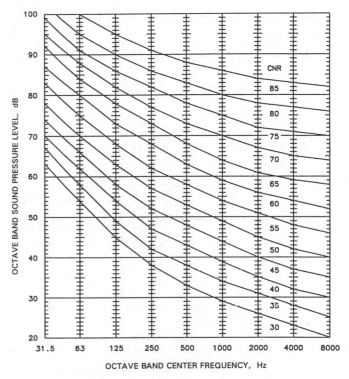

Fig. 7   Non-Normalized Composite Noise Rating (CNR$_{nn}$) Curves

signals are different in contents and levels, separate sound measurements should be taken for the two periods.

The non-normalized composite noise rating associated with a noise equals the highest penetration of any of the octave band sound pressure levels into the curves. If the highest penetration falls between two curves, the non-normalized CNR is the interpolated value between the two curves. The non-normalized CNR must then be normalized or corrected for the background noise conditions that exist in the absence of the intruding noise and for time-of-day and seasonal factors, noise intermittency, noise characteristics, and previous community exposure to similar noise. The correction for background noise can be accomplished in one of two ways.

If it is possible to measure the octave band sound pressure levels associated with the ambient or background noise in the absence of any intruding noise source, the levels should be measured and plotted on Figure 8. The zone into which the major portion of the octave band spectrum falls designates the correction to be applied for the background noise. The correction that should be used is associated with the curve that has a point of tangency closest to the octave band ambient sound pressure level curve. It is not necessary to interpolate between curves. Daytime ambient noise levels should be recorded for daytime intruding noise, and nighttime ambient levels should be used for nighttime intruding noise.

If it is not possible to measure the octave band ambient sound pressure levels, the background sound level corrections given in Table 3 can be used to estimate the correction for background or ambient sound levels. The corrections in Table 3 are based on the general type of community area and nearby traffic activity.

The normalized CNR, corrected for background noise level, is obtained by adding the number (keeping track of the sign in front of number) obtained from either Figure 8 or Table 3 to the non-normalized composite noise rating obtained from Figure 7. The final correction is associated with time-of-day and seasonal factors, noise intermittency, noise characteristics, and previous community exposure to similar noise. The correction factors are obtained from Table 4. The

total correction for these factors is the sum of the corrections associated with each individual factor.

The normalized CNR is calculated by taking the non-normalized CNR obtained from Figure 7 and adding to it the correction number for the background noise obtained from either Figure 8 or Table 3 and the total correction number obtained from Table 4. Once the normalized CNR has been calculated, the anticipated community reaction to the intruding noise is obtained from Figure 9.

The composite noise rating procedure is generally a reliable method of determining community reaction to outdoor noise from mechanical and electrical equipment. However, the procedure may not be reliable when dealing with equipment that generates strong, pure tones (e.g., high pressure blowers, diesel generators, gas turbines). An

Table 3   Background Noise Correction Numbers

Condition	Background Correction Number
Nighttime, rural; no nearby traffic of concern	+15
Daytime, rural; no nearby traffic of concern	+10
Nighttime, suburban; no nearby traffic of concern	+10
Daytime, suburban; no nearby traffic of concern	+5
Nighttime, urban; no nearby traffic of concern	+5
Daytime, urban; no nearby traffic of concern	0
Nighttime, business or commercial area	0
Daytime, business or commercial area	−5
Nighttime, industrial or manufacturing area	−5
Daytime, industrial or manufacturing area	−10
Within 90 m of intermittent light traffic	0
Within 90 m of continuous light traffic	−5
Within 90 m of continuous medium-density traffic	−10
Within 90 m of continuous heavy-density traffic	−15
90 to 300 m from intermittent light traffic	+5
90 to 300 m from continuous light traffic	0
90 to 300 m from continuous medium-density traffic	−5
90 to 300 m from continuous heavy-density traffic	−10
300 to 600 m from intermittent light traffic	+10
300 to 600 m from continuous light traffic	+5
300 to 600 m from continuous heavy-density traffic	−5
600 to 1200 m from intermittent light traffic	+15
600 to 1200 m from continuous light traffic	+10
600 to 1200 m from continuous medium-density traffic	+5
600 to 1200 m from continuous heavy-density traffic	0

Table 4   Correction Numbers

Correction for time-of-day and seasonal factors	
(for full-time operation, total correction is 0)	
Daytime only	−5
Nighttime (2200 to 0700 hrs)	0
Winter only	−5
Winter and summer	0
Correction for intermittency	
(ratio of source "on" time to reference time period)	
1.00 to 0.57	0
0.56 to 0.18	−5
0.17 to 0.06	−10
0.05 to 0.018	−15
0.017 to 0.0057	−20
0.0057 to 0.0018	−25
Correction for character of noise	
Noise is very low frequency (peak level at 1/1 octave center frequency of 125 Hz or lower)	+5
Noise contains tonal components	+5
Impulsive sound	+5
Correction for previous exposure and community attitude	
No prior exposure	+5
Some previous exposure but poor community relations	+5
Some previous exposure and good community relations	0
Considerable previous exposure and good community relations	−5

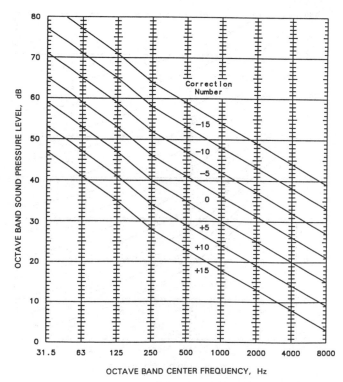

Fig. 8   Correction for Composite Noise Rating (CNR) Associated with Ambient Noise

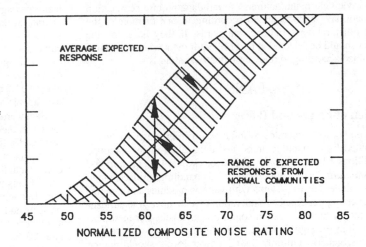

Fig. 9   Estimated Community Reaction to Intruding Noise Versus Normalized Composite Noise Rating (CNR$_n$)

acoustical consultant should be consulted when dealing with these sound sources.

Example 2. The octave band sound pressure levels associated with a cooling tower are listed below:

	Octave Band Center Frequency, Hz							
	63	125	250	500	1000	2000	4000	8000
L_p, dB	64	64	62	60	56	53	51	43

The cooling tower runs 24 hours a day. The location at which the sound pressure levels were measured is a business area. Assume there is previous exposure to similar noise and community relations are good. Determine the composite noise rating associated with the cooling tower noise, and make some statement relative to the anticipated community reaction to the noise.

Solution:

The non-normalized composite noise rating (CNR$_{nn}$) is obtained by plotting the above octave band sound pressure levels on Figure 7. The resulting plot is shown in Figure 10. An examination of the plot indicates that the CNR$_{nn}$ is CNR$_{nn}$ 58.

Because the cooling tower runs 24 hours a day, the normalized composite noise rating (CNR$_n$) for both daytime and nighttime use must be determined. The background noise correction numbers are obtained from Table 3. The numbers for a business area are

> Daytime: −5
> Nighttime: 0

The correction numbers from Table 4 are

> Time-of-day: 0
> Intermittency: 0
> Character of noise: +5
> Previous exposure: 0

Thus, the CNR$_n$ values are

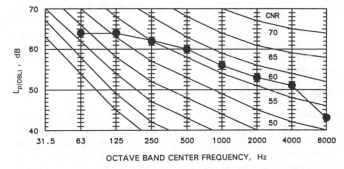

Fig. 10   CNR$_{nn}$ Value for Example 2

Daytime: $\text{CNR}_n = 58 - 5 + 0 + 0 + 5 + 0 = 58$
Nighttime: $\text{CNR}_n = 58 + 0 + 0 + 0 + 5 + 0 = 63$

An examination of Figure 9 indicates there will be no complaints during the day, but there will be some sporadic complaints during the night.

EQUIPMENT SOUND LEVELS

Reliable equipment sound data are important when conducting acoustic analyses of HVAC systems. These data are often available from equipment manufacturers in the form of sound pressure levels L_p at a specified distance from the equipment or equipment sound power levels L_w. To ensure the reliability of equipment sound data, care must be taken to ascertain that equipment manufacturers use the most current versions of appropriate industry standards to obtain data. With respect to sound measurements, the following standards apply (Addresses for these organizations are in Chapter 51.)

ANSI S12.31, Precision Methods for the Determination of Sound Power Levels of Broadband Noise Sources in Reverberation Rooms.

ANSI S12.32, Precision Methods for the Determination of Sound Power Levels of Discrete-Frequency and Narrow-Band Noise Sources in Reverberation Rooms.

ANSI S12.34, Engineering Methods for the Determination of Sound Power Levels of Noise Sources for Essentially Free-Field Conditions over a Reflecting Plane.

AMCA 300, Reverberant Room Method for Sound Testing of Fans.

ASHRAE 68/AMCA 330, Laboratory Method of Testing In-Duct Sound Power Measurement Procedure for Fans.

ASHRAE 70, Method of Testing for Rating the Performance of Air Outlets and Inlets.

ARI 260, Sound Rating of Ducted Air Moving and Conditioning Equipment.

ARI 270, Sound Rating of Outdoor Unitary Equipment.

ARI 275, Application of Sound Rated Outdoor Unitary Equipment.

ARI 300, Rating the Sound Level and Sound Transmission Loss of Packaged Terminal Equipment.

ARI 350, Sound Rating of Non-Ducted Indoor Air-Conditioning Equipment.

ARI 370, Sound Rating of Large Outdoor Refrigerating and Air-Conditioning Equipment.

ARI 530, Method of Measuring Sound and Vibration of Refrigerant Compressors.

ARI 575, Method of Measuring Machinery Sound Within an Equipment Space.

ARI 880, Air Terminals.

ARI 885, Procedure for Estimating Occupied Space Sound Levels in the Application of Air Terminals and Air Outlets.

ARI 890, Rating of Air Diffusers and Air Diffuser Assemblies.

ASTM E 477, Standard Test Method for Measuring Acoustical and Airflow Performance of Duct Liner Materials and Prefabricated Silencers.

When reviewing manufacturers' sound data, require certification that the data have been obtained according to one or more of the above or other relevant industry standards. If they have not, the equipment should be rejected in favor of equipment whose data have been obtained according to relevant industry standards.

FANS

Prediction of Fan Sound Power

The sound power generated by a fan performing at a given duty is best obtained from manufacturers' test data taken under approved test conditions (AMCA *Standard* 300 or ASHRAE *Standard* 68/AMCA *Standard* 330). Applications of air-handling products range from stand alone fans to systems with various modules and appurtenances. These appurtenances and modules can have a significant effect on the air handler system sound power levels. For these reasons, as well as the significant acoustic differences between various similar fans, generic fan-only sound power levels should not be relied on in selecting or comparing acoustic levels between air handler systems. Manufacturers are in the best position to supply information about the effect of acoustic appurtenance on their products. Proper testing to determine the sound power levels associated with configured air-handling units, and not just fans, should be made in accordance with ARI *Standard* 260.

Point of Fan Operation

The point of fan operation has a major effect on the acoustic output of fans. Fan selection at the calculated point of maximum efficiency is common practice to ensure minimum power consumption. However, actual system losses often exceed expectations, causing the fan to operate at a higher total static pressure than originally calculated. This higher loss can cause the fan to run in the stall region, which may produce low-frequency rumble. This problem can be avoided by selecting a fan to operate at a lower static pressure than the point of maximum efficiency (Figure 11).

Centrifugal Fans

Forward curved (FC) fans are commonly used in a wide range of standard air handler products. FC fans are relatively insensitive to inlet flow distortions, including modulation devices such as inlet guide vanes; and they have a wide useful operating range without a sharply defined stall onset. FC fans typically do not have a significant

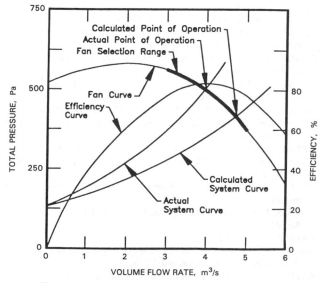

**Fig. 11 Suggested Selection of Calculated Fan
Point of Operation**

blade passage frequency amplitude. FC fans should be selected to operate in the region just to the right of peak efficiency for lowest sound levels. Modulation of inlet guide vanes along a VAV system curve does not cause an increase in sound at 63 Hz and above. The most distinguishing acoustic concern of FC fans is the prevalent occurrence of low-frequency rumble. FC fans are commonly thought to have 31.5 and 63 Hz rumble in applications. This rumble can occur over a wide operating range, but it is worse in the stall region.

Backward inclined (BI) fans and airfoil (AF) fans are often louder in the mid- and upper-frequency range than a given FC fan selected for the same duty. Some BI and AF fans have the additional disadvantage of being much more sensitive to inlet flow obstructions, including modulation devices such as inlet guide vanes. AF and BI fans typically have a significant tonal level at the blade passage frequency that increases in prominence with increasing static pressure and fan speed. The blade pass frequency tone is typically in a frequency range that is difficult to attenuate in an application. BI and AF fans have a narrower useful operating range and a more sharply defined stall region than FC fans. These type of fans must not be run in stall due to the severe increase in noise levels and fan vibration. Outside of the stall region, AF and BI fans are generally perceived to have lower sound amplitude at low frequency (below 100 Hz) than FC fans. Within the usable operating range, the sound level of AF and BI wheels is a strong function of blade tip speed and does not vary much with point of operation at a fixed speed.

Plug and Plenum Fans

Airfoil and backward inclined fan wheels can also be used in a plenum or plug fan configuration. Plenum fans have the advantage of flexibility of discharge location. Plenum fans have less turbulence and lower pressure fluctuations entering the discharge duct than housed AF and BI fans. Unhoused plug and plenum fans can literally be plugged into an air-handling system. This unhoused centrifugal fan may be installed horizontally or vertically in a plenum chamber. Air flows into the fan wheel through an inlet bell located in the chamber wall. The fan, which has no housing around the fan wheel, discharges directly into the chamber. The discharge pressurizes the plenum chamber and forces air through the attached ductwork. These fans can be provided with shaped inlet cones, nested inlet vanes, and extended shafts.

Plug fans may require slightly more power than housed centrifugal fans, but they can result in substantially lower sound power levels if the fan plenum is appropriately sized and acoustically tested.

Vaneaxial Fans

Axial fans are generally thought to have the lowest amplitudes at low frequency of any of the fan types, and for that reason they are often used in applications where the higher frequency noise can be managed with attenuation devices. Axial fans are the most sensitive fan type to inlet flow obstructions. Inlet flow obstructions, including inlet guide vanes for flow modulation, cause a significant increase in axial fan blade pass frequency tone prominence as well as increased low-frequency noise. In the useful operating range, the noise from axial fans is a strong function of blade tip speed and does not vary much with point of operation at a fixed speed. Most axial fans have a sharply defined stall line. In the stall region, sound levels at all frequencies increase significantly.

Propeller Fans

Sound from propeller fans generally has a low-frequency dominated spectrum shape, and the blade passage frequency is typically prominent and occurs in the low-frequency bands due to the small number of blades. Propeller fan blade passage frequency noise is very sensitive to inlet obstructions. In a bad, but fairly typical, application the noise of a propeller fan is often described as sounding like a helicopter. For some propeller fan designs, the shape of the fan inlet bell is a very important parameter that affects sound levels.

General Discussion of Fan Sound

To minimize the required air distribution system sound attenuation, proper selection and installation of the fan (or fans) is vitally important. The following factors should be considered:

- Design the air distribution system for minimum airflow resistance. High airflow resistance results in high fan static pressures, which in turn cause increased fan sound power levels.
- Examine the sound power levels of different types of fan designs for any given job. Different fans generate different sound levels and produce different sound spectra. Select a fan (or fans) that will generate the lowest possible sound power levels and corresponding sound spectra that are commensurate with other fan selection requirements.
- Fans with relatively few blades (fewer than 15) tend to generate fan tones that can dominate the fan sound spectrum. These tones occur at the blade passage frequency and its harmonics. The amplitude of these tones can be affected by resonances within the duct system, fan design, and inlet flow distortions.
- Design duct connections at both the fan inlet and outlet for uniform and straight airflow. Avoid unstable, turbulent, and swirling inlet airflow. Deviation from acceptable practice can severely degrade both the aerodynamic and acoustic performance of any fan and invalidate manufacturers' ratings or other performance predictions.

CHILLERS AND ABSORPTION MACHINES

Few data are available to accurately estimate the sound levels associated with chillers and absorption machines. This is partly associated with the size of these machines. Chillers and absorption machines are usually large, and as a result, the sound levels of these machines must often be measured in the equipment rooms in which they are installed. The acoustic properties of these rooms are sometimes difficult to accurately characterize, and other sound sources are often present.

Chillers have several components that generate and radiate sound. The primary sound sources are usually the compressor or compressors associated with the chiller and the drive units used to run the compressors. The drive units are usually electric motors and sometimes steam turbines. Large chillers may be made with assemblies that have two or more smaller compressors. Whenever possible, manufacturers' sound data, obtained in accordance with ARI *Standard* 575, should be used.

Sound from other pieces of equipment generally masks the sound emitted by an absorption machine. Sound generated by an absorption machine usually includes sound from one or more small pumps; sound from steam flow and steam valves may also be present.

VARIABLE AIR VOLUME SYSTEMS

General Design Considerations

Because VAV systems modulate air capacity, they can provide significant energy savings; but they can be the sources of fan noise problems that are very difficult to mitigate. To avoid these potential problems, the designer should take great care in the design of the ductwork and static pressure control systems, as well as in the selection of the fan/air-handling unit (AHU) and its air modulation device.

As in other aspects of HVAC system design, the duct system should be designed for the lowest feasible static pressure loss, especially in ductwork closest to the fan/AHU. High air velocities and convoluted duct routing can cause excessive pressure drop, as

well fan instabilities that are responsible for excessive noise and/or fan stall.

Many VAV system noise complaints have been traced to problems with the control system. While most of the problems are associated with improper system installation, many are caused by poor system design. The designer should specify high-quality fans/AHUs that will operate in their optimum ranges and not at the edge of their operation ranges, where low system tolerances can lead to inaccurate flow capacity control. Also, the in-duct static pressure sensors should be placed in duct sections having the lowest possible air turbulence (i.e., at least three equivalent duct diameters from any elbow, takeoff, transition, offset, damper, etc.).

VAV system noise problems have been traced to improper air balancing. For example, air balance contractors commonly balance an air distribution system by setting all damper positions without considering the possibility of reducing the fan speed. The end result is a duct system in which no damper is completely open and the fan is delivering air at a higher static pressure than would otherwise be necessary. If the duct system is balanced with at least one balancing damper wide open, the fan speed could be reduced, with a corresponding reduction in fan noise. Lower sound levels occur if most balancing dampers are wide open or eliminated.

Fan Selection

Fans for constant-volume systems should operate near maximum efficiency at the fan design airflow rate. However, VAV systems must operate efficiently and stably throughout their range of modulation. For example, a fan selected for peak efficiency at full output may aerodynamically stall at an operating point of 50% of full output, resulting in significantly increased low-frequency noise. Similarly, a fan selected to operate at the 50% output point may be very inefficient at full output, resulting in substantially increased fan noise at all frequencies. In general, a fan selected for a VAV system should be selected for a peak efficiency at an operating point of around 70 to 80% of the maximum required system capacity. This usually means selecting a fan that is one size smaller than that required for a peak efficiency at 100% of maximum required system capacity (Figure 12). When the smaller fan is operated at higher capacities, it will produce up to 5 dB more noise. This occasional increase in sound level is usually more tolerable than the stall-related sound problems that can occur with a larger fan operating at less than 100% of design capacity most of the time.

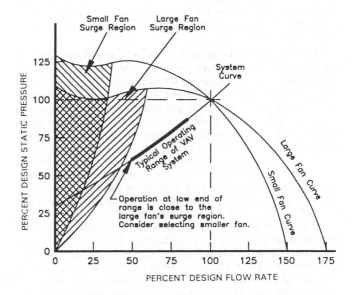

Fig. 12 Basis for Fan Selection in VAV Systems

Air Modulation Devices

The control method selected to vary the air capacity of a VAV system is important. Variable capacity control methods can be divided into three general categories: (1) variable inlet vanes (sometimes called inlet guide vanes) or discharge dampers, which yield a new fan system curve at each vane or damper setting; (2) adjustable pitch fan blades (usually used on inline axial fans), which adjust the blade angle for optimum efficiency at varying capacity requirements; and (3) variable-speed motor drives, which vary speed by modulation of the power line frequency (0 to 60 Hz) or by mechanical means such as gears or continuous belt adjustment. While inlet- vane and discharge-damper volume controls can add noise to a fan system at reduced capacities, variable-speed motor drives and variable pitch fan blade systems are quieter at reduced air output than at full output.

Variable Inlet Vanes. Variable inlet vanes vary airflow capacity by closing off the inlet air to a fan wheel. This type of air modulation varies the total air volume and pressure at the fan while the fan speed remains constant. Pressure and air volume reductions at the fan reduce duct system noise by reducing air velocities and pressures in the ductwork. However, an associated increase in fan noise is caused by the obstructing inlet vanes, which increase turbulence and flow distortions.

Fan manufacturers' test data have shown that as nested inlet vanes close on airfoil type centrifugal fans (vanes mounted inside the fan inlet), the sound level at the blade passing frequency of the fan increases by 2 to 8 dB, depending on the percent of total air volume restricted. For external mounted inlet vanes the increase is on the order of 2 to 3 dB. Forward curved fan wheels with inlet vanes increase noise levels 1 to 2 dB less than airfoil fan wheels. Inline axial fans with inlet vanes generate increased noise levels of 2 to 8 dB in the low-frequency octave bands for a 25 to 50% closed vane position.

Discharge Damper Capacity Control. Discharge dampers are typically located immediately downstream of the supply air fan and reduce airflow and increase pressure drop across the fan while the fan speed remains constant. Because of the air turbulence and flow distortions created by the high pressure drop across discharge dampers, duct rumble often occurs near the damper location. If the dampers are throttled to a very low flow, a stall condition can occur at the fan, also resulting in an increase in low-frequency noise.

Variable Pitch Fan Blades for Capacity Control. In order to reduce the overall airflow through the fan, variable pitch fan blade controls vary the fan blade angle. This type of capacity control system is used predominantly in axial type fans. As air volume and pressure are reduced at the fan, the corresponding noise reduction is usually 2 to 5 dB in the 125 through 4000 Hz octave bands for an 80% to 40% air volume reduction.

Electronic Variable-Speed Controlled Fans. Three types of electronic variable-speed control units are used with fans: (1) current source inverter, (2) voltage source inverter, and (3) pulse-width modulation (PWM). The current source inverter and third generation PWM control units are usually the quietest of the three systems, while the voltage source inverter control unit is usually the noisiest. In all three systems, the matching of motors to control units and the quality of the motor windings determine the noise output of the motor. The motor typically emits a pure tone whose amplitude depends on the smoothness of the waveform from the line current. The frequency of the motor tone depends on the motor type, windings, and speed. Both the inverter control units and motors should be enclosed in areas, such as mechanical rooms or electrical rooms, where the noise impact on surrounding rooms is minimal. The primary acoustic advantage of variable-speed controlled fans is the reduction of fan speed which translates into reduced noise (dB reduction equals $50 \times \log$ [higher speed/lower speed]). Because this speed reduction generally follows the fan system curve, a fan selected at optimum efficiency initially (lowest noise) will not lose that efficiency as the speed is reduced.

The use of variable-speed controllers requires observing the following guidelines:

- Select fan vibration isolators on the basis of the lowest practical speed of the fan. For example, the lowest speed might be 600 rpm for a 1000 rpm fan in a typical commercial system.
- Select a controller with a "critical frequency jump band." This feature allows a user to program the controller to avoid certain fan or motor speed settings that might excite vibration isolation system or building structure resonance frequencies.
- When selecting a fan that will be controlled by a variable-speed motor controller, check the intersection of the fan's various speed curves with the duct system curve, keeping in mind that the system curve does not go to zero static pressure at no flow. Instead, the system curve is asymptotic at the static pressure control setpoint, typically 250 to 370 Pa. An improperly selected fan may be forced to operate in its stall range at slower fan speeds.

Terminal Units

Fans and pressure reducing valves in VAV units should have manufacturer published sound data that indicate the sound power levels (1) that are discharged from the low pressure end of the unit and (2) that radiate from the exterior shell of the unit. These sound power levels vary as a function of valve position and fan point of operation. Sound data for VAV units should be obtained according to ARI *Standard* 880.

If the VAV unit is located away from critical areas (e.g., above a storeroom or corridor), the sound radiated from the shell of the unit may be of no concern. If, however, the unit is located above a critical space and separated from the space by a suspended acoustical ceiling that has little or no sound transmission loss at low frequencies, the sound radiated from the shell may exceed the noise criterion for the room below. In this case, the unit may need to be moved to a noncritical area or enclosed in a construction having a high transmission loss. In general, fan-powered VAV units should not be placed above or near any room with a required rating of less than RC 40(N).

Systems that permit the use of VAV units that will not completely shut off should be considered. If too many units in a system shut off simultaneously, excessive duct system pressure at low flow can occur. This condition can sometimes cause a fan go into a stall condition, resulting in accompanying roar, rumble, and surge. Using minimum airflow instead of shutoff VAV units will help prevent this condition from occurring.

ROOFTOP AIR HANDLERS

Rooftop air handlers have unique noise control requirements because these units are often integrated into low-density roof construction. Large roof openings are often required for supply and return air duct connections. These ducts run directly from noise-generating rooftop air handlers to the building interior. Generally, there is insufficient space or distance between the roof-mounted equipment and the closest occupied spaces below the roof to apply adequate sound control treatments. Rooftop units should be located above spaces that are not acoustically sensitive and should be placed as far as possible from the nearest occupied space. This measure can reduce the amount of sound control treatment necessary to achieve an acoustically acceptable installation.

Four common sound transmission paths are associated with rooftop air handlers: (1) airborne through the bottom of the rooftop unit to spaces below, (2) structure-borne from vibrating equipment in the rooftop unit to the building structure, (3) duct-borne through the supply air duct from the air handler, and (4) duct-borne through the return air duct to the air handler (Figure 13). Airborne paths are associated with casing-radiated sound that passes through the air handler enclosure and roof structure to the spaces below. Airborne

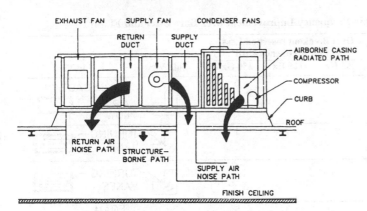

Fig. 13 Sound Paths for Typical Rooftop Installations

sound can be either air handler noise or sound from other components in the rooftop unit.

When a rooftop unit is placed over openings through which the supply and return air ducts pass, the openings should be limited to two openings sized to accommodate only the supply and return air ducts. These openings should be properly sealed after the installation of the ducts. If a large single opening exists under the rooftop unit, it should be structurally sealed with one or more layers of gypsum board or other similar material around the supply and return air ducts. Airborne sound transmission to spaces below a rooftop unit can be greatly reduced by placing a rooftop unit on a structural support extending above the roof structure and running the supply and return air ducts horizontally along the roof for several duct diameters before the ducts turn to penetrate the roof. The roof deck/ceiling system below the unit can be constructed to adequately attenuate the sound radiated from the bottom of the unit.

Structure-borne sound and vibration from vibrating equipment in a rooftop unit can be minimized by using proper vibration isolation. Special curb mounting bases are available to support and provide vibration isolation for rooftop units. For roofs constructed with open web joists, thin long span slabs, wooden construction, and any unusually light construction, all equipment with a mass of more than 140 kg should be evaluated to determine the additional deflection of the structure caused by the equipment. Isolator deflection should be a minimum of 15 times the additional deflection. If the required spring isolator deflection exceeds commercially available products, stiffen the supporting structure or change the equipment location.

The transmission of duct-borne sound through the supply air duct consists of two components: sound transmitted from the air handler through the supply air duct system to occupied areas and sound transmitted via duct breakout through a section or sections of the supply air duct close to the air handler to occupied areas. The transmission of airborne sound through supply air ducts can be minimized by using the design procedures discussed in the section on Acoustical Design of HVAC Systems.

Experience has indicated that sound transmission below 250 Hz via duct breakout is a major acoustical limitation for many rooftop installations. Excessive low-frequency noise associated with fan noise and air turbulence in the region of the discharge section of the fan and the first duct elbow results in duct rumble, which is difficult to attenuate. This problem is often made worse by the presence of a duct with a high aspect ratio at the discharge section of the fan. Rectangular ducts with duct lagging are often ineffective in reducing duct breakout noise.

Duct breakout can be controlled by using either a single- or dual-wall circular duct with a radius elbow coming off the discharge section of the fan. If space does not allow for the use of a single duct, the duct can be split into several parallel circular ducts. Another method that is effective is the use of an acoustic plenum chamber constructed of a minimum 50-mm thick dual-wall plenum panel, which is filled

with fiberglass and which has a perforated inner liner, at the discharge section of the fan. Either circular or rectangular ducts can be taken off the plenum as necessary for the rest of the supply air distribution system. Table 5 illustrates 12 possible rooftop discharge duct configurations with their associated low-frequency noise reduction potential (Harold 1986, Beatty 1987, Harold 1991).

The transmission of duct-borne sound through the return air duct of a rooftop unit is often a problem because generally only one short return air duct section runs from the plenum space above a ceiling to the return air section of the air handler. This short duct does not adequately attenuate the sound between the fan inlet and the spaces below the air handler. The sound attenuation through the return air duct system can be improved by adding at least one (more if possible) branch division where the return air duct splits into two sections that extend several duct diameters before they terminate into the plenum space above the ceiling. The inside surfaces of all the return air ducts should be lined with minimum 25-mm thick duct liner. If conditions permit, duct silencers in the duct branches or an acoustic plenum chamber at the air handler inlet section will give better sound conditions.

AERODYNAMICALLY GENERATED SOUND IN DUCTS

Although fans are a major source of sound in HVAC systems, they are not the only sound source. Aerodynamic sound is generated at duct elbows, dampers, branch takeoffs, air modulation units, sound attenuators, and other duct elements. The sound power levels in each octave frequency band depend on the duct element geometry and the turbulence and velocity of the airflow near the duct element. Duct-related aerodynamic noise problems can be avoided by

- Sizing ductwork or duct configurations so that air velocities are low (See Tables 6 and 7)
- Avoiding abrupt changes in duct cross-sectional area
- Providing for smooth transitions at duct branches, takeoffs, and bends
- Attenuating sound generated at duct fittings with sufficient sound attenuation elements between a fitting and corresponding air-terminal device

Duct Velocities

The amplitude of aerodynamically generated sound in ducts is generally proportional to between the fifth and sixth power of the air velocity in the vicinity of a duct fitting. Tables 6 and 7 give guidelines for recommended velocities in duct sections and duct outlets that are necessary to avoid problems associated with aerodynamically generated sound in ducts (Egan 1988, Schaffer 1991).

Dampers

Dampers are used to balance and control the airflow in a duct system and from duct terminal devices. Depending on its location relative to a duct terminal device, a damper can generate sound that is radiated as unwanted noise into an occupied area of a building.

Volume dampers should not be placed close to an air outlet. For good design, a volume damper should be no closer than 1.5 m from an air outlet. When a volume control damper is installed close to an air outlet to achieve system balance, the acoustic performance of the air outlet must be based not only on the air volume handled, but also on the magnitude of the pressure drop across the damper. The sound level to be added to the diffuser sound rating is proportional to the pressure ratio (PR) of the throttled pressure drop across the damper divided by the minimum pressure drop across the damper. Table 8 shows the effect of damper location on the decibels to be added to the diffuser sound rating.

Balancing dampers, equalizers, and so forth should not be placed directly upstream of air devices or open-ended ducts in

Table 5 Duct Breakout Insertion Loss—Potential Low-Frequency Improvement over Bare Duct and Elbow

Discharge Duct Configuration, 3660 mm of Horizontal Supply Duct	Duct Breakout Insertion Loss at Low Frequencies, dB			Side View	End View
	63 Hz	125 Hz	250 Hz		
Rectangular duct: no turning vanes (reference)	0	0	0		22 GAGE, AIR FLOW
Rectangular duct: one-dimensional turning vanes	0	1	1		TURNING VANES
Rectangular duct: two-dimensional turning vanes	0	1	1		TURNING VANES
Rectangular duct: wrapped with foam insulation and two layers of lead	4	3	5	SEE END VIEW	FOAM INSUL. W/ 2 LAYERS LEAD
Rectangular duct: wrapped with glass fiber and one layer 16-mm gypsum board	4	7	6	SEE END VIEW	GLASS FIBER PRESSED FLAT AGAINST DUCT.
Rectangular duct: wrapped with glass fiber and two layers 16-mm gypsum board	7	9	9	SEE END VIEW	GYPSUM BOARD SCREWED TIGHT
Rectangular plenum drop (12 ga.): three parallel rectangular supply ducts (22 ga.)	1	2	4	12 GAGE	22 GAGE
Rectangular plenum drop (12 ga.): one round supply duct (18 ga.)	8	10	6	12 GAGE	18 GAGE
Rectangular plenum drop (12 ga.): three parallel round supply ducts (24 ga.)	11	14	8	12 GAGE	24 GAGE
Rectangular (14 ga.) to multiple drop: round mitered elbows with turning vanes, three parallel round supply ducts (24 ga.)	18	12	13	24 GAGE	14 GAGE
Rectangular (14 ga.) to multiple drop: round mitered elbows with turning vanes, three parallel round lined double-wall, 560-mm OD supply ducts (24 ga.)	18	13	16	24 GAGE	14 GAGE
Round drop: radius elbow (14 ga.), single 940-mm diameter supply duct	15	17	10	14 GAGE, 18 GAGE	

acoustically critical spaces such as concert halls. They should be located 5 to 10 duct diameters from the opening, and acoustically lined duct should be placed between the damper and duct termination device.

Linear diffusers are often installed in distribution plenums so that the damper may be installed at the plenum entrance. The further a damper is installed from the outlet, the lower the resultant sound level will be (Table 8).

Air Devices

Room air terminal devices such as grilles, registers, diffusers, air-handling light fixtures, and air-handling suspension bars are always rated for noise generation. Manufacturers' test data should be obtained in accordance with ASHRAE *Standard* 70 and/or ARI *Standard* 890.

The room duct termination device should be selected to meet the noise criterion required or specified for the room, bearing in mind that the manufacturer's sound power rating is obtained with a uniform velocity distribution throughout the diffuser neck or grille collar. If a duct turn precedes the entrance to the diffuser or if a balancing damper is installed immediately before the diffuser, the airflow will be turbulent and the noise generated by the device will be substantially higher than the manufacturer's published data. This turbulence can be substantially reduced by placing an equalizer grid in the neck of the diffuser. The equalizer grid provides a uniform velocity gradient within the neck of the diffuser, and the sound power will be close to that listed in the manufacturer's catalog. If the equalizer grid is omitted, the sound power levels associated with the diffuser can increase by as much as 12 dB.

Table 6 Maximum Recommended Duct Airflow Velocities Needed to Achieve Specified Acoustic Design Criteria

Main Duct Location	Design RC(N)	Maximum Airflow Velocity, m/s	
		Rectangular Duct	Circular Duct
In shaft or above drywall ceiling	45	17.8	25.4
	35	12.7	17.8
	25	8.6	12.7
Above suspended acoustic ceiling	45	12.7	22.9
	35	8.9	15.2
	25	6.1	10.2
Duct located within occupied space	45	10.2	19.8
	35	7.4	13.2
	25	4.8	8.6

Notes:
1. Branch ducts should have airflow velocities of about 80% of the values listed.
2. Velocities in final runouts to outlets should be 50% of the values or less.
3. The presence of elbows and other fittings can increase airflow noise substantially, depending on the type of elbow or fitting. Thus, duct airflow velocities should be reduced accordingly.

Table 7 Maximum Recommended "Free" Supply Outlet and Return Air Opening Velocities Needed to Achieve Specified Acoustic Design Criteria

Type of Opening	Design RC(N)	"Free" Opening Airflow Velocity, m/s
Supply air outlet	45	3.2
	40	2.8
	35	2.5
	30	2.2
	25	1.8
Return air opening	45	3.8
	40	3.4
	35	3.0
	30	2.5
	25	2.2

Note: The presence of diffusers or grilles can increase sound levels a little or a lot, depending on how many diffusers or grilles are installed and on their design, construction, installation, etc. Thus, allowable outlet or opening airflow velocities should be reduced accordingly.

Table 8 Decibels to Be Added to Diffuser Sound Rating to Allow for Throttling of Volume Damper

Location of Volume Damper	Damper Pressure Ratio					
	1.5	2	2.5	3	4	6
	dB to Be Added to Diffuser Sound Rating					
In neck of linear diffuser	5	9	12	15	18	24
In inlet of plenum of linear diffuser	2	3	4	5	6	9
In supply duct at least 1.5 m from inlet plenum of linear diffuser	0	0	0	2	3	5

Table 9 Normalized Diffuser Sound Pressure Levels

$f_p/32$	$f_p/16$	$f_p/8$	$f_p/4$	$f_p/2$	f_p	$2f_p$	$4f_p$	$8f_p$	$16f_p$	$32f_p$
−28	−17	−10	−4	−1	0	−1	−5	−11	−19	−29

A flexible duct connection between the diffuser and the air supply duct or VAV unit provides a convenient means of aligning the diffuser with respect to the ceiling grid. A misalignment in this connection that exceeds one-fourth of the diffuser diameter over a length equal to two times the diffuser diameter can cause a significant increase in the diffuser sound power levels relative to the levels provided by the manufacturer. If the diffuser offset is less than one-eighth the length of the connection diameter, the sound power levels will not increase appreciably. If the offset is equal to or greater than the diffuser diameter over a connection length equal to two times the

diffuser diameter, the sound power levels associated with the diffuser can increase as much as 12 dB.

Manufacturers usually supply sound data in terms of NC levels and the pressure drops across their diffusers at specified airflows. To estimate the corresponding sound pressure levels associated with a specific diffuser, perform the following steps (Beranek 1971, Reynolds and Bledsoe 1991, Reynolds and Bevirt 1994a, SMACNA 1990):

1. First determine the mean air velocity in the duct prior to the diffuser.

$$u = Q/S \qquad (1)$$

where

u = mean airflow velocity in the duct prior to the diffuser, m/s
Q = volume airflow, m³/s
S = duct cross-sectional area prior to the diffuser, m²

2. The shape of the octave band sound spectrum for a typical generic diffuser is similar to that shown in Figure 14. The frequency f_p at which the curve in Figure 14 is a maximum can be approximated by

$$f_p = 160u \qquad (2)$$

The octave band sound pressure levels in Figure 14 normalized relative to f_p are given in Table 9. Calculate f_p and determine which octave frequency band contains f_p.

3. Using the sound pressure level values in Table 9 normalized to f_p, specify the normalized octave band sound pressure levels from 63 to 8000 Hz.

4. Position the normalized octave band sound pressure level curve associated with the diffuser on a set of NC curves to where it is tangent to the NC curve that corresponds to the NC level specified by the manufacturer for the design volume air flow through the diffuser and the corresponding pressure drop across the diffuser.

5. Note the octave sound pressure level of the NC curve for the octave frequency band at which the normalized diffuser sound pressure level curve is tangent to the NC curve. Add this octave band sound pressure level to the absolute value of the corresponding octave band sound pressure level for the normalized diffuser sound pressure level curve to obtain the overall sound pressure level value.

6. Add the overall sound pressure level value to the normalized octave band sound pressure levels obtained in Step 3 to obtained the corresponding sound pressure levels associated with the diffuser.

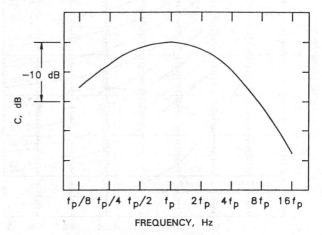

Fig. 14 Generalized Octave Band Spectrum Shape Associated with Diffuser Noise

Manufacturers normally subtract a 10 dB room correction from measured octave band sound power levels of a diffuser to determine the octave band sound pressure levels from which a corresponding NC level is obtained for a specific point of operation for the diffuser. Thus, if it is desired to estimate the octave band sound power levels associated with a specific diffuser, add 10 dB to the octave band sound pressure levels obtained in Step 6 above.

Example 3. The volume flow through a 300-mm by 400-mm duct, which is ahead of a rectangular diffuser, is $Q = 0.55$ m³/s. The specified NC level for the diffuser is NC 32. Determine the octave band sound pressure levels and the corresponding octave band sound power levels associated with the diffuser.

Solution:

The cross-sectional area S and flow velocity u are

$$S = 0.300 \times 0.400 = 0.12 \text{ m}^2$$
$$u = 0.55/0.12 = 4.58 \text{ m/s}$$

The frequency f_p is

$$f_p = 160 \times 4.58 = 733 \text{ Hz}$$

733 Hz is in the 1000 Hz octave frequency band. Thus, the normalized sound pressure levels are

	Octave Band Center Frequency, Hz							
	63	125	250	500	1000	2000	4000	8000
$L_{p(normalized)}$, dB	−17	−10	−4	−1	0	−1	−5	−11

Figure 15 shows that the normalized curve for this example is tangent to the NC 32 curve in the 2000 Hz octave frequency band. The value of the NC 32 curve in the 2000 Hz octave frequency band is 31 dB. Thus, the overall sound pressure level is

$$L_{p(overall)} = 31 + |{-1}| = 32$$

The resulting values for octave band sound pressure levels L_p and corresponding sound power levels L_w are

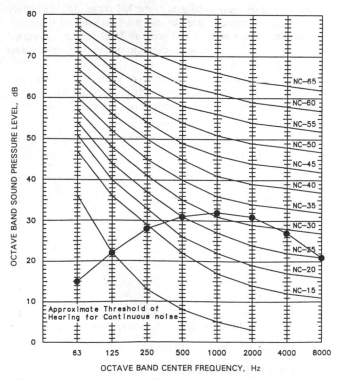

Fig. 15 Plot of Diffuser Sound Pressure Levels for Example 3

	Octave Band Center Frequency, Hz							
	63	125	250	500	1000	2000	4000	8000
$L_{p(normalized)}$, dB	−17	−10	−4	−1	0	−1	−5	−11
$L_{p(overall)}$, dB	32	32	32	32	32	32	32	32
L_p, dB	15	22	28	31	32	31	27	21
Room Corr., dB	10	10	10	10	10	10	10	10
L_w, dB	25	32	38	41	42	41	37	31

DUCT ELEMENT SOUND ATTENUATION

The duct elements covered in this section include sound plenums, unlined rectangular ducts, acoustically lined rectangular ducts, unlined circular ducts, acoustically lined circular ducts, elbows, acoustically lined circular radiused elbows, duct silencers, duct branch power division, duct end reflection loss, and terminal volume regulation units. Simplified tabular procedures for obtaining the sound attenuation associated with these elements are presented.

Plenum Chambers

Plenum chambers are often used to smooth out turbulent airflow associated with air as it leaves the outlet section of a fan and before it enters the ducted air distribution system of a building. The plenum chamber is usually placed between the discharge section of a fan and the main duct of the air distribution system. These chambers are usually lined with acoustically absorbent material to reduce fan and other types of noise. Plenum chambers are usually large rectangular enclosures with an inlet and one or more outlet sections. The transmission loss TL associated with a plenum chamber can be expressed as follows (Beranek 1971, Reynolds and Bledsoe 1991, Reynolds and Bevirt 1994a, SMACNA 1990, Wells 1958):

$$TL = -10 \log \left\{ S_{out} \left(\frac{Q \cos \theta}{4 \pi r^2} + \frac{1 - \alpha_A}{S \alpha_A} \right) \right\} \quad (3)$$

where (refer to Figure 16)

- S_{out} = area of outlet section of plenum, m²
- S = total inside surface area of plenum minus inlet and outlet areas, m²
- r = distance between centers of inlet and outlet sections of plenum, m
- Q = directivity factor, which may be taken as 4
- α_A = average absorption coefficient of plenum lining
- θ = angle of vector representing r to long axis l of duct [see Equation (5)]

The average absorption coefficient α_A of plenum lining is given by

$$\alpha_A = \frac{S_1 \alpha_1 + S_2 \alpha_2}{S} \quad (4)$$

where

- α_1 = sound absorption coefficient of any bare or unlined inside plenum surfaces
- S_1 = surface area of any bare or unlined inside plenum surfaces, m²
- α_2 = sound absorption coefficient of acoustically lined inside plenum surfaces
- S_2 = surface area of acoustically lined inside plenum surfaces, m²

In many situations, 100% of the inside surfaces of a plenum chamber are lined with a sound-absorbing material. For these situations, $\alpha_A = \alpha_2$. Table 10 gives the values of sound absorption coefficients for selected common plenum materials.

The value of $\cos \theta$ is obtained from

$$\cos \theta = \frac{l}{(l^2 + r_v^2 + r_h^2)} \quad (5)$$

Table 10 Sound Absorption Coefficients of Selected Plenum Materials

	Octave Band Center Frequency, Hz						
	63	125	250	500	1000	2000	4000
Non-Sound-Absorbing Materials							
Concrete	0.01	0.01	0.01	0.02	0.02	0.02	0.03
Bare sheet metal	0.04	0.04	0.04	0.05	0.05	0.05	0.07
Sound-Absorbing Materials (Fiberglass Insulation Board)							
25-mm, 48 kg/m^3	0.05	0.11	0.28	0.68	0.90	0.93	0.96
50-mm, 48 kg/m^3	0.10	0.17	0.86	1.00	1.00	1.00	1.00
75-mm, 48 kg/m^3	0.30	0.53	1.00	1.00	1.00	1.00	1.00
100-mm, 48 kg/m^3	0.50	0.84	1.00	1.00	1.00	1.00	0.97

Note: The 63 Hz values are estimated from higher frequency values.

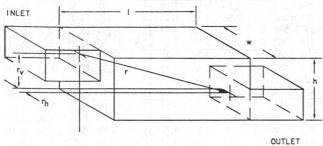

Fig. 16 Schematic of Plenum Chamber

where (refer to Figure 16)

l = length of plenum, m
r_v = vertical offset between axes of plenum inlet and outlet, m
r_h = horizontal offset between axes of plenum inlet and outlet, m

Equation (3) treats a plenum as if it is a large enclosure. Thus, Equation (3) is valid only for the case where the wavelength of sound is small compared to the characteristic dimensions of the plenum. For frequencies that correspond to plane wave propagation in the duct, the results predicted by Equation (3) are usually not valid. Plane wave propagation in a duct exists at frequencies as follows:

$$f_{co} = c_o/2a \qquad (6)$$

or

$$f_{co} = 0.586 c_o/d \qquad (7)$$

where

f_{co} = cutoff frequency, Hz
c_o = speed of sound in air, m/s
a = larger cross-sectional dimension of rectangular duct, m
d = diameter of circular duct, m

The cutoff frequency f_{co} is the frequency above which plane waves no longer propagate in a duct. At these higher frequencies the waves that propagate in the duct are referred to as cross or spinning modes. At frequencies below f_{co}, the actual attenuation usually exceeds the values given by Equation (3) by 5 to 10 dB. Equation (3) usually applies at frequencies of 1000 Hz and higher.

Example 4. A plenum chamber is 1.8 m high, 1.2 m wide, and 1.8 m long. The configuration of the plenum is similar to that shown in Figure 16. The inlet and outlet are each 0.9 m wide by 0.6 m high. The horizontal distance between centers of the plenum inlet and outlet is 0.3 m. The vertical distance is 1.2 m. The plenum is lined with 25-mm thick 48 kg/m^3 density fiberglass insulation board. All of the inside surfaces of the plenum are lined with the fiberglass insulation. Determine the transmission loss associated with this plenum. See Table 10 for the values of the absorption coefficients.

Solution:
The areas of the inlet section, outlet section, and plenum cross-section are

$$S_{in} = 0.9 \times 0.6 = 0.54 \text{ m}^2$$

$$S_{out} = 0.9 \times 0.6 = 0.54 \text{ m}^2$$

$$S_{pl} = 1.8 \times 1.2 = 2.16 \text{ m}^2$$

$l = 1.8$ m; $r_v = 1.2$ m; and $r_h = 0.3$ m. The values of r and $\cos\theta$ are

$$r^2\sqrt{1.8^2 + 1.2^2 + 0.3^2} = 2.184 \text{ m}$$
$$\cos\theta = 1.8/2.184 = 0.824$$

The total inside surface area of the plenum is

$$S = 2(1.2 \times 1.8) + 2(1.2 \times 1.8) + 2(1.8 \times 1.8) - 1.1 = 14.4 \text{ m}^2$$

The value of f_{co} is

$$f_{co} = 343/(2 \times 0.9) = 190 \text{ Hz}$$

The results using Equation (3) are tabulated below.

	Octave Band Center Frequency, Hz						
	63	125	250	500	1000	2000	4000
$Q\cos\theta/4\pi r^2$ ($\times 10^3$)			53.3	53.3	53.3	53.3	53.3
$(1-\alpha A)/S\alpha A$ ($\times 10^3$)			244.3	30.9	6.82	2.881	0.696
TL, dB	—	—	8	13	15	15	15

Unlined Rectangular Sheet Metal Ducts

Straight unlined rectangular sheet metal ducts provide a small amount of airborne sound attenuation. At low frequencies, the attenuation is significant, and it tends to decrease as frequency increases. Table 11 shows the tabulated results of selected unlined rectangular sheet metal ducts (Cummings 1983, Reynolds and Bledsoe 1989b, Ver 1978, Woods 1973). The attenuation values shown in Table 11 apply only to rectangular sheet metal ducts that have gages selected according to SMACNA duct construction standards.

Sound energy attenuated at low frequencies in rectangular ducts may manifest itself as breakout noise elsewhere along the duct. Low-frequency breakout noise should therefore be checked.

Acoustically Lined Rectangular Sheet Metal Ducts

Fiberglass internal duct lining for rectangular sheet metal ducts can be used to attenuate sound in ducts and to thermally insulate ducts. The thickness of duct linings associated with thermal insulation usually varies from 13 to 50 mm. For fiberglass duct lining to be effective for attenuating fan sound, it must have a minimum thickness of 25 mm. Tables 12 and 13 give tabulated attenuation values of selected rectangular sheet metal ducts for 25- and 50-mm duct lining, respectively (Kuntz 1986, Kuntz and Hoover 1987, Machen and Haines 1983, Reynolds and Bledsoe 1989b). The density of the fiberglass lining used in lined rectangular sheet metal ducts usually varies between 24 and 48 kg/m^3.

The insertion loss values given in Tables 12 and 13 are the difference between the sound pressure level measured in a reverberation

Table 11 Sound Attenuation in Unlined Rectangular Sheet Metal Ducts

Duct Size, mm × mm	P/A 1/mm	Attenuation, dB/m Octave Band Center Frequency, Hz			
		63	125	250	>250
150 × 150	0.26	0.98	0.66	0.33	0.33
305 × 305	0.13	1.15	0.66	0.33	0.20
305 × 610	0.10	1.31	0.66	0.33	0.16
610 × 610	0.07	0.82	0.66	0.33	0.10
1220 × 1220	0.03	0.49	0.33	0.23	0.07
1830 × 1830	0.02	0.33	0.33	0.16	0.07

Table 12 Insertion Loss for Rectangular Sheet Metal Ducts with 25-mm Fiberglass Lining

Dimensions, mm × mm	Insertion Loss, dB/m Octave Band Center Frequency, Hz							
	63	125	250	500	1000	2000	4000	8000
102 × 102	6.56	6.76	7.15	12.01	33.14	33.99	15.75	9.42
102 × 152	4.89	5.35	6.20	10.93	29.20	29.36	14.47	8.99
102 × 203	4.20	4.69	5.74	10.37	27.13	26.97	13.81	8.76
102 × 254	3.81	4.33	5.45	10.01	25.59	25.52	13.35	8.63
152 × 152	3.54	4.07	5.25	9.78	25.00	24.54	13.06	8.53
152 × 254	2.69	3.15	4.43	8.73	21.39	20.54	11.78	8.07
152 × 305	2.53	2.95	4.23	8.43	20.44	19.49	11.45	7.94
152 × 457	2.26	2.59	3.87	7.94	18.83	17.75	10.86	7.74
203 × 203	2.53	2.95	4.23	8.43	20.44	19.49	11.45	7.94
203 × 305	2.13	2.43	3.67	7.68	18.01	16.83	10.53	7.61
203 × 457	1.97	2.20	3.41	7.28	16.73	15.49	10.04	7.41
203 × 610	1.84	1.97	3.15	6.86	15.42	14.07	9.51	7.22
254 × 254	2.07	2.33	3.58	7.51	17.52	16.31	10.33	7.55
254 × 406	1.80	1.94	3.08	6.76	15.16	13.81	9.38	7.19
254 × 508	1.74	1.80	2.92	6.50	14.34	12.93	9.06	7.05
254 × 762	1.67	1.67	2.69	6.14	13.19	11.78	8.60	6.86
305 × 305	1.84	1.97	3.15	6.86	15.42	14.07	9.51	7.22
305 × 457	1.67	1.71	2.79	6.23	13.58	12.17	8.76	6.92
305 × 610	1.64	1.57	2.59	5.94	12.63	11.19	8.33	6.76
305 × 914	1.31	1.41	2.43	5.58	11.61	10.17	7.91	6.56
381 × 381	1.67	1.67	2.69	6.14	13.19	11.78	8.60	6.86
381 × 559	1.31	1.44	2.46	5.61	11.71	10.24	7.94	6.59
381 × 762	1.18	1.28	2.23	5.28	10.79	9.35	7.51	6.43
381 × 1143	1.05	1.12	2.03	4.99	9.94	8.50	7.12	6.23
457 × 457	1.31	1.41	2.43	5.58	11.61	10.17	7.91	6.56
457 × 635	1.08	1.15	2.10	5.05	10.14	8.69	7.22	6.27
457 × 914	0.98	1.05	1.94	4.82	9.51	8.07	6.92	6.14
457 × 1372	0.89	0.92	1.77	4.53	8.76	7.35	6.56	5.97
610 × 610	0.98	1.05	1.94	4.82	9.51	8.07	6.92	6.14
610 × 914	0.82	0.85	1.67	4.40	8.37	6.99	6.36	5.91
610 × 1219	0.75	0.79	1.54	4.17	7.78	6.40	6.07	5.77
610 × 1829	0.69	0.69	1.41	3.94	7.19	5.84	5.74	5.61
762 × 762	0.79	0.82	1.61	4.30	8.14	6.76	6.27	5.84
762 × 1143	0.69	0.69	1.41	3.94	7.19	5.84	5.74	5.61
762 × 1524	0.62	0.62	1.28	3.71	6.66	5.35	5.48	5.48
914 × 914	0.69	0.69	1.41	3.94	7.19	5.84	5.74	5.61
914 × 1372	0.59	0.56	1.21	3.58	6.33	5.05	5.28	5.38
914 × 1829	0.52	0.52	1.12	3.38	5.87	4.63	5.05	5.25
1067 × 1067	0.59	0.59	1.25	3.64	6.43	5.15	5.35	5.41
1067 × 1626	0.52	0.49	1.05	3.31	5.64	4.43	4.92	5.18
1067 × 2134	0.46	0.46	0.98	3.15	5.28	4.10	4.69	5.09
1219 × 1219	0.52	0.52	1.12	3.38	5.87	4.63	5.05	5.25
1219 × 1829	0.46	0.43	0.95	3.08	5.18	4.00	4.66	5.05

Table 13 Insertion Loss for Rectangular Sheet Metal Ducts with 50-mm Fiberglass Lining

Dimensions, mm × mm	Insertion Loss, dB/m Octave Band Center Frequency, Hz							
	63	125	250	500	1000	2000	4000	8000
102 × 102	11.61	12.50	14.83	24.97	33.14	33.99	15.75	9.42
102 × 152	8.43	9.81	12.83	22.77	29.20	29.36	14.47	8.99
102 × 203	7.05	8.53	11.78	21.59	27.13	26.97	13.81	8.76
102 × 254	6.30	7.81	11.15	20.87	25.85	25.52	13.35	8.63
152 × 152	5.84	7.32	10.73	20.34	25.00	24.54	13.06	8.53
152 × 254	4.17	5.54	8.99	18.18	21.39	20.54	11.78	8.07
152 × 305	3.81	5.12	8.56	17.59	20.44	19.49	11.45	7.94
152 × 457	3.28	4.43	7.81	16.57	18.83	17.75	10.86	7.74
203 × 203	3.81	5.12	8.56	17.59	20.44	19.49	11.45	7.94
203 × 305	3.05	4.10	7.41	16.04	18.01	16.83	10.53	7.61
203 × 457	2.72	3.64	6.82	15.22	16.73	15.49	10.04	7.41
203 × 610	2.43	3.22	6.23	14.34	15.42	14.07	9.51	7.22
254 × 254	2.89	3.94	7.19	15.72	17.52	16.31	10.33	7.55
254 × 406	2.36	3.12	6.14	14.17	15.16	13.81	9.38	7.19
254 × 508	2.23	2.85	5.77	13.62	14.34	12.93	9.06	7.05
254 × 762	2.03	2.56	5.28	12.83	13.19	11.78	8.60	6.86
305 × 305	2.43	3.22	6.23	14.34	15.42	14.07	9.51	7.22
305 × 457	2.10	2.66	5.45	13.09	13.58	12.17	8.76	6.92
305 × 610	1.97	2.40	5.02	12.40	12.63	11.19	8.33	6.76
305 × 914	1.57	2.10	4.66	11.68	11.61	10.17	7.91	6.56
381 × 381	2.03	2.56	5.28	12.83	13.19	11.78	8.60	6.86
381 × 559	1.57	2.13	4.69	11.75	11.71	10.24	7.94	6.59
381 × 762	1.38	1.87	4.27	11.09	10.79	9.35	7.51	9.71
381 × 1143	1.21	1.61	3.87	10.47	9.94	8.50	7.12	6.23
457 × 457	1.57	2.10	4.66	11.68	11.61	10.17	7.91	6.56
457 × 635	1.25	1.67	3.97	10.60	10.14	8.69	7.22	6.27
457 × 914	1.12	1.51	3.67	10.10	9.51	8.07	6.92	6.14
457 × 1372	0.98	1.31	3.35	9.55	8.76	7.35	6.56	5.97
610 × 610	1.12	1.51	3.67	10.10	9.51	8.07	6.92	6.14
610 × 914	0.92	1.21	3.18	9.22	8.37	6.99	6.36	5.91
610 × 1219	0.85	1.08	2.92	8.76	7.78	6.40	6.07	5.77
610 × 1829	0.75	0.95	2.66	8.23	7.19	5.84	5.74	5.61
762 × 762	0.89	1.15	3.08	9.06	8.14	6.76	6.27	5.84
762 × 1143	0.75	0.95	2.66	8.23	7.19	5.84	5.74	5.61
762 × 1524	0.69	0.85	2.43	7.81	6.66	5.35	5.48	5.48
914 × 914	0.75	0.95	2.66	8.23	7.19	5.84	5.74	5.61
914 × 1372	0.62	0.79	2.30	7.51	6.33	5.05	5.28	5.38
914 × 1829	0.56	0.69	2.10	7.15	5.87	4.63	5.05	5.25
1067 × 1067	0.66	0.79	2.33	7.64	6.43	5.15	5.35	5.41
1067 × 1626	0.56	0.66	2.00	6.96	5.64	4.43	4.92	5.18
1067 × 2134	0.49	0.56	1.77	6.46	5.12	3.97	4.63	5.02
1219 × 1219	0.56	0.69	2.10	7.15	5.87	4.63	5.05	5.25
1219 × 1829	0.49	0.56	1.80	6.53	5.18	4.00	4.66	5.05

room with sound propagating through an unlined section of rectangular duct and the corresponding sound pressure level measured when the unlined section of rectangular duct is replaced with a similar section of acoustically lined rectangular duct. As was mentioned in the section on Unlined Rectangular Ducts, the sound attenuation associated with unlined rectangular duct can be significant at low frequencies. This attenuation is, in effect, subtracted during the process of calculating the insertion loss from measured data. Although not proven at this time, many designers add this attenuation to the insertion loss of correspondingly sized acoustically lined rectangular ducts to obtain the total sound attenuation of acoustically lined rectangular ducts as follows:

$$\text{ATTN}(T) = \text{ATTN} + IL \qquad (8)$$

where

ATTN(T) = total sound attenuation of acoustically lined rectangular duct

ATTN = are duct sound attenuation obtained from Table 11

IL = lined duct insertion loss obtained from Table 12 or 13

Because of structure-borne sound that is transmitted in and through the duct wall, ATTN(T) usually does not exceed 40 dB. Thus, the maximum allowable total sound attenuation in Equation (8) is 40 dB. Insertion loss and attenuation values discussed in this section apply only to rectangular sheet metal ducts that have gages selected according to SMACNA duct construction standards.

Unlined Circular Sheet Metal Ducts

As with unlined rectangular ducts, unlined circular ducts provide some sound attenuation, which should be considered when designing a duct system. In contrast with rectangular ducts, circular ducts are much more rigid and, therefore, do not resonate or absorb as much sound energy. Because of this, circular ducts only provide about one-tenth the sound attenuation at low frequencies as that from rectangular ducts. Table 14 lists sound attenuation values for unlined circular ducts (Woods 1973; Chapter 52, 1987 ASHRAE Handbook).

Acoustically Lined Circular Sheet Metal Ducts

Very few data, except for product data available from manufacturers, are available for the insertion loss of acoustically lined circular

Table 14   Sound Attenuation in Straight Circular Ducts

Diameter, mm	Attenuation, dB/m Octave Band Center Frequency, Hz						
	63	125	250	500	1000	2000	4000
$D \le 180$	0.10	0.10	0.16	0.16	0.33	0.33	0.33
$180 < D \le 380$	0.10	0.10	0.10	0.16	0.23	0.23	0.23
$380 < D \le 760$	0.07	0.07	0.07	0.10	0.16	0.16	0.16
$760 < D \le 1520$	0.03	0.03	0.03	0.07	0.07	0.07	0.07

Table 15   Insertion Loss for Acoustically Lined Circular Ducts with 25-mm Lining

Diameter, mm	Insertion Loss, dB/m Octave Band Center Frequency, Hz							
	63	125	250	500	1000	2000	4000	8000
152	1.25	1.94	3.05	5.02	7.12	7.58	6.69	4.13
203	1.05	1.77	2.92	4.92	7.19	7.12	6.00	3.87
254	0.89	1.64	2.79	4.86	7.22	6.69	5.38	3.67
305	0.75	1.51	2.66	4.76	7.15	6.27	4.86	3.44
356	0.62	1.38	2.53	4.69	7.02	5.87	4.40	3.28
406	0.52	1.25	2.40	4.59	6.82	5.48	3.97	3.12
457	0.43	1.15	2.26	4.49	6.59	5.12	3.61	2.95
508	0.36	1.02	2.13	4.40	6.30	4.76	3.28	2.85
559	0.26	0.92	2.00	4.30	5.97	4.40	3.02	2.72
610	0.23	0.82	1.87	4.20	5.61	4.07	2.79	2.62
660	0.16	0.72	1.74	4.07	5.22	3.74	2.59	2.53
711	0.10	0.62	1.61	3.94	4.79	3.41	2.43	2.43
762	0.07	0.52	1.48	3.81	4.36	3.12	2.26	2.33
813	0.03	0.46	1.38	3.67	3.94	2.85	2.17	2.26
864	0	0.36	1.25	3.51	3.51	2.59	2.07	2.17
914	0	0.26	1.15	3.35	3.05	2.33	1.97	2.10
965	0	0.20	1.02	3.15	2.62	2.10	1.90	2.00
1016	0	0.10	0.92	2.99	2.23	1.87	1.80	1.90
1067	0	0.03	0.82	2.76	1.84	1.64	1.74	1.80
1118	0	0	0.75	2.56	1.48	1.44	1.67	1.71
1168	0	0	0.66	2.33	1.15	1.28	1.57	1.57
1219	0	0	0.59	2.07	0.85	1.12	1.48	1.44
1270	0	0	0.49	1.80	0.62	0.95	1.35	1.31
1321	0	0	0.46	1.51	0.43	0.82	1.21	1.12
1372	0	0	0.39	1.21	0.30	0.72	1.02	0.95
1422	0	0	0.33	0.92	0.26	0.59	0.82	0.72
1473	0	0	0.30	0.56	0.26	0.52	0.59	0.49
1523	0	0	0.26	0.20	0.33	0.46	0.30	0.23

Table 16   Insertion Loss for Acoustically Lined Circular Ducts with 50-mm Lining

Diameter, mm	Insertion Loss, dB/m Octave Band Center Frequency, Hz							
	63	125	250	500	1000	2000	4000	8000
152	1.84	2.62	4.49	7.38	7.12	7.58	6.69	4.13
203	1.67	2.46	4.36	7.32	7.19	7.12	6.00	3.87
254	1.51	2.33	4.23	7.22	7.22	6.69	5.38	3.67
305	1.38	2.20	4.10	7.15	7.15	6.27	4.86	3.44
356	1.25	2.07	3.97	7.05	7.02	5.87	4.40	3.28
406	1.15	1.94	3.84	6.96	6.82	5.48	3.97	3.12
457	1.05	1.84	3.71	6.89	6.59	5.12	3.61	2.95
508	0.95	1.71	3.58	6.79	6.30	4.76	3.28	2.85
559	0.89	1.61	3.44	6.66	5.97	4.40	3.02	2.72
610	0.82	1.51	3.31	6.56	5.61	4.07	2.79	2.62
660	0.79	1.41	3.18	6.43	5.22	3.74	2.59	2.53
711	0.72	1.31	3.05	6.33	4.79	3.41	2.43	2.43
762	0.69	1.21	2.95	6.17	4.36	3.12	2.26	2.33
813	0.66	1.12	2.82	6.04	3.94	2.85	2.17	2.26
864	0.62	1.05	2.69	5.87	3.51	2.59	2.07	2.17
914	0.59	0.95	2.59	5.71	3.05	2.33	1.97	2.10
965	0.56	0.89	2.49	5.54	2.62	2.10	1.90	2.00
1016	0.52	0.79	2.40	5.35	2.23	1.87	1.80	1.90
1067	0.49	0.72	2.30	5.15	1.84	1.64	1.74	1.80
1118	0.43	0.66	2.20	4.92	1.48	1.44	1.67	1.71
1168	0.39	0.56	2.10	4.69	1.15	1.28	1.57	1.57
1219	0.36	0.49	2.03	4.46	0.85	1.12	1.48	1.44
1270	0.30	0.39	1.97	4.20	0.62	0.95	1.35	1.31
1321	0.23	0.33	1.90	3.90	0.43	0.82	1.21	1.12
1372	0.16	0.26	1.84	3.61	0.30	0.72	1.02	0.95
1422	0.07	0.16	1.80	3.28	0.26	0.59	0.82	0.72
1473	0	0.10	1.74	2.95	0.26	0.52	0.59	0.49
1523	0	0	1.74	2.59	0.33	0.46	0.30	0.23

ducts. Tables 15 and 16 give the insertion loss values for dual-wall circular sheet metal ducts with 25- and 50-mm acoustical lining (Reynolds and Bledsoe 1989a). The acoustical lining for the ducts is a 12 kg/m³ density fiberglass blanket, which is covered with an internal liner of perforated galvanized sheet metal that has on open area of 23%. Because the sound attenuation of unlined circular ducts is generally negligible, it is not necessary to include it when calculating the total sound attenuation of lined circular ducts. Because the duct wall transmits structure-borne sound, the total sound attenuation of lined circular ducts usually does not exceed 40 dB.

Nonmetallic Insulated Flexible Ducts

Nonmetallic insulated flexible ducts can reduce airborne noise significantly. Insertion loss values for specified duct diameters and lengths are given in Table 17. Duct lengths should normally be from 1 to 2 m. Care should be taken to keep flexible ducts straight; bends should have as long a radius as possible. While an abrupt bend may provide some additional insertion loss, the noise associated with the airflow in the bend may be unacceptably high. Because of potentially high breakout sound levels associated with flexible ducts, care should be exercised when using flexible ducts above sound-sensitive spaces.

Rectangular Sheet Metal Duct Elbows

Table 18 displays insertion loss values for unlined and lined square elbows without turning vanes (Beranek 1960). For lined square elbows,

the duct lining must extend at least two duct widths w beyond the elbow, and the thickness of the total lining thickness should be at least 10% of the duct width. Table 18 applies only for the situation where the duct is lined before and after the elbow. Table 19 gives the insertion loss values associated with round elbows. Table 20 gives the insertion loss values for unlined and lined square elbows with turning vanes. The symbol fw in Tables 18 through 20 is the center frequency f in kilohertz of the octave frequency band times the width of elbow in millimetres as shown in Figure 17 (Beranek 1960; Chapter 7, 1993 *ASHRAE Handbook*; Chapter 35, 1976 *ASHRAE Handbook*; Chapter 32, 1984 *ASHRAE Handbook*).

Duct Silencers

Duct silencers are used to attenuate sound that is transmitted through HVAC systems, particularly duct systems. They can add pressure and energy losses. Therefore, when selecting silencers, the following parameters should be considered:

- **Airflow pressure drop**, including system effects for less than ideal flow conditions
- **Insertion loss**, which is the reduction in sound power level, in decibels, at a given location due solely to the placement of a sound-attenuating device in the path of transmission between the sound source and the given location

Airflow-generated noise is created as the air flows into, through, and out of the silencer. In the majority of installations the airflow-generated noise is much less than, and does not contribute to, the silenced noise level on the quiet side of the silencer. In general, airflow-generated noise should be evaluated if static pressure drops exceed 87 Pa, the noise criterion is below RC 35, or the silencer is located very close to or in the occupied space.

There are three types of HVAC duct silencers: dissipative (with acoustic media), reactive (no-media), and active silencers.

Table 17   Lined Flexible Duct Insertion Loss

Diameter, mm	Length, m	Insertion Loss, dB Octave Band Center Frequency, Hz						
		63	125	250	500	1000	2000	4000
102	3.7	6	11	12	31	37	42	27
	2.7	5	8	9	23	28	32	20
	1.8	3	6	6	16	19	21	14
	0.9	2	3	3	8	9	11	7
127	3.7	7	12	14	32	38	41	26
	2.7	5	9	11	24	29	31	20
	1.8	4	6	7	16	19	21	13
	0.9	2	3	4	8	10	10	7
152	3.7	8	12	17	33	38	40	26
	2.7	6	9	13	25	29	30	20
	1.8	4	6	9	17	19	20	13
	0.9	2	3	4	8	10	10	7
178	3.7	9	12	19	33	37	38	25
	2.7	6	9	14	25	28	29	19
	1.8	4	6	10	17	19	19	13
	0.9	2	3	5	8	9	10	6
203	3.7	8	11	21	33	37	37	24
	2.7	6	8	16	25	28	28	18
	1.8	4	6	11	17	19	19	12
	0.9	2	3	5	8	9	9	6
229	3.7	8	11	22	33	37	36	22
	2.7	6	8	17	25	28	27	17
	1.8	4	6	11	17	19	18	11
	0.9	2	3	6	8	9	9	6
254	3.7	8	10	22	32	36	34	21
	2.7	6	8	17	24	27	26	16
	1.8	4	5	11	16	18	17	11
	0.9	2	3	6	8	9	9	5
305	3.7	7	9	20	30	34	31	18
	2.7	5	7	15	23	26	23	14
	1.8	3	5	10	15	17	16	9
	0.9	2	2	5	8	9	8	5
356	3.7	5	7	16	27	31	27	14
	2.7	4	5	12	20	23	20	11
	1.8	3	4	8	14	16	14	7
	0.9	1	2	4	7	8	7	4
406	3.7	2	4	9	23	28	23	9
	2.7	2	3	7	17	21	17	7
	1.8	1	2	5	12	14	12	5
	0.9	1	1	2	6	7	6	2

Note: The 63 Hz insertion loss values are estimated from higher frequency insertion loss values. Also, a proposed revision to ARI *Standard* 885 includes these values.

Table 18   Insertion Loss of Unlined and Lined Square Elbows Without Turning Vanes

	Insertion Loss, dB	
	Unlined Elbows	Lined Elbows
$fw < 48$	0	0
$48 \leq fw < 96$	1	1
$96 \leq fw < 190$	5	6
$190 \leq fw < 380$	8	11
$380 \leq fw < 760$	4	10
$fw > 760$	3	10

Note: $fw = f \times w$ where f = center frequency, kHz, and w = width, mm

Table 19   Insertion Loss of Round Elbows

	Insertion Loss, dB
$fw < 48$	0
$48 \leq fw < 96$	1
$96 \leq fw < 190$	2
$fw > 190$	3

Note: $fw = f \times w$ where f = center frequency, kHz, and w = width, mm

Table 20   Insertion Loss of Unlined and Lined Square Elbows with Turning Vanes

	Insertion Loss, dB	
	Unlined Elbows	Lined Elbows
$fw < 48$	0	0
$48 \leq fw < 96$	1	1
$96 \leq fw < 190$	4	4
$190 \leq fw < 380$	6	7
$fw > 380$	4	7

Note: $fw = f \times w$ where f = center frequency, kHz, and w = width, mm

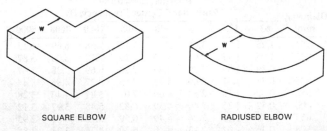

SQUARE ELBOW   RADIUSED ELBOW

Fig. 17   Rectangular Duct Elbows

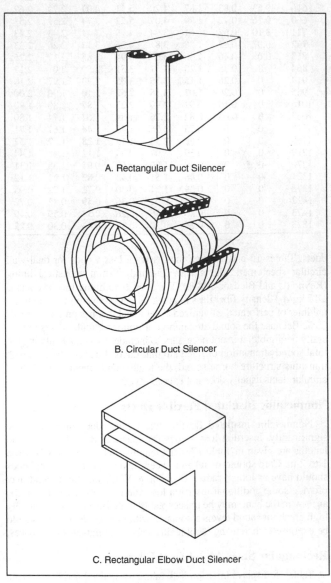

A. Rectangular Duct Silencer

B. Circular Duct Silencer

C. Rectangular Elbow Duct Silencer

Fig. 18   Dissipative Duct Silencer

- *Dissipative silencers* (Figure 18) generally use perforated metal surfaces covering acoustic-grade fiberglass to attenuate sound over a broad range of frequencies. Airflow does not significantly affect the insertion loss if duct approach velocities are under 10 m/s.

- *Reactive silencers* use tuned perforated metal facings covering tuned chambers void of any fibrous material. The outside physical appearance of reactive silencers is similar to that of dissipative silencers (Figure 18). Because of tuning, broadband insertion loss is more difficult to achieve than with dissipative silencers. Longer lengths may be required to achieve similar insertion loss performance. Airflow generally increases the insertion loss of reactive silencers.

- *Active duct silencers* (Figure 19) reduce noise at lower frequencies by producing inverse sound waves that cancel the unwanted noise. An input microphone measures the noise in the duct and converts it to electrical signals. These signals are processed by a digital computer that generates exact opposite "mirror-image" sound waves of equal amplitude. This secondary noise source destructively interferes with the noise and cancels a significant portion of the unwanted sound. An error microphone measures the residual sound beyond the silencer and provides feedback to adjust the computer model to increase performance. Because the components are mounted outside the airflow, there is no pressure loss or generated noise. Performance is limited, however, by the presence of excessive turbulence in the airflow detected by the microphones. Manufacturers recommend using active silencers where duct velocities are less than 7.6 m/s and where the duct configurations are conducive to smooth, evenly distributed airflow.

Data for dissipative and reactive silencers should be obtained from tests in a manner consistent with the procedures outlined in the most current version of ASTM *Standard* E 477. This standard has not been verified for determining performance of active silencers. Since insertion loss measurements use a substitution technique, reasonable (±3 dB) insertion loss values can be achieved down to 63 Hz. Airflow-generated noise, however, cannot be measured accurately below 100 Hz due to the lack of a standard means of qualifying the reverberant rooms for sound power measurements at lower frequencies (a function of room size). A recent series of round robin tests has shown that airflow-generated sound power data has an expected standard deviation of ±3 to 6 dB over the octave band frequency range of 125 to 8000 Hz.

Care should be taken in applying test data to actual project installations. Adverse system effects can have a significant impact on the performance of all standard silencers. Standard silencers should be located at least three duct diameters from a fan, coil, elbow, branch takeoff, or other duct element. Locating a standard silencer closer than three duct diameters can result in a significant increase in both the pressure loss across the silencer and the generated noise. Active silencers may be located in this region without pressure loss, but the acoustic insertion loss may be limited by turbulence.

All silencers come in varying shapes and sizes to fit the project ductwork. Straight silencers with both rectangular and circular cross-sections are available with or without center splitters and pods. Elbow silencers are available as both dissipative and reactive silencers for use where there is insufficient space for a straight silencer.

Elbow silencers effectively attenuate the noise with splitters that aerodynamically turn the air to minimize system pressure drop. Special fan inlet and fan discharge silencers, including cone silencers and inlet box silencers, minimize aerodynamic system effects, maximize acoustic system effects, and contain noise at the source.

Duct Branch Sound Power Division

When sound traveling in a duct encounters a junction, the sound power contained in the incident sound waves in the main feeder duct is distributed between the branches associated with the junction (Ver 1982, 1984a). This division of sound power is referred to as the branch sound power division. The corresponding attenuation of sound power that is transmitted down each branch of the junction is comprised of two components. The first is associated with the reflection of the incident sound wave if the sum of the cross-sectional areas of the individual branches $\Sigma\, S_{Bi}$ differs from the cross-sectional area S_M of the main feeder duct. The second component is associated with the energy division according to the ratio of the cross-sectional area S_{Bi} of an individual branch divided by the sum of the cross-sectional areas of the individual branches, $\Sigma\, S_{Bi}$. The second component is the dominant component. Table 21 provides values for the attenuation of sound power ΔL_{Bi} at a junction that are related to the sound power transmitted down an individual branch of the junction.

Duct End Reflection Loss

When low-frequency plane sound waves interact with openings into a large room, a significant amount of the sound energy incident on this interface is reflected back into the duct. The sound attenuation values ΔL associated with duct end reflection losses for ducts terminated in free space are given in Table 22, and those for ducts terminated flush with a wall are given in Table 23 (Sandbakken et al. 1981, AMCA *Standard* 300). Diffusers that terminate in a suspended lay-in acoustic ceiling can be treated as terminating in free space. If a duct terminating into a diffuser is rectangular, the duct diameter D is given by

$$D = \sqrt{4A/\pi} \qquad (9)$$

where A = area of the rectangular duct, m².

Tables 22 and 23 have some limitations. The tests on which these equations are based were conducted with straight sections of circular ducts. These ducts directly terminated into a reverberation chamber either with no restriction on the end of the duct or with a circular orifice constriction placed over the end of the duct. Diffusers can be either round or rectangular. They usually have a restriction associated with them, which may be a damper, guide vanes to direct airflow, a perforated metal facing, or a combination of these elements. Currently, no data are available to indicate whether these elements react similarly to the orifices used in the test chamber. As a result, the effects of an orifice placed over the end of a duct are not included in Tables 22 and 23. One can assume that using Equation (9) to calculate D will yield reasonable results with diffusers that have low aspect ratios (length/width). However, many diffusers (particularly

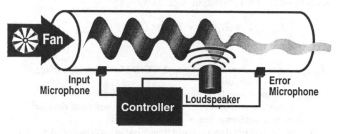

Fig. 19   Active Duct Silencers

Table 21   Duct Branch Sound Power Division

$S_i/\Sigma S_{Bi}$	ΔL_{Bi}	$S_i/\Sigma S_{Bi}$	ΔL_{Bi}
1.00	0	0.10	10
0.80	1	0.08	11
0.63	2	0.063	12
0.50	3	0.050	13
0.40	4	0.040	14
0.32	5	0.032	15
0.25	6	0.025	16
0.20	7	0.020	17
0.16	8	0.016	18
0.12	9	0.012	19

Table 22 Duct End Reflection Loss—Duct Terminated in Free Space

Duct Diameter, mm	End Reflection Loss, dB Octave Band Center Frequency, Hz					
	63	125	250	500	1000	2000
150	20	14	9	5	2	1
200	18	12	7	3	1	0
250	16	11	6	2	1	0
300	14	9	5	2	1	0
400	12	7	3	1	0	0
510	10	6	2	1	0	0
610	9	5	2	1	0	0
710	8	4	1	0	0	0
810	7	3	1	0	0	0
910	6	3	1	0	0	0
1220	5	2	1	0	0	0
1830	3	1	0	0	0	0

Table 23 Duct End Reflection Loss—Duct Terminated Flush with Wall

Duct Diameter, mm	End Reflection Loss, dB Octave Band Center Frequency, Hz				
	63	125	250	500	1000
150	18	13	8	4	1
200	16	11	6	2	1
250	14	9	5	2	1
300	13	8	4	1	0
400	10	6	2	1	0
510	9	5	2	1	0
610	8	4	1	0	0
710	7	3	1	0	0
810	6	2	1	0	0
910	5	2	1	0	0
1220	4	1	0	0	0
1830	2	1	0	0	0

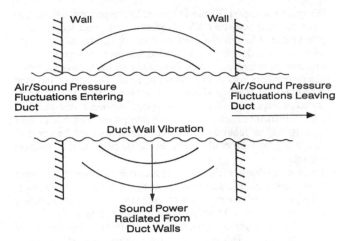

Fig. 20 Airflow-Generated Duct Rumble

slot diffusers) have high aspect ratios. Currently, it is not known whether Tables 22 or 23 can be accurately used with these diffusers.

Finally, many diffusers do not have long straight sections (greater than three duct diameters) before they terminate into a room. Many duct sections between a main feed branch and a diffuser may be curved or may be short and stubby. The effects of these configurations on the duct end reflection loss are not known. While Tables 22 and 23 can probably be used with reasonable accuracy for many diffuser configurations, caution should be exercised when a configuration differs drastically from the test conditions used to derive these tables.

SOUND RADIATION THROUGH DUCT WALLS

Airflow-Generated Duct Rumble

An HVAC fan and its connected ductwork can act as a semi-closed, compressible-fluid pumping system to transmit both acoustic and aerodynamic air pressure fluctuations at the fan to other locations in the duct system. Air pressure fluctuations can be caused by variations in the speed of the fan, motor, or fan belt or by airflow instabilities transmitted to the fan housing or nearby connected ductwork. When the air pressure fluctuations encounter large, flat, unreinforced duct surfaces that have resonance frequencies near or equal to the disturbing frequencies, the duct surfaces vibrate (Ebbing et al. 1978). This vibration occurs in the stiffness- or resonance-controlled regions of the duct. In typical HVAC duct systems, duct wall vibration can produce sound pressure levels of the order of 65 to 95 dB at frequencies that range from 10 to 100 Hz. This type of duct-generated sound is generally called duct rumble (Figure 20).

Figure 21 shows typical duct configurations near a centrifugal fan. Configurations labeled "fair" or "bad" can cause duct rumble. "Good" to "optimum" design of fan inlet and discharge transitions minimizes the potential for duct rumble during the design phase of a project, but this may not completely eliminate duct rumble.

Several methods can be used to eliminate or reduce duct rumble. One method is to alter the fan, motor, or fan belt speed. Changing speed changes the frequency of the air pressure fluctuations so that they differ from the duct wall resonance frequencies, and duct rumble may not occur. Another measure applies rigid materials, such as duct reinforcements and drywall, directly to the duct wall to change the wall resonance frequencies (Figure 22). Noise reductions of 5 to 11 dB in the 31.5 Hz and 63 Hz octave frequency bands have been recorded using this treatment.

The application of mass-loaded materials in combination with absorptive materials does not alleviate duct rumble noise into a space unless both materials are completely decoupled from the vibrating duct wall. An example of this type of construction, using two layers of drywall, is shown in Figure 23. Because the treatment is decoupled from the duct wall, it can provide the greatest amount of noise reduction. Mass-loaded materials combined with absorptive materials that are directly attached to the duct wall are not effective in reducing duct rumble (Figure 24).

Sound Breakout and Breakin in Ducts

Sound associated with fan or airflow noise that radiates through the duct walls into the surrounding area is called breakout (Figure 25). Breakout can be a problem if it is not adequately attenuated before the duct runs over an occupied space (Cummings 1983, Lilly 1987). Sound transmitted into a duct from the surrounding area is known as breakin (Figure 26).

The transmission loss characteristics of ducts presented here ignore sound energy attenuation along the duct and can be applied to ducts where the critical length of duct for noise radiation is the first 6 to 9 m.

Breakout Sound Transmission from Ducts. The sound power level associated with breakout is given by

$$L_{w(out)} = L_{w(in)} + 10 \log (S/A) - TL_{out} \qquad (10)$$

where

$L_{w(out)}$ = sound power level of sound radiated from outside surface of duct walls, dB
$L_{w(in)}$ = sound power level of sound inside duct, dB
S = surface area of outside sound-radiating surface of duct, m²
A = cross-sectional area of inside of duct, m²
TL_{out} = normalized duct breakout transmission loss (independent of S and A), dB

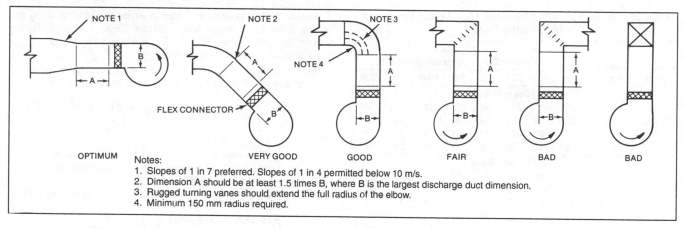

Fig. 21 Various Outlet Configurations for Centrifugal Fans and Their Possible Rumble Conditions

OPTIMUM   VERY GOOD   GOOD   FAIR   BAD   BAD

Notes:
1. Slopes of 1 in 7 preferred. Slopes of 1 in 4 permitted below 10 m/s.
2. Dimension A should be at least 1.5 times B, where B is the largest discharge duct dimension.
3. Rugged turning vanes should extend the full radius of the elbow.
4. Minimum 150 mm radius required.

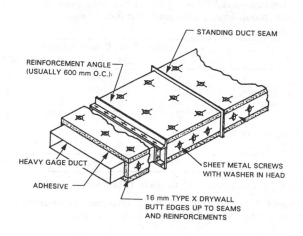

Fig. 22 Drywall Lagging on Duct for Duct Rumble

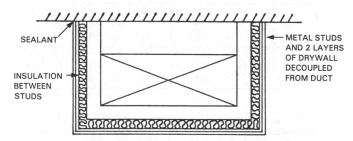

Fig. 23 Decoupled Drywall Enclosure for Duct Rumble

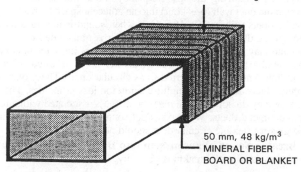

Fig. 24 Rectangular Duct with External Lagging

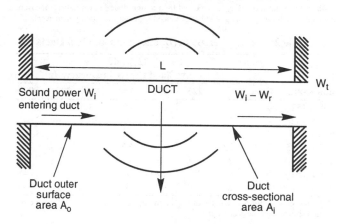

Fig. 25 Duct Breakout

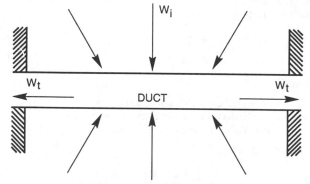

Fig. 26 Duct Breakin

Table 24 TL_{out} Versus Frequency for Various Rectangular Ducts

Duct Size, mm × mm	Gage	TL_{out}, dB Octave Band Center Frequency, Hz							
		63	125	250	500	1000	2000	4000	8000
305 × 305	24	21	24	27	30	33	36	41	45
305 × 610	24	19	22	25	28	31	35	41	45
305 × 1220	22	19	22	25	28	31	37	43	45
610 × 610	22	20	23	26	29	32	37	43	45
610 × 1220	20	20	23	26	29	31	39	45	45
1220 × 1220	18	21	24	27	30	35	41	45	45
1220 × 2440	18	19	22	25	29	35	41	45	45

Note: The data are for duct lengths of 6.1 m, but the values may be used for the cross-section shown regardless of length.

Table 25 Experimentally Measured TL_{out} Versus Frequency for Circular Ducts

Diameter, mm	Length, m	Gage	TL_{out}, dB Octave Band Center Frequency, Hz						
			63	125	250	500	1000	2000	4000
Long Seam Ducts									
203	4.57	26	>45	(53)	55	52	44	35	34
356	4.57	24	>50	60	54	36	34	31	25
559	4.57	22	>47	53	37	33	33	27	25
813	4.57	22	(51)	46	26	26	24	22	38
Spiral Wound Ducts									
203	3.05	26	>48	>64	>75	72	56	56	46
356	3.05	26	>43	>53	55	33	34	35	25
660	3.05	24	>45	50	26	26	25	22	36
660	3.05	16	>48	53	36	32	32	28	41
813	3.05	22	>43	42	28	25	26	24	40

Note: In cases where background sound swamped the sound radiated from the duct walls, a lower limit on TL_{out} is indicated by a > sign. Parentheses indicate measurements in which background sound has produced a greater uncertainty than usual.

Table 26 TL_{out} Versus Frequency for Flat Oval Ducts

Duct Size, mm × mm	Gage	TL_{out}, dB Octave Band Center Frequency, Hz						
		63	125	250	500	1000	2000	4000
305 × 152	24	31	34	37	40	43	—	—
610 × 152	24	24	27	30	33	36	—	—
610 × 305	24	28	31	34	37	—	—	—
1220 × 305	22	23	26	29	32	—	—	—
1220 × 610	22	27	30	33	—	—	—	—
2440 × 610	20	22	25	28	—	—	—	—
2440 × 1220	18	28	31	—	—	—	—	—

Note: The data are for duct lengths of 6.1 m, but the values may be used for the cross-section shown regardless of length.

Table 24 provides values for TL_{out} for rectangular ducts, Table 25 for round ducts, and Table 26 for flat oval ducts (Cummings 1983, 1985). For rectangular ducts,

$$S = 2L(a + b) \tag{11}$$

$$A = ab \tag{12}$$

where

a = larger duct cross-sectional dimension, m
b = smaller duct cross-sectional dimension, m
L = length of duct sound-radiating surface, m

For circular ducts,

$$S = L\pi d \tag{13}$$

$$A = \pi d^2/4 \tag{14}$$

where

d = duct diameter, m
L = length of duct sound-radiating surface, m

For flat oval ducts,

$$S = L[2(a - b) + \pi b] \tag{15}$$

$$A = b(a - b) + \pi b^2/4 \tag{16}$$

where

a = length of larger major axis, m
b = length of minor duct axis, m
L = length of duct sound-radiating surface, m

Equation (10) assumes no interior sound attenuation along the length of the duct sound-radiating surface. Thus, it is valid only for unlined ducts. It is generally valid for duct lengths up to 9 m. $L_{w(out)}$ must always be equal to or less than $L_{w(in)}$.

For most applications, the sound pressure level in an occupied space as a result of duct sound breakout can be obtained from

$$L_p = L_{w(out)} - 10 \log(\pi r L) \tag{17}$$

where

L_p = sound pressure level at specified point in the space, dB
$L_{w(out)}$ = sound power level of sound radiated from outside surface of duct walls [given by Equation (10)], dB
r = distance between duct and position at which L_p is being calculated, m
L = length of duct sound-radiating surface, m

Example 5. A 305-mm by 1220-mm by 4.57-m long rectangular supply duct above a mineral fiber lay-in tile ceiling is constructed of 22 gage (0.853-mm) sheet metal. Given the sound power levels in the duct, what are the sound pressure levels at a listener 1.52 m from the duct?

Solution:

	Octave Band Center Frequency, Hz						
	63	125	250	500	1000	2000	4000
$L_{w(in)}$	90	85	80	75	70	65	60
− TL_{out} (Table 24)	−19	−22	−25	−28	−31	−37	−43
10 log (S/A)	16	16	16	16	16	16	16
$L_{w(out)}$	87	79	71	63	55	44	33
− Ceiling tile (Table 30)	−4	−5	−6	−8	−10	−12	−14
− 10 log ($\pi r L$)	−14	−14	−14	−14	−14	−14	−14
L_p, dB	69	60	51	41	31	18	5

The RC level associated with the above L_p values is RC 30(R).

Example 6. Repeat Example 5 using a circular duct of equivalent airflow area: 696 mm in diameter, 22 gage (0.853 mm), 4.57 m long.

Solution:

	Octave Band Center Frequency, Hz						
	63	125	250	500	1000	2000	4000
$L_{w(in)}$	90	85	80	75	70	65	60
− TL_{out} (Table 25)	−49	−50	−32	−30	−29	−25	−31
10 log (S/A)	14	14	14	14	14	14	14
$L_{w(out)}$	55	49	62	59	55	54	43
− Ceiling tile (Table 30)	−4	−5	−6	−8	−10	−12	−14
− 10 log ($\pi r L$)	−14	−14	−14	−14	−14	−14	−14
L_p, dB	37	30	42	37	31	28	15

The RC level associated with the above L_p values is RC 32(N). The use of the circular duct resulted in a slight increase in the RC level. However, it eliminated the annoying low-frequency rumble present in Example 5.

Examples 5 and 6 demonstrate that when there is a sound breakout problem associated with ducts, circular ducts are much more effective in reducing the low-frequency sound that is transmitted through the duct wall (breakout) into an adjacent space than are rectangular ducts (Figure 27). However, when sound is not transmitted through the duct wall of a circular duct, it is transmitted down the duct and may become a problem at some other point in the duct system. Flexible and rigid fiberglass ducts often come in "circular" configurations and may be referred to as circular ducts. These types of circular ducts do not have high transmission loss properties. This is because they do not have the mass or stiffness associated with circular sheet-metal ducts. Whenever duct sound breakout is a concern, fiberglass or flexible circular duct should not be used.

Breakin Sound Transmission into Ducts. The sound power level associated with breakin is given by

$$L_{w(in)} = L_{w(out)} - TL_{in} - 3 \tag{18}$$

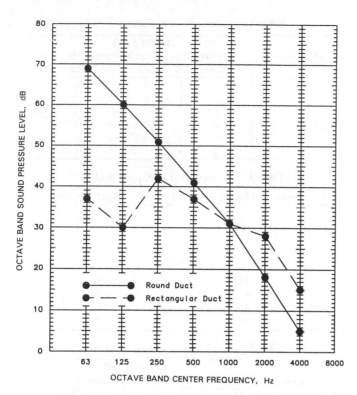

Fig. 27 Duct Breakout Sound Pressure Levels from Examples 5 and 6

Table 27 TL_{in} Versus Frequency for Various Rectangular Ducts

Duct Size, mm × mm	Gage	TL_{out}, dB Octave Band Center Frequency, Hz							
		63	125	250	500	1000	2000	4000	8000
305 × 305	24	16	16	16	25	30	33	38	42
305 × 610	24	15	15	17	25	28	32	38	42
305 × 1220	22	14	14	22	25	28	34	40	42
610 × 610	22	13	13	21	26	29	34	40	42
610 × 1220	20	12	15	23	26	28	36	42	42
1220 × 1220	18	10	19	24	27	32	38	42	42
1220 × 2440	18	11	19	22	26	32	38	42	42

Note: The data are for duct lengths of 6.1 m, but the values may be used for the cross-section shown regardless of length.

Table 28 Experimentally Measured TL_{in} Versus Frequency for Circular Ducts

Diameter, mm	Length, m	Gage	TL_{in}, dB Octave Band Center Frequency, Hz						
			63	125	250	500	1000	2000	4000
Long Seam Ducts									
203	4.57	26	>17	(31)	39	42	41	32	31
356	4.57	24	>27	43	43	31	31	28	22
559	4.57	22	>28	40	30	30	30	24	22
813	4.57	22	(35)	36	23	23	21	19	35
Spiral Wound Ducts									
203	3.05	26	>20	>42	>59	>62	53	43	26
356	3.05	26	>20	>36	44	28	31	32	22
660	3.05	24	>27	38	20	23	22	19	33
660	3.05	16	>30	>41	30	29	29	25	38
813	3.05	22	>27	32	25	22	23	21	37

Note: In cases where background sound swamped the sound radiated from the duct walls, a lower limit on TL_{in} is indicated by a > sign. Parentheses indicate measurements in which background sound has produced a greater uncertainty than usual.

Table 29 TL_{in} Versus Frequency for Flat Oval Ducts

Duct Size, mm × mm	Gage	TL_{in}, dB Octave Band Center Frequency, Hz						
		63	125	250	500	1000	2000	4000
305 × 152	24	18	18	22	31	40	—	—
610 × 152	24	17	17	18	30	33	—	—
610 × 305	24	15	16	25	34	—	—	—
1220 × 305	22	14	14	26	29	—	—	—
1220 × 610	22	12	21	30	—	—	—	—
2440 × 610	20	11	22	25	—	—	—	—
2440 × 1220	18	19	28	—	—	—	—	—

Note: The data are for duct lengths of 6.1 m, but the values may be used for the cross-section shown regardless of length.

where

$L_{w(in)}$ = sound power level of sound transmitted into duct and then transmitted upstream or downstream of point of entry, dB
$L_{w(out)}$ = sound power level of sound incident on the outside of duct walls, dB
TL_{in} = duct breakin transmission loss, dB

Table 27 gives values for TL_{in} for rectangular ducts, Table 28 for round ducts, and Table 29 for flat oval ducts (Cummings 1983, 1985).

SOUND TRANSMISSION IN RETURN AIR SYSTEMS

The fan return air system provides a sound path (through ducts or through an unducted ceiling plenum) between a fan and occupied rooms. Often a direct opening passes through the ceiling plenum to the mechanical equipment room. This condition can result in high sound levels in spaces adjacent to the equipment room. The high sound levels are caused by the close proximity of the fan and other sound sources in the equipment room and the low system attenuation between the equipment room and the adjacent spaces.

Sound in ducted return air systems is controlled by the fan intake sound power levels. Unducted plenum return air systems are impacted by the sound power levels of the fan intake and casing-radiated noise components. In certain installations, sound from other equipment located in the mechanical equipment room may also radiate through the wall opening and into adjacent spaces. Good design practices yield room return air system sound levels that are approximately 5 dB below the corresponding room supply air system sound levels.

When sound levels in spaces adjacent to mechanical equipment rooms are too high, noise control measures must be provided. Before specifying modifications, the controlling sound paths between the mechanical equipment room and adjacent spaces must be identified. Ducted return air systems can be modified using methods applicable to ducted supply systems. Good planning is required with unducted ceiling plenum systems so that the mechanical equipment room is located away from noise-sensitive spaces.

Unducted plenum systems may still require several modifications to satisfy design goals. Prefabricated sound attenuators can be effective when installed at the mechanical equipment room wall opening or at the suction side of the fan. Improvements in the ceiling transmission loss are often limited by typical ceiling penetrations and lighting fixtures. Modifications to the mechanical equipment room wall can also be effective for some constructions. Adding acoustical absorption in the mechanical equipment room reduces the buildup of reverberant sound energy in this space; however, this typically reduces noise only slightly in the low-frequency range in nearby areas.

SOUND TRANSMISSION THROUGH CEILING SYSTEMS

When terminal units, fan-coil units, air-handling units, ducts, or return air openings to mechanical equipment rooms are located in a ceiling plenum above an occupied room, sound transmission through the ceiling system can be high enough to cause excessive noise levels

Table 30 Transmission Loss Values Associated with Typical Integrated Ceiling Systems

Ceiling System	Transmission Loss, dB Octave Band Center Frequency, Hz						
	63	125	250	500	1000	2000	4000
9.5-mm thick gypsum board	6	11	17	21	26	28	23
12.7-mm thick gypsum board	9	14	20	23	27	27	25
15.9-mm thick gypsum board	10	15	21	25	27	26	27
38.1-mm thick gypsum board	13	18	25	27	27	26	29
Mineral fiber lay-in ceiling	4	5	6	8	10	12	14
Mineral fiber concealed-spline ceiling	6	12	12	13	14	15	16
Fiberglass lay-in ceiling	2	3	4	5	7	9	11

in that room. There are no standard test procedures associated with measuring the direct transmission of sound through ceilings. As a result, ceiling product manufacturers rarely publish data that can be used in calculations. This problem is further complicated by the fact that ceiling systems are rarely homogeneous surfaces; they usually have light fixtures, diffusers, grilles, speakers, and so forth, which substantially reduce the transmission loss of the ceiling and thus must be considered.

To estimate the sound levels in a room associated with sound transmission through the ceiling, the sound power levels in the ceiling plenum must be reduced by the transmission loss of the ceiling system before converting from sound power levels to corresponding sound pressure levels in the room. In the absence of a recognized test standard, the transmission loss and insertion loss values in Table 30 may be used.

RECEIVER ROOM SOUND CORRECTION

The sound pressure level at a given location in a room associated with a particular sound source is a function of the sound power level and sound radiation characteristics of the sound source, the acoustic properties of the room (e.g., surface treatments, furnishings), the room volume, and the distance between the sound source and the point of observation. Two types of sound sources are typically encountered in HVAC system applications—point sources and line sources. Point sources are usually associated with sound radiated from grilles, registers, and diffusers; air-valve and fan-powered air terminal units and fan-coil units located in ceiling plenums; and return air openings. Line sources are usually associated with sound breakout from air ducts.

With respect to point sound sources in an enclosed space, diffuse-field theory predicts that as the distance between the sound source and point of observation is increased, the sound pressure level decreases at the rate of 6 dB per doubling of distance to the region in the room where the reverberant sound field begins to dominate. The sound pressure level in the reverberant field approaches a constant level. Diffuse-field theory applies only to rooms in which there is no furniture or other objects that can scatter sound.

Investigators have found that diffuse-field theory does not apply in real rooms with furniture or other sound-scattering objects (Schultz 1985, Thompson 1981). In real rooms, the sound pressure levels decrease at the rate of around 3 dB per doubling of distance between the sound source and the point of observation. Generally, a reverberant sound field does not exist in small rooms (room volume less than 420 m³). In large rooms (room volume greater than 420 m³), reverberant fields usually exist, but usually at distances from the sound sources that are significantly greater than those predicted by diffuse-field theory.

Point Sound Sources

Most normally furnished rooms with regular proportions have acoustic characteristics that range from average to medium-dead. These

Table 31 Values for A in Equation (19)

Room Volume, m³	Value for A, dB Octave Band Center Frequency, Hz						
	63	125	250	500	1000	2000	4000
42	4	3	2	1	0	−1	−2
71	3	2	1	0	−1	−2	−3
113	2	1	0	−1	−2	−3	−4
170	1	0	−1	−2	−3	−4	−5
283	0	−1	−2	−3	−4	−5	−6
425	−1	−2	−3	−4	−5	−6	−7

Table 32 Values for B in Equation (19)

Distance from Sound Source, m	Value for B, dB
0.9	5
1.2	6
1.5	7
1.8	8
2.4	9
3.0	10
4.0	11
4.9	12
6.1	13

Table 33 Values for C in Equation (20)

Distance from Sound Source, m	Value for C, dB Octave Band Center Frequency, Hz						
	63	125	250	500	1000	2000	4000
0.9	5	5	6	6	6	7	10
1.2	6	7	7	7	8	9	12
1.5	7	8	8	8	9	11	14
1.8	8	9	9	9	10	12	16
2.4	9	10	10	11	12	14	18
3.0	10	11	12	12	13	16	20
4.0	11	12	13	13	15	18	22
4.9	12	13	14	15	16	19	24
6.1	13	15	15	16	17	20	26
7.6	14	16	16	17	19	22	28
9.8	15	17	17	18	20	23	30

usually include carpeted rooms that have sound-absorptive ceilings. If a normally furnished room has a room volume less than 420 m³ and the sound source is a single point source, the sound pressure levels associated with the sound source can be obtained from

$$L_p = L_w + A - B \tag{19}$$

where

L_p = sound pressure level at specified distance from sound source, dB
L_w = sound power level of sound source, dB

Values for A and B are given in Tables 31 and 32. If a normally furnished room has a room volume greater than 420 m³ and the sound source is a single point source, the sound pressure levels associated with the sound source can be obtained from

$$L_p = L_w - C - 5 \tag{20}$$

Values for C are given in Table 33. Equation (20) can be used for room volumes of up to 4250 m³. The accuracy of the above equations is typically within 2 to 5 dB.

Distributed Array of Ceiling Sound Sources

In many office buildings, air supply outlets are located flush with the ceiling of the conditioned space and constitute an array of distributed sound sources in the ceiling. The geometric pattern depends on the floor area served by each outlet, the ceiling height, and the

Table 34 Values for D in Equation (21)

Floor Area per Diffuser, m²	Value for D, dB Octave Band Center Frequency, Hz						
	63	125	250	500	1000	2000	4000
Ceiling height 2.4 to 2.7 m							
9.3 to 14	2	3	4	5	6	7	8
18.5 to 23	3	4	5	6	7	8	9
Ceiling height 3.0 to 3.7 m							
14 to 18.5	4	5	6	7	8	9	10
23 to 28	5	6	7	8	9	10	11
Ceiling height 4.3 to 4.9 m							
23 to 28	7	8	9	10	11	12	13
32.5 to 37	8	9	10	11	12	13	14

thermal load distribution. In the interior zones of a building where the thermal load requirements are essentially uniform, the air delivery per outlet is usually the same throughout the space; thus, the sound sources tend to have nominally equal sound power levels. One method of calculating the sound pressure levels in a room associated with such a distributed array is to (1) use Equation (19) or (20) to calculate the sound pressure levels associated with each individual air outlet at specified locations in the room and then (2) logarithmically add the sound pressure levels associated with each diffuser at each observation point. This calculation procedure can be very tedious and time consuming when there are a large number of ceiling air outlets.

For a distributed array of ceiling sound sources (air outlets) of nominally equal sound power, the room sound pressure levels tend to be uniform in a plane parallel to the ceiling. Although the sound pressure levels decrease with distance from the ceiling along a vertical axis, the sound pressure level along any selected horizontal plane is nominally constant. The calculation for a distributed ceiling array can be greatly simplified by using Equation (21) instead of Equation (19) or (20). For this case, it is desirable to use a reference plane of 1.5 m above the floor (the average distance between seated and standing head height):

$$L_{p(1.5\,m)} = L_{w(s)} - D \qquad (21)$$

where

$L_{p(1.5\,m)}$ = sound pressure level 1.5 m above floor, dB
$L_{w(s)}$ = sound power level of single diffuser in array, dB

Values for D are given in Table 34.

Nonstandard Rooms

Equations (19) through (21) are based on the assumption that the acoustic characteristics of a room range from average to medium dead, which is generally true of most rooms. However, some rooms may be acoustically medium-live to live (they have little sound absorption). These rooms may be sports or athletic areas, concert halls, or other rooms that are designed to be live, or they may be rooms that are improperly designed from an acoustic standpoint. For such rooms, Equations (19) through (21) can overestimate the decrease in the sound pressure levels associated with the room sound correction by as much as 10 to 15 dB. Thus, they should not be used for acoustically live rooms. When these or other nonstandard rooms are encountered, the calculation procedures referred to by Reynolds and Bledsoe (1991) and Reynolds and Bevirt (1994a) should be used.

MECHANICAL EQUIPMENT ROOM SOUND ISOLATION

Mechanical equipment is inherently noisy. Pumps, chillers, cooling towers, boilers, and fans create noise that is difficult to contain. Noise-sensitive spaces must be isolated from these sources. One of the best precautions is to locate mechanical spaces away from acoustically

critical spaces. Buffer zones (storage rooms, corridors, or less noise-sensitive spaces) can be placed between mechanical equipment rooms and rooms requiring quiet. Sound transmission through roofs and exterior walls is usually less of a problem, making corner rooms and top-floor spaces reasonably good for housing mechanical equipment.

Construction enclosing a mechanical equipment room should be poured concrete or masonry units with enough surface mass to provide adequate sound transmission loss. Walls must be airtight and caulked at the edges to prevent sound leaks. Floors and ceilings should be concrete slabs, except for the ceiling or roof deck above a top-floor mechanical space.

Penetrations of the mechanical equipment room enclosure create potential paths for sound to escape into adjacent spaces. Therefore, wherever ducts, pipes, conduits, and the like penetrate the walls, floor, or ceiling of a mechanical equipment room, it is necessary to acoustically treat the opening for adequate noise control. A 12- to 16-mm clear space should be left all around the penetrating element and filled with fibrous material for the full depth of the penetration. Both sides of the penetration should be sealed airtight with a non-hardening resilient sealant (Figure 28).

Doors in mechanical equipment rooms are frequently the weak link in the enclosure. Where noise control is important, they should be as heavy as possible, be gasketed around the perimeter, have no grilles or other openings, and be self-closing. If such doors lead to sensitive spaces, two doors separated by a 1- to 3-m corridor may be necessary. Mechanical room doors should open out, not in, so that the negative pressure in the room will keep the door sealed against the jamb instead of holding it against the latch.

Penetration of Walls by Ducts

Ducts passing through the mechanical equipment room enclosure pose an additional problem. Sound can be transmitted to either side of the wall via the duct walls. Airborne sound in the mechanical room can be transmitted into the duct (breakin) and enter an adjacent space by reradiating (breakout) from the duct walls, even if the duct contains no grilles, registers, diffusers, or other openings.

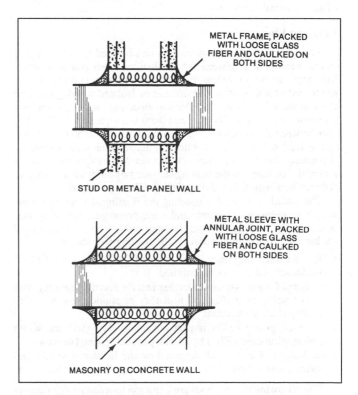

Fig. 28 Typical Duct, Conduit, and Pipe Wall Penetration

Sound levels in ducts close to fans are usually high. Sound can come not only from the fan but also from pulsating duct walls, excessive air turbulence, and air buffeting caused by tight or restricted fan airflow entrance or exit configurations. Controlling low-frequency noise propagation from these sources can be difficult. Thus, duct layout and airflow conditions should be carefully considered to avoid low-frequency sound generation as much as possible.

Mechanical Chases

Mechanical chases and shafts should be treated the same way as mechanical equipment rooms, especially if they contain noise-producing equipment. The shaft should be closed at the mechanical equipment room, and shaft walls should have a surface mass sufficient to reduce sound transmission to noise-sensitive areas to acceptable levels. Chases should not be allowed to become speaking tubes between spaces requiring different acoustical environments. Any vibrating piping, duct conduits, or equipment should be isolated so that vibration is not transmitted to the shaft walls and the general building construction.

If mechanical equipment rooms are to be used as supply or return plenums, all openings into the equipment room plenum space may require noise control treatment. This is especially true if the ceiling space just outside the equipment room is used as a return air plenum and the ceiling is an acoustical ceiling. Most acoustical ceilings are almost acoustically transparent at low frequencies.

Special Wall Construction

When mechanical equipment rooms must be placed adjacent to offices, conference rooms, or other noise-sensitive areas, the building construction around the equipment space must reduce the noise enough to satisfy the acoustical requirements of the nearby spaces. Depending on the degree of noise sensitivity, it may be necessary to consider high transmission loss walls. Tables 35, 36, and 37 give the transmission loss values for selected mechanical equipment room single wall configurations. If simple walls are inadequate, double walls, isolated from each other, should be considered with the advice of an acoustical consultant.

Floating Floors

Typical floating floor construction can be used to further reduce sound transmission between the mechanical room and noise-sensitive areas above or below it. This construction consists of two reinforced concrete slabs: the floating or isolated wearing slab and the structural floor slab. The floating floor slab is designed with a minimum thickness of 100 mm and there is a separation of 25 to 100 mm between the structural and the floating slabs. A 50-mm separation is most common. The connection between the slabs is made by supporting the floating slab on permanent load-bearing resilient material. The edges of the floating slab are isolated from the structure with resilient material, and the joint is resiliently caulked.

The isolation material supporting the floating slab must be resilient, have permanent dynamic and static properties, and safely support both the floating slab and the imposed live load for the life of the building. Isolation material for the floating slab should

- Have a natural frequency in the range of 7 to 15 Hz under the load conditions that exist in the building
- Be tested for load versus deflection and for natural frequency with known aging properties and a history of applications in similar floating floor installations
- Be tested in service for impact insulation class (IIC) and sound transmission class (STC) by an independent testing laboratory
- Be designed for the loads imposed on the isolator in service and during construction

The following two methods are common for constructing floating floors:

Table 35 Transmission Loss Values of Drywall Configurations

	Transmission Loss, dB Octave Band Center Frequency, Hz						
	63	125	250	500	1000	2000	4000

1. 92-mm metal studs; 15.9-mm gypsum wallboard screwed to both sides of studs; all joints taped and finished

	63	125	250	500	1000	2000	4000
	11	20	30	37	47	40	44

2. Same as (1) except 50-mm, 48 kg/m³ fiberglass blanket placed between studs

| | 14 | 23 | 40 | 45 | 53 | 47 | 48 |

3. 92 mm metal studs; two layers 15.9-mm gypsum wallboard screwed to both sides of studs; all joints taped and finished

| | 19 | 27 | 40 | 46 | 52 | 48 | 48 |

4. Same as (3) except 75-mm, 48 kg/m³ fiberglass blanket placed between studs

| | 24 | 32 | 43 | 50 | 52 | 49 | 50 |

5. 64-mm metal studs; two layers 15.9-mm gypsum wallboard screwed to both sides of studs; all exposed joints taped and finished

| | 18 | 26 | 37 | 45 | 51 | 47 | 49 |

6. Same as (5) except 50-mm, 48 kg/m³ fiberglass blanket placed between studs

| | 22 | 31 | 44 | 50 | 53 | 49 | 51 |

7. 92-mm metal studs; two layers 15.9-mm gypsum wallboard screwed to horizontal resilient channels on one side; three layers 15.9-mm gypsum wallboard screwed to studs on other side; 75-mm, 48 kg/m³ fiberglass between studs

| | 27 | 40 | 54 | 61 | 65 | 63 | 67 |

Note: The 63 Hz *TL* values are estimated from higher frequency *TL* values.

Table 36 Transmission Loss Values of Masonry/Floor/Ceiling Configurations

	Transmission Loss, dB Octave Band Center Frequency, Hz						
	63	125	250	500	1000	2000	4000

1. 152 × 203 × 457 mm 3-cell light concrete block (9.5 kg/block)

| | 24 | 28 | 34 | 39 | 44 | 50 | 53 |

2. 152 × 203 × 457 mm 3-cell dense concrete block (16.3 kg/block)

| | 27 | 31 | 36 | 42 | 49 | 50 | 56 |

3. 203 × 203 × 457 mm 3-cell light concrete block (12.7 kg/block)

| | 31 | 33 | 37 | 40 | 46 | 51 | 55 |

4. 203 × 203 × 457 mm 3-cell light concrete block (15.4 kg/block)

| | 29 | 33 | 39 | 45 | 49 | 56 | 60 |

5. 203 × 203 × 457 mm 3-cell dense concrete block (21.8 kg/block)

| | 30 | 34 | 40 | 49 | 52 | 59 | 57 |

6. 76-mm thick, 2560 kg/m³ concrete (195 kg/m²)

| | 30 | 35 | 39 | 43 | 52 | 58 | 64 |

7. 102-mm thick, 2300 kg/m³ concrete (235 kg/m²)

| | 32 | 34 | 35 | 37 | 42 | 49 | 55 |

8. 127-mm thick, 2480 kg/m³ concrete (315 kg/m²)

| | 37 | 41 | 43 | 47 | 54 | 59 | 63 |

9. 152-mm thick, 2560 kg/m³ concrete (390 kg/m²)

| | 33 | 38 | 43 | 50 | 58 | 64 | 68 |

10. 203-mm thick, 2300 kg/m³ concrete (470 kg/m²)

| | 40 | 44 | 47 | 54 | 58 | 63 | 67 |

11. Prefabricated 76-mm deep trapezoidal concrete channel slabs mortared together on 508-mm centers (137 kg/m²)

| | 33 | 34 | 34 | 38 | 45 | 55 | 61 |

Note: The 63 Hz *TL* values are estimated from higher frequency *TL* values.

Table 37 Transmission Loss Values of Painted Masonry Block Walls and Painted Block Walls with Resiliently Mounted Gypsum Wallboard

		Transmission Loss, dB Octave Band Center Frequency, Hz				
63	**125**	**250**	**500**	**1000**	**2000**	**4000**

1. 152 × 203 × 457 mm 3-cell light concrete block (9.5 kg/block), both sides painted

| 32 | 36 | 36 | 41 | 45 | 54 | 58 |

2. 152 × 203 × 457 mm 3-cell dense concrete block (16.3 kg/block), both sides painted

| 32 | 36 | 38 | 42 | 49 | 53 | 59 |

3. 203 × 203 × 457 mm 3-cell light concrete block (12.7 kg/block), both sides painted

| 34 | 38 | 38 | 40 | 46 | 54 | 58 |

4. Same as (1) except 12.7-mm gypsum wallboard screwed to resilient channels on one side

| 32 | 36 | 43 | 48 | 61 | 66 | 66 |

5. Same as (3) except 15.9-mm gypsum wallboard screwed to resilient channels on one side

| 35 | 39 | 40 | 48 | 59 | 61 | 66 |

Note: The 63 Hz *TL* values are estimated from higher frequency *TL* values.

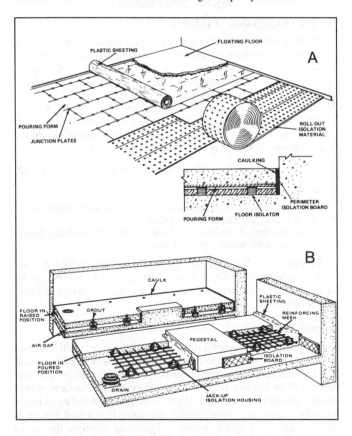

Fig. 29 Two Typical Floating Floor Constructions

1. Individual pads are spaced on 300- to 600-mm centers each way and covered with plywood or sheet metal. Low-density absorption material can be placed between the pads to reduce the tunneling effect. Reinforced concrete is placed directly on the waterproofed panels and cured; then the equipment is set (Figure 29A).
2. Cast-in-place canisters are placed on 600- to 1200-mm centers each way on the structural floor. Reinforced concrete is poured over the canisters; after curing, the entire slab is raised into operating position, the canister access holes are grouted, and the equipment is set (Figure 29B).

Table 38 Sound Transmission Loss Through Typical Floor Construction

	Transmission Loss, dB Octave Band Center Frequency, Hz				
125	**250**	**500**	**1000**	**2000**	**4000**

1. 150-mm concrete structural floor

| 35 | 36 | 40 | 46 | 53 | 58 |

2. 150-mm floor slab, 50-mm airspace, 100-mm floating floor, equipment noise impinging on walls and ceiling

| 44 | 47 | 58 | 68 | 75 | 87 |

3. 150-mm floor slab, 50-mm airspace, 100-mm floating floor, equipment noise contained to avoid flanking

| 52 | 64 | 78 | 90 | 96 | 73 |

With either method, mechanical equipment with vibration isolators can be placed on the floating slab unless the equipment operates at shaft rotation speeds of 0.7 to 1.4 times the resonant frequency of the floating slab system. In that case, the equipment should be supported with isolators on structural slab extensions that penetrate through the floating slab. This type of installation requires careful attention to detail to avoid acoustical flanking associated with improper direct contact between the floating slab and structural penetrations. Floating floors primarily control airborne sound transmission; they are not intended to be used in place of vibration isolators and/or inertia bases.

The actual resonant frequency of the floating slab is determined by both the stiffness of the resilient elements used to support the slab and the stiffness of the airspace between the structural and floating slabs. With a 50-mm airspace and a 100-mm thick floating slab, the resonant frequency can be expected to be near 18 Hz. Thus, equipment operating between about 12.5 and 25 Hz (750 and 1500 rpm) should be supported on structural slab extensions, unless the equipment is rated at about 4 kW or less.

Resilient material inserted between the floating and structural slabs results in an extremely resilient upper concrete slab. This slab must be designed to operate within the safe bending limits of the concrete. While floating floor tests have shown transmission loss ratings exceeding STC 75, these can only be realized in field installations where all flanking sound transmission paths or short circuits have been minimized. Table 38 shows improvement of sound transmission loss by floating floor slabs, with or without flanking noise control.

Enclosed Air Cavity

If acoustically critical spaces are located over a mechanical equipment room, it may be necessary to (1) install a floating floor in the space above it or (2) form a totally enclosed air cavity between the spaces by installing a resiliently suspended dense plaster or gypsum board ceiling in the equipment room.

Ideally, such a ceiling can be positioned between building beams, leaving the beam bottoms available for hanging equipment, piping, ducts, and so forth, without penetrating the ceiling with hanger rods or other devices. In the case of bar joist construction, it may be easier to resiliently attach the ceiling to the underside of the joists and then support all equipment, ducts, and pipes from the mechanical room floor or from wall supports that are properly vibration isolated.

FUME HOOD DUCT SYSTEM DESIGN

Fume hood exhaust systems are often the major sound source in a laboratory and, as such, require noise control. The exhaust system may consist of either individual exhaust fans ducted to separate fume hoods or a central exhaust fan connected through a distribution system to a large number of hoods. In either case, the sound levels produced in the laboratory space can be estimated using procedures described in this section. Table 2 lists noise level design criteria for laboratory spaces using fume hoods.

To minimize static pressure loss and fan power consumption within a duct system, fume hood ducts should either be large enough to permit the rated flow of air through the duct at a velocity no greater than 10 m/s or be consistent with regulatory requirements. Duct velocities in excess of 10 m/s should be avoided for acoustical reasons and to conserve energy, unless the resulting increases in sound levels, static pressure, and fan power requirements are deemed acceptable for the spaces served.

Noise control measures that have been successfully applied to a variety of fume hood systems include the following:

1. Use backward inclined or forward curved rather than radial blade fans where conditions permit.
2. Select the fan to operate at a low tip speed and maximum efficiency.
3. Use prefabricated duct attenuators or sections of lined ducts where conditions permit.

All potential noise control measures should be carefully evaluated for compliance with applicable codes, safety requirements, and corrosion resistance requirements of the specific fume hood system. In addition, vibration isolation for fume hood exhaust fans is generally required. However, for some laboratory facilities, particularly those having electron microscopes, vibration control can be critical, and a vibration specialist should be consulted.

SOUND CONTROL FOR OUTDOOR EQUIPMENT

Outdoor mechanical equipment should be carefully selected, installed, and maintained to prevent sound radiated by the equipment from annoying people outdoors or in nearby buildings and from violating local noise codes. The difference between sound levels at the listening location with the equipment operating and the ambient levels with the equipment not operating will determine whether the noise from the equipment is annoying. Equipment with strong tonal components will more likely evoke complaints than equipment with a broadband noise spectrum.

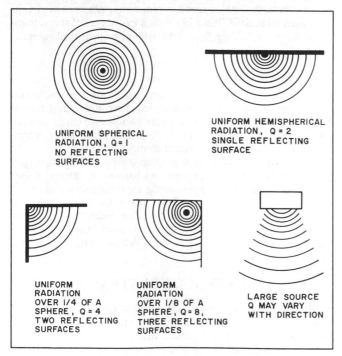

Fig. 30   Directivity Factors for Various Radiation Patterns

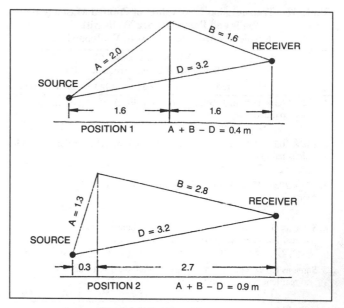

Fig. 31   Noise Barrier

Sound Propagation Outdoors

If the equipment sound power level spectrum and ambient sound pressure level spectrum are known, the contribution of the equipment to the sound level at any location can be estimated by analyzing the sound transmission paths involved. Outdoors, when there are no intervening barriers, the principal factors are reflections from buildings near the equipment and the distance to the specific location (Figure 30).

The following equation may be used to estimate the decibel differences between the sound power level of the equipment and the sound pressure level at any distance from it, at any frequency:

$$L_p = L_w + 10 \log Q - 20 \log d \qquad (22)$$

where

L_p = sound pressure level at distance d (m) from sound source, dB
L_w = sound power level of sound source, dB
Q = directivity factor associated with the way sound radiates from the sound source (refer to Figure 31)

Equation (22) does not apply where d is less than twice the maximum dimension of the sound source. The value of L_p may be low by up to 5 dB where d is between two and five times the maximum sound source dimension.

Sound Barriers

A sound barrier is a solid structure that intercepts the direct sound path from a sound source to a receiver. It reduces the sound pressure level within its shadow zone. Figure 31 illustrates the geometrical aspects of an outdoor barrier where no extraneous surfaces reflect sound into the protected area. Here the barrier is treated as an intentionally constructed noise control structure. If a sound barrier is placed between a sound source and receiver location, the sound pressure level L_p in Equation (22) is reduced by the insertion loss IL associated with the barrier.

Table 39 gives the insertion loss of an outdoor ideal solid barrier when (1) no surfaces reflect sound into the shadow zone, and (2) the sound transmission loss of the barrier wall or structure is at least 10 dB greater at all frequencies than the insertion loss expected of the barrier.

The path-length difference referred to in Table 39 is given by

$$\text{Path-length difference} = A + B - D \qquad (23)$$

where A, B, and D are as specified in Figure 31.

Table 39   Insertion Loss Values of an Ideal Solid Barrier

Path-Length Difference, m	Insertion Loss, dB Octave Band Center Frequency, Hz							
	31	63	125	250	500	1000	2000	4000
0.003	5	5	5	5	5	6	7	8
0.006	5	5	5	5	5	6	8	9
0.015	5	5	5	5	6	7	9	10
0.03	5	5	5	6	7	9	11	13
0.06	5	5	6	8	9	11	13	16
0.15	6	7	9	10	12	15	18	20
0.3	7	8	10	12	14	17	20	22
0.6	8	10	12	14	17	20	22	23
1.5	10	12	14	17	20	22	23	24
3.0	12	15	17	20	22	23	24	24
6.1	15	18	20	22	23	24	24	24
15.2	18	20	23	24	24	24	24	24

The limiting value of about 24 dB is caused by scattering and refraction of sound into the shadow zone formed by the barrier. For large distances outdoors, this scattering and bending of sound waves into the shadow zone reduces the effectiveness of the barrier. For a conservative estimate, the height of the sound source location should be taken as the topmost part of the sound source, and the height of the sound receiver should be taken as the topmost location of the receiver, such as the top of the second-floor windows in a two-floor house or at a height of 1.5 m for a standing person.

Reflecting Surfaces. There should be no other surfaces that can reflect sound around the ends or over the top of the barrier into the barrier shadow zone. Figure 32 shows examples of reflecting surfaces that can reduce the effectiveness of a barrier wall. These situations should be avoided.

Width of Barrier. Each end of the barrier should extend horizontally beyond the line of sight from the outer edge of the source to the outer edge of the receiver position by a distance of at least three times the value of the path-length distance. Near the end of the barrier, the barrier effectiveness is reduced because some sound is diffracted over the top of the barrier, some sound is diffracted around the end of the barrier, and some sound is reflected or scattered from various nonflat surfaces along the ground near the end of the barrier. In critical situations, the barrier should extend around the sound source so that it completely encloses the source. This is necessary to eliminate or reduce the effects of reflecting surfaces that may exist between the sound source and receiver.

Reflection from Barrier. A large, flat reflecting surface, such as a barrier wall, may reflect more sound in the opposite direction than there would have been with no wall present. If there is no special focusing effect, the wall may produce about 2 or 3 dB higher levels at most in the direction of the reflected sound.

SYSTEM DESIGN PROCEDURES

In order to deal effectively with each of the different sound sources and related sound transmission paths associated with an HVAC system, the following design procedures are suggested:

1. Determine the design goal for HVAC system noise for each critical area according to its use and construction. Choose desirable RC levels from Table 2.

2. Relative to equipment that radiates sound directly into a room, select equipment that will be quiet enough to meet the desired design goal.

3. If central or roof-mounted mechanical equipment is used, complete an initial design and layout of the HVAC system, using acoustical treatment where it appears appropriate.

4. Starting at the fan, appropriately add the sound attenuation and sound power levels associated with the central fan(s), fan-powered terminal units (if used), and duct elements between the central fan(s) and the room of interest, and convert to the corresponding sound pressure levels in the room. Be sure to investigate both the supply and return air paths. Investigate and control possible duct sound breakout when fans are adjacent to the room of interest or roof-mounted fans are above the room of interest.

5. If the mechanical equipment room is adjacent to the room of interest, determine the sound pressure levels in the room of interest that are associated with sound transmitted through the mechanical equipment room wall.

6. Combine on an energy basis the sound pressure levels in the room of interest that are associated with all of the sound paths between the mechanical equipment room or roof-mounted unit and the room of interest.

7. Determine the corresponding RC level associated with the calculated total sound pressure levels in the room of interest.

8. If the RC level exceeds the design goal, determine the octave frequency bands in which the corresponding sound pressure levels are exceeded and the sound paths that are associated with these octave frequency bands.

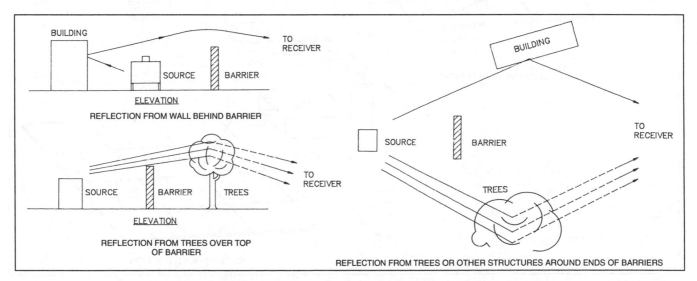

Fig. 32   Examples of Surfaces That Can Reflect Sound Around or Over a Barrier Wall

9. Redesign the system, adding additional sound attenuation to the paths which contribute to the excessive sound pressure levels in the room of interest.
10. Repeat Steps 4 through 9 until the desired design goal is achieved.
11. Repeat Steps 3 through 10 for every room that is to be analyzed.
12. Make sure that noise radiated by outdoor equipment will not disturb adjacent properties or interfere with criteria established in Step 1.

Example 7. Individual examples have been given in the preceding sections that demonstrate how to calculate equipment and airflow-generated sound power levels and sound attenuation values associated with the system elements of HVAC air distribution systems. This is an example of a complete HVAC system, which shows how the information can be combined to determine the sound pressure levels associated with a specific HVAC system. Complete calculations for each system element are not provided. Only a summary of the tabulated results is shown.

Air is supplied to the HVAC system in this example by the rooftop unit shown in Figure 33. The receiver room is directly below the unit. The room has the following dimensions: length = 6.1 m, width = 6.1 m; and height = 2.75 m. For this example, the roof penetrations associated with the supply and return air ducts are assumed to be well sealed, and there are no other roof penetrations. The supply side of the rooftop unit is ducted to a VAV terminal control unit that serves the room in question. A return air grille conducts air to a common ceiling return air plenum. The return air is then directed to the rooftop unit through a short rectangular return air duct.

Three sound paths are examined:

Path 1. Fan airborne supply air sound that enters the room from the supply air system through the ceiling diffuser.

Path 2. Fan airborne supply air sound that breaks out through the wall of the main supply air duct into the plenum space above the room.

Path 3. Fan airborne return air sound that enters the room from the inlet of the return air duct.

The sound power levels associated with the supply air and return air sides of the fan in the rooftop unit are specified by the manufacturer as follows:

	Octave Band Center Frequency, Hz						
	63	125	250	500	1000	2000	4000
Rooftop supply air = 3.3 m³/s at 620 Pa	92	86	80	78	78	74	71
Rooftop return air = 3.3 m³/s at 620 Pa	82	79	73	69	69	67	59

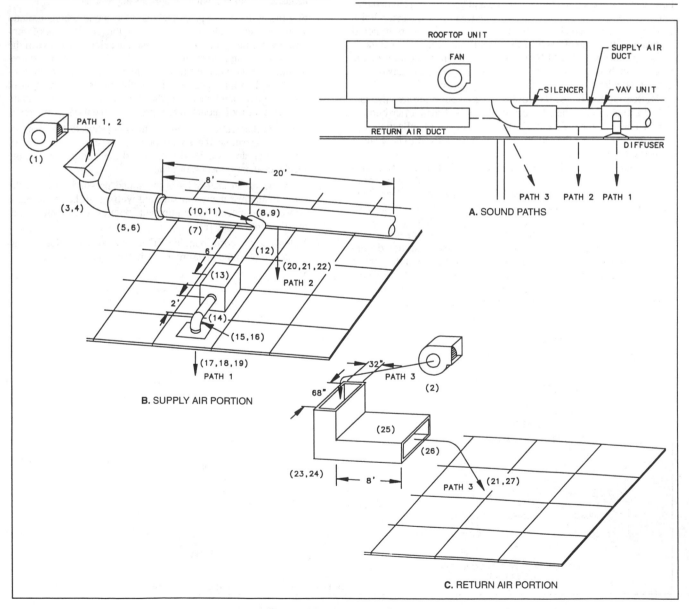

Fig. 33 Layout for Example 7

Path 1 in Example 7

No.	Description	63	125	250	500	1000	2000	4000
		Octave Band Center Frequency, Hz						

No.	Description	63	125	250	500	1000	2000	4000
1	Fan—Supply air, 3.3 m³/s, 620 Pa s.p.	92	86	80	78	78	74	71
3	560-mm wide (dia.) unlined radius elbow	0	−1	−2	−3	−3	−3	−3
	Sum with noise reduction values	92	85	78	75	75	71	68
4	90° bend without turning vanes, 320-mm radius	56	54	51	47	42	37	29
	Sum sound power levels	92	85	78	75	75	71	68
5	560-mm dia. by 1120-mm high pressure silencer	−4	−7	−19	−31	−38	−38	−27
	Sum with noise reduction values	88	78	59	44	37	33	41
6	Register noise from above silencer	68	79	69	60	59	59	55
	Sum sound power levels	88	82	69	60	59	59	55
7	560-mm dia. by 2.44-m unlined circular duct	0	0	0	0	0	0	0
10	Branch pwr. div., M-560 mm dia., B-250 mm dia.	−8	−8	−8	−8	−8	−8	−8
	Sum with noise reduction values	80	74	61	52	51	51	47
11	Duct 90° branch takeoff, 50-mm radius	56	53	50	47	43	37	31
	Sum sound power levels	80	74	61	53	52	51	47
12	250-mm dia. by 1830-mm unlined circular duct	0	0	0	0	0	0	0
13	Terminal volume register unit (gen. attn.)	0	−5	−10	−15	−15	−15	−15
14	250-mm dia. by 610-mm unlined circular duct	0	0	0	0	0	0	0
15	250-mm wide (dia.) unlined radius elbow	0	0	−1	−2	−3	−3	−3
	Sum with noise reduction values	80	69	50	36	34	33	29
16	90° bend without turning vanes, 50-mm radius	49	45	41	37	31	24	16
	Sum sound power levels	80	69	51	40	36	34	29
17	250-mm dia. diffuser end ref. loss	−16	−10	−6	−2	−1	0	0
	Sum with noise reduction values	64	59	45	38	35	34	29
18	380-mm by 380-mm rectangular diffuser	31	36	39	40	39	36	30
	Sum sound power levels	64	59	46	42	40	38	33
19	ASHRAE room correction, 1 ind. sound source	−5	−6	−7	−8	−9	−10	−11
	Sound pressure levels—receiver room	59	53	39	34	31	28	22

Path 2 in Example 7

No.	Description	63	125	250	500	1000	2000	4000
1	Fan—Supply air, 3.3 m³/s, 620 Pa s.p.	92	86	80	78	78	74	71
3	560-mm wide (dia.) unlined radius elbow	0	−1	−2	−3	−3	−3	−3
	Sum with noise reduction values	92	85	78	75	75	71	68
4	90° bend without turning vanes, 320-mm radius	56	54	51	47	42	37	29
	Sum sound power levels	92	85	78	75	75	71	68
5	560-mm dia. by 1120-mm high pressure silencer	−4	−7	−19	−31	−38	−38	−27
	Sum with noise reduction values	88	78	59	44	37	33	41
6	Register noise from above silencer	68	79	69	60	59	59	55
	Sum sound power levels	88	82	69	60	59	59	55
7	560-mm dia. by 2.44-m unlined circular duct	0	0	0	0	0	0	0
8	Branch pwr. div., M-560 mm dia., B-560 mm dia.	−1	−1	−1	−1	−1	−1	−1
	Sum with noise reduction values	87	81	68	59	58	58	54
9	Duct 90° branch takeoff, 50-mm radius	63	60	57	54	50	44	34
	Sum sound power levels	87	81	68	60	59	58	54
20	560-mm dia. by 6.1-m, 26 ga. duct breakout	−29	−29	−21	−11	−9	−7	−5
21	610-mm × 1220-mm × 16-mm lay-in ceiling	−4	−8	−8	−12	−14	−15	−15
22	Line source—Medium-dead room	−6	−5	−4	−6	−7	−8	−9
	Sound pressure levels—receiver room	48	39	35	31	29	28	25

Path 3 in Example 7

No.	Description	63	125	250	500	1000	2000	4000
2	Fan—Return air, 3.3 m³/s, 620 Pa s.p.	82	79	80	78	78	74	71
23	810-mm wide lined square elbow w/o turning vanes	−1	−6	−11	−10	−10	−10	−10
	Sum with noise reduction values	81	73	69	68	68	64	61
24	90° bend w/o turning vanes; 12.7-mm radius	77	73	68	62	55	48	38
	Sum sound power levels	82	76	72	69	68	64	61
25	810-mm × 1730-mm × 2440-mm lined duct	−2	−2	−6	−18	−15	−12	−13
26	810-mm × 1730-mm diffuser end ref. loss	−4	−2	0	0	0	0	0
21	610-mm × 1220-mm × 16-mm lay-in ceiling	−4	−8	−8	−12	−14	−15	−15
27	ASHRAE room corr., 1 ind. sound source	−8	−9	−10	−11	−12	−13	−14
	Sound pressure levels—receiver room	64	55	48	28	27	24	19

Total Sound Pressure Levels from All Paths in Example 7

Description	63	125	250	500	1000	2000	4000
Sound pressure levels Path 1	59	53	39	34	31	28	22
Sound pressure levels Path 2	48	39	35	31	29	28	25
Sound pressure levels Path 3	64	55	48	28	27	24	19
Total sound pressure levels—All paths	65	57	49	37	34	32	28

Paths 1 and 2 are associated with the supply air side of the system. Figure 33B shows a layout of the part of the supply air system that is associated with the receiver room. The main duct is a 560-mm diameter, 26 gage (0.551-mm), unlined, circular sheet metal duct. The flow volume in the main duct is 3.3 m³/s. The silencer after the radiused elbow is a 560-mm diameter by 1120-mm long, high pressure, circular silencer.

The branch junction that occurs 2.44 m from the silencer is a 45° wye. The branch duct between the main duct and the VAV control unit is a 250-mm diameter, unlined, circular sheet metal duct. The flow volume in the branch duct is 0.37 m³/s.

The straight section of duct between the VAV control unit and the diffuser is a 250-mm diameter, unlined circular sheet metal duct. The diffuser is 380 mm by 380 mm square. Assume a typical distance between the diffuser and a listener in the room is 1.5 m.

With regard to the duct breakout sound associated with the main duct, the length of the duct that runs over the room is 6.1 m. The ceiling of the room is comprised of 610-mm by 1220-mm by 16-mm lay-in ceiling tiles that have a density of 2.9 to 3.4 kg/m². The ceiling has integrated lighting and diffusers.

Path 3 is associated with the return air side of the system. Figure 33C shows a layout of the part of the return air system that is associated with the receiver room. The rectangular return air duct is lined with 50-mm thick 48 kg/m³ density fiberglass duct liner. For the return air path, assume the typical distance between the inlet of the return air duct and a listener is 3 m.

Solution:

The analysis associated with each path begins at the rooftop unit (fan) and proceeds through the different system elements to the receiver room. The system element numbers in the tables correspond to the element numbers contained in brackets in Figures 33B and 33C.

The first table is associated with Path 1. The first entry in the table is the manufacturer's values for supply air fan sound power levels (1). The second entry is the sound attenuation associated with the 560-mm diameter unlined radius elbow (3). Because the next entry is associated with the regenerated sound power levels associated with the elbow (4), the results associated with the elbow attenuation must be tabulated to determine the sound power levels at the exit of the elbow. These sound power levels and the elbow regenerated sound power levels are then added logarithmically. In a like fashion, the dynamic insertion loss values of the duct silencer (5) and the silencer regenerated sound power levels (6) are included in the table and tabulated.

Next, the attenuation associated with the 2440-mm section of 560-mm diameter duct (7) and the branch power division (10) associated with sound propagation in the 250-mm diameter branch duct are included in the table. Then the sound power levels that exist in the branch duct after the branch takeoff are calculated so that the regenerated sound power levels (11) in the branch duct associated with the branch takeoff can be logarithmically added to the results.

The sound attenuation values associated with the 1830-mm section of 250-mm diameter unlined duct (12), the terminal volume regulation unit (13), the 610-mm section of 250-mm diameter unlined duct (14), and 250-mm diameter radius elbow (15) are included in the table. The sound power levels that exist at the exit of the elbow are then calculated so that the regenerated sound power levels (16) associated with the elbow can be logarithmically added to the results. The diffuser end reflection loss (17) and the diffuser regenerated sound power levels (18) are appropriately included in the table. The sound power levels that are tabulated after element 18 are the sound power levels that exist at the diffuser in the receiver room. The final entry in the table is the room correction, which converts the sound power levels at the diffuser to their corresponding sound pressure levels at the point of interest in the receiver room.

Elements 1 through 7 in Path 2 are the same as in Path 1. Elements 8 and 9 are associated with the branch power division (8) and the corresponding regenerated sound power levels (9) associated with sound that propagates down the main duct beyond the duct branch. The next three entries in the table are the sound transmission loss associated with the duct breakout sound (20), the sound transmission loss associated with the ceiling (21), and the room correction (22), converting the sound power levels at the ceiling to corresponding sound pressure levels in the room.

The first element in Path 3 is the manufacturer's values for return air fan sound power levels (2). The next two elements are the sound attenuation associated with a 810-mm wide lined square elbow without turning vanes (23) and the regenerated sound power levels associated with the square elbow (24). The final four elements are the insertion loss associated with a 810-mm by 1730-mm by 2440-mm long rectangular sheet metal duct lined with 50-mm thick, 48 kg/m³ fiberglass duct lining (26), the diffuser end reflection loss (27), the transmission loss through the ceiling (21), and the room correction

(27) converting the sound power levels at the ceiling to corresponding sound pressure levels in the room.

The total sound pressure levels in the receiver room from the three paths are obtained by logarithmically adding the individual sound pressure levels associated with each path. From the total sound pressure levels for all three paths, the NC value in the room is NC 42, and the RC value is RC 34(R-H).

USE OF FIBERGLASS PRODUCTS IN HVAC SYSTEMS

Fiberglass duct liner continues to be the most cost-effective solution to noise control in most HVAC air duct systems. Recently, however, the use of fiberglass in several institutional, educational, and medical projects has been banned or severely limited. The decisions are based on concerns that the fibers may be carcinogenic and that the products may promote microbial growth.

ASHRAE's technical committee (TC 2.6) on sound and vibration control reviewed available research and studies and concluded the following:

1. At present, there is no clear evidence that any form of man-made mineral fibers are carcinogenic (Chapter 37, 1993 *ASHRAE Handbook*). The International Agency for Research on Cancer (IARC) has performed extensive research in regard to the carcinogenicity of fiberglass materials and found inadequate evidence to link inhalation of fiberglass wool with cancer in humans (World Health Organization 1986). IARC does indicate a possible link to cancer from glass wool based on heavy-dosage direct-injection into animals. As a result of this, the U.S. Department of Health and Human Services (USDHHS) has added glass wool to its list of substances "reasonably anticipated" to be carcinogenic (USDHHS 1994).

   The Canadian Environmental Protection Act (Environment Canada 1994) indicated that studies have failed to show any evidence that fiberglass is carcinogenic and classified the material as "unlikely to be carcinogenic to humans." In addition, the report indicated that fiberglass is not entering the environment in quantities or conditions that may constitute a danger to human life or health.

2. The best available evidence indicates that both moisture and dirt are required for microbial growth and that microbial growth can occur on any surface, including sheet metal, sound traps, filters, and turning vanes—in addition to duct lining (Morey and Williams 1991). Microbial growth is a system effect not related to one individual component. Thus, removal of fiberglass duct lining by itself will not eliminate microbial growth. Control of moisture and dirt in duct systems is the best method of reducing the potential microbial growth. Duct design for access, proper filtration, humidification, and condensate systems; methods to prevent moisture and dirt accumulation during installation and commissioning; and regular maintenance procedures can reduce the occurrence of microbial growth. ASHRAE *Standard* 62 and the North American Insulation Manufacturers Association (NAIMA), Alexandria, VA, have additional information.

Owners and designers considering removal of fiberglass duct lining from any building must understand the acoustic impact of that decision. Compensating for the removal of fiberglass duct liner may require larger and longer duct runs, larger fans and fan plenums, and the use of alternative noise control devices such as removable lined duct sections, foil-coated liner, no-fill sound traps, and active noise control, some of which may be less effective and more expensive.

Certain health care facility spaces such as operating rooms and critical care spaces should follow duct liner guidelines issued by the American Institute of Architects Committee on Architecture (AIA 1992-93). The complete removal of fiberglass duct liner from duct systems in spaces such as theaters, concert halls, recording and TV studios, and educational facilities will result in extremely poor acoustic qualities and may render the spaces unusable.

VIBRATION ISOLATION AND CONTROL

Mechanical vibration and vibration-induced noise are often major sources of occupant complaints in modern buildings. Lighter construction has made buildings more susceptible to vibration and vibration-related problems. Increased interest in energy conservation has resulted in many new buildings being designed with variable air volume systems. This often results in the location of mechanical equipment in penthouses on the roof, in the use of roof-mounted HVAC units, and in the location of mechanical equipment rooms on intermediate level floors. These trends have resulted in an increase in the number of pieces of mechanical equipment located in a building, and they often have resulted in the location of mechanical equipment adjacent to or above occupied areas.

Occupant complaints associated with building vibration typically take one of three forms:

1. The level of vibration perceived by building occupants is of sufficient magnitude to cause concern or alarm.
2. Vibration energy from mechanical equipment, which is transmitted to the building structure, is transmitted to various parts of the building and then is radiated as structure-borne noise.
3. The vibration present in a building interferes with proper operation of sensitive equipment or instrumentation.

The following sections present basic information to properly select and specify vibration isolators and to analyze and correct field vibration problems. Chapter 7 of the 1993 *ASHRAE Handbook—Fundamentals* and Reynolds and Bevirt (1994a) provide more detailed information.

Equipment Vibration

Vibration can be isolated or reduced to a fraction of the original force with resilient mounts between the equipment and the supporting structure. To determine the excessive forces that must be isolated or that adversely affect the performance or life of the equipment, criteria should be established for equipment vibration. Figures 34 and 35

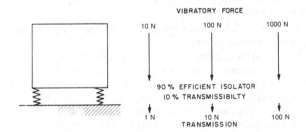

Fig. 34 Transmission to Structure Varies as Function of Magnitude of Vibration Forces

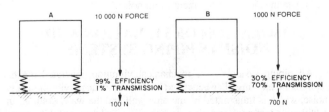

Fig. 35 Interrelationship of Equipment Vibration, Isolation Efficiency, and Transmission

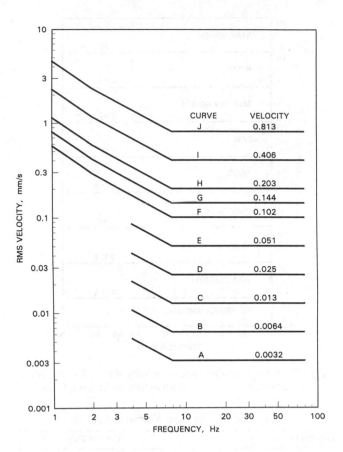

Fig. 36 Building Vibration Criteria for Vibration Measured on Building Structure

show the relation between equipment vibration levels and vibration isolators that have a fixed vibration isolation efficiency. In this case, the magnitude of transmission to the building is a function of the magnitude of the vibration force.

Vibration Criteria

Vibration criteria can be specified relative to three areas: (1) human response to vibration, (2) vibration levels associated with potential damage to sensitive equipment in a building, and (3) vibration severity of a vibrating machine. Figure 36 and Table 40 present recommended acceptable vibration criteria for vibration that can exist in a building structure (Ungar et al. 1990). The occupant vibration criteria are based on guidelines specified by ANSI *Standard* S3.29 and by ISO *Standard* 2631-2. For sensitive equipment, acceptable vibration values specified by equipment manufacturers should be used.

If acceptable vibration values are not available from equipment manufacturers, the values specified in Figure 37 can be used. The figure gives recommended equipment vibration severity ratings based on measured root-mean-square (rms) values (IRD 1988). The vibration values associated with Figure 37 are measured by vibration transducers (usually accelerometers) mounted directly on equipment, equipment structures, or bearing caps.

Vibration levels measured on equipment and its components can be affected by equipment unbalance, misalignment of components, and resonance interaction between a vibrating piece of equipment and the structural floor system on which it is placed. If a piece of equipment is balanced within acceptable tolerances and excessive vibration levels still exists, the equipment and its installation should be checked for possible resonance conditions. Table 41 gives maximum allowable rms velocity levels for selected equipment.

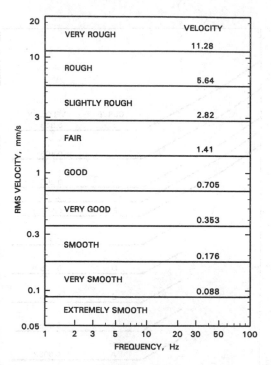

Fig. 37 Equipment Vibration Severity Rating for Vibration Measured on Equipment Structure or Bearing Caps

Table 40 Equipment Vibration Criteria

Human Occupancy	Time of Day	Curve[a]
Workshops	All	J
Office areas	All[b]	I
Residential (good environmental standards)	0700-2200[b]	H-I
	2200-0700[b]	G
Hospital operating rooms and critical work areas	All	F

Equipment Requirements	Curve[a]
Computer areas	H
Bench microscopes up to 100× magnification; laboratory robots	F
Bench microscopes up to 400× magnification; optical and other precision balances; coordinate measuring machines; metrology laboratories; optical comparators; microelectronics manufacturing equipment—Class A[c]	E
Microsurgery, eye surgery, neurosurgery; bench microscope at magnification greater than 400×; optical equipment on isolation tables; microelectronic manufacturing equipment—Class B[c]	D
Electron microscopes up to 30 000× magnification; microtomes; magnetic resonance imagers; microelectronics manufacturing equipment—Class C[c]	C
Electron microscopes at magnification greater than 30 000×; mass spectrometers; cell implant equipment; microelectronics manufacturing equipment—Class D[c]	B
Unisolated laser and optical research systems; microelectronics manufacturing equipment—Class E[c]	A

[a]See Figure 36 for corresponding curves.
[b]In areas where individuals are sensitive to vibration, use curve H.
[c]Classes of microelectronics manufacturing equipment:
  Class A: Inspection, probe test, and other manufacturing support equipment.
  Class B: Aligners, steppers, and other critical equipment for photolithography with line widths of 3 μm or more.
  Class C: Aligners, steppers, and other critical equipment for photolithography with line widths of 1 μm.
  Class D: Aligners, steppers, and other critical equipment for photolithography with line widths of 0.5 μm; includes electron-beam systems.
  Class E: Aligners, steppers, and other critical equipment for photolithography with line widths of 0.25 μm; includes electron-beam systems.

Table 41 Equipment Vibration Criteria

Equipment	Allowable rms Velocity, mm/s
Pumps	3.3
Centrifugal compressors	3.3
Fans (vent sets, centrifugal, axial)	2.3

The vibration levels measured on equipment structures should be in the "Good" region or below (Figure 37). Vibration in the "Fair" or "Slightly Rough" regions may indicate potential problems requiring maintenance. Vibration in these regions should be monitored to ensure problems do not arise. Machine vibration levels in the "Rough" and "Very Rough" regions indicate a potentially serious problem, and immediate action should be taken to identify the problem and correct it.

Specification of Vibration Isolators

Vibration isolators are selected to compensate for floor stiffness. Longer spans also allow the structure to be more flexible, permitting the building to be more easily set into vibration-type motion. Building spans, equipment operating speeds, equipment power, damping, and other factors are considered in the vibration isolator selection guide given in Table 42.

By specifying isolator deflection rather than isolation efficiency, transmissibility, or other theoretical parameters, a designer can compensate for floor stiffness and building resonances by selecting isolators that provide minimum vibration transmission and that have more deflection than the supporting floor. To apply the information from Table 42, base type, isolator type, and minimum deflection columns are added to the equipment schedule. These isolator specifications are then incorporated into mechanical specifications for the project.

The minimum deflections listed in Table 42 are based on the experience of acoustical and mechanical consultants and vibration control manufacturers. Recommended isolator type, base type, and minimum static deflection are reasonable and safe recommendations for 70 to 80% of HVAC equipment installations. The selections are based on concrete equipment room floors 100 to 300 mm thick with typical floor stiffness. The type of equipment, proximity to noise-sensitive areas, and type of building construction may alter these choices.

The following approach is suggested to develop isolator selections for specific applications:

1. Use Table 42 for floors specifically designed to accommodate mechanical equipment.
2. Use recommendations for the 6-m span column for equipment on ground-supported slabs adjacent to noise-sensitive areas.
3. For roofs and floors constructed with open web joists, thin long span slabs, wooden construction, and any unusual light construction, evaluate all equipment with a mass of more than 140 kg to determine the additional deflection of the structure caused by the equipment. Isolator deflection should be 15 times the additional deflection or the deflection shown in Table 42, whichever is greater. If the required spring isolator deflection exceeds commercially available products, consider air springs, stiffen the supporting structure, or change the equipment location.
4. When mechanical equipment is adjacent to noise-sensitive areas, isolate mechanical equipment room noise.

ISOLATION OF VIBRATION AND NOISE IN PIPING SYSTEMS

All piping systems have mechanical vibration generated by the equipment and impeller-generated and flow-induced vibration and noise, which is transmitted by the pipe wall and the water column. In addition, equipment installed on vibration isolators exhibits some motion or movement from pressure thrusts during operation. Vibration isolators have even greater movement during start-up and shutdown,

Table 42 Selection Guide for Vibration Isolation

Equipment Type	Shaft Power, kW and Other	Rpm	Slab on Grade Base Type	Iso-lator Type	Min. Defl., mm	Up to 6 m Floor Span Base Type	Iso-lator Type	Min. Defl., mm	6- to 9-m Floor Span Base Type	Iso-lator Type	Min. Defl., mm	9- to 12-m Floor Span Base Type	Iso-lator Type	Min. Defl., mm	Reference Notes
Refrigeration Machines and Chillers															
Bare compressors	All	All	A	2	6	C	3	20	C	3	45	C	4	65	2,3,12
Reciprocating	All	All	A	2	6	A	4	20	A	3	45	A	4	65	2,3,12
Centrifugal	All	All	A	1	6	A	4	20	A	3	45	A	3	45	2,3,4,12
Open centrifugal	All	All	C	1	6	C	4	20	C	3	45	C	3	45	2,3,12
Absorption	All	All	A	1	6	A	4	20	A	3	45	A	3	45	
Air Compressors and Vacuum Pumps															
Tank-mounted	Up to 7.5	All	A	3	20	A	3	20	A	3	45	A	3	45	3,13,15
	11 and over	All	C	3	20	C	3	20	C	3	45	C	3	45	3,13,15
Base-mounted	All	All	C	3	20	C	3	20	C	3	45	C	3	45	3,13,14,15
Large reciprocating	All	All	C	3	20	C	3	20	C	3	45	C	3	45	3,13,14,15
Pumps															
Closed coupled	Up to 5.6	All	B	2	6	C	3	20	C	3	20	C	3	20	16
	7.5 and over	All	C	3	20	C	3	20	C	3	45	C	3	45	16
Large inline	3.7 to 19	All	A	3	20	A	3	45	A	3	45	A	3	45	
	22 and over	All	A	3	45	A	3	45	A	3	45	A	3	65	
End suction and split case	Up to 30	All	C	3	20	C	3	20	C	3	45	C	3	45	16
	37 to 93	All	C	3	20	C	3	20	C	3	45	C	3	65	10,16
	110 and over	All	C	3	20	C	3	45	C	3	45	C	3	65	10,16
Cooling Towers	All	Up to 300	A	1	6	A	4	90	A	4	90	A	4	90	5,8,18
		301 to 500	A	1	6	A	4	65	A	4	65	A	4	65	5,18
		500 and over	A	1	6	A	4	20	A	4	20	A	4	45	5,18
Boilers—Fire-tube	All	All	A	1	6	B	4	20	B	4	45	B	4	65	4
Axial Fans, Fan Heads, Cabinet Fans, and Fan Sections															
Up to 560 mm dia.	All	All	A	2	6	A	3	20	A	3	20	C	3	20	4,9
610 mm dia. and over	Up to 500 Pa s.p.	Up to 300	B	3	65	C	3	90	C	3	90	C	3	90	9
		300 to 500	B	3	20	B	3	45	C	3	65	C	3	65	9
		501 and over	B	3	20	B	3	45	B	3	45	B	3	45	9
	501 Pa s.p. and over	Up to 300	C	3	65	C	3	90	C	3	90	C	3	90	3,9
		300 to 500	C	3	45	C	3	45	C	3	65	C	3	65	3,8,9
		501 and over	C	3	20	C	3	45	C	3	45	C	3	65	3,8,9
Centrifugal Fans															
Up to 560 mm dia	All	All	B	2	6	B	3	20	B	3	20	C	3	45	9,19
610 mm dia. and over	Up to 30	Up to 300	B	3	65	B	3	90	B	3	90	B	3	90	8,19
		300 to 500	B	3	45	B	3	45	B	3	65	B	3	65	8,19
		501 and over	B	3	20	B	3	20	B	3	20	B	3	45	8,19
	37 and over	Up to 300	C	3	65	C	3	90	C	3	90	C	3	90	2,3,8,9,19
		300 to 500	C	3	45	C	3	45	C	3	65	C	3	65	2,3,8,9,19
		501 and over	C	3	25	C	3	45	C	3	45	C	3	65	2,3,8,9,19
Propeller Fans															
Wall-mounted	All	All	A	1	6	A	1	6	A	1	6	A	1	6	
Roof-mounted	All	All	A	1	6	A	1	6	B	4	45	D	4	45	
Heat Pumps	All	All	A	3	20	A	3	20	A	3	20	A/D	3	45	
Condensing Units	All	All	A	1	6	A	4	20	A	4	45	A/D	4	45	
Packaged AH, AC, H and V Units															
All	Up to 7.5	All	A	3	20	A	3	20	A	3	20	A	3	20	19
	11 and over, up to 1 kPa s.p.	Up to 300	A	3	20	A	3	90	A	3	90	C	3	90	2,4,8,19
		301 to 500	A	3	20	A	3	65	A	3	65	A	3	65	4,19
		501 and over	A	3	20	A	3	45	A	3	45	A	3	45	4,19
	11 and over, 1 kPa s.p. and over	Up to 300	B	3	20	C	3	90	C	3	90	C	3	90	2,3,4,8,9
		301 to 500	B	3	20	C	3	45	C	3	65	C	3	65	2,3,4,9
		501 and over	B	3	20	C	3	45	C	3	45	C	3	65	2,3,4,9
Packaged Rooftop Equipment	All	All	A/D	1	6	D	3	20	See Note 17						5,6,8,17
Ducted Rotating Equipment															
Small fans, fan-powered boxes	Up to 283 L/s	All	A	3	13	A	3	13	A	3	13	A	3	13	7
	284 L/s and over	All	A	3	20	A	3	20	A	3	20	A	3	20	7
Engine-Driven Generators	All	All	A	3	20	C	3	45	C	3	65	C	3	90	2,3,4

Base Types:
A. No base, isolators attached directly to equipment (Note 27)
B. Structural steel rails or base (Notes 28 and 29)
C. Concrete inertia base (Note 30)
D. Curb-mounted base (Note 31)

Isolator Types:
1. Pad, rubber, or glass fiber (Notes 20 and 21)
2. Rubber floor isolator or hanger (Notes 20 and 25)
3. Spring floor isolator or hanger (Notes 22, 23, and 25)
4. Restrained spring isolator (Notes 22 and 24)
5. Thrust restraint (Note 26)

NOTES FOR VIBRATION ISOLATOR SELECTION GUIDE (TABLE 42)

The notes in this section are keyed to the numbers listed in the column titled "Reference Notes" in Table 42 and to other reference numbers throughout the table. While the guide is conservative, cases may arise where vibration transmission to the building is still excessive. If the problem persists after all short circuits have been eliminated, it can almost always be corrected by increasing isolator deflection, using low-frequency air springs, changing operating speed, reducing vibratory output by additional balancing or, as a last resort, changing floor frequency by stiffening or adding more mass.

Note 1. Isolator deflections shown are based on a floor stiffness that can be reasonably expected for each floor span and class of equipment.

Note 2. For large equipment capable of generating substantial vibratory forces and structure-borne noise, increase isolator deflection, if necessary, so isolator stiffness is at least 0.10 times the floor stiffness.

Note 3. For noisy equipment adjoining or near noise-sensitive areas, see the text section on Mechanical Equipment Room Sound Isolation.

Note 4. Certain designs cannot be installed directly on individual isolators (Type A), and the equipment manufacturer or a vibration specialist should be consulted on the need for supplemental support (Base Type).

Note 5. Wind load conditions must be considered. Restraint can be achieved with restrained spring isolators (Type 4), supplemental bracing, or limit stops.

Note 6. Certain types of equipment require a curb-mounted base (Type D). Airborne noise must be considered.

Note 7. See the text section on Resilient Pipe Hangers and Supports for hanger locations adjoining equipment and in equipment rooms.

Note 8. To avoid isolator resonance problems, select isolator deflection so that resonance frequency is 40% or less of the lowest operating speed of equipment.

Note 9. To limit undesirable movement, thrust restraints (Type 5) are required for all ceiling-suspended and floor-mounted units operating at 50 mm and more total static pressure.

Note 10. Pumps over 55 kW may require extra mass and restraining devices.

Isolation for Specific Equipment

Note 12. Refrigeration Machines: Large centrifugal, hermetic, and reciprocating refrigeration machines generate very high noise levels, and special attention is required when such equipment is installed in upper stories or near noise-sensitive areas. If such equipment is to be located near extremely noise-sensitive areas, confer with an acoustical consultant.

Note 13. Compressors: The two basic reciprocating compressors are (1) single- and double-cylinder vertical, horizontal or L-head, which are usually air compressors; and (2) Y, W, and multihead or multicylinder air and refrigeration compressors. Single- and double-cylinder compressors generate high vibratory forces requiring large inertia bases (Type C) and are generally not suitable for upper-story locations. If such equipment must be installed in upper stories or on grade locations near noise-sensitive areas, unbalanced forces should be obtained from the equipment manufacturer, and a vibration specialist should be consulted for design of the isolation system.

Note 14. Compressors: When using Y, W, and multihead and multicylinder compressors, obtain the magnitude of unbalanced forces from the equipment manufacturer so that the necessity for an inertia base can be evaluated.

Note 15. Compressors: Base-mounted compressors through 4 kW and horizontal tank-type air compressors through 8 kW can be installed directly on spring isolators (Type 3) with structural bases (Type B) if required, and compressors 10 to 75 kW on spring isolators (Type 3) with inertia bases (Type C) with a mass of one to two times the compressor mass.

Note 16. Pumps: Concrete inertia bases (Type C) are preferred for all flexible-coupled pumps and are desirable for most close-coupled pumps, although steel bases (Type B) can be used. Close-coupled pumps should not be installed directly on individual isolators (Type A) because the impeller usually overhangs the motor support base, causing the rear mounting to be in tension. The primary requirements for Type C bases are strength and shape to accommodate base elbow supports. Mass is not usually a factor, except for pumps over 55 kW where extra mass helps limit excess movement due to starting torque and forces. Concrete bases (Type C) should be designed for a thickness of one-tenth the longest dimension with minimum thickness as follows: (1) for up to 20 kW, 150 mm; (2) for 30 to 55 kW, 200 mm; and (3) for 75 kW and higher, 300 mm.

Pumps over 55 kW and multistage pumps may exhibit excessive motion at start-up; supplemental restraining devices can be installed if necessary. Pumps over 90 kW may generate high starting forces, so a vibration specialist should be consulted for installation recommendations.

Note 17. Packaged Rooftop Air-Conditioning Equipment: This equipment is usually on light structures that are susceptible to sound and vibration transmission. The noise problem is further compounded by curb-mounted equipment, which requires large roof openings for supply and return air.

The table shows Type D vibration isolator selections for all spans up to 6 m, but extreme care must be taken for equipment located on spans of over 6 m, especially if construction is open web joists or thin low-density slabs. The recommended procedure is to determine the additional deflection caused by equipment in the roof. If additional roof deflection is 6 mm or under, the isolator can be selected for 15 times the additional roof deflection. If additional roof deflection is over 6 mm, supplemental stiffening should be installed or the unit should be relocated.

For units, especially large units, capable of generating high noise levels, consider (1) mounting the unit on a platform above the roof deck to provide an air gap (buffer zone) and (2) locating the unit away from the roof penetration, thus permitting acoustical treatment of ducts before they enter the building.

Some rooftop equipment has compressors, fans, and other equipment isolated internally. This isolation is not always reliable because of internal short circuiting, inadequate static deflection, or panel resonances. It is recommended that rooftop equipment be isolated externally, as if internal isolation were not used.

Note 18. Cooling Towers: These are normally isolated with restrained spring isolators (Type 4) directly under the tower or tower dunnage. Occasionally, high deflection isolators are proposed for use directly under the motor-fan assembly, but this arrangement must be used with extreme caution.

Note 19. Fans and Air-Handling Equipment: The following should be considered in selecting isolation systems for fans and air-handling equipment:

Fans with wheel diameters of 560 mm and under and all fans operating at speeds to 300 rpm do not generate large vibratory forces. For fans operating under 300 rpm, select isolator deflection so that the isolator natural frequency is 40% or less of the fan speed. For example, for a fan operating at 275 rpm, an isolator natural frequency of 110 rpm (1.8 Hz) or lower is required (0.4 × 275 = 110 rpm). A 75-mm deflection isolator (Type 3) can provide this isolation.

Flexible duct connectors should be installed at the intake and discharge of all fans and air-handling equipment to reduce vibration transmission to air ducts.

Inertia bases (Type C) are recommended for all Class 2 and 3 fans and air-handling equipment because extra mass permits the use of stiffer springs, which limit movement.

Thrust restraints (Type 5) that incorporate the same deflection as isolators should be used for all fan heads, all suspended fans, and all base-mounted and suspended air-handling equipment operating at 500 Pa and over total static pressure.

Vibration Isolators: Materials, Types, and Configurations

Notes 20 through 31 are useful for evaluating commercially available isolators for HVAC equipment. The isolator selected for a particular application depends on the required deflection, but life, cost, and suitability must also be considered.

RUBBER PADS (Type 1)

RUBBER
MOUNTS
(Type 2)

Note 20. Rubber isolators are available in pad (Type 1) and molded (Type 2) configurations. Pads are used in single or multiple layers. Molded isolators come in a range of 30 to 70 durometer (a measure of stiffness). Material in excess of 70 durometer is usually ineffective as an isolator. Isolators are designed for up to 13-mm deflection, but are used where 8-mm or less deflection is required. Solid rubber and composite fabric and rubber pads are also available. They provide high load capacities with small deflection and are used as noise barriers under columns and for pipe supports. These pad types work well only when they are properly loaded and the weight load is evenly distributed over the entire pad surface. Metal loading plates can be used for this purpose.

GLASS FIBER PADS (Type 1)

Note 21. Precompressed glass fiber isolation pads (Type 1) constitute inorganic inert material and are available in various sizes in thicknesses of 25 to 100 mm, and in capacities of up to 3.4 MPa. Their manufacturing process assures long life and a constant natural frequency of 7 to 15 Hz over the entire recommended load range. Pads are covered with an elastomeric coating to increase damping and to protect the glass fiber. Glass fiber pads are most often used for the isolation of concrete foundations and floating floor construction.

SPRING ISOLATOR (Type 3)

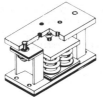

Note 22. Steel springs are the most popular and versatile isolators for HVAC applications because they are available for almost any deflection and have a virtually unlimited life. All spring isolators should have a rubber acoustical barrier to reduce transmission of high-frequency vibration and noise that can migrate down the steel spring coil. They should be corrosion-protected if installed outdoors or in a corrosive environment. The basic types include

1. **Note 23.** Open spring isolators (Type 3) consist of a top and bottom load plate with an adjustment bolt for leveling. Springs should be designed with a horizontal stiffness at least 100% of the vertical stiffness to assure stability, 50% travel beyond rated load and safe solid stresses.

RESTRAINED SPRING ISOLATOR (Type 4)

Note 24. Restrained spring isolators (Type 4) have hold-down bolts to limit vertical movement. They are used with (a) equipment with large variations in mass (boilers, refrigeration machines) to restrict movement and prevent strain on piping when water is removed, and (b) outdoor equipment, such as cooling towers, to prevent excessive movement because of wind load. Spring criteria should be the same as for open spring isolators, and restraints should have adequate clearance so that they are activated only when a temporary restraint is needed.

3. Housed spring isolators consist of two telescoping housings separated by a resilient material. Depending on design and installation, housed spring isolators can bind and short circuit. Their use should be avoided.

AIR SPRINGS

ROLLING LOBE

BELLOWS

Air springs can be designed for any frequency but are economical only in applications with natural frequencies of 1.33 Hz or less (150-mm or greater deflection). Their use is advantageous in that they do not transmit high-frequency noise and are often used to replace high deflection springs on problem jobs. Constant air supply is required, and there should be an air dryer in the air supply.

RUBBER HANGER (Type 2)
SPRING HANGER (Type 3)

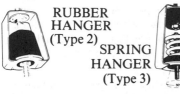

THRUST RESTRAINT (Type 5)

Note 25. Isolation hangers (Types 2 and 3) are used for suspended pipe and equipment and have rubber, springs, or a combination of spring and rubber elements. Criteria should be the same as for open spring isolators. To avoid short circuiting, hangers should be designed for 20 to 35° angular hanger rod misalignment. Swivel or traveler arrangements may be necessary for connections to piping systems subject to large thermal movements.

Note 26. Thrust restraints (Type 5) are similar to spring hangers or isolators and are installed in pairs to resist the thrust caused by air pressure.

DIRECT ISOLATION (Type A)

Note 27. Direct isolation (Type A) is used when equipment is unitary and rigid and does not require additional support. Direct isolation can be used with large chillers, packaged air-handling units, and air-cooled condensers. If there is any doubt that the equipment can be supported directly on isolators, use structural bases (Type B) or inertia bases (Type C), or consult the equipment manufacturer.

STRUCTURAL BASES (Type B)

Note 28. Structural bases (Type B) are used where equipment cannot be supported at individual locations and/or where some means is necessary to maintain alignment of component parts in equipment. These bases can be used with spring or rubber isolators (Types 2 and 3) and should have enough rigidity to resist all starting and operating forces without supplemental hold-down devices. Bases are made in rectangular configurations using structural members with a depth equal to one-tenth the longest span between isolators, with a minimum depth of 100 mm. Maximum depth is limited to 300 mm, except where structural or alignment considerations dictate otherwise.

STRUCTURAL RAILS (Type B)

Note 29. Structural rails (Type B) are used to support equipment that does not require a unitary base or where the isolators are outside the equipment and the rails act as a cradle. Structural rails can be used with spring or rubber isolators and should be rigid enough to support the equipment without flexing. Usual industry practice is to use structural members with a depth one-tenth of the longest span between isolators with a minimum depth of 100 mm. Maximum depth is limited to 300 mm, except where structural considerations dictate otherwise.

CONCRETE BASES (Type C)

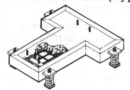

Note 30. Concrete bases (Type C) consist of a steel pouring form usually with welded-in reinforcing bars, provision for equipment hold-down, and isolator brackets. Like structural bases, concrete bases should be rectangular or T-shaped and, for rigidity, have a depth equal to one-tenth the longest span between isolators, with a minimum of 150 mm. Base depth need not exceed 300 mm unless it is specifically required for mass, rigidity, or component alignment.

CURB ISOLATION (Type D)

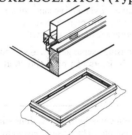

Note 31. Curb isolation systems (Type D) are specifically designed for curb-supported rooftop equipment and have spring isolation with a watertight and airtight curb assembly. The roof curbs are narrow to accommodate the small diameter of the springs within the rails, with static deflection in the 25- to 75-mm range to meet the design criteria described for Type 3.

when the equipment goes through the isolators' resonant frequency. The piping system must be flexible enough to (1) reduce vibration transmission along the connected piping, (2) permit equipment movement without reducing the performance of vibration isolators, and (3) accommodate equipment movement or thermal movement of the piping at connections without imposing undue strain on the connections and equipment.

Flow noise in piping can be minimized by sizing pipe so that velocities are 1.2 m/s maximum for pipe 50 mm and smaller and using a pressure drop limitation of 400 Pa per metre of pipe length with a maximum velocity of 3 m/s for larger pipe sizes. Flow noise and vibration can be reintroduced by turbulence, sharp pressure drops, and entrained air. Care should be taken to avoid these conditions.

Resilient Pipe Hangers and Supports

Resilient pipe hangers and supports are necessary to prevent vibration and noise transmission from the piping system to the building structure and to provide flexibility in the piping.

Suspended Piping. Isolation hangers described in Note 25 of Table 42 should be used for all piping in equipment rooms or for 15 m from vibrating equipment, whichever is greater. To avoid reducing the effectiveness of equipment isolators, at least three of the first hangers from the equipment should provide the same deflection as the equipment isolators, with a maximum limitation of 50-mm

deflection; the remaining hangers should be spring or combination spring and rubber with 20-mm deflection.

Good practice requires that the first two hangers adjacent to the equipment be the positioning or precompressed type, to prevent load transfer to the equipment flanges when the piping system is filled. The positioning hanger aids in installing large pipe, and many engineers specify this type for all isolated pipe hangers for piping 200 mm and over.

While isolation hangers are not often specified for branch piping or piping beyond the equipment room for economic reasons, they should be used for all piping over 50 mm in diameter and for any piping suspended below or near noise-sensitive areas. Hangers adjacent to noise-sensitive areas should be the spring and rubber combination Type 3.

Floor-Supported Piping. Floor supports for piping in equipment rooms and adjacent to isolated equipment should use vibration isolators as described in Note 24 of Table 42. They should be selected according to the guidelines for hangers. The first two adjacent floor supports should be the restrained spring type, with a blocking feature that prevents load transfer to equipment flanges as the piping is filled or drained. Where pipe is subjected to large thermal movement, a slide plate (Teflon, graphite, or steel) should be installed on top of the isolator, and a thermal barrier should be used

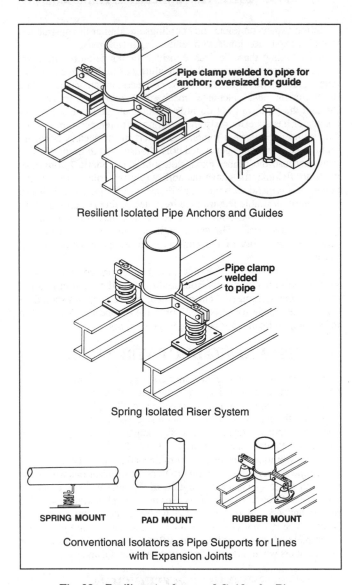

Resilient Isolated Pipe Anchors and Guides

Spring Isolated Riser System

SPRING MOUNT   PAD MOUNT   RUBBER MOUNT

Conventional Isolators as Pipe Supports for Lines
with Expansion Joints

Fig. 38   Resilient Anchors and Guides for Pipes

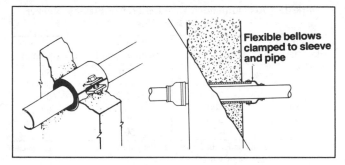

Fig. 39   Acoustical Pipe Penetration Seals

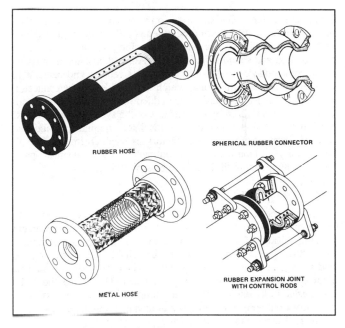

Fig. 40   Flexible Pipe Connectors

when rubber products are installed directly beneath steam or hot water lines.

Riser Supports, Anchors, and Guides. Many piping systems have anchors and guides, especially in the risers, to permit expansion joints, bends, or pipe loops to function properly. Anchors and guides are designed to eliminate or limit (guide) pipe movement and must be rigidly attached to the structure; this is inconsistent with the resiliency required for effective isolation. The engineer should try to locate the pipe shafts, anchors, and guides in noncritical areas, such as next to elevator shafts, stairwells, and toilets, rather than adjoining noise-sensitive areas. Where concern about vibration transmission exists, some type of vibration isolation support or acoustical support is required for the pipe support, anchors, and guides.

Because anchors or guides must be rigidly attached to the structure, the isolator cannot deflect in the sense previously discussed, and the primary interest is that of an acoustical barrier. Such acoustical barriers can be provided by heavy-duty rubber and duck and rubber pads that can accommodate large loads with minimal deflection. Figure 38 shows some arrangements for resilient anchors and guides. Similar resilient-type supports can be used for the pipe.

Resilient supports for pipe, anchors, and guides can attenuate noise transmission, but they do not provide the resiliency required to isolate vibration. Vibration must be controlled in an anchor guide system by designing flexible pipe connectors and resilient isolation hangers or supports.

Completely spring-isolated riser systems that eliminate the anchors and guides have been used successfully in many instances and give effective vibration and acoustical isolation. In this type of isolation system, the springs are sized to accommodate thermal growth as well as to guide and support the pipe. Such systems require careful engineering to accommodate the movements encountered not only in the riser but also in the branch takeoff to avoid overstressing the piping.

Piping Penetrations. Most HVAC systems have many points at which piping must penetrate floors, walls, and ceilings. If such penetrations are not properly treated, they provide a path for airborne noise, which can destroy the acoustical integrity of the occupied space. Seal the openings in the pipe sleeves between noisy areas, (e.g., equipment rooms) and occupied spaces with an acoustical barrier such as fibrous material and caulking or with engineered pipe penetration seals as shown in Figure 39.

Flexible Pipe Connectors

Flexible pipe connectors (1) provide piping flexibility to permit isolators to function properly, (2) protect equipment from strain from misalignment and expansion or contraction of piping, and (3) attenuate noise and vibration transmission along the piping (Figure 40). Connectors are available in two configurations: (1) the hose type, a straight or slightly corrugated wall construction of either rubber or

Table 43 Recommended Live Length[a] of Flexible Rubber and Metal Hose

Nominal Diameter, mm	Length,[b] mm	Nominal Diameter, mm	Length,[b] mm
20	300	100	450
25	300	125	600
40	300	150	600
50	300	200	600
65	300	250	600
80	450	300	900

[a]Live length is end-to-end length for integral flanged rubber hose and is end-to-end less total fitting length for all other types.
[b]Based on recommendations of Rubber Expansion Division, Fluid Sealing Association.

metal; and (2) the arched or expansion joint type, a short length connector of rubber, polytetrafluoroethylene (Teflon), or metal with one or more large radius arches. Metal expansion joints are seldom used for vibration and sound isolation in HVAC systems, and their use is not recommended. All flexible connectors require end restraint to counteract the pressure thrust, which is (1) added to the connector, (2) incorporated by its design, (3) added to the piping system (anchoring), or (4) built in by the stiffness of the system. Connector extension caused by pressure thrust on isolated equipment should also be considered when flexible connectors are used. Overextension will cause failure. Manufacturers' recommendations on restraint, pressure, and temperature limits should be strictly followed.

Hose Connectors

Hose connectors accommodate lateral movement perpendicular to the length; these connectors have very limited or no axial movement capability. Rubber hose connectors are wire reinforced and of molded or handwrapped construction. They are available with metal-threaded end fittings or integral rubber flanges. Threaded fittings should be limited to 75-mm and smaller pipe diameter. The fittings should be the mechanically expanded type to minimize the possibility of pressure thrust blowout. Flanged types are available in larger pipe sizes. Table 43 lists recommended lengths.

Metal hose is constructed with a corrugated inner core and a braided cover that helps attain a pressure rating and provides end restraints that eliminate the need for supplemental control assemblies. Short lengths of metal hose or corrugated metal bellows, or pump connectors, are available without braid and have built-in control assemblies. Metal hose is used to control misalignment and vibration rather than noise and is used primarily where temperature or the pressure of flow media preclude the use of other material. Table 43 lists recommended lengths.

Expansion Joint or Arched-Type Connectors

Expansion joint or arched-type connectors have one or more convolutions or arches and can accommodate all modes of axial, lateral, and angular movement and misalignment. These connectors are available in flanged rubber and Teflon construction. When made of rubber, they are commonly referred to as expansion joints, spool joints, or spherical connectors, and in Teflon, as couplings or expansion joints.

Rubber expansion joints or spool joints are available in two basic types: (1) handwrapped with wire and fabric reinforcing and (2) molded with fabric and wire or with high-strength fabric only (instead of metal) for reinforcing. The handmade type is available in a variety of materials and lengths for special applications. Rubber spherical connectors are molded with high-strength fabric or tire cord reinforcing instead of metal. Their distinguishing characteristic is a large radius arch. The shape and construction of some designs permit use without control assemblies in systems operating to 1.0 MPa. Where thrust restraints are not built in, they must be used as described for rubber hose joints.

Teflon expansion joints and couplings are similar in construction to rubber expansion joints with reinforcing metal rings.

In evaluating these devices, consider the temperature, pressure, and service conditions as well as the ability of each device to attenuate vibration and noise. Metal hose connections can accommodate misalignment and attenuate mechanical vibration transmitted through the pipe wall but do little to attenuate noise. This type of connector has superior resistance to long-term temperature effects. Rubber hose, expansion joints, and spherical connectors attenuate vibration and impeller-generated noise transmitted through the pipe wall. Because the rubber expansion joint and spherical connector walls are flexible, they have the ability to grow volumetrically and attenuate noise and vibration at blade passage frequencies. This is a particularly desirable feature in uninsulated piping systems, such as condenser water and domestic water, which may run adjacent to noise-sensitive areas. However, high pressure has a detrimental effect on the ability of the connector to attenuate vibration and noise.

Because none of the flexible pipe connectors control flow or velocity noise or completely isolate vibration and noise transmission to the piping system, resilient pipe hangers and supports should be used; these are shown in Note 25 of Table 42 and are described in the section on Resilient Pipe Hangers and Supports.

ISOLATING DUCT VIBRATION

Flexible canvas and rubber duct connections should be used at fan intake and discharge. However, they are not completely effective since they become rigid under pressure and allow the vibrating fan to pull on the duct wall. To maintain a slack position of the flexible duct connections, thrust restraints (see Note 26, Table 42) should be used on all equipment as indicated in Table 42.

While vibration transmission from ducts that are isolated by flexible connectors is not a common problem, flow pulsations within the duct can cause mechanical vibration in the duct walls that can be transmitted through rigid hangers. Spring or combination spring and rubber hangers are recommended wherever ducts are suspended below or near a noise-sensitive area. These hangers are especially recommended for large ducts with velocities above 7.5 m/s and for all size ducts when duct static pressures are 500 Pa and over.

SEISMIC PROTECTION

Seismic restraint requirements are specified by applicable building codes that define the design forces to be resisted by the mechanical system, depending on the building location and occupancy, location of the system within the building, and whether it is used for life safety. Where required, seismic protection of resiliently mounted equipment poses a unique problem because resiliently mounted systems are much more susceptible to earthquake damage, due to overturning forces and to resonances inherent in vibration isolators.

As a deficiency in seismic restraint design or anchorage would not become apparent until an earthquake occurs, with possible catastrophic consequences, the adequacy of the restraint system and anchorage to resist design forces must be verified before the event. This verification should be by equipment tests, calculations, or dynamic analysis, depending on the item, with calculations or dynamic analysis performed under the direction of a professional engineer. The verification is often supplied as a package by the vibration isolation vendor.

The restraints for floor-mounted equipment should be designed with adequate clearances so that they are not engaged during normal operation of the equipment. Contact surfaces should be protected with resilient pad material to limit shock during an earthquake, and the restraints should be sufficiently strong to resist the forces in any direction. Due to the difficulty of analyzing these devices, their integrity is normally verified by independent laboratory tests.

Chapter 50 covers seismic restraint design. Calculations or dynamic analysis should have an engineer's seal to verify that input forces are obtained in accordance with code or specification requirements. The anchorage calculations should also be made by a professional engineer in accordance with accepted standards.

VIBRATION INVESTIGATIONS

Theoretically, a vibration isolation system can be selected to isolate vibration forces of extreme magnitude. However, isolators should not be used to mask a condition that should be corrected before it damages the equipment and its operation. A high level of transmitted vibration can indicate faulty equipment in need of correction, or it can be resonance interaction between vibrating equipment and the floor system on which it is placed.

Vibration investigations should include

1. Measurement of the imbalance of reciprocating or rotating equipment components. Refer to ANSI *Standard* 2.19 and ARI *Guideline* G for balance guidelines.
2. Measurement of the vibration levels on vibrating equipment. Refer to Figure 37 for severity ratings of vibrating equipment.
3. Measurement of the vibration levels in buildings on which vibrating equipment is placed. Refer to Figure 36 and Table 40 for recommended building vibration criteria.
4. Examination of equipment vibration generated by such components as bearings and drives.
5. Examination of equipment installation factors, such as equipment alignment and vibration isolator placement. Refer to Table 42.

TROUBLESHOOTING

In spite of all efforts taken by specifying engineers, consultants, and installing contractors, situations arise where there is disturbing noise and vibration. Fortunately, many problems can be readily identified and corrected by

1. Determining which equipment or system is the source of the problem
2. Determining whether the problem is one of airborne sound, vibration and structure-borne noise, or a combination of both
3. Applying appropriate solutions

Troubleshooting is time-consuming, expensive, and often difficult. In addition, once a transmission problem exists, the occupants become more sensitive and require greater reduction of the sound and vibration levels than would initially have been satisfactory. Therefore, the need for troubleshooting should be avoided by carefully designing, installing, and testing the system as soon as it is operational and before the building is occupied.

Determining Problem Source

The system or equipment that is the source of the problem can often be determined without instrumentation. Vibration and noise levels are usually well above the sensory level of perception and are readily felt or heard. A simple and accurate method of determining the problem source is to turn individual pieces of equipment on and off until the vibration or noise is eliminated. Since the source of the problem is often more than one piece of equipment or the interaction of two or more systems, it is always good practice to double check by shutting off the system and operating the equipment individually. Reynolds and Bevirt (1994b) provide information on the measurement and assessment of sound and vibration in buildings.

Determining Problem Type

The next step is to determine whether the problem is one of noise or vibration. If vibration is perceptible, vibration transmission is usually the major cause of the problem. The possibility that light wall or ceiling panels are excited by airborne noise should be considered. If vibration is not perceptible, the problem may still be one of vibration

transmission causing structure-borne noise, which can be checked by the following procedure:

1. If a sound level meter is available, check C-scale and overall scale readings. If the difference is greater than 6 dB, or if the slope of the curve is greater than 5 to 6 dB/octave in the low frequencies, vibration is probably the cause.
2. If the affected area is remote from source equipment, no problem appears in intermediary spaces, and noise does not appear to be coming from the duct system or diffusers, structure-borne noise is probably the cause.

Noise Problems

Noise problems are more complex than vibration problems and usually require the services of an acoustical engineer or consultant. If the affected area adjoins the room where the source equipment is located, structure-borne noise must be considered as part of the problem, and the vibration isolation systems should be checked. A simple but reasonably effective test is to have one person listen in the affected area while another shouts loudly in the equipment room. If the voice cannot be heard, the problem is likely one of structure-borne noise. If the voice can be heard, check for openings in the wall or floor separating the areas. If no such openings exist, the structure separating the areas does not provide adequate transmission loss. In such situations, refer to the section on Mechanical Equipment Room Sound Isolation for possible solutions.

If duct-borne sound (i.e., noise from grilles or diffusers or duct breakout noise) is the problem, measure the sound pressure levels and compare them with the design goal RC curves. Where the measured curve differs from the design goal RC curve, the potential noise source(s) can identified. Once the noise sources have been identified, the engineer can determine whether sufficient attenuation has been provided by analyzing each sound source using the procedures presented in this chapter.

If the sound source is a fan, pump, or similar rotating equipment, determine whether it is operating at the most efficient part of its operating curve. This is the point at which most equipment operates best. Excessive vibration and noise can occur if a fan or pump is trying to move too little or too much air or water. In this respect, check that vanes, dampers, and valves are in the correct operating position and that the system has been properly balanced.

Vibration Problems

Vibration and structure-borne noise problems can occur from

- Equipment operating with excessive levels of vibration, usually caused by unbalance
- Lack of vibration isolators
- Improperly selected or installed vibration isolators that do not provide the required isolator deflection
- Flanking transmission paths such as rigid pipe connections or obstructions under the base of vibration-isolated equipment
- Floor flexibility
- Resonances in equipment, the vibration isolation system, or the building structure

Most field-encountered problems are the result of improperly selected or installed isolators and flanking paths of transmission, which can be simply evaluated and corrected. Floor flexibility and resonance problems are sometimes encountered and usually require analysis by experts. However, the following information identifies such problems. If the equipment lacks vibration isolators, isolators recommended in Table 40 can be added by using structural brackets without altering connected ducts or piping.

Testing Vibration Isolation Systems. Improperly functioning vibration isolation systems are the cause of most field-encountered problems and can be evaluated and corrected by the following procedures.

1. Ensure that the system is free-floating by bouncing the base, which should cause the equipment to move up and down freely and easily. On floor-mounted equipment, check that there are no obstructions between the base and the floor, which would short-circuit the isolation system. This is best accomplished by passing a rod under the equipment. A small obstruction might permit the base to "rock," giving the impression that it is free-floating when it is not. On suspended equipment, make sure that rods are not touching the hanger box. Rigid connections such as pipes and ducts can prevent equipment from floating freely, prohibit isolators from functioning properly, and provide flanking paths of transmission.

2. Determine whether the isolator deflection is as specified or required, changing it if necessary as recommended in Table 42. A common problem is inadequate deflection caused by underloaded isolators. Overloaded isolators are not generally a problem as long as the system is free-floating and there is space between the spring coils.

With the most commonly used spring isolators, determine the spring deflection by measuring the operating height and comparing it to the free height information available from the manufacturer. Once the actual isolator deflection is known, determine its adequacy by comparing it with the recommended deflection in Table 42.

If the natural frequency of the isolator is 25% or less of the disturbing frequency (usually considered the operating speed of the equipment), the isolators should be amply efficient except in situations where heavy equipment is installed on extremely long-span floors or very flexible floors. If a transmission problem exists, it may be caused by (1) excessively rough equipment operation, (2) the system's not being free-floating or flanking path transmission, or (3) a resonance or floor stiffness problem, as described below.

While it is easy to determine the natural frequency of spring isolators by height measurements, such measurements are difficult with pad and rubber isolators and are not accurate in determining their natural frequencies. Although such isolators can theoretically provide natural frequencies as low as 4 Hz, they actually provide higher natural frequencies and generally do not provide the desired isolation efficiencies for upper floor equipment locations.

Generally, field vibration measurements can not be used to determine whether design vibration isolation efficiencies have been achieved in an installation. However, vibration measurements can be made on vibrating equipment, on equipment supports, on floors supporting vibration-isolated equipment, and on floors in adjacent areas to determine whether vibration criteria specified in Table 38 or in Figures 36 and 37 have been achieved in field installations.

Floor Flexibility Problems. Floor flexibility is not a problem with most equipment and structures; however, such problems can occur with heavy equipment installed on long-span floors or thin slabs and with rooftop equipment installed on light structures of open web joist construction. If floor flexibility is suspected, the isolators should be one-tenth or less as stiff as the floor to eliminate the problem. Floor stiffness can be determined by calculating the additional deflection in the floor caused by a specific piece of equipment.

For example, if a 5000 kg piece of equipment causes floor deflection of an additional 2.5 mm, floor stiffness is 19.6 MN/m (2 Gg/m), and an isolator combined stiffness of 1.96 MN/m or less must be used. Note that the floor stiffness or spring rate, not the total floor deflection, is determined. In this example, the total floor deflection might be 25 mm, but if the problem equipment causes 2.5 mm of that deflection, 2.5 mm is the important figure, and floor stiffness k is 19.6 MN/m.

Resonance Problems. These problems occur when the operating speed of the equipment is the same as or close to the resonance frequency of (1) an equipment component such as a fan shaft or bearing support pedestal, (2) the vibration isolation system, or (3) the floor or other building component, such as a wall. Vibration resonances can cause excessive equipment vibration levels, as well as

objectionable and possibly destructive vibration transmission in a building. These conditions must always be identified and corrected.

When vibrating mechanical equipment is mounted on vibration isolators on a flexible floor, there are two resonance frequencies that must be considered. The lower frequency is associated with and primarily controlled by the stiffness (and consequently the static deflection) of the vibration isolators. This frequency is generally significantly lower than the operating speed (or frequency) of the mechanical equipment and is generally not a problem. The higher resonance frequency is associated with and primarily controlled by the stiffness of the floor. This resonance frequency is usually not affected by increasing or decreasing the static deflection of the mechanical equipment vibration isolators. Sometimes when the floor on which mechanical equipment is located is flexible (this occurs with some long-span floor systems and with roof systems supporting rooftop packaged units), the operating speed of the mechanical equipment can coincide with the higher resonance frequency. When this occurs, changing the static deflection of the vibration isolators will not solve the problem.

Vibration Isolation System Resonance. Always characterized by excessive equipment vibration, vibration isolation system resonance usually results in objectionable transmission to the structure. However, transmission might not occur if the equipment is on grade or on a stiff floor. Vibration isolation system resonance can be measured with instrumentation or, more simply, by determining the isolator natural frequency, as described in the section on Testing Vibration Isolation Systems, and comparing this figure to the operating speed of the equipment.

When vibration isolation system resonance exists, the isolator natural frequency must be changed using the following guidelines:

1. If the equipment is installed on pad or rubber isolators, isolators with the deflection recommended in Table 42 should be installed.
2. If the equipment is installed on spring isolators and there is objectionable vibration or noise transmission to the structure, determine whether the isolator is providing maximum deflection. For example, an improperly selected or installed nominal 50-mm deflection isolator could be providing only 3-mm deflection, which would be in resonance with equipment operating at 500 rpm. If this is the case, the isolators should be replaced with ones having enough capacity to provide 50-mm deflection. Since there was no transmission problem with the resonant isolators, it is not necessary to use greater deflection isolators than can be conveniently installed.
3. If the equipment is installed on spring isolators and there is objectional noise or vibration transmission, replace the isolators with spring isolators with the deflection recommended in Table 42.

Building Resonances. These problems occur when some part of the structure has a resonance frequency the same as the disturbing frequency or the operating speed of some of the equipment. These problems can exist even if the isolator deflections recommended in Table 42 are used. The resulting objectionable noise or vibration should be evaluated and corrected. Often, the resonance problem is in the floor on which the equipment is installed, but it can also occur in a remotely located floor, wall, or other building component. If a noise or vibration problem has a remote source that cannot be associated with piping or ducts, resonance must be suspected.

Building resonance problems can often be resolved by applying the following corrections:

1. Reduce the vibration force by balancing the equipment. This is not a practical solution for a true resonance problem; however, it is viable when the disturbing frequency equals the floor natural frequency, as evidenced by the equal displacement of the floor and the equipment, especially when the equipment is operating with excessive vibration.
2. Change the isolator resonance frequency by increasing or decreasing the static deflection of the isolator. Only small changes are

necessary to "detune" the system. Generally, increasing the deflections is preferred. If the initial deflection is 25 mm, a 50- or 75-mm deflection isolator should be installed. However, if the initial isolator deflection is 100 mm, it may be more practical and economical to replace it with a 75- or 50-mm deflection isolator.

3. Change the structure stiffness or the structure resonance frequency. A change in structure stiffness changes the structure resonance frequency; the greater the stiffness, the higher the resonance frequency. However, the structure resonance frequency can also be changed by increasing or decreasing the floor deflection without changing the floor stiffness. While this approach is not recommended, it may be the only solution in certain cases.

4. Change the disturbing frequency by changing the equipment operating speed. This is practical only for belt-driven equipment, such as fans.

REFERENCES

AIA. 1992-93. *Guidelines for construction and equipment of hospital and medical facilities.* The American Institute of Architects Committee on Architecture for Health with Assistance from the U.S. Department of Health and Human Services. AIA Press, Washington, DC.

Beatty, J. 1987. Discharge duct configurations to control rooftop sound. *Heating/Piping/Air Conditioning* (July).

Beranek, L.L. 1960. *Noise reduction.* McGraw-Hill, New York.

Beranek, L.L. 1971. *Noise and vibration control.* McGraw-Hill, New York.

Blazier, W.E., Jr. 1981. Revised noise criteria for design and rating of HVAC systems. *ASHRAE Transactions* 87(1).

Cummings, A. 1983. Acoustic noise transmission through the walls of air-conditioning ducts. *Final Report.* Department of Mechanical and Aerospace Engineering, University of Missouri-Rolla.

Cummings, A. 1985. Acoustic noise transmission through duct walls. *ASHRAE Transactions* 91(2A).

Ebbing, C.E., D. Fragnito, and S. Inglis. 1978. Control of low frequency duct-generated noise in building air distribution systems. *ASHRAE Transactions* 84(2).

Egan, M.D. 1988. *Architectural acoustics.* McGraw-Hill, New York.

Environment Canada. 1994. Mineral fibres: Priority substances list assessment report, Canadian Environmental Protection Act, Ottawa.

Harold, R.G. 1986. Round duct can stop rumble noise in air-handling installations. *ASHRAE Transactions* 92(2).

Harold, R.G. 1991. Rooftop installation sound and vibration considerations. *ASHRAE Transactions* 97(1).

Harris, C.M., ed. 1979. *Handbook of noise control,* 2nd ed. McGraw-Hill, New York.

Hirschorn, M. 1981. Acoustic and aerodynamic characteristics of duct silencers for air-handling installations. *ASHRAE Transactions* 87(1).

IRD. 1988. *Vibration technology-1.* IRD Mechanalysis, Columbus, OH.

Kuntz, H.L. 1986. The determination of the interrelationship between the physical and acoustical properties of fibrous duct liner materials and lined duct sound attenuation. *Report No.* 1068. Hoover Keith and Bruce, Houston, TX.

Kuntz, H.L. and R.M. Hoover. 1987. The interrelationships between the physical properties of fibrous duct lining materials and lined duct sound attenuation. *ASHRAE Transactions* 93(2).

Lilly, J. 1987. Break-out in HVAC duct systems. *Sound & Vibration* (October).

Machen, J. and J.C. Haines. 1983. Sound insertion loss properties of linacoustic and competitive duct liners. *Report No.* 436-T-1778, Johns-Manville Research and Development Center, Denver, CO.

Morey, P.R. and C.M. Williams. 1991. Is porous insulation inside an HVAC system compatible with healthy building? *ASHRAE IAQ Symposium.*

Reynolds, D.D. and W.D. Bevirt. 1994a. *Sound and vibration design and analysis.* National Environmental Balancing Bureau, Rockville, MD.

Reynolds, D.D. and W.D. Bevirt. 1994b. *Procedural standards for the measurement and assessment of sound and vibration.* National Environmental Balancing Bureau, Rockville, MD.

Reynolds, D.D. and J.M. Bledsoe. 1989a. Sound attenuation of acoustically lined circular ducts and radiused elbows. *ASHRAE Transactions* 95(1).

Reynolds, D.D. and J.M. Bledsoe. 1989b. Sound attenuation of unlined and acoustically lined rectangular ducts. *ASHRAE Transactions* 95(1).

Reynolds, D.D. and J.M. Bledsoe. 1991. *Algorithms for HVAC acoustics.* ASHRAE, Atlanta.

Sandbakken, M., L. Pande, and M.J. Crocker. 1981. Investigation of end reflection coefficient accuracy problems with AMCA *Standard* 300-67. HL 81-16. Ray W. Herrick Laboratories, Purdue University, West Lafayette, IN.

Schaffer, M.E. 1991. *A practical guide to noise and vibration control for HVAC systems.* ASHRAE, Atlanta.

Schultz, T.J. 1985. Relationship between sound power level and sound pressure level in dwellings and offices. *ASHRAE Transactions* 91(1).

SMACNA. 1990. *HVAC systems duct design,* 3rd ed. Sheet Metal and Air Conditioning Contractors' National Association, Vienna, VA.

Stevens, K.N., W.A. Rosenblith, and R.H. Bolt. 1955. A community's reaction to noise: Can it be forecast? *Noise Control* (January).

Thompson, J.K. 1981. The room acoustics equation: Its limitation and potential. *ASHRAE Transactions* 87(2).

Ungar, E.E., D.H. Sturz, and C.H. Amick. 1990. Vibration control design of high technology facilities. *Sound and Vibration* (July).

Ver, I.L. 1978. A review of the attenuation of sound in straight lined and unlined ductwork of rectangular cross section. *ASHRAE Transactions* 84(1).

Ver, I.L. 1982. A study to determine the noise generation and noise attenuation of lined and unlined duct fittings. *Report No.* 5092. Bolt, Beranek and Newman, Boston.

Ver, I.L. 1984a. Noise generation and noise attenuation of duct fittings—A review: Part II. *ASHRAE Transactions* 90(2A).

Ver, I.L. 1984b. Prediction of sound transmission through duct walls: Break-out and pickup. *ASHRAE Transactions* 90(2A).

Wells, R.J. 1958. Acoustical plenum chambers. *Noise Control* (July).

Woods Fan Division. 1973. *Design for sound.* The English Electric Company.

World Health Organization. 1986. *International Symposium on Man-Made Mineral Fibers in the Working Environment.* International Agency for Research on Cancer, Copenhagen.

BIBLIOGRAPHY

Armstrong Cork Company. 1981. Single-pass sound transmission loss data.

Beranek, L.L. 1954. *Acoustics.* McGraw-Hill, New York.

Beranek, L.L. 1957. Revised criteria for noise in buildings. *Noise Control* (January).

Chaddock, J.B. et al. 1959. Sound attenuation in straight ventilation ducting. *Refrigeration Engineering* (January).

Cummings, A. 1979. The effects of external lagging on low frequency sound transmission through the walls of rectangular ducts. *Journal of Sound Vibration* 67(2):187-201.

Departments of the Army, the Air Force, and the Navy. 1983. *Noise and Vibration Control for Mechanical Equipment.* Army TM 5-805-4, Air Force AFM 88-37, Navy NAVFAC DM-3.10.

Diehl, G.M. 1973. *Machinery acoustics.* John Wiley and Sons, New York.

Faulkner, L.L., ed. 1976. *Handbook of industrial noise control,* pp. 419-424. Industrial Press.

Fry, A., ed. 1988. *Noise control in building services.* Pergamon Press, Oxford, UK.

Galloway, W.J., and T.J. Schultz. 1979. Noise assessment guidelines —1979 Technical background. *Report No.* 4024. Bolt, Beranek, and Newman, Boston.

Goodfriend, L.S. 1980. Indoor sound rating criteria. *ASHRAE Transactions* 86(2).

Kahn, Greenberg, and Essert. 1987. Break-out noise from lined air-conditioning ducts. *Noise-Con 87.*

Office of Noise Control. 1981. Catalog of STC and IIC ratings for wall and floor/ceiling assemblies. California Department of Health Services, Sacramento.

Owens-Corning Fiberglass Corp. 1970. Sound attenuation of fiberglass ducts and duct liners. Product Testing Laboratories *Report No.* 32433–Final.

Owens-Corning Fiberglass Corp. 1981. *Noise Control Manual,* 4th ed.

Pallet, D.S., R.D. Kilmer, and T.L. Quindry. 1978. *Design guide for reducing transportation noise in and around buildings.* Building Science *Series* 84. U.S. Department of Commerce, National Bureau of Standards, Gaithersburg, MD.

Reynolds, D.D. and J.M. Bledsoe. 1989. Sound transmission through mechanical equipment room walls, floor, or ceiling. *ASHRAE Transactions* 95(1).

Reynolds, D.D. and W.P. Zeng. 1994. New relationship between sound power level and sound pressure level in rooms. *Report No.* URP-93001-1. Ventilation & Acoustic Systems Technology Laboratory, University of Nevada, Las Vegas.

Sessler, S.M. 1973. Acoustical and mechanical considerations for the evaluation of chiller noise. *ASHRAE Journal* 15(10).

CHAPTER 44

CORROSION CONTROL AND WATER TREATMENT

THIS chapter covers the fundamentals of corrosion, its prevention and control, and some of the common problems arising from corrosion in heating and air-conditioning equipment. Further information can be obtained from the references listed in the bibliography and in publications of the National Association of Corrosion Engineers and the American Water Works Association.

DEFINITIONS

Terms commonly used in the field of water treatment and corrosion control are defined as follows:

Alkalinity. The sum of the bicarbonate, carbonate, and hydroxide ions in water. Other ions, such as borate, phosphate, or silicate, can also contribute to alkalinity.

Anion. A negatively charged ion of an electrolyte that migrates toward the anode under the influence of a potential gradient.

Anode. The electrode of an electrolytic cell at which oxidation occurs.

Biological Deposits. Water-formed deposits of biological organisms or the products of their life processes (e.g., barnacles, algae, or slimes).

Cathode. The electrode of an electrolytic cell at which reduction occurs.

Cation. A positively charged ion of an electrolyte that migrates toward the cathode under the influence of a potential gradient.

Corrosion. The deterioration of a material, usually a metal, through reaction with its environment.

Corrosivity. The capacity of an environment to bring about destruction of a specific metal by the process of corrosion.

Electrolyte. A solution through which an electric current can be made to flow.

Filtration. The process of passing a liquid through a suitable porous material in such a manner as to effectively remove the suspended matter from the liquid.

Galvanic Corrosion. Corrosion resulting from the contact of two dissimilar metals in an electrolyte or from the contact of two similar metals in an electrolyte of nonuniform concentration.

Hardness. The sum of the calcium and magnesium ions in water.

Inhibitor. A chemical substance that reduces the rate of corrosion, scale formation, or slime production.

Ion. An electrically charged atom or group of atoms.

Passivity. The tendency of a metal to become inactive in a given environment.

pH. The logarithm of the reciprocal of the hydrogen ion concentration of a solution. pH values below 7 are increasingly acidic; those above 7 are increasingly alkaline.

The preparation of this chapter is assigned to TC 3.6, Corrosion and Water Treatment.

Polarization. The deviation from the open circuit potential of an electrode resulting from the passage of current.

Scale. The formation at high temperature of thick corrosion product layers on a metal surface; the deposition of water-insoluble constituents on a metal surface.

Sludge. A sedimentary water-formed deposit, either of biological origin or suspended particles from the air.

Tuberculation. The formation over a surface of scattered, knob-like mounds of localized corrosion products.

Water-Formed Deposit. Any accumulation of insoluble material derived from water or formed by reaction with water on surfaces in contact with it.

CORROSION THEORY—MECHANISM

Corrosion is destruction of a metal or alloy by chemical or electrochemical reaction with its environment. In most instances, this reaction is electrochemical in nature, much like that in an electric dry cell. The basic background for corrosion is almost always the same: a flow of electricity between certain areas of a metal surface through a solution capable of conducting an electrical current. This electrochemical action causes destructive alteration (eating away) of a metal at areas (called anodes) where the electric current enters the solution. This destruction is the critical step in the series of reactions associated with corrosion.

For corrosion to occur, there must be a release of electrons and a formation of metal ions through oxidation and disintegration of the metal at the anode. At the cathode, there must be a simultaneous acceptance of electrons or formation of negative ions. These actions do not occur independently. According to Faraday's law, the two reactions must be at equivalent rates at the same time. Corrosion, or disintegration of the metal, occurs only at anode areas. Anode and cathode areas may shift from time to time as corrosion proceeds, resulting in uniform corrosion.

Reactions at the cathode surface most often control the rate of corrosion. Depending on the nature of the electrolyte, the hydrogen generated at the cathode surface may (1) accumulate to coat the surface and slow down the reaction (*cathodic polarization*); (2) form bubbles and be swept from the surface, thus allowing the reaction to proceed; or (3) react with oxygen in the electrolyte to form water or a hydroxyl ion. At the anode, the metal ion entering the solution may react with a constituent in the electrolyte to form a corrosion product. With iron or steel, the ferrous ion may react with the hydroxyl ion in water to form ferrous hydroxide and then with oxygen to produce ferric hydroxide (rust). These corrosion products may accumulate on the anode surface and slow down the reaction rate (*anodic polarization*).

ACCELERATING OR INTENSIFYING FACTORS

Moisture

Corrosion does not occur in dry air. However, some moisture is present as water vapor in most natural atmospheres. In pure air, almost no iron corrosion occurs at relative humidities up to 99%; however, when contaminants such as sulfur dioxide or solid particles of charcoal are present, corrosion can proceed at relative humidities of 50% or above. Pure air is seldom encountered. The corrosion reaction proceeds on surfaces of exposed metals such as iron and unalloyed steel as long as the metal remains wet. Many alloys develop thin corrosion product films or oxide coatings and are thus unaffected by moisture.

Oxygen

In electrolytes consisting of water solutions of salts or acids, the presence of dissolved oxygen accelerates the corrosion rate of ferrous metals by depolarizing the cathodic areas through reaction with hydrogen generated at the cathode. This allows the anodic reaction to proceed. In many ferrous systems used for handling water, such as boiler systems or water heating systems, oxygen is removed from the water to reduce the corrosion rate. This is done by adding oxygen-scavenging chemicals, such as sulfites, or by using deaeration equipment to expel the dissolved oxygen.

The presence of oxygen in the medium does not affect all alloys in the manner just described. In alloys that develop protective oxide films (e.g., stainless steels), oxygen can reduce corrosion by maintaining the oxide film. Free of oxygen, the medium might cause some corrosion of the alloy.

Solutes

In ferrous materials such as iron and steel, mineral acids accelerate the corrosion rate, whereas alkalies reduce it. The likelihood of cathodic polarization, which slows down corrosion, decreases as the concentration of hydrogen ions (i.e., the acidity of the medium) increases. Alkaline solutions, which contain a much higher concentration of hydroxyl ions than of hydrogen ions, promote cathodic polarization, thus reducing the rate of dissolution at the anode. Relative acidity or alkalinity of a solution is defined as pH; a neutral solution has a pH of 7. Solutions increase in acidity as the pH decreases and increase in alkalinity as pH increases.

The corrosivity of most salt solutions depends on their pH. For example, aluminum sulfate is the salt of a strong acid (sulfuric acid) and a weak alkali (aluminum hydroxide). A solution of aluminum sulfate is acidic in nature and reacts with iron as would a dilute acid. Because alkaline solutions are generally less corrosive to ferrous systems, it is practical in many closed water systems to minimize corrosion by adding alkali or alkaline salt to raise the pH to 9 or higher.

Differential Solute Concentration

For the corrosion reaction to proceed, there must be a potential difference between anode and cathode areas. This potential difference can be established between different locations on a metal surface because of differences in solute concentration in the medium at these locations. Corrosion caused by such circumstances is called *concentration cell corrosion*. These cells can be metal ion or oxygen concentration cells.

In the metal ion cell, the metal surface in contact with the higher concentration of dissolved metal ion becomes the cathodic area, and the surface in contact with the lower concentration becomes the anode. The metal ions involved may be a constituent of the medium or may come from the corroding surface itself. The concentration differences may be caused by the sweeping away of the dissolved metal ions at one location and not at another. Such differences could occur at crevices or be caused by deposits of one sort or another. The anodic area is outside the crevice or deposit, where the metal ion concentration is least.

In the oxygen concentration cell, the surface area in contact with the medium of higher oxygen concentration becomes the cathodic area, and the surface in contact with the medium of lower oxygen concentration becomes the anode. Crevices or foreign deposits on the metal surface can produce conditions conducive to corrosion. The anodic area, where corrosion proceeds, is in the crevice or under the deposit. Although they are actually manifestations of concentration cell corrosion, *crevice corrosion* and *deposit attack* are sometimes referred to as separate forms of corrosion.

Dissimilar Metals

Another factor that can accelerate the corrosion process is the difference in potential of dissimilar metals coupled together and immersed in an electrolyte. The following factors control the severity of the *galvanic corrosion* resulting from such a coupling of dissimilar metals:

1. The difference in potential. The greater the difference, the greater the driving force of the reaction. The galvanic series for metals in flowing aerated seawater is shown in Table 1.
2. The relative area relationship between anode and cathode areas. Because the amount of current flow and, therefore, total metal loss is determined by the potential difference and resistance of the circuits, a small anodic area corrodes more rapidly than a large anodic area.
3. Polarization of either the cathodic or the anodic area. This can reduce the potential difference and thus the rate of attack of the anode.
4. The mineral content of the water. If the water has a low mineral content (low electrical conductivity), the corrosion rate will be less severe than with water containing high mineral concentrations.

Table 1 Galvanic Series of Metals and Alloys in Flowing Aerated Seawater at 4 to 27°C

Corroded End (Anodic or Least Noble)
Magnesium alloys
Zinc
Beryllium
Aluminum alloys
Cadmium
Mild steel, wrought iron
Cast iron, flake or ductile
Low alloy high strength steel
Ni-Resist, types 1 & 2
Naval Brass (CA464), yel. brass (CA268), Al brass (CA687), red brass (CA230), admlty brass (CA443), Mn Bronze
Tin
Copper (CA102, 110), Si Bronze (CA655)
Lead-tin solder
Tin bronze (G & M)
Stainless steel, 12 to 14% Cr (AISI types 410, 416)
Nickel silver (CA 732, 735, 745, 752, 764, 770, 794)
90/10 Copper-nickel (CA706)
80/20 Copper-nickel (CA710)
Stainless steel, 16 to 18% Cr (AISI type 430)
Lead
70/30 Copper-nickel (CA715)
Nickel-aluminum bronze
Silver braze alloys
Nickel 200
Silver
Stainless steel, 18% Cr, 8% Ni (AISI types 302, 304, 321, 347)
Stainless steel, 18% Cr, 12% Ni-Mo (AISI types 316, 317)
Titanium
Graphite, graphitized cast iron
Protected End (Cathodic or Most Noble)

It is inadvisable to couple a small exposed area of a less noble (anodic) metal with a large area of a more noble (cathodic) metal in media in which the less noble material may tend to corrode by itself. If such couples cannot be avoided, but one of the dissimilar metals can be painted or coated with a nonmetallic coating, the cathodic material should be coated.

In piping systems, the galvanic couple can be avoided if the two materials can be insulated from each other with an insulated joint. Where this is not possible, a *waster* (heavy wall nipple section of the anodic material) can be installed and readily replaced when it fails. In piping systems handling natural waters, galvanic corrosion is not likely to extend more than three to five pipe diameters down the inside of the less noble pipe material. In metal components exposed to the atmosphere, galvanic effects are likely to be confined to the area immediately adjacent to the joint.

Stray Current

Stray current corrosion is a form of galvanic corrosion. The electrical potential driving the corrosion reaction comes from stray electrical currents of an electric generator. This phenomenon can occur on buried or submerged metallic structures. The soil or other medium provides the electrolyte. The anode, or structure suffering accelerated corrosion, may be located some distance from the cathode structure. The corrosion attack is usually restricted to the structure surface in contact with the soil or other medium.

Stress

Stresses in metallic structures rarely have significant effects on the uniform corrosion resistance of metals and alloys. There have been a few instances with some materials where stress has accelerated corrosion; the more severely stressed areas (e.g., pipe threads) were usually anodic to less stressed areas. Stresses in specific metals and alloys can cause corrosion cracking when the metals are exposed to specific corrosive environments. The cracking can have catastrophic effects on the usefulness of the particular metal.

Almost all metals and alloys exhibit susceptibility to stress-corrosion cracking in one or more specific environments. Common examples are steels in hot caustic solutions, high zinc content brasses in ammonia, and stainless steels in hot chlorides. Metal producers have more details on specific materials. Stress-corrosion cracking can often be prevented by using the susceptible alloy in the annealed condition or by selecting a material resistant to attack by the specific medium.

Temperature

According to studies of chemical reaction rates, corrosion rates double for every 10 K rise in temperature. However, such a ratio is not necessarily valid for nonlaboratory corrosion reactions; the effect of temperature on a particular system is difficult to predict without specific knowledge of the characteristics of the metals involved and the constituents of the medium.

In systems where the presence of oxygen in the medium promotes corrosion, an increase in temperature may increase the corrosion rate to a point. Oxygen solubility decreases as temperature increases and, in an open system, may approach zero as the medium boils. Beyond a critical temperature level, therefore, the corrosion rate may decrease due to a decrease in oxygen solubility. However, in a closed system, from which oxygen cannot escape, the corrosion rate may continue to increase with temperature rise.

For those alloys that depend on oxygen in the medium for maintaining a protective oxide film, the reduction in oxygen content due to an increase in temperature can accelerate the corrosion rate by preventing oxide film formation.

Temperature can affect corrosion behavior by causing a salt dissolved in the medium to precipitate on the surface as scale, which can be protective. One example is calcium carbonate scale in hard waters. Temperature can also affect the nature of the corrosion product, which may be relatively stable and protective in certain temperature ranges and unstable and nonprotective in others. An example of this is zinc in distilled water; the corrosion product is nonprotective between 60 and 80°C but reasonably protective at other temperatures.

Pressure

As with temperature, it is difficult to predict the effects of pressure on corrosion. Where dissolved gases such as oxygen and carbon dioxide affect the corrosion rate, pressure on the system may increase their solubility and thus increase corrosion. Similarly, a vacuum on the system reduces the solubility of the dissolved gas, thus reducing corrosion. In a heated system, pressure may rise with temperature. It is seldom possible or practical to control metallic system corrosion by pressure control alone.

In centrifugal pumps, cavitation occurs when the net positive suction pressure is not high enough. This causes either the release of gas (air) bubbles from the water or the generation of steam. The bubbles of gas or steam expand rapidly as the pressure in the pump decreases. They cavitate (explode) on the metal surface and cause erosion. The attack is also speed related; it is most severe at the outer edge of the pump impeller.

Flow Velocity

The effects of the flow velocity of the medium depend on the characteristics of the particular metal or alloy. In systems where oxygen increases the corrosion rate (e.g., iron or steel in water), increased flow velocity can increase the corrosion rate by making more oxygen available to the metal surface for reaction. In systems where alloys (e.g., stainless steels) depend on thin oxide films for corrosion resistance, increased flow velocity increases corrosion resistance up to high velocities.

In metal systems where corrosion products retard corrosion by acting as a physical barrier, high flow velocities may cause these products to be swept away, permitting corrosion to proceed at its initial rate. In specific media and for specific metals, there may a critical velocity below which the corrosion product film remains adherent and protective.

Turbulent medium flow may cause uneven attack, involving localized erosion and corrosion. This attack is called *erosion-corrosion*; it can occur in piping systems at sharp bends if the flow velocity is high. Copper and some of its alloys are subject to this type of attack.

Low velocities can allow suspended solids to precipitate on a metal surface, which initiates concentration cell corrosion.

PREVENTIVE AND PROTECTIVE MEASURES

Materials Selection

Almost any piece of heating or air-conditioning equipment can be constructed with materials that will not corrode significantly under service conditions. However, because of economic and physical limitations, this is rarely an option.

When selecting construction materials for a piece of equipment, consider (1) the corrosion resistance of each proposed metal or other material to the service environment, (2) the nature of the corrosion products that may be formed and their effects on equipment operation, (3) the suitability of the materials to handling by standard fabrication methods, and (4) the effects of design and fabrication limitations on tendencies toward local corrosion (as discussed in the section on Accelerating or Intensifying Factors). The overall economic balance during the projected life of the equipment should also be considered. It may be less expensive in the long run to pay more for a corrosion-resistant material and avoid regular painting or other corrosion control measures than to use a less expensive material that requires regular corrosion control maintenance.

The following are considerations for materials selection:

1. Use of dissimilar metals should be avoided or minimized. Where dissimilar materials are used, insulating gaskets and/or organic coatings must be used to prevent galvanic couples.
2. Temperature- and pressure-measuring equipment using mercury should be avoided because accidental breakage and mercury spillage into the system can cause severe damage.
3. Stagnation or sluggish fluid flow, unnecessarily high fluid velocities, impingement, and erosion should be avoided through proper design of equipment and accessories.
4. Corrosion inhibitors and biological control additives must be compatible with the materials chosen for any given system.

Protective Coatings

The environment in which system components operate largely determines the type of coating system needed. Even with a coating suited to the environment, the protection provided depends on the adhesion of the coating to the base material, which itself depends on the quality of surface preparation and application technique. Consult several reliable manufacturers of industrial coatings before choosing a coating system for any exposure, and follow the manufacturer's recommendations for both surface preparation and application technique.

Shop-applied coatings offer the best results because better conditions for controlled surface preparation and coating application usually exist in the shop.

Field-applied coatings can be the same types as shop-applied coatings. The limiting factor is accessibility; once equipment is in place, obstructions can make it difficult to prepare the surface optimally.

Maintenance. Defects in a coating are virtually inescapable. These defects can be either flaws introduced into the coating during application or mechanical damage sustained after application; the defects must be repaired to prevent premature failure at these points. Depending on the severity of the service, an inspection should be scheduled after installation to detect mechanical damage. Damaged areas should be noted and repaired.

Cathodic Protection

Cathodic protection reduces or prevents corrosion by using direct impressed or galvanic current to transform the metal in a conducting medium into the cathode. It is widely used to control corrosion in (1) water storage tanks; (2) heat exchanger water boxes; (3) water and chemical processing equipment such as filters, reactors, and clarifiers; and (4) the external surfaces of submerged and underground tanks, piping, and piling.

Although cathodic protection can be applied to bare structures underground, it is most widely used to provide corrosion control for metal exposed at the inevitable coating flaws. The rate of metal penetration at coating flaws is often greater than that on bare surfaces.

Cathodic protection may be used with iron, aluminum, lead, and stainless steel. On new steel structures, optimum corrosion control is often achieved by applying cathodic protection in combination with coatings, environment conditioning, or both.

Cathodic protection is unique in that protective effects can be directed from distantly positioned anode current sources onto existing submerged or buried structures. Cathodic protection can be achieved without taking the facilities out of service, exposing the surfaces for coating, or specially treating the surrounding environment.

Because this process superimposes an applied current onto an existing electrochemical corrosion system, its design must be adapted to the needs of the specific corrosion problem. The electrochemical corrosion mechanisms to which cathodic protection can be applied fall into the following two broad classifications:

1. Corrosion of a metal surface by *stray* direct currents, which flow in circuits grounded at more than one point or in subsurface structures purposely made part of a direct current circuit.
2. Galvanic corrosion of a metal surface in an electrolyte between discrete areas of oxidation and reduction reactions.

Complete corrosion control by cathodic protection is approached when the net current at any point on the metal surface either measures zero or is flowing from the corroding medium into the metal. It is not generally feasible to measure current flow directly at all points on a metal surface. Full cathodic protection requires polarizing the cathode areas to the open circuit potential of the anodes. The criterion for cathodic protection of iron is met when all points on the metal surface are polarized to a potential of −0.85 V or more negative, as measured against a copper sulfate reference electrode positioned on the metal surface.

Protective current requirements generally vary with factors influencing corrosion rates. Increasing oxygen concentration, temperature, and velocity increases protective current requirements. Resistive coatings, precipitated calcareous salts, adherent zinc or aluminum flocs, silt, and electrophoretically deposited particles all reduce the protective current requirements.

Sacrificial Anodes. Coupling a more active metal to a structure results in galvanic current flow through the corroding electrolyte, providing a protection effect on the cathode surface. In providing the galvanic protective current flow, the more active metal is electrochemically consumed (sacrificed); it must be replaced. No outside power is required for protection.

The rate of sacrificial anode consumption is greater than the theoretical electrochemical equivalent of its protective current output (Table 2). This is attributable to the self-corrosion current flow superimposed on the protective current output. Sacrificial anode system design is limited by the available driving voltage between the sacrificial anode and the structure to be protected. The resistance between the anodes and the structure mainly determines the protective current flow. A high-resistivity electrolyte requires a larger number of anodes than a more conductive electrolyte to obtain the same amount of protective current flow.

Table 2 Rate of Sacrificial Anode Consumption

Material (and Specification)	Theoretical, kg/(A·yr)	Actual, kg/(A·yr)	Potential (Cu-CuSO$_4$), V
Magnesium (AS-63A)	3.9	7.7	−1.55
Zinc (MIL-A-18001)	10.4	11.4	−1.10[a]
Aluminum (B-605)	3.0	5.4	−1.05[a]
Aluminum (ERP-HP7)	3.0	3.6	−1.20[a]

[a]In seawater

Sacrificial anodes are cast in various forms and weights for optimum design and service life. Special backfills around the anodes are used in soils to maintain moisture at the anode surface and to increase efficiency. In condenser water boxes and similar systems, sacrificial anodes are sometimes encased in perforated plastic containers for optimum service life and performance.

The zinc on galvanized iron protects the underlying metal until it is consumed in the protective process. Zinc and aluminum anodes can usually be used to protect well-coated surfaces without accelerating coating damage.

Impressed Current. An external voltage source can be used to impress protective current flow from anodes through the conducting medium onto the corroding surface. Alternating current power, converted to direct current by an adjustable output selenium or silicon rectifier, is most commonly used.

Sacrificial anodes such as scrap iron or aluminum are sometimes used with impressed current. In impressed current systems, *nonsacrificial anodes* (those not consumed by the electrochemical process) such as graphite, high silicon cast iron alloy, platinum, platinum plated or clad on titanium or tantalum, and silver-lead alloys in seawater are more widely used. The distribution of protective current on the cathode, power costs, and stray current effects on neighboring structures should be considered when selecting the number, form, and arrangement of anodes.

One method of impressed current cathodic protection is *automatic potential control* (Sudrabin 1963). In this system, the potential of the structure is continuously measured against a reference electrode by a monitoring system that regulates the protective current applied to maintain the structure at a preselected protective potential value. Because the protective current requirements vary with conditions, an automatic potential control system protects at all times and prevents excessive protection, which would accelerate (1) coating damage by electroendosmotic effect, (2) cathodic attack on amphoteric metals such as aluminum, and (3) power wastage.

Limitations. Full cathodic protection depends on adequate current flow reaching all surfaces to be protected. On bare surfaces, the relation between anode and cathode configuration determines, in large part, the distribution of protective current. For example, relatively simple anode systems distribute protective current adequately on the outside of storage tanks, well-coated pipelines, tank interiors, and water boxes of heat exchangers. However, the protective current received on the interior of a condenser tube from anodes in the water box diminishes rapidly within a very short distance of the tube entrance.

Rapid attenuation of the protective effect also occurs on the exterior of bare pipelines in conductive media. Therefore, protective current sources must be placed at closer intervals along the pipe. Harmful effects of stray currents from a cathodic protection system on neighboring utilities or other isolated metallic systems must be considered. All underground cathodic protection installations should be reported to, and examined in cooperation with, local electrolysis committees.

Economics. Because the cathodic protection principle must be adapted to meet the specific needs of each corrosion problem, costs vary.

Environmental Change

Another approach to corrosion protection is to change the environment so that it is less aggressive to the equipment. A familiar example is the use of solid desiccants such as silica gel to maintain a low moisture level in refrigerant lines. Desiccants are also applied to minimize atmospheric corrosion. The *mothballing* technique for preserving decommissioned ships and other heavy equipment is an example of this. These methods are employed for dry lay-up. Vapor phase corrosion inhibitors can minimize atmospheric corrosion in certain cases.

Adding alkali to water to raise the pH and using corrosion inhibitors, such as phosphates, are other examples of reducing corrosive attack by changing the environment. Corrosion inhibition in water systems is discussed more fully in the section on Water-Side Corrosion and Deposits. Inhibitors for corrosion control are added to glycol antifreeze solutions, lubricants, lithium bromide absorption refrigeration brines, and other liquids.

Adding filming amines to steam to protect condensate lines and certain additives to fuel oil to reduce fireside corrosion in boilers can make the local environment at the metal surface less corrosive.

Equipment design modifications that reduce the likelihood of corrosion are also considered environmental changes. They include eliminating crevices and providing weep holes to prevent water accumulation.

UNDERGROUND CORROSION

Corrosion activity must always be anticipated on subsurface structures. Corrosion in underground facilities can cause not only economic loss but also loss of valuable fluids, interruption of service, and creation of hazards to life and property. Corrosion control measures must be adapted to the specific conditions in which the structure will exist. A corrosion survey is usually necessary before corrosion control measures can be designed.

Types of Soils

The corrosivity of a soil is affected by its porosity (aeration), electrical resistivity, dissolved salts (depolarizers and inhibitors), moisture, and acidity or alkalinity. Although corrosion rates and characteristics cannot be related exactly to an individual factor, the attack on a metallic structure buried in nonuniform soil will be greatest on those surfaces in contact with the least porous (air-free), least resistive, most alkaline, and most acidic soil. Romanoff (1957) conducted a 45-year study of underground corrosion that covers soil corrosivity, materials, coatings, cathodic protection, and stray currents.

Bacterial Activity

High rates of corrosion in some air-free environments are associated with the presence of sulfate-reducing bacteria (*Desulfovibrio* spp). Bacterial activity is not a special form of corrosion. The conventional use of well-applied coatings and cathodic protection will control corrosion on underground metallic structures in the presence of these bacteria. Coatings that disbond, allowing water and corrodent to reach the metal surface, shield the exposed metal from the effects of applied cathodic protection.

Insulation Failures

Most thermal insulation does not protect against corrosive soil or water. Catastrophic corrosion occurs on hot pipe surfaces that are intermittently contacted by water (Sudrabin 1956). This corrosion process cannot be controlled by cathodic protection. Insulation such as glass fiber or magnesia fiber in underground runs must be kept dry through conduits. Severe corrosion has been experienced within a few months after construction on magnesia-asbestos-insulated snow-melting pipe manifolds extending below the slab into the soil (Sudrabin and LeFebvre 1953). Natural asphalt materials used for underground pipe heat insulation exhibit some protective effect against corrosion. However, they are relatively permeable to water, and repeated heating and cooling initiates attack under the insulation. Similarly, corrosion has occurred at locations where the water table fluctuates.

Radiant-heating and snow-melting pipe embedded in slabs constructed with open expansion joints and vermiculite-filled concrete, brick, or concrete block supports have suffered severe corrosion (Sudrabin and LeFebvre 1953). Severe corrosion of radiant-heating pipe embedded in sand or porous concrete under terrazzo flooring has also been experienced. In such heating systems, good construction practice requires a minimum thickness of 40 mm of cement-rich concrete surrounding the piping, a waterproof membrane under the slab, and steel saddle pipe supports. The piping should be coated at expansion joints, and pipe should not contact the reinforcing steel. Lightweight or insulating concrete produces local cell activity at the nonuniformities in contact with the piping. The piping should be insulated from all metallic structures. Magnesium anodes may be buried in the adjoining soil and attached to the piping for protection.

Cathodic Protection

Although cathodic protection has been applied to radiant-heating piping, it is not always possible to direct adequate amounts of protective current to corroding surfaces. The best corrosion control measures are those integrated into the original design and construction of the heating system.

Galvanic anodes (magnesium or zinc) or impressed current systems can be used for cathodic protection of underground pipelines and tanks. Impressed current may be applied from anodes distributed alongside the structure, from remote anodes, or from deep well anodes.

The method used depends on the economics and the site conditions. Full cathodic protection is achieved when enough current is applied to the structure to prevent current flow into the soil from any and all points on the surface of the structure. Specifically, this protective current must be applied to the structure in an amount sufficient to maintain its external surface negative at every point by at least 0.85 V to a copper-saturated copper sulfate half-cell in its immediate proximity.

Electrically Insulating Protective Coatings

Protective coatings applied to underground structures isolate them from the soil environment and insulate them from electrical effects. Such coating materials must possess long-term qualities of high electrical resistance, inertness to the environment, low water absorption, high resistance to deformation by soil pressures, resistance to the system operating temperatures, and good adhesion. Most coatings require reinforcing and/or shielding to resist soil stresses. The performance of all coating systems, even those of the best material specifications, depends on (1) the metal surface preparation; (2) application procedure and conditions; (3) backfilling; and (4) the physical, chemical, biological, and electrical stresses that occur in service.

The most common coatings for underground structures include (1) hot-applied coal tar enamel; (2) hot-applied petroleum-base coatings, such as asphalt enamels and waxes; (3) polyethylene tape, polyvinyl tape, or extruded polyethylene; (4) coal tar epoxy resin; and (5) cold-applied bituminous and asphalt emulsions, cutback solvents, and greases. Cold-applied coatings are formulated by manufacturers for specific application and service conditions. The National Association of Corrosion Engineers (NACE 1975a, 1975b, 1976, 1985, 1990) publishes standard recommended practices describing acceptable requirements for underground coatings.

WATER-SIDE CORROSION AND DEPOSITS

Because the occurrence and correction of corrosion in water systems are so closely related to the occurrence and correction of other water-caused problems, it is impossible to consider them separately. The most common water problems in heating and cooling systems are (1) corrosion, (2) scale formation, (3) biological growths, and (4) suspended solid matter. Some knowledge of these problems is important because each of them can reduce the cooling or heating efficiency of a system and can lead to premature equipment failure and widespread harm to people in the vicinity.

Controlling water-side problems, particularly in heating and cooling systems, is complex. Handling these problems involves water chemistry, engineering, economics, and personnel administration during each stage of system development, design, construction, installation, and operation.

WATER CHARACTERISTICS

Between the time that water falls as rain, sleet, or snow and the time that it is pumped into a user's premises, it dissolves a small amount of almost every gas and solid substance with which it comes in contact. These dissolved impurities, rather than the water itself, are the primary cause of various water problems.

Table 3 indicates the complexity of water chemistry. Water received at a given location can vary widely over time, either because different supplies are being used or because the composition of a single supply fluctuates, as in the case of river water.

Another often overlooked influence on water-caused problems is the change in composition of the water added to a system after the

system starts to operate. Evaporation, aeration, corrosion, and scale formation all cause changes in chemical composition. These chemical changes, along with temperature changes, formation of biological growths, and the accumulation of suspended matter, all tend to produce operating results that can differ from those predicted based on the chemical analysis of the makeup water.

Chemical Characteristics

The type and amount of dissolved inorganic materials, including gases, define the chemical characteristics of any water. Typical water analyses, such as those shown in Table 3, are not complete but give the major important constituents for municipal and average industrial water use. The many minor constituents present in most water supplies have little or no importance in most water uses.

In water analyses, values are usually given in milligrams per litre; in addition, the chemical species must also be given. Thus, the same calcium concentration in a single water sample might be variously expressed as 100 mg/L $CaCO$, 56 mg/L CaO, or 40 mg/L Ca. Generally, lower pH water tends to be more corrosive and higher pH water tends to be more scale-forming, although many other factors affect both properties. In water chemistry, factors that increase scale formation decrease corrosion, and vice versa.

Most water analyses include only dissolved solids and omit the dissolved gases that are also present. Certain gases, such as nitrogen, have virtually no effect on any water use; others, such as oxygen, carbon dioxide, and hydrogen sulfide, produce important effects in water systems. Carbon dioxide can be either measured directly or estimated from the pH and total alkalinity.

Oxygen and hydrogen sulfide must be measured at the time the sample is collected. When air contacts water, oxygen readily dissolves into the water. Thus, the water chemist generally assumes that any water supply that contacts air will be saturated with oxygen according to its partial pressure in air and the water temperature.

Total hardness is the most commonly recognized chemical constituent in water analysis. In most cases, it corresponds to the calcium and magnesium content of the water. The hardness, particularly the

Table 3 Analyses of Typical Public Water Supplies[a]

Substance		(1)	(2)	(3)	(4)	(5)	(6)	(7)	(8)	(9)
		Concentration, ppm								
Silica	SiO_2	2	6	12	37	10	9	22	14	—
Iron	Fe_2	0	0	0	1	0	0	0	2	—
Calcium	Ca	6	5	36	62	92	96	3	155	400
Magnesium	Mg	1	2	8	18	34	27	2	46	1 300
Sodium	Na	2	6	7	44	8	183	215	78	11 000
Potassium	K	1	1	1	—	1	18	10	3	400
Bicarbonate	HCO_3	14	13	119	202	339	334	549	210	150
Sulfate	SO_4	10	2	22	135	84	121	11	389	2 700
Chloride	Cl	2	10	13	13	10	280	22	117	19 000
Nitrate	NO_3	1		0	2	13	0	1	3	—
Dissolved solids		31	66	165	426	434	983	564	948	35 000
Carbonate hardness	$CaCO_3$	12	11	98	165	287	274	8	172	125
Noncarbonate hardness	$CaSO_4$	5	7	18	40	58	54	0	295	5 900

Source: Collins (1944).

[a]Numbers indicate location or area as follows:
(1) Catskill supply—New York City
(2) Swamp water (colored)—Black Creek, Middleburg, FL
(3) Niagara River (filtered)—Niagara Falls, NY
(4) Missouri River (untreated)—average
(5) Well waters—public supply—Dayton, OH—9 to 18 m
(6) Well water—Maywood, IL—637 m
(7) Well water—Smithfield, VA—100 m
(8) Well water—Roswell, NM
(9) Ocean water—average

calcium, is one of the factors influencing scale formation. The scaling potential increases with increasing hardness.

Alkalinity is a measure of the capacity of a water to neutralize strong acids. In natural waters, the alkalinity almost always consists largely of bicarbonate, although there may also be some carbonate present. Borate, hydroxide, phosphate, and other constituents, if present, will be included in the alkalinity measurement in treated waters. Alkalinity also contributes to scale formation.

Alkalinity is measured using two different end-point indicators. The *phenolphthalein alkalinity* (P alkalinity) measures the strong alkali present; the *methyl orange alkalinity* (M alkalinity), or *total alkalinity*, measures all of the alkalinity present in the water. Note that the total alkalinity includes the phenolphthalein alkalinity. For most natural waters, in which the concentration of phosphates, borates, and other noncarbonated alkaline materials is small, the actual chemical species present can be estimated from the two alkalinity measurements (Table 4).

Alkalinity or acidity is often confused with pH. Such confusion may be avoided by keeping in mind that the pH is a measure of hydrogen ion concentration, in moles per litre, expressed as the logarithm of its reciprocal.

Table 4   Alkalinity Interpretation for Waters[a]

P Alk	Carbonate	Bicarbonate	Free Carbon Dioxide
0	0	M Alk	Present
< 0.5 M	2 P Alk	M Alk − 2 P Alk	0
= 0.5 M Alk	2 P Alk = M Alk	0	0
> 0.5 M Alk[b]	2 (M Alk − P Alk)	0	0

[a]P Alk = Phenolphthalein alkalinity.
M Alk = Methyl orange (total) alkalinity.
[b]Treated waters only. Hydroxide also present.

Dissolved solids. These also affect corrosion and scale formation. Low-solids waters are generally corrosive because they have less tendency to deposit scale. If a high-solids water is nonscaling, it tends to produce more intensive corrosion because of its high conductivity. Dissolved solids are often referred to as total dissolved solids (TDS).

Specific conductance measures the ability of a water to conduct electricity. Conductivity increases with the TDS. Specific conductance can be used to estimate dissolved solids.

Sulfates also contribute to scale formation in high-calcium waters. Calcium sulfate scale, however, forms only at much higher concentrations than the more common calcium carbonate scale. High sulfates also contribute to increased corrosion because of their high conductivity.

Chlorides have no effect on scale formation but do contribute to corrosion because of their conductivity and because the small size of the chloride ion permits the continuous flow of corrosion current when surface films are porous. Chlorides are a useful measuring tool in evaporative systems. Virtually all other constituents in the water increase or decrease as a result either of the addition of common treatment chemicals or of chemical changes that take place in the normal operation of the water system. With few exceptions, chloride concentrations are affected only by evaporation, so that the ratio of chlorides in a water sample from an operating system to those in the makeup water provides a measure of how much the water has been concentrated in the system. (*Note*: Chloride levels will change if the system is continuously chlorinated.)

Soluble iron in the water can originate from metallurgy corrosion within the cooling water systems or as a contaminant in the makeup water supply. The iron can form heat-insulating deposits by precipitation as iron hydroxide or iron phosphate (if a phosphate-based water treatment product is used or if phosphate is present in the makeup water).

Silica can form particularly hard-to-remove scales if permitted to concentrate sufficiently. Fortunately, silicate scales are far less common than others.

Suspended Solids. In addition to dissolved solids, waters (particularly unpurified waters from surface sources or those that have been circulating in open equipment) frequently contain suspended solids, both organic and inorganic. Organic matter in surface supplies may be present as colloidal solutions. Natural coloring matter usually occurs in this form. At high velocities, hard suspended particles can abrade equipment. Settled suspended matter of all types can contribute to concentration cell corrosion.

Biological Characteristics

Bacteria, algae, fungi, and protozoa are present in water systems, and their excessive growth can cause operating problems. Microorganisms are directly affected by the temperature of their habitat. Thus, for every 10 K increase in temperature the microorganism can multiply two or three times, as long as that temperature does not exceed the optimum for the organism. There are established procedures for detecting the presence or absence of selected pathogenic microorganisms. In cooling systems, nonpathogenic microorganisms are the more important cause of operating difficulties. The vast majority of environmental microorganisms are mesophiles, which are middle-temperature-loving microorganisms. When temperatures exceed their optimum growth temperature (generally 38°C), mesophiles are rarely associated with operating problems. Useful microbiological tests have been developed for application in industrial water problems (APHA 1989).

HEATING AND COOLING WATER SYSTEMS

To evaluate the probable effects on a system of the water supply, the boiler water, or the recirculating water, the treatment specialist needs data on the size, construction, operating pattern, and other characteristics of the system. All heating and cooling systems involve a temperature change, which influences the rates at which water-caused problems can develop.

In steam heating systems, the concentration of dissolved solids in the boiler water increases as boiler water is converted to steam. In the ordinary space-heating system, this change is limited because all the steam is condensed and returned to the boiler. On the other hand, if the percentage of condensate returning to the boiler is reduced, the nature of the water changes taking place within the boiler is drastically altered. The condensate has characteristics quite different from those of the boiler water because it is substantially free of dissolved solids but can create different problems due to the presence of dissolved oxygen or carbon dioxide.

Water heating and other closed circulating systems generally have few water-caused problems because there should in theory be virtually no makeup water and no opportunity for significant changes in the composition of the water within the system. Experience has shown, however, that most of these systems do not operate according to this theory (Sussman and Fullman 1953); thus engineers accept the need for water treatment in both ordinary water heating systems and high-temperature hot water systems.

Like water heating systems, closed cooling systems (chilled water, chilled water/hot water, brine, or glycol) require more makeup in practice than in theory. As a result, treatment of the circulating water or solution in these systems is desirable. It is advisable to monitor these systems to ensure that treatment levels and corrosion rates meet specifications.

Condensers and other heat exchangers in refrigeration and air-conditioning systems can be cooled by water passing through the equipment and going to waste. In such cases, water-caused problems are directly related to the chemical composition of the cooling water. Due to increasing water shortages, such once-through cooling is becoming less common and is being replaced by open circulating systems that use cooling towers, evaporative condensers, or spray ponds. In these systems, the makeup water composition is drastically changed by evaporation, aeration, and other chemical and physical processes depending on the contaminants in the air to which the water is exposed.

With humidification and dehumidification equipment, several types of water problems are common. During humidification, water circulating in the equipment is subjected to the same exposure conditions and composition changes that affect open circulating systems. During dehumidification, however, water is removed from the surrounding air. Dissolved solids are low, except for any materials that may be scrubbed from the air. These dissolved solids may not contribute to conductivity, and the condensate may be relatively corrosive.

Materials of Construction

The seriousness of corrosion in heating and cooling systems depends, in part, on the nature of the construction materials, their proximity, their relative areas, and other physical factors.

WATER-CAUSED PROBLEMS

The operator of a heating or cooling system can recognize water-caused problems by the appearance of one or more of the following three symptoms, each of which can be produced by causes other than the water: (1) reduction in heat transfer rate, in which the formation of an insulating deposit on a heat transfer surface significantly reduces the cooling or heating efficiency of the equipment; (2) reduced water flow, which results from a partial or complete blockage of pipelines, condenser tubes, or other openings; and (3) damage to, or destruction of, the equipment, which can result from corrosion of metals, deterioration of wood or plastics, or excessively rapid wear of moving parts such as pumps, shafts, or seals. Several different conditions can act together in the production of any given symptom.

Corrosion contributes to all three symptoms. Deposits of corrosion products can reduce heat transfer rates and create water flow blockages. Damage or destruction of equipment is most often the result of metallic corrosion or the somewhat parallel phenomenon of wood delignification.

Scale formation also contributes to all three symptoms. Even a small buildup of scale on a heat exchange surface reduces water flow. Scale may continue to build up in boilers until heat transfer is so low that the metal overheats, permitting the tubes to rupture under the operating pressure. Scale particles can also accelerate the wear of moving parts.

Biological growths such as slime can act as insulators to reduce cooling efficiency. Bacterial and algae growths often accumulate sufficiently to interfere with water flow in cooling towers. The poultice effect created by accumulations of organic matter can cause localized corrosive attack, resulting in premature equipment failure. Fungus action can also cause destruction of cooling tower wood.

Suspended solid matter, such as dirt scrubbed from the air or finely divided mill scale, can also contribute to all three symptoms. Depending on where they accumulate, these deposits can reduce heat transfer or water flow. Like deposits of organic matter, dirt deposits tend to localize corrosion, and suspended solids cause rapid wear of moving parts.

WATER TREATMENT

General Considerations

Due to the complexity of correcting water-caused problems in heating and cooling systems, it is important to consult a water treatment specialist early in the design stage of any system and regularly thereafter during design, construction, and operation. The improper use of water treatment chemicals can cause problems more serious than those which would have occurred without treatment.

Frequently, there is more than one possible solution for a water treatment problem. The selection may be dictated by such considerations as economics, available space, or labor. The selected treatment program must be followed diligently because it is only as effective as the consistency and control with which it is applied.

Operating and maintenance personnel are not ordinarily familiar with safety precautions pertaining to water treatment chemicals. It is important that (1) the potential hazards associated with any particular water treatment chemical or program be well understood; (2) suitable safety rules be formulated; (3) appropriate safety equipment be supplied; and (4) the safety program be enforced at all times to avoid injury and equipment damage. Material safety data sheets (MSDSs) must be readily available to employees who handle chemicals.

Contamination of drinking water by nonpotable treated or untreated water must be prevented through the elimination of cross-connections between systems or the installation of approved backflow preventers. Disposal of waters treated with certain chemicals into municipal sewers or into streams or lakes may be restricted. Pollution control regulations should be consulted when selecting water treatment. In some cases, removal of treatment chemicals prior to discharging blowdown or drainage can be economically justified.

With very few exceptions, the proper control of chemical water treatment programs depends on proportional feeding of the chemicals to maintain a desired concentration level at all times. Intermittent batch or slug feeding of water treatment chemicals cannot be relied on to produce satisfactory results, particularly in systems that have appreciable makeup rates.

Sound water treatment programs require care and consistent attention, yet devices appear on the market that allegedly prevent scale and corrosion without requiring the operator's attention. Various natural forces, such as electricity, magnetism, or catalysis, generally behaving in some new way, are said to be responsible for the effects claimed. However, independent investigations of these devices have found them to produce no significant effect in preventing or correcting corrosion and scale formation (Meckler 1974; Rosa 1988-1989).

Corrosion Control

Corrosion damage to water systems can be minimized by using corrosion-resistant construction materials, providing protective coatings to separate the water from the metal surfaces of the equipment, removing oxygen from the water, or adding corrosion inhibitors and pH-control chemicals. Two or more of these methods are often used in the same system.

The selection of corrosion-resistant construction materials is the responsibility of the equipment or system designer. Although it is technically possible to build equipment that shows no significant corrosion under almost any operating conditions, economic limitations usually make this impossible. On the other hand, investigation of corrosion failures in air-conditioning and heating equipment sometimes reveals design errors proving that elementary principles of corrosion control have been ignored.

Oxygen removal is effective for closed systems in which opportunities for the pickup of additional oxygen are small. Thus, boiler feedwater can be deaerated mechanically in an open, or deaerating, heater. This process is based on the reduced solubility of oxygen in water at high temperatures (Table 5) and is made more effective by equipment that reduces the partial pressure of oxygen in the gas above the water. The last traces of oxygen are removed chemically by adding sodium sulfite or, at higher temperatures, hydrazine or organic oxygen scavengers. (*Note*: Hydrazine is a suspect carcinogen and must be handled only with skin and respiratory protective equipment.)

Table 5 Solubility of Oxygen from Air in Water at Different Temperatures

Temperature, °C	Millilitres per Litre (mL/L)				
	Air	=	Oxygen	+	Nitrogen
0	28.64	=	10.19	+	18.45
5	25.21	=	8.91	+	16.30
10	22.37	=	7.87	+	14.50
15	20.11	=	7.04	+	13.07
20	18.26	=	6.35	+	11.91
25	16.71	=	5.75	+	10.96
30	15.39	=	5.24	+	10.15
40	13.15	=	4.48	+	8.67
50	11.40	=	3.85	+	7.55
60	9.78	=	3.28	+	6.50
80	6.00	=	1.97	+	4.03
100	0.00	=	0.00	+	0.00

Source: Nordell (1961).

$$\text{Sodium sulfite: } 2Na_2SO_3 + O_2 \rightarrow 2Na_2SO_4$$

$$\text{Hydrazine: } N_2H_4 + O_2 \rightarrow N_2 + 2H_2O$$

Oxygen is less often removed chemically from cold water circuits because of the slow rate of reaction of the sodium sulfite with the dissolved oxygen; however, a small amount of cobalt salt acts as a catalyst to speed the reaction (Pye 1947). In open-spray systems, chemical removal of oxygen would be too expensive because the circulating water is thoroughly oxygenated again with each passage through the spray equipment.

Oxygen may be removed by vacuum deaeration to minimize corrosion on once-through cooling systems. Vaccum deaeration is particularly applicable for waters having an appreciable carbon dioxide concentration because the carbon dioxide as well as the oxygen is removed.

Corrosion control treatment of heating and cooling waters is most often a combination of pH control and the addition of a corrosion inhibitor. Adjustment of the pH to 7.0 is not sufficient to stop corrosion. When the oxygen has been removed, control of pH at certain levels is frequently adequate, as in the case of low-pressure heating boilers maintained at a pH above 10.5. In most other cases, both an inhibitor and pH control are required.

Historically, chromate was used as a highly effective and inexpensive corrosion inhibitor at sodium chromate concentrations from 200 to 2000 mg/L. Cooling system pH was controlled at 6.5 to 7.5 by sulfuric acid addition in order to minimize formation of calcium carbonate scale. Calcium carbonate and other scaling ions are more soluble at low pH than at high pH. Disposal of chromate-treated waters is subject to increasingly stringent pollution control regulations. In the United States, state and federal regulations require that chromate be removed before waters are discharged to sewers or public waterways. Regional offices of the Environmental Protection Agency (EPA) provide these guidelines. Chromate is a suspect carcinogen; because it leaves the cooling tower as part of drift, the EPA has banned its use for comfort cooling systems in the United States.

Alternatives to chromate treatment for comfort cooling systems include blends of phosphate, phosphonate, molybdate, zinc, silicate, and various polymers for scale and corrosion control. Tolyltriazole or benzotriazole are added to these blends to protect copper and copper alloys from corrosion. When selecting a treatment product, consult local, state, and federal environmental guidelines.

In closed systems, chromates are becoming increasingly unacceptable because of their yellow color, disposal problems, and potential hazard if used carelessly. Sodium nitrite has been used as an inhibitor instead of chromate. With ferrous metals, it is nearly as effective as chromate but must be maintained at a minimum concentration of about 500 mg/L and at a pH above 7.0 to avoid breakdown. It is also necessary to monitor nitrite, nitrate, and ammonia concentrations at frequent intervals because certain bacteria may use the nitrite as an alternate electron acceptor, converting it to nitrogen gas; other bacteria may convert it to ammonia or nitrate. As the nitrite disappears, corrosion protection decreases. Sodium nitrite has little or no protective effect on nonferrous metals, and other inhibitors must be included in the blend to provide protection. A commonly used nitrite-base inhibitor includes borax as a pH buffer and sodium tolyltriazole as an inhibitor for nonferrous metals. Historically, hydrazine was used in closed loops, but it, like chromate, is a suspect carcinogen. Organic inhibitors are likely to see more use as chromates and hydrazine are used less.

Under the proper conditions, polyphosphates reduce tuberculation and pitting. For this purpose, polyphosphate concentrations higher than those used for scale control must be provided, as must close pH control, in the 6.0 to 7.0 range. To control pitting corrosion of copper and steel, polyphosphates must be combined with other inhibitors.

Corrosion control treatment of once-through cooling water is practical only with very inexpensive chemicals. Sodium silicate (water glass) or phosphate-silicate mixtures control corrosion in once-through systems, including potable water systems. Sufficient silicate is fed to increase the silica content of the water by about 8 mg/L.

Delignification of wood is the loss of the primary structural component of wood, lignin. Over time, lignin in cooling towers can be dissolved by excessive doses of chlorine (greater than 1 mg/L) or by high alkalinity. Prevention of fungal deterioration is more effective than attempted cures. Manufacturers of cooling towers can provide wood that has been pressure treated with solutions of copper, chromates, and arsenic (CCA).

Scale Control

The methods used for scale control in heating and cooling systems include a variety of internal and external treatment procedures. *Internal treatment*, where chemicals are added directly to the water in the systems, is frequently used in smaller systems. *External treatment* means that the water received some modifications prior to entering the system; an example of this is softening the water prior to sending it to the boiler. The selection of an appropriate method for any one system requires the evaluation of many factors.

Scale control methods attempt to minimize the likelihood of the precipitation of calcium carbonate, the least soluble common constituent of water. The solubility of calcium carbonate depends on the pH, temperature, and total solids content of water, in addition to the calcium and alkalinity (bicarbonate or carbonate). Using these items and one of several nomographs (such as that in Figure 1), the pH_s (i.e., the pH at which any given water is in equilibrium with calcium carbonate) can be calculated. This pH_s can be used with the actual pH of the water in either of two calculations that indicate whether the water tends to precipitate calcium carbonate or to dissolve it. The older of these is the Langelier Saturation Index (Langelier 1936).

$$\text{Saturation Index} = pH - pH_s$$

A positive saturation index shows scale-forming tendency only. The larger the index, the greater this tendency. Other factors may inhibit scale formation under some circumstances. Usually, calcium carbonate precipitates as a scale when the saturation index exceeds +0.5 to +1.0. A negative saturation index indicates that calcium carbonate will dissolve and that bare metal will remain bare and thus accessible for corrosion.

Ryznar (1944) suggested a modified method for predicting calcium carbonate scale formation based on operating performance: the stability index.

$$\text{Stability Index} = 2\,pH_s - pH$$

The stability index is always positive. When it falls below 6.0, scale formation is possible; it becomes more probable as the numerical value of the index decreases.

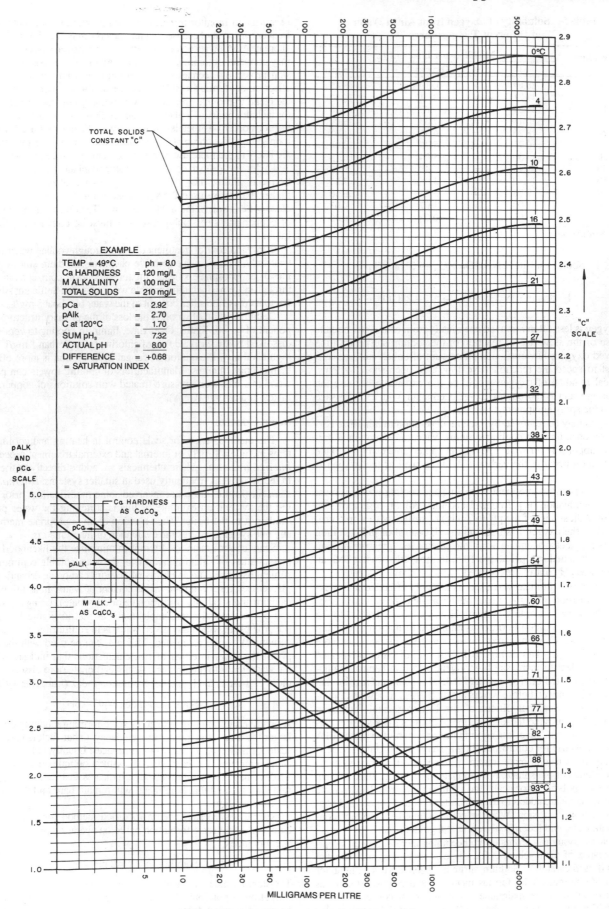

Fig. 1   Langelier Saturation Index Nomograph

Similar but more complex methods for estimating the scale-forming tendencies of calcium sulfate (Denman 1961) and calcium phosphate (Green and Holmes 1947) have been published but are not as commonly used. At cooling water temperatures and lower pressure boiler temperatures, calcium sulfate is far more soluble than calcium carbonate.

An effective method for preventing scale formation is to soften the water by passing it through an ion exchange water softener. This external treatment process removes all but 2 to 5 mg/L of hardness. It is usually carried out in a closed vertical tank about two-thirds filled with small beads of a cation exchange resin. The resin preferentially removes calcium and magnesium ions from the water, while adding sodium ions.

The calcium and magnesium content of the resin ultimately rises to the point that these elements are no longer completely absorbed from the water passing through the resin. For restoration, the resin is backwashed to remove any dirt particles and then regenerated by having a much higher concentration salt solution passed through it. The sodium ions of the strong salt solution displace the absorbed calcium and magnesium ions, restoring the resin to its initial sodium form. After the excess salt is rinsed out, the resin is ready for another softening cycle.

$$\underset{\text{Regenerated Resin}}{Na_2R + Ca^{++}\,(or\;Mg^{++})} \underset{\text{Regeneration}}{\overset{\text{Softening}}{\rightleftharpoons}} \underset{\text{Exhausted Resin}}{CaR\,(or\;MgR) + 2Na^+}$$

When it is possible to accept and control the presence of suspended matter in the water, hardness may be eliminated by precipitation within the operating equipment. This is commonly done in lower pressure process steam boilers. The following are equations for widely used precipitation reactions:

$$Ca(HCO_3)_2 + 2NaOH \rightarrow CaCO_3 + Na_2CO_3 + 2H_2O \quad (1)$$

$$3Ca(HCO_3)_2 + 6NaOH + 2Na_3PO_4 \rightarrow Ca_3(PO_4)_2 \quad (2)$$
$$+ 6Na_2CO_3 + 6H_2O$$

$$Mg(HCO_3)_2 + 4NaOH \rightarrow Mg(OH)_2 + 2Na_2CO_3 + 2H_2O \quad (3)$$

Alkalies are used to precipitate calcium carbonate in accordance with Equation (1). Phosphates are added to remove the last of the hardness, as shown in Equation (2), by virtue of the lower solubility of calcium phosphate. Magnesium is most commonly precipitated as the hydroxide, as shown in Equation (3).

There are situations in which it is economically or otherwise impossible to remove the hardness from water. In these cases, other measures can be taken to control scale formation. One common method is to reduce the alkalinity of the water. This substantially reduces scale formation because the solubility of calcium carbonate is much less than that of other calcium salts.

Most of the alkalinity can be removed by an anion exchange resin in a process called dealkalizing, which is analogous to softening. In this case, the alkalinity (bicarbonate) is retained by the resin, which gives up to the water equivalent amounts of chloride. When the resin begins to pass larger amounts of alkalinity than desired, it is backwashed and regenerated by a salt solution, as described above. Usually, this salt solution contains a small percentage of alkali.

$$\underset{\text{Regenerated Resin}}{R'Cl + HCO_3^-} \underset{\text{Regeneration}}{\overset{\text{Dealkalizing}}{\rightleftharpoons}} \underset{\text{Exhausted Resin}}{R'HCO_3 + Cl^-}$$

Particularly in large cooling towers, sulfuric acid is commonly used to eliminate most of the alkalinity. The solubility of calcium sulfate is about 2000 mg/L, in contrast to 35 mg/L for calcium carbonate; this permits a much higher hardness in the circulating water before scale can form. The acid feeding procedure involves a considerable risk in the absence of careful controls. The alkalinity of the circulating water is reduced to such a low figure that a slight overdose of acid produces corrosive circulating water. Accordingly, it is good practice to (1) restrict acid feed to systems that operate under conditions of constant makeup water composition and evaporation or (2) use automatic pH-control equipment.

For cooling systems in which the above measures cannot be used, two other measures are used together to minimize scale formation. First, the total dissolved solids in the circulating water are controlled by a continuous bleed or bleedoff to a maximum value proportional to the concentration of hardness, silica, or other limited-solubility constituent. Second, scale control adjuncts are added to the water to increase the apparent solubility of the calcium carbonate. Adjuncts include low concentrations (2 to 5 mg/L) of sodium polyphosphates; organic dispersing agents such as various lignin derivatives; synthetic polymer polyelectrolytes; organic phosphates; or mixtures of these. These measures may also be used with acid feeding.

Control of Biological Growths

Algae, bacterial slimes, and fungi can interfere with the operation of cooling systems if they are able to clog distribution holes on the deck, clog spray nozzles, impede water flow and heat transfer in heat exchangers, or weaken the structural integrity of cooling tower wood. Heating systems operate at temperatures outside biological limits and therefore have no microbial problems.

Algae use the energy from the sun to convert bicarbonate or carbon dioxide into biomass. Masses of algae can block piping, distribution holes, and nozzles when allowed to grow in sufficient profusion. A covered distribution deck is one of the most cost-effective additions to a cooling tower. Biocides are used to assist in the control of algae.

Most waters contain organisms capable of producing biological slime, but significant amounts of slime are produced only when conditions are favorable. Optimal conditions are poorly understood, but include sufficient nutrients and appropriate environmental (temperature and pH) conditions. Equipment located near sources of nutrients is particularly susceptible to slime formation. Two common examples are air washers in printing plants, where there is fine paper dust in the air, and cooling towers located in food-packing plants, which receive large amounts of contamination. Slimes can be formed by bacteria, algae, yeasts, or molds and frequently consist of a mixture of these organisms and organic and inorganic debris. Wherever slimes or other microbiological growths threaten to interfere with the efficient functioning of a cooling system, the use of biocidal materials is indicated.

The effective control of slime and algae frequently requires a combination of mechanical and chemical treatments. For example, when a system already contains a considerable accumulation of slime, a preliminary mechanical or detergent cleaning will make the subsequent application of a microbicidal chemical more effective in killing the growths and more enduring in the prevention of further growths. Similarly, periodic mechanical removal of slimes from readily accessible areas reduces the chemical microbicide requirements and makes them more effective.

In large systems, particularly once-through cooling systems using river water, estuarine water, or seawater, macroorganisms such as barnacles and mussels may accumulate. Although antifouling paints can prevent the growth of such macroorganisms on large-diameter pipe surfaces, the paints must be renewed at frequent intervals and are not applicable to inaccessible areas such as the insides of smaller diameter piping. Generally, chlorine is used to destroy these organisms. In once-through systems, intermittent feeding of chlorine in the form of chlorine gas or hypochlorite solutions is the most effective and economical method of controlling microorganisms and slimes. Tin-containing paint and trihalomethane formation from the use of chlorine are both under increasing scrutiny from an environmental standpoint.

Microbicides. Chlorine and chlorine-yielding compounds, such as sodium hypochlorite and calcium hypochlorite, are among the most effective microbicidal chemicals. However, they are not always appropriate for the control of organic materials in cooling systems. In air washers, for instance, their odor may become offensive; in wooden cooling towers, excessive concentrations of chlorine can cause rapid deterioration of wood construction; and in metal equipment, higher concentrations of chlorine can accelerate corrosion. In systems large enough to justify equipment for the controlled feeding of chlorine, its use can be both safe and economical. Some of the disadvantages can be avoided by combining the chlorine with isocyanuric acid, which minimizes the free chlorine concentration. Most chlorine programs can benefit from surfactant (chlorine helper) products or nonoxidizing, organic microbicides.

When selecting a microbicide, consider the pH of the circulating water and the chemical compatibility with the corrosion/scale inhibitor product. The availability of many organic microbicides allows flexibility. Typical products include quaternary ammonium compounds, tributyl tin oxide, methylene bis(thiocyanate), isothiazolones, dibromonitrilopropionamide (DBNPA), glutaraldehyde, bromochlorohydantoin, and many proprietary blends. All microbicides must be handled with care to ensure personal safety. Cooling water microbicides are approved and regulated through the EPA and, by law, must be handled in accordance with labeled instructions. Maintenance staff handling the biocides should read the material safety data sheets and be provided with all the appropriate safety equipment to handle the substance. An automatic biocide feed system, such as a pump and timer, should be installed to minimize the exposure of maintenance staff, who would otherwise dose the cooling system using a bucket.

The manner in which biocides are fed to a system is important. Often, the continuous feeding of low dosages is neither effective nor economical. Better results can be obtained by shock feeding larger concentrations to the system to achieve a toxic level of the chemical in the water for a sufficient time to kill the organisms present. Alternate shock feeding of two different types of microbicides usually gives the most effective results. Much larger dosages are required when biomass has been permitted to accumulate than when it has been kept under control. At times, organisms appear to build up an immunity to the particular chemical being used, making it necessary to change the biocidal chemical to keep the organic growths under control.

Legionnaire's Disease

Legionella pneumophila, the bacterium that causes Legionnaire's disease (legionellosis), first received public attention when it affected persons attending an American Legion Convention at the Bellevue-Stratford Hotel in Philadelphia, Pennsylvania in 1976. Subsequent examination of clinical specimens that had been stored revealed that the disease was not new; various species of *Legionella* had been responsible for cases and epidemics of pneumonia as far back as 1943.

Like other living things, *Legionella* bacteria require moisture for survival. They are widely distributed in natural water systems and are present in many drinking water supplies. Potable hot water systems at temperatures between 27 and 49°C, cooling towers, humidifiers, whirlpools and spas, and the various components of air-conditioning systems are considered to be amplifiers. These bacteria are killed at temperatures above 60°C.

Legionellosis can be acquired by inhalation of *Legionella* organisms in aerosols. Aerosols can be produced by cooling towers, evaporative condensers, decorative fountains, showers, and misters. The aerosol from cooling towers can be transmitted over a distance of only 200 to 300 m. If the air inlet ducts of near-by air-conditioning systems draw in the aerosol from contaminated cooling towers, the building air distribution system itself can transmit the disease. When an outbreak of Legionnaire's disease occurs, cooling towers are often the suspected source. However, other water systems that may produce an aerosol should not be neglected.

Humidifiers. Units that generate a water aerosol for humidity control can become amplifiers of bacteria if their reservoirs are poorly maintained. Manufacturers' recommendations on cleaning and maintenance should be followed closely. Steam humidifiers should not pose a bacterial problem because the steam will not contain bacteria.

Whirlpools and Spas. These units pose a potential hazard to users if not properly maintained. The complexity of some units makes cleaning difficult. It may be necessary to employ a firm that specializes in cleaning these devices. Manufacturers' instructions should be followed carefully.

Decorative Fountains. If these systems become contaminated, *Legionella* may multiply and pose a hazard to those nearby. The recirculating water should be kept clean and clear with proper filtration. Continuous chlorination or use of another biocide with EPA approval for decorative fountains is recommended. Indoor decorative fountains with ponds containing fish and other aquatic life cannot be chlorinated or treated; steps should be taken to ensure that water does not become airborne via sprays or cascading waterfalls.

Roof Ponds. These are used to lower the roof temperature and thus the demand on the air-conditioning system. If contaminated, such ponds can pose a risk to the staff and general public. Roof ponds should be monitored and treated in the same manner as recirculating cooling water systems (see the section on Selection of Water Treatment).

Safety Showers. Safety showers are mandated by code for safety reasons, but they are used infrequently. Standing reservoirs of nonsterile water at room temperature should be avoided. A preventive maintenance program that includes flushing is suggested.

Vegetable Misters or Sprays. Vegetable misters that use water from a reservoir have been implicated in one or two cases of this disease, so bacterial activity should be monitored. Once-through systems using potable water do not appear to be a problem.

Machine Shop Cooling Systems. Water used in machining operations may be mixed with oil or other ingredients to improve cooling and cutting efficiency. It is typically stored in holding tanks that are subject to bacterial contamination, including by *Legionella*. Bacterial activity should be monitored and controlled.

Ice-Making Machines. These are prone to bacterial amplification like any water-using device. Exposure to *Legionella* is not a risk to the person drinking a beverage containing the ice. *Legionella* bacteria are not likely to amplify in this cold environment.

Prevention and Control. The *Legionella* count required to cause illness has not been firmly established because many factors are involved, including (1) virulence and number of *Legionella* in the air, (2) rate at which the aerosol dries, (3) wind direction, and (4) susceptibility to the disease of the person breathing the air. The organism is often found in sites not associated with an outbreak of the disease. Likewise, the value of routine bacterial monitoring and maintenance in preventing epidemics of legionellosis has not been shown conclusively.

From the standpoint of liability, however, it may be appropriate to periodically monitor circulating water for total bacteria count and *Legionella* count by culture methods or direct fluorescent antibody microscopic analysis. It is also important to monitor system cleanliness and use a microbial control agent that has laboratory efficacy or is generally regarded as effective in controlling *Legionella* populations. Other measures to decrease risk include optimizing cooling tower design to minimize drift and locating the tower so that drift is not injected into the air-handling system.

Mechanical Filtration

Suspended solids are undesirable in heating and cooling water systems. Settling out on heat exchange surfaces, they can be as effective as scale or slime in reducing heat transfer. They may accumulate enough to interfere with water flow, and localized corrosive attack can occur beneath such accumulations. Abrasive dirt particles can cause excessively rapid wear of such moving parts as pump shafts and mechanical seals. Treatment with polyelectrolytes, together with blowdown and filtration, helps control most harmful effects of suspended solids.

Strainers, filters, and separators may be used to reduce the suspended solids to an acceptably low level. Generally, if the screen is 200 mesh (74-μm opening), it is called a strainer; if it is finer than 200 mesh, it is a filter.

Strainers. A strainer is a closed vessel with a cleanable screen element designed to remove from various flowing fluids foreign particles down to 25-μm diameter. They extract material that is not wanted in the fluid; this can sometimes be a valuable product that may be saved. Strainers are available as single-basket or duplex, manual or automatic cleaning units, and may be made of cast iron, bronze, stainless steel, alloys, or plastic. Magnetic inserts are available where microscopic iron or steel particles are present in the fluid.

Cartridge Filters. These are typically used as final polishing filters to remove suspended particles from about 100 μm in size down to 1 μm or less. Cartridge filters are, with a few exceptions, disposable (i.e., once plugged, they must be replaced). The frequency of replacement, and thus the economic feasibility of their use, depends on the concentration of suspended solids in the fluid, the size of the smallest particles to be removed, and the removal efficiency level of the selected filter.

In general, cartridge filtration is favored in systems where contamination levels are less than 0.01% by mass (<100 mg/L). They are available in a number of different materials and configurations. Filter media include yarns, felts, papers, nonwovens, resin-bonded fibers, woven wire cloths, sintered metal, and ceramics. The standard configuration is a cylinder having an outside diameter of about 65 to 70 mm, an overall length of about 250 mm, and an internal core, where the filtered fluid collects, with an inside diameter of about 25 mm. Filters having overall lengths of 75 to 1000 mm are readily available.

Cartridges made of yarns or resin bonded fibers normally increase in density toward the center. These depth-type filters capture particles throughout the thickness of the medium. Thin-media filters, such as pleated paper types, have a narrow pore size distribution designed to capture particles at or near the surface of the filter. Surface-type filters can normally handle higher flow rates and provide higher removal efficiency than equivalent depth filters. Cartridge filters are rated in accordance with manufacturer's guidelines. Surface-type filters have an "absolute" rating, while depth-type filters have a "nominal" rating that reflects their general clarification function.

Multimedia Downflow (Sand) Filters. A downflow filter is used to remove suspended solids from a water stream for various applications. The degree of removal depends on the combinations and grades of the media used in the vessel. During filtration, water enters at the top of the filter vessel, passes through a flow impingement plate, and enters the quiescent (calm) freeboard area above the medium.

In multimedia downflow vessels, various grain sizes and types of media provide depth filtration primarily to increase the capacity of the system to hold suspended solids. This increases the length of time between backwashing. Multimedia vessels may also be applied for low suspended solids applications, where chemical additives are required. In the multimedia vessel, the fluid enters the top layer of anthracite, which has an effective size of 1 mm. This relatively coarse layer removes the larger particles, a substantial portion of the smaller particles, and small quantities of free oil. Flow continues downward through the next layer of fine garnet material, which has an effective size of 0.3 mm. A more finely divided range of suspended solids is removed in this polishing layer. The fluid continues into the final layer, a coarse garnet material that has an effective size of 2 mm. Within this layer is the header/lateral assembly, which collects the filtered water.

When the vessel has retained enough suspended solids to cause a substantial pressure drop, the unit must be backwashed. This is accomplished by reversing the direction of flow, which carries the accumulated solids out through the top of the vessel.

Bag-Type Filters. These filters comprise a bag of mesh or felt supported by a removable perforated metal basket in a closed housing having an inlet and an outlet.

The housing is a welded, tubular pressure vessel with a hinged cover on top for access to the bag and basket. Housings are made of carbon or stainless steel. The inlet can be in the cover, in the side (above the bag), or in the bottom (and internally piped to the bag). The side inlet is the simplest type. In any case, the liquid enters the top of the bag. The outlet is located at the bottom of the side (below the bag). Pipe connections can be threaded or flanged. Single-basket housings can handle up to 14 L/s, multibaskets up to 220 L/s.

The support basket is usually of 304 stainless steel perforated with 3-mm holes. (Heavy wire mesh baskets also exist.) The baskets can be lined with fine wire mesh and used without a filter bag as strainers. Some manufacturers offer a second, inner basket (and bag) that fits inside the primary basket. This provides for two-stage filtering: first a coarse filtering stage, then a finer one. The benefits are longer service time and possible elimination of a second housing.

The filter bags may be made of many materials (cotton, nylon, polypropylene, and polyester) and have particle-removal ratings from 1 to 840 μm. Felted materials are most common because of their depth-filtering quality, which provides high dirt-loading capability, and their fine pores. Mesh bags are generally coarser, but they are reusable and therefore less costly. The bags have a metal ring sewn into their opening; this holds the bag open and seats it on top of the basket rim.

During filtration, the liquid enters the bag from above, flows out through the basket, and exits the housing free of particulates down to the desired size. The contaminant is trapped inside the bag, making it easy to remove without spilling any downstream.

Centrifugal-Gravity Separators. In this type of separator, liquids/solids are drawn into the unit through tangential slots and accelerated into the chamber. Centrifugal action tosses the particles heavier than the liquid to the perimeter of the separation chamber. Solids gently drop along the perimeter and into the separator's quiescent collection chamber. Solids-free liquid is drawn to the vortex (low-pressure area) of the separator and up through the outlet. Solids are either purged periodically or bled continuously from the separator with an appropriate valve system.

Special Methods. Localized areas can frequently be protected by special methods. Pump-packing glands or mechanical shaft seals can be protected by using fresh water makeup or by circulating water from the pump casing, through a cyclone separator or filter, and into the lubricating chamber.

In smaller equipment, a good dirt-control measure is the installation of backflush connections and shutoff valves on all condensers and heat exchangers so that they can be readily backflushed with makeup water or detergent solutions to remove accumulated settled dirt. These connections can also be used for acid cleaning to remove calcium carbonate scale.

SELECTION OF WATER TREATMENT

Many methods are available for the correction of almost any water-caused problem. However, no one method of treatment applies to all cases. The selection of the proper water treatment method and the details of the chemicals and equipment necessary to apply that method depend on many factors. The chemical characteristics of the water, which change with the operation of the equipment, are important. Other factors contributing to a lesser degree to the selection of proper water treatment are (1) economics; (2) other nonchemical influences, such as the design of individual major system components (e.g., the cooling tower or boiler); (3) equipment operation; and (4) human factors, such as the quantity and quality of operating personnel available.

Once-Through Systems

Economics is the overriding consideration in the treatment of water in once-through systems. The quantities of water to be treated are usually so large that any treatment other than simple filtration or the addition of a few milligrams per litre of a polyphosphate, silicate, or other inexpensive chemical may not be feasible. Intermittent treatment with polyelectrolytes can help to maintain clean conditions when the cooling water is sediment-laden. In such systems, it is generally less expensive to invest more in corrosion-resistant construction materials than to attempt to treat the water.

Open Recirculating Systems

The size of the system has a considerable effect on the selection of water treatment for open recirculating systems. At one extreme, large industrial cooling tower systems use such huge quantities of water that it is economically necessary to minimize the concentration of any treatment chemical used. In these systems, sizable expenditures for chemical treatment and control equipment can be justified, as can the use of specially trained operating personnel. At the other extreme, small air-conditioning and refrigeration cooling towers use such a small amount of water that original, replacement, and chemical costs for the system are low enough that automatic control equipment and specially trained personnel are not economically justifiable.

Therefore, a typical water treatment scheme for a large industrial cooling tower system operating at a circulation rate above 630 L/s might include scale control by a controlled bleed and alkalinity reduction by automatically pH-controlled sulfuric acid feed combined with corrosion control. Cooling tower controllers can monitor acid feed by pH, mineral content by conductivity, inhibitor concentration by direct or indirect measurement, and corrosion rate by test probes and corrosion coupons (small metal strips).

On the other hand, in a small air-conditioning or refrigeration cooling tower circulating at 6.3 L/s, treatment might consist of a controlled bleed to minimize scale formation, maintenance of 200 to 500 mg/L of an inhibitor for corrosion control, and occasional application of a shock dose of microbicide.

Air Washers and Sprayed Coil Units

A water treatment program for an air washer or a sprayed coil unit is usually complex and depends on the purpose and function of the system. Some systems, such as sprayed coils in office buildings, are used primarily to control temperature and humidity, while other systems are intended also to remove dust, oil vapor, and other airborne contaminants from an airstream. Unless the water is properly treated from the beginning, the fouling characteristics of the contaminants removed from the air will cause operational problems.

Scale control is important in air washers or sprayed coil systems providing humidification because the minerals present in the water may become concentrated (through the process of water evaporation) to such a degree that they cause problems. Inhibitor/dispersant treatments commonly employed in cooling towers are often used in air washers to control scale formation and corrosion.

Suitable dispersants and surfactants are often needed to control oil and dust removed from the airstream. The type of dispersant depends on the nature of the contaminant and the degree of system contamination. For maximum operating efficiency, dispersants should produce a minimal amount of foam in the system.

Control of slime and bacterial growth is also important in the treatment of air washers and sprayed coil systems. Because these systems are constantly exposed to airborne microorganisms, the potential for biological growth is enhanced, especially if the water

contains contaminants that are nutrients for the microorganisms. Because of the variations in conditions and applications of air-washing installations and the possibility of toxicity problems, individual cases should be studied by a chemical consultant before a water treatment program is begun. All microbicides applied in air washers must have specific EPA approval.

Ice Machines

Lime scale formation, cloudy or "milky" ice, objectionable taste and odor, and sediment are the most frequently encountered water problems in ice machines. Lime scale is probably the most serious problem because it interferes with the harvest cycle by forming on the freezing surfaces, preventing the smooth release of ice from the surface to the harvest bin.

Scale is caused by dissolved minerals in the water. Because water tends to freeze in a pure state, these dissolved minerals concentrate in the unfrozen water, and some eventually deposit on the machine's freezing surfaces as a lime scale. Two major factors contributing to the problem are calcium hardness (carbonate) and total alkalinity (bicarbonate).

The probability of scale formation in an ice machine varies directly with the concentrations of carbonate and bicarbonate in the water circulating in the machine. Table 6 shows how the problem can vary with calcium hardness and alkalinity.

Dirty or scaled-up ice makers should be thoroughly cleaned before the water treatment program is started. Water distributor holes should be cleared, and all loose sediment and other material flushed from the system. Existing scale formations can be removed by circulating an acceptable acid solution through the system. Several such solutions are available from air-conditioning and refrigeration parts wholesalers.

Slowly soluble food-grade polyphosphates inhibit scale formation on freezing surfaces during normal operation by keeping hardness minerals in solution. Although polyphosphates can inhibit lime scale in the ice-making section and help prevent sludge deposits of loose particles of lime scale in the sump, they do not eliminate the soft, milky, or white ice caused by high concentrations of dissolved minerals in the water.

Even with proper chemical treatment, recirculating water having a mineral content above 500 to 1000 mg/L cannot produce clear ice. Increasing bleedoff or reducing the thickness of the ice slab or the size of the cubes may mitigate the problem. However, demineralizing or distillation equipment is needed to prevent white ice production. This equipment is expensive and is usually not economical for use with small ice machines.

Another problem frequently encountered with ice machines is objectionable taste or odor. When water containing a material having an offensive taste or odor is used in an ice machine, the taste or odor is trapped in the ice. An activated carbon filter on the makeup water line can remove the objectionable material from the water. Carbon filters need to be serviced or replaced regularly to avoid organic buildup in the carbon bed.

Table 6 Scale Formation in an Ice Machine

Total Alkalinity as Bicarbonate, mg/L	Hardness as Calcium Carbonate, mg/L			
	0 to 49	50 to 99	100 to 199	200 and up
0 to 49	No scale	Very light scale	Very light scale	Very light scale
50 to 99	Very light scale	Moderate scale	Moderate scale	Moderate scale
100 to 199	Very light scale	Troublesome scale	Troublesome scale	Heavy scale
200 and up	Very light scale	Troublesome scale	Heavy scale	Very heavy scale

Occasionally, slime growth is the cause of an odor problem in an ice machine. This problem can be controlled by regularly cleaning the machine with a food-grade acid. If the slime deposits persist, sterilization of the machine may be helpful.

Feedwater often contains suspended solids such as mud, rust, silt, and dirt. To remove these contaminants, a sediment filter of appropriate size can be installed in the feed lines. In addition to improving the quality of the ice, a suspended solids filter protects the solenoid valves in the machine.

Closed Recirculating Systems

These are often defined as systems requiring less than 5% makeup per year. The need for water treatment in such systems (i.e., water heating, chilled water, combined cooling and heating, and closed loop condenser water systems) is often ignored based on the rationalization that the total amount of scale from the water initially filling the system would be insufficient to interfere significantly with heat transfer and that corrosion would not be serious. However, leakage losses are common, and corrosion products can accumulate sufficiently to foul heat transfer surfaces. Therefore, all systems should be adequately treated for corrosion control, and scale-forming waters should be softened before use as makeup. Many closed loop systems contain glycol or alcohol solutions; the inhibitor products chosen for corrosion control must be chemically compatible. Monitoring water use, chemical consumption and concentration, and corrosion rates (with corrosion coupons) can be beneficial.

The selection of a treatment program for closed systems is often influenced by factors other than degree of corrosion protection offered. Chromate and hydrazine treatments used in the past are progressively less acceptable from a safety standpoint. Alternatives include buffered nitrite, molybdate, and organic blends. Nonconductive couplings of different metals should be employed throughout the construction of the system. The presence of aluminum or its alloys along with other metals makes corrosion prevention more difficult and usually necessitates supplementary additives such as nitrites or silicates. With buffered sodium nitrite inhibitors, a minimum of 500 mg/L as sodium nitrite is required; the pH must be maintained in the range of 8.0 to 10.0.

Before new systems are treated, they must be cleaned and flushed. Grease, oil, construction dust, dirt, and mill scale are always present in varying degrees and must be removed from the metallic surfaces to ensure adequate heat transfer and to reduce the opportunity for localized corrosion. Detergent cleaners with organic dispersants are available for proper cleaning and preparation of new closed systems.

Water Heating Systems

Secondary and Low-Temperature. Closed chilled water systems that are converted to secondary water heating during winter and primary low-temperature hot water heating systems, both of which usually operate in the temperature range of 60 to 120°C, require sufficient inhibitors to control corrosion to less than 0.13 mm per year. Proprietary organic inhibitors containing no chromates or nitrites are available for such systems.

Medium- and High-Temperature. Medium-temperature water heating systems (120 to 175°C) and high-temperature, high-pressure hot water systems (above 175°C) require careful consideration of treatment for corrosion and deposit control. Makeup water for such systems should be demineralized or softened to prevent scale deposits. For corrosion control, oxygen scavengers such as sodium sulfite should be introduced to remove dissolved oxygen, and neutralizing amines and caustic soda should be used to maintain the pH in the 8.0 to 10.0 range.

Electrode boilers are sometimes used to supply low- or high-temperature hot water. Such systems use heat generated due to the electrical resistance of the water between electrodes. The conductivity of the recirculating water must be in a specific range depending on the voltage used. Treatment of this type of system for corrosion and deposit control varies. In some cases, sodium sulfite and caustic soda are used to maintain the desired specific conductance and sulfite residual and a pH value from 7.0 to 9.0. In other applications, neutralizing amines (which do not add to the conductivity) can remove dissolved oxygen and maintain the pH value at 7.0 to 9.0. The choice depends on the required conductivity of the electrode boiler and of the makeup water. Before selecting the water treatment for an electrode boiler, consult the boiler manufacturer or operating manual.

Brine Systems

Brine systems must be treated to control corrosion and deposits. The standard chromate treatment program is the most effective. Calcium chloride brines require a minimum of 1800 mg/L of sodium chromate with a pH of 6.5 to 8.5. Sodium chloride brines require a minimum of 3600 mg/L of sodium chromate and also a pH of 6.5 to 8.5. Sodium nitrite at 3000 mg/L in calcium brines or 4000 mg/L in sodium brines and pH between 7.0 and 8.5 should provide adequate protection. Organic inhibitors are available that may provide adequate protection where neither chromates nor nitrites can be used.

Ethylene glycol or propylene glycol are used instead of brine as antifreeze in secondary hot water systems. Such glycols are available commercially, with inhibitors such as sodium nitrite, potassium phosphate, and organic inhibitors for nonferrous metals added by the manufacturer. These require no further treatment, but softened water should be used for all filling and makeup requirements. Samples from the systems should be checked periodically to ensure that the inhibitor has not been depleted. Analytical services are available from the glycol manufacturers and others for this purpose.

Boilers

There is a wide range of water treatment procedures for boilers; the method selected must depend on the composition of the makeup water, the operating pressure of the boiler, and the makeup rate.

Minimum makeup water pretreatment for low-pressure boilers should consist of ion exchange softening, with further consideration given to dealkalizers, desilicizers, or demineralizers. In low-pressure systems where steam is used for cooking or humidification, some makeup is required, and the use of inhibitors is ruled out. In these cases, it is necessary to remove dissolved oxygen from the feedwater by using a deaerator or feedwater heater. This is then followed by treatment to raise the pH value above 10.5 and sodium sulfite to remove the last traces of dissolved oxygen (Blake 1968). Scale control may also be required. In the United States, steam used for cooking and humidification is under increasing scrutiny by the FDA, EPA, and USDA; chemicals used in steam for these applications must follow guidelines.

As operating pressures and makeup rates go up, the corrosion problem does not decrease, but the scale problem becomes more important. Corrosion is controlled by (1) using an open heater (at higher operating pressures, a deaerating heater) to remove oxygen from the boiler feedwater; (2) the maintenance in the boiler water of a sulfite or hydrazine residual as an oxygen scavenger; and (3) the maintenance of a sufficiently high pH in the boiler water. Scale is controlled by external softening and by internal treatment, usually with phosphates and organic dispersants, to precipitate residual traces of calcium, generally as a phosphate (hydroxyapatite), and magnesium (as magnesium hydroxide or a basic silicate). The ASME standards for boiler water quality are given in Table 7.

Other chemicals used for corrosion control in closed heating and cooling systems include volatile amines (such as morpholine and cyclohexylamine) and filming amines (such as octadecylamine) used for steam condensate line protection. Various proprietary chemical mixtures are used for glycol or alcohol antifreeze solutions in chilled water or snow-melting systems, where the inhibitor must be

Table 7   ASME Standards for Boiler Water Quality

Drum Pressure, MPa (gage)	Silica, mg/L SiO$_2$	Total Alkalinity,[a] mg/L CaCO$_3$	Neutralized Specific Conductance, micromhos/cm	Suspended Solids,[d] mg/L
0-2.1	150	700[b]	7000	300
2.1-3.1	90	600[b]	6000	250
3.1-4.1	40	500[b]	5000	150
4.1-5.2	30	400[b]	4000	100
5.2-6.2	20	300[b]	3000	60
6.2-6.9	8	200[b]	2000	40
6.9-10.3	2	0[c]	150	20
10.3-13.8	1	0[c]	100	10

Source: ASME (1979).

[a]Minimum level of OH alkalinity in boilers below 7 MPa must be individually specified with regard to silica solubility and other components of internal treatment.

[b]Alkalinity not to exceed 10% of specific conductance.

[c]Zero in these cases refers to free sodium or potassium hydroxide alkalinity. Some small variable amount of total alkalinity will be present and measurable with the assumed coordinated control or volatile treatment employed at these high-pressure ranges.

[d]American Boiler and Affiliated Industries Manufacturers Association's maximum limits for boiler water concentrations in units with a steam drum.

chemically compatible with the antifreeze agent. (*Note*: Volatile amines may present health hazards.)

Treatment of the boiler water is often determined by the end use of the steam. Treatment may range from reduction of alkalinity to minimize condensate line corrosion by keeping the concentration of carbon dioxide in the steam low, to silica removal for the protection of steam turbines against siliceous turbine blade deposits. Antifoam agents are used for improved operation and steam quality.

Return Condensate Systems

Three approaches are used to minimize corrosion in condensate systems: (1) eliminating alkalinity from all water entering the boiler to minimize the amount of carbon dioxide in the condensate lines; (2) using of volatile amines, such as morpholine and cyclohexylamine, to neutralize carbon dioxide and thus raise the pH of the condensate; and (3) introducing filming amines, such as octadecylamine, into the steam to form a thin, hydrophobic film on the condensate line surfaces. The third approach is particularly effective in minimizing corrosion by oxygen. The need for chemical treatment can be minimized by designing return systems so that the condensate is still very hot when it reaches the boiler feed pump.

FIRE-SIDE CORROSION AND DEPOSITS

The surfaces of flues and boilers that are contacted by combustion products are seldom corroded while the equipment is operating, unless halogenated hydrocarbons, such as certain degreasing solvents, are present in the combustion air. Chimney connectors, smoke hoods, and canopies in contact with flue gas may, however, be subject to attack during warm-up periods or when the rate of operation is so low that the temperature of the flue gas is below its dew point. In stack sections where flue gas temperatures drop below the dew point during operations, corrosion is inevitable.

Cast iron or acid-resistant, vitreous enameled steel are used in flue gas connections to appliances. Metal surfaces with temperatures below 200°C can be protected by periodic applications of paints. Protective coatings with organic binders are destroyed rapidly at temperatures above 200°C because of the decomposition of the organic materials. Stacks and other surfaces that reach higher temperatures can be protected with special coatings, such as silicones, that are resistant to these higher temperatures.

Corrosion on the fire side of boilers is common, and certain precautions must be taken to prevent it. It has been reported that 15% of boiler tube failures have been caused by fire-side attack (Hinst 1955). The most common cause of fire-side corrosion is the condensation of sulfuric acid at cold ends of the furnace. All fossil fuels contain varying amounts of sulfur. Even low-sulfur fuel oil containing less than 1.0% sulfur forms gaseous oxides of sulfur when burned. Sulfur dioxide has no particularly corrosive effect on the boiler when it remains in the gaseous state and is emitted from the stack, but it pollutes the atmosphere. However, in boilers that operate with fluctuating loads and therefore varying amounts of excess air in the combustion chamber, the sulfur dioxide can further oxidize to sulfur trioxide, which is corrosive to the metallic surfaces of the boiler. Metals such as vanadium and iron act as catalysts to this reaction.

Sulfur trioxide combines with moisture to form sulfuric acid. The condensation point of sulfuric acid is 165°C. If a boiler has a variable load, the metal surface temperature in the furnace may drop below 165°C at low firing rates and condense sulfuric acid, causing serious corrosion. This may happen regularly throughout the normal operating period, but it is most likely to occur when the boiler is shut down for some period. Furthermore, moisture may condense on the metallic surfaces during shutdown due to the high relative humidity. In the presence of moisture and acidic soot deposits, fire-side corrosion will continue.

Fire-side corrosion can be reduced by modifying the firing cycle to minimize the number of times the flue gas temperature drops below its dew point. Other methods include (1) using low-sulfur fuel oils, (2) regulating excess air, and (3) using fuel oil additives that neutralize acidic condensate and deposits on fire-side surfaces. Good maintenance practices are essential. A boiler should be fired to maintain a constant temperature. A boiler let down at the end of a day and left at low is conducive not only to condensation of sulfuric acid, which causes corrosion, but also to soot and slag formation and stress-corrosion cracking around the tube ends.

Additives such as magnesium and calcium oxide slurries have effectively neutralized sulfuric acid formed on fire sides of boilers. Certain manganese-bearing compounds inhibit the formation of sulfur trioxide significantly and are thereby effective in reducing sulfuric acid corrosion. These compounds also reduce smoke, soot, and deposits (Belyea 1966). Some additives contain detergents (surfactants) and dispersants that prevent fuel from adhering to burner nozzles, a cause of sticking, fuel dribbling, improper spray pattern, smoking, and soot deposits. These deposits absorb acid gases and cause serious corrosion and combustion inefficiency.

Care after shutdown is equally important in preventing fire-side corrosion. The fire side of a boiler should always be thoroughly brushed and cleaned to remove accumulations of soot and other deposits (Hinst 1955). This should be followed by air drying.

Moisture absorbents such as silica gel and quicklime can be spread on trays on top of the tubes or in the bottom of the boiler drum or shell. This will reduce moisture condensation and the resulting corrosion during idle periods.

REFERENCES

APHA. 1989. *Standard methods for the examination of water and wastewater*, 17th ed. American Public Health Association, Washington, D.C.

ASME. 1979. Consensus on operating practices for the control of feedwater and boiler water quality in modern industrial boilers. Research Committee on Water in Thermal Power Systems, Industrial Boiler Subcommittee. American Society of Mechanical Engineers, New York.

Belyea, A.R. 1966. Manganese additive reduces SO$_3$. *Power* (November).

Blake, R.T. 1968. Cure for pitting in low pressure boiler systems. *Air Conditioning, Heating and Ventilating* (November):45.

Collins, W.D. 1944. Typical water analyses for classification with reference to industrial use. *Proceedings of the American Society for Testing and Materials* 44:953.

Greene, J. and J.A. Holmes. 1947. Calculation of the pH of saturation of tricalcium phosphate. *Journal of the American Waterworks Association* 39:1090.

Hinst, H.F. 1955. Eleven ways to avoid boiler tube corrosion. *Heating, Piping and Air Conditioning* (January).

Langelier, W.F. 1936. The analytical control of anticorrosion water treatment. *Journal of the American Water Works Association* 28:1500.

Meckler, M. 1974. Corrosion and scale formation. *Heating, Piping and Air Conditioning* (August).

NACE. 1975a. Application and handling of wax-type protective coatings and wrapper systems for underground pipelines. RP0375-75. National Association of Corrosion Engineers, Houston, TX.

NACE. 1975b. Application of organic coatings to the external surface of steel pipe for underground service. RP0275-75. National Association of Corrosion Engineers, Houston, TX.

NACE. 1976. Extruded asphalt mastic type protective coatings for underground pipelines. RP02760-76. National Association of Corrosion Engineers, Houston, TX.

NACE. 1985. Extruded polyolefin resin coating systems for underground or submerged pipe. RP0185-85. National Association of Corrosion Engineers, Houston, TX.

NACE. 1990. External protective coatings for joints, fittings, and valves on metallic underground or submerged pipelines and piping systems. RP0190-90. National Associating of Corrosion Engineers, Houston, TX.

Nordell, E. 1961. *Water treatment for industrial and other uses.* Reinhold Publishing Co., New York.

Pye, D. 1947. Chemical fixation of oxygen. *Journal of the American Water Works Association* 39:1121.

Romanoff, M. 1957. Underground corrosion. *Circular* No. 570. National Bureau of Standards.

Rosa, F. 1988-1989. A historical perspective of devices to control scale, corrosion and bio-growths in HVAC systems. *National Engineer* 92(3-12), 93(1-5).

Ryznar, J.W. 1944. A new index for determining amount of calcium carbonate scale formed by a water. *Journal of the American Water Works Association* 36:472.

Sudrabin, L.P. 1956. An anomaly in pipe line corrosion diagnosis. *Corrosion* 12(3):17.

Sudrabin, L.P. 1963. Designing automatic controls for cathodic protection. *Materials Protection* 2(2):64.

Sudrabin, L.P. and F.J. LeFebvre. 1953. External corrosion of piping in radiant heating and snow melting systems. Heating, Piping and Air Conditioning Contractors National Association [currently Mechanical Contractors Association of America, Rockville, MD] *Official Bulletin* 60 (July).

Sussman, S. and J.B. Fullman. 1953. Corrosion in closed circulating water systems. *Heating and Ventilation* (October):77.

BIBLIOGRAPHY

ARI. 1958. *Corrosion and its prevention.* Air-Conditioning and Refrigeration Institute, Arlington, VA.

ASTM. 1978. *Manual on water.* STP 442A. American Society for Testing and Materials, Philadelphia.

Berk, A.A. 1962. *Handbook—Questions and answers on boiler feed water conditioning.* Bureau of Mines, U.S.Department of Interior, Washington, D.C.

Carrier Corp. 1963. *System design manual,* Part 5, Water conditioning. Syracuse, NY.

Evans, U.R. 1960. *The corrosion and oxidation of metals.* Edward Arnold, London.

Evans, U.R. 1963. *An introduction to metallic corrosion,* 2nd ed. Edward Arnold, London.

Hamer, P., J. Jackson, and E.F. Thurston. 1961. *Industrial water treatment practice.* Butterworth & Co., London.

McCoy, J.W. 1974. *The chemical treatment of cooling water.* Chemical Publishing Company, New York.

McCoy, J.W. 1980. *Microbiology of cooling water.* Chemical Publishing Company, New York.

NACE. 1969. Control of external corrosion on underground or submerged metallic piping systems. NACE-RP-1-69. National Association of Corrosion Engineers, Houston, TX.

NACE. 1980. *NACE corrosion engineer's reference book.* National Association of Corrosion Engineers, Houston, TX.

Nalco. 1979. *Nalco water handbook.* The Nalco Company, Oak Brook, IL.

Peabody, A.W. 1963. Pipeline corrosion survey techniques. *Materials Protection* 2(4):62.

Powell, S.T. 1954. *Water conditioning for industry.* McGraw-Hill, New York.

Rosa, F. 1985. *Water treatment specification manual.* McGraw-Hill, New York

Sudrabin, L.P. 1963. A review of cathodic protection theory and practice. *Materials Protection* 2(5):8.

CHAPTER 45

SERVICE WATER HEATING

A SERVICE water heating system has (1) a heat energy source, (2) heat transfer equipment, (3) a distribution system, and (4) terminal hot water usage devices.

Heat energy sources may be (1) fuel combustion, (2) solar energy collection, (3) electrical conversion, and/or (4) recovered waste heat from such sources as flue gases, ventilation and air-conditioning systems, refrigeration cycles, and process waste discharge.

Heat transfer equipment is either of the direct or indirect type. For direct equipment, heat is derived from combustion of fuels or direct conversion of electrical energy into heat and is applied within the water heating equipment. For indirect heat transfer equipment, heat energy is developed from remote heat sources, such as boilers, solar energy collection, cogeneration, refrigeration, or waste heat, and is then transferred to the water in a separate piece of equipment. Storage tanks may be part of or associated with either type of heat transfer equipment.

Distribution systems transport the hot water produced by the water heating equipment to plumbing fixtures and other terminal points. The water consumed must be replenished from the building water service main. For locations where constant supply temperatures are desired, circulation piping or a means of heat maintenance must be provided. This is common for remote fixtures.

Terminal hot water usage devices are plumbing fixtures and equipment requiring hot water. They typically have periods of irregular flow, constant flow, and no flow. These patterns and their related water usage requirements vary with different buildings, process applications, and personal preference.

In this chapter, it is assumed that an adequate supply of service water is available. If this is not the case, alternate strategies such as water accumulation, pressure control, and flow restoration should be considered.

SYSTEM PLANNING

Flow rate and temperature are the primary factors to be determined in the hydraulic and thermal design of water heating and piping systems. Operating pressures and water quality are also mandatory considerations. Separate procedures are used to select water heating equipment and to design the piping system.

Water heating equipment, storage facilities, and piping should (1) have sufficient capacity to provide the required hot water while minimizing the waste of energy or water and (2) allow economical system installation, maintenance, and operation.

Hot water can be provided in many ways. The service water heating system for a given application must be based on the overall design and energy demand of the building's hot water and other mechanical systems.

Water heating equipment types and designs are based on (1) the energy source, (2) the application of the developed energy to heating the water, and (3) the control method used to deliver the necessary hot water at the required temperature under varying water demand conditions. Application of water heating equipment within the overall design of the hot water system is based on (1) location of the equipment within the system, (2) related temperature requirements, and (3) the volume of water to be used.

Energy Sources

The choice among available energy sources is interrelated with the choices among equipment types and locations. These decisions should be made only after evaluating purchase, installation, operating, and maintenance costs. A life-cycle analysis is highly recommended.

In making energy conservation choices, current editions of the following energy conservation guides should be consulted: the ANSI/ASHRAE/IES *Standard* 90 series or the sections on Service Water Heating of the ANSI/ASHRAE/IES *Standard* 100 series (see also the section on Design Considerations in this chapter).

HEAT TRANSFER EQUIPMENT

Gas-Fired or Oil-Fired

Residential water heating equipment is usually the automatic storage type. For industrial and commercial applications, commonly used types of heaters are (1) automatic storage, (2) circulating tank, (3) instantaneous, (4) hot water supply boilers, and (5) immersion storage.

Installation guidelines for gas-fired water heaters can be found in the National Fuel Gas Code, NFPA *Standard* 54 (ANSI Z223.1). This code also covers the sizing and installation of venting equipment and controls. Installation guidelines for oil-fired water heaters can be found in NFPA *Standard* 31, Installation of Oil-Burning Equipment (ANSI Z95.1).

Automatic storage heaters incorporate the burner(s), storage tank, outer jacket, insulation, and controls in a single unit. They are normally installed independent of other hot water storage equipment.

Circulating tank heaters are classified in two types: (1) automatic, in which the thermostat is located in the water heater, and (2) nonautomatic, in which the thermostat is located within an associated storage tank.

Instantaneous heaters have minimal water storage capacity. They usually include a flow switch as part of the control system. Instantaneous heaters may have a modulating fuel valve that varies fuel flow as water flow changes.

The preparation of this chapter is assigned to TC 6.6, Service Water Heating.

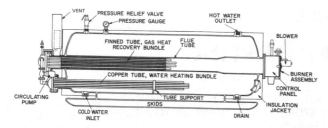

Fig. 1   Indirect Immersion-Fired Storage Type Heater

Hot water supply boilers are capable of providing service hot water. They are typically installed with separate storage tanks and applied as an alternative to circulating tank or instantaneous heaters.

Immersion storage-type heaters (Figure 1) have a power burner firing into a horizontal tube containing a finned-tube bundle. An intermediate heat transfer fluid, usually water, is pumped through the finned bundle and then to the water heating bundle located below the fire tube in the shell or storage tank. The heat transfer fluid, continuously circulated at a controlled velocity, is the basic source of heat to the stored water, even though the firing tube is immersed in the water and serves as an additional heat surface. The heat transfer fluid is circulated at a maximum temperature of 120°C, and the closed system is operated under pressure to prevent boiling.

Electric

Electric water heaters are generally of the automatic storage type, consisting of a tank with one or more immersion heating elements attached to line voltages of 120, 208, 240, 277, 480, or 600 V. Element wattages are selected to meet recovery requirements or electrical demand considerations. Electric water heating elements consist of resistance wire embedded in refractories having good heat conduction properties and electrical insulating values. Heating elements are generally sheathed in copper or alloy tubes and fitted into a threaded or flanged mounting for insertion into a tank. Thermostats controlling heating elements may be of the immersion or surface-mounted type.

Residential storage water heaters range up to 450 L with input up to 12 kW. They have a primary resistance heating element near the bottom and often a secondary element located in the upper portion of the tank. Each element is controlled by its own thermostat. In twin-element heaters, the thermostats are usually interlocked so that the lower heating element cannot operate if the top of the tank is cold.

Commercial storage water heaters are available in many combinations of element number, wattage, voltage, and storage capacity. Storage tanks may be horizontal or vertical. Compact, low-volume models are used in point-of-use applications to reduce hot water piping length. Location of the water heater near the point of use makes recirculation loops unnecessary.

Instantaneous electric water heaters are commonly used for dishwasher rinse, hot tub, whirlpool bath, and swimming pool applications.

Heat pump water heaters (HPWHs) use a vapor-compression refrigeration cycle to extract energy from an air or water source to heat water. Most HPWHs are air-to-water units. As the HPWH collects heat, it provides a potentially useful cooling effect and dehumidifies the air. HPWHs typically have a maximum output temperature of 60°C. Where higher delivery temperatures are required, a conventional storage-type or booster water heater downstream of the heat pump storage tank should be used. HPWH systems function most efficiently where the inlet water temperature is low and the entering air is warm and humid. Systems should be sized to allow high HPWH run time. Control systems should allow the HPWH to operate whenever possible without conflicting with the conventional equipment. The effect of HPWH cooling output on the building's energy balance should be considered. Cooling output

should be directed to provide occupant comfort and avoid interfering with temperature-sensitive equipment (EPRI 1990).

Demand controlled water heating can significantly reduce the cost of heating water electrically. Demand controllers operate on the principle that a building's peak electrical demand exists for a short period, during which heated water can be supplied from storage rather than hot water recovery. Avoiding the use of electricity for service water heating during peak demand periods allows water heating at the lowest electric energy cost in many electric rate schedules. The building electrical load must be detected and compared with peak demand data. When the load is below peak demand, the control device allows the water heater to operate. Some controllers can program deferred loads in steps as capacity is available. The priority sequence may involve each of several banks of elements in (1) a water heater, (2) multiple water heaters, or (3) water heating and other equipment having a deferrable load, such as pool heating and snow melting. When load controllers are used, hot water storage must be used.

Instantaneous and hot water supply boilers described in the section on Gas-Fired or Oil-Fired water heating equipment are also commonly available with electric heating elements.

Electric off-peak storage water heating is a water heating equipment load management strategy whereby electrical demand to a water heating system is time-controlled, primarily in relation to the utility electrical load profile. This approach usually requires an increase in tank storage capacity and/or stored water temperature to accommodate water use during peak periods.

Sizing recommendations in this chapter apply only to water heating systems without demand or off-peak controlled systems. When demand control devices are used, the storage and recovery rate must be increased to supply all the hot water needed during the peak period and during the ensuing recovery period. Manian and Chackeris (1974) include a detailed discussion on load-limited storage heating system design.

Indirect

In indirect water heating, the heating medium is steam, hot water, or another fluid that has been heated in a separate generator or boiler. The water heater extracts heat through an external or internal heat exchanger.

When the heating medium is at a higher pressure than the service water, the service water may be contaminated by leakage of the heating medium through a damaged heat transfer surface. In the United States, some national, state, and local codes require double-wall, vented tubing in indirect water heaters to reduce the possibility of cross-contamination. When the heating medium is at a lower pressure than the service water, other jurisdictions allow single-wall tubing heaters because any leak would be into the heating medium.

If the heating medium is steam, high rates of condensation occur, particularly when a sudden demand causes an inflow of cold water. The steam pipe and condensate return pipes should be of ample size. Condensate should drain by gravity without lifts to a vented condensate receiver located below the level of the heater. Otherwise, water hammer, reduced capacity, or heater damage may result. The condensate may be cooled by preheating the cold water supply to the heater.

Corrosion is minimized on the heating medium side of the heat exchanger because no makeup water, and hence no oxygen, is brought into that system. The metal temperature of the service water side of the heat exchanger is usually less than that in direct-fired water heaters. This minimizes scale formation from hard water.

Storage water heaters are designed for service conditions where hot water requirements are not constant, i.e., where a large volume of heated water must be held in storage for periods of peak load. The amount of storage required depends on the nature of the load and the recovery capacity of the water heating system. Residential equipment is usually a single tank with an immersed heat exchanger. For commercial applications, either a number of units joined by manifold or an individual large tank may be used.

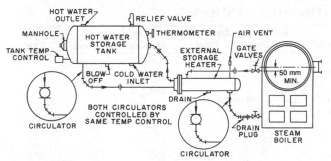

Fig. 2 Indirect, External Storage Water Heater

External storage water heaters are designed for connection to a separate tank (Figure 2). The boiler water circulates through the heater shell, while service water from the storage tank circulates through the tubes and back to the tank. Circulating pumps are usually installed in both the boiler water piping circuit and the circuits between the heat exchanger and the storage tank. Steam can also be used as the heating medium in a similar scheme.

Instantaneous indirect water heaters (tankless coils) are best used for a steady, continuous supply of hot water. In these units, the water is heated as it flows through the tubes. Because the heating medium flows through a shell, the ratio of hot water volume to heating medium volume is small. As a result, variable flow of the service water causes uncertain temperature control unless a thermostatic mixing valve is used to maintain the hot water supply to the plumbing fixtures at a more uniform temperature.

Some indirect instantaneous water heaters are located inside a boiler. The boiler is provided with a special opening through which the coil can be inserted. While the coil can be placed in the steam space above the water line of a steam boiler, it is usually placed below the water line. The water heater transfers heat from the boiler water to the service water. The gross output of the boiler must be sufficient to serve all loads.

Semi-Instantaneous

These water heaters have limited storage, determined by manufacturers to meet the average momentary surges of hot water demand. They usually consist of a heating element and control assembly devised for close control of the temperature of the leaving hot water.

Circulating Tank

These water heaters are instantaneous or semi-instantaneous types used with a separate storage tank and a circulating pump. The storage acts as a flywheel to accommodate variations in the demand for hot water.

Blending Injection

These water heaters inject steam or hot water directly into the process or volume of water to be heated. This is often associated with point-of-use applications, e.g., certain types of commercial laundry, food, and process equipment. *Caution*: Cross-contamination of potable water is possible.

Solar

The availability of solar energy at the building site, the efficiency and cost of solar collectors, and the availability and cost of other fuels determine whether solar energy collection units should be used as a primary heat energy source. Solar energy equipment can also be included to supplement other energy sources and conserve fuel or electrical energy.

The basic elements of a solar water heater include solar collectors, a storage tank, piping, controls, and a transfer medium. The system may use natural convection or forced circulation. Auxiliary heat energy sources may be added, if needed.

Collector design must allow operation in below-freezing conditions, where applicable. Antifreeze solutions in a separate collector piping circuit arrangement are often used, as are systems that allow water to drain back to heated areas when low temperatures occur. Uniform flow distribution in the collector or bank of collectors and stratification in the storage tank are important for good system performance.

The application of solar water heaters depends on (1) auxiliary energy requirements; (2) collector orientation; (3) temperature of the cold water; (4) general site, climatic, and solar conditions; (5) installation requirements; (6) area of collectors; and (7) amount of storage. For a more detailed discussing, see Chapter 30, Solar Energy Utilization.

Waste Heat Use

Waste heat can be recovered from equipment or processes by using appropriate heat exchangers in the hot gaseous or liquid streams. Heat recovered from this separate equipment is frequently used to preheat the water entering the service water heater. Refrigeration heat reclaim systems use refrigerant-to-water heat exchangers connected to the refrigeration circuit between the compressor and condenser of a host refrigeration or air-conditioning system to extract heat for service water heating. Water is heated only when the host is operating at the same time there is a call for water heating. A conventional water heater is typically required to augment the output of the refrigeration heat reclaim system and provide hot water during periods when the host system is not in operation.

Waste heat recovery is an energy conservation method that can reduce (1) energy costs, (2) energy requirements of the building heating and service water heating equipment, or (3) cooling requirements of applicable air-conditioning systems.

Refrigeration Heat Reclaim

These systems heat water with heat that would otherwise be rejected through a refrigeration system condenser. Because many simple systems reclaim only superheat energy from the refrigerant, they are often called desuperheaters. However, some units are also designed to provide partial condensing. The refrigeration heat reclaim heat exchanger is generally of vented, double-wall construction to isolate potable water from refrigerant. Some heat reclaim devices are designed for use with multiple refrigerant circuits. Controls are required to limit high water temperature, prevent low condenser pressure, and provide for freeze protection. Refrigeration systems with higher run time and lower efficiency provide more heat reclaim potential. Most systems are designed with a preheat water storage tank connected in series with a conventional water heater (EPRI 1992). In all installations, care must be taken to prevent the inappropriate venting of refrigerants.

DISTRIBUTION SYSTEMS

Piping System Material

Traditional piping materials have included galvanized steel used with galvanized cast iron or galvanized malleable iron screwed fittings. Copper piping and copper tube types K, L, or M have been used with brass, bronze, or wrought copper solder fittings. Legislation or plumbing code changes have banned the use of lead in solders or pipe-jointing compounds in potable water piping because of possible lead contamination of the water supply. See ASHRAE *Standard* 90 series for pipe insulation requirements.

Today, most potable water supplies require treatment before distribution; this may cause the water to become more corrosive. Therefore, depending on the water supply, traditional galvanized steel piping or copper tube may no longer be satisfactory, due to accelerated corrosion.

Table 1 Hot Water Demand in Fixture Units (60°C Water)

	Apartments	Club	Gymnasium	Hospital	Hotels and Dormitories	Industrial Plant	Office Building	School	YMCA
Basin, private lavatory	0.75	0.75	0.75	0.75	0.75	0.75	0.75	0.75	0.75
Basin, public lavatory	—	1	1	1	1	1	1	1	1
Bathtub	1.5	1.5	—	1.5	1.5	—	—	—	—
Dishwasher	1.5	Five fixture units per 250 seating capacity							
Therapeutic bath	—	—	—	5	—	—	—	—	—
Kitchen sink	0.75	1.5	—	3	1.5	3	—	0.75	3
Pantry sink	—	2.5	—	2.5	2.5	—	—	2.5	2.5
Service sink	1.5	2.5	—	2.5	2.5	2.5	2.5	2.5	2.5
Shower[a]	1.5	1.5	1.5	1.5	1.5	3.5	—	1.5	1.5
Circular wash fountain	—	2.5	2.5	2.5	—	4	—	2.5	2.5
Semicircular wash fountain	—	1.5	1.5	1.5	—	3	—	1.5	1.5

[a]In applications where the principal use is showers, as in gymnasiums or at end of shift in industrial plants, use conversion factor of 1.00 to obtain design water flow rate in L/s.

Galvanized steel piping is particularly susceptible to corrosion (1) when hot water is between 60 and 80°C and (2) where repairs have been made using copper tube without a nonmetallic coupling.

Before selecting any water piping material or system, consult the local code authority. The local water supply authority should also be consulted about any history of water aggressiveness causing failures of any particular material.

Alternate piping materials that may be considered are (1) stainless steel tube and (2) various plastic piping and tubes. Particular care must be taken to make sure that the application meets the design limitations set by the manufacturer and that the correct materials and methods of jointing are used. These precautions are easily taken with new projects but become more difficult to control during repairs of existing work. The use of incompatible piping, fittings, and jointing methods or materials must be avoided, as they can cause severe problems.

Pipe Sizing

Sizing of hot water supply pipes involves the same principles as sizing of cold water supply pipes (see Chapter 33 of the 1993 *ASHRAE Handbook—Fundamentals*). The water distribution system must be correctly sized for the total hot water system to function properly. Hot water demand varies with the type of establishment, usage, occupancy, and time of day. The piping system should be capable of meeting peak demand at an acceptable pressure loss.

Supply Piping System

Table 1, Figures 23 and 24, and manufacturers' specifications for fixtures and appliances can be used to determine hot water demands. These demands, together with procedures given in Chapter 33 of the 1993 *ASHRAE Handbook—Fundamentals*, are used to size the mains, branches, and risers.

Allowance for pressure drop through the heater should not be overlooked when sizing hot water distribution systems, particularly where instantaneous water heaters are used and where the available pressure is low.

Pressure Differentials

Sizing of both cold and hot water piping requires that pressure differentials at the point of use of blended hot and cold water be kept to a minimum. This required minimum differential is particularly important for tubs and showers, since sudden changes in flow at fixtures cause discomfort and a possible scalding hazard. Pressure-compensating devices are available.

Return Piping System

Return piping systems are commonly provided for hot water supply systems in which it is desirable to have hot water available continuously at the fixtures. This includes cases where the hot water piping system exceeds 30 m.

The water circulation pump may be controlled by a thermostat (in the return line) set to start and stop the pump over an acceptable temperature range. An automatic time switch or other control should turn the water circulation off when hot water is not required. Because hot water is corrosive, circulating pumps should be made of corrosion-resistant material.

For small installations, a simplified pump sizing method is to allow 3 mL/s for every fixture unit in the system, or to allow 30 mL/s for each 20- or 25-mm riser; 60 mL/s for each 32- or 40-mm riser; and 130 mL/s for each riser 55 mm or larger.

Werden and Spielvogel (1969) and Dunn et al. (1959) cover heat loss calculations for large systems. For larger installations, piping heat losses become significant. A quick method to size the pump and return for larger systems is as follows:

1. Determine the total length of all hot water supply and return piping.
2. Choose an appropriate value for piping heat losses from Table 2 or other engineering data (usually supplied by insulation companies, etc.). Multiply this value by the total length of piping involved.

   A rough estimation can be made by multiplying the total length of covered pipe by 30 W/m or uninsulated pipe by 60 W/m. Table 2 gives actual heat losses in pipes at a service water temperature of 60°C and ambient temperature of 21°C. The values of 30 or 60 W/m are only recommended for ease in calculation.
3. Determine pump capacity as follows:

$$Q_p = q / (\rho c_p \Delta t) \qquad (1)$$

where

Q_p = pump capacity, L/s
q = heat loss, W
ρ = density of water = 1.00 kg/L
c_p = specific heat of water = 4180 J/(kg·K)
Δt = allowable temperature drop, K

Table 2 Heat Loss of Pipe
(at 60°C Inlet, 21°C Ambient Temperatures)

Nominal Pipe Size, mm	Bare Copper Tubing, W/m	13-mm Glass Fiber Insulated Copper Tubing, W/m
20	28.8	17.0
25	36.5	19.5
32	43.2	22.5
40	50.9	24.4
50	63.4	28.4
65	76.9	32.5
80	90.4	38.0
100	115.4	46.5

For a 10 K allowable temperature drop,

$$Q_p \text{(L/s)} = q/(1 \times 4180 \times 10) = q/42\,000 \quad (2)$$

Caution: This assumes that a 10 K temperature drop is acceptable at the last fixture.

4. Select a pump to provide the required flow rate, and obtain from the pump curves the pressure created at this flow.

5. Check that the pressure does not exceed the allowable friction loss per metre of pipe.

6. Determine the required flow in each circulating loop, and size the hot water return pipe based on this flow and the allowable friction loss.

Where multiple risers or horizontal loops are used, balancing valves with means of testing are recommended in the return lines. A swing-type check valve should be placed in each return to prevent entry of cold water or reversal of flow, particularly during periods of high hot water demand.

Three common methods of arranging circulation lines are shown in Figure 3. Although the diagrams apply to multistory buildings, arrangements (A) and (B) are also used in residential designs. In circulation systems, air venting, pressure drops through the heaters and storage tanks, balancing, and line losses should be considered. In Figures 3A and 3B, air is vented by connecting the circulating line below the top fixture supply. With this arrangement, air is eliminated from the system each time the top fixture is opened. Generally, for small installations, a 15- or 20-mm hot water return is ample.

All storage tanks and piping on recirculating systems should be insulated as recommended by the ASHRAE *Standard* 90 series and *Standard* 100 series.

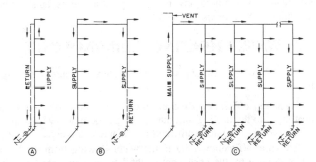

Fig. 3   Arrangements of Hot Water Circulation Lines

Heat Traced, Nonreturn Piping System

In this system, the fixtures can be as remote as in the return piping system. The hot water supply piping is heat traced with electric resistance heating cable preinstalled under the pipe insulation. Electrical energy input is self-regulated by the cable's construction to maintain the required water temperature at the fixtures. No return piping system or circulation pump is required.

Special Piping—Commercial Dishwashers

Adequate flow rates and pressures must be maintained for automatic dishwashers in commercial kitchens. To reduce operating difficulties, piping for automatic dishwashers should be installed according to the following recommendations:

1. The cold water feed line to the water heater should be no smaller than 25-mm ID.
2. The supply line that carries 82°C water from the water heater to the dishwasher should not be smaller than 20-mm ID.
3. No auxiliary feed lines should connect to the 82°C supply line.
4. A return line should be installed if the source of 82°C water is more than 1.5 m from the dishwasher.

5. Forced circulation by a pump should be used if the water heater is installed on the same level as the dishwasher, if the length of return piping is more than 18 m, or if the water lines are trapped.

6. If a circulating pump is used, it is generally installed in the return line. It may be controlled by (a) the dishwasher wash switch, (b) a manual switch located near the dishwasher, or (c) an immersion or strap-on thermostat located in the return line.

7. A pressure-reducing valve should be installed in the low-temperature supply line to a booster water heater, but external to a recirculating loop. It should be adjusted, with the water flowing, to the value stated by the washer manufacturer (typically 140 kPa).

8. A check valve should be installed in the return circulating line.

9. If a check valve type of water meter or a backflow prevention device is installed in the cold water line ahead of the heater, it is necessary to install a properly sized diaphragm-type expansion tank between the water meter or prevention device and the heater.

10. National Sanitation Foundation (NSF) standards require the installation of an 8-mm IPS connection for a pressure gage mounted adjacent to the supply side of the control valve. They also require a water line strainer ahead of any electrically operated control valve (Figure 4).

11. NSF standards do not allow copper water lines that are not under constant pressure, except for the line downstream of the solenoid valve on the rinse line to the cabinet.

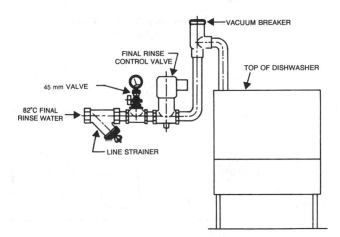

Fig. 4   National Sanitation Foundation (NSF) Plumbing Requirements for Commercial Dishwasher

Water Pressure—Commercial Kitchens

Proper flow pressure must be maintained to achieve efficient dishwashing. NSF standards of for dishwasher water flow pressures are 100 kPa (gage) minimum, 170 kPa (gage) maximum, and 140 kPa (gage) ideal. Flow pressure is the line pressure measured when water is flowing through the rinse arms of the dishwasher.

Low flow pressure can be caused by undersized water piping, stoppage in piping, or excess pressure drop through heaters. Low water pressure causes an inadequate rinse, resulting in poor drying and sanitizing of the dishes. If flow pressure in the supply line to the dishwasher is below 100 kPa (gage), a booster pump or other means should be installed to provide supply water at 140 kPa (gage).

A flow pressure in excess of 170 kPa (gage) causes atomization of the 82°C rinse water, resulting in an excessive temperature drop. The temperature drop between the rinse nozzle and the dishes can be as much as 8 K. A pressure regulator should be installed in the supply water line adjacent to the dishwasher and external to the return circulating loop (if used). The regulator should be set to maintain a pressure of 140 kPa (gage).

Two-Temperature Service

Where multiple temperature requirements are met by a single system, the system temperature is determined by the maximum temperature needed. Where the bulk of the hot water is needed at the higher temperature, lower temperatures can be obtained by mixing hot and cold water. Automatic blending to reduce the temperature of the hot water available at certain outlets may be necessary to prevent injury or damage (Figure 5). Applicable codes should be consulted for blending valve requirements.

Where predominant usage is at lower temperatures, the common approach is to heat all water to the lower temperature and then use a separate booster heater to further heat the water for the higher temperature services (Figure 6). This method offers better protection against scalding.

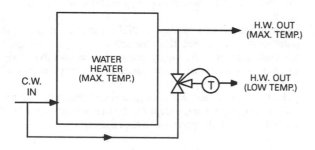

Fig. 5 Two-Temperature Service with Three-Way Control Valve

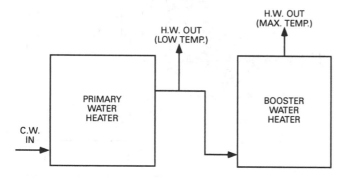

Fig. 6 Two-Temperature Service with Primary Heater and Booster Heater in Series

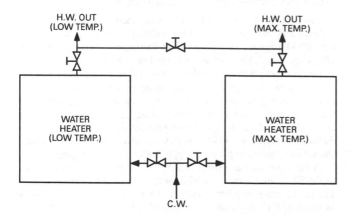

Fig. 7 Two-Temperature Service with Separate Heater for Each Service

A third method uses separate heaters for the higher temperature service (Figure 7). It is common practice to cross-connect the two heaters, so that one heater can serve the complete installation temporarily while the other is valved off for maintenance. Each heater should be sized for the total load unless hot water consumption can be reduced during maintenance periods.

Manifolding

Where one heater does not have sufficient capacity, two or more water heaters may be installed in parallel. If blending is needed in such an installation, a single mixing valve of adequate capacity should be used. It is difficult to obtain even flow through parallel mixing valves.

Heaters installed in parallel should be identical, i.e., have the same input and storage capacity, with inlet and outlet piping arranged so that an equal flow is received from each heater under all demand conditions.

An easy way to get parallel flow is to use reverse/return piping (Figure 8). The unit having its inlet closest to the cold water supply is piped so that its outlet will be farthest from the hot water supply line. Quite often this results in a hot water supply line that reverses direction to bring it back to the first unit in line; hence the name reverse/return.

TERMINAL HOT WATER USAGE DEVICES

Details on the vast number of devices using service hot water are beyond the scope of this chapter. Nonetheless, they are important to a successful overall design. Consult the manufacturer's literature for information on required flow rates, temperature limits, and/or other operating factors for specific items.

WATER HEATING TERMINOLOGY

The following terms apply to water heaters or water heating systems:

Recovery efficiency. Heat absorbed by the water divided by the heat input to the heating unit during the period that water temperature is raised from inlet temperature to final temperature.

Thermal efficiency. Heat in the water flowing from the heater outlet divided by the heat input of the heating unit over a specific period of steady-state conditions.

Energy factor. An indicator of the overall efficiency of a residential storage water heater representing heat in the estimated daily quantity of delivered hot water divided by the daily energy consumption of the water heater.

First hour rating. The amount of hot water that the water heater can supply in one hour of operation, starting with the tank full of heated water.

Standby loss. The amount of heat lost from the water heater during periods of nonuse of service hot water.

Hot water distribution efficiency. Heat contained in the water at points of use divided by the heat delivered at the heater outlet at a given flow.

Heater/system efficiency. Heat contained in the water at points of use divided by the heat input to the heating unit at a given flow rate (thermal efficiency times distribution efficiency).

Overall system efficiency. Heat in the water delivered at points of use divided by the heat supplied to the heater for any selected time period.

System standby loss. The amount of heat lost from the water heating system and the auxiliary power consumed during periods of nonuse of service hot water.

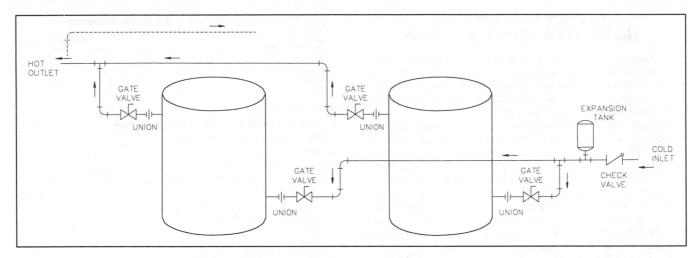

Fig. 8 Reverse/Return Manifold Systems

DESIGN CONSIDERATIONS

Hot water system design should consider the following:

- Water heaters of different sizes and insulation may have different standby losses, recovery efficiency, thermal efficiency, or energy factors.
- A distribution system should be properly designed, sized, and insulated to deliver adequate water quantities at temperatures satisfactory for the uses served. This reduces system standby loss and improves hot water distribution efficiency.
- Heat traps between recirculation mains and infrequently used branch lines reduce convection losses to these lines and improve heater/system efficiency.
- Controlling circulating pumps to operate only as needed to maintain proper temperature at the end of the main reduces losses on return lines. This improves overall system efficiency.
- Provision for shutdown of circulators during building vacancy reduces system standby losses.

WATER QUALITY, SCALE, AND CORROSION

A complete water analysis and an understanding of system requirements are needed to protect water heating systems from scale and corrosion. Analysis shows whether water is hard or soft. Hard water may cause scale (fouling or liming of heat transfer and water storage surfaces); soft water may aggravate corrosion problems.

Scale formation is also affected by system requirements and equipment. As shown in Figure 9, the rate of scaling increases with temperature and usage; this is because calcium carbonate and other scaling compounds lose solubility at higher temperatures. In water tube-type equipment, scaling problems can be offset by increasing the water flow over the heat transfer surfaces, which reduces tube surface temperatures. Also, the turbulence of the flow, if high enough, works to keep any scale that does precipitate off the surface. When water hardness is over 140 mg/L, water softening or other water treatments are often recommended.

Corrosion problems increase with temperature because corrosive oxygen and carbon dioxide gases are released from the water. Electrical conductivity also increases with temperature, enhancing electrochemical reactions such as rusting (Toaborek et al. 1972). A deposit of scale provides some protection from corrosion; however, this deposit also reduces the heat transfer rate, and it is not under the control of the system designer (Talbert et al. 1986).

Steel vessels can be protected to varying degrees by galvanizing them, nickel plating them, or lining them with copper, glass, cement,

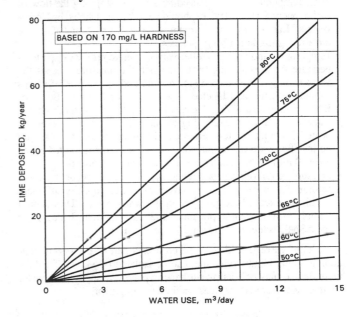

Fig. 9 Lime Deposited Versus Temperature and Water Use
(Purdue University Bulletin No. 74)

or other corrosion-resistant material. Glass-lined vessels are almost always supplied with electrochemical protection. Typically, one or more rods of magnesium alloy (the anode) are installed in the vessel by the manufacturer. This electrochemically active material sacrifices itself to reduce or prevent corrosion of the tank (the cathode). Higher temperatures, softened water, and high water usage may lead to rapid anode consumption. Manufacturers recommend periodic replacement of the rod(s) to prolong the life of the vessel. Some waters have very little electrochemical activity. In this instance, a standard anode shows little or no activity, and the vessel is not adequately protected. If this condition is suspected, the equipment manufacturer should be consulted on the possible need for a high potential anode, or the use of vessels made of nonferrous materials should be considered.

Water heaters and hot water storage tanks constructed of stainless steel, copper, or other nonferrous alloys are protected against oxygen corrosion. However, care must still be taken, as stainless steels may be adversely affected by chlorides, and copper may be attacked by ammonia or carbon dioxide.

SAFETY DEVICES FOR HOT WATER SUPPLY SYSTEMS

Regulatory agencies differ as to the selection of protective devices and methods of installation. It is therefore essential to check and comply with the manufacturer's instructions and the applicable local codes. In the absence of such instructions and codes, the following recommendations may be used as a guide:

- Thermal expansion control devices limit the pressure that results when the water in the tank is heated and expands in a closed system. Although the water heating system is under service pressure, the pressure will rise rapidly if backflow is prevented by devices such as a check valve, pressure-reducing valve, or backflow preventer in the cold water line or by temporarily shutting off the cold water. When backflow is prevented, the pressure rise during heating may rupture the tank or cause other damage. Systems having this potential problem must be protected by a properly sized and located diaphragm-type expansion tank.
- Temperature limiting devices (energy cutoff/high limit) prevent water temperatures from exceeding 99°C by stopping the flow of fuel or energy. These devices should be listed and labeled by Underwriters' Laboratories, the American Gas Association (AGA), or other recognized certifying agencies.
- Temperature and pressure relief valves open to prevent water temperature from exceeding 99°C or when the pressure exceeds the valve setting. Combination temperature and pressure relief valves should be listed and labeled by the AGA or National Board of Boiler and Pressure Vessel Inspectors and have a water discharge capacity equal to or exceeding the heat input rating of the water heater. Combination temperature and pressure relief valves should be installed with the temperature-sensitive element located in the top 150 mm of the tank, i.e., where the water is hottest.
- Pressure relief valves open when the pressure exceeds the valve setting. These valves should have a discharge capacity sufficient to relieve excess fluid pressure in the water heater. They should comply with current applicable American National Standards or the ASME Boiler and Pressure Vessel Code.

A relief valve should be installed in any part of the system containing a heat input device that can be isolated by valves. The heat input device may be solar water heating panels, desuperheater water heaters, heat recovery devices, or similar equipment.

SPECIAL CONCERNS

Legionella pneumophila (Legionnaire's Disease)

The bacteria that causes Legionnaire's disease when inhaled has been discovered in the service water systems of various buildings in the United States and abroad. Infection has often been traced to *Legionella pneumophila* colonies in shower heads. Ciesielki et al. (1984) determined that *Legionella pneumophila* can colonize in hot water systems maintained at 46°C or lower. Segments of service water systems in which the water stagnates (e.g., shower heads, faucet aerators, and certain sections of storage-type water heaters) provide ideal breeding locations.

Service water temperatures in the 60°C range are recommended to limit the potential for *Legionella pneumophila* growth. This high temperature increases the potential for scalding, so care must be taken. Supervised periodic flushing of fixture heads with 77°C water is recommended in hospitals and health care facilities because the already weakened patients are generally more susceptible to infection.

Temperature Requirements

Typical temperature requirements for some services are shown in Table 3. In some cases, slightly lower temperatures may be satisfactory. When gas- or oil-fired storage-type water heaters are used, temperatures below 60°C are usually obtained by blending hot and cold

Table 3 Representative Hot Water Temperatures

Use	Temperature, °C
Lavatory	
Hand washing	40
Shaving	45
Showers and tubs	43
Therapeutic baths	35
Commercial or institutional laundry, based on fabric	up to 82
Residential dish washing and laundry	60
Surgical scrubbing	43
Commercial spray-type dish washing[a]	
Single- or multiple-tank hood or rack type	
Wash	65 minimum
Final rinse	82 to 90
Single-tank conveyor type	
Wash	71 minimum
Final rinse	82 to 90
Single-tank rack or door type	
Single-temperature wash and rinse	74 minimum
Chemical sanitizing types[b]	60
Multiple-tank conveyor type	
Wash	65 minimum
Pumped rinse	71 minimum
Final rinse	82 to 90
Chemical sanitizing glass washer	
Wash	60
Rinse	24 minimum

[a]As required by NSF.
[b]See manufacturer for actual temperature required.

water at point of use. A 60°C water temperature minimizes flue gas condensation in the equipment.

Hot Water from Tanks and Storage Systems

With storage systems, 60 to 80% of the hot water in a tank is assumed to be usable before dilution by cold water lowers the temperature below an acceptable level. However, better designs can exceed 90%. Thus, the hot water available from a self-contained storage heater is usually considered to be

$$V_t = Rd + MS_t \qquad (3)$$

where

V_t = available hot water, L
R = recovery rate at the required temperature, L/s
d = duration of peak hot water demand, s
M = ratio of usable water to storage tank capacity
S_t = storage capacity of the heater tank, L

Usable hot water from an unfired tank is

$$V_a = MS_a \qquad (4)$$

where

V_a = usable water available from an unfired tank, L
S_a = capacity of unfired tank, L

Note: Assumes tank water at required temperature.

Hot water obtained from a water heating system using a storage heater with an auxiliary storage tank is

$$V_z = V_t + V_a = Rd + M(S_t + S_a) \qquad (5)$$

where V_z = total hot water available from system during one peak, L.

Placement of Water Heaters

Many types of water heaters may be expected to leak at the end of their useful life. They should be placed where such leakage will not cause damage. Alternatively, suitable drain pans piped to drains should be provided.

Water heaters not requiring combustion air may generally be placed in any suitable location, as long as relief valve discharge pipes open to a safe location.

Water heaters requiring ambient combustion air must be located in areas with air openings large enough to admit the required combustion/dilution air (see NFPA *Standard* 54).

For gas- or oil-fired water heaters located in areas where flammable vapors are likely to be present, additional precautions should be taken to eliminate the probable ignition of flammable vapors (see sections 5.1.9 through 5.1.12 of NFPA *Standard* 54). However, use, spillage, or release of combustible gases or liquids (e.g., liquefied petroleum gases, paint thinner, or gasoline) near gas- or oil-fired water heaters should be avoided.

HOT WATER REQUIREMENTS AND STORAGE EQUIPMENT SIZING

Methods for sizing storage water heaters vary. Those using recovery versus storage curves are based on extensive research. All methods provide adequate hot water if the designer allows for unusual conditions.

Table 4   Typical Residential Usage of Hot Water per Task

Use	High Flow, L	Low Flow (Water Savers Used), L
Food preparation	19	11
Hand dish washing	15	15
Automatic dishwasher	57	57
Clothes washer	121	80
Shower or bath	76	57
Face and hand washing	15	8

RESIDENTIAL

Table 4 shows typical hot water usage in a residence. In its Minimum Property Standards for One- and Two-Family Living Units, No. 4900.1-1982, HUD-FHA established minimum permissible water heater sizes (Table 5). Storage water heaters may vary from the sizes shown in the table if combinations of recovery and storage are used that produce the required 1-h draw.

The first hour rating (FHR) is the amount of hot water that the water heater can supply in one hour of operation (DOE 1990); it is gaining recognition as a sizing criterion for selecting water heaters. The procedure for determining the first hour rating specifies that the tank initially be full of heated water (DOE 1993). The linear regression lines shown in Figure 10 represent the FHR for 2071 electric heaters and 3379 gas heaters (GAMA 1993). Regression lines are not

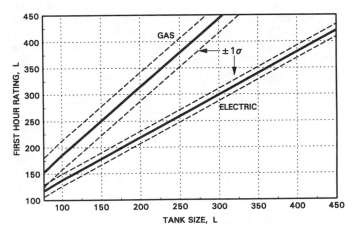

Fig. 10   First Hour Rating Relationships for Residential Water Heaters

Table 5   HUD-FHA Minimum Water Heater Capacities for One- and Two-Family Living Units

Number of Baths	1 to 1.5			2 to 2.5				3 to 3.5			
Number of Bedrooms	1	2	3	2	3	4	5	3	4	5	6
GAS[a]											
Storage, L	76	114	114	114	150	150	190	150	190	190	190
kW input	7.9	10.5	10.5	10.5	10.5	11.1	13.8	11.1	11.1	13.8	14.6
1-h draw, L	163	227	227	227	265	273	341	273	311	341	350
Recovery, mL/s	24	32	32	32	32	36	42	34	34	42	44
ELECTRIC[a]											
Storage, L	76	114	150	150	190	190	250	190	250	250	300
kW input	2.5	3.5	4.5	4.5	5.5	5.5	5.5	5.5	5.5	5.5	5.5
1-h draw, L	114	167	220	220	273	273	334	273	334	334	387
Recovery, mL/s	10	15	19	19	23	23	23	23	23	23	23
OIL[a]											
Storage, L	114	114	114	114	114	114	114	114	114	114	114
kW input	20.5	20.5	20.5	20.5	20.5	20.5	20.5	20.5	20.5	20.5	20.5
1-h draw, L	337	337	337	337	337	337	337	337	337	337	337
Recovery, mL/s	62	62	62	62	62	62	62	62	62	62	62
TANK-TYPE INDIRECT[b,c]											
I-W-H-rated draw, L in 3 h, 55 K rise		150	150		250	250[e]	250	250	250	250	250
Manufacturer-rated draw, L in 3 h, 55 K rise		186	186		284	284[e]	284	284	284	284	284
Tank capacity, L		250	250		250	250[e]	310	250	310	310	310
TANKLESS-TYPE INDIRECT[c,d]											
I-W-H-rated draw, mL/s, 55 K rise		170	170		200	200[e]	240	200	240	240	240
Manufacturer-rated draw, L in 5 min, 55 K rise		57	57		95	95[e]	133	95	133	133	133

[a]Storage capacity, input, and recovery requirements indicated in the table are typical and may vary with each individual manufacturer. Any combination of these requirements to produce the stated 1-h draw will be satisfactory.

[b]Boiler-connected water heater capacities (82°C boiler water, internal, or external connection).

[c]Heater capacities and inputs are minimum allowable. Variations in tank size are permitted when recovery is based on 4.2 mL/s·kW at 55 K rise for electrical, AGA recovery ratings for gas, and IBR ratings for steam and hot water heaters.

[d]Boiler-connected heater capacities (93°C boiler water, internal, or external connection).

[e]Also for 1 to 1.5 baths and 4 bedrooms for indirect water heaters.

included for oil-fired and heat-pump water heaters because of limited data. The FHR represents water heater performance characteristics that are similar to those represented by the one-hour draw values listed in Table 5.

Another factor to consider when sizing water heaters is the set point temperature, as lower hot water temperatures may increase the volume of hot water used. Currently, manufacturers are shipping residential water heaters with a recommendation that the initial set point be about 49°C to minimize the potential for scalding. Reduced set points generally lower standby losses and increase the water heater's efficiency and recovery capacity.

Over the last decade, the structure and life-style of a typical family has altered the household's hot water consumption. Due to variations in family size, age of family members, presence and age of children, hot water use volumes and temperatures, and other factors, demand patterns fluctuate widely in both magnitude and time distribution.

Perlman and Mills (1985) developed the overall and peak average hot water use volumes shown in Table 6. Average hourly patterns and 95% confidence level profiles are illustrated in Figures 11 and 12. Samples of results from the analysis of similarities in hot water use are given in Figures 13 and 14.

Table 6 Overall (OVL) and Peak Average Hot Water Use Volumes

| | Average Hot Water Use, L | | | | | | | |
| | Hourly | | Daily | | Weekly | | Monthly | |
Group	OVL	Peak	OVL	Peak	OVL	Peak	OVL	Peak
All families	9.8	17.3	236	254	1652	1873	7178	7700
"Typical" families	9.9	21.9	239	252	1673	1981	7270	7866

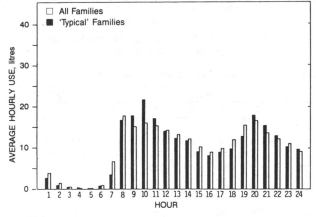

Fig. 11 Residential Average Hourly Hot Water Use

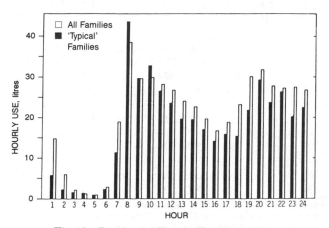

Fig. 12 Residential Hourly Hot Water Use—95% Confidence Level

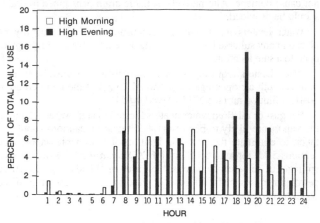

Fig. 13 Residential Hourly Water Use Pattern for Selected High Morning and High Evening Users

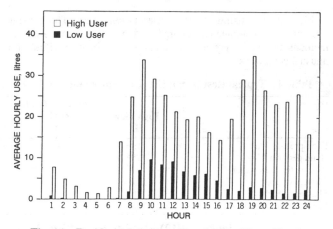

Fig. 14 Residential Average Hourly Hot Water Use for Low and High Users

COMMERCIAL AND INSTITUTIONAL

Most commercial and institutional establishments use hot or warm water. The specific requirements vary in total volume, flow rate, duration of peak load period, and temperature. Water heaters and systems should be selected based on these requirements.

This section covers sizing recommendations for central storage water heating systems. Hot water usage data and sizing curves for dormitories, motels, nursing homes, office buildings, food service establishments, apartments, and schools are based on EEI-sponsored research (Werden and Spielvogel 1969). Caution must be taken in applying these data to small buildings. Also, within any given category there may be significant variation. For example, the motel category encompasses standard, luxury, resort, and convention motels.

When additional hot water requirements exist, a designer should increase the recovery and/or storage capacity accordingly. For example, if there is food service in an office building, the recovery and storage capacities required for each additional hot water use should be added when sizing a single central water heating system.

Peak hourly and daily demands for various categories of commercial and institutional buildings are shown in Table 7. These demands for central storage-type hot water systems represent the maximum flows metered in this 129-building study, excluding extremely high and very infrequent peaks. Table 7 also shows average hot water consumption figures for these buildings. Averages for schools and food service establishments are based on actual days of operation, while all others are based on total days. These averages can be used to estimate monthly consumption of hot water.

Table 7 Hot Water Demands and Use for Various Types of Buildings

Type of Building	Maximum Hour	Maximum Day	Average Day
Men's dormitories	14.4 L/student	83.3 L/student	49.7 L/student
Women's dormitories	19 L/student	100 L/student	46.6 L/student
Motels: Number of units[a]			
20 or less	23 L/unit	132.6 L/unit	75.8 L/unit
60	20 L/unit	94.8 L/unit	53.1 L/unit
100 or more	15 L/unit	56.8 L/unit	37.9 L/unit
Nursing homes	17 L/bed	114 L/bed	69.7 L/bed
Office buildings	1.5 L/person	7.6 L/person	3.8 L/person
Food service establishments:			
Type A—full meal restaurants and cafeterias	5.7 L/max meals/h	41.7 L/max meals/h	9.1 L/average meals/day[b]
Type B—drive-ins, grilles, luncheonettes, sandwich and snack shops	2.6 L/max meals/h	22.7 L/max meals/h	2.6 L/average meals/day[b]
Apartment houses: Number of apartments			
20 or less	45.5 L/apartment	303.2 L/apartment	159.2 L/apartment
50	37.9 L/apartment	276.7 L/apartment	151.6 L/apartment
75	32.2 L/apartment	250 L/apartment	144 L/apartment
100	26.5 L/apartment	227.4 L/apartment	140.2 L/apartment
200 or more	19 L/apartment	195 L/apartment	132.7 L/apartment
Elementary schools	2.3 L/student	5.7 L/student	2.3 L/student[b]
Junior and senior high schools	3.8 L/student	13.6 L/student	6.8 L/student[b]

[a]Interpolate for intermediate values.
[b]Per day of operation.

Research conducted for ASHRAE (Becker et al. 1991, Thrasher and DeWerth 1994) and others (Goldner 1993, 1994) included a compilation and review of service hot water use information in commercial and multifamily structures along with new monitoring data. Some of this work found consumption levels comparable to those shown in Table 7; however, many of the studies showed higher consumption levels.

Dormitories

Hot water requirements for college dormitories generally include showers, lavatories, service sinks, and clothes washers. Peak demand usually results from the use of showers. Load profiles and hourly consumption data indicate that peaks may last 1 or 2 h and then taper off substantially. Peaks occur predominantly in the evening, mainly around midnight. The figures do not include hot water used for food service.

Military Barracks

Design criteria for military barracks are available from the engineering departments of the U.S. Department of Defense. Some measured data exist for hot water use in these facilities. For published data, contact the U.S. Army Corps of Engineers or Naval Facilities Engineering Command.

Motels

Domestic hot water requirements are for tubs and showers, lavatories, and general cleaning purposes. Recommendations are based on tests at low- and high-rise motels located in urban, suburban, rural, highway, and resort areas. Peak demand, usually from shower use, may last 1 or 2 h and then drop off sharply. Food service, laundry, and swimming pool requirements are not included.

Nursing Homes

Hot water is required for tubs and showers, wash basins, service sinks, kitchen equipment, and general cleaning. These figures include hot water for kitchen use. When other equipment, such as that for heavy laundry and hydrotherapy purposes, is to be used, its hot water requirement should be added.

Office Buildings

Hot water requirements are primarily for cleaning and lavatory use by occupants and visitors. Hot water use for food service within office buildings is not included.

Food Service Establishments

Hot water requirements are primarily for dish washing. Other uses include food preparation, cleaning pots and pans and floors, and hand washing for employees and customers. The recommendations are for establishments serving food at tables, counters, booths, and parked cars. Establishments that use disposable service exclusively are not covered in Table 7.

Dish washing, as metered in these tests, is based on the normal practice of dish washing after meals, not on indiscriminate or continuous use of machines irrespective of the flow of soiled dishes. The recommendations include hot water supplied to dishwasher booster heaters.

Apartments

Hot water requirements for both garden-type and high-rise apartments are for one- and two-bath apartments, showers, lavatories, kitchen sinks, dishwashers, clothes washers, and general cleaning purposes. Clothes washers can be either in individual apartments or centrally located. These data apply to central water heating systems only.

Elementary Schools

Hot water requirements are for lavatories, cafeteria and kitchen use, and general cleaning purposes. When showers are used, their additional hot water requirements should be added. The recommendations include hot water for dishwashers but not for extended school operation such as evening classes.

High Schools

Senior high schools, grades 9 or 10 through 12, require hot water for showers, lavatories, dishwashers, kitchens, and general cleaning. Junior high schools, grades 7 through 8 or 9, have requirements similar to those of the senior high schools. Junior high schools having no showers follow the recommendations for elementary schools.

Requirements for high schools are based on daytime use. Recommendations do not take into account hot water usage for additional activities such as night school. In such cases, the maximum hourly demand remains the same, but the maximum daily and the average daily usage increases, usually by the number of additional people using showers and, to a lesser extent, eating and washing facilities.

Additional Data

Fast Food Restaurants. Hot water is used for food preparation, cleanup, and rest rooms. Dish washing is usually not a significant load. In most facilities, peak usage occurs during the cleanup period, typically soon after opening and immediately prior to closing. Hot water consumption varies significantly among individual facilities. Fast food restaurants typically consume 950 to 1900 L per day (EPRI 1994).

Supermarkets. The trend in supermarket design is to incorporate food preparation and food service functions, substantially increasing the usage of hot water. Peak usage is usually associated with cleanup periods, often at night, with a total consumption of 1100 to 3800 L per day (EPRI 1994).

Apartments. Table 8 differs from Table 7 in that it represents low-medium-high (LMH) guidelines rather than specific singular volumes. The values presented in Table 8 are based on monitored domestic hot water consumption found in the studies by Becker et al. (1991), Thrasher and DeWerth (1994), and Goldner (1993, 1994).

Table 8 Hot Water Demand and Use Guidelines for Apartment Buildings (Litres per Person)

Guideline	Maximum Hourly	Peak 15 Minutes	Maximum Daily	Average Daily
Low	11	4	76	53
Medium	19	7.5	18	114
High	34	11.5	340	204

"Med" is the overall average. "Low" is the lowest peak value. Such values are generally associated with apartment buildings having such occupant demographics as (1) all occupants working, (2) seniors, (3) couples with children, (4) middle income, or (5) higher population density. "High" is the maximum recurring value. These values are generally associated with (1) high percentage of children, (2) low income, (3) public assistance, or (4) no occupants working.

In applying these guidelines, the designer should note that a building may outlast its current use. This may be a reason to increase the design capacity of the domestic hot water system or allow for future enhancement of the service hot water system. Building management practices, such as the explicit prohibition (in the lease) of apartment clothes washers or the existence of bath/kitchen hookups, should be factored into the design process. A diversity factor that lowers the probability of coincident consumption may also be employed in larger buildings.

SIZING EXAMPLES

Figures 15 through 22 show the relationships between recovery and storage capacity for the various building categories. Any combination of storage and recovery rates that falls on the proper curve will satisfy the building requirements. Using the minimum recovery rate and the maximum storage capacity on the curves yields the smallest hot water capacity capable of satisfying the building requirement. The higher the recovery rate, the greater the 24-h heating capacity and the smaller the storage capacity required.

These curves can be used to select recovery and storage requirements accommodate water heaters that have fixed storage or recovery rates. Where hot water demands are not coincident with peak electric, steam, or gas demands, greater heater inputs can be selected if they do not create additional energy system demands, and the corresponding storage tank size can be selected from the curves.

Ratings of gas-fired water heating equipment are based on sea level operation and apply for operation at elevations up to 600 m. For operation at elevations above 600 m, and in the absence of specific recommendations from the local authority, equipment ratings should be reduced at the rate of 4% for each 300 m above sea level before selecting appropriately sized equipment.

Recovery rates in Figures 15 through 22 represent the actual hot water required without considering system heat losses. Heat losses from storage tanks and recirculating hot water piping should be calculated and added to the recovery rates shown. Storage tanks and hot water piping must be insulated.

The storage capacities shown are net usable requirements. Assuming that 60 to 80% of the hot water in a storage tank is usable, the actual storage tank size should be increased by 25 to 66% to compensate for unusable hot water.

Examples

Example 1. Determine the required water heater size for a 300-student women's dormitory for the following criteria:

a. Storage system with minimum recovery rate.

b. Storage system with recovery rate of 2.6 mL/s per student.

c. With the additional requirement for a cafeteria to serve a maximum of 300 meals per hour for minimum recovery rate, combined with item *a*; and for a recovery rate of 1.0 mL/s per maximum meals per hour, combined with item *b*.

Solution:

a. The minimum recovery rate from Figure 15 for women's dormitories is 1.1 mL/s per student or 330 mL/s total. At this rate, storage required is 45 L per student or 13.5 m^3 total. On a 70% net usable basis, the necessary tank size is 13.5/0.7 = 19.3 m^3.

b. The same curve shows 19 L storage per student at 2.6 mL/s recovery, or $300 \times 19 = 5700$ L storage with recovery of $300 \times 2.6 = 780$ mL/s. The tank size will be 5700/0.7 = 8140 L.

c. The requirements for a cafeteria can be determined from Figure 19 and added to those for the dormitory. For the case of minimum recovery rate, the cafeteria (Type A) requires $300 \times 0.47 = 140$ mL/s recovery rate and $300 \times 26.5/0.7 = 11.4$ m^3 of additional storage. The entire building then requires $330 + 140 = 470$ mL/s recovery and $19.3 + 11.4 = 30.7$ m^3 of storage.

With 1.0 mL/s recovery at the maximum hourly meal output, the recovery required is 300 mL/s, with $300 \times 7.5/0.7 = 3260$ mL/s of additional storage. Combining this with item *b*, the entire building requires $780 + 300 = 1080$ mL/s recovery and $8140 + 3260 = 11\ 400$ L = 11.4 m^3 of storage.

Note: Recovery capacities shown are for heating water only. Additional capacity must be added to offset the system heat losses.

Example 2. Determine the water heater size and monthly hot water consumption for an office building to be occupied by 300 people:

a. Storage system with minimum recovery rate.

b. Storage system with 3.8 L per person storage.

c. Additional minimum recovery rate requirement for a luncheonette open 5 days a week, serving a maximum of 100 meals per hour and an average of 200 meals per day.

d. Monthly hot water consumption.

Solution:

a. With minimum recovery rate of 0.1 mL/s per person from Figure 18, 30 mL/s recovery is required, while the storage is 6 L per person, or $300 \times 6 = 1800$ L. If 70% of the hot water is usable, the tank size will be 1800/0.7 = 2570 L.

b. The curve also shows 3.8 L storage per person at 0.18 mL/s per person recovery, or $300 \times 0.18 = 54$ mL/s. The tank size will be $300 \times 3.8/0.7 = 1630$ L.

c. The hot water requirements for a luncheonette (Type B) are contained in Figure 19. With a minimum recovery capacity of 0.26 mL/s per maximum meals per hour, 100 meals per hour would require 26 mL/s recovery, while the storage would be 7.6 L per maximum meals per hour, or $100 \times 7.6/0.7 = 1090$ L storage. The combined requirements with item *a* would then be 56 mL/s recovery and 3660 L storage.

Combined with item *b*, the requirement is 80 mL/s recovery and 2720 L storage.

d. Average day values are found in Table 7. The office building will consume an average of 3.8 L per person per day × 30 days per month × 300

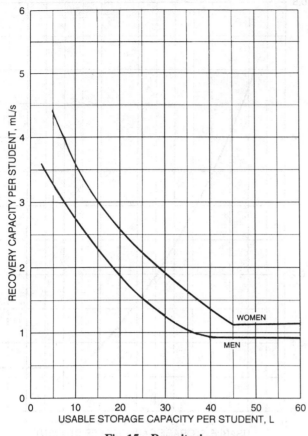

Fig. 15   Dormitories

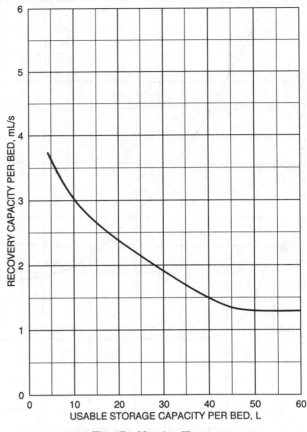

Fig. 17   Nursing Homes

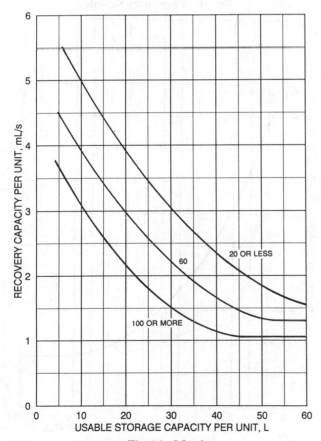

Fig. 16   Motels

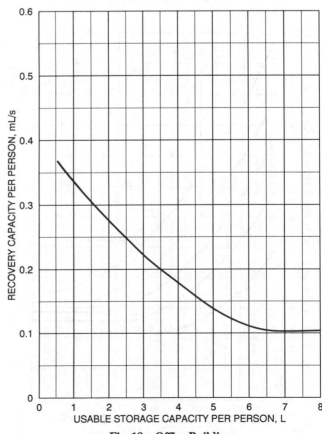

Fig. 18   Office Buildings

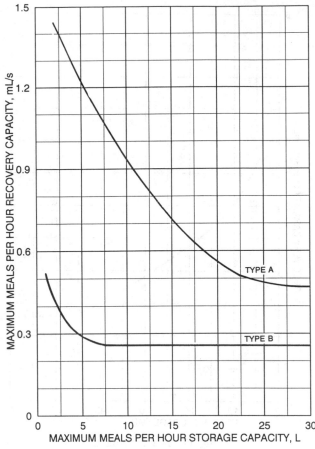

Fig. 19 Food Service

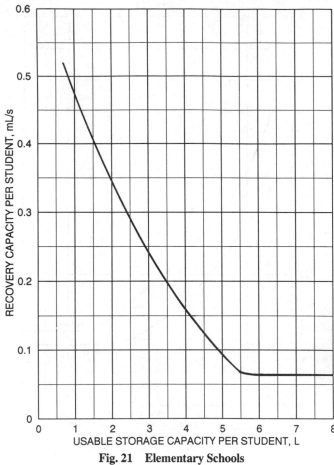

Fig. 21 Elementary Schools

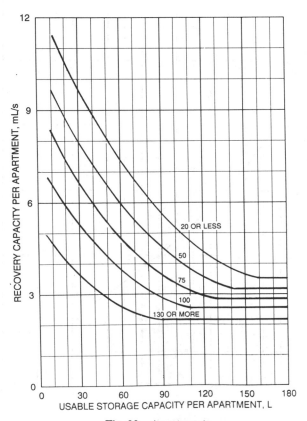

Fig. 20 Apartments

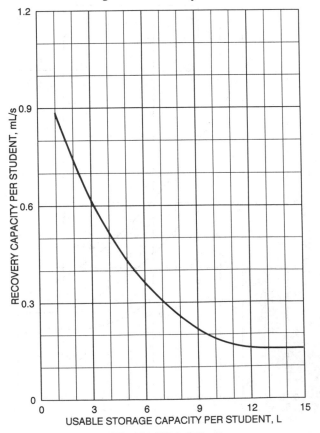

Fig. 22 High Schools

people = 34.2 m³ per month, while the luncheonette will consume 2.6 L per meal × 200 meals per day × 22 days per month = 11.4 m³ per month, for a total of 45.6 m³ per month.

Note: Recovery capacities shown are for heating water only. Additional capacity must be added to offset the system heat losses.

Example 3. Determine the water heater size for a 200-unit apartment house:

  a. Storage system with minimum recovery rate.

  b. Storage system with 4.2 mL/s per apartment recovery rate.

  c. Storage system for each of two 100-unit wings.

    1. Minimum recovery rate.

    2. Recovery rate of 4.2 mL/s per apartment.

Solution:

  a. The minimum recovery rate, from Figure 20, for apartment buildings with 200 apartments is 2.2 mL/s per apartment, or a total of 440 mL/s. The storage required is 90 L per apartment, or 18 m³. If 70% of this hot water is usable, the necessary tank size is 18/0.7 = 25.7 m³.

  b. The same curve shows 19 L storage per apartment at a recovery rate of 4.2 mL/s per apartment, or 200 × 4.2 = 840 mL/s. The tank size will be 200 × 19/0.7 = 5.43 m³.

  c. Solution for a 200-unit apartment house having two wings, each with its own hot water system.

    1. With minimum recovery rate of 2.6 mL/s per apartment (see Figure 20), a 260 mL/s recovery is required, while the necessary storage is 106 L per apartment, or 100 × 106 = 10.6 m³. The required tank size is 10.6/0.7 = 15.2 m³ for each wing.

    2. The curve shows that for a recovery rate of 4.2 mL/s per apartment the storage would be 50 L per apartment, or 100 × 50 = 5 m³, with recovery of 100 × 4.2 = 420 mL/s. The necessary tank size is 5/0.7 = 7.2 m³ in each wing.

Note: Recovery capacities shown are for heating water only. Additional capacity must be added to offset the system heat losses.

Example 4. Determine the water heater size and monthly hot water consumption for a 2000-student high school.

  a. Storage system with minimum recovery rate.

  b. Storage system with 15 m³ maximum storage capacity.

  c. Monthly hot water consumption.

Solution:

  a. With the minimum recovery rate of 0.16 mL/s per student (from Figure 22) for high schools, 320 mL/s recovery is required. The storage required is 11.5 L per student, or 2000 × 11.5 = 23 m³. If 70% of the hot water is usable, the tank size is 23/0.7 = 32.9 m³.

  b. The net storage capacity will be 0.7 × 15 = 10.5 m³, or 5.25 L per student. From the curve, a recovery capacity of 0.41 mL/s per student or 2000 × 0.41 = 820 mL/s is required.

  c. From Table 7, monthly hot water consumption is 2000 students × 6.8 L per student per day × 22 days = 299 m³.

Note: Recovery capacities shown are for heating water only. Additional capacity must be added to offset the system heat losses.

Table 9 can be used to determine the size of water heating equipment from the number of fixtures. To obtain the probable maximum demand, multiply the total quantity for the fixtures by the demand factor in line 19. The heater or coil should have a water heating capacity equal to this probable maximum demand. The storage tank should have a capacity equal to the probable maximum demand multiplied by the storage capacity factor in line 20.

Example 5. Determine heater and storage tank size for an apartment building from a number of fixtures.

Solution:

60 lavatories	×	7.6 L/h =	456 L/h
30 bathtubs	×	76 L/h =	2280 L/h
30 showers	×	114 L/h =	3420 L/h
60 kitchen sinks	×	38 L/h =	2280 L/h
15 laundry tubs	×	76 L/h =	1140 L/h
Possible maximum demand		=	9576 L/h
Probable maximum demand	=	9576 × 0.30 =	2870 L/h
Heater or coil capacity	=	2870/3600 =	0.80 L/s
Storage tank capacity	=	2870 × 1.25 =	3590 L/s

Showers

In many housing installations such as motels, hotels, and dormitories, the peak hot water load is usually due to the use of showers. Tables 1 and 9 indicate the probable hourly hot water demand and

Table 9 Hot Water Demand per Fixture for Various Types of Buildings
(Litres of water per hour per fixture, calculated at a final temperature of 60°C)

	Apartment House	Club	Gymnasium	Hospital	Hotel	Industrial Plant	Office Building	Private Residence	School	YMCA
1. Basin, private lavatory	7.6	7.6	7.6	7.6	7.6	7.6	7.6	7.6	7.6	7.6
2. Basin, public lavatory	15	23	30	23	30	45.5	23	—	57	30
3. Bathtub[c]	76	76	114	76	76	—	—	76	—	114
4. Dishwasher[a]	57	190-570	—	190-570	190-760	76-380	—	57	76-380	76-380
5. Foot basin	11	11	46	11	11	46	—	11	11	46
6. Kitchen sink	38	76	—	76	114	76	76	38	76	76
7. Laundry, stationary tub	76	106	—	106	106	—	—	76	—	106
8. Pantry sink	19	38	—	38	38	—	38	19	38	38
9. Shower	114	568	850	284	284	850	114	114	850	850
10. Service sink	76	76	—	76	114	76	76	57	76	76
11. Hydrotherapeutic shower				1520						
12. Hubbard bath				2270						
13. Leg bath				380						
14. Arm bath				130						
15. Sitz bath				114						
16. Continuous-flow bath				625						
17. Circular wash sink				76	76	114	76		114	
18. Semicircular wash sink				38	38	57	38		57	
19. DEMAND FACTOR	0.30	0.30	0.40	0.25	0.25	0.40	0.30	0.30	0.40	0.40
20. STORAGE CAPACITY FACTOR[b]	1.25	0.90	1.00	0.60	0.80	1.00	2.00	0.70	1.00	1.00

[a]Dishwasher requirements should be taken from this table or from manufacturers' data for the model to be used, if this is known.

[b]Ratio of storage tank capacity to probable maximum demand/h. Storage capacity may be reduced where an unlimited supply of steam is available from a central street steam system or large boiler plant.

[c]Whirlpool baths require specific consideration based on their capacity. They are not included in the bathtub category.

the recommended demand and storage capacity factors for various types of buildings. Hotels could have a 3- to 4-h peak shower load. Motels require similar volumes of hot water, but the peak demand may last for only a 2-h period. In some types of housing, such as barracks, fraternity houses, and dormitories, all occupants may take showers within a very short period. In this case, it is best to find the peak load by determining the number of shower heads and the rate of flow per head; then estimate the length of time the shower will be on. It is estimated that the average shower time per individual is 1 to 3 min.

The flow rate from a shower head varies depending on type, size, and water pressure. At 280 kPa water pressure, available shower heads have nominal flow rates of blended hot and cold water from about 160 to 380 mL/s. In multiple shower installations, flow control valves on shower heads are recommended because they reduce the flow rate and maintain it regardless of fluctuations in water pressure. Flow can usually be reduced to 50% of the manufacturer's maximum flow rating without adversely affecting the spray pattern of the shower head. Flow control valves are commonly available with capacities from 1.5 to 4.0 gpm.

If the manufacturer's flow rate for a shower head is not available, and no flow control valve is used, the following average flow rates may serve as a guide for sizing the water heater:

Small shower head	160 mL/s
Medium shower head	280 mL/s
Large shower head	380 mL/s

Food Service

In a restaurant, bacteria are usually killed by rinsing the washed dishes with 82 to 90°C water for several seconds. In addition, an ample supply of general-purpose hot water, usually 60 to 65°C, is required for the wash cycle of dishwashers. Although a water temperature of 60°C is reasonable for dish washing in private dwellings, in public places, the NSF or local health departments require 82 to 90°C water in the rinsing cycle. However, the NSF allows lower temperatures when certain types of machines and chemicals are used. The two-temperature hot water requirements of food service establishments present special problems. The lower temperature water is distributed for general use, but the 82°C water should be confined to the equipment requiring it and should be obtained by

Table 10   NSF Final Rinse Water Requirements for Dish Washing Machines[a]

Type and Size of Dishwasher	Flow Rate, mL/s	82 to 95°C Hot Water Requirements	
		Heaters Without Internal Storage,[b] mL/s	Heaters with Internal Storage to Meet Flow Demand,[c] mL/s
Door type:			
410 × 410 mm	400	440	73
460 × 460 mm	550	550	91
510 × 510 mm	660	660	109
Undercounter	320	320	74
Conveyor type:			
Single tank	440	440	440
Multiple tank (dishes flat)	360	360	360
Multiple tank (dishes inclined)	290	290	290
Silver washers	440	440	47
Utensil washers	500	500	80
Makeup water requirements	150	150	146

Note: Values are extracted from NSF *Standard* 5-92.
[a]Flow pressure at dishwashers is assumed to be 140 kPa (gage).
[b]Based on the flow rate in L/s.
[c]Based on dishwasher operation at 100% of mechanical capacity.

boosting the temperature. It would be dangerous to distribute 82°C water for general use. NSF *Standard* 26-80 covers the design of dishwashers and water heaters used by restaurants. The American Gas Association (Dunn et al. 1959) has published a recommended procedure for sizing water heaters for restaurants that consists of determining the following:

1. Types and sizes of dishwashers used (manufacturers' data should be consulted to determine the initial fill requirements of the wash tanks)
2. Required quantity of general-purpose hot water
3. Duration of peak hot water demand period
4. Inlet water temperature
5. Type and capacity of existing water heating system
6. Type of water heating system desired

After the quantity of hot water withdrawn from the storage tank each hour has been taken into account, the following equation may be used to size the required heater(s). The general-purpose and 82 to 90°C water requirements are determined from Tables 10 and 11.

$$q_i = Q_h c_p \rho \Delta t / \eta \qquad (6)$$

where

q_i = heater input, W
Q_h = flow rate, mL/s
c_p = specific heat of water = 4.18 kJ/(kg·K)
ρ = density of water = 1.0 kg/L
Δt = temperature rise, K
η = heater efficiency

To determine the quantity of usable hot water from storage, the duration (in hours) of consecutive peak demand must be estimated. This peak demand period usually coincides with the dishwashing period during and after the main meal and may last from 1 to 4 h.

Any hour in which the dishwasher is used at 70% or more of capacity should be considered a peak hour. If the peak demand lasts for 4 h or more, the value of a storage tank is reduced, unless especially large tanks are used. Some storage capacity is desirable to meet momentary high draws.

NSF *Standard* 5 recommendations for hot water rinse demand are based on 100% operating capacity of the machines, as are the data provided in Table 10. NSF *Standard* 5 states that 70% of operating rinse capacity is all that is normally attained, except for rackless-type conveyor machines.

Examples 6, 7, and 8 demonstrate the use of Equation (6) in conjunction with Tables 10 and 11. The calculations assume a heater efficiency of 75%. Higher efficiency units have lower input requirements.

Table 11   General-Purpose Hot Water (60°C) Requirement for Various Kitchens Uses[a,b]

Equipment	mL/s
Vegetable sink	47
Single pot sink	32
Double pot sink	63
Triple pot sink	95
Prescrapper (open type)	189
Preflush (hand-operated)	47
Preflush (closed type)	252
Recirculating preflush	42
Bar sink	32
Lavatories (each)	5.3

Source: Dunn et al. (1959).
[a]Supply water pressure at equipment is assumed to be 140 kPa (gage).
[b]Dishwasher operation at 100% of mechanical capacity.

Example 6. Determine the hot water demand for a new water heating system in a cafeteria kitchen with one vegetable sink, five lavatories, one prescrapper, one utensil washer, and one two-tank conveyor dishwasher (dishes inclined) with makeup device. The initial fill requirement for the tank of the utensil washer is 90 mL/s at 60°C. The initial fill requirement for the dishwasher is 21 mL/s for each tank, or a total of 42 mL/s, at 60°C. The maximum period of consecutive operation of the dishwasher at or above 70% capacity is assumed to be 2 h. The supply water temperature is 15°C.

Solution: The required quantities of general purpose (60°C) and rinse (82°C) water for the equipment, from Tables 10 and 11 and given values, are shown in the following tabulation:

Item	Quantity Required at 60°C, mL/s	Quantity Required at 82°C, mL/s
Vegetable sink	47	—
Lavatories (5)	26.5	—
Prescrapper	189	—
Dishwasher	—	290
Initial tank fill	42	—
Makeup water	—	146
Utensil washer	—	80
Initial tank fill	90	—
Total requirements	395	516

The total consumption of 60°C water is 395 mL/s. The total consumption at 82°C depends on the type of heater to be used. For a heater that has enough internal storage capacity to meet the flow demand, the total consumption, based on the recommendation of the NSF, is 70% of the total requirements taken from Table 10, or approximately 360 mL/s ($0.70 \times 516 = 361$ mLs). For an instantaneous heater without internal storage capacity, the total quantity of 82°C water consumed must be based on the flow demand. From Table 10, the quantity required for the dishwasher is 290 mL/s; for the makeup, 146 mL/s; and for the utensil washer, 500 mL/s. The total consumption of 82°C water is $290 + 146 + 500 = 936$ mL/s, or approximately 940 mL/s.

Example 7. Determine gas input requirements for heating water in the cafeteria kitchen described in Example 6, by the following systems, which are among many possible solutions:

a. Separate, self-contained, storage-type heaters.

b. Single instantaneous-type heater having no internal storage, to supply both 82°C and 60°C water through a mixing valve.

c. Separate instantaneous-type heaters having no internal storage.

d. Combination of heater and external storage tank for 60°C water, plus a booster heater for 82°C water. The heater and external storage are to supply 60°C water for both the general-purpose requirement and the booster heater. The booster heater is to have sufficient storage capacity to meet the flow demand of 82°C rinse water.

Solution:

a. The temperature rise for 60°C water is $60 - 15 = 45$ K. From Equation (6), the gas input required to produce 395 mL/s of 60°C water with a 45 K temperature rise at 75% efficiency is about 100 kW. One or more heaters may be selected to meet this total requirement.

From Equation (6), the gas input required to produce 360 mL/s of 82°C water with a temperature rise of $82 - 15 = 67$ K at 75% efficiency is 135 kW. One or more heaters with this total requirement may be selected from manufacturers' catalogs.

b. The correct sizing of instantaneous-type heaters depends on the flow rate of the 82°C rinse water. From Example 6, the consumption of 82°C water based on the flow rate is 940 mL/s; consumption of 60°C water is 395 mL/s.

Gas input required to produce 940 mL/s of 82°C water with a 67 K temperature rise is 350 kW. Gas input to produce 395 mL/s of 60°C water with a temperature rise of 45 K is 100 kW. Total heater requirement is $350 + 100 = 450$ kW. One or more heaters meeting this total input requirement can be selected from manufacturers' catalogs.

c. Gas input required to produce 60°C water is the same as for Solution *b*, 100 kW. One or more heaters meeting this total requirement can be selected.

Gas input required to produce 82°C water is also the same as in Solution *b*, 350 kW. One or more heaters meeting this total requirement can be selected.

d. The *net* hourly hot water requirement must be determined in order to size the heater required to supply 60°C water. Assuming a 2000 L tank is used for storage, from Equation (4), the quantity of usable hot water in storage V_a is $0.7 \times 2000 = 1400$ L. The total quantity of 60°C water required for the 2 h is 2 h (395 mL/s + 360 mL/s) = 5530 L. The heater must therefore supply $(5530 \text{ L} - 1400 \text{ L})/2$ h = 575 mL/s.

From Equation (6), the gas input required to produce 575 mL/s of 60°C water with a 45 K temperature rise is 144 kW.

For systems involving storage tanks, it is assumed that water in the tanks has been brought up to temperature prior to the peak dish washing period and that enough time will elapse before the next peak period to permit recovery of the water temperature in the storage tank.

The booster heater is sized to heat 375 mL/s from 60 to 82°C, a 22 K rise. From Equation (6), the gas input required is 46 kW.

Example 8. A luncheonette has purchased a door-type dishwasher that will handle 410×410-mm racks. The existing hot water system is capable of supplying the necessary 60°C water to meet all requirements for general-purpose use and for the booster heater that is to be installed. Determine the size of the following booster heaters operating at 75% thermal efficiency required to heat 60°C water to provide sufficient 82°C rinse water for the dishwasher:

a. Booster heater with no storage capacity.

b. Booster heater with enough storage capacity to meet flow demand.

Solution:

a. Because the heater is the instantaneous type, it must be sized to meet the 82°C water demand at a rated flow. From Table 10, this rated flow is 440 mL/s. From Equation (6), the required gas input with a 22 K temperature rise is 54 kW. A heater meeting this input requirement can be selected from manufacturers' catalogs.

b. In designing a system with a booster heater having storage capacity, the dishwasher's hourly flow demand can be used instead of the flow demand used in Solution *a*. The flow demand from Table 10 is 73 mL/s when the dishwasher is operating at 100% mechanical capacity. However, the NSF states that 70% of operating rinse capacity is all that is normally attained for this type of dishwasher. Therefore, the hourly flow demand is $0.70 \times 73 = 51$ mL/s. From Equation (6), with a 22 K temperature rise, the gas input required is 6300 W. A booster heater with this input can be selected from manufacturers' catalogs.

Estimating Procedure. Hot water requirements for kitchens are sometimes estimated on the basis of the number of meals served (assuming eight dishes per meal). Demand for 82°C water for a dishwasher is

$$D_1 = C_1 N / \theta \qquad (7)$$

where

D_1 = water for dishwasher, mL/s
N = number of meals served
θ = hours of service
C_1 = 0.84 for single-tank dishwasher
= 0.53 for two-tank dishwasher

Demand for water for a sink with gas burners is

$$D_2 = C_2 V \qquad (8)$$

where

D_2 = water for sink, mL/s
C_2 = 3.2
V = sink capacity (380 mm depth), L

Demand for general-purpose hot water at 60°C is

$$D_3 = C_3 N / (\theta + 2) \qquad (9)$$

where

D_3 = general-purpose water, mL/s
C_3 = 1.3

Total demand is

$$D = D_1 + D_2 + D_3$$

For soda fountains and luncheonettes, use 75% of the total demand. For hotel meals or other elaborate meals, use 125%.

Schools

Service water heating in schools is needed for janitorial work, lavatories, cafeterias, shower rooms, and sometimes swimming pools.

Hot water used in cafeterias is about 70% of that usually required in a commercial restaurant serving adults and can be estimated by the method used for restaurants. Where NSF sizing is required, follow *Standard 5*.

Shower and food service loads are not ordinarily concurrent. Each should be determined separately, and the larger load should determine the size of the water heater(s) and the tank. Provision must be made to supply 82°C sanitizing rinse. The booster must be sized according to the temperature of the supply water. If feasible, the same water heating system can be used for both needs. If the distance between the two points of need is great, separate water heating systems should be used.

A separate water heating system for swimming pools can be sized as outlined in the section on Swimming Pools/Health Clubs.

Domestic Coin-Operated Laundries

Small domestic machines in coin laundries or apartment house laundry rooms have a wide range of draw rates and cycle times. Domestic machines provide wash water temperatures (normal) as low as 50°C. Some manufacturers recommend temperatures of 70°C; however, the average appears to be 60°C. Hot water sizing calculations must assure a supply to both the instantaneous draw requirements of a number of machines filling at one time and the average hourly requirements.

The number of machines that will be drawing at any one time varies widely; the percentage is usually higher in smaller installations. One or two customers starting several machines at about the same time has a much sharper effect in a laundry with 15 or 20 machines than in one with 40 machines. Simultaneous draw may be estimated as follows:

1 to 11 machines	100% of possible draw
12 to 24 machines	80% of possible draw
25 to 35 machines	60% of possible draw
36 to 45 machines	50% of possible draw

Possible peak draw can be calculated from

$$F = 1000 \, NPV_f/T \tag{10}$$

where

> F = peak draw, mL/s
> N = number of washers installed
> P = number of washers drawing hot water divided by N
> V_f = quantity of hot water supplied to machine during hot wash fill, L
> T = wash fill period, s

Recovery rate can be calculated from

$$R = (1000NPV_f) / [60(\theta + 10)] = 16.7NPV_f / (\theta + 10) \tag{11}$$

where

> R = total hot water (per machine) used for entire cycle (machine adjusted to hottest water setting), mL/s
> θ = actual machine cycle time, min

Note: $(\theta + 10)$ is the cycle time plus 10 min for loading and unloading.

Commercial Laundries

Commercial laundries generally use a storage water heating system. The water may be softened to reduce soap use and improve quality. The trend is toward installation of high-capacity washer-extractor wash wheels, resulting in high peak demand.

Sizing Data. Laundries can normally be divided into five categories. The required hot water is determined by the mass of the material processed. Average hot water requirements at 82°C are

Institutional	4.6 mL/(kg·s)
Commercial	4.6 mL/(kg·s)
Linen supply	5.8 mL/(kg·s)
Industrial	5.8 mL/(kg·s)
Diaper	5.8 mL/(kg·s)

Total mass of the material times these values give the average hourly hot water requirements. The designer must consider peak requirements; for example, a 270-kg machine may have a 1.25 L/s average requirement, but the peak requirement could be 22 L/s.

In a multiple-machine operation, it is not reasonable to fill all machines at the momentary peak rate. Diversity factors can be estimated by using 1.0 of the largest machine plus the following balance:

	Total number of machines				
	2	**3 to 5**	**6 to 8**	**9 to 11**	**12 and over**
1.0 +	0.6	0.45	0.4	0.35	0.3

For example, four machines would have a diversity factor of 1.0 + 0.45 = 1.45.

Types of Systems. Service water heating systems for laundries are pressurized or vented. The pressurized system uses city water pressure, and the full peak flow rates are received by the softeners, reclaimer, condensate cooler, water heater, and lines to the wash wheels. The flow surges and stops at each operation in the cycle. A pressurized system depends on an adequate water service.

The vented system uses pumps from a vented (open) hot water heater or tank to supply hot water. The tank's water level fluctuates from about 150 mm above the heating element to a point 300 mm from the top of the tank; this fluctuation defines the working volume. The level drops for each machine fill, while makeup water runs continuously at the average flow rate under water service pressure during the complete washing cycle. The tank is sized to have full working volume at the beginning of each cycle. Lines and softeners may be sized for the average flow rate from the water service to the tank, not the peak machine fill rate as with a closed, pressurized system.

The waste heat exchangers have a continuous flow across the heating surface at a low flow rate, with continuous heat reclamation from the wastewater and flash steam. Automatic flow-regulating valves on the inlet water manifold control this low flow rate. Rapid fill of machines increases production; i.e., more batches can be processed.

Heat Recovery. Commercial laundries are ideally suited for heat recovery because 58°C wastewater is discharged to the sewer. Fresh water can be conservatively preheated to within 8 K of the wastewater temperature for the next operation in the wash cycle. Regions with an annual average temperature of 13°C can increase to 50°C the initial temperature of fresh water going into the hot water heater. For each litre or kilogram per second of water preheated 37 K (13 to 50°C), heat reclamation will be 155 kW. This saves 18.5 m³ of natural gas or 14.8 litres of oil per hour.

Flash steam from a condensate receiving tank is often wasted to the atmosphere. The heat in this flash steam can be reclaimed with a suitable heat exchanger. Makeup water to the heater can be preheated 5 to 10 K above the existing makeup temperature with the flash steam.

Swimming Pools/Health Clubs

The desirable temperature for swimming pools is 27°C. Most manufacturers of water heaters and boilers offer specialized models for pool heating; these include a pool temperature controller and a water bypass to prevent condensation. The water heating system is usually installed prior to the return of treated water to the pool. A circulation rate to generate a change of water every 8 h for residential pools and 6 h for commercial pools is acceptable. An indirect heating system, in which piping is embedded in the walls or floor of the pool, has the advantage of reduced corrosion, scaling, and condensation because pool water does not flow through the pipes. The disadvantage of this type of system is the high initial installation cost.

The installation should have a pool temperature control and a water pressure or flow safety switch. The temperature control should be installed at the inlet to the heater, while the pressure or flow switch can be installed at either the inlet or outlet, depending on the manufacturer's instructions. It affords protection against inadequate water flow.

Characteristics of pool use are of prime importance in sizing the water heater, and, for economical reasons, the heater should be on only during and just prior to the period of pool use. Sizing should be based on four considerations:

1. Conduction through the pool walls
2. Convection from the pool surface
3. Radiation from the pool surface
4. Evaporation from the pool surface

Except in aboveground pools and in rare cases where cold groundwater flows past the pool walls, conduction losses are small and can be ignored. Because convection losses depend on temperature differentials and wind speed, these losses can be greatly reduced by the installation of windbreaks such as hedges, solid fences, or buildings.

Radiation losses occur when the pool surface is subjected to temperature differentials; these frequently occur at night, when the sky temperature may be as much as 45 K below ambient air temperature. This usually occurs on clear, cool nights. During the daytime, however, an unshaded pool will receive a large amount of radiant energy, often as much as 30 kW. These losses and gains may offset each other. An easy method of controlling nighttime radiation losses is to use a floating pool cover; this also substantially reduces evaporative losses.

Evaporative losses constitute the greatest heat loss from the pool—50 to 60% in most cases. If it is possible to cut evaporative losses drastically, the heating requirement of the pool may be cut by as much as 50%. The floating pool cover can accomplish this reduction.

A pool heater with an input great enough to provide a heat-up time of 24 h would be the ideal solution. However, it may not be the most economical system for pools that are in continuous use during an extended swimming season. In this instance, a less expensive unit providing an extended heat-up period of as much as 48 h can be used. Pool water may be heated by several methods. Fuel-fired water heaters and boilers, electric boilers, tankless electric circulation water heaters, air-source heat pumps, and solar heaters have all been used successfully. Air-source heat pumps and solar heating systems would probably be used to extend a swimming season rather than to allow intermittent use with rapid pickup.

At one time, pools were essentially rectangular in shape, but now the shape is limited only by the imagination of the pool designer. The following equations provide some assistance in determining the area and volume of pools:

Elliptical

Area = 3.14 AB
A = short radius
B = long radius
Volume = Area × Average Depth

Kidney Shape

Area = 0.45 $L(A+B)$ (approximately)
L = length
A = width at one end
B = width at other end
Volume = Area × Average Depth

Oval (for circular, set L = 0)

Area = 3.14 $R + LW$
L = length of straight sides
W = width or $2R$
R = radius of ends
Volume = Area × Average Depth

Rectangular

Area = LW

L = length
W = width
Volume = Area × Average Depth

The following is an effective method for heating outdoor pools. Additional equations can be found in Chapter 4.

1. Obtain pool water capacity, in cubic metres, from the above.
2. Determine the desired heat pickup time in hours.
3. Determine the desired pool temperature—if not known, use 27°C.
4. Determine the average temperature of the coldest month of use.

The required heater output q_t can now be determined by the following equations:

$$q_1 = \rho c_p V (t_f - t_i) / \theta \qquad (12)$$

where

q_1 = pool heat-up rate, kW
ρ = density of water = 998 kg/m^3
c_p = specific heat of water = 4.18 kJ/(kg·K)
V = pool volume, m^3
t_f = desired temperature (usually 27°C)
t_i = initial temperature of pool, °C
θ = pool heat-up time, h

$$q_2 = UA (t_p - t_a) \qquad (13)$$

where

q_2 = heat loss from pool surface, kW
U = surface heat transfer coefficient = 0.060 kW/(m^2·K)
A = pool surface area, m^2
t_p = pool temperature, °C
t_a = ambient temperature, °C

$$q_t = q_1 + q_2 \qquad (14)$$

Notes: These heat loss equations assume a wind velocity of 5 to 8 km/h. For pools sheltered by nearby fences, dense shrubbery, or buildings, an average wind velocity of less than 5.6 km/h can be assumed. In this case, use 75% of the values calculated by Equation (13). For a velocity of 8 km/h, multiply by 1.25; for 16 km/h, multiply by 2.0.

Because Equation (13) applies to the coldest monthly temperatures, the calculated results may not be economical. Therefore, a value of one-half the surface loss plus the heat-up value yields a more viable heater output figure. The heater input then equals the output divided by the efficiency of the fuel source.

Whirlpools and Spas

Hot water requirements for whirlpool baths and spas depend on temperature, fill rate, and total volume. Water may be stored separately at the desired temperature or, more commonly, regulated at the point of entry by blending. If rapid filling is desired, provide storage at least equal to the volume needed; fill rate can then be varied at will. An alternative is to establish a maximum fill rate and provide an instantaneous water heater that will handle the flow.

Industrial Plants

Hot water is used in industrial plants for cafeterias, showers, lavatories, gravity sprinkler tanks, and industrial processes. If the same hot water system is used only for the cafeteria, employee cleanup, laundry, and small miscellaneous uses, the water heater can be sized to meet employee cleanup load (with additional provision for the sanitizing rinse needs of the cafeteria). Employee cleanup load is usually heaviest and not concurrent with other uses. The other loads should be checked before sizing, however, to be certain that this is true.

The employee cleanup load consists of one or more of the following: (1) wash troughs or standard lavatories, (2) multiple wash sinks, and (3) showers. Hot water requirements for employees using standard wash fixtures can be estimated at 3.8 L of hot water for each clerical and light-industrial employee per work shift and 7.6 L for each heavy-industrial worker.

For sizing purposes, the number of workers using multiple wash fountains is disregarded. Hot water demand is based on full flow for the entire cleanup period. This usage over a 10-min period is indicated in Table 12. The shower load depends on the flow rate of the shower heads and their length of use. Table 12 may be used to estimate flow based on a 15-min period.

Table 12 Hot Water Usage for Industrial Wash Fountains and Showers

	Multiple Wash Fountains		Showers	
Type	L of 60°C Water Required for 10-min Period[a]		Flow Rate, mL/s	L of 60°C Water Required for 15-min Period[b]
910-mm Circular	150		190	110
910-mm Semicircular	83		250	150
1370-mm Circular	250		320	185
1370-mm Semicircular	150		380	220

[a]Based on 43°C wash water and 5°C cold water at average flow rates.
[b]Based on 40°C shower water and 5°C cold water.

Water heaters used to prevent freezing in gravity sprinkler tanks or water storage tanks should be part of a separate system. The load depends on tank heat loss, tank capacity, and winter design temperature.

Process hot water load must be determined separately. Volume and temperature vary with the specific process. If the process load occurs at the same time as the shower or cafeteria load, the system must be sized to reflect this total demand. Separate systems can also be used, depending on the size of the various loads and the distance between them.

Ready-Mix Concrete

In cold weather, ready-mix concrete plants need hot water to mix the concrete so that it will not be ruined by freezing before it sets. Operators prefer to keep the mix at about 21°C by adding hot water to the cold aggregate. Usually, water at about 65°C is considered proper for cold weather. If the water temperature is too high, some of the concrete will flash set.

Generally, 150 L of hot water per cubic metre of concrete mix is used for sizing. To obtain the total hot water load, this number is multiplied by the number of trucks loaded each hour and the capacity of the trucks. The hot water is dumped into the mix as quickly as possible at each loading, so ample hot water storage or large heat exchangers must be used. Table 13 shows a method of sizing water heaters for concrete plants.

Table 13 Water Heater Sizing for Ready-Mix Concrete Plant
(Input and Storage Tank Capacity to Supply 65°C Water at 4°C Inlet Temperature)

Truck Capacity, m³	Water Heater Storage Tank Volume, L	Time Interval Between Trucks, min[a]					
		50	35	25	10	5	0
		Water Heater Capacity, kW					
4.6	1630	134	179	230	403	536	809
5.7	1860	154	205	264	463	615	923
6.9	2130	174	232	299	524	697	1049
8.4	2430	201	268	344	604	803	1207

[a]This table assumes that there is 10-min loading time for each truck. Thus, for a 50-min interval between trucks, it is assumed that 1 truck/h is served. For 0 min between trucks, it is assumed that one truck loads immediately after the truck ahead has pulled away. Thus, 6 trucks/h are served.

It is also assumed that each truck carries a 450-L storage tank of hot water for washing down at the end of dumping the load. This hot water is drawn from the storage tank and must be added to the total hot water demands. This has been included in the table.

Part of the heat may be obtained by heating the aggregate bin by circulating hot water through pipe coils in the walls or sides of the bin. If the aggregate is warmed in this manner, the temperature of the mixing water may be lower, and the aggregate will flow easily from the bins. When aggregate is not heated, it often freezes into chunks, which must be thawed before they will pass through the dump gates. If hot water is used for thawing, too much water accumulates in the aggregate, and control of the final product may vary beyond allowable limits. Therefore, jets of steam supplied by a small boiler and directed on the large chunks are often used for thawing.

SIZING INSTANTANEOUS AND SEMI-INSTANTANEOUS WATER HEATERS

The methods for sizing storage water heating equipment should not be used for instantaneous and semi-instantaneous heaters. The following is based on the Hunter method for sizing hot and cold water piping, with diversity factors applied for hot water and various building types.

Fixture units (Table 1) are assigned to each fixture using hot water and totalled. Maximum hot water demand is obtained from Figure 23 or Figure 24 by matching total fixture units to the curve for the type of building. Hot water for fixtures and outlets requiring constant flows should be added to the demand.

The heater can then be selected for the total demand and the total temperature rise required. For critical applications such as hospitals, multiple heaters with 100% standby are recommended. Consider multiple heaters for buildings in which continuity of service is

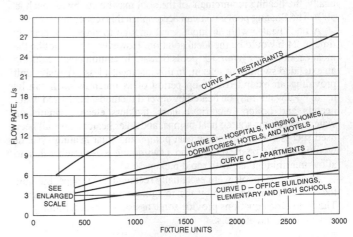

Fig. 23 Modified Hunter Curve for Calculating Hot Water Flow Rate

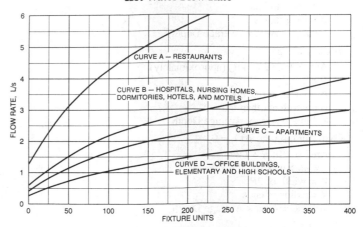

Fig. 24 Enlarged Section of Figure 23 (Modified Hunter Curve)

important. The minimum recommended size for semi-instantaneous heaters is 0.65 L/s, except for restaurants, in which it is 0.95 L/s. When the flow for a system is not easily determined, the heater may be sized for the full flow of the piping system. Heaters with low flows must be sized carefully, and care should be taken in the estimation of diversity factors. Unusual hot water requirements should be analyzed to determine whether additional capacity is required. One example is a dormitory in a military school, where all showers and lavatories are used simultaneously when students return from a drill. In this case, the heater and piping should be sized for the full flow of the system.

While the fixture count method bases heater size of the diversified system on hot water flow, hot water piping should be sized for the full flow to the fixtures. Recirculating hot water systems are adaptable to instantaneous heaters. When recirculating systems are installed, the heater capacity should be checked and increased if necessary to offset heat losses of the recirculating system.

To make preliminary estimates of hot water demand when the fixture count is not known, use Table 14 with Figure 23 or Figure 24. The result will usually be higher than the demand determined from the actual fixture count. Actual heater size should be determined from Table 1. Hot water consumption over time can be assumed the same as that in the section on Storage Equipment Sizing.

Table 14   Preliminary Hot Water Demand Estimate

Type of Building	Fixture Units
Hospital or nursing home	2.50 per bed
Hotel or motel	2.50 per room
Office building	0.15 per person
Elementary school	0.30 per student[a]
Junior and senior high school	0.30 per student[a]
Apartment house	3.00 per apartment

[a]Plus shower load.

Example 9. A 600-student elementary school has the following fixture count: 60 public lavatories, 6 service sinks, 4 kitchen sinks, 6 showers, and 1 dishwasher at 0.5 L/s. Determine the hot water flow rate for sizing a semi-instantaneous heater based on the following:

   a. Estimating the number of fixture units.

   b. Actual fixture count.

Solution:

   a. Use Table 14 to find the estimated fixture count: 600 students × 0.3 fixture units per student = 180 fixture units. As showers are not included, Table 1 shows 1.5 fixture units per shower × 6 showers = 9 additional fixture units. The basic flow is determined from curve D of Figure 24, which shows that the total flow for 189 fixture units is 1.4 L/s.

   b. To size the unit based on actual fixture count and Table 1, the calculation is as follows:

60 public lavatories	×	1.0	FU	=	60 FU
6 service sinks	×	2.5	FU	=	15 FU
4 kitchen sinks	×	0.75	FU	=	3 FU
6 showers	×	1.5	FU	=	9 FU
Subtotal					87 FU

At 87 fixture units, curve D of Figure 22 shows 1.0 L/s, to which must be added the dishwasher requirement of 0.5 L/s. Thus, the total flow is 1.5 L/s.

Comparing the flow based on actual fixture count to that obtained from the preliminary estimate shows the preliminary estimate to be slightly lower in this case. It is possible that the preliminary estimate could have been as much as twice the final fixture count. To prevent such oversizing of equipment, use the actual fixture count method to select the unit.

SIZING REFRIGERANT-BASED WATER HEATING SYSTEMS

Refrigerant-based water heating systems such as heat pump water heaters and refrigeration heat reclaim systems cannot be sized like conventional systems to meet peak loads. The seasonal and instantaneous efficiency and output of these systems vary greatly with operating conditions. Microcomputer software programs that perform detailed performance simulations taking these factors into account are recommended for sizing and analysis.

The capacities of these systems and any related supplemental water heating equipment should be selected to achieve high average daily run time for the system and the lowest combination of operating and equipment costs. For heat pump water heaters, the adequacy of the heat source and the potential effect of the cooling output must be addressed.

BOILERS FOR INDIRECT WATER HEATING

When service water is heated indirectly by a space heating boiler, Figure 25 may be used to determine the additional boiler capacity required to meet the recovery demands of the domestic water heating load. Indirect systems include immersion coils in boilers as well as heat exchangers with space heating media.

Because the boiler capacity must meet not only the water supply requirement but also the space heating loads, Figure 25 indicates the reduction of additional heat supply for water heating if the ratio of water heating load to space heating load is low. This reduction is possible because

1. Maximum space heating requirements do not occur at the time of day when the maximum peak hot water demands occur.
2. Space heating requirements are based on the lowest outdoor design temperatures, which may occur for only a few days of the total heating season.
3. An additional heat supply or boiler capacity to compensate for pickup and radiation losses is usual. The pickup load cannot occur at the same time as the peak hot water demand because the building must be brought to a comfortable temperature before the occupants will be using hot water.

The factor obtained from Figure 25 is multiplied by the peak water heating load to obtain the additional boiler output capacity required.

For reduced standby losses in summer and improved efficiency in winter, step-fired modular boilers may be used. Units not in operation cool down and reduce or eliminate jacket losses. The heated

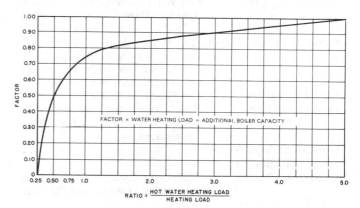

Fig. 25   Sizing Factor for Combination Heating and Water Heating Boilers

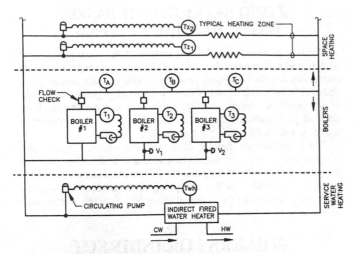

Fig. 26 Typical Modular Boiler for Combined Space and Water Heating

boiler water should not pass through an idle boiler. Figure 26 shows a typical modular boiler combination space and water heating arrangement.

Typical Control Sequence

1. Any control zone or indirectly fired water heater thermostat starts its circulating pump and supplies power to boiler No. 1 control circuit.
2. If T1 is not satisfied, burner is turned on, boiler cycles as long as any circulating pump is on.
3. If after 5 min TA is not satisfied, V1 opens and boiler No. 2 comes on line.
4. If after 5 min TB is not satisfied, V2 opens and boiler No. 3 comes on line.
5. If TC is satisfied and two boilers or fewer are firing for a minimum of 10 min, V2 closes.
6. If TB is satisfied and only one boiler is firing for a minimum of 10 min, V1 closes.
7. If all circulating pumps are off, boiler No. 1 shuts down.

The ASHRAE/IES *Standard* 90 series discusses combination service water heating/space heating boilers and establishes restrictions on their use. The ASHRAE/IES *Standard* 100 series section on Service Water Heating also has information on this subject.

CODES AND STANDARDS

ANSI/AGA Z21.10.1-1993	Gas Water Heaters, Volume I: Storage Water Heaters with Input Ratings of 75,000 Btu per Hour or Less
ANSI/AGA Z21.10.3-1993	Gas Water Heaters, Volume III: Storage, with Input Ratings Above 75,000 Btu per Hour, Circulating, and Instantaneous Water Heaters
ANSI Z21.56-1991	Gas-Fired Pool Heaters
ANSI Z21.22-1986	Relief Valve and Automatic Gas Shutoff Devices for Hot Water Supply Systems
ASHRAE *Standard* 90 series and 100 series	
ASHRAE 118.1-1993	Methods of Testing for Rating Commercial Gas, Electric, and Oil Water Heaters
ASHRAE 118.2-1993	Methods of Testing for Rating Residential Water Heaters

ASME Boiler and Pressure Vessel Code	Section IV-92: Rules for Construction of Heating Boilers
	Section VIII D1-92: Rules for Construction of Pressure Vessels
HUD-FHA 4900.1	Minimum Property Standards for One- and Two-Family Living Units
NFPA 54-92 (ANSI Z223.1-1992)	National Fuel Gas Code
NFPA 31-92	Installation of Oil-Burning Equipment
NSF 26-80	Pot, Pan, and Utensil Washers
UL 174-1989	Household Electric Storage Tank Water Heaters
UL 732-1988	Oil-Fired Unit Heaters
UL 1261-1992	Electric Water Heaters for Pools and Tubs
UL 1453-1988	Electric Booster and Commercial Storage Tank Water Heaters

REFERENCES

AGA. Sizing and equipment data for specifying swimming pool heaters. Catalog No. R-00995. American Gas Association, Cleveland, OH.

Becker, B.R., W.H. Thrasher, and D.W. DeWerth. 1991. Comparison of collected and compiled existing data on service hot water use patterns in residential and commercial establishments. *ASHRAE Transactions* 97(2): 231-39.

Ciesielki, C.A. et al. 1984. Role of stagnation and obstruction of water flow in isolation of *Legionella pneumophila* from hospital plumbing. *Applied and Environmental Microbiology* (November):984-87.

DOE. 1990. Final rule regarding test procedures and energy conservation standards for water heaters. 10 CFR Part 430, Federal Register 55(201). U.S. Department of Energy.

DOE. 1993. Uniform test method for measuring the energy consumption of water heaters. 10 CFR Part 430, Subpart B, Appendix E. U.S. Department of Energy.

Dunn, T.Z., R.N. Spear, B.E. Twigg, and D. Williams. 1959. Water heating for commercial kitchens. *Air Conditioning, Heating and Ventilating* (May):70. Also published as a bulletin, *Enough hot water—hot enough.* American Gas Association (1959).

EPRI. 1990. *Commercial heat pump water heaters applications handbook.* CU-6666. Electric Power Research Institute, Palo Alto, CA.

EPRI. 1992. *Commercial water heating applications handbook.* TR-100212. Electric Power Research Institute, Palo Alto, CA.

EPRI. 1994. High-efficiency electric technology fact sheet: Commercial heat pump water heaters. BR-103415. Electric Power Research Institute, Palo Alto, CA.

GAMA. 1993. Consumers' directory of certified efficiency rating for residential heating and water heating equipment. Gas Appliance Manufacturers Association, Arlington, VA.

Goldner, F.S. 1994. Energy use and domestic hot water consumption: Final Report—Phase 1. Report No. 94-19. New York State Energy Research and Development Authority, Albany, NY.

Goldner, F.S. 1994. DHW system sizing criteria for multifamily buildings. *ASHRAE Transactions* 100(1):963-77.

Manian, V.S. and W. Chackeris. 1974. Off peak domestic hot water systems for large apartment buildings. *ASHRAE Transactions* 80(1):147-65.

NSF. 1992. Water heaters, hot water supply boilers, and heat recovery equipment. *Standard* 5-92. National Sanitation Foundation, Ann Arbor, MI.

Perlman, M. and B. Mills. 1985. Development of residential hot water use patterns. *ASHRAE Transactions* 91(2A):657-79.

Talbert, S.G., G.H. Stickford, D.C. Newman, and W.N. Stiegelmeyer. 1986. The effect of hard water scale buildup and water treatment on residential water heater performance. *ASHRAE Transactions* 92(2B):433-47.

Toaborek, J. et al. 1972. Fouling—The major unresolved problem in heat transfer. *Chemical Engineering Progress* (February):59.

Thrasher, W.H. and D.W. DeWerth. 1994. New hot-water use data for five commercial buildings (RP-600). *ASHRAE Transactions* 100(1):935-47.

Werden, R.G. and L.G. Spielvogel. 1969. Sizing of service water heating equipment in commercial and institutional buildings, Part 1. *ASHRAE Transactions* 75:81.

BIBLIOGRAPHY

AGA. Comprehensive on commercial and industrial water heating. Catalog No. R-00980. American Gas Association, Cleveland, OH.

AGA. 1962. Water heating application in coin operated laundries. Catalog No. C-10540. American Gas Association, Cleveland, OH.

AGA. 1965. *Gas engineers handbook.* American Gas Association, Cleveland, OH.

Brooks, F.A. Use of solar energy for heating water. Smithsonian Institution, Washington, D.C.

Carpenter, S.C. and J.P. Kokko. 1988. Estimating hot water use in existing commercial buildings. *ASHRAE Transactions* 94(2):3-12

Coleman, J.J. 1974. Waste water heat reclamation. *ASHRAE Transactions* 80(2):370.

EPRI. 1992. WATSIM 1.0: Detailed water heating simulation model user's manual. TR-101702. Electric Power Research Institute, Palo Alto, CA.

EPRI. 1993. HOTCALC 2.0: Commercial water heating performance simulation tool, Version 2.0. SW-100210-R1. Electric Power Research Institute, Palo Alto, CA.

Hebrank, E.F. 1956. Investigation of the performance of automatic storage-type gas and electric domestic water heaters. *Engineering Experiment Bulletin* No. 436. University of Illinois.

Jones, P.G. 1982. The consumption of hot water in commercial building. *Building Services Engineering, Research and Technology* 3:95-109.

Olivares, T.C. 1987. Hot water system design for multi-residential buildings. Report No. 87-239-K. Ontario Hydro Research Division.

Schultz, W.W. and V.W. Goldschmidt. 1978. Effect of distribution lines on stand-by loss of service water heater. *ASHRAE Transactions* 84(1): 256-65.

Smith, F.T. 1965. Sizing guide for gas water heaters for in-ground swimming pools. Catalog No. R-00999. American Gas Association, Cleveland, OH.

Vine, E., R. Diamond, and R. Szydlowski. 1987. Domestic hot water consumption in four low income apartment buildings. *Energy* 12(6).

Wetherington, T.I., Jr. 1975. Heat recovery water heating. *Building Systems Design* (December/January).

SNOW MELTING

THE practicality of melting snow with heat has been demonstrated in a large number of installations, including sidewalks, roadways, ramps, and runways. Melting eliminates the need for snow removal, provides greater safety for pedestrians and vehicles, and reduces the labor of slush removal.

There are three types of snow-melting systems:

1. Hot fluid circulated in embedded pipes (hydronic)
2. Embedded electric heating resistance cable or wire (electric)
3. Overhead high intensity infrared radiant heating (infrared)

In this chapter, embedded and infrared systems are both covered in the section on Electric System Design.

Components of the system design include (1) heating requirement, (2) pavement design, (3) control, and (4) hydronic or electric system design.

HEATING REQUIREMENT (HYDRONIC AND ELECTRIC)

The heating requirement for snow melting depends on four atmospheric factors: (1) rate of snowfall, (2) air temperature, (3) relative humidity, and (4) wind velocity. The effects of these factors can be evaluated by considering the fate of snow falling on a warmed surface.

The first flakes fall on a dry, warm surface, where they are warmed to 0°C and melted. The water from the melted snow forms a film over the entire area and starts to evaporate. This evaporation is a mass and heat transfer from the surface to the atmosphere. There is also heat transfer from the film to the ambient air and surfaces.

Sensible Heat and Heat of Fusion. The rate of snowfall determines the heat required to warm the snow to 0°C and to melt it.

Mass Transfer by Evaporation. The evaporation rate of the melted snow from the snow-melting slab is affected by the wind speed and by the difference in vapor pressure between the air and the melted snow. The air vapor pressure, however, is fixed by the relative humidity and temperature of the air. Thus, if the slab surface temperature is fixed, the evaporation rate varies with changes in air temperature, relative humidity, and wind speed.

Heat Transfer by Convection and Radiation. The combined convection and radiation loss from a wetted surface, such as the film of melted snow on the slab, to the air depends on the film coefficient and the difference in temperature between the surface and the air. The coefficient is a function of wind speed alone. Because the surface temperature is fixed, convection and radiation losses vary with changes in air temperature and wind speed.

Free Area Ratio

Before equations giving quantitative values for the effects of the four climatic factors can be derived, the insulating effect of the

unmelted snow must be considered. While the flakes are being warmed and before they are completely melted, they act as tiny blankets or insulators. The effect of this insulation can be large. As the snowflakes cover a fraction of the surface area, it is convenient to think of the insulating effect in terms of an area ratio. The *free area ratio* (A_r) is the ratio of the uncovered, or free, area to the total area:

$$A_r = A_f / A_t \qquad (1)$$

where

A_r = free area ratio
A_f = free area, m^2
A_t = total area, m^2

Therefore,

$$0 \le A_r \le 1$$

To maintain $A_r = 1$, the system must melt the snow so rapidly that accumulation is zero. This is impossible, but for purposes of design, it is permissible to assume $A_r = 1$ as a maximum. For $A_r = 0$, the surface must be completely covered with snow to a depth sufficient to prevent evaporation and heat transfer losses. Research on the insulating effects of snow indicates that in practice free area ratio can be adequately expressed by one of three values—0, 0.5, and 1.

Heating Equations

Chapman (1952) derives and explains equations for the heating requirement of a snow-melting system. Chapman and Katunich (1956) derive the general equation for the required slab heat output q_o:

$$q_o = q_s + q_m + A_r(q_e + q_h) \qquad (2)$$

where

q_s = sensible heat transferred to snow, W/m^2
q_m = heat of fusion, W/m^2
A_r = ratio of snow-free area to total area, dimensionless
q_e = heat of evaporation, W/m^2
q_h = heat transfer by convection and radiation, W/m^2

The sensible heat q_s to bring the snow to 0°C is

$$q_s = s c_p \rho (0 - t_a) / c_1 \qquad (3)$$

where

s = rate of snowfall, mm of water equivalent per hour
ρ = density of water equivalent of snow, 998 kg/m^3
c_p = specific heat of snow, 2090 $J/(kg \cdot K)$
t_a = air temperature, °C
c_1 = 1000 mm/m $\times$ 3600 s/h = 3.6×10^6

$$q_s = -0.58 s t_a \qquad (3a)$$

The preparation of this chapter is assigned to TC 6.1, Hydronic and Steam Equipment and Systems.

Table 1 Data for Determining Operating Characteristics of Snow-Melting Systems

	Period of No Snowfall				Period of Snowfall[b]												
	% of Hours with No Snowfall[a]		Mean Air Temp. During Freezing Period,[d] °C	Wind Speed During Freezing Period,[d] km/h	Hours of Snowfall[a,b]		Free Area Ratio, A_r	Required Output,[c] W/m²									Maximum Output,[c] W/m²
City	Air Temp. >0°C	Air Temp. ≤0°C			%	per Year		0 to 156	157 to 313	314 to 471	472 to 629	630 to 787	788 to 944	945 to 1102	1103 to 1259	1260 up	
								Frequency Distribution of Snowfall Hours at Above Outputs, %[e]									
Albuquerque, NM	74.7	24.7	−3.3	13.7	0.6	22	1	62.0	25.4	7.6	4.2	—	0.8	—	—	—	817
							0	94.1	5.9	—	—	—	—	—	—	—	259
Amarillo, TX	73.1	26.0	−4.1	21.4	0.9	33	1	33.7	35.4	15.4	10.7	3.0	1.8	—	—	—	820
							0	88.1	19.1	1.8	—	—	—	—	—	—	451
Boston, MA	64.6	31.4	−4.1	22.9	4.0	145	1	51.5	30.0	12.3	4.3	1.2	0.6	0.1	—	—	1009
							0	83.2	14.0	2.0	0.3	0.3	0.1	—	0.2	—	1167[f]
Buffalo-Niagara Falls, NY	46.5	46.9	−4.5	17.4	6.6	240	1	50.7	32.6	11.2	3.7	1.4	0.2	0.2	—	—	975
							0	95.9	3.4	0.2	0.5	—	—	—	—	—	606
Burlington, VT	39.0	54.5	−6.9	17.4	6.5	236	1	53.7	29.9	13.2	2.5	0.6	0.1	—	—	—	883
							0	91.8	7.6	0.6	—	—	—	—	—	—	448
Caribou-Limestone, ME	21.4	70.6	−8.6	16.1	8.0	290	1	35.0	39.7	16.0	5.7	2.0	1.0	0.5	0.1	—	1192
							0	92.0	7.5	0.5	—	—	—	—	—	—	435
Cheyenne, WY	46.4	49.8	−5.9	24.6	3.8	138	1	16.5	26.2	19.4	13.1	8.6	4.7	4.2	4.7	2.6	1574
							0	94.3	5.4	0.3	—	—	—	—	—	—	498
Chicago, IL	45.4	50.9	−5.9	18.5	3.7	134	1	45.8	37.4	11.4	3.1	1.4	0.6	0.2	0.1	—	1161
							0	91.5	8.1	0.3	0.1	—	—	—	—	—	520
Colorado Springs, CO	54.3	43.6	−5.5	18.5	2.1	76	1	26.8	36.3	19.0	7.5	4.4	5.5	0.5	—	—	981
							0	98.4	1.6	—	—	—	—	—	—	—	199
Columbus, OH	59.0	38.1	−4.2	16.1	2.9	105	1	65.8	22.4	8.0	1.7	1.7	0.4	—	—	—	823
							0	97.7	2.3	—	—	—	—	—	—	—	227
Detroit, MI	47.0	49.3	−4.4	17.1	3.7	134	1	60.4	27.7	9.3	1.5	0.8	0.3	—	—	—	877
							0	95.9	3.5	0.6	—	—	—	—	—	—	442
Duluth, MN	12.6	80.5	−9.8	19.3	6.9	250	1	23.7	32.9	20.6	13.7	4.3	2.5	1.7	0.6	—	1205
							0	94.8	4.7	—	0.3	0.2	—	—	—	—	650
Flamouth, MA	68.5	29.5	−3.6	20.6	2.0	73	1	50.0	33.9	14.2	1.6	0.3	—	—	—	—	643
							0	91.5	7.4	1.1	—	—	—	—	—	—	454
Great Falls, MT	49.0	46.2	−8.6	23.2	4.8	174	1	26.2	27.6	16.7	16.4	7.5	4.6	0.3	0.5	0.2	1422
							0	94.6	4.8	0.6	—	—	—	—	—	—	435
Hartford, CT	56.4	38.9	−4.3	13.2	4.7	171	1	48.4	34.6	11.2	4.3	0.8	0.7	—	0.1	—	1249
							0	80.4	16.7	2.2	0.5	—	0.1	—	0.1	—	1208
Lincoln, NB	45.0	52.5	−6.3	16.3	2.5	91	1	32.7	26.2	20.0	13.9	5.7	1.5	—	—	—	924
							0	97.2	2.6	—	—	0.2	—	—	—	—	637
Memphis, TN	87.2	12.5	−2.8	18.5	0.3	11	1	48.4	28.3	6.7	13.3	3.3	—	—	—	—	716
							0	85.0	8.3	6.7	—	—	—	—	—	—	454
Minneapolis-St. Paul, MN	23.6	70.8	−8.4	17.9	5.6	203	1	28.4	31.4	21.7	14.1	3.5	0.6	0.3	—	—	987
							0	96.5	3.1	0.3	0.1	—	—	—	—	—	489
Mt. Home, ID	56.3	42.6	−4.0	15.3	1.1	40	1	74.2	21.9	3.9	—	—	—	—	—	—	451
							0	98.1	1.9	—	—	—	—	—	—	—	284
New York, NY	55.7	42.2	−4.4	19.0	2.1	76	1	53.1	31.8	9.4	2.2	1.5	1.7	—	0.3	—	1214
							0	87.6	9.6	1.5	0.7	0.3	0.3	—	—	—	940
Ogden, UT	50.0	45.6	−4.3	15.1	4.4	160	1	64.6	29.2	5.8	0.3	0.1	—	—	—	—	681
							0	88.8	9.4	1.4	0.3	0.1	—	—	—	—	681[f]
Oklahoma City, OK	79.0	19.8	−4.1	25.4	1.2	44	1	27.8	18.7	17.0	12.6	14.3	5.9	2.7	1.0	—	1243
							0	95.7	4.3	—	—	—	—	—	—	—	255
Philadelphia, PA	75.8	22.6	−3.0	15.6	1.6	58	1	62.3	23.6	10.4	2.3	0.9	0.5	—	—	—	934
							0	84.3	14.0	1.1	0.2	0.4	—	—	—	—	722
Pittsburgh, PA	55.2	39.8	−4.3	18.7	5.0	182	1	53.6	30.8	8.4	4.6	1.9	0.7	—	—	—	889
							0	93.3	5.9	0.7	0.1	—	—	—	—	—	495
Portland, OR	92.9	6.1	−1.8	13.5	1.0	36	1	78.0	16.9	5.1	—	—	—	—	—	—	394
							0	91.5	8.5	—	—	—	—	—	—	—	306
Rapid City, SD	45.2	51.6	−7.1	20.8	3.2	116	1	29.7	29.0	16.0	8.4	6.3	3.6	1.9	2.0	3.1	1832
							0	97.6	2.2	0.2	—	—	—	—	—	—	322
Reno, NV	56.0	41.6	−4.3	9.0	2.4	87	1	82.6	15.4	1.8	0.2	—	—	—	—	—	479
							0	90.2	8.0	1.6	0.2	—	—	—	—	—	486[f]
St. Louis, MO	68.7	30.4	−3.9	18.5	0.9	33	1	42.9	31.4	16.7	7.1	1.9	—	—	—	—	710
							0	85.2	11.6	2.6	0.6	—	—	—	—	—	479
Salina, KS	60.0	38.5	−4.9	17.5	1.5	54	1	44.9	31.9	12.7	7.6	2.2	0.7	—	—	—	902
							0	93.5	6.2	0.3	—	—	—	—	—	—	378[f]
Sault Ste. Marie, MI	21.3	69.2	−7.5	15.1	9.5	345	1	45.7	32.8	14.3	5.7	1.4	0.1	—	—	—	826
							0	97.9	2.0	0.1	—	—	—	—	—	—	454
Seattle-Tacoma, WA	88.0	10.8	−2.0	9.5	1.2	44	1	86.3	12.3	1.4	—	—	—	—	—	—	432
							0	91.0	8.1	0.9	—	—	—	—	—	—	404
Spokane, WA	48.5	46.1	−3.5	17.2	5.4	196	1	62.6	28.7	7.4	1.1	2.0	—	—	—	—	647
							0	92.0	7.8	0.2	—	—	—	—	—	—	401
Washington, D.C.	77.9	21.2	−2.9	15.5	0.9	33	1	59.0	29.8	10.6	0.6	—	—	—	—	—	486
							0	85.7	11.8	2.5	—	—	—	—	—	—	382

Source: Air Conditioning, Heating and Ventilating, August, 1957, p. 87.

Note: The period covered by this table is from November 1 to March 31, with February taken as a 28.25-day month. Total hours in period = 3630.

[a]The percentages in Columns 2, 3, and 6 total 100%. Note that Hours of Snowfall per Year does not include idling time and is not actual operating time. See text.

[b]Snowfalls of trace amounts are not included; hence, only those hours of 0.25 mm or more water equivalent snowfall per hour are listed.

[c]Output does not include allowance for back or edge losses because these depend on slab construction.

[d]Freezing Period is that during No Snowfall when the air temperature is 0°C or below.

[e]Percentages total 100% of the number of hours of snowfall.

[f]When heat output for $A_r = 0$ equals or exceeds the heat output for $A_r = 1$, the heat transfer q_h is from the air *to* the slab. This occurs when snowfall is at temperatures above 0°C.

The heat of fusion q_m to melt the snow is

$$q_m = s h_f \rho / c_1 \tag{4}$$

where h_f = enthalpy of fusion for water, 334 kJ/kg

$$q_m = 92.6s \tag{4a}$$

The heat of evaporation q_e (mass transfer) is

$$q_e = h_{fg}(0.005V + 0.022)(0.625 - p_{av}) \tag{5}$$

where

h_{fg} = heat of evaporation at the film temperature, kJ/kg
V = wind speed, km/h
p_{av} = vapor pressure of moist air, kPa

The heat transfer q_h (convection and radiation) is

$$q_h = 64.74(0.0125V + 0.055)(t_f - t_a) \tag{6}$$

where t_f = water film temperature, °C, usually taken as 0.5°C.

The designer can use Equations (2) through (6) to determine the heating requirement of a snow-melting system. The solution to Equation (2), however, requires the simultaneous consideration of all four climatic factors (wind speed, air temperature, relative humidity, and rate of snowfall). Annual averages or maximums for the climatic factors should not be used because there is no assurance that they will ever occur simultaneously. It is necessary, therefore, to perform a frequency analysis of the solutions to Equation (2) for all the occurrences of snowfall over several years.

Weather Data

Table 1 shows analyses of 33 cities and operating information for applicable snow-melting systems. In freezing temperatures (0°C and below) without snowfall, the system may be idling, which means that some heat is supplied to the slab so that melting begins immediately when snow starts to fall. Column 4 of Table 1 gives the mean temperature during freezing periods. This temperature, together with wind speed, is used to calculate the *idling load*. The column headed Hours of Snowfall indicates the number of hours that snow is falling at rates equal to or greater than 0.25 mm of water equivalent per hour. There are snowfalls of trace quantities about twice as often as there are of measurable quantities. These light falls can normally be handled by the idling slab.

The remaining columns of Table 1 represent the frequency distribution of required heat output. This distribution is based on the solution to Equation (2) for two values of the free area ratio A_r and is the basis for Tables 2 and 3.

Performance Classification

Snow-melting installations are classified in Table 2 according to type as Class I, II, or III. Chapman (1957) discusses these classes, which are defined in more detail in the footnotes to Table 2. Snow-melting systems are generally classified, by urgency of melting, as follows:

- **Class I (minimum):** Residential walks or driveways; interplant ways or paths.
- **Class II (moderate):** Commercial sidewalks and driveways; steps of hospitals.
- **Class III (maximum):** Toll plazas of highways and bridges; aprons and loading areas of airports; hospital emergency entrances.

The difference between a Class I system and a Class II system is the required ability of each system to melt snow. For example, the accumulation of 25 mm of snow in an hour during a heavy storm might not be objectionable for a residential system. On the other hand, a store manager might consider the system inadequate if

Table 2 Design Data for Three Classes of Snow-Melting System

| City | Design Output, W/m² | | |
	Class I System[a]	Class II System[b]	Class III System[c]
Albuquerque, NM	224	259	527
Amarillo, TX	309	451	760
Boston, MA	338	729	804
Buffalo-Niagara Falls, NY	252	606	968
Burlington, VT	284	448	770
Caribou-Limestone, ME	281/293	435	968
Cheyenne, WY	262	407	1340
Chicago, IL	281	520	1104
Colorado Springs, CO	154/199	199	924
Columbus, OH	164	227	798
Detroit, MI	218	442	804
Duluth, MN	262/360	650	1180
Falmouth, MA	293	454	520
Great Falls, MT	265/353	435	1173
Hartford, CT	363	801	820
Lincoln, NB	202/211	637	776
Memphis, TN	423	454	669
Minneapolis-St. Paul, MN	199/300	489	807
Mt. Home, ID	158	284	442
New York, NY	382	940	1079
Ogden, UT	309	681	684
Oklahoma City, OK	208	256	1104
Philadelphia, PA	306	722	830
Pittsburgh, PA	281	495	867
Portland, OR	271	306	350
Rapid City, SD	183/271	322	1410
Reno, NV	309	486	489
St. Louis, MO	385	479	624
Salina, KS	268	378	719
Sault Ste. Marie, MI	164/246	454	672
Seattle-Tacoma, WA	290	404	420
Spokane, WA	274	401	596
Washington, DC	369	382	454

Source: Air Conditioning, Heating and Ventilating, August 1957, p. 92.
General note: Output does not include allowance for back or edge losses because these depend on slab construction.
[a]For Class I (residential) systems, the design output is set at the required heat output (see Table 1) at the 98th percentile of the frequency distribution for $A_r = 0$; that is, where 98% of the hours have this output or less. Where idling rate is greater than Class I design rate, idling rate value follows slash and should be used as Class I design output.
[b]For Class II (commercial) systems, the design output is the maximum output for $A_r = 0$ in Table 1 (see last column).
[c]For Class III (industrial) systems, the design output is determined by the following four requirements: (1) output is never exceeded for two consecutive hours; (2) output for $A_r = 1$, Table 1, is at least 3.15 W/m² greater than maximum output for $A_r = 0$; (3) A_r must be greater than or equal to 0.5 with respect to the maximum requirement shown in Table 1 for $A_r = 1$; that is, $q_o = q_s + q_m + 0.5(q_e + q_h)$ for the conditions where $q_s + q_m + q_h + q_e$ are a maximum; and (4) the free area ratio A_r is unity for at least 98% of the hours listed in Table 1.

13 mm of snow accumulated on the sidewalk in front of the store. In a residential system where initial cost must be kept at a minimum, more frequent snow accumulations may have to be accepted.

Table 2 contains the design heat requirements for all three classes of snow-melting systems. Design outputs for Class I and II systems are based on an A_r value of 0. For details and for an explanation of the selection of design output for Class III systems, see the footnotes of Table 2. The designer can alter design rates if a particular job has different design criteria from those given in these footnotes. Any change in the design conditions used should be based on the frequency distribution given in Table 1.

Back and Edge Losses

Back and edge losses, which increase the required slab output found by Equation (2), must also be considered. Adlam (1950) demonstrates that these losses vary from 4 to 50%, depending on factors such as pavement construction, operating temperature, ground temperature, or back exposure. With construction such as that shown in Figure 1 or Figure 3 and ground temperature of 5°C at a depth of 600 mm, back losses are approximately 20%. Higher losses would result from (1) colder ground temperature, (2) more cover over pipe or cable, or (3) exposed back, as on bridges or parking decks.

Heating Requirement Example

Example 1. Snow-melting systems are to be designed for the service areas of a turnpike running from the eastern edge of the Wisconsin-Illinois border northwest to the Wisconsin-Minnesota border just east of St. Paul, Minnesota. It is decided that Chicago, Illinois, data will be adequate for the southern terminus and that Minneapolis-St. Paul data will be adequate for the northern terminus. Determine the heating requirements of the systems for service areas between Chicago and St. Paul.

Solution: For the walkways to the restaurant from the parking area, a Class I design rate could be used. Table 2 gives the design rate at 281 for Chicago and idling rate at 300 for Minneapolis. For points in between, this rate could be estimated at 284 W/m².

The lanes leading from the turnpike to the gasoline pumps and parking areas should be rated as Class II areas. A look at Tables 1 and 2 indicates that 505 W/m² would be adequate.

If an emergency area for a wrecking truck, ambulance, or police garage is included, it would be wise to consider a Class III rating for this area. An inspection of Table 1 shows that for $A_r = 1$, a rate of 867 W/m² would be adequate 99.4% [45.8 + 37.4 + 11.4 + 3.1 + 1.4 + (0.612) = 99.4%] of the time in both Chicago and St. Paul. Therefore, 867 W/m² seems sufficient for the emergency areas. The Class III column in Table 2 lists 1104 W/m² for Chicago and 801 W/m² for St. Paul, but for the uses in this example, 867 W/m² should be adequate.

The above figures should be adjusted upward by 20 to 30% to compensate for back and edge losses.

Operating Cost Example

Example 2. A Class III snow-melting system of 200 m² is installed in Chicago. Table 3 shows that annual output is 100 MJ/m². Assuming a fossil fuel cost of $9/GJ, an electric cost of $0.07/kWh, and back loss at 30%, find the annual operating cost of the system.

Solution: Operating cost O may be expressed as follows:

$$O = A q_a F / [1 - (B/100)]$$

where

O = operating cost, $/yr
A = area, m²
q_a = annual output, J/m²·yr or kWh/m²·yr
F = fuel cost, $/J or $/kWh
B = back loss, %

Operating cost for fossil fuel is

$$O = (200 \times 100 \times 9 \times 10^{-3}) / [1 - (30/100)]$$
$$= \$257/yr$$

Operating cost for electric heat is

$$O = 200 (100/3.6) (0.07) / [1 - (30/100)]$$
$$= \$556/yr$$

PAVEMENT DESIGN (HYDRONIC AND ELECTRIC)

Either concrete or asphalt pavement may be used for snow-melting systems. The thermal conductivity of asphalt is less than that of concrete; pipe or cable spacing and temperatures are thus different. Hot asphalt may be damaging to plastic or electric snow-melting systems unless adequate precautions are taken. For specific recommendations, refer to the sections on Hydronic and Electric System Design.

Concrete slabs containing hydronic or electric snow-melting apparatus must be designed and constructed with a subbase, expansion-contraction joints, reinforcement, and drainage to prevent slab cracking; otherwise, crack-induced shearing or tensile forces would break the pipe or cable. The pipe or cable must not run through expansion-contraction joints, keyed construction joints, or control joints (dummy grooves); however, the pipe or cable may be run under 3-mm score marks (block and other patterns). Control joints must be placed wherever the slab changes size or incline. The maximum distance between control joints for ground-supported slabs should be less than 4.6 m, and the length should be no greater than twice the width, except for ribbon driveways or sidewalks. In ground-supported slabs, most cracking occurs during the early cure. Depending on the amount of water used in a concrete mix, shrinkage during cure may be up to 60 mm per 100 m. If the slab is more than 4.6 m long, the concrete does not have sufficient strength to overcome friction between it and the ground while shrinking during the cure period.

If the slabs are poured in two separate layers, the top layer, which contains the snow-melting apparatus, does not usually contribute toward total slab strength; therefore, the lower layer must be designed to provide the total strength.

The concrete mix of the top layer should give maximum weatherability. The compressive strength should be 28 to 34 MPa; recommended slump is 75 mm maximum, 50 mm minimum. Aggregate size and air content should be as follows:

Maximum Size Crushed Rock Aggregate, mm	Air Content, %
64	5 ± 1
25	6 ± 1
13	7.5 ± 1

Note: Do not use river gravel or slag.

The pipe or cable may be placed in contact with an existing sound pavement (either concrete or asphalt) and then covered as described in the sections on Hydronic and Electric System Design. If there are signs of cracking or heaving, the pavement should be replaced. Pipe or cable should not be placed over existing expansion-contraction, control, or construction joints. The finest grade of asphalt is best for the top course; stone diameter should not exceed 10 mm.

A moisture barrier should be placed between any insulation and the fill. The joints in the barrier should be mopped and the fill made smooth enough to eliminate holes or gaps for moisture transfer. Also, the edges of the barrier should be flashed to the surface of the pavement to seal the ends.

Snow-melting systems should have good surface drainage. When the ambient air temperature is 0°C or below, runoff from melting snow freezes immediately on leaving the heated area. Any water that is able to get under the pavement also freezes when the system is deenergized, causing extreme frost heaving. Runoff should be piped away in drains that are heated or below the frost line.

The area to be protected by the snow-melting system must first be measured and planned. For total snow removal, hydronic or electric heat must cover the entire area. In larger installations, it may be desirable to melt snow and ice from only the most frequently used areas, such as walkways and wheel tracks for trucks and autos. Planning for separate circuits should be considered so that areas within the system can be heated individually, as required.

Where snow-melting apparatus must be run around obstacles (e.g., a storm sewer grate), the pipe or cable spacing should be uniformly reduced. Because some drifting will occur adjacent to walls or vertical surfaces, extra heat should be added in these areas, if possible also in the vertical surface. Drainage flowing through the area expected to be drifted tends to wash away some snow.

CONTROL
(HYDRONIC AND ELECTRIC)

Snow-melting systems can be controlled either manually or automatically.

Manual Control

Manual operation is strictly by two-position control; an operator must activate and deactivate the system when snow falls. If the system is not turned off after snowfall, operating cost is increased.

Automatic Control

If the snow-melting system is not turned on until snow starts falling, it may not melt snow effectively for several hours, giving additional snowfall a chance to accumulate and increasing the time needed to melt the area. Automatic controls provide satisfactory operation because they turn on the system when light snow starts, allowing adequate warm-up before heavy snowfall develops. Operating costs are reduced with automatic turn-off.

Snow Detectors. Snow detectors monitor precipitation and temperature. They allow operation only when snow is present and may incorporate a delay-off timer. Snow detectors located in the heated area activate the snow-melting system when precipitation (snow) occurs at a temperature below the preset pavement temperature (usually 4°C). Another type of snow detector is mounted above ground, adjacent to the heated area, without cutting into the existing system; however, it does not detect tracked or drifting snow. Both types of sensors should be located so that they are not affected by overhangs, trees, blown snow, or other local conditions.

Pavement Temperature Sensor. To limit energy waste during normal and light snow conditions, it is common to include a remote temperature sensor installed midway between two pipes or cables in the pavement; the setting is adjusted between 5 and 15°C. Thus, during mild weather snow conditions, the system is automatically modulated or cycled on and off to keep the pavement temperature at the sensor at set point.

Outdoor Thermostat. The control system may include an outdoor thermostat that turns the system off when the outdoor ambient temperature rises above 2 to 5°C as automatic protection against accidental operation in summer.

Control Selection

For optimum operating convenience and minimum operating cost, all of the aforementioned controls should be incorporated in the snow-melting system.

Operating Cost

To evaluate operating cost during idling or melting, use the annual output data from Table 3. Idling and melting data is based on pavement surface temperature control at 0°C, which requires a pavement temperature sensor. Without the pavement temperature sensors, operating costs will be substantially higher.

HYDRONIC SYSTEM DESIGN

Hydronic system design includes selection of the following components: (1) heat transfer fluid, (2) pipe system, (3) fluid heater, (4) pump(s) to circulate the fluid, and (5) controls. In concrete pavement, thermal stress is also a design consideration.

Heat Transfer Fluid

There are a variety of fluids suitable for transferring the heat from the fluid heater to the pavement, including brine, oils, and glycol-water. Freeze protection is essential because most systems will not be operated continuously in subfreezing weather. Without freeze protection, power loss or pump failure could cause freeze damage to the pipe system and pavement.

Brine is the least costly heat transfer fluid, but it has a lower specific heat than glycol. The use of brine may be discouraged because of the cost of heating equipment that can resist its corrosive potential.

While heat transfer oils are not corrosive, they are more expensive than and have a lower specific heat and higher viscosity than brine or glycol. Petroleum distillates used as fluids in snow-melting systems are classified as nonflammable but have fire points between 150 and 180°C. When oils are used as heat transfer fluids, any oil dripping from the seals on the pump should be collected. It is good practice to place a barrier between the oil lines and the boiler so that a flashback from the boiler cannot ignite a possible oil leak. Other nonflammable fluids, such as those used in some transformers, can be used as antifreeze.

Glycols (ethylene and propylene) are used the most often in snow-melting systems because of their moderate cost, high specific heat, and low viscosity; ease of corrosion control is another advantage. Automotive glycols containing silicates are not recommended because they can cause fouling, pump seal wear, fluid gelation, and reduced heat transfer. The pipe system should be designed for periodic addition of inhibitor. Glycols should be tested annually to determine any change in reserve alkalinity and freeze protection. Only inhibitors obtained from the manufacturer of the glycol should be added. Heat exchanger surfaces should be kept below 140°C, which corresponds to about 280 kPa (gage) steam. Temperatures above 150°C accelerate the deterioration of the inhibitors.

Because ethylene glycol and petroleum distillates are toxic, no permanent connection should be installed between the snow-melting system and the drinking water supply. Gordon (1950) discusses precautions concerning internal corrosion, flammability, toxicity, cleaning, joints, and hook-up that should be taken during the installation of hydronic pipe systems. The properties of brine and glycol are discussed in Chapter 18 of the 1993 *ASHRAE Handbook—Fundamentals*. The effect of glycol on system performance is detailed in Chapter 12 of the 1992 *ASHRAE Handbook—Systems and Equipment*.

Pipe System

Pipe systems may be metal or plastic. Steel, iron, and copper pipes have long been used. Steel and iron may corrode rapidly if the pipe is not shielded by a coating and/or cathodic protection. The use of salts for deicing and the elevated temperature accelerate corrosion of metallic components. NACE (1978) states that the corrosion rate roughly doubles for each 10 K rise in temperature.

Chapman (1952) derived the equation for the fluid temperature required to provide an output q_o. For construction similar to that shown in Figure 1, the equation is

$$t_m = 0.088q_o + t_f \tag{7}$$

where t_m = mean fluid (antifreeze solution) temperature, °C. Equation (7) applies to 25-mm as well as 20-mm IPS pipe (Figure 1).

For specific conditions or for cities other than those given in Tables 1 and 2, Equations (2) and (7) are used. Table 4 gives solutions to these equations at a relative humidity of 80%. Equations (8) and (9), which are derived from Equations (2), (5), and (7), can be used to determine the corrections for other values of relative humidity.

$$dq_o/dp_{av} = -A_r(0.005V + 0.022) h_{fg} \tag{8}$$

$$dt_m/dp_{av} = -(A_r/2)(0.005V + 0.022) h_{fg} \tag{9}$$

Appropriate values for the climatic variables must be determined before Equation (2) can be solved. The best procedure is to contact

Table 3   Yearly Operating Data

City	Idling			Melting			
	Time,[a] h/yr	Rate,[b] W/m²	Annual Output,[c] kWh/m²	Time,[d] h/yr	Annual Output,[e,f] kWh/m²		
					Class I	Class II	Class III
Albuquerque, NM	897	102.5	91.9	22	2.9	3.1	3.6
Amarillo, TX	944	161.2	152.2	33	6.8	7.9	8.7
Boston, MA	1140	164.3	187.3	145	25.2	28.6	28.7
Buffalo-Niagara Falls, NY	1702	158.3	269.4	240	37.2	46.0	47.0
Burlington, VT	1978	242.5	479.7	236	37.2	42.6	43.5
Caribou-Limestone, ME	2563	293.3	751.7	290	56.1[g]	65.0	71.0
Cheyenne, WY	1808	245.1	443.1	138	30.7	41.6	63.7
Chicago, IL	1848	213.8	395.1	134	22.7	26.5	27.4
Colorado Springs, CO	1583	200.0	316.6	76	12.5[g]	12.5	23.3
Columbus, OH	1383	141.9	196.2	105	11.3	13.2	16.9
Detroit, MI	1790	154.5	276.6	134	17.5	21.6	22.3
Duluth, MN	2922	358.9	1048.7	250	104.7[g]	120.2	124.6
Falmouth, MA	1071	139.4	149.3	73	12.1	13.4	13.5
Great Falls, MT	1677	352.0	590.3	174	43.2[g]	80.1	60.2
Hartford, CT	1412	132.2	186.7	171	31.0	34.1	34.1
Lincoln, NB	1906	211.9	403.9	91	15.0[g]	26.3	26.9
Memphis, TN	454	100.9	45.8	11	2.2	2.3	2.5
Minneapolis-St. Paul, MN	2570	299.9	770.7	203	44.8[g]	55.5	58.0
Mt. Home, ID	1546	132.2	204.4	40	4.0	4.8	5.0
New York, NY	1532	159.9	245.0	76	13.2	14.8	14.9
Ogden, UT	1655	140.7	232.9	160	22.2	23.2	23.2
Oklahoma City, OK	719	177.3	127.5	44	7.5	8.8	17.0
Philadelphia, PA	820	99.0	81.2	58	8.5	9.8	9.8
Pittsburgh, PA	1445	156.1	225.6	182	28.5	33.7	35.0
Portland, OR	221	54.9	12.1	36	4.1	4.2	4.3
Rapid City, SD	1873	272.5	510.4	116	23.5[g]	26.0	42.3
Reno, NV	1510	116.4	175.8	87	9.4	9.6	9.6
St. Louis, MO	1104	141.3	156.0	33	6.9	7.2	7.5
Salina, KS	1398	170.0	237.7	54	9.2	10.6	12.0
Sault Ste. Marie, MI	2512	245.1	615.7	345	55.5[g]	70.0	73.2
Seattle-Tacoma, WA	392	54.2	21.2	44	4.4	4.5	4.5
Spokane, WA	1673	123.3	206.3	196	27.3	29.5	30.2
Washington, DC	770	96.5	74.3	33	5.2	5.2	5.3

Source: Air Conditioning, Heating and Ventilating, August 1957, p. 94.
[a]Column 3, Table 1 × 3630 winter hours per year.
[b]Rate when idling, W/m² = $(0.96V + 18.7)(0 - t)$, where V = wind speed from Column 5, Table 1, and t = air temperature from Column 4, Table 1.
[c]Product of the preceding columns.
[d]Hours of Snowfall, Column 7, Table 1.

[e]See footnotes for Table 2 for explanation of classes.
[f]Based on the condition that surface temperature is maintained at 0.5°C until required output exceeds designed output, at which time design output is used regardless of required output. Distribution of required output based on Table 1.
[g]Based on idling rate rather than design rate.

Table 4   Heat Output and Fluid Temperature for Snow-Melting System

s, Rate of Snowfall, mm/h	A_r		$t_a = -18°C$ Wind Speed V, km/h			$t_a = -12°C$ Wind Speed V, km/h			$t_a = -6.5°C$ Wind Speed V, km/h			$t_a = -1°C$ Wind Speed V, km/h		
			8	16	24	8	16	24	8	16	24	8	16	24
2.0	1.0	q_o	476	647	820	401	530	659	322	404	486	237	265	296
		t_m	42.2	57.2	72.2	36.1	47.2	58.9	29.4	36.1	43.3	21.1	23.9	26.1
	0.0	q_o	208	208	208	202	202	202	196	196	196	189	189	189
		t_m	18.9	18.9	18.9	18.3	18.3	18.3	17.8	17.8	17.8	17.2	17.2	17.2
4.2	1.0	q_o	686	861	1031	609	735	864	520	602	684	426	454	486
		t_m	61.1	76.1	92.2	53.4	65.0	76.6	47.2	53.9	61.1	37.8	40.5	42.8
	0.0	q_o	419	419	419	407	407	407	394	394	394	382	382	382
		t_m	37.2	37.2	37.2	36.1	36.1	36.1	35.0	35.0	35.0	33.9	33.9	33.9
6.4	1.0	q_o	921	1094	1265	836	962	1091	741	823	905	640	669	697
		t_m	81.7	96.7	112.2	73.9	85.6	96.7	66.1	72.8	80.0	56.1	59.4	62.2
	0.0	q_o	656	656	656	637	637	637	615	615	615	593	593	593
		t_m	58.3	58.3	58.3	56.7	56.7	56.7	55.0	55.0	55.0	52.8	52.8	52.8

Note: Table developed from Equations (2) and (7). Relative humidity is assumed to be 80% at all air temperatures.

A_r = free area ratio
q_o = slab output, W/m²

t_a = air temperature, °C
t_m = fluid temperature, °C. Based on construction as shown in Figure 1.

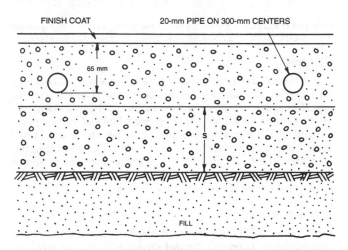

F = depth of finish coat—assumed as 12 mm of concrete. Finish coat may be asphalt, but then cover slab should be reduced from 75 mm. Depth of slab should always keep thermal resistance equal to 75 mm of concrete.

S = depth required by structural design (should be a minimum of 50 mm of concrete).

Fig. 1   Detail of Hot Fluid Panel for Melting Snow

the local National Weather Service office and examine the Local Climatological Summary. An estimate of required slab output q_o can be taken from Table 4, using the appropriate value for snowfall rate s from Table 5 and the following parameters for the different system performance classes.

1. When designing a Class I system, use

   $A_r = 1.0$, $t_a = -1°C$, and $V = 24$ km/h

2. When designing a Class II system, use

   $A_r = 1.0$, $t_a = -6.5°C$, and $V = 24$ km/h

3. When designing a Class III system, use

   $A_r = 1.0$, $t_a = -18°C$, and $V = 24$ km/h

It is satisfactory to use 20-mm pipe or tube on 300-mm centers as a standard coil. If pumping loads require reduced friction, the pipe size can be increased to 25 mm, but the pavement depth must be increased accordingly.

The piping should be supported by a minimum of 50 mm of concrete above and below. This requires a 130-mm pavement for 20-mm pipe and 140-mm for 25-mm pipe.

Plastic pipe (polyethylene, polybutylene) is popular due to its lower material cost, lower installation cost, and corrosion resistance. Considerations when using plastic pipe include stress crack resistance, temperature limitations, and thermal conductivity. Heat transfer oils should not be used with plastic pipe. Polyethylene and polybutylene pipe is furnished in coils. The smaller sized pipe can be bent to form a variety of heating panel designs without elbows or joints. Mechanical compression connections can be used to connect the heating panel pipe to the larger supply and return piping leading to the pump and fluid heater. Plastic pipe may be fused using the appropriate fittings and fusion equipment. Fusion joining eliminates metallic components and thus the possibility of corrosion in the pipe system; it does, however, require considerable installation training.

When plastic pipe is used, the system must be designed so that the fluid temperature required will not damage the pipe. If a design requires a temperature above the tolerance of plastic pipe, the heat output will never meet design requirements. A logical solution is to decrease the pipe spacing. Adlam (1950) addresses the parameter of pipe size and the effect on heat output of pipe spacing. A typical solution is summarized in Table 6, which shows a way of designing

Table 5   Snowfall Data for Various Cities

City	Number of Readings at Various Snowfall Rates[a] Snowfall Rate in Equivalent Millimetres of Water per 6 h 0.0 to 6.3	6.4 to 12.6	12.7 to 19	19.1 to 25.1	Total Readings Taken[b]	Assumed Design Rate of Snowfall,[c] mm/h
Albany, NY	2052	29	5	1	3720	4.2
Asheville, NC	463	5	1	0	3536	2.0
Billings, MT	1640	4	0	0	3532	2.0
Bismarck, ND	2838	0	0	0	3720	2.0
Cincinnati, OH	1045	3	0	0	3720	2.0
Cleveland, OH	1569	2	0	0	3720	2.0
Evansville, IN	916	5	1	1	3720	2.0
Kansas City, MO	1189	12	2	1	3720	4.2
Madison, WI	2370	5	2	0	3720	2.0
Portland, ME	2054	33	4	1	3720	4.2

Note: Data from U.S. Weather Bureau. Based on readings taken at 1:30 a.m., 7:30 a.m., 1:30 p.m., and 7:30 p.m. daily from November 15 to February 15, 1940 to 1949.
[a]Readings are of 6-h periods with a maximum temperature below freezing.
[b]Where the total readings are less than 3720, the period of record is less than 10 years. The difference between Column 6 and the sum of readings in Columns 2, 3, 4, and 5 is the number of readings with a maximum temperature (in the 6-h period) above freezing.
[c]The design rate is found as follows: Proceed to left (on line from any city) from Column 5 until the column containing the tenth reading is found. Assume that the larger value in the heading of the selected column is an *average maximum* value, and should be multiplied by 2 to obtain the maximum rate for a 6-h period. This maximum rate divided by 6 is the design rate per hour. This is equivalent to dividing the larger value in the heading of the selected column by 3.

For example, for Albany, Columns 5 and 4 total six readings, and consequently the tenth reading is in Column 3, which has the larger value of 12.6 in the column heading. Dividing 12.6 by 3, the design water equivalent of 4.2 mm/h is found, as listed in Column 7.

Table 6   Plastic Pipe Specifications
(Mean fluid temperature = 55°C)

Class	Heat Input,[a] W/m^2	Snow-Melting Rate,[b] mm/h	Minimum Air Temp.,[c] °C	Pipe Circuit Spacing,[d] mm
I	630	25	-12	300
II	790	40	-15	230
III	950	65	-18	150
Steps	1260	25	-12	100

[a]Heat input per unit area. Includes 30% back loss. Increase to 50% for bridges and structures with exposed back.
[b]Snow-melting rate with -6.5°C air, 16 km/h wind.
[c]Minimum air temperature at which surface can be maintained above 0°C with 16 km/h wind, no snow.
[d]Space pipes 25 mm closer for each 25 mm of concrete cover over 50 mm. Space pipes 50 mm closer for each 25 mm of brick paver and mortar.

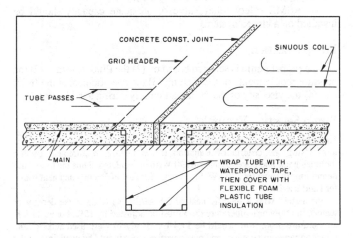

Fig. 2   Piping Detail for Concrete Construction Joints

pipe spacing according to heating requirements. This table also shows adjustments for the effect of more than 50 mm of concrete or paver over the pipe.

It is good design practice to avoid passing any embedded piping through a concrete expansion joint; otherwise, the pipe may be stressed and possibly ruptured. A method of protecting piping that must pass through a concrete expansion joint from stress under normal conditions is shown in Figure 2.

Because introduction of air causes deterioration of the antifreeze, the pipe system should not be vented to the atmosphere. It should be divided into smaller zones to facilitate filling and allow isolation when service is necessary.

After pipe installation, but before pavement installation, all piping should be air tested to about 700 kPa (gage). This pressure should be maintained until all welds and connections have been checked for leaks. Testing should not be done with water for the following reasons: (1) small leaks may not be observed during pavement installation; (2) water leaks may damage the concrete during installation; (3) the system may freeze before antifreeze is added; and (4) it is difficult to add antifreeze when the system is filled with water.

Fluid Heater and Air Control

The heat transfer fluid can be heated using any of a variety of energy sources, depending on availability at the location of the snow-melting system. A fluid heater can use steam, hot water, gas, oil, or electricity. In some applications, heat may be available from other secondary sources, such as engine generators, computers, the sun, and condensate.

The design capacity of the fluid heater can be established by evaluating the data in the section on Heating Requirements; it is usually 630 to 950 W/m^2, which includes back and edge losses.

Design of the fluid heater should follow standard practice, with adjustments for the film coefficient. Consideration should be given to flue gas condensation and thermal shock in boilers due to low fluid temperatures. Bypass flow controls may be necessary to maintain recommended boiler temperatures. Boilers should be derated for high-altitude applications.

Air can be eliminated from the piping during the initial filling by pumping the antifreeze from an open container into isolated zones of the pipe system. A properly sized pump and piping system that maintains an adequate fluid velocity, together with an air separator and expansion tank, will keep air from entering the system during operation.

A strainer, sediment trap, or other means for cleaning the piping system may be provided. It should be placed in the return line ahead of the heat exchanger and must be cleaned frequently during the initial system operation to remove scale and sludge. A strainer should be checked and cleaned, if necessary, at the start of and periodically during each snow-melting season.

An ASME safety relief valve of adequate capacity should be installed on a closed system.

Pump Selection

The proper pump is selected based on (1) the fluid flow rate; (2) the pressure requirements of the system; (3) the specific heat of the fluid; and (4) the viscosity of the fluid, particularly during a cold startup.

Pump Selection Example

Example 3. An area of 1000 m^2 is to be designed with a snow-melting system. Using the criteria of Equation (2), a heating requirement (which includes back and edge losses) of 790 W/m^2 is selected. Thus the total fluid heater output must be 790 kW. Size heater for fuel efficiency and heat transfer fluid used.

For a fluid with a specific heat of 3.5 kJ/(kg·K) (0.85 × the density of water), the flow or temperature drop must be adjusted by 15%. For an 11.5 K temperature drop, flow would be 17.4 L/s. If glycol were used at 55°C, the effect on pipe pressure loss and pump performance would be negligible. If the system pressure were determined to be 120 kPa and pump efficiency 60%,

pump power would be 3.5 kW. Centrifugal pump design criteria may be found in Chapter 39 of the 1992 *ASHRAE Handbook—Systems and Equipment.*

Controls

The controls discussed earlier provide convenience and operating economy but are not required for operation. Hydronic systems require fluid temperature control for safety and for component longevity. Pavement stress and the temperature limits of the heat transfer fluid, pipe components, and fluid heater need to be considered. Certain plastic pipe materials should not be subjected to temperatures above 60°C. A secondary fluid temperature sensor should deactivate the snow-melting system if the primary control fails, and possibly activate an alarm.

Thermal Stress

Chapman (1955) discusses the problems of thermal stress in concrete pavement. In general, thermal stresses will cause no problems if the following installation and operation rules are observed.

1. Minimize the temperature difference between the fluid and the slab surface by maintaining (a) close pipe spacing (see Figure 1), (b) a low temperature differential in the fluid (less than 10 K), and (c) continuous operation (if economically feasible).
2. Install pipe within about 50 mm of the surface.
3. Use reinforcing steel designed for thermal stress if high structural loads are expected (such as on highways).

Thermal shock to the pavement may occur if heated fluid is introduced from a large source of residual heat such as a storage tank, a large pipe system, or another snow-melting area. The pavement should be brought up to temperature by maintaining the fluid temperature differential at less than 10 K.

ELECTRIC SYSTEM DESIGN

General

Snow-melting systems using electricity as an energy source have heating elements in the form of (1) mineral insulated (MI) cable, (2) a resistance wire assembly embedded in the paving materials, or (3) high-intensity infrared lamps.

Heat Density

The basic load calculations for electric systems are the same as presented earlier in this chapter. However, because electric system output is determined by the resistance installed and the voltage impressed, it cannot be altered by fluid flow rates or temperatures. Consequently, neither safety factors nor marginal capacity systems are design considerations.

Heat intensity within a slab can be varied by altering the cable or wire spacing to compensate for anticipated drift areas or other high heat loss areas. Watt density should not exceed 1300 W/m^2 (NFPA 1993).

Switchgear and Conduit

Double-pole, single-throw switches or tandem circuit breakers should be used to open both sides of the line. The switchgear may be in any protected, convenient location. It is also advisable to include a pilot lamp on the load side of each switch so that there is a visual indication when the system is energized.

The power supply conduit is run underground, outside the slab, or in a prepared base. With concrete pavement, this conduit should be installed before the reinforcing mesh.

Mineral Insulated Cable

Mineral insulated (MI) heating cable is a magnesium oxide (MgO)-filled, die-drawn cable with one or two copper or copper alloy conductors and a seamless copper or stainless steel alloy sheath. The metal outer sheath is protected from salts and other chemicals by a high-density polyethylene jacket that is important whenever MI cable is embedded in a medium. Although it is heavy-duty cable, MI cable is practical in any snow-melting installation.

Cable Layout. To determine the characteristics of the MI heating cable needed for a specific area, the following must be known:

- Heated area size
- Watt density required
- Voltage(s) available
- Approximate cable length needed

To find the approximate MI cable length, estimate 6 linear metres of cable per square metre of concrete. This corresponds to a 150-mm on-center spacing. Actual cable spacing will vary between 75 and 230 mm to provide the proper watt density.

Cable spacing is dictated primarily by the heat-conducting ability of the material in which the cable is embedded. Concrete has a higher heat transmission coefficient than asphalt, permitting wider cable spacing. The following is a procedure to select the proper MI heating cable:

1. Determine total wattage.

$$W = Aw$$

2. Determine total resistance.

$$R = E^2/W$$

3. Calculate cable resistance per metre.

$$r_1 = R/L_1$$

where

W = total wattage needed, W
A = heated area of each heated slab, m^2

w = desired watt density input, W/m^2
E = voltage available, V
r_1 = calculated cable resistance, Ω per metre of cable
L_1 = estimated cable length, m
R = total resistance of cable, Ω
L = actual cable length needed, m
r = actual cable resistance, Ω/m
S = cable center to center spacing, mm
I = total current per MI cable, A

Commercially available mineral insulated heating cables have actual resistance values (if there are two conductors, the value is the total of the two resistances) ranging from 0.005 to 2 Ω/m. Manufacturing tolerances are ±10% on these values. MI cables are die-drawn, with the internal conductor drawn to size indirectly via pressures transmitted through the mineral insulation. Special cables are not economical unless the quantity needed is 30 000 m or more.

4. From manufacturers' literature, choose a cable with a resistance r closest to the calculated r_1. Table 7 illustrates typical cable resistances.

5. Determine the actual cable length needed to give the wattage desired.

$$L = R/r$$

6. Determine cable spacing within the heated area.

$$S = 1000A/L$$

Cable spacing for optimum performance should be within the following limits: in concrete, 75 mm to 230 mm; in asphalt, 75 mm to 150 mm.

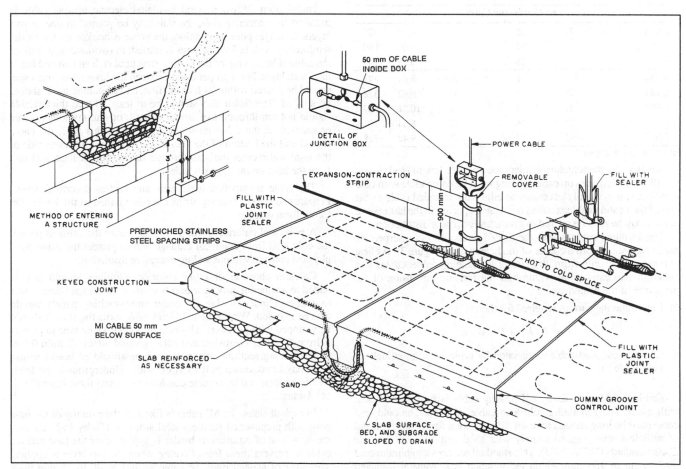

Fig. 3   Typical Mineral Insulated Heating Cable Installation in Concrete Slab

Table 7 Typical Heating Cable Resistances

Ω/m at 25°C	Volt Rating	Conductor			Sheath		
		Material[a]	American Wire Gage		Material[a]	OD, mm	kg/m
Single-Conductor Cable							
2.001	300	B	26		B	3.05	22
0.119			25			3.25	30
0.984			23			3.68	51
0.656			21			3.81	57
0.328			18			4.19	67
0.164			15			4.44	83
0.098			13			5.08	107
0.066			11			5.46	127
0.033			8			6.25	179
0.656	600	B	21		B	4.65	79
0.344			18			5.21	103
0.197			16			5.33	110
0.115			16			5.46	127
0.082			12			6.10	149
0.066			11			6.43	167
0.043			9			7.04	213
0.020			6			6.10	301
0.021	600	B	18		B	5.06	97
0.013			16			5.46	109
0.008			14			5.84	134
0.005			12			6.25	156
0.003			10			7.04	201
Two-Conductor Cable							
2.625	300	B	25		B	4.19	75
1.312			21			4.65	100
0.410			18			6.25	143
0.230	600	B	16		B	8.64	271
0.144		B	14		B	9.42	328
0.092		B	12		B	10.21	387
0.027		C	16		C	8.64	271
0.017		C	14		C	9.42	328

[a]B = Bronze; C = Copper

Table 8 Mineral Insulated Cold Lead Cables (Maximum Voltage—600 V)

Single-Conductor Cable		Two-Conductor Cable	
Current Capacity, A	American Wire Gage	Current Capacity, A	American Wire Gage
30	14	25	14/2
40	12	30	12/2
55	10	40	10/2
70	8	50	8/2
100	6	70	6/2
135	4	90	4/2
155	3		
180	2		
210	1		

Because the manufacturing tolerance on cable length is ±1%, and installation tolerances on cable spacing must be compatible with field conditions, it is usually necessary to adjust the installed cable as the end of the heating cable is rolled out. Cable spacing in the last several passes may have to be altered to give a uniform heat distribution.

The installed cable within the heated areas follows a serpentine path originating from a corner of the heated area (Figure 3). As heat is conducted evenly from all sides of the heating cable, cables in a concrete slab can be run within half the spacing dimension of the perimeter of the heated area.

7. Determine the current required for the cable.

$$I = E/R, \text{ or } I = W/E$$

8. Choose cold lead cable as dictated by typical design guidelines (see Table 8).

Cable Cold Lead. Every MI heating cable is factory fabricated with a non-heat-generating cold lead cable attached. The cold lead cable must be long enough to reach a dry location for termination, and of sufficient wire gage to comply with local and *National Electric Code* standards (NFPA 1993). This standard requires a minimum cold lead length of 150 mm within the junction box. Mineral insulated cable junction boxes must be located such that the box remains dry,

and at least 0.9 m of cold lead cable is available at the end for any future service (Figure 3). Preferred junction box locations are indoors; on the side of a building, utility pole, or wall; or inside a manhole on the wall. Boxes should have a hole in the bottom to drain condensation. Outdoor boxes should be completely watertight except for the condensation drain hole. Where junction boxes are mounted below grade, the cable end seals must be coated with an epoxy to prevent moisture entry. Cable end seals should extend into the junction box far enough to allow the end seal to be removed if necessary.

Although magnesium oxide, the insulation in MI cable, is hygroscopic, the only vulnerable part of the cable is the end seal. However, should moisture penetrate the seal, it can easily be detected with a megohmmeter and driven out by applying a torch 0.6 to 0.9 m from the end and working the flame toward the end.

Installation. When mineral insulated electric heating cable is installed in a concrete slab, the slab may be poured in one or two layers. In single-pour application, the cable is hooked on top of the reinforcing mesh before the pour is started. In two-layer application, the cable is laid on top of the bottom structural slab and embedded in the finish layer. For a proper bond between the layers, the finish slab should be poured within 24 h of the first, and a bonding grout should be applied. The finish slab should be at least 50 mm thick. Cable should not run through expansion, control, or dummy joints (score or groove). If the cable must cross a joint of this type, the cable should exit the bottom of the slab at least 100 mm from one side of the joint and reenter the slab through the bottom at least 100 mm from the joint on the opposite side.

The cable is uncoiled from reels and laid as described above. Prepunched copper spacing strips are often nailed to the lower slab for uniform spacing.

A polyvinyl chloride jacket is extruded over the cable to protect the cold lead from chemical damage and to protect the cable from physical damage without adding excessive insulation.

Calcium chloride or other chloride additives should not be added to a concrete mix in winter because chlorides are destructive to copper. Cinder or slag fill under snow-melting panels should also be avoided. Where the cold lead cable exits the slab, it should be wrapped with polyvinyl chloride or polyethylene tape to protect it from fertilizer corrosion and other ground attack. Within 0.6 m of the heating section, only polyethylene should be used because heat may break down polyvinyl chloride. Underground, the leads should be installed in suitable conduits to protect them from physical damage.

In asphalt slabs, the MI cable is fixed in place on top of the base pour with prepunched stainless steel strips or 150 by 150 mm wire mesh. A coat of bituminous binder is applied over the base and the cable to prevent them from floating when the top layer is applied. The layer of asphalt over the cable should be 40 to 75 mm thick (Figure 4).

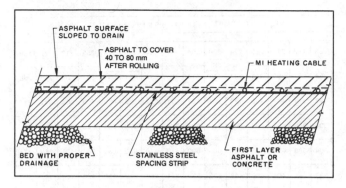

Fig. 4 Typical Section, Mineral Insulated Heating Cable in Asphalt

Testing. Mineral insulated heating cables should be thoroughly tested before, during, and after installation to ensure they have not been damaged either in transit or during installation.

Because of the hygroscopic nature of the magnesium oxide insulation, damage to the cable sheath is easily detectable with a 500-V field megohmmeter. Cable insulation resistance should be measured on arrival of the cable. Cable with insulation resistance of less than 200 MΩ should not be used. Cable with insulation resistance of less than 5 MΩ between the conductor(s) and the metal sheath should not be used. Cable that shows a marked loss of insulation resistance after installation should be investigated for damage. Cable should also be checked for electrical continuity.

Embedded Wire Systems

In an embedded wire system, the resistance elements may consist of a length of copper wire or alloy with a given amount of resistance. When energized, these elements will produce the required amount of heat. Witsken (1965) describes this system in further detail.

Elements are either solid-strand conductors or conductors wrapped in a spiral around a nonconducting fibrous material core. Both types are covered with a layer of insulation such as polyvinyl chloride or silicone rubber.

The heat-generating portion of an element is the conductive core. The resistance is specified in ohms per linear metre of core. Alternately, a manufacturer may specify the wire in terms of watts per metre of core, where the power is a function of the resistance of the core, the applied voltage, and the total length of core.

Considerations in the selection of insulating materials for heating elements are watt density, chemical inertness, application, and end use. Polyvinyl chloride is the least expensive insulation and is widely used because it is inert to oils, hydrocarbons, and alkalies. An outer covering of nylon is often added to increase its physical strength and to protect it from abrasion. The heat output of embedded polyvinyl chloride is limited to 16 W per linear metre. Silicone rubber is not inert to oils or hydrocarbons. It requires an additional covering—metal braid, conduit, or fiberglass braid—for protection. This material can dissipate up to 30 W per linear metre.

Lead can be used to encase resistance elements insulated with glass fiber. The lead sheath is then covered with a vinyl material. Output is limited to approximately 30 W/m by the polyvinyl chloride jacket.

Teflon has good physical and electrical properties and can be used at temperatures up to 260°C.

Low watt density (less than 30 W/m) resistance wires may be attached to plastic or fiber mesh to form a mat unit. Prefabricated factory-assembled mats are available in a variety of watt densities for embedding in specified paving materials to match desired snow-melting capacities. Mats of lengths up to 18 m are available for installation in asphalt sidewalks and driveways.

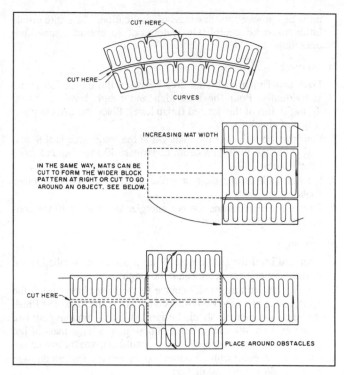

Fig. 5 Shaping Mats Around Curves and Obstacles

Preassembled mats of appropriate widths are also available for stair steps. Mats are seldom made larger than 5.6 m^2, since larger ones are more difficult to install, both mechanically and electrically. With a series of cuts, mats can be tailored to follow contours of curves and fit around objects, as shown in Figure 5. Extreme care should be exercised to prevent damage to the heater wire (or lead) insulation during this operation.

The mats should be installed from 40 to 75 mm below the finished surface of asphalt or concrete. Installing the mats deeper decreases the snow-melting efficiency. Only mats that can withstand hot asphalt compaction should be used for asphalt paving.

Layout. Heating wires should be long enough to fit between the concrete slab dummy groove control or construction joints. Because concrete forms may be inaccurate, 50 to 100 mm should be allowed between the edge of the concrete and the heating wire for clearance. Approximately 100 mm should be allowed between the adjacent heating wires at the control or construction joints.

For asphalt, the longest wire or largest heating mat that can be used on straight runs should be selected. The mats must be placed at least 300 mm in from the pavement edge. Adjacent mats must not overlap. Junction boxes should be located so that each accommodates the maximum number of mats. Wiring must conform to requirements of the *National Electric Code*. It is best to position junction boxes adjacent to or above the slab.

Installation

General

1. Check the wire or mats with an ohmmeter before, during, and after installation.
2. Temporarily lay the mats in position and install conduit feeders and junction boxes. Leave enough slack in the lead wires to permit temporary removal of the mats during the first pour. Carefully ground all leads, using the grounding braids provided.
3. Secure all splices with approved crimped connectors or setscrew clamps. Tape all of the power splices with plastic tape to make them waterproof. All junction boxes, fittings, and snug bushings

must be approved for this class of application. The entire installation must be completely waterproof to ensure trouble-free operation.

In Concrete

1. Pour and finish each slab area between the expansion joints individually. Pour the base slab and rough level to within 40 to 50 mm of the desired finish level. Place the mats in position and check for damage.
2. Pour the top slab over the mats while the rough slab is still wet, and cover the mats to a depth of at least 40 mm, but not more than 50 mm.
3. Do not walk on the mats or strike them with shovels or other tools.
4. Except for brief testing, do not energize the mats until the concrete is completely cured.

In Asphalt

1. Pour and level the base course. If units are to be installed on an existing asphalt surface, clean it thoroughly.
2. Apply a bituminous binder course to the lower base, install the mats, and apply a second binder coating over the mats. The finish topping over the mats should be applied in a continuous pour to a depth of 30 to 40 mm. *Note*: Do not dump a large mass of hot asphalt on the mats because the heat could damage the insulation.
3. Check all circuits with an ohmmeter to be sure that no damage occurred during the installation.
4. Do not energize the system until the asphalt has completely hardened.

Infrared Snow-Melting Systems

While overhead infrared systems can be designed specifically for snow-melting and pavement drying, they are usually installed for the additional features they offer. Infrared systems provide comfort heating, which can be particularly useful at the entrances of plants, office buildings, and hospitals or on loading docks. Infrared lamps can improve the security, safety, and appearance of a facility. These additional benefits may justify the somewhat higher cost of infrared systems.

Infrared fixtures can be installed under entrance canopies, along building facades, and on freestanding poles. Approved equipment is available for recess, surface, and pendant mounting.

Infrared Fixture Layout. The same infrared fixtures used for comfort heating installations (as described in Chapter 15 of the 1992 *ASHRAE Handbook—Systems and Equipment*) can be used for snow-melting systems. The major differences in fixture selection result from the difference in the orientation of the target area. Whereas in comfort applications the *vertical* surfaces of the human body constitute the target of irradiation, in snow-melting applications it is a *horizontal* surface that is targeted. When snow melting is the primary design concern, fixtures with narrow beam patterns confine the radiant energy within the target area for more efficient operation. Asymmetric reflector fixtures, which aim the radiation primarily to one side of the fixture centerline, are often used near the periphery of the target area.

Infrared fixtures usually have a longer energy pattern parallel to the long dimension of the fixture than at right angles to it (Frier 1965). Therefore, fixtures should be mounted in a row parallel to the longest dimension of the area. Where the target area is 2.4 m or more in width, it is best to locate the fixtures in two or more parallel rows. This arrangement also provides better comfort heating because radiation is directed across the target area from both sides at a more favorable incident angle.

Radiation Spill. In theory, the most desirable energy distribution would be uniform throughout the snow-melting target area at a density equal to the design requirement. The design of heating fix-

INTENSITY ON PAVEMENT FROM FOUR INFRARED FIXTURES, W/m²														
158	172	213	255	274	296	304	301	304	296	274	255	213	172	158
				HEATER					HEATER					
255	267	309	341	384	411	414	424	414	411	384	341	309	267	255
277	341	404	460	499	532	562	571	562	532	499	460	404	341	277
304	369	461	503	551	600	630	678	630	600	551	503	461	369	304
304	369	461	503	551	600	630	678	630	600	551	503	461	369	304
277	341	404	460	499	532	562	571	562	532	499	460	404	341	277
255	267	309	341	384	411	414	424	414	411	384	341	309	267	255
				HEATER					HEATER					
158	172	213	255	274	296	304	301	304	296	274	255	213	172	158

Fig. 6 Typical Energy Levels per Unit Area for Infrared Snow-Melting System

ture reflectors determines the percentage of the total fixture radiant output scattered outside of the target area design pattern.

Even the best controlled beam fixtures do not produce a completely sharp cutoff at the beam edges. Therefore, if uniform distribution is maintained for the full width of the area, a considerable amount of radiant energy falls outside the target area. For this reason, infrared snow-melting systems are designed so that the intensity on the pavement begins to decrease near the edge of the area (Frier 1964). This design procedure minimizes stray radiant energy losses.

Figure 6 shows the watts per square metre values obtained in a sample snow-melting problem (Frier 1965). The sample design average is 480 W/m². It is apparent that the incident watt density is above the design average value at center of the target area and below the average at the periphery. Figure 6 shows how the watt density and distribution in the snow-melting area depends on the number, wattage, beam pattern, mounting height, and position relative to the pavement of the heaters (Frier 1964).

With distributions similar to the one in Figure 6, snow begins to collect at the edges of the area as the energy requirements for snow melting approach or exceed system capacity. As the snowfall lessens, the snow at the edges of the area and possibly beyond is the melted if the system continues to operate.

Target Area Watt Density. Theoretical target area wattage densities for snow melting with infrared systems are the same as those for commercial applications of embedded element systems; however, it should be emphasized that theoretical density values are for radiation incident on the pavement surface, not that emitted from the lamps. Merely multiplying the recommended snow-melting wattage density by the pavement area to obtain the total power input for the system does not result in good performance. Experience has shown that multiplying this product by a correction factor of 1.6 gives a more realistic figure for the total required power input. The resulting wattage compensates not only for the radiant inefficiency involved, but also for the radiation falling outside the target area. For small areas, or when the fixture mounting height exceeds 5 m, the multiplying factor can be as large as 2.0; large areas with sides of approximately equal length can have a factor of about 1.4.

The point-by-point method is the best way to calculate the fixture requirements for an installation. This method involves dividing the target area into 1-m squares and adding the radiant energy from each infrared fixture incident on each square (Figure 6). The radiant energy distribution of a given infrared fixture can be obtained from

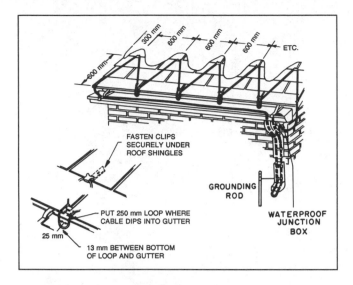

Fig. 7 Typical Insulated Wire Layout to Protect Roof Edge and Downspout

the equipment manufacturer and should be followed for that fixture size and placement.

System Operation. With infrared energy, the target area can be preheated to snow-melting temperatures in 20 to 30 min, unless the air temperature is well below −7°C, or wind velocity is high (Frier 1965). This short warm-up time makes it unnecessary to turn on the system before snow begins to fall. The equipment can be turned on either manually or with a snow detector. A timer is sometimes used to turn the system off 4 to 6 h after the snow stops falling, allowing time for the pavement to be dried completely.

If the snow is allowed to accumulate before the infrared system is turned on, there will be a delay in clearing the pavement, as is the case with embedded systems. Since the infrared energy is absorbed in the top layer of snow rather than by the pavement surface, the length of time needed depends on the snow depth and on atmospheric conditions. Generally, a system that maintains a clear pavement by melting 25 mm of snow an hour as it falls requires 1 h to clear 25 mm of accumulated snow under the same conditions.

To ensure maximum efficiency, fixtures should be cleaned at least once a year, preferably at the beginning of the winter season. Other maintenance requirements are minimal.

Snow Melting in Gutters and Downspouts

Both MI cable and insulated wire are used to prevent heavy snow and ice accumulation on roof overhangs and to prevent ice dams from forming in gutters and downspouts (Lawrie 1966). Figure 7 shows a typical insulated wire layout for protecting a roof edge and downspout. Wire or MI cable for this purpose is generally rated at approximately 20 to 50 W/m, and about 750 mm of wire is installed per linear metre of roof edge. One metre of heated wire per linear metre of gutter or downspout is usually adequate.

If the roof edge or gutters (or both) are heated, downspouts that carry away melted snow and ice must also be heated. A heated length of wire (weighted, if necessary) is dropped inside the downspout to the bottom, even if it is underground.

Lead wires should be spliced or plugged into the main power line in a waterproof junction box, and a ground wire should be installed from the downspout or gutter.

Manual switch control is generally used, although a protective thermostat that senses outdoor temperature should be used to prevent system operation at ambient temperatures above 5°C.

REFERENCES

Adlam, T.N. 1950. *Snow melting*. The Industrial Press, New York; University Microfilms, Ann Arbor, MI.

Chapman, W.P. 1952. Design of snow melting systems. *Heating and Ventilating* (April):95 and (November):88.

Chapman, W.P. 1955. Are thermal stresses a problem in snow melting systems? *Heating, Piping and Air Conditioning* (June):92 and (August):104.

Chapman, W.P. 1957. Calculating the heat requirements of a snow melting system. *Air Conditioning, Heating and Ventilating* (September through August).

Chapman, W.P. and S. Katunich. 1956. Heat requirements of snow melting systems. *ASHAE Transactions* 62:359.

Frier, J.P. 1964. Design requirements for infrared snow melting systems. *Illuminating Engineering* (October):686. Also discussion, December.

Frier, J.P. 1965. Snow melting with infrared lamps. *Plant Engineering* (October):150.

Gordon, P.B. 1950. Antifreeze protection for snow melting systems. *Heating, Piping and Air Conditioning Contractors National Association Official Bulletin* (February):21.

Lawrie, R.J. 1966. Electric snow melting systems. *Electrical Construction and Maintenance* (March):110.

NACE. 1978. Basic corrosion course text (October). National Association of Corrosion Engineers, Houston, TX.

NFPA. 1993. National electrical code. National Fire Protection Association, Quincy, MA.

Witsken, C.H. 1965. Snow melting with electric wire. *Plant Engineering*. (September):129.

BIBLIOGRAPHY

Chapman, W.P. 1955. Snow melting system hydraulics. *Air Conditioning, Heating and Ventilating* (November).

Hydronics Institute. 1994. Snow melting calculation and installation guide. Berkeley Heights, NJ.

Kilkis, i.B. 1994. Design of embedded snow-melting systems: Part 1, Heat requirements—An overall assessment and recommendations. *ASHRAE Transactions* 100(1):423-33.

Kilkis, i.B. 1994. Design of embedded snow-melting systems: Part 2, Heat transfer in the slab—A simplified model. *ASHRAE Transactions* 100(1): 434-41.

EVAPORATIVE AIR COOLING

EVAPORATIVE cooling can be energy-efficient, environmentally benign, and cost-effective if current technology and equipment are used. Applications are found for comfort cooling in commercial and institutional buildings, in addition to the traditional industrial applications for improvement of worker comfort in mills, foundries, power plants, and other hot environments. Several types of apparatus cool by evaporating water directly in the airstream, including (1) evaporative coolers, (2) spray-filled and wetted-surface air washers, (3) sprayed coil units, and (4) humidifiers. Interest has increased in indirect evaporative cooling equipment that combines the evaporative cooling effect in a secondary airstream with heat exchange to produce cooling without adding moisture to the primary airstream.

In the past, indirect evaporative cooling equipment was considered too expensive compared with assembly line refrigerated equipment, but energy conservation, increasing energy costs, concern for indoor air quality, and environmental concerns regarding chlorofluorocarbons have revived interest in both indirect and direct evaporative cooling systems.

Evaporative cooling reduces the dry-bulb temperature and provides a better environment for human occupancy and farm livestock. Evaporative cooling is also used to improve products grown or manufactured by controlling dry-bulb temperatures and/or relative humidity levels.

When temperature or humidity must be controlled within narrow limits, mechanical refrigeration can be combined with evaporative cooling in stages; it can also be used as a backup system. Evaporative cooling equipment, including unitary equipment and air washers, is covered in Chapter 19 of the 1992 *ASHRAE Handbook—Systems and Equipment*.

GENERAL APPLICATIONS

Cooling

Evaporative cooling is commonly used to improve (1) the environment for people, animals, or processes, without attempting to control ambient temperature or humidity (spot cooling); and (2) ambient conditions in a space (area cooling).

In a hot environment, where ambient heat control is difficult or impractical, cooling is accomplished by passing air that is cooler than the skin over the body. Evaporative coolers are suited to this purpose. The performance of evaporative cooling is directly related to climatic conditions. The entering wet-bulb temperature governs the final dry-bulb temperature of the air discharged from a direct evaporative cooler. The capability of the direct evaporative cooler is determined by the amount by which the dry-bulb temperature exceeds the wet-bulb temperature. The performance of indirect evaporative coolers is also limited by the wet-bulb temperature of the secondary airstream.

Indirect applications using room exhaust as secondary air or incorporating precooled air in the secondary airstream may produce leaving dry-bulb temperatures that approach the wet-bulb temperature of the secondary airstream.

The direct evaporative cooling process is an adiabatic exchange of heat. For water to evaporate, heat must be added. The heat is supplied by the air into which water is evaporated. The dry-bulb temperature is lowered, and sensible cooling results. The amount of heat removed from the air equals the amount of heat absorbed by the water evaporated as heat of vaporization. If water is recirculated in the evaporative cooling apparatus, the water temperature in the reservoir will approach the wet-bulb temperature of the air entering the process. By definition, no heat is added to, or extracted from, an adiabatic process. The initial and final conditions of an adiabatic process fall on a line of constant total heat (enthalpy), which nearly coincides with a line of constant wet-bulb temperatures.

The maximum reduction in dry-bulb temperature is the difference between the entering air dry- and wet-bulb temperatures. If the air is cooled to the wet-bulb temperature, it becomes saturated and the process is 100% effective. System effectiveness is the depression of the dry-bulb temperature of the air leaving the apparatus divided by the difference between the dry- and wet-bulb temperatures of the entering air. Evaporative cooling is less than 100% effective, although systems may be 85 to 90% or even more effective.

When a direct evaporative cooling unit cannot provide the desired conditions, several alternatives can satisfy application requirements and still be energy effective and economical to operate. The recirculating water supplying the evaporative cooling unit can be chilled by mechanical refrigeration to provide lower leaving wet-and dry-bulb temperatures and lower humidity. Compared to the cost of using mechanical refrigeration only, this arrangement reduces operating costs by as much as 25 to 40%. Indirect evaporative precooling applied as a first stage, upstream from a second direct evaporative stage, makes it possible to reduce both the entering dry- and wet-bulb temperatures before the air enters the direct evaporative cooling unit. Indirect evaporative cooling systems may save as much as 60 to 75% or more of the total cost of operating a mechanical refrigeration system to produce the same cooling effect. Systems may combine indirect evaporative cooling, direct evaporative cooling, and mechanical refrigeration, or any two of these processes.

The psychrometric chart in Figure 1 illustrates what happens when air is passed through a direct evaporative cooling unit. In the example shown, assume an entering condition of 35°C db and 24°C wb. The initial difference is 35 − 24 = 11 K. If the effectiveness is 80%, the depression is 0.80 × 11 = 8.8 K. The dry-bulb temperature leaving the evaporative cooler is 35 − 8.8 = 26.2°C. In the adiabatic evaporative cooler, only a portion of the water recirculated is assumed to evaporate and the water supply is recirculated. The recirculated water will reach an equilibrium temperature that is approximately the same as the wet-bulb temperature of the entering air.

The preparation of this chapter is assigned to TC 5.7, Evaporative Cooling.

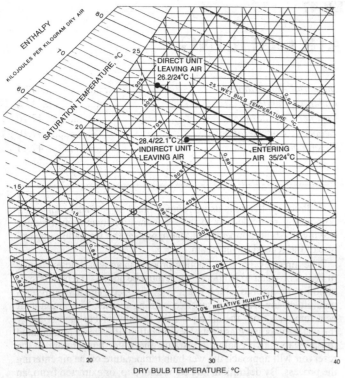

Fig. 1 Psychrometrics of Evaporative Cooling

The performance of an indirect evaporative cooling system can also be shown on a psychrometric chart. Many manufacturers of indirect evaporative cooling equipment use a similar definition of effectiveness as is used for a direct evaporative cooler. The term performance factor (PF) is also used. In indirect evaporative cooling, the cooling process in the primary airstream follows a line of constant moisture content (constant dew point). Performance factor (or effectiveness) is the dry-bulb depression in the primary airstream divided by the difference between the entering dry-bulb temperature of the primary airstream and the entering wet-bulb temperature of the secondary air. Depending on the heat exchanger design and relative air quantities of primary and secondary air, effectiveness ratings may be as high as 85%.

Continuing the example, assuming an effectiveness of 60%, and assuming both primary air and secondary air enter the apparatus at the outdoor condition of 35°C db and 24°C wb, the dry-bulb depression is 0.60 (35 − 24) = 6.6 K. The dry-bulb temperature leaving the indirect evaporative cooling process is 35 − 6.6 = 28.4°C. Because the process cools without adding moisture, the wet-bulb temperature is also reduced. Plotting on the psychrometric chart shows that the final wet-bulb temperature is 22.1°C. Because both the wet- and the dry-bulb temperatures in the indirect evaporative cooling process are reduced, indirect evaporative cooling can be used as a substitute for a portion of the refrigeration load in many applications.

Humidification

Air can be humidified with an evaporative cooler by three methods: (1) using recirculated water without prior treatment of the air, (2) preheating the air and treating it with recirculated water, or (3) using heated water. In any evaporative cooler installation, the air should not enter with a wet-bulb temperature of less than 4°C; otherwise, the water may freeze.

Recirculated Spray Water

Except for both the small amount of outside energy added by the recirculating pump in the form of shaft work and the small amount

of heat leakage into the apparatus from outside (including through the pump and its connecting piping), evaporative cooling is strictly adiabatic. Evaporation occurs from the recirculated liquid. Its temperature should adjust to the thermodynamic wet-bulb temperature of the entering air.

The whole airstream is not brought to complete saturation, but its state point should move along a line of constant thermodynamic wet-bulb temperature. The extent to which the leaving air temperature approaches the thermodynamic wet-bulb temperature of the entering air is expressed by a saturation effectiveness ratio, often called the humidifying effectiveness in humidifiers. The representative saturation, or humidifying effectiveness, of a spray-type air washer with various spray arrangements is listed in Table 1.

The degree of saturation depends on the extent of the contact between air and water. Other conditions being equal, a low-velocity airflow is conducive to higher humidifying effectiveness.

Table 1 Effectiveness of Spray Arrangements in a Spray-Type Air Washer

Bank	Arrangement	Length, m	Effectiveness, %
1	Downstream	1.2	50 to 60
1	Downstream	1.8	60 to 75
1	Upstream	1.8	65 to 80
2	Downstream	2.4 to 3.0	80 to 90
2	Opposing	2.4 to 3.0	85 to 95
2	Upstream	2.4 to 3.0	90 to 98

Preheating Air

Preheating the air increases both the dry- and wet-bulb temperatures and lowers the relative humidity, but it does not alter the humidity ratio (i.e., the mass ratio of water vapor to dry air). At a higher wet-bulb temperature, but with the same humidity ratio, more water can be absorbed per unit mass of dry air in passing through the evaporative cooler (if the humidifying effectiveness of the evaporative cooler is not adversely affected by operation at the higher wet-bulb temperature). The analysis of the process that occurs in the evaporative cooler is the same as that for recirculated water. The final preferred conditions are achieved by adjusting the amount of preheating to give the required wet-bulb temperature at the entrance to the evaporative cooler.

Heated Recirculated Water

Even if heat is added to the recirculated water, the mixing in the evaporative cooler may still be regarded as adiabatic. The state point of the mixture should move toward the specific enthalpy of the heated water. By elevating the water temperature, it is possible to raise the air temperature (both dry and wet bulb) above the dry-bulb temperature of the entering air.

The relative humidity of the leaving air may be controlled by (1) bypassing some of the air around the evaporative cooler and remixing the two airstreams downstream or (2) automatically reducing the number of operating spray nozzles or sections of media wetted by operating valves in the different recycle header branches.

Dehumidification and Cooling

Evaporative coolers are also used to cool and dehumidify air. Heat and moisture removed from the air raise the water temperature. If the entering water temperature is below the entering wet-bulb temperature, both the dry- and wet-bulb temperatures are lowered. Dehumidification results if the leaving water temperature is below the entering dew-point temperature. Moreover, the final water temperature is determined by the sensible and latent heat pickup and the amount of water circulated. However, this final temperature must not exceed the final required dew point, with one or two degrees below dew point being common.

The air leaving an evaporative cooler that is being used as a dehumidifier is substantially saturated. Usually, the spread between dryand wet-bulb temperatures is less than 0.5 K. The spread between leaving air and leaving water depends on the difference between entering dry- and wet-bulb temperatures and on certain design features, such as the length and height of a spray chamber, the cross-sectional area and depth of the media being used, air velocity, quantity of water, and the spray pattern. The rise in water temperature is usually between 3 and 7 K, although higher increases have been used successfully. Lower temperature increases are usually selected when the water is chilled by mechanical refrigeration because of possible higher refrigerant temperatures. It is often desirable to make an economic analysis of the effect of higher refrigerant temperature compared to the benefits of a greater increase in water temperature. For systems receiving water from a well or other source at an acceptable temperature, it may be desirable to design on the basis of a high temperature increase and a minimum water flow.

Air Cleaning

Evaporative coolers of all types perform some air cleaning. Drip-type coolers are the least effective, removing particulates down to about 10 μm in size. Air washers can be highly effective air cleaners, provided they are equipped with high-pressure nozzles and are followed by well-designed, efficient entrainment separators.

The dust removal efficiency of evaporative coolers depends largely on the size, density, wettability, and solubility of the dust particles. Larger, more wettable particles are the easiest to remove. Separation is largely a result of the impingement of particles on the wetted surface of the eliminator plates or on the surface of the media. Because the force of impact increases with the size of the solid, the impact (together with the adhesive quality of the wetted surface) determines the cooler's usefulness as a dust remover. The standard low-pressure spray is relatively ineffective in removing most atmospheric dusts.

Evaporative coolers are of little use in removing soot particles because their greasy surface will not adhere to the wet plates or media. Evaporative coolers are also ineffective in removing smoke, because the small particles (less than 1 μm) do not impinge with sufficient impact to pierce the water film that covers the plates and be held on the media or the wet plates. Instead, the particles follow the air path between the media surfaces or the plates.

Control of Gaseous Contaminants. When used in a makeup air system comprised of a mixture of outside air and recirculated air, evaporative coolers function as scrubbers and are effective in reducing the gaseous contaminants found in urban atmospheres. For more information on controlling gaseous contaminants, see Chapter 41.

INDIRECT EVAPORATIVE PRECOOLING

Outdoor Air Systems

Because there is no increase in absolute humidity in the primary airstream, indirect evaporative cooling is well suited to precooling the air entering a refrigerated coil. The cooling effect provided by the upstream indirect evaporative equipment is a sensible cooling load reduction to the downstream refrigerated coil and compression apparatus. This reduces the size of the required refrigeration system as well as energy and operating costs. By contrast, direct evaporative cooling equipment, when used with a refrigerated coil, will exchange latent heat for sensible heat, increasing the latent load on the coil in proportion to the sensible cooling achieved. The enthalpy of the air entering the coil is not changed.

The power input per kilowatt of cooling effect is substantially lower with indirect evaporative cooling than with conventional refrigerated equipment. Experimental observations by Peterson and Hunn (1991) for Dallas, Texas, show that (1) the seasonal energy efficiency ratio

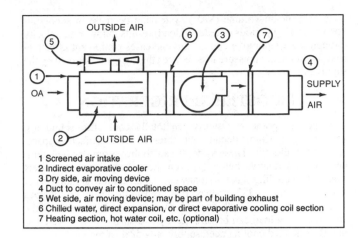

Fig. 2 Indirect Evaporative Cooling Configuration

1 Screened air intake
2 Indirect evaporative cooler
3 Dry side, air moving device
4 Duct to convey air to conditioned space
5 Wet side, air moving device; may be part of building exhaust
6 Chilled water, direct expansion, or direct evaporative cooling coil section
7 Heating section, hot water coil, etc. (optional)

(SEER) of an indirect evaporative cooler can be 70% higher than that of a conventional air conditioner and (2) nearly 12% of the air-conditioning capacity can be displaced by the indirect evaporative cooler. The indirect equipment selected must result in only a minimal addition of static pressure loss in the primary air system. The added static pressure loss increases the primary air fan motor power, and the total effect on the system must be considered, even when continuous cooling is not required. Static pressure loss for a nominal selection may be as low as 50 Pa, which represents a minimal addition to the supply fan power required. Furthermore, the equipment may also require additional energy for water pumping and for moving secondary air across the evaporative surfaces.

The cooling configuration is shown in Figure 2. The primary air side of the indirect unit is positioned at the intake to the refrigerated cooling coil. The secondary air to the unit can come from outdoor ambient air or from room exhaust air. Exhaust air from the space may have a lower wet-bulb temperature than outdoor ambient, depending on climate, time of year, and space latent load. Latent cooling may be possible in the primary airstream using room exhaust air as secondary air. This may occur if the dew-point temperature of the primary air is above the exhaust (secondary air) wet-bulb temperature. If this is possible, provision to drain the water condensed from the primary airstream may be necessary.

In many areas, an indirect precooler can satisfy more than one-half of the annual cooling load. For example, Supple (1982) showed that 30% of the annual cooling load for Chicago can be accomplished by indirect evaporative precooling. Indirect evaporative precooling systems using exhaust air for secondary air may be as effective in warm, humid climates as in drier areas.

Mixed Air Systems

Indirect evaporative cooling can also save energy in systems that use a mixture of return and outside air. The apparatus configuration is similar to that shown in Figure 2, except that a mixing section with outside- and return-air dampers is added upstream of the indirect precooling section. Outdoor air would be used as secondary air for the indirect evaporative cooler.

A typical indirect evaporative precooling stage can reduce the dry-bulb temperature by as much as 60 to 80% of the difference between the entering dry-bulb temperature and the wet-bulb temperature of the secondary air. When the dry-bulb temperature of the mixed air is more than a few degrees above the wet-bulb temperature of the secondary airstream, indirect evaporative precooling of the mixed airstream may reduce the amount of refrigerated cooling required.

The precooling contribution depends on the differential between mixed air dry-bulb temperature and secondary air wet-bulb temperature. As mixture temperature increases, or as secondary air wet-bulb decreases, the precooling contribution becomes more significant.

In variable-air-volume (VAV) systems, a decrease in supply air volume (during periods of reduced load) results in lower air velocity through the evaporative cooler; this increases equipment effectiveness. Lower static pressure loss reduces the energy consumed by the supply fan motor.

BOOSTER REFRIGERATION

Staged evaporative systems can totally cool office buildings, schools, gymnasiums, department stores, restaurants, factory space, and other buildings. These systems can control room dry-bulb temperature and relative humidity, even though one stage is a direct evaporative cooling stage. In many cases, booster refrigeration is not required. Supple (1982) showed that even in higher humidity areas with a 1% mean wet-bulb design temperature of 24°C, 42% of the annual cooling load can be satisfied by two-stage evaporative cooling. Refrigerated cooling need supply only 58% of the load.

Figure 3 shows indirect/direct two-stage system performance for 16 cities in the United States. Performance is based on 60% effectiveness of the indirect stage and 90% for the direct stage. Supply air temperatures (leaving the direct stage) at the 1% design dry-bulb/ mean wet-bulb condition range from 11.5 to 21.5°C. Energy use ranges from 17 to 39%, compared to conventional refrigerated equipment.

Booster mechanical refrigeration allows the designer to provide indoor design comfort conditions regardless of the outdoor wet-bulb temperature without having to size the mechanical refrigeration equipment for the total cooling load. If the indoor design condition becomes questionable, the quantity of moisture introduced into the airstream must be limited in order to control room humidity. Where the upper relative humidity design level is critical, a life-cycle cost analysis would favor a system design composed of an indirect cooling stage and a mechanical refrigeration stage.

RESIDENTIAL OR COMMERCIAL COOLING

In dry climates, evaporative cooling is effective, with lower air velocities than those required in humid climates. This makes it suitable for use in applications where low air velocity is desirable. Packaged evaporative coolers are commonly used for residential and commercial application. Cooler capacity requirements may be determined from standard heat gain calculations (see Chapters 25 and 26 of the 1993 *ASHRAE Handbook—Fundamentals*).

Detailed calculation of heat load, however, is usually not economically justified. Instead, one of several estimates, with proper consideration, will give satisfactory results.

In one method of calculating heat load, the difference between dry-bulb design temperature and coincident wet-bulb temperature multiplied by 10 is equated with the number of seconds needed for each air change. This or any other arbitrary rules for equating cooling capacity with airflow depends on an evaporative cooler effectiveness of 70 to 80%. Obviously, the rule must be modified for unusual conditions such as large unshaded glass areas, uninsulated roof exposure, or high internal heat gain.

Such empirical methods make no attempt to predict air temperature at specific points in the system; they merely establish an air quantity for use in sizing equipment.

Example 1. An indirect evaporative cooling system is to be installed in a 15- by 24.4-m one-story office building with a 3-m ceiling and a flat roof. Outdoor design conditions are assumed to be 35°C dry bulb and 18.3°C wet bulb. The following heat gains are to be used in the cooling system design:

	Gains, kW
All walls, doors, and roof	23.0
Glass area	1.7
Occupants (sensible load)	5.0
Lighting	18.4
Total sensible heat load	48.1
Total latent load (occupants)	6.2
Total heat load	54.3

Find the required air quantity, the temperature and humidity ratio of the air leaving the cooler (entering the office), and the temperature and humidity ratio of the air leaving the office.

Solution: A temperature rise of 5 K in the cooling air is assumed. The airflow rate that must be supplied by the evaporative cooler may be found from the following equation:

$$Q_{ra} = q_s / 1.2 (t_1 - t_s) \tag{1}$$

$$Q_{ra} = 48.1 / (1.2 \times 5) = 8.0 \text{ m}^3/\text{s}$$

City	Outside Air Design db/wb, °C	Indirect/Direct Performance (Supply Air = 1.55 W per L/s)			
		Indirect db/wb, °C	Supply Air db, °C	EER	EUC, %
Los Angeles, CA	34/21	26.2/18.6	19.4	7.7	30
San Francisco, CA	28/18	22.0/15.8	16.2	10.2	23
Seattle, WA	29/20	23.6/18.3	18.8	7.1	33
Albuquerque, NM	36/16	24.0/11.4	12.7	12.9	18
Denver, CO	34/15	22.6/10.3	11.5	13.9	17
Salt Lake City, UT	36/17	24.6/12.7	13.9	12.4	19
Phoenix, AZ	43/22	30.4/18.1	19.3	8.1	29
El Paso, TX	38/18	26.0/13.7	14.9	11.1	21
Santa Rosa, CA	37/20	26.8/16.6	17.6	9.3	25
Spokane, WA	34/18	24.4/14.5	15.5	11.0	21
Boise, ID	36/18	25.2/14.2	15.3	10.6	22
Billings, MT	34/18	24.4/14.5	15.5	11.0	21
Portland, OR	32/20	24.8/17.7	18.4	8.7	27
Sacramento, CA	38/21	27.8/17.9	18.9	8.3	28
Fresno, CA	39/21	28.2/17.6	18.7	8.3	28
Austin, TX	38/23	29.0/20.7	21.5	6.0	39

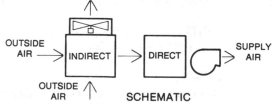

INDIRECT/DIRECT SYSTEM PERFORMANCE

Outdoor air design condition: 1% dry bulb/mean coincident wet bulb.

EER = Energy Efficiency Ratio = watts cooling output per watt of electrical input. Comparison base to conventional system with 15.6°C supply air and 13.9 K temperature drop.

EUC = Energy Use Comparison to a conventional refrigeration system with EER = 2.3.

I/D effectiveness: Indirect = 60% or 0.6 (db − wb); Direct = 90% or 0.9 (db − wb).

Note: Sea level psychrometric chart used. 1500-m elevation will increase supply air temperature 3 to 4%.

Fig. 3   Indirect/Direct Two-Stage System Performance

where

Q_{ra} = required airflow rate, m³/s
q_s = instantaneous sensible heat load, kW
t_1 = indoor air dry-bulb temperature, °C
t_s = room supply air dry-bulb temperature, °C

This air volume represents a 137-s [15 × 24.4 × 3/8.0] air change for a building of this size. The evaporative air cooler is assumed to have a saturation effectiveness of 80%. This is the ratio of the reduction of the dry-bulb temperature to the wet-bulb depression of the entering air. The dry-bulb temperature of the air leaving the evaporative cooler is found from the following equation:

$$t_2 = t_1 - e_h (t_1 - t') / 100 \qquad (2)$$
$$= 35 - 80 (35 - 18.3) / 100 = 21.6°C$$

where

t_2 = dry-bulb temperature of leaving air, °C
t_1 = dry-bulb temperature of entering air, °C
e_h = humidifying or saturating effectiveness, %
t' = thermodynamic wet-bulb temperature of entering air, °C

From the psychrometric chart, the humidity ratio W_2 of the cooler discharge air is 11.85 g/kg dry air. The humidity ratio W_3 of the air leaving the space being cooled is found from the following equation:

$$W_3 = q_e / (3.010 \ Q_{ra}) + W_2 \qquad (3)$$
$$= 6.2 / (3.010 \times 8.0) + 11.85$$
$$= 12.11 \text{ g/kg dry air}$$

where q_e = latent heat load, kW.

The remaining values of wet-bulb temperature and relative humidity for the problem may be found from the psychrometric chart. Figure 4 illustrates the various relationships of outdoor air, supply air to the space, and discharge air.

The wet-bulb depression (WBD) method to estimate airflow gives the following result:

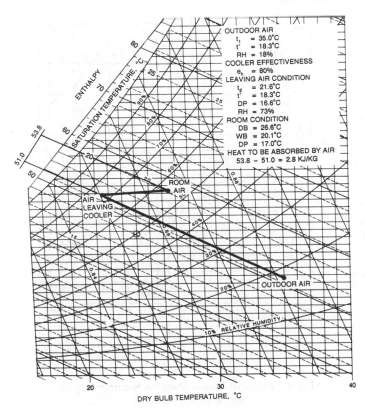

OUTDOOR AIR
t_1 = 35.0°C
t' = 18.3°C
RH = 18%
COOLER EFFECTIVENESS
e_h = 80%
LEAVING AIR CONDITION
t_2 = 21.6°C
t' = 18.3°C
DP = 16.6°C
RH = 73%
ROOM CONDITION
DB = 26.6°C
WB = 20.1°C
DP = 17.0°C
HEAT TO BE ABSORBED BY AIR
53.8 − 51.0 = 2.8 KJ/KG

ROOM AIR
AIR LEAVING COOLER
OUTDOOR AIR

DRY BULB TEMPERATURE, °C

Fig. 4   Solution for Example 1

$$\text{WBD} \times 10 = (35 - 18.3) \times 10 = 167 \text{ s per air change}$$

$$Q_{ra} = \text{volume/air change rate} = 15 \times 24.4 \times 3 / 167 = 6.6 \text{ m}^3/\text{s}$$

While not exactly alike, these two air volume calculations are close enough to select cooler equipment of the same size.

EXHAUST REQUIRED

If air is not exhausted freely, the static pressure buildup reduces airflow through the coolers. The result is a marked increase in the moisture and the heat absorbed per unit mass of air leaving the cooler, and a reduction in the room air velocity. These effects combine to produce almost certain discomfort. A good residential application includes attic exhausters and thermostatic operation.

For normal commercial systems with no internal heat emission, the exhaust should nearly equal the quantity of evaporatively cooled air introduced. Where heat emission occurs, additional exhaust capacity must be added. For most commercial and industrial installations, both power and gravity exhaust should be used. The power exhaust should be controlled so that it increases as the outdoor relative humidity increases. This reduces the buildup of excessive humidity in the space being cooled.

TWO-STAGE COOLING

Two-stage systems for commercial applications can extend the range of atmospheric conditions under which comfort requirements can be met, as well as increase the energy savings. (For the same design conditions, two-stage cooling provides lower cool air temperatures, thereby reducing the required airflow rate.)

High internal load areas are similar to industrial environment conditions and are discussed in the following section.

INDUSTRIAL APPLICATIONS

In a factory with a large internal heat load, it is difficult to approach outdoor conditions during the summer without using an extremely large quantity of outdoor air. Using a more reasonable quantity of outdoor air, evaporative cooling alleviates this heat problem, increasing worker efficiency and improving employee morale. If the heat problem is not alleviated, increased absenteeism, high labor turnover, and danger to health and safety can be expected during the summer months. On hot days, production decreases in uncooled plants may range from 25 to 40%.

An examination of the effective temperature chart reveals that a dry-bulb temperature reduction due to the evaporation of water always results in a lower effective temperature, regardless of the relative humidity level. Figure 5 shows an effective temperature chart for air velocities ranging from 0.1 to 3.5 m/s. Although the maximum velocity shown on the chart is 3.5 m/s, workers exposed to high heat-producing operations often request air movement up to 20 m/s. Because the working range of the chart is approximately midway between the vertical dry- and wet-bulb scales, each of these temperatures is equally responsible for the effective temperature. A reduction in either one will have the effect of decreasing the effective temperature by about one-half of the reduction. This is graphically illustrated by lines ED and CD on the chart.

Conditions of 35°C dry-bulb and 24°C wet-bulb were chosen as the original state, because this set of conditions is usually considered the summer design criteria in most areas. Reducing the temperature 8 K by evaporating water adiabatically provides an effective temperature reduction of 3 K for air moving at 0.1 m/s and a reduction of 5 K for air moving at 3.5 m/s—an improvement of 2 K.

Furthermore, the reduction in dry-bulb temperature through water evaporation increases the effectiveness of the cooling power of moving air in this example by over 100%. On line ED, the effective temperature varies from 28.5°C at 0.1 m/s to 26.5°C at 3.5 m/s,

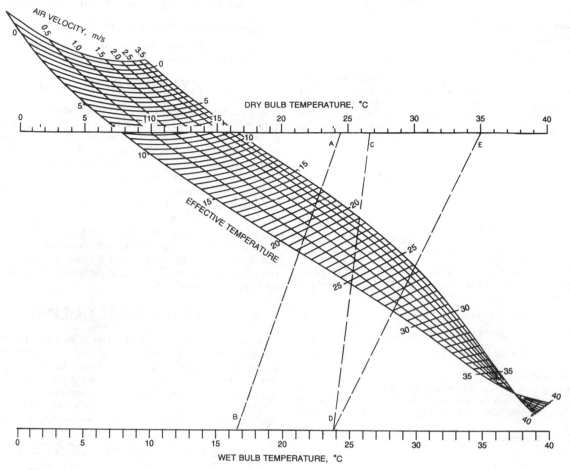

Fig. 5   Effective Temperature Chart

whereas line CD indicates an effective temperature of 25.5°C at 0.1 m/s and 21.5°C at 3.5 m/s. In the former case, increasing the air velocity from 0.1 to 3.5 m/s resulted in only a 2 K decrease in effective temperature. This contrasts with a 4 K decrease in effective temperature for the same range of air movement when the dry-bulb temperature was lowered by water evaporation. These relationships between water evaporative air cooling and effective temperatures can provide relief cooling of factories, almost regardless of geographical location.

Several methods can be used to evaluate the environmental improvement that may be achieved with evaporative coolers. One method is shown in Figure 6, in which temperature is plotted on the vertical axis and time of day on the horizontal. Curve A is plotted from maximum dry-bulb temperature recordings at the hours indicated. Curve B is a plot of the corresponding wet-bulb temperatures.

Curve C depicts the effective temperature when air is moved over a person at 1.5 m/s with all temperature conditions as stated previously. Note its relationship to the suggested maximum effective temperature of 27°C. While the maximum suggested effective temperature is exceeded, both the differential and the total hours are substantially reduced from the conditions in still air.

If the air is passed through an 80% efficient evaporative cooler before being projected over the person at 1.5 m/s, the effective temperature would, in all cases, be below the suggested maximum as shown by Curve D.

Curve E shows the additional decrease in effective temperature when air velocities of 3.5 m/s are provided. Higher velocities lower the effective temperature proportionally.

In spite of the high wet-bulb temperatures, the in-plant environment can be maintained below the suggested upper limit of 27°C

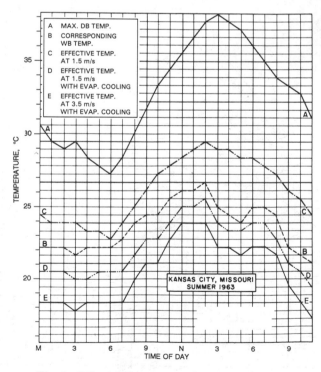

Fig. 6   Effective Temperature for Summer Day in Kansas City, Missouri
(Worst-Case Basis)

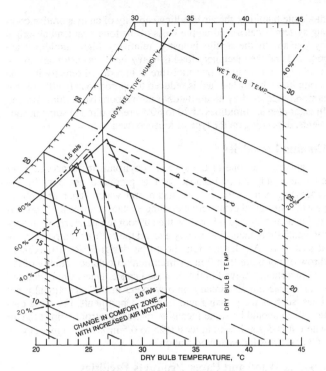

**Fig. 7 Change in Human Comfort Zone as
Air Movement Increases**

effective temperature; this should allow for full production without undue stress on individuals. This assumes that the combination of air velocity, duct length, and insulation between the cooler and the duct outlet is such that there is little heat transfer between air in the ducts and warmer air under the roof.

Another method of determining the effect of using evaporative coolers plots ambient wet- and dry-bulb temperatures on an ASH-RAE Effective Temperature Chart, as in Figure 7 (Crow 1972). The dashed lines show the improvement to expect when using an 80% efficient evaporative cooler. In these examples, a satisfactory effective temperature is provided when the air velocity around the occupant is 3 m/s.

Area Cooling

Evaporative cooling of industrial buildings may be on an area basis or by spot cooling. Evaporative coolers can be either automatically or manually controlled. The air-handling capacity of these systems can also be used to supply tempered fresh air during the fall, winter, and spring. The spent air is exhausted by gravity and/or power roof ventilators. Ducts should be designed to distribute air in the lower 3 m of the space to ensure that air is supplied to the workers.

Because cooling requirements vary from season to season (in fact, from day to day), adjustable discharge grilles should be provided. If the incoming air temperature drops to a point where the cooling airstream becomes an annoying draft, the horizontal blades in the grille may be adjusted so that the air is discharged above the workers' heads and does not blow directly on them. In some cases, provision should be made for air volume control, either at the individual outlet or for the entire system. It may be necessary to vary the exhaust volume accordingly. In a typical arrangement for area cooling, the evaporatively cooled air is brought in and down to the 3-m level, where it is distributed from the duct in opposite directions.

Spot Cooling

Area cooling lends itself well to applications in which personnel move about within a cooled space, but spot cooling yields a more efficient use of the equipment when the personnel are relatively stationary. In these instances, cool air can be brought in at levels even lower than 3 m; in some cases, it can even be introduced from ducts located under the floor. Selection of the height depends on the location of other equipment in the area. For best results, the throw should be kept to a minimum. Spot cooling is especially applicable in hot areas where cooling for individuals is needed, such as in chemical plants and die casting shops, and near glass-forming machines, billet furnaces, and pig and ingot casting. Controls may be automatic or manual, with the fan often operating throughout the year.

The volume of cooling air supplied to each worker depends on the throw and the amount of worker activity and heat to be overcome. Usually, the air supply ranges from 0.9 to 2.4 m³/s per worker with a target velocity of 1 to 20 m/s. If possible, air outlets should be no more than 1.2 to 3 m from the workstations to avoid entrainment of warm air and to effectively blanket workers with cooler air. Provisions should be made so that workers can control the direction of discharge, because air motion that is appropriate for hot weather may be too severe for cool weather or even cool mornings. Volume controls should be considered for some installations to prevent overcooling of the building and to minimize excessive grille-blade adjustment.

These cooling systems are usually installed in rooms that have elevated temperatures; there are no climatic or geographical boundaries. When the dry-bulb temperature of the air is below skin temperature, a body is cooled by convection rather than evaporation. A 27 to 30°C airstream feels so good that its relative humidity is inconsequential.

Cooling Large Motors

The rating of electrical generators and motors is generally based on a maximum ambient temperature of 40°C. If this temperature is exceeded, excessive temperatures develop in the electrical windings unless the load on the motor or generator is reduced. If air is supplied to the windings from an evaporative cooler, this equipment may be safely operated without reducing the load, and in many cases, an overload may be carried. Likewise, transformer capacity can be increased using evaporative cooling.

The heat emitted by high-capacity electrical equipment may also be sufficient to raise the ambient condition to an uncomfortable level. With mill drive motors, an additional problem is often encountered with the commutator. If the air used for motor ventilation is dry, the temperature rise through the motor results in a still lower relative humidity. Destruction of brush film with resultant brush and commutator wear as well as dusting occurs at low humidities. General design considerations require approximately 60 L/s of ventilating air per kilowatt hour of losses with a temperature rise through the motor of 14 K. Assuming 35°C inlet air, the air leaving the motor would be 49°C, which represents an average motor temperature of over 42°C. This temperature is 2 K higher than it should be for the normal 40°C ambient. An evaporative cooling system with a 97% saturation effectiveness, the same quantity of air, and a 24°C wet-bulb area would give an average motor temperature of 31°C. This avoids the necessity of using special high-temperature insulation and improves the ability of the motor to absorb temporary overloads. By comparison, an air quantity of 90 L/s would be required if supplied by a cooler with 80% saturation effectiveness. Figure 8 shows three basic arrangements for motor cooling systems.

The air from the cooler may be directed on the motor windings or into the room, which requires increased air volume to compensate for the building heat load. Operation of these evaporative cooling systems should be keyed to the motor operation to provide the following safeguards: (1) saturated or nearly saturated air should never be introduced into a motor until it has had time to warm up, and (2) if more than one motor is served by a single system, air circulation through idle motors should be prevented.

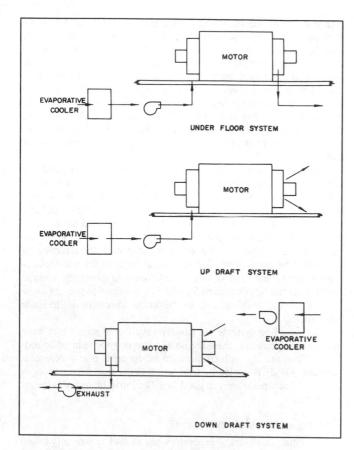

Fig. 8   Arrangements for Cooling Large Motors

Cooling Gas Turbine Engines and Generators

Gas turbines that drive electric generators require large quantities of clean cool air, generally on the order of 36 to 40 kg of air per kilowatt hour. For example, a 5000-kW generator requires about 45 m³/s. Within limits, the temperature of the air supplied directly affects gas turbine output. If air is supplied at 38°C db, the output will be about 10% less than with air at 27°C. Similarly, dropping the air temperature to 16°C results in an increase of 10% over the performance at 27°C. In an installation of this type, the following precautions for evaporative cooling must be taken: (1) the eliminators must be designed to stop the entrainment of all free moisture droplets, (2) a temperature difference of at least 1.5 to 2 K between the dry- and wet-bulb temperatures must be maintained to prevent condensation due to the pressure drop at the compressor inlet, and (3) coolers must be turned off at temperatures below 7°C to prevent icing. In addition, airflow should be considered when designing evaporative coolers for large gas turbines to avoid dry spots in media and/or water carryover.

Evaporative cooling has been used to provide relief from the heat, increased generator capacity, and improved engine performance in primary aluminum reduction plants, where gas engine-generators are used to supply the heavy power requirements. Evaporative coolers with 80% effectiveness and a capacity of 23 m³/s have been provided for each 1200-kW engine-generator set; 4.2 m³/s is supplied to the scavenging air blower of the engine, 8.5 m³/s is drawn through the 1000-kW generator windings, and 10.3 m³/s is supplied to the engine room to provide relief to operators stationed at the control panels.

Process Cooling

In the manufacture of tobacco and textiles and in processes such as spray coating, accurate humidities are required, and comfort cooling can be provided by evaporative cooling. For conditions of moderate humidity, the air-handling capacity of an evaporative cooling system is calculated on the basis of the total heat load absorbed by the air. In the textile industry, relatively high humidities are required and the machinery load is heavy, and it is customary to use a split system whereby free moisture is introduced directly into the room; here, the air handled is reduced to approximately 60% of that normally required by an all-outdoor air evaporative cooling system. In cigar plants, humidities of 70 to 80% are required in some departments, depending on the type of tobacco used.

Cooling Laundries

Laundries are one of the severest environments evaporative air cooling is applied in because heat is produced not only by the processing equipment, but by steam and water vapor as well. A properly designed evaporative cooling system reduces the temperature in a laundry 3 to 6 K below the outdoor temperature. With only fan ventilation, laundries usually exceed the outdoor temperature by at least 6 K. Air distribution should be designed for a maximum throw of not more than 9 m. A minimum circulated velocity of 0.5 to 1 m/s should prevail in the occupied space. Ducts can be located to discharge the air directly onto workers in the exceptionally hot areas, such as the pressing and ironing departments. For these outlets, there should be some means of manual control to direct the air where it is desired, with at least 0.25 to 0.5 m³/s at a target velocity of 3 to 4.5 m/s for each workstation.

Cooling Wood and Paper Products Facilities

Because of gases and particulates present in most paper plant atmospheres, water-cooled systems are preferred over air-cooled systems. The most prevalent contaminants are chlorine gas, caustic soda, and sulfur compounds. With more efficient air cleaning, the air quality in and about most mills is becoming adequate for properly placed air-cooled chillers or condensing units with well-analyzed and properly applied coil and housing coatings. Phosphor-free brazed coil joints are recommended in areas where sulfur compounds are present.

Heat is readily available from the processing operations and should be used whenever possible. Most plants have quality hot water and steam, which can be readily geared to unit heater, central station, or reheat use. Evaporative cooling should not be ignored. Newer plant air-conditioning methods that use energy conservation techniques (such as temperature stratification) lend themselves well to this type of large structure. As in most industrial applications, absorption systems can be considered for pulp and paper plants, because they afford some degree of energy recovery from high-temperature steam processes. Refer to Chapter 22 for further information on air conditioning paper product facilities.

OTHER APPLICATIONS

Cooling Power Generation Facilities

An appropriate air-conditioning system can be selected once the preliminary heating and cooling loads are determined and the criteria are established for temperature, humidity, pressure, and airflow control. The same considerations for system selection apply to power generation facilities and industrial facilities in terms of ventilation—natural or forced heating and adiabatic cooling or refrigeration cooling.

Cooling Mines

Chapter 25 describes various evaporative cooling methods developed for cooling mines.

Cooling Animals

The design criteria for farm animal environments and the need for cooling animal shelters are discussed in Chapter 20. Evaporative cooling is ideally suited to farm animal shelters because all outdoor air is used and, therefore, tight construction is not required. The fresh air also removes odors and reduces the harmful effects of ammonia fumes from urine and excrement. At night and in the spring and fall, the evaporative cooling system can also be used for ventilation.

These systems should be sized to change the air within the shelter in 90 to 120 s, assuming the ceiling height does not exceed 3 m. This flow rate will keep the shelter at 30°C or lower. In a typical milking parlor, packaged air coolers can improve working conditions for the operator, which makes handling the cows easier. Evaporative cooling is also used to cool farrowing and gestation houses to improve production.

In a poultry house, the production efficiency of egg layers and breeders can be enhanced in some areas by using evaporative coolers. Most applications require an air change every 45 to 90 s, with the majority at 60 s. Placement of the fans at the ends or the center of the house with the evaporative cooling system strategically placed, creates a tunnel ventilation system. The fans are generally selected for a total pressure drop of 30 Pa, which means that the evaporative cooling media cannot have a pressure drop in excess of 20 Pa. Thus, to guard against an inadequate volume of air being pulled through the poultry house, the designer must know the pressure drop through the evaporative media being selected.

Experiments with evaporative cooling of broiler chicken houses have shown that, while the feed conversion ratio is not necessarily improved, the average bird weighs more at the conclusion of the growth cycle. Many broiler houses lower the side curtains to decrease natural draft and reduce the temperature within the house. But, when the ambient outside temperature exceeds 38°C, evaporative cooling is often the only way to keep the flock of chickens alive.

Produce Storage Cooling

Potatoes. Evaporative cooling systems for potato storage should pass air directly through the potato pile. The capacity of the cooling system should be based on a 180-s air change to give quick and even cooling. The total volume of the storage should be considered in calculating the size of the system to provide sufficient cooling if the depth of storage is increased at a future time. Above-floor ducts can be used, but underfloor ducts are preferred as they can also be used for handling the potatoes with bin unloaders. Ducts are typically 510 mm wide and at least 360 mm deep to allow room for inserting the bin unloader conveyors. Duct tops must be removable and should be able to support the weight of a loaded truck. Boards should be spaced to provide 15-mm slots. Because distributing and delivery ducts act as extended plenums, it is unnecessary to reduce duct depth as the far end of the ducts is approached. Delivery ducts should extend to within 1.8 m of the walls to provide uniform airflow through the storage. Since friction loss in the duct system is negligible, the resistance of the potato pile is the principal load on the fan in the evaporative cooler. The resistance is approximately 60 Pa of static pressure. The additional resistance of the ducts, inlets, dampers, and evaporative cooler will add another 60 Pa for a total of 120 Pa that the fan must overcome. In large storages, dividing the house into two sections keeps the cooling system to a reasonable size. If main and distributing ducts are to be of reasonable size, a single system should not handle over 5.7 m³/s of air.

Apples. Evaporative cooling systems for apple storage without refrigeration should distribute cool air to all parts of the storage. The evaporative cooler may be floor-mounted or located near the ceiling in a fan room. The system should be designed to discharge air horizontally at the ceiling level. Because the degree of cooling is limited by the prevailing wet-bulb temperature, a system with the maximum reasonable size should be installed to reduce the storage temperature rapidly and as close to the wet-bulb temperature as possible. Generally, a system with a 180-s air change capacity is the largest system that can be installed. This capacity results in a 60- to 90-s air change when the storage is loaded.

For further information on apple storage, see Chapter 15 of the 1994 *ASHRAE Handbook—Refrigeration*.

Citrus. The chief purpose of evaporative cooling as it is applied to fruits and vegetables is to provide an effective, yet inexpensive, means of improving common storage. However, it also serves a special function in the case of oranges, grapefruit, and lemons. Although mature and ready for harvest, citrus fruits are often still green. Color change (degreening) is achieved through a sweating process in special rooms equipped with evaporative cooling. Air with a high relative humidity and a moderate temperature is circulated continuously during the operation. Ethylene gas, the concentration depending on the variety and intensity of green pigment in the rind, is discharged into the rooms. Ethylene destroys the chlorophyll in the rind, allowing the yellow or orange color to become evident. During the degreening operation, a temperature of 21°C and a relative humidity of 88 to 90% are maintained in the sweat room. (In the Gulf States, 28 to 30°C with 90 to 92% rh is recommended.) The evaporative cooling system is designed to deliver 1.9 L/s per field box of fruit (0.11 L/s per kilogram).

Evaporative cooling is also used as a supplement to refrigeration in the storage of citrus fruit. Citrus storage requires refrigeration in the summer, but the required conditions can often be obtained using evaporative cooling during the fall, winter, and spring when the outdoor wet-bulb temperature is low. For further information, see Chapter 16 of the 1994 *ASHRAE Handbook—Refrigeration*.

Cooling Greenhouses

Proper regulation of greenhouse temperatures during the summer is essential for developing high-quality crops. The principal load on a greenhouse is solar radiation, which at sea level at about noon in the temperate zone is approximately 630 W/m². Smoke, dust, or heavy clouds reduce the radiation load. Table 2 gives solar radiation loads for representative cities in the United States. Note that the values cited are average solar heat gains, not peak loads. Temporary rises in temperature inside a greenhouse can be tolerated; an occasional rise above design conditions is not likely to cause damage.

Not all the solar radiation that reaches the inside of the greenhouse becomes a cooling load. About 2% of the total solar radiation is used in photosynthesis. Transpiration of moisture varies from crop to crop, but typically uses about 48% of the solar radiation. This leaves 50% to be removed by the cooling system. Example 2 shows a method for calculating the size of a greenhouse evaporative cooling system.

Example 2. An evaporative cooling system is to be installed in a 15- by 30-m greenhouse. Design conditions are assumed to be 34°C db and 23°C wb, and the average solar radiation is 435 W/m². An indoor temperature of 32°C db must not be exceeded at design conditions.

Solution: The evaporative air cooler is assumed to have a saturation effectiveness of 80%. Equation (2) may be used to determine the dry-bulb temperature of the air leaving the evaporative cooler:

$$t_2 = 34 - 80(34 - 23)/100 = 25.2°C$$

The following equation may be used to calculate the airflow rate that must be supplied by the evaporative cooler:

$$Q_{ra} = A I_t / [2400(t_1 - t_2)] \qquad (4)$$

where

A = greenhouse floor area, m²
I_t = total incident solar radiation, W per m² of receiving surface

For the conditions of this problem, Equation (4) becomes

$$Q_{ra} = (15 \times 30 \times 435)/[2400(32 - 25.2)] = 12.0 \text{ m}^3/\text{s}$$

Horizontal illumination due to direct rays from a noonday summer sun with a clear sky is as much as 100 kilolux (klx); under clear glass, this is approximately 90 klx. Crops such as chrysanthemums and carnations grow best in full sun, but many foliage plants, such as

Table 2 Three-Year Average Solar Radiation for Horizontal Surface During Peak Summer Month

City	W/m²	City	W/m²
Albuquerque, NM	625	Lemont, IL	448
Apalachicola, FL	536	Lexington, KY	536
Astoria, OR	416	Lincoln, NE	473
Atlanta, GA	498	Little Rock, AR	467
Bismarck, ND	442	Los Angeles, CA	511
Blue Hill, MA	404	Madison, WI	435
Boise, ID	489	Medford, OR	536
Boston, MA	394	Miami, FL	483
Brownsville, TX	552	Midland, TX	558
Caribou, ME	363	Nashville, TN	486
Charleston, SC	480	Newport, RI	435
Cleveland, OH	480	New York, NY	442
Columbia, MO	483	Oak Ridge, TN	467
Columbus, OH	401	Oklahoma City, OK	521
Davis, CA	581	Phoenix, AZ	631
Dodge City, KS	581	Portland, ME	420
East Lansing, MI	416	Prosser, WA	555
East Wareham, MA	416	Rapid City, SD	480
El Paso, TX	615	Richland, WA	432
Ely, NV	552	Riverside, CA	555
Fort Worth, TX	555	St. Cloud, MN	416
Fresno, CA	593	San Antonio, TX	555
Gainesville, FL	492	Santa Maria, CA	593
Glasgow, MT	480	Sault Ste. Marie, MI	435
Grandby, CO	470	Sayville, NY	467
Grand Junction, CO	546	Schenectady, NY	369
Great Falls, MT	473	Seabrook, NJ	426
Greensboro, NC	489	Seattle, WA	369
Griffin, GA	517	Spokane, WA	439
Hatteras, NC	558	State College, PA	445
Indianapolis, IN	442	Stillwater, OK	527
Inyokern, CA	688	Tallahassee, FL	423
Ithaca, NY	457	Tampa, FL	527
Lake Charles, LA	505	Upton, NY	467
Lander, WY	558	Washington, D.C.	448
Las Vegas, NV	615		

gloxinias and orchids, do not need more than 16 to 22 klx. If shade is needed, it should be noted that solar radiation is approximately proportional to light intensity. Thus, the greater the amount of shade, the smaller the cooling system capacity required. A value of 1 klx per hour is approximately equivalent to 9 W/m². Although atmospheric conditions, such as clouds and haze, affect the relationship, this is a safe conversion factor to use. This relationship should be used instead of the information given in Table 2 for greenhouses where illumination can be determined by design or by actual observation.

Evaporative cooling systems for greenhouses may be either the supply or extraction type. Regardless of the type of system used, the length of air travel should not exceed 50 m. The rise in temperature of the cool air limits the throw to this value. Air movement must be kept low to prevent mechanical damage to the plants; it should generally not exceed 0.5 m/s in the crop area.

WEATHER CONSIDERATIONS

The effectiveness of evaporative cooling depends on weather conditions. System design is affected by the prevailing outdoor dry-and wet-bulb temperatures as well as by the application. For example, a simple residential direct evaporative cooling system, with an effectiveness of 80%, will provide satisfactory room conditions (given an adequate quantity of outdoor supply air and a well-designed exhaust

system) throughout the cooling season in areas such as Reno, Nevada. Here the 1% design dry-bulb and mean coincident wet-bulb temperatures are 35.5 and 16°C, respectively. In the same location, additional cooling effect can be gained by the addition of an indirect evaporative precooling stage, which lowers the temperature (both dry- and wet-bulb) entering the direct evaporative cooling stage and, consequently, lowers the supply air temperature.

In a geographic location such as Atlanta, Georgia, with design temperatures of 35 and 23.5°C, the same direct evaporative cooler could supply only 25.5°C. This could be reduced to 22.5°C by the addition of an indirect evaporative precooling stage that is 65% effective (Supple 1982).

In a busy restaurant in Atlanta, a well-designed direct evaporative cooling system, including adequate powered exhaust and an air velocity of 0.5 to 1.5 m/s, has provided comfort. This type of approach uses effective temperature and carefully designed exhaust and supply systems to provide a comfort level that is generally thought unattainable in a high wet-bulb design area. Long-term benefits to the owners include 20 to 40% of the utility costs compared to mechanical refrigeration (Watt 1988).

ECONOMIC CONSIDERATIONS

Design of evaporative cooling systems and sizing of equipment should be based on the load requirements of the application and on the local dry- and wet-bulb design conditions, which may be found in Chapter 24 of the 1993 *ASHRAE Handbook—Fundamentals*.

Total energy use for a specific application during a set period may be forecast by using annual weather data. Dry-bulb and mean coincident wet-bulb temperatures, with the hours of occurrence, can be summarized and used in a modified bin procedure. The calculations must reflect the hours of occupancy. Results may vary for different types of buildings, even though the same weather station data may be used. When comparing systems, the cost analysis should include the annual energy reduction at the applicable electrical rate, plus the anticipated energy cost escalation over the expected life of the system.

Reducing air-conditioning kilowatt demand is even more important in areas with ratcheted demand rates (Scofield and Deschamps 1980). In these areas, the demand rate is not set monthly. A summer month with a heavy cooling peak energy demand can set the demand rate for the entire year.

Many areas have time-of-day electrical metering as an incentive to use energy during off-peak hours when rates are lowest. Thermal storage using ice banks or chilled water storage vessels may be used as part of a multistage evaporative-refrigerated cooling system to combine the energy-saving advantages of evaporative cooling; off-peak savings are obtained using thermal storage. In addition to the cost savings due to off-peak energy rates, thermal storage systems may actually save energy because the refrigeration equipment operates at a time when the ambient temperature and resulting condensing temperatures are lower (Eskra 1980).

Direct System Energy Savings

Residences in arid or semiarid regions can use direct or indirect evaporative coolers. Direct, open systems are usually sized to meet the full sensible load using once-through airflows (no recirculation of return air). In such cases, the monthly or seasonal energy savings, compared to conventional vapor compression cooling, can be estimated using the bin method.

Indirect System Energy Savings

Commercial building sensible cooling loads can be met by direct or indirect evaporative cooling. Indirect cooling, or a combination of direct and indirect cooling, is customarily used with the central fan system because of humidity control requirements in commercial buildings. With indirect evaporative cooling, the wet-side (secondary) air can consist of full outside air, full return air, or any combination of the two. The monthly or seasonal energy savings resulting from, or the cooling produced by, an indirect evaporative system can be calculated using the bin method.

PSYCHROMETRICS

Figure 9 shows the two-stage (indirect/direct) process applied to nine western cities in the United States. The entering conditions to the first-stage indirect unit are at the recommended 1% design dry- and wet-bulb temperatures from Chapter 24 of the 1993 *ASHRAE Handbook—Fundamentals*. The effectiveness ratings are 60% for the first (indirect) stage and 90% for the second (direct) stage. The leaving air temperatures range from 11 to 21°C and the relative humidities are 80% and higher. Figure 10 projects these second-stage supply temperatures (based on a 95% room sensible heat factor, i.e., room sensible heat/room total heat) to a room condition at

25.5°C db. Given a 1% entering condition in the cities shown, room conditions are maintained within the comfort zone without a refrigerated third stage. However, Figures 9 and 10 indicate the need to consider the following factors:

1. As the room sensible heat factor increases, the required supply air temperature decreases to maintain a given room condition.

2. As the supply air temperature increases, the supply air quantity increases, resulting in higher air-side system initial cost and increased supply air fan power.

3. A decrease in the required room dry-bulb temperature causes an increase in the supply air quantity. At a given room sensible heat factor, a decrease in room dry-bulb temperature may cause the relative humidity to exceed the comfort zone.

4. The suggested 1% entering air (dry-bulb/mean wet-bulb) conditions to the system are only one concern. Partial load conditions

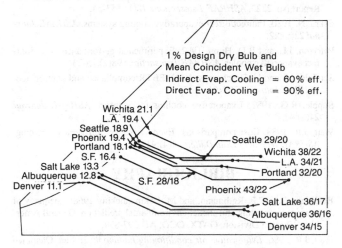

Fig. 9   Two-Stage Evaporative Cooling at 1% Design Condition in Various Cities

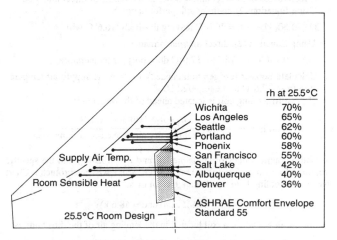

Fig. 10   Final Room Design Conditions after Two-Stage Evaporative Cooling

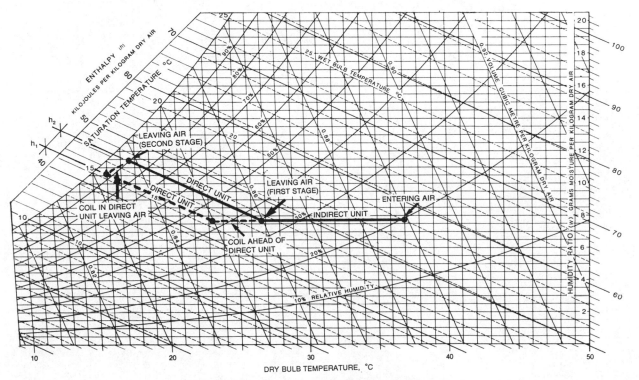

Fig. 11   Three-Stage Evaporative Cooling for Example 3

must also be considered, along with the effect (extent and duration) of spike wet-bulb temperatures. Mean wet-bulb temperatures can be used to determine energy use of the indirect/direct system accurately. However, the higher wet-bulb temperature spikes should be considered to determine their effect on room temperatures.

An ideal condition for the maximum use with minimum energy consumption of a two- and three-stage indirect/direct system is a room sensible heat factor of 90% and higher, a supply air temperature of 16°C, and a dry-bulb room temperature of 26°C. In many cases, three-stage refrigeration is required to ensure satisfactory dry-bulb temperature and relative humidity. Example 3 shows a method for determining the refrigeration capacity for three-stage cooling. The calculations are illustrated in Figure 11.

Example 3. Assume the following:

1. Supply air quantity = 11.3 m/s; supply air temperature = 16°C
2. Design condition = 37°C db and 20°C wb
3. Effectiveness of indirect unit = 60%; effectiveness of direct unit = 90%
4. Using Equation (2), indirect unit performance is

   $37 - 0.60(37 - 20) = 26.8°C$ leaving dry-bulb (16.6°C wb)

   Using Equation (2), direct unit performance is

   $26.8 - 0.90(26.8 - 16.6) = 17.6°C$ db supply air temperature

5. Calculate booster refrigeration capacity to drop the supply air temperature from 17.6°C to the required 16°C.

   If the refrigerating coil is located ahead of the direct unit,

$$\text{kW cooling} = \frac{(h_1 - h_2) \text{ (supply air, L/s)}}{\text{specific volume dry air at leaving air condition}}$$

With numeric values of enthalpies h_1 and h_2 (in kJ/kg) and the specific volume of air (in m³/kg dry air) taken from ASHRAE Psychrometric Chart No. 1, the cooling load is calculated as follows:

$$(46.5 - 42.8)11.3/0.860 = 48.6 \text{ kW}$$

The cooling load for a coil located in the leaving air of the direct unit is

$$(46.6 - 42.9)11.3/0.838 = 49.9 \text{ kW}$$

Depending on the location of the booster coil, the above calculations can be used to determine third-stage refrigeration capacity and to select a cooling coil.

Using this example, refrigeration sizing can be compared to a conventional refrigerated system without staged evaporative cooling. Assuming mixed air conditions to the coil of 27°C db and 19.1°C wb, and the same 16°C db supply air as shown in Figure 13, the refrigerated capacity is

$$(54.1 - 42.9)11.3/0.833 = 152 \text{ kW}$$

This represents an increase of 103 kW. The staged evaporative effect reduces the required refrigeration by 68%.

REFERENCES

ASHRAE. 1992. Thermal environmental conditions for human occupancy. *Standard* 55-1992.

Crow, L.W. 1972. Weather data related to evaporative cooling. Research Report No. 2223. *ASHRAE Transactions* 78(1):153-64.

Eskra, N. 1980. Indirect/direct evaporative cooling systems. *ASHRAE Journal* 22(5):22.

Peterson, J.L. and B.D. Hunn. 1992. Experimental performance of an indirect evaporative cooler. *ASHRAE Transactions* 98(2):15-23.

Scofield, M. and N. Deschamps. 1980. EBTR compliance and comfort too. *ASHRAE Journal* 22(6):61.

Supple, R.G. 1982. Evaporative cooling for comfort. *ASHRAE Journal* 24(8):42.

Watt, J.R. 1988. Cost comparisons: Evaporative vs. refrigerative cooling. *ASHRAE Technical Data Bulletin* 4(4):41.

BIBLIOGRAPHY

Baschiere, R.J., C.E. Rathmann, and M. Lokmanhekim. 1968. Adequacy of evaporative cooling and shelter environmental prediction. General American Research Division, GATX, OCD, AD 697-874.

Watt, J.R. 1986. *Evaporative air conditioning handbook*, 2nd ed. Chapman & Hall, New York.

SMOKE MANAGEMENT

I N building fires, smoke often flows to locations remote from the fire, threatening life and damaging property. Stairwells and elevators frequently become smoke-filled, thereby blocking or inhibiting evacuation. Smoke causes the most deaths in fires. Smoke is defined as the airborne solid and liquid particulates and gases evolved when a material undergoes pyrolysis or combustion, together with the quantity of air that is entrained or otherwise mixed into the mass.

In the late 1960s, the idea of using pressurization to prevent smoke infiltration of stairwells began to attract attention. This concept was followed by the idea of the "pressure sandwich," i.e., venting or exhausting the fire floor and pressurizing the surrounding floors. Frequently, a building's ventilation system is used for this purpose. Smoke management systems use pressurization produced by mechanical fans to limit smoke movement in fire situations.

This chapter discusses fire protection and smoke control systems in buildings as they relate to the HVAC field. For a more complete discussion, refer to *Design of Smoke Management Systems* (Klote and Milke 1992). The *Guide for Smoke and Heat Venting* (NFPA 204M) provides information about venting of large industrial and storage buildings. For further information, refer to *Recommended Practice for Smoke Control Systems* (NFPA 92A-93) and the *Guide for Smoke Management Systems in Malls, Atria, and Large Areas* (NFPA 92B).

The objective of fire safety is to provide some degree of protection for a building's occupants, the building and the property inside it, and neighboring buildings. Various forms of system analysis have been used to help quantify protection. Specific life safety objectives differ with occupancy; for example, nursing home requirements are different from those for office buildings.

Two basic approaches to fire protection are to prevent fire ignition and to manage fire impact. Figure 1 shows a decision tree for fire protection. The building occupants and managers have the primary role in preventing fire ignition. The building design team may incorporate features into the building to assist the occupants and managers in this effort. Because it is impossible to prevent fire ignition completely, managing fire impact has become significant in fire protection design. Examples of fire impact management include compartmentation, suppression, control of construction materials, exit systems, and smoke management. The *Fire Protection Handbook* (NFPA 1991a) and *Smoke Movement and Control in High-Rise Buildings* (Tamura 1994) contain detailed fire safety information.

Historically, fire safety professionals have considered the HVAC system as a potentially dangerous penetration of natural building membranes (wall, floors, and so forth) that can readily transport smoke and fire. For this reason, the systems have traditionally been shut down when fire is discovered. Although shutting

---

The preparation of this chapter is assigned to TC 5.6, Control of Fire and Smoke.

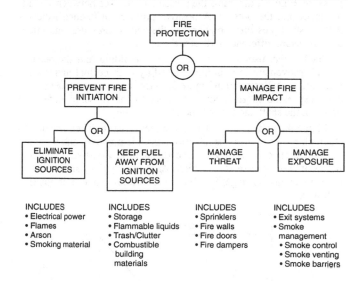

Fig. 1 Simplified Fire Protection Decision Tree

down the system prevents fans from forcing smoke flow, it does not prevent smoke movement through ducts due to smoke buoyancy, stack effect, or the wind. To solve the problem of smoke movement, methods of smoke control have been developed; it should be viewed as only one part of the overall building fire protection system.

SMOKE MOVEMENT

A smoke control system must be designed so that it is not overpowered by the driving forces that cause smoke movement, which include stack effect, buoyancy, expansion, the wind, and the heating, ventilating, and air-conditioning system. In a fire, smoke is generally moved by a combination of these forces.

Stack Effect

When it is cold outside, air often moves upward within building shafts, such as stairwells, elevator shafts, dumbwaiter shafts, mechanical shafts, or mail chutes. This normal stack effect occurs because the air in the building is warmer and less dense than the outside air. Normal stack effect is great when outside temperatures are low, especially in tall buildings. However, normal stack effect can exist even in a one-story building.

When the outside air is warmer than the building air, a downward airflow, or reverse stack effect, frequently exists in shafts. At standard atmospheric pressure, the pressure difference due to either normal or reverse stack effect is expressed as

$$\Delta p = 3460\left(\frac{1}{T_o} - \frac{1}{T_i}\right)h \qquad (1)$$

where

Δp = pressure difference, Pa
T_o = absolute temperature of outside air, K
T_i = absolute temperature of air inside shaft, K
h = distance above neutral plane, m

For a building 60-m tall with a neutral plane at the midheight, an outside temperature of −18°C, and an inside temperature of 21°C, the maximum pressure difference due to stack effect would be 55 Pa. This means that at the top of the building, a shaft would have a pressure 55 Pa greater than the outside pressure. At the bottom of the shaft, the shaft would have a pressure 55 Pa less than the outside pressure. Figure 2 diagrams the pressure difference between a building shaft and the outside. A positive pressure difference indicates that the shaft pressure is higher than the outside pressure, and a negative pressure difference indicates the opposite.

Stack effect usually exists between a building and the outside. The air movement in buildings caused by both normal and reverse stack effect is illustrated in Figure 3. In this case, the pressure difference expressed in Equation (1) refers to the pressure difference between the shaft and the outside of the building.

Figure 4 can be used to determine the pressure difference due to stack effect. For normal stack effect, $\Delta p/h$ is positive, and the pressure difference is positive above the neutral plane and negative below it. For reverse stack effect, $\Delta p/h$ is negative, and the pressure difference is negative above the neutral plane and positive below it.

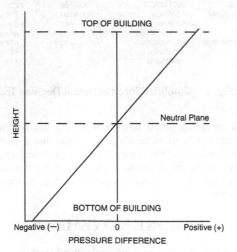

Fig. 2   Pressure Difference Between a Building Shaft and the Outside Due to Normal Stack Effect

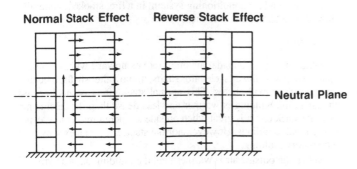

Note: Arrows Indicate Direction of Air Movement

Fig. 3   Air Movement Due to Normal and Reverse Stack Effect

In unusually tight buildings with exterior stairwells, Klote (1980) observed reverse stack effect even with low outside air temperatures. In this situation, the exterior stairwell temperature is considerably lower than the building temperature. The stairwell represents the cold column of air, and other shafts within the building represent the warm columns of air.

If the leakage paths are uniform with height, the neutral plane is near the midheight of the building. However, when the leakage paths are not uniform, the location of the neutral plane can vary considerably, as in the case of vented shafts. McGuire and Tamura (1975) provide methods for calculating the location of the neutral plane for some vented conditions.

Smoke movement from a building fire can be dominated by stack effect. In a building with normal stack effect, the existing air currents (as shown in Figure 3) can move smoke considerable distances from the fire origin. If the fire is below the neutral plane, smoke moves with the building air into and up the shafts. This upward smoke flow is enhanced by buoyancy forces due to the temperature of the smoke. Once above the neutral plane, the smoke flows from the shafts into the upper floors of the building. If the leakage between floors is negligible, the floors below the neutral plane, except the fire floor, are relatively smoke-free until the quantity of smoke produced is greater than can be handled by stack effect flows.

Smoke from a fire located above the neutral plane is carried by the building airflow to the outside through exterior openings in the building. If the leakage between floors is negligible, all floors other than the fire floor remain relatively smoke-free until the quantity of smoke produced is greater than can be handled by stack effect flows. When the leakage between floors is considerable, the smoke flows to the floor above the fire floor.

The air currents caused by reverse stack effect (Figure 3) tend to move relatively cool smoke down. In the case of hot smoke, buoyancy forces can cause smoke to flow upward, even during reverse stack effect conditions.

Buoyancy

High-temperature smoke from a fire has a buoyancy force due to its reduced density. The pressure difference between a fire compartment and its surroundings can be expressed as follows:

$$\Delta p = 3460\left(\frac{1}{T_o} - \frac{1}{T_f}\right)h \qquad (2)$$

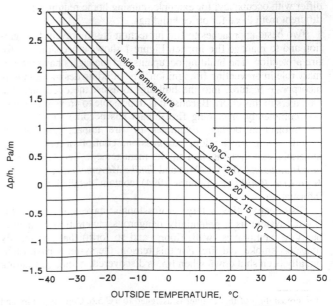

Fig. 4   Pressure Difference Due to Stack Effect

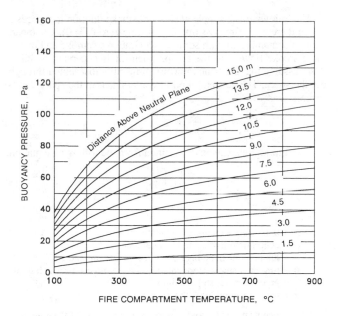

Fig. 5 Pressure Difference Due to Buoyancy

where

Δp = pressure difference, Pa
T_o = absolute temperature of surroundings, K
T_f = absolute temperature of fire compartment, K
h = distance above neutral plane, m

The pressure difference due to buoyancy can be obtained from Figure 5 for the surroundings at 20°C. The neutral plane is the plane of equal hydrostatic pressure between the fire compartment and its surroundings. For a fire with a fire compartment temperature at 800°C, the pressure difference 1.5 m above the neutral plane is 13 Pa. Fang (1980) studied pressures caused by room fires during a series of full-scale fire tests. During these tests, the maximum pressure difference reached was 16 Pa across the burn room wall at the ceiling.

Much larger pressure differences are possible for tall fire compartments where the distance h from the neutral plane can be larger. If the fire compartment temperature is 700°C, the pressure difference 10.7 m above the neutral plane is 90 Pa. This is a large fire, and the pressures it produces are beyond present smoke control methods. However, the example illustrates the extent to which Equation (2) can be applied.

In sprinkler-controlled fires, the temperature in the fire room remains at that of the surroundings except for a short time before sprinkler activation. Sprinklers are activated by a thin (50 to 100 mm) layer of hot gas under the ceiling. This layer is called the *ceiling jet*. The maximum temperature of the ceiling jet depends on the fire location, the activation temperature the sprinkler, and the thermal lag of the sprinkler link. For most residential and commercial applications, the ceiling jet is between 80 and 150°C. In Equation (2), T_f is the average temperature of the fire compartment. For a sprinkler-controlled fire

$$T_f = [T_o (H - H_j) + T_j H_j]/H$$

where

H = floor to ceiling height
H_j = thickness of ceiling jet
T_j = absolute temperature of ceiling jet

For example, for H = 2.5 m, H_j = 0.1 m, T_o = 20 + 273 = 293 K, and T_j = 150 + 293 = 423 K,

$$T_f = [293 (2.5 - 0.1) + 423 \times 0.1]/2.5 = 298.2 \text{ K or } 25.2°C$$

In Equation (2), this results in a pressure difference of 0.5 Pa, which is insignificant for smoke control applications.

Expansion

In addition to buoyancy, the energy released by a fire can move smoke by expansion. In a fire compartment with only one opening to the building, building air will flow in, and hot smoke will flow out. Neglecting the added mass of the fuel, which is small compared to the airflow, the ratio of volumetric flows can be expressed as a ratio of absolute temperatures:

$$\frac{Q_{out}}{Q_{in}} = \frac{T_{out}}{T_{in}}$$

where

Q_{out} = volumetric flow rate of smoke out of fire compartment, m³/s
Q_{in} = volumetric flow rate of air into fire compartment, m³/s
T_{out} = absolute temperature of smoke leaving fire compartment, K
T_{in} = absolute temperature of air into fire compartment, K

For a smoke temperature of 700°C (973 K) and an entering air temperature of 20°C (293 K), the ratio of volumetric flows is 3.32. Note that absolute temperatures are used in the calculation. In such a case, if the air flowing into the fire compartment is 1.5 m³/s, the smoke flowing out of the fire compartment would be 5.0 m³/s, with the gas expanding to more than three times its original volume.

For a fire compartment with open doors or windows, the pressure difference across these openings due to expansion is negligible. However, for a tightly sealed fire compartment, the pressure differences due to expansion may be important.

Wind

In many instances, wind can have a pronounced effect on smoke movement within a building. The pressure the wind exerts on a surface can be expressed as

$$p_w = 0.5 C_w \rho_o V^2 \qquad (3)$$

where

C_w = dimensionless pressure coefficient
ρ_o = outside air density, kg/m³
V = wind velocity, m/s

The pressure coefficients C_w are in the range of –0.8 to 0.8, with positive values for windward walls and negative values for leeward walls. The pressure coefficient depends on building geometry and varies locally over the wall surface. In general, wind velocity increases with height from the surface of the earth. Sachs (1972), Houghton and Carruther (1976), Simiu and Scanlan (1978), and MacDonald (1975) give detailed information concerning wind velocity variations and pressure coefficients. Shaw and Tamura (1977) have developed specific information about wind data with respect to air infiltration in buildings.

A 15.6 m/s wind produces a pressure on a structure of 120 Pa with a pressure coefficient of 0.8. The effect of wind on air movement within tightly constructed buildings with all exterior doors and windows closed is slight. However, the effects of wind can be important for loosely constructed buildings or for buildings with open doors or windows. Usually, the resulting airflows are complicated, and computer analysis is required.

Frequently in fire situations, a window breaks in the fire compartment. If the window is on the leeward side of the building, the negative pressure caused by the wind vents the smoke from the fire compartment. This reduces smoke movement throughout the building. However, if the broken window is on the windward side, the wind forces the smoke throughout the fire floor and to other floors, which endangers the lives of building occupants and hampers fire fighting. Pressures induced by the wind in this situation can be large and can dominate air movement throughout the building.

HVAC Systems

Before methods of smoke control were developed, HVAC systems were shut down when fires were discovered because the systems frequently transported smoke during fires.

In the early stages of a fire, the HVAC system can aid in fire detection. When a fire starts in an unoccupied portion of a building, the system can transport the smoke to a space where people can smell it and be alerted to the fire. However, as the fire progresses, the system transports smoke to every area it serves, thus endangering life in all those spaces. The system also supplies air to the fire space, which aids combustion. Although shutting the system down prevents it from supplying air to the fire, it does not prevent smoke movement through the supply and return air ducts, air shafts, and other building openings due to stack effect, buoyancy, or wind.

SMOKE MANAGEMENT

In this chapter, smoke management includes all methods that can be used singly or in combination to modify smoke movement for the benefit of occupants or firefighters or for the reduction of property damage. The use of barriers, smoke vents, and smoke shafts are traditional methods of smoke management. The effectiveness of barriers is limited by the extent to which they are free of leakage paths. Smoke vents and smoke shafts are limited by the fact that smoke must be sufficiently buoyant to overcome any other driving forces that could be present. In the last few decades, fans have been used with the intent of overcoming the limitations of traditional approaches. The mechanisms of compartmentation, dilution, airflow, pressurization, and buoyancy are used by themselves or in combination to manage smoke conditions in fire situations. These mechanisms are discussed in the following sections.

Compartmentation

Barriers with sufficient fire endurance to remain effective throughout a fire exposure have a long history of providing protection against fire spread. In such fire compartmentation, the walls, partitions, floors, doors, and other barriers provide some level of smoke protection to spaces remote from the fire. This section discusses the use of passive compartmentation. The use of compartmentation in conjunction with pressurization is discussed later. Many codes, such as NFPA 101 (1994), provide specific criteria for the construction of smoke barriers (including doors) and smoke dampers in these barriers. The extent to which smoke leaks through such barriers depends on the size and shape of the leakage paths in the barriers and the pressure difference across the paths.

Dilution Remote from Fire

Dilution of smoke is sometimes referred to as smoke purging, smoke removal, smoke exhaust, or smoke extraction. Dilution can be used to maintain acceptable gas and particulate concentrations in a compartment subject to smoke infiltration from an adjacent space. It can be effective if the rate of smoke leakage is small compared to either the total volume of the safeguarded space or the rate of purging air supplied to and removed from the space. Also, dilution can be beneficial to the fire service for removing smoke after a fire has been extinguished. Sometimes, when doors are opened, smoke flows into areas intended to be protected. Ideally, the doors are only open for short periods during evacuation. Smoke that has entered spaces remote from the fire can be purged by supplying outside air to dilute the smoke.

The following is a simple analysis of smoke dilution for spaces in which there is no fire. Assume that at time zero ($\theta = 0$), a compartment is contaminated with some concentration of smoke and that no more smoke flows into the compartment or is generated within it. Also, assume that the contaminant is uniformly distributed throughout the space. The concentration of contaminant in the space can be expressed as

$$\frac{C}{C_o} = e^{-a\theta} \qquad (4)$$

The dilution rate can be determined from the following equation:

$$a = \frac{1}{\theta}\ln\left(\frac{C_o}{C}\right) \qquad (5)$$

where

C_o = initial concentration of contaminant
C = concentration of contaminant at time θ
a = dilution rate in number of air changes per minute
θ = time after smoke stops entering space or time after which smoke production has stopped, min
e = base of natural logarithm (approximately 2.718)

The concentrations C_o and C must be expressed in the same units, and they can be any units appropriate for the particular contaminant being considered. McGuire et al. (1971) evaluated the maximum levels of smoke obscuration from a number of fire tests and a number of proposed criteria for tolerable levels of smoke obscuration. Based on this evaluation, they state that the maximum levels of smoke obscuration are greater by a factor of 100 than those relating to the limit of tolerance. Thus, they indicate that a space can be considered "reasonably safe" with respect to smoke obscuration, if the concentration of contaminants in the space is less than about 1% of the concentration in the immediate fire area. This level of dilution will increase visibility by about a factor of 100 (for example from 0.15 m to 15 m). Such dilution would also reduce the concentrations of toxic smoke components. Toxicity is a more complex problem, and no parallel statement has been made regarding dilution needed to obtain a safe atmosphere with respect to toxic gases.

In reality, it is impossible to ensure that the concentration of the contaminant is uniform throughout the compartment. Because of buoyancy, it is likely that higher concentrations are near the ceiling. Therefore, exhausting smoke near the ceiling and supplying air near the floor probably dilutes smoke even faster than indicated by Equation (4). Caution should be exercised in placing the supply and exhaust points to prevent the supply air from blowing into the exhaust inlet, which short circuits the dilution operation.

Example 1. A space is isolated from a fire by smoke barriers and self-closing doors, so that no smoke enters the compartment when the doors are closed. When a door is opened, smoke flows through the open doorway into the space. If the door is closed when the contaminant in the space is 20% of the burn room concentration, what dilution rate is required so that six minutes later the concentration will be 1% of that in the burn room?

The time $\theta = 6$ min. and $C_o/C = 20$. From Equation (5), the dilution rate is about 0.5 changes per minute or 30 air changes per hour.

Caution About Dilution Near Fire

Many people have unrealistic expectations about what dilution can accomplish in the fire space. No theoretical or experimental evidence indicates that using a building's heating, ventilation or air-conditioning (HVAC) system for smoke dilution will significantly improve tenable conditions within the fire space. Because HVAC systems promote a considerable degree of air mixing within the spaces they serve and because very large quantities of smoke can be produced by building fires, it is generally believed that dilution of smoke by an HVAC system in the fire space will not result in any practical improvement in the tenable conditions in that space. Thus smoke purging systems should not be used to improve hazard conditions within the fire space or in spaces connected to the fire space by large openings.

Pressurization

Systems that pressurize an area using mechanical fans are referred to as smoke control in this chapter and by NFPA (1993b). A pressure difference across a barrier can control smoke movement as illustrated in Figure 6. Within the barrier is a door. The high pressure side of the door can be either a refuge area or an egress route. The low pressure side is exposed to smoke from a fire. Airflow through the gaps around the door and through construction cracks prevents smoke infiltration to the high pressure side.

For smoke control analysis, the orifice equation can be used to estimate the flow through building flow paths:

$$Q = CA\sqrt{2\Delta p/\rho} \tag{6}$$

where

Q = volumetric airflow rate, m^3/s
C = flow coefficient
A = flow area (also called leakage area), m^2
Δp = pressure difference across flow path, Pa
ρ = density of air entering flow path, kg/m^3

The flow coefficient depends on the geometry of the flow path, as well as on turbulence and friction. In the present context, the flow coefficient is generally in the range of 0.6 to 0.7. For $\rho = 1.2$ kg/m^3 and $C = 0.65$, the flow equation above can be expressed as

$$Q = 0.839\,A\sqrt{\Delta p} \tag{7}$$

The flow area is frequently the same as the cross-sectional area of the flow path. A closed door with a crack area of 0.01 m^2 and a pressure difference of 2.5 Pa has an air leakage rate of approximately 0.013 m^3/s. If the pressure difference across the door is increased to 75 Pa, the flow is 0.073 m^3/s.

Frequently, in field tests of smoke control systems, pressure differences across partitions or closed doors have fluctuated by as much as 5 Pa. These fluctuations have generally been attributed to wind, although they could have been due to the HVAC system or some other source. To control smoke movement, the pressure differences produced by a smoke control system must be sufficiently large so they are not overcome by pressure fluctuations, stack effect, smoke buoyancy, and the forces of wind. However, the pressure difference should not be so large that the door is difficult to open.

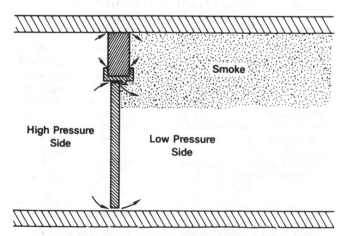

Fig. 6   Pressure Difference Across a Barrier of a Smoke Control System Preventing Smoke Infiltration to the High Pressure Side of the Barrier

Airflow

Airflow has been used extensively to manage smoke from fires in subway, railroad, and highway tunnels. Large flow rates of air are needed to control smoke flow, and these flow rates can supply additional oxygen to the fire. Because of the need for complex controls, airflow is not used so extensively in buildings. The control problem consists of having very small flows when a door is closed and then having those flows increase significantly when that door is opened. Furthermore, it is a major concern that the airflow supplies oxygen to the fire. This section presents the basics of smoke control by airflow and demonstrates why this technique is not recommended, except when the fire is suppressed or in the rare cases when fuel can be restricted with confidence.

Thomas (1970) determined that airflow in a corridor in which there is a fire can almost totally prevent smoke from flowing upstream of the fire. Molecular diffusion is believed to result in the transfer of trace amounts of smoke, which are not hazardous but are detectable as the smell of smoke upstream. Based on work by Thomas, the critical air velocity for most applications can be approximated as

$$V_k = 0.0292\left(\frac{E}{W}\right)^{1/3} \tag{8}$$

where

V_k = critical air velocity to prevent smoke backflow, m/s
E = energy release rate into corridor, W
W = corridor width, m

This relation can be used when the fire is located in the corridor or when the smoke enters the corridor through an open doorway, air transfer grille, or other opening. While the critical velocities calculated from this equation are approximate, they are indicative of the kind of air velocities required to prevent smoke backflow from fires of different sizes.

While Equation (8) can be used to estimate the airflow rate necessary to prevent smoke backflow through an open door, the oxygen supplied is a concern. Huggett (1980) evaluated the oxygen consumed in the combustion of numerous natural and synthetic solids. He found that for most materials involved in building fires, the energy released per unit of mass of air is approximately 13.1 Mg/kg. Air is 23.3% oxygen by mass. Thus, if all the oxygen in a kilogram of air is consumed, 3.0 MJ of heat is liberated. If all the O_2 is consumed by fire in 1 m^3/s of air with a density of 1.2 kg/m^3, 3.6 MW of heat will be liberated.

As can be seen from Examples 2 and 3, the air needed to prevent smoke backflow can support an extremely large fire. In most commercial and residential buildings, sufficient fuel (paper, cardboard, furniture, etc.) is present to support very large fires. Even when the amount of fuel is normally very small, short-term fuel loads (during building renovation, material delivery, etc.) can be significant. Therefore, the use of airflow for smoke control is not recommended, except when the fire is suppressed or in the rare cases when fuel can be restricted with confidence.

Example 2. What airflow at a doorway is needed to stop smoke backflow from a room fully involved in fire, and how large a fire can this airflow support?

A room fully involved in fire can have an energy release rate on the order of 2.4 MW. Assume the door is 0.9 m wide and 2.1 m high. From Equation (8), $V_k = 0.0292 (2.4 \times 10^6/0.9)^{1/3} = 4.9$ m/s. A flow through the doorway of $4 \times 0.9 \times 2.1 = 7.6$ m^3/s is needed to prevent smoke backflow.

If all the O_2 in this airflow is consumed in the fire, the heat liberated is 7.6 $m^3/s \times 3.6$ MW per m^3/s of air = 27 MW. This is many times greater than the fully involved room fire and indicates why airflow is generally not recommended for smoke control in buildings.

Example 3. What airflow is needed to stop smoke backflow from a wastebasket fire and how large a fire can this airflow support?

A wastebasket fire can have an energy release rate on the order of 150 kW. As in Example 2, $V_k = 0.0292 (150 \times 10^3/0.9)^{1/3} = 1.6$ m/s. A flow through the doorway of $1.6 \times 0.9 \times 2.1 = 3.0$ m³/s is needed to prevent smoke backflow.

If all the O_2 in this airflow is consumed in the fire, the heat liberated is 3 m³/s × 3.6 MW per m³/s of air = 10.8 MW. This is still many times greater than the fully involved room fire and further indicates why airflow is generally not recommended for smoke control in buildings.

Buoyancy

The buoyancy of hot combustion gases is employed in both fan-powered and non-fan-powered venting systems. Such fan-powered venting for large spaces is commonly employed for atriums and covered shopping malls, and non-fan-powered venting is commonly used for large industrial and storage buildings. There is a concern that the sprinkler flow will cool the smoke, reducing buoyancy and thus the system effectiveness. Research is needed in this area. Refer to Klote and Milke (1992) and NFPA (1991b, c) for detailed design information about these systems.

DOOR-OPENING FORCES

The door-opening forces resulting from the pressure differences produced by a smoke control system must be considered. Unreasonably high door-opening forces can result in occupants having difficulty or being unable to open doors to refuge areas or escape routes.

The force required to open a door is the sum of the forces to overcome the pressure difference across the door and to overcome the door closer. This can be expressed as

$$F = F_{dc} + \frac{WA\Delta p}{2(w-d)} \qquad (9)$$

where

F = total door-opening force, N
F_{dc} = force to overcome door closer, N
W = door width, m
A = door area, m²
Δp = pressure difference across door, Pa
d = distance from doorknob to edge of knob side of door, m

This relation assumes that the door-opening force is applied at the knob. Door-opening forces due to pressure difference can be determined from Figure 7 for a value of $d = 75$ mm. The force to overcome the door closer is usually greater than 13 N and, in some cases,

can be as great as 90 N. For a door that is 2.1 m high and 0.9 m wide and subject to a pressure difference of 75 Pa, the total door-opening force is 133 N, if the force to overcome the door closer is 53 N.

FLOW AREAS

In designing smoke control systems, airflow paths must be identified and evaluated. Some leakage paths are obvious, such as cracks around closed doors, open doors, elevator doors, windows, and air transfer grilles. While construction cracks in building walls are less obvious, they are equally important.

The flow area of most large openings, such as open windows, can be calculated easily. However, flow areas of cracks are more difficult to evaluate. The area of these leakage paths depends on the workmanship, for example, how well a door is fitted or how well weather stripping is installed. A door that is 0.9 m by 2.1 m with an average crack width of 3 mm has a leakage area of 0.018 m². However, if this door is installed with a 20 mm undercut, the leakage area is 0.033 m²—a significant difference. The leakage area of elevator doors is in the range of 0.051 to 0.065 m² per door.

For open stairwell doorways, Cresci (1973) found that complex flow patterns exist and that the resulting flow through open doorways was considerably below the flow calculated by using the geometric area of the doorway as the flow area in Equation (6). Based on this research, it is recommended that the design flow area of an open stairwell doorway be *half* that of the geometric area (door height times width) of the doorway. An alternate approach for open stairwell doorways is to use the geometric area as the flow area and use a reduced flow coefficient. Because it does not allow the direct use of Equation (6), this alternate approach is not used here.

Typical leakage areas for walls and floors of commercial buildings are tabulated as area ratios in Table 1. These data are based on a relatively small number of tests performed by the National Research Council of Canada (Tamura and Shaw 1976a, 1976b, 1978; Tamura and Wilson 1966). Actual leakage areas depend primarily on the workmanship rather than construction materials, and, in some cases, the flow areas in particular buildings may vary from the values listed. Data concerning air leakage through building components is also provided in Chapter 23, Infiltration and Ventilation, of the 1993 *ASHRAE Handbook—Fundamentals*.

Because a vent surface is usually covered by a louver and screen, the flow area of a vent is less than its area (vent height times width). Since the slats in louvers are frequently slanted, calculation of the flow area is further complicated. Manufacturer's data should be used for specific information.

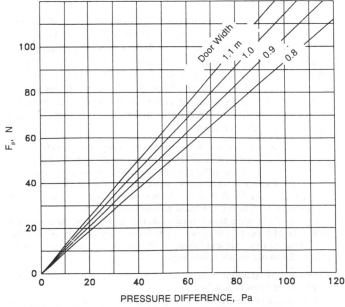

Fig. 7   Door-Opening Force Due to Pressure Differences

Table 1   Typical Leakage Areas for Walls and Floors of Commercial Buildings

Construction Element	Wall Tightness	Area Ratio A/A_w
Exterior building walls[a]	Tight	0.70×10^{-4}
(includes construction cracks, cracks	Average	0.21×10^{-3}
around windows and doors)	Loose	0.42×10^{-3}
	Very Loose	0.13×10^{-2}
Stairwell walls[a]	Tight	0.14×10^{-4}
(includes construction cracks, but not	Average	0.11×10^{-3}
cracks around windows or doors)	Loose	0.35×10^{-3}
Elevator shaft walls[a]	Tight	0.18×10^{-3}
(includes construction cracks, but not	Average	0.84×10^{-3}
cracks around doors)	Loose	0.18×10^{-2}
		A/A_f
Floors[b]	Average	0.52×10^{-4}
(includes construction cracks and areas around penetrations)		

A = leakage area; A_w = wall area; A_f = floor area
[a]Flow areas evaluated at 75 Pa
[b]Flow areas evaluated at 25 Pa

Effective Flow Areas

The concept of effective flow areas is useful for analyzing smoke control systems. The paths in the system can be in parallel with one another, in series, or a combination of parallel and series. The effective area of a system of flow areas is the area that results in the same flow as the system when it is subjected to the same pressure difference over the total system of flow paths. This is similar to the effective resistance of a system of electrical resistances. The effective area for parallel leakage areas is the sum of the individual leakage paths:

$$A_e = \sum_{i=1}^{n} A_i \qquad (10)$$

where n is the number of flow areas A_i in parallel.

For example, the effective area A_e for the three parallel leakage areas in Figure 8 is

$$A_e = A_1 + A_2 + A_3 \qquad (11)$$

If A_1 is 0.10 m^2 and A_2 and A_3 are each 0.05 m^2, then the effective flow area A_e is 0.20 m^2.

Three leakage areas in series from a pressurized space are illustrated in Figure 9. The effective flow area of these paths is

$$A_e = \left(\frac{1}{A_1^2} + \frac{1}{A_2^2} + \frac{1}{A_3^2} \right)^{-0.5} \qquad (12)$$

The general rule for any number of leakage areas in series is

$$A_e = \left[\sum_{i=1}^{n} \frac{1}{A_1^2} \right]^{-0.5} \qquad (13)$$

where n is the number of leakage areas A_i in series. In smoke control analysis, there are frequently only two paths in series and the effective leakage area is

$$A_e = \frac{A_1 A_2}{\sqrt{A_1^2 + A_2^2}} \qquad (14)$$

Example 4. Calculate the effective leakage area of two equal flow paths in series.

Let $A = A_1 = A_2 = 0.02$ m^2. From Equation (14),

$$A_e = \frac{A^2}{\sqrt{2A^2}} = 0.014 \text{ m}^2$$

Example 5. Calculate the effective area of two flow paths in series, where $A_1 = 0.02$ m^2 and $A_2 = 0.2$ m^2. From Equation (14),

$$A_e = \frac{A_1 A_2}{\sqrt{A_1^2 + A_2^2}} = 0.0199 \text{ m}^2$$

This example illustrates that when two areas are in series, and one is much larger than the other, the effective area is approximately equal to the smaller area.

The method of developing an effective area for a system of both parallel and series paths is to combine groups of parallel paths and series paths systematically. The system illustrated in Figure 10 is analyzed as an example. The figure shows that A_2 and A_3 are in parallel; therefore, their effective area is

$$(A_{23})_{eff} = A_2 + A_3$$

Areas A_4, A_5, and A_6 are also in parallel, so their effective area is

$$(A_{456})_{eff} = A_4 + A_5 + A_6$$

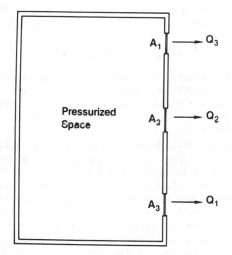

Fig. 8   Leakage Paths in Parallel

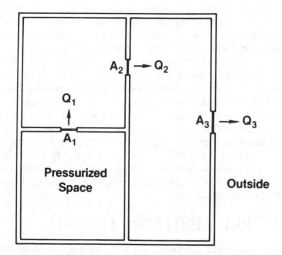

Fig. 9   Leakage Paths in Series

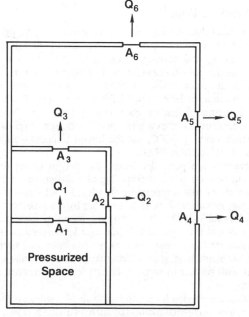

Fig. 10   Combination of Leakage Paths in Parallel and Series

These two effective areas are in series with A_1. Therefore, the effective flow area of the system is given by

$$A_e = \left[\frac{1}{A_1^2} + \frac{1}{(A_{23})_{eff}^2} + \frac{1}{(A_{456})_{eff}^2} \right]^{-0.5}$$

Example 6. Calculate the effective area of the system in Figure 10, if the leakage areas are $A_1 = A_2 = A_3 = 0.02$ m^2 and $A_4 = A_5 = A_6 = 0.01$ m^2.

$$(A_{23})_{eff} = 0.04 \text{ m}^2$$
$$(A_{456})_{eff} = 0.03 \text{ m}^2$$
$$A_e = 0.015 \text{ m}^2$$

Symmetry

The concept of symmetry is useful in simplifying problems. Figure 11 illustrates the floor plan of a multistory building that can be divided in half by a plane of symmetry. Flow areas on one side of the plane of symmetry are equal to corresponding flow areas on the other side. For a building to be treated in this manner, every floor of the building must be such that it can be divided in the same manner by the plane of symmetry. If wind effects are not considered in the analysis, or if the wind direction is parallel to the plane of symmetry, the airflow is only one-half of the total for the building analyzed. It is not necessary that the building be geometrically symmetric, as shown in Figure 11; it must be symmetric only with respect to flow.

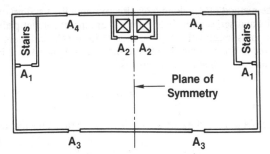

Fig. 11   Building Floor Plan Illustrating Symmetry Concept

Design Weather Data

Little weather data has been developed specifically for the design of smoke control systems. A designer may use the design temperatures for heating and cooling found in Chapter 24 of the 1993 *ASHRAE Handbook—Fundamentals*. In a normal winter, approximately 22 h are at or below the 99% design value, and approximately 54 h are at or below the 97.5% design value. Furthermore, extreme temperatures can be considerably lower than the winter design temperatures. For example, the 99% design temperature for Tallahassee, Florida, is –3°C, but the lowest temperature observed there was –19°C (NOAA 1979).

Temperatures are generally below the design values for short periods, and because of the thermal lag of building materials, these short intervals of low temperature usually do not cause problems with heating. However, there is no time lag for a smoke control system; thus it is subjected to all the extreme forces of stack effect that exist the moment it is operated. If the outside temperature is below the winter design temperature for which a smoke control system was designed, problems from stack effect may result. A similar situation can result with respect to summer design temperatures and reverse stack effect.

Wind data are needed for a wind analysis of a smoke control system. At present, no formal method of such an analysis exists, and the approach most generally taken is to design a smoke control system that minimizes any effects of wind.

Design Pressure Differences

Both the maximum and minimum allowable pressure differences across the boundaries of smoke control should be considered. The maximum allowable pressure difference should not result in excessive door-opening forces.

The minimum allowable pressure difference across a boundary of a smoke control system might be that difference such that no smoke leakage occurs during building evacuation. In this case, the smoke control system must produce sufficient pressure differences to overcome forces of wind, stack effect, or buoyancy of hot smoke. The pressure differences due to wind and stack effect can be large in the event of a broken window in the fire compartment. Evaluation of these pressure differences depends on evacuation time, rate of fire growth, building configuration, and the presence of a fire-suppression system. NFPA 92A (1993b) suggests values of minimum and maximum design pressure difference.

Open Doors

Another design concern is the number of doors that could be opened simultaneously when the smoke control system is operating. A design that allows all doors to be open simultaneously may ensure that the system always works, but it probably adds to the cost of the system.

Deciding how many doors will be open simultaneously depends largely on building occupancy. For example, in a densely populated building, it is likely that all doors will be open during evacuation. However, if a staged evacuation plan or refuge area concept is incorporated in the building fire emergency plan, or if the building is sparsely occupied, only a few of the doors may be open during a fire.

FIRE AND SMOKE DAMPERS

Openings for ducts in walls and floors with fire resistance ratings should be protected by fire dampers and ceiling dampers, as required by local codes. Air transfer openings should also be protected. These dampers should be classified and labeled in accordance with UL *Standard* 555 (1990).

A smoke damper can be used for either traditional smoke management (smoke containment) or for smoke control. In smoke management, a smoke damper inhibits the passage of smoke under the forces of buoyancy, stack effect, and wind. Generally, for smoke containment, smoke dampers should have low leakage characteristics at elevated temperatures. However, smoke dampers are only one of many elements (partitions, floors, doors) intended to inhibit smoke flow. For smoke management, the smoke dampers selected should have leakage characteristics similar to those of the other system elements.

In a smoke control system, a smoke damper inhibits the passage of air that may or may not contain smoke. Low leakage characteristics of a damper are not necessary when outside (fresh) air is on the high pressure side of the damper, as is the case for dampers that shut off supply air from a smoke zone or that shut off exhaust air from a nonsmoke zone. In these cases, moderate leakage of smoke-free air through the damper does not adversely affect the control of smoke movement. It is best to design smoke control systems so that only smoke-free air is on the high pressure side of a smoke damper.

Smoke dampers should be classified and listed in accordance with UL *Standard* 555S (1993). At locations requiring both smoke and fire dampers, combination dampers meeting the requirements of both can be used. Fire, ceiling, and smoke dampers should be installed in accordance with the manufacturer's instructions. NFPA *Standard* 90-A (1993a) gives general guides regarding locations requiring these dampers.

PRESSURIZED STAIRWELLS

Many pressurized stairwells have been designed and built to provide a smoke-free escape route in the event of a building fire. They also provide a smoke-free staging area for fire fighters. On the fire floor, a pressurized stairwell must maintain a positive pressure difference across a closed stairwell door so that smoke infiltration is prevented.

During building fire situations, some stairwell doors are opened intermittently during evacuation and fire fighting, and some doors may even be blocked open. Ideally, when the stairwell door is opened on the fire floor, airflow through the door should be sufficient to prevent smoke backflow. Designing such a system is difficult because of the many combinations of open stairwell doors and weather conditions affecting airflow.

Stairwell pressurization systems may be single and multiple injection systems. A single injection system has pressurized air supplied to the stairwell at one location—usually at the top. Associated with this system is the potential of smoke entering the stairwell through the pressurization fan intake. Therefore, automatic shutdown during such an event should be considered.

For tall stairwells, single injection systems can fail when a few doors are open near the air supply injection point. Such a failure is especially likely when a ground-level stairwell door is open in bottom injection systems.

For tall stairwells, supply air can be supplied at a number of locations over the height of the stairwell. Figures 12 and 13 show two examples of multiple injection systems that can be used to overcome

the limitations of single injection systems. In these figures, the supply duct is shown in a separate shaft. However, systems have been built that have eliminated the expense of a separate duct shaft by locating the supply duct in the stairwell itself. In such a case, care must be taken that the duct does not become an obstruction to orderly building evacuation.

STAIRWELL COMPARTMENTATION

Compartmentation of the stairwell into a number of sections is one alternative to multiple injection (Figure 14). When the doors between compartments are open, the effect of compartmentation is lost. For this reason, compartmentation is inappropriate for densely populated buildings where total building evacuation by the stairwell is planned in the event of fire. However, when a staged evacuation plan is used and when the system is designed to operate successfully when the maximum number of doors between compartments is open, compartmentation can be an effective means of providing stairwell pressurization for tall stairwells.

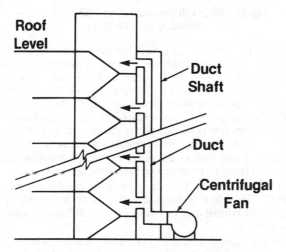

Fig. 12   Stairwell Pressurization by Multiple Injection with Fan Located at Ground Level

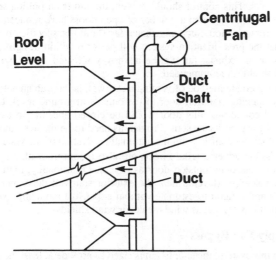

Fig. 13   Stairwell Pressurization by Multiple Injection with Roof-Mounted Fan

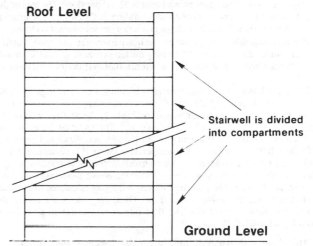

Note: Each four-floor compartment has at least one supply air injection point.

Fig. 14   Compartmentation of Pressurized Stairwell

STAIRWELL ANALYSIS

This section presents an analysis for a pressurized stairwell in a building without vertical leakage. The performance of pressurized stairwells in buildings without elevators may be closely approximated by this method. It is also useful for buildings with vertical leakage in that it yields conservative results. Only one stairwell is considered in the building; however, the analysis can be extended to any number of stairwells by the concept of symmetry. For evaluation of vertical leakage through the building or with open stairwell doors, computer analysis is recommended. The analysis is for buildings where the leakage areas are the same for each floor of the building and where the only significant driving forces are the stairwell pressurization system and the temperature difference between the indoors and outdoors.

The pressure difference Δp_{sb} between the stairwell and the building can be expressed as

$$\Delta p_{sb} = \Delta p_{sbb} + By / [1 + (A_{sb}/A_{bo})^2] \tag{15}$$

where

Δp_{sbb} = pressure difference from stairwell to building at stairwell bottom, Pa

y = distance above stairwell bottom, m

A_{sb} = flow area between stairwell and building (per floor), m^2

A_{bo} = flow area between building and outside (per floor), m^2
B = 3460 $[1/(273 + t_o) - 1/(273 - t_s)]$
t_o = temperature of outside air, °C
t_s = temperature of stairwell air, °C

For a stairwell with no leakage directly to the outside, the flow rate of pressurization air is

$$Q = 0.559 \, N \, A_{sb} \left(\frac{\Delta p_{sbt}^{3/2} - \Delta p_{sbb}^{3/2}}{\Delta p_{sbt} - \Delta p_{sbb}} \right) \quad (16)$$

where

Q = volumetric flow rate, m^3/s
N = number of floors
Δp_{sbt} = pressure difference from stairwell to building at stairwell top, Pa

Example 7. Each story of a 20-story stairwell is 3.3 m high. The stairwell has a single-leaf door at each floor leading to the occupant space and one ground-level door to the outside. The exterior of the building has a wall area of 560 m^2 per floor. The exterior building walls and stairwell walls are of average leakiness. The stairwell wall area is 52 m^2 per floor. The area of the gap around each stairwell door to the building is 0.024 m^2. The exterior door is well gasketed, and its leakage can be neglected when it is closed.

For this example, the following design parameters are used: outside design temperature, t_o = −10°C; stairwell temperature, t_s = 21°C; maximum design pressure differences when all stairwell doors are closed of 137 Pa.

Using the leakage ratios for an exterior building wall of average tightness from Table 1, A_{bo} = 560(0.21 × 10^{-3}) = 0.118 m^2. Using leakage ratios for a stairwell wall of average tightness from Table 1, the leakage area of the stairwell wall is 52(0.11 × 10^{-3}) = 0.006 m^2. The value of A_{sb} equals the leakage area of the stairwell wall plus the gaps around the closed doors. A_{sb} = 0.006 + 0.024 = 0.030 m^2. The temperature factor B is calculated at 1.39 Pa/m. The pressure difference at the stairwell bottom is selected as Δp_{sbb} = 20 Pa to provide an extra degree of protection above the minimum allowable value of 13 Pa. The pressure difference Δp_{sbt} is calculated from Equation (15) at 106 Pa, using y = 66.1 m. Thus, Δp_{sbt} does not exceed the maximum allowable pressure. The flow rate of pressurization air is calculated from Equation (16) at 3.9 m^3/s.

The flow rate depends highly on the leakage area around the closed doors and on the leakage area that exists in the stairwell walls. In practice, these areas are difficult to evaluate and even more difficult to control. If the flow area A_{sb} in Example 7 were 0.050 m^2 rather than 0.030 m^2, a flow rate of pressurization air of 6.5 m^3/s would be calculated from Equation (16). A fan with a sheave allows adjustment of supply air to offset for variations in actual leakage from the values used in design calculations.

STAIRWELL PRESSURIZATION AND OPEN DOORS

The simple pressurization system discussed previously has two limitations regarding open doors. First, when a stairwell door to the outside and doors to the building are open, the simple system cannot provide sufficient airflow through doorways to the building to prevent smoke backflow. Second, when stairwell doors are open, the pressure difference across the closed doors can drop to low levels. Two systems used to overcome these problems are overpressure relief (Tamura 1990) and supply fan bypass.

Overpressure Relief

The total airflow rate is selected to provide the minimum air velocity when a specific number of doors are open. When all the doors are closed, part of this air is relieved through a vent to prevent excessive pressure buildup, which could cause excessive door-opening forces. This excess air can be vented either to the building or to the outside. Since exterior vents can be subject to adverse effects of the wind, wind shields are recommended.

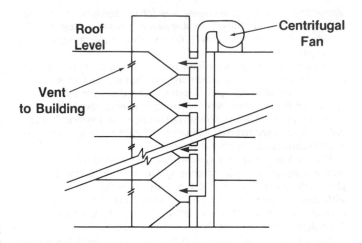

Notes:
1. Vents to the building have a barometric damper and a fire damper in series.
2. A roof-mounted supply fan is shown; however, the fan may be located at any level.
3. A manually operated damper may be located at the stairwell top for smoke purging by the fire department.

Fig. 15   Stairwell Pressurization with Vents to the Building at Each Floor

Barometric dampers that close when the pressure drops below a specified value can minimize the air losses through the vent when doors are open. Figure 15 illustrates a pressurized stairwell with overpressure relief vents to the building at each floor. In systems with vents between the stairwell and the building, the vents typically have a fire damper in series with the barometric damper. As an energy conservation feature, these fire dampers are normally closed, but they open when the pressurization system is activated. This arrangement also reduces the possibility of the annoying damper chatter that frequently occurs with barometric dampers.

An exhaust duct can provide overpressure relief in a pressurized stairwell. This system is designed so that the normal resistance of a nonpowered exhaust duct maintains pressure differences within the design limits.

Exhaust fans can also relieve excessive pressures when all stairwell doors are closed. An exhaust fan should be controlled by a differential pressure sensor, so that it will not operate when the pressure difference between the stairwell and the building falls below a specific level. This control should prevent the fan from pulling smoke into the stairwell when a number of open doors have reduced stairwell pressurization. Such an exhaust fan should be specifically sized so that the pressurization system will perform within design limits. Because an exhaust fan can be adversely affected by the wind, a wind shield is recommended.

An alternate method of venting a stairwell is through an automatically opening stairwell door to the outside at ground level. Under normal conditions, this door would be closed and, in most cases, locked for security reasons. Provisions need to be made so that this lock does not conflict with the automatic operation of the system.

Possible adverse wind effects are also a concern with a system that uses an open outside door as a vent. Occasionally, high local wind velocities develop near the exterior stairwell door, and such winds are difficult to estimate without expensive modeling. Nearby obstructions can act as wind breaks or wind shields.

Supply Fan Bypass

In this system, the supply fan is sized to provide at least the minimum air velocity when the design number of doors are open. Figure 16 illustrates such a system. The flow rate of air into the stairwell is

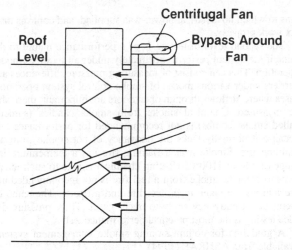

Notes:
1. Fan bypass controlled by one or more static pressure sensors located between the stairwell and the building.
2. A roof-mounted supply fan is shown; however, the fan may be located at any level.
3. A manually operated damper may be located at the stairwell top for smoke purging by the fire department.

Fig. 16 Stairwell Pressurization with Bypass Around Supply Fan

varied by modulating bypass dampers, which are controlled by one or more static pressure sensors that sense the pressure difference between the stairwell and the building. When all the stairwell doors are closed, the pressure difference increases and the bypass damper opens to increase the bypass air and decrease the flow of supply air to the stairwell. In this manner, excessive stairwell pressures and excessive pressure differences between the stairwell and the building are prevented.

ELEVATORS

Elevator smoke control systems intended for use by fire fighters should keep elevator cars, elevator shafts, and elevator machinery rooms smoke-free. Small amounts of smoke in these spaces are acceptable, provided that the smoke is nontoxic and that the operation of the elevator equipment is not affected. Elevator smoke control systems intended for fire evacuation of the handicapped or other building occupants should also keep elevator lobbies smoke-free or nearly smoke-free. The long-standing obstacles to fire evacuation by elevators include (1) the logistics of evacuation, (2) the reliability of electrical power, (3) the jamming of elevator doors, and (4) fire and smoke protection. All these obstacles, except smoke protection, can be addressed by existing technology (Klote 1984).

Klote and Tamura (1986) studied conceptual elevator smoke control systems for handicapped evacuation. The major problem was maintaining pressurization with open doors, especially doors on the ground floor. Of the systems evaluated, only one with a supply fan bypass with feedback control maintains adequate pressurization with any combination of open or closed doors. There are probably other systems capable of providing adequate smoke control, and the procedure used by Klote and Tamura can be viewed as an example of a method of evaluating the performance of a system to determine whether it meets the particular characteristics of a building under construction.

The transient pressures due to *piston effect* when an elevator car moves in a shaft have been a concern with regard to elevator smoke control. Piston effect is not a concern for slow-moving cars in multiple car shafts. However, for fast cars in single car shafts, the piston effect can be considerable.

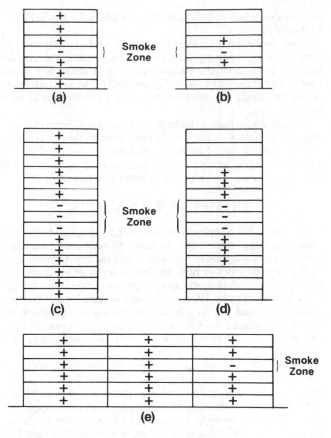

Note:
In the above figures, the smoke zone is indicated by a minus sign, and pressurized spaces are indicated by a plus sign. Each floor can be a smoke control zone as in (a) and (b), or a smoke zone can consist of more than one floor as in (c) and (d). All the nonsmoke zones in a building may be pressurized as in (a) and (c) or only nonsmoke zones adjacent to the smoke zone may be pressurized as in (b) and (d). A smoke zone can also be limited to a part of a floor as in (e).

Fig. 17 Some Arrangements of Smoke Control Zones

ZONE SMOKE CONTROL

Klote (1990) conducted a series of tests on full-scale fires, which demonstrated that zone smoke control can restrict smoke movement to the zone where a fire starts.

Pressurized stairwells are intended to prevent smoke infiltration into stairwells. However, in a building with only stairwell pressurization, smoke can flow through cracks in floors and partitions and through shafts to damage property and threaten life at locations remote from the fire. The concept of zone smoke control is intended to limit such smoke movement.

A building is divided into a number of smoke control zones, each zone separated from the others by partitions, floors, and doors that can be closed to inhibit smoke movement. In the event of a fire, pressure differences and airflows produced by mechanical fans limit the smoke spread from the zone in which the fire initiated. The concentration of smoke in this zone goes unchecked; thus, in zone smoke control systems, the occupants should evacuate the smoke zone as soon as possible after fire detection.

A smoke control zone can consist of one floor, more than one floor, or a floor can be divided into more than one smoke control zone. Sprinkler zones and smoke control zones should be coordinated so that sprinkler water flow will activate the zone's smoke control system. Some arrangements of smoke control zones are illustrated in Figure 17. All the nonsmoke zones in the building may be pressurized as in Figure 17. The term *pressure sandwich* describes cases where only adjacent zones to the smoke zone are pressurized, as in Figures 17 (b) and (d).

Zone smoke control is intended to limit smoke movement to the smoke zone by the two principles of smoke control. Pressure differences in the desired direction across the barriers of a smoke zone can be achieved by supplying outside (fresh) air to nonsmoke zones, by venting the smoke zone, or by a combination of these methods.

Venting smoke from a smoke zone prevents significant overpressures due to thermal expansion of gases caused by the fire. However, venting results in only slight reduction of smoke concentration in the smoke zone. This venting can be accomplished by exterior wall vents, smoke shafts, and mechanical venting (exhausting).

COMPUTER ANALYSIS

Because of the complexity of airflow in buildings, network computer programs were developed to model the airflow with pressurization systems. These models represent rooms and shafts by nodes, and airflow is from nodes of high pressure to nodes of lower pressure. Some programs calculate steady-state airflow and pressures throughout a building (Sander 1974, Sander and Tamura 1973). Other programs go beyond this to calculate the smoke concentrations that would be produced throughout a building in the event of a fire (Yoshida et al. 1979, Evers and Waterhouse 1978, Wakamatsu 1977, and Rilling 1978).

The ASCOS program was developed specifically for analysis of pressurization smoke control systems (Klote 1982). ASCOS is the most widely used program for smoke control analysis (Said 1988), and it has been validated against field data from flow experiments at an eight-story tower in Champs sur Marne, France (Klote and Bodart 1985). ASCOS and the other network models have been used extensively for design and for parametric analysis of the performance of smoke control systems. However, ASCOS was originally intended as a research tool for application to 10- and 20-story buildings. It is not surprising that convergence failures have been encountered with applications to much larger buildings.

Wray and Yuill (1993) evaluated several flow algorithms to find the most appropriate one for analysis of smoke control systems. The AIRNET flow routine developed by Walton (1989) was selected as the best algorithm based on computational speed and use of computer memory. None of the algorithms from this study takes advantage of the repetitive nature of building flow networks, so data entry is difficult. However, Walton (1994) developed the public domain program, CONTAM93, with an improved version of the AIRNET flow routine and an easier method of input.

These models are appropriate for analyzing systems that use pressurization to control smoke flow. For systems that rely on buoyancy of hot smoke (such as atrium smoke exhaust), zone fire models are appropriate for design calculations. The concepts behind zone fire modeling are discussed by Bukowski (1991), Jones (1983), and Mitler (1985). Some frequently used zone models are ASET (Cooper 1985), CCFM (Cooper and Forney 1987), CFAST (Peacock et al. 1993). Milke and Mowrer (1994) have enhanced the CCFM model for atrium applications.

ACCEPTANCE TESTING

Regardless of the care, skill, and attention to detail with which a smoke control system is designed, an acceptance test is needed as assurance that the system, as built, operates as intended.

An acceptance test should be composed of two levels of testing. The first is of a functional nature to determine if everything in the system works as it is supposed to, that is, an initial check of the system components. The importance of the initial check has become apparent because of the problems encountered during tests of smoke control systems. These problems include fans operating backward,

fans to which no electrical power was supplied, and controls that did not work properly.

The second level of testing is of a performance nature to determine if the system performs adequately under all required modes of operation. This can consist of measuring pressure differences across barriers under various modes of smoke control system operation. In cases where airflows through open doors are important, these should be measured. Chemical smoke from smoke candles (sometimes called smoke bombs) is not recommended for performance testing because it normally lacks the buoyancy of hot smoke from a real building fire. Smoke near a flaming fire has a temperature in the range of 540 to 1100°C. Heating chemical smoke to such temperatures to emulate smoke from a real fire is not recommended unless precautions are taken to protect life and property. The same comments about buoyancy apply to tracer gases. Thus, pressure difference testing is the most practical performance test.

A guideline for commissioning smoke management systems is available from ASHRAE (1994).

REFERENCES

ASHRAE. 1994. Commissioning smoke management systems. *Guideline* 5.
Bukowski, R.W. 1991. Fire models, the future is now! *NFPA Journal* 85(2): 60-69.
Cooper, L.Y. 1985. ASET—A computer program for calculating available safe egress time. *Fire Safety Journal* 9:29-45.
Cooper, L.Y. and G.P. Forney. 1987. Fire in a room with a hole: A prototype application of the consolidated compartment fire model (CCFM) computer code. Presented at the 1987 Combined Meetings of Eastern Section of Combustion Institute and NBS Annual Conference on Fire Research.
Cresci, R.J. 1973. Smoke and fire control in high-rise office buildings—Part II, Analysis of stair pressurization systems. Symposium on Experience and Applications on Smoke and Fire Control, ASHRAE Annual Meeting, June.
Evers, E. and A. Waterhouse. 1978. A computer model for analyzing smoke movement in buildings. Building Research Est., Fire Research Station, Borehamwood, Herts, England.
Houghton, E.L. and N.B. Carruther. 1976. *Wind forces on buildings and structures.* John Wiley & Sons, New York.
Jones, W.W. 1983. A review of compartment fire models. NBSIR 83-2684. U.S. National Bureau of Standards.
Klote, J.H. 1980. Stairwell pressurization. *ASHRAE Transactions* 86(1): 604-73.
Klote, J.H. 1982. A computer program for analysis of smoke control systems. NBSIR 82-2512. U.S. National Bureau of Standards.
Klote, J.H. 1984. Smoke control for elevators. *ASHRAE Journal* 26(4):23-33.
Klote, J.H. 1990. Fire experiments of zoned smoke control at the Plaza Hotel in Washington, DC. *ASHRAE Transactions* 96(2):399-416.
Klote, J.H. and X. Bodart. 1985. Validation of network models for smoke control analysis. *ASHRAE Transactions* 91(2B):1134-45.
Klote, J.H. and J.A. Milke. 1992. *Design of smoke management systems.* ASHRAE.
Klote, J.H. and G.T. Tamura. 1986. Smoke control and fire evacuation by elevators. *ASHRAE Transactions* 92(1).
MacDonald, A.J. 1975. *Wind loading on buildings.* John Wiley & Sons, New York.
McGuire, J.H. and G.T. Tamura. 1975. Simple analysis of smoke flow problems in high buildings. *Fire Technology* 11(1):15-22 (February).
McGuire, J.H., G.T. Tamura, and A.G. Wilson. 1970. Factors in controlling smoke in high buildings. Symposium on Fire Hazards in Buildings. ASHRAE Winter Meeting.
Milke, J.A. and F.W. Mowrer. 1994. Computer aided design for smoke management in atriums and covered malls. *ASHRAE Transactions* 100(2).
Mitler, H.E. 1985. Comparison of several compartment fire models: An interim report. NBSIR 85-3233. U.S. National Bureau of Standards.
NFPA. 1991a. *Fire protection handbook*, 17th ed. National Fire Protection Association, Quincy, MA.
NFPA. 1991b. Guide for smoke management systems in malls, atria, and large areas. NFPA 92B-91. National Fire Protection Association, Quincy, MA.
NFPA 1991c. Guide for smoke and heat venting. NFPA 204M-91. National Fire Protection Association, Quincy, MA.

NFPA. 1993a. Installation of air-conditioning and ventilating systems. NFPA 90A-93. National Fire Protection Association, Quincy, MA.

NFPA. 1993b. Recommended practice for smoke control systems. NFPA 92A-93. National Fire Protection Association, Quincy, MA.

NFPA. 1994. Code for safety to life from fire in buildings and structures. NFPA 101-94, National Fire Protection Association, Quincy, MA.

NOAA. 1979. Temperature extremes in the United States. National Oceanic and Atmospheric Administration (U.S.), National Climatic Center, Asheville, NC.

Peacock, R.D., G.P. Forney, P. Reneke, R. Portier, and W.W. Jones. 1993. CFAST, the consolidated model of fire growth and smoke transport. *NIST Technical Note* 1299. National Institute of Standards and Technology, Gaithersburg, MD.

Rilling, J. 1978. Smoke study, 3rd Phase, Method of calculating the smoke movement between building spaces. Centre Scientifique et Technique du Batiment (CSTB), Champs sur Marne, France.

Sachs, P. 1972. *Wind forces in engineering.* Pergamon Press, New York.

Said, M.N.A. 1988. A review of smoke control models. *ASHRAE Journal* 30(4):36-40.

Sander, D.M. and G.T. Tamura. 1973. FORTRAN IV program to simulate air movement in multi-story buildings. DBR Computer Program No. 35. National Research Council, Ottawa, Canada.

Sander, D.M. 1974. FORTRAN IV program to calculate air infiltration in buildings. DBR Computer Program No. 37. National Research Council, Ottawa, Canada (May).

Shaw, C.Y. and G.T. Tamura. 1977. The calculation of air infiltration rates caused by wind and stack action for tall buildings. *ASHRAE Transactions* 83(2):145-58.

Simiu, E. and R.H. Scanlan. 1978. *Wind effects on structures: An introduction to wind engineering.* John Wiley & Sons, New York.

Tamura, G.T. 1994. Smoke movement and control in high-rise buildings. National Fire Protection Association, Quincy, MA.

Tamura, G.T. 1990. Field tests of stair pressurization systems with overpressure relief. *ASHRAE Transactions* 96(1):951-58.

Tamura, G.T. and C.Y. Shaw. 1976a. Studies on exterior wall air tightness and air infiltration of tall buildings. *ASHRAE Transactions* 83(1):122-34.

Tamura, G.T. and C.Y. Shaw. 1976b. Air leakage data for the design of elevator and stair shaft pressurization systems. *ASHRAE Transactions* 83(2): 179-90.

Tamura, G.T. and C.Y. Shaw. 1978. Experimental studies of mechanical venting for smoke control in tall office buildings. *ASHRAE Transactions* 86(1):54-71.

Tamura, G.T. and A.G. Wilson. 1966. Pressure differences for a 9-story building as a result of chimney effect and ventilation system operation. *ASHRAE Transactions* 72(1):180-89.

Thomas, P.H. 1970. Movement of smoke in horizontal corridors against an airflow. *Institution of Fire Engineers Quarterly* 30(77):45-53.

UL. 1990. UL standard for safety fire dampers, 4th ed. UL 555-90. Underwriters Laboratories, Northbrook, IL.

UL. 1993. UL standard for safety leakage rated dampers for use in smoke control systems, 2nd ed. UL 555S-93. Underwriters Laboratories, Northbrook, IL.

Wakamatsu, T. 1977. Calculation methods for predicting smoke movement in building fires and designing smoke control systems. Fire *Standards and Safety,* ASTM STP 614, pp. 168-93. American Society for Testing and Materials, Philadelphia.

Walton, G.N. 1989. AIRNET—A computer program for building airflow network modeling. National Institute of Standards and Technology, Gaithersburg, MD.

Walton, G.N. 1994. CONTAM93 user manual. NISTIR 5385. National Institute of Standards and Technology, Gaithersburg, MD.

Wray, C.P. and G.K. Yuill. 1993. An evaluation of algorithms for analyzing smoke control systems. *ASHRAE Transactions* 99(1):160-174.

Yoshida, H., C.Y. Shaw, and G.T. Tamura. 1979. A FORTRAN IV program to calculate smoke concentrations in a multi-story building. DBR Computer Program No. 45. National Research Council, Ottawa, Canada.

CHAPTER 49

RADIANT HEATING AND COOLING

RADIANT heating and cooling applications are classified as panel heating or cooling if the surface temperature is below 150°C and are classified as low, medium, or high intensity if the surface or source temperature range exceeds 150°C.

Radiant energy is transmitted by electromagnetic waves that travel in straight lines, can be reflected, and heat solid objects but do not heat the air through which the energy is transmitted. Because of these characteristics, radiant systems are effective both for spot heating and for entire building space heating or cooling requirements.

LOW, MEDIUM, AND HIGH INTENSITY INFRARED

Low, medium, and high intensity infrared heaters are compact, self-contained direct-heating devices used in hangars, warehouses, factories, greenhouses, and gymnasiums, as well as in areas such as loading docks, racetrack stands, outdoor restaurants, animal breeding areas, swimming pool lounge areas, and under marquees. Infrared heating is also used for snow melting (e.g., on stairs and ramps) and in process heating (e.g., paint baking and drying). An infrared heater may be electric, gas-fired, or oil-fired and is classified by the source temperature as follows:

- Low intensity for source temperatures to 650°C
- Medium intensity for source temperatures to 980°C
- High intensity for source temperatures to 2800°C

The source temperature is determined by such factors as the source of energy, the configuration, and the size. Reflectors can be used to direct the distribution of radiation in specific patterns.

As floors and solid objects are heated by infrared radiation, they reradiate at lower source temperatures. Furthermore, air in contact with the warm surfaces is heated, and the ambient temperature is raised by convection.

Additional information on radiant equipment is available in Chapter 15 of the 1992 *ASHRAE Handbook—Systems and Equipment.*

PANEL HEATING AND COOLING

Panel heating and cooling systems include the following:

- Metal ceiling panels
- Embedded hydronic tubing or attached piping in ceilings, walls, or floors
- Air-heated floors or ceilings
- Electric ceiling or wall panels
- Electric heating cable in ceilings or floors
- Deep heat, a modified storage system using electric heating cable or embedded hydronic tubing in ceilings or floors

---

The preparation of this chapter is assigned to TC 6.5, Radiant Space Heating and Cooling.

In these systems, generally more than 50% of the heat transfer from the controlled temperature surface to other surfaces is by radiation. Panel heating and cooling systems are used in residences, office buildings for perimeter heating, classrooms, hospital patient rooms, swimming pool areas, repair garages, and industrial and warehouse applications. Additional information is available in Chapter 6 of the 1992 *ASHRAE Handbook—Systems and Equipment.*

Some radiant panel systems combine controlled heating and cooling surfaces with central station air conditioning and are referred to as hybrid conditioning systems. They are used more for cooling than for heating (Wilkins and Kosonen 1992). The controlled temperature surfaces may be in the floor, walls, or ceiling, with the temperature maintained by electric resistance or by circulating water or air. The central station can be a basic, one-zone, constant-temperature, or constant-volume system, or it can incorporate some or all the features of dual-duct, reheat, multizone, or variable-volume systems. When used in combination with other water/air systems, radiant panels provide zone control of temperature and humidity.

Metal ceiling panels may be integrated into the central heating and cooling system to provide individual room or zone heating and cooling. These panels can be designed as small units to fit the building module, or they can be arranged as large continuous areas for economy. Room thermal conditions are maintained primarily by direct transfer of radiant energy, normally using four-pipe hot and chilled water. These metal ceiling panel systems have generally been used in hospital patient rooms.

The application of metal ceiling panel systems is discussed in Chapter 6 of the 1992 *ASHRAE Handbook—Systems and Equipment.*

ELEMENTARY DESIGN RELATIONSHIPS

When considering radiant heating or cooling for human comfort, the following five terms can be used to describe the temperature and energy characteristics of the total radiant environment:

- *Mean radiant temperature* (MRT) $\bar{t}_r$ is the temperature of an imaginary isothermal black enclosure in which an occupant would exchange the same amount of heat by radiation as in the actual nonuniform environment.
- *Ambient temperature* t_a is the temperature of the air surrounding the occupant.
- *Operative temperature* t_o is the temperature of a uniform isothermal black enclosure in which the occupant would exchange the same amount of heat by radiation and convection as in the actual nonuniform environment.

For air velocities less than 0.4 m/s and mean radiant temperatures less than 50°C, the operative temperature is approximately equal to the adjusted dry-bulb temperature, which is the average of the air and mean radiant temperatures.

- *Adjusted dry-bulb temperature* is the average of the air temperature and the mean radiant temperature at a given location. The adjusted dry-bulb temperature is approximately equivalent to the operative temperature for air motions less than 0.4 m/s and mean radiant temperatures less than 50°C.

- *Effective radiant flux* (ERF) is defined as the *net radiant heat exchanged* at the ambient temperature t_a between an occupant, whose surface is hypothetical, and all enclosing surfaces and directional heat sources and sinks. Thus, ERF is the net radiant energy received by the occupant from all surfaces and sources whose temperatures *differ* from t_a. ERF is particularly useful in high intensity radiant heating applications.

The relationship between the aforementioned terms can be shown for an occupant at surface temperature t_{sf}, exchanging sensible heat H_m in a room with ambient air temperature t_a and mean radiant temperature $\bar{t}_r$. The linear radiative and convective heat transfer coefficients are h_r and h_c, respectively; the latter coefficient is a function of the air movement V. The heat balance equation is

$$H_m = h_r(t_{sf} - \bar{t}_r) + h_c(t_{sf} - t_a) \tag{1}$$

During thermal equilibrium, H_m is equal to metabolic heat minus work minus evaporative cooling by sweating. By definition of operative temperature,

$$H_m = (h_r + h_c)(t_{sf} - t_o) \tag{2}$$

Using Equations (1) and (2) to solve for t_o yields

$$t_o = (h_r\bar{t}_r + h_c t_a) / (h_r + h_c) \tag{3}$$

Thus, t_o is an average of $\bar{t}_r$ and t_a, weighted by their respective heat transfer coefficients; it represents how people sense the thermal level of their total environment as a single temperature.

The combined heat transfer coefficient is h, where $h = h_r + h_c$. Rearranging Equation (1) and substituting $h - h_r$ for h_c,

$$H_m + h_r(\bar{t}_r - t_a) = h(t_{sf} - t_a) \tag{4}$$

where $h_r(\bar{t}_r - t_a)$ is, by definition, the effective radiant flux (ERF) and represents the radiant energy absorbed by the occupant from all sources whose temperatures differ from t_a.

The principal relationships between $\bar{t}_r$, t_a, t_o, and ERF are as follows:

$$ERF = h_r(\bar{t}_r - t_a) \tag{5}$$

$$ERF = h(t_o - t_a) \tag{6}$$

$$\bar{t}_r = t_a + ERF/h_r \tag{7}$$

$$t_o = t_a + ERF/h \tag{8}$$

$$\bar{t}_r = t_a + (h/h_r)(t_o - t_a) \tag{9}$$

$$t_o = t_a + (h_r/h)(\bar{t}_r - t_a) \tag{10}$$

In Equations (1) through (10), the radiant environment is treated as a blackbody with temperature $\bar{t}_r$. The effect of the emittance of the source, radiating at absolute temperature in kelvins, and the absorptance of the skin and clothed surfaces is reflected in the effective values of $\bar{t}_r$ or ERF and not in the coefficient h_r, which is given in general by

$$h_r = 4\sigma f_{eff}[(\bar{t}_r + t_a)/2 + T]^3 \tag{11}$$

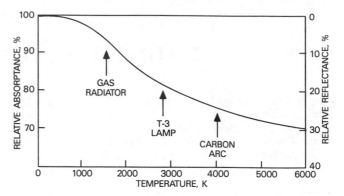

Fig. 1 Relative Absorptance and Reflectance of Skin and Typical Clothing Surfaces at Various Color Temperatures

where

h_r = linear radiative heat transfer coefficient, W/(m²·K)
f_{eff} = ratio of radiating surface of the human body to its total DuBois surface area A_D = 0.71
σ = Stefan-Boltzmann constant = 5.67×10^{-8} W/(m²·K⁴)
T = 273 for temperatures in °C

The convective heat transfer coefficient for an occupant is

$$h_c = 8.5\ V^{0.5} \tag{12}$$

where h_c is in W/(m²·K) and V is air movement in m/s.

When $\bar{t}_r > t_a$, ERF adds heat to the body system; when $t_a > \bar{t}_r$, heat is lost from the body system by radiant cooling. ERF is independent of the surface temperature of the occupant and can be measured directly by a black globe thermometer or any blackbody radiometer or flux meter using the ambient air t_a as its heat sink.

In the above definitions and for radiators below 925°C (1200 K), the body clothing and skin surface are treated as blackbodies, exchanging radiation with an imaginary blackbody surface at temperature $\bar{t}_r$. The effectiveness of a radiating source on human occupants is governed by the absorptance α of the skin and clothing surface for the color temperature (in K) of that radiating source. The relationship between α and temperature is illustrated in Figure 1. The values for α are those expected relative to the matte black surface normally found on globe thermometers or radiometers for measuring radiant energy. A gas radiator usually operates at 925°C (1200 K); a quartz lamp, for example, radiates at 2200°C (2475 K) with 240 V; and the sun's radiating temperature is 5530°C (5800 K). The use of α in estimating the ERF and t_o caused by sources radiating at temperatures above 925°C (1200 K) is discussed in the section on Test Instrumentation for Radiant Heating.

DESIGN CRITERIA FOR ACCEPTABLE RADIANT HEATING

Perceptions of comfort, temperature, and thermal acceptability are related to activity, the transfer of body heat from the skin to the environment, and the resulting physiological adjustments and body temperature. Heat transfer is affected by the ambient air temperature, thermal radiation, air movement, humidity, and clothing worn. Thermal sensation is described by feelings of hot, warm, slightly warm, neutral, slightly cool, cool, and cold. An acceptable environment is defined as one in which at least 80% of the occupants perceive a thermal sensation between "slightly cool" and "slightly warm." Comfort is associated with a neutral thermal sensation during which the human body regulates its internal temperature with minimal physiological effort for the activity concerned. In contrast, warm discomfort is related primarily to physiological strain necessary to maintain the body's thermal equilibrium rather

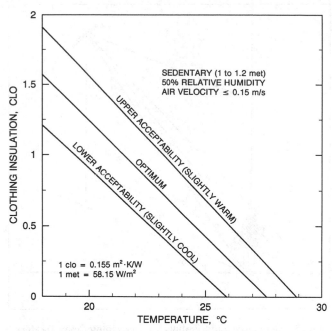

Fig. 2   Range of Thermal Acceptability for Various Clothing Insulations and Operative Temperatures

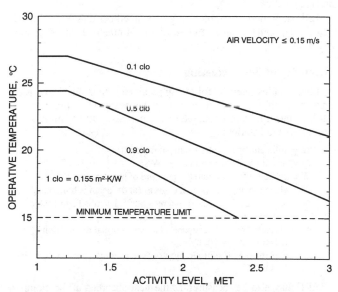

Fig. 3   Optimum Operative Temperatures for Active People in Low Air Movement Environments

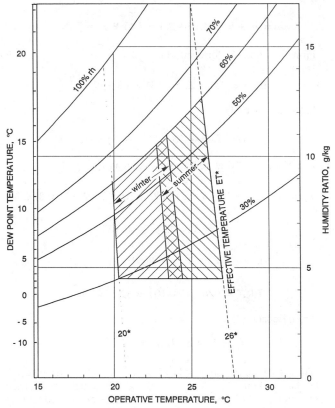

Fig. 4   ASHRAE Comfort Chart Modified for Radiant Heating and Cooling

than to the temperature sensation experienced. For a full discussion on the interrelation of physical, psychological, and physiological factors, refer to Chapter 8 of the 1993 *ASHRAE Handbook—Fundamentals*.

ASHRAE studies show a linear relationship between the clothing insulation worn and the operative temperature t_o for comfort. This is illustrated for a sedentary subject in Figure 2. Figure 3 shows the effect of both activity and clothing on the t_o for comfort. Figure 4 shows the effects of high and low humidity on a sedentary person wearing average clothing.

A comfortable t_o at 50% rh is perceived as slightly warmer as the humidity increases or is perceived as slightly cooler as the humidity decreases. Changes in humidity have a much greater effect on "warm" and "hot" discomfort. In contrast, "cold" discomfort is only

slightly affected by humidity and is very closely related to a "cold" thermal sensation. Figures 2, 3, and 4 are adapted from ANSI/ASHRAE *Standard* 55, Thermal Environmental Conditions for Human Occupancy.

Determining the specifications for a radiant heating installation designed for human occupancy and acceptability must involve the following steps:

1. Define the probable activity (metabolism) level of and clothing worn by the occupant and the air movement in the occupied space. The following are two examples:

Case 1: Sedentary (65 W/m^2 or 1.1 met)

Clothing insulation = 0.09 m^2·K/W or 0.6 clo; air movement = 0.15 m/s

Case 2: Light work (116 W/m^2 or 2 met)

Clothing insulation = 0.14 m^2·K/W or 0.9 clo; air movement = 0.5 m/s

2. From Figure 2 or 3, determine the optimum t_o for comfort and acceptability:

$$\text{Case 1: } t_o = 23.5°C; \quad \text{Case 2: } t_o = 17°C$$

3. For the ambient air temperature t_a, calculate the mean radiant temperature $\bar{t}_r$ and/or ERF necessary for comfort and thermal acceptability.

Case 1: For $t_a = 15°C$ and assuming $\bar{t}_r = 30°C$.

Solve for h_r from Equation (11):

$$h_r = 4 \times 5.67 \times 10^{-8} \times 0.71 \left[(30 + 15)/2 + 273 \right]^3 = 4.15$$

Solve for h_c from Equation (12):

$$h_c = 8.5 (0.15)^{0.5} = 3.3$$

Then,

$$h = h_r + h_c = 4.15 + 3.3 = 7.45 \ W/(m^2 \cdot K)$$

From Equation (6), for comfort,

$$ERF = 7.45 \ (23.5 - 15) = 63.3 \ W/m^2$$

From Equation (9),

$$\bar{t}_r = 15 + (7.45/4.15) \ (23.5 - 15) = 30.2°C$$

Case 2: For $t_a = 10°C$ at 50% rh and assuming $\bar{t}_r = 25°C$.

$$h_r = 4 \times 5.67 \times 10^{-8} \times 0.71 \ [\ (25 + 10) \ /2 + 273]^3 = 3.95$$

$$h_c = 8.5 \ (0.5)^{0.5} = 6.01$$

$$h = 3.95 + 6.01 = 9.96 \ W/(m^2 \cdot K)$$

$$ERF = 9.96 \ (17 - 10) = 69.7 \ W/m^2$$

From Equation (7),

$$\bar{t}_r = 10 + 69.7/3.95 = 27.6°C$$

The t_o for comfort, predicted by Figure 2, is on the "slightly cool" side when the humidity is low; for very high humidities, the predicted t_o for comfort is "slightly warm." This small effect on comfort can be seen in Figure 4. For example, for high humidity at $t_{dp} = 13°C$, the t_o for comfort is

Case 1: $t_o = 22.2°C$, compared to 23.5°C at 50% rh

Case 2: $t_o = 16°C$, compared to 17°C at 50% rh

When thermal acceptability is the primary consideration in an installation, humidity can sometimes be ignored in preliminary design specifications. However, for conditions where radiant heating and the work level cause sweating and high heat stress, humidity is a major consideration.

Equations (3) to (12) can also be used to determine the ambient air temperature t_a required when the mean radiant temperature MRT is to be maintained by a specified radiant system.

When making heat loss calculation, t_a must be determined. For a radiant system that is to maintain a MRT of $\bar{t}_r$, the operative temperature t_o can be determined from Figure 4. Then, t_a can be calculated by recalling that t_o is approximately equal to the average of t_a and t_r. For example, for a t_o of 23°C and a radiant system designed to maintain an MRT of 26°C, the t_a would be 20°C.

DESIGN FOR BEAM RADIANT HEATING

Spot beam radiant heat can improve comfort at a specific location in a large, poorly heated work area. The design problem is specifying the type, capacity, and orientation of the beam heater.

Using the same reasoning used for Equations (1) through (10), the effective radiant flux ΔERF that must be added to an unheated work space with an operative temperature t_{uo} to result in a t_o for comfort (as given by Figure 2 or 3) is

$$\Delta ERF = h \ (t_o - t_{uo}) \tag{13}$$

or

$$t_o = t_{uo} + (\Delta ERF) \ /h \tag{14}$$

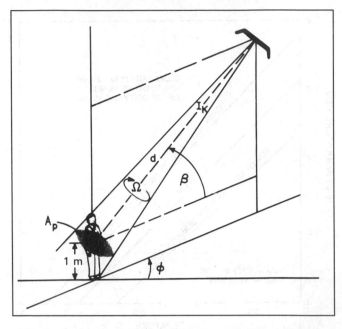

Fig. 5 Geometry and Symbols for Describing Beam Heaters

This equation is unaffected by air movement. The heat transfer coefficient h for the occupant in Equation (13) is given by Equations (11) and (12), with $h = h_r + h_c$.

ERF is, by definition, the energy absorbed per unit of total body surface A_D (DuBois area) and *not* the total effective radiating area A_{eff} of the body.

Geometry of Beam Heating

Figure 5 illustrates the following parameters that must be considered in specifying a beam radiant heater designed to produce the ERF, or mean radiant temperature $\bar{t}_r$, necessary for comfort at an occupant's workstation:

Ω = solid angle of heater beam, steradians (sr)
I_K = irradiance from beam heater, W/sr
K = subscript for absolute temperature of beam heater, K
β = elevation angle of heater, degrees (at 0°, beam is horizontal)
Φ = azimuth angle of heater, degrees (at 0°, beam is facing subject)
d = distance from beam heater to center of occupant, m
A_p = projected area of occupant on a plane normal to direction of heater beam (Φ, β), m²
α_K = absorptance of skin-clothing surface at emitter temperature (see Figure 1)

ERF may also be measured as the heat absorbed at the clothing and skin surface of the occupant from a beam heater at absolute temperature:

$$ERF = [\alpha_K \cdot I_K \cdot (A_p/d^2)] \ /A_D \tag{15}$$

where ERF is in W/m² and (A_p/d^2) is the solid angle subtended by the projected area of the occupant from the radiating beam heater I_K, which is treated here as a point source. A_D is the DuBois area:

$$A_D = 0.202 W^{0.425} \times H^{0.725}$$

where

W = occupant mass, kg
H = occupant height, m

For additional information on radiant flux distribution patterns and sample calculations of radiation intensity I_K and ERF, refer to Chapter 15 of the 1992 *ASHRAE Handbook—Systems and Equipment*.

FLOOR RERADIATION

In most low, medium, and high intensity radiant heater installations, local floor areas are strongly irradiated. The floor absorbs most of this energy and warms to an equilibrium temperature t_f, which is higher than that of the ambient air temperature t_a and the unheated room enclosure surfaces. The temperature of the unheated areas (walls, ceiling, and floor that are distant from heaters) can be assumed to be the same as the air temperature. Part of the energy directly absorbed by the floor is transmitted by conduction to the cooler underside (or, for slabs-on-grade, to the ground), part is transferred by natural convection to room air, and the remainder is reradiated. The warmer floor will raise ERF or $\bar{t}_r$ over that caused by the heater alone.

For a person standing on a large flat floor that has a temperature raised by direct radiation t_f, the linearized $\bar{t}_r$ due to the floor and unheated walls is

$$\bar{t}_{rf} = F_{p-f}\,t_f + (1 - F_{p-f})\,t_a \qquad (16)$$

where the unheated walls, ceiling, and ambient air are assumed to be at t_a, and F_{p-f} is the angle factor governing the radiation exchange between the heated floor and the person.

The ERF_f from the floor affecting the occupant, which is due to the $(t_f - t_a)$ difference, is

$$ERF_f = h_r\,(\bar{t}_{rf} - t_a)$$
$$= h_r F_{p-f}\,(t_f - t_a) \qquad (17)$$

where h_r is the linear radiative heat transfer coefficient for a person as given by Equation (11). Figures 50 and 53 in Fanger (1973) show F_{p-f} is 0.44 for a standing or sitting subject when the walls are farther than 5 m away. For an average size 5 m by 5 m room, a value of F_{p-f} of 0.35 is suggested. For detailed information on floor reradiation, see Chapter 15 in the 1992 *ASHRAE Handbook—Systems and Equipment*.

In summary, when radiant heaters warm occupants in a selected area of a poorly heated space, the radiation heat necessary for comfort consists of two additive components: (1) ERF directly caused by the heater and (2) reradiation ERF_f from the floor. The effectiveness of floor reradiation can be improved by choosing flooring with a low specific conductivity. Flooring with high thermal inertia may be desirable during radiant transients, which may occur as the heaters are cycled by a thermostat set to the desired operative temperature t_o.

Asymmetric Radiant Fields

In the past, comfort heating has required flux distribution in occupied areas to be uniform, which is not possible with beam radiant heaters. Asymmetric radiation fields, such as those experienced when lying in the sun on a cool day or when standing in front of a warm fire, can be pleasant. Therefore, a limited amount of asymmetry, which is allowable for comfort heating, is referred to as "reasonable uniform radiation distribution" and is used as a design requirement.

To develop criteria for judging the degree of asymmetry allowable for comfort heating, Fanger et al. (1980) proposed defining radiant temperature asymmetry as the difference between the plane radiant temperature of two opposing surfaces. Plane radiant temperature is the equivalent $\bar{t}_{r1}$ caused by radiation on one side of the subject, compared with the equivalent $\bar{t}_{r2}$ caused by radiation on the opposite side. Gagge et al. (1967) conducted a study of subjects (eight clothed and eight unclothed) seated in a chair and heated by two lamps. Unclothed subjects found a $(\bar{t}_r - t_a)$ asymmetry as high as 11 K to be comfortable, but clothed subjects were comfortable for an asymmetry as high as 17 K.

For an unclothed subject lying on an insulated bed under a horizontal bank of lamps, neutral temperature sensation occurred for a t_o of 22°C, which corresponds to a $(t_o - t_a)$ asymmetry of 11 K or a $(\bar{t}_r - t_a)$ asymmetry of 15 K, both averaged for eight subjects (Stevens et al. 1969). In studies on heated ceilings, 80% of eight male and eight female clothed subjects voted conditions as comfortable and acceptable for asymmetries as high as 11 K. The latter study compared the floor and heated ceilings. The asymmetry in the MRTs for direct radiation from three lamps and for floor reradiation is about 0.5 K, which is negligible.

In general, the human body has a great ability to sum sensations from many hot and cold sources. For example, Australian aborigines sleep unclothed next to open fires in the desert at night, where t_a is 6°C. The $\bar{t}_r$ caused directly by three fires alone is 77°C, and the cold sky $\bar{t}_r$ is −1°C; the resulting t_o is 28°C, which is acceptable for human comfort (Scholander 1958).

According to the limited field and laboratory data available, an allowable design radiant asymmetry of 12 ± 3 K should cause little discomfort over the comfortable t_o range used by ANSI/ASHRAE *Standard* 55 and in Figures 2 and 3. Increased clothing insulation allows increases in the acceptable asymmetry, but increased air movement reduces it. Increased activity also reduces human sensitivity to changing $\bar{t}_r$ or t_o and, consequently, increases the allowable asymmetry. The design engineer should use caution with an asymmetry greater than 15 K, as measured by a direct beam radiometer or estimated by calculation.

RADIATION PATTERNS

Figure 6 indicates the basic radiation patterns commonly used in design for radiation from point or line sources (Boyd 1962). A point source radiates over an area that is proportional to the square of the

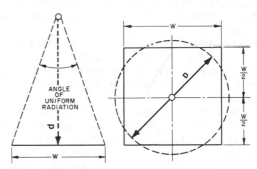

(A) PATTERN OF RADIATION FROM A POINT SOURCE

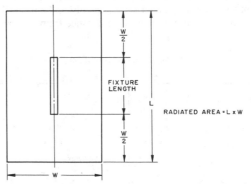

(B) PATTERN OF RADIATION FROM A LINE SOURCE

Note: The projected area W^2 normal to a beam that is Ω steradians wide at distance d is Ωd^2. The floor area irradiated by a beam heater at an angle elevation β is $W^2/\sin\beta$. Fixture length L increases the area irradiated by the factor $(1 + L/W)$.

Fig. 6 Basic Radiation Patterns for System Design
(Boyd 1962)

distance from the source. The area for a (short) line source also varies substantially as the square of the distance, with about the same area as the circle actually radiated at that distance. For line sources, the width of the pattern is determined by the reflector shape and position of the element within the reflector. The rectangular area used for installation purposes as the pattern of radiation from a line source assumes a length equal to the width plus the fixture length. This assumed length is satisfactory for design, but is often two or three times the pattern width.

Electric infrared fixtures are often identified by their beam pattern (Rapp and Gagge 1967), which is the radiation distribution normal to the line source element. The beam of a high intensity infrared fixture may be defined as that area in which the intensity is at least 80% of the maximum intensity encountered anywhere within the beam. This intensity is measured in the plane in which maximum control of energy distribution is exercised.

The beam size is usually designated in angular degrees and may be symmetrical or asymmetrical in shape. For adaptation to their design specifications, some manufacturers indicate beam characteristics based on 50% maximum intensity.

The control used for an electric system affects the desirable maximum end-to-end fixture spacing. Actual pattern length is about three times the design pattern length, so control in three equal stages is achieved by placing every third fixture on the same circuit. If all fixtures are controlled by input controllers or variable voltage to electric units, end-to-end fixture spacing can be nearly three times the design pattern length. Side-to-side minimum spacing is determined by the distribution pattern of the fixture and is not influenced by the method of control.

Low intensity equipment typically consists of a steel tube hung near the ceiling and parallel to the outside wall. Circulation of the products of combustion inside the tube elevates the tube temperature and radiant energy is emitted. The tube is normally provided with a reflector to direct the radiant energy down into the space to be conditioned.

Radiant ceiling panels for heating only are installed in a narrow band around the perimeter of an occupied space and are usually the primary heat source for the space. The radiant source is a (long) linear source. The flux density is inversely proportioned to the distance from the source.

The rate of radiation exchange between a panel and a particular object depends on their temperatures, emissivities, and geometrical orientation (i.e., shape factor). It also depends on the temperatures and configurations of all the other objects and walls within the space.

The energy flux for an ideal line source is shown in Figure 7. All objects (except the radiant source) are at the same temperature, and the radiant source is suspended symmetrically in the room. The flux density is inversely proportional to the distance from the source.

DESIGN FOR TOTAL SPACE HEATING

Radiant heating differs from conventional heating by a moderately elevated ERF, $\bar{t}_r$, or t_o over the ambient temperature t_a. Standard methods of design are normally used, although informal studies indicate that radiant heating requires a lower heating capacity than convection heating (Zmeureanu et al. 1987). Buckley and Seel (1987) demonstrated that a combination of elevated floor temperature, higher mean radiant temperature, and reduced ambient temperature results in lower thermostat settings and a reduced temperature differential across the building envelope, and thus, a lower heat loss and lower heating load for the structure. In addition, the peak load may be decreased due to heat (cool) stored in the structure (Kilkis 1990 and 1992).

Most gas radiation systems for full building heating concentrate the bulk of capacity at mounting heights of 3 to 5 m at the perimeter, directed at the floor near the walls. Units can be mounted considerably higher. Successful application depends on supplying the proper amount of heat in the occupied area. Heaters should be located to take maximum advantage of the pattern of radiation produced. Exceptions to perimeter placement include walls with high transmission losses and extreme height, as well as large roof areas where roof heat losses exceed perimeter and other heat losses.

Electric infrared systems installed indoors for complete building heating have used layouts that uniformly distribute the radiation throughout the area used by people, as well as layouts that emphasize perimeter placement, such as in ice hockey rinks. Some electric radiant heaters emit a significant amount of visible radiation and provide both heating and illumination.

The orientation of equipment and people is less important for general area heating (large areas within larger areas) than it is for spot heating. With reasonably uniform radiation distribution in work or living areas, the exact orientation of the units is not important. Higher intensities of radiation may be desirable near walls with outside exposure. Radiation shields (reflective to infrared) fastened a few centimetres from the wall to allow free air circulation between the wall and shield are effective for frequently occupied work locations close to outside walls.

In full building heating, units should be placed where their radiant and convective output best compensates for the structure's heat losses. The objective of a complete heating system is to provide a warm floor with low conductance to the heat sink beneath the floor. This thermal storage may permit cycling of units with standard controls.

TEST INSTRUMENTATION FOR RADIANT HEATING

In designing a radiant heating system, the calculation of radiant heat exchange may involve some untested assumptions. During installation of the radiant heating equipment in the field, the designer must test and adjust the equipment to ensure that it provides acceptable comfort conditions. The black globe thermometer and directional radiometer can be used to evaluate the installation.

Black Globe Thermometer

The classic (Bedford) globe thermometer is a thin-walled, matte-black, hollow sphere with a thermocouple, thermistor, or thermometer placed at the center. It can directly measure $\bar{t}_r$, ERF, and t_o. When a black globe is in thermal equilibrium with its environment, the gain in radiant heat from various sources is balanced by the convective loss to ambient air. Thus, in terms of the globe's linear radiative and convective heat transfer coefficients, h_{rg} and h_{cg}, respectively, the heat balance at equilibrium is

$$h_{rg}(\bar{t}_{rg} - t_g) = h_{cg}(t_g - t_a) \qquad (18)$$

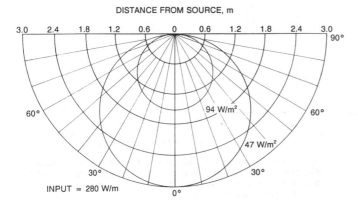

DISTANCE FROM SOURCE, m

Fig. 7   Lines of Constant Radiant Flux for a Line Source

where $\bar{t}_{rg}$ is the mean radiant temperature measured by the globe and t_g is the temperature in the globe.

In general, the $\bar{t}_{rg}$ of Equation (18) equals the $\bar{t}_r$ affecting a person when the globe is placed at the center of the occupied space and when the radiant sources are distant from the globe.

The effective radiant flux measured by a black globe is

$$ERF_g = h_{rg}(\bar{t}_{rg} - t_a) \qquad (19)$$

which is analogous to Equation (5) for occupants. From Equations (18) and (19), it follows that

$$ERF_g = (h_{rg} + h_{cg})(t_g - t_a) \qquad (20)$$

If the ERF_g of Equation (20) is modified by the factors that describe the skin-clothing absorptance α_K and shape f_{eff} of an occupant relative to the black globe, the corresponding ERF affecting the occupant is

$$ERF \text{ (for a person)} = f_{eff}\alpha_K ERF_g \qquad (21)$$

where α_K is defined in the section on Geometry of Beam Heating and f_{eff}, which is defined after Equation (11), is approximately 0.71 and equals the ratio h_r/h_{rg}. The definition of t_o affecting a person, in terms of t_g and t_a, is given by

$$t_o = Kt_g + (1 - K)t_a \qquad (22)$$

where the coefficient K is

$$K = \alpha_K f_{eff}(h_{rg} + h_{cg}) / (h_r + h_c) \qquad (23)$$

Ideally, when K is unity, the t_g of the globe would equal the t_o affecting a person.

For an average comfortable equilibrium temperature of 25°C and noting that f_{eff} for the globe is unity, Equation (11) yields

$$h_{rg} = 6.01 \text{ W}/(\text{m}^2 \cdot \text{K}) \qquad (24)$$

and Equation (12) yields

$$h_{cg} = 6.32D^{-0.4}V^{0.5} \qquad (25)$$

where

D = globe diameter, m
V = air velocity, m/s

Equation (25) is Bedford's convective heat transfer coefficient for a 150-mm globe's convective loss, modified for D. For any radiating source below 1200 K, the ideal diameter of a sphere that makes $K = 1$ and that is independent of air movement is 200 mm (see Table 1). Table 1 shows the value of K for various values of globe diameter D and ambient air movement V. The table shows that the uncorrected temperature of the traditional 150-mm globe would overestimate the true $(t_o - t_a)$ difference by 6% for velocities up to 1 m/s, and the probable error of overestimating t_o by t_g uncorrected would be less than 0.5 K. Globe diameters between 150 and 200 mm are optimum for using the uncorrected t_g measurement for t_o. The exact value for K may be used for the smaller sized globes when estimating t_o from t_g and t_a measurements. The value of $\bar{t}_r$ may be found by substituting Equations (24) and (25) in Equation (18), because $\bar{t}_r$ (person) is equal to $\bar{t}_{rg}$. The smaller the globe, the greater the variation in K caused by air movement. Globes with D greater than 200 mm will overestimate the importance of radiation gain versus convection loss.

For sources radiating at high temperature (1000 to 5800 K), the ratio α_m/α_g may be set near unity by using a pink-colored globe surface, whose absorptance for the sun is 0.7, a value similar to that of human skin and normal clothing (Madsen 1976).

Table 1 Value of K for Various Air Velocities and Globe Diameters ($\alpha_g = 1$)

Air Velocity, m/s	Approximate Globe Diameter, mm			
	50	100	150	200
0.25	1.35	1.15	1.05	0.99
0.5	1.43	1.18	1.06	0.99
1.0	1.49	1.21	1.07	1.00
2.0	1.54	1.23	1.08	1.00
4.0	1.59	1.26	1.09	1.00

In summary, the black globe thermometer is simple and inexpensive and may be used to determine $\bar{t}_r$ [Equation (18)] and the ERF [Figure 1 and Equations (20) and (21)]. When the radiant heater temperature is less than 1200 K, the uncorrected t_g of a 150- to 200-mm black globe is a good estimate of the t_o affecting the occupants. A pink globe extends its usefulness to sun temperatures (5800 K). A globe with a low mass and low thermal capacity is more useful because it reaches thermal equilibrium in less time.

Using the heat exchange principles described, many instruments of various shapes, heated and unheated, have been designed to measure acceptability in terms of t_o, $\bar{t}_r$, and ERF, as sensed by their own geometric shape. Madsen (1976) developed an instrument that can determine the predicted mean vote (PMV) from the $(t_g - t_a)$ difference, as well as correct for clothing insulation, air movement, and activity (ISO 1984).

Directional Radiometer

The angle of acceptance (in steradians) in commercial radiometers allows the engineer to point the radiometer directly at a wall, floor, or high-temperature source and read the average temperature of that surface. Directional radiometers are calibrated to measure either the radiant flux accepted by the radiometer or the equivalent blackbody radiation temperature of the emitting surface. Many are collimated to sense small areas of either body, clothing, wall, or floor surface. A directional radiometer allows rapid surveys and analyses of important radiant heating factors such as the temperature of skin, clothing surfaces, and walls and floors, as well as the radiation intensity I_K of heaters on the occupants. One radiometer for direct measurement of the equivalent radiant temperature has an angle of acceptance of 2.8° or 0.098 sr, so that at 1 m, it measures the average temperature over a projected circle about 30 mm in diameter.

APPLICATION CONSIDERATIONS

All bodies with a surface temperature above absolute zero radiate energy with wavelengths that depend on the body surface temperature. Each facet of a surface emits rays in straight lines at right angles to the facet. When examined under a microscope, the surface of concrete or rough plaster is seen to be covered with numerous facets, each giving off radiant energy. Polished steel or similar surfaces show no such facets. Thus, a rough surface radiates heat more efficiently than a polished surface.

The effect of radiant heat is experienced when the body is exposed to the sun's rays on a cool but sunny day. Some of the rays incident on the body come directly from the sun and include a range of wavelengths. Other rays come from other objects that absorbed radiant energy from the sun and reradiate the energy at longer wavelengths and lower temperatures, producing a comfortable feeling of warmth. When a cloud briefly obscures the sun, it instantly creates a sensation of cold, although in such a short interval, the air temperature does not vary at all.

To create and maintain conditions that satisfy the physiological demands of the human body, a system must consider the three main factors controlling heat loss from the human body: radiation, convection, and evaporation. Radiant heating systems may be thought to be applicable only for certain buildings and only in certain climates. However, wherever people live, the three factors of heat loss must be considered, making comfort conditions as important in very cold climates as they are in moderate climates. Low-temperature radiation can be used to maintain comfort conditions even in the most severe weather conditions.

Panel heating and cooling systems provide a comfortable environment by controlling surface temperatures and minimizing air motion within a space. Thermal comfort, as defined by ANSI/ASHRAE *Standard* 55, is "that condition of mind which expresses satisfaction with the thermal environment." A person who is comfortable may not be not aware that the environment is being heated or cooled. The mean radiant temperature (MRT) strongly influences the feeling of comfort.

Radiant energy is also applied to control condensation on surfaces such as the large glass exposures in the Chicago airport terminal.

When the surface temperature of outside walls, particularly those with large areas of glass, deviates too much from the room air temperature and from the temperature of other surfaces, simplified calculations for the load and the operative temperature may lead to errors in sizing and locating the panels. In such cases, more detailed radiant exchange calculations may be required, with separate estimation of heat exchange between the panels and each surface. A large window area may lead to significantly lower mean radiant temperatures than expected. For example, Athienitis and Dale (1987) reported an MRT 3 K lower than room air temperature for a room with a glass area equivalent to 22% of its floor area.

Other factors to consider when placing radiant heaters in specific applications include the following:

- Gas and electric high-temperature infrared heaters must not be placed where they could ignite flammable dust or vapors, or decompose vapors into toxic gases.

- Fixtures must be located with recommended clearances to ensure proper heat distribution. Stored materials must be kept far enough from the fixtures to avoid hot spots. Manufacturers' recommendations must be followed.

- Unvented gas heaters inside tight, poorly insulated buildings may cause excessive humidity with condensation on cold surfaces. Proper insulation, vapor barriers, and ventilation prevent these problems.

- Combustion-type heaters in tight buildings may require makeup air to ensure proper venting of combustion gases. Some infrared heaters are equipped with induced draft fans to relieve this problem.

- Some transparent materials may break due to uneven application of high intensity infrared. Infrared energy is transmitted without loss from the radiator to the absorbing surfaces. The system must produce the proper temperature distribution at the absorbing surfaces. Problems are rarely encountered with glass 6 mm or less in thickness.

- Comfort heating with infrared heaters requires a reasonably uniform flux distribution in the occupied area. While thermal discomfort can be relieved in warm areas with high air velocity, such as on loading docks, the full effectiveness of a radiant heater installation is reduced by the presence of high air velocity. At specific locations in large work areas, direct radiant heating may be one means of providing comfort for the occupants.

APPLICATIONS

Low, Medium, and High Intensity Infrared Applications

Low, medium, and high intensity infrared are used extensively in industrial, commercial, and military applications. This equipment is particularly effective in large areas with high ceilings, such as in buildings with large air volumes and in areas with high infiltration rates, large access doors, or large ventilation requirements.

Factories. Low intensity radiant equipment suspended near the ceiling around the perimeter of facilities with high ceilings enhances the comfort of employees because it warms floors and equipment in the work area. For older uninsulated buildings, the energy cost for low intensity radiant equipment is less than that of other heating systems. High intensity infrared is particularly effective for spot heating in large unheated facilities.

Warehouses. Low and high intensity infrared are used for heating warehouses, which usually have a large volume of air, are often poorly insulated, and have high infiltration. Low intensity infrared equipment is installed near the ceiling around the perimeter of the building. High intensity infrared equipment, also suspended near the ceiling, is arranged to control radiant intensity and provide uniform heating at the working level.

Garages. Low intensity infrared provides comfort for mechanics working near or on the floor. With elevated MRT in the work area, comfort is provided at a lower ambient temperature.

In winter, opening the large overhead doors to admit equipment for service allows a substantial entry of cold outdoor air. On closing the doors, the combination of reradiation from the warm floor and radiant heat warming the occupants (not the air) provides rapid recovery of comfort. Radiant energy rapidly warms the cold (perhaps snow-covered) vehicles. Radiant floor panel heating systems are also effective in garages.

Low intensity equipment is suspended near the ceiling around the perimeter, often with greater concentration near overhead doors. High intensity equipment is also used to provide additional heat near doors.

Aircraft Hangars. Equipment suspended near the roof of hangars, which have high ceilings and large access doors, provides uniform radiant intensity throughout the working area. A heated floor is particularly effective in restoring comfort after an aircraft has been admitted. As in garages, the combination of reradiation from the warm floor and radiation from the radiant heating system provides rapid regain of comfort. Radiant energy also heats aircraft moved into the work area.

Greenhouses. In greenhouse applications, a uniform flux density must be maintained throughout the facility to provide acceptable growing conditions. In a typical application, low intensity units are suspended near and running parallel to the peak of the greenhouse.

Outdoor Applications. Applications include loading docks, racetrack stands, outdoor restaurants, and under marquees. Low, medium, and high intensity infrared are used in these facilities, depending on their layout and requirements.

Other Applications. Radiant heat may be used in a variety of large, high-ceilinged facilities, including churches, gymnasiums, swimming pools, enclosed stadiums, and facilities that are open to the outdoors.

Low, medium, and high intensity infrared are also used for other industrial applications, including process heating for component or paint drying ovens, humidity control for corrosive metal storage, and snow control for parking or loading areas.

Panel Heating and Cooling

Residences. Embedded pipe coil systems, electric resistance panels, and forced warm-air panel systems have all been used in residences. The embedded pipe coil system is most common, using plastic or rubber tubing in the floor slab or copper tubing systems in older plaster ceilings. These systems are suitable for conventionally constructed residences with normal amounts of glass. Lightweight hydronic metal panel ceiling systems have also been applied to residences, and prefabricated electric panels are advantageous, particularly in rooms that have been added on.

Office Buildings. A panel system is usually applied as a perimeter heating system. Panels are typically piped to provide exposure control with one riser on each exposure and all horizontal piping incorporated in the panel piping. In these applications, the air system provides individual room control. Perimeter radiant panel systems have also been installed with individual zone controls. However, this type of installation is usually more expensive and, at best, provides minimal energy savings and limited additional occupant comfort. Radiant panels can be used for cooling as well as heating. Cooling installations are generally limited to retrofit or renovation jobs where ceiling space is insufficient for the required ductwork sizes. In these installations, the central air supply system provides ventilation air, dehumidification, and some sensible cooling. Water distribution systems using the two- and four-pipe concept may be used. Hot water supply temperatures are commonly reset by outside temperature, with additional offset or flow control to compensate for solar load. Panel systems are readily adaptable to accommodate most changes in partitioning. Electric panels in lay-in ceilings have been used for full perimeter heating.

Schools. In all areas except gymnasiums and auditoriums, panels are usually selected for heating only, and may be used with any type of approved ventilation system. The panel system is usually sized to offset the transmission loads plus any reheating of the air. If the school is air conditioned by a central air system and has perimeter heating panels, a single-zone piping system may be used to control the panel heating output, and the room thermostat modulates the supply air temperature or volume. Heating and cooling panel applications are similar to those in office buildings. Panel heating and cooling for classroom areas has no mechanical equipment noise to interfere with instructional activities.

Hospitals. The principal application of heating and cooling radiant panel systems has been for hospital patient rooms. Perimeter radiant heating panel systems are typically applied in other areas of hospitals. Compared to conventional systems, radiant heating and cooling systems are well suited to hospital patient rooms because they (1) provide a draft-free, thermally stable environment, (2) have no mechanical equipment or bacteria and virus collectors, and (3) do not take up space in the room. Individual room control is usually achieved by throttling the water flow through the panel. The supply air system is often a 100% outdoor air system; minimum air quantities delivered to the room are those required for ventilation and exhaust of the toilet room and soiled linen closet. The piping system is typically a four-pipe design. Water control valves should be installed in corridors so that they can be adjusted or serviced without entering the patient rooms. All piping connections above the ceiling should be soldered or welded and thoroughly tested. If cubicle tracks are applied to the ceiling surface, track installation should be coordinated with the radiant ceiling. Security panel ceilings are often used in areas occupied by mentally disturbed patients so that equipment cannot be damaged by a patient or used to inflict injury.

Swimming Pools. A partially clothed person emerging from a pool is very sensitive to the thermal environment. Panel heating systems are well suited to swimming pool areas. Floor panel temperatures must be controlled so they do not cause foot discomfort. Ceiling panels are generally located around the perimeter of the pool, not directly over the water. Panel surface temperatures are higher to compensate for the increased ceiling height and to produce a greater radiant effect on partially clothed bodies. Ceiling panels may also be placed over windows to reduce condensation.

Apartment Buildings. For heating, pipe coils are embedded in the masonry slab. The coils must be carefully positioned so as not to overheat one apartment while maintaining the desired temperature in another. The slow response of embedded pipe coils in buildings with large glass areas may be unsatisfactory. Installations for heating and cooling have been made with pipes embedded in hung plaster ceilings. A separate minimum-volume dehumidified air system provides the necessary dehumidification and ventilation for each apartment. The application of electric resistance elements embedded in floors or

behind a skim coat of plaster at the ceiling has increased. Electric panels are easy to install and simplify individual room control.

Industrial Applications. Panel systems are widely used for general space conditioning of industrial buildings in Europe. For example, the walls and ceilings of an internal combustion engine test cell are cooled with chilled water. Although the ambient air temperature in the space reaches up to 35°C, the occupants work in relative comfort when 13°C water is circulated through the ceiling and wall panels.

Other Building Types. Metal panel ceiling systems can be operated as heating systems at elevated water temperatures and have been used in airport terminals, convention halls, lobbies, and museums, especially those with large glass areas. Cooling may also be applied. Because radiant energy travels through the air without warming it, ceilings can be installed at any height and remain effective. One particularly high ceiling installed for a comfort application is 15 m above the floor, with a panel surface temperature of approximately 140°C for heating. The ceiling panels offset the heat loss from a single-glazed, all-glass wall.

The high lighting levels in television studios make them well suited to panel systems that are installed for cooling only and are placed above the lighting system to absorb the radiation and convection heat from the lights and normal heat gains from the space. The panel ceiling also improves the acoustical properties of the studio.

Metal panel ceiling systems are also installed in minimum and medium security jail cells and in facilities where disturbed occupants are housed. The ceiling construction is made more rugged by increasing the gage of the ceiling panels, and security clips are installed so that the ceiling panels cannot be removed. Part of the perforated metal ceiling can be used for air distribution.

New Techniques

With the introduction of thermoplastic and rubber tubing and new design techniques, interest has developed in radiant panel heating and cooling. The systems are energy efficient and use low water temperatures available from solar collector systems and heat pumps (Kilkis 1993).

Metal radiant panels can be integrated into the ceiling design to provide a narrow band of radiant heating around the perimeter of the building. Compared to baseboard or overhead air, radiant systems have better appearance, comfort, operating efficiency and cost, and product life.

SYMBOLS

A_D	total DuBois surface area of person, m²
A_{eff}	effective radiating area of person, m²
A_p	projected area of occupant normal to the beam, m²
clo	unit of clothing insulation equal to 0.155 m²·K/W
D	diameter of globe thermometer, m
d	distance of beam heater from occupant, m
ERF	effective radiant flux (person), W/m²
ERF_f	radiant flux caused by heated floor on occupant, W/m²
ERF_g	effective radiant flux (globe), W/m²
F_{p-f}	angle factor between occupant and heater floor
f_{eff}	ratio of radiating surface (person) to its total area (DuBois)
H_m	net metabolic heat loss from body surface, W/m²
h	combined heat transfer coefficient (person), W/(m²·K)
h_c	convective heat transfer coefficient for person, W/(m²·K)
h_{cg}	convective heat transfer coefficient for globe, W/(m²·K)
h_r	linear radiative heat transfer coefficient (person), W/(m²·K)
h_{rg}	linear radiative heat transfer coefficient for globe, W/(m²·K)
I_K	irradiance from beam heater, W/sr
K	coefficient that relates t_a and t_g to t_o [Equation (22)]
K	subscript indicating absolute irradiating temperature of beam heater, K
L	fixture length, m
met	unit of metabolic energy equal to 58.2 W/m²
t_a	ambient air temperature near occupant, °C
t_f	floor temperature, °C

t_g temperature in globe, °C
t_o operative temperature, °C
$\bar{t}_r$ mean radiant temperature affecting occupant, °C
$\bar{t}_{rf}$ linearized $\bar{t}_r$ caused by floor and unheated walls on occupant, °C
t_{sf} exposed surface temperature of occupant, °C
t_{uo} operative temperature of unheated workspace, °C
V air velocity, m/s
W width of a square equivalent to the projected area of a beam of angle Ω steradians at a distance d, m
α relative absorptance of skin-clothing surface to that of matte black surface
α_g absorptance of globe
α_K absorptance of skin-clothing surface at emitter temperature
α_m absorptance of skin-clothing surface at emitter temperatures above 925°C
β elevation angle of beam heater, degrees
Ω radiant beam width, sr
Φ azimuth angle of heater, degrees
σ Stefan-Boltzmann constant = 5.67×10^{-8} W/(m$^2 \cdot$ K^4)

REFERENCES

ASHRAE. 1992. Thermal environmental conditions for human occupancy. *Standard* 55-1992.

Athienitis, A.K. and J.D. Dale. 1987. A study of the effects of window night insulation and low emissivity coating on heating load and comfort. *ASHRAE Transactions* 93(1A):279-94.

Boyd, R.L. 1962. Application and selection of electric infrared comfort heaters. *ASHRAE Journal* 4(10):57.

Buckley, N.A. and T.P. Seel. 1987. Engineering principles support an adjustment factor when sizing gas-fired low-intensity infrared equipment. *ASHRAE Transactions* 93(1):1179-91.

Fanger, P.O. 1973. *Thermal comfort.* McGraw-Hill, New York.

Fanger, P.O., L. Banhidi, B.W. Olesen, and G. Langkilde. 1980. Comfort limits for heated ceiling. *ASHRAE Transactions* 86(2):141-56.

Gagge, A.P., G.M. Rapp, and J.D. Hardy. 1967. The effective radiant field and operative temperature necessary for comfort with radiant heating. *ASHRAE Transactions* 73(1):I.2.1-9; and *ASHRAE Journal* 9(5):63-66.

ISO. 1984. Moderate thermal environments—Determination of the PMV and PPD indices and specifications of the conditions for thermal comfort. *Standard* 7730-1984. International Standard Organization, Geneva.

Kilkis, B. 1990. Panel cooling and heating of buildings using solar energy. ASME Winter Meeting: Solar Energy in the 1990s (Nov. 25-30, Dallas), vol. 10:1-7.

Kilkis, B. 1992. Enhancement of heat pump performance using radiant floor heating systems. ASME Winter Meeting: Advanced Energy Systems, Recent Research in Heat Pump Design, Analysis, and Application (Nov. 8-13, Anaheim), vol. 28:119-27.

Kilkis, B. 1993. Radiant ceiling cooling with solar energy: Fundamentals, modeling, and a case design. *ASHRAE Transactions* 99(2):521-33.

Madsen, T.L. 1976. Thermal comfort measurements. *ASHRAE Transactions* 82(1):60-70.

Rapp, G.M. and A.P. Gagge. 1967. Configuration factors and comfort design in radiant beam heating of man by high temperature infrared sources. *ASHRAE Transactions* 73(3):1.1-1.8.

Scholander, P.E. 1958. Cold adaptation in the Australian aborigines. *Journal of Applied Physiology* 13:211-18.

Stevens, J.C., L.E. Marks, and A.P. Gagge. 1969. The quantitative assessment of thermal comfort. *Environmental Research* 2:149-65.

Wilkins, C.K. and R. Kosonen. 1992. Cool ceiling system: A European air-conditioning alternative. *ASHRAE Journal* 34(8):41-5.

Zmeureanu, R., P.P. Fazio, and F. Haghighat. Thermal performance of radiant heating panels. *ASHRAE Transactions* 94(2):13-27.

BIBLIOGRAPHY

AGA. 1960. Literature review of infra-red energy produced with gas burners. *Research Bulletin* 83, Catalog No. 35/IR, American Gas Association.

Boyd, R.L. 1959. High intensity infrared radiant heating. *Heating, Piping and Air Conditioning* (November):140.

Boyd, R.L. 1961. How to apply high intensity infrared heaters for comfort heating. *Heating, Piping and Air Conditioning* (January):230.

Boyd, R.L. 1963. Control of electric infrared energy distribution. *Electrical Engineering* (February):103.

Boyd, R.L. 1964. Beam characteristics of high intensity infrared heaters. IEEE Paper CP 64-123, Institute of Electrical and Electronics Engineers (February).

Frier, J.P. and W.R. Stephens. 1962. Design fundamentals for space heating with infrared lamps. *Illuminating Engineering* (December).

Griffiths, I.S. and D.A. McIntyre. 1974. Subjective response to overhead thermal radiation. *Human Factors* 16(3):415-22.

Hardy, J.D., A.P. Gagge, and J.A.J. Stolwijk. 1970. *Physiological and behavioral temperature regulation.* CC. Thomas, Springfield, IL.

McIntyre, D.A. 1974. The thermal radiating field. *Building Science* (9):247-62.

McNall, P.E., Jr. and R.E. Biddison. 1970. Thermal and comfort sensations of sedentary persons exposed to asymmetric radiant fields. *ASHRAE Transactions* 76(1):123-36.

Olesen, S., P.O. Fanger, P.B. Jensen, and O.J. Nielsen. 1972. Comfort limits for man exposed to asymmetric thermal radiation. Building Research Est. Report, *Thermal comfort and moderate heat stress*.

Walker, C.A. 1962. Control of high intensity infrared heating. *ASHRAE Journal* 4(10):66.

CHAPTER 50

SEISMIC RESTRAINT DESIGN

EARTHQUAKE damage to inadequately restrained heating, ventilating, air-conditioning, and refrigerating (HVAC&R) equipment can be extensive. The cost of restraining the equipment properly is relatively small compared to the high costs incurred replacing or repairing damaged equipment. It is also possible that the building will be unfit for occupation if the ventilation and/or cooling capacities become inadequate. The earthquake-resistive design of nonstructural building elements (components attached to the structural frame) did not receive engineering analysis until the 1964 Alaska earthquake, which measured 8.4 on the Richter scale. Damage reports from that destructive earthquake and from subsequent earthquakes stimulated the development of practical design methodologies.

This chapter covers restraint design to limit the movement of HVAC&R equipment during an earthquake. Large or conservative safety factors are applied to reduce the complexity of earthquake response analysis and evaluation. However, the additional cost of higher capacity restraints is small.

Local building officials must be contacted for specific requirements that may be more stringent than those presented in this chapter. Nearly all local codes in the United States and Canada are based on model building codes developed by four organizations: the International Conference of Building Officials (ICBO); Building Officials and Code Administrators International (BOCAI); the Southern Building Code Conference, Inc. (SBCCI); and the National Building Code of Canada (NBCC). Most seismic requirements adopted by local jurisdictions are based on the Uniform Building Code (UBC), which is developed by ICBO, and on the requirements of the National Earthquake Hazards Reduction Program (NEHRP).

The Sheet Metal and Air Conditioning Contractors' National Association publishes *HVAC Systems—Duct Design* (SMACNA 1991), which includes seismic restraint guidelines for ductwork and piping. The National Fire Protection Association (NFPA) has developed standards on restraint design for fire protection systems. Restraint design for nuclear facilities, which is not discussed in this chapter, is covered in U.S. Department of Energy DOE 6430.1A and in the American Society of Mechanical Engineers' publication ASME AG-1.

In seismically active areas where governmental agencies carefully regulate the earthquake-resistive design of buildings (e.g., California), the HVAC engineer usually does not prepare the code-required seismic restraint calculations. The HVAC engineer selects all the heating and cooling equipment and, with the assistance of the acoustical engineer (if there is one on the project), selects the required vibration isolation devices. The HVAC engineer specifies these devices and calls for shop drawing submittals from the contractors, but the manufacturer employs a registered engineer to design and detail the installation. The HVAC engineer reviews the shop design and details the installation, reviews the shop drawings

and calculations, and obtains the approval of the architect and structural engineer before issuance to the contractors for installation. Anchors for tanks, brackets, and other equipment supports that do not require vibration isolation are designed by the building's structural engineer based on layout drawings prepared by the HVAC engineer. The building officials maintain the code-required quality control over the design by requiring that all building design professionals are registered (licensed).

TERMINOLOGY

Base plate thickness. Thickness of the equipment bracket fastened to the floor.

Effective shear force (V_{eff}). Maximum shear force of one seismic restraint or tie-down bolt.

Effective tension force (T_{eff}). Maximum tension force or pullout force on one seismic restraint or tie-down bolt.

Equipment. Any HVAC&R component that must be restrained from movement during an earthquake.

Fragility level. Maximum lateral acceleration force that the equipment is able to withstand. This data may be available from the equipment manufacturer and is generally on the order of four times the force of gravity ($4g$) for mechanical equipment.

Resilient support. Active seismic device (such as a spring with a bumper) to prevent equipment from moving more than a specified amount.

Response spectra. Relationships between the acceleration response of the ground and the peak acceleration of the earthquake in a damped single degree of freedom at various frequencies. The ground motion response spectrum varies with different soil conditions.

Rigid support. Passive seismic device used to restrict any movement.

Shear force (V). Force generated at the plane of the seismic restraints, acting to cut the restraint at the base.

Seismic restraint. Device designed to withstand an earthquake.

Snubber. Device made of steel-housed, resilient bushings arranged to prevent equipment from moving beyond an established gap.

Tension force (T). Force generated by overturning moments at the plane of the seismic restraints, acting to pull out the bolt.

UBC 94 seismic zones. The geographical location of a facility determines its seismic zone, as given in the Uniform Building Code (ICBO 1994). See Figure 1 and Table 3.

CALCULATIONS

The calculations presented in this chapter assume that the equipment support is an integrated resilient support and restraint device. When the two functions of resilient support and motion restraint are separate or act separately, additional spring loads may need to be added to the anchor load calculation for the restraint device. Internal loads within integrated devices are not addressed in this chapter. Such devices must be designed to withstand the full anchorage loads plus any internal spring loads.

The preparation of this chapter is assigned to the Task Group for Seismic Restraint Design.

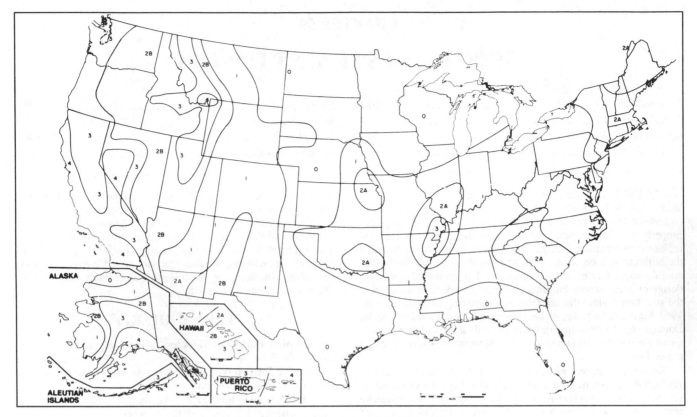

Fig. 1   Seismic Zone Map of the United States
(Reproduced from the 1994 edition of the *Uniform Building Code*™,
copyright 1994, with the permission of the publisher, the International Conference of Building Officials.)

Both static and dynamic analyses reduce the force generated by an earthquake to an equivalent static force, which acts in a horizontal direction at the component's center of gravity. The resulting overturning moment is resisted by shear and tension (pullout) forces on the tie-down bolts. Static analysis is used for both rigid-mounted and resilient-mounted equipment.

Dynamic Analysis

A dynamic analysis is based on site-specific ground motions developed by a geotechnical or soils engineer. A common approach assumes an elastic response spectrum. The results of the dynamic analysis are then scaled up or down as a percentage of the total lateral force obtained from the static analysis performed on the building. The scaling coefficient is established by the UBC or by the governing building official. The scaled acceleration calculated by the structural engineer at any level in the structure can be determined and compared to the force calculated in Equation (1). The greater of the two should be used in the anchor design. The horizontal force factor C_p should be multiplied by a factor of two in either case, as shown in Table 2.

Static Analysis

The total design lateral seismic force is given by the following equation:

$$F_p = Z I_p C_p W_p \qquad (1)$$

where

F_p = total design lateral seismic force
Z = seismic zone factor
I_p = importance factor
C_p = horizontal force factor
W_p = mass of equipment

Table 1   Seismic Zone Factor Z

Zone	Z
1	0.075
2A	0.15
2B	0.20
3	0.30
4	0.40

Table 2   Horizontal Force Factor C_p

Equipment or nonstructural components	C_p
Mechanical equipment, plumbing, and electrical equipment and associated piping rigidly mounted	0.75
All equipment resiliently mounted (maximum 2.0)	$2C_p$

Note: Stacks and tanks should be evaluated for compliance with the applicable codes by a qualified engineer.

Figure 1 and Table 3 may be used to determine the seismic zone. The seismic zone factor Z can then be determined form Table 1. The *Technical Manual—Seismic Design for Buildings* (Army, Navy, and Air Force 1992) can also be used to determine the seismic zone.

The importance factor I_p from the UBC (ICBO 1994) ranges from 1 to 1.5, depending on the building occupancy and hazard level. For equipment, I_p should be conservatively set at 1.5.

The horizontal force factor C_p is determined from Table 2 based on the type of equipment, the tie-down configuration, and the type of base.

The mass W_p of the equipment should include all the items attached or contained in the equipment.

Table 3 International Seismic Zones

Country	City	Seismic Zone
Albania	Tirana	3
Algeria	Algiers	4
	Oran	4
Angola	Luanda	0
Antigua and Barbuda	St. Johns	3
Argentina	Buenos Aires	0
Armenia	Yerevan	3
Australia	Brisbane	1
	Canberra	1
	Melbourne	1
	Perth	1
	Sydney	1
Austria	Salzburg	2B
	Vienna	2B
Azerbaijan	Baku	3
Bahamas	Nassau	0
Bahrain	Manama	0
Bangladesh	Dhaka	3
Barbados	Bridgetown	3
Belarus	Minsk	1
Belgium	Antwerp	1
	Brussels	1
Belize	Belize City	1
Benin	Cotonou	0
Bermuda	Hamilton	0
Bolivia	La Paz	3
Botswana	Gaborone	0
Brazil	Belo Horizonte	0
	Brasilia	0
	Porto Alegre	0
	Recife	0
	Rio de Janeiro	0
	Sao Paulo	1
Brunei	Bandar Seri Begawan	1
Bulgaria	Sofia	3
Burkina Faso	Ouagadougou	0
Burma	Mandalay	3
	Rangoon	2B
Burundi	Bujumbura	3
Cameroon	Douala	0
	Yaounde	0
Canada	Calgary	1
	Halifax	1
	Montreal	2A
	Ottawa	2A
	Quebec	3
	Toronto	1
	Vancouver	3
Cape Verde	Praia	0
Central African Republic	Bangui	0
Chad Republic	N'Djamena	0
Chile	Santiago	4
China	Beijing (Peking)	3
	Chengdu	3
	Guangzhou (Canton)	2B
	Shanghai	2B
	Shenyang (Mukden)	4
Colombia	Barranquilla	2B
	Bogota	3
Congo	Brazzaville	0
Costa Rica	San Jose	3
Cuba	Havana	1
Cyprus	Nicosia	3
Czech Republic	Prague	1
Denmark	Copenhagen	1
Djibouti	Djibouti	3
Dominican Republic	Santo Domingo	3
Ecuador	Guayaquil	3
	Quito	4
Egypt	Alexandria	2B
	Cairo	2B
El Salvador	San Salvador	4
Equatorial Guinea	Malabo	0
Estonia	Tallinn	2A
Ethiopia	Addis Ababa	3
	Asmara	3
Fiji Islands	Suva	3
Finland	Helsinki	1
France	Bordeaux	2B
	Lyon	1
	Marseille	2B
	Paris	0
	Strasbourg	2B
Gabon	Libreville	0
Gambia	Banjul	0
Georgia	Tbilisi	3
Germany	Berlin	0
	Bonn	1
Germany (Continued)	Dusseldorf	1
	Frankfurt	1
	Hamburg	0
	Munich	1
	Stuttgart	2B
Ghana	Accra	3
Greece	Athens	3
	Thessaloniki	4
Grenada	St. George's	3
Guatemala	Guatemala City	4
Guinea	Conakry	0
Guinea-Bissau	Bissau	0
Guyana	Georgetown	0
Haiti	Port-au-Prince	3
Honduras	Tegucigalpa	3
Hong Kong	Hong Kong	2B
Hungary	Budapest	2B
Iceland	Reykjavik	4
India	Bombay	3
	Calcutta	2B
	Madras	1
	New Delhi	2B
Indonesia	Jakarta	3
	Medan	3
	Surabaya	3
Iraq	Baghdad	2B
Ireland	Dublin	0
Israel	Jerusalem	3
	Tel Aviv	1
Italy	Florence	3
	Genoa	2B
	Milan	2B
	Naples	2B
	Palermo	4
	Rome	2B
Ivory Coast	Abidjan	3
Jamaica	Kingston	3
Japan	Fukuoka	3
	Kobe	3
	Naha	3
	Okinawa	3
	Osaka	3
	Sapporo	3
	Tokyo	4
Jordan	Amman	3
Kazakhstan	Alma-Ata	4
Kenya	Nairobi	2B
Korea	Seoul	2A
Kuwait	Kuwait	1
Kyrgyzstan	Bishkek	4
Laos	Vientiane	1
Latvia	Riga	1
Lebanon	Beirut	3
Lesotho	Maseru	2B
Liberia	Monrovia	1
Lithuania	Vilnius	1
Luxembourg	Luxembourg	1
Madagascar	Antananarivo	0
Malawi	Lilongwe	3
Malaysia	Kuala Lumpur	1
Mali Republic	Bamako	0
Malta	Valletta	2B
Martinique	Martinique	3
Mauritania	Nouakchott	0
Mauritius	Port Louis	0
Mexico	Cuidad Juarez	2B
	Guadalajara	3
	Hermosillo	3
	Matamoros	0
	Merida	0
	Mexico City	3
	Monterrey	0
	Nuevo Laredo	0
	Tijuana	3
Moldova	Kishinev	2B
Morocco	Casablanca	2B
	Rabat	2B
Mozambique	Maputo	2B
Nepal	Kathmandu	3
Netherlands	Amsterdam	0
	The Hague	0
Netherlands Antilles	Curacao	3
New Zealand	Auckland	2B
	Wellington	4
Nicaragua	Managua	4
Niger Republic	Niamey	0
Nigeria	Kaduna	0
	Lagos	0
Norway	Oslo	2B
Oman	Muscat	2B
Pakistan	Islamabad	4
	Karachi	2B
	Lahore	2B
	Peshawar	3
Panama	Panama City	2B
Papua New Guinea	Port Moresby	3
Paraguay	Asuncion	0
Peru	Lima	4
Philippines	Baguio	3
	Cebu	4
	Manila	4
Poland	Krakow	2B
	Poznan	1
	Warsaw	1
Portugal	Azor	3
	Lisbon	3
	Oporto	2B
	Ponta Delgada	3
Qatar	Doha	0
Romania	Bucharest	3
Russia	Khabarovsk	1
	Moscow	1
	St. Petersburg	0
	Vladivostok	1
Rwanda	Kigali	3
Saudi Arabia	Dhahran	0
	Jeddah	2B
	Riyadh	0
Senegal Republic	Dakar	0
Seychelles Islands	Victoria	0
Sierra Leone	Freetown	0
Singapore	Singapore	1
Slovakia	Bratislava	2B
Somalia	Mogadishu	0
South Africa	Cape Town	2B
	Durban	2B
	Johannesburg	2B
	Pretoria	2B
Spain	Barcelona	2B
	Bilbao	2B
	Madrid	0
Sri Lanka	Colombo	0
Sudan	Khartoum	2B
Suriname	Paramaribo	0
Swaziland	Mbabane	2B
Sweden	Stockholm	0
Switzerland	Bern	2B
	Geneva	1
	Zurich	2B
Syrian Arab Republic	Damascus	3
Taiwan	Taipei	4
Tajikistan	Dushanbe	4
Tanzania	Dar Es Salaam	2B
	Zanzibar	2B
Thailand	Bangkok	1
	Chiang Mai	2B
	Songkhla	0
Togo	Lome	1
Trinidad and Tobago	Port of Spain	3
Tunisia	Tunis	3
Turkey	Adana	2B
	Ankara	2B
	Istanbul	4
	Izmir	4
Turkmenistan	Ashkhabad	4
Uganda	Kampala	2B
Ukraine	Kiev	1
United Arab Emirates	Abu Dhabi	0
	Dubai	0
United Kingdom	Belfast	0
	Edinburgh	1
	London	1
Uruguay	Montevideo	0
Uzbekistan	Tashkent	3
Vatican City	Vatican City	2B
Venezuela	Caracas	3
	Maracaibo	2B
Vietnam	Ho Chi Minh City	0
Yemen Arab Republic	Aden City	3
	Sana	3
Yugoslavia	Belgrade	2A
	Zagreb	3
Zaire	Kinshasa	0
	Lubumbashi	2B
Zambia	Lusaka	2B
Zimbabwe	Harare	2B

APPLYING STATIC ANALYSIS

The forces acting on the equipment are the lateral and vertical forces resulting from the earthquake, the force of gravity, and the forces of the restraint holding the equipment in place. The analysis assumes the equipment does not move during an earthquake, thus, the sum of the forces and moments must be zero. When calculating the overturning moment, the vertical component F_{pv} at the center of gravity is given by the following equation:

$$F_{pv} = F_p /3 \qquad (1a)$$

The forces of the restraint holding the equipment in position include shear and tension forces. It is important to determine the number of bolts that are affected by the earthquake forces. The direction of the lateral force should be evaluated in both horizontal directions as shown in Figure 2. All bolts or as few as a single bolt may be affected.

Figure 2 shows a typical rigid floor mount installation of a piece of equipment. To calculate the shear force, the sum of the forces in the horizontal plane is given by the following equation:

$$0 = F_p - V \qquad (2)$$

The effective shear force V_{eff} is given by the following equation:

$$V_{eff} = F_p /N_{bolt} \qquad (3)$$

where N_{bolt} = the number of bolts in shear.

The restraints shown in Figure 2 have two bolts on each side, so that four bolts are in shear. To calculate the tension force, the sum of the moments for overturning are as follows:

$$F_p h_{cg} - (W_p - F_p /3)(D_1 /2) - TD_1 = 0$$

Thus,

$$T = [F_p h_{cg} - (W_p - F_p /3)(D_1 /2)]/D_1 \qquad (4)$$

For the example shown in Figure 2, two bolts are in tension. The effective tension force T_{eff}, where the overturning affects only one side, is given by the following equation:

$$T_{eff} = T/N_{bolt} \qquad (5)$$

For the example in Figure 2, the shear and tension forces (V and T) should be calculated independently for both axes as shown in the front and side views. The worst case governs seismic restraint design; however, the direction of seismic loading that will govern the design is not always obvious. For example, if three bolts were installed on each side, the lateral force applied as shown in the side view of Figure 2 affects six bolts in shear and a minimum of two in tension. The lateral force applied as shown on the front view results

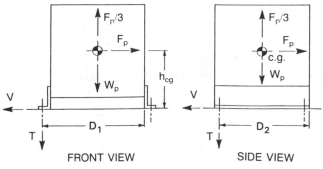

Diameter, mm	T_{allow}, kN	V_{allow}, kN
13	1.3	3.9
16	2.0	9.8
19	3.0	13.3

Notes:
1. The allowable tensile forces are for installations without special inspection (torque test) and may be doubled if the installation is inspected.
2. Additional tension and shear values may be obtained from published ICBO reports.

Table 4   Typical Allowable Loads for Wedge-Type Anchors

in six bolts affected in shear and three in tension. Also, D_1 and D_2 are is different for each axis.

Equations (1), (2), and (3) may be applied to ceiling-mounted equipment. Equation (4) must be modified to Equation (6), because the mass of the equipment adds to the overturning moment. By summing the moments the effective tension force may be determined from the following equation:

$$T = [F_p h_{cg} + (W_p + F_p /3)(D/2)]/D \qquad (6)$$

Interaction Formula. To evaluate the combined effective tension and shear forces that act simultaneously on the bolt, the following equation applies:

$$(T_{eff} / T_{allow}) + (V_{eff} / V_{allow}) \le 1.0 \qquad (7)$$

The allowable forces T_{allow} and V_{allow} are the generic allowable capacities given in Table 4 for wedge-type anchor bolts.

ANCHOR BOLTS

Several types of anchor bolts are manufactured. Wedge and sleeve anchors perform better than self-drilling or drop-in types. Epoxy-type anchors are stronger than other anchors, but lose their strength at elevated temperatures, for example, on rooftops and in areas damaged by fires.

Wedge-type anchors have a wedge on the end with a small clip around the wedge. After a hole is drilled, the bolt is inserted and the external nut tightened. The wedge expands the small clip, which bites into the concrete.

A self-drilling anchor is basically a hollow drill bit. The anchor is used to drill the hole and is then removed. A wedge is then inserted on the end of the anchor, and the assembly is drilled back into place; the drill twists the assembly fully in place. The self-drilling anchor is weaker than other types because it forms a rough hole.

Drop-in expansion anchors are hollow cylinders with a tapered end. After they are inserted in a hole, a small rod is driven through the hollow portion, expanding the tapered end. These anchors are recommended only for shallow installations because they have no reserve expansion capacity.

A sleeve anchor is a bolt covered by a threaded, thin-wall, split tube. As the bolt is tightened, the thin wall expands. Additional load tends to further expand the thin wall. The bolt must be properly preloaded or the friction force will not develop the required holding force.

Adhesive anchors may be in glass capsules or installed with various tools. Pure epoxy, polyester, or vinyl ester resin adhesives are used with a threaded rod supplied by the contractor or the adhesive manufacturer. Some adhesives have a problem with shrinkage; others are degraded by heat. However, some adhesives have been tested without protection to 590°C before they fail—all mechanical anchors will fail at this temperature. Where required, or if there is a concern, anchors should be protected with fire retardants similar to those applied to steel decks in high-rise buildings.

The manufacturer's instructions for installing the anchor bolts should be followed. Performance test data published by manufacturers should include shock, fatigue, and seismic resistance. For allowable

Fig. 2   Equipment with Rigidly Mounted Structural Bases

forces used in the design, refer to the International Conference of Building Officials (ICBO) reports. Add a safety factor of two if the installation has not been inspected by a qualified firm or individual.

WELD CAPACITIES

Weld capacities may be calculated to determine the size of welds needed to attach equipment to a steel plate or to evaluate raised support legs and attachments. A static analysis provides the effective tension and shear forces. The capacity of a weld is given per unit length of weld based on the shear strength of the weld material. For steel welds, the allowable shear strength capacity is 110 MPa on the throat section of the weld. The section length is 0.707 times the specified weld size.

For a 1.5-mm weld, the length of shear in the weld is 0.707 (1.5) = 1.06 mm. The allowable weld force $(F_w)_{allow}$ for a 1.5-mm weld is

$$(F_w)_{allow} = 1.06 \times 110 = 117 \text{ N per millimetre of weld}$$

For a 3-mm weld, the capacity is 233 N/mm.

The effective weld force is the sum of the vectors calculated in Equations (3) and (5). Because the vectors are perpendicular, they are added by the method of the square root of the sum of the squares (SRSS):

$$(F_w)_{eff} = \sqrt{(T_{eff})^2 + (V_{eff})^2}$$

The length of weld required is given by the following equation:

$$\text{Length} = (F_w)_{eff} / (F_w)_{allow} \qquad (8)$$

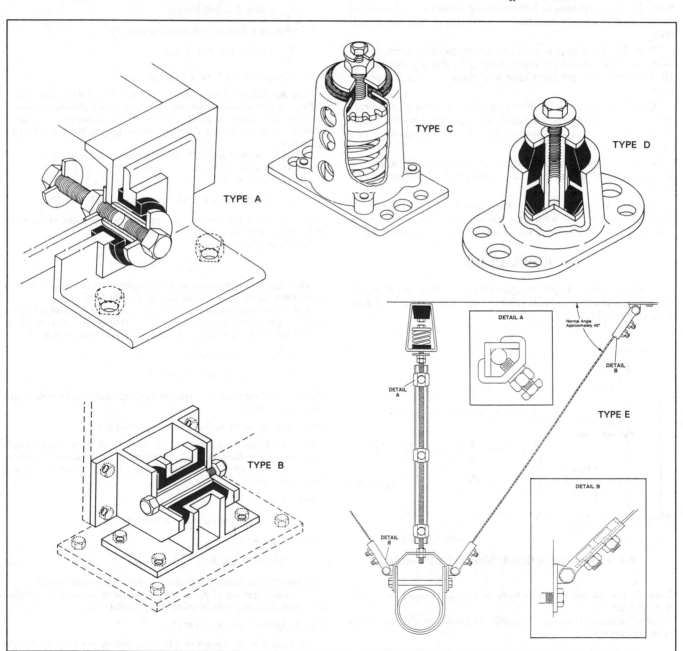

Fig. 3   Seismic Snubbers

SEISMIC SNUBBERS

Several types of snubbers are manufactured or field fabricated. All snubber assemblies should meet the following minimum requirements to avoid imparting excessive accelerations to the HVAC&R equipment: (1) the impact surface should have a high-quality elastomeric surface that is not cemented in place, (2) the resilient material should be easy to inspect for damage and be replaceable if necessary, (3) the assembly must provide restraint in all directions, and (4) snubbers should be tested by an independent test laboratory and analyzed by a registered engineer to ensure the stated load capacity and to avoid serious design flaws. Typical snubber types available are classified as Types A through E (see Figure 3). Many devices are presently approved with Office of Statewide Health Planning Development (OSHPD) ratings.

Type A. All-directional with molded, replaceable neoprene element. Neoprene element of bridge-bearing quality is a minimum of 5 mm thick. Snubber must have a minimum of two anchor bolt holes.

Type B. All-directional with molded, replaceable neoprene element. Neoprene element of bridge-bearing quality is a minimum of 20 mm thick. Snubber must have a minimum of two anchor bolt holes.

Type C. Snubber built into a resilient mounting. All-directional, molded bridge-bearing quality neoprene element is a minimum of 3 mm thick. Mounting must have a minimum of two anchor bolt holes.

Type D. Neoprene mount capable of sustaining seismic loads in all directions. Mounting must have a minimum of two anchor bolt holes.

Type E. Aircraft wire rope with galvanized end connections that avoid bending the wire rope across sharp edges. This type of snubber is mainly used with suspended pipe duct and equipment.

EXAMPLES

The following examples are provided to assist in the design of equipment anchorage to resist seismic forces. Assume seismic zone 4 for all examples.

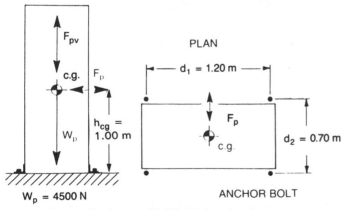

Fig. 4 Equipment Rigidly Mounted to Structure

Example 1. Anchorage design for equipment rigidly mounted to the structure (see Figure 4).

From Equations (1) and (1a), calculate the lateral seismic force and its vertical component:

$$F_p = 0.4 \times 1.5 \times 0.75 \times 4500 = 2025 \text{ N}$$

$$F_{pv} = F_p/3 = 2025/3 = 675 \text{ N}$$

Calculate the overturning moment (OTM):

$$\text{OTM} = F_p h_{cg} \tag{9}$$
$$= (2025 \times 1.0) = 2025 \text{ N} \cdot \text{m}$$

Calculate the resisting moment (RM):

$$\text{RM} = (W_p \pm F_{pv}) \, d_{min}/2 \tag{10}$$
$$= (4500 \pm 675) \, 0.70/2 = 1811 \text{ or } 1339 \text{ N} \cdot \text{m}$$

Calculate the tension force T, using RM_{min} to determine the maximum tension force:

$$T = (\text{OTM} - \text{RM}_{min}) / d_{min} \tag{11}$$
$$= (2025 - 1339) / 0.70 = 980 \text{ N}$$

This force is the same as that obtained using Equation (4).

Calculate T_{eff} per bolt from Equation (5):

$$T_{eff} = 980/2 = 490 \text{ N/bolt}$$

Calculate shear force per bolt from Equation (3):

$$V_{eff} = 2025/4 = 506 \text{ N/bolt}$$

Case 1. Equipment attached to a timber structure.

From the *National Design Specification for Wood Construction* (NDS) (NFPA 1986). Selected fasteners must be secured to solid lumber, not to plywood or other similar material. The following calculations are made to determine whether a 13-mm diameter, 100-mm long lag screw will hold the required load.

From Table 8.1A in the NDS, which is inch-pound units, for Group IV (mixed species), $G = 0.35$, and from Table 8.6A, the allowable withdrawal load is 203 lb/in. $\times$ 4.45 N/lb $\times$ (1 in./25.4 mm) = 35.6 N/mm, and

$$T_{allow} = (35.6 \text{ N/mm} \times 88 \text{-mm penetration}) \, 2/3 = 2090 \text{ N}$$

where the factor 2/3 accounts for the fact that about one-third of the length of a lag screw or bolt has no threads on the shank.

From Table 8.6C in the NDS,

$$V_{allow} = 800 \text{ N}$$

Other types of wood may be used with appropriate factors from Table 8 and/or other reductions as specified in Part II of the NDS.

In timber construction, the interaction formula given in Equation (7) does not apply per Section 8.6.8 of the NDS. The ratios of the calculated shear and tension values to the allowable values should each be less than 1.0:

$$T/T_{allow} = 490/2090 = 0.23 < 1.0$$

$$V/V_{allow} = 506/800 = 0.63 < 1.0$$

Therefore, a 13-mm diameter, 100-mm long lag screw can be used at each corner of the equipment.

Case 2: Equipment attached to concrete with drill-in bolts.

It is good design practice to specify a minimum of 13-mm diameter bolts to attach roof or floor-mounted equipment to the structure. Determine whether 13-mm wedge anchors without special inspection provisions will hold the required load.

From Table 4,

$$T_{allow} = 1300 \text{ N and } V_{allow} = 3900 \text{ N}$$

From Equation (7),

$$490/1300 + 506/3900 = 0.51 < 1.0$$

Therefore, 13-mm diameter, 100-mm long drill-in bolts can be used.

If special inspection of the anchor installation is provided by qualified personnel, T_{allow} only may be increased by a factor of 2.

Case 3: Equipment attached to steel.

For the case where equipment is attached directly to a steel member, the analysis is the same as that shown in Case 1 above. The allowable values for the attaching bolts are given in the *Manual of Steel Construction* (AISC 1989). Values for A307 bolts are given in Table 5.

Table 5   Allowable Loads for A307 Bolts

Diameter, mm	T_{allow}, kN	V_{allow}, kN	A_b, mm^2
13	17.3	8.7	126
16	27.1	13.8	198
19	39.1	19.6	285
25	69.8	35.1	506

The interaction formula given in Equation (7) does not apply to steel-to-steel connections. Instead, the allowable tension load must be modified as in the following equation:

$$(T_{allow})_{mod} = F_t A_b$$

where

$$F_t = 116 - 1.8(V/N_{bolt})A_b \leq 140(4/3) = 187$$

V/N_{bolt} is in kilonewtons and F_t is in kilopascals. The 33% stress increase (4/3) is allowed for short-term loads such as wind or earthquakes.

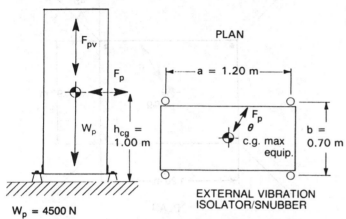

$W_p = 4500$ N

Fig. 5   Equipment Supported by External Spring Mounts

Example 2. Anchorage design for equipment supported by external spring mounts (see Figure 5).

A mechanical or acoustical consultant should choose the type of isolator or snubber or combination of the two. Then the product vendor should select the actual spring snubber.

Assume that the center of gravity (CG) of the equipment coincides with the CG of the isolator group.

If T = maximum tension on isolator,

C = maximum compression on isolator, and

$F_{pv} = F_p/3$, then

$$T = (-W_p + F_{pv})/4 + F_p h_{cg} \cos\theta/2b + F_p h_{cg} \sin\theta/2a$$
$$= (-W_p + F_{pv})/4 + (F_p h_{cg}/2)(\cos\theta/b + \sin\theta/a)$$

To find maximum T or C, set $dT/d\theta = 0$:

$$dT/d\theta = (F_p h_{cg}/2)(-\sin\theta/b + \cos\theta/a) = 0$$

$$\theta_{max} = \tan^{-1}(b/a) \tag{12}$$
$$= \tan^{-1}(28/48) = 30.26°$$

$$T = (-W_p + F_{pv})/4 + (F_p h_{cg}/2)(\cos\theta_{max}/b + \sin\theta_{max}/a) \tag{13}$$

$$C = (-W_p - F_{pv})/4 - (F_p h_{cg}/2)(\cos\theta_{max}/b + \sin\theta_{max}/a) \tag{14}$$

From Equations (1) and (1a),

$$F_p = 0.4 \times 1.5 \times 2 \times 0.75 \times 4500 = 4050 \text{ N}$$

$$F_{pv} = F_p/3 = 4050/3 = 1350 \text{ N}$$

From Equations (13) and (14),

$$T = -788 + 3349 = 2561 \text{ N}$$

$$C = -1463 - 3349 = -4812 \text{ N}$$

Calculate the shear force per isolator:

$$V = F_p/N_{iso} \tag{15}$$
$$= 4050/4 = 1013 \text{ N}$$

This shear force is applied at the operating height of the isolator. Uplift tension T on the vibration isolator provides the worst condition for the design of the anchor bolts. The compression force C must be evaluated to check the adequacy of the structure to resist the loads (Figure 6).

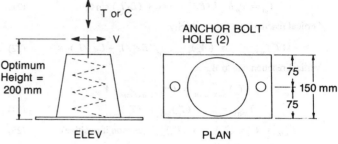

Fig. 6   Spring Mount Detail

$$(T_1)_{eff} \text{ per bolt} = T/2 = 2561/2 = 1280 \text{ N}$$

The value of $(T_2)_{eff}$ per bolt due to overturning on the isolator is

$$(T_2)_{eff} = V \times (\text{op. ht.})/0.85 d N_{bolt}$$

where d = distance from edge of isolator base plate to center of bolt hole.

$$(T_2)_{eff} = 1013 \times 200/(0.85 \times 75 \times 2) = 1590 \text{ N}$$

$$(T_{max})_{eff} = (T_1)_{eff} + (T_2)_{eff} = 1280 + 1590 = 2870 \text{ N}$$

$$V_{eff} = 1013/2 = 507 \text{ N}$$

Determine whether 16-mm drill-in bolts with special inspection will handle this load. From Equation (7) and Table 4,

$$2870/4050 + 507/9800 = 0.76 < 1.0$$

Therefore, 16-mm drill-in bolts will carry the load.

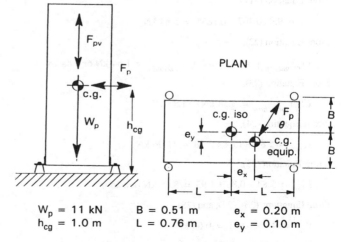

$W_p = 11$ kN $B = 0.51$ m $e_x = 0.20$ m
$h_{cg} = 1.0$ m $L = 0.76$ m $e_y = 0.10$ m

Fig. 7   Equipment with Different Center of Gravity than Isolator Group (in Plan View)

Example 3. Anchorage design for equipment with a center of gravity different from that of the isolator group (see Figure 7).

Anchor properties: $\quad I_x = 4B^2; \quad I_y = 4L^2$

Angles:

$$\theta = \tan^{-1}(B/L) \tag{16}$$

$$\alpha = \tan^{-1}(e_x/e_y) \tag{17}$$

$$\beta = 180 - |\alpha - \theta| \tag{18}$$

$$\phi = \tan^{-1}(LI_x/BI_y) \tag{19}$$

Vertical reactions:

$$(W_n)_{max/min} = W_p \pm F_{pv} \tag{20}$$

Vertical reaction due to overturning moment:

$$T_m = F_p h_{cg}[(B/I_x)\cos\phi + (L/I_y)\sin\phi] \tag{21}$$

Vertical reaction due to eccentricity:

$$(T_e)_{max/min} = (W_n)_{max/min}(Be_y/I_x + Le_x/I_y) \tag{22}$$

Vertical reaction due to W_p:

$$(T_w)_{max/min} = (W_n)_{max/min}/4 \tag{23}$$

$$T_{max} = T_m + (T_e)_{max} + (T_w)_{max} \tag{24}$$

$$T_{min} = T_m + (T_e)_{min} + (T_w)_{min} \quad \text{(tension if positive)} \tag{25}$$

Horizontal reactions:

Horizontal reaction due to rotation:

$$V_{rot} = F_p[(e_x^2 + e_y^2)/16(B^2 + L^2)]^{0.5} \tag{26}$$

$$V_{dir} = F_p/4 \tag{27}$$

$$V_{max} = (V_{rot}^2 + V_{dir}^2 - 2V_{rot}V_{dir}\cos\beta)^{0.5} \tag{28}$$

From Equations (1) and (1a),

$$F_p = 0.4 \times 1.5 \times 2 \times 0.75 \times 11 = 9.9 \text{ kN}$$

$$F_{pv} = 9.9/3 = 3.3 \text{ kN}$$

$$I_x = 4(0.51)^2 = 1.04 \text{ m}^2; \quad I_y = 4(0.76)^2 = 2.31 \text{ m}^2$$

From Equations (16) through (19),

$$\theta = 33.86° \qquad \alpha = 63.43°$$

$$\beta = 150.43° \qquad \phi = 33.86°$$

From Equation (20),

$$(W_n)_{max/min} = 14.3 \text{ kN or } 7.7 \text{ kN}$$

From Equation (21),

$$T_m = 9.9(0.407 + 0.183) = 5.84 \text{ kN}$$

From Equation (22),

$$(T_e)_{max/min} = 0.1148(W_n)_{max/min} = 1.64 \text{ kN or } 0.88 \text{ kN}$$

From Equation (23),

$$(T_w)_{max/min} = 3.58 \text{ kN or } 1.93 \text{ kN}$$

From Equation (24),

$$T_{max} = 5.84 + 1.64 + 3.58 = 11.06 \text{ kN}$$

From Equation (25),

$$T_{min} = 5.84 + 0.88 + 1.93 = 8.65 \text{ kN (tension)}$$

From Equations (26), (27), and (28),

$$V_{rot} = 9.9 \times 0.061 = 0.60 \text{ kN}$$

$$V_{dir} = 9.9/4 = 2.48 \text{ kN}$$

$$V_{max} = 3.02 \text{ kN}$$

The values of T_{min} and V_{max} are used to design the anchorage of the isolators and/or snubbers, and T_{max} is used to verify the adequacy of the structure to resist the vertical loads.

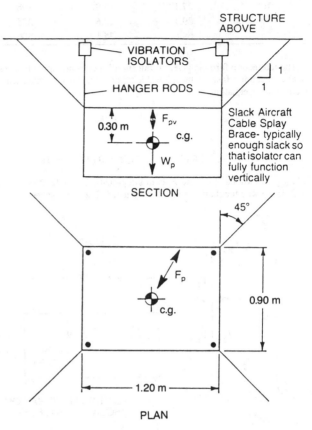

Fig. 8 Supports and Bracing for Suspended Equipment

Example 4. Anchorage design for equipment with supports and bracing for suspended equipment (see Figure 8).

Because drill-in bolts may not withstand published allowable static loads when subjected to vibratory loads, vibration isolators should be used between the equipment and the structure to dampen vibrations generated by the equipment.

$$W_p = 2200 \text{ N}$$

From Equations (1) and (1a),

$$F_p = 0.4 \times 1.5 \times 2 \times 0.75 \times 2200 = 1980 \text{ N}$$

$$F_{pv} = 1980/3 = 660 \text{ N}$$

From Equation (9),

$$\text{OTM} = 1980 \times 0.30 = 594 \text{ N} \cdot \text{m}$$

From Equation (10),

$$\text{RM} = (2200 \pm 660)0.90/2 = 1287 \text{ or } 693 \text{ N} \cdot \text{m}$$

Because RM is greater than OTM, overturning is not critical.

Force to the hanger rods:

$$T_{eff} = (W_p + F_{pv})/4$$

$$= (2200 + 660)/4 = 715 \text{ N}$$

Force in the splay brace $= \sqrt{2}\, F_p = 2800$ N at a 1:1 slope

Due to the force being applied at the critical angle, as in Example 2, only one splay brace is effective in resisting the lateral load F_p. If eccentricities occur, as in Example 3, a similar method of analysis must be done to obtain the design forces.

Design of hanger rod/vibration isolator and connection to the structure.

When installing drill-in anchors in the underside of a concrete beam or slab, the allowable tension loads on the anchors must be reduced to account for the cracking of the concrete. A general rule is to use half the allowable load.

Determine whether a 13-mm wedge anchor with special inspection provisions will hold the required load.

$$T_{allow} = 2600 \times 0.5 = 1300 \text{ N} > T_{eff} = 715 \text{ N}$$

Therefore, a 13-mm rod and drill-in bolt should be used at each corner of the unit.

For anchors installed without special inspection,

$$T_{allow} = 1300 \times 0.5 = 650 \text{ N} < T_{eff} = 715 \text{ N}$$

Therefore, a larger anchor would have to be chosen.

Design of splay brace and connection to the structure.

Force in the slack cable = 2800 N.

Force in the connection to the structure:

$$V_{max} = 2800/\sqrt{2} = 1980 \text{ N} \qquad T_{max} = F_p = 1980 \text{ N}$$

Determine whether a 19-mm wedge-type anchor with special inspection provisions will hold the required load. From Table 4,

$$T_{allow} = 6000/2 = 3000 \text{ N} \qquad V_{allow} = 13\,300 \text{ N}$$

From Equation (7),

$$1980/3000 + 1980/13\,300 = 0.81 < 1.0$$

Therefore, it is permissible to use a 19-mm anchor or multiple anchors of a smaller size bolted through a clip and to the structure.

Because the cable forces are relatively small, a 9.5-mm aircraft cable attached to clips with cable clamps should be used. The clips, in turn, may be attached to either the structure or to the equipment.

INSTALLATION PROBLEMS

The following should be considered when installing seismic restraints.

- Anchor location affects the required strengths. Concrete anchors should be located away from edges, stress joints, or existing fractures. ASTM *Standard* E488 should be followed as a guide for edge distances and center-to-center spacing.
- Concrete anchors should not be too close together. Epoxy-type anchors can be closer together than expansion-type anchors. Expansion-type anchors (self-drilling and drop-in) can crush the concrete where they expand and impose internal stresses in the concrete. Spacing of all anchor bolts should be carefully reviewed. (See manufacturer's recommendations.)
- Supplementary steel bases and frames, concrete bases, or equipment modifications may void some manufacturer's warranties. Snubbers, for example, should be properly attached to a subbase. Bumpers may be used with springs.
- Static analysis does not account for the effects of resonant conditions within a piece of equipment or its components. Because all equipment has different resonant frequencies during operation and nonoperation, the equipment itself might fail even if the restraints do not. Equipment mounted inside a housing should be seismically restrained to meet the same criteria as the exterior restraints.
- Snubbers used with spring mounts should withstand motion in all directions. Some snubbers are only designed for restraint in one direction; sets of snubbers or snubbers designed for multidirectional purposes should be used.
- Equipment must be strong enough to withstand the high deceleration forces developed by resilient restraints.
- Flexible connections should be provided between equipment that is braced and piping and ductwork that need not be braced.
- Flexible connections should be provided between isolated equipment and restraint piping and ductwork.

- Bumpers installed to limit horizontal motion should be outfitted with resilient neoprene pads to soften the potential impact loads of the equipment.
- Anchor installations should be inspected; in many cases, damage occurs because bolts were not properly installed. To develop the rated restraint, bolts should be installed according to manufacturer's recommendations.
- Brackets in structural steel attachments should be matched to reduce bending and internal stresses at the joint. Rigid seismic restraints should not have slotted holes.

REFERENCES

AISC. 1989. *Manual of steel construction,* 9th ed. American Institute of Steel Construction, Chicago.

Army, Navy, and Air Force. 1992. Technical manual seismic design for buildings. TM5-809-10, NAVFAC P-355, AFN 88-3, Chapter 13 (October).

ASTM. 1990. Standard test methods for strength of anchors in concrete and masonry elements. *Standard* E488-90. American Society for Testing and Materials, Philadelphia, P.A.

ICBO. 1994. *Uniform building code.* International Conference of Building Officials, Whittier, California.

NFPA. 1986. National design specification, Wood construction. National Forest Products Association, Washington, D.C.

BIBLIOGRAPHY

Associate Committee on the National Building Code. 1985. *National Building Code of Canada* 1985, 9th ed. National Research Council of Canada, Ottawa.

Associate Committee on the National Building Code. 1986. *Supplement to the National Building Code of Canada* 1985, 2nd ed. National Research Council of Canada, Ottawa. First errata, January.

Astaneh, A., V.V. Bertero, B.A. Bolt, S.A. Mahin, J.P. Moehle, and R.B. Seed. 1989. Preliminary *report* on the seismological and engineering aspects of the October 17, 1989, Santa Cruz (Loma Prieta) Earthquake. Report No. UCB/EERC-89/14. Earthquake Engineering Research Center, College of Engineering, University of California at Berkeley (October).

Ayrers, J.M. and T.Y. Sun. 1971. Nonstructural damage. In *The San Fernando earthquake of February 9, 1971,* vol. 1B, p. 735. National Oceanic and Atmospheric Administration, Washington, D.C.

Ayrers, J.M. and T.Y. Sun. 1973. Criteria for building services and furnishings. In *Building practices* for *disaster mitigation, Building Science Series* 46, p. 253. U.S. Department of Commerce, National Bureau of Standards.

Ayrers, J.M., T.Y. Sun, and F.R. Brown. 1973. *Nonstructural damage of buildings,* p. 346. National Academy of Sciences, Washington, D.C.

BOCA. 1993. *The BOCA National Building Code.* Building Officials & Code Administrators International, Inc. Country Club Hills, Illinois.

Bolt, B.A. 1987, 1988. *Earthquakes.* W.H. Freeman, New York. [*Note: Earthquakes* is a revision of *Earthquakes: A primer,* 1978.]

DOE. 1989. General design criteria. DOE Order 6430.1A. U.S. Department of Energy, Washington, D.C.

Guidelines for seismic restraints of mechanical systems and plumbing & piping systems, 1st ed. 1982. The Sheet Metal Industry Fund of Los Angeles and the Plumbing & Heating Industry Council, Inc.

Jones, R.S. 1984. *Noise and vibration control in buildings.* McGraw-Hill, New York.

Kennedy, R.P., S.A. Short, J.R. McDonald, M.W. McCann, and R.C. Murray. 1989. Design and evaluation guidelines for the Department of Energy facilities subjected to natural phenomena hazards.

Lew, H.S., E.V. Leyendecker, and R.D. Dikkers. 1971. *Engineering aspects of the 1971 San Fernando earthquake, Building Science Series* 40. U.S. Department of Commerce, National Bureau of Standards (December).

Maley, R., A. Acosta, F. Ellis, E. Etheredge, L. Foote, D. Johnson, R. Porcella, M. Salsman, and J. Switzer. 1989. Department of the Interior, U.S. geological survey. U.S. geological survey strong-motion records from the Northern California (Loma Prieta) earthquake of October 17, 1989. Open-file *report* 89-568.

Naeim, F. 1989. *The seismic design handbook.* Van Nostrand Reinhold International Company Ltd., London, England.

SBCCI. 1991. *Standard Building Code.* Southern Building Code Congress International, Birmingham, Alabama.

SMACNA. 1991. *HVAC Systems—Duct Design.* Sheet Metal and Air Conditioning Contractors' National Association, Chantilly, VA.

Weigels, R.L. 1970. *Earthquake engineering,* 10th ed. Prentice-Hall, Englewood Cliffs, New Jersey.

CHAPTER 51

CODES AND STANDARDS

THE Codes and Standards listed in Table 1 represent practices, methods, or standards published by the organizations indicated. They are valuable guides for the practicing engineer in determining test methods, ratings, performance requirements, and limits applying to the equipment used in heating, refrigerating, ventilating, and air conditioning. *Copies can usually be obtained from the organization listed in the Publisher column.* These listings represent the most recent information available at the time of publication.

Table 1   Codes and Standards Published by Various Societies and Associations

Subject	Title	Publisher	Reference
Air Conditioners	Room Air Conditioners	CSA	C22.2 No. 117-1970 (R 1992)
Room	Room Air Conditioners	AHAM	ANSI/AHAM (RA C-1)
	Method of Testing for Rating Room Air Conditioners and Packaged Terminal Air Conditioners	ASHRAE	ANSI/ASHRAE 16-1983 (RA 88)
	Method of Testing for Rating Room Air Conditioners and Packaged Terminal Air Conditioner Heating Capacity	ASHRAE	ANSI/ASHRAE 58-1986 (RA 90)
	Methods of Testing for Rating Room Fan-Coil Air Conditioners	ASHRAE	ANSI/ASHRAE 79-1984 (RA 91)
	Heating and Cooling Equipment	CSA/UL	CAN/CSA-C22.2 No. 236.M90
	Performance Standard for Room Air Conditioners	CSA	CAN/CSA-C368.1-M90
	Room Air Conditioners (1993)	UL	UL 484
Packaged Terminal	Packaged Terminal Air Conditioners	ARI	ARI 310-90
	Packaged Terminal Heat Pumps	ARI	ARI 380-90
	Standards for Packaged Terminal Air-Conditioners and Heat Pumps	CSA	C744-93
Transport	Air Conditioning of Aircraft Cargo (1978)	SAE	SAE AIR806A
	Nomenclature, Aircraft Air-Conditioning Equipment (1978)	SAE	SAE ARP147C
Unitary	Load Calculation for Commercial Summer and Winter Air Conditioning, 4th ed. (1988)	ACCA	ACCA Manual N
	Application of Sound Rated Outdoor Unitary Equipment	ARI	ARI 275-84
	Commercial and Industrial Unitary Air-Conditioning Equipment	ARI	ANSI/ARI 360-86
	Sound Rating of Outdoor Unitary Equipment	ARI	ARI 270-84
	Unitary Air-Conditioning and Air-Source Heat Pump Equipment	ARI	ANSI/ARI 210/240-89
	Methods of Testing for Rating Heat Operated Unitary Air-Conditioning Equipment for Cooling	ASHRAE	ANSI/ASHRAE 40-1986 (RA 92)
	Methods of Testing for Rating Unitary Air-Conditioning and Heat Pump Equipment	ASHRAE	ANSI/ASHRAE 37-1988
	Methods of Testing for Seasonal Efficiency of Unitary Air Conditioners and Heat Pumps	ASHRAE	ANSI/ASHRAE 116-1983
	Performance Standard for Split-System Central Air Conditioners and Heat Pumps	CSA	CAN/CSA-C273.3-M91
	Performance Standard for Single Package Central Air Conditioners and Heat Pumps	CSA	CAN/CSA-C656-M92
	Performance Standard for Rating Large Air Conditioners and Heat Pumps	CSA	CAN/CSA-C746-93
	Air Conditioners, Central Cooling (1982)	UL	ANSI/UL 465-1984
Air Conditioning	Commercial Low Pressure, Low Velocity Duct System Design	ACCA	Manual Q
	Duct Design for Residential Buildings	ACCA	ACCA Manual D
	Load Calculation for Residential Winter and Summer Air Conditioning, 7th ed. (1986)	ACCA	ACCA Manual J
	Gas-Fired Absorption Summer Air Conditioning Appliances (with 1982 addenda)	AGA	ANSI Z21.40.1-1981; Z21.40.1a-1982
	Heating and Cooling Equipment (1990)	CSA/UL	ANSI/UL 1995-1992
	Environmental System Technology (1984)	NEBB	NEBB
	Automotive Air-Conditioning Hose (1989)	SAE	ANSI/SAE J51 MAY89
	HVAC Systems—Applications, 1st ed. (1986)	SMACNA	SMACNA
	HVAC Systems—Duct Design (1990)	SMACNA	SMACNA
	Installation Standards for Residential Heating and Air Conditioning Systems (1988)	SMACNA	SMACNA
	Requirements for Gas-Fired, Engine-Driven Air Conditioning Appliances	AGA	4-89
	Requirements for Gas-Fired Desiccant Type Dehumidifiers and Air Conditioners	AGA	9-90
	Air Conditioning Equipment, General Requirements for Subsonic Airplanes (1961)	SAE	SAE ARP85E
	General Requirements for Helicopter Air Conditioning (1970)	SAE	SAE ARP292B
	Testing of Commercial Airplane Environmental Control Systems (1973)	SAE	SAE ARP217B

Table 1 Codes and Standards Published by Various Societies and Associations (*Continued*)

Subject	Title	Publisher	Reference
Air Conditioning (Continued)	Service Hose for Automatic Air Conditioning	SAE	SAE J22196
	Standard of Purity for use in Mobile Air Conditioning Systems	SAE	SAE J1991
	Extraction and Recycle Equipment for Mobile Automotive Air Conditioning Systems	SAE	SAE J1990
	Guide to the Application and Use of Passenger Car Air Conditioning Compressor Face Seals	SAE	SAE J1954
	Rating Air Conditioner Evaporator Air Delivery and Cooling Capacities	SAE	SAE J1487
	Information Relating to Duty Cycles and Average Power Requirements of Truck and Bus Engine Accessories	SAE	SAE J1343
	Test Method for Measuring Power Consumption of Air Conditioning and Brake Compressors for Trucks and Buses	SAE	SAE J1340
	Design Guidelines for Air Conditioning Systems for Off-Road Operator Enclosures	SAE	SAE J169
	Aircraft Ground Air Conditioning Service Connection	SAE	SAE AS4262
	Recommended Practice for the Design of Tubing Installations for Aerospace Fluid Power Systems	SAE	SAE ARP994
	Control of Excess Humidity in Avionics Cooling	SAE	SAE ARP987
	Guide for Qualification Testing of Aircraft Air Valves	SAE	SAE ARP986
	Air Cycle Air Conditioning Systems for Military Air Vehicles	SAE	SAE ARP4073
	Engine Bleed Air Systems for Aircraft	SAE	SAE ARP1796
	Air Conditioning of Subsonic Aircraft at High Altitudes	SAE	SAE AIR795
	Aircraft Fuel Weight Penalty Due to Air Conditioning	SAE	SAE AIR1168/8
Unitary	Method of Rating Computer and Data Processing Room Unitary Air Conditioners	ASHRAE	ANSI/ASHRAE 127-1988
	Method of Rating Unitary Spot Air Conditioners	ASHRAE	ANSI/ASHRAE 128-1988
Air Curtains	Air Curtains for Entranceways in Food and Food Service Establishments	NSF	NSF-37
	Air Distribution Basics for Residential and Small Commercial Buildings	ACCA	ACCA Manual T
	Residential Equipment Selection	ACCA	ACCA Manual S
	Test Code for Grilles, Registers and Diffusers	ADC	ADC 1062:GRD-84
	Metric Units and Conversion Factors	AMCA	AMCA 99-0100-76
	Test Methods for Air Curtain Units	AMCA	AMCA 220-91
	Air Terminals	ARI	ANSI/ARI 880-89
	Method of Testing for Rating the Performance of Air Outlets and Inlets	ASHRAE	ANSI/ASHRAE 70-1991
	Standard Methods for Laboratory Air Flow Measurement	ASHRAE	ANSI/ASHRAE 41.2-1987 (RA 92)
	Rating the Performance of Residential Mechanical Ventilating Equipment	CSA	CAN/CSA-C260-M90
	Rating the Performance of Residential Mechanical Ventilating Equipment	CSA	CAN/CSA-C260-M90
	High Temperature Pneumatic Duct Systems for Aircraft (1981)	SAE	ANSI/SAE ARP699D
	Direct Gas-Fired Door Heaters	AGA	ANSI Z83.17-1990; Z21.17a-1991
Air Ducts and Fittings	Commercial Low Pressure, Low Velocity Duct Systems	ACCA	Manual Q
	Duct Design for Residential Winter and Summer Air Conditioning	ACCA	Manual D
	Flexible Air Duct Test Code	ADC	ADC FD-72R1-1979
	Flexible Duct Performance and Installation Standards	ADC	ADC-91
	Pipes, Ducts and Fittings for Residential Type Air Conditioning Systems	CSA	B228.1-1968
	Installation of Air Conditioning and Ventilating Systems (1993)	NFPA	ANSI/NFPA 90A-1993
	Installation of Warm Air Heating and Air-Conditioning Systems (1993)	NFPA	ANSI/NFPA 90B-1993
	Ducted Electric Heat Guide for Air Handling Systems (1971)	SMACNA	SMACNA
	HVAC Air Duct Leakage Test Manual (1985)	SMACNA	SMACNA
	HVAC Duct Construction Standards—Metal and Flexible, 1st ed. (1985)	SMACNA	SMACNA
	HVAC Duct Systems Inspection Guide (1989)	SMACNA	SMACNA
	Rectangular Industrial Duct Construction (1980)	SMACNA	SMACNA
	Round Industrial Duct Construction (1977)	SMACNA	SMACNA
	Thermoplastic Duct (PVC) Construction Manual (Rev. A, 1974)	SMACNA	SMACNA
	Closure Systems for Use with Rigid Air Ducts and Air Connectors (1991)	UL	UL 181A
	Factory-Made Air Ducts and Air Connectors (1990)	UL	UL 181
	Marine Rigid and Flexible Air Ducting (1986)	UL	ANSI/UL 1136-1986
Air Filters	Method for Measuring Performance of Portable Household Electrical Cord-Connected Room Air Cleaners	AHAM	ANSI/AHAM AC-1
	Commercial and Industrial Air Filter Equipment	ARI	ARI 850-84
	Residential Air Filter Equipment	ARI	ARI 680-86
	Gravimetric and Dust Spot Procedures for Testing Air Cleaning Devices Used in General Ventilation for Removing Particulate Matter	ASHRAE	ANSI/ASHRAE 52.1-1992
	Method for Sodium Flame Test for Air Filters	BSI	BS 3928
	Particulate Air Filters for General Ventilation—Requirements, Testing Marking	BSI	BS EN 779:1993
	Electrostatic Air Cleaners (1989)	UL	ANSI/UL 867-1988
	High-Efficiency, Particulate, Air Filter Units (1990)	UL	ANSI/UL 586-1990
	Test Performance of Air Filter Units (1987)	UL	ANSI/UL 900-1987

Table 1 Codes and Standards Published by Various Societies and Associations (*Continued*)

Subject	Title	Publisher	Reference
Air-Handling Units	Commercial Low Pressure, Low Velocity Duct Systems	ACCA	Manual Q
	Duct Design for Residential Winter and Summer Air Conditioning	ACCA	Manual D
	Central Station Air-Handling Units	ARI	ANSI/ARI 430-89
	Direct Gas-Fired Make-up Air Heaters	AGA	ANSI Z83.4-1991; Z83.4a-1992
Air Leakage	Air Leakage Performance for Detached Single-Family Residential Buildings	ASHRAE	ANSI/ASHRAE 119-1988
Boilers	A Guide to Clean and Efficient Operation of Coal Stoker-Fired Boilers	ABMA	ABMA
	Boiler Water Limits and Steam Purity Recommendations for Watertube Boilers	ABMA	ABMA
	Boiler Water Requirements and Associated Steam Purity—Commercial Boilers	ABMA	ABMA
	Fluidized Bed Combustion Guidelines	ABMA	ABMA
	Guidelines for Industrial Boiler Performance Improvement	ABMA	ABMA
	Matrix of Recommended Quality Control Requirements	ABMA	ABMA
	Operation and Maintenance Safety Manual	ABMA	ABMA
	Recommended Design Guidelines for Stoker Firing of Bituminous Coals	ABMA	ABMA
	(Selected) Summary of Codes and Standards of the Boiler Industry	ABMA	ABMA
	Thermal Shock Damage to Hot Water Boilers as a Result of Energy Conservation Measures	ABMA	ABMA
	Commercial Applications Systems and Equipment	ACCA	Manual CS
	Boiler and Pressure Vessel Code (11 sections) (1989)	ASME	ASME
	Boiler, Pressure Vessel, and Pressure Piping Code	CSA	B51-M1986
	Heating, Water Supply, and Power Boilers—Electric (1991)	UL	ANSI/UL834-1991
Cast-Iron	Ratings for Cast-Iron and Steel Boilers (1993)	HYDI	IBR
	Testing and Rating Heating Boilers (1989)	HYDI	IBR
Gas or Oil	Gas-Fired Low-Pressure Steam and Hot Water Boilers	AGA	ANSI Z21.13-1991; Z21.13a-1993
	Gas Utilization Equipment in Large Boilers (with 1972 and 1976 addenda; R-1983, 1989)	AGA	ANSI Z83.3-1971; Z83.3a-1972; Z83.3b-1976
	Requirements for High Pressure Steam Boilers	AGA	3-89
	Control and Safety Devices for Automatically Fired Boilers	ASME	ANSI/ASME CSD.1-1988
	Oil-Fired Steam and Hot-Water Boilers for Residential Use	CSA	B140.7.1-1976 (R 1991)
	Oil-Fired Steam and Hot-Water Boilers for Commercial and Industrial Use	CSA	B140.7.2-1967 (R 1991)
	Single Burner Boiler Operations	NFPA	ANSI/NFPA 8501-1992
	Prevention of Furnace Explosions/Implosions in Multiple Burner Boiler-Furnaces (1991)	NFPA	NFPA 85C-1991
	Commercial-Industrial Gas Heating Equipment (1973)	UL	UL 795
	Oil-Fired Boiler Assemblies (1990)	UL	ANSI/UL 726-1990
	Standards and Typical Specifications for Deaerators, 5th ed. (1992)	HEI	HEI
	Method and Procedure for the Determination of Dissolved Oxygen, 2nd ed. (1963)	HEI	HEI
Building Codes	ASTM Standards Used in Building Codes	ASTM	ASTM
	National Building Code, 11th ed. (1993)	BOCA	BOCA
	National Property Maintenance Code, 3rd ed. (1990)	BOCA	BOCA
	One- and Two-Family Dwelling Code (1992)	CABO	CABO
	Model Energy Code (1992)	CABO	CABO
	Uniform Building Code (1991)	ICBO	ICBO
	Uniform Building Code Standards (1991)	ICBO	ICBO
	Directory of Building Codes and Regulations (1993 ed.)	NCSBCS	NCSBCS
	Standard Building Code (1991 ed. with 1992/1993 Revisions)	SBCCI	SBCCI
Mechanical	Safety Code for Elevators and Escalators (plus two yearly supplements)	ASME	ANSI/ASME A 17.1-1990
	BOCA National Mechanical Code, 7th ed. (1993)	BOCA	BOCA
	Uniform Mechanical Code (1991) (with Uniform Mechanical Code Standards)	ICBO/ IAPMO	ICBO/IAPMO
	Standard Gas Code (1991 ed. with 1992/1993 revisions)	SBCCI	SBCCI
	Standard Mechanical Code (1991 ed. with 1992/1993 revisions)	SBCCI	SBCCI
Burners	Guidelines for Burner Adjustments of Commercial Oil-Fired Boilers	ABMA	ABMA
	Domestic Gas Conversion Burners	AGA	Z21.17-1991; Z21.17a-1993
	Installation of Domestic Gas Conversion Burners	AGA	ANSI Z21.8-1984; Z21.8a-1990
	General Requirements for Oil Burning Equipment	CSA	CAN/CSA-B140.0-M87 (R 1991)
	Installation Code for Oil Burning Equipment	CSA	CAN/CSA-B139-M91
	Oil Burners; Atomizing Type	CSA	CAN/CSA-B140.2.1-M90
	Pressure Atomizing Oil Burner Nozzles	CSA	B140.2.2-1971 (R 1991)
	Replacement Burners and Replacement Combustion Heads for Residential Oil Burners	CSA	B140.2.3-M1981 (R 1991)
	Vaporizing-Type Oil Burners	CSA	B140.1-1966 (R 1991)
	Commercial-Industrial Gas Heating Equipment (1973)	UL	UL 795
	Oil Burners (1989)	UL	ANSI/UL 296-1988

Table 1   Codes and Standards Published by Various Societies and Associations (*Continued*)

Subject	Title	Publisher	Reference
Capillary Tubes	Capillary Tubes Method of Testing Flow Capacity of Refrigerant	ASHRAE	ANSI/ASHRAE 28-1988
Chillers	Methods of Testing Liquid Chilling Packages	ASHRAE	ASHRAE 30-1978
	Absorption Water-Chilling and Water Heating Packages	ARI	ARI 560-92
	Centrifugal and Rotary Screw Water-Chilling Packages	ARI	ARI 550-92
	Positive Displacement Compressor Water-Chilling Packages	ARI	ARI 590-92
	Performance Standard for Rating Packaged Water Chillers	CSA	C743-93
Chimneys	Design and Construction of Masonry Chimneys and Fireplaces	CSA	CAN/CSA-A405-M87
	Chimneys, Fireplaces, and Vents, and Solid Fuel Burning Appliances	NFPA	ANSI/NFPA 211-1992
	Chimneys, Factory-Built, Medium Heat Appliance (1986)	UL	ANSI/UL 959-1992
	Chimneys, Factory-Built, Residential Type and Building Heating Appliance (1989)	UL	ANSI/UL 103-1988
Cleanrooms	Procedural Standards for Certified Testing of Cleanrooms (1988)	NEBB	NEBB-1988
Coils	Forced-Circulation Air-Cooling and Air-Heating Coils	ARI	ARI 410-91
	Methods of Testing Forced Circulation Air Cooling and Air Heating Coils	ASHRAE	ASHRAE 33-1978
Comfort Conditions	Thermal Environmental Conditions for Human Occupancy	ASHRAE	ANSI/ASHRAE 55-1992
Compressors	Compressors and Exhausters (reaffirmed 1986)	ASME	ANSI/ASME PTC 10-1965 (R 1986)
	Displacement Compressors, Vacuum Pumps and Blowers	ASME	ANSI/ASME PTC9-1974 (R 1992)
	Safety Standard for Air Compressor Systems	ASME	ANSI/ASME B19.1-1990
	Safety Standard for Compressor for Process Industries	ASME	ASME/ANSI B19.3-1991
	Compressed Air and Gas Handbook, 5th ed. (1988)	CAGI	CAGI
Refrigeration	Ammonia Compressor Units	ARI	ANSI/ARI 510-87
	Method for Presentation of Compressor Performance Data	ARI	ARI 540-91
	Positive Displacement Refrigerant Compressors, Compressor Units and Condensing Units	ARI	ANSI/ARI 520-90
	Method for Presentation of Compressor Performance Data	ARI	ARI 540-91
	Methods of Testing for Rating Positive Displacement Refrigerant Compressors and Condensing Units	ASHRAE	ANSI/ASHRAE 23-1993
	Hermetic Refrigerant Motor-Compressors (1991) (Harmonized with CAN/CSA-C22.2 No.140.2-M91)	CSA/UL	UL 984
Computers	Protection of Electronic Computer/Data Processing Equipment	NFPA	ANSI/NFPA 75-1992
Condensers	Commercial Applications Systems and Equipment (for equipment selection only)	ACCA	Manual CS
	Remote Mechanical Draft Air-Cooled Refrigerant Condensers	ARI	ARI 460-87
	Water-Cooled Refrigerant Condensers, Remote Type	ARI	ARI 450-87
	Methods of Testing for Rating Remote Mechanical-Draft Air-Cooled Refrigerant Condensers	ASHRAE	ASHRAE 20-1970
	Methods of Testing Remote Mechanical-Draft Evaporative Refrigerant Condensers	ASHRAE	ANSI/ASHRAE 64-1989
	Methods of Testing for Rating Water-Cooled Refrigerant Condensers	ASHRAE	ANSI/ASHRAE 22-1992
	Addendum I, Standards for Steam Surface Condensers (1989)	HEI	HEI
	Standards for Steam Surface Condensers, 8th ed. (1984)	HEI	HEI
	Standards for Direct Contact Barometric and Low Level Condensers, 5th ed. (1970)	HEI	HEI
Condensing Units	Commercial Applications Systems and Equipment	ACCA	Manual CS
	Residential Equipment Selection	ACCA	Manual S
	Commercial and Industrial Unitary Air-Conditioning Condensing Units	ARI	ANSI/ARI 365-87
	Methods of Testing for Rating Positive Displacement Refrigerant Compressors and Condensing Units	ASHRAE	ANSI/ASHRAE 23-1993
	Standard for Safety Heating and Cooling Equipment (Harmonized with CAN/CSA C22.2 No. 236-M90)	UL	UL 1995-90
	Refrigeration and Air-Conditioning Condensing and Compressor Units (1987)	UL	ANSI/UL303-1988
Contactors	Definite Purpose Contactors for Limited Duty	ARI	ANSI/ARI 790-86
	Definite Purpose Magnetic Contactors	ARI	ANSI/ARI 780-86
Controls	Quick-Disconnect Devices for Use with Gas Fuel	AGA	ANSI Z21.41-1989; Z21.41a-1990; Z21.41b-1992
	Energy Management Control Systems Instrumentation	ASHRAE	ANSI/ASHRAE 114-1986
	Temperature-Indicating and Regulating Equipment	CSA	C22.2 No. 24-1993
	Control Centers for Changing Message Type Electric Signals (1991)	UL	UL 1433
	Industrial Control Equipment (1993)	UL	UL 508
	Limit Controls (1989)	UL	ANSI/UL 353-1988
	Primary Safety Controls for Gas- and Oil-Fired Appliances (1985)	UL	ANSI/UL 372-1985

Table 1   Codes and Standards Published by Various Societies and Associations (*Continued*)

Subject	Title	Publisher	Reference
Controls (continued)	Solid State Controls for Appliances (1986)	UL	ANSI/UL244A-1987
	Temperature-Indicating and -Regulating Equipment (1988)	UL	ANSI/UL 873-1987
	Tests for Safety-Related Controls Employing Solid-State Devices (1989)	UL	UL 991
Commercial and Industrial	General Standards for Industrial Control and Systems	NEMA	NEMA ICS 1-1988
	Industrial Control Devices, Controllers and Assemblies	NEMA	NEMA ICS 2-1988
	Instructions for the Handling, Installation, Operation and Maintenance of Motor Control Centers	NEMA	NEMA ICS 2.3-1983 (R 1990)
	Maintenance of Motor Controllers after a Fault Condition	NEMA	NEMA ICS 2.2-1983 (1988)
	Preventive Maintenance of Industrial Control and Systems Equipment	NEMA	NEMA ICS 1.3-1986
Residential	Automatic Gas Ignition Systems and Components	AGA	ANSI Z21.20-1989; Z21.20a-1991; Z21.20b-1992
	Gas Appliance Pressure Regulators	AGA	ANSI Z21.18-1987; Z21.18-1989; Z21.18-1992
	Gas Appliance Thermostats	AGA	ANSI Z21.23-1989; Z21.23a-1991
	Manually Operated Gas Valves for Appliances, Appliance Connector Valves and Hose End Valves	AGA	ANSI Z21.15-1992
	Manually-Operated Piezo Electric Spark Gas Ignition Systems and Components	AGA	ANSI Z21.77-1989
	Hot Water Immersion Controls	NEMA	NEMA DC-12-1985 (R 1991)
	Line Voltage Integrally-Mounted Thermostats for Electric Heaters	NEMA	NEMA DC 13-1979 (R 1985)
	Quick Connect Terminals	NEMA	ANSI/NEMA DC 2-1982 (R 1988)
	Residential Controls—Class 2 Transformers	NEMA	NEMA DC 20-1986 (R 1992)
	Residential Controls—Surface Type Controls for Electric Storage Water Heaters	NEMA	NEMA DC 5-1989
	Safety Guidelines for the Application, Installation, and Maintenance of Solid State Controls	NEMA	NEMA ICS 1.1-1984 (R 1988)
	Temperature Limit Controls for Electric Baseboard Heaters	NEMA	NEMA DC 10-1083 (R 1989)
	Wall-Mounted Room Thermostats	NEMA	NEMA DC 3-1989
	Electrical Quick-Connect Terminals (1991)	UL	UL 310
Coolers Air	Refrigeration Equipment	CSA	CAN/CSA-C22.2 No. 120-M91
	Unit Coolers for Refrigeration	ARI	ANSI/ARI 420-89
	Methods of Testing Forced Convection and Natural Convection Air Coolers for Refrigeration	ASHRAE	ANSI/ASHRAE 25-1990
	Commercial Bulk Milk Dispensing Equipment	NSF	NSF 20
Bottled Beverage	Methods of Testing and Rating Bottled and Canned Beverage Vendors and Coolers	ASHRAE	ANSI/ASHRAE 32-1986 (RA 90)
	Refrigerated Vending Machines (1989)	UL	ANSI/UL 541-1988
Drinking Water	Application and Installation of Drinking Fountains and Drinking Water Coolers	ARI	ANSI/ARI 1020-84
	Drinking Fountains and Self-Contained, Mechanically Refrigerated Drinking Water Coolers	ARI/ANSI	ANSI/ARI 1010-84
	Methods of Testing for Rating Drinking-Water Coolers with Self-Contained Mechanical Refrigeration Systems	ASHRAE	ANSI/ASHRAE 18-1987 (RA 91)
	Drinking-Water Coolers (1993)	UL	ANSI/UL 399-1992
	Manual Food and Beverage Dispensing Equipment	NSF	ANSI/NSF 18
Liquid	Refrigerant-Cooled Liquid Coolers, Remote Type	ARI	ANSI/ARI 480-87
	Methods of Testing for Rating Liquid Coolers	ASHRAE	ANSI/ASHRAE 24-1989
Cooling Towers	Commercial Applications Systems and Equipment	ACCA	Manual CS
	Atmospheric Water Cooling Equipment	ASME	ANSI/ASME PTC 23-1986 (R 1991)
	Water-Cooling Towers	NFPA	ANSI/NFPA 214-1992
	Acceptance Test Code for Spray Cooling Systems (1985)	CTI	CTI ATC-133-1985
	Acceptance Test Code for Water Cooling Towers: Mechanical Draft, Natural Draft Fan Assisted Types, Evaluation of Results, and Thermal Testing of Wet/Dry Cooling Towers (1990)	CTI	CTI ATC-105-1990
	Certification Standard for Commercial Water Cooling Towers (1991)	CTI	CTI STD-201-1991
	Code for Measurement of Sound from Water Cooling Towers	CTI	CTI ATC-128-1981
	Fiberglass-Reinforced Plastic Panels for Application on Industrial Water Cooling Towers	CTI	CTI STD-131-1986
	Nomenclature for Industrial Water-Cooling Towers	CTI	CTI NCL-109-1983
	Recommended Practice for Airflow Testing of Cooling Towers	CTI	CTI-1993
Dehumidifiers	Commercial Applications Systems and Equipment	ACCA	Manual CS
	Dehumidifiers	AHAM	ANSI/AHAM DH 1
	Dehumidifiers	CSA	C22.2 No. 92-1971 (R 1992)
	Dehumidifiers (1987)	UL	ANSI/UL 474-1987
Desiccants	Method of Testing Desiccants for Refrigerant Drying	ASHRAE	ANSI/ASHRAE 35-1992
Driers	Method of Testing Liquid Line Refrigerant Driers	ASHRAE	ANSI/ASHRAE 63.1-1988
	Liquid Line Driers	ARI	ANSI/ARI 710-86

Table 1   Codes and Standards Published by Various Societies and Associations (*Continued*)

Subject	Title	Publisher	Reference
Electrical	Voltage Ratings for Electrical Power Systems and Equipment	ANSI	ANSI C84.1-1989
	Canadian Electrical Code, Part 1 (16th ed.)	CSA	C22.1-1990
	Application Guide for Ground Fault Interrupters	NEMA	NEMA 280-1990
	Application Guide for Ground Fault Protective Devices for Equipment	NEMA	NEMA PB 2.2-1988
	Enclosures for Electric Equipment	NEMA	NEMA 250-1991
	Enclosures for Industrial Control and Systems	NEMA	NEMA ICS 6-1988
	General Requirements for Wiring Devices	NEMA	NEMA WD 1-1983 (R 1989)
	Low Voltage Cartridge Fuses	NEMA	NEMA FU 1-1986
	Molded Case Circuit Breakers	NEMA	NEMA AB 1-1986
	Terminal Blocks for Industrial Use	NEMA	NEMA ICS 4-1983 (R 1988)
	National Electric Code (1990)	NFPA	ANSI/NFPA 70-1993
	Compatibility of Electrical Connectors and Wiring (1988)	SAE	SAE AIR1329 A
	Manufacturers' Identification of Electrical Connector Contacts, Terminals and Splices (1982)	SAE	SAE AIR1351 A
	Class T Fuses (1988)	UL	ANSI/UL 198H-1987
	Enclosures for Electrical Equipment (1992)	UL	UL 50
	Fuseholders (1993)	UL	ANSI/UL 512-1992
	High-Interrupting-Capacity Fuses, Current-Limiting Types (1986)	UL	ANSI/UL 198C-1986
	Molded-Case Circuit Breakers and Circuit-Breaker Enclosures (1991)	UL	UL 489
	Terminal Blocks (1993)	UL	UL 1059
	Thermal Cutoffs for Use in Electrical Appliances and Components (1983)	UL	ANSI/UL 1020-1986
Energy	Air Conditioning and Refrigerating Equipment Nameplate Voltages	ARI	ARI 110-90
	Energy Conservation in Existing Buildings—High Rise Residential	ASHRAE	ANSI/ASHRAE/IES 100.2-1991
	Energy Conservation in Existing Buildings—Commercial	ASHRAE	ANSI/ASHRAE/IES 100.3-1985
	Energy Conservation in Existing Facilities—Industrial	ASHRAE	ANSI/ASHRAE/IES 100.4-1984
	Energy Conservation in Existing Buildings—Institutional	ASHRAE	ANSI/ASHRAE/IES 100.5-1991
	Energy Conservation in Existing Buildings—Public Assembly	ASHRAE	ANSI/ASHRAE/IES 100.6-1991
	Energy Conservation in New Building Design—Residential only	ASHRAE	ANSI/ASHRAE/IES 90A-1980
	Energy Efficient Design of New Buildings Except Low Rise Residential Buildings	ASHRAE	ASHRAE/IES 90.1-1989
	Model Energy Code (MEC) (1992)	CABO	BOCA/ICBO/SBCCI
	Uniform Solar Energy Code (1991)	IAPMO	IAPMO
	Energy Directory (1991)	NCSBCS	NCSBCS
	Energy Management Guide for the Selection and Use of Polyphase Motors	NEMA	NEMA MG 10-1983 (R 1988)
	Energy Management Guide for the Selection and Use of Single Phase Motors	NEMA	MEA MG 11-1977 (R 1992)
	Total Energy Management Handbook, 3rd ed.	NEMA	NEMA 05101-1986
	Energy Conservation Guidelines (1984)	SMACNA	SMACNA
	Energy Recovery Equipment and Systems, Air-to-Air (1991)	SMACNA	SMACNA
	Retrofit of Building Energy Systems and Processes (1982)	SMACNA	SMACNA
	Energy Management Equipment (1984)	UL	ANSI/UL 916-1987
Exhaust Systems	Commercial Low Pressure, Low Velocity Duct Systems	ACCA	Manual Q
	Fundamentals Governing the Design and Operation of Local Exhaust Systems	ANSI	ANSI/AIHA Z9.2-1991
	Laboratory Ventilation	ANSI	ANSI/AIHA Z9.5-1992
	Open-Surface Tanks—Ventilation and Operation	ANSI	ANSI/AIHA Z9.1-1991
	Safety Code for Design, Construction, and Ventilation of Spray Finishing Operations (reaffirmed 1971)	ANSI	ANSI/AIHA Z9.3-1985
	Ventilation and Safe Practices of Abrasives Blasting Operations	ANSI	ANSI/AIHA Z9.4-1985
	Method of Testing Performance of Laboratory Fume Hoods	ASHRAE	ANSI/ASHRAE 110-1985
	Compressors and Exhausters	ASME	ANSI/ASME PTC 10-1974 (R 1986)
	Mechanical Flue-Gas Exhausters	CSA	CAN 3-B255-M81
	Exhaust Systems for Air Conveying of Materials	NFPA	ANSI/NFPA 91-1992
	Draft Equipment (1993)	UL	UL 378
Expansion Valves	Thermostatic Refrigerant Expansion Valves	ARI	ANSI/ARI 750-87
	Method of Testing for Capacity Rating of Thermostatic Refrigerant Expansion Valves	ASHRAE	ANSI/ASHRAE 17-1986
Fan Coil Units	Room Fan-Coil Air Conditioners	ARI	ARI 440-89
	Methods of Testing for Rating Room Fan-Coil Air Conditioners	ASHRAE	ANSI/ASHRAE 79-1984 (R 1991)
	Fan-Coil Units and Room Fan-Heater Units (1986)	UL	ANSI/UL 883-1986
Fans	Commercial Low Pressure, Low Velocity Duct Systems	ACCA	Manual Q
	Duct Design for Residential Winter and Summer Air Conditioning	ACCA	Manual D
	Designation for Rotation and Discharge of Centrifugal Fans	AMCA	AMCA 99-2406-83
	Drive Arrangements for Centrifugal Fans	AMCA	AMCA 99-2404-78
	Drive Arrangements for Tubular Centrifugal Fans	AMCA	AMCA 99-2410-82
	Inlet Box Positions for Centrifugal Fans	AMCA	AMCA 99-2405-83

Table 1   Codes and Standards Published by Various Societies and Associations (*Continued*)

Subject	Title	Publisher	Reference
Fans (continued)	Laboratory Methods of Testing Fans for Rating	AMCA	ANSI/AMCA 210-85
	Motor Positions for Belt or Chain Drive Centrifugal Fans	AMCA	AMCA 99-2407-66
	Site Performance Test Standard Power Plant and Industrial Fans	AMCA	AMCA 803-87
	Standards Handbook	AMCA	AMCA 99-86
	Fans and Blowers	ARI	ARI 670-90
	Laboratory Methods of Testing Fans for Rating	ASHRAE	ANSI/ASHRAE 51-1985 ANSI/AMCA 210-85
	Methods of Testing Dynamic Characteristics of Propeller Fans— Aerodynamically Excited Fan Vibrations and Critical Speeds	ASHRAE	ANSI/ASHRAE 87.1-1992
	Fans	ASME	ANSI/ASME PTC 11-1984 (R 1990)
	Fans and Ventilators	CSA	C22.2 No. 113-M1984 (R 1993)
	Rating the Performance of Residential Mechanical Ventilating Equipment	CSA	CAN/CSA C260-M90
	Electric Fans (1991)	UL	UL 507
Ceiling	AC Electric Fans and Regulators	ANSI	ANSI-IEC Pub. 385
Filters	Flow-Capacity Rating and Application of Suction-Line Filters and Filter Driers	ARI	ANSI/ARI 730-86
	Exhaust Hoods for Commercial Cooking Equipment (1990)	UL	ANSI/UL 710-1992
	Grease Filters for Exhaust Ducts (1979)	UL	UL 1046
Fire Dampers	Fire Dampers (1990)	UL	ANSI/UL 555-1989
Fireplaces	Factory-Built Fireplaces (1988)	UL	ANSI/UL 127-1992
Fire Protection	Standard Method for Fire Tests of Building Construction and Materials	ASTM	ASTM E 119-88
	Method of Test of Surface Burning Characteristics of Building Materials	ASTM/ NFPA	ASTM E 84-89a; NFPA 255-1990
	BOCA National Fire Prevention Code, 8th ed. (1990)	BOCA	BOCA
	Uniform Fire Code (1991)	IFCI	IFCI
	Uniform Fire Code Standards (1991)	IFCI	IFCI
	Interconnection Circuitry of Non-Coded Remote Station Protective Signalling Systems	NEMA	NEMA SB 3-1969 (R 1989)
	Fire Doors and Windows	NFPA	ANSI/NFPA 80-1992
	Fire Prevention Code	NFPA	ANSI/NFPA 1-1992
	Fire Protection Handbook, 17th ed.	NFPA	NFPA
	Flammable and Combustible Liquids Code	NFPA	ANSI/NFPA 30-1990
	Life Safety Code	NFPA	ANSI/NFPA 101-1991
	National Fire Codes (issued annually)	NFPA	NFPA
	Smoke Control Systems	NFPA	NFPA 92A-1988
	Smoke Management Systems	NFPA	NFPA 92B-1993
	Fire Tests of Door Assemblies	NFPA	ANSI/NFPA 252-1990
	Standard Fire Prevention Code (1991 ed. with 1992/1993 revisions)	SBCCI	SBCCI
	Fire Tests of Building Construction and Materials (1992)	UL	UL 263
	Fire Tests of Through-Penetration Firestops	UL	UL 1479-1984
	Heat Responsive Links for Fire-Protection Service (1987)	UL	ANSI/UL 33-1987
Fireplace Stoves	Fireplace Stoves (1988)	UL	ANSI/UL 737-1988
Flow Capacity	Method of Testing Flow Capacity of Suction Line Filters and Filter Driers	ASHRAE	ANSI/ASHRAE 78-1985 (RA 90)
Freezers	Refrigeration Equipment	CSA	CAN/CSA-C22.2 No. 120-M91
Household	Household Refrigerators, Combination Refrigerator-Freezers, and Household Freezers	AHAM	ANSI/AHAM; HRF 1
	Capacity Measurement and Energy Consumption Test Methods for Refrigerators, Combination Refrigerator-Freezers, and Freezers	CSA	CAN/CSA-C300-M91
	Household Refrigerators and Freezers	CSA	C22.2 No. 63-M1987
	Household Refrigerators and Freezers (1983)	UL	ANSI/UL 250-1991
Commercial	Dispensing Freezers	NSF	ANSI/NSF-6-1989
	Food Service Refrigerators and Storage Freezers	NSF	NSF-7
	Ice Cream Makers (1993)	UL	ANSI/UL 621-1992
	Commercial Refrigerators and Freezers (1992)	UL	ANSI/UL 471-1991
	Ice Makers (1992)	UL	ANSI/UL 563-1991
Furnaces	Commercial Applications Systems and Equipment	ACCA	Manual CS
	Residential Equipment Selection	ACCA	Manual S
	Direct Vent Central Furnaces	AGA	ANSI Z21.64-1990; Z21.64a-1992
	Gas-Fired Central Furnaces (except Direct Vent)	AGA	ANSI Z21.47-1990; Z21.47a-1990; Z21.47b-1992
	Gas-Fired Duct Furnaces	AGA	ANSI Z83.9-1990; Z83.9a-1992
	Gas-Fired Gravity and Fan Type Direct Vent Wall Furnaces	AGA	ANSI Z21.44-1991; Z21.44a-1992
	Gas-Fired Gravity and Fan Type Floor Furnaces	AGA	Z21.48-1992
	Gas-Fired Gravity and Fan Type Vented Wall Furnaces	AGA	Z21.49-1992

Table 1   Codes and Standards Published by Various Societies and Associations (*Continued*)

Subject	Title	Publisher	Reference
Furnaces (continued)	Methods of Testing for Annual Fuel Utilization Efficiency of Residential Central Furnaces and Boilers	ASHRAE	ANSI/ASHRAE 103-1993
	BOCA National Mechanical Code	BOCA	BOCA
	Installation Code for Solid-Fuel-Burning Appliances and Equipment	CSA	CAN/CSA-B365-M91
	Solid Fuel-Fired Central Heating Appliances	CSA	CAN/CSA-B366.1-M91
	Heating and Cooling Equipment	CSA/UL	CAN/CSA C22.2 No.236-M90
	Oil Burning Stoves and Water Heaters	CSA	B140.3-1962 (R 1991)
	Oil-Fired Warm Air Furnaces	CSA	B140.4-1974 (R 1991)
	Installation of Oil Burning Equipment	NFPA	NFPA 31-1992
	Standard Gas Code (1991 ed. with 1992/1993 revisions)	SBCCI	SBCCI
	Standard Mechanical Code (1991 ed. with 1992/1993 revisions)	SBCCI	SBCCI
	Commercial-Industrial Gas Heating Equipment (1973)	UL	UL 795
	Oil-Fired Central Furnaces (1986)	UL	UL 727-1986
	Oil-Fired Floor Furnaces (1987)	UL	ANSI/UL 729-1987
	Oil-Fired Wall Furnaces (1987)	UL	ANSI/UL 730-1986
	Residential Gas Detectors (1991)	UL	UL 1484
	Single and Multiple Station Carbon Monoxide Detectors (1992)	UL	UL 2034
	Solid-Fuel and Combination-Fuel Central and Supplementary Furnaces (1991)	UL	ANSI/UL 391-1991
Heat Exchangers	Remote Mechanical-Draft Evaporative Refrigerant Condensers	ARI	ANSI/ARI 490-89
	Method of Testing Air-to-Air Heat Exchangers	ASHRAE	ANSI/ASHRAE 84-1991
	Standard Methods of Test for Rating the Performance of Heat-Recovery Ventilators	CSA	CAN/CSA-C439-88
	Standards for Power Plant Heat Exchangers, 2nd ed. (1990)	HEI	HEI
	Standards of Tubular Exchanger Manufacturers Association, 7th ed. (1988)	TEMA	TEMA
Heat Meters	Method of Testing Thermal Energy Heat Meters for Liquid Streams in HVAC Systems	ASHRAE	ANSI/ASHRAE 125-1992
Heat Pumps	Commercial Applications Systems and Equipment	ACCA	Manual CS
	Heat Pump Systems: Principles and Applications (Commercial and Residence)	ACCA	Manual H
	Residential Equipment Selection	ACCA	Manual S
	Commercial and Industrial Unitary Heat Pump Equipment	ARI	ANSI/ARI 340-86
	Ground Water-Source Heat Pumps	ARI	ARI 325-85
	Water-Source Heat Pumps	ARI	ANSI/ARI 320-86
	Methods of Testing for Rating Unitary Air-Conditioning and Heat Pump Equipment	ASHRAE	ANSI/ASHRAE 37-1988
	Heating and Cooling Equipment	CSA/UL	CAN/CSA C22.2 No. 236-M90
	Installation Requirements for Air-to-Air Heat Pumps	CSA	C273.5-1980 (R 1991)
	Performance Standard for Split System Central Air Conditioners and Heat Pumps	CSA	CAN/CSA-C273.3-M91
	Heat Pumps (1985)	UL	ANSI/UL 559-1985
Gas-Fired	Requirements for Gas-Fired, Absorption and Adsorption Heat Pumps	AGA	10-90
Heat Recovery	Gas Turbine Heat Recovery Steam Generators	ASME	ANSI/ASME PTC 4.4-1981 (R 1992)
	Energy Recovery Equipment and Systems, Air-to-Air (1991)	SMACNA	SMACNA
	Requirements for Heat Reclaimer Devices for Use with Gas-Fired Appliances	AGA	1-80
Heaters	Direct Gas-Fired Industrial Air Heaters	AGA	ANSI Z83.18-1990; Z83.18a-1991; Z83.18b-1992
	Gas-Fired Construction Heaters	AGA	ANSI Z83.7-1990; Z83.7a-1991
	Gas-Fired Infrared Heaters	AGA	ANSI Z83.6-1990; Z83.6a-1992
	Gas-Fired Pool Heaters	AGA	ANSI Z21.56-1991
	Gas-Fired Room Heaters, Vol. I, Vented Room Heaters (with 1989 and 1990 addenda)	AGA	ANSI Z21.11.1-1991
	Gas-Fired Room Heaters, Vol. II, Unvented Room Heaters (with 1990 addenda)	AGA	ANSI Z21.11.2-1992
	Requirements for Unvented Room Heaters Equipped with Oxygen Depletion Safety Shutoff Systems	AGA	2-79
	Gas-Fired Unvented Commercial and Industrial Heaters (with 1984 and 1989 addenda)	AGA	ANSI Z83.16-1982; Z83.16a-1984; Z83.16b-1989
	Requirements for Gas-Fired Vented Catalytic Type Room Heaters	AGA	1-81
	Requirements for High Pressure LP Infrared Poultry and Livestock Heating Systems	AGA	4-87
	Requirements for Direct Gas-Fired Circulating Heaters for Agricultural Buildings	AGA	5-88
	Requirements for Residential Radiant Tube Heaters	AGA	7-89
	Requirements for Gas-Fired Infrared Patio	AGA	5-90

Table 1   Codes and Standards Published by Various Societies and Associations (*Continued*)

Subject	Title	Publisher	Reference
Heaters (continued)	Gas-Fired Unvented Catalytic Room Heaters for Use with Liquefied Petroleum (LP) Gases	AGA	ANSI Z21.76-1991; Z21.76a-1992
	Desuperheater/Water Heaters	ARI	ARI 470-87
	Air Heaters	ASME	ANSI/ASME PTC 4.3-1968 (R 1991)
	Standards for Closed Feedwater Heaters, 5th ed. (1992)	HEI	HEI
	Fuel-Fired Heaters—Air Heating—for Construction and Industrial Machinery (1989)	SAE	SAE J1024 MAY89
	Motor Vehicle Heater Test Procedure (1982)	SAE	SAE J638 JUN82
	Electric Air Heaters (1980)	UL	ANSI/UL 1025-1991
	Electric Central Air Heating Equipment (1986)	UL	ANSI/UL 1096-1985
	Electric Dry Bath Heaters (1989)	UL	ANSI/UL 875-1989
	Electric Heaters for Use in Hazardous (Classified) Locations (1991)	UL	ANSI/UL 823-1990
	Electric Heating Appliances (1987)	UL	ANSI/UL 499-1987
	Electric Oil Heaters (1991)	UL	ANSI/UL 574-1990
	Fixed and Location-Dedicated Electric Room Heaters (1992)	UL	UL 2021
	Commercial-Industrial Gas Heating Equipment (1973)	UL	UL 795
	Movable and Wall- or Ceiling-Hung Electric Room Heaters (1992)	UL	UL 1278 (1992)
	Oil-Fired Air Heaters and Direct-Fired Heaters (1975)	UL	UL 733
	Oil-Burning Stoves (1973)	UL	UL 896
	Room Heaters, Solid Fuel-Type (1988)	UL	ANSI/UL 1482-1988
	Unvented Kerosene-Fired Room Heaters and Portable Heaters (1993)	UL	UL 647
Combination	Requirements for Gas-Fired Combination Space Heating/Water Heating Appliances	AGA	11-90
Heating	Commercial Applications Systems and Equipment	ACCA	Manual CS
	Residential Equipment Selection	ACCA	Manual S
	Determining the Required Capacity of Residential Space Heating and Cooling Appliances	CSA	CAN/CSA-F280-M90
	Automatic Flue-Pipe Dampers for Use with Oil-Fired Appliances	CSA	B140.14-M1979 (R 1991)
	Electric Duct Heaters	CSA	C22.2 No. 155-M1986 (R 1992)
	Heater Elements	CSA	C22.2 No.72-M1984 (R 1992)
	Oil-Fired Service Water Heaters and Swimming Pool Heaters	CSA	B140.12-1976 (R 1991)
	Performance Requirements for Electric Heating Line-Voltage Wall Thermostats	CSA	C273.4-M1978 (R 1992)
	Portable Industrial Oil-Fired Heaters	CSA	B140.8-1967 (R 1991)
	Portable Kerosine-Fired Heaters	CSA	CAN 3-B140.9.3 M86
	Electric Air Heaters	CSA	C22.2 No.46-M1988
	Advanced Installation Guide for Hydronic Heating Systems, 1991	HYDI	IBR 250
	Comfort Conditioning Heat Loss Calculation Guide	HYDI	IBR H-21 1984, IBR H-22 (1989)
	Installation Guide for Residential Hydronic Heating Systems, 6th ed. (1988)	HYDI	IBR 200
	Radiant Floor Heating (1993)	HYDI	IBR400
	Environmental System Technology (1984)	NEBB	NEBB
	Pulverized Fuel Systems	NFPA	ANSI/NFPA 8503-1992
	Aircraft Electrical Heating Systems (1965) (reaffirmed 1983)	SAE	SAE AIR860
	Performance Test for Air-Conditioned, Heated, and Ventilated Off-Road Self-Propelled Work Machines	SAE	SAE J1503
	Heating Value of Fuels	SAE	SAE J1498
	Selection and Application Guidelines for Diesel, Gasoline, and Propane Fire Liquid Cooled Engine Pre-Heaters	SAE	SAE J1350
	Electric Engine Preheaters and Battery Warmers for Diesel Engines	SAE	SAE J1310
	Heater, Aircraft, Internal Combustion Heat Exchanger Type	SAE	SAE AS8040
	Total Temperature Measuring Instruments (Turbine Powered Subsonic Aircraft)	SAE	SAE AS793
	Heaters, Aircraft, Internal Combustion Heat Exchanger Type	SAE	SAE AS143
	Heater, Airplane, Engine Exhaust Gas to Air Heat	SAE	SAE ARP86
	Installation, Heaters, Airplane, Internal Combustion Heater Exchange Type	SAE	SAE ARP266
	Performance of Low Pressure Ratio Ejectors for Engine Nacelle Cooling	SAE	SAE AIR1191
	HVAC Systems—Applications, 1st ed. (1986)	SMACNA	SMACNA
	Installation Standards for Residential Heating and Air Conditioning Systems (1988)	SMACNA	SMACNA
	Electric Baseboard Heating Equipment (1987)	UL	ANSI/UL 1042-1986
	Electric Central Air Heating Equipment (1986)	UL	ANSI/UL 1096-1985
	Heating and Cooling Equipment (1990)	CSA/UL	ANSI/UL 1995-1992
Humidifiers	Appliance Humidifiers	AHAM	ANSI/AHAM HU 1
	Central System Humidifiers	ARI	ANSI/ARI 610-89
	Self-Contained Humidifiers	ARI	ANSI/ARI 620-89
	Humidifiers and Evaporative Coolers	CSA	C22.2 No. 104-M1983
	Humidifiers (1987)	UL	ANSI/UL 998-1985

Table 1   Codes and Standards Published by Various Societies and Associations (*Continued*)

Subject	Title	Publisher	Reference
Ice Makers	Automatic Commercial Ice Makers	ARI	ARI 810-91
	Ice Storage Bins	ARI	ANSI/ARI 820-88
	Methods of Testing Automatic Ice Makers	ASHRAE	ANSI/ASHRAE 29-1988
	Refrigeration Equipment	CSA	CAN/CSA-C22.2 No. 120-M91
	Automatic Ice-Making Equipment	NSF	NSF-12
	Ice Makers (1992)	UL	ANSI/UL 563-1991
Incinerators	Incinerators, Waste and Linen Handling Systems and Equipment	NFPA	ANSI/NFPA 82-1990
	Residential Incinerators (1993)	UL	UL 791
Induction Units	Room Air-Induction Units	ARI	ANSI/ARI 445-87
	Frame Assignments for Alternating Current Integral-Horsepower Induction Motors	NEMA	NEMA MG 13-1984 (R 1990)
Industrial Duct	Rectangular Industrial Duct Construction (1980)	SMACNA	SMACNA
	Round Industrial Duct Construction (1977)	SMACNA	SMACNA
Insulation	Specification for Adhesives for Duct Thermal Insulation	ASTM	ASTM C916
	Specification for Thermal and Acoustical Insulation (Mineral Fiber, Duct Lining Material)	ASTM	ASTM C1071-86
	Test Method for Steady-State Heat Flux Measurements and Thermal Transmission Properties by Means of the Guarded Hot Plate Apparatus	ASTM	ASTM C177-85
	Test Method for Steady-State Heat Flux Measurements and Thermal Transmission Properties by Means of the Heat Flow Meter Apparatus	ASTM	ASTM C518-85
	Test Method for Steady-State Heat Transfer Properties of Horizontal Pipe Insulations	ASTM	ASTM C335-89
	Test Method for Steady-State and Thermal Performance of Building Assemblies by Means of a Guarded Hot Box	ASTM	ASTM C236-89
	Thermal Insulation, Mineral Fibre, for Buildings	CSA	A101-M1983
	National Commercial and Industrial Insulation Standards	MICA	MICA 1993
Louvers	Test Method for Louvers, Dampers, and Shutters	AMCA	AMCA 500-89
Lubricants	Method of Testing the Floc Point of Refrigeration Grade Oils	ASHRAE	ANSI/ASHRAE 86-1983
	Practice for Calculating Viscosity Index from Kinematic Viscosity at 40 and 100°C	ASTM	ASTM D2270-91
	Practice for Conversion of Kinematic Viscosity to Saybolt Universal Viscosity or to Saybolt Furol Viscosity	ASTM	ASTM D2161-87
	Method for Estimation of Molecular Weight of Petroleum Oils from Viscosity Measurements	ASTM	ASTM D2502-87
	Method for Separation of Representative Aromatics and Nonaromatics Fractions of High-Boiling Oils by Elution Chromatography	ASTM	ASTM D2549-91
	Classification for Viscosity System for Industrial Fluid Lubricants	ASTM	ASTM D2422-86
	Test Method for Carbon-Type Composition of Insulating Oils of Petroleum Origin	ASTM	ASTM D2140-86
	Test Method for Dielectric Breakdown Voltage of Insulating Liquids Using Disk Electrodes	ASTM	ASTM D877-87
	Test Method for Dielectric Breakdown Voltage of Insulating Oils of Petroleum Origin Using VDE Electrodes	ASTM	ASTM D1816-84a (90)
	Test Method for Mean Molecular Weight of Mineral Insulating Oils by the Cryoscopic Method	ASTM	ASTM D2224-78 (1983)
	Test Method for Molecular Weight of Hydrocarbons by Thermoelectric Measurement of Vapor Pressure	ASTM	ASTM D2503-82 (1987)
	Test Methods for Pour Point of Petroleum Oils	ASTM	ASTM D97-87
	Semiconductor Graphite	NEMA	NEMA CB 4-1989
Measurements	Procedure for Bench Calibration of Tank Level Gaging Tapes and Sounding Rules	ASME	ANSI MC88.2-1974 (R 1987)
	Standard Methods of Measuring and Expressing Building Energy Performance	ASHRAE	ANSI/ASHRAE 105-1984 (RA 90)
	Engineering Analysis of Experimental Data	ASHRAE	ASHRAE Guideline 2-1986 (RA 90)
	Standard Method for Temperature Measurement	ASHRAE	ANSI/ASHRAE 41.1-1986 (RA 91)
	Standard Method for Pressure Measurement	ASHRAE	ANSI/ASHRAE 41.3-1989
	Standard Method for Measurement of Proportion of Oil in Liquid Refrigerant	ASHRAE	ANSI/ASHRAE 41.4-1984
	Standard Method for Measurement of Moist Air Properties	ASHRAE	ANSI/ASHRAE 41.6-1982
	Standard Method for Measurement of Flow of Gas	ASHRAE	ANSI/ASHRAE 41.7-1984 (RA 91)
	Standard Methods of Measurement of Flow of Liquids in Pipes Using Orifice Flowmeters	ASHRAE	ANSI/ASHRAE 41.8-1989
	A Standard Calorimeter Test Method for Flow Measurement of a Volatile Refrigerant	ASHRAE	ANSI/ASHRAE 41.9-1988
	Glossary of Terms Used in the Measurement of Fluid Flow in Pipes	ASME	ANSI/ASME MFC-1M-1991
	Guide for Dynamic Calibration of Pressure Transducers	ASME	ANSI MC88-1-1972 (R 1987)

Table 1 Codes and Standards Published by Various Societies and Associations (*Continued*)

Subject	Title	Publisher	Reference
Measurements (*continued*)	Measurement of Fluid Flow in Pipes Using Orifice, Nozzle, and Venturi	ASME	ASME MFC-3M-1989
	Measurement of Fluid Flow in Pipes Using Vortex Flow Meters	ASME	ASME/ANSI MFC-6M-1987
	Measurement of Gas Flow by Means of Critical Flow Venturi Nozzles	ASME	ASME/ANSI MFC-7M-1987
	Measurement of Gas Flow by Turbine Meters	ASME	ANSI/ASME MFC-4M-1986 (R 1990)
	Measurement of Industrial Sound	ASME	ANSI/ASME PTC 36-1985
	Measurement of Liquid Flow in Closed Conduits Using Transit-Time Ultrasonic Flowmeters	ASME	ANSI/ASME MFC-5M-1985 (R 1989)
	Measurement of Rotary Speed	ASME	ANSI/ASME PTC 19.13-1961 (R 1986)
	Measurement Uncertainty	ASME	ANSI/ASME PTC 19.1-1985 (R 1990)
	Measurement Uncertainty for Fluid Flow in Closed Conduits	ASME	ANSI/ASME MFC-2M-1983 (R 1988)
	Pressure Measurement	ASME	ASME/ANSI PTC 19.2-1987
	Temperature Measurement	ANSI	ANSI/ASME PTC 19.3-1974 (R 1986)
Mobile Homes and Recreational Vehicles	Load Calculation for Residential Winter and Summer Air Conditioning	ACCA	Manual J
	Recreational Vehicle Cooking Gas Appliances (with 1989 addenda)	AGA	ANSI Z21.57-1990; Z21.57a-1991
	Mobile Homes	CSA	CAN/CSA-Z240 MH Series-92
	Mobile Home Parks	CSA	Z240.7.1-1972
	Oil-Fired Warm-Air Heating Appliances for Mobile Housing and Recreational Vehicles	CSA	B140.10-1974 (R 1991)
	Recreational Vehicle Parks	CSA	Z240.7.2-1972
	Recreational Vehicles	CSA	CAN/CSA-Z240 RV Series-M86 (R 1992)
	Gas Supply Connectors for Manufactured Homes	IAPMO	IAPMO TSC 9-1992
	Manufactured Home Installations	NCSBCS	ANSI A225.1-93
	Recreational Vehicles	NFPA	NFPA 501C-1993
	Plumbing System Components for Manufactured Homes and Recreational Vehicles	NSF	ANSI/NSF-24
	Gas Burning Heating Appliances for Mobile Homes and Recreational Vehicles (1965)	UL	UL 307B
	Gas-Fired Cooking Appliances for Recreational Vehicles (1993)	UL	UL 1075
	Liquid Fuel-Burning Heating Appliances for Manufactured Homes and Recreational Vehicles (1990)	UL	ANSI/UL 307A-1989
	Low Voltage Lighting Fixtures for Use in Recreational Vehicles (1993)	UL	UL 234
	Roof Jacks for Manufactured Homes and Recreational Vehicles (1990)	UL	UL 311
	Roof Trusses for Manufactured Homes (1990)	UL	UL 1298
	Shear Resistance Tests for Ceiling Boards for Manufactured Homes (1992)	UL	UL 1296
Motors and Generators	Steam Generating Units	ASME	ANSI/ASME PTC 4.1-1964 (R 1991)
	Testing of Nuclear Air-Treatment Systems	ASME	ANSI/ASME N510-1989
	Nuclear Power Plant Air Cleaning Units and Components	ASME	ANSI/ASME N509-1989
	Energy Efficiency Test Methods for Three-Phase Induction Motors (Efficiency Quoting Method and Permissible Efficiency Tolerance)	CSA	C390-93
	Motors and Generators	CSA	CAN/CSA-C22.2 No. 100-92
	Guide for the Development of Metric Standards for Motors and Generators	NEMA	NEMA 10407-1980
	Motion/Position Control Motors and Controls	NEMA	NEMA MG 7-1987
	Motors and Generators	NEMA	NEMA MG 1-1987 (R 1992)
	Renewal Parts for Motors and Generators—Performance, Selection, and Maintenance	NEMA	NEMA RPI-1981 (R 1987)
	Electric Motors (1989)	UL	ANSI/UL 1004-1988
	Electric Motors and Generators for Use in Hazardous (Classified) Locations (1989)	UL	UL 674
	Impedance-Protected Motors (1982)	UL	ANSI/UL 519-1984 (R 1989)
	Thermal Protectors for Motors (1991)	UL	ANSI/UL 547-1991
Operation and Maintenance	Preparation of Operating and Maintenance Documentaion for Building Systems	ASHRAE	ASHRAE Guideline 4P-1993
Outlets and Inlets	Method of Testing for Rating the Performance of Air Outlets and Inlets	ASHRAE	ANSI/ASHRAE 70-1991
Pipe, Tubing, and Fittings	Power Piping	ASME	ANSI/ASME B31.1-1989
	Refrigeration Piping	ASME	ASME/ANSI B31.5-1987
	Scheme for the Identification of Piping Systems	ASME	ANSI/ASME A13.1-1981 (R 1985)
	Specification for Acrylonitrile-Butadiene-Styrene (ABS) Plastic Pipe, Schedules 40 and 80	ASTM	ASTM D1527-89
	Specification for Polyethylene (PE) Plastic Pipe, Schedule 40	ASTM	ASTM D2104-90
	Specification for Polyvinyl Chloride (PVC) Plastic Pipe, Schedules 40, 80, and 120	ASTM	ASTM D1785-91
	Specification for Seamless Copper Pipe, Standard Sizes	ASTM	ASTM B42-89

Table 1 Codes and Standards Published by Various Societies and Associations (*Continued*)

Subject	Title	Publisher	Reference
Pipe, Tubing, and Fittings (continued)	Specification for Seamless Copper Tube for Air Conditioning and Refrigeration Field Service	ASTM	ASTM B280-88
	Specification for Welded Copper and Copper Alloy Tube for Air Conditioning and Refrigeration Service	ASTM	ASTM B640-90
	Standards of the Expansion Joint Manufacturers Association, Inc., 6th ed. (1993)	EJMA	EJMA
	Corrugated Polyolefin Coilable Plastic Utilities Duct	NEMA	NEMA TC 5-1990
	Corrugated Polyvinyl-Chloride (PVC) Coilable Plastic Utilities Duct	NEMA	NEMA TC 12-1991
	Electrical Nonmetallic Tubing (ENT)	NEMA	NEMA TC 13-1986
	Electrical Plastic Tubing (EPT) and Conduit Schedule EPC-40 and EPC-80	NEMA	NEMA TC 2-1990
	Extra-Strength PVC Plastic Utilities Duct for Underground Installation	NEMA	NEMA TC 8-1990
	Filament Wound Reinforced Thermosetting Resin Conduit and Fittings	NEMA	NEMA TC 14-1984 (R 1986)
	Fittings, Cast Metal Boxes, and Conduit Bodies for Conduit and Cable Assemblies	NEMA	NEMA FB 1-1988
	Fittings for ABS and PVC Plastic Utilities Duct for Underground Installation	NEMA	NEMA TC 9-1990
	Polyvinyl-Chloride (PVC) Externally Coated Galvanized Rigid Steel Conduit and Intermediate Metal Conduit	NEMA	NEMA RN 1-1989
	PVC and ABS Plastic Utilities Duct for Underground Installations	NEMA	NEMA TC 6-1990
	Smooth Wall Coilable Polyethylene Electrical Plastic Duct	NEMA	NEMA TC 7-1990
	National Fuel Gas Code	NFPA/ AGA/ANSI	ANSI/NFPA 54-1992/ ANSI Z223.1-1992
	Plastics Piping Components and Related Materials	NSF	ANSI/NSF-14
	Refrigeration Tube Fittings (1977)	SAE	ANSI/SAE J513 OCT77
	Seismic Restraint Manual Guidelines for Mechanical Systems (1991)	SMACNA	SMACNA
	Rubber Gasketed Fittings for Fire-Protection Service (1993)	UL	UL 213
	Tube Fittings for Flammable and Combustible Fluids, Refrigeration Service, and Marine Use (1993)	UL	UL 109
Plumbing	BOCA National Plumbing Code, 8th ed. (1990)	BOCA	BOCA
	Uniform Plumbing Code (1991) (with IAPMO Installation Standards)	IAPMO	IAPMO
	National Standard Plumbing Code (NSPC)	NAPHCC	NSPC 1993
	Standard Plumbing Code (1991 ed. with 1992/1993 revisions)	SBCCI	SBCCI
Pumps	Centrifugal Pumps	ASME	ASME PTC 8.2-1990
	Displacement Compressors, Vacuum Pumps and Blowers	ASME	ANSI/ASME PTC 9-1974 (R 1992)
	Liquid Pumps	CSA	CAN/CSA C.22.2 No.108-M89
	Performance Standard for Liquid Ring Vacuum Pumps, 1st ed. (1987)	HEI	HEI
	Centrifugal Pump	HI	HI 1.1-1.6 (1993)
	Vertical Pump	HI	HI 2.1-2.6 (1993)
	Rotary Pump	HI	HI 3.1-3.6 (1993)
	Sealless Rotary Pump	HI	HI 4.1-4.6 (1993)
	Sealless Centrifugal Pump	HI	HI 5.1-5.6 (1993)
	Reciprocating Power Pump	HI	HI 6.1-6.6 (1993)
	Controlled Volume Pump	HI	HI 7.1-7.5 (1993)
	Direct Acting (Steam) Pump	HI	HI 8.1-8.5 (1993)
	General Pump	HI	HI 9.1-9.6 (1993)
	Hydraulic Institute Engineering Data Book, 2nd Edition (1991)	HI	HI
	Circulation System Components and Related Materials for Swimming Pools, Spas/Hot Tubs	NSF	ANSI/NSF-50
	Swimming Pool Pumps, Filters and Chlorinators (1986)	UL	ANSI/UL 1081-1985
	Motor-Operated Water Pumps (1991)	UL	ANSI/UL 778-1991
	Pumps for Oil-Burning Appliances (1993)	UL	ANSI/UL 343-1992
Radiation	Ratings for Baseboard and Fin-Tube Radiation (1993)	HYDI	IBR
	Testing and Rating Code for Baseboard Radiation, 6th ed. (1990)	HYDI	IBR
	Testing and Rating Code for Finned-Tube Commercial Radiation (1990)	HYDI	IBR
Receivers	Refrigerant Liquid Receivers	ARI	ANSI/ARI 495-85
Refrigerant-Containing Components	Refrigerant-Containing Components for Use in Electrical Equipment	CSA	C22.2 No.140.3-M1987 (R 1993)
	Refrigerant-Containing Components and Accessories, Non-Electrical (1993)	UL	ANSI/UL 207-1993
Refrigerants	Performance of Refrigerant Recovery, Recycling, and/or Reclaim Equipment	ARI	ARI 740-91
	Specifications for Fluorocarbon Refrigerants	ARI	ANSI/ARI 700-88
	Methods of Testing Discharge Line Refrigerant-Oil Separators	ASHRAE	ANSI/ASHRAE 69-1990
	Number Designation and Safety Classification of Refrigerants	ASHRAE	ANSI/ASHRAE 34-1992
	Reducing Emission of Fully Halogenated Chlorofluorocarbon (CFC) Refrigerants in Refrigeration and Air-Conditioning Equipment and Applications	ASHRAE	ASHRAE Guideline 3-1990

Table 1 Codes and Standards Published by Various Societies and Associations (*Continued*)

Subject	Title	Publisher	Reference
Refrigerants (continued)	Refrigeration Oil Description	ASHRAE	ANSI/ASHRAE 99-1981 (RA 87)
	Sealed Glass Tube Method to Test the Chemical Stability of Material for Use Within Refrigerant Systems	ASHRAE	ANSI/ASHRAE 97-1983 (RA 89)
	Refrigerant Recovery/Recycling Equipment (1989)	UL	ANSI/UL 1963-1991
	Recommended Service Procedure for the Containment of HFC-134a	SAE	SAE J2211
	HFC-134a Recycling Equipment for Mobile Air Conditioning Systems	SAE	SAE J2210
	CFC-12 (R-12) Extraction Equipment for Mobile Automotive Air Conditioning Systems	SAE	SAE J2209
	HFC-134a (R-134a) Service Hose Fittings for Automotive Air Conditioning Service Equipment	SAE	SAE J2197
	Standard of Purity for Recycled HFC-134a for Use in Mobile Air Conditioning Systems	SAE	SAE J2099
	Recommended Service Procedure for the Containment of R-12	SAE	SAE J1989
Refrigeration	Safety Code for Mechanical Refrigeration	ASHRAE	ANSI/ASHRAE 15-1992
	Refrigeration Equipment	CSA	CAN/CSA-C22.2 No.120-M91
	Equipment, Design and Installation of Ammonia Mechanical Refrigeration Systems	IIAR	ANSI/IIAR 2-1984
	Refrigerated Medical Equipment (1978)	UL	UL 416
Refrigeration Systems Steam Jet	Ejectors	ASME	ASME PTC 24-1976 (R 1982)
	Standards for Steam Jet Vacuum Systems, 4th ed. (1988)	HEI	HEI
Transport	Mechanical Transport Refrigeration Units	ARI	ARI 1110-92
	Mechanical Refrigeration Installations on Shipboard	ASHRAE	ANSI/ASHRAE 26-1978 (RA 85)
	General Requirements for Application of Vapor Cycle Refrigeration Systems for Aircraft (1973) (reaffirmed 1983)	SAE	SAE ARP731A
	Safety Practices for Mechanical Vapor Compression Refrigeration Equipment or Systems Used to Cool Passenger Compartment of Motor Vehicles (1987)	SAE	SAE J639 JAN87
Refrigerators	Method of Testing Open Refrigerators for Food Stores	ASHRAE	ANSI/ASHRAE 72-1983
	Methods of Testing Closed Refrigerators	ASHRAE	ANSI/ASHRAE 117-1992
Commercial	Food Carts	ANSI/NSF	NSF 59
	Food Equipment	ANSI/NSF	NSF-2
	Food Service Refrigerators and Storage Freezers	ANSI/NSF	NSF 7
	Soda Fountain and Luncheonette Equipment	NSF	NSF 1
	Commercial Refrigerators and Freezers (1992)	UL	ANSI/UL 471-1991
	Refrigerating Units (1989)	UL	ANSI/UL 427-1989
	Refrigeration Unit Coolers (1980)	UL	ANSI/UL 412-1984
Household	Refrigerators Using Gas Fuel	AGA	ANSI Z21.19-1990; Z21.19a-1992
	Household Refrigerators and Household Freezers	AHAM	AHAM HRF 1
	Capacity Measurement and Energy Consumption Test Methods for Refrigerators, Combination Refrigerator-Freezers, and Freezers	CSA	CAN/CSA C300-M91
	Household Refrigerators and Freezers (1983)	UL	ANSI/UL 250-1984
Roof Ventilators	Commercial Low Pressure, Low Velocity Duct Systems	ACCA	Manual Q
	Power Ventilators (1984)	UL	ANSI/UL 705-1984
Solar Equipment	Method of Measuring Solar-Optical Properties of Materials	ASHRAE	ANSI/ASHRAE 74-1988
	Method of Testing to Determine the Thermal Performance of Flat-Plate Solar Collectors Containing a Boiling Liquid Materials	ASHRAE	ANSI/ASHRAE 109-1986 (RA 90)
	Method of Testing to Determine the Thermal Performance of Solar Collectors	ASHRAE	ANSI/ASHRAE 93-1986 (RA 91)
	Methods of Testing to Determine the Thermal Performance of Solar Domestic Water Heating Systems	ASHRAE	ASHRAE 95-1981 (RA 87)
	Methods of Testing to Determine the Thermal Performance of Unglazed Flat-Plate Liquid-Type Solar Collectors	ASHRAE	ANSI/ASHRAE 96-1980 (RA 90)
Solenoid Valves	Solenoid Valves for Use with Volatile Refrigerants	ARI	ANSI/ARI 760-87
Sound Measurement	Methods for Calculating Fan Sound Ratings from Laboratory Test Data	AMCA	AMCA Standard 301-90
	Laboratory Method of Testing—In-Duct Sound Power Measurement Procedure for Fans	AMCA	ANSI/AMCA 330-86
	Reverberant Room Method for Sound Testing of Fans	AMCA	AMCA 300-85
	Application of Sound Rated Outdoor Unitary Equipment	ARI	ARI 275-84
	Method of Measuring Machinery Sound within Equipment Rooms	ARI	ARI 575-87
	Method of Measuring Sound and Vibration of Refrigerant Compressors	ARI	ANSI/ARI 530-89
	Rating the Sound Levels and Sound Transmission Loss of Packaged Terminal Equipment	ARI	ANSI/ARI 300-88
	Sound Rating of Large Outdoor Refrigerating and Air-Conditioning Equipment	ARI	ARI 370-86
	Sound Rating of Non-Ducted Indoor Air-Conditioning Equipment	ARI	ARI 350-86
	Sound Rating of Outdoor Unitary Equipment	ARI	ARI 270-84

Table 1 Codes and Standards Published by Various Societies and Associations (*Continued*)

Subject	Title	Publisher	Reference
Sound Measurement *(continued)*	Guidelines for the Use of Sound Power Standards and for the Preparation of Noise Test Codes	ASA	ASA 94; ANSI S12.30-1990
	Method for the Calibration of Microphones (reaffirmed 1986)	ASA	ANSI S1.10-1966 (R 1986)
	Specification for Sound Level Meters (reaffirmed 1986)	ASA	ASA 47; ANSI S1.4-1983; ANSI S1.4A-1985
	Measurement of Industrial Sound	ASME	ASME/ANSI PTC 36-1985
	Procedural Standards for Measuring Sound and Vibration	NEBB	NEBB-1977
	Sound and Vibration in Environmental Systems	NEBB	NEBB-1977
	Sound Level Prediction for Installed Rotating Electrical Machines	NEMA	NEMA MG 3-1974 (R 1990)
Space Heaters	Electric Air Heaters	CSA	C22.2 No. 46-M1988
	Electric Air Heaters (1980)	UL	ANSI/UL 1025-1991
	Fixed and Location-Dedicated Electric Room Heaters (1992)	UL	UL 2021
	Movable and Wall- or Ceiling-Hung Electric Room Heaters (1992)	UL	UL 1278
	Gas-Fired Room Heaters, Vol. I, Vented Room Heaters	AGA	ANSI Z21.11.1-1991
	Gas-Fired Room Heaters, Vol. II, Unvented Room Heaters	AGA	ANSI Z21.11.2-1992
Symbols	Graphic Electrical Symbols for Air-Conditioning and Refrigeration Equipment	ARI	ARI 130-88
	Graphic Symbols for Electrical and Electronic Diagrams	IEEE	ANSI/IEEE 315-1975
	Graphic Symbols for Heating, Ventilating, and Air Conditioning	ASME	ANSI/ASME Y32.2.4-1949 (R 1984)
	Graphic Symbols for Pipe Fittings, Valves and Piping	ASME	ANSI/ASME Y32.2.3-1949 (R 1988)
	Graphic Symbols for Plumbing Fixtures for Diagrams used in Architecture and Building Construction	ASME	ANSI/ASME Y32.4-1977 (R 1987)
	Symbols for Mechanical and Acoustical Elements as used in Schematic Diagrams	ASME	ANSI/ASME Y32.18-1972 (R 1985)
Testing and *Balancing*	Site Performance Test Standard—Power Plant and Industrial Fans	AMCA	AMCA 803-87
	General Pump (Including "Measurement of Airborne Sound")	HI	HI 9.1-9.6 (1993)
	Procedural Standards for Certified Testing of Cleanrooms (1988)	NEBB	NEBB-1988
	Procedural Standards for Testing, Adjusting, Balancing of Environmental Systems, 5th ed. (1991)	NEBB	NEBB-1991
	HVAC Systems—Testing, Adjusting and Balancing (1993)	SMACNA	SMACNA
Terminals, Wiring	Quick Connect Terminals	NEMA	NEMA DC 2-1982 (R 1988)
	Electrical Quick-Connect Terminals (1991)	UL	UL 310
	Equipment Wiring Terminals for Use with Aluminum and/or Copper Conductors (1988)	UL	ANSI/UL 486E-1987
	Splicing Wire Connectors (1991)	UL	ANSI/UL 486C-1990
	Wire Connectors and Soldering Lugs for Use with Copper Conductors (1991)	UL	ANSI/UL 486A-1990
	Wire Connectors for Use with Aluminum Conductors (1991)	UL	ANSI/UL 486B-1990
Thermal Storage	Commissioning of HVAC Systems	ASHRAE	ASHRAE Guideline 1-1989
	Metering and Testing Active Sensible Thermal Energy Storage Devices Based on Thermal Performance	ASHRAE	ANSI/ASHRAE 94.3-1986 (RA 90)
	Method of Testing Active Latent Heat Storage Devices Based on Thermal Performance	ASHRAE	ANSI/ASHRAE 94.1-1985 (RA 91)
	Methods of Testing Thermal Storage Devices with Electrical Input and Thermal Output Based on Thermal Performance	ASHRAE	ANSI/ASHRAE 94.2-1981 (RA 89)
	Practices for Measurement, Testing and Balancing of Building Heating, Ventilation, Air-Conditioning, and Refrigeration Systems	ASHRAE	ANSI/ASHRAE 111-1988
Turbines	Land Based Steam Turbine Generator Sets	NEMA	NEMA SM 24-1991
	Steam Turbines for Mechanical Drive Service	NEMA	NEMA SM23-1991
Unit Heaters	Gas Unit Heaters	AGA	ANSI Z83.8-1990; Z83.8a-1990
	Oil-Fired Unit Heaters (1988)	UL	ANSI/UL 731-1987
Valves	Automatic Gas Valves for Gas Appliances	AGA	ANSI Z21.21-1987; Z21.21a-1989; Z21.21b-1992
	Manually Operated Gas Valves for Appliances, Appliance Connection Valves, and Hose End Valves	AGA	ANSI Z21.15-1992
	Relief Valves and Automatic Gas Shutoff Devices for Hot Water Supply Systems	AGA	ANSI Z21.22-1986; Z21.22a-1990
	Combination Gas Controls for Gas Appliances	AGA	ANSI Z21.78-1992
	Requirements for Manually Operated Gas Valves for Use in House Piping Systems	AGA	3-88
	Requirements for Automatic Non-Shutoff Modulating Gas Valves	AGA	1-92
	Requirements for Manually Operated Valves for High Pressure Natural Gas	AGA	2-93
	Requirements for Gas Operated Valves for High Pressure Natural Gas	AGA	3-93
	Refrigerant Access Valves and Hose Connectors	ARI	ANSI/ARI 720-88
	Refrigerant Pressure Regulating Valves	ARI	ARI 770-84

Table 1   Codes and Standards Published by Various Societies and Associations (*Continued*)

Subject	Title	Publisher	Reference
Valves (continued)	Solenoid Valves for Use with Volatile Refrigerants	ARI	ANSI/ARI 760-87
	Thermostatic Refrigerant Expansion Valves	ARI	ANSI/ARI 750-87
	Methods of Testing Nonelectric, Nonpneumatic Thermostatic Radiator Valves	ASHRAE	ANSI/ASHRAE 102-1983 (RA 89)
	Face-to-Face and End-to-End Dimensions of Valves	ASME	ASME/ANSI B16.10-1986
	Large Metallic Valves for Gas Distribution (Manually Operated, NPS-2 1/2 to 12, 125 psig Maximum)	ASME	ANSI/ASME B16.38-1985
	Manually Operated Metallic Gas Valves for Use in Gas Piping Systems up to 125 psig	ASME	ANSI B16.33-1990
	Manually Operated Thermoplastic Gas Shutoffs and Valves in Gas Distribution Systems	ASME	ANSI/ASME B16.40-1985
	Safety and Relief Valves	ASME	ANSI/ASME PTC25.3-1988
	Valves—Flanged Threaded, and Welding End	ASME	ANSI/ASME B16.34-1988
	Electrically Operated Valves (1982)	UL	ANSI-UL 429-1988
	Pressure Regulating Valves for LP-Gas (1985)	UL	ANSI/UL 144-1985
	Safety Relief Valves for Anhydrous Ammonia and LP-Gas (1993)	UL	UL 132
	Valves for Anhydrous Ammonia and LP-Gas (Other than Safety Relief) (1993)	UL	UL 125
	Valves for Flammable Fluids (1980)	UL	UL 842
Vending Machines	Methods of Testing Pre-Mix and Post-Mix Soft Drink Vending and Dispensing Equipment	ASHRAE	ANSI/ASHRAE 91-1976 (RA 91)
	Vending Machines	CSA	CAN/CSA-C22.2 No.128-M90
	Vending Machines for Food and Beverages	NSF	ANSI/NSF-25
	Refrigerated Vending Machines (1989)	UL	ANSI/UL 541-1988
Vent Dampers	Automatic Vent Damper Devices for Use with Gas-Fired Appliances	AGA	ANSI Z21.66-1988; Z21.66a-1991; Z21.66b-1991
	Vent or Chimney Connector Dampers for Oil-Fired Appliances (1988)	UL	ANSI/UL 17-1988
Venting	Draft Hoods	AGA	ANSI Z21.12-1990; Z21.12a-1993
	National Fuel Gas Code	AGA	ANSI Z223.1-1992
	Requirements for Electrically Operated Automatic Combustion and Ventilation Air Control Devices for Use with Gas-Fired Appliances	AGA	1-88
	Requirements for Mechanical Venting Systems	AGA	6-90
	Chimneys, Fireplaces, Vents and Solid Fuel Burning Appliances	NFPA	ANSI/NFPA 211-1992
	Explosion Prevention Systems	NFPA	ANSI/NFPA 69-1992
	Guide for Steel Stack Design and Construction (1983)	SMACNA	SMACNA
	Draft Equipment (1993)	UL	UL 378
	Gas Vents (1991)	UL	ANSI/UL 441-1991
	Low-Temperature Venting Systems, Type L (1986)	UL	ANSI/UL 641-1985
Ventilation	Commercial Low Pressure, Low Velocity Duct Systems	ACCA	Manual Q
	Guide for Testing Ventilation Systems	ACGIH	ACGIH
	Industrial Ventilation (1992)	ACGIH	ACGIH
	Method of Determining Air Change Rates in Detatched Dwellings	ASHRAE	ANSI/ASHRAE 136-1993
	Method of Testing for Room Air Diffusion	ASHRAE	ANSI/ASHRAE 113-1990
	Ventilation for Acceptable Indoor Air Quality	ASHRAE	ANSI/ASHRAE 62-1989
	Residential Mechanical Ventilation Systems	CSA	CAN/CSA F326-M91
	Ventilation Directory (1990)	NCSBCS	NCSBCS
	Parking Structures; Repair Garages	NFPA	ANSI/NFPA 88A-1991; 88B-1991
	Removal of Smoke and Grease-Laden Vapors from Commercial Cooking Equipment	NFPA	ANSI/NFPA 96-1991
	Food Equipment	NSF	ANSI/NSF-2
	Class II (Laminar Flow) Biohazard Cabinetry	NSF	NSF-49
	Performance Test for Air-Conditioned, Heated, and Ventilated Off-Road Self-Propelled Work Machines	SAE	SAE J1503
	Test Procedure for Battery Flame Retardant Venting Systems	SAE	SAE J1495
	Hose, Air Duct, Flexible Nonmetallic, Aircraft	SAE	SAE AS1501
	High Pressure Oxygen System Filler Valve	SAE	SAE AS1225
	Heater, Airplane, Engine Exhaust Gas to Air Heat	SAE	SAE ARP86
	Aerothermodynamic Systems Engineering and Design	SAE	SAE AIR1168/3
Water Heaters	Gas Water Heaters, Vol. I, Storage Water Heaters with Input Ratings of 75,000 Btu per Hour or Less	AGA	ANSI Z21.10.1-1990; Z21.10.1a-1991; Z21.10.1b-1992
	Gas Water Heaters, Vol. III, Storage, with Input Ratings Above 75,000 Btu per Hour, Circulating and Instantaneous Water Heaters	AGA	ANSI Z21.10.3-1990; Z21.10.3a-1990; Z21.10.3b-1992
	Requirements for Non-Metallic Dip Tubes for Use in Gas-Fired Water Heaters	AGA	1-89
	Requirements for Indirect Water Heaters for Use with External Heat Source	AGA	1-91

Table 1   Codes and Standards Published by Various Societies and Associations (*Continued*)

Subject	Title	Publisher	Reference
Water Heaters (continued)	Methods of Testing for Rating Commercial Gas, Electric and Oil Water Heaters	ASHRAE	ANSI/ASHRAE 118.1-1993
	Methods of Testing for Rating Residential Water Heaters	ASHRAE	ANSI/ASHRAE 118.2-1993
	Methods of Testing to Determine the Thermal Performance of Solar Domestic Water Heating Systems	ASHRAE	ANSI/ASHRAE 95-1981 (RA 87)
	Methods of Testing for Rating Combination Space-Heating and Water-Heating Appliances	ASHRAE	ANSI/ASHRAE 124-1991
	Construction and Test of Electric Storage-Tank Water Heaters	CSA	CAN/CSA-C22.2 No. 110-M90
	Performance of Electric Storage Tank Water Heaters	CSA	CAN/CSA-C191-SERIES-M90
	Oil Burning Stoves and Water Heaters	CSA	B140.3-1962 (R 1991)
	Oil-Fired Service Water Heaters and Swimming Pool Heaters	CSA	B140.12-1976 (R 1991)
	Water Heaters, Hot Water Supply Boilers, and Heat Recovery Equipment	NSF	NSF-5
	Commercial-Industrial Gas Heating Equipment (1973)	UL	UL 795
	Electric Booster and Commercial Storage Tank Water Heaters (1988)	UL	ANSI/UL 1453-1987
	Household Electric Storage Tank Water Heaters (1989)	UL	ANSI/UL 174-1989
	Oil-Fired Storage Tank Water Heaters (1988)	UL	ANSI/UL 732-1987
Woodburning Appliances	Method of Testing for Performance Rating of Woodburning Appliances	ASHRAE	ANSI/ASHRAE 106-1984
	Installation Code for Solid Fuel Burning Appliances and Equipment	CSA	CAN/CSA-B365-M91
	Solid-Fuel-Fired Central Heating Appliances	CSA	CAN/CSA-B366.1-M91
	Chimneys, Fireplaces, Vents and Solid Fuel Burning Appliances	NFPA	ANSI/NFPA 211-1992
	Commercial Cooking, Rethermalization and Powered Hot Food Holding and Transport Equipment	NSF	ANSI/NSF-4
	Room Heaters, Solid-Fuel Type (1988)	UL	ANSI/UL 1482-1988

ABBREVIATIONS AND ADDRESSES

ABMA	American Boiler Manufacturers Association, 950 N. Glebe Road, Suite 160, Arlington, VA 22203
ACCA	Air Conditioning Contractors of America, 1513 16th Street, NW, Washington, D.C. 20036
ACGIH	American Conference of Governmental Industrial Hygienists, 6500 Glenway Avenue, Building D-7, Cincinnati, OH 45211
ADC	Air Diffusion Council, Suite 200, 111 E. Wacker Drive, Chicago, IL 60601
AGA	American Gas Association, 1515 Wilson Boulevard, Arlington, VA 22209
AHAM	Association of Home Appliance Manufacturers, 20 N. Wacker Drive, Chicago, IL 60606
AIHA	American Industrial Hygiene Association, 2700 Prosperity Avenue, Suite 250, Fairfax, VA 22031
AMCA	Air Movement and Control Association, Inc., 30 W. University Drive, Arlington Heights, IL 60004-1893
ANSI	American National Standards Institute, 11 West 42nd Street, New York, NY 10036
ARI	Air-Conditioning and Refrigeration Institute, 4301 North Fairfax Drive, Suite 425, Arlington, VA 22203
ASA	Acoustical Society of America, Standards Secretariat, 120 Wall Street, New York, NY 10005
ASHRAE	American Society of Heating, Refrigerating and Air-Conditioning Engineers, Inc., 1791 Tullie Circle, NE, Atlanta, GA 30329
ASME	The American Society of Mechanical Engineers, 345 E. 47 Street, New York, NY 10017
	For ordering publications: ASME Marketing Department, Box 2350, Fairfield, NJ 07007-2350
ASTM	American Society for Testing and Materials, 1916 Race Street, Philadelphia, PA 19103
BOCA	Building Officials and Code Administrators International, Inc., 4051 W. Flossmoor Road, Country Club Hills, IL 60478-5795
BSI	British Standards Institution, 2 Park Street, London, W1A 2BS, England
CABO	Council of American Building Officials, 5203 Leesburg Pike, Suite 708, Falls Church, VA 22041
CAGI	Compressed Air and Gas Institute, 1300 Sumner Avenue, Cleveland, OH 44115
CSA	Canadian Standards Association, 178 Rexdale Boulevard, Rexdale (Toronto), Ontario M9W 1R3, Canada
CTI	Cooling Tower Institute, P.O. Box 73383, Houston, TX 77273
EJMA	Expansion Joint Manufacturers Association, Inc., 25 N. Broadway, Tarrytown, NY 10591
HEI	Heat Exchange Institute, 1300 Sumner Ave., Cleveland, OH 44115
HI	Hydraulic Institute, 9 Sylvan Way, Suite 180, Parsippany, NJ 07054-3802
HYDI	Hydronics Institute, 35 Russo Place, Berkeley Heights, NJ 07922
IAPMO	International Association of Plumbing and Mechanical Officials, 20001 Walnut Drive South, Walnut, CA 91789-2825
ICBO	International Conference of Building Officials, 5360 Workman Mill Road, Whittier, CA 90601
IFCI	International Fire Code Institute, 5360 Workman Mill Road, Whittier, CA 90601-2298
IIAR	International Institute of Ammonia Refrigeration, 111 East Wacker Drive, Chicago, IL 60601
MICA	Midwest Insulation Contractors Association, 2017 South 139th Circle, Omaha, NE 68144
NAPHCC	National Association of Plumbing-Heating-Cooling Contractors, P.O. Box 6808, Falls Church, VA 22040
NCSBCS	National Conference of States on Building Codes and Standards, 505 Huntmar Park Drive, Suite 210, Herndon, VA 22070
NEBB	National Environmental Balancing Bureau, 1385 Piccard Drive, Rockville, MD 20850
NEMA	National Electrical Manufacturers Association, 2101 L Street, NW, Suite 300, Washington, D.C. 20037-1526
NFPA	National Fire Protection Association, 1 Batterymarch Park, P.O. Box 9101, Quincy, MA 02269-9101
NSF International	National Sanitation Foundation, P.O. Box 130140, Ann Arbor, MI 48113-0140
SAE	Society of Automotive Engineers, 400 Commonwealth Drive, Warrendale, PA 15096
SBCCI	Southern Building Code Congress International, Inc., 900 Montclair Road, Birmingham, AL 35213-1206
SMACNA	Sheet Metal and Air Conditioning Contractors' National Association, 4201 Lafayette Center Drive, Chantilly, VA 22021
TEMA	Tubular Exchanger Manufacturers Association, Inc., 25 N. Broadway, Tarrytown, NY 10591
UL	Underwriters Laboratories Inc., 333 Pfingsten Road, Northbrook, IL 60062-2096

COMPOSITE INDEX
ASHRAE HANDBOOK SERIES

This index covers the current Handbook series published by ASHRAE. The four volumes in the series are identified as follows:

A = 1995 Applications

R = 1994 Refrigeration

F = 1993 Fundamentals

S = 1992 Systems and Equipment

The index is alphabetized in a letter-by-letter system; for example, **Airborne** is before **Air cleaners** and **Heaters** is before **Heat exchangers**.

The page reference for an index entry includes the book letter and chapter number, which may be followed by a decimal point and the beginning page or page range within the chapter. For example, A31.4-6 means the information may be found in the 1995 Applications volume, Chapter 31, pages 4 to 6.

Each Handbook is revised and updated on a four-year cycle. Because technology and the interests of ASHRAE members change, the following topics are not included in the current Handbook series but may be found in the Handbooks cited.

- Degree days
  Average Monthly and Yearly Degree Days for Cities in the United States and Canada
  1981 Fundamentals, Chapter 24, Table 4, pp. 23-28
- Steam-jet refrigeration
  1983 Equipment, Chapter 13, Steam-Jet Refrigeration Equipment
- Survival shelters
  1991 Applications, Chapter 11, Environmental Control for Survival
- Thermoelectric cooling
  1981 Fundamentals, Chapter 1, pp. 27-33
